Geologic Time Scale 2020

Volume 2

Geologic Time Scale 2020

Volume 2

Edited by

Felix M. Gradstein

James G. Ogg

Mark D. Schmitz

Gabi M. Ogg

ELSEVIER

Elsevier
Radarweg 29, PO Box 211, 1000 AE Amsterdam, Netherlands
The Boulevard, Langford Lane, Kidlington, Oxford OX5 1GB, United Kingdom
50 Hampshire Street, 5th Floor, Cambridge, MA 02139, United States

Notices
Knowledge and best practice in this field are constantly changing. As new research and experience broaden our
understanding, changes in research methods, professional practices, or medical treatment may become necessary.

Practitioners and researchers must always rely on their own experience and knowledge in evaluating and using any
information, methods, compounds, or experiments described herein. In using such information or methods they
should be mindful of their own safety and the safety of others, including parties for whom they have a professional
responsibility.

To the fullest extent of the law, neither the Publisher nor the authors, contributors, or editors, assume any liability for
any injury and/or damage to persons or property as a matter of products liability, negligence or otherwise, or from any
use or operation of any methods, products, instructions, or ideas contained in the material herein.

British Library Cataloguing-in-Publication Data
A catalogue record for this book is available from the British Library

Library of Congress Cataloging-in-Publication Data
A catalog record for this book is available from the Library of Congress

Volume 1 ISBN: 978-0-12-824362-6
This Volume (2) ISBN: 978-0-12-824363-3
Set ISBN: 978-0-12-824360-2

For Information on all Elsevier publications
visit our website at https://www.elsevier.com/books-and-journals

Front cover of Volume 2: Geologic excursion along the superbly exposed Kimmeridge Clay Formation (Upper Jurassic)
at Kimmeridge Bay, Dorset, UK. Photograph by F.M. Gradstein.

Publisher: Candice Janco
Acquisitions Editor: Amy Shapiro
Editorial Project Manager: Susan Ikeda
Production Project Manager: Kiruthika Govindaraju
Cover Designer: Mark Rogers

Typeset by MPS Limited, Chennai, India

Printed in Great Britain

Last digit is the print number: 10 9 8 7 6 5 4

Quotes

To place all the scattered pages of Earth history in their proper chronological order is by no means an easy task.

Arthur Holmes

The fascination in creating a new geologic time scale is that it evokes images of creating a beautiful carpet by many skilled hands. All stitches must conform to a pre-determined pattern, in this case the pattern of physical, chemical and biological events on Earth aligned along the arrow of time.

This book—Foreword

Contents

Volume 1

Contributors   xi
Editors' Biographies   xv
Preface   xvii
Abbreviations and acronyms   xix

Part I
Introduction   1

1. **Introduction**   3
   F.M. Gradstein

2. **The Chronostratigraphic Scale**   21
   F.M. Gradstein and J.G. Ogg

Part II
Concepts and Methods   33

3. **Evolution and Biostratigraphy**   35
   Coordinated by F.M. Gradstein

   3A. Trilobites   36
   S. Peng, L.E. Babcock and P. Ahlberg

   3B. Graptolites   43
   J. Zalasiewicz, M. Williams and A. Rushton

   3C. Chitinozoans   50
   A. Butcher

   3D. Conodonts   56
   C.M. Henderson

   3E. Ammonoidea   61
   A.S. Gale, D. Korn, A.J. McGowan,
   J. Cope and C. Ifrim

   3F. Calcareous nannofossils   69
   D.K. Watkins and I. Raffi

   3G. Planktonic foraminifera   74
   M.R. Petrizzo, B.S. Wade and
   F.M. Gradstein

   3H. Larger benthic foraminifera   88
   M.D. Simmons

   3I. Dinoflagellates   99
   R.A. Fensome and D.K. Munsterman

   3J. Plants, spores, and pollen   109
   H. Kerp, G. Mangerud and
   S.R. Gradstein

   3K. Cretaceous microcrinoids   122
   A.S. Gale

   3L. Three major mass extinctions and
   evolutionary radiations in
   their aftermath   125
   S. Esmeray-Senlet

4. **Astrochronology**   139
   J. Laskar

5. **Geomagnetic Polarity Time
   Scale**   159
   J.G. Ogg

6. **Radioisotope Geochronology**   193
   M.D. Schmitz, B.S. Singer and A.D. Rooney

7. **Strontium Isotope Stratigraphy**   211
   J.M. McArthur, R.J. Howarth, G.A. Shields
   and Y. Zhou

8. **Osmium Isotope Stratigraphy**   239
   B. Peucker-Ehrenbrink and G.E. Ravizza

9. **Sulfur Isotope Stratigraphy**   259
   A. Paytan, W. Yao, K.L. Faul and E.T. Gray

10. **Oxygen Isotope Stratigraphy**   279
    E.L. Grossman and M.M. Joachimski

viii   Contents

11. Carbon Isotope Stratigraphy   309

B.D. Cramer and I. Jarvis

12. Influence of Large Igneous Provinces 345

R.E. Ernst, D.P.G. Bond and S.H. Zhang

13. Phanerozoic Eustasy   357

*M.D. Simmons, K.G. Miller, D.C. Ray,
A. Davies, F.S.P. van Buchem and B. Gréselle*

14. Geomathematics   401

14A. Geomathematical and Statistical
Procedures   402
*F.P. Agterberg, A-C. Da Silva and
F.M. Gradstein*

14B. Global Composite Sections
and Constrained Optimization   425
P. Sadler

**Part III
Geologic Periods: Planetary and
Precambrian**   441

15. The Planetary Time Scale   443

H. Hiesinger and K. Tanaka

16. Precambrian (4.56–1 Ga)   481

*R. Strachan, J.B. Murphy, J. Darling, C. Storey
and G. Shields*

17. The Tonian and Cryogenian Periods 495

G. Halverson, S. Porter and G. Shields

18. The Ediacaran Period   521

S.H. Xiao and G.M. Narbonne

Volume 2

Abbreviations and acronyms   xi

Part IV
Geologic Periods: Phanerozoic   **563**

19. **The Cambrian Period**   565

    S.C. Peng, L.E. Babcock and P. Ahlberg

20. **The Ordovician Period**   631

    D. Goldman, P.M. Sadler and S.A. Leslie

21. **The Silurian Period**   695

    M.J. Melchin, P.M. Sadler and B.D. Cramer

22. **The Devonian Period**   733

    R.T. Becker, J.E.A. Marshall and A.-C. Da Silva

23. **The Carboniferous Period**   811

    M. Aretz, H.G. Herbig and X.D. Wang

24. **The Permian Period**   875

    C.M. Henderson and S.Z. Shen

25. **The Triassic Period**   903

    J.G. Ogg and Z.-Q. Chen

26. **The Jurassic Period**   955

    S.P. Hesselbo, J.G. Ogg and M. Ruhl

27. **The Cretaceous Period**   1023

    A.S. Gale, J. Mutterlose and S. Batenburg

28. **The Paleogene Period**   1087

    R.P. Speijer, H. Pälike, C.J. Hollis, J.J. Hooker and J.G. Ogg

29. **The Neogene Period**   1141

    I. Raffi, B.S. Wade and H. Pälike

30. **The Quaternary Period**   1217

    P.L. Gibbard and M.J. Head

31. **The Anthropocene**   1257

    J. Zalasiewicz, C. Waters and M. Williams

Appendix 1: Recommended color
              coding of stages   1281

Appendix 2: Radioisotopic ages used
              in GTS2020   1285

Index   1351

ORGANIZATIONS

CGMW Commission for the Geological Map of the World
DNAG Decade of North American Geology
DSDP Deep Sea Drilling Project
GSC Geological Survey of Canada
ICS International Commission of Stratigraphy
IODP International Ocean Drilling Project
IGC International Geological Congress
IGCP International Geological Correlation Project
INQUA International Quaternary Association
IUGS International Union of Geological Sciences
IUPAC International Union of Pure and Applied Chemistry
ODP Ocean Drilling Project
SNS Subcommission (of ICS) on Neogene Stratigraphy
PGS Subcommission (of ICS) on Paleogene Stratigraphy
SQS Subcommission (of ICS) on Quaternary Stratigraphy
STS Subcommission (of ICS) on Triassic Stratigraphy
SOS Subcommission (of ICS) on Ordovician Stratigraphy
SCS Subcommission (of ICS) on Cambrian Stratigraphy
UNESCO United Nations Education, Scientific, and Cultural Organization
USGS United States Geological Survey

TIME SCALE PUBLICATIONS

NDS82 *Numerical Dating in Stratigraphy* (Odin et al., 1982)
GTS82 *A Geologic Time Scale* (Harland et al., 1982)
DNAG83 *Geologic Time Scale, Decade of North American Geology* (Palmer, 1983)
KG85 Kent and Gradstein (1985)
EX88 Exxon 1988 (Haq et al., 1988)
GTS89 *A Geologic Time Scale 1989* (Harland et al., 1990)
OB93 Obradovich (1993)
JGR94 *Journal of Geophysical Research* 1994 (Gradstein et al., 1994)
SEPM95 Society for Sedimentary Geology 1995 (Gradstein et al., 1995)
GO96 Gradstein and Ogg (1996)
GTS2004 Gradstein, Ogg, and Smith (2004)
GTS2008 Ogg, Ogg, and Gradstein (2008)
GTS2012 Gradstein, Ogg, Schmitz, and Ogg (2012)
GTS2016 Ogg, Ogg, and Gradstein (2016)

GEOSCIENTIFIC CONCEPTS

CA-TIMS Chemical abrasion-thermal ionization mass spectrometry (in U−Pb dating)
FAD First appearance datum
FOD First occurrence datum
FCT (FCs) Fish Canyon Tuff sanidine monitor standard (in Ar−Ar dating)
GPTS Geomagnetic polarity time scale
GSSP Global Boundary Stratotype Section and Point
GSSA Global Standard Stratigraphic Age (in Precambrian)
HO Highest occurrence level
HR-SIMS High-resolution secondary-ion mass spectrometry (in U−Pb dating)
ID-TIMS Isotope dilution-thermal ionization mass spectrometry (in U−Pb dating)
LAD Last appearance datum
LA-ICPMS Laser ablation-inductively coupled plasma mass Spectrometry (in U−Pb dating)
LO Lowest occurrence level
LOD Last occurrence datum
LA2004 Laskar 2004 numerical solution of orbital periodicities
LA2010 Laskar 2010 numerical solution of orbital periodicities (Laskar et al., 2011)
MMhb-1 McClure Mountain hornblende monitor standard (in Ar−Ar dating)
SL13 Sri Lanka 13 monitor zircon standard (in HR-SIMS dating)
TCs Taylor Creek Rhyolite sanidine monitor standard (in Ar−Ar dating)

SYMBOLS

ka 10^3 years ago (kilo annum)
kyr 10^3 years duration
Ma 10^6 years ago (mega annum)
Myr 10^6 years duration
Ga 10^9 years ago (giga annum)
Gyr 10^9 years duration
SI Le Système Internationale d'Unités
a annus (year)
s second

Part IV

Geologic Periods: Phanerozoic

S.C. Peng, L.E. Babcock and P. Ahlberg

Chapter 19

The Cambrian Period

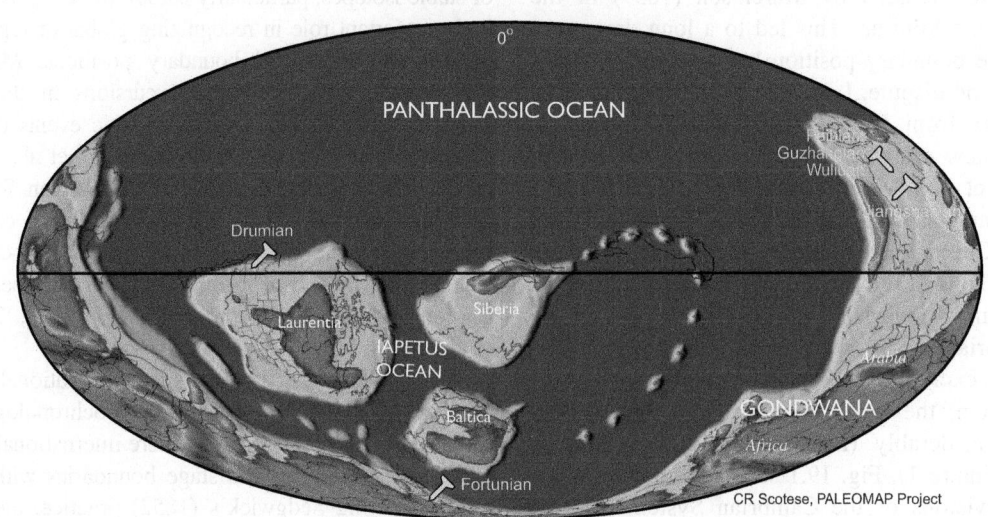

510 Ma Cambrian

CR Scotese, PALEOMAP Project

Chapter outline

19.1 History and subdivisions		**566**
19.1.1 Terreneuvian Series		573
19.1.2 Series 2 (undefined)		574
19.1.3 Miaolingian Series		575
19.1.4 Furongian Series		579
19.1.5 Regional Cambrian stage suites		585
19.2 Cambrian stratigraphy		**599**
19.2.1 Faunal provinces		599
19.2.2 Trilobite zones		599
19.2.3 Archaeocyathan zones		600
19.2.4 Small shelly fossil zones		600

19.2.5 Conodont zones		601
19.2.6 Magnetostratigraphy		601
19.2.7 Chemostratigraphy		602
19.2.8 Sequence stratigraphy		605
19.2.9 Cambrian evolutionary events		606
19.3 Cambrian time scale		**607**
19.3.1 Age of the Ediacaran–Cambrian boundary		607
19.3.2 Age of internal boundaries		609
Acknowledgments		**613**
Bibliography		**613**

Abstract

Appearance of metazoans with mineralized skeletons, "explosion" in biotic diversity and disparity, infaunalization of the substrate, occurrence of metazoan Konservat Fossil-lagerstätten, establishment of most invertebrate phyla, strong faunal provincialism, dominance of trilobites, generally warm climate but with possible glacial–interglacial cycles in the later part, opening of the Iapetus Ocean, progressive equatorial drift and separation of Laurentia, Baltica, Siberia, and Avalonia from Gondwana characterize the Cambrian Period.

Geologic Time Scale 2020. DOI: https://doi.org/10.1016/B978-0-12-824360-2.00019-X

19.1 History and subdivisions

The name "Cambrian" is derived from *Cambria*, the classical name for Wales. *Cambria* is latinized from the Welsh name *Cymru*, which refers to the Welsh people. The term "Cambrian" was first used by Adam Sedgwick (in Sedgwick and Murchison, 1835) for the "Cambrian successions" in North Wales and Cumberland (northwestern England). These successions are now known to include Neoproterozoic, Cambrian, Ordovician, and early Silurian rocks. Sedgwick divided strata of the area into three groups, Lower, Middle, and Upper Cambrian. Some strata originally defined as Cambrian were included by Murchison (1839) in the lower part of the Silurian. This led to a long drawn-out conflict on the boundary position between the two systems. To end the dispute, Lapworth (1879) excluded the disputed strata from both systems and proposed the interval as a new system, the Ordovician (Stubblefield, 1956; Cowie et al., 1972; Bassett, 1985). In 1960, the term Cambrian was officially accepted at the 21st International Geological Congress (IGC) in Copenhagen, Denmark, as the lowest system of the Paleozoic. As used today, the name Cambrian applies only to part of the "Lower Cambrian" as refined by Sedgwick in 1852. The modern usage excludes the Tremadoc and Arenig slates from the system; the lower boundary has been shifted downward considerably (Peng et al., 2006; Babcock et al., 2011: Figure 1). Fig. 19.1 shows the chronostratigraphic subdivisions of the Cambrian System adopted by the Cambrian Subcommission of the International Commission on Stratigraphy (ICS), global standard stages (ages), series (epochs), and Global Boundary Stratotype Sections and Points (GSSPs; also referred to as Global Standard Stratotype-sections and Points) ratified by the ICS and International Union of Geological Sciences (IUGS) through August, 2020. Other provisional subdivisions are identified with numbers, and potential GSSP levels are indicated. Also shown on this figure are the carbon isotope curve and its (named) excursions, a schematic sea-level curve, geomagnetic polarity reversal trends, and Cambrian regional subdivisions of China, Laurentia, and Russia (revised from Babcock et al., 2005; Peng et al., 2012a; Ogg et al., 2016).

The Cambrian marks an important phase in the history of life on earth. The system is characterized by the appearance of numerous animals (metazoans) bearing mineralized or chitinous skeletons, and by a rapid diversification of animals commonly referred to as the "Cambrian explosion." Nearly all animal phyla known from the fossil record appeared during the Cambrian Period. The biostratigraphically most useful fossil group is the trilobites, especially agnostoid trilobites, which show a remarkable evolutionary diversification, particularly in the upper half of the Cambrian. The phylogenetic

relationship between polymerid trilobites and agnostoids is unresolved, but for the sake of simplicity the agnostoids are herein treated as a specialized "trilobite" group (see for discussion Babcock et al., 2017). Inarticulate brachiopods, archaeocyaths, conodonts, acritarchs, and diverse skeletal remains referred to as "small shelly fossils" also provide good biostratigraphic control in appropriate facies. Trace fossils have been used to zone the lowermost part of the system and to identify its base. The principal regional biostratigraphic zonal schemes of the Cambrian are shown in Fig. 19.2.

Among nonbiostratigraphic correlation criteria, excursions of stable isotopes, particularly carbon ($\delta^{13}C$), play an increasingly important role in recognizing global or regional stratigraphic tie points and boundary positions. Many widely recognizable carbon isotopic excursions in the Cambrian seem to correlate with important biotic events (Peng et al., 2004a, 2012a; Zhu et al., 2006; Babcock et al., 2007, 2015, 2017). For example, the base of the Paibian Stage, which coincides with the first appearance of the cosmopolitan agnostoid trilobite *Glyptagnostus reticulatus*, is closely associated with the onset of the SPICE (Steptoean Positive Carbon isotopic Excursion; Saltzman et al., 2000; Peng et al., 2004a; Figs. 19.1 and 19.2).

For many years, there was no international agreement on standard chronostratigraphic or geochronologic subdivisions of the Cambrian nor was there international agreement on positions of series and stage boundaries within the system. Following Sedgwick's (1852) practice, the Cambrian has traditionally been divided into lower, middle, and upper parts (corresponding to series/epochs) in most parts of the world. Unfortunately, due to the absence of an internationally accepted standard, the boundary of each series (epoch) was placed at a chronostratigraphic level that varied from region to region. The placement of the base of the Middle Cambrian was especially subject to widely differing interpretations (Geyer, 1990, 1998, 2005; Geyer et al., 2000; Sundberg et al., 2016). Numerous different nomenclatures for regional stages were established throughout the years, and there was little uniformity from region to region. Sometimes, terminology even varied within individual regions or according to stratigraphic practice (Babcock et al., 2011).

Through the years, differing stratigraphic philosophies have been used for definition of series and stages. Most older definitions of series and stages were based on the unit stratotype concept (see Salvador, 1994), in which a unit is defined and characterized with reference to a type section. The lower and upper boundaries of a unit are normally specified by reference to a type section. Some of the more recently established definitions, however, have been based on the boundary stratotype concept (see Salvador, 1994), in which a point in strata is used to define the base of a series or stage, and in

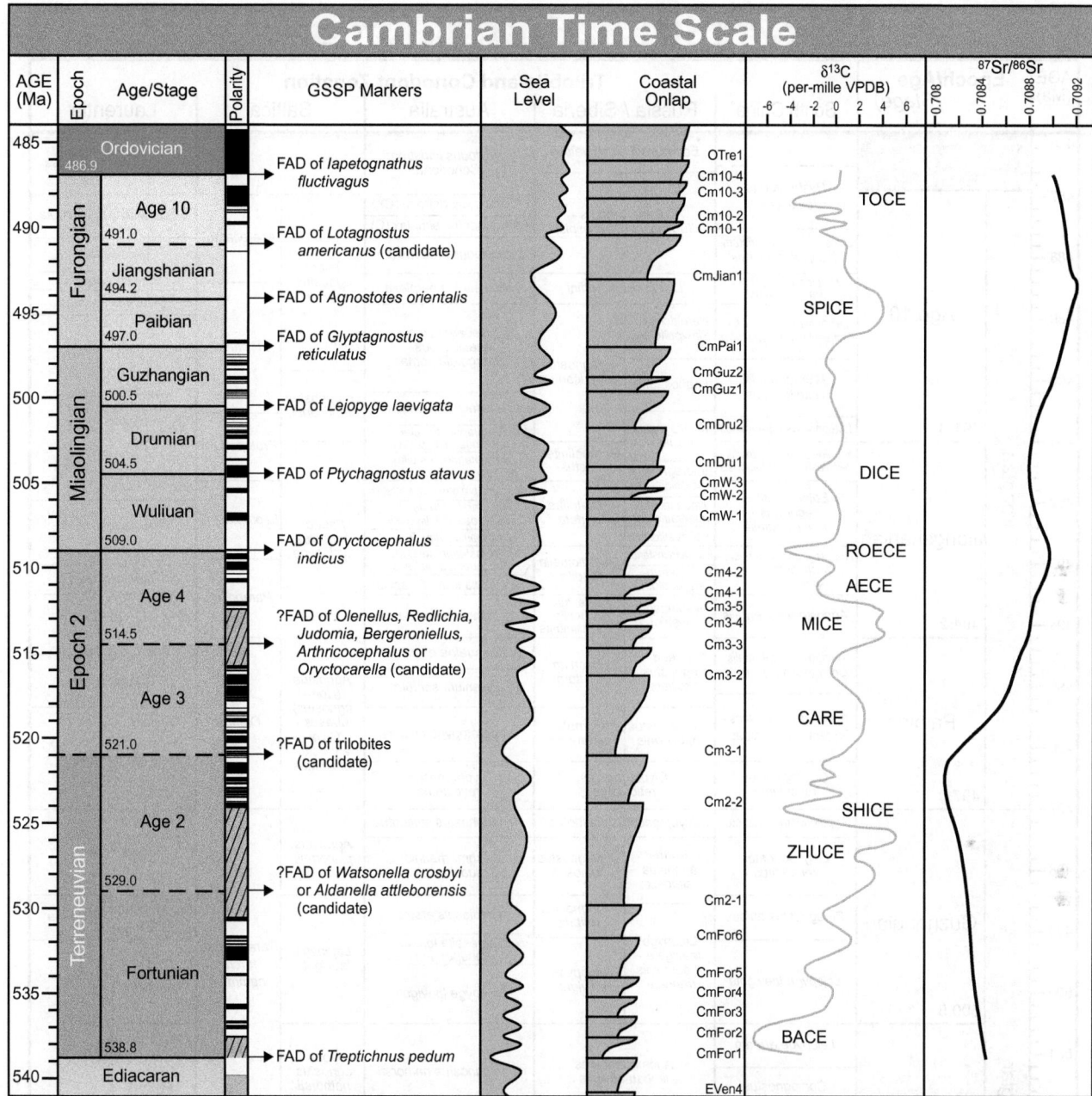

FIGURE 19.1 Cambrian overview. The main markers for the currently (as of August 2020) ratified GSSPs of Cambrian stages are the trace fossil *Treptichnus pedum* for the base of the Cambrian, FADs of cosmopolitan agnostoid trilobite taxa in late Cambrian and of the polymerid trilobite *Oryctocephalus indicus* for the base of the Wuliuan; see text for details. ("Age" is the term for the time equivalent of the rock-record "stage.") Other provisional subdivisions are identified with numbers, and potential GSSP levels for these units are indicated. Magnetic polarity scale is a composite by Peng et al. (2012a), which included a Furongian pattern modified from Kouchinsky et al. (2008) and an early Cambrian modified from a Siberian compilation by Varlamov et al. (2008a, 2008b), but most of the polarity pattern awaits verification. Schematic sea-level, coastal onlap curves and sequence nomenclature are modified from Haq and Schutter (2008) following advice of Bilal Haq (pers. comm., 2008, to J. Ogg); although Babcock et al. (2015) have a slightly different sea-level version that emphasizes that the FADs of the GSSP-marker agnostoid trilobites coincide with rapid regional coastal onlaps. See also review in (Simmons et al. (2020), Ch. 13, Phanerozoic eustasy, this book). The nomenclature of the major widespread events on the δ13Ccarb curve are modified from Zhu et al. (2006) (see text for explanations of their acronyms); see (Cramer and Jarvis (2020), Ch. 11, Carbon isotope stratigraphy, this book) for an enhanced compilation. Strontium isotope curve is modified from (McArthur et al. (2020), Ch.7, Strontium isotope stratigraphy, this book) through Wuliuan with extension to base of Cambrian from Peng et al. (2012a). The vertical scale of this diagram is standardized to match the vertical scales of the period-overview figure in all other Phanerozoic chapters. *FAD*, First appearance datum; *GSSPs*, Global Boundary Stratotype Sections and Points (also referred to as Global Standard Stratotype-sections and Points); *VPDB*, Vienna PeeDee Belemnite δ13C standard (modified from Peng et al., 2012a; Ogg et al., 2016).

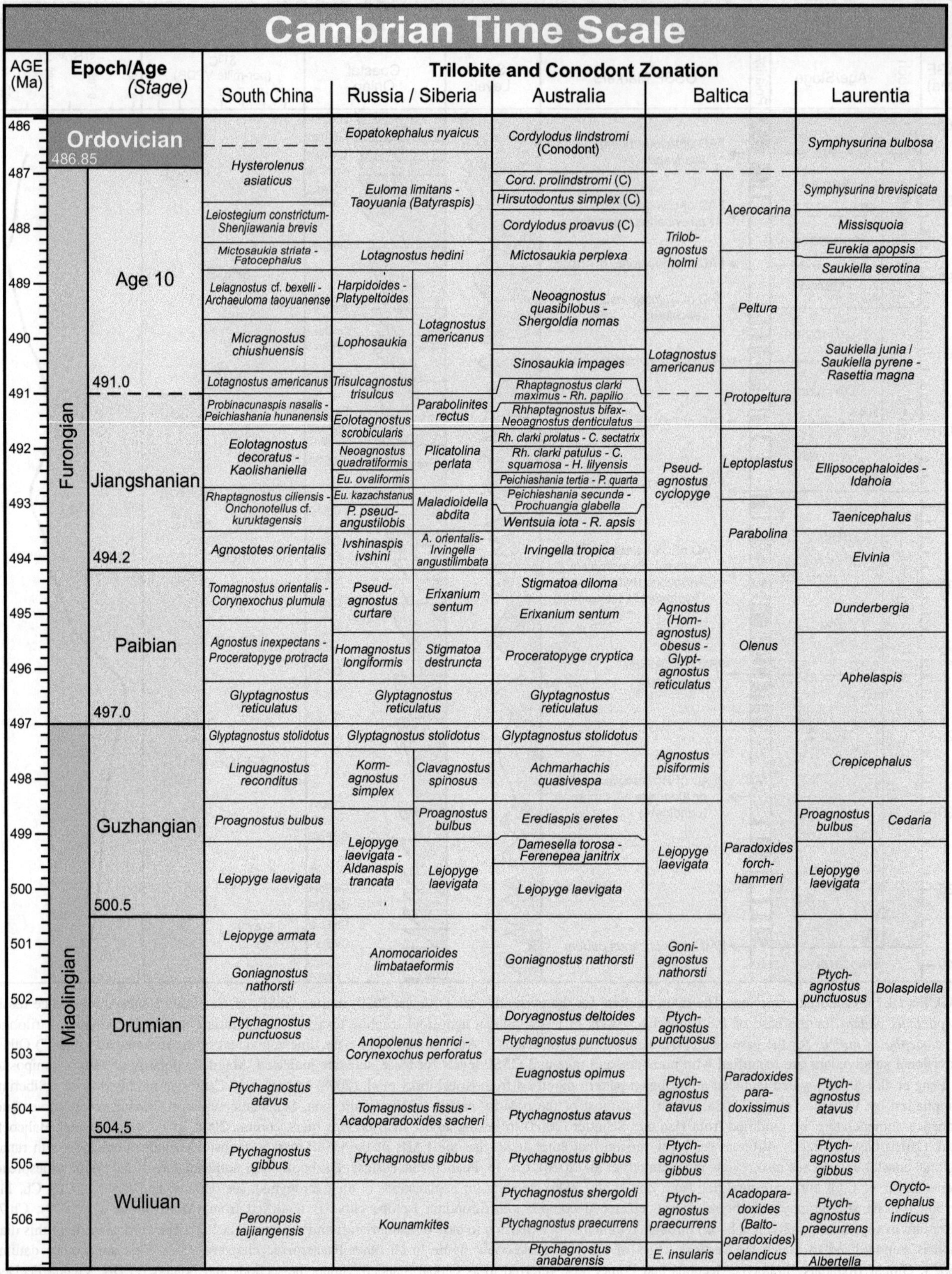

Cambrian Time Scale

AGE (Ma)	Epoch/Age (Stage)		Trilobite and Conodont Zonation				
			South China	Russia / Siberia	Australia	Baltica	Laurentia
486	Ordovician 486.85			Eopatokephalus nyaicus	Cordylodus lindstromi (Conodont)		Symphysurina bulbosa
487		Age 10	Hysterolenus asiaticus	Euloma limitans - Taoyuania (Batyraspis)	Cord. prolindstromi (C)		Symphysurina brevispicata
					Hirsutodontus simplex (C)	Acerocarina	
488			Leiostegium constrictum- Shenjiawania brevis		Cordylodus proavus (C)	Trilob- agnostus holmi	Missisquoia
			Mictosaukia striata - Fatocephalus	Lotagnostus hedini	Mictosaukia perplexa		Eurekia apopsis
							Saukiella serotina
489			Leiagnostus cf. bexelli - Archaeuloma taoyuanense	Harpidoides - Platypeltoides	Neoagnostus quasibilobus - Shergoldia nomas	Peltura	
490	Furongian		Micragnostus chiushuensis	Lophosaukia	Lotagnostus americanus		Saukiella junia / Saukiella pyrene - Rasettia magna
					Sinosaukia impages	Lotagnostus americanus	
491		491.0	Lotagnostus americanus	Trisulcagnostus trisulcus	Rhaptagnostus clarki maximus - Rh. papilio		
		Jiangshanian	Probinacunaspis nasalis - Peichiashania hunanensis	Eolotagnostus scrobicularis	Parabolinites rectus	Rhhaptagnostus bifax- Neoagnostus denticulatus	Protopeltura	
492			Eolotagnostus decoratus - Kaolishaniella	Neoagnostus quadratiformis	Plicatolina perlata	Rh. clarki prolatus - C. sectatrix	Leptoplastus	Ellipsocephaloides - Idahoia
						Rh. clarki patulus - C. squamosa - H. lilyensis	Pseud- agnostus cyclopyge	
				Eu. ovaliformis		Peichiashania tertia - P. quarta		
493			Rhaptagnostus ciliensis Onchonotellus cf. kuruktagensis	Eu. kazachstanus	Maladioidella abdita	Peichiashania secunda - Prochuangia glabella		Taenicephalus
				P. pseud- angustilobis		Wentsuia iota - R. apsis	Parabolina	
494		494.2	Agnostotes orientalis	Ivshinaspis ivshini	A. orientalis- Irvingella angustilimbata	Irvingella tropica		Elvinia
495			Tomagnostus orientalis - Corynexochus plumula	Pseud- agnostus curtare	Erixanium sentum	Stigmatoa diloma	Agnostus (Hom- agnostus) obesus - Glypt- agnostus reticulatus	Dunderbergia
		Paibian				Erixanium sentum		
496			Agnostus inexpectans - Proceratopyge protracta	Homagnostus longiformis	Stigmatoa destruncta	Proceratopyge cryptica	Olenus	Aphelaspis
497		497.0	Glyptagnostus reticulatus	Glyptagnostus reticulatus	Glyptagnostus reticulatus		
			Glyptagnostus stolidotus	Glyptagnostus stolidotus	Glyptagnostus stolidotus	Agnostus pisiformis	Crepicephalus
498		Guzhangian	Linguagnostus reconditus	Korm- agnostus simplex	Clavagnostus spinosus	Achmarhachis quasivespa		
499			Proagnostus bulbus	Proagnostus bulbus	Erediaspis eretes	Lejopyge laevigata	Paradoxides forch- hammeri	Proagnostus bulbus	Cedaria
				Lejopyge laevigata - Aldanaspis trancata	Lejopyge laevigata	Damesella torosa - Ferenepea janitrix		Lejopyge laevigata
500	Miaolingian	500.5	Lejopyge laevigata			Lejopyge laevigata		
501			Lejopyge armata	Anomocarioides limbataeformis	Goniagnostus nathorsti	Goni- agnostus nathorsti	Ptych- agnostus punctuosus	
502			Goniagnostus nathorsti					Bolaspidella
		Drumian	Ptychagnostus punctuosus	Anopolenus henrici - Corynexochus perforatus	Doryagnostus deltoides	Ptych- agnostus punctuosus		
503					Ptychagnostus punctuosus			
					Euagnostus opimus	Ptych- agnostus atavus	Paradoxides para- doxissimus	Ptych- agnostus atavus
504			Ptychagnostus atavus	Tomagnostus fissus - Acadoparadoxides sacheri	Ptychagnostus atavus			
		504.5						
505			Ptychagnostus gibbus	Ptychagnostus gibbus	Ptychagnostus gibbus	Ptych- agnostus gibbus		Ptych- agnostus gibbus
		Wuliuan			Ptychagnostus shergoldi	Ptych- agnostus praecurrens	Acadopara- doxides (Balto- paradoxides) oelandicus	Orycto- cephalus indicus
506			Peronopsis taijiangensis	Kounamkites	Ptychagnostus praecurrens		Ptych- agnostus praecurrens	
					Ptychagnostus anabarensis	E. insularis		Albertella

FIGURE 19.2 Principal regional biostratigraphic zonal schemes of the Cambrian.

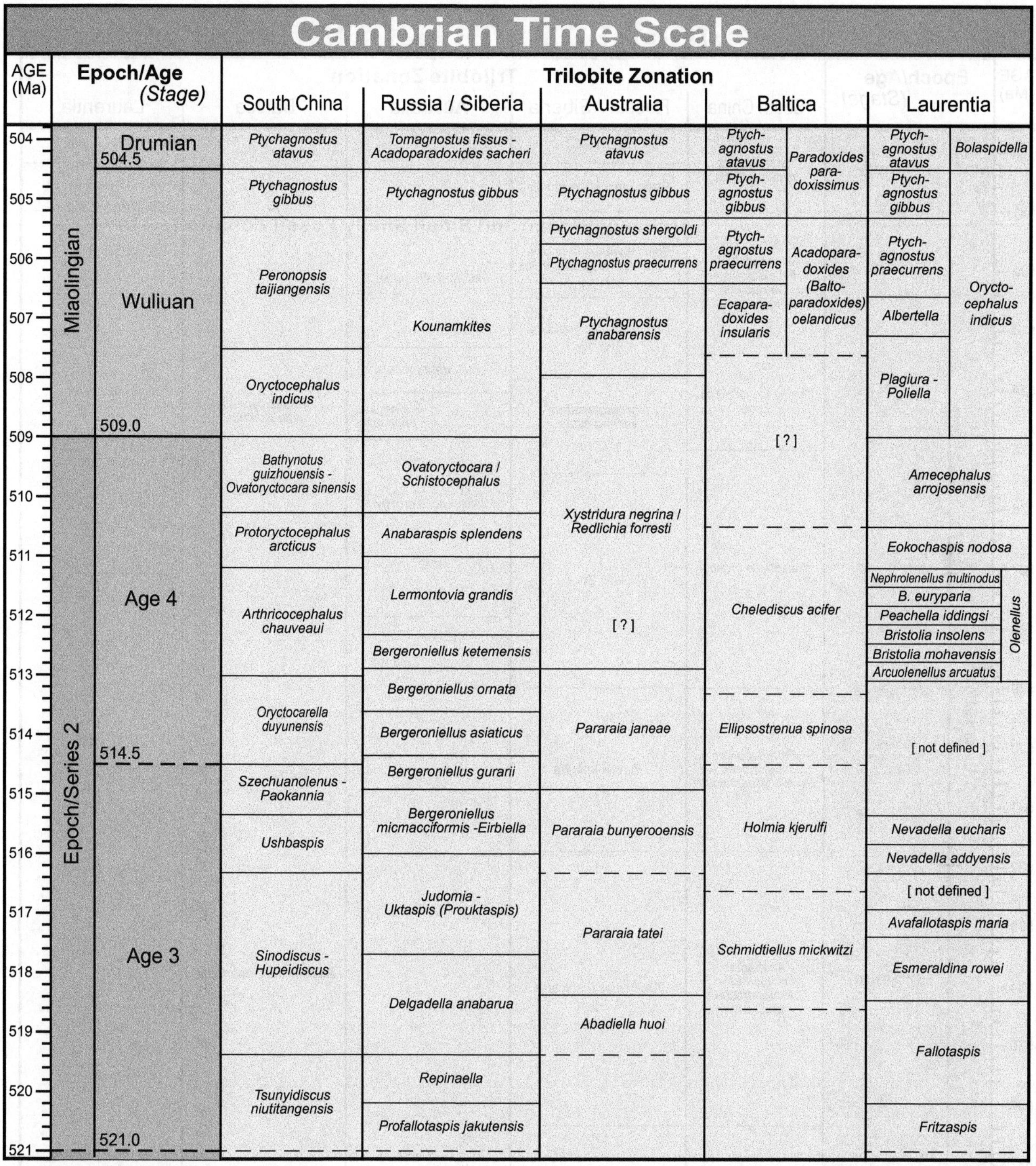

Cambrian Time Scale

AGE (Ma)	Epoch/Age (Stage)		Trilobite Zonation				
			South China	Russia / Siberia	Australia	Baltica	Laurentia
504	Miaolingian	Drumian 504.5	Ptychagnostus atavus	Tomagnostus fissus - Acadoparadoxides sacheri	Ptychagnostus atavus	Ptychagnostus atavus / Paradoxides paradoxissimus	Ptychagnostus atavus / Bolaspidella
505		Wuliuan	Ptychagnostus gibbus	Ptychagnostus gibbus	Ptychagnostus gibbus	Ptychagnostus gibbus	Ptychagnostus gibbus
506			Peronopsis taijiangensis	Kounamkites	Ptychagnostus shergoldi / Ptychagnostus praecurrens	Ptychagnostus praecurrens / Acadoparadoxides (Baltoparadoxides) oelandicus	Ptychagnostus praecurrens / Oryctocephalus indicus
507					Ptychagnostus anabarensis	Ecaparadoxides insularis	Albertella
508			Oryctocephalus indicus				Plagiura - Poliella
509	Epoch/Series 2	509.0				[?]	
510		Age 4	Bathynotus guizhouensis - Ovatoryctocara sinensis	Ovatoryctocara / Schistocephalus	Xystridura negrina / Redlichia forresti		Amecephalus arrojosensis
511			Protoryctocephalus arcticus	Anabaraspis splendens			Eokochaspis nodosa
512			Arthricocephalus chauveaui	Lermontovia grandis	[?]	Chelediscus acifer	Nephrolenellus multinodus / B. euryparia / Peachella iddingsi (Olenellus)
513				Bergeroniellus ketemensis			Bristolia insolens / Bristolia mohavensis / Arcuolenellus arcuatus
514			Oryctocarella duyunensis	Bergeroniellus ornata	Pararaia janeae	Ellipsostrenua spinosa	[not defined]
		514.5		Bergeroniellus asiaticus			
515			Szechuanolenus - Paokannia	Bergeroniellus gurarii			
516			Ushbaspis	Bergeroniellus micmacciformis -Eirbiella	Pararaia bunyerooensis	Holmia kjerulfi	Nevadella eucharis / Nevadella addyensis
517		Age 3		Judomia - Uktaspis (Prouktaspis)	Pararaia tatei	Schmidtiellus mickwitzi	[not defined] / Avafallotaspis maria
518			Sinodiscus - Hupeidiscus				Esmeraldina rowei
519				Delgadella anabarua	Abadiella huoi		
520			Tsunyidiscus niutitangensis	Repinaella			Fallotaspis
521		521.0		Profallotaspis jakutensis			Fritzaspis

FIGURE 19.2   (Continued).

which the upper limit of each chronostratigraphic unit is automatically defined by the base of the overlying chronostratigraphic unit. The recently introduced global chronostratigraphic units all have definitions based on the boundary stratotype concept.

The first steps toward achieving internationally acceptable subdivisions of the Cambrian, and toward definition of those subdivisions, were taken at the 1960 IGC, with the inception of the idea of forming a subcommission on Cambrian stratigraphy. Since its foundation in 1964, the International Subcommission on Cambrian Stratigraphy (ISCS) has worked to develop an internationally applicable, standard chronostratigraphic scale for the Cambrian System. From 1964 onward, extensive studies of Cambrian stratigraphy have been carried out throughout the world to resolve correlation problems in various facies and

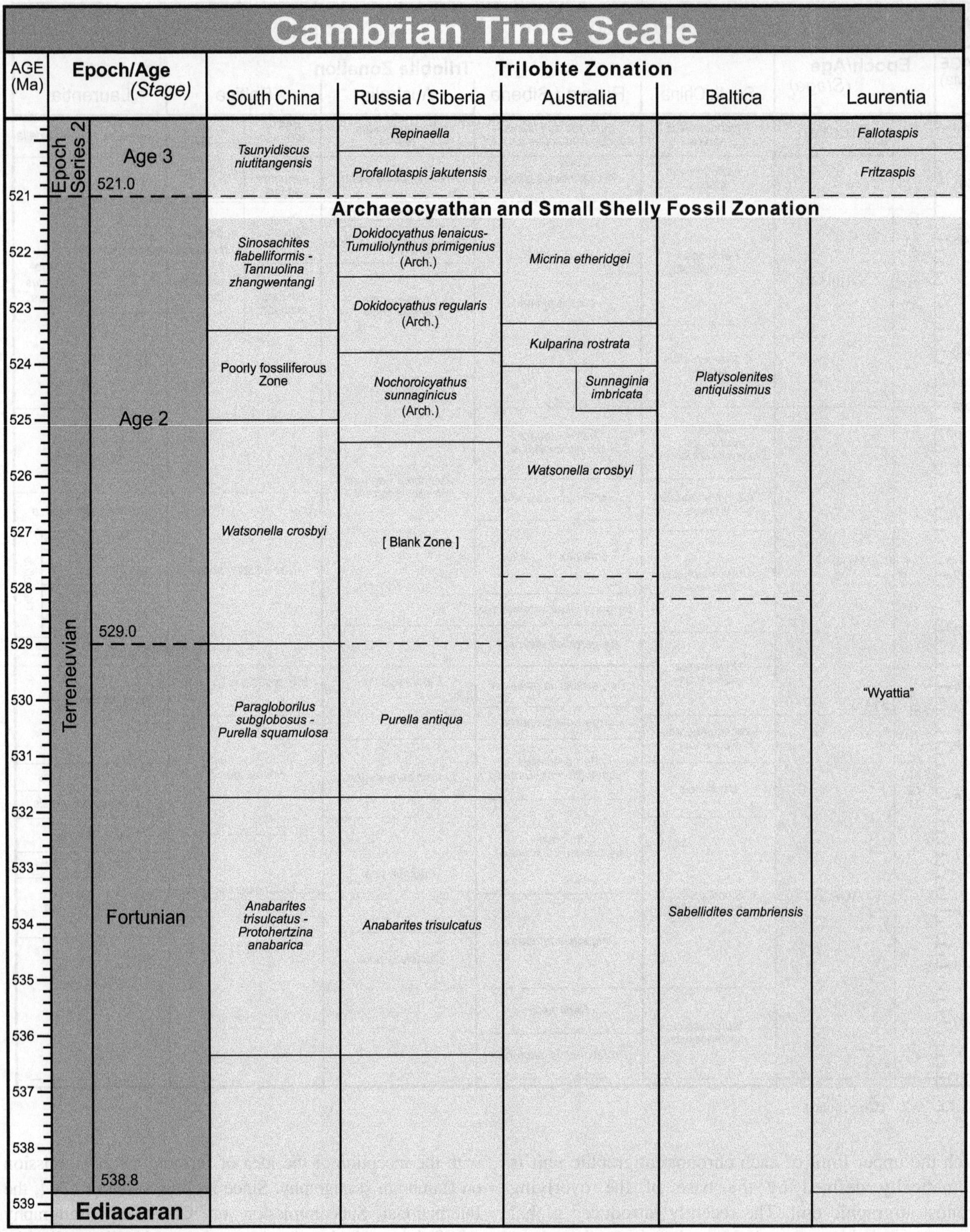

Cambrian Time Scale

AGE (Ma)	Epoch/Age (Stage)	Trilobite Zonation				
		South China	Russia / Siberia	Australia	Baltica	Laurentia
	Epoch Series 2 — Age 3 (521.0)	Tsunyidiscus niutitangensis	Repinaella			Fallotaspis
521			Profallotaspis jakutensis			Fritzaspis

Archaeocyathan and Small Shelly Fossil Zonation

AGE (Ma)	Epoch/Age (Stage)	South China	Russia / Siberia	Australia	Baltica	Laurentia
522	Age 2	Sinosachites flabelliformis - Tannuolina zhangwentangi	Dokidocyathus lenaicus-Tumuliolynthus primigenius (Arch.)	Micrina etheridgei	Platysolenites antiquissimus	
523			Dokidocyathus regularis (Arch.)			
524		Poorly fossiliferous Zone	Nochoroicyathus sunnaginicus (Arch.)	Kulparina rostrata / Sunnaginia imbricata		
525						
526		Watsonella crosbyi		Watsonella crosbyi		
527			[Blank Zone]			
528						
529	(529.0)					
530	Fortunian	Paragloborilus subglobosus - Purella squamulosa	Purella antiqua			"Wyattia"
531						
532						
533		Anabarites trisulcatus - Protohertzina anabarica			Sabellidites cambriensis	
534			Anabarites trisulcatus			
535						
536						
537						
538						
539	(538.8) Ediacaran					

(Terreneuvian labeled along left margin)

FIGURE 19.2 (Continued).

biogeographic realms, to identify the stratigraphic horizons having the best correlation potential on intracontinental and global scales and to establish GSSPs of formal chronostratigraphic units. In 1968 the ISCS decided that resolving the problem of the "base of the Cambrian System" should be one of its first tasks toward a precise definition of the system. In 1972 the first working group of the Cambrian Subcommission (called the Working Group on the Precambrian–Cambrian Boundary) was established. After two decades of study a GSSP for the base of Cambrian, the first GSSP for the system, was erected in the Fortune Head Section, eastern Newfoundland, Canada, in 1992 (Narbonne et al., 1987; Landing, 1991, 1994; Brasier et al., 1994). The boundary position is identified by a significant change in trace fossil associations (see Fig. 19.3C). This criterion supplanted the other historic criteria for marking the boundary level, namely the appearance of trilobites (Brøgger, 1886; Walcott, 1889a, 1889b, 1890a, 1890b; Wheeler, 1947), and the appearance of pretrilobitic skeletal faunas (Rozanov, 1967; Cowie, 1978). It also extended the base of the system to a level well below both trilobites and the earliest small shelly fossil (SSF) assemblages. With formal definition of the base of the Cambrian in the section at Fortune Head, Newfoundland, Canada, the system "added" a thick pretrilobite interval bearing trace fossils of Phanerozoic aspect and SSFs. Prior to ratification of the Cambrian base in 1992, this "added" section was correlated with the upper part of the Proterozoic.

The Cambrian–Ordovician boundary was defined in 1997 with ratification of the Ordovician base (Cooper and Nowlan, 1999; Cooper et al., 2001). This was followed by an acceleration in the pace of the work of the ISCS toward subdividing the Cambrian System. A detailed, region-by-region correlation chart of the Cambrian was published by Geyer et al. (2000). On this chart, 14 stratigraphic levels were recognized as having strong correlation potential. Voting members of the ISCS identified six of those levels as being recognizable on global or intercontinental scales (see GSSP levels in Fig. 19.1), and therefore potentially useful for defining global stages (Geyer and Shergold, 2000). This led to the establishment of a number of working groups charged with further, detailed study of these levels. Following further investigation, five intra-Cambrian GSSPs have so far been erected in northwestern Hunan, western Zhejiang, and eastern Guizhou, China, and the Great Basin, United States (Peng et al., 2004a, 2009, 2012b; Babcock et al., 2007; Peng and Zhao, 2018; Zhao et al., 2019). Other GSSPs will be decided upon in the near future.

Addition of a thick pretrilobitic interval to the traditional Lower Cambrian opened discussions on the possibility of dividing the system (period) into four series (epochs) (see Fig. 19.1). Important to the discussion were two significant facts:

1. The suggestion from geochronologic dating that the expanded Early Cambrian Epoch represents a duration of time that is longer than the traditional Middle and Late Cambrian combined.
2. Longstanding recognition that the traditional Early Cambrian bears a clear and important bioevent, the appearance of trilobites. This event has long been viewed as useful for defining an epoch boundary (Fig. 19.1).

Fourfold regional subdivisions of the Cambrian were proposed for Laurentia by Palmer (1998) and South China by Peng (2000a, 2000b). In 2004 it was the unanimous opinion of participants in a Cambrian Subcommision workshop held in South Korea that four series should be established for the global chronostratigraphic scale of the Cambrian System. Subsequently, the subcommission approved a subdivision of the system with one subtrilobitic series and three trilobite-dominated series (Peng and Babcock, 2005a; Babcock et al., 2005; Figs. 19.1 and 19.2).

In the current conceptual model of the Cambrian the system is further subdivided into 10 global stages (Fig. 19.1). Because of strong faunal provincialism in the earlier part of the Cambrian Period, the lower half of the Cambrian System (i.e., the lower two series) bears only a few levels that have potential for global or intercontinental correlation. For this reason, these two series may each be subdivided into two stages. More diverse faunas in the upper half of the Cambrian System enable subdivision of each series into three stages. As illustrated in Fig. 19.1, the levels used for defining stages in the upper half of the Cambrian are based on the first appearance datum (FAD) horizons of key species of agnostoid trilobites. Apart from the base of the system, no agreement has yet been reached on the criteria to be used for defining stages of the lower half of the Cambrian, but attention is now focusing on several horizons deemed to have intercontinental correlation potential.

In accordance with ICS standards, units of the new Cambrian chronostratigraphic time scale are to be defined by GSSPs. To avoid any possible confusion with regional series and stage concepts applied previously, the Cambrian Subcommission decided to introduce a set of new globally applicable names for all series (epochs) and stages (ages) as new GSSPs are established. The new terms are based on geographic features, preferably ones associated with the GSSP-bearing sections, and all are defined according to the boundary stratotype concept. By the end of 2018, six Cambrian stages (Fortunian, Wuliuan, Drumian, Guzhangian, Paibian, and Jiangshanian) and three series (Terreneuvian, Miaolingian, and Furongian) had received formal names (Fig. 19.1; Babcock et al., 2007; Peng et al., 2004a, 2009, 2012a; Landing et al., 2007; Zhao et al., 2019). Other series and stages are as yet undefined and have received provisional numerical designations (see Fig. 19.16 at the end of this chapter).

Base of the Fortunian Stage of the Cambrian System
at Fortune Head, southeastern Newfoundland, Canada

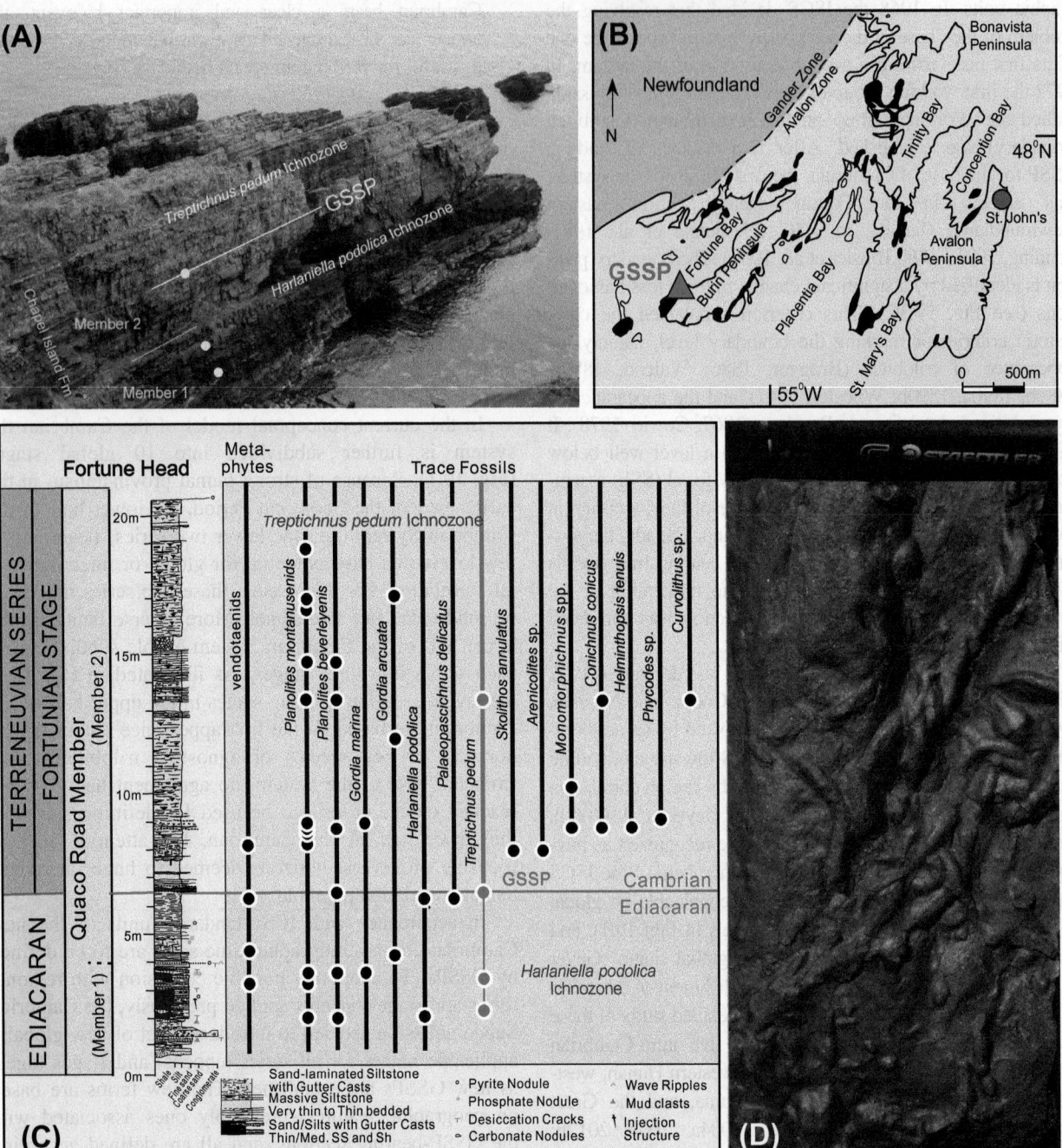

FIGURE 19.3   (A) and (B) GSSP at the base of the Phanerozoic Eonothem, Paleozoic Erathem, Cambrian System, Terreneuvian Series, and Fortunian Stage; Fortune Head section, Newfoundland, Canada. The yellow dots in (A) are occurrences of *Treptichnus pedum* in the stratotype section. Note that the two dots in Member 1 are sub-GSSP occurrences at 3.11 and 4.41 m below the GSSP. A fault, 6.5 m below the GSSP lies just beyond the bottom ridge of the image; (C) zonation of trace fossils and body fossils associated with the basal Cambrian GSSP with thick vertical range lines recognized at the time of boundary stratotype ratification (1992), and thin line for *T. pedum* subsequently reported by Gehling et al. (2001); (D) the trace fossil *T. pedum* (Seilacher, 1955), whose FAD at the time of boundary ratification coincides with the base of the Fortunian Stage. *FAD*, First appearance datum; *GSSP*, Global Boundary Stratotype Section and Point.

19.1.1 Terreneuvian Series

The GSSP for the base of the Terreneuvian Series is located in the Quaco Road Member of the Chapel Island Formation in the Fortune Head Section of the Burin Peninsula, eastern Newfoundland, Canada [(Fig. 19.3B); Landing, 1994, 1996, 2004]. The Terreneuvian Series is a fully subtrilobite succession, characterized in its lower part by complex, substrate-penetrating (i.e., Phanerozoic-type) trace fossils, and in its higher part by diverse, biomineralized (calcareous and phosphatic) or secondarily phosphatized SSFs (Landing et al., 1989).

The name "Terreneuvian" evokes "Terre Neuffve," which, prior to a spelling reform, was the formal name for the 17th century French colony in Newfoundland that essentially corresponds to the Burin Peninsula. The base of the series is conterminous with the base of the Phanerozoic Eonothem, the Paleozoic Erathem, the Cambrian System, and the Fortunian Stage. The GSSP for the Cambrian System was ratified by the ICS and the IUGS at the 29th IGC at Kyoto, Japan, in 1992 (Brasier et al., 1994; Landing, 1994). The Cambrian's lowest series and stage were not formally named at that time. Afterward, they were provisionally termed Series 1 and Stage 1 of the Cambrian System, respectively. In 2006, the Cambrian Subcommission voted to apply the name "Terreneuvian" to the lowermost series and "Fortunian" to the lowermost stage of the series. The new terms were approved by the ICS and ratified by the IUGS in 2007 (Landing et al., 2007).

The current concept is that the Terreneuvian Series comprises two stages, the Fortunian Stage and an overlying stage provisionally called Cambrian Stage 2. The Terreneuvian Series is overlain by a series provisionally termed "Cambrian Series 2." The base of Series 2 has not been defined yet, but is expected to be close to the horizon marking the first appearance of trilobites in Gondwana, Siberia, and Laurentia. That horizon and its GSSP, once they are ratified, will automatically define the top of the Terreneuvian Series.

19.1.1.1 Fortunian Stage and the base of the Cambrian System

The Fortunian Stage was named for the Fortune Head section, which contains the GSSP, on the southern Burin Peninsula, eastern Newfoundland, Canada. The Fortunian is the lowest stage of the Terreneuvian Series and the Cambrian System. The GSSP is a point that lies 2.4 m above the base of what was earlier referred to as "Member 2" of the Chapel Island Formation (Narbonne et al., 1987). This horizon is exposed in coastal cliffs low in the 440-m-thick Fortune Head section, and just above the transition to a storm-influenced facies (Narbonne et al., 1987; Landing, 1994; Brasier et al., 1994; Fig. 19.3A and C). The units earlier termed members 1 and 2 of the Chapel Island

Formation (Narbonne et al., 1987) constitute the lower part of what is now called the Quaco Road Member of the Chapel Island Formation (Landing, 1996). The GSSP coincides with the FAD of the ichnofossil *Phycodes pedum* at the time of boundary ratification. The trace fossil (Fig. 19.3D) also has been referred to by various authors as *Treptichnus pedum*, *Trichophycus pedum*, or *Manykodes pedum*. As intended, the FAD of *Treptichnus pedum* defines the base of the *Treptichnus pedum* Ichnozone, an assemblage-zone based on trace fossils. It reflects the appearance of complex sediment-disturbing behavior by multiple epifaunal and infaunal animals. Since the time of its ratification, it has been shown that the GSSP level was not placed at the FAD of *Treptichnus pedum* but at a level 4.41 m above the lowest occurrence of *Treptichnus* (Gehling et al., 2001) in the stratotype section. Jensen et al. (2000) argued that *Treptichnus* includes a range of expressions and that some traces identified as *Treptichnus* traces in the Ediacaran may represent part of the range of variation of *T. pedum*. The GSSP has been challenged as posing an ambiguous correlation level as well (see for review Babcock et al., 2014). Difficulties have been encountered in precisely correlating the horizon coinciding with the GSSP to strata on most paleocontinents, but especially to Siberia and South China (Gondwana).

Babcock et al. (2014) and the International Commission on Cambrian Stratigraphy (ISCS), in conjunction with the International Subcommission on Ediacaran Stratigraphy (ISES), proposed to reassess the Cambrian base in order to develop a well-reasoned, practical solution that will stand the test of time. A working group has been established by the ICS to refine criteria for its recognition and correlation.

Strata below the GSSP in the lower part of the Quaco Road Member include uppermost Proterozoic (Ediacaran) layers assigned to the *Harlaniella podolica* Ichnozone. The highest observed occurrence of *H. podolica* is 0.2 m below the GSSP, and the lowest observed occurrence of a low-diversity SSF assemblage is about 400 m higher in the succession (Narbonne et al., 1987; Landing et al., 1989). The lowest trilobite-bearing strata (assigned to the *Callavia broeggeri* Zone) lie approximately 1400 m above the GSSP (Landing, 1996). However, a major regional unconformity (a type-1 sequence boundary) separates the Terreneuvian Series from the overlying trilobite-bearing strata (provisional Series 2) across the Avalon paleocontinent, including areas both in eastern North America (such as eastern Newfoundland) and in the United Kingdom (Landing, 1996, 2004).

19.1.1.2 Stage 2 (undefined)

The base of the second stage of the Cambrian has not yet been defined. Likewise, a criterion for defining the boundary has not yet been determined. Few biostratigraphically defined

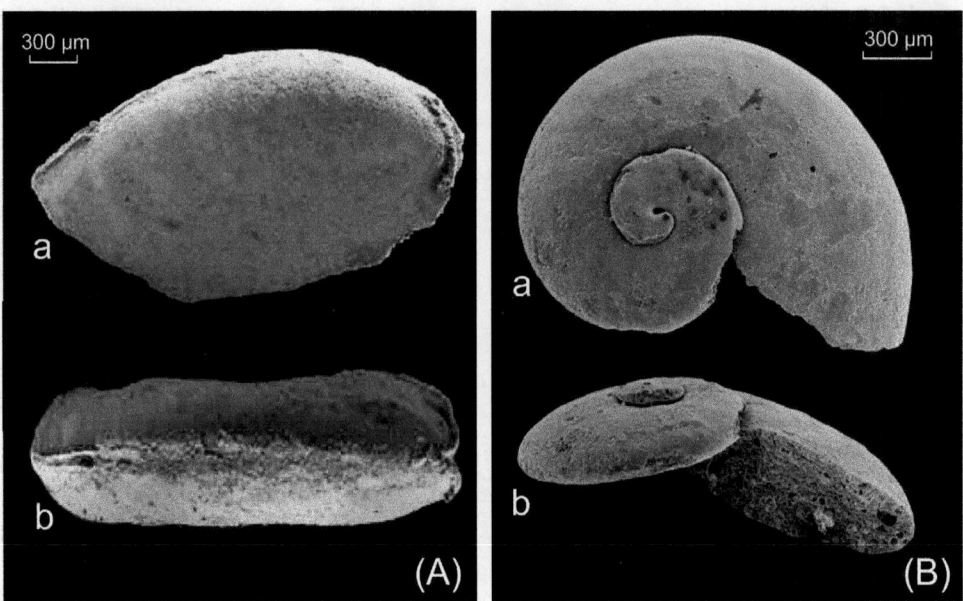

FIGURE 19.4 Key small shelly guide fossils associated with the base of provisional Stage 2. (A) *Watsonella crosbyi* Grabau, 1900, lateral (a) and dorsal (b) views; (B) *Aldanella attleborensis* (Shaler and Foerste, 1888), apical (a) and oblique apertural (b) views.

levels within the Terreneuvian Series have good potential for global or even intercontinental correlation. Ultimately, the FAD of an SSF or an archaeocyath may be used to define the base (Babcock et al., 2005). SSFs of the Terreneuvian Series have been extensively studied during the last two decades especially in Siberia, South China, Avalonia, and Laurentia. Although most seem to be highly endemic, a few species have potential in long-distance correlation (Li et al., 2007). The micromollusk (rostroconch) *Watsonella crosbyi* (Fig. 19.4A), for example, has a wide distribution in the subtrilobitic strata of Siberia, South China, Australia, Mongolia, Kyrgyzstan, Avalonian North America (eastern Newfoundland and Massachusetts), and France (Demidenko, 2001; Jacquet et al., 2017). It ranges through a narrow stratigraphic interval and occurs in both carbonate and siliciclastic successions. The mollusk *Aldanella attleborensis* (Fig. 19.4B), which usually co-occurs with *W. crosbyi*, is also widely distributed. Specimens are known from the subtrilobitic strata of Siberia, South China, Tarim, Iran, Laurentia, Avalonia, Baltica, W. Mongolia, and possibly Spain (Parkhaev and Karlova, 2011). Archaeocyaths have excellent application in biostratigraphic zonation of the lower part of the traditional Lower Cambrian of Siberia, but high faunal provincialism and endemism limit the potential of this group for interregional correlation (Zhuravlev, 1995; Kruse and Shi, 2000).

19.1.2 Series 2 (undefined)

Series 2 is the first trilobite-dominated series. It has fewer levels that can be used for intercontinental correlation

potential than the two younger series because of strong provicialism and endemism among trilobite faunas. The series is likely to be divided into two stages. The base of the series is expected to be placed at a horizon close to the first appearance of trilobites.

During Epoch 2 of the Cambrian Period the world oceans experienced an explosive diversification of metazoans, but the faunas, particularly the polymerid trilobites and archaeocyaths, were highly provincial (Kobayashi, 1971, 1972; Debrenne, 1992; Palmer and Repina, 1993; Kruse and Shi, 2000; Álvaro et al., 2013; Zhang et al., 2017). Olenellid and redlichiid trilobites, which characterize two separate faunal provinces, diversified during Epoch 2. The olenellids and most redlichiids became extinct close to the end of the epoch, a significant bioevent recognized as the earliest mass extinction of trilobites.

19.1.2.1 Stage 3 (undefined)

The base of Cambrian Stage 3, which is also the base of Cambrian Series 2, has not been defined. This boundary is expected to be marked by a significant and widely recognizable bioevent that will divide the lower half of the Cambrian as equally as possible. The FAD of trilobites, with its historic aspects, is a possibility for the boundary position, although the FAD of an SSF could eventually be selected as the primary marker of the boundary. In any case, definition of the base of provisional Stage 3 will divide the lower half of the Cambrian subequally into a subtrilobitic series and a trilobite-dominated series. Such a distinction of two series has been introduced in Laurentia as the Begadean and Waucobian series (Palmer, 1998), in South China as the Diandongian and Qiandongian series (Peng, 2000a, 2000b, 2003), in western Gondwana (Morocco and Iberia) as the Cordubian and Atlashian series (Geyer and Landing, 2004), and in Avalonia as the Placentian and Branchian series (Landing, 1992).

Precise correlation of a horizon marked solely by the FAD of trilobites will be hard to achieve because trilobites appear in strata at slightly different positions in different regions (Zhang et al., 2017). Differences in the point of appearance among different regions are due to diachroneity, and to the appearance of the earliest trilobites immediately above unconformities in some regions. If the first appearance of were accepted as a marker for the lower boundary of provisional Stage 3 (and Series 2), the most likely regions for defining the GSSP are Siberia, western Gondwana (Morocco or Iberia), and western Laurentia. The earliest occurrences of trilobites as recognized currently on other paleocontinents appear to be younger in age. The earliest trilobites include *Profallotaspis jakutensis* in Siberia, *Hupetina antiqua* in Morocco, and *Fritzaspis* in Laurentia (see Zhang et al., 2017 for a review). In Siberia, *P. jakutensis* occurs 10 m above the base of the Atdabanian Stage (Astashkin et al., 1990, 1991; Shergold et al., 1991; Varlamov et al., 2008a); in Morocco, the FAD of *H. antiqua* coincides with the base of the Issendealenian Stage (Geyer, 1990, 1995); and in western Laurentia *Fritzaspis*, *Profallotaspis?*, *Amplifallotaspis*, and *Repinaella* all occur below the lower boundary of the Montezuman Stage, and the Waucoban Series (Palmer, 1998; Hollingsworth, 2007, 2011).

To sidestep issues associated with use of a trilobite as the correlation tool coinciding with the base of Stage 3, it is possible that an SSF such as *Pelagiella subangulata* or *Microdictyon effusum* might be selected. The micromollusk *P. subangulata* has a narrow stratigraphic range and is widely distributed through Siberia, South China, Australia, Antarctica, Iran, India, continental Europe (Germany and Sardinia), the United Kingdom, Canada (Nova Scotia), and Kazakhstan. Sclerites of the armored lobopod *M. effusum* are known from Siberia, South China, Australia, the United Kingdom, Baltica, Kazakhstan, and Laurentia. At least eight species of *Microdictyon* have been named, and some may be synonyms, so a taxonomic overhaul of the genus is required.

19.1.2.2 Stage 4 (undefined)

The base of provisional Stage 4 has not been formally defined. One level, the base of the interval bearing the eodiscoid trilobites *Hebediscus, Calodiscus, Serrodiscus,* and *Triangulaspis*, referred to as the HCST band (Geyer, 2005), was proposed as a potential stage marker in the upper part of the traditional Lower Cambrian (Geyer and Shergold, 2000), but it has not received broad support from

members of the Cambrian Subcommission. Instead, the subcommission has favored placing the base of Stage 4 at a level coinciding with the FAD of a single trilobite species. Possibilities include a species of *Olenellus* (s.l.), *Redlichia* (s.l.), *Judomia, Bergeroniellus,* or *Oryctocarella* (Babcock et al., 2011; Peng et al., 2017). Such a position would be at a level roughly corresponding to the base of the Dyeran Stage of Laurentia (Palmer, 1998), the base of the Duyunian Stage of South China (Peng, 2000a, 2000b, 2003), and the base of the Botoman Stage of Siberia (Repina et al., 1964; Khomentovsky and Repina, 1965).

Stage 4 spans almost the entire range of *Olenellus* (s.l.) and *Redlichia* (s.l.), the representative forms of the "olenellid" and "redlichiid" faunal realms. Both olenellid and redlichiid trilobites became extinct near the end of the stage.

19.1.3 Miaolingian Series

The Miaolingian Series is the second trilobite-dominated series. Its name is derived from the Miaoling Mountains, which traverse the southeastern part of Guizhou Province, South China, and are inhabited largely by the Miao ethnic minority. Previously, the series was provisionally termed Series 3, and its lowermost stage, the Wuliuan Stage, was provisionally termed Stage 5. The interval that contains their conterminous base has been extensively referred to in literature as the "Lower-Middle Cambrian boundary." The traditional usage of this interval as the boundary stems from separate works by Brøgger (1878, 1882, 1886) and Walcott (1889a, 1889b, 1890a, 1890b), who developed concepts of the "Early Cambrian" and "Middle Cambrian" that were of different duration and that were applied in different faunal provinces (Robison et al., 1977; Fletcher, 2003; Geyer, 2005). Until the time of boundary ratification in 2018, there was considerable variance in the horizon interpreted as the "Lower-Middle Cambrian boundary" (Geyer and Shergold, 2000; Shergold and Cooper, 2004).

The GSSP for the conterminous base of the Miaolingian Series and the Wuliuan Stage coincides with the FAD of the oryctocephalid trilobite *Oryctocephalus indicus*. This level is close to the traditional, regional usage of the "Lower-Middle Cambrian boundary" in Laurentia and eastern Gondwana. It automatically defines the top of provisional Cambrian Series 2 and its uppermost stage (Stage 4). The top of the Miaolingian Series is defined by the base of the Furongian Series. That horizon coincides with the FAD of the agnostoid trilobite *Glyptagnostus reticulatus*, a level that lies two full biozones (the *Linguagnostus reconditus* and *Glyptagnostus stolidotus* Zones) above the traditional "Middle-Upper Cambrian Boundary" (Peng et al., 2004a). The Miaolingian Series, therefore, does not correspond to the traditional Middle Cambrian in concept, and it is of considerably greater stratigraphic thickness than the traditional Middle Cambrian.

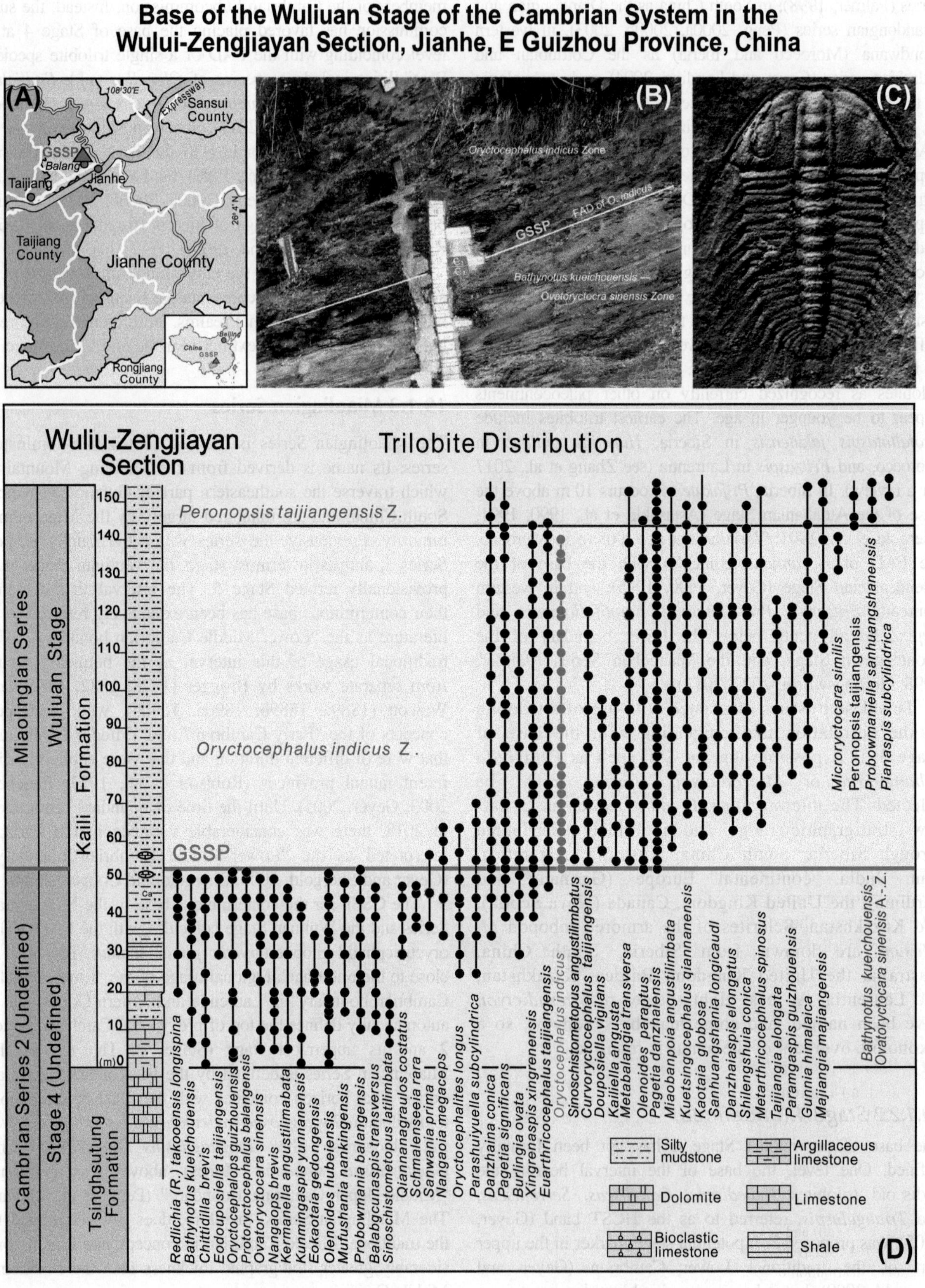

Base of the Wuliuan Stage of the Cambrian System in the Wului-Zengjiayan Section, Jianhe, E Guizhou Province, China

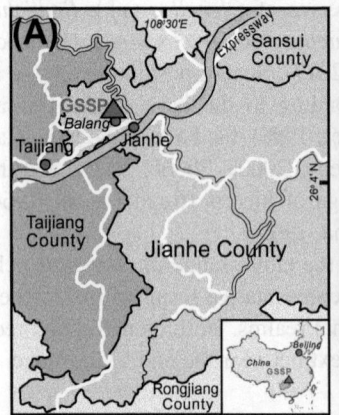

(A) (B) (C) (D)

Wuliu-Zengjiayan Section / **Trilobite Distribution**

- Miaolingian Series — Wuliuan Stage — Kaili Formation
  - *Peronopsis taijiangensis* Z.
  - *Oryctocephalus indicus* Z.
  - GSSP
- Cambrian Series 2 (Undefined) — Stage 4 (Undefined) — Tsinghsutung Formation

Trilobite taxa (listed):

Redlichia (R.) takooensis longispina, *Bathynotus kueichouensis*, *Chittidilla brevis*, *Eodouposiella taijiangensis*, *Oryctocephalops guizhouensis*, *Protoryctocephalus balangensis*, *Ovatoryctocara sinensis*, *Nangaops brevis*, *Kermanella angustilimabata*, *Kunmingaspis yunnanensis*, *Eokaotaia gedongensis*, *Olenoides hubeiensis*, *Mufushania nankingensis*, *Probowmania balangensis*, *Balabgcunaspis transversus*, *Sinoschistometopus latilimbata*, *Qiannanagraulos costatus*, *Schmalenseeia rara*, *Sanwania prima*, *Nangaoia megaceps*, *Oryctocephalites longus*, *Parashuiyuella subcylindrica*, *Danzhaina conica*, *Pagetia significans*, *Burlingia ovata*, *Xingrenaspis xingrenensis*, *Euarthricocephalus taijiangensis*, *Oryctocephalus indicus*, *Sinoschistometopus angustilimbatus*, *Curvoryctocephalus taijiangensis*, *Douposiella vigilans*, *Kailiella angusta*, *Metabalangia transversa*, *Olenoides paraptus*, *Pagetia danzhaiensis*, *Miaobanpoia angustilimbata*, *Kuetsingocephalus qiannanensis*, *Kaotaia globosa*, *Sanhuangshania longicaudl*, *Danzhaiaspis elongatus*, *Shilengshuia conica*, *Metarthricocephalus spinosus*, *Taijiangia elongata*, *Paramgaspis guizhouensis*, *Gunnia himalaica*, *Majiangia majiangensis*, *Microryctocara similis*, *Peronopsis taijiangensis*, *Probowmaniella sanhuangshanensis*, *Pianaspis subcylindrica*, *Bathynotus kueichouensis - Ovatoryctocara sinensis A.-Z.*

Legend:
- Silty mudstone
- Argillaceous limestone
- Dolomite
- Limestone
- Bioclastic limestone
- Shale

The Miaolingian Series is divided into three stages, all of which have been formally defined. The Miaolingian Epoch was a time in which trilobite faunas recovered from the mass extinction near the end of Epoch 2 (Álvaro et al., 2013). Trilobite diversity increased dramatically, so there is a major change in faunal composition on Cambrian continents. The series is characterized by the presence of highly diversified metazoans and the appearance of agnostoids in the lowermost stage of the series; by the predominance of widespread ptychagnostid and diplagnostid agnostoids in the succession, and by endemic paradoxidid and ptychopariid polymerids. In addition, oryctocephalid and bathynotid polymerids predominate in the lower part of the series, whereas anomocarid and damesellid polymerids are more common in the upper part of the series. Ceratopygid polymerids appear in the uppermost stage of the series, and the damesellids became extinct near the end of the series. The extinction of damesellids is regarded as an important bioevent in eastern Gondwana.

19.1.3.1 Wuliuan Stage

The Wuliuan Stage is the lowermost stage of the Miaolingian Series. The conterminous base of the Wuliuan Stage and Miaolingian Series is defined by a GSSP coinciding with the widely distributed oryctocephalid trilobite *Oryctocephalus indicus* (Fig. 19.5C and D). The stage name is derived from Wuliu, a small hill near Balang Village, Jianhe County, about 2.5 km northwest of Jianhe township, eastern Guizhou, China. The GSSP, which was ratified in 2018, is 58.2 m above the base of the Kaili Formation in the Wuliu–Zengjiayan section (Fig. 19.5B and D). The stratotype section is exposed in natural outcrops situated along a ridge between Wuliu and Zengjiayan, which is about 0.5 km north of the Balang Village (Zhao et al., 2012, 2019).

Oryctocephalus indicus has an intercontinental distribution, having been reported from the Indian Himalaya, northeastern Siberia, the Great Basin (United States), North Greenland, North Korea, and South China (e.g., Reed, 1910; Saito, 1934; Jell and Hughes, 1997; Sundberg and McCollum, 1997, 2003; Yuan et al., 1997, 2002; Sundberg et al., 1999; Geyer and Shergold, 2000; Peng et al., 2000b, 2012a; Zhao et al., 2001, 2004; Korovnikov, 2001; Fletcher, 2007; Shabanov et al., 2008; Geyer and Peel, 2011). In South China, Laurentia, and the Indian Himalaya (Zhao et al., 2001; Sundberg and McCollum, 2003; Hughes, 2016; Singh et al., 2016, 2017), *O. indicus* has been in use as zonal

fossil. In these areas and elsewhere, the base of the Wuliuan Stage can be closely constrained by additional stratigraphic criteria. In South China, the western United States, Australia, and Siberia, the first appearance of *O. indicus* lies slightly above the last occurrence datum (LAD) of the cosmopolitan polymerid trilobite *Bathynotus*, and above the LAD of olenellid or redlichiid trilobites. Another intercontinentally distributed oryctocephalid trilobite, *Ovatoryctocara granulata* (see Fletcher, 2003; Geyer, 2005; Sundberg et al., 2016), helps to constrain the base of the Wuliuan Stage. This species first appears slightly below the base of the Wuliuan Stage in Siberia, eastern and western Gondwana, Avalonia (Newfoundland), and North Greenland. With the aid of Siberian taxa associated with *Acadoparadoxides* and *Eccaparadoxides*, the base of Wuliuan Stage can be recognized in the Mediterranean region.

The base of the Wuliuan Stage nearly coincides with a high-amplitude negative shift in carbon isotopic (δ^{13}C) values (c. −4‰). This excursion is referred to as the ROECE (Redlichiid−Olenellid Extinction Carbon isotope Excursion) and is one of the largest negative carbon isotope excursions known from the Cambrian (Zhu et al., 2006; Fig. 19.1 and Section 19.2.7). The ROECE begins just below the traditional "Lower-Middle Cambrian boundary" as recognized in Laurentia and suggests the onset of major paleoceanographic and paleoclimatic changes associated with the extinction of olenellid trilobites in Laurentia and redlichiid trilobites in Gondwana (Montañez et al., 2000; Zhao et al., 2014; Faggetter et al., 2017).

19.1.3.2 Drumian Stage

The Drumian Stage is the middle stage of the Miaolingian Series. The name, which was ratified in 2006, is derived from the Drum Mountains in northern Millard County, western Utah, United States (Fig. 19.6C). The base of the stage is defined by a GSSP coinciding with the FAD of a cosmopolitan agnostoid trilobite *Ptychagnostus atavus* (Fig. 19.6B and D). The GSSP is located 62 m above the base of the Wheeler Formation in the Stratotype Ridge section of the Drum Mountains (Babcock et al., 2004, 2007; Fig. 19.6A and D). *P. atavus* has been identified from nearly all major Cambrian regions of the world (e.g., Robison et al., 1977; Rowell et al., 1982; Robison, 1999; Peng and Robison, 2000; Geyer and Shergold, 2000; Shergold and Geyer, 2003; Babcock et al., 2004, 2007; Ahlberg et al., 2007) and, even long before definition of the boundary level, had been in use as a zonal guide fossil in deposits of Baltica, Gondwana, Kazakhstania, and

---

FIGURE 19.5 (A) and (B) GSSP of the Miaolingian Series and Wuliuan Stage in the Wuliu–Zengjiayan section, Guizhou, China; (C) *Oryctocephalus indicus* (Reed, 1910), an oryctocephalid (polymerid) trilobite whose FAD coincides with the base of the Wuliuan Stage; (D) Stratigraphic distribution of trilobites close to the base of the Miaolingian Series and the Wuliuan Stage in the Wuliu–Zengjiayan section, Guizhou, China. *FAD*, First appearance datum; *GSSP*, Global Boundary Stratotype Section and Point (modified from Zhao et al., 2019).

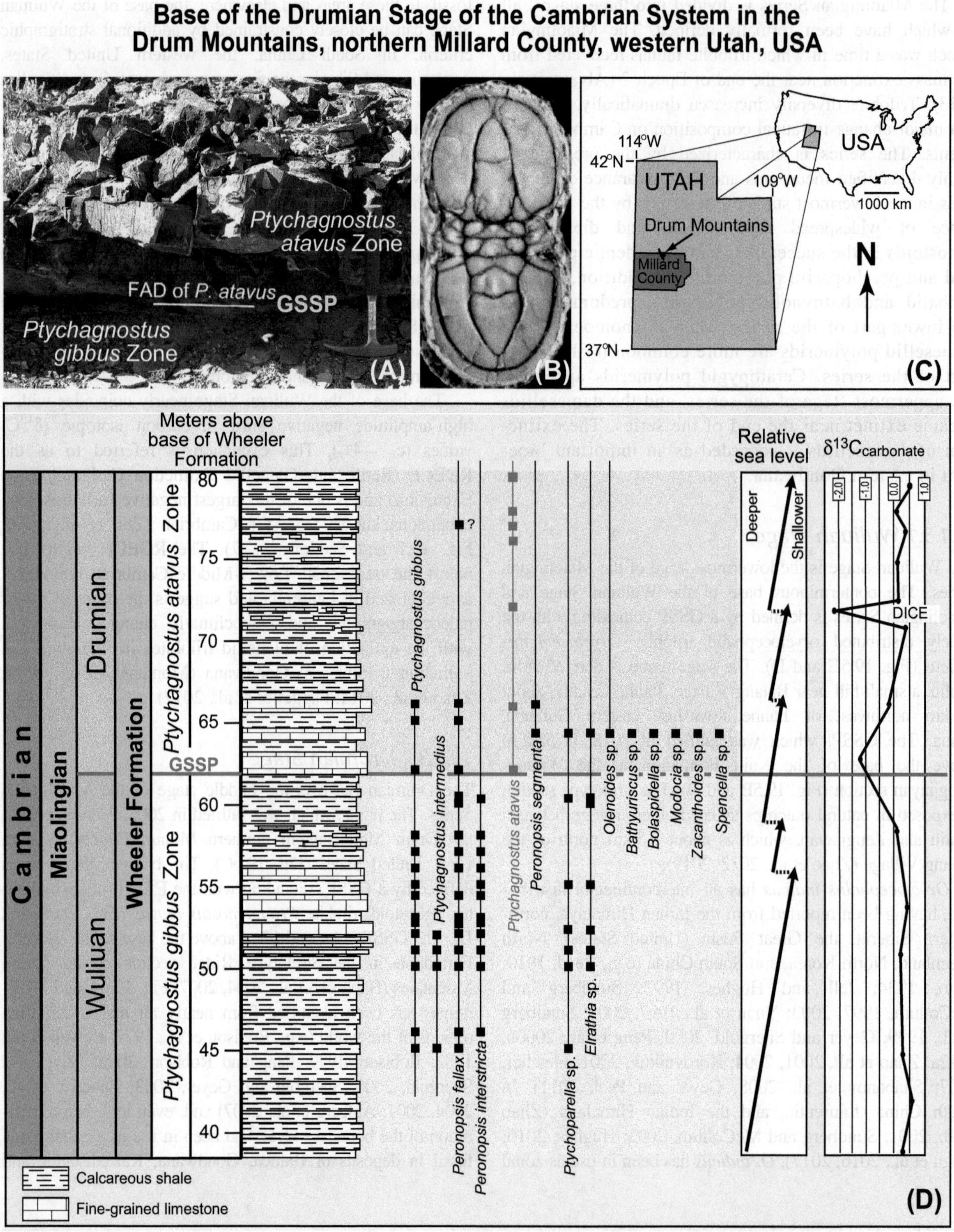

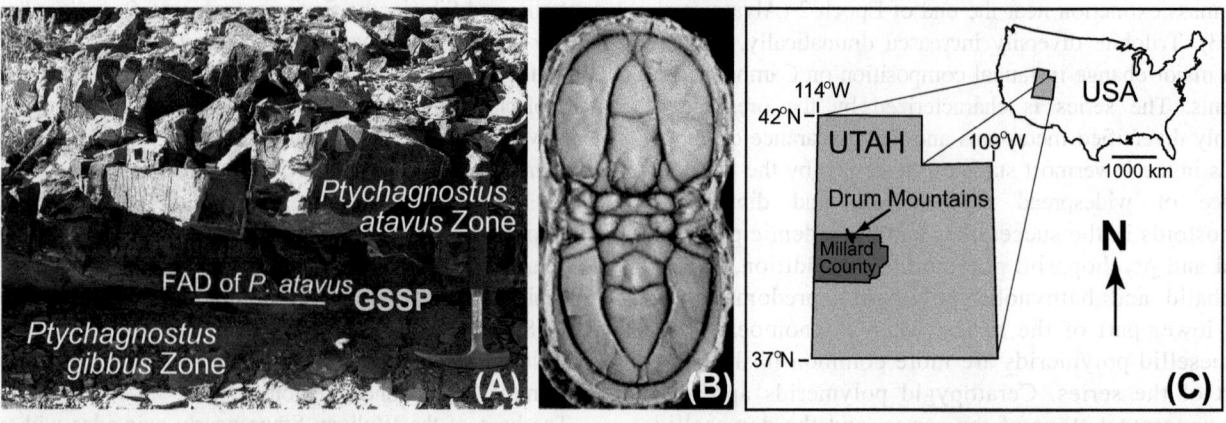

Base of the Drumian Stage of the Cambrian System in the Drum Mountains, northern Millard County, western Utah, USA

FIGURE 19.6 (A) and (C) Drumian Stage GSSP in the Stratotype Ridge section, Drum Mountains, Utah, United States; (B) *Ptychagnostus atavus* (Tullberg, 1880), an agnostoid trilobite whose FAD coincides with the base of the Drumian Stage; (D) Stratigraphic distribution of trilobites close to the base of the Drumian Stage in the Stratotype Ridge section, Utah, United States (redrawn after Babcock et al., 2007). *FAD*, First appearance datum; *GSSP*, Global Boundary Stratotype Section and Point.

Laurentia (e.g., Westergård, 1946; Robison, 1976, 1984; Öpik, 1979; Ergaliev, 1980; Geyer and Shergold, 2000; Peng and Robison, 2000). In addition to the FAD of *P. atavus*, two other related, cosmopolitan agnostoid species help constrain the base of the Drumian Stage. The first appearance of *P. atavus* always succeeds the first appearance of *Ptychagnostus gibbus*, and it precedes the first appearance of *Ptychagnostus punctuosus*. Both *P. gibbus* and *P. punctuosus* have been widely used as zonal guide fossils (Geyer and Shergold, 2000).

The base of the Drumian Stage can be recognized on a global scale not only by the FAD of *Ptychagnostus atavus* but also by significant changes in polymerid trilobites faunas, conodont faunas, and by a carbon isotope excursion. The position coincides closely with the bases of two biozones based on polymerid trilobites: the *Bolaspidella* Zone of Laurentia (Robison, 1976; Palmer, 1998, 1999; Babcock et al., 2004, 2007) and the *Dorypyge richthofeni* Zone of South China (Peng et al., 2004b). It also corresponds closely to a turnover in conodont faunas. The base of the Drumian Stage occurs just below the base of the *Gapparodus bisulcatus−Westergaardodina brevidens* Assemblage-Zone (Dong and Bergström, 2001a, 2001b). The base of the stage corresponds to the onset of a medium-scale negative carbon excursion (Babcock et al., 2007; Brasier and Sukhov, 1998), referred to as the DICE (DrumIan Carbon isotope Excursion; Zhu et al., 2006; Figs. 19.1 and 19.6 and see also Section 19.2.7).

19.1.3.3 Guzhangian Stage

The Guzhangian Stage is the third and last stage of the Miaolingian Series. The name is derived from Guzhang County in the Wuling Mountains of northwestern Hunan Province, China. The base of the Guzhangian Stage is defined by a GSSP that coincides with the FAD of the cosmopolitan agnostoid trilobite *Lejopyge laevigata* (Fig. 19.7C and D). The GSSP, which was ratified in 2007, is 121.3 m above the base of the Huaqiao Formation in the Luoyixi section, exposed along a roadcut on the south bank of the Youshui River, 4 km northwest of the town of Luoyixi (Peng et al., 2009; Fig. 19.7A).

Lejopyge laevigata, which is the primary correlation tool for the base of the Guzhangian Stage, has been recognized from all Cambrian paleocontinents (e.g., Westergård, 1946; Cowie et al., 1972; Robison, 1976, 1984; Öpik, 1979; Geyer and Shergold, 2000; Peng and Robison, 2000; Axheimer et al., 2006; Peng et al., 2006). The species has been used as a zonal guide fossil in Baltica, Gondwana, Kazakhstania, Siberia, Laurentia, and eastern Avalonia. In addition to the FAD of *L. laevigata*, two other congeneric species, *Lejopyge armata* and *Lejopyge calva*, can be used to help constrain the position of the Guzhangian base. *L. laevigata* consistently appears slightly above the first appearance of either *L. armata*

(in Australia, South China, North China, Kazakhstan, Siberia, and Sweden) or *L. calva* (in Laurentia).

The Guzhangian GSSP can be tightly constrained by stratigraphic criteria other than just the ranges of *Lejopyge* species. The Guzhangian base is above the LAD of the agnostoid *Goniagnostus nathorsti*, and below the FAD of the agnostoid *Proagnostus bulbus*. The first appearance of *L. laevigata* closely corresponds to the first appearance of damesellid trilobites (e.g., *Palaeodotes, Parablackwelderia, Blackwelderia*) in South China and Kazakhstan (Ergaliev, 1980; Ergaliev and Ergaliev, 2004; Peng et al., 2004b, 2006). It also approximates the position of a faunal change associated with the base of the *Paradoxides forchhammeri* Zone in western Avalonia (Geyer and Shergold, 2000).

The range of *Lejopyge laevigata* has long been used regionally as a major stratigraphic marker. The FAD of *L. laevigata* defines the base of the Ayusokkanian Stage in Kazakhstan, the base of the Boomerangian Stage in Australia (Öpik, 1967; Geyer and Shergold, 2000), and the base of the *Aldanaspis* Zone in Siberia (Egorova et al., 1982). By using the first appearance of *L. laevigata*, rather than its local abundance, the base of the Scandinavian *L. laevigata* Zone has been moved downward to the base of the *Solenopleura? brachymetopa* Zone (Axheimer et al., 2006). This change makes the base of the Guzhangian Stage identical with the base of the revised *L. laevigata* Zone in Scandinavia (a zone that now embraces the traditional *S.? brachymetopa* Zone; Ahlberg et al., 2009).

The Guzhangian Stage embraces the interval of the traditional "Middle-Upper Cambrian boundary." In Sweden, the traditional "Upper Cambrian" was marked by the base of the *Agnostus pisiformis* Zone (Westergård, 1946). Peng and Robison (2000) stated that the "Middle-Upper Cambrian boundary" in Sweden is commonly drawn at a position marking the local abundance of *A. pisiformis* rather than its first appearance. Recently, *Linguagnostus reconditus* was reported from the *A. pisiformis* Zone in Sweden, and it is associated with abundant *A. pisiformis* (Ahlberg and Ahlgren, 1996; Ahlberg, 2003). This suggests that the traditional "Middle-Upper Cambrian boundary" of Sweden is close to, if not identical with, the base of the *L. reconditus* Zone. The *L. reconditus* Zone is the third of four agnostoid zones recognized within Guzhangian Stage strata of South China.

19.1.4 Furongian Series

The Furongian Series is the uppermost series of the Cambrian System. Its name is derived from Furong, which means lotus, in reference to Hunan Province, the Lotus State of China. The GSSP for the conterminous base of the Furongian Series and its lowermost stage, the Paibian Stage, is located in the Wuling Mountains of northwestern Hunan, China. The GSSP coincides with the FAD of *Glyptagnostus reticulatus*, a level that is stratigraphically much higher than

Base of the Guzhangian Stage of the Cambrian System in the Luoyixi Section in the Wuling Mountains, NW Hunan Province, China

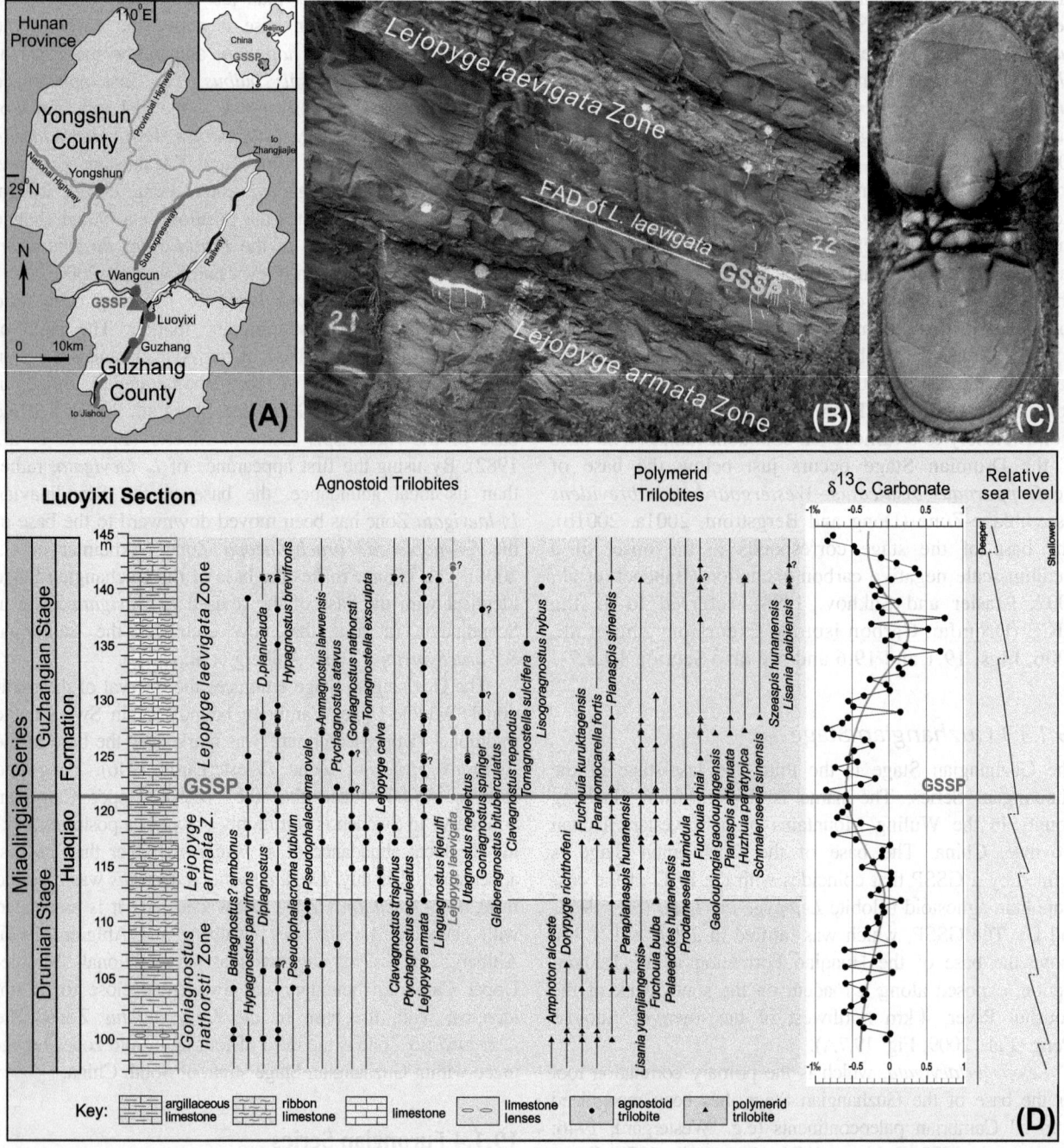

FIGURE 19.7   (A) and (B) Guzhangian Stage GSSP in the Luoyixi section, Hunan, China; (C) *Lejopyge laevigata* (Linnarsson, 1869), an agnostoid trilobite whose FAD coincides with the base of the Guzhangian Stage; (D) Stratigraphic distribution of trilobites close to the base of the Guzhangian Stage in the Luoyixi section, Hunan, China (redrawn after Peng et al., 2009). *FAD*, First appearance datum; *GSSP*, Global Boundary Stratotype Section and Point.

the traditional "Middle-Upper Cambrian boundary" as recognized earlier in Sweden. Definition of a GSSP for the base of the Furongian Series resulted in an "Upper Cambrian" that was restricted in length compared to that of traditional usage.

This is because the *A. pisiformis* Zone of Sweden, which corresponds to two biozones as used in South China, was excluded (Ahlberg, 2003; Axheimer et al., 2006; see also preceding discussion of the Guzhangian Stage).

The upper boundary of the Furongian Series is automatically defined by the base of the Tremadocian Series (Lower Ordovician). The basal Tremadocian GSSP coincides with the FAD of the conodont *Iapetognathus fluctivagus*. Currently, the majority opinion within the Cambrian Subcommission is that the Furongian should be subdivided into three subequal stages. The lowermost stage (Paibian) and the middle stage (Jiangshanian) have already been ratified. The base of the upper stage is expected to coincide with, or be close to, the FAD of an agnostoid, *Lotagnostus americanus* (Babcock et al., 2005).

The Furongian Epoch marks a time of great faunal turnover among polymerid trilobites. Stepwise extinction of polymerids in Gondwana and Laurentia at the end of the Miaolingian Epoch was followed by recovery during the early part of the Furongian. In eastern Gondwana, damesellid trilobites were replaced by leiostegiid (e.g., *Chuangia* and *Prochuangia*) and ceratopygid trilobites. In Laurentia, the Furongian Series corresponds to the Pterocephaliid Biomere, and strata are zoned according to the record of evolution in pterocephaliid, elviniid, and saukiid polymerids. In Baltica, evolutionary changes in olenid trilobites allow for fine regional zonation of the series. Agnostid and pseudagnostid agnostoids are important for regional and global subdivision of the series.

19.1.4.1 Paibian Stage

The base of the Paibian Stage, and the base of the Furongian Series, is defined by a GSSP coinciding with the first appearance of the cosmopolitan agnostoid trilobite *Glyptagnostus reticulatus* (Fig. 19.8B and D). The stage is named for Paibi, a village in Huayuan County, about 35 km west of Jishou, northwestern Hunan, China. The GSSP, which was ratified in 2003, is 369 m above the base of the Huaqiao Formation along a south-facing hill in the Paibi Section (Fig. 19.8A and D).

The FAD of *G. reticulatus* is one of the most widely recognizable stratigraphic horizons in the Cambrian. Even before its selection as the criterion for marking the base of the Paibian Stage and Furongian Series, *G. reticulatus* was used as a zonal guide fossil in Siberia, Kazakhstan, South China, Australia, and Laurentia. The interval containing the FAD of *Glyptagnostus reticulatus* marks a time of significant faunal change, and that change has been formalized in various regional stratigraphic schemes. The position corresponds to the base of the Pterocephaliid Biomere or the base of the Steptoean Stage in Laurentia, the base of the Idamean Stage in Australia, the base of the Sakian Stage in Kazakhstan and in Siberia (previously the base of the Maduan Horizon in Siberia) (Ergaliev and Ergaliev, 2004; Pegel et al., 2016; Rozanov, 2016), and the base of the Furongian Series in Scandinavia (Terfelt et al., 2008). The level is also near the base of a large positive shift in $\delta^{13}C$ values, referred to as the SPICE (Saltzman et al., 2000; Fig. 19.1 and see also Section 19.2.7).

19.1.4.2 Jiangshanian Stage

The base of the Jiangshanian Stage is defined by a GSSP coinciding with the first appearance of the cosmopolitan agnostoid trilobite *Agnostotes orientalis* (Fig. 19.9C and D). The stage is named for Jiangshan County, Zhejiang Province, China. The GSSP, which was ratified in 2011, is 108.12 m above the base of the Huayansi Formation in the Duibian B section (Fig. 19.9A and D). The stratotype section is exposed in natural outcrops situated at the base of Dadoushan Hill, west of Duibian Village (Peng et al., 2012b).

The holotype of *A. orientalis* (Kobayashi, 1935) is poorly preserved, and because the species was poorly characterized originally, a number of junior synonyms have been proposed. In some regions, the species is better known by junior synonyms such as *Pseudoglyptagnostus clavatus*, *Agnostotes (Pseudoglyptagnostus) clavatus*, *Agnostotes clavata*, and *Glyptagnostotes elegans*. Currently the species is recognized from South China (Hunan and Zhejiang), South Korea, Siberia (Kharaulakh Ridge, northeastern Siberian Platform and the Chopko River of the Norilsk Region, northwestern Siberian Platform), southern Kazakhstan (Malyi Karatau), and Laurentia (Nevada, United States and Mackenzie Mountains and southeastern British Columbia, Canada).

In many regions *A. orientalis* co-occurs with the polymerid trilobite *Irvingella* (Lazarenko, 1966; Öpik, 1967; Ergaliev, 1980; Peng, 1992; Pratt, 1992; Chatterton and Ludvigsen, 1998; Choi, 2004; Hong et al., 2003; Varlamov et al., 2005; Ergaliev and Ergaliev, 2008; Varlamov and Rozova, 2009; Peng et al., 2012b; Chatterton and Gibb, 2016; Pegel et al., 2016), and both trilobites can be used to constrain the base of the Jiangshanian Stage. Together, the two trilobites have been used as zonal guide fossils in South China, northeastern Siberia, northwestern Siberia, and South Korea. *Irvingella*, however, has a wider paleogeographic distribution (Geyer and Shergold, 2000), and it allows close correlation into Australia, Baltica, Avalonia, eastern, and western Laurentia, Argentina, and probably Antarctica. In South China and Canada, *A. orientalis* and *Irvingella angustilimbata* make their first appearances at the same stratigraphic level (Peng, 1992; Pratt, 1992; Peng et al., 2012b). This is true in the Duibian B section, which contains the GSSP. The horizon corresponds to the base of the Taoyuanian Stage as used previously in South China (Peng and Babcock, 2001, 2008; Shergold and Cooper, 2004) and the base of the *Proceratopyge rectispinata* Zone in the Mackenzie Mountains, Canada (Pratt, 1992). The base of the *P. rectispinata* Zone lies somewhat below the base of the Sunwaptan Stage of Laurentia, but above the base of the *Elvinia* Zone, the uppermost zone of the stage. The FAD of *A. orientalis* corresponds closely to the base of the Iverian Stage in Australia, the base of the *Parabolina spinulosa* Zone of Sweden and England, the base of the *Pseudagnostus vastulus–Irvingella tropica* Zone of Kazakhstan (Ergaliev

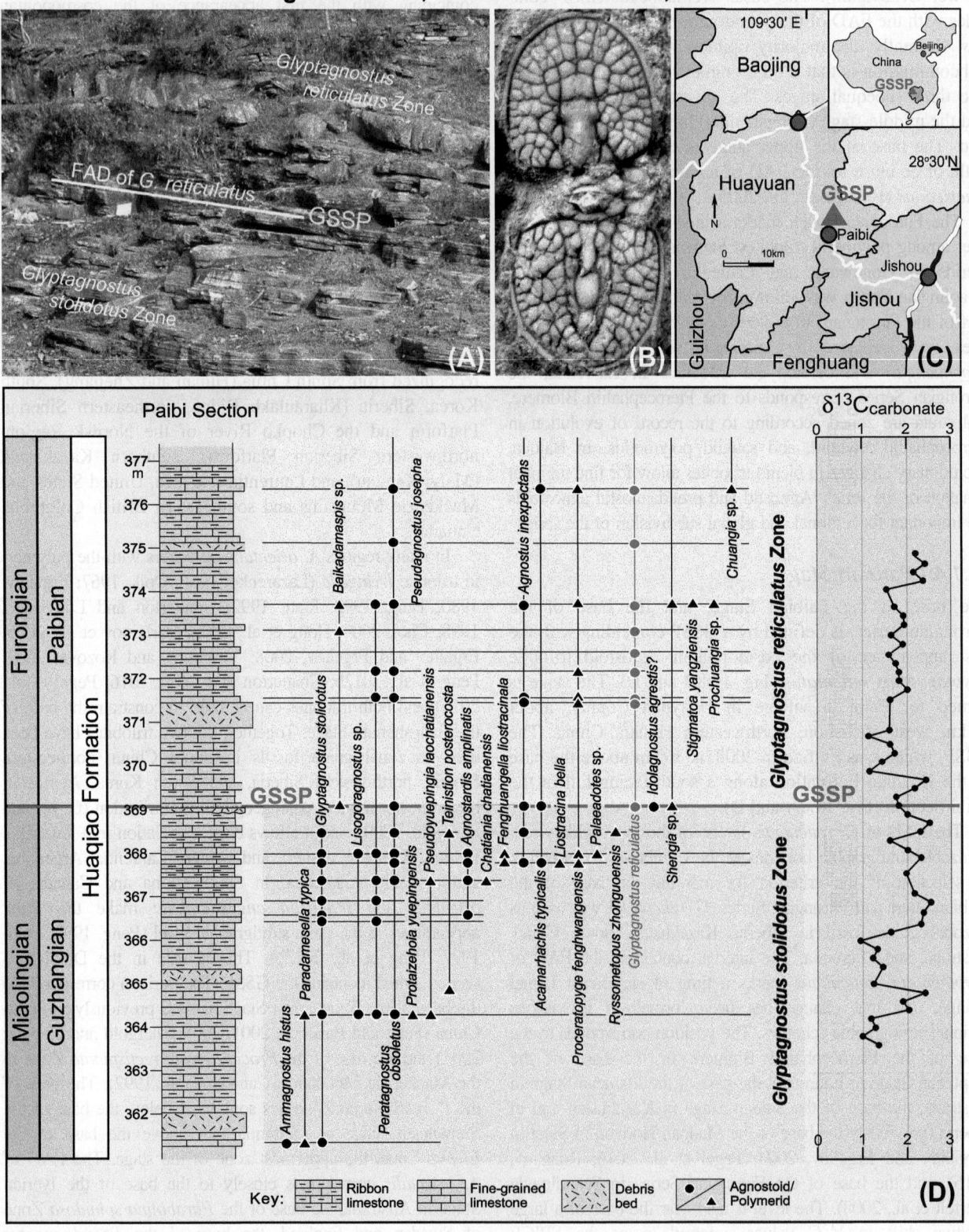

FIGURE 19.8 (A) and (C) GSSP of the Furongian Series and Paibian Stage in the Paibi section, Hunan, China; (B) *Glyptagnostus reticulatus* (Angelin, 1851), an agnostoid trilobite whose FAD coincides with the base of the Furongian Series and Paibian Stage; (D) Stratigraphic distribution of trilobites close to the base of the Furongian Series and Paibian Stage in the Paibi section, Hunan, China (redrawn after Peng et al., 2004a). *FAD*, First appearance datum; *GSSP*, Global Boundary Stratotype Section and Point.

Base of the Jiangshanian Stage of the Cambrian System in the Duibian B Section, W Zhejiang Province, China

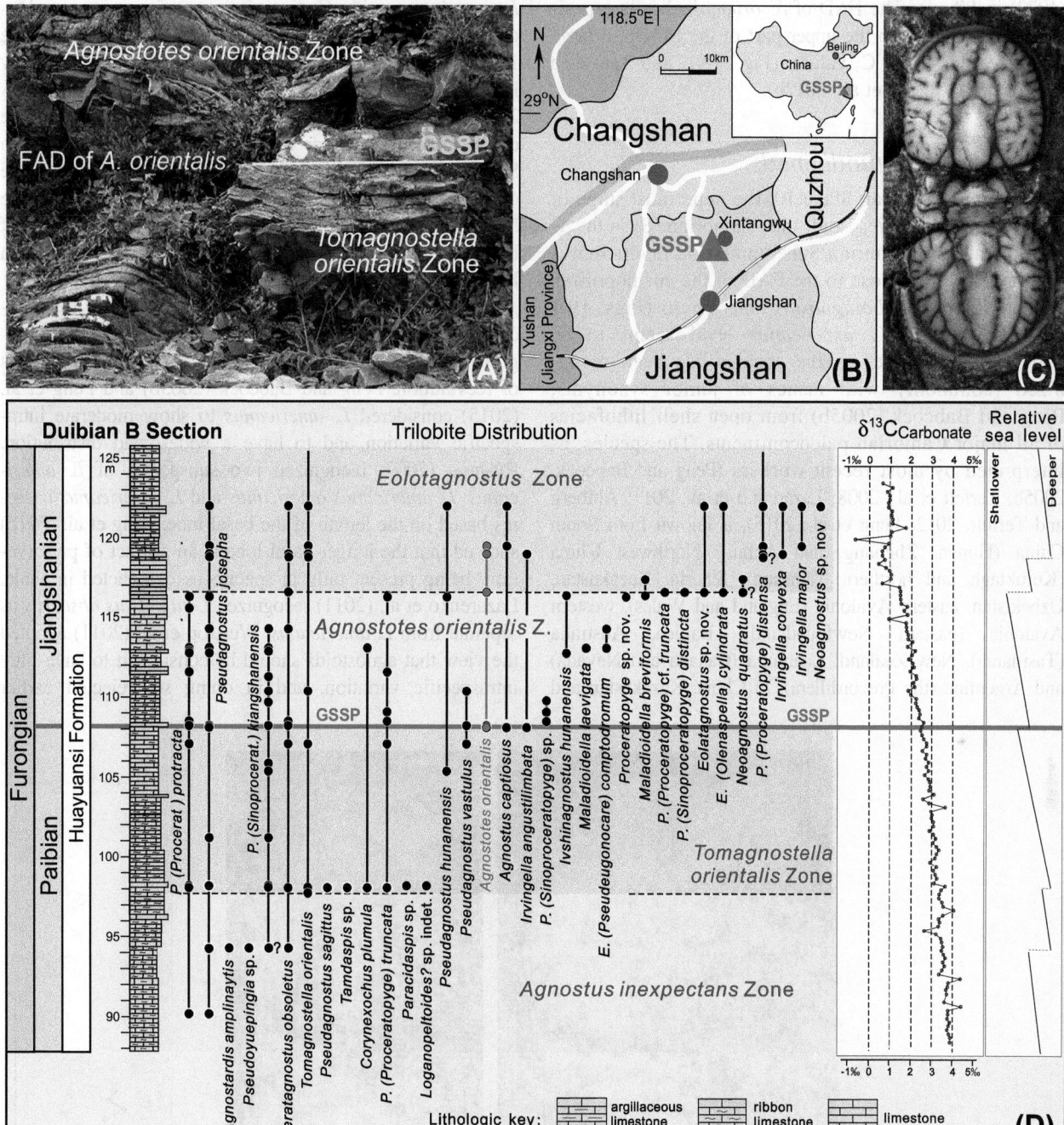

FIGURE 19.9 (A) and (B) GSSP of the Jiangshanian Stage in the Duibian B section, Zhejiang, China; (C) *Agnostotes orientalis* (Kobayashi, 1935), an agnostoid trilobite whose FAD coincides with the base of the Jiangshanian Stage; (D) Stratigraphic distribution of trilobites close to the base of the Jiangshanian Stage in the Duibian B section, Zhejiang, China (redrawn after Peng et al., 2012b). *FAD*, First appearance datum; *GSSP*, Global Boundary Stratotype Section and Point.

and Ergaliev, 2008), and a level that is somewhat below the base of the "Tukalandyan Stage" of Rozova (1963) or within the Chekurovian Stage of Lazarenko and Nikiforiv (1972) in Siberia. The FAD of *A. orientalis* lies in a position corresponding to the upper part of the SPICE, a large positive shift in δ^{13}C values (Fig. 19.1 and see also Section 19.2.7; Peng et al., 2012b).

19.1.4.3 Stage 10 (undefined)

The base of Cambrian Stage 10, the uppermost stage of the Furongian Series and the Cambrian System, is undefined. The Cambrian Subcommission favors marking the base at or close to the FAD of the cosmopolitan agnostoid trilobite *Lotagnostus americanus* (Figs. 19.1 and 19.10). The *L. americanus* level seems to be widely recognizable, as the species has been recognized (commonly with names of junior synonyms; Peng and Babcock, 2005b) from open shelf lithofacies of all major Cambrian paleocontinents. The species, as interpreted by most recent workers (Peng and Babcock, 2005b; Terfelt et al., 2008; Lazarenko et al., 2011; Ahlberg and Terfelt, 2012; Peng et al., 2015), is known from South China (Hunan, Zhejiang, and Anhui), Northwest China (Kuruktagh and northern Tianshan), Siberia, Kazakhstan, Uzbekistan, eastern Avalonia (England and Wales), western Avalonia (eastern Newfoundland), Baltica, Australia (Tasmania), New Zealand, Laurentia (Canada and Nevada) and Argentina (the Precordillera). In China, Kazakhstan and

Siberia, *L. americanus* is used as a zonal guide fossil (Xiang and Zhang, 1985; Lu and Lin, 1989; Peng, 1992; Ergaliev, 1992; Lazarenko et al., 2008, 2011). The co-occurrence of *L. americanus* with *Hedinaspis* in certain regions enables correlations to be extended even further. If a GSSP for Stage 10 is established at the FAD of *L. americanus*, the boundary level will correspond to the base of the *Ctenopyge spectabilis* Zone of the *Protopeltura* Superzone or alternatively the base of the *Ctenopyge tumida* Zone of the *Peltura* Superzone in Sweden (Terfelt et al., 2008; Babcock et al., 2017), the base of the Aksayan Stage in Kazakhstan, the base of the Niuchehean Stage in South China, and possibly the bases of both the Payntonian Stage in Australia and the Ketyan Horizon in Siberia.

On account of the importance of *Lotagnostus* for intercontinental correlation, species concepts and the paleogeographic distribution of included species are in the process of reevalution. Peng and Babcock (2005b) and Peng et al. (2015) considered *L. americanus* to show moderate intraspecific variation and to have a widespread distribution. Rushton (2009) recognized two subspecies of *L. americanus*: *L. americanus americanus* and *L. americanus trisectus* based on the length of the basal lobe. Peng et al. (2015) showed that the longer basal lobe is an artifact of preservation, being present only in specimens compacted in shale. Lazarenko et al. (2011) recognized *Lotagnostus obscurus* as separate from *L. americanus*. Westrop et al. (2011) adopted the view that agnostoids should be considered to have little intraspecific variation, and in doing so, rejected earlier

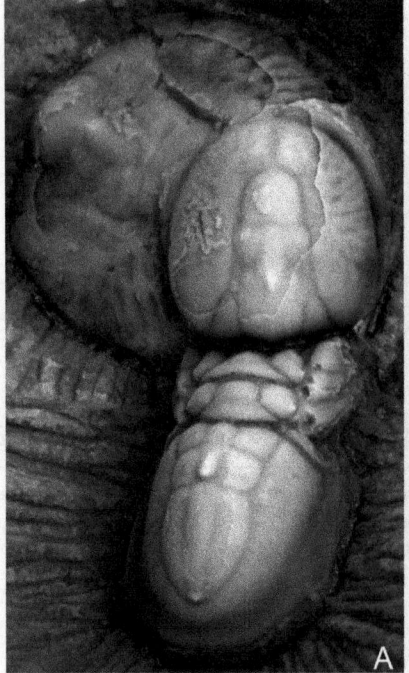

FIGURE 19.10 *Lotagnostus americanus* (Billings, 1860), a key agnostoid trilobite for recognizing the base of provisional Stage 10. (A) A specimen showing weak scrobiculation; (B) a specimen showing well-developed scrobiculation.

interpretations of intraspecific variability in *L. americanus*. As a consequence of this view, which is at odds with most recent interpretations of species concepts in agnostoids (e.g., Pratt, 1992; Robison, 1984, 1994; Ahlberg and Ahlgren, 1996; Peng and Robison, 2000; Ahlberg et al., 2004; Lazarenko et al., 2011; Chatterton and Gibb, 2016), including statistical studies based on large populations (Rowell et al., 1982), Westrop et al. (2011) and Westrop and Landing (2017) recommended recognition of several described forms including *L. trisectus* as species distinct from *L. americanus*. In the view of Westrop et al. (2011), *L. americanus* should be restricted to specimens from a single locality in Quebec, Canada.

Two conodont species, *Eoconodontus notchpeakensis* and *Cordylodus andresi*, have been suggested as possible markers for the base of the uppermost Cambrian stage (Miller et al., 2006, 2011). One of these is the euconodont *E. notchpeakensis*, which is recognizable in western Utah, United States (Miller et al., 2006, 2011; Landing et al., 2010a, 2010b, 2011) and elsewhere. Transport of the elements of this and other euconodont species allows for the identification of *E. notchpeakensis* through peritidal, platform, and slope deposits of Laurentia, Gondwana, Baltica, and Kazakhstan (Dubinina, 2009; Landing et al., 2010, 2011; Dong and Zhang, 2017; Bagnoli et al., 2017), but questions remain whether the first known occurrences of the species in these different places are isochronous or nearly so. This species has its first observed appearance in Laurentia at the base of the *E. notchpeakensis* Subzone of the *Eoconodontus* Zone (conodonts), equivalent to the middle of the *Saukia* Zone (*Saukiella junia* Subzone, polymerid trilobites). The FAD coincides with the onset of the TOCE (Top of Cambrian carbon isotope Excursion), a negative δ^{13}C excursion alternatively referred to as the HERB Event (Ripperdan, 2002; Miller et al., 2006, 2011, 2015; Li et al., 2017; but see Landing et al., 2011, who recognized two separate excursions/events). *E. notchpeakensis* has a long stratigraphic range, extending into the *I. fluctivagus* Zone of the Lower Ordovician. The onset of the TOCE (Figs. 19.1 and 19.12) is at present regarded as occurring about halfway through provisional Stage 10, and the return to positive δ^{13}C values occurs a little below the Ordovician base. A stage whose base would be identified by the FAD of *E. notchpeakensis* would be about half the stratigraphic thickness of Stage 10 as currently envisioned. The FAD of the euconodont *C. andresi* at the base of the *Cordylodus proavus* Zone in Utah, United States, marks another horizon that could serve as the base of the uppermost Cambrian stage (Miller et al., 2006). This position is recognizable intercontinentally, and occurs above the first appearance of *E. notchpeakensis*. A stage whose base would be identified by the FAD of *C. andresi* would represent a thickness less than half that currently envisioned for Stage 10. If either *E. notchpeakensis* or *C. andresi* were selected as the marker for the base of provisional Stage 10, this stage

would represent a much shorter span of time than any other Cambrian stage/age. If the *E. notchpeakensis* level were selected, the terminal Cambrian stage would represent a time duration of about 2 Myr, and if the *C. andresi* level were selected, the stage would represent even less time.

19.1.5 Regional Cambrian stage suites

Regional stages and series have been erected for many parts of the Cambrian world. Examples of intensively studied areas are South China, Australia, Siberia, Laurentia, and Baltica (Fig. 19.11).

19.1.5.1 Cambrian Stages of South China

A chronostratigraphic model for the Cambrian of South China with boundary stratotype-based stages and series has been developed (Peng et al., 1998, 1999, 2000a, 2000b; Peng, 2000a, 2000b, 2003, 2008; Peng and Babcock, 2001; Fig. 19.11, column 2). Since 2012 this model has been used for the national standard stratigraphic scale of China by the China Geological Survey and the National Commission on Stratigraphy of China (Wang et al., 2013). Apart from the lowermost two stages, the stages are based on sections in the Jiangnan Slope Belt, where the Cambrian succession yields rich agnostoids having significance for global or intercontinental correlation. The lower boundary of the Cambrian System in South China has been difficult to identify using *Treptichnus pedum* (Zhu et al., 2001; Steiner et al., 2007). Instead, the BACE (BAse of Cambrian isotope Excursion) δ^{13}C excursion is a more reliable stratigraphic marker. South China is home to the GSSPs of four global stages, the Wuliuan, Guzhangian, Paibian, and Jiangshanian. Once these global stages were established, it became advantageous to replace regional stage names having essentially the same concept and content.

19.1.5.1.1 Jinningian

The lowest stage of the Cambrian System in South China was proposed by Peng (2000a, 2000b). It refers to the strata in the Yuhucun Formation below the "Marker B" (or "China B Point") in the Meishucun section, near Meishucun town, Jinning County, eastern Yunnan Province (Luo et al., 1984, 1990, 1991, 1992, 1994; Cowie et al., 1989; Brasier et al., 1994). The "Marker B" was one of the levels proposed in the late 1980s as a possibility for the global Precambrian–Cambrian boundary. The Jinningian Stage embraces the oldest SSF zone, the *Anabarites trisulcatus–Protohertzina anabarica* Zone, which is characterized by the predominance of simple, low-diversity hyolithids. The base of the SSF zone was originally defined at "Marker A," the lowest observed occurrence of SSFs in the Meishucun section. In the Meishucun section an unconformity lies slightly below "Marker A," and a relatively thick

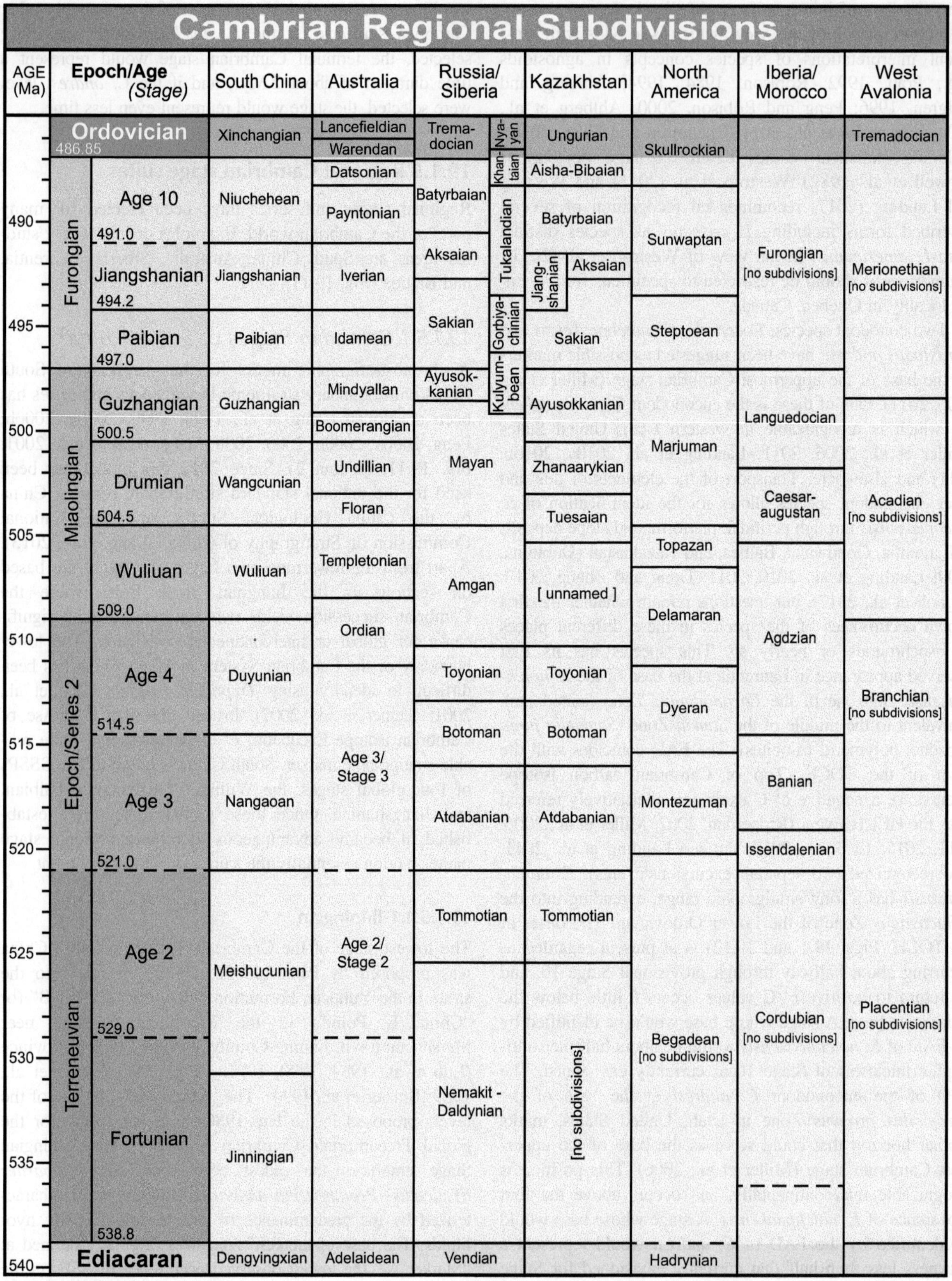

Cambrian Regional Subdivisions

AGE (Ma)	Epoch/Age (Stage)	South China	Australia	Russia/Siberia		Kazakhstan	North America	Iberia/Morocco	West Avalonia
	Ordovician 486.85	Xinchangian	Lancefieldian	Trema-docian	Khan-taian Nya-yan	Ungurian	Skullrockian		Tremadocian
490	Age 10 491.0	Niuchehean	Warendan			Aisha-Bibaian		Furongian [no subdivisions]	Merionethian [no subdivisions]
			Datsonian	Batyrbaian	Tukalandian	Batyrbaian	Sunwaptan		
			Payntonian						
	Jiangshanian 494.2	Jiangshanian	Iverian	Aksaian		Jiang-shanian / Aksaian			
495	Paibian 497.0	Paibian	Idamean	Sakian	Gorbiya-chinian	Sakian	Steptoean		
	Guzhangian 500.5	Guzhangian	Mindyallan	Ayusok-kanian	Kulyum-bean	Ayusokkanian	Marjuman	Languedocian	
500			Boomerangian	Mayan		Zhanaarykian		Caesar-augustan	Acadian [no subdivisions]
	Drumian 504.5	Wangcunian	Undillian						
505			Floran			Tuesaian	Topazan		
	Wuliuan 509.0	Wuliuan	Templetonian	Amgan		[unnamed]	Delamaran	Agdzian	
510	Age 4	Duyunian	Ordian						Branchian [no subdivisions]
	514.5			Toyonian		Toyonian			
515			Age 3/ Stage 3	Botoman		Botoman	Dyeran	Banian	
	Age 3	Nangaoan		Atdabanian		Atdabanian	Montezuman		
520	521.0							Issendalenian	
	Age 2	Meishucunian	Age 2/ Stage 2	Tommotian		Tommotian			Placentian [no subdivisions]
525									
	529.0						Begadean [no subdivisions]	Cordubian [no subdivisions]	
530				Nemakit - Daldynian		[no subdivisions]			
	Fortunian	Jinningian							
535									
	538.8								
540	Ediacaran	Dengyingxian	Adelaidean	Vendian			Hadrynian		

FIGURE 19.11 Principal regional stage schemes of the Cambrian, and the four-fold division of the system into series adopted by the International Commission on Stratigraphy (ICS). See text for discussion of caveats in applying the numerical scale to stage boundaries.

interval, corresponding to the Daibu Member of northeastern Yunnan (Zhu et al., 2001), is missing. The Daibu Member is regarded as part of the Cambrian System (Li et al., 2001). *Treptichnus pedum* was reported as occurring within the *A. trisulcatus−P. anabarica* Zone in the Meishucun section, much higher than "Marker A," but this is apparently not its lowest occurrence. As originally defined, the base of the Jinningian Stage is the base of the Cambrian System. The horizon equivalent to the basal Cambrian GSSP, however, remains unknown in South China (Zhu et al., 2001).

19.1.5.1.2 Meishucunian

The name of the stage was originally proposed as a lithologic unit, the Meishucun Formation (Jiang et al., 1964), which comprises subtrilobite sequences of "phosphate-bearing beds" in the basal part of the Meishucun section. Qian (1997) regarded the formation as the first stage of the Cambrian System of China but failed to define its base at that time. Subsequently, Luo et al. (1994) revised the stage concept, drawing the base of the stage at the base of the *Paragloborites−Siphogonuchites* Zone (i.e., "Marker B"), which coincides with a major change in SSF fauna marked by the abrupt appearances of phosphatized micromollusks and problematica. It is also the base of Bed 7 in the section. "Marker B" was selected as a possibility for the base of the Cambrian System by the Precambrian−Cambrian Boundary Working Group (Cowie, 1985; Xing et al., 1991; Luo et al., 1994). The point is in the Yuhucun Formation in the Meishucun section and is currently identified by the FAD of the hyolithid *Paragloborilus subglobosus*. The Meishucunian Stage covers an interval occupied by three biozones with abundant and diverse SSFs and an interzone that is poorly fossiliferous (Luo et al., 1994; Peng and Babcock, 2001; Steiner et al., 2007).

19.1.5.1.3 Nangaoan

As originally defined (Peng, 2000a), the base of the stage is placed at the FAD of trilobites, as this level represents an important event in biotic development. This criterion has also been provisionally adopted by the Cambrian Subcommission to define the base of provisional global Stage 3 (Peng and Babcock, 2005a, 2008; Babcock et al., 2005; Babcock and Peng, 2007). In practice, the base of the Nangaoan Stage is drawn at the lowest observed occurrence of *Tsunyidiscus niutitangensis* in the middle part of Bed 5 of the Xiaosai section near Xiaosai, Yuqing County, eastern Guizhou Province (Zhang et al., 1979). The bed belongs to the Niutitang Formation, which is composed of black shale. Detailed work is expected to define the base of the stage more precisely. The Nangaoan Stage is the oldest trilobite-bearing stage in South China, with its lower part characterized by the occurrences of diverse eodiscoids (*Tsunyidiscus*, *Neocobboldia*, *Sinodiscus*, and *Hupeidiscus*), which are

associated with protolenids (*Paraichangia*) in eastern Guizhou and the occurrence of the Chengjiang Biota in eastern Yunnan. The redlichiid *Parabadiella* has been reported from South China (Zhang, 1987). It is one of the earliest trilobites, but the redlichiids as a whole are regarded as having evolved from fallotaspidoids (Jell, 2003). As defined, the base of the stage corresponds closely to base of the traditional Chiungchussuan Stage of Yunnan.

19.1.5.1.4 Duyunian

Peng (2000a) defined the base of the Duyunian Stage by the FAD of *Arthricocephalus duyunensis*, a species that was for some time considered to be a junior synonym of *Arthricocephalus chauveaui* (Lane et al., 1988; Blaker and Peel, 1997; McNamara et al., 2003; Geyer and Peel, 2011). Work by Peng et al. (2017) clarified the concept of *A. chauveaui*, showed *Arthricocephalus duyunensis* to be a separate species, and referred *A. duyunensis* to the genus *Oryctocarella* (i.e., *O. duyunensis*). The lower boundary of the stage lies in the Niutitang Formation, about 25 m above the base of Bed 10 in the Jiumenchong section near Nangao, Danzhai, eastern Guizhou (Zhang et al., 1979). There are indications that the first occurrence of *O. duyunensis* in the Jiumenchong section is at a lower position than in other sections of eastern Guizhou (Zhou Zhiyi, pers. comm.). *O. duyunensis* is widely distributed in South China and is also recorded from Greenland (Lane et al., 1988; Blaker and Peel, 1997; Geyer and Peel, 2011). In the Jiumenchong section, its observed lowest occurrence is near the base of the Balang Formation. The Duyunian Stage occupies an interval with four trilobite zones (Peng, 2000a, 2000b), which are characterized by the early development of oryctocephalids and the flourishing of redlichiids. The cosmopolitan trilobites *Bathynotus* and *Ovatoryctocara*, plus various primitive ptychopariids, all occur near the top of the stage. They comprise much of the first assemblage of the Kaili Biota. There is a medium-scale faunal extinction event at the end of the Duyunian Age (Yuan et al., 2002; Zhen and Zhou, 2008), and only about 25% of the trilobite genera present in the upper Duyunian (regional stage) range upward into the overlying Wuliuan Stage.

19.1.5.1.5 Wuliuan

The global Wuliuan Stage, defined in South China, replaces the regional Taijiangian Stage because they both share the same concept and the same stratotype section (Peng and Zhao, 2018). The base of the stage is defined by the FAD of the intercontinentally distributed trilobite *Oryctocephalus indicus*, which occurs in the lower part of the Kaili Formation, close to the traditional Lower−Middle Cambrian boundary as recognized in China (Fig. 19.5). The boundary interval is exposed in a hillside section between Wuliu and Zengjiayan, near Balang Village, Jianhe County, eastern

Guizhou. The FAD of *O. indicus* lies at the base of Bed 10, which is 52.8 m above the base of the Kaili Formation in the stratotype (Zhao et al., 2001, 2012, 2019). Originally the Taijiangian Stage was represented by three zones, the *O. indicus* Zone (lowermost), the *Ptychagnostus gibbus* Zone, and the *P. atavus* Zone (uppermost). Various other zonations, based on a mix of agnostoid and polymerid trilobites, have been proposed, but for uniformity with the global standard, the stage was restricted to the *O. indicus* (polymerid zonation) and *P. gibbus* (agnostoid) zones. Strata of the *P. atavus* Zone (agnostoid zonation) are assigned to the overlying Wangcunian Stage (see Peng, 2009; Peng et al., 2012a). The base of the Wangcunian Stage (South China usage), marked by the base of the *P. atavus* Zone, coincides with the base of the Drumian Stage (global µsage).

The Wuliuan Stage is characterized by an abundance of oyctocephalid trilobites (a polymerid group) in the lower part, and by the initial diversification of agnostoids in the upper part. On account of differences in the ecology of agnostoid and polymerid trilobites, it has been useful to recognize separate zonation schemes for these groups upward from the Wuliuan (e.g., Robison, 1976; Terfelt et al., 2008, 2011; Robison and Babcock, 2011; Ahlberg and Terfelt, 2012; Nielsen et al., 2014; Robison et al., 2015; Babcock et al., 2017), including in South China (e.g., Peng and Robison, 2000; Peng et al., 2004a, 2004b). Yao et al. (2009) recognized an additional agnostoid zone, the *Peronopsis taijiangensis* Zone, in strata of South China that are equivalent to the upper part of the *O. indicus* Zone (s. l.).

The Kaili Biota, a Burgess Shale-type assemblage yielding numerous fossil groups, occurs in Wuliuan strata of eastern Guizhou Province (e.g., Zhao et al., 2002, 2005b, 2011, 2012).

19.1.5.1.6 Wangcunian

The base of the Wangcunian Stage, defined originally at the FAD of the cosmopolitan agnostoid trilobite *Ptychagnostus punctuosus* (Peng et al., 1998) in the Huaqiao Formation near Wangcun, Yongshun County, northwestern Hunan Province, was shifted downward to the level marked by the FAD of *P. atavus* (Peng, 2009). This level, which coincides with the base of the global Drumian Stage, is 1.2 m above the base of the Huaqiao Formation in the Wangcun section (Peng and Robison, 2000; Peng et al., 2004b). This revision conforms with the global correlation standard for the Drumian Stage. As revised, the Wangcunian Stage in South China is characterized by the diversification of ptychagnostid and hypagnostid agnostoids, the diversification of corynexochid and proasaphiscid trilobites, the occurrence of lisaniids (*Lisania, Qiandonaspis*) in the upper part, and the first appearance of the agnostoid trilobite *Lejopyge armata* and the damesellid *Palaeodotes* near the top of the stage. More than 90 trilobite taxa, of which 25 are agnostoids (Peng and Robison, 2000; Peng et al., 2004b), occur in the type area of the Wangcunian Stage.

19.1.5.1.7 Guzhangian

The Guzhangian Stage is a global stage that replaces the regional Youshuian Stage (as revised, with the base shifted downward from the FAD of *Linguagnostus reconditus* to coincide with the FAD of *Lejopyge laevigata*). The base of the stage is defined by a GSSP in the Luoyixi Section, a roadcut on the south bank of the Youshui River, opposite the Wangcun section. The Wangcun section is the stratotype for the abandoned Youshuian Stage. As recognized by Peng and Robison (2000), the stage is represented in its type area by four successive agnostoid zones: the *L. laevigata* Zone, the *Proagnostus bulbus* Zone, the *L. reconditus* Zone, and the *Glyptagnostus stolidotus* Zone. The base of the *L. reconditus* Zone is closely correlative with the base of the traditional Upper Cambrian as defined in Sweden by Westergård (1922, 1947) (see also Peng and Robison, 2000; Ahlberg, 2003; Ahlberg et al., 2004; Axheimer et al., 2006). The Guzhangian Stage is characterized by a high abundance and diversification of trilobites, especially ones assigned to the families Clavagnostidae, Damesellidae, and Lisaniidae, and by the genera *Lejopyge, Linguagnostus, Proagnostus, Torifera,* and *Fenghuangella*. More than 150 taxa have been described from the Guzhangian Stage, of which 45 are agnostoid trilobites (Peng and Robison, 2000; Peng et al., 2004b; 2020). There is a major faunal extinction event near the end of the Guzhangian Age, an event resulting in the extinction of more than 90 percent of taxa. In the type area, all but one of the damesellid trilobites is confined to the stage; only one species ranges upward. The extinction event is recognized in other regions of the world as the faunal crisis at the beginning of the Idamean Stage in Australia (Öpik, 1966; Shergold and Cooper, 2004), and the top of the Marjumiid Biomere in Laurentia (Palmer, 1979, 1984, 1998; Ludvigsen and Westrop, 1985; Saltzman et al., 2000).

19.1.5.1.8 Paibian

The Paibian Stage is a global stage that is also applied now as a regional stage in South China because of its origin. It replaces the abandoned Waergangian Stage as used previously for South China. The GSSP that defines the base of the stage coincides with the FAD of the cosmopolitan agnostoid *Glyptagnostus reticulatus* in the upper part of the Huaqiao Formation of the Paibi section, near Paibi, Huayuan, northwestern Hunan. A remarkable faunal turnover occurs at the beginning of the stage with the first coappearances of a number of taxa—the leiostegiid *Chuangia*, the pagodiid *Prochuangia*, the eulomiid *Stigmatoa*, the olenid *Olenus*, the lisaniid *Shengia quadrata*, the agnostoid *G. reticulatus*, and diverse species of *Pseudagnostus* and *Proceratopyge*, of which only a few species range upward from the underlying Guzhangian Stage (Peng, 1992; Peng et al., 2004b). The Paibian trilobite fauna marks a recovery period following the end-Guzhangian extinction. Only about 30 taxa occur in three successive zones.

19.1.5.1.9 Jiangshanian

The Jiangshanian Stage is a global chronostratigraphic unit applied also as a regional stage in South China. It replaces the revised Taoyuanian Stage (Peng, 2008), which is the equivalent of the lower part of the Taoyuanian Stage as originally defined (i.e., the interval below the FAD of *Lotagnostus americanus*). The upper part of the original Taoyuanian Stage was proposed as a new stage, the Niuchehean Stage (Peng, 2008).

The GSSP that defines the base of the Jiangshanian Stage coincides with the FAD of the cosmopolitan agnostoid *Agnostotes orientalis* in the upper part of the Huayansi Formation of the Duibian B section, at Duibian, western Zhejiang. In the boundary stratotype section, and also in the Wa'ergang section, northwestern Hunan, the FAD of *A. orientalis* coincides with the FAD of *Irvingella angustilimbata* (Peng, 1992; Peng et al., 2012b). As a cosmopolitan polymerid trilobite, *I. angustilimbata* can also be used to constrain the base of the Jiangshanian Stage. The stage is characterized by the diversification of the agnostoid subfamily Pseudagnostinae, with successive separation of *Pseudagnostus*, *Rhaptagnostus*, and *Neoagnostus*; the diversification of the superfamily Leiostegioidae; and the first occurrences of Dikelocephaloidea, Saukiidea, Hapalopleuridae, Shumardiidae, and Macropygiinae. In northwestern Hunan the Jiangshanian Stage contains 70 trilobite taxa in four biozones.

19.1.5.1.10 Niuchehean

The Niuchehean Stage was proposed by Peng (2008) for the upper part of the abandoned Taoyuanian Stage as originally defined. It is the uppermost stage of the Cambrian System in South China. The base of the stage is defined at the base of the *Lotagnostus americanus* Zone in the Wa'ergang section, Taoyuan, northwestern Hunan. The top of the stage in the stratotype section is the level marked by the first appearance of the conodont *Iapetognathus fluctivagus* within the conodont *Cordylodus lindstromi* Zone (Dong et al., 2004; Dong and Zhang, 2017). This level also lies within the *Hysterolenus* Zone (polymerid trilobite zonation), and coincides with the base of the global Tremadocian Stage of the Ordovician System (Peng, 2009; Dong and Zhang 2017). The stage is named for Niuchehe, a township that governs Wa'ergang Village. The base of the uppermost global stage of the Cambrian System will probably be defined at or close to the FAD of *L. americanus* (Peng and Babcock, 2005a; Babcock et al., 2005). The stage embraces four and a half assemblage-zones, collectively containing more than 80 trilobite taxa in its type area (Peng, 1984, 1990, 1992). It is characterized by the abundance and diversification of ceratopygids (*Charchaqia*, *Diceratopyge*, *Hedinaspsis*, *Hunanopyge*, *Macropyge*, *Promacropyge*, *Yuepingia*, *Hysterolenus*, *Onychopyge*); the diversification of eulomids (*Archaeuloma*, *Proteuloma*, *Euloma*, *Karataspis*, *Ketyna*); the

appearance of remopleuriids (*Fatocephalus*, *Ivshinanspis*), nileids (*Troedssonia*, *Shenjiawania*), pilekids (*Parapilekia*), and harpidiids (*Eotrinucleus*); the separation of true asaphids from ceratopygids; and the occurrence of leiostegiids and saukiids. Agnostoids of the stage are dominated by members of the family Agnostidae. Apart from trilobites, the Niuchehean Stage in the type area is also characterized by the occurrences of euconodonts, including *Eoconodontus notchpeakensis*, *Cordylodus proavus*, *C. intermedius*, *C. lindstromi*, and by primitive nautiloids (Peng and Chen, 1983; Peng, 1984; Bagnoli et al., 2017; Dong and Zhang, 2017).

19.1.5.2 Australian Cambrian Stages

Australian stages were summarized by Shergold (1995), Young and Laurie (1996), and Kruse et al. (2009), on which the following outline is based (Fig. 19.11, column 3). The stages are described as "biochronological" units and are defined in terms of their contained fauna (Shergold, 1995). Boundary stratotypes therefore have not been designated. Apart from the Ordian, the stages discussed below were all erected in the Georgina Basin of western Queensland. Recently Betts et al. (2016, 2017, 2018) recognized "Stage 2 (upper part)," "Stage 3," and a possible "Stage 4" (basal part) in the Arrowie and Stansbury basins, South Australia, which, as regional "stages," remain unnamed.

19.1.5.2.1 Pre-Ordian

Stages have not yet been designated for most of the traditional Lower Cambrian of Australia except for those recognized by Betts et al. (2016, 2017, 2018). Correlations based on archaeocyaths, SSFs, and trilobites, as well as on carbon isotope curves, indicate that the global stage 2−4, equivalent to the upper Nemakit-Daldynian to Toyonian stages of the Siberian Platform and the Altay−Sayan Foldbelt of Russia, can be recognized through southern and central Australia. Ichnofaunas in central and southern Australia possibly correlate with the Tommotian and Nemakit-Daldynian of Siberia (Walter et al., 1989; Bengtson et al., 1990; Shergold, 1996; Betts et al., 2016, 2017, 2018).

19.1.5.2.2 Ordian

The Ordian Stage was originally proposed by Öpik (1968) as a time and time-rock division of the Cambrian characterized by the occurrence of the *Redlichia chinensis* faunal assemblage. The Templetonian Stage, "a liberal interpretation of Whitehouse's (1936) Templetonian Series" (Öpik, 1968), was originally conceived by Öpik as containing the *Xystridura templetonensis* assemblage of Western Queensland, followed by faunas of the *Ptychagnostus gibbus* (under the name of *Triplagnostus gibbus*) Zone. In practice, it is difficult to distinguish the *Redlichia* and *Xystridura* faunas because four species of *Xystridura*, similar eodiscoid and ptychoparioid trilobites, some bradoriid arthropods, and chancelloriids occur in rocks of both Ordian and early Templetonian ages. Accordingly Shergold (1995) regarded the Ordian—early Templetonian as a single standard regional unit. Later, Laurie (2004, 2006) redefined the base of the Templetonian Stage, and thereby defining the top of the Ordian Stage, based on recognition of three pre-*P. gibbus* agnostoid zones in southern Georgina Basin drillholes. The Ordian Stage had long been regarded as the earliest traditional Middle Cambrian Stage in Australia even though it apparently correlates with the traditional Longwangmiaoan Stage of China (Shergold, 1997; Chang, 1998; Geyer et al., 2000; 2003; Peng, 2003) and with the Toyonian Stage of the Siberian Platform (Zhuravlev, 1995), both of which were long regarded as terminal "Lower" Cambrian. The Longwangmiaoan and Toyonian regional Stages are now considered to be part of global Stage 4. Kruse et al. (2009) considered the Ordian Stage to be at least partly (and possibly entirely) equivalent to global Stage 4 (or the uppermost part of the traditional Lower Cambrian).

19.1.5.2.3 Templetonian

Laurie (2004, 2006) redefined the base of the Templetonian Stage, fixing its base at the base of the agnostoid trilobite *Ptychagnostus anabarensis* (as *Pentagnostus anabarensis*) Zone. As revised by Laurie (2004, 2006), the Templetonian Stage includes four agnostoid zones: the *P. anabarensis*, *Ptychagnostus praecurrens* (as *Pentagnostus praecurrens*), *Ptychagnostus shergoldi* (as *Pentagnostus shergoldi*), and *P. gibbus* zones. The *P. praecurrens*, *P. shergoldi*, and *P. gibbus* zones equal the interval recognized earlier as the·early Templetonian Stage by Shergold (1995), who divided the Templetonian Stage into lower and upper portions. Shergold (1995) assigned the lower portion to the Ordian—Lower Templetonian Stage and assigned the upper portion to the Upper Templetonian—Floran Stage.

The Templetonian is important because it contains trilobites that are useful for intercontinental correlation (Shergold, 1969), including oryctocephalids and cosmopolitan agnostoids (e.g., *P. praecurrens* and *P. gibbus*).

19.1.5.2.4 Floran

As originally defined (Öpik, 1979), the Floran Stage contained the agnostoid trilobite zones of *P. atavus* (under the name *Acidusus atavus*) and *Euagnostus opimus*. This concept was revised by Shergold (1995) to include subjacent strata of the late Templetonian zone of *Ptychagnostus gibbus*, argued on the grounds of faunal continuity (overlap in the ranges of *P. gibbus* and *P. atavus* in the Georgina Basin, Western Queensland) and sequence stratigraphy (Southgate and Shergold, 1991). By redefining the Templetonian Stage, the base of the Floran has been restored to its original level (Laurie, 2004, 2006; Kruse et al., 2009), and it again contains two agnostoid zones, the *P. atavus* and *Euagnostus opimus* zones. The base of the Floran coincides

with the base of the global Drumian Stage, and the base of the revised Wangcunian Stage of South China.

19.1.5.2.5 Undillan

The Undillan Stage, defined by Öpik (1979), contains the fauna of two agnostoid zones, the *Ptychagnostus punctuosus* Zone and the succeeding *Goniagnostus nathorsti* Zone. A third zone, based on *Doryagnostus deltoides* and containing 15 agnostoid species including *P. punctuosus* and *G. nathorsti*, was recognized by Öpik (1979) in the Undilla region of the Georgina Basin. The agnostoid fauna of the Undillan Stage has a cosmopolitan distribution. Polymerid trilobites of the Undillan include ptychoparioids, anomocarids, mapaniids, damesellids, conocoryphids, corynexochids, nepeiids, and dolichometopids, all of widespread distribution.

19.1.5.2.6 Boomerangian

The Boomerangian Stage (Öpik, 1979) is essentially the *Lejopyge laevigata* Zone divided into three. A *Ptychagnostus cassis* Zone at the base is overlain by zones defined by the polymerid trilobites *Proampyx agra* and *Holteria arepo*. Boomerangian agnostoids are accompanied by a range of polymerid trilobites including species of *Centropleura*, dolichometopids, olenids, mapaniids, corynexochids, and damesellids. A "Zone of Passage," characterized by the occurrence of *Damesella torosa* and *Ascionepea janitrix*, was interposed by Öpik (1966, 1967) between the Boomerangian (uppermost Middle Cambrian) and Mindyallan (considered at that time to mark the beginning of the Upper Cambrian) stages. Subsequently, Daily and Jago (1975) restricted this zone to the Boomerangian, and placed the Middle—Upper Cambrian boundary within the early Mindyallan Stage.

19.1.5.2.7 Mindyallan

Originally, Öpik (1963) defined the Mindyallan Stage to include a *Glyptagnostus stolidotus* Zone (above) and a "prestolidotus" Zone (below). Subsequently Öpik (1966, 1967) revised the stage, placing the *G. stolidotus* Zone in the upper Mindyallan and dividing the underlying strata into an initial Mindyallan *Erediaspis eretes* Zone and an overlying *Acmarhachis quasivespa* (under the name *Cyclagnostus quasivespa*) Zone. The *E. eretes* Zone contains 45 trilobites, including 18 agnostoid genera. The polymerid trilobites belong to a variety of families; anomocarids, asaphiscids, catillicephalids, damesellids, leiostegiids?, lonchocephalids, menomoniids, nepeiids, norwoodiids, rhyssometopids, and tricrepicephalids are represented. The *A. quasivespa* Zone has 18 species of trilobites confined to it, but many other species range upward from lower zones. Daily and Jago (1975) subdivided the *A. quasivespa* Zone into two assemblages based on the occurrence of *Leiopyge cos* and *Blackwelderia sabulosa*. Because *L. cos* appears to be synonymous with *L. armata*, a late Middle Cambrian taxon, they drew the Middle–Upper Cambrian boundary between these two assemblages. Only eight species range from Öpik's (1966, 1967) *A. quasivespa* Zone into the overlying *G. stolidotus* Zone. The *G. stolidotus* Zone contains 75 species; some (asaphiscids, auritamids, catillicephalids, norwoodiids, and raymondinids) have Laurentian biogeographic affinities, and others (damesellids and liostracinids) have Chinese biogeographic affinities.

19.1.5.2.8 Idamean

The Idamean Stage, as introduced by Öpik (1963), encompassed five successive assemblage-zones: the *Glyptagnostus reticulatus* with *Olenus ogilviei* Zone, the *G. reticulatus* with *Proceratopyge nectans* Zone, the *Corynexochus plumula* Zone, the *Erixanium sentum* Zone, and the *I. tropica* with *Agnostotes inconstans* Zone. Henderson (1976, 1977) proposed an alternative zonation in which the two zones with *Glyptagnostus* were united into a single *G. reticulatus* Zone, the *Corynexochus* Zone was renamed the *Proceratopyge cryptica* Zone, and the *E. sentum* Zone was subdivided into a zone of *E. sentum* followed by a zone of *Stigmatoa diloma*. The name *Irvingella tropica–Agnostotes inconstans* Zone was changed to *I. tropica* Zone. Henderson's scheme was adopted by Shergold (1982). The *I. tropica* Zone is now regarded as the lowermost zone of the succeeding Iverian Stage (see Shergold, 1982, for justification; Shergold, 1993).

There was a major faunal crisis at the beginning of the Idamean. Few Mindyallan trilobite genera and no Mindyallan species survived the extinction (Öpik, 1966). There was also a major reorganization of trilobite families as outer shelf communities dominated by agnostoids, olenids, pterocephaliids, leiostegiids, eulomids, and ceratopygids abruptly replaced shallow shelf taxa of the Mindyallan biota. Shergold (1982) recorded 69 Idamean taxa; they permit a highly resolved biochronology enabling precise international correlations.

19.1.5.2.9 Iverian

The Iverian Stage (Shergold, 1993) was proposed for the concept of a post-Idamean–pre-Payntonian interval in the eastern Georgina Basin, western Queensland, the only region where a reasonably complete sequence has so far been described (Shergold, 1972, 1975, 1980, 1982, 1993). Paleontologically the Iverian Stage is distinct. On the basis of trilobites, it is characterized by (1) the occurrence of the cosmopolitan genus *Irvingella* in Australia; (2) the diversification of the agnostoid subfamily Pseudagnostinae during which *Pseudagnostus*, *Rhaptagnostus*, and *Neoagnostus* separate and become biostratigraphically important; (3) diversification of the Leiostegioidea, especially the families Kaolishaniidae and Pagodiidae; (4) the first occurrence of the Dikelocephaloidea, Remopleuridoidea, and Shumardiidae; (5) the separation of the true asaphids from ceratopygids.

Nine trilobite assemblage-zones have been recognized in the Iverian based on successive species of *Irvingella*, *Peichiashania*, *Hapsidocare*, and *Lophosaukia* (Shergold, 1993). In subsequent work (Shergold and Cooper, 2004; Kruse et al., 2009), the *Hapsidocare lilyensis* and the *Rhaptagnostus clarki patulus–Caznaia squamosa* assemblage-zones were united into a single *R. clarki patulus–C. squamosa–H. lilyensis* Assemblage-zone. More than 160 trilobite taxa occur in the type area of the Iverian Stage.

19.1.5.2.10 Payntonian

As defined by Jones et al. (1971), the Payntonian Stage is recognized on the basis of its trilobite assemblages (Shergold, 1975), its base lying at the point in its type section (Black Mountain, western Queensland), where the comingled Laurentian–Asian assemblages of the Iverian are replaced by others of only Asian biogeographic affinity. These are dominated by tsinaniid, leiostegioidean, saukiid, ptychaspidid, dikelocephaloidean, and remopleuridoidean trilobites. A tripartite zonal scheme has emerged from biostratigraphic revisions by Nicoll and Shergold (1991), Shergold and Nicoll (1992), and Shergold (1993, 1995). In ascending order, these zones are based on *Sinosaukia impages*, *Neoagnostus quasibilobus* with *Shergoldia nomas*, and *Mictosaukia perplexa*. These zones are calibrated by conodont biostratigraphy (Nicoll, 1990, 1991; Shergold and Nicoll, 1992). The Payntonian Stage contains a total of 30 trilobite taxa.

19.1.5.2.11 Datsonian

The concept of the Datsonian Stage remains as defined by Jones et al. (1971), with its base located at the FAD of the conodont *Cordylodus proavus*. Only rare trilobites, *Onychopyge* and leiostegiids, occur and these are insufficient for the establishment of a trilobite biostratigraphy. Accordingly, the Datsonian Stage is defined solely on the basis of conodonts and embraces three successive zones: the *C. proavus* Zone, the *Hirsutodontus simplex* Zone, and the *Cordylodus prolindstromi* Zone.

19.1.5.2.12 Warendan

The Warendan (corrected from Warendian by Kruse et al., 2009) Stage was originally defined at the base of the conodont *Cordylodus prion–Scolopodus* Assemblage-zone (Jones et al., 1971). Revision of the cordylodids by Nicoll (1990, 1991) resulted in introduction of a single zone of *Cordylodus lindstromi* to replace the assemblage-zone as the lowermost zone of the Warendan Stage because *C. prion* was recognized as a part of the septimembrate apparatus of the eponymous species (Shergold and Nicoll, 1992).

The Warendan Stage of Australia is inferred to be Cambrian in the basal part, and Ordovician through most of its extent. In the Ordovician stratotype at Green Point, Newfoundland, Canada, the FAD of the conodont *Iapetognathus fluctivagus*, which is the guide event for the base of the Ordovician, lies within the interval containing the co-ranging species *Cordylodus lindstromi* and *C. prion* (Cooper et al., 2001). The Ordovician base thus correlates to a level above the base of the *C. lindstromi* Zone in the Green Point section. At Wa'ergang, South China, *C. prion* also first occurs within the *C. lindstromi* Zone (Dong et al., 2004). Although *I. fluctivagus* has not been identified from Warenden Stage strata of Australia, correlations based on co-occuring taxa from the sections in Canada and China suggest that the lowermost part of the stage is correlative to the uppermost Cambrian.

19.1.5.3 Siberian Cambrian Stages

The first attempts to develop a Cambrian Chronostratigraphy in the former USSR were based on sections of the Siberian Platform (Pokrovskaya, 1954; Suvorova, 1954). A scale with four Lower and two Middle Cambrian stages was accepted as the national standard of the Soviet Union at the All-Union Stratigraphic Meeting in 1982. In addition to the Lower–Middle Cambrian stages, the Upper Cambrian standard scale for the Soviet Union adopted the stages established by Ergaliev (1980, 1981) in the Malyi Karatau Range, Kazakhstan, which then was a republic of the USSR. The updated General Stratigraphic Scale (GSS) of Russia, adopted by the Commission of Cambrian System, Russian National Stratigraphic Committee (RNSC) in 2015 (Rozanov, 2016), remains a tripartite subdivision. Siberian stages are used for the Lower Cambrian and the lower–middle Middle Cambrian, and Kazakhstanian stages (with modification) are used for the uppermost Middle Cambrian and the Upper Cambrian. The base of the Ayusokkanian Stage on the GSS, which is the uppermost stage of the Middle Cambrian as recognized in Russia, is placed at a higher level than it is in Kazakhstan. The GSS uses the stage names Sakskian (Sakian), Aksaian, and Batyrbaian, all derived from Kazakhstan, for the Upper Cambrian. The base of the Batyrbaian Stage on the GSS, however, is placed one zone above its position in Kazakhstan (Fig. 19.11).

Traditionally, the regional chronostratigraphic standard for the uppermost Middle through Upper Cambrian in Siberia has been subdivided into four stages (superhorizons), each of which includes two regional "horizons." These units were based on inner shelf sections with endemic faunas, along the Kulymbe River in the Koyuy-Igarka region, northwestern Siberia (Rozova, 1963, 1964, 1967, 1968, 1970). The lower boundary of the Cambrian in Siberia is commonly drawn between the Tommotian Stage and the underlying Nemakit-Daldynian Stage. In Russia, the Nemakit-Daldynian Stage usually has been regarded as Precambrian (Vendian) (Khomentovsky, 1974, 1976, 1984; Khomentovsky and Karlova, 1993, 2005; Rozanov et al., 2008; Varlamov et al., 2008b), although on biostratigraphic and carbon isotopic evidence it has been widely regarded internationally as the lowermost Cambrian. All traditional Siberian stages (Fig. 19.11, column 4) are unit stratotype-based, with the stage boundaries defined at the base of either a biozone or a lithologic unit.

19.1.5.3.1 Nemakit-Daldynian

The Nemakit-Daldynian Stage was named for the Nemakit-Daldyn River, Kotuy River Basin, northwestern Siberia, Russia. The stage name was introduced directly from a lithologic unit, the Nemakit-Daldyn "horizon." As originally proposed by Savitsky (1962), the horizon comprises a carbonate succession cropping out on the upper reaches of the Nemakit-Daldyn River, and both its lower and upper boundaries are defined at disconformities. The stage was proposed by Khomentovsky (1976, 1984), who recognized the Nemakit-Daldynian Stage as the uppermost of three successive stages (horizons) that he proposed for the Vendian System (Yudomian Series) of the Yudoma-Anabar facies region. The Nemakit-Daldynian Stage is characterized by the appearance of the first skeletal fossils belonging to the *Anabarites trisulcatus* Zone, the oldest SSF assemblage, and by the fauna of the succeeding *Purella antiqua* Zone (Khomentovsky and Karlova, 1993). No lower boundary can be reliably defined at the type section of the Manukai (Nemakit-Daldyn) Formation of the eastern Pre-Anabar Area, which is regarded as the stratotype of the stage (V.V. Khomentovsky, pers. comm., 2007). Practically, the lower boundary of the stage is recognized by the change of strata bearing Ediacaran fossils to the strata bearing SSFs of the *A. trisulcatus* Zone; the upper boundary is placed at the boundary between the *P. antiqua* Zone and the overlying *Nochoroicyathus sunnaginicus* Zone.

19.1.5.3.2 Tommotian

The Tommotian Stage was named for the town of Tommot on the Aldan River, Russia. As originally defined (Rozanov, 1966; Rozanov and Missarzhevsky, 1966; Rozanov et al., 1969) the Tommotian was the lowermost stage of the Cambrian, represented by a stage stratotype with 85-m-thick successions in the middle reaches of the Aldan River, cropping out from Dvortsy to Ulakhan-Sulugur Creek. The stage is characterized by the abrupt appearance of SSFs including hyoliths, gastropods, inarticulate brachiopods, and problematic fossils, and by the occurrence of primitive archaeocyath sponges having simple systems of porous walls and septae (the first evolutionary stage). Trilobites have not been found in the Tommotian Stage. This pre-trilobite stage embraces three successive zones based on archaeocyath assemblages:

the *Nochoroicyathus sunnaginicus* Zone at the base, followed by the *Dokidocyathus regularis* Zone and the *Dokidocyathus lenaicus* Zone (previously *D. lenaicus—Tumuliolynthus primigenius* Zone).

19.1.5.3.3 Atdabanian

The Atdabanian Stage, named for Atdaban Village on the Lena River, Russia, was named by Zhuravleva et al. (1969) for strata cropping out between the mouth of the Negyurchyune River and the mouth of Achchagyy—Kyyry—Taas Creek (Zhuravlev and Repina, 1990). The lower half of the stage is not exposed in the stratotype area, and the lower boundary of the stage is defined elsewhere (at the base of Bed 4 in the Zhurinsky Mys section, along the Lena River; Varlamov et al., 2008a). *Profallotaspis*, the oldest trilobite known from Siberia, first appears 2.6 m above the boundary in Bed 4. The stage is characterized by the first appearance and the early development of trilobites, dominated by fallotaspidids, in the lower half, and an increasing diversity of trilobites in the upper half. The stage embraces four trilobite zones; in ascending order they are the *Profallotaspis jakutensis* Zone, the *Fallotaspis* Zone, the *Pagetiellus anabarus* Zone, and the *Judomia—Uktaspis (Prouktaspis)* Zone. The stage is also characterized by a sharp increase in archaeocyaths that bear compound skeletal elements. The Atdabanian contains the second stage in the evolution of archaeocyaths. Four archaeocyath-based zones have been established for the Atdabanian (Rozanov and Sokolov, 1984). Mollusk and SSF diversity are rather low in the lower part of the stage, and the upper part of the stage is characterized by SSFs having intercontinental distributions.

19.1.5.3.4 Botoman

The Botoman Stage (alternatively referred to in literature as Botomian Stage) was named by Repina et al. (1964) for the Botoma River, a tributary in the middle reaches of the Lena River, Russia. The stage stratotype section lies on the left bank of the Botoma River (Rozanov and Sokolov, 1984), and the lower boundary of the stage is at the base of unit III, within the Perekhod Formation, in the Ulakhan—Kyyry—Taas section located 1.5 km downstream from the Ulakhan—Kyyry—Taas Creek mouth (Rozanov and Sokolov, 1984). The stage base corresponds to the conterminous base of a trilobite biozone (the *Bergeroniellus micmacciformis—Erbiella* Zone) and an archaeocyathan biozone (the *Porocyathus squamosus—Botomocyathus zelenovi* Zone). The stage is characterized by a high diversity of trilobites, archaeocyaths, brachiopods, rare mollusks, and various hyoliths. Trilobites first appearing above the base of the stage include *Neocobboldia*, *Protolenus*, *Bergeroniellus*, *Bergeroniaspis*, *Micmaccopsis*, *Erbiella*, and *Judomiella*. The trilobites are numerous and diverse, especially in the lower half, in which protolenids

predominate. In contrast, trilobite diversity in the upper half of the stage is greatly reduced. Archaeocyaths are abundant and diverse but restricted to the basal part of the Botoman in the stratotype. The archaeocyaths are characterized mainly by forms having complex walls, the third stage in their evolutionary history.

19.1.5.3.5 Toyonian

The Toyonian Stage was introduced as the national standard stage of the uppermost Lower Cambrian in the former USSR (Spizharsky et al., 1983). The name is derived from Ulakhan-Toyon Island on the Lena River, Russia. The stage stratotype is in a carbonate succession that crops out in the middle reaches of the Lena River between Tit-Ary and Elanka villages. Previously this interval of strata was named the Lenan Stage by Repina et al. (1964) and the Elankan Stage by Rozanov (1973). The lower boundary of the stage is drawn at the base of the Keteme Formation, which is also the base of the *Bergeroniellus ketemensis* Zone. However, in the stratotype, the principal guide fossil, *B. ketemensis*, and other trilobites first appear 6 m above the lower boundary of the *B. ketemensis* Zone (Zhuravlev and Repina, 1990). The stage embraces three trilobite zones: the *B. ketemensis* Zone, the *Lermontovia grandis* Zone, and the *Anabaraspis splendens* Zone. Trilobites of the Toyonian Stage include edelsteinaspidids, dorypygids (*Kootenia*, *Kooteniella*), and dinesidians (*Erdia*), and these forms predominate in the Anabar-Sinsk facies region. In the deeper water Yudoma-Olenek facies, menneraspidid, lermontoviine, and paramicmaccine trilobites are more common. In general, archaeocyaths occur throughout the stage, but in the stratotype section they are confined to the *L. grandis* Zone. The Toyonian contains the fourth (and last) evolutionary stage of archaeocyaths. Before the end of the Toyonian, archaeocyaths became extinct. Hyolithids are represented by forms having shells with polygonal cross sections.

19.1.5.3.6 Amgan

The Amgan is the lowermost stage of the Russian Middle Cambrian Series. The Amgan Stage was named for the Amga River, a tributary of the Aldan River, Russia. Its stratotype is in the middle reaches of the Amga River (Chernysheva, 1961). The base of the Amgan Stage is placed at the base of the *Schistocephalus* Zone, 27 m above the base of the Elanka Formation on the Lena River, near Elanka Village. More than 100 trilobite taxa of Amgan age have been documented from the stratotype area (Chernysheva, 1961; Egorova et al., 1976). The stage is characterized by a turnover of faunas and includes the first appearances of paradoxidids (*Paradoxides*, *Schistocephalus*), oryctocephalids (*Oryctocara*, *Ovatoryctocara*, *Oryctocephalus*, *Tonkinella*), ptychopariids (*Kounamkites*, *Ptychoparia*), and agnostoids after the extinction of the "Lower Cambrian" ellipsocephalids (*Bergeroniellus, Lermontovia, Paramicmacca, Protolenus, Protolenellus*), and redlichiids (*Redlichia, Redlichina*).

In the Yudoma-Olenek facies region (Egorova et al., 1976), the *Schistocephalus* Zone, at the base of the Amgan Stage, corresponds to the *Oryctocara* Zone (also published as either the *Ovatoryctocara* Zone or the *Ovatoryctocara–Schistocephalus* Zone). The *Ovatoryctocara* Zone is identified by the first appearance of the oryctocephalid *O. granulata* (Rozanov, 2016). By implication the base of this series and stage is somewhat lower than the base of the global Wuliuan Stage and the Miaolingian Series. In the Yudoma-Olenek region the *Ovatoryctocara* Zone is succeeded by the *Kounamkites* Zone, the *Ptychagnostus gibbus* Zone, and the *Tomagnostus fissus–Paradoxides sacheri* Zone.

19.1.5.3.7 Mayan

The Mayan Stage was named for the Maya River, a tributary of the Aldan River, Russia, by Egorova et al. (1982). The stage stratotype comprises a number of outcrops on the Yudoma and Maya rivers. The lower boundary of the stage was previously drawn at the base of the *Anopolenus henrici* Zone (Chernysheva, 1967), but Egorova et al. (1982) advocated shifting the boundary downward to the base of the underlying *Tomagnostus fissus–Paradoxides sacheri* Zone. This change was approved for the Russian standard by the Commission on the Cambrian System of the RNSC (Rozanov, 2016). Originally, the upper boundary of this regional stage was defined, but imprecisely, by the disappearance of *Lejopyge, Goniagnostus*, etc., and by the appearance of abundant "Late Cambrian" trilobites such as *Homagnostus fecundus, Buttsia pinga*, and *Toxotis* in the Kharaulakh area. In northwestern Siberia, the top of the stage is characterized by the disappearance of *Maiaspis, Aldanaspis, Buitella*, etc., and by the appearance of *Pauciella prima, Nganasanella*, and *Homagnostus paraobesus* (Egorova et al., 1982). More than 230 trilobites species were documented by Egorova et al. (1982), including various agnostoids, most of which are widely distributed. As adopted for the Russian standard scale (Rozanov, 2016), the upper boundary of the Mayan is defined at the base of an agnostoid zone, the *Kormagnostus simplex* Zone. This zone is lowermost in the Russian Ayusokkanian Stage, and its position is above the base of the Ayusokkanian Stage as used in Kazakhstan. In South Kazakhstan, the base of the Ayusokkanian Stage coincides with the base of the *L. laevigata* Zone (Ergaliev and Ergaliev, 2004, 2008), which occupies a position below the *K. simplex* Zone.

19.1.5.3.8 Kulyumbean

The Kulyumbean Stage is named for the middle reaches of the Kulyumbe River (Rozova, 1963), on the northwestern Siberian Platform, Russia. The stage is subdivided into two "horizons," the Nganasanian Horizon (below) and the

Tavgian Horizon (above). The lower boundary of the stage is drawn at the base of a 10-m-thick limestone breccia that occurs at the bottom of the Nganasanian Horizon. The trilobite fauna in the Kulyumbean Stage is characterized by a high diversity of polymerid trilobites comprising acrocephalitids, eoacidaspidids, lonchocephalids, crepicephalids, pterocephalids, catillicephalids, etc., all of which are basically endemic forms. A few agnostoids, such as the widely distributed *Nahannagnostus nganasanicus*, are also present in the stage. With wide distributions, the agnostoids suggest a correlation to the upper Marjuman Stage in the Mackenzie Mountains, Canada (Pratt, 1992), and to the upper Guzhangian Stage of South China and Northwest China (Zhou et al., 1996; Peng and Robison, 2000).

19.1.5.3.9 Gorbiyachinian

The Gorbiyachinian Stage was named for the Gorbiyachin River, a tributary of the Kulyumbe River, Russia (Rozova, 1963, 1968). The stage stratotype directly overlies the Kulyumbean Stage stratotype, both of which are in the middle reaches of the Kulyumbe River. The lower boundary of the Gorbiyachinian Stage is identified by an abundance of the plethopeltid trilobite *Koldinia mino*. As originally defined by Rozova (1963), the Gorbiyachinian Stage embraces two "horizons," the Maduan Horizon, and the overlying Entsian Horizon. The upper boundary of the stage is drawn at a level where the illaenurid trilobite *Yurakia yurakiensis* occurs in abundance. The stage is characterized by a turnover of trilobite faunas, with only a single species ranging upward from the underlying Kulyumbean Stage in the stratotype section (Rozova, 1968). The trilobite fauna comprises various endemic polymerids, and characteristic forms are *Acidaspidina, Maduiya, Kulyumbopeltis, Taenicephalus*, and *Parakoldinia*. The equivalent of the stage in the Chopko River section (Varlamov et al., 2005) is characterized by the occurrence of the agnostoid *A. orientalis* and the polymerid *Irvingella*, both of which have cosmopolitan distributions.

19.1.5.3.10 Tukalandian

The Tukalandian Stage derives its name from the Tukalandy River, a tributary of the Khantai River, Russia (Rozova, 1963, 1968). The base of the Tukalandian Stage is defined by the local abundance of the illaenurid trilobite *Yurakia yurakiensis* in the stratotype section, where Tukalandian strata lie in succession over strata of the Gorbiyachinian Stage. The upper boundary of the stage is defined by the occurrence of the eulomid trilobite *Dolgeuloma abunda*. The Tukalandian is characterized by the presence of various endemic polymerid trilobites, primarily eulomids (*Kujandaspis = Ketyna*), aphelaspidids (*Amorphella*), advanced eoacidaspidids (*Eoacidaspis*), lonchocephalids (*Graciella, Monosulcatina, Nordia*), and illaenurids (*Polyariella, Yurakia*), and by a turnover of trilobite faunas. There are no species known to range upward from the underlying Gorbiyachinian Stage (Rozova, 1968). The stage is subdivided into two horizons, the Yurakian Horizon below and the Ketyan Horizon above.

19.1.5.3.11 Khantaian

The Khantaian Stage is named for the Khantai River, a tributary of the Yenisey River, in northwestern Siberia, Russia. Its lower boundary with the Tukalandian Stage is defined by the appearance of the polymerid trilobite *Dolgeuloma abunda* (Rozova, 1963, 1968).

The Khantaian Stage of Russia is inferred to be uppermost Cambrian in the lower part and Lower Ordovician (Tremadocian Stage) in the upper part. Originally, Rozova (1963) did not subdivide the Khataian Stage, but in later work she (Rozova, 1967, 1968) subdivided the stage into two "horizons," the Mansian Horizon and the Loparian Horizon, assigning both to the Upper Cambrian. Present in the upper part of the Loparian Horizon is the graptolite *Rhabdinopora flabelliformis* (a cosmopolitan species that elsewhere first occurs in the Tremadocian Stage). This inference of an Ordovician age for that part of the Loparian Horizon is further supported by the presence of the polymerid trilobite *Plethopeltides*, which has an Ordovician aspect. The trilobites that characterize the stage include *Pseudoacrocephalites*, *Kaninia*, *Dolgeuloma*, and *Mansiella*, all of which are endemic and of low diversity.

19.1.5.4 Laurentian Cambrian Stages

The use of Cambrian stage and series nomenclature in Laurentia was reviewed by Babcock et al. (2011). Two sets of regional names have been used, but neither extends through the entire system. The first stages introduced, Dresbachian, Franconian, and Trempealeauan, were based on trilobite assemblages from formations representing shallow shelf lithofacies in the Upper Mississippi Valley area (e.g., Lochman-Balk and Wilson, 1958).

A second set of regional names were influenced by the concept of the biomere (segment of life) introduced by Palmer (1965a) and subsequently discussed by Stitt (1975), Palmer (1979, 1984), Taylor (1997, 2006), and others. As originally defined (Palmer, 1965a), a biomere is a regional biostratigraphic unit bounded by an abrupt extinction event on the shallow cratonic shelf. When this happens to trilobite faunas, an evolving shelf fauna is replaced by a new, low-diversity fauna dominated by simple ptychoparioid trilobites invading from the outer shelf or shelf break. The new fauna then evolves until another extinction event occurs. Six such cycles were suggested by Palmer (1981), but the lower two are as yet undefined. In ascending order they are the "Olenellid," "Corynexochid," Marjumiid, Pterocephaliid, Ptychaspid, and Symphysurinid biomeres.

Ludvigsen and Westrop (1985) considered biomeres to be stages because they were based on an aggregate of trilobite zones and subzones. They named three stages in the "Upper" Cambrian, Marjuman, Steptoean, and Sunwaptan; sections in western North America showing a biomere pattern were selected for reference. The new stage names were intended to replace the terms Dresbachian, Franconian, and Trempealeauan, but these terms continue to be used in certain circumstances, particularly in subsurface studies. Palmer (1998) considered biomeres to be retained as units subtly different from stages and extended Ludvigsen and Westrop's (1985) proposed sequence of stages for the Laurentian Cambrian based on trilobites (Fig. 19.11, column 6). The portion of the Cambrian named by Palmer (1998) as the Begadean Series, which he regarded as pretrilobitic, still lacks defined stages. However, Hollingsworth (2007, 2011) subsequently reported a small assemblage of polymerid trilobites from the upper part of this interval.

19.1.5.4.1 Montezuman

The Montezuman Stage (Palmer, 1998) was named for the Montezuma Range, Nevada, United States. Its base was originally defined by the appearance of characteristic fallotaspidid trilobites. As in Morocco and Siberia, the fallotaspidids are followed by nevadiids and holmiids. At least three families of Olenellina, which are different from olenellines of the succeeding stage (see generic range charts in Palmer and Repina, 1993), are present in the Montezuman Stage. The Montezuman Stage also contains the oldest Laurentian archaeocyaths.

19.1.5.4.2 Dyeran

The Dyeran Stage (Palmer, 1998) was named for the town of Dyer, Nevada, United States, and covers the biostratigraphic interval that for many years was assigned to the *Olenellus* Zone. The *Olenellus* Zone of historic usage has subsequently been regarded as multizonal (Palmer and Repina, 1993; Palmer, 1998; Webster, 2011). The base of the stage coincides with a major change in the olenelloid fauna following the nevadiid-bearing late Montezuman. A similar change was documented by Fritz (1992) from British Columbia, Canada. Olenellid trilobites are characteristic of the Dyeran Stage. Webster (2011) proposed six olenelloid trilobite zones for the upper part of the stage. In ascending order, the zones are the *Arcuolenellus arcuatus*, *Bristolia mohavensis*, *Bristolia insolens*, *Peachella iddingsi*, *Bolbolenellus euryparia*, and *Nephrolenellus multinodus* zones.

19.1.5.4.3 Delamaran

The stratotype of the Delamaran Stage (Palmer, 1998) is the Oak Spring Summit section, Delamar Mountains, Nevada, United States. The stage embraces the "Corynexochid" biomere and the *Plagiura−Poliella*, *Albertella*, and

Glossopleura zones in restricted shelf environments (Palmer and Halley, 1979; Eddy and McCollum, 1998). It is characterized by ptychoparioid, corynexochid, zacanthoidid, dolichometopid, and oryctocephalid trilobites.

19.1.5.4.4 Topazan

The Topazan Stage was erected (Sundberg, 2005) by restriction of the Marjuman Stage. Recognition of the Topazan Stage resulted from restoration of the base of the overlying Marjuman Stage to its original proposed level (Ludvigsen and Westrop, 1985; *contra* Palmer, 1998), the FAD of the cosmopolitan agnostoid *Ptychagnostus atavus* (see Marjuman Stage, Section 19.1.5.4.5). The Topazan Stage is defined as the interval between the top of the Delamaran Stage and the base of the Marjuman Stage (as restricted; Sundberg, 2005). The stage was named for the Topaz Internment Camp (active during World War II), located some 25 km to the southwest of the stratotype. The base of the stage is defined within a shale sequence, 2.6 m above the base of the upper shale member of the Chisholm Formation at section Do2 of Sundberg (1990, 1994) in the Drum Mountains, Utah, United States. The basal 10 cm of the stage contains the FAD of the polymerid trilobite *Proehmaniella basilica*.

The Topazan Stage embraces only a single polymerid zone, the *Ehmaniella* Zone, which is subdivided into four subzones: the *Proehmaniella*, the *Elrathiella*, the *Ehmaniella*, and the *Altiocculus*. In outer shelf facies the agnostoid *Ptychagnostus praecurrens* and *P. gibbus* zones characterize the Topazan Stage. Nearly 70 trilobite taxa of Topazan age have been documented from Nevada and Utah (Sundberg, 1994, 2005).

19.1.5.4.5 Marjuman

The Marjuman Stage (Ludvigsen and Westrop, 1985; emended by Palmer, 1998; restricted by Sundberg, 2005) takes its name from Marjum Pass in the House Range, Utah, United States, and was intended to replace the Marjumian Biomere. Ludvigsen and Westrop (1985) originally defined the base of the Marjuman Stage at the base of the *P. atavus* Zone (which closely corresponds to the base of the *Bolaspidella* Zone based on polymerid trilobites) but this was not a major extinction event according to Palmer (1998). Ludvigsen and Westrop (1985) equated the Marjuman Stage with the Marjumiid biomere (Palmer, 1981), but the biomere event occurred earlier on the inner shelf with a major change from trilobites of the *Glossopleura* Zone to those of the *Ehmaniella* Zone (*Proehmaniella* Subzone; Sundberg, 1994). In open shelf environments, this event corresponds to the base of the *Bathyuriscus–Elrathina* Zone. Palmer (1998) moved the base of the Marjuman Stage downward to coincide with the lowest occurrence of *P. basilica*, which marks the base of the Marjumiid Biomere as he (Palmer, 1981) envisioned it. The revised stage embraced three polymerid trilobite zones;

in ascending order they are the *Ehmaniella* Zone (with the *P. basilica* Subzone at the base), the *Bolaspidella* Zone, and the *Crepicephalus* Zone (Palmer, 1999).

Sundberg (2005) restored the original concept of Ludvigsen and Westrop's (1985) Marjuman and proposed a new Topazan Stage for the interval between the top of the underlying Delamaran Stage and the base of the Marjuman Stage as defined at the base of the *Ptychagnostus atavus* Zone. Subsequently, the base of the *P. atavus* Zone, which coincides with the FAD of the eponymous species in the Drum Mountains of northern Millard County, Utah, was designated as the primary stratigraphic marker coinciding with the GSSP for the Drumian Stage of global chronostratigraphy (Babcock et al., 2007). The Marjum Pass, Utah, section, for which the Marjuman Stage was named, shows considerable structural complications, and the true first appearance of *P. atavus* there is unknown.

The Marjuman is characterized in open shelf environments by cedariid trilobites, four zones of which were documented by Pratt (1992). Cedariid and crepicephalid trilobites characterize inner shelf facies of the Marjuman Stage. The Marjuman Stage, as conceptualized by Ludvigsen and Westrop (1985) and Sundberg (2005), embraces agnostoid zones from the *P. atavus* Zone to the *Glyptagnostus stolidotus* Zone, and corresponds to the Drumian through Guzhangian stages of global usage.

19.1.5.4.6 Steptoean

The Steptoean Stage (Ludvigsen and Westrop, 1985) was named for Steptoe Valley, in the Duck Creek Range, near McGill, eastern Nevada, United States, and was intended to replace the Pterocephaliid Biomere. The base of the stage is defined at the base of the polymerid trilobite *Aphelaspis* Zone, which also corresponds to the base of the Pterocephaliid Biomere (Palmer, 1965b). The *Aphelaspis* Zone contains the cosmopolitan agnostoid *Glyptagnostus reticulatus*, which allows precise correlation globally. Above the *Aphelaspis* Zone, the Steptoean Stage embraces the *Dicanthopyge*, *Prehousia*, *Dunderbergia*, and lower *Elvinia* zones in restricted shelf environments. The agnostoid *Agnostotes orientalis* occurs in the uppermost part of the *Elvinia* Zone (Westrop and Adrain, 2013; Chatterton and Gibb, 2016), which suggests that the base of the global Jiangshanian Stage lies near the top of the Steptoean. The *G. reticulatus*, *Olenaspella regularis* and *Olenaspella evansi* zones characterize the Steptoean Stage in outer shelf facies. The *Parabolinoides calvilimbata* and *Proceratopyge rectispinata* faunas, documented by Pratt (1992), are typical of open shelf environments.

19.1.5.4.7 Sunwaptan

The Sunwaptan Stage was named for Sunwapta Creek, Wilcox Peak, Jasper National Park, in southern Alberta, Canada (Ludvigsen and Westrop, 1985), and was intended to replace the Ptychaspid Biomere (see Longacre, 1970; Stitt, 1975). The base of the Sunwaptan Stage is at the base of the *Irvingella major* Subzone of the *Elvinia* Zone, which Chatterton and Ludvigsen (1998) argued should be regarded as a separate zone. This is succeeded by the *Taenicephalus* Zone, the *Stigmacephalus oweni* fauna, and the *Ellipsocephaloides* Zone in the lower Sunwaptan, and the *Illaenurus* Zone, and most of the *Saukia* Zone in the upper Sunwaptan. More than 130 trilobite taxa of Sunwaptan age have been documented from Alberta by Westrop (1986), and from the District of Mackenzie, Northwest Territories, Canada, by Westrop (1995). Characteristics are dikelocephalid, ptychaspidid, parabolinoidid, saukiid, ellipsocephaloid, illaenurid, and elviniid trilobites.

19.1.5.4.8 Skullrockian

The Skullrockian Stage was named for Skull Rock Pass in the House Range, Utah, United States (Ross et al., 1997). It was originally conceived of as the lowermost stage of the Ibexian Series, which at the time was considered to be Lower Ordovician. The base of the Skullrockian is defined by conodonts at the base of the *Hirsutodontus hirsutus* Subzone of the *Cordylodus proavus* Zone. On the polymerid trilobite zonal scale, this level corresponds to the base of the *Eurekia apopsis* Zone (Ross et al., 1997; Miller et al., 2006). The *E. apopsis* Zone has a limited trilobite fauna, as does the overlying *Mississquoia* Zone, and the primary group used for high-resolution correlation is conodonts.

The Skullrockian Stage of Laurentia is Cambrian in the lower part, and Lower Ordovician through most of its stratigraphic extent. Following definition of the Ordovician GSSP at the FAD of the euconodont *Iapetognathus fluctivagus*, a position that is partway up through the Skullrockian Stage, the lower part of the stage (through the *Cordylodus lindstromi* Zone of conodont zonation and the *Symphysurina brevispicata* Subzone of the *Symphysurina* Zone of polymerid trilobite zonation) was automatically reassigned to the Cambrian. Most of the stage remains assigned to the Ordovician, however (Miller et al., 2003, 2006). Difficulties in achieving precise correlation between the horizon containing *I. fluctivagus* from the Ordovician stratotype at Green Point, Newfoundland, Canada, and western Utah, United States, where the Skullrockian was defined, were discussed by Miller et al. (2003). As a result, redefinition of the Skullrockian Stage has not taken place, nor has a replacement stage whose base corresponds to the base of the Ordovician, been proposed.

19.1.5.5 Baltica Cambrian "Stages" and Biostratigraphy

Regional stages have not been widely adopted for the Cambrian of Baltica. A substantial part of the Cambrian of eastern Europe has, however, been subdivided into acritarch zones or "horizons" that have been considered as regional stages by some local authors (see Nielsen and Schovsbo, 2011, and references therein). Most of these regional "stages" are not strict chronostratigraphic units and cannot be easily recognized outside of eastern Europe (e.g., Eklund, 1990; Moczydłowska, 1991). Hence, the eastern European regional "stages" generally have not been accepted and used by authors working on the Cambrian of Scandinavia. Nielsen and Schovsbo (2011, 2015), however, adopted a slightly emended version of the eastern European "stage" classification and changed the names by adding the suffix "ian" (e.g., Dominopolian, Vergalian–Rausvian, and Kibartian). Nielsen and Schovsbo (2015) also introduced two new regional stages in the traditional Middle Cambrian, the Bödan Stage for the *Acadoparadoxides oelandicus* Superzone, and the Almbackenian Stage for the *Paradoxides paradoxissimus* Superzone. As most of the regional "stages" (alternatively, zones or horizons; see Moczydłowska, 1991), originally recognized from stratigraphically incomplete successions in Estonia, Russia, Ukraine, and Lithuania, are not well defined and have not gained wide acceptance, this section focuses on the biostratigraphy of Baltica, and Scandinavia in particular.

The Terreneuvian-to-Series 2 succession of Baltica is generally unfossiliferous or poorly fossiliferous except in its uppermost part. By contrast, the Miaolingian and Furongian strata of Scandinavia are generally richly fossiliferous with faunas commonly dominated by polymerid and agnostoid trilobites. These trilobites provide a firm basis for the biostratigraphic classification (e.g., Westergård, 1922, 1946, 1947; Henningsmoen, 1957; Ahlberg, 2003; Axheimer and Ahlberg, 2003; Axheimer et al., 2006; Høyberget and Bruton, 2008). Recent efforts to produce a high-resolution trilobite zonation of the Miaolingian and Furongian in Scandinavia, especially in southern Sweden and southern Norway, have resulted in new zonal nomenclature. Because of significant differences in ecologic and geographic distributions, separate zonal schemes are now being used for the polymerid trilobites and the agnostoids of Scandinavia (e.g., Terfelt et al., 2008, 2011; Ahlberg and Terfelt, 2012; Nielsen et al., 2014; Rasmussen et al., 2015; Babcock et al., 2017).

19.1.5.5.1 Terreneuvian and Series 2

The biostratigraphy and regional correlation of the Terreneuvian and Cambrian Series 2 of Baltica have been reviewed by Ahlberg (1981, 1985, 1991), Bergström (1981), Ahlberg et al., 1986, Moczydłowska (1991), Moczydłowska et al. (2001), Ebbestad et al. (2003), Nielsen and Schovsbo (2007, 2011), Axheimer et al. (2007), Zhang et al. (2017), and Meidla (2017), among others. Zonation and regional correlations commonly have been based on polymerid trilobites, acritarchs, trace fossils, and SSFs. The zones can be categorized as assemblage-zones and are generally vaguely defined. Conventionally, the lowermost zone, the *Platysolenites antiquissimus* Zone of the upper Terreneuvian, is characterized by small shelly or tubular fossils, such as the eponymous species, *Spirosolenites spiralis*, and *Aldanella kunda* (e.g., Føyn and Glaessner, 1979). The four successive Cambrian Series 2 trilobite zones as usually recognized are, in ascending order, the *Schmidtiellus mickwitzi*, *Holmia inusitata*, *Holmia kjerulfi*, and *Ornamentaspis? linnarssoni* zones. Nielsen and Schovsbo (2011) suggested abandoning use of the *H. inusitata* Zone and also recommended subdividing the *O.? linnarssoni* Zone into a lower subzone characterized by *O.? linnarssoni* and an upper subzone characterized by *Comluella? scanica* and *Ellipsocephalus? lunatus*.

Recent work has revealed important new information on the distribution and stratigraphic significance of trilobites in Cambrian Series 2 strata of Scandinavia, and it is evident that some zonal assemblages reflect biofacies more than temporal differences. For the purpose of simplicity and to minimize ambiguity of the biozonation, Ahlberg et al. (2016) proposed a subdivision of the series into four interval-zones: the *S. mickwitzi*, *H. kjerulfi*, *Strenuaeva? (=Ellipsostrenua) spinosa*, and *Chelediscus acifer* zones (Fig. 19.2). Each zone is defined by the first appearance of the eponymous species and delimited at the top by the base of the succeeding zone. The species are distinctive, easily recognizable and, except for *S. mickwitzi*, widespread.

The first appearance of *Strenuaeva primaeva*, a distinctive ellipsocephalid trilobite, nearly coincides with the first appearance of *H. kjerulfi* and can be used as a secondary marker for the base of the *H. kjerulfi* Zone. *Calodiscus lobatus* is an eodiscoid trilobite that has been reported from Baltica, Avalonia, West Gondwana, Laurentia, and Siberia. Its first appearance in Scandinavia nearly coincides with the first appearance of *Ellipsostrenua spinosa* (Cederström et al., 2009, 2012). The eodiscoid *Chelediscus acifer* is known from the Protolenid–strenuellid Zone (*Protolenus* Zone) of eastern Avalonia (England), the *Geyerorodes howleyi* Zone of western Avalonia (southeastern Newfoundland), and northernmost Sweden (Rushton, 1966; Fletcher, 2003; Axheimer et al., 2007). Strata with *C. acifer* are inferred to be younger than those yielding eodiscoid trilobites of the *Serrodiscus bellimarginatus–Triangulaspis annio–Hebediscus attleborensis* assemblage, and older than, or partly correlative with, strata of the *Morocconus notabilis* Zone of Newfoundland and Morocco (Axheimer et al., 2007). The *C. acifer* Zone may correlate with the *C.? scanica–E.? lunatus* Zone of southern Sweden.

19.1.5.5.2 Miaolingian

In Scandinavia the base the Miaolingian generally rests on a prominent unconformity ascribed to nondeposition and erosion during a eustatic sea-level fall that correlates partly with the regressive "Hawke Bay Event" (Nielsen and

Schovsbo, 2015). The Miaolingian of Scandinavia is currently subdivided into three superzones, which in ascending order are the *Acadoparadoxides oelandicus*, *Paradoxides paradoxissimus*, and *P. forchhammeri* superzones. These superzones were considered to be regional stages by Westergård (1946). The *A. oelandicus* Superzone corresponds to the Bödan Regional Stage, and the *P. paradoxissimus* Superzone corresponds to the Almbackenian Stage (Nielsen and Schovsbo, 2015).

With minor modifications, the global agnostoid zonation proposed by Peng and Robison (2000) can be applied to the Miaolingian succession of Scandinavia (e.g., Axheimer and Ahlberg, 2003; Weidner et al., 2004; Axheimer et al., 2006; Høyberget and Bruton, 2008; Weidner and Nielsen, 2014; Weidner and Ebbestad, 2014). Noteworthy is that the *Proagnostus bulbus* Zone of Peng and Robison (2000) is represented by the upper *Lejopyge laevigata* Zone in Scandinavia, and the *Agnostus pisiformis* Zone forms the uppermost zone in the Miaolingian of Scandinavia (Axheimer et al., 2006). Beginning in strata correlated with the Wuliuan Stage, the recognized agnostoid zonation includes the *Ptychagnostus praecurrens*, *P. gibbus*, *P. atavus*, *P. punctuosus*, *Goniagnostus nathorsti*, *Lejopyge laevigata*, and *Agnostus pisiformis* zones. The first appearance of *P. atavus* corresponds to the base of the global Drumian Stage, and the first appearance of *L. laevigata* corresponds to the base of the global Guzhangian Stage.

19.1.5.5.3 Furongian

The Furongian biostratigraphy of Scandinavia is largely based on the succession of olenid trilobites. The rate of faunal turnover is high, which enabled Westergård (1922, 1947) and Henningsmoen (1957) to establish a high-resolution biostratigraphy. Their biostratigraphic scheme has been modified (Terfelt et al., 2008; Terfelt et al., 2011; Høyberget and Bruton, 2012; Weidner and Nielsen, 2013; Rasmussen et al., 2015) and the Furongian of Scandinavia is now subdivided into six superzones and 26 polymerid (olenid) trilobite zones that can be linked to four parallel agnostoid zones (Nielsen et al., 2014; Rasmussen et al., 2017; Babcock et al., 2017; Nielsen et al. 2020). In ascending order, the Furongian includes the *Olenus*, *Parabolina*, *Leptoplastus*, *Protopeltura*, *Peltura*, and *Acerocarina* superzones (Fig. 19.2). Recognized agnostoid zones are the *Glyptagnostus reticulatus–Agnostus (Homagnostus) obesus*, *Pseudagnostus cyclopyge*, *Lotagnostus americanus*, and *Trilobagnostus holmi* zones. The conterminous base of the Furongian Series and the Paibian Stage is placed at the lowest occurrence of *Olenus gibbosus*, which coincides with the first appearance of the agnostoid *G. reticulatus* (see Peng et al., 2004a; Ahlberg and Terfelt, 2012; Nielsen et al., 2014). Polymerid trilobites and agnostoids from near the base of the *Parabolina* Superzone suggest a correlation with the base of the Jiangshanian Stage

(Ahlberg and Terfelt, 2012). The first appearance of *L. americanus* in the upper part of the *Protopeltura* Superzone suggests a correlation with the base of provisional Stage 10.

19.2 Cambrian Stratigraphy

19.2.1 Faunal provinces

The Cambrian Period is noteworthy from a biologic standpoint because it marks the appearance of most multicellular phyla that have populated the Earth. Faunal provincialism tended to be strong, and biostratigraphic zonal schemes based on benthic and nektobenthic taxa generally cannot be applied beyond their provincial boundaries.

Álvaro et al. (2013) summarized the history of studies on Cambrian trilobite biogeography and provided a comprehensive review based on an updated database of Cambrian genera. Most authors have recognized biogeographic differentiation into two main provinces during Cambrian Epoch 2 (Cowie, 1971; Kobayashi, 1972; Palmer, 1973; Lu et al., 1974; Lu, 1981; Chang, 1989; Palmer and Repina, 1993; Álvaro et al., 2013). One faunal province, the Redlichiid Province of Gondwana, is characterized by endemic redlichiids, pandemic ellipsocephaloids, and eodiscids. The other faunal province, the Olenellid Province, comprising much of Baltica, Laurentia, and Siberia, is characterized by endemic olenellids, pandemic ellipsocephaloids, and eodiscids. An overlap in the geographic ranges of taxa characteristic of both major provinces in some peri-Gondwanan margins led Pillola (1991) to erect the intermediate Bigotinid Province.

For trilobites of the Miaolingian and Furongian epochs, Palmer (1973) and others (e.g., Sdzuy, 1972; Jell, 1974; Chang, 1989) have recognized more complicated biogeographic schemes. Terms such as Pacific (or North American) and Atlantic (or Acado-Baltic) have often been used to distinguish biogeographic units. Chang (1989) characterized the Pacific Province using an assemblage of centropleurid, xystridurid, and olenid trilobites and characterized the Atlantic Province using an assemblage of paradoxidid and olenid trilobites. The Acado-Baltic Province (*sensu* Sdzuy, 1972), characterized by a paradoxidid-solenopleurid-conocoryphid assemblage, was widespread through Avalonia, the Mediterranean and central-European areas, and Baltica (Álvaro and Vizcaïno, 2003). Babcock (1994a, 1994b) showed that this trilobite assemblage was widely distributed in cool marine waters of various latitudes, including in deep water surrounding tropical Laurentia. Analysis of a large data set of Cambrian genera led Jell (1974) to recognize three trilobite provinces: (1) Columban in North and South America; (2) Viking in Europe, maritime North America, and northwestern Africa; and (3) Tollchuticook in Asia, Australia, and Antarctica. Palmer (1973), Robison (1976), Pegel (2000), Robison and

Babcock (2011), among others, recognized differentiation between trilobite faunas of outer shelf areas and inner shelf areas of low-latitude continents such as Laurentia and Siberia.

Biogeographic studies on Cambrian trilobites have played an integral role in the recognition of tectonostratigraphic terranes. In general, the juxtaposition of trilobites representing two distinct faunal units in neighboring strata has been used to help infer the boundary of an accreted terrane (e.g., Secor et al., 1983; Samson et al., 1990). However, Babcock (1994b) advised caution in such interpretations, as warm water shelf faunas can occur in close association with cooler water faunas of adjacent deepwater in tropical regions. Stratigraphic and structural/tectonic evidence must be used to supplement biogeographic information to arrive at a conclusion as to a terrane's provenance. Today, a complex mosaic of tectonostratigraphic terranes is recognized, particularly for areas such as Europe, Asia, and the margins of North America. Álvaro et al. (2013) analyzed the biogeographic affinities of trilobites among all Cambrian continents and numerous terranes. These results are in general agreement with more classical interpretations of biogeographic provinces but provide considerable additional information about biogeographic links between regions.

Archaeocyaths also showed provincialism during the Cambrian Period. Debrenne (1992) identified three archaeocyathan faunal provinces that existed in the early half of the Cambrian: (1) an Afro-European Province, which possibly extends to China, and characterized by Anthomorphidae; (2) an Australo-Antarctica Province characterized by Flindersicyathidae, Metacyathidae, and Syringocnemidae; and (3) a Siberian Province characterized by genera belonging to all of these families. Kruse and Shi (2000), who analyzed the distributions of archaeocyaths statistically, recognized five provinces:

1. Siberia—Mongolia
2. Europe—Morocco
3. Central Asia—East Asia
4. Australia—Antarctica
5. North America—Koryakia

19.2.2 Trilobite zones

The most widely used fossil group for biostratigraphic zonation of the Cambrian is trilobites, the best known group of Paleozoic arthropods. Beginning in provisional Series 2, they enable fine stratigraphic subdivision and good correlation reliability. Babcock et al. (2017) provided a review of Cambrian trilobite biostratigraphy and its role in calibrating other chronostratigraphic techniques. In general, polymerid and agnostoid trilobites have different biogeographic distributions and correlation value. Polymerid species and genera tend to be endemic to individual regions or paleocontinents and are thus of greatest use in correlating deposits of the continental shelf and platform (Robison, 1976; Babcock, 1994a; Peng et al., 2004b; Babcock et al., 2007, 2011, 2017; Robison and Babcock, 2011). Agnostoid species tend to be much more widespread, and many are cosmopolitan. They are of great value in correlating open shelf- deposits to shelf margin deposits intercontinentally (Westergård, 1946; Robison, 1976, 1984, 1994; Öpik, 1979; Peng and Robison, 2000; Ahlberg, 2003; Ahlberg et al., 2004; Peng et al., 2004a; Babcock et al., 2007, 2011, 2017; Robison and Babcock, 2011; Ahlberg and Terfelt, 2012).

In the late half of the Cambrian, trilobite diversification and evolutionary turnover were extreme. For this reason, fine zonations of polymerids and agnostoids have been established on the major paleocontinents (Fig. 19.2). Cosmopolitan agnostoids enable global correlation of Miaolingian and Furongian strata (Westergård, 1946; Öpik, 1979; Peng and Robison, 2000; Peng and Babcock, 2005b; Ergaliev and Ergaliev, 2008; Babcock et al., 2011, 2017; Robison and Babcock, 2011; Terfelt et al., 2011; Ahlberg and Terfelt, 2012). Thirteen agnostoid zones have been defined beginning in the upper part of the Wuliuan Stage, with the zonal bases being placed at the first appearances of eponymous species. In ascending order, these are the *Ptychagnostus gibbus, P. atavus, P. punctuosus, Goniagnostus nathorsti, Lejopyge armata, L. laevigata, Proagnostus bulbus, Linguagnostus reconditus, Glyptagnostus stolidotus, G. reticulatus, Agnostotes orientalis,* and *Lotagnostus americanus* zones (Robison, 1984; Peng and Robison, 2000; Peng and Babcock, 2005b). Some zones, such as the *G. nathorsti, L. armata, P. bulbus, L. reconditus,* and *G. stolidotus* zones are not recognized on all paleocontinents (Robison and Babcock, 2011; Babcock et al., 2011, 2017). In the Wuliuan Stage below the first appearance of *P. gibbus,* a globally consistent zonation of agnostoids has not been established. Babcock et al. (2011) demonstrated the synchroneity of most agnostoid appearances in shelf areas of separate paleocontinents associated with the initial stages of eustatic sealevel rises.

China, Russia, North America, Scandinavia, and Australia have the most complete Cambrian trilobite zonal successions (Fig. 19.2). Those of South China, Siberia, North America, Scandinavia, and Australia are shown in Fig. 19.2. Historically, differing biostratigraphic philosophies have been applied in different regions, and these have resulted in differing concepts of trilobite zones (see review in Babcock et al., 2017). In North America, for example, the concept of a zone was commonly based on the range of a characteristic species or genus (an interval-zone; Robison, 1994). In China, Russia, Scandinavia, and Australia, species-zones or assemblage-zones were most commonly applied. In Scandinavia, the pre-Furongian agnostoid zones were, until recently (Terfelt et al., 2008, 2011; Ahlberg et al., 2009), based on local abundance of eponymous species (Westergård, 1946; Peng and Robison,

2000; Axheimer et al., 2006). Increasingly in recent years, zones based on the first appearances of characteristic species have been replacing the older, regional concepts of zones. Particularly where widespread species, such as agnostoids, are the characteristic species of zones, this practice has led to precise correlation regionally and intercontinentally.

19.2.3 Archaeocyathan zones

More than 300 genera of regular Archaeocyatha and Radiocyatha are known from carbonate platforms in the lower half of the Cambrian (Kruse and Shi, 2000). Archaeocyaths have been used extensively for biostratigraphy in certain regions. The most detailed archaeocyathan biostratigraphy has been developed in Siberia, where the Tommotian Stage embraces three successive assemblage-zones, the Atdabanian four, the Botoman three, and the Toyonian three (Debrenne and Rozanov, 1983; Zhuravlev, 1995). Archaeocyathan zones have also been established in South Australia (5), Laurentia (9), Spain (11), Morocco (4), and South China (4) (Zhuravlev, 1995; Yang et al., 2005). Problems associated with correlation of these areas on the basis of archaeocyaths are due primarily to regional endemism. For example, Kruse and Shi (2000) noted that of the 240 archaeocyathan species in Australia and Antarctica, only 26 are shared between the two continents, and only genera with wide stratigraphic distributions are common to Australia and Siberia (Zhuravlev and Gravestock, 1994).

19.2.4 Small shelly fossil zones

The primarily and secondarily phosphatic skeletonized microfossils, termed small shelly fossils, occur in differing levels of abundance in the lower half of the Cambrian. In South China, Siberia, and Australia, SSFs are usually used in regional biostratigraphy. A detailed regional SSF biostratigraphy with three assemblage-zones embracing eight subzones has been developed for the Diandongian Series of South China with the Jinningian Stage embracing five subzones and the Meishucunian Stage (s.s.) three (Luo et al., 1984). Qian et al. (1999) subsequently recognized only four assemblage-zones for the series but added four assemblage-zones for the overlying Qiandongian Series. In Siberia, an SSF biostratigraphy with two zones has been developed for the Nemakit-Daldynian Stage, although a diverse SSF assemblage occurs in the basal part of the succeeding Tommotian Stage (Khomentovsky and Karlova, 1993). In Australia, three informal SSF zones have been established for the mid-Stage 3 through basal Stage 4 interval (middle and upper "Atdabanian" through lower "Toyonian") in the Arrowie and Stansbury basins (Demidenko et al., 2001; Jago et al., 2006). More recently, Betts et al. (2016; 2017, 2018) proposed a new lower Cambrian biostratigraphy, also based on sections of the Arrowie and Stansbury basins, with three new SSF zones (the *Kulparina rostrata*, *Micrina etheridgei*, and *Dailyatia odyssey* zones), in the upper Stage 2 through lower Stage 4 interval. Two additional horizons, one yielding *Sunnaginia imbricata*, and the other yielding *Watsonella crosbyi*, underlie these zones (Betts et al., 2016; Jacquet et al., 2017). An SSF biostratigraphy or succession has also been developed for the Terreneuvian Series of England, Poland, Iran, southern France, and Mongolia (Brasier, 1984; Keber, 1988; Orłowski, 1992; Hamdi et al., 1989; Khomentovsky and Gibsher, 1996). Because of apparent regional endemism, SSF correlation is more or less limited and problematic. It is not certain at present how much of the apparent endemism of SSFs is related to the development of separate taxonomic nomenclature in separate regions of Cambrian exposure.

19.2.5 Conodont zones

Conodont elements, including slender, simple cones referred to as protoconodonts, range through Cambrian strata beginning about the middle of the Terreneuvian Series (Bengtson, 1976). In both Siberia and South China, the protoconodont *Protohertzina* has been used as an eponymous genus of the *Anabarites—Protohertzina* Zone that occurs in the basal part of a regional stage (Nemakit-Daldynian in Siberia; Jinningian in South China). Conodonts began to diversify in the middle of the Miaolingian Epoch, and in the Furongian, they had become sufficiently common and differentiated to be biostratigraphically useful.

Conodonts of the Furongian Series, including protoconodonts, paraconodonts, and euconodonts, have been most intensively studied in the Great Basin (e.g., Miller, 1980, 1988; Landing et al., 2011; Miller et al., 2011, 2018), Western Queensland, Australia (Black Mountain; Druce and Jones, 1971; Nicoll and Shergold, 1991; Shergold and Nicoll, 1992), and South China (Dong and Bergström, 2001a, 2001b; Dong et al., 2004; Dong and Zhang, 2017). In Utah, 11 subzones have been defined through the interval of the upper Sunwaptan Stage through the lower Skullrockian Stage (*Saukia junia* Subzone of the *Saukia* Zone through the *Symphysurina bulbosa* Subzone of the *Symphysurina* Zone) (Miller, 1980; Miller et al., 2006). The subzones are named for *Proconodontus posterocostatus*, *Proconodontus muelleri*, *Eoconodontus notchpeakensis*, *Cambrooistodus minutus*, *Hirsutodontus hirsutus*, *Fryxellodontus inornatus*, *Clavohamulus elongatus*, *Hirsutodontus simplex*, and *Clavohamulus hintzei*. Above this interval lies the *Cordylodus lindstromi* Zone, which has lower and upper subzones. In Australia, at least seven conodont assemblages have been recognized through the Jiangshanian and provisional Stage 10. These assemblage-zones are, in ascending order, based on *Teridontus nakamurai*, *Hispidodontus resimus*, *Hispidodontus*

appressus, *Hispidodontus discretus*, *Cordylodus proavus*, *Hirsutodontus simplex*, *Cordylodus prolindstromi*, and *C. lindstromi* (basal part only) (Shergold and Nicoll, 1992; Kruse et al., 2009). Dong and Bergström (2001a, 2001b), Dong et al. (2004), and Dong and Zhang (2017) developed a comprehensive conodont biostratigraphy with 12 and one-half zones ranging through the interval of the Drumian Stage through the Furongian Series in Hunan, China. In ascending order, the zones are the *Gapparodus bisulcatus—Westergaardodina brevidens* Zone, *Shandongodus priscus—Hunanognathus tricuspidatus* Zone, *Westergaardodina quadrata* Zone, *W. matsushitai—W. grandidens* Zone, *W. lui—W. ani* Zone, *W.* cf. *calix—Prooneotodus rotundatus* Zone, *Proconodontus tenuiserratus* Zone, *Proconodontus* Zone, *Eoconodontus* Zone, *Cordylodus proavus* Zone, *C. intermedius* Zone, *C. lindstromi* Zone, and *Co. angulatus* Zone (lower part).

19.2.6 Magnetostratigraphy

Two types of magnetostratigraphic information have been used for correlation of Cambrian strata, magnetic polarity studies and magnetic susceptibility studies. Most work has involved development of a magnetic polarity time scale (Fig. 19.1). To the present, such a time scale remains incomplete, for reasons summarized by Trench (1996). Detailed studies, however, have been undertaken through parts of all four series, with the most intense research being concentrated on the Cambrian—Ordovician boundary interval. So far, magnetic susceptibility work has been applied only to strata within the Drumian Stage and to strata near the base of the stage.

Kirschvink and Rozanov (1984), Kirschvink et al. (1991), and Varlamov et al. (2008b) provided a detailed magnetostratigraphic polarity scale for the uppermost Terreneuvian Series and lower part of Series 2 derived from studies along the Lena River of Siberia. In the Tommotian and Atdabanian stages, as used regionally on the Siberian Platform, Kirschvink and Rozanov (1984) and Kirschvink et al. (1991) found many polarity reversals. The Tommotian correlates approximately to the upper part of Stage 2, and the Atdabanian correlates approximately to the lower to middle part of Stage 3. Pavlov and Gallet (2001) challenged the interpretation of Kirschvink and Rozanov's (1984) results because the paleomagnetic pole they obtained differs significantly from other pole positions obtained from the Siberian Platform and because a predominant reversed polarity is most often observed for this time interval (Khramov and Rodionov, 1980; Pisarevsky et al., 1997). Nevertheless, the information from Siberia is essentially in agreement with results obtained from Morocco and South China near the equivalents of the Tommotian—Atdabanian boundary (Kirschvink et al., 1991, 1997), where several magnetic reversals were discovered. Magnetostratigraphic information, calibrated with chemostratigraphic results, can be used in a general way to correlate among these three areas (Kirschvink et al., 1991).

Rudimentary magnetostratigraphic polarity results are available for the Botoman and Toyonian stages of Siberian usage (middle to upper part of Series 2), derived from studies along the Lena River of Siberia (Kirschvink and Rozanov, 1984; Kirschvink et al., 1991; Varlamov et al., 2008b). These results show two long episodes each of normal and reversed polarity. The Botoman correlates approximately to upper Stage 3 through lowermost Stage 4, and the Toyonian correlates approximately to lower–middle Stage 4. In contrast, magnetostratigraphic data from the Yuanshan Member of the Chiungchussu Formation (also known as Maotianshan Shale) from the Chengjiang area, eastern Yunnan, China, reveal a relatively high frequency of magnetic pole reversals (Yin, 2002). In the Yuanshan Member, which is the unit containing the Chengjiang Biota, at least 29 magnetic polarity intervals are recognized. This member correlates to the lowermost part of Series 2, Stage 3 (lower Nangaoan Stage of South China regional usage).

Magnetic polarity studies have been conducted on several paleocontinents in the Cambrian–Ordovician boundary interval, and studies from the middle of the Miaolingian Series through the lower Furongian have been reported from Siberia. Early, and rather limited, investigations in strata adjacent to the Ordovician base were made by Kirschvink (1978a, 1978b) and Klootwijk (1980) in central and South Australia. More detailed information comes from studies across the Cambrian–Ordovician boundary interval at Black Mountain in western Queensland, Australia (Ripperdan and Kirschvink, 1992; Ripperdan et al., 1992), at Batyrbai, southern Kazakhstan (Apollonov et al., 1992), at Dayangcha and Tangshan, North China (Ripperdan et al., 1993; Yang et al., 2002), and along the Kulyumbe River, northwestern Siberian Platform (Pavlov and Gallet, 2001, 2005; Kouchinsky et al., 2008). Geomagnetic results for the Drumian through Paibian stages were reported from the Kulyumbe River section of Siberia by Pavlov and Gallet (1998, 2001, 2005), Kirschvink and Raub (2003), Pavlov et al. (2008), and Kouchinsky et al. (2008). Combining results, a composite magnetic polarity time scale is now available from the Drumian Stage upward into the Tremadocian Stage of the Ordovician System (Kouchinsky et al., 2008).

According to Kouchinsky et al. (2008), the "Middle Cambrian" (presumably approximating the Miaolingian Series) has up to 100 geomagnetic reversals, although data were presented only for the Drumian and Guzhangian stages. A total of 100 geomagnetic intervals corresponds to a reversal frequency of 10 per million years, an extremely high rate. In the "Upper Cambrian" (closely equivalent to the Furongian Series), only 10–11 magnetic intervals were recognized, and this corresponds to a reversal rate of about 1 per million years. This rate is an order of magnitude lower than that in the Miaolingian Series. The Furongian is dominated by

intervals of reversed polarity, most of them relatively long, and mostly short intervals of normal polarity. Except for a short interval of normal polarity in the lower part of the stage, the Paibian shows a long interval of continuous reversed polarity. The longest interval of normal polarity in the Furongian Series is close to the top of the series. The initial Ordovician is dominated by periods of normal polarity with a couple of intervals of reversed polarity (Ripperdan et al., 1993; Yang et al., 2002; Pavlov and Gallet, 2005; Wang et al., 2019). Only two to three geomagnetic intervals were recognized from the Tremadocian Stage (Ordovician) by Kouchinsky et al. (2008).

Studies of magnetic susceptibility recently have been applied to limited intervals of the Drumian Stage, and to strata adjacent to the base of the stage. Magnetic susceptibility shows excellent potential for high-resolution correlation and adds to the magnetostratigraphic information obtained through magnetic polarity studies. A detailed profile across the interval containing the Drumian Stage GSSP in the Drum Mountains, Utah, United States, shows a positive deflection in magnetic susceptibility that correlates precisely to the beginning of the DICE $\delta^{13}C$ excursion (Babcock et al., 2009). Halgedahl et al. (2009) showed that higher in the Drumian Stage peaks in magnetic susceptibility can be matched among sections within the same general area of Utah.

19.2.7 Chemostratigraphy

A significant and stratigraphically important body of chemostratigraphic information now exists. Particularly important are stable isotopes of carbon ($\delta^{13}C$) and strontium ($^{87}Sr/^{86}Sr$) (Fig. 19.12).

Zhu et al. (2006) synthesized previous studies of $\delta^{13}C$ isotopic values in Cambrian deposits (e.g., Brasier, 1993; Derry et al., 1994; Zhang et al., 1997; Brasier and Sukhov, 1998; Saltzman et al., 1998; Montañez et al., 2000; Corsetti and Hagadorn, 2001; Buggisch et al., 2003; Peng et al., 2004a, 2006; Zhu et al., 2004; Babcock et al., 2005; Guo et al., 2005; Maloof et al., 2005; Kouchinsky et al., 2005) and added new information from carbonates of South China to develop a generalized curve encompassing the entire Cambrian System. They recognized 10 distinct isotopic excursions, many of which coincide with important biotic events such as evolutionary radiations and extinctions, or with times of taphonomic windows (Fig. 19.12). Three positive $\delta^{13}C$ excursions recorded in the Terreneuvian Series correspond to times when faunas of SSFs radiated on the Yangtze Platform. Only the last of these three excursions has been named (the ZHUCE or ZHUjiaqing Carbon isotope Excursion; Zhu et al., 2006). The CARE (Cambrian Arthropod Radiation isotope Excursion; Zhu et al., 2006) is a positive shift in $\delta^{13}C$ values associated with the appearance of a wide variety of arthropod fossils (the Cambrian Arthropod Radiation Event), particularly in major Konservat-

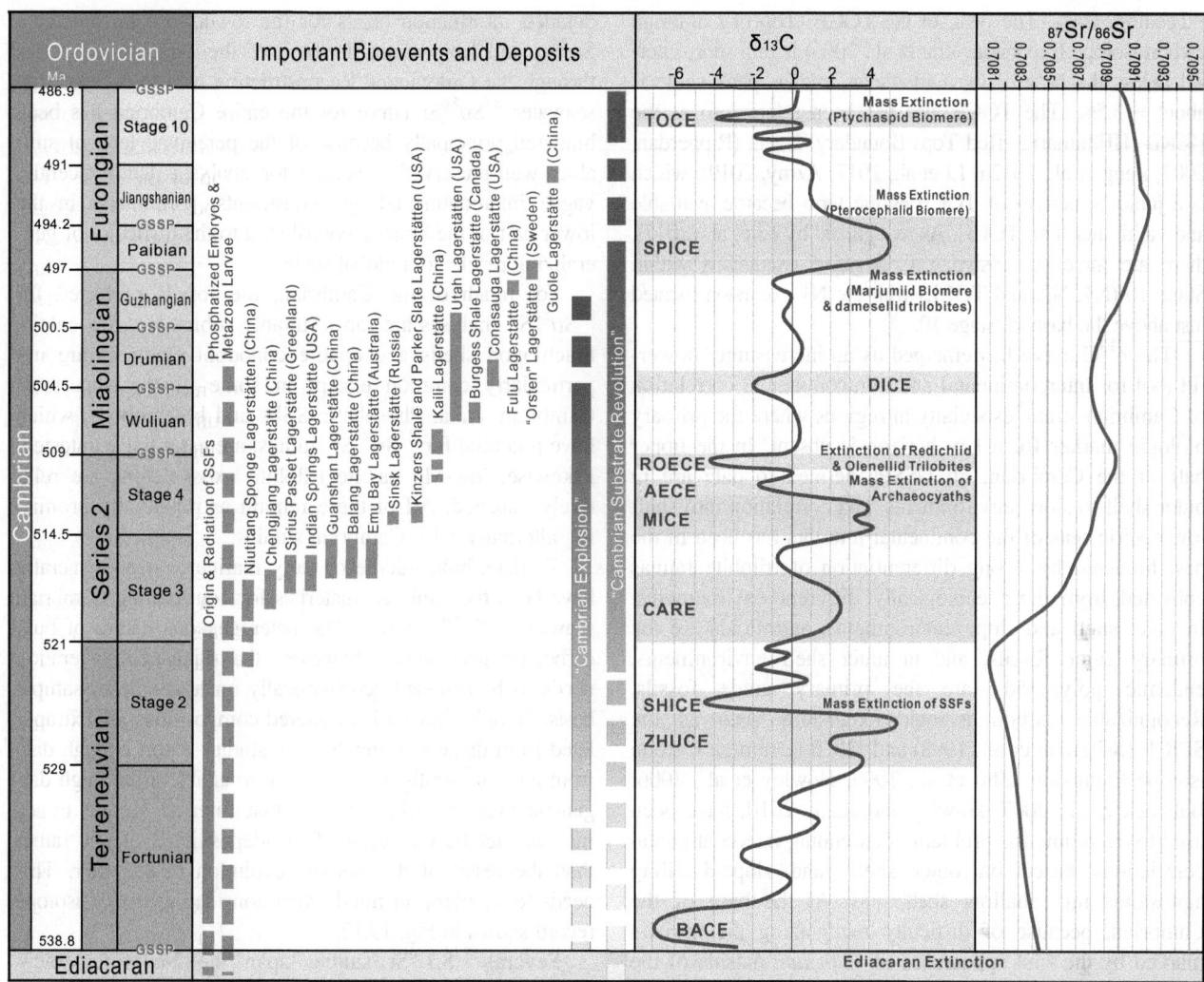

FIGURE 19.12 Carbon isotope (δ^{13}C) and strontium isotope (^{87}Sr/^{86}Sr) chemostratigraphy of the Cambrian System and comparison to biotic events. The carbon isotope chemostratigraphy and comparison to biotic events is modified from Zhu et al. (2006), Miller et al. (2006), Howley and Jiang (2010), and Li et al. (2017). The strontium curve is a composite derived from curves for the upper Terreneuvian Series and lower Series 2 (Derry et al., 1994); upper Series 2 through the Miaolingian Series (Montañez et al., 2000); and much of the Furongian Series (Saltzman et al., 1995; Kouchinsky et al., 2008), except for the uppermost part of Stage 10 (Ebneth et al., 2001).

lagerstätten. The extinction of acritarchs and other organisms near the end of the Ediacaran Period corresponds to a strong negative shift in δ^{13}C values recorded in carbonate sediments (BACE excursion, or Basal Cambrian Carbon isotope Excursion; Zhu et al., 2006). Mass extinctions of SSFs during Cambrian Stage 2 (SHICE, or SHIyantou Carbon isotope Excursion; Zhu et al., 2006), extinction of archaeocyaths in Stage 4 (AECE excursion, or Archaeocyathid Extinction Carbon isotope Excursion; Zhu et al., 2006), and extinction of redlichiid and olenellid trilobites at the end of Stage 4 (ROECE; Zhu et al., 2006) all are associated with strong shifts toward negative δ^{13}C values. The SPICE (Steptoean Positive Carbon isotopic Excursion, Saltzman et al., 1998) corresponds to the Pterocephaliid Biomere of Laurentia (Saltzman et al., 1998; Peng et al., 2004a; Zhu et al., 2006). Onset of this large positive excursion marks the extinction of

marjumiid trilobites (at the top of the Marjumiid Biomere) in Laurentia. A similar biotic turnover with the extinction of damesellid trilobites is recognized at an equivalent horizon in eastern Gondwana (Yang, 1992; Peng et al., 2004a, 2004b; Peng, 2018). Subsequent extinction of pterocephaliid trilobites in Laurentia (at the top of the Pterocephaliid Biomere) is reflected in a return of δ^{13}C values to near 0.

Each Cambrian series is bracketed by a pair of distinct carbon isotopic excursions. The base of the Cambrian corresponds to the onset of the BACE excursion, which reaches a peak value exceeding -6‰. The base of Series 2 is associated with the onset of the CARE, which reaches a peak value of about $+2.5$‰. The base of the Miaolingian Series is associated with the ROECE, which reaches a peak value exceeding -4‰. The base of the Furongian Series is associated with the onset of the SPICE, which reaches a peak value exceeding +4‰. The peak of the TOCE (Top of Cambrian carbon isotope Excursion; Zhu et al., 2006) is in a short interval below the base of the Ordovician, and its peak value is about −3.5‰. The TOCE also has been referred to as the HERB (HEllnmaria−Red Tops Boundary) Event (Ripperdan, 2002; Peng et al., 2012a; Li et al., 2017; Azmy, 2019), which is a junior synonym of TOCE as the term became available and valid later than TOCE. As recognized by Li et al. (2017), there are three successive $\delta^{13}C$ negative excursions within Stage 10 (N1, N2, and TOCE) with the N1 excursion located just above the base of Stage 10.

The $\delta^{13}C$ curve has emerged as an increasingly powerful tool for intercontinental and intracontinental correlation of Cambrian strata, especially in regions where the primary biologic marker for a key horizon is absent. In the upper half of the Cambrian, all GSSPs defined to date are in outer shelf to slope environments, and correlation into shallow epeiric seas of the continental interiors has been in the past hindered by strong differentiation of trilobite faunas collected from these ecologically different environments. In outer shelf and slope environments, agnostoids are the primary guide fossils, and in inner shelf environments, endemic polymerids are the primary guide fossils. Recognizable carbon isotopic excursions such as the SPICE (Saltzman et al., 1998) and DICE (DrumIan Carbon isotope Excursion; Zhu et al., 2006; Howley et al., 2006; Babcock et al., 2007; Howley and Jiang, 2010) have been used to overcome the problem of extending intercontinental correlations based on outer shelf- and slope-dwelling agnostoids into shallow shelf seas. At the base of the Cambrian, because of difficulty recognizing the horizon marked by the first appearance of *T. pedum* outside of the Avalonian paleocontinent, constraining the base of the system intercontinentally is more commonly by means of the BACE excursion (e.g., Corsetti and Hagadorn, 2001; Zhu et al., 2001, 2006; Amthor et al., 2003; Babcock et al., 2011; Peng et al., 2012a; Zhu et al., 2019).

Studies suggest a relationship between eustatic sea-level history and the carbon isotopic curve for the Cambrian (Babcock et al., 2015). Fine-scale eustatic information is not yet available for the entire system, but well-resolved interpretations (e.g., Miller et al., 2003; Peng et al., 2004a; Babcock et al., 2005; Howley et al., 2006; Howley and Jiang, 2010) suggest a close correspondence in the timing of sea-level change and changes in $\delta^{13}C$ values. The base of the Drumian Stage, which is in the lower phase of a eustatic rise, is associated with the onset of the DICE (Babcock et al., 2007, 2015; Howley et al., 2006; Howley and Jiang, 2010). The base of the Paibian Stage is also in the lower part of a eustatic rise, and it is closely associated with the onset of the SPICE (Peng et al., 2004a; Babcock et al., 2015).

Data on the temporal variation of strontium isotopes ($^{87}Sr/^{86}Sr$) are now available through most of the Cambrian, although the scale of resolution is variable. The most

detailed information exists for the lowermost Terreneuvian Series, and from about the base of the Miaolingian Series through the Ordovician. Reconstruction of a high-resolution seawater $^{87}Sr/^{86}Sr$ curve for the entire Cambrian has been hindered principally because of the perceived lack of suitable, well-preserved materials for analysis. Until recently, vague intercontinental age constraints, particularly in the lower half of the system, contributed to the difficulty of generalizing results to a global scale.

For much of the Cambrian, the fossils preferred for $^{87}Sr/^{86}Sr$ analyses are rare. Suitably robust low-Mg calcite brachiopod shells and apatite conodont elements are not particularly common except in some of the uppermost Cambrian strata. Phosphatic SSFs and brachiopods, which have potential for $^{87}Sr/^{86}Sr$ studies, are essentially untested. Likewise, low-Mg calcite trilobite exoskeletons are relatively untested. All of these remain as potentially promising alternatives for Cambrian studies.

To date, bulk micrite or early marine cements generally have been the preferred materials for constraining Cambrian seawater $^{87}Sr/^{86}Sr$ ratios. The heterogeneous nature of bulk carbonate necessitates, however, that diagenetic alteration needs to be assessed geochemically on a sample-by-sample basis. Samples having least-altered compositions are extrapolated from diagenetic trends. Few studies report enough data from any one stratigraphic horizon to enable a thorough diagenetic analysis, which means that inferred secular trends may in fact be the result of postdepositional effects rather than the result of the isotopic evolution of seawater. This needs to be borne in mind when considering the Sr isotope record shown in Fig. 19.12.

Several $^{87}Sr/^{86}Sr$ studies span the Neoproterozoic−Cambrian boundary interval. The most comprehensive work is that of Brasier et al. (1996). The results of that study and others (Derry et al., 1994; Kaufman et al., 1996; Nicholas, 1996; Valledares et al., 2006; Jiang et al., 2007; Sawaki et al., 2008; Li et al., 2013) constrain latest Ediacaran and earliest Cambrian $^{87}Sr/^{86}Sr$ to about 0.70845 ± 0.0005. Least-altered samples from Mongolia and Siberia (Brasier et al., 1996; Kaufman et al., 1996) reveal a decreasing trend to a low of 0.70805 ± 0.0005 by the end of the Terreneuvian, before rising through Series 2.

Four studies provide data from close to the base of the Miaolingian Series. Values from least-altered samples reported by Montañez et al. (2000) from the Great Basin (United States), Wotte et al. (2007) from France and Spain, and Wang et al. (2011) from Tarim Basin (China) are mutually consistent, whereas high Mg/Ca ratios indicate that the slightly lower values reported by Derry et al. (1994) from Siberia, Russia, developed at the time of dolomitization. The values of 0.70891−0.70898 on least-altered samples are not significantly different from those reported by Kouchinsky et al. (2008) for samples from higher in the Miaolingian Series (Drumian and Guzhangian stages).

Published data show an increasing trend of $^{87}Sr/^{86}Sr$ values from the middle to upper part of the Miaolingian Series through most of the Furongian Series. The trend is abruptly reversed with a decrease in $^{87}Sr/^{86}Sr$ values near the top of the series. Values near the base of the Guzhangian Stage are 0.70893 ± 0.0002, and they increase until close to the top of the Paibian Stage (Furongian Series). Values on least-altered samples in the SPICE interval (Paibian Stage) reach 0.70910 ± 0.0001 (Montañez et al., 2000; Kouchinsky et al., 2008). Isotopic data reported from western North America (Saltzman et al., 1995) for the *Elvinia—Taenicephalus* Biozone boundary (lower part of the Jiangshanian Stage) are internally consistent and imply that seawater $^{87}Sr/^{86}Sr$ rose to its highest ever value (0.70925) in the Jiangshanian Stage, before falling sharply to 0.70914 near the top of the Jiangshanian Stage, and to 0.70910–0.70911 in Stage 10. A study by Ebneth et al. (2001) on samples from conodont elements confirms that

this decrease continued to 0.70900 near the base of the Ordovician.

19.2.8 Sequence stratigraphy

Mei et al. (2007) provided a summary of second- and third-order eustatic changes in South China during the Cambrian (Fig. 19.13). They recognized two second-order sequences that correspond to (1) the Terreneuvian Series plus Series 2 and (2) the Miaolingian Series plus the Furongian Series. Within the Terreneuvian Series 2 sequence, five third-order sequences were recognized, a large one terminating in the upper part of Stage 2, another large one extending to near the top of Stage 3, and three cycles of short duration through upper Stages 3 and 4. Seven third-order cycles of short to moderately long duration occupy the Miaolingian—Furongian sequence.

More detailed sea-level histories have been provided by, among others, Babcock et al. (2005, 2007, 2015), Howley

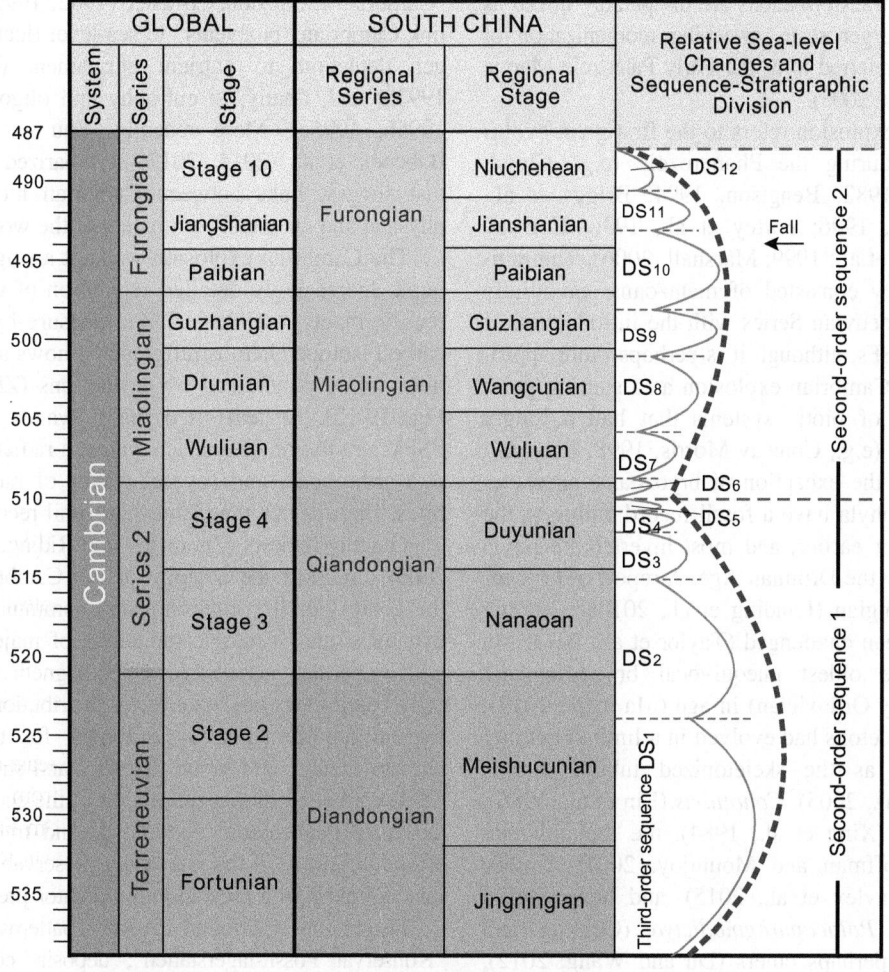

FIGURE 19.13 Cambrian sequence stratigraphy, showing second- and third-order cycles recognized from Guizhou, South China (redrawn after Mei et al. 2007).

et al. (2006), Miller et al. (2003, 2006), Jago et al. (2006), Peng et al. (2009, 2012b), Howley and Jiang (2010), and Nielsen and Schovsbo (2011, 2015). Babcock et al. (2005, 2007, 2015) and Peng et al. (2009, 2012b) showed that the first appearances of some agnostoid guide fossils in the upper half of the Cambrian closely follow small-scale eustatic rises of sea level. Babcock et al. (2015) summarized stratigraphic and paleontologic evidence suggesting that rapid sea-level fluctuation in the Miaolingian and Furongian epochs may have been related to glacial−interglacial cycles, with the likely center of continental glaciation being in high southern latitudes of Gondwana. Agnostoid biozones begin in the lower parts of transgressive systems tracts.

19.2.9 Cambrian evolutionary events

The Cambrian records two important, and evidently linked evolutionary events, the "Cambrian explosion" (Cloud, 1968) and the "Cambrian substrate revolution" (e.g., Bottjer et al., 2000). These biotic transformations are inseparably linked as components of a larger scale, sweeping reorganization of marine ecosystems referred to as the Early Paleozoic Marine Revolution (Babcock, 2003).

The Cambrian explosion refers to the first great evolutionary radiation during the Phanerozoic (e.g., Cloud, 1968; Runnegar, 1982; Bengtson, 1994; Briggs et al., 1994; Chen et al., 1996; Fortey et al., 1996; Conway Morris, 1998; Hou et al., 1999; Marshall, 2006). The radiation, which largely consisted of metazoans, essentially began in the Terreneuvian Series with the introduction of the first biota of SSFs, although it is perhaps more appropriate to view the Cambrian explosion as a starting point for reorganization of biotic systems that had a longer Proterozoic history (e.g., Conway Morris, 1998; Babcock, 2003, 2005). With the exception of bryozoans, all skeletonized metazoan phyla have a fossil record dating to the Furongian Epoch or earlier, and most invertebrate phyla were established by the Drumian Age. A report of bryozoans from the Furongian (Landing et al., 2010a; Landing et al., 2015) has been challenged (Taylor et al., 2013; Ma et al., 2013); the oldest unequivocal bryozoans are Tremadocian (Early Ordovician) in age (Ma et al., 2013). Biomineralized skeletons had evolved in a limited number of animals, such as the skeletonized tubular fossils *Cloudina* (Hua et al., 2005), *Conotubus* (Lin et al., 1986), and *Shaanxilithes* (Xing et al., 1984), the lophophorate *Namacalathus* (Hoffman and Mountjoy, 2001; Amthor et al., 2003; Zhuravlev et al., 2015), and hexactinellid sponges such as *Palaeophragmodictya* (Gehling and Rigby, 1996) and perhaps others (Du and Wang, 2012), during the Ediacaran Period but most clades that evolved biomineralized skeletons did so during the Cambrian Period (Terreneuvian Epoch or Epoch 2). Through the Terreneuvian, Series 2, and Miaolingian Series, there was

a spectacular burst in diversity (number of species and genera) and in disparity (number of distinct body plans). Brasier (1979) gave an extensive review of the fossil record during the early half of the Cambrian, and Kouchinsky et al. (2012) summarized the first appearances of major animal groups with mineralized skeletons and the chronology of biomineralization.

Fossil groups involved in the Cambrian explosion include prokaryotes; eukaryotic protoctists, acritarchs, and chitinozoans; larger algae and vascular plants; Parazoa (Porifera, Chancelloriida, Radiocyatha, Archaeocyatha, Stromatoporoidea); Radiata; Bilateria (Priapulida, Sipunculida, Mollusca, Annelida, Arthropoda including Lobopoda and Tardigrada, Pogonophora, Brachiopoda, Ectoprocta, Phoronida, Mitrosagophora, Tommotiida, mobergellids, Echinodermata, Hemichordata, Chaetognatha, Conodontophorida, and Chordata). Brasier (1979) also commented on phyletic changes, skeletal changes, niche changes, size changes, and environment-related changes. In attempting to explain the "Cambrian Explosion," Brasier (1982, 1995a) attempted to link Cambrian "bioevents" to sea-level fluctuation and oxygen depletion, to nutrient enrichment (Brasier, 1992a, 1992b) and, finally, to eutrophy and oligotrophy (Brasier, 1995b, 1995c). More recently, Zhu et al. (2006) and Babcock et al. (2015, 2017) summarized integrated data that suggest links between Cambrian biotic history and physical and chemical parameters of the world ocean.

The Cambrian explosion was not a single evolutionary burst. Increasingly detailed resolution of the stratigraphic record, made possible in large measure by application of carbon isotope chemostratigraphy, shows a series of radiations, often punctuated by extinctions (Zhu et al., 2006; Fig. 19.12). At least two major waves of radiation of SSFs, a radiation of archaeocyaths, a radiation of nontrilobite arthropods, and several waves of radiation of trilobites, characterize the Cambrian fossil record.

Lengthy reviews (Zhuravlev and Riding, 2001; Marshall, 2006) discussed the ecology of the Cambrian radiation in the context of life environments, community patterns and dynamics, and "ecologic radiation" of major fossil groups, with important chapters on paleomagnetically and tectonically based maps of global facies distribution, and supercontinental amalgamation as a trigger for the "explosion," climate change, and biotic diversity and structure. Babcock (2003) provided further discussion of the ecologic context of the early Phanerozoic body fossil and trace fossil record, attributing much of the increasing preservability of fossils to factors linked with escalation in predator-prey systems.

Insight into Cambrian diversity patterns is provided by "Konservat Fossil-lagerstätten"; deposits containing exquisitely well-preserved fossils, particularly of nonbiomineralized (soft) body parts. They are known globally from approximately 40 localities (Conway Morris, 1985; Babcock et al., 2001; Gaines et al., 2012) if the "Orsten"-type preservation style of Sweden, China, and elsewhere (e.g., Maas et al., 2006; Eriksson et al., 2016) is included. The richest and most spectacular Lagerstätten are in the Buen Formation of North Greenland (e.g., Conway Morris et al., 1987; Conway Morris and Peel, 1990; Budd, 1997; Budd and Peel, 1998; Babcock and Peel, 2007; Zhuravlev et al., 2015; Hammarlund et al., 2018), the Yuanshan Member of the Chiungchussu Formation or Maotianshan Shale (Chengjiang Biota) of Yunnan, China (e.g., Chen et al., 1996; Hou et al., 1999, 2017; Luo et al., 1999; Babcock et al., 2001), the Burgess Shale of British Columbia (e.g., Conway Morris, 1977, 1985; Whittington, 1977; Briggs et al., 1994), the Kaili Formation of Guizhou, China (e.g., Zhao et al., 2002, 2005a, 2005b, 2011), and the Sandu Formation (Guole Biota) of Guangxi, China (e.g., Han and Chen, 2008; Zamora et al., 2013; Zhu et al., 2016). Collectively, deposits of the Great Basin (e.g., Gunther and Gunther 1981; Robison, 1991; Briggs et al., 2005, 2008; English and Babcock, 2010; Hollingsworth and Babcock, 2011; Stein et al., 2011; Robison and Babcock, 2011; Robison et al., 2015; Butler et al., 2015) have also produced spectacular material, but in lower numbers than these other localities.

The Cambrian Substrate Revolution (e.g., Seilacher and Pfluger, 1994; Bottjer et al., 2000) has been used to describe the changes in marine substrates through Cambrian time, changes that were evidently coupled to biotic radiation. Marine substrates of the Ediacaran Period were largely stabilized by microbial mat communities (Gehling, 1999), or matgrounds. The relatively few known Ediacaran traces fossils were essentially surface traces. Matgrounds continued into the Cambrian (Buatois et al., 2014), but increasingly through the period, more penetrative traces appear (Mángano and Buatois, 2014). In the upper Furongian traces penetrating several centimeters are not unusual, and some sedimentary layers are well bioturbated (Droser and Bottjer, 1989). Thus the Cambrian substrate revolution involves a transition from mat-dominated substrates to more fluidized, bioturbated substrates. Correlated with this change is a decline in helicoplacoids (Domke and Dornbos, 2010), edrioasteroids (Zamora et al., 2017; Wen et al., 2019), and other organisms having a "mat sticking" life habit, or other life habit dependent on microbial mats (Bottjer et al., 2000). Among the available bioturbators of the Cambrian were a variety of priapulid and other worms (Conway Morris, 1977). Babcock (2003) showed that polymerid trilobites increasingly burrowed into Cambrian substrates, sometimes in search of prey, which were themselves burrowers.

19.3 Cambrian Time Scale

19.3.1 Age of the Ediacaran–Cambrian boundary

The base of the Cambrian System/Period has received considerable attention from geochronologists, in part because

of major biotic changes that occurred during the late Ediacaran–Cambrian interval. For this reason, calibration of the lower boundary of the Cambrian, and thus of the concurrent Fortunian Stage/Age, Terreneuvian Series/Epoch, Paleozoic Erathem/Era, and Phanerozoic Eonothem/Eon, is relatively well constrained (Fig. 19.14). Further refinement in calibration of the boundary probably depends more on advances in biostratigraphy than in geochronology, particularly on the discovery of new stratigraphic sections and fossil occurrences that help in the identification and correlation of the boundary. Dated samples used for calibration of the base and top of the Cambrian Period and the Cambrian stages are listed in Appendix 2 of this volume. The age and two-sigma error range given in the original cited references are adjusted here, following the guidelines and procedures outlined in Schmitz et al. (2020, Ch. 6: Radioisotope geochronology, this book). In the following account the adjusted ages are used.

Until 2016 the base of the Cambrian was inferred to have an age close to 541 Ma, but more recent evidence places it at about 538.8 Ma (Linnemann et al., 2019). The older estimate was based on a U–Pb date of 540.61 ± 0.88 (originally given as 543.3 ± 1) Ma on zircon from a volcanic ash in the upper Spitskopf Member of the Schwarzrand Subgroup in Namibia (Grotzinger et al., 1995; Schmitz, 2012). This ash bed, the lowermost of eight in the terminal Ediacaran–earliest Cambrian succession in the upper Spitskopf Member, is assigned to the latest Ediacaran as it lies in an interval containing the calcified *Namacalathus* and *Cloudina*, and below the last appearance of the characteristic nonbiomineralized Ediacaran organisms *Swarpuntia* and *Pteridinium* (Grotzinger et al., 1995; Linnemann et al., 2017). As such, this date provides a maximum age constraint on the base of the Cambrian. Linnemann et al. (2017, 2019) reported new, high-resolution age dates from ashes 1 to 5 in the Spitskopf Member of the Urusis Formation (Fig. 19.14). All of these ash beds are below the last occurrences of *Swarpuntia germsi* and *Pteridinium simplex* of the Ediacaran biota. The youngest age (538.99 ± 0.21 Ma) was obtained from ash bed 5 and suggests that the base of the Cambrian is younger than previously estimated (Linnemann et al., 2019). Ash bed 6 is in the overlying Nomtsas Formation and above the lowest occurrences of trace fossils of Fortunian and Phanerozoic aspect, including *T.* cf. *pedum* and *Streptichnus narbonnei*. It yielded an age of 538.58 ± 0.19 Ma. This age provides a minimum age for the base of the Cambrian. Consequently, the Cambrian base should lie between ash beds 5 and 6, which have calculated ages of 538.99 ± 0.21 Ma and 538.58 ± 0.19 Ma, respectively. At present, the best estimated radioisotopic date for the base of the Cambrian is 538.8 ± 0.60 Ma (Linnemann et al., 2019; and see Appendix 2, this book).

Previous estimates of the age of the base of the Cambrian are consonant with the newly calculated age

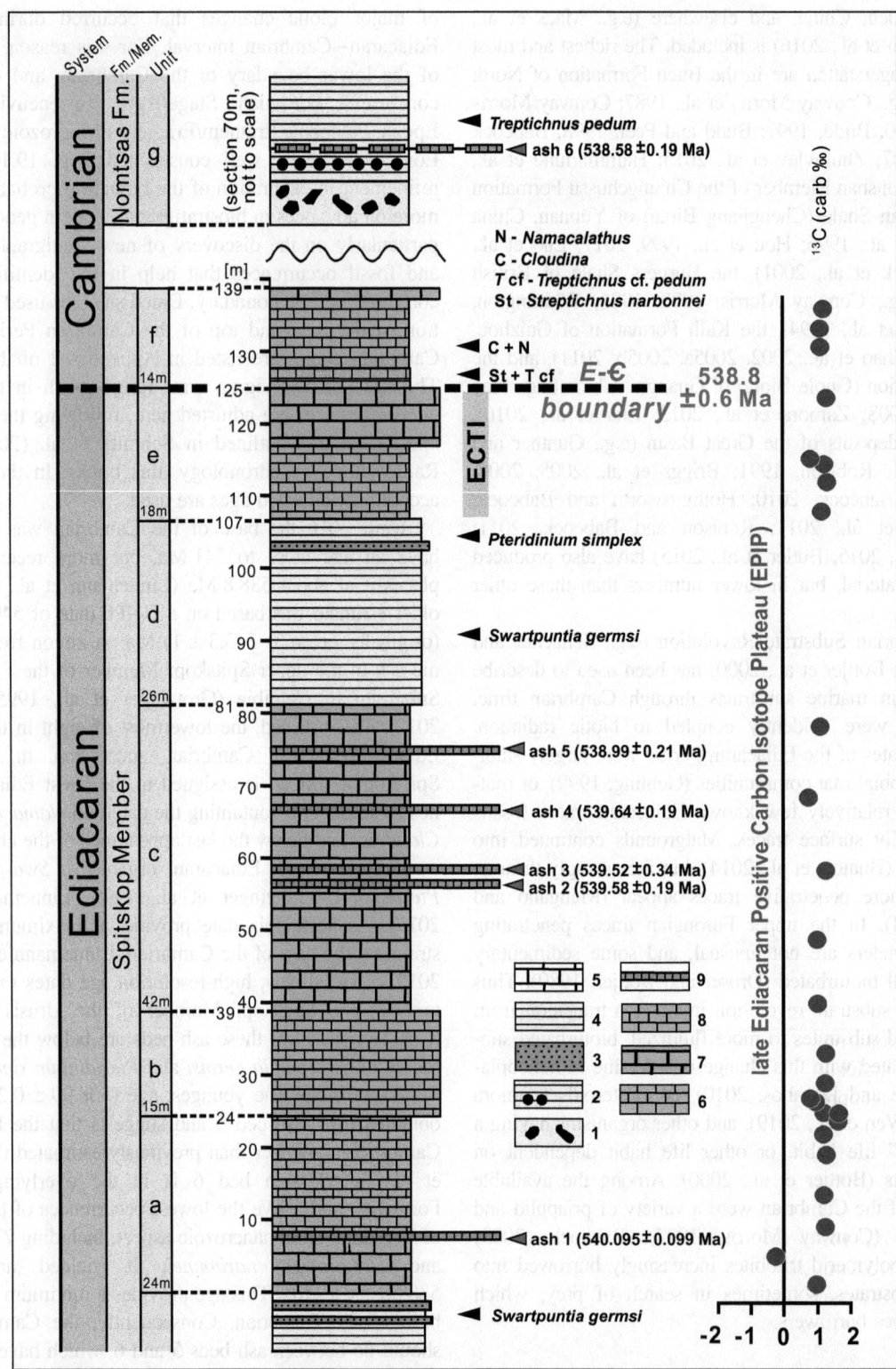

FIGURE 19.14 Ediacaran−Cambrian boundary interval in the Swartpunt section (units a−f) and Swartkloofberg section (unit g), southern Namibia. Precise U−Pb ages obtained by CA-ID-TIMS techniques with uncertainties given at 2-sigma level, carbon isotope values and fossil horizons. ECTI = Ediacaran−Cambrian transition interval; (1) debris flow, shale, olistoliths; (2) shale, sandstone, conglomerate; (3) gray-green sandstone; (4) greenish shale; (5) gray thick-bedded micrite; (6) gray thin-bedded micrite; (7) black thick-bedded micrite; (8) black thin-bedded micrite; (9) ash bed (redrawn after Linnemann et al., 2019).

estimate. A U−Pb date of 538.18 ± 1.24 Ma was obtained (Grotzinger et al., 1995) from the Nomtsas Formation, which overlies the Spitskopf Member in Namibia. The basal contact of the Nomtsas Formation with the Spitskopf Formation is disconformable, and the basal beds of the Nomtsas contain the trace fossil *Triptichnus pedum*. Previously reported data from Namibia suggest that the adjusted age of the base of the Cambrian is between 538 and 540 Ma (Brasier et al., 1994; Grotzinger et al., 1995). These data, plus the new estimates (Linnemann et al., 2019), approximate other zircon dates, stratigraphically less constrained, from Siberia (Bowring et al., 1993) and from the upper part of the Ediacaran (Grotzinger et al., 1995; Tucker and McKerrow, 1995).

Some elements of the globally distributed Ediacaran biota occur stratigraphically above ash bed 1 in the Spitskopf Member of Namibia, and this has been used as evidence that Ediacaran-grade organisms range into the earliest Cambrian (Grotzinger et al., 1995). Linnemann et al. (2019) reported nonbiomineralized Ediacaran-type fossils above ash 5, but below bilaterian trace fossils of Cambrian aspect near the top of the Spitskopf Member. Similar faunal relationships have been reported from South Australia (Jensen et al., 1998). Other reports of putative nonbiomineralized Ediacaran-grade organisms ranging well into Cambrian strata of other regions include Conway Morris (1993), Jensen et al. (1998), Hagadorn et al. (2000), Zhang and Babcock (2001), Lin et al. (2006), Shu et al. (2006), and Babcock and Ciampaglio (2007). Vidal et al. (1994), Zhuravlev et al. (2009), and Zhu et al. (2017) reported an overlap of *Cloudina* with SSFs of Cambrian aspect in the Ust'-Yudoma Formation of Siberia, Russia. Together, these reports confirm that not all members of the biota became extinct prior to the end of the Ediacaran Period.

In the Ara group of Oman, chemostratigraphic and paleontologic data on subsurface samples were originally interpreted to indicate the simultaneous occurrence of an extinction of Neoproterozoic biomineralized organisms (*Namacalathus* and *Cloudina*) and a large-magnitude negative excursion in carbon isotopes (Amthor et al., 2003; Bowring et al., 2003, 2007), the BACE excursion (Zhu et al., 2006), which is widely equated with the boundary (Grotzinger et al., 1995; Bartley et al., 1998; Corsetti and Hagadorn, 2001; Kimura and Watanabe, 2001; Zhu et al., 2006, 2019). However, Geyer and Landing (2016) reexamined the core from which samples were derived, and expressed skepticism that the age dates correspond closely to the Cambrian base. The radioisotopic dates, which were likely obtained from the upper Ediacaran System, should be interpreted as maximum constraints on the age of the Ediacaran−Cambrian boundary. Following Bowring et al. (2007) the ash bed at the peak of the isotope excursion, 1 m above the base of the A4C carbonate unit, yielded an adjusted age of 541.00 ± 0.63 Ma (2-sigma). Adjusted ages obtained from two ash layers in the next carbonate unit in succession, the A3C carbonate unit, are 542.37 ± 0.63 Ma and 542.9 ± 0.63 Ma. Based on these data, the peak of the BACE negative carbon isotope excursion in the Oman sequence has an age of at least 541 Ma. The top of the excursion likely has an age closer to 538 or 539 Ma. Geyer and Landing (2016) considered 539 Ma a better estimate of the age of the Cambrian base based on the data from Oman.

The Namibian sequence, which has the best paleontological constraints, has an age calculated as 538.8 ± 0.60 Ma for base of the Cambrian, and data from the Oman sequence suggests that such as position is equivalent to the upper part of the BACE excursion. Pending further information, we regard 538.8 ± 0.60 Ma, as outlined above to be the optimum age for the base of the Cambrian, Paleozoic, and Phanerozoic. The difference between the age of the top of the Cambrian Period (i.e., of the beginning of the Ordovician) at 486.9 ± 0.8 Ma, and the base at 538.8 ± 0.60 Ma gives 51.9 Myr for its duration.

19.3.2 Age of Internal Boundaries

Ages of internal Cambrian boundaries are generally poorly constrained, especially in the lower part of the Cambrian, and caution should be exercised when using the numerical scale in Figs. 19.1, 19.2, 19.11−19.15 and Table 19.1.

Fossil diversity and abundance are generally low in the Terreneuvian Series and in provisional Series 2. As a result, the biostratigraphic framework is rather vague. In the lower part of the Terreneuvian Series, resolution of the time scale is limited in part by a paucity of biostratigraphically useful fossils, inconsistent identification and correlation of the level corresponding to the base of the series (and system), and a lack of radioisotopic data except for the interval bracketing the Ediacaran−Cambrian boundary. Our estimates of stage/age durations are correspondingly intuitive and the estimated ages of stage/age boundaries in the Terreneuvian Series/Epoch and Series/Epoch 2 shown in Figs. 19.2, 19.12, and 19.15 should be regarded as highly approximate.

The diversity and abundance of fossils increase upward from the lower part of the Miaolingian Series. The Miaolingian and Furongian series are finely zoned by trilobite biostratigraphy, and stages are proportioned approximately according to the number of trilobite zones they contain. The biostratigraphic zonation of South China is used as a reference standard. In the Miaolingian of South China, 11 zones are recognized, and in the Furongian, 13 zones are recognized. Through most of the Miaolingian, agnostoids provide reliable zonation and intercontinental correlation. In the Furongian, agnostoid zonation is used through the base of provisional Stage 10. The method of equating trilobite zonation to time assumes a more or less

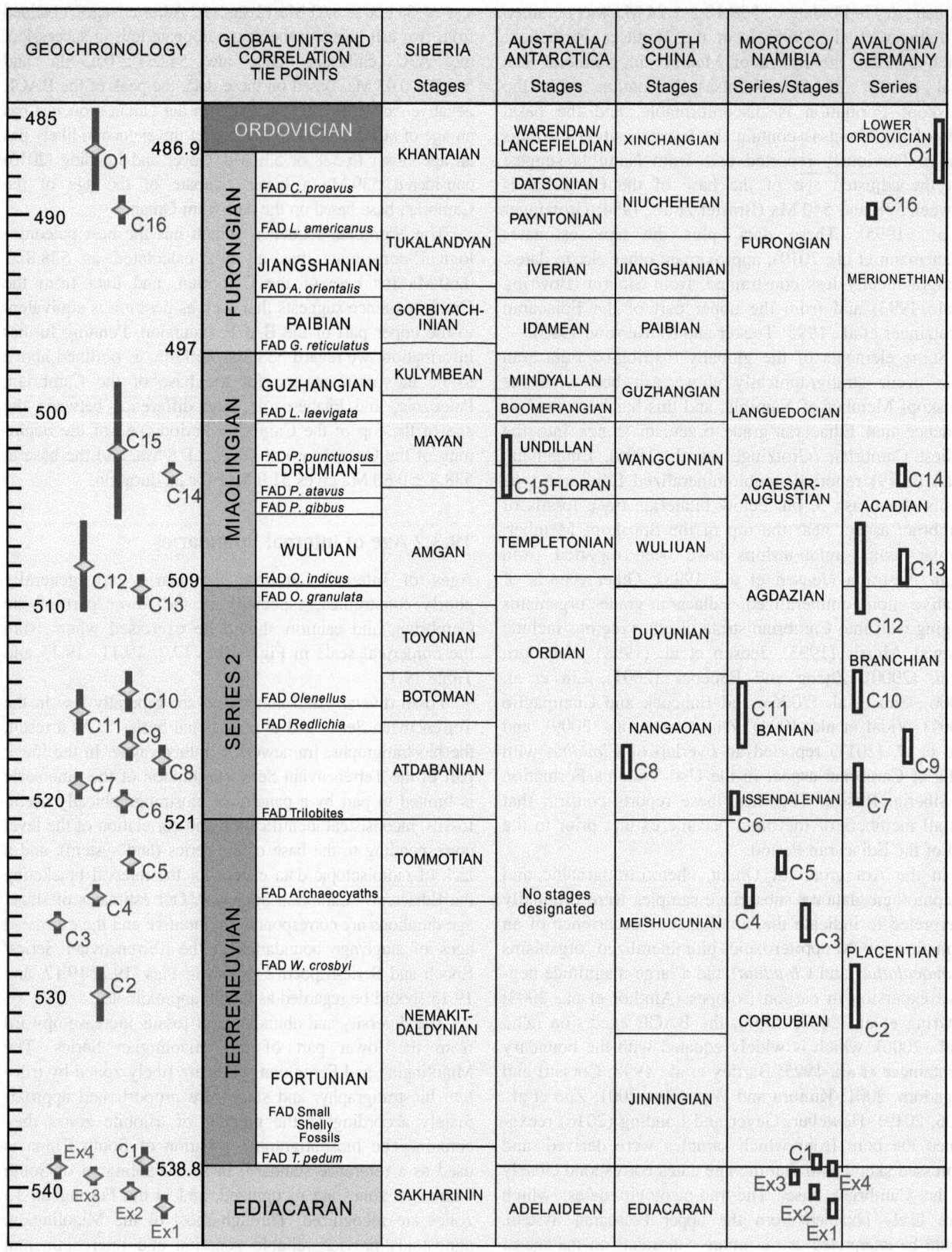

FIGURE 19.15 Distribution of the numerical age dates used for Cambrian time scale based on $^{206}Pb/^{238}U$ and $^{207}Pb/^{206}Pb$ radioisotopic analyses (see text of Section 19.3 and Appendix 2 for stratigraphic and/or radioisotopic method details).

TABLE 19.1 Cambrian epoch and stage boundary interpolated ages, with uncertainty estimates where feasible, stage durations and comparison to GTS2012.

Chronostratigraphic unit			GTS2012		GTS2020	
Period	Epoch	Stage/Age	Base (Ma)	Stage duration (Myr)	Base (Ma)	Stage duration (Myr)
Ordovician		Tremadocian	485.4 ± 1.9		486.9 ± 1.5	
Cambrian	Furongian	Stage 10	489.5	4.0	491.0	4.1
		Jiangshanian	494.0	4.5	494.2	3.2
		Paibian	497.0	3.0	497.0	2.8
	Miaolingian	Guzhangian	500.5	3.5	500.5	3.5
		Drumian	504.5	4.0	504.5	4.0
		Wuliuan	509.0	4.5	509.0	4.5
	Epoch 2	Stage 4	514.5	5.0	514.5	5.5
		Stage 3	521.0	7.0	521.0	6.5
	Terreneuvian	Stage 2	529.0	8.0	529.0	8.0
		Fortunian	541.0 ± 0.6	12.0	538.8 ± 0.6	9.8

For details see text.

constant rate of evolutionary turnover and a uniformity in paleontologic practice in zonal designation, both of which are unlikely to be completely true.

Radiometric dates bracketing the base of the Terreneuvian Series and the conterminant Fortunian Stage are discussed in Section 19.3.1. An age of approximately 539 Ma is inferred.

Ash beds in New Brunswick, Canada, with an age of 530.02 ± 1.20 Ma, from the upper part of the *Rusophycus avalonensis* Zone in the "Placentian Series" of Avalon, are regarded as equivalent to the Nemakit-Daldynian Stage of Siberia (Isachsen et al., 1994), and provide a rather loose numerical calibration for the Nemakit-Daldynian (equivalent to the Fortunian Stage plus the lower half of provisional Stage 2). An ash bed in the upper Adoudou Formation of the Anti-Atlas Mountains of Morocco gives a good U–Pb weighted mean zircon age of 525.23 ± 0.61 Ma (Maloof et al., 2005). On the basis of the correlation of a $+7‰$ $\delta^{13}C$ excursion at the dated level with a global excursion correlative with the base of the Tommotian of Siberia (near the middle of provisional Stage 2), an age of 525 Ma for the position can be inferred.

The age of the base of Series 2 (and Stage 3) is poorly constrained and is here taken to be about 521 Ma.

In Pembrokeshire, South Wales, an ash bed at Cwm Bach in the Caerfai Bay Shales has yielded an age of 519.38 ± 0.28 Ma, and this is consistent with an age of 519 ± 1 Ma for the same formation reported by Landing et al. (1998). The dated ash bed is correlated with a horizon within the Lower Comley Sandstone on the presence of the bradoriid arthropod *Indiana lentiformis* (Siveter and

Williams, 1995; Harvey et al., 2011), which suggests a position in provisional Stage 3. Although biostratigraphic control is not good, the radioisotopic age is well corroborated.

An ash bed in the Lower Comley Sandstone of Shropshire, United Kingdom, has yielded a weighted mean $^{206}Pb/^{238}U$ age of 514.5 ± 0.81 (Harvey et al., 2011). The ash bed lies within the *Callavia* Trilobite zone, in the upper part of provisional Stage 3 or possibly in the lower part of Stage 4. The age is generally consistent with the Moroccan ash bed ages of Landing et al. (1998).

Ash beds from Morocco are taken as equivalent to the "middle Botomian to Toyonian" of Siberia (Landing et al., 1998), and therefore are equivalent to the upper part of Stage 3 through the lower half of Stage 4. Five single-grain zircon analyses provided an adjusted weighted mean age of 515.56 ± 1.16 Ma.

Based on reasonably well-corroborated ages, the base of the Miaolingian Series and the conterminant Wuliuan Stage is inferred to be close to 509 Ma. Ash beds associated with "upper Lower Cambrian" trilobites of the *O. granulata–Protolenus howleyi* zones (upper Stage 4) in New Brunswick, Canada (Landing et al., 1998), have yielded zircons that give a recalculated weighted mean age of 508.05 ± 2.75 Ma. An ash bed in the Upper Comley Sandstone of Shropshire, United Kingdom, has given a weighted mean $^{206}Pb/^{238}U$ age of 509.02 ± 0.79 Ma (Harvey et al., 2011). Beds immediately overlying this position yield trilobites including *Acadoparadoxides harlani*, which indicate that the *A. harlani* Zone of Newfoundland,

GSSPs of the Cambrian Stages, with location and primary correlation criteria

Stage	GSSP Location	Latitude, Longitude	Boundary Level	Correlation Events	Reference
Stage 10				*Agnostoid trilobite, FAD of* Lotagnostus americanus	
Jiangshanian	Duibian B Section, Jiangshan County, W. Zhejiang Province, S. China	28°48.977'N 108°36.887'E	108.12 m above base of the Huayansi Formation	Agnostoid trilobite, FAD of *Agnostotes orientalis*	Episodes **35**/4, 2012
Paibian	Paibi, Huayuan County, NW Hunan Province, S. China	28°23.37'N 109°31.54'E	at 396 m above the base of the Huaqiao Formation	Agnostoid trilobite, FAD of *Glyptagnostus reticulatus*	Lethaia **37**/4, 2004
Guzhangian	Luoyixi, Guzhang County, NW Hunan Province, S. China	28°43.20'N 109°57.88'E	121.3 m above the base of the Huaqiao Formation	Agnostoid trilobite, FAD of *Lejopyge laevigata*	Episodes **32**/1, 2009
Drumian	Drum Mountains, Millard County, Utah, USA	39°30.705'N 112°59.489'W	at the base of a dark- gray thinly laminated calcisiltite layer, 62 m above the base of the Wheeler Formation	Agnostoid trilobite, FAD of *Ptychagnostus atavus*	Episodes **30**/2, 2007
Wuliuan	Balang, Jianhe County E. Guizhou Province, S. China	26°44.843'N 108°24.830'E	at the base of a silty mud-stone layer 52.8 m above the base of the Kaili Formation	Polymerid trilobite, FAD of *Oryctocephalus indicus*	Episodes **42**/2, 2019
Stage 4				*Trilobite, FAD of* Olenellus, Redlichia, Judomia, Bergeroniellus, Arthrico-cephalus, *or* Oryctocarella	
Stage 3				*FAD of trilobites*	
Stage 2				*Small shelly fossils, FAD of* Watsonella crosbyi *or* Aldanella attleborensis	
Fortunian	Fortune Head, Burin Peninsula, E. Newfoundland, Canada	47°4'34.47"N 55°49'51.71"W*	2.4 m above the base of Member 2 in the Chapel Island Formation	Trace fossil, previously considered FAD of *Treptichnus pedum*	Episodes **17**/1-2, 1994; Episodes **30**/3, 2007

* according to Google Earth

FIGURE 19.16 GSSPs of the Cambrian stages, with location and primary correlation criteria (status as of August 2020). *GSSP*, Global Boundary Stratotype Section and Point.

Canada, is correlative with the upper *O. indicus* Zone (Geyer, 2005; Fletcher, 2006; Sundberg et al., 2016). The ash bed thus helps to constrain the age of the base of Miaolingian Series. Although it conflicts with the age of the stratigraphically lower New Brunswick ash bed (508.05 ± 2.75 Ma) reported by Landing et al. (1998), the conflict is easily accommodated within the error ranges for the two dates (Fig. 19.15).

The Taylor Formation in Antarctica (Encarnación et al., 1999) has yielded zircons with a recalculated weighted mean age of 502.1 ± 3.50 Ma on ashes interbedded with trilobite-bearing limestones correlated with the Undillan Stage of Australia (equivalent to the middle−upper Drumian Stage). Based on the number of agnostoid zones, we can estimate the age of the base of the Guzhangian Stage and of the overlying Furongian Series. This base of the Undillan Stage of Australia approximates the base of the *G. nathorsti* Zone of South China. From this position upward there are two agnostoid zones recognized in South China. If each agnostoid zone

equals about 1 Myr, the base of the Guzhangian Stage is likely to be about 500 Ma; we take 500.5 Ma to be the best estimate. From the Undillan base upward, there are six agnostoid zones recognized in Australia, and there are six agnostoid zones recognized in South China above the base of the *G. nathorsti* Zone. Again, assuming that each agnostoid zone equals about 1 Myr, the base of the Furongian Series (and Paibian Stage), is likely to be about 496 or 497 Ma; we take 497 Ma to be the best estimated age.

The base of the traditional Upper Cambrian as commonly applied in most regions is at a position below the base of the Furongian Series. It is equivalent to the base of the *A. pisiformis* Zone, which is close to the base of the *L. reconditus* Zone of South China (Peng and Robison, 2000; Ahlberg, 2003; Ahlberg et al., 2004) or to the base of its equivalent *A. quasivespa* Zone of Australia. This position is in the mid-Guzhangian Stage and has an inferred age of about 499 Ma.

The base of the Jiangshanian Stage was previously estimated to be about 494 Ma, as it lies three agnostoid zones above the base of the conterminous Furongian Series and Paibian Stage, suggesting that the Paibian had a duration of 3 Myr. Since the base of provisional Stage 10 is re-estimated to be 491 Ma (see below), which gives a 3 Myr duration also for the Jiangshanian Stage. However, the Jiangshanian Stage is slightly thicker than the Paibian Stage in both Siberia and South China. This may suggest that the Jiangshanian is of longer duration than the Paibian. Therefore the base of Jiangshanian is set at about 494.2 Ma, making a duration of 2.8 Myr for the Paibian and 3.4 Myr for the Jiangshanian.

In the uppermost Furongian a volcanic sandstone in North Wales yielded a maximum age for the *Ctenopyge bisulcata* Subzone of the middle *Peltura scarabaeoides* Zone, of 488.71 ± 2.78 Ma (Davidek et al., 1998; Landing et al., 2000) that was redated as 490.1 ± 0.6 Ma. In Sweden, the *C. bisulcata* Zone overlies the agnostoid *L. americanus* Zone, which corresponds to two elevated polymerid zones. The base of the *L. americanus* Zone coincides with the base of provisional Stage 10. This level lies a single agnostoid zone, or two polymerid zones, below the *C. bisulcata* Zone. If we assume the average duration of an agnostoid zone is about 1 Myr (Peng and Robison, 2000), or slightly less, the base of Stage 10 is likely to be about 491 Ma.

To summarize, the duration of the Cambrian is regarded as approximately 51.9 Myr, ranging from 538.8 to 486.9 Ma (Table 19.1). The base of Series 2 is approximately 521 Ma, the base of the Miaolingian Series is near 509 Ma, and the base of the Furongian Series is near 497 Ma. The Terreneuvian Epoch lasted for approximately 18 Myr, Epoch 2 lasted for approximately 12 Myr, the Miaolingian Epoch lasted for approximately 12 Myr, and the Furongian Epoch lasted for approximately 10 Myr.

Acknowledgments

We thank all colleagues and friends who have helped to guide our thinking about Cambrian stratigraphy over the past decades. The research leading to this chapter was supported by long-term funding from the National Natural Science Foundation of China (Grants no. 47072003, 48702077, 40023002, 40332018, and 41330101), Chinese Academy of Sciences (Grant no. XDB26000000), China Geological Survey (Grant no. DD20160120, DD20190009), and State Key Laboratory of Palaeobiology and Stratigraphy (Grant no. 0926040116; 20191101) to Shanchi Peng; from the Subsurface Energy Research Center, The Ohio State University to Loren E. Babcock; and from the Crafoord Foundation to P. Ahlberg. We thank I. Gogin (St. Petersburg) and T. V. Pegel (Novosibirsk) for access to Russian literature.

Bibliography

Ahlberg, P., 1981, Ptychopariid trilobites in the Lower Cambrian of Scandinavia. *In* Taylor, M.E. (ed.), *Short Papers for the Second International Symposium on the Cambrian System.* U.S. Geological Survey Open-File Report, **81−743**: pp. 5−7.

Ahlberg, P., 1985, Lower Cambrian trilobite faunas from Scandinavian Caledonides − a review. *In* Gee, D.G., and Sturt, B.A. (eds), *The Caledonide Orogen − Scandinavia and Related Areas.* Chichester: Wiley, 339−346.

Ahlberg, P., 1991, Trilobites in the Lower Cambrian of Scandinavia. *Geologiska Föreningens i Stockholm Förhandlingar*, **113**: 74−75.

Ahlberg, P., 2003, Trilobites and intercontinental tie points in the Upper Cambrian of Scandinavia. *Geologica Acta*, **1**: 127−134.

Ahlberg, P., and Ahlgren, J., 1996, Agnostids from the Upper Cambrian of Västergötland, Sweden. *GFF*, **118**: 129−140.

Ahlberg, P., Bergström, J., and Johansson, J., 1986, Lower Cambrian olenellid trilobites from the Baltic Faunal Province. *Geologiska Föreningens i Stockholm Förhandlingar*, **108**: 39−56.

Ahlberg, P., Axheimer, N., and Robison, R.A., 2007, Taxonomy of *Ptychagnostus atavus*: a key trilobite in defining a global Cambrian stage boundary. *Geobios*, **40**: 709−714.

Ahlberg, P., Axheimer, N., Babcock, L.E., Eriksson, M.E., Schmitz, B., and Terfelt, F., 2009, Cambrian high-resolution biostratigraphy and carbon isotope chemostratigraphy in Scania, Sweden: first record of the SPICE and DICE excursions in Scandinavia. *Lethaia*, **42**: 2−16.

Ahlberg, P., Axheimer, N., Eriksson, M., and Terfelt, F., 2004, Agnostoids and intercontinental tie points in the middle and upper Cambrian of Scandinavia. *GFF*, **126**: 108.

Ahlberg, P., Cederström, P., and Babcock, L.E., 2016, Cambrian Series 2 biostratigraphy and chronostratigraphy of Scandinavia: a reappraisal. *In* Laurie, J.R., Kruse, P.D., García-Bellido, D.C., and Holmes, J.D. (eds), *Palaeo Down Under 2, Adelaide, 11−15 July 2016.* Geological Society of Australia Abstracts, **117**: p. 16.

Ahlberg, P., and Terfelt, F., 2012, Furongian (Cambrian) agnostoids of Scandinavia and their implications for intercontinental correlation. *Geological Magazine*, **149**: 1001−1012.

Álvaro, J.J., Ahlberg, P., Babcock, L.E., Bordonaro, O.L., Choi, D.K., Cooper, R.A., et al., 2013, Global Cambrian trilobite palaeobiogeography assessed using parsimony analysis of endemicity. *Geological Society of London, Memoirs*, 38. pp. 273−296.

Álvaro, J.J., and Vizcaíno, D., 2003, The conocoryphid biofacies, a benthic assemblage of normal-eyed and blind trilobites. *Special Papers in Palaeontology*, **70**: 127−140.

Amthor, J.E., Grotzinger, J.P., Schröder, S., Bowring, S.A., Ramezani, J., Martin, M.W., et al., 2003, Extinction of *Cloudina* and *Namacalathus* at the Precambrian boundary in Oman. *Geology*, **31**: 431−434.

Angelin, N.P., 1851, *Palaeontologia Suecica, pars I: Iconographia Crustaceorum Formationis Transitionis.* Weigel, Lund, p. 24.

Apollonov, M.K., Bekteleuov, A.K., Kirschvink, J.L., Ripperdan, R.L., Tursunov, B.N., and Tynbaev, K.L., 1992, Paleomagnetic scale of Upper Cambrian and Lower Ordovician in Batyrbai section (Lesser Karatau, southern Kazakhstan). *Izvestiya Akademii Nauk Kazakhskoy SSR, Seriya Geologicheskaya*, 4 (326): 51−57 (in Russian).

Astashkin, V.A., Pegel, T., Shabanov, Y.Y., Sukhov, S.S., Sundukov, V.M., Repina, L.N., et al., 1991, *The Cambrian System of the Siberian Platform*, Vol. 27. IUGS Publication, p. 133.

Astashkin, V.A., Varlamov, A.I., Esakova, N.V., Zhuravlev, A.Y., Repina, L.N., Rozanov, A.Y., et al., 1990, Guide on Aldan and Lena Rivers, Siberian Platform. Third International Symposium on Cambrian System (in Russian). Novosibirsk: Institute of Geology and Geophysics, Siberian Branch, Academy of Sciences of USSR, p. 115.

Axheimer, N., and Ahlberg, P., 2003, A core drilling through Cambrian strata at Almbacken, Scania, S. Sweden: trilobites and stratigraphical assessment. *GFF*, **125**: 139−156.

Axheimer, N., Ahlberg, P., and Cederström, P., 2007, A new lower Cambrian eodiscoid trilobite fauna from Swedish Lapland and its implications for intercontinental correlation. *Geological Magazine*, **144**: 953–961.

Axheimer, N., Eriksson, M.E., Ahlberg, P., and Bengtsson, A., 2006, The middle Cambrian cosmopolitan key species *Lejopyge laevigata* and its biozone: new data from Sweden. *Geological Magazine*, **143**: 447–455.

Azmy, K., 2019, Carbon-isotope stratigraphy of the uppermost Cambrian in eastern Laurentia: implications for global correlation. *Geological Magazine*, **156**: 759–771.

Babcock, L.E., 1994a, Systematics and phylogenetics of polymeroid trilobites from the Henson Gletscher and Kap Stanton formations (Middle Cambrian), North Greenland. *Grønland Geologiske Undersøgelse Bulletin*, **169**: 79–127.

Babcock, L.E., 1994b, Biogeography and biofacies patterns of Middle Cambrian polymeroid trilobites from North Greenland: palaeogeographic and palaeo-oceanographic implications. *Grønland Geologiske Undersøgelse Bulletin*, **169**: 129–147.

Babcock, L.E., 2003, Trilobites in Paleozoic predator-prey systems, and their role in reorganization of early Paleozoic ecosystems. *In* Kelley, P.A., Kowalewski, M., and Hansen, T.A. (eds), *Predator-Prey Interactions in the Fossil Record*. New York: Kluwer Academic/ Plenum Publishers, 55–92.

Babcock, L.E., 2005, Interpretation of biological and environmental changes across the Neoproterozoic-Cambrian boundary: developing a refined understanding of the radiation and preservational record of early multicellular organisms. *Palaeogeography, Palaeoclimatology, Palaeoecology*, **220**: 1–5.

Babcock, L.E., and Ciampaglio, C.N., 2007, Frondose fossil from the Conasauga Formation (Cambrian: Drumian Stage) of Georgia, USA. *Memoirs of the Association of Australasian Palaeontologists*, **34**: 555–562.

Babcock, L.E., and Peel, J.S., 2007, Palaeobiology, taphonomy and stratigraphic significance of the trilobite *Buenellus* from the Sirius Passet Biota, Cambrian of North Greenland. *Memoirs of the Association of Australasian Palaeontologists*, **34**: 401–418.

Babcock, L.E., and Peng, S.C., 2007, Cambrian chronostratigraphy: current state and future plans. *Palaeogeography, Palaeoclimatology, Palaeoecology*, **254**: 62–66.

Babcock, L.E., Peng, S.C., and Ahlberg, P., 2017, Cambrian trilobite biostratigraphy and its role in developing an integrated history of the Earth system. *Lethaia*, **50**: 381–399.

Babcock, L.E., Peng, S.C., Brett, C.E., Zhu, M.Y., Ahlberg, P., Bevis, M., et al., 2015, Global climate, sea level cycles, and biotic events in the Cambrian Period. *Palaeoworld*, **24**: 5–15.

Babcock, L.E., Peng, S.C., and Ellwood, B.B., 2009, Progress toward completion of a Cambrian chronostratigraphic scale. *In* Ergaliev, G.K., Nikitina, O.I., Zhemchuzhnikov, V.G., Popov, L.E., and Bassett, M.G. (eds), *Stratigraphy, Fossils and Progress of International Stratigraphic Scale of Cambrian System. Materials of the 14th International Field Conference of Excursion of the Cambrian Stage Subdivision Working Group*. Almaty: Gylym, 7–8.

Babcock, L.E., Peng, S.C., Geyer, G., and Shergold, J.H., 2005, Changing perspectives on Cambrian chronostratigraphy and progress toward subdivision of the Cambrian System. *Geosciences Journal*, **9**: 101–106.

Babcock, L.E., Peng, S.C., Zhu, M.Y., Xiao, S.H., and Ahlberg, P., 2014, Proposed reassessment of the Cambrian GSSP. *Journal of African Earth Sciences*, **98**: 3–10.

Babcock, L.E., Rees, M.N., Robison, R.A., Langenburg, E.S., and Peng, S.C., 2004, Potential Global Standard Stratotype-section and Point

(GSSP) for a Cambrian stage boundary defined by the first appearance of the trilobite *Ptychagnostus atavus*, Drum Mountains, Utah, USA. *Geobios*, **37**: 149–158.

Babcock, L.E., Robison, R.A., and Peng, S.C., 2011, Cambrian stage and series nomenclature of Laurentia and the developing global chronostratigraphic scale. *Museum of Northern Arizona Bulletin*, **67**: 12–26.

Babcock, L.E., Robison, R.A., Rees, M.N., Peng, S.C., and Saltzman, M.R., 2007, The Global boundary Stratotype Section and Point (GSSP) of the Drumian Stage (Cambrian) in the Drum Mountains, Utah, USA. *Episodes*, **30**: 85–95.

Babcock, L.E., Zhang, W.T., and Leslie, S.A., 2001, The Chengjiang Biota: record of the Early Cambrian diversification of life and clues to exceptional preservation. *GSA Today*, **11** (2): 4–9.

Bagnoli, G., Peng, S.C., Qi, Y.P., and Wang, C.Y., 2017, Conodonts from the Wa'ergang section, China, a potential GSSP for the uppermost stage of the Cambrian. *Rivista Italiana di Paleontologia e Stratigrafia*, **123** (1): 1–10.

Bartley, J.K., Pope, M., Knoll, A.H., Semikhatov, M.A., and Petrov, P.Y., 1998, A Vendian-Cambrian Boundary succession from the northwestern margin of the Siberian Platform: stratigraphy, paleontology, and correlation. *Geological Magazine*, **135**: 473–494.

Bassett, M., 1985, Towards a "common language" in stratigraphy. *Episodes*, **8**: 87–92.

Bengtson, S., 1976, The structure of some Middle Cambrian conodonts, and the early evolution of conodont structure and function. *Lethaia*, **9**: 185–206.

Bengtson, S., 1994, The advent of animal skeletons. *In* Bengtson, S. (ed.), *Early Life on Earth*. New York: Columbia University Press, 412–425.

Bengtson, S., Conway Morris, S., Cooper, B.J., Jell, P.A., and Runnegar, B.N., 1990, Early Cambrian Fossils from South Australia. *Memoirs of the Association of Australasian Palaeontologists*, **9**: 1–364.

Bergström, J., 1981, Lower Cambrian shelly faunas and biostratigraphy in Scandinavia. *In* Taylor, M.E. (ed.), *Short Papers for the Second International Symposium on the Cambrian System*. U.S. Geological Survey Open-File Report, 22–25 81–743.

Betts, M.J., Paterson, J.R., Jacquet, S.M., Andrew, A.S., Hall, P.A., Jago, J.B., et al., 2018, Early Cambrian chronostratigraphy and geochronology of South Australia. *Earth-Science Reviews*, **185**: 498–543.

Betts, M.J., Paterson, J.R., Jago, J.B., Jacquet, S.M., Skovsted, C.B., Topper, T.P., et al., 2016, A new lower Cambrian shelly fossil biostratigraphy for South Australia. *Gondwana Research*, **36**: 163–195.

Betts, M.J., Paterson, J.R., Jago, J.B., Jacquet, S.M., Skovsted, C.B., Topper, T.P., et al., 2017, Global correlation of the early Cambrian of South Australia: shelly fauna of the *Dailyatia odyssey* Zone. *Gondwana Research*, **46**: 240–279.

Billings, E., 1860, On some new species of fossils from the limestone near Point Levi, opposite Quebec. *Canadian Naturalist and Geologist*, **5**: 301–324.

Blaker, M.R., and Peel, J.S., 1997, Lower Cambrian trilobites from North Greenland. *Meddelelser om Grønland Geoscience*, **35**: 1–145.

Bottjer, D.J., Hagadorn, J.W., and Dornbos, S.Q., 2000, The Cambrian substrate revolution. *GSA Today*, **10** (9): 1–7.

Bowring, S.A., Grotzinger, J.P., Condon, D.J., Ramezani, J., Newall, M.J., and Allen, P.A., 2007, Geochronologic constraints on the chronostratigraphic framework of the Neoproterozoic Huqf Supergroup, Sultanate of Oman. *American Journal of Science*, **307**: 1097–1145.

Bowring, S.A., Grotzinger, J.P., Isachsen, C.E., Knoll, A.H., Pelechaty, S.M., and Kolosov, P., 1993, Calibrating rates of Early Cambrian evolution. *Science*, **261**: 1293–1298.

Bowring, S.A., and Schmitz, M.D., 2003, High-precision U-Pb Zircon geochronology and the stratigraphic record. *Reviews in Mineralogy and Geochemistry*, **53**: 305–326.

Brasier, M.D., 1979, The Cambrian radiation event. *In* House, M.R. (ed.), *The Origin of Major Invertebrate Groups*. Systematics Association Special Volume, **12**: 103–159.

Brasier, M.D., 1982, Sea-level changes, facies changes and the late Precambrian-early Cambrian evolutionary explosion. *Precambrian Research*, **17**: 105–123.

Brasier, M.D., 1984, Microfossils and small shelly fossils from the Lower Cambrian Hyolithes Limestone at Nuneaton, English Midlands. *Geological Magazine*, **121**: 229–253.

Brasier, M.D., 1992a, Paleoceanography and changes in the biological cycling of phosphorus across the Precambrian-Cambrian boundary. *In* Lipps, J.H., and Signor, P.W. (eds), *Origin and Early Evolution of the Metazoa*. New York: Plenum Press, 483–523.

Brasier, M.D., 1992b, Nutrient-enriched waters and the early skeletal fossil record. *Journal of the Geological Society, London*, **149**: 621–629.

Brasier, M.D., 1993, Towards a carbon isotope stratigraphy of the Cambrian System: potential of the Great Basin succession. *In* Hailwood, E.A., and Kidd, R.B. (eds), *High Resolution Stratigraphy*. Geological Society of London Special Publication, **70**: 341–350.

Brasier, M.D., 1995a, Fossil indicators of nutrient levels. 1: Eutrophication and climate change. *In* Bosence, D.W.J., and Allison, P.A. (eds), *Marine Palaeoenvironmental Analysis from Fossils*. Geological Society of London Special Publication, **83**: 113–132.

Brasier, M.D., 1995b, Fossil indicators of nutrient levels. 2: Evolution and extinction in relation to oligotrophy. *In* Bosence, D.W.J., and Allison, P.A. (eds), *Marine Palaeoenvironmental Analysis from Fossils*. Geological Society of London Special Publication, **83**: 133–150.

Brasier, M.D., 1995c, The basal Cambrian transition and Cambrian bio-events (from terminal Proterozoic extinctions to Cambrian bio-meres). *In* Walliser, O.H. (ed.), *Global Events and Event Stratigraphy in the Phanerozoic*. Berlin: Springer, 113–118.

Brasier, M.D., Cowie, J., and Taylor, M., 1994, Decision on the Precambrian-Cambrian boundary stratotype. *Episodes*, **17**: 3–8.

Brasier, M.D., Shields, G., Kuleshov, V.N., and Zhegallo, E.A., 1996, Integrated chemo- and biostratigraphic calibration of early animal evolution of southwest Mongolia. *Geological Magazine*, **133**: 445–485.

Brasier, M.D., and Sukhov, S.S., 1998, The falling amplitude of carbon isotopic oscillations through the Lower to Middle Cambrian: northern Siberia data. *Canadian Journal of Earth Sciences*, **35**: 353–373.

Briggs, D.E.G., Erwin, D.H., and Collier, F.J., 1994, *The Fossils of the Burgess Shale*. Washington, DC: Smithsonian Institution Press 238 pp.

Briggs, D.E.G., Lieberman, B.S., Halgedahl, S.L., and Jarrard, R.D., 2005, A new metazoan from the Middle Cambrian of Utah and the nature of the Vetulicolia. *Palaeontology*, **48**: 681–686.

Briggs, D.E.G., Lieberman, B.S., Hendricks, J.R., Halgedahl, S.L., and Jarrard, R.D., 2008, Middle Cambrian arthropods from Utah. *Journal of Paleontology*, **82**: 238–254.

Brøgger, W.C., 1878, Om Paradoxidesskifrene ved Krekling. *Nyt Magazin Naturvidenskap*, **24**: 18–88.

Brøgger, W.C., 1882, Die silurischen Etagen 2 und 3 im Kristianiagebiet und auf Eker, ihre Gliederung, Fossilien, Schichtenstörungen und Kontactmetamorfosen. *Kristiania Universitäts-Programm*, **32**: 1–376.

Brøgger, W.C., 1886, Om alderen af Olenelluszonen i Nordamerika. *Geologiska Föreningens i Stockholm Förhandlingar*, **8**: 182–213.

Buatois, L.A., Narbonne, G.M., Mángano, M.G., Carmona, N.B., and Myrow, P., 2014, Ediacaran matground ecology persisted into the earliest Cambrian. *Nature Communications*, **5**: 3544.

Budd, G.E., 1997, Stem group arthropods from the Lower Cambrian Sirius Passet fauna of North Greenland. *In* Fortey, R.A., and Thomas, R.H. (eds), *Arthropod Relationships*. Systematics Association Special, **Vol. 55**: 125–138.

Budd, G.E., and Peel, J.S., 1998, A new xenusiid lobopod from the Early Cambrian Sirius Passet fauna of North Greenland. *Palaeontology*, **41**: 1201–1213.

Buggisch, W., Keller, M., and Lehnert, O., 2003, Carbon isotope record of Late Cambrian to Early Ordovician carbonates of the Argentine Precordillera. *Palaeogeography, Palaeoclimatology, Palaeoecology*, **195**: 357–373.

Butler, A.D., Streng, M., Holmer, L.E., and Babcock, L.E., 2015, Exceptionally preserved *Mickwitzia* from the Indian Springs Lagerstätte (Cambrian Stage 3), Nevada. *Journal of Paleontology*, **89**: 933–955.

Cederström, P., Ahlberg, P., Babcock, L.E., Ahlgren, J., Høyberget, M., and Nilsson, C.H., 2012, Morphology, ontogeny and distribution of the Cambrian Series 2 ellipsocephalid trilobite *Strenuaeva spinosa* from Scandinavia. *GFF*, **134**: 157–171.

Cederström, P., Ahlberg, P., Clarkson, E.N.K., Nilsson, C.H., and Axheimer, N., 2009, The Lower Cambrian eodiscoid trilobite *Calodiscus lobatus* from Sweden: morphology, ontogeny and distribution. *Palaeontology*, **52**: 491–539.

Chang (Zhang), W.T., 1989, *World Cambrian Biogeography*. Beijing: Science Press, pp. 209–220 Developments in Geoscience, Chinese Academy of Sciences.

Chang (Zhang), W.T., 1998, Cambrian biogeography of the Perigondwana Faunal Realm. *In* Gámez Vintaned, J.A., Palacios, T., Liñán, E., Gozalo, R., and Martínez Chacón (eds), *II Field Conference of the Cambrian Stage Subdivision Working Groups (Spain, 13–21 September 1996)*. Revista Española de Paleontología, 35–49 Extra no. Homenaje al Prof. Gonzalo Vidal.

Chatterton, B.D.E., and Gibb, S., 2016, Furongian (Upper Cambrian) trilobites from the McKay Group, Bull River Valley, near Cranbrook, southeastern British Columbia, Canada. *Palaeontographica Canadiana*, **35**: 1–275.

Chatterton, B.D.E., and Ludvigsen, R., 1998, Upper Steptoean (Upper Cambrian) trilobites from the McKay Group of southeastern British Columbia, Canada. *Paleontological Society Memoir*, **49**: 1–43.

Chen, J.Y., Zhou, G.Q., Zhu, M.Y., and Yeh, K.Y., 1996, *The Chengjiang Biota, A Unique Window of the Cambrian Explosion*. Taichun: National Museum of Natural Sciences, p. 222 (in Chinese).

Chernysheva, N.E., 1961, Stratigraphy of the Cambrian of the Aldan Anticline and the palaeontological basis for separation of the Amgan Stage. *Trudy Vsesoyuznogo Nauchno-Issledobatelskogo Geologicheskogo Instituta (VSEGEI), New Series*, **49**: 1–347.

Chernysheva, N.E., 1967, Stratigraphical section of Mayan Stage of Middle Cambrian. *Bulletin of Science and Technology information, ONTI VIEMS, Series Geology, Ore Deposits, Useful Mineral, and Regional Geology*, **7**: 38–42 (in Russian).

Choi, D.K., 2004, Stop 12, Machari Formation at the Gonggiri section. *In* Choi, D.K., Chough, S.K., Fitches, W.R., Kwon, Y.K., Lee, S.B., Kang, I., et al., (eds), *Korea, 2004, Cambrian in the Land of Morning Calm*. Seoul: Seoul National University, 51–56.

Cloud, P.E., 1968, Pre-metazoan evolution and the origins of the Metazoa. *In* Drake, E.T. (ed.), *Evolution and Environment*. New Haven, CT: Yale University Press, 1–72.

Conway Morris, S., 1977, Fossil priapulid worms. *Special Papers in Palaeontology*, **20**: 1–159.

Conway Morris, S., 1985, Cambrian Lagerstätten: their distribution and significance. *Philosophical Transactions of the Royal Society of London, Series B*, **311**: 49–65.

Conway Morris, S., 1993, Ediacaran fossils in Cambrian Burgess Shale-type faunas of North America. *Palaeontology*, **36**: 539–635.

Conway Morris, S., 1998, *The Crucible of Creation. The Burgess Shale and the Rise of Animals*. Oxford: Oxford University Press, p. 276.

Conway Morris, S., and Peel, J.S., 1990, Articulated halkieriids from the Lower Cambrian of North Greenland. *Nature*, **345**: 802–805.

Conway Morris, S., Peel, J.S., Higgins, A.K., Soper, N.J., and Davis, N.C., 1987, A Burgess Shale-like fauna from the Lower Cambrian of Greenland. *Nature*, **326**: 181–183.

Cooper, R.A., and Nowlan, G.S., 1999, Proposed global stratotype section and point for base of the Ordovician System. *Acta Universitatis Carolinae Geologica*, **43**: 61–64.

Cooper, R.A., Nowlan, G.S., and Williams, H.S., 2001, Global Stratotype Section and Point for base of the Ordovician System. *Episodes*, **24**: 19–28.

Corsetti, F.A., and Hagadorn, J.W., 2001, The Precambrian–Cambrian transition in the southern Great Basin, USA. *The Sedimentary Record*, **1** (1): 4–8.

Cowie, J.W., 1971, Lower Cambrian faunal provinces. *In* Middlemiss, F.A., Rawson, P.F., and Newall, G. (eds), *Faunal provinces in space and time* (Proceedings of the 17th Inter-university Geological Congress held in Queen Mary College (University of London), 17, 18, 19 December 1969). Seel House Press, 31–46.

Cowie, J.W., 1978, I.U.G.S./I.G.C.P. Project 29, Precambrian-Cambrian Boundary Working Group in Cambridge, 1978. *Geological Magazine*, **115**: 151–152.

Cowie, J.W., 1985, Continuing work on the Precambrian-Cambrian boundary. *Episodes*, **8**: 93–97.

Cowie, J.W., Rushton, A.W.A., and Stubblefield, C.J., 1972, A correlation of Cambrian rocks in the British Isles. *Geological Society of London, Special Report*, **2**: 1–42.

Cowie, J.W., Ziegler, W., and Remane, J., 1989, Stratigraphic Commission accelerates progress, 1984–1989. *Episodes*, **12**: 79–83.

Cramer, B.D., and Jarvis, I., 2020, Chapter 11 - Carbon isotope stratigraphy. *In* Gradstein, F.M., Ogg, J.G., Schmitz, M.D., and Ogg, G.M. (eds), *The Geologic Time Scale 2020*. Elsevier, Boston, MA, Vol. 1 (this book).

Daily, B., and Jago, J.B., 1975, The trilobite *Leiopyge* Hawle and Corda and the Middle-Upper Cambrian boundary. *Palaeontology*, **18**: 527–550.

Davidek, K., Landing, E., Bowring, S.A., Westrop, S.R., Rushton, A.W.A., Fortey, R.A., et al., 1998, New uppermost Cambrian U-Pb date from Avalonian Wales and age of the Cambrian-Ordovician boundary. *Geological Magazine*, **135**: 305–309.

Debrenne, F., 1992, Diversification of Archaeocyatha. *In* Lipps, J.H., and Signor, P.W. (eds), *Origin and Early Evolution of the Metazoa*. New York: Plenum Press, 425–443.

Debrenne, F., and Rozanov, A.Y., 1983, Paleogeographic and stratigraphic distribution of regular Archaeocyatha (Lower Cambrian fossils). *Geobios*, **16**: 727–736.

Demidenko, Y.E., 2001, Small shelly fossils. *In* Alexander, E.M., Jago, J.B., Rozanov, A.Y., and Zhuravlev, A.Y. (eds), *The Cambrian Biostratigraphy of the Stansbury Basin, South Australia*. Transactions of the Palaeontological Institute, Russian Academy of Sciences, **282**: 85–117.

Demidenko, Y.E., Jago, J.B., Lin, T.R., Parkhaev, P., Rozanov, A.Y., Ushatinskaya, G.T., et al., 2001, Conclusions. *In* Alexander, E.M., Jago, J.B., Rozanov, A.Y., and Zhuravlev, A.Y. (eds), *The Cambrian Biostratigraphy of the Stansbury Basin, South Australia*. Transactions of the Palaeontological Institute, Russian Academy of Sciences, **282**: 66–73.

Derry, L.A., Brasier, M.D., Corfield, R.M., Rozanov, A.Y., and Zhuravlev, A.Y., 1994, Sr and C isotopes in Lower Cambrian carbonates from the Siberian craton: a paleoenvironmental record during the "Cambrian explosion". *Earth and Planetary Science Letters*, **128**: 671–681.

Domke, K.L., and Dornbos, S.Q., 2010, Paleoecology of the middle Cambrian edrioasteroid echinoderm *Totiglobus*: implications for unusual Cambrian morphologies. *Palaios*, **25**: 209–214.

Dong, X.P., and Bergström, S.M., 2001a, Middle and Upper Cambrian protoconodonts and paraconodonts from Hunan, South China. *Palaeontology*, **44**: 949–985.

Dong, X.P., and Bergström, S.M., 2001b, Stratigraphic significance of Middle and Upper Cambrian protoconodonts and paraconodonts from Hunan, South China. *Palaeoword*, **13**: 307–309.

Dong, X.P., Repetski, J.E., and Bergström, S.M., 2004, Conodont biostratigraphy of the Middle Cambrian through lowermost Ordovician in Hunan, South China. *Acta Geologica Sinica*, **78**: 1185–1260.

Dong, X.P., and Zhang, H.Q., 2017, Middle Cambrian through lowermost Ordovician conodonts from Hunan, South China. *Journal of Paleontology*, **91** (Supplement S73): 1–89.

Droser, M.L., and Bottjer, D.J., 1988, Trends in depth and extent of bioturbation in Cambrian carbonate marine environments, western United States. *Geology*, **16**: 233236.

Droser, M.L., and Bottjer, D.J., 1989, Ichnofabric of sandstones deposited in the high energy nearshore environments: measurement and utilization. *Palaios*, **4**: 598–604.

Druce, E.C., and Jones, P.-J., 1971, Cambro-Ordovician conodonts from the Burke River Structural Belt, Queensland. *Bulletin of the Bureau of Mineral Resources of Australia*, **110**: 1–158.

Du, W., and Wang, X.L., 2012, Hexactinellid sponge spicules in Neoproterozoic dolostone from South China. *Paleontological Research*, **16**: 199–207.

Dubinina, S.V., 2009, Proposed new terminal stage of the Cambrian System in Kazakhstan. *In* Ergaliev, G.K., Nikitina, O.I., Zhemchuzhnikov, V.G., Popov, L.E., and Bassett, M.G. (eds), *Stratigraphy, Fossils and Progress of International Stratigraphic Scale of Cambrian System. Materials of the 14th International Field Conference of Excursion of the Cambrian Stage Subdivision Working Group*. Almaty: Gylym, 18–19.

Ebbestad, J.O.R., Ahlberg, P., and Høyberget, M., 2003, Redescription of *Holmia inusitata* (Trilobita) from the Lower Cambrian of Scandinavia. *Palaeontology*, **46**: 1039–1054.

Ebneth, S., Shields, G.A., Veizer, J., Miller, J.F., and Shergold, J.H., 2001, High resolution strontium isotope stratigraphy across the Cambrian-Ordovician transition. *Geochimica et Cosmochimica Acta*, **65**: 2273–2292.

Eddy, J.D., and McCollum, L.B., 1998, Early Middle Cambrian *Albertella* Biozone trilobites of the Pioche Shale, Southeastern Nevada. *Journal of Paleontology*, **72**: 864–887.

Egorova, L.I., Shabanov, Y.Y., Pegel, T.V., Savitsky, V.E., Sukhov, S.S., and Chernysheva, N.E., 1982, Type section of the Mayan Stage (Middle Cambrian of the southeastern Siberian Platform. *Academy of Sciences of the USSR, Ministry of Geology of the USSR, Interdepartmental Stratigraphic Committee of the USSR, Transactions*, **8**: 1−146 (in Russian).

Egorova, L.I., Shabanov, Y.Y., Rozanov, A.Y., Savitsky, V.E., Chernysheva, N.E., and Shishkin, B.B., 1976, Elansk and Kuonamsk facies stratotypes in the lower part of the Middle Cambrian of Siberia. *Sibirskogo Nauchno-issledovatel'skogo Instituta Geologii, Geofiziki i Mineral'nogo Syr'ya, Seriya Paleontologiya i Stratigrafiya*, **211**: 1−167 (in Russian).

Eklund, C., 1990, Lower Cambrian acritarch stratigraphy of the Bårstad 2 core, Östergötland, Sweden. *Geologiska Föreningens i Stockholm Förhandlingar*, **112**: 19−44.

Encarnación, J., Rowell, A.J., and Grunow, A.M., 1999, A U-Pb age for the Cambrian Taylor Formation, Antarctica: implications for the Cambrian time scale, *Journal of Geology*, 107. pp. 497−504.

English, A.M., and Babcock, L.E., 2010, Census of the Indian Springs Lagerstätte, Poleta Formation (Cambrian), western Nevada, USA. *Palaeogeography, Palaeoclimatology, Palaeoecology*, **295**: 236−244.

Ergaliev, G.K., 1980, *Middle and Upper Cambrian Trilobites From Malyi Karatau*. Alma-Ata: Akademiya Nauk Kazakhskoi SSR 211 pp. (in Russian).

Ergaliev, G.K., 1981, Upper Cambrian biostratigraphy of the Kyrshabakty section, Malyi Karatau, southern Kazakhstan. *In* Taylor, M.E. (ed.), *Short Papers for the Second International Symposium on the Cambrian System*. U.S. Geological Survey Open-file Report, 82−88 81−743.

Ergaliev, G.K., 1992, On palaeogeography of Karatau (southern Kazakhstan) in Middle Cambrian to Early Ordovician. *Proceedings of Academy of Sciences, Republic of Kazakhstan, Series Geology*, **6**: 51−56.

Ergaliev, G.K., and Ergaliev, F.G., 2004, Stages and zones of the Middle and Upper portions of Cambrian of Malyi Karatau for the Project of International Stratigraphic Chart. *In* Daukeev, S.Z. (ed.), *Geology of Kazakhstan*. Almaty: Kazakhstan Geological Society, pp. 36−52 (in Russian).

Ergaliev, G.K., and Ergaliev, F.G., 2008, *Middle and Upper Cambrian Agnostida of the Aksai National Geological Reserve, South Kazakhstan (Kyrshabakty River, Malyi Karatau Range), Part 1.* Almaty: Gylym Press, p. 376.

Eriksson, M.E., Terfelt, F., Elofsson, R., Maas, A., Marone, F., Lindskog, A., et al., 2016, Baring it all: undressing Cambrian 'Orsten' phosphatocopine crustaceans using synchotron radiation X-ray tomographic microscopy. *Lethaia*, **49**: 312−326.

Faggetter, L.E., Wignall, P.B., Pruss, S.B., Newton, R.J., Sun, Y., and Crowley, S., 2017, Trilobite extinctions, facies changes and the ROECE carbon isotope excursion at the Cambrian Series 2−3 boundary, Great Basin, western USA. *Palaeogeography, Palaeoclimatology, Palaeoecology*, **478**: 53−66.

Fletcher, T.P., 2003, *Ovatoryctocara granulata*: the key to a global Cambrian stage boundary and the correlation of the olenellid, red-lichiid and paradoxidid realms. *Special Papers in Palaeontology*, **70**: 73−102.

Fletcher, T.P., 2006. Bedrock Geology of the Cape St. Mary's Peninsula, Southwest Avalon Peninsula, Newfoundland (includes parts of NTS map sheets 1M/1, 1N/4, 1L/6 and 1K/13). Government of Newfoundland and Laborador, Geological Survey,

Department of Natural Resources, St. John's, Report 06-02, pp. 1−117.

Fletcher, T.P., 2007, The base of Cambrian Series 3: the global significance of key oryctocephalid trilobite ranges in the Kaili Formation of South China. *Memoirs of the Association of Australasian Palaeontologists*, **33**: 29−33.

Fortey, R.A., Briggs, D.E.G., and Wills, M.A., 1996, The Cambrian evolutionary 'explosion': decoupling cladogenesis from morphological disparity. *Biological Journal of the Linnean Society*, **57**: 13−33.

Føyn, S., and Glaessner, M.F., 1979, *Platysolenites*, other animal fossils, and the Precambrian−Cambrian transition in Norway. *Norsk Geologisk Tidsskrift*, **59**: 25−46.

Fritz, W.H., 1992, Walcott's Lower Cambrian olenellid trilobite collection 61K, Mount Robson area, Canadian Rocky Mountains. *Geological Survey of Canada*, **432**: 1−65.

Gaines, R.R., Droser, M.L., Orr, P.J., Garson, D., Hammarlund, E., Qi, C.S., et al., 2012, Burgess shale-type biotas were not entirely burrowed away. *Geology*, **40**: 283−286.

Gehling, J.G., 1999, Microbial mats in Proterozoic siliciclastics: Ediacaran death masks. *Palaios*, **14**: 40−57.

Gehling, J.G., Jensen, S., Droser, M.L., Myrow, P.M., and Narbonne, G. M., 2001, Burrowing below the basal Cambrian GSSP, Fortune Head, Newfoundland. *Geological Magazine*, **138**: 213−218.

Gehling, J.G., and Rigby, L.K., 1996, Long expected sponges from the Neoproterozoic Ediacara fauna of South Australia. *Journal of Paleontology*, **70**: 185−195.

Geyer, G., 1990, Correlation along the Lower/Middle Cambrian boundary − a puzzling story with an elusory end? *In* Repina, L.N., and Zhuravlev, A.Y. (eds), *Third International Symposium on the Cambrian System*. Novosibirsk: Institute of Geology and Geophysics, 100−102.

Geyer, G., 1995, The Fish River Subgroup in Namibia: stratigraphy, depositional environments and the Proterozoic-Cambrian boundary problem revisited. *Geological Magazine*, **142**: 465−498.

Geyer, G., 1998, Intercontinental, trilobite-based correlation of the Moroccan early Middle Cambrian. *Canadian Journal of Earth Sciences*, **35**: 374−401.

Geyer, G., 2005, The base of a revised Middle Cambrian: are suitable concepts for a series boundary in reach? *Geosciences Journal*, **9**: 81−99.

Geyer, G., and Landing, E., 2004, A unified Lower-Middle Cambrian chronostratigraphy for West Gondwana. *Acta Geologica Polonica*, **54**: 179−218.

Geyer, G., and Landing, E., 2016, The Precambrian−Phanerozoic and Ediacaran−Cambrian boundaries: a historical approach to a dilemma, Special Publications. *In* Brasier, A.T., McIlroy, D., and McLoughlin, N. (eds), *Earth System Evolution and Early Life: A Celebration of the Work of Martin Brasier*. London: Geological Society, **448**: pp. 311−349. http://doi.org/10.1144/SP448.10.

Geyer, G., and Peel, J.S., 2011, The Henson Gletscher Formation, North Greenland, and its bearing on the global Cambrian Series 2-Series 3 boundary. *Bulletin of Geosciences*, **86**: 465−534.

Geyer, G., Peng, S.C., Shergold, J., 2000. Correlation chart for major Cambrian areas. In: Geyer, G., Shergold, J. (eds.), The Quest for Internationally Recognized Divisions of Cambrian time. *Episodes*, **23**. pp. 190−191.

Geyer, G., Peng, S.C., Shergold, J., 2003. Correlation chart for major Cambrian areas (revised version, 2002). In: Shergold, J., Geyer, G.

(eds.), The Subcommission on Cambrian Stratigraphy: The Status Quo. *Geologica Acta*, **1**: pp. 6–7.

Geyer, G., and Shergold, J., 2000, The quest for internationally recognized divisions of Cambrian time. *Episodes*, **23**: 188–195.

Grabau, A.W., 1900, Palaeontology of the Cambrian terranes of the Boston Basin. *Occasional Papers of the Boston Society of Natural History*, **4**: 601–694.

Grotzinger, J.P., Bowring, S.A., Saylor, B.Z., and Kaufman, A.J., 1995, Biostratigraphic and geochronologic constraints on early animal evolution. *Science*, **270**: 598–604.

Gunther, L.F., and Gunther, V.G., 1981, Some Middle Cambrian fossils of Utah. *Brigham Young University Geology Studies*, **28**: 1–81.

Guo, Q.J., Strauss, H., Liu, C.Q., Zhao, Y.L., Pi, D.H., Fu, P.Q., et al., 2005, Carbon and oxygen isotopic composition of Lower to Middle Cambrian sediments at Taijiang, Guizhou Province, China. *Geological Magazine*, **142**: 723–733.

Hagadorn, J.W., Fedo, C.M., and Waggoner, B.M., 2000, Early Cambrian Ediacaran-type fossils from California. *Journal of Paleontology*, **74**: 731–740.

Halgedahl, S.L., Jarrard, R.D., Brett, C.E., and Allison, P.A., 2009, Geophysical and geological signatures of relative sea level change in the upper Wheeler Formation, Drum Mountains, west-central Utah: a perspective into exceptional preservation of fossils. *Palaeogeography, Palaeoclimatology, Palaeoecology*, **277**: 34–56.

Hamdi, B., Brasier, M.D., and Jiang, Z., 1989, Earliest skeletal fossils from Precambrian-Cambrian boundary strata, Elburz Mountains, Iran. *Geological Magazine*, **126**: 283–289.

Hammarlund, E.U., Smith, M.P., Rasmussen, J.A., Nielsen, A.T., Canfield, D.E., and Harper, D.A.T., 2018, The Sirius Passet Lagerstätte of North Greenland – a geochemical window on early Cambrian low-oxygen environments and ecosystems. *Geobiology*, **2018**: 1–15.

Han, N.R., and Chen, G.Y., 2008, The discovery of Stylophora in China: a new species of Phyllocysits from the Furongian (upper Cambrian) of western Guangxi, South China. *Science in China Series D*, **38**: 1–9 (in Chinese).

Haq, B.U., and Schutter, S.R., 2008, A chronology of Paleozoic sealevel changes. *Science*, **322**: 6468.

Harvey, T.H.P., Williams, M., Condon, D.J., Wilby, P.R., Siveter, D.J., Rushton, A.W.A., et al., 2011, A refined chronology for the Cambrian succession of southern Britain. *Journal of the Geological Society*, **168**: 705–716.

Henderson, R.A., 1976, Upper Cambrian (Idamean) trilobites from western Queensland, Australia. *Palaeontology*, **19**: 325–364.

Henderson, R.A., 1977, Stratigraphy of the Georgina Limestone and a revised zonation for the early Upper Cambrian Idamean Stage. *Journal of the Geological Society of Australia*, **23**: 423–433.

Henningsmoen, G., 1957, *The trilobite family Olenidae with description of Norwegian material and remarks on the Olenid and Tremadocian Series.* Series: Norske videnskaps-akademi i Oslo.; I. Matematisk-Naturvidenskapelig Klasse; Skrifter. Publisher: Oslo, I kommisjon hos H. Aschehoug,: 303 pp.

Hoffman, H.J., and Mountjoy, E.W., 2001, *Namacalathus-Cloudina* assemblage in Neoproterozoic Miette Group (Byng Formation), British Columbia: Canada's oldest shelly fossils. *Geology*, **29**: 1091–1094.

Hollingsworth, J.S., 2007, Fallotaspidoid trilobite assemblage (Lower Cambrian) from the Esmeralda Basin (western Nevada, U.S.A.): the

oldest trilobites from Laurentia. *Memoirs of the Association of Australasian Palaentologists*, **33**: 123–140.

Hollingsworth, J.S., 2011, Lithostratigraphy and biostratigraphy of Cambrian Stage 3 in western Nevada and eastern California. *Museum of Northern Arizona Bulletin*, **67**: 26–42.

Hollingsworth, J.S., and Babcock, L.E., 2011, Base of Dyeran Stage (Cambrian Stage 4) in the middle member of the Poleta Formation, Indian Springs Canyon, Miontezuma Range, Nevada. *Museum of Northern Arizona Bulletin*, **67**: 256–262.

Hong, P., Lee, J.G., and Choi, D.K., 2003, Late Cambrian trilobite *Irvingella* from the Machari Formation, Korea: evolution and correlation. *Special Papers in Palaeontology*, **70**: 175–196.

Hou, X.G., Bergström, J., Wang, H.F., Feng, X.H., and Chen, A.L., 1999, *The Chengjiang Fauna: Exceptionally Well-Preserved Animals From 530 Million Years Ago.* Kunming: Yunnan Science and Technology Press 170 pp. (in Chinese).

Hou, X.G., Siveter, D.J., Siveter, D.J., Aldridge, R.J., Cong, P.Y., Gabbott, S.E., et al., 2017, *The Cambrian Fossils of Chengjiang, China: The Flowering of Early Animal Life, second ed.* New York: Wiley 316 pp.

Howley, R.A., and Jiang, G.Q., 2010, The Cambrian Drumian carbon isotope excursion (DICE) in the Great Basin, western United States. *Palaeogeography, Palaeoclimatology, Palaeoecology*, **296**: 138–150.

Howley, R.A., Rees, M.N., and Jiang, G.Q., 2006, Significance of Middle Cambrian mixed carbonate-siliciclastic units for global correlation: southern Nevada, USA. *Palaeoworld*, **15**: 360–366.

Høyberget, M., and Bruton, D.L., 2008, Middle Cambrian trilobites of the suborders Agnostina and Eodiscina from the Oslo Region, Norway. *Palaeontographica Abteilung A*, **286**: 1–87.

Høyberget, M., and Bruton, D.L., 2012, Revision of the trilobite genus *Sphaerophthalmus* and relatives from the Furongian (Cambrian) Alum Shale Formation, Oslo Region, Norway. *Norwegian Journal of Geology*, **92**: 433–450.

Hua, H., Chen, Z., Yuan, X.L., Zhang, L.Y., and Xiao, S.H., 2005, Skeletogenesis and asexual reproduction in the earliest biomineralizing animal Cloudina. *Geology*, **33**: 277–280.

Hughes, N.C., 2016, The Cambrian palaeontological record of the Indian subcontinent. *Earth-Science Reviews*, **159**: 428–451.

Isachsen, C.E., Bowring, S.A., Landing, E., and Samson, S.D., 1994, New constraint on the division of Cambrian time. *Geology*, **22**: 496–498.

Jacquet, S.M., Brougham, T., Skovsted, C.B., Jago, J.B., Laurie, J.R., Betts, M.J., et al., 2017, *Watsonella crosbyi* from the lower Cambrian (Terreneuvian, Stage 2) Normanville Group in South Australia. *Geological Magazine*, **154**: 1088–1104.

Jago, J.B., Zang, W., Sun, X., Brock, G.A., Paterson, J.R., and Skovsted, C.B., 2006, A review of the Cambrian biostratigraphy of South Australia. *Palaeoworld*, **15**: 406–423.

Jell, P.A., 1974, Faunal provinces and possible planetary reconstruction of the Middle Cambrian. *Journal of Geology*, **82**: 319–350.

Jell, P.A., 2003, Phylogeny of Early Cambrian trilobites. *Special Papers in Palaeontology*, **70**: 45–57.

Jell, P.A., and Hughes, N.C., 1997, Himalayan Cambrian trilobites. *Special Papers in Palaeontology*, **58**: 7–113.

Jensen, S., Gehling, J.G., and Droser, M.L., 1998, Ediacara-type fossil in Cambrian sediments. *Nature*, **393**: 567–569.

Jensen, S., Saylor, B.Z., Gehling, J.G., and Germs, G.J.B., 2000, Complex trace fossils from the terminal Neoproterozoic of Namibia. *Geology*, **28**: 143–146.

Jiang, G.Q., Kaufman, A.J., Christie-Blick, N., Zhang, S.H., and Wu, H.C., 2007, Carbon isotope variability across the Ediacaran Yangtze platform in South China: implications for a large surface-to-deep ocean $\delta^{13}C$ gradient. *Earth and Planetary Science Letters*, **261**: 303−320.

Jiang, N.R., Wang, Z.Z., and Chen, Y.G., 1964, Cambrian stratigraphy of eastern Yunnan. *Acta Geologica Sinica*, **44**: 137−155 (in Chinese).

Jones, P.J., Shergold, J.H., and Druce, E.C., 1971, Late Cambrian and Early Ordovician stages in western Queensland. *Journal of the Geological Society of Australia*, **18**: 1−32.

Kaufman, A.J., Knoll, A.H., Semikhatov, M.A., Grotzinger, J.P., Jacobsen, S.B., and Adams, W., 1996, Integrated chronostratigraphy of Proterozoic-Cambrian boundary beds in the western Anabar region, northern Siberia. *Geological Magazine*, **133**: 509−533.

Keber, M., 1988, Mikrofossilien aus unterkambrischen Gesteinen der Montagne Noire, Frankreich. *Palaeontographica Abteilung A*, **202**: 127−203.

Khomentovsky, V.V., 1974, Principles for the recognition of the Vendian as a Paleozoic system. *In* Yanshin, A.L. (ed.), *Essays on Stratigraphy*. Moscow: Nauka, pp. 33−70 (in Russian).

Khomentovsky, V.V., 1976, Vendian. *Transactions of the Institute of Geology and Geophysics, Siberian Branch of Russian Academy of Sciences*, **234**: 1−271 (in Russian).

Khomentovsky, V.V., 1984, Vendian, (in Russian). *In* Yanshin, A.L. (ed.), *Phanerozoic of Siberia*. Novosibirsk: Nauka, **Vol. 1**: pp. 5−35.

Khomentovsky, V.V., and Gibsher, A.S., 1996, The Neoproterozoic-Lower Cambrian in northern Govi-Altai, western Mongolia: regional setting, lithostratigraphy and biostratigraphy. *Geological Magazine*, **133**: 371−390.

Khomentovsky, V.V., and Karlova, G.A., 1993, Biostratigraphy of the Vendian-Cambrian beds and the Lower Cambrian boundary in Siberia. *Geological Magazine*, **130**: 29−45.

Khomentovsky, V.V., and Karlova, G.A., 2005, The Tommotian Stage base as the Cambrian lower boundary in Siberia. *Stratigraphy and Geological Correlation*, **13**: 21−34.

Khomentovsky, V.V., and Repina, L.N., 1965, *The Lower Cambrian Stratotype Sections in Siberia*. Moscow: Nauka 200 pp (in Russian).

Khramov, A., and Rodionov, V., 1980, The geomagnetic field during the Palaeozoic time. *Journal of Geomagnetism and Geoelectricity*, **32** ((Suppl. III): 99−115.

Kimura, H., and Watanabe, Y., 2001, Oceanic anoxia at the Precambrian-Cambrian boundary. *Geology*, **29**: 995−998.

Kirschvink, J.L., 1978a, The Precambrian-Cambrian boundary problem: magnetostratigraphy of the Amadeus Basin, central Australia. *Geological Magazine*, **115**: 139−150.

Kirschvink, J.L., 1978b, The Precambrian-Cambrian boundary problem: palaeomagnetic directions from the Amadeus Basin, central Australia. *Earth and Planetary Science Letters*, **40**: 91−100.

Kirschvink, J.L., Margaritz, M., Ripperdan, R.L., Zhuravlev, A.Y., and Rozanov, A.Y., 1991, The Precambrian-Cambrian boundary: magnetostratigraphy and carbon isotopes resolve correlation problems between Siberia, Morocco and South China. *GSA Today*, **1**: 61−91.

Kirschvink, J.L., and Raub, T.D., 2003, A methane fuse for the Cambrian explosion: carbon cycles and true polar wander. *Geoscience*, **335**: 65−78.

Kirschvink, J.L., Ripperdan, R.L., and Evans, D., 1997, Evidence for a large scale reorganisation of Early Cambrian continental masses by inertial interchange true polar wander. *Science*, **227**: 541−545.

Kirschvink, J.L., and Rozanov, A.Y., 1984, Magnetostratigraphy of Lower Cambrian strata from the Siberian Platform: a palaeomagnetic pole and a preliminary polarity time-scale. *Geological Magazine*, **121**: 189−203.

Klootwijk, C.T., 1980, Early Palaeozoic magnetism in Australia. *Tectonophysics*, **64**: 249−332.

Kobayashi, T., 1935, The Cambro-Ordovician formations and faunas of South Chosen, Palaeontology, Part 3, Cambrian faunas of South Chosen with a special study on the Cambrian trilobite genera and families. *Journal of the Faculty of Science, Imperial University of Tokyo, Section II*, **4**: 49−344.

Kobayashi, T., 1971, The Cambro-Ordovician faunal provinces and the inter-provincial correlation discussed with special reference to the trilobites in eastern Asia. *Journal of the Faculty of Science, University of Tokyo, Section II*, **18** (1): 129−299.

Kobayashi, T., 1972, Three faunal provinces in the Early Cambrian Period. *Proceedings of the Japan Academy*, **48**: 242−247.

Korovnikov, I.V., 2001, Lower and Middle Cambrian boundary and trilobites from Northeast Siberian Platform. *Palaeoworld*, **13**: 270−275.

Korovnikov, I.V., 2005, Lower and Middle Cambrian boundary in open shelf facies of the Siberian Platform. *Acta Micropaleontologica Sinica*, **22** (Suppl.): 80−84.

Kouchinsky, A., Bengtson, S., Gallet, Y., Korovnikov, I., Pavlov, V., Runnegar, B., et al., 2008, The SPICE carbon isotope excursion in Siberia: a combined study of the upper Middle Cambrian-lowermost Ordovician Kulyumbe River section, northwestern Siberian Platform. *Geological Magazine*, **145**: 609−622.

Kouchinsky, A., Bengtson, S., Pavlov, V., Runnergar, B., Val'kov, A., and Young, E., 2005, Pre-Tommotian age of the lower Pestrotsvet Formation in the Selinde section on the Siberian Platform: carbon isotope evidence. *Geological Magazine*, **142**: 319−325.

Kouchinsky, A., Bengtson, S., Runnegar, B., Skovsted, C., Steiner, M., and Vendrasco, M., 2012, Chronology of early Cambrian biomineralization. *Geological Magazine*, **149**: 221−251.

Kruse, P., Jago, J.B., and Laurie, J.R., 2009, Recent developments in Australian Cambrian biostratigraphy. *Journal of Stratigraphy*, **33**: 35−47.

Kruse, P., Shi, G.R., 2000. Archaeocyaths and radiocyaths. In: Brock, G., Engelbresten, M.J., Jago, J.B., Kruse, P.D., Laurie, J.R., Shergold, J.H., et al. (eds.), Palaeo-Biogeographic Affinities of Australian Cambrian Faunas, *Memoirs of the Association of Australasian Palaeontologists*, **23**: pp. 13−20.

Landing, E., 1991, A unified uppermost Precambrian through Lower Cambrian lithostratigraphy for Avalonian North America. *Geological Society of America Abstracts with Programs*, **23**: 56.

Landing, E., 1992, Lower Cambrian of southeastern Newfoundland: Epeirogeny and Lazarus faunas, lithofacies-biofacies linkages, and the myth of a global chronostratigraphy. *In* Lipps, J.H., and Signor, P.W. (eds), *Origin and Early Evolution of Metazoa*. New York: Plenum Press, 283−309.

Landing, E., 1994, Precambrian-Cambrian boundary global stratotype ratified and a new perspective of Cambrian time. *Geology*, **22**: 179−182.

Landing, E., 1996, Avalon: insular continent by the latest Precambrian, Special Paper. *In* Nance, R.D., and Thompson, M.D. (eds), *Avalon and Related Peri-Gondwana Terranes of the Circum-North Atlantic*. Geological Society of America, **304**: 29−63.

Landing, E., 2004, Precambrian-Cambrian boundary interval deposition and the marginal platform of the Avalon microcontinent. *Journal of Geodynamics*, **37**: 411–435.

Landing, E., Antcliffe, J.B., Brasier, M.D., and English, A.B., 2015, Distinguishing Earth's oldest known bryozoan (*Pywackia*, late Cambrian) from pennatulacean octocorals (Mesozoic–Recent). *Journal of Paleontology*, **89** (2): 292–317.

Landing, E., Bowring, S.A., Davidek, K., Westrop, S.R., Geyer, G., and Heldmaier, W., 1998, Duration of the Early Cambrian: U-Pb ages of volcanic ashes from Avalon and Gondwana. *Canadian Journal of Earth Sciences*, **5**: 329–338.

Landing, E., Bowring, S.A., Davidek, K.L., Rushton, A.W.A., Fortey, R.A., and Wimbledon, W.A.P., 2000, Cambrian-Ordovician boundary age and duration of the lowest Ordovician Tremadoc Series based on U-Pb zircon dates from Avalonian Wales. *Geological Magazine*, **137**: 485–494.

Landing, E., English, A.M., and Keppie, J.D., 2010a, Cambrian origin of all skeletalized metazoan phyla–discovery of Earth's oldest bryozoans (Upper Cambrian, southern Mexico). *Geology*, **38**: 547–550.

Landing, E., Myrow, P., Benus, A.P., and Narbonne, G.M., 1989, The Placentian Series: appearance of the oldest skeletalized faunas in southeastern Newfoundland. *Journal of Paleontology*, **63**: 739–769.

Landing, E., Peng, S.C., Babcock, L.E., Geyer, G., and Moczydłowska-Vidal, M., 2007, Global standard names for the lowermost Cambrian Series and Stage. *Episodes*, **30**: 287–299.

Landing, E., Westrop, S.R., and Adrain, J.M., 2011, The Lawsonian Stage – the *Eoconodontus notchpeakensis* (Miller, 1969) FAD and HERB carbon isotope excursion define a globally correlatable terminal Cambrian stage. *Bulletin of Geosciences*, **86**: 621–640.

Landing, E., Westrop, S.R., and Miller, J.F., 2010b, Globally practical base for the uppermost Cambrian (Stage 10) — FAD of the conodont *Eoconodontus notchpeakensis* and the Housian Stage. *In* Fatka, P., and Budil, P. (eds), *Prague 2010, The 15th Field Conference of the Cambrian Stage Subdivision Working Group, International Subcommission on Cambrian Stratigraphy*. Prague: Czech Geological Survey, p. 18.

Lane, P.D., Blaker, M.R., and Zhang, W.T., 1988, Redescription of the Early Cambrian trilobite *Arthricocephalus chauveaui* Bergeron, 1899. *Acta Palaeontologica Sinica*, **27**: 553–560 (in Chinese and English).

Lapworth, C., 1879, On the tripartite classification of the Lower Paleozoic rocks. *Geological Magazine, New Series*, **6**: 1–15.

Laurie, J.R., 2004, Early Middle Cambrian trilobite faunas from NTGS Elkedra 3 corehole, southern Georgina Basin, Northern Territory. *Memoirs of the Association of Australasian Palaeontologists*, **30**: 221–260.

Laurie, J.R., 2006, Early Middle Cambrian trilobites from Pacific Oil 8L Gas Baldwin 1 well, southern Georgina Basin, Northern Territory. *Memoirs of the Association of Australasian Palaeontologists*, **32**: 127–204.

Lazarenko, N.P., 1966, Biostratigraphy and some new trilobites of the Upper Cambrian Olenek Rise and Kharaulakh Mountains. *Uchenie Zapisky Paleontologiya i Biostratigrafiya (NIIGA)*, **11**: 33–78.

Lazarenko, N.P., and Nikiforiv, N.I., 1972, *Middle and Upper Cambrian of the Northern Siberian Platform and Adjacent Folded Areas. Stratigraphy, Paleogeography, and Mineral Deposits of the North Arctic*. Leningrad: NIIGA Press, pp. 4–9.

Lazarenko, N.P., Gogin, I.Y., Pegel, T.V., Sukhov, S.S., Sukhov, S.S., Abaimova, G.P., et al., 2008, Excursion 1b. Cambrian stratigraphy of the northeastern Siberian Platform and potential stratotypes of lower boundaries of the proposed Upper Cambrian Chekurovkian and Nelegerian stages in the Ogon'or Formation section at the Khos-Nelege River; the boundaries are defined by FAD of *Agnostotes orientalis* and *Lotagnostus americanus*. *In* Rozanov, A.Y., Varlamov, A.I. (eds). *The Cambrian System of the Siberian Platform, Part 2: North-east of the Siberian Platform*. Moscow and Novosibirsk: PIN, RAS, 60–106 (in Russian and English).

Lazarenko, N.P., Gogin, I.Y., Pegel, T.V., Sukhov, S.S., and Abaimova, G.P., 2011, The Khos-Nelege section of the Ogon'or Formation: a potential candidate for the GSSP of Stage 10, Cambrian System. *Bulletin of Geosciences*, **86**: 555–568.

Li, D., Ling, H.F., Shields-Zhou, G.A., Chen, X., Cremonese, L., Och, L., et al., 2013, Carbon and strontium isotope evolution of seawater across the Ediacaran–Cambrian transition: evidence from the Xiaotan section, NE Yunnan, South China. *Precambrian Research*, **225**: 128–147.

Li, D.D., Zhang, X.L., Chen, K.F., Chen, X.Y., Huang, W., Peng, S.C., et al., 2017, High-resolution carbon isotope chemostratigraphic record from the uppermost Cambrian stage (Stage 10) in South China, implications for defining the base of Stage 10 and palaeoenvironmental change. *Geological Magazine*, **154**: 1232–1243.

Li, G.X., Steiner, M., Zhu, M.Y., Zhu, X.J., and Erdtmann, B.-D., 2007, Early Cambrian fossil record of metazoans in South China: generic diversity and radiation patterns. *Palaeogeography, Palaeoclimatology, Palaeoecology*, **254**: 226–246.

Li, G.X., Zhang, J.M., and Zhu, M.Y., 2001, Litho- and biostratigraphy of the Lower Cambrian Meishucunian Stage in the Xiaotan section, Eastern Yunnan. *Acta Palaeontologica Sinica*, **40** (Suppl.): 40–53.

Li, G.X., Zhao, X., Gubanov, A., Zhu, M.Y., and Na, L., 2011, Early Cambrian mollusk *Watsonella crosbyi*: a potential GSSP index fossil for the base of Cambrian Stage 2. *Acta Geologica Sinica*, **85**: 309–319.

Lin, J.P., Gon III, S.M., Gehling, J.G., Babcock, L.E., Zhao, Y.L., Zhang, X.L., et al., 2006, A *Parvancorina*-like arthropod from the Cambrian of South China. *Historical. Biology*, **18** (1): 33–45.

Lin, S., Zhang, Y., Zhang, L., Tao, X., and Wang, M., 1986, Body and trace fossils of metazoa and algal macrofossils from the upper Sinian Gaojiashan Formation in southern Shaanxi. *Geology of Shaanxi*, **4**: 9–17.

Linnarsson, J.G.O., 1869, Om Vestergötlands cambriska och siluriska aflagringar. *Kongliga Svenska Vetenskaps-Akademiens Handlingar*, **8** (2): 1–89.

Linnemann, U., Ovtcharova, M., Schaltegger, U., Gärtner, A., Hautmann, M., Geyer, G., et al., 2019, New high-resolution age data from the Ediacaran–Cambrian boundary indicate rapid, ecologically driven onset of the Cambrian explosion. *Terra Nova*, **31**: 49–58.

Linnemann, U., Ovtcharova, M., Schaltegger, U., Vickers-Rich, P., Gärtner, A., Hofmann, M., et al., 2017. New geochronological and stratigraphic constraints on the Precambrian-Cambrian boundary (Swartpunt section, South Namibia). In: McIlroy, D. (Ed.), International Symposium on the Ediacaran-Cambrian Transition. Abstract Volume, p. 64.

Lochman-Balk, C., and Wilson, J.L., 1958, Cambrian biostratigraphy in North America. *Journal of Paleontology*, **32**: 312–350.

Longacre, S.A., 1970, Trilobites of the Upper Cambrian Ptychaspid Biomere, Wilberns Formation, central Texas. *Journal of Paleontology*, **44** (Suppl. 2): 1–70.

Lu, Y.H., 1981, Provincialism, dispersal, development, and phylogeny of trilobites. *Geological Society of America, Special Paper,* **187**: 143–150.

Lu, Y.H., and Lin, H.L., 1989, The Cambrian trilobites of western Zhejiang. *Palaeontologia Sinica,* **25**: 1–287.

Lu, Y.H., Zhu, Z.L., Qian, Y.Y., Lin, H.L., Zhou, Z.Y., and Yuan, K.X., 1974, The bio-environmental control hypothesis and its application to the Cambrian biostratigraphy and palaeozoo-geography. *Memoir of Nanjing Institute of Geology and Palaeontology, Academia Sinica,* **5**: 27–110.

Ludvigsen, R., and Westrop, S.R., 1985, Three new Upper Cambrian stages for North America. *Geology,* **13**: 139–143.

Luo, H.L., Hu, S.X., Chen, L.Z., Zhang, S.S., and Tao, Y.H., 1999, *Early Cambrian Chengjiang Fauna From Kunming Region, China.* Kunming: Yunnan Science and Technology Press 129 pp. (in Chinese).

Luo, H.L., Jiang, Z.W., Wu, X.C., Ou, Y.L., Song, X.L., and Xue, X.F., 1992, A further research on the Precambrian-Cambrian boundary at Meishucun section of Jinning, Yunnan, China. *Acta Geologica Sinica,* **5**: 197–207.

Luo, H.L., Jiang, Z.W., Wu, X.C., Song, X.L., and Ou, Y.L., 1990, The global biostratigraphic correlation of the Meishucunian Stage and the Precambrian-Cambrian boundary. *Science in China, Series B,* **1990**: 313–318 (in Chinese).

Luo, H.L., Jiang, Z.W., Wu, X.C., Song, X.L., Ou, Y.L., Xin, Y.S., et al., 1984, *Sinian-Cambrian Boundary Stratotype Section at Meishucun, Jinning, Yunnan, China.* Kunming: Yunnan People's Publishing House 154 pp. (in Chinese).

Luo, H.L., Jiang, Z.W., and Tang, L.D., 1994, *Stratotype Sections for Lower Cambrian Stages in China.* Kunming: Yunnan Science and Technology Press 183 pp. (in Chinese).

Luo, H.L., Wu, X.C., and Ou, Y.L., 1991, Facies changes and transverse correlation of the Sinian-Cambrian boundary strata in eastern Yunnan. *Sedimentary Facies and Palaeogeography,* **1991** (4): 27–35 (in Chinese).

Ma, J.Y., Taylor, P.D., and Xia, F.S., 2013, New observations on the skeletons of the earliest bryozoans from the Fengsiang Formation (Tremadocian, Lower Ordovician), Yichang, China. *Palaeoworld,* **23**: 25–30.

Maas, A., Braun, A., Dong, X.P., Donoghue, P.C.J., Müller, K.J., Olempska, E., et al., 2006, The 'Orsten'—more than a Cambrian Konservat-lagerstätte yielding exceptional preservation. *Palaeoworld,* **15**: 266–282.

Mángano, M.G., and Buatois, L.A., 2014, Decoupling of body-plan diversification and ecological structuring during the Ediacaran-Cambrian transition: evolutionary and geobiological feedbacks. *Proceedings of the Royal Society B, Biological Sciences,* **281**: 20140038. https://doi.org/10.1098/rspb.2014.0038.

Maloof, A.C., Schrag, D.P., Crowley, J.L., and Bowring, S.A., 2005, An expanded record of Early Cambrian carbon cycling from the Anti-Atlas Margin, Morocco. *Canadian Journal of Earth Sciences,* **42**: 2195–2216.

Marshall, C.R., 2006, Explaining the Cambrian "Explosion" of animals. *Annual Review of Earth and Planetary Sciences,* **34**: 355–384.

McArthur, J.M., Howarth, R.J., Shields, G.A., and Zhou, Y., 2020, Chapter 7 - Strontium isotope stratigraphy. *In* Gradstein, F.M., Ogg, J.G., Schmitz, M.D., and Ogg, G.M. (eds), *The Geologic Time Scale 2020.* Elsevier, Boston, MA, Vol. 1 (this book).

McCollum, L.B., and Sundberg, F.A., 1999, Stop 9. Biostratigraphy of the traditional Lower-Middle Cambrian boundary interval in the outer shelf Emigrant Formation, Split Mountain East section, Esmeralda County, Nevada. *In* Palmer, A.R. (ed.), *Laurentia 99, V Field Conference of the Cambrian Stage Subdivision Working Group.* Boulder, CO: Institute for Cambrian Studies, 29–34.

McNamara, K.J., Yu, F., and Zhou, Z.Y., 2003, Ontogeny and heterochrony in the oryctocephalid trilobite *Arthricocephalus* from the Early Cambrian of China. *Special Papers in Palaeontology,* **70**: 103–126.

Mei, M.X., Ma, Y.S., Zhang, H., Meng, X.Q., and Chen, Y.H., 2007, Sequence-stratigraphic framework for the Cambrian of the Upper Yangtze region: ponder on the sequence stratigraphic background of the Cambrian biological diversity events. *Journal of Stratigraphy,* **31**: 68–78 (in Chinese).

Meidla, T., 2017, Ediacaran and Cambrian stratigraphy in Estonia: an updated review. *Estonian Journal of Earth Sciences,* **66**: 152–160.

Miller, J.F., 1980, Taxonomic revisions of some Upper Cambrian and Lower Ordovician conodonts with comments on their evolution. *University of Kansas Paleontological Contributions, Paper,* **99**: 1–43.

Miller, J.F., 1988, Conodonts as biostratigraphic tools for redefinition and correlation of the Cambrian-Ordovician boundary. *Geological Magazine,* **125**: 349–362.

Miller, J.F., Ethington, R.L., Evans, K.R., Holmer, L.E., Loch, J.D., Popov, L.E., et al., 2006, Proposed stratotype for the base of the highest Cambrian stage at the first appearance datum of *Cordylodus andresi,* Lawson Cove section, Utah, USA. *Palaeoworld,* **15**: 384–405.

Miller, J.F., Evans, K.R., Freeman, R.L., Loch, J.D., Ripperdan, R.L., and Taylor, J.F., 2018, Combining biostratigraphy, carbon isotope stratigraphy and sequence stratigraphy to define the base of Cambrian Stage 10. *Australasian Palaeontological Memoirs,* **51**: 19–64.

Miller, J.F., Evans, K.R., Freeman, R.L., Ripperdan, R.L., and Taylor, J.F., 2011, Proposed stratotype for the base of the Lawsonian Stage (Cambrian Stage 10) at the first appearance datum of *Eoconodontus notchpeakensis* (Miller) in the House Range, Utah, USA. *Bulletin of Geosciences,* **86**: 595–620.

Miller, J.F., Evans, K.E., Loch, J.D., Ethington, R.L., Stitt, J.H., Holmer, L.E., et al., 2003, Stratigraphy of the Sauk III interval (Cambrian-Ordovician) in the Ibex area, western Millard County, Utah. *Brigham Young University Geology Studies,* **47**: 23–118.

Miller, J.F., Ripperdan, R.L., Loch, J.D., Freeman, R.L., Evans, K.R., Taylor, J.F., et al., 2015, Proposed GSSP for the base of Cambrian Stage 10 at the lowest occurrence of *Eoconodontus notchpeakensis* in the House Range, Utah, USA. *Annales de Paléontologie,* **101**: 199–211.

Moczydłowska, M., 1991, Acritarch biostratigraphy of the Lower Cambrian and the Precambrian–Cambrian boundary in southeastern Poland. *Fossils and Strata,* **29**: 1–127.

Moczydłowska, M., Jensen, S., Ebbestad, J.O.R., Budd, G.E., and Martí-Mus, M., 2001, Biochronology of the autochthonous Lower Cambrian in the Laisvall-Storuman area, Swedish Caledonides. *Geological Magazine,* **138**: 435–453.

Montañez, I.P., Osleger, D.A., and Banner, J.L., 2000, Evolution of the Sr and C isotope composition of Cambrian oceans. *GSA Today,* **10** (5): 1–7.

Murchison, R.I., 1839, *The Silurian System, Founded on Geological Researches in the Counties of Salop, Hereford, Radnor, Montgomery, Caermarthen, Brecon, Pembroke, Monmouth, Gloucester, Worcester, and Stafford; With Descriptions of the Coal-Fields and Overlying Formations,* Vol. 1. London: John Murray, pp. 1–576 Vol. 2: pp. 577–768.

Naimark, E., Shabanov, Y., and Korovnikov, I., 2011, Cambrian trilobite *Ovatoryctocara* Tchernysheva, 1962 from Siberia. *Bulletin of Geosciences,* **86**: 405–422.

Narbonne, G.M., Myrow, P., Landing, E., and Anderson, M.M., 1987, A candidate stratotype for the Precambrian-Cambrian boundary,

Fortune Head, Burin Peninsula, southeastern Newfoundland. *Canadian Journal of Earth Sciences*, **24**: 277–293.

Nicholas, C.J., 1996, The Sr isotopic evolution of the oceans during the "Cambrian explosion". *Journal of the Geological Society, London*, **153**: 243–254.

Nicoll, R.S., 1990, The genus *Cordylodus* and a latest Cambrian-earliest Ordovician conodont biostratigraphy. *Bureau of Mineral Resources of Australia, Journal of Australian, Geology and Geophysics*, **11**: 529–558.

Nicoll, R.S., 1991, Differentiation of Late Cambrian–Early Ordovician species of *Cordylodus* (Conodonta) with biapical basal cavities. *Bureau of Mineral Resources of Australia, Journal of Australian Geology and Geophysics*, **12**: 223–244.

Nicoll, R.S., and Shergold, J.H., 1991, Revised Late Cambrian (pre-Payntonian-Datsonian) conodont stratigraphy at Black Mountain, Georgina Basin, western Queensland, Australia. *Bureau of Mineral Resources of Australia, Journal of Australian Geology and Geophysics*, **12**: 93–118.

Nielsen, A.T., and Schovsbo, N.H., 2007, Cambrian to basal Ordovician lithostratigraphy in southern Scandinavia. *Bulletin of the Geological Society of Denmark*, **53**: 47–92.

Nielsen, A.T., Høyberget, M., and Ahlberg, P., 2020, The Furongian (upper Cambrian) Alum Shale of Scandinavia: revision of zonation. *Lethaia*. https://doi.org/10.1111/let.12370.

Nielsen, A.T., and Schovsbo, N.H., 2011, The Lower Cambrian of Scandinavia: depositional environment, sequence stratigraphy and palaeogeography. *Earth-Science Reviews*, **107**: 207–310.

Nielsen, A.T., and Schovsbo, N.H., 2015, The regressive Early-Mid Cambrian 'Hawke Bay Event' in Baltoscandia: Epeirogenic uplift in concert with eustasy. *Earth-Science Reviews*, **151**: 288–350.

Nielsen, A.T., Weidner, T., Terfelt, F., and Høyberget, M., 2014, Upper Cambrian (Furongian) biostratigraphy in Scandinavia revisited: definition of superzones. *GFF*, **136**: 193–197.

Ogg, J.G., Ogg, G.M., and Gradstein, F.M., 2016, *A Concise Geologic Time Scale 2016*. Amsterdam: Elsevier Publisher, p. 234.

Öpik, A.A., 1963, Early Upper Cambrian fossils from Queensland. *Bureau of Mineral Resources of Australia Bulletin*, **64**: 1–133.

Öpik, A.A., 1966, The early Upper Cambrian crisis and its correlation. *Journal and Proceedings of the Royal Society of New South Wales*, **100**: 9–14.

Öpik, A.A., 1967, The Mindyallan fauna of north-western Queensland. *Bureau of Mineral Resources of Australia Bulletin*, **74** (1): 404 and (2): 167.

Öpik, A.A., 1968, The Ordian Stage of the Cambrian and its Australian Metadoxididae. *Bureau of Mineral Resources of Australia Bulletin*, **92**: 133–170.

Öpik, A.A., 1979, Middle Cambrian agnostids: systematics and biostratigraphy. *Bureau of Mineral Resources of Australia Bulletin*, **172** (1): 188 and (2): 67.

Orłowski, S., 1992, Cambrian stratigraphy and stage subdivision in the Holy Cross Mountains, Poland. *Geological Magazine*, **129**: 471–474.

Palmer, A.R., 1965a, Biomere – a new kind of biostratigraphic unit. *Journal of Paleontology*, **39**: 149–153.

Palmer, A.R., 1965b, Trilobites of the Late Cambrian Pterocephaliid Biomere in the Great Basin, United States. *U.S. Geological Survey Professional Paper*, **493**: 1–105.

Palmer, A.R., 1973, Cambrian trilobites. *In* Hallam, A. (ed.), *Atlas of Palaeobiogeography*. New York: Elsevier, 3–18.

Palmer, A.R., 1979, Biomere boundaries re-examined. *Alcheringa*, **3**: 33–41.

Palmer, A.R., 1981. Subdivision of the Sauk Sequence. In: Taylor, M.E. (ed.), *Short Papers for the Second International Symposium on the Cambrian System*, U.S. Geological Survey Open-file Report, pp. 81–743, 160–162.

Palmer, A.R., 1984, The biomere problem: evolution of an idea. *Journal of Paleontology*, **58**: 599–611.

Palmer, A.R., 1998, A proposed nomenclature for stages and series for the Cambrian of Laurentia. *Canadian Journal of Earth Sciences*, **35**: 323–328.

Palmer, A.R., 1999, Introduction. *In* Palmer, A.R. (ed.), *Laurentia 99, V Field Conference of the Cambrian Stage Subdivision Working Group*. Boulder, CO: International Subcommission on Cambrian Stratigraphy. Institute for Cambrian Studies, 1–4.

Palmer, A.R., and Halley, R.B., 1979, Physical stratigraphy and trilobite biostratigraphy of the Carrara Formation (Lower and Middle Cambrian) in the southern Great Basin. *U.S. Geological Survey, Professional Paper*, **1047**: 1–131.

Palmer, A.R., and Repina, L.N., 1993, Through a glass darkly: taxonomy, phylogeny, and biostratigraphy of the Olenellina. *University of Kansas Paleontological Contributions, New Series*, **3**: 1–35.

Parkhaev, P.Yu., and Karlova, G.A., 2011, Taxonomic revision and evolution of Cambrian mollusks of the genus *Aldanella* Vostokova, 1962 (Gastropoda: Archaeobranchia). *Paleontological Journal*, **45** (10): 1145–1205.

Parkhaev, P.Yu., Karlova, G.A., and Rozanov, A.Y., 2011, Taxonomy, stratigraphy and biogeography of *Aldanella attleborensis* – a possible candidate for defining the base of Cambrian Stage 2. *Museum of Northern Arizona Bulletin*, **67**: 298–300.

Pavlov, V., Bachtadse, V., and Mikhailov, V., 2008, New Middle Cambrian and Middle Ordovician palaeomagnetic data from Siberia: Llandelian magnetostratigraphy and relative rotation between the Aldan and Anabar-Angara blocks. *Earth and Planetary Science Letters*, **276**: 229–242.

Pavlov, V., and Gallet, Y., 1998, Upper Cambrian to Middle Ordovician magnetostratigraphy from the Kulumbe River section (northwestern Siberia). *Physics of the Earth and Planetary Interiors*, **108**: 49–59.

Pavlov, V., and Gallet, Y., 2001, Middle Cambrian high magnetic reversal frequency (Kulumbe River section, northwestern Siberia) and reversal behaviour during the Early Palaeozoic. *Earth and Planetary Science Letters*, **185**: 173–183.

Pavlov, V., and Gallet, Y., 2005, Third superchron during the early Paleozoic. *Episodes*, **28**: 78–84.

Pegel, T.V., 2000, Evolution of trilobite biofacies in Cambrian basins of the Siberian Platform. *Journal of Paleontology*, **74**: 1000–1019.

Pegel, T.V., Shabanov, Y.Y., Fedorov, A.B., and Sukhov, S.S., 2016, Chapter 4. Regional Cambrian stratigraphic scheme of the Siberian Platform. *In* Shabanov, Y.Y. (ed.), *Stratigraphy of Oil and Gas Basins of Siberia. Cambrian of Siberian Platform, Volume 1 Stratigraphy*. Novosibirsk: IPGG SB RAS, 213–324 (In Russian with English summary).

Peng, S.C., 1984, Cambrian-Ordovician boundary in the Cili-Taoyuan border area, northwestern Hunan with descriptions of relative trilobites. *In* Nanjing Institute of Geology and Palaleontology, Academia Sinica (ed.), *Stratigraphy and Palaeontology of Systemic Boundaries in China, Cambrian-Ordovician Boundary (1)*. Hefei: Anhui Science and Technology Publishing House, 285—405.

Peng, S.C., 1990, Tremadoc stratigraphy and trilobite faunas of northwestern Hunan. *Beringeria*, **2**: 1—171.

Peng, S.C., 1992, Upper Cambrian biostratigraphy and trilobite faunas of Cili-Taoyuan area, northwestern Hunan, China. *Memoirs of the Association of Australasian Palaeontologists*, **13**: 1—119.

Peng, S.C., 2000a, Cambrian of slope facies (of China). *In* Nanjing Institute of Geology and Palaeontology (ed.), *Stratigraphical Studies in China (1979—1999)*. Hefei: Anhui Science and Technology Publishing House, pp. 23—38 (in Chinese).

Peng, S.C., 2000b, A new chronostratigraphic subdivision of Cambrian for China. *In* Aceñolaza, G.F., and Peralta, S. (eds), *Cambrian From the Southern Edge*. INSUGEO, **Vol. 6**: 119—122.

Peng, S.C., 2003, Chronostratigraphic subdivision of the Cambrian of China. *Geologica Acta*, **1** (1): 135—144.

Peng, S.C., 2008, Revision on Cambrian chronostratigraphy of South China and related remarks. *Journal of Stratigraphy*, **32**: 239—245 (in Chinese).

Peng, S.C., 2009, The newly developed Cambrian biostratigraphic succession and chronostratigraphic scheme for South China. *Chinese Science Bulletin*, **54**: 2691—2698.

Peng, S.C., 2018, Cambrian System. *In* National Commission on Stratigraphy of China (ed.), *Explanatory Notes for the Stratigraphic Chart of China (2014)*. Beijing: Geological Press, pp. 71—100 (in Chinese).

Peng, S.C., and Chen, J.Y., 1983, Late Cambrian nautiloids from Cili and Taoyuan, Hunan. *Geology of Hunan*, **2** (2): 13—17, 25.

Peng, S.C., and Babcock, L.E., 2001, Cambrian of the Hunan-Guizhou Region, South China. In: Peng, S.C., Babcock, L.E., and Zhu, M.Y. (eds), Cambrian System of South China. *Palaeoworld*, **13**: 3—51.

Peng, S.C., and Babcock, L.E., 2005a, Towards a new global subdivision of the Cambrian System. *Journal of Stratigraphy*, **29**: 171—177, 204 (in Chinese).

Peng, S.C., and Babcock, L.E., 2005b, Two Cambrian agnostoid trilobites, *Agnostotes orientalis* (Kobayashi, 1935) and *Lotagnostus americanus* (Billings, 1860): key species for defining global stages of the Cambrian System. *Geosciences Journal*, **9**: 107—115.

Peng, S.C., and Babcock, L.E., 2008, Cambrian Period. *In* Ogg, J.G., Ogg, G.J., and Gradstein, F.M. (eds), *The Concise Geological Timescale*. Cambridge: Cambridge University Press, 37—46.

Peng, S.C., Babcock, L.E., and Cooper, R.A., 2012a, The Cambrian Period. *In* Gradstein, F.M., Ogg, J.G., Schmitz, M.D., and Ogg, G.M. (eds), *The Geologic Time Scale 2012*. Elsevier Publisher, 437—488 <https://doi.org/10.1016/B978-0-444-59425-9.00016-0>.

Peng, S.C., Babcock, L.E., Geyer, G., and Moczydłowska, M., 2006, Nomenclature of Cambrian epochs and series based on GSSPs— comments on an alternative proposal by Rowland and Hicks. *Episodes*, **29**: 130—132.

Peng, S.C., Babcock, L.E., and Lin, H.L., 2004b, *Polymerid Trilobites From the Cambrian of Northwestern Hunan, China. Volume 1. Corynexochida, Lichida, and Asaphida*. Beijing: Science Press 333 pp.

Peng, S.C., Babcock, L.E., Robison, R.A., Lin, H.L., Ress, M.N., and Saltzman, M.R., 2004a, Global Standard Stratotype-section and Point (GSSP) of the Furongian Series and Paibian Stage (Cambrian). *Lethaia*, **37**: 365—379.

Peng, S.C., Babcock, L.E., Zhu, X.J., Ahlberg, P., Terfelt, F., and Dai, T., 2015, Intraspecific variation and taphonomic alteration in the Cambrian (Furongian) agnostoid *Lotagnostus americanus*: new information from China. *Bulletin of Geosciences*, **90**: 281—306.

Peng, S.C., Babcock, L.E., Zhu, X.J., Lei, Q.P., and Dai, T., 2017, Revision of the oryctocephalid trilobite genera *Arthricocephalus* Bergeron and *Oryctocarella* Tomashpolskaya and Karpinski (Cambrian) from South China and Siberia. *Journal of Paleontology*, **91**: 933—959.

Peng, S.C., Babcock, L.E., Zuo, J.X., Lin, H.L., Zhu, X.J., Yang, X.F., et al., 2012b, Global standard stratotype—section and point (GSSP) for the base of the Jiangshanian Stage (Cambrian: Furongian) at Duibian, Jiangshan, Zhejiang, Southeast China. *Episodes*, **35**: 1—16.

Peng, S.C., Babcock, L.E., Zuo, J.X., Lin, H.L., Zhu, X.J., Yang, X.F., et al., 2009, The Global boundary Stratotype Section and Point of the Guzhangian Stage (Cambrian) in the Wuling Mountains, northwestern Hunan, China. *Episodes*, **32**: 41—55.

Peng, S.C., and Robison, R.A., 2000, Agnostoid biostratigraphy across the Middle-Upper Cambrian Boundary in China. *Journal of Paleontology 74, Memoir*, **53**: 1—104.

Peng, S.C., Yang, X.F., Liu, Y., Zhu, X.J., Sun, H.J., Zamora, S., et al., 2020, Fulu biota, a new exceptionally-preserved Cambrian fossil assemblage from the Longha Formation in southeastern Yunnan. *Palaeoworld*. https://doi.org/10.1016/j.palwor.202.

Peng, S.C., Yuan, J.L., and Zhao, Y.L., 2000b, Taijiangian Stage: a new chronostratigraphic unit for the traditional Lower Middle Cambrian in South China. *Journal of Stratigraphy*, **24**: 56—57 (in Chinese).

Peng, S.C., and Zhao, Y.L., 2018, The proposed global standard stratotype-section and point (GSSP) for the conterminous base of Miaolingian Series and Wuliuan Stage at Balang, Jianhe, Guizhou, China was ratified by IUGS. *Journal of Stratigraphy*, **42**: 325—327.

Peng, S.C., Zhou, Z.Y., and Lin, T.R., 1998, Late Middle—late Upper Cambrian chronostratigraphy of China. *In* Ahlberg, P., Eriksson, M., and Olsson, I. (eds), *IV Field Conference of the Cambrian Stage Subdivision Working Group, International Subcommission on Cambrian Stratigraphy, Abstracts*. Lund: Lund University, p. 20.

Peng, S.C., Zhou, Z.Y., and Lin, T.R., 1999, A proposal of the Cambrian chronostratigraphic scale in China. *Geosciences*, **31** (2): 242 (in Chinese).

Peng, S.C., Zhou, Z.Y., Lin, T.R., and Yuan, J.L., 2000a, Research on Cambrian chronostratigraphy: present and tendency. *Journal of Stratigraphy*, **24**: 8—17 (in Chinese).

Pillola, G.L., 1991, Trilobites du Cambrien inférieur du SW de la Sardaigne, Italie. *Palaeontographica Italica*, **78**: 1—174.

Pisarevsky, S., Gurevich, E., and Khramov, A., 1997, Paleomagnetism of the Lower Cambrian sediments from the Olenek River section (northern Siberia): paleopoles and the problem of the magnetic polarity in the Early Cambrian. *Geophysics Journal International*, **130**: 746—756.

Pokrovskaya, N.V., 1954, Stratigraphy of the Cambrian sediments in the south of the Siberian Platform. *In* Shatsky, N.S. (ed.), *Issue of Geology in Asia, Vol. 1*. Moscow: Academy of Sciences Press of USSR, 444—465.

Pratt, B.R., 1992, Trilobites of the Marjuman and Steptoean stages (Upper Cambrian), Rabbitkettle Formation, southern Mackenzie Mountains, northwest Canada. *Palaeontographica Canadiana*, **9**: 1—179.

Qian, Y., 1997, Hyolitha and some problematica from the Lower Cambrian Meishucun Stage in Central and SW China. *Acta Palaeontologica Sinica*, **16**: 255—278 (in Chinese).

Qian, Y., Chen, M.E., He, T.G., Zhu, M.Y., Yin, G.Z., Fen, W.M., et al., 1999, *Taxonomy and Biostratigraphy of Small Shelly Fossils in China*. Beijing: Science Press, p. 247 (in Chinese).

Rasmussen, B.W., Nielsen, A.T., and Schovsbo, N.H., 2015, Faunal succession in the upper Cambrian (Furongian) *Leptoplastus* Superzone at Slemmestad, southern Norway. *Norwegian Journal of Geology*, **95**: 1–22.

Rasmussen, B.W., Rasmussen, J.A., and Nielsen, A.T., 2017, Biostratigraphy of the Furongian (upper Cambrian) Alum Shale Formation at Degerhamn, Öland, Sweden. *GFF*, **139**: 92–118.

Reed, F.A., 1910, The Cambrian fossils of Spiti. *Memoirs of Geological Survey of India, Palaeontologia India*, **15**: 1–70.

Repina, L.N., Khomentovsky, V.V., Zhuravleva, I.T., and Rozanov, A. Y., 1964, *Lower Cambrian Biostratigraphy of the Saya-Altay Folded Zone*. Moscow: Nauka, p. 378 (in Russian).

Ripperdan, R.L., 2002, The HERB Event: end of Cambrian carbon cycle paradigm? *Geological Society of America Abstracts With Programs*, **34** (6): 413.

Ripperdan, R.L., and Kirschvink, J.L., 1992, Paleomagnetic results from the Cambrian-Ordovician boundary section at Black Mountain, Georgina Basin, western Queensland, Australia. *In* Webby, B.D., and Laurie, J.R. (eds), *Global Perspectives on Ordovician Geology*. Rotterdam: A.A. Balkema, 381–394.

Ripperdan, R.L., Magaritz, M., and Kirschvink, J.L., 1993, Magnetic polarity and carbon isotope evidence for non-depositional events within the Cambrian-Ordovician boundary section near Dayangcha, Jilin Province, China. *Geological Magazine*, **130**: 443–452.

Ripperdan, R.L., Magaritz, M., Nicoll, R.S., and Shergold, J.H., 1992, Simultaneous changes in carbon isotopes, sea level, and conodont biozones within the Cambrian-Ordovician boundary interval at Black Mountain, Australia. *Geology*, **20**: 1039–1042.

Robison, R.A., 1976, Middle Cambrian trilobite biostratigraphy of the Great Basin. *Brigham Young University Geology Studies*, **23**: 93–109.

Robison, R.A., 1984, Cambrian Agnostida of North America and Greenland, part 1, Ptychagnostidae. *University of Kansas Paleontological Contributions, Paper*, **109**: 1–59.

Robison, R.A., 1991, Middle Cambrian biotic diversity: examples from four Utah Lagerstätten. *In* Simonetta, A.M., and Conway Morris, S. (eds), *The Early Evolution of Metazoa and the Significance of Problematic Taxa*. Cambridge: Cambridge University Press, 77–98.

Robison, R.A., 1994, Agnostoid trilobites from the Henson Gletscher and Kap Stanton formations (Middle Cambrian), North Greenland. *Grønland Geologiske Undersøgelse Bulletin*, **169**: 25–77.

Robison, R.A., 1999, Base of *Ptychagnostus atavus* Zone, candidate stratotype for base of unnamed international series. *In* Palmer, A.R. (ed.), *Laurentia 99. V Field Conference of the Cambrian Stage Subdivision Working Group, International Subcommission on Cambrian Stratigraphy*. Boulder, CO: Institute for Cambrian Studies, 15–17.

Robison, R.A., and Babcock, L.E., 2011, Systematics, paleobiology, and taphonomy of some exceptionally preserved trilobites from Cambrian Lagerstätten of Utah. *Paleontological Contributions*, **5**: 1–47.

Robison, R.A., Babcock, L.E., and Gunther, V.G., 2015, Exceptional Cambrian Fossils from Utah: a Window into the Age of Trilobites. *Utah Geological Survey Miscellaneous Publication*, **15-1**: 1–97.

Robison, R.A., Rosova, A.V., Rowell, A.J., and Fletcher, T.P., 1977, Cambrian boundaries and divisions. *Lethaia*, **10**: 257–262.

Ross Jr., R.J., Hintze, L.F., Ethington, R.L., Miller, J.F., Taylor, M.E., Repetski, J.E., et al. 1997, The Ibexian, lowermost series in the North American Ordovician. In: Taylor, M.E. (Ed.), Early Paleozoic Biochronology of the Great Basin, Western United States, U.S. Geological Survey Professional Paper, pp. 1–50.

Rowell, A.J., Robison, R.A., and Strickland, D.K., 1982, Aspects of Cambrian agnostoid phylogeny and chronocorrelation. *Journal of Paleontology*, **56**: 161–182.

Rozanov, A.Y., 1966, The problem of the lower boundary of the Cambrian. *In* Keller, B.M. (ed.), *Results of Science, General Geology, Stratigraphy*. Moscow: VINTTI Press, pp. 92–111 (in Russian).

Rozanov, A.Yu., 1967, The Cambrian lower boundary problem. *Geological Magazine*, **104** (5): 425–434.

Rozanov, A.Y., 1973, Regularities of the morphological evolution of archaeocyaths and the issues of stage subdivision of Lower Cambrian. *Transactions of the Geological Institute, Academy of Sciences SSSR*, **241**: 1–164 (in Russian).

Rozanov, A.Y., 2016, Commission on Cambrian System on revised stage subdivision of Middle and Upper Series of Cambrian System of the General Stratigraphic Scale of Russia. *Decision of Intrabranch Division of Stratigraphic Commission and its Standing Committee*, **44**: 25–29 (in Russian).

Rozanov, A.Y., Khomentovsky, V.V., Shabanov, Y.Y., Karlova, G.A., Varlamov, A.I., Parkhaev, P., et al., 2008, To the problem of stage subdivision of the Lower Cambrian. *Stratigraphy and Geological Correlation*, **16** (1): 3–21.

Rozanov, A.Y., and Missarzhevsky, V.V., 1966, Biostratigraphy and fauna of the Cambrian lower horizons. *Transactions of Geological Institute, Academy of Sciences of USSR*, **148**: 1–127 (in Russian).

Rozanov, A.Y., Missarzhevsky, V.V., Volkova, N.A., Voronova, L.G., Krylov, L.N., Keller, B.M., et al., 1969, *The Tommotian Stage and the problem of the lower boundary of the Cambrian, Transactions of Geological Institute, Academy of Sciences of USSR*, 206. pp. 1–380 [in Russian; English translation: Raaben, M.E. (ed), 1981. New Delhi: Amerind Publishing.

Rozanov, A.Y., and Sokolov, B.S. (eds), 1984. Lower Cambrian Stage Subdivision. Stratigraphy. Nauka. Moscow, 184 pp. (in Russian).

Rozova, A.V., 1963, Biostratigraphic chart of the zonation of the Upper Cambrian and the terminal Middle Cambrian of the northwestern Siberian Platform and New Upper Cambrian trilobites from the Kulyumbe River. *Geology and Geophysics*, **1963** (9): 3–19 (in Russian).

Rozova, A.V., 1964, *Biostratigraphy and Description of Trilobites From the Middle and Upper Cambrian of the Northwestern Siberian Platform*. Moscow: Nauka 148 pp. (in Russian).

Rozova, A.V., 1967, The Cambrian lower boundary problem. *Geological Magazine*, **104**: 415–434.

Rozova, A.V., 1968, *Cambrian and Lower Ordovician of the Northwestern Siberian Platform*. Moscow: Nauka 196 pp. (in Russian).

Rozova, A.V., 1970, On the biostratigraphic charts of the Upper Cambrian and Lower Ordovician of the northwestern Siberian Platform. *Geology and Geophysics*, **1970** (5): 26–31 (in Russian).

Runnegar, B., 1982, The Cambrian explosion: animals or fossils? *Journal of the Geological Society of Australia*, **29**: 395–411.

Rushton, A.W.A., 1966, The Cambrian trilobites from the Purley Shales of Warwickshire. *Monographs of the Palaeontographical Society*, **120**: 1–55.

Rushton, A.W.A., 2009, Revision of the Furongian agnostoid *Lotagnostus trisectus* (Salter). *Memoirs of the Association of Australasian Palaeontologists*, **37**: 273–279.

Saito, K., 1934, Older Cambrian trilobite and Conchostraca from north-western Korea. *Japanese Journal of Geology and Geography*, **11**: 211–237.

Saltzman, M.R., Davidson, J.P., Holden, P., Runnegar, B., and Lohmann, K.C., 1995, Sea-level-driven changes in ocean chemistry at an upper Cambrian extinction horizon. *Geology*, **23**: 893–896.

Saltzman, M.R., Ripperdan, R.L., Brasier, M.D., Lohmann, K.C., Robison, R.A., Chang, W.T., et al., 2000, Global carbon isotope excursion (SPICE) during the Late Cambrian: relation to trilobite extinctions, organic-matter burial and sea level. *Palaeogeography, Palaeoclimatology, Palaeoecology*, **162**: 211–223.

Saltzman, M.R., Runnegar, B., and Lohmann, K.C., 1998, Carbon isotope stratigraphy of Upper Cambrian (Steptoean Stage) sequences of the eastern Great Basin: record of a global oceanographic event. *Geological Society of America Bulletin*, **110**: 285–297.

Salvador, A. (ed), 1994. International Stratigraphic Guide. second ed. International Union of Geological Sciences and the Geological Society of America. Trondheim, 214 pp.

Samson, S., Palmer, A.R., Robison, R.A., and Secor Jr., D.T., 1990, Biogeographical significance of Cambrian trilobites from the Carolina slate belt. *Geological Society of America Bulletin*, **102**: 1459–1470.

Savitsky, V.E., 1962, Relationships between Upper Precambrian and Cambrian in the Anabar Shield. *Proceedings of the Conference on Upper Precambrian Stratigraphy in Siberia and Far East (Abstract)*. Novosibirsk: Institute of Geology and Geophysics, pp. 53–54.

Sawaki, Y., Ohno, T., Fukushi, Y., Komiya, T., Ishikawa, T., Hirata, T., et al., 2008, Sr isotope excursion across the Precambrian-Cambrian boundary in the Three Gorges area, South China. *Gondwana Research*, **14**: 134–147.

Schmitz, M.D., 2012, Radioisotopic isotope geochronology. *In* Gradstein, F.M., Ogg, J.G., Schmitz, M.D., and Ogg, G.J. (eds), *The Geologic Timescale 2012*. Amsterdam: Elsevier BV, **Vol. 1**: 115–126.

Schmitz, M.D., Singer, B.S., and Rooney, A.D., 2020. Chapter 6 - Radioisotope geochronology. *In* Gradstein, F.M., Ogg, J.G., Schmitz, M.D., and Ogg, G.J. (eds), *The Geologic Timescale 2020*. Elsevier, Boston, MA, Vol. 1 (this book).

Scotese, C.R., 2014. Atlas of Cambrian and Early Ordovician Paleogeographic Maps (Mollweide Projection), Maps 81-88, The Early Paleozoic, PALEOMAP Atlas for ArcGIS, PALEOMAP Project. **vol. 5**. Evanston, IL. https://www.academia.edu/16785571/Atlas_of_Cambrian_and_Early_Ordovician_Paleogeographic_Maps.

Sdzuy, K., 1972, Das Kambrium der Acadobaltischen Faunenprovinz – Gegenwärtiger Kenntnisstand und Probleme. *Zentralblatt für Geologie und Paläontologie (II)*, **1972**: 1–91.

Secor Jr., D.T., Samson, S.L., Snoke, A.W., and Palmer, A.R., 1983, Confirmation of the Carolina slate belt as an exotic terrane. *Science*, **221**: 649–650.

Sedgwick, A., 1852, On the classification and nomenclature of the Lower Palaeozoic rocks of England and Wales. *Quarterly Journal of the Geological Society, London*, **8**: 136–168.

Sedgwick, A., and Murchison, R.I., 1835, On the Silurian and Cambrian Systems, exhibiting the order in which the older sedimentary strata

succeed each other in England and Wales. *The London and Edinburgh Philosophical Magazine and Journal of Science*, **7**: 483–535.

Seilacher, A., 1955, Spuren und Fazies im Unterkambrium. *Akademie der Wissenschaften und der Literatur zur Mainz, Mathematisch-Naturwissenschaftliche Klasse, Abhandlungen*, **10**: 373–399.

Seilacher, A., and Pfluger, E., 1994, From biomats to benthic agriculture: a biohistoric revolution. *In* Krumbein, W.S., Paterson, D.M., and Stal, L.J. (eds), *Biostabilization of Sediments*. Oldenberg: Bibliotheks und Informationsystem der Universität Oldenberg, 97–105.

Shabanov, Y.Y., Korovnikov, I.V., Pereladov, V.S., and Fefelov, A.F., 2008, Excursion 1a. The traditional Lower-Middle Cambrian boundary in the Kuonamka Formation of the Molodo River section (the southeastern slope of the Olenek Uplift of the Siberian Platform) proposed as a candidate for GSSP of the lower boundary of the Middle Cambrian and its basal (Molodian) stage, defined by the FAD of *Ovatoryctocara granulata*. *In* Rozanov, A.Y., and Varlamov, A.I. (eds), *The Cambrian System of the Siberian Platform. Part 2: North-East of the Siberian Platform*. Moscow and Novosibirsk: PIN, RAS, pp. 8–59 (in Russian and English).

Shaler, N.S., and Foerste, A.F., 1888, Preliminary description of North Attleborough fossils. *Bulletin of the Museum of Comparative Zoology*, **16**: 27–41.

Shergold, J.H., 1969, Oryctocephalidae (Trilobita: Middle Cambrian) of Australia. *Bureau of Mineral Resources of Australia Bulletin*, **104**: 1–66.

Shergold, J.H., 1972, Late Upper Cambrian trilobites from the Gola Beds, western Queensland. *Bulletin of the Bureau of Mineral Resources of Australia*, **112**: 1–126.

Shergold, J.H., 1975, Late Cambrian and Early Ordovician trilobites from the Burke River structural belt, western Queensland, Australia. *Bureau of Mineral Resources of Australia Bulletin*, **153** (1): 1–251.

Shergold, J.H., 1980, Late Cambrian trilobites from the Chatsworth Limestone, western Queensland. *Bulletin of the Bureau of Mineral Resources of Australia*, **186**: 1–111.

Shergold, J.H., 1982, Idamean (Late Cambrian) trilobites, Burke River structural belt, western Queensland. *Bulletin of the Bureau of Mineral Resources of Australia*, **187**: 1–69.

Shergold, J.H., 1993, The Iverian, a proposed Late Cambrian stage, and its subdivision in the Burke River structural belt, western Queensland. Bureau of Mineral Resources of Australia. *Journal of Australian Geology and Geophysics*, **13**: 345–358.

Shergold, J.H., 1995. Timescales 1. Cambrian. Australian Phanerozoic timescales, biostratigraphic charts and explanatory notes. Second series. Australian Geological Survey Organisation, Record 1995/30: pp. 1–32.

Shergold, J.H., 1996, Cambrian (Chart 1). *In* Young, G.C., and Laurie, J. R. (eds), *An Australian Phanerozoic Timescale*. Melbourne: Oxford University Press, 63–76.

Shergold, J.H., 1997, Explanatory notes for the Cambrian correlation chart. *In* Whittington, H.B., Chatterton, B.D.E., Speyer, S.E., Fortey, R.A., Owens, R.M., Chang, W.T., et al., (eds), *Treatise on Invertebrate Paleontology, Part O Arthropoda 1, Trilobita (Revised), (Volume 1) Introduction, Order Agnostida, Order Redlichiida*. Boulder, CO; and Lawrence, KS: Geological Society of America and University of Kansas, 303–311.

Shergold, J.H., and Cooper, R.A., 2004, The Cambrian Period. *In* Gradstein, F.M., Ogg, J.G., and Smith, A.G. (eds), *(Coordinators)*,

A Geologic Time Scale 2004. Cambridge: Cambridge University Press, 147–164.

Shergold, J.H., and Geyer, G., 2003, The Subcommission on Cambrian Stratigraphy: the status quo. *Geologica Acta*, 1–5–9.

Shergold, J.H., and Nicoll, R.S., 1992, Revised Cambrian-Ordovician boundary biostratigraphy, Black Mountain, western Queensland. *In* Webby, B.D., and Laurie, J.R. (eds), *Global Perspectives on Ordovician Geology*. Rotterdam: Balkema, 81–92.

Shergold, J.H., Rozanov, A.Y., and Palmer, A.R., 1991, *The Cambrian System on the Siberian Platform*. Trondheim: International Union of Geological Sciences Publications, p. 133.

Shu, D.G., Conway Morris, S., Han, J., Li, Y., Zhang, X.L., Hua, H., et al., 2006, Lower Cambrian vendobionts from China and early diploblast evolution. *Science*, 312: 731–734.

Simmons, M.S., Miller, K.G., Ray, D.C., Davies, A., van Buchem, F.S.P., and Gréselle, B., 2020, Chapter 13 - Phanerozoic eustasy. *In* Gradstein, F.M., Ogg, J.G., Schmitz, M.D., and Ogg, G.M. (eds), *The Geologic Time Scale 2020*. Elsevier, Boston, MA, Vol. 1 (this book).

Singh, B.P., Chaubey, R.S., Bhargava, O.N., Prasad, S.K., and Negia, R.S., 2017, The Cambrian trilobite fauna from the Shian (Saybang) section, Pin Valley (Spiti) and its biostratigraphic significance. *Palaeoworld*, 26: 25–36.

Singh, B.P., Virmani, N., Bhargava, O.N., Negia, R.S., Kishore, N., and Gill, A., 2016, Trilobite fauna of basal Cambrian Series 3 (Stage 5) from the Parahio Valley (Spiti), Northwest Himalaya, India and its biostratigraphic significance. *Annales de Paléontologie*, 102 (1): 59–67.

Siveter, D.J., and Williams, M., 1995, An Early Cambrian assignment for the Caerfai Group of South Wales. *Journal of the Geological Society*, 152: 221–224.

Southgate, P.N., and Shergold, J.H., 1991, Application of sequence stratigraphic concepts to Middle Cambrian phosphogenesis, Georgina Basin, Australia. *Journal of Australian Geology and Geophysics*, 12: 119–144.

Spizharsky, T.N., Ergaliev, G.K., Zhuravleva, I.T., Repina, L.N., and Rozanov, A.Y., 1983, Stage scale of the Cambrian System. *Soviet Geology*, 1983 (8): 57–72.

Stein, M., Church, S.B., and Robison, R.A., 2011, A new Cambrian arthropod, *Emeraldella brutoni*, from Utah. *Paleontological Contributions*, 3: 1–9.

Steiner, M., Li, G.X., Qian, Yi, Zhu, M.Y., and Erdtmann, B.-D., 2007, Neoproterozoic to early Cambrian small shelly fossil assemblages and a revised biostratigraphic correlation of the Yangtze Platform (China). *Palaeogeography, Palaeoclimatology, Palaeoecology*, 254: 67–99.

Stitt, J.H., 1975, Adaptive radiation, trilobite paleoecology and extinction, Ptychaspidid Biomere, Late Cambrian of Oklahoma. *Fossils and Strata*, 4: 381–390.

Stubblefield, G.J., 1956. Cambrian palaeogeography in Britian. In: Rodgers, J. (Ed.), El Sistema Cámbrico, su Paleogeografiá y el Problema de su Base. 20th International Geological Congress, Mexico City 1, pp. 1–43.

Sundberg, F.A., 1990. Morphological Diversification of Ptychopariid Trilobites in the Marjumiid Biomere (Middle to Upper Cambrian). Ph.D. dissertation. Blacksburg: Virginia Polytechnic Institute and State University, p. 674.

Sundberg, F.A., 1994, Corynexochida and Ptychopariida (Trilobita, Arthropoda) of the *Ehmaniella* Biozone (Middle Cambrian), Utah and Nevada. *Natural History Museum of Los Angeles County, Contributions in Science*, 446: 1–137.

Sundberg, F.A., 2005, The Topazan Stage, a new Laurentian stage (Lincolnian Series—"Middle" Cambrian). *Journal of Paleontology*, 79: 63–71.

Sundberg, F.A., Geyer, G., Kruse, P.D., McCollum, L.B., Pegel, T.V., Żylińska, A., et al., 2016, International correlation of the Cambrian Series 2-3, Stages 4-5 boundary interval. *Australasian Palaeontological Memoirs*, 49: 83–124.

Sundberg, F.A., and McCollum, L.B., 1997, Oryctocephalids (Corynexochida: Trilobita) of the Lower-Middle Cambrian boundary interval from California and Nevada. *Journal of Paleontology*, 71: 1065–1090.

Sundberg, F.A., and McCollum, L.B., 2003, Early and Middle Cambrian trilobites from the outer-shelf deposits of Nevada and California, USA. *Palaeontology*, 46: 945–986.

Sundberg, F.A., Yuan, J.L., McCollum, L.B., and Zhao, Y.L., 1999, Correlation of Lower-Middle Cambrian boundary of South China and Western United States of America. *Acta Palaeontoloica Sinica*, 38 (Suppl.): 102–107.

Suvorova, N.P., 1954, On the Lenian Stage of the Lower Cambrian in Yakutia, (in Russian). *In* Shatsky, N.S. (ed.), *Problem of the Geology of Asia*. Moscow: Academy of Sciences Press of USSR, Vol. 1: pp. 466–483.

Taylor, J.F., 1997. Upper Cambrian biomeres and stages, two distinctly different and equally vital stratigraphic units. Second International Trilobite Conference, August 1997, St. Catharines, Ontario, Abstracts with Program, p. 47.

Taylor, J.F., 2006, History and status of the biomere concept. *Memoirs of the Association of Australasian Palaeontologists*, 32: 247–265.

Taylor, P.D., Berning, B., and Wilson, M.A., 2013, Reinterpretation of the Cambrian 'bryozoan' *Pywackia* as an octocoral. *Journal of Paleontology*, 87: 984–990.

Terfelt, F., Ahlberg, P., and Eriksson, M.E., 2011, Complete record of Furongian polymerid trilobites and agnostoids of Scandinavia – a biostratigraphical scheme. *Lethaia*, 44: 8–14.

Terfelt, F., Eriksson, M.E., Ahlberg, P., and Babcock, L.E., 2008, Furongian (Cambrian) biostratigraphy of Scandinavia – a revision. *Norwegian Journal of Geology*, 88: 73–87.

Trench, A., 1996, Magnetostratigraphy: Cambrian to Silurian. *In* Young, G.C., and Laurie, J.R. (eds), *An Australian Phanerozoic Timescale*. Melbourne: Oxford University Press, 23–29.

Tucker, R.D., and McKerrow, W.S., 1995, Early Paleozoic chronology: a review in light of new U-Pb zircon ages from Newfoundland and Britain. *Canadian Journal of Earth Sciences*, 32: 368–379.

Tullberg, S.A., 1880, Om Agnostus-arterna i de kambriska aflagringarne vid Andrarum. *Sveriges Geologiska Undersökning*, C42: 1–38.

Valledares, M.I., Ugidos, J.M., Barba, P., Fallick, A.E., and Ellam, R.M., 2006, Oxygen, carbon and strontium isotope records of Ediacaran carbonates in central Iberia (Spain). *Precambrian Research*, 147: 354–365.

Varlamov, A.I., and Rozova, A.V., 2009, New Upper Cambrian (Evenyiskie) Regional Stage of Siberia. *In* Efimov, A.C. (ed.), *New Data on Lower Palaeozoic Stratigraphy and Palaeontology of Siberia*. Novosibirsk: SNIIGGiMS, 3–61.

Varlamov, A.I., Pak, K.L., and Rosova, A.V., 2005, *The Upper Cambrian of the Chopko River Section, Norilsk Region, Northwestern Siberian Platform: Stratigraphy and Trilobites*. Novosibirsk: Nauka, p. 84 [in Russian; English edition: *Paleontological Journal*, 40 (Suppl. 1): 1–56].

Varlamov, A.I., Rozova, A.V., Khamentovsky, Y.Y., Shabnov, Y.Y., Abaimova, G.P., Demidenko, Y.E., et al., 2008a, The Zhurinsky

Mys section. *In* Rozanov, A.Y., and Varlamov, A.I. (eds), *The Cambrian System of the Siberian Platform. Part 1: The Aldan-Lena Region*. Moscow and Novosibirsk: Paleontological Institute, Russian Academy of Sciences, pp. 71–92 (in Russian and English).

Varlamov, A.I., Rozova, A.V., Khamentovsky, Y.Y., Shabnov, Y.Y., Abaimova, G.P., Demidenko, Y.E., et al., 2008b, Introduction. *In* Rozanov, A.Y., and Varlamov, A.I. (eds), *The Cambrian System of the Siberian Platform. Part 1: The Aldan-Lena Region*. Moscow and Novosibirsk: Paleontological Institute, Russian Academy of Sciences, pp. 6–11 (in Russian and English).

Vidal, G., Palacios, T., Gámez-Vintaned, J.A., Díez Balda, M.A., and Grant, S.W., 1994, Neoproterozoic-early Cambrian geology and palaeontology of Iberia. *Geological Magazine*, **131**: 729–765.

Walcott, C.D., 1889a, Stratigraphic position of the *Olenellus* fauna in North America and Europe. *American Journal of Science*, **37**: 374–392.

Walcott, C.D., 1889b, Stratigraphic position of the *Olenellus* fauna in North America and Europe (continued). *American Journal of Science*, **38**: 29–42.

Walcott, C.D., 1890a. The fauna of the Lower Cambrian or *Olenellus* Zone. Tenth Annual Report of the Director, 1888–1889. Part 1. United States Geological Survey, pp. 509–774.

Walcott, C.D., 1890b, Descriptive notes on new genera and species from the Lower Cambrian or *Olenellus* Zone of North America. *Proceedings of the U.S. National Museum*, **12**: 33–46.

Walter, M.R., Elphinstone, R., and Heys, G.R., 1989, Proterozoic and Early Cambrian trace fossils from the Amadeus and Georgina Basins, central Australia. *Alcheringa*, **13**: 209–256.

Wang, X.L., Hua, W.X., Yao, S.P., Chen, Q., and Xie, X.M., 2011, Carbon and strontium isotopes and global correlation of Cambrian Series 2–Series 3 carbonate rocks in the Keping area of the northwestern Tarim Basin, NW China. *Marine and Petroleum Geology*, **28**: 992–1002.

Wang, X.F., Stouge, S., Maletz, J., Bagnoli, G., Qi, Y.P., Raevskaya, E.G., et al., 2019, Correlating the global Cambrian–Ordovician boundary: precise comparison of the Xiaoyangqiao section, Dayangcha, North China with the Green Point GSSP section, Newfoundland, Canada. *Palaeoworld*, **28**: 243–275. <https://doi.org/10.1016/j.palwor.2019.01.003>.

Wang, Z.J., Huang, Z.G., Yao, J.X., and Ma, X.L., 2013, Characteristics and main progress of the Stratigraphic Chart of China and directions. *Acta Geoscientica Sinica*, **35**: 271–276.

Webster, M., 2011, Upper Dyeran trilobite biostratigraphy in the southern Great Basin, U.S.A. *Museum of Northern Arizona Bulletin*, **67**: 121–154.

Weidner, T., Ahlberg, P., Axheimer, N., and Clarkson, E.N.K., 2004, The middle Cambrian *Ptychagnostus punctuosus* and *Goniagnostus nathorsti* zones in Västergötland, Sweden. *Bulletin of the Geological Society of Denmark*, **51**: 39–45.

Weidner, T., and Ebbestad, J.O.R., 2014, The early middle Cambrian agnostid *Pentagnostus praecurrens* (Westergård 1936) from Sweden. *Memoirs of the Association of Australasian Palaeontologists*, **45**: 403–419.

Weidner, T., and Nielsen, A.T., 2013, The late Cambrian (Furongian) *Acerocarina* Superzone (new name) on Kinnekulle, Västergötland, Sweden. *GFF*, **135**: 30–44.

Weidner, T., and Nielsen, A.T., 2014, A highly diverse trilobite fauna with Avalonian affinities from the Middle Cambrian *Acidusus atavus* Zone (Drumian Stage) of Bornholm, Denmark. *Journal of Systematic Palaeontology*, **12**: 23–92.

Wen, R.Q., Babcock, L.E., Peng, J., and Robison, R.A., 2019, New edrioasteroid (Echinodermata) from the Spence Shale (Cambrian), Idaho, USA: further evidence of attachment in the early evolutionary history of edrioasteroids. *Bulletin of Geosciences*, **94**: 115–124.

Westergård, A.H., 1922, Sveriges Olenidskiffer. *Sveriges Geologiska Undersökning*, **Ca18**: 1–205.

Westergård, A.H., 1946, Agnostidea of the Middle Cambrian of Sweden. *Sveriges Geologiska Undersökning*, **C477**: 1–141.

Westergård, A.H., 1947, Supplementary notes on the Upper Cambrian trilobites of Sweden. *Sveriges Geologiska Undersökning*, **C489**: 1–34.

Westrop, S.R., 1986, Trilobites of the Upper Cambrian Sunwaptan Stage, southern Canadian Rocky Mountains, Alberta. *Palaeontographica Canadiana*, **3**: 1–179.

Westrop, S.R., 1995, Sunwaptan and Ibexian (Upper Cambrian-Lower Ordovician) trilobites of the Rabbitkettle Formation, Mountain River region, northern Mackenzie Mountains, northwest Canada. *Palaeontographica Canadiana*, **12**: 1–75.

Westrop, S.R., and Adrain, J.M., 2013, Biogeographic shifts in a transgressive succession: the Cambrian (Furongian, Jiangshanian; latest Steptoean–Earliest Sunwaptan) agnostoid arthropods *Kormagnostella romanenko* and *Biciragnostus ergaliev* in North America. *Journal of Paleontology*, **87**: 804–817.

Westrop, S.R., Adrain, J.M., and Landing, E., 2011, The Cambrian (Sunwaptan, Furongian) agnostoid arthropod *Lotagnostus* Whitehouse, 1936, in Laurentian and Avalonian North America: systematics and biostratigraphic significance. *Bulletin of Geosciences*, **86**: 569–594.

Westrop, S.R., and Landing, E., 2017, The agnostoid arthropod *Lotagnostus* Whitehouse, 1936 (late Cambrian; Furongian) from Avalonian Cape Breton Island (Nova Scotia, Canada) and its significance for international correlation. *Geological Magazine*, **154**: 1001–1021.

Wheeler, H.E., 1947, Base of the Cambrian System. *Journal of Geology*, **55**: 153–159.

Whitehouse, F.W., 1936, The Cambrian faunas of northeastern Australia. Parts 1 and 2. *Memoirs of the Queensland Museum*, **11**: 59–112.

Whittington, H.B., 1977, The Middle Cambrian trilobite *Naraoia*, Burgess Shale, British Columbia. *Philosophical Transactions of the Royal Society of London, Series B*, **280**: 409–443.

Wotte, T., Álvaro, J.J., Shields, G.A., Brown, B., Brasier, M.D., and Veizer, J., 2007, C-, O- and Sr-isotope stratigraphy across the Lower-Middle Cambrian transition of the Cantabrian Zone (Spain) and the Montagne Noire (France), West Gondwana. *Palaeogeography, Palaeoclimatology, Palaeoecology*, **256**: 47–70.

Xiang, L.W., and Zhang, T.R., 1985, Stratigraphy and trilobite faunas of the Cambrian in the western part of northern Tianshan, Xinjiang. *Ministry of Geology and Mineral Resources, Geological Memoirs, Series 2*, **4**: 1–243.

Xing, Y.S., Ding, Q.X., Luo, H.L., He, T.G., and Wang, Y.G., 1984, The Sinian-Cambrian boundary of China. *Bulletin of the Institute of Geology Chinese Academy of Geological Sciences*, **10**: 1–262 (in Chinese).

Xing, Y.S., Luo, H.L., Jiang, Z.W., and Zhang, S.S., 1991, A candidate Global Stratotype Section and Point for the Precambrian-Cambrian boundary at Meishucun, Yunnan, China. *Journal of China University of Geosciences*, **2** (1): 47–57.

Yang, A.H., Zhu, M.Y., Debrenne, F., Yuan, K.X., Vannier, J., Zhang, J.M., et al., 2005, Early Cambrian archaeocyathan zonation of the Yangtze

Platform and biostratigraphic implications. *Acta Macropalaeontologica Sinica*, **22** (Suppl.): 205–210.

Yang, J.L., 1992, Evolution of the Family Damesellidae (Trilobita) and boundary between the Middle and Upper Cambrian. *Earth Science*, **17** (3): 251–260 (in Chinese).

Yang, Z.Y., Otofuji, Y., Sun, Z.M., and Huang, B.C., 2002, Magnetostratigraphic constraints on the Gondwanan origin of North China: Cambrian/Ordovician boundary results. *Geophysical Journal International*, **151**: 1–10.

Yao, L., Peng, J., Fu, X.P., and Zhao, Y.L., 2009, Ontogenesis of *Tuzoia bispinosa* (Arthropoda) from the Middle Cambrian Kaili biota, Guizhou, China. *Acta Palaeontologica Sinica*, **48**: 56–64 (in Chinese).

Yin, J.Y., 2002, Research of Palaeomagenetism. *In* Chen, L.X., Luo, H.L., Hu, S.X., Yin, J.Y., Jiang, Z.W., Wu, Z.L., et al., (eds), *Early Cambrian Chengjiang Fauna in Eastern Yunnan, China*. Yunnan: Science and Technology Press, pp. 98–113 (in Chinese).

Young, G.C., and Laurie, J.R., 1996, *An Australian Phanerozoic Timescale*. Melbourne: Oxford University Press, p. 279.

Yuan, J.L., Zhao, Y.L., Li, Y., and Huang, Y.Z., 2002, *Trilobite Fauna of the Kaili Formation (Uppermost Lower Cambrian–Lower Middle Cambrian) From Southeastern Guizhou, South China*. Shanghai: Shanghai Science and Technology Press, p. 422 (in Chinese).

Yuan, J.L., Zhao, Y.L., Wang, Z.Z., Zhou, Z.Y., and Cheng, X.Y., 1997, A preliminary study on Lower-Middle Cambrian boundary and trilobite fauna at Balang, Taijiang, Guizhou, South China. *Acta Palaeontonlgica Sinica*, **36**: 494–524 (in Chinese).

Zamora, S., Deline, B., Álvaro, J.J., and Rahman, I.A., 2017, The Cambrian substrate revolution and the early evolution of attachment in suspension-feeding echinoderms. *Earth-Science Reviews*, **171**: 478–491.

Zamora, S., Zhu, X.J., and Lefebvre, B., 2013, A new Furongian (Cambrian) echinoderm-Lagerstätte from the Sandu Formation (South China). *Cahiers de Biologie Marine*, **54**: 565–569.

Zhang, J.M., Li, G.X., Zhou, C.M., Zhu, M.Y., and Yu, Z.Y., 1997, Carbon isotope profiles and their correlation across the Neoproterozoic-Cambrian boundary interval on the Yangtze Platform, China. *Bulletin of National Museum of Natural Sciences*, **10**: 107–116.

Zhang, W.T., 1987, World's oldest Cambrian trilobites from eastern Yunnan. *In* Nanjing Institute of Geology and Palaeontology (ed.), *Stratigraphy and Palaeontology of Systemic Boundaries in China. Precambrian-Cambrian Boundary (1)*. Nanjing: Nanjing University Publishing House, 1–16.

Zhang, W.T., and Babcock, L.E., 2001, New extraordinarily preserved fossils, possibly with Ediacaran affinities, from the Lower Cambrian of Yunnan, China. *Acta Palaeontologica Sinica*, **40** (Suppl.): 201–213.

Zhang, W.T., Yuan, K.X., Zhou, Z.Y., Qian, Y., and Wang, Z.Z., 1979, Cambrian of southwestern China. *In* Nanjing Institute of Geology and Palaeontology (ed.), *Carbonate Biostratigraphy of Southwest China*. Beijing: Science Press, pp. 39–107 (in Chinese).

Zhang, X.L., Ahlberg, P., Babcock, L.E., Choi, D.K., Geyer, G., Gozalo, R., et al., 2017, Challenges in defining the base of Cambrian Series 2 and Stage 3. *Earth-Science Reviews*, **172**: 124–139.

Zhao, Y.L., Peng, J., Yuan, J.L., Guo, Q.J., Tai, T.S., Yin, L.M., et al., 2012, Stop 5: The Kaili Formation and Kaili Biota at the Wuliu-Zengjiayan section of Guizhou Province, China and proposed Global Standard Stratotype-section and Point (GSSP) of the unnamed

Cambrian Series 3, Stage 5. *Journal of Guizhou University (Natural Sciences)*, **29** (Suppl. 1): 108–124.

Zhao, Y.L., Yang, R.D., Yuan, J.L., Zhu, M.Y., Guo, Q.J., Yang, X.L., et al., 2001, Cambrian stratigraphy at Balang, Guizhou Province, China: candidate section for a global unnamed series and stratotype section for the Taijiangian Stage. *Palaeoworld*, **13**: 189–208.

Zhao, Y.L., Yuan, J.L., Babcock, L.E., Guo, Q.J., Peng, J., Yin, L.M., et al., 2019, Global Standard Stratotype-Section and Point (GSSP) for the conterminous base of the Miaolingian Series and Wuliuan Stage (Cambrian) at Balang, Jianhe, Guizhou, China. *Episodes*, **42**: 165–184.

Zhao, Y.L., Yuan, J.L., Guo, Q.J., Peng, J., Yin, L.M., Yang, Y.L., et al., 2014, Comments on some important issues concerning the establishment of a GSSP for Cambrian Stage 5. *GFF*, **136**: 333–336.

Zhao, Y.L., Yuan, J.L., Peng, S.C., Guo, Q.L., Zhu, L., Peng, J., et al., 2004, Proposal and prospects for the global Lower–Middle Cambrian boundary. *Progress in Natural Science*, **14**: 1033–1038.

Zhao, Y.L., Yuan, J.L., Zhu, M.Y., Babcock, L.E., Peng, J., Wang, Y., et al., 2005a, Balang section, Guizhou, China: stratotype section for the Taijiangian Stage and candidate for GSSP of an unnamed Cambrian series. *In* Peng, S.C., Babcock, L.E., and Zhu, M.Y. (eds), *Cambrian System of China and Korea, Guide to Field Excursions*. Hefei: University of Science and Technology of China Press, 62–83.

Zhao, Y.L., Yuan, J.L., Zhu, M.Y., Yang, R.D., Guo, Q.J., Peng, J., et al., 2002, Progress and significance in research on the early Middle Cambrian Kaili biota, Guizhou Province, China. *Progress in Natural Science*, **12**: 649–654.

Zhao, Y.L., Zhu, M.Y., Babcock, L.E., and Peng, J., 2011, *The Kaili Biota: Marine Organisms From 508 Million Years Ago*. Guiyang: Guizhou Publishing Group 251 pp.

Zhao, Y.L., Yuan, J.L., Zhu, M.Y., Yang, R.D., Guo, Q.J., and Qian, Y., 1999, A progress report on research on the early Middle Cambrian Kaili Biota, Guizhou, PRC. *Acta Palaeontologica Sinica*, **38** (Suppl.): 1–14.

Zhao, Y.L., Zhu, M.Y., Babcock, L.E., Yuan, J.L., Parsley, R.L., Peng, J., et al., 2005b, *Kaili Biota: a taphonomic window on diversification of metazoans from the basal Middle Cambrian: Guizhou, China, Acta Geologica Sinica*, 79. pp. 51–765.

Zhen, Y.Y., and Zhou, Z.Y., 2008, History of trilobite biodiversity: a Chinese perspective. *In* Zhou, Z.Y., and Zhen, Y.Y. (eds), *Trilobite Record of China*. Beijing: Science Press, 301–330.

Zhou, Z.Q., Cao, X.D., Hu, Y.X., and Zhao, J.T., 1996, Early Palaeozoic stratigraphy and sedimentary-tectonic evolution in eastern Qilian Mountains, China. *Northwest Geosciences*, **17**: 1–58 (in Chinese).

Zhu, M.Y., Babcock, L.E., and Peng, S.C., 2006, Advances in Cambrian stratigraphy and paleontology: integrating correlation techniques, paleobiology, taphonomy and paleoenvironmental reconstruction. *Palaeoworld*, **15**: 217–222.

Zhu, M.Y., Li, G.X., Zhang, J.M., Steiner, M., Qian, Y., and Jiang, Z.W., 2001, Early Cambrian stratigraphy of East Yunnan, southwestern China: a synthesis. *Acta Palaeontologica Sinica*, **40** (Suppl.): 4–39.

Zhu, M.Y., Yang, A.H., Yuan, J.L., Li, G.X., Zhang, J.M., Zhao, F.C., et al., 2019, Cambrian integrative stratigraphy and timescale of China. *Science China Earth Sciences*, **62**: 25–60.

Zhu, M.Y., Zhang, J.M., Li, G.X., and Yang, A.H., 2004, Evolution of C isotopes in the Cambrian of China: implications for Cambrian subdivision and trilobite mass extinctions. *Geobios*, **37**: 287−310.

Zhu, M.Y., Zhuravlev, A.Y., Wood, R.A., Zhao, F., and Sukhov, S.S., 2017, A deep root for the Cambrian explosion: implications of new bio-and chemostratigraphy from the Siberian Platform. *Geology*, **45**: 459−462.

Zhu, X.J., Peng, S.C., Zamora, S., Lefebvre, B., and Chen, G.Y., 2016, Furongian (upper Cambrian) Guole Konservat-Lagerstätte from South China. *Acta Geologica Sinica*, **90**: 30−37.

Zhuravlev, A.Y., 1995. Preliminary suggestions on the global Early Cambrian zonation. In: Landing, E., Geyer, G. (Eds.), Morocco '95: The Lower-Middle Cambrian Standard of Western Gondwana. Beringeria, 2 (Special Issue): 147−160.

Zhuravlev, A.Y., Gámex Vintaned, J.A., and Ivantsov, A.Y., 2009, First finds of problematic Ediacaran fossil *Gaojiashania* in Siberia and its origin. *Geological Magazine*, **146**: 775−780.

Zhuravlev, A.Y., and Gravestock, D.I., 1994, Archaeocyaths from Yorke Peninsula, South Australia and archaeocyathan Early Cambrian zonation. *Alcheringa*, **18**: 1−54.

Zhuravlev, A.Y., and Repina, L.N. (eds), 1990. Guidebook for Excursion on the Aldan and Lena Rivers, Siberian Platform. Institute of Geology and Geophysics, Siberian Branch of Academy of Sciences of USSR. Novosibirsk, 115 pp.

Zhuravlev, A.Y., and Riding, R. (eds), 2001. The ecology of the Cambrian Radiation. *Perspectives in Paleobiology and Earth History* **7**: 1−525.

Zhuravlev, A.Y., Wood, R.A., and Penny, A.M., 2015, Ediacaran skeletal metazoan interpreted as a lophophorate. *Proceedings of the Royal Society B, Biological Sciences*, **282** (1818): 20151860. https://doi.org/10.1098/rspb.2015.1860.

Zhuravleva, I.T., Korshunov, V.I., and Rozanov, A.Y., 1969, Atdabanian Stage and its archaeocyathan basis in the stratotype section. *In* Zhuravleva, I.T. (ed.), *Lower Cambrian Biostratigraphy and Palaeontology of Siberia and Far East*. Moscow: Nauka, 5−59.

D. Goldman, P.M. Sadler and S.A. Leslie
With contributions by M.J. Melchin, F.P. Agterberg and F.M. Gradstein

Chapter 20

The Ordovician Period

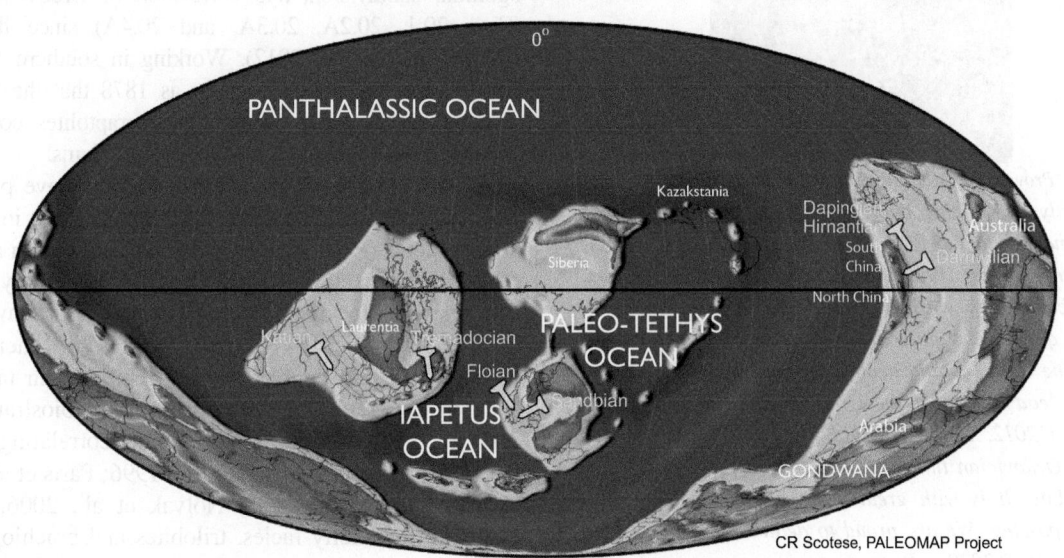

456 Ma Ordovician

CR Scotese, PALEOMAP Project

Chapter outline

20.1 **History and subdivisions** 632
  20.1.1 Stages of the Lower Ordovician 635
  20.1.2 Stages of the Middle Ordovician 643
  20.1.3 Stages of the Upper Ordovician 646
20.2 **Regional subdivisions** 651
  20.2.1 Australasian chronostratigraphic units 651
  20.2.2 East Baltic chronostratigraphic units 651
  20.2.3 Bohemo-Iberian chronostratigraphic units 653
  20.2.4 North and South China chronostratigraphic units 653
  20.2.5 North American chronostratigraphic units 653

20.2.6 Ordovician stage slices 654
20.3 **Ordovician stratigraphy** 655
  20.3.1 Biostratigraphy 655
  20.3.2 Chemostratigraphy 657
  20.3.3 Cyclostratigraphy 661
  20.3.4 Biotic and climatic events 662
20.4 **Ordovician time scale** 664
  20.4.1 Radioisotopic dates 664
  20.4.2 Building the Ordovician and Silurian Composite Standard 667
  20.4.3 Age of stage boundaries 674
Acknowledgments 682
Bibliography 682

Abstract

The Ordovician Period (486.9—443.1 Ma) encompasses two extraordinary biological events in the history of life on the Earth. The first, the "Great Ordovician Biodiversification Event," is a great evolutionary radiation of marine life and the second is a catastrophic Late Ordovician extinction. Understanding the duration, rate, and magnitude of these events requires an increasingly precise time scale. The Ordovician time scale is based on the subdivision of a Lower Paleozoic CONOP9 composite graptolite range chart derived from 837 stratigraphic sections and 2651 graptolite taxa with interpolated radioisotopic dates. Thirty-seven new radio-isotope dates are used in the scaling of the new Ordovician time scale. The base of the Ordovician Period is defined at the level of the first appearance of the conodont *Iapetognathus fluctivagus* at the Green Point Newfoundland section. Its top, the base of the Silurian Period, is set as the level of the first appearance of the graptolite *Akidograptus ascensus* at Dob's Linn, Scotland. For the first time an independently time-scaled CONOP9 composite cono-dont range chart is presented to facilitate the application of the time scale to carbonate facies sections.

Provided by Margaret Low, GNS Science

Shortly before this book went to press, our generous friend and colleague, Roger Cooper, passed away. Roger, the former Chief Paleontologist of the New Zealand Geological Survey, worked at the forefront of biostratigra-phy and quantitative paleobiology. He headed the compila-tion of the New Zealand geological time scale (2004), and was the lead author of the Ordovician chapters for GTS 2004 and 2012. His substantial and innovative contribu-tions to Ordovician time scales guided us through the cur-rent update. It is with great sadness that we learned of Roger's passing. We are proud to dedicate the Ordovician Chapter of GTS2020 to Roger Cooper.

20.1 History and subdivisions

The Ordovician System was proposed by Lapworth in 1879 as a compromise solution to the great Cambrian—Silurian controversy over strata in North Wales. Rocks that had been assigned by Adam Sedgwick to his Cambrian System (Sedgwick, 1847, 1852) were also included in the lower part of the overlying Silurian System by Roderick Murchison (Murchison, 1839). In reality, Lapworth's (1879) solution was more than a compromise, as it was based on the identification of three stratigraphically successive Lower Paleozoic faunas that had widespread geographical distribu-tion in the Northern Hemisphere (Bassett, 1979). Barrande had already recognized the existence of three distinct Lower Paleozoic faunas in Bohemia (Watts, 1939), and Lapworth was convinced that a tripartite division of Lower Paleozoic strata would greatly facilitate correlations between Britain, the European Continent, and across the Atlantic to North America. Lapworth (1879) noted that it would be fitting for the new middle division to be named after the *Ordovices*, a Welsh tribe whose geographical location was midway between the southern and northern tribes from which the

Cambrian and Silurian systems derived their names. Although it was initially slow to be accepted in Britain, where it was instead generally called Lower Silurian well into the 20th cen-tury, the Ordovician was soon recognized and used elsewhere, such as in the Baltic region and Australia. The name "Ordovician" was officially adopted at the 1960 International Geological Congress in Copenhagen.

Dark graptolite-bearing shales are common in Ordovician sedimentary successions around the world. Lapworth (1879—1880) described the stratigraphic distribution of British graptolites at the same time that he proposed the Ordovician System, and graptolites have played a major role in the rec-ognition, subdivision, and correlation of Ordovician rocks (Figs. 20.1, 20.2A, 20.3A, and 20.4A) since that time (Cooper and Sadler, 2012). Working in southern Scotland, Lapworth demonstrated as early as 1878 that the fine bio-stratigraphic zonation possible with graptolites could help solve both stratigraphic and structural problems.

In the last several decades, conodonts have proved to be of comparable global biostratigraphic value in the car-bonate facies (Figs. 20.2B, 20.3B, and 20.4B). In addition, conodont elements have proved to be extremely durable and resistant to diagenesis, properties that have made them extraordinarily valuable sources of chemostrati-graphic data. Substantial work on chitinozoan taxonomy and biostratigraphy has provided a third biostratigraphic data set that is extremely useful correlating Lower Paleozoic strata (e.g., Paris, 1990, 1996; Paris et al., 2004; Nõlvak and Grahn, 1993; Nõlvak et al., 2006; Achab, 1989). In the shelly facies, trilobites and brachiopods are used extensively for local and regional zonation, and cor-al—stromatoporoid communities enable biostratigraphic subdivision in the Upper Ordovician. Many other biotic groups became established and diversified during the Ordovician and enable biostratigraphic subdivision in local regions. Radiolarian and acritarch zonations are use-ful as correlation tools in certain intervals and facies. Correlation between disparate biofacies remains a difficult problem, but ongoing work focused on integrating grapto-lite, conodont, and chitinozoan biostratigraphies with new chemostratigraphic data is providing ever more precise and widespread correlations (e.g., Bergström et al., 2009, 2010 a,b,c, 2018; Ainsaar et al., 2010; Saltzman et al., 2014; Wu et al., 2017).

Subdivision of the Ordovician into Upper and Lower, or Upper, Middle, and Lower parts has been very inconsis-tent (Jaanusson, 1960; Webby, 1998). The International Subcommission on Ordovician Stratigraphy voted to recog-nize a threefold subdivision of the system (Webby, 1995), now accepted by the International Commission on Stratigraphy (ICS) as the Lower, Middle, and Upper Ordovician series. Because of marked faunal provincial-ism and facies differentiation throughout most of the Ordovician, no existing regional suite of stages or series

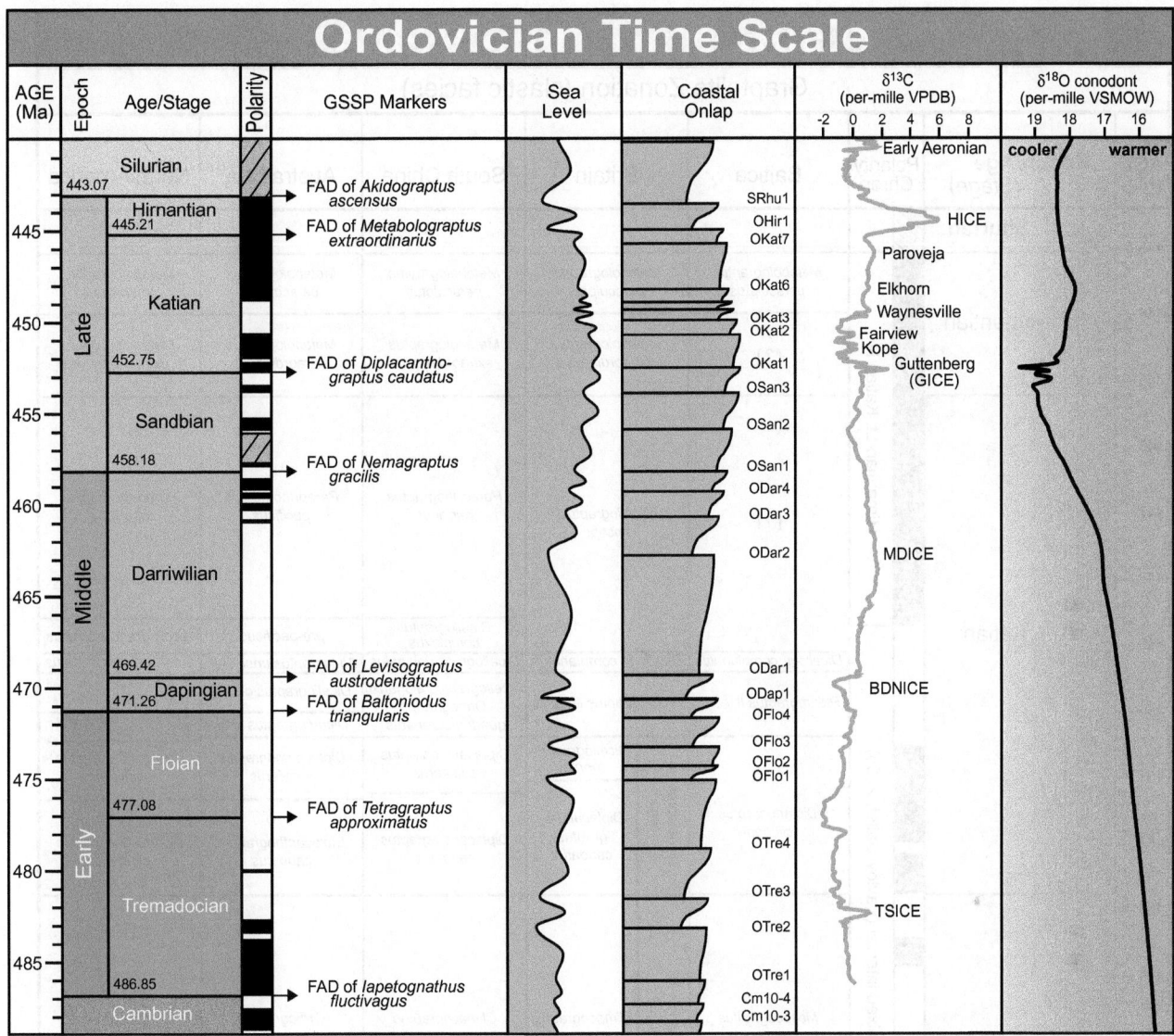

Ordovician Time Scale

FIGURE 20.1 Ordovician overview. Main markers for GSSPs of Ordovician stages are FADs of graptolite taxa (except conodont FADs for the bases of the Tremadocian and Dapingian stages). ("Age" is the term for the time equivalent of the rock-record "stage.") Magnetostratigraphy from Pavlov and Gallet (2005); but many intervals have not yet been studied or verified. Schematic sea-level, coastal onlap curves and sequence nomenclature are modified from Haq and Schutter (2008) following advice of Haq (pers. comm., 2008, to J. Ogg) and from Dronov et al. (2011). See also review in Simmons et al. (2020, Ch. 13: Phanerozoic eustasy, this book). The carbon isotope curve with major excursions is from Cramer and Jarvis (2020, Ch. 11: Carbon isotope stratigraphy, this book), who had compiled numerous studies, including Edwards and Saltzman (2014), Wu et al. (2017), Bauert et al. (2014a), Bergström et al. (2010a,b,c), and Young et al. (2010). The oxygen isotope curve and implied sea-surface temperature trends is the statistical-mean curve by Grossman and Joachimski (2020, Ch. 10: Oxygen isotope stratigraphy, this book) from a synthesis of numerous conodont apatite studies, including Trotter et al. (2008). The vertical scale of this diagram is standardized to match the vertical scales of the period-overview figure in all other Phanerozoic chapters. *FADs*, First appearance datums; *VPDB*, Vienna PeeDee Belemnite δ^{13}C standard; *VSMOW*, Vienna Standard Mean Ocean Water δ^{18}O standard.

is entirely satisfactory for global application. Therefore the Ordovician subcommission decided the best way to correlate between regions was to identify the best fossil-based levels, wherever they are found, and to use these for definition of global chronostratigraphic (and chronologic) units (Webby, 1995, 1998). In this respect, it deviated from the course followed by the Silurian and Devonian subcommissions, both of which recommended

the adoption of preexisting (regional) stage or series schemes for global use. During the early 1990s the Ordovician subcommission established a number of working groups to investigate and recommend levels within the period that were suitable for international correlation, and therefore for defining international stages (Webby, 1995, 1998). Seven general chronostratigraphic levels were certified as primary correlation levels for the seven

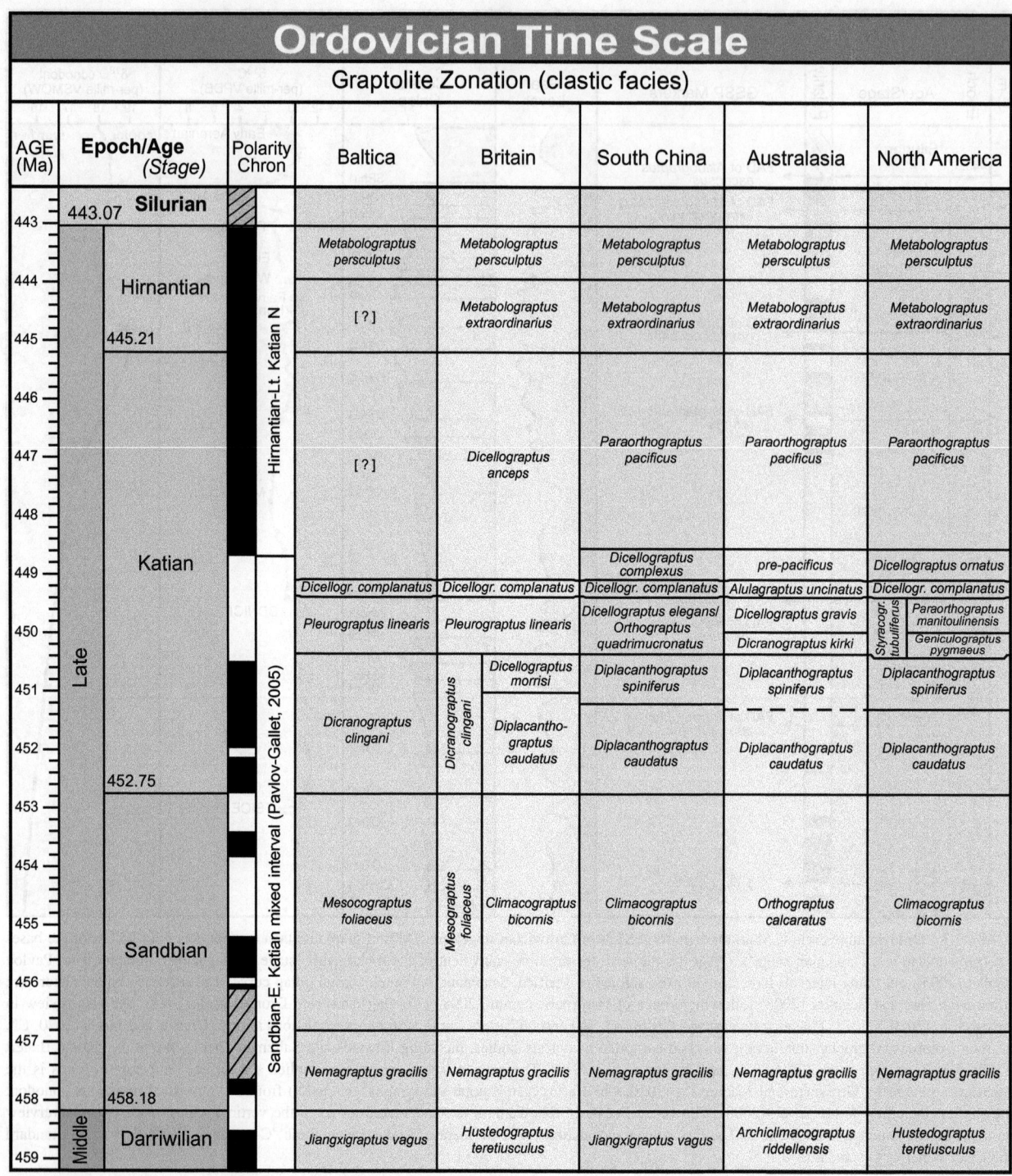

FIGURE 20.2   (A and B) Late Ordovician time scale, schematic geomagnetic polarity pattern, and graptolite, conodont, and chitinozoan zonal schemes. Zonations and correlations for graptolites are modified from Loydell (2011), Maletz and Ahlberg (2018), VandenBerg and Cooper (1992), Cooper and Sadler (2012), Goldman et al. (2007a), and Zalasiewicz et al. (2009); for conodonts modified from Bergström and Wang (1995), Wang et al. (2018), and Zhang et al. (2019); and for chitinozoans from Nõlvak et al. (2006) and Paris et al. (2004). Complete citations for biostratigraphic schemes can be found in text Section 20.3.1.

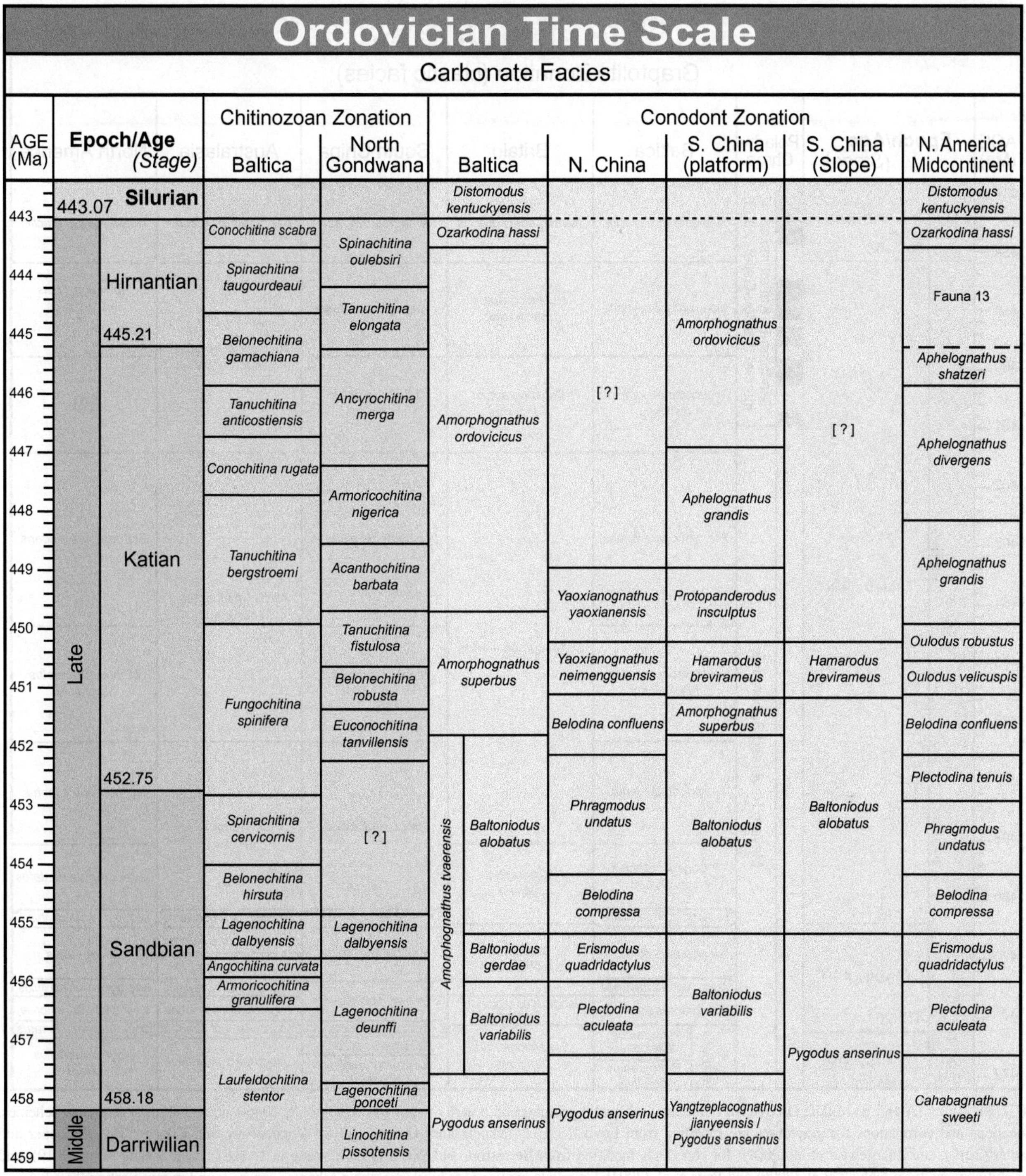
Ordovician Time Scale
Carbonate Facies
AGE (Ma)
Epoch/Age (Stage)
Chitinozoan Zonation
Baltica
North Gondwana
Conodont Zonation
Baltica
N. China
S. China (platform)
S. China (Slope)
N. America Midcontinent
443
444
445
446
447
448
449
450
451
452
453
454
455
456
457
458
459
443.07
Silurian
445.21
452.75
458.18
Hirnantian
Katian
Sandbian
Darriwilian
Late
Middle
Conochitina scabra
Spinachitina taugourdeaui
Belonechitina gamachiana
Tanuchitina anticostiensis
Conochitina rugata
Tanuchitina bergstroemi
Fungochitina spinifera
Spinachitina cervicornis
Belonechitina hirsuta
Lagenochitina dalbyensis
Angochitina curvata
Armoricochitina granulifera
Laufeldochitina stentor
Spinachitina oulebsiri
Tanuchitina elongata
Ancyrochitina merga
Armoricochitina nigerica
Acanthochitina barbata
Tanuchitina fistulosa
Belonechitina robusta
Euconochitina tanvillensis
[?]
Lagenochitina dalbyensis
Lagenochitina deunffi
Lagenochitina ponceti
Linochitina pissotensis
Distomodus kentuckyensis
Ozarkodina hassi
Amorphognathus ordovicicus
Amorphognathus superbus
Amorphognathus tvaerensis
Baltoniodus alobatus
Baltoniodus gerdae
Baltoniodus variabilis
Pygodus anserinus
[?]
Yaoxianognathus yaoxianensis
Yaoxianognathus neimengguensis
Belodina confluens
Phragmodus undatus
Belodina compressa
Erismodus quadridactylus
Plectodina aculeata
Pygodus anserinus
Amorphognathus ordovicicus
Aphelognathus grandis
Protopanderodus insculptus
Hamarodus brevirameus
Amorphognathus superbus
Baltoniodus alobatus
Baltoniodus variabilis
Yangtzeplacognathus jianyeensis / Pygodus anserinus
[?]
Hamarodus brevirameus
Baltoniodus alobatus
Pygodus anserinus
Distomodus kentuckyensis
Ozarkodina hassi
Fauna 13
Aphelognathus shatzeri
Aphelognathus divergens
Aphelognathus grandis
Oulodus robustus
Oulodus velicuspis
Belodina confluens
Plectodina tenuis
Phragmodus undatus
Belodina compressa
Erismodus quadridactylus
Plectodina aculeata
Cahabagnathus sweeti

FIGURE 20.2 (Continued)

international stages. They are based on the first appearance of key graptolite or conodont species. All boundaries have been formally voted on and are defined by a Global Boundary Stratigraphic Section and Point (GSSP). From the bottom upward they are Tremadocian, Floian, Dapingian, Darriwilian, Sandbian, Katian, and Hirnantian (Figs. 20.1 and 20.18).

20.1.1 Stages of the Lower Ordovician

20.1.1.1 The Cambrian–Ordovician Boundary and the Tremadocian Stage

The Global Boundary Stratotype Section and Point (GSSP) for the base of the Tremadocian Stage, the Lower Ordovician

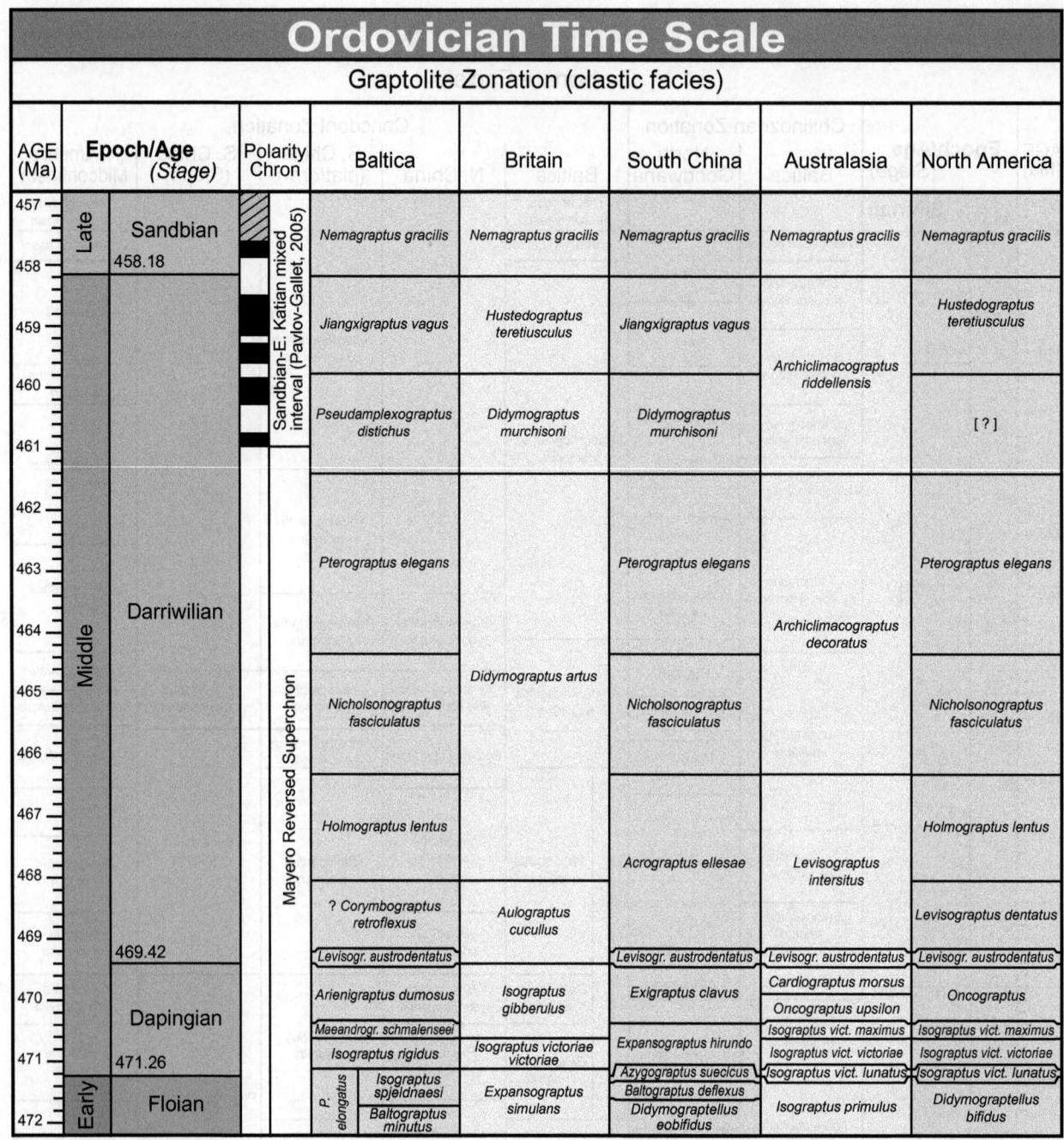

Ordovician Time Scale

Graptolite Zonation (clastic facies)

AGE (Ma)	Epoch/Age (Stage)	Polarity Chron	Baltica	Britain	South China	Australasia	North America
457–458	Late — Sandbian 458.18	Sandbian–E. Katian mixed interval (Pavlov-Gallet, 2005)	Nemagraptus gracilis	Nemagraptus gracilis	Nemagraptus gracilis	Nemagraptus gracilis	Nemagraptus gracilis
459	Middle — Darriwilian		Jiangxigraptus vagus	Hustedograptus teretiusculus	Jiangxigraptus vagus	Archiclimacograptus riddellensis	Hustedograptus teretiusculus
460–461			Pseudamplexograptus distichus	Didymograptus murchisoni	Didymograptus murchisoni		[?]
462–463		Mayero Reversed Superchron	Pterograptus elegans		Pterograptus elegans	Archiclimacograptus decoratus	Pterograptus elegans
464				Didymograptus artus			
465–466			Nicholsonograptus fasciculatus		Nicholsonograptus fasciculatus		Nicholsonograptus fasciculatus
467			Holmograptus lentus		Acrograptus ellesae	Levisograptus intersitus	Holmograptus lentus
468							
469			? Corymbograptus retroflexus	Aulograptus cucullus			Levisograptus dentatus
469.42			Levisogr. austrodentatus		Levisogr. austrodentatus	Levisogr. austrodentatus	Levisogr. austrodentatus
470	Early — Dapingian		Arienigraptus dumosus	Isograptus gibberulus	Exigraptus clavus	Cardiograptus morsus	Oncograptus
						Oncograptus upsilon	
			Maeandrogr. schmalenseei		Expansograptus hirundo	Isograptus vict. maximus	Isograptus vict. maximus
471	471.26		Isograptus rigidus	Isograptus victoriae victoriae		Isograptus vict. victoriae	Isograptus vict. victoriae
					Azygograptus suecicus	Isograptus vict. lunatus	Isograptus vict. lunatus
	Early — Floian	P. elongatus — Isograptus spjeldnaesi	Expansograptus simulans	Baltograptus deflexus	Isograptus primulus	Didymograptellus bifidus
472			Baltograptus minutus		Didymograptellus eobifidus		

FIGURE 20.3 (A and B) Middle Ordovician time scale, schematic geomagnetic polarity pattern, and graptolite, conodont, and chitinozoan zonal schemes. Zonations and correlations for graptolites are modified from Loydell (2011), Maletz and Ahlberg (2018), VandenBerg and Cooper (1992), Cooper and Sadler (2012), and Zalasiewicz et al. (2009); for conodonts modified from Bergström and Wang (1995), Wang et al. (2018), and Zhang et al. (2019); and for chitinozoans from Nõlvak et al. (2006) and Paris et al. (2004). Complete citations for biostratigraphic schemes can be found in text Section 20.3.1.

Series, and the Cambrian−Ordovician Boundary is in the Green Point section of western Newfoundland (Fig. 20.5). The boundary is defined as the level in Bed 23, lower Broom Point Member of the Green Point Formation that is marked by the first appearance of the conodont species *Iapetognathus fluctivagus*, and is 4.8 m below the first appearance of *Rhabdinopora praeparabola* (Bed 25,

Cooper et al., 2001a). The appearance of *I. fluctivagus* and the first planktonic graptolites within a narrow stratigraphic interval at the Green Point section facilitates the use of both conodonts and graptolites in the global correlation of the boundary and allows for the precise correlation of the Cambrian−Ordovician boundary into shale sections devoid of conodonts. Radiolarians of the

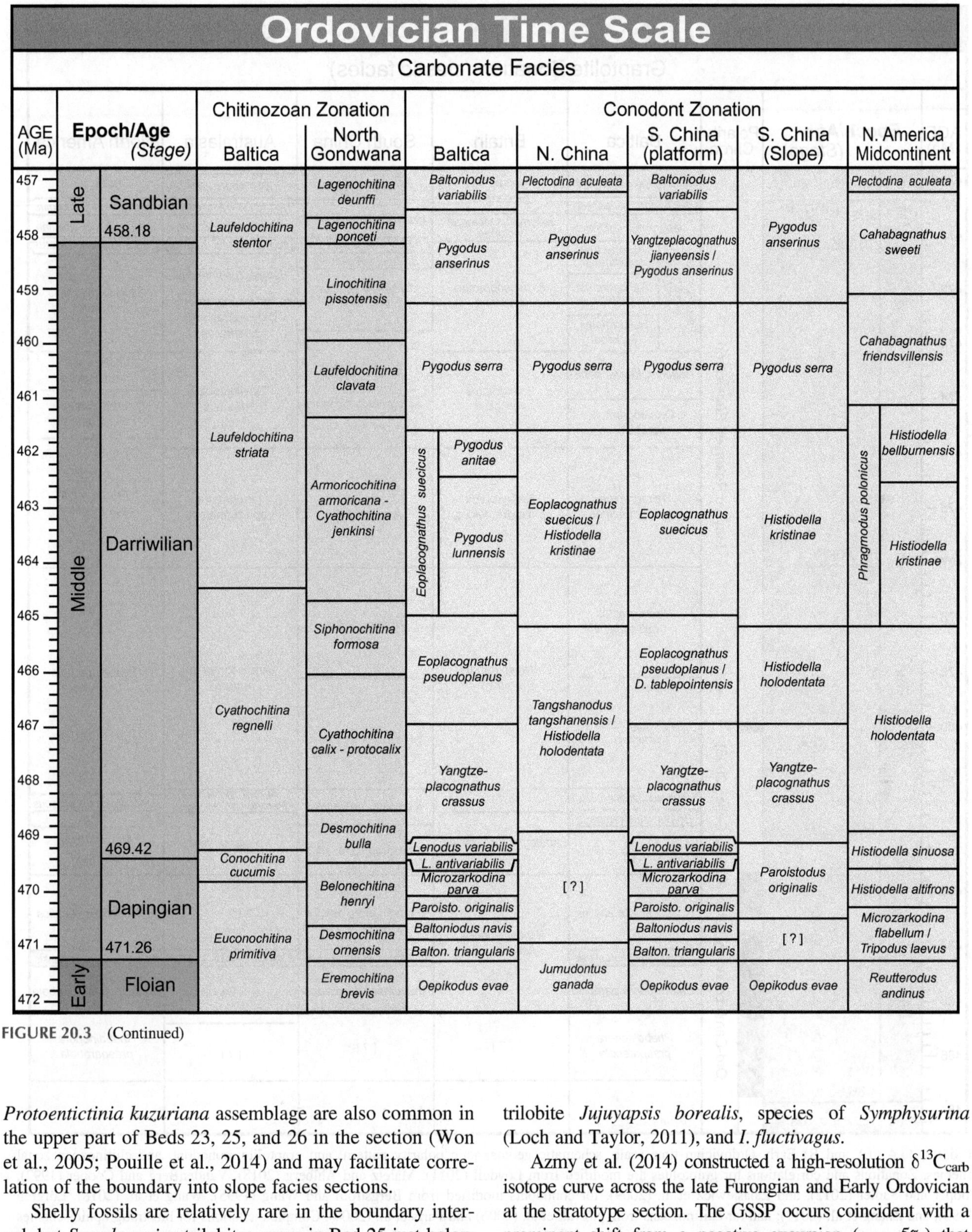

Ordovician Time Scale

Carbonate Facies

AGE (Ma)	Epoch/Age (Stage)		Chitinozoan Zonation		Conodont Zonation				
			Baltica	North Gondwana	Baltica	N. China	S. China (platform)	S. China (Slope)	N. America Midcontinent
457	Late	Sandbian	Laufeldochitina stentor	Lagenochitina deunffi	Baltoniodus variabilis	Plectodina aculeata	Baltoniodus variabilis	Pygodus anserinus	Plectodina aculeata
458		458.18		Lagenochitina ponceti	Pygodus anserinus	Pygodus anserinus	Yangtzeplacognathus jianyeensis / Pygodus anserinus		Cahabagnathus sweeti
459				Linochitina pissotensis					
460		Darriwilian	Laufeldochitina striata	Laufeldochitina clavata	Pygodus serra	Pygodus serra	Pygodus serra	Pygodus serra	Cahabagnathus friendsvillensis
461	Middle								
462				Armoricochitina armoricana - Cyathochitina jenkinsi	*Eoplacognathus suecicus* — Pygodus anitae		Eoplacognathus suecicus / Histiodella kristinae	Eoplacognathus suecicus	Histiodella kristinae	Histiodella bellburnensis
463					Pygodus lunnensis	Eoplacognathus suecicus / Histiodella kristinae			Histiodella kristinae
464									
465			Cyathochitina regnelli	Siphonochitina formosa	Eoplacognathus pseudoplanus		Eoplacognathus pseudoplanus / D. tablepointensis	Histiodella holodentata	*Phragmodus polonicus*
466						Tangshanodus tangshanensis / Histiodella holodentata			Histiodella holodentata
467				Cyathochitina calix - protocalix					
468				Yangtze-placognathus crassus	Yangtze-placognathus crassus		Yangtze-placognathus crassus	Yangtze-placognathus crassus	
469		469.42	Conochitina cucumis	Desmochitina bulla	Lenodus variabilis / L. antivariabilis Microzarkodina parva	[?]	Lenodus variabilis / L. antivariabilis Microzarkodina parva	Paroistodus originalis	Histiodella sinuosa
470		Dapingian		Belonechitina henryi	Paroisto. originalis		Paroisto. originalis		Histiodella altifrons
471		471.26	Euconochitina primitiva	Desmochitina ornensis	Baltoniodus navis / Balton. triangularis		Baltoniodus navis / Balton. triangularis	[?]	Microzarkodina flabellum / Tripodus laevus
472	Early	Floian		Eremochitina brevis	Oepikodus evae	Jumudontus ganada	Oepikodus evae	Oepikodus evae	Reutterodus andinus

FIGURE 20.3 (Continued)

Protoenctictinia kuzuriana assemblage are also common in the upper part of Beds 23, 25, and 26 in the section (Won et al., 2005; Pouille et al., 2014) and may facilitate correlation of the boundary into slope facies sections.

Shelly fossils are relatively rare in the boundary interval, but *Symphysurina* trilobites occur in Bed 25 just below the FAD of *R. praeparabola* (Stouge et al., 2017). These fossils provide a straightforward correlation into the western Laurentian platform sections, which also contain the trilobite *Jujuyapsis borealis*, species of *Symphysurina* (Loch and Taylor, 2011), and *I. fluctivagus*.

Azmy et al. (2014) constructed a high-resolution $\delta^{13}C_{carb}$ isotope curve across the late Furongian and Early Ordovician at the stratotype section. The GSSP occurs coincident with a prominent shift from a negative excursion ($\sim -5\permil$) that begins in Bed 22 below the boundary to a positive one ($\sim +2\permil$) just above the FAD of *I. fluctivagus* in Bed 23 (Azmy et al., 2014). This double inflection also lies just

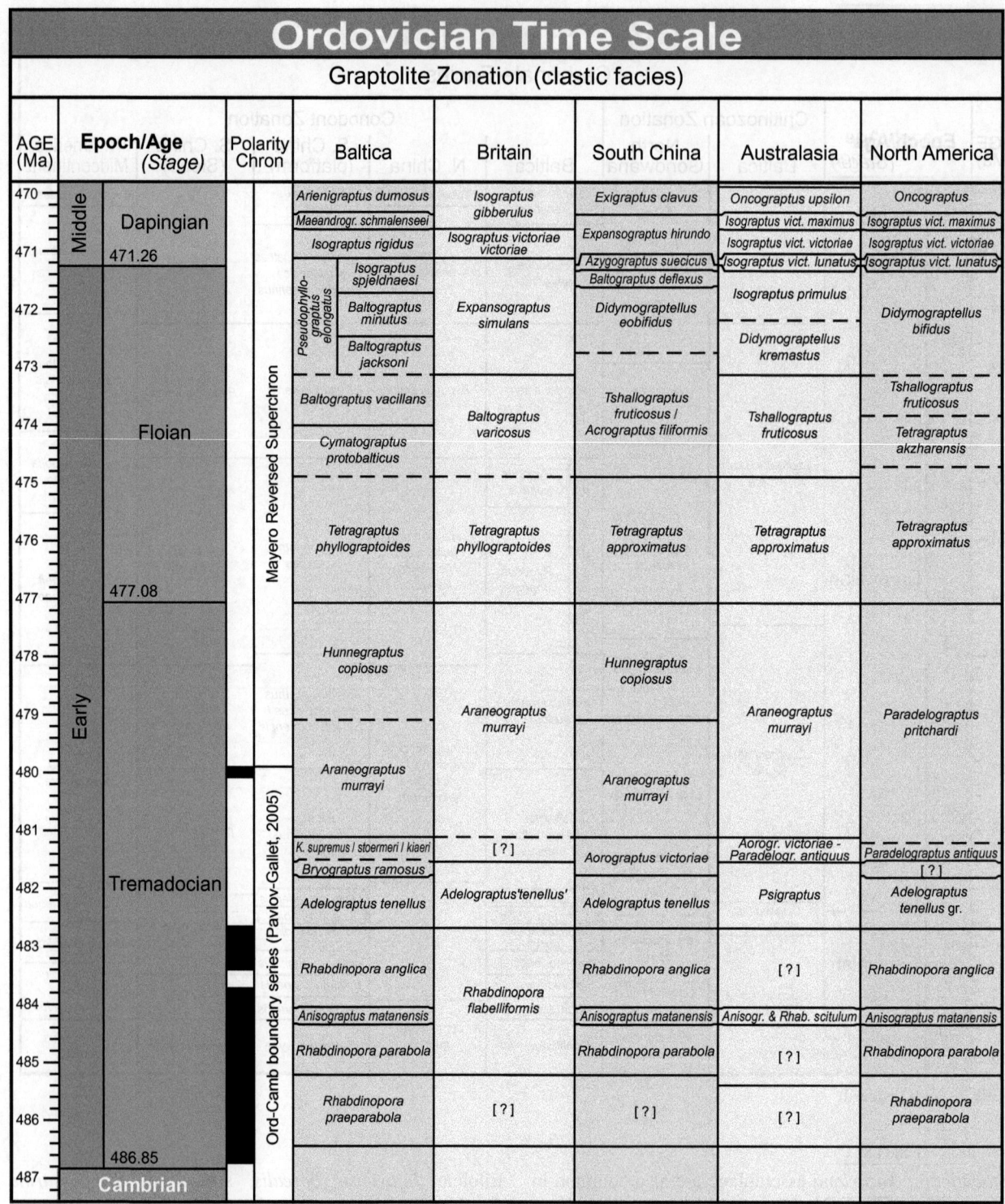

AGE (Ma)	Epoch/Age (Stage)		Polarity Chron		Baltica	Britain	South China	Australasia	North America

Ordovician Time Scale
Graptolite Zonation (clastic facies)

FIGURE 20.4 (A and B) Early Ordovician time scale, schematic geomagnetic polarity pattern, and graptolite, conodont, and chitinozoan zonal schemes. Zonations and correlations for graptolites are modified from Loydell (2011), Maletz and Ahlberg (2018), VandenBerg and Cooper (1992), Cooper and Sadler (2012), and Zalasiewicz et al. (2009); for conodonts modified from Bergström and Wang (1995), Wang et al. (2018), Löfgren (1993), and Zhang et al. (2019); and for chitinozoans from Nõlvak et al. (2006) and Paris et al. (2004). Complete citations for biostratigraphic schemes can be found text Section 20.3.1.

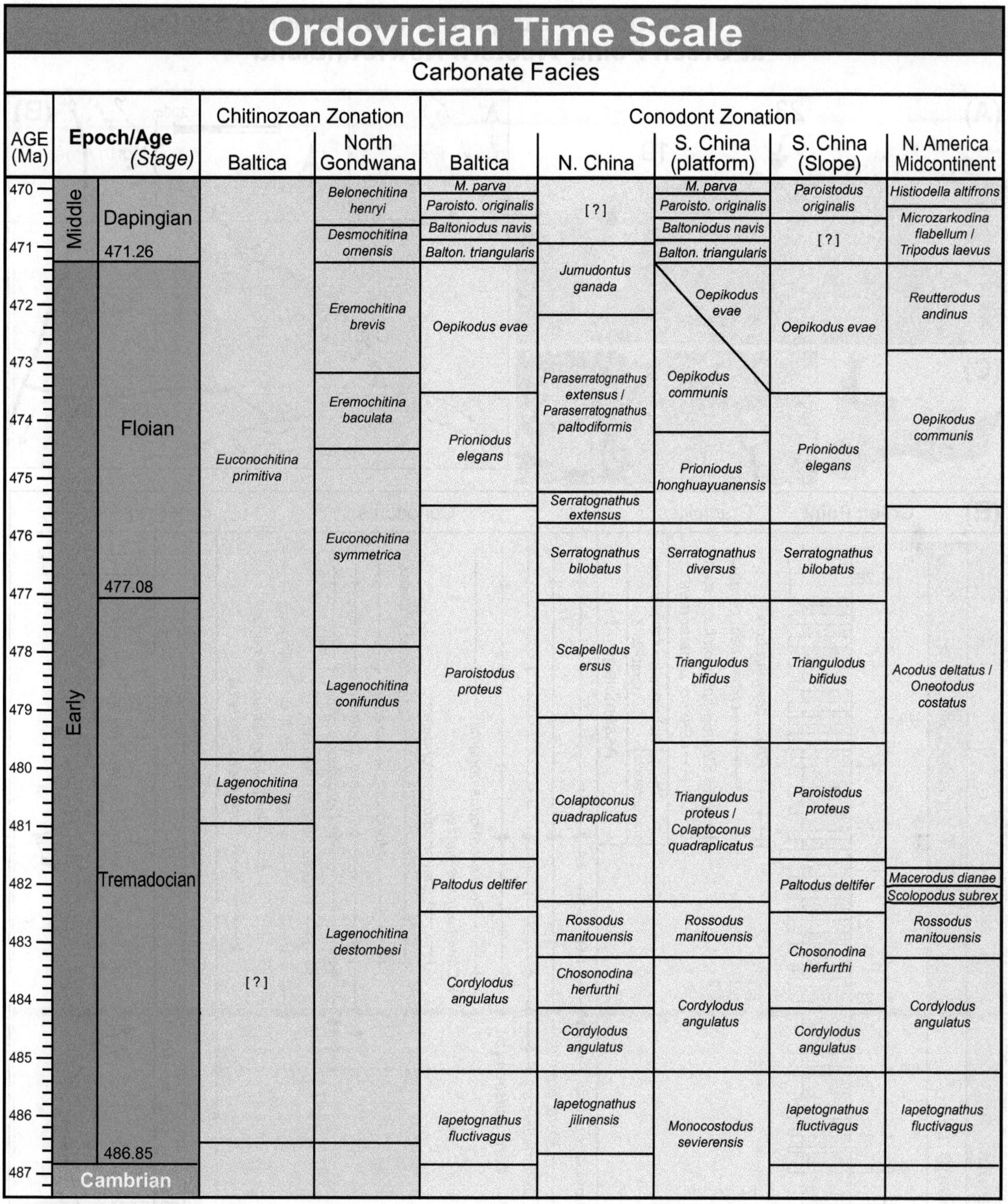

Ordovician Time Scale

Carbonate Facies

AGE (Ma)	Epoch/Age (Stage)		Chitinozoan Zonation		Conodont Zonation				
			Baltica	North Gondwana	Baltica	N. China	S. China (platform)	S. China (Slope)	N. America Midcontinent
470	Middle	Dapingian 471.26	Euconochitina primitiva	Belonechitina henryi	M. parva	[?]	M. parva	Paroistodus originalis	Histiodella altifrons
					Paroisto. originalis		Paroisto. originalis		Microzarkodina flabellum / Tripodus laevus
471				Desmochitina ornensis	Baltoniodus navis		Baltoniodus navis	[?]	
					Balton. triangularis		Balton. triangularis		
472	Early	Floian 477.08		Eremochitina brevis	Oepikodus evae	Jumudontus ganada	Oepikodus evae	Oepikodus evae	Reutterodus andinus
473						Paraserratognathus extensus / Paraserratognathus paltodiformis	Oepikodus communis		
474				Eremochitina baculata	Prioniodus elegans			Prioniodus elegans	Oepikodus communis
475							Prioniodus honghuayuanensis		
476				Euconochitina symmetrica		Serratognathus extensus			
477						Serratognathus bilobatus	Serratognathus diversus	Serratognathus bilobatus	
478				Lagenochitina conifundus	Paroistodus proteus	Scalpellodus ersus	Triangulodus bifidus	Triangulodus bifidus	Acodus deltatus / Oneotodus costatus
479									
480		Tremadocian 486.85	Lagenochitina destombesi			Colaptoconus quadraplicatus	Triangulodus proteus / Colaptoconus quadraplicatus	Paroistodus proteus	
481									
482					Paltodus deltifer			Paltodus deltifer	Macerodus dianae / Scolopodus subrex
483			[?]	Lagenochitina destombesi		Rossodus manitouensis	Rossodus manitouensis	Chosonodina herfurthi	Rossodus manitouensis
484					Cordylodus angulatus	Chosonodina herfurthi	Cordylodus angulatus		Cordylodus angulatus
485						Cordylodus angulatus		Cordylodus angulatus	
486					Iapetognathus fluctivagus	Iapetognathus jilinensis	Monocostodus sevierensis	Iapetognathus fluctivagus	Iapetognathus fluctivagus
487	Cambrian								

FIGURE 20.4 (Continued)

above the well-documented uppermost Cambrian positive $\delta^{13}C_{carb}$ excursion in Bed 22 ($= Hirsutodontus simplex$ spike). The distinctive shifts in $\delta^{13}C_{carb}$ isotope values should aid in the correlation of the GSSP to other Cambrian−Ordovician boundary sections.

Unfortunately, the placement of the Cambrian−Ordovician boundary at the FAD of *I. fluctivagus* in the Green Point section is not without controversy. Terfelt et al. (2012) have questioned the taxonomic identification of the specimens of *I. fluctivagus* in Bed 23 at Green Point suggesting

Base of the Tremadocian Stage of the Ordovician System at Green Point, Western Newfoundland

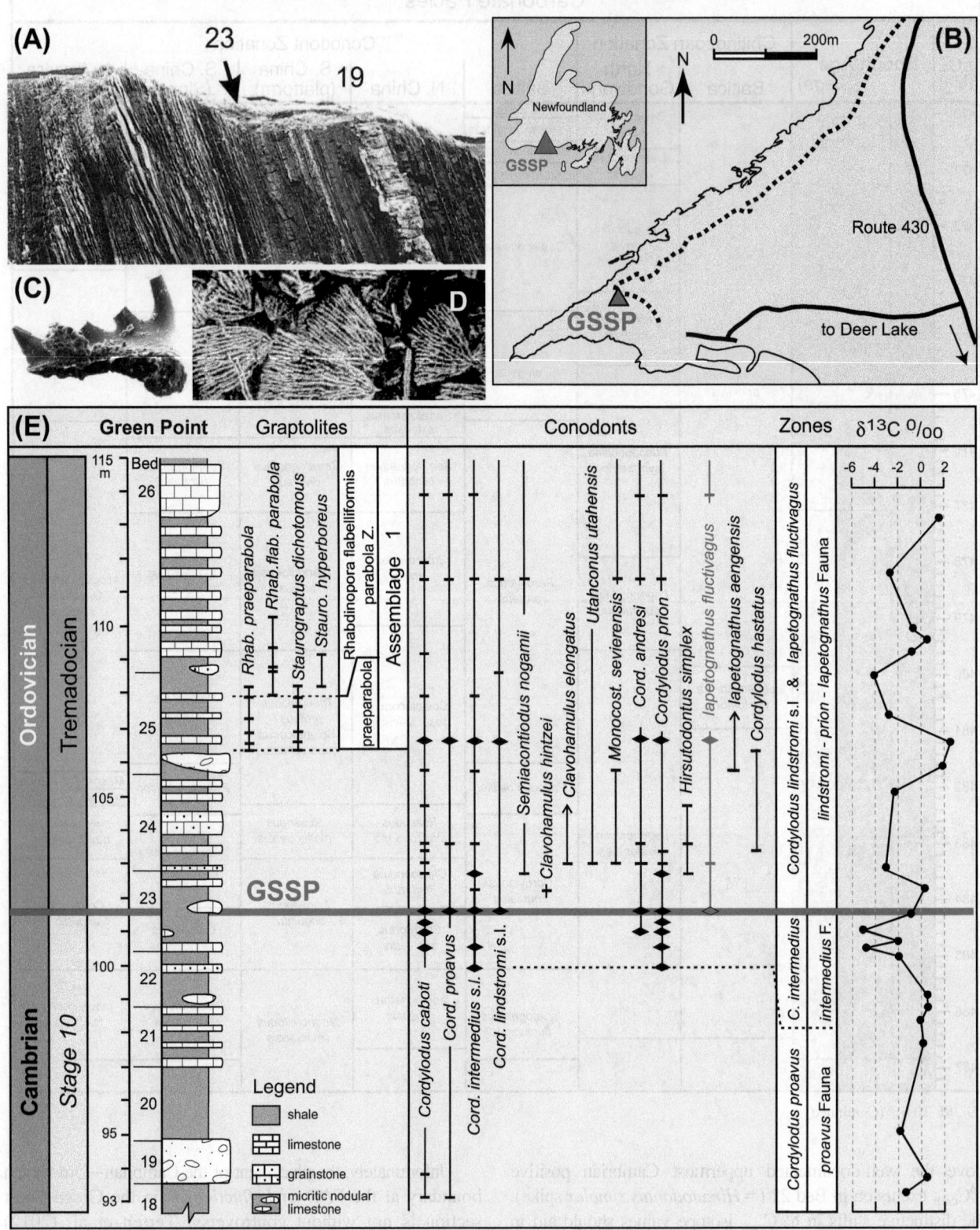

instead that they belong to *Iapetognathus praeangensis*, and that Bed 23 is well above the FAD of that latter taxon at Green Point. Reevaluation of the conclusions of Terfelt et al. (2012) regarding the taxonomy of *Iapetognathus*, and the level at which the various species of *Iapetognathus* appear is ongoing (Stouge et al., 2017).

The upper and lower boundaries of the Tremadocian closely match those of the British Tremadoc Series (Rushton, 1982), and the name was approved by the ICS in 1999 (Cooper and Sadler, 2012). The Tremadocian Stage includes the evolution of planktonic graptolites, which become a major component of marine macroplankton and represent the first abundant and widespread fossil record of the Earth's zooplankton.

The Tremadocian base is one of two Ordovician stage boundaries recognized by a conodont first appearance datum. The early Tremadocian contains two widespread conodont zones (Fig. 20.4B): the zones of *I. fluctivagus* and *Cordylodus angulatus*, which correlate with the graptolite zones of *Rhabdinopora praeparabola* through *R. anglica*. The middle and late Tremadocian comprises the zones of *Paltodus deltifer* and *Paroistodus proteus* (lower) in Baltica, and the zones of *Rossodus manitouensis*, *Macerodus dianae*, and *Acodus deltatus* in Laurentia. This interval comprises the *Adelograptus* "*tenellus*," *Araneograptus murrayi*, and *Hunnegraptus copiosus* graptolite zones in shale successions. Recent work by Wang et al. (2018) has provided important new zonations for North and South China (Fig. 20.4B).

The graptolite fauna of the early Tremadocian is dominated by species of *Rhabdinopora*, and other members of the Anisograptidae (Cooper et al., 1998; Erdtmann, 1988). In the middle and late Tremadocian other anisograptids, particularly *Paradelograptus*, *Paratemnograptus*, *Kiaerograptus*, *Aorograptus*, *Araneograptus*, *Hunnegraptus*, and *Clonograptus* become abundant (Maletz, 1999; Maletz and Egenhoff, 2001). Cooper (1999a) recognized nine global graptolite chronozones in upward sequence (Fig. 20.4A), the zones of *R. praeparabola*, *Rhabdinopora flabelliformis parabola*, *Anisograptus matanensis*, *R. flabelliformis anglica*, *Adelograptus*, *Paradelograptus antiquus*, *Kiaerograptus*, *Araneograptus murrayi*, and *Hunnegraptus copiosus* (see also Maletz, 1999).

Correlations between conodont and graptolite zones are facilitated by the conodont- and graptolite-bearing sections of western Newfoundland (e.g., Cooper et al., 2001a,b), southwest Sweden (Löfgren, 1993, 1996), the eastern

Cordillera of Argentina (e.g., Ortega and Albanesi, 2005; Zeballo et al., 2008), and Dayangcha, China (Chen, 1986, Chen et al., 1988; Zhang and Erdtmann, 2004). Integrated graptolite and conodont biostratigraphy together with recent studies in stable isotope geochemistry (e.g., Azmy et al., 2014) provide a global correlation framework of high precision for the Tremadocian Stage.

20.1.1.2 The Floian Stage

The Floian Stage is named for the village of Flo, which lies about 80 km NNE from Göteborg and 12 km ESE from Vänersborg, in the province of Västergötland, southern Sweden (Fig. 20.6). The base of the Floian Stage, the second stage in the Ordovician, is defined at the stratigraphic level of the first appearance of the graptolite *Tetragraptus approximatus* in a section of lower Tøyen Shale exposed in the Diabasbrottet quarry on the northwestern slope of Mount Hunneberg (Maletz et al., 1996; Bergström et al., 2004) adjacent to Lake Vänern, about 5 km northwest of Flo. This biostratigraphic level can be widely recognized throughout low- to middle-paleolatitude regions where it commonly occurs with *Tetragraptus phyllograptoides*. The latter taxon, *T. phyllograptoides*, also debuts in the GSSP horizon at Diabasbrottet and is well known from high-latitude localities (e.g., Britain). Hence, a precise correlation of the boundary level can be traced through graptolite successions around the world. The GSSP was ratified by the ICS in 2002 and the stage named in 2005. This biostratigraphic level was also adopted for the base of the revised British Arenig Series (Fortey et al., 1995).

The Tøyen Shale at the Diabasbrottet quarry also contains a rich conodont fauna (Löfgren, 1993; Maletz et al., 1996). The GSSP horizon occurs just above the base of the conodont subzone of *Oelandodus elongatus/Acodus deltatus*, the highest subzone of the *P. proteus* Zone in the Baltic (mid-paleolatitude) sequence (Bergström et al., 2004). It lies at, or very close to, the base of the *A. deltatus/O. costatus* Zone of the North American midcontinent (low paleolatitude) succession, and the *Serratognathus bilobatus* Zone in North China (Wang et al., 2018) and the Canning Basin of Australia (Zhen et al., 2017). The boundary lies within the *Megistaspis* (*Paramegistaspis*) *planilimbata* trilobite zone.

Bergström et al. (2019a) published the first $\delta^{13}C$ chemostratigraphy from the Floian GSSP at Diabasbrottet. The Diabasbrottet $\delta^{13}C_{org}$ curve (Fig. 20.6) does not exhibit any

---

FIGURE 20.5 GSSP for base of the Ordovician (base of Tremadocian Stage) at the Green Point section, Newfoundland, eastern Canada. Strata in the cliff exposure and extensive shore platform (A) are overturned; therefore the sequence becomes younger from right to left; and the prominent massive limestone conglomerate bed at right is Bed 19 at the bottom of stratigraphic column (E). The GSSP level at Bed 23 in the section was intended to coincide with the lowest occurrence of conodont *Iapetognathus fluctivagus* (Nicoll et al., 1992). Conodont *I. fluctivagus* (C) is 0.5 mm long. The GSSP is 4.8 m below the lowest appearance of planktonic graptolites, and the zonal graptolite taxa *Rhabdinopora praeparabola* and *Rhabdinopora parabola* [bottom specimen in (D) is 17 mm long; from Cooper and Sadler, 2001a]. $\delta^{13}C_{carb}$ isotope curve is from Azmy et al. (2014). Stratigraphic column and ranges of taxa are modified from the original GSSP publication (Cooper and Sadler, 2001a), images of fossils are from Cooper and Sadler (2012), and the outcrop photograph is by S.H. Williams.

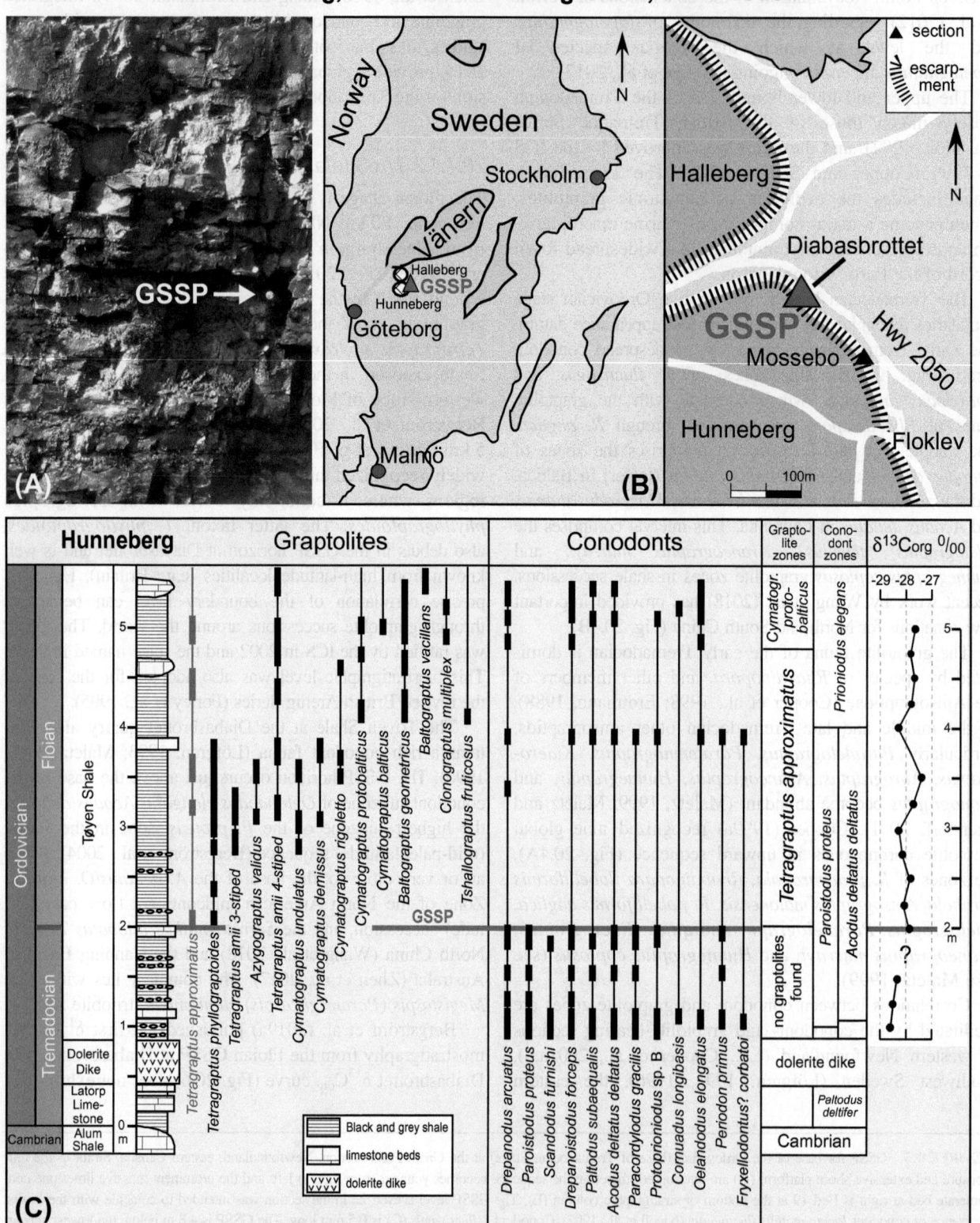

FIGURE 20.6 GSSP for base of the Floian Stage in the Diabasbrottet section, Mt Hunneberg, Sweden, with stratigraphic ranges of graptolites and conodonts. The boundary point coincides with the lowest appearance of graptolite *Tetragraptus approximatus* (*arrow*) in this section. After Bergström et al. (2004: Figure 7). Organic carbon isotope curve is from Bergström et al. (2019a).

conspicuous excursions with most values between -28.5 and $-27.5‰$. Due to the lack of any prominent excursion and the few $\delta^{13}C_{org}$ studies through this interval, carbon isotope chemostratigraphy does not currently facilitate or expand the global correlation of the base of the Floian.

Cooper et al. (2014) noted that a sustained period of moderately elevated graptolite origination rates in the Floian resulted in a significant increase in the diversity and abundance of planktonic graptolites. Although anisograptid graptolites become rare, the exponential radiation of the Dichograptidae and Sinograpta drives species richness to one of its highest levels in the Ordovician.

20.1.2 Stages of the Middle Ordovician

20.1.2.1 The Dapingian Stage

The base of the Dapingian Stage and the Middle Ordovician Series is defined at the stratigraphic level of the first appearance datum of the conodont *Baltoniodus triangularis*, 10.57 m above the base of the Dawan Formation at the base of Bed SHOD-16 in the Huanghuachang Section, 22 km NE of Yichang City, Hubei Province, South China (Fig. 20.7) (Wang et al., 2009). This level is 0.2 m below the FAD of the conodont *Microzarkodina flabellum*, is close to the boundary between the lower and upper *Azygograptus suecicus* graptolite zone (as defined by the first appearance of *Azygograptus ellesi*), and nearly coincides with the base of the *Belonechitina* cf. *henryi* chitinozoan biozone. The graptolite fauna from the Huanghuachang section indicates that the GSSP level is close to the first appearance of the biostratigraphically important graptolite, *Isograptus victoriae victoriae*. It also lies within the *Ampullulae—Barakella felix* acritarch assemblage zone.

Carbon isotope chemostratigraphy of the Dapingian GSSP at Huanghuachang has been discussed by Wang et al. (2009). A steadily increasing trend in the $\delta^{13}C_{carb}$ isotope curve occurs through most of the Lower Dawan Formation (Lower Ordovician *Oepikodus evae* conodont zone). The $\delta^{13}C_{carb}$ curve begins with a value of approximately 0‰ in the lowest Dawan Formation and then follows a steady rise of $\delta^{13}C_{carb}$ values until the maximum value of $+0.68‰$, which occurs approximately 1.4 m above the GSSP horizon in Bed SHod21, and 1 m below the base of the middle Dawan Formation (Wang et al., 2009).

Several important speciation events occur in the *Baltoniodus*, *Gothodus*, *Microzarkodina*, and *Periodon* conodont lineages across the Lower and Middle Ordovician Series boundary. This succession of new taxa facilitates the correlation of the GSSP into conodont successions where *B. triangularis* is uncommon. The Dapingian Stage also includes the evolutionary radiation of the well-known graptolite genus *Isograptus* and closely related genera, particularly the *I. victoriae* and *Parisograptus caduceus* groups, providing fine zonal

subdivision and correlation. Two Australasian stages and five zones are represented, which from oldest to youngest are, the zones of *I. v. victoriae*, *I. victoriae maximus*, *I. victoriae maximodivergens* (Castlemainian Stage), and *Oncograptus upsilon* and *Cardiograptus morsus* (Yapeenian Stage).

Dapingian graptolites and conodonts also occur together in the Cow Head Group of western Newfoundland (Williams and Stevens, 1988; Stouge, 2012), and the Trail Creek region of Idaho, western United States (Goldman et al., 2007b). Hence, the base of the Dapingian Stage and the Middle Ordovician can be easily recognized and globally correlated with high precision in both relatively shallow-water carbonate facies and in deep-water graptolite facies. The Dapingian Stage was ratified by the ICS in 2007.

20.1.2.2 The Darriwilian Stage

The GSSP for the base of the Darriwilian Stage is defined at the first appearance of the graptolite species *Levisograptus austrodentatus* at the base of Bed AEP 184, 22 m below the top of the Ningkuo Formation at the Huangnitang section, near Changshan, Zhejiang Province, south-east China (Mitchell et al., 1997). This level lies within a well-controlled succession of graptolite first appearances, including *Levisograptus sinodentatus*, *Arienigraptus zhejianensis*, *Undulograptus formosus*, and *Undulograptus primus* (Fig. 20.8). The base of the Darriwilian Stage correlates with the *Aulograptus cucullus* graptolite zone in high-latitude successions (Britain).

The Ningkuo Formation at Huangnitang is dominantly clastic, but limestone interbeds yield diagnostic conodonts (Mitchell et al., 1997). Samples from below the GSSP level (beginning with sample AEP 167) are referable to the *Paroistodus originalis* Zone, and samples from Bed AEP 250, approximately 13 m above the GSSP contain *Yangtzeplacognathus crassus*. By interpolation the stage base lies close to the base of the conodont zone of *Lenodus antivariabilis*, which marks the appearance of *L. antivariabilis* and *Baltoniodus norrlandicus* in the Baltoscandian succession. It lies slightly above the appearance of the zone index *Histiodella sinuosa* in the North American (midcontinent) conodont succession (Chen and Bergström, 1995). In the Argentine Precordillera and Laurentia species of the evolving conodont genus *Histiodella* provide a biostratigraphically useful set of biozones that span the lower and middle Darriwilian. From oldest to youngest these are the *H. sinuosa*, *Histiodella holodentata*, *Histiodella kristinae*, and *Histiodella bellburnensis* zones (Stouge, 2012; Serra et al, 2019; Fig. 20.3B, herein).

The boundary between the Dapingian and Darriwilian stages marks a major faunal turnover in graptoloids. A fauna previously dominated by dichograptids and isograptids is replaced by one dominated by diplograptids and glossograptids. The rapid evolutionary radiation of the

Base of the Dapingian Stage of the Ordovician System at Huanghuachang, near Yichang City, Hubei Province, China

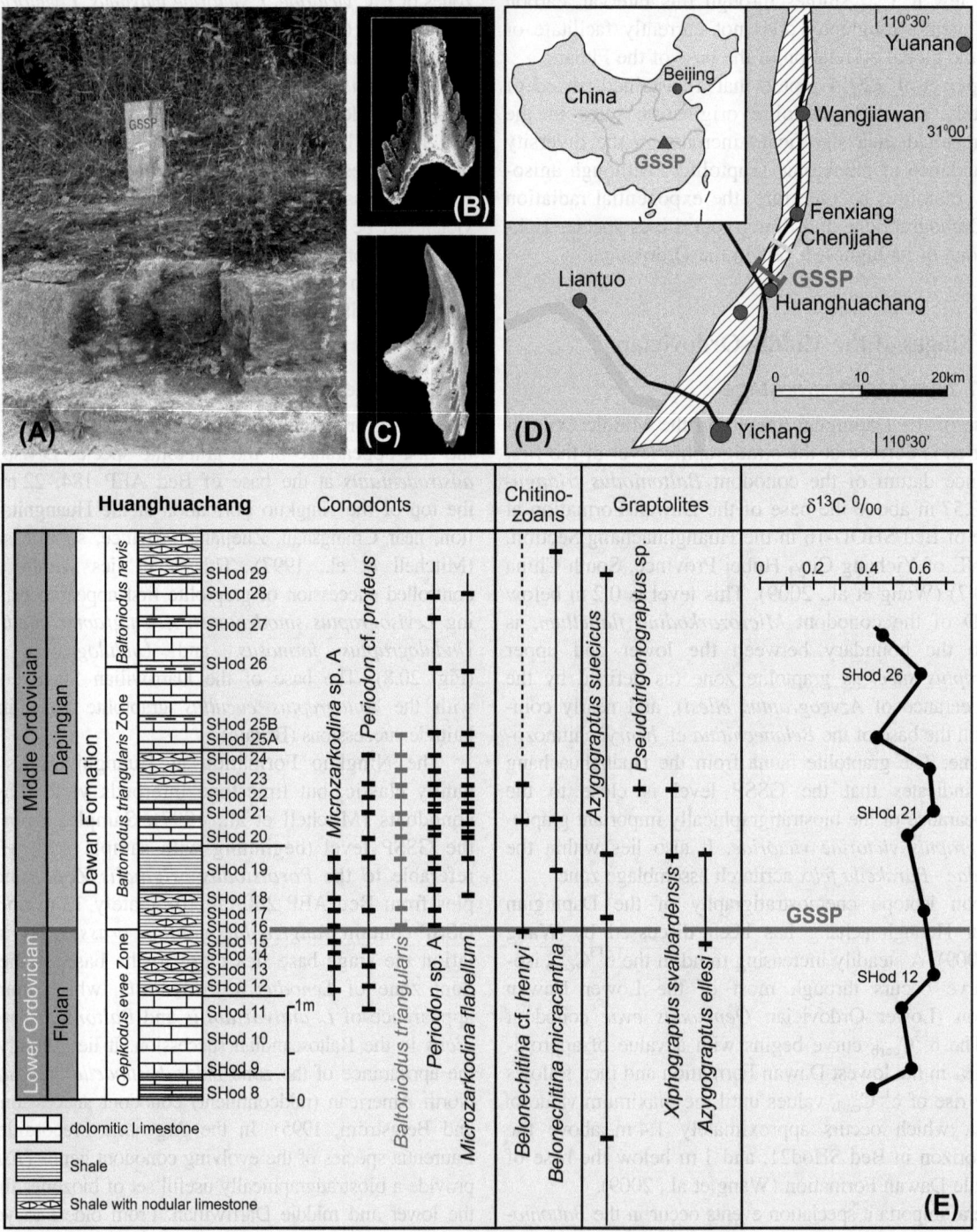

FIGURE 20.7   GSSP for base of the Middle Ordovician (base of Dapingian Stage) at the Huanghuachang section, Yichang, Hubei, South China, showing stratigraphic ranges of conodonts, chitinozoans and graptolites, and the δ¹³Ccarb isotope curve. The GSSP level coincides with the lowest occurrence of the conodont *Baltoniodus triangularis*. Images of *B. triangularis*: (C) Pa element, lateral view, specimen is 0.23 mm in height; (B) Sa element, posterior view, specimen is 0.16 mm in height. Modified from Cooper and Sadler (2012) and Wang et al. (2005: Figure 4).

Base of the Darriwilian Stage of the Ordovician System at Huangnitang, Changshan County, Zhejiang Province, southeast China

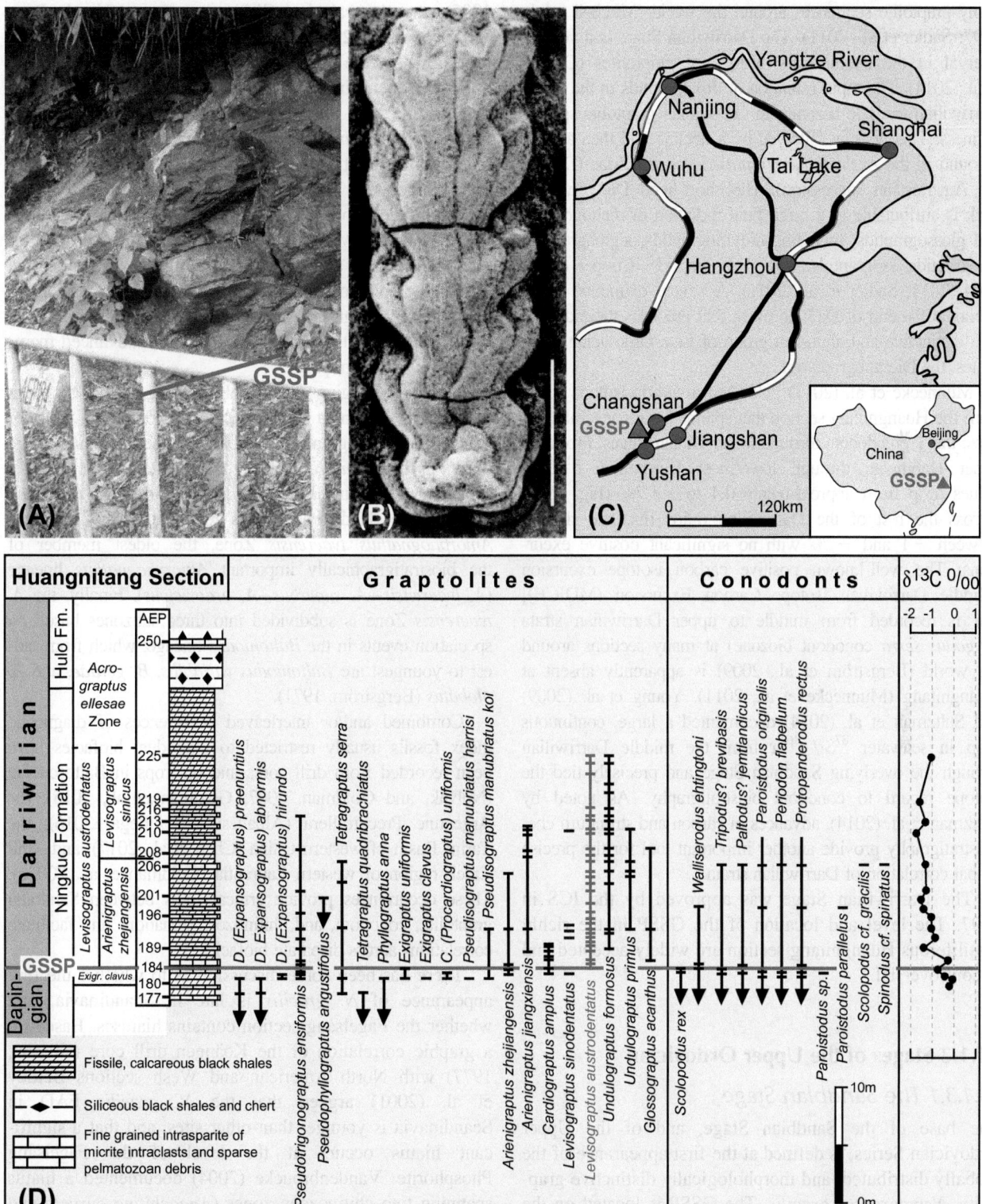

FIGURE 20.8 GSSP for base of the Darriwilian Stage at Huangnitang, Changshan, China, showing stratigraphic ranges of graptolites and conodonts. *Levisograptus austrodentatus* and adjacent zones are from Mitchell et al. (1997: Figure 5) and δ13Ccarb isotope curve is from Munnecke et al. (2011). (B) Scanning electron micrograph of three-dimensionally preserved, pyritized graptolite *Levisograptus austrodentatus* from the Ningkuo Formation, internal mold. Its first appearance coincides with the GSSP. Scale bar 1 mm. Figure supplied by Zhang Yuandong.

Diplograptacea along with the appearance of several distinctive pseudisograptid and glossograptid species represents a distinctive faunal transition that is well documented from many graptolite sequences around the world (Mitchell et al., 1997; Sadler et al., 2011). The Darriwilian Stage is a volatile interval in the evolutionary history of graptolites (Cooper et al., 2014). The rapid radiation of diplograptids in the lower Darriwilian leads to a zenith in Ordovician graptolite species richness. This peak is followed by a decline and then a short rebound in the *Archiclimacograptus decoratus* Zone (Da3 of the Australasian succession). The short-lived Da3 diversity peak is attributable to a brief, final radiation of dichograptids and glossograptids, including didymograptids, sigmagraptids, sinograptids, isograptids, and glossograptids (Cooper et al., 2004, 2014; Sadler et al., 2011). A severe extinction event occurs at the end of Da3, an event that precedes the radiation of a distinctive and abundant group of Late Ordovician graptolites, the Dicranograptacea.

Munnecke et al. (2011) published a carbon isotope curve from the Huangnitang section that spanned, with several large gaps, the Tremadocian through lower Katian stages. From the upper Dapingian through lowermost Darriwilian $\delta^{13}C_{carb}$ values drop from approximately 0.4 to $-1.7‰$ (Fig. 20.8). Across the rest of the Darriwilian ratios fluctuate slightly between -1 and $-2‰$ with no significant positive excursions. The well-known positive carbon isotope excursion [Middle Darriwilian Isotope Carbon Excursion (MDICE)] that is recorded from middle to upper Darriwilian strata (*Pygodus serra* conodont biozone) at many sections around the world (Bergström et al., 2009) is apparently absent at Huangnitang (Munnecke et al., 2011). Young et al. (2009) and Saltzman et al. (2014) documented a large, continuous drop in seawater $^{87}Sr/^{86}Sr$ from the middle Darriwilian through the overlying Sandbian Stage and precisely tied the isotope record to conodont biostratigraphy. As noted by Saltzman et al. (2014), advances in carbon and strontium chemostratigraphy provide another important tool for the precise global correlation of Darriwilian strata.

The Darriwilian Stage was approved by the ICS in 1997. The level and location of the GSSP in the richly fossiliferous Huangnitang section are widely accepted and uncontroversial.

20.1.3 Stages of the Upper Ordovician

20.1.3.1 The Sandbian Stage

The base of the Sandbian Stage, and of the Upper Ordovician Series, is defined at the first appearance of the globally distributed, and morphologically distinctive graptolite, *Nemagraptus gracilis*. The GSSP is located on the south bank of Sularp Brook, in outcrop E14b, at the locality known as Fågelsång in the Province of Scania, Sweden, 1.4 m below the Fågelsång Phosphorite marker bed in the *Dicellograptus* Shale (Fig. 20.9) (Bergström et al., 2000). The FAD of *N. gracilis* is also used to set the base of the British Caradoc Series (Fortey et al., 1995), the Australasian Gisbornian Stage (VandenBerg and Cooper, 1992), and the Chinese Hanjiang Series. The base lies near the middle of the *Pygodus anserinus* conodont zone (near the base of the *Amorphognathus inaequalis* Subzone) and also corresponds to the middle part of the *Laufeldochitina stentor* chitinozoan zone (approximately coincident with the base of the *Eisenackitina rhenana* Subzone). The stage name is derived from the community of South Sandby, where the GSSP lies. The GSSP was approved by the ICS in 2002 and the name in 2005 (Nõlvak et al., 2006). Following an abrupt decline in the later part of the *Darriwilian*, graptolites expanded in diversity through the middle Sandbian, an increase driven by a radiation of the Dicranograptidae and advanced members of the Diplograptina.

Species belonging to several evolving Baltic conodont lineages are important zonal indices and correlation tools for Sandbian biostratigraphy. The base of the Sandbian Stage occurs in the *P. anserinus* Zone, the youngest of the member of the *Pygodus* lineage (*P. anitae−P. serra−P. aserinus*). Most of the Sandbian Stage is referable to the overlying *Amorphognathus tvaerensis* Zone, the oldest member of the biostratigraphically important *Amorphognathus* lineage (*A. tvaerensis−A. superbus−A. ordovicicus*). Finally, the *A. tvaerensis* Zone is subdivided into three subzones based on speciation events in the *Baltoniodus* lineage, which from oldest to youngest are *Baltoniodus variabilis*, *B. gerdae*, and *B. alobatus* (Bergström, 1971).

Combined and/or interleaved occurrences of diagnostic index fossils usually restricted to individual biofacies have been recorded from drill cores and outcrops in Baltoscandia (Nõlvak, and Goldman, 2007; Goldman et al., 2015), the Argentine Precordillera (Albanesi and Ortega, 2016), the Tarim Basin of western China (Chen et al., 2017), and Trail Creek region of western Laurentia (Goldman et al., 2007b). These occurrences provide critical links between Sandbian graptolite, conodont, and chitinozoan zonations and facilitate correlations across disparate biofacies.

There has been some discussion as to whether the first appearance of *N. gracilis* is late in Scandinavia, and whether the Fågelsång section contains hiatuses. Based on a graphic correlation of the Koängen drill core (Nilsson, 1977) with North American and Wesh sections Bettley et al. (2001) argued that the *N. gracilis* FAD in Scandinavia is younger than other sites, and that a significant hiatus occurs at the level of the Fågelsång Phosphorite. Vandenbroucke (2004) documented a hiatus spanning two chitinozoan zones (*Angochitina curvata* and *Armoricochitina granulifera*) at the level of the Fågelsång Phosphorite. Further work using quantitative correlation and multiple fossil groups is needed to resolve these issues.

Base of the Sandbian Stage of the Ordovician System at Fågelsång, Southern Sweden

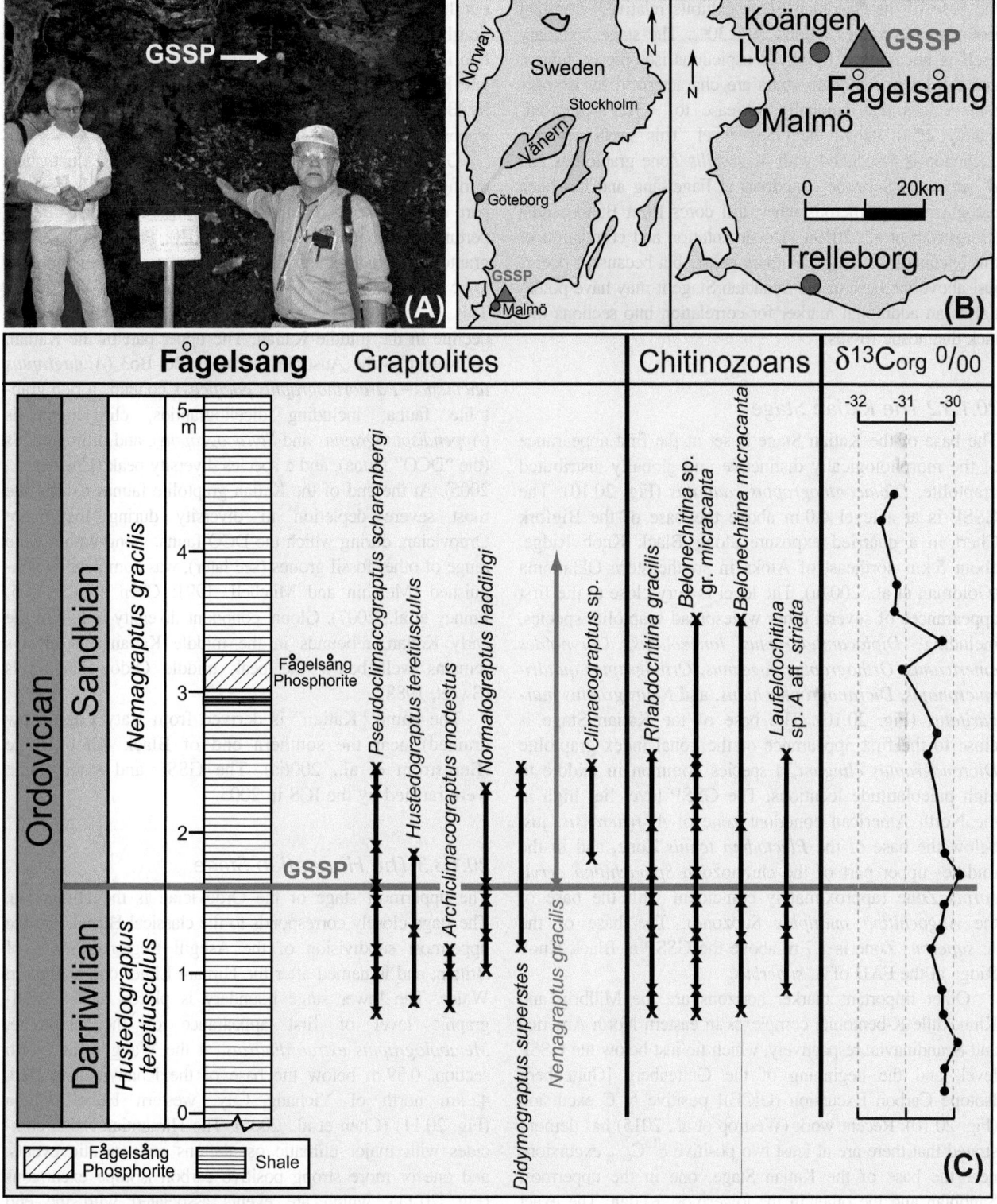

FIGURE 20.9   GSSP for base of the Upper Ordovician (base of Sandbian Stage) at the E14b outcrop, Fågelsång, Sweden with ranges of graptolites and chitinozoans from Bergström et al. (2000: Figure 5). The GSSP level coincides with the lowest occurrence of graptolite *Nemagraptus gracilis*. Position of samples are marked with *x* along the range bars. Photograph and stratigraphic column are from Cooper and Sadler (2012) and organic carbon isotope curve is modified from Bergström et al. (2019a).

Bergström et al. (2019a) provided a $\delta^{13}C_{org}$ curve across the Middle and Upper Ordovician boundary at the Fågelsång GSSP. The lower part of the Fågelsång $\delta^{13}C_{org}$ curve across the base of the Sandbian Stage exhibits relatively constant isotopic ratio values around $\approx -30\text{‰}$. The stage boundary itself is not marked by any conspicuous isotopic excursion, and the lower Sandbian strata are characterized by isotopic ratio values that gradually decrease to -31.5‰ approximately 2.5 m above the GSSP level. This small negative excursion is associated with *N. gracilis* Zone graptolites and *B. variabilis* Subzone conodonts at Fågelsång and has been recognized in numerous other drill cores from Baltoscandia (Bergström et al., 2019a). Documentation and correlation of this excursion is in a preliminary phase, but because it occurs just above the base of the Sandbian Stage it may have potential as an additional marker for correlation into sections that lack diagnostic fossils.

20.1.3.2 The Katian Stage

The base of the Katian Stage is set at the first appearance of the morphologically distinctive and globally distributed graptolite, *Diplacanthograptus caudatus* (Fig. 20.10). The GSSP is at a level 4.0 m above the base of the Bigfork Chert in a quarried exposure along Black Knob Ridge, about 5 km northeast of Atoka in southeastern Oklahoma (Goldman et al., 2007a). The level is very close to the first appearances of several other widespread graptolite species, including, *Diplacanthograptus lanceolatus*, *Corynoides americanus*, *Orthograptus pageanus*, *Orthograptus quadrimucronatus*, *Dicranograptus hians*, and *Neurograptus margaritatus* (Fig. 20.10). The base of the Katian Stage is close to the first appearance of the zonal index graptolite *Dicranograptus clingani*, a species common in middle to high paleolatitude locations. The GSSP level lies high in the North American conodont zone of *A. tvaerensis*, just below the base of the *Plectodina tenuis* Zone, and in the middle—upper part of the chitinozoan *Spinachitina cervicornis* Zone (approximately coincident with the base of the *Angochitina multiplex* Subzone). The base of the *A. superbus* Zone is 1.7 m above the GSSP at Black Knob Ridge at the FAD of *A. superbus*.

Other important marker horizons are the Millbrig and Kinnekulle K-bentonite complexes in eastern North America and Scandinavia, respectively, which lie just below the GSSP level, and the beginning of the Guttenberg [Guttenberg Isotope Carbon Excursion (GICE)] positive $\delta^{13}C$ excursion (Fig. 20.10). Recent work (Westrop et al., 2015) has demonstrated that there are at least two positive $\delta^{13}C_{carb}$ excursions near the base of the Katian Stage, one in the uppermost Sandbian and the other in the lowermost Katian. The exact age (Sandbian or Katian) of the type excursion from the Guttenberg Member of the Decorah Formation in the Upper Mississippi Valley of the United States is still unclear, as is

the precise correlation between the $\delta^{13}C_{org}$ curve from Black Knob Ridge and the better known $\delta^{13}C_{carb}$ curves from other localities. Further investigation into these issues is ongoing, but there is some current uncertainty regarding the exact relationship and correlation of the GSSP level with positive carbon isotope excursions of a *similar* age. Several middle and late Katian positive carbon isotope excursions (summarized by Bergström et al., 2009; and Fig. 20.1 herein) are also important stratigraphic markers.

Graptolite diversity exhibits a pattern of strong fluctuation during the Katian (Cooper et al., 2014). In the lowermost part of the Katian, perhaps in response to global climate perturbations (Goldman and Wu, 2010; Pohl et al., 2016), graptolite faunas exhibit a distinct decline in both morphologic and taxonomic diversity (Goldman and Wu, 2010). This is followed by a brief diversity peak, and another decline in the middle Katian. The upper part of the Katian, equivalent to the Australasian zones Bol-Bo3 (*Alulagraptus uncinatus—Paraorthograptus pacificus*), contains a rich graptolite fauna, including dicellograptids, climacograptids (*Appendispinograptus* and *Styracograptus*), and orthograptids (the "DCO" fauna), and a species diversity peak (Chen et al., 2005). At the end of the Katian graptolite faunas exhibit the most severe depletion in diversity during the entire Ordovician, during which the DCO fauna, along with a wide range of other fossil groups (see later), was completely extinguished (Melchin and Mitchell, 1991; Chen et al., 2005; Finney et al., 2007). Global conodont diversity is low in the early Katian, rebounds in the middle Katian, but always remains well below the peak middle Ordovician levels (Sweet, 1988).

The name "Katian" is derived from Katy Lake (now drained) near the southern end of Black Knob Ridge (Bergström et al., 2006a). The GSSP and stage name were ratified by the ICS in 2005.

20.1.3.3 The Hirnantian Stage

The uppermost stage of the Ordovician is the Hirnantian. The stage closely corresponds to the classical Hirnantian, the uppermost subdivision of the Ashgill Regional Series of Britain, and is named after the Hirnant Limestone at Bala in Wales. The lower stage boundary is placed at the stratigraphic level of first appearance of the graptolite, *Metabolograptus extraordinarius*, at the Wangjiawan North section, 0.39 m below the base of the Kuanyinchiao Bed, 42 km north of Yichang City, western Hubei, China (Fig. 20.11) (Chen et al., 2006). The Hirnantian Stage coincides with major climatic oscillations and eustatic events, and one or more strong positive carbon isotope excursions (Fig. 20.11), which are events associated with the end Ordovician glaciation (Brenchley et al., 2003; Melchin et al., 2013). The Late Ordovician Mass Extinction (LOME) occurred in the *M. extraordinarius* and *Metabolograptus*

Base of the Katian Stage of the Ordovician System at Black Knob Ridge, Southeastern Oklahoma, USA

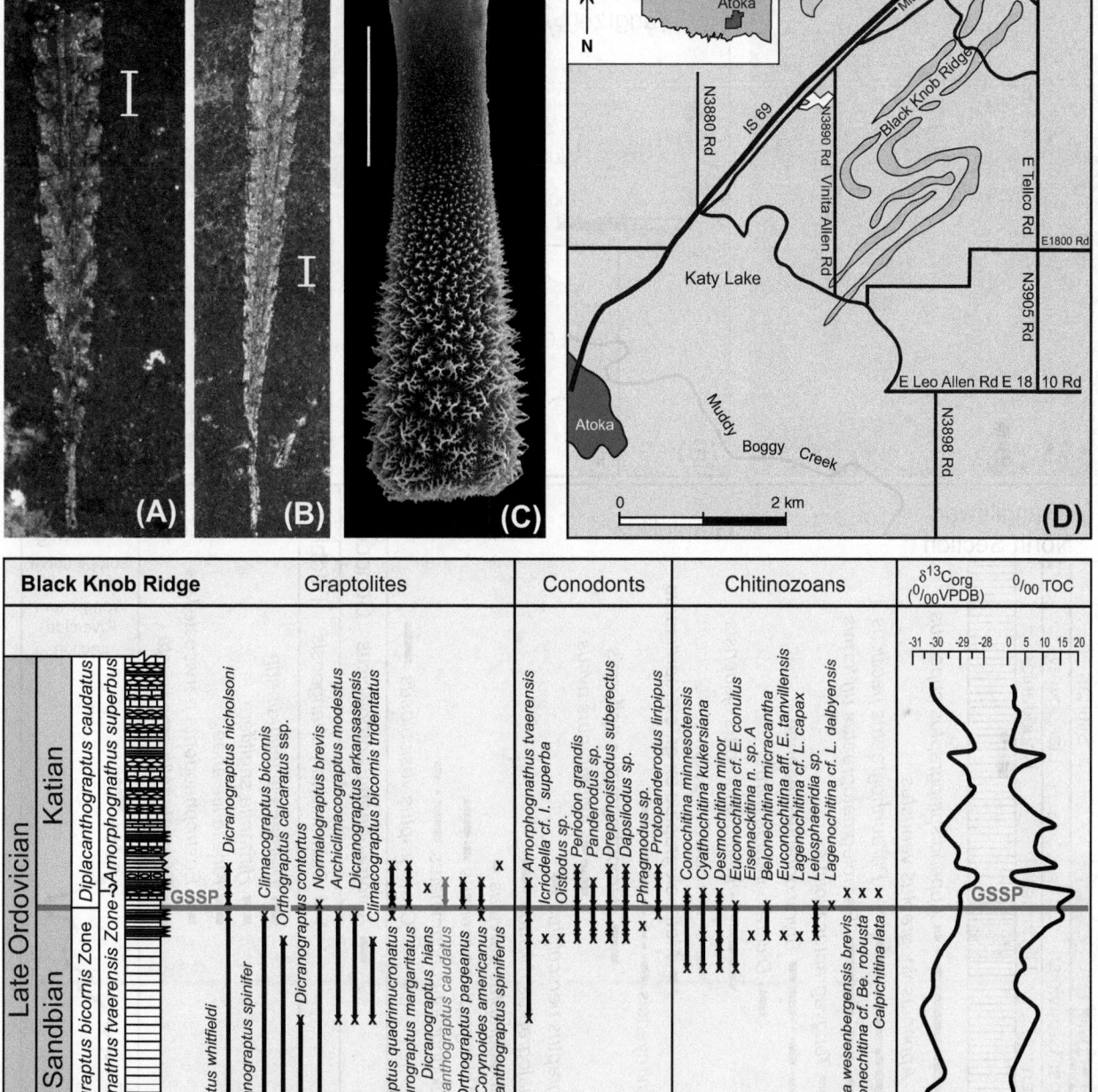

FIGURE 20.10   GSSP for base of the Katian Stage in the Big Fork Chert at Black Knob Ridge section, Oklahoma, United States (D), with a range chart (E) of graptolites, conodonts, and chitinozoans (from Goldman et al., 2007a: Figure 5). The GICE positive excursion in the carbon isotope ($\delta^{13}C_{org}$) stratigraphy of the Sandbian–Katian transition is considered to have interregional, possibly global, correlation value (see also Fig. 20.14). The GSSP for the base of the Katian Stage coincides with the lowest occurrence of graptolite *Diplacanthograptus caudatus* [subfigures (A) and (B); photograph by D. Goldman; scale bar is 1 mm). Subfigure (C) is a specimen of the chitinozoan index *Belonechitina robusta* (Eisenack) from the lowermost Viola Springs Formation, Sect. D near Fittstown, Oklahoma, United States (figure supplied by J. Nõlvak). Length of scale bar is 0.1 mm. *GICE*, Guttenberg isotope carbon excursion.

Base of the Hirnantian Stage of the Ordovician System at Wangjiawan North Section, Yichang, Western Hubei, China

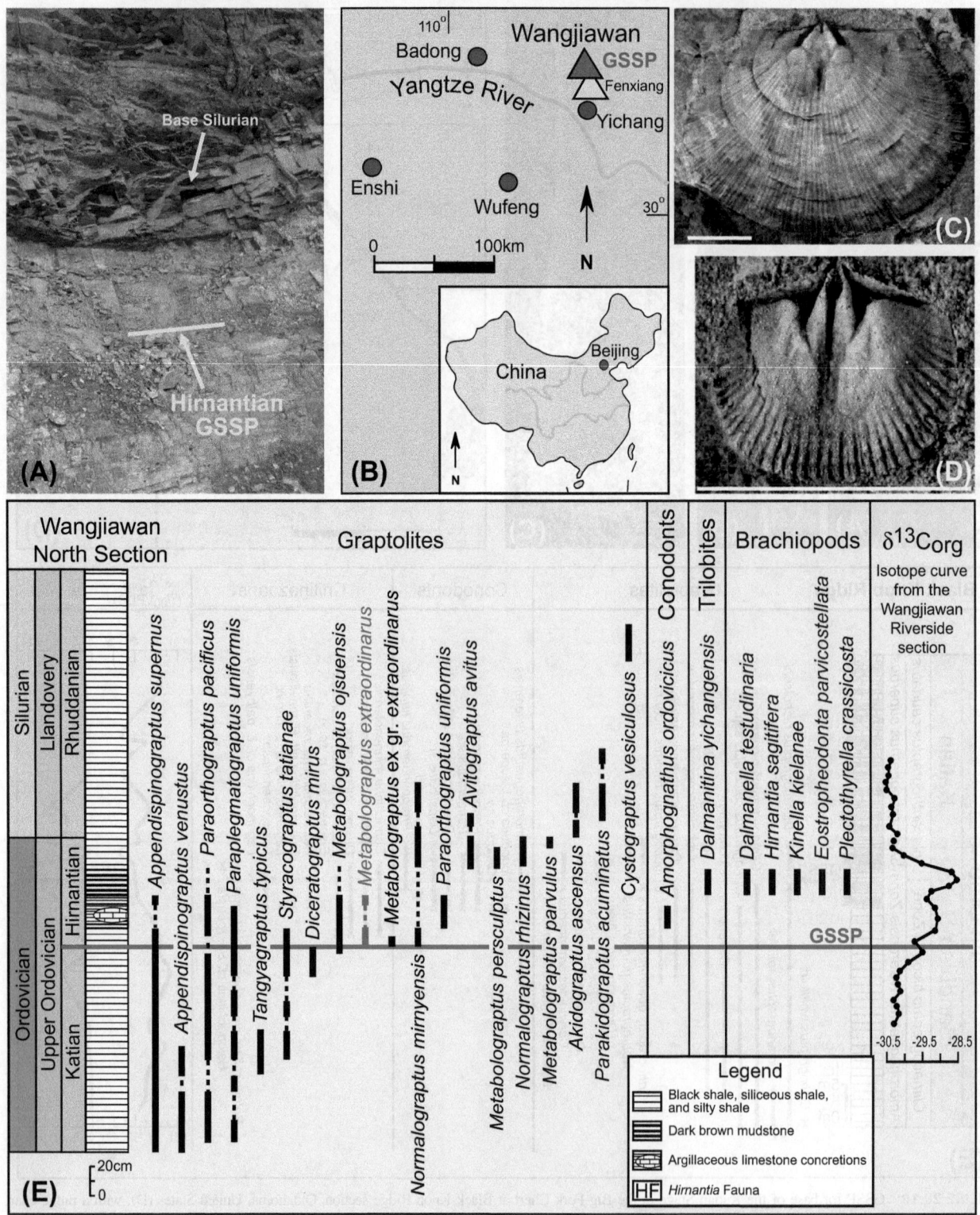

FIGURE 20.11 GSSP for base of the Hirnantian Stage and the Ordovician–Silurian boundary interval (E) at the Wangjiawan North section, Yichang, western Hubei, China (B), with ranges of graptolites and shelly faunas (modified from Chen et al., 2006: Figure 6). The organic carbon isotopic curve of isolated kerogen with the HICE positive excursion (E) is from the nearby Wangjiawan Riverside section (from Chen et al., 2006: Figure 11). The GSSP level in the Wangjiawan outcrop section [*yellow arrow* in (A)] concides with the lowest occurrence of graptolite

persculptus graptolite zones, although some diversity decline began in the latest Katian. The first of the Phanerozoic's five great extinction events, the LOME, apparently driven by climatic and associated sea-level cycles, drastically reduced the diversity of the Earth's marine biota, including graptolites (e.g., Crampton et al., 2018; Harper et al., 2014).

A distinctive shelly fossil association, the Hirnantia—Dalmanitina fauna (commonly referred to as the "Hirnantia fauna") (Fig. 20.11), is widely distributed around the world in strata of latest Ordovician (generally Hirnantian) age (Rong and Harper, 1988). There is some controversy regarding the correlation of the Hirnantian GSSP into carbonate facies sections (compare for instance, Holmden et al., 2013 and Bergström et al., 2014), but recent works from Baltoscandia and Anticosti Island place the boundary in the *Belonechitina gamachiana* chitinozoan zone (Melchin, 2008; Mauviel and Desrochers, 2016; Amberg et al., 2017). The top of the stage is defined by the base of the Silurian System, which occurs at the stratigraphic level of first appearance of the graptolite *Akidograptus ascensus* in the Dob's Linn section of southern Scotland (Melchin et al., 2012).

20.2 Regional subdivisions

The new global stages—Tremadocian, Floian, Dapingian, Darriwilian, Katian, and Hirnantian—have gained widespread acceptance in Ordovician stratigraphy, and their boundaries, while still subject to reevaluation and refinement, are relatively controversy-free. Precise correlation of regional Ordovician subdivisions with the global stages is a critical and ongoing task. Prior to the ratification of the Ordovician GSSPs by the ICS, the most widely used chronostratigraphic units were the British series (Tremadoc, Arenig, Llanvirn, Llandeilo, Caradoc, and Ashgill), which were established in North Wales and England. Indeed, since the 19th century the succession of British series and stages had been used as the global standard in Ordovician stratigraphy. However, unlike the Silurian and Devonian subcommissions, the Ordovician subcommission chose to establish new chronostratigraphic units based on the most reliable fossil-based bioevents, and not to accept with redefinition preexisting regional stages or series (Cooper and Sadler, 2012).

Cooper and Sadler (2012) reviewed and summarized the history, type location, and boundary definitions of the classic British units, and readers are referred to Whittington et al. (1984), Fortey et al. (1991, 1995, 2000), and Cooper and Sadler (2012) for a full review of the British series and a discussion of their correlation with the global Ordovician stages. Other regional subdivisions have received less attention historically but have become increasingly important in studies of Ordovician biostratigraphy, chemostratigraphy, sequence stratigraphy, climate, and biodiversity. These subdivisions are briefly reviewed next, and a comparison of Ordovician chronostratigraphic units, including the British, Australasian, North American, East Baltic, North and South China, and Bohemo-Iberian subdivisions, is presented in Fig. 20.12.

20.2.1 Australasian chronostratigraphic units

Graptolite-based stages were established in Victoria, Australia, in the late 19th and early 20th centuries (Hall, 1895; Harris and Keble, 1932; Harris and Thomas, 1938). A set of 9 stages has been used for the Ordovician of Australia and New Zealand for over 70 years (Harris and Thomas, 1938; Thomas, 1960; VandenBerg and Cooper, 1992). From oldest to youngest they are the Lancefieldian, Bendigonian, Chewtonian, Castlemainian, Yapeenian, Darriwilian, Gisbornian, Eastonian, and Bolindian stages. These units are widely applicable in graptolite successions around the world, particularly those representing low paleolatitude regions (30°N—30°S of the paleo-equator), such as western North America, Cordilleran South America, Greenland, and Spitsbergen. As originally defined and used, the 9 stages were, in effect, groupings of graptolite zones (see next) and their lower boundaries were taken at zone boundaries (see Figs. 20.2A, 20.3A, and 20.4A). Only one stage, the Lancefieldian, has a lower boundary formally defined by a boundary stratotype (Cooper and Stewart, 1979). VandenBerg and Cooper (1992) defined the stage boundaries with bioevents and established reference sections in an effort to create formal chronostratigraphic units. One stage, the Darriwilian, has since been adopted for international use, with a lower boundary stratotype (GSSP) established in China (see earlier). Cooper and Sadler (2004, 2012) and Sadler et al. (2009) used the Australasian stages for calibration of the Ordovician time scale because they are readily correlated with other regional stage successions and with the international stage GSSPs.

20.2.2 East Baltic chronostratigraphic units

The cratonic Baltoscandic Ordovician succession is richly fossiliferous, very condensed, and remarkably complete stratigraphically. Dominated by carbonates, these strata

---

Metabolograptus extraordinarius. The Hirnantia fauna brachiopods from the Kuanyinchiao Bed (upper *M. extraordinarius* Zone to lower *Metabolograptus persculptus* Zone), Wangjiawan, western Hubei, central China, are part of a distinctive Hirnantia shelly fauna that is known worldwide. (C) Brachiopod *Hirnantia sagittifera* (M'Coy), dorsal internal mold. (D) Brachiopod *Kinnella kielanae* (Temple), dorsal internal mold. Scale bar is 5 mm for both figures (supplied and identified by Rong Jiayu). *HICE*, Hirnantian isotope carbon excursion.

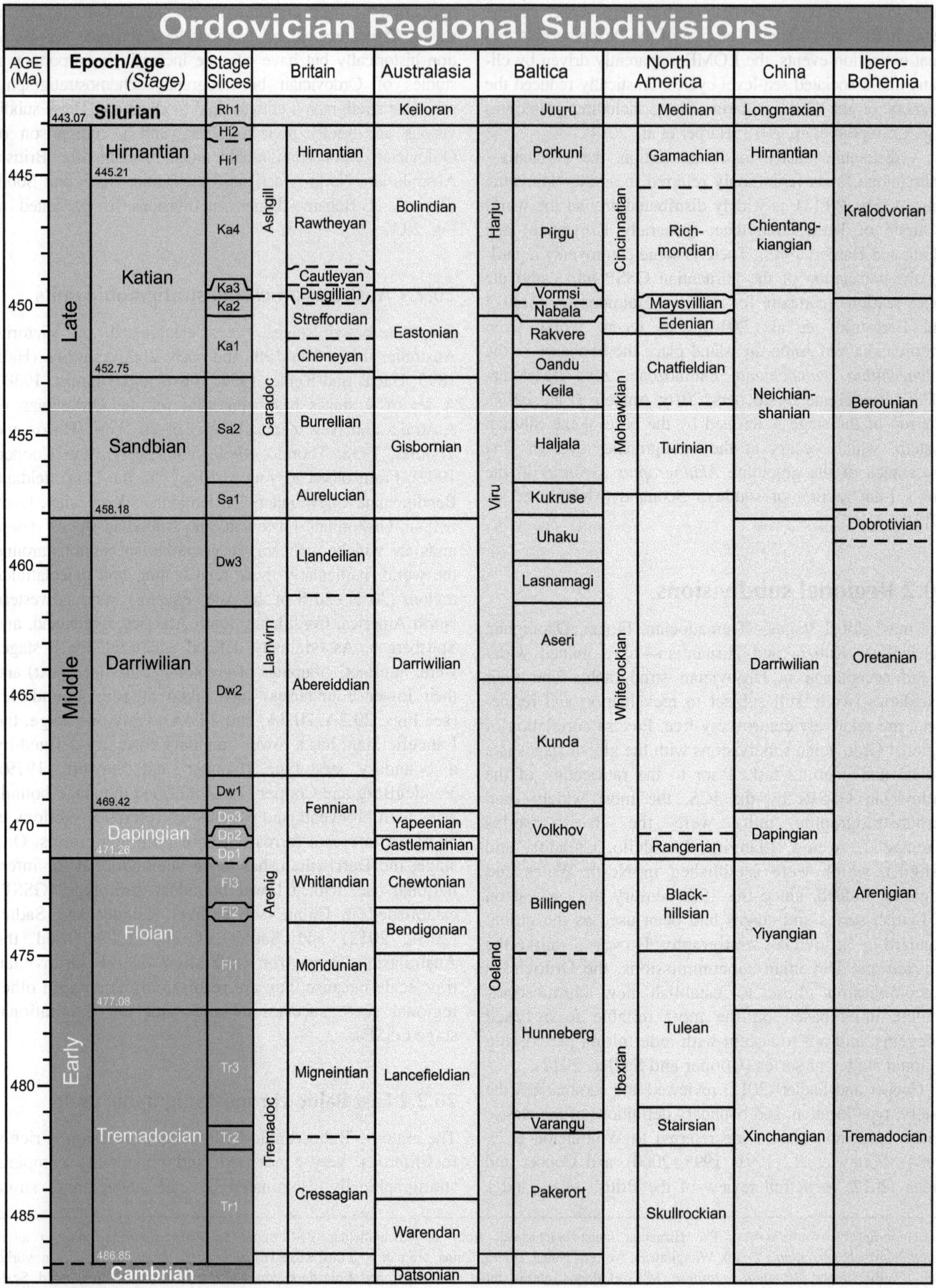

Ordovician Regional Subdivisions

AGE (Ma)	Epoch/Age (Stage)		Stage Slices	Britain		Australasia	Baltica		North America		China	Ibero-Bohemia
443.07	**Silurian**		Rh1			Keiloran		Juuru		Medinan	Longmaxian	
445	Hirnantian 445.21		Hi2					Porkuni		Gamachian	Hirnantian	
			Hi1	Hirnantian		Bolindian	Harju					Kralodvorian
		Ashgill	Ka4	Rawtheyan				Pirgu	Cincinnatian	Rich-mondian	Chientang-kiangian	
450	Katian		Ka3	Cautleyan				Vormsi		Maysvillian		
			Ka2	Pusgillian				Nabala		Edenian		
				Streffordian		Eastonian		Rakvere				
		Caradoc	Ka1	Cheneyan				Oandu		Chatfieldian	Neichian-shanian	Berounian
	452.75							Keila	Mohawkian			
455	Sandbian		Sa2	Burrellian		Gisbornian		Haljala		Turinian		
			Sa1	Aurelucian			Viru	Kukruse				
	458.18							Uhaku				Dobrotivian
460			Dw3	Llandeilian				Lasnamagi				
		Llanvirn						Aseri			Darriwilian	Oretanian
	Darriwilian		Dw2	Abereiddian		Darriwilian			Whiterockian			
465								Kunda				
	469.42		Dw1	Fennian								
470	Dapingian 471.26		Dp3			Yapeenian		Volkhov			Dapingian	
			Dp2			Castlemainian				Rangerian		
			Dp1									
		Arenig	Fl3	Whitlandian		Chewtonian		Billingen		Black-hillsian		Arenigian
	Floian		Fl2			Bendigonian					Yiyangian	
475			Fl1	Moridunian			Oeland		Ibexian			
	477.08							Hunneberg		Tulean		
480			Tr3	Migneintian		Lancefieldian						Tremadocian
	Tremadocian	Tremadoc	Tr2					Varangu		Stairsian	Xinchangian	
				Cressagian				Pakerort				
485			Tr1							Skullrockian		
	486.85					Warendan						
	Cambrian					Datsonian						

FIGURE 20.12 International Ordovician stages and selected regional suites of Ordovician stages and series. The calibration, taken from Figs. 20.1–20.4, applies to the Australasian stage boundaries. Other regions are calibrated by correlation with the Australasian stages. References for regional stages are cited in the text, Section 20.2. Stage slices are from Bergström et al. (2009).

have been extensively sampled in numerous outcrops and in drill cores, and large collections of conodont and chitinozoan fossils were obtained by digesting rock samples in acids. Although originally defined as hybrid bio/lithostratigraphic units (Jaanusson, 1976), many of the East Baltic stage boundaries have been redefined using chitinozoan first appearances (Nõlvak and Grahn, 1993; Nõlvak et al., 2006) and can be employed in precise regional correlations. The Ordovician East Baltic succession is subdivided into 18 regional stages, which in ascending order are—the Pakerort, Varangu, Hunneberg, Billingen, Volkov, Kunda, Aseri, Lasnamägi, Uhaku, Kukruse, Haljala, Keila, Oandu, Rakvere, Nabala, Vormsi, Pirgu, and Porkuni stages (Nõlvak et al., 2006). Apart from the lower Tremadocian, which is represented by a clastic lithofacies, the taxonomy and biostratigraphy of Ordovician conodonts and chitinozoans in Baltoscandia are as well controlled as in almost any other region in the world. In addition, the broad, shallow North Estonian Platform, which extends eastward into the St. Petersburg region of NW Russia, has been the subject of several sequence stratigraphic studies (e.g., Harris et al., 2004; Dronov et al., 2011), which indicate that the regional stages generally correspond to sequence boundaries.

20.2.3 Bohemo-Iberian chronostratigraphic units

The Ordovician GSSPs can be difficult to correlate into the high-paleolatitude settings of southern peri-Gondwana, which tend to be characterized by highly endemic shelly faunas. Gutiérrez-Marco et al. (2017) have recently discussed the advantages of using an alternative Bohemo-Iberian regional scheme that comprises five regional stages (in order, the Arenigian, Oretanian, Dobrotivian, Berounian, and Kralodvorian; refined from Havlíček and Marek, 1973) for the interval between the global Tremadocian and Hirnantian for the "Mediterranean" Ordovician (Fig. 20.12). These chronostratigraphic units are defined on the occurrence of endemic shelly fossils, chitinozoans, sparse graptolites, and palynology (Gutiérrez-Marco et al., 2017).

Given the importance of high paleolatitude biostratigraphy in understanding Late Ordovician climate change and its associated mass extinction, precise regional and global correlation of peri-Gondwanan strata is essential. The Bohemo-Iberian regional Ordovician scheme facilitates the potential for detailed correlations within southwestern and central Europe (Ibero-Armorica, Sardinia, Bohemia, Bulgaria) and can be extended around the Mediterranean region. Sparse occurrences of graptolites and shelly faunas of Baltic or Avalonian affinities also allow for indirect correlation with the global stages through their own regional subdivisions (Gutiérrez-Marco et al., 2017). For example, recent revisions to the taxonomy and distribution of Katian and earliest Hirnantian graptolites from the Prague Basin by

Kraft et al. (2015) have substantially improved correlations between the Upper Ordovician of Bohemia and the international chronostratigraphic units.

20.2.4 North and South China chronostratigraphic units

Three of the seven international stages, the Dapingian, Darriwilian, and Hirnantian, have their stratotypes in China. The Ordovician System is subdivided in China into three series and seven stages: Lower (Xinchangian, Yiyangian), Middle (Dapingian, Darriwilian), and Upper (Neichiashanian, Chientangkiangian, Hirnantian), respectively. This scheme is largely congruent with the standard international classification, which is easily applied to the Ordovician of China (Zhang et al., 2019). Detailed graptolite, conodont, and chitinozoan biostratigraphies have been developed from South China, North China, and Tarim (e.g., Zhang et al., 2019; Wang et al., 2018; Zhen et al., 2011; Chen et al., 2005, 2017; Tang et al., 2016; Liang and Tang, 2016a,b) allowing for precise correlation both among major Chinese paleoplates and with the international subdivisions.

20.2.5 North American chronostratigraphic units

The Ordovician of North America is subdivided into four series, which from oldest to youngest are the Ibexian, Whiterockian, Mohawkian, and Cincinnatian. The history of North American stratigraphic nomenclature is long and complicated (see discussion in Ross et al., 1982; and Sweet, 1995) and as with all regional subdivisions, the North American units follow the complexities of local geology and biostratigraphy. The lower two series are defined in the great Lower Ordovician carbonate platform of western Laurentia, and the upper two in the more clastic-rich facies of the midcontinent and Appalachian Basin. Accordingly, the bases of the Ibexian and Whiterockian series are coincident with the first appearances of conodonts characteristic of shelf assemblages, *I. fluctivagus* and *Tripodus laevis*, respectively (Miller et al., 2003, 2014). Miller et al. (2016) proposed that the FAD of *I. fluctivagus* at the Lawson Cove section, Utah could also serve as the auxiliary GSSP for the Cambrian–Ordovician boundary. The top of the Ibexian Series in its type area coincides with the correlated base of the Whiterockian Series, which is drawn at the base of the *T. laevis* conodont zone, a level just below the base of the Dapingian global stage (Bergström et al., 2009). The Ibexian Series approximately correlates with the Tremadocian and Floian stages, and the Whiterockian Series with the Dapingian, Darriwilian, and lower Sandbian stages.

The base of the Mohawkian Series was defined as the FAD of the conodont taxon *B. gerdae*, a level at the base (or very low) in the Elway Formation at the Lay School section in Tennessee (Bergström, in Ross et al., 1982). Sweet (1984, 1988) constructed a high-resolution, post-Whiterockian graphic correlation network in which the Mohawkian—Cincinnatian composite was subdivided into 80 chronostratigraphic units. Sweet (1984) noted that the base the Elway Formation and the FAD of *B. gerdae* were separated by only 1 m in the CSS and placed the base of the Mohawkian Series at the projected level of the FAD of *B. gerdae*, 809 m in the Composite Standard Section. The critical interval of the stratotype is now largely covered, but it is exposed in the Porterfield Quarry near Saltville, Virginia (Bergström et al., 1988), and Leslie and Bergström (1995) suggested this as the series boundary reference section. The Mohawkian Series boundary lies within the upper part of the Sandbian Stage (*Climacograptus bicornis* graptolite zone).

Historically, the base of the overlying Cincinnatian Series corresponds to the base of the Kope Formation at the Clifton Avenue section in Cincinnati, Ohio, a level that is unfortunately no longer exposed (Bergström, in Ross et al., 1982). In Sweet's (1984, 1988) graphic correlation framework the base of the Cincinnatian Series was coincident with the level of the base of the Kope Formation at the Moffett Road section, Kentucky, at 1063 m in the Composite Standard Section. The lowermost Kope Formation contains graptolites that are characteristic of the *Diplacanthograptus spiniferus* Zone and correlates with a level just above the base of the *A. superbus* North Atlantic conodont zone and the *Belodina confluens* Midcontinent conodont zone (Bergström and Mitchell, 1986, 1990, 1994). However, more recent work suggests these relations may be complicated by facies dependence of the critical fossils (see, e.g., Brett et al., 2004 and Sell et al., 2015).

In its type region the top of the Cincinnatian Series is marked by a prominent unconformity. The scope of this hiatus varies regionally, but the top of the youngest Cincinnatian stage, the Richmondian, is well below the Ordovician—Silurian boundary (Bergström, in Ross et al., 1982). Following Twenhofel's (1928) work on the more complete Ordovician—Silurian succession on Anticosti Island, several authors (e.g., Twenhofel et al., 1954; McCracken and Barnes, 1981) have suggested using the term "Gamachian" for the post-Richmond—pre-Silurian interval that is missing around Cincinnati. The base of the Gamachian Stage is placed at the base of the Ellis Bay Formation on Anticosti Island, coincident with an abrupt change to a conodont fauna dominated by the species *Gamachignathus ensifer* and *G. hastatus* (McCracken and Barnes, 1981; McCracken and Nowlan, 1988). This "Gamachignathus fauna" is referred to as Fauna 13 because it overlies Sweet's (1984, 1988) faunal interval 12 (McCracken and Barnes, 1981; McCracken and Nowlan, 1988) (see Fig. 20.2B, herein).

There is some disagreement regarding the correlation of the base Hirnantian Stage into the Anticosti Island succession. Bergström et al. (2009) noted that the base of the Ellis Bay Formation is somewhat below the beginning of the Hirnantian $\delta^{13}C$ excursion (HICE) and suggested that the Gamachian may correlate with a pre-Hirnantian level in the uppermost Katian Stage. Alternatively, Jin and Copper (2008) and Jin and Zhan (2008) argued that the brachiopod faunal succession on Anticosti Island indicated that the Hirnantian Stage spanned a much greater portion of the Late Ordovician succession there. Melchin and Holmden (2006b) correlated the HICE as recorded in the uppermost strata of the Ellis Bay Formation with the mid to late Hirnantian (*M. persculptus* graptolite zone) and noted that the excursion does not encompass much of the lower to mid Hirnantian at this site in contrast to the latest Ordovician carbon isotopic history elsewhere. This latter interpretation, which approximately correlates the base of the Hirnantian with the base of the Ellis Bay and hence the base of the Gamachian, is corroborated by recent work on Ellis Bay graptolite biostratigraphy (Melchin, 2008), chitinozoan biostratigraphy (Achab et al., 2011; Amberg et al., 2017), and carbon isotope stratigraphy (Mauviel and Desrochers, 2016).

20.2.6 Ordovician stage slices

In an attempt to define correlation units finer than stages for the analysis of biodiversification, Webby et al. (2004a) proposed a set of what they called "time slices." They constructed 6 primary divisions (labeled 1—6, essentially equivalent to the 6 Ordovician stages) and 19 secondary divisions (labeled 1a—d, 2 a—d, 3a—d, 4a—d, 5a—d, and 6a—c) were listed. As Bergström et al. (2009) pointed out, the designation "time slice" implies that these are chronostratigraphic units; however, their original definitions were rather broad and their status is not entirely clear. They have nevertheless proved to be useful for infra stage-level correlation and have become increasingly common in the literature (e.g. Trotter et al., 2008; Shields et al., 2003; Nõlvak et al., 2006). Bergström et al. (2009) revised, redefined, and renamed the units, calling them "stage slices," and recognizing 20 in all. They stated that, in terms of stratigraphic scope, "a stage slice falls between a stage and a faunal zone, that is, it corresponds to a substage or superzone" (Bergström et al., 2009). Stage slices are stage subdivisions based on the FAD of the presumed most reliable index fossil, either graptolite or conodont, for a particular interval. The exception is the Hirnantian Hi2 stage slice, the lower boundary of which is at the top of the global $\delta^{13}C$ excursion (HICE). Although stage slices are referred to as "informal, but defined chronostratigraphic units" (Bergström et al., 2009), it remains debatable as to whether they are essentially chronostratigraphic or biostratigraphic in nature. In terms of definition, the stage slices are directly equivalent to the Australasian graptolite zones, which have been used widely for interregional correlation of graptolite sequences.

20.3 Ordovician stratigraphy

20.3.1 Biostratigraphy

The two fossil groups most used for correlation in the Ordovician are graptolites and conodonts. Although these two groups have generally been considered the most reliable and cosmopolitan fossils for building widely applicable biostratigraphic zonations, recent work has demonstrated the importance and utility of a third group, chitinozoans (Achab, 1989; Paris, 1990, 1996; Paris et al., 2004; Nõlvak and Grahn, 1993; Grahn and Nõlvak, 2007). Graptolites are most abundant in shale sections, particularly those of the outer continental shelf, slope, and ocean settings, whereas conodonts and chitinozoans are most abundant in carbonate sections of the shelf and platform. Graptolites isolated from limestone successions, and conodonts and chitinozoans identified from shale surfaces have facilitated zonal ties between otherwise disparate biofacies (e.g., Bergström, 1986; Vandenbroucke et al., 2010; Goldman et al., 2007a,b). When precisely tied together, these fossil zonations represent a biostratigraphic correlation framework that can be applied with confidence across a wide range of facies and latitudinal zones (Bergström, 1986; Cooper, 1999b). Other fossil groups that are useful for regional and interregional correlation include trilobites, brachiopods, and, in upper ordovician carbonate facies, corals, and stromatoporoids (Webby et al., 2000, 2004b). Radiolarians and acritarchs are increasingly being used for global correlation in some facies and intervals (Pouille et al., 2014; Webby et al., 2004b).

20.3.1.1 Graptolite zones

Graptolites (Phylum Hemichordata) were a major component of the Ordovician macroplankton and represent the first substantial fossil record of zooplankton. They lived at various depths in the ocean waters (Chen, 1990; Cooper et al., 1991, 2012; Chen et al., 2001), were particularly abundant along continental margins in upwelling zones (Fortey and Cocks, 1986; Finney and Berry, 1997), and are found in a wide range of sedimentary facies, although are most abundant in distal shelf/slope deposits. Many graptolite species dispersed rapidly, are widespread globally, and have relatively short stratigraphic durations, lasting for 2 Myr or less. It is estimated (Foote et al., 2019) that the published graptoloid record captures up to 75% of original species richness and, on average, 85% of the original durations for species known from more than one collection. These attributes combine to make graptolites extremely valuable fossils for parsing and correlating Ordovician and Silurian strata (Skevington, 1963; Cooper

and Lindholm, 1990; Webby et al., 2004a; Cooper and Sadler, 2012).

Graptolite faunas exhibit a relatively strong latitudinal association, and zonations, which have been erected from several regions around the world (Loydell, 2011), tend to fall into high- and low-latitude groups. In low-latitude environs ("Pacific Province," 30°N−30°S), one of the most detailed and best established regional zonal schemes, spanning almost the entire Ordovician, is that of Australasia (Figs. 20.2A, 20.3A, and 20.4A; Harris and Thomas, 1938; Thomas, 1960; VandenBerg and Cooper, 1992). Thirty zones, two of which are divided into subzones, are recognized giving an average duration of 1.5 Myr each. Zones are based on the stratigraphic ranges of species, most zonal boundaries being tied to first appearance events. The Australasian zonation is applicable in most Pacific Province graptolite successions, but in some regions and intervals local zonations may provide a finer resolution for correlation (e.g., the endemic lower to middle Katian zonation from the Appalachian Basin of eastern North America). Sadler et al. (2009) calibrated the zonal boundaries of the Australasian succession using the CONOP (constrained optimization) method and a data set of over 400 stratigraphic sections from around the world. This method seeks the earliest occurrence of each taxon in any section and builds a composite that can differ in detail from the regional successions of first and last appearance events. Radioisotopic age calibration of the graptolite CONOP composite is the basis of the Ordovician and Silurian time scale (Cooper and Sadler, 2012; Sadler et al., 2009).

The most representative zonal scheme for middle to high paleolatitudes (Atlantic Province) is that of southern Britain (Zalasiewicz et al., 2009) where some 16 zones span the Tremadoc to middle Caradoc (Tremadocian through Sandbian global stages), averaging 2.3 Myr each. Recent work on the taxonomy and biostratigraphy of central European graptolite faunas (Kraft et al., 2015, Gutiérrez-Marco et al., 2017) has helped refine a peri-Gondwanan graptolite zonation (Figs. 20.2A, 20.3A, 20.4A). Although many species were cosmopolitan and correlation between high- and low-paleolatitude regions is well controlled throughout most of the period, recognizing the global stages in peri-Gondwana successions remains a work in progress. Other important graptolite zonal schemes for correlation are those of Cordilleran North America, eastern North America, Scandinavia, South China, and Tarim (western China).

20.3.1.2 Conodont zones

Conodont animals are widely recognized as an important group of early vertebrates with tooth-like structures that comprised a feeding apparatus that represents the earliest expression of mineralized skeletons among vertebrates

(Aldridge et al., 1993, 2013; Donoghue et al., 2000, 2008). The individual tooth-like structures of the feeding apparatus of the conodont animal are composed of calcium phosphate and referred to as either conodonts or conodont elements. The soft body of the conodont animal is rarely preserved, although from the few localities that have yielded soft parts, a basic understanding of the conodont animal body plan has emerged. There is a general consensus that they were small eel-like animals that occupied a variety of niches in early Cambrian to late Triassic seas (Aldridge and Briggs, 1989; Aldridge et al., 1993; Briggs et al., 1983). They were most abundant on continental shelves and were readily preserved in shelf carbonates, although bedding plane occurrences on shale are common. Because conodonts are composed of calcium phosphate they can be extracted relatively easily from carbonate rock by acid digestion. This led to extensive sampling of carbonate platforms for isolated conodont elements from acid residues and resulted in detailed biostratigraphic zonation of carbonate facies (Bergström, 1971). Some conodont species are found in a wide range of sedimentary environments and geographical regions, making them valuable fossils for long-range correlation. Conodonts range from the early Cambrian through the Triassic and are used as zone fossils in all these periods.

In the Ordovician, conodont faunas, like graptolite faunas, are distributed in two major biogeographic provinces (Sweet and Bergström, 1976, 1984). A warm-water province ranged about 30 degrees N and S of the Equator, and a cool-water province extended poleward from 30 to 40 degrees latitudes. The warm-water province, typified by the North American midcontinent region (Sweet and Bergström, 1976), contains a diverse and rich fauna that can be finely zoned into 26 zones, some of which are further divided into subzones (Figs. 20.2B, 20.3B, and 20.4B). It is important to note that the zones of the midcontinent region were initially based on traditional biostratigraphic zonal practices (Sweet and Bergström, 1976 and references therein). Sweet (1984, 1995) used graphic correlation to construct and formally name midcontinent conodont chronozones, using many of the same taxon names that were used for the biostratigraphic zones. Sweet (1988) provided a useful range chart for many taxa in his midcontinent conodont chronozonation as well as for the North Atlantic biozonation, but in a practice that would later prove to be confusing, referred to the named midcontinent conodont chronozones as biozones. Although there is a history of the named midcontinent conodont zones to be based on chronozones defined through graphic correlation, they are often incorrectly used as biozones, generally based on the first appearance of the biozone name-bearing taxon. This practice together with the apparent control of many Laurentian conodont species ranges by local habitat conditions related to water depth has led to conflicting or misleading correlations. The cold-water North Atlantic province is best known from the Baltic region in which 17 zones and several subzones are recognized. These two zonal successions provide good resolution through the Lower and Middle Ordovician. In the Upper Ordovician, there are 11 warm-water zones and only 3 cold-water zones. New zonations derived from partially endemic faunas in North and South China (Zhang et al., 2019; Wang et al., 2018) have added substantially to our understanding of conodont provincialism and biostratigraphy, but additional work to establish precise zonal ties is required.

For the first time radioisotopic dates have been interpolated into a global CONOP conodont composite derived from nearly 200 stratigraphic sections, and conodont zonal boundaries have been age dated. In Section 20.4.3 we present an independently calibrated time scale derived from carbonate facies sections that can be compared with the calibrated graptolite composite from mainly dark shale sections.

20.3.1.3 Chitinozoan zones

Chitinozoans are an extinct group of organic-walled microfossils that are abundant in the Lower to Middle Paleozoic strata. Their taxonomic affinity is unknown but several workers have suggested that they are the fossilized eggs of ancient marine metazoans (e.g., Paris and Nõlvak, 1999). However, a new study by Liang et al. (2019) on population variation in chitinozoans has questioned that hypothesis. Along with graptolites and conodonts, chitinozoans are an important tool in high-resolution Ordovician–Silurian biostratigraphy (Achab, 1989; Paris, 1990, 1996; Paris et al., 2004; Nõlvak and Grahn, 1993; Grahn and Nõlvak, 2007). Chitinozoans are most commonly isolated from carbonates by acid dissolution, but recent work by Vandenbroucke et al. (2005, 2013) on extracting specimens from black shale has improved correlation across biofacies and added important data to some long-standing stratigraphic problems in Ordovician stratigraphy. In the Ordovician, two regional zonations, Baltic and North Gondwana, are well developed and used extensively to correlate Ordovician strata (Nõlvak and Grahn, 1993; Paris, 1990). The Baltic zonation includes 18 zones and 10 subzones but lacks resolution in the Lower Ordovician (Nõlvak et al., 2006). The North Gondwana succession is finely divided into 25 zones. Achab (1989) provided a partial zonation for Ordovician rocks of eastern Laurentia, and a substantial amount of new work is currently underway in South China (e.g., Liang and Tang, 2016a,b; Liang et al., 2018; Wang et al., 2013). Figs. 20.2B, 20.3B, and 20.4B list the succession of Baltic and North Gondwana chitinozoan zones.

20.3.1.4 Integration of graptolite, conodont, and chitinozoan zones

Precise correlation between disparate biofacies remains a difficult problem in Ordovician biostratigraphy. In a pioneering study, Bergström (1986) comprehensively reviewed the literature for co-occurrences of Ordovician graptolite and conodont and recognized 78 individual tie points between the various zonations. Bergström et al. (in Chen et al., 2017) added 27 new zonal ties from combined fossil occurrences in Darriwilian and Sandbian strata. Exhaustive searches for fossil occurrences outside their common biofacies association, detailed sampling of regions that have shelf to slope transects, and use of quantitative stratigraphic methods to interleave data sets have further advanced the integration of graptolite, conodont, and chitinozoan zones.

The most common types of zonal ties are produced by graptolite occurrences in carbonate successions and conodonts derived from limestone interbeds in graptolite-bearing dark shales. Examples of the former are three-dimensionally preserved, Sandbian to Katian graptolites from the Viola Group of Oklahoma, United States (e.g., Finney, 1986); from the Darriwilian to Katian carbonates in the Kandava drill core of Latvia (Goldman et al., 2015), and the Athens Shale of Alabama, United States (Finney, 1977) (Darriwilian to Sandbian). There are numerous examples of the latter situation, but the integrated biostratigraphy of the Cow Head Group in western Newfoundland (e.g., Cooper et al., 2001a; Stouge, 2012) and the Dawangou section in Tarim, western China (auxiliary GSSP for the base of the Upper Ordovician and Sandbian Stage), are particularly noteworthy (Chen et al., 2006, 2017). Finding diagnostic conodonts on shale surfaces and extracting chitinozoans from clastics with hydrofluoric acid dissolution has also resulted in more precise zonal ties (e.g., Vandenbroucke et al., 2013; Vandenbroucke, 2004, 2008a,b; Goldman et al., 2007a,b; Leslie et al., 2008).

In some regions Ordovician strata can be sampled from outcrops and/or drill cores in nearly continuous shelf to slope transects. Interfingering of lithofacies in sections at the margins of facies belts may result in co-occurrences of fossils that are not commonly found together. The Middle and Upper Ordovician rocks of Baltoscandia have been divided into spatially distinct, composite litho- and biofacies units called "confacies" belts (Jaanusson, 1976, 1995). These confacies belts—the Scanian (slope, black shale), Central Baltoscandian (outer shelf, argillaceous limestones), and North Estonian (carbonate platform) belts—record an offshore to onshore transition along the eastern to southeastern margin of the East European Craton. Using CONOP methodology Goldman et al. (2012, 2013) constructed a fully integrated multiclade range chart (graptolites, conodonts, chitinozoans, and ostracods) from Middle and Upper Ordovician strata in outcrops and drill cores that spanned all three confacies belts. Serra et al. (2019) and Bryan et al. (2018) have also used quantitative correlation to integrate conodont and graptolite biostratigraphy in the Argentine Precordillera.

20.3.2 Chemostratigraphy

20.3.2.1 Carbon isotope stratigraphy

Chemostratigraphy of stable isotopes as a means of correlation of Ordovician rocks has become general practice for both regional and global correlations. Observed $\delta^{13}C_{carb}$ records are interpreted to represent changes in the global seawater-dissolved inorganic carbon (DIC) values (or globally synchronized during early diagenesis) only if trends are documented from multiple sections on a global scale. This can be rigorously evaluated by generating similar $\delta^{13}C_{carb}$ trends from a wide range of depositional settings with differing diagenetic histories and are shown to be demonstrably synchronous by independent data (Saltzman and Thomas, 2012). Therefore, to construct meaningful correlations of global trends, it is necessary to compare a large number of carbon isotope records from various paleoenvironmental settings based on independent dating methods such as a precise biostratigraphic framework. The trends in $\delta^{13}C_{carb}$ from Ordovician marine carbonates ($\delta^{13}C_{carb}$) have been intensely studied for chemostratigraphic correlation and potential links between the global carbon cycle and the biosphere have been established (e.g., Bergström et al., 2006b, 2010a,b; Melchin and Holmden, 2006b; Panchuk et al., 2006; Saltzman and Thomas, 2012; Melchin et al., 2013; Metzger et al., 2014).

Correlations using $\delta^{13}C_{carb}$ data sets focus on rapid changes in isotope values, or carbon isotope excursions, commonly shorthanded as an initial with the extension "ICE" added (e.g., Hirnantian isotopic carbon excursion, HICE) or "CIE" added (e.g., Ireviken CIE). There is little variability in Lower to lower Middle Ordovician $\delta^{13}C_{carb}$ records. Edwards and Saltzman (2014) and Lehnert et al. (2014) described the Lower Ordovician $\delta^{13}C_{carb}$ excursions albeit using different names for the same excursions. Of those excursions the TSICE (=C1 of Edwards and Saltzman, 2014) has the greatest potential for correlation. Major perturbations to the global carbon cycle (i.e., ICEs or CIEs) become more widespread and recurrent during the Middle to Late Ordovician, roughly coincident with the start of secular shifts in $^{87}Sr/^{86}Sr$ (Qing et al., 1998; Shields et al., 2003; Young et al., 2009; Saltzman et al., 2014). There are many locally named ICEs that are used primarily for regional correlation (e.g., Ludvigson et al., 2004), but that are not recognizable between basins, likely because these represent local processes affecting carbon cycling only within a particular region (e.g., Saltzman and Edwards, 2017). In addition to local processes affecting $\delta^{13}C_{carb}$ values, significant isotopic heterogeneity (up to 2‰ in $\delta^{13}C_{carb}$) within single hand samples can be superimposed over stratigraphic intervals where carbon isotope excursions have been recorded (e.g., Metzger and Fike, 2013), and without petrographic and geochemical screening such isotopic variation could artificially enhance or

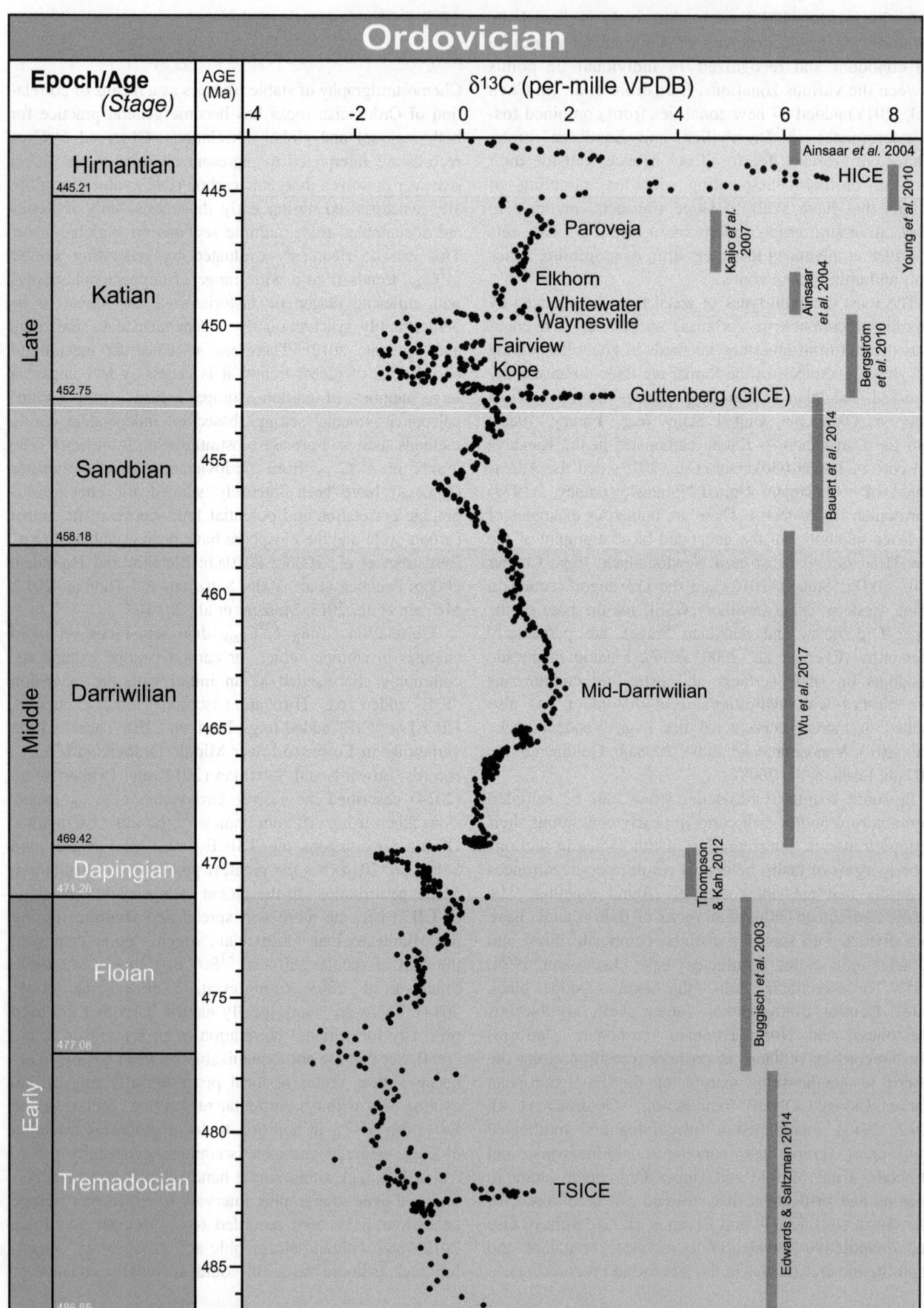

FIGURE 20.13   Ordovician δ^{13}Carbon isotope curve. From Cramer and Jarvis (2020, Ch. 11: Carbon isotope stratigraphy, this book). See Fig. 20.1 explanation and text Section 20.3.2.1 for complete references to the Ordovician $\delta^{13}C_{carb}$ isotope curve. *BDNICE*, Basal Dapingian Negative Isotope Carbon Excursion; *GICE*, Guttenberg Isotope Carbon Excursion; *HICE*, Hirnantian Isotope Carbon Excursion; *MDICE*, Middle Darriwilian Isotope Carbon Excursion.

mask the primary $\delta^{13}C_{carb}$ signal. The uppermost Sandbian—basal Katian Guttenberg (GICE, Bergström et al., 2010a; Hatch et al., 1987; Young et al., 2005), Hirnantian (HICE, Bergström et al., 2006b, 2014; Brenchley et al., 2003; Fanton and Holmden, 2007; Marshall and Middleton, 1990), mid Darriwilian (MDICE, Kaljo et al., 2007; Schmitz et al., 2010; Wu et al., 2017), and to a lesser extent the Elkhorn (EICE, Bergström et al., 2010b) are the best documented ICEs from multiple basins/continents, whereas the mid Katian Paroveja (Estonia), Moe (Estonia), Whitewater (Laurentia), Waynesville (Laurentia), Saunja (Estonia), Fairview, (Laurentia), Rakvere (Estonia), and Kope (Laurentia) are less well constrained (Smith et al., 2011, Ainsaar et al., 2010) but have potential for regional—global correlation (Bergström et al., 2019b).

Efforts to construct a global $\delta^{13}C_{carb}$ curve have produced a general framework that estimates the magnitudes of the named excursions. Bergström et al. (2009) constructed a generalized $\delta^{13}C_{carb}$ curve and placed it next to what they termed "stage slices" (see Section 20.2.5). Smith et al. (2011), Saltzman and Thomas (2012), and Lehnert et al. (2014) also assembled similar $\delta^{13}C_{carb}$ curves for the Ordovician by linking data sets together from the same regions and from different areas utilizing established biostratigraphy. The newly compiled global $\delta^{13}C_{carb}$ curve in Fig. 20.13 is from Fig. 11.5 in Jarvis and Cramer (2020, Ch. 11: Carbon isotope stratigraphy, this book). Detailed comparisons between System-spanning $\delta^{13}C_{carb}$ curves become more problematic when expanding the scope to the global scale, although some general patterns and excursions are recognizable. The Smith et al. (2011) and Saltzman and Thomas (2012) data sets, taken together with data from Bergström et al. (2010a,b,c) and $\delta^{13}C_{carb}$ records from Baltica (Ainsaar et al., 2010; Bergström et al., 2016; Lindskog et al., 2019; Jarvis and Cramer, 2020, this book) convincingly demonstrate that the MDICE, GICE, EICE, and HICE are recognizable on a near global scale. Although these generalized and composited curves depict the MDICE as prominent and the GICE as a single uncomplicated excursion, it is important to note that caution is needed in carrying out such correlations (see Fanton and Holmden, 2007; Schiffbauer et al., 2017). For example, $\delta^{13}C_{carb}$ data sets from the Great Basin in the western United States and the Precordillera of Argentina do not show a clearly discernible MDICE record (Albanesi et al., 2013; Edwards and Saltzman, 2014; Saltzman and Edwards, 2017; Henderson et al., 2018). Both Argentinian and North American successions typically show distinctly lower $\delta^{13}C_{carb}$ values for the MDICE interval compared to those from Baltoscandia (Lindskog et al., 2019), and these differences are likely due to local processes (e.g., organic matter remineralization, sluggish seawater circulation, sediment reworking) that are superimposed upon a global

perturbation of the carbon cycle (e.g., Saltzman and Edwards, 2017, and references therein). Zhang and Munnecke (2016) published a $\delta^{13}C_{carb}$ data set representing the entire Ordovician of Tarim Basin of the Tarim Plate and correlated it with South China Plate $\delta^{13}C_{carb}$ records (Munnecke et al., 2011) and the global composite of Bergström et al. (2010b). The correlation with the HICE interval is fairly straightforward; however, correlation of the timing and magnitude of other excursions are less clear.

The GICE has been recognized in sections across North America (Hatch et al., 1987; Ludvigson et al., 1996, 2000, 2004; Patzkowsky et al., 1997; Finney et al., 1999; Pancost et al., 1999; Bergström et al., 2009; Young et al., 2005; Barta et al., 2007; Goldman et al., 2007a; Jones et al., 2016) in Baltoscandia (e.g., Ainsaar et al., 1999, 2004, 2010; Meidla et al., 1999; Kaljo et al., 2004, 2007; Ainsaar and Meidla, 2008; Bergström et al., 2010a,b, 2012), South China (Young et al., 2008; Bergström et al., 2009; Munnecke et al., 2011; Fan et al., 2015), and Malaysia (Bergström et al., 2010c). Recent work by Westrop et al. (2015) has raised the possibility that there may be two or more excursions that have been referred to as the GICE, occurring within Sandbian—Katian boundary strata, and are contributing to erroneous interregional correlations of this interval of time. In addition, $\delta^{13}C_{carb}$ records from the Sandbian—Katian boundary intervals in China appear to show a delayed and prolonged GICE record relative to the sequences from North America (Zhang et al., 2019). Although the chronostratigraphic significance of the GICE base and peak are debated, the timing of the GICE roughly matches with deposition of upwelling-driven phosphatic deposition in Laurentia (Kolata et al., 2001; Pope and Steffen, 2003; Herrmann et al., 2004) and is one of the most widespread and globally documented ICEs in the Ordovician demonstrating its truly global nature.

There are a relatively large number named post-GICE to pre-HICE excursions that occur regionally (e.g., Bergström et al., 2010b). For some of these excursions, attempts to correlate them outside their region have been made, but there is work to be done to better constrain many of these correlations. Ainsaar et al. (2010) and Bergström et al. (2015) suggested that several excursions correlate from Baltoscandia into the North American midcontinent (e.g., Moe = Whitewater, Saunja = Waynesville, and Rakvere = Kope). However, there is substantial disagreement regarding the global correlation of these mid to late Katian excursions (e.g. Ainsaar et al., 2010; Jones et al., 2016; Holmden et al., 2013; Myrow et al., 2018). Problems with recognition and correlation of these Katian ICEs or CIEs are in many cases due to poor biostratigraphic constraint, and some issues are inherent in trying to match small magnitude (<2‰) shifts in $\delta^{13}C_{carb}$ records that can be influenced by local processes (see previous discussion) and or not preserved in certain regions due to cryptic unconformities/hiatuses.

The HICE (Hirnantian $\delta^{13}C_{carb}$ excursion) is a distinctive, globally documented, positive excursion that is commonly used for regional and global correlations of late Katian and Hirnantian strata, although the absolute values and the magnitude of the ICE record vary depending on the section and the depositional environment (e.g., Ainsaar et al., 2010; Bergström et al., 2010b, 2011; Brenchley et al., 2003; Demski et al., 2015; Desrochers et al., 2010; Fan et al., 2009; Finney et al., 1999; Jones et al., 2011; Kaljo et al., 2012; Kump et al., 1999; LaPorte et al., 2009; Melchin and Holmden, 2006b; Melchin et al., 2013; Underwood et al., 1997; Young et al., 2010; Mauviel and Desrochers, 2016; Zhang et al., 2019; Holmden et al., 2013; Jones et al., 2015, 2016; Myrow et al., 2018). In carbonate shelf environments the excursion may be a much as 7‰, whereas in outer shelf and slope settings it is often about 3‰, possibly as a result of restricted water mass exchange with the global carbon reservoir (Melchin and Holmden, 2006a). Geochemical modeling suggests the actual change in open oceanic DIC may have been only 1‰−2‰ during the HICE (Ahm et al., 2017) and that the isotopic signature of the syn-glacial tropical carbonate platforms may have been substantially effected by early diagenetic alteration (Holmden et al., 2012; Jones et al., 2016; Kimmig and Holmden, 2017).

Mauviel and Desrochers (2016) identified a two-peaked HICE on Anticosti Island that displays a distinct long-term trend with a long sustained lower HICE followed abruptly by an upper HICE and return to baseline values of +0.5‰ prior to the Rhuddanian. The global correlation of the dual-peak Anticosti HICE is still a matter of debate, and the biostratigraphy of the HICE interval (Ellis Bay Formation) has proven to be contentious (e.g., Bergström et al., 2006b; Melchin and Holmden, 2006b; Kaljo et al., 2001, 2008; Achab et al., 2011; Hints et al., 2014). Furthermore, Hirnantian $\delta^{13}C_{carb}$ records from Nevada, Missouri, Yukon Territory, Arctic Canada, South China, and Baltica show a singular-peak HICE (Bergström et al., 2006b; Brenchley et al., 2003; Kump et al., 1999; LaPorte et al., 2009; Melchin and Holmden, 2006b; Yan et al., 2009). The upper HICE as recorded from Anticosti Island is correlated with the final deglaciation of Hirnantian glaciation (Ghienne et al., 2014; Mauviel and Desrochers, 2016) and is the most pronounced and well correlated of all Ordovician $\delta^{13}C_{carb}$ excursions.

20.3.2.2 Paired carbon isotopes

Paired carbon isotopic analysis of $\delta^{13}C_{carb}$ and $\delta^{13}C_{org}$ and their difference ($\Delta^{13}C = \delta^{13}C_{carb} - \delta^{13}C_{org}$) can be used to estimate changes in isotopic fractionation associated with photosynthetic reduction pathways via marine

primary producers through time as a function of O_2/CO_2 in the atmosphere (Young et al., 2008; Metzger et al., 2014; Edwards and Saltzman, 2016). These inferences about relative changes in atmospheric pO_2 and pCO_2 can really only be attempted once $\delta^{13}C_{org}$ records have been demonstrated to be largely reflective of global changes in primary production rather than local factors. Edwards and Saltzman (2016) compiled a composite $\delta^{13}C_{org}$ curve and compared it with published $\delta^{13}C_{carb}$ to generate a $\Delta^{13}C$ curve for the Ordovician. Although there is a fair amount of scatter in $\delta^{13}C_{org}$, a rather clear trend shows that Lower Ordovician $\delta^{13}C_{org}$ data range from ca. −26‰ to −28‰, decreasing throughout the Lower−Middle Ordovician to ca. −29‰ to −31‰ (Edwards and Saltzman, 2016). $\delta^{13}C_{org}$ values remain at their low throughout the Sandbian and then in the Katian display prominent excursions for both the GICE and HICE. $\Delta^{13}C$ values from well-preserved intervals generally vary between +26‰ to +28‰ throughout the lower to middle Ordovician and increase to +31‰ during the mid−late Darriwilian and mid Sandbian. This ~3‰ increase coincides with a previously interpreted period of ocean cooling and some of the earliest pulses of global biodiversity of marine invertebrates and planktonic organisms (Edwards and Saltzman, 2016).

20.3.2.3 Strontium isotope stratigraphy

Strontium isotope stratigraphy (SIS) can be used as a means of global correlation when $^{87}Sr/^{86}Sr$ secular curves are calibrated against biostratigraphy and geochronology (McArthur et al., 2012). The Ordovician seawater $^{87}Sr/^{86}Sr$ curve exhibits $^{87}Sr/^{86}Sr$ values of ~0.7090 in the Tremadocian falling to ~0.7078 in the Katian (Denison et al., 1998; Qing et al., 1998; Shields et al., 2003; Young et al., 2009; McArthur et al., 2012; Saltzman et al., 2014). A growing consensus based on climate modeling and $^{87}Sr/^{86}Sr$ analyses is that the Ordovician icehouse was triggered by a pCO_2 drawdown resulting from increased silicate weathering of the juvenile Taconic arc terranes (Buggisch et al., 2010; Young et al., 2009; Swanson-Hysell and Macdonald, 2017) which would have driven down oceanic $^{87}Sr/^{86}Sr$ values (Smith et al, 2011). A recent calibration of the Ordovician seawater $^{87}Sr/^{86}Sr$ curve with conodont biostratigraphy and geochronology found a maximum resolution of SIS at ~0.5 to 1.0 Myr in the Darriwilian and Sandbian (Shields et al., 2003; Young et al., 2009, Saltzman et al., 2014).

Edwards et al. (2015) demonstrated that although $^{87}Sr/^{86}Sr$ conodont measurements represent the best estimate of Ordovician $^{87}Sr/^{86}Sr$ seawater, bulk carbonate can also faithfully record $^{87}Sr/^{86}Sr$ seawater and provide a high-resolution record of seawater $^{87}Sr/^{86}Sr$ when either (1) Sr concentrations are >300 ppm, even at relatively high burial temperatures (e.g., CAI = 5), or (2) CAI values of associated conodonts are low (≤ 2), even at low [Sr] in bulk carbonate. For parts of the Ordovician in which rates of change in $^{87}Sr/^{86}Sr$ are relatively high ($\sim 5.0-10.0 \times 10^{-5}$ per Myr), Sr isotope stratigraphy is likely to be useful as a high-resolution correlation tool that in some cases is on par with, or potentially better than, conodont biostratigraphy (Saltzman et al., 2014). Sr isotope stratigraphy is particularly useful in strata that only preserve long-ranging conodonts or species that have poorly constrained age ranges.

The resolution of Sr isotope stratigraphy in Lower Ordovician stages (Tremadocian–Floian) is relatively low based on a $^{87}Sr/^{86}Sr$ rate of fall of $\sim 1.6 \times 10^{-5}$ per Myr, and this trend continues into the Middle Ordovician Dapingian Stage, where $^{87}Sr/^{86}Sr$ falls from 0.70880 to 0.70875, and the rate of change is 1.9×10^{-5} per Myr (Saltzman et al., 2014). Rates of $^{87}Sr/^{86}Sr$ change in the Darriwilian and Sandbian stages are the greatest and provide the highest potential for geochronologic resolution, with rates of $^{87}Sr/^{86}Sr$ change as high as 8.9×10^{-5} per Myr in the late Darriwilian *Cahabagnathus friendsvillensis* conodont zone (Saltzman et al., 2014). Rates of change slow dramatically in the late Sandbian through the Katian stages until the curve reverses course and $^{87}Sr/^{86}Sr$ values rise from the upper part of the Katian through the terminal Ordovician Hirnantian Stage (Saltzman et al., 2014).

20.3.2.4 Oxygen isotope stratigraphy

Although many studies of $\delta^{13}C_{carb}$ include $\delta^{18}O_{carb}$ data sets, these data sets are generally less useful for stratigraphic correlations because $\delta^{18}O_{carb}$ is relatively more sensitive to lithologic variations and diagenetic alteration than $\delta^{13}C_{carb}$ (see e.g., Kah, 2000; Melim et al., 2001). Factors responsible for observed $\delta^{18}O_{carb}$ trends in ancient carbonates are likely complex (e.g., Saltzman and Edwards, 2017) and therefore $\delta^{18}O_{carb}$ is generally not used as a primary means of correlation.

Conodont $\delta^{18}O$ records have provided a wealth of data for reconstructing Ordovician climate history, albeit some of it inconsistent. Trotter et al. (2008) recorded a steady cooling trend through the Early and Middle Ordovician and hypothesized that the attainment of a more moderate climate played an important role in the remarkable marine diversification that occurred during the Ordovician Period. Elrick et al. (2013) collected relatively sparse $\delta^{18}O$ data from parasequence scale lithologic cycles in Katian age strata on Anticosti Island (Vaureal Formation) and the North American Midcontinent (Lexington Limestone and Kope Formations), and interpreted them as orbitally paced, pre-Hirnantian glacioeustatic cycles that were associated with large magnitude water depth and sea-surface temperature fluctuations. More detailed conodont $\delta^{18}O$ records from locations throughout Laurentia (Trotter et al., 2008; Buggisch et al., 2010; Herrmann et al., 2010; Rosenau et al., 2012; Quinton and

MacLeod, 2014; Quinton et al., 2017) exhibit more constant $\delta^{18}O$ values through the Katian and indicate a relatively stable climate in this interval.

The growing number of conodont $\delta^{18}O$ records have allowed for the testing of some hypotheses about the relationship between carbon isotope excursions and ice sheet formation. The HICE is associated with glaciation (Ghienne et al, 2014; Mauviel and Desrochers, 2016), and the GICE is interpreted to represent global-scale enhanced organic carbon burial (Ainsaar et al., 1999; Saltzman and Young, 2005; Young et al., 2005). The GICE, therefore, was interpreted as evidence for a long-lived late Ordovician glaciation beginning in the uppermost Sandbian–lower Katian (e.g., Frakes et al., 1992; Pope and Steffen, 2003; Saltzman and Young, 2005). A short cooling interval inferred from $\delta^{18}O$ values is coincident with the GICE and was tentatively attributed to initiation of Ordovician icehouse conditions (Rosenau et al., 2012). However, a detailed study across the Sandbian–Katian boundary and into the GICE interval based on published and new conodont $\delta^{18}O$ data from Laurentia is consistent with a diachronous warming for the late Sandbian–early Katian and shows no evidence for cooling (Quinton et al., 2017). While the existence of Katian ice sheets remains an open question (Jones et al., 2017), the majority of conodont $\delta^{18}O$ data (Trotter et al., 2008; Buggisch et al., 2010; Herrmann et al., 2010; Rosenau et al., 2012; Quinton and MacLeod, 2014; Quinton et al., 2017) do not support long-term cooling through the Katian.

Unfortunately, there is a scarcity of published conodont $\delta^{18}O$ data from the Hirnantian Stage, a critical interval in understanding the relationship between Ordovician climate and extinction. Limited conodont $\delta^{18}O$ (Trotter et al., 2008) and clumped isotope data (Finnegan et al., 2011) suggest the presence of a significant positive $\delta^{18}O$ excursion in the Hirnantian. However, a smoothed Ordovician $\delta^{18}O$ curve derived from a locfit regression of 513 published conodont $\delta^{18}O$ values along with isotopic sea-surface temperatures does not depict this excursion (Fig. 20.1, adapted from Fig. 10.5 in Grossman and Joachimski, 2020, Ch. 10: Oxygen isotope stratigraphy, this book). The paucity of Hirnantian data hampers climate interpretations and highlights the need to widen the search for Hirnantian sections that could yield conodonts for isotopic analysis.

20.3.3 Cyclostratigraphy

Cyclostratigraphy is the rhythmic variation in sedimentary deposits resulting from astronomically forced paleoclimatic changes (Hinnov, 2013). Recent studies have related cyclical sedimentary variation in Lower Paleozoic strata (e.g., lithologic, biotic, and magnetic susceptibility cycles) to astronomical (Milankovitch) cycles and, hence, have expanded the geologic toolbox for fine-tuning the Ordovician time scale. In the following paragraphs we provide a few examples of the most recent work on Ordovician astrochronology.

Zhong et al. (2019) conducted a detailed cyclostratigraphic analysis of magnetic susceptibility logs from Middle Ordovician strata of the Huangnitang Darriwilian GSSP section and the nearby CJ-3 drillcore. They detected Milankovitch periodicities, in particular a 405-kyr eccentricity cycle that is considered to be stable throughout geologic time (Laskar et al., 2004). Zhong et al. (2019) used these data to construct an integrated 405-kyr-tuned floating astronomical time scale of the Middle Ordovician for the Huangnitang area, which gave the durations of 8.38 ± 0.4 Myr for the Darriwilian Stage and 1.97 ± 0.7 Myr for the Dapingian Stage. The duration of the Dapingian compares well with the GTS2020 time scale (1.97 ± 0.7 Myr compared to 1.84 ± 0.49 Myr, herein), but the Darriwilian is considerably shorter (8.38 ± 0.4 Myr compared to 11.24 ± 0.51 Myr, herein).

Fang et al. (2016) conducted a cyclostratigraphic analysis on the Pingliang Formation, in the Ordos Basin of North China. They detected astronomical cycles with periods consistent with long orbital eccentricity (≈ 405 kyr), short orbital eccentricity (≈ 100 kyr), obliquity (≈ 30.6 kyr), and precession cycles (19.2 and 16.2 kyr) in both lithological (limestone-shale couplets) and magnetic susceptibility series. Fang et al. (2016) hypothesized that precession-forced climate changes controlled the production of lime mud in the basin and was the main influence on the formation of the limestone-shale couplets, although both precession and obliquity controlled the production of biogenic silica. The Pingliang Formation at the Guanzhuang section is well known for its rich *N. gracilis* and *C. bicornis* age graptolite faunas (Chen et al., 2017), but unfortunately the zones are incomplete at the bottom and top, respectively, making precise calculations for the duration of the zones impossible at that locality.

Svensen et al. (2015) matched rhythmic variations in magnetic susceptibility logs of Late Ordovician strata at the Vollen section in the Oslo Region of southern Norway to long- (400-kyr) and short- (100-kyr) eccentricity cycles, and 30-kyr obliquity cycles. They then used both the radioisotopic age difference and cyclostratigraphy to calculate sedimentation rates between two age-dated Late Sandbian K-bentonites, the Kinnekulle and the upper Grimstorp (454.52 ± 0.50 and 453.91 ± 0.37 Ma, respectively). Extrapolating the cyclostratigraphy sedimentation rate of 0.92 cm/kyr to the interval above the Grimstorp K-bentonite, Svensen et al. (2015) calculated an age difference of 2.03 Myr from the upper Grimstorp bentonite up to the estimated Sandbian−Katian boundary position. Assuming a constant sedimentation rate, this provided an estimated age for the base of the Katian Stage of 451.88 ± 0.37 Ma. This calculation is somewhat younger

than the age of the Sandbian−Katian boundary in the age model provided herein (452.75 Ma). The Sandbian−Katian boundary age is well constrained by age-dated K-bentonites in the uppermost Sandbian and lowermost Katian (Sell et al., 2011 and MacDonald et al., 2017, respectively), constraints that likely make the Svensen et al. (2015) calculation about a million years too young (see Section 20.4.1).

Ballo et al. (2019) dated a suite of Sandbian K-bentonites at the Sinsen section, a locality very close to the Svensen et al. (2015) Vollen locality. Using a cyclostratigraphy-derived sedimentation rate of 1.12 cm/kyr and the onset of GICE as a proxy for the Sandbian−Katian stage boundary, Ballo et al. (2019) estimated the age of the base of the Katian Stage at 452.62 ± 0.39 Ma. This calculation is very close to, and within the standard error of the Sandbian−Katian boundary in the age model provided herein.

Lu et al. (2019) undertook a cyclostratigraphic analysis of a late Katian through Hirnantian succession in South China. Analyzing mainly variations of Fe^{3+} and total organic carbon concentrations, they were able to identify what they interpreted to be astronomically forced cycles in the succession over a range of time scales. The most significant were ~1.2-Myr (long-period obliquity) and ~405-kyr (long eccentricity) cycles. They related these to interpreted cycles of sea-level change and processes governing organic-rich shale formation. They also used the results to make relatively precise estimates of the durations of the biozones of the late Katian and Hirnantian. Their estimate of the duration of the Hirnantian (1.74 Myr ± 0.4) is approximately 0.36 Myr shorter than the calculation presented here (2.1 Myr). It is worth noting that the late Katian and Hirnantian have the fewest radioisotopic dates of any interval in the Ordovician.

Finally, Crampton et al. (2018) used a global graptolite composite range chart constructed using CONOP (see Sadler, 2020, Ch. 14B in Geomathematics, this book) to examine Lower Paleozoic graptoloid species turnover (speciation and extinction probabilities). They found that the time series of graptolite species turnover contained an ≈ 1.3 Myr cyclicity early in their clade history followed by a strong 2.6-Myr rhythm, with the transition between the two occurring between 460 and 453 Ma. These cycles are close to that of the modern day orbital obliquity and eccentricity grand cycles, respectively. Hence, these Milankovitch grand cycles may have played a significant role in Lower Paleozoic climate variation, which in turn influenced species turnover in an important group of Early Paleozoic zooplankton (Crampton et al., 2018).

20.3.4 Biotic and climatic events

The Ordovician Period encompasses two extraordinary biological events in the history of life on the Earth. The first is a great evolutionary radiation of marine life, which is known as the "Great Ordovician Biodiversification Event" or GOBE (Webby, 2004), and the second a catastrophic Late Ordovician extinction that wiped out nearly 85% of all marine species (Brenchley, 1989; Jablonski, 1991; Sepkoski, 1995; Sheehan, 2001; Bambach, 2006). These events are also related to significant changes that occurred in the Ordovician environment, providing important insight into the coevolution of the Earth and life.

The "GOBE" occurred over an interval of approximately 30 million years (from ca. 485 to 455 Ma), and estimates of the increase in marine generic diversity range from 3-fold (Webby, 2004) to over 10-fold (Kröger and Lintulaakso, 2017). However, the GOBE was not merely an increase in species richness, as it also included a substantial ecological restructuring that resulted in the establishment of modern aspect marine ecosystems (Droser and Sheehan, 1997; Servais et al., 2010; Servais and Harper, 2018). In addition, as noted by Webby (2004) and more recently Servais and Harper (2018), the GOBE is not a single event, but the sum of multiple radiations that occurred at different regional and temporal scales, encompassed both taxonomic and morphologic diversification (Miller, 2004), and included significant biotic immigration events (called BIME's by Stigall et al., 2017).

In a recent review of the duration, structure, and drivers of the GOBE, Servais and Harper (2018) noted that Ordovician diversification comprised three distinct phases—a plankton revolution that began in the late Cambrian and continued through the Early Ordovician (Servais et al., 2008, 2010), a diversification of level-bottom communities that spanned the late Tremadocian through Sandbian, and a radiation of Middle to Late Ordovician reef-building organisms (Webby, 2004). The plankton revolution included the radiation of graptolites, the first major group of zooplankton to be preserved in the fossil record. The most dramatic increase in Ordovician biodiversity occurred in the Darriwilian Stage among benthic organisms, particularly brachiopods, well after the onset of the plankton revolution, and culminated in a peak of diversity in the early to middle Late Ordovician (Sepkoski, 1995; Webby, 2004; Harper, 2006; Harper et al., 2013, 2015; Kröger and Lintulaakso, 2017). Trilobites, corals, echinoderms, bryozoans, gastropods, bivalves, nautiloids, graptolites, and conodonts also exhibit significant generic increases through the Middle Ordovician (Sepkoski, 1995; Webby, 2004). These benthic and planktic radiations lead to the recognition of the second of Sepkoski's (1995) statistically based marine evolutionary faunas, the Paleozoic Evolutionary Fauna, which replaced the trilobite-dominated communities of the Cambrian Evolutionary Fauna.

There has been a good deal of published discussion regarding the driving mechanisms of the GOBE. Miller and

Mao (1995) suggested a link between biological diversification and increased Ordovician volcanism and orogenesis. Trotter et al. (2008) related Middle Ordovician diversification to a reduction in sea-surface temperatures. Schmitz et al. (2008) noted that the major Middle Ordovician phase of the GOBE was coincident with asteroid breakup and an increase in bombardment by extraterrestrial material. Harper (2006) stressed the importance of parsing the global diversity signal into α-diversity (intracommunity), β-diversity (intercommunity), and γ-diversity (interprovince) components in order to understand the ecological and biogeographical dimensions of the GOBE.

Thus the Great Ordovician Biodiversification event was a complicated evolutionary transition that encompassed a substantial interval of geologic time and occurred differently at local, regional, and global scales. As noted by Servais and Harper (2018), the complexity of the GOBE likely requires an equally complex and varied set of driving mechanisms and enabling events.

The second biological event that dominates the pattern of Lower Paleozoic biodiversity is a catastrophic LOME. The LOME is considered to be the oldest of five great Phanerozoic extinction events (Sheehan, 2001; Bambach, 2006), which resulted in the loss of almost half of marine invertebrate genera and an estimated 85% of species. The LOME is also closely associated with major Late Ordovician climatic and oceanic changes, providing an important deep time analog for the effects of global environmental change on the Earth's biota. Like the GOBE, the LOME was a complex event that most studies indicate occurred in at least two pulses (latest Katian and mid-Hirnantian), and although clearly associated with climate and sea-level events, may have been driven by multiple mechanisms (see e.g., summaries in Sheehan, 2001; Melchin et al., 2013; Harper et al., 2014). Most recently, a number of authors have suggested that a major event involving the formation of a large igneous province was a key driving mechanism for the environmental changes that occurred in this interval (Jones et al., 2017; Bond and Grasby, 2017; Gong et al., 2017).

Chemostratigraphic and lithostratigraphic data suggest that Ordovician greenhouse—icehouse transitions and resultant Gondwanan glaciations probably began in the Sandbian or even earlier (Buggisch et al., 2010; Melchin et al., 2013; Pohl et al., 2016). However, the major Gondwanan ice sheet formation, as evidenced by oxygen isotope, carbon isotope and sequence stratigraphic records, occurs at the Katian—Hirnantian boundary near the base of the *M. extraordinarius* graptolite zone (Le Heron et al., 2010; Loi et al., 2010; Melchin et al., 2013; Ghienne et al., 2014). Most studies (e.g., Brenchley et al., 2001; Sheehan, 2001; Rong and Zhan, 2004; Rong et al., 2006; Harper et al., 2014) describe two phases of Hirnantian mass extinction that coincide

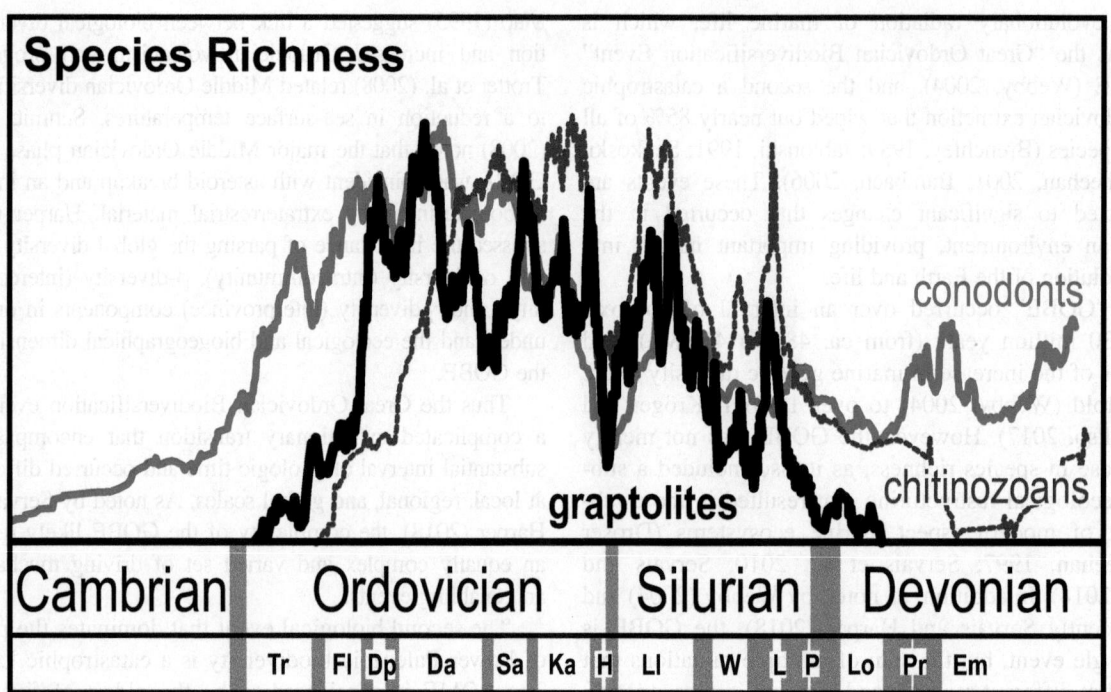

FIGURE 20.14 Taxon richness trajectories of three clades used for Ordovician biozonation. By rescaling all three curves to their Lower Paleozoic richness maxima, the entire graptolite trajectory (*solid black line* based on 2291 taxa from 619 sections) is compared with Lower Paleozoic portions of longer clade trajectories for conodonts (*red line*; 3788 taxa from 1221 sections) and chitinozoans (*blue dashed line*; 231 taxa from 1210 sections). The three curves were composited separately then aligned by interpolated age. They summarize a somewhat earlier state of the database than that used to calibrate the Ordovician and Silurian periods in this book, but without culling of rare taxa known from few sections. Epoch and stage boundaries are positioned with gray vertical bands to represent 2-sigma age uncertainty.

with rapid, climate-driven changes in oceanic anoxia. The first extinction occurred at the onset of glaciation and with the loss of anoxic conditions at the end of the Katian. The second extinction occurred at the termination of glaciation and coincided with the return of anoxic conditions during the late Hirnantian−early Rhuddanian. In an interesting alternative hypothesis, Holland and Patzkowsky (2015) suggested that the apparent two-phase extinction could also be an artifact of a stratigraphic and paleontological record controlled by eustatic sea-level changes. A single extended extinction interval could be preserved as two distinct phases because the last occurrences of fossil species will tend to cluster at sequence-bounding subaerial unconformities, major flooding surfaces, and surfaces of forced regression and stratigraphic condensation. Recently, Wang et al. (2019) also suggested that the LOME comprised just one phase of mass extinction at the Katian−Hirnantian boundary, which was followed by a lengthy three-phase recovery.

Among clades used in the construction of the Ordovician time scale, graptolites suffered some of the greatest extinction intensities during the LOME. The species-rich and morphologically diverse Diplograptina were driven to extinction by the Hirnantian Earth system changes and were replaced by the Neograptina, a far smaller and morphologically

less complex group during the Ordovician (Bapst et al., 2012). Lower Paleozoic species richness curves that illustrate the Ordovician diversification and extinction for graptolites, conodonts, and chitinozoans are provided in Fig. 20.14.

20.4 Ordovician time scale

20.4.1 Radioisotopic dates

Cooper and Sadler (in Gradstein et al., 2012) summarized the history, current standards, and biostratigraphic framework for the radioisotopic dates used in GTS2012. The distribution of those dates in the Ordovician Period was uneven, with the majority coming from the Middle and Upper Ordovician. During the following 8 years to our GTS2020 compilation, 37 new dates have been calculated, primarily from zircons in K-bentonite beds that occur within fossiliferous strata. Importantly, nearly 25% of these new dates come from Lower Ordovician strata and are well-constrained biostratigraphically. All the new dates, except where noted later, are calculated from the $^{206}Pb/^{238}U$ decay system by the chemical abrasion-isotope dilution thermal ionization mass spectrometry (CA-IDTIMS) method from analyses composed of single zircon grains or fragments of grains (Mattinson, 2005;

Mattinson, 2010). Dates previously reported under various uranium and lead isotopic decay systems ($^{206}Pb/^{238}U$, $^{207}Pb/^{235}U$, and $^{207}Pb/^{206}Pb$), which are retained from GTS2012 and used in the current Ordovician time scale, have, where possible, here been mapped onto the $^{206}Pb/^{238}U$ decay system (see Schmitz et al., 2020, Ch. 6: Radioisotope geochronology, this book). Similarly, error propagation of analytical, tracer/standard, and decay constant errors has been treated consistently and rigorously for each published age. All dates have been reappraised for analytical and biostratigraphic reliability. The radioisotopic ages (and their biostratigraphic association) used in the current calibration of the Ordovician time scale are listed in Appendix 2.

Of 37 new radioisotopic dates, 15 are biostratigraphically associated with conodonts, 13 with graptolites, 1 with chitinozoans, and 8 can be directly related to zonations from at least two different fossil groups. With respect to the Ordovician series and stages, 19 new dates are from Upper Ordovician strata (1 from Hirnantian Strata, 6 from lower Katian successions, and 12 from the Sandbian), 7 dates are from Middle Ordovician rocks (5 from Darriwilian successions and 2 from Dapingian units), and 11 dates are from the Lower Ordovician (10 from the Floian and 1 from the upper Tremadocian). The following comments explain the details of the biostratigraphic constraints on the dated samples.

Landing et al. (2000) published a date of 486.78 ± 2.57 Ma from a crystal rich tuff in the Dolgellau Formation in North Wales. This level was 5 m below a horizon that contained *R. praeparabola*, indicating a level near the Cambrian/Ordovician boundary. Schmitz et al. (2020, Ch. 6: Radioisotope geochronology, this book) reanalyzed the zircons from that tuff using modern state-of-the-art techniques and calculated a date of 490.10 ± 0.57 Ma. Hence, the Cambrian/Ordovician boundary is now calculated at 486.85 Ma, 1.45 Myr older than the age in GTS2012.

Normore et al. (2018) reviewed the three Lower Ordovician dates used in the GTS2012, and noted that two of them from the Tremadocian (481.13 ± 2.76 Ma, recalculated from Landing et al., 1997) and Floian (472.98 ± 0.79 Ma, from Cooper et al., 2008) need to be treated with caution due to small numbers of zircons or outdated analytical techniques. Normore et al. (2018) published seven new CA-IDTIMS U−Pb zircon ages from a sequence of K-bentonite beds in Olympic 1 drill core from the Canning Basin, western Australia. The bentonites occur in the Nambeet and Willara formations, which span the late Tremadocian through early Dapingian stages, and are well constrained within a conodont biostratigraphic framework. The oldest bed has a weighted mean U−Pb dates range from 479.37 ± 0.16 Ma within the late Tremadocian *P. proteus* conodont biozone, followed by two beds just above the base of the *S. bilobatus* conodont zone that

yielded early Floian zircon dates (477.07 ± 0.21 and 477.03 ± 0.16 Ma). This conodont association also suggests a level very close to the base of the *T. approximatus* graptolite zone. These beds are succeeded by three dates in the middle Floian *Oepikodus communis* Zone (472.82 ± 0.13, 471.78 ± 0.13, and 471.32 ± 0.11 Ma), one near the boundary between the Floian and Dapingian stages within the *Jumudontus gananda* conodont zone ($=O. evae$ Zone, 470.18 ± 0.13 Ma).

Thompson et al. (2012) provided 4 new ID-TIMS U−Pb zircon ages of K-bentonites from measured sections of the San Juan Formation (Talacasto and Cerro La Chilca sections) in the Argentine Precordillera. These dates provide stratigraphically consistent ages that range from 473.45 ± 0.70 to 469.53 ± 0.62 Ma, and are tied to samples taken for carbon isotope chemostratigraphy, but unfortunately lack a direct biostratigraphic framework. Thompson et al. (2012) and Thompson and Kah (2012) plotted carbon isotope samples from several Precordilleran sections (including those that bracketed their dated K-bentonites) along with other samples from the Table Head Group at Table Point, western Newfoundland. All these samples were then correlated with a negative pre-MDICE excursion described by Buggisch et al. (2003). The Buggisch et al. (2003) carbon isotope curve was tied to conodont biostratigraphy, and Thompson et al. (2012) and Thompson and Kah's (2012) chemostratigraphic cycle "match" correlated this part of the San Juan Formation with *Baltoniodus navis/Microzarkodina parva* conodont zones and with the Table Head Group. It also forces the Table Head Group to be of Dapingian age. As noted by Albanesi et al. (2013), this chemostratigraphic correlation is likely to be erroneous because it results in incorrect conodont ages for both the San Juan Formation and the Table Head Group. Baldo et al. (2003) published a date (469.8 ± 2.3 Ma; U−Pb SHRIMP) from a bentonite low in the Talacasto section, which Albanesi et al. (2006) correlated with the base of *Oepikodus intermedius* Zone. Unfortunately, the exact stratigraphic position of the bed dated by Baldo et al. (2003) is unknown and we are not using SHRIMP dates in our calibration. Hence, Thompson et al.'s (2012) important dates for the Lower to Middle Ordovician still require better biostratigraphic control.

Lindskog et al. (2017) report a zircon U−Pb date of 467.50 ± 0.28 Ma from the "Likhall" bed within the meteorite-bearing interval of the Kinnekulle section, southern Sweden. The "Likhall" bed is precisely located at the boundary between the *Lenodus variabilis* and *Yangtzeplacognathus crassus* conodont zones. This is close to the boundary between the *Aulograptus cucullus* and *Holmograptus lentus* (or, *Didymograptus artus*) graptolite zones, as well as that between the regional *Asaphus expansus* and *Asaphus raniceps* trilobite zones (Lindskog et al., 2017). This level is likely just above the base of the *Levisograptus dentatus* graptolite zone and within the lower Darriwilian Da2 Zone of the Australasian zonal scheme. Lindskog et al. (2017) then used this U–Pb date and cosmic-ray exposure ages of co-occurring meteoritic materials to calculate sedimentation rates at Kinnekulle, and extrapolate an age of ≈469.6 Ma for the base of the Darriwilian. This date is consistent with the age of the Darriwilian base in the GTS2020 time scale, 469.42 ± 0.93 Ma.

Sell et al. (2013) analyzed single zircon crystals from K-bentonites in two sections of graptolite-rich strata on the Port au Port Peninsula of western Newfoundland. One bed occurs at 41 m, in the Mainland Section (Cape Cormorant) and one, bed D, from the West Bay Centre quarry. The samples from the Mainland Section at 41 m and West Bay Centre Quarry D yielded weighted mean $^{206}Pb/^{238}U$ ages of 464.5 ± 0.4 and 464.57 ± 0.95 and are associated with grapto-lites of the *Pterograptus elegans* and *Holmograptus spinosus* graptolite zones (Darriwilian), respectively. The K-bentonite from the West Bay Centre Quarry is also associated with cono-donts from the Upper *H. holodentata* conodont zone (Klebold et al., 2019) providing an important link of a radioisotopic date with both graptolite and conodont biostratigraphy.

Sell et al. (2013), Bauert et al. (2014b), and Svensen et al. (2015) reported dates from the late Sandbian Kinnekulle K-bentonite and stratigraphically nearby beds in Baltoscandia. Sell et al. (2013) analyzed eight zircon crystals from Kinnekulle K-bentonite at the Vasagard Section on the island of Bornholm, Denmark, and five zircon crystals from bed 46 in the Ristikula 174 core in SW Estonia. These yielded ages of 454.41 ± 0.17/0.21/0.53 Ma, and 454.65 ± 0.56/0.58/ 0.75 Ma, respectively. Bauert et al. (2014b) used a laser ablation inductively coupled plasma mass spectrometer to obtain single-grain zircon U–Pb ages for the Kinnekulle K-bentonite sampled at its type locality at Mossen on Kinnekulle in southern Sweden, and from the Paaskiila Hillock at Tallinn in northern Estonia. The weighted mean age for the Mossen zircons is 453.4 ± 6.6 Ma and that for Tallinn zircons is 454.9 ± 4.9 Ma. These ages generally agree with or are within the 2σ error of previously published dates for the Kinnekulle (Sell et al., 2013; Min et al., 2001; Tucker, 1992, recalculated by Cooper and Sadler, 2012; and Tucker and McKerrow, 1995), although their 2σ error is rather large for current standards. Svensen et al. (2015) dated the Kinnekulle K-bentonite and the uppermost tephra layer (the upper Grimstorp K-bentonite) in the Upper Ordovician of the Oslo region, Norway. The bentonites were sampled from the upper part of the Arnestad Formation (Sandbian) south of Oslo and gave ages of 454.52 ± 0.50 and 453.91 ± 0.37 Ma for the Kinnekulle and Grimstorp, respectively.

The Kinnekulle K-bentonite is biostratigraphically assigned to the upper *Amorphognathus tvaerensis*

conodont zone, which correlates with the upper *Diplograptus foliaceus* graptolite zone (=*Climacograptus bicornis* Zone) (Bergström et al., 2011) and the *Spinachitina cervicornis* chitinozoan zone (Nõlvak and Grahn, 1993). The short-ranged and stratigraphically valu-able chitinozoan species *Angochitina multiplex* (Schallreuter) appears just above the Kinnekulle K-bentonite bed in Baltic drill cores. The Kinnekulle, Grimstorp, and Ristikula Bed 46 K-bentonites are strati-graphically below but close to both the GICE and the Sandbian–Katian boundary (Bergström et al., 2011), and these new and increasingly precise dates place an impor-tant lower constraint on the age of the stage boundary.

Ballo et al. (2019) analyzed a suite of five Sandbian K-bentonites, including the Kinnekulle and the upper Grimstorp K-bentonites, from the Arnestad Formation at the Sinsen section in the Oslo region of southern Norway. Four of these were used in the Ordovician time scale cali-bration. Ballo et al. (2019) bed 1 (EB16-S-0) yielded a date of 456.81 ± 0.30/0.38/0.61 Ma, and bed 16 (EB16-S-28.63) has a calculated date of 456.13 ± 0.39/0.45/ 0.66 Ma. Bed 30 (EB16-S-39.60) is the Kinnekulle K-bentonite and yielded a date of 454.06 ± 0.43/0.49/ 0.68 Ma. Bed 32 (EB16-S-42.71) is the upper Grimstorp K-bentonite and is dated at 453.48 ± 0.74/0.78/0.91 Ma. Ballo et al. (2019) did not provide biostratigraphic infor-mation in association with these dated K-bentonites, but as noted above the Kinnekulle K-bentonite is referable to the upper *A. tvaerensis* conodont zone and the *C. bicornis* graptolite zone. Hence, the Sinsen section K-bentonite suite spans the middle and upper Sandbian.

In North America, Sell et al. (2013) and Oruche et al. (2018) calculated ages from two bentonites, the Deike and Millbrig that are similar in biostratigraphic age to the Kinnekulle. Sell et al. (2013) recorded a date of 453.74 ± 0.20/0.28/0.56 Ma from the Deicke K-bentonite and an age of 452.86 ± 0.29/0.34/0.59 Ma from the Millbrig, both at Shakertown, Kentucky, United States. Oruche calculated a date of 453.36 ± 0.38 Ma for the Millbrig, sampled from the top of the L'Orignal Formation in the Ottawa Embayment. These new U–Pb zircon ages show an improvement in analytical precision over previously published radioisotopic age determinations on the same ben-tonites (Cooper and Sadler, 2012). These dates are within the error of other U–Pb zircon age determinations for the Millbrig and Deicke K-bentonites (Tucker, 1992; Tucker and McKerrow, 1995; Schoene et al., 2006) and are similar to recent Ar–Ar biotite and sanidine ages for the Millbrig and Deicke (Min et al., 2001; Smith et al., 2011) when accounting for the K–Ar primary standard and decay con-stant errors as shown by Schoene et al. (2006).

Sell et al. (2013) also provided dates for two K-bentonites in the upper Womble Shale at Black Knob Ridge, the GSSP for the base of the Katian Stage

(Goldman et al., 2007a,b). These beds yield ages of 453.98 ± 0.33/0.38/0.62 Ma and 453.16 ± 0.24/0.31/0.57 Ma and are stratigraphically 5.5 and 6.6 m below the Katian GSSP level. They are directly associated with *C. bicornis* Zone graptolites and *A. tvaerensis* Zone conodonts. The Womble K-bentonites are of similar age but have different geochemical signatures than the Deike and Millbrig (Sell et al., 2015), which also occur in the upper Sandbian (Upper *C. bicornis* graptolite zone and lower *Phragmodus undatus* conodont zone). All these beds are comparable in age to the Kinnekulle in Europe and have a similar stratigraphic relationship (just below) to the Sandbian−Katian stage boundary and GICE excursion.

MacDonald et al. (2017) calculated exceedingly precise dates for several K-bentonites from the Flat Creek and Indian Castle Members of the Utica Shale in the Mohawk Valley of New York State, United States. These beds yield dates of 452.63 ± 0.06 Ma, 451.71 ± 0.13 Ma, 451.26 ± 0.11 Ma, and 450.71 ± 0.20 Ma and occur within the lower Katian *Corynoides americanus* (two older dates) and *Diplacanthograptus spiniferus* (two younger dates) graptolite zones, respectively. In conjunction with the upper Sandbian dates discussed in the previous paragraph, the MacDonald et al. (2017) dates bracket and tightly constrain the age of the Sandbian−Katian boundary and the GICE, allowing for global correlation of the stage boundary into disparate biofacies.

A paper by Ling et al. (2019) that contains four new radioisotope dates from upper Katian and Hirnantian strata in northeastern Yunnan, South China, was published after this chapter and age model was finalized. Their calculated duration of the Hirnantian Stage differs substantially from our estimate. We think that the Ling et al.'s (2019) article suffers from a number of methodological and analytical problems, including geochronologic (insufficient number of zircons analyzed for the main horizons), biostratigraphic (the source stratigraphic article by Tang et al. (2017) did not illustrate the key index graptolites or provide a range chart to indicate uncertainties, including for the limits of the Hirnantian Stage), and sedimentologic (the Hirnantian duration was based on unrealistic extrapolation of sediment accumulation rates that do not adjust for the substantial environmental and sea-level changes that occur during this time interval). Until these issues are resolved we cannot utilize these four dates in the CONOP graptolite-date composite used for the construction of the Ordovician time scale for GTS2020.

Altogether, 51 radioisotopic-dated samples are regarded as having sufficient analytical quality and biostratigraphic constraint to be used for calibration of the Ordovician and Silurian time scales (Appendix 2). They range from latest Cambrian to earliest Devonian in age. In an important advance in Ordovician geochronology, 11 new lower Ordovician dates have been added to the time scale

calibration and the 42 intra-Ordovician dates are now evenly distributed. All Ordovician samples are U−Pb zircon ages that are analyzed by the CA-IDTIMS method.

20.4.2 Building the Ordovician and Silurian Composite Standard

Although the number of reliably dated Ordovician and Silurian tuff beds has increased significantly since the calibration of lower Paleozoic periods for Geologic time scale 2012, three fundamental and challenging tasks remain. First, because units of the time scale are recognized by their fossil content, the position of dated beds must be determined as precisely as possible relative to fossil taxon ranges. This task demands much of the raw data collection: either the fossils and bentonites must be collected from the same stratigraphic sections or the ash falls must be characterized in sufficient detail that they can be traced from the dated locality to others from which fossils have been described. Carbon isotope excursions assist with some of these linkages, although as the number of described early Paleozoic excursions has increased so has the uncertainty of their correlation. Next, a proxy time scale must be developed on which to interpolate ages from the dated beds to unit boundaries that lie between them. This scale should approximate the relative durations of stages, zones, and subzones without recourse to radioisotopic dates. The third task tracks uncertainties that accumulate at each step in the process. Age interpolation typically adds significant uncertainty to Paleozoic time scales, because the proxy scale requires weak assumptions about the likely distribution of time across strata. The following paragraphs describe our approach to developing a numerical proxy scale on which to interpolate age.

The options for an interpolation scale are limited. Assumptions are required that, explicitly or implicitly, simplify the relationship between elapsed time and either biotic turnover, or rock thickness, or depositional cycles (Sadler et al., 2009). For example, several authors have made interpolations based upon the assumption that the relative durations of units are proportional to the number of biozones that they contain (Boucot, 1975; McKerrow et al., 1985; Harland et al., 1990) or to their thicknesses (Churkin et al., 1977). Recently, the durations of two Middle Ordovician stages (Zhong et al., 2018) and an early Silurian stage (Gambacorta et al., 2018) have been estimated by tuning lithological cycles to periodic orbital fluctuations. These simplifying assumptions require judicious selection of reference sections.

The notion of enhancing a reference section by projecting information into it from other locations is more than half a century old (Shaw, 1964). Shaw's graphic correlation technique has often been used to build composite sections of Ordovician and Silurian taxon ranges (Sweet, 1984, 1988, 1995; Finney et al., 1996; Kleffner, 1989; Fordham, 1992; Cooper, 1999b; Fan et al., 2013). Traditional graphic correlation projects information, one section at a time, into a reference section and must then mitigate bias introduced by the order in which sections were added. The final composite section has the thickness scale of the reference section, but can resolve far more event levels than any one contributing location. It would be a good proxy section for age interpolation. Unfortunately, the iterative compositing process becomes unwieldy as the number of sections and events increases. Additional assumptions are needed when the time span of the problem exceeds the time span of the longest section.

We employ a CONOP approach (Kemple et al., 1995; Sadler, 2003, 2012; Sadler et al., 2011, 2014) that uses evolutionary programming to perform a graphic correlation that is multidimensional in the sense that it considers all the local stratigraphic sections simultaneously. The order of addition of information is not an issue. The heuristic optimization algorithm uses iterative trial-and-error analogous to those that search for most parsimonious cladograms. It avoids the need to select a standard reference section by searching first for a composite sequence of all range-end events with minimal misfit to all the field observations. Misfit between local sections and the composite is determined by the net number of event horizons through which observed range ends must be extended in order to make all the observed range charts fit the same composite sequence. Chemostratigraphic excursions, tuff beds, and time-stratigraphic marker events can all be included in the optimization process (Sadler, 2012). At this stage, however, the composite is a purely ordinal sequence of events, based only on superposition of events in local sections. Assumptions for scaling relative time intervals between events are selected *after* the best fit sequence has been determined. In this way, scaling assumptions do not compromise the sequencing task—an advantage over traditional graphic correlation, adapted from Edwards' (1978) no-space graph variant.

The process of scaling the optimized composite sequence for time scale interpolation is summarized by Sadler and Cooper (2003) and described in detail by Sadler et al. (2009). Briefly, the total thickness of each section can be rescaled according to the number of events that it spans in the composite sequence. This assumes that net biologic change is a more reasonable guide to relative duration, in the long term, than raw stratigraphic thickness. The spacing of every successive pair of events in the composite can be replaced by averages of the raw or rescaled spacing of events in the sections. This assumes that relative thickness is a better guide to relative duration in the short term. The proxy scale is, therefore, derived from all of the sections and based on ratios of rock thickness, not absolute thickness. The goal is to minimize the influence of aberrant sections, incomplete preservation, and nonuniform, unsteady depositional rates. Where biotic turnover is rapid, many range-end events fall at the same horizon in measured sections; these "zero spacings" are included in the averaging process and prevent high diversity from being misinterpreted as long time intervals. The procedure is most vulnerable to intervals of extraordinarily low diversity, especially those near the origination and extinction of a clade, for example. As the number of taxa and other events increases, these concerns diminish. As the number of dated events increases, the scaling method becomes less critical and even the uniformly spaced ordinal composite sequence may support age interpolation. For the Ordovician–Silurian graptolite clade, the entire supporting database has now grown to 837 sections, yielding 2651 taxa, and 385 other events. That amounts to 33,480 local records of taxon range ends, marker beds, and isotopic excursion intervals that inform the optimization. The Cambrian through early Devonian portion of the conodont database consists of 1040 sections, 1289 taxa, and 440 other events (36,630 local records).

Cooper and Sadler (2012) interpolated the ages of Ordovician and Silurian unit boundaries on a proxy scale of first and last appearances of graptolite taxa, composited from 512 contributing sections. Some of the radioisotopic-dated bentonites and intervals of carbon isotope excursion that are associated with these fossils are more robustly linked to conodont-bearing limestone sections than to graptolite-rich shale sections. For this chapter, therefore, composite sequences were developed separately from graptolite- and conodont-bearing sections. Both clades are pandemic and their decades of use in biostratigraphy ensures adequate taxonomic stability. The conodont composite sequence provides better constraint on late Cambrian and earliest Ordovician events that precede the rich graptolite record. The dual approach is made possible by a much enlarged database.

20.4.2.1 Calibration of zone and stage boundaries by composite standard optimization

Events in the scaled composites are spaced along an axis of arbitrary "composite units" to provide proxy time scales. Graptolite and conodont zonal boundaries and stage boundaries are located in the scaled composite to produce a relative time scale. The zones thus recognized are chronozones rather than biozones in the sense that they are tied to estimates of global range-end datums, rather than local or regional bioevents. Zone and stage boundary ages are estimated, as before, by interpolation between dated events in the composite. With more dates available than in previous calibration exercises, it has become necessary to improve the way that dated events are incorporated in the optimization process.

When there were few dated events, each could be tagged with its analytical mean age as a direct indicator of the true sequence of events dated in separate sections. With more dated events, many pairs of age determinations now overlap within their 2-sigma analytical uncertainties. Mean ages may no longer always reveal the true order of events. Accordingly, dated events are now represented as uncertainty intervals (Sadler, 2012). Age-scaled "pseudo-sections" are added in which dated events are located as pairs of values; that is, paired maximum and minimum values determined by the 2-sigma analytical uncertainties. Separate age-scaled sections are used to distinguish dates based on the U−Pb system and the Ar−Ar system. Internal comparison of Ar−Ar dates may provide a trustworthy indicator of relative age, while only U−Pb system dating is used for calibration.

In the sampled sections the uncertainty interval for a dated event is typically clamped down to the sampled horizon but may be wider if the sample position is uncertain. Other events preserved in the sampled sections may establish that the true order of dated events is the opposite of their mean age estimates. The simplest example would be two imprecisely dated events in the same section; if the two uncertainty intervals overlap, superposition may reliably contradict the order of mean ages. When the optimizing algorithms are seeking the best fit composite sequence, local uncertainty intervals may be shrunk to fit (minimized to preserve the correct stratigraphic order), unlike taxon ranges, which are stretched to fit. This matches the logic that true taxon range ends may lie *outside* the locally preserved range while true mineral ages most likely lie *within* analytical uncertainty. The uncertainty interval may shrink in the age-scaled pseudo-section, in response to other stratigraphic evidence (superposition) of relative age. One result of combining large amounts of time-stratigraphic information may therefore be a reduction in the uncertainty of age of a dated event. This feature allows dates with large uncertainties to be included in the data set without compromising the results, because the uncertainty limits are more trustworthy than the analytical mean age.

Coding dated events as uncertainty intervals also simplifies the determination of uncertainty in the stratigraphic placement of these events. When dated events were tagged with a single (mean) age, stratigraphic uncertainty was determined by comparing the placement of a dated event across several best fit sequences that resulted from different data sets or different measures of fit (e.g., Cooper and Sadler, 2012). Now that dated events are tagged with both a maximum and a minimum likely age, these paired ends of the uncertainty interval land separately in the best fit sequence. Because the shrinkage of a local uncertainty interval is treated as a measure of misfit, equivalent to the stretching of a local taxon range, the maximum and minimum positions in the composite will remain apart unless forced closer together in response to other information. Their final separation in the optimal sequence is the stratigraphic uncertainty, which can now be determined in a single pass of the optimization routine. A bivariate plot of the scaled composite against the age-scaled pseudo-section displays both analytical uncertainty and stratigraphic uncertainty. Ages may be interpolated, by regression, for all levels in the scaled composite sequence. Regressions (global polynomials, locally weighted polynomials, or splines) may be determined separately for the older and younger ends of the uncertainty intervals. Of several scaling options in CONOP, one is chosen that leads to a simple yet well-fit regression. The distribution of plotted ages need not be linear; assumptions underlying the proxy time scaling are weak. LOESS regressions are nonparametric (Jacoby, 2000), requiring no assumptions about the relationship between age and position in the composite but, for age calibration, the local sliding window must be wide enough to ensure that the slope is monotonic. Wider windows dampen the influence of outlier points derived from less precise dates. Fig. 20.15B compares the LOESS regressions calculated with smoothing factors of 0.2, 0.33, and 0.5; and Table 20.1 provides age estimates for Ordovician stage and graptolite zone bases and durations using polynomial, LOESS, and cubic spline scaling (also compared graphically in Fig. 20.15A−C).

Unit boundaries may be calibrated three ways. Most often, biozone boundaries must be located in the composite sequence, using definitive range ends and characteristic taxa; that is, treating the composite as one would a real stratigraphic section. Local biozone boundary placements are deliberately *not* included among the events that are sequenced (Cody et al., 2008).

Second, GSSPs have been established for some stage boundaries in sections that yield conodonts and/or graptolites. These GSSPs are part of the optimized database and can be projected from the scaled composite into the age-scaled section. Like a taxon found at only one level in a single section, however, the constraints on GSSP location in the composite sequence are not necessarily tight. Uncertainty in the age calibration of each GSSP depends, in part, upon the graptolite and conodont ranges preserved in the stratotype section.

Third, many published sections with richly populated range charts include their authors' best estimates of the position of stage boundaries. Sometimes the boundary position has been based on fossil groups other than graptolites and conodonts. These local boundary estimates have been included in the compositing exercise as very conservative (wide) uncertainty intervals. Like dated

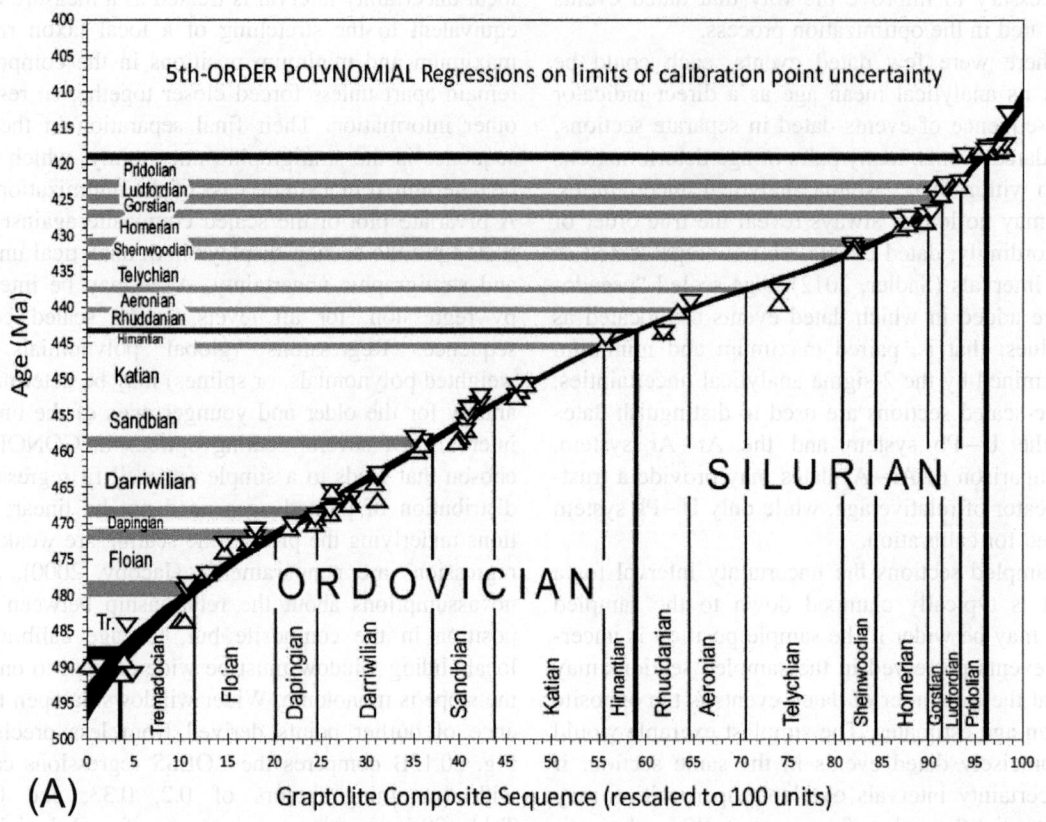

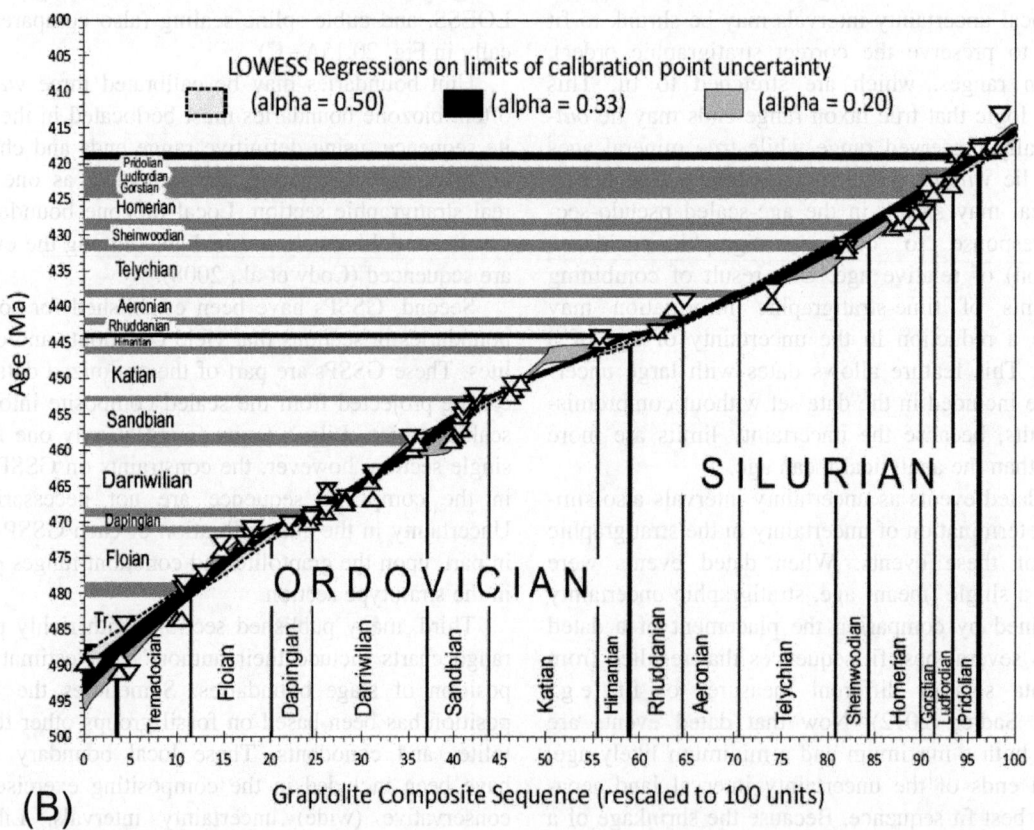

FIGURE 20.15 Geochronology of the Ordovician and Silurian stages with boundary ages interpolated from the CONOP graptolite composite using: (A) fifth-order polynomial regression; (B) LOESS regression comparing smoothing factors of alpha = 0.2, 0.33, and 0.5; and (C) cubic spline. The curves are fitted

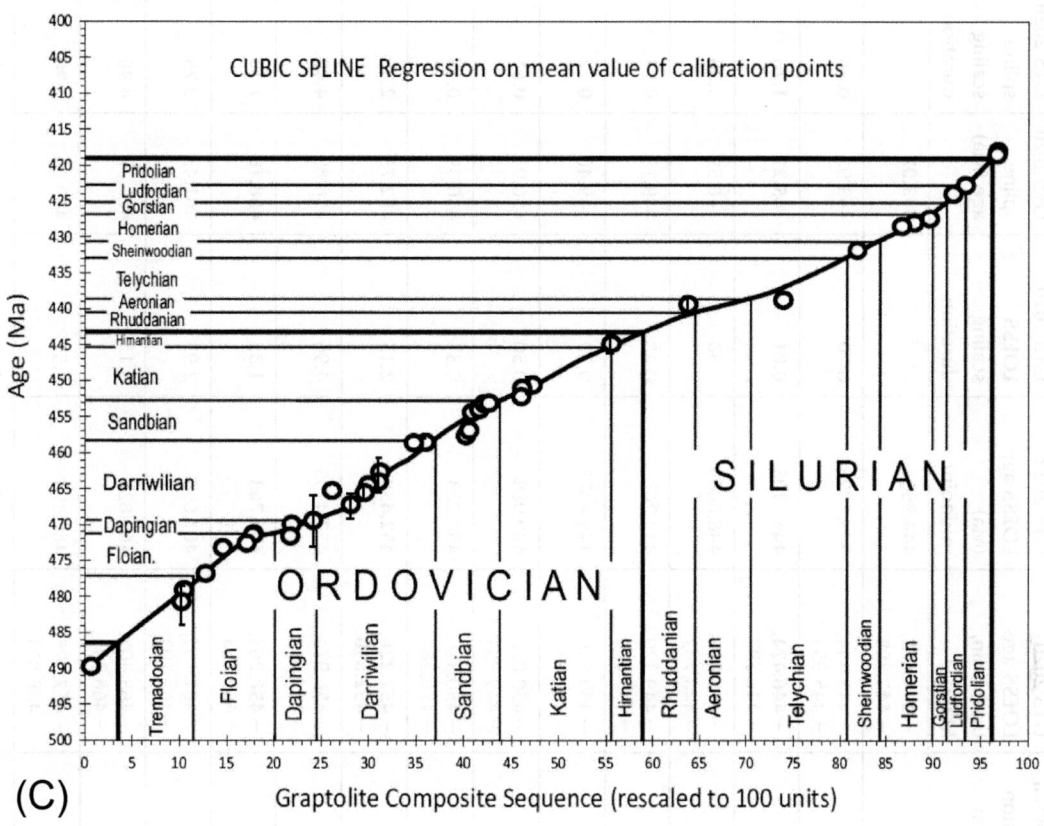

(C)

Graptolite Composite Sequence (rescaled to 100 units)

FIGURE 20.15 (Continued)

through the 49 radioisotopic-dated ash beds used for calibrating the Ordovician and Silurian scales (Appendix 2) plotted on the CONOP graptolite composite (X-axis rescaled to 100 percentage units) against their analytical age (Y-axis). Uncertainty bands are achieved for parts (A) and (B) by regressing separately the upper and lower 2-sigma analytical uncertainties. Compared to GTS2012, the base Ordovician is 1.45 Myr older (486.85 ± 1.53 vs 485.4 ± 1.9 Ma), the base of the Silurian is 0.7 Myr younger (443.1 ± 0.9 vs 443.8 ± 1.8 Ma), and the base Devonian remains nearly identical in age (419.0 ± 1.8 vs 419.2 ± 3.2 Ma).

events, they facilitate the earliest trials in the optimization process by guiding the coarse seriation of sections that do not overlap in age. Although they lose influence before the fine-tuning end of the optimization process, their position and size of these uncertainty intervals in the optimal composite section give an indication of common practice and consensus in stage boundary placement.

Note: Technical information on the CONOP method and applications is given in the website: https://drive.google.com/drive/folders/1xj5uZATaANSvhfauB-Je_YXsjSM468OH.

20.4.2.2 Uncertainty in the composite standard sequence

When Sadler et al. (2009) first described the calibration process, they discussed four sources of uncertainty. One is analytical uncertainty in radioisotopic age estimation, usually expressed as 2-sigma. This contribution to the calibration process was considered to be minimized because dated events are used as an ordered ensemble and boundary ages are derived by regression. The number of dated events in the ensemble has

increased and, as explained previously, the treatment of dates as "shrink-to-fit" intervals reduces some of the uncertainty.

A second uncertainty ("best fit" intervals of Sadler and Cooper, 2003) derives from nonuniqueness in the composite sequence that best fits the input data; the biostratigraphic sequencing problem is underdetermined. Although event placement is relatively stable across repeated optimization of the same data, there are irresolvable knots of events and some predictable anomalies in the composite placement of poorly constrained taxa. Where the order of two events is not resolvable, they emerge at the same level in the scaled composite. The resolving power of the composite section increases with the number of well-constrained taxon ranges.

For taxa known from only a single section or from minimally fossiliferous sections, there may be very little superpositional information with which to constrain the position of their range-end events in the composite sequence. Few observed coexistences with other taxa are likely. Such taxa tend to be placed somewhat arbitrarily or in low-diversity parts of the composite sequence, where

TABLE 20.1 Comparison of Ordovician age scaling using different methods of statistical regression: Columns are stages, graptolite zones, Level in CONOP composite, polynomial regression age, LOESS regression age, GTS (Spline) age, 2-sigma max−min, and interval durations.

Zone	Level in 2020 composite (%)	GTS2020 Polynomial age (Ma) (Min, Max)[1]	Polynomial age (Ma)[2]	Polynomial regression duration	GTS 2020 LOESS age (Ma) (Min, Max)[3]	GTS 2020 LOESS age (Ma) midpoint	GTS 2020 LOESS scaling duration	GTS 2020 spline age (Ma)	GTS 2020 spline scaling duration
Silurian GSSP level	**58.967**	**− 443.362, − 443.442**	**− 443.50**		**− 444.683, − 445.308**	**444.9955**		**443.07**	
Metabolograptus persculptus Zone Bo5	57.466	− 444.227, − 444.337	− 444.37	0.87	− 445.444, − 445.954	445.699	0.70	443.95	0.88
Metabolograptus extraordinarius Zone, Bo4—Hirnantian Base	55.478	**− 445.417, − 445.579**	**− 445.57**	1.20	**− 446.424, − 446.597**	**446.5105**	**0.81**	**445.21**	1.26
Paraorthograptus pacificus pacificus Zone, Bo3	50.555	− 448.575, − 448.901	− 448.76	3.19	− 448.885, − 448.732	448.8085	2.30	448.59	3.38
Dicellograptus complexus Zone (≈ Bo2)	49.768	− 449.105, − 449.460	− 449.30	0.54	− 449.320, − 449.200	449.26	0.45	449.13	0.54
Dicellograptus complanatus Zone (≈ Bo1)	49.365	− 449.379, − 449.749	− 449.57	0.28	− 449.540, − 449.435	449.4875	0.23	449.40	0.27
Dicellograptus gravis Zone, Ea4	48.458	− 450.001, − 450.406	− 450.20	0.63	− 450.022, − 449.954	449.988	0.50	450.01	0.61
Styracograptus tubuliferus Zone, (≈ Ea3)	47.880	− 450.401, − 450.828	− 450.61	0.40	− 450.320, − 450.286	450.303	0.31	450.39	0.38
Diplacanthograptus caudatus Zone (≈ Ea1) – GSSP Katian Base	**43.830**	**− 452.709, − 453.259**	**− 452.94**	2.33	**− 452.226, − 452.690**	**452.458**	2.15	**452.75**	2.36
Climacograptus bicornis bicornis Zone (≈ Gi2)	38.582	− 457.151, − 457.897	− 457.43	4.49	− 456.974, − 457.853	457.4135	4.96	456.80	4.05
Nemagraptus gracilis gracilis Zone—GSSP Sandbian Base, Gi1	**37.020**	**− 458.322, − 459.106**	**− 458.61**	1.18	**− 458.263, − 459.263**	**458.763**	1.35	**458.18**	1.38
Pseudamplexograptus distichus Zone (≈ Da4)	33.675	− 460.841, − 461.689	− 461.15	2.54	− 461.193, − 462.193	461.693	2.93	461.43	3.25
Archiclimacograptus decoratus Zone, Da3	28.765	− 464.555, − 465.453	− 464.89	3.74	− 465.307, − 466.345	465.826	4.13	466.33	4.90
Levisograptus dentatus Zone (≈ Da2)	24.866	− 467.522, − 468.440	− 467.87	2.98	− 467.975, − 468.813	468.394	2.57	469.23	2.90
Levisograptus austrodentatus Zone, Da1—GSSP Darriwilian Base	**24.539**	**− 468.511, − 469.437**	**− 468.86**	0.99	**− 468.860, − 469.711**	**469.2855**	**0.89**	**469.42**	0.19

Graptolite Zone	[3]	[1]	[2]		[2]	[3]			
Oncograptus upsilon Zone, Ya1–2	22.523	−469.325, −470.261	−469.68	0.82	−469.601, −470.518	470.0595	0.77	470.39	0.97
Isograptus victoriae maximus Zone, Ca3	21.910	−469.801, −470.744	−470.16	0.48	−470.048, −470.995	470.5215	0.46	470.64	0.24
Isograptus victoriae victoriae Zone, Ca2	20.386	−470.996, −471.961	−471.36	1.20	−471.255, −472.298	**471.7765**	1.26	471.15	0.52
Isograptus victoriae lunatus Zone—Graptolite Dapingian Base	**20.042**	**−471.268, −472.240**	**−471.64**	**0.27**	**−471.540, −472.613**	**472.0765**	**0.30**	**471.26**	**0.11**
Didymograptellus bifidus Zone, Ch1,2	16.350	−474.271, −475.365	−474.67	3.03	−474.640, −476.077	475.3585	3.28	473.14	1.87
Tshallograptus fruticosus Zone, Be1–4	14.365	−475.970, −477.184	−476.39	1.72	−476.293, −477.919	477.106	1.75	474.90	1.76
Tetragraptus approximatus Zone—GSSP Floian Base	**12.356**	**−477.777, −479.166**	**−478.24**	**1.84**	**−477.956, −479.771**	**478.8635**	**1.76**	**477.08**	**2.18**
Araneograptus murrayi Zone, La2b	8.769	−481.300, −483.186	−481.86	3.62	−480.929, −483.090	482.0095	3.15	481.12	4.05
Aorograptus victoriae Zone, La2a	8.179	−481.925, −483.922	−482.50	0.65	−481.421, −483.642	482.5315	0.52	481.78	0.66
Adelograptus tenellus Zone	7.367	−482.811, −484.974	−483.42	0.92	−482.101, −484.404	483.2525	0.72	482.69	0.90
Rhabdinopora flabelliformis anglica Zone	6.117	−484.234, −486.691	−484.90	1.48	−483.153, −485.585	484.369	1.12	484.07	1.39
Anisograptus matanensis Zone	5.876	−484.519, −487.039	−485.20	0.30	−483.358, −485.815	484.5865	0.22	484.34	0.27
Rhabdinopora parabola Zone	5.077	−485.483, −488.223	−486.20	1.01	−484.037, −486.577	485.307	0.72	485.22	0.88
Rhabdinopora praeparabola Zone—Graptolite Base of Tremadocian/Ordovician	**3.960**	**−486.895, −489.980**	**−487.68**	**1.48**	**−484.993, −487.650**	**486.3215**	**1.01**	**486.45**	**1.23**

1—separate 5th order polynomial regressions through max and min 2σ uncertainty limits of radioisotope dates
2—5th order polynomial regression through 2σ uncertainty limits of radioisotope dates
3—separate loess regression with 0.50 smoothing parameter on 2σ uncertainty limits of radioisotope dates
Ages in bold font are bases of stages.

they add the fewest new coexistences to those proven in local range charts (a secondary measure of misfit that is minimized). This overfitting of the model can readily be mitigated by omitting rare taxa now that the database is much larger. We perform an initial optimization that is limited to taxa known from at least four or five sections. If more of the rarer taxa are needed to more finely divide the composite sequence, they are added in a second optimization run, in which the result of the first optimization is treated as one of the sections, but with a high enough weight that its sequence of events emerges intact within the enlarged optimal sequence.

The number of local occurrences is an imperfect indicator of the usefulness of taxa. Taxa found only in the top or bottom sample at a local section are almost unconstrained with regard to upward or downward adjustment, respectively. Such occurrences might offer no useful constraint on the global range. A simple solution to this problem is to treat those samples as a thin interval and, thus, separate the first and last local occurrences of the taxon by a minute distance. This empowers the preference given to composite sequences that add fewer new coexistences to the list know from field observation.

The third uncertainty derives from the range of positions in the composite within which a stage or zone boundary could equally be placed well. The age regression converts this important uncertainty to a time interval. Boundary placement is subjective and depends on matching a local succession, to which the zone applies, with the global composite. For several reasons the composite may not match the expectations for any one real section. Although the optimizing algorithms minimize the implied diachronism of range ends, the result is a conservative estimate of maximum known ranges, not a prediction for a single section (Cooper et al., 2001a,b). The database combines ranges from different provinces, studied by various authors. Uncertainties in the placement of biozone boundaries in a sequence of composited taxon ranges likely exceed those associated with both the radioisotopic dating and the optimization. The uncertainty of boundary placement cannot easily be mitigated. Fortunately, it is not cumulative.

The fourth source of uncertainty is no longer as large as in earlier calibrations. This uncertainty is derived from the need to compare successive composite sequences, built as the database expanded and as alternative optimizing rules were tried (Sadler et al., 2009). The database now includes more sections with correspondingly better constraint on more taxa. Best settings for the CONOP algorithms have been established. In particular, the length of the trial-and-error process is now self-regulating, based on diminishing returns monitored at run time. Multiple optimization runs with different data sets are no longer needed to estimate stratigraphic

uncertainty for the placement of dated events (see earlier). The database now provides more insightful comparison by supporting separate compositing of conodont- and graptolite-bearing sections.

20.4.3 Age of stage boundaries

Statistical treatment of GTS2020 age data for Paleozoic periods is essentially similar to the methods described in Sections 14.2.2 for treatment of Paleozoic age estimates in GTS2004, and in subsection 14.3.5 for treatment of the GTS2012 age data.

The authors of the Ordovician and Silurian period chapters of GTS2004 and GTS2012 selected a CONOP graptolite composite, described previously, as the primary scale for interpolating boundary ages. This practice is followed for GTS2020. For the first time, however, we also provide an independently scaled, global CONOP conodont composite (Fig 20.16) because more than half the radioisotopic dates used to calibrate the Ordovician time scale come from carbonate sections and are biostratigraphically associated with conodont fossils. That conodont composite is part of a longer Cambrian−Triassic CONOP sequence. The approximately 68 Myr (Ordovician−Early Devonian) portion used here is considered less reliable because it has had less critical evaluation by expert biostratigraphers than has been stimulated in the 15 years since the introduction of a graptolite composite by IGCP 410 (Sadler and Cooper, 2004). In addition, composite values for graptolites and conodonts in the Ordovician−Silurian are not directly comparable with one another. They are outcomes of independent sequencing exercises using different taxa for intervals of somewhat different duration. Although both were rescaled to 100 percentage units, their differences are evident from the shapes of the spline curves in Figs. 14.6 and 14.7 of Agterberg et al. (2020, Ch. 14A in Geomathematics, this book). Spline-fit stage ages using the conodont composite have larger 2-sigma values than that using the graptolites, mirroring the choice of the graptolite composite for the GTS2020 time scale. The Ordovician stage ages derived from the scaled conodont composite and their comparison with the standard graptolite-based time scale are provided in Table 20.2.

As for GTS2012, smoothing splines were obtained by relating the radioisotopic dates for both successive periods with the CONOP composite values. The age dates are listed in Appendix 2 of this volume. Note that the number of radioisotopic dates listed in Appendix 2 is slightly larger than used here, the result of averaging a few clusters in composite placement that were essentially identical and have insignificant error. Numbers of radioisotopic dates used increased from 26 in GTS2012 to 49 for GTS2020. Both graptolite and conodont composite sequences used all dates, even though many were far better constrained in one sequence or the other. In addition

Ordovician Time Scale - Conodont

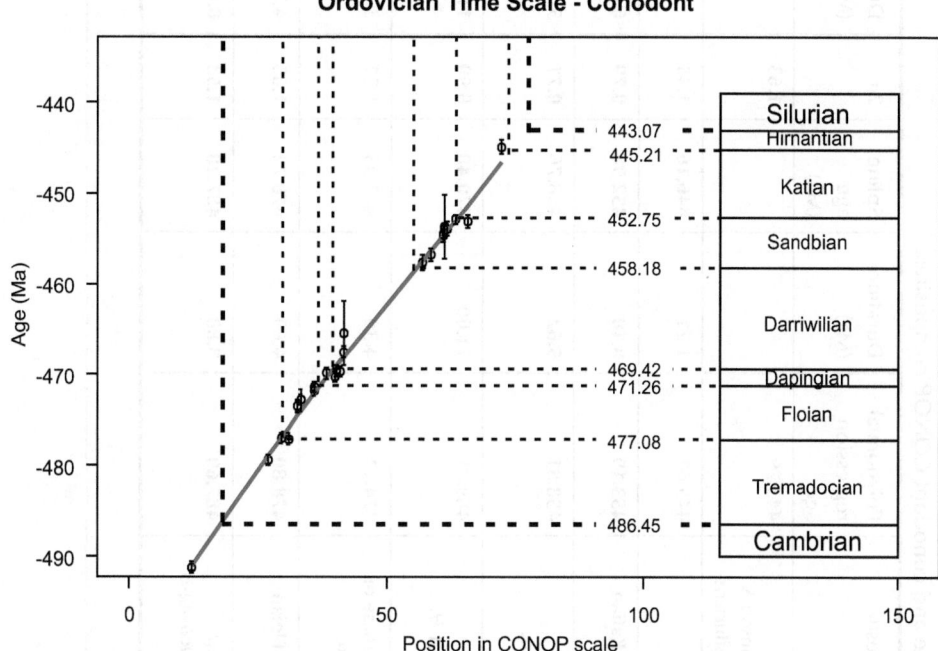

FIGURE 20.16 Geochronology of the Ordovician stages derived from a calibration of the CONOP conodont composite. Boundary ages and error bars using a smoothing spline fitted through 24 radioisotopic-dated ash beds used for calibrating the Ordovician time scale (Appendix 2) plotted on the CONOP composite (*X*-axis) against their analytical age (*Y*-axis).

to associated conodont or graptolite range ends, dates may be constrained in their composite placement by superposition with other dates, carbon isotope excursions and named bentonites. The Ordovician, which lasted 43.8 Myr, has five times as many radioisotopic dates as the Silurian, which is 24.1 Myr long. Two-sigma, external error of the dates is between 0.7 and 1.8 Myr, with the tail ends of the spline having larger error bars. Silurian stage boundaries have larger uncertainty of interpolated age than in the Ordovician, because the Silurian age constraints are sparser. A comparison of Ordovician stage boundary ages in successive time scales since 2004 is shown in Fig. 20.17, and the GSSP definition and locations for the Ordovician stage are shown in Fig. 20.18.

The best fitting spline curves for graptolites and conodonts are shown in Figs. 20.15C and 20.16 (and in Figs. 14.6 and 14.7 of Agterberg et al., 2020, Ch. 14A in Geomathematics, this book) with rather smooth fits. The primary reason that the graptolite spline deviates from a linear fit is fluctuating taxon richness. We note, however, that the fit between the composite best fits using a polynomial and the GTS2020 spline is virtually linear; that is, the age model is relatively insensitive to the choice of regression/interpolation technique.

Age estimates with their 2-sigma uncertainties for the 16 Ordovician stages and biozone boundaries are shown in Table 20.3 for the graptolite spline. Numerically, the age estimates for the stages using the GTS2020 data set are not particularly different from GTS2012, except for the Tremadocian, Dapingian, and Darriwilian, the result of more and better age constraints available since 2012.

The finalized time scale using the splined graptolite composite of the Ordovician and Silurian are in Tables 20.3 and 20.4, which compare the stage base ages to those of GTS2012 also. Using the defined level of the FAD of the conodont species *I. fluctivagus*, the base Tremadocian age is 486.9 ± 1.5 Ma, which is 0.4 Myr older than the 486.45 Ma value in the spline for base Tremadocian, which is actually the FAD of the graptolite species *R. praeparabola*. Hence, the Cambrian−Ordovician boundary is 486.9 ± 1.5 Ma.

The Tremadocian is estimated to have lasted from 486.9 ± 1.5 to 477.1 ± 1.2 Ma, for a duration of 15.6 Myr. The Floian duration is 5.8 Myr, from 477.1 ± 1.2 to 471.3 ± 1.0 Ma. The Dapingian Stage from 471.3 ± 1.0 to 469.4 ± 0.9 Ma with a duration of 1.9 Myr. The Darriwilian Stage is 11.2 Myr long, from 469.4 ± 0.9 to 458.2 ± 0.7 Ma, and is thus the longest stage in the Ordovician and Silurian, followed by Tremadocian. The Sandbian Stage is estimated to have started at 458.2 ± 0.7 and ended at 452.8 ± 0.7 Ma, lasting 5.4 Myr. The Katian Stage is estimated to have run from 452.8 ± 0.7 to 445.2 ± 0.9 Ma for a 7.6 Myr duration, and the Hirnantian to have had a 2.1 Myr duration from its base at 445.2 ± 0.9 Ma to the Rhuddanian (base Silurian) at 443.1 ± 0.9 Ma. Both the Katian and Sandbian stages have the smallest (0.7 Myr) stage base uncertainty. The Early Ordovician is estimated to have lasted 15.6 Myr, the

TABLE 20.2 Comparison of GTS2020 stage base ages from the independently calibrated graptolite and conodont CONOP composites.

Graptolite composite stages	Polynomial regression age	Duration (Ma)	Spline age (Ma)	2σ	Duration (Ma)	Conodont composite stages	Polynomial regression age	Duration (Ma)	Spline age (Ma)	2σ	Duration (Ma)
Base of Silurian/Rhuddanian/A. ascensus	443.50		443.07	0.91		FAD Rexroadus kentuckyensis (Branson & Branson)/Base of Silurian?	443.79			1.63	
GSSP Level Base of Hirnantian	445.57	2.07	445.21	0.86	2.14	Conodont Base of Hirnantian	445.01	1.21	446.16	1.45	
GSSP Level Base of Katian	452.94	7.37	452.75	0.73	7.54	Conodont Base of Katian	453.19	8.18	452.98	0.79	6.82
GSSP Level Base of Sandbian	458.61	5.67	458.18	0.73	5.43	Conodont Base of Sandbian; base of A. tvaerensis Zone	458.81	5.62	458.76	0.71	5.78
GSSP Level Base of Darriwilian	468.86	10.25	469.42	0.93	11.24	Conodont Base of Darriwilian; Base of H. sinuosa Zone	469.81	11.00	470.40	0.90	11.64
Graptolite base of Dapingian/base lunatus Zone	471.64	2.77	471.26	0.98	1.84	Base of Dapingian/base of B. triangularis Zone	474.11	4.30	473.93	1.21	3.53
GSSP Level Base of Floian	478.24	6.60	477.08	1.17	5.82	Conodont Base of Floian	478.80	4.69	478.72	1.83	4.79
Base of Ordovician/Tremadocian	487.68	9.44	486.85	1.52	9.77	Base of Ordovician/Tremadocian/I. fluctivagus Zone	487.60	8.80	487.20	1.53	8.48

FAD, First appearance datum.

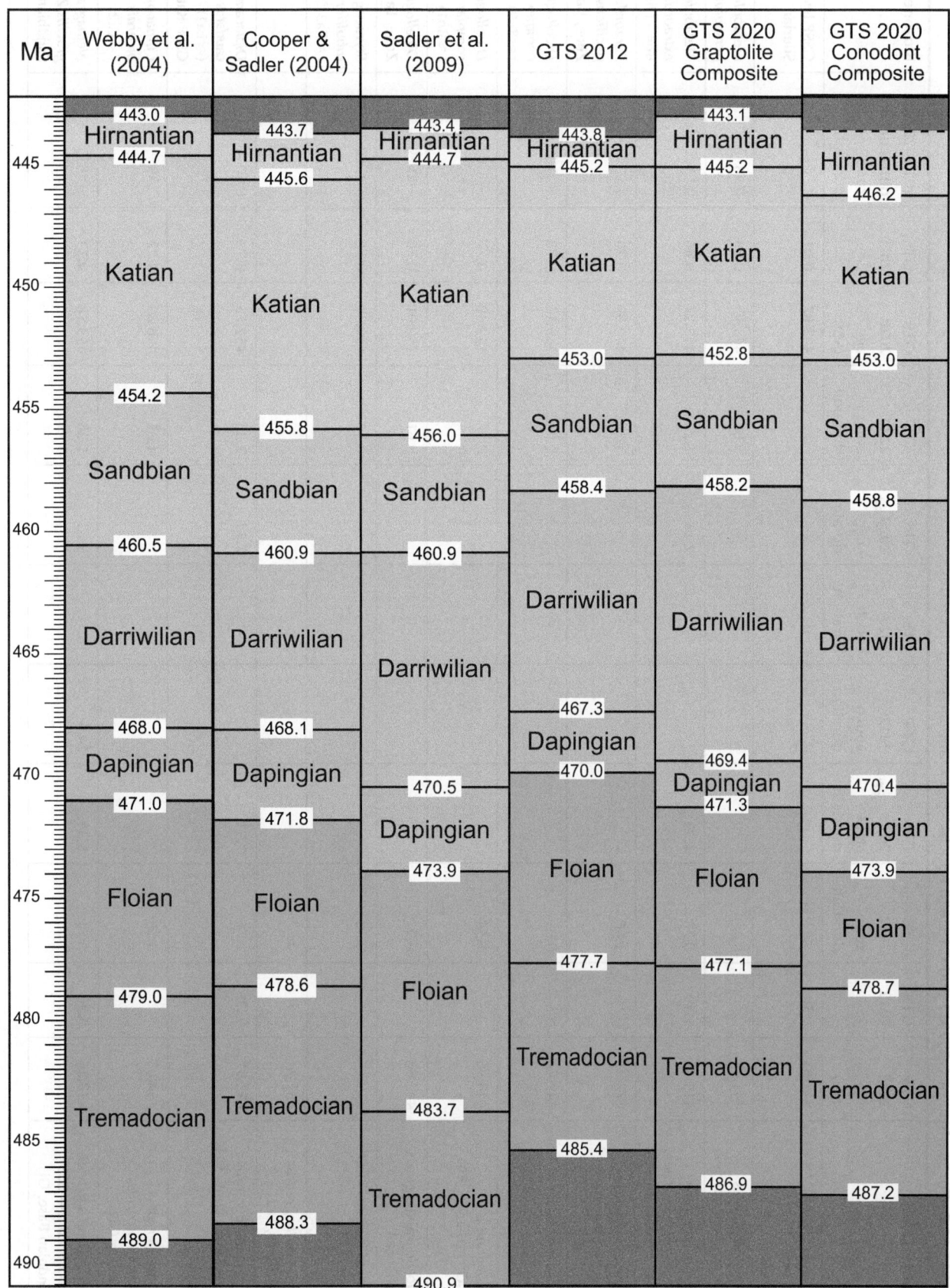

FIGURE 20.17 Comparison of stage base ages for the independently scaled GTS2020 graptolite and conodont CONOP composites with previous time scales; Cooper and Sadler (2004) is GTS2004 (see text for discussion).

TABLE 20.3 Comparison of GTS2012 and GTS2020 epochs/stage/zone ages and durations.

Zone	GTS 2012 age (Ma)	GTS 2012 2σ	GTS 2012 zone duration	GTS 2012 stage duration	GTS 2012 Epoch duration	GTS 2020 Epoch duration	GTS 2020 stage duration	GTS 2020 zone duration	GTS 2020 age (Ma)	GTS 2020 2σ	2012–2020 difference in zone duration	Biozone
Akidograptus ascensus Base of Silurian	443.83	1.50							443.07	0.91		GSSP level Base of Silurian
Metabolograptus persculptus Zone Bo5	444.43		0.6					0.88	443.95	0.89	0.28	Metabolograptus persculptus Zone Bo5
Metabolograptus extraordinarius Zone, Bo4 - Hirnantian Base	445.16	1.40	0.73	1.33			2.14	1.26	445.21	0.86	0.53	Metabolograptus extraordinarius Zone, Bo4 - Hirnantian Base
Paraorthograptus pacificus pacificus Zone, Bo3	447.02		1.86					3.38	448.59	0.79	1.52	Paraorthograptus pacificus pacificus Zone, Bo3
pre-pacificus Zone, Bo2	447.62		0.6					0.54	449.13	0.78	−0.06	Dicellograptus complexus Zone (≈ Bo2)
Alulagraptus uncinatus Zone, Bo1	448.96		1.34					0.27	449.40	0.77	−1.07	Dicellograptus complanatus Zone (≈ Bo1)
Dicellograptus gravis Zone, Ea4	449.86		0.9					0.61	450.01	0.76	−0.29	Dicellograptus gravis Zone, Ea4
Dicranograptus kirki Zone, Ea3	450.44		0.58					0.38	450.39	0.76	−0.20	Styracograptus tubuliferus Zone, (≈Ea3)
Diplacanthograptus spiniferus, Ea2	451.21		0.77									
Diplacanthograptus lanceolatus Zone, Ea1	452.97	0.70	1.76	7.81			7.54	2.36	452.75	0.73	0.60	Diplacanthograptus caudatus Zone (≈Ea1) - GSSP Katian Base
Climacograptus bicornis bicornis Zone (= Gi2)	456.63		3.66					4.05	456.80	0.72	0.39	Climacograptus bicornis bicornis Zone (≈ Gi2)
Nemagraptus gracilis gracilis Zone—GSSP Sandbian Base, Gi1	458.36	0.90	1.73	5.39	14.5	15.1	5.43	1.38	458.18	0.73	−0.35	Nemagraptus gracilis gracilis Zone GSSP Sandbian Base, Gi1

Zone												Zone
Pseudamplexograptus distichus Zone (≈ Da4)	− 0.34	0.76	461.43	3.25					3.59		461.95	*Archiclimacograptus riddellensis*, Da4
Archiclimacograptus decoratus Zone, Da3	2.88	0.85	466.33	4.90					2.02		463.97	*Archiclimacograptus decoratus* Zone, Da3
Levisograptus dentatus Zone (≈ Da2)	1.21	0.92	469.23	2.90					1.69		465.66	*Undulograptus intersitus* Zone, Da2
Levisograptus austrodentatus Zone, Da1—GSSP Darriwilian Base	− 1.40	0.93	469.42	0.19	11.24			8.89	1.59	1.10	467.25	*Levisograptus austrodentatus* Zone, Da1—GSSP Darriwilian Base
Oncograptus upsilon Zone, Ya1—2	− 0.09	0.96	470.39	0.97					1.06		468.31	*Oncograptus upsilon* Zone, Ya1—2
									0.3		468.61	*Isograptus victoriae maximodivergens* Zone, Ca4
Isograptus victoriae maximus Zone, Ca3	0.10	0.96	470.64	0.24					0.14		468.75	*Isograptus victoriae maximus* Zone, Ca3
Isograptus victoriae victoriae Zone, Ca2	− 0.69	0.98	471.15	0.52					1.21		469.96	*Isograptus victoriae victoriae* Zone, Ca2
Isograptus victoriae lunatus Zone - Graptolite Dapingian Base	− 0.35	0.98	471.26	0.11	1.84	13.1	11.6	3.17	0.46	1.40	470.42	*Isograptus victoriae lunatus* Zone - Graptolite Dapingian Base
Didymograptellus bifidus Zone, Ch1,2	− 0.12	1.04	473.14	1.87					1.99		472.41	*Didymograptellus protobifidus* Zone, Ch1,2
Tshallograptus fruticosus Zone, Be1—4	− 1.80	1.10	474.90	1.76					3.56		475.97	*Tshallograptus fruticosus* Zone, Be1—4

(Continued)

TABLE 20.3 (Continued)

Zone	GTS 2012 age (Ma)	GTS 2012 2σ	GTS 2012 zone duration	GTS 2012 stage duration	GTS 2012 Epoch duration	GTS 2020 Epoch duration	GTS 2020 stage duration	GTS 2020 zone duration	GTS 2020 age (Ma)	GTS 2020 2σ	2012–2020 difference in zone duration	Biozone
Tetragraptus approximatus Zone, La3 - GSSP Floian Base	477.72	**1.40**	1.75	7.3			**5.82**	**2.18**	**477.08**	**1.17**	0.43	*Tetragraptus approximatus* Zone - GSSP Floian Base
Araneograptus murrayi Zone, La2b	479.86		2.14					4.05	481.12	1.32	1.91	*Araneograptus murrayi* Zone, La2b
Aorograptus victoriae Zone, La2a	480.55		0.69					0.66	481.78	1.34	– 0.03	*Aorograptus victoriae* Zone, La2a
Psigraptus Zone, La1b	481.67		1.12					0.90	482.69	1.37	– 0.22	*Adelograptus tenellus* Zone
								1.39	484.07	1.43		*Rhabdinopora flabelliformis anglica* Zone
R. scitulum & *Anisograptus* Zone, La1b	484.28		2.61					0.27	484.34	1.44		*Anisograptus matanensis* Zone
Rhabdinopora parabola Zone	484.97		0.69					0.88	485.22	1.47	0.19	*Rhabdinopora parabola* Zone
								1.23	486.45	**1.52**	1.23	*Rhabdinopora praeparabola* Zone
Iapetognathus fluctivagus Zone - Base of Tremadocian/Ordovician	485.37	**1.90**	**0.4**	7.65	15.4	15.6	9.7741	**0.4**	**486.85**		0.00	*Iapetognathus fluctivagus* Zone - Base of Tremadocian/Ordovician

Note that for GTS2012 the standard graptolite zonation used for calibration was the Australasian Succession, but for GTS2020 a slightly different set of zones was used. Ages in bold font are at bases of stages.

TABLE 20.4 Interpolated ages of Ordovician and Silurian epoch and stage boundaries, with uncertainty estimates, stage durations, and comparison to GTS2012 (for details see text).

Chronostratigraphic unit			GTS2012				GTS2020			
Period	Epoch	Stage/age	Base Ma	Est. ± Myr (2σ)	Epoch duration	Stage duration	Base Ma	Est. ± Myr (2σ)	Epoch duration	Stage duration
Devonian	Early	Lochkovian	419.2	3.2			419.0	1.8		
Silurian	Pridoli	Pridolian	423.0	2.3	3.8	3.8	422.7	1.6	3.7	3.7
	Ludlow	Ludfordian	425.6	0.9	4.4	2.6	425.0	1.5	4	2.3
		Gorstian	427.4	0.5		1.8	426.7	1.5		1.7
	Wenlock	Homerian	430.5	0.7	6.0	3.1	430.6	1.3	6.2	3.9
		Sheinwoodian	433.4	0.8		2.9	432.9	1.2		2.3
	Llandovery	Telchyan	438.5	1.1	10.5	5.1	438.6	1.1	10.2	5.7
		Aeronian	440.8	1.2		2.3	440.5	1.0		1.9
		Rhuddanian	443.8	1.5		3.1	443.1	0.9		2.6
Ordovician	Late	Hirnantian	445.2	1.4	14.5	1.3	445.2	0.9	15.1	2.1
		Katian	453.0	0.7		7.8	452.8	0.7		7.6
		Sandbian	458.4	0.9		5.4	458.2	0.7		5.4
	Middle	Darriwillian	467.3	1.1	11.6	8.9	469.4	0.9	13.1	11.2
		Dapingian	470.0	1.4		2.7	471.3	1.0		1.9
	Early	Floian	477.7	1.4	15.4	7.8	477.1	1.2	15.6	5.8
		Tremadocian	485.4	1.9		7.7	486.9	1.5		9.8

GSSPs of the Ordovician Stages, with location and primary correlation criteria

Stage	GSSP Location	Latitude, Longitude	Boundary Level	Correlation Events	Reference
Hirnantian	Wangjiawan North section, N of Yichang city, Western Hubei Province, China	30°59'2.68"N 111°25'10.76"E	0.39 m below the base of the Kuanyinchiao Bed	Graptolite, FAD of *Metabolograptus extraordinarius*	Episodes **29/3**, 2006
Katian	Black Knob Ridge section, Atoka, Oklahoma (USA)	34°25.829'N 96°4.473'W	4.0 m above the base of the Bigfork Chert	Graptolite, FAD of *Diplacanthograptus caudatus*	Episodes **30/4**, 2007
Sandbian	Sularp Brook, Fågelsång, Sweden	55°42'49.3"N 13°19'31.8"E*	1.4 m below a phosphorite marker bed in the E14b outcrop	Graptolite, FAD of *Nemagraptus gracilis*	Episodes **23/2**, 2000
Darriwilian	Huangnitang section, Changshan, Zhejiang Province, SE China	28°51'14"N 118°29'23"E*	base of Bed AEP 184	Graptolite, FAD of *Levisograptus austrodentatus*	Episodes **20/3**, 1997
Dapingian	Huanghuachang section, NE of Yichang city, Hubei Province, S China	30°51'37.8"N 110°22'26.5"E	10.57 m above base of the Dawan Formation	Conodont, FAD of *Baltoniodus triangularis*	Episodes **32/2**, 2009
Floian	*Diabasbrottet, Hunneberg, Sweden*	58°21'32.2"N 12°30'08.6"E	in the lower Tøyen Shale, 2.1 m above the top of the Cambrian	Graptolite, FAD of *Tetragraptus approximatus*	Episodes **27/4**, 2004
Tremadocian	Green Point Section, western Newfoundland	49°40'58.5"N 57°57'55.09"W*	at the 101.8 m level,within Bed 23, in the measured section	Conodont, FAD of *Iapetognathus fluctivagus*	Episodes **24/1**, 2001

* according to Google Earth

FIGURE 20.18   GSSPs of the Ordovician stages, with location and primary correlation criteria.

Middle Ordovician 13.1, and the Late Ordovician 15.1 Myr. The Ordovician Period lasted 43.8 Myr, which is 2.3 Myr longer than estimated in GTS2012, where its estimate spanned 41.5 Myr.

Compared to GTS2012, the GTS2020 scale for the Ordovician–Silurian is more reliable, the result of a biostratigraphic composite based on substantial new data and almost double the number of age dates, a majority of which have less analytical uncertainty than in GTS2012. For the Ordovician the largest differences from the GTS2012 time scale are a 1.5 Myr older base Tremadocian boundary, a 1.3 Myr older base Dapingian, and a 2.1 Myr older base Darriwilian boundary (see Tables 20.3 and 20.4 for comparison).

Acknowledgments

For help with correlation charts, GSSP charts, and photos of fossil sand outcrops we thank Y-Y. Zhen, S. Stouge, J. Maletz, S.M. Bergström, J. Repetski, I. Percival, G. Albanesi, Y. Liang, and J. Nõlvak. Assistance for figures retained wholly or in part from 2012 were provided by G. Ogg, G. Nowlan, S.M. Bergström, J. Nõlvak, Rong Jiayu, Chen Xu, C.E. Mitchell, Zhang Yuandong, and Zhou Zhiyi. C.E. Mitchell and R.A. Cooper kindly reviewed the text.

Bibliography

Achab, A., 1989, Ordovician chitinozoan zonation of Quebec and western Newfoundland. *Journal of Paleontology*, **63**: 14–24.

Achab, A., Asselin, E., Desrochers, A., Riva, J.F., and Farley, C., 2011, Chitinozoan contribution to the development of a new Upper Ordovician stratigraphic framework for Anticosti Island. *Geological Society of America Bulletin*, **123**: 186–205.

Agterberg, F.P., da Silva, A.-C., and Gradstein, F.M., 2020, Chapter 14A - Geomathematical and statistical procedures. *In* Gradstein, F.M., Ogg, J.G., Schmitz, M.D., and Ogg, G.M. (eds), *The Geologic Time Scale 2020*. **Vol. 1** (this book). Elsevier, Boston, MA.

Ahm, A.-S.C., Bjerrum, C.J., and Hammarlund, E.U., 2017, Disentangling the record of diagenesis, local redox conditions, and global seawater chemistry during the latest Ordovician glaciation. *Earth and Planetary Science Letters*, **459**: 145–156.

Ainsaar, L., and Meidla, T., 2008, Ordovician carbon isotopes. *In* Põldvere, A. (ed), *Männamaa (F-367) Drill Core*. Tallinn: Geological Survey of Estonia, pp. 27–29.

Ainsaar, L., Meidla, T., and Martma, T., 1999, Evidence for a widespread carbon isotopic event associated with late Middle Ordovician sedimentological and faunal changes in Estonia. *Geological Magazine*, **136**: 49–62.

Ainsaar, L., Meidla, T., and Martma, T., 2004, The middle Caradoc facies and faunal turnover in the Late Ordovician Baltoscandian paleobasin. *Palaeogeography, Palaeoclimatology, Palaeoecology*, **210**: 119–133.

Ainsaar, L., Kaljo, D., Martma, T., Meidla, T., Mannik, J., Nolvak, J., et al., 2010, Middle and Upper Ordovician carbon isotope chemostratigraphy in Baltoscandia: a correlation standard and clues to environmental history. *Palaeogeography, Palaeoclimatology, Palaeoecology*, **294**: 189–201.

Albanesi, G.L., Carrera, M.G., Cañas, F.L., and Saltzman, M.R., 2006, A proposed Global Boundary Stratotype Section and Point for the base of the Middle Ordovician Series: the Niquivil section, Precordillera of San Juan, Argentina. *Episodes*, **29**: 1–15.

Albanesi, G.L., Bergström, S.M., Schmitz, B., Serra, F., Feltes, N.A., Voldman, G.G., et al., 2013, Darriwilian (Middle Ordovician) $\delta^{13}C_{carb}$ chemostratigraphy in the Precordillera of Argentina: documentation of the middle Darriwilian Isotope Carbon Excursion (MDICE) and its use for intercontinental correlation. *Palaeogeography, Palaeoclimatology, Palaeoecology*, **389**: 48–63.

Albanesi, G.L., and Ortega, G., 2016, Conodont and graptolite biostratigraphy of the Ordovician System of Argentina. *In* Montenari, M. (ed), *Stratigraphy and Timescales*. Elsevier Academic Press, pp. 61–121.

Aldridge, R.J., and Briggs, D.E.G., 1989, A soft body of evidence. *Natural History*, **5/89**: 6–11.

Aldridge, R.J., Briggs, D.E.G., Smith, M.P., Clarkson, E.N.K., and Clark, N.D.L., 1993, The anatomy of conodonts. *Philosophical Transactions of the Royal Society, B: Biological Sciences*, **340** (1294), 405–442.

Aldridge, R.J., Murdock, D.J.E., Gabbott, S.E., and Theron, J.N., 2013, A 17-element conodont apparatus from the Soom Shale Lagerstätte (Upper Ordovician), South Africa. *Palaeontology*, **36**: 261–276.

Alroy, J., 2008, Phanerozoic trends in the global diversity of marine invertebrates. *Science*, **321**: 97–100.

Amberg, C.E.A., Vandenbroucke, T.R.A., Nielsen, A.T., Munnecke, A., and McLaughlin, P.I., 2017, Chitinozoan biostratigraphy and carbon isotope stratigraphy from the Upper Ordovician Skogerholmen Formation in the Oslo Region. A new perspective for the Hirnantian lower boundary in Baltica. *Review of Palaeobotany and Palynology*, **246**: 109–119.

Azmy, K., Stouge, S., Brand, U., Bagnoli, G., and Ripperdan, R., 2014, High-resolution chemostratigraphy of the Cambrian–Ordovician GSSP: enhanced global correlation tool. *Palaeogeography, Palaeoclimatolology, Palaeoecology*, **409**: 135–144.

Baldo, E.G., Fanning, C.M., Rapela, C.W., Pankhurst, R.J., Casquet, C., and Galindo, C., 2003, U–Pb shrimp dating of rhyolite volcanism in the Famatinian belt and K-bentonites in the Precordillera. *In* Albanesi, G.L., Beresi, M.S., and Peralta, S.H. (eds), *Ordovician from the Andes, Tucumán*: INSUGEO. Serie Correlación Geológica, **17**: 185–190.

Ballo, E.G., Augland, L.E., Hammer, Ø., and Svensen, H.H., 2019, A new age model for the Ordovician (Sandbian) K-bentonites in Oslo, Norway. *Palaeogeography, Palaeoclimatology, Palaeoecology*, **520**: 203–213.

Bambach, R.K., 2006, Phanerozoic biodiversity mass extinctions. *Annual Review of Earth and Planetary Sciences*, **34**: 127–155.

Bambach, R.K., Knoll, A.H., and Wang, S.C., 2004, Origination, extinction, and mass depletions of marine diversity. *Paleobiology*, **30**: 522–542.

Bapst, D.W., Bullock, P.C., Melchin, M.J., Sheets, H.D., and Mitchell, C.E., 2012, Graptoloid diversity and disparity became decoupled during the Ordovician mass extinction. *Proceedings of the National Academy of Science*, **109**: 3428–3433.

Barta, N.C., Bergström, S.M., Saltzman, M.R., and Schmitz, B., 2007, First record of the Ordovician Guttenberg $\delta^{13}C$ excursion (GICE) in New York State and Ontario: local and regional chronostratigraphic implications. *Northeastern Geology and Environmental Sciences*, **29**: 276–298.

Bassett, M.G., 1979, 100 years of Ordovician geology. *Episodes*, **2**: 18–21.

Bauert, H., Ainsaar, L., Põldsaar, K., and Sepp, S., 2014a, $\delta^{13}C$ chemostratigraphy of the Middle and Upper Ordovician succession in the Tartu-453 drillcore, southern Estonia, and the significance of the HICE. *Estonian Journal of Earth Sciences*, **63**: 195–200.

Bauert, H., Isozaki, Y., Holmer, L.E., Aoki, K., Sakata, S., and Hirata, T., 2014b, New U-Pb zircon ages of the Sandbian (Upper Ordovician) "Big K-bentonite" in Baltoscandia (Estonia and Sweden) by LA-ICPMS. *GFF*, **136**: 30–33.

Bergström, S.M., 1971, Conodont biostratigraphy of the Middle and Upper Ordovician of Europe and eastern North America. *Memoirs— Geological Society of America*, **127**: 83–157.

Bergström, S.M., 1986, Biostratigraphic integration of Ordovician graptolite and conodont zones - a regional review. *In* Rickards, R.B., and Hughes, C.P. (eds), *Paleobiology and Biostratigraphy of Graptolites*. Oxford: Blackwell Scientific, pp. 61–78.

Bergström, S.M., and Mitchell, C.E., 1986, The graptolite correlation of the North American Upper Ordovician Standard. *Lethaia*, **19**: 247–266.

Bergström, S.M., and Mitchell, C.E., 1990, Trans-Pacific graptolite faunal relations: the biostratigraphic position of the base of the Cincinnatian Series (Upper Ordovician) in the standard Australian graptolite zone succession. *Journal of Paleontology*, **64**: 992–997.

Bergström, S.M., and Mitchell, C.E., 1994, Regional relationships between late Middle and early Late Ordovician standard successions in New York and Quebec and the Cincinnati region in Ohio, Indiana, and Kentucky. *Bulletin of the New York State Museum*, **481**: 5–10.

Bergström, S.M., and Wang, Z., 1995, Global correlation of Castlemainian to Darriwilian conodont faunas and their relation to the graptolite zone succession. *In* Chen, X., and Bergström, S.M. (eds), *The Base of the Austrodentatus Zone as a Level for Global Subdivision of the Ordovician System*. Nanjing: Nanjing University Press, pp. 92–98.

Bergström, S.M., Carnes, J.B., Hall, J.C., Kurapkat, W. and O'Neil, B.E. 1988. Conodont biostratigraphy of some Middle Ordovician stratotypes in the southern and central Appalachians. In: Landing, E. (ed.), The Canadian Paleontology and Biostratigraphy Symposium. *New York State Museum Bulletin*, 462: 20–31.

Bergström, S.M., Finney, S.C., Xu, C., Palsson, C., Zhi-Gao, W., and Grahn, Y., 2000, A proposed global boundary stratotype for the base of the Upper Series of the Ordovician System: the Fågelsång section, Scania, southern Sweden. *Episodes*, **23** (2), 102–109.

Bergström, S.M., Löfgren, A., and Maletz, J., 2004, The GSSP of the second (upper) stage of the Lower Ordovician Series: diabasbrottet at Hunneberg, Province of Vaastergötland, southwest Sweden. *Episodes*, **27** (4), 265–272.

Bergström, S.M., Finney, S.M., Chen, X., Goldman, D., and Leslie, S.A., 2006a, Three new Ordovician global stage names. *Lethaia*, **39**: 287–288.

Bergström, S.M., Saltzman, M.R., and Schmitz, B., 2006b, First record of the Hirnantian (Upper Ordovician) $\delta^{13}C$ excursion in the North American Midcontinent and its regional implications. *Geological Magazine*, **143**: 657–678.

Bergström, S.M., Xu, C., Guteirrez-Marco, J.-C., and Dronov, A., 2009, The new chronostratigraphic classification of the Ordovician System and its relations to major regional series and stages and to $\delta^{13}C$ chemostratigraphy. *Lethaia*, **42**: 97–107.

Bergström, S.M., Schmitz, B., Saltzman, M.R., and Huff, W.D., 2010a, The Upper Ordovician Guttenberg excursion (GICE) in North America and Baltoscandia: Occurrence, chronostratigraphic significance, and paleoenvironmental relationships. *In* Finney, S.C., and Berry, W.B.N. (eds), The Ordovician Earth System. *Geological Society of America Special Paper*, **466**: pp. 37–67.

Bergström, S.M., Young, S.A., and Schmitz, B., 2010b, Katian (Upper Ordovician) δ13C chemostratigraphy and sequence stratigraphy in the United States and Baltoscandia: a regional comparison. *Palaeogeography, Palaeoclimatology, Palaeoecology*, **296**: 217–234.

Bergström, S.M., Agematsu, S., and Schmitz, B., 2010c, Global Upper Ordovician correlation by means of $\delta^{13}C$ chemostratigraphy: implications of the discovery of the Guttenberg $\delta^{13}C$ excursion (GICE) in Malaysia. *Geological Magazine*, **147**: 641–651.

Bergström, S.M., Kleffner, M., Schmitz, B., Cramer, B.D., and Dix, G., 2011, Revision of the position of the Ordovician-Silurian boundary in southern Ontario: regional chronostratigraphic implications of $\delta^{13}C$ chemostratigraphy of the Manitoulin Formation and associated strata. *Canadian Journal of Earth Sciences*, **48**: 1447–1470.

Bergström, S.M., Lehnert, O., Calner, M., and Joachimski, M.M., 2012, A new upper Middle Ordovician-Lower Silurian drillcore standard succession from Borenshult in Östergötland, southern Sweden: 2. Significance of $\delta^{13}C$ chemostratigraphy. *Geologiska Föreningens Förhandlingar*, **134**: 39–63.

Bergström, S.M., Eriksson, M.E., Young, S.A., Ahlberg, P., and Schmitz, B., 2014, Hirnantian (latest Ordovician) $\delta^{13}C$ chemostratigraphy in southern Sweden and globally: a refined integration with the graptolite and conodont zone successions. *Geologiska Föreningens Förhandlingar*, **136**: 355–386.

Bergström, S.M., Saltzman, M.R., Leslie, S.A., Young, S.A., and Feretti, A., 2015, Trans-Atlantic application of the Baltic Middle and Upper Ordovician carbon isotope zonation. *Estonian Journal of Earth Sciences*, **64** (1), 8–12.

Bergström, S.M., Eriksson, M.E., Schmitz, B., Young, S.A., and Ahlberg, P., 2016, Upper Ordovician $\delta^{13}C$ chemostratigraphy, K-bentonite stratigraphy, and biostratigraphy in southern Scandinavia: a reappraisal. *Palaeogeography, Palaeoclimatology, Palaeoecology*, **454**: 175–188.

Bergström, S.M., Ahlberg, P., Maletz, J., Lundberg, F., and Joachimski, M., 2018, Darriwilian (Middle Ordovician) chemostratigraphy linked to graptolite, conodont, and trilobite biostratigraphy in the Fågelsång-3 drill core, Scania, Sweden. *Geologiska Föreningens Förhandlingar*, **140** (3), 229–240.

Bergström, S.M., Schmitz, B., Terfelt, F., Eriksson, M.E., and Ahlberg, P., 2019a, The $\delta^{13}C$ chemostratigraphy of Ordovician global stage stratotypes: geochemical data from the Floian and Sandbian GSSPs in Sweden. *Geologiska Föreningens Förhandlingar*, 1–10, https://doi.org/10.1080/11035897.2019.1631883.

Bergström, S.M., Kleffner, M., and Eriksson, M.E., 2019b, Upper Katian (Upper Ordovician) trans-Atlantic $\delta^{13}C$ chemostratigraphy: the geochronological equivalence of the ELKHORN and PAROVEJA excursions and its implications: upper Katian (upper Ordovician) chemostratigraphy. *Lethaia*, **52**: 1–18.

Bettley, R.M., Fortey, R.A., and Siveter, D.J., 2001, High-resolution correlation of Anglo-Welsh Middle to Upper Ordovician sequences and its relevance to international chronostratigraphy. *Journal of the Geological Society*, **158** (6), 937–952.

Bond, D.P.G., and Grasby, S.E., 2017, On the causes of mass extinctions. *Palaeogeography, Palaeoclimatology, Palaeoecology*, **478**: 3–29.

Bottjer, D.J., Droser, M.L., Sheehan, P.M., and McGhee, G.R., 2001, The ecological architecture of major events in the Phanerozoic history of marine invertebrate life. *In* Allmon, W.D., and Bottjer, D.J. (eds), *Evolutionary Paleoecology: The Ecological Context of Macroevolutionary Change*. New York: Columbia University Press, pp. 35–61.

Boucot, A.J., 1975, *Evolution and Extinction Rate Controls*. Amsterdam: Elsevier, p. 427.

Brenchley, P.J., 1989, The Late Ordovician extinction. *In* Donovan, S.K. (ed), *Mass Extinctions: Processes and Evidence*. London: Belhaven Press, pp. 104–132.

Brenchley, P.J., Marshall, J.D., and Underwood, C.J., 2001, Do all mass extinctions represent an ecological crisis? Evidence from the Late Ordovician. *Geological Journal*, **36**: 329–340.

Brenchley, P.J., Carden, G.A., Hints, L., Kaljo, D., Marshall, J.D., Martma, T., et al., 2003, High-resolution stable isotope stratigraphy of Upper Ordovician sequences: constraints on the timing of bioevents and environmental changes associated with mass extinction and glaciation. *Geological Society of America Bulletin*, **115** (1), 89–104.

Brett, C.E., McLaughlin, P.I., Cornell, S.R., and Baird, G.C., 2004, Comparative sequence stratigraphy of two classic Upper Ordovician successions, Trenton Shelf (New York-Ontario) and Lexington Platform (Kentucky-Ohio): implications for eustasy and local tectonism in eastern Laurentia. *Palaeogeography, Palaeoclimatology, Palaeoecology*, **210**: 295–329.

Briggs, D.E.G., Clarkson, E.N.K., and Aldridge, R.J., 1983, The conodont animal. *Lethaia*, **16**: 1–14.

Bryan, A., Goldman, D., Albanesi, G.L., Ortega, G., and Serra, F., 2018, Computer-assisted graphic correlation of Ordovician conodonts and graptolites from the Argentine Precordillera and wetern Newfoundland using constrained optimization (CONOP9). *In International Geoscience Programme Project 653, 3rd Annual Meeting, Program, and Abstracts*, Athens, OH, p. 25.

Buggisch, W., Keller, M., and Lehnert, O., 2003, Carbon isotope record of Late Cambrian to Early Ordovician carbonates of the Argentine Precordillera. *Palaeogeography, Palaeoclimatology, Palaeoecology*, **195**: 357–373.

Buggisch, W., Joachimski, M.M., Lehnert, O., Bergström, S.M., Repetski, J. E., and Webers, G.F., 2010, Did intense volcanism trigger the first Late Ordovician icehouse? *Geology*, **38**: 327–330.

Chen, J.Y., 1986, *Aspects of Cambrian–Ordovician Boundary in Dyangcha, China*. Beijing: China Prospect Publishing House, pp. 1–410.

Chen, X., 1990, Graptolite depth zonation. *Acta Palaeontologica Sinica*, **29** (5), 507–526.

Chen, X., and Bergström, S.M., 1995, *The Base of the Austrodentatus Zone as a Level for Global Subdivision of the Ordovician System*. Nanjing: Nanjing University Press, p. 117.

Chen, J.Y., Qian, Y.Y., Zhang, J.M., Lin, Y.K., Yin, L.M., Wang, Z.H., et al., 1988, The recommended Cambrian-Ordovician global boundary stratotype of the Xiaoyangqiao section Dayangcha, Jilin Province, China. *Geological Magazine*, **125** (4), 415–444.

Chen, X., Zhang, Y., and Mitchell, C.E., 2001, Early Darriwilian graptolites from central and western China. *Alcheringa*, **25**: 191–210.

Chen, X., Melchin, M.J., Sheets, H.D., Mitchell, C.E., and Fan, J.-X., 2005, Patterns and processes of latest Ordovician graptolite extinction and recovery based on data from south China. *Journal of Paleontology*, **79**: 842–861.

Chen, Xu, Zhang, Yuandong, Bergström, S.M., Xu, Honggen, 2006, Upper Darriwilian graptolite and conodont zonation in the global stratotype section of the Darriwilian stage (Ordovician) at Huangnitang, Changshan, Zhejiang, China. *Palaeoworld*, **15**: 150–170.

Chen, X., Rong, J., Fan, J., Zhan, R., Mitchell, C.E., Harper, D.A.T., et al., 2006, The global boundary stratotype section and point (GSSP) for the base of the Hirnantian Stage (the uppermost of the Ordovician System). *Episodes*, **29**: 183–196.

Chen, X., Zhang, Y.-D., Goldman, D., Bergström, S.M., Fan, J.-X., Wang, Z.-H., et al., 2017, *Darriwilian to Katian (Ordovician) Graptolites From Northwest China (in Chinese)*. Hangzhou: Zhejiang University Press, pp. 1–321.

Churkin, M., Carter, C., and Johnson, B.R., 1977, Subdivision of Ordovician and Silurian time scale using accumulation rates of graptolitic shale. *Geology*, **5**: 452–456.

Cody, R.D., Levy, R.H., Harwood, D.M., and Sadler, P.M., 2008, Thinking outside the zone: high resolution quantitative diatom biochronology for the Antarctic Neogene. *Palaeogeography, Palaeoclimatology, Palaeoecology*, **260**: 92–121.

Cooper, R.A., 1999a, Ecostratigraphy, zonation and global correlation of earliest planktic graptolites. *Lethaia*, **32**: 1–16.

Cooper, R.A., 1999b, The Ordovician Time Scale – calibration of graptolite and conodont zones. *Acta Universitatis Carolinae—Geologica*, **43** (1/2), 1–4.

Cooper, R.A., and Lindholm, K., 1990, A precise worldwide correlation of Early Ordovician graptolite sequences. *Geological Magazine*, **127**: 497–525.

Cooper, R.A., and Sadler, P.M., 2004, Ordovician. *In* Gradstein, F.M., Ogg, J.G., and Smith, A.G. (eds), *A Geologic Time Scale 2004*. Cambridge: Cambridge University Press, pp. 165–187.

Cooper, R.A., and Sadler, P.M., 2012, The Ordovician Period. *In* Gradstein, F.M., Ogg, J.G., Schmitz, M.D., and Ogg, G.M. (eds), *The Geologic Time Scale 2012*. Amsterdam: Elsevier Press, pp. 489–523.

Cooper, R.A., and Stewart, I.R., 1979, The Tremadoc graptolite sequence of Lancefield, Victoria. *Palaeontology*, **22**: 767–797.

Cooper, R.A., Fortey, R.A., and Lindholm, K., 1991, Latitudinal and depth zonation of early Ordovician graptolites. *Lethaia*, **24**: 199–218.

Cooper, R.A., Maletz, J., Wang, H., and Erdtmann, B.-D., 1998, Taxonomy and evolution of the earliest Ordovician graptolites. *Norsk Geologisk Tidsskrift*, **78**: 3–32.

Cooper, R.A., Nowlan, G.S., and Williams, H.S., 2001a, Global Stratotype Section and Point for base of the Ordovician System. *Episodes*, **24**: 19–28.

Cooper, R.A., Crampton, J.S., Raine, J.I., Gradstein, F.M., Morgans, H.E.G., Sadler, P.M., et al., 2001b, Quantitative biostratigraphy of the Taranaki basin, New Zealand: a deterministic and probabilistic approach. *The American Association of Petroleum Geologists Bulletin*, **85** (8), 1469–1498.

Cooper, R.A., Maletz, J., Taylor, L., and Zalasiewicz, J.A., 2004, Estimates of Ordovician mean standing diversity in low, middle and high paleolatitudes. *In* Webby, B.D., Paris, F., Droser, M.L., and Percival, I.G. (eds), *The Great Ordovician Biodiversification Event*. New York: Columbia University Press, pp. 281–293.

Cooper, M.R., Crowley, Q.G., and Rushton, A.W.A., 2008, New age constraints for the Ordovician Tyrone Volcanic Group, Northern Ireland. *Journal of the Geological Society*, **165**: 333–339.

Cooper, R.A., Rigby, S., Loydell, D.K., and Bates, D.E.B., 2012, Palaeoecology of the Graptoloidea. *Earth-Science Reviews*, **112**: 23–41.

Cooper, R.A., Sadler, P.M., Munnecke, A., and Crampton, J.S., 2014, Graptoloid evolutionary rates track Ordovician–Silurian global climate change. *Geological Magazine*, **151** (2), 349–364.

Cramer, B.D., and Jarvis, I., 2020, Chapter 11 - Carbon isotope stratigraphy. *In* Gradstein, F.M., Ogg, J.G., Schmitz, M.D., and Ogg, G.M. (eds), *The Geologic Time Scale 2020*. **Vol. 1** (this book). Elsevier, Boston, MA.

Crampton, J.S., Meyers, S.R., Cooper, R.A., Sadler, P.M., Foote, M., and Harte, D., 2018, Pacing of Paleozoic macroevolutionary rates by Milankovitch grand cycles. *Proceedings of the National Academy of Sciences of the United States of America*, **115** (22), 5686–5691.

Demski, M.W., Wheadon, B., Stewart, L.A., Elias, R.J., Young, G.A., Nowlan, G.S., et al., 2015, Hirnantian strata identified in major intracratonic basins of central North America: implications for uppermost Ordovician stratigraphy. *Canadian Journal of Earth Sciences*, **52**: 68–76.

Denison, R.E., Koepnick, R.B., Burke, W.H., and Hetherington, E.A., 1998, Construction of the Cambrian and Ordovician seawater $^{87}Sr/^{86}Sr$ curve. *Chemical Geology*, **152**: 325–340.

Desrochers, A., Farley, C., Achab, A., Asselin, E., and Riva, J.F., 2010, A far-field record of the end Ordovician glaciation: the Ellis Bay Formation, Anticosti Island, Eastern Canada. *Palaeogeography, Palaeoclimatology, Palaeoecology*, **296**: 248–263.

Donoghue, P.C.J., Forey, P.L., and Aldridge, R.J., 2000, Conodont affinity and chordate phylogeny. *Biological Reviews*, **75**: 191–251.

Donoghue, P.C.J., Purnell, M.A., Aldridge, R.J., and Zhang, S., 2008, The interrelationships of 'complex' conodonts (Vertebrata). *Journal of Systematic Palaeontology*, **6** (2), 119–153.

Dronov, A.V., Ainsaar, L., Kaljo, D., Meidla, T., Saadre, T., and Einasto, R., 2011, Ordovician of Baltoscandia: facies, sequences and sea-level changes. *In* Gutiérrez-Marco, J.C., Rábano, I., and García-Bellido, D. (eds), *Ordovician of the World*. Madrid: Cuadernos del Museo Geominero, Instituto Geológico y Minero de Espanã, pp. 143–150.

Droser, M.L., and Sheehan, P.M., 1997, Palaeoecology of the Ordovician radiation: resolution of large-scale patterns with individual clade histories, palaeogeography and environments. *Geobios, Mémoire Spécial*, **20**: 221–229.

Edwards, C.T., and Saltzman, M.R., 2014, Carbon isotope ($\delta^{13}C_{carb}$) stratigraphy of the Lower–Middle Ordovician (Tremadocian–Darriwilian) in the Great Basin, western United States: implications for global correlation. *Palaeogeography, Palaeoclimatology, Palaeoecology*, **399**: 1–20.

Edwards, C.T., and Saltzman, M.R., 2016, Paired carbon isotopic analysis of Ordovician bulk carbonate ($\delta^{13}C_{carb}$) and organic matter ($\delta^{13}C_{org}$) spanning the Great Ordovician Biodiversification Event. *Palaeogeography, Palaeoclimatology, Palaeoecology*, **458**: 102–117.

Edwards, C.T., Saltzman, M.R., Leslie, S.A., Bergström, S.M., Sedlacek, Howard, A., et al., 2015, Strontium isotope ($^{87}Sr/^{86}Sr$) stratigraphy of Ordovician bulk carbonate: implications for preservation of primary seawater values. *Geological Society of America Bulletin*, **127**: 1275–1289.

Edwards, L.E., 1978, Range charts and no-space graphs. *Computers & Geosciences*, **4**: 247–255.

Elrick, M., Reardon, D., Labor, W., Martin, J., Desrochers, A., and Pope, M., 2013, Orbital-scale climate change and glacioeustasy during the early Late Ordovician (pre-Hirnantian) determined from $\delta^{18}O$ values in marine apatite. *Geology*, **41**: 775–778.

Erdtmann, B.-D., 1988, The earliest Ordovician nematophorous graptolites: taxonomy and correlation. *Geological Magazine*, **125**: 327–348.

Fan, J.X., Peng, P.A., and Melchin, M.J., 2009, Carbon isotopes and event stratigraphy near the Ordovician-Silurian boundary, Yichang, South China. *Palaeogeography, Palaeoclimatology, Palaeoecology*, **276**: 160–169.

Fan, J., Chen, Q., Melchin, M.J., Sheets, H.D., Chen, Z., Zhang, L., et al., 2013, Quantitative stratigraphy of the Wufeng and Lungmachi black shales and graptolite evolution during and after the Late Ordovician mass extinction. *Palaeogeography, Palaeoclimatology, Palaeoecology*, **389**: 96–114.

Fan, R., Bergström, S.M., Lu, Y.Z., Zhang, X.L., Zhang, S.B., Li, X., and Deng, S.H., 2015, Upper Ordovician carbon isotope chemostratigraphy on the Yangtze Platform, Southwestern China: implications for the correlation of the Guttenberg $\delta^{13}C$ excursion (GICE) and paleoceanic change. *Palaeogeography, Palaeoclimatology, Palaeoecology,* **433**: 81−90.

Fang, Q., Wu, H., Hinnov, L.A., Wang, X., Yang, T., Li, H., et al., 2016, A record of astronomically forced climate change in a Late Ordovician (Sandbian) deep marine sequence, Ordos Basin, North China. *Sedimentary Geology,* **341**: 163−174.

Fanton, K.C., and Holmden, C., 2007, Sea-level forcing of carbon isotope excursions in epeiric seas: implications for chemostratigraphy. *Canadian Journal of Earth Sciences,* **44**: 807−818.

Fearnsides, W.G., 1905, On the geology of Arenig Fawr and Moel Llyfnant. *The Quarterly Journal of the Geological Society of London,* **61**: 608−637.

Fearnsides, W.G., 1910, The Tremadoc Slates of south-east Carnarvonshire. *The Quarterly Journal of the Geological Society of London,* **66**: 142−188.

Finnegan, S., Bergmann, K.D., Eiler, J.M., Jones, D.S., Fike, D., Eisenman, I., et al., 2011, The magnitude and duration of Late Ordovician-Early Silurian glaciation. *Science,* **331**: 903−906.

Finnegan, S., Bergmann, K.D., Eiler, J.M., Jones, D.S., Fike, D., Eisenman, I., et al., 2011, The magnitude and duration of Late Ordovician-Early Silurian glaciation. *Science,* **331**: 903−906, https://doi.org/10.1126/science.1200803.

Finney, S.C., 1977, Graptolites of the Middle Ordovician Athens Shale, Alabama. *Unpublished Ph.D. dissertation.* The Ohio State University, Columbus, OH. 585 p.

Finney, S.C., 1986, Graptolite biofacies and correlation of eustatic, subsidence, and tectonic events in the Middle to Upper Ordovician of North America. *Palaios,* **1**: 435−461.

Finney, S.C., and Berry, W.B.N., 1997, New perspectives on graptolite distributions and their use as indicators of platform margin dynamics. *Geology,* **25**: 919−922.

Finney, S.C., Grubb, B.J., and Hatcher, R.D., 1996, Graphic correlation of Middle Ordovician graptolite shale, southern Appalachians: an approach for examining the subsidence and migration of a Taconic foreland basin. *Geological Society of America Bulletin,* **108** (3), 355−371.

Finney, S.C., Berry, W.B.N., Cooper, J.D., Ripperdan, R.L., Sweet, W. C., Jacobson, S.R., et al., 1999, Late Ordovician mass extinction: a new perspective from stratigraphic sections in central Nevada. *Geology,* **27**: 215−218.

Finney, S.C., Berry, W.B.N., and Cooper, J.D., 2007, The influence of denitrifying seawater on graptolite extinction and diversification during the Hirnantian (latest Ordovician) mass extinction event. *Lethaia,* **40**: 281−291.

Foote, M., Sadler, P.M., Cooper, R.A., and Crampton, J.S., 2019, Completeness of the known graptoloid paleontological record. *Journal of the Geological Society,* 18 pp. https://doi.org/10.1144/jgs2019-061.

Fordham, B.G., 1992, Chronometric calibration of mid-Ordovician to Tournaisian conodont zones: a compilation from recent graphic correlation and isotope studies. *Geological Magazine,* **129**: 709−721.

Fortey, R.A., and Cocks, L.R.M., 1986, Marginal faunal belts and their structural implications, with examples from the Lower Palaeozoic. *Journal of the Geological Society,* **143**: 151−160.

Fortey, R.A., Bassett, M.G., Harper, D.A.T., Hughes, R.A., Ingham, J.K., Molyneux, A.W., et al., 1991, Progress and problems in the selection of stratotypes for the bases of series in the Ordovician System of the historical type area in the U.K. *In* Barnes, C.R., and Williams, S.H. (eds), Advances in Ordovician Geology. *Geological Survey of Canada Paper,* **90-9**: pp. 5−25.

Fortey, R.A., Harper, D.A.T., Ingham, J.K., Owen, A.W., and Rushton, A.W.A., 1995, A revision of Ordovician series and stages from the historical type area. *Geological Magazine,* **132**: 15−30.

Fortey, R.A., Harper, D.A.T., Ingham, J.K., Owen, A.W., Parkes, M.A., Rushton, A.W.A., et al., 2000, A revised correlation of the Ordovician rocks of the British Isles. *Geological Society Special Publication,* **24**: 83.

Frakes, L.A., Francis, J.E., and Syktus, J.I., 1992, *Climate Modes of the Phanerozoic.* Cambridge: Cambridge University Press, pp. 15−26.

Gambacorta, G., Menichetti, E., Trinianti, E., and Torricelli, S., 2018, Orbital control on cyclical productivity and benthis anoxia: astronomical tuning of the Telychian Stage (Early Silurian), Palaeoclimatology, Palaeogeography. *Palaeoecology,* **495**: 152−162.

Ghienne, J.-F., Desrochers, A., Vandenbroucke, T.R.A., Achab, A., Asselin, E., Dabard, M.-P., et al., 2014, A Cenozoic-style scenario for the end-Ordovician glaciation. *Nature Communications,* **5**: 4485.

Goldman, D., and Wu, S.-Y., 2010, Paleogeographic, paleoceanographic, and tectonic controls on early Late Ordovician graptolite diversity patterns. *In* Finney, S.C., and Berry, W.B.N. (eds), The Ordovician Earth System. Geological Society of America Special Paper. *Denver, CO: The Geological Society of America,* **466**: 149−161.

Goldman, D., Leslie, S.A., Nõlvak, J., Young, S., Bergström, S.M., and Huff, W.D., 2007a, The Global Stratotype Section and Point (GSSP) for the base of the Katian Stage of the Upper Ordovician Series at Black Knob Ridge, southeastern Oklahoma, USA. *Episodes,* **30**: 258−270.

Goldman, D., Mitchell, C.E., Maletz, J., Riva, J.F.V., Leslie, S.A., and Motz, G.J., 2007b, Ordovician graptolites and conodonts of the Trail Creek Region of central Idaho: a revised, integrated biostratigraphy. *Acta Palaeontologica Sinica,* **46** (Suppl): 155−162.

Goldman, D., Podhalańska, T., Sheets, H., David, Bergström, S.M., Nõlvak, J., et al., 2012, High resolution stratigraphic correlation and biodiversity dynamics of Middle and Late Ordovician marine fossils from Poland and Baltoscandia, In: *GeoShale 2012: Recent Advances in Geology of Fine Grained Sediments. Abstracts and Field Trip Guidebook.* Warsaw, Poland: Polish Geological Institute and NationalResearch Institute, p. 35.

Goldman, D., Sheets, H.D., Bergström, S.M., Nõlvak, J., Podhalańska, T., Mitchell, C.E., et al., 2013, High-resolution stratigraphic correlation and biodiversity dynamics of Middle and Late Ordovician marine fossils from Baltoscandia and Poland. *Abstracts with Programs—Geological Society of America,* **45** (7), 236.

Goldman, D., Nõlvak, J., and Maletz, J., 2015, Mid to Late Ordovician graptolite and chitinozoan biostratigraphy of the Kandava-25 drill core in Western Latvia. *Geologiska Föreningens Förhandlingar,* **137**: 197−211.

Gong, Q., Wang, X., Zhao, L., Grasby, S.E., Chen, Z.-Q., Zhang, L., et al., 2017, Mercury spikes suggest volcanic driver of the Ordovician-Silurian mass extinction. *Nature, Scientific Reports,* **7**: article #5304.

Grahn, Y., and Nõlvak, J., 2007, Ordovician Chitinozoa and biostratigraphy from Skåne and Bornholm, southernmost Scandinavia - an overview and update. *Bulletin of Geosciences,* **82**: 11−26.

Grossman, E.L., Joachimski, M.M., 2020, Chapter 10 - Oxygen isotope stratigraphy. *In* Gradstein, F.M., Ogg, J.G., Schmitz, M.D., and

Ogg, G.M. (eds), *The Geologic Time Scale 2020*. Vol. 1 (this book). Elsevier, Boston, MA.

Gutiérrez-Marco, J.C., Sá, A.A., García-Bellido, D.C., and Rábano, I., 2017, The Bohemo-Iberian regional chronostratigraphical scale for the Ordovician System and palaeontological correlations within South Gondwana. *Lethaia*, **50**: 258–295.

Hall, T.S., 1895, The geology of Castlemaine, with a subdivision of part of the Lower Silurian rocks of Victoria, and a list of minerals. *The Proceedings of the Royal Society of Victoria*, **7**: 55–88.

Haq, B.U., and Schutter, S.R., 2008, A chronology of Paleozoic sea-level changes. *Science*, **322**: 64–68.

Harland, W.B., Armstrong, R.L., Cox, A.V., Craig, L.E., Smith, A.G., and Smith, G., 1990, *A Geologic Time Scale 1989*. Cambridge: Cambridge University Press, p. 263.

Harper, D.A.T., 2006, The Ordovician biodiversification: setting an agenda for marine life. *Palaeogeography, Palaeoclimatology, Palaeoecology*, **232**: 148–166.

Harper, D.A.T., Rasmussen, C.M.Ø., Liljeroth, M., Blodgett, R.B., Candela, Y., Jin, J., et al., 2013, Biodiversity, biogeography and phylogeography of Ordovician rhynchonelliform brachiopods. *Geological Society, London, Memoirs*, **38**: 127–144.

Harper, D.A.T., Hammarlund, E.U., and Rasmussen, C.M.Ø., 2014, End Ordovician extinctions: a coincidence of causes. *Gondwana Research*, **25**: 1294–1307.

Harper, D.A.T., Zhan, R.B., and Jin, J., 2015, The Great Ordovician Biodiversification Event: reviewing two decades of research on diversity's big bang illustrated by mainly brachiopod data. *Palaeoworld*, **24**: 75–85.

Harris, T.M., Sheehan, P.M., Ainsaar, L., Hints, L., Männik, P., Nõlvak, J., et al., 2004, Upper Ordovician sequences of western Estonia. *Palaeogeography, Palaeoclimatology, Palaeoecology*, **210**: 135–148.

Harris, W.J., and Keble, R.A., 1932, Victorian graptolite zones, with correlations and description of species. *The Proceedings of the Royal Society of Victoria*, **44**: 25–48.

Harris, W.J., and Thomas, D.E., 1938, A revised classification and correlation of the Ordovician graptolite beds of Victoria. *Mining and Geological Journal*, **1**: 62–72.

Hatch, J.R., Jacobson, S.R., Witzke, B.J., Risatti, J.B., Anders, D.E., Wattney, W.L., et al., 1987, Possible late Middle Ordovician organic carbon isotope excursion: evidence from Ordovician oils and hydrocarbon source rocks, mid-continent and east-central United States. *The American Association of Petroleum Geologists Bulletin*, **71**: 1342–1354.

Havlíček, V., and Marek, L., 1973, Bohemian Ordovician and its international correlation. *Časopis pro Mineralogii a Geologii*, **18**: 225–232.

Henderson, M.A., Serra, F., Feltes, N.A., Albanesi, G.A., and Kah, L.C., 2018, Paired isotope records of carbonate and organic matter from the Middle Ordovician of Argentina: intrabasinal variation and effects of the marine chemocline. *Palaeogeography, Palaeoclimatology, Palaeoecology*, **490**: 107–130.

Herrmann, A.D., Haupt, B.J., Patzkowsky, M.E., Seidov, D., and Slingerland, R.L., 2004, Response of Late Ordovician paleoceanography to changes in sea level, continental drift, and atmospheric pCO_2: potential causes for long-term cooling and glaciations. *Palaeogeography, Palaeoclimatology, Palaeoecology*, **210**: 385–401.

Herrmann, A.D., MacLeod, K.G., and Leslie, S.A., 2010, Did a volcanic mega-eruption cause global cooling during the Late Ordovician? *Palaios*, **25**: 831–836.

Hinnov, L.A., 2013, Cyclostratigraphy and its revolutionizing applications in the earth and planetary sciences. *Geological Society of America Bulletin*, **125** (11-12), 1703–1734.

Hints, O., Martma, T., Mannik, P., Nõlvak, J., Põldvere, A., Shen, Y., et al., 2014, New data on Ordovician stable isotope record and conodont biostratigraphy from the Viki reference drill core, Saaremaa Island, western Estonia. *Geologiska Föreningens Förhandlingar*, **136**: 100–104.

Holland, S.M., and Patzkowsky, M.E., 2015, The stratigraphy of mass extinction. *Palaeontology*, **58** (5), 903–924.

Holmden, C., Panchuk, K., and Finney, S.C., 2012, Tightly coupled records of Ca and C isotope changes during the Hirnantian glaciation event in an epeiric sea setting. *Geochimica et Cosmochimica Acta*, **98**: 94–106.

Holmden, C., Mitchell, C.E., LaPorte, D.F., Patterson, W.P., Melchin, M.J., and Finney, S.C., 2013, Nd isotope records of late Ordovician sealevel change: implications for glaciation frequency and global stratigraphic correlation. *Palaeogeography, Palaeoclimatology, Palaeoecology*, **386**: 131–144.

Jaanusson, V., 1960, On series of the Ordovician System. Reports of the 21st International Geological Congress. *Copenhagen*, **7**: 70–81.

Jaanusson, V., 1976, Faunal Dynamics in the Middle Ordovician (Viruan) of Balto-Scandia. *In The Ordovician System: Proceedings of a Palaeontological Association Symposium Birmingham, September 1974*. University of Wales Press and National Museum of Wales, Cardiff, pp. 301–326.

Jaanusson, V., 1995, Confacies differentiation and upper Middle Ordovician correlation in the Baltoscandian basin. *Proceedings of the Estonian Academy of Sciences. Geology*, **44**: 73–86.

Jablonski, D., 1991, Extinctions: a paleontological perspective. *Science*, **253**: 754–757.

Jacoby, W.G., 2000, Loess: a non-parametric, graphical tool for depicting relationships between variables. *Electoral Studies*, **19**: 577–613.

Jin, J., and Copper, P., 2008, Response of brachiopod communities to environmental change during the Late Ordovician mass extinction interval, Anticosti Island, eastern Canada. *Fossils and Strata*, **54**: 41–51.

Jin, J., and Zhan, R.-b, 2008, *Late Ordovician Orthide and Billingsellide Brachiopods From Anticosti Island, Eastern Canada: Diversity Change Through Mass Extinction*. Ottawa, ON, Canada: NRC Research Press.

Jones, D.S., Fike, D.A., Finnegan, S., Fischer, W.W., Schrag, D.P., and McCay, D., 2011, Terminal Ordovician carbon isotope stratigraphy and glacioeustatic sealevel change across Anticosti Island (Québec, Canada). *Geological Society of America Bulletin*, **123**: 1645–1664.

Jones, D.S., Creel, R.C., Rios, B., and Ramos, D., 2015, Chemostratigraphy of an Ordovician–Silurian carbonate platform: $\delta^{13}C$ records below glacioeustatic exposure surfaces. *Geology*, **43**: 59–62.

Jones, D.S., Creel, R.C., and Rios, B.A., 2016, Carbon isotope stratigraphy and correlation of depositional sequences in the Upper Ordovician Ely Springs Dolostone, eastern Great Basin, USA. *Palaeogeography, Palaeoclimatology, Palaeoecology*, **458**: 85–101.

Jones, D.S., Martini, A.M., Fike, D.A., and Kaiho, K., 2017, A volcanic trigger for the Late Ordovician mass extinction? Mercury data from south China and Laurentia. *Geology*, **45**: 631–634.

Kah, L.C., 2000, Depositional $\delta^{18}O$ signatures in Proterozoic dolostones: constraints on seawater chemistry and early diagenesis. *In* Grotzinger, J.P., and James, N.P. (eds), Carbonate Sedimentation and Diagenesis in the Evolving Precambrian World. *SEPM Special Publication*, **67**: 345–360.

Kaljo, D.L., Hints, L., Martma, T., and Nõlvak, J., 2001, Carbon isotope stratigraphy in the latest Ordovician of Estonia. *Chemical Geology*, **175**: 49–59.

Kaljo, D.L., Hints, L., Martma, T., Nõlvak, J., and Oraspõld, A., 2004, Late Ordovician carbon isotope trend in Estonia, its significance in stratigraphy and environmental analysis. *Palaeogeography, Palaeoclimatology, Palaeoecology*, **210** (2–4), 165–185.

Kaljo, D.L., Martma, T., and Saadre, T., 2007, Post-Hunnebergian Ordovician carbon isotope trend in Baltoscandia, its environmental implications and some similarities with that of Nevada. *Palaeogeography, Palaeoclimatology, Palaeoecology*, **245**: 138–145.

Kaljo, D., Hints, L., Männik, P., and Nõlvak, J., 2008, The succession of Hirnantian events based on data from Baltica: brachiopods, chitinozoans, conodonts, and carbon isotopes. *Estonian Journal of Earth Sciences*, **57**: 197–218.

Kaljo, D., Männik, P., Martma, T., and Nõlvak, J., 2012, More about the Ordovician-Silurian transition beds at Mirny Creek, Omulev Mountains, NE Russia: carbon isotopes and conodonts. *Estonian Journal of Earth Sciences*, **61**: 277–294.

Kemple, W.G., Sadler, P.M., and Strauss, D.J., 1995, Extending graphic correlation to many dimensions: Stratigraphic correlation as constrained optimisation. *In* Mann, K.O., Lane, H.R., and Scholle, P.A. (eds), Graphic Correlation. *SEPM Special Publication*, **53**: pp. 65–82.

Kimmig, S.R., and Holmden, C., 2017, Multi-proxy geochemical evidence for primary aragonite precipitation in a tropical-shelf 'calcite sea' during the Hirnantian glaciation. *Geochimica et Cosmochimica Acta*, **206**: 254–272.

Klebold, C.J., Leslie, S.A., Goldman, D., and Stouge, S., 2019, Conodont biostratigraphy of the Middle Ordovician (Darriwilian) succession at West Bay Centre Piccadilly Head Quary, Port au Port Peninsula, Newfoundland. *Abstracts with Programs, Geological Society of America*, online: https://gsa.confex.com/gsa/2019AM/webprogram/Paper339107.html.

Kleffner, M.A., 1989, A conodont-based Silurian chronostratigraphy. *Geological Society of America Bulletin*, **101**: 904–912.

Kolata, D.R., Huff, W.D., and Bergström, S.M., 2001, The Ordovician Sebree trough: an oceanic passage to the Midcontinent United States. *Geological Society of America Bulletin*, **113** (8), 1067–1078.

Kraft, P., Storch, P., and Mitchell, C.E., 2015, Graptolites of the Králův Dvůr Formation (mid Katian to earliest Hirnantian, Czech Republic). *Bulletin of Geosciences*, **90**: 195–225.

Kröger, B., and Lintulaakso, K., 2017, RNames, a stratigraphical database designed for the statistical analysis of fossil occurrences – the Ordovician diversification as a case study. *Pakaeontologia Electronica*, **20** (1), 1T, 1–12.

Kump, L.R., Arthur, M.A., Patzkowsky, M.E., Gibbs, M.T., Pinkus, D. S., and Sheehan, P.M., 1999, A weathering hypothesis for glaciation at high atmospheric pCO_2 during the Late Ordovician. *Palaeogeography, Paleoclimatology, Palaeoecology*, **152**: 173–187.

Landing, E., Bowring, S.A., Fortey, R.A., and Davidek, K., 1997, U-Pb zircon date from Avalonian Cape Breton Island and geochronological calibration of the Early Ordovician. *Canadian Journal of Earth Sciences*, **34**: 724–730.

Landing, E., Bowring, S.A., Davidek, K.L., Rushton, A.W.A., Fortey, R. A., and Wimbledon, W.A.P., 2000, Cambrian-Ordovician boundary age and duration of the lowest Ordovician Tremadoc Series based on U-Pb zircon dates from Avalonian Wales. *Geological Magazine*, **137**: 485–494.

LaPorte, D.F., Holmden, C., Patterson, W.P., Loxton, J.D., Melchin, M.J., Mitchell, C.E., et al., 2009, Local and global perspectives on carbon and nitrogen cycling during the Hirnantian glaciation. *Palaeogeography, Palaeoclimatology, Palaeoecology*, **276**: 182–195.

Lapworth, C., 1879, On the tripartite classification of the Lower Paleozoic rocks. *Geological Magazine*, **6**: 1–15.

Lapworth, C., 1879–80, On the geological distribution of the Rhabdophora. *Annals and Magazine of Natural History*, series **5** (3): 245–257. 449–455; (4): 333–341, 423–431; (5): 45–62, 273–285, 359–369; (6): 16–29, 185–207 [vols. 3–4, 1879, vols. 5–6, 1880].

Laskar, J., Robutel, P., Joutel, F., Gastineau, M., Correia, A.C.M., and Levrard, B., 2004, A long term numerical solution for the insolation quantities of the Earth. *Astronomy & Astrophysics*, **428**: 261–285.

Le Heron, D.P., Armstrong, H.A., Wilson, C., Howard, J.P., and Gindre, L., 2010, Glaciation and deglaciation of the Libyan Desert: the Late Ordovician record. *Sedimentary Geology*, **223** (1–2), 100–125.

Lehnert, O., Meinhold, G., Wu, R., Calner, M., and Joachimski, M., 2014, $\delta^{13}C$ chemostratigraphy in the upper Tremadocian through lower Katian (Ordovician) carbonate succession of the Siljan district, central Sweden. *Estonian Journal of Earth Sciences*, **63** (4), 277–286.

Leslie, S.A., and Bergström, S.M., 1995, Timing of the Trenton Transgression and revision of the North American late Middle Ordovician stage classification based on K-bentonite bed correlation. *In* Cooper, J.D., Droser, M.L., and Finney, S.C. (eds), *Ordovician Odyssey: Short Papers for the Seventh International Symposium on the Ordovician System*. Pacific Section SEPM, Vol. 77, pp. 49–54.

Leslie, S.A., Goldman, D., Williams, N.A., Derose, L.M., and Hawkins, A. D., 2008, Bedding plane co-occurrence of biostratigraphically useful conodonts and graptolites in Ordovician shale sequences. *Abstracts with Programs—Geological Society of America*, **40** (6), 475.

Liang, Y., and Tang, P., 2016a, Early-Middle Ordovician chitinozoan biostratigraphy of the Upper Yangtze region, South China. *Journal of Stratigraphy*, **40**: 136–150.

Liang, Y., and Tang, P., 2016b, Early Ordovician to early Late Ordovician chitinozoan biodiversity of the Upper Yangtze region, South China. *Scientia Sinica Terrae*, **46**: 809–823.

Liang, Y., Hints, O., Luan, X., Tang, P., Nõlvak, J., and Zhan, R., 2018, Lower and Middle Ordovician chitinozoans from Honghuayuan, South China: Biodiversity patterns and response to environmental changes. *Palaeogeography, Palaeoclimatology, Palaeoecology*, **500**: 95–105.

Liang, Y., Bernardo, J., Nõlvak, J., Goldman, D., Tang, P., and Hints, O., 2019, Morphological variation suggests that chitinozoans may be fossils of individual microorganisms rather than metazoan eggs. *Proceedings of the Royal Society – Biological Sciences*, **286**: article #20191270. https://doi.org/10.1098/rspb.2019.1270.

Lindskog, A., Costa, M., Rasmussen, C.Ø., Connelly, J., and Eriksson, M.E., 2017, Refined Ordovician timescale reveals no link between asteroid breakup and biodiversification. *Nature Communications*, **8**: article #14066, 1–8.

Lindskog, A., Eriksson, M.E., Bergström, S.M., and Young, S.A., 2019, Lower–Middle Ordovician carbon and oxygen isotope chemostratigraphy at Hällekis, Sweden: implications for regional to global correlation and palaeoenvironmental development. *Lethaia*, **52**: 204–219.

Ling, M.X., Zhan, R.B., Wang, G.X., Wang, Y., Amelin, Y., Tang, P., et al., 2019, An extremely brief end Ordovician mass extinction linked to abrupt onset of glaciation. *Solid Earth Sciences*, **4** (4), 190–198.

Loch, J.D., and Taylor, J.F., 2011, New symphysurinid trilobites from the Cambrian-Ordovician boundary interval in the western United States. *Memoirs of the Association of Australasian Palaeontologists*, **42**: 417–436.

Loi, A., Ghienne, J.F., Dabard, M.P., Paris, F., Botquelen, A., Christ, N., et al., 2010, The Late Ordovician glacio-eustatic record from a high-latitude storm-dominated shelf succession: the Bou Ingarf section (Anti-Atlas, Southern Morocco). *Palaeogeography, Palaeoclimatology, Palaeoecology*, **296**: 332–358.

Loydell, D.K., 2011, Graptolite biozone correlation charts. *Geological Magazine*, **149** (1), 124–132.

Löfgren, A., 1993, Conodonts from the lower Ordovician at Hunneberg, south-central Sweden. *Geological Magazine*, **130**: 215–232.

Löfgren, A., 1996, Lower Ordovician conodonts, reworking, and biostratigraphy of the Orreholmen quarry, Vastergötland, south-central Sweden. *Lethaia*, **118**: 169–183.

Lu, Y., Huang, C., Jiang, S., Zhang, J., Lu, Y., and Liu, Y., 2019, Cyclic late Katian through Hirnantian glacioeustasy and its control of the development of the organic-rich Wufeng and Longmaxi shales, South China. *Palaeogeography, Palaeoclimatology, Palaeoecology*, **526**: 96–109.

Ludvigson, G.A., Jacobson, S.R., Witzke, B.J., and Gonzalez, L.A., 1996, Carbonate component chemostratigraphy and depositional history of the Ordovician Decorah Formation, Upper Mississippi Valley. *In* Witzke, B. J., Ludvigson, G.A., and Day, J. (eds), *Paleozoic Sequence Stratigraphy: Views from the North American Craton*. Geological Society of America, Special Paper 306, pp. 67–86.

Ludvigson, G.A., Witzke, B.J., Schneider, C.L., Smith, E.A., Emerson, N.R., Carpenter, S.J., et al., 2000, A profile of the mid-Caradoc (Ordovician) carbon isotope excursion at the McGregor Quarry, Clayton County, Iowa, Geological Society of Iowa, Guidebook. *In* Anderson, R.R. (ed), The Natural History of Pikes Peak State Park. *County, IA: Clayton*, **70**: pp. 25–31.

Ludvigson, G.A., Witzke, B.J., Schneider, C.L., Smith, E.A., Emerson, N.R., Carpenter, S.J., et al., 2004, Late Ordovician (Turinian-Chatfieldian) carbon isotope excursions and their stratigraphic and paleoceanic significance. *Paleogeography, Paleoclimatology, Paleoecology*, **210**: 187–214.

MacDonald, F.A., Karabinos, P.M., Crowley, J.L., Hodgin, E.B., Crockford, P.W., and Delano, J.W., 2017, Bridging the gap between the foreland and hinterland ii: geochronology and tectonic setting of Ordovician magmatism and basin formation on the Laurentian margin of New England and Newfoundland. *American Journal of Science*, **317**: 555–596.

Maletz, J., 1999, Late Tremadoc graptolites and the base of the *Tetragraptus approximatus* Zone. *Acta Universitatis Carolinae Geologica*, **43**: 25–28.

Maletz, J., and Egenhoff, S.O., 2001, Late Tremadoc to Early Arenig graptolite faunas of southern Bolivia and their implications for a worldwide biozonation. *Lethaia*, **34**: 47–62.

Maletz, J., and Ahlberg, P., 2018, The Lower Ordovician Tøyen Shale succession in the Fågelsång-3 drill core, Scania, Sweden. *GFF*, **140**: 293–305.

Maletz, J., Löfgren, A., and Bergström, S.M., 1996, The base of the *Tetragraptus approximatus* Zone at Mt. Hunneberg, S.W. Sweden: a

proposed Global Stratotype for the Base of the Second Series of the Ordovician System. *Newsletters on Stratigraphy*, **34**: 129–159.

Marshall, J.D., and Middleton, P.D., 1990, Changes in marine isotopic composition and the late Ordovician glaciation. *Journal of the Geological Society of London*, **147**: 1–4.

Mattinson, J.M., 2005, Zircon U–Pb chemical abrasion ('CA-TIMS') method: combined annealing and multi-step partial dissolution analysis for improved precision and accuracy of zircon ages. *Chemical Geology*, **220**: 47–66.

Mattinson, J.M., 2010, Analysis of the relative decay constants of 235U and 238U by multi-step CA-IDTIMS measurements of closed-system natural samples. *Chemical Geology*, **275**: 1–13.

Mauviel, A., and Desrochers, A., 2016, A high-resolution, continuous δ13C record spanning the Ordovician–Silurian boundary on Anticosti Island, eastern Canada. *Canadian Journal of Earth Sciences*, **53**: 795–801.

McArthur, J.M., Howarth, R.J., and Shields, G.A., 2012, Strontium isotope stratigraphy. *In* Gradstein, F.M., Ogg, J.G., Schmitz, M.D., and Ogg, G.M. (eds), *The Geologic Time Scale 2012*. Amsterdam, The Netherlands: Elsevier Science, 127–144.

McCracken, A.D., and Barnes, C.R., 1981, Conodont biostratigraphy and paleoecology of the Ellis Bay Formation, Anticosti Island, Quebec, with special reference to Late Ordovician-Early Silurian chronostratigraphy and the system boundary. *Geological Survey of Canada Bulletin*, **329**: 51–134.

McCracken, A.D., and Nowlan, G.S., 1988, The Gamachian Stage and Fauna 13. *Bulletin of the New York State Museum*, **462**: 71–79.

McKerrow, W.S., Lambert, R.St.J., and Cocks, L.R.M., 1985, The Ordovician, Silurian and Devonian Periods. *In* Snelling, N.J. (ed), The Chronology of the Geological Record. *Memoir of the Geological Society of London*, **10**: 73–80.

Meidla, T., Ainsaar, L., Hints, L., Hints, O., Martma, T., and Nõlvak, J., 1999, The mid-Caradocian biotic and isotopic event in the Ordovician of the East Baltic. *Acta Universitatis Carolinae – Geologica*, **43** (1), 503–506.

Melchin, M.J., 2008, Restudy of some Ordovician–Silurian boundary graptolites from Anticosti Island, Canada, and their biostratigraphic significance. *Lethaia*, **41**: 155–162.

Melchin, J.M., and Mitchell, C.E., 1991, Late Ordovician extinction in the Graptoloidea. *In* Barnes, C.R., and Williams, S.H. (eds), *Advances in Ordovician Geology*. Geological Survey of Canada, 143–156 Paper, 90-9.

Melchin, M.J., and Holmden, C., 2006a, Carbon isotope chemostratigraphy of the Llandovery in Arctic Canada: implications for global correlation and sea-level change. *Geologiska Föreningens Förhandlingar*, **128**: 173–180.

Melchin, M.J., and Holmden, C., 2006b, Carbon isotope chemostratigraphy in Arctic Canada. Sea-level forcing of carbonate platform weathering and implications for Hirnantian global correlation. *Palaeogeography, Palaeoclimatology, Palaeoecology*, **234**: 186–200.

Melchin, M.J., Sadler, P.M., and Cramer, B.D., 2012, The Silurian Period (Chapter 21). *In* Gradstein, F.M., Ogg, J.G., Schmitz, M.D., and Ogg, G.M. (eds), *The Geologic Time Scale 2012*. Amsterdam: Elsevier Press, pp. 525–558.

Melchin, J.M., Mitchell, C.E., Holmden, C., and Štorch, P., 2013, Environmental changes in the Late Ordovician – Early Silurian: review and new insights from black shales and nitrogen isotopes. *Geological Society of America Bulletin*, **125**: 1635–1670.

Melim, L.A., Swart, P.K., and Maliva, R., 2001, *In* Ginsburg, R.N. (ed), Meteoric and Marine Burial Diagenesis in the Subsurface of Great Bahama Bank. *SEPM Special Publication*, **70**: 137–162.

Metzger, J.G., and Fike, D.A., 2013, Techniques for assessing spatial heterogeneity of carbonate δ¹³C: implications for cratonwide isotope gradients. *Sedimentology*, **60**: 1405–1431.

Metzger, J.G., Fike, D.A., and Smith, L.B., 2014, Applying carbon-isotope stratigraphy using well cuttings for high-resolution chemostratigraphic correlation of the subsurface. *Bulletin of the American Association of Petroleum Geologists*, **98**: 1551–1576.

Miller, A.I., 2004, The Ordovician radiation. *In* Webby, B.D., Droser, M.L., Paris, F., and Percival, I. (eds), *The Great Ordovician Biodiversification Event*. New York, NY: Columbia University Press, 72–76.

Miller, A.I., and Mao, S., 1995, Association of orogenic activity with the Ordovician radiation of marine life. *Geology*, **23**: 305–308.

Miller, J.F., Evans, K.R., Loch, J.D., Ethington, R.L., Stitt, J.H., Holmer, L., et al., 2003, Stratigraphy of the Sauk III interval (Cambrian–Ordovician) in the Ibex area, western Millard County, Utah and central Texas. *Brigham Young University Geology Studies*, **47**: 23–118.

Miller, J.F., Repetski, J.E., Nicoll, R.S., Nowlan, G., and Ethington, R.L., 2014, The conodont *Iapetognathus* and its value for defining the base of the Ordovician System. *Geologiska Föreningens Förhandlingar*, **136**: 185–188.

Miller, J.F., Evans, K.R., Ethington, R.L., Freeman, R.L., Loch, J.D., Repetski, J.E., et al., 2016, Proposed Auxiliary Boundary Stratigraphic Section and Point (ASSP) for the base of the Ordovician System at Lawson Cove, Utah, USA. *Stratigraphy*, **12** (3–4), 219–236.

Min, K., Renne, P.R., and Huff, W.D., 2001, ⁴⁰Ar/³⁹Ar dating of Ordovician K-bentonites in Laurentia and Baltoscandia. *Earth and Planetary Science Letters*, **185**: 121–134.

Mitchell, C.E., Chen, X., Bergström, S.M., Zhang, Y., Wang, Z., Webby, B.D., et al., 1997, Definition of a global boundary stratotype for the Darriwilian Stage of the Ordovician System. *Episodes*, **20**: 158–166.

Munnecke, A., Zhang, Y., Liu, X., and Cheng, J., 2011, Stable carbon isotope stratigraphy in the Ordovician of South China. *Palaeogeography, Palaeoclimatology, Palaeoecology*, **307**: 17–43.

Murchison, R.I., 1839, *The Silurian System Founded on Geological Researches in the Counties of Salop, Hereford, Radnor, Montgomery, Caermarthen, Brecon, Pembroke, Monmouth, Gloucester, Worcester, and Stafford: With Descriptions of the Coal-Fields, and Overlying Formations*, 2. London: John Murray, vols.

Myrow, P.M., Fike, D.A., Malmskog, E., Leslie, S.A., Singh, B.P., Chaubey, R.S., et al., 2018, Ordovician–Silurian boundary strata of the Indian Himalaya: record of the latest Ordovician Boda event. *Geological Society of America Bulletin*, **131** (5–6), 881–898.

Nicoll, R.S., Thorshøj Nielsen, A., Laurie, J.R., and Shergold, J.H., 1992, Preliminary correlation of latest Cambrian to Early Ordovician sea level events in Australia and Scandinavia. *In* Webby, B.D., and Laurie, J.R. (eds), *Global Perspectives on Ordovician Geology: Proceedings of the Sixth International Symposium on the Ordovician System, University of Sydney, Australia, 15–19 July 1991*. Balkema Press, Rotterdam, pp. 381–394.

Nilsson, R., 1977, A boring through middle and upper Ordovician strata at Koängen in western Scania, southern Sweden. *Sveriges Geologiska Undersökning*, **71** (8), 1–58.

Nõlvak, J., and Goldman, D., 2007, Biostratigraphy and taxonomy of three-dimensionally preserved nemagraptids from the Middle and Upper Ordovician of Baltoscandia. *Journal of Paleontology*, **81**: 254–260.

Nõlvak, J., and Grahn, Y., 1993, Ordovician chitinozoan zones from Baltoscandia. *Review of Palaeobotany and Palynology*, **79**: 245–269.

Nõlvak, J., Hints, O., and Männik, P., 2006, Ordovician time scale in Estonia: recent developments. *Proc. Estonian Acad. Sciences, Geol.*, **55**: 98–108.

Normore, L.S., Zhen, Y.Y., Dent, L.M., Crowley, J.L., Percival, I.G., and Wingate, M.T.D., 2018, Early Ordovician CA-IDTIMS U–Pb zircon dating and conodont biostratigraphy, Canning Basin, Western Australia. *Australian Journal of Earth Sciences*, **65** (1), 61–73.

Ortega, G., and Albanesi, G.L., 2005, Tremadocian Graptolite-Conodont Biostratigraphy of the South American Gondwana margin (Eastern Cordillera, NW Argentina). *Geologica Acta*, **3** (4), 355–371.

Oruche, N.E., Dix, G.R., and Kamo, S.L., 2018, Lithostratigraphy of the upper Turinian – lower Chatfieldian (Upper Ordovician) foreland succession, and a U–Pb ID–TIMS date for the Millbrig volcanic ash bed in the Ottawa Embayment. *Canadian Journal of Earth Sciences*, **55**: 1079–1102.

Panchuk, K.M., Holmden, C.E., and Leslie, S.A., 2006, Local controls on carbon cycling in the Ordovician midcontinent region of North America, with implications for carbon isotope secular curves. *Journal of Sedimentary Research*, **76**: 200–211.

Pancost, R.D., Freeman, K.H., and Patzkowsky, M.E., 1999, Organic matter source variation and the expression of a late-Middle Ordovician carbon isotope excursion. *Geology*, **27**: 1015–1018.

Paris, F., 1990, The Ordovician chitinozoan biozones of the Northern Gondwana domain. *Review of Palaeobotany and Palynology*, **66**: 181–209.

Paris, F., 1996, Chitinozoan biostratigraphy and paleoecology. *In* Jansonius, J., and McGregor, D.C. (eds), Palynology: Principles and Applications. *American Association of Stratigraphic Palynologists Foundation*, **2**: 531–552.

Paris, F., and Nõlvak, J., 1999, Biological interpretationand paleobiodiversity of a cryptic fossil group: The "chitinozoan animal". *Geobios*, **32**: 315–324.

Paris, F., Achab, A., Asselin, E., Chen, X.-H., Grahn, Y., Nõlvak, J., et al., 2004, Chitinozoans. *In* Webby, B.D., Paris, F., Droser, M.L., and Percival, I.G. (eds), *The Great Ordovician Biodiversification Event*. New York: Columbia University Press, 281–293.

Patzkowsky, M.E., Slupik, L.M., Arthur, M.A., Pancost, R.D., and Freeman, K.H., 1997, Late Middle Ordovician environmental change and extinction: Harbinger of the Late Ordovician or continuation of Cambrian patterns? *Geology*, **25**: 911–914.

Pavlov, V., and Gallet, Y., 2005, A third superchron during the Early Paleozoic. *Episodes*, **28**: 78–84.

Pohl, A., Donnadieu, Y., Le Hir, G., Ladant, J.B., Dumas, C., Alvarez-Solas, J., et al., 2016, Glacial onset predated Late Ordovician climate cooling. *Paleoceanography*, **31**: 800–821.

Pope, M.C., and Steffen, J.B., 2003, Widespread, prolonged late Middle to Late Ordovician upwelling in North America: a proxy record of glaciation? *Geology*, **31** (1), 63–66.

Pouille, L., Danelian, T., and Maletz, J., 2014, Radiolarian diversity changes during the Late Cambrian–Early Ordovician transition as recorded in the Cow Head Group of Newfoundland (Canada). *Marine Micropaleontology*, **110**: 25–41.

Qing, H., Barnes, C.R., Buhl, D., and Veizer, J., 1998, The strontium isotopic composition of Ordovician and Silurian brachiopods and conodonts: relationships to geological events and implications for coeval seawater. *Geochimica et Cosmochimica Acta*, **62**: 1721–1733.

Quinton, P.C., and MacLeod, K.G., 2014, Oxygen isotopes from conodont apatite of the midcontinent US: implications for Late Ordovician climate evolution. *Palaeogeography, Palaeoclimatology, Palaeoecology*, **404**: 57–66.

Quinton, P.C., Law, S., MacLeod, K.G., Herrmann, A.D., Haynes, J.T., and Leslie, S.A., 2017, Testing the early Late Ordovician cool-water hypothesis with oxygen isotopes from conodont apatite. *Geological Magazine*, **155**: 1727–1741.

Rong, J.Y., and Harper, D.A.T., 1988, A global synthesis of latest Ordovician Hirnantian brachiopod faunas. *Transactions of the Royal Society of Edinburgh Earth Sciences*, **79**: 383–402.

Rong, J.Y., and Zhan, R.B., 2004, Late Ordovician brachiopod mass extinction of South China. *In* Rong, J.Y., and Fang, Z.J. (eds), *Mass Extinction and Recovery—Evidences from the Palaeozoic and Triassic of South China*. University of Science and Technology of China Press, 71–96 (in Chinese with English summary).

Rong, J.Y., Boucot, A.J., Harper, D.A.T., Zhan, R.B., and Neuman, R.B., 2006, Global analyses of brachiopod faunas through the Ordovician and Silurian transition: reducing the role of the Lazarus effect. *Canadian Journal of Earth Sciences*, **43**: 23–39.

Rong, J.Y., Huang, B., Zhan, R.B., and Harper, D.A.T., 2008, Latest Ordovician brachiopod and trilobite assemblage from Yuhang, northern Zhejiang, East China: a window on Hirnantian deep-water benthos. *Historical Biology*, **20**: 137–148.

Rosenau, N.A., Herrmann, A.D., and Leslie, S.A., 2012, Conodont apatite $\delta^{18}O$ values from a platform margin setting, Oklahoma, USA: implications for initiation of Late Ordovician icehouse conditions. *Palaeogeography, Palaeoclimatology, Palaeoecology*, **315–316**: 172–180.

Ross, R.J., Adler, F.J., Amsden, T.W., Bergstrom, D., Bergström, S.M., Carter, C., et al., 1982, *The Ordovician System in the United States, Correlation Chart and Explanatory Notes*. International Union of Geological Sciences Publication, p. 73.

Rushton, A.W.A., 1982, The biostratigraphy and correlation of the Merioneth-Tremadoc Series boundary in North Wales. *In* Bassett, M.G., and Dean, W.T. (eds), The Cambrian-Ordovician Boundary: Sections, Fossil Distributions, and Correlations. *National Museum of Wales Geological Series*, **3**: 41–59.

Sadler, P.M., 2003, Constrained optimization – approaches to the paleobiologic correlation and seriation problems: a users' guide and reference manual to the CONOP program family. Unpublished, Riverside, CA, 92 pp. https://extras.springer.com/2003/978-94-017-5087-5/MAIN/Documentation/GUIDE9.pdf

Sadler, P.M., 2012, Integrating carbon isotope excursions into automated stratigraphic correlation: an example from the Silurian of Baltica. *Bulletin of Geosciences*, **87** (4), 681–694.

Sadler, P.M., 2020, Chapter 14B - Global composite sections and constrained optimization. *In* Gradstein, F.M., Ogg, J.G., Schmitz, M.D., and Ogg, G.M. (eds), *The Geologic Time Scale 2020*. **Vol. 1** (this book). Elsevier, Boston, MA.

Sadler, P.M., and Cooper, R.A., 2003, Best-fit intervals and consensus sequences; comparison of the resolving power of traditional biostratigraphy and computer-assisted correlation. *In* Harries, P. (ed), *High Resolution Approaches in Stratigraphic Paleontology*. Dordrecht: Kluwer Academic, 49–94.

Sadler, P.M., and Cooper, R.A., 2004, Calibration of the Ordovician time scale. *In* Webby, B.D., Paris, F., Droser, M.L., and Percival, I.G. (eds), *The Great Ordovician Biodiversification Event*. New York: Columbia University Press, 48–51.

Sadler, P.M., Cooper, R.A., and Melchin, M.R., 2009, High resolution, early Paleozoic (Ordovician-Silurian) time scales. *Geological Society of America Bulletin*, **121**: 887–906.

Sadler, P.M., Cooper, R.A., and Melchin, M.J., 2011, Sequencing the graptolite clade: building a global diversity curve from local range-charts, regional composites and global time-lines. *Proceedings of the Yorkshire Geological Society*, **58** (4), 329–343.

Sadler, P.M., Cooper, R.A., and Crampton, J.S., 2014, High-resolution geobiologic time-lines: progress and potential, fifty years after the advent of graphic correlation. *The Sedimentary Record*, **12** (3), 4–9.

Saltzman, M.R., and Edwards, C.T., 2017, Gradients in the carbon isotopic composition of Ordovician shallow water carbonates: a potential pitfall in estimates of ancient CO_2 and O_2 estimates. *Earth and Planetary Science Letters*, **464**: 46–54.

Saltzman, M.R., and Thomas, E., 2012, Carbon isotope stratigraphy. *In* Gradstein, F.M., Ogg, J.G., Schmitz, M.D., and Ogg, G.M. (eds), *The Geologic Time Scale 2012*. Amsterdam: Elsevier Press, 207–232.

Saltzman, M.R., and Young, S.A., 2005, Long-lived glaciation in the Late Ordovician? Isotopic and sequence-stratigraphic evidence from western Laurentia. *Geology*, **33**: 109–112.

Saltzman, M.R., Edwards, C.T., Leslie, S.A., Dwyer, G.S., Bauer, J.A., Repetski, J.E., et al., 2014, Calibration of a conodont apatite-based Ordovician $^{87}Sr/^{86}Sr$ curve to biostratigraphy and geochronology: implications for stratigraphic resolution. *Geological Society of America Bulletin*, **126**: 1551–1568.

Schiffbauer, J.D., Huntley, J.W., Fike, D.A., Jeffrey, M.J., Gregg, J.M., Kevin, L., et al., 2017, Decoupling biogeochemical records, extinction, and environmental change during the Cambrian SPICE event. *Science Advances*, **3**: e1602158.

Schmitz, B., Harper, D., Peucker-Ehrenbrink, B., Stouge, S., Alwmark, C., Cronholm, A., et al., 2008, Asteroid breakup linked to the Great Ordovician Biodiversification Event. *Nature Geoscience*, **1**: 49–53. https://doi.org/10.1038/ngeo.2007.37.

Schmitz, B., Bergström, S.M., and Wang, X., 2010, The middle Darriwilian (Ordovician) $\delta 13C$ excursion (MDICE) discovered in the Yangtze Platform succession in China; implications of its first recorded occurrence outside Baltoscandia. *Journal of the Geological Society of London*, **167**: 249–259.

Schmitz, M.D., Singer, B.S., and Rooney, A.D., 2020, Chapter 6 - Radioisotope geochronology. *In* Gradstein, F.M., Ogg, J.G., Schmitz, M.D., and Ogg, G.M. (eds), *The Geologic Time Scale 2020*. **Vol. 1** (this book). Elsevier, Boston, MA.

Schoene, B., Crowley, J.L., Condon, D.J., Schmitz, M.D., and Bowring, S.A., 2006, Reassessing the uranium decay constants for geochronology using ID-TIMS U–Pb data. *Geochimica et Cosmochimica Acta*, **70**: 426–445.

Scotese, C.R., 2014, Atlas of Silurian and Middle-Late Ordovician Paleogeographic Maps (Mollweide Projection), Maps 73–80, Volumes 5, The Early Paleozoic, PALEOMAP Atlas for ArcGIS, PALEOMAP Project, Evanston, IL.

Sedgwick, A., 1847, On the classification of the fossiliferous slates of North Wales, Cumberland, Westmoreland and Lancashire. *The Quarterly Journal of the Geological Society of London*, **3**: 133–164.

Sedgwick, A., 1852, On the classification and nomenclature of the Lower Paleozoic rocks of England and Wales. *Quarterly Journal of the Geological Society of London*, **8**: 136–168.

Sell, B.K., and Samson, S.D., 2011, Apatite phenocryst compositions demonstrate a miscorrelation between the Millbrig and Kinnekulle K-bentonites of North America and Scandinavia. *Geology*, **39**: 303–306.

Sell, B.K., Ainsaar, L., and Leslie, S.A., 2013, Precise timing of the Late Ordovician (Sandbian) supereruptions and associated environmental, biological, and climatological events. *Journal of the Geological Society, London*, **170**: 711–714.

Sell, B.K., Samson, S.D., Mitchell, C.E., McLaughlin, P.I., Koenig, A.E., and Leslie, S.A., 2015, Stratigraphic correlations using trace elements in apatite from Late Ordovician (Sandbian-Katian) K-bentonites of eastern North America. *Geological Society of America Bulletin*, **127**: 1259–1274.

Sepkoski, J.J., 1995, The Ordovician radiations: diversification and extinction shown by global genus-level taxonomic data. *In* Cooper, J.D., Droser, M.L., and Finney, S.F. (eds), *Ordovician Odyssey: Short Papers for the Seventh International Symposium on the Ordovician System. The Pacific Section for the Society for Sedimentary Geology (SEPM)*, Fullerton, CA, pp. 393–396.

Serra, F., Feltes, N.A., Albanesi, G.L., and Goldman, D., 2019, High-resolution conodont biostratigraphy from the Darriwilian Stage (Middle Ordovician) of the Argentine Precordillera and biodiversity analyses: a CONOP9 approach. *Lethaia*, **52**: 188–203.

Servais, T., and Harper, D.A.T., 2018, The Great Ordovician Biodiversification Event (GOBE): definition, concept and duration. *Lethaia*, **51**: 151–164.

Servais, T., Lehnert, O., Li, J., Mullins, G.L., Munnecke, A., Nützel, A., et al., 2008, The Ordovician Biodiversification: revolution in the oceanic trophic chain. *Lethaia*, **41**: 99–109.

Servais, T., Owen, A.W., Harper, D.A.T., Kröger, B., and Munnecke, A., 2010, The Great Ordovician Biodiversification Event (GOBE): the palaeoecological dimension. *Palaeogeography, Palaeoclimatology, Palaeoecology*, **294**: 99–119.

Shaw, A.B., 1960, *Time in Stratigraphy*. New York: McGraw-Hill, 365 pp.

Sheehan, P.M., 2001, The Late Ordovician mass extinction. *Annual Review of Earth and Planetary Sciences*, **29**: 331–364.

Shields, G.A., Carden, G.A.F., Vezier, J., Meidla, T., and Jia-yu, R., 2003, Sr, C, and O isotope geochemistry of Ordovician brachiopods: a major isotopic event around the Middle-Late Ordovician transition. *Geochimica et Cosmochimica Acta*, **67**: 2005–2035.

Simmons, M.S., Miller, K.G., Ray, D.C., Davies, A., van Buchem, F.S.P., and Gréselle, B., 2020, Chapter 13 - Phanerozoic eustasy. *In* Gradstein, F.M., Ogg, J.G., Schmitz, M.D., and Ogg, G.M. (eds), *The Geologic Time Scale 2020*. Vol. 1 (this book). Elsevier, Boston, MA.

Skevington, D., 1963, A correlation of Ordovician graptolite-bearing sequences. *Geologiska Föreningens Förhandlingar*, **85**: 298–319.

Smith, M.E., Singer, B.S., and Simo, T., 2011, A time like our own? Radioisotopic calibration of the Ordovician greenhouse to icehouse transition. *Earth and Planetary Science Letters*, **311**: 364–374.

Stigall, A.L., Bauer, J.E., Lam, A.R., and Wright, D.F., 2017, Biotic immigration events, speciation, and the accumulation of biodiversity in the fossil record. *Global and Planetary Change*, **148**: 242–257.

Stouge, S., 2012, Middle Ordovician (late Dapingian–Darriwilian) conodonts from the Cow Head Group and Lower Head Formation, western Newfoundland, Canada. *Canadian Journal of Earth Science*, **49**: 59–90.

Stouge, S., Bagnol, G., and McIlroy, D., 2017, Cambrian-Middle Ordovician Platform-Slope Stratigraphy, Palaeontology and Geochemistry of Western Newfoundland. *International Symposium on the Ediacaran-Cambrian Transition Field Guidebook*. St. John's, Newfoundland, June 2017, pp. 1–106.

Svensen, H.H., Hammer, Ø., and Corfu, F., 2015, Astronomically forced cyclicity in the Upper Ordovician and U–Pb ages of interlayered tephra, Oslo Region, Norway. *Palaeogeography, Palaeoclimatology, Palaeoecology*, **418**: 150–159.

Swanson-Hysell, N.L., and Macdonald, F.A., 2017, Tropical weathering of the Taconic orogeny as a driver for Ordovician cooling. *Geology*, **45**: 719–722.

Sweet, W.C., 1984, Graphic correlation of upper Middle and Upper Ordovician rocks, North American Midcontinental Province, USA. *In* Bruton, D.L. (ed), Aspects of the Ordovician System. *Paleontological Contributions of the University of Oslo*, **295**: 23–35.

Sweet, W.C., 1988, Mohawkian and Cincinnatian chronostratigraphy. *Bulletin of the New York State Museum*, **462**: 84–90.

Sweet, W.C., 1995, A conodont-based composite standard for the North American Ordovician: progress report. *In* Cooper, J.D., Droser, M.L., and Finney, S.F. (eds), *Ordovician Odyssey: Short Papers for the Seventh International Symposium on the Ordovician System. The Pacific Section for the Society for Sedimentary Geology (SEPM)*, Fullerton, CA, pp. 15–20.

Sweet, W.C., and Bergström, S.M., 1976, Conodont biostratigraphy of the Middle and Upper Ordovician of the United Staes Midcontinent. *In* Bassett, M.G. (ed), *The Ordovician System: Proceedings of a Palaeontological Association Symposium, Birmingham*, University of Wales Press and National Museum of Wales, Cardiff, pp. 121–151.

Sweet, W.C., and Bergström, S.M., 1984, Conodont provinces and biofacies of the Late Ordovician. *Geological Society of America Special Paper*, **196**: 69–87.

Tang, P., Wang, Y., Xu, H.-H., Jiang, Q., Yang, Z.-L., Zhan, J.-Z., et al., 2016, Late Ordovician (late Katian) cryptospores and chitinozoans from the Mannan-1 borehole, south Tarim Basin, China. *Palaeoworld*, **26**: 50–63.

Tang, P., Huang, B., Wu, R.C., Fan, J.X., Yan, K., Wang, G.X., et al., 2017, On the Upper Ordovician Daduhe Formation of the Upper Yangtze Region. *Journal of Stratigraphy*, **41**: 119–133.

Terfelt, F., Bagnoli, G., and Stouge, S., 2012, Re-evaluation of the conodont *Iapetognathus* and implications for the base of the Ordovician System GSSP. *Lethaia*, **45**: 227–237.

Thomas, D.E., 1960, The zonal distribution of Australian graptolites. *Journal and Proceedings of the Royal Society of New South Wales*, **94**: 1–58.

Thompson, C.K., and Kah, L.C., 2012, Sulfur isotope evidence for widespread euxinia and a fluctuating oxycline in Early to Middle Ordovician greenhouse oceans. *Palaeogeography, Palaeoclimatology, Palaeoecology*, **313–314**: 189–214.

Thompson, C.K., Kah, L.C., Astini, R., Bowring, S.A., and Buchwaldt, R., 2012, Bentonite geochronology, marine geochemistry, and the Great Ordovician Biodiversification Event (GOBE). *Palaeogeography, Palaeoclimatology, Palaeoecology*, **321–322**: 88–101.

Trotter, J.A., Williams, I.S., Barnes, C.R., Lecuyer, C., and Nicoll, R.S., 2008, Did cooling oceans trigger Ordovician biodiversification? Evidence from conodont thermometry. *Science*, **321**: 550–554.

Tucker, R.D., 1992, U-Pb dating of Plinian-eruption ashfalls by the isotope dilution method: a reliable and precise tool for time-scale calibration and biostratigraphic correlation. *Geological Society of America, Abstracts with Programs*, **24** (7), A198.

Tucker, R.D., and McKerrow, W.S., 1995, Early Paleozoic chronology: a review in light of new U-Pb zircon ages from Newfoundland and Britain. *Canadian Journal of Earth Sciences*, **32**: 368–379.

Twenhofel, W.H., 1928, The Geology of Anticosti Island. *Memoirs of the Geological Survey of Canada*, **154**: 1–481.

Twenhofel, W.H., Bridge, J., Cloude Jr., P.E., Cooper, B.N., Cooper, G.A., Cumings, E.R., et al., 1954, Correlation of the Ordovician Formations of North America. *Geological Society of America Bulletin*, **65**: 247–298.

Underwood, C.J., Crowley, S.F., Marshall, J.D., and Brenchley, P.J., 1997, High-Resolution carbon isotope stratigraphy of the basal Silurian Stratotype (Dob's Linn, Scotland) and its global correlation. *Journal of the Geological Society, London*, **154**: 709–718.

VandenBerg, A.H.M., and Cooper, R.A., 1992, The Ordovician graptolite sequence of Australasia. *Alcheringa*, **16**: 33–65.

Vandenbroucke, T.R.A., 2004, Chitinozoan biostratigraphy of the Upper Ordovician Fågelsång GSSP, Scania, southern Sweden. *Review of Palaeobotany and Palynology*, **130**: 217–239.

Vandenbroucke, T.R.A., 2008a, Upper Ordovician chitinzoans from the historical type area in the UK. London: The Geological Society. *Monograph of the Palaeontographical Society*, **161** (628), 113.

Vandenbroucke, T.R.A., 2008b, An Upper Ordovician Chitinozoan biozonation in British Avalonia (England, and Wales). *Lethaia*, **41**: 275–294.

Vandenbroucke, T.R.A., Rickards, B., and Verniers, J., 2005, Upper Ordovician Chitinozoan biostratigraphy from the type Ashgill Area (Cautley district) and the Pus Gill section (Dufton district, Cross Fell Inlier), Cumbria, Northern England. *Geological Magazine*, **142**: 783–807.

Vandenbroucke, T.R.A., Armstrong, H.A., Williams, M., Paris, F., Sabbe, K., Zalasiewicz, J.A., et al., 2010, Epipelagic chitinozoan biotopes map a steep latitudinal temperature gradient for earliest Late Ordovician seas: implications for a cooling Late Ordovician climate. *Palaeogeography, Palaeoclimatology, Palaeoecology*, **294**: 202–219.

Vandenbroucke, T.R.A., Recourt, P., Nõlvak, J., and Nielsen, A.T., 2013, Chitinozoan biostratigraphy of the Late Ordovician *D. clingani* and *P. linearis* graptolite Biozones on the Island of Bornholm, Denmark. *Stratigraphy*, **10** (4), 281–301.

Wang, X., Stouge, S., Erdtmann, B.-D., Chen, X., Li, Z., Wang, C., et al., 2005, A proposed GSSP for the base of the Middle Ordovician Series: the Huanghuachang section, Yichang, China. *Episodes*, **28** (2), 105–117.

Wang, G.X., Zhan, R.B., and Percival, I.G., 2019, The end-Ordovician mass extinction: A single-pulse event? *Earth-Science Reviews*, **192**: 15–33.

Wang, X.F., Stouge, S., Chen, X.H., Li, Z.H., Wang, C.S., Finney, S.C., et al., 2009, The Global Stratotype Section and Point for the Middle Ordovician Series and the Third Stage (Dapingian). *Episodes*, **32** (2), 96–114.

Wang, Z.-H., Bergström, S.M., Zhen, Y.Y., Chen, X., and Zhang, Y.D., 2013, On the integration of Ordovician conodont and graptolite biostratigraphy: new examples from the Ordos Basin and Inner Mongolia in China. *Alcheringa*, **37** (4), 510–528.

Wang, Z.-H., Zhen, Y.Y., Bergström, S.M., Wu, R.C., Zhang, Y., and Ma, X., 2018, A new conodont biozone classification of the Ordovician System in South China. *Palaeoworld*, **28** (1–2), 173–186.

Watts, W.W., 1939, The author of the Ordovician System; Charles Lapworth, M.Sc., LL.D., F.R.S., F.G.S. *Proceedings of the Geologists' Association*, **50** (2), 235–286.

Webby, B.D., 1995, Towards an Ordovician time scale. *In* Cooper, J.D., Droser, M.L., Finney, S.F. (eds), *Ordovician Odyssey: Short Papers for the Seventh International Symposium on the Ordovician System, The Pacific Section for the Society for Sedimentary Geology (SEPM)*, Fullerton, CA, pp. 5–9.

Webby, B.D., 1998, Steps towards a global standard for Ordovician stratigraphy. *Newsletters in Stratigraphy*, **36**: 1–33.

Webby, B.D., 2004, Introduction. *In* Webby, B.D., Paris, F., Droser, M.L., and Percival, I.G. (eds), *The Great Ordovician Biodiversification Event*. New York: Columbia University Press, 1–38.

Webby, B.D., Cooper, R.A., Bergström, S.M., and Paris, F., 2004, Stratigraphic framework and time slices. *In* Webby, B.D., Paris, F., Droser, M.L., and Percival, I.G. (eds), *The Great Ordovician Biodiversification Event*. New York: Columbia University Press, 41–47.

Webby, B.D., Paris, F., Droser, M.L., and Percival, I.G., 2004, *The Great Ordovician Biodiversification Event*. New York: Columbia University Press, p. 496.

Westrop, S., Amati, L., Brett, C.E., Swisher, R.E., Carlucci, J.R., Goldman, D., et al., 2015, The more the merrier? Reconciling sequence stratigraphy, chemostratigraphy and multiple biostratigraphic indices in the correlation of the Katian Reference Section, central Oklahoma. *Stratigraphy* **12** (2), 139. In Leslie, S.A., Goldman, D., and Orndorff, R.C. (eds), *12th International Symposium on the Ordovician System, Short Papers and Abstracts*.

Whittington, H.B., Dean, W.T., Fortey, R.A., Rickards, R.B., Rushton, A.W.A., and Wright, A.D., 1984, Definition of the Tremadoc series and the series of the Ordovician System in Britain. *Geological Magazine*, **121**: 17–33.

Williams, S.H., and Stevens, R.K., 1988, Early Ordovician (Arenig) graptolites from the Cow Head Group, western Newfoundland. *Palaeontographica Canadiana*, **5**: 1–167.

Won, M.-Z., Iams, W.J., and Reed, K., 2005, Earliest Ordovician (Early To Middle Tremadocian) radiolarian faunas of the Cow Head Group, Western Newfoundland. *Journal of Paleontology*, **79** (3), 433–459.

Woodcock, N.H., 1990, Sequence stratigraphy of the Palaeozoic Welsh Basin. *Journal of the Geological Society*, **147**: 537–547.

Wu, R., Calner, M., and Lehnert, O., 2017, Integrated conodont biostratigraphy and carbon isotope chemostratigraphy in the Lower–Middle Ordovician of southern Sweden reveals a complete record of the MDICE. *Geological Magazine*, **154** (2), 334–353.

Yan, D., Chen, D., Wang, Q., Wang, J., and Wang, Z., 2009, Carbon and sulfur isotopic anomalies across the Ordovician–Silurian boundary on the Yangtze Platform, South China. *Palaeogeography, Palaeoclimatology, Palaeoecology*, **274**: 32–39.

Young, S.A., Saltzman, M.R., and Bergström, S.M., 2005, Upper Ordovician (Mohawkian) carbon isotope ($\delta^{13}C$) stratigraphy in eastern and central North America: regional expression of a perturbation of the global carbon cycle. *Palaeogeography, Palaeoclimatology, Palaeoecology*, **222**: 53–76.

Young, S., Saltzman, M.R., Bergström, S.M., Leslie, S.A., and Chen, X., 2008, Paired $\delta^{13}C_{carb}$ and $\delta^{13}C_{org}$ records of Upper Ordovician (Sandbian-Katian) carbonates in North America and China: implications for paleoceanographic change. *Palaeogeography, Palaeoclimatology, Palaeoecology*, **270**: 166–178.

Young, S.A., Saltzman, M.R., Foland, K.A., Linder, J.S., and Kump, L., 2009, A major drop in seawater $^{87}Sr/^{86}Sr$ during the middle Ordovician (Darriwilian): links to volcanism and climate? *Geology*, **37**: 951–954.

Young, S.A., Saltzman, M.R., Ausich, W.I., Desrochers, A., and Kaljo, D., 2010, Did changes in atmospheric CO_2 coincide with latest Ordovician glacial–interglacial cycles? *Palaeogeography, Palaeoclimatology, Palaeoecology*, **296**: 376–388.

Zalasiewicz, J.A., Taylor, L., Rushton, A.W.A., Loydell, D.K., Rickards, R.B., and Williams, M., 2009, Graptolites in British stratigraphy. *Geological Magazine*, **146**: 785–850.

Zeballo, F.J., Albanesi, G.L., and Ortega, G., 2008, New late Tremadocian (Early Ordovician) conodont and graptolite records from the southern South American Gondwana margin (Eastern Cordillera, Argentina). *Geologica Acta*, **6** (2), 131–145.

Zhang, Y.-D., and Erdtmann, B.D., 2004, Tremadocian (Ordovician) biostratigraphy and graptolites at Dayangcha (Baishan, Jilin, NE China). *Paläontologische Zeitschrift*, **78** (2), 323–354.

Zhang, Y.-D., and Munnecke, A., 2016, Ordovician stable carbon isotope stratigraphy in the Tarim Basin, NW China. *Palaeoclimatology, Palaeogeography, Palaeoecology*, **458**: 154–175.

Zhang, Y.-D., Zhan, R.-B., Zhen, Y.-Y., Wang, Z.-H., Yuan, W.-W., Fang, X., et al., 2019, Ordovician integrative stratigraphy and time scale of China. *Science China Earth Sciences*, **62**: 61–88.

Zhen, Y.Y., Wang, Z.H., Zhang, Y.D., Bergström, S.M., Percival, I.G., and Chen, J.F., 2011, Middle to Late Ordovician (Darriwilian-Sandbian) conodonts from the Dawangou section, Kalpin area of the Tarim Basin, northwestern China. *Records of the Australian Museum*, **63** (3), 203–266.

Zhen, Y.-Y., Percival, I.G., Normore, L.S., and Dent, L.M., 2017, Floian (Early Ordovician) conodonts of the Canning Basin, Western Australia—biostratigraphy and palaeobiogeographic affinities with Chinese faunas. *In* Zhang, Y.-D., Zhan, R.-B., Fan, J.-X., and Muir, L.A. (eds), *Filling the Gap Between the Cambrian Explosion and the GOBE—IGCP Project 653 Annual Meeting 2017, Extended Summaries*. Hangzhou, China: Zhejiang University Press, 233–239.

Zhong, Y., Wu, H., Zhang, Y., Zhang, S., Yang, T., Li, H., et al., 2018, Astronomical calibration of the Middle Ordovician of the Yangtze Block, South China, Palaeoclimatology, Palaeogeography. *Palaeoecology*, **505**: 86–99.

Zhong, Y., Chen, D., Fan, J., Wu, H., Fang, Q., and Shi, M., 2019, Cyclostratigraphic calibration of the Upper Ordovician (Sandbian–Katian) Pagoda and Linhsiang Formations in the Yichang Area, South China. *Acta Geologica Sinica (English Edition)*, **93**: 177–180.

M.J. Melchin, P.M. Sadler and B.D. Cramer

Chapter 21

The Silurian Period

425.6 Ma Silurian

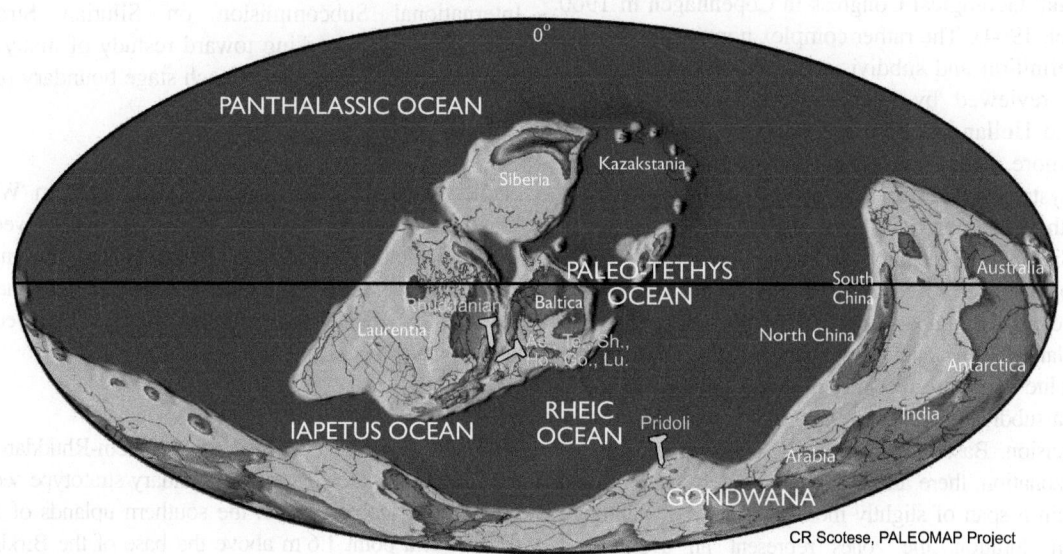

0°

PANTHALASSIC OCEAN

Siberia

Kazakstania

PALEO-TETHYS OCEAN

South China

Australia

Rhuddanian

Baltica

Laurentia

Ta. Sh.,
Tomo., Lu.

North China

Antarctica

IAPETUS OCEAN

RHEIC OCEAN

Pridoli

India

Arabia

GONDWANA

CR Scotese, PALEOMAP Project

Chapter outline

21.1 **History and subdivisions** 696
  21.1.1 Llandovery Series 696
  21.1.2 Wenlock Series 703
  21.1.3 Ludlow Series 706
  21.1.4 Pridoli Series 709
  21.1.5 Regional stage classifications 709
  21.1.6 Silurian stage slices 712
21.2 **Silurian stratigraphy** 712
  21.2.1 Biostratigraphy 712
  21.2.2 Magnetostratigraphy 713
  21.2.3 Chemostratigraphy 714
  21.2.4 Eustasy 716
  21.2.5 Climatic events 717
  21.2.6 Volcanism and K-bentonite stratigraphy 717
  21.2.7 Astrochronology 718
21.3 **Silurian time scale** 718
  21.3.1 Radioisotopic dates 718
  21.3.2 Ages of stage boundaries 722
Acknowledgments 724
Bibliography 724

Abstract

The Silurian Period (443.1−419.0 Ma) was a time of general convergence of continental plates, strong fluctuations in global sea level, and the early stages of colonization of land. The base of the Silurian System is defined at the level of the first appearance of the graptolite species *Akidograptus ascensus* at Dob's Linn, Scotland. Silurian time can be finely resolved using integrated graptolite, conodont, and isotope biochemostratigraphy. The Silurian time scale is based on a CONOP9 composite of graptolite range data derived from 837 stratigraphic sections and 2651 graptolite taxa, with interpolated radioisotope dates, spanning the Ordovician into the Lower Devonian. There is a succession of at least seven globally recognizable positive carbon-isotope excursions, most of which are associated with important bioevents and environmental changes indicated by other geochemical proxies. These data show that the Silurian was a time of dramatic changes in climate, ocean chemistry, and biodiversity.

Geologic Time Scale 2020. DOI: https://doi.org/10.1016/B978-0-12-824360-2.00021-8

21.1 History and subdivisions

The Silurian System was erected by Murchison (1839) and named after the Silures, a Welsh borderland tribe. As originally conceived, the Silurian embraced rocks that were claimed as Cambrian by Sedgwick, leading to a protracted debate. The disputed rocks were separated out as the new Ordovician System by Lapworth (1879) but the debate lingered and Lower Silurian was used in Britain in parallel with Ordovician for many years, while Gotlandian was used in parallel with upper Silurian in some regions. Eventually, the name Silurian was officially adopted in its restricted (upper Silurian) sense at the International Geological Congress in Copenhagen in 1960 (Sorgenfrei, 1964). The rather complex nomenclatural history of definition and subdivision of the Silurian System has been reviewed by Whittard (1961), Cocks et al. (1971), and Holland (1989). Melchin et al. (2004, 2012) provided more recent reviews of our understanding of the Silurian System and its time scale.

As with the Ordovician, black shales are widely developed in Silurian sedimentary successions around the world and graptolites have proved to be valuable fossils for correlation. However, as compared with the Ordovician, it was a time of relatively low faunal provincialism. Over 30 successive graptolite zones are recognized widely around the world providing a subdivision and correlation framework of extraordinary precision. Based on the present scale and generalized graptolite zonation, there are 34 globally widely recognizable zones within a span of slightly more than 24 Myr. Although variable in duration, the zones represent an average of approximately 0.71 Myr. Conodonts have proved to be of considerable global biostratigraphic value in shallow-water carbonate facies and the conodont biostratigraphic scale is rapidly becoming increasingly well resolved. Chitinozoan and acritarch zonations have been developed for several regions and are proving to be increasingly useful for correlation in many circumstances. A rich and diverse fauna is present in the shelly facies where trilobites and brachiopods are used extensively for zonation, and coral-stromatoporoid communities enable local biostratigraphic subdivision and correlation. Vertebrate microfossil, sporomorph, and radiolarian zonations are also being developed for the Silurian. The zonation based on sporomorphs is applicable in terrestrial and sometimes shallow marine facies.

Recently, variations and excursions in the record of stable isotopes, particularly carbon, have proven to be an important tool for correlation, as well as for understanding the record of paleoenvironmental changes through Silurian time. Most importantly, many recent studies have combined biostratigraphic data, often from two or more fossil groups, together with carbon-isotope chemostratigraphic data into integrated biochemostratigraphic analyses to improve the resolution in regional and global correlation.

The Silurian comprises four series, the Llandovery, Wenlock, Ludlow, and Pridoli, in upward sequence. All series and their constituent stages (Fig. 21.1) have designated lower boundary Stratotype Sections and Points (Bassett, 1985; Cocks, 1985). All of the Global Boundary Stratotype Section and Points (GSSPs) (as shown in Fig. 21.14 at the end of this chapter) were biostratigraphically marked with reference to standard graptolite zones (Fig. 21.2), although in some cases the index taxa for these zones (Fig. 21.3) are not found at the GSSP localities. Many of the GSSPs of the Silurian stages have been shown to have significant shortcomings as levels for precise global correlation (see review in Ray, 2011) and the International Subcommission on Silurian Stratigraphy (ISSS) has been working toward restudy of many of these intervals (see discussions of each stage boundary below).

21.1.1 Llandovery Series

Named from the type area in Dyfed, southern Wales, the Llandovery Series comprises three stages, approved by the ICS (Bassett, 1985): the Rhuddanian, Aeronian, and Telychian stages. The Aeronian Stage is approximately, but not exactly, equivalent to the previously employed Idwian and Fronian stages.

21.1.1.1 Rhuddanian Stage

Although the stage is named for the Cefn-Rhuddan Farm in the Llandovery area, its lower boundary stratotype section and point are at Dob's Linn in the southern uplands of Scotland, defined at a point 1.6 m above the base of the Birkhill Shale in the Linn Branch Trench section (Fig. 21.4). This point was previously regarded as coincident with the local base of the *Parakidograptus acuminatus* Zone (Cocks, 1985). More recent resampling and systematic revisions have shown, however, that *P. acuminatus* has its first occurrence datum 1.6 m above this level and that the succession can be readily subdivided, both at this section and globally, into a lower *Akidograptus ascensus* Zone and upper *P. acuminatus* Zone (Melchin and Williams, 2000). Melchin and Williams (2000) therefore proposed that the base *A. ascensus* Zone, marked by the first occurrences of *A. ascensus* (Fig. 21.3A) and *Parakidograptus praematurus* (the latter was identified by Williams, 1983 as *P. acuminatus* sensu lato), be regarded as the biostratigraphic horizon that is locally coincident with the base of the Silurian System. This proposed revision was ratified by the International Union of Geological Sciences (Rong et al., 2008). Thus redefined, the Rhuddanian Stage spans the *A. ascensus* through *Coronograptus cyphus* zones.

21.1.1.2 Aeronian Stage

The Aeronian Stage is named for the Cwm-coed-Aeron Farm in the Llandovery area. The stratotype section and point are located in the Trefawr Formation, in the Trefawr

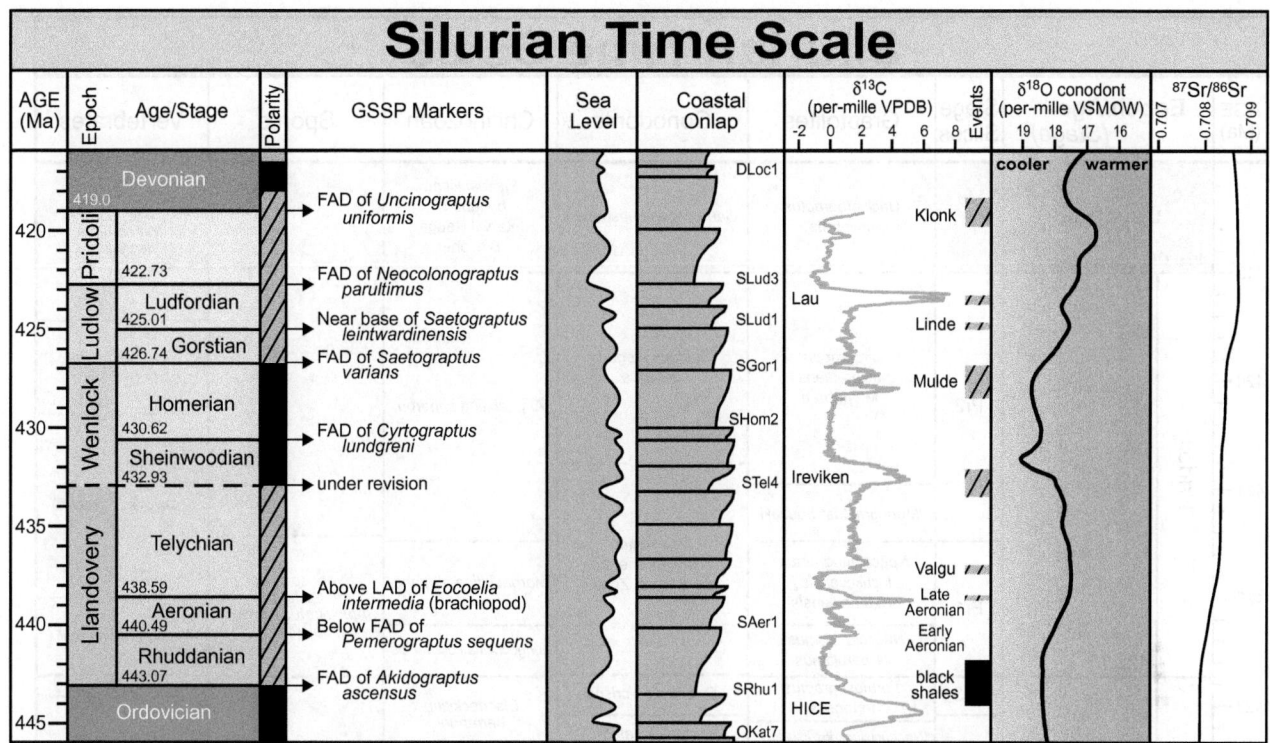

Silurian Time Scale

AGE (Ma)	Epoch	Age/Stage	Polarity	GSSP Markers	Sea Level	Coastal Onlap	δ¹³C (per-mille VPDB) -2 0 2 4 6 8	Events	δ¹⁸O conodont (per-mille VSMOW) 19 18 17 16	⁸⁷Sr/⁸⁶Sr 0.707 0.708 0.709
420	Pridoli	Devonian 419.0		FAD of *Uncinograptus uniformis*		DLoc1		Klonk	cooler warmer	
		422.73		FAD of *Neocolonograptus parultimus*		SLud3	Lau			
425	Ludlow	Ludfordian 425.01		Near base of *Saetograptus leintwardinensis*		SLud1		Linde		
		Gorstian 426.74		FAD of *Saetograptus varians*		SGor1		Mulde		
430	Wenlock	Homerian 430.62		FAD of *Cyrtograptus lundgreni*		SHom2				
		Sheinwoodian 432.93		under revision		STel4		Ireviken		
435	Llandovery	Telychian						Valgu		
		438.59		Above LAD of *Eocoelia intermedia* (brachiopod)				Late Aeronian		
440		Aeronian 440.49		Below FAD of *Pemerograptus sequens*		SAer1		Early Aeronian		
		Rhuddanian 443.07		FAD of *Akidograptus ascensus*		SRhu1		black shales		
445		Ordovician				OKat7		HICE		

FIGURE 21.1 Silurian Overview. Main markers for GSSPs of Silurian stages are intended to be FADs of graptolite taxa, but some of the current GSSPs do not allow a precise calibration (see text). ("Age" is the term for the time equivalent of the rock-record "stage."). Magnetostratigraphy is essentially unstudied within the Silurian. Schematic sea-level, coastal onlap curves, and sequence nomenclature are modified from Haq and Schutter (2008) following the advice of Bilal Haq (pers. comm., 2008, to J. Ogg), but there are several other published curves with different details (see Fig. 21.12; and Simmons et al., 2020, Ch. 13: Phanerozoic eustacy, this book). The δ¹³C curve and the placement of major widespread biotic events (extinction episodes, etc.) are from Melchin et al. (2012, 2013) and Cramer and Jarvis (2020, Ch. 11: Carbon isotope stratigraphy, this book). The oxygen-isotope curve and implied sea-surface temperature trends is the statistical-mean curve from a synthesis from numerous conodont apatite studies, including Trotter et al. (2016) (see Grossman and Joachimski, 2020, Ch. 10: Oxygen isotope stratigraphy, this book). Strontium isotope curve is modified from McArthur et al. (2020, Ch. 7: Strontium isotope stratigraphy, this book). The vertical scale of this diagram is standardized to match the vertical scales of the period-overview figure in all other Phanerozoic chapters. *FADs*, First-appearance datums; *GSSPs*, Global Boundary Stratotype Section and Points; *VPDB*, Vienna PeeDee Belemnite δ¹³C standard; *VSMOW*, Vienna Standard Mean Ocean Water δ¹⁸O standard.

track section 500 m north of the Cwm-coed-Aeron Farm, between Locality 71 and 72 of Cocks et al. (1984). The stratotype point is just below the level of occurrence of *Pernerograptus sequens* (=*Monograptus austerus sequens*), which indicates the *Demirastrites triangulatus* Zone (Bassett, 1985; Cocks, 1989) (Fig. 21.5). However, *P. sequens* has previously been reported from very few other localities, all within Europe (Sudbury, 1958; Hutt, 1974; Bjerreskov, 1975), where its level of first occurrence is within but significantly higher than the base of the *D. triangulatus* Zone, although Štorch et al. (2018) questionably identified this species in Bohemia, first occurring at a level just above the lowest appearance of *D. triangulatus* (Fig. 21.3B). In addition, at the stratotype section, the *D. triangulatus* Zone is confidently represented by only a single fossil horizon (Davies et al., 2011, 2013). Thus although the stratotype point can be shown to occur between levels representing the *C. cyphus* and *D. triangulatus* zones, it cannot be shown to correlate precisely with the boundary between those zones and it probably represents a level within the *D. triangulatus* Zone.

A working group of the ISSS is currently restudying this boundary, focusing on the base of the *D. triangulatus* Zone as the preferred level. A formal proposal for a new GSSP for this boundary has been published for the Hlásná Třebaň section in the Prague Synform of the Czech Republic (Štorch et al., 2018). The Liangshan Danangou section of Nanzheng, southern Shaanxi, China, has also been proposed as a possible stratotype section for this boundary (Shao et al., 2018), although Štorch and Melchin (2018) noted that the only illustrated specimen assigned to *D. triangulatus* from that section does not belong to that species. Preliminary results of study of a second candidate section at Rheidol Gorge, Wales, have been presented by Melchin et al. (2018), and the chitinozoan record from that section has been described by De Weirdt et al. (2019). Study of another possible candidate section in China is also underway. Given the uncertainty in the correlation of the current GSSP with current biostratigraphic zonations, and in keeping with current usage, the base of the *D. triangulatus* Zone is here used to mark the base of the Aeronian.

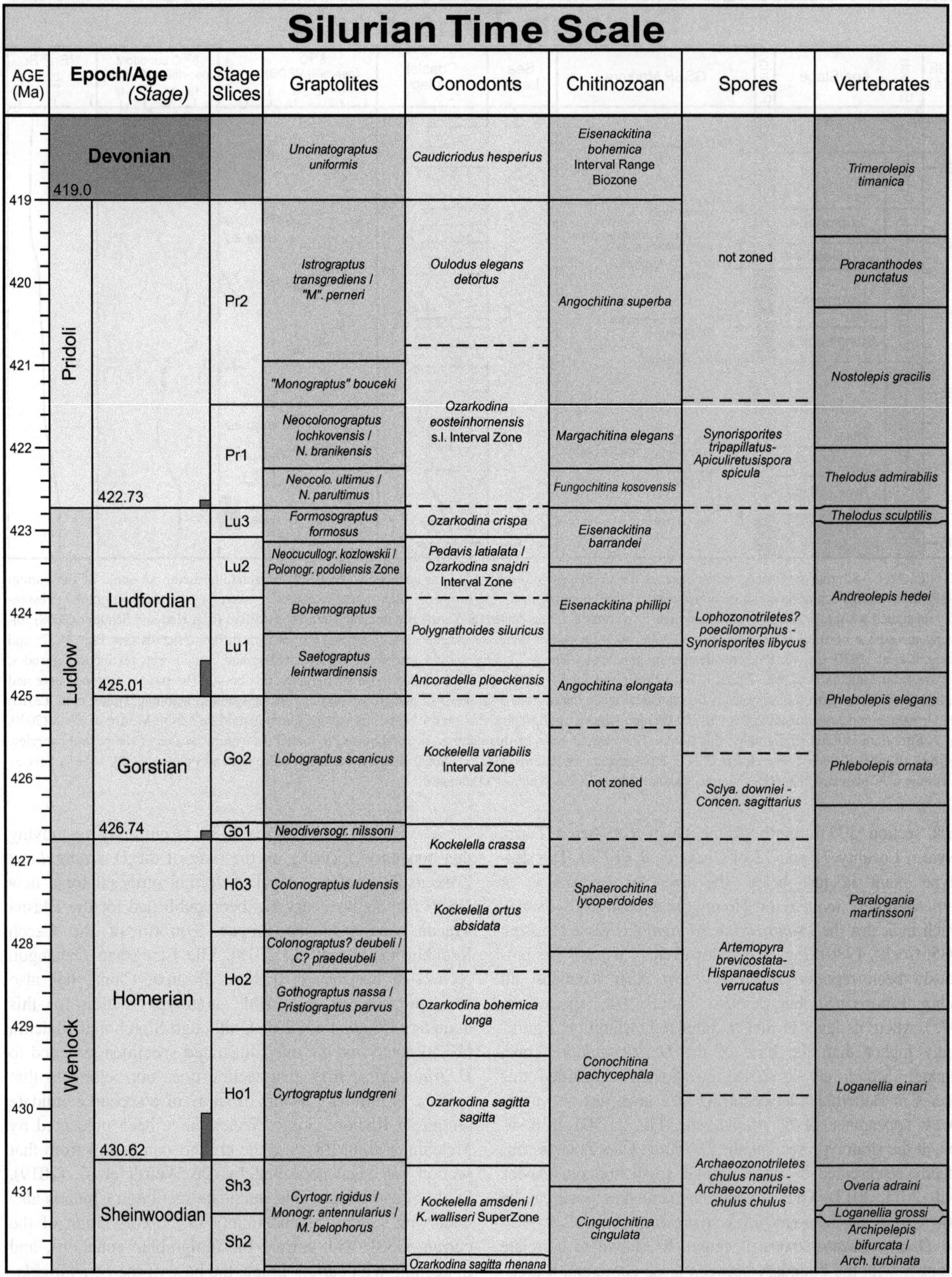

Silurian Time Scale

AGE (Ma)	Epoch/Age (Stage)	Stage Slices	Graptolites	Conodonts	Chitinozoan	Spores	Vertebrates
419.0	Devonian		Uncinatograptus uniformis	Caudicriodus hesperius	Eisenackitina bohemica Interval Range Biozone	not zoned	Trimerolepis timanica
420	Pridoli	Pr2	Istrograptus transgrediens / "M". perneri	Oulodus elegans detortus	Angochitina superba		Poracanthodes punctatus
421			"Monograptus" bouceki				Nostolepis gracilis
422		Pr1	Neocolonograptus lochkovensis / N. branikensis	Ozarkodina eosteinhornensis s.l. Interval Zone	Margachitina elegans	Synorisporites tripapillatus-Apiculiretusispora spicula	Thelodus admirabilis
422.73			Neocolo. ultimus / N. parultimus		Fungochitina kosovensis		
423	Ludfordian	Lu3	Formosograptus formosus	Ozarkodina crispa	Eisenackitina barrandei		Thelodus sculptilis
		Lu2	Neocucullogr. kozlowskii / Polonogr. podoliensis Zone	Pedavis latialata / Ozarkodina snajdri Interval Zone	Eisenackitina phillipi	Lophozonotriletes? poecilomorphus - Synorisporites libycus	Andreolepis hedei
424		Lu1	Bohemograptus	Polygnathoides siluricus			
425.01			Saetograptus leintwardinensis	Ancoradella ploeckensis	Angochitina elongata		Phlebolepis elegans
426	Gorstian	Go2	Lobograptus scanicus	Kockelella variabilis Interval Zone	not zoned	Sclya. downiei - Concen. sagittarius	Phlebolepis ornata
426.74		Go1	Neodiversogr. nilssoni	Kockelella crassa			
427	Homerian	Ho3	Colonograptus ludensis	Kockelella ortus absidata	Sphaerochitina lycoperdoides	Artemopyra brevicostata-Hispanaediscus verrucatus	Paralogania martinssoni
428		Ho2	Colonograptus? deubeli / C? praedeubeli				
			Gothograptus nassa / Pristiograptus parvus	Ozarkodina bohemica longa			
429		Ho1	Cyrtograptus lundgreni	Ozarkodina sagitta sagitta	Conochitina pachycephala		Loganellia einari
430							
430.62	Sheinwoodian	Sh3	Cyrtogr. rigidus / Monogr. antennularius / M. belophorus	Kockelella amsdeni / K. walliseri SuperZone		Archaeozonotriletes chulus nanus - Archaeozonotriletes chulus chulus	Overia adraini
431		Sh2		Ozarkodina sagitta rhenana	Cingulochitina cingulata		Loganellia grossi / Archipelepis bifurcata / Arch. turbinata

FIGURE 21.2   Silurian Epoch and Age time scale, stage slices, graptolite, conodont, chitinozoan, spore, and vertebrate zonal schemes. Stage slices from Cramer et al. (2011a); graptolite zonation slightly modified after Melchin et al. (2012); conodont zonation based on McAdams et al. (2019); chitinozoan zonation from Verniers et al. (1995), updated based on Nestor (2012); spore zonation from Subcommission on Silurian Stratigraphy (1995),

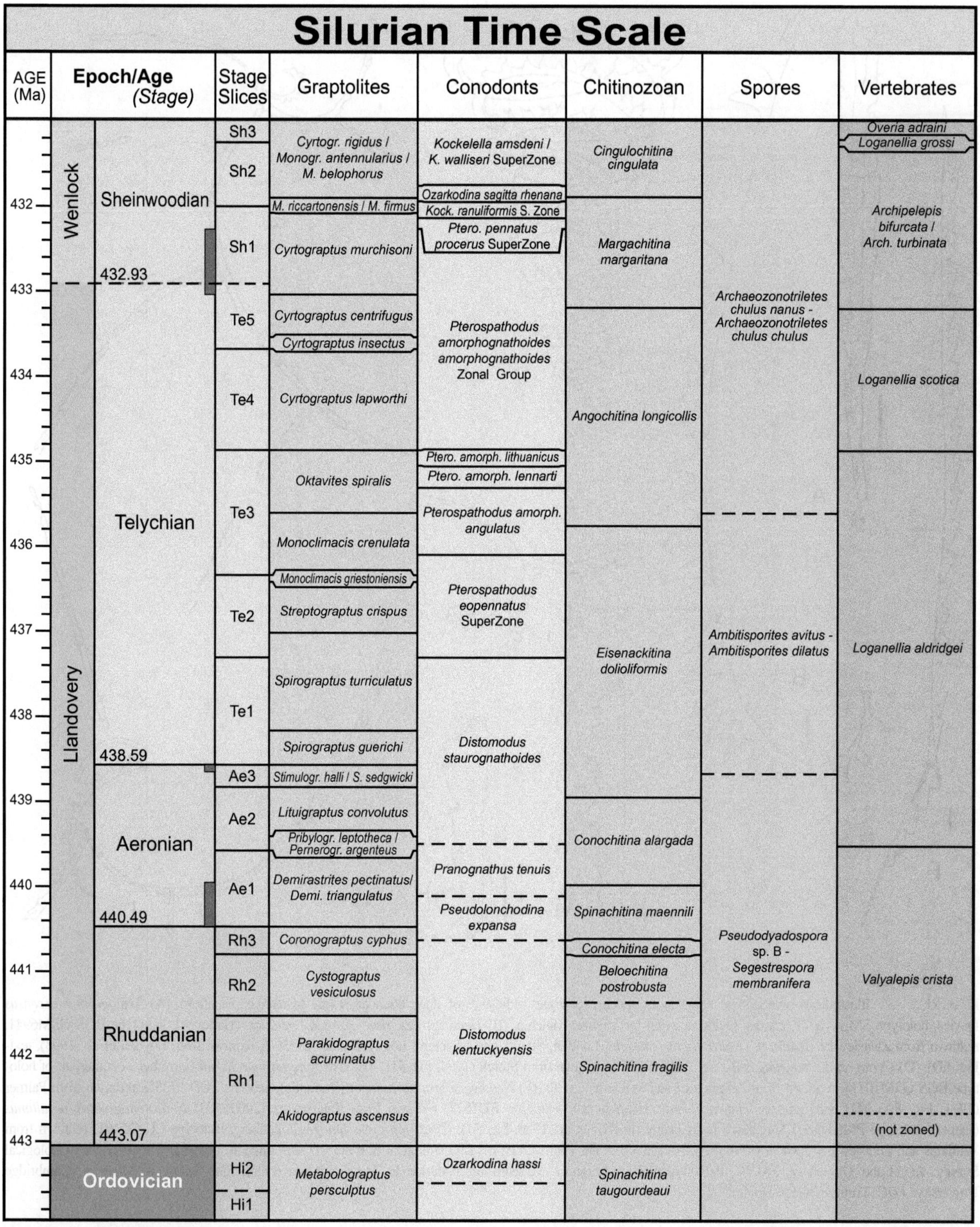

Silurian Time Scale

AGE (Ma)	Epoch/Age (Stage)		Stage Slices	Graptolites	Conodonts	Chitinozoan	Spores	Vertebrates
432	Wenlock	Sheinwoodian	Sh3	Cyrtogr. rigidus / Monogr. antennularius / M. belophorus	Kockelella amsdeni / K. walliseri SuperZone	Cingulochitina cingulata	Archaeozonotriletes chulus nanus - Archaeozonotriletes chulus chulus	Overia adraini
			Sh2					Loganellia grossi
				M. riccartonensis / M. firmus	Ozarkodina sagitta rhenana			Archipelepis bifurcata / Arch. turbinata
			Sh1	Cyrtograptus murchisoni	Kock. ranuliformis S. Zone	Margachitina margaritana		
433		432.93			Ptero. pennatus procerus SuperZone			
	Llandovery	Telychian	Te5	Cyrtograptus centrifugus	Pterospathodus amorphognathoides amorphognathoides Zonal Group	Angochitina longicollis		Loganellia scotica
434				Cyrtograptus insectus				
			Te4	Cyrtograptus lapworthi				
435			Te3	Oktavites spiralis	Ptero. amorph. lithuanicus			Loganellia aldridgei
					Ptero. amorph. lennarti			
436					Pterospathodus amorph. angulatus		Ambitisporites avitus - Ambitisporites dilatus	
				Monoclimacis crenulata				
437			Te2	Monoclimacis griestoniensis	Pterospathodus eopennatus SuperZone	Eisenackitina dolioliformis		
				Streptograptus crispus				
438			Te1	Spirograptus turriculatus	Distomodus staurognathoides			
				Spirograptus guerichi				
		Aeronian	438.59					
439			Ae3	Stimulogr. halli / S. sedgwicki				
			Ae2	Lituigraptus convolutus		Conochitina alargada		
440				Pribylogr. leptotheca / Pernerogr. argenteus	Pranognathus tenuis			
			Ae1	Demirastrites pectinatus/ Demi. triangulatus	Pseudolonchodina expansa	Spinachitina maennili		
		Rhuddanian	440.49				Pseudodyadospora sp. B - Segestrespora membranifera	
			Rh3	Coronograptus cyphus		Conochitina electa		
441			Rh2	Cystograptus vesiculosus		Beloechitina postrobusta		Valyalepis crista
442			Rh1	Parakidograptus acuminatus	Distomodus kentuckyensis	Spinachitina fragilis		
443				Akidograptus ascensus				(not zoned)
		443.07						
	Ordovician		Hi2	Metabolograptus persculptus	Ozarkodina hassi	Spinachitina taugourdeaui		
			Hi1					

FIGURE 21.2   (Continued)

modified after Burgess and Richardson (1995); and vertebrate zonation from Märss and Männik (2013). Gray boxes at stage boundaries indicate the interval of uncertainty in correlation between stratotype points and the graptolite zonation (see text for explanation). Dashed lines at zonal boundaries indicate significant uncertainties in placement or correlation of the zonal boundaries relative to the composite scale.

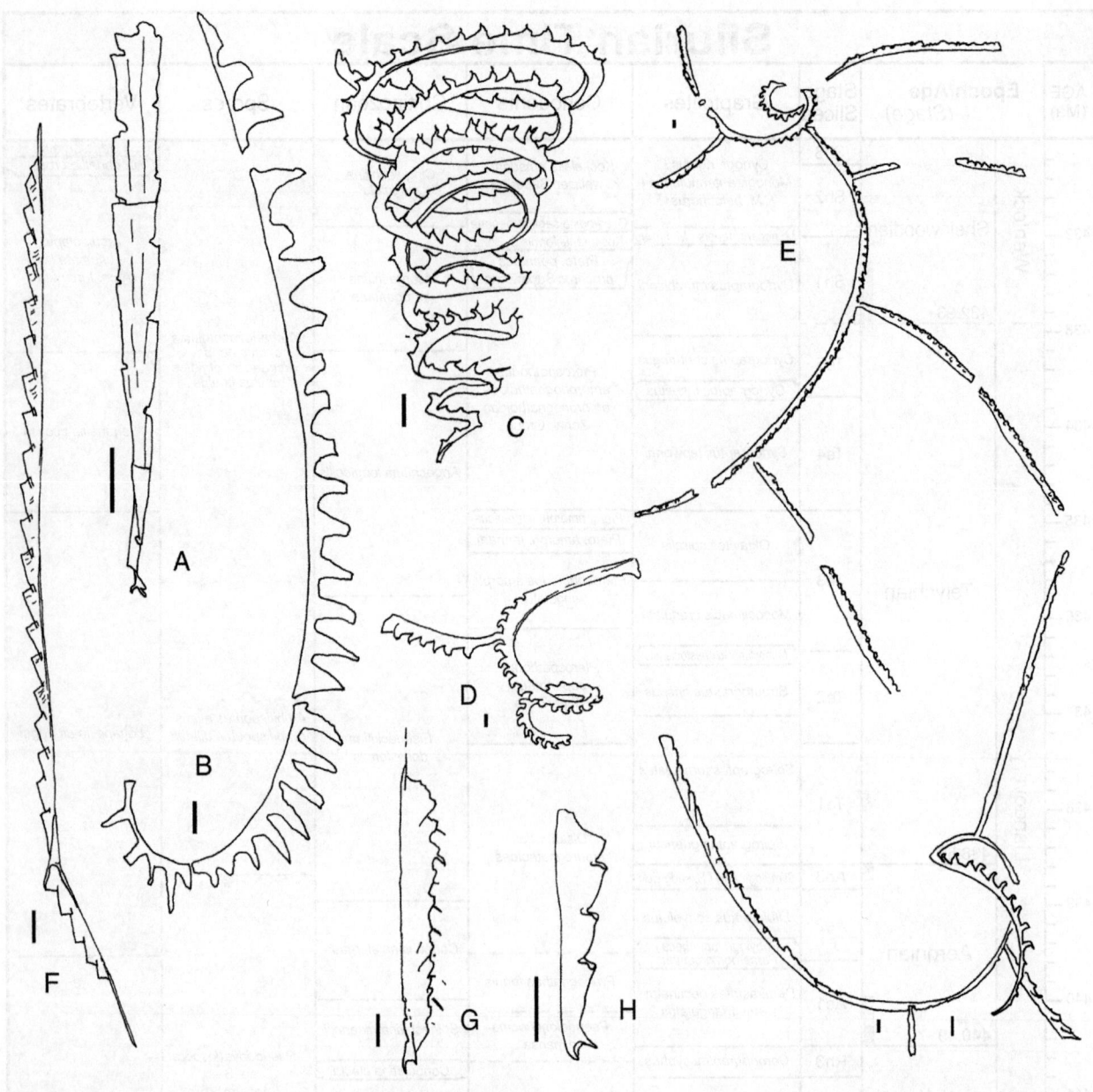

FIGURE 21.3 Illustrations of graptolite zonal index species for zones indicative of stage boundaries (see text for explanation). (A) *Akidograptus ascensus* Davies, holotype SM A10021, camera lucida drawing by Michael Melchin. (B) *Demirastrites triangulatus triangulatus* (Harkness), lectotype BGS GSM6941, redrawn from Zalasiewicz (2008). (C) *Spirograptus guerichi* (Loydell, Štorch, and Melchin), holotype CGS PŠ359/1, redrawn from Loydell et al. (1993), text-Fig. 6D). (D) *Cyrtograptus centrifugus* Bouček, CGS PŠ572, redrawn from Štorch (1994, Fig. 7H). (E) *Cyrtograptus murchisoni* Carruthers, counterpart of holotype BGS GSM10718, redrawn from Zalasiewicz and Williams (2008). (F) *Neodiversograptus nilssoni* (Barrande), neotype TCD 9735D, redrawn from Palmer (1971, Fig. 8A). (G) *Saetograptus leintwardinensis* (Hopkinson), lectotype BU1527, redrawn from Zalasiewicz (2000). (H) *Neocolonograptus parultimus* (Jaeger), holotype PMHUg607.1, redrawn from Jaeger (in Kříž et al., 1986, Fig. 31). (I) *Cyrtograptus lundgreni* Tullberg, lectotype LU O546T, redrawn from Williams and Zalasiewicz (2004, text-Fig. 8A). All scale bars are 1 mm. *BGS*, British Geological Survey; *BU*, Birmingham University; *CGS*, Czech Geological Survey; *LU*, Lund University; *PMHU*, Paläontologisches Museum, Museum für Naturkunde, Humboldt-Universität; *SM*, Sedgwick Museum, Cambridge University; *TDC*, Trinity College Dublin.

The Aeronian Stage has been previously regarded as extending through the *Stimulograptus sedgwickii* Zone, although in several regions, the interval included within this zone, in its broad sense, can be subdivided into a lower *S. sedgwickii* Zone and upper *Stimulograptus halli* Zone (e.g., Loydell, 1991; Zalasiewicz et al., 2009; Loydell et al., 2015). In this case, the latter is regarded as the uppermost graptolite zone of the Aeronian Stage.

Base of the Rhuddinian Stage of the Silurian System in Dob's Linn near Moffat in the Southern Uplands of Scotland, U.K.

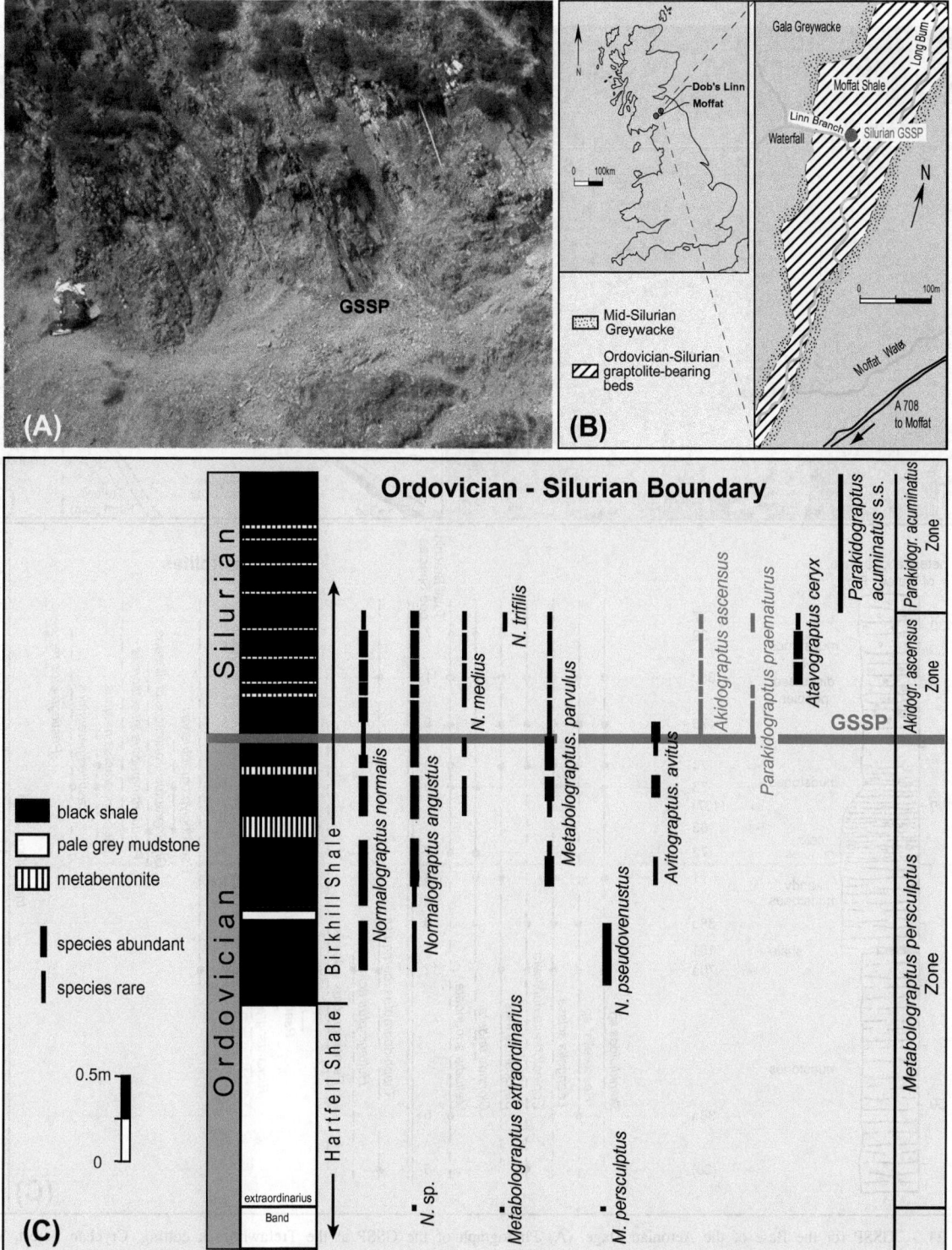

FIGURE 21.4 GSSP for the Base of the Silurian System (base of Rhuddanian Stage). (A) Photograph of the GSSP at Dob's Linn, Scotland. Yellow stick is 1 m in length. Strata are overturned. Photo by Michael Melchin. (B) Location map of the GSSP. (C) Stratigraphy and ranges of selected graptolite taxa at the GSSP. Ranges after Williams (1983), Melchin et al. (2003), and Rong et al. (2008). Taxon names updated after Melchin et al. (2011). *GSSPs*, Global Boundary Stratotype Section and Points.

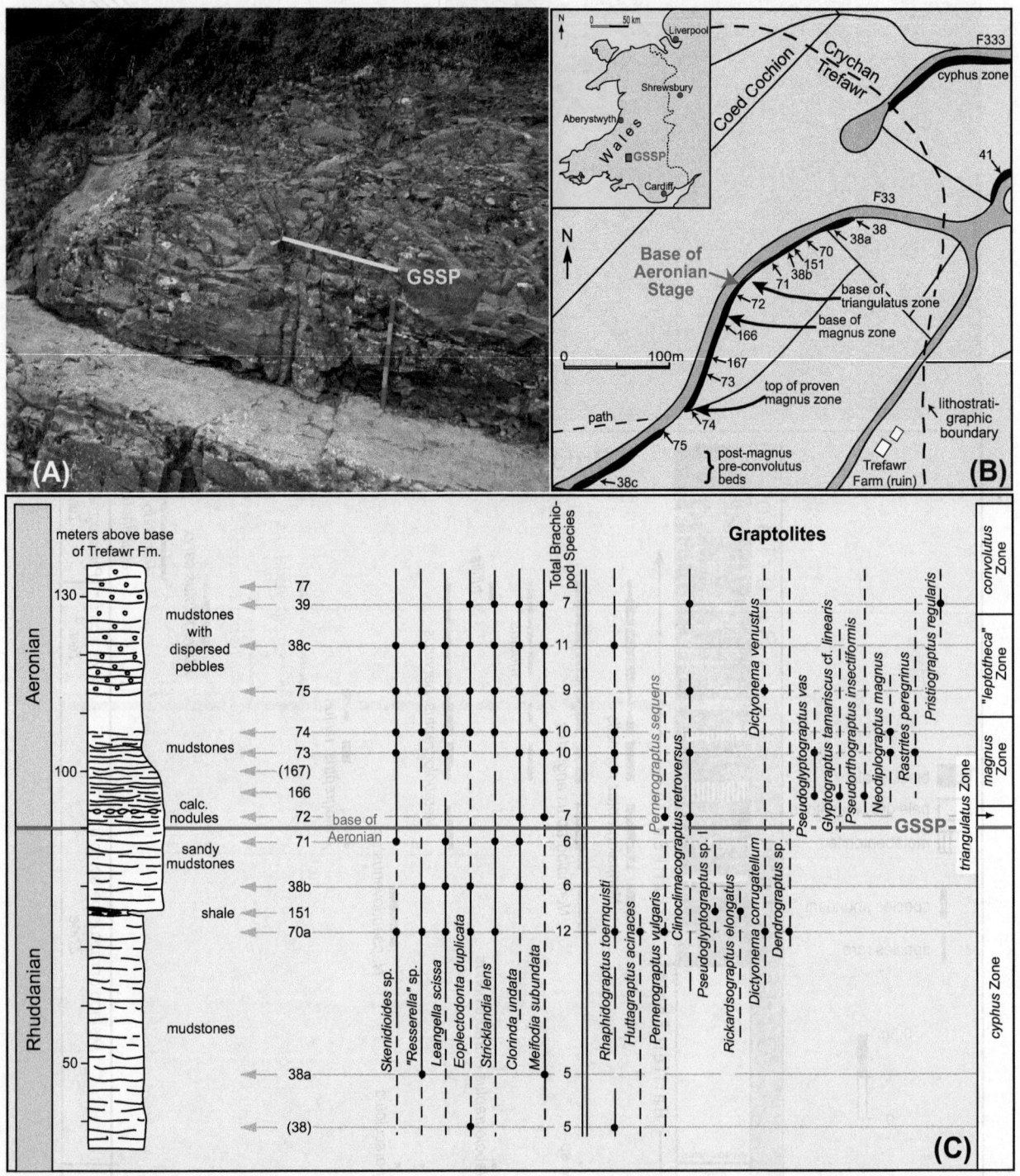

FIGURE 21.5 GSSP for the Base of the Aeronian Stage. (A) Photograph of the GSSP at the Trefawr track cutting, Crychan Forest, Wales. (B) Location map of the GSSP. Exposures are shown in dense black. The localities numbers correspond to the numbers in the profile. (C) Stratigraphy and ranges of key taxa at the GSSP (from Cocks et al., 1984). Graptolite taxon names updated after Štorch et al. (2018). *GSSPs*, Global Boundary Stratotype Section and Points. (A) Photo by Michael Melchin.

21.1.1.3 Telychian Stage

The Telychian Stage is named for the Pen-lan-Telych Farm. The stratotype section and point are located in an abandoned quarry that forms part of the Cefn-cerig Road section, at locality 162 of Cocks et al. (1984) and Cocks (1989), approximately 31 m below the top of the Wormwood Formation. Biostratigraphically, it is marked at a level above the highest occurrence of the brachiopod *Eocoelia intermedia* and below the first appearance of *Eocoelia curtisi* (Bassett, 1985) (Fig. 21.6). This was regarded as corresponding with the base of the *Spirograptus turriculatus* Zone and this was considered to be supported by the occurrence of *Paradiversograptus runcinatus* in the beds above the stratotype level, although not at the stratotype section.

Loydell et al. (1993) revised the species of the genus *Spirograptus* and found that the stratigraphically lower specimens previously assigned to *S. turriculatus* and *S. turriculatus minor* actually belong to a distinct, new species, *S. guerichi* (Fig. 21.3C). Accordingly, those authors found that the strata that had previously been assigned to the lower part of the *S. turriculatus* Zone (the *S. turriculatus minor* Zone of some authors and the *Rastrites linnaei* Zone of others) could be regarded as belonging to a globally correlatable *S. guerichi* Zone. This zone is now regarded as the lowest graptolite zone of the Telychian. However, it should be noted that *P. runcinatus*, the only identifiable graptolite that had been found in the lowest Telychian beds in the vicinity of the stratotype, is known to have its first occurrence in upper Aeronian strata elsewhere in Wales (Loydell, 1991), although it reaches its acme in the lower *S. guerichi* Zone. In addition, Doyle et al. (1991) showed that elsewhere in Britain, the last occurrence of *E. intermedia* occurs within the upper part of the *S. sedgwickii* Zone rather than at the base of the *S. guerichi* Zone. Thus there remains uncertainly regarding the precise correlation of the stratotype point with the graptolite zonation, but it appears to occur within the upper part of the *S. sedgwickii* Zone.

Recent work has documented the occurrence of *S. guerichi* in strata above the level of the GSSP, at approximately the same level of the previous record of *P. runcinatus* (Davies et al., 2013). However, the same study (see also Davies et al., 2011) showed that the strata in between the last occurrence of *E. intermedia* and below the first appearance of *E. curtisi* at the GSSP locality are structurally disrupted, and mudrocks within that interval and in higher strata contain graptolites and chitinozoa indicative of a mid-Sheinwoodian age, intermixed with sediments containing typically Telychian brachiopods, in a sedimentary melange. This evidence shows that the Telychian stratotype locality does not record a continuous stratigraphic succession through the upper Aeronian and lower Telychian.

A working group of the Subcommission on ISSS is currently restudying this boundary, focusing on the base of the *S. guerichi* Zone as the preferred level. The results of biochemostratigraphic study of a candidate GSSP section at El Pintado, Spain, were published by Loydell et al. (2015). Given the uncertainty in the correlation of the current GSSP with current biostratigraphic zonations, and in keeping with current usage, the base of the *S. guerichi* Zone is here used to mark the base of the Telychian.

The Telychian has been most commonly regarded as extending from the *S. guerichi* Zone through the *Cyrtograptus insectus* Zone, although recent restudy of the GSSP for the base of the overlying Sheinwoodian Stage suggests that the *Cyrtograptus centrifugus* Zone and the lowest part of the *Cyrtograptus murchisoni* Zone are also within the uppermost Telychian (see discussion of the Sheinwoodian Stage in Section 21.1.2.1).

21.1.2 Wenlock Series

The Wenlock Series is named for the type area, Wenlock Edge, in the Welsh borderlands of England. It has been divided into two stages, the Sheinwoodian Stage below and the Homerian Stage above.

21.1.2.1 Sheinwoodian Stage

The type locality for the Sheinwoodian Stage occurs in Hughley Brook, 200 m southeast of Leasowes Farm and 500 m northeast of Hughley Church. The stratotype point is the base of the Buildwas Formation as described by Bassett et al. (1975) and Bassett (1989). The stratotype point occurs within the *Pterospathodus amorphognathoides amorphognathoides* Conodont Zonal Group (*sensu* Jeppsson, 1997a), between the base of acritarch zone 5 and the last occurrence of *P. am. amorphognathoides* (Mabillard and Aldridge, 1985) (Fig. 21.7). This level was considered to be approximately correlative with the base of the *C. centrifugus* graptolite zone, although no graptolites are known to occur in the boundary interval at the stratotype section.

The occurrence of *Monoclimacis* aff. *vomerina* and *Pristiograptus watneyae* higher in the Buildwas Formation, together with species indicative of the *M. greistoniensis* and *M. crenulata* zones in the underlying Purple Shales Formation were regarded as indicating that the stratotype point was near the *centrifugus–crenulata* zonal boundary. However, it has since been demonstrated that three additional graptolite zones can be identified between these zones (e.g., Zalasiewicz et al., 2009).

Loydell (2011a) reviewed the data pertaining to the biostratigraphic position of the GSSP for the base of the Wenlock. He noted that conodont data suggested that the GSSP was within a few centimeters of the base of the Upper *Pseudooneotodus bicornis* Zone (Jeppsson, 1997a), or Datum 2 of the Ireviken Event (Jeppsson, 1998). He also noted that data from Estonia and Latvia (Loydell et al., 1998,

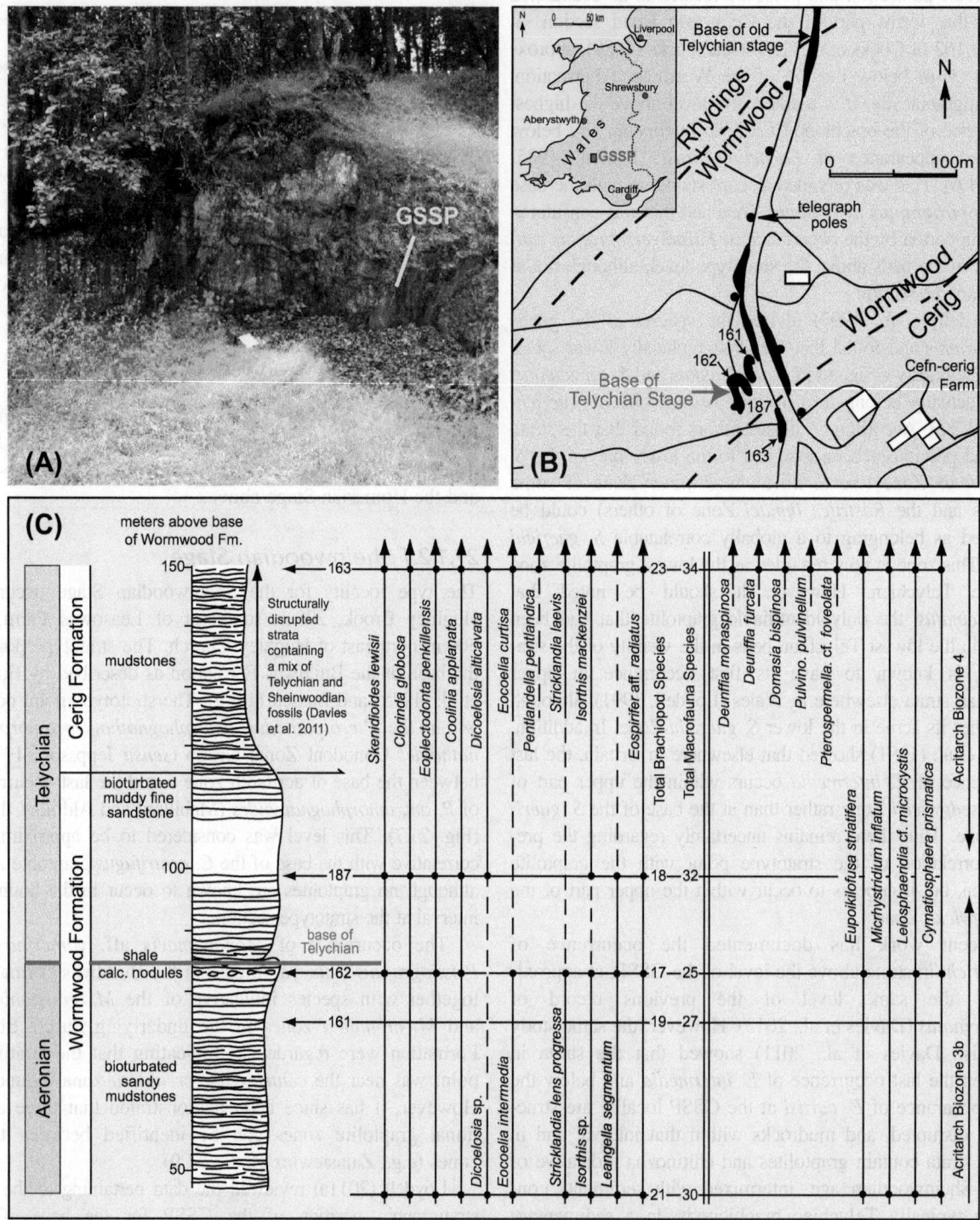

Base of the Telychian Stage of the Silurian System in the area of Dyfed, Wales, Great Britain

FIGURE 21.6 GSSP for the Base of the Telychian Stage. (A) Photograph of the GSSP for the Base of the Telychian Stage at Cefn-Cerig Quarry, near Llandovery, Wales. Geologic hammer indicates the scale. Strata are not overturned (B) Map of the GSSP. (C) Stratigraphy and ranges of key taxa at the GSSP (from Cocks, 1989). Note that level thought to represent Wormwood–Cerig formational contact has been recently reinterpreted as a regional synsedimentary slide surface (Davies et al., 2011; Davies et al., 2013) and the overlying strata as a sedimentary mélange contain a mix of sediments yielding both Telychian and Sheinwoodian fossils. *GSSPs,* Global Boundary Stratotype Section and Points. (A) Photo by Michael Melchin.

Base of the Sheinwoodian Stage (and Wenlock Series) in Hughley Brook, southeast of Leasows Farm, Great Britain

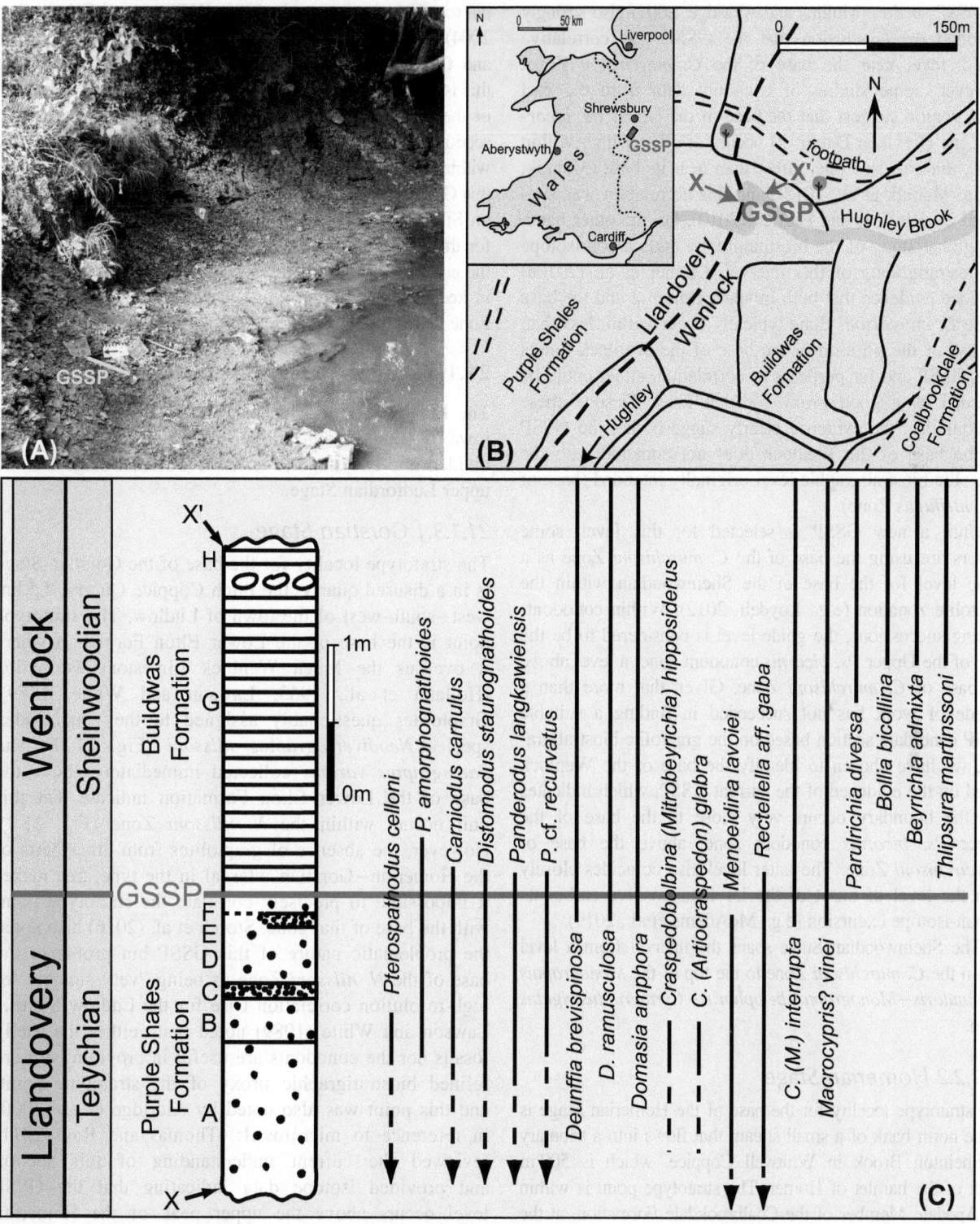

FIGURE 21.7 GSSP for the Base of the Wenlock Series and Sheinwoodian Stage. (A) Photograph of the GSSP at Hughley Brook, Shropshire. White scale card is 15 cm in length. (B) Location map of the GSSP. (C) Stratigraphy and ranges of key taxa at the GSSP (from Bassett, 1989). *GSSPs*, Global Boundary Stratotype Section and Points. (A) Photo by Michael Melchin.

2003) indicated that this conodont datum was correlative with a level near the base of the *C. murchisoni* Zone, the graptolite zone above the *C. centrifugus* Zone. Chitinozoan data from the GSSP section (Mullins and Aldridge, 2004) also strongly supported the conclusion that the GSSP was correlative with a level near the base of the *C. murchisoni* Zone. However, some studies of conodont data from the east Baltic region suggest that the base of the Upper *Ps. bicornis* Zone (Ireviken Datum 2) occurs at a level high within the *C. murchisoni* Zone, rather than near its base (Männik, 2007a; Männik et al., 2015), and this correlation was supported by Kleffner and Barrick (2010). On the other hand, in a global study of the biostratigraphy and carbon-isotope chemostratigraphy of this interval, Cramer et al. (2010a) provided evidence that both Ireviken Datum 2 and the base of the *C. murchisoni* Zone typically occur within less than 200 kyr of the position of the base of the Wenlock Series at its GSSP and for purposes of correlation, either could be regarded as a good proxy for that level. Despite these uncertainties, the evidence clearly suggests that the GSSP for the base of the Wenlock does not coincide with the graptolite biostratigraphic level originally intended (base of *C. centrifugus* Zone).

Until a new GSSP is selected for this level, some authors are using the base of the *C. murchisoni* Zone as a guide level for the base of the Sheinwoodian within the graptolite zonation (e.g., Loydell, 2012). Within conodont-bearing successions, the guide level is considered to be the base of the Upper *Ps. bicornis* conodont zone, a level above the base of *C. murchisoni* Zone. Given that more than a decade of work has not succeeded in finding a suitable GSSP candidate section based on the graptolite biostratigraphy, we have chosen to identify the base of the Wenlock based on the evidence of the current GSSP, which indicates that the boundary occurs very close to the base of the Upper *Ps. bicornis* conodont zone, above the base of *C. murchisoni* Zone. The latter level also coincides closely with the level of onset of the Ireviken positive carbonate carbon-isotope excursion (e.g., McAdams et al., 2019).

The Sheinwoodian Stage spans the interval from a level within the *C. murchisoni* Zone to the top of the *Monograptus antennularis—Monograptus belophorus—Cyrtograptus rigidus* Zone.

21.1.2.2 Homerian Stage

The stratotype locality for the base of the Homerian Stage is in the north bank of a small stream that flows into a tributary of Sheinton Brook in Whitwell Coppice, which is 500 m north of the hamlet of Homer. The stratotype point is within the Apedale Member of the Coalbrookdale Formation, at the point of first appearance of a graptolite fauna containing *Cyrtograptus lundgreni* (Figs. 21.3I and 21.8). Underlying strata contain graptolites of the *Cyrtograptus ellesae* Zone (Bassett et al., 1975; Bassett, 1989). Although the faunas of the *C. ellesae* and *C. lundgreni* zones seem to be

stratigraphically and taxonomically distinct in this and some other regions, recent work in Wales has shown a succession where the first occurrence of *C. lundgreni* is below that of *C. ellesae* (Zalasiewicz et al., 1998; Williams and Zalasiewicz, 2004). In addition, Loydell (2011b) suggested that *C. ellesae* and *C. lundgreni* may be the same species. Whether or not this is true, the currently available data suggest that the ranges of the zonal index taxa may be incomplete at the stratotype section and that the stratotype point for the Homerian is likely within the *C. lundgreni* Zone rather than at its base. Although this GSSP is in need of restudy, the International Commission on Silurian Stratigraphy has not yet struck a working group for this boundary. Given the uncertainty in the correlation of the current GSSP with current biostratigraphic zonations, and in keeping with current usage, the base of the *C. lundgreni* Zone is here used to mark the base of the Homerian.

21.1.3 Ludlow Series

The Ludlow Series is named for the type area near the town of Ludlow, in Shropshire, England. It has been divided into two stages, the lower Gorstian Stage and the upper Ludfordian Stage.

21.1.3.1 Gorstian Stage

The stratotype locality for the base of the Gorstian Stage is in a disused quarry, the Pitch Coppice Quarry, 4.5 km west—south-west of the town of Ludlow. The stratotype point is the base of the Lower Elton Formation where it overlies the Much Wenlock Limestone Formation (Holland et al., 1963; Lawson and White, 1989). Graptolites questionably assigned to the zonal index species *Neodiversograptus nilssoni* (Fig. 21.3F) and *Saetograptus varians* collected immediately above the base of the Lower Elton Formation indicate that this unit occurs within the *N. nilssoni* Zone (Fig. 21.9). However, the absence of graptolites from other parts of the Homerian—Gorstian interval in the type area makes it impossible to precisely correlate the stratotype point with the base of that zone. Štorch et al. (2016) also noted the problematic nature of this GSSP but promoted the base of the *N. nilssoni* Zone as being "very suitable for high-resolution correlation base for the Ludlow Series." Lawson and White (1989) noted that neither the shelly fossils nor the conodonts are useful in providing a more refined biostratigraphic proxy of the stratotype point, and this point was also noted by Aldridge et al. (2000) in reference to microfossils. Thomas and Ray (2011) reviewed the current understanding of this section and provided isotope data indicating that the GSSP level occurs above the upper peak of the Homerian (Mulde) positive carbon-isotope excursion. The conodont *Kockelella crassa* has been used to approximate the position of this boundary; however, recent biochemostratigraphic data (McAdams et al., 2019) indicates that the first appearance of *K. crassa* is likely just below the

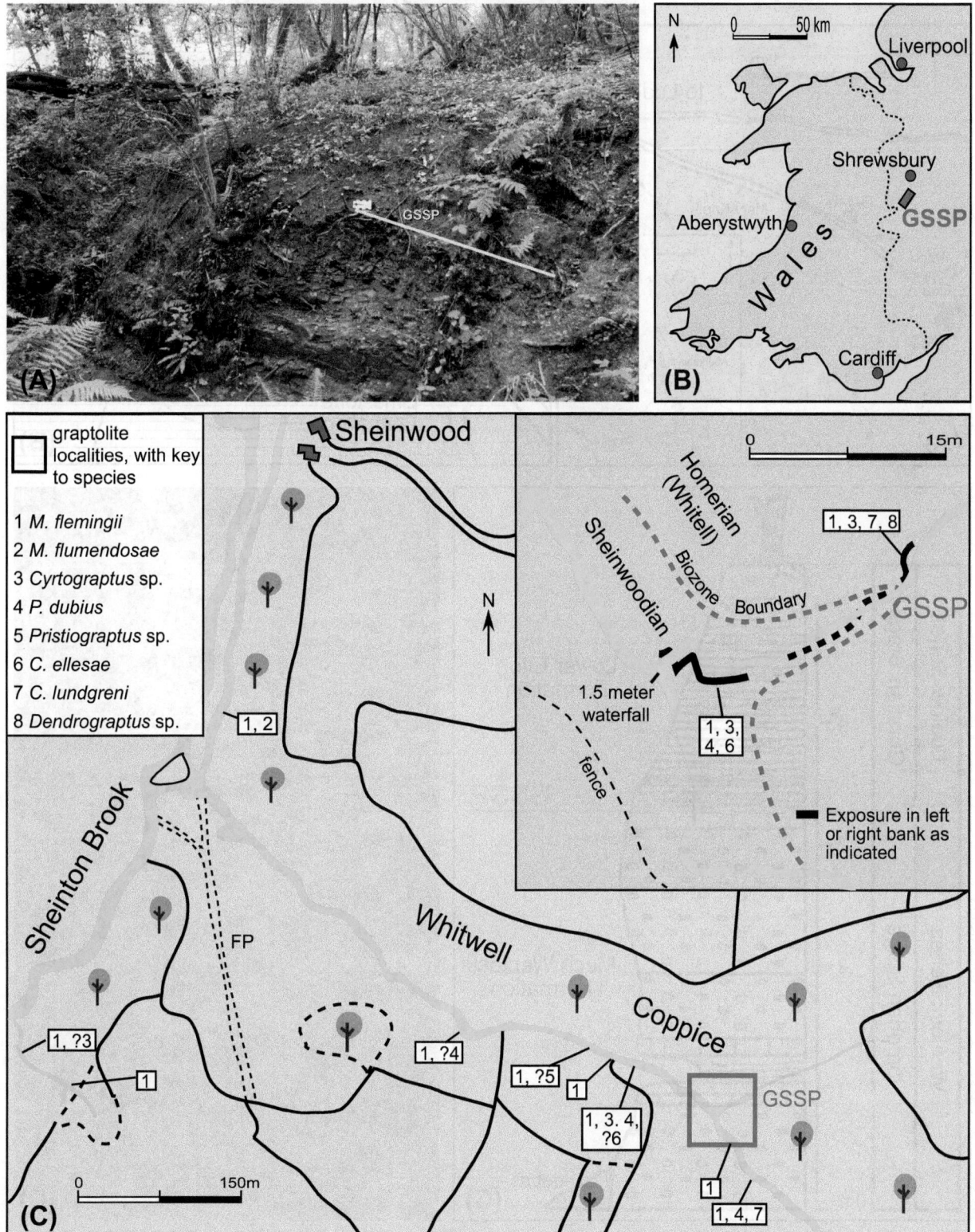

Base of the Homerian Stage of the Silurian System, in a Tributary to the Sheinton Brook, near the Hamlet of Homer, United Kingdom

FIGURE 21.8 GSSP for the Base of the Homerian Stage. (A) Photograph of the GSSP at Whitwell Coppice, Shropshire. White scale card is 15 cm in length. (B, C) Location maps of the GSSP, including positions of key biostratigraphic levels (from Bassett, 1989). *GSSPs*, Global Boundary Stratotype Section and Points. (A) Photo by Michael Melchin.

Base of the Gorstian Stage of the Silurian System in Pitch Coppice Quarry, United Kingdom.

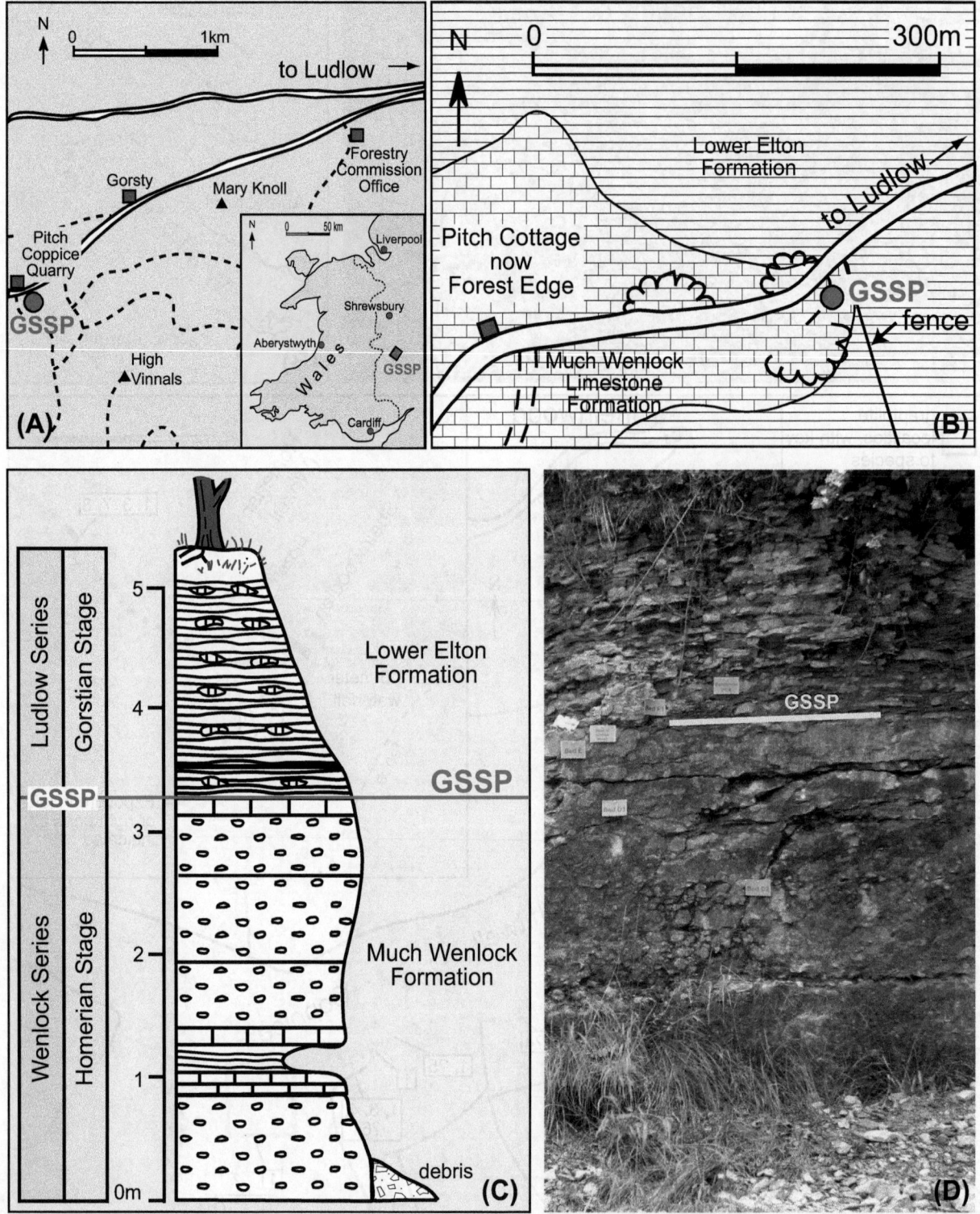

FIGURE 21.9   Base of the Ludlow Series and Gorstian Stage. (A, B) Location maps of the GSSP at Pitch Coppice near Ludlow, Shropshire, Great Britain. (C) Lithologic log of GSSP section (from Lawson and White, 1989). (D) Photograph of the GSSP. White scale card is 15 cm in length. (D) Photo by Michael Melchin.

position of the GSSP. Although this GSSP is in need of restudy, the International Commission on Silurian Stratigraphy has not yet struck a working group for this boundary. Given the uncertainty in the correlation of the current GSSP with current biostratigraphic zonations, and in keeping with current usage, the base of the *N. nilssoni* Zone is here used to mark the base of the Gorstian.

21.1.3.2 Ludfordian Stage

The locality for the stratotype of the Ludfordian Stage is at Sunnyhill Quarry, approximately 2.5 km southwest of the town of Ludlow. The level coincides with the contact between the Upper Bringewood Formation and the Lower Leintwardine Formation (Lawson and White, 1989, Fig. 21.10). The graptolite *Saetograptus leintwardinensis* (Fig. 21.3G) occurs in the basal beds of the Lower Leintwardine Formation, and it becomes common higher in the formation. The underlying Upper Bringewood Formation is devoid of identifiable graptolites, although the Lower Bringewood Formation contains graptolites indicative of the *Pristograptus tumescens/Saetograptus incipiens* Zone. Therefore the stratotype point is considered to approximate the base of the *S. leintwardinensis* Zone (Lawson and White, 1989), although it may occur within the lower part of that zone. Štorch et al. (2014) noted the synonymy of *S. leintwardinensis* and *Saetograptus linearis*, another species commonly used to identify the base of the Ludfordian and, given the shortcomings of the current GSSP, promoted the use of the base of the *S. leintwardinensis* Zone for recognition of the base of this stage.

The stratotype level is also marked by the disappearance of a number of distinctive brachiopod taxa as well as changes in relative abundances among others. No distinctive conodont taxa appear at the stratotype point. However, there are significant changes in the palynological assemblages at or near the formational contact (Lawson and White, 1989). Cherns (2011) reviewed the current understanding of this section and provided carbon-isotope data that indicate a weak (~ 1‰) positive shift beginning approximately 50 cm below the GSSP level. This may represent the onset of a weak positive excursion recognized at other localities, which has been related to the Linde extinction event by some authors (e.g., Cramer et al., 2011a). Although this GSSP is in need of restudy, the International Commission on Silurian Stratigraphy has not yet struck a working group for this boundary. Given the uncertainty in the correlation of the current GSSP with current biostratigraphic zonations, and in keeping with current usage, the base of the *S. leintwardinensis* Zone is here used to mark the base of the Ludfordian.

21.1.4 Pridoli Series

The Pridoli Series is named for the Přídolí area, near Prague, Bohemia, Czech Republic. This series has not been subdivided into stages. The stratotype section is the Požáry section in the Daleje Valley, near Řeporyje, Prague (Kříz et al.,

1986; Kříz, 1989). The stratotype point is in bed 96, approximately 2 m above the base of the Požáry Formation, marked by the first appearance of *Neocolonograptus parultimus* (Figs. 21.3H and 21.11). Graptolites are absent from the immediately underlying strata, however, so it is possible that the stratotype point occurs within the lower part of that zone. A number of other fossil groups are common in the type area of the Pridoli besides the graptolites, but only chitinozoa show potential for more detailed biostratigraphic correlation in this region. The base of the *Fungochitina kosovensis* chitinozoan zone occurs approximately 20 cm above the stratotype point. The disappearance of the conodont *Ozarkodina crispa* has been used to approximate the position of the base of the Pridoli Series, however, the presence of the graptolite *Neocolonograptus parultimus* below the last appearance of *O. crispa* in the Cellon section demonstrates that *O. crispa* extends at least into the lowermost part of the Pridoli Series (Corradini et al., 2015).

21.1.5 Regional stage classifications

Although a number of regions of the world have had regional series and stage classifications, the majority of these have fallen out of usage since the global standard series and stage stratotype sections and points were defined. The generally high degree of faunal cosmopolitanism has greatly facilitated the global usage of the standard time scale. However, there remain significant intervals in which correlation between the graptolite biostratigraphic scale and that of conodonts and carbonate shelf facies are imprecise. As a result, in the East Baltic Region, which has a long history of detailed faunal and stratigraphic study in mainly carbonate strata, workers continue to refer to a scale of regional stages (East Baltic Regional Stages, Fig. 21.12) based on faunal and facies changes in that area (Bassett et al., 1989).

The interior of Laurentia, including the midcontinent regions of the United States and Canada, has also seen some continued usage of regional series and stages. Unlike the Baltic, however, the terminology has not been consistently employed between different regions within North America, and many of the series and stage boundaries have not been clearly defined. Cramer et al. (2011a) reviewed the current state of understanding of the North American series and stage terminology and its correlation with the global standard and identified a number of intervals where there remains considerable uncertainty in this correlation.

It is worth noting that in many Russian stratigraphic studies, the Silurian is formally divided into two subsystems. The Lower Silurian includes the Llandovery and Wenlock series and the Upper Silurian includes the Ludlow and Pridoli series (e.g., Tesakov, 2015), although in some Russian studies, the Upper and Lower Silurian are regarded as series (Zhamoida, 2015) and the Llandovery, Wenlock, Ludlow, and Pridoli as stages. The usage of "lower" and "upper" Silurian is often

Base of the Ludfordian Stage of the Silurian System at Sunnyhill Quarry, United Kingdom.

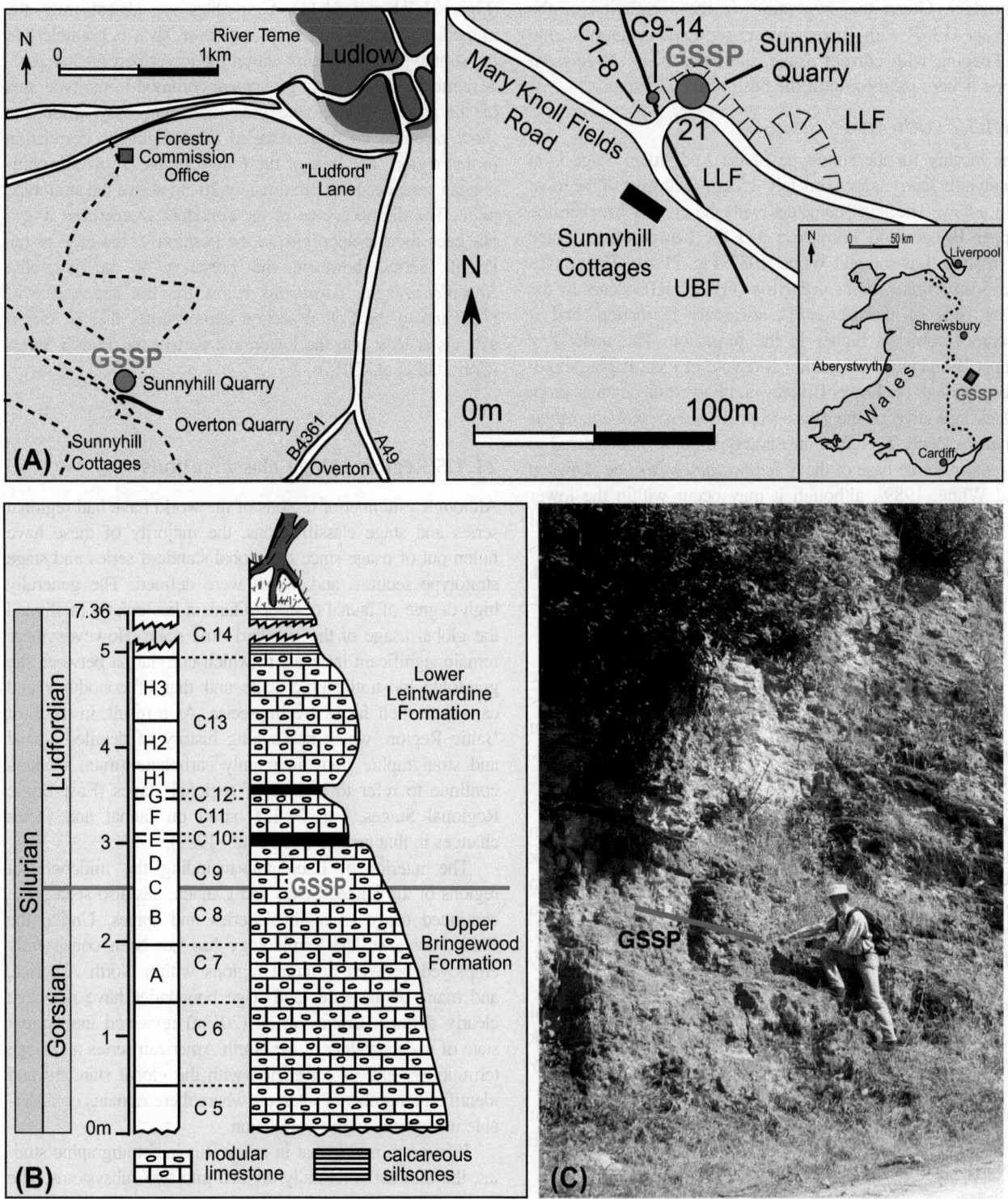

FIGURE 21.10   GSSP for the Base of the Ludfordian Stage. (A) Location maps of the GSSP at Sunnyhill Quarry near Ludlow, Shropshire, Great Britain. (B) Lithologic log of GSSP section (from Lawson and White, 1989). (C) Photograph of the GSSP. *GSSPs*, Global Boundary Stratotype Section and Points. (C) Photo by Michael Melchin.

Base of the Pridolí Series of the Silurian System in the Daleje Valley, Prague, Czech Republic

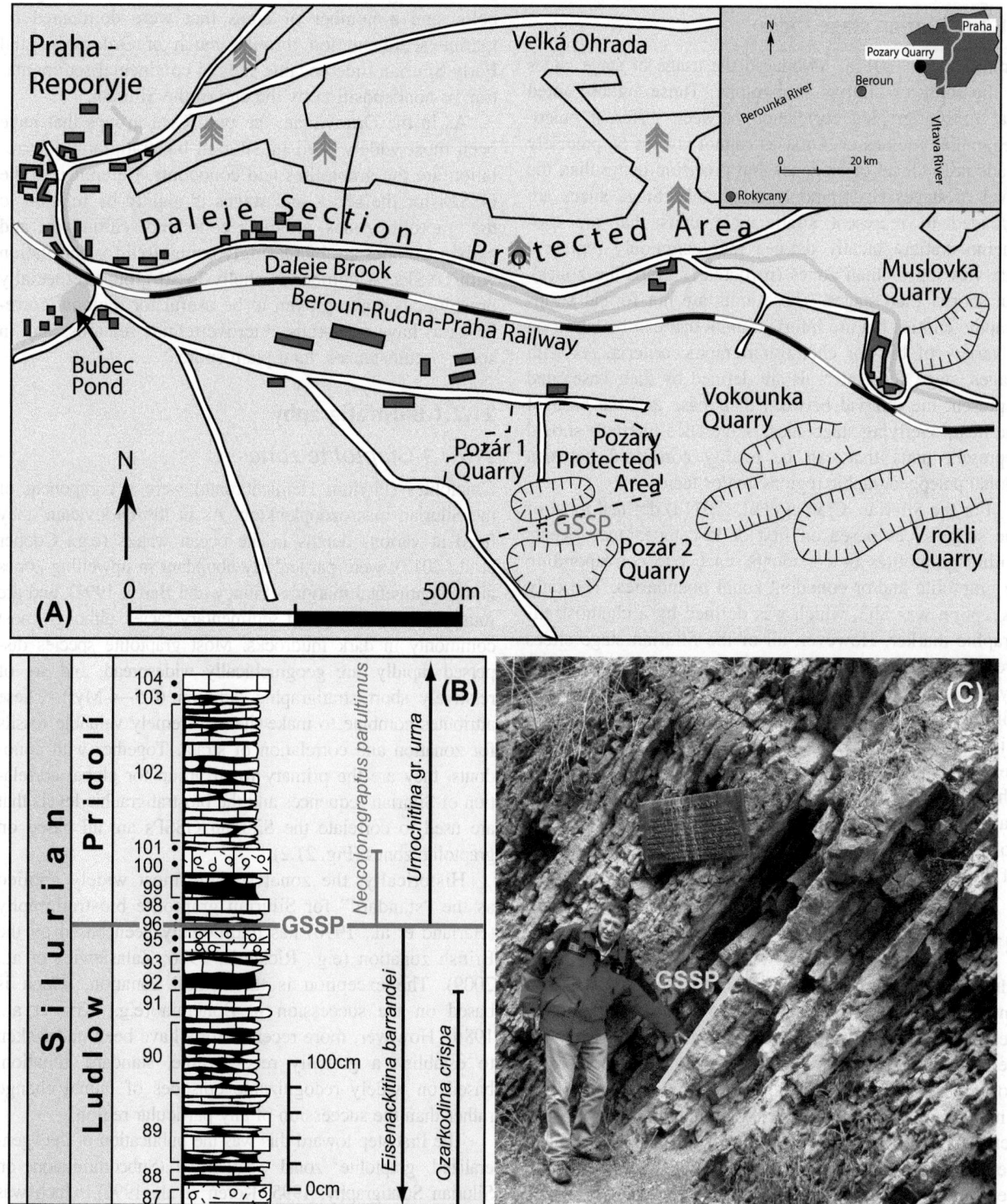

FIGURE 21.11 GSSP for the Base of the Pridoli Series. (A) Location map of the GSSP in the Daleje Valley, Prague, Czech Republic. (B) Lithologic log of the GSSP section showing ranges of some key taxa (from Kříž, 1989). (C) Photograph of the GSSP. *GSSPs*, Global Boundary Stratotype Section and Points. (C) Photo by Michael Melchin.

applied informally above the series level in stratigraphic studies in other parts of the world.

21.1.6 Silurian stage slices

Cramer et al. (2011a) introduced the usage of stage slices to the study of Silurian stratigraphy. Those authors noted that biostratigraphic correlation between different paleogeographic regions or biofacies cannot always be precisely made at the level of biozones, but precision better than the level of stages is commonly achievable. Stage slices are intended to represent stratigraphic units that are bio-chemo-stratigraphically defined and commonly represent intervals longer than zones (particularly graptolite zones), but shorter than stages. These units are not formal stratigraphic entities but are informal units that can be based on a variety of bio- or chemostratigraphic criteria. As with stages, stage slice intervals are defined by their bases and represent the interval between their base and the base of the next, overlying stage slice. Stage slice intervals should represent units that can be readily correlated between global paleogeographic regions and/or facies.

For the Silurian, Cramer et al. (2011a) defined most of the stage slices based on first or last appearance events within graptolites or conodonts, each one corresponding to graptolite and/or conodont zonal boundaries. The only exception was Sh3, which was defined by a chemostratigraphic marker. However, all of the Silurian stage slices were chosen and described by Cramer et al. and employed by McAdams et al. (2019) in reference to a generalized Silurian δ^{13}C curve. The succession of 24 Silurian stage slices is shown in Fig. 21.2. All of the stage slices are defined as in Cramer et al. (2011a), except for the base of Sh1, which we have slightly revised to coincide with the base of the Upper *Ps. bicornis* Zone (Ireviken Datum 2), so that it remains approximately at the level of the base of the Sheinwoodian Stage.

21.2 Silurian Stratigraphy

Much of the sedimentary record of the Silurian in basinal and continental margin settings is represented by graptolite-bearing mudrocks. In the paleotropical regions, epicontinental settings are dominated by carbonate successions, commonly with well-developed reefs. In later Silurian time, large evaporite basins developed in some epeiric settings. Terrestrial alluvial deposits of Silurian age have been reviewed by Davies and Gibling (2010). Glacial deposits, mainly of Llandovery age, have been recognized in South America (e.g., Caputo, 1998; Díaz-Martínez, Grahn 2007; Cuervo et al., 2018).

The Silurian Period was generally a time of relative convergence of continental landmasses and narrowing and closing of ocean basins (Torsvik and Cocks, 2017). One of the results of this is the generally low degree of provincialism seen in marine faunas. However, another result of this convergence is tectonic uplift in several orogenic belts, and a number of areas that were dominated by marine sedimentation through much of Ordovician and Early Silurian time became sites of continental sedimentation or nondeposition by the end of the Silurian.

As in the Ordovician, the two fossil groups that have been most widely used in Silurian biostratigraphic correlation are the graptolites and conodonts. Reference to the GSSPs for the series and stages is mainly by relation to the graptolite zones, although conodonts, chitinozoa, and carbon isotopes are increasingly being used for correlation with GSSPs. A number of shelly fossil groups, especially brachiopods, have proven to be useful for regional correlation as have acritarchs, microvertebrate remains, and, in some circumstances, land plant spores.

21.2.1 Biostratigraphy

21.2.1.1 Graptolite zones

Graptolites (Phylum Hemichordata) were a component of the Silurian macrozooplankton. As in the Ordovician, they lived at various depths in the ocean waters (e.g., Cooper et al., 2012), were particularly abundant in upwelling zones along continental margins (Finney and Berry, 1997), and are found in a wide range of sedimentary facies, although most commonly in dark mudrocks. Most graptolite species dispersed rapidly, are geographically widespread, and are of relatively short stratigraphic duration (0.5–4 Myr). These attributes combine to make them extremely valuable fossils for zonation and correlation of strata. Together with conodonts, they are the primary fossil group for global correlation of Silurian sequences and the biostratigraphic levels that are used to correlate the Silurian GSSPs are all based on graptolite zones (Fig. 21.2).

Historically, the zonal scheme most widely applied as the "standard" for Silurian graptolite biostratigraphy (Harland et al., 1990) has traditionally been based on the British zonation (e.g., Rickards, 1976; Zalasiewicz et al., 2009). The exception is the Pridoli zonation, which is based on the succession in Bohemia (e.g., Kříž et al., 1986). However, more recent efforts have been undertaken to establish a globally recognizable, standard zonation, based on widely recognizable episodes of faunal change rather than the succession of any particular region.

The first step toward this was the publication of the "generalized graptolite zonal sequence" (Subcommission on Silurian Stratigraphy, 1995; Koreń et al., 1996), which was assembled for the purpose of a coordinated study of global paleogeography, but was also used for a study of patterns of global diversity and survivorship in Silurian graptolites (Melchin et al., 1998). In the course of the latter study, it was found that a number of the generalized zones recognized by

Koreń et al. (1996) could be readily subdivided and still recognized in several different paleogeographic regions of the world. Further refinements of the Generalized Graptolite Zonation were presented in Melchin et al. (2004), Sadler et al. (2009), Cramer et al. (2011a, b), and Melchin et al. (2012). The graptolite zonation used in this chapter is slightly modified from that of Melchin et al. (2012). The criterion used in this zonation, as was the case for the original Generalized Graptolite Zonation (Koreń et al., 1996), was that zonal base should be recognizable in at least three different paleogeographic regions of the world. Loydell (2012) provided correlations for several of the regional graptolite zonations for the Ordovician and Silurian.

Fig. 21.2 shows the remarkable precision in correlation obtainable using Silurian graptolites. The 34 graptolite zonal divisions of the Silurian span 24.1 Myr and thus average approximately 709,000 years each in duration. At the regional level, where more zones are recognized in some intervals, the precision is even better.

21.2.1.2 Conodont zones

Conodonts are tooth-like structures in the feeding apparatus of small lamprey-like primitive chordates (Aldridge and Briggs, 1989; Purnell et al., 2000). Although they are most readily preserved in and extracted from carbonate facies, the free-swimming conodont animal roamed widely and some species are found in a broad variety of lithologies and depositional environments, making them an important tool for global stratigraphic correlation. The fluorapatite (calcium phosphate) conodont elements are typically extracted via acid digestion using either acetic or (preferably) formic acid. Particularly in dolomitic Silurian carbonate rocks, the use of buffered formic acid has greatly improved the yield (number of elements/kg of rock) during digestion.

Exceptionally detailed conodont biozonations have been constructed for portions of the Silurian and tremendous improvements to both taxonomy and biostratigraphy of Silurian conodonts have taken place over the past two decades (e.g., Jeppsson, 1997a; Männik, 1998, 2007a,b; Corradini and Serpagli, 1999; Murphy et al., 2004; Jeppsson et al., 2006; Carls et al., 2007; Corriga and Corradini, 2009; Cramer et al., 2010a, 2010b; Slavik, 2014; Männik et al., 2015; Corradini et al., 2015; Waid and Cramer, 2017a; McAdams et al., 2019; Bancroft and Cramer, 2020). At the finest scale of resolution (<500 kyr), it is becoming clear that the first appearances of some zonally important conodont species are not synchronous globally (Cramer et al., 2010b; Slavik, 2014; Waid and Cramer, 2017a; Bancroft and Cramer, 2020). Similar difficulties can be found among Silurian graptolites and, these diachroneities must be addressed when working at extremely high resolution (i.e., approaching the shorter astrochronological cycle periods). The conodont zonation of the Baltic Basin remains the most highly refined of any Silurian paleocontinent, and the recognition of these zones

elsewhere has begun to demonstrate the fine-scale differences in timing of first appearances. In the comparatively few studies published from outside of Baltica that have been conducted at sufficient resolution and using suitable digestion techniques, the vast majority of the Baltic zones appear to be recognizable. The composite biozonation used here (Fig. 21.2) is from McAdams et al. (2019), which was constructed specifically to include the entire Silurian and to be more globally applicable than previously published zonations.

21.2.1.3 Chitinozoan zones

Chitinozoa are organic-walled microfossils of unknown biological affinities. Many accept the hypothesis that they were the planktonic egg capsules of some metazoan (Servais et al., 2013), although recently this hypothesis has been challenged (Liang et al., 2019). They occur in a variety of marine facies and many species were geographically widespread and relatively short-lived. Verniers et al. (1995) proposed a global biozonation for Silurian chitinozoa, based on correlation of well-known successions in Laurentia, Avalonia, Baltica, and Gondwana (Fig. 21.2). Global biozonal levels are defined by well-established taxa whose first appearances are regarded as synchronous in two or more distinct paleogeographic regions. Many of these biozonal levels have been defined or recognized in direct reference to Global Boundary Stratotype Sections and Points. Although some authors have more recently described more refined regional zonations for all or part of the Silurian (e.g., Nestor, 2012; Steeman et al., 2016), these are still correlated in reference to the global biozonation of Verniers et al. (1995).

21.2.1.4 Other zonal groups

The Subcommission on Silurian Stratigraphy (1995), which produced generalized zonations for graptolites, conodonts, and chitinozoa, also provided Silurian zonations for spores and vertebrates (Fig. 21.2). The former, which has been slightly revised by Burgess and Richardson (1995) (see also Melchin et al., 2012), is particularly important; in that it provides the possibility for biostratigraphic correlation in terrestrial strata and between the terrestrial and marine realms. The vertebrate zonation, based mainly on disarticulated remains (ichthyoliths), was refined by Märss and Männik (2013).

21.2.2 Magnetostratigraphy

Understanding of the magnetostratigraphic scale for the Silurian System is still in a very preliminary state and is based on incomplete data from only a few localities. Based on the presently available data, it appears that much of the Silurian is characterized by a mixed polarity, with a predominantly normal phase through much of the Wenlock (Fig. 21.1; Trench et al., 1993).

21.2.3 Chemostratigraphy

Chemostratigraphic analyses are now available at fairly high levels of stratigraphic resolution for all or most of the Silurian System for the $\delta^{13}C$, $\delta^{18}O$, and Sr isotopic systems (Fig. 21.1, Fig. 21.12).

21.2.3.1 Carbon-isotope stratigraphy

The carbon isotopic ratio of Silurian seawater, measured by proxy either in carbonate ($\delta^{13}C_{carb}$) or organic matter ($\delta^{13}C_{org}$), was highly variable through time and indicates that the global carbon cycle and global climate system were less stable during the Silurian than almost any other period during the Phanerozoic. As is the case with most of the Paleozoic, the $\delta^{13}C_{carb}$ record is better known than the $\delta^{13}C_{org}$ record, and throughout the rest of this chapter, we will be referring to the $\delta^{13}C_{carb}$ record unless otherwise indicated.

Significant positive excursions to the carbon isotopic ratio of Silurian seawater occurred during the early and late Aeronian, early Telychian, early to middle Sheinwoodian, middle to late Homerian, middle Ludfordian, early Pridoli and during the latest Pridoli—crossing the base of the Devonian (see Cramer et al., 2011a; Cramer and Jarvis, 2020, Ch. 11: Carbon isotope stratigraphy, this book; and Fig. 21.12). The late Aeronian (Wenzel, 1997; Waid and Cramer, 2017b; Braun, 2018), Sheinwoodian (Samtleben et al., 1996; McAdams et al., 2019), Ludfordian (Samtleben et al., 1996; Frýda and Manda, 2013), and Silurian—Devonian boundary (Saltzman, 2002) positive $\delta^{13}C_{carb}$ excursions have all been shown to approach or exceed +5.0‰ (Vienna PeeDee Belemnite $\delta^{13}C$ standard). The Late Aeronian excursion had been among the least studied of Silurian carbon events but it has been demonstrated that this excursion reaches $> +4.5‰$ in Baltica (Oslo Graben, Wenzel, 1997) and Laurentia (Iowa, Waid and Cramer, 2017b; Anticosti, Braun 2018) and appears to be of similar magnitude as the larger and better studied events. This excursion was well documented in the $\delta^{13}C_{org}$ record by Melchin and Holmden (2006), and Melchin et al. (2015) showed that it may be resolvable into several distinct peaks. The Homerian excursion tends to be of lower amplitude than those that occurred during the Sheinwoodian or Ludfordian in any given basin; however, values greater than +4.0‰ have been recorded from Homerian strata (e.g., Danielsen et al., 2019) and this relationship is not always the case in every basin (e.g., Podolia; Kaljo et al., 2007). The Ludfordian excursion is exceptional in its amplitude (typically $> +8.0‰$, e.g., Frýda and Manda, 2013) in that it is likely the largest post-Cambrian excursion of the entire Phanerozoic. By the middle Ludfordian, it had been more than 100 million years since the carbon isotopic ratio of the ocean had reached such an elevated level. The presence of other lower magnitude positive carbon-isotope excursions was debated at the time of publication of GTS2012, but there appears to be increasing evidence that there were other low-magnitude positive carbon-isotope events during the Silurian that are only beginning to be recognized. The lower Aeronian excursion is now known from Laurentia and Baltica (Kaljo et al., 2003; Bancroft et al., 2015; Waid and Cramer, 2017b; Braun, 2018), and its occurrence in the $\delta^{13}C_{org}$ record (e.g., Melchin and Holmden, 2006; Li et al., 2019) indicates that it may represent an important marker for the base of the Aeronian Stage (Štorch et al., 2018). The Gorstian—Ludfordian boundary excursion remains difficult to define even when it may be present in a dataset (e.g., McAdams et al., 2019) and a growing body of literature is beginning to demonstrate that there is a separate carbon-isotope excursion within the Pridoli Series, above the Ludfordian excursion and below the Silurian—Devonian boundary excursion (e.g., Kaljo et al., 2012). In addition, Hammarlund et al. (2019) recognized a positive excursion within the Telychian *Oktavites spiralis* Zone in Baltica. Although, as yet, this event has not been widely recognized elsewhere, there was a positive shift documented in the mid-Telychian in Wisconsin (McLaughlin et al., 2013) that may be correlative. It is also becoming clearer that there is significant structure to the carbon isotope curve of the Pridoli Series (e.g., Oborny et al., 2020) that warrants further investigation.

Carbon-isotope chemostratigraphy can be used to investigate changes in the global carbon cycle and the ocean—atmosphere—biosphere system and, in addition, can be used as a purely stratigraphic tool. It is in the latter capacity that the Silurian $\delta^{13}C_{carb}$ and $\delta^{13}C_{org}$ records have become among the most highly resolved of any period of the Paleozoic. Because $\delta^{13}C$ chemostratigraphy can be recovered from strata that contain conodonts or graptolites (i.e., from carbonate or clastic settings), carbon-isotope chemostratigraphy has become an invaluable tool for comparing and refining the correlation between biostratigraphically useful groups, such as graptolites and conodonts, which are infrequently found together in the stratigraphic record (Kaljo et al., 2015; Cramer et al., 2015a).

Whereas the utility of carbon-isotope chemostratigraphy is widely accepted in its application to correlation, the relationship between the carbon isotopic record and Silurian biotic/global climate events remains a highly contentious issue (e.g., Cramer and Saltzman, 2007; Lehnert et al., 2007; Loydell, 2007, 2008; Cramer and Munnecke, 2008; Munnecke et al., 2010; Lehnert et al., 2010; Cramer et al., 2012; Frýda and Manda, 2013; Cooper et al., 2014; Radzevičius et al., 2014; Kozłowski, 2015; Husson et al., 2016; Jarochowska et al., 2016; Spiridonov et al., 2016; Trotter et al., 2016; Crampton et al., 2016; Mergl et al., 2018; Rose et al., 2019). As is the case with the majority of the positive carbon-isotope excursions during the Paleozoic, paleobiodiversity events occurred immediately prior to and/or during the onset of positive excursions, whereas much of the stratigraphic/temporal extent of any given excursion coincides with the postextinction/recovery phase of the biotic event. Reconstruction of Silurian sea levels remains problematic (Munnecke et al., 2010). It is

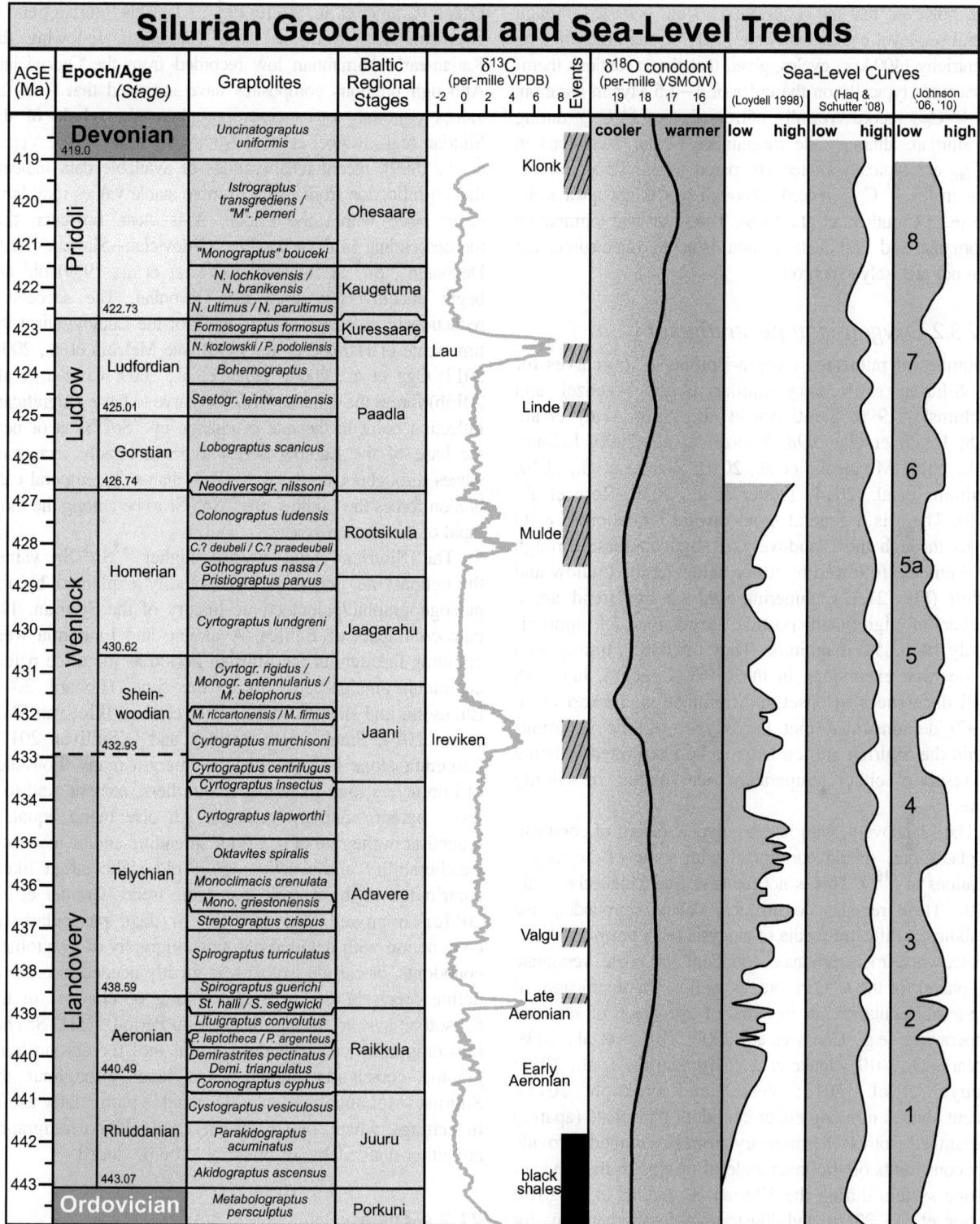

FIGURE 21.12 Silurian geochemical and sea-level trends. The GTS2020 age model for Silurian chronostratigraphy and graptolite biozonation was used to rescale the Baltic regional stages, the carbon-isotope curve with major widespread graptolite and conodont bioevents (e.g., Jaeger, 1991; Jeppsson, 1998; Jeppsson et al., 2006; Melchin et al., 1998, 2012, 2013), the oxygen-isotope curve with implied sea-surface temperature trends, and the relative sea-level curves of Loydell (1998), Haq and Schutter (2008) and Johnson (2006, 2010). The δ¹³C curve and the placement of major widespread biotic events (extinction episodes, etc.) are from Melchin et al. (2012, 2013) and Cramer and Jarvis (2020, Ch. 11: Carbon isotope stratigraphy, this book). The oxygen-isotope curve and implied sea-surface temperature trends is the statistical-mean curve from a synthesis of numerous conodont apatite studies, including Trotter et al. (2016) (see Grossman and Joachimski, 2020, Ch. 10: Oxygen isotope stratigraphy, this book). *VPDB*, Vienna PeeDee Belemnite δ¹³C standard; *VSMOW*, Vienna Standard Mean Ocean Water δ¹⁸O standard.

clear, however, that the positive excursions persisted through several sea-level cycles that were likely on the order of long eccentricity (400 kyr) cycles, given that the excursions themselves were typically on the order of 1−1.5 Myr in duration. The $\delta^{13}C_{org}$ record typically mirrors that of $\delta^{13}C_{carb}$ during the Silurian, although the magnitudes of the excursions in $\delta^{13}C_{org}$ are typically muted compared to the same isotopic event in the $\delta^{13}C_{carb}$ record, although this is not always the case (e.g., Caruthers et al., 2018). The onset and termination of organic and carbonate carbon isotopic fluctuations are often not precisely synchronous.

21.2.3.2 Oxygen-isotope stratigraphy

A number of published oxygen-isotope ($\delta^{18}O$) curves for the Silurian show very similar trends (Wenzel and Joachimski, 1996; Samtleben et al., 1996; Azmy et al., 1998; Heath et al., 1998; Wenzel et al., 2000; Lehnert et al., 2010; Munnecke et al., 2010; Žigaitė et al., 2010; Cummins et al., 2014; Trotter et al., 2016; Rose et al., 2019). There is a general trend toward reduction in $\delta^{18}O$ values through the Llandovery, a slight increase through the Wenlock, followed by lower values in the Ludlow and Pridoli (Fig. 21.12). Superimposed on this trend are a number of significant positive excursions of approximately 1‰−3.5‰ magnitude. They overlap in timing with the positive excursions in the $\delta^{13}C_{carb}$ record, but with small differences in onset and termination. Bickert et al. (1997) demonstrated that the oxygen-isotope variations within the Silurian are too large to be interpreted as being the result of either temperature, ice-volume, or salinity alone.

The Llandovery, with its unequivocal record of continental glaciations, should correspond with some of the largest variations in $\delta^{18}O$. This is not the case (see Munnecke et al., 2010). There remains considerable debate surrounding the reliability of different media of analysis (e.g., brachiopod carbonate, whole-rock carbonate, conodont phosphate, vertebrate phosphate) for the $\delta^{18}O$ record, as well as the significance of long-term, secular trends in terms of estimation of seawater temperatures (e.g., Came et al., 2007; Trotter et al., 2008; Pucéat et al., 2010; Žigaitė et al., 2010; Nardin et al., 2011; Lécuyer et al., 2013; Veizer and Prokoph, 2015). Recent studies utilizing either conodont phosphate (apatite) or clumped isotope thermometry from brachiopods provide new constraints on the magnitude of change in the oxygen-isotope system during the Silurian (Cummins et al, 2014; Trotter et al., 2016) and illustrate a future pathway for improved Silurian oxygen-isotope records.

21.2.3.3 Strontium isotope stratigraphy

The strontium isotopic ratio ($^{87}Sr/^{86}Sr$) of Silurian seawater shows a generally increasing trend from about 0.70795 during the early Rhuddanian to about 0.70870 during the latest Pridoli (Cramer et al., 2011c; Fig. 21.1). This Silurian trend is a continuation from the latest Hirnantian, following the Katian-early Hirnantian low recorded from the Ordovician. Although previous composites have indicated that $^{87}Sr/^{86}Sr$ values continue to increase throughout the whole of the Silurian (e.g., Ruppel et al., 1996; Azmy et al., 1999; Veizer et al., 1999), recent recalibrations of available data indicate that an inflection from rising to more stable values may have taken place within the Pridoli. It is clear, however, that the descending limb of the Late Ordovician−Silurian−Early Devonian $^{87}Sr/^{86}Sr$ fluctuation (Veizer et al., 1999) did not begin unequivocally until the Devonian. The successive reduction in estimates of the length of the Ludlow since the time scale of Harland et al. (1990) (see Melchin et al., 2004, 2012; Ogg et al., 2008; Sadler et al., 2009; Cramer et al., 2015b) forces the Silurian $^{87}Sr/^{86}Sr$ curve to have a significant inflection point in the rate of change in $^{87}Sr/^{86}Sr$ at or near the base of the Ludlow Series to more rapidly increasing values throughout the Ludlow. This change in temporal calibration forces the Ludlow rise $^{87}Sr/^{86}Sr$ to be among the most rapid of the Phanerozoic.

The Silurian trend toward higher $^{87}Sr/^{86}Sr$ values throughout the period is most readily explained by the paleogeographic/paleotectonic history of the Silurian. The paleocontinents of Baltica, Avalonia, and Laurentia were colliding throughout the Silurian Period as the three paleocontinents amalgamated during this time (Hibbard, 2000; Ettensohn and Brett, 2002; Cramer et al., 2011c; van Staal et al., 2014; Pinet, 2016; Bradley and O'Sullivan 2017). Laurentia alone had a collisional margin many thousands of kilometers long including its northern, eastern, and to a lesser extent, southern margins. All else being equal, a reduction in the rate of sea-floor spreading and/or increased weatherability and weathering of old sialic strata likely contributed to the overall increasing trend (Cramer et al., 2011c). Improved Silurian $^{87}Sr/^{86}Sr$ data, particularly in conjunction with detailed chronostratigraphy of graptolites, conodonts, or carbon isotopes, is greatly needed in order to define precisely the rates and timing of changes in the Sr isotope system during the Silurian Period. $^{87}Sr/^{86}Sr$ chemostratigraphy can provide a useful tool for chronostratigraphic correlation at the stage level throughout the Silurian (McLaughlin et al., 2013) and is particularly useful in settings where biostratigraphy and chemostratigraphy are either unavailable or unlikely to be produced.

21.2.4 Eustasy

A considerable body of literature exists in the effort to construct a eustatic curve for the Silurian Period. There have been several different approaches to the problem of estimating eustatic changes. Johnson (1996) and Munnecke et al. (2010) summarized a series of earlier papers in which the sea-level histories of individual regions have been reconstructed mainly based on the use of benthic assemblages and sedimentary structures. Sea-level curves for each region are then correlated and compared to identify global signals. Johnson et al. (1998) added to this the study of submergence of paleotopographic features as measures of absolute sea-level change. Ross and Ross (1996) also employed biofacies data, but additionally incorporated some lithofacies information; their curve is based almost entirely on sections in Laurentia. Johnson (2006) provided an update of the Silurian eustatic record, based on this approach (Fig. 21.12), and included a discussion of the relationship between sea-level change and the patterns of extinction and oceanic events.

In contrast, Loydell (1998; see Fig. 21.12) defined episodes of sea-level rise and fall based primarily on identification of the oxidation state of the strata under investigation and the graptolite fauna contained therein. For example, he argued that transgressive intervals are recognized as being generally condensed and organic-rich shales with diverse graptolite faunas, whereas regressive intervals are thicker and less organic-rich with more depauperate graptolite assemblages. This method has some advantages; in that the graptolite biostratigraphy provides a more precise and globally correlatable temporal control. In addition, it is also based mainly on deeper water successions, which are less susceptible to truncation by subaerial exposure and erosion. On the other hand, this method provides no means of deriving quantitative estimates of magnitudes of sea-level change, and the oxidation state of the sea-floor needs not necessarily be related to sea level.

A paper that interpreted global eustatic patterns during the Silurian by Haq and Schutter (2008) was based upon sequence stratigraphic interpretation of records from cratonic basin successions. Intriguingly, Haq and Schutter's record proposes a higher number of short-term global sea-level fluctuations than any of the previous studies (Fig. 21.12). Johnson (2010) provided a detailed discussion and comparison of the methodologies and results of the Johnson (2006) and Haq and Schutter (2008) sea-level studies.

Reconstruction of Silurian sea levels is complicated especially by two issues: (1) the interpretation of the stratigraphic record and (2) the chronostratigraphic correlation of the studied sections. It is often the case that the inability to sufficiently correlate stratigraphic sections either with each other, or with the Silurian time scale, produces diametrically opposing curves (e.g., Fig. 21.12; Munnecke et al., 2010, Fig. 2).

21.2.5 Climatic events

There is considerable debate surrounding the various means used to estimate climatic conditions during the Silurian (e.g., Landing and Johnson, 1998; Munnecke et al., 2003, 2010; Lehnert et al., 2007, 2010; Loydell, 2007, 2008; Cramer and Munnecke, 2008; Cramer et al., 2012; Frýda and Manda, 2013; Cooper et al., 2014; Radzevičius et al., 2014;

Kozłowski, 2015; Veizer and Prokoph, 2015; Husson et al., 2016; Jarochowska et al., 2016; Spiridonov et al., 2016; Trotter et al., 2016; Crampton et al., 2016; Mergl et al., 2018; Rose et al., 2019). Nevertheless, it is now clear that the history of Silurian climate was extremely dynamic, as recorded in the high frequency and high-magnitude fluctuations in C and O isotopic values (Munnecke et al., 2010; Bachan et al., 2017; McAdams et al., 2019) as well as depositional patterns indicating pronounced changes in ocean state (e.g., Jeppsson, 1990; Brunton et al., 1998; Page et al., 2007; Melchin et al., 2013).

Following the major glacial event of the Late Ordovician (e.g., Brenchley et al., 1994), there were several episodes of glacial advance in the Llandovery (e.g., Caputo, 1998; Díaz-Martínez, Grahn 2007; Cuervo et al., 2018) and the Hirnantian glacial interval is perhaps better thought of as the Late Ordovician to Early Silurian glacial interval (e.g., the "Early Palaeozoic Icehouse" interval of Page et al., 2007). As noted above, the existence of significant excursions in the record of $\delta^{18}O$ and $\delta^{13}C$ that persist through several sea-level cycles and, in many ways, resemble the glacially-related events of the Late Ordovician (Munnecke et al., 2003; Žigaitė et al., 2008) indicates that it is possible that there were later Silurian glacial episodes that were yet to be documented in the physical stratigraphic record (e.g., Lehnert et al., 2010; Kozłowski, 2015; Trotter et al., 2016; Radzevičius et al., 2017). Regardless, the frequency and magnitude (Bachan et al., 2017) of several of the Silurian isotopic shifts clearly indicate a time interval of rapidly evolving climatic and environmental conditions. Many of these events are closely associated with intervals of biotic change (Cooper et al., 2014; Crampton et al., 2016), most clearly seen in the record of graptolites and conodonts (Fig. 21.12).

21.2.6 Volcanism and K-bentonite stratigraphy

Volcanic ash beds or K-bentonites have been widely reported through the Silurian, particularly from Europe and eastern North America (e.g., Bergström et al., 1998a). Geochemical studies of these bentonites suggest that those distributed over Laurentia, Avalonia, and Baltica can be attributed to at least three different volcanic centers within the closing Iapetus and Rheic Ocean basins (Bergström et al., 1997; Huff et al., 2000). However, some of these individual K-bentonite units have been shown by geochemical fingerprinting to be geographically very widespread and serve as excellent marker beds for high-resolution, regional correlations (e.g., Bergström, et al, 1998b; Batchelor et al, 2003; Ray, 2007; Kiipli et al., 2008a, 2008b, 2012, 2014, 2016; Ray et al., 2011). Bentonites also provide the principal source of crystals for radioisotope dates within stratigraphic successions for calibration of the Silurian time scale (see later).

21.2.7 Astrochronology

Astrochronologic tuning of the Silurian time scale remains in its infancy. Whereas fine-scale cyclic sedimentation has been long recognized within the Silurian System (e.g., Goodwin et al., 1986; Yolkin et al., 1997; Marshall, 2000), caution has been advised in making an *a priori* assumption that observed cyclicity must be astronomical in origin (Westphal et al., 2010). Recent publications have only just begun the process of exploring for astrochonologic signatures in the Silurian stratigraphic record and have employed a variety of techniques, stratigraphic information, and analytical tools. Stratigraphic information has included magnetic susceptibility (Crick et al., 2001), conodont abundance (Spiridonov et al., 2016), gamma-ray logs (Radzevičius et al., 2017), and gray-scale imagery of cores as a proxy for organic productivity (Gambacorta et al., 2018). Crampton et al. (2018) reported that a significant component of the variability in graptoloid turnover rates in the Silurian could be explained by long-period astronomical cycles, specifically 2.6 Myr eccentricity grand cycles. These cycles in graptoloid turnover patterns, however, were not directly tied to series analysis of any climate proxy data. At present, there has been no consistency in the analytical tools and statistical thresholds for astronomical signal detection limits (e.g., Meyers et al., 2012; Hinnov, 2013, 2018).

21.3 Silurian time scale

As with the Ordovician, there have been many published radioisotopic dates for calibrating the Silurian time scale (Gale, 1985; Kunk et al., 1985; McKerrow et al., 1985; Odin, 1985; Snelling, 1985; Harland et al., 1990; Tucker et al., 1990; Compston and Williams, 1992; Tucker and McKerrow, 1995; Compston, 2000a,b; Cramer et al., 2012, 2015b). Unfortunately, few of the dates prior to 2000 are of sufficient analytical reliability and precision and/or biostratigraphic constraint to meet modern standards for time scale calibration. Following the advent of the CA-ID-TIMS method for U−Pb dating (Mattinson, 2005, 2010; Condon et al., 2007) and improvements to the Ar−Ar geochronometer (Kuiper et al., 2008), the current iteration of Silurian time scale calibration is based solely on U−Pb dates from zircons.

21.3.1 Radioisotopic dates

Under the new standardized treatment (see Schmitz et al., 2020, Ch. 6: Radioisotope geochronology, this book), error propagation of analytical, tracer/standard, and decay constant errors has been treated consistently and rigorously for each published age. All dates used in our calibration have been reappraised since the publication of Gradstein et al.

(2012) for analytical and biostratigraphic reliability. The radioisotopic ages for some of the dated samples, therefore, may differ from those given in the original publications. Details of the dated samples used here are given in Appendix 2. The following comments refer mainly to the reliability of the biostratigraphic constraints on the dated samples.

Of the six Silurian radioisotopic dates used for calibration by Tucker and McKerrow (1995), only one is accepted here (their dated item 22). Date No. 22 is from the Birkhill Shales, Scotland (^{206}Pb/^{238}U age of 439.57 ± 1.33 Ma, Toghill, 1968; Ross et al., 1982; Tucker et al., 1990). Although the exact level of the bentonite in the section is uncertain (Cooper and Sadler, 2004; Bergström et al., 2008), we accept the assignment of Ross et al. (1982) of this bentonite to the *C. cyphus* Zone.

Five high-quality zircon dates were utilized in the construction of the 2012 Silurian time scale of Melchin et al. (2012). The first is from the Osmundsberget K-bentonite of Dalarna, Sweden (Bergström et al., 2008). Four out of five zircons yielded a weighted mean ^{206}Pb/^{238}U age of 438.74 ± 1.20 Ma; one zircon was affected by inheritance. The graptolites *S. turriculatus* and *Streptograptus johnsoni* are found in the section about 0.5 m below the K-bentonite, and *S. turriculatus* is also present 0.5 m above it (Loydell and Maletz, 2002). We take the overlap in range of *S. turriculatus* and *S. johnsoni* in the CONOP composite as the stratigraphic uncertainty range and the midpoint in this range as the level. A second K-bentonite from Dalarna, in the Kallholn Shale, gave a ^{206}Pb/^{238}U age of 437.0 ± 0.5 Ma; however, uncertainty about its biostratigraphic age and correlation led Bergström et al. (2008) to reject it for time scale calibration, and we have not included it here.

The other four of the five new high-quality zircon ^{206}Pb/^{238}U dates used in GTS2012 came from Cramer et al. (2012). Preliminary dating of the Ireviken Bentonite from Gotland, Sweden, was reported by Jeppsson et al. (2005), but the complete analytical details remained unavailable until recently. Cramer et al. (2012), using the CA-ID-TIMS method of Mattinson (2005) and Condon et al. (2007), show that 11 out of 15 zircons yielded a weighted mean ^{206}Pb/^{238}U age of 431.80 ± 0.71 Ma for the Ireviken Bentonite from the Lusklint 1 outcrop (for locality guide see Eriksson and Calner, 2005). At this well-studied section (e.g., Jeppsson, 1997a, 1997b), the Ireviken Bentonite occurs 2.85 m above the Lusklint Bentonite (the local reference level), which is roughly 8.28 m above sea level. The Ireviken Bentonite occurs within the Lower Visby Formation and is within the Upper *Ps. bicornis* conodont zone between Datums 2 and 3 of the Ireviken Event (Jeppsson, 1997a, 1997b). Based upon our present knowledge of the base Wenlock GSSP, Datum 2 of the Ireviken Event is likely within a few centimeters of the "golden spike" at Hughley Brook (see above; Mabillard and Aldridge, 1985; Jeppsson, 1997a; Cramer et al., 2010a), and the Ireviken Bentonite is 14 cm above Datum 2 in the Lusklint 1 outcrop. Based upon biostratigraphic correlation of the conodont zonations with the graptolite zonations (e.g., Cramer et al., 2010a), as well as the correlation of bentonites throughout the Baltic where they co-occur with graptolites (e.g., Aizpute core, Kiipli et al., 2008b), the Ireviken Bentonite occurs within the *C. murchisoni* graptolite zone and is below the onset of the Sheinwoodian (Ireviken) positive δ^{13}C excursion (Munnecke et al., 2003).

The second date from the Swedish Island of Gotland comes from the outcrop Hörsne 3 (Laufeld, 1974; Kiipli et al., 2008b). The name "Grötlingbo Bentonite" was first given to a 38-cm-thick bentonite recognized in the Grötlingbo drillcore (Jeppsson and Calner, 2003), where it occurs within the *Gothograptus nassa/Pristiograptus parvus* graptolite zone (Calner et al., 2006). The only identified exposure of this bentonite on Gotland is the Hörsne 3 locality where it is 30 cm thick (Kiipli et al., 2008b). At Hörsne 3, the Grötlingbo Bentonite is within strata equivalent to the Mulde Brickclay Member of the Halla Formation and within the *Ozarkodina bohemica longa* conodont zone (Jeppsson et al., 2006; Kiipli et al., 2008b). A bentonite from the Hörsne 3 locality was previously dated using K−Ar at 425.8 ± 6.0 Ma by Odin et al. (1986, sample B126). The Grötlingbo Bentonite has been correlated more broadly around the East Baltic by Kiipli et al. (2008a; 2008b), who demonstrated that the identification and correlation based on the mineralogy of magmatic phenocrysts are consistent with its biostratigraphic position throughout the Baltic Basin. This sample (G05−335LJ of Kiipli et al., 2008b) comes from the upper part of the 30 cm-thick exposure and five out of nine zircons yielded a weighted mean ^{206}Pb/^{238}U age of 428.47 ± 0.72 Ma (Cramer et al., 2012). The Grötlingbo Bentonite occurs within the rising limb of the first peak of the Homerian "Mulde" positive δ^{13}C excursion (compare Calner et al., 2006; Kaljo et al., 2007; Kiipli et al., 2008b).

The third date from Gotland comes from the Djupvik 1 locality (Jeppsson, 1982; Eriksson and Calner, 2005) and was previously studied by Odin et al. (1986, sample B108) and Batchelor and Jeppsson (1999). Odin et al. (1986) determined a K−Ar age of 427.4 ± 6.0 Ma from this locality. The lower of two bentonites, which are separated by 16 cm, was sampled (G03−340LJ of Kiipli et al., 2008b) by Cramer et al. (2012) and six out of nine zircons yielded a weighted mean ^{206}Pb/^{238}U age of 428.06 ± 0.68 Ma. This bentonite occurs within the Djupvik Member of the Halla Formation, within the *Kockelella ortus absidata* conodont zone (Kiipli et al., 2008b) and likely correlates to a position near the boundary between the *Colonograptus praedeubeli* and *Colonograptus deubeli* graptolite zones. The Djupvik Member, and the bentonites contained therein, roughly marks the position of the low point between the two peaks of the Homerian "Mulde" positive δ^{13}C excursion (Calner et al., 2006).

The final date from Cramer et al. (2012) comes from Wren's Nest Hill, Dudley, England. This bentonite occurs in Lion's Mouth Cavern within the uppermost part of the Upper Quarried Limestone Member of the Much Wenlock Limestone Formation, near the base of PS12 (Parasequence 12) of Ray et al. (2010). By correlation to the base Ludlow GSSP, this bentonite (Wren's Nest Hill 15—Ray et al., 2011) likely corresponds with a level only a few centimeters below the golden spike at Pitch Coppice Quarry (see above). There is little direct biostratigraphic information available from this outcrop; however, through the detailed correlation of Ray et al. (2010), and the position of the bentonite with respect to the base of the Lower Elton Formation, this bentonite can be correlated to a position very near the top of the *Colonograptus ludensis* graptolite zone and the *K. o. absidata* conodont zone, just above the final conclusion of the Homerian "Mulde" positive δ^{13}C excursion (Ray et al., 2010). Cramer et al. (2012) reported that 6 out of 11 zircons yielded a weighted mean ^{206}Pb/^{238}U age of 427.77 ± 0.68 Ma.

Two new Silurian high-precision CA-ID-TIMS dates have become available since the publication of GTS2012 and both come from the Silurian section in Podolia, Ukraine. The first is the M12 bentonite from the Malynivtsi 150 section, Podolia, southwestern Ukraine within the upper Hrynchuk Formation just below the onset of the Ludfordian "Lau" positive carbon-isotope excursion (Cramer et al., 2015b), placing it within the upper part of the *Polygnathoides siluricus* conodont zone of McAdams et al. (2019), although the nominative species has not been recovered from this section. Seven zircons yielded concordant and equivalent dates with a weighted mean ^{206}Pb/^{238}U age of 424.08 ± 0.53 Ma.

The second date from Podolia, Ukraine, comes from the C6 bentonite from the very top of the Pryhorodok Formation at the Ataky-117 outcrop. Cramer et al. (2015b) correlated this bentonite to a position within the upper part of the global *O. crispa* conodont zone and therefore close to the Ludlow−Pridoli boundary in the uppermost part of the Ludfordian Stage of the Ludlow Series. Based upon the revised understanding of the position of the LAD of *O. crispa* with respect to the graptolite zonation discussed above in Section 21.1.4, and if the identification of the bentonites preserved in the Gushcha-4015 drillcore is correct (Tsegelnyuk et al., 1983), the position of this bentonite is best placed not only within the uppermost part of the *Ozarkodina crispa* conodont zone but also the *Neocolonograptus parultimus* graptolite zone of the lowermost part of the Pridoli Series. In the Gushcha-4015 drillcore, bentonite C6 was not identified, and only two of the C-series bentonites were identified (C4 and C5) causing uncertainty in their identification. However, *N. parultimus* occurs below bentonite C4 and *N. ultimus* first occurs above the bentonite identified as C5. Therefore if these identifications are correct, it is most likely that the dated bentonite from the Ataky-117 section belongs in the *N. parultimus* graptolite zone, the uppermost part of

TABLE 21.1 Comparison of ages and durations of Silurian graptolite zones and stages recognized in the CONOP composite in GTS2012 (Melchin et al., 2012) and that of GTS2020, obtained by spline interpolation. Also shown are the ages and durations resulting from using polynomial and LOWESS regression methods.

Stage	Zone	GTS2012 Spline Age (Ma)	GTS2012 Duration	GTS2020 2-sigma of Spline Age (Myr)	GTS2020 Spline Age (Ma)	GTS2020 Spline Duration (Myr)	GTS2020 Polynomial Age (Ma)	GTS2020 Polynomial Duration (Myr)	GTS2020 Polynomial Age (Ma) (Min, Max)[a]	GTS2020 LOWESS Age (Ma) (Min, Max)[a]
Pridoli	"Monograptus" perneri–Neocolonograptus? transgrediens	421.26	2.06	1.69	420.95	1.94	420.37	2.31	421.282, 420.708	419.860, 421.091
Pridoli	"Monograptus" bouceki	422.13	0.87	1.67	421.48	0.53	420.98	0.6	421.833, 421.347	420.189, 421.421
Pridoli	Neocolonograptus brankensis–N. lochkovensis	422.83	0.69	1.64	422.25	0.77	421.83	0.85	422.610, 422.244	420.672, 421.903
Pridoli	Neocolonograptus parultimus–N. ultimus	422.96	0.13	1.62	422.73	0.48	422.36	0.53	423.094, 422.800	420.983, 422.214
Ludfordian	Formosograptus formosus	423.69	0.73	1.61	423.14	0.41	422.8	0.44	423.496, 423.260	421.250, 422.480
Ludfordian	Neocolonograptus kozlowskii–Polonograptus podoliensis	424.96	1.27	1.59	423.64	0.5	423.33	0.54	423.987, 423.819	421.584, 422.813
Ludfordian	Bohemograptus (not used as a distinct zone in 2012)			1.56	424.25	0.61	423.99	0.65	424.586, 424.498	422.007, 423.234
Ludfordian	Saetograptus leintwardinensis	425.57	0.61	1.53	425.01	0.76	424.79	0.81	425.325, 425.329	422.552, 423.776
Gorstian	Lobograptus scanicus	426.93	1.36	1.47	426.54	1.53	426.42	1.62	426.816, 426.983	423.745, 424.959
Gorstian	Neodiversograptus nilssoni	427.36	0.43	1.47	426.74	0.2	426.63	0.22	427.016, 427.201	423.915, 425.127
Homerian	Colonograptus ludensis	427.92	0.56	1.42	427.86	1.12	427.89	1.26	428.182, 428.465	424.964, 426.162
Homerian	Colonograptus praedeubeli–Colonograptus deubeli	428.18	0.26	1.41	428.34	0.48	428.4	0.5	428.648, 428.964	425.414, 426.604
Homerian	Pristiograptus parvus–Gothograptus nassa	429.26	1.08	1.38	429.04	0.71	429.17	0.78	429.370, 429.728	426.145, 427.321
Homerian	Cyrtograptus lundgreni	430.45	1.19	1.32	430.62	1.58	430.82	1.65	430.912, 431.319	427.874, 429.008
Sheinwoodian	Monograptus belophorus–M. antenularius–Cyrtograptus rigidus	432.36	1.59	1.27	431.91	1.29	432.03	1.21	432.056, 432.464	429.319, 430.407
Sheinwoodian	Monograptus firmus–M. riccartonensis	432.59	0.23	1.27	432.1	0.19	432.19	0.17	432.215, 432.619	429.531, 430.610

TABLE 21.1 (Continued)

Stage	Zone								
Base of Sheinwoodian		433.35		432.93	1.24		432.89	432.875, 433.263	430.441, 431.485
Telychian–Sheinwoodian	Cyrtograptus murchisoni	433.35	0.76	433.06	1.23	0.79	432.98	432.968, 433.353	430.573, 431.611
Telychian	Cyrtograptus centrifugus	433.46	0.11	433.56	1.22	0.39	433.38	433.345, 433.715	431.118, 432.132
Telychian	Cyrtograptus insectus	433.63	0.17	433.7	1.21	0.1	433.48	433.445, 433.810	431.264, 432.271
Telychian	Cyrtograptus lapworthi	435.15	1.52	434.89	1.17	0.88	434.36	434.297, 434.618	432.547, 433.489
Telychian	Oktavites spiralis	439.15	1	435.63	1.14	0.55	434.91	434.832, 435.120	433.377, 434.271
Telychian	Monoclimacis crenulata	436.51	0.36	436.36	1.12	0.56	435.48	435.384, 435.635	434.244, 435.083
Telychian	Monoclimacis griestoniensis	436.61	0.1	436.44	1.12	0.07	435.54	435.448, 435.695	434.345, 435.178
Telychian	Streptograptus crispus	437.2	0.59	437.04	1.1	0.5	436.04	435.942, 436.155	435.127, 435.907
Telychian	Spirograptus turriculatus	438.13	0.93	438.19	1.05	1.17	437.21	437.099, 437.238	436.970, 437.623
Telychian	Spirograptus guerichi	438.49	0.36	438.59	1.04	0.51	437.73	437.607, 437.718	437.787, 438.405
Aeronian	Stimulograptus sedgwickii–Stimulograptus halli	438.76	0.27	438.85	1.03	0.36	438.09	437.967, 438.061	438.307, 438.974
Aeronian	Lituigraptus convolutus	439.21	0.45	439.43	1.02	0.87	438.96	438.835, 438.895	439.460, 440.233
Aeronian	Pernerograptus argentus–Pribylograptus leptotheca	439.47	0.26	439.6	1.01	0.26	439.22	439.096, 439.149	439.751, 440.656
Aeronian	Demirastrites triangulatus–Demirastrites pectinatus	440.77	1.3	440.49	0.99	0.78	440	439.876, 439.913	440.662, 441.695
Rhuddanian	Coronograptus cyphus	441.57	0.8	440.82	0.98	0.99	440.99	440.859, 440.891	441.967, 442.848
Rhuddanian	Cystograptus vesiculosus	442.47	0.9	441.54	0.96	0.87	441.86	441.729, 441.769	442.954, 443.778
Rhuddanian	Parakidograptus acuminatus	443.4	0.93	442.39	0.93	0.94	442.8	442.664, 442.723	444.006, 445.736
Rhuddanian	Akidograptus ascensus	443.83	0.43	443.07	0.91	0.4	443.2	443.058, 443.128	444.399, 445.063

aMax and min refer to separate regressions for the older and younger limits, respectively, of 2-sigma uncertainty intervals on radioisotopic dates, for which initial analytical uncertainty may have shrunk in response to stratigraphic constraints (see Section 20.4.2.1 of Goldman et al. (Ch. 20: The Ordovician Period, this volume) for further details).

the *O. crispa* conodont zone, and the lowermost part of the Pridoli Series. Eight zircons yielded concordant and equivalent dates with a weighted mean $^{206}Pb/^{238}U$ age of 422.91 ± 0.49 Ma.

21.3.2 Ages of stage boundaries

The methods and procedures used to build the composite sequence for the Ordovician and Silurian, to calibrate the zone and stage boundaries, and to determine the uncertainty within the composite standard sequence, all of which were used to estimate the ages of the Ordovician–Silurian stages and zones, are outlined in Sections 20.4.2 and 20.4.3 of Goldman et al. (2020, Ch. 20: The Ordovician Period, this volume), and with a more technical explanation in Section 14A.2 of Agterberg et al. (Ch. 14: Geomathematics, this book).

Although the various sources of uncertainty inherent in deriving these time scales are described in (Section 14.2), those chapters of this book, it is worthwhile to reiterate the most important sources of error here. The first source of uncertainty is the 2-sigma analytical error on the radioisotopic dates that are used to calibrate the scale. The second source of error is uncertainty within the algorithmic sequencing of events in the compositing process—an underdetermined problem with too little information to yield a unique answer. The third source of uncertainty is the final, expert placement of stage and zonal boundaries within the composite. Since the sequence of events (e.g., taxon range ends, ash bed levels, levels of onset or end of isotope excursion events) in the composite is a global composite and does not match the sequences at individual sections, expert judgment is challenged to select the most appropriate levels for stage and zonal boundaries within the composite (Sadler et al., 2009). As noted in Goldman et al. (2020, Ch. 20: The Ordovician Period, this volume), the placement of GSSPs within the composite is less well constrained than global taxon FADs, because the GSSPs represent only a single horizon from a single section, whereas taxon range ends are commonly constrained by the compilation of data from many sections. This problem with GSSPs is worse for the stages of the Silurian, most of which, as noted above, are not well marked by a set

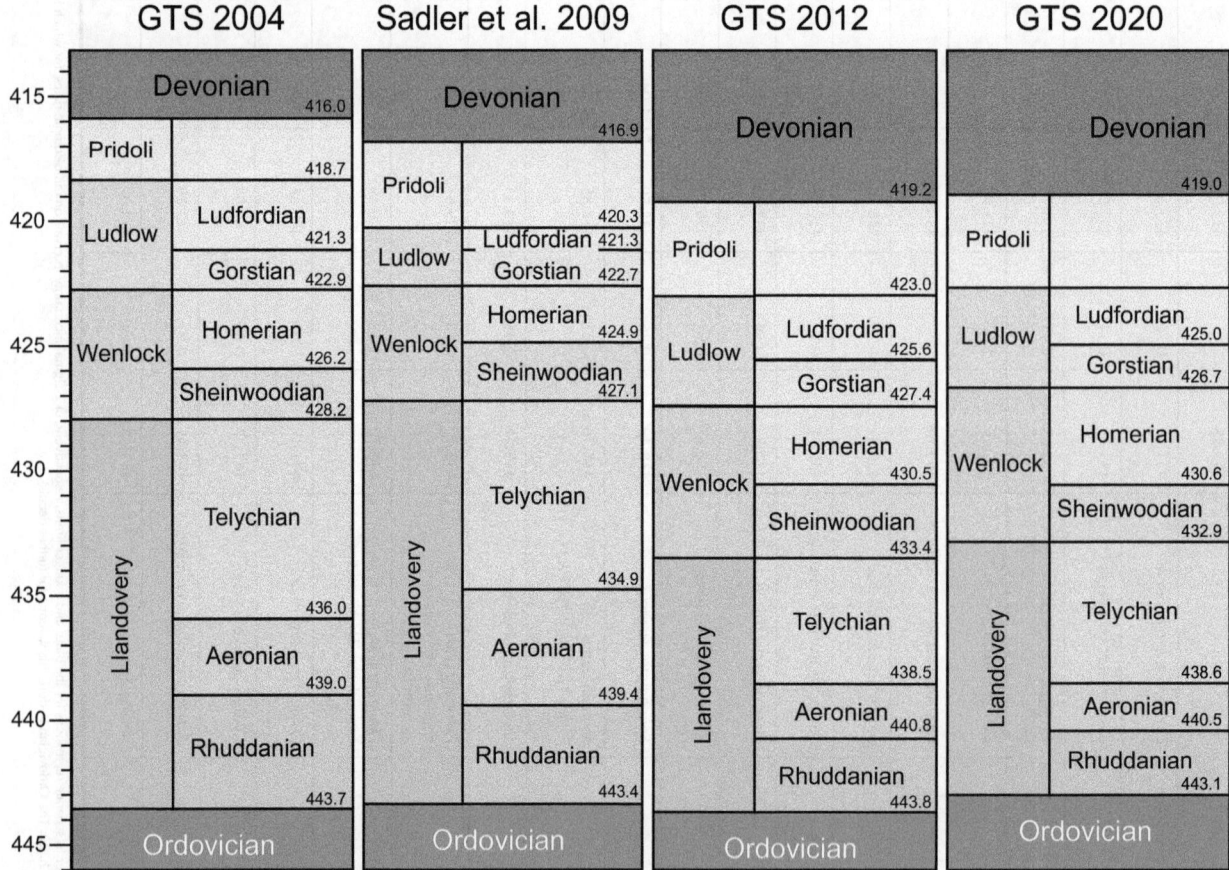

FIGURE 21.13 Comparison of the Silurian time scales of Melchin et al. (2004, GTS2004), Sadler et al. (2009), Melchin et al. (2012, GTS2012), and present GTS2020, showing series and stages. Note that in Melchin et al. (2004) and Sadler et al. (2009), the base of the Wenlock was drawn at a level correlative with the base of the *Cyrtograptus centrifugus* Zone, whereas in GTS2012 it is drawn at the level of the base of the overlying *Cyrtograptus murchisoni* Zone. In the present work, it is drawn at the level of the GSSP, which is correlated with a level slightly above the base of the overlying *C. murchisoni* Zone (see text for explanation).

TABLE 21.2 Interpolated ages of Ordovician and Silurian epoch and stage boundaries, with uncertainty estimates, stage durations, and comparison to GTS2012.

Chronostratigraphic unit			GTS2012				GTS2020			
Period	Epoch	Stage/Age	Base Ma	Est. ± Myr (2σ)	Epoch duration	Stage duration	Base Ma	Est. ± Myr (2σ)	Epoch duration	Stage duration
Devonian	Early	Lochkovian	419.2	3.2			419.0	1.8		
Silurian	Pridoli		423.0	2.3	3.8	3.8	422.7	1.6	3.7	3.7
	Ludlow	Ludfordian	425.6	0.9		2.6	425.0	1.5		2.3
		Gorstian	427.4	0.5	4.4	1.8	426.7	1.5	4	1.7
	Wenlock	Homerian	430.5	0.7		3.1	430.6	1.3		3.9
		Sheinwoodian	433.4	0.8	6.0	2.9	432.9	1.2	6.2	2.3
	Llandovery	Telchyan	438.5	1.1		5.1	438.6	1.1		5.7
		Aeronian	440.8	1.2	10.5	2.3	440.5	1.0	10.2	1.9
		Rhuddanian	443.8	1.5		3.1	443.1	0.9		2.6
Ordovician	Late	Hirnantian	445.2	1.4		1.3	445.2	0.9		2.1
		Katian	453.0	0.7	14.5	7.8	452.8	0.7	15.1	7.6
		Sandbian	458.4	0.9		5.4	458.2	0.7		5.4
	Middle	Darriwillian	467.3	1.1		8.9	469.4	0.9		11.2
		Dapingian	470.0	1.4	11.6	2.7	471.3	1.0	13.1	1.9
	Early	Floian	477.7	1.4		7.8	477.1	1.2		5.8
		Tremadocian	485.4	1.9	15.4	7.7	486.9	1.5	15.6	9.8

of globally useful biostratigraphic datums, at, below, and above the GSSP levels. Therefore most of the stage boundaries for the Silurian are here recognized by a biostratigraphic level within the composite, rather than by the level of the GSSP. A judgment of the uncertainty in the level of the GSSPs relative to the time scale is shown as gray boxes adjacent to the stage boundaries in Fig. 21.2. Expert judgment in the placement of stage and zone boundary levels is subjective and difficult to quantify in the probabilistic manner of uncertainties in dating and regression. This judgment uncertainty is not included in the numerical calculations of errors on the stage boundaries.

Another source of uncertainty, which is taken care of in the geomathematical analysis, is in the interpolation process (see Chapter 14A: Geomathematical and Statistical Procedures, this book). The main process used for interpolation in the age scaling in this chapter (and throughout this book) is spline fitting, having optimal uncertainty analysis of the data. However, for the Ordovician and Silurian time scales two alternative methods of interpolation scaling were also used internally: fifth-order polynomial regression, and LOWESS curve fitting. To illustrate the degree of uncertainty that can result from the choice of interpolation method, we compare outcomes for each of these three methods in Table 21.1. Graphs showing these results are presented in Fig. 20.15 of Goldman et al. (Ch. 20: The Ordovician Period,

this volume). In the LOWESS and polynomial regressions the Ordovician−Silurian boundary is constrained by comparably rich data on both sides, but the Silurian−Devonian boundary lies closer to the limits of the data and is less well constrained. The spline fitting can be adjusted for information in successive chapters. The ages adopted for this chapter are those produced by spline fitting.

Compared to GTS2012, the GTS2020 scale for the Ordovician−Silurian should be more reliable (Table 21.2). It is the product of a composite sequence that includes more recent publications and almost double the number of age dates, a majority of which are also more reliable. Although a majority of these new, precise dates are from Ordovician strata, the calibration regressions use the whole ensemble of dates from both systems.

The Rhuddanian Stage is now estimated to have extended from 443.1 ± 0.9 to 440.5 ± 1.0 Ma, for a duration of 2.6 Myr. The Aeronian estimate is 1.9 Myr, from 440.5 ± 1.0 to 438.6 ± 1.1 Ma. The Telychian Stage ranges from an estimated 438.6 ± 1.1 to 432.9 ± 1.2 Ma, for a duration of 5.7 Myr, and is thus the longest stage in the Silurian Period. The Sheinwoodian estimate is 2.3 Myr, from 432.9 ± 1.2 to 430.6 ± 1.3 Ma. The Homerian Stage is estimated to have started at 430.6 ± 1.3 and ended at 426.7 ± 1.5 Ma, lasting 3.9 Myr. The Gorstian estimate runs from 426.7 ± 1.5 to 425 ± 1.5 Ma for 1.7 Myr and is thus the shortest stage

GSSPs of the Silurian Stages, with location and primary correlation criteria

Stage	GSSP Location	Latitude, Longitude	Boundary Level	Correlation Events	Reference
Přídolí (Series)	Požáry Section, Řeporyjie, Prague, Czech Republic	50°01'39.82"N 14°19'29.56"E*	within Bed 96	Graptolite, FAD of *Neocolonograptus parultimus*	Episodes 8/2, 1985; Geol. Ser., Nat. Mus. Wales **9**, 1989
Ludfordian	Sunnyhill Quarry, near Ludlow, UK	52°21'33"N 2°46'38"W*	coincident with the base of the Leintwardine Formation	Near base of graptolite, *Saetograptus leintwardinensis*	Lethaia **14**; Episodes 5/3, 1982; Geol. Ser.,Nat. Mus. Wales **9**, 1989
Gorstian	Pitch Coppice Quarry near Ludlow, UK	52°21'33"N 2°46'38"W*	coincident with the base of the Lower Elton Formation	Imprecise, just below FAD of graptolite questionably assigned to *Saetograptus varians*	Lethaia **14**; Episodes 5/3, 1982; Geol. Ser., Nat. Mus. Wales **9**, 1989
Homerian	Sheinton Brook, Homer, UK	52°36'56"N 2°33'53"W*	within upper part of the Apedale Member of the Coalbrookdale Formation	Graptolite, FAD of *Cyrtograptus lundgreni*	Lethaia **14**; Episodes 5/3, 1982; Geol. Ser., Nat. Mus. Wales **9**, 1989
Sheinwoodian (under revision)	Hughley Brook, UK	52°34'52"N 2°38'20"W*	base of the Buildwas Formation	Imprecise, between the base of acritarch biozone 5 and LAD of conodont *Pterospathodus amorphognathoides*	Lethaia **14**; Episodes 5/3, 1982; Geol. Ser., Nat. Mus. Wales **9**, 1989
Telychian	Cefn-cerig Road Section, Wales, UK	51.97°N 3.79°W**	approximately 31 m below the top of the Wormwood Formation	Just above LAD of Brachiopod *Eocoelia intermedia* and below FAD of *Eocoelia curtisi*	Episodes 8/2, 1985; Geol. Ser., Nat. Mus. Wales **9**, 1989
Aeronian	Trefawr Track Section, Wales, UK	52.03°N 3.70°W**	within Trefawar Formation	Imprecise, just below occurrence of graptolite *Pernerograptus sequens*	Geol. Series, Nat. Mus. Wales **9**, 1989
Rhuddanian	Dob's Linn, Scotland	55.44°N 3.27°W**	1.6 m above the base of the Birkhill Shale Formation	Graptolite, FAD of *Akidograptus ascensus*	Episodes 8/2, 1985; Episodes **31**/3 2008

* according to Google Earth, ** derived from map

FIGURE 21.14 Silurian GSSP locations, boundary levels, primary marker fossils and correlation events, and key references. *GSSPs*, Global Boundary Stratotype Section and Points.

in the Silurian. The Ludfordian estimate is 2.3 Myr, from 425.0 ± 1.5 Ma to the base of the Pridoli at 422.7 ± 1.6 Ma. The Pridoli Series estimate is 3.7 Myr, from 422.7 ± 1.6 to 419.0 ± 1.8 Ma at the base of the Devonian.

The Llandovery Epoch has an estimated duration of 10.2 Myr, the Wenlock 6.2 Myr, the Ludlow 4.0 Myr, and the Pridoli 3.7 Myr. The total Silurian estimate of 24.1 Myr, from 443.1 to 419 Ma, is 0.6 Myr shorter than was estimated in GTS2012—a duration difference that is smaller than the uncertainty on many of the calibration ages.

The differences in age and duration estimates of the 2020 Silurian time scale compared to those of the GTS2012, Sadler et al. (2009) and GTS2004 are illustrated in Fig. 21.13. For the Silurian the largest differences between the 2020 and 2012 time scales are 0.7 Myr younger bases for the Rhuddanian and the Gorstian, with a correspondingly longer duration for the Homerian and a 0.6 Myr younger age for the lower boundary of the Ludfordian Stage (see Table 21.2). Note, however, that none of the differences in ages of stage

boundaries between the 2020 and 2012 time scales is greater than the 2σ uncertainties on those boundary ages.

Acknowledgments

The research leading to this chapter was made possible by the following sources of research funding: a Natural Sciences and Engineering Research Council of Canada Discovery Grant to MJM; National Science Foundation Grants no. EAR-0948277 and CAREER-1455030 to BDC. Valuable discussions with many collaborators have led to the refined correlations and zonations presented in this chapter. We particularly thank Dan Goldman for his close collaboration on the production of the overall Ordovician–Silurian time scale, and Rosemary Melchin for editorial assistance.

Bibliography

Agterberg, F.P., da Silva, A.-C., and Gradstein, F.M., 2020, Chapter 14A - Geomathematical and statistical procedures. *In* Gradstein, F.M., Ogg, J.G.,

Schmitz, M.D., and Ogg, G.M. (eds), *The Geologic Time Scale 2020*. **Vol. 1** (this book). Elsevier, Boston, MA.

Aldridge, R.J., and Briggs, D.E.G., 1989, A soft body of evidence: a puzzle that has confused paleontologists for more than a century has finally been solved. *Natural History*, **89**: 6–9.

Aldridge, R.J., Siveter, D.J., Siveter Derek, J., Lane, P.D., Palmer, D., and Woodcock, N.H., 2000, *British Silurian Stratigraphy*. Peterborough: Geological Conservation Review Series, Joint Nature Conservation Committee 542 pp.

Azmy, K., Veizer, J., Bassett, M.G., and Copper, P., 1998, Oxygen and carbon isotopic composition of Silurian brachiopods: implications for coeval seawater and glaciations. *Geological Society of America Bulletin*, **110**: 1499–1512.

Azmy, K., Veizer, J., Wenzel, B., Bassett, M.G., and Copper, P., 1999, Silurian strontium isotope stratigraphy. *Geological Society of America Bulletin*, **111**: 475–483.

Bachan, A., Lau, K.V., Saltzman, M.R., Thomas, E., Kump, L.R., and Payne, J.L., 2017, A model for the decrease in amplitude of carbon isotope excursions across the Phanerozoic. *American Journal of Science*, **317**: 641–676.

Bancroft, A.M., and Cramer, B.D., 2020, Silurian conodont biostratigraphy of the east-central Appalachian Basin (eastern USA): Re-examination of the C.T. Helfrich Collection. *Bulletin of Geosciences*, **95**: 1–22. https://doi.org/10.3140/bull.geosci.1748.

Bancroft, A.M., Brunton, F.R., and Kleffner, M.A., 2015, Silurian conodont biostratigraphy and carbon (d^{13}C$_{carb}$) isotope stratigraphy of the Victor Mine (V-03-270-AH) core in the Moose River Basin. *Canadian Journal of Earth Sciences*, **52**: 1169–1181.

Bassett, M.G., 1985, Towards a "common language" in stratigraphy. *Episodes*, **8**: 87–92.

Bassett, M.G., 1989, The Wenlock Series in the type area. *In*: Holland, C.H., and Bassett, M.G. (eds), A Global Standard for the Silurian System. National Museum of Wales. **Vol. 9**. Geological Series 51–73 pp.

Bassett, M.G., Cocks, L.R.M., Holland, C.H., Rickards, R.B., and Warren, P.T., 1975, The type Wenlock Series. *Report of the Institute of Geological Sciences*, **75/13**: 1–19.

Bassett, M.G., Kaljo, D., and Teller, L., 1989, The Baltic region. *In*: Holland, C.H., and Bassett, M.G. (eds), A Global Standard for the Silurian System. National Museum of Wales. **Vol. 9**. Geological Series, 158–170 pp.

Batchelor, R.A., and Jeppsson, L., 1999, Wenlock metabentonites from Gotland, Sweden: geochemistry, sources and potential as chemostratigraphic markers. *Geological Magazine*, **136**: 661–669.

Batchelor, R.A., Harper, D.A.T., and Anderson, T.B., 2003, Geochemistry and potential correlation of Silurian (Telychian) metabentonites from Ireland and SW Scotland. *Geological Journal*, **38**: 161–174.

Bergström, S.M., Huff, W.D., Kolata, D.R., and Melchin, M.J., 1997, Occurrence and significance of Silurian K-bentonite beds at Arisaig, Nova Scotia, eastern Canada. *Canadian Journal of Earth Sciences*, **34**: 1630–1643.

Bergström, S.M., Huff, W.D., and Kolata, D.R., 1998a. Silurian K-bentonites in North America and Europe. *Instituto Technológico Geominero de España, Temas Geologico-Mineros*, **23**: 54–56.

Bergström, S.M., Huff, W.D., and Kolata, D.R., 1998b, The Lower Silurian Osmundsberg K-bentonite. Part I: Stratigraphic position, distribution, and paleogeographic significance. *Geological Magazine*, **135**: 1–13.

Bergström, S.M., Toprak, F.O., Huff, W.D., and Mundil, R., 2008, Implications of a new, biostratigraphically well-controlled, radio-isotopic

age for the lower Telychian Stage of the Llandovery Series (Lower Silurian, Sweden). *Episodes*, **31**: 309–314.

Bickert, T., Pätzold, J., Samtleben, C., and Munnecke, A., 1997, Paleoenvironmental changes in the Silurian indicated by stable isotopes in brachiopod shells from Gotland, Sweden. *Geochimica et Cosmochimica Acta*, **61**: 2717–2730.

Bjerreskov, M., 1975, Llandoverian and Wenlockian graptolites from Bornholm. *Fossilium Strata*, **8**: 94.

Bradley, D.C., and O'Sullivan, P., 2017, Detrital zircon geochronology of pre- and syncollisional strata, Acadian orogen, Maine Appalachians. *Basin Research*, **29**: 571–590.

Braun, M., 2018, High Resolution Chemostratigraphy and Cyclostratigraphy of Lower Silurian Neritic Carbonates from Anticosti Island, Quebec, Canada. *Unpublished M.Sc. thesis*, University of Ottawa, 104 p.

Brenchley, P.J., Marshall, J.D., Carden, G.A.F., Robertson, D.B.R., Long, D.G.F., Meidla, T., et al., 1994, Bathymetric and isotopic evidence for a short-lived Late Ordovician glaciation in a greenhouse period. *Geology*, **22**: 295–298.

Brunton, F.R., Smith, L., Dixon, O.A., Copper, P., Nestor, H., and Kershaw, S., 1998, Silurian reef episodes, changing seascapes, and paleobiogeography. *In*: Landing, E., and Johnson, M.E. (eds), *Silurian Cycles: Linkages of Dynamic Stratigraphy With Atmospheric, Oceanic, and Tectonic Changes*. Albany: New York State Museum. James Hall Centennial Volume. 265–282 pp.

Burgess, N.D., and Richardson, J.B., 1995, Late Wenlock to early Přídolí cryptospores and miospores from south and southwest Wales, Great Britain. *Palaeontographica Abteilung B*, **236**: 1–44.

Calner, M., Kozlowska, A., Masiak, M., and Schmitz, B., 2006, A shoreline to deep basin correlation chart for the middle Silurian coupled extinction-stable isotopic event. *GFF*, **128** (2), 79–84.

Came, R.E., Eiler, J.M., Veizer, J., Azmy, K., Brand, U., and Weidman, C.R., 2007, Coupling of surface temperatures and atmospheric CO2 concentrations during the Palaeozoic era. *Nature*, **449** (7159), 198.

Caputo, M.V., 1998, Ordovician-Silurian glaciations and global sea-level changes. *In*: Landing, E., and Johnson, M.E. (eds), Silurian Cycles: Linkages of Dynamic Stratigraphy With Atmospheric, Oceanic and Tectonic Changes. *New York State Museum Bulletin*, **Vol. 491**: pp. 15–25.

Carls, P., Slavik, L., and Valenzuela-Rios, J.I., 2007, Revisions of conodont biostratigraphy across the Silurian-Devonian boundary. *Bulletin of Geosciences*, **82**: 145–164.

Caruthers, A.H., Gröcke, D.R., Kaczmarek, S.E., Rine, M.J., Kuglitsch, J., and Harrison III, W.B., 2018, Utility of organic carbon isotope data from the Salina Group halite (Michigan Basin): A new tool for stratigraphic correlation and paleoclimate proxy resource. *Geological Society of America Bulletin*, **130**: 1782–1790. https://doi.org/10.1130/B31972.1.

Cherns, L., 2011, The GSSP for the base of the Ludfordian Stage, Sunnyhill Quarry. In: Ray, D.C. (ed), *Siluria Revisited: A Field Guide. International Subcommission on Silurian Stratigraphy, Field Meeting 2011*, 75–81 pp.

Cocks, L.R.M., 1985, The Ordovician-Silurian boundary. *Episodes*, **8**: 98–100.

Cocks, L.R.M., 1989, The Llandovery Series in the Llandovery area. *In*: Holland, C.H., and Bassett, M.G. (eds), A Global Standard for the Silurian System. National Museum of Wales. **Vol. 9**. Geological Series, 36–50 pp.

Cocks, L.R.M., Holland, C.H., Rickards, R.B., and Strachan, I., 1971, A correlation of the Silurian rocks on the British Isles. *Journal of the Geological Society, London*, **127**: 103−136.

Cocks, L.R.M., Woodcock, N.H., Rickards, R.B., Temple, J.T., and Lane, P.D., 1984, The Llandovery Series of the type area. *Bulletin of the British Museum Natural History (Geology)*, **38**: 131−182.

Compston, W., 2000a, Interpretations of SHIMP and isotope dilution zircon ages for the geological time-scale: 1. The early Ordovician and late Cambrian. *Mineralogical Magazine*, **64**: 43−57.

Compston, W., 2000b, Interpretations of SHIMP and isotope dilution zircon ages for the Palaeozoic time-scale: II. Silurian to Devonian. *Mineralogical Magazine*, **64**: 1127−1146.

Compston, W., and Williams, I.S., 1992, Ion probe ages for the British Ordovician and Silurian stratotypes. *In* Webby, B.D., and Laurie, J. (eds), *Global Perspectives on Ordovican Geology*. Rotterdam: Balkema, pp. 59−67.

Condon, D., Schoene, B., Bowring, S., Parrish, R., McLean, N., Noble, S., et al., 2007, EARTHTIME: isotopic tracers and optimized solutions for high-precision U-Pb ID-TIMS geochronology. *Eos, Transactions, American Geophysical Union 88, Full Meeting Supplement, Abstract V41E-06*.

Cooper, R.A., and Sadler, P.M., 2004, Chapter 15, Ordovician. *In*: Gradstein, F., Ogg, J.G., and Smith, A.G. (eds), *A Geologic Time Scale*. Cambridge University Press, pp. 165−187.

Cooper, R.A., Rigby, S., Loydell, D.K., and Bates, D.E.B., 2012, Palaeoecology of the Graptoloidea. *Earth-Science Reviews*, **112**: 23−41.

Cooper, R.A., Sadler, P.M., Munnecke, A., and Crampton, J.S., 2014, Graptoloid evolutionary rates track Ordovician-Silurian global climate change. *Geological Magazine*, **151**: 349−364.

Corradini, C., and Serpagli, E., 1999, A Silurian conodont zonation from late Llandovery to end Přídolí in Sardinia. *In*: Serpagli, E. (ed), Studies on Conodonts, Proceedings of the Seventh European Conodont Symposium. *Bollettino della Societé Paleontologica Italiana*, **37**: 255−283.

Corradini, C., Corriga, M.G., Männik, P., and Schonlaub, H.P., 2015, Revised conodont stratigraphy of the Cellon section (Silurian, Carnic Alps). *Lethaia*, **48**: 56−71.

Corriga, M.G., and Corradini, C., 2009, Upper Silurian and Lower Devonian conodonts from the Monte Cocco II section (Carnic Alps, Italy). *Bulletin of Geosciences*, **84**: 155−168.

Cramer, B.D., and Jarvis, I., 2020, Chapter 11 - Carbon isotope stratigraphy. *In* Gradstein, F.M., Ogg, J.G., Schmitz, M.D., and Ogg, G.M. (eds), *The Geologic Time Scale 2020*. **Vol. 1** (this book). Elsevier, Boston, MA.

Cramer, B.D., and Munnecke, A., 2008, Early Silurian positive delta C-13 excursions and their relationship to glaciations, sea-level changes and extinction events: discussion. *Geological Journal*, **43**: 517−519.

Cramer, B.D., and Saltzman, M.R., 2007, Fluctuations in epeiric sea carbonate production during Silurian positive carbon isotope excursions: a review of proposed paleoceanographic models. *Palaeogeography, Palaeoclimatology, Palaeoecology*, **245**: 37−45.

Cramer, B.D., Loydell, D.K., Samtleben, C., Munnecke, A., Kaljo, D., Männik, P., et al., 2010a, Testing the limits of Paleozoic chronostratigraphic correlation via high-resolution (<500,000yrs) integrated conodont, graptolite, and carbon isotope (δ13Ccarb) biochemostratigraphy across the Llandovery-Wenlock (Silurian) boundary: is a unified Phanerozoic timescale achievable? *Geological Society of America Bulletin*, **122**: 1700−1716. https://doi.org/10.1130/B26602.1.

Cramer, B.D., Kleffner, M.A., Brett, C.E., McLaughlin, P.I., Jeppsson, L., Munnecke, A., et al., 2010b, Paleobiogeography, high-resolution stratigraphy, and the future of Paleozoic biostratigraphy: fine-scale diachroneity of the Wenlock (Silurian) conodont *Kockelella walliseri*. *Palaeogeography, Palaeoclimatology, Palaeoecology*, **294**: 232−241. https://doi.org/10.1016/j.palaeo2010.01.002.

Cramer, B.D., Brett, C.E., Melchin, M.J., Männik, P., Kleffner, M.A., McLaughlin, P.I., et al., 2011a, Revised chronostratigraphic correlation of the Silurian System of North America with global and regional chronostratigraphic units and δ¹³C_carb chemostratigraphy. *Lethaia*, **44**: 185−202. https://doi.org/10.1111/j.1502-3931.2010.00234.x.

Cramer, B.D., Davies, J.R., Ray, D.C., Thomas, A.T., and Cherns, L., 2011b. In Ray, D.C. (ed), *Siluria Revisited: A Field Guide. International Subcommission on Silurian Stratigraphy, Field Meeting 2011*, 6−27 pp.

Cramer, B.D., Munnecke, A., Schofield, D.I., Hasse, K.M., and Haase-Schramm, A., 2011c, A revised ⁸⁷Sr/⁸⁶Sr curve for the Silurian: implications for global ocean chemistry and the Silurian timescale. *Journal of Geology*, **119**: 335−349. https://doi.org/10.1086/660117.

Cramer, B.D., Condon, D.J., Söderlund, U., Marshall, C., Worton, G.J., Thomas, A.T., et al., 2012, U-Pb (zircon) age constraints on the timing and duration of Wenlock (Silurian) paleocommunity collapse and recovery during the "Big Crisis". *Geological Society of America Bulletin*, **124**: 1841−1857.

Cramer, B.D., Vandenbroucke, T.R.A., and Ludvigson, G.A., 2015a, High-Resolution Event Stratigraphy (HiRES) and the quantification of stratigraphic uncertainty: Silurian examples of the quest for precision in stratigraphy. *Earth-Science Reviews*, **141**: 136−153.

Cramer, B.D., Schmitz, M.D., Huff, W.D., and Bergstrom, S.M., 2015b, High-precision U-Pb zircon age constraints on the duration of rapid biogeochemical events during the Ludlow Epoch (Silurian Period). *Journal of the Geological Society*, **172**: 157−160.

Crampton, J.S., Cooper, R.A., Sadler, P.M., and Foote, M., 2016, Greenhouse-icehouse transition in the Late Ordovician marks a step change in extinction regime in the marine plankton. *Proceedings of the National Academy of Sciences of the United States of America*, **113**: 1498−1503.

Crampton, J.S., Meyers, S.R., Cooper, R.A., Sadler, P.M., Foote, M., and Harte, D., 2018, Pacing of Paleozoic macroevolutionary rates by Milankovitch grand cycles. *Proceedings of the National Academy of Sciences of the United States of America*, **115**: 5686−5691.

Crick, R.E., Ellwood, B.B., Hladil, J., El Hassani, A., Hrouda, F., and Chlupac, I., 2001, Magnetostratigraphy susceptibility of the Pridolian-Lochkovian (Silurian-Devonian) GSSP (Klonk, Czech Republic) and a coeval sequence in Anti-Atlas Morocco. *Palaeogeography, Palaeoclimatology, Palaeoecology*, **167**: 73−100.

Cuervo, H.D.R., Soares, E.A.A., Caputo, M.V., and Dino, R., 2018, Sedimentology and stratigraphy of new outcrops of Silurian glaciomarine strata in the Presidente Figueiredo region, northwestern margin of the Amazonas Basin. *Journal of South American Earth Sciences*, **85**: 43−56.

Cummins, R.C., Finnegan, S., Fike, D.A., Eiler, J.M., and Fischer, W. W., 2014, Carbonate clumped isotope constraints on Silurian ocean temperature and seawater delta O-18. *Geochimica et Cosmochimica Acta*, **140**: 241−258.

Danielsen, E.M., Cramer, B.D., and Kleffner, M.A., 2019, Identification of a global sequence boundary within the upper Homerian (Silurian) Mulde Event: High-resolution chronostratigraphic correlation of the midcontinent United States with Sweden and the United Kingdom. *Geosphere*, **15**: 839−855. https://doi.org/10.1130/GES01685.1.

Davies, J.R., Molyneux, S.G., Vandenbroucke, T.R.A., Verniers, J., Waters, R.A., Williams, M., et al., 2011, Pre-conference field trip to the Type Llandovery area. In: Ray, D.C. (ed), *Siluria Revisited: A Field Guide. International Subcommission on Silurian Stratigraphy, Field Meeting 2011*, 29−72 pp.

Davies, J.R., Waters, R.A., Molyneux, S.G., Williams, M., Zalasiewicz, J.A., Vandenbroucke, T.R.A., et al., 2013, A revised sedimentary and biostratigraphical architecture for the Type Llandovery area, Central Wales. *Geological Magazine*, **150**: 300−332.

Davies, N.S., and Gibling, M.R., 2010, Cambrian to Devonian evolution of alluvial systems: the sedimentological impact of the earliest land plants. *Earth-Science Reviews*, **98**: 171−200.

De Weirdt, J., Vandenbroucke, T.R.A., Cocq, J., Russell, C., Davies, J.R., Melchin, M., et al., 2019, Chitinozoan biostratigraphy of the Rheidol Gorge section, central Wales, UK: a GSSP replacement candidate section for the Rhuddanian-Aeronian boundary. *Papers in Palaeontology*. https://doi.org/10.1002/spp2.1260.

Díaz-Martínez, E., and Grahn, Y., 2007, Early Silurian glaciation along the western margin of Gondwana (Peru, Bolivia, and northern Argentina): Palaeogeographic and Geodynamic setting. *Palaeogeography, Palaeoclimatology, Palaeoecology*, **245**: 62−81.

Doyle, E.N., Hoey, A.N., and Harper, D.A.T., 1991, The rhynchonellide brachiopod *Eocoelia* from the Upper Llandovery of Ireland and Scotland. *Palaeontology*, **34**: 439−454.

Eriksson, M.E., and Calner, M., 2005, The Dynamic Silurian Earth. *Sveriges Geologiska Undersökning, Rapporter och Meddelanden*, **121**: 1−99.

Ettensohn, F.R., and Brett, C.E., 2002, Stratigraphic evidence from the Appalachian Basin for continuation of the Taconian orogeny into Early Silurian time. *Physics and Chemistry of the Earth*, **27**: 279−288.

Finney, S.C., and Berry, W.B.N., 1997, New perspectives on graptolite distributions and their use as indicators of platform margin dynamics. *Geology*, **25**: 919−922.

Frýda, J., and Manda, S., 2013, A long-lasting steady period of isotopically heavy carbon in the late Silurian ocean: evolution of the delta C-13 record and its significance for an integrated delta C-13, graptolite and conodont stratigraphy. *Bulletin of Geosciences*, **88**: 463−482.

Gale, N.H., 1985, Numerical calibration of the Paleozoic time-scale; Ordovician, Silurian and Devonian periods. *In*: Snelling, N.J. (ed), *The Chronology of the Geological Record. Memoir 10, Geological Society*. Oxford: Blackwell Scientific Publications, pp. 81−88.

Gambacorta, G., Menichetti, E., Trincianti, E., and Torricelli, S., 2018, Orbital control on cyclical primary productivity and benthic anoxia: astronomical tuning of the Telychian Stage (Early Silurian). *Palaeogeography, Palaeoclimatology, Palaeoecology*, **495**: 152−162.

Goldman, D., Sadler, P.M., and Leslie, S.A., 2020, Chapter 20 - The Ordovician Period. *In* Gradstein, F.M., Ogg, J.G., Schmitz, M.D., and Ogg, G.M. (eds), *The Geologic Time Scale 2020*. **Vol. 2** (this book). Elsevier, Boston, MA.

Goodwin, P.W., Anderson, E.J., Goodman, W.M., and Saraka, L.J., 1986, Punctuated aggradational cycles: implications for stratigraphic analysis. *Palaeoceanography Paleoclimatology*, **2**: 417−429.

Gradstein, F.M., Ogg, J.G., and Smith, A.G. (eds), 2012. A Geologic Time Scale. Elsevier. Amsterdam, 1144 p.

Grossman, E.L., and Joachimski, M.M., 2020, Chapter 10 - Oxygen isotope stratigraphy. *In* Gradstein, F.M., Ogg, J.G., Schmitz, M.D., and Ogg, G.M. (eds), *The Geologic Time Scale 2020*. **Vol. 1** (this book). Elsevier, Boston, MA.

Hammarlund, E.U., Loydell, D.K., Nielsen, A.T., and Schovsbo, N.H., 2019, Early Silurian δ^{13}Corg excursions in the foreland basin of Baltica, both familiar and surprising. *Palaeogeography Palaeoclimatology Palaeoecology*, **526**: 126−135.

Harland, W.B., Armstrong, R.L., Cox, A.V., Craig, L.E., Smtih, A.G., and Smith, D.G., 1990, *A Geologic Time Scale*. Cambridge University Press 263 p.

Haq, B.U., and Schutter, S.R., 2008, A chronology of Paleozoic sea-level changes. *Science*, **322**: 64−68.

Heath, R.J., Brenchley, P.J., and Marshall, J.D., 1998, Early Silurian carbon and oxygen stable-isotope stratigraphy of Estonia: implications for climate change. *In*: Landing, E., and Johnson, M.E. (eds), Silurian Cycles: Linkages of Dynamic Stratigraphy With Atmospheric, Oceanic and Tectonic Changes. *New York State Museum Bulletin*, **Vol. 491**: pp. 313−327.

Hibbard, J., 2000, Docking Carolina: Mid-Paleozoic accretion in the southern Appalachians. *Geology*, **28**: 127−130.

Hinnov, L.A., 2013, Cyclostratigraphy and its revolutionizing applications in the earth and planetary sciences. *Geological Society of America Bulletin*, **125**: 1703−1734.

Hinnov, L.A., 2018, Chapter 1 − Cyclostratigraphy and astrochronology in 2018. *In*: Montenari, M. (ed), Stratigraphy & Timescales. *Amsterdam: Elsevier*, **Vol. 3**: pp. 1−80.

Holland, C.H., 1989, History of classification of the Silurian System. *In*: Holland, C.H., and Bassett, M.G. (eds), A Global Standard for the Silurian System. *National Museum of Wales*. Geological Series, 7−26 pp.

Holland, C.H., Lawson, J.D., and Walmsley, V.G., 1963, The Silurian rocks of the Ludlow District, Shropshire. *Bulletin of the British Museum Natural History (Geology)*, **8**: 93−171.

Huff, W.D., Bergström, S.M., and Kolata, D.R., 2000, Silurian K-bentonites of the Dnestr Basin, Podolia, Ukraine. *Journal of the Geological Society, London*, **157**: 493−504.

Husson, J.M., Schoene, B., Bluher, S., and Maloof, A.C., 2016, Chemostratigraphic and U-Pb geochronologic constraints on carbon cycling across the Silurian-Devonian boundary. *Earth and Planetary Science Letters*, **436**: 108−120.

Hutt, J.E., 1974, A new group of Llandovery biform monograptids. *In* Rickards, R.B., Jackson, D.E., and Hughes, C.P. (eds), *Graptolite Studies in Honour of O. M. B. Bulman*. Special Papers in Paleontology, The Palaeontological Association, **Vol. 13**: pp. 108−120.

Jaeger, H., 1991, Neue Standard-Graptolithenzonenfolge nach der 'Grossen Krise' an der Wenlock/Ludlow Grenze (Silurs). *Neues Jarhbuch für Geologie und Palaontologie Abhandlungen*, **182**: 303−354.

Jarochowska, E., Munnecke, A., Frisch, K., Ray, D.C., and Castagner, A., 2016, Faunal and facies changes through the mid Homerian (late Wenlock, Silurian) positive carbon isotope excursion in Podolia, western Ukraine. *Lethaia*, **49**: 170−198.

Jeppsson, L., 1982, *Third European Conodont Symposium (ECOS III). Guide to Excursion*, Vol. 239. Sweden: Publications of Mineralogy, Paleontology and Quaternary Geology, University of Lund, 32 p.

Jeppsson, L., 1990, An oceanic model for lithological and faunal changes tested on the Silurian record. *Journal of the Geological Society, London*, **147**: 663−674.

Jeppsson, L., 1997a, A new latest Telychian, Sheinwoodian and early Homerian (early Silurian) standard conodont zonation. *Transactions of the Royal Society of Edinburgh Earth Science,* **88**: 91–114.

Jeppsson, L., 1997b, Recognition of a probable secundo-primo event in the early Silurian. *Lethaia,* **29**: 311–315.

Jeppsson, L., 1998, Silurian oceanic events: summary of general characteristics. *In:* Landing, E., and Johnson, M.E. (eds), Silurian Cycles: Linkages of Dynamic Stratigraphy With Atmospheric, Oceanic and Tectonic Changes. *New York State Museum Bulletin,* **Vol. 491:** pp. 239–257.

Jeppsson, L., and Calner, M., 2003, The Silurian Mulde Event and a scenario for secundo-secundo events. *Transactions of the Royal Society of Edinburgh Earth Science,* **93**: 135–154.

Jeppsson, L., Eriksson, M.E., and Calner, M., 2005, Locality descriptions. *In:* Eriksson, M.E., and Calner, M. (eds), The Dynamic Silurian Earth. Subcommission on Silurian Stratigraphy Field Meeting 2005. Field Guide and Abstracts: Geological Survey of Sweden. *Rapporter och Meddelanden,* **Vol. 121:** pp. 22–56.

Jeppsson, L., Eriksson, M.E., and Calner, M., 2006, A latest Llandovery to latest Ludlow high-resolution biostratigraphy based on the Silurian of Gotland–a summary. *GFF,* **128**: 109–114.

Johnson, M.E., 1996, Stable cratonic sequences and a standary for Silurian eustacy. *In:* Witzke, B.J., Ludvigson, G.A., and Day, J. (eds), Paleozoic sequence stratigraphy: views From the North American Craton. Geological Society of America Special Paper, **306**: 203–211.

Johnson, M.E., 2006, Relationship of Silurian sea-level fluctuations to oceanic episodes and events. *GFF,* **128**: 123–129.

Johnson, M.E., 2010, Tracking Silurian eustasy: alignment of empirical evidence or pursuit of deductive reasoning? *Palaeogeography, Palaeoclimatology, Palaeoecology,* **296**: 276–284.

Johnson, M.E., Rong, J.Y., and Kershaw, S., 1998, Calibrating Silurian eustacy against the erosion and burian of coastal paleotopography. *In:* Landing, E., and Johnson, M.E. (eds), Silurian Cycles: Linkages of Dynamic Stratigraphy With Atmospheric, Oceanic and Tectonic Changes. *New York State Museum Bulletin,* **Vol. 491:** pp. 3–13.

Kaljo, D., Martma, T., Männik, P., and Viira, V., 2003, Implications of Gondwana glaciations in the Baltic late Ordovician and Silurian and a carbon isotopic test of environmental cyclicity. *Bulletin de la Société Géologique de France,* **174**: 59.66.

Kaljo, D., Grytsenko, V., Martma, T., and Mõtus, M.-A., 2007, Three global carbon isotope shifts in the Silurian of Podolia (Ukraine): stratigraphical implications. *Estonian Journal of Earth Sciences,* **56**: 205–220.

Kaljo, D., Martma, T., Grytsenko, V., Brazauskas, A., and Kaminskas, D., 2012, Pridoli carbon isotope trend and upper Silurian to lowermost Devonian chemostratigraphy based on sections in Podolia (Ukraine) and the East Baltic area. *Estonian Journal of Earth Sciences,* **61**: 162–180.

Kaljo, D., Einasto, R., Martma, T., Märss, T., Nestor, V., and Viira, V., 2015, A bio-chemostratigraphical test of the synchroneity of biozones in the upper Silurian of Estonia and Latvia with some implications for practical stratigraphy. *Estonian Journal of Earth Sciences,* **64**: 267–283.

Kiipli, T., Orlova, K., Kiipli, E., and Kallaste, T., 2008a, Use of immobile trace elements for the correlation of Telychian bentonites on Saaremaa Island, Estonia, and mapping of volcanic ash clouds. *Estonian Journal of Earth Sciences,* **57**: 39–52.

Kiipli, T., Radzevičius, S., Kallaste, T., Motuza, V., Jeppsson, L., and Wickström, L.M., 2008b, Wenlock bentonites in Lithuania and correlation with bentonites from sections in Estonia, Sweden and Norway. *GFF,* **130**: 203–210.

Kiipli, T., Radzevičius, S., Kallaste, T., Kiipli, E., Siir, S., Soesoo, A., et al., 2012, The Geniai Tuff in the southern East Baltic area—a new correlation tool near the Aeronian/Telychian stage boundary, Llandovery, Silurian. *Bulletin of Geosciences,* **87**: 695–704.

Kiipli, T., Radzevičius, S., and Kallaste, T., 2014, Silurian bentonites in Lithuania: correlations based on sanidine phenocryst composition and graptolite biozonation—interpretation of volcanic source regions. *Estonian Journal of Earth Sciences,* **63**: 18–29.

Kiipli, E., Kiipli, T., Kallaste, T., and Märss, T., 2016, Chemical weathering east and west of the emerging Caledonides in the Silurian—Early Devonian, with implications for climate. *Canadian Journal of Earth Sciences,* **53**: 774–780.

Kleffner, M.A., and Barrick, J.E., 2010, Telychian-early Sheinwoodian (early Silurian) conodont-, graptolite-, chitinozoan- and event-based chronostratigraphy developed using the graphic correlation method. *Memoirs of the Association of Australasian Palaeontologists,* **39**: 191–210.

Koreń, T.N., Lenz, A.C., Loydell, D.K., Melchin, M.J., Štorch, P., and Teller, L., 1996, Generalized graptolite zonal sequence defining Silurian time intervals for global paleogeographic studies. *Lethaia,* **29**: 59–60.

Kozłowski, W., 2015, Eolian dust influx and massive whitings during the *kozlowski*/Lau Event: carbonate hypersaturation as a possible driver of the mid-Ludfordian Carbon Isotope Excursion. *Bulletin of Geosciences,* **90**: 807–840.

Kříž, J., 1989, The Přídolí Series in the Prague Basin (Barrandian area, Bohemia). *In:* Holland, C.H., and, and Bassett, M.G. (eds), *A Global Standard for the Silurian System.* National Museum of Wales. Vol. 9. Geological Series, 90–100 pp.

Kříž, J., Jaeger, H., Paris, F., and Schönlaub, H.-P., 1986, Přídolí—the Fourth Series of the Silurian. *Jahrbuch der Geologischen Bundesanstalt, Wein,* **129**: 291–360.

Kuiper, K.F., Deino, A., Hilgen, F.J., Krijgsman, W., Renne, P.R., and Wijbrans, J.R., 2008, Synchronizing rock clocks of Earth history. *Science,* **320** (5875), 500–504. https://doi.org/10.1126/science.1154339.

Kunk, M.J., Sutter, J., Obradovitch, J.D., and Lanphere, M.A., 1985, Age of biostatigraphic horizons within the Ordovician and Silurian Systems. *In* Snelling, N.J. (ed), *The Chronology of the Geological Record. Memoir 10, Geological Society.* Oxford: Blackwell Scientific Publications, pp. 89–92.

Lapworth, C., 1879, On the tripartite classification of Lower Palaeozoic rocks. *Geological Magazine,* **6**: 1–15.

Landing, E., and Johnson, M.E., 1998, Silurian cycles: linkages of dynamic stratigraphy with atmospheric, oceanic, and tectonic changes. *New York State Museum Bulletin, James Hall Centennial Volume,* Vol. 49. Albany, NY: New York State Museum, 327 p.

Laufeld, S., 1974, Silurian Chitinozoa from Gotland. *Fossilium Strata,* **5**: 1–130.

Lawson, J.D., and White, D.E., 1989, The Ludlow Series in the type area. *In* Holland, C.H., and, and Bassett, M.G. (eds), *A Global Standard for the Silurian System.* National Museum of Wales, **Vol. 9:** pp. 1–15.

Lécuyer, C., Amiot, R., Touzeau, A., and Trotter, J., 2013, Calibration of the phosphate delta O-18 thermometer with carbonate-water oxygen isotope fractionation equations. *Chemical Geology,* **347**: 217–226.

Lehnert, O., Frýda, J., Buggisch, W., Munnecke, A., Nützel, A., Kříž, J., et al., 2007, Delta C-13 records across the late Silurian Lau event: new data from middle palaeo-latitudes of northern peri-Gondwana (Prague Basin, Czech Republic). *Palaeogeography, Palaeoclimatology, Palaeoecology,* **245**: 227–244.

Lehnert, O., Männik, P., Joachimski, M.M., Calner, M., and Frýda, J., 2010, Palaeoclimate perturbations before the Sheinwoodian glaciation: a trigger for extinctions during the 'Ireviken Event'. *Palaeogeography, Palaeoclimatology, Palaeoecology*, **296**: 320−331.

Li, Y.F., Zhang, T.W., Shao, D.Y., and Shen, B.J., 2019, New U-Pb zircon age and carbon isotope records from the Lower Silurian Longmaxi Formation on the Yangtze Platform, South China: implications for stratigraphic correlation and environmental change. *Chemical Geology*, **509**: 249−260.

Liang, Y., Bernardo, J., Goldman, D., Nõlvak, J., Tang, P., Wang, W., et al., 2019, Morphological variation suggests that chitinozoans may be fossils of individual microorganisms rather than metazoan eggs. *Proceedings of the Royal Society B*, **286**: 20191270. <https://doi.org/10.1098/rspb.2019.1270>.

Loydell, D.K., 1991, The biostratigraphy and formational relationships of the upper Aeronian and lower Telychian (Llandovery, Silurian) formations of western mid-Wales. *Geological Journal*, **26**: 209−244.

Loydell, D.K., 1998, Early Silurian sea-level changes. *Geological Magazine*, **135**: 447−471.

Loydell, D.K., 2007, Early Silurian positive δ13C excursions and their relationship to glaciations, sea-level changes and extinction events. *Geological Journal*, **42**: 531−546.

Loydell, D.K., 2008, Reply to 'Early Silurian positive δ13C excursions and their relationship to glaciations, sea-level changes and extinction events: discussion' by Bradley D. Cramer and Axel Munnecke. *Geological Journal*, **43**: 511−515.

Loydell, D.K., 2011a, The GSSP for the base of the Wenlock Series, Hughley Brook. In: Ray, D.C. (ed), *Siluria Revisited: A Field Guide. International Subcommission on Silurian Stratigraphy, Field Meeting 2011*, 91−99 pp.

Loydell, D.K., 2011b, The GSSP for the base of the Homerian Stage, Whitwell Coppice. In: Ray, D.C. (ed), *Siluria Revisited: A Field Guide. International Subcommission on Silurian Stratigraphy, Field Meeting 2011*, 85−90 pp.

Loydell, D.K., 2012, Graptolite biozone correlation charts. *Geological Magazine*, **149**: 124−132.

Loydell, D.K., and Maletz, J., 2002, Isolated *'Monograptus' gemmatus* from the Silurian of Osmundsberget, Sweden. *GFF*, **124**: 193−196.

Loydell, D.K., Štorch, P., and Melchin, M.J., 1993, Taxonomy, evolution and biostratigraphical importance of the Llandovery graptolite *Spirograptus. Palaeontology*, **36**: 909−926.

Loydell, D.K., Kaljo, D., and Männik, P., 1998, Integrated biostratigraphy of the lower Silurian of the Ohesaare core, Saaremaa, Estonia. *Geological Magazine*, **135**: 769−783.

Loydell, D.K., Männik, P., and Nestor, V., 2003, Integrated biostratigraphy of the lower Silurian of the Aizpute-41 core, Latvia. *Geological Magazine*, **140**: 205−229.

Loydell, D.K., Frýda, J., and Gutiérrez-Marco, J.C., 2015, The Aeronian/Telychian (Llandovery, Silurian) boundary, with particular reference to sections around the El Pintado reservoir, Seville Province, Spain. *Bulletin of Geosciences*, **90**: 743−794.

Mabillard, J.E., and Aldridge, R.J., 1985, Microfossil distribution across the base of the Wenlock Series in the type area. *Palaeontology*, **28**: 89−100.

Männik, P., 1998, Evolution and taxonomy of Silurian conodont Pterospathodus. *Palaeontology*, **41**: 1001−1050.

Männik, P., 2007a, Some comments on Telychian-early Sheinwoodian conodont faunas, events and stratigraphy. *Acta Palaeontologica Sinica*, **46** (Suppl.), 305−310.

Männik, P., 2007b, An updated Telychian (Late Llandovery, Silurian) conodont zonation based on Baltic faunas. *Lethaia*, **40**: 45−60.

Männik, P., Loydell, D.K., Nestor, V., and Nolvak, J., 2015, Integrated Upper Ordovician-lower Silurian biostratigraphy of the Grotlingbo-1 core section, Sweden. *GFF*, **137**: 226−244.

Marshall, J., 2000, Palynofacies-defined cyclicity and the recognition of transgressive-regressive cycles using palynomorph ratios: an example from the Silurian of Saudi Arabia. *GeoArabia*, **5**: 136.

Märss, T., and Männik, P., 2013, Revision of Silurian vertebrate biozones and their correlation with the conodont succession. *Estonian Journal of Earth Sciences*, **62**: 181−204.

Mattinson, J.M., 2005, Zircon U-Pb chemical abrasion ("CA-TIMS") method: combined annealing and multi-step partial dissolution analysis for improved precision and accuracy of zircon ages. *Chemical Geology*, **220**: 47−66.

Mattinson, J.M., 2010, Analysis of the relative decay constants of U-235 and U-238 by multi-step CA-TIMS measurements of closed-system natural zircon samples. *Chemical Geology*, **275**: 186−198.

McAdams, N.E.B., Cramer, B.D., Bancroft, A.M., Melchin, M.J., Devera, J.A., and Day, J.E., 2019, Integrated δ13Ccarb, conodont, and graptolite biochemostratigraphy of the Silurian from the Illinois Basin and stratigraphic revision of the Bainbridge Group. *Geological Society of America Bulletin*, **131**: 335−352. doi.org/10.1130/B32033.1.

McArthur, J.M., Howarth, R.J., Shields, G.A., and Zhou, Y., 2020, Chapter 7 - Strontium isotope stratigraphy. *In* Gradstein, F.M., Ogg, J.G., Schmitz, M.D., and Ogg, G.M. (eds), *The Geologic Time Scale 2020.* **Vol. 1** (this book). Elsevier, Boston, MA.

McKerrow, W.S., St Lambert, R.J., and Cocks, L.R.M., 1985, The Ordovician, Silurian and Devonian Periods. *In:* Snelling, N.J. (ed), *The Chronology of the Geological Record. Memoir 10, Geological Society*. Oxford: Blackwell Scientific Publications, pp. 73−80.

McLaughlin, P.I., Mikulic, D.G., and Kluessendorf, J., 2013, Correlation of Silurian rocks in Sheboygan, Wisconsin, using integrated stable carbon isotope chemostratigraphy and facies analysis. *Geoscience Wisconsin*, **21**: 15−38.

Melchin, M.J., and Holmden, C., 2006, Carbon isotope chemostratigraphy of the Llandovery in Arctic Canada: implications for global correlation and sea-level change. *GFF*, **128**: 173−180.

Melchin, M.J., and Williams, S.H., 2000, A restudy of the akidograptine graptolites from Dob's Linn and a proposed redefined zonation of the Silurian Stratotype. *Palaeontology 2000, Geological Society of Australia*, **61**: 63.

Melchin, M.J., Koren', T.N., and Štorch, P., 1998, Global diversity and survivorship patterns of Silurian graptoloids. *In:* Landing, E., and Johnson, M.E. (eds), Silurian Cycles: Linkages of Dynamic Stratigraphy With Atmospheric, Oceanic and Tectonic Changes. *New York State Museum Bulletin,* **Vol. 491**: pp. 165−182.

Melchin, M.J., Holmden, C., and Williams, S.H., 2003, Correlation of graptolite biozones, chitinozoan biozones, and carbon isotope curves through the Hirnantian. *In:* Albanesi, G.L., Beresi, M.S., and Peralta, S.H. (eds), Ordovician From the Andes. *Tucumán: Comunicarte Editorial,* **Vol. 17**: pp. 101−104.

Melchin, M.J., Cooper, R.A., and Sadler, P.M., 2004, The Silurian Period. *In:* Gradstein, F.M., Ogg, J.G., and Smith, A.G. (eds), *A Geological Time Scale 2004.* Cambridge: Cambridge University Press, pp. 188−201.

Melchin, M.J., Mitchell, C.E., Naczk-Cameron, A., Fan, J.-X., and Loxton, J., 2011, Phylogeny and adaptive radiation of the Neograpta (Graptoloida) during the hirnantian mass extinction and Silurian recovery. *Proceedings of the Yorkshire Geological Society*, **58**: 281–309.

Melchin, M.J., Sadler, P.M., and Cramer, B.D., 2012, The Silurian Period. *In* Gradstein, F.M., Ogg, J.G., and Smith, A.G. (eds), *A Geologic Time Scale 2012*. Amsterdam: Elsevier, 525–558.

Melchin, M.J., Mitchell, C.E., Holmden, C., and Štorch, P., 2013, Environmental changes in the Late Ordovician-early Silurian: review and new insights from black shales and nitrogen isotopes. *Geological Society of America Bulletin*, **125**: 1635–1670.

Melchin, M.J., MacRae, K.-D., and Bullock, P., 2015, A multi-peak organic carbon isotope excursion in the late Aeronian (Llandovery, Silurian): evidence from Arisaig, Nova Scotia, Canada. *Palaeoworld*, **24**: 191–197.

Melchin, M.J., Davies, J.R., De Weirdt, J., Russell, C., Vandenbroucke, T.R.A., and Zalasiewicz, J.A., 2018, *Integrated Stratigraphic Study of the Rhuddanian-Aeronian (Llandovery, Silurian) Boundary Succession at Rheidol Gorge, Wales: A Preliminary Report*. Nottingham: British Geological Survey 16 p. (OR/18/139).

Mergl, M., Frýda, J., and Kubajko, M., 2018, Response of organophosphatic brachiopods to the mid-Ludfordian (late Silurian) carbon isotope excursion and associated extinction events in the Prague Basin (Czech Republic). *Bulletin of Geosciences*, **93**: 369–400.

Meyers, S.R., Siewert, S.E., Singer, B.S., Sageman, B.B., Condon, D.J., Obradovich, J.D., et al., 2012, Intercalibration of radioisotopic and astrochronologic time scales for the Cenomanian-Turonian boundary interval, Western Interior Basin, USA. *Geology*, **40**: 7–10.

Murphy, M.A., Valenzuela-Ríos, J.I., and Carls, P., 2004, *On Classification of Pridoli (Silurian) — Lochkovian (Devonian) Spathognathodontidae (Conodonts)*. Riverside, CA: University of California Campus Museum Contribution 6, 27 p.

Mullins, G.L., and Aldridge, R.J., 2004, Chitinozoan biostratigraphy of the basal Wenlock Series (Silurian) Global Stratotype Section and Point. *Palaeontology*, **47**: 745–773.

Munnecke, A., Samtleben, C., and Bickert, T., 2003, The Ireviken Event in the lower Silurian of Gotland, Sweden — relation to similar Palaeozoic and Proterozoic events. *Palaeogeography, Palaeoclimatology, Palaeoecology*, **195**: 99–124.

Munnecke, A., Calner, M., Harper, D.A.T., and Servais, T., 2010, Ordovician and Silurian sea-water chemistry, sea level, and climate: a synopsis. *Palaeogeography, Palaeoclimatology, Palaeoecology*, **296**: 389–413.

Murchison, R.I., 1839, *The Silurian System, Founded on Geological Researches in the Counties of Salop, Hereford, Radnor, Montgomery, Caermarthen, Brecon, Pembroke, Monmouth, Gloucester, Worcester and Stafford; With Descriptions of the Coal-Fields and Overlying Formations*. London: John Murray 768 p.

Nardin, E., Godderis, Y., Donnadieu, Y., Le Hir, G., Blakey, R.C., Puceat, E., et al., 2011, Modeling the early Paleozoic long-term climatic trend. *Geological Society of America Bulletin*, **123**: 1181–1192.

Nestor, V., 2012, A summary and revision of the East Baltic Silurian chitinozoan biozonation. *Estonian Journal of Earth Sciences*, **61**: 242–260.

Odin, G.S., 1985, Remarks on the numerical scale of Ordovician to Devonian times. *In*: Snelling, N.J. (ed), *The Chronology of the Geological Record. Memoir 10, Geological Society*. Oxford: Blackwell Scientific Publications, pp. 93–97.

Odin, G.S., Hunziker, J.C., Jeppsson, L., and Spjeldnaes, N., 1986, Ages radiometriques K-Ar de biotites pyroclatiques sedimentées dans le Wenlock de Gotland (Suède). *In*: Odin, G.S. (ed), Calibration of the Phanerozoic Time Scale. *Chemical Geology, 59*: pp. 117–125.

Ogg, J.G., Ogg, G., and Gradstein, F.M. (eds), 2008. The Concise Geologic Time Scale. Cambridge University Press. Cambridge, 177 p.

Oborny, S.C., Cramer, B.D., Brett, C.E., Alyssa, M., and Bancroft, A.M., 2020, Integrated Silurian conodont and carbonate carbon isotope stratigraphy of the east-central Appalachian Basin. *Palaeogeography, Palaeoclimatology, Palaeoecology*, **554**. https://doi.org/10.1016/j.palaeo.2020.109815.

Page, A.A., Zalasiewicz, J.A., Williams, M., and Popov, L.E., 2007, Were transgressive black shales a negative feedback modulating glacioeustasy during the Early Palaeozoic Icehouse? *In* Williams, M., Haywood, A.M., Gregory, F.J., and Schmidt, D.N. (eds), *Deep Time Perspectives on Climate Change: Marrying the Signal From Computer Models and Biological Proxies*. London: The Micropalaeontological Society Special Publications. The Geological Society, 123–156.

Palmer, D., 1971, The Ludlow graptolites *Neodiversograptus nilssoni* and *Cucullograptus (Lobograptus) progenitor*. *Lethaia*, **4**: 357–394.

Pinet, N., 2016, Far-field effects of Appalachian orogenesis: a view from the craton. *Geology*, **44**: 83–86.

Pogson, D.J., 2009, The Siluro-Devonian time scale: a critical review and interim revision. *Geol. Surv. NSW, Q. Notes*, **130**: 1–11.

Pucéat, E., Joachimski, M.M., Bouilloux, A., Monna, A., Bonin, A., Motreuil, S., et al., 2010, Revised phosphate-water fractionation equation reassessing paleotemperatures derived from biogenic apatite. *Earth and Planetary Science Letters*, **298**: 135e142.

Purnell, M.A., Donoghue, P.C.J., and Aldridge, R.J., 2000, Orientation and anatomical notation in conodonts. *Journal of Paleontology*, **74**: 113–122.

Radzevičius, S., Spiridonov, A., and Brazauskas, A., 2014, Integrated middle-upper Homerian (Silurian) stratigraphy of the Viduklė-61 well, Lithuania. *GFF*, **136**: 218–222.

Radzevičius, S., Tumakovaite, B., and Spiridonov, A., 2017, Upper Homerian (Silurian) high-resolution correlation using cyclostratigraphy: an example from western Lithuania. *Acta Geologica Polonica*, **67**: 307–322.

Ray, D.C., 2007, The correlation of Lower Wenlock Series (Silurian) bentonites from the Lower Hill Farm and Eastnor Park boreholes, Midland Platform, England. *Proceedings of the Geologists' Association*, **118**: 175–185.

Ray, D.C. (ed), 2011, *Siluria Revisited: A Field Guide. International Subcommission on Silurian Stratigraphy, Field Meeting 2011*, 166 p.

Ray, D.C., Brett, C.E., Thomas, A.T., and Collings, A.V.J., 2010, Late Wenlock sequence stratigraphy in central England. *Geological Magazine*, **147**: 123–144.

Ray, D.C., Collings, A.V.J., Worton, G.J., and Jones, G., 2011, Upper Wenlock bentonites from Wren's Nest Hill, Dudley; comparisons with prominent bentonites along Wenlock Edge, Shropshire, England. *Geological Magazine*, **148**: 670–681. https://doi.org/10.1017/S0016756811000288.

Rickards, R.B., 1976, The sequence of Silurian graptolite zones in the British Isles. *Geological Journal*, **11**: 153–188.

Rong, J.Y., Melchin, M.J., Williams, S.H., Koren, T.N., and Verniers, J., 2008, Report of the restudy of the defined global stratotype of the base of the Silurian System. *Episodes*, **31**: 315–318.

Rose, C.V., Fischer, W.W., Finnegan, S., and Fike, D.A., 2019, Records of carbon and sulfur cycling during the Silurian Ireviken Event in Gotland, Sweden. *Geochimica et Cosmochimica Acta*, **246**: 299–316.

Ross, C.A., and Ross, J.P.R., 1996, Silurian sea-level fluctuations. *In* Witzke, B.J., Ludvigson, G.A., and Day, J. (eds), *Paleozoic Sequence Stratigraphy: Views From the North American Craton.* Geological Society of America Special Paper 306. Geological Society of America, pp. 83–100.

Ross, R.J., Naeser, C.W., Izett, G.A., Obradovich, J.D., Bassett, M.J., Hughes, C.P., et al., 1982, Fission track dating of British Ordovician and Silurian stratotypes. *Geological Magazine*, **119**: 135–153.

Ruppel, S.C., James, E.W., Barrick, J.E., Nowlan, G., and Uyeno, T. T., 1996, High-resolution ^{87}Sr/^{86}Sr chemostratigraphy of the Silurian: implications for event correlation and strontium flux. *Geology*, **24**: 831–834.

Sadler, P.M., Cooper, R.A., and Melchin, M.J., 2009, High-resolution, early Paleozoic (Ordovician-Silurian) time scales. *Geological Society of America Bulletin*, **121**: 887–906.

Saltzman, M.R., 2002, Carbon isotope δ^{13}C stratigraphy across the Silurian-Devonian transition in North America: evidence for a perturbation of the global carbon cycle. *Palaeogeography, Palaeoclimatology, Palaeoecology*, **187**: 83–100.

Samtleben, C., Munnecke, A., Bickert, T., and Pätzold, J., 1996, The Silurian of Gotland (Sweden): facies interpretation based on stable isotopes in brachiopod shells. *Geologische Rundeschau*, **85**: 278–292.

Schmitz, M.D., Singer, B.S., and Rooney, A.D., 2020, Chapter 6 - Radioisotope geochronology. *In* Gradstein, F.M., Ogg, J.G., Schmitz, M.D., and Ogg, G.M. (eds), *The Geologic Time Scale 2020.* **Vol. 1** (this book). Elsevier, Boston, MA.

Scotese, C.R., 2014, *Atlas of Silurian and Middle-Late Ordovician Paleogeographic Maps (Mollweide Projection), Maps 73–80, Volumes 5, The Early Paleozoic, PALEOMAP Atlas for ArcGIS,* PALEOMAP Project, Evanston, IL.

Servais, T., Achab, A., and Asselin, E., 2013, Eighty years of chitinozoan research: from Alfred Eisenack to Florentin Paris. *Review of Palaeobotany and Palynology*, **197**: 205–217.

Shao, T.Q., Jia, C.H., Liu, Y.H., Fu, L.P., Zhang, Y.A., Qin, J.C., et al., 2018, The Llandovery graptolite zonation of the Danangou section in Nanzheng, Shaanxi Province, central China, and comparisons with those of other regions. *Geological Journal*, **53**: 414–428.

Simmons, M.S., Miller, K.G., Ray, D.C., Davies, A., van Buchem, F.S.P., and Gréselle, B., 2020, Chapter 13 - Phanerozoic eustacy. *In* Gradstein, F.M., Ogg, J.G., Schmitz, M.D., and Ogg, G.M. (eds), *The Geologic Time Scale 2020.* **Vol. 1** (this book). Elsevier, Boston, MA.

Slavik, L., 2014, Revision of the conodont zonation of the Wenlock-Ludlow boundary in the Prague Synform. *Estonian Journal of Earth Sciences*, **63**: 305–311.

Snelling, N.J. (ed), 1985. The chronology of the geological record. *Memoir 10, Geological Society.* Blackwell Scientific Publications, Oxford, 343 p.

Sorgenfrei, T. (ed), 1964, *Report of the 21st International Geological Congress, Copenhagen, Norden, 1960, Part 28: General Proceedings,* 277 p.

Spiridonov, A., Brazauskas, A., and Radzevičius, S., 2016, Dynamics of abundance of the Mid- to Late Pridoli Conodonts from the Eastern part of the Silurian Baltic Basin: multifractals, state shifts, and oscillations. *American Journal of Science*, **316**: 363–400.

Steeman, T., Vandenbroucke, T.R.A., Williams, M., Verniers, J., Perrier, V., Siveter, D.J., et al., 2016, Chitinozoan biostratigraphy of the Silurian

Wenlock-Ludlow boundary succession of the Long Mountain, Powys, Wales. *Geological Magazine*, **153**: 110–111.

Štorch, P., 1994, Graptolite biostratigraphy of the Lower Silurian (Llandovery and Wenlock) of Bohemia. *Geological Journal*, **29**: 137–165.

Štorch, P., and Melchin, M.J., 2018, Lower Aeronian triangulate monograptids of the genus *Demirastrites* Eisel, 1912: biostratigraphy, palaeobiogeography, anagenetic changes and speciation. *Bulletin of Geosciences*, **93**: 513–537.

Štorch, P., Manda, S., and Loydell, D.K., 2014, The early Ludfordian *leintwardinensis* graptolite event and the Gorstian-Ludfordian boundary in Bohemia (Silurian, Czech Republic). *Palaeontology*, **57**: 1003–1043.

Štorch, P., Manda, S., Slavik, L., and Tasaryova, Z., 2016, Wenlock-Ludlow boundary interval revisited: new insights from the offshore facies of the Prague Synform, Czech Republic. *Canadian Journal of Earth Sciences*, **53**: 666–673.

Štorch, P., Manda, S., Tasáryová, Z., Frýda, J., Chadimová, L., and Melchin, M.J., 2018, A proposed new global stratotype for Aeronian Stage of the Silurian System: Hlásná Třebaň section, Czech Republic. *Lethaia*, **51**: 357–388.

Subcommission on Silurian Stratigraphy, 1995, Left hand column for correlation charts. *Silurian Times*, **2**: 7–8.

Sudbury, M., 1958, Triangulate monograptids from the Monograptus gregarius Zone (Lower Llandovery) in the Rheidol Gorge (Cardiganshire). *Philosophical Transactions of the Royal Society of London, Series B*, **241**: 485–554.

Tesakov, Y.I., 2015, Correlation of chronostratigraphic and biostratigraphic units (example of the Silurian System). *Russian Geology and Geophysics*, **56**: 631–651.

Thomas, A.T., and Ray, D.C., 2011, Pitch Coppice: GSSP for the base of the Ludlow Series and Gorstian Stage, Whitwell Coppice. *In*: Ray, D.C. (ed), *Siluria Revisited: A Field Guide. International Subcommission on Silurian Stratigraphy, Field Meeting 2011,* 80–84 pp.

Toghill, P., 1968, The graptolite assemblages and zones of the Birkhill Shales (Lower Silurian) at Dobb's Linn. *Palaeontology*, **11**: 654–668.

Torsvik, T.H., and Cocks, L.R.M., 2017, *Earth History and Palaeogeography.* Cambridge: Cambridge University Press 317 p.

Trench, A., McKerrow, W.S., Torsvik, T.H., Li, X., and McCracken, S.R., 1993, The polarity of the Silurian magnetic field: indications from global data compilation. *Journal of the Geological Society, London*, **150**: 823–831.

Trotter, J.A., Williams, I.S., Barnes, C.R., Lécuyer, C., and Nicoll, R.S., 2008, Did cooling oceans trigger Ordovician biodiversification? Evidence from conodont thermometry. *Science*, **321**: 550–554. https://doi.org/10.1126/science.1155814.

Trotter, J.A., Williams, I.S., Barnes, C.R., Männik, P., and Simpson, A., 2016, New conodont δ^{18}O records of Silurian climate change: implications for environmental and biological events. *Palaeogeography, Palaeoclimatology, Palaeoecology*, **443**: 34–48.

Tsegelnyuk, P.D., Gritsenko, V.P., et al., 1983, *The Silurian of Podolia: The Guide to Excursion. International Union of Geological Sciences, Subcommission on Silurian Stratigraphy.* Kiev: Naukova Dumka 224 p.

Tucker, R.D., and McKerrow, W.S., 1995, Early Paleozoic chronology: a review in light of new U-Pb zircon ages from Newfoundland and Britain. *Canadian Journal of Earth Sciences*, **32**: 368–379.

Tucker, R.D., Krogh, T.E., Ross, R.J., and Williams, S.H., 1990, Time-scale calibration by high-precision U-Pb zircon dating of interstratified volcanic ashes in the Ordovician and lower Silurian stratotypes of Britain. *Earth and Planetary Science Letters*, **100**: 51–58.

van Staal, C.R., Zagorevski, A., McNicoll, V.J., and Rogers, N., 2014, Time-transgressive Salinic and Acadian orogenesis, magmatism and Old Red Sandstone sedimentation in Newfoundland. *Geoscience Canada*, **41**: 138–164 <https://doi.org/10.12789/geocanj.2014.41.031>.

Veizer, J., and Prokoph, A., 2015, Temperatures and oxygen isotopic composition of Phanerozoic oceans. *Earth-Science Reviews*, **146**: 92–104.

Veizer, J., Ala, D., Azmy, K., Bruckschen, P., Buhl, D., Bruhn, F., et al., 1999, $^{87}Sr/^{86}Sr$, $\delta^{13}O$ evolution of Phanerozoic seawater. *Chemical Geology*, **161**: 59–88.

Verniers, J., Nestor, V., Paris, F., Dufka, P., Sutherland, S., and van Grootel, G., 1995, A global Chininozoa biozonation for the Silurian. *Geological Magazine*, **132**: 651–666.

Villeneuve, M., 2004, Radiogenic isotope geochronology. *In*: Gradstein, F.M., Ogg, J.G., and Smith, A.G. (eds), *A Geological Time Scale 2004*. Cambridge: Cambridge University Press, pp. 87–95.

Waid, C.B.T., and Cramer, B.D., 2017a, Telychian (Llandovery, Silurian) conodonts from the LaPorte City Formation of eastern Iowa, USA (East-Central Iowa Basin) and their implications for global Telychian conodont biostratigraphic correlation. *Palaeontologia Electronica*, **20**.

Waid, C.B.T., and Cramer, B.D., 2017b, Global chronostratigraphic correlation of the Llandovery Series (Silurian System) in Iowa, USA, using high-resolution carbon isotope (delta(13) C-carb) chemostratigraphy and brachiopod and conodont biostratigraphy. *Bulletin of Geosciences*, **92**: 373–390.

Wenzel, B., 1997, Isotopenstratigraphische Untersuchungen an silurischen Abfolgen und deren paläozeanographische Interpretation. *Erlanger Geologische Abhandlungen*, **129**: 1–117.

Wenzel, B., and Joachimski, M.M., 1996, Carbon and oxygen isotopic composition of Silurian brachiopods (Gotland/Sweden): palaeoceanographic implications. *Palaeogeography, Palaeoclimatology, Palaeoecology*, **122**: 143–166.

Wenzel, B., Lécuyer, C., and Joachimski, M.M., 2000, Comparing oxygen isotope records of Silurian calcite and phosphate – $\delta^{18}O$ compositions of brachiopods and conodonts. *Geochimica et Cosmochimica Acta*, **64**: 1859–1872.

Westphal, H., Hilgen, F., and Munnecke, A., 2010, An assessment of the suitability of individual rhythmic carbonate successions for astrochronological application. *Earth-Science Reviews*, **99**: 19–30.

Whittard, W.F., 1961, *Lexique Stratigraphique Internationale, Vol. 1: Europe. Fasc. 3a, Pays de Galles, Ecosse, V., Silurien*. Paris: Centre National de la Recherche Scientifique, 273 p.

Williams, M., and Zalasiewicz, J.A., 2004, The Wenlock *Cyrtograptus* species of the Builth Wells district, central Wales. *Palaeontology*, **47**: 223–263.

Williams, S.H., 1983, The Ordovician-Silurian boundary graptolite fauna of Dob's Linn, southern Scotland. *Palaeontology*, **26**: 605–639.

Yolkin, E.A., Sennikov, N.V., Bakharev, N.L., Izokh, N.G., and Yazikov, A.Y., 1997, Periodicity of deposition in the Silurian and relationships of global geological events in the middle Paleozoic of the southwestern margin of the Siberian continent. *Russian Geology and Geophysics*, **38**: 636–647.

Zalasiewicz, J.A., 2000, *Saetograptus (Saetograptus) leintwardinensis leintwardinensis* (Hopkinson MS, Lapworth, 1880). *In*: Zalasiewicz, J.A., Rushton, A.W.A., Hutt, J.E., and Howe, M.P.A. (eds), *Atlas of Graptolite Type Specimens, Folio 1.90*. Palaeontographical Society.

Zalasiewicz, J.A., 2008, *Monograptus triangulatus triangulatus* (Harkness, 1851). *In*: Zalasiewicz, J.A., and Rushton, A.W.A. (eds), *Atlas of Graptolite Type Specimens, Folio 2.92*. Palaeontographical Society.

Zalasiewicz, J.A., and Williams, M., 2008, Cyrtograptus murchisoni (Carruthers, 1867). *In*: Zalasiewicz, J.A., and Rushton, A.W.A. (eds), *Atlas of Graptolite Type Specimens. Folio 2.60*. Palaeontographical Society.

Zalasiewicz, J., Williams, M., Verniers, J., and Jachowicz, M., 1998, A revision of the graptolite biozonation and calibration with the chitinozoa and acritarch biozonations for the Wenlock succession of the Builth Wells district, Wales, U.K. *In* Gutiérrez-Marco, J. C., Rábano, I. (eds), *Proceedings of the Sixth International Graptolite Conference of the GWG (IPA) and the 1998 Field Meeting of the International Subcommission on Silurian Stratigraphy (ICS-IUGS)*, Instituto Tecnológico Geominero de España, *Temas Geológico-Mineros*, **23**: 141.

Zalasiewicz, J.A., Taylor, L., Rushton, A.W.A., Loydell, D.K., Rickards, R.B., and Williams, M., 2009, Graptolites in British stratigraphy. *Geological Magazine*, **146**: 785–850.

Zhamoida, A.I., 2015, General Stratigraphic Scale of Russia: state of the art and problems. *Russian Geology and Geophysics*, **56**: 511–523.

Žigaitė, Ž., Joachimski, M.M., Lehnert, O., and Brazauskas, A., 2008, $\delta^{18}O$ composition of conodont apatite indicates climatic cooling during the Middle Pridoli. *Palaeogeography, Palaeoclimatology, Palaeoecology*, **294**: 242–247.

Žigaitė, Ž., Joachimski, M.M., Lehnert, O., and Brazauskas, A., 2010, $\delta^{18}O$ composition of conodont apatite indicates climatic cooling during the Middle Pridoli. *Palaeogeography, Palaeoclimatology, Palaeoecology*, **294**: 242–247.

R.T. Becker, J.E.A. Marshall and A.-C. Da Silva
With contributions by F.P. Agterberg, F.M. Gradstein and J.G. Ogg

Chapter 22

The Devonian Period

388.2 Ma Devonian

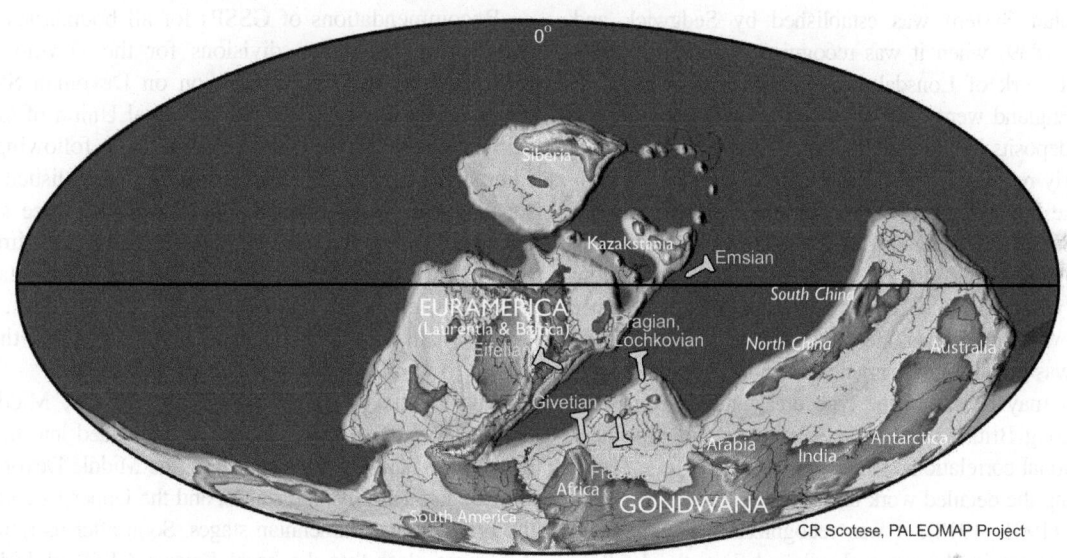

CR Scotese, PALEOMAP Project

Chapter outline

22.1 History and chronostratigraphic subdivisions 734
   22.1.1 Lower Devonian Series 737
   22.1.2 Middle Devonian Series 741
   22.1.3 Upper Devonian Series 744
22.2 Devonian stratigraphy 749
   22.2.1 Biostratigraphy 749
   22.2.2 Physical and chemical stratigraphy 762
22.3 Devonian time scale 772
   22.3.1 Previous scales 772
   22.3.2 Zonal scaling and biostratigraphic comments on the radioisotopic dates 773
   22.3.3 Radioisotopic data 778
   22.3.4 Age of stage boundaries 779
Bibliography 781

Abstract

All seven Devonian stages have been defined by Global Boundary Stratotype Sections and Points (GSSPs), but revisions of the base of the Emsian and of the Devonian—Carboniferous boundary are ongoing. Most of the Devonian Period was a time of exceptionally high sea-level stand and inferred widespread equable climates, but glaciations occurred immediately before its end in the south polar areas of Gondwana (South America, Central, and South Africa). There is even evidence for contemporaneous mountain glaciers in tropical latitudes (in the Appalachians of eastern North America). The cold-water Malvinocaffric Province of southern Gondwana

existed throughout the Early Devonian but disappeared stepwise in the Middle Devonian. Most present-day continental areas and shelves were grouped in one hemisphere, creating a giant "Proto-Pacific" or Panthalassa Ocean, whose margins are poorly preserved in allochthonous terrains. Following the tectonic events of the Caledonian Orogeny of Laurasia, many "Old Red Sandstone" terrestrial deposits formed. After the closure of the narrow Rheic Ocean early in the Devonian, Eovariscan tectonic movements affected Middle and Late Devonian strata in the western Proto-Tethys of Europe and North Africa. Other active fold belts existed in western North America (early Antler Orogeny), polar Canada,

Geologic Time Scale 2020. DOI: https://doi.org/10.1016/B978-0-12-824360-2.00022-X

733

in the Appalachians, in the Urals, along the southern margin of Siberia, in NW China, and in eastern Australia. The Devonian is the time of greatest carbonate production, with a peak of reef growth, and of the greatest diversity of marine fauna in the Paleozoic. Forests became established near the end of the Middle Devonian. Aquatic tetrapods appeared early in the Middle Devonian and diversified in the Upper Devonian.

22.1 History and chronostratigraphic subdivisions

The Devonian System was established by Sedgwick and Murchison (1839) when it was recognized through the then unpublished work of Lonsdale (1840) that marine rocks in southwest England were the equivalent of terrestrial Old Red Sandstone deposits in Wales, the north of England, and Scotland: an early recognition of facies change. Murchison's definition of the boundary between the Silurian System and the Old Red Sandstone in Wales and the Welsh Borders has some ambiguities, but general opinion is that the Ludlow Bone Bed was very close to the intention. However, other boundaries were used over the next century (White, 1950). The result was that there was no clear definition of the boundary in what may be called the type area, and no consistent practice among British geologists. All this was of little help for international correlations.

Following the detailed work on British graptolites by Elles and Wood (1901–1918), it was recognized that graptolites were last present in the upper Ludlow below the Ludlow Bone Bed. Therefore the extinction of graptolites was considered to be the major guide to the position of the base of the Devonian elsewhere in the world. It was not until 1960 that it became clear from evidence outside the British Isles that megaplanktonic graptolites continued long after the time equivalent of the Ludlow Bone Bed. The last monograptids occur in an important transgressive black shale near the base of the (traditional) Emsian (Slavík, 2004b), which was named as *atopus* Event in Becker et al. (2012). After a long period of global, intensive research (Martinsson, 1977), it was, therefore, decided to define the base of the Devonian by evolutionary changes within monograptids at Klonk, Bohemia. The new definition raised the Silurian–Devonian boundary to beneath the *Psammosteus* Limestone in Wales (see review by Marshall and House, 2000).

Sedgwick and Murchison (1840) placed the top of the Devonian at a fairly unambiguous boundary in North Devon, but faunal and floral studies were not then precise enough for accurate correlation. As a result, stratigraphic levels were taken that were subsequently demonstrated to be inaccurate, and other names were used for strata where there was some uncertainty of assignment. In the latter category were names such as the Kinderhookian in North America, and Etroeungt

and Strunian in continental Europe. Following a proposal by Paeckelmann and Schindewolf at the Second Heerlen Carboniferous Congress in 1935 (published in 1937), the Devonian–Carboniferous (D/C) boundary became the first formally decided chronostratigraphic level and was fixed by a type section (Oberrödinghausen railway cut, northern Rhenish Massif, Germany). However, subsequent research (Alberti et al., 1974) showed that the latter contains an unconformity right at the chosen boundary, which resulted in the long search by an international working group for a new stratotype section. Finally, a new basal Carboniferous GSSP (Global Boundary Stratotype Sections and Point) was fixed at La Serre in southern France (Paproth et al., 1991).

Recommendations of GSSPs for all boundaries of system, series, and stage divisions for the Devonian were completed by the Subcommission on Devonian Stratigraphy (SDS) and ratified by International Union of Geological Sciences (IUGS) in 1996. In the following year, summary accounts of all decisions were published in two special volumes (Bultynck, 2000a, 2000b). Since some of the stages are relatively long, and since there are important second/third order global extinctions and sedimentary perturbations, giving natural breaks within stages (e.g., House, 1985; Walliser, 1996; Becker et al., 2016a), the SDS started to work on formal substage definitions.

The Devonian is divided into the Lower, Middle, and Upper Series. The Lower Devonian is divided into the Lochkovian, Pragian, and Emsian stages, the Middle Devonian into the Eifelian and Givetian stages, and the Upper Devonian into the Frasnian and Famennian stages. Soon after its ratification, it became clear that the basal Emsian GSSP of Uzbekistan lies far below the base of the classical Emsian of German usage and that it correlates with a level in the lower half of the classical Pragian of Bohemia (Carls et al., 2008). Therefore the SDS decided in 2008 to revise the Emsian base, a task that has not yet been completed. Kaiser (2009) provided new conodont data from the La Serre Devonian-Carboniferous boundary stratotype section and demonstrated that basal Carboniferous index conodont occurs below the GSSP level. In the absence of any alternative marker for a precise GSSP correlation (Kaiser and Becker, 2007; Kaiser and Corradini, 2008), a new international task group was set up in order to revise the D/C boundary for a second time. This work is ongoing (e.g., Becker et al., 2016b; Aretz and Group, 2016).

The current standard international chronostratigraphic divisions for the Devonian are given in Figs. 22.1 and 22.2. There are also many terms widely used as local and regional stages, for example, for the neritic Lower Devonian of Europe the Gedinnian and Siegenian (for new data see Jansen, 2016). Conodont workers use a subdivision of the Lochkovian into three substages (e.g., Valenzuela-Ríos and Murphy, 1997). Fig. 22.1 shows proposed conodont levels that have not yet been voted on by the SDS. The current basal Emsian GSSP shall be used in future to define an Upper Pragian Substage.

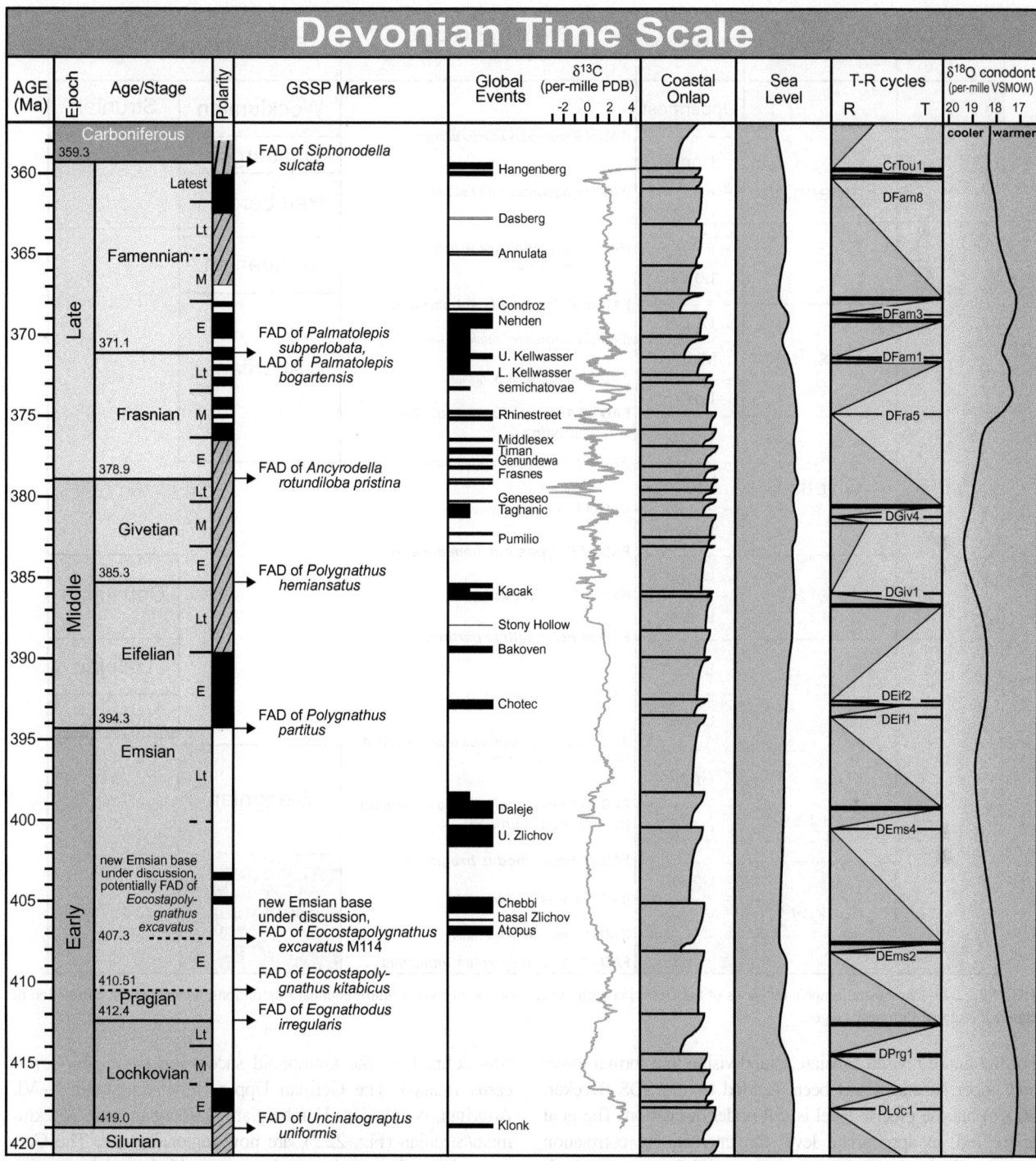

FIGURE 22.1 Devonian Overview. Most main markers for GSSPs of Devonian stages are FADs of conodont taxa as detailed in the text. ("Age" is the term for the time equivalent of the rock-record "stage".) See Carboniferous chapter for discussion on potential revised definition of the Devonian/Carboniferous boundary. The magnetic polarity scale is from Becker et al. (2012) with revised Frasnian–Famennian from Hansma et al. (2015). The carbon isotope (δ^{13}C) curve from Cramer and Jarvis (2020, Ch. 11: Carbon isotope stratigraphy, this book) is a composite of numerous studies. Global events (often associated with widespread anoxic events, as indicated in black) are modified from Becker et al. (2012). The schematic sea-level curve, coastal-onlap curve and the main transgression (T)–regression (R) trends with sequence nomenclature are modified from Haq and Schutter (2008) following advice of Bilal Haq (pers. comm., 2008, to J. Ogg) with additional revised age calibrations by this study (see detailed schematic coastal-onlap curve in Fig. 22.11). The oxygen isotope curve and implied sea-surface temperature trends is a statistical-mean curve by Grossman and Joachimski (2020, Ch. 10: Oxygen isotope stratigraphy, this book) from a synthesis of numerous conodont apatite studies, including Joachimski (2006). The vertical scale of this diagram is standardized to match the vertical scales of the period-overview figure in all other Phanerozoic chapters. *FADs*, First-appearance datums; *GSSP*, Global Boundary Stratotype Section and Point; *PDB*, PeeDee Belemnite $\delta13$C standard; *VSMOW*, Vienna Standard Mean Ocean Water δ^{18}O standard.

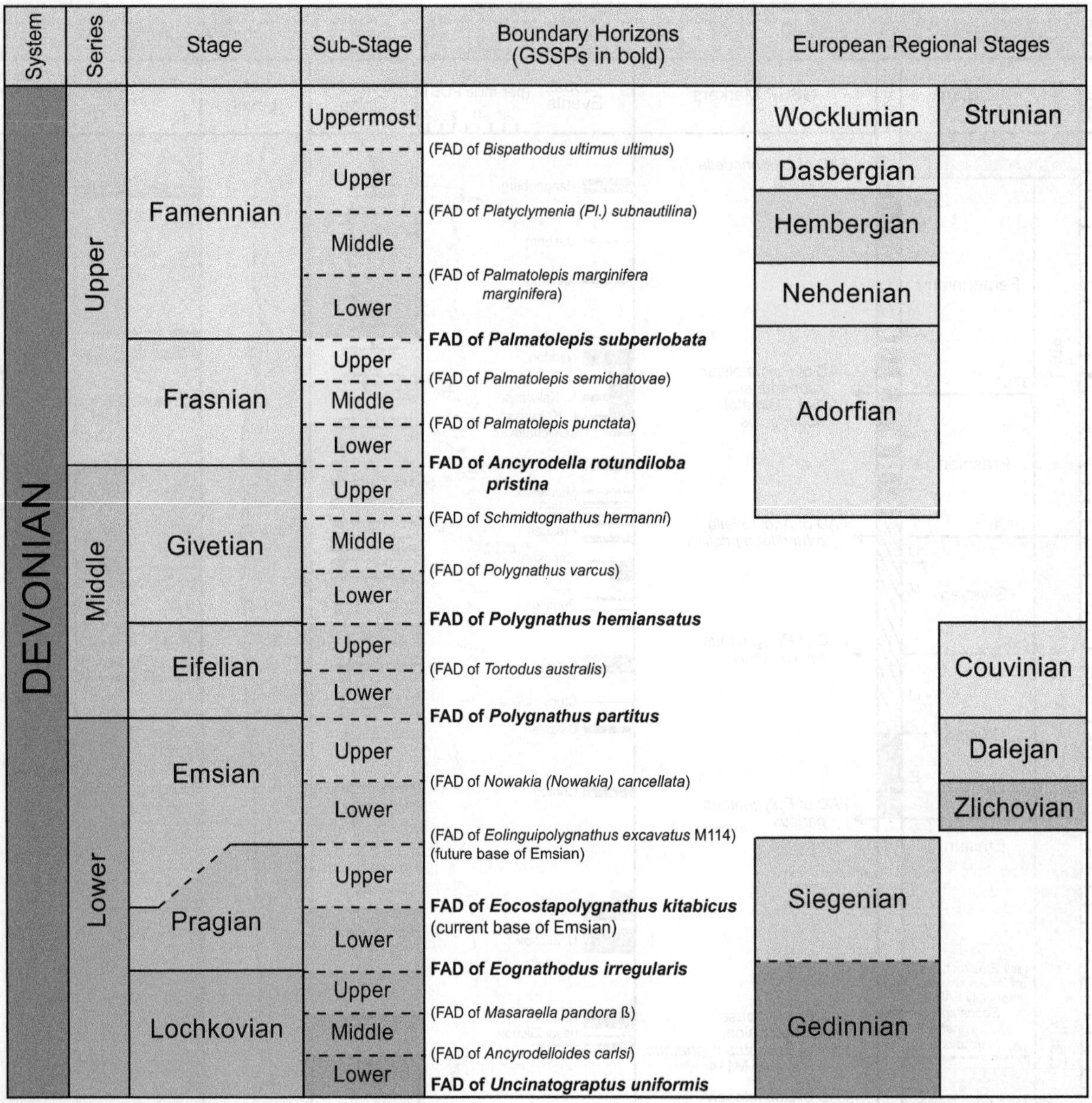

System	Series	Stage	Sub-Stage	Boundary Horizons (GSSPs in bold)	European Regional Stages	
DEVONIAN	Upper	Famennian	Uppermost		Wocklumian	Strunian
				(FAD of *Bispathodus ultimus ultimus*)	Dasbergian	
			Upper	(FAD of *Platyclymenia (Pl.) subnautilina*)		
			Middle		Hembergian	
				(FAD of *Palmatolepis marginifera marginifera*)	Nehdenian	
			Lower	**FAD of *Palmatolepis subperlobata***		
		Frasnian	Upper	(FAD of *Palmatolepis semichatovae*)	Adorfian	
			Middle	(FAD of *Palmatolepis punctata*)		
			Lower	**FAD of *Ancyrodella rotundiloba pristina***		
	Middle	Givetian	Upper	(FAD of *Schmidtognathus hermanni*)		
			Middle	(FAD of *Polygnathus varcus*)		
			Lower	**FAD of *Polygnathus hemiansatus***		
		Eifelian	Upper	(FAD of *Tortodus australis*)	Couvinian	
			Lower	**FAD of *Polygnathus partitus***		
	Lower	Emsian	Upper	(FAD of *Nowakia (Nowakia) cancellata*)	Dalejan	
			Lower	(FAD of *Eolinguipolygnathus excavatus* M114) (future base of Emsian)	Zlichovian	
		Pragian	Upper	**FAD of *Eocostapolygnathus kitabicus*** (current base of Emsian)	Siegenian	
			Lower	**FAD of *Eognathodus irregularis***		
		Lochkovian	Upper	(FAD of *Masaraella pandora* ß)	Gedinnian	
			Middle	(FAD of *Ancyrodelloides carlsi*)		
			Lower	**FAD of *Uncinatograptus uniformis***		

FIGURE 22.2 Chronostratigraphic divisions of the Devonian with the current and proposed future boundary definitions, as well as the correlation of classical regional European stages.

For the future revised Emsian, a subdivision into formal lower and upper substages has been decided by the SDS (Becker, 2007c) but the precise level is still under discussion. The goal is to find an appropriate level for international correlation near the Zlíchovian/Dalejan boundary of Bohemian terminology, which is characterized by the entry of the dacryoconarid *Nowakia (Now.) cancellata*. In the past, it was the miscorrelation of the Dalejan deepening with the Eifelian deepening that led to many complications in the definition of the Lower/Middle Devonian boundary. The term Couvinian (Fig. 22.2) is now a regional term of the Ardennes only. In 2006 the SDS decided to subdivide the Givetian into Lower, Middle, and Upper substages, with the latter positioned close to the old base of the Upper Devonian of German tradition, which

was defined by the ammonoid succession (base of *Pharciceras* faunas). The German Upper Devonian Stufen I—VI, Adorfian, Nehdenian, Hembergian, Dasbergian, and Wocklumian/Strunian (Fig. 22.2), are now regional terms. The German Devonian Subcommission decided to harmonize boundaries with international stage boundaries (Schindler et al., 2018). The Frasnian was decided by the SDS in 2006 to be formally subdivided into lower, middle, and upper substages and precise levels have been selected (Becker, 2007a; Fig. 22.2) but are not yet ratified by the International Commission on Stratigraphy (ICS). The rather long Famennian will include four substages named as Lower, Middle, Upper, and Uppermost Famennian. The latter will roughly equal the Strunian of Belgian regional chronostratigraphy

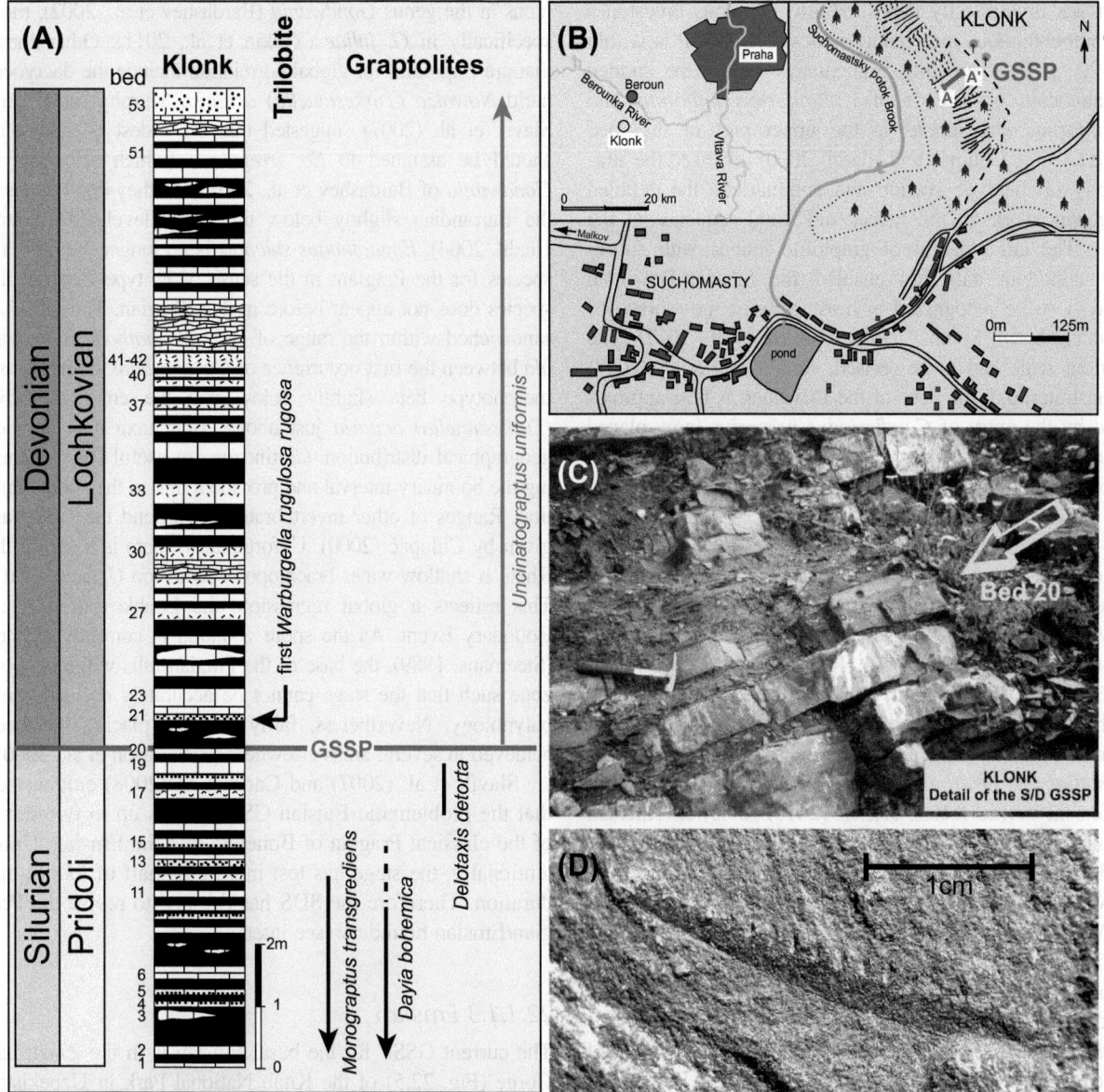

FIGURE 22.3 Basal Devonian (basal Lochkovian) GSSP at Klonk near Suchomasty, Barrandian Czech Republic. (A) Lithological log and ranges of important biostratigraphic markers in relation to the GSSP level; (B) geographic position of the GSSP; (C) photo (by L. Slavík) of GSSP section, with the GSSP positioned within the upper part of Bed 20; (D) the basal Devonian index taxon, *Uncinatograptus uniformis uniformis* (photo by L. Slavík).

(Streel et al., 1998). It is important to note that in several cases (sub)stage names in the past have been used differently by individual authors.

22.1.1 Lower Devonian Series

The GSSP for the Silurian-Devonian boundary, Lower Devonian Series, and Lochkovian Stage is at Klonk in the Czech Republic as documented in Martinsson (1977), in

which D. J. McLaren recounted the scientific and political issues in reaching a decision on its placement. Important faunal characters used in the initial definition are the entry of the graptolite *Uncinatograptus uniformis uniformis* and of the *Warburgella rugulosa* group of trilobites.

22.1.1.1 Lochkovian

The basal-Devonian and Lochkovian GSSP is situated southwest of Prague in the Paleozoic area known as the

Barrandian, where the Klonk section (Fig. 22.3) near Sucho-masty is a natural 34 m cliff section embracing the latest Silurian (Pridoli) and the lower Lochkovian. The sequence comprises rhythmically deposited allochthonous limestones with autochthonous intervening shales. The GSSP is within Bed 20, a 7–10 cm unit immediately below the sudden and abundant occurrence of *U. uniformis uniformis* and *U. uniformis angustidens* in the upper part of that bed (Jaeger, 1977). Chlupáč and Hladil (2000) reviewed the stratigraphy of the type section and summarized the detailed subsequent work on the faunal and floral sequence at the GSSP. The combination of graptolite faunas with subsequent conodont data has enabled the Silurian-Devonian boundary to be recognized in most parts of the world, for example, now in South America (Mestre et al., 2017). The conodont scale had to be revised, since Carls et al. (2007) demonstrated that the base of the Devonian is best approximated by the entry of *Caudicriodus hesperius* (now placed in the genus *Cypricriodus*; Murphy et al., 2018a, 2018b), not by the traditional and widely quoted *woschmidti* Zone. True *Caud. woschmidt* (in the sense of its type) enter somewhat higher in the Lochkovian, locally even later than its previously supposed descendent *Caud. postwoschmidti*. However, some recent publications still show the entry of *Caud. woschmidti* (s.l.) near the base of the Devonian (e.g., Corradini and Corriga, 2012). Richardson and McGregor (1986) and Richardson et al. (2000) provided spore evidence for the correlation of Lochkovian areas and for facies not covered by conodonts and graptolites. Trilobite lineages (*Acastella* and *Warburgella*) provide precise correlation markers in neritic facies. Jansen (2016) reviewed Rhenish brachiopod ranges across the Silurian-Devonian boundary, which enable the distinction of a "lowermost Gedinnian" interval that falls in the Pridoli. Saltzman (2002), Buggisch and Mann (2004), Kleffner et al. (2009), Zhao et al. (2011), Racki et al. (2012), and others showed that the Silurian-Devonian boundary can be approximated by a major positive carbon isotope excursion, the so-called Klonk Event. It can be separated from a topmost Silurian isotopic peak and extinction interval (*transgrediens* Event) in the Barrandian (Manda and Frýda, 2010) or in Podolia (Malkowski et al., 2009). The positive isotope spike has been recognized both in carbonate and organic carbon (e.g., Racki et al., 2012; Zhao et al., 2015). Fig. 22.2 shows proposed Lochkovian substage boundaries defined by the conodonts *Ancyrodelloides carlsi* (alternatively placed in the genus *Lanea*) and *Masaraella pandora* morphotype ß (e.g., Slavík et al., 2012).

22.1.1.2 Pragian

The base of this stage is defined by the GSSP at Velka Chuchle near Prague (Fig. 22.4), Czech Republic (Chlupáč and Oliver, 1989; reviewed by Chlupáč, 2000). The primary correlation marker for this boundary used to be the first occurrence of the conodont *Eognathodus sulcatus*. However, subsequent taxonomic restrictions placed early Pragian specimens in the genus *Gondwania* (Bardashev et al., 2002), more specifically in *G. juliae* (Yolkin et al., 2011). Other forms that are important for global correlation include the dacryoconarid *Nowakia (Turkestanella) acuaria*. Murphy (2005) and Slavík et al. (2007) suggested that the oldest eognathodids should be assigned to *Eo. irregularis* (which also falls in *Gondwania* of Bardashev et al., 2002) and they commence in the Barrandian slightly below the GSSP level (Slavík and Hladil, 2004). *Eognathodus sulcatus* is no longer the defining species for the Pragian; in the sense of its type material the species does not appear before middle Pragian. The GSSP is sandwiched within the range of early *Eognathodus/Gondwania* between the first occurrence of *Caudicriodus steinachensis* morphotype beta slightly below and the entry of *Now. (Turkestanella) acuaria* just above. Both taxa have a wide geographical distribution. Chitinozoa are useful for recognizing the boundary interval and provide a link to the spore zonation. Ranges of other invertebrate taxa around the GSSP are given by Chlupáč (2000). Unfortunately, there is a gap in the Rhenish shallow-water brachiopod succession (Jansen, 2016). This reflects a global regression, the Lochkovian–Pragian Boundary Event. As the spore zonation is currently defined (Steemans, 1989), the base of the Pragian falls within a spore zone such that the stage cannot be accurately defined using palynology. Nevertheless, fairly accurate placing has been achieved in several areas elsewhere (Richardson et al., 2000).

Slavík et al. (2007) and Carls et al. (2008) emphasized that the problematic Emsian GSSP places up to two-thirds of the classical Pragian of Bohemia into the Emsian. Unintentionally, the stage has lost more than half of its original duration. Therefore the SDS has decided to revise the Pragian/Emsian boundary (see later).

22.1.1.3 Emsian

The current GSSP for the basal Emsian is in the Zinzil'ban Gorge (Fig. 22.5) of the Kitab National Park in Uzbekistan (Yolkin et al., 1998). The key conodont used for the stage definition is *Polygnathus kitabicus*, which falls in the genus *Eocostapolygnathus* of Bardashev et al. (2002) (see comments in Aboussalam et al., 2015). As noted above, the "*kitabicus* boundary" lies much lower than the base of the Emsian in the classical region of Germany (Carls and Valenzuela-Ríos, 2007; Jansen, 2008; Carls et al., 2008) and it correlates with a position in the lower half of the Praha Limestone of Bohemia (Slavík et al., 2007; Da Silva et al., 2016). Consequently, and after completion of the required 10-year moratorium, the SDS decided in 2008 to revise the Emsian base. The entry of *Eoc. kitabicus* and the current GSSP shall define the base of a future formal Upper Pragian Substage (Fig. 22.2). The revised Emsian Base has been

Base of the Pragian Stage of the Devonian System in the Velká Chuchle Quarry in the southwest part of Prague, Czech Republic

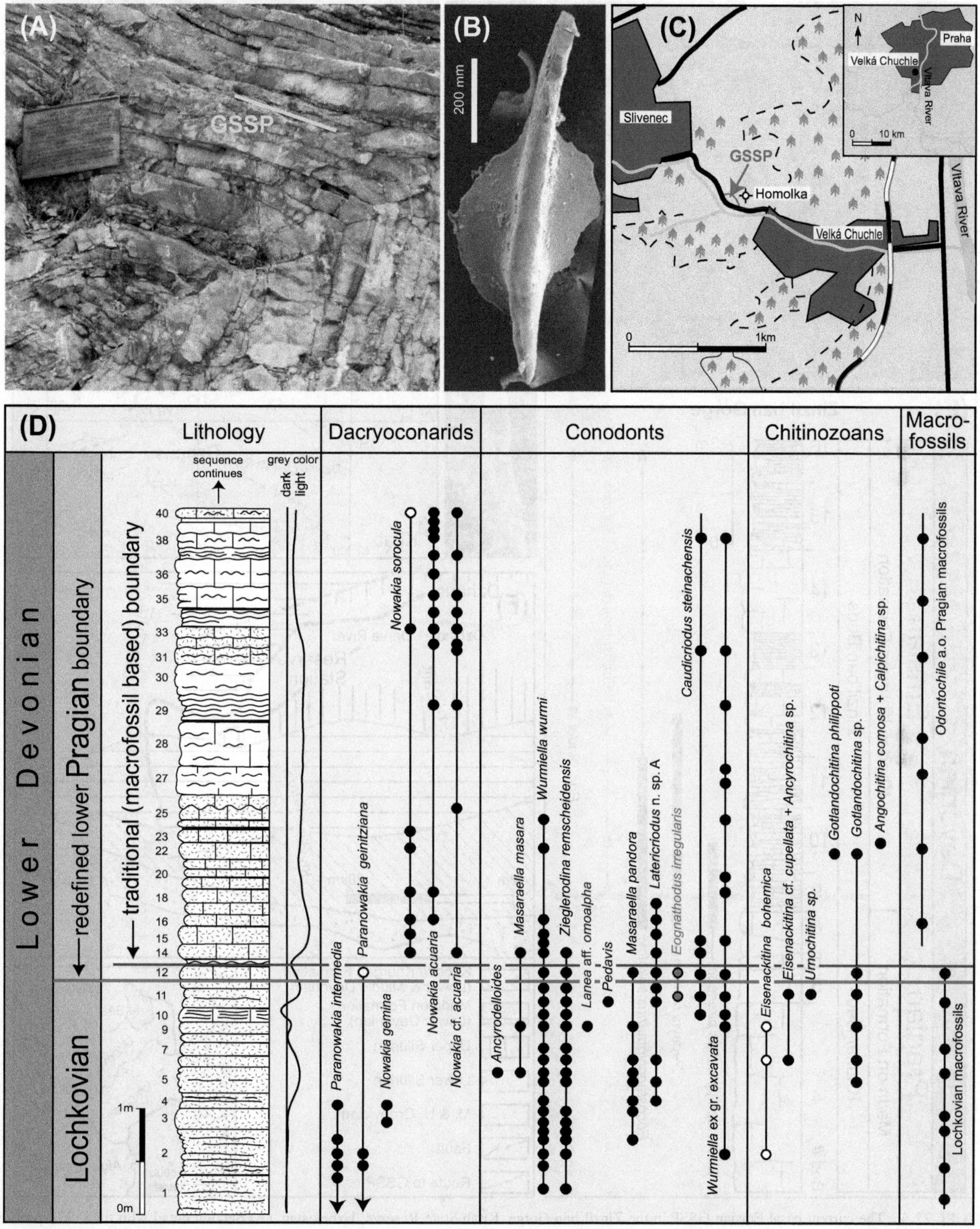

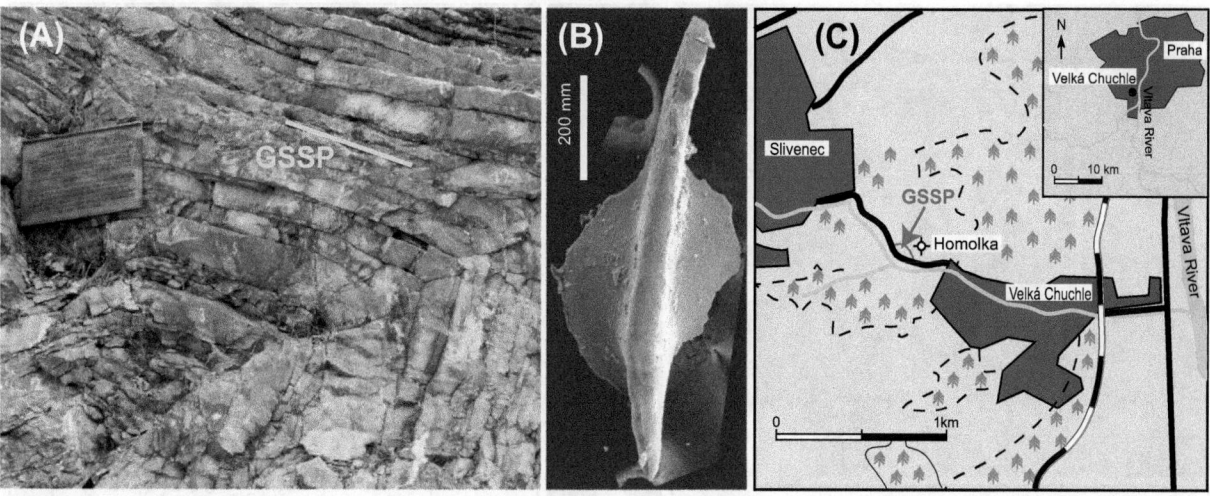

FIGURE 22.4 Basal Pragian GSSP at Velka Chuchle south of Prague, Barrandian, Czech Republic. (A) Section detail around the GSSP level at the base of Bed 12, with the plate explaining the stratotype (photo by L. Slavík); (B) photo of an "*Eognathodus sulcatus* s.l." (*Gondwania irregularis* Gp.) from the GSSP Bed (Slavík and Hladil, 2004, pl. 1, fig. 9); (C) locality map showing the GSSP position; (D) lithological log with ranges of important index fossils in relation to the GSSP (based on data from Chlupáč and Hladil, 2000, and Slavík and Hladil, 2004; with updated taxonomy).

Base of the Emsian Stage of the Devonian System at Zinzil'ban Gorge, Uzbekistan.

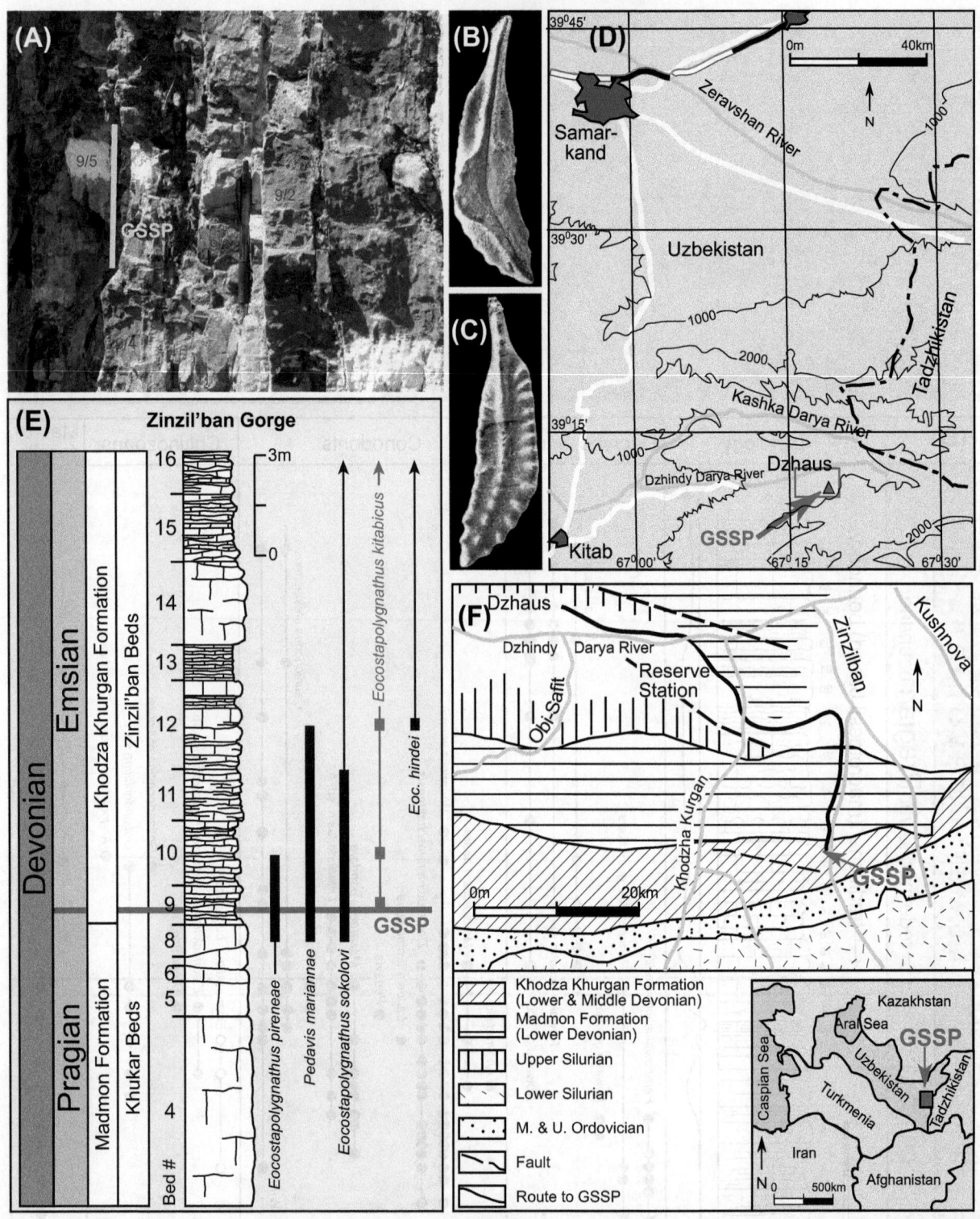

FIGURE 22.5 The current basal Emsian GSSP in the Zinzil'ban Gorge, Kitab State Reserve, Uzbekistan. (A) Section detail with the GSSP position at the base of Bed 9-5; (B/C) lower and upper views of the defining taxon, *Eocostapolygnathus kitabicus*, from the GSSP Bed (Yolkin et al., 1994, pl. 1, figs. 1−2); (D) geographic position of the Kitab region and GSSP location in eastern Uzbekistan; (E) lithological log with ranges of marker conodonts in relation to the GSSP (based on data in Yolkin et al., 2008; generic taxonomy updated); (F) geographic position of the GSSP within the Zinzil'ban Gorge.

proposed to be raised to a level near the entry of *Eolinguipo-lygnathus excavatus* Morphotype 114 sensu (Carls and Valenzuela-Ríos, 2002; Becker, 2009). Based on brachiopod-conodont correlations in Celtiberia (Spain), this new conodont level will be close to the traditional Emsian Base. A review of relevant Rhenish brachiopod faunas was provided by Jansen (2016). While it was originally hoped to place the new basal Emsian GSSP also in the Zinzilban area, resampling campaigns yielded only sparse polygnathids from the critical interval and these are separated by faunal gaps (Slavík et al., 2017). This significantly slowed the revision process and led to the conclusion that other Emsian regions, such as the Pyrenees (Valenzuela-Ríos et al., 2018), need to be reconsidered and better investigated.

A major paleoecological changeover in pelagic areas is shown by the gradual reduction and eventual loss of the Grap-toloidea, with the uniserial monograptids becoming extinct within the basal Emsian (of any definition), at the end of the transgressive *atopus* Event (Chlupáč, Lukes, 1999; Becker et al., 2012). Closely above, coiled ammonoids enter (Becker et al., 2019), diversified rapidly, and spread with the transgressive Chebbi Event (Klug et al., 2008; Aboussalam et al., 2015). They became a dominant group in marine facies until their extinction at the close of the Cretaceous. The Czech terms Zlíchovian and Dalejan are currently used informally for regional substages (Fig. 22.2). A formal Lower/Upper Emsian substage boundary shall be placed close to the traditional Zlíchovian—Dalejan boundary (Chlupác and Lukes, 1999), marked by the extinction of typical lower Emsian goniatites (*Anetoceras* faunas), and close to the entry of the widespread dacryoconarid *Now. cancellata*. Problems of this substage definition have been summarized by Becker (2007c). Ammonoid (Becker et al., 2010) and dacryoconarid data (Kim, 2011) suggest a short overlap of typical lower Emsian goniatites with the first *Now. cancellata* in Uzbeki-stan. The Daleje Shale of Bohemia marks a transgressive interval that is widely recognized in Europe and North Africa and gave rise to the term Daleje Event (sensu Chlupáč and Kukal, 1988). Other authors (Ferrová et al., 2012) have used the term in a wider sense, including the older, separate transgressive phase of the Upper Zlíchov Event sensu García-Alcalde (1997; see discussion in Aboussalam et al., 2015 and Tonarová et al., 2017).

22.1.2 Middle Devonian Series

22.1.2.1 Eifelian

The base of the Middle Devonian Series and of the Eifelian Stage is drawn at a GSSP at Wetteldorf (Fig. 22.6) in the Prüm Syncline of the Eifel Mountains of Germany (Ziegler and Werner, 1982; Ziegler, 2000). The recommendation of the SDS was ratified at meetings of IUGS held with the ICS

in Moscow in 1984. The primary correlation marker for this boundary is the junction of the *Polygnathus patulus* and *P. partitus* conodont zones (Weddige, 1982), which lies somewhat below the anoxic pulse of the global Choteč Event that occurred near the end of the *partitus* Zone (e.g., Berkyová, 2009; Koptíková, 2011; Becker and Aboussalam, 2013a; Brocke et al., 2016). The GSSP is in a trench (Richtschnitt) and the locality is protected by a building (the Happel Hut) erected in 1990 by the Senckenbergische Naturforschende Gesellschaft. As this boundary is in the Eifel Hills, its definition involved only a minor change from historical usage of the term Eifelian in the area. There are informal subdivisions of the Eifelian into lower and upper substages. Currently, it is discussed within the SDS to use the base of the *Tortodus australis* Zone as the future substage (see Fig. 22.2) boundary but there has been no formal vote on this yet. The multiphase, global Kačák Event (e.g., Budil, 1995; House, 1996a; Schöne, 1997; DeSantis et al., 2007; Walliser and Bultynck, 2011; Königshof et al., 2016; Kabanov, 2018) occurred at the end of the Eifelian. It includes a major transgressive pulse (e.g., House, 1985; Chan et al, 2017) and has been traced in terrestrial environments of the Old Red Continent (Marshall et al., 2007).

22.1.2.2 Givetian

The base of the Givetian Stage is drawn at a GSSP in southern Morocco, at Jebel Mech Irdane (Fig. 22.7) in the western Tafilalt Area of the eastern Anti-Atlas, 12 km southwest of Rissani (Walliser et al., 1995; Walliser, 2000; Ellwood et al., 2003, 2011a; Schmitz et al., 2006; Walliser and Bultynck, 2011; Becker et al., 2018c). The primary correlation marker is the base of the *Polygnathus hemiansatus* conodont zone, corresponding to an upper part of the former *Polygnathus ensensis* Zone. *Icriodus obliquimarginatus* is an additional important marker conodont, especially in neritic successions. The oldest maenioceratid ammonoids (*Bensaidites koeneni*), the goniatite group that traditionally coined the term *Maenioceras* Stufe, commence one bed below the boundary. Based on evidence from the Eifel region, the spore *Geminospora lemurata* (known in situ from the progymnosperm *Archaeopteris*, an important component of Middle and Late Devonian forests) enters slightly above the boundary (Streel et al., 2000a).

The SDS's recommendation was ratified at an IUGS meeting in London in 1994. The original Givet Limestone and "Assisse de Givet" of the Ardennes gave the name to the stage, which had been defined in several different ways, based either on local neritic characters or on inferred correlation with fauna in pelagic areas. The lower boundary of the present stage is not too far below the base of the Givet Limestone and very close to the entry of the classical Givetian brachiopod marker *Stringocephalus* (Bultynck

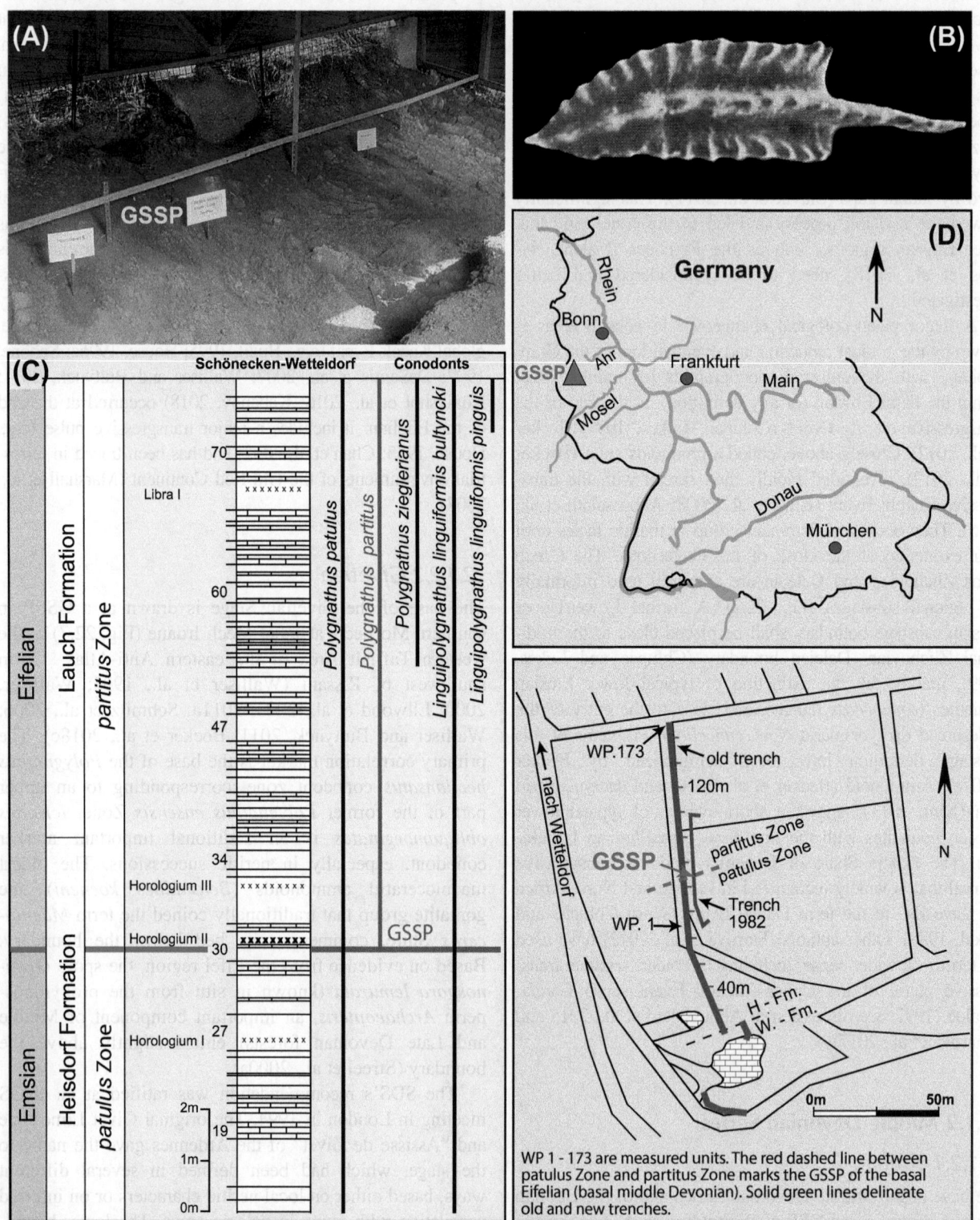

Base of the Eifelian Stage (Base of the Middle Devonian Series) of the Devonian System near Schönecken-Wetteldorf, Germany

FIGURE 22.6   Basal Eifelian (basal Middle Devonian) GSSP in the sheltered trench at Wetteldorf, Eifel Mountains, Germany. (A) Section details within the Happel Hut, with the marked GSSP position (photo by K. Weddige); (B) upper view of a specimen of the index conodont, *Polygnathus partitus*, from the GSSP bed (photo by K. Weddige); (C) lithological log showing the range of marker conodonts in relation to the GSSP; (D) geographic position of the trench close to Wetteldorf.

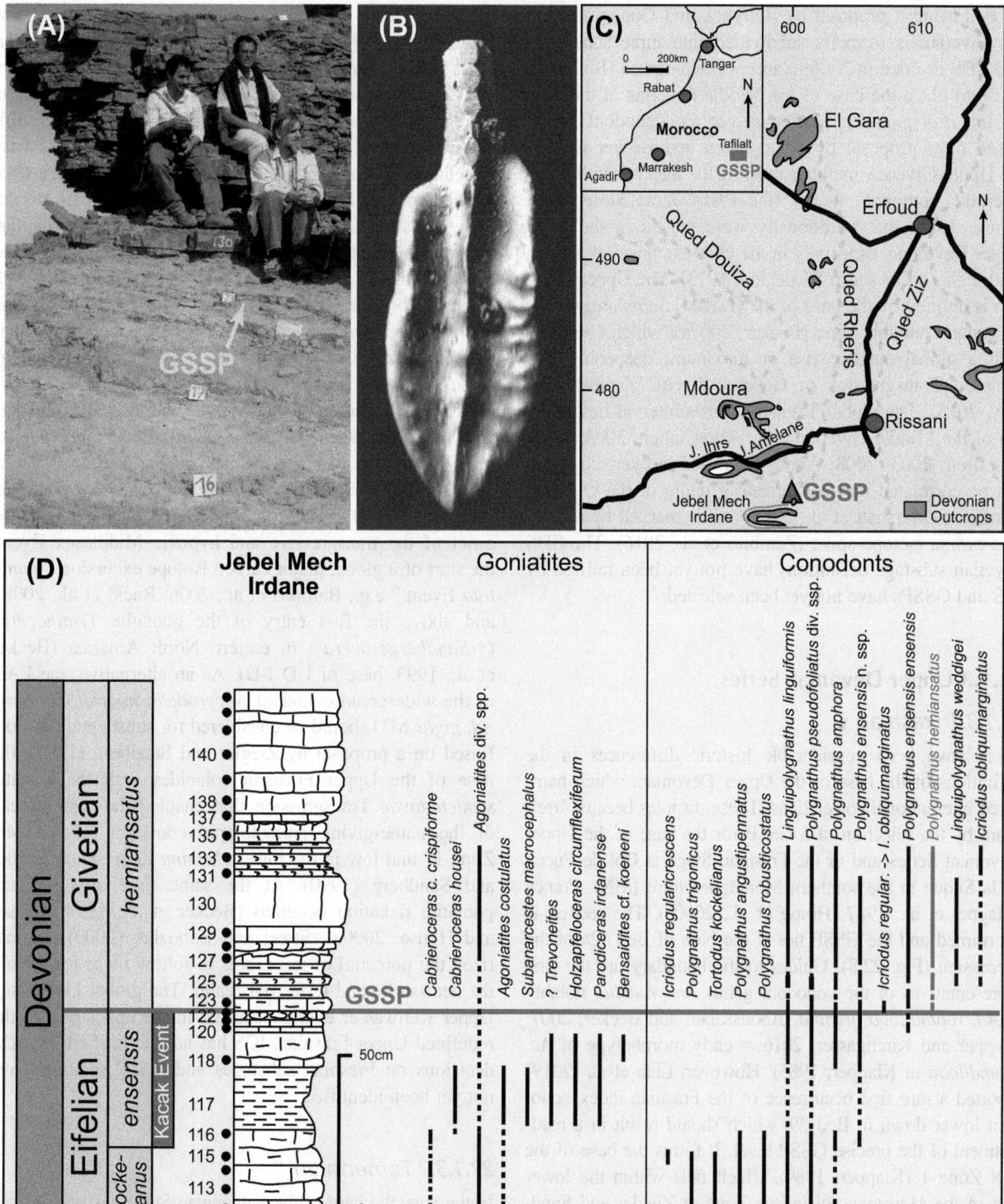

FIGURE 22.7 Basal Givetian GSSP in the western part of Jebel Mech Irdane, western Tafilalt, Anti-Atlas, SE Morocco. (A) Section details showing the position of the GSSP near the base of the upper cliff; (B) upper view of a specimen of the index conodont, *Polygnathus hemiansatus*, from the GSSP Bed (Walliser, 2000, fig. 3C3); (C) geographic position of Jebel Mech Irdane within the northern Tafilalt; (D) lithological log showing the range of marker goniatites and conodonts around the GSSP (taxonomy updated).

and Hollevoet, 1999). In Morocco, the boundary falls in the top part of the polyphase Kačák Event interval (Walliser and Bultynck, 2011).

Following a proposal by Bultynck and Gouwy (2002), the Givetian is formally subdivided into three substages. The SDS decided in 2006 to accept the proposal (Bultynck, 2005) to place the base of the Middle Givetian at the base of the *Polygnathus rhenanus–varcus* conodont zone. Based on a proposal by Aboussalam and Becker (2002), the Upper Givetian includes most of the interval with pharciceratid ammonoid faunas (the *Pharciceras* Stufe sensu House, 1985) that traditionally were placed in the basal Upper Devonian, especially in the German classical ammonoid scale (do Iα sensu Wedekind, 1913). The Upper Givetian is defined by the base of the (Lower) *Schmidtognathus hermanni* conodont zone (Becker, 2007a), which coincides with a global eustatic rise or maximum deepening, the Geneseo Transgression or Geneseo Event (Zambito and Day, 2015). The global Taghanic Crisis interval lies at the top of the Middle Givetian (e.g., Aboussalam, 2003; Baird and Brett, 2003, 2008; Aboussalam and Becker, 2011). It can be correlated into the terrestrial realm of the Old Red Continent (Marshall et al., 2011) and is marked by a positive carbon isotope spike (Zambito et al., 2016). The SDS Givetian substage definitions have not yet been ratified by ICS and GSSPs have not yet been selected.

22.1.3 Upper Devonian Series

22.1.3.1 Frasnian

There have been considerable historic differences in the definitions of the base of the Upper Devonian, which hampered international correlations. These disputes became irrelevant by the selection of a GSSP for the base of the Upper Devonian Series and of the Frasnian Stage at Col de Puech de la Suque in the southern Montagne Noire (MN), France (Klapper et al., 1987; House et al., 2000a). The section is overturned and the GSSP lies at the base of Bed 42a of the succession (Fig. 22.8). Guides to the boundary are the first representatives of the conodont genus *Ancyrodella*, notably of *Ad. rotundiloba pristina* (Aboussalam and Becker, 2007; Klapper and Kirchgasser, 2016; = early morphotype of *Ad. rotundiloba* in Klapper, 1985). However, Liao et al. (2019) reported a rare first occurrence of the Frasnian index conodont lower down in Bed 39, which should result in a readjustment of the precise GSSP level. It forms the base of the MN Zone 1 (Klapper, 1989), which falls within the lower part of the *Mesotaxis falsiovalis* Zone of Ziegler and Sandberg (1990). The zonal index species of the latter has been regarded as a subjective junior synonym of the Chinese *Mesotaxis guanwushanensis* (Aboussalam and Becker, 2007) but this was rejected by Bardashev and Ovnatanova (2018). Both taxa are closely related and the type-level of

M. falsiovalis lies much above the Frasnian base, which questions the widespread use of the term *falsiovalis* Zone. In the type area for the naming of the stage (Frasnes, Belgium), the GSSP level almost coincides with the base of the Frasnes Group. The entry of the goniatite genus *Neopharciceras* occurs immediately above the GSSP in Bed 43 (Korn, 1992 cited in House et al., 2000a), and this provides a useful correlation with some Asian successions. There are no sufficient brachiopod or palynomorph data that enable an easy correlation of the GSSP level into neritic and terrestrial clastic deposits. The polyphase Frasnes Events (e.g., Becker and Aboussalam, 2004) ranges from the topmost Givetian (*norrisi* Zone) into the MN Zone 2, with a maximum spread of black shales and dacryoconarid mass occurrences well above the boundary (Aboussalam and Becker, 2007). Within the extended event interval, the base of the Frasnian is marked by a minor transgression–regression couplet (e.g., Aboussalam and Becker, 2007).

The SDS decided to formally subdivide the Frasnian into three substages (Becker, 2007a). Becker and House (1999) proposed to place the base of the Middle Frasnian at the base of the MN 5 or *Palmatolepis punctata* conodont zone (Fig. 22.2). However, this level lies just above the onset of the transgressive and hypoxic Middlesex Event, the start of a global major carbon isotope excursion ("*punctata* Event," e.g., Balinski et al., 2006; Racki et al., 2008), and above the first entry of the goniatite *Triainoceras* (=*Sandbergeroceras*) in eastern North America (Becker et al., 1993, base of UD I-D). As an alternative, the FAD of the widespread conodont *Ancyrodella nodosa* (formerly *Ad. gigas* M1) should be considered for substage definition. Based on a proposal by Ziegler and Sandberg (1997), the base of the Upper Frasnian coincides with the eustatic *semichatovae* Transgression that enabled the wide spread of the name-giving palmatolepid conodont low in MN Zone 11 and low in the lower *rhenana* Zone sensu Ziegler and Sandberg (1990). At the same time, a significant goniatite radiation occurred (Becker et al., 1993; Becker and House, 2000a). Streel and Loboziak (2000) summarized the potential of miospores to follow these levels into the near-shore and terrestrial realm. The global Lower and Upper Kellwasser events occurred in the upper part of the redefined Upper Frasnian. ICS has not yet ratified the SDS decisions on Frasnian substages and GSSP sections have not yet been identified.

22.1.3.2 Famennian

In the past, the base of the Famennian Stage has been placed at different levels, reflecting problems of correlation between faunal groups. The definition for the base of the Famennian proposed by the SDS was ratified by the IUGS in 1993. The GSSP is very close to the base of the Famennian as formerly used in the Famenne Area in Belgium (Bultynck and Martin,

Base of the Frasnian Stage of the Devonian System at Col du Puech de la Suque, Montagne Noire, France.

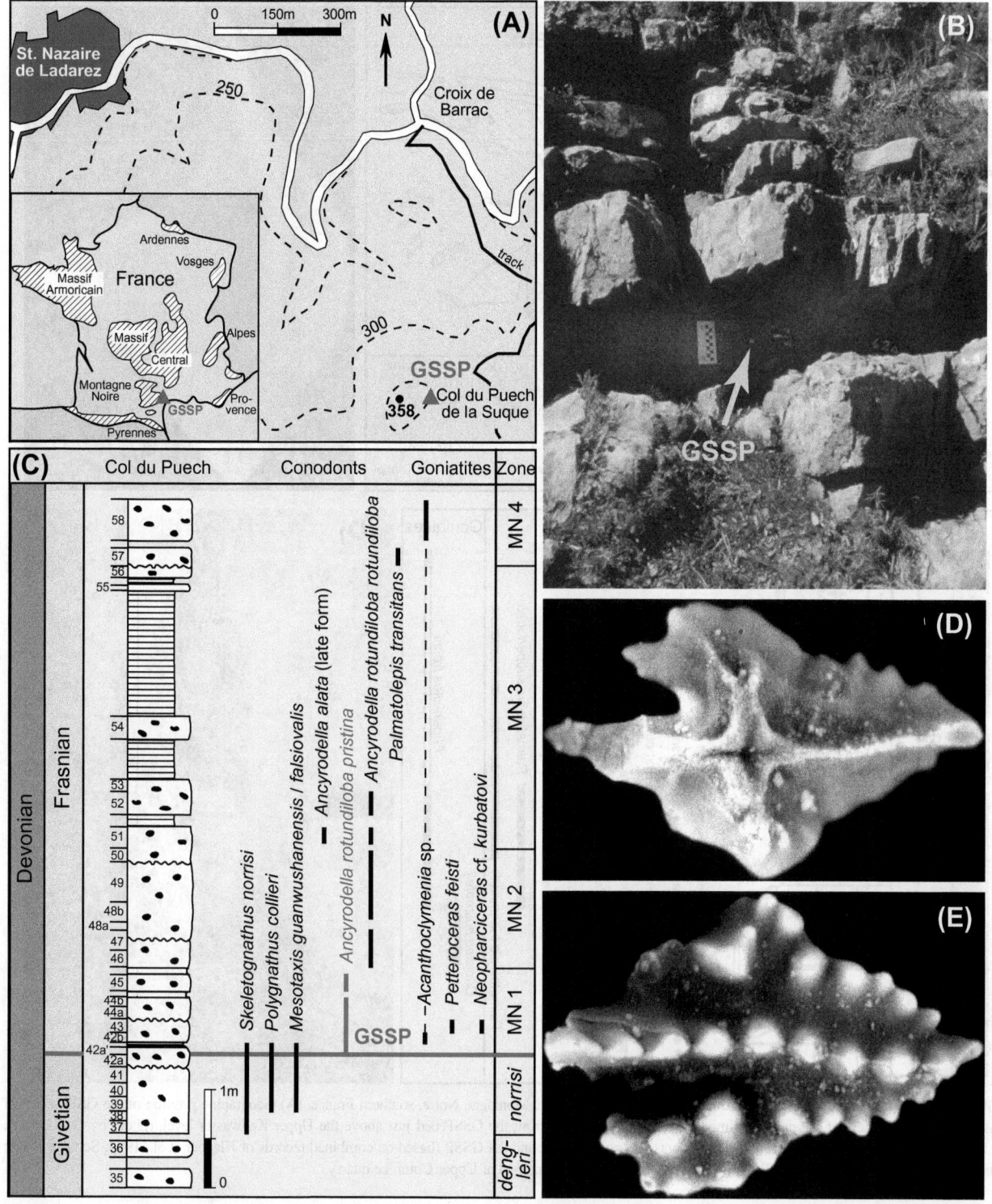

FIGURE 22.8 Basal Frasnian (basal-Upper Devonian) GSSP at Col de Puech de la Suque, Montagne Noire, southern France. (A) Geographic position of the GSSP within the Mont Peyroux Nappe SE of St. Nazaire de Ladarez; (B) section details with the (current) GSSP position in a recessive part of the succession (photo by R. Feist); (C) lithological log with the ranges of marker conodonts and goniatites around the GSSP (with updated taxonomy; note that Liao et al., 2019 recently reported a lower range extension for *Ad. rotundiloba pristina* down into Bed 39', and a lower range extension for *Skeletognathus norrisi* down into Bed 35); (D/E) representative adult specimen of the index conodont, *Ancyrodella rotundiloba pristina*, from the (current) GSSP bed (photos by G. Klapper).

Base of the Famennian Stage of the Devonian System near the Upper Coumiac Quarry in Montagne Noire, France

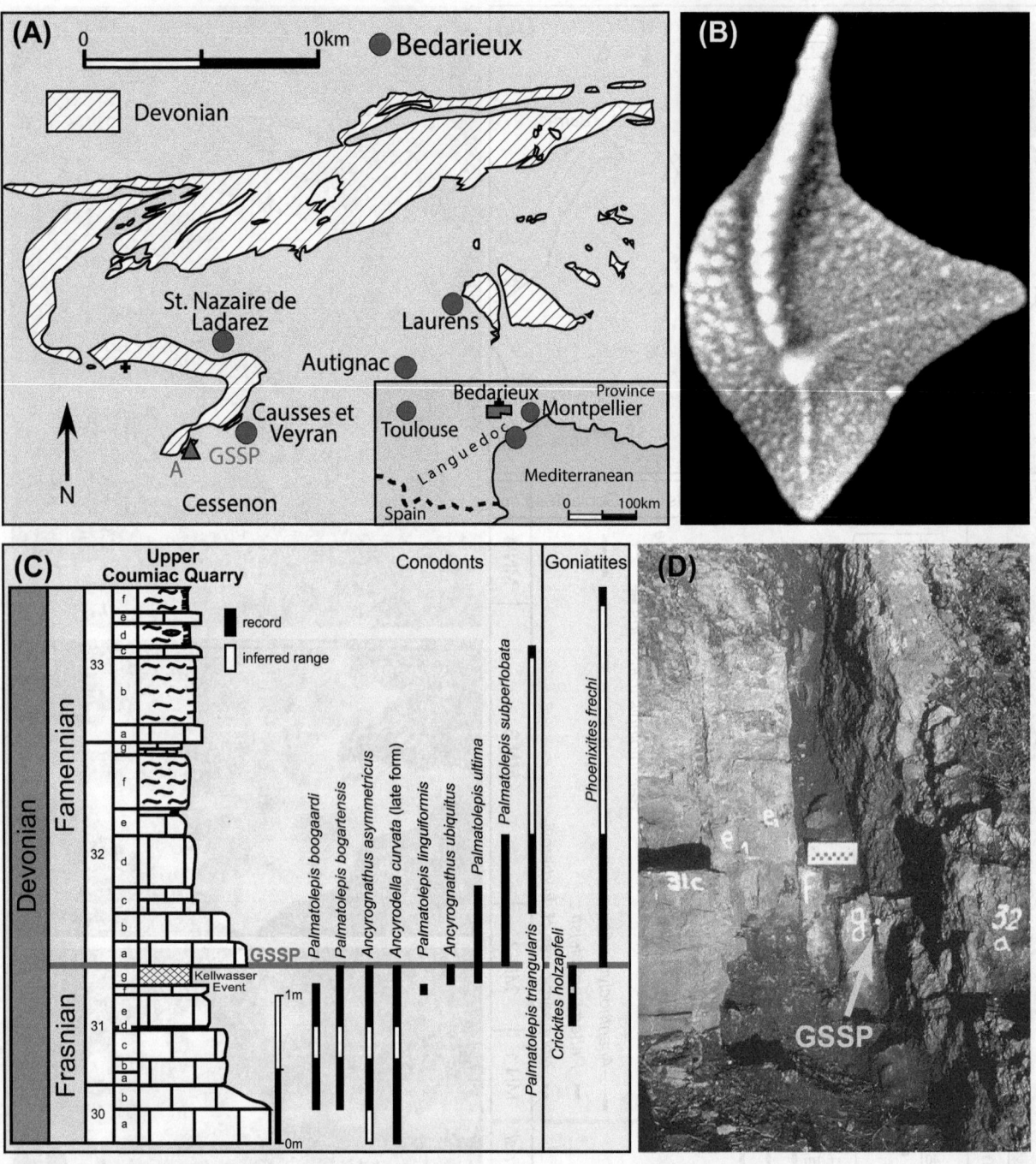

FIGURE 22.9 Basal Famennian GSSP in the Upper Quarry at Coumiac, Montagne Noire, southern France. (A) Geographic position of the GSSP north of Cessenon; (B) upper view of a typical specimen of *Palmatolepis ultima* from the GSSP bed just above the Upper Kellwasser level, (photo by G. Klapper); (C) lithological log with the ranges of marker conodonts and goniatites around the GSSP (based on combined records of Klapper et al., 1993; Schülke, 1995; House et al., 2000b; Girard et al., 2005); (D) outcrop photo of the GSSP interval at Upper Coumiac quarry.

1995). It lies above the Upper Coumiac Quarry (Fig. 22.9), N of Cessenon, in the MN, southern France (Becker et al., 1989; Klapper et al., 1993; House et al., 2000b). The boundary falls at the junction of the *Palmatolepis linguiformis* (or top of MN Zone 13c, Girard et al., 2005) and traditional lower *Palmatolepis triangularis* conodont zones (Klapper, 2000; Klapper et al., 2004). The GSSP does not coincide with the entry (FAD) of *Pa. triangularis*, but with the extinction of Frasnian palmatolepids (especially of *Palmatolepis bogartensis*), ancyrodellids, and ancyrognathids, and the FAD of *Pa. subperlobata* in beds with a sudden flood occurrence of *Palmatolepis ultima*. Therefore the basal Famennian conodont zone was renamed by Spalletta et al. (2017) as *subperlobata* Zone, with the entry of *Pa. triangularis* s.str. as the index taxon of a slightly higher level (second Famennian conodont zone or subzone). The Gephuroceratidae and Beloceratidae, the last families of the order Agoniatitida (suborder Gephuroceratina), became extinct at the Frasnian-Famennian (F/F) boundary. Most tornoceratid lineages became Lazarus Taxa, only the opportunistic genus *Phoenixites* flourished (Becker, 1993a). The Lower Famennian marker genus *Cheiloceras* enters significantly higher, which proved a long delay of post-event recovery. The old terms *Cheiloceras* Stufe (do II) or the regional Nehden-Stufe of Germany, therefore, did not include the basal Famennian but it has been suggested to harmonize both terms and to reassign the so-called post Iδ interval or the *Ph. frechi* Zone to the do II (Becker 1993a; Schindler et al., 2018).

At the GSSP section, both anoxic pulses of the Lower and Upper Kellwasser events are present; the GSSP level is immediately above the latter. It marks the main extinction of nektonic organisms, such as ammonoids, conodonts, and various fish groups (placoderms, agnathans) that bloomed in general in the hypoxic Upper Kellwasser Beds. The boundary falls within a global positive carbon isotope excursion (e.g., Joachimski et al., 2001, 2002; Buggisch and Joachimski, 2006). It is marked by a sudden climatic cooling indicating by a positive oxygen isotope spike of conodont apatite (e.g., Joachimski and Buggisch, 2002; Balter et al., 2008; Le Houedec et al., 2013; Joachimski et al., 2009; discussion in Hartenfels et al., 2016). There was no increase of the micrometeoritic flux at the time, only an eccentricity-related increase of cosmic dust deposition just below the Upper Kellwasser Limestone (Schmitz et al., 2019). The F/F boundary extinction occurred separately from the basal Upper Kellwasser extinction. It is a precise and global stratigraphic marker, independent from the development of organic-rich Kellwasser Beds (e.g., Becker et al., 1991; George et al., 2014; Hartenfels et al., 2016). It has been recognized variably in shallow neritic (e.g., Sandberg et al., 1988; Wang and Geldsetzer, 1995; Kirilishina, 2006; Tagarieva, 2012), post-reefal (e.g., Ji, 1989a; Klapper, 2007b; Becker et al., 2016c; Weiner et al., 2017), pelagic platform/ramp/seamount (e.g., Schindler, 1990, 1993; Girard et al., 2005; Savage et al., 2006; Gereke and Schindler, 2012; Chang et al., 2017; Becker et al., 2018d), slope (e.g., Ziegler and Sandberg, 1990; Yudina et al., 2002; Huang and Gong, 2016), or argillaceous deep-water facies (e.g., Over, 1997; Klapper et al., 2004; Gereke, 2007; Hartkopf-Fröder et al., 2007; Zhang et al., 2008; Hartenfels et al., 2013; Becker et al., 2016g; Huang et al., 2018). Right at the F/F boundary there are widespread unconformities caused by a major regression and/or

high-energy depositional events triggered by synsedimentary tectonics (e.g., Racki, 1998; Becker, 2018). The regression caused also a sharp signal in magnetic susceptibility (MS) studies (e.g., Riquier et al., 2007; Weiner et al., 2017).

Based on a proposal by Streel et al. (1998) and after a prolonged discussion, SDS decided in 2003 to formally subdivide the Famennian into four substages. The old German classification with UD II (Nehdenian), III/IV (Hembergian), V (Dasbergian), and VI (Wocklumian) (Fig. 22.2) was not adopted internationally but has been used in pelagic settings of widely distant basins of several continents (e.g., in Australia and in Russian basins). There are several proposals for the precise position of substage boundaries (Becker, 1998; Sandberg and Ziegler, 1999; Streel and Loboziak, 1998; Streel, 2005; Hartenfels et al., 2009; review by Becker, 2013), but the SDS has not yet formally voted on these. There are good arguments to place the base of the future Middle and Upper Famennian substages at transgressive levels that enabled the global spread of marker conodonts and ammonoids (e.g., slightly above the base of the *Palmatolepis marginifera marginifera* Zone, base of global Lower *Annulata* Event = base of Famennian IV of German regional chronostratigraphy).

It has been proposed to place the future Uppermost Famennian at the base of the former Upper *Palmatolepis gracilis expansa* conodont zone (Streel, 2005), which now should be called *Bispathodus ultimus ultimus* Zone (Söte et al., 2017). It allows an approximation with the widely used but not formally recognized Strunian stage of regional Ardennes chronostratigraphy (Streel et al., 2006; Fig. 22.2). *Bispathodus ultimus ultimus*, *Palmatolepis gracilis gonioclymeniae*, and *Pseudopolygnathus trigonicus* are the main index conodonts but the documentation of their precise entries in potential GSSP sections is still ongoing. With outstanding precision in strictly cyclic successions, Bartzsch and Weyer (2012) and Kononova and Weyer (2013) published relevant data from Thuringia. Streel (2009) provided an update of Famennian miospore−conodont correlation; the proposed Uppermost Famennian base would fall within the LL Zone, below a morphometric change within the index miospore *Retispora lepidophyta* (from *l. lepidophyta* to *l. minor*; Maziane et al., 2002; Streel, 2015). Just slightly lower within the *ultimus ultimus* Zone, the widespread index foraminifers *Quasiendothyra kobeitusana* and *Q. konensis*, the markers of DFZ 7, enter in the Ardennes (e.g., Poty et al., 2006) or in the Urals (Kulagina, 2013). They enable a correlation of shallow-water carbonates as far as Russia and South China. Becker et al. (2016b) provided an update of Uppermost Famennian ammonoid zones and their conodont correlation.

22.1.3.3 Base of the Carboniferous

Following the recommendations of a special working group set up by ICS, the IUGS, accepted in 1989 a GSSP for the

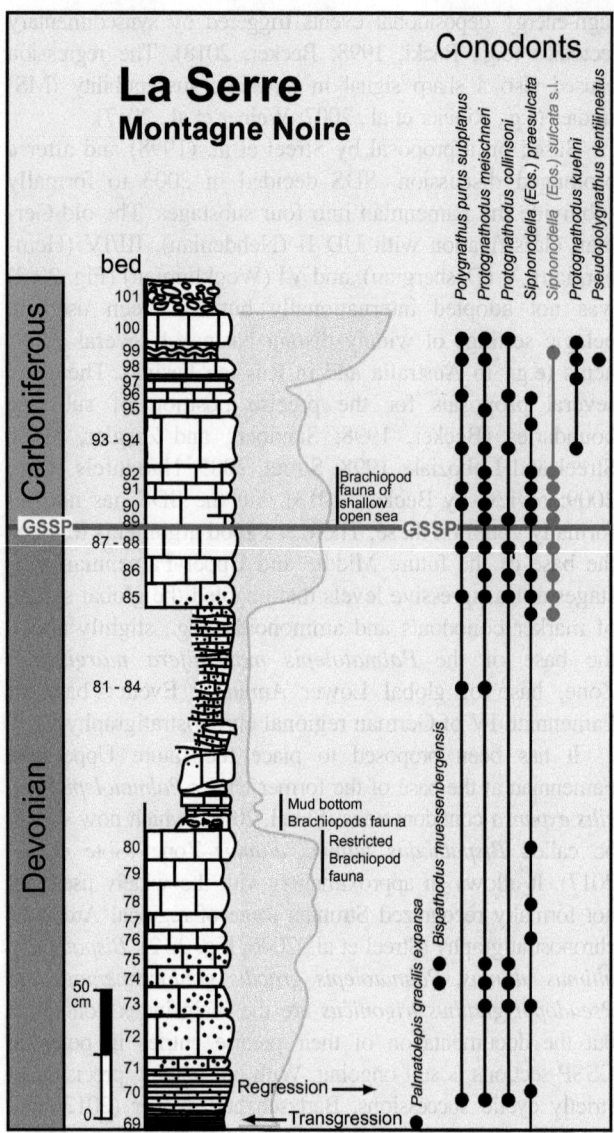

FIGURE 22.10 Lithological log and ranges of index conodonts in relation to the current Devonian-Carboniferous boundary GSSP level at La Serre, Trench E, Montagne Noire, southern France. More section details are given in Aretz et al. (2020, Ch. 23: The Carboniferous Period, this volume).

Devonian-Carboniferous (D/C) boundary at La Serre (Fig. 22.10) near Cabrières, in the MN, southern France (Paproth et al., 1991; Feist et al., 2000). The GSSP guide was the entry of the conodont *Siphonodella (Eosiphonodella) sulcata* in a supposed gradual transition from *S. (E.) praesulcata* (Flajs and Feist, 1988). However, the taxonomy of the group is currently problematic and subjective (Kaiser and Corradini, 2011). The type of the Carboniferous index species has been lost, and it is possibly from a much younger Lower Tournaisian level (Evans et al., 2013). Even the *S. (E.) praesulcata* holotype is from a level above the main Hangenberg extinction (top of Unit 5 of the Montana Hangenberg Crisis interval, which postdates the LN Zone; see Kaiser et al., 2016; Rice

et al., 2017), not from the preextinction *praesulcata* Zone. Kaiser (2009) showed that *S. (E.) sulcata* morphotypes identical with those that were used to define the GSSP occur significantly lower at La Serre and that the assumed *S. praesulcata-sulcata* lineage is not preserved in partly reworked faunas from oolites. The current GSSP level lies high in the *sulcata* Zone (s.l.) and cannot be correlated with precision into any other section but most likely with a level above the entry of the classical Carboniferous goniatite *Gattendorfia* elsewhere. Hence, a GSSP revision became inevitable (Kaiser and Becker, 2007; Kaiser and Corradini, 2008). During the D/C Boundary Task Group Meeting in Montpellier in autumn 2016, it was decided to focus the ongoing search for a new GSSP level on the interval of the early postglacial transgression, above the siliciclastic interval of the Hangenberg Regression (Hangenberg Sandstones and equivalents), on the *Protognathodus kockeli* Zone, early re-radiation of fossil groups, and Lower Stockum Limestones. This is the Upper Hangenberg Crisis Interval sensu Becker (2016b) below the entry of Carboniferous-type siphonodellids and below the current GSSP level. However, the taxonomy of *Pr. kockeli*, especially its gradual evolution from the ancestral *Protognathodus collinsoni* and the type level of its Rhenish types, requires revision and refinements (Corradini et al., 2011; Hartenfels and Becker, 2018, 2019). Calcareous foraminifers, especially the onset of the *Tornayellina pseudobeata* Zone (DFZ 8), are important for the correlation of the interval into neritic carbonates (Poty et al., 2006; Kulagina, 2013; Kalvoda et al., 2013).

Palynologically, the D/C transition is well marked by the last appearance of the globally distributed *Retispora lepidophyta*, which disappears together with other taxa (e.g., all the bifurcate tipped spores) at the LN/VI zone boundary just beneath the first occurrence of *Siphonodella sulcata* (Higgs et al., 1993) in the Hasselbachtal auxiliary stratotype section (Becker and Paproth, 1993; Becker, 1996). The LN Zone enters below with the Hangenberg Black Shale (Higgs and Streel, 1994). A second auxiliary D/C stratotype was designated in more neritic facies at Nanbiancun, NW of Guilin, Guangxi Province, South China (Yu, 1988), but there is a dispute concerning its conodont stratigraphy and correlation with the La Serre GSSP (e.g., Ji, 1989b; Gong et al., 1991). The main global extinction (main Hangenberg Event), especially of goniatites, clymeniids, trilobites, and corals, occurred below the boundary and coincided with the initial anoxic pulse (see review by Kaiser et al., 2016), a peak of seawater temperatures, and with the onset of a positive carbon isotope excursion (Kaiser et al., 2006, 2011; Kumpan et al., 2013). The short (50–100 kyr, Myrow et al., 2013) but significant main glacial phase (e.g., Streel et al., 2000a; Caputo et al., 2008; Lakin et al., 2016) and major sea-level fall (Hangenberg Regression) followed the pelagic mass extinction and black shale event. It forms the Middle Crisis

Interval sensu Becker et al. (2016b). Isaacson et al. (2008) and Wicander et al. (2011) suggested several upper/uppermost Famennian glacial episodes in Bolivia but this was rejected by Streel et al. (2013); for a detailed discussion see Lakin et al. (2016). It is still a matter of debate whether the extinction of neritic faunas occurred later than in the pelagic realm, with the major Hangenberg Regression (e.g., Denayer et al., 2019). The gradual decline of *R. lepidophyta* in miospore assemblages led to the recognition of an upper "atypical LN Zone" with rare *lepidophyta* but in an assemblage of increased spore diversity (Streel, 2015). The Upper Crisis Interval, the postglacial topmost Famennian Transgression and peak of isotopic excursion (e.g., Cramer et al., 2008; Day et al., 2011; Kumpan et al., 2014; Qie et al., 2015; Cole et al., 2015) in the *P. kockeli* Zone (=Upper *praesulcata* Zone), saw the initial radiation of Carboniferous-type ostracods, conodonts, ammonoids, trilobites, brachiopods, and corals (e.g., Korn et al., 1994; Kaiser et al., 2016). The International D/C Boundary Task Group decided in August 2019 to use this interval to define a future, revised Devonian-Carboniferous boundary.

22.2 Devonian stratigraphy

22.2.1 Biostratigraphy

There are refined zonations in pelagic facies using especially conodonts, ammonoids, entomozoid ostracods, dacryoconarids, and, for parts of the Lower Devonian, monograptid graptolites. While the original aim was to establish globally applicable "standard zones," it has become more evident in recent years that it is equally important to work out variably less or more detailed regional zonations. These may make use of endemic taxa and can be combined in complex global correlation schemes. The SDS has devoted special attention to the correlation of chronostratigraphic levels across facies boundaries and from tropical areas toward higher latitudes (e.g., Becker and Kirchgasser, 2007). Biozonations based on different faunal groups can be combined with physical stratigraphy levels in holostratigraphic charts (e.g., Becker, 1996 for the D/C boundary).

The acritarch zonation is less detailed than the conodont or ammonoid scales but of high importance in shelf siliciclastics. For neritic facies, brachiopods, trilobites, and ostracods of the Eifelian ecotype are regionally important (e.g., Jansen, 2016), but faunas tend to have many endemic characteristics. Chitinozoa occur both in the pelagic and in the neritic facies and both in tropical and boreal climates. In the terrestrial facies, miospores, macroplants, and various fish groups are stratigraphically useful. In near-shore facies, miospores provide an important tool for the correlation with open marine zonations. A listing of biozones and their general correlation is given in Figs. 22.11 and 22.12, which elaborate data published in Bultynck (2000b).

22.2.1.1 Conodont zonations

Starting with the proposal for a conodont-based stage tripartition and with a new zonation for the middle part by Valenzuela-Ríos and Murphy (1997), the Lochkovian conodont zonation has been much revised and refined (Fig. 22.11). Following the revision by Carls et al. (2007), the Lower Lochkovian comprises now mostly the *Cypricriodus hesperius* and *Caudicriodus postwoschmidti* zones (Corradini and Corriga, 2012; Becker et al., 2012, 2016a), with *Caudicriodus transiens* as a marker for highest parts. The new *Zieglerodina petrea* Hušková and Slavík (2019) is an alternative basal Devonian marker conodont for limestone succession. There is a different regional subdivision for Bohemia (Slavík et al., 2012) and small disagreement whether the middle Lochkovian shall begin with *Lanea omoalpha* (Valenzuela-Ríos and Murphy, 1997; Becker et al., 2012) or *Ancyrodelloides/Lanea carlsi* (Slavík et al., 2012; Corradini and Corriga, 2012). The highest time-resolution is reached in the middle Lochkovian, with up to seven zones/subzones defined by members of the evolving *Lanea/Ancyrodelloides* lineages (e.g., Van Hengstum and Gröcke, 2008; Mavrinskaya and Slavík, 2013; Valenzuela-Ríos et al., 2015). The upper Lochkovian has only two zones defined by the entries of *Masaraella pandora* morphotype beta and *Pedavis gilberti*.

The Pragian conodont zonation requires further revision after the three traditional zones had to be discarded (Slavík, 2004a; Murphy, 2005; Slavík et al., 2007). Currently, there is no good correlation between Europe, North America, and Asian regions. For example, the endemic *Gagievodus−Vjaloviodus* lineage provides a regional upper Lochkovian−Pragian zonation for Arctic regions (Baranov, 2012). Increasing evidence (e.g., Schönlaub et al., 2017) suggests that *Pelekysgnathus serratus* is a widespread and useful zone fossil, better than *Gondwania kindlei* or *Gondwania profunda*. The oldest polygnathids originated from *Eognathodus* in the higher *kindlei* Zone and this evolutionary event may be used for a zonal subdivision (e.g., Yolkin et al., 2011). As noted above, the next higher *Eoc. kitabicus* Zone shall define in future a formal Upper Pragian Substage.

Principles of the lower Emsian polygnathid zonation were established by Klapper and Johnson (1975). Aboussalam et al. (2015) reviewed the subsequent changes introduced by Mawson (1987, 1995), Bultynck (1989), Yolkin et al. (1994), Bardashev et al. (2002), and Martínez-Pérez et al. (2011). There is a different regional zonation for NE Asia published by Baranov et al. (2014) and Baranov and Blodgett (2016). Weddige (1977) and Klapper et al. (1978) established the currently used upper Emsian polygnathid succession. Vodrážková et al. (2011) granted the index conodonts at the Emsian−Eifelian boundary, *Po. patulus* and *Po. partitus*, full species status. Emsian icriodids are

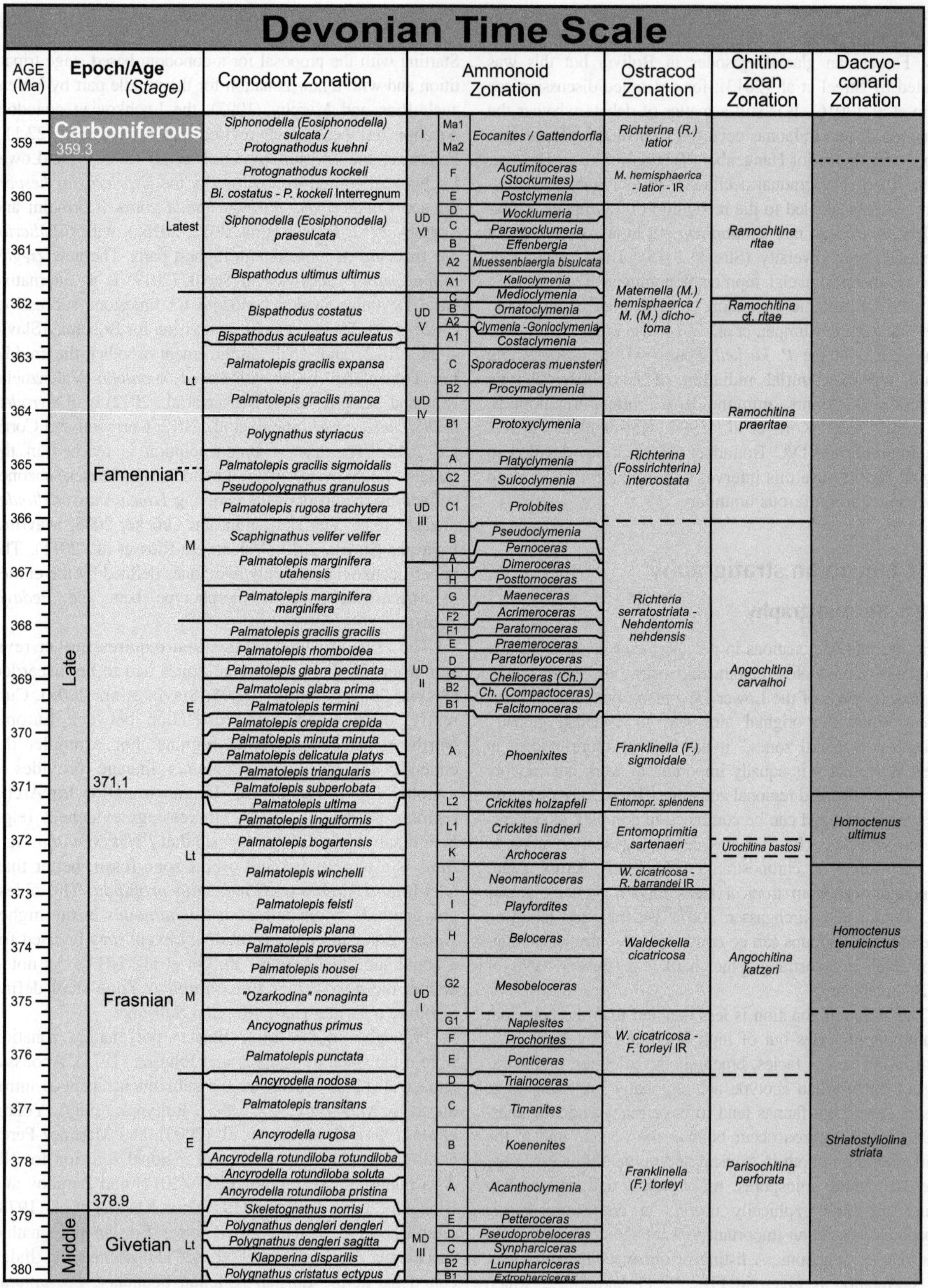

Devonian Time Scale

AGE (Ma)	Epoch/Age (Stage)	Conodont Zonation	Ammonoid Zonation			Ostracod Zonation	Chitinozoan Zonation	Dacryoconarid Zonation
359	**Carboniferous** 359.3	Siphonodella (Eosiphonodella) sulcata / Protognathodus kuehni	I	Ma1 - Ma2	Eocanites / Gattendorfia	Richterina (R.) latior	Ramochitina ritae	
360	Latest	Protognathodus kockeli		F	Acutimitoceras (Stockumites)	M. hemisphaerica - latior - IR		
		Bi. costatus - P. kockeli Interregnum		E	Postclymenia			
		Siphonodella (Eosiphonodella) praesulcata	UD VI	D	Wocklumeria			
361				C	Parawocklumeria			
				B	Effenbergia			
		Bispathodus ultimus ultimus		A2	Muessenbiaergia bisulcata			
362				A1	Kalloclymenia			
		Bispathodus costatus	UD V	C	Medioclymenia	Maternella (M.) hemisphaerica / M. (M.) dicho-toma	Ramochitina cf. ritae	
				B	Ornatoclymenia			
363	Lt	Bispathodus aculeatus aculeatus		A2	Clymenia -Gonioclymenia			
				A1	Costaclymenia			
		Palmatolepis gracilis expansa		C	Sporadoceras muensteri			
364	Famennian	Palmatolepis gracilis manca	UD IV	B2	Procymaclymenia		Ramochitina praeritae	
				B1	Protoxyclymenia			
365		Polygnathus styriacus		A	Platyclymenia	Richterina (Fossirichterina) intercostata		
		Palmatolepis gracilis sigmoidalis		C2	Sulcoclymenia			
366		Pseudopolygnathus granulosus	UD III	C1	Prolobites			
		Palmatolepis rugosa trachytera		B	Pseudoclymenia			
	M	Scaphignathus velifer velifer		A	Pernoceras			
367		Palmatolepis marginifera utahensis		I	Dimeroceras	Richteria serratostriata - Nehdentomis nehdensis		
				H	Posttornoceras			
368		Palmatolepis marginifera marginifera		G	Maeneceras		Angochitina carvalhoi	
				F2	Acrimeroceras			
		Palmatolepis gracilis gracilis		F1	Paratornoceras			
		Palmatolepis rhomboidea		E	Praemeroceras			
369		Palmatolepis glabra pectinata	UD II	D	Paratorleyoceras			
		Palmatolepis glabra prima		C	Cheiloceras (Ch.)			
	E	Palmatolepis termini		B2	Ch. (Compactoceras)			
370		Palmatolepis crepida crepida		B1	Falcitornoceras			
		Palmatolepis minuta minuta						
		Palmatolepis delicatula platys		A	Phoenixites	Franklinella (F.) sigmoidale		
		Palmatolepis triangularis						
371	371.1	Palmatolepis subperlobata						
		Palmatolepis ultima		L2	Crickites holzapfeli	Entomopr. splendens		Homoctenus ultimus
		Palmatolepis linguiformis		L1	Crickites lindneri	Entomoprimitia sartenaeri		
372	Lt	Palmatolepis bogartensis		K	Archoceras		Urochitina bastosi	
		Palmatolepis winchelli		J	Neomanticoceras	W. cicatricosa - R. barrandei IR		
373		Palmatolepis feisti		I	Playfordites			
		Palmatolepis plana		H	Beloceras			Homoctenus tenuicinctus
374		Palmatolepis proversa				Waldeckella cicatricosa	Angochitina katzeri	
		Palmatolepis housei		G2	Mesobeloceras			
375	Frasnian M	"Ozarkodina" nonaginta	UD I					
		Ancyognathus primus		G1	Naplesites			
				F	Prochorites	W. cicatricosa - F. torleyi IR		
376		Palmatolepis punctata		E	Ponticeras			
		Ancyrodella nodosa		D	Triainoceras			
377		Palmatolepis transitans		C	Timanites			
	E	Ancyrodella rugosa		B	Koenenites			Striatostyliolina striata
378		Ancyrodella rotundiloba rotundiloba				Franklinella (F.) torleyi	Parisochitina perforata	
		Ancyrodella rotundiloba soluta		A	Acanthoclymenia			
	378.9	Ancyrodella rotundiloba pristina						
379	Skeletognathus norrisi		E	Petteroceras			
	Middle	Polygnathus dengleri dengleri		D	Pseudoprobeloceras			
	Givetian Lt	Polygnathus dengleri sagitta	MD III	C	Synpharciceras			
380		Klapperina disparilis		B2	Lunupharciceras			
		Polygnathus cristatus ectypus		B1	Extropharciceras			

(Left margin shows "Late" spanning the Famennian and Frasnian stages.)

FIGURE 22.11 Devonian chronostratigraphy with conodont, ammonoid, pelagic ostracod, chitinozoan, and dacryoconarid zonations (left sides) and the carbon isotope curve, global event succession, sea-level history (coastal onlap, second order sea-level trends, former major transgression-regression cycles) and global temperature trends based on conodont-phosphate oxygen isotopes (right sides). For details, especially for the extensive background literature, see the main text.

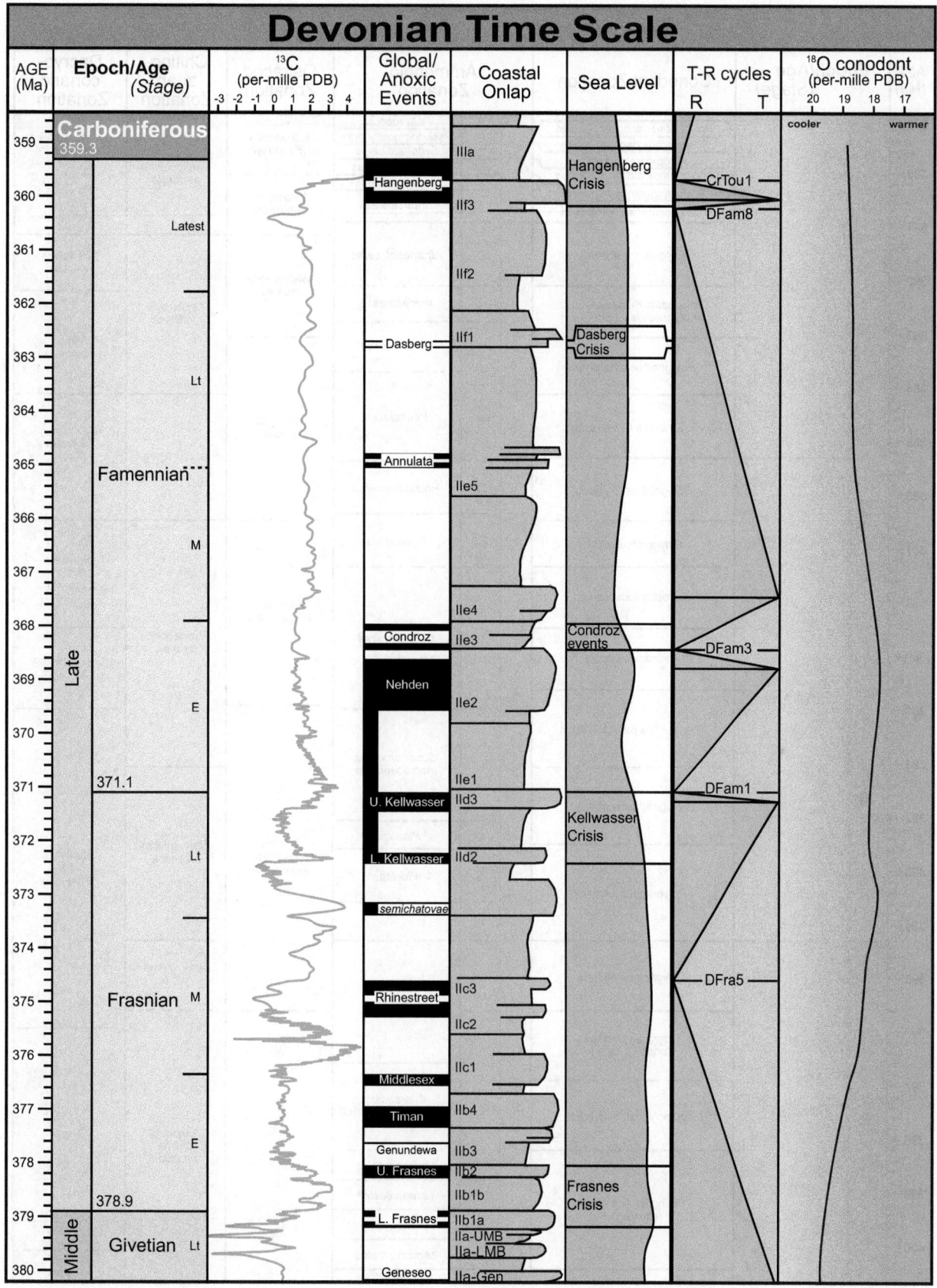

FIGURE 22.11   (Continued)

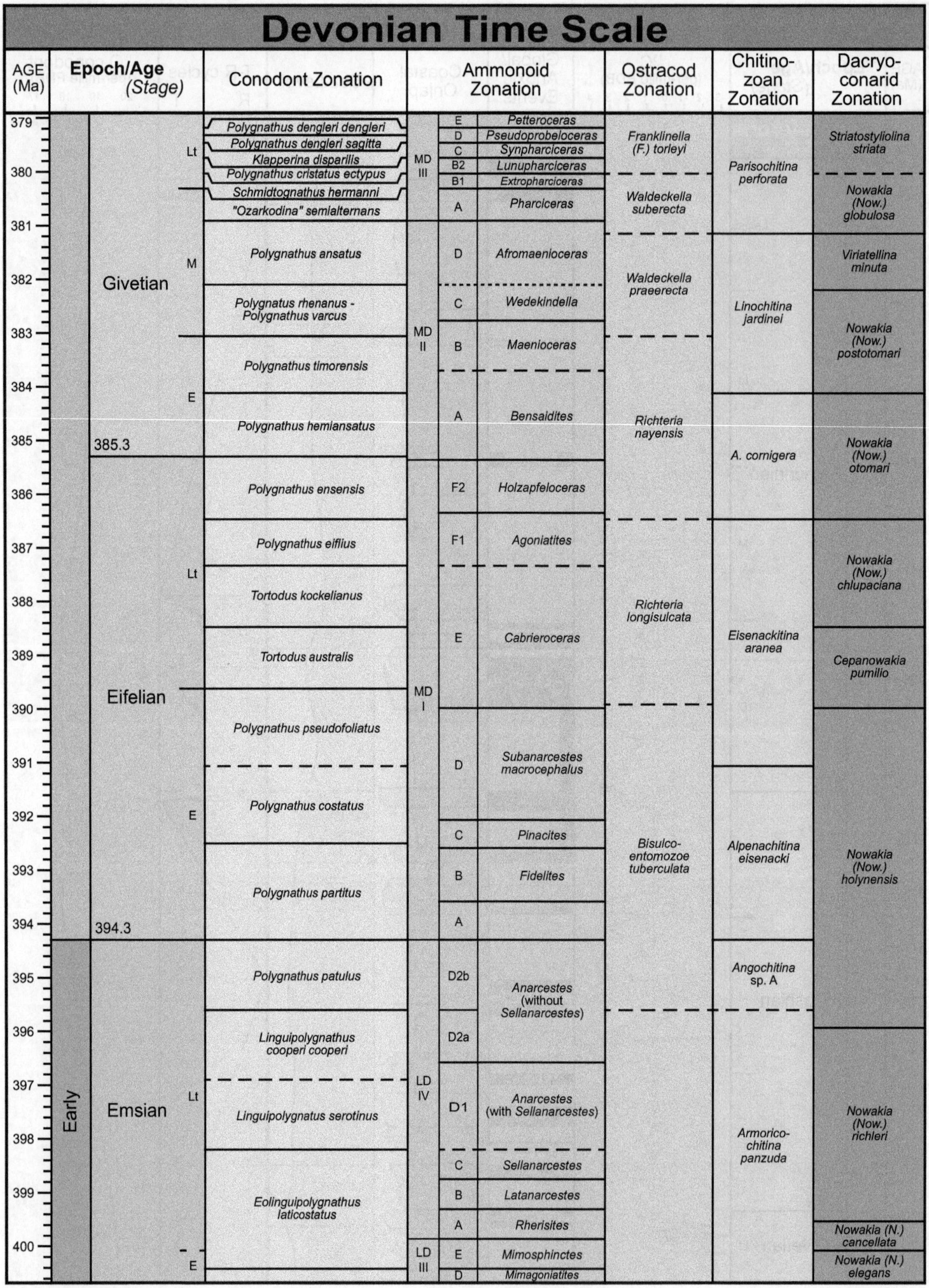

Devonian Time Scale

AGE (Ma)	Epoch/Age (Stage)			Conodont Zonation	Ammonoid Zonation			Ostracod Zonation	Chitino-zoan Zonation	Dacryo-conarid Zonation
379		Givetian	Lt	Polygnathus dengleri dengleri	MD III	E	Petteroceras	Franklinella (F.) torleyi	Parisochitina perforata	Striatostyliolina striata
380				Polygnathus dengleri sagitta		D	Pseudoprobeloceras			
				Klapperina disparilis		C	Synpharciceras			
				Polygnathus cristatus ectypus		B2	Lunupharciceras			Nowakia (Now.) globulosa
				Schmidtognathus hermanni		B1	Extropharciceras	Waldeckella suberecta		
381				"Ozarkodina" semialternans		A	Pharciceras			
382			M	Polygnathus ansatus	MD II	D	Afromaenioceras	Waldeckella praeerecta	Linochitina jardinei	Viriatellina minuta
383				Polygnatus rhenanus - Polygnathus varcus		C	Wedekindella			Nowakia (Now.) postotomari
						B	Maenioceras			
384			E	Polygnathus timorensis		A	Bensaidites	Richteria nayensis	A. cornigera	
385		385.3		Polygnathus hemiansatus						Nowakia (Now.) otomari
386				Polygnathus ensensis	MD I	F2	Holzapfeloceras			
387		Eifelian	Lt	Polygnathus eiflius		F1	Agoniatites			Nowakia (Now.) chlupaciana
388				Tortodus kockelianus				Richteria longisulcata	Eisenackitina aranea	
389				Tortodus australis		E	Cabrieroceras			Cepanowakia pumilio
390										
391			E	Polygnathus pseudofoliatus		D	Subanarcestes macrocephalus		Alpenachitina eisenacki	
392				Polygnathus costatus				Bisulco-entomozoe tuberculata		Nowakia (Now.) holynensis
393				Polygnathus partitus		C	Pinacites			
						B	Fidelites			
394		394.3				A				
395	Early			Polygnathus patulus	LD IV	D2b	Anarcestes (without Sellanarcestes)		Angochitina sp. A	
396				Linguipolygnathus cooperi cooperi		D2a				
397		Emsian	Lt	Linguipolygnatus serotinus		D1	Anarcestes (with Sellanarcestes)		Armorico-chitina panzuda	Nowakia (Now.) richleri
398						C	Sellanarcestes			
399				Eolinguipolygnathus laticostatus		B	Latanarcestes			
						A	Rherisites			Nowakia (N.) cancellata
400			E		LD III	E	Mimosphinctes			Nowakia (N.) elegans
						D	Mimagoniatites			

FIGURE 22.11   (Continued)

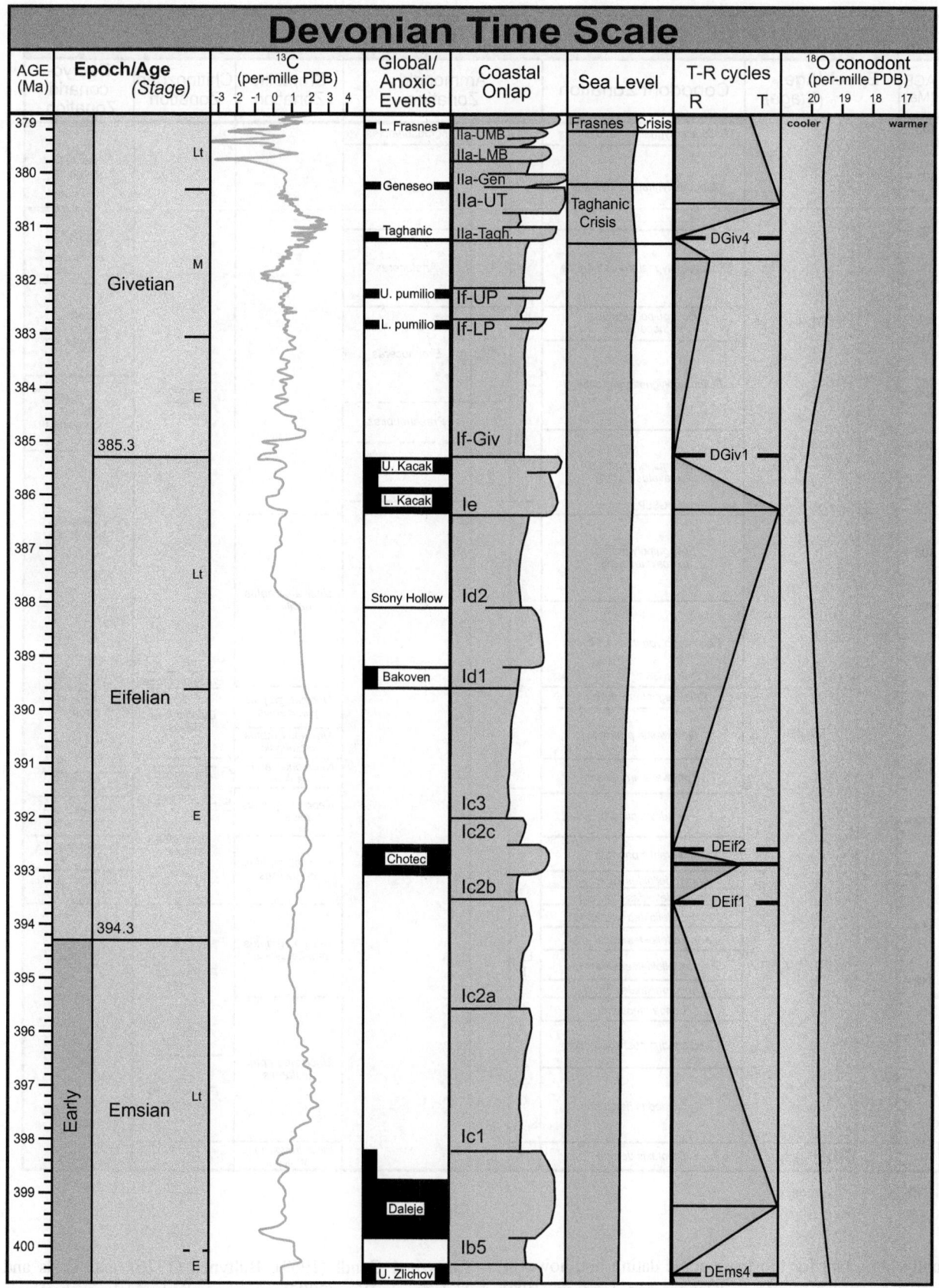

FIGURE 22.11 (Continued)

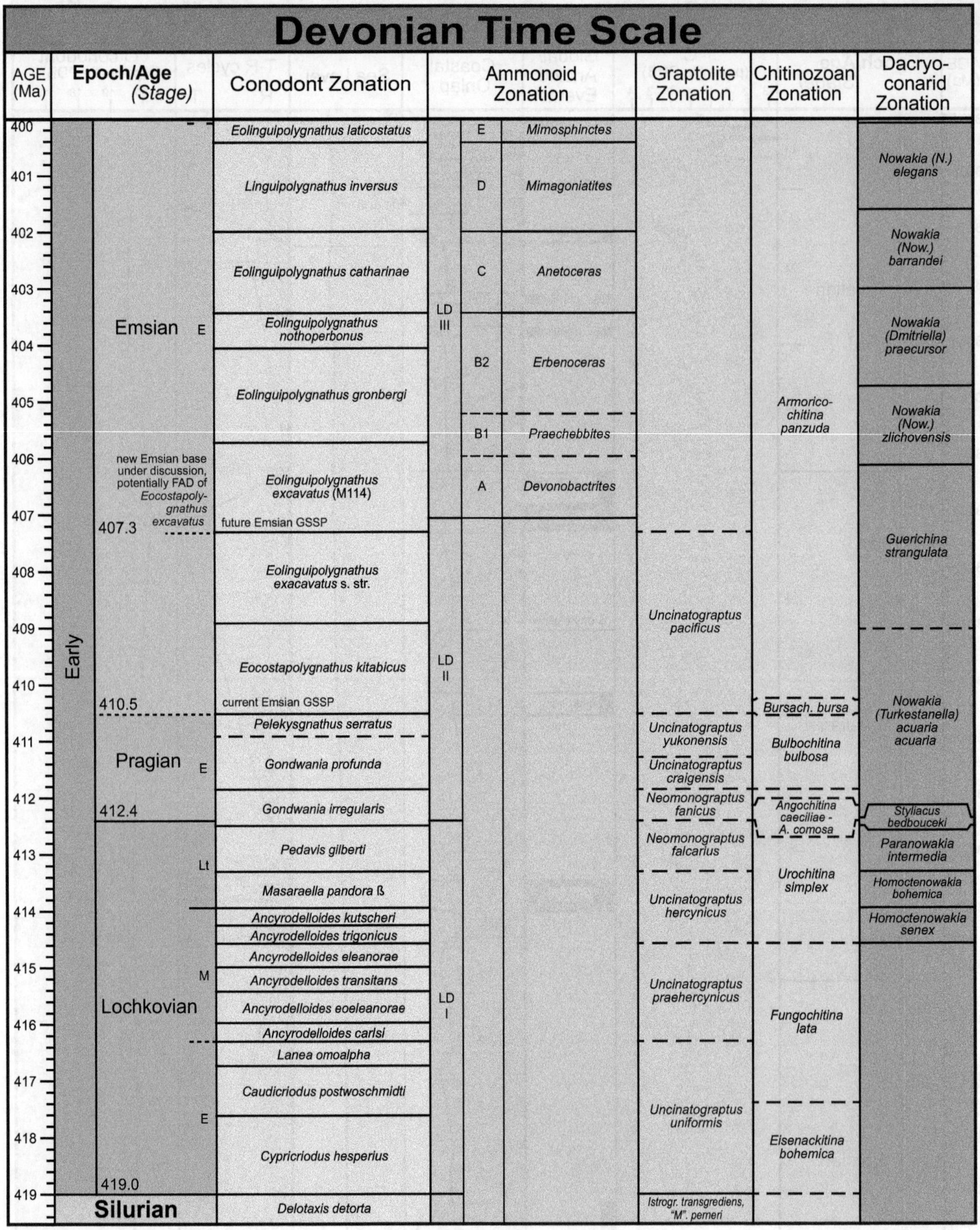

Devonian Time Scale

AGE (Ma)	Epoch/Age *(Stage)*		Conodont Zonation	Ammonoid Zonation			Graptolite Zonation	Chitinozoan Zonation	Dacryo-conarid Zonation
400			*Eolinguipolygnathus laticostatus*	E	Mimosphinctes				
401			*Linguipolygnathus inversus*	D	Mimagoniatites				Nowakia (N.) elegans
402				C	Anetoceras				Nowakia (Now.) barrandei
403	Emsian	E	*Eolinguipolygnathus catharinae*			LD III			
404			*Eolinguipolygnathus nothoperbonus*	B2	Erbenoceras				Nowakia (Dmitriella) praecursor
405			*Eolinguipolygnathus gronbergi*					Armorico-chitina panzuda	
				B1	Praechebbites				Nowakia (Now.) zlichovensis
406	new Emsian base under discussion, potentially FAD of *Eocostapolygnathus excavatus*		*Eolinguipolygnathus excavatus (M114)*	A	Devonobactrites				
407	407.3		future Emsian GSSP						Guerichina strangulata
408			*Eolinguipolygnathus exacavatus s. str.*						
409	Early						Uncinatograptus pacificus		
410			*Eocostapolygnathus kitabicus*			LD II			Nowakia (Turkestanella) acuaria acuaria
	410.5		current Emsian GSSP					Bursach. bursa	
411			*Pelekysgnathus serratus*				Uncinatograptus yukonensis	Bulbochitina bulbosa	
	Pragian	E	*Gondwania profunda*				Uncinatograptus craigensis		
412	412.4		*Gondwania irregularis*				Neomonograptus fanicus	Angochitina caeciliae - A. comosa	Styliacus bedbouceki
413		Lt	*Pedavis gilberti*				Neomonograptus falcarius		Paranowakia intermedia
414			*Masaraella pandora ß*				Uncinatograptus hercynicus	Urochitina simplex	Homoctenowakia bohemica
			Ancyrodelloides kutscheri						Homoctenowakia senex
			Ancyrodelloides trigonicus						
415		M	*Ancyrodelloides eleanorae*				Uncinatograptus praehercynicus		
	Lochkovian		*Ancyrodelloides transitans*			LD I			
416			*Ancyrodelloides eoeleanorae*					Fungochitina lata	
			Ancyrodelloides carlsi						
			Lanea omoalpha						
417			*Caudicriodus postwoschmidti*				Uncinatograptus uniformis		
418		E	*Cypricriodus hesperius*					Eisenackitina bohemica	
419	419.0								
	Silurian		*Delotaxis detorta*				Istrogr. transgrediens, "M". perneri		

FIGURE 22.11   (Continued)

equally important for biostratigraphic dating but, however, affected by endemism (e.g., Klapper and Johnson, 1980; Bultynck, 2003). The Spanish succession established by Carls and Gandl (1969), Bultynck (1976), and Carls and Valenzuela-Ríos (2002) can be applied widely in the European—North African realm (e.g., Bultynck, 1985, 1989;

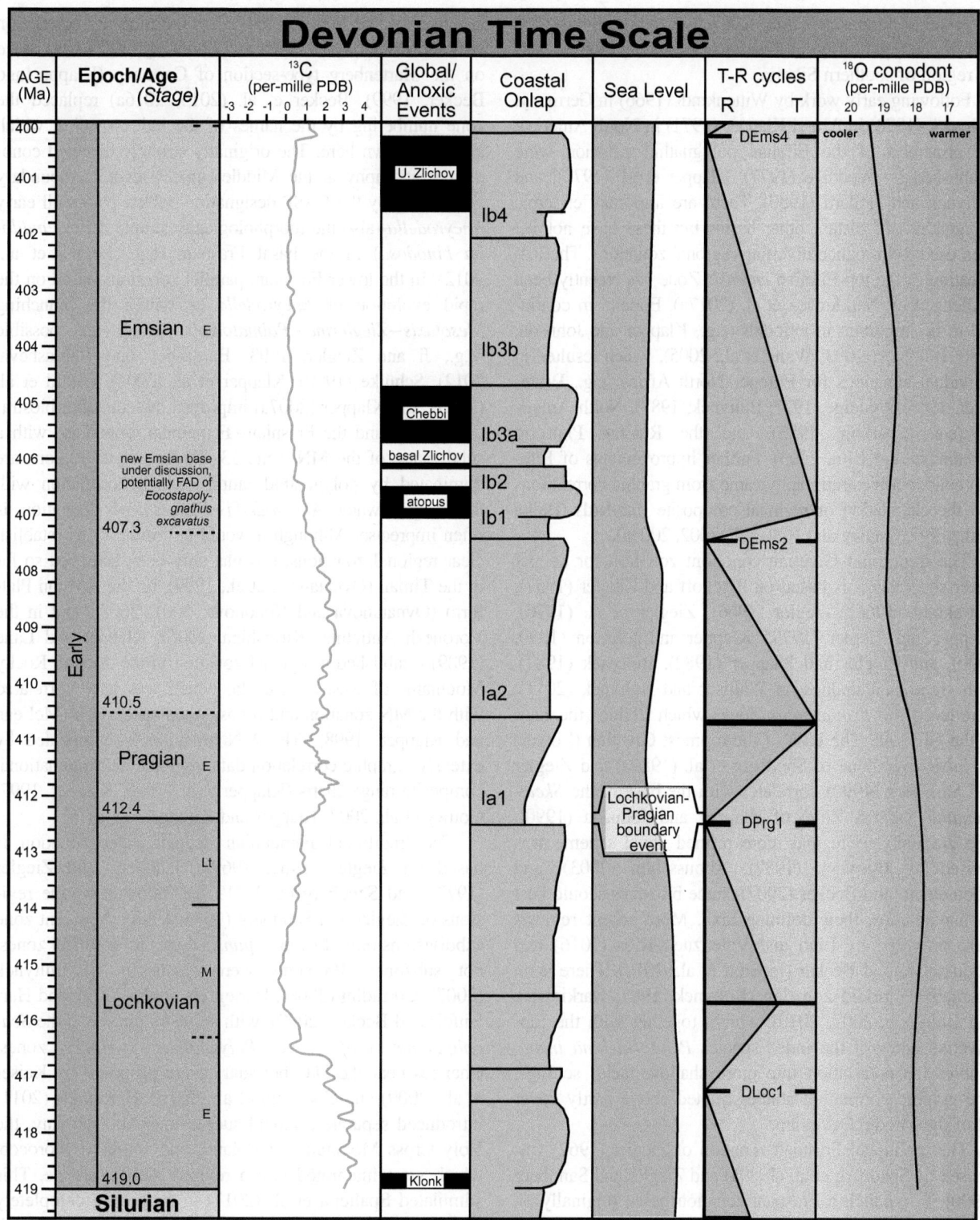

FIGURE 22.11 (Continued)

Bultynck and Hollard, 1980; Aboussalam et al., 2015) but not in North America (Johnson et al., 1980; Johnson and Klapper, 1981). It is of special importance for the correlation with neritic brachiopod faunas (Jansen et al., 2007).

The *Icriodus fusiformis* Zone of Ziegler (1971) and Carls et al. (1972), which first was thought to indicate the Lower to Middle Devonian transition, enables today the recognition of the basal upper Emsian, which is not marked in polygnathid faunas (Becker, 2007c). Izokh (1990, 1998) established a yet different *Caudicriodus* lineage for the Salair region of southern Siberia.

Following early work by Wittenkindt (1966) in Germany, Bultynck (1970) in 3, and Klapper (1971) in North America, the principles of the Eifelian polygnathid zonation were established by Weddige (1977), Klapper et al. (1978), and Bultynck and Hollard (1980). There are important endemic polygnathids of distant other basins but these have not yet been used to introduce diverting regional zonations. The justification of the top-Eifelian *ensensis* Zone has recently been challenged by Narkiewicz et al. (2017a). Endemism continued to be important in icriodids (e.g., Klapper and Johnson, 1980; Bultynck, 2003; Wang et al., 2005), which resulted in individual zonations for Europe−North Africa (e.g., Wittekindt, 1966; Weddige, 1977; Bultynck, 1987), North America (e.g., Sparling, 1983), and the Russian Platform (Kononova and Kim, 2003). Further improvements of Eifelian conodont biostratigraphy came from graphic correlations and the elucidation of regional composite standards (Belka et al., 1997; Gouwy and Bultynck, 2002, 2003a).

The traditional Givetian conodont zonation for deeper water successions is based on Bischoff and Ziegler (1957), Wittekindt (1966), Ziegler (1966), Ziegler et al. (1976), Klapper and Ziegler (1979), Klapper and Johnson (1980, 1990), and Ziegler and Klapper (1982). Bultynck (1987), with significant updates in Walliser and Bultynck (2011), introduced the *hemiansatus* Zone, which defines the base of the Givetian. The base of the topmost Givetian (Lower) *M. falsiovalis* Zone of Sandberg et al. (1989a) and Ziegler and Sandberg (1990) correlates with the base of the *Skeletognathus norrisi* Zone of Klapper and Johnson (1990). The gradually more and more refined zonal scheme was revised by Bultynck (1987), Aboussalam (2003), and Aboussalam and Becker (2007), since biozones should best be named after their defining taxa. More recent reviews were provided by Liao and Valenzuela-Ríos (2016) and Aboussalam and Becker (in Brett et al., 2018). There is an alternative icriodid zonation (Bultynck, 1987; Narkiewicz and Bultynck, 2007, 2010), which, together with the top-Givetian entry of the index species *Pandorinellina insita*, enables the correlation into more shallow facies settings. The graphic correlation studies quoted above partly cover all of the Middle Devonian.

The traditional Frasnian zonation of Ziegler (1962) was revised by Sandberg et al. (1989a) and Ziegler and Sandberg (1990). In parallel, a Frasnian zonation based originally on the MN succession was worked out by Klapper (1989) and numbered, giving MN Zones 1−13. It is more detailed than the so-called standard zonation and can be applied globally (e.g., Klapper et al., 1996: Timan, northern Russia; Klapper et al., 2004: Moose River Basin, Canada; Klapper, 2007b: Australia; Becker and Aboussalam, 2013a, 2013b: southern Morocco; Klapper and Kirchgasser, 2016: Appalachian

Basin; Day and Witzke, 2017: Iowa Basin). Its correlation with the zonation of Ziegler and Sandberg (1990) was based on the Martenberg type-section of Germany (Klapper and Becker, 1999). Becker et al. (2013, 2016a) replaced the zone numbering by the names of the index species, which are also shown here. The originally strongly disputed conodont stratigraphy at the Middle-Upper Devonian boundary was settled by the GSSP designation, which places all early *Ancyrodella* (also the morphologically simple *Ad. rotundiloba binodosa*) in the basal Frasnian (e.g., Becker et al., 2012). In the lower Frasnian, parallel zonations based on the rapid evolution of *Ancyrodella* or within the branching *Mesotaxis−Zieglerina−Palmatolepis* Group are possible (e.g., Ji and Ziegler, 1993; Bardashev and Bardasheva, 2012). Schülke (1999), Klapper et al. (2004), Girard et al. (2005), and Klapper (2007a) improved the conodont biostratigraphy around the Frasnian−Famennian boundary, with a subdivision of the MN Zone 13. Shallow-water regions are dominated by polygnathid faunas, which correlation with the deeper water *Ancyrodella−Palmatolepis* Zonation is often imprecise. Although it would be fortuitous to establish clear regional zonations, this has only been achieved so far in the Timan (Ovnatanova et al., 1999), on the Russian Platform (Ovnatanova and Kononova, 2001, 2008), and in the Voronezh Anticline (Kirilishina, 2006). Klapper and Lane (1989) established a regional zonation in the Alberta Rocky Mountains of western Canada, which was later correlated with the MN zonation and rugose coral succession (McLean and Klapper, 1998). The MN zonation is augmented by extensive graphic correlation data, regional and international composite range charts (Klapper et al., 1995; Klapper, 1997; Gouwy et al., 2007; Klapper and Kirchgasser, 2016).

The traditional Famennian deeper water zonation is based on Ziegler (1962, 1969), Sandberg and Ziegler (1973), and Sandberg et al. (1978), followed by the revisions of Ziegler and Sandberg (1984, 1990). Apparent zone subdivisions (e.g., Lower *expansa* Zone) refer to full zones, not subzones. Problems were discussed by Bultynck (2007), Corradini (2008), Hartenfels et al. (2009), and Hartenfels and Becker (2009), with some emphasis on the *Scaphignathus velifer* and *Polygnathus styriacus* zones. Changes near the D/C boundary were proposed by Kaiser et al. (2009) and Corradini et al. (2016). Hartenfels (2011) introduced separate regional successions for Germany, the Holy Cross Mountains of Poland, and southern Morocco, which were integrated into a revised global scheme. This stimulated Spalletta et al. (2017) to propose a completely new zonal nomenclature for all of the Famennian, with all zones named after the lowest occurrence of index taxa. However, this included a minor imprecision (e.g., for the *Bispathodus ultimus ultimus* Zone; discussion in Söte et al., 2017) and led to some ignorance toward the major global conodont extinction associated with the Hangenberg Black Shale (Kaiser et al., 2016). Therefore the D/C boundary zonal scheme of Kaiser et al. (2009) is kept here. For the often very different and more endemic shallow-water successions, alternative icriodid zones (Sandberg and Dreesen, 1984) or regional successions with polygnathids and other genera have been proposed (e.g., Lipnjagow, 1979: Donetz Basin; Aristov, 1984: Russian basins; Shilo et al., 1984: Omolon, Russian Far East; Kuźmin, 1992 and Vorontzova, 1993: Kazakhstan; Wang et al., 2016: Junggar Basin, NW China).

22.2.1.2 Ammonoid zonations

The coiled ammonoids appeared low in the Emsian of its classical sense (e.g., Klug, 2001; Becker et al, 2019), above a level with the first abundant bactritids (Klug et al., 2008) (Fig. 22.11). Building on Bultynck and Hollard (1980) and Chlupáč and Turek (1983), the current Emsian zonation, with LD III for the lower Emsian, and LD IV for the upper Emsian, was introduced by Becker and House (2000a) and Becker and House (1994b). There are important additional lower Emsian data for southern Morocco (Klug, 2001; De Baets et al., 2010; Aboussalam et al., 2015; Klug, 2017; Becker et al., 2018b, 2019), Uzbekistan (Becker et al., 2010; Naglik et al., 2019), and South China (Ruan, 1996). These are currently the most important regions to study early ammonoid evolution in a time frame, at least until more precision can be given to the various Russian faunas. Klug et al. (2000), Klug (2002), and Ebbighausen et al. (2011) provided detailed upper Emsian data for several Anti-Atlas regions of southern Morocco, which is currently the best ammonoid region on a global scale for that interval (Becker et al., 2018a).

Building on Wedekind (1920), Hollard (1974, 1978), the review by House (1978), Bultynck and Hollard (1980), and Chlupáč and Turek (1983), a revised global Eifelian zonation was introduced by Becker and House (1994b, 2000a). There are subsequent updates for Morocco by Klug et al. (2000), Klug (2002), Ebbighausen et al. (2011), and Klug and Pohle (2018), especially for the Eifelian−Givetian boundary GSSP (Walliser et al., 1995; Becker et al., 2018c). Other important Eifelian ammonoid regions are Algeria (Göddertz, 1987) and the Cantabrian Mountains (Montesinos, 1990). The more individual and episodic faunas from many other regions can be tied to the Moroccan "standard succession."

Goniatite records are surprisingly rare and of very low diversity on a global scale in the lower Givetian. The poorly used Givetian zonations of Wedekind (1920) and Schmidt (1926) were revised by Becker and House (1994b, 2000a). It includes the *Afromaenioceras* Zone (MD II-D) discovered by Göddertz (1987) in southern Algeria. Becker (2005, 2007b) provided short reviews of the global record, with a focus on correlations with the proposed Givetian substage levels. The most detailed middle Givetian succession is known from southern Morocco (Bensaid, 1974; Becker et al., 2000b, 2004b, 2018b; Aboussalam and Becker, 2011) but

many details and taxa are still unpublished. Ebbighausen et al. (2007) correlated a new local Rhenish succession with the Dra Valley zonation of the Moroccan Anti-Atlas. All upper Givetian ammonoid faunas were previously included in the lower Frasnian *Pharciceras lunulicosta* Zone (do I α) of Wedekind (1913), which became the *Pharciceras* Stufe of House (1985), and then the MD III (Becker and House, 1994b). Aboussalam and Becker (2001, 2002) proposed a future substage boundary that can be recognized both by conodonts and goniatites. House et al. (1985) established a subdivision of the *Pharciceras* Stufe in the MN, which provided the base for the refined zonation, with MD-A to E, of Becker and House (1994b, 2000a, 2000b). The eastern Anti-Atlas comprises on a global scale the richest upper Givetian ammonoid faunas (e.g., Bockwinkel et al., 2009, 2013, 2015, 2017). Recent investigations suggested that an additional faunal level characterized by unusually large-sized pharciceratids can be recognized as an upper subdivision of MD III-C (Aboussalam et al., 2018).

The traditional Frasnian ammonoid zonation of Germany is based on Wedekind (1913) and Matern (1931). However, ammonoid-conodont correlations by House and Ziegler (1977) showed that large parts of the lower/middle Frasnian were not zoned in the Rhenish Massif. This suspicion gained fundamental support from the much more detailed succession of the Appalachian Basin (House, 1978; revised by House and Kirchgasser, 2008). The comparison of the Canning Basin and New York State faunal sequences with the global record, led Becker et al. (1993) and House and Kirchgasser (1993) to establish a completely new Frasnian zonation, assigned to UD I-A to L (compare Becker and House, 2000a). Subsequently, regional zonations were described from the MN (Becker and House, 1994b; House et al., 2000a, 2000b), Timan, northern Russia (Yatskov and Kuźmin, 1992; Becker et al., 2000), and the Tafilalt of southern Morocco (Becker and House, 2000b; new data in Becker et al., 2004c, 2018b; Becker and Aboussalam, 2013a, 2013b; Aboussalam et al., 2018). Nikolaeva et al. (2009) demonstrated that it is possible to correlate isolated Russian faunas with the Timan zonal scheme; this still has to be applied to the varied Frasnian faunas from the Urals and southern Siberia described by Bogoslovskiy (1969, 1971).

The principles of Famennian ammonoid biostratigraphy go back for more than a century and are primarily based on the early studies of German (Rhenish Massif and Saxothuringia) localities (e.g., Frech, 1902; Wedekind, 1908, 1914; Schindewolf, 1923, 1937; Schmidt, 1922, 1924). Much later, the analysis of the global lower Famennian (UD II) ammonoid record enabled a refined zonation (Becker, 1993a). Studies on younger (UD III−VI) German faunas (e.g., Korn, 1981, 1991, 2002b, 2004; Korn and Price, 1987; Price and Korn, 1989; Clausen et al., 1989; Becker, 1992, 1988, 1996; Korn et al., 1994; Korn and Ziegler, 2002) were incorporated in zonal reviews by Korn and Luppold (1987), Becker

(1993b, 1998), Becker and House (2000a), and Korn (2002a). Additional high-resolution stratigraphic data were more recently added by Hartenfels (2011), Bartzsch and Weyer (2012), Hartenfels et al. (2016), and Becker et al. (2016e, 2016f). While there are distinctive regional zonations for the Famennian of the far distant Canning Basin of Western Australia (Becker and House, 2009) and for the Junggar Basin of NW China (Zong et al., 2015), very rich middle to uppermost Famennian faunas of Russia and Kazakhstan (e.g., Bogoslovskiy, 1981; Nikolaeva and Bogoslovskiy, 2005; Nikolaeva, 2007) and Poland (Czarnocki, 1989; Dzik, 2006) have not yet been treated similarly. The equally diverse Famennian faunas of southern Morocco form another important regional succession (Korn, 1999; Korn et al., 2000, 2004, 2015b; Becker et al., 2002, 2018a, 2018b, 2018e; Hartenfels and Becker, 2009, 2016a, 2016b, 2018; Hartenfels et al., 2013; Klug et al., 2016). Differences to Europe at species-level become more and more evident by morphometric studies but the very similar successive entries of marker genera enable a good correlation. Korn et al. (2014, 2015) introduced an alternative zonation for southern Morocco that is only based on endemic species of the Prionoceratidae. A similar approach, with parallel zonations based on different goniatite and clymeniid lineages, can be recognized at the D/C transition (Becker et al., 2016b). The Famennian ammonoid succession of southern Algeria (Korn et al., 2015a) resembles closely that of southern Morocco.

22.2.1.3 Graptolite zonations

From the uppermost Silurian (Pridoli) to the basal Emsian, megaplanktonic graptolites, the Graptoloidea, consisted mostly of monograptids, which included several lineages/genera (Fig. 22.11). More complex genera with many more thecae, such as *Linograptus* and *Abiesgraptus*, ranged only into the lower Lochkovian (e.g., Jaeger, 1978, 1979). Monograptids are very important for global Lower Devonian stratigraphy, for example, for the definition of the basal Devonian GSSP at Klonk, Czech Republic. But there was a steady decline of the group, coupled with a strongly decreasing number of sections and regions, especially after the (traditional) Pragian (e.g., Jaeger, 1978; Koren, 1979). The Lochkovian to basal Emsian Zonation was reviewed by Jaeger (1979). Subsequent studies enlarged the number of regions with stratigraphically significant occurrences and improved the correlation with other fossil groups. Examples are the conodont–graptolite correlations in Nevada, Alaska, and Yukon (Murphy and Berry, 1983; Savage et al., 1985), or the investigation of Chinese regions (e.g., Ni and Jiao, 1983; Mu and Ni, 1983, 1985; Wang, 1980; Wang and Zao, 1984). During the Second International Symposium on the Devonian System in Calgary, new reviews were presented by Jaeger (1989) and Lenz (1989). The first included a new *Monograptus kayseri* Zone in the topmost Lochkovian. Subsequent contributions

significant for graptolite stratigraphy were the precise dating of the entry of *Monograptus hercynicus* (now a species of *Uncinatograptus*) within the Lochkovian (Springer and Murphy, 1994), the review of western European faunas (Lenz et al., 1996), the correlation of graptolite and geochemical events in Poland (Porebska and Sawlowicz, 1997), the review of faunas from Australia (Rickards and Wright, 2000), and the review of the graptolite succession of Arctic Islands of Canada (Lenz, 2013). The latter includes the modern generic assignment of zonal index species. For example, the basal Devonian index taxon, *Monograptus uniformis uniformis*, also belongs to the genus *Uncinatograptus*. Based on Tian Shan conodont data, Koren et al. (2007) showed that the supposed lower Pragian *Neomonograptus falcarius* Zone falls in the upper Lochkovian. Chen et al. (2015) studied a new Lochkovian graptolite succession of Guangxi, which mostly agrees with the global zonal scale.

The knowledge of the youngest, lower Emsian Graptoloidea is still restricted. It is was long unclear how to correlate the regional *U. pacificus* Zone of Alaska (Churkín et al., 1970), based on an endemic species, with the successions of Europe and Asia. Data by N.V. Sennikov (in Yolkin et al., 2008) show that *U. pacificus pacificus* enters in the higher part of the *Eoc. kitabicus* Zone in the Zinzil'ban Emsian GSSP section. The youngest graptolite level of Bohemia with *Neomonograptus atopus* (Bouček, 1966) falls below the base of the Zlíchov Limestone (or type Zlíchovian) but in the basal Emsian of any definition, more precisely in the *Latericriodus bilatericrescens gracilis* Zone (Slavík, 2004a, 2004b) and in the *Guerichina strangulata* dacryoconarid zone. It marks a transgressive interval, the *atopus* Event of Aboussalam et al. (2015). However, *Neom. atopus* has a lower, Pragian, range in Asia (Jaeger, 1983; Li and Wang, 1983), overlapping with the *Uncinatograptus yukonensis* Zone (Jaeger, 1989). Koren and Sennikov (in Yolkin et al., 2008) claim that monograptids range much higher in the Zinzilban section of Uzbekistan, with a last rich fauna (Bed 44: seven taxa) above the entry of advanced lower Emsian polygnathids (Bed 42/24, conodont data of Slavík et al., 2017) and with last specimens even above the first *Eolinguipolygnathus nothoperbonus*. This would mean an overlap with early ammonoids, a feature unknown from anywhere else. Since both groups have not been found together in sections of the Zeravshan Range, new data on the precise conodont age of these possibly youngest Graptoloidea are required.

22.2.1.4 Dacryoconarid zonations

Planktonic tentaculitoids, especially members of the Nowakiidae (Dacryoconarida; see review by Lüttke, 1979) and Homoctenida, are of high importance for global biostratigraphy until their extinction at the Frasnian/Famennian boundary (Fig. 22.11). Most fundamental for the Lower and

Middle Devonian is the work of Bouček (1964), Alberti (e.g., 1981, 1982, 1993, 1998, 2000). Gessa (1996) suggested a restriction and subdivision of forms that are usually included in the Pragian index species *Nowakia (Turkestanella) acuaria*. Kim (2011) provided new data for the Emsian stratotype region of Uzbekistan. Ferrová et al. (2012) restudied nowakiid biostratigraphy across the Daleje Event (s.l.) in Bohemia and discussed its significance for Emsian substage definition. A further revision of *Now. cancellata*, its ancestors, representatives outside Bohemia, and of possible homeomorphs, will be crucial since the species could become the upper Emsian index taxon (Fig. 22.2). Higher up, the role of the lower Eifelian global Choteč Event for dacryoconarid extinctions and biogeography was recently discussed by Brocke et al. (2016). Givetian and Frasnian ranges of stratigraphically important taxa were summarized in Liashenko (1967) and Sauerland (1983). Bond (2006) reviewed the sudden extinction in the course of the Kellwasser Events.

22.2.1.5 Ostracod zonations

The fast evolution of planktonic and outer shelf Entomozoacea provides a fine, globally significant biostratigraphic tool (Fig. 22.11). Major initial studies were conducted by Rabien (1954) and Blumenstengel (1965). Wang (1989) investigated the more complete Chinese entomozoid succession, which was compared with the German sequence by Gross-Uffenorde and Wang (1989). Groos-Uffenorde et al. (2000) published a general review for all Devonian groups. Benthic ostracods are excellent indicators of ecotypes, but a global biozonation has not been established. In general, ecologically controlled assemblages dominate successions, which, however, can be used for regional ecostratigraphy, for example, on the East European platform (e.g., Tschigova, 1977; Orlov, 1990), or in southern Siberia (e.g., Bakharev, 2005a, 2005b). Braun (1978, 1990) published a rather detailed assemblage zonation for the Middle and Upper Devonian of Western Canada. Lethiers (1974, 1984) introduced a simple zonation for the neritic upper Givetian and Frasnian of the Ardennes to Boulonnais regions (Belgium/Northern France), which was complemented by an upper Frasnian *Svantovites lethiersi* Zone (Casier, 1979), and, most recently, by a new *Ovatoquassilites avesnellensis* Zone for the lower Famennian (Casier, 2018). Through all of the Devonian, the evolution of the genus *Polyzygia* is useful for stratigraphy (e.g., Crasquin-Soleau et al., 1994). In the Lower Devonian of southern Siberia, a similar "phylozone" concept was based on the genus *Miraculum* (Bakharev and Bazarova, 2004).

22.2.1.6 Radiolarian zonations

Radiolaria can be abundant in Devonian outer shelf to oceanic siliceous sediments. Their zonation developed gradually in the two last decades of the 20th century (e.g., Holdsworth and Jones, 1980; Nazarov and Ormiston, 1985; Schwartzapfel and Holdsworth, 1996; Aitchison et al., 1999; Wang et al., 2000). Radiolarian zones for the Lower and Middle Devonian of Japan (e.g., Umeda, 1998; Kurihara, 2004) are, unfortunately, not well correlated with chronostratigraphy or other biostratigraphic scales. In general, there has been some focus on the Upper Devonian. For example, Afanasieva (2000) distinguished three zones in the Frasnian, representing roughly the three substages, and a lower Famennian *Tetrentactinia barysphaera— Ceratoikiscum famennium* Zone. Afanasieva (2009) and Afanasieva and Amon (2011, 2012) added many new faunas from the Urals and other Russian regions. Especially useful for correlation are joint radiolarian-conodont assemblages (e.g., Zhang et al., 2008; Obut and Shcherbanenko, 2008; Izokh et al., 2008), for example, at the Frasnian—Famennian boundary. The recent review by Aitchison et al. (2017) showed up to 30 successive assemblages for all of the Devonian, 17 of which are thought to have more than regional value. Since many zones are not sufficiently correlated to the conodont scale, they are not shown in Fig. 22.11.

22.2.1.7 Miospore zonations

Two spore zonations have been proposed that cover the entirety of the Devonian (Fig. 22.12). That of Richardson and McGregor (1986) also included the Silurian and had 14 Devonian zones with the nominate taxa occurring at or close to the base of each zone. The zones were each also defined on a characteristic assemblage. These zones were defined on sections on, and peripheral to, the Old Red Sandstone Continent in Arctic Canada, the United States, the British Isles, and the classic Ardennes-Rhenish Area. The other zonation is that of Streel et al. (1987) that was based entirely on these classic Ardennes-Rhenish sections, which have stratigraphic ties to marine faunas. The Streel et al. (1987) zonation included 51 Oppel zones based on inceptions of single taxa. These zones are variously subdivided into interval zones and lineage subzones to give 75 levels of correlation. Some of these zones, key taxa, and sections are in common with those in Richardson and McGregor (1986). Intervals of both zonations have been refined and better calibrated to conodont zonations. This particularly includes the Givetian to Frasnian Interval as refined by Matyja and Turnau (2008), Turnau and Narkiewicz (2011), and Tel'nova (2008). Other refinements that provide better ties to the Famennian marine faunas and sequence stratigraphy, particularly in the D/C boundary interval, are by Maziane et al. (1999, 2002) and Streel (2000, 2009).

The scheme of Streel et al (1987) has been adapted outside the Ardennes-Rhenish region although with differences in the inception order of the zone defining taxa. These differences are clearly evident when the scheme was applied to

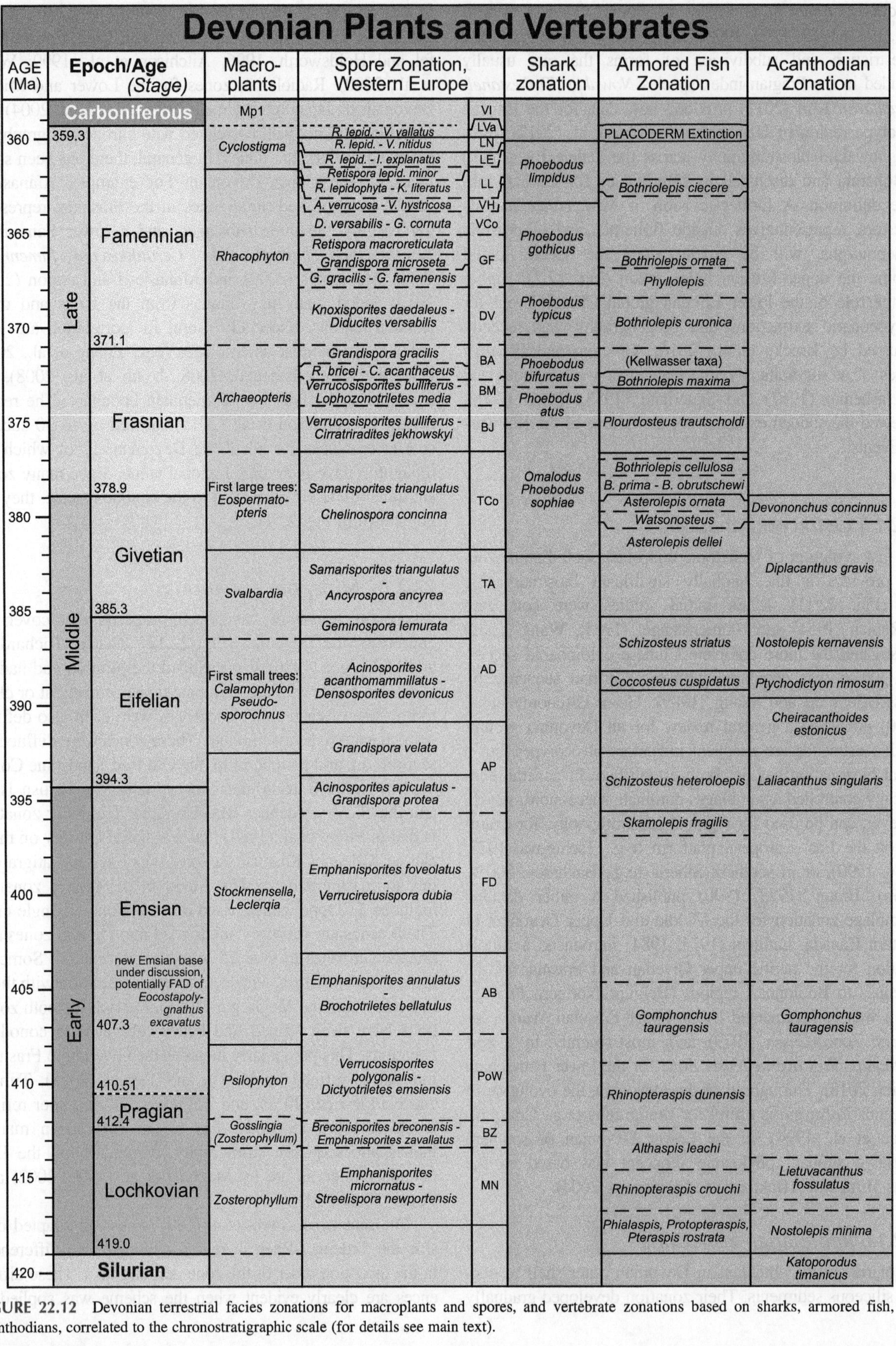

Devonian Plants and Vertebrates

AGE (Ma)	Epoch/Age (Stage)		Macro-plants	Spore Zonation, Western Europe		Shark zonation	Armored Fish Zonation	Acanthodian Zonation
360	Carboniferous		Mp1		VI			
	359.3			R. lepid. - V. vallatus	LVa		PLACODERM Extinction	
			Cyclostigma	R. lepid. - V. nitidus	LN	Phoebodus limpidus		
				R. lepid. - I. explanatus	LE		Bothriolepis ciecere	
		Late		Retispora lepid. minor	LL			
				R. lepidophyta - K. literatus				
365	Famennian		Rhacophyton	A. verrucosa - V. hystricosa	VH			
				D. versabilis - G. cornuta	VCo	Phoebodus gothicus		
				Retispora macroreticulata	GF		Bothriolepis ornata	
				Grandispora microseta			Phyllolepis	
				G. gracilis - G. famenensis				
370				Knoxisporites daedaleus - Diducites versabilis	DV	Phoebodus typicus	Bothriolepis curonica	
	371.1							
			Archaeopteris	Grandispora gracilis	BA	Phoebodus bifurcatus	(Kellwasser taxa)	
				R. bricei - C. acanthaceus	BM		Bothriolepis maxima	
				Verrucosisporites bulliferus - Lophozonotriletes media		Phoebodus atus		
375	Frasnian			Verrucosisporites bulliferus - Cirratriradites jekhowskyi	BJ		Plourdosteus trautscholdi	
						Omalodus Phoebodus sophiae	Bothriolepis cellulosa	
	378.9		First large trees: Eospermato-pteris	Samarisporites triangulatus - Chelinospora concinna	TCo		B. prima - B. obrutschewi	
380							Asterolepis ornata	Devononchus concinnus
							Watsonosteus	
	Givetian		Svalbardia	Samarisporites triangulatus - Ancyrospora ancyrea	TA		Asterolepis dellei	
	385.3	Middle						Diplacanthus gravis
385				Geminospora lemurata				
			First small trees: Calamophyton Pseudo-sporochnus	Acinosporites acanthomammillatus - Densosporites devonicus	AD		Schizosteus striatus	Nostolepis kernavensis
	Eifelian						Coccosteus cuspidatus	Ptychodictyon rimosum
390								Cheiracanthoides estonicus
				Grandispora velata	AP			
	394.3			Acinosporites apiculatus - Grandispora protea			Schizosteus heteroloepis	Laliacanthus singularis
395							Skamolepis fragilis	
				Emphanisporites foveolatus - Verruciretusispora dubia	FD			
400	Emsian	Early	Stockmensella, Leclerqia					
	new Emsian base under discussion, potentially FAD of Eocostapoly-gnathus excavatus			Emphanisporites annulatus - Brochotriletes bellatulus	AB		Gomphonchus tauragensis	Gomphonchus tauragensis
405								
	407.3							
410			Psilophyton	Verrucosisporites polygonalis - Dictyotriletes emsiensis	PoW		Rhinopteraspis dunensis	
	410.51							
	Pragian		Gosslingia (Zosterophyllum)	Breconisporites breconensis - Emphanisporites zavallatus	BZ		Althaspis leachi	
	412.4							Lietuvacanthus fossulatus
415	Lochkovian		Zosterophyllum	Emphanisporites micrornatus - Streelispora newportensis	MN		Rhinopteraspis crouchi	
							Phialaspis, Protopteraspis, Pteraspis rostrata	Nostolepis minima
	419.0							
420	**Silurian**							Katoporodus timanicus

FIGURE 22.12 Devonian terrestrial facies zonations for macroplants and spores, and vertebrate zonations based on sharks, armored fish, and acanthodians, correlated to the chronostratigraphic scale (for details see main text).

the Early and Middle Devonian of Bolivia (Troth et al., 2011) and Saudi Arabia (Breuer and Steemans, 2013). A spore zonation for Gondwana has also been devised in the Amazon Basin (Loboziak and Melo, 2002) using shallow boreholes and, in the absence of any accompanying cono-donts or goniatites, was calibrated against the zonation of Streel et al. (1987). This gives some 12 Devonian zones with known gaps in the section. Importantly, the section is cali-brated against the Gondwana chitinozoan zonation (Grahn and Melo, 2005).

An independently defined zonation (Avkhimovitch et al., 1993) was devised for the upper Emsian to Upper Devonian from Eastern Europe. It has 14 zones and 21 subzones and was based on the efforts of the many palynologists who worked in Russia, Belarus, the Baltic Republics, and the Ukraine. This has been calibrated with the zonation of Streel et al. (1987) and now, more importantly, it can be calibrated with the increasingly refined conodont zonation that is avail-able from recent research in Russia (e.g., Ovnatanova et al., 1999).

22.2.1.8 Acritarch zonation

Acritarchs are organic-walled microfossils of unknown and probably varied biological affinities, but their stratigraphi-cal use, especially in the Middle and Upper Devonian has increased. At present, the current zonation falls behind that of some other groups in resolving power. Reviews of zona-tions are given by Molyneux et al. (1996) and Le Hérissé et al. (2000).

22.2.1.9 Chitinozoan zonation

Progress toward a global Devonian chitinozoan zonation was last summarized by Paris et al. (2000) (Fig. 22.11A–C). Based on detailed studies in South America (e.g., Grahn and Melo, 2004; Grahn, 2005), there have been significant improvements in subsequent years, at least in terms of a within-Gondwana Zonation. External correlations remain, as ever, a problem in the absence of conodonts and goniatites. Recently there has been progress in recognizing elements of the Devonian sea-level curve and other Devonian events, par-ticularly near the Kačák Event (e.g., Brocke et al., 2016). These correlations rely on integrated palynostratigraphy, which utilizes spores, acritarchs, and chitinozoa within models for transgressions. They have been applied in Boli-via (Troth et al., 2011), Brazil (Grahn et al., 2010), and the Falkland Islands (Marshall, 2016). Importantly by recog-nizing Devonian events outside Euramerica and northern Gondwana, they can tell us much about underlying mechanisms.

22.2.1.10 Plant megafossil zonation

Vascular plants, which began well before the Devonian, rise in dominance during the period (Fig. 22.12). Among the

oldest currently known tree-sized forests is the cladoxylopsid forest from the Eifelian of Lindlar, Germany (Giesen and Berry, 2013). These plants are subsequently found across Euramerica and into the upper Givetian of New York State. *Archaeopteris* forests have attracted much interest as a speculative driver of Devonian change. The first archaeopteridalean progymnosperms appear in the earli-est Givetian (as evidenced by their microspore *G. lemur-ata*) with the fossils becoming increasingly abundant in the Givetian of Euramerica (Stein et al., 2012), forming mixed forests with the cladoxylopsid *Wattieza* in New York State. The third forest type is that with the tree-sized lycopod *Protolepidodendropsis* from the paleo-equatorial Devonian and known in situ from the Givetian and lower Frasnian of Svalbard, where it grew in dense thickets (Berry and Marshall, 2015). Seed plants became increasingly abundant in the later Devonian and are largely of the *Elkinsia* type. However, there is no real understanding of their abundance or how they might respond to, or still less drive the series of mass extinc-tions that occur through this interval.

A broad division of the Devonian into seven zones was suggested by Banks (1980) and this has been further refined by Edwards et al. (2000), whose zonation, extend-ing Banks' numbers, is shown in Fig. 22.12.

22.2.1.11 Vertebrate zonations

The Devonian is marked by the origin and early evolution of tetrapods; vertebrates with limbs and digits, which are only of very broad stratigraphic use due to the rarity of specimens and relative isolation of most known genera (Blom et al., 2004; Fig. 22.12). Tetrapod trackways have recently been discovered in the Eifelian of Poland (Niedzwiedzki et al., 2010). The ear-liest body fossils are known from the Frasnian of Latvia, Scot-land, and China (Clack, 2002; latest global reviews in Blieck et al., 2007, 2010). During the Devonian these vertebrates are in many ways fish-like and aquatic, and fully terrestrial tetra-pods are not known before the earliest Carboniferous (Tour-naisian). In terrestrial facies, fish are helpful in age determination (Fig. 22.12). Several zonations have developed using various fish groups and teeth and other microvertebrate remains. Major zonations include those for thelodonts and het-erostracans and for placoderm and acanthodian fish and are of importance for the difficult problem of marine to nonmarine correlation. A summary of achievements in Paleozoic verte-brate biostratigraphy was published by Blieck and Turner (2000). It includes summaries for the Old Red Sandstone, Bal-tica and Belarus, Gondwana regions, China, and an impressive literature compilation. In open shelf environments, shark teeth assemblages give a widely applicable zonation (e.g., Ginter and Ivanov, 2000; Ginter et al., 2002; Trinajstic and George, 2009), which, however, is less detailed than that of the co-occurring conodonts or ammonoids.

22.2.2 Physical and chemical stratigraphy

22.2.2.1 Global event stratigraphy

Devonian global events are complex and numerous (Figs. 22.1 and 22.11). They include extinctions of variable magnitude, with two mass extinctions at the Frasnian—Famennian (Upper Kellwasser Event, e.g., Racki, 2005) and near the Devonian-Carboniferous boundary (Hangenberg Crisis, e.g., Kaiser et al., 2016). Most "events" are staged and characterized by rapid eustatic changes, pulses of eutrophication and black shale formation, isotopic excursions, other geochemical anomalies, opportunistic blooms, sudden migrations, and radiations. Rapid global climate change in association with major pulses of volcanism and tectonism is widely assumed primary trigger mechanisms. The role of cosmic impacts is ambiguous.

The principles of Devonian event stratigraphy were published by Walliser (1984, 1996), House (1985, 2002), Becker (1993b), and Sandberg et al. (2002). There are regional special volumes or compilations for the event succession of Belgium (Denayer and Mottequin, 2015), the Appalachian Basin (Brett et al., 2009), the Great Basin of western North America (Sandberg et al., 1989b), southern Morocco (Bultynck and Walliser, 2000), the Polar Urals (Sobolev and Soboleva, 2018), Australia (Talent et al., 1993), Vietnam (preliminary data of Königshof et al., 2017), and South China (Bai et al., 1994; Ma et al., 2017; Qie et al., 2018). The comprehensive global review by Becker et al. (2016a) emphasized the distinction between individual global events (short-termed, within one biozone or at a biozone boundary) and crises (polyphase, stretching over two or several biozones, consisting of a distinctive event succession; e.g., Kellwasser and Hangenberg Crises). Most Devonian events/crises can be assigned to orders of magnitude, from first (the two mass extinctions, with the loss of complete ecosystems) to fourth order (sudden global extinction of only a few but very widespread fossil groups). Some other events are defined by the fast radiation, sudden blooms, or the global spread of index fossils (e.g., Givetian *pumilio* Events, Nehden Event), or by rapid eustatic pulses (e.g., *semichatovae* Event).

The third-order Klonk Event (Jeppson, 1998) at the base of the Devonian is characterized by some faunal overturn (conodonts, graptolites; e.g., Chlupáč and Kukal, 1988; Chlupáč and Hladil, 2000) and a globally significant carbon isotope spike (see isotope stratigraphy). It can be correlated with the Hüinghausen Event of shallow-water facies (Jansen, 2016). A neglected fourth-order extinction and faunal overturn associated with regression occurred at the end of the lower Lochkovian (e.g., Valenzuela-Ríos and García-López, 1998; papers in Becker et al., 2013). Based on the first description in the Gerri de la Sol Area of the central Pyrenees (Valenzuela-Ríos, 1990), the term Gerri Event is proposed here. It ended black shale deposition in several regions (e.g., Pyrenees, Catalonia, southern Morocco), led to the final

disappearance of the globally distributed, megaplanctonic scyphocrinitids (see Haude et al., 2014), and a replacement of various icriodids by the rapidly radiating ancyrodelloid conodonts. The event will be useful for future substage recognition. The sudden eustatic regression at the end of the Lochkovian has been named as end-*pesavis* Event by Talent et al. (1993) and as Lochkovian—Pragian Boundary Event by Walliser (1996). It caused moderate extinctions, for example, in conodonts Ziegler and Lane, 1989, an isotope spike, and local massive reworking (Suttner and Kido, 2016) or unconformities in near-shore settings (Rhenish Gap, Jansen, 2016). Within the Pragian, there are no known global events in outer shelf facies apart from the marked conodont radiation associated with the current basal Emsian GSSP level (Yolkin et al., 2011; ? = the Gensberg Event of Jansen, 2016).

The globally youngest Graptoloidea occur locally (Bohemia) in black shales of the short, transgressive *atopus* Event (e.g., Slavík, 2004a, 2004b). This eustatic pulse was significant in Morocco and for the spread of early Emsian bactritoids (Cephalopoda), in the sense of the upcoming revised Emsian Stage (Klug et al., 2008; Aboussalam et al., 2015). Extinction patterns at this level still have to be worked out in regions apart from Bohemia and Morocco. As pointed out by Aboussalam et al. (2015), the Basal Zlíchov Event of Chlupáč and Kukal (1988) refers to a slightly younger and more regional facies change at the base of the Zlíchov Limestone in Bohemia. The term Chebbi Event was introduced by Becker and Aboussalam (2011) for the transgression that enabled in southern Morocco the diversification of the oldest Ammonoidea in black shale facies (e.g., Klug, 2001; Klug et al., 2008). García-Alcalde (1997) named the next phase of eustatic rise in the lower Emsian [c. base of *Nowakia (Now.) elegans* Zone] as upper Zlíchov Event (compare Tonarová et al., 2017). Correlations with transgressive phases of the Yujiang Event recognized in South China (Yu et al., 2018) are still equivocal. The subsequent third-order Daleje Event sensu House (1985) and Chlupáč and Kukal (1988) was a major but somewhat gradual transgressive phase in the European-North African realm. Its base coincided with the extinction of the last Mimosphinctaceae (e.g., Chlupáč and Turek, 1983; House, 2002) and of benthic fauna (see the neritic Berlé Event of Jansen, 2016). However, Becker (2007c) noted that the Daleje Deepening is difficult to trace in North America (? = base of sequence Ems-4 of Ver Straeten, 2007), which hampers the search for a suitable intra-Emsian substage boundary. Ma et al. (2017) correlated the southern Chinese Dale sea-level rise with the Daleje Event.

The (Basal) Choteč Event of House (1985) and Chlupáč and Kukal (1988) equals the *Pinacites* or *jugleri* Event of Walliser (1984). It refers to the sudden interruption of well-oxygenated outer shelf facies of Bohemia (Koptíková et al., 2008; Koptíková, 2011; Vodrážková et al., 2013), Morocco (Becker and House, 1994b; Walliser, 2000; Becker and Aboussalam, 2013a; Becker et al., 2018c), southern Siberia

(Bakharev and Sobolev, 2011), and South China (Qie et al., 2018) by a thin interval of pelagic black shales and styliolinites. Associated are moderate extinctions in several organism groups (e.g., Pedder, 2010), short-termed faunal blooms (Brocke et al., 2009), and migration pulses (e.g., Brocke et al., 2016). Higher in the Eifelian, the small-scale (fourth order) Bakoven and Stony Hollow events were first recognized in eastern North America (DeSantis and Brett, 2011) but they seem to represent eustatic pulses; the first may assist a possible substage subdivision at the base of the *australis* Zone. The top-Eifelian, third-order Kačák (House, 1985) or *otomari* Event (Walliser, 1984) was polyphase (e.g., Walliser, 1996; Walliser and Bultynck, 2011), with the basal GSSP having been placed in its upper part (Walliser et al., 1995). There are summaries for Bohemia by Budil (1995) and, with a more global approach, by House (1996a). Becker and House (1994b) and Ward et al. (2013) published additional data for Morocco, Schöne (1997) and Königshof et al. (2016) provided correlations into neritic successions of the Rhenish Massif, DeSantis et al. (2007) into eastern Laurentia, Troth et al. (2011) and Horodyski et al. (2013) into regions of the cold-water Malvinokaffric areas of South America, and Marshall et al. (2007) into the terrestrial successions of the Old Red Continent. There are corresponding deepening pulses in several regions of southern Siberia (spread of top-Eifelian *Agoniatites* faunas; e.g., Bakharev and Sobolev, 2011), in the Arctic parts of Siberia (Sennikov et al., 2018), in the Belarusian Basin (Narkiewicz et al., 2017b), in northern Spain (Liao and García-López, 2018; Askew and Wellman, 2018), in the Ardennes (Gouwy and Bultynck, 2003b; Denayer and Mottequin, 2015), and in the Northwest Territories of Canada (Gouwy and Uyeno, 2017; Kabanov, 2018). Conodont and trilobite extinctions occurred at the base, ammonoid extinction at the top of the main black shale interval (e.g., Becker and House, 1994b; House, 1996b).

Two small-scale bioevents, the Lower and Upper *pumilio* Events (Lottmann, 1990), fall in the basal and middle part of the Middle Givetian. The name stems from the sudden bloom of minute brachiopods (e.g., *Ense pumilio*), also of dacryoconarids, probably as the result of short-termed eutrophication, not due to tsunami deposition, as originally proposed. The events are best developed in central Europe to North Africa but can probably be correlated with short-termed deepening pulses in the Appalachian Basin (Brett et al., 2011; Aboussalam and Becker, 2011) and South China (Ma et al., 2017). The second-order Taghanic Crisis (Aboussalam, 2003) consists of several retransgressive pulses at the end of the middle Givetian (e.g., Baird and Brett, 2003, 2008; Brett et al., 2011, 2018), including the long-known Taghanic Onlap of Johnson (1970), followed by the basal Upper Givetian Geneseo Transgression or Event (Aboussalam and Becker, 2011; Zambito and Day, 2015), which is the Yidade Transgression of South China (Ma et al., 2014; for Hunan see Ma and Zong, 2010). The

Taghanic Crisis was related to global climatic overheating (Joachimski et al., 2004), which finally ended the Malvinocaffric Province of South America and South Africa (Melo, 1989; Boucot, 1989). In low latitudes, it caused major staged ammonoid (e.g., House, 1985; Aboussalam and Becker, 2001, 2002, 2011), trilobite (e.g., Feist, 1991; Aboussalam, 2003), ostracod (Maillet et al., 2016), brachiopod (e.g., Dutro, 1981; Talent et al., 1993), coral (Oliver and Pedder, 1989; Schröder, 2002), and reef extinctions.

The term Frasnes Events (House, 1985; better Frasnes Crisis; *Manticoceras* Event of Walliser, 1996) includes three distinctive transgressive pulses in the top-Givetian *norrisi* Zone (the Ense Event of Ebert, 1993), basal Frasnian (*Ancyrodella rotundiloba pristina* = MN 1 Zone), and in the *Ad. rotundiloba soluta* = lower MN 2 Zone (see brief review by Becker and Aboussalam, 2004). The maximum of black shale formation occurred in the last phase (Aboussalam and Becker, 2007; Becker et al., 2018c; Ma et al., 2017). This significant second-order crisis, with 100 % regional ammonoid extinctions right at the Middle-Upper Devonian Boundary, has so far insufficiently been studied. It caused reef extinctions (e.g., Eichholt and Becker, 2016; Stichling et al., 2017) or shifts of reef growth (e.g., Becker et al., 2016c; Grudev et al., 2016). There are recent data for the Ardennes type area, where the Fromelennes carbonate platform drowned (e.g., Devleeschouwer et al., 2010; Pas et al., 2015; Denayer and Mottequin, 2015). The subsequent, minor Genundewa Event (House and Kirchgasser, 1993) is based on the bipartite maximum transgression of the Genundewa Limestone of New York State, around the MN 2/3 zone boundary. Elsewhere, it enabled a spread of pelagic ammonoid facies as far away as in the Polar Urals (Soboleva et al., 2018) and Western Australia (Becker and House, 1997). The Timan Event (Becker and House, 1997) refers to the sudden spread of ammonoids (especially of *Timanites*) with the next younger maximum transgression in the upper *Ancyrodella rugosa* (upper MN 3 Zone with *Ad. pramosica* and *Ad. africana* and the *Ad. pramosica* Zone of Pizarzowska et al., 2020) to lower *Palmatolepis transitans* Zone (lower MN 4 Zone). Apart from its northern Russian type-region, the event can be recognized as far as Western Canada or Western Australia (House et al., 2000c; Soboleva et al., 2018). In southern Morocco, it triggered a second major interval of black shale and styliolinite deposition (Aboussalam and Becker, 2007; Aboussalam et al., 2018). The next higher, fourth-order Middlesex Event (Becker et al., 1993; = *punctata* Event of Śliwiński et al., 2011) was named after the strong deepening of the New York State Middlesex Shale (House and Kirchgasser, 1993; precise dating as top MN 4 Zone with *Ad. nodosa* = *gigas* M1 in Over et al., 2003). On other continents, it led to the onset of the organic-rich Domanik deposition on wide parts of the Russian Platform and in the Polar Urals (Soboleva et al., 2018), or it drowned the F2d reefs of the Ardennes (e.g., House et al., 2000c). There were possibly relationships with the Alamo Impact of western North America, which occurred within the interval with *Ad. nodosa* (=*gigas* M1) and below the first local record of *Pa. punctata* (see biostratigraphy data in Morrow et al., 2009). A special volume (Balinski et al., 2006) was devoted to the "eventful" lower/middle Frasnian transition (Timan to Middlesex Event interval), with some focus on Polish successions. Recently, the Middlesex level was also recognized within anoxic facies of southern Morocco (Becker et al., 2018b; Aboussalam et al., 2018).

The (Basal) Rhinestreet Event (Becker et al., 1993) is based on the onset of the thickest black shale package of eastern North America (e.g., House and Kirchgasser, 1993; Blood and Lash, 2018). It can be correlated with the transgression at the base of the Upper Domanik of the Russian Platform (House et al., 2000c) and with a thin, organic-rich styliolinite of southern Morocco (Becker and Aboussalam, 2013b; Aboussalam et al., 2018). Ma et al. (2017) suggested a possible correlation with the Longkouchong Transgression of South China. Aboussalam and Becker (2013) and Aboussalam et al. (2018) noted a second, slightly younger (upper MN Zone 7 = *nonaginta* Zone) styliolinite interval in southern Morocco, which was named as Upper Rhinestreet Event. The *semichatovae* Event (Sandberg et al., 1992) was named after the sudden spread of its index conodont (*Palmatolepis semichatovae*) with the major transgression that shall define in future the base of the Upper Frasnian (Ziegler and Sandberg, 1997). This sudden eustatic rise has been recognized in many parts of the world and led to strong radiations in deeper water organism groups, such as conodonts and ammonoids.

Most famous for Devonian event stratigraphy are the hypoxic Lower and Upper Kellwasser Events of the polyphase Kellwasser Crisis (Schindler, 1990). There are too many publications to be quoted here and these cover most regions of the world, apart from the cold-water regions of South America, South Africa, and Antarctica. The last review by Racki (2005) seems now to be outdated but even earlier models (e.g., Joachimski and Buggisch, 1993; Becker and House, 1994a) include many still valid observations, processes, and patterns. Most significant are volumes dedicated to the Frasnian-Famennian (F/F) boundary (Balinski et al., 2002; Racki and House, 2002; Over et al., 2005) and summaries for specific regions, such as the Rhenohercynian type region of Germany (Gereke and Schindler, 2012), the Ardennes (Mottequin and Poty, 2016), southern France (Becker and House, 1994), the Polish–Moravian Basin (Matyja and Narkiewicz, 1992; Racki et al., 2002), the Russian Platform (e.g., Kirilishina, 2006), the South Urals (Abramova and Artyushkova, 2004; Veimarn et al., 2004; Artyushkova et al., 2011; Tagarieva, 2012), Afghanistan (Farsan, 1986), South China (e.g., Ji, 1989a; Ma and Bai, 2002; Wang and Ziegler, 2002; Ma et al.,

2015), central to eastern (Over, 2002; Day and Witzke, 2017), and western United States (Morrow, 2000; Morrow and Sandberg, 2003), Ontario, Canada (e.g., Levman and von Bitter, 2002; Klapper et al., 2004), and the Canadian Rocky Mountains (e.g., Mountjoy and Becker, 2000; Whalen et al., 2002, 2016). For a precise understanding of the mass extinction, it is important to separate the two organic-rich Kellwasser intervals as well as the benthos extinctions at the base of the hypoxic Upper Kellwasser level from the sharp extinction of pelagic biota right at the F/F boundary (e.g., Becker et al., 1989; Schindler, 1990; Hartenfels et al., 2016). The latter coincided with a very peculiar interval of widespread seismic events causing synchronous mass flows on several continents (e.g., Sandberg et al., 1988; Hou et al., 1989; Racki, 1998; Whalen et al., 2002; Racki et al., 2002; Becker et al., 2016e). Only in very rare cases, diverse benthic faunas occur in time equivalents of the Upper Kellwasser Limestone (e.g., Hartenfels et al., 2016).

The "Nehden Event" (House, 1985) is not a true event but refers to the stepwise, long-lasting lower Famennian eustatic rise, which culminated in episodes of maximum flooding and maxima of black shale deposition, notably in the higher *termini* (=Middle *crepida*) to *glabra pectinata* (=Uppermost *crepida*) zones, especially in the German type region (Becker et al., 2016f). The deepening enabled the gradual recovery of conodonts (e.g., Schülke, 1999) and goniatites (Becker, 1993a, 1993b) from the F/F boundary mass extinction. The two pulses of the following global Condroz Events (Becker, 1993a), named after the biphasic onset of the Condroz Sandstone Group of Belgium, ended this long phase of global high sea level. The short-termed regressions led to a significant global ammonoid faunal overturn in the (Lower) *rhomboidea* and *gracilis* (=Upper *rhomboidea*) zones (Becker, 1993a). Many lineages disappeared with the termination of the formerly widespread eutrophic black shale habitats. The Condroz sea-level fall is well expressed in South China (Ma et al., 2017). The term Enkeberg Event (House, 1985) refers to an interval of regional overturn in Rhenish pelagic faunas and cannot be regarded as a global event. Not identical, and much more important as a phase of global radiation and for global chronostratigraphy (future Middle Famennian Substage definition), is the rapid eustatic rise slightly above the base of the *marginifera marginifera* Zone (compare the Maguano Transgression of South China, Ma et al., 2017).

Three global hypoxic and transgressive events interrupted the overall shallowing trend of the middle/upper Famennian, the Lower and Upper *Annulata* Events (House, 1985) high in the *Palmatolepis rugosa trachytera* conodont zone and the biphasic Dasberg Crisis around the boundary of the Lower and Middle *expansa* zones (now *gracilis expansa* and *aculeatus aculeatus* zones; Becker, 1993a; Becker et al., 2004a; Hartenfels et al., 2009; Hartenfels and Becker, 2009; Hartenfels, 2011; Racka et al., 2010). The *Annulata* Events were named after two thin marker black shales with *Platyclymenia annulata* first recognized by Schmidt (1924) in the Rhenish Massif (see reviews by Becker, 1992; Korn, 2004). Their international significance for correlation and future chronostratigraphy were emphasized by Hartenfels and Becker (2016a), supplemented by new data for NW China (Zong et al., 2015), Moravia (Weiner and Kalvoda, 2016), and Bulgaria (Boncheva et al., 2015). In South China, a corresponding Tieshan Event has been noted (Ma et al., 2017). The transgressive phase is marked by the sudden global spread of marker clymeniids and goniatites; severe ammonoid extinction occurred in a prolonged interval just before, possibly due to global extreme oligotrophy (Becker and Hartenfels, 2010; Becker et al., 2018e). The Dasberg Crisis is characterized by a strange short time difference between black shale units of European blocks and of North Africa. It is also characterized by blooms, radiations, and the sudden spread of taxa; extinction occurred during the preceding regression, which is also characterized by extreme oligotrophy (top of Famennian III). Marynowski et al. (2010) documented for Poland (Holy Cross Mountains) the characteristics of palynology and organic geochemistry of Dasberg Black Shales, supporting previous ideas of anoxia and sudden eutrophication. A higher regional black shale (Kowala Shale) and the transgressive Epinette Event of the Ardennes named by Dreesen et al. (1989) have not yet been sufficiently correlated globally (see discussion in Hartenfels and Becker, 2016b). This is also true for the Chinese Shaodong Event (Hou, 2008; Ma et al., 2017) and for two thin anoxic peaks between the *Annulata* and Dasberg Black Shales of southern Morocco (Korn, 1999; Becker et al., 2018e). All these so far local occurrences indicate minor additional upper Famennian events, which are significant for the general understanding of event processes and any potential cyclicity.

The global Hangenberg Crisis near the Devonian-Carboniferous (D/C) Boundary has been reviewed by Becker et al. (2016b) and Kaiser et al. (2016). A summary of the timing of glaciations leading to the major top-Famennian Regression was provided in the same volume by Lakin et al. (2016). Therefore there is no reason to recycle here the wealth of publications on the topic. Important for correlation and the redefinition of the D/C boundary is the distinction of a regressive prelude (Drewer Sandstone), followed in the Lower Crisis Interval by the main Hangenberg Extinction at the base of globally widespread anoxia and black shales (Hangenberg Black Shale), by the Middle Crisis Interval defined by staged glaciogenic regression (Hangenberg Shale/Sandstone), and by the polyphase Upper Crisis Interval, which embraces both the initial recovery of Carboniferous-type marine taxa during rewarming and new transgression, and the main terrestrial extinction. In the frame of the ongoing D/C boundary revision, new studies on the Hangenberg Crisis continue to be published and there are forthcoming special D/C boundary volumes. For example, Kalvoda et al. (2015) improved the conodont-foraminifer correlation in Moravia, Yao et al. (2016) published on the postcrisis proliferation of microbial carbonates, Bábek et al. (2016) on geochemical aspects of D/C boundary sequence stratigraphy, Liu et al. (2018) on the Hangenberg Black Shale in Tibet, Kalvoda et al. (2018) on the geochemistry of Middle Crisis Interval microbial laminites from Moravia, Plotitsyn et al. (2018) on new data for sections in the northern Ural, and Zhang et al. (2018) on the first record of Hangenberg Black Shale with survivor cymaclymeniids in South China. Martinez et al. (2018) located for the first time the crisis interval and Hangenberg Black Shale equivalents in the thick, hypoxic Ohio succession. Denayer et al. (2019) pointed out that extinctions in the pelagic and neritic realm (e.g., on the Ardennes Shelf) might not have been synchronous.

22.2.2.2 Sequence stratigraphy/eustatic changes

In a broad sense, sea-level changes in the Devonian have been commented upon at least since the collative work of French geologists in the 19th century (Figs. 22.1 and 22.11). Much later, a more systematic comparison between New York (United States) and Europe was published by House (1983) as an initial attempt to create a Devonian eustatic curve. This was improved using detailed conodont evidence for the Laurussian area, leading to the sea-level curve and depophase terminology of Johnson et al. (1985), which soon became a widely quoted standard that is still recycled in many publications. However, the "Johnson et al. curve" is not based on a sequence stratigraphic approach and concentrates on the correlation of times of fastest and maximum transgression. Another problem is the lack of precise definitions for some of the depophases. They were based on examples from different regions that are now known not to correlate at all, creating ambiguity, which level should be used for a refined definition.

Many subsequent publications have summarized regional relative sea-level curves, often with some focus on short-termed eustatic fluctuations reflected by sudden hypoxic events. Regional reviews correlated against the standardized conodont time scale were bundled in a thematic volume by

House and Ziegler (1997). Other examples are the Middle/ Upper Devonian sea-level histories for central and western North America (Johnson and Sandberg, 1989; Day et al., 1996), the Iowa Basin (e.g., Witzke et al., 1989; last review in Day and Witzke, 2017), eastern North America (Ver Straeten, 2007; Bartholomew and Brett, 2007; Brett et al., 2011), the Timan of northern Russia (House et al., 2000c), the Polar Urals (Sobolev and Soboleva, 2018), Australia (Talent and Yolkin, 1987; Talent, 1989; Becker et al., 1993; Becker and House, 1997), and South China (Ma et al., 2009). Most recently, Bábek et al. (2018) analyzed Lower and Middle Devonian sea-level changes and sequence stratigraphy of Bohemia. For the numerous papers on individual eustatic events see the previous chapter. Prominent and regular sea-level oscillations in thick shallow-water carbonate platforms and siliciclastics, unfortunately, mostly lack a precise age control. The clear identification of sequence boundaries and maximum flooding episodes combined with a biostratigraphic-tuned correlation of the diverse regional trends should result in a revised eustatic curve with third and fourth order global sequences.

Several authors have published refinements for the "Johnson et al. curve," including depophase subdivisions. These include Klapper (1992), García-Alcalde (1997), and Ver Straeten (2009) for the Emsian, Baird and Brett (2003, 2008), Aboussalam (2003), and Aboussalam and Becker (2011) for the middle/upper Givetian, Becker (1993a) for all of the Upper Devonian, Day et al. (1996) for the Givetian/Frasnian, House et al. (2000c) for the Frasnian, Hartenfels and Becker (2009, 2016a, 2016b) for the middle/ upper Famennian, and Bless et al. (1993), Becker (1996), and Kaiser et al. (2016) for the D/C boundary interval.

22.2.2.3 Stable isotope stratigraphy

The stratigraphic use of stable isotope data has increased significantly in the Devonian in the last two decades (Figs. 22.1 and 22.11). Especially the use of carbon isotopes has become a standard procedure. Short-timed peaks are excellent markers for global correlation and to identify changes in the global carbon reservoirs in association with global events. Early studies by Talent et al. (1993) and Hladíková et al. (1997) gave way to attempts to establish complete carbon isotope curves, for example, for the Devonian of Europe or of the Great Basin of western North America (Saltzman, 2005). This was either based on whole rock carbonate data or on well-preserved brachiopod shells (Buggisch and Mann, 2004; Buggisch and Joachimski, 2006; Van Geldern et al., 2006; Saltzman and Thomas, 2012). Positive $\delta^{13}C$ spikes can also be recognized in organic carbon (e.g., Joachimski, 1997) and in individual biomarkers (Joachimski et al., 2002). There are also studies that employ carbon isotopes to assist sequence stratigraphy in the correlation of

complete carbonate complexes, such as the Canning Basin (Playton et al., 2013; Hillbun et al., 2016).

Significant isotope spikes are now known for many of the Devonian Events discussed above. Records are as follows:

Silurian-Devonian boundary or *Klonk Event* (Schönlaub et al., 1994; Hladíková et al., 1997; Buggisch and Joachimski, 2006: Bohemia and Carnic Alps; Andrew et al., 1994: Australia; Saltzman, 2002: eastern to western North America; Buggisch and Mann, 2004: central and Southern Europe; Kleffner et al., 2009; Jacobi et al., 2009; Hasson et al., 2016: Appalachian Basin; Malkowski et al., 2009; Racki et al., 2012; Podolia, Ukraine; Manda and Frýda, 2010: Bohemia; Zhao et al., 2011; Zhao et al., 2015: South China).

Gerri Event, basal middle Lochkovian (Ma et al., 2017; Qie et al., 2018: Three Rivers region, China)

Lochkovian—Pragian Boundary Event (Buggisch and Joachimski, 2006; Qie et al., 2018)

Kitab Reserve Emsian GSSP level and conodont radiation phase (future mid-Pragian; Izokh, 2011a)

Basal Zlíchov Event (Buggisch and Mann, 2004)

(Basal) Choteč Event (Vodrážková et al., 2013: Bohemian Type Region)

Kačák Crisis (Hladíková et al., 1997: Sageman et al., 2003: Appalachian Basin; Bohemia; Buggisch and Joachimski, 2006: MN; Van Hengstum and Gröcke, 2008: Ontario, Canada; Qie et al., 2018: South China)

Pumilio Events (Buggisch and Joachimski, 2006: Germany and southern France)

Taghanic Crisis (Aboussalam, 2003: southern France and Morocco; Day et al., 2010: Iowa Basin; Zambito and Day, 2015; Zambito et al., 2016: Appalachian Basin; Qie et al., 2018: South China)

Frasnes Events (Wang and Bai, 2002: Guangxi; Sageman et al., 2003: Appalachian Basin; Buggisch and Joachimski, 2006; Becker et al., 2016d: southern France, Rhenish Massif).

Timan and Middlesex Events (Sageman et al., 2003: Appalachian Basin; Buggisch and Joachimski, 2006: Rhenish Massif, Carnic Alps, and Western Australia; Pisarzowska et al., 2006; Pisarzowska and Racki, 2012: Holy Cross Mountains, Poland; Yans et al., 2007: Belgium; Morrow et al., 2009: Nevada, Great Basin; Śliwiński et al., 2011; Labounty and Whalen, 2016: Western Canada; Becker and Aboussalam, 2013b: southern Morocco; Izokh et al., 2015: Rudny Altai, Siberia; Tulipani et al., 2015: Canning Basin, Western Australia)

Lower Rhinestreet Event (Sageman et al., 2003; Blood and Lash, 2018: New York State; Becker and Aboussalam, 2013a, 2013b: southern Morocco)

Kellwasser Crisis (Goodfellow et al., 1989; Joachimski et al., 2002; Stephens and Sumner, 2003: Canning Basin, Western Australia; Goodfellow et al., 1989; Geldsetzer et al., 1993: Western Canada; Halas et al., 1992;

Joachimski et al., 2001; Joachimski in Racki et al., 2002; Trela and Malec, 2007; Malec, 2014; Rakociński et al., 2016: Holy Cross Mountains, Poland; Joachimski and Buggisch, 1993, 2002; Buggisch and Joachimski, 2006: Rhenish Massif, Germany; Joachimski et al., 1994; Buggisch and Joachimski, 2006: Carnic Alps; Joachimski et al., 2002: Great Basin, western United States, and Moroccan Meseta; Yudina et al., 2002: northern Urals; Chen et al., 2002, 2005; Gong et al., 2005; Xu et al., 2012; Chang et al., 2017; Song et al., 2017: Guangxi, South China; Sageman et al., 2003; Uveges et al., 2018; Blood and Lash, 2018: Illinois and Appalachian basins; Buggisch and Joachimski, 2006: Thuringia, Germany; Savage et al., 2006; Königshof et al., 2012: western Thailand; De la Rue et al., 2007: Indiana; Izokh et al., 2009: Kuznetsk Basin, Siberia; Izokh, 2011b: South Urals; Azmy et al., 2012: Ardennes; Suttner et al., 2014; Wang et al., 2016: Junggar Basin, NW China; Hedhli et al., 2016: Alberta and Montana; Day and Witzke, 2017: Iowa Basin; Lash, 2017)

Basal middle Famennian transgression (Buggisch and Joachimski, 2006: Franconia)

Upper—uppermost Famennian Epinette or *Strunian events* (Saltzman, 2005; Myrow et al., 2011: Great Basin, western North America; Buggisch and Joachimski, 2006; Kumpan et al., 2014: Ardennes; Kaiser et al., 2008, 2017: Franconia)

Hangenberg Crisis (Xu et al., 1986: Guangxi, South China; Schönlaub et al., 1992; Kaiser et al., 2006, 2008: Carnic Alps; Kaiser et al., 2006: Rhenish Massif, Germany; Bai et al., 1994: Hunan, China; Brand et al., 2004; Buggisch and Joachimski, 2006: MN, southern France; Cramer et al., 2008; Clark et al., 2009; Day et al., 2011: Illinois and Missouri; Kaiser et al., 2008; Bojar et al., 2013: Graz Paleozoic, Austria; Kumpan et al., 2013: Moravia; Azmy et al., 2009; Kumpan et al., 2014: Ardennes; Myrow et al., 2011; Hagadorn et al, 2016: Colorado; Qie et al., 2015: several regions of China; Cole et al., 2015: Utah/Montana, Great Basin; Martinez et al., 2018: Ohio)

Diagenetic overprinting and locally variable primary producers prevent the use of the magnitude of isotopic excursions for correlation (e.g., Da Silva and Boulvain, 2008; Qie et al., 2015). This is especially relevant for C_{org} measurements (King Phillips et al., 2017). The diagenetic recycling of C_{org} during carbonate recrystallization may cause pronounced local negative spikes in sharp contrast to the global trend. Examples are Wang and Bai (1989) for the F/F boundary of Hunan, Wei and Ji (1989) for the D/C boundary of Guizhou, Talent et al. (1993) for the Kačák and Taghanic Crises of Eastern Australia, Zheng et al. (1993) for the Kellwasser Crisis of Guangxi, Bai et al. (1994) for the Hangenberg Crisis of Guangxi, Becker and Aboussalam (2013b) for the Timan and Lower Rhinestreet Events of southern Morocco, and Becker et al. (2013) for the Upper Hangenberg Crisis Interval of southern Morocco.

Widespread diagenetic alteration of carbonates prevents also the use of whole rock oxygen isotope data (e.g., Halas et al., 1992; 1997; Chen et al., 2005; Bojar et al., 2013) for correlation and paleotemperature reconstructions (Brand, 2004). At least in the Middle and Upper Devonian, oxygen data from conodont phosphate (Elrick et al., 2009; Joachimski et al., 2009) seem to be superior to measurements of supposed pristine brachiopod calcite (Joachimski et al., 2004; Van Geldern et al., 2006) as a tool to reconstruct Devonian seawater temperatures. However, Pucéat et al. (2010) emphasized the difficulty to calculate absolute temperatures from measured isotopic values. Elrick et al. (2009) found evidence for cooling intervals in the lower and upper Eifelian. There are negative $\delta^{18}O$ peaks in the upper Givetian (Joachimski et al., 2009: warming) and strong fluctuations associated with the Kellwasser Events in the Rhenish Massif (Joachimski and Buggisch, 2002), MN, and in the Moroccan Meseta (Balter et al., 2008; Le Houedec et al., 2013). In epeiric basins with restricted water circulations (Belarus Basin, Narkiewicz et al., 2017; Illinois Basin, Gouwy et al., 2012), isotopic signatures may have been overprinted by other factors, such as salinity fluctuations. Kaiser et al. (2006) used conodont isotope data to reconstruct climatic changes associated with the Hangenberg Crisis. The available data set is still much too incomplete to use oxygen isotope data for improvements of the Devonian time scale.

The marine $^{87}Sr/^{86}Sr$ ratio through the Devonian (Diener et al., 1996; Denison et al., 1997; Van Geldern et al., 2006) shows a broad trough, centered on the basal Eifelian to middle Givetian. It reaches 0.7078 at its minimum between a Lower Devonian high of 0.7087 and a Lower Carboniferous high of 0.7083. This range is large enough for using Sr isotope stratigraphy in the interval (see also McArthur et al., 2020, Ch. 7: Strontium isotope stratigraphy, this book). Reliable data (Brand, 2004) are best obtained from pristine brachiopod calcite, but measurements of conodont phosphate (Kürschner et al., 1993; Ebneth et al., 1997; Veizer et al., 1997) or bulk rock samples (e.g., Carpenter et al., 1991; Chen et al., 2005; Carmichael et al., 2015; Hagadorn et al., 2016; Wang et al., 2018) may be meaningful, too. The Sr isotopic record is studied in order to track episodic changes in detrital Sr discharge or of mid-oceanic ridge activity. Recently, it was also tried to reconstruct changing weathering regimes at the Frasnian-Famennian (F/F) boundary with the help of Mg (Huang et al., 2017) and Zn isotopes (Wang et al., 2018).

Based on fluctuations in oceanic circulation, productivity, bacterial sulfate reduction, and oxygenation, sulfur isotope data have used as event stratigraphic markers around the Eifelian/Givetian (Sageman et al., 2003), F/F (e.g., Geldsetzer et al., 1987; Halas et al., 1992; Wang et al., 1996; Joachimski et al., 2001; Levman and von Bitter, 2002; Chen et al., 2013), and D/C boundaries (Bojar et al., 2013).

Simon et al. (2007) used a still limited data set to reconstruct a curve for the sulfur isotope composition of Devonian marine sulfate, which shows a significant depression centered in the middle Emsian and with a broad high in the lower/middle Famennian (see also Paytan et al., 2020, Ch. 7: Sulfur isotope stratigraphy, this book). Bingham-Koslowski et al. (2016) found a possible short, negative F/F spike and showed how a marked positive $\delta^{34}S$ excursion characterizes an upper lower Famennian episode of improved oxygenation in Ontario, which may reflect local effects of the Condroz Regressions.

Only a few studies have so far utilized neodymium isotopic changes for Devonian stratigraphy. Water mass movements and sea-level fluctuations are well recorded in values measured from conodont phosphate. Dopieralska et al. (2006) showed that the basal upper Frasnian *semichatovae* Transgression can be recognized by the sudden incursion of oceanic water with more radioisotopic values on the northern Gondwana shelf, while the eustatic pre-Hangenberg shallowing is recorded in the Sudetes by an unusual fall of εNd values. Dopieralska et al. (2016) expanded the Moroccan and Polish data to German and French F/F boundary sections and used the Nd isotope record for sea-level reconstructions. Dopieralska et al. (2012) applied Upper Devonian conodont Nd data for paleogeographic reconstructions in SE Asia. In a case study involving two Frasnian Sections of the Great Basin (Nevada), Theiling et al. (2017) suggested that Nd and Sm isotopes from whole rock samples indicate complex transport patterns, resulting in fluctuating εNd values with time.

A rather new development is the use of $\delta^{15}N$ fluctuations to reconstruct nitrogen fixation or recycling during the black shales of the Kellwasser (Uveges et al., 2019) and Hangenberg Crises (Liu et al., 2016; Martinez et al., 2018). Uranium ($\delta^{238}U$) isotopic changes also correlate with carbon isotope trends near the F/F boundary (Song et al., 2017; White et al., 2018) but their interpretation is still equivocal since there are strong contradictions to other geochemical parameters.

22.2.2.4 Organic geochemistry stratigraphy

Measurements of total organic carbon contents of carbonates and shales aid the recognition of various Devonian anoxic events (e.g., Joachimski et al., 2001; Sageman et al., 2003; Mahmudy-Gharaie et al., 2004; Gong et al., 2005; Hartkopf-Fröder et al., 2007; Song et al., 2017 for the Kellwasser Crisis; Kaiser et al., 2006; Marynowski and Filipiak, 2007; Bojar et al., 2013 for the Hangenberg Crisis). Based on the mostly high thermal maturity of Devonian sediments, organic geochemistry analyses, including biomarkers and sterane/hopane ratios, are of geographically restricted use (e.g., Schwark and Empt, 2006). However, they can document short-termed changes in primary producers and can be used for regional correlation and event interpretation (e.g., Joachimski et al., 2001; Gong et al., 2007; Chen et al., 2013; Haddad et al., 2016; Martinez et al., 2018). For example, biomarkers of green sulfur bacteria give evidence for anoxia reaching the open marine photic zone in Upper Kellwasser time (Hartkopf-Fröder et al., 2007), in the subsequent lower Famennian (Marynowski et al., 2011), and during the *Annulata* (Racka et al., 2010) and Dasberg Events (Marynowski et al., 2010). Tulipani et al. (2015) documented biomarker spikes for an anoxic/euxinic setting around the Givetian/Frasnian boundary (level of the Frasnes Events) below the onset of reef growth in the Canning Basin of Western Australia.

22.2.2.5 Inorganic geochemistry stratigraphy

Many publications have used stratigraphically organized inorganic geochemistry data to fingerprint, interpret, and correlate Devonian events, especially black shale units (e.g., Robl and Barron, 1989; Rimmer et al., 2004). A focus lay on the reconstruction of fluctuating detrital influxes (erosion and sea-level proxies), oxygenation levels (redox-sensitive elements), changing nutrient availability, and recycling (paleoproductivity proxies), which all may be used for correlation. Studies cover longer intervals of complete regions (Lower/Middle Devonian of Bohemia, Bábek et al., 2018; Middle/Upper Devonian of the Appalachian Basin, Sageman et al., 2003) or yielded significant results individually for the Kačák Crisis (Ellwood et al., 2011a; Kabanov, 2018), Middlesex/*punctata* (Śliwiński et al., 2011; Labounty and Whalen, 2016; Kabanov, 2018), and *Annulata* Events (Racka et al., 2010). But many more studies focused on the Kellwasser Crisis (McGhee et al., 1996; Pujol et al., 2006: Rhenish Massif, Germany; Goodfellow et al., 1989: global record; Geldsetzer et al., 1993: NW Canada, Bratton et al., 1999: Great Basin, western North America; Yudina et al., 2002: northern Urals; Ma and Bai, 2002; Gong et al., 2005; Ma et al., 2009: South China; Tribovillard et al., 2004; Averbuch et al., 2005; Pujol et al., 2006: southern France; Riquier et al., 2005, 2007; Averbuch et al., 2005: Moroccan Meseta; Riquier et al., 2006: Harz Mountains, Germany; Formolo et al., 2014: Michigan Basin; George et al., 2014: Western Australia; Rakociński et al., 2016: Poland; Weiner et al., 2017: Moravia; Lash, 2017; Blood and Lash, 2018: New York State). Numerous similar research dealt with the Hangenberg Crises (Wang and Yang, 1988; Bai and Ning, 1989; Wei and Ji, 1989; Wang and Xia, 2002; Zeng et al., 2010, 2011: South China; Hao, 2001: Tarim Basin, NW China; Bojar et al., 2013: Graz Paleozoic; Kumpan et al., 2013; Kalvoda et al., 2018: Moravia; Kumpan et al., 2014: Ardennes; Kumpan et al., 2015; Becker et al., 2016e: Rhenish Massif; Carmichael et al., 2015: Junggar Basin, NW China; Bábek et al., 2016: various European regions).

Nothdurft et al. (2004) proved for the thermally weakly overprinted reef complexes of the Canning Basin that pristine REE patterns can be preserved in Upper Devonian limestones. The search for volcanism signatures stimulated investigations of whole rock REE patterns at the F/F and D/C boundaries (Wang and Yang, 1988; Wang and Xia, 2002: South China; Kalvoda et al., 2018: Moravia). Without conclusive trends, F/F boundary REE contents of conodonts have also been investigated (Grandjean-Lécuyer et al., 1993; Girard and Albarède, 1996; Girard and Lécuyer, 2002). Zircon fingerprinting enabled the precise correlation of thin volcaniclastic layers in the Frasnian of central Europe, which proved that major eruptions took place in now eroded areas, leading to a repeated very wide spread of ash (Winter, 2015). Most recently, short and distinctive spikes of mercury enrichments added geochemical evidence for pulses of major F/F volcanism (Moreno et al., 2018; Racki et al., 2018), but the correlation with the Viluy Traps of eastern Siberia (e.g., Ricci et al., 2013) is ambiguous.

Supposed impact signatures near the Eifelian−Givetian (Ellwood et al., 2003; contra: Racki and Koeberl, 2004) and Frasnian−Famennian boundaries, including microspherules and iridium anomalies (e.g., Playford et al., 1984; Wang et al., 1991), could not be substantiated in subsequent studies (McGhee et al., 1996; Claeys et al., 1996; Girard et al., 1997; Schmitz et al., 2006), are so far regional features (Bai et al., 1994; Ma and Bai, 2002), or do not coincide with extinction levels (Wang, 1992; Wang et al., 1994; Marini et al., 1997). The well-dated Frasnian Flynn Creek (e.g., Schieber and Over, 2005) and Alamo Impacts (e.g., Warme and Sandberg, 1996; Morrow et al., 2009) were regional events in the Frasnian of North America; they do not correspond closely to extinction levels. Mason and Caffee-Cooper (2016) reported on new F/F microtectites from Indiana. The role of possibly D/C boundary impacts in South China (Bai et al., 1987, 1994; Bai and Ning, 1989) and Australia (Glikson et al., 2005, 2013) requires further studies (discussion in Kaiser et al., 2016).

22.2.2.6 Cyclostratigraphy

Due to the chaotic diffusion of the inner solar system, the accurate theoretical calculation of precession, obliquity, and their phase relationship with eccentricity is limited to the last 50 Myr (Laskar et al., 2004; Fig. 22.13). However, eccentricity orbital variations are considered as stable during time, and their durations (mostly 405 kyr) were proposed to be used as a geologic chronometer for successions beyond 50 Myr (Laskar et al., 2011). The counting of the 405-kyr eccentricity cycles provides progress for the estimate of stages or conodont zone durations during the Devonian.

The first cyclostratigraphic studies focusing on the Devonian were mostly based on direct observations and cycle counting on outcrop, by allocating main observed cycles to precession or eccentricity. In some cases,

uncertainties of absolute durations prevented the authors to assign fifth-order sea-level fluctuations estimated at 35−85 kyr clearly to Milankovitch cyclicity (McLean and Mountjoy, 1994). Elrick (1995) found peritidal and subtidal cycles of c. 50- and c. 130-kyr duration, respectively, in the Great Basin of the western United States. Bai et al. (1995) estimated eustatic cycles form the Middle/Upper Devonian of South China to represent 100-kyr cycles. House (1995) linked small-scale microrhythms to precession cycles and estimated the Givetian to have lasted about 6.5 Myr. Early work includes the exploratory study of Chlupáč (2000), who worked on the Barrandian (Czech) historical stratotypes and identified ∼410−450 bedding couplets within the Lochkovian and estimated about 350−380 couplets within the historical Pragian (roughly equivalent to the Praha Formation), pointing to a longer duration of the Lochkovian compared to the historical Pragian. Chen and Tucker (2003) counted cycle bundling around the Frasnian/Famennian boundary in China and estimated the timing of the biotic crisis to be constrained at ∼450 kyr in duration.

Since the 2012 Devonian time scale (Becker et al., 2012), recent efforts in constructing precise floating astronomical time scales for portions of the Devonian have been carried out, through the application of spectral analysis and statistical techniques (synthesis in Fig. 22.13). This lead to an estimated duration of the Lochkovian of 7.7 Myr based on various spectral techniques applied on magnetic susceptibility (MS) and gamma-ray signal (GRS) data from the Pod Barrandovem section, Czech Republic (Da Silva et al., 2016). The same study provided a duration of 1.7 Myr for the (restricted) Pragian (using the definition of GTS2012) and 5.7 Myr for the Praha Formation, based on the MS and GRS from the Požár and Pod Barrandovem outcrops of the Czech Republic (Da Silva et al., 2016). Ellwood et al. (2015) proposed a duration of 6.2 Myr for the Eifelian based on the MS signal of Bou Tchrafine section in the Tafilalt of SE Morocco. Combining the MS signals from outcrops in Morocco, France, and New York State (United States), allowed to build an astronomical time scale with a duration of ∼5.6 Myr for the Givetian (Ellwood et al., 2011b). On the other hand, De Vleeschouwer et al. (2015) proposed a duration of 4.35 ± 0.45 Myr for the Givetian, based on four overlapping MS records from Belgium, and Pas et al. (2020) interpreted a duration for the Eifelian of c. 5 Myr from sections in the Appalachian Basin of New York. De Vleeschouwer et al. (2012a) calculated a duration of 6.5 ± 0.4 Myr for the Frasnian, based on five overlapping MS records from Western Canada. Pas et al. (2018) focused on three overlapping cores from the Illinois Basin of North-America, going through the upper Frasnian to Lower Carboniferous. Spectral analysis on the MS records of these cores suggested a duration of 13.5 Myr for the Famennian and a placement of the Frasnian−Famennian boundary at 372.4 Ma. A cyclostratigraphy study of a conodont-zoned

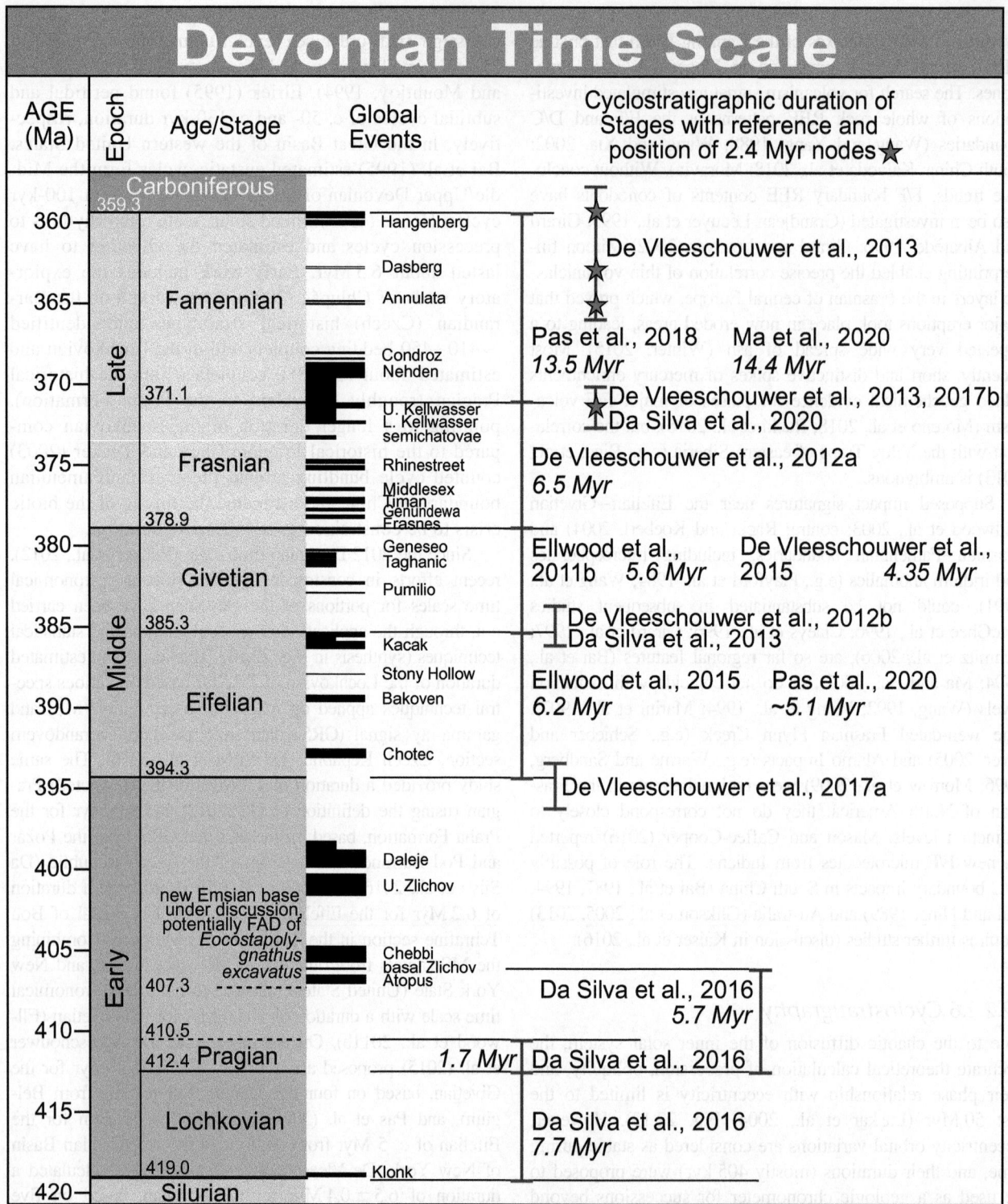

FIGURE 22.13 Devonian time frame from cyclostratigraphy studies with correlation to the succession of global events. Relevant cyclostratigraphic studies are shown with their time intervals (blue bars), corresponding references; and, when the study covers a complete Stage, its estimated duration for that Stage (in italics). The positions of 2.4-Myr long eccentricity cycle nodes are indicated through red stars. The four 2.4-Myr nodes are all associated with important anoxic events.

carbonate reference section in the Guizhou Province of South China indicated a slightly longer duration for the Famennian of 14.4 ± 0.3 Myr (Ma et al., 2020).

On top of the calculation of these long-lasting stage durations, some strategic short intervals around specific stage boundaries were also studied. The Emsian−Eifelian boundary at the GSSP section at Wetteldorf, Germany, is interpreted as lasting ~250 kyr (De Vleeschouwer et al., 2017a). The Eifelian—Givetian boundary was studied in Belgium on two different sections (Monts de Baileux section in Da Silva et al., 2013; La Couvinoise in De Vleeschouwer et al., 2012b), allowing to develop a floating time scale for this interval. An orbitally calibrated chronology across the Lower and Upper Kellwasser over six globally distributed Frasnian—Famennian sections stipulates that 600 kyr separates the Lower and Upper Kellwasser positive $\delta^{13}C$ excursions and give a span of 800 kyr between the onset of the positive Lower Kellwasser carbon isotope excursion and the Frasnian—Famennian boundary (De Vleeschouwer et al., 2017b). The Famennian anoxic shales of the *Annulata*, Dasberg and Hangenberg events were studied in Poland based on the facies alternation and the high-resolution cyclostratigraphic framework; it indicates that those shales were deposited at 2.2 and 2.4 Myr intervals respectively (De Vleeschouwer et al., 2013).

The development of these cyclostratigraphic frameworks also allows to develop paleoclimatic interpretations for their respective stratigraphic interval. De Vleeschouwer et al. (2015) noticed a strong imprint of precession and obliquity in Belgian Givetian sections deposited at paleotropical latitudes. This development of obliquity is relatively unexpected since obliquity classically has a limited impact on the incoming solar radiations at tropical latitudes. This imprint at low latitude is interpreted as related to a climatic teleconnection, which could be related to Givetian ice sheets at high latitude (De Vleeschouwer et al., 2015) for which, however, there is so far no geological evidence.

The cyclostratigraphic framework of the Frasnian—Famennian boundary interval (De Vleeschouwer et al., 2017b; Da Silva et al., 2020) also points to a strong pacing of obliquity during the anoxic Upper Kellwasser Event, similar to what occurred during the Cretaceous Ocean-Anoxic-Event-2 (OAE-2), influencing the development of anoxia. For this interval the strong development of obliquity is interpreted as associated with a minimum eccentricity configuration, leading to lower seasonal extremes allowing organic carbon to accumulate. Furthermore, the formation of the upper/uppermost Famennian *Annulata*, Dasberg, and Hangenberg anoxic shales are considered as separated by a 2.4-Myr eccentricity cycle (De Vleeschouwer et al., 2013). This astronomical configuration, with high eccentricity, would lead to sea-level rise by triggering the collapse of small continental Gondwanan ice-sheets.

The position of these eccentricity extremes (at the Frasnian Upper Kellwasser, Famennian *Annulata*, Dasberg, and Hangenberg Black Shales) is important (Figs. 22.11 and 22.13) because they probably had a strong impact on the climatic system and possibly have helped to pass some threshold, enhancing an already critical situation. Therefore they can also be used as climatically triggered correlation markers.

22.2.2.7 Magnetic susceptibility

Magnetic susceptibility (MS) measurements are bulk, nonoriented measurements, classically normalized against sample mass and represent the sum of all individual contributions of various magnetic minerals present in the sample. MS of sedimentary rock samples is interpreted as reflecting changes in detrital inputs from fluvial or aeolian sources (e.g., Ellwood et al., 2000; Hladil et al., 2010a). MS records are commonly relatively robust and can be correlated within and between different sedimentary basins. However, caution should be taken when interpreting MS signals, which can carry not only convolved information, mixtures of climatic, sea level, tectonic primary information but also diagenetic secondary overprinting. Decrypting the origin of the magnetic minerals carrying the MS signal is then essential for a meaningful interpretation of MS signals as a paleoenvironmental proxy, as a correlation tool, or for cyclostratigraphy. Protocols have been developed to screen the origin of the MS signal, through (1) comparison with other proxies influenced by paleoenvironmental parameters; (2) magnetic property measurements to obtain information on the magnetic mineralogy and grain size; and (3) direct observation of the magnetic minerals via dissolution (Riquier et al., 2010; Da Silva et al., 2013; synthesis in Da Silva et al., 2015). When the MS record is carefully checked and acknowledged to be relatively well preserved, it can be used as a correlation tool, as a proxy for sea level or paleoclimatic changes, or to build cyclostratigraphic scale.

The application of MS on Devonian sedimentary rocks started in the 1990s, mostly with the work of B.B. Ellwood and coauthors (e.g., Crick et al., 1997; Ellwood et al., 1999). MS curves were produced with a focus on specific interval such as the Silurian-Devonian boundary interval (Crick et al., 2001; Vacek et al., 2010), Lochkovian—Pragian boundary (Slavík et al., 2016), Choteč Event/Emsian—Eifelian boundary (Ellwood et al., 2006; Machado et al., 2010), Kačák Event/Eifelian—Givetian boundary (Crick et al., 2000; Hladil et al., 2002), Givetian—Frasnian boundary (Devleeschouwer et al., 2010), *punctata* Event (=Middlesex Event) of the Middle Frasnian (Da Silva et al., 2010; Śliwiński et al., 2012, 2010), and the Frasnian—Famennian interval (Crick et al., 2002; Racki et al., 2002; Riquier et al., 2010; Whalen et al., 2015).

MS records were also produced through much longer interval, such as the Lochkovian to Mid-Emsian (Hladil et al., 2011, 2010b; Koptíková et al., 2010a, 2010b), Emsian (Ellwood et al., 2001; Michel et al., 2010), Eifelian (Mabille et al., 2008), Givetian (Boulvain et al., 2010; Casier et al., 2011a, 2011b), Frasnian (Da Silva and Boulvain, 2012; Da Silva et al., 2009; Pas et al., 2015; Weiner et al., 2017) to Famennian (Pas et al., 2014; Whalen and Day, 2010). An UNESCO IGCP project (IGCP-580), on the "Application of magnetic susceptibility as a paleoclimate proxy on Paleozoic sedimentary rocks and characterization of the magnetic signal", with a main focus on Devonian, permitted many new results to be published (two special issues: Da Silva and Boulvain, 2010; Da Silva et al., 2015). Further MS records were also produced in the last few years and most of them are used to build astrochronologic time scales and are mentioned in Section 22.1.1.5.

22.2.2.8 (Classical) Magnetostratigraphy

The widespread remagnetization of Devonian rocks in the course of subsequent Variscan and Alpidian orogenies has prevented so far the introduction of a zonation based on magnetic reversals. A recent attempt to overcome the situation was published for the weakly deformed Upper Devonian of the Canning Basin (Hansma et al., 2015), which was positioned at c. 10°S.

22.2.2.9 Gamma-ray spectroscopy

Correlations based on outcrop gamma-ray spectrometry have been introduced in the Devonian stratigraphic literature only since 2010. They provide cyclic signals that are correlated with MS data and interpreted in terms of sea-level changes, sometimes also with color analysis (spectral reflectance, Koptíková et al., 2008). Several studies have contributed to event and cyclostratigraphy in the Lower Devonian (Slavík et al., 2016) and Middle Devonian to Frasnian (Gersl and Hladil, 2004; Hladil et al., 2006; Bábek et al., 2007; Koptíková, 2011; Chan et al., 2017). Bábek et al. (2018) investigated the complete Lochkovian to Eifelian succession of Bohemia. High-resolution data are only moderately distinctive across the Frasnian—Famennian boundary (Bond et al., 2004; Harris et al., 2013; Weiner et al., 2017) but significant for precise correlation within the Hangenberg Crisis Interval (Kumpan et al., 2013, 2014, 2015; Bábek et al., 2016). Bond and Zaton (2003) used gamma-ray logs to reconstruct the oxygenation history of the Upper Devonian of the Holy Cross Mountains, with anoxic peaks at the Frasnian—Famennian boundary and during the basal upper Famennian *Annulata* Events. Most recently, Blood and Lash (2018) demonstrated that gamma-ray logging in wells of the Appalachian Basin is very useful to identify various Upper Devonian black shale events, such as the Middlesex, Lower Rhinestreet, and Lower Kellwasser events, and the Dunkirk Shale as expression of the basal Famennian transgression.

22.3 Devonian time scale

22.3.1 Previous scales

Devonian geologic time scales have been composed by Tucker et al. (1998), Compston (2000), Williams et al. (2000), and more recently by GTS2004 (House and Gradstein, 2004) and GTS2012 (Becker et al., 2012). Noticeable in reviews of the Devonian time scale are the limited number of good-quality chronostratigraphically fixed radioisotopic age dates within the period. As a result,

various assumptions were made to arrive at estimates of duration and boundary ages of the stages, such as equal duration of stages or of zones or subzones, particularly for conodonts; even though the available cyclostratigraphic information suggests that these assumptions have no logical base. Despite these stratigraphic misgivings, one does not have to look far back in literature to realize that the relative number of subzones per stage was often utilized as scaler for successions of stages (see, e.g., chapters on the Mesozoic periods in GTS2004). The finer the subdivisions on this stratigraphic ruler, and the more successive age dates, the better the scaling. In those cases, there simply was no alternative scaling method available for time scale calculation.

Older time scales that assumed constant sedimentation rates, or relied on the subjective views of specialists on probable zonal durations are now generally discarded. A comparison of the age of the base and/or top of the Devonian Period with the corresponding duration from selected time scales published during the last two decades is given in Table 22.1. Relative to the GTS2020 results, the GTS2012 Devonian time scale is 400 kyr shorter.

Although a substantial number of key dates were published after Harland et al. (1990), Young and Laurie (1996), and House and Gradstein (2004) presented their limited evidence and interpolations, the Devonian time scale (internally) was not well established in 2012. A review of procedures followed in GTS2012 is provided below.

For the Devonian and lowermost Carboniferous segments of GTS2012, 18 radioisotopic dates were used as listed in Appendix 2 of that volume. The three Re—Os dates D11, D12, and D17 were not included. Their inclusion produced undesirable curvature in the spline through the lower Famennian, which segment had no other age

TABLE 22.1 Comparison of age and duration of the Devonian Period in selected published time scales.

	Base	Top	Duration
	(Ma)	(Ma)	(Myr)
Harland et al. (1990)	408.5	362.5	46
Young and Laurie (1996)	410	354	56
Tucker et al. (1998)	418	362	56
Compston (2000) (1)	411.6	359.6	52
Compston (2000) (2)	418	362	56
Williams et al. (2000a)	418	362	56
Gradstein et al. (2004)	416 ± 2.8	359.2 ± 2.5	56.8
Becker et al. (2012)	419.2 ± 3.2	358.9 ± 0.4	60.3
De Vleeshouwer and Parnell (2014)	418.8 ± 2.9	359.1 ± 0.4	59.7
This study	419.0 ± 1.8	359.3 ± 0.3	59.7

dates to corroborate such curvature. Stratigraphic positions were assigned to all radioisotopic dates according to the relative chronostratigraphic age scale in the Devonian chapter of GTS2012 ("Becker Scale") and are summarized in Table 22.2. All data points are expressed with age and stratigraphic uncertainties, creating a rectangular error-box. The zonal position of the conodont zonation in the Devonian stages was based on initial scaling estimate of stages and zones (see also discussion of cyclostratigraphy in Becker et al., 2012). The age of the base Devonian, with error bar, was specified according to the boundary age and error of the Ordovician−Silurian spline of GTS2012.

In a first run of the smoothing spline procedure outlined in Chapter 14 (Statistical Procedures) of GTS2012 the cross-validation suggested an optimal smoothing factor of 1.85. With such a high-smoothing factor, six of the data points did not pass the χ^2 test (D4, $\chi^2 = 8.12$, $P = .004$; D5, $\chi^2 = 21.7$, $P = .000$; D6, $\chi^2 = 5.45$, $P = .020$; D8, $\chi^2 = 7.85$, $P = .005$; D9, $\chi^2 = 4.86$, $P = .032$; and D15, $\chi^2 = 4.43$, $P = .035$). For each of these points the combined stratigraphic and radioisotopic error s_{xy} was therefore increased to enforce a P value of .5.

The adjusted errors were used for a second computation of the spline. In this second run the optimal smoothing factor was 0.45. The resulting final spline is seen in the Devonian chapter GTS2012 (Becker et al., 2012).

The Givetian did, *and still does not*, contain radioisotopic dates, and the shape of the spline is relatively unconstrained in this interval, as long as it is monotonic. To estimate the upper and lower boundaries of the Givetian a pragmatic exception was made to the splining procedure, reverting to straight-line interpolation between the closest points along the scale (D7 and D9). As for the spline, error bars for this straight line were estimated by a Monte Carlo procedure, randomly moving D7 and D9 within their stratigraphic and radioisotopic errors. This method gives a duration for the Givetian of 5.0 Myr, which is closer to the 6.5 Myr of House (1995) based on assumed precessional cycles, than the 4.3 Myr resulting from the spline. Interestingly, Ellwood et al. (2011b) recently suggested 5.6 Myr, based on the analysis of cyclic magneto-susceptibility data.

Since the Middle Devonian time scale is relatively unconstrained, it is useful to include cyclicity data, as was attempted by De Vleeshouwer and Parnell (2014), also using the BChron R software package of Haslett and Parnell (2018) to derive at a modified Devonian time scale using the original GTS2012 input dates. In this modified Devonian time scale the stage boundaries of Eifelian, Givetian and Frasnian are over 2 Myr younger and have a larger uncertainty. An issue is that current cyclicity information is at variance with the stage durations in De Vleeshouwer and Parnell (2014) and also with those in GTS2020, with the latter having a superior set of radioisotopic age dates.

22.3.2 Zonal scaling and biostratigraphic comments on the radioisotopic dates

Relative numerical scaling of the Devonian zonation, as a pre-requisite to calculate the GTS2020 Devonian time scale is an update of that scaling proposed by Becker et al. (2012). That Devonian zonation was used as a template to create an updated "composite" zonal column. The latter was achieved by generating a detailed new linear scale along the zonal column, with scaling itself based on a combination of relative thickness of ammonite subzones and number of cycles. Spread in Table 22.2 refers to the relative "thickness" of the units in this table, with linear stratigraphic level, spread, and uncertainty generated from interpolation. Relative linear position of the eight Devonian stage boundary levels is also shown in Table 22.2. The Pragian–Emsian boundary (scale level 404) is taken at the base of the Eol. excavatus M114 Zone, in line with current chronostratigraphic recommendation for the potentially revised stage boundary position.

Ideally, the Devonian zonal composite column in Table 22.2 should be based on quantitative methodology as employed for the Ordovician–Silurian, Carboniferous–Permian, and Cretaceous Periods, but such is not (yet) done for Devonian. In principle such quantitative scaling would be achieved by creating a master dataset of, let say, the 25 stratigraphically most complete and long sections with their fossil and isotopes and other vital occurrences. Using relatively simple compositing methods like applied for the Ordovician–Silurian and Carboniferous–Permian scales, a Devonian composite standard could have been calculated. On this composite standard stage boundary levels would have been assigned from appropriate markers. Next, this quantitative zonal composite could be directly linked to the radioisotopic and cyclic data with their uncertainties and thus interpolated with the smoothing cubic spline method. This way a more objective and more detailed Devonian time scale could have been generated than presented now, but, as outlined, we do not (yet) have a suitable database of many outcrop sections with many fossil and other observed data points.

Below are brief summaries on the sixteen zonal levels accompanying the 30 + radioisotopic dates listed in Table 22.2.

D1−D7—Husson et al. (2016) and McAdams et al. (2017) provided new geochronological data for ash beds in the basal Devonian of the Helderberg Group of New York State and West Virginia. Unfortunately, the biostratigraphy of the region is far from satisfactory based on the rarity of zonally diagnostic conodonts. McAdams et al. (2017) redated the "Kalkberg" (now New Scotland Fm.) or better Judds Falls Bentonite as middle Lochkovian, based on supposed Pb elements of the index genus *Ancyrodelloides*, which does not occur in the lower Lochkovian. J.I.

TABLE 22.2 Listing of the Zonal level composite values and their spread for the latest Silurian through earliest Carboniferous radioisotopic dates with their 2-sigma uncertainty for calculation of the Devonian geologic time scale.

Stage	No.	Age in Ma	± 2σ total	Zone or Assemblage	Zonal level	± spread	Reference
Tournaisian p.p.	CB3	357.26	0.42	MFZ23 (foraminifer)	356.9	0.5	Davydov et al. (2010)
	CB2	358.43	0.42	lower *S. (S.) duplicata*	357.9	0.4	Davydov et al. (2011)
	CB1	358.71	0.42	upper *S. (Eos.) sulcata*	358.5	0.2	„ „
					Tournaisian 359.2		
Famennian							
	D27	358.89	0.48	middle/upper *costatus–kockeli* interregnum	360.1	0.3	Myrow et al. (2013)
	D26	358.97	0.43	upper part of *S. (Eos.) praesulcata*, UD VI-D	360.7	0.3	„ „
	D25	359.25	0.42	„ „	360.7	0.3	Davydov et al. (2011)
	D24	361.3	2.4	upper *Bi. costatus* to lower *costatus–kockeli* Interregnum	361.5	1.6	Selby and Creaser (2005)
	D23	362.87	0.96	*hystricosus* to *lepidophyta*, VH to basal LL (spores), middle/upper *Bi. costatus*	363	1.4	Tucker et al. (1998)
	D22	364.08	2.25	„ „	363	1.4	„ „
	D21	367.7	2.8	*Pa. delicatula platys* (Middle *Pa. triangularis*)	375.5	0.9	Turgeon et al. (2007)
					Famennian 376.4		
Frasnian							
	D20	374.2	4.2	lower *Pa. linguiformis* (lower MN 13b)	376.7	0.3	„ „
	D19	372.36	0.41	middle part of *Pa. bogartensis* (middle MN 13a)	377.5	0.5	Percival et al. (2018)
	D18	375.14	0.45	*Pa. housei* (MN 8)	380.3	0.3	Lanik et al. (2016)
	D17	375.25	0.45	„ „	380.3	0.3	„ „
	D16	375.55	0.44	„ „	380.3	0.3	„ „
					Frasnian 385		
Givetian							
					Givetian 392		
Eifelian							
	D15	390.14	0.47	*Po. costatus*	395.9	0.8	Harrigan et al. (in review)
	D14	390.82	0.48	„ „	395.9	0.8	„ „
					Eifelian 397		
Emsian							
	D13	394.29	0.47	uppermost *Po. patulus*	397.1	0.2	„ „
	D12	407.75	1.16	*N. (D.) praecursor* (dacryoconarid) and LD III-B (ammonoid)	402	0.4	Kaufmann et al. (2005)
					Emsian 404		

TABLE 22.2 (Continued)

Stage	No.	Age in Ma	± 2σ total	Zone or Assemblage	Zonal level	± spread	Reference
Pragian							
	D11	411.7	1.2	below *N. (T.) acuaria* (dacryoconarid), lower/middle Pragian	408	1	Bodorkos et al. (2017)
	D10	411.5	1.3	*polygonalis−emsiensis*, PoW (spores), Pragian	408	2.5	Parry et al. (2011)
					Pragian 408.7		
Lochkovian							
	D9	415.6	0.8	*eurekaensis−delta* Zones, lower Lochkovian	412	2	Bodorkos et al. (2017)
	D8	415.6	0.8	*hesperius−eurekaensis* Zones, lower Lochkovian	415	2	Bodorkos et al. (2017)
	D7	417.22	0.5	*Caudicriodus postwoschmidti*	415	1.5	Husson et al. (2016)
	D6	417.61	0.5	" "	415	1.5	McAdams et al. (2017)
	D5	417.68	0.52	" "	415	1.5	Husson et al. (2016)
	D4	417.56	0.51	" "	415	1.5	" "
	D3	417.73	0.53	" "	415	1.5	" "
	D2	417.85	0.54	" "	415	1.5	" "
	D1	418.42	0.53	" "	415	1.5	" "
					Lochkovian 417		
basal Pridoli	S8	422.91	0.49	basal *N. parultimus* Zone; upper *O. crispa*	421	1.5	Cramer et al. (2015)
lower Ludfordian	S7	424.08	0.53	lower *Bohemograptus* Zone; upper *P. siluricus*	423	1.5	" "

Spread refers to the estimated "thickness" of the composite value for each age date in the "Becker scale" (Becker et al., 2012). Over half of these 32 radioisotopic dates are new since GTS2012, with a majority having less than 0.5-Myr uncertainties, thus providing a good estimate for the interpolated ages of the Devonian stage boundaries. *UD*, Upper Devonian ammonoid zone.

Valenzuela-Ríos (Valencia), one of the leading experts on ancyrodelloids, rejected this identification in an email of mid-January 2019. Since the associated *Ozarkodina plani-lingua* is long-ranging in the lower/middle Lochkovian, the decisive conodont from the Kalkberg/New Scotland Fm. is the *C. postwoschmidti* Group (Kleffner et al., 2009) that indicates the higher part of the lower Lochkovian; *C. post-woschmidti* is not known to range into the middle Lochko-vian (Corradini and Corriga, 2012). Further revisions of the New York caudicriodids may provide better constraints in the future. Currently, the stratigraphic error bar for the dated ash beds should be the c. upper half of the lower Lochkovian.

D8—The biostratigraphic age of volcanics from the lower Turondale Formation of the Hills Ends Trough in New South Wales, dated by Jagodzinski and Black (1999),

is also difficult to assess. They are older than the middle Lochkovian *Ancyrodelloides transitans* Zone since the brec-ciated limestone at the top of the Turondale Formation yielded *Zieglerodina remscheidensis* in one sample (Pack-ham et al., 2001). This species does not range above the *A. transitans* Zone (see range chart of Corradini and Corriga, 2012) or just into the basal *Ancyrodelloides eleanorae* Zone in Nevada. The associated *Amydrotaxis praejohnsoni* (reported as *Amydrotaxis johnsoni* alpha morphotype in Packham et al., 2001) is a middle Lochkovian species, which occurs typically in the *Kimognathus delta* Zone of western North America (Klapper and Murphy, 1980). In Nevada, it may enter just slightly earlier than *A. transitans* but it ranges to the top of the middle Lochkovian (Valenzuela-Ríos et al., 2015). *Flajsella stygia* occurs in possibly related olistolites of the Palmers Oakey area (Bischoff and Ferguson, 1982), but these are not well correlated with the main Hills End Trough lithostratigraphy (Talent and Mawson, 1999). *F. stygia* does not enter before the upper *A. eleanorae* Zone (Corradini and Corriga, 2012) but it occurs also in the North American *K. delta* Zone (Klapper and Murphy, 1980). In any case the volcanite-bearing lower part of the Turondale Formation is significantly older than the limestones that fall in middle to upper parts of the middle Lochkovian.

Quoted brachiopod evidence from the area and its indirect conodont correlation include many uncertainties. Wright in Packham (1969) reported brachiopods from Paling Yards in the Limekilns Group of the Mudgee District, which is adjacent to but not identical with the Hills End Trough succession that provided the radioisotopic ages. There is only an outdated list of genera insufficient to assign it to the regional brachiopod zonation of Garratt and Wright (1989). In that publication the Paling Yards fauna is not specified but the authors mention that the *Boucotia australis* Zone is recognized in the Waterbeach Formation near Limekilns, which obviously refers to the Paling Yards fauna. The Waterbeach Fm. overlies the Turondale Fm. and its brachiopods, therefore, should be younger than the conodont faunas discussed above. However, Jagodzinski and Black (1999) reported that new maps showed that there is no Waterbeach Fm. at Palin Yards. The brachiopods either derived from the top Turondale Fm. or the Waterbeach Fm. has not been separated during mapping due to poor outcrop conditions. Packham et al. (2001), therefore, assign the Paling Yards fauna to the Turondale Fm. but add that "there is some uncertainty as to the correct formation assignment of this fossil locality." Packham (2003) then referred to an incompletely studied Lochkovian brachiopod fauna from the Turondale Fm. at Limekilns, which was correlated by Garratt and Wright (1989, not 1982) with a "Mandagery Park Formation fauna" that occurs further to the west.

How does the *B. australis* brachiopod zone that postdates the dated volcanics correlate with the conodont zonation? Data in Savage (1973) for conodonts from the Lower "Mandagery Park Limestone," which yielded brachiopods, are moderately conclusive since all reported taxa are long-ranging in the lower/middle Lochkovian. The Mandagery Park Area was shown by Mawson et al. (1989) to have three successive limestone units. The lower two (including the one with the brachiopods) were reassigned to the Garra Formation and yielded associated *Z. remscheidensis*, *Amydrotaxis sexidentata*, and *Pandorinellina optima*. The first species does not range above the basal *A. eleanorae* Zone (see above), the second always predates *Amy. praejohnsoni* and co-occurs in middle to upper parts of the lower Lochkovian with *Cypricriodus hesperius* and *Eurekadonta eurekaensis* (Murphy, 2018, *eurekaensis* Zone), the third defines in Bohemia a zone c. a third above the base of the lower Lochkovian (Slavík et al., 2012) but ranges much higher. The Mandagery Park fauna falls in higher parts of the lower Lochkovian but

Garratt and Wright (1989) also showed that there are other faunas of the *B. australis* Zone that are much younger, up to the upper Lochkovian *Pedavis pesavis* Zone. The current evidence gives no reason why the undescribed Paling Yards brachiopod fauna should be from the lower, not from the middle or upper parts of the long-ranging *B. australis* Zone, resulting in a larger biostratigraphic uncertainty than for the conodont faunas. Consequently, the biostratigraphic error bar for the lower Turondale Fm. zircons has to be much of the lower Lochkovian, even if the thick Turondale siliciclastics deposited quickly.

D10—Parry et al. (2011) assigned an andesitic lava from the Rhynie Outlier in NE Scotland to the *polygonalis-emsiensis* or PoW miospore zone, which occupies most of the Pragian, creating a large biostratigraphic error bar.

D11—In the Hills End Trough succession of New South Wales, Jagodzinski and Black (1999) published a higher geochronological age for the Merrion Formation. A record of the Pragian index dacryoconarid *Nowakia (Turkestanella) acuaria* is from the overlying Limekiln Formation, but from an adjacent, different area. The Cunningham Formation that alternatively overlies the Merrion Fm. has a diachronous base from west to east. A brachiopod fauna from the lower Cunningham Fm. in the region that yielded the zircon ages (Packham et al., 2001) is also no older than the Pragian based on a ?*Nadiastrophia* (see regional brachiopod zonation of Garratt and Wright, 1989). The zircons theoretically could predate the Pragian or come from any level within it. If, however, the Merrion volcanism ended at the same time as the Riversdale volcanics of New South Wales, as discussed by Pogson (2009), then it is not younger than the top *G. kindlei* Zone. This gives a stratigraphic error bar for the Merrion Fm. age stretching c. the lower/middle Pragian.

D12—There are no conodonts in the famous Hunsrück Slate of the SW Rhenish Massif of Germany but there is now a refined dacryoconarid-ammonoid-conodont correlation (Aboussalam et al., 2015) based on the Moroccan sequence. The dated Hunsrück Slate bentonite (Hans-Platte) is now known as the Kühstabl Tuffite since it falls in the Kühstabl Member of the Kaub Formation (De Baets et al., 2013). The dacryoconarid from Eschenbach, possibly from above the bentonite, is *Nowakia (Dmitriella) praecursor* (Alberti, 1982), index of the *praecursor* Zone. The same interval was quoted in Kaufmann et al. (2005) as the "upper part of the *zlichovensis* Zone." The Kühstabl Member goniatites belong to *Ivoites* (formerly included in *Anetoceras*, De Baets et al., 2013), and, based on *N. (D.) praecursor*, correlate with the higher part of the Lower Devonian LD III-B of Morocco (*Metabactrites−Erbenoceras* Zone; Becker et al., 2019, in press). The Kühstabl Member is still missing true *Anetoceras* (the LD III-C marker), which is characteristic for the *Eol. nothoperbonus* and *Eolinguipolygnathus catharinae* subzones (Upper *gronbergi* Zone of other authors; Abous-salam et al., 2015). The biostratigraphic error bar for the Hunsrück Slate age is, therefore, very short within the lower Emsian (lower/middle *N. (D). praecursor* Zone and main *Eolinguipolygnathus gronbergi* Zone or Subzone).

D13–D15—*Polygnathus patulus* and *Polygnathus costatus* were granted full species status by Vodrážková et al. (2011).

D16–D18—Our Frasnian biostratigraphic chart uses the Montagne Noire (MN) Zonation, which has been correlated by Klapper and Becker (1999) with the less detailed zonation of Ziegler and Sandberg (1990). Becker et al. (2016a) added the names of index species to the MN zones. The biostratigraphic and geochronological age of the Belpre Ash Suite of Tennessee was revised by Lanik et al. (2013, 2016). It falls in the *Palmatolepis housei* or MN Zone 8 (upper part of Lower *hassi* Zone). However, the Rhinestreet Shale Ash of New York State was placed in the (upper) *Ozarkodina nonaginta* Zone (upper MN 7 Zone), which equals the c. "middle part of the Lower *hassi* Zone."

D19—The bentonite bed (Bed 36) of Steinbruch Schmidt, Kellerwald, eastern Rhenish Massif, Germany, was first analyzed by Kaufmann et al. (2004) and recently redated by Percival et al. (2018). It falls in the middle part of the *Pa. bogartensis* Zone or MN Zone 13a (middle part of the Upper *rhenana* Zone sensu Ziegler and Sandberg, 1990).

D20—Turgeon et al. (2007) provided Re–Os depositional ages for dark-gray shales from around the Frasnian–Famennian boundary in the top Hanover Formation and basal Dunkirk Formation of a core in western New York State. No conodonts have been described from the West Valley core, which means that the Frasnian conodont ages are based on the supposed correlation with other Hanover Shale sections. D20 is sample WVC785 from 2.9 m below the assumed top of an Upper Kellwasser equivalent and F/F boundary. At Irish Gulf (Over, 1997), the oldest evidence for the *Pa. linguiformis* Zone (MN 13b Zone) comes from a concretion 1.75 m below the F/F boundary. But based on graphic correlation, Klapper and Kirchgasser (2016) suggested that a fauna from Beaver Creek at c. 9.5 m below the F/F boundary projects already into the (basal) *Pa. linguiformis* Zone. A position of D20 within the lower part of the *Pa. linguiformis* Zone (MN Zone 13b) is accepted with some reservation since the sedimentation rate at the West Valley core could be significantly different than in the named other surface outcrops.

D21—WVC754 from the lower Dunkirk Fm. of the West Valley core (Turgeon et al., 2007) falls in the *Palmatolepis delicatula platys* Zone, which replaced terminologically the former Middle *triangularis* Zone (Spalletta et al., 2017). Over (1997) found the zonal index species in the top Hanover Shale and *Palmatolepis clarki*, the original index species of the Middle *triangularis* Zone, in the basal Dunkirk Fm.

D22/D23—The geochronological dating of zircons by Tucker et al. (1998) included samples from intrusive volcanics in the Carrow Formation of New Brunswick, eastern United States. The Famennian miospore-conodont correlation has been revised by Streel (2009). Based on McGregor and McCutcheon (1988), the Carrow Fm. yielded among many other taxa *Vallatisporites pusillites* s.l. (=*Cirratriradites hystricosus*) and a single specimen of *Retispora lepidophyta* (an unusually large form). *Vallatisporites hystricosus* defines the European VH zone, *R. lepidophyta* the next higher LL Zone. Both taxa range higher but in the absence of index species for higher zones, and with respect to the rarity of *R. lepidophyta*, a level near the VH/LL boundary is likely. This correlates with a position within the middle to upper parts of the *Bispathodus costatus* Subzone sensu Hartenfels (2011), a middle/upper subdivision of the former Middle *expansa* Zone. The biostratigraphic error bar is relatively short although the VH and basal LL zones cover much of the Famennian V of German definition.

D24—Selby and Creaser (2005) provided Re–Os ages for black shales from the Exshaw Formation of the Canadian Rocky Mountains. Their sample was from the upper 10 cm of the noncalcareous black shale subunit of the Lower Member. The base of the latter yielded *Bi. costatus* M1, the index species of the *Bi. costatus* Subzone that begins c. a third above the base of the *Bispathodus aculeatus aculeatus* Zone (=former Middle *expansa* Zone; Hartenfels, 2011). Its upper range is the main Hangenberg Extinction at the top of the *S. (E.) praesulcata* Zone in the revised sense of Kaiser et al. (2009). *Pa. gracilis sigmoidalis* occurs at 3.1 and 5.6 m above the Exshaw base. In sections with calm deposition and without reworking (which applies to the black shale), this form also disappears at the top of the *S. (Eos.) praesulcata* Zone. Two other subspecies of *Pa. gracilis*, which also did not survive the main Hangenberg Extinction, are found within the noncalcareous black shale at other Exshaw localities (Johnston et al., 2010).

Since there is continuous black shale from the last conodont level to the sampled geochronology interval, it is very unlikely that the level of the major Hangenberg Regression and its sequence boundary has been reached or crossed. Therefore the Re–Os date comes from an interval stretching from the higher *Bi. costatus* Subzone to the basal *costatus–kockeli*–Interregnum (*ck*I) sensu Kaiser et al., 2009 (Hangenberg Black Shale level of the Lower Hangenberg Crisis Interval, Kaiser et al., 2016). Richards and Higgins (1989) identified a discontinuity surface at the contact between the two subunits of the Lower Member, which was used to place arbitrarily the Devonian-Carboniferous boundary. The discontinuity surface may represent the Hangenberg Regression (Middle Hangenberg Crisis Interval, upper *ck*I), followed by the calcareous black shale subunit of possibly topmost Famennian (top *ck*I to *kockeli* Zone) to basal Tournaisian age. Contrary to reports in the literature, there is no biostratigraphic information for the calcareous black shale subunit. A fauna with *S. (S.) cooperi* found by (Macqueen and Sandberg, 1970) from the "upper 2.5 ft. of the black shale unit" has been misunderstood by subsequent authors. As clearly noted by (Macqueen and Sandberg, 1970), the conodonts come from the distinctive unit with abundant goniatites (see House, 1993), which is the basal 60–70 cm of the Upper Member of the Exshaw Formation. The goniatite level is also marked in Richards and Higgins (1989) but yielded to them only the long-ranging *Neopolygnathus communis communis*. It is the basal "Siltstone Member" that clearly falls in higher parts of the Lower Tournaisian, not the calcareous black shale. Therefore it is possible to restrict the upper uncertainty range for the Re−Os age of Selby and Creaser (2005) to a level below the Hangenberg Regression.

Ekhoff et al. (2013) dated four ash layers from within the upper calcareous black shale of the Lower Exshaw Member as 359–360 Ma. These dates should be assigned to the Upper Hangenberg Crisis Interval [top *ck*I to *S. (E.) sulcata* Zone] sensu Becker et al. (2016b). Based on the significant age differences of various Exshaw levels, Ekhoff et al. (2013) emphasize the significance of unconformities within the formation.

D26—Myrow et al. (2013) dated ash beds from around the Devonian-Carboniferous (D/C) boundary at Kowala in the Holy Cross Mountains, Poland. The D/C boundary conodont stratigraphy has been revised by Kaiser et al. (2009). The Middle *praesulcata* Zone should not be used any more since it was very poorly defined by the locally diachronous disappearance of *Pa. gracilis gonioclymeniae*. Instead, the (revised) *praesulcata* Zone ranges to the main Hangenberg Extinction at the base of the Hangenberg Black Shale. Unfortunately, the D/C conodont stratigraphy of Kowala has not yet been fully investigated. The so far best data were published by Dzik (1997) and Malec (2014). Both lower ash beds fall in the upper part of the (revised) *S. (E.) praesulcata* Zone and in the higher *Wocklumeria sphaeroides* Zone (UD VI-D) of the ammonoid zonation. The upper ash bed lies more precisely at the top of the *S. (E.) praesulcata* Zone and in UD VI-D2 (*Epiwocklumeria applanata* was recorded by Czarnocki, 1989).

D27—So far there is no good documentation for the *kockeli* Zone (=old Upper *praesulcata* Zone) at Kowala. The thin limestone above the Hangenberg Black Shale (Dzik, 1997: Sample 73) and the subsequent tuffaceous shales (Samples 72-71) yielded a restricted conodont fauna that is characteristic for the *costatus*−*kockeli*−Interregnum (*ck*I). Especially typical is a small opportunistic bloom of *N. communis communis*. Associated squashed *Acutimitoceras (Stockumites)* are widely known from this level (Middle Hangenberg Crisis Interval) and fall in the lower UD VI-F (Becker et al., 2016b). The next thin calcareous bed has acutimitoceratids but no conodonts so far. A set of three subsequent limestones (more precisely Samples 51 and 24 of Dzik, 1997; shown as one bed in Myrow et al., 2013) yielded "early" protognathodids but not yet *Pr. kockeli*;

therefore they still fall in the (upper) *ck*I. The next thin limestone was placed by Myrow et al. (2013) in the basal Carboniferous although no siphonodellids have been reported from it. Malec (2014) noted *Po. purus purus* in the corresponding Sample 177, which in the past was regarded as an alternative marker for the *S. (E.) sulcata* Zone. However, a range into the *kockeli* Zone cannot be excluded (Becker et al., 2016b). In any case, the dated upper ash bed at Kowala does not fall in the *kockeli* Zone but in the immediately underlying middle/upper *ck*I.

22.3.3 Radioisotopic data

The detailed and high-resolution conodont−ammonoid zonation for the Devonian, with over 75 zones (Fig. 22.11) is linked to 30 + radioisotopic age dates employed in Devonian time scale building, a majority of which have 0.5 Myr or less uncertainty (see Table 22.2). Over half of the radioisotopic dates are new since GTS2012, and several dates employed in GTS2012 have been revised. The best radioisotopic coverage is in the upper and lower parts of the Devonian, with six dates only in the Pragian through Givetian interval. Thanks to Harrigan et al. (in review), the lower Eifelian Tioga Ashes (first dated by Roden et al., 1990) and top-Emsian Hercules bentonites (see Kaufmann et al., 2005) now have high-resolution age dates, and appear to have good stratigraphic control. No dates are currently available for Givetian, a situation not changed since GTS2012. Thanks to the new D11, D9, and D8 dates in Pragian and Lochkovian by Bodorkos et al. (2017), these two stages have better and more reliable constraints. The Devonian radioisotopic age dates are listed in detail in Appendix 2.

In order to provide constraints on the age of the base and top of the Devonian, we included three lowermost Carboniferous and two uppermost Silurian age dates in Table 22.2. Those zonal level and their spread were interpolated by extending the "Becker scale" (Becker et al., 2012) in these zones from the zonal tables of Carboniferous and Silurian in GTS2012.

For construction of the Devonian geologic time scale in GTS2020, radioisotopic ages, almost exclusively generated with the U/Pb ID-TIMS method, were plotted against the relative (sub)zonal columns of Fig. 22.11 in this chapter.

The best-fit line for the two-way plot (Fig. 22.14) of radioisotopic ages against the zones and stages was calculated with a cubic-spline-fitting method that also combines stratigraphic uncertainty estimates with the 2-sigma error bars of the radioisotopic data. A special routine was developed to generate reliable tiepoints for those intervals where clustering of age dates prevents use of a simple mean value. Details of the best fitting method and error analysis to arrive at best age estimates with uncertainty for the Devonian stage boundaries are outlined in Chapter 14, Geomathematics, this book.

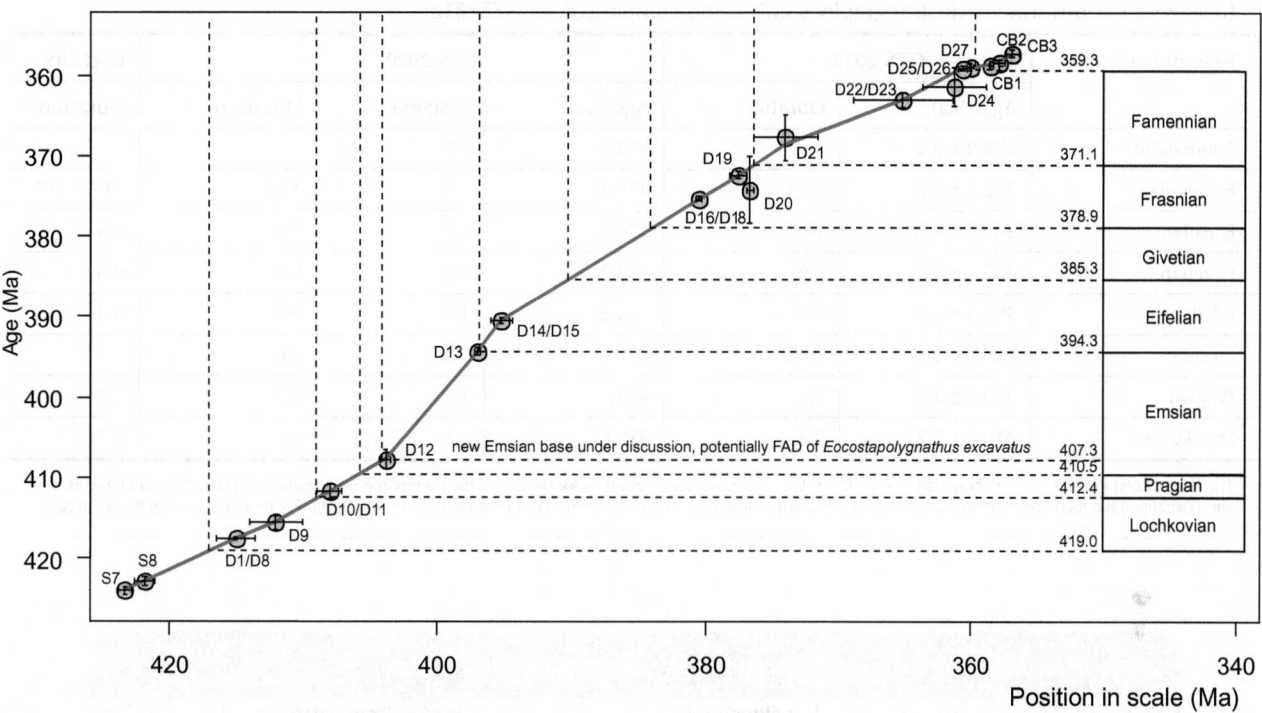

FIGURE 22.14 Two-way graphics plot of the scaled zonation in Fig. 22.11 and Table 22.2 versus radioisotopic dates for construction of the Devonian time scale, using clustering technique, cubic splining and error analysis. The plotted data are in Table 22.2, with details in Section 22.3 of this chapter. Note that the Silurian—Devonian boundary age in GTS2020 is based on the more detailed and updated Ordovician—Silurian spline (see Table 20.3 of the Ordovician chapter of this volume).

22.3.4 Age of stage boundaries

For the Devonian and lowermost Carboniferous segments of GTS2020, we used the radioisotopic dates discussed in the previous section, including their 2-sigma error as listed in Table 22.2. Stratigraphic positions with their rectangular error bars were assigned for all age dates according to the chronostratigraphic scaling. The best-fit line for the two-way plot of radioisotopic ages against the zones and stages, as outlined in the previous section and visualized in Fig. 22.14, yields the age for the eight Devonian stage boundaries. The age and error derived with the Devonian spline for the base and the top of Devonian were aligned with the age and error for those boundaries with respectively the Ordovician—Silurian and Carboniferous—Permian splines (Chapter 21: The Silurian Period, and Chapter 23: The Carboniferous Period, this volume). The new Silurian-Devonian boundary value (419.0 ± 1.8 Ma) is taken from the Ordovician—Silurian data and spline fit, using a more updated data set for Ludfordian—Pridoli stage intervals than used in the slightly earlier "crunched" Devonian ages (B. Cramer, pers. comm. to F. Gradstein, August 2019).

The Devonian-Carboniferous boundary age of 359.3 ± 0.3 Ma is the average of the ages for this boundary

calculated with the spline fits for the Devonian and for the Carboniferous—Permian. This age of 359.3 ± 0.3 Ma is realistic with focus on a potential GSSP level for a base Tournaisian within the higher Hangenberg Event Interval, shown in Fig. 22.11. It agrees well with the U—Pb zircon age of 359.6 ± 1.9 Ma, which was derived from an ash bed in the *kockeli* Zone of Dapaoushang, Guizhou, South China (Liu et al., 2012).

The Givetian does not contain any radioisotopic dates, and the shape of the spline is relatively unconstrained in this interval, but provides an acceptable and likely fit. The new Pragian—Emsian boundary age of 407.3 ± 1.1 Ma provides an estimate for the age of the *Eol. excavatus* M114 zonal base, in line with chronostratigraphic recommendation. The base *kitabicus* interpolates at 410.5 ± 1.2 Ma.

The interpolated Devonian stage boundaries with their confidence intervals are given in Table 22.3. Base of the Devonian is at 419.0. ± 1.8 Ma, and top is 359.3 ± 0.3 Ma, both ages respectively 0.6 and 0.2 Myr younger than in GTS2012, with even larger numeric changes in the age of base Pragian and Frasnian. Base Pragian is almost 2 Myr older and base Frasnian over 4 Myr younger than in GTS2012. All stages significantly change in duration. The

TABLE 22.3 Interpolated ages of Devonian stage boundaries from Fig. 22.2, with uncertainty estimates, comparison to stage durations from cyclostratigraphy studies, and comparison to GTS2012.

Base of Stage	GTS 2012		GTS 2020			Cyclicity
	Age (Ma)	Duration	Age (Ma)	2-Sigma	Duration	Duration
Tournaisian	358.9 ± 0.4		359.3	0.3		
Famennian	372.2 ± 1.6	13.3	371.1	1.1	11.8	13.5 ± 0.5
Frasnian	382.7 ± 1.0	10.5	378.9	1.2	7.8	6.5
Givetian	387.7 ± 0.8	5.0	385.3	1.2	6.4	5.6
Eifelian	393.3 ± 1.2	5.6	394.3	1.1	9.0	6.2
Emsian	407.6 ± 2.6	14.2	407.3	1.1	13	
Pragian	410.8 ± 2.8	3.2	412.4	1.1	5.1	1.7
Lochkovian	419.2 ± 3.2	8.4	419.0	1.8	6.6	7.7

The interpolated Pragian–Emsian boundary age of 407.3 ± 1.1 Ma provides an estimate for the age of the base of the *Eol. excavatus* M114 zone, in line with chronostratigraphic recommendation for the future GSSP marker. The current Emsian GSSP has a projected age of 410.5 Ma. For more details, see the main text.

GSSPs of the Devonian Stages, with location and primary correlation criteria

Stage	GSSP Location	Latitude, Longitude	Boundary Level	Correlation Events	Reference
Famennian	Coumiac Quarry, near Cessenon, Montagne Noire, France	43°27'40.6"N 3°02'25"E*	base of Bed 32a	Conodont, FAD of *Palmatolepis subperlobata*, Conodont, LAD of *Palmatolepis bogartensis*	Episodes **16**/4, 1993
Frasnian	Col du Puech de la Suque, Montagne Noire, France	43°30'11.4"N 3°05'12.6"E*	base of Bed 42a at Col du Puech de la Suque section E	Conodont, FAD of *Ancyrodella rotundiloba pristina*	Episodes **10**/2, 1987
Givetian	Jebel Mech Irdane, Morocco	31°14'14.7"N 4°21'14.8"W*	base of Bed 123	Conodont, FAD of *Polygnathus hemiansatus*	Episodes **18**/3, 1995
Eifelian	Wetteldorf, Eifel Hills, Germany	50°08'58.6"N 6°28'17.6"E*	21.25 m above the base of the exposed section, base of unit WP30	Conodont, FAD of *Polygnathus partitus*	Episodes **8**/2, 1985
Emsian	Zinzil'ban Gorge, Uzbekistan	39°12'N 67°18'20"E	base of Bed 9/5 in the Zinzil'ban Gorge in the Kitab State Geological Reserve	Conodont, FAD of *Eocostapolygnathus kitabicus* New Emsian base under discussion, potentially FAD of *Eolinguipolygnathus excavatus* M114	Episodes **20**/4, 1997
Pragian	Velká Chuchle, Prague, Czech Republic	50°00'53"N 14°22'21.5"E*	base of Bed 12 in Velká Chuchle Quarry	Conodont, just above FAD of *Eognathodus irregularis*	Episodes **12**/2, 1989
Lochkovian	Klonk, near Prague, Czech Republic	48.855°N 13.792°E**	within Bed 20	Graptolite, FAD of *Uncinatograptus uniformis*	IUGS Series A, **5**, 1977

* according to Google Earth, ** derived from map

FIGURE 22.15 Devonian GSSP locations with precise stratigraphic positions, the primary marker fossils and correlation events, and the main references. *GSSP*, Global Stratotype Section and Point.

anomalous short duration from cycle scaling of the Pragian Stage is due to the fact that only part of it is scaled this way, as outlined in Da Silva et al. (2016). We note that future revision of the Pragian–Emsian boundary definition will shorten duration of the Emsian Stage.

The duration of the three epochs and seven stages in the Devonian is as follows:

The Early Devonian lasted 24.7 Myr, with the Lochkovian Stage lasting 6.6 Myr (base at 419.0 ± 1.8 Ma), the Pragian Stage 5.1 Myr (base at 412.4 ± 1.1 Ma) and the Emsian Stage (using its pending redefinition of the Pragian/Emsian boundary) 13.0 Myr, starting at 407.3 ± 1.1 Ma. The Pragian is the shortest stage in the Devonian at 5.1 Myr. The Emsian is the longest stage in the Devonian (13.0 Myr), followed by the Famennian (11.8 Myr).

The Middle Devonian has a duration of 15.4 Myr, with the Eifelian lasting 9.0 Myr (starting at 394.3 ± 1.1 Ma), and the Givetian 6.4 Myr (starting at 385.3 ± 1.2 Ma). The Late Devonian started at 378.9 ± 1.2 Ma and ended at 359.3 ± 0.3 Ma, lasting 19.6 Myr, being over 5 Myr shorter than the Early Devonian. The Frasnian Stage lasted 7.8 Myr and the Famennian Stage 11.8 Myr; the latter starts at 371.1 ± 1.1 Ma. Error bars increase, as expected, with scarcity of spread out radioisotopic dates.

As shown in Table 22.3 stage duration from cycles substantially mismatches such from current spline fit scaling. A detailed study by Pas et al. (2018) on cyclicity using MS measurements on deeper marine late Frasnian−Famennian strata of the Illinois Basin, United States, implied a duration for the Famennian of 13.5 ± 0.5 Myr. This is several 400-kyr cycles longer than calculated using 15 Frasnian−Famennian−Tournaisian radioisotopic dates, the relevant ones and majority with less than 0.5-Myr uncertainty. An age of 372.4 ± 0.9 Ma for base Famennian in Pas et al. (2018), combining the Illinois Basin cycle data with the GTS2012 time scale, almost overlaps with the current estimate of 371.1 ± 1.1 Ma.

The ages for several GTS2020 stage boundaries are outside the GTS2012 spline error bars for these levels, reflecting new chronostratigraphic information and more and more accurate age dates. We consider that part of Devonian scale is maturing, although more infra-stage dates are desirable, and more stage duration estimation from cycles is essential. The new and total duration of the Devonian is 59.7 Myr (whereas it was 60.3 Myr in GTS2012 and 56.8 Myr in GTS2004).

Bibliography

Aboussalam, Z.S., 2003, Das «Taghanic-Event» im höheren Mittel-Devon von West-Europa und Marokko. *Münstersche Forschungen zur Geologie und Paläontologie*, **97**: 1−332.

Aboussalam, Z.S., and Becker, R.T., 2001, Prospects for an upper Givetian substage. Mitteilungen aus dem Mus. *Für Naturkunde Berlin, Geowissenschaftliche Reihe*, **4**: 83−99.

Aboussalam, Z.S., and Becker, R.T., 2002, The base of the *hermanni* Zone as the base of an Upper Givetian substage. *Subcommission on Devonian Stratigraphy Newsletter*, **19**: 25−34.

Aboussalam, Z.S., and Becker, R.T., 2007, New upper Givetian to basal Frasnian conodont faunas from the Tafilalt (Anti-Atlas, Southern Morocco). *Geological Quarterly*, **51** (4): 345−374.

Aboussalam, Z.S., and Becker, R.T., 2011, The global Taghanic Biocrisis (Givetian) in the eastern Anti-Atlas, Morocco. *Palaeogeography, Palaeoclimatology, Palaeoecology*, **304**: 136−164.

Aboussalam, Z.S., Becker, R.T., and Bultynck, P., 2015, Emsian (Lower Devonian) conodont stratigraphy and correlation of the Anti-Atlas (Southern Morocco). *Bulletin of Geosciences*, **90** (4): 893−980.

Aboussalam, Z.S., Becker, R.T., Mayer, O., Gibb, A., and Hüneke, H., 2018, Emsian to middle Famennian bioevents and cephalopod faunas

at Mdoura-East (western Tafilalt Platform). In: Hartenfels, S., Becker, R.T., El Hassani, A., and Lüddecke, F. (Eds.), 10th International Symposium "Cephalopoda − Present and Past", Fes, March 26−April 3, 2018, Field Guidebook. *Münstersche Forschungen zur Geologie und Paläontologie*, **110** (online): 312−326.

Abramova, A.N., and Artyushkova, O.V., 2004, The Frasnian-Famennian boundary in the Southern Urals. *Geological Quarterly*, **48** (3): 217−232.

Afanasieva, M., 2000, Upper Devonian radiolarian zones of the Timan-Pechora Basin. *Ichthyolith Issues, Special Publication*, **6**: 8−14.

Afanasieva, M.S., 2009, New radiolariens from the Devonian of the southern Ural Mountains: 2. Middle-Late Devonian. *Paleontologicheskiy Zhurnal*, **2009** (1): 33−46.

Afanasieva, M.S., and Amon, E.O., 2011, Devonian Radiolarians of Russia. *Paleontological Journal*, **45**: 1313−1532.

Afanasieva, M.S., and Amon, E.O., 2012, Biostratigraphy and Paleobiogeography of Devonian Radiolarians of Russia. *Russian Academy of Sciences, Borissiak Paleontological Institute*, 280 pp.

Aitchison, J.C., Davis, A.M., Stratford, J.M.C., and Spiller, F.C.P., 1999, Lower and Middle Devonian radiolarian biozonation of the Gamilaroi terrane, New England Orogen, eastern Australia. *Micropaleontology*, **45** (2): 138−162.

Aitchison, J.C., Suzuki, N., Caridroit, M., Danelian, T., and Noble, P., 2017, Paleozoic radiolarian biostratigraphy. *Geodiversitas*, **39** (3): 503−531.

Alberti, G.K.B., 1981, Paläontologische Daten zur Stratigraphischen Verbreitung der Nowakiidae (Dacryoconarida) im Devon von Nord-Afrika (Marokko, Algerien). *Senckenbergiana lethaea*, **62** (2/6): 205−216.

Alberti, G.K.B., 1982, Nowakiidae (Dacryoconarida) aus dem Hunsrückschiefer von Bundenbach (Rheinisches Schiefergebirge). *Senckenbergiana lethaeae*, **63** (5/6): 451−463.

Alberti, G.K.B., 1993, Dacryoconaride und homoctenide Tentaculiten des Unter- und Mittel-Devons. Teil I. *Courier Forschungs-Institut Senckenberg*, **158**: 1−229.

Alberti, G.K.B., 1998, Planktonische Tentakuliten des Devon. III. Dacryoconarida Fisher 1962 aus dem Unter-Devon und oberen Mitteldevon. *Palaeontographica, Abteilung A*, **250** (1−3): 1−46.

Alberti, G.K.B., 2000, Planktonische Tentakuliten des Devon. IV. Dacryoconarida FISHER 1962 aus dem Unter-Devon. *Palaeontographica, Abteilung A*, **254**: 1−23.

Alberti, H., Groos-Uffenorde, H., Streel, M., Uffenorde, H., and Walliser, O.H., 1974, The stratigraphical significance of the *Protognathodus* fauna from Stockum (Devonian/Carboniferous boundary, Rhenish Schiefergebirge). *Newsletters on Stratigraphy*, **3**: 263−276.

Andrew, A.S., Hamilton, P.J., Mawson, R., Talent, J.A., and Whitford, D.J., 1994, Isotopic correlation tools in the mid-Palaeozoic and their relation to extinctions events. *Australian Petrology Exploration Association*, **34**: 268−277.

Aretz, M., and Group, Task, 2016, Report of the joint Devonian-Carboniferous Boundary GSSP Reappraisal Task Group. *Newsletter on Carboniferous Stratigraphy*, **32**: 26−29.

Aristov, V.A., 1984, Devonian and Lower Carboniferous conodonts of Eurasia: assemblages, zonal subdivision, correlation of different facies sediments. *Russian Academy of Sciences, Geological Institute, Transactions*, **484**: 1−192 (in Russian).

Aretz, M., Herbig, H.-G., and Wang, X.D., 2020, Chapter 23 - The Carboniferous Period. *In* Gradstein, F.M., Ogg, J.G., Schmitz, M.D., and Ogg, G.M. (eds), *The Geologic Time Scale 2020*. Vol. 2 (this book). Elsevier, Boston, MA.

Artyushkova, O.V., Maslov, V.A., Pazukhin, V.N., Kulagina, E.I., Tagarieva, R.C., Mizens, L.I., et al., 2011, Devonian and Lower Carboniferous type sections of the Western South Urals. *International Conference "Biostratigraphy, Paleogeography and Events in the Devonian and Lower Carboniferous"*, In: *Memory of Evgeny A. Yolkin (SDS/IGCP 596 Joint Field Meeting), Ufa, Novosibirsk, Russia, July 20−August 10, 2011, Pre-Conference Field Excursion Guidebook, Ufa*, 91 pp.

Averbuch, O., Tribovillard, N., Devleeschouwer, X., Riquier, L., Mistiaen, B., and Van Vliet-Lanoe, B., 2005, Mountain building-enhanced continental weathering and organic carbon burial as major causes for climatic cooling at the Frasnian−Famennian boundary (c. 376 Ma)? *Terra Nova*, **17**: 25−34.

Askew, A.J., and Wellman, C.H., 2018, An endemic flora of dispersed spores from the Middle Devonian of Iberia. *Papers in Palaeontology*, **2018**: 1−45.

Avkhimovitch, V.I., Tchibrikova, E.V., Obukhovskaya, T.G., Nazarenko, A.M., Umnova, V.T., Raskatova, L.G., et al., 1993, Middle and Upper Devonian miospore zonation of eastern Europe. *Bulletin des Centres Recherches Exploration-Production Elf-Aquitaine*, **17**: 79−147.

Azmy, K., Poty, E., and Mottequin, B., 2012, Biochemostratigraphy of the Upper Frasnian in the Namur-Dinant Basin, Belgium: implications for a global Frasnian-Famennian pre-event. *Palaeogeography, Palaeoclimatology, Paleoecology*, **313/314**: 93−106.

Azmy, K., Poty, E., and Brand, U., 2009, High-resolution isotope stratigraphy of the Devonian-Carboniferous boundary in the Namur-Dinant Basin, Belgium. *Sedimentary Geology*, **2016**: 117−124.

Bábek, O., Prikryl, T., and Hladil, J., 2007, Progressive drowning of carbonate platform in the Moravo-Silesian Basin (Czech Republic) before the Frasnian/Famennian event: facies, compositional variation and gamma-ray spectrometry. *Facies*, **53**: 293−316.

Bábek, O., Kumpan, T., Kalvoda, J., and Grygar, T.M., 2016, Devonian/Carboniferous boundary glacioeustatic fluctuations in a platform-to-basin direction: a geochemical approach of sequence stratigraphy in pelagic settings. *Sedimentary Geology*, **337**: 81−99.

Bábek, O., Faměra, M., Šimíček, D., Weinerová, H., and Hladil, J., 2018, Sea-level changes vs. organic productivity as controls on Early and Middle Devonian bioevents: facies- and gamma-ray based sequence-stratigraphic correlation of the Prague Basin, Czech Republic. *Global and Planetary Change*, **160**: 75−95.

Bai, S.L., 1995, Milankovitch Cyclicity and Time Scale of the Middle and Upper Devonian. *International Geology Review*, **37**: 1109−1114.

Bai, S.L., and Ning, Z.S., 1989, Faunal change and events across the Devonian-Carboniferous boundary of Huangmao section, Guangxi, South China, (imprint 1988). *In* McMillan, N.J., Embry, A.F., and Glass, D.J. (eds), Devonian of the World. *Canadian Society of Petroleum Geology, Memoir*, **14(III)**: 147−157.

Bai, S.L., Ning, Z.S., Chai, Z.F., Ma, S.L., Mao, X.Y., Mai, J.G., et al., 1987, Zonation and Geochemical Anomaly of the Devonian/Carboniferous Boundary Beds of Huangmao, Guangxi. *Acta Scientiarum Naturalis, Universitatis Pekinensis*, **1987** (4): 105−111.

Bai, S.L., Bai, Z.Q., Ma, X.P., Wang, D.R., and Sun, Y.L., 1994, *Devonian Events and Biostratigraphy of South China*. Peking University Press, 303 pp., 45 pls.

Bai, S.L., Bai, Z.Q., Ma, X.P., Wang, D.R., and Sun, Y.L., 1995, *Devonian Events and Biostratigraphy of South China*. Peking University Press, 303 pp.

Baird, G.C., and Brett, C.E., 2003, Shelf and off-shelf deposits of the Tully Formation in New York and Pennsylvania: faunal incursions, eustasy and tectonics. *Courier Forschungsinstitut Senckenberg*, **242**: 141−156.

Baird, G.C., and Brett, C.E., 2008, Late Givetian Taghanic bioevents in New York State: new discoveries and questions. *Bulletin of Geosciences*, **83** (4): 357−370.

Bakharev, N.K., 2005a, Lower Devonian (Emsian) ostracod associations from different facies in the vicinity of Gur'evsk, Salair (south of West Siberia). In: Yolkin, E.A., Izokh, N.G., Obut, O.T., and Kipriyanova, T.P. (Eds.), *"Devonian Terrestrial and Marine Environments: From Continent to Shelf" IGCP 499 Project/SDS joint field meeting, Novosibirsk, Russia, July 25−August 9, 2005, Contributions*, pp. 22−26.

Bakharev, N.K., 2005b, Devonian ostracods from the Salair and Kuznetsk Basin: taxonomic composition and stratigraphic distribution. In: Yolkin, E.A., Izokh, N.G., Obut, O.T., and Kipriyanova, T.P. (Eds.), *"Devonian Terrestrial and Marine Environments: From Continent to Shelf" IGCP 499 Project/SDS Joint Field Meeting, Novosibirsk, Russia, July 25−August 9, 2005, Contributions*, pp. 26−27.

Bakharev, N.K., and Bazarova, L.S., 2004, Silurian and Devonian ostracodes of the genus *Miraculum*: new species and phylozones. *News on Paleontology and Stratigraphy, Supplement to Geologiya i Geofisika*, **6/7**: 75−87.

Bakharev, N.K., and Sobolev, E.S., 2011, Ammonoidea and Middle Devonian biostratigraphy of Salair. In: Obut, O.T., and Kipriyanova, T.P. (Eds.), *"Biostratigraphy, Palaeogeography and Events in Devonian and Lower Carboniferous" (SDS/IGCP 596 Joint Field Meeting), July 20−August 10, 2011, Ufa, Novosibirsk, Contributions of International Conference*. Novosibirsk: Publishing House SB RAS, pp. 27−33.

Balinski, A., Olempska, E., and Racki, G. (eds), 2002. Biotic responses to the Late Devonian global events. *Acta Palaeontologica Polonica* **47** (2): 186−404.

Balinski, A., Olempska, E., and Racki, G. (eds), 2006. Biotic aspects of the Early-Middle Frasnian eventful transition. *Acta Palaeontologica Polonica* **51** (4): 606−832.

Balter, V., Renaud, S., Girard, C., and Joachimski, M.M., 2008, Record of climate-driven morphological changes in 376 Ma Devonian fossils. *Geologie*, **36**: 907−910.

Banks, H.P., 1980, Floral assemblages in the Siluro−Devonian. *In* Dilcher, D.L., and Taylor, T.N. (eds), *Biostratigraphy of Fossil Plants*. Stroudsville: Dowden, Hutchinson & Ross Inc, pp. 1−24.

Baranov, V.V., 2012, Lower Devonian Conodont Zonation in Arctic regions of Eurasia. *Stratigraphy Geological Correlations*, **20** (2): 179−198.

Baranov, V.V., and Blodgett, R.B., 2016, Some Emsian (Lower Devonian) polygnathids of Northeast Eurasia. *New Mexico Museum of Natural History and Science, Bulletin*, **74**: 13−24.

Baranov, V.V., Slavík, L., and Blodgett, R.B., 2014, Early Devonian polygnathids of Northeast Asia and correlation of Pragian/Emsian strata of the marginal seas of Angarida. *Bulletin of Geosciences*, **89**: 645−678.

Bardashev, I.A., and Bardasheva, 2012, Platform Conodonts from the Givetian-Frasnian boundary (Middle-Upper Devonian). Dushanbe: Academy of Science, Republic of Tadjikistan, "Donish" Publishing House, pp. 91.

Bardasheva, I.A., and Ovnatanova, N.S., 2018, On the validity of conodont species Mesotaxis falsiovalis Sandberg, Ziegler et Bultynck, 1989. *Paleontological Journal*, **52**: 39−41.

Bardashev, I.A., Weddige, K., and Ziegler, W., 2002, The Phylomorphogenesis of some Early Devonian Platform Conodonts. *Senckenbergiana lethaea*, **82** (2): 375−451.

Bartholomew, A.J., and Brett, C.E., 2007, Correlation of Middle Devonian Hamilton Group-equivalent strata in east-central North America: implications for eustasy, tectonics and faunal provinciality. *In* Becker,

R.T., and Kirchgasser, W.T. (eds), Devonian Events and Correlations. *The Geological Society of London Special Publication,* **278**: 105–131.

Bartzsch, K., and Weyer, D., 2012, Zur Stratigraphie des Breternitz-Members (Obere Clymenien-Schichten, Oberdevon) von Saalfeld (Schwarzburg-Antiklinorium, Thüringisches Schiefergebirge). *Freiberger Foschungshefte,* **C542** (20): 1–54.

Becker, R.T., 1988, Ammonoids from the Devonian-Carboniferous Boundary in the Hasselbach Valley (Northern Rhenish Slate Mountains). *Courier Forschungsinstitut Senckenberg,* **100**: 193–213 pls. 1–2.

Becker, R.T., 1992, Zur Kenntnis von Hemberg-Stufe und *Annulata*-Event im Nordsauerland (Oberdevon, Rheinisches Schiefergebirge, GK 4611 Hohenlimburg). *Berliner geowissenschaftliche Abhandlungen,* **E3**: 3–41.

Becker, R.T., 1993a, Stratigraphische Gliederung und Ammonoideen-Fauna im Nehdenium (Oberdevon II) von Europe und Nord-Afrika. *Courier Forschungsinstitut Senckenberg,* **155**: 1–405.

Becker, R.T., 1993b, Anoxia, eustatic changes, and Upper Devonian to lowermost Carboniferous global ammonoid diversity. *Systematics Association Special,* **47**: 115–163.

Becker, R.T., 1996, New faunal records and holostratigraphic correlation of the Hasselbachtal D/C-Boundary Auxiliary Stratotype. *Annales de la Société géologique de Belgique,* **117** (1): 19–45.

Becker, R.T., 1998, Prospects for an International Substage subdivision of the Famennian. *Subcommission on Devonian Stratigraphy Newsletter,* **15**: 14–17.

Becker, R.T., 2005, Ammonoids and substage subdivisions in the Givetian open shelf facies. In: Yolkin, E.A., Izokh, N.G., Obut, O.T., and Kipriyanova, T.P. (Eds.), *Devonian Terrestrial and Marine Environments: From Continent to Shelf. Contributions From the IGCP 499 Project/SDS Joint Field Meeting, Novosibirsk, Russia, July 25–August 9, 2005,* pp. 29–31.

Becker, R.T., 2006, Myths and facts concerning the Frasnian/Famennian boundary mass extinction. *In* Yang, Q., Wang, Y., and Weldon, E.A. (eds), *Ancient Life and Modern Approaches Abstracts of the Second International Palaeontological Congress, June 17-21, 2006, Beijing, China.* Peking University of Science and Technology of China Press, p. 352.

Becker, R.T., 2007a, Results of the voting on Givetian and Frasnian substages. *Subcommission on Devonian Stratigraphy Newsletter,* **22**: 2.

Becker, R.T., 2007b, Correlation of the proposed middle Givetian substage with the global ammonoid record. *Subcommission on Devonian Stratigraphy Newsletter,* **22**: 17–23.

Becker, R.T., 2007c, Emsian substages and the Daleje Event – a consideration of conodont, dacryoconarid, ammonoid and sea level data. *Subcommission on Devonian Stratigraphy Newsletter,* **22**: 29–32.

Becker, R.T., 2009, Minutes of the SDS Business Meeting, Kitab State Geological Reserve, Uzbekistan. *Subcommission on Devonian Stratigraphy Newsletter,* **24**: 12–15.

Becker, R.T., 2013, Towards the formal voting on Famennian substages. *Subcommission on Devonian Stratigraphy Newsletter,* **28**: 24–29.

Becker, R.T., 2018, The Frasnian-Famennian boundary mass extinction – widespread seismic events, the timing of climatic pulses, "pelagic death zones", and opportunistic survivals. *The Fossil Week, Fifth International Palaeontological Congress, July 9–13, 2018, France, Abstract Book,* p. 107.

Becker, R.T., Aboussalam, Z.S., 2004, The Frasnes Event – a phased 2nd order global crisis and extinction period around the Middle-Upper Devonian boundary. In: El Hassani, A. (Ed.), *Devonian neritic-pelagic correlation and events in the Dra Valley (Western Anti-Atlas, Morocco), International Meeting on Stratigraphy, Rabat, March 1–10, 2004, Abstracts,* pp. 8–10.

Becker, R.T., and Aboussalam, Z.S., 2011, Emsian chronostratigraphy – preliminary new data and a review of the Tafilalt (SE Morocco). *Subcommission on Devonian Stratigraphy Newsletter,* **26**: 33–43.

Becker, R.T., and Aboussalam, Z.S., 2013a, The global Chotec Event at Jebel Amelane (western Tafilalt Platform) – preliminary data. In: Becker, R.T., El Hassani, A., and Tahiri, A. (Eds.), *International Field Symposium "The Devonian and Lower Carboniferous of northern Gondwana", March 22–29, 2013, Field Guidebook,* Vol. 27. Rabat: Document de Í Institut Scientifique, pp. 129–134.

Becker, R.T., Aboussalam, Z.S., 2013b, Middle Givetian – middle Frasnian event stratigraphy at Mdoura-East (western Tafilalt). In: Becker, R.T., El Hassani, A., and Tahiri, A. (Eds.), *International Field Symposium "The Devonian and Lower Carboniferous of northern Gondwana", March 22–29, 2013, Field Guidebook,* Vol. 27. Rabat: Document de l' Institut Scientifique, pp. 143–150.

Becker, R.T., and Hartenfels, S., 2010, Der Faunenwechsel bei Ammonoideen des mittleren Famenniums (Oberdevon III/IV) – Folge einer globalen trophischen Krise in Außenschelfen? *Zitteliana, Reihe B,* **29**: 20–21.

Becker, R.T., and House, M.R., 1994a, Kellwasser Events and goniatite succession in the Devonian of the Montagne Noire with comments on possible causations. *Courier Forschungsinstitut Senckenberg,* **169**: 45–77.

Becker, R.T., and House, M.R., 1994b, International Devonian goniatite zonation, Emsian to Givetian, with new records from Morocco. *Courier Forschungsinstitut Senckenberg,* **169**: 79–135.

Becker, R.T., and House, M.R., 1997, Sea-level changes in the Upper Devonian of the Canning Basin, Western Australia. *Courier Forschungsinstitut Senckenberg,* **199**: 129–146.

Becker, R.T., and House, M.R., 1999, Proposals for an international substage subdivision of the Frasnian. *Subcommission on Devonian Stratigraphy Newsletter,* **15**: 17–22.

Becker, R.T., and House, M.R., 2000a, Devonian ammonoid zones and their correlation with established series and stage boundaries. *Courier Forschungsinstitut Senckenberg,* **220**: 113–151.

Becker, R.T., and House, M.R., 2000b, Late Givetian and Frasnian ammonoid succession at Bou Tchrafine (Anti-Atlas, Southern Morocco). *Notes et Mémoires du Service Géologique,* **399**: 27–36.

Becker, R.T., and House, M.R., 2009, Devonian ammonoid biostratigraphy of the Canning Basin. *Geological Survey of Western Australia, Bulletin,* **145**: 415–439.

Becker, R.T., and Kirchgasser, W.T. (eds), 2007. *Devonian Events and Correlations.* pp. 1–280.

Becker, R.T., and Paproth, E., 1993, Auxiliary stratotype section for the Global Stratotype Section and Point (GSSP) for the Devonian-Carboniferous boundary: Hasselbachtal. *Annales de la Société Géologique de Belgique,* **115** (2): 703–706.

Becker, R.T., Feist, R., Flajs, G., House, M.R., and Klapper, G., 1989, Frasnian-Famennian extinction events in the Devonian at Coumiac, southern France. *Comptes rendus de l'Academie des Sciences Paris, Series II,* **309**: 259–266.

Becker, R.T., House, M.R., Kirchgasser, W.T., and Playford, P.E., 1991, Sedimentary and faunal changes across the Frasnian/Famennian boundary in the Canning Basin of Western Australia. *Historical Biology,* **5**: 183–196.

Becker, R.T., House, M.R., and Kirchgasser, W.T., 1993, Devonian goniatite biostratigraphy and timing of facies movements in the Frasnian of the Canning Basin, Western Australia. *In* Hailwood, E.A., and Kidd, R.B. (eds), High Resolution Stratigraphy. *Geological Society of London Special Publication,* **70**: 293–321.

Becker, R.T., House, M.R., Menner, V.V., and Ovnatanova, N.S., 2000b, Revision of ammonoid biostratigraphy in the Frasnian (Upper Devonian) of the Southern Timan (Northeast Russian Platform). *Acta Geologica Polonica*, **50** (1): 67–97.

Becker, R.T., House, M.R., Bockwinkel, J., Ebbighausen, V., and Aboussalam, Z.S., 2002b, Famennian ammonoid zones of the eastern Anti-Atlas (southern Morocco). *Münstersche Forschungen zur Geologie und Paläontologie*, **93**: 159–205.

Becker, R.T., Ashouri, A.R., and Yazdi, M., 2004a, The Upper Devonian *Annulata* Event in the Shotori Range (eastern Iran). *Neues Jahrbuch für Geologie und Paläontologie, Abhandlungen*, **231** (1): 119–143.

Becker, R.T., Jansen, U., Plodowski, G., Schindler, E., Aboussalam, Z.S., and Weddige, K., 2004b, Devonian litho- and biostratigraphy of the Dra Valley area – an overview. *Documents de l'Institut Scientifique*, **19**: 3–18.

Becker, R.T., Aboussalam, Z.S., Bockwinkel, J., Ebbighausen, V., El Hassani, A., and Nübel, H., 2004c, The Givetian and Frasnian at Oued Mzerreb (Tata region, eastern Dra Valley). *Document de Í Institut Scientifique*, **19**: 29–43.

Becker, R.T., De Baets, K., and Nikolaeva, S., 2010, New ammonoid records from the lower Emsian of the Kitab Reserve (Uzbekistan) – preliminary results. *Subcommission on Devonian Stratigraphy Newsletter*, **25**: 20–28.

Becker, R.T., Gradstein, F.M., and Hammer, O., 2012, The Devonian Period. *In* Gradstein, F.M., Ogg, J.G., Schmitz, M.D., and Ogg, G.M. (eds), *The Geologic Time Scale 2012*. Elsevier Publisher, **2**: pp. 559–601.

Becker, R.T., El Hassani, A., and Tahiri, A. (eds), 2013. International Field Symposium "The Devonian and Lower Carboniferous of Northern Gondwana". Field Guidebook. pp. 1–150.

Becker, R.T., Königshof, P., and Brett, C.E., 2016a, Devonian climate, sea level and evolutionary events: an introduction, Special Publications. *In* Becker, R.T., Königshof, P., and Brett, C.E. (eds), Devonian Climate, Sea Level and Evolutionary Events. *London: Geological Society*, **423**: 1–10.

Becker, R.T., Kaiser, S.I., and Aretz, M., 2016b, Review of chrono-, litho- and biostratigraphy across the global Hangenberg Crisis and Devonian-Carboniferous Boundary, Special Publications. *In* Becker, R.T., Königshof, P., and Brett, C.E. (eds), Devonian Climate, Sea Level and Evolutionary Events. *London: Geological Society*, **423**: 355–386.

Becker, R.T., Aboussalam, Z.S., Hartenfels, S., Nowak, H., Juch, D., and Drozdzewski, G., 2016c, Drowning and sedimentary cover of Velbert Anticline reef complexes (northwestern Rhenish Massif). *In* Becker, R.T., Hartenfels, S., Königshof, P., and Helling, S. (eds), Middle Devonian to Lower Carboniferous Stratigraphy, Facies, and Bioevents in the Rhenish Massif, Germany, An IGCP 596 Guidebook. *Münstersche Forschungen zur Geologie und Paläontologie*, **108**: 76–101.

Becker, R.T., Aboussalam, Z.S., Stichling, S., May, A., and Eichholt, S., 2016d, The Givetian-Frasnian Hönne Valley Reef Complex (northern Sauerland) – an outline of stratigraphy and facies development. *In* Becker, R.T., Hartenfels, S., Königshof, P., and Helling, S. (eds), Middle Devonian to Lower Carboniferous Stratigraphy, Facies, and Bioevents in the Rhenish Massif, Germany, An IGCP 596 Guidebook. *Münstersche Forschungen zur Geologie und Paläontologie*, **108**: 126–140.

Becker, R.T., Hartenfels, S., Weyer, D., and Kumpan, T., 2016e, The Famennian to Lower Viséan at Drewer (northern Rhenish Massif). *In* Becker, R.T., Hartenfels, S., Königshof, P., and Helling, S. (eds), Middle Devonian to Lower Carboniferous Stratigraphy, Facies, and Bioevents in the Rhenish Massif, Germany, An IGCP 596 Guidebook. *Münstersche Forschungen zur Geologie und Paläontologie*, **108**: 158–178.

Becker, R.T., Hartenfels, S., Helling, S., and Schreiber, G., 2016f, The "Nehden Goniatite Shale" (lower Famennian, Brilon Reef Complex, NE Rhenish Massif). *In* Becker, R.T., Hartenfels, S., Königshof, P., and Helling, S. (eds), Middle Devonian to Lower Carboniferous Stratigraphy, Facies, and Bioevents in the Rhenish Massif, Germany, An IGCP 596 Guidebook. *Münstersche Forschungen zur Geologie und Paläontologie*, **108**: 179–195.

Becker, R.T., Piecha, M., Gereke, and Spellbrink, K., 2016g, The Frasnian/Famennian boundary in shelf basin facies north of Diemelsee-Adorf. *In* Becker, R.T., Hartenfels, S., Königshof, P., and Helling, S. (eds), Middle Devonian to Lower Carboniferous Stratigraphy, Facies, and Bioevents in the Rhenish Massif, Germany, An IGCP 596 Guidebook. *Münstersche Forschungen zur Geologie und Paläontologie*, **108**: 220–231.

Becker, R.T., El Hassani, A., Aboussalam, Z.S., Hartenfels, S., Baidder, L., 2018a, The Devonian and Lower Carboniferous of the eastern Anti-Atlas: introduction to a "cephalopod paradise". In: Hartenfels, S., Becker, R.T., El Hassani, A., and Lüddecke, F. (Eds.), 10th International Symposium "Cephalopoda – Present and Past", Fes, March 26–April 3, 2018, Field Guidebook. *Münstersche Forschungen zur Geologie und Paläontologie*, **110**: 145–157.

Becker, R.T., Aboussalam, Z.S., Hartenfels, S., El Hassani, A., Baidder, L., 2018b, Bou Tchrafine – central Tafilalt reference section for Devonian stratigraphy and cephalopod succession. In: Hartenfels, S., Becker, R.T., El Hassani, A., and Lüddecke, F. (Eds.), 10th International Symposium "Cephalopoda – Present and Past", Fes, March 26–April 3, 2018, Field Guidebook. *Münstersche Forschungen zur Geologie und Paläontologie*, **110**: 158–187.

Becker, R.T., Aboussalam, Z.S., El Hassani, A., 2018c, Jebel Mech Irdane – the Eifelian/Givetian boundary GSSP and an important cephalopod locality. In: Hartenfels, S., Becker, R.T., El Hassani, A., and Lüddecke, F. (Eds.), 10th International Symposium "Cephalopoda – Present and Past", Fes, March 26–April 3, 2018, Field Guidebook. *Münstersche Forschungen zur Geologie und Paläontologie*, **110**: 214–228.

Becker, R.T., Aboussalam, Z.S., Hartenfels, S., Gibb, A., Mayer, O., Hüneke, H., 2018d, Emsian events, Frasnian-Famennian boundary, and *Gonioclymenia* Limestone at Jebel Ihrs (western Tafilalt Platform). In: Hartenfels, S., Becker, R.T., El Hassani, A., and Lüddecke, F. (Eds.), 10th International Symposium "Cephalopoda – Present and Past", Fes, March 26–April 3, 2018, Field Guidebook. *Münstersche Forschungen zur Geologie und Paläontologie*, **110**: 229–243.

Becker, R.T., Hartenfels, S., Klug, C., Aboussalam, Z.S., Afhüppe, L., 2018e, The cephalopod-rich Famennian and Tournaisian of the Aguelmous Syncline (southern Maïder). In: Hartenfels, S., Becker, R.T., El Hassani, A., and Lüddecke, F. (Eds.), 10th International Symposium "Cephalopoda – Present and Past", Fes, March 26–April 3, 2018, Field Guidebook. *Münstersche Forschungen zur Geologie und Paläontologie*, **110**: 273–306.

Becker, R.T., Aboussalam, Z.S., and Hartenfels, S., 2018f, The Frasnian-Famennian boundary mass extinction – widespread seismic events, the timing of climatic pulses, "pelagic death zones", and opportunistic survivals. *5th International Paleontological Congress – Paris, 10th-13th July 2018, Abstract Book*, 107.

Becker, R.T., Klug, C., Söte, T., Hartenfels, S., Aboussalam, Z.S., and El Hassani, A., 2019, The oldest ammonoids of Morocco. *Swiss Journal of Palaeontology*, **138**: 9–25.

Belka, Z., Kaufmann, B., and Bultynck, P., 1997, Conodont-based quantitative biostratigraphy for the Eifelian of the eastern Anti-Atlas, Morocco. *Geological Society of America, Bulletin*, **109** (6): 643–651.

Bensaid, M., 1974, Etude sur des Goniatites a la limite du Dévonien Moyen et Supérieur, du Sud Marocain. *Notes de Service Carte géologique du Maroc*, **36** (2364): 81−140.

Berkyová, S., 2009, Lower-Middle Devonian (upper Emsian-Eifelian, *serotinus-kockelianus* zones) conodont faunas from the Prague Basin, Czech Republic. *Bulletin of Geosciences*, **84** (4): 667−686.

Berry, C.M., and Marshall, J.E.A., 2015, Lycopsid forests in the early Late Devonian paleoequatorial zone of Svalbard. *Geology*, **43**: 1043−1046.

Bingham-Koslowski, N., Tsujita, C., Jin, J., and Azmy, K., 2016, Widespread Late Devonian marine anoxia in eastern North America: a case study of the Kettle Point Formation black shale, southwestern Ontario. *Canadian Journal of Earth Sciences*, **53**: 837−855.

Bischoff, G., and Ziegler, W., 1957, Die Conodontenchronologie des Mitteldevons und des tiefsten Oberdevons. *Abhandlungen des Hessischen Geologischen Landesamtes für Bodenforsch*, **22**: 1−136.

Bischoff, G.C.O., and Ferguson, C.L., 1982, Conodont distributions, and ages of Silurian and Devonian limestones in the Palmers Oakey district, N.S. W. *Journal oft he Geological Society of Australia*, **29**: 469−476.

Bless, M.J.M., Becker, R.T., Higgs, K., Paproth, E., and Streel, M., 1993, Eustatic cycles around the Devonian-Carboniferous Boundary and the sedimentary and fossil record in Sauerland (Federal Republic of Germany). *Annales de la Société géologique de Belgique*, **115** (2): 689−702.

Blieck, A., and Turner, S. (eds), 2000. Palaeozoic Vertebrate Biochronology and Global Marine/Non-Marine Correlation − final report of IGCP 328 (1991-1996). *Courier Forschungsinstitut Senckenberg*, **223**: 1−575.

Blieck, A., Clément, G., Blom, H., Lelievre, H., Luksevics, E., Streel, M., et al., 2007, The biostratigtaphical and palaeogeographical framework of the earliest diversification of tetrapods (Late Devonian). *In* Becker, R.T., and Kirchgasser, W.T. (eds), Devonian events and correlations. *The Geological Society, London, Special Publication*, **278**: 219−235.

Blieck, A., Clément, G., and Streel, M., 2010, The biostratigraphical distribution of earliest tetrapods (Late Devonian) − a revised version with comments on biodiversification. *In* Vecolli, M., Clément, G., and Meyer-Berthaud, B. (eds), The Terrestrialization Process: Modelling Complex Interactions at the Biosphere-Geosphere Interface. *The Geological Society, London, Special Publication*, **339**: 129−138.

Blom, H., Clack, J.A., and Ahlberg, P.E., 2005, Localities, distribution and stratigraphical context of the Late Devonian tetrapods of East Greenland. *Meddelelser om Grønland Geoscience*, **43**: 1−50.

Blood, D., Lash, G.G., 2018, FT-02 Stratigraphy and Sedimentology of the Upper Devonian Black Shales of Western New York State with Considerations for Unconventional Reservoir Development. *Eastern Section AAPG, 47th Annual Meeting, October 7, 2018, Field Trip Guidebook*, 106 pp.

Blumenstengel, H., 1965, Zur Taxionomie und Biostratigraphie verkieselter Ostracoden aus dem Thiiringer Oberdevon. *Freiberger Forschungshefte*, **C183**: 1−127.

Bockwinkel, J., Becker, R.T., and Aboussalam, Z.S., 2017, Ammonoids from the late Givetian *Taouzites* Bed of Ouidane Chebbi (eastern Tafilalt, SE Morocco). *Neues Jahrbuch für Geologie und Paläontologie, Abhandlungen*, **284** (3): 307−354.

Bockwinkel, J., Becker, R.T., and Ebbighausen, V., 2009, Upper Givetian ammonoids from Dar Kaoua (Tafilalt, SE Anti-Atlas, Morocco). *Berliner pläobiologische Abhandlungen*, **10**: 61−128.

Bockwinkel, J., Becker, R.T., and Ebbighausen, V., 2013, Late Givetian ammonoids from Hassi Nebech (Tafilalt Basin, Anti-Atlas, southern Morocco). *Fossil Record*, **16** (1): 5−65 58 pp. online supplement.

Bockwinkel, J., Becker, R.T., and Ebbighausen, V., 2015, Late Givetian ammonoids from Äit Ou Amar (northern Maider, Anti-Atlas, southeastern Morocco). *Neues Jahrbuch für Geologie und Paläontologie, Abhandlungen*, **278** (2): 123−158.

Bodorkos, S., Blevin, P.L., Eastlake, M.A., Downes, P.M., Campbell, L.M., Gilmore, P.J., et al. 2015. New SHRIMP U-Pb zircon ages from the central and eastern Lachlan Orogen, New South Wales: July 2013−June 2014. *Geoscience Australia, Canberra; Report* **GS2015/0002**. Geological Survey of New South Wales, Maitland. http://doi.org/10.11636/Record.2015.002

Bodorkos, S., Pogson, D.J., Friedman, R.M., 2017, Zircon U-Pb dating of biostratigraphically constrained felsic volcanism in the Lachlan Orogen via SHRIMP and CA-IDTIMS: implications for the division of Early Devonian time. *Granites 2017, September 2017, Benalla*, pp. 11−14.

Bogoslovskiy, B.I., 1969, Devonskie ammonoidey. I. Agoniatity. *Trudy Paleontologicheskogo Instituta, Akademiya Nauk SSSR*, **124**: 1−341, 29 pls.

Bogoslovskiy, B.I., 1971, Devonskie ammonoidey. II. Goniatity. *Trudy Paleontologicheskogo Instituta, Akademiya Nauk SSSR*, **127**: 1−228, 19 pls.

Bogoslovskiy, N.I., 1981, Devonskie ammonoidey. III. Klimenii (Podotryad Gonioclymeniina). *Trudy Paleontologicheskogo Instituta, Akademiya Nauk SSSR*, **191**: 1−123, 16 pls.

Bojar, A.-V., Neubauer, F., and Koeberl, C., 2013, Geochemical record of Late Devonian to Early Carboniferous events, Palaeozoic of Graz, Eastern Alps, Austria, Special Publications. *In* Gasiewicz, A., and Slowakiewicz, M. (eds), Palaeozoic Climate Cycles: Their Evolutionary and Sedimentologicalb Impact. *London: Geological Society*, **376**: pp. 87−108.

Boncheva, I., Sachanski, V., and Becker, R.T., 2015, Sedimentary and faunal evidence for the Late Devonian Kellwasser and *Annulata* events in the Balkan Terrane (Bulgaria). *Geologica Balcanica*, **44** (1/3): 17−24.

Bond, D., 2006, The fate of the homoctenids (Tentaculitoidea) during the Frasnian-Famennian mass extinction (Late Devonian). *Geobiology*, **4**: 167−177.

Bond, D., and Zaton, M., 2003, Gamma-ray spectrometry across the Upper Devonian basin succession at Kowala in the Holy Cross Mountains (Poland). *Acta Geologica Polonica*, **53** (2): 93−99.

Bond, D., Wignall, P.B., and Racki, G., 2004, Extent and duration of marine anoxia during the Frasnian−Famennian (Late Devonian) mass extinction in Poland, Germany, Austria and France. *Geological Magazine*, **141** (2): 173−193.

Bouček, B., 1964, The tentaculites of Bohemia; their morphology, taxonomy, ecology, phylogeny and biostratigraphy. *Czechoslovak Academy of Sciences, Prague*, 215 pp, 40 pls.

Bouček, B., 1966, Eine neue und bisher jüngste Graptolithenfauna aus dem böhmischen Devon. *Neues Jahrbuch für Geologie und Paläontologie, Monatshefte*, **1966** (3): 161−168.

Boucot, A.J., 1989, Devonian Biogeography: an update, (imprint 1988). *In* McMillan, N.J., Embry, A.F., and Glass, D.J. (eds), Devonian of the World. *Canadian Society of Petroleum Geology, Memoir*, **14(III)**: 211−227.

Boulvain, F., Da Silva, A.C., Mabille, C., Hadil, J., Gersl, M., Koptíková, L., et al., 2010, Magnetic susceptibility correlation of km-thick Eifelian-Frasnian sections (Ardenne-Moravian karst). *Geologica Belgica*, **13** (4): 309−318.

Brand, U., 2004, Carbon, oxygen and strontium isotopes in Paleozoic carbonate components: an evaluation of original seawater-chemistry proxies. *Chemical Geology*, **204**: 23−24.

Brand, U., Legrand-Blain, M., and Steel, M., 2004, Biochemostratigraphy of the Devonian-Carboniferous boundary global stratotype section and point, Griotte Formation, La Serre, Montagne Noire, France. *Palaeogeography, Palaeoclimatology, Palaeoecology*, **195**: 99–124.

Bratton, J.F., Berry, W.B.N., and Morrow, J.R., 1999, Anoxia pre-dates Frasnian-Famennian boundary mass extinction horizon in the Great Basin, USA. *Palaeogeography, Palaeoclimatology, Palaeoecology*, **154**: 275–292.

Braun, W.K., 1978, Devonian ostracodes and biostratigraphy of western Canada, Special Paper. *In* Stelck, C.R., and Chatterton, B.D.E. (eds), Western and Arctic Canadian Biostratigraphy. *Geological Association of Canada*, **18**: 259–288.

Braun, W.K., 1990, Devonian ostracod faunas of western Canada; their evolution, biostratigraphical potential and environmental settings. *In* Whatley, R., and Maybury, C. (eds), *Ostracoda and Global Events*. London, New York: Chapman and Hall, 59–70.

Brett, C.E., Ivany, L.C., Bartholomew, A.J., DeSantis, M.K., and Baird, G.C., 2009, Devonian ecological-evolutionary subunits in the Appalachian Basin: a revision and a test of persistence and discreteness. *In* Königshof, P. (ed), Devonian Change: Case Studies in Palaeogeography and Palaeoecology. *The Geological Society, London, Special Publications, 314*: 7–36.

Brett, C.E., Baird, G.C., Bartholomew, A., DeSantis, M., and Ver Straeten, C., 2011, Sequence stratigraphy and revised sea level curve for the Middle Devonian of Eastern North America. *Palaeogeography, Palaeoclimatology, Palaeoecology*, **304**: 21–53.

Brett, C.E., Zambito, J.J., Schindler, E., and Becker, R.T., 2012, Diagenetically-enhanced trilobite obrution deposits in concretionary limestones: the paradox of "Rhythmic Event Beds". *Palaeogeography, Palaeoclimatology, Palaeoecology*, **367/368**: 30–43.

Brett, C.E., Zambito, J.J., Baird, G.C., Aboussalam, Z.S., Becker, R.T., and Bartholomew, A.J., 2018, Litho-, bio-, and sequence stratigraphy of the Boyle-Portwood Succession (Middle Devonian, Central Kentucky, USA). *Palaeobiodiversity Palaeoenvironments*, **98** (2): 331–368.

Breuer, P., and Steemans, P., 2013, Devonian spore assemblages from northwestern Gondwana: taxonomy and biostratigraphy. *Special Papers in Palaeontology*, **89**: 1–163.

Brezinski, D.K., Cecil, C.B., and Skema, V.W., 2010, Late Devonian glacigenic and associated facies from the central Appalachian Basin, eastern United States. *GSA Bulletin*, **122** (1/2): 265–281.

Brocke, R., Berkyová, S., Budil, P., Fatka, O., Fryda, J., Schindler, E., 2009, Phytoplankton bloom in the early Middle Devonian (Eifelian): The Basal Choteč Event in the Barrandian area (Czech Republic). In: Röhling, H.-G., Linnemann, U., and Lange, J.-M. (Eds.), GeoDresden 2009, Geologie der Böhmischen Masse – Regionale und Angewandte Geowissenschaften in Mitteleuropa, Dresden, September 30–October 2, Abstracts. *Schriftenreihe der Deutschen Gesellschaft für Geowissenschaften*, **63**: 63.

Brocke, R., Fatka, O., Lindemann, R.H., Schindler, E., and Ver Straeten, C.A., 2016, Palynology, dacryoconarids and the lower Eifelian (Middle Devonian) Basal Choteč Event: case studies from the Prague and Appalachian basins, Special Publications. *In* Becker, R. T., Königshof, P., and Brett, C.E. (eds), Devonian Climate, Sea Level and Evolutionary Events. *London: Geological Society,* **423**: 123–169.

Budil, P., 1995, Demonstrations of the Kačák event (Middle Devonian, uppermost Eifelian) at some Barrandian localities. *Vestnik Ceského geologeckého ústavu*, **70**: 1–24.

Buggisch, W., and Mann, W., 2004, Carbon isotope stratigraphy of Lochkovian to Eifelian limestones from the Devonian of central and southern Europe. *International Journal of Earth Sciences*, **93**: 521–541.

Buggisch, W., and Joachimski, M.M., 2006, Carbon isotope stratigraphy of the Devonian of Central and Southern Europe. *Palaeogeography, Palaeoclimatology, Palaeoecology*, **240**: 68–88.

Bultynck, P., 1970, Révision stratigraphique et paléontologique de la coupe type du Couvinien. *Mémoires de l'Institut Géologie de l'Université de Louvain*, **26**: 1–152, 39 pls.

Bultynck, P., 1976, Le Silurien superieur et le Dévonien inferieur de la Sierra de Guadarrama (Espagne Centrale). Troisième partie: elements icriodiformes, pelekysgnathiformes et polygnathiformes. *Bulletin du Institut royal des Sciences naturelles Belgique, Sciences de la Terre*, **49** (5): 1–74.

Bultynck, P., 1985, Lower Devonian (Emsian)–Middle Devonian (Eifelian and lowermost Givetian) conodont successions from the Ma'der and the Tafilalt, southern Morocco. *Courier Forschungsinstitut Senckenberg*, **75**: 261–286.

Bultynck, P., 1987, Pelagic and neritic conodont successions from the Givetian of pre-Sahara Morocco and the Ardennes. *Bulletin de l'Institut Royal des Sciences Naturelles de Belgique, Sciences de la Terre*, **57**: 149–181.

Bultynck, P., 1989, Conodonts from the La Grange Limestone (Emsian), Armorican Massif, North-Western France. *Courier Forschungsinstitut Senckenberg*, **117**: 173–203.

Bultynck, P., 2000a, Fossil groups important for boundary definition. *Courier Forschungsinstitut Senckenberg*, **220**: 1–205.

Bultynck, P., 2000b, Recognition of Devonian series and stage boundaries in geological areas. *Courier Forschungsinstitut Senckenberg*, **225**: 1–347.

Bultynck, P., 2003, Devonian Icriodontidae: biostratigraphy, classification and remarks on palaeoecology and dispersal. *Revista Española de Paleontología*, **35** (3): 295–314.

Bultynck, P., 2005, The Givetian Working Group. *Subcommission on Devonian Stratigraphy Newsletter*, **21**: 20–22.

Bultynck, P., 2007, Limitations on the application of the Devonian standard conodont zonation. *Geological Quarterly*, **51** (4): 339–344.

Bultynck, P., Gouwy, S., 2002, Towards a standardization of global Givetian substages. In: Yushkin, N.P., Tsyganko, V.S., and Männik, P. (Eds.), *Geology of the Devonian System: Proceedings of the International Symposium, July 9–12 2002, Syktyvkar, Komi Republic*, pp. 142–144.

Bultynck, P., and Hollard, H., 1980, Distribution comparée de Conodonts et Goniatites dévoniens de plaines du Dra, du Ma'der et du Tafilalt (Maroc). *Aardkundige Mededlingen*, **1**: 73.

Bultynck, P., and Hollevoet, C., 1999, The Eifelian-Giverian boundary and Struve's Middle Devonian Great Cap in the Couvin area (Ardennes, southern Belgium). *Senckenbergiana lethaea*, **79** (1): 3–11.

Bultynck, P., and Martin, F., 1995, Assessment of an old stratotype: the Frasnian/Famennian boundary at Senzeilles, Southern Belgium. *Bulletin de L'Institut Royal des Sciences Naturelles de Belgique*, **65**: 5–34.

Bultynck, P., and Walliser, O.H., 2000, Devonian Boundaries in the Moroccan Anti-Atlas. *Courier Forschungsinstitut Senckenberg*, **225**: 211–226.

Caplan, M.L., and Bustin, R.M., 1999, Devonian-Carboniferous Hangenberg mass extinction event, widespread organic-rich mudrock and anoxia; causes and consequences. *Palaeogeography, Palaeoclimatology, Palaeoecology*, **148** (4): 187–207.

Caputo, M.V., Melo, J.H.G., Steel, M., and Isbell, J.L., 2008, Late Devonian and Early Carboniferous glacial records of South America. *Geological Society of America, Special*, **441**: 161–173 Paper.

Carls, P., and Gandl, J., 1969, Stratigraphie und Conodonten des Unter-Devons des Östlichen Iberischen Ketten (NE-Spanien). *Neues Jahrbuch für Geologie und Paläontologie, Abhandlungen*, **132**: 155–218.

Carls, P., Gandl, J., Groos-Uffenorde, H., Jahnke, H., and Walliser, O.H., 1972, Neue Daten zur Grenze Unter-/Mittel-Devon. *Newsletters on Stratigraphy*, **2** (3): 115–147.

Carls, P., and Valenzuela-Ríos, J.I., 2002, Early Emsian conodonts and associated shelly faunas of the Mariposas Fm (Iberian Chains, Aragón, Spain). *Cuadernos del Museo Geominero*, **1**: 315–333.

Carls, P., and Valenzuela-Ríos, J.I., 2007, From the Emsian GSSP to the early late-Emsian correlations with historical boundaries. *Subcommission on Devonian Stratigraphy Newsletter*, **22**: 24–28.

Carls, P., Slavík, L., and Valenzuela-Ríos, J.I., 2007, Revisions of conodont biostratigraphy across the Silurian-Devonian boundary. *Bulletin of Geosciences*, **82** (2): 145–164.

Carls, P., Slavík, L., and Valenzuela-Ríos, J.I., 2008, Comments on the GSSP for the basal Emsian stage boundary: the need for its redefinition. *Bulletin of Geosciences*, **83** (4): 383–390.

Carmichael, S.K., Waters, J.A., Batchelor, C.J., Coleman, D.M., Suttner, T.J., Kido, E., et al., 2015, Climate instability and tipping points in the Late Devonian: detection of the Hangenberg Event in an open oceanic island arc in the Central Asian Orogenic Belt. *Gondwana Research*, **32**: 213–231.

Carpenter, S.J., Lohmann, K.C., Holden, P., Walter, L.M., Huston, T.J., and Haliday, A.N., 1991, $\delta^{18}O$ values, $^{87}Sr/^{86}Sr$ and Sr/Mg ratios of Late Devonian abiotic calcite: implications for the composition of ancient seawater. *Geochimica et Cosmochimica Acta*, **55**: 1991–2010.

Casier, J.-G., 1979, La Zone à *Svantovites lethiersi*, zone nouvelle d'ostracodes de la fin du Frasnien et du début du Famennien des bassins de Namur et de Dinant. *Bulletin de l'Institut royal des Sciences naturelles de Belgiqie, Sciences de la Terre*, 51.

Casier, J.-G., 2018, Ostracods across the Frasnian/Famennian boundary (Devonian) in the Hony railway section (southern border of the Dinant Synclinorium, Belgium)-geochemical consequences. *Palaeobiodiversity Palaeoenvironments*, **98** (3): 431–439.

Casier, J.G., Devleeschouwer, X., Moreau, J., Petitclerc, E., and Préat, A., 2011a, Ostracods, rock facies and magnetic susceptibility records from the stratotype of the Terres d'Haurs Formation (Givetian) at the Mont d'Haurs (Givet, France). *Bulletin de l'Institut Royal des Sciences Naturelles de Belgique*, **81**: 97–128.

Casier, J.G., Devleeschouwer, X., Petitclerc, E., and Préat, A., 2011b, Ostracods, rock facies and magnetic susceptibility of the Hanonet Formation/Trois-Fontaines Formation boundary interval (Early Givetian) at the Mont d'Haurs (Givet, France). *Bulletin de l'Institut Royal des Sciences Naturelles de Belgique*, **81**: 63–96.

Chan, W.C., Kabanov, P., Gouwy, S., 2017, Capturing an Eifelian-Givetian drowning event in the northwestern Canadian sub-arctic mainland. In: Liao, J.-C., and Valenzuela-Ríos, J.I. (Eds.), Fourth International Conodont Symposium. ICOS IV "Progress on Conodont Investigations". *Cuadernos del Museo Geominero*, **22**: 141–145.

Chang, J.J., Bai, Z.Q., Sun, Y.L., Peng, Y.B., Qin, S., and Shen, B., 2017, High resolution bio- and chemostratigraphic framework at the Frasnian-Famennian boundary: implications for regional stratigraphic correlation between different sedimentary facies in South China (online) *Palaeogeography, Palaeoclimatology, Palaeoecology*, <https//doi.org/10.1016/j.palaeo.2017.05.019>.

Chen, D.Z., and Tucker, M.E., 2003, The Frasnian-Famennian mass extinction: insights from high-resolution sequence stratigraphy and cyclostratigraphy in South China. *Palaeogeography, Palaeoclimatology, Palaeoecology*, **193**: 87–111.

Chen, D.Z., Tucker, M.E., Shen, Y., Yans, J., and Preat, A., 2002, Carbon isotope excursions and sea-level change: implications for the Frasnian-Famennian biotic crisis. *Journal of the Geological Society, London*, **159**: 623–626.

Chen, D.Z., Qing, H.R., and Li, R.W., 2005, The Late Devonian Frasnian-Famennian (F/F) biotic crisis: insights from $\delta^{13}C_{carb}$, $\delta^{13}C_{org}$ and $^{87}Sr/^{86}Sr$ isotopic systematics. *Earth and Planetary Science Letters*, **235**: 151–166.

Chen, D.Z., Wang, J.G., Racki, G., Li, H., Wang, C.Y., Ma, X.P., et al., 2013, Large sulphur isotopic perturbations and oceanic changes during the Frasnian-Famennian transition of the Late Devonian. *Journal of the Geological Society, London*, **170**: 465–476.

Chen, X., Ni, Y., Lenz, A.C., Zhang, L., Chen, Z., and Tang, L., 2015, Early Devonian graptolites from the Qinzhou–Yulin region, southeast Guangxi, China. *Canadian Journal of Earth Sciences*, **52**: 1–14.

Chlupáč, I., 2000, Cyclicity and duration of Lower Devonian stages: observations from the Barrandian area, Czech Republic. *Neues Jahrbuch für Geologie und Paläotologie, Abhandlungen*, **215** (1): 97–124.

Chlupáč, I., and Hladil, J., 2000, The global stratotype section and point of the Silurian-Devonian boundary. *Courier Forschungsinstitut Senckenberg*, **225**: 1–7.

Chlupáč, I., and Kukal, Z., 1988, Possible global events and the stratigraphy of the Palaeozoic of the Barrandian (Cambrian-Middle Devonian, Czechoslovakia). *Sborník geologických věd, Geologie*, **43**: 86–146.

Chlupáč, I., and Lukes, P., 1999, Pragian/Zlichovian and Zlichovian/Dalejan boundary sections in the Lower Devonian of the Barrandian area, Czech Republic. *Newsletters on Stratigraphy*, **37** (1/2): 75–100.

Chlupáč, I., and Oliver Jr., W.A., 1989, Decision on the Lochkovian–Pragian Boundary Stratotype (Lower Devonian). *Episodes*, **12**: 109–113.

Chlupáč, I., and Turek, V., 1983, Devonian goniatites from the Barrandian area, Czechoslovakia. *Rozpravy Ùstrédnho ústavu geologického*, **46**: 1–159.

Churkín, M., Jaeger, H., and Eberlein, G.G., 1970, Lower Devonian graptolites from southern Alaska. *Lethaia*, **3**: 183–202.

Clack, J.A., 2002, *Gaining Ground: The Origin and Evolution of Tetrapods*. Bloomington, IN: Indiana University Press, 369 pp.

Claeys, P., Kyte, F.T., Herbosch, A., and Casier, J.-G., 1996, Geochemistry of the Frasnian-Famennian boundary in Belgium: mass extinction, anoxic oceans and microtektite layer, but not much iridium. *Geological Scoiety of America*, **307**: 491–504 Special Paper.

Clark, S., Day, J., Ellwood, B., Harry, R., and Tomkin, J., 2009, Astronomical Tuning of Integrated Upper-Famennian-Early Carboniferous Faunal, Carbon Isotope and High Resolution Magnetic Susceptibility Records: Western Illinois Basin. *Subcommission on Devonian Stratigraphy Newsletter*, **24**: 27–35.

Clausen, C.-D., Lorn, D., Luppold, F.W., and Stoppel, D., 1989, Untersuchungen zur Devon/Karbon-Grenze auf dem Müssenberg (Nördliches Rheinisches Schiefergebirge). *Bulletin de la Société belge de Géologie*, **98** (3/4): 353–369.

Cole, D., Myrow, P.M., Fike, D.A., Hakim, A., and Gehrels, G.E., 2015, Uppermost Devonian (Famennian) to Lower Mississippian events of the western U.S.: stratigraphy, sedimentology, chemostratigraphy, and detrital zircon geochronology. *Palaeogeography, Palaeoclimatology, Palaeoecology*, **427**: 1–19.

Compston, W., 2000, Interpretations of SHRIMP and isotope dilution zircon ages for the Palaeozoic time-scale: II. Silurian to Devonian. *Mineralogical Magazine*, **64** (6): 1127–1146.

Compston, W., 2004, SIMS U–Pb zircon ages for the Upper Devonian Snobs Creek and Cerberean Volcanics from Victoria, with age uncertainty based on UO2/UO v. UO/U precision. *Journal of the Geological Society, London*, **161**: 223–228.

Corradini, C., 2008, Revision of Famennian-Tournaisian (Late Devonian–Early Carboniferous) conodont biostratigraphy of Sardinia, Italy. *Revue de Micropaléonologie*, **51**: 123–132.

Corradini, C., and Corriga, M.G., 2012, A Přídoli-Lochkovian conodont zonation in Sardinia and the Carnic Alps: implications for a global zonation scheme. *Bulletin of Geosciences*, **87** (4): 635–650.

Corradini, C., Kaiser, S.I., Perri, M.C., and Spalletta, C., 2011, Conodont genus *Protognathodus* and its potential tool for defining the Devonian/Carboniferous boundary. *Rivista Italiana di Paleontologia e Stratigrafia*, **117**: 15–28.

Corradini, C., Spalletta, C., Mossoni, A., Matyja, H., and Over, J.D., 2016, Conodonts across the Devonian/Carboniferous boundary: a review and implication for the redefinition of the boundary and a proposal for an updated conodont zonation (online) *Geological Magazine*, **154** (4): 888–902.

Cramer, B.D., Saltzman, M.R., Day, J., and Witzke, B.J., 2008, Record of the Late Devonian Hangenberg global positive carbon isotope excursion in epeiric sea setting: carbonate production, organic carbon burial, and paleoceanography during the Late Famennian, Paper. *In* Pratt, B., and Holmden, C. (eds), Epeiric Seas. *Geological Association of Canada*, **48**: 103–118.

Cramer, B.D., Schmitz, M.D., Huff, W.D., and Bergstrom, S.M., 2015, High-precision U-Pb zircon age constraints on the duration of rapid biogeochemical events during the Ludlow Epoch (Silurian Period). *Journal of The Geological Society*, **172**: 157–160.

Cramer, B.D., and Jarvis, I., 2020, Chapter 11 - Carbon isotope stratigraphy. *In* Gradstein, F.M., Ogg, J.G., Schmitz, M.D., and Ogg, G.M. (eds), *The Geologic Time Scale 2020*. Vol. 1 (this book). Elsevier, Boston, MA.

Crasquin-Soleau, S., Lethiers, F., and Tassy, P., 1994, Évolution des espèces du genre *Polyzygia* (Ostracoda, Dévonien). *Compte Rendu de l'Academie des Sciences Paris, série II*, **319**: 365–371.

Crick, R.E., Ellwood, B.B., Feist, R., El Hassani, A., Feist, R., and Hladil, J., 1997, Magnetosusceptibility Event and Cyclostratigraphy (MSEC) of the Eifelian-Givetian GSSP and associated boundary sequences in North Africa and Europe. *Episodes*, **20**: 167–174.

Crick, R.E., Ellwood, B.B., Feist, R., El Hassani, A., and Feist, R., 2000, Proposed Magnetostratigraphy Susceptibility Stratotype for the Eifelian-Givetian GSSP (Anti-Atlas Morocco). *Episodes*, **23**: 93–101.

Crick, R.E., Ellwood, B.B., El Hassani, A., Hladil, J., Hrouda, F., and Chlupáč, I., 2001, Magnetostratigraphy susceptibility of the Pridoli-Lochkovian (Silurian-Devonian) GSSP (Klonk, Czech Republic) and a coeval sequence in Anti-Atlas Morocco. *Palaeogeography, Palaeoclimatology, Palaeoecology*, **167**: 73–100.

Crick, R.E., Ellwood, B.B., Feist, R., El Hassani, A., Schindler, E., Dreesen, R., et al., 2002, Magnetostratigraphy susceptibility of the Frasnian/Famennian boundary. *Palaeogeography, Palaeoclimatology, Palaeoecology*, **181**: 67–90.

Czarnocki, J., 1989, Klimenie Gór Świetokrzyskich. *Prace Pánstwowego Instytutu Geologicznego*, **127**: 1–91 50 pls.

Da Silva, A.-C., and Boulvain, F., 2002, Sedimentology, Magnetic susceptibility and isotopes of a Middle Frasnian carbonate platform: Talifer Section, Belgium. *Facies*, **46**: 89–102.

Da Silva, A.-C., and Boulvain, F., 2006, Upper Devonian carbonate platform correlations and sea level variations recorded in magnetic susceptibility. *Palaeogeography, Palaeoclimatology, Palaeoecology*, **240**: 373–388.

Da Silva, A.-C., and Boulvain, F., 2008, Carbon isotope lateral variability in a Middle Frasnian carbonate platform (Belgium): significance of facies, diagenesis and sea-level history. *Palaeogeography, Palaeoclimatology, Palaeoecology*, **269**: 189–204.

Da Silva, A.C., and Boulvain, F. (eds), 2010. Application of magnetic susceptibility, correlations and paleoenvironments. *Geologica Belgica, Special issue* **13** (4).

Da Silva, A.C., and Boulvain, F., 2012, Analysis of the Devonian (Frasnian) platform from Belgium: a multi-faceted approach for basin evolution reconstruction. *Basin Research*, **24**: 338–356.

Da Silva, A.C., Potma, K., Weissenberger, J.A., Whalen, M.T., Humblet, M., Mabille, C., and Boulvain, F., 2009, Magnetic susceptibility evolution and sedimentary environments on carbonate platform sediments and atolls, comparison of the Frasnian from Belgium and Alberta, Canada. *Sedimentary Geology*, **214**: 3–18.

Da Silva, A.C., Yans, J., and Boulvain, F., 2010, Early-middle Frasnian (early Late Devonian) sedimentology and magnetic susceptibility of the Ardennes area (Belgium): identification of severe and rapid sea-level fluctuations. *Geologica Belgica*, **13** (4): 319–332.

Da Silva, A.C., De Vleeschouwer, D., Boulvain, F., Claeys, P., Fagel, N., Humblet, M., et al., 2013, Magnetic susceptibility as a high-resolution correlation tool and as a climatic proxy in Paleozoic rocks – merits and pitfalls: examples from the Devonian in Belgium. *Marine and Petroleum Geology*, **46**: 173–189.

Da Silva, A.C., Whalen, M.T., Hladil, J., Chadimova, L., Chen, D., Spassov, S., et al., 2015, Magnetic susceptibility application: a window onto ancient environments and climatic variations: foreword. *Geological Society Special Publication*, **414**: 1–13.

Da Silva, A.C., Hladil, J., Chadimová, L., Slavík, L., Hilgen, F.J.J., Bábek, O., et al., 2016, Refining the Early Devonian time scale using Milankovitch cyclicity in Lochkovian–Pragian sediments (Prague Synform, Czech Republic). *Earth and Planetary Science Letters*, **455**: 125–139.

Da Silva, A.C., Sinnesael, M., Claeys, P., Davies, J.H.F.L., de Winter, N.J., Percival, L.M.E., et al., 2020, Anchoring the Late Devonian mass extinction in absolute time by integrating climatic controls and radio-isotopic dating. *Scientific Reports*, **10**, article #12940: 12 pp. https://doi.org/10.1038/s41598-020-69097-6.

Davydov, V., Crowley, J., Schmitz, M.D., and Poletaev, V., 2010, High-precision U-Pb zircon age calibration of the global Carboniferous time scale and Milankovitch band cyclicity in the Donets Basin, eastern Ukraine. *Geochemistry Geophysics Geosystems*, **11**:Q0AA04, https://doi.org/10.1029/2009GC00273.

Davydov, V.I., Schmitz, M., and Korn, D., 2011, The Hangenberg Event was abrupt and short at the global scale: the quantitative integration and intercalibration of biotic and geochronologic data within the Devonian-Carboniferous transition. *Geological Society of America, Abstracts with Programs*, **43** (5): 128.

Day, J., and Witzke, B.J., 2017, Upper Devonian Biostratigraphy, Event Stratigraphy, and Late Frasnian Kellwasser Extinction Bioevents in the

Iowa Basin: Western Euramerica. *In* Montenari, M. (ed), Stratigraphy and Timescales. *Burlington: Academic Press,* **Vol. 2**: 243–332.

Day, J., Uyeno, T., Norris, W., Witzke, B.J., and Bunker, B.J., 1996, Middle-Upper Devonian relative sea-level histories of central and western North American interior basin. *Geological Society of America, Special Paper,* **306**: 259–275.

Day, J., Witzke, B.I., Bunker, B.J., Holmden, C., Rowe, H., 2010, Epeiric C^{13}carb record from the Middle and Upper Devonian Cedar Valley Group – Iowa Basin of Central North America. *GSA Annual Meeting, Abstracts With Programs,* **42**: 514.

Day, J., Witzke, B., Rowe, H., 2011, Development of an empirical subtropical paleoclimate record from western Euramerica: late Frasnian-earliest Tournaisian stable carbon isotope record from the Yellow Spring-New Albany Groups of the northwestern Illinois Basin. *GSA Annual Meeting, Abstracts With Programs,* **43**: 151.

De Baets, K., Klug, C., and Plusquellec, Y., 2010, Zlíchovian faunas with early ammonoids from Morocco and their use for the correlation of the eastern Anti-Atlas and the western Dra Valley. *Bulletin of Geosciences,* **85** (2): 317–352.

De Baets, K., Klug, C., Korn, D., Bartels, C., and Poschmann, M., 2013, Emsian Ammonoidea and the age of the Hunsrück Slate (Rhenish Mountains, Western Germany). *Palaeontographica, Abteilung A,* **299**: 1–113.

De la Rue, S.R., Rowe, H.D., and Rimmer, S., 2007, Palynological and bulk geochemical constraints on the paleoceanographic conditions across the Frasnian-Famennian boundary, New Albany Shale, Indiana. *International Journal of Coal Geology,* **71**: 72–84.

Denayer, J., Mottequin, B., 2015, Middle and Upper Devonian Events in Belgium: review and new insights. In: Mottequin, B., Denayer, J., Königshof, P., Prestianni, C., and Olive, S. (Eds.), IGCP 596 – SDS Symposium, Climate change and Biodiversity patterns in the Mid-Palaeozoic, September 20–22, 2015, Brussels – Belgium, Abstracts. *Strata, Travaux de Géologie sédimentaire et Paleontologie, Série 1: communications,* **16**: 40–42.

Denayer, J., Prestianni, C., Mottequin, B., Hance, L., Poty, E., 2019, Revision of the Devonian-Carboniferous Boundary in Belgium and surrounding areas: a scenario. In: Hartenfels, S., Herbig, H.-G., and Aretz, M. (Eds.), 19th International Congress on the Carboniferous and Permian, Cologne, July 29–August 2, 2019, Abstracts. *Kölner Forum für Geologie und Paläontologie,* **23**: 75–76.

Denison, R.E., Koepnick, R.B., Burke, W.H., Hetherington, E.A., and Fletcher, A., 1997, Construction of the Silurian and Devonian seawater ^{87}Sr/^{86}Sr curve. *Chemical Geology,* **140**: 109–121.

DeSantis, M.K., and Brett, C.E., 2011, Late Eifelian (Middle Devonian) biocrises: timing and signature of the pre-Kačák Bakoven and Stony Hollow Events in eastern North America. *Palaeogeography, Palaeoclimatology, Palaeoecology,* **304**: 113–135.

DeSantis, M.K., Brett, C.E., and Ver Straeten, C.A., 2007, Persistent depositional sequences and bioevents in the Eifelian (early Middle Devonian) of eastern Laurentia: North American evidence of the Kacák Events? *In* Becker, R.T., and Kirchgasser, W.T. (eds), Devonian Events and Correlations. *The Geological Society of London, Special Publication,* **278**: 83–104.

De Vleeschouwer, D., Whalen, M.T., Day, J.E., and Claeys, P., 2012a, Cyclostratigraphic calibration of the Frasnian (Late Devonian) timescale (Western Alberta, Canada). *Geological Society of America bulletin,* **124**: 928–942.

De Vleeschouwer, D., Da Silva, A.C., Boulvain, F., Crucifix, M., and Claeys, P., 2012b, Precessional and half-precessional climate forcing

of Mid-Devonian monsoon-like dynamics. *Climate of the Past,* **8**: 337–351.

De Vleeschouwer, D., Rakociński, M., Racki, G., Bond, D.P.G.G., Sobień, K., and Claeys, P., 2013, The astronomical rhythm of Late-Devonian climate change (Kowala section, Holy Cross Mountains, Poland). *Earth and Planetary Science Letters,* **365**: 25–37.

De Vleeschouwer, D., and Parnell, A.C., 2014, Reducing time-scale uncertainty for the Devonian by integrating astrochronology and bayesian statistics. *Geology,* **42**: 491–494.

De Vleeschouwer, D., Boulvain, F., Da Silva, A., Pas, D., Labaye, C., and Claeys, P., 2015, The astronomical calibration of the Givetian (Middle Devonian) timescale (Dinant Synclinorium, Belgium, Special Publications. *In* Da Silva, A.C., Whalen, M.T., Hladil, J., Chadimová, L., Chen, D., Spassov, S., Boulwain, F., and Dveleeschouwer, X. (eds), Magnetic Susceptibility Application: A Window onto Ancient Environments and Climatic Variations. *London: Geological Society,* **414**: 245–256.

De Vleeschouwer, D., Königshof, P., and Claeys, P., 2017a, Reading time and paleoenvironmental change in the Emsian–Eifelian boundary GSSP section (Wetteldorf, Germany): a combination of cyclostratigraphy and facies analysis (online) *Newsletters on Stratigraphy,* **41** (2): 209–226.

De Vleeschouwer, D., Da Silva, A.C., Sinnesael, M., Chen, D., Day, J.E., Whalen, M.T., et al., 2017b, Timing and pacing of the Late Devonian mass extinction event regulated by eccentricity and obliquity. *Nature Communications,* **8** (2268): 1–12. https://doi.org/10.1038/s41467-017-02407-1.

Devleeschouwer, X., Petitclerc, E., Spassov, S., and Préat, A., 2010, The Givetian-Frasnian boundary at Nismes parastratotype (Belgium): the magnetic susceptibility signal controlled by ferromagnetic minerals. *Geologica Belgica,* **13** (4): 351–366.

Diener, A., Ebneth, S., Veizer, J., and Buhl, D., 1996, Strontium isotope stratigraphy of the Middle Devonian: brachiopods and conodonts. *Geochimica et Cosmochimica Acta,* **60**: 639–652.

Dopieralska, J., Belka, Z., and Haack, U., 2006, Geochemical decoupling of water masses in the Variscan oceanic system during Late Devonian times. *Palaeogeography, Palaeoclimatology, Palaeoecology,* **240**: 108–119.

Dopieralska, J., Belka, Z., Königshof, P., Racki, G., Savage, N., Lutat, P., et al., 2012, Nd isotopic composition of Late Devonian seawater in western Thailand: Geotectonic implications for the origin of the Sibumasu terrane. *Gondwana Research,* **22** (3/4): 1102–1109.

Dopieralska, J., Belka, Z., and Walczak, A., 2016, Nd isotope composition of conodonts: an accurate proxy of sea-level fluctuations. *Gondwana Research,* **34**: 284–295.

Dreesen, R., Paproth, E., and Thorez, J., 1989. Events documented in Famennian sediments (Ardenne-Rhenish Massif, Late Devonian, NW Europe). *In* McMillan, N.J., Embry, A.F., and Glass, D.J. (eds), Devonian of the World. *Canadian Society of Petroleum Geology, Memoir,* **14(I)**, pp. 295–308. [imprint 1988].

Dutro, J.T., 1981, Devonian brachiopod biostratigraphy of New York State. In: Oliver, W.A., and Klapper, G. (Eds.), *Devonian Biostratigraphy of New York, Part 1, Text.* Subcommission on Devonian Stratigraphy, IUGS, Washington, DC, pp. 57–82.

Dzik, J., 1997, Emergence and succession of Carboniferous conodont and ammonoid communities in the Polish part of the Variscan Sea. *Acta Palaeontologica Polonica,* **42** (1): 57–170.

Dzik, J., 2006, The Famennian "Golden Age" of conodonts and ammonoids in the Polish part of the Variscan sea. *Palaeontologia Polonica,* **63**: 1–360.

Ebbighausen, V., Becker, R.T., Bockwinkel, J., and Aboussalam, Z.S., 2007, Givetian (Middle Devonian) brachiopod-goniatite-correlation in the Dra Valley (Anti-Atlas, Morocco) and Bergisch Gladbach-Paffrath

Syncline (Rhenish Massif, Germany), Special Publications. *In* Becker, R.T., and Kirchgasser, W.T. (eds), Devonian Events and Correlations. *London: Geological Society,* **278**: 157–172.

Ebbighausen, V., Becker, R.T., and Bockwinkel, J., 2011, Emsian and Eifelian Ammonoids from Oufrane, eastern Dra Valley (Anti-Atlas, Morocco) – taxonomy, stratigraphy and correlation. *Neues Jahrbuch für Geologie und Paläontologie,* **259** (3): 313–379.

Ebert, J., 1993, Globale Events im Grenz-Bereich Mittel-/Ober-Devon. *Göttinger Arbeiten zur Geologie und Paläontologie,* **59**: 1–109.

Ebneth, S., Diener, A., Buhl, D., and Veizer, J., 1997, Strontium isotope systematics of conodonts: Middle Devonian, Eifel Mountains, Germany. *Palaeogeography, Palaeoclimatology, Palaeoecology,* **132**: 79–96.

Edwards, D., Fairon-Demaret, M., and Berry, C.M., 2000, Plant megafossils in Devonian stratigraphy: a progress report. *Courier Forschungsinstitut Senckenberg,* **220**: 25–37.

Eichholt, S., and Becker, R.T., 2016, Middle Devonian reef facies and development in the Oued Cherrat Zone and adjacent regions (Moroccan Meseta). *Facies,* **62** (7): 29.

Ekhoff, J., Bundy, K., Schmitz, M., Dacydov, V., Over, D.J., 2013, U-Pb zircon geochronology in the Late Devonian Exshaw Formation: global correlation with the Hangenberg Black Shale and extinction event. *GSA Annual Meeting, Denver, 125th Anniversary of GSA, October 27–30, 2013, Abstract With Program, Session 247, Paper No. 7.*

Elles, G.L., and Wood, E.M.R., 1901, A monograph of British graptolites. *Monograph of the Palaeontographical Society,* 11 parts, 539 pp.

Ellwood, B.B., Crick, R.E., and El Hassani, A., 1999, Magnetosusceptibility event and cyclostratigraphy (MSEC) method used in geological correlation of Devonian rocks from Anti-Atlas Morocco. *The American Association of Petroleum geologists, Bulletin,* **83**: 1119–1134.

Ellwood, B.B., Crick, R.E., El Hassani, A., Benoist, S., and Young, R., 2000, Magnetosusceptibility Event and Cyclostratigraphy (MSEC) in marine rocks and the question of detrital input versus carbonate productivity. *Geology,* **28**: 1135–1138.

Ellwood, B.B., Crick, R.E., García-Alcalde Fernandez, J.L., Soto, F.M., Truyols-Massoni, M., El Hassani, A., et al., 2001, Global correlation using magnetic susceptibility data from Lower Devonian rocks. *Geology,* **29**: 583–586.

Ellwood, B.B., Benoist, S.L., El Hassani, A., Wheeler, C., and Crick, R.E., 2003, Impact ejecta layer from the mid-Devonian: possible connection to global mass extinctions. *Science,* **300**: 1734–1737.

Ellwood, B.B., García-Alcalde, J.L., El Hassani, A., Hladil, J., Soto, F.M., Truyóls-Massoni, M., et al., 2006, Stratigraphy of the Middle Devonian boundary: formal definition of the susceptibility magnetostratigraphy in Germany with comparisons to sections in the Czech Republic, Morocco and Spain. *Tectonophysics,* **418**: 31–49.

Ellwood, B.B., Algeo, T.J., El Hassani, A., Tomkin, J.H., and Rowe, H.D., 2011a, Defining the timing and duration of the Kačak Interval within the Eifelian/Givetian boundary GSSP, Mech Irdane, using geochemical and magnetic susceptibility patterns. *Palaeogeography, Palaeoclimatology, Palaeoecology,* **304**: 74–84.

Ellwood, B.B., Tomkin, J.H., El Hassani, A., Bultynck, P., Brett, C.E., Schindler, E., et al., 2011b, A climate-driven model and development of a floating point time scale for the entire Middle Devonian Givetian Stage: a test using magnetostratigraphy susceptibility as a climate proxy. *Palaeogeography, Palaeoclimatology, Palaeoecology,* **304**: 85–95.

Ellwood, B.B., El Hassani, A., Tomkin, J.H., and Bultynck, P., 2015, A climate-driven model using time-series analysis of magnetic susceptibility (χ) datasets to represent a floating-point high-resolution geological timescale for the Middle Devonian Eifelian stage. *In* Da Silva, A.C., Whalen, M.T., Hladil, J., Chadimová, L., Chen, D., Spassov, S., Boulvain, F., and Dveleeschouwer, X. (eds), Magnetic Susceptibility Application: A Window Onto Ancient Environments and Climatic Variations. *Geological Society, London, Special Publications,* **414**: 209–223.

Elrick, M., 1995, Cyclostratigraphy of Middle Devonian carbonates of the eastern Great Basin. *Journal of Sedimentary Research,* **B65** (1): 61–79.

Elrick, M., Berkyová, S., Klapper, G., Sharp, Z., Joachimski, M., and Frýda, J., 2009, Stratigraphic and oxygen isotope evidence for My-scale glaciation driving eustasy in the Early-Middle Devonian greenhouse world. *Palaeogeography, Palaeoclimatology, Palaeoecology,* **276**: 170–181.

Evans, S.D., Over, D.J., Day, J.E., Hasenmueller, N.R., 2013, The Devonian/Carboniferous boundary and the holotype of *Siphonodella sulcata* (Huddle; 1934) in the upper New Albany Shale, Illinois Basin, southern Indiana. In: El Hassani, A., Becker, R.T., and Tahiri, A. (Eds.), *International Field Symposium "The Devonian and Lower Carboniferous of northern Gondwana", March 22–29, 2013, Abstract Book,* Vol. 26. Rabat: Documents de l'Institut Scientifique, pp. 42–43.

Farsan, N.M., 1986, Faunenwandel oder Faunenkrise? Faunistische Untersuchungen der Grenze Frasnium/Famennium im mittleren Südasien. *Newsletters on Stratigraphy,* **16** (3): 113–131.

Feist, R., 1991, The Late Devonian trilobite crisis. *Historical Biology,* **5**: 197–214.

Feist, R., and Klapper, G., 1985, Stratigraphy and conodonts in pelagic sequences across the Middle–Upper Devonian boundary, Montagne Noire, France. *Palaeontographica,* **A188**: 1–8.

Feist, R., Flajs, G., and Girard, C., 2000, The stratotype section of the Devonian-Carboniferous boundary. *Courier Forschungsinstitut Senckenberg,* **225**: 77–82.

Ferrová, L., Frýda, J., and Lukeš, P., 2012, High-resolution tentaculite biostratigraphy and facies developments across the Early Devonian Daleje Event in the Barrandian (Bohemia): implications for global Emsian stratigraphy. *Bulletin of Geosciences,* **87** (3): 587–624.

Flajs, G., and Feist, R., 1988, Index conodonts, trilobites and environment of the Devonian-Carboniferous Boundary beds at La Serre (Montagne Noire, France). *Courier Forschungsinstitut Senckenberg,* **100**: 53–107.

Formolo, M.J., Riedinger, N., and Gill, B.C., 2014, Geochemical evidence for euxinia during the Late Devonian extinction events in the Michigan Basin (U.S.A.). *Palaeogeography, Palaeoclimatology, Palaeoecology,* **414**: 146–154.

Frech, F., 1902, Über devonische Amoneen. *Beiträge zur Paläontologie Österreich-Ungarns und des Orients,* **14**: 27–112, pls. 2–5.

García-Alcalde, J.L., 1997, North Gondwanan Emsian events. *Episodes,* **20** (4): 241–246.

Garrat, M.J., and Wright, A.J., 1989, Late Silurian to Early Devonian biostratigraphy of southeastern Australia, (imprint 1988). *In* McMillan, N.J., Embry, A.F., Glass, D.J. (eds), Devonian of the World. *Canadian Society of Petroleum Geology, Memoir,* **4(III)**: 647–662. [imprint 1988].

Geldsetzer, H.H.J., Goodfellow, W.D., McLaren, D.J., and Orchard, M.J., 1987, Sulfur-isotope anomaly associated with the Frasnian-Famennian extinction, Medicine Lake, Alberta, Canada. *Geology,* **15**: 393–396.

Geldsetzer, H.H.J., Goodfellow, W.D., and McLaren, D.J., 1993, The Frasnian-Famennian extinction event in a stable cratonic shelf setting: Trout River, Northwest Territories, Canada. *Palaeogeography, Palaeoclimatology, Palaeoecology,* **104**: 81–95.

George, A.D., Chow, N., and Trinajstic, K.M., 2014, Oxic facies and the Late Devonian mass extinction, Canning Basin, Australia. *Geology,* **42** (4): 327–330.

Gereke, M., 2007, Die oberdevonische Kellwasser-Krise in der Beckenfazies von Rhenohercynikum und Saxothuringicum (spätes Frasnium/frühestes Famennium, Deutschland). *Kölner Forum für Geologie und Paläontologie,* **17**: 1–228.

Gereke, M., and Schindler, E., 2012, "Time-Specific Facies" and biological crises – the Kellwasser Event interval near the Frasnian/Famennian boundary (Late Devonian). *Palaeogeography, Palaeoclimatology, Palaeoecology,* **367/368**: 19–29.

Gersl, M., and Hladil, J., 2004, Gamma-ray and magnetic susceptibility correlation across a Frasnian carbonate platform and the search for *punctata* equivalents in stromatopoproid-coral limestone facies of Moravia. *Geological Quarterly,* **48** (3): 283–292.

Gessa, S., 1996, *Nowakia* dacryoconarids in the Pragian of the Praha Basin (Lower Devonian, Czech Republic). *Revue de Micropaléontologie,* **39** (4): 315–337.

Giesen, P., and Berry, C.M., 2013, Reconstruction and growth of the early tree *Calamophyton* (Pseudosporochnales, Cladoxylopsida) based on exceptionally complete specimens from Lindlar, Germany (Mid-Devonian): organic connection of *Calamophyton* branches and *Duisbergia* trunks. *International Journal of Plant Science,* **174**: 665–686.

Girard, C., and Albarède, F., 1996, Trace elements in conodont phosphates from the Frasnian/Famennian boundary. *Palaeogeography, Palaeoclimatology, Palaeoecology,* **126**: 195–209.

Ginter, M., and Ivanov, A., 2000, Stratigraphic distribution of chondrichthyans in the Devonian on the East European Platform margin. *Courier Forschungsinstitut Senckenberg,* **223**: 325–339.

Ginter, M., Hairapetian, V., and Klug, C., 2002, Famennian chondrichthyans from the shelves of North Gondwana. *Acta Geologica Polonica,* **52** (2): 169–215.

Girard, C., and Lécuyer, C., 2002, Variation in Ce anomalies of conodonts through the Frasnian/Famennian boundary of Poland Kowala—Holy Cross Mountains; implications for the redox state of seawater and biodiversity. *Palaeogeography, Palaeoclimatology, Palaeoecology,* **181**: 209–311.

Girard, C., Robin, E., Rocchia, R., Froget, L., and Feist, R., 1997, Search for impact remains at the Frasnian-Famennian boundary in the stratotype area, southern France. *Palaeogeography, Palaeoclimatology, Palaeoecology,* **132**: 391–397.

Girard, C., Klapper, G., and Feist, R., 2005, Subdivision of the terminal Frasnian linguiformis conodont Zone, revision of the correlative interval of Montagne Noire Zone 13, and discussion of stratigraphically significant associated trilobites. *In* Over, D.J., Morrow, J.R., and Wignall, P.B. (eds), Understanding Late Devonian and Permian-Triassic Biotic and Climatic Events: Towards an Integrated Approach. *Developments in Palaeontology & Stratigraphy,* **20**: 181–198.

Glikson, A.Y., Moray, A.J., Iasky, R.P., Pirajno, F., Goldring, S.D., and Uyrsal, I.T., 2005, Woodleigh, Southern Carnarvon Basin, Western Australia: history of discovery, Late Devonian age, and geophysical and morphometric evidence for a 120 km-diameter impact structure. *Australian Journal of Earth Sciences,* **52**: 545–553.

Glikson, A.Y., Uysal, I.T., Gerald, J.D., and Saygin, E., 2013, Geophysical anomalies and quartz microstructures, Eastern Warbzrton Basin, Northeast South Australia: tectonic or impact metamorphic origin? *Tectonophysics,* **589**: 57–76.

Göddertz, B., 1987, Devonische Goniatiten aus SW-Algerien und ihre stratigraphische Einordnung in die Conodonten-Abfolge. *Palaeontographica, Abteilung A,* **197**: 127–220, pls. 1–15.

Gong, X.B., Huang, H.D., Zhang, M.L., and Huang, Q.D., 1991, *The Stratigraphic Classification and Correlation of Carbonate Rocks of Upper Devonian and Lower Carboniferous in Guilin Karst Region.* Guilin: Guangxi Science and Technology Publishing House, 106 pp.

Gong, Y.M., Xu, R., Tang, Z.D., Si, Y.L., and Li, B.H., 2005, Relationships between bacterial-algal proliferating and mass extinction in the Late Devonian Frasnian-Famennian transition: enlightening from carbon isotopes and molecular fossils. *Science in China, Series D, Earth Sciences,* **48** (10): 1656–1665.

Gong, Y.M., Xu, R., Feng, Q., Zhang, L.J., Ma, H., and Zeng, J.W., 2007, Hypersaline and anoxia in the Devonian Frasnian-Famennian transition: molecular fossil and mineralogical evidence from Guangxi, South China. *Frontiers in Earth Sciences China,* **1** (4): 458–469.

Goodfellow, W.D., Geldsetzer, H.H.J., McLaren, D.J., Orchard, M.J., and Klapper, G., 1989, The Frasnian-Famennian extinction: current results and possible causes, (imprint 1988). *In* McMillan, N.J., Embry, A.F., and Glass, D.J. (eds), Devonian of the World. *Canadian Society of Petroleum Geologists, Memoir,* **14(III)**: 9–21.

Gouwy, S.A., and Bultynck, P., 2000, Graphic correlation of Frasnian sections (Upper Devonian) in the Ardennes, Belgium. *Bulletin de l'Institut Royal des Sciences Naturelles de Belgique, Sciences de la Terre,* **70**: 25–52.

Gouwy, S.A., and Bultynck, P., 2002, Graphic correlation of Middle Devonian sections in the Ardenne region (Belgium) and the Mader-Tafilalt region (Morocco): development of a Middle Devonian composite standard. In: Proceedings of the First Geologica Belgica International Meeting, Leuven, September 11–15, 2002. *Aardkundige Mededelingen,* **12**: 105–108.

Gouwy, S.A., and Bultynck, P., 2003a, Conodont based graphic correlation of the Middle Devonian formations of the Ardenne (Belgium): implications for stratigraphy and construction of a regional composite. *Revista Española de Micropaleontologia,* **35** (3): 315–344.

Gouwy, S.A., and Bultynck, P., 2003b, Conodont data across the Eifelian-Givetian boundary at Aisemont, southern Namur Synclinorium, Belgium: correlation and implications. *Courier Forschungsinstitut Senckenberg,* **242**: 239–255.

Gouwy, S.A., Haydukiewicz, J., and Bultynck, P., 2007, Conodont-based graphic correlation of upper Givetian-Frasnian sections of the Eastern Anti-Atlas (Morocco). *Geological Quarterly,* **51** (4): 375–392.

Gouwy, S.A., and Uyeno, T., 2017, A new assessment of the Middle and Upper Devonian conodont biostratigraphy of the Horn River Group in the Powell Creek reference section (northern Mackenzie Mountains, NWT, Canada). In: Liao, J.-C., and Valenzuela-Ríos, J.I. (Eds.), Fourth International Conodont Symposium. ICOS IV "Progress on Conodont Investigations". *Cuadernos del Museo Geominero,* **22**: 153–156.

Gouwy, S., Day, J.E., Macleod, K.G., 2012, Pragian to Givetian Conodont Biostratigraphy and Conodont Apatite δ^{18}O Results of the Grand Tower and Saint Laurent Formations: Southern Illinois Basin-Reel Foot Embayment (Illinois, USA). *Fourth International Geologica Belgica Meeting 2012, Abstracts,* 1 p.

Grahn, Y., 2003, Silurian and Devonian chitinozoan assemblages from the Chaco-Paraña Basin, northeastern Argentina and Central Uruguay. *Revista Española de Paleontología*, **35**: 1−8.

Grahn, Y., 2005, Devonian chitinozoan biozones of western Gondwana. *Acta Geologica Polonica*, **55**: 211−227.

Grahn, Y., and Melo, J.H.G., 2004, Integrated Middle Devonian chitinozoan and miospore zonation of the Amazonas Basin, northern Brazil. *Revue de micropaléontologie*, **47**: 71−85.

Grahn, Y., and Melo, J.H.G., 2005, Middle and Late Devonian Chitinozoa and biostratigraphy of the Parnaiba and Jatoba Basins, northeastern Brazil. *Palaeontographica, Abt. B*, **272**: 1−50.

Grahn, Y., Mauller, P.M., Pereira, E., and Loboziak, S., 2010, Palynostratigraphy of the Chapada Group and its significance in the stratigraphy of the Paraná Basin, south Brazil. *Journal of South American Earth Sciences*, **29**: 354−370.

Grandjean-Lécuyer, P., Feist, R., and Albarède, F., 1993, Rare earth elements in old biogenic apatites. *Geochimica et Cosmochimica Acta*, **57**: 2507−2514.

Groos-Uffenorde, H., and Wang, S.Q., 1989, The entomozoacean succession of China and Germany (Ostrracoda, Devonian). *Courier Forschungsinstitut Seenckenberg*, **110**: 61−79.

Groos-Uffenorde, H., Lethiers, F., and Blumenstengel, H., 2000, Ostracodes and Devonian stage boundaries. *Courier Forschungsinstitut Senckenberg*, **220**: 99−111.

Grossman, E.L., and Joachimski, M.M., 2020, Chapter 10 - Oxygen isotope stratigraphy. *In* Gradstein, F.M., Ogg, J.G., Schmitz, M.D., and Ogg, G.M. (eds), *The Geologic Time Scale 2020*. Vol. 1 (this book). Elsevier, Boston, MA.

Grudev, D.A., Soboleva, M.A., Sobolev, D.B., and Zhuravlev, A.V., 2016, The Frasnian deposits in the Bol'shaya Nadota River region (sub-Polar Urals) − stratigraphy and depositional environment. *Litosfera*, **2016** (6): 97−116 (in Russian with English summary).

Haddad, E.E., Tuite, M.L., Martinez, A.M., Williford, K., Boyer, D.L., Droser, M.L., et al., 2016, Lipid biomarker stratigraphic records through the Late Devonian Frasnian/Famennian boundary: comparison of high- and low-latitude epicontinental marine settings. *Organic Geochemistry*, **98**: 38−53.

Hagadorn, J.W., Bullecks, J., Soar, L.K., Lahey, B.L., Over, D.J., Wistort, Z., et al., 2016, Colorado mass extinction: weird facies and cool fossils from the end-Devonian Dyer Formation. *GSA Annual Meeting, Abstracts with Programs*, **48** (7): 1.

Halas, S., Balínski, A., Gruszczynski, M., Hoffman, A., Malkowski, K., and Narkiewicz, M., 1992, Stable isotope record at the Frasnian/Famennian boundary in southern Poland. *Neues Jahrbuch für Geologie und Paläontologie, Monatshefte*, **1992** (3): 129−138.

Halas, S., Szaran, J., and Niezgoda, H., 1997, Experimental determination of carbon isotope equilibrium fractionation between dissolved carbonate and carbon dioxide. *Geochimica et Cosmochimica Acta*, **61**: 2691−2695.

Harrigan, C.O., Schmitz, M.D., Over, D.J., in review, New high-precision U-Pb zircon dates and age-depth modeling to revise the Devonian time scale. *Geological Society of America Bulletin*: in review.

Hansma, J., Tohver, E., Yan, M.D., Trinajstic, K., Roelofs, B., Peek, S., et al., 2015, Late Devonian carbonate magnetostratigraphy from the Oscar and Horse Spring Ranges, Lennard Shelf, Canning Basin, Western Australia. *Earth and Planetary Science Letters*, **409**: 232−242.

Hao, W.C., 2001, The Devonian-Carboniferous Boundary and Events at Bachu, Xinjiang, Northwestern China. *International Geology Review*, **43**: 276−284.

Haq, B.U., and Schutter, S.R., 2008, A chronology of Paleozoic sea-level changes. *Science*, **322**: 64−68.

Harland, W.B., Armstrong, R.L., Cox, A.V., Craig, L.A., Smith, A.G., and Smith, D.G., 1990, *A Geologic Time Scale 1989*. Cambridge University Press, 263 pp.

Harris, N.B., Mnich, C.A., Selby, D., and Korn, D., 2013, Minor and trace element and Re-Os chemistry of the Upper Devonian Woodford Shale, Permian Basin, west Texas: insights into metal abundance and basin processes. *Chemical Geology*, **356**: 76−83.

Hartenfels, S., 2011, Die globalen *Annulata*-Events und die Dasberg-Krise (Famennium, Oberdevon) in Europa und Nord-Afrika − hochauflösende Conodonten-Stratigraphie, Karbonat −Mikrofazies, Paläoökologie und Paläodiversität. *Münstersche Forschungen zur Geologie und Paläontologie*, **105**: 17−527.

Hartenfels, S., and Becker, R.T., 2009, Timing of the global Dasberg Crisis − implications for Famennian eustasy and chronostratigraphy. *In* Over, D.J. (ed), Studies in Devonian Stratigraphy: Proceedings of the 2007 International Meeting of the Subcommission on Devonian Stratigraphy and IGCP 499. *Palaeontographica Americana*, **63**: 69−95.

Hartenfels, S., and Becker, R.T., 2016a, The global *Annulata* Events: review and new data from the Rheris Basin (northern Tafilalt) of SE Morocco, Special Publications. *In* Becker, R.T., Königshof, P., and Brett, C.E. (eds), Devonian Climate, Sea Level and Evolutionary Events. *London: Geological Society*, **423**: pp. 291−354.

Hartenfels, S., and Becker, R.T., 2016b, Age and correlation of the transgressive *Gonioclymenia* Limestone (Famennian, Tafilalt, eastern Anti-Atlas, Morocco) (online) *Geological Magazine*, **155** (3): 586−629.

Hartenfels, S., and Becker, R.T., 2018, The upper Famennian Dasberg Crisis and Gonioclymenia Limestone at Oum el Jerane (Amessoui Syncline, southern Tafilalt Platform). In: Hartenfels, S., Becker, R.T., El Hassani, A., and Lüddecke, F. (Eds.), 10th International Symposium "Cephalopoda − Present and Past", Fes, March 26−April 3, 2018, Field Guidebook. *Münstersche Forschungen zur Geologie und Paläontologie*, **110**: 261−272.

Hartenfels, S., and Becker, R.T., 2018b, Borkewehr near Wocklum (northern Rhenish Massif, Germany), a possible future Devonian/Carboniferous boundary GSSP section. *GeoBonn 2018, Living Earth, 2-6 September 2018, Bonn, Germany, Abstracts*, 252.

Hartenfels, S., Becker, R.T., 2019, The Devonian/Carboniferous transition in the Rhenish Massif − Borkewehr, a poterntial GSSP section. In: Hartenfels, S., Herbig, H.-G., and Aretz, M. (Eds.), 19th International Congress on the Carboniferous and Permian, Cologne, July 29−August 2, 2019, Abstracts. *Kölner Forum für Geologie und Paläontologie*, **23**: 140−141.

Hartenfels, S., Becker, R.T., and Tragelehn, H., 2009, Marker conodonts around the global *Annulata* Events and the definition of an Upper Famennian substage. *Subcommission on Devonian Stratigraphy Newsletter*, **24**: 40−48.

Hartenfels, S., Becker, R.T., Aboussalam, Z.S., El Hassani, A., Baidder, L., Fischer, T., et al., 2013, The Upper Devonian at El Khraouia (southern Tafilalt). In: Becker, R.T., El Hassani, A., and Tahiri, A. (Eds.), *International Field Symposium "The Devonian and Lower Carboniferous of northern Gondwana", March 22−29, 2013, Field Guidebook*, Vol. 27. Rabat: Document de l'Institut Scientifique, pp. 41−50.

Hartenfels, S., Becker, R.T., and Aboussalam, Z.S., 2016, Givetian to Famennian stratigraphy, Kellwasser, *Annulata* and other events at Beringhauser Tunnel (Messinghausen Anticline, eastern Rhenish Massif). *In* Becker, R.T., Hartenfels, S., Königshof, P., and Helling, S. (eds), Middle Devonian to Lower Carboniferous Stratigraphy, Facies, and

Bioevents in the Rhenish Massif, Germany, An IGCP 596 Guidebook. *Münstersche Forschungen zur Geologie und Paläontologie,* **108**: 196–219.

Hartkopf-Fröder, C., Kloppisch, M., Mann, U., Neumann-Mahlkau, P., Schaefer, R.G., and Wilkes, H., 2007, The end-Frasnian mass extinction in the Eifel Mountains, Germany: new insights from organic matter composition and preservation. *In* Becker, R.T., and Kirchgasser, W.T. (eds), Devonian Events and Correlations. *The Geological Society, London, Special Publication,* **278**: 173–196.

Haslett, A.C. and Parnell, A., 2018. BChron R package. Available at: http://andrewcparnell.github.io/Bchron/

Hasson, J.M., Schoene, B., Bluher, S., and Maloof, A.C., 2016, Chemostratigraphic and U-Pb geochronologic constraints on carbon cycling across the Silurian-Devonian boundary. *Earth and Planetary Science Letters,* **436**: 108–120.

Haude, R., Corriga, M.G., Corradini, C., and Walliser, O.H., 2014, Bojen-Seelilien (Scyphocrinitidae, Echinodermata Geosciences) in neudatierten Schichten vom oberen Silur bis untersten Devon Südost-Marokkos. *Göttingen Contributions Geosciences,* **77**: 129–145.

Hedhli, M., Grasby, S.E., Beauchamp, B., Ardakani, O.H., and Sanei, H., 2016, Latest Devonian demise of carbonate factories caused by basin-to-shelf anoxia on western Laurentia. *GSA Annual Meeting, Abstracts with Programs,* **48** (7): 1. https://doi.org/10.1130/abs/2016AM-284512.

Higgs, K.T., and Streel, M., 1994, Palynological age for the lower part of the Hangenberg Shales in Sauerland, Germany. *Annales de la Société géologique de Belgique,* **115**: 551–557.

Higgs, K.T., Streel, M., Korn, D., and Paproth, E., 1993, Palynological data from the Devonian-Carboniferous boundary beds in the new Stockum trench II and the Hasselbach borehole, Northern Rhenish Massif, Germany. *Annales de la Société Géologique de Belgique,* **115**: 551–557.

Hillbun, K., Playton, T.E., Katz, D.A., Tohver, E., Trinajstic, K., Haines, P.W., et al., 2016, Correlation and sequence stratigraphic interpretation of Upper Devonian carbonate slope facies using carbon isotope chemostratigraphy, Lennard Shelf, Canning Basin, Western Australia. In: *New Advances in Devonian Carbonates: Outcrop Analogs, Reservoirs, and Chronostratigraphy,* SEPM Special Publication **107**, pp. 302–318.

Hladíková, J., Hladil, J., and Kríbek, B., 1997, Carbon and oxygen isotope record across Pridoli to Givetian stage boundaries in the Barrandian basin (Czech Republic). *Palaeogeography, Palaeoclimatology, Palaeoecology,* **132**: 225–241.

Hladil, J., Pruner, P., Venhodová, D., Hladíkova, T., and Man, O., 2002, Toward an exact age of Middle Devonian Celechovice corals – past problems in biostratigraphy and present solutions complemented by new magnetosusceptibility measurements. *Coral Research Bulletin,* **7**: 65–71.

Hladil, J., Gersl, M., Strnad, L., Frana, J., Langrova, A., and Spisiak, J., 2006, Stratigraphic variation of complex impurities in platform limestones and possible significance of atmospheric dust: a study with emphasis on gamma-ray spectrometry and magnetic susceptibility outcrop logging (Eifelian-Frasnian, Moravia, Czech Republic). *International Journal of Earth Sciences,* **95**: 703–723.

Hladil, J., Cejchan, P., Bábek, O., Koptíková, L., Navratil, T., and Kubinova, P., 2010a, Dust—a geology-orientated attempt to reappraise the natural components, amounts, inputs to sediment, and importance for correlation purposes. *Geologica Belgica,* **13** (4): 367–384.

Hladil, J., Vondra, M., Cejchan, P., Vich, R., Koptíková, L., and Slavík, L., 2010b, The dynamic time-warping approach to comparison of magnetic-susceptibility logs and application to lower Devonian calci-turbidites (Prague Synform, Bohemian Massif). *Geologica Belgica,* **13** (4): 385–406.

Hladil, J., Slavík, L., Vondra, M., Koptíková, L., Cejchan, P., Schnabl, P., et al., 2011, Pragian-Emsian successions in Uzbekistan and Bohemia: magnetic susceptibility logs and their dynamic time warping alignment. *Stratigraphy,* **8**: 217–235.

Holdsworth, B.K., and Jones, D.L., 1980, Preliminary radiolarian zonation for Late Devonian through Permian time. *Geology,* **8**: 281–285.

Hollard, H., 1974, Recherches sur la stratigraphie des formations du Dévonien moyen, de l'Emsien supérieur au Frasnien, dans le Sud du Tafilalt et dans le Ma'der (Anti-Atlas oriental). *Notes et Mémoires du Service Géologique du Maroc,* **264**: 7–68.

Horodyski, R.S., Holz, M., Grahn, Y., and Bosetti, E.P., 2013, Remarks on sequence stratigraphy and taphonomy of the Malvinokaffric shelly fauna during the Kačák Event in the Apucarana Sub-basin (Paraná Basin), Brazil. *International Journal of Earth Sciences,* **103** (1): 367–380.

Hou, H.-F., 2008, A comprehensive research report on Shaodong Stage (Devonian). *In* The Third National Stratigraphic Council of China (ed), *The Research Report on the Main Divisions of Stratigraphy Into Stages in China (2001-2005).* Beijing: Geological Publishing House, pp. 324–332 (in Chinese).

Hou, H.F., Ji, Q., and Wang, J.X., 1989, Preliminary report on Frasnian-Famennian Events in South China, (imprint 1988). *In* McMillan, N.J., Embry, A.F., and Glass, D.J. (eds), Devonian of the World. *Canadian Society of Petroleum Geology, Memoir,* **14(III)**: 63–69.

House, M.R., 1978, Devonian ammonoids from the Appalachians and their bearing on international zonation and correlation. *Special Papers in Palaeontology,* **21**: 1–70.

House, M.R., 1983, Devonian eustatic events. *Proceedings of the Ussher Society,* **5**: 396–405.

House, M.R., 1985, Correlation of mid-Palaeozoic ammonoid evolutionary events with global sedimentary perturbations. *Nature,* **313**: 17–22.

House, M.R., 1991, Devonian Period. *Encyclopedia Brittanica,* **19**: 804–814.

House, M.R., 1993, Earliest Carboniferous goniatite recovery after the Hangenberg Event. *Annales de la Société géologique de Belgique,* **115** (2): 559–579.

House, M.R., 1995, Devonian precessional and other signatures for establishing a Givetian timescale. *Geological Society of London Special Publication,* **85**: 37–49.

House, M.R., 1996a, The Middle Devonian Kačak Event. *Proceedings of the Ussher Society,* **9**: 79–84.

House, M.R., 1996b, Juvenile goniatite survival strategies following Devonian extinction events. *In* Hart, M.B. (ed), *Biotic Recovery From Mass Extinction Events. Geological Society Special Publication.* pp. 1063–185.

House, M.R., 2002, Strength, timing, setting and cause of mid-Palaeozoic extinctions. *Palaeogeography, Palaeoclimatology, Palaeoecology,* **181**: 5–25.

House, M.R., and Ziegler, W., 1977, The Goniatite and Conodont sequences in the early Upper Devonian at Adorf, Germany. *Geologica et Palaeontologica,* **11**: 69–108.

House, M.R., and Kirchgasser, W.T., 1993, Devonian goniatite biostratigraphy and timing of facies movements in the Frasnian of eastern North America. *In* Hailwood, E.A., and Kidd, R.B. (eds), High Resolution Stratigraphy. *Geological Society Special Publication,* **70**: 267–292.

House, M.R., and Ziegler, W. (Eds.), 1997. On sea-level fluctuations in the Devonian. *Courier Forschungsinstitut Senckenberg*. **199**: 1–146.

House, M.R., and Gradstein, F.M., 2004, The Devonian Period. *In* Gradstein, F.M., Ogg, J.G., and Smith, A.G. (eds), *A Geologic Time Scale 2004*. Cambridge University Press, pp. 202–221.

House, M.R., and Kirchgasser, W.T., 2008, Late Devonian Goniatites (Cephalopoda, Ammonoidea) from New York State. *Bulletins of American Paleontology*, **374**: 1–288.

House, M.R., Kirchgasser, W.T., Price, J.D., and Wade, G., 1985, Goniatites from Frasnian (Upper Devonian) and adjacent strata of the Montagne Noire. *Hercynica*, **1**: 1–21.

House, M.R., Feist, R., and Korn, D., 2000a, The Middle/Upper Devonian boundary at Puech de la Suque, Southern France. *Courier Forschungsinstitut Senckenberg*, **225**: 49–58.

House, M.R., Becker, R.T., Feist, R., Flajs, G., Girard, C., and Klapper, G., 2000b, The Frasnian/Famennian boundary GSSP at Coumiac, southern France. *Courier Forschungsinstitut Senckenberg*, **225**: 59–75.

House, M.R., Menner, V.V., Becker, R.T., Klapper, G., Ovnatanova, N.S., and Kuźmin, V., 2000c, Reef episodes, anocxia and sea-level changes in the Frasnian of the southern Timan (NE Russian platform), Special Publications. *In* Insalaco, E., Skelton, P.W., and Palmer, T.J. (eds), Carbonate Platform Systems: Components and Interactions. *Geological Society, London, Special Publications*, **178**: 147–176.

Huang, C., and Gong, Y.M., 2016, Timing and patterns of the Frasnian-Famennian event: evidences from high-resolution conodont biostratigraphy and event stratigraphy at the Yangi section, Guangxi, South China. *Palaeogeography, Palaeoclimatology, Palaeoecology*, **448**: 317–338.

Huang, T.Z., Sun, Y.L., Zhao, Z.Q., Li, C., Zhao, H., Nie, T., et al. 2017, Reconstruction the weathering intensity during the early evolution of vascular land plants. *Annual GSA Meeting 2017, Seattle, Abstracts With Programs, 49: Paper 60-1*.

Huang, C., Song, J.J., Shen, J., and Gong, Y.M., 2018, The influence of the Late Devonian Kellwasser events on deep-water ecosystems: evidence from palaeontological and geochemical records from South China. *Palaeogeography, Palaeoclimatology, Palaeoecology*, **504**: 60–64.

Hušková, A., and Slavík, L., 2019, In search of Silurian/Devonian boundary conodont markers in carbonate environments of the Prague Synform (Czech Republic) (online). *Palaeogeography, Palaeoclimatology, Palaeoecology*. https://doi.org/10.1016/j.palaeo.2019.03.027.

Husson, J.N., Schoene, B., Bluher, S., and Maloof, A.C., 2016, Chemostratigraphic and U-Pb geochronologic constraints on carbon cycling across the Silurian-Devonian boundary. *Earth and Planetary Science Letters*, **436**: 108–120.

Isaacson, P.E., Dáz-Martínez, E., Grader, G.W., Kalvoda, J., Babek, O., and Devuyst, F.X., 2008, Late Devonian-earliest Mississippian glaciation in Gondwanaland and its biogeographic consequences. *Palaeogeography, Palaeoclimatology, Palaeoecology*, **268**: 126–142.

Izokh, N., 1990, Ikriodusy telengitskogo nadgorizonta Salaira (konodonty; devon, ems). *In* Yolkin, E.A., and Kanygin, A.V. (eds), *Novoe v paleontologii I biostratigrafii paleozoica aziatskoi chasti SSSR*. Novosibirsk: Nauka, pp. 107–111.

Izokh, N., 1998, Konodonty I zonal'noe raschlenenie emskikh otloszheniy Salaira. [Conodonts and Zonal Division of the Emsian Deposits of the Salair Ridge] Izdatel'stvo, Cand. Sci. (Geol.) Dissertation, Novosibirsk: IGNiG SO RAN. *Novosirbisk*, pp. 1–24 [In Russian].

Izokh, O., 2011a, C isotope evidence of global perturbation in the marine ecosystem at the early Emsian time (base of *kitabicus*

conodont biozone). *In* Yolkin, E.A., Izokh, N.G., Obut, O.T., and Kipriyanova, T.P. (eds), *"Devonian Terrestrial and Marine Environments: From Continent to Shelf" IGCP 499 Project/SDS Joint Field Meeting, Novosibirsk, Russia, July 25–August 9, 2005, Contributions*, pp. 61–62.

Izokh, O., 2011b, Carbon-isotope characteristics of Frasnian-Famennian carbonates in the South Urals. *In* Yolkin, E.A., Izokh, N.G., Obut, O.T., and Kipriyanova, T.P. (eds), *"Devonian Terrestrial and Marine Environments: From Continent to Shelf" IGCP 499 Project/SDS Joint Field Meeting, Novosibirsk, Russia, July 25–August 9, 2005, Contributions*, pp. 63–64.

Izokh, N.G., Obut, O.T., Morrow, J., and Sandberg, C.A., 2008, New findings of Upper Frasnian conodonts and radiolarians. *In* Königshof, P., and Linnemann, U. (eds), *From Gondwana and Laurussia to Pangea: Dynamics of Oceans and Supercontinents. 20th International Senckenberg Conference and Second Geinitz Conference, Final Meeting of IGCP 497 and IGCP 499, Abstracts and Programme*, pp. 186–188.

Izokh, O., Izokh, N.G., Ponomarchuk, V.A., and Semenova, D.V., 2009, Carbon and oxygen isotopes in the Frasnian-Famennian section of the Kuznetsk Basin (southern West Siberia). *Russian Geology of Geophysics*, **50** (7): 610–617.

Izokh, O.P., Izokh, N.D., Saraev, S.V., and Dokukina, G.A., 2015, C isotopic variations in the lower-middle Frasnian (lower Upper Devonian) of the Rudny Altai. *Geological Magazine*, **152** (3): 565–571.

Jacobi, D.J., Barrick, J.E., Kleffner, M.A., Karlsson, H.R., 2009, Stable isotope chemostratigraphy and conodont biostratigraphy across the Silurian-Devonian boundary in southwestern Laurentia. In: Over, D.J. (Ed.), Studies in Devonian Stratigraphy: Proceedings of the 2007 International Meeting of the Subcommission on Devonian Stratigraphy and IGCP 499. *Palaeontographica Americana*, **63**: 9–31.

Jaeger, H., 1977, Graptolites. *In* Martinsson, A. (ed), The Silurian-Devonian boundary. *International Union of Geological Sciences, Series A*, **5**: 337–345.

Jaeger, H., 1978, Late graptoloid faunas and the problem of graptoloid extinction. *Acta Palaeontologica Polonica*, **23** (4): 497–521.

Jaeger, H., 1979, Devonian Graptolithina. *In* House, M.R., Scrutton, C.T., and Bassett, M.G. (eds), The Devonian System. *Special Papers in Palaeontology*, **23**: 335–339.

Jaeger, H., 1983, Unterdevonische Graptolithen aus Burma und zu vergleichende Formen. *Jahrbuch der Geologischen Bundesanstalt*, **126** (2): 245–252.

Jaeger, H., 1989, Devonian Graptoloidea, (imprint 1988). *In* McMillan, N.J., Embry, A.F., and Glass, D.J. (eds), Devonian of World. *Canadian Society of Petroleum Geologists, Memoir*, **14(III)**: 431–438.

Jagodzinski, E.A., and Black, L.P., 1999, U-Pb dating of silicic lavas, sills and syneruptive resedimented volcaniclastic deposits of the Lower Devonian Crudine Group, Hill End Trough, New South Wales. *Australian Journal of Earth Sciences*, **46**: 749–764.

Jansen, U., 2008, Biostratigraphy and correlation of the traditional Emsian stage. In: Kim, A.I., Salimova, F.A., and Meshchankina, N.A. (Eds.), *Global Alignments of Lower Devonian Carbonate and Clastic Sequences: Contributions From the IGCP 499 Project/SDS Joint Field Meeting, Kitab State Geological Reserve, Uzbekistan, August 25–September 3, 2008*, pp. 42–45.

Jansen, U., 2016, Brachiopod faunas, facies and biostratigraphy of the Pridolian to lower Eifelian succession in the Rhenish Massif (Rheinisches Schiefergebirge, Germany, Special Publications. *In* Becker, R.T.,

Königshof, P., and Brett, C.E. (eds), Devonian Climate, Sea Level and Evolutionary Events. *London: Geological Society,* **423**: pp. 45–122.

Jansen, U., Lazreq, N., Plodowski, G., Schemm-Gregory, M., Schindler, E., and Weddige, K., 2007, Neritic-pelagic correlation in the Lower and basal Middle Devonian of the Dra Valley (Southern Anti-Atlas, Moroccan Pre-Sahara). *In* Becker, R.T., and Kirchgasser, W.T. (eds), Devonian Events and Correlations. *Geological Society of London, Special Publications,* **278**: 9–37.

Jeppson, L. 1998. Silurian oceanic events: summary of general characteristics. In: Landing, E., and Johnson, M.E. (Eds.), *Silurian Cycles – Linkeages of Dynamic Stratigraphy with Atmospheric, Oceanic and Tectonic Changes.* New York State Bulletin, 491, pp. 239–257.

Ji, Q., 1989a, On the Frasnian-Famennian Mass Extinction Event in South China. *Courier ForschungsInstitut Senckenberg,* **117**: 275–301.

Ji, Q., 1989b, A comparison between the Dapoushang section and the other candidate sections for the Devonian-Carboniferous boundary stratotype. *In* Ji, Q. (ed), *The Dapoushang Section, An Excellent Section for the Devonian-Carboniferous Boundary Stratotype in China.* Bejing Science Press, pp. 66–79.

Ji, Q., and Ziegler, W., 1993, Lali section: an excellent reference section for Upper Devonian in south China. *Courier Forschungsinstitut Senckenberg,* **157**: 183.

Joachimski, M.M., 1997, Comparisons of organic and inorganic carbon isotope patterns across the Frasnian-Famennian boundary. *Palaeogeography, Palaeoclimatology, Palaeoecology,* **132**: 133–145.

Joachimski, M.M., and Buggisch, W., 1993, Anoxic events in the late Frasnian—causes of the Frasnian-Famennian faunal crisis? *Geology,* **21**: 657–678.

Joachimski, M.M., and Buggisch, W., 2002, Conodont apatite $\delta^{18}O$ signatures indicate climatic cooling as a trigger of the Late Devonian mass extinction. *Geology,* **30** (8): 711–714.

Joachimski, M.M., Buggisch, W., and Anders, T., 1994, Mikrofazies, Conodontenstratigraphie und Isotopengeochemie des Frasne/Famenne-Grenzprofils Wolayer Gletscher (Karnische Alpen). *Abhandlungen der Geologischen Bundes-Anstalt,* **50**: 183–195.

Joachimski, M.M., Ostertag-Henning, C., Pancost, R.D., Strauss, H., Freeman, K.H., Littke, R., et al., 2001, Water column anoxia, enhanced productivity and concomitant changes in $\delta^{13}C$ and $\delta^{34}S$ across the Frasnian-Famennian boundary (Kowala – Holy Cross Mountains/ Poland). *Chemical Geology,* **175**: 109–131.

Joachimski, M.M., Pancrost, R.D., Freeman, K.A., Ostertag-Henning, C., and Buggisch, W., 2002, Carbon isotope geochemistry of the Frasnian–Famennian transition. *Palaeogeography, Palaeoclimatology, Palaeoecology,* **181**: 91–109.

Joachimski, M.M., van Geldern, R., Breisig, S., Buggisch, W., and Day, J., 2004, Oxygen isotope evolution of biogenic calcite and apatite diring the Middle and Late Devonian. *International Journal of Earth Sciences,* **93**: 542–553.

Joachimski, M.M., Breisig, S., Buggisch, W., Talent, J.A., Mawson, R., Gereke, M., et al., 2009, Devonian climate and reef evolution: insights from oxygen isotopes in apatite. *Earth and Planetary Science Letters,* **284**: 599–609.

Johnson, J.G., 1970, Taghanic onlap and the end of North American provinciality. *Geological Society of America Bulletin,* **81**: 2077–2106.

Johnson, D.B., and Klapper, G., 1981, New Early Devonian conodont species of central Nevada. *Journal of Paleontology,* **55** (6): 1237–1250.

Johnson, J.G., Klapper, G., and Trojan, W.R., 1980, Brachiopod and conodont successions in the Devonian of the northern Antelope Range, central Nevada. *Geologica et Paleontologica,* **14**: 77–116.

Johnson, J.G., Klapper, G., and Sandberg, C.A., 1985, Devonian eustatic fluctuations in Euramerica. *Geological Society of America bulletin,* **96**: 567–587.

Johnson, J.G., and Sandberg, C.A., 1989, Devonian eustatic events in the Western United States and their biostratigraphic responses, (imprint 1988). *In* McMillan, N.J., Embry, A.F., and Glass, D.J. (eds), Devonian of the World. *Canadian Society of Petroleum Geology, Memoir,* **14 (III)**: 171–178.

Johnston, D.I., Henderson, C.M., and Schmidt, M.J., 2010, Upper Devonian to Lower Mississppian conodont biostratigraphy of uppermost Wabamun Group and Palliser Formation to lowermost Banff and Lodgepole formations, southern Alberta sand southeastern British Columbia, Canada: implications for correlations and sequence stratigraphy. *Bulletin of American Petroleum Geology,* **58** (4): 295–341.

Kabanov, P., 2018, Devonian (c. 388-375 Ma) Horn River Group of Mackenzie Platform (NW Canada) is an open-shelf succession recording oceanic anoxic events. *Journal of the Geological Society,* 17 pp. <https://doi.org/10.1144/jgs2018-075>.

Kaiser, S.I., 2009, The Devonian/Carboniferous boundary stratotype section (La Serre, France) revisited. *Newsletters on Stratigraphy,* **43** (2): 195–205.

Kaiser, S.I., Becker, R.T., 2007. The required revision of the Devonian-Carboniferous boundary. In: Wang, Y., Zhang, H., Wang, X.J. (eds), XVI International Congress on the Carboniferous and Permian, Abstracts. Journal of Stratigraphy, 31(supplement), p. 95.

Kaiser, S.I., and Corradini, C., 2008, Should the Devonian/Carboniferous boundary be redefined? *Subcommission on Devonian Stratigraphy, Newsletter,* **23**: 55–56.

Kaiser, S.I., and Corradini, C., 2011, The early siphonodellids (Conodonta, Late Devonian-Early Carboniferous): overview and taxonomic state. *Neues Jahrbuch für Geologie und Paläontologie, Abhandlungen,* **261** (1): 19–35.

Kaiser, S.I., Steuber, T., Becker, R.T., and Joachimski, M.M., 2006, Geochemical evidence for major environmental change at the Devonian-Carboniferous boundary in the Carnic Alps and the Rhenish Massif. *Palaeogeography, Palaeoclimatology, Palaeoecology,* **240** (1-2): 146–160.

Kaiser, S.I., Steuber, T., and Becker, R.T., 2008, Environmental change during the Late Famennian and Early Tournaisian (Late Devonian-Early Carboniferous): implications from stable isotopes and conodont biofacies in southern Europe. *Geological Journal,* **43**: 241–260.

Kaiser, S.I., Becker, R.T., and Spaletta, C., 2009, High-resolution conodont stratigraphy, biofacies and extinctions around the Hangenberg Event in pelagic successions from Austria, Italy, and France. *In* Over, D.J. (ed), Studies in Devonian Stratigraphy: Proceedings of the 2007 International Meeting of the Subcommission on Devonian Stratigraphy and IGCP 499. *Palaeontographica Americana,* **63**: 97–139.

Kaiser, S.I., Becker, R.T., Steuber, T., and Aboussalam, Z.S., 2011, Climate-controlled mass extinctions, facies, and sea-level changes around the Devonian-Carboniferous boundary in the eastern Anti-Atlas (SE Morocco). *Palaeogeography, Palaeoclimatology, Palaeoecology,* **310**: 340–364.

Kaiser, S.I., Aretz, M., and Becker, R.T., 2016, The global Hangenberg Crisis (Devonian-Carboniferous transition): review of a first order mass extinction, Special Publications. *In* Becker, R.T., Königshof, P., and Brett, C.E. (eds), Devonian Climate, Sea Level and Evolutionary Events. *London: Geological Society,* **423**: 387–437.

Kaiser, S.I., Joachimski, M.M., Hartenfels, S., 2017, First evidence for a late Famennian carbon isotope excursion in Franconia (Germany). In:

Liao, J.-C., and Valenzuela-Ríos, J.I. (Eds.), Fourth International Conodont Symposium. ICOS IV "Progress on Conodont Investigations". *Cuadernos del Museo Geominero*, **22**: 179–181.

Kalvoda, J., Kumpan, T., and Bábek, O., 2013, Upper Famennian and Lower Tournaisian sections of the Moravian Karst (Moravo-Silesian Zone, Czech Republic): a proposed key area for correlation of the conodont and foraminiferal zonations. *Geological Journal*, **50** (1): 17–38.

Kalvoda, J., Kumpan, T., and Babek, O., 2015, Upper Famennian and Lower Tournaisian sections of the Moravian Karst (Moravio-Silesian Zone, Czech Republic): a proposed key area for correlation of the conodont and foraminiferal zones. *Geological Journal*, **50**: 17–38.

Kalvoda, J., Kumpan, T., Holá, M., Bábek, O., Kanický, V., and Škoda, R., 2018, Fine-scale LA-ICP-MS study of redox oscillations and REEY cycling during the latest Devonian Hangenberg Crisis (Moravian Karst, Czech Republic). *Palaeogeography, Paleoclimatology, Palaeoecology*, **493**: 30–43.

Kaufmann, B., 2006, Calibrating the Devonian Time Scale: a synthesis of U-Pb ID-TIMS ages and conodont stratigraphy. *Earth-Science Reviews*, **76**: 175–190.

Kaufmann, B., Trapp, E., and Mezger, K., 2004, The numerical age of the Upper Frasnian (Upper Devonian) Kellwasser Horizons: a new U-Pb Zircon date from Steinbruch Schmidt (Kellerwald, Germany). *The Journal of Geology*, **112**: 495–501.

Kaufmann, B., Trapp, E., Mezger, K., and Wedige, K., 2005, Two new Emsian (Early Devonian) U-Pb zircon ages from volcanic rocks of the Rhenish Massif (Germany): implications for the Devonian time scale. *Journal of Geological Society, London*, **162**: 363–371.

Kim, A.I., 2011, Devonian tentaculites from the Kitab State Geological Reserve (Zeravshan-Gissar mountains area, Uzbekistan). *News on Paleontology and Stratigraphy, Supplement to Geologiya I Geofizika*, **52** (15): 65–81.

King Phillips, E.J., Cohen, P., Junium, C.K., Uveges, B.T.I., 2017, Examining the Marine Isotope Gradient Through the Late Devonian Using Microfossils From the Kellwasser Horizons. *GSA Annual Meeting 2017, Seattle, Abstracts with Programs, 49(6): Paper 378/278.*

Kirilishina, E.M., 2006, The conodont characteristics of the Frasnian – Famennian boundary interval (Late Devonian) of the Voronezh Anteclise. *Sovremennaia paleontologia klassicheskie i noveisie metodi, PIN PAL*, **2006**: 43–54.

Klapper, G., 1971, Sequence within the conodont *Polygnathus* in the New York lower Middle Devonian. *Geologica et. Palaeontologica*, **5**: 59–79.

Klapper, G., 1985, Sequence in conodont genus *Ancyrodella* in Lower *asymmetricus* Zone (earliest Frasnian, Upper Devonian) of the Montagne Noire, France. *Palaeontographica, Abteilung A.*, **188**: 19–34.

Klapper, G., 1989, The Montagne Noire Frasnian (Upper Devonian) conodont succession, (imprint 1988). *In* McMillan, N.J., Embry, A.F., and Glass, D.J. (eds), Devonian of the World. *Canadian Society of Petroleum Geology, Memoir*, **14(III)**: 449–468.

Klapper, G., 1992, North American Midcontinent Devonian T-R Cycles. *Okla. Geological Survey Bulletin*, **145**: 127–135.

Klapper, G., 1997, Graphic correlation of Frasnian (Upper Devonian) sequences in Montagne Noire, France, and western Canada. *Geological Society of America, Special Paper*, **321**: 113–129.

Klapper, G., 2000, Species of Spathiognathodontidae and Polygnathidae (Conodonta) in the recognition of Upper Devonian stage boundaries. *Courier Forschungsinstitut Senckenberg*, **220**: 153–159.

Klapper, G., 2007a, Conodont taxonomy and the recognition of the Frasnian/Famennian (Upper Devonian) Stage Boundary. *Stratigraphy*, **4** (1): 67–76.

Klapper, G., 2007b, Frasnian (Upper Devonian) conodont succession at Horse Spring and correlative sections, Canning Basin, Western Australia. *Journal of Paleontology*, **81** (3): 513–537.

Klapper, G., and Becker, R.T., 1999, Comparison of Frasnian (Upper Devonian) conodont zonations. *Bolletino della Società Paleontologica Italiana*, **37**: 339–347.

Klapper, G., and Johnson, J.G., 1975, Sequence in the conodont genus *Polygnathus* in Lower Devonian at Lone Mountain, Nevada. *Geologica et Palaeontologica*, **9**: 65–83.

Klapper, G., and Johnson, J.G., 1980, Endemism and dispersal of Devonian conodonts. *Journal of Paleontology*, **54**: 400–455.

Klapper, G., and Johnson, J.G., 1990, Revision of Middle Devonian conodont zones. *Journal of Paleontology*, **54**: 400–455.

Klapper, G., and Kirchgasser, W.T., 2016, Frasnian Late Devonian conodont biostratigraphy in New York: graphic correlation and taxonomy. *Journal of Paleontology*, **90** (3): 525–554.

Klapper, G., and Lane, H.R., 1989, Frasnian (Upper Devonian) conodont sequence at Luscar Mountain and Mount Haultain, Alberta Rocky Mountains, (imprint 1988). *In* McMillan, N.J., Embry, A.F., and Glass, D.J. (eds), Devonian of the World. *Canadian Society of Petroleum Geology, Memoir*, **14(III)**: 469–478.

Klapper, G., and Murphy, M.A., 1980, Conodont zonal species from the *delta* and *pesavis* Zones (Lower Devonian) in central Nevada. *Neues Jahrbuch für Geologie und Paläontologie, Monatshefte*, **1980** (8): 490–504.

Klapper, G., and Ziegler, W., 1979, Devonian conodont biostratigraphy. *Special Papers in Palaeontology*, **23**: 199–224.

Klapper, G., Ziegler, W., and Mashkova, T.V., 1978, Conodonts and correlation of Lower-Middle Devonian boundary beds in the Barrandian area of Czechoslowakia. *Geologica et. Palaeontologica*, **12**: 103–116.

Klapper, G., Feist, R., and House, M.R., 1987, Decision on the Boundary Stratotype for the Middle/Upper Devonian Series Boundary. *Episodes*, **10**: 97–101.

Klapper, G., Feist, R., Becker, R.T., and House, M.R., 1993, Definition of the Frasnian/Famennian Stage boundary. *Episodes*, **16**: 433–441.

Klapper, G., Kirchgasser, W.T., and Baesemann, J.F., 1995, Graphic correlation of a Frasnian (Upper Devonian) composite standard. *SEPM Special Publication*, **53**: 177–184.

Klapper, G., Kuz'min, A., and Ovnatanova, N.S., 1996, Upper Devonian conodonts from the Timan-Pechora region, Russia, and correlation with a Frasnian composite standard. *Journal of Paleontology*, **70** (1): 131–152.

Klapper, G., Uyeno, T.T., Armstrong, D.K., and Telford, P.G., 2004, Conodonts of the Williams Island and Long Rapids Formations (Upper Devonian, Frasnian-Famennian) of the Onakawana B Drillhole, Moose River Basin, Northern Ontario, with a revision of Lower Famennian species. *Journal of Paleontology*, **78** (2): 371–387.

Kleffner, M.A., Barrick, J.E., Ebert, J.R., Matteson, D.K., and Karlsson, H.R., 2009, Conodont biostratigraphy, $\delta^{13}C$ chemostratigraphy, and recognition of Silurian/Devonian boundary in the Cherry Valley, New York region of the Appalachian Basin. *In* Over, D.J. (ed), Conodont Studies Commemorating the 150th Anniversary of the First Conodont Paper (Pander, 1856) and the 40th Anniversary of the Pander Society. *Palaeontographica Americana*, **62**: 57–73.

Klug, C., 2001, Early Emsian ammonoids from the eastern Anti-Atlas (Morocco) and their succession. *Paläontologische Zeitschrift*, **74**: 479–515.

Klug, C., 2002, Quantitative stratigraphy and taxonomy of late Emsian and Eifelian ammonoids of the eastern Anti-Atlas (Morocco). *Courier Forschungsinstitut Senckenberg*, **238**: 1–109.

Klug, C., 2017, First description of the early Devonian ammonoid *Mimosphinctes* from Gondwana and stratigraphical implications. *Swiss Journal of Palaeontology*, **136**: 345–358.

Klug, C., and Pohle, A., 2018, The eastern Amessoui Syncline — a hotspot for Silurian to Carboniferous cephalopod research. In: Hartenfels, S., Becker, R.T., El Hassani, A., and Lüddecke, F. (Eds.), 10th International Symposium "Cephalopoda — Present and Past", Fes, March 26–April 3, 2018, Field Guidebook. *Münstersche Forschungen zur Geologie und Paläontologie*, **110**: 244–260.

Klug, C., Korn, D., Reisdorf, A., 2000, Ammonoid and conodont stratigraphy of the late Emsian to early Eifelian (Devonian) at Jebel Ouaoufilal (near Taouz, Tafilalt, Morocco). In: Tahiri, A., and El Hassani, A. (Eds.), Proceedings of the Subcommission on Devonian Stratigraphy (SDS) — IGCP 421 Morocco Meeting. *Travaux de l'Institut Scientifique, Série Géologie & Géographie Physique*, **20**: 45–56.

Klug, C., Kröger, B., Rücklin, M., Korn, D., Schemm-Gregory, M., De Baets, K., et al., 2008, Ecological change during the early Emsian (Devonian) in the Tafilalt (Morocco), the origin of the Ammonoidea, and the first African pyrgocystid edrioasterioids, machaerids, and phyllocarids. *Palaeontographica, Abteilung A*, **283**: 83–176.

Klug, C., Frey, L., Korn, D., Jattiot, R., and Rücklin, M., 2016, The oldest Gondwanan cephalopod mandibles (Hangenberg Blackshale, Late Devonian) and the mid-Palaeozoic rise of jaws. *Paleontology*, **59** (5): 611–629.

Königshof, P., Savage, N.M., Lutat, P., Sardsud, A., Dopieralska, J., Belka, Z., et al., 2012, Late Devonian sedimentary record of the Paleotethys Ocean — the Mae Sariang section, northwestern Thailand. *Journal of Asian Earth Sciences*, **52**: 146–157.

Königshof, P., Da Silva, A.C., Suttner, T.J., Kido, E., Waters, J., Carmichael, S.K., et al., 2016, Shallow-water facies setting around the Kačák Event: a multidisciplinary approach, Special Publications. *In* Becker, R.T., Königshof, P., and Brett, C.E. (eds), Devonian Climate, Sea Level and Evolutionary Events. *London: Geological Society*, **423**: 171–199.

Königshof, P., Narkiewicz, K., Ta Hoa, P., Carmichael, S., and Waters, J., 2017, Devonian events: examples from the eastern Palaeotethys (Si Phai section, NE Vietnam). *Palaeodiversity Palaeoenvoronments*, **97** (3): 481–496.

Kononova, L.I., and Kim, S.-Y., 2003, Eifelian Conodonts from the Central Russian Platform. *Paleontological Journal*, **39** (Suppl. 2): S55–S134.

Kononova, L.I., and Weyer, D., 2013, Upper Famennian conodonts from the Breternitz Member (Upper Clymeniid Beds) of the Saalfeld region, Thuringia (Germany). *Freiberger Forschungshefte*, **545** (21): 15–97.

Koptíková, L., 2011, Precise position of the Basal Choteč event and evolution of sedimentary environments near the Lower-Middle Devonian boundary: the magnetic susceptibility, gamma-ray spectrometric, lithological, and geochemical record of the Prague Synform. *Palaeogeography, Palaeoclimatology, Palaeoecology*, **304**: 96–112.

Koptíková, L., Berkyova, S., Hladil, J., Slavík, L., Schnabl, P., Frana, J., et al., 2008, Long-distance correlation of basal Choteč Event sections using magnetic susceptibility (Barrandian vs. Nevada) and lateral and vertical variations in fine-grained non-carbonate mineral phases. In: Kim, A.I., Salimova, F.A., and Meshchankina, N.A. (Eds.), *Global Alignments of Lower Devonian Carbonate and Clastic Sequences: Contributions From the IGCP 499 Project/SDS Joint Field Meeting, Kitab State Geological Reserve, Uzbekistan, August 25–September 3, 2008*, pp. 60–62.

Koptíková, L., Bábek, O., Hladil, J., Kalvoda, J., and Slavík, L., 2010a, Stratigraphic significance and resolution of spectral reflectance logs in Lower Devonian carbonates of the Barrandian area, Czech Republic; a correlation with magnetic susceptibility and gamma ray logs. *Sedimentary Geology*, **225**: 83–98.

Koptíková, L., Hladil, J., Slavík, L., Čejchan, P., and Bábek, O., 2010b, Fine-grained non-carbonate particles embedded in neritic to pelagic limestones (Lochkovian to Emsian, Prague Synform, Czech Republic): composition, provenance and Links to magnetic susceptibility and gamma-ray logs. *Geologica Belgica*, **13** (4): 407–430.

Koren, T.N., 1979, Late monograptid faunas and the problem of graptolite extinction. *Acta Palaeontologica Polonica*, **24** (1): 79–106.

Koren, T.N., Kim, A.I., and Walliser, O.H., 2007, Contribution to the biostratigraphy around the Lochkovian-Pragian boundary in Central Asia (graptolites, tentaculites, conodonts). *Palaeobiodiversity Palaeoenvironments*, **87** (2): 187–219.

Korn, D., 1981, *Cymaclymenia* — eine besonders langlebige Clymenien-Gattung (Ammonoidea, Cephalopoda). *Neues Jahrbuch für Geologie und Paläontologie, Abhandlungen*, **161** (2): 172–208.

Korn, D., 1991, Three dimensionally preserved clymeniids from the Hangenberg Black Shale of Drewer (Cephalopoda, Ammonoidea; Devonian-Carboniferous boundary; Rhenish Massif). *Neues Jahrbuch für Geologie und Paläontologie, Monatshefte*, **1991** (9): 553–563.

Korn, D., 1999, Famennian Ammonoid Stratigraphy of the Ma'der and Tafilalt (Eastern Anti-Atlas, Morocco). *Abhandlungen der Geologischen Bundesanst.*, **54**: 147–179.

Korn, D., 2002a, Historical Subdivisions of the Middle and Late Devonian Sedimentary Rocks in the Rhenish Mountains by Ammonoid Faunas. *Senckenbergiana lethaea*, **82** (2): 545–555.

Korn, D., 2002b, Die Ammonoideen-Fauna der Platyclymenia annulata-Zone von Kattensiepen (Oberdevon, Rheinisches Schiefergebirge). *Senckenbergiana lethaea*, **82** (2): 557–608.

Korn, D., 2004, The mid-Famennian ammonoid succession in the Rhenish Mountains: the "annulata Event" reconsidered. *Geological Quarterly*, **48** (3): 245–252.

Korn, D., and Luppold, F.W., 1987, Nach Clymenien und Conodonten gegliederte Profile des oberen Famennium im Rheinischen Schiefergebirge. *Courier Forschungsinstitut Senckenberg*, **92**: 199–223.

Korn, D., and Price, J., 1987, Taxonomy and Phylogeny of the Kosmocly-meniinae subfam. nov. (Cephalopoda, Ammonoidea, Clymeniida). *Courier Forschungsinstitut Senckenberg*, **92**: 5–75.

Korn, D., and Ziegler, W., 2002, The ammonoid and conodont zonation at Enkenberg (Famennian, Late Devonian; Rhenish Mountains). *Senckenbergiana lethaea*, **82** (2): 453–462.

Korn, D., Clausen, C.-D., and Luppold, F.W. (eds), 1994. Die Devon/Karbon-Grenze im Rheinischen Schiefergebirge. *Geologie und Paläontologie in Westfalen* **29**: 1–221.

Korn, D., Klug, C., and Reisdorf, A., 2000, Middle Famennian ammonoid stratigraphy in the Amessoui Syncline (Late Devonian; eastern Anti-Atlas, Morocco). *Trav. de. l'Institut Scientifique, Rabat, Série Géologie & Géographie Phys.*, **20**: 69–77.

Korn, D., Klug, C., Ebbighausen, V., and Bockwinkel, J., 2004, The youngest African clymeniids (Ammonoidea; Late Devonian) — failed survivors of the Hangenberg Event. *Lethaia*, **37**: 307–315.

Korn, D., Bockwinkel, J., and Ebbighausen, V., 2014, Middle Famennian (Late Devonian) ammonoids from the Anti-Atlas of Morocco. 1. Prionoceras. *Neues Jahrbuch für Geologie und Paläontologie, Abhandlungen*, **272** (2): 167–204.

Korn, D., Bockwinkel, J., and Ebbighausen, V., 2015a, The Late Devonian ammonoid *Mimimitoceras* in the Anti-Atlas of Morocco. *Neues Jahrbuch für Geologie und Paläontologie, Abhandlungen*, **275** (2): 125–150.

Korn, D., Bockwinkel, J., and Ebbighausen, V., 2015b, Middle Famennian (Late Devonian) ammonoids from the Anti-Atlas of Morocco. 2. Sporadoceratidae. *Neues Jahrbuch für Geologie und Paläontologie, Abhandlungen*, **278** (1): 47–77.

Korn, D., Zohara, F., Benyoucef, M., Bockwinkel, J., and Ebbighausen, V., 2017, The late Famennian ammonoid succession in the vicinity of Béni Abbès (Saoura Valley, Algeria). *Neues Jahrbuch für Geologie und Paläontologie, Abhandlungen*, **285** (2): 123–159.

Kulagina, E.I., 2013, Taxonomic diversity of foraminifers of the Devonian-Carboniferous boundary interval in the South Urals. *Bulletin of Geosciences*, **88** (2): 265–282.

Kumpan, T., Bábek, O., Kalvoda, J., Frída, J., and Grygar, T.N., 2013, A high-resolution, multiproxy stratigraphic analysis of the Devonian-Carboniferous boundary sections in the Moravian Karst (Czech Republic) and a correlation with the Carnic Alps (Austria). *Geological Magazine*, **151**: 201–215.

Kumpan, T., Bábek, O., Kalvoda, J., Grygar, and Frýda, J., 2014, Sea-level and environmental changes around the Devonian-Carboniferous boundary in the Namur-Dinant Basin (S Belgium, NE France): a multi-proxy stratigraphic analysis of carbonate ramp archives and its use in regional and interregional correlations. *Sedimentary Geological*, **311**: 43–59.

Kumpan, T., Bábek, O., Kalvoda, J., Grygar, T.M., Frýda, J., Becker, R.T., et al., 2015, Petrophysical and geochemical signature of the Hangenberg Events: an integrated stratigraphy of the Devonian-Carboniferous boundary interval in the Northern Rhenish Massif (Avalonia, Germany). *Bulletin of Geosciences*, **90** (3): 667–694.

Kurihara, T., 2004, Silurian and Devonian radiolarian biostratigraphy of the Hida Gaien Belt, central Japan. *Journal of the Geological Society of Japan*, **110**: 620–639.

Kürschner, W., Becker, R.T., Buhl, D., and Veizer, J., 1993, Strontium isotopes in conodonts: Devonian/Carboniferous transition, the northern Rhenish Slate Mountains, Germany. *Annales de la Société géologique de Belgique*, **115** (2): 595–622.

Kuźmin, A.V., 1992, Use of conodonts for dividing Famennian sediments of Tosuy ore region (central Kazakhstan). *Izvestiya Rossia Akademia Nauk, Seria Geologii*, **8**: 30–40 (in Russian).

Labounty, D., and Whalen, M.T., 2016, Microfacies and trace element variation across the Frasnian punctata Event within the Bear Biltmore drill core (Alberta, Canada). *Annual Meeting, Abstracts with Programs*, **48** (7): 1. https://doi.org/10.1130/abs/2016AM-284521.

Lakin, J.A., Marshall, J.E.A., Troth, I., and Harding, I.C., 2016, Greenhouse to icehouse: a biostratigraphic review of latest Devonian-Mississppian glaciations and their global effects, Special Publications. *In* Becker, R.T., Königshof, P., and Brett, C.E. (eds), Devonian Climate, Sea Level and Evolutionary Events. *London: Geological Society*, **423**:439–464.

Lange, W., 1929, Zur Kenntnis des Oberdevons am Enkeberg und bei Balve (Sauerland). *Abhandlungen der Preußischen Geologischen Landesanstalt, Neue Folge*, **119**: 1–132.

Lanik, A., Over, D.J., Schmitz, M.D., Hogencamp, N., 2013, Conodont biostratigraphy and new zircon dates of the Upper Devonian Belpre Ashes, Chattanooga Shale, Tennessee, and lower Rhinestreet Shale, New York, eastern North America. *GSA Annual Meeting, Denver, 125th Anniversary of GSA, October 27–30, 2013, Abstract With Program, Session 376, Paper No. 5*.

Lanik, A., Over, J.D., Schmitz, M., and Kirchgasser, W.T., 2016, Testing the limits of chronostratigraphic resolution in the Appalachian Basin, Late Devonian (middle Frasnian), eastern North America: new U-Pb zircon dates for the Belpre Ash suite. *Bulletin of the Geological Society of America*, **128** (11/12): 1813–1821. https://doi.org/10.1130/B31408.1.

Lash, G.G., 2017, A multiproxy analysis of the Frasnian-Famennian transition in western New York State, U.S.A. *Palaeogeography, Palaeoclimatology, Palaeoecology*, **473**: 108–122.

Laskar, J., Robutel, P., Joutel, F., Gastineau, M., Correia, A.C.M., and Levrard, B., 2004, A long-term numerical solution for the insolation quantities of the Earth. *Astronomy & Astrophysics*, **428**: 261–285.

Laskar, J., Fienga, A., Gastineau, M., and Manche, H., 2011, La2010: a new orbital solution for the long-term motion of the Earth. *Astronomy and Astrophysics*, **532** (A89): 1–15.

Le Hérissé, A., Servais, T., and Wicander, R., 2000, Devonian acritarchs and related forms. *Courier Forschungsinstitut Senckenberg*, **220**: 195–204.

Le Houedec, S., Girard, C., and Balter, V., 2013, Conodont Sr/Ca and $\delta^{18}O$ record seawater changes at the Frasnian-Famennian boundary. *Palaeogeography, Palaeoclimatology, Palaeoecology*, **376**: 114–121.

Lenz, A.C., 1989, Revision of upper Silurian and Lower Devonian graptolite biostratigraphy and morphological variation in Monograptus yukonensis and related Devonian graptolites, Northern Yukon, Canada, (imprint 1988). *In* McMillan, N.J., Embry, A.F., and Glass, D.J. (eds), Devonian of World. *Canadian Society of Petroleum Geologists, Memoir*, **14(III)**: 439–447.

Lenz, A.C., 2013, Early Devonian graptolites and graptolite biostratigraphy, Arctic Islands, Canada. *Canadian Journal of Earth Sciences*, **50** (11): 1097–1115.

Lenz, A.C., Robardet, M., Gutiérrez-Marco, J.C., and Picarra, J., 1996, Devonian graptolites from southwestern Europe: a review with new data. *Geological Journal*, **31**: 349–358.

Lethiers, F., 1974, Biostratigraphie des Ostracodes dans le Dévonien supérieur du Nord de la France et de la Belgique. *Newsletters Stratigraphy*, **3** (2): 73–79.

Lethiers, F., 1984, Zonation du Dévonien supérieur par les ostracodes (Ardenne et Boulonnais). *Revue de Micropaléontologie*, **27**: 30–42.

Levman, B.G., and von Bitter, P.H., 2002, The Frasnian-Famennian (mid-Late Devonian) boundary in the type section of the Long Rapids Formation, James Bay Lowlands, northern Ontario, Canada. *Canadian Journal of Earth Sciences*, **39**: 1795–1818.

Li, D.Y., and Wang, J.W., 1983, A discovery of Early Devonian strata in Changning, Yunnan. *Journal of Stratigraphy*, **7**: 71–73 (in Chinese).

Liao, J.C., and Valenzuela-Ríos, J.I., 2016, Givetian (Middle Devonian) historical bio- and chronostratigraphical subdivision based on conodonts. *Lethaia*, **50** (3): 447–463.

Liao, J.C., and García-López, S., 2018, Capturing the Kačák Episode (Middle Devonian) in the Palentine Domain (Cantabrian Mountains, NW Spain). *The Fossil Week, Fifth International Palaeontological Congress, July 9–13, 2018, France, Abstract Book*, p. 797.

Liao, J.C., Girard, C., Valewnzuela-Ríos, J.I., et al., 2019, New conodont data from the Middle-Upper Devonian boundary stratotype at Col du

Puech de la Suque (Montagne Noire, France). *STRATI 2019, Third International Congress on Stratigraphy, July 2–5, 2019, Milano, Abstract Book*, p. 188.

Liashenko, G.P., 1967, Coniconchia (Tentaculitida, Nowakiida, Styliolinida) and their importance in Devonian Biostratigraphy. In: Oswald, D.H. (Ed.), *International Symposium on the Devonian System*, Vol. II. Calgary, pp. 897–903.

Lipnjagow, O.M., 1979, The conodonts of Ct1a and Ct1b of the Donetz Basin. *Service Géologique de Belgique, Professional Paper*, **161**: 42–49.

Liu, Y.Q., Ji, Q., Kuang, H.W., Jiang, X.J., Xu, H., and Peng, N., 2012, U-Pb zircon age, sedimentary facies, and sequence stratigraphy of the Devonian-Carboniferous boundary, Daposhang Section, Guizhou, China. *Palaeoworld*, **21**: 100–107.

Liu, J.S., Qie, W.K., Algeo, T.J., Yao, L., Huang, J.H., and Luo, G.M., 2016, Changes in marine nitrogen fixation and denitrification rates during the end-Devonian mass extinction. *Palaeogeography, Palaeoclimatology, Palaeoecology*, **448**: 195–206.

Liu, F., Kerp, H., Peng, H.P., Zhu, H.C., and Peng, J.G., 2018, Palynostratigraphy of the Devonian-Carboniferous transition in the Tulong section in South Tibet: a Hangenberg Event sequence analogue in the Himalaya-Tethys zone. *Palaeogeography, Palaeoclimatology, Palaeoecology*, (online). 12 pp, <https//doi.org/10.1016/j.palaeo.2018.03.016>.

Loboziak, S., and Melo, J.H.G., 2002, Devonian miospore successions of Western Gondwana: update and correlation with Southern Euramerican miospore zones. *Review of Palaeobotany Palynology*, **121**: 133–148.

Lonsdale, W., 1840, Notes on the age of the Limestones of South Devonshire. *Transactions of the Geological Society of London, Series 2*, **5**: 721–738.

Lottmann, J., 1990, Die *pumilio*-Events (Mittel-Devon). *Göttinger Arbeiten zur Geologie und Paläontologie*, **44**: 1–98.

Lüttke, F., 1979, Biostratigraphical significance of the Devonian Dacryoconarida. *In* House, M.R., Scrutton, C.T., and Bassett, M.G. (eds), The Devonian System. *Special Papers in Palaeontology, 23*: 281–289.

Ma, K.Y., Hinnov, L.A., Zhang, X.S., and Gong, Y.M., 2020, Astronomical time calibration of the Upper Devonian Lali section, South China. *Global and Planetary Change*, **193**, article #103267: 12 pp. https://doi.org/10.1016/j.gloplacha.2020.103267.

Ma, X.P., and Bai, S.L., 2002, Biological, depositional, microspherule, and geochemical records of the Frasnian/Famennian boundary beds, South China. *Palaeogeography, Palaeoclimatology, Palaeoecology*, **181**: 325–346.

Ma, X.P., and Zong, P., 2010, Middle and Late Devonian brachiopod assemblages, sea level change and paleogeography of Hunan, China. *Science China, Earth Sciences*, **53** (12): 1849–1863.

Ma, X.P., Wang, C.Y., Racki, G., and Racka, M., 2008, Facies and geochemistry across the Early-Middle Frasnian transition (Late Devonian) on South China carbonate shelf: comparison with the Polish reference successions. *Palaeogeography, Palaeoclimatology, Palaeoecology*, **269**: 130–151.

Ma, X.P., Liao, W.H., and Wang, D.M., 2009, The Devonian System of China, with a discussion on sea-level change in South China. *In* Königshof, P. (ed), Devonian Change: Case Studies in Palaeogeography and Palaeoecology. *The Geological Society, London, Special Publications*, **314**: 241–262.

Ma, X.P., Zhang, Y.B., and Zhang, M.Q., 2014, Lithologic and biotic aspects of major Devonian events in South China. *Subcommission on Devonian Stratigraphy Newsletter*, **29**: 21–33.

Ma, X.P., Gong, Y.M., Chen, D., Racki, G., Chen, X.C., and Liao, W.H., 2015, The Late Devonian Frasnian-Famennian Event in South China – patterns

and causes of extinctions, sea level changes, and isotope variations. *Palaeogeography, Palaeoclimatology, Palaeoecology*, **448**: 224–244.

Ma, X.P., Wang, H.H., and Zhang, M.Q., 2017, Devonian event succession and sea level change in South China – with Early and Middle Devonian carbon and oxygen isotopic data. *Subcommission on Devonian Stratigraphy Newsletter*, **32**: 17–24.

Mabille, C., Pas, D., Aretz, M., Boulvain, F., Schröder, S., and Silva, A.C., 2008, Deposition within the vicinity of the Mid-Eifelian High: detailed sedimentological study and magnetic susceptibility of a mixed ramp-related system from the Eifelian Lauch and Nohn formations (Devonian; Ohlesberg, Eifel, Germany). *Facies*, **54**: 597–612.

Machado, G., Hladil, J., Slavík, L., Koptíková, L., Moreira, N., and Fonseca, P., 2010, An Emsian-Eifelian calciturbidite sequence and the possible correlatable pattern of the Basal Choteč event in Western Ossa-Morena Zone, Portugal (Odivelas Limestone). *Geologica Belgica*, **13** (4): 431–446.

Macqueen, R.W., and Sandberg, C.A., 1970, Stratigraphy, age, and interregional correlation of the Exshaw Formation, Alberta Rocky Mountains. *Bulletin of Canadian Petroleum Geology*, **18** (1): 32–66.

Mahmudy-Gharaie, M.H., Matsumoto, R., Kakuwa, Y., and Milroy, P.G., 2004, Late Devonian facies variety in Iran: volcanism as a possible trigger of the environmental perturbation near the Frasnian-Famennian boundary. *Geological Quarterly*, **48** (4): 323–332.

Maillet, S., Milhau, B., Vreulx, M., and Sánchez de Posada, L.-C., 2016, Givetian ostracods of the Candás Formation (Asturias, North-western Spain): taxonomy, stratigraphy, palaeoecology, relationships to global events and palaeogeographical implications. *Zootaxa*, **4068** (1): 1–78.

Malec, J., 2014, The Devonian/Carboniferous boundary in the Holy Cross Mountains (Poland). *Geological Quarterly*, **58**: 217–234.

Malkowski, K., Racki, G., Drygant, D., and Szaniawski, H., 2009, Carbon isotope stratigraphy across the Silurian-Devonian transition in Podolia, Ukraine: evidence for a global biogeochemical perturbation. *Geological Magazine*, **146** (5): 674–689.

Manda, S., and Frýda, J., 2010, Silurian-Devonian boundary events and their influence on cephalopod evolution: evolutionary significance of cephalopod egg size during mass extinction. *Bulletin of Geosciences*, **85** (3): 513–540.

Marini, F., Casier, J.-G., Claude, J.-M., and Théry, J.-M., 1997, Cosmic magnetic sphaerules in the Famennian of the Bad Windsheim borehole (Germany): preliminary study and implications. *Sphaerula*, **1** (1): 4–19.

Mark-Kurik, E., Blieck, A., Lobaziak, S., and Candilier, A.-M., 1999, Miospore assemblage from the Lode member (Gauja Formation) in Estonuia and the Middle-Upper Devonian boundary problem. *Proc. Estonian Acad. Sciences, Geological*, **48**: 86–98.

Marshall, J.E.A., 2016, Palynological calibration of Devonian events at near-polar palaeolatitudes in the Falkland Islands, South Atlantic. *In* Becker, R.T., Königshof, P., and Brett, C.E. (eds) Devonian Climate, Sea Level and Evolutionary Events. *Geological Society of London, Special Publication*, **423**: 25–44.

Marshall, J.E.A., 2019, There was a mass extinction in plants at the Devonian-Carboniferous Boundary. In: *STRATI 2019, Third International Congress on Stratigraphy, July 2–5, 2019, Milano, Abstract Book*, pp. 190.

Marshall, J.E.A., and House, M.R., 2000, Devonian Stage Boundaries in England, Wales and Scotland. *Courier Forschungsinstitut Senckenberg*, **225**: 83–90.

Marshall, J.E.A., Astin, T.R., Brown, J.F., Mark-Kurik, E., and Lazauskiene, J., 2007, Recognizing the Kačák Event in the Devonian terrestrial environment and its implications for understanding land–sea interactions. *In* Becker, R.T., and Kirchgasser, W.T. (eds), Devonian Events and Correlations. *The Geological Society of London, Special Publication*, **278**: 133–155.

Marshall, J.E.A., Brown, J.F., and Astin, T.R., 2011, Recognising the Taghanic Crisis in the Devonian terrestrial environment and its implications for understanding land-sea interactions. *Palaeogeography, Palaeoclimatology, Palaeoecology*, **304**: 165–183.

Martinez, A.M., Boyer, D., Droser, M.L., Barrie, C., and Love, G.D., 2018, A stable and productive marine microbial community was sustained through the end-Devonian Hangenberg Crisis within the Cleveland Shale of the Appalachian Basin, United States. *Geobiology*, **2018**: 1–16.

Martínez-Pérez, C., Valenzuela-Ríos, J.I., Navas-Parejo, P., Liao, J.-C., and Botella, H., 2011, Emsian (Lower Devonian) Polygnathids (Conodont) succession in the Spanish Central Pyrenees. *Journal of Iberian Geology*, **37** (1): 45–64.

Martinsson, A. (ed), 1977. The Silurian-Devonian boundary: final report of the Committee of the Siluro-Devonian Boundary within IUGS Commission on Stratigraphy and a state of the art report for Project Ecostratigraphy. *International Union of Geological Sciences, Series A* **5**: 347 pp.

Marynowski, L., and Filipiak, P., 2007, Water column euxinia and wildfire evidence during deposition of the Upper Famennian Hangenberg event horizon from the Holy Cross Mountains (central Poland). *Geological Magazine*, **144** (3): 569–595.

Marynowski, L., Filipiak, P., and Zaton, M., 2010, Geochemical and palynological study of the Upper Famennian Dasberg event horizon from the Holy Cross Mountains (central Poland). *Geological Magazine*, **147** (4): 527–550.

Marynowski, L., Rakocinski, M., Borcuch, E., Kremer, B., Schubert, B.A., and Jahren, A.H., 2011, Molecular and petrographic indicators of redox conditions and bacterial communities after the F/F mass extinction (Kowala, Holy Cross Mountains, Poland). *Palaeogeography, Palaeoclimatology, Palaeoecology*, **306**: 1–14.

Mason, G.M., and Caffee-Cooper, D.D., 2016, Microtectite-like glassy spherules near the Frasnian-Famennian Boundary, Late Devonian, New Albany Shale, Floyd County, Indiana. *GSA Annual Meeting, Abstracts with Programs*, **48** (7): 1. https://doi.org/10.1130/abs/2016AM-285731.

Matern, H., 1931, Das Oberdevon der Dill-Mulde. *Abhandlungen der Preußischen Geologischen Landesanstalt, Neue Folge*, **134**: 1–139.

Matyja, H., and Narkiewicz, M., 1992, Conodont Biofacies Succession near the Frasnian/Famennian Boundary. Some Polish Examples. *Courier Forschungsinstitut Senckenberg*, **154**: 125–147.

Matyja, H., Turnau, E., 2008, Integrated analyses of miospores and conodonts: a tool for correlation of shallow-water, mixed siliciclastic-carbonate successions of Lower? and Middle Devonian (Pomeranian Basin, NW Poland). In: Kim, A.I., Salimova, F.A., and Meshchankina, N.A. (Eds.), *Global Alignments of Lower Devonian Carbonate and Clastic Sequences: Contributions From the IGCP 499 Project/SDS Joint Field Meeting, Kitab State Geological Reserve, Uzbekistan, August 25–September 3, 2008*, pp. 67–71.

Mavrinskaya, T., and Slavík, L., 2013, Correlation of Early Devonian (Lochkovian-early Pragian) conodont faunas of the South Urals (Russia). *Bulletin of Geosciences*, **88** (2): 283–296.

Mawson, R., 1987, Early Devonian conodont faunas from Buchan and Bindi, Victoria, Australia. *Palaeontology*, **30** (2): 251–297.

Mawson, R., 1995, Early Devonian polygnathid conodont lineages with special reference to Australia. *Corier Forschungsinstitut Senckenberg.*, **182**: 389–398.

Mawson, R., Talent, J.A., Bear, D.S., Brock, G.A., Farrell, J.R., Hyland, K.A., et al., 1989, Conodont data in relation to resolution of stage and zonal boundaries for the Devonian of Australia, (imprint 1988). *In* McMillan, N.J., Embry, A.F., and Glass, D.J. (eds), Devonian of the World. *Canadian Society of Petroleum Geology, Memoir*, **14(III)**: 485–527.

Maziane, N., Higgs, K.T., and Streel, M., 1999, Revision of the late Famennian miospore zonation scheme in eastern Belgium. *Journal of Micropalaeontology*, **18**: 17–25.

Maziane, N., Higgs, K., and Streel, M., 2002, Biometry and paleoenvironment of *Retispora lepidophyta* (Kedo) Playford 1976 and associated miospores in the latest Famennian nearshore marine facies, eastern Belgium. *Review of Palaeobotany and Palynology*, **10**: 170–175.

McAdams, N.E.B., Schmitz, M.D., Kleffner, M.A., Verniers, J., Vandenbroucke, T.R.A., Ebert, J.R., et al., 2017, A new, high-precision CA-ID-TIMS date for the 'Kalkberg' K-bentonite (Judds Falls Bentonite) (online) *Lethaia*, **51** (3): 344–356.

McArthur, J.M., Howarth, R.J., Shields, G.A., and Zhou, Y., 2020, Chapter 7 - Strontium isotope stratigraphy. *In* Gradstein, F.M., Ogg, J.G., Schmitz, M.D., and Ogg, G.M. (eds), *The Geologic Time Scale 2020*. Vol. 1 (this book). Elsevier, Boston, MA.

McGhee Jr., G.R., Orth, C.J., Quintana, L.R., Gilmore, J.S., and Olsen, E. J., 1996, Late Devonian "Kellwasser Event" mass-extinction horizon in Germany: no geochemical evidence for a large-body impact. *Geology*, **14**: 776–779.

McGregor, D.C., and McCutcheon, S.R., 1988, Implications of spore evidence for Late Devonian age of Pskahegan Group, southeastern New Brunswick. *Canadian Journal of Earth Sciences*, **25**: 1349–1364.

McLean, D., and Mountjoy, E.W., 1994, Allocyclic control on late Devonian buildup development, southern Canadian Rocky Mountains. *Journal of Sedimentary Research*, **B64** (3): 326–340.

McLean, R.A., and Klapper, G., 1998, Biostratigraphy of Frasnian (Upper Devonian) strata in western Canada, based on conodonts and rugose corals. *Bulletins of Canadian Petroleum Geology*, **46** (4): 515–563.

McMillan, N.J., Embry, A.F., Glass, D.J. (Eds.), 1989, Devonian Geology of the world. *Canadian Society of Petroleum Geology, Memoir*, **14**, **I**, 795 pp.; **II**, 674 pp.; **III**, 714 pp.

Melo, J.H.G., 1989, The Malvinokaffric Realm in the Devonian of Brazil, (imprint 1988). *In* McMillan, N.J., Embry, A.F., and Glass, D.J. (eds), Devonian of the World. *Canadian Society of Petroleum Geology, Memoir*, **14(I)**: 669–703.

Melo, J.H.G., and d Loboziak, S., 2003, Devonian-Early Carboniferous miospore biostratigraphy of the Amazon Basin, Northern Brazil. *Review of Palaeobotany and Palynology*, **124**: 131–202.

Mestre, A., Gómez, M.J., Garcías, Y., Corradini, C., Heredia, S., 2017, Advances on Silurian-Devonian conodont biostratigraphy in the central Precordillera, Argentina. In: Liao, J.-C., and Valenzuela-Ríos, J.I. (Eds.), Fourth International Conodont Symposium. ICOS IV "Progress on Conodont Investigations". *Cuadernos del Museo Geominero*, **22**: 105–108.

Michel, J., Boulvain, F., Philippo, S., and Da Silva, A.C., 2010, Palaeoenvironmental study and small scale correlations using facies analysis and magnetic susceptibility of the mid-Emsian (Himmelbaach quarry, Luxembourg). *Geologica Belgica*, **13** (4): 447–458.

Molyneux, S.G., Le Hérissé, A., and Wicander, R., 1996, Paleozoic phytoplankton. *In* Jansonius, J., and McGregor, D.C. (eds), *Palynology:*

Principles and Applications. Dallas: American Association of Stratigraphic Palynologists, pp. 493–529.

Montesinos, J.R., 1990, Las Biozonas de Ammonoideos del Devonico palentino (Emsiense inferior – Famenniense inferior): critica as sistema de classification zonal. *Revista Española de Paleontología*, **5**: 3–17.

Moreno, C., Gonzalvez, F., Sáez, R., Melgarejo, J.C., and Suárez-Ruiz, I., 2018, The Upper Devonian Kellwasser event recorded in a regressive sequence from inner shelf to lagoonal pond, Catalan Coastal Range, Spain. *Sedimentology*, **65**: <https//doi.org/10.1111/sed.12457>.

Morrow, J.R., 2000, Shelf-to-Basin Lithofacies and Conodont Paleoecology across Frasnian-Famennian (F-F, mid-Late Devonian) Boundary, Central Great Basin (Western U.S.A.). *Courier Forschungsinstitut Senckenberg*, **219**: 1–57.

Morrow, J.R., and Sandberg, C.A., 2003, Late Devonian Sequence and Event Stratigraphy Across the Frasnian-Famennian (F-F) Boundary, Utah and Nevada. *In* Harries, P.J. (ed), *Approaches in High-Resolution Stratigraphic Paleontology.* Kluwer Adacemic Publishers, pp. 351–419.

Morrow, J.R., Sandberg, C.A., Malkowski, K., and Joachimski, M.M., 2009, Carbon isotope chemostratigraphy and precise dating of middle Frasnian (lower Upper Devonian) Alamo Breccia, Nevada, USA. *Palaeogeography, Palaeoclimatology, Palaeoecology*, **282**: 105–118.

Mottequin, B., Poty, E., 2016, Kellwasser horizons, sea-level changes and brachiopod-coral crises during the late Frasnian in the Namur-Diant Basin (southern Belgium): a synopsis. In: Becker, R.T., El Hassani, A., and Tahiri, A. (Eds.), *International Field Symposium "The Devonian and Lower Carboniferous of northern Gondwana", March 22–29, 2013, Field Guidebook*, Vol. 27. Rabat: Document de l'Institut Scientifique, pp. 235–250.

Mountjoy, E., and Becker, S., 2000, Frasnian to Famennian sea-level changes and the Sassenach Formation, Jasper Basin, Alberta Rocky Mountains. *In* Homewood, P., and Eberli, G. (eds), *Genetic Stratigraphy on the Exploration and the Production Scales: Case Studies From the Pennsylvanian of the Paradox Basin and Upper Devonian of Alberta.* Elf EP – Editions, Memoire, Pau, pp. 181–202.

Mu, E.Z., and Ni, Y.N., 1983, Lower Devonian graptolites from Yunnan. *Acta Paleontologica Sin.*, **22** (3): 295–307 (in Chinese with English summary).

Mu, E.Z., and Ni, Y.N., 1985, Researches on the graptolites of Xizang (Tibet). *Palaeontologia Cathayana*, **2**: 1–15.

Murphy, E.M., Sageman, B.B., and Hollander, D.J., 2000, Eutrophication by decoupling of the marine biogeochemical cycles of C, N, and P: a mechanism for the Late Devonian mass extinction. *Geology*, **28** (5): 427–430.

Murphy, M.A., 2000, Conodonts first occurrences in Nevada. *In* Weddige, K. (ed), Devonian Correlation Table. *Senckenbergiana Lethaea*, **80**:695.

Murphy, M.A., 2005, Pragian conodont zonal classification in Nevada, Western North America. *Revista Española de Paleontología*, **20** (2): 177–206.

Murphy, M.A., and Berry, W.B.N., 1983, Early Devonian conodont-graptolite collation and correlation with brachiopods and coral zones. *Central Nevada. AAPG*, **67**: 371–379.

Murphy, M.A., and Valenzuela-Ríos, J.I., 1999, *Lanea* new genus, lineage of Early Devonian conodonts. *Bolletino della Società Paleontologica Italiana*, **37** (2/3): 321–334.

Murphy, M.A., Carls, P., and Valenzuela-Ríos, J.I., 2018a, Cypricriodus hesperius *(Klapper and Murphy), Taxonomy and Biostratigraphy*, 8. University of California, Riverside Campus Museum Contributions, 23 pp.

Murphy, M.A., Carls, P., and Valenzuela-Ríos, J.I., 2018b, Cypricriodus hesperius *(Klapper and Murphy, 1975), Taxonomy and Biostratigraphy*, 8. University of California, Riverside Campus Museum Contribution, pp. 1–21, 2 pp. in appendix.

Myrow, P.M., Ramezani, J., Hanson, A.E., Bowring, S.A., Racki, G., and Rakociński, M., 2013, High-precision U-Pb age and duration of the latest Devonian (Famennian) Hangenberg event, and its implications. *Terra Nova*, **26**: 222–229.

Myrow, P.M., Strauss, J.V., Creveling, J.R., Sicard, K.R., Ripperdan, R., Sandberg, C.A., et al., 2011, A carbon isotopic and sedimentological record of the latest Devonian (Famennian) from the Western U.S. and Germany. *Palaeogeography, Palaeoclimatology, Palaeoecology*, **306**: 147–159.

Naglik, C., De Baets, K., and Klug, C., 2019, Early Devonian ammonoid faunas in the Zeravshan Mountains (Uzbekistan and Tadjikistan) and the transition from a carbonate platform setting to pelagic sedimentation. *Bulletin of Geosciences*, **94** (3): 32. https://doi.org/10.3140/bull.geosci.1721.

Narkiewicz, K., and Bultynck, P., 2007, Conodont biostratigraphy of shallow marine Givetian deposits from the Radom-Lublin area, SE Poland. *Geological Quarterly*, **51** (4): 419–442.

Narkiewicz, K., and Bultynck, P., 2010, The Upper Givetian (Middle Devonian) *subterminus* conodont zone in North America, Europe and North Africa. *Journal of Paleontology*, **84** (4): 588–625.

Narkiewicz, K., Narkiewicz, M., Bultynck, P., Königshof, P., 2017a, The past, present and future of the upper Eifelian conodont zonation. In: Liao, J.-C., and Valenzuela-Ríos, J.I. (Eds.), Fourth International Conodont Symposium, ICOS IV, "Progress on Conodont Investigation". *Cuadernos del Museo Geominero*, **22**: 137–140.

Narkiewicz, M., Narkiewicz, K., Krzemińska, E., and Kruchek, S., 2017b, Oxygen isotopic composition of conodont apatite in the equatorial epeiric Belarusian Basin (Eifelian) – relationships to fluctuating seawater salinity and temperature. *Palaios*, **32**: 439–447.

Nawrocki, J., Polechonska, O., and Werner, T., 2008, Magnetic susceptibility and selected geochemical-mineralogical data as proxies for Early to Middle Frasnian (Late Devonian) carbonate depositional settings in the Holy Cross Mountains, southern Poland. *Palaeogeography, Palaeoclimatology, Palaeoecology*, **269**: 176–188.

Nazarov, B.B., and Ormiston, A.R., 1985, Evolution of Radiolaria in the Paleozoic and its correlation with the development of other marine fossil groups. *Senckenbergiana lethaea*, **66** (3/5): 203–215.

Ni, Y.N., and Jiao, S.D., 1983, Lower Devonian graptolites from Yunnan. *Acta Palaeontologica Sinica*, **22**: 295–307 (in Chinese with English summary).

Niedzwiedzki, G., Szrek, P., Narkiewicz, M., and Ahlberg, P.E., 2010, Tetrapod trackways from the early Middle Devonian Period of Poland. *Nature*, **463**: 43–48.

Nikolaeva, S.V., 2007, New Data on the Clymeniid Faunas of the Urals and Kazakhstan. *In* Landman, N.H., Davis, R.A., and Mapes, R.H. (eds), *Cephalopods – Present and Past: New Insights and Fresh Perspectives.* New York: Springer, pp. 317–343.

Nikolaeva, S.V., and Bogoslovskiy, B.I., 2005, Devonskie Ammonoidei. IV. Klymenii (Podotryad Clymeniina). *Trudy Paleontologicheskogo Instituta, Akademya Nauk SSSR*, **287**: 1–220, 32 pls.

Nothdurft, L.D., Webb, G.E., and Kamber, B.S., 2004, Rare Earth element geochemistry of Late Devonian reefal carbonates, Canning Basin, Western Australia: confirmation of a seawater REE proxy in ancient limestones. *Geochimica et Cosmochimica Acta*, **68** (2): 263–283.

Nikolaeva, S.V., Kuzmichev, A.B., and Aristov, V.A., 2009, On Frasnian Ammonoids of the New Siberian Islands. *Paleontological Zhurnal*, **43** (2): 134–141.

Obukhovskaya, T.G., Avkhimovitch, V.I., Streel, M., and Loboziak, S., 2000, Miospores from the Frasnian-Famennian Boundary deposits in Eastern Europe (the Pripyat Depression, Belarus and the Timan-Pechora Province, Russia) and comparisons with Western Europe (Northern France). *Review of Palaeobotany and Palynology*, **112**: 229–246.

Obut, O.T., and Shcherbanenko, T.A., 2008, Late Devonian radiolarians from the Rudnyi Altai (SW Siberia). *Bulletin of Geosciences*, **83** (4): 371–382.

Oliver Jr., W.A., and Pedder, A.E.H., 1989, Origins, migrations, and extinctions of Devonian Rugosa on the North American Plate. *Memoirs of the Association of Australasian Palaeontologists*, **8**: 231–237.

Orlov, A.N., 1990, Kompleksy ostracod is verchnedevonskich otloszheniy r. Sharyu (Sewero-Wostok Timano-Petchorskoi Provinzii). *VNIGRI*, **1990**: 22–29.

Over, D.J., 1997, Conodont biostratigraphy of the Java Formation (Upper Devonian) and the Frasnian-Famennian boundary in western New York State, Special Paper. *In* Klapper, G., Murphy, M.A., and Talent, J.A. (eds), Paleozoic Sequence Stratigraphy, Biostratigraphy, and Biogeography: Studies in Honor of J. Granville ("Jess") Johnson. *Geological Society of America, Special Paper,* **321**: 161–177.

Over, D.J., 2002, The Frasnian/Famennian boundary in central and eastern United States. *Palaeogeography, Palaeoclimatology, Palaeoecology*, **181**: 153–169.

Over, J.D., Hopkins, T.H., Brill, A., and Spaziani, A.L., 2003, Age of the Middlesex Shale (Upper Devonian, Frasnian) in New York State. *Courier Forschungsinstitut Senckenberg*, **242**: 217–223.

Over, D.J., Morrow, J.R., and Wignall, P.B. (eds), 2005. Understanding Late Devonian and Permian-Triassic biotic and climatic events — towards an integrated approach. *Dev. Palaeontology & Stratigraphy* **20**: 1–337.

Ovnatanova, N.S., and Kononova, L.I., 2001, Conodonts and Upper Devonian (Frasnian) Biostratigraphy of Central Regions of Russian Platform. *Courier Forschungsinstitut Senckenberg*, **233**: 1–155.

Ovnatanova, N.S., and Kononova, L.I., 2008, Frasnian Conodonts from the Eastern Russian Platform. *Paleontological Journal*, **42** (10): 997–1166.

Ovnatanova, N.S., Kuźmin, A.V., and Menner, V.V., 1999, The Succession of Frasnian Conodont Assemblages in the type sections of the Southern Timan-Pechora Province (Russia). *Bolletino della Società Paleontologica Italiana*, **37** (2/3): 349–360.

Packham, G.H., 1969, Southern and Central Highlands Fold Belt. *Journal of the Geological Society of Australia*, **16** (1): 73–226.

Packham, G.H., 2003, U-Pb SHRIMP ages of volcanic zircons from the Merrions and Turondale Formations, New South Wales, and the Early Devonian time-scale: a biostratigraphic and sedimentological assessment. *Australian Journal of Earth Sciences*, **50**: 169–179.

Packham, G.H., Percival, I.G., Rickards, R.B., and Wright, A.J., 2001, Late Silurian and Early Devonian biostratigraphy in the Hills End Trough and the Limekilns area; New South Wales. *Alcheringa*, **25**: 251–261.

Paeckelmann, W., Schindewolf, O.H., 1937, Die Devon-Karbon-Grenze. *Compte Rendu 2e Congrés Carbonifère, 1935, Heerlen*, **2**: 703–714.

Paproth, E., Feist, R., and Flajs, G., 1991, Decision on the Devonian-Carboniferous boundary stratotype. *Episodes*, **14** (4): 331–336.

Paris, F., Winchester-Seeto, T., Boumendjel, K., and Grahn, Y., 2000, Towards a global biozonation of Devonian chitinozoans. *Courier Forschungsinstitut Senckenberg*, **220**: 39–55.

Parry, S.F., Noble, S.R., Crowley, Q.G., and Wellman, C.H., 2011, A high-precision U-Pb age constraint on the Rhynie Chert Konservat-Lagerstätte: time scale and other implications. *Journal of the Geological Society, London*, **168**: 863–872.

Pas, D., Da Silva, A.-C., Suttner, T., Kido, E., Bultynck, P., Pondrelli, M., et al., 2014, Insight into the development of a carbonate platform through a multi-disciplinary approach: a case study from the Upper Devonian slope deposits of Mount Freikofel (Carnic Alps, Austria/Italy). *International Journal of Earth Sciences*, **103**: 519–538.

Pas, D., Da Silva, A.-C., Devleeschouwer, X., De Vleeschouwer, D., Labaye, C., Cornet, P., et al., 2015, Sedimentary development and magnetic susceptibility evolution of the Frasnian in Western Belgium (Dinant Synclinorium, La Thure section). *Geological Society, London, Special Publications*, **414**: 15–36.

Pas, D., Hinnov, L., Day, J.E., Kodama, K., Sinnesael, M., and Liu, W., 2018, Cyclostratigraphic calibration of the Famennian stage (Late Devonian, Illinois Basin, USA). *Earth and Planetary Science Letters*, **488**: 102–114.

Pas, D., Da Silva, A.-C., Over, D.J., Brett, C.E., Brandt, L., Over, J.-S., et al., 2020, Cyclostratigraphic calibration of the Eifelian Stage (Middle Devonian, Appalachian Basin, Western New York, USA). *Geological Society of America Bulletin*, **132**: in press. https://doi.org/10.1130/B35589.1.

Paytan, A., Yao, W.Q., Faul, K., and Gray, E.T., 2020, Chapter 9 - Sulfur isotope stratigraphy. *In* Gradstein, F.M., Ogg, J.G., Schmitz, M.D., and Ogg, G.M. (eds), *The Geologic Time Scale 2020*. Vol. 1 (this book). Elsevier, Boston, MA.

Pedder, A.E.H., 2010, Lower-Middle Devonian rugose coral faunas of Nevada: contribution to an understanding of the "barren" E Zone and Choteč Event in the Great Basin. *Bulletin of Geosciences*, **85** (1): 1–26.

Percival, L.M.E., Davies, J.H.F.L., Schlategger, U., De Vleeschouwer, D., Da Silva, A.-C., and Föllmi, K.B., 2018, Precisely dating the Frasnian-Famennian boundary: implications for the cause of the Late Devonian mass extinction. *Nature, Scientific Reports*, **2018** (8): 9578, 10. https://doi.org/10.1038/s41598-018-27847-7.

Pisarzowska, A., Sobstel, M., and Racki, G., 2006, Conodont-based event stratigraphy of the Early-Middle Frasnian transition on the South Polish carbonate shelf. *Acta Palaeontologica Polonica*, **51** (4): 609–646.

Pisarzowska, A., and Racki, G., 2012, Isotopic chemostratigraphy across the Early-Middle Frasnian transition (Late Devonian) on the South Polish carbonate shelf: a reference for the global *punctata* Event. *Chemical Geology*, **334**: 199–220.

Pisarzowska, A., Becker, R.T., Aboussalam, Z.S., Szczerba, M., Sobień, K., Kremer, B., et al., 2020, Middlesex/*punctata* event in the Rhenish Basin (Padberg section, Sauerland, Germany) — Geochemical clues to the early-middle Frasnian perturbation of global carbon cycle. *Global and Planetary Change*, **191**, article #103211: 14 pp. https://doi.org/10.1016/j.gloplacha.2020.103211.

Playford, P.E., McLaren, D.J., Orth, C.J., Gilmore, J.S., and Goodfellow, W.D., 1984, Iridium anomaly in the Upper Devonian of the Canning Basin, Western Australia. *Science*, **226**: 437–439.

Playton, T.E., Hocking, R., Montgomery, P., Tohver, E., Hillbun, K., Katz, D., et al., 2013, Development of a regional stratigraphic framework for Upper Devonian Reef Complexes using integrated chronostratigraphy: Lennard Shelf, Canning Basin, Western Australia. In: *West Australian Basin Symposium 2013, Perth, WA, August 18–21, 2013*, 15 pp.

Plotitsyn, A.N., Zhuravlev, A.V., Sobolev, D.B., Vevel', Y.A., and Gruzdev, D.A., 2018, The Devonian-Carboniferous Boundary in the western slope of the North Urals and the Cis-Urals. *Proceedings of the Paleontological Society*, **1**: 90–107 (in Russian).

Pogson, D.J., 2009, The Siluro-Devonian geological time scale: a critical review and interim revision. *Quarterly Notes, Geological Survey of New South Wales*, **130**: 1–13.

Porebska, B., and Sawlowicz, Z., 1997, Palaeoceanographic limits of geochemical and graptolite events around the Silurian-Devonian boundary in Barzkie Mountains (Southwest Poland). *Palaeogeography, Palaeoclimatology, Palaeoecology*, **132**: 343–354.

Poty, E., Devuyst, F.-X., and Hance, L., 2006, Upper Devonian and Mississippian foraminiferal and rugose coral zonation of Belgium and Northern France: a tool for Eurasian correlations. *Geological Magazine*, **143**: 829–857.

Price, J.D., and Korn, D., 1989, Stratigraphically important Clymeniids (Ammonoidea) from the Famennian (Late Devonian) of the Rhenish Massif, West Germany. *Courier Forschungsinstitut Senckenberg*, **110**: 257–294.

Pucéat, E., Joachimski, M.M., Bouilloux, A., Monna, F., Bonin, A., Motreuil, S., et al., 2010, Revised phosphate-water fractionation equation reassessing paleotemperatures derived from biogenic apatite. *Earth and Planetary Science Letters*, **298**: 135–142.

Pujol, F., Berner, Z., and Stüben, D., 2006, Palaeoenvironmetal changes at the Frasnian/Famennian boundary in key European sections: chemostratigraphic constraints. *Palaeogeography, Palaeoclimatology, Palaeoecology*, **240**: 120–145.

Qie, W.K., Liu, J.S., Chen, J.T., Wang, X.D., Mii, H.S., Zhang, X.H., et al., 2015, Local overprints on the global carbonate $\delta^{13}C$ signal in Devonian-Carboniferous boundary successions of South China. *Palaeogeography, Palaeocllimatology, Palaeoecology*, **418**: 290–303.

Qie, W.K., Ma, X.P., Xu, H.H., Qiao, L., Liang, K., Guo, W., et al., 2018, Devonian integrative stratigraphy and timescale of China (online) *Science China, Earth Sciences*, **62** (1): 112–134 <https//doi.org/10.1007/s11430-017-9259-9>.

Rabien, A., 1954, Zur Taxionomie und Chronologie der Oberdevonischen Ostracoden. *Abhandlungen des Hessischen Landesamtes für Bodenforschung*, **9**: 1–268.

Racka, M., Marynowski, L., Filipiak, P., Sobstel, M., Pisarzowska, A., and Bond, D.P.G., 2010, Anoxic *Annulata* Events in the Late Famennian of the Holy Cross Mountains (Southern Poland): geochemical and palaeontological record. *Palaeogeography, Palaeoclimatology, Palaeoecology*, **297** (3/4): 549–575.

Racki, G., 1998, Frasnian-Famennian biotic crisis: unvalued tectonic control? *Palaeogeography, Palaeoclimatology, Palaeoecology*, **141**: 177–198.

Racki, G., 1999, The Frasnian-Famennian biotic crisis: how many (if any) bolide impacts? *Geologische Rundsch.*, **87**: 617–632.

Racki, G., 2005, Toward understanding Late Devonian global events: few answers, many questions. *In* Over, D.J., Morrow, J.R., and Wignall, P.B. (eds), Understanding Late Devonian and Permian-Triassic Biotic and Climatic Events: Towards an Integrated Approach. *Developments in Palaeontology & Stratigraphy*, **20**: 5–36.

Racki, G., House, M.R. (eds), 2002. The Frasnian/Famennian boundary extinction event. Palaeogeography, Palaeoclimatology, Palaeoecology, 181, 374 pp.

Racki, G., and Koeberl, C., 2004, Comment on "Impact ejecta layer from the mid-Devonian: possible connection to global mass extinctions". *Science*, **303**: 471.

Racki, G., Racka, M., Matyja, H., and Devleeschouwer, X., 2002, The Frasnian/Famennian boundary interval in the South Polish-Moravian shelf basins: integrated event-stratigraphical approach. *Palaeogeography, Palaeoclimatology, Palaeoecology*, **181**: 251–297.

Racki, G., Piechota, A., Bond, D., and Wignall, P.B., 2004, Geochemical and ecological aspects of lower Frasnian pyrite-ammonoid level at Kostomloty (Holy Cross Mountains, Poland). *Geological Quarterly*, **48** (3): 267–282.

Racki, G., Joachimski, M.M., Morrow, J.R. (eds), 2008. A major perturbation of the global carbon budget in the Early-Middle Frasnian transition (Late Devonian). Palaeogeography, Palaeoclimatology, Palaeoecology, **269**: 127–204.

Racki, G., Balinski, A., Wrona, R., Malkowski, K., Drygant, D., and Szaniawski, H., 2012, Faunal dynamics across the Silurian-Devonian positive isotope excursion ($\delta^{13}C$, $\delta^{18}O$) in Podolia, Ukraine: comparative analysis of the Ireviken and Klonk events. *Acta Palaeontologica Polonica*, **57** (4): 795–832.

Racki, G., Rakociński, M., Marynowski, L., and Wignall, P.B., 2018, Mercury enrichments and the Frasnian-Famennian biotic crisis: a volcanic trigger proved? *Geology*, **48** (6): 543–546.

Rakociński, M., Pisarzowska, A., Janiszewska, K., and Szrek, P., 2016, Depositional conditions during the Lower Kellwasser Event (Late Frasnian) in the deep-shelf Lysogóry Basin of the Holy Cross Mountains, Poland. *Lethaia*, **49**: 571–590.

Ricci, J., Quidelleur, X., Pavlov, V., Orlov, S., Shatsillo, A., and Courtillot, V., 2013, New $^{40}Ar/^{39}Ar$ and K-Ar ages of the Viluy traps (eastern Siberia): further evidence for a relationship with the Frasnian-Famennian mass extinction. *Palaeogeography, Palaeoclimatology, Paleoecology*, **181**: 251–297.

Rice, B.J., Doughty, P.T., Grader, G.W., di Pasquo, M.M., Isaacson, P.E., 2017, Revision of the type *Siphonodella praesulcata* conodont locality at Lick Creek, Montana. *GSA Annual Meeting in Seattle, Washington, USA, 2017, Final Paper 174-8*, 1 p.

Richards, B.C., and Higgins, A.C., 1989, Devonian-Carboniferous boundary beds of the Palliser and Exshaw Formations at Jura Creek, Rocky mountsains, southwestern Alberta, (imprint 1988). *In* McMillan, N.J., Embry, A.F., and Glass, D.J. (eds), Devonian of the World. *Canadian Society of Petroleum Geology, Memoir*, **1C4(II)**: 399–412.

Richards, B.C., Ross, G.M., and Utting, J., 2002, U−Pb geochronology, lithology and biostratigraphy of tuff in the upper Famennian to Tournaisian Exshaw Formation: evidence for a mid-Paleozoic magmatic arc on the northwestern margin of North America. *Canadian Society of Petroleum Geologists, Memoir*, **19**: 158–207.

Richardson, J.B., and McGregor, D.C., 1986, Silurian and Devonian spore zones of the Old Red Sandstone Continent and adjacent areas. *Geological Survey of Canada Bulletin*, **364**: 1–79.

Richardson, J.B., Rodriguex, R.M., and Sutherland, J.E., 2000, Palynology and recognition of the Silurian/Devonian boundary in some British terrestrial sediments by correlation with other European marine sequences—a progress report. *Courier Forschungsinstitut Senckenberg*, **220**: 1–7.

Rickards, R.B., and Wright, A.J., 2000, Early Devonian graptolites from Limekilns, New South Wales. *Records of the Western Australian Museum*, **58** (Suppl.): 123–131.

Rimmer, S.M., Thompson, J.A., Goodnight, S.A., and Robl, T.L., 2004, Multiple controls on the preservation of organic matter in Devonian-

Mississippian marine black shales: geochemical and petrographic evidence. *Palaeogeography, Palaeoclimatology, Palaeoecology*, **215**: 125–154.

Riquier, L., Tribovillard, N., Averbuch, O., Joachimski, M.M., Racki, G., Dveleeschouwer, X., et al., 2005, Productivity and bottom warer redox conditions at the Frasnian-Famennian boundary on both sides of the Eovariscan Belt: constraints from trace-element geochemistry. *In* Over, D.J., Morrow, J.D., and Wignall, P.B. (eds), Understanding Late Devonian and Permian-Triassic Biotic and Climatic Events: Towards an Integrated Approach. *Developments in Paleontology & Stratigraphy*, **20**: 199–224.

Riquier, L., Tribovillard, N., Averbuch, O., Devleeschouwer, X., and Riboulleau, A., 2006, The Late Frasnian Kellwasser horizons of the Harz Mountains (Germany): two oxygen-deficient periods resulting from different mechanisms. *Chemical Geology*, **233**: 137–155.

Riquier, L., Averbuch, O., Tribovillard, N., El Albani, A., Lazreq, N., and Chakiri, S., 2007, Environmental changes at the Frasnian-Famennian boundary in Central Morocco (Northern Gondwana): integrated rock-magnetic and geochemical studies. *In* Becker, R.T., and Kirchgasser, W.T. (eds), Devonian Events and Correlations. *The Geological Society of London Special Publication*, **278**: 197–217.

Riquier, L., Averbuch, O., Devleeschouwer, X., and Tribovillard, N., 2010, Diagenetic versus detrital origin of the magnetic susceptibility variations in some carbonate Frasnian-Famennian boundary sections from Northern Africa and Western Europe: implications for paleoenvironmental reconstructions. *International Journal of Earth Sciences*, **99**: 57–73.

Robl, T.L., and Barron, L.S., 1989, The geochemistry of Devonian black shales in central Kentucky and its relationship to inter-basinal correlation and depositional environments. *Canadian Society of Petroleum Geologists, Memoir*, **14** (II): 377–392 [Imprint, 1988].

Roden, M.K., Parrish, R.R., and Miller, D.S., 1990, The absolute age of the Eifelian Tioga ash bed, Pennsylvania. *The Journal of Geology*, **98**: 282–285.

Rotondo, K.A., and Over, D.J., 2000, Biostratigraphic age of the Belpre Ash (Frasnian), Chattanooga and Rhinestreet shales in the Appalachian Basin. *Geological Society of America, Abstracts with Programs*, **32** (6): A70.

Ruan, Y., 1996, Zonation and distribution of the early Devonian primitive ammonoids in South China. *In* Wang, H.Z., and Wang, X.L. (eds), *Centennial Memorial Volume of Professor Sun Yunzhu (Y.C. Sun): Palaeontology and Stratigraphy*. Wuhan: China University of Geosciences Press, pp. 104–112.

Sageman, B.B., Murphy, A.E., Werne, J.P., Ver Straeten, C.A., Hollander, D.J., and Lyons, T.W., 2003, A tale of shales: the relative roles of production, decomposition, and dilution in the accumulation of organic-rich strata, Middle-Upper Devonian, Appalachian basin. *Chemical Geology*, **195**: 229–273.

Saltzman, M.R., 2002, Carbon isotope (δ^{13}C) stratigraphy across the Silurian-Devonian transition in North America: evidence for a perturbation of the global carbon cycle. *Palaeogeography, Palaeoclimatology, Palaeoecology*, **187**: 83–100.

Saltzman, M.R., 2005, Phosphorus, nitrogen, and the redox evolution of the Paleozoic oceans. *Geology*, **33** (7): 573–576.

Saltzman, M.R., and Thomas, E., 2012, Carbon isotope stratigraphy. *In* Gradstein, F.M., Ogg, J.G., Schmitz, M.D., and Ogg, G.M. (eds), *The Geologic Time Scale 2012*. Elsevier Publisher, **Vol. 1**: 207–232.

Sandberg, C.A., and Dreesen, R., 1984, Late Devonian icriodontid biofacies models and alternate shallow water conodont zonation, Special

Paper. *In* Clark, D.L. (ed), Conodont Biofacies and Provincialism. *Geological Society of America*, **196**: 143–178.

Sandberg, C.A., and Ziegler, W., 1973, Refinement of standard Upper Devonian conodont zonation based on sections in Nevada and West Germany. *Geologica et. Palaeontologica*, **7**: 97–122.

Sandberg, C.A., and Ziegler, W., 1999, Comments on proposed Frasnian and Famennian Subdivisions. *Subcommission on Devonian Stratigraphy Newsletter*, **15**: 43–46 [imprint 1998].

Sandberg, C.A., Ziegler, W., Leuteritz, K., and Brill, S.M., 1978, Phylogeny, speciation and zonation of *Siphonodella* (Conodonta, Upper Devonian and Lower Carboniferous). *Newsletters Stratigraphy*, **7**: 102–120.

Sandberg, C.A., Ziegler, W., Dreesen, R., and Butler, J.L., 1988, Late Frasnian mass extinction: conodont event stratigraphy, global changes, and possible causes. *Courier Forschungsinstitut Senckenberg*, **102**: 263–307.

Sandberg, C.A., Ziegler, W., and Bultynck, P., 1989a, New Standard Conodont Zones and Early *Ancyrodella* Phylogeny across Middle-Upper Devonian Boundary. *Courier Forschungsinstitut Senckenberg*, **110**: 195–230.

Sandberg, C.A., Poole, F.G., and Johnson, J.G., 1989b, Upper Devonian of western United States, (imprint 1988). *In* McMillan, N.J., Embry, A.F., and Glass, D.J. (eds), Devonian of the World. *Canadian Society of Petroleum Geology, Memoir*, **14(I)**: 183–220.

Sandberg, C.A., Morrow, J.R., and Ziegler, W., 2002, Late Devonian sea-level changes, catastrophic events, and mass extinctions. *In* Koeberl, C., and MacLeod, K.G. (eds), *Catastrophic Events and Mass Extinctions: Impacts and Beyond. Geological Society of America. Special Paper*, **356**, pp. 4783–487.

Sauerland, U., 1983, Dacryoconariden und Homocteniden der Givet- und Adorf-Stufe aus dem Rheinischen Schiefergebirge (Tentaculitoidea, Devon). *Göttinger Arbeiten zur Geologie und Paläontologie*, **25**: 1–86.

Savage, N.M., 1973, Lower Devonian conodonts from New South Wales. *Palaeontology*, **16** (2): 307–333.

Savage, N.M., Blodgett, R.B., and Jaeger, H., 1985, Conodonts and associated graptolites from the late Early Devonian of east-central Alaska and Western Yukon Territory. *Canadian Journal of Earth Sciences*, **22**: 1880–1883.

Savage, N.M., Sardsurd, A., and Buggisch, W., 2006, Late Devonian conodonts and the global Frasnian-Famennian extinction event, Thong Pha Phum, western Thailand. *Palaeoworld*, **15**: 171–184.

Schieber, J., and Over, D.J., 2005, Sedimentary infill of the Late Devonian Flynn Creek crater: a hard target marine impact. *In* Over, D.J., Morrow, J.R., and Wignall, P.B. (eds), Understanding Late Devonian and Permian-Triassic Biotic and Climatic Events: Towards an Integrated Approach. *Developments in Palaeontology and Stratigraphy*, **20**: 51–69.

Schindewolf, O.H., 1923, Beiträge zur Kenntnis des Paläozoikums in Oberfranken, Ostthüringen und dem Sächsischen Vogtlande. I. Stratigraphie und Ammoneenfauna des Oberdevons von Hof a. S. *Neues Jahrbuch für Mineralogie, Geologie und Paläontologie, Beilage-Band*, **49**: 250–357, 393–509, pls. 14-18.

Schindewolf, O.H., 1937, Zur Stratigraphie und Paläontologie der Wocklemer Schichten (Oberdevon). *Abhandlungen der Preussischen geologischen landesanstalt, Neue Folge*, **178**: 1–132.

Schindler, E., 1990, Die Kellwasser-Krise (hohe Frasne-Stufe, Ober-Devon). *Göttinger Arbeiten zur Geologie und Paläontologie*, **46**: 1–115.

Schindler, E., 1993, Event-stratigraphic markers within the Kellwasser crisis near the Frasnian/Famennian boundary (Upper Devonian) in Germany. *Palaeogeography, Palaeoclimatology, Palaeoecology*, **104**: 115–125.

Schindler, E., Brocke, R., Becker, R.T., Buchholz, P., Jansen, U., Luppold, F.W., et al., 2018, The Devonian in the Stratigraphic Table of Germany 2016. *Zeitschrift der Deutschen Gesellschaft für Geowissenschaften*, **168** (4): 447–463.

Schmidt, H., 1922, Das Oberdevon-Culm-Gebiet von Warstein i.W. und Belecke. *Jahrbuch der Preußischen Heologischen Landesanst.*, **41**: 254–339 (for 1920), pls. 12–13.

Schmidt, H., 1924, Zwei Cephalopodenfaunen an der Devon-Carbongrenze im Sauerland. *Jahrbuch der Preußischen Geologischen Landesanst.*, **44**: 98–171 (for 1923), pls. 6–8.

Schmidt, H., 1926, Beobachtungen über mitteldevonische Zonen-Goniatiten. *Senckenbergiana*, **8**: 291–295.

Schmitz, B., Ellwood, B.B., Peucker-Ehrenbrink, B., El Hassani, A., and Bultynck, P., 2006, Platinum group elements and $^{187}Os/^{188}Os$ in a purported impact ejecta layer near the Eifelian-Givetian stage boundary, Middle Devonian. *Earth and Planetary Science Letters*, **249**: 162–172.

Schmitz, B., Feist, R., Meier, M.M.M., Martin, E., Heck, P.R., Lenaz, D., et al., 2019, The micrometeoritic flux to Earth during the Frasnian-Famennian transition reconstructed in the Coumiac GSSP section, France. *Earth and Planetary Science Letters*, **522**: 234–243.

Schöne, B.R., 1997, Der *otomari*-Event und seine Auswirkungen auf die Fazies des Rhenoherzynischen Schelfs (Devon, Rheinisches Schieferge-birge). *Göttinger Arbeiten zur Geologie und Paläontologie*, **70**: 1–140.

Schönlaub, H.P., Attrep, M., Boeckelmann, K., Dreesen, R., Feist, R., Fenninger, A., et al., 1992, The Devonian/Carboniferous boundary in the Carnic Alps—a multidisciplinary approach. *Jahresbericht der Geologischen Bundesanstalt Wien.*, **135**: 57–98.

Schönlaub, H.P., Kreuzer, L., Joachimski, M.M., and Buggisch, W., 1994, Paleozoic boundary sections of the Carnic Alps (Southern Austria). *Erlanger Geologische Abhandlungen*, **122**: 77–103.

Schönlaub, H.P., Corradini, C., Corriga, M.G., and Ferretti, A., 2017, Chrono-, litho- and conodont bio-stratigraphy of the Rauchkofel Boden Section (Upper Ordovician – Lower Devonian), Carnic Alps, Austria. *Newsletters Stratigraphy*, **50** (4): 445–469.

Schröder, S., 2002, Rugose Korallen aus dem hohen Givetium und tiefen Frasnium (Devon) des Messinghäuser Sattels (Rheinisches Schieferge-birge/Sauerland). *Coral Research Bulletin*, **7**: 175–189.

Schülke, I., 1995, *Evolutive Prozesse bei* Palmatolepis *in der frühen Famenne-Stufe (Conodonta, Ober-Devon), Göttinger Arbeiten zur Geologie und Paläontologie*, **67**: 1–108.

Schülke, I., 1999, Conodont multielement reconstructions from the early Famennian (Late Devonian) of the Montagne Noire. *Geologica et Palaeontologica, SB*, **3**: 1–124.

Schwark, L., and Empt, P., 2006, Sterane biomarkers as indicators of Palaeozoic algal evolution and extinction events. *Palaeogeography, Palaeoclimatology, Palaeoecology*, **240**: 225–236.

Schwartzapfel, J.A., and Holdsworth, B.K., 1996, Upper Devonian and Mississippian radiolarian zonation and biostratigraphy of the Woodford, Sycamore, Caney and Goddard formations, Oklahoma. *Cushman Foundation for Foraminiferal Research, Special Publication*, **33**: 1–275.

Scotese, C.R., 2014, *Atlas of Devonian Paleogeographic Maps, PALEO-MAP Atlas for ArcGIS, Volume 4, The Late Paleozoic, Maps 65-72, Mollweide Projection*, PALEOMAP Project. Evanston, IL. https://doi.org/10.13140/2.1.1542.5280.

Sedgwick, A., and Murchison, R.I., 1839, Stratification of the older stratified deposits of Devonshire and Cornwall. *Philosophical Magazine, Series 3*, **14**: 241–260.

Sedgwick, A., and Murchison, R.I., 1840, On the physical structure of Devonshire, and on the subdivisions and geological relations of its older stratified deposits. *Transactions of the Geological Society of London, Series 2*, **5**: 633–704.

Selby, D., and Creaser, R.A., 2005, Direct radiometric dating of the Devonian-Mississippian time-scale boundary using the Re-Os black shale geochronometer. *Geology*, **33** (7): 545–548.

Sennikov, N.V., Shcherbanenko, T.A., Varaksina, I.V., Izokh, N.G., Sobolev, E.S., and Yazikov, A.Y., 2018, Biostratigraphy and Sedimentary Settings of the Middle Devonian Succession of the Yuryung-Tumus Peninsula, Khatanga Gulf of the Laptev Sea. *Stratigraphy Geological Correlation*, **26** (3): 267–282.

Shilo, N.A., Bouckaert, J., Afanasjeva, G.A., Bless, M.J.M., Conil, R., Erlanger, O.A., et al., 1984, Sedimentological and paleontological atlas of the late Famennian and Tournaisian deposits in the Omolon region (NE-USSR). *Annales de. la. Société Géologique de Belgique*, **107**: 137–247.

Simon, L., Goddéris, Y., Buggisch, W., Strauss, H., and Joachimski, M.M., 2007, Modeling the carbon and sulfur isotope compositions of marine sediments: climate evolution during the Devonian. *Chemical Geology*, **246**: 19–38.

Slavík, L., 2004a, A new conodont zonation of the Pragian Stage (Lower Devonian) in the stratotype area (Barrandian, central Bohemia). *Newsletters Stratigraphy*, **40** (1/2): 39–71.

Slavík, L., 2004b, The Pragian-Emsian conodont succession of the Barrandian area: search of an alternative to the GSSP polygnathid-based correlation concept. *Geobios*, **37**: 454–470.

Slavík, L., and Hladil, J., 2004, Lochkovian/Pragian GSSP revisited: evidence about conodont taxa and their stratigraphic distribution. *Newsletters on Stratigraphy*, **40** (3): 137–153.

Slavík, L., Valenzuela-Ríos, J.I., Hladil, J., and Carls, P., 2007, Early Pragian conodont-based correlations between the Barrandian area and the Spanish Central Pyrenees. *Geological Journal*. **42**: 499–512.

Slavík, L., Carls, P., Hladil, J., and Koptíková, L., 2012, Subdivision of the Lochkovian Stage based on conodont faunas from the stratotype area (Prague Synform, Czech Republic). *Geological Journal.*, **47**: 616–631.

Slavík, L., Valenzuela-Ríos, J.I., Hladil, J., Chadimová, L., Liao, J.-C., Hušková, A., et al., 2016, Warming or cooling in the Pragian? Sedimentary record and petrophysical logs across the Lochkovian-Pragian boundary in the Spanish Central Pyrenees. *Palaeogeography, Palaeoclimatology, Palaeoecology*, **449**: 300–320.

Slavík, L., Izokh, N., and Valenzuela-Ríos, J.I., 2017, Final report on results from the SDS fieldwork in Kitab State Geological Reserve, Tien-Shan Mts., Uzbekistan 2015. *Subcommission on Devonian Stratigraphy Newsletter*, **32**: 14–16.

Śliwiński, M.G., Whalen, M.T., and Day, J., 2010, Trace element variations in the Middle Frasnian *punctata* zone (Late Devonian) in the Western Canada sedimentary basin – changes in oceanic bioproductivity and paleoredox spurred by a pulse of terrestrial afforestation? *Geologica Belgica*, **13** (4): 459–482.

Śliwiński, M.G., Whalen, M.T., Newberry, R.J., Payne, J.H., and Day, J.E., 2011, Stable isotope $\delta^{13}C_{carb\ and\ org}$, $\delta^{15}N_{org}$ and trace element anomalies during the Late Devonian 'punctata' Event' in the Western Canada Sedimentary Basin. *Palaeogeography, Palaeoclimatology, Palaeoecology*, **307**: 245–271.

Śliwiński, M.G., Whalen, M.T., Meyer, F.J., and Majs, F., 2012, Constraining clastic input controls on magnetic susceptibility and trace element anomalies during the Late Devonian *punctata* Event in the Western Canada Sedimentary Basin. *Terra Nova*, **24**: 301–309.

Sobolev, D.B., and Soboleva, M.A., 2018, Reflection of global events Frasnian Epoch in the section of the western slope Polar Urals. *Lithosphere (Russia)*, **18** (3): 341–362 (in Russian with English summary).

Soboleva, M.A., Sobolev, D.B., and Matveeva, N.A., 2018, Frasnian section of the Kozhim River (the western slope of Polar urals) – results of biostratigraphic, bio- and lithofacies, isotopic and geochemical studies. *Neftegasovaâ geologiâ. Teoriâ i practika*, **13** (1): 1–55 (in Russian with English summary).

Song, H.Y., Song, H.J., Algeo, T.J., Tong, J.N., Romaniello, S.J., Zhu, Y.Y., et al., 2017, Uranium and carbon isotopes document global-ocean redox-productivity relationships linked to cooling during the Frasnian-Famennian mass extinction. *Geology*, **45** (10): 887–890.

Söte, T., Hartenfels, S., and Becker, R.T., 2017, Uppermost Famennian stratigraphy and facies development of the Reigern Quarry near Hachen (northern Rhenish Massif, Germany), Special Issue(3). *In* Mottequin, B., Slavík, L., and Königshof, P. (eds), Climate Change and Biodiversity Patterns in the Mid-Palaeozoic. *Palaeodiversity and Palaeoenvironments, 97*: 633–654.

Spalletta, C., Perri, M.C., Over, J.D., and Corradini, C., 2017, Famennian (Upper Devonian) conodont zonation: revised global standard. *Bulletin of Geosciences*, **92** (1): 31–57.

Sparling, D.R., 1983, Conodont biostratigraphy and biofacies of lower Middle Devonian limestones, north-central Ohio. *Journal of paleontology*, **57** (4): 825–864.

Springer, K.B., and Murphy, M.A., 1994, Punctuated stasis and collateral evolution in the Devonian lineage of *Monograptus hercynicus*. *Lethaia*, **27** (2): 119–128.

Steemans, P., 1989, Etude palynostratigraphique du Devonien Inferieur dans l'ouest de l'Europe. *Mémoire, Explication des Cartes Géologiques et Mineralogiques de Belgique*, **27**: 1–453.

Stein, W.E., Berry, C.M., Van Aller Hernick, L., and Mannolini, F., 2012, Surprisingly complex community discovered in the mid-Devonian fossil forest at Gilboa. *Nature*, **483**: 78–81.

Stephens, N.P., and Sumner, D.Y., 2003, Late Devonian carbon isotope stratigraphy and sea level fluctuations, Canning Basin, Western Australia. *Palaeogeography, Palaeoclimatology, Palaeoecology*, **191**: 203–219.

Stichling, S., Becker, R.T., Hartenfels, S., Aboussalam, Z.S., 2017, Conodont dating of reef drowning and extinction in the Hönne Valley (northern Rhenish Massif, Germany). In: Liao, J.-C., and Valenzuela-Ríos, J.I. (Eds.), Fourth International Conodont Symposium. ICOS IV "Progress on Conodont Investigations". *Cuadernos del Museo Geominero*, **22**: 163–165.

Streel, M., 2000, The late Famennian and early Frasnian datings given by Tucker and others (1998) are biostratigraphically poorly constrained. *Subcommission on Devonian Stratigraphy Newsletter*, **17**: 59.

Streel, M., 2005, Subdivision of the Famennian stage into four substages and correlation with the neritic and continental miospore zonation. *Subcommission on Devonian Stratigraphy Newsletter*, **21**: 14–17.

Streel, M., 2009, Upper Devonian miospore and conodont zone correlation in western Europe. *In* Königshof, P. (ed), Devonian Change: Case Studies in Palaeogeography and Palaeoecology. *The Geological Society of London, Special Publications*, **314**: 163–176.

Streel, M., 2015, Palynomorphs (miospores, acritarchs, prasinophytes) before and during the Hangenberg crisis. In: Mottequin, B., Denayer, J., Königshof, P., Prestianni, C., and Olive, S. (Eds.), IGCP 596 – SDS Symposium, Climate change and Biodiversity patterns in the Mid-Palaeozoic, September 20–22, 2015, Brussels – Belgium, Abstracts. *Strata, Travaux de Géologie sédimentaire et Paleontologie, Série 1: Communications*, **16**: 140–143.

Streel, M., and Loboziak, S., 1998, Proposal of boundaries for subdivision of the Famennian Stage: miospore implications. *Subcommission on Devonian Stratigraphy Newsletter*, **15**: 46–47.

Streel, M., and Loboziak, S., 2000, Correlation of the proposed conodont based Upper Devonian substage boundary levels into the neritic and terrestrial miospore zonation. *Subcommission on Devonian Stratigraphy Newsletter*, **17**: 12–14.

Streel, M., Higgs, K., Loboziak, S., Riegel, W., and Steemans, P., 1987, Spore stratigraphy and correlation with faunas and floras in the type marine Devonian of the Ardenne–Rhenish regions. *Review Palaeobotany and Palynology*, **50**: 211–229.

Streel, M., Brice, D., Degardin, J.-M., Derycke, C., Dreesen, R., Groessens, E., et al., 1998, Proposal for a Strunian substage and a subdivision of the Famennian Stage into four Substages. *Subcommission on Devonian Stratigraphy Newsletter*, **15**: 47–52.

Streel, M., Caputo, M.V., Loboziak, S., and Melo, J.H.G., 2000a, Late Frasnian-Famennian climates based on palynomorph analyses and the question of the Late Devonian glaciations. *Earth Science Reviews*, **52** (1–3): 121–173.

Streel, M., Loboziak, S., Steemanns, P., and Bultynck, P., 2000b, Devonian miospore stratigraphy and correlation with the global stratotype sections and points. *Courier Forschungsinstitut Senckenberg*, **220**: 9–23.

Streel, M., Brice, D., and Mistiaen, B., 2006, Strunian. *Geologica Belgica*, **9** (1/2): 105–109.

Streel, M., Caputo, M.V., Melo, J.H.G., and Perez-Leyton, M., 2013, What do latest Famennian and Mississippian miospores from South American diamictites tell us? *Palaeobiodiversity Palaeoenvironments*, **93**: 299–316.

Suttner, T.J., and Kido, E., 2016, Distinct sea-level fluctuations and deposition of a megaclast horizon in the neritic Rauchkofel Limestone (Wolayer area, Carnic Alps) correlate with the Lochkov-Prag Event, Special Publications. *In* Becker, R.T., Königshof, P., and Brett, C.E. (eds), Devonian Climate, Sea Level and Evolutionary Events. *London: Geological Society*, **423**: 11–23.

Suttner, T.J., Kido, E., Chen, X., Mawson, R., Waters, J.A., Frýda, J., et al., 2014, Stratigraphy and facies development of the marine Late Devonian near the Boulongour Reservoir, northwest Xinjiang, China. *Journal of Asian Earth Sciences*, **80**: 101–118.

Tagarieva, R.C., 2012, Conodont biodiversity of the Frasnian-Famennian boundary interval (Upper Devonian) in the Southern Urals. *Bulletin of Geosciences*, **88** (2): 297–314.

Talent, J.A., 1989, *Transgression-regression pattern for the Silurian and Devonian of Australia. Edwin Sherborn Hills Memorial Volume*. Carlton: Blackwell, pp. 201–219.

Talent, J.A., and Yolkin, E.A., 1987, Transgression-regression patterns for the Devonian of Australia and southern West Siberia. *Courier Forschungsinstitut Senckenberg*, **92**: 235–249.

Talent, J.A., Mawson, R., Andrew, A.S., Hamilton, P.J., and Whitford, D.J., 1993, Middle Palaeozoic extinction events: faunal and isotopic data. *Palaeogeography, Palaeoclimatology, Palaeoecology*, **104**: 139–152.

Talent, J.A., and Mawson, R., 1999, North-Eastern Molong Arch and Adjacent Hill End Trough (Eastern Australia): Mid-Palaeozoic Conodont Data and Implications. *Abhandlungen der Geologischen Bundesanst.*, **54**: 49–105.

Tel'nova, O.P., 2008, Palynological characterization of the Givetian-Frasnian deposits in the reference borehole Section 1-Balneologicheskaya (Southern Timan). *Stratigraphy and Geological Correlation*, **16**: 41–58.

Theiling, B.P., Elrick, M., and Asmerom, Y., 2017, Constraining the timing and provenance of trans-Laurentian transport using Nd and Sm isotopes from Silurian and Devonian marine carbonates. *Palaeogeography, Palaeoclimatology, Palaeoecology*, **466**: 392–405.

Tonarová, P., Vodrážková, S., Ferrová, L., de la Puente, G.S., Hints, O., Frýda, J., et al., 2017, Palynology, microfacies and biostratigraphy across the Daleje Event (Lower Devonian, lower to upper Emsian): new insights from the offshore facies of the Prague Basin, Czech Republic, (3). *In* Mottequin, B., Slavík, L., and Königshof, P. (eds), Climate Change and Biodiversity Patterns in the Mid-Palaeozoic. *Palaeodiversity and Palaeoenvironments*, **97**: 419–438.

Trapp, E., Kaufmann, B., Mezger, K., Korn, D., and Weyer, D., 2004, Numerical calibration of the Devonian-Carboniferous boundary: two new U-Pb isotope dilution-thermal ionization mass spectrometry single-zircon ages from Hasselbachtal (Sauerland, Germany). *Geology*, **32** (10): 857–860.

Trela, W., and Malec, J., 2007, Zapis $\delta^{13}C$ w osadach pogranicza dewonu I karbonu w poludniowej czesci Gor Swietokrzyskich. *Przeglad Geologiczny*, **55**: 411–415.

Tribovillard, N., Averbusch, O., Devleeschouwer, X., Racki, G., and Riboulleau, A., 2004, Deep-water anoxia over the Frasnian-Famennian boundary (La Serre, France): a tectonically induced oceanic anoxiv event? *Terra Nova*, **16**: 288–295.

Trinajstic, K., and George, A.D., 2009, Microvertebrate biostratigraphy of Upper Devonian (Frasnian) carbonate rocks in the Canning and Carnarvon basins of Western Australia. *Palaeontology*, **52** (3): 641–659.

Troth, I., Marshall, J.E.A., Racey, A., and Becker, R.T., 2011, Devonian sea-level change in Bolivia: a high palaeolatitude biostratigraphical calibration of the global sea-level curve. *Palaeogeography, Palaeoclimatology, Palaeoecology*, **304**: 3–20.

Tschigova, V., 1977, *Stratigrafya I korrelyatsya neftegazonosnych otloszheniy d3evona I karbona evropeckoiv chasti SSSR I zapubezhnich stran.* Moskva: Nedra, 194 pp., 47 pls.

Tucker, R.D., Bradley, D.C., Ver Straeten, C.A., Harris, A.G., Ebert, J.R., and McCutcheon, S.R., 1998, New U. Pb zircon ages and the duration and division of Devonian time. *Earth and Planetary Science Letters*, **158** (3–4): 175–186.

Tulipani, S., Grice, K., Greenwood, P.F., Haines, P.W., Sauer, P.E., Schimmelmann, A., et al., 2015, Changes of palaeoenvironmental conditions recorded in Late Devonian reef systems from the Canning Basin, Western Australia: a biomarker and stable isotope approach. *Gondwana Research*, **28**: 1500–1515.

Turgeon, S.C., Creaser, R.A., and Algeo, T.J., 2007, Re-Os depositional ages and seawater Os estimates for the Frasnian-Famennian boundary: implications for weathering rates, land plant evolution, and extinction mechanisms. *Earth and Planetary Science Letters*, **261**: 649–661.

Turnau, E., and Narkiewicz, K., 2011, Biostratigraphical correlation of spore and conodont zonations within Givetian and Frasnian of the Lublin area (SE Poland). *Review of Palaeobotany and Palynology*, **164**: 30–38.

Umeda, M., 1998, Upper Silurian-Middle Devonian radiolarian zones of the Yokokurayama and Knomori areas in the Kurosegawa Belt, Southwest Japan. *The Island Arc*, **7**: 637–646.

Uveges, B.T., Junium, C.K., Boyer, D.L., Cohen, P.A., and Day, J.E., 2019, Biogeochemical controls on black shale deposition during the Frasnian-Famennian biotic crisis in the Illinois and Appalachian Basins, USA, inferred from stable isotopes of nitrogen and carbon. *Palaeogeography, Palaeoclimatology, Palaeoecology*, **531**, Part A, article #108787: 14 pp. <https//doi.org/10.1016/j.palaeo.2018.05.031>.

Vacek, F., Hladil, J., and Schnabl, P., 2010, Stratigraphic correlation potential of magnetic susceptibility and gamma-ray spectrometric variations in calciturbiditic facies mosaics (Silurian-Devonian boundary, Barrandian area, Czech Republic). *Geologica Carpathica*, **61**: 257–272.

Valenzuela-Ríos, J.I., 1990, Lochkovian conodonts and stratigraphy at Gerri de la Sal (Pyrennes). *Courier Forschunginstitut Senckenberg*, **118**: 53–63.

Valenzuela-Ríos, J.I., and Murphy, M.A., 1997, A new zonation of middle Lochkovian (Lower Devonian) conodonts and evolution of *Flajsella* n. gen. (Conodonta). *Geological Society of America, Special Paper*, **321**: 131–144.

Valenzuela-Ríos, J.I., and García-López, S., 1998, Using conodonts to correlate abiotic events: an example from the Lochkovian (Early Devonian) of NE Spain. *Palaeontologia Polonica*, **58**: 191–199.

Valenzuela-Ríos, J.I., Slavík, L., Liao, J.-C., Calvo, H., Hušková, A., and Chadimová, L., 2015, The middle and upper Lochkovian (Lower Devonian) conodont successions in key peri-Gondwana localities (Spanish Central Pyrenees and Prague Synform) and their relevance for global correlations. *Terra Nova*, **27**: 409–415.

Valenzuela-Ríos, J.I., Carls, P., Martínez-Pérez, C., Liao, J.-C., 2018, Conodont sequences spanning the Pragian/Emsian transition (Lower Devonian) in NE Spain (Pyrenees and Iberian Chains) and its relevance for the redefinition of the base of the Emsian. *Fifth International Palaeontological Congress, July 9–13, 2018, France, Abstract Book*, pp. 812.

Van Geldern, R., Joachimski, M.M., Day, J., Jansen, U., Alvarez, F., Yolkin, E.A., et al., 2006, Carbon, oxygen and strontium isotope records of Devonian brachiopod shell calcite. *Palaeogeography, Palaeoclimatology, Palaeoecology*, **240**: 47–67.

Van Hengstum, P.J., and Gröcke, D.R., 2008, Stable isotope record of the Eifelian-Givetian boundary Kačák-*otomari* Event (Middle Devonian) from Hungry Hollow, Ontario, Canada. *Canadian Journal of Earth Sciences*, **45**: 353–366.

Veimarn, A.N., Puchkov, V.N., Abramova, A.N., Artyushkova, O.V., Baryshev, V.N., Degtyaryov, K.E., et al., 2004, Stratigraphy and geological events at the Frasnian-Famennian boundary in the Southern Urals. *Geological Quarterly*, **48** (3): 233–244.

Veizer, J., Buhl, D., Diener, A., Ebneth, S., Podlaha, O.G., Bruckschen, P., et al., 1997, Strontium isotope stratigraphy: potential resolution and event correlation. *Palaeogeography, Palaeoclimatology, Palaeoecology*, **132**: 65–77.

Ver Straeten, C.A., 2007, Basinwide stratigraphic synthesis and sequence stratigraphy, upper Pragian, Emsian and Eifelian stages (lower to Middle Devonian), Appalachian Basin, Special Publications. *In* Becker, R.T., and Kirchgasser, W.T. (eds), Devonian Events and Correlations. *London: Geological Society*, **278**: 39–81.

Ver Straeten, C.A., 2009, Devonian T-R Cycle 1b: The "Lumping" of Emsian sea-level history. In: Over, D.J. (Ed.), Studies in Devonian Stratigraphy: Proceedings of the 2007 International Meeting of the

Subcommission on Devonian Stratigraphy and IGCP 499. *Palaeontographica Americana*, **63**: 33–47.

Vodrážková, S., Klapper, G., and Murphy, M.A., 2011, Early Middle Devonian conodont faunas (Eifelian, *costatus-kockelianus* zones) from the Roberts Mountains and adjacent areas in central Nevada. *Bulletin of Geosciences*, **86** (4): 737–764.

Vodrážková, S., Frýda, J., Suttner, T.J., Koptíková, L., and Tonarová, P., 2013, Environmental changes close to the Lower-Middle Devonian boundary: the Basal Choteč Event in the Prague Basin (Czech Republic). *Facies*, **59**: 425–449.

Vorontzova, T.N., 1993, The genus *Polygnathus* sensu lato (Conodonta): phylogeny and systematics. *Paleontologicheskiy Zhurnal*, **1993** (3): 66–78, pl. VIII.

Walliser, O.H., 1984, Geologic processes and global events. *Terra Cognita*, **4**: 17–20.

Walliser, O.H., 1996, Global events in the Devonian and Carboniferous. In Walliser, O.H. (ed), *Global Events and Event Stratigraphy in the Phanerozoic*. Berlin: Springer-Verlag, pp. 225–250.

Walliser, O.H., 2000, The Eifelian-Givetian Stage boundary. *Courier Forschungsinstitut Senckenberg*, **225**: 37–47.

Walliser, O.H., and Bultynck, P., 2011, Extinctions, survival and innovations of conodont species during the Kačák Episode (Eifelian-Givetian) in south-eastern Morocco. *Bulletin de l'Institut royal des Sciences naturelles de Belgique, Sciences de la Terre*, **81**: 5–25.

Walliser, O.H., Bultynck, P., Weddige, K., Becker, R.T., and House, M.R., 1995, Definition of the Eifelian-Givetian Stage boundary. *Episodes*, **18**: 107–115.

Wang, X.F., 1980, On the distribution, zonation and correlation of graptolite-bearing Silurian and Early Devonian in China. *Chinese Academy of Geological Sciences, Bulletin, Series VII*, **1**: 23–36 (in Chinese with English Summary).

Wang, S.Q., 1989, Pelagic ostracods from Givetian to Tournaisian in South China. *Bulletin of the Nanjung Institute of Geology and Palaeontology, Academia Sinica*, **9**: 1–80, 16 pls.

Wang, K., 1992, Glassy Microspherules (Microtectites) from an Upper Devonian Limestone. *Science*, **256**: 1547–1550.

Wang, K., and Bai, S.L., 1989, Faunal changes and events near the Frasnian-Famennian boundary of South China. *In* McMillan, N.J., Embry, A.F., and Glass, D.J. (eds), Devonian of the World. *Canadian Society of Petroleum Geologists, Memoir*, **14**(III): 71–78.

Wang, D.-R., and Bai, Z.Q., 2002, Chemostratigraphic characters of the Middle-Upper Devonian boundary in Guangxi, South China. *Journal of Stratigraphy*, **26**: 50–54 (in Chinese).

Wang, K., and Geldsetzer, H.H.J., 1995, Late Devonian Conodonts define the precise horizon of the Frasnian–Famennian boundary at Cinquefoil Mountain, Jasper, Alberta. *Canadian Journal of Earth Sciences*, **32**: 1825–1834.

Wang, C.Y., and Xia, W., 2002, Geochemical Properties and Stratigraphical Correlation of Frasnian-Famennian Transitional Strata in Wuzishan Section. *Journal of China University of Geoscuiences*, **13** (1): 48–52.

Wang, Y.X., and Yang, J.D., 1988, Paleoenvironment deduced from trace element geochemistry. *In* Yu, C.-M. (ed), *Devonian-Carboniferous Boundary in Nanbiancun, Guilin, China – Aspects and Records*. Beijing: Science Press, pp. 68–98.

Wang, X.F., and Zao, Z.G., 1984, Early Devonian graptolite faunas from Yulin, Guangxi. *Geological Reviews*, **30**: 416–429.

Wang, C.Y., and Ziegler, W., 2002, The Frasnian-Famennian Conodont Mass Extinction and Recovery in South China. *Senckenbergiana lethaea*, **82** (2): 463–493.

Wang, K., Orth, C.J., Attrep Jr., M., Chatterton, B.D.E., Hou, H., and Geldsetzer, H.H., 1991, Geochemical evidence for a catastrophic biotic event at the Frasnian-Famennian boundary in south China. *Geology*, **19**: 776–779.

Wang, K., Geldsetzer, H.H.J., and Chatterton, B.D.E., 1994, *A Late Devonian Extraterrestrial Impact and Extinction in Eastern Gondwana: Geochemical, Sedimentological, and Faunal Evidence*, 293. Geological Society of America, pp. 111–120, Special Paper.

Wang, K., Geldsetzer, H.H.J., Goodfellow, W.D., and Krouse, H.R., 1996, Carbon and sulfur isotope anomalies across the Frasnian-Famennian extinction boundary, Alberta, Canada. *Geology*, **24** (2): 187–191.

Wang, C.Y., Chuluun, M., Weddige, K., Ziegler, W., Gonchigdorj, S., Jugdernamjil, M., et al., 2005, Devonian (Emsian-Eifelian) conodonts from South Gobi, Mongolia. *Acta Micropalaeontologica Sinica*, **22** (1): 19–28.

Wang, Y.J., Fang, Z.J., Yang, Q., Zhou, Z.C., Cheng, Y.N., Duan, Y.X., et al., 2000, Middle-Late Devonian strata of cherty facies and radiolarian faunas from West Yunnan. *Acta Micropalaeontologica Sinica*, **17** (3): 235–254.

Wang, Z.H., Becker, R.T., Aboussalam, Z.S., Hartenfels, S., Joachimski, M.M., and Gong, Y.M., 2016, Conodont and carbon isotope stratigraphy near the Frasnian/Famennian (Devonian) boundary at Wulankeshun, Junggar Basin, NW China. *Palaeogeography, Palaeoclimatology, Palaeoecology*, **448**: 279–297.

Wang, X., Liu, S.A., Wang, Z.G., Chen, D.Z., and Zhang, L.Y., 2018, Zinc and strontium isotope evidence for climate cooling and constraints on the Frasnian-Famennian (~372 Ma) mass extinction. *Palaeogeography, Palaeoclimatology, Palaeoecology*, **498**: 68–82.

Ward, D., Becker, R.T., Aboussalam, Z.S., Rytina, M., and Stichling, S., 2013, The Devonian at Oued Ferkla (Tinejdad region, SE Morocco). In: Becker, R.T., El Hassani, A., and Tahiri, A. (Eds.), *International Field Symposium "The Devonian and Lower Carboniferous of northern Gondwana", March 22–29, 2013, Field Guidebook*, Vol. 27. Rabat: Document de l'Institut Scientifique, pp. 23–29.

Warme, J.E., and Sandberg, C.A., 1996, Alamo Megabreccia: record of a Late Devonian impact in southern Nevada. *Geology Today*, **6** (1): 1–7.

Weddige, K., 1977, Die Conodonten der Eifel-Stufe im Typusgebiet und in benachbarten Faziesgebieten. *Senckenbergiana Lethaea*, **58**: 271–419.

Weddige, K., 1982, The Wetteldorf Richtschnitt as boundary stratotype from the view point of conodont stratigraphy. *Courier Forschungsinstitut Senckenberg*, **55**: 26–37.

Weddige, K. (ed), 1996. Devon-Korrelationstabelle. *Senckenbergiana Lethaea* **76**: 267–286.

Wedekind, R., 1908, Die Cephalopodenfauna des höheren Oberdevon am Enkeberge. *Neues Jahrbuch für Mineralogie, Geologie und Paläontologie, Beilage-Band*, **26**: 565–633, pls. 39–45.

Wedekind, R., 1913, Die Goniatitenkalke des unteren Oberdevon von Martenberg bei Adorf. *Sitzungsberichte der Gesellschaft Naturforschender Freunde zu Berlin*, **1913**: 23–77, pls. 4–7.

Wedekind, R., 1914, Monographie der Clymenien des Rheinischen Gebirges. Abhandlungen der Königlichen Ges. *Der Wissenschaften zu Göttingen, mathematisch-Physikalische Klasse, Neue Folgen*, **10** (1): 173, pls. 1–7.

Wedekind, R., 1920, *Geologie des Rheinischen Schiefergebirges. 1. Lieferung*. Marburg: R. Friedrich's Universitäts-Buchdruckerei, 32 pp.

Wei, J.Y., and Ji, Q., 1989, Stable isotopes of carbon and oxygen. *In* Ji, Q. (ed), *The Dapoushang Section, An Excellent Section for the Devonian-*

Carboniferous Boundary Stratotype in China. Beijing: Science Press, pp. 48–52.

Weiner, T., and Kalvoda, J., 2016, Biostratigraphic and sedimentary record of the *Annulata* Events in the Moravian Karst (Famennian, Czech Republic). *Facies*, **62** (6): 25.

Weiner, T., Kalvoda, J., Kumpan, T., Schindler, E., and Šimíček, D., 2017, An Integrated Stratigraphy of the Frasnian-Famennian Boundary Interval (Late Devonian) in the Moravian Karst (Czech Republic) and Kellerwald (Germany). *Bulletin of Geosciences*, **92** (2): 257–281.

Whalen, M.T., and Day, J.E., 2008, Magnetic susceptibility, biostratigraphy, and sequence stratigraphy: insights into Devonian carbonate platform development and basin infilling, western Alberta, Canada. *SEPM Special Publication*, **89**: 291–314.

Whalen, M.T., and Day, J.E., 2010, Cross-basin variations in magnetic susceptibility influenced by changing sea level, paleogeography, and paleoclimate: upper Devonian, Western Canada. *Journal of Sedimentary Research*, **80**: 1109–1127.

Whalen, M.T., Day, J., Eberli, G.P., and Homewood, P.W., 2002, Microbial carbonates as indicators of environmental change and biotic crisis in carbonate systems: examples from the Late Devonian, Alberta basin, Canada. *Palaeogeography, Palaeoclimatology, Palaeoecology*, **181**: 127–151.

Whalen, M.T., Śliwiński, M.G., Payne, J.H., Day, J.E., Chen, D., and Da Silva, A.C., 2015, *Chemostratigraphy and magnetic susceptibility of the Late Devonian Frasnian–Famennian transition in western Canada and southern China. Implications for carbon and nutrient cycling and mass extinction*, Geological Society, London, Special Publications, **414**, pp. 37–72.

Whalen, M.T., De Vleeschouwer, D., Payne, J.H., Day, J.E., Over, D.J., and Claeys, P., 2016, *Pattern and timing of the Late Devonian biotic crisis in Western Canada: insights from carbon isotopes and astronomical calibration of magnetic susceptibility data, New Advances in Devonian Carbonates: Outcrop, Reservoirs, and Chronostratigraphy*, 107. SEPM Special Publication, pp. 185–201.

White, E.I., 1950, The vertebrate faunas of the Lower Old Red sandstone of the Welsh Borders. *Bulletin of the British Museum of Natural History (Geology)*, **1**: 51–67.

White, D.A., Elrick, M., Romaniello, S., and Zang, F., 2018, Global seawater redox trends during the Late Devonian mass extinction detected using U isotopes of marine limestones. *Earth and Planetary Science Letters*, **503**: 68–77.

Wicander, R., Clayton, G., Marshall, J.E.A., Troth, I., and Racey, A., 2011, Was the latest Devonian glaciation a multiple event? New palynological evidence from Bolivia. *Palaeogeography, Palaeoclimatology, Palaeoecology*, **305**: 84–92.

Williams, E.A., Friend, P.F., and Williams, P.J., 2000, A review of Devonian time scales: databases, construction and new data. *In* Friend, P.F., and Williams, B.P.J. (eds), New Perspectives on the Old Red Sandstone. *Geological Society of London, Special Publications*, **180**: 1–21.

Winter, J., 2015, Vulkanismus und Kellwasser-Krise – Zirkon-Tephrostratigraphie, Identifizierung und Herkunft distaler Fallout-Aschenlagen (Oberdevon, Synklinorium von Dinant, Rheinisches Schiefergebirge, Harz). *Zeitschrift der Deutschen Gesellschaft für Geowissenschaften*, **166**: 227–251.

Wittekindt, H., 1966, Zur Conodontenchronologie des Mitteldevons. Fortschritte in der. *Geologie von Rheinland und Westfalen*, **9**: 621–646 (imprint 1965).

Witzke, B.J., Bunker, B.J., and Rogers, F.S., 1989, Eifelian through Lower Frasnian stratigraphy and deposition in the Iowa area, central midcontinent, U.S.A, (imprint 1988). *In* McMillan, N.J., Embry, A.F., and Glass, D.J. (eds), Devonian of the World. *Calgary: Canadian Society of Petroleum Geologists*, **Vol. I**: 221–250.

Xu, D.Y., Yan, Z., Zhang, Q.W., Shen, Z.D., Sun, Y.Y., and Ye, L.F., 1986, Significance of a $\delta^{13}C$ anomaly near the Devonian/Carboniferous boundary at the Muhua section, South China. *Nature*, **321**: 854–855.

Xu, B., Gu, Z.Y., Wang, C.Y., Hao, Q.Z., Han, J.T., Liu, Q., et al., 2012, Carbon isotopic evidence for the association of decreasing atmospheric CO_2 level with the Frasnian-Famennian mass extinction. *Journal of Geophysical Research*, **117**: 1–2. https://doi.org/10.1029/2011JG001847.

Yans, J., Corfield, R., Racki, G., and Préat, A., 2007, Evidence for perturbations of the carbon cycle in the Middle Frasnian punctata Zone (Late Devonian). *Geological Magazine*, **144**: 263–370.

Yao, L., Aretz, M., Chen, J.T., Webb, G.E., and Wang, X.D., 2016, Global microbial carbonate proliferation after the end-Devonian mass extinction: mainly controlled by demise of skeletal bioconstructors. *Scientific Reports*, **6**: 9. https://doi.org/10.1038/srep39694.

Yatskov, S.V., and Kuźmin, A.V., 1992, On the Relationships between the Ammonoid and Conodont Assemblages in the Lower Frasnian Deposits of Southern Timan. *Bulletin MOIP, Otdelenie Geologitscheskoe*, **67** (1): 85–90 (in Russian).

Yolkin, E.A., Weddige, K., Izokh, N.G., and Erina, M.V., 1994, New Emsian conodont zonation (Lower Devonian). *Courier Forschungsinstitut Senckenberg*, **168**: 139–157.

Yolkin, E.A., Kim, A.I., Weddige, K., Talent, J.A., and House, M.R., 1998, Definition of the Pragian/Emsian Stage Boundary. *Episodes*, **20**: 235–240.

Yolkin, E.A., Kim, A.I., and Talent, J.A. (eds), 2008. Devonian Sequences of the Kitab Reserve Area. Field Excursion *Guidebook, International Conference "Global Alignments of Lower Devonian Carbonate and Clastic Sequences" (SDS/IGCP 499 Project Joint Field Meeting), Kitab State Geological Reserve, Uzbekistan*. Publishing House of SB RAS. Novosibirsk, August 25–September 3, 2008. 97 pp.

Yolkin, E.A., Izokh, N.G., Weddige, K., Erina, V., Valenzuela-Rios, J.I., and Apekina, L.S., 2011, Eognathodid and polygnathid lineages from the Kitab State Geological Reserve sections (Zeravshan-Gissar mountainous area, Uzbekistan) as the bases for improvements of the Pragian-Emsian standard conodont zonation. *News on Paleontology and Stratigraphy, Supplement to Geologiya I Geofizika*, **52** (15): 37–47.

Young, G.C., 2000, Flawed timescale, or flawed logic. *Australian Geologist*, **115**: 6.

Young, G.C., and Laurie, J.R., 1996, *An Australian Phanerozoic Timescale*. Oxford: Oxford University Press, 279 pp.

Yu, C.M. (ed), 1988. Devonian-Carboniferous Boundary in Nanbiancun, Guilin, China – Aspects and Records. Science Press. Beijing, 379 pp.

Yu, C.M., Qie, W.K., and Lu, J.F., 2018, Emsian (Early Devonian) Yujiang Event in South China. *Palaeoworld*, **27**: 53–65.

Yudina, A.N., Racki, G., Savage, N.M., Racka, M., and Malkowski, K., 2002, The Frasnian-Famennian events in a deep-shelf succession, Subpolar Urals: biotic, depositional, and geochemical records. *Acta Palaeontologica Polonica*, **47** (2): 355–372.

Zambito, J., and Day, J., 2015, Integrated stratigraphic analysis of the Middle Devonian (late Givetian) Geneseo Event in the Appalachian and Michigan basins. *In* Mottequin, B., Denayer, J., Königshof, P., Prestianni, C., and Olive, S. (eds), IGCP 596 – SDS Symposium, Climate Change and

Biodiversity Patterns in the Mid-Palaeozoic, September 20–22, 2015, Brussels – Belgium, Abstracts. *Strata, Travaux de Géologie sédimentaire et Paleontologie, Série 1: communications,* **16**: 152–153.

Zambito, J.J., Joachimski, M.M., Brett, C.E., Baird, G.C., and Aboussalam, Z.S., 2016, A carbonate carbon isotope record for the late Givetian (Middle Devonian) Global Taghanic Biocrisis in the type region (northern Appalachian Basin, Special Publications. *In* Becker, R.T., Königshof, P., and Brett, C.E. (eds), Devonian Climate, Sea Level and Evolutionary Events. *Geological Society, London, Special Publications,* **423**: 223–233.

Zeng, J.W., Xu, R., and Gong, Y.M., 2010, Geochemistry of the Late Devonian F-F Transitional Rare Earth Elements in the Yangdi Section from Guilin, South China. *Journal of Earth Science,* 21: 96–98.

Zeng, J.W., Xu, R., and Gong, Y.M., 2011, Hydrothermal activities and seawater acidification in the Late Devonian F-F transition: evidence from geochemistry of rare earth elements. *Science China, Earth Sciences,* **54** (4): 540–549.

Zhang, N., Xia, W., Dong, Y., and Shang, H., 2008, Conodonts and radiolarians from pelagic cherts of the Frasnian-Famennian boundary interval at Bancheng, Guangxi, China: global recognition of the upper Kellwasser event. *Marine Micropaleontology,* **67**: 180–190.

Zhang, M., Becker, R.T., and Ma, X.P., 2018, Hangenberg Black Shale with cymaclymeniid ammonoids in the terminal Devonian of South China. *Palaeodiversity Palaeoenvironments,* (online).14 pp. <https//doi.org/10.1007/s12549-018-0348-x>.

Zhao, W.J., Wang, N.Z., Zhu, M., Mann, U., Herten, U., and Lücke, A., 2011, Geochemical stratigraphy and microvertebrate assemblage sequences across the Silurian/Devonian transition in South China. *Acta Geologica Sinica,* **85** (2): 340–353.

Zhao, W.J., Jia, G.D., Zhu, M., and Zhu, Y.A., 2015, Geochemical and palaeontological evidence for the definition of the Silurian/Devonian boundary in the Changwantang Section, Guangxi Province, China. *Estonian Journal of Earth Sciences,* **64** (1): 110–114.

Zheng, Y., Hou, H.F., and Ye, L.F., 1993, Carbon and oxygen isotope event markers near the Frasnian-Famennian boundary, Louxiu section, South China. *Palaeogeography, Palaeoclimatology, Palaeoecology,* **104**: 97–104.

Ziegler, W., 1962, Taxionomie und Phylogenie Oberdevonischer Conodonten un ihre stratigraphische Bedeutung. *Abhandlungen des Hessischen Geologischen Landesamtes für Bodenforschung.,* **38**: 1–166.

Ziegler, W., 1966, Eine Verfeinerung der Conodontengliederung an der Grenze Mittel-/Oberdevon. *Fortschritte in der Geologie von Rheinland und Westfalen,* **9**: 647–676 (imprint 1965).

Ziegler, W., 1969, Eine neue Conodontenfauna aus dem höchsten Oberdevon. *Fortschritte in der Geologie von Rheinland und Westfalen,* **17**: 179–191.

Ziegler, W., 1971. Conodont stratigraphy of the European Devonian. Geological Society of America Memoir, **127**: 227–268

Ziegler, W., 2000, The Lower Eifelian Boundary. *Courier Forschungsinstitut Senckenberg,* **225**: 27–36.

Ziegler, W., and Klapper, G., 1982, The *disparilis* conodont Zone, the proposed level for the Middle-Upper Devonian boundary. *Courier Forschungsinstitut Senckenberg,* **55**: 463–492.

Ziegler, W., and Lane, H.R., 1989, Cycles in conodont evolution from Devonian to mid-Carboniferous. *In* Aldridge, R.J. (ed), *Palaeobiology of Conodonts.* Chichester: Horwood Press, pp. 147–164.

Ziegler, W., and Sandberg, C.A., 1984, *Palmatolepis*-based revision of upper part of standard Late Devonian conodont zonation, Special Paper. *In* Clark, D.L. (ed), Conodont Biofacies and Provincialism. *Geological Society of America,* **196**: 179–194.

Ziegler, W., and Sandberg, C.A., 1990, The Late Devonian Standard Zonation. *Courier Forschungsinstitut Senckenberg,* **121**: 1–115.

Ziegler, W., and Sandberg, C.A., 1997, Proposal of boundaries for a late Frasnian Substage and for subdivision of the Famennian Stage into three Substages. *Subcommission on Devonian Stratigraphy Newsletter,* **14**: 11–12.

Ziegler, W., and Sandberg, C.A., 1998, Comments on proposed Frasnian and Famennian Subdivisions. *Subcommission on Devonian Stratigraphy Newsletter,* **15**: 43–46.

Ziegler, W., and Werner, R. (eds), 1982. On Devonian Stratigraphy and Palaeontology of the Ardenno-Rhenish mountains and related Devonian matters. *Courier Forschungsinstitut Senckenberg* **55**: 1–505.

Ziegler, W., Klapper, G., and Johnson, J.G., 1976, Redefinition and subdivision of the *varcus*-Zone (Conodonts, Middle-?Upper Devonian) in Europe and North America. *Geologica et Palaeontologica,* **10**: 109–140.

Zong, P., Becker, R.T., and Ma, X., 2015, Upper Devonian (Famennian) and Lower Carboniferous (Tournaisian) ammonoids from western Junggar, Xinjiang, northwestern China – stratigraphy, taxonomy and palaeobiogeography. *Palaeobiodiversity Palaeoenvironments,* **95** (2): 159–202.

M. Aretz, H.G. Herbig and X.D. Wang
With contributions by F.M. Gradstein, F.P. Agterberg and J.G. Ogg

Chapter 23

The Carboniferous Period

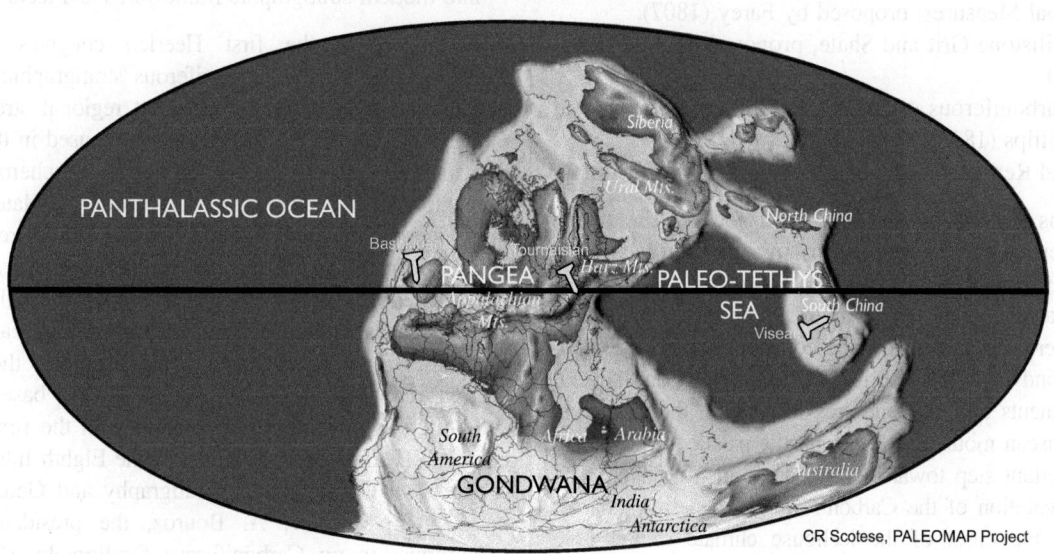

306Ma Carboniferous

CR Scotese, PALEOMAP Project

Chapter outline

23.1 History and subdivisions   811
     23.1.1 Mississippian Subsystem (Lower Carboniferous)   815
     23.1.2 Pennsylvanian Subsystem (Upper Carboniferous)   819
     23.1.3 Regional Carboniferous subsystem, stage, and
           substage units   823
23.2 Carboniferous stratigraphy   836
     23.2.1 Biostratigraphy   836
     23.2.2 Magnetostratigraphy   843
     23.2.3 Chemostratigraphy   843

23.2.4 Sequence stratigraphy and cyclostratigraphy   845
23.2.5 Physical stratigraphy: further tools   847
23.2.6 Paleoclimate and global correlation   847
23.3 Carboniferous time scale   848
     23.3.1 Previous scales   848
     23.3.2 Radioisotopic dates   848
     23.3.3 Carboniferous and Permian composite standard   853
     23.3.4 Ages of stage boundaries   853
Bibliography   855

Abstract

The Carboniferous Period experienced recovery and major diversification in the marine and terrestrial realms after the end-Devonian mass extinctions. Marked evolutions took place in the terrestrial realm with the development of extended forests and the appearance of reptiles and the first flying organisms. Strong faunal provincialism resulted from geodynamic and climatic changes during the Visean. The Carboniferous Period is also characterized by the onset of the Late Paleozoic Ice Age with characteristic glacial—interglacial cycles, the closure of the western Paleotethys Ocean, and the Variscan orogeny.

23.1 History and subdivisions

The Carboniferous is the stratigraphic period associated to "coal." Its name is derived from the Latin *carbo* (charcoal) and *ferronus* (i.e., bearing) or the Italian *Carbonarium* (charcoal producer). The extensive Carboniferous coal formations were fueling the industrial revolution of the late 18th century in Western Europe, which subsequently spread around the globe. It is difficult to exactly pinpoint the beginning of stratigraphic work in Carboniferous strata, but first stratigraphic

Geologic Time Scale 2020. DOI: https://doi.org/10.1016/B978-0-12-824360-2.00023-1

ideas aroused independently at the end of the 18th century in Belgium (de Witry, 1776/1780) and in Great Britain (Whitehurst, 1778). The use of the term Carboniferous as formalized stratigraphic unit was by Farey (1811) who introduced "coral measures or Carboniferous strata" in Great Britain (see Ramsbottom, 1984). In other regions, for example, Belgium, the connection to coals is found at the same time in names as "Bituminiferous Formation" and "Terrain Anthraxifère" by d'Omalius d'Halloy (1808, 1828), respectively. Conybeare and Phillips (1822) introduced the Medal or Carboniferous Order to regroup four stratigraphic units, which were in descending order:

- the Coal Measures, proposed by Farey (1807),
- the Millstone Grit and Shale, proposed by Whitehurst (1778),
- the Carboniferous or Mountain Limestone, first listed by Phillips (1818), and
- the Old Red Sandstone, later assigned to the Devonian.

Phillips (1835) is the first to associate the terms Carboniferous and System following the then new concept of Murchison (1835) for the Silurian.

The Carboniferous was a time of important changes in the biosphere and geosphere. The collision of two large continents, Gondwana and Laurussia (Euramerica), and several microcontinents and volcanic arcs resulted in the formation of the Variscan mountain belt along the paleoequator, which is an important step toward the supercontinent Pangea. The climatic evolution of the Carboniferous is generally characterized by the onset of the icehouse climate of the Late Paleozoic Ice Age (LPIA) with superposed abundant smaller scaled glacial and interglacial cycles. The evolution of the marine biosphere is marked by an important diversification in the aftermath of the end-Devonian extinction events. However, most remarkable is the development of the terrestrial biosphere. The colonization of the landmasses reached a first peak with the development of extensive forests, the origination of reptiles, and the conquest of the air by arthropods.

The Carboniferous is one of the most confusing and complicated geologic periods in terms of stratigraphic classification (Davydov et al., 2012). This is rooted not only in its complex geological, biological, and climatic history, but also in the approaches of the different Carboniferous stratigraphic communities. The major contributing factors include the following:

- The faunal and floral provincialisms are high as consequences of important geographical and climatic boundaries during the Variscan Orogeny and formation of the supercontinent Pangea and an icehouse climate (LPIA).
- The Carboniferous is one of the first established geologic periods, but the discussion of how to subdivide it is almost as old. This is well illustrated in the traditional subdivisions in Europe, Russia, and North

America already on the stratigraphic level today called subsystem.

- The exploration of coal basins has driven long time the stratigraphic work in the Carboniferous, and many detailed regional schemes have been developed. However, these often concern terrestrial or only partly marine records, which have only partly successfully transferred into modern chronostratigraphic subdivisions.
- The decline of the economic activities in the coal basins, especially in Europe, resulted in considerable shrinkage of the Carboniferous stratigraphic community. This still results in massive problems to integrate the older records into modern stratigraphic frameworks and technics.

Starting with the first Heerlen congress in 1927 (Jongmans, 1928), the Carboniferous stratigraphic community attempted to unify the different regional stratigraphic scales (see Section 23.1.3). This work focused in the beginning on the western and central European schemes, but a more global perspective started to develop at latest at the third Heerlen congress in 1951, when the correlation of North American and European stratigraphic classifications was debated. It is interesting to note that what will be later called boundary stratotype concept was used already in the first half of the 20th century; for example, the second Heerlen congress in 1935 defined the base of the Carboniferous in a stratotype section with the first appearance data (FADs) of a goniatite. At the Eighth International Congress on Carboniferous Stratigraphy and Geology held in 1975 in Moscow, A. Bouroz, the president of the Subcommission on Carboniferous Stratigraphy (SCCS) at that time, presented the first attempt of a truly global chronostratigraphic division of the Carboniferous in using elements of the Western European, Eastern European, and North American classifications (Bouroz et al., 1978). This proposal stimulated many discussions inside and outside of SCCS. After intensive debates in the 1970s–1990s, SCCS has come to a consensus for the current subdivision of the Carboniferous Period (Heckel and Clayton, 2006). The Carboniferous is divided into two subsystems, named Mississippian and the Pennsylvanian. These subsystems are both divided into unnamed lower, middle, and upper series. The Mississippian contains the Tournaisian, Visean, and Serpukhovian stages, the Pennsylvanian the Bashkirian, Moscovian, Kasimovian, and Gzhelian stages (Fig. 23.1). So far, only the base of the Tournaisian and Bashkirian stages are defined by fully ratified Global Boundary Stratotype Section and Points (GSSPs) (Paproth et al., 1991; Lane et al., 1999) (see Fig. 23.9 at the end of this chapter). The GSSP of the Visean (Devuyst et al., 2003; Richards and Aretz, 2009) has been ratified by International Commission on Stratigraphy (IUGS-ICS), but the final publication is still missing. Task groups for the definition of the remaining GSSPs at the base of the Serpukhovian, Moscovian,

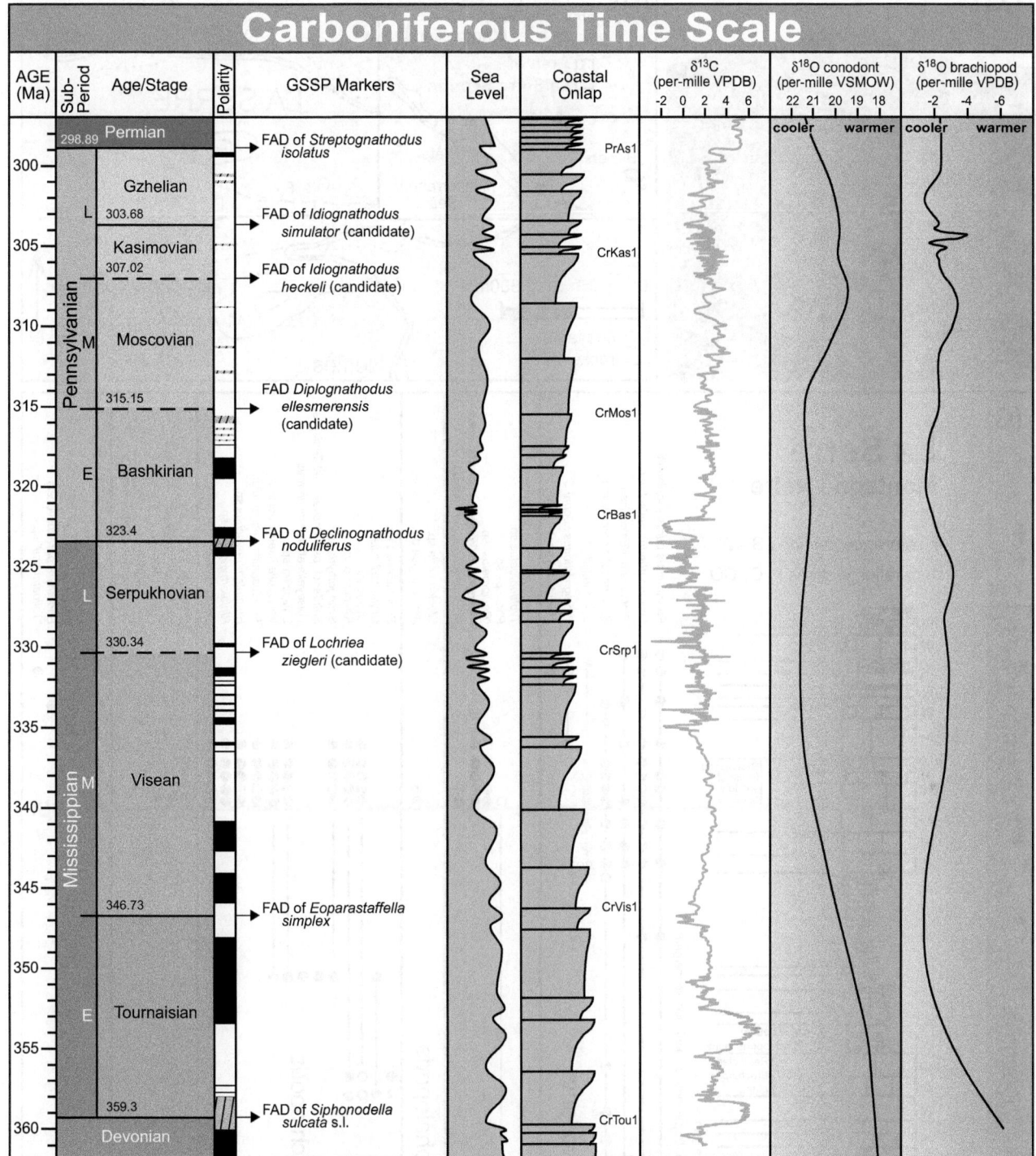

FIGURE 23.1 Carboniferous overview. The majority of the main markers or candidate markers for GSSPs of Carboniferous stages are FADs of conodont taxa as detailed in the text. ("Age" is the term for the time equivalent of the rock-record "stage.") Magnetic polarity scale is from Davydov et al. (2012). Schematic sea-level, coastal onlap curves and sequence nomenclature are modified from Haq and Schutter (2008) following advice of Bilal Haq (pers. comm., 2008) and with the incorporation of revised Mississippian calibrations by Poty (2016) and Herbig (2016), as detailed in the text. See also review in Simmons et al. (2020, Ch. 13, Phanerozoic eustasy, this book). The carbon isotope curve is from Cramer and Jarvis (2020, Ch. 11, this book), who had compiled numerous studies, including Saltzman (2003) and Batt et al. (2007). The oxygen isotope curve and implied sea-surface temperature trends are the statistical mean curves by Grossman and Joachimski (2020, Ch. 10, this book) from a synthesis of numerous conodont apatite and of brachiopod studies. The vertical scale of this diagram is standardized to match the vertical scales of the period-overview figure in all other Phanerozoic chapters. *FAD*, First appearance datum; *GSSP*, Global Boundary Stratotype Section and Point; *VPDB*, Vienna PeeDee Belemnite $\delta^{13}C$ and $\delta^{18}O$ standard; *VSMOW*, Vienna Standard Mean Ocean Water $\delta^{18}O$ standard.

Base of the Tournaisian Stage of the Carboniferous System in the La Serre Section, Montagne Noire, France

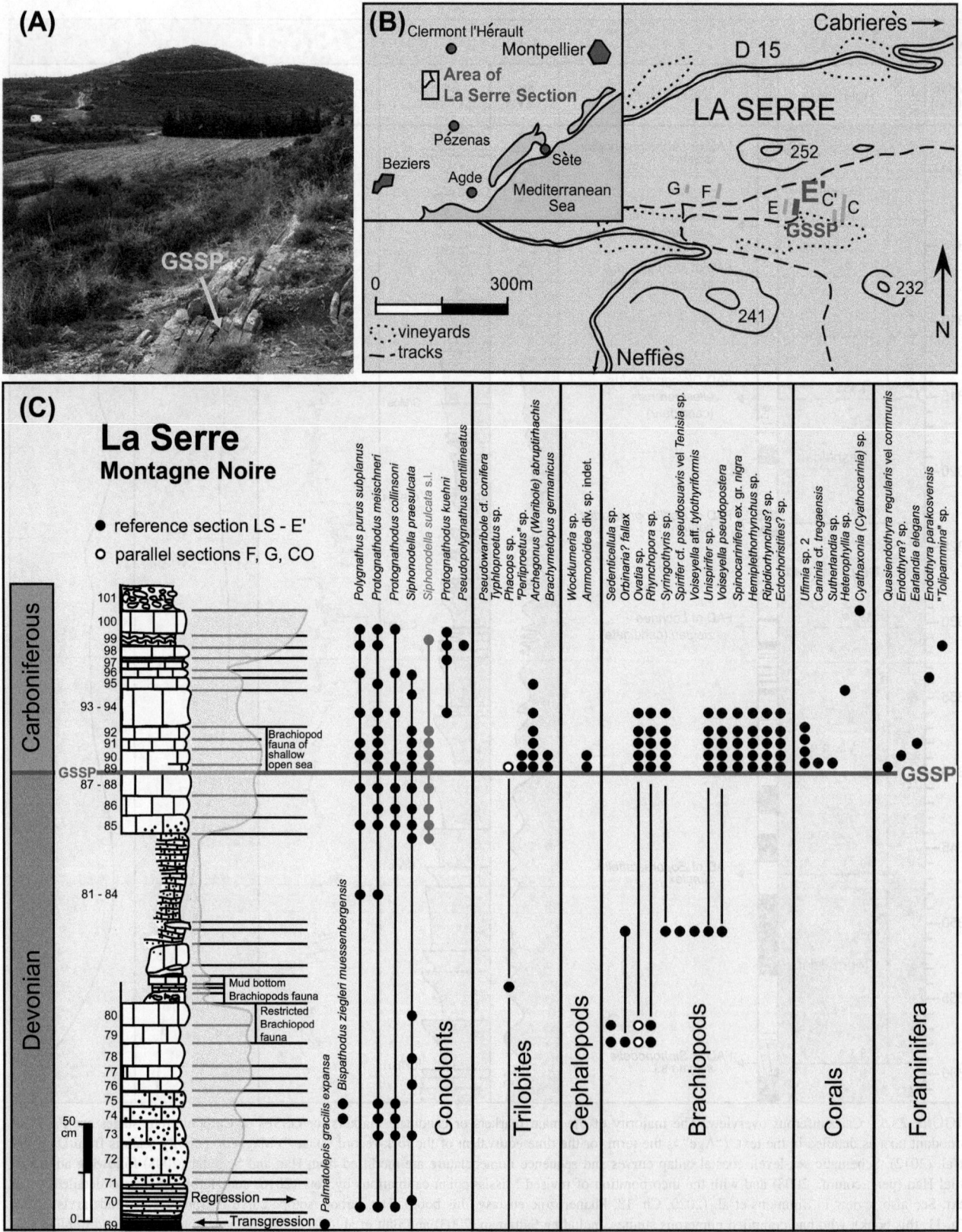

FIGURE 23.2 The GSSP for the base of the Tournaisian Stage, Mississippian Sub-System and Carboniferous System has been defined in trench E' at La Serre section, Montagne Noire, southern France. The section lies on the southern slope of the La Serre hill, 2.5 km southwest of the village of Cabrières, near the "classic" base of the *Gattendorfia* ammonoid Stufe. The GSSP level was intended to coincide with the lowest occurrence of conodont *Siphonodella sulcata* in the lineage *S. praesulcata-S. sulcata*, but this taxon is now found at a lower level. A revised criterion for the basal marker and potentially a different boundary section for the Devonian–Carboniferous boundary is now being sought. Outcrop photograph provided by Markus Aretz; stratigraphic column from Paproth et al. (1991).

Kasimovian, and Gzhelian stages have been active for the last two decades, but so far only the boundary criteria for the Gzhelian has been formally established. The top of the Carboniferous is defined by the GSSP for the base of the Permian (Davydov et al., 1998). A task group is working on the redefinition of the base of the Carboniferous (Devonian−Carboniferous boundary) currently established by the FAD of the conodont *Siphonodella sulcata* (Paproth et al., 1991), which has been found to occur earlier in the GSSP (Kaiser, 2009) and poses taxonomic problems (Kaiser and Corradini, 2011).

The chronostratigraphic subdivisions of the Carboniferous System adopted by the Carboniferous Subcommission with global standard stages (ages), series (epochs), and GSSPs ratified by the ICS and IUGS through July 2019 are shown in Fig. 23.1.

23.1.1 Mississippian Subsystem (Lower Carboniferous)

The Mississippian Subsystem is named after the Mississippi Valley. It is divided into three series corresponding to three stages. The GSSP for the conterminous base of the Carboniferous System, the Mississippian Subsystem, the Lower Mississippian Series, and the Tournaisian Stage coincides with the FAD of the conodont *S. sulcata*. The GSSP for the conterminous base of the Middle Mississippian Series and the Visean Stage coincides with the FAD of the foraminifera *Eoparastaffella simplex*. A GSSP for the conterminous base of the Upper Mississippian Series and the Serpukhovian Stage has not been defined so far.

The Devonian−Carboniferous boundary interval was a time of profound changes in the marine and terrestrial biosphere, and in the climate associated with important sea-level fluctuations (Kumpan et al., 2014; Kaiser et al., 2015; Poty, 2016). A sequence boundary is located slightly below the Devonian−Carboniferous boundary, but it is debated if this is a major sequence boundary (e.g. Paproth et al., 1991; Herbig, 2016) or if it is of minor order (e.g., Poty, 2016). However, the Famennian glaciation (e.g., Brezinski et al., 2010) is a prelude of the so-called Late Paleozoic Ice Age (LPIA) (e.g., Fielding et al., 2008b; Montañez and Poulsen, 2013). In the terrestrial deposits of the paleotropical Greenland, Marshall et al. (2013) demonstrated a change from cold and arid conditions of the Famennian to warm and wet conditions in the Tournaisian. The beginning of the polyphased LPIA is often set into the late Visean, but isotope (e.g., Buggisch et al., 2008) and sedimentological data (e.g., Poty, 2016) demonstrate an earlier onset. The tropical carbonate systems are under influence of glacio-eustatic sequences (global high-frequency regressive−transgressive cycles) at least from the early Visean onward (Poty, 2016) and the major sequence

boundary in the top of the Tournaisian (SB 4−5 of Hance et al., 2001) may indicate the first major impact in the far-field records of glaciations in the southern polar regions. The near-field glacial records from the southern hemisphere (e.g., Fielding et al., 2008a; Gulbranson et al., 2010; Limarino et al., 2014) point to an ice-free southern pole during the Tournaisian.

Important differences in the composition of marine faunal assemblages containing conodonts, foraminifers, rugose corals, and brachiopods in Northern America and Europe contradict the postulate of Ross and Ross (1988) that Mississippian marine faunas were generally worldwide in their distribution and latitudinal differences are more strongly developed than longitudinal differences. The closure of the seaway between Laurussia and Gondwana during the Mississippian strongly influenced the marine communication between the Paleotethys and Panthalassa oceans, and important differences existed between pelagic nektonic and benthic forms (e.g., Korn et al., 2012; Davydov and Cózar, 2019).

23.1.1.1 Tournaisian Stage and the base of the Carboniferous System

The name Tournaisian derives from the town of Tournais in western Belgium. The GSSP for the conterminous base of the Carboniferous System, the Mississippian Subsystem, the Lower Mississippian Series, and the Tournaisian Stage, is situated in the southeastern Montagne Noire, southern France (Paproth et al., 1991) (Fig. 23.2). It is located in a trenched section (trench E') on the southern slope of La Serre Hill about 0.5 km to the E of La Rouquette farmhouse, c. 2.5 km S of the village of Cabrières. The boundary level of the GSSP is near the entry of *Gattendorfia subinvoluta*, which was the classical boundary criteria defined at the second Heerlen Congress in 1935.

The current Devonian-Carboniferous boundary (DCB) is based on the FAD of the conodont *S. sulcata* in the evolutionary lineage *Siphonodella praesulcata−S. sulcata*. In the La Serre GSSP, this FAD was found in bed 89 (Paproth et al., 1991). On top of the latest Famennian compact cephalopod calcilutites, the interval straddling the boundary is a 3.7-m thick unit of well-bedded, graded biodetrital limestones with reworked and/or transported brachiopod and crinoid bioclasts, composed of two oolitic intervals separated by a clay-rich level with reworked upper Famennian blocks and pebbles witnessing of increased dynamic gravitational transport into an intermittently subsiding probably channelized environment. Bed 89 is situated in the upper part of that unit (upper oolitic interval).

Further work on the conodont distribution in the La Serre section (Kaiser, 2009) showed 0.45 cm below the defined boundary the entry of *S. sulcata* already in Bed 85. This level is immediately above a clay-rich level, hence questioning the usefulness and validity of the La Serre GSSP and also the Carboniferous time scale. Proposals for solving the issue can be summarized as solving the taxonomic problems of the index fossil (e.g., Kaiser, 2009; Kaiser and Corradini, 2011), reevaluating other conodont species and phylogenies (e.g., Corradini et al., 2011, 2017) or physical criteria (e.g., Kumpan et al., 2014; Kaiser et al., 2015; Bábek et al., 2016). Becker et al. (2016) listed some possible criteria for establishing a new DCB, which may correspond to a succession of timelines useful in the boundary interval. A joined task group of the Devonian and Carboniferous subcommissions (Aretz, 2014; Aretz and Corradini, 2016; Aretz and Task Group, 2018) working on a redefined DCB currently favors a multiproxy approach and new criteria applicable in different facies realms, which respects stratigraphic stability. This new criterion (Aretz and Corradini, 2019), defined by a timeline corresponding to "the end of the Devonian mass extinction and beginning of the Carboniferous radiation, the base of the *Protognathodus kockeli* conodont zone and the top of a major regression (top of HSS) and end of mass extinction." This criterion will somewhat lower the boundary, but in many sections not significantly. However, for the moment the reference for the Devonian–Carboniferous boundary remains Bed 89 in the La Serre section based on the originally proposed FAD of *S. sulcata*.

Together with the La Serre GSSP, two auxiliary stratotype sections were established for the Devonian–Carboniferous boundary, which are located in the Hasselbachtal, Sauerland area, Germany (Becker and Paproth, 1993), and in Nanbiancun, Guangxi Autonomous Region, southern China (Wang, 1993). The Hasselbachtal section is not accessible anymore, but it was important for the presence of several volcanic ash layers within the DCB interval, which provided radioisotopic dates (Claoué-Long et al., 1995; Davydov et al., 2008a).

The last appearance datum (LAD) of the miospore *Retispora lepidophyta* [*R. lepidophyta/Verrucosisporites nitidus* Zone (LN)] almost coincides with the FAD of the conodont *S. sulcata*. It should be noted that the validity of the LN zone as a biozone has been questioned by Prestianni et al. (2016) and supposed gaps in the stratigraphic record based on the absence of *V. nitidus* may be in fact an ecological or facies problem. The ammonoid genus *Gattendorfia* enters near the base of the Tournaisian, but the genus is not present in the earliest Tournaisian deposits, which are still characterized by *Acutimitoceras* (*Stockumites*) (Becker et al., 2016). In the type area of the Tournaisian in the Namur–Dinant Basin of Belgium, the basal Tournaisian is characterized by the foraminiferal DFZ 8 *Tournayellina beata pseudobeata* and MFZ1 unilocular zones and the rugose coral RC1 *Conilophyllum* Zone (Poty et al., 2006, 2014).

The duration of the Tournaisian is long (12.6 Myr), which is reflected in its former use as series. Poty et al. (2014) proposed to use the Belgian regional substages of Hastarian and Ivorian for further international subdivision.

23.1.1.2 Visean Stage

The name Visean (or Viséan) derives from the town of Visé in northeastern Belgium. The base of the middle Mississippian Series is conterminous to the base of the Visean Stage. At the Sixth International Congress on Carboniferous Stratigraphy and Geology in 1967 a Tournaisian–Visean boundary stratotype was formally adopted in the Belgian Namur–Dinant Basin (SCCS, 1969). It corresponded to the base of Bed 141, the first black limestone intercalation (Calcaires et Marbres Noirs) in the Leffe facies at the Bastion section near Dinant (George et al., 1976; Hance et al., 1997, 2006b). This boundary coincided with the first occurrence (FO) of the foraminiferal genus *Eoparastaffella*. The FO of the conodont *Gnathodus homopunctatus* is about 1 m higher in Bed 143 (Groessens and Noel, 1977). After the adoption in Sheffield, the Visean boundary at Baston failed in various correlation attempts not only across Eurasia but also with section situated nearby. This caused anomalous results (e.g., Conil et al., 1989) and the base of the Visean fluctuated in a time interval covered in modern biostratigraphic nomenclature by the foraminiferal biozones MFZ8–9.

A task group of the Carboniferous Subcommission reexanimated potential levels for the base of the Tournaisian, and decided to maintain stratigraphic stability by keeping the base of the Visean Stage as close as possible to the classical level (Hance et al., 1997). Because the Tournaisian–Visean boundary interval does not contain a significant faunal turnover in the pelagic realm (neither conodonts or ammonoids) conventionally used for defining GSSPs in the Paleozoic, it was decided to use as boundary marker the phylogenetic lineage of the foraminifera genus *Eoparastaffella*. The task group used as boundary criterion the first appearance of *Eoparastaffella* with an outer subangular periphery (morphotype 2 corresponding to *E. simplex*) succeeding forms characterized by a rounded periphery (morphotype 1, *Eoparastaffella*. ex. gr. *ovalis*) (Hance and Muchez, 1995; Hance, 1997; Sevastopulo et al., 2002). Because morphotype 1 does not appear in the Bastion section, the reference section was transferred to a stream section near the Pengchong village, Guangxi Autonomous Region, South China (Fig. 23.3), where the FAD of *E. simplex* is found in bed 83 (Devuyst et al., 2003; Richards and Aretz, 2009). Pengchong is situated about 15 km N/NE of the city of Liuzhou, several 100 m north of the Shantou–Kunming Expressway. At Pengchong, mainly dark gray, thick-bedded limestones with subordinate thin calcareous shale interbeds compose the Tournaisian–Visean

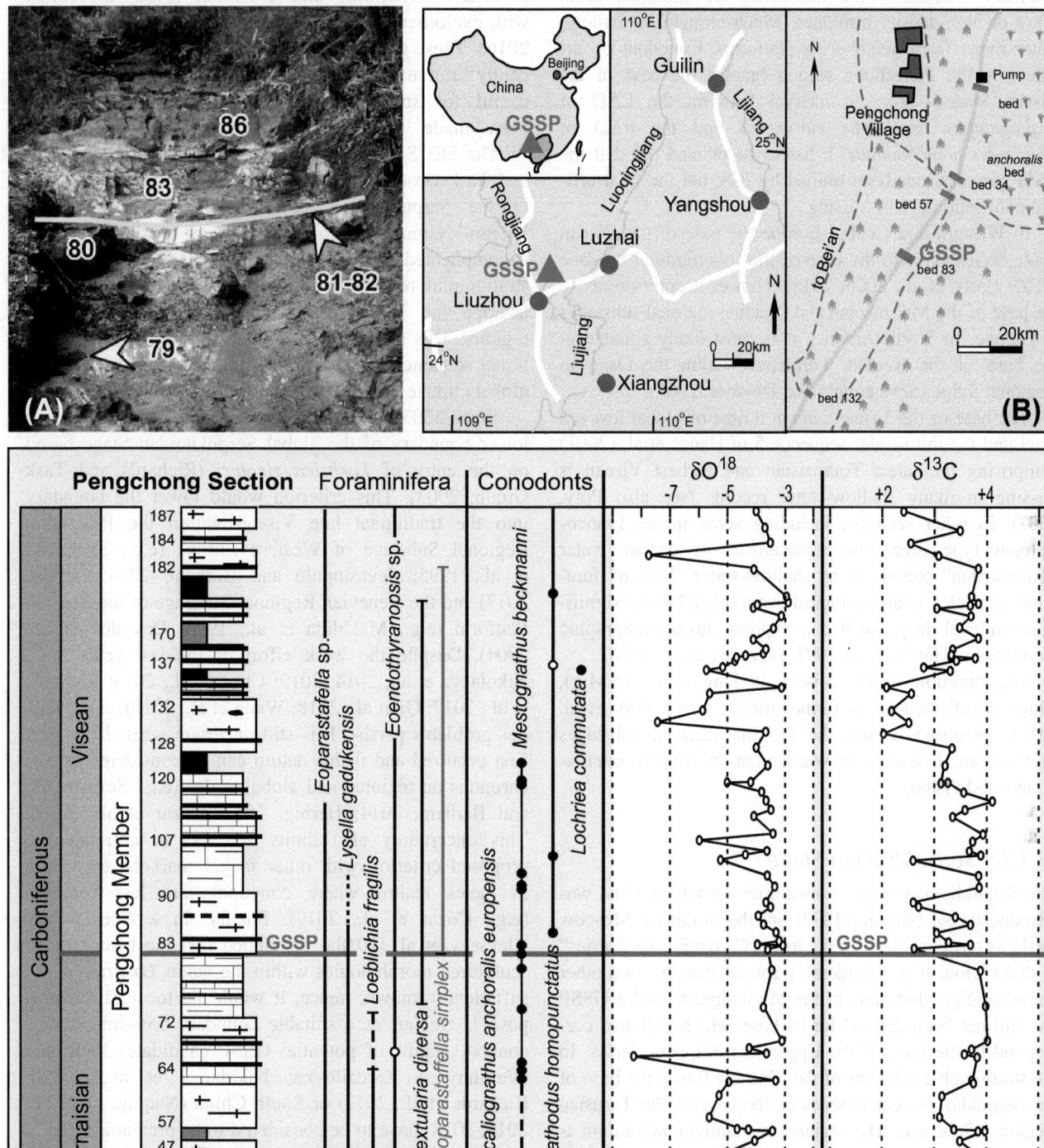

FIGURE 23.3 The GSSP section for the base of the Visean Stage of the Carboniferous System is situated in the Pengchong section, Guangxi Autonomous Region, South China. The Pengchong section is situated about 15 km N/NE of the city of Liuzhou, and several 100 m north of the Shantou-Kunming Expressway. The GSSP section lies south of the village in a small northward flowing creek. The base of the Visean corresponds to the first appearance of the foraminifera genus *Eoparastaffella* with an outer subangular periphery (morphotype 2 corresponding to *E. simplex*) succeeding forms characterized by a rounded periphery (morphotype 1, *E.* ex. gr. *ovalis*). (Photograph, stratigraphic column, and schematic sea-level trends from Davydov et al. 2012).

boundary interval. These limestones are modified grain flows or high-density turbidites, which brought in shallow-water fauna from neighboring platforms. Conodont occurrences in the Pengchong section bracket the base of the Visean Stage within an interval between the LAD of *Scaliognathus anchoralis europensis* and the FAD of *Gnathodus homopuctatus*. It has to be pointed out that the GSSP proposal has been ratified by ICS, but the final official publication is still missing.

In Western and Central Europe the base of the Visean Stage corresponds to the base of the foraminiferal biozone MFZ9 (Poty et al., 2006, 2014). Hence, it corresponds to the base of the Moliniacian and Chadian regional substages in Europe. In North America it is most likely situated at the base of the Keokuk Formation within the Osagean Regional Stage (Sevastopulo and Devuyst, 2005).

The base of the Visean falls in a time of global low sea level and the third-order sequence 5 of Hance et al. (2001) comprising the latest Tournaisian and earliest Visean is missing in many shallow-water records (see also Poty, 2007). In other sections, including some in the Franco-Belgian type area, the replacement of deeper water "Tournaisian" conodonts by shallow-water "Visean" foraminifers is clearly an ecological event related to the significant sea-level drop, but it may support chronostratigraphic correlations (Hance et al., 1997; Davydov et al., 2010).

The duration of the Visean is long (over 16 Myr), which is reflected in its former use as series. Poty et al. (2014) proposed to use the Belgian regional substages Moliniacian, Livian, and Warnantian for further international subdivision.

23.1.1.3 Serpukhovian Stage

The Serpukhovian Stage, named after Serpukhov City, was introduced by Nikitin (1890) in the southern Moscow Basin as terminal stage of the lower Carboniferous "series" in that region. It is considerably shorter than the two other chronostratigraphic units of the Mississippian, and a GSSP has still not been defined for its base, which will also correspond to the base of the Upper Mississippian Series. In the stratigraphic schemes of the Moscow Basin, the base of the Serpukhovian corresponds to the base of the Tarusian Regional Substage. The regional biostratigraphic datum is the base of the local *Neoarchaediscus postrugosus* foraminifera zone. The index taxa of this acme zone and other stratigraphic important taxa range from the late Visean Venevian Regional Substage into the Serpukhovian (Gibshman, 2003; Davydov et al., 2012). Makhlina et al. (1993) highlighted the ecological control on the distribution of benthic faunas in the Russian Platform, which results in many diachronous appearances. In addition, the stratigraphic record of the late Visean and Serpukhovian in the Moscow Basin is incomplete as expressed by common

paleokarstic features and erosional levels associated with cyclothem successions (e.g., Kabanov et al., 2013, 2016). These gaps and resulting "missing fossils" significantly limit attempts to reconstruct phylogenetic lineages useful for stratigraphy, even if such attempts have been made (e.g., Gibshman and Baranova, 2007).

The SCCS has set up a task group in 2003 (Richards and Task Group, 2003) to establish a GSSP for the base of the Serpukhovian Stage close to the traditional Visean–Namurian boundary. The work of this task group is complicated by marine biotic endemism and incomplete stratigraphic records or major facies changes, which characterize the Visean–Serpukhovian transition in many regions. This sharp impact on the biosphere and depositional sequences results from pronounced changes in the global climate and the plate configurations.

Since 2003 the task group is testing a conodont-based lower boundary of the global Serpukhovian Stage based on the entry of *Lochriea ziegleri* (Richards and Task Group, 2003). This criterion would lower the boundary into the traditional late Visean within the Brigantian Regional Substage of Western Europe (e.g., Skompski et al., 1995; Sevastopulo and Barham, 2014; Herbig, 2017) and the Venevian Regional Substage of the Russian Platform (e.g., Makhlina et al., 1993; Davydov et al., 2004). Despite the work effort of the last years (e.g., Nikolaeva et al., 2014, 2019; Chen et al., 2016; Richards et al., 2017; Qi et al., 2018; Wang et al., 2018), some serious problems persist. It is still uncertain when *L. ziegleri* first occurred and if this datum can be considered as isochronous on regional and global scales (e.g., Sevastopulo and Barham, 2014; Herbig, 2017; Cózar et al., 2019). This uncertainty also limits attempts to correlate the proposed criterion with other faunal markers, especially in facies realms where conodonts are less common (e.g., Cózar et al., 2019). Finally, in a recent study Alekseev et al. (2018a) concluded that the taxonomy of sculptured morphologies within the genus *Lochriea* is not sufficiently known; hence, it would be too early to propose *L. ziegleri* as a suitable boundary criterion. In this context, claims of potential GSSP candidates in Russia (Verkhnyaya Kardailovka; Nikolaeva et al., 2009b; Richards et al., 2017) or South China (Naqing; Qi et al., 2013, 2018) have to be considered to be premature.

It has to be underlined that currently no formal decision has been made on the criterion to define the global Serpukhovian Stage. Hence, the anticipation of any decision must be avoided, because it may result in erroneous determinations of the boundary and/or its correlation (Herbig, 2017; Alekseev et al., 2018a, 2018b; Cózar et al., 2019). The Serpukhovian can only be used in the defined stratigraphic frame of the Moscow Basin and the entry of *L. ziegleri* is during the latest Visean in current chronostratigraphy.

23.1.2 Pennsylvanian Subsystem (Upper Carboniferous)

The Pennsylvanian Subsystem is named after the state of Pennsylvania in the United States. It is divided into three series corresponding to four stages. The GSSP for the conterminous base of the Pennsylvanian Subsystem, the Lower Pennsylvanian Series, and the Bashkirian Stage coincides with the FAD of the conodont *Declinognathodus noduliferus* s.l. GSSPs for the conterminous base of the Middle Pennsylvanian Series and the Moscovian Stage, the conterminous base of the Upper Pennsylvanian Series and the Kasimovian Stage, and the base of the Gzhelian Stage within the Upper Pennsylvanian Series have not been defined so far.

Pennsylvanian times are generally characterized by glacial—interglacial cycles of different durations and intensities. The resulting glacio-eustatic sea-level fluctuations are documented globally and result in many places in typical cyclothem sedimentation patterns. On the one side, these cyclic patterns offer important tools for global correlation (e.g., Heckel et al., 2007; Heckel, 2013; Ogg et al., 2016), but they also contain and sometimes mask important sedimentological and stratigraphic gaps. The Pennsylvanian is also a time, when in many regions thick coal deposits formed in paralic and intramountane/limnic basins. The terrestrial record for Pennsylvanian times is better known than that of the Mississippian, due to intensive research related to coal mining.

The Mississippian—Pennsylvanian transition coincides with an important marine extinction (e.g., McGhee et al., 2013), which in its magnitude rivals with the main extinction events of the Phanerozoic (McGhee et al., 2012). The faunal turnover results in many taxonomic groups in distinctive Mississippian and Pennsylvanian faunas or associations. This can be often seen on the family level in, for example, conodonts, ammonoids, and rugose corals. The Serpukhovian—Bashkirian boundary corresponds in many sections to stratigraphic gaps, which resulted from a major sea-level drop as consequence of a major glaciation phase in Gondwana. The main pulse of this glaciation is thought to be of Bashkirian age (Saltzman, 2003; Fielding et al., 2008b; Grossman et al., 2008; Bishop et al., 2009). The closure of the seaway between Laurussia and Gondwana reduced the marine communication between the Paleotethys and Panthalassa oceans, but this exchange was not completely blocked and at least occasionally marine passages did exist.

23.1.2.1 Bashkirian Stage and the Mid-Carboniferous Boundary

At the Tenth International Congress of Carboniferous Stratigraphy and Geology in 1983, it was decided to define a Mid-Carboniferous boundary, today corresponding to the base of the Pennsylvanian Subsystem and Bashkirian Stage, by the FAD of the conodont *D. noduliferus* s.l. This modern chronostratigraphic base of the Bashkirian Stage is much lower than the original definition in Russia, where the FO of the foraminifera *Pseudostaffella antique* was used (Semikhatova, 1934).

The GSSP for the base of the Bashkirian Stage is defined near the northern end of the Arrow Range, in Arrow Canyon, United States, about 75 km NE of Las Vegas (Lane et al., 1999) (Fig. 23.4). *D. noduliferus* s.l. first appears in Unit G in the lower part of the Bird Spring Formation at Sample 61B. This sample was taken 82.90—83.05 m above the contact of the Indian Springs and Battleship Wash formations (Lane et al., 1999). However, Manger (2017) stated that today the FAD of *Declinognathodus inaequalis*, a former subspecies of *D. noduliferus*, should define the GSSP level, which considerably restricts the original definition used for the boundary definition. In the absence of an official decision on this new definition the FAD of *D. noduliferus* s.l. defines the base of the Bashkirian Stage. This can potentially result in stratigraphic miscorrelations; for example, Sanz-Lopez et al. (2006) showed that in Spain *D. noduliferus bernesgae* appears in the strata dated with ammonoids and other conodonts as Serpukhovian.

The Mid-Carboniferous boundary interval at Arrow Canyon represents a shallow-water, mid-ramp to inner-ramp, dominantly carbonate succession divided into several forth-order glacio-eustatic cycles related to the waning and waxing of the ice sheets on Gondwana (e.g., Lane et al., 1999; Barnett and Wright, 2008). The transgressive parts of the cycles are composed of bioclastic and mixed bioclastic—siliciclastic limestones and fine sandstone, whereas mudstone, oolitic, and brecciated limestone intervals represent the regressive cycle parts. The upper cycle parts and cycle boundaries are characterized by subaerial hiatuses as shown in paleokarstic surfaces, paleosols, and calcretes (e.g., Richards et al., 2002; Barnett and Wright, 2008; Bishop et al., 2009).

The GSSP for the base of the Bashkirian Stage is situated 68 cm above the base of the transgressive—regressive cycle 3 of Lane et al. (1999). Barnett and Wright (2008) showed on top of this cycle, about 1 m above the GSSP level, a major second-order shift in the stacking patterns of cycles. Lane et al. (1999) interpreted this as a regionally developed second-order sequence boundary within the Bird Spring Formation.

The presence of subaerial hiatuses in the GSSP section resulted in a debate of the completeness of its stratigraphic record and hence its usefulness as GSSP section. Lane et al. (1999) claimed that times of subaerial exposure were short, whereas Barnett and Wright (2008) showed that these exposures could be rather long (>1 Myr) and that important parts of Western European regional substages could be missing at Arrow Canyon. If

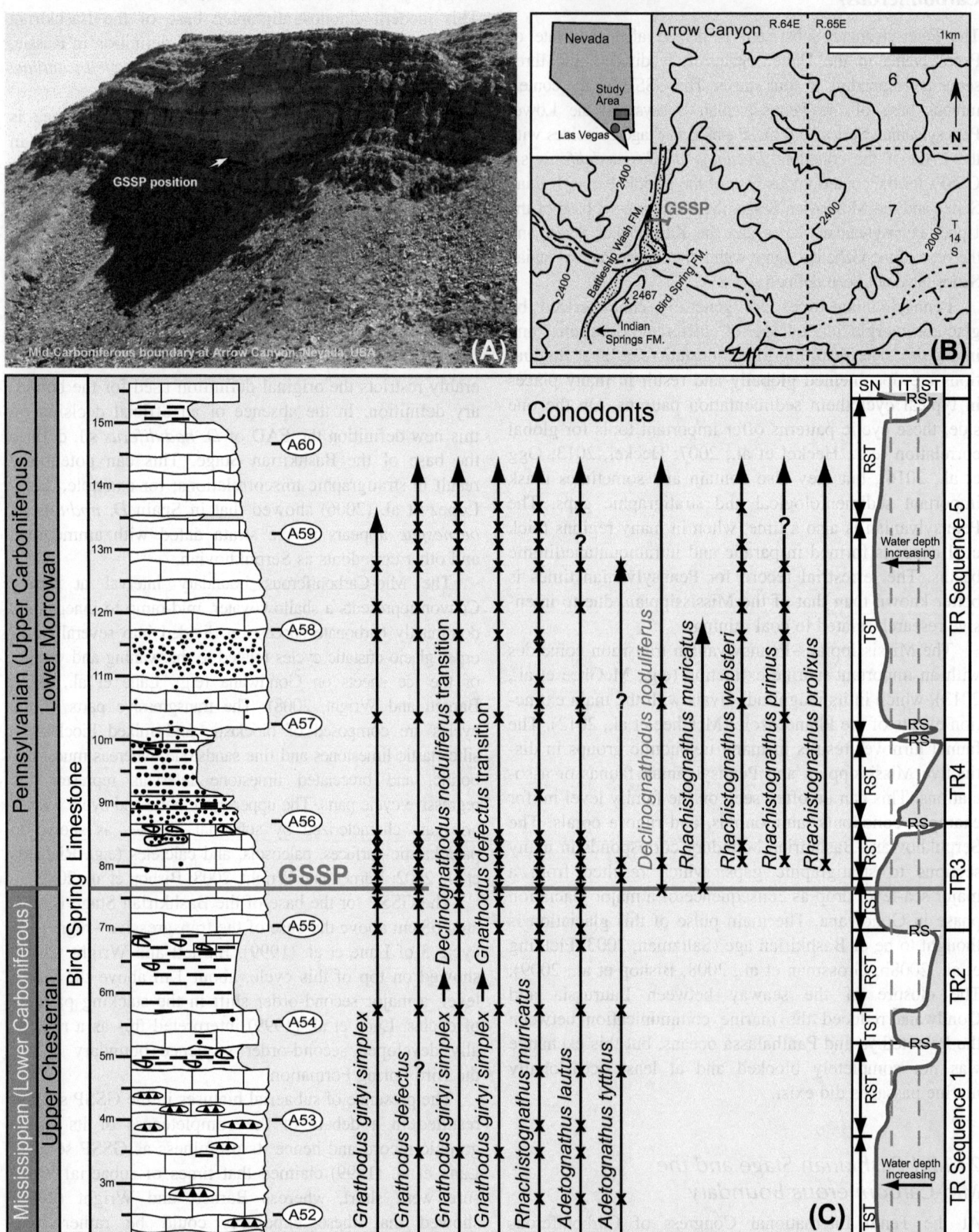

Base of the Pennsylvanian Sub-system of the Carboniferous System at Arrow Canyon, Nevada, U.S.A.

FIGURE 23.4 The GSSP for the base of the Bashkirian Stage and Pennsylvanian Sub-System and is situated in Arrow Canyon, in the Great Basin, Nevada, USA. The type section is located approximately 75 km northwest of Las Vegas. The GSSP level coincides with the first-appearance datum (FAD) of conodont *Declinognathodus noduliferus* (sensu lato). (Photograph, stratigraphic column, and schematic sea-level trends from Davydov et al., 2012).

the latter is correct, an important impact on fine-scaled global correlations has to be taken into considerations. Therefore Smith et al. (2014) recommended redefining the GSSP based on its incompleteness and the facies changes.

The reconstruction of the phylogenetic lineage for *D. noduliferus* s.l. has also been a matter of discussion. The GSSP is based on the original reconstruction that *D. noduliferus* s.l. evolved from *Gnathodus girty simplex* (Lane and Manger, 1985). Later Nemyrovska (1999) showed a chronocline of *Gnathodus postbilineatus—Declinognathodus praenoduliferus—D. noduliferus* in the Donets Basin. This may indicate a polyphyletic origin of the guide taxa, which potentially implies a diachroneity of the boundary between Eastern and Western Hemispheres.

Lane et al. (1999) indicate that the base of the Bashkirian Stage corresponds to the base of the *noduliferus—primus* conodont zone of North America (Baesemann and Lane, 1985). They also showed that the foraminifers *Eosigmoilina robertsoni* and *Brenckleina rugosa* disappear slightly above the base of the Bashkirian Stage (samples 62 and 63 at Arrow Canyon). At the base of cycle 4 at Arrow Canyon appears the foraminifer *Globivalvulina bulloides*. However, this species is known in Japan (Mizuno and Ueno, 1997), Arctic Alaska (Harris et al., 1997; Baesemann et al., 1998), the Pyrenees (Perret, 1993), and possibly the Donets Basin (Vdovenko et al., 1990) already below the boundary. Outside North America the occurrence of species of the foraminiferal genus *Plectostaffella*: *Pl. varvariensis*, *P. jakhensis*, *P. posochovae*, and *P. bogdanovkensis* mark the base of the Bashkirian Stage (Davydov et al., 2012). Waters et al. (2011) showed that in the stratotype for the Chokierian Regional Substage the conodont *D. noduliferus* s.l. occurs 9.4 m above the *Isohomoceras subglobosum* Marine Band. Hence, the base of the Bashkirian Stage may not coincide in the ammonoid zonation with the base of the *Homoceras* Zone or *I. subglobosum* Zone of Great Britain, Nevada, and Central Asia (Ramsbottom and Saunders, 1985; Kullmann and Nikolaeva, 2002). Manger (2017) stated that the occurrence of *I. subglobosum* is earlier in Central Asia and Nevada (Nemirovskaya and Nigmadganov, 1994; Titus et al., 1997) than in western Europe. The duration of the Bashkirian is about 8 Myr with the uncertainty for the placement of the base of the Moscovian Stage.

23.1.2.2 Moscovian Stage

The base of the global Moscovian Stage, which is equivalent to the Middle Pennsylvanian Series, has so far not been defined. The stage is named after the Russian capital Moscow and the Moskva River. In its type area (Nikitin, 1890) in the Moscow Basin the Moscovian is represented by alternations of shallow-marine bioclastic limestones and colorful claystones. They unconformably overlying Mississippian limestones indicate a stratigraphic gap of at least the Bashkirian. The base of the Moscovian is represented by the Vereiskian

Regional Substage (Ivanova and Khvorova, 1955), which in its lower part is dominated by terrestrial, alluvial facies (Alekseev, 2009). However, the fusulinid species *Aljutovella aljutovica* and the conodont species *Declinognathodus donetzianus* were found 3 m above the base of the Aljutovskaya Formation, which today represents the base of the Vereiskian Regional Substage (Makhlina et al., 2001). In more complete marine sections *D. donetzianus* appears significantly earlier than *A. aljutovica* (Ueno and Nemirovska, 2008; Davydov, 2009; Kulagina et al., 2009).

However, the *A. aljutovica* fusulinid zone is commonly accepted as the foraminiferal zone defining the base of the Moscovian in the former Soviet Union. Although it is widely correlated along the northern and eastern margins of Pangea (Kulagina et al., 2009), further correlation is hampered by the strong endemism of fusulinids between the western and eastern hemispheres. This excludes fusulinids as *A. aljutovica*, *Depratina prisca* or *Eofusilina* for the definition of the base of the global stage.

Therefore the task group for the establishment of the Moscovian GSSP, near the traditional Bashkirian—Moscovian boundary, is looking for a conodont species as potential index fossil. Several proposals have been made, using different species, but also placing the boundary in different stratigraphic positions in respect to the historical base of the Moscovian in the Moscow Basin. Nemyrovska (1999) proposed to base the boundary on the FO of *D. donetzianus* in the *Declinognathodus marginodosus—D. donetzianus* conodont lineage. Advantages of this proposal are its stratigraphic position near the traditional base of the Moscovian and its established evolutionary lineage (Nemyrovska et al., 1999; Goreva and Alekseev, 2007). Qi et al. (2016) noted that it is rare and that the originally assumed wide geographical distribution (see, e.g., Davydov et al., 2012) is mainly restricted to the Donets and Moscow basins and Great Britain, due to the lumping of transitional forms into the species. Another proposal is based on the FO of *Diplognathodus ellesmerensis* in the *Diplognathodus* aff. *orphanus—D. ellesmerensis* conodont lineage (Groves and Task Group, 2007). Despite initial skepticism of the suitability of the species for boundary definition (Groves and Task Group, 2008), it has gained support in recent years (Qi et al., 2016). The species is one of the most widely distributed Pennsylvanian conodont species with records in North America, boreal Canada, South America, Eurasia and China. Its taxonomy and phylogeny have been revised, it is easily identifiable, and it has a restricted and well-defined stratigraphic range (see Qi et al., 2016 for details). However, those authors also noted that *Diplognathodus* specimens are a minor faunal constituent and they are usually small. Hence, the FO of the species in many sections may not correspond to its phylogenetic FAD, which will influence the placement of the boundary. Davydov et al. (2010) indicated that a boundary based on this species would be about 0.5 Myr younger than the traditional boundary, but this might be partly related to the problems in finding the species. Alekseev et al. (2018b) reported the cooccurrence of *D. donetzianus* with several morphotypes of early *Diplognathodus*, including primitive forms of *D. ellesmerensis* in the deeper water sections in the southern Urals. This is a first step toward reliable correlations of the Bashkirian−Moscovian boundary interval in the Donets and Moscow basins with the succession of the South China.

An alternative proposal, which would place the boundary significantly higher, is to use the FO of *Neognathodus bothrops* to define the base of the global Moscovian Stage (Kabanov and Alekseev, 2011; Alekseev and Goreva, 2013). This stratigraphic level defines the base of the second Moscovian regional substage, the Kashirian in the Moscow Basin. Such a boundary definition would facilitate correlations with the boundary between the *Winslowoceras−Diaboloceras* and *Wellerites−Paralegoceras* ammonoid zones at the base of the Kashirian. However, this proposal did not gain much support within the task group, since *Neognathodus* is rare in most regions outside the Moscow Basin and North America. Hence, this proposal will not receive further evaluation by the Task Group (Alekseev and Task Group, 2014).

Like for the Serpukhovian Stage, already candidate sections have been proposed before a consensus on the criterion has been achieved (Kulagina et al., 2009; Qi et al., 2016). The base of the Moscovian Stage would be based on the FO *D. donetzianus* supported by the FO of the fusulinid *D. prisca* in the Basu section, Russia, and by the FO of *D. ellesmerensis* supported by the fusulinid *Profusulinella* in the Naqing section, China.

The duration of the Moscovian is about 8 Myr with the uncertainty for the placement of its base and top.

23.1.2.3 Kasimovian Stage

The conterminous base of the upper Pennsylvanian Series and the Kasimovian Stage has so far not been defined. Ivanov (1926) individualized from the Moscovian under the name "Teguliferina" Horizon, the stratigraphic interval, which was later termed Kasimovian after the town of Kasimov along the Oka River. The fusulinid-based definition of its base by Rauser-Chernousova and Reitlinger (1954) was later redefined by Davydov (1990), who placed the Moscovian/Kasimovian boundary between the *Praeobsoletes burkemensis−Protriticites ovatus* and *Obsoletes obsoletes−Protriticites pseudomontiparus* fusulinid zones, slightly lower than the original definition in the Moscow Basin. The boundary interval corresponds to a eustatic sea-level low, which results in disconformities in many sections around the world (see references in Davydov et al., 2012).

Despite the significant progress made by the SCCS task group for the definition of the base of the Kasimovian over more than two decades, the criterion for the definition of the global stage has still not been determined. For the moment the

task group favors a conodont-based boundary. Discussions are centered on three proposals. There are two long-standing proposals to use the FO of *Idiognathodus turbatus* or *Idiognathodus sagittalis*. These dates are close to each other, but they are within the middle Kasimovian (Goreva et al., 2009; Rosscoe and Barrick, 2009), hence about one substage higher than the traditional base of the Kasimovian (Villa and Task Group, 2008). This is a serious disadvantage, since the stage is already rather short as noted by Davydov et al. (2012). If the boundary should be based on a species of *Idiognathodus*, *Idiognathodus heckeli*, a species defined by Rosscoe and Barrick (2013), which is the direct ancestor of *Id. turbatus*, may be an alternative (Ueno and Task Group, 2014). This would bring down the boundary toward the traditional base of the Kasimovian. Russian researchers (in Ueno and Task Group, 2017) proposed to return to the traditional boundary of the Kasimovian, the base of the regional Krevyakinian Substage, in using the FO of *Swadelina subexcelsa* (*Streptognathodus subexcelsus* for some authors). This species is known in many Pennsylvanian basins, including the Moscow and Donets basins, the Midcontinent (Oklahoma) in the United States, and South China. In this light the proposal of Davydov (2007) and Davydov and Khodjanyazova (2009) to use the change in the evolutionary chronocline within the fusulinid genus *Protriticites* from primitive to advanced forms has gained importance, since this change occurs around the traditional Moscovian/Kasimovian boundary. In using a benthic marker, such a boundary definition would resemble that for the Visean Stage, and the neostratotype of the Kasimovian in the Moscow Basin, the Afanasievo section (Goreva et al., 2009), which contains abundant fusulinids, p. ex. *Protriticites* and *Montiparus*, and the conodonts *Id. sagittalis* and *Sw. subexcelsa* would become a candidate. However, the section is composed of cyclothems separated by hiatuses and paleosols, which represent unknown time.

Two candidate sections have been proposed as potential boundary stratotype in deeper water facies and using a conodont marker. The Usolka section in the Russian Urals yields *Id. sagittalis* and fusulinids, but in this section the boundary interval contains some alternations of mudstone and limestone, which may obscure the evolutionary FO. The second candidate is the deep-water carbonate slope section at Naqing in South China. It contains several of the discussed conodont markers (Qi et al., 2012; Hu and Qi, 2017), the *Idiognathodus swadei−Id. heckeli−Id. turbatus* lineage and the *Sw. subexcelsa*, as well as some fusulinids.

The Kasimovian is the shortest stage of the Carboniferous with a duration of 3.3 Myr.

23.1.2.4 Gzhelian Stage

The youngest Carboniferous stage, the Gzhelian Stage, was named after the Gzhel village in the Moscow Region. The biostratigraphic criterion for the base of the Gzhelian Stage is defined by the FAD of the conodont *Idiognathodus simulator* (s.s.) in the chronocline *Idiognathodus eudoraensis—I. simulator* (Villa and Task Group, 2007; Heckel et al., 2008; Villa et al., 2009). This ancestry of *I. simulator* is different to that proposed by Chernykh (2005), who indicated a chronocline *Idiognathodus praenuntiuse—I. simulator—Idiognathodus auratus*. The top of the Gzhelian Stage is defined by the GSSP for the base of the Permian System (Davydov et al., 1998).

Heckel et al. (2008) demonstrated the applicability of the boundary criterion and its precise correlation and dating in the cyclothems typical in many regions at this time interval. They showed that an exposure surfaces is very valuable in continent—continent correlation, but it also makes the search for a complete stratigraphic record suitable for a GSSP difficult. However, in the shallow-water successions, fusulinids are very useful in placing the Kasimovian—Gzhelian boundary. In Eurasia, one example is *Rauserites rossicus*, which first appears together with *I. simulator* in the upper Rusavkino Formation in the Gzhel section (Villa et al., 2009), which had been used previously for boundary identifications (Ivanova and Rozovskaya, 1967; Barskov et al., 1980).

In terms of ammonoid biostratigraphy the base of the *Shumardites—Vidrioceras* Zone has been conventionally placed at the base of the Gzhelian in the former Soviet Union (Bogoslovskaya, 1984). In North America the base of the *Shumardites* Zone is equaled to the base of the global Gzhelian Stage (Boardman and Work, 2013). This is also confirmed by data from Central Asia (Popov et al., 1989; Popov, 1999; Boardman and Work., 2004). However, in the southern Urals, the entry of *Shumardites* is delayed (Popov et al., 1985; Davydov et al., 1994).

The task group for the definition of the Kasimovian—Gzhelian boundary is currently discussing potential stratotype candidates, which are not affected by exposure surfaces and major facies changes. Both candidates were also named candidates for the base of the Kasimovian Stage. The deep-water Usolka section in the southern Urals displaying mixed carbonate—siliciclastic facies is the older candidate section (Chernykh et al., 2006). The conodont chronocline *I. praenuntiuse—I. simulator—I. auratus* is well documented, abundant fusulinids occur in several horizons (including *Rauserites*), numerous volcanic ash beds in the boundary transition have been dated, and a Sr isotope curve based on conodonts is available (Chernykh et al., 2006; Davydov et al., 2008b; Schmitz et al., 2009; Schmitz and Davydov, 2012). However, *I. simulator* first occurs above a 0.16 m thick brownish-gray mudstone with small phosphatic concretions (2—3 cm) (Sungatullina et al., 2016) and silicification of strata influences the extraction of conodonts (Alekseev in Ueno and Task Group, 2017).

The second candidate is the Naqing section in South China. It yields *I. simulator* as well as other abundant transitional conodonts between *I. eudoraensis* and *I. simulator*, which indicate great potential to find the direct ancestor of the latter. The fusulinids are, however, not rich in this section. Chen et al. (2016) provided oxygen isotopes of conodont apatite and Chen et al. (2018) Sr isotope data for the section. The duration of the Gzhelian is 4.8 Myr.

23.1.3 Regional Carboniferous subsystem, stage, and substage units

Regional subsystems, series, stages, and substages have been erected for many parts of the Carboniferous world. Examples of intensively studied areas in the tropical realm are Western and Eastern Europe, North America, and South China, and in higher latitudes Gondwana and Angaraland (Siberia) (Fig. 23.5).

23.1.3.1 Western European Subdivisions

Historically, a twofold subdivision into series is applied to the Carboniferous succession of Western and Central Europe. The lower part, dominated by marine facies, was called Lower Carboniferous or Dinantian and the upper part, dominated by terrestrial facies, was called Upper Carboniferous or Silesian. The boundary between these two series was defined at the Forth Heerlen Congress in 1958 by the FO of the ammonoid *Cravenoceras leion*. This traditional scheme became obsolete with the definition of the mid-Carboniferous boundary, which corresponds to a boundary within the upper series corresponding roughly to the Namurian A—B boundary in traditional nomenclature.

Today, boundary stratotype—based regional substages, stages, and series are available mainly based on the succession in Belgium and Great Britain, with additions form northern Spain in the highest stratigraphic levels.

The subdivision of the Belgian Dinantian started in the 19th century, and the concept of the Tournaisian and Visean stages, as Calcaires de Tournais succeeded by the Calcaire de Visé, can be traced back to Dumont (1832). Both name-giving localities, respectively, in western and eastern Belgium, are famous for their strata rich in marine fauna.

In Belgian, formalization of the stages in the 19th century resulted in their subdivision in the legends of the geologic maps (e.g., Tn1b, V2a), which became a first detailed chronostratigraphic scheme (see Demanet, 1929, 1958). Substages as defined by Conil et al. (1977) are in ascending order Hastarian (lower Tournaisian), Ivorian (upper Tournaisian), Moliniacian (lower Visean), Livian (middle Visean), and Warnantian (upper Visean). Following detailed revisions, these substages are now tied into a framework of modern biostratigraphic, lithostratigraphic, and sequence stratigraphic data (Poty et al., 2014; Poty, 2016).

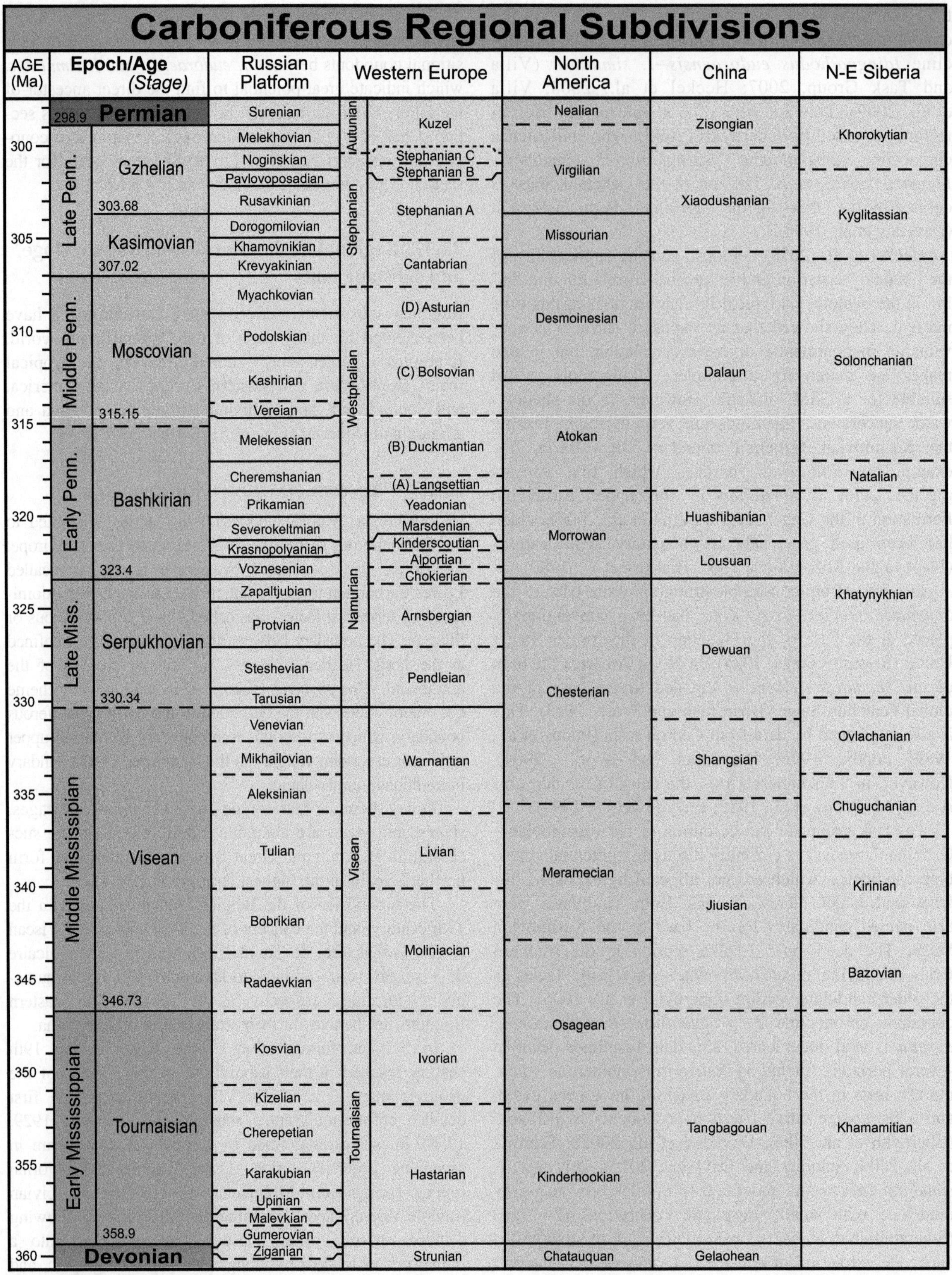

Carboniferous Regional Subdivisions

AGE (Ma)	Epoch/Age (Stage)	Russian Platform	Western Europe	North America	China	N-E Siberia
298.9	**Permian**	Surenian	Autunian — Kuzel	Nealian	Zisongian	Khorokytian
300	Gzhelian (Late Penn.)	Melekhovian				
		Noginskian	Stephanian — Stephanian C / Stephanian B	Virgilian	Xiaodushanian	
		Pavlovoposadian				
303.68		Rusavkinian	Stephanian A			Kyglitassian
305	Kasimovian (Late Penn.)	Dorogomilovian		Missourian		
		Khamovnikian				
307.02		Krevyakinian	Cantabrian			
310	Moscovian (Middle Penn.)	Myachkovian	Westphalian — (D) Asturian	Desmoinesian		Solonchanian
		Podolskian			Dalaun	
		Kashirian	(C) Bolsovian			
315		Vereian				
315.15		Melekessian	(B) Duckmantian	Atokan		
	Bashkirian (Early Penn.)	Cheremshanian	(A) Langsettian			Natalian
320		Prikamian	Yeadonian		Huashibanian	
		Severokeltmian	Namurian — Marsdenian	Morrowan		
		Krasnopolyanian	Kinderscoutian / Alportian		Lousuan	
323.4		Voznesenian	Chokierian			
325	Serpukhovian (Late Miss.)	Zapaltjubian	Arnsbergian			Khatynykhian
		Protvian			Dewuan	
		Steshevian	Pendleian	Chesterian		
330	330.34	Tarusian				
	Visean (Middle Mississippian)	Venevian	Warnantian		Shangsian	Ovlachanian
335		Mikhailovian				
		Aleksinian				Chuguchanian
		Tulian	Visean — Livian	Meramecian		
340					Jiusian	Kirinian
		Bobrikian				
			Moliniacian			
345		Radaevkian				Bazovian
346.73						
	Tournaisian (Early Mississippian)	Kosvian	Ivorian	Osagean		
350						
		Kizelian			Tangbagouan	Khamamitian
		Cherepetian	Tournaisian —			
355		Karakubian	Hastarian	Kinderhookian		
		Upinian				
358.9		Malevkian / Gumerovian				
360	**Devonian**	Ziganian	Strunian	Chatauquan	Gelaohean	

FIGURE 23.5 Correlation chart of the Carboniferous with international subdivisions and selected regional nomenclature (mainly stages and substages).

The oldest substage is named **Hastarian** with the Anseremme section as its neostratotype (Hance and Poty, 2006). Its base coincides with the base of the Tournaisian Stage and the Carboniferous System. It is defined by the FO of the conodont *S. sulcata*. The substage (Poty et al., 2006, 2014) correlates with the foraminifer zones MFZ1−3 and the lower part of MFZ4, the rugose coral zones RC1 and 2 and the base of RC3α, the *Siphonodella* conodont interval (CC1 of Conil et al., 1977), and the *Gattendorfia−Eocanites* and *Goniocyclus−Eocanites* ammonoid genus zones. It comprises the four miospore zones from the *Vallatisporites verrucosus−Retusotriletes incohatus* (VI) to the *Spelaeotriletes pretiosus−Raistrickia clavate* (PC) zones (Higgs et al., 1992). The second substage of the Tournaisian, the **Ivorian**, is named after a disused quarry near to the Yvoir railway station on the right bank of the Meuse River. The biostratigraphic marker (Hance et al., 2006a) is the FO of the conodont *Polygnathus communis carina*. The substage comprises most of the foraminiferal zone MFZ4 and the succeeding zones MFZ5−8, the rugose coral zones RC3α to RC4β1 (except of the base of RC3α, which is Hastarian, the *Polygnathus communis carina* and the *S. anchoralis* conodont zones and the lower part of the CC4 zone (the local *Mestognathus praebeckmanni* Subzone) (Groessens in Conil et al., 1991; Poty et al., 2006). It corresponds to the *Pericyclus−Progoniatites* and the lower part of the *Fascipericyclus−Ammonellipsites* ammonoid genus zones, and the CM (*Schopfites daviger−Auroraspora macra*) Zone and probably the base of the Pu (*Lycospora pusilla*) Miospore zones.

The base of the *Visean* Stage defined by the FO of the foraminifer *E. simplex* is the base of the **Moliniacian** Substage following the revision of Hance et al. (2006b). Its stratotype is situated near Salet in the Molignée Valley (Devuyst et al., 2006), the substage comprises the foraminiferal zones MFZ9−11, the rugose coral zones RC4β2 and the lower part of RC5, the *G. homopunctatus* and lower *Taphrognathus transatlanticus* conodont zones, the upper part of the *Fascipericyclus−Ammonellipsites* and lower part of the *Bollandites−Bollandoceras* ammonoid genus zones, most parts of the *S. daviger−A. macra* (CM) and the basal *Koxisporites triradiatus−Koxisporites stephanephorus* (TS) miospore zones. The middle Visean substage, the **Livian**, appears to represent a rather long time (Poty and Hance, 2006; Poty et al., 2014), in contrast to the estimates by Davydov et al. (2012). The stratotype is situated on the right bank of the Meuse Valley in a cliff and disused quarry near the village of Lives (Poty and Hance, 2006). In the stratotype the base of the Livian corresponds to an argillaceous bed of volcano-sedimentary origin ("Banc d'or de Bachant"; bentonite L1 of Delcambre, 1989), for which Pointon et al. (2018) provide an absolute age of 340.13 ± 0.18 Ma. This bed indicates a third-order sequence boundary; hence, strata and subsequently time is missing in the stratotype section

(Poty et al., 2014). The Livian is characterized by FOs of the foraminifers *Koskinotextularia* and *Pojarkovella nibelis* (Paproth et al., 1983; Conil et al., 1991). The later species appears 14.3 m above the base of the Livian in its stratotype (see Poty et al., 2014). The substage comprises a single foraminifer zone (MFZ12), the upper part of the rugose coral zone RC5 and the entire RC6, and the upper part of the *Bollandites−Bollandoceras* and lower part of the *Entogonites* ammonoid genus zones. The *Taphrognathus transatlanticus* conodont zone and *K. triradiatus−K. stephanephorus* (TS) miospore zone start before and after this substage. The late Visean substage, the Warnantian, has its stratotype in the disused Camp de César quarry in the Thom-Samson Valley. The Warnantian is covered by the MFZ13−15 foraminifer zones (the former Cf6 zone of Conil et al., 1977, 1991), the rugose coral zones RC8−9, the upper *Entogonites* to *Lusitanoceras−Lyrogoniatites* ammonoid genus zones, upper parts of the *T. transatlanticus* and the *Gnathodus bilineatus* conodont zones, and the upper *K. triradiatus−K. stephanephorus* (TS) to lower *Reticulatisporites carnosus−Bellispores nitidus* (CN) miospore zones. The latter corresponds to the *Cingulizonates* cf. *capistratus* (CC) Subzone. A continuous succession from the Warnantian into the Namurian is only known in few outcrops (e.g., Bioul−Anhée, Blaton regions).

The Late Carboniferous or Silesian starts in Belgian with a major facies shift toward siliciclastic facies, which is well documented at the base of the **Namurian** Stage. The stage is named after outcrops along the Citadelle de Namur (Dusar, 2006). As mentioned previously, the Mid-Carboniferous Boundary is placed within this stratigraphic unit. The traditional middle and late Silesian stages, **Westphalian** and **Stephanian**, respectively, were named after coal mining districts in Western Germany (Westphalia) and in central France (town of Saint-Etienne). In the late Upper Carboniferous strata of Belgium, detailed subdivisions, including substages as the **Chokierian** (named after a village in the Liège mining district), follow the Western European chronostratigraphic classification (see later).

In Great Britain the subdivision of the Carboniferous was largely accelerated by the pioneering biostratigraphic work of Vaughan (1905) on corals and brachiopods in SW England and in the deeper water ammonoid-bearing facies by Bisat (1924). The first chronostratigraphic scheme for the British Carboniferous by George et al. (1976) and Ramsbottom et al. (1978) was updated and modified by Waters et al. (2011).

The oldest Carboniferous substage of the British Isles is the **Courceyan** with a stratotype at the Old Head of Kinsale in County Cork, Ireland. The base of the substage in its stratotype equates with the base of the *V. verrucosus−R. incohatus* (VI) miospore zone, which is almost coincident with the Devonian−Carboniferous boundary. The application of the boundary between the

Belgian substages Hastarian and Ivorian has been proven to be difficult in the British Isles (Ramsbottom and Mitchell, 1980). Since the former lower Chadian strata are now being attributed to the Courceyan (Waters et al., 2011), the Courceyan is the only substage of the Tournaisian and represents it in total.

The Visean Stage in the British Isles is represented by five substages, which have basal boundaries corresponding to lithologic changes below the incoming of diagnostic fauna (George et al., 1976). The Chadian and Arundian substages correlate approximately to the lower Visean, the Holkerian to the middle Visean, and the Asbian and Brabantian to the upper Visean. It should be underlined that these boundaries are diachronous (Poty et al., 2014). The **Chadian** is named after Chatburn near Clitheroe, Lancashire. The base of the Chadian should correspond to the base of the Visean Stage (Waters et al., 2011), and hence the FAD of *E. simplex*. This criterion has not been established in the stratotype (see Cossey et al., 2004 for a detailed discussion), and Waters et al. (2011) recommend relocating the stratotype to a more suitable location. The second substage of the Visean, the **Arundian**, is defined by the first lithological change below the appearance of the foraminifer family Archaediscidae (George et al., 1976). In its stratotype at Hobbyhorse Bay, Dyfed (South Wales), it corresponds to the base of the Pen-y-Holt Limestone. This definition was maintained by Waters et al. (2011) although the basal 16 m of the Pen-y-Holt Limestone lacks the Arundian fauna (Ramsbottom, 1981), and Riley (1993) recommended using the entry of the Archaediscidae as boundary criterion. According to Waters et al. (2011), the Chadian–Arundian boundary can be equaled to the boundary between the *Fascipericyclus–Ammonellipsites* and *Bollandites–Bollandoceras* ammonoid genus zones. The third substage, the **Holkerian**, has not the same duration as the Livian in Belgian as previously stated by Davydov et al. (2012) (see Poty et al., 2014; Poty, 2016). In its stratotype at Barker Scar, south Cumbria, the base of the substage corresponds to the boundary between the Dalton and Park Limestone formations. This is the first lithological change below the diagnostic Holkerian fauna composed of the corals *Axophyllum vaughani* and *Lithostrotion vorticale*, the brachiopods *Davidsonia carbonaria*, *Composita ficoides*, and *Linoprotonia corrugatohemispherica*, the foraminifers *Archaediscus* spp. of the *A. concavus* stage and paleotextularidiids (George et al., 1976). It corresponds to the lower and middle part of the foraminifer Cf5 zone (=MFZ12) (Waters et al., 2011). The stratotype may be in need for relocation due to significant nondeposition (Riley, 1993). The **Asbian** Substage is traditionally equaled with the beginning of the late Visean. This assumption is not supported by its stratotype at Little Asby Scar, Ravenstonedale, Cumbria, where

the base of the Asbian corresponds to the boundary between the Ashfell Limestone and Potts Beck Limestone formations. The diagnostic Asbian coral fauna appears higher up in the section (Aretz and Nudds, 2005). This is also backed by although contrasting data on foraminifers (Strank in Ramsbottom, 1981; White, 1992), which do place the boundary either higher or lower. Hence, the fauna indicated by George et al. (1976), the corals *Dibunophyllum bourtonense*, *Siphonodendron pauciradiale*, and *Siphonodendron juncerum*, and the brachiopods *Linoprotonia hemispherica* and *Daviesiella llangollensis*, cannot be used as markers of the substage, but their presence characterizes the substage. The substage corresponds to the upper part of foraminifer Cf5 zone (=MFZ12) and Cf6α − Cf6γ2 zones (Somerville, 2008; Waters et al., 2011). The last substage of the Visean, the **Brigantian**, has its stratotype at Janny Wood near Kirky Stephen, Cumbria. George et al. (1976) placed its base at the base of the Peghorn Limestone, below the income of the typical fauna composed of the corals *Diphyphyllum lateseptatum*, *Lonsdaleia (Actinocyathus) floriformis*, *Nemistium edmondsi*, *Orionastrea* spp. and *Palastrea regia*, and the brachiopods *Productus hispidus*, *Productus productus*, and *Pugilis pugilis*. Riley (1993) noted already problems with the stratotype and recommended its relocation and a redefinition to the base of the *Arnsbergites falcatus* (P_{1b}) ammonoid zone. Work on foraminifers form Janny Wood allowed Cózar and Somerville (2004) to propose the relocation of the substage boundary further down to the base of the Birkdale Limestone. However, the original definition of the substage was maintained by Waters et al. (2011).

The **Namurian** Regional Stage comprises seven substages in ascending order Pendleian (E_1), Arnsbergian (E_2), Chokierian (H_1), Alportian (H_2), Kinderscoutian (R_1), Marsdenian (R_2), and Yeadonian (G_1). Except of the Arnsbergian, all substages have their type localities in northern England in the hemipelagic claystones and mudstones of the Craven Basin. All these substages correspond to ammonoid biozones originally defined by Bisat (1924), and the ammonoid faunas are often from distinctive so-called marine bands. In the global chronostratigraphic scheme the Serpukhovian Stage is represented by the Pendleian and Arnsbergian. The rest of the Namurian belongs to the Bashkirian Stage.

The **Pendleian** has its type locality at Light Clough, Pendle Hill, Lancashire. The substage is defined by the FO of the ammonoid *C. leion*, which has already used at the Heerlen Congress in 1935 to define the base of the Pendleian and the Namurian, at that time considered to be stage and series, respectively. It has to be underlined that the conodont *L. ziegleri*, currently under consideration to be a marker of the base of the Serpukhovian Stage, appears several ammonoid zones below the base of the Pendleian (e.g., Sevastopulo and Barham, 2014; Herbig, 2017). The second substage of the Namurian, the

Arnsbergian, is named after the German town of Arnsberg in the Rhenish Mountains, but the marker of the Arnsbergian, the ammonoid *Cravenoceras cowlingense*, is absent in the Arnsberg region. Hence, the section Stony River, Slieve Anierin, in NW Ireland is often considered to be the boundary stratotype following a proposal of Ramsbottom (1969). The **Chokierian** is named after the village of Chokier, several kilometers west of Liège, Belgium, but its type locality is at Stonehead Beck (formerly Gill Beck) in North Yorkshire, Great Britain. The base of the Chokierian corresponds to the FO of *I. subglobosum*. The Stonehead Beck section was the British candidate for the Mid-Carboniferous boundary and there the conodont *Declinognathodus* occurs first 9.4 m above the base of the Chokierian (Waters et al., 2011). The *Alportian* is defined by the FO of *Hudsonoceras proteus*. The substage is named after the Alport Borehole in Derbyshire and has a type locality at Blake Brook, Staffordshire. The **Kinderscoutian** was named after the Kinder Scout area of Derbyshire, and the occurrence of the ammonoid genus *Reticuloceras*. Today, its type locality is at Samlesbury Bottoms, near Blackpool, Lancashire, where the *Hodsonites magistorum* Marine Band marks the substage base. The **Marsdenian** has its stratotype at Park Clough, Marsden, Yorkshire, and its base corresponds to the FO of *Bilinguites gracilis*. The highest substage of the Namurian, the **Yeadonian**, is named after Yeadon, West Yorkshire. The FO of the ammonoid *Cancellocers cancellatum* marks the base of the substage in its new stratotype at Orchard Farm, Derbyshire.

At the Heerlen Congress in 1935 the **Westphalian** was divided into four parts named A, B, C, and D. Later at the Krefeld Congress in 1971, the Westphalian was considered to be a series and the lower boundaries of the first three stages corresponded to the Sarnsbank, Katharina, and Aegir marine beds (George and Wagner, 1972). In the following three stages, today substages, were formally defined based on stratotypes in the English East Pennines Coalfield: Langsettian (Westphalian A), Duckmantian (Westphalian B), and Bolsovian (Westphalian C) (Owens et al., 1985). For the Westphalian D a proposal was made to name it Asturian (Wagner et al., 2002), but SCCS has not made a formal decision. The boundary between the Bashkirian and Moscovian stages can be placed in the late Duckmantian. That between the Moscovian and Kasimovian stages may be in the late Westphalian D (e.g., Opluštil et al., 2016) or higher up (e.g., Falcon-Lang et al., 2018).

The oldest substage of the Westphalian Regional Stage is the **Langsettian** with a stratotype at Langsett, River Little Don, South Yorkshire. It starts with the FO of *Gastrioceras subcrenatum* in the Subcrenatum Marine Bed. The **Duckmantian** with a stratotype at Duckmanton, Derbyshire, starts with by the FO of *Anthracoceratites*

vanderbeckei in the Vanderbeckei Marine Bed. The **Bolsovian** with its stratotype at the River Doe Lea, Bolsover, Derbyshire, starts with the FO of *Donetzoceras aegiranum* in the Aegiranum Marine Bed. The unratified youngest substage of the Westphalian Regional Stage, **Asturian** (former Westphalian D), is named after the Asturias region in NW Spain. Due to the absence of marine fauna in the paralic basins of West and Central Europe, this substage was defined in the marine-influenced succession of NW Spain based on paleobotanic evidences with the FO of *Neuropteris ovata* Wagner et al. (2002). Knight and Alvarez-Vazquez (2019) indicated that the proposed stratotype in the La Riosa Valley might not be suitable.

The Western European chronostratigraphic classification of the **Stephanian** is originally based on outcrops in the intramontane basins of the French Massif Central and the French–German Lorraine–Saar Basin. It was divided into Stephanian A, B, and C. Work in the Cantabrian Mountains showed an important stratigraphic gap at the base of the Stephanian, which resulted in the introduction of the **Cantabrian** Substage at its base (Wagner and Winkler Prins, 1985). Its base corresponds to the base of the Villanueva Marine Formation in the Guardo-Cervera Coalfield and has its stratotype in the Ojosa section of La Pernia. The Cantabrian comprises the *Odonopteris cantabrica* megafloral zone. The conceptually determined Stephanian A of the French Central Massif was formally renamed **Barruelian** based on strata in the Cantabrian Mountains in northern Spain. Its stratotype is in the Barruelo Coalfield and it comprises the *Lobatopteris lamarina* megafloral zone (Wagner and Winkler Prins, 1985). Wagner and Álvarez-Vázquez (2010) and Knight and Wagner (2014) proposed a **Saberian** Substage, which incorporates the time between the top of the Barruelian and the lower conceptual Stephanian B of the French Central Massif. The Saberian is equated with the *Alethopteris zeilleri* megafloral zone (Knight and Wagner, 2014) and has its stratotype in the in the Sabero Coalfield (León, NW Spain). The stratigraphic division of the remaining time between the top of the Saberian and the base of the Permian has still to be defined, but it comprises strata often called Stephanian B and C. Recent radioisotopic data from the French Autun Basin indicate that the Carboniferous–Permian boundary is located near the base of the **Autunian**, the first substage in the regional subdivision of the Central European Permian (Pellenard et al., 2017).

23.1.3.2 Eastern European Subdivisions

A threefold subdivision into lower, middle, and upper series applied to the Carboniferous succession of Russia and surrounding territories in Eastern Europe and Asia based on the work of Murchison (1845) and Möller (1878) was firmly accepted and formalized in the 20th century. These series were further subdivided into stages (Anonymous, 1944). The Lower Carboniferous was divided into the three stages using the Western European names Tournaisian, Visean, and Namurian. However, the Namurian in the Eastern European sense was shorter than in Western Europe (see Van Leckwijck, 1960). Subsequently, its use was abandoned and this stratigraphic interval was formally renamed Serpukhovian Stage. The Middle Carboniferous was subdivided into the Bashkirian (Nikitin, 1890) and Moscovian (Nikitin, 1890) stages, the Upper Carboniferous into Kasimovian (Theodorovich, 1936) and Gzhelian (Nikitin, 1890).

Detailed stratigraphic work in the former Soviet Union resulted in stratigraphic schemes, which had often chronostratigraphic value (e.g., Aizenverg et al., 1978; Kagarmanov and Donakova, 1990). These schemes have been developed further into often boundary stratotype—based units with robust biostratigraphic control (e.g., Alekseev et al., 1996, 2004, 2013; Menning et al., 2006; Zhamoida, 2015). In this process, lithostratigraphic units (formations, horizons) have been equaled or developed into regional substages and integrated into global stratigraphic scales (e.g., Menning et al., 2006; Davydov et al., 2012; Kabanov et al., 2016). These regional substages are not necessarily thick and fully marine units, and historical boundaries often correspond to facies changes or stratigraphic discontinuities. Some of the substages are equaled to assemblage biozones with FOs of the zonal markers in older strata. The succession of the Russian Platform is divided into 33 substages and that of the Donets Basin into 30 substages (e.g., Einor, 1996; Alekseev et al., 2004; Davydov et al., 2010). The schemes of the Russian Platform are based on the successions of the Moscow Syncline and its shallow-marine and terrestrial records, and the successions at the eastern margin of the platform in the central and southern Urals, where also distal slope and basinal facies occurs (e.g., Alekseev et al., 1996; Makhlina, 1996).

The base of the **Tournaisian** Stage is placed with the marker for the global stage, the FO of the conodont *S. sulcata* within the **Gumerovian** Regional Substage (Pazukhin et al., 2009). The following **Malevkian** Regional Substage can be equaled to the *Earlandia minima* foraminifera zone, which comprises the top of the *S. sulcata* and *Siphonodella duplicata* conodont zones (Kulagina et al., 2003). The **Upian** Regional Substage corresponds to the *Prochenyshinella disputabilis* foraminifera zone and the *Siphonodella belkai* conodont zone (Kulagina et al., 2003). Due to the absence of marine facies in the Moscow Syncline, the **Karabukian** Regional Substage has its type section in the marine facies of the Donets Basin. In the Urals the Upian is followed by the **Kosorechian** Regional Substage, which is equaled with the *Chernysbinella glomformis* foraminifera subzone of the *Palaeospiroplectamina tchernyshinensis* foraminifera zone and the *Siphonodella*

quadruplicata conodont zone (Kulagina et al., 2003). Below a significant late Tournaisian—middle Visean gap in the Moscow Basin, the shallow-marine deposits of the **Chereptian** Regional Substage are equaled to the *S. quadruplicata* conodont zone and the *Chernyshinella glomiformis—Chernyshinella paraglomiformis*, *C. glomiformis—Spinoendothyra krainica—Paleospiroplectammina tchernyshinensis* foraminifera zones (Alekseev et al., 2004). In the Urals (Kulagina et al., 2003) the upper part of the *P. tchernyshinensis* foraminifera zone corresponds to the local *Latiendothyra latispiralis* subzone, which is correlated with the *Siphonodella isosticha* conodont zone. These zones define the base of the **Kizelian** Regional Substage, which together with the following **Kosvian** Regional Substage is considered to represent the late Tournaisian. Much of the Kizilian is represented by the *Spinoendotyra costifera* foraminifera zone, which corresponds to the lower part of the *Gnathodus typicus* conodont zone (Kulagina et al., 2003). The topmost *S. costifera* foraminifera zone is already Kosvian in age. The Kosvian Regional Substage comprises the *Eotextularia diversa* and *Eoparastaffillina rotunda* foraminifera zones and the upper *G. typicus* (*Dollymae bouckaerti* subzone) and *S. anchoralis* zones (Kulagina et al., 2003; Menning et al., 2006), hence a rather long stratigraphic interval of almost the entire Belgian Ivorian.

The **Visean** Stage is divided into the six regional substages. Characteristic for the lower part, comprising approximately the lower two regional substages, are coal-bearing deposits in the central parts of the Russian Platform (Moscow Syncline). This terrestrial facies is gradually replaced by marine facies in the upper part of the Visean. For those parts of the platform, where a marine record is known, Kulagina et al. (2003) place the *E. simplex* foraminifera zone at the base of the Visean and tentatively correlate this zone with the **Radaevkian** Regional Substage. In terms of conodont zones, it is attributed to the lower part of the *Gnathodus texanus* conodont zone. The following *Uralodiscus rotundus* foraminifera zone is equivalent to the **Bobrikovian** Regional Substage. It corresponds to the middle part of the *G. texanus* conodont zone. The Visean succession above the Bobrikovian was called Okian Stage by Makhlina (1996). The base of the **Tulian** Regional Substage is correlated with the Molinacian/Livian boundary, which indicates a middle to late Visean age for the Tulian to Venevian substages, which all corresponds to foraminifera zones. The carbonate—siliciclastic succession of the Tulian Regional Substage comprises the *Endothyranopsis compressa—Archaediscus krestovnikovi* Zone (Makhlina et al., 1993) or *E. compressa—Paraarchaediscus koktjubensis* Zone (Alekseev, 2008). It is a rather long stratigraphic interval corresponding to the Belgian Livian and basal Warnantian (Kulagina et al., 2003). The remaining three regional substages are, especially in their type area, characterized by clean limestones. Historically, the differentiation was based on parallel unconformities or rhizoid limestones, but today biostratigraphic criteria predominate (see lit. in Kabanov et al., 2016). Every substage corresponds to a foraminifera biozone; the **Aleksinian** Regional Substage to the lower part of the *Endothyranopsis crassa–Archaediscus gigas* Zone (Alekseev, 2008), the **Mikhailovian** Regional Substage to the *Eostaffella ikensis* Zone (Makhlina et al., 1993), and the **Venevian** Regional Substage to the *Endothyranopsis sphaerica–Eostaffella tenebrosa* Zone (Kulagina et al., 2003). In terms of conodont biostratigraphy the (upper) Tulian, Aleksinian, and lower Mikhailovian correspond to the *G. bilineatus bilineatus* Zone, the middle Mikhailovian to middle Venevian to the *Lochriea nodosa* Zone and the upper Venevian to the *L. ziegleri* Zone (Makhlina et al., 1993; Alekseev et al., 2004; Menning et al., 2006; Kabanov et al., 2016).

The **Serpukhovian** Stage is divided into four regional substages. The first two represent a lower and the latter two an upper Serpukhovian. The base of the Serpukhovian Stage corresponds to the base of the **Tarusian** Regional Stage, which coincides with the FO of the foraminifers *N. postrugosus* and "*Millerella*" *tortula*. Those foraminifera define the local, short-ranging *N. postrugosus* Zone, which may or may be not included into the succeeding *Pseudoendothyra globosa* Zone (see different approaches in Alekseev et al., 2004; Gibshman et al., 2009; Kabanov et al., 2009, 2016). The **Steshevian** Regional Stage is equaled to the *Eostaffellina decurta* foraminifera zone, but Gibshman et al. (2009) and Kabanov et al. (2009) showed the entry of the zonal markers, below the zone in the upper Taurisan. Kabanov et al. (2016) indicate for the lower Steshevian in the type area of the Serpukhovian the FO of the conodonts *Vogelgnathus campbelli* (rare) and *Cavusgnathus unicornis* and *Cavusgnathus naviculus*. The **Protvian** corresponds to the *Eostaffellina paraprotvae* or *Eostaffella protvae–Eostaffella minifica* foraminifera zone. The stratigraphically important conodonts *Adetognathus unicornis* and *Gnathodus bollandensis* occur in the lower Protvian, but their precise FOs are currently unknown (Kabanov et al., 2016). Upper Serpukhovian–Bashkirian strata are missing in the Moscow Basin (e.g., Alekseev et al., 2004), so the stratigraphic nomenclature is based on the successions in the Donets Basin and the Urals. The **Zapaltyubian** Regional Substage as youngest Serpukhovian substage originates from the Donets Basin. It comprises the *Monotaxinoides transitorius* foraminiferal zone.

With the definition of the global Mid-Carboniferous boundary (Lane et al., 1999; see above), the base of the **Bashkirian** Stage corresponds on the platform to the base of the Voznesenskian Regional Substage and in the deeper marine facies to the Bogdanovkian Regional Substage. Overall, two sets of Bashkirian regional substages are available. In the platform facies the stage is divided into the Voznesenskian, Krasnopolyanian, Severokeltmenian, Prikamian, Chermshankian, and Melekessian regional substages. The oldest stage is based on the strata from the Donets Basin; the other five have their type area on the eastern Russian Platform. In terms of foraminifera biozones the **Voznesenskian** Regional Substage corresponds to the *P. bogdanovkensis* Zone, the **Krasnopolyanian** Regional Substage to the *Semistaffella variabilis* Zone, the **Severokeltmenian** Regional Substage to the *Pseudostaffella antiqua* Zone, the **Prikamian** Regional Substage to *Staffellaeformes staffellaeformis–Pseudostaffella praegorskyi* Zone, the **Cheremshankian** Regional Substage to the *Ozawainella pararhomboidalis–Pseudostaffella gorskyi* Zone, and the **Melekessian** Regional Stage to the *Tikhonovichiella tikhonovichi* Zone. In the foredeep of the southern Urals a second set of regional substages is available. Here the Bashkirian Stage is divided into the Syuranian, Akavasian, Askynbashian, and Arkhangelskian regional substages. In terms of conodont zones (Kulagina et al., 2009) the **Syuranian** Regional Substage corresponds to the *D. noduliferus and Idiognathodus sinuatus* zones, the *Akavasian* to the *Neognathodus askynensis–Neognathodus symmetricus* Zone, the **Askynbashian** Regional Substage to the *Idiognathodus sinuosus* Zone, and the **Arkhangelskian** Regional Substage to the *D. marginodosus* Zone. The Syuranian Regional Substage can be further divided into a Bogdanovkian Horizon (*D. noduliferus* and basal *Id. sinuatus* zones) and a Kamennogorian Horizon (*Id. sinuatus* zones except its basal part). The Arkhangelskian Regional Substage can be divided into the Tashastian and Asatauian Horizons using the *Branneroceras–Gastrioceras* and *Diaboloceras–Axinolobus* ammonoid genozones. The Asatauian also comprises the *Verella spicata–T. tikhonovichi* foraminifera zone.

Around the Bashkirian–Moscovian boundary, marine facies return to the central parts of the Russian Platform. Hence, most regional substages of the middle and late Pennsylvanian have their type area in the cyclothems of the wider Moscow Syncline. The **Moscovian** Stage is divided into four regional substages. The base of the **Vereian** Regional Substage corresponds to the base of the Aljutovo Formation (Makhlina et al., 2001), and the appearance of the biostratigraphic markers *A. aljutovica* (fusulinid) and *D. donetzianus* (conodont), which also give the name to the respective conodont and foraminifera biozones. The upper part of the substage (Ordynka Formation) comprises the *Ovatella arta* fusulinid zone and the *Idiognathoides ouachitensis* and *Streptognathus transitivus* conodont zones. The **Kashirian** Regional Substage is divided into four formations in its type area (Makhlina et al., 2001). In ascending order each formation is characterized by a fusulinid biozone: *Priscoidella priscoidea* (Tsna Formation), *Hemifusulina moelleri–Beedeina* *pseudoelegans* (Nara Formation), *Moellerites praecolania–Fusulina subpulchra* (Lopasnya Formation), and *Hemifusulina vozhgolica* (Smedva Formation) zones (Makhlina et al., 2001; Alekseev et al., 2004). The lower two formations contain the *N. bothrops* conodont zone, whereas the upper two equal, respectively, to the *Neognathodus medadultimus* and *Swadelina coninna – Idiognathodus robustus* zones. The **Podolskian** Regional Substage comprises three fusulinid zones (Baranova et al., 2014): *Putrella brazhnikovae, Fusulinella colanii–Fusulina ulitinensis*, and *Fusulina chernovi*. The boundary between the conodont zones *Neognathodus medexultimus–Idiognathodus podolskens* and *Neognathodus inaequalis* lies within the *P. brazhnikovae* fusulinid zone (Makhlina et al., 2001; Baranova et al., 2014). The **Myachkovian** Regional Stage comprises three fusulinid zones (*Fusulinella bocki, Fusulina cylindrica*, and *P. ovatus*) and the upper part of the *Neognathodus inaequalis* conodont zone and the following *Neognathodus roundyi* conodont zone. It is important to note that the sequence stratigraphic model of Kabanov (2003) and Kabanov and Baranov (2007) impacts the traditional attribution of formations names in the upper Moscovian.

The **Kasimovian** Stage is divided into the Krevyakinian, Khamovnikian, and Dorogomilovian regional stages, which were defined on the eastern Russian Platform. Its base corresponds to a significant eustatic sea-level low (see references in Davydov et al., 2012).

The **Krevyakinian** Regional Substage corresponds to the *O. obsoletes–P. pseudomontiparus* fusulinid zone (Goreva et al., 2009). The conodont *Sw. subexcelsa* first occurs near the base of this substage. The **Khamovnikian** Regional Substage corresponds to the *Montiparus montiparus* fusulinid zone (Alekseev et al., 2004). The two conodont species, discussed for the definition of the global stage, *Id. turbatus* and *Id. sagittalis* first occur in the Khamovnikian. The **Dorogomilovian** Regional Substage corresponds to the *Triticites quasiarcticus–Schwageriniformis mosquensis* and *Tricites irregularis–Tricites actus* fusulinid zones fusulinid zone (Alekseev et al., 2004). Three conodont zones have been identified in this substage: *Idiognathodus mestsherensis, Idiognathodus toretzianus*, and *Streptognathodus firmus*.

The definition of the base of the global Gzhelian Stage moves the traditional base of the **Gzhelian** of the Russian Platform slightly higher (Alekseev et al., 2009) above the FO the fusulinid species *R. rossicus*. In the Moscow Basin the Gzhelian is divided into the Dobryatinian, Pavlovoposadian, Noginskian, and Melekhovian regional substages, which all corresponded to one fusulinid zone and several conodonts zones (Goreva and Alekseev, 2010). In terms of fusulinids the **Dobryatinian** Regional Substage corresponds to the *R. rossicus–Rauserites paraarcticus* Zone, **Pavlovoposadian** Regional Substage to the *Jigulites jigulensis* Zone, **Noginskian** Regional Substage to the *Daixina sokensis* Zone, and the **Melekhovian** Regional Substage to the

Daixina robusta–Daixina bosbytauensis Zone. Based on sections in deeper water and flysch facies, a detailed conodont biostratigraphy can be scaled against the substages (Goreva and Alekseev, 2010). The Dobryatinian comprises the *Streptognathodus simulator* and *Streptognathodus vitali* zones, the Pavlovoposadian most of the *Streptognathodus virgilicus* zone, which reaches into the lower part of the Noginskian. The rest of the Noginskian contains the *Streptognathodus bellus* zone. The *Streptognathodus wabaunsensis* Zone characterizes the Melekhovian.

23.1.3.3 North American Subdivisions

In North America the Carboniferous was divided into the Mississippian (Winchell, 1869) and Pennsylvanian (Williams, 1891; see Manger, 2017). Since the beginning of the 20th century they have been regarded as independent systems by the North American stratigraphers. The Mississippian has its type area in the Mississippi Valley from north of Burlington, Iowa, to southern Illinois, where marine Carboniferous strata covers the interval between the Devonian–Carboniferous and Mid-Carboniferous boundaries (Lane and Brenckle, 2005). The Pennsylvanian was named after the state of Pennsylvania and its coal-producing strata with the obvious resemblance to western European "Coal Measures" or late Carboniferous.

It is important to note that the North American regional substages are not based on boundary stratotypes like their counterparts in other parts of the world, even if they are used in this sense (e.g., Lucas, 2016). The Mississippian is commonly subdivided into Kinderhookian, Osagean, Meramecian, and Chesterian using unconformable boundaries in the succession in the Mississippian type area in the Illinois Basin. They became chronostratigraphic units for North America in the first half of the 20th century, after biostratigraphic characteristics had been added to each unit (Lane and Brenckle, 2005). According to recent stratigraphic classification these four units are regional stages. The correlation with the global middle and upper Mississippian stages has not been fully achieved, mostly due to faunal endemism in North America. The Tournaisian–Visean boundary is situated within the Osagean Regional Stage. Sevastopulo and Devuyst (2005) indicate that its position is most likely at the base of the Keokuk Formation when using conodont data. Hence, the level of uncertainty is reduced to max. 15-m thickness. The position of the Visean–Serpukhovian boundary within the Chesterian Regional Stage is not precisely known. Lane and Brenckle (2005) correlate the base of the Pendleian with the base of the Menard Limestone. However, new ammonoid data (Korn and Titus, 2011; Titus et al., 2015) may help to further constrain the position of the VSB in North America.

The **Kinderhookian** Regional Stage is named after the village of Kinderhook, Illinois. Its base is defined by the FO of the conodonts *S. sulcata* and *Protognathodus kuehni* at the base of the Horton Creek Formation (Conkin and Conkin, 1973). The Kinderhookian comprises the *Sulcata* to *Isosticha* conodont zones (FU1A-FU1G of Lane, 1974). The **Osagean** Regional Stage was named after outcrops along the Osage River, Missouri. In the Mississippian type area the Kinderhookian−Osagean boundary is an unconformity between the Chouteau Formation and the Meppen Limestone. The base of the regional stage corresponding to the FO of *Punctatus* conodont zone (Lane, 1974; Lane and Brenckle, 2005) has been found on the opposite of the Ozark Uplift in southwestern Missouri (Thompson and Fellows, 1970). The Osagean comprises the *Punctatus* to the lower part of the *Texanus* conodont zones (FU2−FU7 of Lane, 1974). The **Meramecian** Regional Stage was defined at the Meramec Highlands Quarry along Meramec River, southwest of St. Louis, Missouri. In the type section the base of the stage is characterized by the FO of the brachiopods *Warsawia lateralis*, *Planalvus densa*, *Crossacanthia perlamellosa*, *Setigerites altonensis*, and *Tetracamera subcuneata*, the blastoid *Pentremites conoideus*, and the foraminifer *Globoendothyra baileyi* (Kammer et al., 1990; Lane and Brenckle, 2005). This level may correspond to the FO of the conodont *Hindeodus penescitulus* within the *Texanus* conodont zone. The top of the Meramecian comprises the *Scitulus−Scalenus* conodont zone. The last regional stage of the Mississippian is the **Chesterian**. It is named after Chester in southern Illinois. Today, it comprises in its type area the strata from the Ste. Genevieve Formation to the Grove Church Shale, but the upper part of the stage is missing. Lane and Brenckle (2005) use the FO of asteroarchaediscin foraminifers (*Asteroarchaediscus* and *Neoarchaediscus*) to define base of the Chesterian. This boundary approximates the base of the *Bilineatus* conodont zone. The Chesterian Regional Stage comprises the *Bilineatus* to Lower *Muricatus* conodont zones (Lane, 1974; Lane and Brenckle, 2005).

The Pennsylvanian is divided into five regional stages. They were defined in cyclic marine sequences of the Midcontinent Basin. The **Morrowan** Regional Stage is named after a town in northwestern Arkansas. It has been considered to correspond to the zone of primitive fusulinid *Millerella* (Moore and Thompson, 1949; today *Millerella−Eostaffella* Genozone, Wahlman, 2013), and it is also characterized by reticuloceratid and schistoceratid ammonoids (Heckel, 2013). Due to a widespread Mississippian−Pennsylvanian unconformity, the base of the Morrowan, including in its type area, is mostly absent in North America. Exceptions are found in Nevada (GSSP for the Mid-Carboniferous Boundary) and more basinal areas like in Arkansas (Lane et al., 1999). The Morrowan foraminiferal assemblages are low diverse and highly provincial. The Morrowan comprises the *D. noduliferus−Rhachistognathus primus* to *Idiognathoides convexus* or *Neognathus higginsi* to lower parts of the *Neognathodus nataliae* conodont zones (Barrick et al., 2013).

The **Atokan** Regional Stage is named after Atoka town in southeastern Oklahoma. Moore (1948) and Moore and Thompson (1949) considered that the Atokan corresponds to the *Profusulinella* and *Fusulinella* fusulinid genozones. Groves (1986) established the current base of the stage with the FO of the fusulinid genus *Eoschubertella*, and hence the *Eoschubertella−Pseudostaffella* fusulinid genozone. Wilde (1990) used these fusulinid genozones to subdivide the Atokan into three parts. In terms of conodont biostratigraphy (Barrick et al., 2013) the Atokan starts with *N. nataliae* conodont zone in which *Idiognathoides incurvus* occurs and ends with the extinction of the *Idiognathoides* in the *Neognathodus colombiensis* conodont zone.

The **Desmoinesian** Regional Stage is named after outcrops along the Des Moines River in Central Iowa. An unconformity exists between the Atokan and Desmoinesian in its type area and more complete successions have been described in the Ardmore Basin in Oklahoma (Clopine, 1992). In general, the Desmoinesian Regional Stage is equaled with the *Beedeina* fusulinid genozone. However, the base of the stage is not precisely defined since the *Fusulinella iowensis* Zone is either included or excluded into this genozone. This is probably rooted in the transitional evolutionary character between *Fusulinella* and *Beedeina* (Davydov et al., 2010). The Desmoinesian comprises the *N. colombiensis* (without *Idiognathoides*) to *N. roundyi* (sensu lato) and the *Idiognathodus amplificus−Idiognathodus obliquus* to *Swadelina nodocarinata* conodont zones (Barrick et al., 2013) and the *Wellerites* and *Eothalassoceras* ammonoid zones (Boardman and Work, 2013). Typical for the Desmoinesian and the following two regional stages is the cyclic sedimentation pattern of shales, limestone, coals, and sandstone, related to glacio-eustatic variations and known as cyclothems, which are the backbone of very detailed stratigraphic correlations within the stages (see summary in Heckel, 2013).

The **Missourian** Stage is named after outcrops along the Missouri River in Iowa and Missouri. Traditionally, it corresponds to the lower part of the *Triticites* fusulinid genozone, although this genus only occurs in the upper two-thirds of the stage (Thompson et al., 1956). The base of the stage has been redefined by Heckel et al. (2002) as corresponding to the base of the Exline cyclothem, which corresponds to the FO of the conodont *Idiognathodus eccentricus*. This boundary is higher than the originally disconformable base of the stage at the base of the Pleasanton Group. The Desmoinesian−Missourian boundary also corresponds to a major extinction event in the North American province (Davydov et al., 2012). Important biostratigraphic markers for the Missourian are the FOs of the ammonoid genus *Pennoceras* (Boardman et al., 1991), and the conodont genus *Streptognathodus* (s.l.) (Barrick et al., 2004). The stage comprises the *I. eccentricus* to *I. eudoraensis* conodont zones (Barrick et al., 2013) and the *Pennoceras* ammonoid zone up to the *Pseudoaktubites newelli* ammonoid subzone (Boardman and Work, 2013). Foraminifer data

(Wahlman, 2013) show a gap in the foraminifer record above the conodont-based base of the stage. Missourian foraminifera zones range from the *Eowaeringella ultimita* to the lower part of the *Triticites secalicus—Triticites oryziformis* zones.

The **Virgilian** Stage is named after a town in east-central Kansas. Moore (1932) used a major unconformity to separate it from the original Missourian. Most fossil groups only changed slightly between the Missourian and Virgilian. The stage is considered to correspond to upper part of the *Triticites* fusulinid genozone and the *Uddenites* ammonoid genozone (Moore and Thompson, 1949). The base of the stage was previously placed at several different levels within, at the top, or above the South Bend cyclothem (see Boardman et al., 1994; Heckel and Watney, 2002). Today, the FO of the conodont *Streptognathodus zethus* in the lower part of the Cass cyclothem is used for the boundary definition (Heckel, 1999; Barrick et al., 2013). In the Midcontinent the first appearance of *Streptognathodus isolatus*, as used by Boardman et al. (2009), indicates the Permian System. Hence, the Virgilian comprises the *S. zethus* to *Streptognathodus binodosus* conodont zones (Barrick et al., 2013), and the upper part of the *T. secalicus—T. oryziformis* to the *Triticites ventricosus—Schwagerina longissimoidea* fusulinid zones (Wahlman, 2013). The stage comprises the interval defined by the *Pseudoaktubites stainbrooki* ammonoid subzone to *Subperrinites* ammonoid zone (Boardman and Work, 2013).

The correlation of the North American regional stages to the global Pennsylvanian stages remains partly problematic since it is hampered by the unaccomplished work of the task groups to precisely define the bases of the global stages, the absence of boundary stratotypes for the North American stratigraphic units and different views on the completeness of the North American succession (e.g., Heckel, 2013; Richards, 2013; Lucas, 2016; Wagner and Winkler Prins, 2016; Wagner, 2017).

23.1.3.4 Chinese Subdivisions

Originally, a twofold subdivision into Fengningian (Mississippian) and Pennsylvanian was applied to the Carboniferous succession of South and North China (Ting and Grabau, 1936). Later, a threefold subdivision into Lower, Middle, and Upper series was also applied (e.g., Yang et al., 1962; Wu et al., 1974; Zhang, 1988). The proposal of Yang et al. (1979) to use the development of biota and the sedimentological history in two distinct phases in South China and other parts of China, for a twofold subdivision of the Carboniferous into Fengningian (Lower) and Hutianian (Upper) series, has been widely accepted (e.g., Wu et al., 1987; Wang and Hou, 1990). Jin et al. (2000) proposed a refined scheme, which consists of the Fengningian and Hutianian subsystems divided into four series (Aikuanian, Tatangian, Weiningian, and

Mapingian) and eight stages (Jiusian, Shangsian, Dewuan, Luosuan, Huashibanian, Dalaun, and Xiaodushanian). Wang et al. (2019) provided the most recent detailed stratigraphic scale of Carboniferous strata in China.

The **Fengningian** Regional Subsystem, named after the old district name "Fengning" in southern Guizhou, South China, was originally described by Ting (1931) as a system, which consists of Kolaoho (Gelaohe) Limestone and Tangpakou (Tangbagou) Sandstone of the Aikuan Group and Chiussu (Jiusi) Sandstone and Shangssu (Shangsi) Limestone of the Tatang Group. Those lithological units later gradually became regional chronostratigraphic units. The historical type area of the Fengningian is located between Shangsi and Jiusi towns, SW Dushan County, and is represented by shallower-water facies. Its reference section is the Baihupo section near Dushan City.

The **Aikuanian** Regional Series is named after Yanguan (=Aikuan) Village, Dushan County Seat, Guizhou (Ting, 1931). Originally considered as a stage, it was elevated to a series by Jin et al. (2000) and Wang and Jin (2000). Ting (1931) included the Gelaohe Formation with its *Cystophrentis* fauna into the Aikuanian, but it was transferred in the middle 1980s, based on microfossils, into the Upper Devonian. Today, the basal boundary is marked by the occurrence of the rugose coral *Pseudouralinia* in the coral—brachiopod-dominated sequences, which are widespread in South China, and by the appearance of *S. sulcata* or *Gattendorfia* in the conodont—cephalopod-dominated sequences with limited distribution. The references sections are the Baihupo and Qilinzhai sections near Dushan City (Wu et al., 1987).

The Aikuanian corresponds to the Lower Series of the Mississippian Subsystem. It comprises a single stage, the **Tangbagouan** Regional Stage (Jin et al., 2000), originally established by Ting (1931) as Tangpakou (Tangbagou) Sandstone with a type locality at Tangbagou (Dushan County Seat). In geological practice, the Tangbagouan had served as a chronostratigraphic standard in China for decades due to widely recognizable presence of the rugose coral *Pseudouralinia* and its associated fossils. Tangbagouan strata in its reference section, the Qilinzhai section, are composed of fossiliferous sandstone and mudstone of inner shelf facies and comprise fossils. In the pelagic realm the Tangbagouan comprises the *S. sulcata* to *S. anchoralis—Gnathodus pseudosemiglaber* conodont zones (Ji and Xiong, 1985; Hu et al., 2019; Wang et al., 2019). The Tangbagouan Regional Stage corresponds to the Tournaisian Stage.

The **Tatangian** Regional Series was originally established in its type locality at Xiguan village, Datang (=Tatang) Town, South Guizhou (Ting, 1931). Since the Tatangian was practically applied as the Upper Fengningian (Mississippian) Subsystem and was correlated with the Visean and Serpukhovian stages, Jin et al. (2000) formalized it to be a series divided into three regional stages: the Jiusian, Shangsian, and Dewuan. The

Tatangian Regional Series corresponds to the Middle and Upper series of the Mississippian Subsystem.

The **Jiusian** Regional Stage was originally established by Ting (1931) as Chiussu (=Jiusi) Sandstone from its type locality at Jiusi Village, near Datang Town. In the type are, it includes the stratigraphic sequence composed of the Xiangbai and Jiusi formations (Jin et al., 2000; Wang and Jin, 2003). The Jiusian comprises the *Pseudognathodus homopunctatus* and *Lochriea commutata* conodont zones in South Guizhou and Guangxi (Wang, 1991; Hou et al., 2011). It may correspond to the lower half of the Visean Stage.

The **Shangsian** Regional Stage was first introduced as the Shangssu (Shangsi) Limestone from its type locality at Shangsi Village, 25 km South of the Dushan County Seat (Ting, 1931). The Shangsian faunas can be grouped into the *Yuanophyllum* rugose coral zone (Yu, 1931), the *G. bilineatus bilineatus* and *L. nodosa* conodont zones (Wang, 1991; Wang and Qi, 2003), the *Vitiliproductus groberi−Pugilis hunanensis* and *Gigantoproductus moderatus* brachiopod zones (Wu et al., 1974), and the *A. krestovnikovi* and *Asteroarchaediscus baschkiricus* foraminifer zones (Wang, 1983; Lin et al., 1990). The Shangsian Regional Stage may correspond to the upper half of the Visean Stage.

The **Dewuan** Regional Stage has its type locality in the Dewu District of Liupanshui City, Guizhou (Yang, 1962). It comprises the *L. ziegleri*, *G. bollandensis*, and *G. postbilineatus* conodont zones (Qi and Wang, 2005; Hu et al., 2019); the *Dombarites−Eumorphoceras* ammonoid zone (Ruan, 1981); the *Gigantoproductus edelburgensis−Gondolina−Striatifera* brachiopod zone (Li, 1987); and the *Janischewskina delicata/Plectomillerella tortula*, *E. paraprotvae*, *Bradyina cribrostomata*, and *M. transitorius* foraminifera zones (Sheng et al., 2018). The Dewuan Regional Stage is approximately correlated with the Serpukhovian Stage.

The **Hutianian** Regional Subsystem, named after its type locality at the Hutian Town, central Hunan, South China, was originally established as Hutian Limestone by Tien and Wang (1932). It now consists of the Weiningian and Mapingian series and corresponds to the Pennsylvanian Subsystem.

The **Weiningian** Regional Series is named after the Weining County Seat. The Weiningian as applied by Chi (1931) and Ting and Grabau (1936) represented the strata overlying the Fengningian. Jin et al. (2000) assigned it to be a series subdivided it into the Luosuan, Dalaun, and Huashibanian stages. The Weiningian Regional Series corresponds to the Lower and Middle series of the Pennsylvanian System. The lower boundary of the Weiningian is marked by the FO of conodont *D. noduliferus* s.l., accordingly.

The **Luosuan** Regional Stage is named after Luosu town, Luodian County in South Guizhou. In its type section the Naqing (Nashui) section (Rui et al., 1987), the Luosuan strata are mainly wackestone beds with planar lamination interbedded with packstones, grainstones, and cherty beds, representing slope facies. The basal boundary of the Luosuan is marked by the FOs of conodont *D. noduliferus* s.l. and fusulinid *Millerella marblensis* (Rui et al., 1987). It comprises the *D. noduliferus* s.l. and *Idiognathoides sulcatus−Id. sinuatus* conodont zones (Hu et al., 2019), and the *Homoceras* and *Reticuloceras* ammonoid zones (Ruan, 1981; Wang et al., 2019). A brachiopod assemblage with characteristic elements such as *Weiningia* dominated the Luosuan strata in shelf facies (Li, 1987). The Luosuan Regional Stage correlates with the lower part of the Bashkirian Stage.

The **Huashibanian** Regional Stage is named after Huashiban Village, 30 km East of Panxian (now Panzhou) City, Guizhou. The lower boundary in its stratotype section was marked by the FO of fusulinid *P. antiqua* (Zhang et al., 2004). According to Zhang et al. (2004), it comprises the *P. antiqua−P. antiqua posterior* and *Pseudostaffella composite−Pseudostaffella paracompressa* fusulinid zones and *Reticuloceras* and *Gastrioceras−Branneroceras* ammonoid zones. In slope section the Huashibanian correlates with the *N. symmetricus*, *Idiognathodus primulus* and *Streptognathodus expansus* M1 conodont zones (Wang et al., 2019; Hu et al., 2019). The Huashibanian Regional Stage can be correlated with the middle part of the Bashkirian Stage.

The **Dalaun** Regional Stage is named after Dala Village, 30 km East of Panxian (=Panzhou) City, Guizhou (Jin et al., 1962). The lower boundary of the Dalaun is marked by the FO of fusulinid *Profusulinella priscoidea* and/or *Profusulinella parva* (Zhang et al., 2005). In the pelagic facies it comprises the *Winslowoceras* and *Owenoceras* ammonoid zones (Wang et al., 2019) and the interval between the *S. expansus* M2 and *Swadelina makhlinae* conodont zones (Wang et al., 2019; Hu et al., 2019). The Dalaun Regional Stage may correlate with the upper part of the Bashkirian and Moscovian stages.

The **Mapingian** Regional Series is named after Maping, former name of Liuzhou City, Guangxi (Ting, in Yoh, 1929). The section at Zhaojiashan near Weining of Guizhou was commonly regarded as the type section (Wu et al., 1987). The Mapingian used to comprise the interval of the *Montiparus*, *Triticites*, and *Pseudoschwagerina* of the fusulinacea genozones (Yang et al., 1979; Wu et al., 1987; Wang and Hou, 1990). Since the traditional Mapingian contains the Carboniferous−Permian boundary, it was subsequently split into Carboniferous (*Montiparus−Triticites* zones) and Permian (*Pseudoschwagerina* Zone) parts. Those divisions got several names, but today the Xiaodushanian Regional Stage is employed (Jin et al., 2000). The Mapingian may correspond to the Upper Series of the Pennsylvanian Subsystem; it consists of only the Xiaodushanian Regional Stage.

The **Xiaodushanian** Regional Stage was designated by Zhou et al. (1987) to represent the stratigraphic interval corresponding to the fusulinacean succession consisting of the *Protriticites subschwagerinoides*, *Triticites montiparus*, *Triticites schwageriniformis*, *Triticites dictyophorus*, and *Triticites shikhanensis compactus* zones in its type section, Xiaodushan section, 5 km East of the Babao Town, Guangnan County, Yunnan. In the pelagic facies this interval comprises the *Id. turbatus* to *S. wabaunsensis* conodont zones (Wang and Qi, 2003; Hu et al., 2019). The Xiaodushanian Regional Stage may correlate with the Kasimovian and Gzhelian stages.

23.1.3.5 Gondwana

Gondwana consists of the modern southern continents, peninsular India, and the Cimmerian terranes that separated from the northern rim of the supercontinent since the beginning of the Permian (Yeh and Shellnutt, 2016). Large intracontinental basins of different tectonic settings show predominant continental, often strongly discontinuous sediment piles.

At least for certain Carboniferous time slices, the European and North American stratigraphic nomenclatures have been directly applied to the marine Carboniferous basins facing the Paleotethys along the northern margin of Gondwana, (e.g., Sahara Basins, Conrad, 1985; Amazon—Solimões basins, Cardoso et al., 2017; Bonaparte Basin, NW Australia, Gorter et al., 2005). For most other parts of Gondwana, chronostratigraphic correlation with the marine tropical realm is strongly hampered. Reasons are (1) the predominance of continental depositional regimes in many regions during extended time intervals, (2) concomitant scarce biostratigraphic data that have to rely on palynomorphs, fewer macrofloristic remains, and subordinate vertebrates, and (3) the paleoclimatic constraints caused by the position of major parts of Gondwana in high latitudes. This caused the well-known Permo-Carboniferous paleofloristic "Gondwana Province" (Gothan, 1937; Chaloner and Lacey, 1973; Archangelsky, 1990, and many others). Further partitioning by differing ecological conditions has to be considered (DiMichele et al., 2001, 2005).

In consequence, lithostratigraphic subdivisions prevail in such intracontinental basins, with supergroups/groups used as parachronostratigraphic units that are roughly correlated with the global chronostratigraphic stages. The paleoclimatic constraints in the marine realm resulted in siliciclastic successions, opposed to the carbonate-dominated paleoequatorial domain with faunal groups prone to global correlation, such as foraminifera, ammonoids, and conodonts. Organism associations adapted to temperate to cool water conditions prevail (e.g., López Gamundi, 1989; Cisterna and Sterren, 2010; Fielding et al., 2008a) and result in regional biozonations based on endemic, basin-wide occurring taxa. As a rule, only few faunal elements allow some correlation along the southern (Panthalassan) margin of Gondwana, and punctually into the paleotropical realm (e.g., González, 1993; Taboada, 2010).

Correlation with the marine realm might be enhanced by radioisotopic dating methods that, however, have to rely on the presence of suitable volcanic material. Except for Australia, data from the Carboniferous are scarce. An excellent and still valuable overview of the Southern continents was given in Wagner et al. (1985).

23.1.3.5.1 Southwestern and northwestern Gondwana

The South African Cape—Karoo Basin has the most complete Carboniferous succession of the large intracontinental basins of southwestern Gondwana. The Givetian to Tournaisian Witteberg Group consists of marginal marine formations of mudstone, siltstone, and sandstone. Glaciogenic deposits of the Visean to Lower Permian Dwyka Group overly with pronounced erosional unconformity. An extraordinary chronostratigraphic spike is the positioning of the Devonian—Carboniferous boundary based on a rich and diversified fish fauna. The Hangenberg Event is known for widespread extinction of fish taxa, among those the global extinction of placoderms and the majority of acanthodians and sarcopterygians, and the spread of the actinopterygians in the lower Carboniferous. The extinction is well documented at the top of the sandstone-dominated Witpoort Formation and the radiation of actinopterygians in mudstone- and siltstone-dominated formations above (Gess, 2016). This confirms the earlier placement of the Devonian—Carboniferous boundary based on the global sea-level curve with regressive sandy Famennian and transgressive muddy to silty Tournaisian strata (Cooper, 1986). Based on a single productive palynological sample, and by correlation with the western European PC miospore zone, respectively, with the Upper *crenulata—isosticha* conodont zone, Streel and Theeron (1999) placed the upper boundary of the Witteberg Group oldest into the mid Tournaisian but did not exclude a later Tournaisian age. From the Dwyka Group biostratigraphic data are missing. A single radioisotopic age from a rhyolitic—andesitic volcanic tuff, 400 m above the base, and 200 m below its top yielded an age close to the Carboniferous—Permian boundary (297 ± 1.8 Ma, Bangert et al., 1999). Similar ages from tuffs intercalated in offshore marine shales from the uppermost Dwyka Group of Namibia cluster at 302 ± 3.0 Ma (Gzhelian) (Stollhofen et al., 2008). Tuffs from shales from the top of the overlying deglaciation sequence III yielded a weighted mean $^{206}Pb/^{238}U$ age of 297.1 ± 1.8 Ma (earliest Asselian). As these shales are considered to be correlatives of the *Eurydesma* transgression, which is also known from Argentina, India, and Australia, this is an important age marker throughout Gondwana. New U—Pb single zircon TIMS ages for the southern Paraná Basin center around 298 Ma (Lowermost Asselian, Griffis et al., 2018) and point to similar deglaciation in South Africa and Argentina.

The adjacent Amazon and Solimões basins are the most important basins from Northwest Gondwana. Due to their intermediate paleogeographic and paleoclimatic position, they are key regions to establish and refine further correlations between the paleoequatorial realm and the glaciogenic-dominated sequences of southwestern Gondwana.

Extended hiatuses during different parts of the Mississippian evoke similarities with the Karoo Basin, but late Carboniferous glaciogenic sediments are missing. Silty to sandy fluvial to shallow-marine formations were replaced by calcareous sediments in the Westphalian. Melo and Loboziak (2003) showed the importance of palynostratigraphy for global chronostratigraphic for the Devonian to lower Westphalian, and Playford and Dino (2000a, 2000b) for the Westphalian. The combination of Euramerican and western Gondwanan palynomorphs as zonal taxa permitted regional biostratigraphic subdivisions, basin-wide correlation, correlations with other Paleozoic basins in West Gondwana, for example, the Parnaíba Basin, and with the marine epicontinental of the Saharan Basins (Loboziak et al., 1999). Lithostratigraphic units were linked to the global Carboniferous stages, although certain intervals were not accurately correlated with definite coeval Euramerican miospore zones due to the scarceness or absence of the eponymous taxa, or of characteristic assemblages (Melo and Loboziak, 2003). The Devonian—Carboniferous boundary is very well discerned by the local *R. lepidophyta—Vallatisporites vallatus* Miospore Zone (LVa Zone). The additional, though scarce, occurrence of *V. nitidus* allows correlation with the upper part, or all, of the western European LN Interval Zone. The earliest to early late Tournaisian palynoflora is distinctly different. Three local zones are correlated with the western European palynozones, from below the undifferentiated VI and HD zones, BP Zone, and PC Zone. According to the conodont zonation, this is the interval from uppermost *praesulcata* to upper *crenulata—isosticha* zones, including the lower part of following *communis carina* Zone (see Melo and Loboziak, 2003 for further references). Embraced by unconformities at base and top, the *Cordylosporites magnidictyus* Palynozone is correlated with the western European undifferentiated TC and NM palynozones and, thus, with the British Asbian Regional Substage.

The Westphalian (Bashkirian—Moscovian) succession of the Amazon Basin (lower part of Tapajós Group) is bound by erosional unconformities at base and top. Playford and Dino (2000a, 2000b) differentiated five palynozones. They suggested correlations with the differing regional zonations of other South American basins, and a rough correlation with other fragments of Gondwana. General problems and short-comings of the zonation were summarized by Playford and Dino (2005). Due to the establishment of warm-water/tropical shallow-marine carbonates in the Amazon and Solimões basins during the Westphalian, the palynozonation could be complemented by fusulinids (Altiner and Savini, 1995). The regional zones were correlated with the upper Morrowan to lower Desmoinesian interval of the US Midcontinent zonation, respectively, with the upper Bashkirian to Moscovian of the Paleotethys realm. Well-studied conodont faunas allowed erection of regional conodont zones and confirm the stratigraphic interval indicated by palynoflora and fusulinids (Scomazzon et al., 2016, and references therein) However, first Cardoso et al. (2017) established four conodont zones and additional subzonal associations derived from the zonal division of the US Midcontinent region, thoroughly discussed correlations with European Pennsylvanian substages, and reconsidered the range of the palynozones. Conodont zones are from below *N. symmetricus* Interval Zone and *Declinognathodus coloradoensis* Interval Zone (both upper Bashkirian = middle Morrowan to lower Atokan = Langsettian to Duckmantian) *D. ellesmerensis* Interval Zone (lowermost Moscovian = Atokan pro parte = Bolsovian), regional *Idiognathodus itaitubensis* Taxon-range Zone [Moscovian pro parte (Kashirian to Podolskian) = upper Atokan to middle Desmoinesian = Asturian]. They also first proved the existence of upper Kasimovian to lower Gzhelian strata by the presence of *Streptognathodus firmus*.

23.1.3.5.2 Eastern Gondwana

The Arabian Peninsula is characterized by the onset of late Westphalian (late Moscovian glaciogenic sediments above a major "Hercynian" unconformity that truncates successions ranging from Cambrian to early Carboniferous (Konert et al., 2001; Osterloff et al., 2004; Melvin and Sprague, 2006). In peninsular India presumably latest Carboniferous (Gzhelian) glaciogenic sediments occur within long-stretched suture zones of Precambrian craton blocks (Barghava, 2008; Mukhopadhyay et al., 2010).

In Australia, sedimentary basins with Carboniferous sedimentary successions are widely distributed (Roberts, 1985). In general, early and mid-Mississippian biota of many Australian basins (and from the Sibumasu Terrane fragments) bear many taxa to place the successions into the global stratigraphic stages. This ended with the later Visean/Serpukhovian climate deterioration (Jones and Metcalfe, 1998).

In eastern Australia (New South Wales, Queensland), shallow-marine successions, including carbonate sediments, occur during the Mississippian. Brachiopod and conodont biozonations were summarized by Fielding et al. (2008a), relying on unpublished data by Roberts et al. A total of 8 regional brachiopod zones based on the revised zonation of Roberts et al. (1993) and 11 conodont zones based on Jenkins et al. (1993) and Roberts et al. (1993) were correlated with western European stages. Especially during the Tournaisian, seven mostly global applicable conodont zones allow well-constrained correlations, whereas more unfavorable facies hamper the correlations during the Visean. Throughout the Pennsylvanian nonsedimentation or discontinuous glaciogenic successions have been recorded and were tabled by Fielding et al. (2008a). Biostratigraphic control relying on palynostratigraphy is scarce (e.g., Jones and Truswell, 1992) and consists of four palynozones, from the cosmopolitan latest Famennian *R. lepidophyta* Assemblage, to the lower Namurian.

Even though there are many datings of Carboniferous rocks of eastern Australia (Roberts et al., 1995a, 1995b), inconsistent ages due to the usage of different reference standards and conflicting with other stratigraphic data have been noted and were recalibrated in Fielding et al. (2008a).

In western Australia, Mississippian successions occur in the Carnarvon, Canning, and Bonaparte basins. Biostratigraphic knowledge of the mixed carbonate−siliciclastic deposits in the Carnarvon and Canning basins is scarce (Mory and Haig, 2011; Mory and Hocking, 2011) and mostly restricted to the Tournaisian, with four local shallow-water conodont assemblage zones (Nicoll and Druce, 1979) and four ostracod assemblage zones (Jones, 1989). Some middle Tournaisian smaller calcareous foraminifers from the Canning Basin were described by Edgell (2004). Recently, Vachard et al. (2014) described a middle Visean (Livian Substage, MFZ12 foraminifer zone of Poty et al., 2006) assemblage of calcareous algae and mostly endemic smaller calcareous foraminifers from the southern Carnarvon Basin and correlated the calcareous interval with time-equivalent limestone from the Bonaparte Basin of northwest Australia.

In spite of frequent hiatuses, the Bonaparte Basin displays the most complete Carboniferous succession (Gorter et al., 2005). This is due to its position in lower latitudes compared to the other western Australian basins (compare Mory et al., 2008 for the earliest Permian), but like else where in Australia, Pennsylvanian (middle Bashkirian to Gzhelian) successions are siliciclastic, and at least partly of glaciogenic origin. Local lowermost to middle Tournaisian (Hastarian) conodont, ostracod, and brachiopod zonations are detailed, but less differentiated during the upper Tournaisian and Visean (Roberts, 1971; Druce, 1969, 1974; Jones, 1989). The mid-Carboniferous boundary is approximated by the occurrence of a floating *Declinognathus noduliferus−Idiognathodus corrugatus* Assemblage Zone (Gorter et al., 2005). Carboniferous palynozones are from Eyles et al. (2002). Most important, however, are Mississippian calcareous smaller foraminifers, which indicate position of the Bonaparte Basin in the Paleotethys realm, and enable quite reliable correlation with the western European substages (Mamet and Belford, 1968; Belford, 1970; Gorter et al., 2005; Vachard et al., 2014).

In all western Australian basins, glaciogenic successions start discontinuously above the Proterozoic to Mississippian substratum, earliest in the latest Visean (northern Perth Basin), in the late Bashkirian (northern and central Canning Basin), or in the Gzhelian (southern Canning Basin, southern Perth Basin) (Mory et al., 2008). Macrofaunas are missing and only a crude palynostratigraphy was established consisting of three regional zones spanning the Tournaisian−Visean interval and three regional zones from the Serpukhovian to Gzhelian (Eyles et al., 2002, cum lit.).

In summary, like the Amazon and Solimões basins in western Gondwana, also the most temperate, northernmost basin of eastern Gondwana, that is, the Bonaparte Basin facing the Prototethys, forms a bridge for improved global correlation also of further Carboniferous Gondwana successions.

23.2 Carboniferous stratigraphy

23.2.1 Biostratigraphy

23.2.1.1 Marine realm

Before microfossils became important stratigraphic tools in the mid-20th century, the stratigraphic calibration of the Carboniferous marine strata was based on invertebrate macrofossils. Pioneer work, mainly in neritic successions of western and eastern Europe, first used corals and brachiopods (e.g., Frech, 1887; Nikitin, 1890; Vaughan, 1905; Delepine, 1911). Today, the temporal distribution of these groups is often considered to be controlled by paleoecologic and paleogeographic limitations (Davydov et al., 2012); hence, their stratigraphic value is mostly reduced to local scales. Ammonoids became the favored chronostratigraphic markers for interregional and global correlations beginning with the first Heerlen Congress in 1927. Their presence in the marine beds of the paralic successions of the Pennsylvanian in Europe has been invaluable for the correlation between marine and nonmarine facies realms.

Starting in the 1960s microfossils, especially conodonts and benthic foraminifers, became important widely used stratigraphic tools for subdivisions in the Carboniferous, and during the last decades they have been the preferred index markers of the International Subcommission for the definition of the base of the global stages. However, it should be underlined that both groups are not free from provincialism and paleoecologic limitations in the light of the formation of Pangea and the LPIA, which influence their temporal and spatial distributions. Good examples are provided by Davydov and Cózar (2019), who showed the impact of the closure of the Rheic-Tethys Seaway on foraminiferal diversity and provincialism, and Cózar et al. (2019) who showed the migration and delayed occurrences of biostratigraphically important foraminiferal markers for the VSB. The different standard zonations that have been developed for conodonts and benthic foraminifers in North America, Europe, and China showcase the impact of such factors, and major differences may exist between the two sides of the developing Pangea (Eurasia vs western America) (Figs. 23.6 and 23.7).

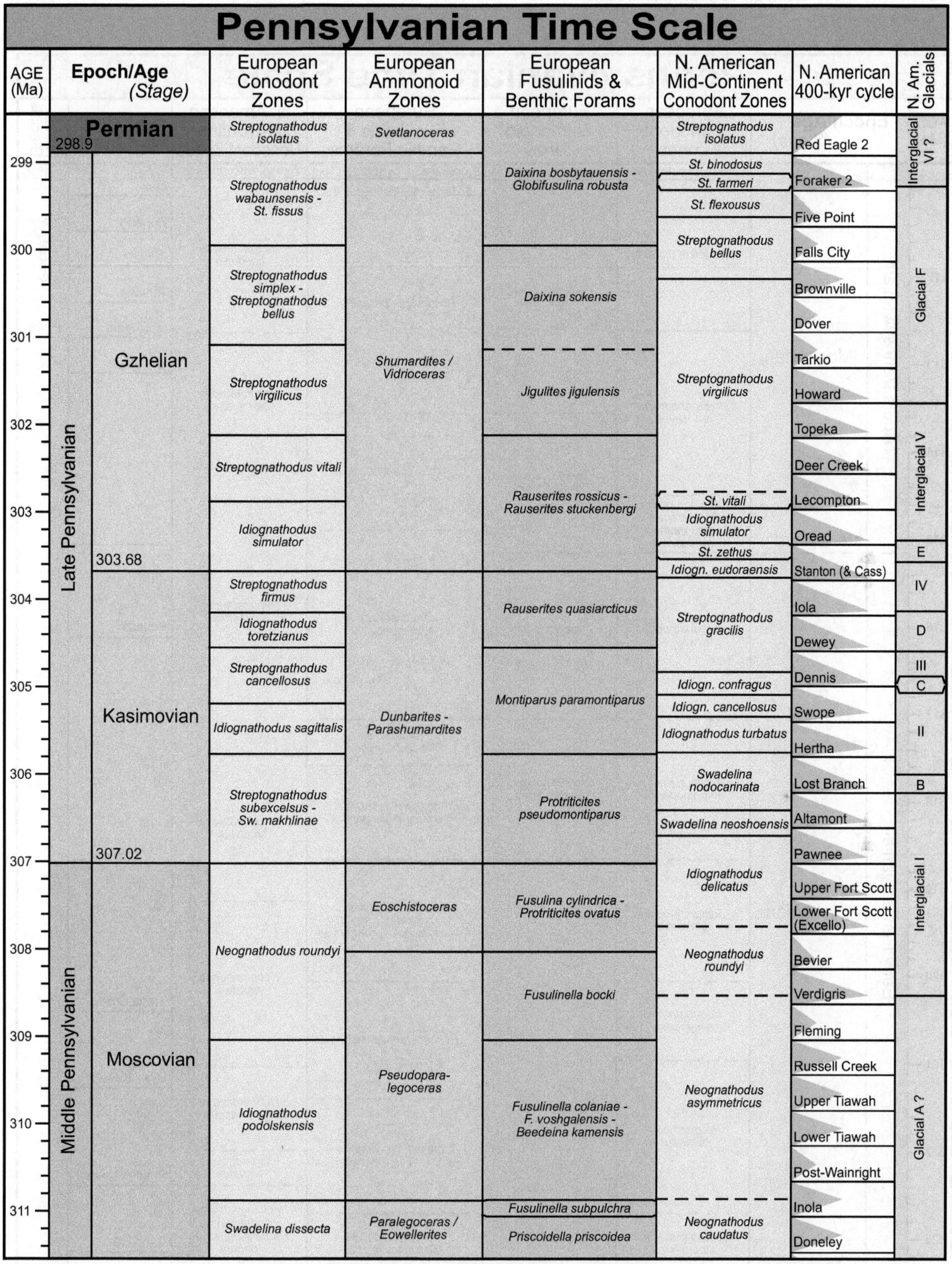

Pennsylvanian Time Scale

AGE (Ma)	Epoch/Age (Stage)	European Conodont Zones	European Ammonoid Zones	European Fusulinids & Benthic Forams	N. American Mid-Continent Conodont Zones	N. American 400-kyr cycle	N. Am. Glacials
298.9	Permian	Streptognathodus isolatus	Svetlanoceras		Streptognathodus isolatus	Red Eagle 2	Interglacial VI ?
299	Gzhelian (Late Pennsylvanian)	Streptognathodus wabaunsensis - St. fissus		Daixina bosbytauensis - Globifusulina robusta	St. binodosus	Foraker 2	
					St. farmeri		
					St. flexousus	Five Point	Glacial F
300		Streptognathodus simplex - Streptognathodus bellus	Shumardites / Vidrioceras	Daixina sokensis	Streptognathodus bellus	Falls City	
						Brownville	
301						Dover	
		Streptognathodus virgilicus		Jigulites jigulensis	Streptognathodus virgilicus	Tarkio	
						Howard	
302		Streptognathodus vitali		Rauserites rossicus - Rauserites stuckenbergi		Topeka	Interglacial V
						Deer Creek	
303					St. vitali	Lecompton	
		Idiognathodus simulator			Idiognathodus simulator	Oread	
303.68					St. zethus		E
	Kasimovian	Streptognathodus firmus		Rauserites quasiarcticus	Idiogn. eudoraensis	Stanton (& Cass)	IV
304		Idiognathodus toretzianus			Streptognathodus gracilis	Iola	
		Streptognathodus cancellosus	Dunbarites - Parashumardites	Montiparus paramontiparus		Dewey	D
305					Idiogn. confragus	Dennis	III
		Idiognathodus sagittalis			Idiogn. cancellosus	Swope	C
					Idiognathodus turbatus		II
306		Streptognathodus subexcelsus - Sw. makhlinae		Protriticites pseudomontiparus	Swadelina nodocarinata	Hertha	
						Lost Branch	B
					Swadelina neoshoensis	Altamont	
307.02	Moscovian (Middle Pennsylvanian)					Pawnee	Interglacial I
307		Neognathodus roundyi	Eoschistoceras	Fusulina cylindrica - Protriticites ovatus	Idiognathodus delicatus	Upper Fort Scott	
						Lower Fort Scott (Excello)	
308					Neognathodus roundyi	Bevier	
				Fusulinella bocki		Verdigris	
309			Pseudopara-legoceras		Neognathodus asymmetricus	Fleming	
						Russell Creek	
		Idiognathodus podolskensis		Fusulinella colaniae - F. voshgalensis - Beedeina kamensis		Upper Tiawah	Glacial A ?
310						Lower Tiawah	
						Post-Wainright	
311				Fusulinella subpulchra	Neognathodus caudatus	Inola	
		Swadelina dissecta	Paralegoceras / Eowellerites	Priscoidella priscoidea		Doneley	

FIGURE 23.6 Selected marine biostratigraphic zonations for the Pennsylvanian Sub-Period with 405-kyr sea-level cycles from eccentricity-forced oscillations of the Gondwana ice sheets. ("Age" is the term for the time equivalent of the rock-record "Stage."). The 405-kyr cycles (cyclothems) and scaling of European and North American Mid-Continent conodont zones to these cycles are a composite from GTS2016 (Ogg et al., 2016, and references therein).

Pennsylvanian Time Scale

AGE (Ma)	Epoch/Age (Stage)		European Conodont Zones	European Ammonoid Zones	European Fusulinids & Benthic Forams	N. American Mid-Continent Conodont Zones	N. American 400-kyr cycle	N. Am. Glacials
311	Middle Pennsylvanian	Moscovian	*Idiogn. podolskensis*	*Pseudoparalegoceras*	*F. colaniae - F. voshg. - B. kam.*	*Neogn. asymmetricus*	Inola	Glacial A ?
					Fusulinella subpulchra		Doneley	
312			*Swadelina dissecta*	*Paralegoceras / Eowellerites*	*Priscoidella priscoidea*	*Neognathodus caudatus*	Spaniard	
							Tamaha	
							McCurtain	
313			*"Streptognathodus" transtivus*			*Neognathodus colombiensis*	unnamed	
314				*Diaboloceras - Winslowoceras*	*Aljutovella aljutovica*			
315		315.15	*Diplognathus ellesmerensis - Declinognathodus donetzianus*					
316	Early Pennsylvanian	Bashkirian	*Neognathodus atokaensis*	*Diaboloceras - Axinolobus*	*Verella spicata - Alj. tikhonovichi*	*Neognathodus atokaensis*	Pounds	
317					*Profusulinella rhombiformis*			
318			*Declinognathodus marginodosus*	*Branneroceras / Gastrioceras*	*Profusulinella primitiva - Pseudostaffella gorskyi*	*Neognathodus nataliae*	Bostwick	
319			*Idiognathodus sinuosus*	*Bilinguites / Cancelloceras*	*Staffellaeformes staffellaeformis - Pseudostaffella praegorskyi*			
320						*Neognathodus bassleri*	Trace Creek	
			Neognathodus askynensis		*Pseudostaffella antiqua*			
321							Dye - Kessler	
			Idiognathodus sinuatus	*Baschkortoceras / Reticuloceras*		*Neognathodus symmetricus*		
322					*Semistaffella variabilis - Semistaffella minuscilaria*	*Idiognathoides sinuatus*	Prairie Grove 2	
323			*Declinognathodus noduliferus*	*Homoceras / Hudsonoceras*	*Plectostaffella bogdanovkensis*	*Declinognathodus noduliferus, Rhachistognathus primus*	Cane Hill	
		323.4						
	Mississipian		*Gnathodus postbilineatus*	*Eumophoceras / Cravenoceratoides*	*Monotaxinoides transitorius*	upper *Rhachistognathus muricatus*		

FIGURE 23.6   continued

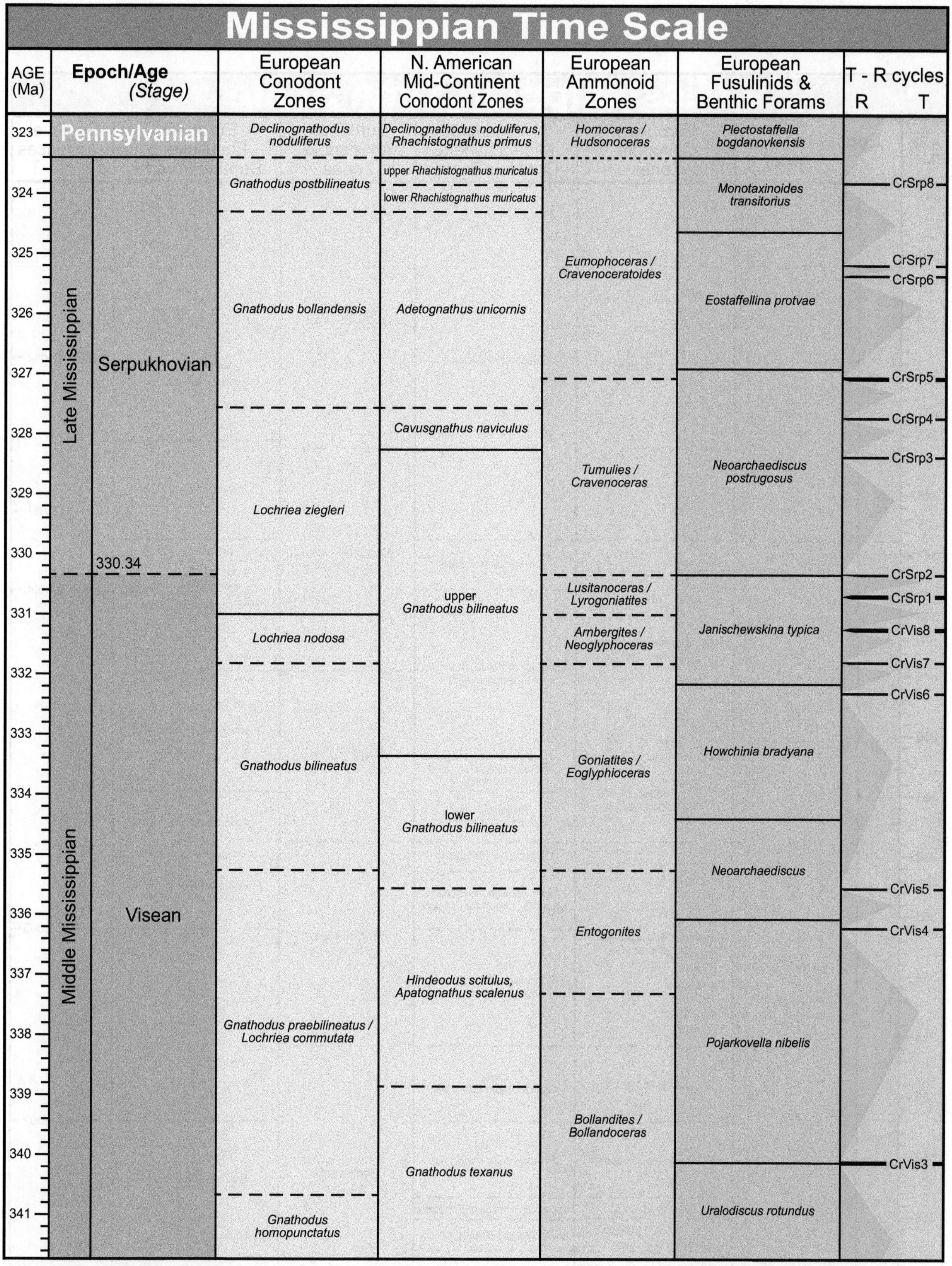

Missippian Time Scale

AGE (Ma)	Epoch/Age (Stage)		European Conodont Zones	N. American Mid-Continent Conodont Zones	European Ammonoid Zones	European Fusulinids & Benthic Forams	T - R cycles R T
323		Pennsylvanian	Declinognathodus noduliferus	Declinognathodus noduliferus, Rhachistognathus primus	Homoceras / Hudsonoceras	Plectostaffella bogdanovkensis	
324	Late Mississippian	Serpukhovian	Gnathodus postbilineatus	upper Rhachistognathus muricatus		Monotaxinoides transitorius	CrSrp8
				lower Rhachistognathus muricatus			
325			Gnathodus bollandensis	Adetognathus unicornis	Eumophoceras / Cravenoceratoides	Eostaffellina protvae	CrSrp7 / CrSrp6
326							
327							CrSrp5
328				Cavusgnathus naviculus	Tumulies / Cravenoceras	Neoarchaediscus postrugosus	CrSrp4 / CrSrp3
329			Lochriea ziegleri				
330		330.34		upper Gnathodus bilineatus	Lusitanoceras / Lyrogoniatites		CrSrp2 / CrSrp1
331	Middle Mississippian	Visean	Lochriea nodosa		Arnbergites / Neoglyphoceras	Janischewskina typica	CrVis8 / CrVis7
332							CrVis6
333			Gnathodus bilineatus		Goniatites / Eoglyphioceras	Howchinia bradyana	
334				lower Gnathodus bilineatus			
335						Neoarchaediscus	
336					Entogonites		CrVis5 / CrVis4
337			Hindeodus scitulus, Apatognathus scalenus				
338			Gnathodus praebilineatus / Lochriea commutata			Pojarkovella nibelis	
339					Bollandites / Bollandoceras		
340				Gnathodus texanus			CrVis3
341			Gnathodus homopunctatus			Uralodiscus rotundus	

FIGURE 23.7 Selected marine biostratigraphic zonations for the Mississippian Sub-Period with T-R cycles are partly based on Davydov et al. (2012) and Poty (2016). ("Age" is the term for the time equivalent of the rock-record "Stage.").

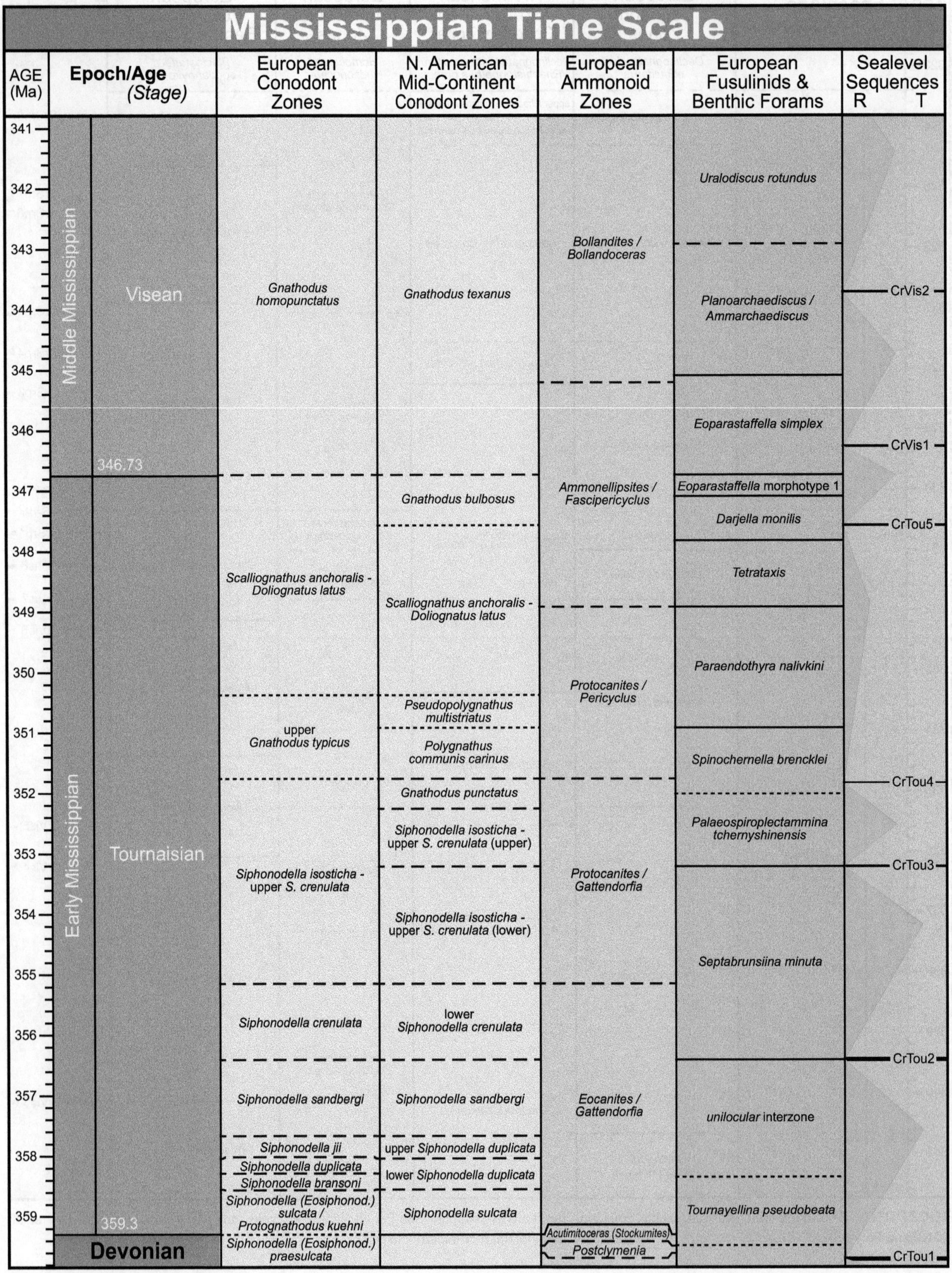

Mississippian Time Scale

AGE (Ma)	Epoch/Age (Stage)	European Conodont Zones	N. American Mid-Continent Conodont Zones	European Ammonoid Zones	European Fusulinids & Benthic Forams	Sealevel Sequences R T
341						
342	Middle Mississippian — Visean	Gnathodus homopunctatus	Gnathodus texanus	Bollandites / Bollandoceras	Uralodiscus rotundus	
343						CrVis2
344					Planoarchaediscus / Ammarchaediscus	
345						
346					Eoparastaffella simplex	CrVis1
346.73						
347	Early Mississippian — Tournaisian		Gnathodus bulbosus	Ammonellipsites / Fascipericyclus	Eoparastaffella morphotype 1	
					Darjella monilis	CrTou5
348		Scalliognathus anchoralis - Doliognatus latus			Tetrataxis	
349			Scalliognathus anchoralis - Doliognatus latus			
350				Protocanites / Pericyclus	Paraendothyra nalivkini	
351		upper Gnathodus typicus	Pseudopolygnathus multistriatus			
			Polygnathus communis carinus		Spinochernella brencklei	CrTou4
352			Gnathodus punctatus			
		Siphonodella isosticha - upper S. crenulata	Siphonodella isosticha - upper S. crenulata (upper)		Palaeospiroplectammina tchernyshinensis	CrTou3
353				Protocanites / Gattendorfia		
354			Siphonodella isosticha - upper S. crenulata (lower)			
355					Septabrunsiina minuta	
356		Siphonodella crenulata	lower Siphonodella crenulata			CrTou2
357		Siphonodella sandbergi	Siphonodella sandbergi	Eocanites / Gattendorfia	unilocular interzone	
358		Siphonodella jii	upper Siphonodella duplicata			
		Siphonodella duplicata	lower Siphonodella duplicata			
		Siphonodella bransoni			Tournayellina pseudobeata	
359		Siphonodella (Eosiphonod.) sulcata / Protognathodus kuehni	Siphonodella sulcata			
359.3				Acutimitoceras (Stockumites)		
	Devonian	Siphonodella (Eosiphonod.) praesulcata		Postclymenia		CrTou1

FIGURE 23.7 (continued).

The methodological base for the use of benthic foraminifers was developed in the Soviet Union in the 1930s and 1940s. Detailed biozonations have been established in various regions, mainly when neritic carbonate facies is present, and those can be found throughout the Carboniferous System and on almost all continents. Detailed Mississippian biozonations were established, for example, in eastern Europe (Lipina and Reitlinger, 1970), western Europe (Conil et al., 1991; Poty et al., 2006; Somerville, 2008), and North America (Mamet and Skipp, 1970; Lane and Brenckle, 2005). The MFZ biozones developed in the Belgian Tournaisian and Visean (Poty et al., 2006) can be applied to many successions within the Paleotethys Ocean (Hance et al., 2011; Poty et al., 2014). However, zonations based on the successions in England Pennines and especially the Moscow Basin are preferred for the late Visean and Serpukhovian (Vachard et al., 2016; Cózar et al., 2019). Detailed biozonations were developed for the Pennsylvanian succession, for example, in eastern Europe (Rauser-Chernousova, 1941, 1949; Rosovskaya, 1950; Solovieva, 1977), western Europe (Villa, 1995), and North America (Ross and Ross, 1988; Wahlman, 2013). A clear provincialism exists between the Eurasian and the American foraminiferal Assemblages, and with the closure of the equatorial seaway faunal exchange between the two realms is limited (Davydov and Cózar, 2019). However, the still-open Franklinian corridor in today's Arctic North America enabled east–west exchange, but in many cases foraminiferal markers occur earlier in Eurasia then in northern America (Davydov, 2014).

First standard conodont zonations have been developed for the Mississippian and Pennsylvanian succession of North America and western and eastern Europe in the 1970s (e.g., Lane et al., 1971, 1980; Barskov and Alekseev, 1975; Groessens, 1975; Higgins, 1975; Sandberg et al., 1978; Barskov et al., 1980). Those zonations were worldwide applied and have been refined and further developed over the last decades (e.g., Perret and Weyant, 1994; Nemyrovska, 1999; Lane and Brenckle, 2005; Alekseev and Goreva, 2007; Hu et al., 2019). Especially the correlation of conodont biozones and global sea-level fluctuations in the Pennsylvanian (e.g., Heckel et al., 2007; Heckel, 2013) improved the correlation between the zonation schemes and contributed to a robust, often very detailed stratigraphic framework. However, the situation in the Mississippian is slightly different. Whereas in Tournaisian times, especially in the lower Tournaisian, conodonts evolved rather fast, which allows the definition of short biozones, the evolutionary changes are considerably slower in many parts of the Visean and Serpukhovian, which results in biozones representing often very long time intervals. A further unresolved issue is the correlation of the zonation

schemes for so-called shallow-water conodonts (e.g., Qie et al., 2014), and the standard zonations mainly based on deeper water outer platform facies.

However, the integration of absolute dates obtained from ash beds into the correlation schemes for conodonts and benthic foraminifers (Schmitz and Davydov, 2012) has helped to improve the Pennsylvanian time scale.

Historically, the stratigraphic divisions in the pelagic facies were based on ammonoids (e.g., Bisat, 1924, 1928; Schmidt, 1925). Ammonoids have been provided very detailed biostratigraphic schemes in western Europe (e.g., Ramsbottom and Saunders, 1985; Korn, 1996, 2006), eastern Europe (e.g., Ruzhenzev, 1965; Ruzhenzev and Bogoslovskaya, 1978; Nikolaeva et al., 2009a), Northern Africa (e.g., Ebbighausen and Bockwinkel, 2007; Korn et al., 2007), and northern America (e.g., Korn and Titus, 2011; Boardman and Work, 2013). Korn and Klug (2015) noted that ammonoid faunas are recorded from many regions in Mississippian and Bashkirian strata, but they become progressively rarer in their distribution and by the end of the Carboniferous rich faunas are only known from the American midcontinent, the southern Urals, and Uzbekistan. Ammonoids have partly lost their importance for global chronostratigraphy due to the development of strong faunal provincialism starting in the late Visean (Korn et al., 2012) and the favoring of conodonts as index fossils for chronostratigraphy. However, it has to be underlined that at least for some time intervals as the late Visean and Serpukhovian, ammonoids still provide the most detailed biozonation schemes, which allow not only very precise dating but also help global correlation (e.g., Korn et al., 2012; Korn and Wang, 2019). In this respect the importance of ammonoids for intercontinental and global correlation should not be neglected and better considered for the definition of chronostratigraphic units in the Carboniferous.

Neritic macrofaunas are still important tools for correlation, and their potential may often be underestimated. Brunton (1984) showed the high potential of Tournaisian and Visean brachiopods for interregional and intercontinental correlation. Good examples for the use of brachiopods for regional stratigraphic schemes are known from North America (Carter, 1991), western Europe (Legrand-Blain, 1991), eastern Europe (Poletaev and Lazarev, 1994), southern China (Li, 1987), and eastern Australia (Roberts et al., 1993). Corals are important index fossils in many marine shallow-water successions during the Carboniferous. Regional scales have been established throughout the tropical belt (e.g., Poty, 1985; Poty et al., 2006; Kossovaya, 1996; Bamber and Fedorowski, 1998) and Sando (1991) even developed a global Mississippian coral zonation. It is interesting to note that in the Belgian Dinantian the same number of coral biozones are known as in the showcased fine-scaled zonation for foraminifera

(Poty et al., 2006, 2014). Many of those coral biozones can be correlated throughout the Paleotethys realm (Poty et al., 2014; Poty, 2016) further showing the importance of this group for interregional correlation in tropical carbonate facies.

Other micro- and macrofaunal groups as radiolarians (e.g., Nazarov and Ormiston, 1985; Gourmelon, 1987), ostracods (e.g., Gorak, 1977; Crasquin, 1985), marine bivalves (e.g., Amler, 2004; Zhang and Yan, 1993), crinoids (e.g., Stukalina, 1988), and trilobites (e.g., Engel and Morris, 1985; Hahn, 1991; Owens and Hahn, 1993) can be important local and regional stratigraphic tools, but in general they have low potential for global chronostratigraphy. This is not only due to paleoecologic limitations but also due to an underrepresentation of stratigraphically well-studied faunas. The use of these faunal groups can be illustrated by the basinal facies of the Variscan foreland basins in Europe. In this so-called Kulm facies trilobites, radiolarians, ostracods, and marine bivalves can be important secondary markers supporting the stratigraphic subdivisions based on ammonoids and conodonts (Brauckmann and Hahn, 2006; Braun and Schmidt-Effing, 1993; Blumenstengel, 2006; Amler, 2006).

The Hangenberg Crisis had a major impact on the diversity and body size of fishes and sharks, which a long recovery time up into the late Mississippian (Sallan and Coates, 2010; Sallan and Galimberti, 2015; Smithson et al., 2016). Occurrences in the marine realm of fishes and sharks are commonly correlated to biostratigraphic zonations established for groups such as conodonts, foraminifera, or palynomorphs.

Due to the transport of floral remains into the marine realm, the terrestrial flora, and especially palynomorphs, can be stratigraphically important in the marine realm. Hence, the mixed marine and nonmarine floras and faunas provide an important tool to correlate the global chronostratigraphic units defined in the marine realm into the nonmarine deposits.

23.2.1.2 Continental realm

Micro- and macrofloras have been important elements to date the Carboniferous coal basins. Very detailed plant biozonations were established for those basins (e.g., Novik, 1952; Read and Mamay, 1964; Josten, 1991; Opluštil et al., 2016). Marked differences between the composition of floras of the equatorial belt and those of higher latitudes resulted in the differentiation of distinct floras provinces (e.g., Gothan, 1937). The (boreal) floras of the higher latitudes in the northern and southern hemispheres are less well known (e.g., Archangelsky et al., 1995; Meyen et al., 1996) and especially their integration into the global chronostratigraphic schemes is still not well known (e.g., Ganelin and Durante, 2002; Wagner and Winkler Prins, 2016). The floras of the tropical belt are much better integrated into the

chronostratigraphic scheme, especially in the context of the Western European regional substages, based on the megafloral zones elaborated by Wagner (1984). He defined 16 floral zones: 5 in the Mississippian and 11 in the Pennsylvanian. Especially the first 3 Mississippian zones are rather long ranging. The uppermost zone may be partly Permian depending on the definition of the Autunian Regional Substage (Wagner, 1984, see also Pellenard et al., 2017; Knight and Alvarez-Vazquez, 2019).

One consequence of the application of the megafloral concept is the questioning of the completeness of the Pennsylvanian records in the Appalachians and the American Midcontinent (e.g., Wagner and Lyons, 1997). Contrasting views expressed on this question (Falcon-Lang et al., 2011; Wagner, 2011, 2017) are an ideal case to showcase the need for a solid cross-continental correlation based on well-established (regional) stratigraphic units.

Biozonations based on spore assemblages have been developed for various regions covering the entire lengths of the Carboniferous System (e.g., Neves et al., 1972; Teteriuk, 1976; Gao, 1984; Playford, 1991; Peppers, 1996; Melo and Loboziak, 2003). Although Carboniferous microflora biozonations have been developed regionally, already in the late 1970s the Commission Internationale de la Microflore du Paléozoïque (e.g., Owens, 1984; Owens et al., 1989) attempted to correlate the different zonations. Clayton (1985), based on earlier work of Sullivan (1965), demonstrated for the Mississippian the presence of regional microfloras related to specific paleoclimatologic and paleogeographic contexts. These microfloras explain the different regional biozonations well.

Nonmarine and limnic bivalves have provided important regional biostratigraphic zonations, especially in the Westphalian strata in the Western European coal basins (e.g., Calver, 1969; Paproth, 1978; Eager, 2005).

At the end of the Pennsylvanian the climate in equatorial Pangea becomes more continental (arid), with increased seasonality and accentuation of meso- and microclimatic effects (Schneider and Werneburg, 2012). Hence, in many continental deposits the classical Carboniferous index fossils of wetter and more balanced climates are not present anymore (e.g., Uhl and Cleal, 2010). Therefore biostratigraphic zonations are based on groups, which are especially important in the continental facies of the Permian. Schneider and Werneburg (2006, 2012) proposed an insect zonation comprising six zones for the Pennsylvanian starting in the uppermost Moscovian. The last zone straddles the Carboniferous—Permian boundary. The conchostracan zonation of Schneider and Scholze (2018) is composed of six assemblage zones for the late Bashkirian to the top of the Gzhelian, in which the top two zones roughly correspond to the Kasimovian and Gzhelian, respectively. In the tetrapod footprint zonation, which is restricted to Euramerica, three biochrons have been defined for the Carboniferous (Lucas, 2003; Fillmore et al., 2013). The *Hylopus/Notalacerta* biochron boundary falls into the Bashkirian and corresponds to the base of the Westphalian Regional Stage. The *Notalacerta/Dromopus* biochron boundary is roughly equaled with the Kasimovian−Gzhelian boundary. Further zonations have been developed for late Carboniferous nonmarine vertebrates, such as fish (Zajic, 2000), Xenacanth sharks (e.g., Schneider et al., 2000), and amphibians (Werneburg, 1996), but they may be more of local or ecological value.

23.2.2 Magnetostratigraphy

The Carboniferous magnetic polarity time scale is poorly known (Davydov et al., 2012; Hounslow et al., 2014; Ogg et al., 2016), and the potential of magnetostratigraphy for global correlation remains largely unexplored. A magnetic polarity pattern correlated to stage, biostratigraphic, or sequence stratigraphic boundaries has yet to be achieved for the Carboniferous. A current view based on Davydov et al. (2012) and Ogg et al. (2016) using composite sections is shown in Fig. 23.1. [Note: Hounslow (2020, in press) presents a detailed synthesis of Carboniferous magnetostratigraphy and polarity-bias records; and his compilation has partially clarified portions of the polarity pattern and calibration to biozones.]

The general pattern of polarity changes during the Carboniferous can be divided into two parts. The Mississippian and basal Pennsylvanian contain magnetozones of both normal and reverse polarity. In the Russian nomenclature this is called the Donetz mixed megazone or hyperchron (Khramov and Rodionov, 1981; Pechersky et al., 2010), which contains the Tikhvinian (late Frasnian−base Serpukhovian) and Debaltzevian (Serpukhovian−Moscovian) superzones. It is topped by a long-time dominance of reversed polarity starting in the Pennsylvanian and reaching into the Late Permian, which is named Permo-Carboniferous Reversed-polarity Superchron (PCRS) or Kiaman Superchron. Although this overall polarity pattern of the Carboniferous is known for several decades (Irving and Parry, 1963; Khramov et al., 1974), many questions still persist like the detailed shorter termed polarity changes within the individual superchrons and the precise dating of the base of the PCRS.

One the fundamental question in establishing precise magnetostratigraphic scales is the quality of the signal, since remagnetization is a widespread problem and rendering many older data obsolete or difficult to interpret (Hounslow et al., 2014; Iosifidi et al., 2018, 2019). In addition, questions about correlation of the data points into modern (bio)stratigraphic frameworks are not uncommon (Hounslow et al., 2014; Kamenikova et al., 2019).

The Davydov et al. (2012) scale uses for the Tournaisian data from the far east of Russia and southern Belgium (Kolesov, 1984, 2001). Stone et al. (2003) suggested contamination by remagnetization for those data of the far east of Russia, and the Belgian data are also questionable (Hounslow et al., 2014). Data from Poland (Kamenikova et al., 2019) may have the potential to fill this gap. The Visean to Bashkirian part of the Davydov scale is based on data from eastern North America (DiVenere and Opdyke, 1991a,1991b), and the Russian Platform (Khramov et al., 1974; Khramov, 2000). The dominance of reverse polarity in the data of Khramov et al. (1974) is in contradiction to the data from North America and western Europe (e.g., Piper. et al., 1991; Kamenikova et al., 2019), which show many short-term polarity changes. Hounslow et al. (2014) mentioned at least 16 normal polarity magnetozones in the Visean and Serpukhovian, which if confirmed and stratigraphically precisely calibrated might be very useful for global correlation.

The Moscovian to Gzhelian part of the Davydov scale is based on data from Australia, the Russian Platform, the southern Urals, North Caucasus, and Central Asia (Irving and Parry, 1963; Khramov and Davydov, 1984, 1993; Davydov and Khramov, 1991; Opdyke et al., 2000). This time interval belongs to the PCRS or Kiaman Superchron. Although it is considered to be one of the most important Paleozoic magnetostratigraphic markers, the precise age of its base is still not precisely known. Radioisotopic dating in Australia, where the Superchron was originally defined, indicates either a beginning at about 318 or 327.2 ± 2.9 Ma. The first date places its base approximately at the base of the Westphalian Regional Stage of Western Europe, the older date would bring it down to lower Serpukhovian (Opdyke and Channell, 1996). This old Australian date might be supported by data from the Sahara Craton (Derder et al., 2001) but strongly contrasts with the normal polarities observed in North America and western and eastern Europe. However, more work is needed to solve this problem.

Within the PCRS, few normal polarity magnetozones, for example, a magnetozone dated as 317−313 Ma in eastern Australia (Opdyke et al., 2000) and three magnetozones in the Donets Basin within the lower/middle Moscovian (Khramov, 1987), have yet to be reported globally. The most pronounced normal polarity magnetozone within the lower PCRS is found in various regions at the top of the Carboniferous. In the GSSP section for the base of the Permian, this magnetozone named Kartamyshian (Davydov and Khramov, 1991) has been constrained within the *Ultradaixina bosbytauensis−Schwagerina robusta* fusulinid zone in the uppermost Carboniferous. Hence, this Kartamyshian magnetozone is an important magnetostratigraphic marker for separation of the Carboniferous and Permian periods.

23.2.3 Chemostratigraphy

Variations in the $^{87}Sr/^{86}Sr$, $^{18}O/^{16}O$, and $^{13}C/^{12}C$ records in the global ocean water through time have been documented through stratigraphic study of marine sediments or fossil skeletons (calcite or phosphate), which are widely used as an important tool for stratigraphic division and correlation.

The Carboniferous $^{87}Sr/^{86}Sr$ curve was originally derived using calcitic brachiopods from North America and Eastern Europe (Bruckschen et al., 1999; Korte et al., 2006). However, due to relatively low temporal resolution, stratigraphic uncertainties between the two regions, and possible diagenetic alteration, the Carboniferous $^{87}Sr/^{86}Sr$ values show great scatter and gaps. Recently, Chen et al. (2018) present an $^{87}Sr/^{86}Sr$ record of conodonts recovered from an open-water carbonate slope succession in the Naqing section, Guizhou, South China, at an unprecedented temporal resolution (10^5 yr) for ~38 Myr of the Late Paleozoic Ice Age. The resulting $^{87}Sr/^{86}Sr$ curve is largely consistent with previously published data (Brand et al., 2009; Henderson et al., 2012), shows significant reduction in data scatter, fills certain stratigraphic gaps, and refines the structure of the middle Mississippian to early Permian seawater $^{87}Sr/^{86}Sr$ records. The Carboniferous $^{87}Sr/^{86}Sr$ curve shows four phases: (1) a gradual fall from high values of ~0.70840 around the Devonian−Carboniferous boundary to ~0.70769 in the middle Visean over a 23-Myr period (McArthur et al., 2012, 2020), which may be related to global cooling or enhanced basalt weathering during this time interval; (2) $^{87}Sr/^{86}Sr$ values increase rapidly (average rate of 0.000035/Myr) from middle Visean to early Bashkirian over a 16-Myr period (334−318 Ma) to 0.70827, as a result of increased continental weathering triggered by the Variscan orogeny, and partly from proliferation of rainforest in low latitude regions (Chen et al., 2018); (3) a ~15-Myr plateau throughout much of the Pennsylvanian (318−303 Ma), which would have been caused by sustained high weathering rate due to westward propagation of the Variscan orogeny from Europe to North America; (4) a decreasing trend in $^{87}Sr/^{86}Sr$ occurred in early Gzhelian throughout early Permian, consistent with the results of Korte et al (2006).

The oxygen isotope ratios of marine carbonate and phosphate fossils are dependent on the oxygen isotopic composition and temperature of ambient seawater and have yield a refined Carboniferous marine stratigraphy (Grossman, 2012; Chen et al., 2016). Grossman et al. (2008) compiled a record of Carboniferous $\delta^{18}O$ data based on well-preserved brachiopod shells from the North American craton and the Russia platform. The oxygen isotope record from the North American craton shows a 3‰ (VPDB) increase in the Tournaisian, a Visean decline to −4‰ to −3‰, and a mid-Carboniferous rise of 1‰−2‰ with relatively constant Pennsylvanian values of −3‰ to −1‰. The brachiopod and conodont $\delta^{18}O$ records for the Carboniferous are similar, but the $\delta^{18}O$

data from conodonts are less variable. Buggisch et al. (2008) identified two major $\delta^{18}O$ positive shifts of 2‰ and 1.5‰ (VSMOW) in the late Tournaisian and Serpukhovian, respectively, as recorded by conodonts in the Mississippian of Europe and North America. In South China, Chen et al. (2016) reported a high-temporal resolution record of late Visean to middle Permian $\delta^{18}O$ based on conodont apatite from the Naqing slope succession. The new data show slight variation in $\delta^{18}O$ values around 22‰ for the late Visean, a minor decrease to 21.2‰ in the early Serpukhovian, and a prominent rise to the maximum value of 23.3‰ across the mid-Carboniferous boundary. The $\delta^{18}O$ values start to decline from the middle Bashkirian to early Moscovian, stay stable around 22‰ from the Moscovian to Kasimovian, and further decrease to 21.5‰ in the late Gzhelian.

Carboniferous $\delta^{13}C$ curves from Euramerica and South China have been established by Buggisch et al. (2008, 2011), Grossman et al. (2008), Saltzman and Thomas (2012), Wang et al. (2019), and Tian et al. (2019), but the isotopic patterns in different sections and regions can be quite different. To better understand the global versus local contribution in $\delta^{13}C$ records, it is necessary to conduct detailed sedimentary facies and diagenesis analyses and compare isotopic data from time-equivalent intervals on a global scale. During the Carboniferous period, increase in $\delta^{13}C_{carb}$ values is generally regarded as an indicator of enhanced organic carbon burial, which would result in lowering of atmospheric pCO_2, climatic cooling and glaciations (Mii et al., 1999; Chen et al., 2018). Resulting carbon isotopic curves in Euramerica and South China sections revealed several major positive shifts in $\delta^{13}C$ values (Buggisch et al., 2011; Saltzman and Thomas, 2012; Wang et al., 2019). The "mid-Tournaisian carbon isotope excursion" (TICE), termed by Yao et al. (2015), coincides with sedimentologic and oxygen isotopic evidence of climatic cooling and glaciation in the *S. isosticha* conodont zone and represents a global event. The carbon isotope excursion was previously reported from North America, Europe, and Russia (Bruckschen and Veizer, 1997; Mii et al., 1999; Saltzman et al., 2004; Buggisch et al., 2008). In South China, carbon isotope ratios show a major positive shift with an amplitude of 3‰ and 6‰ at the Long'an and Malanbian section, respectively (Yao et al., 2015). In Nevada, $\delta^{13}C$ excursion reaches 7.1‰ in the upper part of the Joana Limestone, and the TICE represents one of the largest carbon isotopic excursions in the Phanerozoic (Saltzman et al., 2000). Around the mid-Carboniferous boundary, carbon isotope ratios show a major positive shift, with an amplitude up to 3.0‰ in the Paleotethyan region (Bruckschen et al., 1999). Divergence of $\delta^{13}C$ values between North America and Europe has been documented across the Mississippian−Pennsylvanian transition and indicates significant interoceanic variability due to the closure of oceanic gateway during the assembly of Pangea (Saltzman, 2003). Recently, Tian et al. (2019)

discovered a smaller amplitude $\delta^{13}C$ positive excursion by <1.0‰ in carbonate slope successions in South China, suggesting enhanced upwelling in the eastern Paleotethys Ocean. In the Gzhelian a large, positive carbon isotope excursion with maximum values of 5‰–6‰ was reported in the Naqing section, South China (Buggisch et al., 2011), which was updated with respect to ages of the data in subsequent studies (Chen et al., 2018; Wang et al., 2019). It is worth noting that C-isotope profiles of several Chinese Platform Successions show negative excursions from Gzhelian to Asselian, which are most likely attributed to meteoric diagenetic alteration due to subaerial exposure of the carbonate Platforms during glacio-eustatic fall as a consequence of maximum ice sheets extensions (Buggisch et al., 2011).

23.2.4 Sequence stratigraphy and cyclostratigraphy

During the Late Paleozoic Ice Age (LPIA), extensive continental ice sheets formed in high-latitude southern Gondwana. They waxed and waned in response to orbital forcing leading to potentially Pleistocene-scale fluctuations in sea level (Montañez et al., 2016; Montañez, 2019). In response, cyclic third- to fifth-order transgressive–regressive depositional cycles developed in the tropical to subtropical marine shallow-water successions of epicontinental cratonic seas in Northern America and Eurasia. These have been regarded since long time to be glacio-eustatic and, therefore, global in distribution (e.g., Ross and Ross, 1985, 1987, 1988; Veevers and Powell, 1987; Izart and Vachard, 1994; Heckel, 2002; Izart et al., 2002; Haq and Schutter, 2008). Changes in the Earth's orbital parameters have been hypothesized to be the driving force since the 1930s (e.g., Wanless and Shepard, 1936; Ramsbottom, 1977; Heckel, 1986, 1994; Maynard and Leeder, 1992; Weedon and Read, 1995; Rasbury et al., 1998; Izart et al., 2003; Ueno et al., 2013). Correspondingly, assumed periods of cycles in cyclothemic successions were used to estimate durations of stages of the time scale (e.g., Heckel, 1986, 1991; Algeo and Wilkinson, 1988; Boardman and Heckel, 1989; Klein, 1990, 1991; Connolly and Stanton, 1992; Dickinson et al., 1994; Heckel et al., 2007). Still, these early usages of cycles for global correlation and calibration of the time scale were hampered by uncertainties based on lithostratigraphic resolution, including the problem of stratigraphic gaps and autocyclic developments, transcontinental biostratigraphic correlation, scarcity and inaccuracy of radioisotopic datings, and problems to extend orbital cycles in deep time, as expressed, for example, in Strasser et al. (2006) and Davydov et al. (2012). However, the long-period, 405-kyr eccentricity-driven orbital cycle was recognized as theoretically stable throughout the Phanerozoic

(Berger et al., 1992; Laskar, 1999; Laskar et al., 2004; Hinnov, 2000, 2004, 2013). In consequence, and based on new radioisotopic data, Davydov et al. (2010, 2012) related Moscovian to Gzhelian cyclothems from the Donets Basin, previously recorded by Izart et al. (2002, 2006), to the 405-kyr eccentricity signal. Based on improved radioisotopic and biostratigraphic scaling of cycles by Schmitz and Davydov (2012), Barrick et al. (2013) and Heckel (2013), Ogg et al. (2016) summarized the Donets Basin and the North American Midcontinent 404-kyr cycles, especially with the conodont zonation, and thus created c. 400-kyr time slices for the Late Pennsylvanian. As a result, the Moscovian Stage is subdivided into 18 cycles SM1–SM18 and the Kasimovian Stage in 8 cycles SK1–SK7 and SG1 (formerly included at the base of the Gzhelian by Davydov et al., 2012). The Gzhelian consists of 11 cycles SG2–SG13, and SG14–16, formerly regarded as separate 400-kyr cycles by Davydov et al. (2012). As a consequence, the cyclothem-calibrated ages of conodont markers for the base of the Moscovian, Kasimovian, and Gzhelian stages resulted in slightly younger numerical ages in Ogg et al. (2016) compared to GTS2012 (Davydov et al., 2012). Still, inconsistencies remain that need to be solved. In fact, the number of Moscovian to Gzhelian cycles depicted in Schmitz and Davydov (2012) and relying on the Donets and North American Midcontinent deviates from the number of cycles in Eros et al. (2012) that also have been derived from the Donets Basin and from those of Ueno et al. (2013) from Guizhou, South China, though all authors claim long-term eccentricity control.

Uppermost Serpukhovian and Bashkirian North American Midcontinent cyclothems recognized by Heckel (2013) and depicted in Ogg et al. (2016) do not fit into the 400-kyr scheme. Also Eros et al. (2012) documented longer cycles during the Serpukhovian and Bashkirian of the Donets Basin.

Recently, Wu et al. (2019) presented an astronomical time scale from the latest Visean to Gzhelian Naqing section, South China, including results from Fang et al. (2018). In contrast to the recognition of cyclothems by mainly sedimentologic-facies methods in earlier studies, these authors applied high-resolution magnetic susceptibility (MS) measurements (see, e.g., Da Silva et al., 2013) at 5-cm spacing through the entire ~232-m-long section of almost 34 Myr duration. They were able to discern 3-m-thick sedimentary cycles related to 405-kyr long-term eccentricity cycles as well as shorter cycles related to short orbital eccentricity (136, 122, and 96 kyr); obliquity (31 kyr) and precession (22.9 and 19.7 kyr); and additional long-term modulations of eccentricity and obliquity. Duration of the Serpukhovian, Bashkirian, and Gzhelian stages (7.6, (8.1, and 4.83 Myr, respectively) are consistent with the ages in GTS2012 (Davydov et al., 2012). The duration of the Moscovian will increase to ∼9 Myr, that of the Kasimovian shrink to about 2.5 Myr, due to the revised choice of the marker conodont for the base of the Kasimovian (see discussion in Wu et al., 2019). The numerical ages of the Pennsylvanian stage boundaries are statistically indistinguishable from GTS2012.

The orbital tuning of Pennsylvanian tropical to subtropical successions by eccentricity scale cycles and teleconnections with high-latitude glaciation on the Southern hemisphere was stressed by Eros et al. (2012). Recently, it was underpinned by a high-resolution record of atmospheric pCO_2 using soil carbonate−based and fossil leaf−based proxies that have been derived from upper Moscovian to lower Gzhelian cyclothems of the Illinois Basin, United States and correlated onlap−offlap record of the Donets Basin (Montañez et al., 2016; Montañez, 2019). The authors demonstrated CO_2 variations on the 10^5-year scale with minima comparable to Pleistocene minima and maxima predicted for anthropogenic change for the 21st century. This coincides with the repeated restructuring of Euramerican tropical forests (DiMichele, 2014). Summing up, an increasingly holistic understanding of chronostratigraphic drivers and feedbacks of the Pennsylvanian ice age emerged during the last decade.

Sequence stratigraphic approaches in the Mississippian had not been considered in previous editions of the Geologic Time Scale in spite of their important, not yet fully exploited, potential for chronostratigraphic subdivision and correlation. Already Ramsbottom (1973, 1979) and van Steenwinkel (1990, 1993) recognized the importance of eustatic sea-level variations in the European Tournaisian and Visean. Later, Ross and Ross (1985, 1987, 1988) demonstrated global synchronicity of sea-level changes on the major carbonate platforms in North America, Northwest Europe, and Russia. Based on the platform successions in southern Belgium, Hance et al. (2001) first established a latest Famennian (Strunian) to Visean sequence stratigraphic scheme consisting of nine third-order sequences. Firmly rooted in biostratigraphy, the model was expanded to correlate platform sequences from Great Britain, especially based on sections in southern Wales (Hance et al., 2002), from southern Poland (Poty et al., 2007) and South China (Poty et al., 2011). Strong similarities to the development of the succession in the Bonaparte Basin, Northwest Australia (Gorter et al., 2005), have to be stressed. Later, Poty (2016) redescribed and refined the Belgian sequence, subdividing the late Tournaisian sequence 4 into two sequences: 4A and 4B. A major breakthrough was recognition of the third-order sequences in key sections of the adjacent deeper water foreland basin in western Germany (Rhenish Mountains' Kulm Basin) (Herbig, 2016) in completely differing facies, thus elucidating the isochronous, facies-breaking, nature of eustatic controlled sequences. Herbig (2016) also extended the scheme by adding three latest Visean sequences (10A, 10B, and 11), and two less well-constrained Serpukhovian sequences (12 and 13). It is of major importance that at least certain sequences are recognizable in other European regions, for example, in the Polish Sudetes and in the northern Moravo-Silesian Zone of Czechia (Herbig, 1998), as well as in southern France Montagne Noire (Aretz, 2016). In future the biostratigraphically well-constrained Mississippian third-order sequences might strongly aid in interbasinal correlations in the Paleotethyan realm, even across facies boundaries between platforms and basins. This is underlined by the potential of the Belgian foraminifer and rugose coral biozonations and correspondingly defined Tournaisian and Visean stages to serve as chronostratigraphic standard in this realm (Poty et al., 2006, 2014) See also review in Simmons et al. (2020, Ch. 13, Phanerozoic eustasy, this book).

Poty (2016) evaluated the durations of the Belgian Tournaisian−Visean third-order sequences and superimposed shorter fourth- to sixth-order cycles. The smaller cycles are pluridecimetric to plurimetric strata assemblages that are considered to represent eccentricity, obliquity, and precession cycles according to the scale of Berger et al. (1992). During the Tournaisian, thin precession cycles predominate that are grouped into the glacio-eustatic third-order sequences. At the base of the Visean, cycle architecture strongly changed with the onset of thicker, more variable, shallowing-upward cycles, that is, characteristic parasequences. They are considered to represent, short, eccentricity-driven glacial−interglacial cycles that are integrated into the glacio-eustatic third-order cycles. This indicates the first important glaciations of the LPIA.

In analogy to the duration of three calibrated upper Visean (upper Asbian and Brigantian) third-order cycles from Nova Scotia (Giles, 2009), the duration of the Hastarian Sequence 2 that was based on the count of precession cycles (Poty et al., 2013), and the comparable thickness of other third-order cycles, duration of most Tournaisian and Visean third-order sequences was correlated with the eccentricity period of 2.38 Myr. Only sequence 7 comprising the Livian (4.76 Myr) and sequence 8 spanning the Livian-Warnantian boundary (0.79 Myr) deviate. This results in a duration of the Tournaisian of 10.42 Myr, and of the Visean of 17.45 Myr (Poty et al., 2019). The duration of both stages (27.87 Myr) is almost identical to the duration indicated in GTS2012 (28.0 Myr Davydov et al., 2012) and in Ogg et al. (2016), though stage durations deviate. The Tournaisian is 1.78 Myr shorter and the Visean 1.65 Myr longer according to astrochronology. In this context, new radioisotopic data from the base of the Livian type section are of major importance (Pointon et al., 2019). They indicate an age of 340.13 ± 0.18 Ma, almost 2 Myr younger than the age of 342.01 ± 0.01 Ma from time-equivalent strata of the Donets Basin (Styl'skaya Formation, close to the base of the $C_1^ve_2$ biozone; Davydov et al., 2010). The new Belgian radioisotopic data would shrink the length of the middle and late Visean—and probably of the complete Visean—and give more credit to the astrochronological calibration.

23.2.5 Physical stratigraphy: further tools

The application of extinctions and hypoxic event stratigraphy in combination with time-specific facies, which has been successfully applied in Devonian times (e.g., Becker et al., 2012), has not been proven successful in the Carboniferous. The previously mentioned (Section 23.1.2) severe extinction at the Serpukhovian—Bashkirian Boundary is not well constrained, but it is not correlated to a hypoxic/anoxic event like many extinctions in the Devonian.

The only comparable event in the sense of the Devonian global hypoxic events is the Lower Alum Shale Event in the Tournaisian (e.g., Siegmund et al., 2002). It can be easily traced in the different facies throughout the Paleotethys Ocean and likely beyond. Other event horizons as the *crenistria* Event (Mestermann, 1998) or the *Actinopteria* Shale Event in the sense of Nyhuis et al. (2015) can be marker horizons for regional calibrations in the basinal facies of the late Visean in Northwest Europe. Boy and Schindler (2000) presented the usefulness of bioevents for stratigraphic correlation in the terrestrial realm.

Gamma-ray spectrometry (GRS) and MS logging have been used in various facies for stratigraphic purposes throughout the Carboniferous (e.g., Davies and Elliott, 1996; Ehrenberg and Svånå, 2001; Bábek et al., 2010; Alekseev et al., 2015). In the surface outcrops, these two methods are often integrated into multiproxy approaches with robust biostratigraphic data (e.g., Kalvoda et al., 2011). Both tools often show cyclic signals, which have been correlated to sea-level and climatic changes. As noted previously, Wu et al. (2019) obtained a very detailed MS record for the Naqing section spanning the late Visean to the late Gzhelian. This MS record was used for the construction of an astronomical-tuned time scale, but the reproducibility of these results in other sections has still to be tested. A systematic compilation of GRS and MS trends throughout the Carboniferous is currently lacking; hence, the reliability of these tools for global correlation, as opposed to regional usage, has yet to be evaluated.

23.2.6 Paleoclimate and global correlation

In GTS2012, Davydov et al. (2012) presented an approach to correlate changes in the diversity of benthic foraminifers with global climate variations observed during the Carboniferous. It is not the aim to reproduce or further develop this approach in the GTS2020, but to briefly indicate some examples for the interlinked relationships of paleoclimate trends, sedimentary patterns, and stratigraphy.

The Carboniferous climate is characterized by the transition into the icehouse climate of the Late Paleozoic Ice Age (LPIA). Similar to the climate during the late Cenozoic, glacial and interglacial periods of different intensities alternate with varying periodicities (e.g., Isbell et al., 2012; Montañez, 2016). Several generalized major glaciation phases can be identified in the high latitudes of the southern continents (e.g., Fielding et al., 2008b; Gulbranson et al., 2010; Isbell et al., 2012). These climate alternations result in contraction and dilatation of the equatorial tropical, which can be partly seen in the diversity changes of tropical benthos (e.g., Aretz et al., 2014; Davydov, 2014, Huang et al., 2019). The existence of distinct floral and faunal provinces during almost the entire Carboniferous clearly hampers the application of a universal stratigraphic scheme (see Section 23.1.3). Today's chronostratigraphy is essentially based on tropical marine facies; and especially the boreal or more temperate zones on Gondwana and Angaraland/Siberia are somewhat excluded (e.g., Wagner and Winkler Prins, 1997, 2016). In those regions the stratigraphic zonation remains strictly regional with limited correlations to the chronostratigraphic scheme (see Section 23.1.5). This major shortcoming of global correlation directly impacts research oriented toward global trends and processes during the Carboniferous such as climate evolution.

The onset of the LPIA remains strongly debated, although it is commonly correlated with the cyclothemic sedimentation patterns in the latest Visean/early Serpukhovian (e.g., Wright and Vanstone, 2001; Isbell et al., 2003; Montañez and Poulsen, 2013). Isotope data (e.g., Saltzman, 2003; Buggisch et al., 2008) that suggest cold temperatures in the later Tournaisian are often interpreted as a prelude of the LPIA. An interesting insight on the Tournaisian and Visean climate evolution has been worked out in the tropical carbonates of Belgium (see Section 23.2.5). The latest Tournaisian is marked by the most pronounced sea-level drop of the entire Tournaisian and Visean, and the subsequent third-order sequence 5 (Hance et al., 2001) is missing in most shelf systems (Paleotethys and beyond). It results in important stratigraphic gaps around the Tournaisian—Visean boundary, which explains previous problems to define the base of the Visean and its global correlation. Poty (2016) related this important sea-level drop to major climate change, which should be the first major global glaciation phase of the LPIA. This is backed by the change of the sedimentary patterns around the TVB. Sedimentary patterns during the Tournaisian are characterized by orbital-forced precession cycles. Starting with the basal Visean, recorded shallowing-upward sequences with sharp boundaries correspond to glacio-eustatic parasequences. At the end of the Visean, those sequences start to be expressed in the well-known Carboniferous cyclothems. Hence, isotope, sedimentary, and stratigraphic data point to an onset of the LPIA more than 10 Myr earlier than commonly admitted.

Similar to the TVB, another important sea-level drop, known as Mississippian–Pennsylvanian unconformity (e.g., Siever, 1951) or end-Mississippian regression (e.g., Wagner and Winkler Prins, 2016), is related to a major glaciation phase of the LPIA. This sea-level drop clearly impacts stratigraphic practice. In the GSSP at Arrow Canyon (see Section 23.1.2.1) it corresponds to a major sequence boundary, which correlates to a drastic change in the staking patterns of sedimentary cycles (e.g., Barnett and Wright, 2008). Interestingly the Mid-Carboniferous boundary was defined few decimeters below this major change. This position can facilitate the correlation of the boundary interval or at least narrow it down in sections with poor or absent biostratigraphic control; and even in continental sections, the change of the marine base level may have some potential to be preserved. However, the problem arises that the boundary was defined below the regression. Hence, it is unclear how much time is missing at the GSSP horizon directly above the boundary, and whether attempts at high-resolution global correlation of the basal Bashkirian GSSP might be erroneous. The debate of the potential missing time and the correlation based on the GSSP (e.g., Lane et al., 1999; Richards et al., 2002; Barnett and Wright, 2008; Bishop et al., 2009) resembles the situation before the absence of the basal Visean for many shelf sections could be demonstrated (see above and Section 21.1.1.2).

At its current GSSP the base of the Carboniferous is defined several meters above a major regression related to an interpreted cooling or glacial episode at the end of the Devonian. The other stage boundaries cannot be correlated to such major sea-level changes. However, Heckel et al. (2007) and Davydov et al. (2010) have demonstrated that the smaller scale eustatic sea-level changes are very powerful tools for regional and global correlations and they are essential for the Carboniferous time scale.

23.3 Carboniferous time scale

23.3.1 Previous scales

A quick overview of Mississippian–Pennsylvanian geologic time scales, going back to Harland et al. (1990), shows substantial differences among them (Table 23.1). Actually, the linear time scale was not well established prior to GTS2012, but since about 2010 a significant number of new and reliable U–Pb dates from the Rhenish Mountains, Donets

TABLE 23.1 Comparison of age and duration of the Carboniferous Period in selected published time scales.

Reference	Mississippian			Pennsylvanian	
	Base	Top	Duration	Top	Duration
	(Ma)	(Ma)	(Myr)	(Ma)	(Myr)
Harland et al. (1990)	362.5	322.8	39.7	290	32.8
Young and Laurie (1996)	354	314	40	298	16
Tucker et al. (1998)	362				
Compston (2000)	359.6				
Ross and Ross (1988)	360	320	40	286	34
Jones (1995)	356	317	39	300	17
Menning et al. (2000)	354	320	34	292	28
Menning et al. (2006)	358	320	38	296	24
Menning 2018	361	320	41	296	24
Heckel (2002)		320		290	30
GTS2004	359.2	318.1	41.1	299	19.1
GTS2012	358.9	323.2	35.7	298.9	24.3
GTS2020	359.3	323.4	35.9	298.9	24.5

Basin, and Urals have become available for GTS building; this, together with a new composite standard model for independent scaling of Carboniferous zones and stages. The result is a more detailed geochronologic framework for most of the Carboniferous, which thus contributes to a more stable time scale.

23.3.2 Radioisotopic dates

The detailed high-resolution conodont–foraminifer–ammonoid zonation for the Carboniferous is now constrained in time with 47 radioisotopic dates (Table 23.2). This is a significant improvement over GTS2012, which had only 27 age dates with a greater spread in uncertainties. All but two out of nine stage boundary ages appear to be stable since GTS2012. In the current GTS2020 solution the result of more direct radioisotopic dates indicates that the Tournaisian stage boundary is 0.4 Myr older and the Serpukhovian stage boundary is 0.6 Myr younger.

It should be noted that only four out of nine Carboniferous stage boundaries have ratified GSSP levels with one, the base of the Tournaisian, being under revision to find a suitable marker horizon near the Hangenberg Event. Since nearly all of the Carboniferous radioisotopic dates have a 2-sigma uncertainty of

TABLE 23.2 For the purpose of GTS2020, the composite standard of Davydov et al. (2012) was updated with the new radioisotopic dates.

Stage	Number	Age in Ma	± 2s	Zone or Assemblage	Zonal level	Reference
Induan p.p.						
	T5	251.5	0.29	Meishan: above P/T boundary in lower *H. parvus* Zone		Yin et al. (2001)
	T4	251.58	0.29	Meishan: 175 cm above P/T boundary in lower *H. parvus* Zone		" "
	T3	251.91	0.28	Penglaitan: 50 cm above P/T boundary in lower *H. parvus* Zone		" "
	T2	251.95	0.28	Dongpan: 30 cm above P/T boundary in lower *H. parvus* Zone		" "
	T1	251.88	0.28	Meishan: 8 cm above the P/T boundary (GSSP) and 20 cm above FO of *H. parvus*		" "
Changhsingian				base Induan ~2656.70		
	P37	251.94	0.28	*Meishan: H. latidentatus–C. meishanensis* zones; 19 cm below P/T boundary (GSSP)	2658.97	" "
	P36	252.1	0.28	Meishan: 4.3 m below P/T boundary (=base of Bed 27c)		" "
	P35	252.06	0.28	Penglaitan: 0.3 m below P/T boundary; 0.6 m below FO of *H. parvus*		Baresel et al. (2017)
	P34	252.13	0.29	Penglaitan: 0.6 m 0.9 m		" "
	P33	252.14	0.29	Penglaitan: 1.1 m 1.4 m "		" "
	P32	251.96	0.28	Dongpan: 2.7 m "		" "
	P31	251.98	0.28	Dongpan: 3.2 m "		" "
	P30	252.12	0.28	Dongpan: 6.4 m "		" "
	P29	252.12	0.28	Dongpan: 7.3 m "		" "
	P28	252.17	0.28	Dongpan: 9.7 m "		" "
	P27	252.85	0.29	Meishan: 17.3 m "		" "
	P26	253.45	0.29	Meishan: 36.8 m "		Yin et al. (2001)
	P25	253.49	0.28	Meishan: 38 m "		" "
	P24	253.24	0.45	Shangsi: 7.6 m below FO of *H. parvus*	2627.25	Mundil et al. (2004)
	P23	253.6	0.29	Shangsi: 12.8 m below the major extinction event and 13.6 m below P/T boundary	2616.07	Shen et al. (2011)
Wuchiapingian				base Changhsingian 2610.63		
	P22	257.79	0.31	Shangsi: 27.5 m below the major extinction event and 28.3 m below P/T boundary	2561	Shen et al. (2011)
	P21	258.14	0.36	in *M. tenkensis* bivalve zone, lower Wuchiapingian		Davydov et al. (2018a)
	P20	260.16	0.5	" " "		" "
	P19	260.74	0.9	64 m below the major extinction event and 64.8 m below P/T boundary		Mundil et al. (2004)
	P18	259.1	1.05	76 m 76.8 m "		" "

(Continued)

TABLE 23.2 (Continued)

Stage	Number	Age in Ma	± 2s	Zone or Assemblage	Zonal level	Reference
Capitanian				base Wuchiapingian 2515.97		
	P17	260.57	0.31	upper Capitanian Stage	2493.7	Davydov et al. (2018b)
	P16	262.45	0.37	lower Capitanian Stage	2455.1	"
	P15	262.58	0.53	*J. postserrata* Zone		Nicklen et al. (2011)
Wordian				base Capitanian 2438.82		
	P14	265.46	0.39	*J. aserrata* Zone	2435.07	Glenister et al. (1999)
	P13	266.5	0.37	lower *J. aserrata* Zone	2405	Nicklen et al. (2011)
Roadian				base Wordian 2393.78		
	P12	271.04	0.33	middle *J. nankingensis* Zone		Wu et al. (2017)
	P11	272.95	0.34	"	2374	"
	P10	273.12	0.34	"		Davydov et al. (2018a)
	P9	274	0.34	" plus *S. harkeri*	2354.18	"
Kungurian				base Roadian 2354.18		
Artinskian				base Kungurian 2277.23		
	P8	288.21	0.34	*S. "whitei"* zone, 12.3 m above FAD *S. whitei*	2218.44	Schmitz and Davydov (2012)
	P7	288.36	0.35	" 10.3 m "	2213.92	"
Sakmarian				base Artinskian 2141.53		
	P6	290.81	0.35	*S. anceps* Zone, Sterlitamakian Substage	2133.26	"
	P5	290.5	0.35	"	2138.88	"
	P4	291.1	0.36	"	2104.31	"
Asselian				base Sakmarian 2028.15		
	P3	296.69	0.37	*S. fusus* Zone, Uskalikian Substage	1956.62	C. Henderson (pers. comm., 2018)
	P2	298.05	0.56	*S. cristellaris–St. sigmoidalis* Zone, Surenian Substage	1931.16	Ramezani et al. (2007)
	P1	298.49	0.37	*S. isolatus* Zone	1921.55	"
Carboniferous				base Asselian 1915.96		
Ghzelian						
	CB47	299.22	0.37	*S. waubaunsensis* Zone, 0.65 m below FAD of *S. isolatus*, Noginian Substage	1912.24	Ramezani et al. (2007)
	CB46	300.22	0.35	*S. simplex–S. bellus* Zone, Noginian Substage	1889.68	Schmitz and Davydov (2012)
	CB45	301.29	0.36	*S. virgilicus* Zone, Pavlovoposadian Substage	1876.25	"
	CB44	301.82	0.36	*S. simulator* Zone, Rusavkian Substage	1873.47	"
	CB43	303.1	0.36	"	1842.09	"

TABLE 23.2 (Continued)

Stage	Number	Age in Ma	± 2s	Zone or Assemblage	Zonal level	Reference
Kasimovian				base Ghzelian 1838.22		
	CB42	303.54	0.39	S. firmus Zone, "	1837.87	"
	CB41	304.49	0.36	I. toretzianus Zone, Khamovnikian Substage	1830.75	"
	CB40	304.83	0.36	I. sagittalis Zone, "	1823.94	"
	CB39	305.49	0.36	" "		"
	CB38	305.51	0.37	"	1815.63	"
	CB37	305.95	0.37	S. subexelsus Zone, Krevyakian Substage	1808.17	"
	CB36	305.96	0.36	"	1803.58	"
Moscovian				base Kasimovian 1796.11		
	CB35	307.66	0.37	N. roundyi–S. cancellosus Zone, Peskovian Substage	1787.42	"
	CB34	308	0.37	", Myachkovian Substage		"
	CB33	308.36	0.38	" "	1781.19	"
	CB32	308.5	0.36	" "	1778.65	"
	CB31	307.26	0.38	", base C31 Zone, Donetz Basin		"
	CB30	310.55	0.38	I. podolskiensis Zone, base C2mc Zone,"	1764.26	"
	CB29	312.01	0.37	S. dissectus Zone, base Csmb Zone, "	1749.94	"
	CB28	312.23	0.37	" "	1746.73	"
	CB27	313.16	0.37	N. uralicus Zone, base C2ma Zone, " "	1736.65	"
	CB26	313.78	0.38	10–12 m below Aegiranum marine band and base of Bolsovian Substage gives a latest Duckmantian age		Pointon et al. (2012)
	CB25	314.37	0.53	14 m above base of Aegiranum marine band and base of Bolsovian Substage		Waters and Condon (2012)
	CB24	314.4	0.37	D. donetzianus Zone, base C2ma Zone, Donetz Basin	1719.12	Davydov et al. (2010)
Bashkirian				base Moscovian 1715.12		
	CB23	317.54	0.38	Cheremshanian Substage	1688.24	Schmitz and Davydov (2012)
	CB22	317.63	0.39	Langsettian (Westph.A) Substage		Pointon et al. (2012)
	CB21	318.63	0.4	Prikamian Substage, G1/G12 limestone	1681.38	Schmitz and Davydov (2012)
	CB20	319.09	0.38	", F1/F11 limestone	1670.76	"

(Continued)

TABLE 23.2 (Continued)

Stage	Number	Age in Ma	± 2s	Zone or Assemblage	Zonal level	Reference
Serpukhovian				base Bashkirian 1607.79		
	CB19	324.54	0.46	Namurian E2b2 amm. Subzone, cycle E2b2ii		Pointon et al. (2012)
	CB18	325.64	0.4	Goniatite E2 (*Eumorphoceras bisulcatum*) Zone, Arnsbergian Substage	1563	Jirásek et al. (2018)
	CB17	325.58	0.46	"		"
	CB16	327.58	0.39	*L. ziegleri* Zone, Pendleian Substage		Jirásek et al. (2013)
	CB15	327.35	0.49	Pendleian Substage		
	CB14	328.14	0.4	*L. ziegleri* Zone, middle C1vg2, *Eumorphoceras* Zone	1501.14	Davydov et al. (2010)
	CB13	328.34	0.55	*Eumorphoceras yatesae* Zone (upper part) = E2a3, lower Arnsbergian		Waters and Condon (2012)
	CB12	328.48	0.41	Pendleian Substage, lower Serpukhovian	1472	Gastaldo et al. (2009)
Visean				base Serpukhovian 1445.34		
	CB11	332.5	0.4	middle MFZ14 Zone, upper Asbian Substage		Pointon et al. (2014)
	CB10	333.87	0.39	middle *L. mononodosa* Zone, 1.5 m below *F. ziegleri*	1392.63	Schmitz and Davydov (2012)
	CB9	335.59	0.44	near base MFZ13 Zone, lower Asbian Substage		Pointon et al. (2014)
	CB8	336.22	0.4	near base MFZ12 Zone, top Holkerian Substage		
	CB7	340.05	0.41	*Globoendothyra, Eostafella, A. krestovnikovi, Endothyra*		Jirásek et al. (2014)
	CB6	342.01	0.41	base *E. porikensis*–*A. gigas* Zone; base MFZ12 zone, base Holkerian Substage	1281.96	Davydov et al. (2010)
	CB5	345	0.4	*H. rotundus* Zone, base MFZ10 Zone	1208.19	"
	CB4	345.17	0.41	"	1206.37	"
Tournaisian				base Visean 1187.23		
	CB3	357.26	0.42	MFZ23	1059.84	Davydov et al. (2010)
	CB2	358.43	0.42	lower *S. duplicata*	1037.63	Davydov et al. (2011)
	CB1	358.71	0.42	upper *S. sulcata*	1010.44	"
				base Tournaisian ~ 991.64		

This composite standard now includes 47 Carboniferous and 37 Permian age dates, all ID-TIMS U–Pb dates and almost universally with 0.3–0.4 Myr 2-sigma (external) uncertainties. For details see text. *FO*, First occurrence; GSSP, Global Boundary Stratotype Section and Point. Full names of the biozones are in Figures 23.6 and 23.7, and in Permian chapter Figure 24.10 (this volume).

0.4 Myr, and the composite values are without error, the interpolated GTS2020 stage boundary ages above the Tournaisian are also assigned a 0.4 Myr uncertainty.

For the final determination of ages of stage boundaries for GTS2020, the Carboniferous and Permian radioisotopic dates were combined. This adds 37 dates to the population of useful radioisotopic dates for a total of 84, all of which are high-resolution U−Pb dates and the majority have 2-sigma (total) uncertainty of 0.3−0.4 Myr (Table 23.2). Details on the Permian radioisotopic dates are in Chapter 24, The Permian Period.

23.3.3 Carboniferous and Permian composite standard

A composite standard independently scales stages for building a two-way plot of radioisotopic dates versus biozones and hence enables derivation of the ages for all stage boundaries and for the biozones used in the composite standard. The composite standard of GTS2012 with minor updates was also utilized for GTS2020. This composite standard was generated in GTS2012 for both Carboniferous and Permian, using the CONOP program (Sadler and Cooper, 2003; Sadler et al., 2003) as summarized in Davydov et al. (2012).

For the Donets Basin the GTS2012 composite from late Devonian to Asselian was built with 2641 species of foraminifers, conodonts, ammonoids and algae (5282 paired bioevents, where paired means first and last occurrence of an event in a section), and 357 unpaired events (coals and limestone), including 9 ash beds from 46 wells and sections.

The GTS2012 composite in the Urals is based on 1156 Visean through Kungurian species of foraminifers (including fusulinids), conodonts and ammonoids (2312 paired bioevents) in 27 stratigraphic sections and includes 29 ash beds coded as dated unpaired events.

The Guadalupian composite includes 286 species of conodonts and foraminifera and one volcanic ash. The Lopingian composite was built with 33 sections in South China, Iran, Azerbaijan, Armenia, Pakistan, Italy, Austria, Turkey, Canadian Arctic and Greenland and includes 316 species of conodonts, foraminifera, ammonoids and radiolaria. Eight ash beds from biostratigraphic-constrained sections (Meishan and Shangsi sections) dated with ID-TIMS (Mundil et al., 2004) integrated in the Lopingian composite. Paleomagnetic events turned to be frequently inconsistent with biostratigraphy for the Lopingian composite and were not involved in the compositing process.

Davydov et al., 2012 in GTS2012 details the Carboniferous and Permian composite standard with its main zonal units and incorporated ash beds with their numbered radioisotopic age dates, with 95% confidence limits and external error estimates.

For the purpose of GTS2020 the composite standard of Davydov et al. for 2012 was updated with the new radioisotopic ages for GTS2020. This composite standard now includes 47 Carboniferous and 37 Permian age dates, all ID-TIMS dates, and almost universally with 0.3−0.4 2-sigma (external) uncertainty. On the advice of C. Henderson (senior author of the Permian Chapter for GTS2020), the composite value for base Sakmarian was changed from 1987.05 to 2028.15, making it slightly younger in agreement with the ratified GSSP definition for the Sakmarian Stage. Also, the conodont *Sverdrupites harkeri* was added to the composite at the level where *Jinogondolella nankingensis* appears at composite level 2354.18. This strengthens the best fit in the Artinskian−Roadian interval. The base Visean follows the revised calibration within the Tournaisian−Visean boundary interval by the SCCS in 2014.

23.3.4 Ages of stage boundaries

In the analysis to interpolate the ages of the stage boundaries and durations in the Carboniferous (and Permian), the chronometric and stratigraphic ages of the radiogenic isotope dates were subjected to a cubic-spline fit (Fig. 23.8). The method and its submethods are described in Chapter 14, Geomathematics.

The age and error derived by the Devonian spline for the top of Devonian was aligned with the age and error for that same boundary from the Carboniferous−Permian cubic-spline fit. Thus the final age for the Devonian−Carboniferous boundary of 359.3 ± 0.3 Ma for GTS2020 is the average of both ages for this boundary calculated with the two spline fits for the Devonian and for the Carboniferous−Permian. This age of 359.3 ± 0.3 Ma is also realistic with future focus on a potential GSSP level for the base Tournaisian within the Hangenberg Event.

The Permian−Triassic boundary was interpolated with level T1 at 251.88 ± 0.3 and with level P37 at 251.94 ± 0.3 Ma in the Meishan GSSP section in China to yield 251.9 ± 0.3 Ma.

With the endpoints in place the final best fit line for age versus stage is shown in Fig. 23.8. It shows a close fit of all data points, which is an indication that the GTS2020 Composite was updated well from GTS2012. All points passed a goodness of fit test. Interpolated stage boundary ages with their error bars and durations are given in Table 23.3. Compared to GTS2012 stage boundary ages are identical except for the Tournaisian stage boundary being 0.4 Myr older and the Serpukhovian stage boundary being 0.6 Myr younger, the result of more direct radioisotopic dates. The error of 0.4 for all but one date is larger and more realistic than calculated with less sophisticated methods in GTS2012.

The Tournaisian Stage spanned 359.3 ± 0.3 to 346.7 ± 0.4 Ma for a duration of 12.6 Myr. Visean lasted

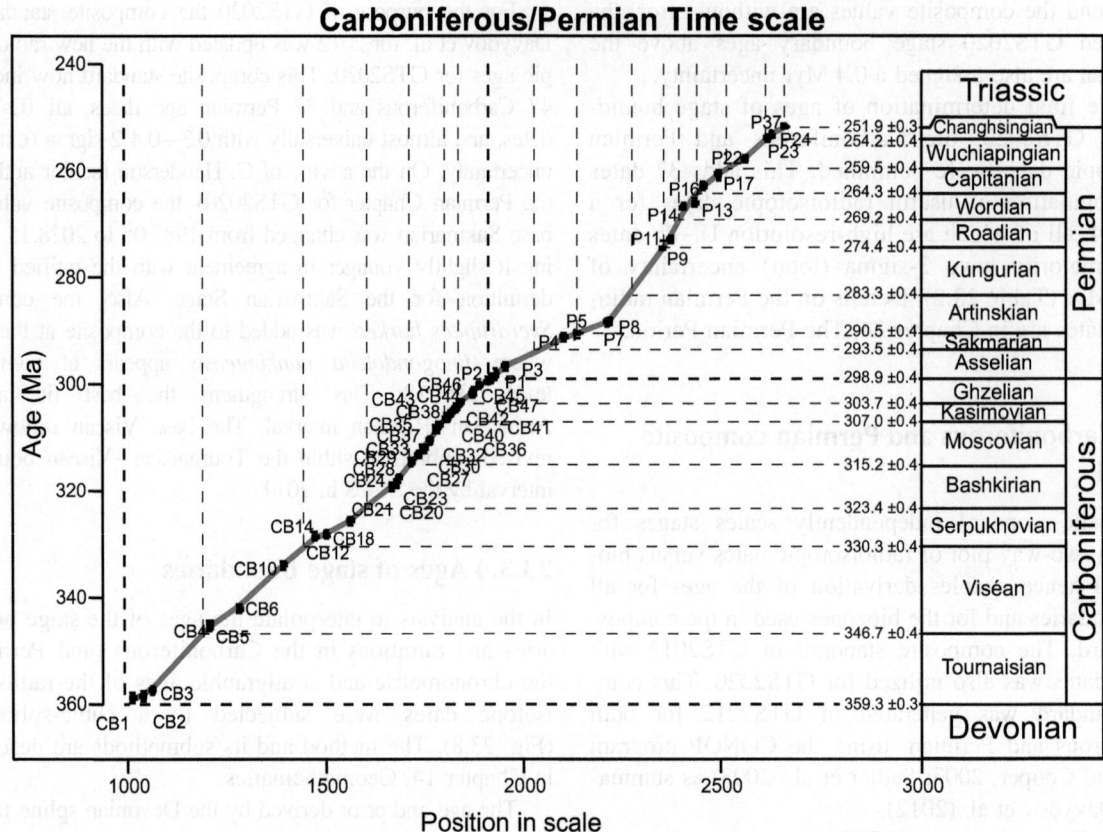

FIGURE 23.8   Best-fitting spline-curve for the Carboniferous-Permian. For details see text.

16.4 Myr from 346.7 ± 0.4 to 330.3 ± 0.4 Ma, thus being the second longest stage in the Phanerozoic, after Norian. The Serpukhovian Stage ranged from 330.3 ± 0.4 to 323.2 ± 0.4 Ma, for a duration of 6.9 Myr. Bashkirian lasted for 8.2 Myr from 323.4 ± 0.4 to 315.2 ± 0.4 Ma. The Moscovian Stage started at 315.2 ± 0.4 and ended at 307 ± 0.4 Ma, lasting 8.2 Myr.

The Kasimovian ran from 307 ± 0.4 to 303.7 ± 0.4 Ma for only 3.3 Myr, thus it is the shortest Carboniferous stage. The Gzhelian lasted 4.8 Myr from 303.7 ± 0.4 Ma to base Permian at 298.9 ± 0.4 Ma. The Mississippian Subperiod had a duration of 35.9 Myr (was 35.7), and the Pennsylvanian 24.5 Myr (was 24.3). The Carboniferous Period lasted from 359.3 to 298.9 Ma, for

TABLE 23.3 GTS2020 scale for the Carboniferous Period with comparison of ages from GTS2012.

Base of Stage	GTS2012		GTS2020		Cyclicity
	Age (Ma)	Duration (Myr)	Age (Ma)	Duration (Myr)	Duration (Myr)
Asselian (Permian)	298.9 ± 0.2		298.9 ± 0.4		
Gzhelian	303.7 ± 0.1	4.8	303.7 ± 0.4	4.8	4.5
Kasimovian	307.0 ± 0.2	3.3	307.0 ± 0.4	3.3	3.3
Moscovian	315.2 ± 0.2	8.2	315.2 ± 0.4	8.2	7.9
Bashkirian	323.4 ± 0.2	8	323.4 ± 0.4	8.2	
Serpukhovian	330.9 ± 0.3	7.7	330.3 ± 0.4	6.9	
Viséan	346.7 ± 0.4	15.8	346.7 ± 0.4	16.4	
Tournaisian	358.9 ± 0.4	12.2	359.3 ± 0.3	12.6	

GSSPs of the Carboniferous Stages, with location and primary correlation criteria

Stage	GSSP Location	Latitude, Longitude	Boundary Level	Correlation Events	Reference
Gzhelian	*Candidates are in southern Urals or Nashui (South China)*			*Conodont, FAD of Idiognathodus simulator (candidate)*	
Kasimovian	*Candidates are in Usolka, Russia, Nashui (South China)*			*Conodont, FAD of Idiognathodus heckeli*	
Moscovian	*Candidates are in southern Urals or Nashui (South China)*			*Conodont, FAD Diplognathodus ellesmerensis (candidate)*	
Bashkirian	Arrow Canyon, Nevada, USA	36°44'00" N, 114°46'40" W**	82.9 m above the top of the Battleship Formation in the lower Bird Spring Formation	Conodont, FAD of *Declinognathodus noduliferus*	Episodes **22**/4, 1999
Serpukhovian	*Candidates are Verkhnyaya Kardailovka (Urals) or Nashui (South China)*			*Conodont, FAD of Lochriea ziegleri (candidate)*	
Visean	Pengchong, South China	24°26'8.88"N, 109°27'19.49"E	base of bed 83 in the Pengchong Section	Foraminifer, FAD of *Eoparastaffella simplex*	Episodes **26**/2, 2003
Tournaisian (GSSP under reevaluation)	La Serre, France	43°33'19.9"N 3°21'26.3"E*	base of Bed 89 in Trench E' at La Serre, (but FAD now known to be at base of Bed 85)	Conodont, FAD of *Siphonodella sulcata* s.l.	Episodes **14**/4, 1991; Kölner Forum Geol. Paläont., **15**, 2006

* according to Google Earth, ** derived from map

FIGURE 23.9 Ratified GSSPs and potential primary markers under consideration for defining the Carboniferous stages (status as of early 2020). Details of each GSSP are available at http://www.stratigraphy.org, at https://timescalefoundation.org/gssp/, and in the Episodes publications.

a duration of 60.4 Myr, which is 0.4 Myr longer than in GTS2012.

Ogg et al. (2016) in GTS2016 aligned the Moscovian, Kasimovian, and Gzhelian stage boundaries with the cycle-derived stratigraphy for those stages (Davydov et al., 2012). With 20 high-resolution ID-TIMS age dates now for this 16-Myr interval, one might expect good convergence between cyclic scaling and the spline fit method. Such is indeed the case here, as shown in Table 23.3, bearing in mind the 0.4-Myr uncertainty of the GTS2020 ages. The important duration indicator shows Kasimovian having the same duration with both methods, and duration of Moscovian and Gzhelian being only 0.3 Myr longer, less than the inherent uncertainty. Aligning the cycle durations 100% would create undesirable distortion of the boundary ages for adjacent stages.

Bibliography

Aizenverg, D.E., Brazhnikova, N.E., Vassilyuk, M.V., Vdovenko, M.V., Gorak, S.V., Dunaeva, N.N., et al., 1978, Carboniferous sequence of the Donets Basin as the standard section of the Carboniferous. *Huitième Congrès International de Stratigraphie et de Géologie du Carbonifère, Compte Rendu*, Moscow: Nauka, **Vol. 1**, pp. 158−169.

Alekseev, A.S., 2008, *Carboniferous system. In: Sostoyane izuchennosti stratigrafii dokembriya i fanerozoya Rossii. Zadachi dalneishikh*

issledovanii. Postanovlenie MSK i ego postoyannykh komissii, **38**. St. Petersburg: VSEGEI, pp. 61−68. (in Russian).

Alekseev, A.S., 2009, Geological setting and Carboniferous stratigraphy of Moscow Basin. *In*: Alekseev, A.S., and Goreva, N.V. (eds), *Type and Reference Carboniferous Sections in the South Part of the Moscow Basin. Field Trip Guidebook of International Field Meeting of the I.U.G.S. Subcommission on Carboniferous Stratigraphy "The Historical Type Sections, Proposed and Potential GSSP of the Carboniferous in Russia," Moscow, August 11−12, 2009.* Moscow: Borissiak Paleontological Institute of RAS, pp. 5−12.

Alekseev, A.S., and Goreva, N.V., 2007, Conodont zonation for the type Kasimovian and Gzhelian stages in the Moscow Basin, Russia. *In*: Wong, T.E. (ed), *Proceedings of the XVth International Congress on Carboniferous and Permian Stratigraphy*. Utrecht: Royal Netherlands Academy of Arts and Sciences, pp. 229−242. August 10−16, 2003.

Alekseev, A.S., and Goreva, N.V., 2013, The conodont *Neognathodus bothrops* Merrill, 1972 as the marker for the lower boundary of the Moscovian Stage (Middle Pennsylvanian). *In*: Lucas, S.G., DiMichele, W.A., Barrick, J.E., Schneider, J.W., and Spielmann, J.A. (eds), The Carboniferous-Permian Transition. *New Mexico Museum of Natural History and Science Bulletin*, **60**: 1−6.

Alekseev, A.S., and Task Group, 2014, Report of the task group to establish a GSSP close to the Existing Bashkirian−Moscovian Boundary. *Newsl. Carbonif. Stratigraphy*, **32**: 32−33.

Alekseev, A.S., Kononova, L.I., and Nikishin, A.M., 1996, The Devonian and Carboniferous of the Moscow Syneclise (Russian Platform): stratigraphy and sea-level changes. *Tectonophysics*, **268**: 149−168.

Alekseev, A.S., Goreva, N.V., Isakova, T.N., Kossovaya, O.L., and Makhlina, M.K., 2004, Biostratigraphy of the Carboniferous in the Moscow Syneclise, Russia. *Newsletter on Carboniferous Stratigraphy*, **22**: 28−35.

Alekseev, A.S., Goreva, N.V., Isakova, T.N., and Kossovaya, O.L., 2009, The stratotype of the Gzhelian Stage (Upper Carboniferous) in Moscow Basin, Russia. *In:* Puchkov, V.N. (ed), The Carboniferous Type Sections in Russia, Potential and Proposed Stratotypes. Proceedings of the International Conference Ufa—Sibai, *13−18 August 2009*. Ufa: Institute of Geology, Bashkirian Academy of Science, pp. 165−177.

Alekseev, A.S., Kossovaya, O.L., Goreva, N.V., 2013. Current state and prospects of improvement of the General Scale of the Carboniferous System of Russia, In: *General Stratigraphic Scale of Russia: Current State and Prospects*, pp. 165−177. Moscow: Geological Institute of the RAS (in Russian).

Alekseev, A.O., Kabanov, P.B., Alekseeva, T.V., and Kalinin, P., 2015, Magnetic susceptibility and geochemical characterization of an upper Mississippian cyclothemic section Polotnyanyi Zavod, (Moscow Basin, Russia). *In:* Da Silva, A.C., Whalen, M.T., Hladil, J., Chadimova, L., Chen, D., Spassov, S., Boulvain, F., and Devleeschouwer, X. (eds), Magnetic Susceptibility Application − A Window onto Ancient Environments and Climatic Variations. *Geological Society, London, Special Publications*, **414**: 181−196. https://doi.org/10.1144/SP414.1.

Alekseev, A.S., Nikolaeva, S.V., Goreva, N.V., Gatovsky, Y.A., and Kulagina, E.I., 2018a, Selection of marker conodont species for the lower boundary of the global Serpukhovian Stage (Mississippian). *Newsl. Carbonif. Stratigraphy*, **34**: 34−39.

Alekseev, A.S., Kulagina, E.I., Kononova, L.I., Zhuravleva, N.D., and Nikolaeva, S.V., 2018b, Joint occurrence of conodonts *Declinognathodus donetzianus* and *Diplognathodus ellesmerensis* close to Bashkirian-Moscovian boundary in the Basu section, South Urals, Russia. *Newsl. Carbonif. Stratigraphy*, **34**: 39−42.

Algeo, T.J., and Wilkinson, B.H., 1988, Periodicity of mesoscale Phanerozoic sedimentary cycles and the role of Milankovitch orbital modulation. *J. Geol.*, **96**: 313−322. https://doi.org/10.1086/629222.

Altiner, D., and Savini, R., 1995, Pennsylvanian Foraminifera and biostratigraphy of the Amazonas and Solimões Basin (North Brazil). *Rev. de. Paleobiologie*, **14**: 417−453.

Amler, M., 2004, Bivalve biostratigraphy of the Kulm Facies (Early Carboniferous, Mississippian) in central Europe. *Newsl. Stratigraphy*, **40**: 183−207. https://doi.org/10.1127/0078-0421/2004/0040-0183.

Amler, M., 2006, Bivalven und Rostroconchien. *In* Deutsche Stratigraphische Kommission, Amler, M.R.W., and Stoppel, D. (eds), Stratigraphie von Deutschland VI. Unterkarbon (Mississippium). *Schriftenreihe der Deutschen Gesellschaft für Geowissenschaften*, **41**: 121−146.

Anonymous, 1944, *Geology of the USSR, Donets Basin*. Moskva: Gosgeolizd., **Vol. 7**. pp. 1−898.

Archangelsky, S., 1990, Plant distribution in Gondwana during the late Paleozoic. *In :* Taylor, T.N., and Taylor, E.L. (eds), *Antarctic Paleobiology* Springer, pp. 102−117. https://doi.org/10.1007/978-1-4612-3238-4-9.

Archangelsky, S., Arrondo, O.G., and Leguizamon, R.R., 1995, Floras Paleozoicas. Contribuciones a la palaeophytologia Argentina y revision y actualizacion de la obra paleobotanica de Kurtz en la Republica Argentina. *Actas de. la. Academia Nacional de Cienc. de la Repub. Argentina*, **11**: 85−125.

Aretz, M., 2014, Redefining the Devonian−Carboniferous boundary: an overview of problems and possible solutions. *In:* Rocha, R., Pais, J.,

Kullberg, J., and Finney, S. (eds), *STRATI 2013*. Springer Geology, pp. 227−231. https://doi.org/10.1007/978-3-319-04364-7_46.

Aretz, M., 2016, The Kulm facies of the Montagne Noire (Mississippian, southern France). *Geologica Belgica*, **19** (1-2), 69−80. https://doi.org/10.20341/gb.2015.018.

Aretz, M., and Corradini, C., 2016, The redefinition of the Devonian-Carboniferous Boundary: recent developments. *Berichte des. Inst. für Geologie und Paläontologie der Karl-Franzens-Universität Graz*, **22**: 6−7.

Aretz, M., and Corradini, C., 2019, The redefinition of the Devonian-Carboniferous Boundary: state of the art. *In:* Hartenfels, S., Herbig, H.-G., Amler, M.R.W., and Aretz, M. (eds), Abstracts, 19th International Congress on the Carboniferous and Permian, Cologne July 29−August 2, 2019. *Kölner Forum für Geologie und Paläontologie*, **23**: 31−32.

Aretz, M., Nardin, E., and Vachard, D., 2014, Diversity patterns and palaeobiogeographical relationships of latest Devonian-Lower Carboniferous foraminifers from South China: What is global, what is local? *J. Palaeogeography*, **3**: 35−59. https://doi.org/10.3724/SP.J.1261.2014.00002.

Aretz, M., and Nudds, J.R., 2005, The coral fauna of the Holkerian/Asbian boundary stratotype section (Carboniferous) at Little Asby Scar (Cumbria, Great Britain) and implications for the boundary. *Stratigraphy*, **2**: 167−190.

Aretz, M., and Task Group, 2018, Report of the joint Devonian/Carboniferous Boundary GSSP Reappraisal Task Group. *Newsl. Carbonif. Stratigraphy*, **34**: 12−13.

Bábek, O., Kalvoda, J., Aretz, M., Cossey, P.J., Devuyst, F.X., Herbig, H.G., et al., 2010, The correlation potential of magnetic susceptibility and outcrop Gamma-Ray logs at Tournaisian−Visean boundary sections in Western Europe. *Geologica Belgica*, **13**: 291−308.

Bábek, O., Kumpan, T., Kalvoda, J., and Grygar, T.M., 2016, Devonian/Carboniferous boundary glacioeustatic fluctuations in a platform-to-basin direction: A geochemical approach of sequence stratigraphy in pelagic settings. *Sediment. Geol.*, **337**: 81−99. https://doi.org/10.1016/j.sedgeo.2016.03.009.

Baesemann, J.F., Brenckle, P.L., and Gruzlovic, P.D., 1998, Composite standard correlation of the Mississippian-Pennsylvanian (Carboniferous) Lisburne Group from Prudhoe Bay to the eastern Arctic National Wildlife Refuge, North Slope, Alaska. *In:* Clough, J.G., and Larson, F. (eds), *Short Notes on Alaska Geology 1997:* Alaska Department of Natural Resources, Division of Geological and Geophysical Surveys, Professional Report. **118**. 23−36. http://doi.org/10.14509/2327.

Baesemann, J.F., and Lane, H.R., 1985, Taxonomy and evolution of the genus Rhachistognathus Dunn (Conodonta; Late Mississippian to early Middle Pennsylvanian). *Cour. Forschungsinstitut Senckenberg.*, **74**: 93−135.

Bamber, E.W., and Fedorowski, J., 1998, Biostratigraphy and systematics of Upper Carboniferous cerioid rugose corals, Ellesmere Island, Arctic Canada. *Geol. Surv. Can. Bull.*, **511**, pp. 127.

Bangert, B., Stollhofen, H., Lorenz, V., and Armstrong, R., 1999, The geochronology and significance of ashfall tuffs in the glaciogenic Carboniferous-Permian Dwyka Group of Namibia and South Africa. *J. Afr. Earth Sci.*, **29**: 33−50. https://doi.org/10.1016/S0899-5362(99)00078-0.

Baranova, D.V., Kabanov, P.B., and Alekseev, A.S., 2014, Fusulinids (Foraminifera), lithofacies and biofacies of the Upper Moscovian (Carboniferous) of the Southern Moscow Basin and Oka−Tsna

Swell. *Paleontological J.*, **48**: 701−849. https://doi.org/10.1134/S0031030114070016.

Baresel, B., Bucher, H., Brosse, M., Cordey, F., Guodun, K., and Schaltegger, U., 2017, Precise age for the Permian−Triassic boundary in South China from high-precision U-Pb geochronology and Bayesian age−depth modeling. *Solid Earth*, **8**: 361−378.

Barghava, O.N., 2008, Paleozoic Successions of the Indian Plate. *Mem. Geol. Soc. India*, **74**: 209−244.

Barnett, A.J., and Wright, V.P., 2008, A sedimentological and cyclostratigraphic evaluation of the completeness of the Mississippian-Pennsylvanian (Mid-Carboniferous) global stratotype section and point, Arrow Canyon, Nevada, USA. *J. Geol. Soc.*, **165**: 859−873. https://doi.org/10.1144/0016-76492007-122.

Barrick, J.E., Lambert, L.L., Heckel, P.H., and Boardman II, D.R., 2004, Pennsylvanian conodont zonation for Midcontinent North America. *Rev. Espanola de. Micropaleontologia*, **36**: 231−250.

Barrick, J.E., Lambert, L.L., Heckel, P.H., Rosscoe, S.J., and Boardman, D.R., 2013, Midcontinent Pennsylvanian conodont zonation. *Stratigraphy*, **10**: 55−72.

Barskov, I.S., and Alekseev, A.S., 1975, Conodonts of the middle and upper Carboniferous of Moscow Basin. *N. Acad. Sci. USSR, Geol. Ser.*, **6**: 84−99.

Barskov, I.S., Alekseev, A.S., and Goreva, N.V., 1980, Conodonts and stratigraphic scale of the Carboniferous. *N. Acad. Sci. USSR, Geol. Ser.*, **3**: 43−45.

Batt, L.S., Pope, M.C., Isaacson, P.E., Montañez, I.M., and Abplanalp, J., 2007, Multi-carbonate component reconstruction of mid-carboniferous (Chesterian) seawater $\delta^{13}C$. *Palaeogeography, Palaeoclimatology, Palaeoecology*, **256**: 298−318. https://doi.org/10.1016/j.palaeo.2007.02.049.

Becker, R.T., Gradstein, F.M., and Hammer, O., 2012, Chapter 22 − The Devonian period. *In* Gradstein, F.M., Ogg, J.G., Schmitz, M.D., and Ogg, G.M. (eds), *The Geologic Time Scale 2012*, Elsevier, pp. 559−601. https://doi.org/10.1016/B978-0-444-59425-9.00022-6.

Becker, R.T., Kaiser, A.I., and Aretz, M., 2016, Review of the chrono-, litho- and biostratigraphy across the global Hangenberg Crisis and Devonian-Carboniferous Boundary. *In*: Becker, R.T., Königshof, P., and Brett, C.E. (eds), Devonian Climate, Sea Level and Evolutionary Events. *Geological Society, London, Special Publications*, **423**: 355−386. https://doi.org/10.1144/SP423.10.

Becker, R.T., and Paproth, E., 1993, Auxiliary stratotype sections for the Global Stratotype Section and Point (GSSP) for the Devonian-Carboniferous boundary. Hasselbachtal. *Annales de. la. Société Géologique de Belgique*, **115**: 703−706.

Belford, D.J., 1970, Upper Devonian and Carboniferous Foraminifera, Bonaparte Gulf Basin, northwestern Australia. *Australian Bur. Miner. Resources, Geol. Geophysics, Bull.*, **108**: 39.

Berger, A., Loutre, M.F., and Laskar, J., 1992, Stability of the astronomical frequencies over the Earth's history for paleoclimate studies. *Science*, **255**: 560−566. https://doi.org/10.1126/science.255.5044.560.

Bisat, W.S., 1924, The Carboniferous goniatites of the north of England and their zones. *Proc. Yorks. Geol. Society, N. Ser.*, **20**: 40−124.

Bisat, W.S., 1928, *The Carboniferous goniatite zones of England and their continental equivalents. Compte Rendue Congrès International de Stratigraphie et de Géologie du Carbonifère (Heerlen 1927)*. Vaillant-Carmanne, pp. 117−133.

Bishop, J.W., Montañez, I.P., Gulbranson, E.L., and Brenckle, P.L., 2009, The onset of mid-Carboniferous glacio-eustasy: sedimentologic and diagenetic

constraints, Arrow Canyon, NV. *Palaeogeography, Palaeoclimatology, Palaeoecology*, **276**: 217−243. https://doi.org/10.1016/j.palaeo.2009.02.019.

Blumenstengel, H., 2006, Ostracoden. In: Deutsche Stratigraphische Kommission. *In*: Amler, M.R.W., and Stoppel, D. (eds), Stratigraphie von Deutschland VI. Unterkarbon (Mississippium). *Schriftenreihe der Deutschen Gesellschaft für Geowissenschaften*, **41**: 178−182.

Boardman II, D.R., and Heckel, P.H., 1989, Glacial-eustatic sea-level curve for early Late Pennsylvanian sequence in north-central Texas and biostratigraphic correlation with curve for Midcontinent North America. *Geology*, **17**: 802−805. https://doi.org/10.1130/0091-7613(1989)017<0802:GESLCF>2.3.CO;2.

Boardman II, D.R., Heckel, P.H., Barrick, J.E., Nestell, M., and Peppers, R.A., 1991, Middle-Upper Pennsylvanian chronostratigraphic boundary in the Midcontinent region of North America. *Cour. Forschungsinstitut Senckenberg.*, **130**: 319−337. (imprinted 1990).

Boardman II, D.R., Work, D.M., Mapes, R.H., and Barrick, J.E., 1994, Biostratigraphy of Middle and Late Pennsylvanian (Desmoinesian-Virgilian) ammonoids. *Kans. Geol. Survey, Rep.*, **232**: 1−121.

Boardman, D.R., and Work, D.M., 2004, Stratigraphic distribution of Stratigraphic distribution of the ammonoids Shumardites and Vidrioceras and implications for the definition and correlation of the global Gzhelian Stage, Upper Pennsylvanian Series. *Newsl. Carbonif. Stratigraphy*, **22**: 23−27.

Boardman, D.R., and Work, D.M., 2013, Pennsylvanian (Desmoinesian-Virgilian) ammonoid zonation for Midcontinent North America. *Stratigraphy*, **10**: 105−116.

Boardman, D.R., Wardlaw, B.R., and Nestell, M.K., 2009, Stratigraphy and conodont biostratigraphy of the uppermost Carboniferous and lower Permian from the North American Midcontinent. *Kans. Geol. Survey, Bull.*, **255**: 1−42.

Bogoslovskaya, M.F., 1984, The main stages of the ammonoid evolution in the Late Carboniferous. *In*: Menner, V.V., and Grigorjeva, A.D. (eds), Upper Carboniferous of the USSR. *Mezhvedomstrennyi stratigraficheskii komitet, Trudy*, **11**, pp. 88−91.

Bouroz, A., Wagner, R.H., and Winkler, P.C., 1978, Report and proceedings of the IUGS Subcomission on Carboniferous Stratigraphy Meeting in Moscow, 8−12 September 1975. *Huitième Congrès International de Stratigraphie et de Géologie Carbonifère, Moscow 1975, Compte Rendu*, **Vol. 1**, pp. 27−35.

Boy, J.A., and Schindler, T., 2000, Ecostratigraphic bioevents at the Stephanian-Autunian transition (uppermost Carboniferous) of the Saar-Nahe Basin (SW Germany) and neighboring regions. *Neues Jahrb. für Geologie und Paläontologie, Abhandlungen*, **216**: 89−152. https://doi.org/10.1127/njgpa/216/2000/89.

Brand, U., Tazawa, J.-I., Sano, H., Azmy, K., and Lee, X., 2009, Is mid-late Paleozoic ocean-water chemistry coupled with epeiric seawater isotope records? *Geology*, **37**: 823−826. https://doi.org/10.1130/G30038A.1.

Brauckmann, C., and Hahn, G., 2006, Trilobiten. In: Deutsche Stratigraphische Kommission. *In*: Amler, M.R.W., and Stoppel, D. (eds), Stratigraphie von Deutschland VI. Unterkarbon (Mississippium). *Schriftenreihe der Deutschen Gesellschaft für Geowissenschaften*, **41**: 171−177.

Braun, A., and Schmidt-Effing, R., 1993, Biozonation, diagenesis and evolution of radiolarians in the Lower Carboniferous of Germany. *Mar. Micropaleontology*, **21**: 369−383. https://doi.org/10.1016/0377-8398(93)90027-U.

Brezinski, D.K., Cecil, C.B., and Skema, V.W., 2010, Late Devonian glacigenic and associated facies from the central Appalachian Basin, eastern United States. *Geol. Soc. Am. Bull.*, **122**: 265−281. https://doi.org/10.1130/B26556.1.

British Lower Carboniferous Stratigraphy. *In* Cossey, P.J., Adams, A.E., Purnell, M.A., Whiteley, M.J., Whyte, M.A., and Wright, V.P. (eds), Geological Conservation Review Series. 29. 1−617.

Bruckschen, P., Oesmann, S., and Veizer, J., 1999, Isotope stratigraphy of the European Carboniferous: Proxy signals for ocean chemistry, climate and tectonics. *Chem. Geol.*, **161**: 127−163. https://doi.org/10.1016/S0009-2541(99)00084-4.

Bruckschen, P., and Veizer, J., 1997, Oxygen and carbon isotopic composition of Dinantian brachiopods: paleoenvironmental implications for the Lower Carboniferous of Western Europe. *Paleogeography, Paleoclimatology, Paleoecology*, **132**: 243−264. https://doi.org/10.1016/S0031-0182(97)00066-7.

Brunton, C.H.C., 1984, The use of brachiopods in Carboniferous stratigraphy. *9ième Congres International de Stratigraphie et de Géologie du Carbonifère, Washington and Champaign-Urbana 1979, Compte Rendu*, Carbondalle and Edwardsville: Southern Illinois University Press, **Vol. 2**, pp. 35−51.

Buggisch, W., Joachimski, M.M., Sevastopulo, G., and Morrow, J.R., 2008, Mississippian $\delta^{13}C$carb and conodont apatite $\delta^{18}O$ records: their relation to the late Palaeozoic glaciation. *Palaeogeography, Palaeoclimatology, Palaeoecology*, **268**: 273−292. https://doi.org/10.1016/j.palaeo.2008.03.043.

Buggisch, W., Wang, X.D., Alekseev, A.S., and Joachimski, M.M., 2011, Late Carboniferous−Permian carbon isotope stratigraphy of successions from China (Yangtze platform), USA (Kansas) and Russia (Moscow Basin and Urals). *Palaeogeography, Palaeoclimatology, Palaeoecology*, **301**: 18−38. https://doi.org/10.1016/j.palaeo.2010.12.015.

Calver, M.A., 1969, Westphalian of Britain. *Sixième Congres International de Stratigraphie et de Géologie du Carbonifère, Sheffield 1967, Compte Rendu*, Maastricht: Ernest van Aelst, **Vol. 1**, pp. 231−254.

Cardoso, C.N., Sanz-López, J., and Blanco-Ferrera, S., 2017, Pennsylvanian conodont zonation of the Tapajós Group (Amazon Basin, Brazil). *Stratigraphy*, **14**: 35−58.

Carter, J.L., 1991, Subdivision of the Lower Carboniferous in North America by means of articulate brachiopod generic ranges. *Cour. Forschungsinstitut Senckenberg.*, **130**: 145−155. (imprinted 1990).

Chaloner, W.G., and Lacey, W.S., 1973, The distribution of late Paleozoic floras. *Spec. Pap. Palaeontology*, **12**: 271−289.

Chen, B., Joachimski, M.M., Wang, X., Shen, S., Qi, Y., and Qie, W., 2016, Ice volume and paleoclimate history of the Late Paleozoic Ice Age from conodont apatite oxygen isotopes from Naqing (Guizhou, China). *Palaeogeography, Palaeoclimatology, Palaeoecology*, **448**: 151−161. https://doi.org/10.1016/j.palaeo.2016.01.002.

Chen, J.T., Montañez, I.P., Qi, Y.P., Wang, X.D., Wang, Q.L., and Lin, W., 2016, Coupled sedimentary and $\delta^{13}C$ records of late Mississippian platform-to-slope successions from South China: Insight into $\delta^{13}C$ chemostratigraphy. *Palaeogeography, Palaeoclimatology, Palaeoecology*, **448**: 162−178. https://doi.org/10.1016/j.palaeo.2015.10.051.

Chen, J., Montañez, I.P., Qi, Y., Shen, S., and Wang, X., 2018, Strontium and carbon isotopic evidence for decoupling of pCO_2 from continental weathering at the apex of the late Paleozoic glaciation. *Geology*, **46**: 395−398. https://doi.org/10.1130/G40093.1.

Chernykh, V.V., 2005, *Zonal Method in Biostratigraphy: Zonal Conodont Scale of the Lower Permian in the Urals*. Ekaterinburg: Institute of Geology and Geochemistry, Uralian Branch of the Russian Academy of Sciencesp. 217. (In Russian).

Chernykh, V.V., Chuvashov, B.I., Davydov, V.I., Schmitz, M.D., and Snyder, W.S., 2006, Usolka section (Southern Urals, Russia): a potential candidate for GSSP to define the base of the Gzhelian Stage in the global chronostratigraphic scale. *Geologija*, **49**: 205−217. https://doi.org/10.5474/geologija.2006.015.

Chi, Y.S., 1931, Weiningian (Middle Carboniferous) Corals of China. *Palaeontologia Sin. Ser. B*, **12** (5), 1−55.

Cisterna, G.A., and Sterren, A.F., 2010, "Levipustula Fauna" in central-western Argentina and its relationship with the Carboniferous glacial event in the southwestern Gondwanan margin. *In*: López-Gamundi, O.R., and Buatois, L.A. (eds), Late Glacial Events and Postglacial Transgressions in Gondwana. *Geological Society of America, Special Paper*, **468**: 133−147. https://doi.org/10.1130/2010.2468(06).

Claoué-Long, J.C., Compston, W., Roberts, J., and Fanning, C.M., 1995, Two Carboniferous ages: a comparison of SHRIMP zircon dating with conventional zircon ages and 40Ar/39Ar analysis. *SEPM Special Publication*, **54**: 3−21. https://doi.org/10.2110/pec.95.04.0003.

Clayton, G., 1985, Dinantian miospores and inter-continental correlation. *10ième Congres International de Stratigraphie et de Géologie du Carbonifère, Madrid 1983, Compte Rendu*, Madrid: IGME, **Vol. 4**, pp. 9−23.

Clopine, W.W., 1992, Lower and Middle Pennsylvanian fusulinid biostratigraphy of southern New Mexico and westernmost Texas. *N. Mexico Bur. Geol. Miner. Resources, Rep.*, **143**: 1−68.

Compston, W., 2000, Interpretation of SHRIMP and isotope dilution zircon ages for the Palaeozoic time-scale, II: Silurian to Devonian. *Mineralogical Magazine*, **64** (6), 1127−1146.

Conil, R., Groessens, E., Laloux, M., and Poty, E., 1989, La limite Tournaisien/Viséen dans la région-type. *Annales de. la. Société géologique de Belgique*, **112**: 177−189.

Conil, R., Groessens, E., and Pirlet, H., 1977, Nouvelle charte stratigraphique du Dinantien type de la Belgique. *Annales de. la. Société géologique du. Nord.*, **96**: 363−371.

Conil, R., Groessens, E., Laloux, M., Poty, E., and Tourneur, F., 1991, Carboniferous guide Foraminifera, corals and conodonts in the Franco-Belgian and Campine Basins: Their potential for widespread correlation. *Cour. Forschungsinstitut Senckenberg.*, **130**: 15−30. (imprinted 1990).

Conkin, J.E., and Conkin, B.M., 1973, The paracontinuity and the determination of the Devonian-Mississippian boundary in the Lower Mississippian type area of North America. *Univ. Louisville Stud. Paleontology Stratigraphy*, **1**: 1−36.

Connolly, W.M., and Stanton Jr., R.J., 1992, Interbasinal cyclostratigraphic correlation of Milankovitch band transgressive-regressive cycles: correlation of Desmoinesian−Missourian strata between southeastern Arizona and the midcontinent of North America. *Geology*, **20**: 999−1002. https://doi.org/10.1130/0091-7613(1992)020<0999:ICCOMB>2.3.CO;2.

Conrad, J., 1985, Northwestern and central Saharan areas: stratigraphic and structural framework. *In*: Martínez Díaz, C., Wagner, R.H., Winkler Prins, C.F., and Granados, L.F. (eds), *The Carboniferous of the World II: Australia, Indian Subcontinent, South America and North Africa*. Instituto Geologico y Minero de España IGME and Empresa Nacional Adaro de Investigaciones Mineras, pp. 315−317.

Conybeare, W.D., and Phillips, W., 1822, *Outlines of the geology of England and Wales, With an Introduction Compendium of the General Principles of That Science, and Comparative Views of the Structure of Foreign Countries. Part 1*. London: William Phillips, p. 470.

The Carboniferous Period **Chapter | 23** **859**

Cooper, M.R., 1986, Facies shifts, sea-level changes and event stratigraphy in the Devonian of South Africa. *South. Afr. J. Sci.*, **82**: 255–258.

Corradini, C., Kaiser, S.I., Perri, M.C., and Spalletta, C., 2011, *Protognathodus* (Conodonta) and its potential as a tool for defining the Devonian/Carboniferous boundary. *Riv. Ital. di Paleontologia e Stratigrafia*, **117**: 15–28. https://doi.org/10.13130/2039-4942/5960.

Corradini, C., Spalletta, C., Mossoni, A., Matyja, H., and Over, D.J., 2017, Conodont across the Devonian/Carboniferous boundary: a review and implication for the redefinition of the boundary and a proposal for an updated conodont zonation. *Geol. Mag.*, **154**: 888–902. https://doi.org/10.13130/2039-4942/5960.

Cózar, P., and Somerville, I.D., 2004, New algal and foraminiferal assemblages and evidence for recognition of the Asbian-Brigantian boundary in northern England. *Proc. Yorks. Geol. Soc.*, **55**: 43–65. https://doi.org/10.1144/pygs.55.1.43.

Cózar, P., Vachard, D., Aretz, M., and Somerville, I.D., 2019, Foraminifers of the Visean–Serpukhovian boundary interval in Western Palaeotethys: a review. *Lethaia*, **52**: 260–284. https://doi.org/10.1111/let.12311.

Cramer, B.D., and Jarvis, I., 2020, Chapter 11 - Carbon isotope stratigraphy. *In* Gradstein, F.M., Ogg, J.G., Schmitz, M.D., and Ogg, G.M. (eds), *The Geologic Time Scale 2020*. Vol. 1 (this book): Elsevier. Boston, MA (this book).

Crasquin, S., 1985, Zonation par les Ostracodes dans le Mississippien de l'Ouest canadien. *Rev. de. Paléobiologie*, **4**: 45–52.

d'Omalius d'Halloy, J.B., 1828, *Mémoires pour servir a la description géologique des Pays-Bas, de la France et de quelques contrées voisines*, Namur: Imprimerie D. Gerard. pp. 307.

Da Silva, A.C., De Vleeschouwer, D., Boulvain, F., Claeys, P., Nagel, N., Humblet, M., et al., 2013, Magnetic susceptibility as a high-resolution correlation tool and as a climatic proxy in Paleozoic rocks – merits and pitfalls: examples from the Devonian in Belgium. *Mar. Pet. Geol.*, **46**: 173–189. https://doi.org/10.1016/j.marpetgeo.2013.06.012.

Davies, S.J., and Elliott, T., 1996, Spectral gamma ray characterization of high resolution sequence stratigraphy: examples from Upper Carboniferous fluvio-deltaic systems, County Clare, Ireland. *In* : Howell, J.A., and Aitken, J.F. (eds), High Resolution Sequence Stratigraphy: Innovations and Applications. *Geological Society London, Special Publications*, **104**: 25–35. https://doi.org/10.1144/GSL.SP.1996.104.01.03.

Davydov, V.I., 1990, On the clarification of the origin and phylogeny of triticitids and the Mid-Upper Carboniferous boundary. *Paleontological J.*, **24**: 13–25. (in Russian).

Davydov, V.I., 2007, Protriticites foraminiferal fauna and its utilization in the Moscovian-Kasomovian boundary definition. *In:* Wong, T.E., (ed), *Proceedings of the XVth International Congress on Carboniferous and Permian Stratigraphy, Utrecht, 10–16 August 2003.* Amsterdam: Royal Netherlands Academy of Arts and Sciences, pp. 456–466.

Davydov, V.I., 2009, Bashkirian-Moscovian transition in Donets Basin: the key for Tethyan-Boreal correlation. *In:* Puchkov, V.N. (ed), *The Carboniferous Type Sections in Russia, Potential and Proposed Stratotypes. Proceedings of the International Conference Ufa—Sibai, 13–18 August 2009.* Ufa: Institute of Geology, Bashkirian Academy of Science, pp. 188–192.

Davydov, V.I., 2014, Warm water benthic Foraminifera document the Pennsylvanian-Permian warming and cooling events – the record from the western Pangea tropical shelves. *Palaeogeography, Palaeoclimatology, Palaeoecology*, **414**: 284–295. https://doi.org/10.1016/j.palaeo.2014.09.013.

Davydov, V.I., Barskov, I.S., Bogoslovskaya, M.F., and Leven, E.Y., 1994, Carboniferous-Permian boundary in the stratotype sections of the Southern Urals and its correlation. *Stratigraphy Geol. Correlations*, **2**: 32–45.

Davydov, V.I., Biakov, A.V., Schmitz, M.D., and Silantiev, V.V., 2018a, Radioisotopic calibration of the Guadalupian (middle Permian) series – Review and updates. *Earth Science Reviews*, **176**: 222–240.

Davydov, V.I., Crowley, J.L., Schmitz, M.D., and Snyder, W.S., 2018b, New U–Pb constraints identify the end-Guadalupian and possibly end-Lopingian extinction events conceivably preserved in the passive margin of North America: implication for regional tectonics. *Geological Magazine*, **155**: 119–131.

Davydov, V.I., Chernykh, V.V., Chuvashov, B.I.M.D.S., and Snyder, W.S., 2008b, Faunal assemblage and correlation of Kasimovian-Gzhelian transition at Usolka section, Southern Urals, Russia (a potential candidate for GSSP to define base of Gzhelian Stage). *Stratigraphy*, **5**: 113–135.

Davydov, V.I., and Cózar, P., 2019, The formation of the Alleghenian Isthmus triggered the Bashkirian glaciation: constraints from warm-water benthic Foraminifera. *Palaeogeography, Palaeoclimatology, Palaeoecology*, **531**: 108403. https://doi.org/10.1016/j.palaeo.2017.08.012. part B.

Davydov, V.I., Crowley, J.L., Schmitz, M.D., and Poletaev, V.I., 2010, High precision U-Pb zircon age calibration of the global Carboniferous time scale and Milankovitch-band cyclicity in the Donets Basin, eastern Ukraine. *Geochemistry, Geophysics, Geosystems*, **11**: Q0AA04. https://doi.org/10.1029/2009GC002736.

Davydov, V.I., Glenister, B.F., Spinosa, C., Ritter, S.M., Chernykh, V.V., Wardlaw, B.R., et al., 1998, Proposal of Aidaralash as GSSP for base of the Permian System. *Episodes*, **21**: 11–18.

Davydov, V.I., and Khodjanyazova, R., 2009, Moscovian-Kasimovian transition in Donets Basin: Fusulinid taxonomy, biostratigraphy correlation and paleobiogeography. *In:* Puchkov, V.N. (ed), *The Carboniferous Type Sections in Russia, Potential and Proposed Stratotypes. Proceedings of the International Conference Ufa—Sibai, August 13–18, 2009.* Ufa: Institute of Geology, Bashkirian Academy of Science, pp. 193–196.

Davydov, V.I., and Khramov, A.N., 1991, Paleomagnetism of Upper Carboniferous and Lower Permian in the Karachatyr region (southern Ferhgana) and the problems of correlation of the Kiama hyperzone. *In:* Khramov, A.N. (ed), *Paleomagnetism and Paleogeodynamics of the Territory of USSR.* Transactions of VNIGRI, pp. 45–53. (in Russian).

Davydov, V.I., Korn, D., Schmitz, M.D., Gradstein, F.M., and Hammer, O., 2012, Chapter 23 – the Carboniferous Period. *In:* Gradstein, F.M., Ogg, J.G., Schmitz, M.D., and Ogg, G.M. (eds), *The Geologic Time Scale 2012*, Elsevier, pp. 603–651. https://doi.org/10.1016/B978-0-444-59425-9.00023-8.

Davydov, V.I., Schmitz, M.D., and Korn, D., 2011, The Hangenberg Event was abrupt and short at the global scale: the quantitative integration and intercalibration of biotic and geochronologic data within the Devonian-Carboniferous transition. *Abstracts with Programs – Geological Society of America*, **43**: 128.

Davydov, V.I., Schmitz, M.D., Chernykh, V.V., Crowley, J., Henderson, C.M., and Korn, D., 2008a, Carboniferous and Permian geologic time scale: state of the art, chronostratigraphic, biostratigraphic and radioisotopic calibration and integration. *33rd International Geological Congress 2008, Oslo, Abstracts on CD-ROM.*

Davydov, V.I., Wardlaw, B.R., and Gradstein, F.M., 2004, The Carboniferous Period. *In* : Gradstein, F.M., Ogg, J.G., and Smith, A.G. (eds), *A Geologic Time Scale 2004.* Cambridge University Press, pp. 222–248.

de Witry, L.H. (abbé d'Everlange), 1780, Mémoire sur les fossiles du Tournaisis et les pétrifications en général, relativement à leur utilité pour la vie civile (lu à la séance du 9 décembre 1776). *Mémoire de. l'Académie impériale et. royale des. Sci. et B.-lett. de Brux.*, **3**: 15–44.

Delcambre, B., 1989, Marqueurs téphrostratigraphiques au passage des calcaires de Neffe vers ceux de Lives. *Bull. de. la. Société Belg. de Géologie*, **98–2**: 163–170.

Delepine, G., 1911, Recherches sur le calcaire carbonifère de la Belgique. *Mémoires et Travaux de la Faculté catholique de Lille*, **Vol. 8**: 1–419.

Demanet, F., 1929, Les lamellibranches du Marbre noir de Dinant (Viséen inférieur). *Mémoire du. Musée R. d'Histoire naturelle de. Belgique*, **40**: 1–79.

Demanet, F., 1958, Contribution à l'étude du Dinantien de la Belgique. *Inst. R. des. Sci. naturelles de. Belgique, Mémoire*, **141**: 1–152.

Derder, M.E.M., Smith, B., Henry, B., Yelles, A.K., Bayou, B., Djellit, H., et al., 2001, Juxtaposed and superimposed palaeomagnetic primary and secondary components from the folded Middle Carboniferous sediments in the Reggane basin (Saharan Craton, Algeria). *Tectonophysics*, **344**: 403–422. https://doi.org/10.1016/S0040-1951(00)00298-5.

Devuyst, F.X., Hance, L., Hou, H., Wu, X., Tian, S., Coen, M., et al., 2003, A proposed Global Stratotype Section and Point for the base of the Visean Stage (Carboniferous): the Pengchong section, Guangxi, South China. *Episodes*, **26**: 105–115.

Devuyst, F.X., Hance, L., and Poty, E., 2006, Moliniacian. *Geologica Belgica*, **9**: 123–131.

Dickinson, W.R., Soreghan, G.S., and Giles, K.A., 1994, Glacio-eustatic origin of Permo-Carboniferous stratigraphic cycles: evidence from the southern Cordilleran foreland region. *In*: Dennison, J.M., and Ettensohn, F.R. (eds), *SEPM, Concepts in Sedimentology and Paleontology*. **4**: 25–34. https://doi.org/10.2110/csp.94.04.0025.

DiMichele, W.A., 2014, Wetland-dryland vegetational dynamics in the Pennsylvanian ice age tropics. *Int. J. Plant. Sci.*, **175**: 123–164. https://doi.org/10.1086/675235.

DiMichele, W., Pfefferkorn, H.W., and Gastaldo, R.A., 2001, Response of Late Carboniferous and Early Permian plant communities to climate change. *Annu. Rev. Earth Planet. Sci.*, **29**: 461–487. https://doi.org/10.1146/annurev.earth.29.1.461.

DiMichele, W., Gastaldo, R.A., and Pfefferkorn, H.W., 2005, Plant biodiversity partitioning in the Late Carboniferous and Early Permian and its implications for ecosystem assembly. *Proceedings of the California Academy of Sciences*, **56** [Suppl. I (4)]: 24–41.

DiVenere, V.J., and Opdyke, N.D., 1991a, Magnetic polarity stratigraphy and Carboniferous paleopole position from the Joggins Section, Cumberland Basin, Nova Scotia. *J. Geophys. Res.*, **96**: 4051–4064. https://doi.org/10.1029/90JB02148.

DiVenere, V.J., and Opdyke, N.D., 1991b, Magnetic polarity stratigraphy in the uppermost Mississippian Mauch Chunk Formation, Pottsville, Pennsylvania. *Geology*, **19**: 127–130. https://doi.org/10.1130/0091-7613(1991)019<0127:MPSITU>2.3.CO;2.

d'Omalius d'Halloy, J.B., 1808, Essai sur la Géologie du Nord de la France. *J. des. Mines, Paris: Bossange & Masson*, **XXIV**: 1–154.

Druce, E.C., 1969, Devonian and Carboniferous conodonts from the Bonaparte Gulf Basin, Northern Australia, and their use in international correlation. *Australian Bur. Miner. Resources, Bull.*, **98**: 243.

Druce, E.C., 1974, Australian Devonian and Carboniferous conodont faunas. *In* Bouckaert, J., and Streel, M. (eds), *International Symposium on Belgian Micropalaeontological Limits From Emsian*

to Visean. *Namur, Bruxelles:* Ministry of Economic Affairs, Adm. Mines-Geological Survey of Belgium, Publication No. 5, p. 18.

Du Toit, A.L., 1937, *Our wandering continents*. Oliver and Boyd, p. 366.

Dumont, A., 1832, Mémoire sur la constitution géologique de la Province de Liège. *Mémoires couronnés et. mémoires des. savants étrangers de. l'Académie des Sci. et B.-lett. de Brux.*, **8**: 1–374.

Dusar, M., 2006, Namurian. *Geologica Belgica*, **9**: 163–175.

Eager, R.M.C., 2005, Non-marine and limnic bivalves. In: Deutsche Stratigraphische Kommission. *In*: Wrede, V. (ed), Stratigraphie von Deutschland, V. Das Oberkarbon (Pennsylvanium) in Deutschland. *Courier Forschungsinstitut Senckenberg*, **254**: 55–86.

Ebbighausen, V., and Bockwinkel, J., 2007, Tournaisian (Early Carboniferous/Mississippian) ammonoids from the Ma'der Basin (Anti-Atlas, Morocco). *Foss. Rec.*, **10**: 125–163. https://doi.org/10.1002/mmng.200700003.

Edgell, H.S., 2004, Upper Devonian and Lower Carboniferous Foraminifera from the Canning Basin, Western Australia. *Micropaleontology*, **50**: 1–26. https://doi.org/10.1661/0026-2803(2004)050[0001:UDALCF]2.0.CO;2.

Ehrenberg, S.N., and Svånå, T.A., 2001, Use of spectral gamma-ray signature to interpret stratigraphic surfaces in carbonate strata: an example from the Finnmark carbonate platform (Carboniferous-Permian), Barents Sea. *AAPG Bull.*, **85**: 295–308. https://doi.org/10.1306/8626C7C1-173B-11D7-8645000102C1865D.

Einor, O.L., 1996, The former USSR. *In* Wagner, R.H., Winkler Prins, C.F., and Granados, L.F. (eds), *The Carboniferous of the World III. The Former USSR, Mongolia, Middle Eastern Platform, Afghanistan and Iran*. Instituto Geologico y Minero de España IGME and Empresa Nacional Adaro de Investigaciones Mineras, pp. 521.

Engel, B.A., and Morris. N., 1985, The biostratigraphy of Carboniferous trilobites in eastern Australia. *10ème Congres International de Stratigraphie et de Géologie du Carbonifère, Madrid 1983, Compte Rendu*, Madrid: IGME, **Vol. 4**, pp. 491–499.

Eros, J.M., Montañez, I.P., Osleger, D.A., Davydov, V.I., Nemyrovska, T.I., Poletaev, V.I., et al., 2012, Sequence stratigraphy and onlap history of the Donets Basin, Ukraine: insight into Carboniferous icehouse dynamics. *Palaeogeography, Palaeoclimatology, Palaeoecology*, **313–314**: 1–25. https://doi.org/10.1016/j.palaeo.2011.08.019.

Eyles, N., Mory, A.J., and Backhouse, J., 2002, Carboniferous–Permian palynostratigraphy of west Australian marine rift basins: resolving tectonic and eustatic controls during Gondwanan glaciations. *Palaeogeography, Palaeoclimatology, Palaeoecology*, **184**: 305–319. https://doi.org/10.1016/S0031-0182(02)00260-2.

Falcon-Lang, H., Heckel, P.H., DiMichele, W.A., Blake, B.M., Easterday, C.R., Eble, C.F., et al., 2011, No major stratigraphic gap exists near the Middle-Upper Pennsylvanian (Desmoinesian-Missourian) Boundary in North America. *Palaios*, **26**: 125–139. https://doi.org/10.2110/palo.2010.p10-049r.

Falcon-Lang, H.J., Nelson, W.J., Heckel, P.H., DiMichele, W.A., and Elrick, S.D., 2018, New insights on the stepwise collapse of the Carboniferous Coal Forests: Evidence from cyclothems and coniferopsid tree-stumps near the Desmoinesian–Missourian boundary in Peoria County, Illinois, USA. *Palaeogeography, Palaeoclimatology, Palaeoecology*, **490**: 375–392. https://doi.org/10.1016/j.palaeo.2017.11.015.

Fang, Q., Wu, H.C., Wang, X.L., Yang, T.S., Li, H.J., and Zhang, S.H., 2018, Astronomical cycles in the Serpukhovian–Moscovian (Carboniferous) marine sequence, South China, and their implications for geochronology and icehouse dynamics. *J. Asian Earth Sci.*, **156**: 302–315. https://doi.org/10.1016/j.jseaes.2018.02.001.

Farey, J., 1807, Coal. *In:* Reese, A. (ed), *The New Cyclopaedia*. London: Longman, Hurst, Rees, Orme & Brown.

Farey, J., 1811, *A General View of the Agriculture and Minerals of Derbyshire. Volume 1*. London: Board of Agriculture, pp. 532.

Fielding, C.R., Frank, T.D., Birgenheier, L.P., Rygel, M.C., Jones, A.T., and Roberts, J., 2008a, Stratigraphic imprint of the Late Paleozoic Ice Age in eastern Australia: a record of alternating glacial and non-glacial climate regime. *J. Geol. Soc. Lond.*, **165**: 129–140. https://doi.org/10.1144/0016-76492007-036.

Fielding, C.R., Frank, T.D., and Isbell, J.L., 2008b, The late Paleozoic ice age—a review of current understanding and synthesis of global climate patterns. *In* Fielding, C.R., Frank, T.D., and Isbell, J.L. (eds), Resolving the Late Paleozoic Ice Age in Time and Space. *Geological Society of America, Special Paper*, **441**: 343–354.

Fillmore, D.L., Lucas, S.G., and Simpson, E.L., 2013, Ichnology of the continental Mississippian Mauch Chunk Formation, eastern Pennsylvania. *In :* Lucas, S.G., DiMichele, W.A., Barrick, J.E., Schneider, J.W., and Spielmann, J.A. (eds), The Carboniferous-Permian Transition. *New Mexico Museum of Natural History and Science Bulletin*, **60**: 81–97.

Frech, F., 1887, Die paläozoischen Bildungen von Cabrières (Languedoc). *Z. der Deutschen Geologischen Ges.*, **39**: 360–487.

Ganelin, V.G., and Durante, M.V., 2002, Biostratigraphy of the Carboniferous of Angaraland. *Newsl. Carbonif. Stratigraphy*, **20**: 23–26.

Gao, L., 1984, Carboniferous spore assemblages in China. *Neuvième Congrès International de Stratigraphie et de Géologie du Carbonifère, Washington and Champaign-Urbana 1979, Compte Rendu*, Carbondalle and Edwardsville: Southern Illinois University Press, **Vol. 2**, pp. 103–108.

Gastaldo, R.A., Purkyňová, E., Simůnek, Z., and Schmitz, M.D., 2009, Ecological persistence in the Late Mississippian (Serpukhovian - Namurian A) megafloral record of the Upper Silesian Basin, Czech Republic. *Palaios*, **24**: 336–350.

George, T.N., Johnson, G.A.L., Mitchell, M., Prentice, J.E., Ramsbottom, W.H.C., Sevastopulo, G.D., et al., 1976, A correlation of Dinantian rocks in the British Isles. *Geological Society of London, Special Report*, **7**: 1–87.

George, T.N., and Wagner, R.H., 1972, International Union of Geological Sciences, Subcommission on Carboniferous Stratigraphy, Proceedings and Report of the General Assembly at Krefeld, *21–22 August. Septième Congrès International de Stratigraphie et de Géologie du Carbonifère, Krefeld 1971, Compte Rendu*, Krefeld: Geologisches Landesamt Nordrhein-Westfalen, **Vol. 1**, pp. 139–147.

Gess, R.W., 2016, Vertebrate biostratigraphy of the Witteberg Group and the Devonian–Carboniferous boundary in South Africa. *In:* Linol, B., and de, Wit (eds), *Origin and evolution of the Cape Mountains and Karoo Basin. Regional Geology Reviews*: pp. 131–140. https://doi.org/10.1007/978-3-319-40859-0-13.

Gibshman, N.B., 2003, Characteristics of foraminifers from the Serpukhovian stratotype, Zabor'ye Quarry; Moscow region. *Stratigrafiya, Geologicheskaya Korrelyatsiya*, **11**: 39–63. (in Russian).

Gibshman, N.B., and Baranova, D.V., 2007, The foraminifers Janischewskina and 'Millerella', their evolutionary patterns and biostratigraphic potential for the Visean-Serpukhovian boundary. *In :* Wong, T.E. (ed), *Proceedings of the XVth International Congress on Carboniferous and Permian Stratigraphy*. Utrecht: Royal Netherlands Academy of Arts and Sciences, 269–281.

Gibshman, N.B., Kabanov, P.B., Alekseev, A.S., Goreva, N.V., and Moshkina, M.A., 2009, Novogurovsky Quarry. Upper Visean and Serpukhovian. *In :* Alekseev, A.S., and Goreva, N.V. (eds), *Type and Reference Carboniferous Sections in the South Part of the Moscow Basin. Field Trip Guidebook of International Field Meeting of the I.U.G. S. Subcommission on Carboniferous Stratigraphy "The Historical Type Sections, Proposed and Potential GSSP of the Carboniferous in Russia," Moscow, August 11–12, 2009*. Moscow: Borissiak Paleontological Institute of RAS, pp. 13–44.

Giles, P.S., 2009, Orbital forcing and Mississippian sea level change: time series analysis of marine flooding events in the Visean Windsor Group of eastern Canada and implications for Gondwana glaciation. *Bull. Can. Pet. Geol.*, **57**: 449–471. https://doi.org/10.2113/gscpgbull.57.4.449.

Glenister, B.F., Wardlaw, B.R., Lambert, L.L., Spinosa, C., Bowring, S.A., Erwin, D.H., et al., 1999, Proposal of Guadalupian and component Roadian, Wordian and Capitanian Stages as International Standards for the Middle Permian Series. *Permophiles*, **34**: 3–11.

González, C.R., 1993, Late Paleozoic faunal succession in Argentina. *Douzième Congrès International de la Stratigraphie et Géologie du Carbonifère et Permien, Buenos Aires 1991, Compte Rendu*, Buenos Aires: Tallares Graficos Curt Latté, **Vol. 1**, pp. 537–550.

Gorak, S.V., 1977, *Carboniferous Ostracods of the Great Donets Basin: Paleoecology, Paleozoogeography and Biostratigraphy* (in Russian). Kiev: Naukova Dumka, p. 148.

Goreva, N.V., and Alekseev, A.S., 2007, Correlation of upper Carboniferous (Pennsylvanian) deposits in the Moscow Syneclise and Donets Basin bu conodonts. *In :* Gozhik, P.F. (ed), *Paleontological Investigations in Ukraine: History, Current Status and Perspectives*. Kiev: Naukova Dumka, pp. 110–114. (in Russian).

Goreva, N.V., and Alekseev, A.S., 2010, Upper Carboniferous Conodont Zones of Russia and their global correlation. *Stratigraphy Geol. Correlation*, **18**: 593–606. https://doi.org/10.1134/S086959381006002X.

Goreva, N., Alekseev, A., Isakova, T., and Kossovaya, O., 2009, Biostratigraphical analysis of the Moscovian-Kasimovian transition at the neostratotype of Kasimovian Stage (Afanasievo section, Moscow Basin, Russia). *Palaeoworld*, **18**: 102–113. https://doi.org/10.1016/j.palwor.2009.04.008.

Gorter, J.D., Jones, P.J., Nicoll, R.S., and Golding, C.J., 2005, A reappraisal of the Carboniferous stratigraphy and the petroleum potential of the southeastern Bonaparte Basin (Petrel Sub-Basin), northwestern Australia. *Australian Pet. Prod. & Exploration Assoc. J.*, **2005**: 275–295. https://doi.org/10.1071/AJ04024.

Gothan, W., 1937, Geobotanische Provinzen im Karbon und Perm. *Deuxième Congrès pour l'Avancement des Études de Stratigraphie Carbonifère (Heerlen 1935), Compte Rendu*, **Vol. 1**, pp. 225–226.

Gourmelon, F., 1987, Les Radiolaires tournaisiens des nodules phosphates de la Montagne Noire et des Pyrenees centrales; systematique; biostratigraphie, paleobiogeographie. *Biostratigraphie du. Paleozoique*, **6**: 1–172.

Griffis, N.P., Mundil, R., Montáñez, I.P., Isbell, J., Fedorchuck, N., Vesely, F., et al., 2018, A new stratigraphic framework built on U-Pb single-zircon TIMS ages and implications for the timing of the penultimate icehouse (Paraná Basin, Brazil). *Geol. Soc. America, Bull.*, **130**: 848–858. https://doi.org/10.1130/B31775.1.

Groessens, E., 1975, Distribution des conodontes dans le Dinantien de la Belgique. *In* Service géologicaue de Belgique (ed), *International Symposium on Belgian Micropaleontological Limits, From Emsian to Visean. Bruxelles: Ministry of Economic Affairs, Adm. Mines-Geological Survey of Belgium*, Publication No. **17**, pp. 1–193.

Groessens, E., Noel, B., 1977, Etude litho- et biostratigraphique du Rocher du Bastion et du Rocher Bayard à Dinant. *In* Service géologicaue de Belgique (ed), *International Symposium on Belgian Micropaleontological limits, From Emsian to Visean. Bruxelles: Ministry of Economic Affairs, Adm. Mines-Geological Survey of Belgium*, Publication No. **15**, pp. 1–17.

Grossman, E.L., 2012, Chapter 10—Oxygen isotope stratigraphy. *In* Gradstein, F.M., Ogg, J.G., Schmitz, M.D., and Ogg, G.M. (eds), *The Geologic Time Scale 2012*, Elsevier, pp. 181–206. https//doi.org/10.1016/B978-0-444-59425-9.00010-X.

Grossman, E.L., Yancey, T.E., Jones, T.E., Bruckschen, P., Chuvashov, B., Mazzullo, S.J., et al., 2008, Glaciation, aridification, and carbon sequestration in the Permo-Carboniferous: the isotopic record from low latitudes. *Palaeogeography, Palaeoclimatology, Palaeoecology*, **268**: 222–233. https://doi.org/10.1016/j.palaeo.2008.03.053.

Grossman, E.L., Joachimski, M.M., 2020, Chapter 10 - Oxygen isotope stratigraphy. *In* Gradstein, F.M., Ogg, J.G., Schmitz, M.D., and Ogg, G.M. (eds), *The Geologic Time Scale 2020*. Vol. 1 (this book): Elsevier. Boston, MA.

Groves, J.R., 1986, Foraminiferal characterization of the Morrowan-Atokan (lower Middle Pennsylvanian) boundary. *Geol. Soc. Am. Bull.*, **97**: 346–353. https://doi.org/10.1130/0016-7606(1986)97<346:FCOTML>2.0.CO;2.

Groves, J.R., and Task Group, 2007, Report of the Task Group to establish a GSSP close to the existing Bashkirian-Moscovian boundary. *Newsl. Carbonif. Stratigraphy*, **25**: 6–7.

Groves, J.R., and Task Group, 2008, Report of the Task Group to establish a GSSP close to the existing Bashkirian-Moscovian boundary. *Newsl. Carbonif. Stratigraphy*, **26**: 10–11.

Gulbranson, E.L., Montañez, I.P., Schmitz, M.D., Limarino, C.O., Isbell, J.L., Marenssi, S.A., et al., 2010, High-precision U–Pb calibration of Carboniferous glaciation and climate history, Paganzo Group, NW Argentina. *Geol. Soc. Am. Bull.*, **122**: 1480–1498. https://doi.org/10.1130/B30025.1.

Hahn, G., 1991, Palaeobiogeographic distribution and biostratigraphic significance of Lower Carboniferous trilobites. *Cour. Forschungsinstitut Senckenberg.*, **130**: 157–171. (imprinted 1990).

Hance, L., 1997, Eoparastaffella, its evolutionary pattern and biostratigraphic potential. *Cushman Found. Foraminifer. Research, Spec. Publ.*, **36**: 59–62.

Hance, L., Brenckle, P.L., Coen, M., Hou, H., Liao, Z., Muchez, P., et al., 1997, The search for a new Tournaisian-Visean boundary stratotype. *Episodes*, **20**: 176–180.

Hance, L., Devuyst, F.-X., and Poty, E., 2002, Sequence stratigraphy of the Belgian Lower Carboniferous - tentative correlation with the British Isles. *Can. Soc. Pet. Geologists, Mem.*, **19**: 41–51.

Hance, L., Hou, H., and Vachard, D., 2011, *Upper Famennian to Visean Foraminifers and Some Carbonate Microproblematica From South China – Hunan, Guangxi and Guizhou*. Beijing: Beijing Geological Publishing House, p. 359.

Hance, L., and Muchez, P., 1995, Study of the Tournaisian-Visean transitional strata in South China (Guangxi). *13th International Congress on Carboniferous-Permian, Krakow, Poland, 1995*. Abstract **Vol. 51**.

Hance, L., and Poty, E., 2006, Hastarian. *Geologica Belgica*, **9**: 111–116.

Hance, L., Poty, E., and Devuyst, F.-X., 2001, Stratigraphie séquentielle du Dinantien type (Belgique) et corrélation avec le Nord de la France (Boulonnais, Avesnois). *Bull. de. la. Société géologique de Fr.*, **172**: 411–426. https://doi.org/10.2113/172.4.411.

Hance, L., Poty, E., and Devuyst, F.-X., 2006a, Ivorian. *Geologica Belgica*, **9**: 117–122.

Hance, L., Poty, E., and Devuyst, F.-X., 2006b, Visean. *Geologica Belgica*, **9**: 55–62.

Haq, B.U., and Schutter, S.R., 2008, A chronology of Paleozoic sea-level changes. *Science*, **322** (5898), 64–68. https://doi.org/10.1126/science.1161648.

Harland, W.B., Armstrong, R.L., Cox, A.V., Craig, L.A., Smith, A.G., and Smith, D.G., 1990, *A Geologic Time Scale 1989*. Cambridge: Cambridge University Press, p. 263.

Harris, A.G., Brenckle, P.L., Baesemann, J.F., Krumhardt, A.P., and Gruzlovic, P.D., 1997, Comparison of conodont and calcareous microfossil biostratigraphy and lithostratigraphy of the Lisburne Group (Carboniferous), Sadlerochit Mountains, Northeast Brooks Range, Alaska. *U. S. Geol. Surv. Professional Pap.*, **1574**: 195–220.

Heckel, P.H., 1986, Sea-level curve for Pennsylvanian eustatic marine transgressive-regressive depositional cycles along Midcontinent outcrop belt, North America. *Geology*, **14**: 330–334. https://doi.org/10.1130/0091-7613(1986)14<330:SCFPEM>2.0.CO;2.

Heckel, P.H., 1991, Comments and replies on "Pennsylvanian time scales and cycle periods. *Geology*, **19**: 405–410. https://doi.org/10.1130/0091-7613(1991)019<0405:CAROPT>2.3.CO;2.

Heckel, P.H., 1994, Evaluation of evidence for glacio-eustatic control over marine Pennsylvanian cyclothems in North America and consideration of possible tectonic events. *SEPM, Concepts Sedimentology Paleontology*, **4**: 65–87. https://doi.org/10.2110/csp.94.04.0065.

Heckel, P.H., 1999, Overview of Pennsylvanian (Upper Carboniferous) stratigraphy in Midcontinent region of North America. *In* Heckel, P.H. (ed), *Guidebook, Fieldtrip #8, XIV International Congress on the Carboniferous-Permian, Kansas Geological Survey. Open File Report, 99–27*, pp. 68–102.

Heckel, P.H., 2002, Overview of Pennsylvanian cyclothems in Midcontinent North America and brief summary of those elsewhere in the world. *Can. Soc. Pet. Geologists, Mem.*, **19**: 79–98.

Heckel, P.H., 2013, Pennsylvanian stratigraphy of Northern Midcontinent Shelf and biostratigraphic correlation of cyclothems. *Stratigraphy*, **10**: 3–39.

Heckel, P.H., Alekseev, A.S., Barrick, J.E., Boardman II, D.R., Goreva, N.V., Nemyrovska, T.I., et al., 2007, Cyclothem ["digital"] correlation and biostratigraphy across the global Moscovian-Kasimovian-Gzhelian stage boundary interval (Middle-Upper Pennsylvanian) in North America and Eastern Europe. *Geology*, **35**: 607–610. https://doi.org/10.1130/G23564A.1.

Heckel, P.H., Alekseev, A.S., Barrick, J.E., Boardman II, D.R., Goreva, N.V., Isakova, T.N., et al., 2008, Choice of conodont *Idiognathodus simulator* (sensu stricto) as the event marker for the base of the global Gzhelian Stage (Upper Pennsylvanian Series, Carboniferous System). *Episodes*, **31**: 319–325.

Heckel, P.H., Boardman, D.R., and Barrick, J.E., 2002, Desmoinesian-Missourian regional stage boundary reference position for North America. *Can. Soc. Pet. Geologists, Mem.*, **19**: 710–724.

Heckel, P.H., and Clayton, G., 2006, The Carboniferous System. Use of the new official names for the subsystems, series, and stages. *Geologica Acta*, **4**: 403–407.

Heckel, P.H., and Watney, W.L., 2002, Revision of stratigraphic nomenclature and classification of the Pleasanton, Kansas City, Lansing, and lower part of the Douglas Groups (lower Upper Pennsylvanian, Missourian) in Kansas. *Kans. Geol. Survey, Bull.*, **246**: 1–69.

Henderson, C.M., Wardlaw, B.R., Vladimir, I.D., Schmitz, M.D., Schiappa, T.A., Tierney, K.E., et al., 2012, Proposal for base-Kungurian GSSP. *Permophiles*, **56**: 8–21.

Herbig, H.-G., 1998, The late Asbian transgression in the central European Culm basins (Late Visean, cd IIIα). *Z. der Deutschen Geologischen Ges.*, **149**: 39–58.

Herbig, H.-G., 2016, Mississippian (Early Carboniferous) sequence stratigraphy of the Rhenish Kulm Basin, Germany. *Geologica Belgica*, **19**: 81–110. https://doi.org/10.20341/gb.2016.010.

Herbig, H.-G., 2017, Taxonomic and stratigraphic problems concerning the conodonts Lochriea senckenbergica Nemirovskaya, Perret & Meischner, 1994 and *Lochriea ziegleri* Nemirovskaya, Perret & Meischner, 1994 – consequences for defining the Visean-Serpukhovian boundary. *Newsl. Carbonif. Stratigraphy*, **33**: 28–35.

Higgins, A.C., 1975, Conodont zonation of the late Visean-early Westphalian strata of the south and central Pennines of northern England. *Geol. Surv. Gt. Br. Bull.*, **53**: 1–90.

Higgs, K.T., Dreesen, R., Dusar, M., and Streel, M., 1992, Palynostratigraphy of the Tournaisian (Hastarian) rocks in the Namur Synclinorium, West Flanders, Belgium. *Rev. Palaeobotany Palynology*, **72**: 149–158. https://doi.org/10.1016/0034-6667(92)90182-G.

Hinnov, L., 2000, New perspectives on orbitally forced stratigraphy. *Annu. Rev. Earth Planet. Sci.*, **28**: 419–475. https://doi.org/10.1146/annurev.earth.28.1.419.

Hinnov, L.A., 2004, Earth's orbital parameters and cyclostratigraphy. *In* Gradstein, F.M., Ogg, J.G., and Smith, A.G. (eds), *A Geologic Time Scale 2004*, Cambridge University Press, 55–62. https://doi.org/10.1017/CBO9780511536045.005.

Hinnov, L.A., 2013, Cyclostratigraphy and its revolutionizing applications in the earth and planetary sciences. *Geol. Soc. America, Bull.*, **125**: 1703–1734. https://doi.org/10.1130/B30934.1.

Hou, H.F., Wu, X.H., and Yin, B.A., 2011, Correlation of the Tournaisian-Visean Boundary Beds. *Acta Geologica Sinica-English Ed.*, **82**: 354–365. https://doi.org/10.1111/j.1755-6724.2011.00404.x.

Hounslow, M.W., Davydov, V.I., Klootwijk, C.T., and Turner, P., 2014, Magnetostratigraphy of the Carboniferous: a review and future Prospects. *Newsl. Carbonif. Stratigraphy*, **22**: 35–40.

Hounslow, M.W., 2020, *A geomagnetic polarity time scale for the Carboniferous. Geological Society.* London: Special Publication. in press.

Hu, K.Y., and Qi, Y.P., 2017, The Moscovian (Pennsylvanian) conodont genus Swadelina from Luodian, southern Guizhou, South China. *Stratigraphy*, **14**: 197–215.

Hu, K.Y., Qi, Y.P., Qie, W.K., and Wang, Q.L., 2019, Carboniferous conodont zonation of China. *Newsl. Stratigraphy*. https://doi.org/10.1127/nos/2019/0498.

Huang, X., Aretz, M., Zhang, X., Du, Y., and Luan, T., 2019, Upper Visean coral biostrome in a volcanic-sedimentary setting from the Eastern Tianshan, Northwest China. *Palaeogeography, Palaeoclimatology, Palaeoecology*, **531**: 108739. https://doi.org/10.1016/j.palaeo.2018.04.014. Part B.

Iosifidi, A.G., Mikhailova, V.A., Popov, V.V., Sergienko, E.S., Danilova, A.V., and Otmas, N.M., 2018, The Carboniferous of the Moscow Syneclise: Paleomagnetic Data. *Izvestiya, Phys. Solid. Earth*, **54**: 163–177. https://doi.org/10.1134/S1069351318010081.

Iosifidi, A.G., Mikhailova, V.A., Popov, V.V., Sergienko, E.S., Danilova, A.V., Otmas, N.M., et al., 2019, Chapter 4. Carboniferous of the Russian Platform: Paleomagnetic Data. *In* Nurgaliev, D., Shcherbakov, V., Kosterov, A., and Spassov, S. (eds), *Recent Advances in Rock Magnetism, Environmental Magnetism and Paleomagnetism*, Springer Geophysics, 37–54. https://doi.org/10.1007/978-3-319-90437-5.

Irving, E., and Parry, L.G., 1963, The magnetism of some Permian rocks from New South Wales. *Geophys. J.*, **7**: 395–411. https://doi.org/10.1111/j.1365-246X.1963.tb07084.x.

Isbell, J.L., Miller, M.F., Wolfe, K.L., and Lenaker, P.A., 2003, Timing of late Paleozoic glaciation in Gondwana: was glaciation responsible for the development of northern hemisphere cyclothems? *In* Chan, M.A., and Archer, A.W. (eds), Extreme Depositional Environments: Mega End Members in Geologic Time. *Geological Society of America Special Paper*, **370**: 5–24. https://doi.org/10.1130/0-8137-2370-1.5.

Isbell, J.L., Henry, L.C., Gulbranson, E.L., Limarino, C.O., Fraiser, M.L., Koch, Z.J., et al., 2012, Glacial paradoxes during the late Paleozoic ice age: evaluating the equilibrium line altitude as a control on glaciation. *Gondwana Res.*, **22**: 1–19. https://doi.org/10.1016/j.gr.2011.11.005.

Ivanov, A.P., 1926, Middle-Upper Carboniferous deposits of the Moscow province. *Bull. Mosc. Soc. Nat. Studies, Geological. Ser.*, **4**: 133–180. (in Russian).

Ivanova, E.A., and Khvorova, I.V., 1955, Stratigraphy of the Middle and Upper Carboniferous of the western part of the Moscow syneclise. *Trans. Paleontological Inst. Acad. Sci. USSR*, **53**: 3–279. (in Russian).

Ivanova, E.A., and Rozovskaya, S.E., 1967, To biostratigraphy of the Upper Carboniferous of the Russian Platform in the scope of investigation of stratotypes. *Bull. Mosc. Soc. Nat. Studies, Geological. Ser.*, **42**: 86–99. (in Russian).

Izart, A., Le Nindre, Y., Stephenson, R., Vaslet, D., and Stovba, S., 2003, Quantification of the control of sequences by tectonics and eustacy in the Dniepr-Donets Basin and on the Russian Platform during Carboniferous and Permian. *Bull. de. la. Société Géologique de Fr.*, **174**: 93–100. https://doi.org/10.2113/174.1.93.

Izart, A., Sachsenhofer, R.F., Privalov, V.A., Elie, M., Panova, E.A., Antsiferov, V.A., et al., 2006, Stratigraphic distribution of macerals and biomarkers in the Donets Basin: implications for paleoecology, paleoclimatology and eustacy. *Int. J. Coal Geol.*, **66**: 69–107. https://doi.org/10.1016/j.coal.2005.07.002.

Izart, A., and Vachard, D., 1994, Subsidence tectonique, eustatisme et contrôle des séquences dans les bassins namuriens et westphaliens de l'Europe de l'ouest, de la CEI et des USA. *Bull. de. la. Société géologique de Fr.*, **165**: 499–514.

Izart, A., Vachard, D., Vaslet, D., Fauvel, P.-J., Süss, P., Kossovaya, O., et al., 2002, Sequence stratigraphy of the Serpukhovian, Bashkirian and Moscovian in Gondwanaland, Western and Eastern Europe and USA. *Can. Soc. Pet. Geologists, Mem.*, **19**: 144–157.

Jenkins, T.B.H., Crane, D.T., and Mory, A.J., 1993, Conodont biostratigraphy of the Visean Series in eastern Australia. *Alcheringa*, **17**: 211–283. https://doi.org/10.1080/03115519308619605.

Ji, Q., and Xiong, J.F., 1985, 1. Conodont biostratigraphy. *In* Hou, H.F., Ji, Q., Wu, X.H., Xiong, J.F., Wang, S.T., Gao, L.D., Sheng, H.B., Wei, J.Y., and Turner, S. (eds), *Muhua Sections of Devonian-*

Carboniferous Boundary Beds. Beijing: Geological Publishing House, pp. 30—38. (in Chinese).

Jin, X.H., Zhai, Z.Q., Li, X.J., and Liu, C.A., 1962, Problems of the subdivision between Lower and Middle Carboniferous in Guizhou (in Chinese). *Annual Conference of the Chinese Geological Society*, **2**: 3—5.

Jin, Y.G., Fan, Y.N., Wang, X.D., and Wang, R.N., 2000, *The Carboniferous System. The Stratigraphic Lexicon of China, No. 8.* Beijing: Geological Publishing Housep. 136. (in Chinese).

Jirásek, J., Hýlová, L., Sivek, M., Jureczka, J., Martínek, K., Sýkorová, I., and Schmitz, M., 2013, The Main Ostrava Whetstone: composition, sedimentary processes, palaeogeography and geochronology of a major Mississippian volcaniclastic unit of the Upper Silesian Basin (Poland and Czech Republic). *International Journal of Earth Sciences*, **102**: 989—1006.

Jirásek, J., Opluštil, S., Sivek, M., Schmitz, M.D., and Abels, H.A., 2018, Astronomical forcing of Carboniferous paralic sedimentary cycles in the Upper Silesian Basin, Czech Republic (Serpukhovian, latest Mississippian): Newradiometric ages afford an astronomical age model for European biozonations and substages. *Earth Science Reviews*, **177**: 715—741.

Jirásek, J., Wlosok, J., Sivek, M., Matýsek, D., Schmitz, M., Sýkorová, I., and Vašiček, Z., 2014, U-Pb zircon age of the Krásné Loučky tuffite: the dating of Visean flysch in the Moravo-Silesian Paleozoic Basin (Rhenohercynian Zone, Czech Republic). *Geological Quarterly*, **58**: 659—672.

Jones, P.J., 1989, Lower Carboniferous Ostracoda (Beyrichicopida and Kirkbyocopa) from the Bonaparte Basin, northwestern Australia. *Australian Bur. Miner. Resources, Geol. Geophysics, Bull.*, **228**: 96.

Jones, P.J., and Metcalfe, I., 1998, Carboniferous biogeography of Australasian faunas and floras. *Newsl. Carbonif. Stratigraphy*, **16**: 16—19.

Jones, M.J., and Truswell, E.M., 1992, Late Carboniferous and Early Permian palynostratigraphy of the Joe Joe Group, southern Galilee Basin, Queensland, and implications for Gondwanan stratigraphy. *Australian Bur. Miner. Resour. J. Australian Geol. Geophysics*, **13**: 143—185.

Jones, C.E., 1995, Time scale 5: Carboniferous. *Australian Geological Survey Organisation Record*, **1995/34**: 3—45.

Jongmans, W.J., 1928, Congrès pour l'étude de la stratigraphie du Carbonifère dans les différents centres houillers de l'Europe. *Compte Rendu Congrès pour l'avancement des études de stratigraphie carbonifère, Heerlen 1927*, pp. i—xlviii.

Josten, K.-H., 1991, Die Steinkohlen-Floren Nordwestdeutschlands. *Fortschritte der Geologie von. Rheinl. und Westfal.*, **31**: 327.

Kabanov, P., 2003, The Upper Moscovian and basal Kasimovian (Pennsylvanian) of central European Russia: facies, sub-aerial exposures and depositional model. *Facies*, **49**: 243—270. https://doi.org/10.1007/s10347-003-0034-x.

Kabanov, P.B., and Alekseev, A.S., 2011, Progress in cyclothem/sequence stratigraphy of type Lower Moscovian succession of Moscow Basin, Russia. *Newsl. Carbonif. Stratigraphy*, **29**: 42—50.

Kabanov, P.B., Alekseev, A.S., Gabdullin, R.R., Gibshman, N.B., Bershov, A., Naumov, S., et al., 2013, Progress in sequence stratigraphy of upper Visean and lower Serpukhovian of southern Moscow Basin, Russia. *Newsl. Carbonif. Stratigraphy*, **30**: 55—65.

Kabanov, P.B., Alekseev, A.S., Gibshman, N.B., Gabdullin, R.R., and Bershov, A.V., 2016, The upper Visean-Serpukhovian in the type area for the Serpukhovian Stage (Moscow Basin, Russia): part 1. Sequences, disconformities and biostratigraphical summary. *Geol. Journal*, **51**: 163—194. https://doi.org/10.1002/gj.2612.

Kabanov, P., and Baranova, D., 2007, Cyclothems and stratigraphy of the Upper Moscovian—basal Kasimovian succession of central and northern European Russia. *In* Wong, T.E. (ed), *Proceedings of the XVth International Congress on Carboniferous and Permian Stratigraphy. Utrecht, 2003.* Royal Netherlands Academy of Arts and Sciences, pp. 147—160.

Kabanov, P.B., Gibshman, N.B., Barskov, I.S., Alekseev, A.S., and Goreva N.V., 2009, Zaborie Section. Lectostratotype of Serpukhovian Stage. *In* Alekseev, A.S., and Goreva, N.V. (eds), *Type and Reference Carboniferous Sections in the South Part of the Moscow Basin. Field Trip Guidebook of International Field Meeting of the I.U.G.S. Subcommission on Carboniferous Stratigraphy "The Historical Type Sections, Proposed and Potential GSSP of the Carboniferous in Russia," Moscow, 11—12 August 2009.* Moscow: Borissiak Paleontological Institute of RAS, pp. 45—64.

Kagarmanov, A.Kh, and Donakova, L.M. (eds), 1990. Decision of Interdepartmental Regional Stratigraphic Conference on Middle and Upper Paleozoic Russian Platform With Regional Stratigraphic Scales, 1988, Leningrad. Vsesoyuznyi Geologisheskii Institut. Leningrad,. pp. 41 with 95 scale sheets. (in Russian).

Kaiser, S.I., 2009, The Devonian/Carboniferous boundary stratotype section (La Serre, France) revisited. *Newsl. Stratigraphy*, **43**: 195—205. https://doi.org/10.1127/0078-0421/2009/0043-0195.

Kaiser, S.I., Aretz, M., and Becker, R.T., 2015, The global Hangenberg Crisis (Devonian—Carboniferous transition): review of a first-order mass extinction. *In* Becker, R.T., Königshof, P., and Brett, C.E. (eds), Devonian Climate, Sea Level and Evolutionary Events. *Geological Society, London, Special Publications*, **423**: 387—437. https://doi.org/10.1144/SP423.9.

Kaiser, S.I., and Corradini, C., 2011, The early siphonodellids (Conodonta, Late Devonian-Early Carboniferous): overview and taxonomic state. *Neues Jahrb. für Geologie und Paläontologie, Abhandlungen*, **261**: 19—35. https://doi.org/10.1127/0077-7749/2011/0144.

Kalvoda, J., Bábek, O., Devuyst, F.X., and Sevastopulo, G.D., 2011, Biostratigraphy, sequence stratigraphy and gamma-ray spectrometry of the Tournaisian-Visean boundary interval in the Dublin Basin. *Bull. Geosci.*, **86**: 683—706. https://doi.org/10.3140/bull.geosci.1265.

Kamenikova, T., Hounslow, M.W., van der Boon, A., Sprain, C., Wójcik, K., Nawocki, J., et al., 2019, Tournaisian-Visean magnetostratigraphy: new data from British and Polish limestones. *In Third International Congress on Stratigraphy, Strati 2019, Milano. Abstract Book.* Scoieta Geologica Italiana, p. 206.

Kammer, T.W., Brenckle, P.L., Carter, J.L., and Ausich, W.I., 1990, Redefinition of the Osagean-Meramecian boundary in the Mississippian stratotype region. *Palaios*, **5**: 414—431. https://doi.org/10.2307/3514835.

Khramov, A.N., 1987, *Paleomagnetology.* Springer-Verlag, p. 308.

Khramov, A.N., 2000, The General magnetostratigraphic scale of the Phanerozoic. *In* Zhamoida, A.I. (ed), *Supplements to the Stratigraphic Code of Russia.* Transactions of VSEGEI, St. Petersburg, 24—45. (in Russian).

Khramov, A.N., and Davydov, V.I., 1984, *Paleomagnetism of Upper Carboniferous and Lower Permian in the South of the U.S.S.R. and the Problems of Structure of the Kiama Hyperzone* (in Russian). St. Petersburg: Transactions of VNIGRI, pp. 55—73.

Khramov, A.N., and Davydov, V.I., 1993, Results of Paleomagnetic investigations. Permian System, Guides to Geological Excursions in the Uralian Type Localities. *Occasional Publ. Earth Sci. Resour. Inst.*, **10**: 34—42.

Khramov, A.N., Goncharov, G.I., Komisssarova, R.A., Osipova, E.P., Pogarskaya, I.A., Rodionov, V., et al., 1974, *Paleozoic Paleomagnetism* (in Russian). St. Petersburg: Transactions of VNIGRI, pp. 238.

Khramov, A.N., and Rodionov, V.P., 1981, The geomagnetic field during Palaeozoic time. *In* McElhinny, M.W., Khramov, A.N., Ozima, M., and Valencio, D.A. (eds), Global Reconstruction and the Geomagnetic Field during the Palaeozoic. *Advances in Earth and Planetary Science*, **10**: 99−115.

Klein, G.D., 1990, Pennsylvanian time scales and cycle periods. *Geology*, **18**: 455−457.

Klein, G.D., 1991, Comments and replies on "Pennsylvanian time scales and cycle periods. *Geology*, **19**: 405−410. https://doi.org/10.1130/0091-7613(1991)019＜0405:CAROPT＞2.3.CO;2.

Knight, J.A., and Alvarez-Vazquez, C., 2019, The West European Regional Stratigraphic Framework for the upper Pennsylvanian, July 29−August 2, 2019. *In* Hartenfels, S., Herbig, H.-G., Amler, M.R.W., and Aretz, M. (eds), Abstracts, 19th International Congress on the Carboniferous and Permian. *Cologne: Kölner Forum für Geologie und Paläontologie*, **23**: pp. 177−178.

Knight, J.A., and Wagner, R.H., 2014, Proposal for the recognition of a Saberian Substage in the mid-Stephanian (West European chronostratigraphic scheme). *Freib. Forschungshefte*, **C548**: 179−195.

Kolesov, E.V., 1984, Paleomagnetic stratigraphy of the Devonian-Carboniferous boundary beds in the Soviet North-East and in the Franco-Belgian Basin. *Annales de. la. Société Géologique de Belgique*, **107**: 135−136.

Kolesov, E.V., 2001, The Paleomagnetism of the Upper Paleozoic volcanogenic-sedimentary sequences of Prikolym uplift. *In* Simakov, K.K. (ed), *Paleomagnetic Studies of Geological Rocks on the North-Eastern of Russia*. Magadan: Far East Branch of the Russian Academy of Sciences, pp. 32−44. (in Russian).

Konert, G., Afifi, A.M., Al-Hajri, S.A., and Droste, H.J., 2001, Paleozoic stratigraphy and hydrocarbon habitat of the Arabian Plate. *GeoArabia*, **6**: 407−442. https://doi.org/10.1306/M74775C24.

Korn, D., 1996, Revision of the Late Visean goniatite stratigraphy. *Annales de. la. Société Géologique de Belgique*, **117**: 205−212.

Korn, D., 2006, Ammonoideen. Deutsche Stratigraphische Kommission. *In* Amler, M.R.W., and Stoppel, D. (eds), Stratigraphie von Deutschland VI. Unterkarbon (Mississippium). *Schriftenreihe der Deutschen Gesellschaft für Geowissenschaften*, **41**: 147−170.

Korn, D., Bockwinkel, J., and Ebbighausen, V., 2007, Tournaisian and Visean ammonoid stratigraphy in North Africa. *Neues Jahrb. für Geologie und Paläontologie, Abhandlungen*, **243**: 127−148. https://doi.org/10.1127/0077-7749/2007/0243-0127.

Korn, D., and Klug, C., 2015, Paleozoic Ammonoid Biostratigraphy. *In* Klug, C., Korn, D., de Baets, K., Kruta, I., and Mapes, R.H. (eds), Ammonoid Paleobiology: From Macroevolution to Paleogeography. Topics in Geobiology. 44. 299−328. https://doi.org/10.1007/978-94-017-9633-0-12.

Korn, D., and Titus, A.L., 2011, Goniatites Zone (middle Mississippian) ammonoids of the Antler Foreland Basin (Nevada, Utah). *Bull. Geosci.*, **86**: 107−196. https://doi.org/10.3140/bull.geosci.1242.

Korn, D., Titus, A.L., Ebbighausen, V., Mapes, R.H., and Sudar, M.N., 2012, Early Carboniferous (Mississippian) ammonoid biogeography. *Geobios*, **45**: 67−77. https://doi.org/10.1016/j.geobios.2011.11.013.

Korn, D., and Wang, Q., 2019, Ammonoids and problems with the correlation of the Visean-Serpukhovian boundary. *In* Hartenfels, S.,

Herbig, H.-G., Amler, M.R.W., and Aretz, M. (eds), Abstracts, 19th International Congress on the Carboniferous and Permian, Cologne 29 July−2 August 2 2019. *Kölner Forum für Geologie und Paläontologie*, **23**: 181−182.

Korte, C., Jasper, T., Kozur, H.W., and Veizer, J., 2006, $^{87}Sr/^{86}Sr$ record of Permian seawater. *Palaeogeography, Palaeoclimatology, Palaeoecology*, **240**: 89−107. https://doi.org/10.1016/j.palaeo.2006.03.047.

Kossovaya, O.L., 1996, Correlation of uppermost Carboniferous and lowermost Permian rugose coral zones from the Urals to Western North America. *Palaios*, **11**: 71−82. https://doi.org/10.2307/3515118.

Kulagina, E.I., Gibshman, N.B., and Pazukhin, V.N., 2003, Foraminiferal zonal standard for Lower Carboniferous of Russia and its correlation with conodont zonation. *Riv. Ital. di Paleontologia e Stratigrafia*, **109**: 173−185. https://doi.org/10.13130/2039-4942/5500.

Kulagina, E.I., Pazukhin, V.N., and Davydov, V.I., 2009, Pennsylvanian biostratigraphy of the Basu River section with emphasis on the Bashkirian-Moscovian transition. *In* Puchkov, V.N. (ed), *The Carboniferous Type Sections in Russia, Potential and Proposed Stratotypes. Proceedings of the International Conference Ufa—Sibai, August 13–18, 2009*. Ufa: Institute of Geology, Bashkirian Academy of Science, pp. 42−63.

Kullmann, J., and Nikolaeva, S.V., 2002, Mid-Carboniferous boundary and the global lower Bashkirian ammonoid biostratigraphy. *Can. Soc. Pet. Geologists, Memoir*, **19**: 780−795.

Kumpan, T., Bábek, O., Kalvoda, J., Fryda, J., and Grygar, M.T., 2014, A high-resolution, multiproxy stratigraphic analysis of the Devonian−Carboniferous boundary sections in the Moravian Karst (Czech Republic) and a correlation with the Carnic Alps (Austria). *Geol. Mag.*, **151**: 201−215. https://doi.org/10.1017/S0016756812001057.

Lane, H.R., 1974, Mississippian of southeastern New Mexico and West Texas: a wedge-on-wedge relation. *Am. Assoc. Pet. Geologists Bull.*, **58**: 269−282. https://doi.org/10.1306/83D913D0-16C7-11D7-8645000102C1865D.

Lane, H.R., and Brenckle, P.L., 2005, *Type Mississippian subdivisions and biostratigraphic succession, Guidebook Series Illinois State Geological Survey, Report*, 34. pp. 76−98.

Lane, H.R., Brenckle, P.L., Baesemann, J.F., and Richards, B., 1999, The IUGS boundary in the middle of the Carboniferous; Arrow Canyon, Nevada, USA. *Episodes*, **22**: 272−283.

Lane, H.R., and Manger, W.L., 1985, Toward a boundary in the middle of the Carboniferous. *Cour. Forschungsinstitut Senckenberg.*, **74**: 15−34.

Lane, H.R., Merrill, G.K., Straka II, J.J., and Webster, G.D., 1971, North American Pennsylvanian conodont biostratigraphy. *Geol. Soc. North. America, Mem.*, **127**: 395−414. https://doi.org/10.1130/MEM127-p395.

Lane, H.R., Sandberg, C.A., and Ziegler, W., 1980, Taxonomy and phylogeny of some Lower Carboniferous conodonts and preliminary standard post-Siphonodella zonation. *Geologica et. Palaeontologica*, **14**: 117−164.

Laskar, J., 1999, The limits of Earth orbital calculations for geological time scale use. *Philos. Trans. R. Society, Ser. A*, **357**: 1735−1759. https://doi.org/10.1098/rsta.1999.0399.

Laskar, J., Robutel, P., Joutel, F., Gastineau, M., Correia, A.C.M., and Levrard, B., 2004, A long-term numerical solution for the insolation quantities of the earth. *Astron. Astrophys.*, **428**: 261−285. https://doi.org/10.1051/0004-6361:20041335.

Legrand-Blain, M., 1991, Brachiopods as potential boundary-defining organisms in the Lower Carboniferous of Western Europe: recent data and productid distribution. *Cour. Forschungsinstitut Senckenberg.*, **130**: 157–171. (imprinted 1990).

Li, S.J., 1987, Late Early Carboniferous to early Late Carboniferous brachiopods from Qixu, Nandan, Guangxi and their palaeoecological significance. *In* Wang, C.Y. (ed), *Carboniferous Boundaries in China*. Beijing: Science Press, 132–150.

Limarino, C.O., Césari, S.N., Spaletti, L.A., Taboada, A.C., Isbell, J.L., Geuna, S., et al., 2014, A paleoclimatic review of Southern South America during the Late Paleozoic: a record from icehouse to extreme greenhouse conditions. *Gondwana Res.*, **25**: 1396–1421. https://doi.org/10.1016/j.gr.2012.12.022.

Lin, J.X., Li, J.X., and Sun, Q.Y., 1990, *Late Palaeozoic Foraminifera From South China*. Beijing: Science Pressp. 297. (in Chinese).

Lipina, O.A., and Reitlinger, E.A., 1970, Stratigraphie zonale et paléozoogéographie du Carbonifère inférieur d'après les foraminifères. *Sixième Congrès International de Stratigraphie et de Géologie du Carbonifère, Sheffield 1967, Compte Rendu*, Maastricht: Ernest van Aelst, **Vol. 6**, pp. 1101–1112.

Loboziak, S., Melo, H.H., and Streel, M., 1999, Latest Devonian and Early Carboniferous palynostratigraphy of Northern Brazil and North Africa – a proposed integration of Western European and Gondwanan miospore biozonations. *Bull. du. Cent. de. Rech. Elf Exploration Prod.*, **22** (1998), 241–259.

López Gamundi, O.R., 1989, Postglacial transgressions in Late Paleozoic basins of western Argentina: a record of glacioeustatic sea level rise. *Palaeography, Palaeoclimatology, Palaeoecology*, **71**: 257–270. https://doi.org/10.1016/0031-0182(89)90054-0.

Lucas, S.G., 2003, Triassic tetrapod footprint biostratigraphy and biochronology. *Albertiana*, **28**: 75–84.

Lucas, S.G., 2016, Defining North American Pennsylvanian Stages. *Newsl. Carbonif. Stratigraphy*, **32**: 42–47.

Makhlina, M.Kh, 1996, Cyclic stratigraphy, facies and fauna of the Lower Carboniferous (Dinantian) of the Moscow Syneclise and Voronezh Anteclise. *In* Strogen, P., Somerville, I.D., and Jones, G.L. (eds), Recent Advances in Lower Carboniferous Stratigraphy. *Geological Society of London, Special Publication*, **107**: 359–364. https://doi.org/10.1144/GSL.SP.1996.107.01.25.

Makhlina, M.K., Vdovenko, M.V., Alekseev, A.S., Byvsheva, T.V., Donakova, L.M., Zhulitova, V.E., et al., 1993, *Lower Carboniferous of the Moscow Syneclise and Voronezh Anteclise*. Moscow: Naukap. 219. (in Russian).

Makhlina, M.K., Alekseyev, A.S., Goreva, N.V., Isakova, T.N., and Drutskoy, S.N., 2001, Stratigraphy. *In* Alekseev, A.S., and Shik, S.M. (eds), *Middle Carboniferous of the Moscow Syneclise, Vol. 1*. Rossiyskaya Akademiya Nauk, Paleontologicheskiy Institut, Moscow, p. 244. (in Russian).

Mamet, B.L., and Belford, D.J., 1968, Carboniferous Foraminifera, Bonaparte Gulf Basin, northwestern Australia. *Micropalaeontology*, **14**: 339–347. https://doi.org/10.2307/1484694.

Mamet, B., and Skipp, B., 1970, Lower Carboniferous calcareous Foraminifera: preliminary zonation and stratigraphic implications for the Mississippian of North America. *Sixième Congrès International de Stratigraphie et de Géologie du Carbonifère, Sheffield 1967, Compte Rendu*, Maastricht: Ernest van Aelst, **Vol. 6**, pp. 1129–1146.

Manger, W.L., 2017, Journey to the Mississippian-Pennsylvanian Boundary GSSP: a long and winding road. *Stratigraphy*, **14**: 247–258.

Marshall, J.E.A., Lakin, J.A., and Finney, S.M., 2013, Terrestrial climate and ecosystem change from the Devonian–Carboniferous boundary to the earliest Visean interval in East Greenland. *Doc. de. lInstitut Scientifique, Rabat*, **26**: 81–82.

Maynard, T.R., and Leeder, M.R., 1992, On the periodicity and magnitude of Late Carboniferous glacio-eustatic sea-level changes. *J. Geol. Soc. Lond.*, **149**: 303–311. https://doi.org/10.1144/gsjgs.149.3.0303.

McArthur, J.M., Howarth, R.J., and Shields, G.A., 2012, Strontium isotope stratigraphy. *In* Gradstein, F.M., Ogg, J.G., Schmitz, M.D., and Ogg, G.M. (eds), *The Geologic Time Scale 2012*, Elsevier, 127–144. https://doi.org/10.1016/B978-0-444-59425-9.00007-X.

McArthur, J.M., Howarth, R.J., Shields, G.A., and Zhou, Y., 2020, Chapter 7 - Strontium isotope stratigraphy. In Gradstein, F.M., Ogg, J.G., Schmitz, M.D., and Ogg, G.M. (eds), *The Geologic Time Scale 2020*. Vol. **1** (this book). Elsevier, Boston, MA.

McGhee, G.R., Sheehan, P.M., Bottjer, D.J., and Droser, M.L., 2012, Ecological ranking of Phanerozoic biodiversity crises: the Serpukhovian (early Carboniferous) crisis had a greater ecological impact than the end-Ordovician. *Geology*, **40**: 147–150. https://doi.org/10.1130/G32679.1.

McGhee, G.R., Clapham, M.E., Sheehan, P.M., Bottjer, D.J., and Droser, M.L., 2013, A new ecological-severity ranking of major Phanerozoic biodiversity crises. *Palaeogeography, Palaeoclimatology, Palaeoecology*, **370**: 260–270. https://doi.org/10.1016/j.palaeo.2012.12.019.

Melo, J.H.G., and Loboziak, S., 2003, Devonian–Early Carboniferous miospore biostratigraphy of the Amazon Basin, Northern Brazil. *Rev. Palaeobotany Palynology*, **124**: 131–202. https://doi.org/10.1016/S0034-6667(02)00184-7.

Melvin, J., and Sprague, R.A., 2006, Advances in Arabian stratigraphy: origin and stratigraphic architecture of glaciogenic sediments in Permian-Carboniferous lower Unayzah sandstones, eastern central Saudi Arabia. *GeoArabia*, **11**: 105–152.

Menning, M., Alekseev, A.S., Chuvashov, B.I., Davydov, V.I., Devuyst, F.-X., Forke, H.C., et al., 2006, Global time scale and regional stratigraphic reference scales of Central and West Europe, East Europe, Tethys, South China, and North America as used in the Devonian–Carboniferous–Permian Correlation Chart 2003 (DCP 2003). *Palaeogeography, Palaeoclimatology, Palaeoecology*, **240**: 318–372. https://doi.org/10.1016/j.palaeo.2006.03.058.

Menning, M., 2018. Die Stratigraphische Tabelle von Deutschland 2016 (STD 2016) / The Stratigraphic Table of Germany 2016 (STG 2016). – *Zeitschrift der Deutschen Gesellschaft für Geowissenschaften* (ZDGG), **169**: 105–128, Stuttgart.

Menning, M., Weyer, D., Drozdzewski, G., van Amerom, H.W.J., and Wendt, I., 2000, A Carboniferous Time Scale 2000. Discussion and use of geological parameters as time indicators from Central and Western Europe. *Geologische Jahresberichte*, **A156**: 3–44.

Menning, M., Weyer, D., Drozdzewski, G., van Amerom, H.W.J., and Wendt, I., 2000, A Carboniferous Time Scale 2000. Discussion and use of geological parameters as time indicators from Central and Western Europe. *Geologische Jahresberichte*, **A156**: 3–44.

Mestermann, B., 1998, Mikrofazies, Paläogeographie und Eventgenese des crenistria-Horizontes (Obervisé, Rhenohercynicum). *Kölner Forum für Geologie und Paläontologie*, **2**: 1–77.

Meyen, S.V., Afanasieva, G.A., Betekhtina, O.A., Durante, M.V., Ganelin, V.G., Gorelova, S.G., et al., 1996, Angara and surrounding marine basins. *In* Martinez Diaz, C., Wagner, R.H., Winkler Prins, C.F., and Granados, L.F. (eds), *The Carboniferous of the World III. The Former USSR, Mongolia, Middle Eastern Platform, Afghanistan, & Iran*. Instituto Geologico y Minero de

España IGME and Empresa Nacional Adaro de Investigaciones Mineras, 180−237.

Mii, H.-S., Grossman, E.L., and Yancey, T.E., 1999, Carboniferous isotope stratigraphies of North America: implications for Carboniferous paleoceanography and Mississippian glaciation. *Geol. Soc. Am. Bull.*, **111**: 960−973. https://doi.org/10.1130/0016-7606(1999) 111 < 0960:CISONA > 2.3.CO;2.

Mizuno, Y., and Ueno, K., 1997, Conodont and foraminiferal changes across the Mid-Carboniferous boundary in the Hina Limestone Group, southwest Japan. *In Proceedings of the XIII International Congress on the Carboniferous and Permian, Krakow 1995, Prace Panstwowego Instytutu Geologicznego*, **157**: 189−205.

Möller, V., 1878, Die Spiral-gewundenen Foraminiferen des russischen Kohlenkalks. *Mémoires de. l'Académie impériale des. Sci. de St.-Pétersbourg série 7*, **25**: 1−147.

Montañez, I.P., 2016, A Late Paleozoic climate window of opportunity. *Proc. Natl Acad. Sci.*, **113**: 2334−2336. https://doi.org/10.1073/pnas.1600236113.

Montañez, I.P., 2019, Understanding feedbacks between climate, pCO_2, and ecosystems in the late Paleozoic Earth System. In: Hartenfels, S., Herbig, H.-G., Amler, M.R.W., and Aretz, M. (eds), Abstracts, 19th International Congress on the Carboniferous and Permian, Cologne, 29 July−2 August 2019. *Kölner Forum für Geologie und Paläontologie*, **23**: 18−19.

Montañez, I.P., McElwain, J.C., Poulsen, C.J., White, J.D., DiMichele, W.A., Wilson, J.P., et al., 2016, Climate, pCO_2 and terrestrial carbon cycle linkages during late Palaeozoic glacial−interglacial cycles. *Nat. Geosci.*, **9**: 824−828. https://doi.org/10.1038/ngeo2822.

Montañez, I.P., and Poulsen, C.J., 2013, The late Paleozoic ice age; an evolving paradigm. *Annu. Rev. Earth Planet. Sci.*, **41**: 629−656. https://doi.org/10.1146/annurev.earth.031208.100118.

Moore, R.C., 1932, Reclassification of the Pennsylvanian System in the northern midcontinent region. Guidebook. *6th Annual Field Conference*. Kansas Geological Society, pp. 79−98.

Moore, R.C., 1948, Classification of Pennsylvanian rocks in Iowa, Kansas, Missouri, Nebraska, and northern Oklahoma. *Bull. Am. Assoc. Pet. Geologists*, **32**: 2011−2040.

Moore, R.C., and Thompson, M.L., 1949, Main divisions of Pennsylvanian period and system. *Bull. Am. Assoc. Pet. Geologists*, **33**: 275−302.

Mory, A.J., and Haig, D.W., 2011, Permian−Carboniferous geology of the northern Perth and southern Carnarvon Basins, Western Australia − a field guide. *Geological Survey of Western Australia, Record 2011/14*, p. 65.

Mory, A.J., and Hocking, R.M., 2011, Permian, Carboniferous and Upper Devonian Geology of the northern Canning Basin, Western Australia − a field guide. *Geological Survey of Western Australia, Record 2011/16*, p. 36.

Mory, A.J., Redfern, J., and Martin, J.R., 2008, A review of Permian−Carboniferous glacial deposits in Western Australia. *In Fielding, C.R., Frank, T.D., and Isbell, J.L. (eds), Resolving the Late Paleozoic Ice Age in Time and Space. *Geological Society of America, Special Paper*, **441**: 29−40. https://doi.org/10.1130/2008.2441(02).

Mukhopadhyay, G., Mukhopadhyay, S.K., Roychowdhury, M., and Parui, P.K., 2010, Stratigraphic correlation between different Gondwana Basins of India. *J. Geol. Soc. India*, **76**: 251−266. https://doi.org/10.1007/s12594-010-0097-6.

Mundil, R., Ludwig, K.R., Metcalfe, I., and Renne, P.R., 2004, Age and timing of the end Permian mass extinctions: U/Pb geochronology on closed system zircons. *Science*, **305**: 1760−1763. https://doi.org/10.1126/science.1101012.

Murchison, R.I., 1835, On the Silurian System of rocks. *London, Edinb. Dublin Philos. Mag. J. Science, 3rd Ser.* **7**: 46−52.

Murchison, R.I., 1845, *Geology of Russia in Europe and the Ural Mountains*. London: John Murray, pp. 700.

Nazarov, B.B., and Ormiston, A.R., 1985, Radiolaria from the late Paleozoic of the Southern Urals, USSR and West Texas, USA. *Micropaleontology*, **31**: 1−54. https://doi.org/10.2307/1485579.

Nemirovskaya, T.I., and Nigmadganov, I.M., 1994, The Mid-Carboniferous conodont event. *Cour. Forschungsinstitut Senckenberg.*, **168**: 319−335.

Nemyrovska, T.I., 1999, Bashkirian conodonts of the Donets Basin, Ukraine. *Scr. Geologica*, **119**: 1−115.

Nemyrovska, T.I., Perret-Mirouse, M.-F., and Alekseev, A., 1999, On Moscovian (Late Carboniferous) conodonts of the Donets Basin, Ukraine. *Neues Jahrb. für Geologie und Paläontologie. Abhandlungen*, **214**: 169−194. https://doi.org/10.1127/njgpa/214/1999/169.

Nemyrovska, T.I., 1999, Bashkirian conodonts of the Donets Basin, Ukraine. *Scripta Geologica*, **119**: 1−115.

Neves, R., Gueinn, K.J., Clayton, G., Ioannides, N.S., and Neville, R.S. W., 1972, A scheme of miospore zones for the British Dinantian. *Septième Congrès International de Stratigraphie et de Géologie du Carbonifère, Krefeld, 1971, Compte Rendu*, Krefeld: Geologisches Landesamt Nordrhein-Westfalen, **Vol. 1**, pp. 347−353.

Nicoll, R.S., and Druce, E.C., 1979, Conodonts from the Fairfield Group, Canning Basin, Western Australia. *Australian Bur. Miner. Resources, Bull.*, **190**: 1−134.

Nicklen, B.L., 2011. *Establishing a Tephrochronologic Framework for the Middle Permian (Guadalupian) Type Area and Adjacent Portions of the Delaware Basin and Northwestern Shelf, West Texas and Southeastern New Mexico, USA*. University of Cincinnati Dissertation, 134 pp.

Nikitin, S.N., 1890, Carboniferous deposits of Moscow Basin and artesian water around Moscow. *Trans. Geol. Comm.*, **5**: 1−182. (in Russian).

Nikolaeva, S.V., Akhmetshina, L.Z., Konovalova, V.A., Korobkov, V.F., and Zainakaeva, G.F., 2009a, The Carboniferous carbonates of the Dombar Hills (western Kazakhstan) and the problem of the Visean-Serpukhovian boundary. *Palaeoworld*, **18**: 80−93. https://doi.org/10.1016/j.palwor.2009.04.004.

Nikolaeva, S.V., Kulagina, E.I., Pazukhin, V.N., Kochetova, N.N., and Konovalova, V.A., 2009b, Paleontology and microfacies of the Serpukhovian in the Verkhnyaya Kardailovka Section, South Urals, Russia: potential candidate for the GSSP for the Visean-Serpukhovian boundary. *Newsl. Stratigraphy*, **43**: 165−193. https://doi.org/10.1127/0078-0421/2009/0043-0165.

Nikolaeva, S.V., Alekseev, A.S., Kulagina, E.I., Gibshman, N.B., Richards, B.C., Gatovsky, Yu.A., et al., 2014, New microfacies and fossil records (ammonoids, conodonts, foraminifers) from the Visean-Serpukhovian boundary beds in the Verkhnyaya Kardailovka section. *Newsl. Carbonif. Stratigraphy*, **31**: 41−51.

Nikolaeva, S.V., Alekseev, A.S., Kulagina, E.I., Gatovsky, Yu.A., Ponomareva, G.Yu, and Gibshman, N.B., 2019, An evaluation of biostratigraphic markers across multiple geological sections in the search for the GSSP of the base of the Serpukhovian Stage (Mississippian). *Palaeoworld*. https://doi.org/10.1016/j.palwor.2019.01.006.

Novik, E.O., 1952, The Carboniferous flora of the European portion of the U.S.S.R. *Akademia Nauk SSSR, Palaeontology, Nov. Ser.*, **1**: 1−468. (in Russian).

Nyhuis, C.R., Amler, M.R.W., and Herbig, H.-G., 2015, Facies and palaeoecology of the late Visean Actinopteria Black Shale Event in the Rhenish Mountains (Germany, Mississippian). *Z. der Deutschen Ges. für Geowissenschaften*, **166**: 55–69. https://doi.org/10.1127/1860-1804/2015/0087.

Ogg, J.G., Ogg, G.M., and Gradstein, F.M., 2016, 9 – Carboniferous. *In* Ogg, J.G., Ogg, G.M., and Gradstein, F.M. (eds), *A Concise Geologic Time Scale*Elsevier, 99–113. https://doi.org/10.1016/B978-0-444-59467-9.00009-1.

Opdyke, N.D., and Channell, J.E.T., 1996, *Magnetic Stratigraphy*. London and San Diego, CA: Academic Press, p. 364.

Opdyke, N.D., Roberts, J., Claoue-Long, J., Irving, E., and Jones, P.J., 2000, Base of Kiaman: its definition and global significance. *Geol. Soc. Am. Bull.*, **112**: 1315–1341. https://doi.org/10.1130/0016-7606(2000)112<1315:BOTKID>2.0.CO;2.

Opluštil, S., Schmitz, M., Cleal, C.J., and Martínek, K., 2016, A review of the Middle–Late Pennsylvanian west European regional substages and floral biozones, and their correlation to the Geological Time Scale based on new U–Pb ages. *Earth-Science Rev.*, **154**: 301–335. https://doi.org/10.1016/j.earscirev.2016.01.004.

Osterloff, P., Penney, R., Aitken, J., Clark, N., and Al-Husseini, M., 2004, Depositional sequences of the Al Khlata Formation, subsurface Interior Oman. *In* Al-Husseini, M.I. (ed), Carboniferous, Permian and Triassic Arabian Stratigraphy. *GeoArabia Special Publication*, **3**: 61–81.

Owens, B., 1984, Miospore zonation of the Carboniferous. *Neuvième Congrès International de Stratigraphie et de Géologie du Carbonifère, Compte Rendu, Carbondalle and Edwardsville*: Southern Illinois University Press, Vol. 2, pp. 90–102.

Owens, R.M., and Hahn, G., 1993, Biogeography of Carboniferous and Permian trilobites. *Geologica et. Palaeontologica*, **27**: 165–180.

Owens, B., Riley, N.J., and Calver, M.A., 1985, Boundary stratotypes and new stage names for the lower and middle Westphalian sequences in Britain. *Dixième Congres International de Stratigraphie et de Géologie du Carbonifère, Madrid: IGME 1983, Compte Rendu*, **Vol. 4**, pp. 461–472.

Owens, B., Clayton, G., Gao, L., and Loboziak, S., 1989, Miospore correlation of the Carboniferous deposits of Europe and China. *Onzième Congres International de Stratigraphie et de Géologie du Carbonifère, Beijing, 1987, Compte Rendu*, Nanjing: Nanjing University Press, **Vol. 3**, pp. 189–210.

Paproth, E., 1978, Nicht-marine Muscheln als Spiegel der Fazies Entwicklung im paralischen Kohlengebiet Northwest-Europas. *Sonderveröffentlichungen des. Geologischen Inst. der Universität Köln*, **33**: 91–100.

Paproth, E., Conil, R., Bless, M.J.M., Boonen, P., Carpentier, N., Coen, M., et al., 1983, Bio- and lithostratigraphic subdivisions of the Dinantian in Belgium, a review. *Annales de. la. Société Géologique de Belgique*, **106**: 185–239.

Paproth, E., Feist, R., and Flajs, G., 1991, Decision on the Devonian-Carboniferous boundary stratotype. *Episodes*, **14**: 331–336.

Pazukhin, V.N., Kulagina, E.I., and Sedaeva, K.M., 2009, The Devonian/Carboniferous boundary on the western slope of the South Urals. *In* Puchkov, V.N., Kulagina, E.I., Nikolaeva, S.V., and Kochetova, N.N. (eds), *Carboniferous Type Sections in Russia and Potential Global Stratotypes: Southern Urals Session. Proceedings of the International Field Meeting "The Historical Type Sections, Proposed and Potential GSSP of the Carboniferous in Russia", Ufa-Sibai, August 13–18, 2009* (in Russian). Ufa: DesignPolygraphService Ltd., pp. 22–33.

Pechersky, D.M., Lyubushin, A.A., and Sharonova, Z.V., 2010, On the synchronism in the events within the core and on the surface of the earth: the changes in the organic world and in the polarity of the geomagnetic field in the Phanerozoic. *Izvestiya, Phys. Solid. Earth*, **46**: 613–623. https://doi.org/10.1134/S1069351310070050.

Pellenard, P., Gand, G., Schmitz, M., Galtier, J., Broutin, J., and Steyer, J. S., 2017, High-precision U-Pb zircon ages for explosive volcanism calibrating the NW European continental Autunian stratotype. *Gondwana Res.*, **51**: 118–136. https://doi.org/10.1016/j.gr.2017.07.014.

Peppers, R.A., 1996, Palynological correlation of major Pennsylvanian (Middle and Upper Carboniferous) chronostratigraphic boundaries in the Illinois and other coal basins. *Geol. Soc. Am. Mem.*, **188**: 111. https://doi.org/10.1130/0-8137-1188-6.1.

Perret, M.-F., 1993, Recherches micropaléontologiques et biostratigraphiques (Conodontes-Foraminifères) dans le Carbonifère Pyrénéen. *Strata*, **21**: 1–597.

Perret, M.-F., and Weyant, M., 1994, Les biozones à conodontes du Carbonifère des Pyrénées. Comparaisons avec d'autres régions du globe. *Geobios*, **27**: 689–715. https://doi.org/10.1016/S0016-6995(94)80056-1.

Phillips, W., 1818, *A Selection of Facts From the Best Authorities Arranged so as to form an Outline of the Geology of England and Wales, With a Map and Sections of the Strata*. London: Published by William Phillips pp. 292. fourth, enlarged ed.

Phillips, J., 1835, *Illustrations of the Geology of Yorkshire; Or, A Description of the Strata and Organic Remains: Accompanied by a Geological Map, Sections, and Plates of the Fossil Plants and Animals*. London: John Murray, pp. 236.

Piper, J.D.A., Atkinson, D., Norris, S., and Thomas, S., 1991, Palaeomagnetic study of the Derbyshire lavas and intrusions, central England: definition of Carboniferous apparent polar wander. *Phys. Earth Planet. Inter.*, **69**: 37–55. https://doi.org/10.1016/0031-9201(91)90152-8.

Playford, G., 1991, Australian Lower Carboniferous miospores relevant to extra-Gondwanic correlations: an evaluation. *Cour. Forschungsinstitut Senckenberg.*, **130**: 85–125. (imprinted 1990).

Playford, G., and Dino, R., 2000a, Palynostratigraphy of upper Palaeozoic strata (Tapajós Group), Amazonas Basin, Brazil: Part one. *Palaeontographica Abt. B.*, **255**: 1–46. https://doi.org/10.1127/palb/255/2000/1.

Playford, G.B., and Dino, R., 2000b, Palynostratigraphy of upper Palaeozoic strata (Tapajós Group), Amazonas Basin, Brazil: Part two. *Palaeontographica Abt. B*, **255**: 87–145. https://doi.org/10.1127/palb/255/2000/1.

Playford, G.B., and Dino, R., 2005, Carboniferous and Permian palynostratigraphy. *In* Koutsoukos, E.A.M. (ed), Applied Stratigraphy. Topics in Geobiology. 23. 101–121. https://doi.org/10.1007/1-4020-2763-X-5.

Pointon, M.A., Chew, D.M., Ovtcharova, M., Sevastopulo, G.D., and Crowley, Q.G., 2012, New high-precision U-Pb dates from western European Carboniferous tuffs; implications for time scale calibration, the periodicity of late Carboniferous cycles and stratigraphical correlation. *Journal of the Geological Society*, **169**: 713–721.

Pointon, M.A., Chew, D.M., Ovtcharova, M., Sevastopulo, G.D., and Delcambre, B., 2014, High-precision UPb zircon CA-ID-TIMS dates from western European late Viséan bentonites. *Journal of the Geological Society*, **171**: 649–658.

Pointon, M.A., Chew, D.M., Delcambre, B., and Sevasqtopulo, G.D., 2018, Geochemistry and origin of Carboniferous (Mississippian; Visean) bentonites in the Namur-Dinant Basin, Belgium: evidence for a Variscan volcanic source. *Geologica Belgica*, **21**: 1–17. https://doi.org/10.20341/gb.2017.011.

Pointon, M.A., Chew, D.M., Delcambre, B. Ovtcharova, M., and Sevastopulo, G., 2019, Hig-precision U-Pb Ca-ID-TIMS dates from the Livian Substage (Mississippian) of Belgium. In: Hartenfels, S., Herbig, H.-G., Amler, M.R.W., and Aretz, M. (eds), Abstracts, 19th International Congress on the Carboniferous and Permian, Cologne, 29 July–2 August 2019. *Kölner Forum für Geologie und Paläontologie*, **23**: 256–257.

Poletaev, V.I., and Lazarev, S.S., 1994, General stratigraphic scale and brachiopod evolution in the Late Devonian and Carboniferous subequatorial belt. *Bull. de. la. Société Belg. de Géologie*, **103**: 99–107.

Popov, A.V., 1999, Gzhelian ammonoids from Karachatyr, Central Asia. *Voprosy Paleontologii (St. Petersburg)*, **11**: 75–87. (in Russian).

Popov, A.V., Davydov, V.I., Donakova, L.M., and Kossovaya, O.L., 1985, Gzhelian stratigraphy in Southern Urals. *Sovetskaya Geologiya*, **3**: 57–67. (in Russian).

Popov, A.V., Davydov, V.I., and Kossovaya, O.L., 1989, Gzhelian stratigraphy in Central Asia. *Sovetskaya Geologiya*, **3**: 64–76. (in Russian).

Poty, E., 1985, A rugose coral biozonation for the Dinantian of Belgium as a basis for a coral biozonation of the Dinantian of Eurasia. *Dixième Congrès International de Stratigraphie et de Géologie du Carbonifère, Madrid 1983, Compte Rendu*, Madrid: IGME, **Vol. 4**, pp. 29–31.

Poty, E., 2007, The Avins event: a remarkable worldwide spread of corals at the end of the Tournaisian (Lower Carboniferous). *Schriftenreihe der Erdwissenschaftlichen Kommissionen, Oesterreichische Akademie der Wissenschaften*, **17**: 231–249.

Poty, E., 2016, The Dinantian (Mississippian) succession of southern Belgium and surrounding areas: stratigraphy improvement and inferred climate reconstruction. *Geologica Belgica*, **19**: 177–200. https://doi.org/10.20341/gb.2016.014.

Poty, E., and Hance, L., 2006, Livian. *Gelogica Belgica*, **9**: 133–138.

Poty, E., Devuyst, F.-X., and Hance, L., 2006, Upper Devonian and Mississippian foraminiferal and rugose coral zonation of Belgium and Northern France: a tool for Eurasian correlations. *Geol. Mag.*, **143**: 829–857. https://doi.org/10.1017/S0016756806002457.

Poty, E., Berkowski, B., Chevalier, E., and Hance, L., 2007, Sequence stratigraphic correlations between the Dinantian deposits of Belgium and southern Poland (Krakow area). *In* Wong, T.E. (ed), *Proceedings of the XVth International Congress on Carboniferous and Permian Stratigraphy. Utrecht, 10–16 August 2003.* Amsterdam: Royal Netherlands Academy of Arts and Sciences, pp. 97–107.

Poty, E., Aretz, M., Hou, H., and Hance, L., 2011, Bio- and sequence stratigraphic correlations between Western Europe and South China: to a global model of the eustatic variations during the Mississippian. *In* Håkansson, E., and Trotter, J.A. (eds), *Programme and Abstracts: The XVII International Congress on the Carboniferous and Permian, Perth 2011. Geological Survey of Western Australia, Record, 2011/20*, p. 104.

Poty, E., Aretz, M., and Hance, L., 2014, Belgian substages as a basis for an international chronostratigraphic division of the Tournaisian and Visean. *Geol. Mag.*, **151**: 229–243. https://doi.org/10.1017/S0016756813000587.

Poty, E., Mottequin, B., and Denayer, J., 2013, An attempt of time calibration of the Lower Tournaisian (Hastarian Substage) based on orbitally forced sequences. *Doc. de. l'Institut Scientifique, Rabat*, **26**: 105–107.

Poty, E., Mottequin, B., and Denayer, J., 2019, An attempt of time calibration of the Tournaisian and Visean stages (Lower and Middle Mississippian) based on long duration orbitally forced sequences. In: Hartenfels, S., Herbig, H.G., Amler, M.R.W., and Aretz, M. (eds), Abstracts, 19th International Congress on the Carboniferous and Permian, Cologne 29 July–2 August 2019. *Kölner Forum für Geologie und Paläontologie*, **23**: 258–259.

Prestianni, C., Denayer, J., and Sautois, M., 2016, Disrupted continental environments around the Devonian-Carboniferous Boundary: introduction of the tener event. *Geologica Belgica*, **19**: 135–145. https://doi.org/10.20341/gb.2016.013.

Qi, Y.P., Hu, K.Y., Barrick, J.E., Wang, Q.L., and Lin, W., 2012, Discovery of the conodont lineage from *Idiognathodus swadei* to *I. turbatus* in South China and its implications. *J. Stratigraphy*, **36**: 551–557. (in Chinese with English abstract).

Qi, Y., Nemyrovska, T., Wang, X.D., Chen, J., Wang, Z., Lane, R., et al., 2013, Late Visean–early Serpukhovian conodont succession at the Naqing (Nashui) section in Guizhou, South China. *Geol. Mag.*, **151**: 254–268. https://doi.org/10.1017/S001675681300071X.

Qi, Y.P., Lambert, L.L., Nemyrovska, T.I., Wang, X.D., Hu, K.Y., and Wang, Q.L., 2016, Late Bashkirian and early Moscovian conodonts from the Naqing section, Luodian, Guizhou, South China. *Palaeoworld*, **25**: 170–187. https://doi.org/10.1016/j.palwor.2015.02.005.

Qi, Y.-P., Nemyrovska, T.I., Wang, Q.-L., Hu, K.-Y., Wang, X.-D., and Lane, H.R., 2018, Conodonts of the genus *Lochriea* near the Visean–Serpukhovian boundary (Mississippian) at the Naqing section, Guizhou Province, South China. *Palaeoworld*, **27**: 423–437. https://doi.org/10.1016/j.palwor.2018.09.001.

Qi, Y.P., and Wang, Z.H., 2005, Serpukhovian conodont sequence and the Visean-Serpukhovian boundary in South China. *Riv. Ital. di Paleontologia e Stratigrafia*, **111**: 3–10. https://doi.org/10.13130/2039-4942/6260.

Qie, W., Zhang, X., Du, Y., Yang, B., Ji, W., and Luo, G., 2014, Conodont biostratigraphy of Tournaisian shallow-water carbonates in central Guangxi, South China. *Geobios*, **47**: 389–401. https://doi.org/10.1016/j.geobios.2014.09.005.

Ramezani, J., Schmitz, M.D., Davydov, V.I., Bowring, S.A., Snyder, W.S., and Northrup, C.J., 2007, High-precision U-Pb zircon age constraints on the Carboniferous-Permian boundary in the Southern Urals stratotype. *Earth and Planetary Science Letters*, **256**: 244–257.

Ramsbottom, W.H.C., 1969, The Namurian of Britain. *Sixième Congres International de Stratigraphie et de Géologie du Carbonifère (Sheffield, 1967), Compte Rendu*, Maastricht: Ernest van Aelst, **Vol. 1**, pp. 219–232.

Ramsbottom, W.H.C., 1973, Transgression and regression in the Dinantian: a new synthesis of British Dinantian stratigraphy. *Proc. Yorks. Geol. Soc.*, **39**: 567–607. https://doi.org/10.1144/pygs.39.4.567.

Ramsbottom, W.H.C., 1977, Major cycles of transgression and regression (mesothems) in the Namurian. *Proc. Yorks. Geol. Soc.*, **41**: 261–291. https://doi.org/10.1144/pygs.41.3.261.

Ramsbottom, W.H.C., 1979, Rates of transgression and regression in the Carboniferous of NW Europe. *J. Geol. Soc. Lond.*, **136**: 147–153. https://doi.org/10.1144/gsjgs.136.2.0147.

Ramsbottom, W.H.C., 1981, Eustacy, sea level and local tectonism, with examples from the British Carboniferous. *Proc. Yorks. Geol. Soc.*, **43**: 473–482. https://doi.org/10.1144/pygs.43.4.473.

Ramsbottom, W.H.C., 1984, The founding of the Carboniferous System. *Neuvième Congres International de Stratigraphie et de Géologie du Carbonifère, Compte Rendu*, Carbondalle and Edwardsville: Southern Illinois University Press, **Vol. 1**, pp. 109–112.

Ramsbottom, W.H.C., Calver, M.A., Eagar, R.M.C., Hodson, F., Holliday, D.W., Stubblefield, C.J., et al., 1978, *A correlation of Silesian rocks in the British Isles*, Geological Society of London, Special Report, **10**. pp. 1–81.

Ramsbottom, W.H.C., and Mitchell, M., 1980, The recognition and division of the Tournaisian series in Britain. *J. Geol. Soc.*, **137**: 61–63. https://doi.org/10.1144/gsjgs.137.1.0061.

Ramsbottom, W.H.C., and Saunders, W.B., 1985, Evolution and evolutionary biostratigraphy of Carboniferous ammonoids. *J. Paleontology*, **59**: 123–139.

Rasbury, T., Hanson, G.N., Meyers, W.J., Holt, W.E., Goldstein, R.H., and Saller, A.H., 1998, U–Pb dates of paleosols: constraints on late Paleozoic cycle durations and boundary ages. *Geology*, **26**: 403–406. https://doi.org/10.1130/0091-7613(1998)026 < 0403:UPDOPC > 2.3.CO;2.

Rauser-Chernousova, D.M., 1941, New Upper Carboniferous stratigraphic data from the Oksko-Tsninskyi Dome. *Rep. Acad. Sci. USSR*, **30**: 434–436. (in Russian).

Rauser-Chernousova, D.M., 1949, Stratigraphy of Upper Carboniferous and Artinskian deposits of Bashkirian Preurals, (35) (in Russian). *In* Rauser-Chernousova, D.M. (ed), Foraminifers of Upper Carboniferous and Artinskian Deposits of Bashkirian Pre-Urals. *Transactions of the Geological Institute of the Academy of Sciences of USSR*, **105**: 3–21.

Rauser-Chernousova, D.M., and Reitlinger, E.A., 1954, Biostratigraphic distribution of foraminifers in Middle Carboniferous deposits of southern limb of Moscow Syneclise. *In* Nalivkin, D.V., and Menner, V.V. (eds), *Regional Stratigraphy of the U.S.S.R. Stratigraphy of Middle Carboniferous Deposits of Central and Eastern parts of Russian Platforms (Based on Foraminifers Study)*, **Vol. 2**. Transactions of the Geological Institute of the Academy of Science of the USSR, Moscow, 7–120. (in Russian).

Read, C.B., and Mamay, S.H., 1964, Upper Paleozoic floral zones and floral provinces of the United States. *U.S. Geol. Survey, Professional Pap.*, **454-K**: 35.

Richards, B.C., 2013, Current status of the international Carboniferous time scale. *In* Lucas, S.G., DiMichele, W.A., Barrick, J.E., Schneider, J.W., and Spielmann, J.A. (eds), The Carboniferous-Permian Transition. *New Mexico Museum of Natural History and Science Bulletin*, **60**: 348–353.

Richards, B.C., and Aretz, M., 2009, Report of the task group to establish a GSSP for the Tournaisian-Visean boundary. *Newsl. Carbonif. Stratigraphy*, **27**: 9–10.

Richards, B.C., Lane, H.R., and Brenckle, P.L., 2002, The IUGS Mid-Carboniferous (Mississippian-Pennsylvanian) global boundary stratotype section and point at Arrow Canyon, Nevada, USA. *Can. Soc. Pet. Geologists, Mem.*, **19**: 802–831.

Richards, B.C., Nikolaeva, S.V., Kulagina, E.I., Alekseev, A.S., Gorozhanina, E.N., Gorozhanin, V.M., et al., 2017, A candidate for the Global Stratotype Section and Point at the base of the Serpukhovian in the South Urals, Russia. *Stratigraphy Geol. Correlation*, **25** (7), 1–62. https://doi.org/10.1134/S0869593817070036.

Richards, B.C., and Task Group, 2003, Report of the Task Group to establish a GSSP close to the existing Visean-Serpukhovian boundary. *Newsl. Carbonif. Stratigraphy*, **21**: 6–10.

Riley, N.J., 1993, Dinantian (Lower Carboniferous) biostratigraphy and chronostratigraphy in the British Isles. *J. Geol. Soc.*, **150**: 427–446. https://doi.org/10.1144/gsjgs.150.3.0427.

Roberts, J., 1971, Devonian and Carboniferous brachiopods from the Bonaparte Gulf Basin, northwestern Australia. *Australian Bur. Miner. Resources, Bull.*, **122**: 317.

Roberts, J., 1985, Australia. *In* Martínez Díaz, C., Wagner, R.H., Winkler Prins, C.F., and Granados, L.F. (eds), *The Carboniferous of the World II: Australia, Indian Subcontinent, South America and North Africa*. Instituto Geologico y Minero de España IGME and Empresa Nacional Adaro de Investigaciones Mineras, 9–145.

Roberts, J., Claoué-Long, J., Jones, P.J., and Foster, C.B., 1995a, SHRIMP zircon age control of Gondwanan sequences in Late Carboniferous and Early Permian Australia. *In* Dunay, R.E., and Hailwood, E.A. (eds), Dating and Correlating Biostratigraphically Barren Strata. *Geological Society, London, Special Publications*, **89**: 145–174. https://doi.org/10.1144/GSL.SP.1995.089.01.08.

Roberts, J., Claoué-Long, J., and Jones, P.J., 1995b, Australian Early Carboniferous time. *In* Berggren, W.A., Kent, D.V., Aubry, M.-P., and Hardenbol, J. (eds), Geochronology, Time Scales and Stratigraphic Correlation. *Society for Sedimentary Geology (SEPM), Special Publications*, **54**: 23–40. https://doi.org/10.2110/pec.95.04.0023.

Roberts, J., Jones, P.J., and Jenkins, T.B.H., 1993, Revised correlations for Carboniferous marine invertebrate zones of eastern Australia. *Alcheringa*, **17**: 353–376. https://doi.org/10.1080/03115519308619598.

Rosovskaya, S.E., 1950, The *Triticites* genus, its development and stratigraphic significance. *Trans. Paleontological Inst. Acad. Sci. USSR*, **26**: 3–79. (in Russian).

Ross, C.A., and Ross, J.R.P., 1985, Late Paleozoic depositional sequences are synchronous and worldwide. *Geology*, **13**: 194–197. https://doi.org/10.1130/0091-7613(1985)13 < 194:LPDSAS > 2.0.CO;2.

Ross, C.A., and Ross, J.R.P., 1987, Late Paleozoic sea levels and depositional sequences. *Cushman Foundation for Foraminiferal Research, Special Publication*, **24**: 137–149.

Ross, C.A., and Ross, J.R.P., 1988, Late Paleozoic transgressive-regressive deposition. *Society of Economic Paleontologists and Mineralogists, Special Publication*, **42**: 227–247. https://doi.org/10.2110/pec.88.01.0227.

Rosscoe, S., and Barrick, J.E., 2009, Revision of Idiognathodus species from the Middle-Upper Pennsylvanian boundary interval in the Midcontinent basin, North America. *Palaeontographica Americana*, **62**: 115–147.

Rosscoe, S., and Barrick, J.E., 2013, North American species of the conodont genus *Idiognathodus* from the Moscovian-Kasimovian boundary composite sequence and correlation of the Moscovian-Kasimovian stage boundary. *In* Lucas, S.G., DiMichele, W.A., Barrick, J.E., Schneider, J.W., and Spielmann, J.A. (eds), The Carboniferous-Permian Transition. *New Mexico Museum of Natural History and Science Bulletin*, **60**: 354–371.

Ruan, Y.P., 1981, Carboniferous ammonoid faunas from Qixu in Nandan of Guangxi. *Mem. Nanjing Inst. Geol. Palaeontology, Acad. Sci.*, **15**: 152–232. (in Chinese).

Rui, L., Wang, Z.H., and Zhang, L.X., 1987, Luosuan—a new chronostratigraphic unit for Lower-Upper Carboniferous. *J. Stratigraphy*, **11**: 103–115. (in Chinese).

Ruzhenzev, V.E., 1965, The major ammonoid assemblages of the Carboniferous Period. *Paleontological J.*, **2**: 3–17. (in Russian).

Ruzhenzev, V.E., and Bogoslovskaya, M.F., 1978, The Namurian Stage in the evolution of the Ammonoidea: Late Namurian Ammonoidea. *Trans. Paleontological Inst. Acad. Sci. USSR*, **167**: 1–338. (in Russian).

Sadler, P.M., and Cooper, R.A., 2003, Best-fit intervals and consensus sequences—a comparison of the resolving power of traditional biostratigraphy and computer assisted correlation. *In* Harries, P.J. (ed), *High Resolution Approaches in Stratigraphic Paleontology*. Topics in Geobiology. 21. 49–94. https://doi.org/10.1007/978-1-4020-9053-0_2.

Sadler, P.M., Kemple, W.G., and Kooser, M.A., 2003, CONOP9 programs for solving the stratigraphic correlation and seriation problems as constrained optimization. *In* Harries, P.J. (ed), *High Resolution Approaches in Stratigraphic Paleontology*. Topics in Geobiology. 21. 461–465. https://doi.org/10.1007/978-1-4020-9053-0_13.

Sallan, L., and Galimberti, A.K., 2015, Body-size reduction in vertebrates following the end-Devonian mass extinction. *Science*, **350**: 812–815. https://doi.org/10.1126/science.aac7373.

Sallan, L.C., and Coates, M.I., 2010, End-Devonian extinction and a bottleneck in the early evolution of modern jawed vertebrates. *Proceedings of the National Academy of Sciences of the United States of America*, **107**: 10131–10135. https://doi.org/10.1073/pnas.0914000107.

Saltzman, M.R., 2003, Late Paleozoic ice age: Oceanic gateway or pCO_2? *Geology*, **31**: 151–154. https://doi.org/10.1130/0091-7613(2003)031<0151:LPIAOG>2.0.CO;2.

Saltzman, M.R., González, L.A., and Lohmann, K.C., 2000, Earliest Carboniferous cooling step triggered by the Antler orogeny? *Geology*, **28**: 347–350. https://doi.org/10.1130/0091-7613(2000)28<347:ECCSTB>2.0.CO;2.

Saltzman, M.R., Groessens, E., and Zhuravlev, A.V., 2004, Carbon cycle models based on extreme changes in $\delta^{13}C$: an example from the lower Mississippian. *Paleogeography, Paleoclimatology, Paleoecology*, **213**: 359–377. https://doi.org/10.1016/j.palaeo.2004.07.019.

Saltzman, M.R., and Thomas, E., 2012, Carbon isotope stratigraphy. *In* Gradstein, F.M., Ogg, J.G., Schmitz, M.D., and Ogg, G.M. (eds), *The Geologic Time Scale 2012*, Elsevier, pp. 207–232. https://doi.org/10.1016/B978-0-444-59425-9.00011-1.

Sandberg, C.A., Ziegler, W., Leuteritz, K., and Brill, S.M., 1978, Phylogeny, speciation, and zonation of Siphonodella (Conodonta, Upper Devonian and Lower Carboniferous). *Newsl. Stratigraphy*, **7**: 102–120. https://doi.org/10.1127/nos/7/1978/102.

Sando, W.J., 1991, Global Mississippian coral zonation. *Cour. Forschungsinstitut Senckenberg*, **130**: 179–187. (imprinted 1990).

Sanz-Lopez, J., Blanco-Ferrera, S., Garcia-Lopez, S., and Sanchez de Posada, L.C., 2006, The Mid-Carboniferous boundary in Northern Spain: difficulties for correlation of the Global Stratotype Section and Point. *Riv. Ital. di Paleontologia e Stratigrafia*, **112**: 3–22. https://doi.org/10.13130/2039-4942/5847.

SCCS, 1969, General discussion of the Subcommission on Carboniferous Stratigraphy. *Sixième Congres International de Stratigraphie et de Géologie du Carbonifère (Sheffield, 1967), Compte Rendu*, Maastricht: Ernest van Aelst, **Vol. 1**, pp. 185–189.

Schmidt, H., 1925, Die carbonischen Goniatiten Deutschlands. *Jahrb. der Preussischen Geologischen Landesanst.*, **45**: 489–609.

Schmitz, M.D., and Davydov, V.I., 2012, Quantitative radioisotopic and biostratigraphic calibration of the Pennsylvanian–Early Permian (Cisuralian) time scale and pan-Euramerican chronostratigraphic correlation. *Geol. Soc. Am. Bull.*, **124**: 549–577. https://doi.org/10.1130/B30385.1. and data-repository item 2012020.

Schmitz, M.D., Davydov, V.I., and Snyder, W.S., 2009, Permo-Carboniferous conodonts and tuffs: High-precision marine Sr isotope geochronology. *Permophiles, ICOS 2009 Abstr.*, **53**: 48.

Schneider, J.W., Hampe, O., and Soler-Gijón, R., 2000, The Late Carboniferous and Permian: aquatic vertebrate zonation in Southern Spain and German basins. *Cour. Forschungsinstitut Senckenberg*, **223**: 543–561.

Schneider, J.W., and Scholze, F., 2018, Late Pennsylvanian–Early Triassic conchostracan biostratigraphy: a preliminary approach. *In* Lucas, S.G., and Shen, S.Z. (eds), The Permian Time scale. *Geological Society London, Special Publications*, **450**: 365–386. https://doi.org/10.1144/SP450.6.

Schneider, J.W., and Werneburg, R., 2006, Insect biostratigraphy of the European continental Late Pennsylvanian and Early Permian. *In* Lucas, S.G., Cassinis, G., and Schneider, J.W. (eds), Non-Marine Permian Biostratigraphy and Biochronology. *Geological Society London, Special Publications*, **265**: 325–336. https://doi.org/10.1144/GSL.SP.2006.265.01.15.

Schneider, J.W., and Werneburg, R., 2012, Biostratigraphie des Rotliegend mit Insekten und Amphibien. In: Deutsche Stratigraphische Kommission. *In* Lützner, H., and Kowalczyk, G. (eds), Stratigraphie von Deutschland X. Rotliegend. Teil I: Innervariscische Becken. *Schriftenreihe der Deutschen Gesellschaft für Geowissenschaften*, **61**: 100–142.

Scomazzon, A.K., Moutinho, L.P., Nascimento, S., Lemos, V.B., and Matsuda, N.S., 2016, Conodont biostratigraphy and paleoecology of the marine sequence of the Tapajós Group, Early-Middle Pennsylvanian of Amazonas Basin, Brazil. *J. South. Am. Earth Sci.*, **65**: 25–42. https://doi.org/10.1016/j.jsames.2015.11.004.

Scotese, C.R., 2014, *Atlas of Permo-Carboniferous Paleogeographic Maps (Mollweide Projection), Maps 53–64, Volumes 4, The Late Paleozoic, PALEOMAP Atlas for ArcGIS, PALEOMAP Project*. Evanston, IL [Available at: https://www.academia.edu/16664729/Atlas_of_Permo-Carboniferous_Paleogeographic_Maps].

Semikhatova, S.V., 1934, The deposits of Moscovian in lower and middle Povolzhie region and the position of Moscovian Stage in the Carboniferous System. *Probl. Sov. Geol.*, **3**: 73–90. (in Russian).

Sevastopulo, G.D., and Barham, M., 2014, Correlation of the base of the Serpukhovian Stage (Mississippian) in NW Europe. *Geol. Mag.*, **152**: 244–253. https://doi.org/10.1017/S0016756813000630.

Sevastopulo, G., and Devuyst, F.X., 2005, Correlation of the base of the Visean Stage in the type Mississippian region of North America. *Newsl. Carbonif. Stratigraphy*, **23**: 12–15.

Sevastopulo, G., Devuyst, F.X., Hance, L., Hou, H.F., Coen, M., Clayton, G., et al., 2002, Progress report of the Working Group to establish a boundary close to the existing Tournaisian-Visean boundary within the Lower Carboniferous. *Newsl. Carbonif. Stratigraphy*, **20**: 6–7.

Shen, S.Z., Crowley, J.L., Wang, Y., Bowring, S.A., Erwin, D.H., Sadler, P.M., et al., 2011, Calibrating the end-Permian mass extinction. *Science*, **334**: 1367–1372.

Sheng, Q.Y., Wang, X.D., Brenckle, P., and Huber, B.T., 2018, Serpukhovian (Mississippian) foraminiferal zones from the Fenghuangshan section, Anhui Province, South China: implications for biostratigraphic correlations. *Geol. J.*, **53**: 45–57. https://doi.org/10.1002/gj.2877.

Siegmund, H., Trappe, J., and Oschmann, W., 2002, Sequence stratigraphic and genetic aspects of the Tournaisian "Liegender Alaunschiefer" and adjacent beds. *Int. J. Earth Sci. (Geologische Rundsch.)*, **91**: 934–949. https://doi.org/10.1007/s00531-001-0252-9.

Siever, R., 1951, The Mississippian-Pennsylvanian unconformity in southern Illinois. *Am. Assoc. Pet. Geologists, Bull.*, **35**: 542–581.

Simmons, M.S., Miller, K.G., Ray, D.C., Davies, A., van Buchem, F.S.P., and Gréselle, B., 2020, Chapter 13 - Phanerozoic eustacy. *In* Gradstein, F.M., Ogg, J.G., Schmitz, M.D., and Ogg, G.M. (eds), *The Geologic Time Scale 2020*. Vol. 1 (this book). Elsevier. Boston, MA.

Skompski, S., Alekseev, A., Meischner, D., Nemirovskaya, T., Perret, M.-F., and Varker, W.J., 1995, Conodont distribution across the Visean/Namurian boundary. *Cour. Forschungsinstitut Senckenberg.*, **188**: 177–209.

Smith, A.G., Barry, T., Bown, P., Cope, J., Gale, A., Gibbard, P., et al., 2014, GSSPs, global stratigraphy and correlation. *In* Smith, D.G., Bailey, R.J., Burgess, P.M., and Fraser, A.J. (eds), Strata and Time: Probing the Gaps in Our Understanding. *Geological Society, London, Special Publications*, **404**: 37–67. https://doi.org/10.1144/SP404.8.

Smithson, T.R., Richards, K.R., and Clack, J.A., 2016, Lungfish diversity in Romer's Gap: reaction to the end-Devonian extinction. *Palaeontology*, **59**: 29–44. https://doi.org/10.1111/pala.12203.

Solovieva, M.N., 1977, Zonal fusulinid stratigraphy of Middle Carboniferous of the USSR. *Quest. Micropaleontology*, **28**: 3–23. (in Russian).

Somerville, I.D., 2008, Biostratigraphical zonation and correlation of Mississippian rocks in Western Europe: some case studies in the late Visean/Serpukhovian. *Geol. J.*, **43**: 209–240. https://doi.org/10.1002/gj.1097.

Stollhofen, H., Werner, M., Stanistreet, I.G., and Armstrong, R.A., 2008, Single-zircon U-Pb dating of Carboniferous-Permian tuffs, Namibia, and the intercontinental deglaciation cycle framework. *In* Fielding, C.R., Frank, T.D., and Isbell, J.L. (eds), Resolving the Late Paleozoic Ice Age in Time and Space. *The Geological Society of America, Special Paper*, **441**: 83–96. https://doi.org/10.1130/2008.2441(06).

Stone, D.B., Minyuk, P., and Kolesev, E., 2003, New palaeomagnetic paleolatitudes for the Omulevka terrane of northeast Russia: a comparison with the Omolon terrane and the eastern Siberian platform. *Tectonophysics*, **377**: 55–82. https://doi.org/10.1016/j.tecto.2003.08.025.

Strasser, A., Hilgen, F., and Heckel, P.H., 2006, Cyclostratigraphy – concepts, definitions, and applications. *Newsl. Stratigraphy*, **42**: 75–114. https://doi.org/10.1127/0078-0421/2006/0042-0075.

Streel, M., and Theeron, J.N., 1999, The Devonian–Carboniferous boundary in South Africa and the age of the earliest episode of the Dwyka glaciation: new palynological result. *Episodes*, **22**: 41–44.

Stukalina, G.A., 1988, Studies in Paleozoic crinoid columnals and stems. *Palaeontographica Abt. A*, **204**: 1–66.

Sullivan, H.J., 1965, Palynological evidence concerning the regional differentiation of Upper Mississippian flora. *Pollen Spores*, **7**: 539–560.

Sungatullina, G.M., Davydov, V.I., Barrick, J.E., and Sungatullin, R.K., 2016, Conodonts of Kasimovian-Gzhelian transition, Usolka section, Southern Urals, Russia: new data. *Newsletters on Carboniferous Stratigraphy*, **32**: 54–57.

Taboada, A.C., 2010, Mississippian–Early Permian brachiopods from western Argentina: tools for middle- to high-latitude correlation, paleobiogeographic and paleoclimatic reconstruction. *Palaeogeography, Palaeoclimatology, Palaeoecology*, **298**: 152–173. https://doi.org/10.1016/j.palaeo.2010.07.008.

Teteriuk, V.K., 1976, Namurian stage analogues in the Carboniferous period of the Donetz Basin (based on palynological data). *Geol. J.*, **36**: 110–122. (in Russian).

Theodorovich, G.I., 1936, On some questions of the stratigraphy of the Carboniferous deposits of the West slope of the Southern Urals. *Probl. Sov. Geol.*, **6** (7), 613–617. (in Russian).

Thompson, T.L., and Fellows, L., 1970, Stratigraphy and conodont biostratigraphy of Kinderhookian and Osagean (lower Mississippian) rocks of southwestern Missouri and adjacent areas. *Mo. Geol. Survey, Rep. Investigations*, **45**: 1–263.

Thompson, M.L., Verville, G.J., and Lokke, D.H., 1956, Fusulinids of the Desmoinesian-Missourian contact. *J. Paleontology*, **30**: 793–811.

Tian, X., Chen, J., Yao, L., Hu, K., Qi, Y., and Wang, X.D., 2019, Glacio-eustasy and δ13C across the Mississippian–Pennsylvanian boundary in the eastern Paleo-Tethys Ocean (South China): implications for mid-Carboniferous major Glaciation. *Geol. Journal*. https://doi.org/10.1002/gj.3551.

Tien, C.C., and Wang, H.C., 1932, A study of the geology of Tsumenchias coal field, Hsianghiang. *Bull. Geol. Soc. China*, **13**: 4–16.

Ting, V.K., 1931, On the stratigraphy of the Fengninian System. *Bull. Geol. Soc. China*, **10** (1), 31–48.

Ting, V.K., and Grabau, A.W., 1936, The Carboniferous of China and its bearing on the classification of the Mississippian and Pennsylvanian. *16th International Geological Congress Report, Washington, DC (1933)*, **I**: 555–571.

Titus, A.L., Webster, G.D., Manger, W.L., and Dewey, C.P., 1997, Biostratigraphic analysis of the mid-Carboniferous boundary at the South Syncline Ridge Section, Nevada Test Site, Nevada, United States. *Proceedings of the XIII International Congress on the Carboniferous and Permian, Krakow 1995*, **3**: 207–213.

Titus, A.L., Korn, D., Harrell, J.E., and Lambert, L.L., 2015, Late Visean (late Mississippian) ammonoids from the Barnett Shale, Sierra Diablo Escarpment, Culberson County, Texas, USA. *Foss. Rec.*, **18**: 81–104. https://doi.org/10.5194/fr-18-81-2015.

Tucker, R.D., Bradley, D.C., Ver Straeten, C.A., Harris, A.G., Ebert, J.R., and McCutcheon, S.R., 1998, New U-Pb zircon ages and the duration and division of Devonian time. *Earth and Planetary Science Letters*, **158**: 175–186.

Ueno, K., Hayakawa, N., Nakazawa, T., Wang, Y., and Wang, X., 2013, Pennsylvanian–Early Permian cyclothemic succession on the Yangtze Carbonate Platform, South China. *Geological Society of London, Special Publication*, **376**: 235–267. 10.1144/SP376.5.

Ueno, K., and Nemirovska, T.I., 2008, Bashkirian-Moscovian (Pennsylvanian, Upper Carboniferous) boundary in the Donets Basin, Ukraine. *J. Geogr.*, **117**: 919–932. https://doi.org/10.5026/jgeography.117.919.

Ueno, K., and Task Group, 2014, Report of the task group to establish a GSSP close to the existing Moscovian–Kasimovian and Kasimovian–Gzhelian Boundaries. *Newsl. Carbonif. Stratigraphy*, **32**: 33–37.

Ueno, K., and Task Group, 2017, Report of the task group to establish a GSSP close to the existing Moscovian–Kasimovian and Kasimovian–Gzhelian Boundaries. *Newsl. Carbonif. Stratigraphy*, **33**: 18–20.

Uhl, D., and Cleal, C.J., 2010, Late Carboniferous vegetation change in lowland and intramontane basins in Germany. *Int. J. Coal Geol.*, **83**: 318–328. https://doi.org/10.1016/j.coal.2009.07.007.

Vachard, D., Cózar, P., Aretz, M., and Izart, A., 2016, Late Visean-Serpukhovian foraminifers in the Montagne Noire (France): biostratigraphic revision and correlation with the Russian substages. *Geobios*, **49**: 469–498. https://doi.org/10.1016/j.geobios.2016.09.002.

Vachard, D., Haig, D.W., and Mory, A.J., 2014, Lower Carboniferous (middle Visean) foraminifers and algae from an interior sea, southern Carnarvon Basin, Australia. *Geobios*, **47**: 57–74. https://doi.org/10.1016/j.geobios.2013.10.005.

Van Leckwijck, W.P., 1960, Report of the Subcommission on Carboniferous Stratigraphy. *Quatrième Congrès pour l'avancement des études de stratigraphie carbonifère, Heerlen, 15–20 Septembre 1958, Compte Rendu*, Maestricht: Ernest van Aelst, **Vol. I**, pp. XXIV–XXVI.

Van Steenwinkel, M., 1990, Sequence stratigraphy from "spot" outcrops—example from a carbonate-dominated setting: Devonian-Carboniferous transition, Dinant synclinorium (Belgium). *Sediment. Geol.*, **69**: 259–280. https://doi.org/10.1016/0037-0738(90)90053-V.

Van Steenwinkel, M., 1993, The Devonian-Carboniferous boundary in southern Belgium: biostratigraphic identification criteria of sequence boundaries. *In* Posamentier, H.W., Summerhayes, C.P., Haq, B.U., and Allen, G.P. (eds), Sequence Stratigraphy and Facies Associations. Special Publications of the International Association of Sedimentologists. 18. 237–246.

Vaughan, A., 1905, The palaeontological sequence in the Carboniferous Limestone of the Bristol area. *Q. J. Geol. Soc. Lond.*, **61**: 181–307. https://doi.org/10.1144/GSL.JGS.1905.061.01-04.13.

Vdovenko, M.V., Aisenverg, D.Y., Nemirovskaya, T.I., and Poletaev, V. I., 1990, An overview of Lower Carboniferous biozones of the Russian platform. *J. Foraminifer. Res.*, **20**: 184–194. https://doi.org/10.2113/gsjfr.20.3.184.

Veevers, J.J., and Powell, M., 1987, Late Paleozoic glacial episodes in Gondwanaland reflected in transgressive-regressive depositional sequences in Euramerica. *Geol. Soc. Am. Bull.*, **98**: 475–487. https://doi.org/10.1130/0016-7606(1987)98<475:LPGEIG>2.0.CO;2.

Villa, E., 1995, Fusulináceos carboníferos del este de Austurias (N de España). *Biostratigraphie du. Paléozoïque*, **13**: 261.

Villa, E., Alekseev, A.S., Barrick, J.E., Boardman, D.R., Djenchuraeva, A. V., Fohrer, B., et al., 2009, Selection of the conodont *Idiognathodus simulator* (Ellison) as the event marker for the base of the global Gzhelian Stage (Upper Pennsylvanian, Carboniferous). *Palaeoworld*, **18**: 114–119. https://doi.org/10.1016/j.palwor.2009.04.002.

Villa, E., and Task Group, 2007, Progress report of the task group to establish the Moscovian-Kasimovian and Kasimovian-Gzhelian boundaries. *Newsl. Carbonif. Stratigraphy*, **25**: 7–8.

Villa, E., and Task Group, 2008, Progress report of the task group to establish the Moscovian-Kasimovian and Kasimovian-Gzhelian boundaries. *Newsl. Carbonif. Stratigraphy*, **26**: 12–13.

Wagner, R.H., 1984, Megafloral zones of the Carboniferous. *Neuvième Congres International de Stratigraphie et de Géologie du Carbonifère, Washington and Champaign-Urbana 1979, Compte Rendu*, Carbondalle and Edwardsville: Southern Illinois University Press, **Vol. 2**, pp. 109–134.

Wagner, R.H., 2011, Comment: no major stratigraphic gap exists near the Middle-Upper Pennsylvanian (Desmoinesian-Missourian) Boundary in North America. *Palaios*, **26**: 669–670. https://doi.org/10.2110/palo.2011.p11-033r.

Wagner, R.H., 2017, The 'global' scheme of Pennsylvanian chronostratigraphic units contrasted with the West European and North American regional classifications: discussion of paleogeographic zones/regions and problems of correlation. *Stratigraphy*, **14**: 405–423.

Wagner, R.H., and Álvarez-Vázquez, C., 2010, The Carboniferous floras of the Iberian Peninsula: a synthesis with geological connotations. *Rev. Palaeobotany Palynology*, **162**: 238–324. https://doi.org/10.1016/j.revpalbo.2010.06.005.

Wagner, R.H., and Lyons, P.C., 1997, A critical analysis of the higher Pennsylvanian megafloras of the Appalachian region. *Rev. Palaeobotany Palynology*, **95**: 255–283. https://doi.org/10.1016/S0034-6667(96)00037-1.

Wagner, R.H., Sánchez De Posada, L.C., Martínez Chacón, M.L., Fernández, L.P., Villa, E., and Winkler Prins, C.F., 2002, The Asturian Stage: a preliminary proposal for the definition of a substitute for Westphalian D. *Can. Soc. Pet. Geologists, Mem.*, **19**: 832–850.

Wagner, R.H., and Winkler Prins, C.F., 1985, The Cantabrian and Barruelian stratotypes: a summary of basin development and biostratigraphic information. In: Lemos de Sousa, M. J., and Wagner, R.H. (eds), *Papers on the Carboniferous of the Iberian Peninsula*, Anais Faculdade de Ciências, Universidade do Porto, Supplement to volume 64 (1983): 359–410.

Wagner, R.H., and Winkler Prins, C.F., 1997, Carboniferous stratigraphy: Quo vadis?. *Proceedings of the XIII International Congress on the Carboniferous and Permian, Kraków, 1995, Part 1*, pp. 187–196.

Wagner, R.H., and Winkler Prins, C.F., 2016, History and current status of the Pennsylvanian chronostratigraphic units: problems of definition and interregional correlation. *Newsl. Stratigraphy*, **49**: 281–320. https://doi.org/10.1127/nos/2016/0073.

Wagner, R.H., Winkler Prins, C.F., and Granados, L.F. (eds), 1985. *The Carboniferous of the World, II. Australia, Indian Subcontinent, South Africa, South America and North Africa*. Instituto Geologico y Minero de España IGME and Empresa Nacional Adaro de Investigaciones Mineras.

Wahlman, G.P., 2013, Pennsylvanian to Lower Permian (Desmoinesian-Wolfcampian) fusulinid biostratigraphy of Midcontinent North America. *Stratigraphy*, **10**: 73–104.

Wang, Z.H., 1991, Conodont zonation of the Lower Carboniferous in South China and phylogeny of some important species. *Cour. Forschungsinstitut Senckenberg.*, **130**: 41–46. (imprinted 1990).

Wang, K.L., 1983, Early Carboniferous Foraminifera from Shaoyang area of Hunan Province and their stratigraphic significance. *Bull. Nanjing Inst. Geol. Palaeontology, Academia Sin.*, **6**: 209–224. (in Chinese).

Wang, C.Y., 1993, Auxiliary stratotype sections for the global stratotype section and point (GSSP) for the Devonian-Carboniferous boundary: Nanbiancun. *Annales de. la. Société Géologique de Belgique*, **115**: 707–708.

Wang, Q., Korn, D., Nemyrovska, T., and Qi, Y., 2018, The Wenne river bank section – an excellent section for the Visean-Serpukhovian boundary based on conodonts and ammonoids (Mississippian; Rhenish Mountains, Germany). *Newsl. Stratigraphy*, **51**: 427–444. https://doi.org/10.1127/nos/2018/0440.

Wang, X.D., Hu, K.Y., Qie, W.K., Sheng, Q.Y., Chen, B., Lin, W., et al., 2019, Carboniferous integrative stratigraphy and time scale of China. *Sci. China Earth Sci.*, **62**: 135–153. https://doi.org/10.1007/s11430-017-9253-7.

Wang, X.D., and Jin, Y.G., 2000, An outline of Carboniferous Chronostratigraphy. *J. Stratigraphy*, **24** (2), 90–98. (in Chinese).

Wang, X.D., and Jin, Y.G., 2003, Carboniferous Biostratigraphy of China. *In* Zhang, W.T., Chen, P.J., and Palmer, A.R. (eds), *Biostratigraphy in China*. Beijing: Science Press, 281–330.

Wang, Z.J., and Hou, H., 1990, *The Carboniferous System of China (Stratigraphy of China, No. 8)*. Beijing: Geological Publishing House. 419 p. (in Chinese).

Wang, Z.H., and Qi, Y.P., 2003, Upper Carboniferous (Pennsylvanian) conodonts from South Guizhou of China. *Riv. Ital. di Paleontologia e Stratigrafia*, **109**: 379–397. https://doi.org/10.13130/2039-4942/5513.

Wanless, H.R., and Shepard, F.P., 1936, Sea level and climatic changes related to late Paleozoic cycles. *Geol. Soc. Am. Bull.*, **47**: 1177–1206. https://doi.org/10.1130/GSAB-47-1177.

Waters, C.N., Somerville, I.D., Jones, N.S., Cleal, C.J., Collinson, J.D., Waters, R.A., et al., 2011, A revised correlation of Carboniferous rocks in the British Isles. *Geol. Soc. Spec. Rep.*, **26**: 1–186. https://doi.org/10.1144/SR26.

Waters, C.N., and Condon, D.J., 2012, Nature and timing of Late Mississippian to Mid-Pennsylvanian glacio-eustatic sea-level changes of the Pennine Basin, UK. *Journal of the Geological Society*, **169**: 37–51.

Weedon, G.P., and Read, W.A., 1995, Orbital-climatic forcing of Namurian cyclic sedimentation from spectral analysis of the Limestone Coal Formation, central Scotland. *In* House, M.R., and Gale, A.S. (eds), Orbital Forcing Time scales and Cyclostratigraphy. *Geological Society London, Special Publication*, **85**: 51–66. https://doi.org/10.1144/GSL.SP.1995.085.01.04.

Werneburg, R., 1996, Temnospondyle Amphibien aus dem Karbon Mitteldeutschlands. *Naturhistorisches Mus. Schloss Bertholdsburg, Schleusingen, Veröffentlichungen*, **11**: 23–64.

White, F.M., 1992, *Aspects of the Palaeoecology and Stratigraphic Significance of Late Dinantian (Early Carboniferous) Foraminifera of Northern England*. Unpublished Ph.D. thesis. University of Manchester, p. 677.

Whitehurst, J., 1778, *An Enquiry Into the Original State and Formation of Earth*. London: Printed for the author and W. Bent by J. Cooper, p. 199.

Wilde, G.L., 1990, Practical fusulinid zonation: the species concept; with Permian basin emphasis. *West. Tex. Geol. Soc.*, **29**: 5–34.

Williams, H.S., 1891, Correlation papers: Devonian and Carboniferous. *U S Geol. Survey, Bull.*, **80**: 1–279.

Winchell, A., 1869, On the geological age and equivalents of the Marshall group. *Proceedings of the American Philosophical Society*, **1**: 57–82, 385–418.

Wright, V.P., and Vanstone, S.D., 2001, Onset of Late Palaeozoic glacio-eustasy and the evolving climates of low latitude areas: a synthesis of current understanding. *J. Geol. Soc.*, **158**: 579–582. https://doi.org/10.1144/jgs.158.4.579.

Wu, W.S., Zhang, L.X., and Jin, Y.G., 1974, The Carboniferous System of western Guizhou. *Mem. Nanjing Inst. Geol. Palaeontology, Academia Sin.*, **6**: 72–98. (in Chinese).

Wu, W.S., Zhang, L.X., Zhao, X.H., Jin, Y.G., and Liao, Z.T., 1987, *Carboniferous stratigraphy of China*. Beijing: Science Press, p. 160.

Wu, Q., Ramezani, J., Zhang, H., Wang, T.-T., Yuan, D.-X., Mu, L., et al., 2017, Calibrating the Guadalupian Series (Middle Permian) of South China. *Palaeogeography Palaeoclimatology Palaeoecology*, **466**: 361–372.

Wu, H., Fang, Q., Wang, X.-D., Hinnov, L., Qi, Y., Shen, S.-Z., et al., 2019, An ~34 m.y. astronomical time scale for the uppermost Mississippian through Pennsylvanian of the Carboniferous System of the Paleo-Tethyan realm. *Geology*, **47**: 83–86. https://doi.org/10.1130/G45461.1.

Yang, S.P., 1962, Lower Carboniferous subdivision based on brachiopods in Guizhou. *Joint Meeting on the Second National Congress of the Palaeontological Society of China and the 10th Annual Conference of PSC, Abstract Volume* (in Chinese), pp. 38–39.

Yang, J.Z., Sheng, J.Z., Wu, W.S., and Lu, L.H., 1962, Proceeding of The National Meeting on Stratigraphy, The Carboniferous in China *(in Chinese)*. Beijing: Science Press, p. 113.

Yang, J.Z., Wu, W.S., Zhang, L.X., Liao, Z.T., and Ruan, Y.P., 1979, New perspectives on the Series level subdivision of Carboniferous. *Acta Stratigraphy Sin.*, **3**: 188–192. (in Chinese).

Yao, L., Qie, W., Luo, G., Liu, J., Alego, T., Bai, X., et al., 2015, The TICE event: Perturbation of carbon–nitrogen cycles during the mid-Tournaisian (Early Carboniferous) greenhouse–icehouse transition. *Chem. Geol.*, **401**: 1–14. https://doi.org/10.1016/j.chemgeo.2015.02.021.

Yeh, M.-W., and Shellnutt, J.G., 2016, The initial break-up of Pangaea elicited by Late Palaeozoic deglaciation. *Sci. Rep.*, **6** (31442), 9. https://doi.org/10.1038/srep31442.

Yin, H.F., Zhang, K.X., Tong, J.N., Yang, Z.Y., and Wu, S.B., 2001, The Global Stratotype Section and Point (GSSP) of the Permian–Triassic Boundary. *Episodes*, **24**: 102–114.

Yoh, S.X., 1929, Preliminary report on the geology and mineral resources of Nan Tan Hsien, Ho Chi Hsien, I Shan Hsien, Ma Ping Hsien, northern Kwangsi Province. *Geol. Survery Kwangtung Kwangsi*, **1**: 97–120.

Young, G.C., and Laurie, J.R., 1996, *An Australian Phanerozoic Time scale*. Melbourne: Oxford University Press. pp. 279.

Yu, C.C., 1931, The correlation of the Fengning System, the Chinese Lower Carboniferous, as based on coral zones. *Bull. Geol. Soc. China*, **10**: 1–30.

Zajic, J., 2000, Vertebrate zonation of the non-marine Upper Carboniferous–Lower Permian basins of the Czech Republic. *Cour. Forschungsinstitut Senckenberg.*, **223**: 563–575.

Zhamoida, A.I., 2015, General Stratigraphic Scale of Russia: state of the art and problems. *Russian Geol. Geophysics*, **56**: 511–523. https://doi.org/10.1016/j.rgg.2015,03.003.

Zhang, Z.Q., 1988, The Carboniferous System in China. *Newsl. Stratigraphy*, **18**: 51–73.

Zhang, L.X., Wang, Z.H., and Zhou, J.P., 2004, The Huashibanian Stage of Upper Carboniferous in China. *J. Stratigraphy*, **28**: 18–34. (in Chinese).

Zhang, R., and Yan, D., 1993, Stratigraphic and paleobiogeographic summary of Carboniferous marine bivalves of China. *J. Paleontology*, **67**: 850–856. https://doi.org/10.1017/S0022336000037100.

Zhang, L.X., Wang, Z.H., and Zhou, J.P., 2005, The Upper Carboniferous Dalaan Stage of China. *J. Stratigraphy*, **29**: 500–511.

Zhou, T.M., Sheng, J.Z., and Wang, Y.J., 1987, Carboniferous-Permian boundary beds and fusulinid zones at Xiaodushan, Guangnan, Eastern Yunnan. *Acta Micropalaeontologica Sin.*, **4**: 123–160. (in Chinese).

C.M. Henderson and S.Z. Shen
With contributions by F.M. Gradstein and F.P. Agterberg

Chapter 24

The Permian Period

255.7 Ma Permian

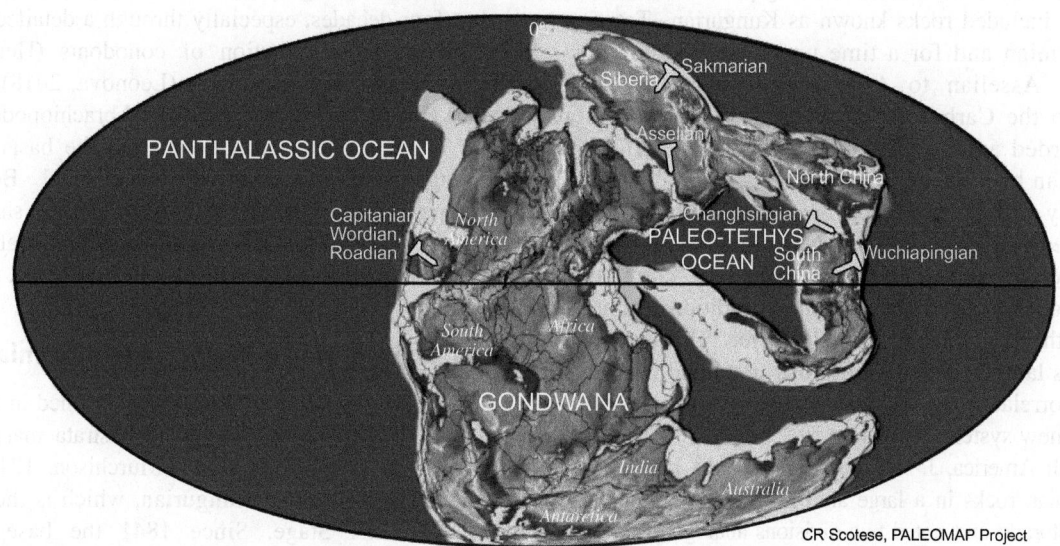

CR Scotese, PALEOMAP Project

Chapter outline

24.1 History and subdivisions	875	
24.1.1 The Cisuralian Series: Lower Permian	876	
24.1.2 The Guadalupian Series: Middle Permian	881	
24.1.3 The Lopingian Series: Upper Permian	886	
24.2 Regional correlations	890	
24.2.1 Russian platform	890	
24.2.2 United States (West Texas)	890	
24.2.3 South China	891	
24.2.4 Western Europe (Germanic Basin)	891	
24.2.5 Salt Range	894	
24.2.6 Pamirs	894	
24.3 Permian stratigraphy	894	
24.3.1 Biostratigraphy	894	
24.3.2 Physical stratigraphy	895	
24.4 Permian time scale	896	
24.4.1 Previous scales	896	
24.4.2 Radioisotopic dates	896	
24.4.3 Permian composite standard	896	
24.4.4 Age of stage boundaries	897	
Bibliography	898	

Abstract

The supercontinent Pangea completes its assembly and moves north during the Permian. The 47-million-year-long period begins with a great ice age and ends with the Earth's greatest extinction. The interval is characterized by icehouse to greenhouse climate transition, major evaporites, changes in internal and external carbonate invertebrate skeletons, major diversification of fusulinacean foraminifers, ammonoids, bryozoans, and brachiopods. It concludes with the major end-Permian extinction of fusulinacean foraminifers, trilobites, rugose and tabulate corals, blastoids, acanthodians, placoderms, and pelycosaurs and dramatic reduction of bryozoans, brachiopods, ammonoids, sharks, bony fish, crinoids, and land vertebrates.

24.1 History and subdivisions

In 1841 Sir Roderick I. Murchison (1792–1871) named the Permian System to take in the "vast series of beds of

Geologic Time Scale 2020. DOI: https://doi.org/10.1016/B978-0-12-824360-2.00024-3

marl, schist, limestone, sandstone, and conglomerate that surmounted the Carboniferous System throughout a great arc stretching from the Volga eastwards to the Urals and from the Sea of Archangel to the southern steppes of Orenburg." He proposed the name "Permian" based on the extensive region that composed the ancient kingdom of Permia and the city of Perm on the western flank of the Urals. Murchison (1872) indicated that "the animals and plants of the Permian era, though chiefly of new species, are generically connected with those of the preceding or Carboniferous epoch, whilst they are almost wholly dissimilar to those of the next succeeding period, the Trias." Murchison included rocks known as Kungurian–Tatarian in the Permian and for a time underlying strata, now known as Asselian to Artinskian, continued to be referred to the Carboniferous; these underlying strata were regarded as intermediate between Carboniferous and Permian by Dunbar (1940).

As early as 1822 (e.g., Conybeare and Phillips, 1822), the Magnesian Limestone and New Red Sandstone of England were well known, as were the correlative German Rotliegendes and Zechstein (a traditional miner's name) with its valuable Kupferschiefer. However, all these rocks lacked richly fossiliferous strata and were difficult to correlate, and unsuitable to justify the development of a new system in western Europe.

In North America, J. Marcou from 1853 to 1867 recognized Permian rocks in a large area from the Mississippi to the Rio Colorado and noted two divisions analogous to those in Western Europe. He accordingly suggested the name Dyassic as more suitable than Permian and proposed a combined Dyas and Trias as a major period. Murchison (1872) summarized Marcou's reports as follows: "the discovery of strata containing many true Permian species in the northwestern part of Texas and along the eastern edge of the Rocky Mountains teaches us that all of the seas of Paleozoic time, even the very last of them, had a very great extension and were inhabited by similar groups of animals over enormously wide areas." Murchison (1872) also rejected the notion of Dyas saying "I resist the introduction of a new name founded upon a local binary division which, though good in Saxony, is inapplicable to many other countries, in several of which the separation is tripartite … and thus I prefer the simpler geographic name 'Permian,' which, like 'Silurian,' involves no theory." For further historical details on the history of the Permian System, see Lucas and Shen (2018).

The Permian divides naturally into three series (Fig. 24.1), but the youngest series is only well represented in China (Jin et al., 1997). In the area of the southern Urals the Lower Permian (Cisuralian) is well represented by marine deposits and abundant biota. This marine dominance disappears during the Kungurian, and overlying Ufimian, Kazanian, and Tatarian are dominated by terrestrial and marginal marine deposits. In other regions the Middle Permian (Guadalupian) is dominated by diversified and well-studied marine fossil assemblages; these deposits are well known in the Guadalupe Mountains of West Texas where they have been the subject of detailed biostratigraphic and sequence stratigraphic study. The Delaware Basin of West Texas becomes evaporitic near the end of the Middle Permian. However, a third series of Upper Permian (Lopingian) marine rocks with abundant fossils are best represented in China, Iran, and the Transcaucasian region.

Permian biostratigraphy has been greatly refined over the last four decades, especially through a detailed understanding of the distribution of conodonts (Henderson, 2018) in relation to ammonoids (Leonova, 2018), fusulinaceans (Zhang and Wang, 2018), and brachiopods (Shen, 2018). These refined correlations form the basis for the following discussion. All of the Global Boundary Stratotype Section and Points (GSSPs, as shown in Fig. 24.11 at the end of this chapter) were biostratigraphically marked with reference to standard conodont zones.

24.1.1 The Cisuralian Series: Lower Permian

The base of the Permian was originally defined in the Ural Mountains of Russia to coincide with strata marking the initiation of evaporite deposition (Murchison, 1841), now recognized as within the Kungurian, which is the uppermost Cisuralian Stage. Since 1841 the base of the Permian has been lowered to include a succession of faunas with post-Carboniferous affinity. Karpinsky (1874) identified clastic successions that Murchison had included in the British Millstone Grit as being younger, transitional between Carboniferous and Permian, and termed them the Artinskian Series. His subsequent study of the abundant ammonoid fauna (Karpinsky, 1889) led him to add the interval to the Permian. Ruzhencev (1936) recognized the Sakmarian as an independent lower subdivision of the Artinskian based on ammonoid biostratigraphy. He later subdivided the Sakmarian and referred the lower interval to the Asselian Stage (Ruzhencev, 1954). The base of the Asselian and of the Permian System was defined by the appearance of the ammonoid families Paragastrioceratidae, Metalegoceratidae, and Popanoceratidae, concurrent with the first inflated fusulinaceans referable to "*Schwagerina*" (i.e., *Pseudoschwagerina* and *Sphaeroschwagerina*).

The Cisuralian was proposed by Waterhouse (1982) to comprise the Asselian, Sakmarian and Artinskian stages; it was termed Uralian by Jin et al. (1994). The Kungurian was later included in the Cisuralian (Jin et al., 1997), so that the series corresponded to the Lower Permian as recognized in Russia (Likharew, 1966; Kotlyar and Stepanov, 1984) and to the Rotliegendes of Harland et al. (1990).

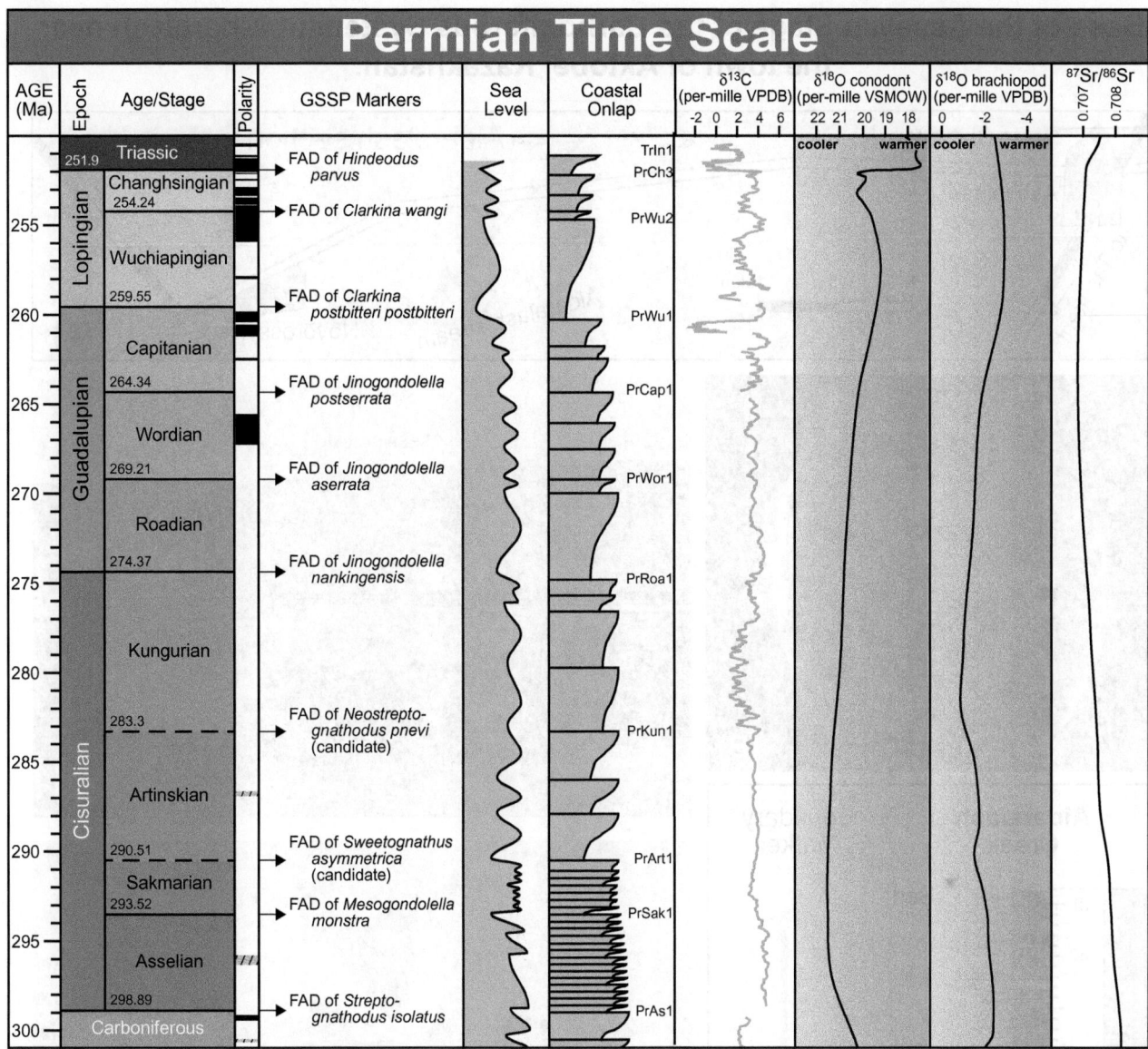

FIGURE 24.1 Permian overview. The main markers or candidate markers for GSSPs of Permian stages are FADs of conodont taxa as detailed in the text. ("Age" is the term for the time equivalent of the rock-record "stage.") Magnetic polarity scale is modified from Steiner (2006), but see Hounslow and Bakabanov (2016) for a revised compilation. Schematic sea-level, coastal onlap curves and sequence nomenclature are modified from Haq and Schutter (2008) following advice of Haq (pers. comm., 2008) and with the incorporation of revised high-frequency glacio-eustatic cycles in the earliest Permian, as detailed in the text. See also review in (Simmons et al. 2020, Ch. 13, Phanerozoic eustacy, this book). The carbon isotope curve is from (Cramer and Jarvis 2020, Ch. 11: Carbon isotope stratigraphy, this book) who had compiled numerous studies, including Buggisch et al. (2011), Cao et al. (2018), and Bagherpour et al. (2018). The oxygen isotope curve and implied sea-surface temperature trends are the statistical-mean curves by (Grossman and Joachimski 2020, Ch.10: Oxygen isotope stratigraphy, this book) from a synthesis of numerous conodont apatite and brachiopod carbonate studies. Strontium isotope curve is modified from (McArthur et al. 2020, Ch. 7: Strontium isotope stratigraphy, this book). The vertical scale of this diagram is standardized to match the vertical scales of the period-overview figure in all other Phanerozoic chapters. *FADs*, First-appearance datums; *VPDB*, Vienna PeeDee Belemnite δ^{13}C and δ^{18}O standard; *VSMOW*, Vienna Standard Mean Ocean Water δ^{18}O standard.

24.1.1.1 Asselian

The base-Permian System is defined at the base-Asselian Stage GSSP located at Aidaralash Creek, Atobe region, northern Kazakhstan (Davydov et al., 1998). The Aidaralash section is approximately 50 km southeast of the city of Atobe where a stone and concrete marker with a plaque has been erected to mark the exact location of the GSSP and the boundary between the Carboniferous and Permian (Fig. 24.2).

Late Paleozoic strata at Aidaralash Creek were deposited on a narrow, shallow-marine shelf that formed the

Base of the Asselian Stage of the Carboniferous System at Aidaralash near the town of Aktobe, Kazakhstan.

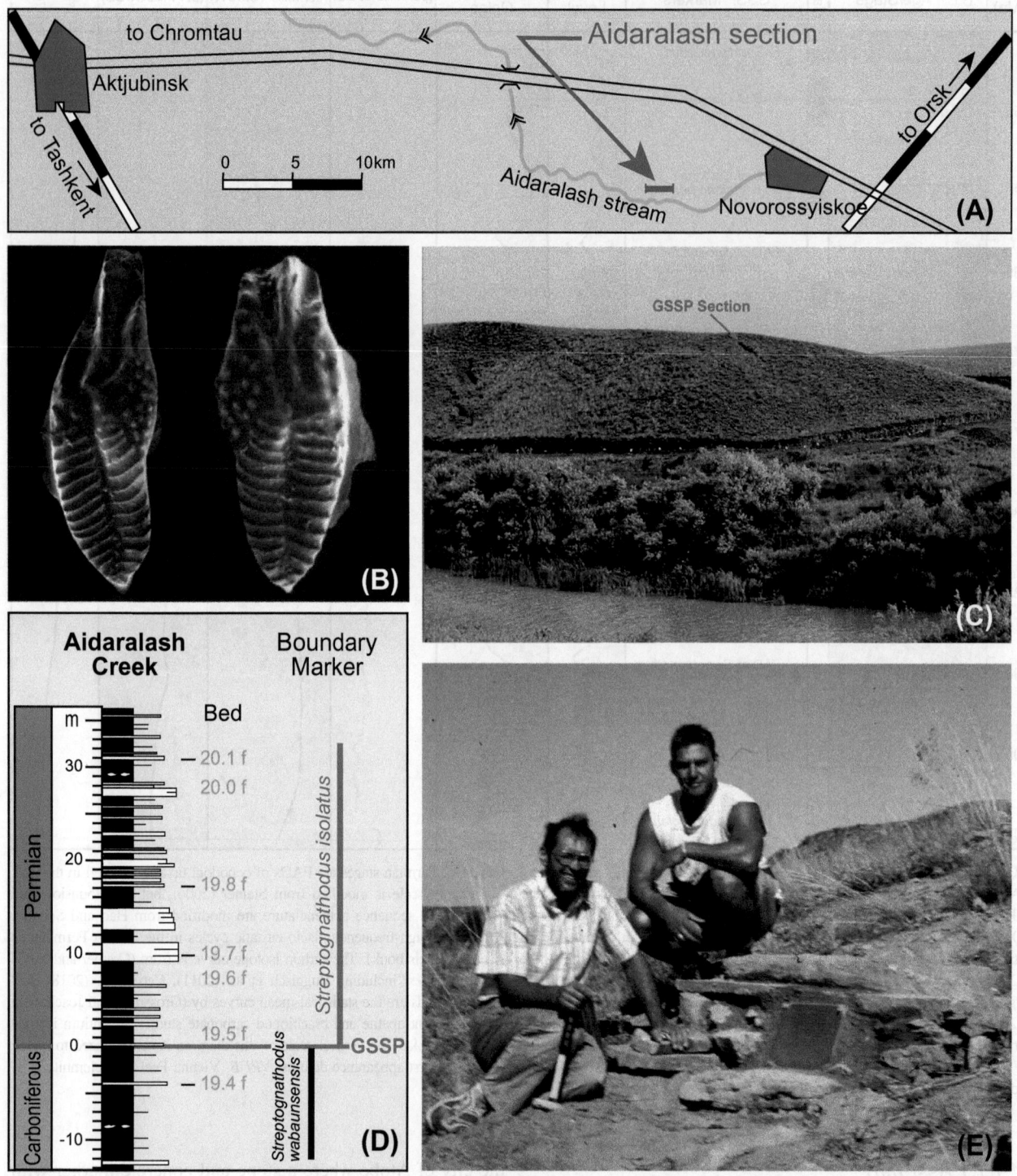

FIGURE 24.2   Carboniferous–Permian GSSP (Global Boundary Stratotype Section and Point) and for the Asselian Stage and Cisuralian Series at Aidaralash, northwest Kazakhstan. Semipermanent plaque marks the boundary; the index conodont shows the holotype of *Streptognathus isolatus* (left, 1.2 mm long and right 0.8 mm long). Outcrop photograph provided by Vladimir Davydov; conodont and section photographs and stratigraphic column from Henderson et al. (2012a).

western boundary of an orogenic zone to the east. A fluvial—deltaic, conglomerate—sandstone succession grades upward into a transgressive, marginal marine systems tract that further grades upward into mudstone, siltstone, and fine sandstone beds with ammonoids, conodonts, and radiolarian representing the deepest facies. This maximum flooding interval is overlain by a regressive sequence grading from offshore to shoreface to delta front and capped by a conglomerate marking an unconformity. The GSSP is defined within the maximum flooding interval and free of disconformities.

The base-Asselian GSSP (Fig. 24.2) is defined by the first occurrence of the conodont *Streptognathodus isolatus* at Aidaralash, which developed from an advanced morphotype in the *S. wabaunsensis* chronomorphocline. This point occurs 27 m above the base of Bed 19 (Bed 19.5) at Aidaralash Creek (Davydov et al., 1998).

The first occurrences of *Streptognathodus invaginatus* and *Streptognathodus nodulinearis* nearly coincide with the first occurrence of *S. isolatus* in many sections and can be used as accessory indicators for the boundary.

The GSSP is 6.3 m below the traditional Russian fusulinacean boundary, which was the base of the *Sphaeroschwagerina vulgaris—Sphaeroschwagerina fusiformis* Zone (Davydov et al., 1998); this position can be widely correlated among Spitsbergen, the Russian Platform, Urals, Central Asia, China and Japan, and thus identifies a Permian age in proximity to the base of the Asselian Stage.

The traditional Russian ammonoid boundary was 26.8 m above the GSSP level; *Prouddenites terminalis*, *Artinskia irinae*, and *Prothalassoceras bashkiricus* occur in lower Bed 19 and *Neopronorites rotundus*, *Daixites antipovi*, *Artinskia kazakhstanica*, and *Prothalassoceras serratum* occur in Bed 20 (Leonova, 2018). Ammonoid taxa are relatively rare in many sections and many taxa are endemic.

Utilization of magnetostratigraphy to assist with recognition and correlation of the Carboniferous—Permian boundary is difficult because it occurs within the reversed polarity Kiaman Superchron (Hounslow and Balabanov, 2018). A tentative normal magnetozone is restricted to the Gzhelian *Daixina bosbytauensis—Globifusulina robusta* fusulinacean zone, just below the Carboniferous—Permian boundary at Aidaralash (Hounslow and Balabanov, 2018). That same stratigraphic polarity relationship is also known elsewhere from the southern Urals, and the northern Caucasus and Donetz Basin, and possibly correlates with the normally polarized magnetic zone in the Manebach Formation of the Thuringian Forest (Menning, 1987).

The conodont succession observed at Aidaralash is displayed in several sections in the southern Urals, especially the basinal reference section at Usolka (Schmitz and Davydov, 2012) where Schmitz et al. (2009) indicate a $^{87}Sr/^{86}Sr$ isotopic value of 0.70801 based on conodont apatite for the base-

Asselian. The lineage is also recognized in the West Texas regional stratotype Wolfcamp Hills (Wardlaw and Davydov, 2000), in South China (Wang, 2000; Mei and Henderson, 2001), and in the Bennett Shale of the Red Eagle cyclothem, the midcontinent United States (Boardman et al., 1998, 2009). The Asselian cyclothem succession in the midcontinent United States also includes *Sweetognathus expansus*, *Sweetognathus merrilli*, and *Sweetognathus whitei* (after Rhodes, 1963); the latter species in the Barneston—Florence limestones of the Chase Group of Kansas (Boardman et al., 2009) in association with *Streptognathodus* spp.

24.1.1.2 Sakmarian

The original Russian base—Sakmarian boundary was proposed by Ruzhencev (1950) at the base of Bed 11 at an unconformable formation break within the Kondurovsky section, based on the occurrence of fusulinacean species *Sakmarella (Pseudofusulina) moelleri*. The first occurrences of *Sweetognathus* aff. *merrilli* and *Mesogondolella uraloceras* were also considered as markers (Chernykh et al., 2016).

The base-Sakmarian GSSP was ratified in 2018 by IUGS and an article is published in Episodes (Chernykh et al., 2020). The base-Sakmarian GSSP (Fig. 24.3) is defined by the first occurrence of the conodont *Mesogondolella monstra* at the Usolka Section in Bed 26/3 (55.4 m) within the lineage *Mesogondolella uralensis* to *M. monstra*. The first occurrence of the conodont *Sweetognathus binodosus* represents an auxiliary marker. The $^{87}Sr/^{86}Sr$ isotopic value based on conodont apatite for the base-Sakmarian is 0.70787 (Chernykh et al., 2020). Tastubian (lower Sakmarian) fusulinaceans, including *Leeina verneuili* are recovered 4 m above the GSSP. The Usolka section also has several ash beds that date the base-Sakmarian as 293.5 Ma ± 0.4 Myr, which is a much younger age compared to GTS2012 because a higher point has been chosen (Chernykh et al., 2013). The higher position was chosen because it nearly coincides with the extinction of the conodont genus *Streptognathus* and because *M. monstra* and *Sweetognathus binododus* are much more widely distributed and the index species is more easily distinguished from its ancestor species.

A significant characteristic of this boundary interval is that it coincides with a major change in stratigraphic cyclicity (Chernykh et al., 2020). The late Carboniferous and Asselian include numerous high-frequency cyclothems attributed to 405-kyr eccentricity cycles. A couple of cyclothems occur in the lower Sakmarian in shallow-water settings, but otherwise the Sakmarian and Artinskian successions represent broad third-order sequences. This signifies the acme and waning of the Late Paleozoic Ice Age.

24.1.1.3 Artinskian

The Artinskian was proposed by Karpinsky in 1874, for the sandstone of the Kashkabash Mountain on the right

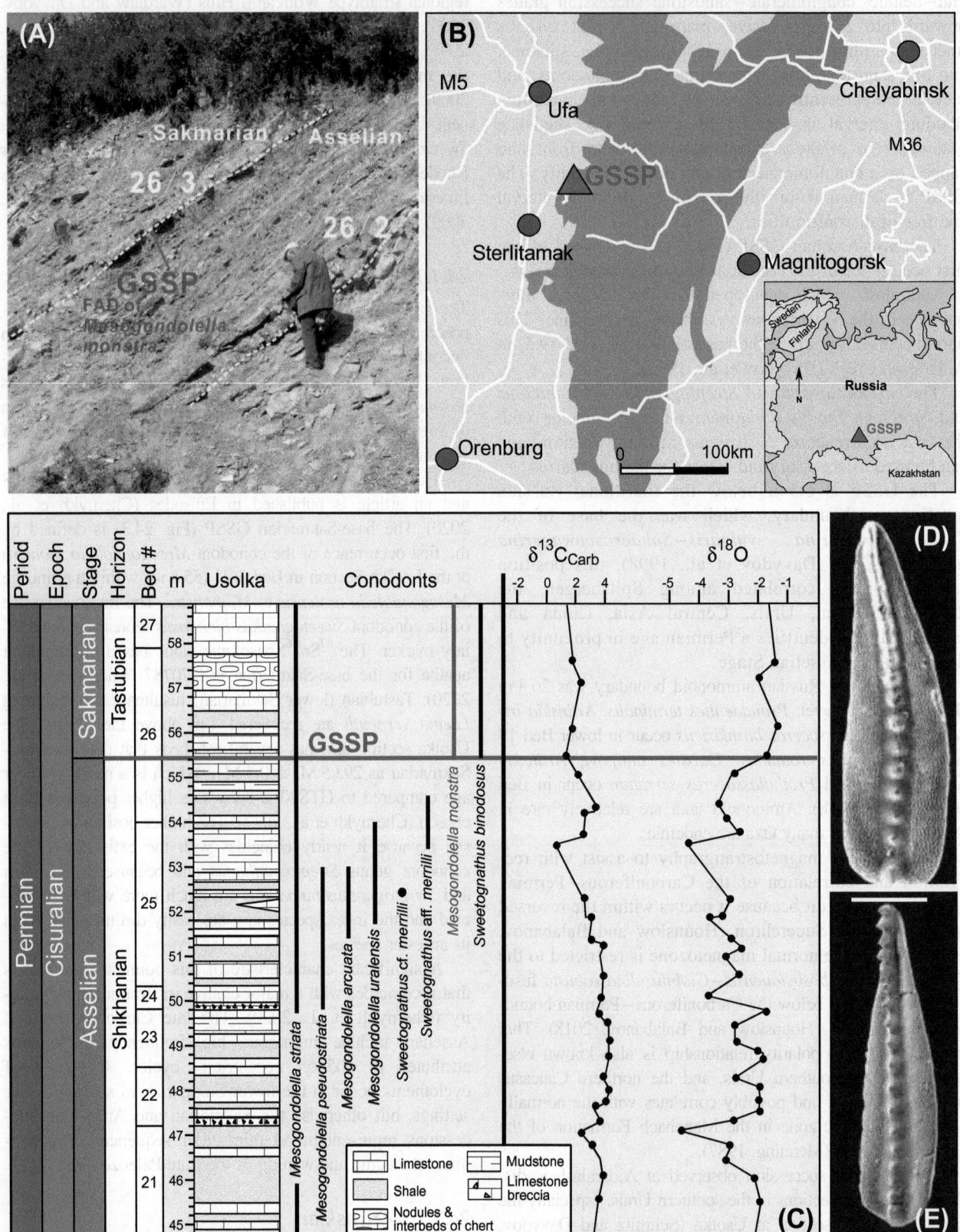

FIGURE 24.3   GSSP for the base of the Sakmarian Stage at Usolka section, Russia. *Mesogondolella monstra* scale bar is 0.2 mm.

bank of the Ufa River, near the village of Arty. Karpinsky (1891) studied abundant and diverse ammonoids in several exposures and small quarries along the Ufa River. Karpinsky (1874) defined two regional belts with ammonoids: the lower at Sakmara River and the upper at Ufa River because the diverse ammonoid assemblage from the Arty area was distinctly more advanced than the Sakmarian assemblage in terms of cephalopod evolution.

Sakmarian–Artinskian boundary deposits are well represented in the Dal'ny Tulkas section, which is near the Usolka section. The upper part of the Sakmarian Stage (Beds 28–31) at the Usolka section and Bed 18 at the Dal'ny Tulkas section are composed of dark-colored marl, argillite, and carbonate mudstone, or less commonly, detrital limestone with fusulinaeans radiolaria, rare ammonoids, and bivalves. The upper part of the Sakmarian includes fusulinaceans characteristic of the Sterlitamakian Horizon, including *Pseudofusulina longa*, *Pseudofusulina fortissima*, *Pseudofusulina plicatissima*, and *Leeina urdalensis*.

There is general consensus that the GSSP will be defined at the Dal'ny Tulkas section in Russia (Chuvashov et al., 2013) and a vote by the Subcommission on Permian Stratigraphy is anticipated in 2020. The point will be defined at the first occurrence of *Sweetognathus asymmetrica* within the chronomorphocline *S. binodosus* to *Sweetognathus anceps* to *S. asymmetrica* (the latter species has previously been referred to *S.* aff. *whitei*) at 2.7 m above bed 4. The succession of *S. binodosus* to *S. asymmetrica* can also be recognized in the Lower Great Bear Cape Formation, southwest Ellesmere Island (Beauchamp and Henderson, 1994; Mei et al., 2002; Chernykh et al., 2020) and in Nevada. Specimens of *S. asymmetrica* in the Canadian Arctic, South China (Sun et al., 2017), and in the Urals are found above high-frequency cyclothems. The occurrence of *S. asymmetrica* in south China is considered near the top of its normal range. The $^{87}Sr/^{86}Sr$ isotopic value based on conodont apatite for the base-Artinskian is approximately 0.70769 (Chernykh et al., 2020) and ash beds at Dal'ny Tulkas section provide an interpolated age of 290.5 ± 0.4 Myr.

24.1.1.4 Kungurian

The stratotype of the Kungurian Stage was not defined when the stage itself was established (Stuckenberg, 1890), but the carbonate–sulfate section exposed along the Sylva River, upstream of the town of Kungur, was later generally accepted. Chuvashov et al. (1999) proposed a new position for the lower Kungurian boundary at the base of the Sarana Horizon, where the section consists of: (1) the Saraninian Horizon, including reefal limestone of the Sylva Formation and its lateral equivalent the Shurtan Formation, composed of marls and clayey limestone; (2) the Filippovian Horizon; and (3) the Irenian Horizon. A disadvantage of the section is the poor fossil

content of the underlying Kamai Formation that contains only small benthic foraminifers, bryozoans and brachiopods, with taxa unsuitable for age determination (Chernykh et al., 2019).

Another section of Artinskian–Kungurian boundary deposits in Russia, located near the Mechetlino settlement on the Yuryuzan River, has a better fossil assemblage both below and above the boundary interval and has been considered as a possible stratotype for the base-Kungurian at the first-appearance datum (FAD) of the conodont *Neostreptognathodus pnevi* in bed 19 (Chuvashov et al., 2002; Chernykh et al., 2019). However, conodonts are rare in most samples and volcanic ash beds provided only reworked zircons. A second section is also considered for the base-Kungurian GSSP. The Rockland section occurs in the Pequop Mountains of northeastern Nevada (Henderson et al., 2012b) and also demonstrates good correlation potential as the chronomorphocline from *Neostreptognathodus pequopensis* to *N. pnevi* (Henderson, 2018) is also present, just as at Mechetlino. Progress has been hampered by the overall poor fossil successions and the lack of additional good markers. The age of the base-Kungurian is interpolated as 283.3 ± 0.4 Myr, but good dates are very widely separated. The overall succession characterizes a lowstand of sea level and better fossil successions are unlikely. The lineage from *N. pequopensis* to *N. pnevi* is found in many regions and remains the best marker for a GSSP between the two longest stages of the Permian.

In summary, the Cisuralian is divided into stages based on the FADs of specific species from three lineages of conodonts. The first, the base of the Permian, and all zones of the Asselian, is based on the widespread occurrence of *Streptognathodus* species, as is most of the Upper Pennsylvanian. Specimens of the genus become progressively rare and finally extinct during the earliest Sakmarian. The Sakmarian is defined within a *Mesogondolella* lineage, and the Artinskian and Kungurian stages will be based on a lineage of *Sweetognathus* and its derived descendant *Neostreptognathodus*.

24.1.2 The Guadalupian Series: Middle Permian

The Guadalupian was first proposed by Girty at the turn of the last century for the spectacular fossils found in the Guadalupe and Glass Mountains of West Texas. These faunas have been well documented and represent an excellent display of an exhumed, well-preserved backreef, reef and basin facies. The West Texas depositional basins include a tropical North American faunal suite, well separated from the more typical tropical Tethyan fauna of Asia and Europe. The Middle Permian was a time of strong provincialism that presents some complexities for correlation. The formal definition of the Guadalupian and its constituent stages is based on the evolution of a single genus of conodont, *Jinogondolella*. The abundant and well-preserved conodont faunas of West Texas show that species of the genus *Jinogondolella* evolved through short-lived transitional morphotypes associated with paedomorphosis. The first species of the genus, *Jinogondolella nankingensis*, is also the marker for the Guadalupian and its basal stage, the Roadian. The species is abundant in West Texas and South China, but occurs rarely in several other sites (i.e., Canadian Arctic, Pamirs); its distribution along the western coast of Pangea represents a geographic cline from the tropical Delaware Basin (West Texas) to the upwelling-influenced Phosphoria Basin (Idaho) to temperate Canadian Arctic and provides excellent correlation globally (Henderson and Mei, 2007).

24.1.2.1 Roadian

Furnish (1973) identified a distinctive ammonoid fauna from the "first limestone member" of the Word Formation, including *Eumedlicottia burckhardti* and *Perrinites hilli*.

The GSSP for the base of the Roadian Stage and Guadalupian Series (Middle Permian) is in Stratotype Canyon, Guadalupe Mountains National Park, Texas, United States (Fig. 24.4); it was ratified in 2001. The marker horizon is the first evolutionary appearance of the conodont *J. nankingensis* from its ancestors *Mesogondolella idahoensis idahoensis* and *M. idahoensis lamberti* (Mei and Henderson, 2002), at 42.7 m above the base of the black, thin-bedded limestone of the Cutoff Formation and 29 cm below a prominent shale band in the upper part of the El Centro Member (Glenister et al., 1999). This member consists of skeletal carbonate mudstone to wackestone with one shale bed, deposited in a basinal setting, proximal to the slope. In terms of magnetostratigraphy, the Cutoff Formation indicates reversed polarity and may fall in the Kiaman reversed superchron. The age for the base-Roadian at 274.4 ± 0.4 Ma is 2.1 Myr older than in GTS2012 because of new ash beds in Chaohu, Anhui Province, southeast China, that suggest slightly younger age, and on Okhotsk Massif, NE Russia (Wu et al., 2017; Davydov et al., 2018).

24.1.2.2 Wordian

Furnish (1973) used Wordian as a stage by considering a distinctive ammonoid fauna from the Word Formation, including the widespread *Waagenoceras*. The ammonoid genus *Waagenoceras* has long been associated with the Guadalupian, and, in particular, the Wordian, but it also occurs within the upper Roadian according to GSSP definitions mentioned previously.

The GSSP for the base of the Wordian Stage in the Guadalupian Series is located in Guadalupe Pass in Texas, a short distance from Stratotype Canyon (Fig. 24.5); it was ratified in 2001. The marker horizon for this stage is the first

evolutionary appearance of *Jinogondolella aserrata* from its ancestor *J. nankingensis* at 7.6 m above the base of the Getaway Ledge outcrop section in Guadalupe Pass, Guadalupe Mountains National Park, Texas, United States. This level is just below the top of the Getaway Limestone Member of the Cherry Canyon Formation, a succession of skeletal carbonate mudstone and wackestone in a slope depositional setting (Glenister et al., 1999). Most of the original work on the Guadalupian GSSPs occurred in the Glass Mountains, which are no longer accessible. As a result the GSSP levels are currently under review in the Guadalupes.

Like the Roadian rocks in the type area, lower Wordian limestone of Guadalupe National Park also display reversed polarity. The end of the reversed polarity Kiamin Superchron occurs during the Wordian and the first normal polarity magnetochron of the Illawarra Superchron (Fig. 24.1) occurs in the middle Wordian (Hounslow and Balabanov, 2018). It is documented at Nipple Hill below the FO of *Jinogondolella postserrata* and below an ash bed at 37.2 m (Bowring et al., 1998; Ramezani and Bowring, 2018), which is dated at 265.46 ± 0.27 Ma. The Illawarra reversal is well known from the lower part of the Tatarian in the Volga region of Russia (Gialanella et al., 1997). It has also been documented from Pakistan. Haag and Heller (1991) show that normal polarity starts at the base of the Wargal Formation in the Nammal Gorge, Salt Range, which is the base of the Illawarra reversal. Peterson and Nairn (1971) record a reversal in West Texas—New Mexico that has been interpreted with additional study by Menning (2000) to occur just below the upper Pinery Limestone Member of the Bell Canyon Formation.

24.1.2.3 Capitanian

The Capitan Limestone was named after the prominent El Capitan reef carbonate forming the southern end of Guadalupe Mountains escarpment. The original biostratigraphy of the Capitanian referred to the *Timorites* ammonoid zone (Glenister and Furnish, 1961).

Like the GSSPs for the Roadian and Wordian stages in the Middle Permian, the GSSP for the Capitanian Stage was also selected in the Guadalupe National Park (Fig. 24.6) and ratified in 2001. The marker horizon for the Capitanian Stage is the first evolutionary appearance of the conodont *J. postserrata* within the lineage *J. nankingensis* to *J. aserrata* to *J. postserrata*. This level is at 4.5 m in the outcrop section at Nipple Hill, in the upper Pinery Limestone Member of the Bell Canyon Formation (Glenister et al., 1999). The GSSP is in a monotonous succession of pelagic carbonate, representing a lower slope depositional setting. Some samples in the Pinery Limestone and overlying Lamar Limestone of the Bell Canyon Formation display normal polarity, with the first

Base of the Roadian Stage of the Guadalupian Series at Stratotype Canyon, Texas, U.S.A.

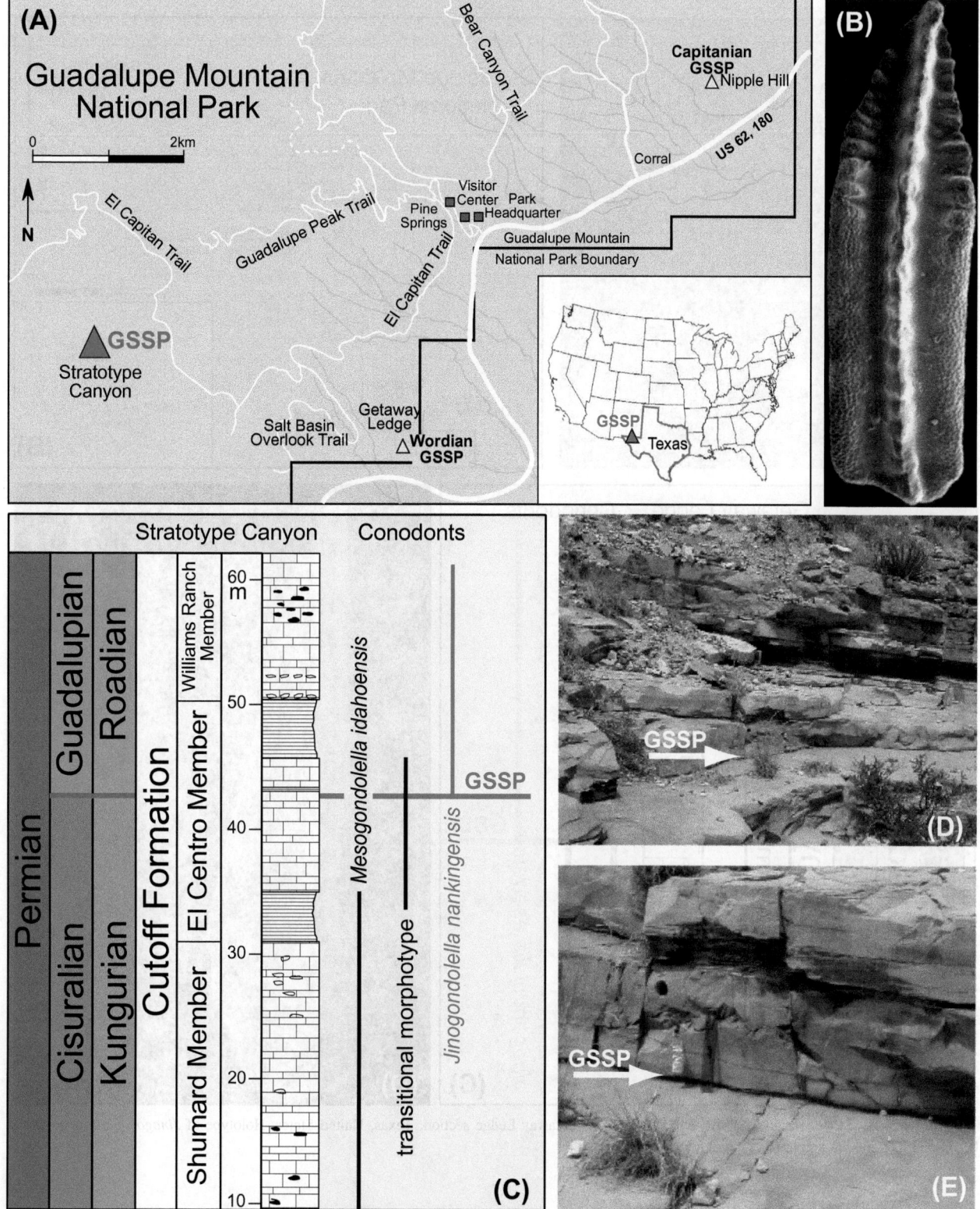

FIGURE 24.4 GSSP for the base of the Roadian Stage and of the Guadalupian Series at Stratotype Canyon, Texas, United States. The GSSP level coincides with the lowest evolutionary occurrence of conodont *Jinogondolella nankingensis*. Holotype of *J. nankingensis* is 1.0 mm long. Photographs and stratigraphic column are from Henderson et al. (2012a).

Base of the Wordian Stage of the Guadalupian Series at Getaway Ledge Section, Texas, U.S.A.

FIGURE 24.5 GSSP for the base of the Wordian Stage at Getaway Ledge section, Texas, United States. Holotype of *Jinogondolella aserrata* is 1.0 mm long on the left and 0.9 mm long on the right.

Base of the Capitanian Stage of the Guadalupian Series at Nipple Hill, Texas, U.S.A.

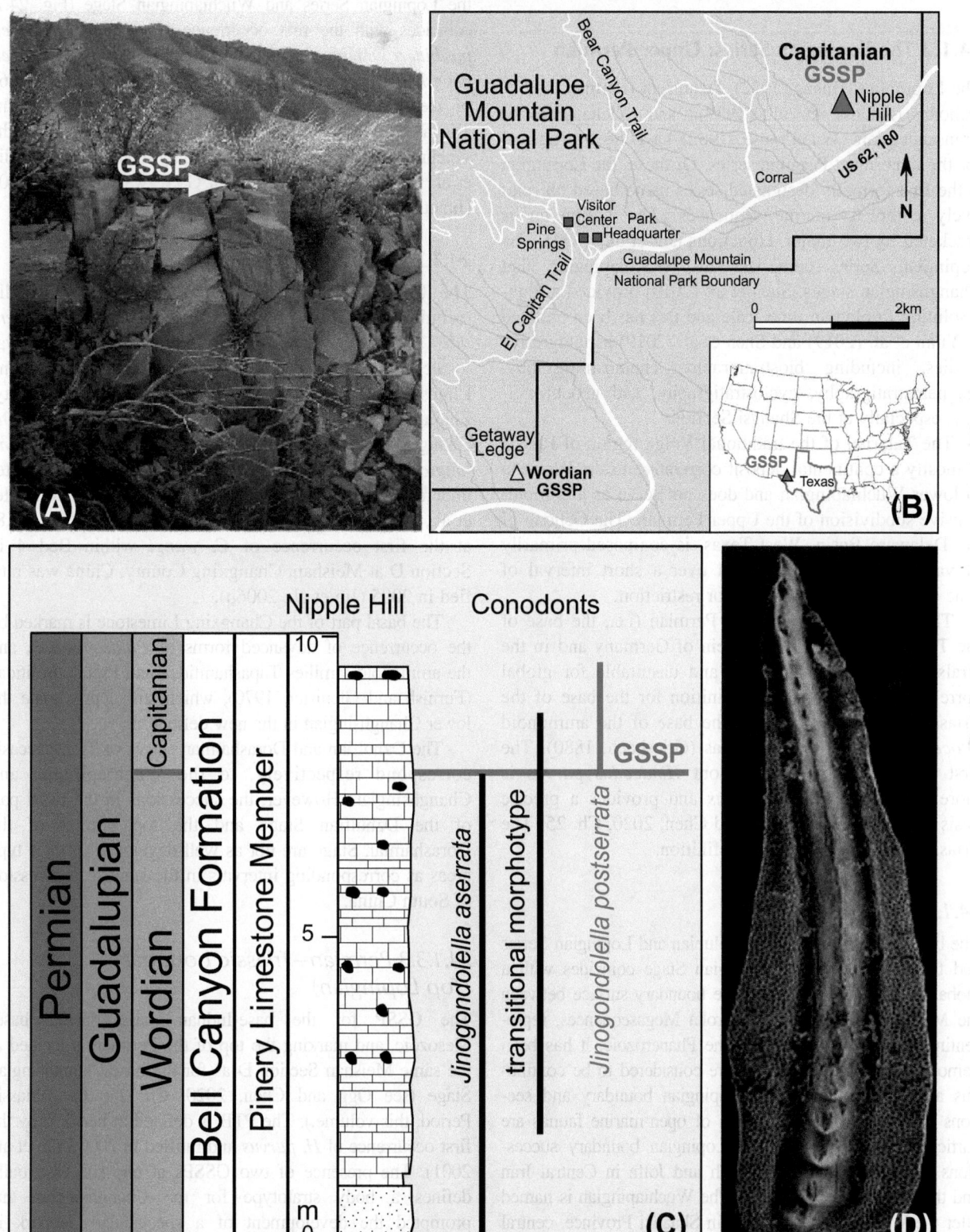

FIGURE 24.6   GSSP for the base of the Capitanian Stage at Nipple Hill, Texas, United States. Holotype of *Jinogondolella postserrata* is 1.1 mm long.

normal polarity indicative of the approximate position of the Illawarra Reversal occurring below the FAD of *J. postserrata* in the middle Wordian.

24.1.3 The Lopingian Series: Upper Permian

The Lopingian (Huang, 1932), Dzhulfian (Furnish, 1973), Ochoan (Adams et al., 1939), and Yichangian or Transcaucasian (Waterhouse, 1982) have been proposed for the uppermost Permian series. Of these, the Lopingian is the first formally designated series name based on relatively complete marine sequences. The Lopingian is bracketed by two major extinctions (Jin et al., 1994). The Lopingian Series comprises the Wuchiapingian and Changhsingian stages. Shen et al. (2010) provided a high-resolution Lopingian time scale and this has been updated in Yuan et al. (2019) and Shen et al. (2019) as integrative studies, including biostratigraphy, chemostratigraphy, magnetostratigraphy, cyclostratigraphy, and geochronology, especially at the Shangsi Section.

The Tatarian of the traditional Volga region of Russia is mostly a continental deposit correlative to the Wordian to lower Wuchiapingian and does not serve as a comprehensive subdivision of the Upper Permian. The Ochoan of the Delaware Basin, West Texas, is composed primarily of varved evaporites deposited over a short interval of time during sea-level lowstand or restriction.

The upper boundary of the Permian (i.e., the base of the Triassic) in the Buntsandstein of Germany and in the Urals of Russia is nonmarine and unsuitable for global correlation. The functional definition for the base of the Triassic was for a long time the base of the ammonoid *Otoceras* Zone of the Himalayas (Griesbach, 1880). The first appearance of the conodont *Hindeodus parvus* is more widespread than *Otoceras* and provides a precise basis for base-Triassic (Ogg and Chen, 2020, Ch. 25: The Triassic Period, this volume) definition.

24.1.3.1 Wuchiapingian

The boundary between the Guadalupian and Lopingian Series and the base of the Wuchiapingian Stage coincides with a global regression, that is, with the boundary surface between the Middle and the Upper Absaroka Megasequences, representing the lowest sea level of the Phanerozoic. It has been demonstrated that few sections are considered to be continuous across the Guadalupian−Lopingian boundary and sections with a complete succession of open-marine faunas are particularly rare. Guadalupian−Lopingian boundary successions were reported from Abadeh and Jolfa in Central Iran and the Salt Range in Pakistan. The Wuchiapingian is named after the Wuchiaping Limestone in Shaanxi Province, central China and traditionally correlated by the fusulinid *Codonofusiella* Zone (Sheng, 1963). The Laibin Syncline in

Guangxi Province, China includes the most complete and interregionally correlatable open-marine succession, including diverse fossils and detailed conodont zonation. The GSSP for the Lopingian Series and Wuchiapingian Stage (Fig. 24.7) coincides with the first occurrence of *Clarkina postbitteri postbitteri* within an evolutionary lineage from *C. postbitteri hongshuiensis* to *Clarkina dukouensis* at the base of Bed 6k of the Penglaitan section (Henderson et al., 2002). The Tieqiao (Iron Rail-Bridge) section on the western slope of the syncline is proposed as a supplementary reference section (Jin et al., 2001). The Wuchiapingian GSSP was ratified in 2004 (Jin et al., 2006a).

24.1.3.2 Changhsingian

The base of the Changhsingian Stage was originally recommended as the horizon between the *Clarkina orientalis* and the *Clarkina subcarinata* zones, which was located at the base of Bed 2 at the base of the Changxing Limestone in Section D at Meishan, Changxing County, Zhejiang Province, China (Zhao et al., 1981). Later the Changhsingian Stage was defined within the *Clarkina longicuspidata−Clarkina wangi* lineage based on a better understanding of conodont taxonomy and evolution (Mei et al., 2004). The GSSP for the Changhsingian (Fig. 24.8) at the first occurrence of *C. wangi* within Bed 4 in Section D at Meishan, Changxing County, China was ratified in 2005 (Jin et al., 2006b).

The basal part of the Changxing Limestone is marked by the occurrence of advanced forms of *Palaeofusulina*, and the ammonoid families Tapashanitidae and Pseudotirolitidae (Furnish and Glenister, 1970), which still approximate the lower Changhsingian in the new definition.

The Dzhulfian and Dorashamian stages of Transcaucasia correspond respectively, to the Wuchiapingian and Changhsingian. However, the successions in the basal part of the Dzhulfian Stage and the top portion of the Dorashamian Stage are not as well developed in their type areas as corresponding intervals in the standard succession of South China.

24.1.3.3 Permian−Triassic boundary (Top Lopingian)

The GSSP for the base-Induan, base-Triassic, base-Mesozoic, and marking the top of the Permian is located at the same Meishan Section D as for the base-Changhsingian Stage (see Ogg and Chen, 2020, Ch. 25: The Triassic Period, this volume.). The PTB is defined at bed 27c by the first occurrence of *H. parvus* and ratified in 2000 (Yin et al., 2001). The presence of two GSSPs at one site essentially defines a body stratotype for the Changhsingian that prompted the development of a spectacular Geopark in Changxing County, Zhejiang Province. The only problem with the GSSP location is condensation, but the boundary

Base of the Wuchiapingian Stage of the Permian System at Penglaitan Section, Southern China

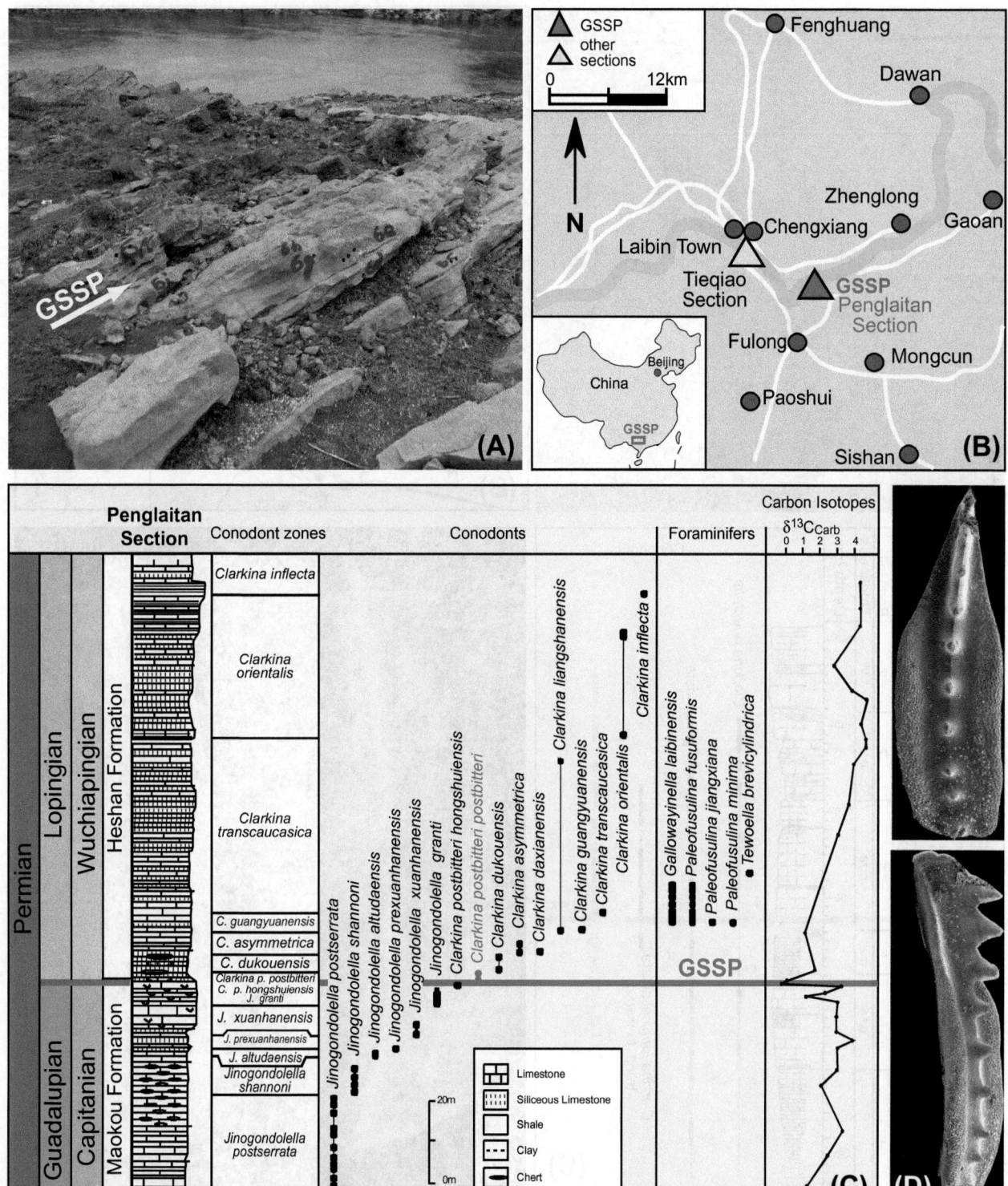

FIGURE 24.7 GSSP for the base of the Wuchiapingian Stage and of the Lopingian Series at Penglaitan section, Guangxi, southern China. The GSSP level coincides with the lowest occurrence of *Clarkina postbitteri postbitteri* within an evolutionary lineage. Holotype of *C. postbitteri* is 1 mm long. Photographs and stratigraphic column are from Henderson et al. (2012a).

Base of the Changhsingian Stage of the Permian System at Meishan Section D, Zhejiang Province, China

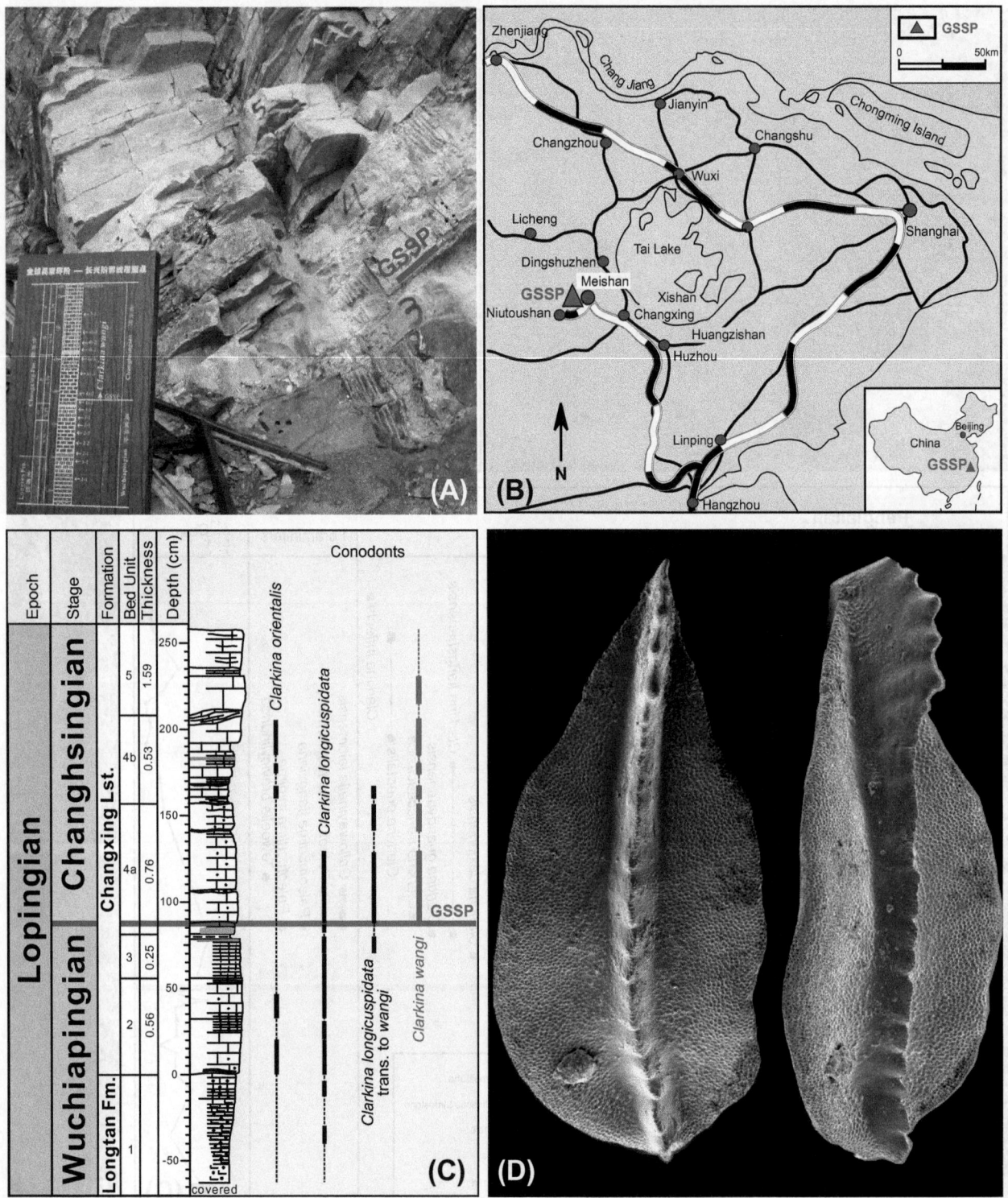

FIGURE 24.8 GSSP for the base of the Changhsingian Stage at Meishan section D, Zhejiang Province, China. Holotype of *Clarkina wangi* is approximately 1 mm long.

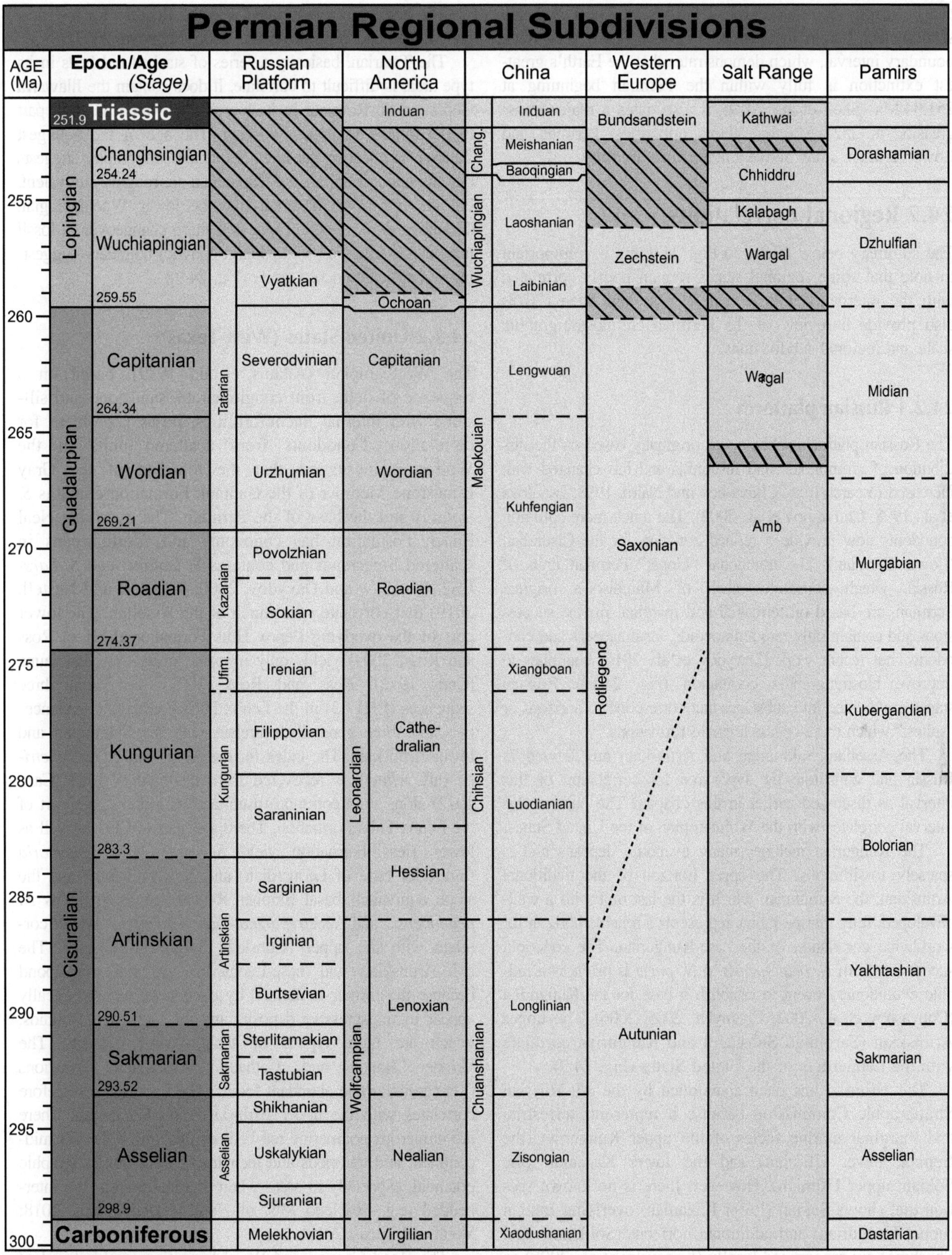

Permian Regional Subdivisions

AGE (Ma)	Epoch/Age (Stage)	Russian Platform	North America	China	Western Europe	Salt Range	Pamirs
251.9	**Triassic**		Induan	Induan	Bundsandstein	Kathwai	
	Lopingian — Changhsingian 254.24	*hiatus*	*hiatus*	Chang. / Meishanian / Baoqingian	*hiatus*	*hiatus* / Chhiddru	Dorashamian
255	Lopingian — Wuchiapingian 259.55	*hiatus* / Vyatkian (Tatarian)	*hiatus* / Ochoan	Wuchiapingian / Laoshanian / Laibinian	Zechstein	Kalabagh / Wargal	Dzhulfian
260	Guadalupian — Capitanian 264.34	Severodvinian (Tatarian)	Capitanian	Lengwuan (Maokouian)	Saxonian	*hiatus* / Wagal	Midian
265	Guadalupian — Wordian 269.21	Urzhumian (Kazanian)	Wordian	Kuhfengian	Saxonian	Amb	Murgabian
270	Guadalupian — Roadian 274.37	Povolzhian / Sokian (Kazanian)	Roadian		Saxonian	Amb	Murgabian
275	Cisuralian — Kungurian 283.3	Irenian (Ufim.) / Filippovian / Saraninian (Kungurian)	Leonardian: Cathedralian / Hessian	Xiangboan / Luodianian (Chihsian)	Rottliegend		Kubergandian / Bolorian
285	Cisuralian — Artinskian 290.51	Sarginian / Irginian / Burtsevian (Artinskian)	Leonardian: Hessian / Wolfcampian: Lenoxian	Longlinian (Chuanshanian)	Autunian		Yakhtashian
290	Cisuralian — Sakmarian 293.52	Sterlitamakian / Tastubian (Sakmarian)	Wolfcampian: Lenoxian	Longlinian	Autunian		Sakmarian
295	Cisuralian — Asselian 298.9	Shikhanian / Uskalykian / Sjuranian (Asselian)	Nealian	Zisongian	Autunian		Asselian
300	**Carboniferous**	Melekhovian	Virgilian	Xiaodushanian			Dastarian

FIGURE 24.9 Correlation chart of the Permian with international subdivisions and selected regional stage and substage nomenclature. Diagonal-lined pattern indicates a widespread regional hiatus.

can be readily correlated around the world. The Penglaitan section near Laibin in Guangxi Province has a very thick boundary interval, which demonstrates that the Earth's greatest extinction is fully within the Permian beginning at 251.94 Ma (Shen et al., 2019); it also dates a major transgression at 252.0 Ma that places uppermost Permian and Lower Triassic strata above a major unconformity.

24.2 Regional correlations

The summary below relates to Fig. 24.9, but it is important to note that some regional stages remain poorly correlated with the international time scale. Lucas and Shen (2018) also provide a review of the Permian chronostratigraphic scale and regional subdivisions.

24.2.1 Russian platform

The Russian platform scheme was originally based on the distribution of ammonoids and fusulinids and has changed with increased research (e.g., Chuvashov and Nairn, 1993; Esaulova et al., 1998; Chuvashov et al., 2002). The much more common conodonts now provide a refined zonation for the Cisuralian (Lower Permian). The traditional "Upper" Permian units of Russia, which constitute much of Murchison's original Permian, are based on terrestrial and marginal marine successions and contain only rare ammonoids, fusulinaceans and conodonts, but recent work (Davydov et al., 2016) continues to improve biostratigraphic correlation (Fig. 24.10). Russian stages are divided into substages that correspond to horizons or "suites," which more-or-less relate to formations.

The Asselian, Sakmarian and Artinskian are defined in Russia and constitute the type area for correlation of this interval as discussed earlier in this chapter. The bulk of this interval correlates with the Wolfcampian of the United States.

The Kungurian includes many evaporite deposits and is sparsely fossiliferous. The upper horizon of the traditional Artinskian, the Saraninian, which is the last unit with a well-developed, fully marine fauna, represents a reliable horizon for worldwide correlation of the base-Kungurian. The conodont succession from N. pequopensis to N. pnevi is taken as a reliable evolutionary event to establish a base for the Kungurian (Chuvashov et al., 2002; Chernykh, 2005, 2006). The upper Artinskian (Sarginian Substage) and Kungurian correlate with the Leonardian of the United States (Fig. 24.9).

The Ufimian has been abandoned by the All Russian Stratigraphic Commission because it represents terrestrial and marginal marine facies of the upper Kungurian (the Irenian, lower Ufimian) and the lower Kazanian (the Sokian, upper Ufimian). However, there is no known section that shows Sokian (lower Kazanian) overlying Irenian (upper Kungurian) and additional horizons (Solikamsk and Sheshma) are recognized locally (Lozovsky et al., 2006) so problems still exist in interpreting this interval.

The Kazanian is correlated with the Roadian for the most part (Leven and Bogoslovskaya, 2006; Leonova, 2007).

The Tatarian, basically a series of stacked red beds in its type area, is difficult to correlate. It does contain the Illawarra Geomagnetic Reversal in its lower part, which ties that part to the middle to upper Wordian. The age of the youngest Tatarian is an open question. Sequence stratigraphy suggests that the upper Tatarian and Capitanian are roughly equivalent, but the uppermost Tatarian may be lower Wuchiapingian (Esaulova et al., 1998). There is a sharp changeover in fossil taxa at the Tatarian—Vetluzhian (Triassic) boundary suggesting a significant unconformity (Fig. 24.9).

24.2.2 United States (West Texas)

The Wolfcampian (Adams et al., 1939), based on a sequence of delta front conglomerate, sandstone and siltstone with internal unconformities, poses a problem for correlation. Conodonts from scattered units in the Wolfcampian suggest that the top bed of the Gray Limestone Member of the Gaptank Formation contains S. isolatus and the base of the Permian. The overlying Neal Ranch Formation has conodonts and fusulinaceans in scattered limestones and contains S. isolatus and S. barskovi (Wardlaw and Davydov, 2000; Wardlaw and Nestell, 2019) that correlate with much of the Asselian. The lower part of the overlying Lenox Hills Formation (LH1 of Ross and Ross, 2009) yields only a sparse upper Asselian fauna (Carr, 1972). Ross and Ross (2003) recognized three sequences (LH1−3) in the Lenox Hills Formation; sequences two and three generally correlate with the Sakmarian and lower Artinskian. The index for the Artinskian, S. asymmetrica (aff. whitei), is recovered from the base of LH3 (Carr, 1972) along with common fusulinaceans in the upper part of the Lenox Hills Formation. The upper part of LH3 as well as lower Hess Formation yield Schwagerina crassitectoria (index for base of Leonardian) and N. pequopensis and the slope equivalent basal Skinner Ranch Formation yields N. pequopensis and Neostreptognathodus exsculptus, which correlate with the upper Artinskian Sarginian Substage. The base-Artinskian and base-Leonardian do not correspond because the former is defined by conodonts, which typically appear in transgressive deposits and the latter by fusulinids, which are more typical of shallow-water deposits. The Skinner Ranch and Cathedral Mountain formations (Leonardian) yield abundant fossils. The Leonardian therefore correlates with the upper Artinskian and Kungurian. There are numerous nonmarine red beds of this age in the US midcontinent and tetropods and their tracks offer biostratigraphic potential, especially in areas where marine deposits are interbedded (e.g., Robledo Mts. of New Mexico; Lucas, 2018; Voigt and Lucas, 2018).

The Road Canyon and Word formations yield abundant and diverse fossils. The Roadian, Wordian and Capitanian

(Guadalupian Series) are defined in west Texas and constitute the type area for correlation of this interval as discussed earlier in this chapter. Slope and basinal equivalents of the Vidrio Formation and Capitan Limestone also yield excellent fauna. Conodonts and radiolaria are documented in the Reef Trail Member of the Bell Canyon Formation that indicate proximity to the Guadalupian−Lopingian boundary (Lambert et al., 2010; Maldonado and Noble, 2010; Nestell et al., 2019).

The Ochoan (Fig. 24.9), based on basin-filling evaporites, poses a problem for correlation. In the Tansill Formation (the shelf equivalent of the Lamar Limestone Member of the Bell Canyon Formation), there is a sparse fauna dominated by a species of *Sweetina* (Croft, 1978). The same species is found overlying the evaporites of the Ochoan, within the Rustler Formation (above the Castille and Salado Formations), implying that the deposition of the basin-filling evaporites occurred in much less than one conodont zone.

24.2.3 South China

Three regional series were recognized in South China (Fig. 24.9), including Chuanshanian, Yangsingian (usually divided into Chihsian and overlying Maokouian), and Lopingian (Sheng and Jin, 1994). There are some issues with correlation of these mostly fusulinid-based series with the mostly conodont-defined international geologic time scale (GTS) as is also the case for the Wolfcampian−Leonardian in the United States. In addition, these series, like formations or groups, may be somewhat diachronous. For example, the Chihsian is typically considered to be transgressive at the base and correlative with the Kungurian, but there are two major transgressions above a maximum regressive surface associated with peak Early Permian glaciation that are dated in other regions as early Sakmarian and early Artinskian; the Kungurian is generally a relative lowstand of sea level.

The Chuanshanian includes two regional stages, including the Zisongian and Longlinian. The Zisongian is based on the fusulinid *Pseudoschwagerina uddeni−Pseudoschwagerina texana* Zone and the *Sphaeroschwagerina* Zone. In many areas of China this unit is correlated with the Maping/ Chuanshan Formations, which includes cyclic, light-colored, thick-bedded limestone deposited on a shallow-water platform and correlates with the Asselian and Sakmarian. At the Tieqiao section, Guangxi Province (Shen et al., 2007), transgressive deposits (carbonaceous shale and organic and phosphate-rich carbonate of the Liangshan Member) overlie the Maping and have been correlated with the Chihsian, but the fusulinid-defined Chihsian base is higher. This transgression is correlated as Artinskian (*S. asymmetrica* ∼34 m above base; Sun et al., 2017), based on the occurrence of *Pamirina*, but the fusulinid genera *Leeina* and *Darvasites* were recognized below and in the basal few meters of the transgression (Shen et al., 2007). These initial transgressive

deposits correlate with the upper Longlinian, which is the upper stage of the Chuanshanian and based on the biostratigraphic sequence between the last *Pseudoschwagerina* and the first *Misellina*. The Zisongian and Longlinian are collectively referred to the Chuanshanian Series and are approximately equal to the Asselian, Sakmarian and part of the Artinskian (Fig. 24.9) (Shen et al., 2019).

The Luodianian Stage traditionally represents the lower Chihsian and is based on the fusulinid *Brevaxina* and *Misellina* zones. At Tieqiao this stage occurs well above the initial transgression and may correlate with the upper Artinskian and much of the Kungurian. The overlying Xiangboan is based on the fusulinid *Cancellina* Zone and correlates with upper Kungurian. The Luodianian and Xiangboan are collectively referred to the Chihsian (lower Yangsingian).

The Maokouan (upper Yangsingian Series) includes the Kuhfengian and Lengwuan (Fig. 24.9). The Kuhfengian is based on the first occurrence of the conodont *Jinogondolella nankingensis*, which correlates with the Roadian, but in some areas of South China the local first occurrence of *J. nankingensis* may be upper Roadian (Henderson and Mei, 2003). The Kuhfengian also includes the range of *J. aserrata*, which correlates with the Wordian. The Lengwuan includes the ranges of the conodonts *J. postserrata*, *J. shannoni* and *J. xuanhanensis* and is correlated with the Capitanian.

The Wuchiapingian and Changhsingian (Lopingian Series) are defined in south China and constitute the type area for correlation of this interval as discussed earlier in this chapter. The Wuchiapingian is based on the first occurrence of the conodont species *C. postbitteri*, just above the origin of this genus. The Changhsingian is based on the first occurrence of the conodont *C. wangi* near the base of the Changxing Limestone.

24.2.4 Western Europe (Germanic Basin)

The Germanic Basin is only briefly dealt with here. The nonmarine succession, including Autunian, Saxonian and Thuringian, are difficult to reconcile with the marine GTS. Cyclostratigraphy and conchostracan biostratigraphy may be among various techniques to offer greater correlation potential (Legler and Schneider, 2013; Schneider and Scholze, 2018). Paleobotanical evidence suggests that the Autunian is at least Sakmarian (Broutin et al., 1999). The Illawarra Reversal is within the upper part of the Rotliegend. Zechstein 1 (Figs. 24.1 and 24.9) contains a fairly diverse marine microfauna that includes the conodonts *Mesogondolella britannica* and *Merrillina divergens*. This assemblage correlates with the early Wuchiapingian (Mei and Henderson, 2001). The duration of the remaining Zechstein units is unknown; the evaporites could represent very short depositional intervals between long hiatuses, or a very short interval similar to that of the

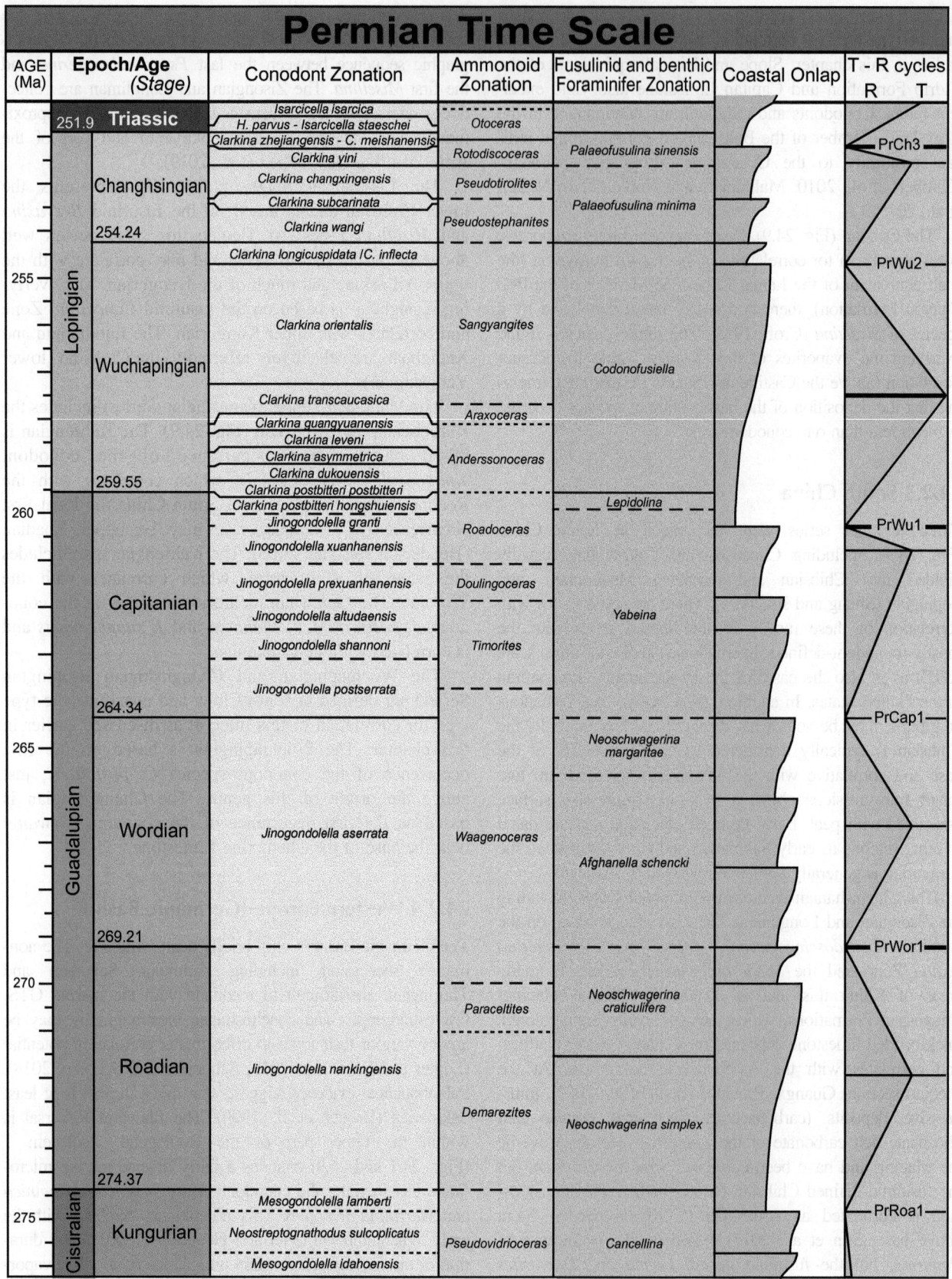

FIGURE 24.10a Permian time scale with conodont and fusulinacean biostratigraphy, coastal onlap and T–R cycles. *T–R*, Transgressive–regressive.

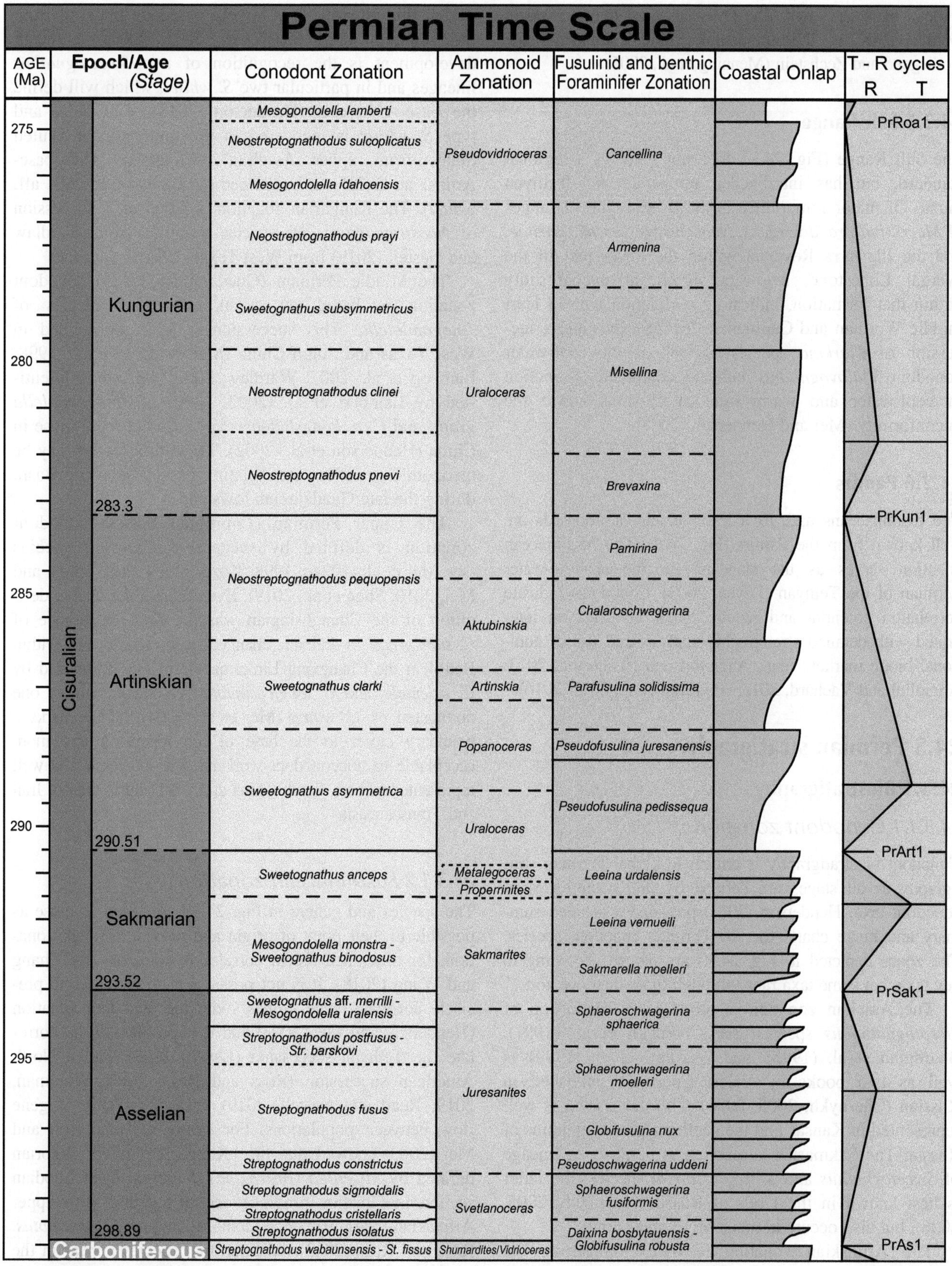

Permian Time Scale

AGE (Ma)	Epoch/Age (Stage)		Conodont Zonation	Ammonoid Zonation	Fusulinid and benthic Foraminifer Zonation	Coastal Onlap	T - R cycles R T
275	Cisuralian	Kungurian	Mesogondolella lamberti	Pseudovidrioceras	Cancellina		PrRoa1
			Neostreptognathodus sulcoplicatus				
			Mesogondolella idahoensis				
			Neostreptognathodus prayi		Armenina		
			Sweetognathus subsymmetricus	Uraloceras			
280			Neostreptognathodus clinei		Misellina		
			Neostreptognathodus pnevi		Brevaxina		
		283.3			Pamirina		PrKun1
285		Artinskian	Neostreptognathodus pequopensis	Aktubinskia	Chalaroschwagerina		
			Sweetognathus clarki	Artinskia	Parafusulina solidissima		
				Popanoceras	Pseudofusulina juresanensis		
			Sweetognathus asymmetrica		Pseudofusulina pedissequa		
290				Uraloceras			
		290.51					PrArt1
		Sakmarian	Sweetognathus anceps	Metalegoceras / Properrinites	Leeina urdalensis		
			Mesogondolella monstra - Sweetognathus binodosus	Sakmarites	Leeina vernueili		
		293.52			Sakmarella moelleri		PrSak1
		Asselian	Sweetognathus aff. merrilli - Mesogondolella uralensis		Sphaeroschwagerina sphaerica		
295			Streptognathodus postfusus - Str. barskovi	Juresanites	Sphaeroschwagerina moelleri		
			Streptognathodus fusus		Globifusulina nux		
			Streptognathodus constrictus		Pseudoschwagerina uddeni		
			Streptognathodus sigmoidalis	Svetlanoceras	Sphaeroschwagerina fusiformis		
			Streptognathodus cristellaris				
		298.89	Streptognathodus isolatus		Daixina bosbytauensis - Globifusulina robusta		PrAs1
	Carboniferous		Streptognathodus wabaunsensis - St. fissus	Shumardites/Vidrioceras			

FIGURE 24.10b (Continued).

Ochoan of West Texas. The Dyas has been reintroduced in Germany as a lithostratigraphic unit comprising the Rotliegend and Zechstein (Menning et al., 2006).

24.2.5 Salt Range

The Salt Range (Fig. 24.9) does not serve as a regional standard, but has interbedded temperate and Tethyan fauna. Of major importance is the overlapping of ranges of *Merrillina praedivergens*, *Neoschwagerina margaritae*, and the Illawarra Reversal within the lower part of the Wargal Limestone, below a significant unconformity within that formation, indicating correlation with at least middle Wordian and Capitanian. *Iranognathus* and a succession of *Clarkina* species as well as the cool-water conodont *Vjalovognathus* indicate continued alternation of cool-water and warm-water successions above the unconformity (Mei and Henderson, 2001).

24.2.6 Pamirs

The fusulinaceans and, to a lesser extent, ammonoids are well known from the Pamirs (Fig. 24.9). The fusulinacean zonation serves as the standard for the shallow-water Permian of the Tethyan (Leven, 2003). Conodonts indicate correlation potential and require major investigation integrated with detailed stratigraphic analysis and use of additional biotic markers (e.g., Angiolini and Henderson, 2013; Angiolini and Vachard, 2015; Angiolini et al., 2015, 2016).

24.3 Permian stratigraphy

24.3.1 Biostratigraphy

24.3.1.1 Conodont zonation

Conodont biostratigraphy is critical to global Permian stratigraphy as all stages are defined by first occurrences of conodont taxa. Henderson (2018) provided a detailed summary and range charts for key Permian conodont species. The zones depicted in Fig. 24.10 are named according to key taxa, but some taxa range through more than one zone.

The Asselian zonation is based on a succession of *Streptognathodus* species; see Chernykh et al. (1997), Boardman et al. (1998), and Wardlaw et al. (1999) as well as two books by Valeri Chernykh published in Russian (Chernykh, 2005, 2006). This succession is well represented in Kansas and the southern Ural Mountains of Russia. The Sakmarian succession is based on a lineage of *Sweetognathus* and *Mesogondolella* species; the latter is best known in the Urals of Russia (Chernykh, 2005, 2006), but also occurs in many other areas now. The Artinskian zonation is based on species of *Sweetognathus* and *Neostreptognathodus* and reflects the major changeover in the forms dominating shelf faunas

during this interval. It is largely based on material from the Southern Urals (Chernykh, 2005, 2006). A recent development is the recognition of two *Sweetognathus* lineages and in particular two *S. whitei*, which will define the base-Artinskian (Henderson, 2018). The older and type *S. whitei* is recovered in the midcontinent United States from upper Asselian cyclothems. The base-Artinskian GSSP will be defined by *S. asymmetrica* (*S.* aff. *whitei*). The Kungurian zonation is based on a succession of *Neostreptognathodus* species modified from Wardlaw and Nestell (2019) from West Texas, USA.

The Middle Permian (Guadalupian Series) conodont zonation is based on a succession of species of *Jinogondolella*. This succession is well represented in West Texas and South China (Mei et al., 1994a,b, 1998; Lambert et al., 2007; Wardlaw, 2000). Specimens identified by Lambert et al. (2002, 2010) of *Jinogondolella granti* and *C. p. hongshuiensis* are different from those in China (Henderson et al., 2002). These differences may be taxonomic or reflect the isolation of the Delaware Basin during the late Guadalupian lowstand.

The Upper Permian (Lopingian Series) conodont zonation is defined by successive *Clarkina* species (see Mei et al., 1994c, 1998; Kozur, 2004, 2005; Shen and Mei, 2010; Shen et al., 2019). Even though the classic definition of the Changhsingian was the first appearance of *C. subcarinata*, a dramatic changeover in fauna occurs within Bed 4 in the Changxing Limestone from one dominated by *C. orientalis* and *C. longicuspidata* (Wuchiapingian) to one dominated by *C. wangi* (Mei et al., 2004), which marks a boundary closer to the base of the formation and more acceptable to nonconodont workers. This succession is well represented in both South China and the Dzhulfa area of Iran and Transcaucasia.

24.3.1.2 Fusulinacean zonation

The species and genera in Fig. 24.10 are listed as close as possible to their point of origin and provide a rough zonation; for a detailed recent correlation summary see Zhang and Wang (2018). It is not possible in many cases to precisely correlate these "zones" with the conodont zonation (Henderson and Mei, 2003) and it is problematic to correlate the Tethyan succession (Leven, 2003) with a North American succession (Ross and Ross, 2003; Wahlman, 2019; Read and Nestell, 2019) because of lack of gene flow between populations. For example, Henderson and Mei (2003) showed that the "Kungurian" lower Bolorian defined by *Misellina (Brevaxina) dyhrenfurthi* at Luodian section in Guizhou province occurs together with upper Artinskian (Sarginian) conodonts, including *N. exsculptus*. Ross and Ross (2003) indicated that species appear in the *M. (Brevaxina) dyhrenfurthi* Zone in Safetdaron Formation of Darvas region that are morphologically similar to lower

Hess fusulinids in N.Amer., associated with *S. crassitectoria* and *S. guembeli*.*S. crassitectoria* defines the base of the Leonardian in the United States and conodonts from the same level include *N. pequopensis* and *N. exsculptus*.

In the southern Urals, sequence boundaries coincide with the bases of several fusulinacean zones and eustatic lowstands correspond to significant fusulinacean extinction events. The base of the Asselian (i.e., the base of the Permian), the base of the Sakmarian, and some fusulinacean zones coincide with highstands. Sea-level lowstands may have been very stressful for global fusulinacean assemblages and a catalyst for both speciation and extinction. Highstands may have created environmental opportunities and appear to be more closely associated with fusulinacean speciation than extinction. Sequence boundaries located within fusulinacean zones, perhaps reflect local tectonism or local climatic changes. Significant turnovers occur in the Kungurian, in particular from schwagerinids to neoschwagerinids. In addition, all of the large fusulinids become extinct during the upper Capitanian. Subsequently the smaller fusulinids *Codonofusiella* (Wuchiapingian) and *Palaeofusulina* (Changhsingian) are all extinct by the end-Permian mass extinction.

24.3.2 Physical stratigraphy

24.3.2.1 Cyclostratigraphy

Waxing and waning of the major continental glaciers generated significant high-frequency and high-magnitude cyclothems that typically follow a long-eccentricity 405-kyr cycle. These cyclothems offer a possible means of high-resolution correlation and stratigraphic tuning in the Asselian (Fig. 24.1) when integrated with conodonts and geochronology (Schmitz and Davydov, 2012; Henderson, 2018).

24.3.2.2 The Illawarra geomagnetic polarity reversal

As mentioned earlier, the Illawarra geomagnetic reversal (Fig. 24.1) is an important tie point within the Guadalupian series at about middle Wordian. The reversal is within the *J. aserrata* Zone in West Texas and near the top of the *Merr. praedivergens* Zone in the Salt Range.

In the Guadalupe Mountains a tuff near the projected position of the Illawarra, within the range of *J. aserrata* and below the first occurrence of *J. postserrata*, the indicator for the Capitanian, yields a date of 265.46 ± 0.27 Ma (Ramezani and Bowring, 2018).

24.3.2.3 Geochemistry

The seawater $^{87}Sr/^{86}Sr$ curve (Fig. 24.1) indicates that a minimum value for all of Phanerozoic time occurred during the Capitanian (Henderson et al., 2012a). The driving

mechanism of the Late Permian variations appears to have been climate change rather than tectonic; Pangea was assembled and fairly stable through this interval with little mid-ocean ridge activity that would normally produce a decrease in the isotopic values. The period of decreasing seawater $^{87}Sr/^{86}Sr$ for the Early−Middle Permian begins with a waning ice age and continues in association with high continental aridity and low external runoff attributed to the huge Pangean landmass. The increase in the isotopic values during the Late Permian and Early Triassic may be related to increased continental weathering by climate changes associated with Siberian Traps volcanism.

The Asselian is marked by high-frequency cyclothems associated with glacial eustasy, and this is consistent with higher values in oxygen isotopes indicating cooler temperature. A fluctuating trend around a mean $\delta^{18}O$ value of −2 begins in the Sakmarian and continues through much of the remaining Permian indicating an overall warmer climate following glaciation. Though not apparent in the oxygen-18 curve, there is faunal evidence for significant cooling after the early Roadian in the Sverdrup Basin continuing for most of the rest of the Permian (Beauchamp, 1994; Henderson, 2002) and cooling in the Capitanian in the Salt Range, indicating bipolarity in cooling during much of the Middle Permian. There is considerable evidence for a runaway greenhouse and significant increase in temperatures in the latest Permian and Early Triassic.

Marine extinctions began as a stepwise function following the Guadalupian (Zhou et al., 1996). This suggests significant temperature fluctuations and the beginning of mass extinction coincided with the increase in $^{87}Sr/^{86}Sr$ (McArthur et al., 2012; Wang et al., 2018). This further suggests an amelioration of the climate, leading to a decrease in continentality, an increase in precipitation, and an overall declining sea level during the Upper Permian, with more area exposed to erosional processes.

There are many excursions in the global carbon cycle that overall show a negative trend, but with significant positive excursions in the Sakmarian, Roadian and early Changhsingian. There are significant negative excursions in the Kungurian, early Wuchiapingian, and the most significant negative carbon excursion is just below the Permian−Triassic boundary. The latter excursion is related to the final extinction event of the Permian indicating a sharp decrease in productivity and burial of organic carbon at the very end of the Permian. Although there is still some uncertainty associated with the causes leading to the Earth's greatest extinction (Erwin, 1993, 1995), there is a general consensus emerging that it is associated with effects related to Siberian Trap volcanism (Burgess and Bowring, 2015; Burgess et al., 2017) and that it occurred very rapidly (Burgess et al., 2014; Shen et al., 2019) and possibly slightly diachronous globally (Algeo et al., 2012).

24.4 Permian time scale

24.4.1 Previous scales

Section 24.4.1 on the geologic time scale in the chapter on the Permian Period in GTS2012 started with the statement that "although precise age constraints are in place for the base and top of the Permian, the Permian time scale is among the least internally constrained in the Phanerozoic" (Henderson et al., 2012a). Fortunately, much has happened since then, not least since new ID-TIMS dates have been generated for the Middle Permian, previously devoid of age dates. Hence, the younger Artinskian through Wordian interval is better and more precisely constrained, with the Artinskian–Kungurian boundary now 4 Myr older than in GTS2012 and the Kungurian–Roadian stage boundary 2.1 Myr older. All new stage ages have 0.4 Myr uncertainties, reflecting uniform uncertainty around this number for a majority of Permian ID-TIMS dates. Ironically, although the Kungurian itself is the longest Permian stage (8.9 Myr), it still has no internal age dates. But the almost linear Middle Permian spline fit provides fair interpolation (Fig. 23.8 in Ch. 23 - The Carboniferous Period (this volume)). In 2012, 7 out of 10 Permian stage boundaries had a ratified stratigraphic definition. This has increased to eight, with the Sakmarian now also having a GSSP, as discussed in Section 24.1.1.2.

24.4.2 Radioisotopic dates

The detailed and high-resolution conodont–foraminifer–ammonoid zonation for the Carboniferous–Permian with over 50 composite standard levels from Tournaisian through Changhsingian is now constrained in time with 84 ID-TIMS radioisotopic dates, as shown in Table 23.2 in (Aretz et al. 2020, Ch.23: The Carboniferous Period, this

volume). This is a significant improvement over GTS2012 with only 45 age dates and more spread in error values. The Permian Period alone now harbors 37 instead of 13 age dates, with 9 new ones by Baresel et al. (2017) for the Wuchiapingian, and 8 important ones for the Sakmarian through Wordian by Schmitz and Davydov (2012), Davydov et al. (2018), Wu et al. (2017), and Nicklen et al. (2015). Fifteen dates alone are in the Changhsingian, reflecting the strong interest in the P/T boundary (see Section 24.1.3.3), with the Permian–Triassic boundary being interpolated precisely at 251.9 ± 0.3 Ma in Table 24.1 (Burgess et al., 2014); the age was 252.2 ± 0.5 Ma in GTS2012 (Shen et al., 2011).

24.4.3 Permian composite standard

The composite standard for the Permian is an extension of that for the Carboniferous and builds on that for GTS2012 (Davydov et al., 2012), as detailed in (Aretz et al. 2020, Ch.23: The Carboniferous Period, this volume). The composite standard derives from the detailed and high-resolution conodont–foraminifer–ammonoid zonation for the Carboniferous–Permian, with over 50 composite standard levels from Tournaisian through Changhsingian listed in Table 23.2.

For the purpose of GTS2020 the composite standard of Davydov et al. (2012) was updated with the 84 radio-isotopic dates for GTS2020. The composite value for base-Sakmarian was changed from 1987.05 to 2028.15, making it slightly younger in agreement with the new ratified GSSP definition for the Sakmarian Stage. Also, the ammonoid *Sverdrupites harkeri* was added to the composite at the level where *J. nankingensis* appears (Davydov et al., 2018) at composite level 2354.18. This strengthens the best fit in the Artinskian–Roadian interval.

TABLE 24.1 GTS2020 scale for the Permian Period with comparison of ages from GTS2012.

Base of stage	GTS2012			GTS2020	
	Age (Ma)		Duration	Age (Ma)	Duration
Induan (Triassic)	252.2 ± 0.5			251.9 ± 0.3	
Changhsingian	254.2 ± 0.3		2	254.2 ± 0.4	2.3
Wuchiapingian	259.8 ± 0.4		5.6	259.5 ± 0.4	5.3
Capitanian	265.1 ± 0.4		5.3	264.3 ± 0.4	4.8
Wordian	268.8 ± 0.5		3.7	269.2 ± 0.4	4.9
Roadian	272.3 ± 0.5		3.5	274.4 ± 0.4	5.2
Kungurian	279.3 ± 0.6		7	283.3 ± 0.4	8.9
Artinskian	290.1 ± 0.2		10.7	290.5 ± 0.4	7.2
Sakmarian	295.5 ± 0.4		5.5	293.5 ± 0.4	3
Asselian	298.9 ± 0.2		3.4	298.9 ± 0.4	5.4

24.4.4 Age of stage boundaries

Analysis of new bentonite samples from Devonian–Carboniferous transition sections provides a revised age estimate for the Devonian–Carboniferous boundary and base-Tournaisian Stage of 359.3 ± 0.3 Ma (was 358.9 ± 0.4 Ma in GTS2012). This age estimate anchors the lower end of the Carboniferous–Permian interval; the anchor for the upper end is the age of the Permian–Triassic boundary of 251.9 Ma. The boundary was interpolated with level T1 at 251.88 ± 0.3 and with level P37 at 251.94 ± 0.3 in the Meishan GSSP section in China to yield 251.9 ± 0.3 Ma.

In the final analysis for the age of the stage boundaries and stage durations in the Permian (and Carboniferous) Periods, the chronometric and stratigraphic ages of the radio-isotopic dates in Table 23.2 (Ch. 23: The Carboniferous Period, this volume), were subjected to a cubic spline fit. The method is described in (Agterberg et al. 2020, Ch. 14A, Geomathematics and statistical procedures, this book), as modified and updated from the spline fitting and error analysis approach used in GTS2012. With the endpoints in place the best fit line for age versus stage is shown in Fig. 23.1.

All of the 50 points are close to or on the best fit line, which provides a simple check on the degree of fit prior to age interpolation. Interpolated stage boundary ages with their error bars and durations are given in Table 24.1; the table also provides age and duration of the stages in GTS2012, with sizable differences for Sakmarian, Kungurian and Roadian.

The base-Permian and base-Asselian interpolate at 298.9 ± 0.4 Ma (Ramezani et al., 2007), an identical age, but with larger uncertainty than in GTS2012. Lower boundary ages of all Permian stages have 0.4 Myr uncertainties; the number will not be repeated below. The Asselian lasted 5.4 Myr (was 3.4 Myr in GTS2012). The Sakmarian lasted 3 Myr (was 5.5 Myr in GTS2012) with the Asselian–Sakmarian boundary being 293.5 Ma in age; it was 295.5 in GTS2012, before the revised GSSP definition. The Artinskian Stage ranges from 290.5 to 283.3 Ma, for a duration of 7.2 Myr (was 10.7 Myr in GTS2012). In GTS2012 the Artinskian started at 290 Ma. The Kungurian lasted for 8.9 Myr (was 7 Myr in GTS2012) from 283.3 to 274.4 Ma. In GTS2012 the lower stage boundary had an age of 279.3 ± 0.6 Ma, 4 Myr younger than in GTS2020.

Stage	GSSP Location	Latitude, Longitude	Boundary Level	Correlation Events	Reference
GSSPs of the Permian Stages, with location and primary correlation criteria					
Changhsingian	Meishan, Zhejiang Province, South China	31°4'55"N 119°42'22.9"E	base of Bed 4a-2, 88 cm above the base of Changxing Limestone at the Meishan D Section	Conodont, FAD of *Clarkina wangi*	Episodes **29/3**, 2006
Wuchiapingian	Penglaitan, Guangxi Province, South China	23°41'43"N 109°19'16"E	base of Bed 6k in the Penglaitan Section	Conodont, FAD of *Clarkina postbitteri postbitteri*	Episodes **29/4**, 2006
Capitanian	Nipple Hill, SE Guadalupe Mountains, Texas, U.S.A	31°54'32.8"N 104°47'21.1"W	4.5 m above the base of the outcrop section of the Pinery Limestone Mbr of the Bell Canyon Formation	Conodont, FAD of *Jinogondolella postserrata*	
Wordian	Guadalupe Pass, Texas, U.S.A	31°51'56.9"N 104°49'58.1"W	17.6 m above the base of the Getaway Ledge outcrop section of the Getaway Limestone Mbr of the Cherry Canyon Formation	Conodont, FAD of *Jinogondolella aserrata*	
Roadian	Stratotype Canyon, Texas, U.S.A	31°52'36.1"N 104°52'36.5"W	42.7 m above the base of the Cutoff Formation	Conodont, FAD of *Jinogondolella nankingensis*	
Kungurian	*candidate Mechetlino Quarry section, Russia*	55°21'42"N 57°59'57"E	Bed 19	*FAD of conodont* Neostreptognathodus pnevi	
Artinskian	*candidate Dal'ny Tulkas section, Russia*	53°53'18.5"N 56°30'58.15"E	2.7 m above the base of Bed 4	*FAD of conodont* Sweetognathus asymmetrica	
Sakmarian	Usolka section, Russia	53°55'28.86"N 56°31'43.38"E	55.4 m above the base of the Usolka section in Bed 26/3	Conodont, FAD of *Mesogondolella monstra*	Episodes accepted 2020
Asselian	Aidaralash Creek, Kazakhstan	50°14'45"N 57°53'29"E*	27m above the base of Bed 19	Conodont, FAD of *Streptognathodus isolatus*	Episodes **21/1**, 1998

* according to Google Earth

FIGURE 24.11 GSSP chart providing information about location, definition, and official publication.

The Roadian Stage started at 274.4 Ma (was 272.3 Ma in GTS2012) and lasted 5.2 Myr (was 3.5 Myr in GTS2012). The Wordian ranges from 269.2 to 264.3 Ma for a duration of 4.9 Myr; in GTS2012 the duration was 3.7 Myr and the lower stage boundary age was 268.8 ± 0.5 Ma. The Capitanian Stage starts at 264.3 Ma for 4.8 Myr; in GTS2012 the stage lasted 5.3 Myr, starting at 265.1 Ma and ending at 259.5 Ma. The Wuchiapingian lasted 5.3 Myr from 259.5 to 254.2 Ma, and Changhsingian started at 254.2 Ma and ended at the Permian–Triassic boundary at 251.9 ± 0.3 Ma. The Changhsingian Stage is the shortest stage in the Permian with only a 2.3 Myr duration, and it is one of the shortest in the Phanerozoic.

The Cisuralian Epoch stretches between 298.9 and 274.4 Ma and has a duration of 24.4 Myr. The Guadalupian Epoch ended at 259.5 Ma and is 14.9 Myr long. The upper epoch of the Permian, the Lopingian, lasted 7.6 Myr, as in GTS2012. The total duration of the Permian Period is 47 Myr, from 298.9 to 251.9 Ma.

The geochronologic scale for the Permian differs substantially from that in GTS2012 for the lower boundary ages of Sakmarian, Kungurian and Roadian stages; and much less so, or not at all for the other six stages. Particularly for the Kungurian, there is a need for stratigraphically meaningful radioisotopic dating information as well as more detailed biostratigraphy.

Bibliography

Adams, J.E., Cheney, M.G., DeFord, R.K., Dickey, R.I., Dunbar, C.O., Hills, J.M., et al., 1939, Standard Permian section of North America. *The American Association of Petroleum Geologists Bulletin*, **23**: 1673–1681.

Algeo, T.J., Henderson, C.M., Ellwood, B., Rowe, H., Elswick, E., Bates, S., et al., 2012, Evidence for a diachronous Late Permian marine crisis from the Canadian Arctic region. *Geological Society of America Bulletin*, **124** (9–10): 1424–1448.

Angiolini, L., Campagna, M., Borlenghi, L., Grunt, T., Vachard, D., Vezzoli, G., et al., 2016, Brachiopods from the Cisuralian-Guadalupian of Darvaz, Tajikistan and implications for Permian stratigraphic correlations. *Palaeoworld*, **25**: 539–568.

Angiolini, L., and Henderson, C.M., 2013, The once and future quest: looking for mid-Permian correlation between the Tethyan and the International Time Scales. *Permophiles*, **57**: 19–20.

Angiolini, L., and Vachard, D., 2015, The Permian sedimentary successions of the Pamir Mountains, Tajikistan. *Permophiles*, **62**: 15–18.

Angiolini, L., Zanchi, A., Zanchetta, S., Nicora, A., Vuolo, I., Berra, F., et al., 2015, From rift to drift in south Pamir (Tajikistan): Permian evolution of a Cimmerian terrane. *Journal of Asian Earth Sciences*, **102**: 146–169.

Agterberg, F.P., da Silva, A.C., and Gradstein, F.M., 2020, Chapter 14A-Geomathematics and statistical procedures. *In* Gradstein, F.M., Ogg, J.G., Schmitz, M.D., and Ogg, G.M. (eds), *The Geologic Time Scale 2020.* **Vol. 1** (this book). Elsevier, Boston, MA.

Aretz, M., Herbig, H.-G., and Wang, X.D., 2020, Chapter 23 - The Carboniferous Period. *In* Gradstein, F.M., Ogg, J.G., Schmitz, M.D., and Ogg, G.M. (eds), *The Geologic Time Scale 2020.* **Vol. 2** (this book). Elsevier, Boston, MA.

Bagherpour, B., Bucher, H., Schneebeli-Hermann, E., Vennemann, T., Chiaradia, M., and Shen, Sz, 2018, Early Late Permian coupled carbon and strontium isotope chemostratigraphy from South China: extended Emeishan volcanism? *Gondwana Research*, **58**: 58–70, https://doi.org/10.1016/j.gr.2018.01.011.

Baresel, B., Bucher, H., Brosse, M., Cordey, F., Guodun, K., and Schaltegger, U., 2017, Precise age for the Permian-Triassic boundary in South China from high precision U-Pb geochronology and Bayesian age depth modelling. *Solid Earth*, **8**: 361–378.

Beauchamp, B., 1994, Permian climatic cooling in the Canadian Arctic. *Geological Society of America, Special Paper*, **288**: 229–246.

Beauchamp, B., and Henderson, C.M., 1994, The Lower Raanes, Great Bear Cape and Trappers Cove formations, Sverdrup Basin, Canadian Arctic: stratigraphy and conodont zonation. *Canadian Society of Petroleum Geologists Bulletin*, **42**: 562–597.

Boardman II, D.R., Nestell, M.K., and Wardlaw, B.R., 1998, Uppermost Carboniferous and lowermost Permian deposition and conodont biostratigraphy of Kansas, USA. *In* Jin, Y., Wardlaw, B.R., and Wang, Y. (eds), Permian Stratigraphy, Environments and Resources, Vol. 2. Stratigraphy and Environments. Palaeoworld. 9: 19–32.

Boardman II, D.R., Wardlaw, B.R., and Nestell, M.K., 2009, Stratigraphy and conodont biostratigraphy of the uppermost Carboniferous and Lower Permian from the North American midcontinent. *Kansas Geological Survey Bulletin*, **255**: 253 pp.

Bowring, S.A., Erwin, D.H., Jin, Y.G., Martin, M.W., Davidek, K., and Wang, W., 1998, U/Pb zircon geochronology and tempo of the end-Permian mass extinction. *Science*, **280**: 1039–1045.

Broutin, J., Chateauneuf, J.J., Galtier, J., and Ronchi, A., 1999, L'Autunian d'Autun reste-t-il une reference pour les depots continentaux du Permien inferieur d'Europe.? Apport des donnees paleobotaniques. *La Géologie de la France*, **2**: 17–31.

Buggisch, W., Wang, Xd, Alekseev, A.S., and Joachimski, M.M., 2011, Carboniferous-Permian carbon isotope stratigraphy of successions from China (Yangtze platform), USA (Kansas) and Russia (Moscow Basin and Urals). *Palaeogeography, Palaeoclimatology, Palaeoecology*, **301**: 18–38, https://doi.org/10.1016/j.palaeo.2010.12.015.

Burgess, S.D., and Bowring, S.A., 2015, High-precision geochronology confirms voluminous magmatism before, during and after Earth's most severe extinction. *Science Advances*, **1**: 1–14.

Burgess, S.D., Bowring, S.A., and Shen, S.Z., 2014, High-precision timeline for Earth's most severe extinction. *Proceedings of the National Academy of Sciences of the United States of America*, **111**: 3316–3321.

Burgess, S.D., Muirhead, J.D., and Bowring, S.A., 2017, Initial pulse of Siberian Traps sills as the trigger of the end-Permian mass extinction. *Nature Communications*, **8**: 164.

Cao, C.Q., Cui, C., Chen, J., Summons, R.E., Shen, Sz, and Zhang, H., 2018, A positive C-isotope excursion induced by sea-level fall in the middle Capitanian of South China. *Palaeogeography, Palaeoclimatology, Palaeoecology*, **505**: 305–316, https://doi.org/10.1016/j.palaeo.2018.06.010.

Carr, T., 1972, Conodont biostratigraphy of the Skinner Ranch and Hess formations (Permian), Glass Mountains, West Texas. *M.Sc. thesis*. Texas Tech University, 43 pp.

Chernykh, V.V., 2005, *Zonal Methods of Biostratigraphy — Conodont Zonal Scheme for the Lower Permian of the Urals*. Russian Academy of Sciences, 217 pp. [In Russian]

Chernykh, V.V., 2006, *Lower Permian Conodonts of the Urals*. Russian Academy of Sciences, 130 pp. [In Russian]

Chernykh, V.V., Chuvashov, B.I., Shen, S.Z., and Henderson, C.M., 2013, Proposal for the Global Stratotype Section and Point (GSSP) for the base-Sakmarian Stage (Lower Permian). *Permophiles*, **58**: 16—26.

Chernykh, V.V., Chuvashov, B.I., Shen, S.Z., and Henderson, C.M., 2016, Proposal for the Global Stratotype Section and Point (GSSP) for the base-Sakmarian Stage (Lower Permian). *Permophiles*, **63**: 4—18.

Chernykh, V.V., Chuvashov, B.I., Shen, S.Z., Henderson, C.M., Yuan, D. X., and Stephenson, M.H., 2020. The Global Stratotype Section and Point (GSSP) for the base-Sakmarian Stage (Cisuralian, Lower Permian). *Episodes*, **43**: in press (19 pp.)

Chernykh, V.V., Kotlyar, G.V., Chuvashov, B.I., Kutygin, R.V., Filimonova, T.V., Sungatullina, G.M., et al., 2019, Multidisciplinary study of the Mechetlino Quarry section (southern Urals, Russia) — the GSSP candidate for the base of the Kungurian Stage (Lower Permian). *Palaeoworld*, https://doi.org/10.1016/j.palwor.2019.05.012.

Chernykh, V.V., Ritter, S.M., and Wardlaw, B.R., 1997, *Streptognathodus isolatus* new species (Conodonta): proposed index for the Carboniferous-Permian boundary. *Journal of Paleontology*, **71**: 162—164.

Chuvashov, B.I., Amon, E.O., Karidrua, M., and Prust, Z.N., 1999, Late Paleozoic radiolarians from the polyfacies formations of Uralian Foreland. *Stratigrafiya, Geologicheskaya Korrelyatsiya*, **7** (1): 41—55.

Chuvashov, B.I., Chernykh, V.V., and Bogoslovskaya, M.F., 2002, Biostratigraphic characteristic of stage stratotypes of the Permian System. *Stratigraphy and Geological Correlation*, **10** (4): 317—333.

Chuvashov, B.I., Chernykh, V.V., Shen, S.Z., and Henderson, C.M., 2013, Proposal for the Global Stratotype Section and Point (GSSP) for the base-Artinskian Stage (Lower Permian). *Permophiles*, **58**: 26—34.

Chuvashov, B.I., and Nairn, A.E.M., 1993, Permian system: guides to geological excursions in the Uralian type localities. *Occasional Publications: Earth Science and Resources Institute*, **10**: 303.

Conybeare, W.D., and Phillips, W., 1822, *Outlines of the Geology of England and Wales, With an Introduction Compendium of the General Principles of That Science, and Comparative Views of the Structure of Foreign Countries. Part 1*. William Phillips, 470 pp.

Cramer, B.D., and Jarvis, I., 2020, Chapter 11 - Carbon isotope stratigraphy. *In* Gradstein, F.M., Ogg, J.G., Schmitz, M.D., and Ogg, G.M. (eds), *The Geologic Time Scale 2020*. **Vol. 1** (this book). Elsevier, Boston, MA.

Croft, J.S., 1978, Upper Permian conodonts and other microfossils from the Pinery and Lamar Limestones Members of the Bell Canyon Formation and from the Rustler Formation, West Texas. M.Sc. thesis. The Ohio State University, Columbus, 176 pp.

Davydov, V.I., Glenister, B.F., Spinosa, C., Ritter, S.M., Chernykh, V.V., Wardlaw, B.R., et al., 1998, Proposal of Aidaralash as GSSP for base of the Permian System. *Episodes*, **21**: 11—18.

Davydov, V.I., Korn, D., and Schmitz, M.D., 2012, The Carboniferous Period. *In* Gradstein, F., et al., (eds), *The Geologic Time Scale 2012*. Elsevier, pp. 603—651.

Davydov, V.I., Biakov, A.S., Isbell, J.L., Crowley, J.L., Schmitz, M.D., and Vedernikov, I.L., 2016, Middle Permian U-Pb zircon ages of the "glacial" deposits of the Atkan Formation, Ayan-Yuryakh

anticlinorium, Magadan province, NE Russia: their significance for global climatic interpretations. *Gondwana Research*, **38**: 74—85.

Davydov, V.I., Crowley, J.L., Schmitz, M.D., and Snyder, W.S., 2018, New U-Pb constraints identify the end-Guadalupian and possibly end-Lopingian extinction events conceivably preserved in the passive margin of North America: implication for regional tectonics. *Geological Magazine*, **155** (1): 119—131.

Dunbar, C.O., 1940, The type Permian: its classification and correlation. *The American Association of Petroleum Geologists Bulletin*, **24** (2): 237—281.

Erwin, D.H., 1993, *The Great Paleozoic Crisis: Life and Death in the Permian*. Columbia University Press, 327 pp.

Erwin, D.H., 1995, The end-Permian mass extinction. *In* Scholle, P.A., Peryt, T.M., and Ulmer-Scholle, D.S. (eds), *The Permian of Northern Pangea: Vol. I, Paleogeography, Paleoclimates, Stratigraphy*. Springer-Verlag, Berlin, pp. 20—34.

Esaulova, N.K., Lozovsky, V.R., and Rozanov, A.Y. (eds), 1998, *Stratotypes and References Sections of the Upper Permian in the Regions of the Volga and Kama Rivers*. Ekotsentr, Kazan, p. 300.

Furnish, W.M., 1973, Permian Stage names. *In* Logan, A., and Hills, L.V. (eds), The Permian and Triassic Systems and Their Mutual Boundary. *Canadian Society of Petroleum Geologists Memoir*, **2**: 522—549.

Furnish, W.M., and Glenister, B.F., 1970, Permian ammonoid *Cyclolobus* from the Salt Range, West Pakistan. *In* Kummel, B., and Teichert, C. (eds), *Stratigraphic Problems: Permian and Triassic of West Pakistan*. *University Press of Kansas*, Lawrence, KS, pp. 153—175.

Gialanella, P.R., Heller, F., Haag, M., Nurgaliev, D., Borisov, A., Burov, B., et al., 1997, Late Permian magnetostratigraphy on the eastern Russian Platform. *In* Dekkers, M.J., Langereis, C.G., and Van der Voo, R. (eds), Analysis of Paleomagnetic Data: A Tribute to Hans Zijderveld. *Geologie en Mijnbouw*, **76**: 145—154.

Glenister, B.F., and Furnish, W.M., 1961, The Permian ammonoids of Australia. *Journal of Paleontology*, **35**: 673—736.

Glenister, B.F., Wardlaw, B.R., Lambert, L.L., Spinosa, C., Bowring, S.A., Erwin, D.H., et al., 1999, Proposal of Guadalupian and component Roadian, Wordian and Capitanian Stages as International Standards for the Middle Permian Series. *Permophiles*, **34**: 3—11.

Griesbach, C.L., 1880, Paleontological notes on the Lower Trias on the Himalayas. *Records of the Geological Survey of India*, **13** (2): 94—113.

Grossman, E.L., and Joachimski, M.M., 2020, Chapter 10 - Oxygen isotope stratigraphy. *In* Gradstein, F.M., Ogg, J.G., Schmitz, M.D., and Ogg, G.M. (eds), *The Geologic Time Scale 2020*. **Vol. 1** (this book). Elsevier, Boston, MA.

Haag, M., and Heller, F., 1991, Late Permian to Early Triassic magnetostratigraphy. *Earth and Planetary Science Letters*, **107** (1): 42—54.

Haq, B.U., and Schutter, S.R., 2008, A chronology of Paleozoic sea-level changes. *Science*, **322**: 64—68.

Harland, W.B., Armstrong, R.L., Cox, A.V., Craig, L.A., Smith, A.G., and Smith, D.G., 1990, *A Geologic Time Scale 1989*. Cambridge University Press, 263 pp.

Henderson, C.M., 2002, Kungurian to Lopingian correlations along western Pangea. *Abstracts with Programs—Geological Society of America*, **34** (3): A—30.

Henderson, C.M., 2010, Update on base-Artinskian GSSP. *Permophiles*, **55**: 15—17.

Henderson, C.M., 2018. Permian conodont biostratigraphy. *In* Lucas, S.G., and Shen, S.Z. (eds). *The Permian Timescale* Geological Society, London, Special Publications, **450**: 119—142.

Henderson, C.M., Wardlaw, B., and Mei, S., 2001, New conodont definitions at the Guadalupian-Lopingian Boundary. *Permophiles*, **38**: 35–36.

Henderson, C.M., and Mei, S.L., 2003, Stratigraphic versus environmental significance of Permian serrated conodonts around the Cisuralian-Guadalupian boundary: new evidence from Oman. *Palaeogeography, Palaeoclimatology, Palaeoecology*, **191**: 301–328.

Henderson, C.M., and Mei, S.L., 2007, Geographical clines in Permian and lower Triassic gondolellids and its role in taxonomy. *Palaeoworld*, **16**: 190–201.

Henderson, C.M., Mei, S.L., and Wardlaw, B.R., 2002, New conodont definitions at the Guadalupian-Lopingian boundary. *In* Hills, L.V., Henderson, C.M., and Bamber, E.W. (eds), *Carboniferous and Permian of the World*. Canadian Society of Petroleum Geologist Memoir, **19**: 725–735.

Henderson, C.M., Davydov, V.I., and Wardlaw, B.R., 2012a, The Permian Period. *In* Gradstein, F., et al., (eds), *The Geologic Time Scale 2012*. Elsevier, pp. 653–679.

Henderson, C.M., Wardlaw, B.R., Davydov, V.I., Schmitz, M.D., Schiappa, T.A., Tierney, K.E., et al., 2012b, Proposal for the base-Kungurian GSSP. *Permophiles*, **56**: 8–21.

Hounslow, M.W., and Balabanov, Y.P., 2016, A geomagnetic polarity timescale for the Permian, calibrated to stage boundaries. *In* Lucas, S.G., Shen, S.-Z. (Eds.), *The Permian Timescale*. Geological Society, London, Special Publications, **450**: 61–103. https://doi.org/10.1144/SP450.8.

Hounslow, M.W., Balabanov, Y.P., 2018, A geomagnetic polarity time-scale for the Permian, calibrated to stage boundaries. *In* Lucas, S.G., and Shen, S.Z. (eds), *The Permian Timescale*. Geological Society, London, Special Publications, **450**: 61–103.

Huang, T.K., 1932, The Permian formations of Southern China. *Memoirs of the Geological Survey of China, Series A*, **10**: 1–40.

Jin, Y.G., Glenister, B.F., Kotlyar, G.V., and Sheng, J.Z., 1994, An operational scheme of Permian chronostratigraphy. *In* Jin, Y., Utting, J., and Wardlaw, B.R. (eds), Permian Stratigraphy, Environments and Resources. Palaeoworld. 4: 1–14.

Jin, Y.G., Wardlaw, B.R., Glenister, B.F., and Kotlyar, G.V., 1997, Permian chronostratigraphic subdivisions. *Episodes*, **20**: 10–15.

Jin, Y.G., Henderson, C.M., Wardlaw, B.R., Glenister, B.F., Mei, S.L., Shen, S.H., et al., 2001, Proposal for the Global Stratotype Section and Point (GSSP) for the Guadalupian-Lopingian boundary. *Permophiles*, **39**: 32–42.

Jin, Y.G., Shen, S.Z., Henderson, C.M., Wang, X.D., Wang, W., Wang, Y., et al., 2006a, The Global Stratotype Section and Point (GSSP) for the boundary between the Capitanian and Wuchiapingian stage (Permian). *Episodes*, **29**: 253–262.

Jin, Y.G., Wang, Y., Henderson, C.M., Wardlaw, B.R., Shen, S.Z., and Cao, C.Q., 2006b, The Global Boundary Stratotype Section and Point (GSSP) for the base of the Changhsingian Stage (Upper Permian). *Episodes*, **29**: 175–182.

Karpinsky, A.P., 1874, Geological investigation of the Orenburg area. *Zapiski Imperatorskargo S. Peterburgskago Mineralogicheskoe Obshchestvo, series 2*, **9**: 212–310. [In Russian].

Karpinsky, A.P., 1889, Über die Ammoneen der Artinsk-Stufe und einige mit denselben verwandte carbonische Forman. *Memoir of the Imperial Academy of Science St. Petersburg, series 7*, **37** (2): 104.

Karpinsky, A.P., 1891, On Artinskian ammonites and some similar Carboniferous forms. *Zapiski Imperatorskargo S. Peterburgskago Mineralogicheskoe Obshchestvo, series 2*, **27**: 1–192.

Kotlyar, G.V., and Stepanov, D.L. (eds), 1984. *Main Features of the Stratigraphy of the Permian System in USSR. Nedra*, Leningrad, p. 233.

Kozur, H.W., 2004, Pelagic uppermost Permian and the Permian–Triassic boundary conodonts of Iran. Part 1: Taxonomy. *Hallesches Jahrbuch Fur Geowissenschaften, Reihe B: Geologie, Palaontologie, Mineralogie*, **18**: 39–68.

Kozur, H.W., 2005, Pelagic uppermost Permian and the Permian-Triassic boundary conodonts of Iran, Part II: Investigated sections and evaluation of the conodont faunas. *Hallesches Jahrbuch Fur Geowissenschaften, Reihe B: Geologie, Palaontologie, Mineralogie*, **19**: 49–86.

Lambert, L.L., Wardlaw, B.R., Nestell, M.K., and Nestell, G.P., 2002, Latest Guadalupian (Middle Permian) conodonts and foraminifers from West Texas. *Micropaleontology*, **48**: 343–364.

Lambert, L.L., Wardlaw, B.R., and Henderson, C.M., 2007, *Mesogondolella* and *Jinogondolella* (Conodonta): multielement definition of the taxa that bracket the basal Guadalupian (Middle Permian Series) GSSP. *Palaeoworld*, **16**: 208–221.

Lambert, L.L., Bell, G.L., Fronimos, J.A., Wardlaw, B.R., and Yisa, M.O., 2010, Conodont biostratigraphy of a more complete Reef Trail Member section near the type section, latest Guadalupian Series type region. *Micropaleontology*, **56** (1–2): 233–253.

Legler, B., and Schneider, J.W., 2013, High-frequency cyclicity preserved in nonmarine and marine deposits (Permian, Germany and North Sea). *In* Lucas, S.G., et al., (eds), *The Carboniferous-Permian Transition. New Mexico Museum of Natural History and Science Bulletin*, **60**: 200–211.

Leonova, T.B., 2007, Correlation of the Kazanian of the Volga-Urals with the Roadian of the global Permian Scale. *Palaeoworld*, **16**: 246–253.

Leonova, T.B., 2018, Permian ammonoid biostratigraphy. *In* Lucas, S.G., and Shen, S.Z. (eds), *The Permian Timescale*. Geological Society, London, Special Publications, **450**: 185–203.

Leven, E.Y., 2003, The Permian stratigraphy and fusulinids of Tethys. *The Rivista Italiana di Paleontologia e Stratigrafia*, **101**: 267–280.

Leven, E.Y., and Bogoslovskaya, M.F., 2006, Roadian stage of the Permian and problems of its global correlation. *Stratigrafiya, Geologicheskaya Korrelyatsiya*, **14**: 67–78.

Likharew, B.K. (ed), 1966. *The Stratigraphy of USSR: The Permian System. Nedra*, Moscow, p. 536.

Lozovsky, V.R., Minikh, M.G., Grunt, T.A., Kukhtinov, D.A., Ponomarenko, A.G., and Sukaceva, I.D., 2006, The Ufimian Stage of the East European scale: status, validity, and correlation potential. *Stratigraphy and Geological Correlation*, **17**: 602–614.

Lucas, S.G., 2018, Permian tetrapod biochronology, correlation and evolutionary events. *In* Lucas, S.G., and Shen, S.Z. (eds), *The Permian Timescale*. Geological Society, London, Special Publications, **450**: 405–444.

Lucas, S.G. and Shen, S.Z., 2018. The Permian chronostratigraphic scale: history, status and prospectus. *In* Lucas, S.G. and Shen, S.Z. (eds). *The Permian Timescale*. Geological Society, London, Special Publications, **450**: 21–50.

Maldonado, A.L., and Noble, P.J., 2010, Radiolarians from the upper Guadalupian (Middle Permian) Reef Trail Member of the Bell Canyon Formation, West Texas and their biostratigraphic implications. *Micropaleontology*, **56** (1–2): 69–115.

McArthur, J.M., Howarth, R.J., and Shields, G.A., 2012, Strontium isotope stratigraphy. *In* Gradstein, F., et al., (eds), *The Geologic Time Scale 2012*. Elsevier, pp. 127–144.

McArthur, J.M., Howarth, R.J., Shields, G.A., and Zhou, Y., 2020, Chapter 7 - Strontium isotope stratigraphy. *In* Gradstein, F.M., Ogg, J.G., Schmitz, M.D., and Ogg, G.M. (eds), *The Geologic Time Scale 2020.* **Vol. 1** (this book). Elsevier, Boston, MA.

Mei, S.L., and Henderson, C.M., 2001, Evolution of Permian conodont provincialism and its significance in global correlation and paleoclimate implication. *Palaeogeography, Palaeoclimatology, Palaeoecology,* **170**: 237–260.

Mei, S.L., and Henderson, C.M., 2002, Conodont definition of the Kungurian (Cisuralian) and Roadian (Guadalupian) Boundary. *In* Hills, L.V., Henderson, C.M., and Bamber, E.W. (eds), *Carboniferous and Permian of the World.* Canadian Society of Petroleum Geologists Memoir, **19**: 529–551.

Mei, S.L., Jin, Y.G., and Wardlaw, B.R., 1994a, Succession of conodont zones from the Permian "Kuhfeng" Formation, Xuanhan, Sichuan and its implications in global correlation. *Acta Palaeontologica Sinica,* **33**: 1–23.

Mei, S.L., Jin, Y.G., and Wardlaw, B.R., 1994b, Succession of Wuchiapingian conodonts from northeastern Sichuan and its worldwide correlation. *Acta Micropalaeontologica Sinica,* **11**: 121–139.

Mei, S.L., Jin, Y.G., and Wardlaw, B.R., 1994c, Zonation of conodonts from the Maokouan-Wuchiapingian boundary strata, South China. *In* Jin, Y.G., Utting, J., and Wardlaw, B.R. (eds), *Permian Stratigraphy, Environments and Resources, Palaeoworld,* **4**: 225–233.

Mei, S.L., Jin, Y.G., and Wardlaw, B.R., 1998, Conodont succession of the Guadalupian-Lopingian boundary strata in Laibin of Guangxi, Chian and West Texas, USA. *Palaeoworld,* **9**: 53–76.

Mei, S.L., Henderson, C.M., and Wardlaw, B.R., 2002, Evolution and distribution of the conodonts *Sweetognathus* and *Iranognathus* and related genera during the Permian, and their implications for climate change. *Palaeogeography, Palaeoclimatology, Palaeoecology,* **180**: 57–91.

Mei, S.L., Henderson, C.M., and Cao, C.Q., 2004, Conodont sample-population approach to defining the base of the Changhsingian Stage, Lopingian Series, Upper Permian. *In* Beaudoin, A., and Head, M. (eds), *The Palynology and Micropalaeontology of Boundaries.* Geological Society Special Publications, **230**: 105–121.

Menning, M., 1987, Problems of stratigraphic correlation: magnetostratigraphy. *In* Lützner, H. (ed), *Sedimentary and Volcanic Rotliegendes of the Saale Depression: Excursion Guidebook.* Academy of Sciences of the GDR, Central Institute for Physics of the Earth, Potsdam, pp. 92–96.

Menning, M., 2000, Magnetostratigraphic results from the Middle Permian type section, Guadalupe Mountains, West Texas. *Permophiles,* **37**: 16.

Menning, M., Aleseev, A.S., et al., 2006, Global Time Scale and regional stratigraphic reference scales of central and west Europe, east Europe, Tethys, south China, and North America as used in the Devonian-Carboniferous-Permian correlation chart 2003 (DCP 2003). *Palaeogeography, Palaeoclimatology, Palaeoecology,* **240**: 318–372.

Murchison, R.I., 1841, First sketch of the principal results of a second geological survey of Russia. *Philosophical Magazine, Series 3,* **19**: 417–422.

Murchison, R.I., 1872, *Siluria: A History of the Oldest Rocks in the British Isles and Other Countries,* 5th ed. John Murray, 566 pp.

Nestell, M. K., Nestell, G. P., and Wardlaw, B. R., 2019, Integrated fusulinid, conodont, and radiolarian biostratigraphy of the Guadalupian (Middle Permian) in the Permian Basin region, USA, *In* Ruppel, S. C., ed., Anatomy of a Paleozoic basin: the Permian Basin, USA (vol. 1, ch. 9): The University of Texas at Austin, Bureau of Economic Geology Report of Investigations 285; *AAPG Memoir,* **118**: 251–291.

Nicklen, B.L., Bell, G.I., Lambert, L.L., and Huff, W.D., 2015, Tephrochronology of the Manzanita Limestone in the Middle Permian (Guadalupian) type area, west Texas and southeastern New Mexico, USA. *Stratigraphy,* **12**: 123–147.

Ogg, J.G., and Chen, Z.Q., 2020, Chapter 25 - The Triassic Period. *In* Gradstein, F.M., Ogg, J.G., Schmitz, M.D., and Ogg, G.M. (eds), *The Geologic Time Scale 2020.* **Vol. 2** (this book). Elsevier, Boston, MA.

Peterson, D.N., and Nairn, A.E.M., 1971, Palaeomagnetism of Permian redbeds from the South-western United States. *Geophysical Journal of the Royal Astronomical Society,* **23** (2): 191–205.

Ramezani, J., Bowring, S.A., 2018. Advances in numerical calibration of the Permian timescale based on radioisotopic geochronology. *In* Lucas, S.G., and Shen, S.Z. (eds). *The Permian Timescale.* Geological Society, London, Special Publications, **450**: 51–60.

Ramezani, J., Schmitz, M.D., Davydov, V.I., Bowring, S.A., Snyder, W.S., and Northrup, C.J., 2007, High-precision U-Pb zircon age constraints on the Carboniferous-Permian boundary in the Southern Urals stratotype. *Earth and Planetary Science Letters,* **256**: 244–257.

Read, M.T., and Nestell, M.K., 2019, Lithostratigraphy and fusulinid biostratigraphy of the Upper Pennsylvanian-Lower Permian Riepe Spring Limestone at Spruce Mountain, Elko County, Nevada, USA. *Stratigraphy,* **16** (4): 195–247.

Rhodes, F.H.T., 1963, Conodonts form the topmost Tensleep Sandstone of the eastern Big Horn Mountains, Wyoming. *Journal of Paleontology,* **37** (2): 401–408.

Ross, C.A., and Ross, J.R.P., 2003, Fusulinid sequence evolution and sequence extinction in Wolfcampian and Leonardian Series (Lower Permian), Glass Mountains, West Texas. *The Rivista Italiana di Paleontologia e Stratigrafia,* **109**: 281–306.

Ross, C.A., and Ross, J.R.P., 2009, Paleontology a tool to resolve late Paleozoic structural and depositional histories. *SEPM Special Publication,* **93**: 95–109.

Ruzhencev, V.E., 1936, New data on the stratigraphy of the Carboniferous and Lower Permian deposits of the Orenburg and Aktyubinsk regions. *Principles of Geochemical Prospecting,* **6**: 470–506.

Ruzhencev, V.E., 1950, Type section and biostratigraphy of the Sakmarian Stage. *Doklady Akademiya Nauk USSR, Novaya Seriya,* **71** (6): 1101–1104.

Ruzhencev, V.E., 1954, Asselian Stage of the Permian System. *Doklady Akademiya Nauk USSR, Seriya Geologicheskaya,* **99** (6): 1079–1082.

Schmitz, M.D., and Davydov, V.I., 2012, Quantitative radiometric and biostratigraphic calibration of the Pennsylvanian – Early Permian (Cisuralian) time scale, and pan-Euramerican chronostratigraphic correlation. *Geological Society of America Bulletin,* **124**: 549–577.

Schmitz, M.D., Davydov, V.I., and Snyder, W.S., 2009, Permo-Carboniferous Conodonts and Tuffs: high precision marine Sr isotope geochronology. *Permophiles,* **53** (Suppl. 1): 48.

Schneider, J.W., and Scholze, F., 2018. Late Pennsylvanian – Early Triassic conchostracan biostratigraphy: a preliminary approach.

In Lucas, S.G., and Shen, S.Z. (eds). *The Permian Timescale* Geological Society, London, Special Publications, **450**: 365–386.

Scotese, C.R., 2014, *Atlas of Middle & Late Permian and Triassic Paleogeographic Maps, Maps 43–52, Volumes 3 & 4, PALEOMAP PaleoAtlas for ArcGIS*. PALEOMAP Project, Evanston, IL.

Sheng, J.Z., 1963, Permian fusulinids of Kwangsi, Kueichouw and Szechuan. *Palaeontologia Sinica, New Series*, **10**: 1–119.

Shen, S.Z., 2018. Global Permian brachiopod biostratigraphy: an overview. *In* Lucas, S.G., and Shen S.Z. (eds). *The Permian Timescale*. Geological Society, London, Special Publications, **450**: 289–320.

Sheng, J.Z., and Jin, Y.G., 1994, Correlation of Permian deposits of China. *In* Jin, Y.G., Utting, J., and Wardlaw, B.R. (eds), *Permian stratigraphy, environments and resources, vol. 1: Palaeontology and Stratigraphy. Palaeoworld*, **4**: 14–113.

Shen, S.Z., and Mei, S.L., 2010, Lopingian (Late Permian) high-resolution conodont biostratigraphy in Iran with comparison to South China zonation. *Geological Journal*, **45**: 135–161.

Shen, S.Z., Wang, Y., Henderson, C.M., Cao, C.Q., and Wang, W., 2007, Biostratigraphy and lithofacies of the Permian System in the Laibin-Heshan area of Guangxi, South China. *Palaeoworld*, **16**: 120–139.

Shen, S.Z., Henderson, C.M., Bowring, S.A., Cao, C.Q., Wang, Y., Wang, W., et al., 2010, High Resolution Lopingian (Late Permian) timescale of South China. *Geological Journal*, **45**: 122–134.

Shen, S.Z., Crowley, J.L., Wang, Y., Bowring, S.A., Erwin, D.H., Sadler, P.M., et al., 2011, Calibrating the end-Permian mass extinction. *Science*, **334**: 1367–1372.

Shen, S.Z., Zhang, H., Zhang, Y.C., Yuan, D.X., Chen, B., He, W.L., et al., 2019, Permian integrative stratigraphy and timescale of China. *Science China Earth Sciences*, **62**: 154–188.

Simmons, M.S., Miller, K.G., Ray, D.C., Davies, A., van Buchem, F.S.P., and Gréselle, B., 2020, Chapter 13 - Phanerozoic eustacy. *In* Gradstein, F.M., Ogg, J.G., Schmitz, M.D., and Ogg, G.M. (eds), *The Geologic Time Scale 2020*. **Vol. 1** (this book). Elsevier, Boston, MA.

Steiner, M.B., 2006, The magnetic polarity time scale across the Permian-Triassic boundary. *In* Lucas, S.G., Cassinis, G., Schmeider, J.W. (eds). *Non-Marine Permian Biostratigraphy and Biochronology*. Geological Society, London, Special Publications, **265**: 15–38.

Stuckenberg, A.A., 1890, Geological map of Russia: sheet 138, geological studies in the northwestern part of sheet 138. *Trudy Geologicheskogo Komiteta*, **4**: 1–115.

Sun, Y.D., Liu, X.T., Yan, J.X., Li, B., Chen, B., Bond, D.P.G., et al., 2017, Permian (Artinskian to Wuchiapingian) conodont biostratigraphy in the Tieqiao section, Laibin area, South China. *Palaeogeography, Palaeoclimatology, Palaeoecology*, **465**: 42–63.

Voigt, S., Lucas, S.G., 2018. Outline of a Permian tetrapod footprint ichnostratigraphy. *In* Lucas, S.G., and Shen, S.Z. (eds). *The Permian Timescale*. Geological Society, London, Special Publications, **450**: 387–404.

Wahlman, G.P., 2019, Pennsylvanian and Lower Permian fusulinid biostratigraphy of the Permian Basin region, southwestern USA, (v.1,

ch.7). *In* Ruppel, S.C. (ed), Anatomy of a Paleozoic Basin: The Permian Basin, USA. *The University of Texas at Austin, Bureau of Economic Geology Reort of Investigations 285, AAPG Memoir*, **118**: 167–227.

Wang, C.Y., 2000, The base of the Permian System in China defined by *Streptognathodus isolatus*. *Permophiles*, **36**: 14–15.

Wang, W.Q., Garbelli, C., Zheng, Q.F., Chen, J., Liu, X.C., Wang, W., et al., 2018, Permian 87Sr/86Sr chemostratigraphy from carbonate sequences in South China. *Palaeogeography, Palaeoclimatology, Palaeoecology*, **500**: 84–94.

Wardlaw, B.R., 2000, Guadalupian Conodont biostratigraphy of the Glass and Del Norte Mountains. *In* Wardlaw, B.R., Grant, R.E., and Rohr, D.M. (eds), *The Guadalupian symposium. Smithsonian Contribution to the Earth Sciences*, **32**: 37–87.

Wardlaw, B.R., and Davydov, V.I., 2000, Preliminary placement of the International Lower Permian Working Standard to the Glass Mountains, Texas. *Permophiles*, **36**: 11–14.

Wardlaw, B.R., and Nestell, M.K., 2019, Conodont biostratigraphy of Lower Permian (Wolfcampian–Leonardian) stratotype sections of the Glass and Del Norte Mountains, West Texas, USA. *In* Ruppel, S. C. (ed.), *Anatomy of a Paleozoic Basin: the Permian Basin, USA* (vol. 1). The University of Texas at Austin, Bureau of Economic Geology Report of Investigations 285; *AAPG Memoir*, **118**: 229–249.

Wardlaw, B.R., Leven, E.Y., Davydov, V.I., Schiappa, T.A., and Snyder, W.S., 1999, The base of the Sakmarian Stage: call for discussion (Possible GSSP in the Kondurovsky Section, Southern Urals, Russia). *Permophiles*, **34**: 19–26.

Waterhouse, J.B., 1982, An early Djulfian (Permian) brachiopod faunule from Upper Shyok Valley, Karakorum Range, and the implications for dating of allied faunas from Iran and Pakistan. *Contribution to Himalayan Geology*, **2**: 188–233.

Wu, Q., Ramezani, J., Zhang, H., Wang, T.T., Yuan, D.X., Mu, L., et al., 2017, Calibrating the Guadalupian Series (Middle Permian) of South China. *Palaeogeography, Palaeoclimatology, Palaeoecology*, **466**: 361–372.

Yin, H.F., Zhang, K.X., Tong, J.N., Yang, Z.Y., and Wu, S.B., 2001, The Global Stratotype Section and Point (GSSP) of the Permian–Triassic Boundary. *Episodes*, **24**: 102–114.

Yuan, D.X., Shen, S.Z., Henderson, C.M., Chen, J., Zhang, H., Zheng, Q.F., et al., 2019, Integrative timescale for the Lopingian (Late Permian): a review and update from Shangsi, South China. *Earth-Science Reviews*, **188**: 190–209.

Zhang, Y.C., and Wang, Y., 2018, Permian fusuline biostratigraphy. *Geological Society, London, Special Publications*, **450**: 253–288.

Zhao, J.K., Sheng, J.Z., Yao, Z.Q., Liang, X.L., Chen, C.Z., Rui, L., et al., 1981, The Changhsingian and Permian-Triassic boundary of South China. *Bulletin of Nanjing Institute of Geology and Palaeontology, Academia Sinica*, **2**: 1–112.

Zhou, Z.R., Glenister, B.F., Furnish, W.M., and Spinosa, C., 1996, Multi-episodal extinction and ecological differentiation of Permian ammonoids. *Permophiles*, **29**: 52–62.

J.G. Ogg and Z.-Q. Chen
With contributions by M.J. Orchard and H.S. Jiang

Chapter 25

The Triassic Period

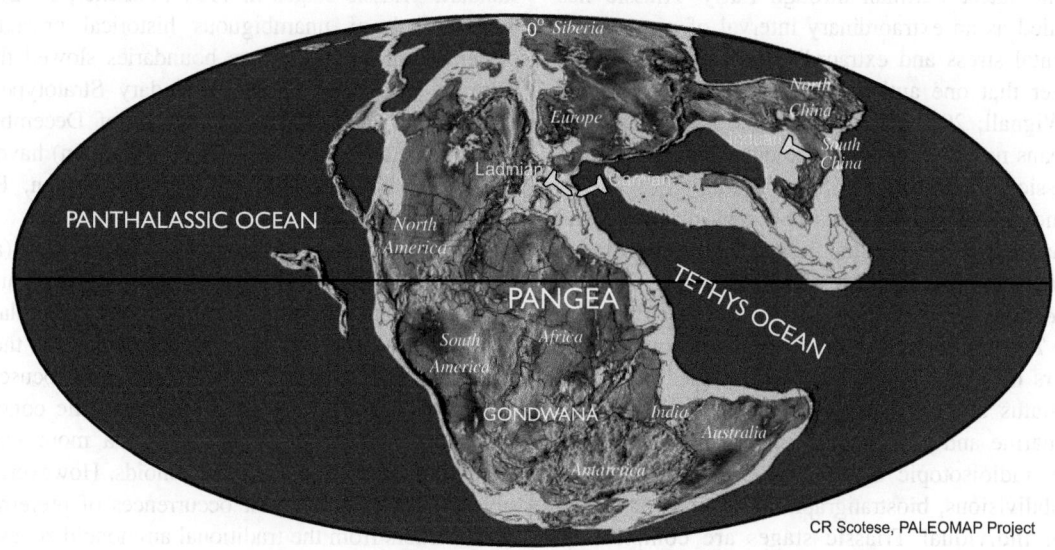

222.6 Ma Triassic

CR Scotese, PALEOMAP Project

Chapter outline

25.1 History and subdivisions 904
   25.1.1 The Permian−Triassic Boundary 904
   25.1.2 Subdivisions of the Lower Triassic 908
   25.1.3 Subdivisions of the Middle Triassic 909
   25.1.4 Subdivisions of the Upper Triassic 912
25.2 Triassic stratigraphy 915
   25.2.1 Marine biostratigraphy 916
   25.2.2 Terrestrial biostratigraphy 918
   25.2.3 Physical stratigraphy 920

25.3 Triassic time scale 926
   25.3.1 Overview 926
   25.3.2 Early Triassic through Anisian age model 926
   25.3.3 Ladinian through Carnian age model 932
   25.3.4 Norian through Rhaetian age model 932
   25.3.5 Estimated uncertainties and future enhancements
      of Triassic age model 933
25.4 Summary 937
Acknowledgments 938
Bibliography 938

Abstract

The Triassic is bound by two mass extinctions that coincide with vast outpourings of volcanic flood basalts. The Mesozoic begins with a gradual recovery of plant and animal life after the end-Permian mass extinction. Conodonts and ammonoids are the main correlation tools for marine deposits. The Pangea supercontinent has no known glacial episodes during the Triassic, but the modulation of its monsoonal climate by Milankovitch cycles left sedimentary signatures useful for high-resolution scaling. Dinosaurs begin to dominate the terrestrial ecosystems in latest Triassic. In contrast to the rapid evolution and pronounced environmental changes that characterize the Early Triassic through Carnian, the Norian−Rhaetian of the Late Triassic was an unusually long interval of stability in Earth history.

Geologic Time Scale 2020. DOI: https://doi.org/10.1016/B978-0-12-824360-2.00025-5

Our understanding of the Triassic underwent a revolution during the past two decades. Collaborations among geochemists, paleomagnetists, paleontologists, and other stratigraphers have concentrated on its exciting boundary intervals but have also enabled compilation of a comparatively detailed bio-mag-geochem-cyclostratigraphy scale for much of the Triassic (Fig. 25.1). High-precision radioisotopic dates and cyclostratigraphic studies, coupled with enhanced methods for global correlation and detailed compilations of geochemical oscillations, have revealed a startling inequality in duration of Triassic subdivisions and in the paces of evolutionary and environmental change. The latest Permian through Early Triassic has been revealed as an extraordinary interval of pronounced environmental stress and extraordinarily rapid evolutionary turnover that one author has termed "The Worst of Times" (Wignall, 2015). In contrast, the Norian of Late Triassic spans nearly three times the duration of the entire Early Triassic without any significant events.

This knowledge and the inevitable new questions and debates have been partially summarized in review articles and special volumes. In particular, the books *Triassic Time Scale* (Lucas, 2010a) and *The Late Triassic World: Earth in a Time of Transition* (Tanner, 2018) contain separate papers on every main stratigraphic topic (e.g., history and status of chronostratigraphy, biostratigraphy of different marine and terrestrial groups, magnetic polarity time scale, radioisotopic age database). Reviews of the history, subdivisions, biostratigraphic zonations, and correlation of individual Triassic stages are compiled in several sources, including *Albertiana* newsletters of The International Commission on Triassic Stratigraphy (https://albertiana-sts.org/), Tozer (1967, 1984), Lucas (2018a), and Tong et al. (2019).

Portions of the following text, especially on the historical background for subdivisions, have been modified from the chapters on Triassic in GTS2012 and GTS2016 (Ogg, 2012; Ogg et al., 2016). As with any review and synthesis, this one compiled in Sept 2019 will quickly become outdated as new major breakthroughs are published.

25.1 History and subdivisions

The *Trias* of von Alberti (1834) united a trio of formations widespread in southern Germany—a lower Buntsandstein (colored sandstone), Muschelkalk (clam limestone), and an upper Keuper (nonmarine reddish beds). These continental and shallow-marine formations were difficult to correlate beyond Germany; therefore most of the traditional stages (Anisian, Ladinian, Carnian, Norian, Rhaetian) were named from ammonoid-rich successions of the Northern Calcareous Alps of Austria. However, the stratigraphy of these Austrian tectonic slices proved unsuitable for establishing formal boundary stratotypes, or even deducing the sequential order

of the stages (Tozer, 1984). For example, the Norian was originally considered to underlie the Carnian stage, but after a convolute scientific-political debate (reviewed in Tozer, 1984), the Norian was established as the younger stage. Over 50 different stage names have been proposed for subdividing the Triassic (tabulated in Tozer, 1984). European stratigraphers commonly use substages with geographic names, whereas North American stratigraphers prefer a generic lower/middle/upper nomenclature (e.g., Fassanian Substage versus Lower Ladinian).

The Subcommission on Triassic Stratigraphy (International Commission on Stratigraphy) adopted seven standard Triassic stages in 1991 (Visscher, 1992), but the general lack of unambiguous historical precedents for placement of Triassic stage boundaries slowed the establishment of formal Global Boundary Stratotype Section and Points (GSSPs) (Fig. 25.1). As of December 2019, only three stages (Induan, Ladinian, Carnian) have GSSPs; and four stages (Olenekian, Anisian, Norian, Rhaetian) lack an international definition.

The traditional definitions for these stages (and substages) were based on lowest occurrences of ammonoid genera within exposures in the Alpine, Himalayan and Mediterranean regions. Current candidates for the GSSPs for the yet-to-be-defined global stages have focused on utilizing the phosphatic teeth of the enigmatic conodonts as the primary markers, which provide a more widespread method of correlation than ammonoids. However, the possible offsets of the lowest occurrences of preferred conodont markers from the traditional ammonoid zones that had defined the stages, disagreements on the taxonomy of the potential marker conodonts along lineages, and their possible diachroneity in interregional appearances are among the factors that have contributed to the delays in formalizing the GSSPs for the remaining half of the Triassic stages and for standardized recognition of substages. Other factors include difficulties in finding suitable sections that have multiple methods for global correlation and in achieving reliable correlations between tropical Tethyan and cooler Boreal marine realms (e.g., review by Konstantinov and Klets, 2009). As with the majority of the stage boundaries in the geologic time scale that have not yet been formally defined, the controversies lie in the details on how to achieve useful and reliable global correlations.

25.1.1 The Permian—Triassic Boundary

25.1.1.1 End-Permian ecological catastrophes and defining the base of the Mesozoic

The Paleozoic terminated in a complex environmental catastrophe and mass extinction of life. This sharp evolutionary division led Phillips (1840, 1841) to introduce Mesozoic (*middle life*, with Triassic at the base) between the Paleozoic

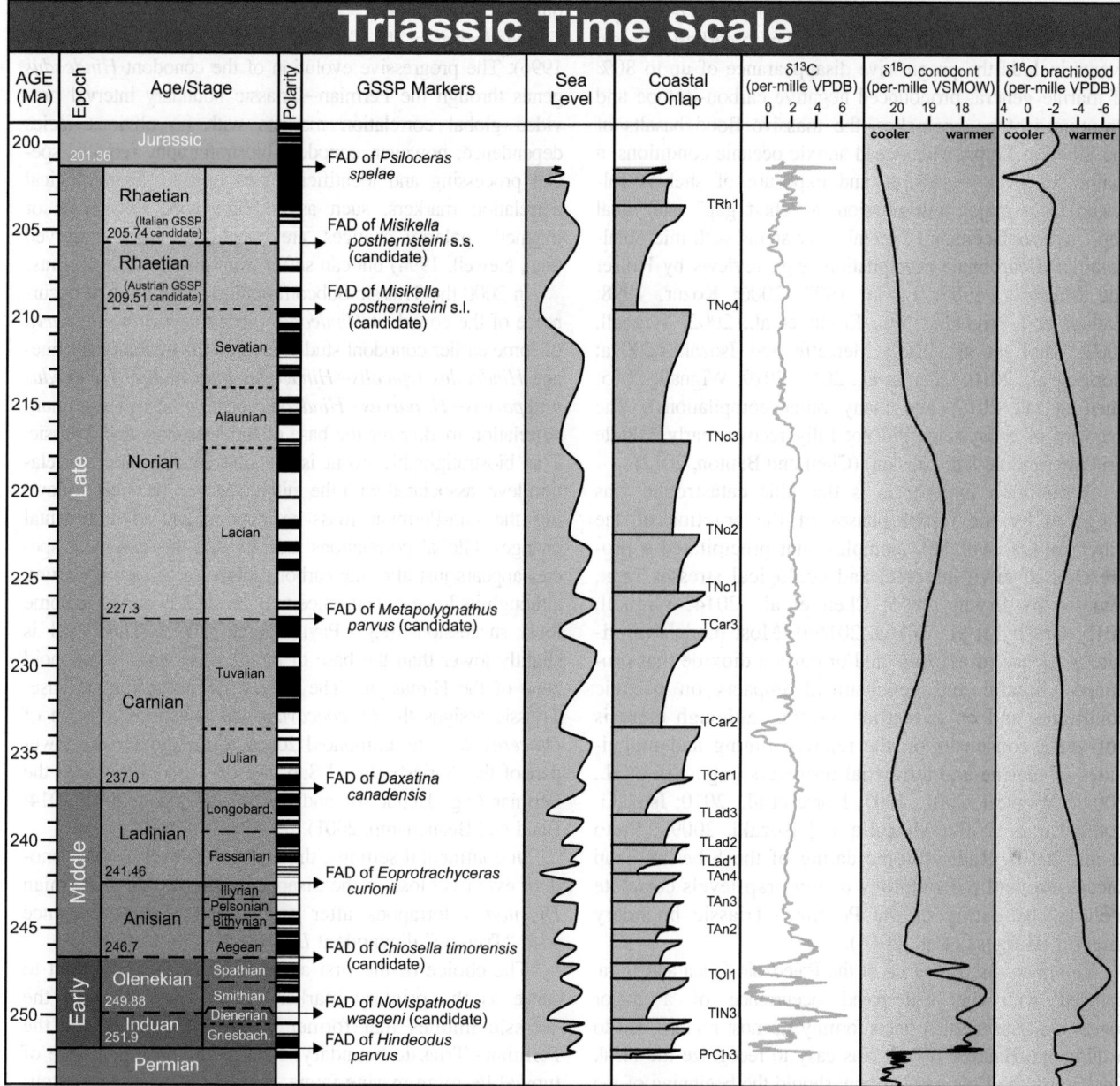

Triassic Time Scale

FIGURE 25.1 Triassic overview. Main markers or candidate markers for GSSPs of Triassic stages are a FAD of ammonoid or conodont taxa as detailed in the text. ("Age" is the term for the time equivalent of the rock-record Stage.) See text for Rhaetian stage boundary options. Magnetic polarity scales from marine sediments are a composite of several syntheses, including Hounslow and Muttoni (2010) and Maron et al. (2015, 2019) with estimated calibrations to the magneto-cyclostratigraphy from thick lacustrine strata in the Newark Basin that span the late Carnian through Rhaetian (Kent and Olsen, 1999; Kent et al., 2017). These magnetic polarity scales are enlarged and detailed in Figs. 25.5– 25.7. Coastal onlap and schematic sea-level curve with labels for selected major sequence boundaries are modified from Haq (2018). See also review in Simmons et al. (2020, Ch. 13: Phanerozoic eustasy, this book). The δ^{13}C curve is a merger of generalized trends with relative magnitudes (see Cramer and Jarvis 2020, Ch. 11: Carbon isotope stratigraphy, this book), including Early Triassic from Sun et al (2012) and portions of Ladinian through Rhaetian from Muttoni et al. (2014). The Triassic δ^{18}O curve (inverted scale) is a statistical fit of conodont-apatite data (Grossman and Joachimski, 2020, Ch. 10: Oxygen isotope stratigraphy, this book) including Early Triassic from Sun et al (2012) and Middle and Late Triassic trend from Trotter et al. (2015) and of brachiopod carbonate data. The vertical scale of this diagram is standardized to match the vertical scales of the stratigraphic summary figure at the beginning of each Phanerozoic chapter. The vertical scale of this diagram is standardized to match the vertical scales of the first stratigraphic summary figure in all other Phanerozoic chapters. *FAD*, First-appearance datum; *GSSP*, Global Boundary Stratotype Section and Point; *VPDB*, Vienna PeeDee Belemnite δ^{13}C and δ^{18}O standard; *VSMOW*, Vienna Standard Mean Ocean Water δ^{18}O standard.

(*old life*, ending with the Permian) and Kainozoic (*recent life*, after the Cretaceous). The latest Permian to earliest Triassic events include the progressive disappearance of up to 80% of marine genera, pronounced negative carbon-isotope and strontium-isotope anomalies, the massive flood basalts of the Siberian Traps, widespread anoxic oceanic conditions, a major sea-level regression, and exposure of shelves followed by a major transgression, a "chert gap" and "coal gap," and replacement of reefal ecosystems with microbial-dominated carbonate precipitation (e.g., reviews by Holser and Magaritz, 1987; Erwin, 1993, 2006; Kozur, 1998; Hallam and Wignall, 1999; Erwin et al., 2002; Wignall, 2007; Knoll et al., 2007; Metcalfe and Isozaki, 2009a; Korte et al., 2010; Chen et al., 2014, 2019; Wignall, 2015; Shen et al., 2019; and many other compilations). The majority of ecosystems did not fully recover early Middle Triassic (middle-late Anisian) (Chen and Benton, 2012).

A common hypothesis is that this catastrophe was triggered by the initial phases of the eruption of the Siberian Trap volcanic complex that precipitated a progression of environmental and ecological stresses (e.g., reviews by Erwin, 2006; Chen et al., 2014; Wignall, 2015; Grasby et al., 2016a, 2016b). Most models implicate a release of aerosols and/or carbon dioxide that produced climatic and geochemical impacts on oceanic conditions and on terrestrial systems, although there is not yet a consensus on the relative timing and magnitudes of marine and terrestrial turnovers (e.g., Yin et al., 2007a; Wignall, 2001, 2007; Korte et al., 2010; Isozaki, 2009; Lucas, 2009; Metcalfe and Isozaki, 2009b; Preto et al., 2010). Radioisotopic dating of the Siberian Trap succession and paleontology of intertrap levels correlate well to the dating of the Permian—Triassic boundary interval (Burgess et al., 2014).

The mass disappearance of the Paleozoic fauna and flora, coupled with the widespread occurrence of a major regression-transgression unconformity in most regions, led to a dilemma (Baud, 2014). It was easy to recognize the bleak final act of the Permian, but how should the beginning of the Mesozoic be defined? The base of the Buntsandstein in SW Germany defined the original Trias concept (von Alberti, 1834), but it is a diachronous boundary within continental beds, now assigned to the upper Permian. Similarly, the base of the Werfen Group (base of Tesero Oolite) in the Italian Alps is a diachronous facies boundary. Ammonoids are the common biostratigraphic tool throughout the Mesozoic, and the *Otoceras* ammonoid genus was long considered to be the first "Triassic" form. Therefore Griesbach (1880) assigned the Triassic base to the base of the *Otoceras woodwardi* Zone in the Himalayan region, but this species is only known from the Perigondwana paleomargin of eastern Tethys (e.g., Iran to Nepal). The first occurrence of *Otoceras* species in the Arctic realm (*Otoceras concavum* Zone) was used by Tozer (1967, 1986, 1994) for a Boreal marker of the base of the Triassic

but is now known to appear significantly prior to *O. woodwardi* in the Tethyan realm (Krystyn and Orchard, 1996). The progressive evolution of the conodont *Hindeodus* genus through the Permian—Triassic boundary interval provided global correlation markers with no obvious facies dependence; however, conodont biostratigraphy requires special processing and identification experience. Nonbiological correlation markers, such as carbon-isotope excursions or magnetic polarity changes, are conclusive when preserved (e.g., Newell, 1994) but can suffer from diagenetic overprints.

In 2000 the Triassic Subcommission chose the first occurrence of the conodont *Hindeodus parvus* (=*Isarcicella parva* of some earlier conodont studies) within the evolutionary lineage *Hindeodus typicalis—Hindeodus latidentatus—Hindeodus praeparvus—H. parvus—Hindeodus postparvus* as the primary correlation marker for the base of the Mesozoic and Triassic. This biostratigraphic event is the first cosmopolitan correlation level associated with the initial stages of recovery following the end-Permian mass extinctions and environmental changes. Global correlations indicate that this conodont species appears just after the carbon-isotope ($\delta^{13}C_{carb}$) minimum, although its lowest occurrence may be slightly earlier in some local successions (e.g., Payne et al., 2009). This level is slightly lower than the base of the *O. woodwardi* ammonoid zone of the Himalayas. The revised definition for the base-Triassic assigns the *O. concavum* and lowermost portion of *Otoceras boreale* ammonoid zones of the Arctic (the lower part of the "Griesbachian" Substage of Tozer, 1967) into the Permian (e.g., Henderson and Baud, 1997; Baud, 2001, 2014; Baud and Beauchamp, 2001).

In continental settings, the correlated level to this conodont event is close to the disappearance of typical Permian *Dicynodon* tetrapods after an interval of co-occurrence with "Triassic" dicynodont *Lystrosaurus* (Kozur, 1998).

The choice of the first appearance of this conodont to serve as the primary marker for the beginning of the Triassic implies that former traditional concepts of the Permian—Triassic boundary, such as the disappearance of typical Permian marine fauna, rapid facies changes, extensive volcanism, and onset of isotope anomalies, are now assigned to the latest Permian.

25.1.1.2 Paleozoic—Mesozoic boundary stratotype (base of Triassic)

The GSSP for the base of the Mesozoic Erathem, the Triassic System, and the Induan Stage is at the base of Bed 27c at a section near Meishan, Zhejiang Province, southern China (Fig. 25.2). This level coincides with the lowest occurrence of conodont *H. parvus* (Yin et al., 2001, 2005). This level was preceded by a pronounced brief negative excursion in carbon isotopes (up to $-6‰$ $\delta^{13}C_{carb}$ relative to the late-Changhsingian *Clarkina yini* conodont zone) and is within a rapid $\sim9°C$ increase in local sea-surface

Base of the Induan Stage of the Triassic System at Meishan, China

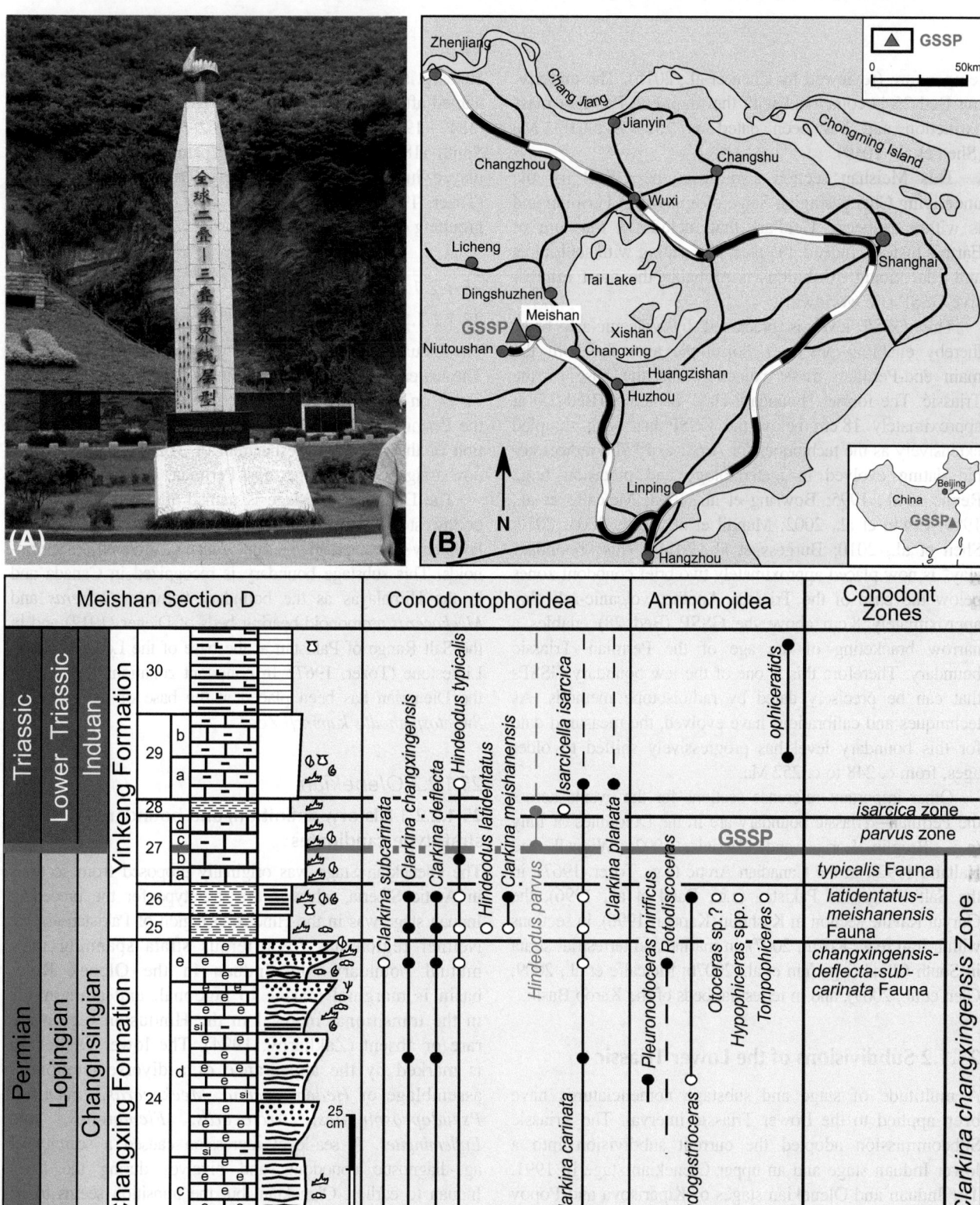

FIGURE 25.2 GSSP for the base of the Triassic (base of Mesozoic; base of Induan Stage) at the Meishan section, South China. The GSSP level coincides with the lowest occurrence of conodont *Hindeodus parvus* and is within a major negative excursion in carbon isotopes. Note that vertical scale is in centimeters to show details of this boundary interval. Conodont ranges are from the ratified GSSP document (Yin et al., 2001), but later studies have revised the assignments (see Chen et al., 2015, and Fig. 24.6 in Henderson and Shen, 2020, Ch. 24: The Permian Period, this volume). Zircons from the altered volcanic-ash components in Bed 25 and Bed 28 have been dated (e.g., Burgess et al., 2014; Shen et al., 2019). The main end-Permian mass extinction event is at the base of Bed 25. The GSSP is preserved within a GeoPark, with the GSSP site located at the top of the stairs on the right side of the decorative wall behind the monument topped by a statue of the conodont. The GSSP for the base of the Changhsingian Stage (uppermost Permian) is in the lower quarry wall beyond the right side of the photo. *GSSP*, Global Boundary Stratotype Section and Point.

temperature (reviewed by Chen et al., 2015). The underlying Bed 25 is coincident with the main end-Permian mass extinction and has been dated as 251.939 ± 0.031 Ma (Shen et al., 2019).

This Meishan section also hosts the GSSP for the underlying Changhsingian Stage of uppermost Permian and is within a special GeoPark that includes a museum of Earth's history. Indeed, the park-like setting with sculptures and educational exhibitions is probably the most impressive GSSP site worldwide.

This GSSP level is bracketed by volcanic-ash beds, thereby enabling precise radioisotopic ages for both the main end-Permian mass extinction and the base of the Triassic. The former "boundary clay" bentonite (Bed 25) at approximately 18 cm below the GSSP has been sampled extensively as the techniques for Ar/Ar and U/Pb radioisotopic dating evolved in methodology and precision (e.g., Renne et al., 1995; Bowring et al., 1998; Metcalfe et al., 1999; Erwin et al., 2002; Mundil et al., 2001, 2004, 2010; Shen et al., 2010; Burgess et al., 2014). This "boundary clay" is now placed approximately two brief conodont zones below the base of the Triassic. Another volcanic-ash clay approximately 8 cm above the GSSP (Bed 28) enables a narrow bracketing of the age of the Permian−Triassic boundary. Therefore this is one of the few boundary GSSPs that can be precisely dated by radioisotope methods. As techniques and calibrations have evolved, the measured date for this boundary level has progressively shifted to older ages, from c. 248 to c. 252 Ma.

Other important reference sections for the events across the Permian−Triassic boundary are in the Dolomites of Italy (e.g., Broglio Loriga and Cassinis, 1992; Wignall and Hallam, 1992), in the Canadian Arctic (e.g., Tozer, 1967), in the Salt Ranges of Pakistan (e.g., Baud et al., 1996), the Guryul Ravine section in Kashmir (Kapoor, 1996), in sections within Iran (e.g., Kozur, 2007), in marine and terrestrial strata in South China (e.g., Yin et al., 2007b; Metcalfe et al., 2009; Glen et al., 2009), and in terrestrial beds of the Karoo Basin.

25.1.2 Subdivisions of the Lower Triassic

A multitude of stage and substage nomenclatures have been applied to the Lower Triassic interval. The Triassic Subcommission adopted the current subdivision into a lower Induan stage and an upper Olenekian stage in 1991. The Induan and Olenekian stages of Kiparisova and Popov (1956, revised in 1964) were named after exposures in the Indus River basin in the Hindustan region of Asia and in the lower reaches of the Olenek River basin of northeast Siberia, respectively.

A suite of four substages is widely used. In an imaginative procedural twist, these Griesbachian, Dienerian, Smithian, and Spathian substages are named after exposures along associated small creeks on Ellesmere and Axel

Heiberg islands in the Canadian Arctic, which in turn were named after the Triassic paleontologists—Carl L. Griesbach (1847−1907), Carl Diener (1862−1928), James Perrin Smith (1864−1931), and Leon Spath (1888−1957)—who played important roles in Lower Triassic biostratigraphy (Tozer, 1965). These substages were originally defined by grouping of ammonoid zones.

25.1.2.1 Induan

25.1.2.1.1 Griesbachian and Dienerian substages

The Induan Stage is informally divided into two substages. The lower substage, Griesbachian, is named after Griesbach Creek on northwest Axel Heiberg Island. The definition of the Permian−Triassic boundary implies that the lower portion of the original Griesbachian of Tozer (1965, 1967) is now assigned to the uppermost Permian.

The Dienerian Substage is named after Diener Creek of northwest Ellesmere Island. The Griesbachian/Dienerian boundary is marked by the appearance of Gyronitidae ammonoids. This substage boundary is recognized in Canada and in the Himalayas as the boundary between *Otoceras* and *Meekoceras* ammonoid-bearing beds of Diener (1912) and in the Salt Range of Pakistan at the base of the Lower Ceratite Limestone (Tozer, 1967). In conodont zonations the base of the Dienerian has been placed at the base of the conodont *Sweetospathodus kummeli* Zone.

25.1.2.2 Olenekian

25.1.2.2.1 History, definition, and boundary stratotype candidates

The Olenekian Stage was originally proposed from sections in Arctic Siberia, whereas the stratotype for the preceding Induan stage was in the Hindustan region of Pakistan−India. Neither region has fossiliferous strata spanning their mutual boundary—the Induan in the Olenek River basin is marginal marine to lagoonal, and ammonoids in the transitional interval in the Hindustan region are rare or absent (Zakharov, 1994). The lower Olenekian is marked by the appearance of a diverse ammonoid assemblage of *Hedenstroemia*, *Meekoceras*, *Juvenites*, *Pseudoprospingites*, *Arctoceras*, *Flemingites*, and *Euflemingites*. A sea-level regression caused a scarcity of age-diagnostic conodonts and bivalves during the latest Induan to earliest Olenekian, but the transition seems to be within the lower portion of the *Novispathodus pakistanensis* conodont zone (Zakharov, 1994; Paull, 1997; Orchard and Tozer, 1997). Proposed ammonoid-based biostratigraphic definitions of the stage boundary were the highest occurrence of the ammonoid *Gyronites subdharmus* and the lowest occurrence of the representatives of the *Meekoceras* or *Hedenstroemia* ammonoid genera (Zakharov et al., 2000, 2002). The base of its lower Smithian Substage was originally defined as the base of a broad *Euflemingites romunderi* ammonoid zone (Tozer, 1965, 1967); then the biostratigraphy was revised to add a *Hedenstroemia hedenstroemi* ammonoid zone (e.g., Orchard and Tozer, 1997). Conodonts were undergoing a pronounced evolutionary radiation at the beginning of the Olenekian, and, even though some taxonomic details remain to be resolved, the widespread distribution and resolution of *Neospathodus* species provide the main method for interregional correlations (Orchard, 2010).

Therefore the base-Olenekian task group selected the lowest occurrence of the conodont *Novispathodus waageni sensu lato* as the primary boundary marker. Correlation of ammonoid and conodont events among paleogeographic provinces indicates the first occurrence of *Neospathodus waageni* is in the lowermost part of the *Rohillites rohilla* ammonoid zone in the Spiti region of the Tethyan realm, is slightly below the lowest occurrence of *Flemingites* and *Euflemingites* ammonoid genera in South China of Tethyan realm, and is in the lower part of *Lepiskites kolyhmensis* Zone of Siberia in the Boreal realm which is just above the regional *Hedenstroemia hedenstromia* Zone (Zakharov et al., 2009). This conodont level is just prior to the peak of the first Triassic positive excursion in $\delta^{13}C_{carb}$, at or just below an upward change from a long-duration reversed-polarity to a brief normal-polarity magnetic zone (LT2r of Hounslow and Muttoni, 2010) at Chaohu, and is just above a widely recognizable sequence boundary (e.g., Krystyn et al., 2007a).

There are two leading candidates for the base-Olenekian GSSP. The West Pingdingshan roadside outcrop near quarries at Chaohu city in the Anhui Province of eastern China has biostratigraphy (with a local *Nov. waageni eowaageni* as the lowest subspecies), carbon isotopes, magnetostratigraphy, and cyclostratigraphy (Tong et al., 2004; Sun et al., 2007; Chinese Triassic Working Group, 2007; Zhao et al., 2007, 2008; Li et al., 2016; Lyu et al., 2018, 2020). A section near Mud (Muth) village in the Spiti valley of northwest India has better ammonoid constraints but lacks the magnetostratigraphy and cyclostratigraphy (Krystyn et al., 2007a, 2007b). A preliminary vote to select Mud as the GSSP was put on hold in 2008 when the *Nov. waageni* conodont marker was identified in strata lower than the proposed GSSP level. An integrated magneto-bio-cyclostratigraphy of another boundary reference section at Daxiakou near the Three Gorges Dam in South China seemed to hint that the lowest occurrence of the conodont marker was ~50 kyr earlier than at the Chaohu GSSP candidate (Zhao et al., 2013; Li et al., 2016). A decision on the base-Olenekian GSSP also requires verifying if these Tethyan-based events can be correlated to the Boreal realm.

25.1.2.2.2 Smithian and Spathian substages

A major environmental and evolutionary event at a global scale occurs within the Olenekian (Algeo et al., 2019a, 2019b; Zhang et al., 2019). The boundary between its two substages, the Smithian and the succeeding Spathian, has been termed the "biggest crisis in Triassic conodont history" (Orchard, 2007a; Goudemand et al., 2008). This boundary is marked by a sudden reduction of ammonoid diversity back to initial Triassic conditions and shift from latitudinal to cosmopolitan distributions (Brayard et al., 2009a), and coincides with a major positive peak in Carbon-13 ($\delta^{13}C_{carb}$) and a cooling climatic shift (e.g., Galfetti et al., 2007a; Goudemand et al., 2019; Algeo et al., 2019a, 2019b).

These two informal substages of the Olenekian Stage were named after the Smith and Spath creeks on Ellesmere Island of the Canadian Arctic. The Spathian Substage is characterized by *Tirolites*, *Columbites*, *Subcolumbites*, *Prohungarites*, and *Keyserlingites* ammonoid genera. In these original stratotypes, the Smithian–Spathian boundary was placed at the base of the *Olenekites pilaticus* ammonoid zone, but there appears to be a missing biostratigraphic interval in the type region (Tozer, 1967; Orchard and Tozer, 1997). The ammonoids recovered in early Spathian with a dramatic evolutionary radiation accompanied by the development of a pronounced latitudinal gradient of diversity (Brayard et al., 2009a). In South China, the FAD of conodont *Nov. pingdingshanensis* was selected as the marker of the base of Spathian (Zhao et al., 2007, 2013; Lyu et al., 2019). The exciting discovery of major geochemical, climatic, and paleontological events that coincide with this substage boundary has led to speculations of concurrent volcanic release of carbon dioxide and/or enhanced storage and preservation of organic matter in the ocean (e.g., Payne and Kump, 2007; Galfetti et al., 2007a, 2007b, 2007c; Algeo et al., 2019b; Grasby et al., 2016b).

25.1.3 Subdivisions of the Middle Triassic

25.1.3.1 Anisian

25.1.3.1.1 History, definition, and boundary stratotype candidates

The Anisian Stage was named after limestone formations near the Enns (=*Anisus*) River at Grossreifling, Austria (Waagen and Diener, 1895). The original Anisian stratotype lacks ammonoids in the lower portion, and lower limit was later clarified in the Mediterranean region (Assereto, 1974). The appearance of a number of ammonoid genera, including *Aegeiceras*, *Japonites*, *Paracrochordiceras*, and *Paradanubites*, may be used to define the base of the Anisian within different regions (e.g., Gaetani, 1993). However, other markers are suggested that may provide a more global correlation value. The lowest occurrence of the *Chiosella timorensis* (or its *sensu stricto* morphotype on its early lineage) conodont slightly precedes the ammonoid level and can be correlated to North American and Asian stratigraphy (Orchard and Tozer, 1997; Orchard, 2010; Chen et al., 2020). Therefore if the lowest occurrence of *Ch. timorensis* is selected as the global marker, then the uppermost part of the ammonoid *Neopopanoceras haugi* Zone of "latest Olenekian" in the Tethyan realm will slightly overlap the basal Anisian. The boundary interval is close to a peak in carbon-13 ($\delta^{13}C_{carb}$) values. Sea-surface temperatures were reported to have cooled by 4°C during this rise in $\delta^{13}C_{carb}$ (Sun et al., 2012), suggesting a sequestration of carbon dioxide. A shift from reversed-polarity- to normal-polarity-dominated magnetostratigraphy (base of normal-polarity magnetozone MT1n of Hounslow and Muttoni (2010) has been proposed as a primary global boundary marker that can be unambiguously correlated between Boreal and Tethyan faunal realms (Hounslow et al., 2007). However, some stratigraphers wish to retain an ammonoid-based assignment for an Anisian GSSP (and some other Triassic stages/substages) (e.g., discussions posted at Ogg et al., 2020).

Published candidates for the base-Anisian GSSP include locations in Romania, in Albania and two in South China. At Deşli-Caira Hill in north Dobrogea, Romania, the Olenekian−Anisian boundary interval is within a condensed Hallstatt limestone facies with ammonites (Gradinaru et al., 2007). The Kçira section in Albania of reddish nodular limestone has magnetostratigraphy and detailed conodont ranges (Muttoni et al., 1996b, 2019), but has not yet been studied for stable isotope stratigraphy.

In South China, the succession of *Ch. timorensis* and other conodont datums within the boundary interval are interbedded with volcanic-ash beds. Although the individual zircons within each of those beds have a widespread of U−Pb radioisotopic dates, the suites enable estimates of the approximate ages for the various datums with a mean near 247 Ma (Lehrmann et al., 2015; Ovtcharova et al., 2015). The Guandao section in the Nanpanjiang Basin (Guizhou province, South China) has an impressive conodont-magnetic-isotope-cyclo stratigraphy with radioisotopic dates (Lehrmann et al., 2006, 2015; Li et al., 2016, 2018); but the critical Olenekian−Anisian boundary interval is distorted by debris flows and did not yet yield the brief M1n normal-polarity subchron. The Wantou roadcut at Jinya, Fengshan County, Guangxi province in South China has been studied for ammonoid, conodont, and carbon-isotope stratigraphy (Galfetti et al., 2007, 2008), and the main events and trends are bracketed by a succession of a dozen volcanic ashes that have yielded ID-TIMS U−Pb ages (Ovtcharova et al., 2006, 2015). Additional work on that Wantou section has yielded a detailed biomagnetostratigraphy that verified that the M1n polarity subchron is near the first *Ch. timorensis* conodont (Chen et al., 2020).

25.1.3.1.2 Anisian substages

The Anisian Stage has three to four informal substages. Assereto (1974) proposed a stratotype for the Lower

Anisian (also called "Aegean" or "Egean") in beds with *Paracrochordiceras* ammonoids at Mount Marathovouno on Chios Island (Aegean Sea, Greece). The Middle Anisian is sometimes subdivided into two substages: a lower "Bithynian," named by Assereto (1974) after the Kokaeli Peninsula (Bithynia) of Turkey, and an upper "Pelsonian," from the Latin name for the region around Lake Balaton in Hungary (Pia, 1930) spanning the *Balatonites balatonicus* ammonoid zone (Assereto, 1974). The Upper Anisian is also called "Illyrian" after the Latin term for Bosnia (Pia, 1930). Southwestern China has continuous Anisian carbonate successions with complete conodont and ammonoid zonations (Enos et al., 2006; Benton et al., 2013; Tong et al. 2019) that have high potential in assisting future decisions on practical inter-regional boundaries of Anisian substages.

25.1.3.2 Ladinian

25.1.3.2.1 History, definition, and boundary stratotype candidates

The Ladinian Stage arose after a heated semantic argument of "Was ist norisch?" (Bittner, 1892), when it was realized that most of the strata that had defined a "pre-Carnian" Norian Stage (Mojsisovics, 1869) were actually deposited *after* the Carnian (Mojsisovics, 1893). This debate and the emergence of the Ladinian Stage split the Vienna geological establishment (vividly reviewed by Tozer, 1984). The Ladinian, named after the Ladini inhabitants of the Dolomites region of northern Italy, encompassed the Wengen and Buchenstein beds (Bittner, 1892).

This historical major revision and even partial inversion of the upper Triassic stratigraphy, coupled with uncertainties about correlation potentials and definition of ammonoid zones, contributed to discussions in assigning the basal limit of the Ladinian Stage (e.g., Gaetani, 1993; Brack and Rieber, 1994; Mietto and Manfrin, 1995; Brack et al., 1995; Vörös et al., 1996; Muttoni et al., 1996a; Orchard and Tozer, 1997; Pálfy and Vörös, 1998). The ammonoid contenders for the primary correlation markers were distributed over at least two zones; including the lowest occurrence of representatives of the *Kellnerites* genus, of the *Nevadites* genus, of the *Eoprotrachyceras* genus, of the *Reitziites reitzi* species, and of the *Aplococeras avisianum* species. In addition, the lowest occurrence of the *Budurovignathus* conodont genus was considered.

The base of *Eoprotrachyceras curionii* Zone (lowest occurrence of *Eoprotrachyceras* ammonoid genus, which is the onset of the Trachyceratidae ammonoid family) was eventually preferred. The Bagolino section (eastern Lombardian Alps, Province of Brescia, Northern Italy) was selected for its multiple stratigraphic records, including the bracketing of the boundary interval by dated volcanic ashes (Brack et al., 2005). The Ladinian GSSP at Bagolino is located at the top

Base of the Ladinian Stage of the Triassic System at Bagolino, Northern Italy

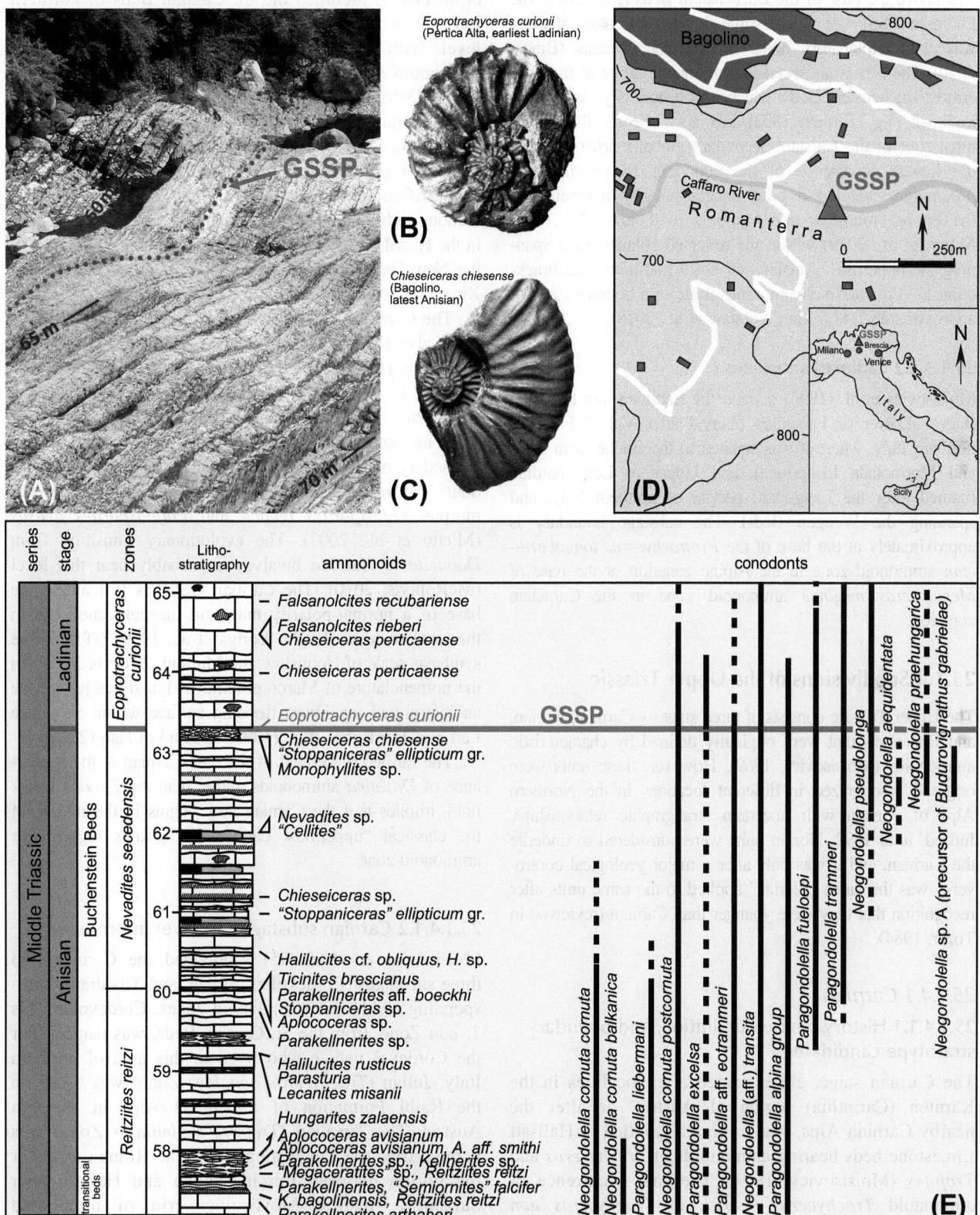

FIGURE 25.3 GSSP for the base of the Ladinian Stage at Bagolino, Italy. The GSSP level coincides with the lowest occurrence of *Eoprotrachyceras* ammonoid genus (base of *Eoprotrachyceras curionii* Zone). *GSSP*, Global Boundary Stratotype Section and Point. *Photos provided by Peter Brack.*

of a distinct 15−20-cm-thick interval of limestone nodules in a shaly matrix (Chiesense groove), located at approximately 5 m above the base of the Buchenstein Beds (Fig. 25.3). The Ladinian GSSP site is accessible through a geological pathway with explanatory notes and ammonoid casts (Brack, 2010). The *Nevadites secedensis* ammonoid zone of the lowermost Buchenstein Beds, which was historically assigned as Ladinian (e.g., Bittner, 1892), has now become the uppermost zone of the Anisian. Secondary global markers include the lowest occurrence of the conodont *Budurovignathus praehungaricus* and a brief normal-polarity magnetozone (MT8n of Hounslow and Muttoni, 2010; and SC2r.1n of Maron et al., 2019) within the reversed-polarity zone spanning the uppermost Anisian and basal Ladinian. The bracketing U−Pb-dated volcanic ashes indicate a boundary age of approximately 241.5 Ma (Wotzlaw et al., 2018).

25.1.3.2.2 Ladinian substages

Mojsisovics et al. (1895) divided the Ladinian into two substages—Lower or Fassanian (named after Val di Fassa in northern Italy, where it was equated to the Buchenstein Beds and Marmolada Limestone), and Upper or Longobardian (named after the Langobard people of northern Italy, and spanning the Wengen Beds). The substage boundary is approximately at the base of the *Protrachyceras longobardicum* ammonoid zone in the Alpine zonation or the base of *Meginoceras meginae* ammonoid zone in the Canadian zonation.

25.1.4 Subdivisions of the Upper Triassic

The Upper Triassic consists of three stages—Carnian, Norian, and Rhaetian—that were originally defined by characteristic ammonoids (Mojsisovics, 1869). However, these units were originally recognized in different locations in the Northern Alps of Austria with uncertain stratigraphic relationships. Indeed, until 1892, Norian units were considered to underlie the Carnian, and it was only after a major geological controversy was the name "Norian" applied to the same units after recognition that they were younger than Carnian (reviewed in Tozer, 1984).

25.1.4.1 Carnian

25.1.4.1.1 History, revised definition, and boundary stratotype candidates

The Carnian stage, either named after localities in the Kärnten (Carinthia) region of Austria, or after the nearby Carnian Alps, was originally applied to Hallsatt Limestone beds bearing ammonoids of *Trachyceras* and *Tropites* (Mojsisovics, 1869). The first occurrence of ammonoid *Trachyceras* (=base of *Trachyceras aon* Zone in Tethys or *Trachyceras desatoyense* in Canada) was the traditional base, although it appears that a

Trachyceras datum would be asynchronous and not cosmopolitan (e.g., Mietto and Manfrin, 1999). Mojsisovics et al. (1895) included the St. Cassian Beds of northern Italy in a revised Carnian subdivision, therefore the level with lowest occurrence of the cosmopolitan ammonoid *Daxatina* at the Prati di Stuores type locality in the Dolomites (northern Italy) was proposed for the base-Carnian GSSP (Broglio Loriga et al., 1998). This section has relatively rapid sedimentation and proved suitable for multiple types of stratigraphy; therefore it was ratified 10 years later (2008). Three other reference sections with multiple biostratigraphic successions are in Spiti in the Himalaya of northwest India (Balini et al., 1998, 2001), the New Pass section of Nevada in western USA, and the Xingyi section of Guizhou in southwestern China.

The Carnian GSSP at Prati di Stores/Stuores Wiesen is 45 m above the base of the St. Cassian (San Cassiano) Formation (Fig. 25.4) (Mietto et al., 2012). This level was selected to coincide with the lowest occurrence of the ammonoid *Daxatina* (base of *Daxatina canadensis* Subzone, lowest subzone of a broad *Trachyceras* Zone). Secondary markers are the lowest occurrences of the conodont "*Paragondolella*" *polygnathiformis* and the palynomorphs *Vallasporites ignacii* and *Patinasporites densus* (Mietto et al., 2007). The evolutionary transition from *Daonella* to *Halobia* bivalves is probably near this level (McRoberts, 2010). The Carnian GSSP is just above the base of a normal-polarity magnetic magnetozone (S2n in the local scale of Broglio Loriga et al., 1999; UT1n in the synthesis scale of Hounslow and Muttoni, 2010; or MA5n in the nomenclature of Maron et al., 2019), and lies just above an interpreted maximum flooding surface within Sequence Lad 3 of Hardenbol et al. (1998) or TLa3 of Haq (2018).

The ratified placement of the base-Carnian at the appearance of *Daxatina* ammonoids, rather than the *T. aon* ammonoid, implies that the Carnian now begins in the middle of the classical "uppermost Ladinian" *Frankites regoledanus* ammonoid zone.

25.1.4.1.2 Carnian substages and wet intermezzo

Mojsisovics et al. (1895) subdivided the Carnian into three substages (Cordevolian, Julian, and Tuvalian) corresponding to his three ammonoid zones. Cordevolian (his *T. aon* Zone, from the St. Cassian Beds, was named after the Cordevol people who lived in this area of northern Italy. Julian (*Trachyceras aonoides* Zone) was based on the Raibl Formation of the Julian Alps in southern Austria. The Tuvalian (*Tropites subbullatus* Zone) was named after the Tuval mountains, the Roman term for the region between Berchtesgaden and Hallein near Salzburg, Austria. Mojsisovics trio of ammonoid zones was later split into additional zones; but his main divisions can be correlated among regions.

Base of the Carnian Stage of the Triassic System in the Prati di Stuores/ Stuores Wiesen Section, near San Cassiano, Italy

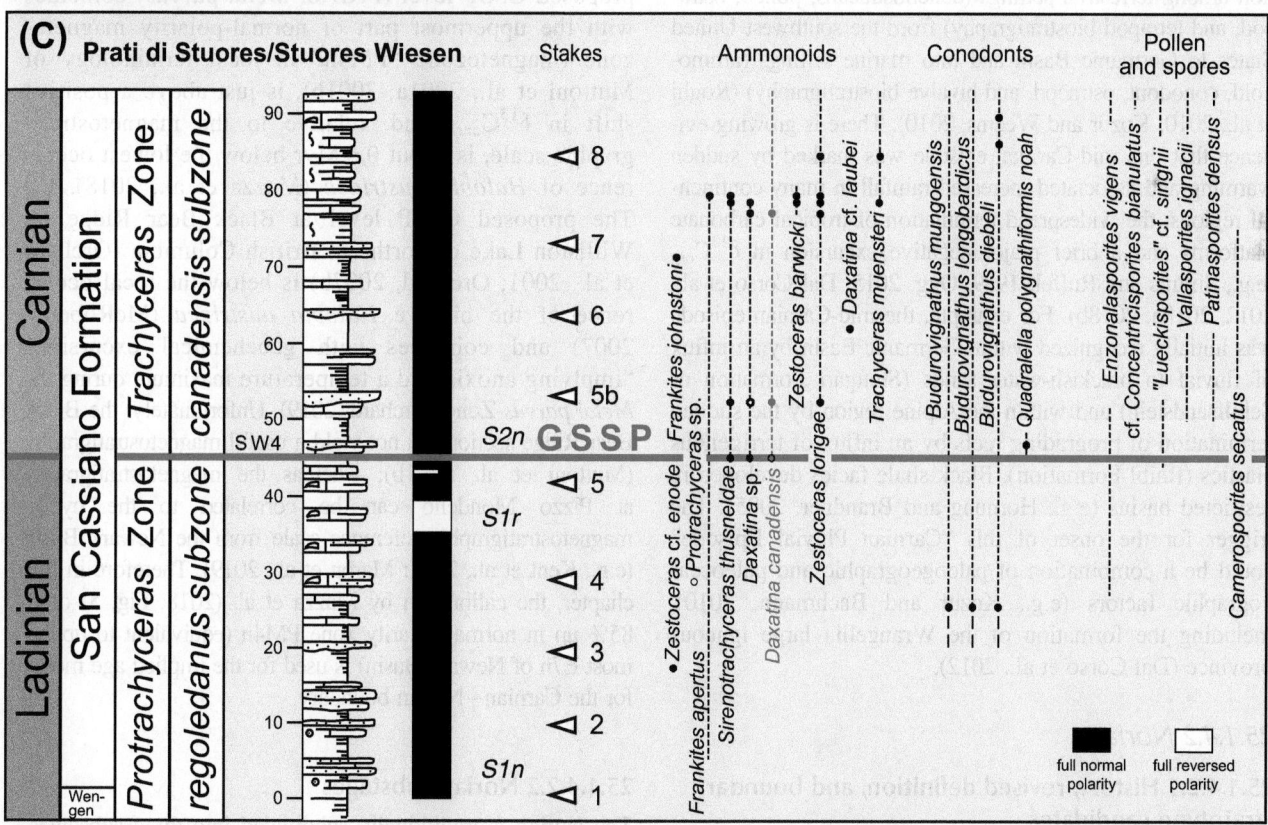

FIGURE 25.4 GSSP for the base of the Upper Triassic (base of Carnian Stage) at the Prati di Stuores (Stuores Wiesen) section, Dolomites region of north Italy. The GSSP level coincides with the lowest occurrence of the cosmopolitan ammonoid *Daxatina* (base of *Daxatina canadensis* Subzone, lowest subzone of *Trachyceras* Zone), and is near the lowest occurrence of conodont *Quadrella polygnathiformis* (inset "B," formerly assigned to the *Paragondolella* genus). Photographs of outcrop and conodont provided by Manuel Rigo. *GSSP*, Global Boundary Stratotype Section and Point.

Stratigraphers often combine Mojsisovics's Cordevolian and Julian into a single Julian Substage, with the Julian/Tuvalian (Lower/Upper Carnian) boundary traditionally assigned as the first occurrence of *Tropites* ammonoids (base of the *T. subbullatus* ammonoid zone of Tethys and *Tropites dilleri* Zone of Canada). Across this substage boundary, the change in ammonoid assemblages is more significant than at the bases of either the Carnian or the Norian stage (Tozer, 1984), the conodont diversity may have been reduced to two genera (Mazza et al., 2010), and there were major changes in radiolarians and other faunal groups (Kozur and Bachmann, 2010).

A dramatic event in the latest Julian that immediately preceded this Julian/Tuvalian boundary is considered to be "the most distinctive climate change within the Triassic" (Preto et al., 2010). This global disruption of the Earth's land-ocean-biological system has various regional names, for example, "Reingraben turnover" (Schlager and Schöllnberger, 1974), "Raibl Event," "Carnian Pluvial Episode" (Simms and Ruffel, 1989; Dal Corso et al., 2018a, 2018b), "Middle Carnian Wet Intermezzo" (Kozur and Bachmann, 2010), or "Carnian Humid Episode" (Ruffell et al., 2016). The distinct fossil assemblages within this "wet intermezzo" interval provide an important means of calibration among terrestrial settings (conchostracans, pollen, ostracod, and tetrapod biostratigraphy) from the southwest United States to Germanic Basin and into marine settings (ammonoid, conodont, ostracod, and bivalve biostratigraphy) (Roghi et al., 2010; Kozur and Weems, 2010). There is growing evidence that this mid-Carnian episode was marked by sudden warming and associated increased rainfall in many continental regions, the widespread termination of tropical carbonate platforms and a brief major negative excursion in $\delta^{13}C_{org}$ (e.g., Simms and Ruffel, 1989; Ogg, 2015; Dal Corso et al., 2012, 2018a, 2018b). For example, the mid-Carnian episode was initially recognized within Germanic Basin by an influx of fluvial to brackish-water sands (Stuttgart Formation or Schilfsandstein) and within the Alpine region by the sudden termination of prograding reefs by an influx of terrigenous clastics (Raibl Formation). Black shale facies developed in restricted basins (e.g., Hornung and Brandner, 2005). The trigger for the onset of this "Carnian Pluvial Episode" could be a combination of paleogeographic and paleoceanographic factors (e.g., Kozur and Bachmann, 2010), including the formation of the Wrangellia large igneous province (Dal Corso et al., 2012).

25.1.4.2 Norian

25.1.4.2.1 History, revised definition, and boundary stratotype candidates

Norian derives its name from the Roman province of Noria, south of the Danube and including the area of Hallstatt, Austria (Mojsisovics, 1869). The stratigraphic extent of strata assigned as "Norian" had a contorted history (reviewed in Tozer, 1984).

Ammonoid successions in Nevada and British Columbia led to a proposal that the base of the Norian is assigned to the base of the *Stikinoceras kerri* ammonoid zone, which is overlying the *Klamathites macrolobatus* Zone (Silberling and Tozer, 1968). This level is approximately coeval with a Tethyan placement between the *Anatropites* and *Guembelites jandianus* ammonoid zones (Krystyn, 1980; Orchard et al., 2000; Jenks et al., 2015). Examinations of conodont lineages have indicated that the first appearance of *Metapolygnathus echinatus* (reclassified as *Metapolygnathus parvus* by Orchard, 2014) is at approximately the beginning of the *S. kerri* ammonoid zone and coincides with a major faunal turnover (Orchard, 2010, 2014) and is approximately coincident with the FAD of the widespread *Halobia austriaca* bivalve. Therefore the *Meta. parvus* conodont level has become favored for assigning the base of the Norian (Mazza et al., 2018; Orchard, 2019).

There are two candidates for the Norian GSSP based on the FAD of *Meta. parvus* conodont: (1) The candidate of Pizzo Mondello in Sicily (Muttoni et al., 2001a; Nicora et al., 2007; Mazza et al., 2012; Balini et al., 2012) has a detailed magnetostratigraphy in which the proposed GSSP level (FAD of *Meta. parvus*) coincides with the uppermost part of normal-polarity magnetozone (magnetozone "PM4n" in local terminology of Muttoni et al., 2001a, 2001b), is just above a positive shift in $\delta^{13}C_{carb}$ and, relative to the magnetostratigraphic scale, is about 0.6 Myr below the lowest occurrence of *Halobia austriaca* (Mazza et al., 2018). (2) The proposed GSSP level at Black Bear Ridge on Williston Lake of northeast British Columbia (Orchard et al., 2001; Orchard, 2007b) is below the local occurrence of the bivalve *Halobia austriaca* (McRoberts, 2007) and coincides with geochemical excursions "implying anoxia and a temperature maximum during the *Meta. parvus* Zone (Orchard, 2019). Unfortunately, the Black Bear Ridge section did not yield a useful magnetostratigraphy (Muttoni et al., 2001b); whereas the magnetostratigraphy at Pizzo Mondello can be correlated to the cyclo-magnetostratigraphic reference scale from the Newark Basin (e.g., Kent et al., 2017: Maron et al., 2019). Therefore in this chapter, the calibration by Mazza et al. (2018; Fig. 5) of c. 85% up in normal-polarity zone PM4n (equivalent to uppermost E7n of Newark Basin) is used for the implied age model for the Carnian—Norian boundary.

25.1.4.2.2 Norian substages

The Norian is traditionally subdivided into three substages, following Mojsisovics et al. (1895). The boundary between the lower Norian (or "Lacian," after the Roman name for the Salzkammergut region of the northern Austrian Alps) and middle Norian (or "Alaunian," named for the Alauns, who lived in the Hallein region of Austria during Roman times) is the base of the Tethyan *Cyrtopleurites bicrenatus* ammonoid zone. The base of the upper Norian (or "Sevatian," after the Celtic tribe who lived between the Inn and Enns rivers of Austria) is generally assigned as the base of the North American *Gnomohalorites cordilleranus* ammonoid zone or the Tethyan *Sagenites quinquepunctatus* ammonoid zone; however, there has not been a consistent usage of this Sevatian Substage and some include the underlying *Halorites macer* ammonoid zone within it (e.g., Kozur, 1999).

25.1.4.3 Rhaetian

The Rhaetian was the first Triassic stage to be established, when von Gümbel (1861) applied the term to strata containing the pteriid bivalve *Rhaetavicula contorta*, such as the Kössen Beds of Austria. This distinctive bivalve is found in shallow-marine facies from the western Tethys and across northwestern Europe. His "Rhätische Gebilde" name was derived from either the Rhätische Alpen or the Roman province of Rhaetium. For a while, it appeared that the Rhaetian interval would be incorporated into the Jurassic (and perhaps renamed as a "Bavarian Stage) or incorporated into the Norian Stage (reviewed in Lucas, 2010a). For example, the Rhaetian was eliminated in some Triassic time scales (e.g., Zapfe, 1974; Palmer, 1983; Tozer, 1984, 1990). In 1991 the Subcommission on Triassic Stratigraphy decided to retain the Rhaetian as an independent stage. Many options were considered for the primary biostratigraphic marker for the lower boundary.

In 2010 the Norian–Rhaetian boundary working group decided that a GSSP level should coincide with lowest occurrence of the conodont *Misikella posthernsteini* (Krystyn, 2010). This conodont is a phylogenetic descendent of *Misikella hernsteini* but is very rare at the beginning of its range. However, even though this morphogenesis is seen worldwide, it appears that the distinctions between the component taxa are not standardized (e.g., discussions in Rigo et al., 2016; Lucas, 2016; Orchard, 2016). Therefore secondary markers should also be employed to assign the base of the Rhaetian (Krystyn, 2010) including the possible: (1) lowest occurrence of conodont *Epigondolella mosheri* (morphotype B sensu Orchard), (2) lowest occurrence of ammonoid *Paracochloceras suessi* and the closely allied genus *Cochloceras* and other taxa, (3) disappearance of ammonoid genus *Metasibirites*, (4) lowest occurrence of radiolarian *Proparvicingula moniliformis* and other species (Carter and Orchard, 2007), (5) disappearance of *Monitis* bivalves, except for continuation by dwarf *Monotis* species in parts of the Tethys (McRoberts et al., 2008), and especially a (6) magnetostratigraphy which would allow a correlation to the cyclo–magnetic reference scale from the Newark Basin.

In the candidate GSSP section at Steinbergkogel near Hallstatt in Austria (Krystyn et al., 2007c, 2007d), the interpreted base of *Mi. posthernsteini* is just above a change from a major normal-polarity magnetozone upward to a reversed-polarity-dominated magnetozone (Gallet et al., 2007; Muttoni et al., 2010; Hounslow and Muttoni, 2010; Hüsing et al., 2011). In contrast, at the candidate section of Pignola-Abriola in Sicily, the interpreted base of the *Mi. posthernsteini* conodont is very high within a reversed-polarity-dominated magnetozone (Maron et al, 2015; Rigo et al., 2016). It appears that the finer features of both of these reversed-dominated magnetozones can be reliably correlated between Austria and Sicily; therefore the proposed correlation of these Rhaetian magnetostratigraphies to the astronomical-cycle-scaled magnetic polarity reference pattern of the Newark Basin implies nearly a 4-million-year offset of the interpreted bases of *Mi. posthernsteini* between those two candidate GSSPs (Maron et al., 2015; Rigo et al., 2016). To explain this discrepancy, Maron et al. (2015) proposed that the interpreted earliest form of *Mi. posthernsteini* at the Austria GSSP candidate is an initial transitional form (sensu lato) at c. 209.5 Ma, whereas the interpreted lowest *Mi. posthernsteini* at the Sicily GSSP is the developed form (sensu stricto) with an estimated age of 205.7 Ma. The proponents of the Pignola-Abriola section consider that the higher level (and the implied shorter-duration Rhaetian) to be more consistent with the traditional recognition of the base of the Rhaetian in other regions, and note that the proposed GSSP level is near a negative isotope excursion in organic-carbon (Rigo et al., 2016). A negative excursion in $\delta^{13}C_{org}$ at the same cyclo-magnetostratigraphic level is observed in terrestrial deposits in Sichuan, China (Li et al., 2017a, 2017b). However, in the Austrian GSSP candidate, the two conodont morphotypes are only separated by one bed (Galbrun et al., 2020).

Both options for the Rhaetian GSSP depending on the eventual morphotype for *Mi. posthernsteini* for the Rhaetian GSSP—the sensu latu (s.l.; Austria candidate) and sensu stricto (s.str.; Italian candidate) are shown on the figures in this chapter; but the Italian "short Rhaetian" is used for the Phanerozoic synthesis.

25.2 Triassic stratigraphy

Ammonoids dominate the historical zonation of the Triassic, but conodonts have become the major tool for global correlation. Thin-shelled bivalves (e.g., *Daonella*, *Halobia*) provide important regional markers. During much of the Triassic, the sedimentary record across the Pangea Supercontinent was dominated by terrestrial deposits, therefore widespread conchostracan, tetrapod, and plant remains are important for global correlation.

Other biostratigraphic, magnetostratigraphic, chemostratigraphic, and other events are typically calibrated to these standard ammonoid or conodont zones. Extensive compilations and intercorrelation of Triassic stratigraphy of European basins were coordinated by Hardenbol et al. (1998), and a suite of detailed Triassic stratigraphic scales are in *The Triassic Timescale* (Lucas, 2010a).

25.2.1 Marine biostratigraphy

25.2.1.1 Ammonoids

The ammonoid successions of the Alps and Canada have historically served as global primary standards for the Triassic (reviews in Tozer, 1967, 1984; Balini et al., 2010). Ammonoids were nearly extinguished at the end of the Permian, with termination of three of the four major Permian clades. However, the rare surviving ammonoids diversified much faster than other marine groups following this catastrophe (Brayard et al., 2009a; Marshall and Jacobs, 2009). The entire Triassic ceratitid group of Ammonoidea is usually considered to have been derived from the morphologically simple *Ophiceras* genus survivor. This rapid surge in diversity was interrupted by major waves of extinction at the end of the Smithian Substage (mid-Olenekian), base of Anisian and end of Julian Substage (mid-Carnian) (e.g., Brayard et al., 2009a, 2009b; Jenks et al., 2015). Peak diversity was attained in middle Norian. A decline in diversity through the latest Triassic culminated in apparently only a single genus, *Psiloceras*, surviving the end-Triassic mass extinctions and rapidly evolving to conquer the Jurassic seas.

Despite their historical importance in subdividing the Triassic, there is not yet a standardized ammonoid zonation (or nomenclature) for the alpine regions. For example, Mietto and Manfrin (1995) proposed a generalized standard for the Middle Triassic of the Tethyan realm that utilized first appearances of widespread genera to define zones and major species to define subzones. But this zonal scheme was immediately rejected by some alpine workers (e.g., critiques by Brack and Rieber, 1996; Vörös et al., 1996). Contributing to this situation is the lack of consensus definitions of species and genera, historical confusions with taxonomy and relative chronostratigraphic placement of taxa, distinct latitudinal and endemic assemblages, use of assemblage or association zones, omission by many authors to provide clarity in their definition of zonal boundaries, and a current reduction in the number of active ammonoid specialists (Balini et al., 2010).

A selection of Triassic zonations and their main index ammonoids are figured in Balini et al. (2010) and in Jenks et al. (2015). The ammonoid zonal columns in the charts in this chapter are generalized versions for the Tethyan and for a blend of the West Canadian Sedimentary Basin and Canadian Arctic realms and are derived mainly from Jenks et al. (2015) and from advice from M. Orchard, respectively.

25.2.1.2 Conodonts

Conodonts are phosphatic feeding apparatuses of an enigmatic pelagic swimmer. The conodont taxa are based on variability of these jaw-like elements, and these evolving features enable widespread correlation of Triassic strata. After surviving the end-Permian mass extinctions and explosively diversifying in the early Olenekian (Orchard, 2007a), the conodonts tragically vanished shortly before the end of the Triassic. Conodonts are generally more widespread than ammonites both in paleogeography and in different marine facies and are approaching a well-defined taxonomy and biostratigraphy (e.g., Anisian through Carnian synthesis by Chen et al., 2016; Carnian through Rhaetian synthesis by Rigo et al., 2018; discussions between Orchard, 2016, and Lucas, 2016). However, there remains a lack of standardization of taxonomic groupings among conodont specialists which implies that publications commonly place species-level taxa under different genera.

Compilations of the calibrations among conodont zones and ammonoids have been proposed for several realms, including Canada (e.g., Orchard and Tozer, 1997; Orchard, 2010, 2014, 2018; Golding, 2019; Golding et al., 2014, 2017), Tethys region (e.g., Krystyn et al., 2002; Kozur, 2003; Rigo et al., 2018), China (Tong et al., 2019), and European basins (Vrielynck, 1998). The generalized columns in the charts in this chapter are composites for the Tethyan (Kozur, 2003; with Carnian through Rhaetian of Moix et al., 2007, as shown in Rigo et al., 2018), for eastern Tethys (Rigo et al., 2018), for China (Tong et al., 2019, with selected enhancements by Haishui Jiang, pers. commun.) and for the Western Canadian Sedimentary Basin or "Arctic/Panthalassa" (contributed by Michael Orchard and Martyn Golding for inclusion in a future synthesis chart by Robert Fensome and Manuel Bringué at the Geological Survey of Canada). Names of genera were standardized to the taxonomy preferred by Michael Orchard.

25.2.1.3 Bivalves

The end-Permian mass extinctions terminated the dominance by the communities of brachiopods, crinoids (and other pelmatozoan echinoderms), and bryozoans that had been typical of Paleozoic marine seabeds. Bivalve and gastropod mollusk communities are typical of Early Triassic shelves, with a later increase in the importance of scleractinian corals. Among the Triassic bivalves, a succession of pelagic forms of thin-shelled "flat clams" (pectinacean pteriid bivalves) was largely restricted to deeper water settings and tend to occur in high densities on certain bedding planes (Hallam, 1981). The distinctive thin-shelled bivalve genera (*Claria*, *Enteropleura*, *Daonella*, *Halobia*, *Monotis*, etc.), which have no modern counterparts, are valuable for global Triassic correlations because their species have both widespread distributions and a mean duration of only 1−2 Myr. Their Jurassic and Cretaceous relatives (*Bositra*, *Buchia*, inoceramids) occupied the same settings. The first comprehensive summary of the ranges of these Triassic pelagic bivalves relative to ammonoid zones in different biogeographic provinces was by McRoberts (2010), who defined 30 discrete zones and regional variants.

25.2.1.4 Radiolarians

The sand-sized opaline skeletons of radiolarians are typically found in Triassic deep-marine facies. After the end-Permian, the most severe extinction event for radiolarians, diversity slowly increased through the Early Triassic, then surged in the Anisian to reach a maximum in early Carnian. Diversity declined through the Late Triassic until the end-Triassic mass extinctions again decimated the radiolarian genera (O'Dogherty et al., 2010).

Radiolarian datums and zonations have been compiled for different regions, including alpine exposures (e.g., De Wever, 1982, 1998; Kozur and Mostler, 1994; Kozur, 2003), Japan (e.g., Sugiyama, 1997), and western North America (e.g., Blome, 1984; Carter, 1993; Carter and Orchard, 2007). A correlation among these regional zonations and a summary for the ranges of the main 281 genera (including schematic images) relative to Triassic substages are compiled by O'Dogherty et al. (2010).

25.2.1.5 Other microfossils

Except for radiolarians, marine microfossil biostratigraphy has not yet been developed as a widespread correlation tool within the Triassic. In contrast to Permian and Jurassic syntheses, the benthic foraminifer stratigraphy of the Triassic has not been compiled on a global scale. Regional scales include a stratigraphic summary of larger benthic foraminifera of the Tethyan realm (Peybernes, 1998) and zonations proposed for the Caucasus (e.g., Vuks, 2000, 2007).

Calcareous nannofossils are only known from Carnian and younger strata, and the first "real coccoliths" appear in the Norian (von Salis, 1998; Bown, 1998).

Records of dinoflagellates are rare, and the oldest representative of this group may be middle Triassic (Hochuli, 1998). However, the latest Triassic species, *Rhaetogonyaulax rhaetica*, is an important means for correlating Rhaetian strata.

25.2.1.6 Reefs

Although not biostratigraphy, per se, there are broad trends with global shallow-water carbonate systems during the Triassic. The metazoan communities that built Permian reefs were completely destroyed in the end-Permian mass extinction, and a "reef gap" ensued that spanned much of the Early Triassic. Simultaneously, there was an Early Triassic "chert gap" following the termination of the extensive Permian chert

accumulations (Beauchamp and Baud, 2002). The accompanying extinction of many benthic grazers probably contributed to the distinctive microbial-dominated carbonates (stromatolites, wavy-laminated micrites, oolites) that are characteristic of the earliest Triassic (Flügel, 2002; Chen and Benton, 2012; Chen et al., 2019). Elevated carbon dioxide levels and/or a pulse of carbonate saturation following CO_2-induced dissolution in an ocean that lacked deep-water carbonate buffering may have contributed to this latest Permian and earliest Triassic interval of microbial and oolitic limestones (e.g., Payne et al., 2007, 2009; Xie et al., 2010). This Early Triassic "reef gap" was mainly caused by a lack of reef-building biota, rather than a crisis in carbonate production/accumulation (Preto et al., 2010). Microbial-sponge and *Tubiphytes* reef communities with a modest contribution from the oldest crown-type scleractinian corals are characteristic of the middle and late Anisian (Kiessling, 2010). Growth rates for some late Anisian through Ladinian reefs were phenomenal, with radioisotopic dates suggesting up to 500 m/Myr for some platforms in the Dolomites of northern Italy (e.g., Brack et al., 2007; Meyers, 2008). The Ladinian-earliest Carnian peak of these microbial-sponge reefs was terminated by the events associated with the mid-Carnian pluvial event (wet intermezzo) (e.g., Stefani et al., 2010). Scleractinian corals contributed to the second phase of reef expansion, which peaked in the Late Norian (Kiessling, 2010). A rapid sea-level fall near the time of the Norian/Rhaetian boundary contributed to the end-Norian Cessation of the spectacular Dolomia Principale/Hauptdolomit platforms of the Alps and other regions (e.g., Berra et al., 2010). The end-Rhaetian wave of extinctions led to a reef crisis that mirrored the end-Permian episode, and reefs were again rare through much of the Early and Middle Jurassic (Kiessling, 2010).

25.2.1.7 Marine reptiles

Beginning in the late Early Triassic, the seas became inhabited by large marine reptiles. Ichthyopterygia ("fish flippers"; or Ichthyosaurs) with their dolphin-like morphology and Sauropterygia (lizard flippers) that retained a more crocodile-like form rapidly diversified during the late Early Triassic (e.g., Motani, 2010). The adaptation of the ichthyosaurs to the open seas, perhaps including a warm-blooded anatomy, enabled their expansion in the Middle and Late Triassic as coastal habitats for other marine reptiles declined (Motani, 2010; Renesto and Dalla Vecchio, 2018). Most of the Sauropterygia clade vanished during the Late Triassic prior to the end-Triassic mass extinction of other marine fauna. The open-ocean long-necked plesiosaurs of the Sauropterygia clade and the parvipelvian group of ichthyosaurs appeared during the Late Triassic and underwent radiation during the Jurassic (Renesto and Dalla Vecchio, 2018).

25.2.2 Terrestrial biostratigraphy

The terrestrial successions of tetrapods (amphibians, reptiles, earliest mammals), plants (spores/pollen, macroflora), and lacustrine organisms (ostracods, and especially conchostracans) are broadly intercorrelated in different regions of Pangea. The calibration of the evolving terrestrial ecosystems to marine-based Triassic stages and substages is fairly well established for the Lower Triassic via a distinctive cycle-scaled magnetostratigraphy in the Germanic Basin, but ties are much less certain and controversial through the Middle and Upper Triassic. An unambiguous correlation of the cycle-scaled magnetostratigraphy of the Carnian–Norian lacustrine-rich strata of the Newark group from the rift valleys of easternmost North America to the magnetostratigraphy of Tethyan marine zones would enable a precise Late Triassic time scale. The interregional calibration of the pollen-tetrapod-conchostracan occurrences within those Newark beds is a work in progress (e.g., Kozur and Weems, 2010; Irmis et al., 2010; Lucas, 2010c, 2010d).

Other nonmarine biostratigraphy, including charophyte cysts (gyrogonites or calcified spores of charophyte green algae), macrofossil plant assemblages, ostracods, bivalves, and fishes, can be useful on a regional scale, but have not yet been developed for use in global correlations (e.g., reviews by Lucas, 2010c, 2018c).

25.2.2.1 Tetrapods and dinosaurs

The Mesozoic is popularly known as the "Age of Dinosaurs," but these famed reptiles did not diverge from other tetrapods until the mid-Triassic. The suite of proposed global zones for tetrapod stratigraphy for Pangea is largely based on widespread occurrences of select semiaquatic tetrapods (both reptiles and amphibians) whose thick skulls, body armor, or other resistant body parts were commonly preserved in the sedimentary record. The Pangean configuration enabled many of these forms to spread across most of the world's land area, therefore the lowest occurrences of distinctive widespread genera enable subdivision of the Triassic into eight "land-vertebrate faunachrons" (LVFs) (Lucas, 1998, 1999, 2010d, and references therein). These "LVFs" have an intercontinental correlation potential that is approximately equivalent to standard Triassic stages; although the placement relative to marine-based substages has not yet been firmly established for all LVFs (e.g., Irmis et al., 2010; Lucas, 2018a, 2018b). When skeletal material is not available, distinctive footprints of tetrapods are a secondary, but less precise, method of correlation (Klein and Lucas, 2010).

The predominant terrestrial or semiaquatic forms in the Early Triassic were therapsids, a major lineage of synapsids (fused arch) that are considered to be "mammal-like reptiles" (e.g., Fraser, 2006). Three groups of therapsids survived the end-Permian mass extinction—cynodonts (dog-toothed), dicynodonts ("two dog-toothed," with two tusks in its upper jaw), and large-skulled therocephalian (beast head) carnivores. The therapsids diminished in importance through the Triassic, but true mammals, of which the oldest record is a mouse-sized *Adelobasileus* of Late Triassic (e.g., Lucas, 2008), are interpreted to be derived from this group. The type-area for the Early Triassic LVFs is the Karoo Basin in South Africa. The Lootsbergian LVF begins in latest Permian with the first appearance of *Lystrosaurus*, a squat, dog-sized, dicynodont herbivore that may have been semiaquatic. The extinction of the *Dicynodon* dicynodont is approximately at the end of the Permian (Lucas, 2009). The lowest occurrence of the robustly built *Cynognathus* cynodont defines the base of the Nonesian LVF, which is approximately equivalent to the marine Olenekian Stage.

The Middle Triassic LVFs (Perokvan and Berdyankian) are derived from the lowest occurrences of massive temnospondyl amphibians (*Eocyclotosaurus* and *Mastodonsaurus giganteus*, respectively) in the Ural foreland basin in Russia.

Diapsid (two arches) reptiles had diverged into different major orders during Early Triassic, and its Archosauria (ruling reptiles) group (dinosaurs, pterosaurs, crocodiles, etc.) would dominate the terrestrial and aerial ecosystems of the Jurassic through Cretaceous. The first occurrences of types of crocodile-like phytosaurs and of large armored herbivore aetosaurs, which resembled elongate armadillos with stout claws, define most of the Late Triassic LVFs. The fossil-rich deposits of the Chinle basin of southwest United States are the main reference sections for four Late Triassic LVFs.

The Otischalkian and Adamanian LVFs of Carnian are defined by FADs of phytosaurs *Parasuchus* (=*Paleorhinus*) and *Rutiodon*, respectively. The beginning of the Revueltian LVF that begins with the aetosaur *Typothorax coccinarium* is estimated to begin near the base of the Norian by Lucas (2018b), but he also places that base at c. 220 Ma, which corresponds to lower-middle Norian and is used on the charts in this chapter. However, Ramezani et al. (2009) place the Adamanian–Revueltian faunal turnover between 219 and 213 Ma based on an array of radiometric ages from the Chinle Formation of southwest United States. The Apachean LVF spans approximately the Late Norian through Rhaetian and begins with by FAD of the phytosaur *Redondasaurus* (e.g., Lucas, 2010d, 2018a, 2018b, 2018c). The Jurassic begins with the FAD of the crocodylomorph *Protosuchus* that defines the base of the Wassonian LVF.

Within this tetrapod-LVF framework, the main groups of dinosaurs seem to have radiated during the late-Middle Triassic (e.g., Heckert and Lucas, 1999; Fraser, 2006). The major dinosaur lineages (Sauropodomorpha, Theropoda, and Ornithischia divisions) became established during the Carnian (Nesbitt et al., 2009; Lucas, 2010a, 2010b, 2010c, 2010d; Benton et al., 2014). But these dinosaurs did not become the dominant land reptiles until the end-Triassic mass extinctions removed their competition from other terrestrial reptiles.

25.2.2.2 Conchostracans (now renamed as Spinicaudata)

Brackish-to-freshwater crustaceans of clam shrimp that were formerly known as conchostracans have a chitinous bivalve carapace of two lateral valves that are typically 2−40 mm in length. However, the group that was called "conchostracans" is now considered to be paraphyletic, therefore this group within the Class Branchiopida have been renamed Spinicaudata (Olesen, 2009, as cited in Geyer and Kelber, 2018). This clade once also lived in the sea during the Paleozoic and early Triassic, then became restricted to non-marine environments.

Their tiny drought-resistant eggs were easily dispersed by wind and water, and rapidly hatched upon exposure to suitable environments. These characteristics and their brief life cycle of only 1−3 weeks enabled conchostracans to be widespread in lakes, shallow seas and temporary pools throughout Pangaea. Their distinctive carapaces were preserved in lacustrine, shallow sea, salt flat, and floodplain deposits. Even through conchostracans are present in pre-Permian through modern sediments, their potential for biostratigraphy and interregional correlation has only been fully developed in Triassic strata (e.g., Kozur and Weems, 2010, and references therein). The application of conchostracan biostratigraphy for other time intervals awaits careful studies of taxonomy, recognition of distinctive taxa, and calibration to other biostratigraphic scales.

Approximately 30 conchostracans zones and regional variants have been defined within the Triassic. These zones have been recognized and intercorrelated within the Germanic Basin, through the southwest United States, in the Newark Supergroup of easternmost North America and within other regions (summarized in Kozur and Weems, 2010, 2011; see also Kozur and Bachmann, 2008). Correlations to marine-based Triassic substages are partially constrained by levels containing ostracods and palynology markers, distinctive facies shifts (e.g., the mid-Carnian "Pluvial Event") and cycle-scaled magnetic magnetozones. The framework established by Kozur and coworkers in the Germanic and Newark basins has been enhanced and modified for Carnian−Norian (Weems and Lucas, 2015; Geyer and Kelber, 2018; Franz et al., 2018) and for the late Permian through Early Triassic (Scholze et al., 2016; Schneider and Scholze, 2018). In particular, the occurrence of *Euestheria gutta* beginning at the base of the Triassic has enabled examination of the response and recovery of terrestrial environments to the catastrophic end-Permian climatic events in basins and terrestrial-marine transitional settings from Germany to southwestern and northern China (e.g., Chu et al., 2019; Zhu et al., 2019; Scholze et al., 2019). The zonal detail of conchostracans is also important for the calibration of the tetrapod LVFs to marine substages (e.g., Weems and Lucas, 2015) and aids in projecting magnetostratigraphy and radioisotopic ages between terrestrial and marine strata.

25.2.2.3 Plants, pollen, and spores

Spores and pollen are important for correlation of marine and terrestrial strata. However, most taxa have relatively long ranges, and changes in assemblages may indicate local climatic-ecosystem shifts rather than a useful temporal marker. Major compilations of Triassic palynology and plant ecosystem evolution are by Wing and Sues (1992), Warrington (2002), Traverse (2007), Yu et al., 2015; Nowak et al., 2018; and Kustatscher et al., 2018. Palynoflora zonations for Triassic strata have been compiled for the Alpine and Germanic regions (e.g., Visscher and Brügman, 1981; Visscher et al., 1994; Hochuli, 1998; Herngreen, 2005; Kürschner and Herngreen, 2010), Australia (Helby et al., 1987), southwest United States (Litwin et al., 1991; Cornet, 1993), Newark basins of eastern United States (e.g., Cornet, 1977; Cornet and Olsen, 1985), and Arctic (Van Veen et al., 1998). However, only the Late Triassic and early Jurassic have a detailed interregional compilation among different continents (Cirilli, 2010; Kustatscher et al., 2018).

Regional diversity of palynomorphs declined at the time of the end-Permian mass extinctions (Kürschner and Herngreen, 2010). Latest Permian plant ecosystems may have undergone a short-lived explosion in abundance in lycopods that interrupted the dominance by gymnosperms to create a c. 10-kyr "spore-spike" within the major negative C-13 isotope decline, but the plant communities rapidly recovered before the main catastrophic marine extinction (Hochuli et al., 2010). Widespread marine clastics of lower Triassic (Induan to lower Olenekian) record a uniquely cosmopolitan "acritarch spike" assemblage of lycopsid spores, small acanthomorph acritarchs, and *Lunatisporites* coniferalean pollen (e.g., Balme and Foster, 1996). Diversity of both pollen and spores peaked in the Germanic Basin during Ladinian−Carnian (Kürschner and Herngreen, 2010; Kustatscher et al., 2018). Globally during this time, there is a rapid diversification of Circumpolloid genera. This broad episode is followed, at least in the Tethyan realm, by a gradual change in palynofloral assemblages without abrupt changes from Carnian into earliest Hettangian (e.g., Cirilli, 2010).

Even though the palynology shifts are generally gradual and mainly important at regional scales, the trends can help identify major hiatuses and correlations in terrestrial records. For example, the strata slightly below flood basalts in the Newark rift basins in the eastern North America display a relatively sharp transition from diverse assemblages of monsaccate and bisaccate pollen to an overlying assemblage containing 60%−90% *Corollina meyeriana* spores, and this level was considered as a regional marker for the Triassic/Jurassic boundary (e.g., Cornet, 1977; Cornet and Olsen, 1985; Fowell and Olsen, 1993, 1995). However, similar palynological changes are recorded near the base of the typical Rhaetian of Europe (e.g., Schuurman, 1979; Orbell, 1983; Van Veen, 1995), therefore some palynologists have interpreted this level to be a major hiatus that includes most of the Rhaetian stage (e.g., Cirilli et al., 2009), as was initially also suggested by the conchostracan assemblages (Kozur and Weems, 2005, 2010). This interpretation of the absence of the uppermost Norian and most of the Rhaetian between the radioisotopic-dated flood basalts and the underlying cycle-scaled magnetozones of the Newark Supergroup lacustrine strata contributed to a pair of Late Triassic age models in GTS2012 (Ogg, 2012); but it now seems from later magnetostratigraphic correlations to marine deposits and to levels dated by radioisotopes in a reference borehole on the Colorado Plateau that there is no significant hiatus in the uppermost Rhaetian of the Newark Basin (Kent et al., 2017, 2018).

25.2.3 Physical stratigraphy

25.2.3.1 Magnetostratigraphy

The compilation of the magnetic polarity time scale for the Triassic is better developed than for the Jurassic or for any of the Paleozoic systems. A concentrated effort by paleomagnetists working closely with biostratigraphers and cyclostratigraphers during the 1990s and first decade of the 21st century has revealed approximately 50 main magnetozones and twice as many lesser polarity subzones (Hounslow and Muttoni, 2010). The composite polarity scale from marine strata has been calibrated to ammonoid and conodont datums in different regions. A parallel polarity scale from terrestrial settings has been partly correlated to conchostracan zones and scaled with cycle-stratigraphy. Magnetic reversals are globally synchronous; therefore the placement of biostratigraphic datums relative to distinctive polarity patterns has been used to determine diachroneity or local distortions in relative timing of markers and a polarity boundary may potentially be used as the primary marker for at least one Triassic GSSP (e.g., GSSP for Anisian by Hounslow et al., 2007). Within Triassic substages, the average of about four major and five minor polarity intervals commonly display a characteristic pattern. However, the sheer abundance of magnetic polarity chrons and lack of broad fingerprints in the patterns at the stage-level implies that utilization of Triassic magnetostratigraphy for interregional or interfacies correlations requires adequate biostratigraphic constraints.

A total of the 133 validated magnetozones in the Triassic have a mean reversal rate of 2.6/Myr, which is similar to the average Cenozoic rate of geomagnetic reversals (Hounslow and Muttoni, 2010). The apparent reversal rate of the Early

and Middle Triassic (4/Myr) is twice the average rate within the Late Triassic.

Each magnetostratigraphy study employed a different system for labeling the observed magnetozones. The verification of these studies by demonstrating reproducibility within the biostratigraphic constraints among different regions enables a nomenclature for the main and minor polarity chrons. In the milestone synthesis by Hounslow and Muttoni (2010), the subjective groupings into main magnetozones are systematically numbered upward within each Triassic series (e.g., MT7 is the seventh "main" polarity episode in the Middle Triassic, although it may contain several brief polarity subchrons, with a normal-polarity-dominated "MT7n" portion followed by a reversed-polarity-dominated "MT7r" portion). An alternative would be to label each magnetozone according to its placement relative to Triassic substages (e.g., Illy-N2; for the second cluster of normal-polarity in the Illyrian Substage of Anisian; which is the same as "MT7n," thereby enabling a user to know the approximate geologic-age relationship (e.g., scales in Ogg et al., 2008). This "stage-abbreviation-then-number" system is the same nomenclature philosophy as used for Phanerozoic sequences (e.g., Hardenbol et al., 1998; Haq, 2018). A third option used by Maron et al. (2019) was to select a single detailed magneto-biostratigraphic reference section (e.g., Seceda in Italy for lower Ladinian) and use its polarity zones ("SC1n," "SC2r," etc.) and their relative thicknesses as the nomenclature and scaling for the global polarity pattern. It was not convenient to show all three systems in the summary figures in this chapter; therefore a hybrid set was used that partly depended upon a subjective decision on the relative documentation of the polarity zones. In any case, there is a caveat that portions of these composite polarity patterns are not well calibrated to boundaries of ammonoid (or conodont) zones and the scaling of the patterns within zones is probably distorted by variable sedimentation rates in the reference sections.

The Triassic magnetic polarity reference scales derived from biostratigraphic-dated marine successions (e.g., composite syntheses by Hounslow and Muttoni, 2010, and Maron et al., 2019) and for astronomically scaled terrestrial basins (e.g., Early Triassic by Szurlies, 2007; Late Triassic by Kent and Olsen, 1999) have been verified by extensive conodont-dated magnetostratigraphy in European and Chinese sections (e.g., Muttoni et al., 2014; Maron et al., 2015, 2019; Lehrmann et al., 2015) including cycle-scaling of significant intervals (e.g., Li et al, 2016, 2017a, 2018; Zhang et al., 2015). The comparison between the magnetostratigraphies of marine sections and of cycle-scaled terrestrial sections appears to have resolved some of the uncertainties about the age models for the Late Triassic and for the Early Triassic. For example, the "long Rhaetian-short Tuvalian" option that was preferred for the GTS2012 scale (Ogg, 2012; Ogg et al., 2014) is consistent with those later compilations; although that Rhaetian appears to correspond to the *"sensu lato"* interpretation of the marker conodont taxon rather than the shorter *sensu stricto* version adapted for the composite Phanerozoic scale (Maron et al., 2015).

The following summary is mainly from Hounslow and Muttoni (2010) and Maron et al. (2019), and only a selection of their main reference sections for biomagnetic calibrations and conclusions are given here. In addition to placing the generalized patterns onto the relevant biostratigraphic scales, they have standardized the stratigraphy of the different regional sections onto a common Tethyan ammonoid zone or substage scale.

25.2.3.1.1 Early Triassic magnetic polarity scales

The Induan stage has two pairs and the Olenekian has seven pairs of polarity chrons that are correlated to Boreal ammonoid and conodont satums in the Arctic (e.g., Ogg and Steiner, 1991; Hounslow et al., 2008a, 2008b) and to conodont ranges in several Tethyan sections in China (e.g., Steiner et al., 1989; Heller et al., 1995; Glen et al., 2009; Li et al., 2016), Iran, and Italy. The Permian—Triassic boundary is near the base of a relatively long normal-magnetozone ("LT1n" or "Gries-N1"; Steiner, 2006). This feature can be identified in the composite magnetostratigraphy from terrestrial deposits within the Germanic Basin (e.g., Nawrocki, 1997; Szurlies, 2004, 2007; Scholze et al., 2016), implying that the base-Triassic is approximately at the transition from the Zechstein evaporite-dominated strata to the Buntsandstein Formation.

The correlation of the marine-based magnetic polarity scale to this Germanic Basin polarity pattern is well established. There is a distinctive dominance of reversed-polarity in the upper Induan and lowermost Olenekian followed by predominantly normal polarity in upper Lower Triassic in both scales, which enables a one-to-one correlation of the individual cycle-scaled Buntsandstein magnetic polarity chrons to those in the marine composite (e.g., Szurlies, 2007; Hounslow and Muttoni, 2010). The same change in polarity dominance and other features are tentatively used to assign geologic stages/substages to terrestrial deposits in the Karoo Basin, southwestern United States, and Russia (Hounslow and Muttoni, 2010; Fig. 4). Therefore the proposed astronomical cyclicity can be projected via these biomagnetostratigraphic correlations to estimate the time spans for Lower Triassic stages and substages (Li et al., 2016).

25.2.3.1.2 Middle Triassic magnetic polarity scales

The seven main polarity pairs of the Anisian display dominance by normal polarity in the lower substages (Aegean, Bithynian) followed by a relative dominance by reversed-polarity through the lowermost Ladinian. This pattern is derived from sections containing Boreal ammonoid zones in Spitsbergen (arctic Norway; Hounslow et al.,

2008a, 2008b) and from those with conodont ranges in China (e.g., Lehrmann et al., 2006, 2015), Albania (e.g., Muttoni et al., 1996b), Romania (e.g., Gradinaru et al., 2007), Austria (Muttoni et al., 1996a), and other regions. The distinctive switch in dominance of polarity is the first-order constraint on projecting ages onto terrestrial facies in the Germanic Basin (e.g., Nawrocki and Szulc, 2000; Szurlies, 2007) and the Catalan and Iberian basins of Spain (e.g., Dinarès-Turell et al., 2005), and to England and China (Hounslow and Muttoni, 2010; Fig. 4).

The main reference sections for the eight main polarity pairs of the Ladinian are in the alpine region of Italy and Austria (Seceda, Mayerling, Stuores sections; e.g., Gallet et al., 1998; Broglio Loriga et al., 1999; Muttoni et al., 2004a; Maron et al., 2019).

25.2.3.1.3 Late Triassic magnetic polarity scales

The alpine Mayerling and Stuores sections, plus the Wayao cyclo-magnetostratigraphic section in Guizhou province of South China (Zhang et al., 2015) and the Bolücektasi Tepe in Turkey (Gallet et al., 1992) are the main reference sections for the Julian Substage (lower Carnian). There had been no adequate biostratigraphic controls on magnetostratigraphic studies that might span the upper Carnian (Tuvalian Substage); therefore Hounslow and Muttoni (2010) assigned a "holding" nomenclature of "UT5−UT12." This interval in boreholes through nonmarine strata in the Germanic Basin is dominated by normal polarity above a lowermost Tuvalian reversed-polarity zone (Zhang et al., 2020).

The uppermost Carnian through Norian pattern is mainly derived from the GSSP candidate for the base-Norian at Pizzo Mondello in Sicily (Muttoni et al., 2001, 2004b) and the Silickà Brezovà section of Turkey (Channell et al., 2003). The uppermost Norian to lowermost Rhaetian is calibrated to conodont datums in Austria (Scheiblkogel, Gallet et al., 1998; and the Austrian candidate for the Rhaetian GSSP at Steinbergkogel; Krystyn et al., 2007c, 2007d; Hüsing et al., 2011) and the Italian candidate for the Rhaetian GSSP at Pignola-Abriola (Rigo et al., 2016). A pair of thick overlapping sections in the southern Alps (Brumano and Italcementi Quarry; Muttoni et al., 2010) spans the middle Rhaetian through lowermost Hettangian, although minor faults complicate the stratigraphy.

In general, the magnetozones of the lower Norian (Lacian and Alaunian substages) are dominated by reversed-polarity; whereas the upper Norian (Sevalian) has approximately equal proportions of normal and reversed-polarity. The uppermost Norian to lowermost Rhaetian (depending upon eventual placement of the Rhaetian GSSP) is mainly reversed-polarity; followed by a prevalence of normal polarity that continues into the Hettangian.

During the Late Triassic, a series of rift valleys along the western margin of the future Central Atlantic accumulated very thick successions of lacustrine deposits that recorded climatic responses to Milankovitch orbital cycles. Drilling of these Newark Basin strata has yielded a complete 30-Myr cycle-scaled pattern of the magnetic reversal history during the Late Triassic (Kent et al., 1995). A few meters above the brief reversed-polarity E23r zone is a palynological turnover event, and a few centimeters higher the lacustrine deposits are overlain by the Orange Mountain basalts, part of the onset of a regional Central Atlantic Magmatic Province (CAMP) dated at approximately 201.5 Ma (e.g., Mundil et al., 2010). Therefore Kent and Olsen (1999; and web page update of 2002) assigned an age of 202 Ma to the top of E23 and tuned the cyclic stratigraphy using the 405-kyr eccentricity cycle and a 1.75-Myr modulating cycle to project the ages of the Late Triassic polarity pattern. Based on palynology (e.g., Cornet, 1977), the base of the Norian had been tentatively assigned to Newark magnetozone E13 at the base of the Passiac Formation (Olsen et al., 1996), but projected to a much lower level (E8r) by correlation to the proposed Norian GSSP at Pizzo Mondello (e.g., option 2 of Muttoni et al., 2004b; Kent et al., 2017). Radioisotopic age control on lower Norian magnetozones in a borehole on the Colorado Plateau supports the cyclostratigraphic age model for the Newark polarity pattern (Kent et al., 2018).

There are many published versions of how this cycle-scaled Newark polarity pattern might correlate to the upper Carnian, Norian and Rhaetian magnetostratigraphy derived from marine sections and composites (e.g., options 1 and 2 of Muttoni et al., 2004b; Ogg, 2004; Ogg et al., 2008; options A, B, and C of Hounslow and Muttoni, 2010; Gallet et al., 2007; Muttoni et al., 2010; Ogg, 2012; Ogg et al., 2016; Maron et al, 2019). Radioisotopic age control and/or cyclostratigraphic scaling of the polarity zones in Norian–Rhaetian marine reference sections is not yet available to provide unambiguous constraints. Once there is an established correlation of the terrestrial to the marine polarity patterns, then the extensive cycle-scaled Newark suite will provide a precise duration and timing for most of the events in the Late Triassic.

25.2.3.2 Chemical stratigraphy

In addition to the major disruptions of climate accompanying the end-Permian and end-Triassic mass extinctions, there are at least two major climatic events indicated by simultaneous excursions in oxygen (temperature) and carbon isotopes (e.g., compilations and reviews by Tanner, 2010a; Preto et al., 2010; Muttoni et al., 2014; Trotter et al., 2015), which are summarized within the geochemical stratigraphy chapters of this book. The anomalous "lethal" tropical temperatures of the Early Triassic (Sun et al., 2012) include an abrupt drop in ammonoid and conodont diversity that was coincident a negative excursion in $\delta^{13}C_{carb}$ near the end of the Smithian Substage. A sudden warming and humid event in the middle

of the Carnian stage is considered to be "the most distinctive climate change within the Triassic" (Preto et al., 2010) and disrupted the global land-ocean-biological system (summarized above with the Carnian substages). The current hypotheses for the triggers for these environmental perturbations may have focused on volcanic activity—eruptive phases from the Siberian Traps and the emplacement of the Wrangellia large igneous province, respectively.

25.2.3.2.1 Carbon-isotope trends and major excursions

The $\delta^{13}C$ curve of Fig. 25.1 is a merger of generalized trends from several publications, including a compilation of Early Triassic to earliest Ladinian from Sun et al. (2012) and a synthesis of Late Ladinian through Rhaetian from Muttoni et al. (2014). Five major carbon-isotope excursions are currently important for global correlations among marine and terrestrial settings. In contrast to the common coincidence of widespread anoxic events with negative carbon-excursions during the Devonian and the Jurassic-Cretaceous, there are not yet any identified widespread black-shale episodes associated with these events, although there are indications that low-oxygen deep-sea waters did impinge onto the continental shelves.

1. End-Permian—The main set of end-Permian marine mass extinctions occurs abruptly within a broader major negative excursion in carbon isotopes in both marine carbonates and organic-carbon. High-resolution studies indicate that the stepwise downward trend in negative carbon isotopes was interrupted by a minor positive-isotope excursion that approximately coincided with this main marine extinction level, followed by a pronounced minimum ^{13}C-values immediately above the basal Triassic (e.g., Korte and Kozur, 2010; Hermann et al., 2010; Luo et al., 2011). Cramer and Jarvis (2020, Ch. 11: Carbon isotope stratigraphy; this book) called the minimum peak the "EGE" for Early-Griesbachian Event. This pronounced trend toward negative carbon-isotope values may have been caused by a combination of decreased marine productivity and influx of light carbon from volcanic, soil-carbon or methane sources (e.g., Holser and Magaritz, 1987; Baud et al., 1989, 1996; Erwin et al., 2002; Korte et al., 2010; Wignall, 2015; Grasby et al., 2016a, 2016b; Chen et al., 2016).

2. Early Triassic—Two positive peaks in carbon-13 within the lower Triassic have been suggested as secondary markers for the bases of the Olenekian (Smithian Substage) and of the Spathian Substage. Therefore Cramer and Jarvis (2020, Ch. 11: Carbon isotope stratigraphy; this book) suggest an abbreviated terminology of DSBE and SSpBE for these two positive excursions. An abrupt drop in ammonoid and conodont diversity is associated with

the onset of the positive carbon-13 excursion at the base of the Spathian Substage (Galfetti et al., 2007a; Algeo et al., 2019a). One hypothesis for the origin of the intervening relatively high-amplitude negative carbon-isotope excursions during the Dienerian through early Spathian is that there were multiple pulses of carbon release during phases of eruption and intrusion of the Siberian Trap flood basalts (e.g., Payne and Kump, 2007; Wignall, 2015; Grasby et al., 2016a, 2016b).

3. Earliest Anisian—The last of these early Triassic positive carbon-isotope excursions is a gradual rise that spans the Olenekian−Anisian boundary. This "Early Anisian Event" or EAnE in the suggested terminology of Cramer and Jarvis (2020, Ch. 11: Carbon isotope stratigraphy; this book) will also be a secondary marker for the yet-to-be-formalized base-Anisian GSSP (Chen et al., 2020).

4. Middle Carnian Event—Carbon isotopes remain relatively constant within a c. 2 per-mil band for c. 30 Myr from middle Anisian to middle Norian, except for a brief set of pronounced negative-excursions that coincide with the middle Carnian "Pluvial Episode" (e.g., Dal Corso et al., 2012, 2018a,b). This excursion may have been caused by the eruption of the Wrangellia large igneous province, which was later accreted to become part of the Alaska-Canada coast.

5. Norian−Rhaetian boundary—A brief positive $\delta^{13}C_{org}$ peak reported from in organic-carbon components at the Norian−Rhaetian boundary (Sephton et al., 2002b; Rigo et al., 2016) has been tentatively attributed to widespread oceanic stagnation coincident with extinction of deep-water invertebrate fauna, but this feature requires verification in additional sections.

6. End-Triassic—The Late Rhaetian (end-Triassic) mass extinctions coincide with a negative carbon-isotope excursion, which, like the end-Permian Event, may be linked to widespread volcanism, oceanic productivity collapse, and a release of methane (e.g., Pálfy et al., 2001; Ward et al., 2001; Hesselbo et al., 2002; Ruhl et al., 2009; Zaffani et al., 2018). However, the interpretation of these carbon-isotope records may be distorted by facies variations in some sections (Zaffani et al., 2018).

25.2.3.2.2 Oxygen isotopes and temperature excursions and trends

The Triassic $\delta^{18}O$ curve from conodont apatite of Fig. 25.1 is a synthesis of several studies, especially for the Early Triassic by Sun et al. (2012) and Middle and Late Triassic trends from Trotter et al. (2015). These imply that tropical temperature during much of the Early Triassic may have exceeded levels that were lethal to larger land-dwelling vertebrates, which may partly explain the dearth of their skeletal remains

from these latitudes until the Middle Triassic (Sun et al., 2012; Wignall, 2015).

The $\delta^{18}O$ curve of Fig. 25.1 from Grossman and Joachimski (2020, Ch. 10: Oxygen isotope stratigraphy; this book) implies a general warming trend from Middle Triassic to mid-Norian, followed by a cooling trend to the end of the Triassic. The resolution of this generalized curve-fit to conodont-apatite data has apparently smoothed over the brief global warm and humid episode in the mid-Carnian (e.g., Preto et al., 2010; Dal Corso et al., 2018a).

25.2.3.2.3 Sulfur-isotope trends and excursions

The marine record of sulfur isotopes ($\delta^{34}S$; Paytan et al., 2020, Ch. 9: Sulfur isotope stratigraphy; this book) display low values of c. 12‰ prior to the end-Permian mass extinction, followed by a brief excursion to approximately 25‰−30‰ near the end-Permian, with potentially a continuation into a broad peak near the Induan−Olenekian boundary (e.g., Kampschulte and Strauss, 2004; Newton et al., 2004; Horacek et al., 2010a, 2010b; Luo et al., 2010; Song et al., 2013; Bernasconi et al., 2017). One interpretation for the rate and magnitude of these excursions is that sequestered hydrogen-sulfide in a latest Permian anoxic ocean was released by oceanic overturning (e.g., review in Paytan et al., 2020, Ch. 9: Sulfur isotope stratigraphy; this book) and that the oceanic sulfate reservoir was anomalously low, perhaps less than 15% of modern size (Luo et al., 2010). Sulfur-isotope stratigraphy within the rest of the Triassic is currently at a low-precision reconnaissance status.

25.2.3.2.4 Strontium and osmium-isotope trends and excursions

The curve of marine $^{87}Sr/^{86}Sr$ through the latest Permian through Triassic (Fig. 25.1) begins with a slow increase from the major trough (0.7068) near the end of the Middle Permian, which suddenly undergoes a sharp increase in the rate of rise at the Permian−Triassic boundary to peak (0.7081) near the base of the Middle Triassic (e.g., review in McArthur et al., 2020, Ch. 7: Strontium isotope stratigraphy; this book). The rapid rise in $^{87}Sr/^{86}Sr$ through the Early Triassic may be a response to increased continental weathering during the anomalous warm temperatures of the Early Triassic (e.g., Martin and Macdougall, 1995; Sedlacek et al., 2014). A relatively rapid decline a minimum (0.7076) at the end of the Middle Triassic is followed by a gradual rise through the Norian to a second peak (c. 0.7079) in latest Triassic followed by a continuous decline through the Rhaetian and Early Jurassic to another major low (to c. 0.7071) in latest Pliensbachian (e.g., Koepnick et al., 1990; Korte et al., 2003; Cohen and Coe, 2007).

Osmium-isotope stratigraphy has not yet been systematically compiled for the Triassic (Peucker-Ehrenbrink and Ravizza, 2020, Ch. 8: Osmium isotope stratigraphy; this book). Analysis of black shales of latest Anisian from Svalbard yielded initial $^{187}Os/^{188}Os$ ratio of 0.83 ± 0.03, which is one of the highest recorded ratios for global seawater between the earliest Cambrian and the Late Early Jurassic (Xu et al., 2009). This peak remains to be verified and delimited in additional sections. In general, the trends in Osmium mirrors those of the Strontium curve, including a gradual decline through the Rhaetian from an initial $^{187}Os/^{188}Os$ ratio from a Late Norian peak of 0.75 that was interrupted by a brief excursion to high (radiogenic) ratios upon eruption of the CAMP at the end of the Triassic (e.g., Cohen and Coe, 2007; Peucker-Ehrenbrink and Ravizza, 2020, Ch. 8: Osmium isotope stratigraphy; this book).

25.2.3.3 Cycle stratigraphy

Numerous cyclostratigraphy analyses have been applied to continental and marine deposits of Triassic age. Huang (2018) compiled over 30 of these Triassic studies to select seven major ones that both encompassed a relatively long time span without major stratigraphic breaks and published the original datasets in a form applicable to a standardized retuning to the 405-kyr long-eccentricity cycle. The merger of these seven cyclostratigraphic datasets enabled an initial estimate of a full astronomical-tuned time scale for the entire Triassic. Those selected studies were the Lower to Middle Triassic carbonates of South China (Li et al., 2016, 2018), Middle to Upper Triassic radiolarian-rich pelagic sediments accreted to Japan (Ikeda et al., 2014; Ikeda et al., 2017), Carnian carbonates of South China (Zhang et al., 2015), and Upper Triassic Lacustrine Sediments of the Newark Basin (Olsen and Kent, 1999; Kent et al., 2017). This suite demonstrates the potential for a full astronomical-tuned Triassic time scale, but this goal is currently hindered by uncertainties in merging the magnetic polarity scales, radiolarian zones, and conodont datums into a global composite.

In continental settings, the monsoon-dominated climate of the Pangea megacontinent was sensitive to Milankovitch orbital-climate cycles, especially to the precession-eccentricity components. Extended and quasi-continuous deposits of continental facies having both excellent magnetostratigraphy and unambiguous cycles are present in central Europe (Lower Triassic) and eastern North America (Upper Triassic). In theory, these successions should be the Rosetta stone to project cycle–scaled durations onto marine sequences for a precise relative time scale, similar to what has been developed for the Cenozoic. In practice, as noted previously, there is a lack of a unique pattern match for correlation of these extended intervals of cycle-scaled magnetostratigraphies with marine-based composite polarity patterns. The interpretations and controversies concerning these Triassic cyclic deposits are critically examined by Tanner (2010b).

Variations in clastic input and the extent of lacustrine influences in the Buntsandstein basins of central Europe during the Early Triassic provide a detailed regional stratigraphy that is applicable to surface exposures and downhole logs (e.g., reviews in Röhling, 1991; Bachmann and Kozur, 2004; Szurlies, 2004; Menning et al., 2005; Feist-Burkhardt et al., 2008). The cycles, spanning about 10–20 m with sandstones fining upward into more clay-rich sediments, are generally interpreted as oscillations between more arid and more humid conditions. Constraints from terrestrial biostratigraphy (conchostracan, pollen-spores) combined with radioisotope ages on the span of the Early Triassic indicate that the depositional sequences appear to coincide with the 100-kyr short-eccentricity cycle (e.g., Bachmann and Kozur, 2004; Menning et al., 2005). However, the expected 405-kyr long-eccentricity has not been unambiguously resolved. The magnetostratigraphy from the Buntsandstein, especially within the lower portion, which has relatively longer duration polarity zones and biostratigraphic constraints, is fairly well correlated to other Early Triassic reference sections (Szurlies, 2007; Hounslow and Muttoni, 2010). Even though a monotonic 100-kyr periodicity is not expected for short-eccentricity and there is a possibility of "missing beats" at possible exposure horizons within this Buntsandstein succession, the projected cycle-scaling of the marine zonation and associated Early Triassic substages via this magnetostratigraphy is a close fit to radioisotopic ages and most of the polarity pattern correlates with the independent cyclo-magnetostratigraphy from Lower Triassic and Anisian Limestones of South China (Li et al., 2016, 2018).

Interbedded marls and limestones of shallow-marine origin spanning the Permian–Triassic boundary interval in the Austrian Alps display cycles with ratios matching Milankovitch periodicities and have been interpreted to imply that the latest Permian extinction and negative carbon-isotope spike spanned less than 30 kyr (Rampino et al., 2000, 2002), which agrees with radioisotopic constraints at the Triassic GSSP at Meishan, China (Burgess et al., 2014).

The Latemar massif in the Italian Dolomites was an atoll-like feature with a core of flat-lying Anisian and Ladinian platform carbonates. Oscillations in sea level were created over 500 thin depositional cycles (Goldhammer et al., 1987). Stacking patterns and spectral analysis of the sea-level oscillations had been interpreted as representing precession modulated by short-term (100 kyr) eccentricity, therefore yielding an implication that the Latemar deposit spans approximately 10 Myr (Goldhammer et al., 1990; Hinnov and Goldhammer, 1991). In contrast, U–Pb ages from coeval tuff-bearing basinal deposits appear to constrain the Latemar platform to span only 2–4 Myr (e.g., Brack et al., 1996, 1997; Mundil et al., 1996; Hardie and Hinnov, 1997; and extended review in Tanner, 2010a). A possible solution to this disparity is that an extremely rapid rate of platform construction (c. 500 m/Myr or greater) enabled recording of sub-Milankovitch sea-level oscillations with misleading similarity in ratios to precession-eccentricity (e.g., Kent et al., 2004, 2006; Hinnov, 2006; Meyers, 2008). This ambiguity in cyclostratigraphic interpretation demonstrates that any cycle-stratigraphic analysis based only on a single section requires verification from other independent basins and facies. Studies of similar oscillating Lofer facies within upper Triassic platform carbonates of the Austrian Alps played an important role in developing fundamental concepts of cyclostratigraphy (e.g., Fischer, 1964), but the reality of regular cyclicity in these deposits has also been debated (e.g., Satterley, 1996, versus Schwarzacher, 2005, and Cozzi et al., 2005; reviewed in Tanner, 2010b).

Radiolarian-rich pelagic chert successions from Japan spanning the Middle Triassic are characterized by ribbon bedding. These chert-clay couplets have been interpreted as productivity fluctuations induced by 20-kyr precession cycles, and the longer term trends in bed thickness correspond to 100- and 405-kyr eccentricity cycles (Ikeda et al., 2010, 2017). These cyclostratigraphic interpretations, the tentative correlation of radiolarian taxa to geologic stages, a potential long-term modulation of c. 3.6 Myr, and the continuity of the bedded-chert sections await further verification.

During the late-Middle Triassic through Early Jurassic, a set of rift basins formed as Pangea underwent an initial phase of breakup. The thick Newark Group of lacustrine sediments from these tropical basins are characterized by oscillations between semistagnant deep lakes and arid playas as the intensity of monsoonal rains responded to Earth's precession modulated by short- (c. 100 kyr) and long-term (c. 400 kyr) eccentricity cycles. Spectral analysis of sediment facies successions in a series of deep-drilling cores enabled compilation of a cycle-scaled stratigraphic record, including a detailed polarity pattern that is unprecedented in its 30-Myr temporal span (e.g., Kent et al., 1995; Olsen et al., 1996; Kent and Olsen, 1999). As discussed previously, the comparison of the cycle-scaled terrestrial polarity signature to the un-scaled marine magnetostratigraphy does not always provide a unique match. However, uppermost Triassic lacustrine deposits with alternating red-to-green coloration at St. Audrie's Bay have yielded both a magnetostratigraphy (Hounslow et al., 2004) and an interpreted 3.7 Myr cyclostratigraphy (Kemp and Coe, 2007) that have a polarity scaling resembling the upper Newark interval of polarity zones E19n–E16n. Cyclostratigraphy of a composite Rhaetian succession in Austria indicates that this stage (using the Austrian GSSP candidate) has a minimum duration of 6.7 Myr (Galbrun et al., 2020).

Details and discussion of these and other Triassic cyclostratigraphic studies are given in Huang (2018).

25.2.3.4 Sequence stratigraphy

Triassic sea-level Trends and sequences of different relative magnitudes have been compiled for Boreal basins (e.g., Embry, 1988; Mørk et al., 1989; Skjold et al., 1998), the classic Germanic Trias (e.g., Aigner and Bachmann, 1992; Geluk and Röhling, 1997), the Dolomites and Italian Alps regions (e.g., De Zanche et al., 1993; Gaetani et al., 1998; Gianolla et al., 1998, Gianolla and Jacquin, 1998), and other regions. Some of these sea-level trends appear to correlate on an interbasin to global scale (e.g., Haq et al., 1988; Haq, 2018; Hallam, 1992; Embry, 1997; Gianolla and Jacquin, 1998). The significant disparity in some proposed major global features for the Triassic (e.g., Hardenbol et al., 1998, compared to Simmons et al., 2007) was difficult to resolve, because the supporting details of reference sections and biostratigraphic control are generally not adequately published.

The major sequences in Figs. 25.1 and 25.5–25.7 are from a Triassic synthesis by Haq (2018) that enhanced a previous compilation and systematic numbering system by Jacquin and Vail (1998). In this compilation the main first-order Triassic sea-level trend is dominated by a single cycle—a progressive transgression that began in the early Anisian, peaks near the Carnian–Norian boundary, followed by a very gradual regression through the middle Rhaetian that turns into a faster regression in the latest Rhaetian through early Hettangian. A lesser transgression–regression cycle began in the latest Permian with a minor peak in latest Induan. Superimposed on these main cycles are several second-order facies cycles with major sequence boundaries at the Induan–Olenekian boundary, Anisian–Ladinian boundary, mid-Carnian (Julian-Tuvalian boundary), latest Carnian, and latest Norian (or the proposed Norian–Rhaetian boundary candidate in Austria). In contrast, Simmons et al. (2007) published a general Triassic scale with seven main sequences (*Tr10* to *Tr80*). See also review in Simmons et al. (2020, Ch. 13: Phanerozoic eustasy, this book). These proposed Triassic suites of sequences and sea-level trend await published documentation of the different compilations and a community consensus on the regional versus global components.

25.2.3.5 Other major stratigraphic events
25.2.3.5.1 Large igneous provinces

The Triassic is delimited by two major volcanic provinces: the Siberian Traps at the base and the CAMP at the top.

1. End-Permian—The Siberian Traps exposed mainly on the Siberian craton was one of the most voluminous volcanic eruption provinces of the Phanerozoic, with an estimated volume greater than 2 million cubic kilometers of basalt flows and volcaniclastic rocks (e.g., reviews by Reichow et al., 2004; Ernst, 2014; Wignall, 2015; in Ernst et al., 2020; and at the Large Igneous Provinces Commission website). The volcanic province on the Siberian craton is

generally subdivided into four distinct geographic regions: Noril'sk, Putorana, Nizhnaya Tunguska, and Maimecha-Kotuy. The main pulses of the voluminous Siberian Trap flood basalts are approximately coeval with the latest Permian mass extinctions, and the waning stages of this volcanic activity continued into the earliest Triassic (e.g., Renne et al., 1995; Erwin et al., 2002, and references therein; Reichow et al., 2009; Burgess and Bowring, 2015; Burgess et al., 2014, 2017; Wang et al., 2018).

2. Middle Carnian—The Wrangellia terrane, which is accreted to British Columbia and Alaska, contains a major episode of tholeiitic flood volcanism in submarine and subaerial environments. Radiometric dating (c. 227 and 232 Ma) and the overlying fossiliferous strata indicate an eruption during Carnian. The original volcanic volume is estimated as 1 million cubic kilometers (e.g., Greene et al., 2008, 2010, 2011).

3. End-Triassic—The CAMP has ages clustering at c. 201 Ma just prior to Triassic—Jurassic boundary and is considered to have been a major causal factor in the end-Triassic extinctions (e.g., Marzoli et al., 1999; Jourdan et al., 2009a; Schoene et al., 2010; Blackburn et al., 2013; Ernst, 2014). The total extrusive volume may have been even greater than the Siberian Traps.

25.2.3.5.2 Major bolide impacts

The 100-km diameter Manicouagan impact structure of Quebec has melt rock directly dated by modern U/Pb methods (214.56 ± 0.05 Ma; Ramezani et al., 2005; Hodych and Dunning, 1992, as reevaluated by Jourdan et al., 2009b). This impact event and associated environmental catastrophe may have contributed to the large-scale turnover of continental tetrapods during mid-Norian, during which dinosaurs attained dominance over competing families (e.g., Benton, 1986, 1993; Lucas, 2010d). However, no record of this impact or major environmental catastrophe has yet been discovered within the extensive lacustrine deposits of the Newark Group of northeastern North America.

The only other significant impact that is assigned to the Triassic is the Saint Martin crater with a c. 40 km diameter in Manitoba, Canada, has a vague estimated mid-Triassic age of 220 ± 32 Ma (Earth Impact Database, 2018, and references therein).

25.3 Triassic time scale

25.3.1 Overview

After the publication of GTS2004 the Triassic was the focus of extensive sampling and application of ultra-high-resolution methods by different geochronology laboratories. Triassic and uppermost Permian strata enabled testing enhanced techniques for single-zircon dating, provided sets for intercalibration of U—Pb and Ar—Ar methods and the standardization

of procedures among laboratories. These new age sets had replaced or called into question nearly all of the Triassic radiogenic isotope ages published before 2004 (reviewed by Mundil et al., 2010). The initiation and termination of the Triassic Period and many of its GSSP-defined stages have been constrained by a remarkably extensive suite of high-precision radioisotopic dates (Appendix 2, this volume), implying a span from 251.9 to 201.4 Ma (50 Myr). The radioisotopic details and stratigraphic placement of these dates are compiled in Appendix 2, and only those dates that constrain stage boundaries will be highlighted below.

The merger of astronomical cycles with magnetostratigraphy is the framework for scaling the majority of the Triassic. The cyclo-magnetostratigraphy is tied to the subset of high-precision radiometric U—Pb dates obtained by ID-TIMS methods. The ages for many of the biozone zones and some of the geochemical curves are derived either from direct correlations to the astronomical scaling in reference sections or from their relative position within magnetic polarity chrons. The age models for all other biostratigraphic datums and geochemical curves are from their estimated placement relative to these primary biostratigraphic or magnetostratigraphic scales. Therefore this composite cyclo—magnetic—biostratigraphic age model that has been developed by different international teams is similar to what has been accomplished for the cyclo—magnetic scaling of the Cenozoic Era.

The precision of the direct radiometric age control on the cyclostratigraphic or magnetic polarity scales is generally less than 0.1 Myr, and the placement of biozones within cyclostratigraphic reference sections is usually to an accuracy of one short-eccentricity cycle of a similar 0.1 Myr duration. As with all other age models for intervals within the Paleozoic—Mesozoic, some caveats include that the published placement of biostratigraphic datums within the main reference sections may not represent global first occurrences and may be influenced by uncertain taxonomic assignments (especially for some conodonts), that some of the cyclo—magnetic—biostratigraphic scales have not yet been verified from independent basins, and that the assignment of some regional biostratigraphic zones is not well calibrated to the primary age scales. Some of these less-certain scalings are indicated by dashed or dotted lines in the synthesis charts (Figs. 25.5—25.7).

25.3.2 Early Triassic through Anisian age model

For the Early Triassic through early Anisian, the magnetic polarity pattern that was calibrated to short-eccentricity (100-kyr) changes in monsoon intensity recorded by clastic cycles in the Germanic Basin (e.g., Szurlies, 2007) has been duplicated in the cyclo-magnetostratigraphy of conodont-bearing carbonate deposits on the margins of the paleogeographic Yangtze Platform of South China (Li et al., 2016, 2018). This cycle-magnetic compilation of South China includes the GSSPs for the Induan (base-Triassic) at Meishan (Zhejiang province), the candidate for the Olenekian GSSP at West Pingdingshan near Chaohu (Anhui province), and a candidate for the Anisian GSSP candidate at Guandao (Guizhou province). The astronomical tuning is tied to the base-Triassic **Induan** GSSP at Meishan, which has an interpolated age of **251.902 ± 0.024 Ma** based on close-spaced U−Pb ID-TIMS dates (Burgess et al., 2014). *[Note that an external uncertainty of c. 0.29 Myr should be included when comparing this version of EARTHTIME-standardized dates to previous versions, as explained in Burgess* et al. (2014)].

The cyclo-bio-magnetostratigraphic reference section at West Pingdingshan near Chaohu in South China is a candidate for the **Olenekian** GSSP. According to the astronomical tuning, the Induan/Olenekian boundary level is 2.0 Myr later than the base-Triassic therefore has an age of **249.9 Ma** (Li et al, 2016). This 2.0-Myr span is independently verified by the 1.9 ± 0.1 Myr duration for the Induan stage derived from cyclo-biostratigraphy of conodont zones in the Montney Formation of British Columbia (Shen et al., 2017; Moslow et al., 2018; Henderson et al, 2018; Shen, 2018).

If the Griesbachian−Dienerian substage boundary is assigned as the base of the conodont *S. kummeli* Zone in these astronomically tuned sections of South China, then the base of **Dienerian** has a projected age of **250.5 Ma** (Li et al., 2016). The 1.4-Myr duration of the Griesbachian Substage is also consistent with a U−Pb date of 251.50 ± 0.6 Ma from the lower-middle Griesbachian at Meishan (Burgess et al., 2014). A reported date of 251.22 ± 0.2 Ma on a volcanic ash from lower Smithian (Galfetti et al., 2007b) was based on pre-EARTHTIME standards, and therefore is anomalously old and should be reanalyzed (Burgess et al., 2014).

There is ID-TIMS radiometric dating of the volcanic-ash beds in the Olenekian−Anisian boundary interval at the GSSP candidate at Wantou (Guangxi province) section (Ovtcharova et al., 2015) and at the Guandao section (Guizhou province) (Lehrmann et al., 2015). These dates,

when coupled with the composite cyclostratigraphy for Early Triassic through Anisian (Li et al., 2016), indicate that the lowest occurrence of the conodont *Ch. timorensis s.str.* as the preferred marker for the base of **Anisian** at approximately **246.7 Ma** (Chen et al., 2020). U−Pb-dated volcanic-ash beds bracketing this base-Anisian marker at Wantou by Ovtcharova et al. (2015) had been interpreted by them to imply a slightly older age (c. 247.3 Ma) for a slightly different conodont marker for the base-Anisian, but they caution that there are inconsistencies in the progression of interpreted dates from the zircon populations in successive layers and the statistical derivation of a mean date from each ash bed was influenced by the decisions of which individual zircons within the distribution of ID-TIMS dates should be included. Therefore the cyclostratigraphic age of 246.7 Ma relative to the base-Triassic control date is used here.

If the Smithian−Spathian substage boundary is placed at the lowest occurrence of conodont *Novispathodus pingdingshanensis* in these sections, then the base of the **Spathian** has a projected age from the astronomical tuning of **248.1 Ma** (Li et al., 2016).

The magnetostratigraphy and interpreted 405-kyr eccentricity-driven cycles at the conodont-rich Guandao section span the Anisian through lowest Ladinian (Lehrmann et al., 2015; Li et al., 2018). This cyclo-magnetostratigraphy age model was applied to estimate the ages for the bases of Anisian substages according to their Tethyan ammonoid-based correlations to magnetic polarity zones (e.g., Hounslow and Muttoni, 2010). The corresponding cyclo-magneto-ammonoid age model projects the base of **Bithynian** at **245.0 Ma** (base of *Kocaella* ammonoid zone), the base of **Pelsonian** at **244.2 Ma** (base of *B. balatonicus* ammonoid zone), and the base of **Illyrian** at **243.3 Ma** (base of *Paraceratites trinodosus* ammonoid zone).

In some cases, these ammonoid-defined substages are different from their conodont-based placements relative to polarity zones in South China (e.g., Lehrmann et al., 2015), for example, the basal substage of Aegean includes 5 ammonoid zones (Jenks et al., 2015), but its conodont-based duration at Guandao would have been only 0.3 Myr, compared to a c. 1.7-Myr duration based on ammonoid correlation to the

FIGURE 25.5   Selected marine and terrestrial biostratigraphic zonations of the Late Triassic. ("*Age*" is the term for the time equivalent of the rock-record *stage. Note that the Ma-age scale is more compact than* in Figs. 25.6 or 25.7). Marine-facies magnetostratigraphy zones are labeled in Rhaetian after Gallet et al. (2007), in Norian after Hounslow and Muttoni (2010), in Carnian−Norian boundary interval after Muttoni et al. (2004b), in Carnian after Zhang et al. (2020) and in Ladinian−Carnian boundary interval after Maron et al. (2019). Those of late Carnian through Rhaetian are shown with one option for the calibrations to the astronomical-scaled polarity chrons of the Newark Basin (Kent et al., 2017). [Note that the durations of Newark Basin polarity zones E1 through E6, which are within noncyclic fluvial sediments, were projected by Kent and Olsen (1999) based on the average accumulation rate for the overlying polarity zones E9 through E14 preserved in lacustrine sediments, and E1 through E4 are uncertain.] Tethyan and Arctic/Panthalassan ammonoid zones are modified from Jenks et al. (2015) and Orchard and Tozer (1997). Tethyan conodont zones are modified from Moix et al. (2007), Kozur (2003) and Rigo et al. (2018) with genera standardized by Michael Orchard, and Tethyan bivalve zones are from McRoberts (2010). Arctic/Panthalassan conodont zones are from Michael Orchard and Martyn Golding (see text). Spinicaudatun (Conchostracan) zones of the Germanic Basin are modified from Kozur and Weems (2010), Weems and Lucas (2015), Geyer and Kelber (2018), and Franz et al. (2018). Land vertebrate zones and markers are from Lucas (2010d, 2018). Main sea-level trends are from Haq (2018). Additional Triassic (and all Phanerozoic) zonations, biostratigraphic markers, geochemical trends, sea-level curves, and details on calibrations are compiled in the internal datasets within the *TimeScale Creator* visualization system (free at www.tscreator.org).

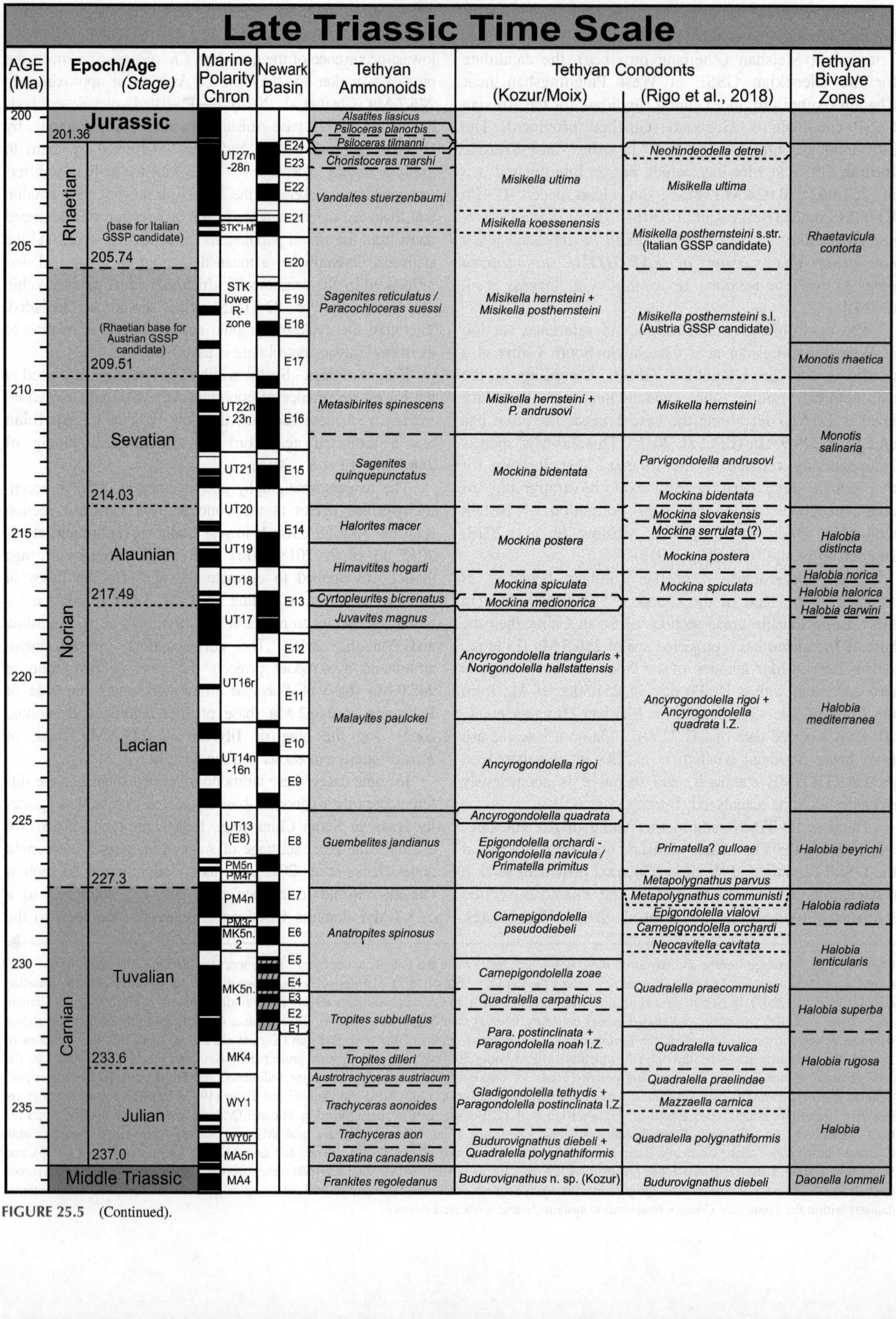

Late Triassic Time Scale

AGE (Ma)	Epoch/Age (Stage)	Marine Polarity Chron	Newark Basin	Tethyan Ammonoids	Tethyan Conodonts (Kozur/Moix)	(Rigo et al., 2018)	Tethyan Bivalve Zones
200 — 201.36	**Jurassic**			Alsatites liasicus / Psiloceras planorbis / Psiloceras tilmanni			
	Rhaetian	UT27n-28n	E24			Neohindeodella detrei	
			E23	Choristoceras marshi	Misikella ultima	Misikella ultima	
			E22				
	(base for Italian GSSP candidate)	STK"I-M"	E21	Vandaites stuerzenbaumi	Misikella koessenensis	Misikella posthernsteini s.str. (Italian GSSP candidate)	Rhaetavicula contorta
205 — 205.74			E20				
		STK lower R-zone	E19	Sagenites reticulatus / Paracochloceras suessi	Misikella hernsteini + Misikella posthernsteini	Misikella posthernsteini s.l. (Austria GSSP candidate)	
	(Rhaetian base for Austrian GSSP candidate) 209.51		E18				Monotis rhaetica
			E17				
210 —	Sevatian	UT22n-23n	E16	Metasibirites spinescens	Misikella hernsteini + P. andrusovi	Misikella hernsteini	Monotis salinaria
		UT21	E15	Sagenites quinquepunctatus	Mockina bidentata	Parvigondolella andrusovi	
214.03		UT20				Mockina bidentata	
215 —	Alaunian	UT19	E14	Halorites macer	Mockina postera	Mockina slovakensis / Mockina serrulata (?)	Halobia distincta
		UT18		Himavitites hogarti		Mockina postera	Halobia norica
217.49			E13	Cyrtopleurites bicrenatus	Mockina spiculata / Mockina medionorica	Mockina spiculata	Halobia halorica / Halobia darwini
		UT17		Juvavites magnus			
			E12		Ancyrogondolella triangularis + Norigondolella hallstattensis		
220 —	Lacian	UT16r	E11	Malayites paulckei		Ancyrogondolella rigoi + Ancyrogondolella quadrata I.Z.	Halobia mediterranea
			E10		Ancyrogondolella rigoi		
		UT14n-16n	E9				
225 —		UT13 (E8)	E8	Guembelites jandianus	Ancyrogondolella quadrata / Epigondolella orchardi - Norigondolella navicula / Primatella primitus	Primatella? gulloae	Halobia beyrichi
227.3		PM5n / PM4r				Metapolygnathus parvus	
	Tuvalian	PM4n	E7	Anatropites spinosus	Carnepigondolella pseudodiebeli	Metapolygnathus communisti / Epigondolella vialovi	Halobia radiata
		PM3r / MK5n.2	E6			Carnepigondolella orchardi / Neocavitella cavitata	Halobia lenticularis
230 —			E5		Carnepigondolella zoae		
		MK5n.1	E4		Quadralella carpathicus	Quadralella praecommunisti	
			E3	Tropites subbullatus			Halobia superba
			E2 / E1		Para. postinclinata + Paragondolella noah I.Z.	Quadralella tuvalica	
233.6		MK4		Tropites dilleri			Halobia rugosa
	Julian			Austrotrachyceras austriacum	Gladigondolella tethydis + Paragondolella postinclinata I.Z.	Quadralella praelindae	
235 —		WY1		Trachyceras aonoides		Mazzaella carnica	Halobia
		WY0r		Trachyceras aon	Budurovignathus diebeli + Quadralella polygnathiformis	Quadralella polygnathiformis	
237.0		MA5n		Daxatina canadensis			
	Middle Triassic	MA4		Frankites regoledanus	Budurovignathus n. sp. (Kozur)	Budurovignathus diebeli	Daonella lommeli

FIGURE 25.5 (Continued).

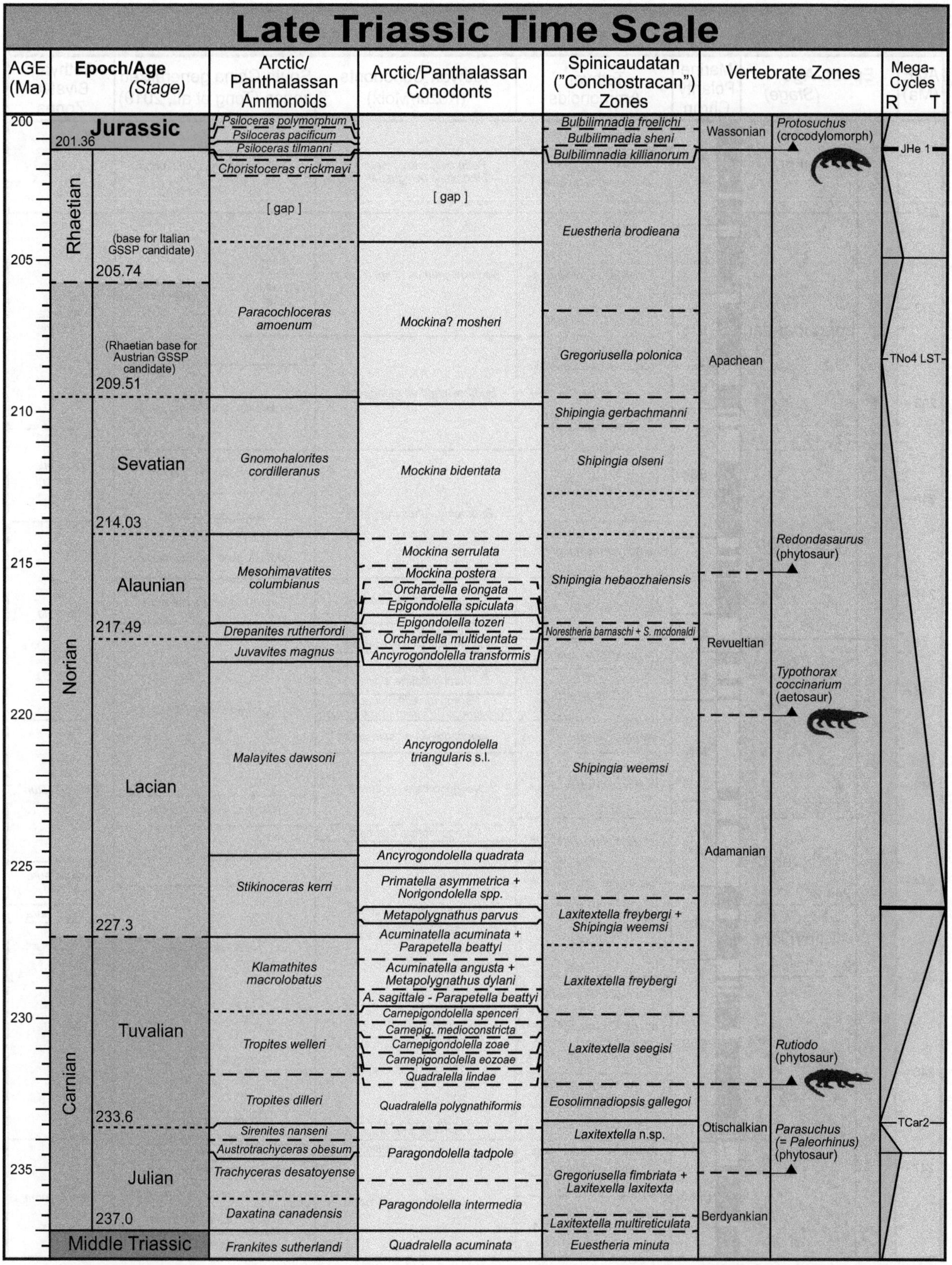

Late Triassic Time Scale

AGE (Ma)	Epoch/Age (Stage)		Arctic/Panthalassan Ammonoids	Arctic/Panthalassan Conodonts	Spinicaudatan ("Conchostracan") Zones	Vertebrate Zones		Mega-Cycles R T
200 — 201.36	**Jurassic**		*Psiloceras polymorphum* / *Psiloceras pacificum* / *Psiloceras tilmanni*		*Bulbilimnadia froelichi* / *Bulbilimnadia sheni* / *Bulbilimnadia killianorum*	Wassonian	*Protosuchus* (crocodylomorph)	JHe 1
	Rhaetian		*Choristoceras crickmayi*	[gap]	*Euestheria brodieana*			
205		[gap]						
		(base for Italian GSSP candidate) 205.74						
			Paracochloceras amoenum	*Mockina? mosheri*	*Gregoriusella polonica*	Apachean		TNo4 LST
		(Rhaetian base for Austrian GSSP candidate) 209.51						
210	Norian				*Shipingia gerbachmanni*			
		Sevatian	*Gnomohalorites cordilleranus*	*Mockina bidentata*	*Shipingia olseni*			
		214.03						
215		Alaunian	*Mesohimavatites columbianus*	*Mockina serrulata* / *Mockina postera* / *Orchardella elongata* / *Epigondolella spiculata*	*Shipingia hebaozhaiensis*		*Redondasaurus* (phytosaur)	
		217.49	*Drepanites rutherfordi*	*Epigondolella tozeri* / *Orchardella multidentata*	*Norestheria barnaschi + S. mcdonaldi*	Revueltian		
			Juvavites magnus	*Ancyrogondolella transformis*				
220		Lacian	*Malayites dawsoni*	*Ancyrogondolella triangularis s.l.*	*Shipingia weemsi*		*Typothorax coccinarium* (aetosaur)	
225				*Ancyrogondolella quadrata*		Adamanian		
			Stikinoceras kerri	*Primatella asymmetrica + Norigondolella spp.*				
		227.3		*Metapolygnathus parvus*	*Laxitextella freybergi + Shipingia weemsi*			
				Acuminatella acuminata + Parapetella beattyi				
	Carnian	Tuvalian	*Klamathites macrolobatus*	*Acuminatella angusta + Metapolygnathus dylani*	*Laxitextella freybergi*			
230				*A. sagittale - Parapetella beattyi*				
			Tropites welleri	*Carnepigondolella spenceri* / *Carnepig. medioconstricta* / *Carnepigondolella zoae* / *Carnepigondolella eozoae* / *Quadralella lindae*	*Laxitextella seegisi*		*Rutiodo* (phytosaur)	
			Tropites dilleri	*Quadralella polygnathiformis*	*Eosolimnadiopsis gallegoi*			
		233.6	*Sirenites nanseni*		*Laxitextella n.sp.*	Otischalkian	*Parasuchus* (= *Paleorhinus*) (phytosaur)	TCar2
			Austrotrachyceras obesum	*Paragondolella tadpole*				
235		Julian	*Trachyceras desatoyense*		*Gregoriusella fimbriata + Laxitexella laxitexta*			
		237.0	*Daxatina canadensis*	*Paragondolella intermedia*	*Laxitextella multireticulata*	Berdyankian		
	Middle Triassic		*Frankites sutherlandi*	*Quadralella acuminata*	*Euestheria minuta*			

FIGURE 25.5 (Continued)

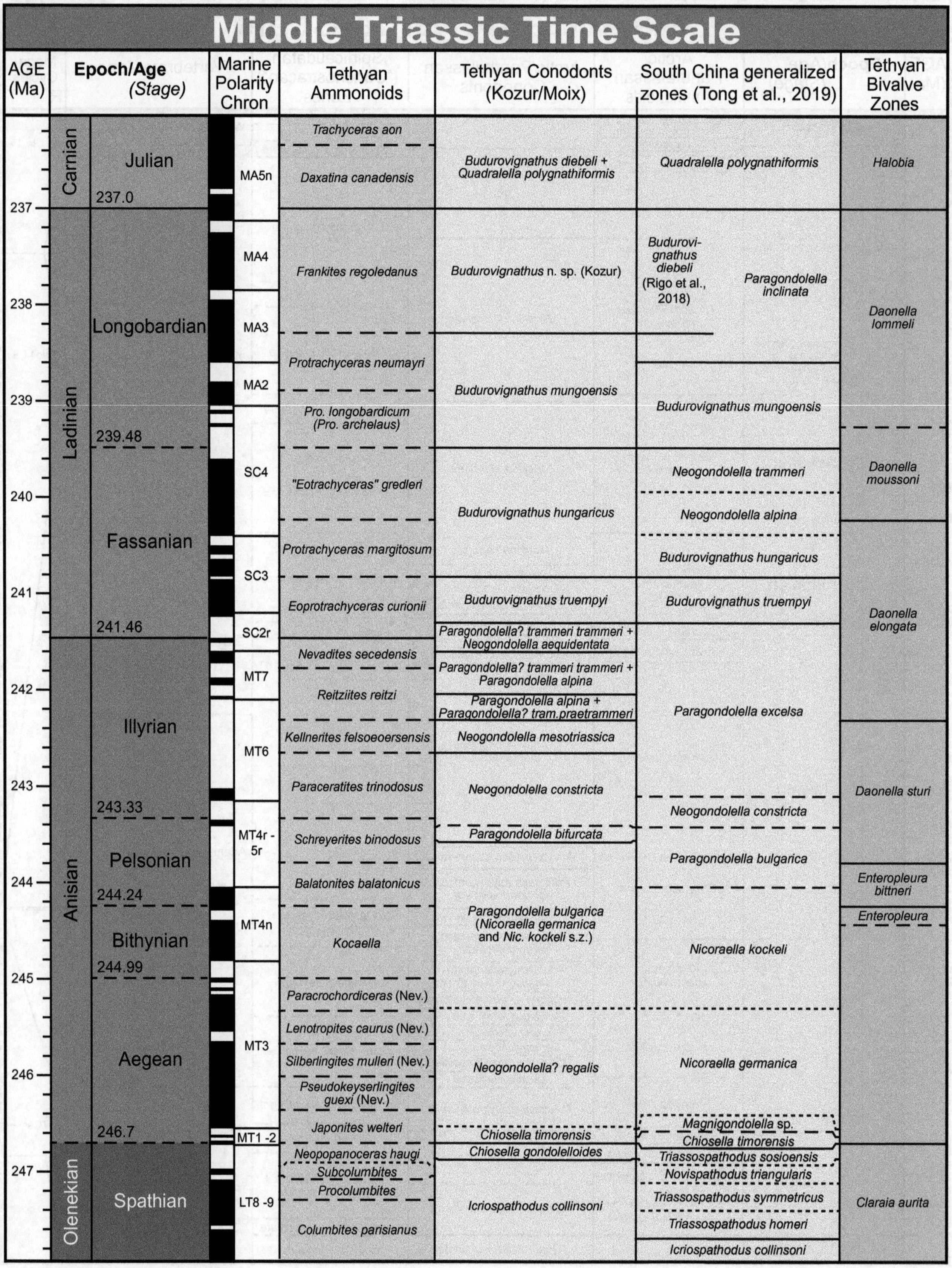

Middle Triassic Time Scale

AGE (Ma)	Epoch/Age (Stage)	Marine Polarity Chron	Tethyan Ammonoids	Tethyan Conodonts (Kozur/Moix)	South China generalized zones (Tong et al., 2019)	Tethyan Bivalve Zones
237 — 237.0	Carnian — Julian	MA5n	Trachyceras aon; Daxatina canadensis	Budurovignathus diebeli + Quadralella polygnathiformis	Quadralella polygnathiformis	Halobia
238 — 239	Ladinian — Longobardian	MA4; MA3; MA2	Frankites regoledanus; Protrachyceras neumayri	Budurovignathus n. sp. (Kozur); Budurovignathus mungoensis	Budurovignathus diebeli (Rigo et al., 2018) / Paragondolella inclinata; Budurovignathus mungoensis	Daonella lommeli
239.48			Pro. longobardicum (Pro. archelaus)			
240 — 241 — 241.46	Ladinian — Fassanian	SC4; SC3; SC2r	"Eotrachyceras" gredleri; Protrachyceras margitosum; Eoprotrachyceras curionii	Budurovignathus hungaricus; Budurovignathus truempyi	Neogondolella trammeri; Neogondolella alpina; Budurovignathus hungaricus; Budurovignathus truempyi	Daonella moussoni; Daonella elongata
242 — 243 — 243.33	Anisian — Illyrian	MT7; MT6	Nevadites secedensis; Reitziites reitzi; Kellnerites felsoeoersensis; Paraceratites trinodosus	Paragondolella? trammeri trammeri + Neogondolella aequidentata; Paragondolella? trammeri trammeri + Paragondolella alpina; Paragondolella alpina + Paragondolella? tram.praetrammeri; Neogondolella mesotriassica; Neogondolella constricta	Paragondolella excelsa; Neogondolella constricta	Daonella sturi
244 — 244.24	Anisian — Pelsonian	MT4r - 5r	Schreyerites binodosus; Balatonites balatonicus	Paragondolella bifurcata	Paragondolella bulgarica	Enteropleura bittneri
244.99	Anisian — Bithynian	MT4n	Kocaella	Paragondolella bulgarica (Nicoraella germanica and Nic. kockeli s.z.)	Nicoraella kockeli	Enteropleura
245 — 246 — 246.7	Anisian — Aegean	MT3; MT1 -2	Paracrochordiceras (Nev.); Lenotropites caurus (Nev.); Silberlingites mulleri (Nev.); Pseudokeyserlingites guexi (Nev.); Japonites welteri	Neogondolella? regalis; Chiosella timorensis	Nicoraella germanica; Magnigondolella sp.; Chiosella timorensis	
247	Olenekian — Spathian	LT8 -9	Neopopanoceras haugi; Subcolumbites; Procolumbites; Columbites parisianus	Chiosella gondolelloides; Icriospathodus collinsoni	Triassospathodus sosioensis; Novispathodus triangularis; Triassospathodus symmetricus; Triassospathodus homeri; Icriospathodus collinsoni	Claraia aurita

FIGURE 25.6   (Continued)

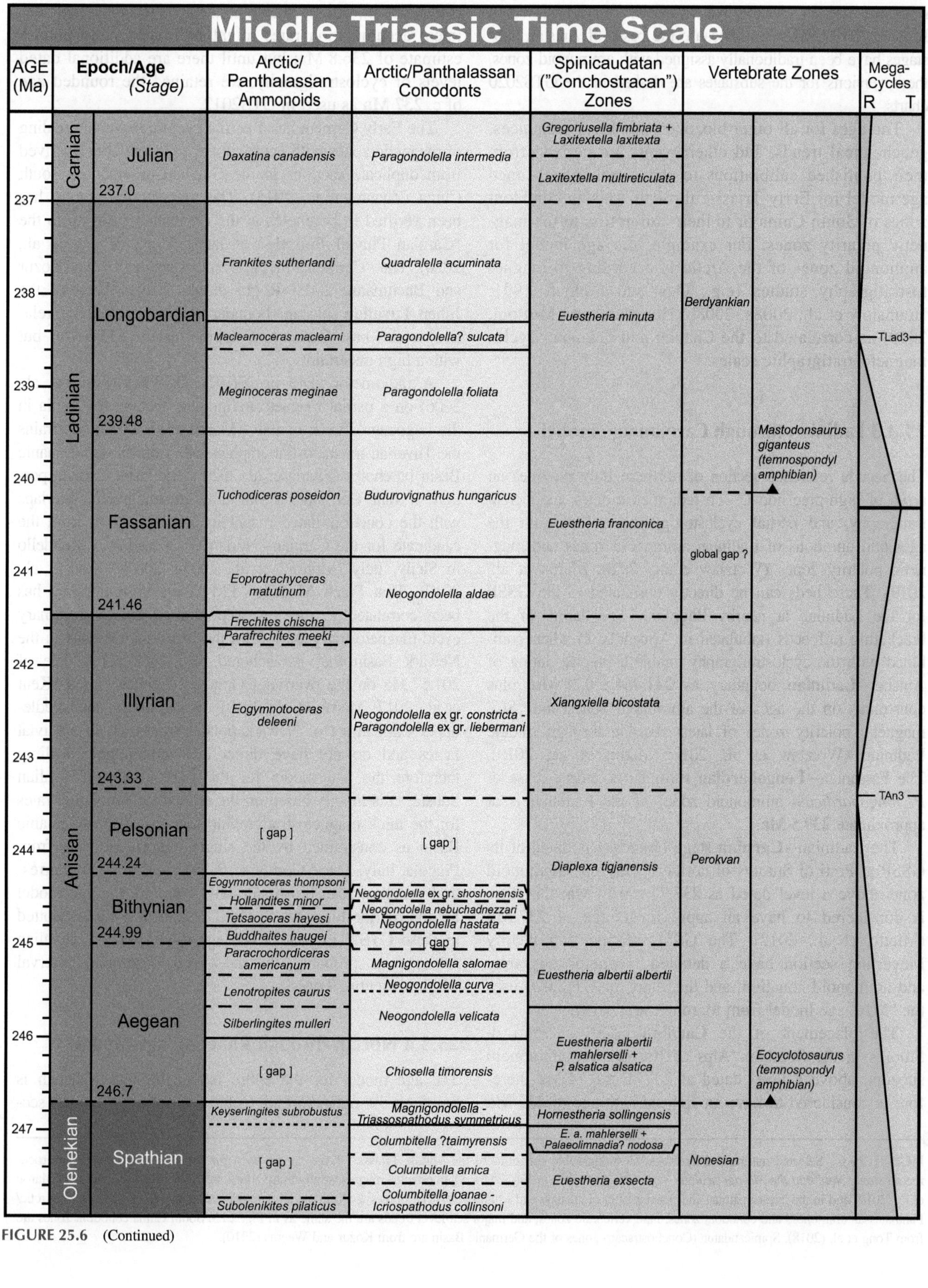

Middle Triassic Time Scale

AGE (Ma)	Epoch/Age (Stage)	Arctic/Panthalassan Ammonoids	Arctic/Panthalassan Conodonts	Spinicaudatan ("Conchostracan") Zones	Vertebrate Zones	Mega-Cycles R / T
237 —	Carnian — Julian — 237.0	Daxatina canadensis	Paragondolella intermedia	Gregoriusella fimbriata + Laxitextella laxitexta — Laxitextella multireticulata		TLad3
238 —	Ladinian — Longobardian	Frankites sutherlandi	Quadralella acuminata	Euestheria minuta (Berdyankian)		
239 —		Maclearnoceras maclearni	Paragondolella? sulcata			
	239.48	Meginoceras meginae	Paragondolella foliata		Mastodonsaurus giganteus (temnospondyl amphibian)	
240 —	Fassanian	Tuchodiceras poseidon	Budurovignathus hungaricus	Euestheria franconica		
241 —		Eoprotrachyceras matutinum	Neogondolella aldae	global gap ?		
	241.46	Frechites chischa — Parafrechites meeki				
242 —	Illyrian	Eogymnotoceras deleeni	Neogondolella ex gr. constricta - Paragondolella ex gr. liebermani	Xiangxiella bicostata		TAn3
243 —	243.33					
244 —	Pelsonian — 244.24	[gap]	[gap]	Diaplexa tigjanensis (Perokvan)		
	Bithynian	Eogymnotoceras thompsoni — Hollandites minor — Tetsaoceras hayesi — 244.99 — Buddhaites haugei	Neogondolella ex gr. shoshonensis — Neogondolella nebuchadnezzari — Neogondolella hastata — [gap]			
245 —		Paracrochordiceras americanum	Magnigondolella salomae — Neogondolella curva	Euestheria albertii albertii		
	Aegean	Lenotropites caurus — Silberlingites mulleri	Neogondolella velicata			
246 —		[gap]	Chiosella timorensis	Euestheria albertii mahlerselli + P. alsatica alsatica	Eocyclotosaurus (temnospondyl amphibian)	
247 —	246.7 — Olenekian — Spathian	Keyserlingites subrobustus	Magnigondolella - Triassospathodus symmetricus	Hornestheria sollingensis		
		[gap]	Columbitella ?taimyrensis	E. a. mahlerselli + Palaeolimnadia? nodosa (Nonesian)		
			Columbitella amica	Euestheria exsecta		
		Subolenikites pilaticus	Columbitella joanae - Icriospathodus collinsoni			

FIGURE 25.6 (Continued)

reference magnetic polarity pattern elsewhere (Hounslow and Muttoni, 2010). Therefore because the bases of Anisian aubstages have been traditionally assigned with ammonoid zones, these versions for the substages are dashed in the GTS2020 charts.

The ages for all other biozones, sea-level sequences, geochemical trends, and other events are derived from their published calibrations to the astronomical-tuned age model for Early Triassic through Anisian conodont zones of South China or to their calibration to the magnetic polarity zones. For example, the age model for ammonoid zones of the Arctic is according to magnetostratigraphy studies (e.g., Ogg and Steiner, 1991; Hounslow et al., 2008a, 2008b; Hounslow and Muttoni, 2010) as correlated to the Chaohu and Guandao cyclomagnetostratigraphic scales.

25.3.3 Ladinian through Carnian age model

The Seceda reference section of northern Italy provides an array of high-precision U−Pb radiometric dates, magnetostratigraphy, and partial cyclostratigraphy to constrain the ages and durations of Ladinian ammonoid zones and magnetic polarity zones (Wotzlaw et al., 2018; Maron et al., 2019). These beds can be directly correlated to the GSSP for the Ladinian at nearby Bagolino. The dating of the bracketing tuff beds (tabulated in Appendix 2) when combined with the cyclostratigraphy enable a precise dating of Anisian−**Ladinian** boundary as **241.464 ± 0.28 Ma**, plus constraints on the ages of the ammonoid zones and "SC" magnetic polarity zones of latest Anisian through middle Ladinian (Wotzlaw et al., 2018; Maron et al., 2019). The Fassanian−**Longobardian** substage boundary (base of *P. longobardicum* ammonoid zone) of the Ladinian is at approximate **239.5 Ma**.

The Ladinian−**Carnian** stage boundary is dated at its GSSP at Prati di Stuores of northern Italy one ammonoid zone above a level dated as 237.77 ± 0.14 Ma; therefore is considered to have an approximate age of **237 Ma** (Mietto et al., 2012). The GSSP section and nearby Mayerling section have a detailed magnetostratigraphy and ammonoid zonation, and the chart for GTS2020 uses the "MA" age model from Maron et al. (2019).

The placement of the **Carnian** GSSP at Prati di Stuores in the Southern Alps of Italy is one ammonoid subzone above a level dated at 237.77 ± 0.14 Ma therefore is considered to have an approximate age of **237 Ma**

(Mietto et al., 2012). Maron et al. (2019) suggest that sediment accumulation rates might indicate a more precise estimate of 236.8 Ma; but until there are additional dated levels or cyclostratigraphy, we retained the rounded age of c. **237 Ma** as used in GTS2012.

The Early Carnian has a partial cyclostratigraphic scaling of magnetic zones with approximate conodont zones derived from duplicate sections in the Guizhou province of South China (Zhang et al., 2015). The magnetostratigraphy has been verified in boreholes in the Germanic Basin up to the "Carnian Pluvial Episode" of latest Julian (Zhang et al., 2020). The "Carnian Pluvial Event" spans c. 1 Myr (Kozur and Bachmann, 2008; Roghi et al., 2010), therefore the Julian-**Tuvalian** substage boundary has a projected age relative to the base-Carnian of approximately **233.6 Ma**; but with a high uncertainty.

A radiometric date of 230.91 ± 0.33 Ma (Furin et al., 2006) on a partial magnetostratigraphic section at Pignola in the Lagonegro Basin of Italy (Maron et al., 2017) constrains the Tuvalian magnetostratigraphic scale from those Germanic Basin boreholes (Zhang et al, 2020). The latest Carnian portion of that Germanic Basin magnetostratigraphy overlaps with the conodont-dated magnetic polarity pattern from the candidate for the Carnian−Norian GSSP at Pizzo Mondello in Sicily, Italy (Muttoni et al., 2004a, 2004b; Kent et al., 2017). That Pizzo Modello "PM" magnetostratigraphy has been correlated to the lowest portion of the extraordinary cyclo-magnetostratigraphy from the lacustrine deposits in the Newark Basin that is anchored to a radiometric date of 201.57 Ma on the overlying Orange Mountain Basalt (Kent et al., 2017; Maron et al., 2019). Unfortunately, the middle-upper Carnian set of Newark polarity zones are in a fluvial facies and do not have direct cyclostratigraphic scaling; therefore the age model for the Late Carnian (Tuvalian Substage) is mainly based on the relative accumulation rates for the thick magnetostratigraphic sections in the Germanic Basin as constrained by the single radiometric date from Pignola, Italy. These Germanic Basin strata have only terrestrial conchostracan biostratigraphy; therefore the age model for the marine biozones in the Tuvalian Substage is dotted in these GTS2020 charts. The age model for the Tuvalian substage is probably the least-well-constrained interval within the entire Triassic.

25.3.4 Norian through Rhaetian age model

The age model for the entire Norian through Rhaetian is based on the magnetostratigraphic correlation of marine sec-

FIGURE 25.6 Selected marine and terrestrial biostratigraphic zonations of the Middle Triassic. ("*Age*" is the term for the time equivalent of the rock-record *stage*. *Note that the Ma-age scale is more expanded than in* Fig. 25.5*).* Marine-facies magnetostratigraphy zones are labeled in Ladinian after Maron et al. (2019) and in the latest Spathian and Anisian after Hounslow and Muttoni (2010). Sources for Tethyan ammonoid, conodont and bivalve zones, Arctic/Panthalassan ammonoid and conodont zones, land vertebrate zones, and major sea-level trends are the same as in Fig. 25.5. South China conodont zones are from Tong et al. (2018). Spinicaudatun (Conchostracan) zones of the Germanic Basin are from Kozur and Weems (2010).

tions to the astronomical-tuned magnetic polarity pattern from the boreholes in the lacustrine strata of the Newark Basin of eastern North America (e.g., Hounslow and Muttoni, 2010; Muttoni et al., 2014; Maron et al., 2015, 2019; Kent et al., 2017). The astronomical tuning of the Newark magnetic polarity time scale has been partially verified by correlation to U−Pb detrital-zircon dates that constrain the magnetostratigraphy from boreholes into the Chinle Formation of southwestern United States (Kent et al., 2018).

The proposed placement of the Carnian−Norian boundary is the lowest occurrence of conodont *Meta. parvus*. In the detailed conodont lineage study of the PM GSSP candidate, this conodont begins at a level 5/6ths up in polarity zone "PM4n" (Mazza et al., 2018), which is equivalent to chron E7n of this astronomical-tuned Newark polarity scale. Therefore the base-**Norian**, based on this working definition and Kent et al. (2017) chron ages, has a projected age of **227.3 Ma**.

This age assignment is partly supported by the U−Pb CA-TIMS radioisotopic dates of 223.81 ± 0.78 Ma and 224.52 ± 0.22 Ma reported from volcanic tuffs in British Columbia that bracket the lower/middle Norian substage boundary, as placed by adjacent conodont assemblages as used in North America (Diakow et al., 2010, abstract; and Mike Orchard, pers. commun., Sept 2015). Unfortunately, when this GTS2020 chapter was being prepared, these and other Triassic ages from Canada still had not yet been officially published.

However, the Carnian−Norian boundary is not yet formalized, and other ages have been promoted that may be partly dependent upon the implicit working definitions. For example, a base-Norian age estimate of c. 220 Ma has been proposed that is partly based on terrestrial correlations and implied potential problems with the Newark cycle−based age model (e.g., Lucas, 2018a, 2018b; Lucas et al., 2012). In GTS2012, two different options for a Norian−Rhaetian age Model were presented of "long Norian" and "short Norian" end members with potential different correlations of the Newark magnetic scale to the magnetostratigraphic records assembled (by Hounslow and Muttoni, 2010) from sedimentary sections that had marine biostratigraphy (Ogg et al., 2012, 2014). The construction of a Norian time scale is also difficult because many portions of the Newark magnetic polarity pattern lack an adequate "fingerprint" to correlate with compilations of magnetostratigraphy of marine-zoned strata (e.g., two options of Muttoni et al., 2004b, and in Ogg et al., 2008; three options discussed in detail within Hounslow and Muttoni, 2010). However, based on the apparent correlations to the candidate for the Norian GSSP at Pizzo Mondello, the c. 231 Ma date from mid-Tuvalian strata in

Sicily and the supporting correlations with the Chinle Magnetostratigraphy with detrital-zircon dating, then only a variation of the "long Norian" option is proposed in this GTS2020.

Using the suggested minimal set of a few magnetostratigraphic correlations shown in Fig. 25.5, then the base of the **Alaunian** Substage (base of Tethyan *C. bicrenatus* ammonoid zone) has a projected age of **217.5 Ma** and the base of the **Sevatian** Substage (if using the base of Tethyan *S. quinquepunctatus* ammonoid zone) is **214.0 Ma**. Visually, there are several other potential magnetostratigraphic correlations within individual biozones or pairs of biozones, but we tried to preserve the approximate scaling of polarity zones as compiled by Hounslow and Muttoni (2010). Obviously, the entire Norian age model for marine biostratigraphy needs to be enhanced and verified with astronomical tuning and radiometric dating of magnetostratigraphic sections having well-defined marine biostratigraphy.

The Norian−Rhaetian boundary will be assigned in a GSSP at a level corresponding to the lowest occurrence of the conodont marker *Mi. posthernsteini*. However, there are different concepts of when that taxon can be differentiated on the *Misikella* morphology lineage (Rigo et al., 2016; Orchard, 2016). There are two main candidate GSSP sections, each having magnetostratigraphy that can be correlated to the astronomically tuned Newark magnetic polarity reference time scale (Maron et al., 2015): (1) Steinbergkogel in Austria that uses a broader *Mi. posthernsteini* sensu lato (Austria; Krystyn et al., 2007c, 2007d; Hüsing et al., 2011) with an implied age of 209.5 Ma and (2) Pignola-Abriola in Sicily that uses a *Mi. posthernsteini* sensu stricto (Maron et al, 2015; Rigo et al., 2016) with an implied age of 205.7 Ma. Both options are shown in Fig. 25.5, but the *Mi. posthernsteini* s.str. (**205.7 Ma**) is used as the base-**Rhaetian** for the summary scales in this GTS2020. The younger option for the base of the Rhaetian is near the extinction of the bivalve *Monotis*, and this level has been dated in Peru as between 205.7 ± 0.15 and 205.3 ± 0.14 Ma (Wotzlaw et al., 2014).

The **Triassic−Jurassic boundary** dated by Schoene et al. (2010) was revised by them (in Wotzlaw et al., 2014) using updated EARTHTIME tracers. This changed the interpolated boundary age from 201.31 ± 0.18 Ma (used in GTS2012) to a slightly older **201.36 ± 0.17 Ma**.

25.3.5 Estimated uncertainties and future enhancements of Triassic age model

The few high-precision radioisotopic dates with well-constrained biostratigraphic ages and astronomical tuning of stratigraphic zonations that constrain the Triassic time scale

---

FIGURE 25.7 Selected marine and terrestrial biostratigraphic zonations of the Early Triassic. ("*Age*" is the term for the time equivalent of the rock-record *stage. Note that the Ma-age scale is more expanded than in* Figs. 25.5 or 25.6). Marine-facies magnetostratigraphy zones are labeled after Hounslow and Muttoni (2010). Sources for Tethyan ammonoid, conodont and bivalve zones, South China conodont zones, Arctic/Panthalassan ammonoid and conodont zones, land vertebrate zones, and major sea-level trends are the same as in Figs. 25.5 and 25.6. Spinicaudatun (Conchostracan) zones of the Germanic Basin are from Kozur and Weems (2010), Scholze et al. (2016), and Schneider and Scholze (2018).

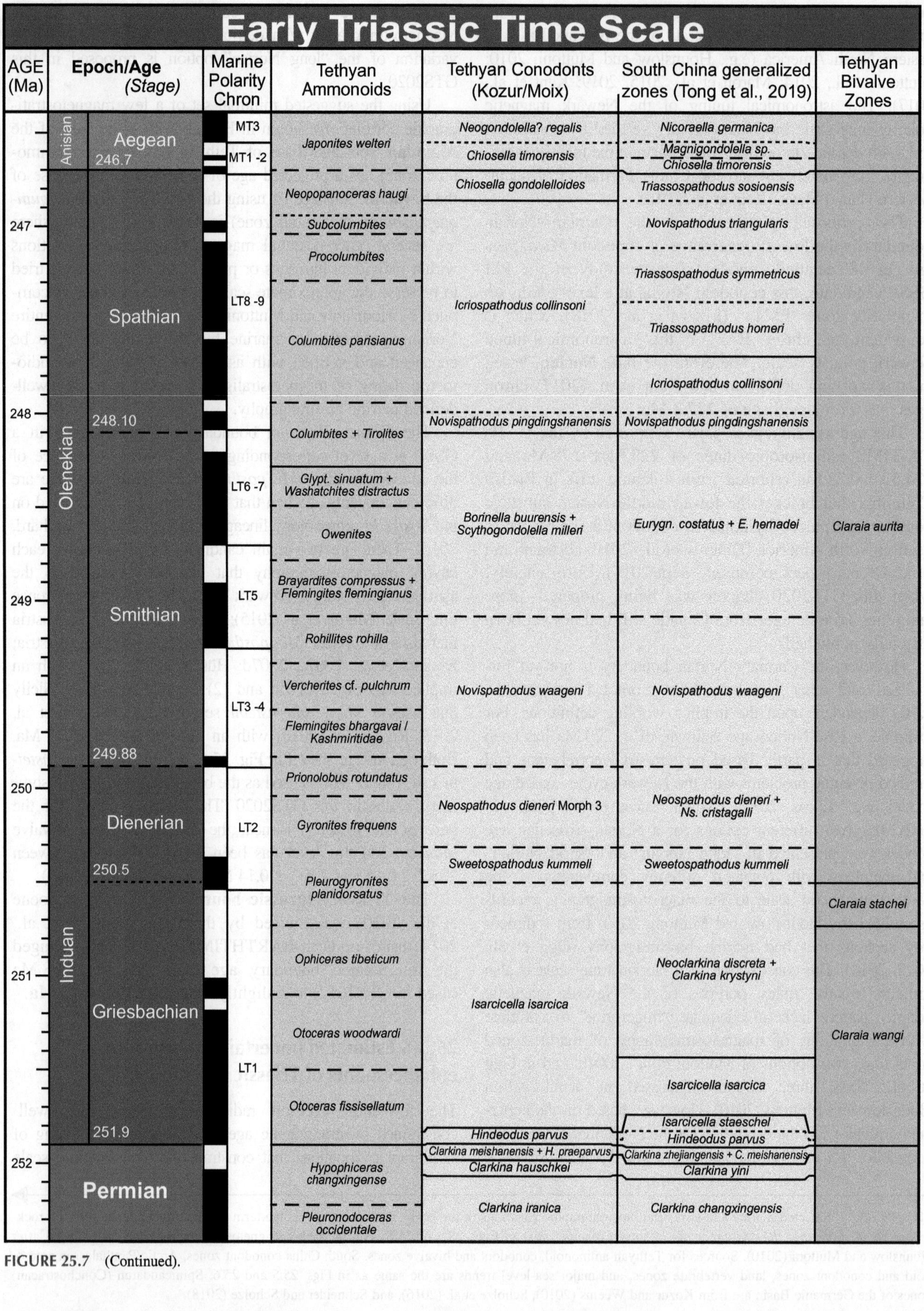

Early Triassic Time Scale

AGE (Ma)	Epoch/Age (Stage)	Marine Polarity Chron	Tethyan Ammonoids	Tethyan Conodonts (Kozur/Moix)	South China generalized zones (Tong et al., 2019)	Tethyan Bivalve Zones
	Anisian — Aegean 246.7	MT3 / MT1-2	Japonites welteri	Neogondolella? regalis / Chiosella timorensis	Nicoraella germanica / Magnigondolella sp. / Chiosella timorensis	
247			Neopopanoceras haugi	Chiosella gondolelloides	Triassospathodus sosioensis	
	Olenekian — Spathian	LT8-9	Subcolumbites	Icriospathodus collinsoni	Novispathodus triangularis	
			Procolumbites		Triassospathodus symmetricus	
			Columbites parisianus		Triassospathodus homeri	
248	248.10		Columbites + Tirolites	Novispathodus pingdingshanensis	Icriospathodus collinsoni / Novispathodus pingdingshanensis	
	Smithian	LT6-7	Glypt. sinuatum + Washatchites distractus	Borinella buurensis + Scythogondolella milleri	Eurygn. costatus + E. hemadei	Claraia aurita
			Owenites			
249		LT5	Brayardites compressus + Flemingites flemingianus			
			Rohillites rohilla	Novispathodus waageni	Novispathodus waageni	
		LT3-4	Vercherites cf. pulchrum			
	249.88		Flemingites bhargavai / Kashmiritidae			
250	Induan — Dienerian	LT2	Prionolobus rotundatus	Neospathodus dieneri Morph 3	Neospathodus dieneri + Ns. cristagalli	
			Gyronites frequens			
	250.5		Pleurogyronites planidorsatus	Sweetospathodus kummeli	Sweetospathodus kummeli	
251	Griesbachian		Ophiceras tibeticum	Isarcicella isarcica	Neoclarkina discreta + Clarkina krystyni	Claraia stachei
						Claraia wangi
		LT1	Otoceras woodwardi		Isarcicella isarcica	
			Otoceras fissisellatum		Isarcicella staeschel	
	251.9			Hindeodus parvus	Hindeodus parvus	
252	Permian		Hypophiceras changxingense	Clarkina meishanensis + H. praeparvus / Clarkina hauschkei	Clarkina zhejiangensis + C. meishanensis / Clarkina yini	
			Pleuronodoceras occidentale	Clarkina iranica	Clarkina changxingensis	

FIGURE 25.7 (Continued).

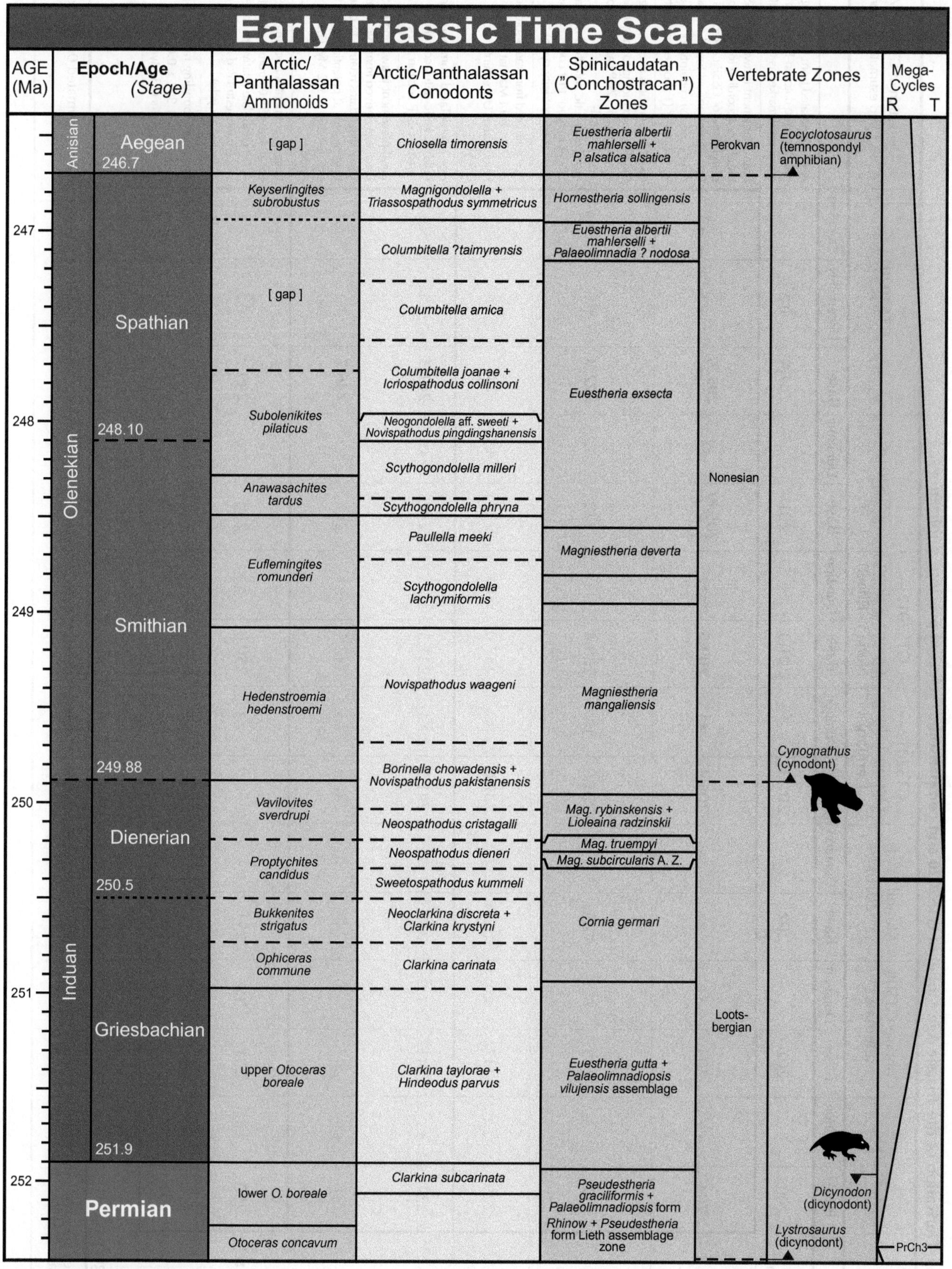

Early Triassic Time Scale

AGE (Ma)	Epoch/Age (Stage)	Arctic/ Panthalassan Ammonoids	Arctic/Panthalassan Conodonts	Spinicaudatan ("Conchostracan") Zones	Vertebrate Zones		Mega-Cycles R T
	Anisian — Aegean 246.7	[gap]	Chiosella timorensis	Euestheria albertii mahlerselli + P. alsatica alsatica	Perokvan	Eocyclotosaurus (temnospondyl amphibian)	
247	Olenekian — Spathian	Keyserlingites subrobustus	Magnigondolella + Triassospathodus symmetricus	Hornestheria sollingensis			
			Columbitella ?taimyrensis	Euestheria albertii mahlerselli + Palaeolimnadia ? nodosa			
		[gap]	Columbitella amica				
248			Columbitella joanae + Icriospathodus collinsoni	Euestheria exsecta			
	248.10	Subolenikites pilaticus	Neogondolella aff. sweeti + Novispathodus pingdingshanensis		Nonesian		
			Scythogondolella milleri				
		Anawasachites tardus	Scythogondolella phryna				
		Euflemingites romunderi	Paullella meeki	Magniestheria deverta			
249			Scythogondolella lachrymiformis				
	Smithian	Hedenstroemia hedenstroemi	Novispathodus waageni	Magniestheria mangaliensis			
	249.88		Borinella chowadensis + Novispathodus pakistanensis			Cynognathus (cynodont)	
250	Dienerian	Vavilovites sverdrupi	Neospathodus cristagalli	Mag. rybinskensis + Lioleaina radzinskii			
		Proptychites candidus	Neospathodus dieneri	Mag. truempyi / Mag. subcircularis A. Z.			
	250.5		Sweetospathodus kummeli				
	Induan — Griesbachian	Bukkenites strigatus	Neoclarkina discreta + Clarkina krystyni	Cornia germari			
		Ophiceras commune	Clarkina carinata		Loots-bergian		
251		upper Otoceras boreale	Clarkina taylorae + Hindeodus parvus	Euestheria gutta + Palaeolimnadiopsis vilujensis assemblage			
	251.9					Dicynodon (dicynodont)	
252	Permian	lower O. boreale	Clarkina subcarinata	Pseudestheria graciliformis + Palaeolimnadiopsis form		Lystrosaurus (dicynodont)	PrCh3
		Otoceras concavum		Rhinow + Pseudestheria form Lieth assemblage zone			

FIGURE 25.7 (Continued).

TABLE 25.1 Derivation of the Triassic Age Model of GTS2020 and comparison to selected previous publications.

Stage	GTS2004		Concise GTS08 alternate		Mundil et al. (2010)		GTS2012—Option 1 (long Carnian)		GTS2012—Option 2 (long Norian–Rhaet)		GTS2016		GTS2020			Derivation of GTS2020 Age Model and estimated uncertainty
	Base	Duration	Base	Duration	Base	Duration	Base	Duration	Base	Duration	Base	Duration	Base	Uncertainty	Duration	
Jurassic (Hettangian)	200		201		202		201.3		201.3		201		**201.4**	0.2		Age interpolated from bounding radioisotopic dates in Peru with a similar ammonoid marker as GSSP is 201.36±0.17 Ma.
Rhaetian	204	4.0	205	3.6	x		205.4	4.1	209.5	8.2	205.8/209.6	4.4	**205.7/209.5**	0.1	4.3	Basal placement is awaiting GSSP decision. Basal age according to magnetostratigraphy correlations of GSSP candidates to Newark Basin astronomical-tuned magnetic polarity time scale. Uncertainty for that correlation is c. 0.1 Myr (1 short-eccentricity cycle).
Norian	217	12.9	229	24.3	<230		221.0	15.7	228.4	18.9	229	22.7	**227.3**	0.1	21.6	Basal placement is also awaiting GSSP decision. Basal age according to magnetostratigraphy correlation of GSSP candidate to Newark Basin astronomical-tuned magnetic polarity time scale. Uncertainty for that correlation is ± 0.1 Myr (1 short-eccentricity cycle).
Carnian	229	12.2	237	8.0	<236		237.0	16.0	237.0	8.7	237	8.5	**237**	0.5	9.7	Age estimated from radioisotopic date of 237.77 ± 0.14 Ma that is one ammonoid subzone below base. Uncertainty = c. 0.5 Myr.
Ladinian	237	8.3	241	3.7	242		241.5	4.5	241.5	4.5	242	4.5	**241.5**	0.3	4.5	Age interpolated from bounding radioisotopic dates correlated to the nearby GSSP section is 241.464 ± 0.28 Ma.
Anisian	246	8.9	247	6.9	247	5.2	247.1	5.6	247.1	5.6	247	5.3	**246.7**	0.2	5.2	Basal placement is awaiting GSSP decision. Basal age derived from cycles at South China GSSP candidates relative to base-Triassic, plus U–Pb dates at potential GSSP. Uncertainty estimated as ± 0.2 Myr (1/2 long-eccentricity cycle).
Olenekian	250	3.6	251	3.6	251	4.1	250.0	2.9	250.0	2.9	250	3.0	**249.9**	0.2	3.2	Basal placement is awaiting GSSP decision. Basal age derived from cycles at South China GSSP candidate relative to base-Triassic. Uncertainty estimated as ± 0.2 Myr (1/2 long-eccentricity cycle).
Induan	251	1.5	253	1.5	252	1.0	252.2	2.2	252.2	2.2	252	2.1	**251.9**	0.3	2	Age from bounding radioisotopic dates at GSSP is 251.902 ± 0.024 Ma. Total uncertainty is 0.3 Myr.
Permian (Changhsingian)																

Ages for bases are in Ma; durations and uncertainties are in Myr. GSSP, Global Boundary Stratotype Section and Point. Radioisotopic dates on stage boundaries do not include the estimated 0.3 Myr external uncertainty as explained in Burgess et al., (2014).

typically have a published uncertainty less than 0.2 Myr. However, an external uncertainty of c. 0.3 Myr should be included if comparing these Triassic EARTHTIME-standardized dates to other dating methods, as explained in Burgess et al. (2014). These dates anchor the cyclostratigraphic scaling for much of the Early and the Late Triassic and for the GSSPs and current GSSP candidates (Table 25.1). There is probably an additional c. 0.2-Myr uncertainty in the assignment of any event within a 405-kyr long-eccentricity cycle. Therefore most of the age assignments for Triassic stage boundaries have a relatively low uncertainty (Table 25.1) when compared to other Paleozoic—Mesozoic stages. However, the uncertainties are greater for the base of the Carnian and for many of the biozone boundaries within the individual stages which have age models that required additional extrapolation from the nearest radio-isotopic dates or from the estimated calibrations to other reference scales.

The main future enhancements to the age model for the Triassic stages will be the decisions on GSSP definition for the Olenekian, Anisian, Norian, and Rhaetian. Other essential developments will be the verification of the cycle-scaling and interregional correlation of the magnetostratigraphy of each stage, and obtaining an astronomical tuning for the zonations within the Ladinian and Carnian stages. It is anticipated that the integration of magnetostratigraphy, astronomical cycles coupled with periodic major changes in sea level, and distinctive stable isotopic excursions will enable a more reliable interregional correlation of biostratigraphic datums for both marine (Tethyan through Boreal realms) and terrestrials settings.

25.4 Summary

During the past decade, Triassic workers have defined most of the stages (Fig. 25.8), greatly enhanced the intercorrelation of biostratigraphic zones, enabled compilation of a nearly complete magnetic polarity pattern calibrated to marine biostratigraphic datums, discovered excursions in stable isotopes (especially major carbon-isotope and oxygen-isotope

GSSPs of the Triassic Stages, with location and primary correlation criteria					
Stage	**GSSP Location**	**Latitude, Longitude**	**Boundary Level**	**Correlation Events**	**Reference**
Rhaetian	*Candidates are Pizzo Mondello, Sicily, Italy, and Steinbergkogel, Austria*			*Near FADs of conodont* Misikella posthernsteini *s.s. or* Misikella posthernsteini *s.l.*	
Norian	*Candidates are Black Bear Ridge in British Columbia (Canada) and Pizzo Mondello, Sicily, Italy*			*FAD of conodont Metapolygnathus parvus. Near base of Stikinoceras kerri ammonoid zone and FAD of bivalve Halobia austriaca*	
Carnian	Section at Prati di Stuores, Dolomites, Italy	46°31'37"N 11°55'49"E	GSSP is base of marly limestone bed SW4, 45 m from base of San Cassiano Formation	FAD of ammonoid *Daxatina canadensis,* conodont *Quadralella polygnathiformis* and *Halobia* bivalves	Episodes **35**/3, 2012
Ladinian	Bagolino, Province of Brescia, Northern Italy	45°49'09.5"N 10°28'15.5"E	base of a 15 – 20 cm thick limestone bed overlying a distinctive groove ("Chiesense groove") of limestone nodules in a shaly matrix, located about 5 m above the base of the Buchenstein Beds	Ammonoid, FAD of *Eoprotrachyceras curionii*	Episodes **28**/4, 2005
Anisian	*Candidates are Desli Caira (Romania), Kçira (Albania), Wantou (Guangxi Province, S. China) and Guandao (Guizhou Province, S. China)*			*FAD of conodont Chiosella timorensis or base of magnetic normal- polarity chronozone MT1n*	
Olenekian	*Candidates are Chaohu, China and Mud (Muth) village, Spiti Valley, India*			*FAD of conodont Novispathodus waageni, near base of Flemingites ammonoid genera*	
Induan	Meishan, Zhejiang Province, China	31°4'47.28"N 119°42'20.90"E	base of Bed 27c in the Meishan Section	Conodont, FAD of *Hindeodus parvus*	Episodes **24**/2, 2001
* according to Google Earth					

FIGURE 25.8 Ratified GSSPs and potential primary markers under consideration for defining the Triassic stages *(status as of January 2020)*. Details of each GSSP are available at http://www.stratigraphy.org, https://timescalefoundation.org/gssp/, and in the *Episodes* publications. *GSSP,* Global Boundary Stratotype Section and Point.

excursions within the lower Triassic), and achieved or rejected cycle-stratigraphic scaling of several intervals. A generalized synthesis of selected Triassic stratigraphic scales is compiled in Figs. 25.5–25.7; and additional geochemical trends are summarized in the geochemical chapters of this book.

Extensive radioisotopic dating using advanced techniques and the astronomical tuning of reference sections in China and Italy (including GSSPs or candidate GSSPs) have replaced much of the radioisotopic-date dataset and extrapolations that were used in GTS2012. The combination of these methods has established a well-constrained age model for the stages and many of the biozones within the Lower and Middle Triassic (Induan GSSP, Olenekian and Anisian candidates, Ladinian and Carnian GSSPs) and for the top of the Triassic (base of Hettangian Stage). There are lingering major uncertainties on the age model and durations for biozones within the yet-to-be-defined Upper Triassic stages of Norian and Rhaetian. Establishing a robust Late Triassic time scale for these zones and other events requires both definitive radioisotopic dates and cyclostratigraphy on marine sections that have standard biostratigraphy. Further advances in formalizing GSSPs, zonal schemes, interregional correlations, and eventual consensus on the best numerical age model interpolations will be found in the *Albertiana* newsletters of the Subcommission on Triassic Stratigraphy.

Acknowledgments

The compilation of the Triassic chronostratigraphy and understanding the disputes on correlations and the age models was also greatly aided over the years by discussions with several colleagues, of which a few are, in alphabetical order: Marco Balini, Charles Henderson (who also reviewed this chapter), Chunju Huang, Mark Hounslow, the late Heinz Kozur (including an intensive 3-day working session at his home), Leopold Krystyn, Spencer Lucas, Manfred Menning, Paul Olsen, Lawrence Tanner, and Hongfu Yin. Mingsong Li and Yang "Wendy" Zhang provided their important cycle-scaling of Early Triassic, Anisian, and Carnian stratigraphy. None of them entirely agree with the selected taxonomic nomenclature for zones used in the figures or with the age models, but all agree that further international efforts will soon resolve all of the disputed calibrations. This work was partly supported by three NSFC grants (41821001, 41830323, 41930322). James Ogg's participation was partly enabled by distinguished visiting professorships at Chengdu University of Technology (Chengdu) and at the China University of Geoscience (Wuhan), China. Additional and alternative zonal schemes are available through the *TimeScale Creator* visualization datapacks (https://timescalecreator.org).

Bibliography

Aigner, T., and Bachmann, G.H., 1992, Sequence-stratigraphic framework of the German Triassic. *Sedimentary Geology*, **80**: 115–135.

Algeo, T., Brayard, A., and Richoz, S. (eds), 2019a. The Smithian-Spathian boundary: a critical juncture in the early Triassic recovery of marine ecosystems. *Earth-Science Reviews*, **Special volume**, 195: 212 pp.

Algeo, T., Brayard, A., and Richoz, S., 2019b, The Smithian-Spathian boundary: a critical juncture in the Early Triassic recovery of marine ecosystems. *Earth-Science Reviews*, **195**: 1–6. https://doi.org/10.1016/j.earscirev.2019.102877.

Assereto, R., 1974, Aegean and Bithynian: proposal for two new Anisian substages. *Schriftenreihe der Erdwissenschaftlichen Kommissionen, Österreichische Akademie der Wissenschaften*, **2**: 23–39.

Bachmann, G.H., and Kozur, H.W., 2004, The Germanic Triassic: correlations with the international scale, numerical ages and Milankovitch cyclicity. *Hallesches Jahrbuch fur Geowissenschaften*, **B26**: 17–62.

Balini, M., Drystyn, L., and Torti, V., 1998, In search of the Ladinian/Carnian boundary: perspectives from Spiti (Tethys Himalaya). *Albertiana*, **21**: 26–32.

Balini, M., Krystyn, L., Nicora, A., and Torti, V., 2001, The Ladinian-Carnian boundary succession in Spiti (Tethys Himalaya) and its bearing to the definition of the GSSP for the Carnian stage (Upper Triassic). *Journal of Asian Earth Sciences*, **19** (3A), 3–4.

Balini, M., Lucas, S.G., Jenks, J.F., and Spielmann, J.A., 2010, Triassic ammonoid biostratigraphy: an overview. In: Lucas, S.G. (ed), The Triassic Timescale, *The Geological Society of London Special Publications*, **334**: 221–262.

Balini, M., Krystyn, L., Levera, M., and Tripodo, A., 2012, Late Carnian-early Norian ammonoids from the GSSP candidate section Pizzo Mondello (Sicani Mountains, Sicily). *Rivista Italiana di Paleontologia e Stratigrafia*, **118**: 47–84.

Balme, B.E., and Foster, C.B., 1996, Triassic, explanatory notes on Biostratigraphic charts. *In* Young, G.C., and Laurie, J.R. (eds), *Australian Phanerozoic Timescales*. Oxford: Oxford University Press, Chart 7.

Baud, A., 2001, The new GSSP, base of the Triassic: some consequences. *Albertiana*, **26**: 4–6.

Baud, A., 2014, The global marine Permian-Triassic boundary: over a century of adventures and controversies (1880-2001). *Albertiana*, **42**: 1–21 Available at: < https://albertiana-sts.org/ > .

Baud, A., and Beauchamp, B., 2001, Proposals for the redefinition of the Griesbachian Substage and for the base of the Triassic in the Arctic regions. *In* Yan, J., and Peng, Y. (eds), *Proceedings of the International Symposium on the Global Stratotype of the Permian-Triassic boundary and the Paleozoic-Mesozoic Events*. Changxing: University of Geosciences Press, 26–28.

Baud, A., Magaritz, M., and Holser, W.T., 1989, Permian-Triassic of the Tethys: carbon isotope studies. *Sonderdruck aus Geologische Rundschau*, **78**: 649–677.

Baud, A., Atudorei, V., and Sharp, Z., 1996, Late Permian and early Triassic evolution of the northern Indian margin: carbon isotope and sequence stratigraphy. *Geodinamica Acta*, **9**: 57–77.

Benton, M.J., 1986, The Late Triassic tetrapod extinction events. *In* Padian, K. (ed), *The Beginning of the Age of Dinosaurs. Faunal Change Across the Triassic-Jurassic Boundary*. Cambridge: Cambridge University Press, 303–320.

Benton, M.J., 1993, Late Triassic extinctions and the origin of the dinosaurs. *Science*, **260**: 769–770.

Benton, M.J., Zhang, Q.Y., Hu, S.X., Chen, Z.Q., Wen, W., Liu, J., et al., 2013, Exceptional vertebrate biotas from the Triassic of China, and the expansion of marine ecosystems after the Permo-Triassic mass extinction. *Earth-Science Reviews*, **125**: 199–243. https://doi.org/10.1016/j.earscirev.2013.05.014.

Benton, M.J., Forth, J., and Langer, M.C., 2014, Models for the rise of the dinosaurs. *Current Biology*, **24**: R87–R95 < https://doi.org/10.1016/j.cub.2013.11.063 >.

Bernasconi, S.M., Meier, I., Wohlwend, S., Brack, P., Hochuli, P.A., Bläsi, H., et al., 2017, An evaporite-based high-resolution sulfur isotope record of Late Permian and Triassic seawater sulfate. *Geochimica et Cosmochimica Acta*, **204**: 331–349.

Berra, F., Jadoul, F., and Anelli, A., 2010, Environmental control on the end of the Dolomia Principale/Hauptdolomit depositional system in the central Alps: coupling sea-level and climatic changes. In: Kustatscher, E., Preto, N., Wignall, P. (eds), Triassic climates, *Palaeogeography, Palaeoclimatology, Palaeoecology*, **290**: 138–150.

Beauchamp, B., and Baud, A., 2002, Growth and demise of Permian biogenic chert along northwest Pangea: evidence for end-Permain collapse of thermohaline circulation. *Palaeogeography, Palaeoclimatology, Palaeoecology*, **184**: 37–63.

Bittner, A., 1892, Was ist norisch? *Jahrbuch Geologischen Reichsanstalt*, **42**: 387–396.

Blackburn, T.J., Olsen, P.E., Browning, S.A., McLean, N.M., Kent, D.V., Puffer, J., et al., 2013, Zircon U-Pb geochronology links the End-Triassic extinction with the Central Atlantic Magmatic Province. *Science*, **340**: 941–945. https://doi.org/10.1126/science.1234204.

Blome, C.D., 1984, Upper Triassic Radiolaria and radiolarian zonation from western North America. *Bulletins of American Paleontology*, **85**: 88 pp.

Bown, P.R. (ed), 1998. Calcareous nanofossils biostratigraphy. Chapman and Hall. London, 314 pp.

Bowring, S.A., Erwin, D.H., Jin, Y.G., Martin, M.W., Davidek, K., and Wang, W., 1998, U/Pb zircon geochronology and tempo of the end-Permian mass extinction. *Science*, **280**: 1039–1045.

Brack, P., 2010, The "golden spike" for the Ladinian is set!. *Albertiana*, **38**: 8–10.

Brack, P., and Rieber, H., 1994, The Anisian/Ladinian boundary: retrospective and new constraints. *Albertiana*, **13**: 25–36.

Brack, P., and Rieber, H., 1996, The new 'High-resolution Middle Triassic ammonoid standard scale' proposed by Triassic researchers from Padova—a discussion of the Anisian/Ladinian boundary interval. *Albertiana*, **17**: 42–50.

Brack, P., Rieber, H., and Mundil, R., 1995, The Anisian/Ladinian boundary interval at Bagolino (Southern Alps, Italy): I. Summary and new results on ammonoid horizons and radiometric dating. *Albertiana*, **15**: 45–56.

Brack, P., Mundil, R., Oberli, F., Meier, M., and Rieber, H., 1996, Biostratigraphic and radiometric age data question the Milankovitch characteristics of the Latemar cycles (Southern Alps, Italy). *Geology*, **24**: 371–375.

Brack, P., Mundil, R., Oberli, F., Meier, M., and Rieber, H., 1997, Biostratigraphic and radiometric age data question the Milankovitch characteristics of the Latemar cycles (Southern Alps, Italy) – reply. *Geology*, **25**: 471–472.

Brack, P., Rieber, H., Nicora, A., and Mundil, R., 2005, The Global Boundary Stratotype Section and Point (GSSP) of the Ladinian Stage (Middle Triassic) at Bagolino (Southern Alps, Northern Italy) and its implications for the Triassic time scale. *Episodes*, **28** (4), 233–244.

Brack, P., Rieber, H., Mundil, R., Blendinger, W., and Maurer, F., 2007, Geometry and chronology of growth and drowning of Middle Triassic carbonate platforms (Cernera and Bivera/Clapsavon) in the Southern Alps (northern Italy). *Swiss Journal of Geosciences*, **10**: 327–347.

Brayard, A., Escarguel, G., Bucher, H., and Brühwiler, 2009a, Smithian and Spathian (Early Triassic) ammonoid assemblages from terranes: paleoceanographic and paleogeographic implications. In: Metcalfe, I., and Isozaki, Y. (eds), End-Permian mass extinction: events & processes, age & timescale, causative mechanism(s) & recovery, *Journal of Asian Earth Sciences*, **36**(6): 420–433.

Brayard, A., Escarguel, G., Bucher, H., Monnet, C., Brühwiler, Goudemand, N., et al., 2009b, Good genes and good luck: ammonoid diversity and the end-Permian mass extinction. *Science*, **325**: 1118–1121.

Broglio Loriga, C., and Cassinis, G., 1992, The Permo–Triassic boundary in the Southern Alps (Italy) and in adjacent Periadriatic regions. *In* Sweet, W.C., Yang, Z., Dickins, J.M., and Yin, H. (eds), *Permo–Triassic Events in the Eastern Tethys. Stratigraphy, Classification, and Relations With the Western Tethys*. Cambridge: Cambridge University Press, 78–97.

Broglio Loriga, C., Cirilli, S., De Zanche, V., di Bari, D., Gianolla, P., Laghi, G.R., et al., 1998, A GSSP candidate for the Ladinian-Carnian boundary: the Prati di Stuores/Stuores Wiesen section (Dolomites, Italy). *Albertiana*, **21**: 2–18.

Broglio Loriga, C., Cirilli, S., De Zanche, V., di Bari, D., Gianolla, P., Laghi, G.R., et al., 1999, The Prati di Stuores/Stuores Wiesen section (Dolomites, Italy): a candidate Global Stratotype Section and Point for the base of the Carnian stage. *Rivista Italiana di Paleontologia e Stratigrafia*, **105**: 37–78.

Brühwiler, T., Hochuli, P., Mundil, R., Schatz, W., and Brack, P., 2007, Bio- and chronostratigraphy of the Middle Triassic Reifling Formation of the westernmost Northern Calcareous Alps. *Swiss Journal of Geosciences*, **100**: 443–455.

Burgess, S.D., and Bowring, S.A., 2015, High-precision geochronology confirms voluminous magmatism before, during, and after Earth's most severe extinction. *Science Advances*, **1** (7), article # e1500470: 14 pp. http://dx.doi.org/10.1126/sciadv.1500470.

Burgess, S.D., Bowring, S., and Shen, Z.Q., 2014, High-precision timeline for Earth's most severe extinction. *Proceedings of the National Academy of Sciences of the United States of America*, **111**: 3316–3321.

Burgess, S.D., Muirhead, J.D., and Bowring, S.A., 2017, Initial pulse of Siberian Traps sills as the trigger of the end-Permian mass extinction. *Nature Communications*, **8**: 164–169. https://doi.org/10.1038/s41467-017-00083-9.

Carter, E.S., 1993, Biochronology and paleontology of uppermost Triassic (Rhaetian) radiolarians, Queen Charlotte Islands, British Columbia, Canada. *Mémoires Géologie (Lausanne)*, **11**: 175 pp.

Carter, E.S., and Orchard, M.J., 2007, Radiolarian–conodont–ammonoid intercalibration around the Norian-Rhaetian Boundary and implications for trans-Panthalassan correlation. *Albertiana*, **36**: 149–163.

Cao, C.Q., Love, G.D., Hays, L.E., Wang, W., Shen, S.Z., and Summons, 2009, Biogeochemical evidence for euxinic oceans and ecological disturbance presaging the end-Permian mass extinction. *Earth and Planetary Science Letters*, **281**: 188–201.

Channell, J.E.T., Kozur, H.W., Sievers, T., Mocl, R., Aubrecht, R., and Sykora, M., 2003, Carnian–Norian biomagnetostratigraphy at Silickà Brezovà (Slovakia): correlation to other Tethyan sections and to the Newark Basin. *Palaeogeography, Palaeoclimatology, Palaecology*, **191**: 65–109.

Chen, Z.Q., and Benton, M.J., 2012, The timing and pattern of biotic recovery following the end-Permian mass extinction. *Nature Geoscience*, **5** (6), 375–383. https://doi.org/10.1038/NGEO1475.

Chen, Y.L., Krystyn, L., Orchard, M.J., Lai, X.L., and Richoz, S., 2016, A review of the evolution, biostratigraphy, provincialism and diversity of Middle and early Late Triassic conodonts. *Papers in Palaeontology*, **2** (2), 235–263. http://dx.doi.org/10.1002/spp2.1038 [and data at Dryad Digital Repository, < https://doi.org/10.5061/dryad.34r55 > .

Chen, Z.Q., Zhao, L.S., Wang, X.D., Luo, M., and Guo, Z., 2018, Great Paleozoic-Mesozoic biotic turnings and paleontological education in China: A tribute to the achievements of Professor Zunyi Yang. *Journal of Earth Science*, **29**: 721–732. https://doi.org/10.1007/s12583-018-0797-1.

Chen, Z.Q., Tu, C.Y., Pei, Y., Ogg, J., Fang, Y.H., Wu, S.Q., et al., 2019, Biosedimentological features of major microbe-metazoan transitions (MMTs) from Precambrian to Cenozoic. *Earth-Science Reviews*, **189**: 21–50. https://doi.org/10.1016/j.earscirev.2019.01.015.

Chen, Y., Jiang, H.S., Ogg, J.G., Zhang, Y., Gong, Y.F., and Yan, C.B., 2020, Early-Middle Triassic boundary interval: integrated chemo-bio-magneto-stratigraphy of potential GSSPs for the base of the Anisian Stage in South China. *Earth and Planetary Science Letters*, **530**, article #115863: 13 pp. https://doi.org/10.1016/j.epsl.2019.115863.

Chen, Z.Q., Algeo, T.J., and Bottjer, D.J., 2014, Global review of the Permian–Triassic mass extinction and subsequent recovery: part I. *Earth-Science Reviews*, **137**: 1–5.

Chen, Z.Q., Yang, H., Luo, M., Benton, J.J., Kaiho, K., Zhao, L., et al., 2015, Complete biotic and sedimentary records of the Permian–Triassic transition from Meishan section, South China: ecologically assessing mass extinction and its aftermath. *Earth-Science Reviews*, **149**: 67–107. http://dx.doi.org/10.1016/j.earscirev.2014.10.005.

Chinese Triassic Working Group, 2007, Final report of the GSSP candidate for the I/O boundary at West Pingdingshan Section in Chaohu, Southeastern China. *Albertiana*, **36**: 10–21.

Chu, D.L., Tong, J.N., Benton, M.J., Yu, J.X., and Huang, Y.F., 2019, Mixed continental-marine biotas following the Permian-Triassic mass extinction in South and North China. *Palaeogeography, Palaeoclimatology, Palaeoecology*, **519**: 95–107.

Cirilli, S., 2010, Upper Triassic–lowermost Jurassic palynology and palynostratigraphy: a review. In: Lucas, S.G. (eds), *The Triassic Timescale*. The Geological Society of London Special Publications, **334**: 285–314.

Cirilli, S., Marzoli, A., Tanner, L., Bertrand, H., Buratti, N., Jourdan, F., et al., 2009, Latest Triassic onset of the Central Atlantic Magmatic Province (CAMP) volcanism in the Fundy Basin (Nova Scotia): new stratigraphic constraints. *Earth and Planetary Science Letters*, **286**: 514–525.

Cohen, A.S., and Coe, A.L., 2007, The impact of the Central Atlantic Magmatic Province on climate and on the Sr- and Os-isotope evolution of seawater. *Palaeogeography, Palaeoclimatology, Palaeoecology*, **244**: 374–390.

Cornet, B., 1977, The Palynology and Age of the Newark Supergroup. *Ph. D. thesis*. Pennsylvania State University, College Park, PA, 505 pp.

Cornet, B., 1993, Applications and limitations of palynology in age, climatic, and paleoenvironmental analyzes of Triassic sequences in North America. In: Lucas, S.G., Morales, M. (eds), The Nonmarine Triassic, *New Mexico Museum of Natural History & Science Bulletin*, **3**: 75–93.

Cornet, B., and Olsen, P.E., 1985, A summary of the biostratigraphy of the Newark Supergroup of Eastern North America with comments on early Mesozoic provinciality. *In* Weber, R. (ed), III Congreso Latinoamericano de Paleontologia Mexico, Simposio Sobre Floras

del Triasico Tardio, su Fitogeograpfia y Paleoecologia, Memoria. Mexico: Universidad Nacional Autónoma de México, 67–81.

Cozzi, A., Hinnov, L.A., and Hardie, L.A., 2005, Orbitally forced Lofer cycles in the Dachstein Limestone of the Julian Alps (northeastern Italy). *Geology*, **33**: 789–792.

Cramer, B.D., and Jarvis, I., 2020, Chapter 11 - Carbon isotope stratigraphy. *In* Gradstein, F.M., Ogg, J.G., Schmitz, M.D., and Ogg, G.M. (eds), *The Geologic Time Scale 2020*. **Vol. 1** (this book). Elsevier, Boston, MA.

Dal Corso, J., Mietto, P., Newton, R.J., Pancost, R.D., Preto, N., Roghi, G., et al., 2012, Discovery of a major negative $\delta^{13}C$ spike in the Carnian (Late Triassic) linked to the eruption of Wrangellia flood basalts. *Geology*, **40** (1), 79–82. https://doi.org/10.1130/G32473.1.

Dal Corso, J., Benton, M.J., Bernardi, M., Franz, M., Gianolla, P., Hohn, S., et al., 2018a, First workshop on the Carnian Pluvial Episode (Late Triassic): a report. *Albertiana*, **44**: 49–57.

Dal Corso, J., Gianolla, P., Rigo, M., Franceschi, M., Roghi, G., Mietto, P., et al., 2018b, Multiple negative carbon-isotope excursions during the Carnian Pluvial Episode (Late Triassic). *Earth-Science Reviews*, **185**: 732–750. https://doi.org/10.1016/j.earscirev.2018.07.004.

De Wever, P., 1982, Radiolaires du Trias et du Lias de la Téthys: systématique, stratigraphy. *Societé Géologique du Nord, Lille*, **7**: 599 pp.

De Wever, P., 1998, Radiolarians. Column for Triassic chart of Mesozoic and Cenozoic sequence chronostratigraphic framework of European basins, by Hardenbol, J., Thierry, J., Farley, M.B., Jacquin, Th., de Graciansky, P.-C., and Vail, P.R. (coordinators). *In* de Graciansky, P.-C., Hardenbol, J., Jacquin, T., and Vail, P.R. (eds), *Mesozoic-Cenozoic Sequence Stratigraphy of European Basins*. SEPM Special Publication, p. 60 Chart 8.

De Zanche, V., Gianolla, P., Mietto, P., Siorpaes, C., and Vail, P.R., 1993, Triassic sequence stratigraphy in the Dolomites. *Memorie di Scienze Geologiche*, **45**: 1–27.

Deenen, M.H.L., 2010, A new chronology for the late Triassic to early Jurassic. PhD thesis, Utrecht University, Faculty of Geosciences, Department of Earth Sciences. *Geologica Ultraiectina*, **323**.

Deenen, M.H.L., Langereis, C.G., Krijgsman, W., El Hachimi, H., and El Hassane, C., 2010a, Paleomagnetic research in the Argana basin, Morocco: Trans-Atlantic correlation of CAMP volcanism and implications for the late Triassic geomagnetic polarity time scale. In: Deenen, M.H.L. (ed), A new chronology for the late Triassic to early Jurassic, *Geologica Ultraiectina*, **323**: 43–64.

Deenen, M.H.L., Ruhl, M., Bonis, N.R., Krijgsman, W., Kuerschner, W.M., Reitsma, M., et al., 2010b, A new chronology for the end-Triassic mass extinction. *Earth and Planetary Science Letters*, **291**: 113–125.

Diakow, L., Orchard, M.J., and Friedman, R., 2010, Absolute ages for the Norian Stage: a contribution from southern British Columbia, Canada. *21st Canadian Paleontological Conference. University of British Columbia, 19–20 Aug 2011, Abstracts and additional details sent by M.J. Orchard to J. Ogg, July 2011 and Sept 2015*.

Diener, C., 1912, The Trials of the Himalayas. *Memoirs of the Geological Survey of India*, **36** (3), 159 pp.

Dinarès-Turell, J., Diez, J.D., Rey, D., and Arnal, I., 2005, 'Buntsandstein' magnetostratigraphy and biostratigraphic reappraisal from eastern Iberia: Early and Middle Triassic stage boundary definitions through correlation to Tethyan sections. *Palaeogeography, Palaeoclimatology, Palaeoecology*, **229**: 158–177.

Dunning, G.R., and Hodych, J.P., 1990, U/Pb zircon and baddeleyite ages for the Palisades and Gettysburg sills of the northeastern

United States: implications for the age of the Triassic/Jurassic boundary. *Geology*, **18**: 795−798.

Earth Impact Database, 2018, *Maintained by the Planetary and Space Science Centre.* University of New Brunswick. Available at: www.passc.net/EarthImpactDatabase/.

Embry, A.F., 1988, Triassic sea-level changes: evidence from the Canadian Arctic Archipelago. In: Wilgus, C. (ed), *Sea level changes—an integrated approach.* SEPM Special Publication, **42**: 249−259.

Embry, A.F., 1997, Global sequence boundaries of the Triassic and their identification in the Western Canada Sedimentary Basin. *Bulletin of Canadian Petroleum Geology*, **45**: 415−533.

Enos, P., Lehrmann, D.J., Wei, J.Y., Yu, Y.Y., Xiao, J.F., Chaikin, et al., 2006, Triassic Evolution of the Yangtze Platform in Guizhou Province, People's Republic of China. *Geological Society of America Special Paper*, **417**: 1−105. https://doi.org/10.1130/SPE417.

Ernst, R.E., 2014, *Large Igneous Provinces.* Cambridge University Press 653 pp.

Ernst, R.E., Dickson, A.J., and Bekker, A. (eds), 2020. *Large Igneous Provinces: A Driver of Global Environmental and Biotic Changes.* AGU Geophysical Monograph. **255** (in press).

Erwin, D.H., 1993, *The Great Paleozoic Crisis: Life and Death in the Permian.* New York: Columbia University Press 327 pp.

Erwin, D.H., 2006, *Extinction: How Life on Earth Nearly Ended 250 Million Years Ago.* Princeton, NJ: Princeton University Press 320 pp.

Erwin, D.H., Bowring, S.A., and Yugan, J., 2002, End-Permian mass extinctions: a review. In: Koeberl, C., MacLeod, K.G. (eds), Catastrophic events and mass extinctions: impacts and beyond, *Geological Society of America Special Paper*, **356**: 363−383.

Feist-Burkhardt, S., Götz, A.E., Szulc, J., Borkhataria, R., Geluk, M., Haas, J., et al., 2008, Triassic. *In* McCann, T. (ed), *The Geology of Central Europe. Vol. 2: Mesozoic and Cenozoic.* London: The Geological Society, 749−821.

Fischer, A.G., 1964, The Lofer cyclothems of the Alpine Triassic. In: Merriam, D.F. (ed), Symposium on Cyclic Sedimentation, *Kansas Geological Survey Bulletin*, **169**: 107−149.

Flügel, E., 2002, Triassic reef patterns. In: Kiessling, W., Flügel, E., Golonka, J. (eds), Phanerozoic Reef Patterns. *SEPM Special Publication*, **72**: 691−733.

Fowell, S.J., and Olsen, P.E., 1993, Time calibration of Triassic/Jurassic microfloral turnover, eastern North America. *Tectonophysics*, **222**: 361−369.

Fowell, S.J., and Olsen, P.E., 1995, Time calibration of Triassic/Jurassic microfloral turnover, eastern North America—reply. *Tectonophysics*, **245**: 96−99.

Franz, M., Bachmann, G.H., Barnasch, J., Heunisch, C., and Röhling, H.-G., 2018, Der Keuper in der Stratigraphischen Tabelle von Deutschland 2016 − kontinuierliche Sedimentation in der norddeutschen Beckenfazies (Variante B)/The Keuper Group in the Stratigraphic Table of Germany 2016 − continuous sedimentation in the North German Basin (variant B). *Zeitschrift der Deutschen Gesellschaft für Geowissenschaften (German Journal of Geology)*, **169** (2), 203−224. https://doi.org/10.1127/zdgg/2018/0114.

Fraser, N.D., 2006, *Dawn of the Dinosaurs: Life in the Triassic.* Bloomington, IN: Indiana University Press 310 pp.

Furin, S., Preto, N., Rigo, M., Roghi, G., Gianolla, P., Crowley, J.L., et al., 2006, High-precision U-Pb zircon age from the Triassic of Italy:

implications for the Triassic time scale and the Carnian origin of calcareous nannoplankton and dinosaurs. *Geology*, **34**: 1009−1012.

Gaetani, M., 1993, Anisian/Ladinian boundary field workshop, Southern Alps—Balaton Highlands, 27 June−4 July 1993. *Albertiana*, **12**: 5−9.

Gaetani, M., Gnaccolini, M., Jadoul, Fl., and Garzanti, E., 1998, Multiorder sequence stratigraphy in the Triassic system of the western Southern Alps. In: de Graciansky, P.-C., Hardenbol, J., Jacquin, T., Vail, P.R. (Eds.), Mesozoic-Cenozoic Sequence Stratigraphy of European Basins, *SEPM Special Publication*, **60**: 701−717.

Galbrun, B., Boulila, S., Krystyn, L., Richoz, S., Gardin, S., Bartolini, A., and Maslo, M., 2020, "Short" or "long" Rhaetian? Astronomical calibration of Austrian key sections. *Global and Planetary Change*, **192**, article #103253: 20 pp. https://doi.org/10.1016/j.gloplacha.2020.103253.

Galfetti, T., Hochuli, P.A., Brayard, A., Bucher, H., Weissert, H., and Vigran, J.O., 2007a, The Smithian/Spathian boundary event: a global climatic change in the wake of the end-Permian biotic crisis. Evidence from palynology, ammonoids and stable isotopes. *Geology*, **35**: 291−294.

Galfetti, T., Bucher, H., Ovtcharova, M., Schaltegger, U., Brayard, A., Brühwiler, T., et al., 2007b, Timing of the Early Triassic carbon cycle perturbations inferred from new U-Pb ages and ammonoid biochronozones. *Earth and Planetary Science Letters*, **258**: 593−604.

Galfetti, T., Bucher, H., Brayard, A., Hochuli, P.A., Weissert, H., Guodun, K., et al., 2007c, Late Early Triassic climate change: insights from carbonate carbon isotopes, sedimentary evolution and ammonoid paleobiogeography. *Palaeogeography, Palaeoclimatology, Palaeoecology*, **243**: 394−411.

Galfetti, T., Bucher, H., Martini, R., Hochuli, P.A., Weissert, H., Crasquin-Soleua, S., et al., 2008, Evolution of Early Triassic outer platform paleoenvironments in the Nanpanjiang Basin (South China) and their significance for the biotic recovery. *Sedimentary Geology*, **204**: 36−60.

Gallet, Y., Besse, J., Krystyn, L., Marcoux, J., and Théveniaut, H., 1992, Magnetostratigraphy of the late Triassic Bolücektasi Tepe section (southwestern Turkey): implications for changes in magnetic reversal frequency. *Physics of the Earth and Planetary Interiors*, **73**: 85−108.

Gallet, Y., Krystyn, L., and Besse, J., 1998, Upper Anisian to Lower Carnian magnetostratigraphy from the Northern Calcareous Alps (Austria). *Journal of Geophysical Research*, **103**: 605−621.

Gallet, Y., Krystyn, L., Marcoux, J., and Besse, J., 2007, New constraints on the End-Triassic (Upper Norian−Rhaetian) magnetostratigraphy. *Earth and Planetary Science Letters*, **255**: 458−470.

Gehrels, G.E., Saleeby, J.B., and Berg, H.C., 1987, Geology of Annette, Gravina, and Duke islands, southeastern Alaska. *Canadian Journal of Earth Sciences*, **24**: 866−881.

Geluk, M.C., and Röhling, H.-G., 1997, High-resolution sequence stratigraphy of the Lower Triassic 'Buntsandstein' in the Netherlands and northwestern Germany. *Geologie en Mijnbouw*, **76**: 227−246.

Geyer, G., and Kelber, K.-P., 2018, Spinicaudata ("Conchostraca", Crustacea) from the Middle Keuper (Upper Triassic) of the southern Germanic Basin, with a review of Carnian−Norian taxa and suggested biozones. *Paläontologische Zeitschrift (PalZ)*, **92**: 1−34.

Gianolla, P., and Jacquin, T., 1998, Triassic sequence stratigraphic framework of western European basins. In: de Graciansky, P.-C., Hardenbol, J., Jacquin, Th., Vail, P.R. (eds), Mesozoic-Cenozoic Sequence Stratigraphy of European Basins, *SEPM Special Publication*, **60**: 643−650.

Gianolla, P., De Zanche, V., and Mietto, P., 1998, Triassic sequence stratigraphy in the Southern Alps (northern Italy): definition of sequences and basin evolution. In: de Graciansky, P.-C., Hardenbol, J., Jacquin, Th., Vail, P.R. (eds.), Mesozoic-Cenozoic Sequence Stratigraphy of European Basins, *SEPM Special Publication*, **60**: 719–747.

Glen, J.M.G., Nomade, S., Lyons, J.J., Metcalfe, I., Mundil, R., and Renne, J.R., 2009, Magnetostratigraphic correlations of Permian–Triassic marine-to-terrestrial sections from China. In: Metcalfe, I., Isozaki, Y. (eds), End-Permian mass extinction: events & processes, age & timescale, causative mechanism(s) & recovery, *Journal of Asian Earth Sciences*, **36**(6): 520–540.

Goldhammer, R.K., Dunn, P.A., and Hardie, L.A., 1987, High frequency glacio-eustatic oscillations with Milankovitch characteristics recorded in northern Italy. *American Journal of Science*, **287**: 853–892.

Goldhammer, R.K., Dunn, P.A., and Hardie, L.A., 1990, Depositional cycles, composite sea level changes, cycle stacking patterns, and the hierarchy of stratigraphic forcing: examples from the Alpine Triassic platform carbonates. *Geological Society of America Bulletin*, **102** (535), 562.

Golding, M.L., 2019, Evaluating tectonic models for the formation of the North American Cordillera using multivariate statistical analysis of Late Triassic conodont faunas. *Palaeobiodiversity Palaeoenvironments*. https://doi.org/10.1007/s12549-019-00393-4.

Golding, M.L., Orchard, M.J., and Zonneveld, J.-P., 2014, A summary of new conodont biostratigraphy and correlation of the Anisian (Middle Triassic) strata in British Columbia, Canada. *Albertiana*, **42**: 33–40.

Golding, M.L., Orchard, M.J., and Zagorevski, A., 2017, Conodonts from the Stikine Terrane in northern British Columbia and southern Yukon. *Geological Survey of Canada, Open File*, **8278**: 23 pp.

Goudemand, N., Orchard, M., Bucher, H., Brayard, A., Brühwiler, T., Galfetti, T., et al., 2008, Smithian-Spathian boundary: the biggest crisis in Triassic conodont history. *Geological Society of America Abstracts With Programs*, **40** (6), 505.

Goudemand, N., Romano, C., Leu, M., Bucher, H., Trotter, J.A., and Williams, I.S., 2019, Dynamic interplay between climate and marine biodiversity upheavals during the Early Triassic Smithian-Spathian biotic crisis. *Earth-Science Reviews*, **195**: 169–178. https://doi.org/10.1016/j.earscirev.2019.01.013.

Gradinaru, E., Orchard, M.J., Nicora, A., Gallet, Y., Besse, J., Krystyn, L., et al., 2007, The Global Boundary Stratotype Section and Point (GSSP) for the base of the Anisian Stage: Desli Caira Hill, North Dobrogea, Romania. *Albertiana*, **36**: 54–71.

Grasby, S.E., Beauchamp, B., and Knies, J., 2016a, Early Triassic productivity crises delayed recovery from world's worst mass extinction. *Geology*, **44** (9), 779–782. https://doi.org/10.1130/abs/2016am-279493.

Grasby, S.E., Beauchamp, B., Bond, D.P., Wignall, P.B., and Sanei, H., 2016b, Mercury anomalies associated with three extinction events (Capitanian crisis, latest Permian extinction and the Smithian/Spathian extinction) in NW Pangea. *Geological Magazine*, **153**: 285–297. https://doi.org/10.1017/S0016756815000436.

Greene, A., Scoates, J., and Weis, D., 2008, The Accreted Wrangellia Oceanic Plateau in Alaska, Yukon, and British Columbia. *Internet Article From Large Igneous Provinces Commission, Large Igneous Province of the Month*: December 2008. Available at: < www.largeigneousprovinces.org/ >

Greene, A.R., Scoates, J.S., Weis, D., Katvala, E.C., Israel, S., and Nixon, G.T., 2010, The architecture of oceanic plateaus revealed by the volcanic stratigraphy of the accreted Wrangellia oceanic plateau. *Geosphere*, **6**: 47–73.

Greene, A., Scoates, J., and Weis, D., 2011, *The Accreted Wrangellia Oceanic Plateau in Alaska, Yukon, and British Columbia*. Available at: < www.eos.ubc.ca/research/wrangellia > [Last update was 2010 when viewed Feb., 2011].

Griesbach, C.L., 1880, Paleontological notes on the Lower Trias on the Himalayas. *Records of the Geological Survey of India*, **13** (2), 94–113.

Grossman, E.L., and Joachimski, M.M., 2020, Chapter 10 - Oxygen isotope stratigraphy. *In* Gradstein, F.M., Ogg, J.G., Schmitz, M.D., and Ogg, G.M. (eds), *The Geologic Time Scale 2020*. **Vol. 1** (this book). Elsevier, Boston, MA.

Hallam, A., 1981, The end-Triassic bivalve extinction event. *Palaeogeography, Palaeoclimatology, Palaeoecology*, **35**: 1–44.

Hallam, A., 1992, *Phanerozoic Sea-Level Changes*. New York: Columbia University Press 266 pp.

Hallam, A., and Wignall, P.B., 1999, Mass extinctions and sea-level changes. *Earth-Science Reviews*, **48**: 217–250.

Haq, B.U., 2018, Triassic eustatic variations reexamined. *GSA Today*, **28**: 4–9 https://doi.org/10.1130/GSATG381A.1; with supplement online at < www.geosociety.org/datarepository/2018/ >.

Haq, B.U., and Al-Qahtani, A.M., 2005, Phanerozoic cycles of sea-level change on the Arabian Platform. *GeoArabia*, **10** (2), 127–160.

Haq, B.U., Hardenbol, J., and Vail, P.R., 1988, Mesozoic and Cenozoic chronostratigraphy and eustatic cycles. In: Wilgus, C.K., Hastings, B.S., Kendall, G.St.C., Posamentier, H.W., Ross, C.A., van Wagoner, J.C. (eds), Sea-Level Changes: An Integrated Approach, *SEPM Special Publication*, **42**: 71–108.

Hardenbol, J., Thierry, J., Farley, M.B., Jacquin, Th., de Graciansky, P.-C., Vail, P.R., et al., 1998, Mesozoic and Cenozoic sequence chronostratigraphic framework of European basins. In: de Graciansky, P.-C., Hardenbol, J., Jacquin, Th., and Vail, P.R. (eds), Mesozoic-Cenozoic Sequence Stratigraphy of European Basins, *SEPM Special Publication*, **60**: 3–13, 763–781.

Hardie, L.A., and Hinnov, L.A., 1997, Biostratigraphic and radiometric age data question the Milankovitch characteristics of the Latemar cycles (Southern Alps, Italy) – comment. *Geology*, **25**: 470–471.

Hayes, J.M., Strauss, H., and Kaufman, 1999, The abundance of ^{13}C in marine organic matter and isotopic fractionation in the global biogeochemical cycle of carbon during the past 800 Ma. *Chemical Geology*, **161**: 103–125.

Heckert, A.B., and Lucas, S.G., 1999, Global correlation and chronology of Triassic theropods (Archosauria: Dinosauria). *Albertiana*, **23**: 22–35.

Heckert, A.B., Lucas, S.G., Dickinson, W.R., and Mortensen, J.K., 2009, New ID-TIMS U-Pb ages for Chinle Group strata (Upper Triassic) in New Mexico and Arizona, correlation to the Newark Supergroup, and implications for the "long Norian". *Geological Society of America Abstracts With Programs*, **41** (7), 123.

Helby, R., Morgan, R., and Partridge, A.D., 1987, A palynological zonation of the Australian Mesozoic. In Jell, P.A. (ed), Studies in Australian Mesozoic Palynology, *Memoir Association Australaisian Paleontologists*, **4**: 1–94.

Heller, F., Haihong, C., Dobson, J., and Haag, M., 1995, Permian-Triassic magnetostratigraphy – new results from south China. *Physics of the Earth and Planetary Interiors*, **89**: 281–295.

Henderson, C., and Baud, A., 1997, Correlation of the Permian- Triassic boundary in Arctic Canada and comparaison with Meishan, China.

In Naiwen, W., and Remane, J. (eds), *Stratigraphy, 11, Proceedings of the 30th IGC*. Beijing: VSP, 143–152.

Henderson, C.M., Golding, M.L., and Orchard, M.J., 2018, Conodont sequence biostratigraphy of the Lower Triassic Montney Formation. In: Euzen, T., Moslow, T.F., Caplan, M. (eds.), The Montney Play: Deposition to Development, *Bulletin of Canadian Petroleum Geology*, Special Volume, **66**: 7–22.

Henderson, C.M., and Shen, S.-Z., 2020, Chapter 24 - The Permian Period. *In* Gradstein, F.M., Ogg, J.G., Schmitz, M.D., and Ogg, G.M. (eds), *The Geologic Time Scale 2020*. **Vol. 2** (this book). Elsevier, Boston, MA.

Hermann, E., Hochuli, P.A., Bucher, H., Vigran, J.O., Weissert, H., and Bernasconi, S.M., 2010, A close-up view of the Permian–Triassic boundary based on expanded organic carbon isotope records from Norway (Trøndelag and Finnmark Platform). *Global and Planetary Change*, **74**: 156–167.

Herngreen, G.F.W., 2005, Triassic sporomorphs of NW Europe: taxonomy, morphology and ranges of marker species with remarks on botanical relationship and ecology and comparison with ranges in the Alpine Triassic. *Kenniscentrum Biogeology, Nederlands Instituut voor Toegepaste Geowetenschappen TNO, Utrecht, TNO report, NITG 04-176-C*.

Hesselbo, S.P., Robinson, S.A., Surlyk, F., and Piasecki, S., 2002, Terrestrial and marine extinctions at the Triassic-Jurassic boundary synchronized with major carbon-cycle perturbation: a link to initiation of massive volcanism? *Geology*, **30**: 251–254.

Hinnov, L.A., 2006, Discussion of "Magnetostratigraphic confirmation of a much faster tempo for sea-level change for the Middle Triassic Latemar platform carbonates" by D.V. Kent, G. Muttoni and P. Brack [Earth Planet. Sci. Lett. 228 (2004), 369–377]. *Earth and Planetary Science Letters*, **243**: 841–846.

Hinnov, L.A., and Goldhammer, R.K., 1991, Spectral analysis of the Middle Triassic Latemar Limestone. *Journal of Sedimentary Research*, **61**: 1173–1193.

Hochuli, P., 1998. Dinoflagellate cysts and Spore pollen. Columns for Triassic chart of Mesozoic and Cenozoic sequence chronostratigraphic framework of European basins, by Hardenbol, J., Thierry, J., Farley, M.B., Jacquin, Th., de Graciansky, P.-C., and Vail, P.R. (coordinators). In: de Graciansky, P.-C., Hardenbol, J., Jacquin, Th., Vail, P.R. (eds), Mesozoic-Cenozoic Sequence Stratigraphy of European Basins, *SEPM Special Publication*, **60**: Chart 8.

Hochuli, P.A., Hermann, E., Vigran, J.O., Bucher, H., and Weissert, W., 2010, Rapid demise and recovery of plant ecosystems across the end-Permian extinction event. *Global and Planetary Change*, **74**: 144–155.

Hodych, J.P., and Dunning, G.R., 1992, Did the Manicouagan impact trigger end-of-Triassic mass extinction? *Geology*, **20**: 51–54.

Holser, W.T., and Magaritz, M., 1987, Events near the Permian–Triassic boundary. *Modern Geology*, **11**: 155–180.

Horacek, M., Brandner, R., Richoz, S., and Povoden-Karadeniz, E., 2010a, Lower Triassic sulphur isotope curve of marine sulphates from the Dolomites, N-Italy. *Palaeogeography, Palaeoclimatology, Palaeoecology*, **290**: 65–70.

Horacek, M., Povoden, E., Richoz, S., and Brandner, R., 2010b, High-resolution carbon isotope changes, litho- and magnetostratigraphy across Permian-Triassic Boundary sections in the Dolomites, N-Italy. New constraints for global correlation. *Palaeogeography, Palaeoclimatology, Palaeoecology*, **290**: 58–64.

Hornung, T., and Brandner, R., 2005, Biochronostratigraphy of the Reingraben Turnover (Hallstatt Facies Belt): local black shale events

controlled by regional tectonics, climatic change and plate tectonics. *Facies*, **51**: 460–479.

Hounslow, M.K., and Muttoni, G., 2010, The geomagnetic polarity timescale for the Triassic: linkage to stage boundary definitions. In: Lucas, S.G. (ed), The Triassic Timescale, *The Geological Society of London Special Publications*, **334**: 61–102.

Hounslow, M.W., Posen, P.E., and Warrington, G., 2004, Magnetostratigraphy and biostratigraphy of the Upper Triassic and lowermost Jurassic succession, St. Audrie's Bay, UK. *Palaeogeography, Palaeoclimatology, Palaeoecology*, **213**: 331–358.

Hounslow, M.K., Szurlies, M., Muttoni, G., and Nawrocki, J., 2007, The magnetostratigraphy of the Olenekian-Anisian boundary and a proposal to define the base of the Anisian using a magnetozone datum. *Albertiana*, **36**: 72–77.

Hounslow, M.K., Peters, C., Mørk, A., Weitschat, W., and Vigran, J.O., 2008a, Biomagnetostratigraphy of the Vikinghøgda Formation, Svalbard (arctic Norway) and the geomagnetic polarity timescale for the Lower Triassic. *Geological Society of America Bulletin*, **120**: 1305–1325.

Hounslow, M.K., Hu, M., Mørk, A., Weitschat, W., Vigran, J.O., Karloukovski, V., et al., 2008b, Intercalibration of Boreal and Tethyan timescales: the magneto-biostratigraphy of the Middle Triassic and the latest Early Triassic, central Spitsbergen (arctic Norway). *Polar Research*, **27**: 469–490.

Huang, C.J., 2018. Astronomical time scale for the Mesozoic. *Stratigraphy and Time Scales*, Elsevier, **Vol. 3**, p. 81–150. https://doi.org/10.1016/bs.sats.2018.08.005.

Hüsing, S.K., Deenen, M.H.L., Koopmans, J.G., and Krijgsman, W., 2011, Magnetostratigraphic dating of the proposed Rhaetian GSSP at Steinbergkogel (Upper Triassic, Austria): implications for the Late Triassic time scale. *Earth and Planetary Science Letters*, **302**: 203–216.

Ikeda, M., and Tada, R., 2014, A 70 million year astronomical time scale for the deep-sea bedded chert sequence (Inuyama, Japan): implications for Triassic-Jurassic geochronology. *Earth and Planetary Science Letters*, **399**: 30–43 <https://doi.org/10.1016/j.epsl.2014.04.031>.

Ikeda, M., Tada, R., and Sakuma, H., 2010, Astronomical cycle origin of bedded chert: a middle Triassic bedded chert sequence, Inuyama, Japan. *Earth and Planetary Science Letters*, **297**: 369–378.

Ikeda, M., Tada, R., and Ozaki, K., 2017, Astronomical pacing of the global silica cycle recorded in Mesozoic bedded cherts. *Nature Communications*, **8**: 15532. https://doi.org/10.1038/ncomms15532.

Irmis, R.B., Martz, J.W., Parker, W.G., and Nesbitt, S.J., 2010, Re-evaluating the correlation between Late Triassic terrestrial vertebrate biostratigraphy and the GSSP-defined marine stages. *Albertiana*, **38**: 40–53.

Isozaki, Y., 2009, Integrated "plume winter" scenario for the double-phased extinction during the Paleozoic–Mesozoic transition: the G-LB and P-TB events from a Panthalassan perspective. In: Metcalfe, I., Isozaki, Y. (eds), End-Permian mass extinction: events & processes, age & timescale, causative mechanism(s) & recovery, *Journal of Asian Earth Sciences*, **36**(6): 459–480.

Jacquin, Th., and Vail, P.R. (coordinators), 1998, Sequence chronostratigraphy. Columns for Triassic chart, Mesozoic and Cenozoic sequence chronostratigraphic framework of European basins by Hardenbol, J., Thierry, J., Farley, M.B., Jacquin, Th., de Graciansky, P.-C., and Vail, P.R. In: de Graciansky, P.-C., Hardenbol, J., Jacquin, Th., Vail, P.R. (eds), Mesozoic-Cenozoic Sequence Stratigraphy of European Basins, *SEPM Special Publication*, **60**: Chart 8.

Jenks, J.F., Monnet, C., Balini, M., Brayard, A., Meier, M., 2015, Chapter 13. Biostratigraphy of Triassic ammonoids. In: Klug, C.,

Korn, D., De Baets, K., Kruta, I., Mapes, R.H. (eds), *Ammonoid Paleobiology: From Macroevolution to Paleogeography*, Topics in Geobiology. Springer Publication, Vol. **44**: 329–371 https://doi.org/10.1007/978-94-017-9633-0_13.

Jenkyns, H.C., Jones, C.E., Gröcke, D.R., Hesselbo, S.P., and Parkinson, D.N., 2002, Chemostratigraphy of the Jurassic system: applications, limitations and implications for palaeoceanography. *Journal of the Geological Society, London*, **159**: 351–378.

Jones, C.E., and Jenkyns, H.C., 2001, Seawater strontium isotopes, oceanic anoxic events, and seafloor hydrothermal activity in the Jurassic and Cretaceous. *American Journal of Science*, **301**: 112–149.

Jourdan, F., Marzoli, A., Bertrand, H., Cirilli, S., Tanner, L.H., Kontak, D.J., et al., 2009a, ^{40}Ar/^{39}Ar ages of CAMP in North America: implications for the Triassic–Jurassic boundary and the ^{40}K decay constant bias. *Lithos*, **110**: 167–180.

Jourdan, F., Renne, P.R., and Reimold, W.U., 2009b, An appraisal of the ages of terrestrial impact structures. *Earth and Planetary Science Letters*, **286**: 1–13.

Kamo, S.L., Czamanske, G.K., and Drogh, T.E., 1996, A minimum U–Pb age for Siberian flood-basalt volcanism. *Geochimica et Cosmochimica Acta*, **60**: 3505.

Kamo, S.L., Czamanske, G.K., Amelin, Y., Fedorenko, V.A., Davis, D.W., and Trofimov, V.R., 2003, Rapid eruption of Siberian flood-volcanic rocks and evidence for coincidence with the Permian-Triassic boundary and mass extinction at 251 Ma. *Earth and Planetary Science Letters*, **214**: 75–91.

Kampschulte, A., and Strauss, H., 2004, The sulfur isotopic evolution of Phanerozoic sea water based on the analysis of structurally substituted sulfate in carbonates. *Chemical Geology*, **204**: 255–286.

Kapoor, H.M., 1996, The Guryul Ravine section, candidate of the global Stratotype and point (GSSP) of the Permian-Triassic boundary (PTB). In: Yin, H.-f. (Ed.), *The Paleozoic-Mesozoic Boundary, Candidates of the Global Stratotype Section and Point of the Permian-Triassic Boundary*. China University of Geosciences Press, Wuhan, pp. 99–110.

Kemp, D.B., and Coe, A.L., 2007, A nonmarine record of eccentricity forcing through the Upper Triassic of southwest England and its correlation with the Newark Basin astronomically calibrated geomagnetic polarity time scale from North America. *Geology*, **35**: 991–994.

Kent, D.V., and Olsen, P.E., 1999, Astronomically tuned geomagnetic polarity timescale for the Late Triassic. *Journal of Geophysical Research*, **104**: 12831–12841 Web page update (2002) posted at Newark Basin Coring Project website: < www.ldeo.columbia.edu/~polsen/nbcp/nbcp.timescale.htm > [Accessed 3 July 2010].

Kent, D.V., Olsen, P.E., and Witte, W.K., 1995, Late Triassic–earliest Jurassic geomagnetic polarity sequence and paleolatitudes from drill cores in the Newark rift basin, eastern North America. *Journal of Geophysical Research*, **100**: 14965–14998.

Kent, D.V., Muttoni, G., and Brack, P., 2004, Magnetostratigraphic confirmation of a much faster tempo for sea-level change for the Middle Triassic Latemar platform carbonates. *Earth and Planetary Science Letters*, **228**: 369–377.

Kent, D.V., Olsen, P.E., and Muttoni, G., 2017, Astrochronostratigraphic polarity time scale (APTS) for the Late Triassic and Early Jurassic from continental sediments and correlation with standard marine stages. *Earth-Science Reviews*, **166**: 153–180. https://doi.org/10.1016/j.earscirev.2016.12.014.

Kent, D.V., Olsen, P.E., Rasmussen, C., Lepre, C., Mundil, R., Irmis, R.B., et al., 2018, Empirical evidence for stability of the 405-kiloyear Jupiter-Venus eccentricity cycle over hundreds of million years. *Proceedings of the National Academy of Sciences of the United States of America*, **115** (24), 6153–6158. https://doi.org/10.1073/pnas.1800891115.

Kiessling, W., 2010, Reef expansion during the Triassic: spread of photosymbiosis balancing climatic cooling. In: Kustatscher, E., Preto, N., Wignall, P. (eds), Triassic climates, *Palaeogeography, Palaeoclimatology, Palaeoecology*, **290**: 11–19.

Kiparisova, L.D., and Popov, Y.N., 1956, Subdivision of the Lower series of the Triassic system into stages. *Doklady Akademiya Nauk USSR*, **109**: 842–845 (in Russian).

Kiparisova, L.D., and Popov, Y.N., 1964. The project of subdivision of the Lower Triassic into stages. *XXII International Geological Congress, Reports of Soviet geologists, Problem 16a* (In Russian) (pp. 91–99).

Klein, H., and Lucas, S.G., 2010, Tetrapod footprints – their use in biostratigraphy and biochronology of the Triassic. In: Lucas, S.G. (ed), The Triassic Timescale, *The Geological Society of London Special Publications*, **334**: 419–446.

Knoll, A.H., Bambach, R.K., Payne, J.L., Pruss, S., and Fischer, W.W., 2007, Paleophysiology and end-Permian mass extinction. *Earth and Planetary Science Letters*, **256**: 295–313.

Koepnick, R.B., Denison, R.E., Burke, W.H., Hetherington, E.A., and Dahl, D.A., 1990, Construction of the Triassic and Jurassic portion of the Phanerozoic curve of seawater ^{87}Sr/^{86}Sr. *Chemical Geology*, **80**: 327–349.

Konstantinov, A.G., and Klets, T.V., 2009, Stage boundaries of the Triassic in northeast Asia. *Stratigraphy and Geological Correlation*, **17**: 173–191. https://doi.org/10.1134/S0869593809020063.

Korte, C., and Kozur, H.W., 2010, Carbon-isotope stratigraphy across the Permian-Triassic boundary: a review. *Journal of Asian Earth Sciences*, **39**: 215–235.

Korte, C., Kozur, H.W., Bruckschen, P., and Veizer, J., 2003, Strontium isotope evolution of late Permian and Triassic seawater. *Geochimica et Cosmochimica Acta*, **67**: 47–62.

Korte, C., Kozur, H.W., and Veizer, J., 2005, ∂^{13}C and ∂^{18}O values of Triassic brachiopods and carbonate rocks as proxies for coeval seawater and palaeotemperature. *Palaeogeography, Palaeoclimatology, Palaeoecology*, **226**: 287–306.

Korte, C., Pande, P., Kalia, P., Kozur, H.W., Joachimski, M.M., and Oberhänsli, 2010, Massive volcanism at the Permian–Triassic boundary and its impact on the isotopic composition of the ocean and atmosphere. *Journal of Asian Earth Sciences*, **37**: 293–311.

Kozur, H.W., 1998, Some aspects of the Permian–Triassic boundary (PTB) and of the possible causes for the biotic crisis around this boundary. *Palaeogeography, Palaeoclimatology, Palaeoecology*, **143**: 227–272.

Kozur, H.W., 1999, Remarks on the position of the Norian-Rhaetian boundary. Proceedings of the Epicontinental Triassic International Symposium, Halle (Germany), 21–23 Sept., 1998. *Zentralblatt für Geologie und Paläontologie*, **I** (7–8): 523–535.

Kozur, H.W., 2003, Integrated ammonoid, conodont and radiolarian zonation of the Triassic. *Hallesches Jahrbuch fur Geowissenschaften*, **B25**: 49–79.

Kozur, H.W., 2007, Biostratigraphy and event stratigraphy in Iran around the Permian–Triassic boundary (PTB): implications for the causes of the PTB biotic crisis. *Global and Planetary Change*, **55**: 155–176.

Kozur, H.W., and Bachmann, G.H., 2005, Correlation of the Germanic Triassic with the international scale. *Albertiana*, **32**: 21−35.

Kozur, H.W., and Mostler, H., 1994, Anisian to Middle Carnian radiolarian zonation and description of some stratigraphically important radiolarians. *Geologisch-Paläontologische Mitteilungen Innsbruck, Sonderband*, **3**: 39−255.

Kozur, H.W., and Bachmann, G.H., 2008, Updated correlation of the Germanic Triassic with the Tethyan scale and assigned numeric ages. *Berichte der Geologischen Bundesanstalt*, **76**: 53−58.

Kozur, H.W., and Bachmann, G.H., 2010, The Middle Carnian Wet Intermezzo of the Stuttgart Formation (Schilfsandstein), Germanic Basin. In: Kustatscher, E., Preto, N., and Wignall, P. (eds), Triassic climates, *Palaeogeography, Palaeoclimatology, Palaeoecology*, **290**: 107−119.

Kozur, H.W., and Weems, R.E., 2005, Conchostracan evidence for a late Rhaetian to early Hettangian age for the CAMP volcanic event in the Newark Supergroup, and a Sevatian (late Norian) age for the immediately underlying beds. *Hallesches Jahrbuch für Geowissenschaften, Reihe B: Geologie, Paläontologie, Mineralogie*, **27**: 21−51.

Kozur, H.W., and Weems, R.E., 2010, The biostratigraphic importance of conchostracans in the continental Triassic of the northern hemisphere. In: Lucas, S.G. (ed), The Triassic Timescale, *The Geological Society of London Special Publications*, **334**: 315−417.

Kozur, H.W., and Weems, R.E., 2011, Detailed correlation and age of continental late Changhsingian and earliest Triassic beds: implications for the role of the Siberian Trap in the Permian-Triassic biotic crisis. *Palaeogeography, Palaeoclimatology, Palaeoecology*, **308**: 22−40 < https://doi.org/10.1016/j.palaeo.2011.02.020 >.

Krull, E.S., and Retallack, G.J., 2000, $\partial^{13}C$ depth profiles from paleosols across the Permian-Triassic boundary: evidence for methane release. *Geological Society of America Bulletin*, **112**: 1459−1472.

Krystyn, L., 1980, Triassic conodont localities of the Salzkammergut region (northern Calcareous Alps). In: Schonlaub, H.P. (ed.), Second European Conodont Symposium ECOS II: Guidebook and Abstracts, *Abhandlungen der Geologischen Bundesanstalt*, **35**: 61−98.

Krystyn, L., 2010, Decision report on the defining event for the base of the Rhaetian stage. *Albertiana*, **38**: 11−12.

Krystyn, L., and Orchard, M.J., 1996, Lowermost Triassic ammonoid and conodont biostratigraphy of Spiti, India. *Albertiana*, **17**: 10−21.

Krystyn, L., Gallet, Y., Besse, J., and Marcoux, J., 2002, Integrated Upper Carnian to Lower Norian biochronology and implications for the Upper Triassic magnetic polarity time scale. *Earth and Planetary Science Letters*, **203**: 343−351.

Krystyn, L., Richoz, S., and Bhargava, O.N., 2007a, The Induan-Olenekian Boundary (IOB) in Mud − an update of the candidate GSSP section M04. *Albertiana*, **36**: 33−49.

Krystyn, L., Bhargava, O.N., and Richoz, S., 2007b, A candidate GSSP for the base of the Olenekian Stage: mud at Pin Valley; district Lahul & Spiti, Himachal Pradesh (Western Himalaya, India). *Albertiana*, **35**: 5−29.

Krystyn, L., Boquerel, H., Kuerschner, W., Richoz, S., and Gallet, Y., 2007c, Proposal for a candidate GSSP for the base of the Rhaetian Stage. *New Mexico Museum of Natural History and Science Bulletin*, **41**: 189−199.

Krystyn, L., Richoz, S., Gallet, Y., Bouquerel, H., Kürschner, W.M., and Spötl, C., 2007d, Updated bio- and magnetostratigraphy from the Steinbergkogel (Austria), candidate GSSP for the base of the Rhaetian stage. *Albertiana*, **36**: 164−173.

Kürschner, W.M., and Herngreen, G.F.W., 2010, Triassic palynology of central and northwestern Europe: a review of palynofloral diversity patterns and biostratigraphic subdivisions. In: Lucas, S.G. (Ed.), The Triassic Timescale, *The Geological Society of London Special Publications*, **334**: 263−283.

Kustatscher, E., Ash, S.R., Karasev, E., Pott, C., Vajda, V., Yu, J.X., et al., 2018, Flora of the Late Triassic. In: Tanner, L. (ed), *The Late Triassic World: Earth in a Time of Transition*. Topics in Geobiology, Springer Publication, Vol. 44: 545−622. https://doi.org/10.1007/978-3-319-68009-5_13

Lehrmann, D.J., Ramezani, J., Bowring, S.A., Martin, M.W., Montgomery, P., Enos, P., et al., 2006, Timing of recovery from the end-Permian extinction: geochronologic and biostratigraphic constraints from south China. *Geology*, **34**: 1053−1056.

Large Igneous Provinces Commission (International Association of Volcanology and Chemistry of the Earth's Interior), 2020, *LIP record*. Internet database available at: < www.largeigneousprovinces.org >.

Lehrmann, D.J., Stepchinski, L., Altiner, D., Orchard, M.J., Montgomery, P., Enos, P., et al., 2015, An integrated biostratigraphy (conodonts and foraminifers) and chronostratigraphy (paleomagnetic reversals, magnetic susceptibility, elemental chemistry, carbon isotopes and geochronology) for the Permian-Upper Triassic strata of Guandao section, Nanpanjiang Basin, south China. *Journal of Asian Earth Sciences*, **108**: 117−135. http://dx.doi.org/10.1016/j.jseaes.2015.04.030.

Li, M.S., Ogg, J.G., Zhang, Y., Huang, C.J., Hinnov, L.A., Chen, Z.-Q., et al., 2016, Astronomical tuning of the end-Permian extinction and the Early Triassic Epoch of South China and Germany. *Earth and Planetary Science Letters*, **441**: 10−25.

Li, M.S., Zhang, Y., Huang, C.J., Ogg, J.G., Hinnov, L.A., and Wang, Y.D., 2017a, Astronomical tuning and magnetostratigraphy of the Upper Triassic Xujiahe Formation of South China and Newark Supergroup of North America: implications for the Late Triassic time scale. *Earth and Planetary Science Letters*, **475**: 207−223.

Li, M.S., Zhang, Y., Huang, C.J., Ogg, J.G., Hinnov, L.A., Wang, Y.D., et al., 2017b, Astrochronology and magnetostratigraphy of the Xujiahe Formation of South China and Newark Supergroup of North America: implications for the Late Triassic time scale. *AGU Fall Meeting, New Orleans, LA*.

Li, M.S., Huang, C.J., Hinnov, L.A., Chen, W.Z., Ogg, J.G., and Tian, W., 2018, Astrochronology of the Anisian Stage (Middle Triassic) at the Guandao reference section, South China. *Earth and Planetary Science Letters*, **482**: 591−606.

Litwin, R.J., Traverse, A., and Ash, S.R., 1991, Preliminary palynological zonation of the Chinle Formation, southwestern U.S.A., and its correlation to the Newark Supergroup (eastern U.S.A.). *Review of Palaeobotany and Palynology*, **68**: 269−287.

Lucas, S.G., 1998, Global Triassic tetrapod biostratigraphy and biochronology. *Palaeogeography, Palaeoclimatology, Palaeoecology*, **143**: 345−382.

Lucas, S.G., 1999, A tetrapod-based Triassic timescale. *Albertiana*, **22**: 31−40.

Lucas, S.G., 2008, *Triassic New Mexico: Dawn of the Dinosaurs*. Albuquerque: New Mexico Museum of Natural History and Science 48 pp.

Lucas, S.G., 2009, Timing and magnitude of tetrapod extinctions across the Permo-Triassic boundary. In: Metcalfe, I., Isozaki, Y. (eds), End-Permian mass extinction: events & processes, age & timescale, causative mechanism(s) & recovery, *Journal of Asian Earth Sciences*, **36**(6): 491–502.

Lucas, S.G., 2010a, The Triassic Timescale. *The Geological Society of London Special Publications*, **334**, 500 pp.

Lucas, S.G., 2010b, The Triassic chronostratigraphy scale: history and status. In: Lucas, S.G. (ed.), The Triassic Timescale, *The Geological Society of London Special Publications*, **334**: 447–500.

Lucas, S.G., 2010c, The Triassic timescale: an introduction. In: Lucas, S. G. (ed), The Triassic Timescale, *The Geological Society of London Special Publications*, **334**: 1–16.

Lucas, S.G., 2010d, The Triassic timescale based on nonmarine tetrapod biostratigraphy and biochronology. In: Lucas, S.G. (ed), The Triassic Timescale, *The Geological Society of London Special Publications*, **334**: 17–39.

Lucas, S.G., 2016, Base of the Rhaetian and a critique of Triassic conodont-based biostratigraphy. *Albertiana*, **43**: 24–27 Available at: < http://paleo.cortland.edu/albertiana/ > [and his "Base of the Rhaetian and a critique of Triassic conodont-based biostratigraphy: reply". *Albertiana*, **43**: 32].

Lucas, S.G., 2018a, The Late Triassic Timescale. In: Tanner, L.H. (ed) *The Late Triassic World: Earth in a Time of Transition*, Topics in Geobiology. Springer Publications, Vol. **46**: 351–405. https://doi. org/10.1007/978-3-319-68009-5_10.

Lucas, S.G., 2018b, Late Triassic terrestrial tetrapods: biostratigraphy, biochronology and biotic events. In: Tanner, L.H. (ed), *The Late Triassic World: Earth in a Time of Transition*, Topics in Geobiology. Springer Publications, Vol. **46**: 351–405. https://doi.org/10.1007/978-3-319-68009-5_10.

Lucas, S.G., 2018c, Permian-Triassic charophytes: distribution, biostratigraphy and biotic events. *Journal of Earth Science*, **29**: 778–793. https://doi.org/10.1007/s12583-018-0786-4.

Lucas, S.G., and Tanner, L.H., 2014, Triassic timescale based on tetrapod biostratigraphy and biochronology. *In* Rocha, R., et al. (eds), *STRATI 2013*. Springer Geology, pp. 1013–1016. https://doi.org/ 10.1007/978-3-319-04364-7_192.

Lucas, S.G., Tanner, L.H., Kozur, H.W., Weems, R.E., and Heckert, A. B., 2012, The Late Triassic timescale: age and correlation of the Carnian-Norian boundary. *Earth-Science Reviews*, **114**: 1–18.

Luo, G.M., Kump, L.R., Wang, Y.B., Tong, J.N., Arthur, M.A., Yang, H., et al., 2010, Isotopic evidence for an anomalously low oceanic sulfate concentration following end-Permian mass extinction. *Earth and Planetary Science Letters*, **300**: 101–111.

Luo, G.M., Wang, Y.B., Yang, H., Algeo, T.J., Kump, L.R., Huang, J.H., et al., 2011, Stepwise and large-magnitude negative shift in delta C-13 (carb) preceded the main marine mass extinction of the Permian-Triassic crisis interval. *Palaeogeography, Palaeoclimatology, Palaeoecology*, **299**: 70–82.

Lyu, Z.Y., Orchard, M.J., Chen, Z.Q., Zhao, L.S., Zhang, L., Zhang, X.M., 2018, A taxonomic re-assessment of the *Novispathodus waageni* group and its role in defining the base of the Olenekian (lower Triassic). *Journal of Earth Science*, **29**: 824–836.

Lyu, Z.Y., Orchard, M.J., Chen, Z.Q., Wang, X.D., Zhao, L.S., and Han, C., 2019, Uppermost Permian to Lower Triassic conodont successions from the Enshi area, western Hubei Province, South China. *Palaeogeography, Palaeoclimatology, Palaeoecology*, **519**: 49–64.

Lyu, Z.Y., Orchard, M.J., Chen, Z.Q., Henderson, C.M., and Zhao, L.S., 2020. A proposed ontogenesis and evolutionary lineage of conodont *Eurygnathodus costatus* and its role in defining the base of the Olenekian (Lower Triassic). *Palaeogeography, Palaeoclimatology, Palaeoecology*, article #109916: 18 pp. https://doi.org/10.1016/j. palaeo.2020.109916.

Maron, M., Rigo, M., Bertinelli, A., Katz, M.E., Godfrey, L., Zaffani, M., et al., 2015, Magnetostratigraphy, biostratigraphy, and chemostratigraphy of the Pignola-Abriola section: new constraints for the Norian-Rhaetian boundary. *Geological Society of America Bulletin*, **127** (7–8), 962–974. https://doi.org/10.1130/B31106.1.

Maron, M., Muttoni, G., Dekkers, M.J., Mazza, M., Roghi, G., Breda, A., et al., 2017, Contribution to the magnetostratigraphy of the Carnian: new magneto-biostratigraphic constraints from Pignola-2 and Dibona marine sections, Italy. *Newsletters on Stratigraphy*, **50** (2), 187–203. https://doi.org/10.1127/nos/2017/0291.

Maron, M., Muttoni, G., Rigo, M., Gianolla, P., and Kent, D.V., 2019, New magnetobiostratigraphic results from the Ladinian of the Dolomites and implications for the Triassic geomagnetic polarity timescale. *Palaeogeography, Palaeoclimatology, Palaeoecology*, **517**: 52–73. https://doi.org/10.1016/j.palaeo.2018.11.024.

Marshall, C.R., and Jacobs, D.K., 2009, Flourishing after the end-Permian mass extinction. *Science*, **325**: 1079–1080.

Martin, E.E., and Macdougall, J.D., 1995, Sr and Nd isotopes at the Permian/Triassic boundary: a record of climate change. *Chemical Geology*, **125**: 73–95.

Marzoli, A., Renne, P.R., Piccirillo, E.M., Ernesto, M., Bellieni, G., and De Min, A., 1999, Extensive 200-million-year-old continental flood basalts of the Central Atlantic Magmatic Province. *Science*, **284**: 616–618.

Mazza, M., Furin, S., Spöti, C., and Rigo, M., 2010, Generic turnovers of Carnian/Norian conodonts: climatic control or competition?. In: Kustatscher, E., Preto, N., Wignall, P. (eds), Triassic climates, *Palaeogeography, Palaeoclimatology, Palaeoecology*, **290**: 120–137.

Marzoli, A., Jourdan, F., Puffer, J.H., Cuppone, T., Tanner, L.H., Weems, R.E., et al., 2011, Timing and duration of the Central Atlantic magmatic province in the Newark and Culpeper basins, eastern U.S.A. *Lithos*, **122**: 175–188.

Mazza, M., Rigo, M., and Gullo, M., 2012, Taxonomy and biostratigraphic record of the Upper Triassic conodonts of the Pizzo Mondello section (western Sicily, Italy). *Rivista Italiana di Paleontologia e Stratigrafia*, **118**: 85–130.

Mazza, M., Nicora, A., and Rigo, M., 2018, *Metapolygnathus parvus* Kozur, 1972 (Conodonta): a potential primary marker for the Norian GSSP (Upper Triassic). *Bollettino della Società Paleontologica Italiana*, **57** (2), 81–101. https://doi.org/10.4435/BSPI.2018.06.

McArthur, J.M., 2008, Comment on 'The impact of the Central Atlantic Magmatic Province on climate and on the Sr- and Os-isotope evolution of seawater' by Cohen, A.S. and Coe, A.L. 2007. Palaeogeography, Palaeoclimatology, Palaeoecology, 244. 374–390. *Palaeogeography, Palaeoclimatology, Palaeoecology*, **263**: 146–149.

McArthur, J.M., 2010, Correlation and dating with strontium-isotope stratigraphy. In: Whitaker, J.E., Hart, M.B. (eds), Micropalaeontology, sedimentary environments and stratigraphy: a tribute to Dennis Curry (1912–2001), *The Micropalaeontological Society Special Publication*, **TSM004**: 133–145.

McArthur, J.M., Howarth, R.J., and Bailey, T.R., 2001, Strontium isotope stratigraphy: LOWESS Version 3: best fit to the marine Sr-isotope curve for 0–509 Ma and accompanying look-up table for deriving numerical age. *Journal of Geology*, **109**: 155–170.

McArthur, J.M., Howarth, R.J., Shields, G.A., and Zhou, Y., 2020, Chapter 7 - Strontium isotope stratigraphy. *In* Gradstein, F.M., Ogg, J.G., Schmitz, M.D., and Ogg, G.M. (eds), *The Geologic Time Scale 2020*. **Vol. 1** (this book). Elsevier, Boston, MA.

McRoberts, C.A., 2007, The halobid bivalve succession across a potential Carnian/Norian GSSP at Black Bear Ridge, Williston Lake, northeast British Columbia, Canada. *Albertiana*, **36**: 142–145.

McRoberts, C.A., 2010, Biochronology of Triassic pelagic bivalves. In: Lucas, S.G. (Ed.), The Triassic Timescale, *The Geological Society of London Special Publications*, **334**: 201–219.

McRoberts, C.A., Krystyn, L., and Shea, A., 2008, Rhaetian (Late Triassic *Monotis* (Bivalvia: Pectinacea) from the Northern Calcareous Alps (Austria) and the end-Norian crisis in pelagic faunas. *Journal of Paleontology*, **51**: 721–735.

Menning, M., Gast, R., Hagdorn, H., Kading, K.-C., Simon, T., Szurlies, M., et al., 2005, Zeitskala für Perm und Trias in der Stratigraphischen Tabelle von Deutschland 2002. Zyklostratigraphische Kalibrierung von höherer Dyas und Germanischer Trias und das Alter der Stufen Roadium bis Rhaetium 2005. In: Menning M., Hendrich, A. (eds), Erläuterungen zur Stratigraphischen Tabelle von Deutschland, *Newsletters of Stratigraphy*, **41**(1/3): 173–210.

Metcalfe, I., and Isozaki, Y. (eds), 2009a. End-Permian mass extinction: events & processes, age & timescale, causative mechanism(s) & recovery. *Journal of Asian Earth Sciences* 36 (6), 407–540.

Metcalfe, I., and Isozaki, Y., 2009b, Current perspectives on the Permian–Triassic boundary and end-Permian mass extinction: preface. In: Metcalfe, I., Isozaki, Y. (eds), End-Permian mass extinction: events & processes, age & timescale, causative mechanism(s) & recovery, *Journal of Asian Earth Sciences*, **36**(6): 407–412.

Metcalfe, I., Nicoll, R.S., Black, L.P., Mundil, R., Renne, P., Jagodzinski, E.A., et al., 1999, Isotope geochronology of the Permian–Triassic boundary and mass extinction in South China. *In* Yin, H.F., and Tong, J.N. (eds), *Pangea and the Paleozoic-Mesozoic Transition*. Wuhan: China University of Geosciences Press, 134–137.

Metcalfe, I., Foster, C.B., Afonin, S.A., Nicoll, R.S., Wang X., and Lucas, S.G., 2009, Stratigraphy, biostratigraphy and C-isotopes of the Permian–Triassic non-marine sequence at Dalongkou and Lucaogou, Xinjiang Province, China. In: Metcalfe, I., Isozaki, Y. (eds), End-Permian mass extinction: events & processes, age & timescale, causative mechanism(s) & recovery, *Journal of Asian Earth Sciences*, **36**(6): 407–412.

Meyers, S.R., 2008, Resolving Milankovitchian controversies: the Triassic Latemar Limestone and the Eocene Green River Formation. *Geology*, **36**: 319–322.

Mietto, P., and Manfrin, S., 1995, A high resolution Middle Triassic ammonoid standard scale in the Tethys Realm: a preliminary report. *Bulletin de la Société Géologique de France*, **166** (5), 539–563.

Mietto, P., and Manfrin, S., 1999, A debate on the Ladinian-Carnian boundary. *Albertiana*, **22**: 23–27.

Mietto, P., Andreetta, R., Broglio Loriga, C., Buratti, N., Cirilli, S., De Zanche, V., et al., 2007, A candidate of the Global Boundary Stratotype Section and Point for the base of the Carnian Stage (Upper Triassic): GSSP at the base of the *canadensis* Subzone (FAD of *Daxatina*) in the Prati di Stuores/Stuores Wiesen section (Southern Alps, NE Italy). *Albertiana*, **36**: 78–97.

Mietto, P., Manfrin, S., Preto, N., Rigo, M., Roghi, G., Furin, S., et al., 2012, A candidate of the Global Boundary Stratotype Section and Point for the base of the Carnian Stage (Upper Triassic): GSSP at the base of the *canadensis* Subzone (FAD of *Daxatina*) in the Prati di Stuores/Stuores Wiesen section (Southern Alps, NE Italy). *Episodes*, **35**: 414–430.

Moix, P., Kozur, H.W., Stampfli, G.M., and Mostler, H., 2007, New paleontological, biostratigraphic and paleogeographic results from the Triassic of the Mersin Mélange, SE Turkey. In: Lucas, S.G., Spielmann, J.A., (eds) The Global Triassic, *New Mexico Museum of Natural History and Sciences Bulletin*, **41**: 282–311.

Mojsisovics, E.V., 1869, Über die Gliederung der oberen Triasbildungen der östlichen Alpen. *Jahrbuch Geologischen Reichsanstalt*, **19**: 91–150.

Mojsisovics, E.V., 1893, Faunistische Ergebnisse aus der Untersuchung der Ammoneen-faunen der Mediterranen Trias. *Abhandlungen der Geologischen Bundesanstalt*, **6**: 810.

Mojsisovics, E., von, Waagen, W., and Diener, C., 1895, Entwurf einer Gliederung der pelagischen Sedimente des Trias-Systems. *Sitzungberichte der Akademie der Wissenschaften in Wien*, **104**: 1271–1302.

Mørk, A., Embry, A.F., and Weitschat, W., 1989, Triassic transgressive-regressive cycles in the Sverdrup Basin, Svalbard and the Barents Shelf. *In* Collinson, J.D. (ed), *Correlation in Hydrocarbon Exploration*. London: Graham and Trotman, 113–130.

Mortensen, J.K., and Hulbert, L.J., 1992, A U-Pb zircon age for a Maple Creek gabbro sill, Tatamagouche Creek area, southwest Yukon Territory. In *Radiogenic Age and Isotopic Studies, Report 5: Geological Survey of Canada, Paper, 91-2* (pp. 175–179).

Moslow, T.F., Haverslew, B., and Henderson, C.M., 2018, Sedimentary facies, petrology, reservoir characteristics, conodont biostratigraphy and sequence stratigraphic framework of a continuous (395 m) full diameter core of the Lower Triassic Montney Formation, northeastern British Columbia. In: Euzen, T., Moslow, T.F., Caplan, M. (eds), The Montney Play: Deposition to Development, *Bulletin of Canadian Petroleum Geology*, Special Volume, **66**: 259–287.

Motani, R., 2010, Warm blooded sea dragons? *Science*, **328**: 1361–1362.

Mundil, R., and Irmis, R., 2008, New U-Pb age constraints for terrestrial sediments in the Late Triassic: implications for faunal evolution and correlations with marine environments. *33rd International Geological Congress, Oslo 2008, Abstract*. Online at: < http://www.cprm.gov.br/33IGC/1342538.html >

Mundil, R., Brack, P., Meier, M., Rieber, H., and Oberli, F., 1996, High resolution U–Pb dating of Middle Triassic volcaniclastics: time-scale calibration and verification of tuning parameters for carbonate sediments. *Earth and Planetary Science Letters*, **141**: 137–151.

Mundil, R., Metcalfe, I., Ludwig, K.R., Renne, P.R., Oberli, F., and Nicoll, R.S., 2001, Timing of the Permian–Triassic biotic crisis: implications from new zircon U/Pb age data (and their limitations). *Earth and Planetary Science Letters*, **187**: 131–145.

Mundil, R., Ludwig, K.R., Metcalfe, I., and Renne, P.R., 2004, Age and timing of the Permian mass extinctions: U/Pb dating of closed-system zircons. *Science*, **305**: 1760–1763.

Mundil, R., Pálfy, J., Renne, P.R., and Brack, P., 2010, The Triassic timescale: new constraints and a review of geochronological data. In: Lucas, S.G. (ed), The Triassic Timescale, *The Geological Society of London Special Publications*, **334**: 41–60.

Muttoni, G., Kent, D.V., Nicora, A., Rieber, H., and Brack, P., 1996a, Magneto-biostratigraphy of the 'Buchenstein Beds' at Frötschbach (western Dolomites, Italy). *Albertiana*, **17**: 51–56.

Muttoni, G., Kent, D.V., Meço, S., Nicora, A., Gaetani, M., Balini, M., et al., 1996b, Magnetobiostratigraphy of the Spathian to Anisian

(Lower to Middle Triassic) Kçira section, Albania. *Geophysical Journal International*, **127**: 503–514 < https://doi.org/10.1111/j.1365-246x.1996.tb04736.x > .

Muttoni, G., Kent, D.V., Meço, S., Balini, M., Nicora, A., Rettori, R., et al., 1998, Towards a better definition of the Middle Triassic magnetostratigraphy and biostratigraphy in the Tethyan realm. *Earth and Planetary Science Letters*, **164**: 285–302.

Muttoni, G., Kent, D.V., DiStefano, P., Gullo, M., Nicora, A., Tait, J., et al., 2001a, Magnetostratigraphy and biostratigraphy of the Carnian/Norian boundary interval from the Pizzo Mondello section (Sicani Mountains, Sicily). *Palaeogeography, Palaeoclimatology, Palaeoecology*, **166**: 383–399.

Muttoni, G., Kent, D.V., and Orchard, M.J., 2001b, Paleomagnetic reconnaissance of early Mesozoic carbonates from Williston Lake, northeastern British Columbia, Canada: evidence for late Mesozoic remagnetization. *Canadian Journal of Earth Sciences*, **38**: 1157–1168.

Muttoni, G., Nicora, A., Brack, P., and Kent, D.V., 2004a, Integrated Anisian−Ladinian boundary chronology. *Palaeogeography, Palaeoclimatology, Palaeoecology*, **208**: 85–102.

Muttoni, G., Kent, D.V., Olsen, P.E., DiStefano, P., Lowrie, W., Bernasconi, S.M., et al., 2004b, Tethyan magnetostratigraphy from Pizzo Mondello (Sicily) and correlation to the Late Triassic Newark astrochronological polarity time scale. *Geological Society of America Bulletin*, **116**: 1043–1058.

Muttoni, G., Kent, D.V., Flavio, J., Olsen, P., Rigo, M., Galli, M.T., et al., 2010, Rhaetian magnetostratigraphy from the Southern Alps (Italy): constraints on Triassic chronology. *Palaeogeography, Palaeoclimatology, Palaeoecology*, **285**: 1–16.

Muttoni, G., Mazza, M., Mosher, D., Katz, M.E., Kent, D.V., and Balini, M., 2014, A Middle-Late Triassic (Ladinian-Rhaetian) carbon and oxygen isotope record from the Tethyan Ocean. *Palaeogeography, Palaeoclimatology, Palaeoecology*, **399**: 246–259 (and on-line data repository item 2015069).

Muttoni, G., Nicora, A., Balini, M., Katz, M., Schaller, M., Kent, D., et al., 2019, A candidate GSSP for the base of the Anisian from Kçira, Albania. *Albertiana*, **45**: 39–49.

Nawrocki, J., 1997, Permian to Early Triassic magnetostratigraphy from the Central European Basin in Poland: implications on regional and worldwide correlations. *Earth and Planetary Science Letters*, **152**: 37–58.

Nawrocki, J., and Szulc, J., 2000, The Middle Triassic magnetostratigraphy from the Peri-Tethys basin in Poland. *Earth and Planetary Science Letters*, **182**: 77–92.

Nesbitt, S.J., Smith, N.D., Irmis, R.B., Turner, A.H., Downs, A., and Norell, M.A., 2009, A complete skeleton of a Late Triassic Saurischian and the early evolution of dinosaurs. *Science*, **326**: 1530–1533.

Newell, N.D., 1994, Is there a precise Permian-Triassic boundary? *Permophiles*, **24**: 46–48.

Newton, R.J., Pevitt, E.L., Wignall, P.B., and Bottrell, S.H., 2004, Large shifts in the isotopic composition of seawater sulphate across the Permo-Triassic boundary in northern Italy. *Earth and Planetary Science Letters*, **218**: 331–345.

Nicora, A., Balini, M., Bellanea, A., Bowring, S.A., Di Stefano, P., Dumitrica, P., et al., 2007, The Carnian/Norian boundary interval at Pizzo Mondello (Sicani Mountains, Sicily) and its bearing for the definition of the GSSP of the Norian Stage. *Albertiana*, **36**: 102–129.

Nowak, H., Schneebeli-Hermann, E., and Kustatscher, E., 2018, Correlations of Lopingian to Middle Triassic palynozones. *Journal of Earth Sciences*, **29**: 755–777. https://doi.org/10.1007/s12583-018-0790-8.

O'Dogherty, L., Carter, E.S., Gorican, S., and Dumitrica, P., 2010, Triassic radiolarian biostratigraphy. In: Lucas, S.G. (ed), The Triassic Timescale, *The Geological Society of London Special Publications*, **334**: 163–200.

Ogg, J.G., 2004, The Triassic Period. *In* Gradstein, F.M., Ogg, J.G., and Smith, A.L. (eds), *A Geologic Time Scale 2004*. Cambridge: Cambridge University Press, 271–306.

Ogg, J.G., 2012, Triassic. *In* Gradstein, F.M., Ogg, J.G., Schmitz, M., and Ogg, G. (eds), *The Geologic Time Scale 2012*. Elsevier Publ., 681–730.

Ogg, J.G., 2015, The mysterious Mid-Carnian "Wet Intermezzo" global event. *Journal of Earth Science*, **26** (2), 181–191.

Ogg, J.G., and Steiner, M.B., 1991, Early Triassic magnetic polarity time scale—integration of magnetostratigraphy, ammonite zonation and sequence stratigraphy from stratotype sections Canadian Arctic Archipelago). *Earth and Planetary Science Letters*, **107**: 69–89.

Ogg, J.G., Ogg, G.M., and Gradstein, F.M., 2008, Triassic Period. *In* Ogg, J.G., Ogg, G.M., and Gradstein, F.M. (eds), *Concise Geologic Time Scale*. Cambridge: Cambridge University Press, 95–106.

Ogg, J.G., Huang, C., and Hinnov, L., 2014, Triassic timescale status: a brief overview. *Albertiana*, **41**: 3–30.

Ogg, J.G., Ogg, G.M., and Gradstein, F.M., 2016, *A Concise Geologic Time Scale 2016*. Elsevier Publ. 234 pp.

Ogg, J.G., Gradinaru, E., Chen, Y., Bucher, F.R., Hounslow, M.W., 2020, Comments on defining the base of Anisian Stage. *ResearchGate*: https://www.researchgate.net/publication/336767153_Early-Middle_Triassic_boundary_interval_Integrated_chemo-bio-magneto-stratigraphy_of_potential_GSSPs_for_the_base_of_the_Anisian_Stage_in_South_China/comments.

Olesen, J., 2009, Phylogeny of Branchiopoda (Crustacea)—character evolution and contribution of uniquely preserved fossils. *Arthropod Systematics & Phylogeny*, **67** (1), 3–39.

Olsen, P.E., and Kent, D.V., 1999, Long-period Milankovitch cycles from the Late Triassic and Early Jurassic of eastern North America and their implications for the calibration of the Early Mesozoic time-scale and the long-term behaviour of the planets. *Philosophical Transactions of the Royal Society of London, Series A: Mathematical, Physical and Engineering Science*, **357**: 1761–1786 < https://doi.org/10.1098/rsta.1999.0400 > .

Olsen, P.E., Kent, D.V., Cornet, B., Witte, W.K., and Schlische, R.W., 1996, High-resolution stratigraphy of the Newark rift basin (early Mesozoic, eastern North America). *Geological Society of America Bulletin*, **108**: 40–77.

Orbell, G., 1983, Palynology of the British Rhaetian. *Bulletin of the Geological Survey of Great Britain*, **44**: 1–44.

Orchard, M.J., 2007a, Conodont diversity and evolution through the latest Permian and Early Triassic upheavals. *Palaeogeography, Palaeoclimatology, Palaeoecology*, **252**: 93–117.

Orchard, M.J., 2007b, A proposed Carnian-Norian Boundary GSSP at Black Bear Ridge, northeast British Columbia, an a new conodont framework for the boundary interval. *Albertiana*, **36**: 130–141.

Orchard, M.J., 2010, Triassic conodonts and their role in stage boundary definitions. In: Lucas, S.G. (ed), The Triassic Timescale, *The Geological Society of London Special Publications*, **334**: 139–161.

Orchard, M.J., 2014, Conodonts from the Carnian-Norian boundary (Upper Triassic) of Black Bear Ridge, northeastern British Columbia, Canada. *New Mexico Museum of Natural History and Science Bulletin*, **64**: 139 pp.

Orchard, M.J., 2016, Base of the Rhaetian and a critique of Triassic conodont-based biostratigraphy: comment. *Albertiana*, **43**: 28–32.

Orchard, M.J., 2018, The lower-middle Norian (Upper Triassic) boundary: new conodont taxa and a refined biozonation. *Bulletins of American Paleontology*, **395-396**: 165–193.

Orchard, M.J., 2019, The Carnian-Norian GSSP candidate at Black Bear Ridge, British Columbia, Canada: update, correlation and conodont taxonomy. *Albertiana*, **45**: 50–68.

Orchard, M.J., and Tozer, E.T., 1997, Triassic conodont biochronology, its intercalibration with the ammonoid standard, and a biostratigraphic summary for the Western Canada Sedimentary Basin. *Bulletin of Canadian Petroleum Geology*, **45**: 675–692.

Orchard, M.J., Carter, E.S., and Tozer, E.T., 2000, Fossil data and their bearing on defining a Carnian-Norian (upper Triassic) boundary in western Canada. *Albertiana*, **24**: 43–50.

Orchard, M.J., Zonneveld, J.P., Johns, M.J., McRoberts, C.A., Sandy, M.R., Tozer, E.T., et al., 2001, Fossil succession and sequence stratigraphy of the Upper Triassic of Black Bear Ridge, northeast British Columbia, a GSSP prospect for the Carnian-Norian boundary. *Albertiana*, **25**: 10–22.

Ovtcharova, M., Bucher, H., Schaltegger, U., Galfetti, T., Brayard, A., and Guex, J., 2006, New Early to Middle Triassic U-Pb ages from South China: calibration with ammonoid biochronozones and implications for the timing of the Triassic biotic recovery. *Earth and Planetary Science Letters*, **243**: 463–475.

Ovtcharova, M., Bucher, H., Goudemand, N., Schaltegger, U., Brayard, A., and Galfetti, T., 2010, New U/Pb ages from Nanpanjiang Basin (South China): implications for the age and definition of the Early-Middle Triassic boundary. *Geophysical Research Abstracts*, **12**: **EGU2010-12505-3**.

Ovtcharova, M., Goudemand, N., Hammer, Ø., Guodun, K., Cordey, F., Galfetti, T., et al., 2015, Developing a strategy for accurate definition of a geological boundary through radio-isotopic and biochronological dating: the Early-Middle Triassic boundary (South China). *Earth-Science Reviews*, **146**: 65–76 <https://doi.org/10.1016/j.earscirev.2015.03.006>.

Pálfy, J., and Vörös, A., 1998, Quantitative ammonoid biochronological assessment of the Anisian–Ladinian (Middle Triassic) stage boundary proposals. *Albertiana*, **21**: 19–26.

Pálfy, J., Smith, P.L., and Mortensen, J.K., 2000a, A U–Pb and ^{40}Ar/^{39}Ar time scale for the Jurassic. *Canadian Journal of Earth Sciences*, **37**: 923–944.

Pálfy, J., Mortensen, J.K., Carter, E.S., Smith, P.L., Friedman, R.M., and Tipper, H.W., 2000b, Timing the end-Triassic mass extinction: first on land, then in the sea? *Geology*, **28**: 39–42.

Pálfy, J., Demeny, A., Haas, J., Hetenyi, M., Orchard, M.J., and Veto, I., 2001, Carbon isotope anomaly and other geochemical changes at the Triassic-Jurassic boundary from a marine section in Hungary. *Geology*, **29**: 1047–1050.

Pálfy, J., Parrish, R.R., David, K., and Voros, A., 2003, Mid-Triassic integrated U/Pb geochronology and ammonoid biochronology from the Balaton Highland (Hungary). *Journal of the Geological Society*, **160**: 271–284.

Palmer, A.R., 1983, *Geologic time scale, Decade of North American Geology (DNAG)*. Boulder, NV: Geological Society of America.

Parrish, R.R., and McNicoll, V.J., 1992. U/Pb age determinations from the southern Vancouver Island area, British Columbia. In *Radiogenic Age and Isotopic Studies: Geological Survey of Canada, Paper, 91-2* (pp. 79–86).

Paull, R.K., 1997, Observations on the Induan-Olenekian boundary based on conodont biostratigraphic studies in the Cordillera of the western United States. *Albertiana*, **20**: 31–32.

Payne, J.L., and Kump, L.R., 2007, Evidence for recurrent Early Triassic massive volcanism from quantitative interpretation of carbon isotope fluctuations. *Earth and Planetary Science Letters*, **256**: 264–277.

Payne, J.L., Lehrmann, D.J., Wei, J., Orchard, M.J., Schrag, D.P., and Knoll, A.H., 2004, Large perturbations of the carbon cycle during recovery from the end-Permian extinction. *Science*, **305**: 506–509.

Payne, J.L., Lehrmann, D.J., Follet, D., Seibel, M., Kump, L.R., Riccardi, A., et al., 2009, Erosional truncation of uppermost Permian shallow-marine carbonates and implications for Permian-Triassic boundary events: reply. *Geological Society of America Bulletin*, **121**: 957–959.

Paytan, A., Yao, W.Q., Faul, K., and Gray, E.T., 2020, Chapter 9 - Sulfur isotope stratigraphy. *In* Gradstein, F.M., Ogg, J.G., Schmitz, M.D., and Ogg, G.M. (eds), *The Geologic Time Scale 2020*. **Vol. 1** (this book). Elsevier, Boston, MA.

Peucker-Ehrenbrink, B., and Ravizza, G.E., 2020, Chapter 8 - Osmium isotope stratigraphy. *In* Gradstein, F.M., Ogg, J.G., Schmitz, M.D., and Ogg, G.M. (eds), *The Geologic Time Scale 2020*. **Vol. 1** (this book). Elsevier, Boston, MA.

Peybernes, B., 1998, Larger benthic foraminifer. Column for Triassic chart of Mesozoic and Cenozoic sequence chronostratigraphic framework of European basins, by Hardenbol, J., Thierry, J., Farley, M.B., Jacquin, T., de Graciansky, P.-C., and Vail, P.R. (coordinators). In: de Graciansky, P.-C., Hardenbol, J., Jacquin, T., Vail, P.R. (eds.), Mesozoic-Cenozoic Sequence Stratigraphy of European Basins, *SEPM Special Publication*, **60**: Chart 8.

Phillips, J., 1840, Palaeozoic Series. *In* Long, G. (ed), The Penny Cyclopaedia of the Society for the Diffusion of Useful Knowledge. *London: Charles Knight*, **Vol. 17**: 153–154.

Phillips, J., 1841, *Figures and Descriptions of the Palaeozoic Fossils of Cornwall, Devon and east Somerset*. London: Longman, Brown, Green and Longmans.

Pia, J., 1930, *Grundbegriffe der Stratigraphie mit ausführlicher anwendung auf die Europäische Mitteltrias*. Leipzig and Wien: Deuticke.

Preto, N., Kustatscher, E., and Wignall, P.B., 2010, Triassic climates – state of the art and perspectives. In: Kustatscher, E., Preto, N., Wignall, P. (eds), Triassic climates, *Palaeogeography, Palaeoclimatology, Palaeoecology*, **290**: 1–10.

Ramezani, J., Bowring, S.A., Pringle, M.S., Winslow III, F.D., and Rasbury, E.T. 2005. The Manicouagan impact melt rock: a proposed standard for the intercalibration of U–Pb and 40Ar/39Ar isotopic systems. *15th V.M. Goldsmidt Conference Abstract Volume, Moscow, ID.* Available online at: <http://goldschmidt.info/2005_cd/Abstract%20Volume/abs_vol.pdf>

Ramezani, J., Bowring, S.A., Martin, M.W., Lehrmann, D.J., Montgomery, P., Enos, P., et al., 2007, Timing of recovery from the end-Permian extinction: geochronologic and biostratigraphic constraints from south China: comment and reply: reply. *Geology*, **35**: e137 Online Forum.

Ramezani, J., Bowring, S.A., Fastovsky, D.E., and Hoke, G.D., 2009, U-Pb-ID-TIMS geochronology of the Late Triassic Chinle Formation, Petrified Forest National Park, Arizona. *Geological Society of America Abstracts With Programs*, **41** (7), 421 [plus copy of summary slide sent to H. Kozur].

Ramezani, J., Bowring, S.A., Fastovsky, D.E., and Hoke, G.D., 2010, Depositional history of the Late Triassic Chinle fluvial system at the Petrified Forest National Park: U-Pb geochronology, regional correlation and insights into early dinosaur evolution. *American Geophysical Union Fall meeting 2010, Abstract #V31A-2313*.

Rampino, M.,R., Prokoph, A., and Adler, A.C., 2000, Tempo of the end-Permian event: high-resolution cyclostratigraphy at the Permian-Triassic Boundary. *Geology*, **28**: 643–646.

Rampino, M.,,R., Prokoph, A., Adler, A.C., and Schwindt, D.M., 2002, Abruptness of the end-Permian mass extinction as determined from biostratigraphic and cyclostratigraphic analysis of European western Tethyan sections. In: Koeberl, C., MacLeod, K.G. (eds), Catastrophic Events and mass extinctions: Impacts and Beyond, *Geological Society of America Special Paper*, **356**: 415–427.

Reichow, M.K., Saunders, A.D., Ivanov, A.V., and Puchkov, V.N., 2004, The Siberian large igneous province. *Internet Article From Large Igneous Provinces Commission, Large Igneous Province of the Month: March 2004*. Available at < www.largeigneousprovinces.org/ >

Reichow, M.K., Pringle, M.S., Al'Mukhamedov, A.I., Allen, M.B., Andreichev, V.L., Buslov, M.M., et al., 2009, The timing and extent of the eruption of the Siberian Traps large igneous province: implications for the end-Permian environmental crisis. *Earth and Planetary Science Letters*, **27**: 9–20.

Renesto, S., and Dalla Vecchio, F.M., 2018, Late Triassic marine reptiles. In: Tanner, L.H. (ed), *The Late Triassic World: Earth in a Time of Transition*, Topics in Geobiology. Springer Publication, pp. 263–313. https://doi.org/10.1007/978-3-319-68009-5_8

Renne, P.R., and Basu, A.R., 1991, Rapid eruption of the Siberian traps flood basalts at the Permo-Triassic boundary. *Science*, **253**: 176–179.

Renne, P.R., Zhang, Z., Richards, M.A., Black, M.T., and Basu, A.R., 1995, Synchrony and causal relations between Permian-Triassic boundary crisis and Siberian flood volcanism. *Science*, **269**: 1413–1416.

Renne, P.R., Mundil, M., Balco, G., Min, K., and Ludwig, K.R., 2010, Joint determination of ^{40}K decay constants and $^{40}Ar*/^{40}K$ for the Fish Canyon sanidine standard, and improved accuracy for $^{40}Ar/^{39}Ar$ geochronology. *Geochimica et Cosmochimica Acta*, **74**: 5349–5367.

Rigo, M., Preto, N., Roghi, G., Tateo, F., and Mietto, P., 2007, A rise in the carbonate compensation depth of western Tethys in the Carnian (Late Triassic): deep-water evidence for the Carnian pluvial event. *Palaeogeography, Palaeoclimatology, Palaeoecology*, **246**: 188–205.

Rigo, M., Bertinelli, A., Concheri, G., Gattolin, G., Godfrey, L., Katz, M.E., et al., 2016, The Pignola-Abriola section (southern Apennines, Italy): a new GSSP candidate for the base of the Rhaetian Stage. *Lethaia*, **49** (3), 287–306. https://doi.org/10.1111/let.12145.

Rigo, M., Mazza, M., Karádi, V., and Nicora, A., 2018, New Upper Triassic conodont biozonation of the Tethyan Realm. In: Tanner, L.H. (ed), *The Late Triassic World: Earth in a Time of Transition*, Topics in Geobiology. Springer Publication, **Vol. 46**: 189–235. https://doi.org/10.1007/978-3-319-68009-5_6.

Rogers, R.R., Swisher III, C.C., Sereno, P.C., Monetta, A.M., Forster, C.A., and Martinez, R.N., 1993, The Ischigualasto tetrapod assemblage (Late Triassic, Argentina) and $^{40}Ar/^{39}Ar$ dating of dinosaur origins. *Science*, **260**: 794–797.

Roghi, G., Gianolla, P., Minarelli, L., Pilati, C., and Preto, N., 2010, Palynological correlation of Carnian humid pulses throughout western Tethys. In: Kustatscher, E., Preto, N., Wignall, P. (eds), Triassic climates, *Palaeogeography, Palaeoclimatology, Palaeoecology*, **290**: 89–106.

Röhling, H.-G., 1991, A lithostratigraphic subdivision of the Lower Triassic in the northwest German Lowlands and the German Sector of the North Sea, based on Gamma-Ray and Sonic Logs. *Geologisches Jahrbuch Reihe A*, **119**: 3–24.

Ruffell, A., Simms, M.J., and Wignall, P.B., 2016, The Carnian Humid Episode of the late Triassic: a review. *Geological Magazine*, **153** (Special Issue 2 (mass extinctions)), 271–284. https://doi.org/10.1017/S0016756815000424.

Ruhl, M., Kuerschner, W.M., and Krystyn, L., 2009, Triassic−Jurassic organic carbon isotope stratigraphy of key sections in the western Tethys realm (Austria). *Earth and Planetary Science Letters*, **281**: 169–187.

Satterley, A.K., 1996, The interpretation of cyclic successions of the Middle and Upper Triassic of the Northern and Southern Alps. *Earth-Science Reviews*, **40**: 181–207.

Schlager, W., and Schöllnberger, W., 1974, Das Prinzip stratigraphischer Wenden in der Schichtenfolge der Nördlichen Kalkalpen. *Mitteilungen der Österreichischen Geologischen Gesellschaft, Wien*, **66/67**: 165–193.

Schaltegger, U., Guex, J., Bartolini, A., Schoene, B., and Ovtcharova, M., 2008, Precise U-Pb age constraints for end-Triassic mass extinction, its correlation to volcanism and Hettangian post-extinction recovery. *Earth and Planetary Science Letters*, **267**: 266–275.

Schneider, J.W., and Scholze, F., 2018, Late Pennsylvanian−Early Triassic conchostracan biostratigraphy: a preliminary approach. *In* Lucas, S.G., and Shen, S.Z. (eds), *The Permian Timescale*. The Geological Society of London Special Publications, **450**: 365–386.

Schoene, B., Guex, J., Bartolini, A., Schaltegger, U., and Blackburn, T.J., 2010, Correlating the end-Triassic mass extinction and flood basalt volcanism at the 100 ka level. *Geology*, **38**: 387–390 and 13-page supplement.

Scholze, F., Schneider, J.W., and Werneburg, R., 2016, Conchostracans in continental deposits of the Zechstein-Buntsandstein transition in central Germany: taxonomy and biostratigraphic implications for the position of the Permian−Triassic boundary within the Zechstein Group. *Palaeogeography, Palaeoclimatology, Palaeoecology*, **449**: 174–193.

Scholze, F., Shen, S.Z., Backer, M., Wei, H.B., Hübner, M., Cui, Y.Y., et al., 2019, Reinvestigation of conchostracans (Crustacea: Branchiopoda) from the Permian−Triassic transition in Southwest China. *Palaeoworld*. https://doi.org/10.1016/j.palwor.2019.04.007.

Schuurman, W.M.N., 1979, Aspects of Late Triassic palynology. 3. Palynology of the latest Triassic and earliest Jurassic deposits of the Northern Limestone Alps in Austria and southern Germany, with special reference to a palynological characterization of the Rhaetian stage in Europe. *Review of Palaeobotany and Palynology*, **27**: 53–75.

Schwarzacher, W., 2005, The stratification and cyclicity of the Dachstein Limestone in Lofer, Leogang and Steinernes Meer (Northern Calcareous Alps, Austria). *Sedimentary Geology*, **181**: 93–106.

Scotese, C.R., 2014, *Atlas of Middle & Late Permian and Triassic Paleogeographic Maps, Maps 43−52, Volumes 3 & 4, PALEOMAP PaleoAtlas for ArcGIS, PALEOMAP Project, Evanston, IL, 2014.*

Sedlacek, A.R., Saltzman, M.R., Algeo, T.J., Horacek, M., Brandner, R., Foland, K., et al., 2014, $^{87}Sr/^{86}Sr$ stratigraphy from the Early Triassic of Zal, Iran: linking temperature to weathering rates and the tempo of ecosystem recovery. *Geology*, **42**: 779−782.

Sephton, M.A., Looy, C.V., Veefkind, R.J., Brinkhuis, H., De Leeuw, J. W., and Visscher, H., 2002a, Synchronous record of $\partial^{13}C$ shifts in the oceans and atmosphere at the end of the Permian. In: Koeberl, C., MacLeod, K.G. (eds), Catastrophic Events and mass extinctions: Impacts and Beyond, *Geological Society of America Special Paper*, **356**: 455−462.

Sephton, M.A., Amor, K., Franchi, I.A., Wignall, P.B., Newton, R., and Zonneveld, J.-P., 2002b, Carbon and nitrogen isotope disturbances and an end-Norian (Late Triassic) extinction event. *Geology*, **30**: 1119−1122.

Shen, C., 2018, Astronomical tuning, astronomical forcing and environmental conditions of the Lower Triassic Montney Formation, northeastern British Columbia, Canada. *Thesis*. University of Calgary, Graduate Program in Geology and Geophysics, 106 pp.

Shen, C., Shoepfer, S.D., and Henderson, C.M., 2017, Astronomical tuning of the Early Triassic Induan Stage in the distal Montney Formation, NE British Columbia, Canada. *Geological Society of America Abstracts With Programs*, **(6)**: 49. https://doi.org/10.1130/abs/2017AM-300972.

Shen, S.Z., Henderson, C.M., Bowring, S.A., Cau, C.Q., Wang, Y., Wang, W., et al., 2010, High-resolution Lopingian (Late Permian) timescale of South China. *Geological Journal*, **45**: 122−134.

Shen, S.Z., Ramezani, J., Chen, J., Cao, C.Q., Erwin, D.H., Zhang, H., et al., 2019, A sudden end-Permian mass extinction in South China. *Geological Society of America Bulletin*, **131**: 205−223. https://doi.org/10.1130/B31909.1.

Silberling, N.J., and Tozer, E.T., 1968, Biostratigraphic classification of the marine Triassic in North America. *Geological Society of America Special Papers*, **10**: 1−63.

Simms, M.J., and Ruffel, A.H., 1989, Synchroneity of climate change and extinctions in the Late Triassic. *Geology*, **17**: 265−268.

Simmons, M.D., Sharland, P.R., Casey, D.M., Davies, R.B., and Sutcliffe, O.E., 2007, Arabian Plate sequence stratigraphy: potential implications for global chronostratigraphy. *GeoArabia*, **12** (4), 101−130.

Simmons, M.S., Miller, K.G., Ray, D.C., Davies, A., van Buchem, F.S.P., and Gréselle, B., 2020, Chapter 13 - Phanerozoic eustasy. *In* Gradstein, F.M., Ogg, J.G., Schmitz, M.D., and Ogg, G.M. (eds), *The Geologic Time Scale 2020*. **Vol. 1** (this book). Elsevier, Boston, MA.

Skjold, L.J., Van Veen, P.M., Kristensen, S.-E., and Rasmussen, A.R., 1998, Triassic sequence stratigraphy of the southwestern Barents Sea. In: de Graciansky, P.-C., Hardenbol, J., Jacquin, T., Vail, P.R. (eds), Mesozoic-Cenozoic Sequence Stratigraphy of European Basins, *SEPM Special Publication*, **60**: 651−666.

Song, H.Y., Tong, J.N., Algeo, T.J., Horacek, M., Qiu, H.O., Song, H.J., et al., 2013, Large vertical delta C-13(DIC) gradients in Early Triassic seas of the South China craton: Implications for oceanographic changes related to Siberian Traps volcanism. *Global and Planetary Change*, **105**: 7−20. https://doi.org/10.1016/j.gloplacha.2012.10.023.

Stefani, M., Furin, S., and Gianolla, P., 2010, The changing climate framework and depositional dynamics of Triassic carbonate platforms from the Dolomites. In: Kustatscher, E., Preto, N., Wignall, P. (eds.), Triassic climates, *Palaeogeography, Palaeoclimatology, Palaeoecology*, **290**: 43−57.

Steiner, M.B., 2006, The magnetic polarity time scale across the Permian−Triassic boundary. In: Lucas, S.G., Cassinis, G., and Schneider, J.W. (eds), Non-Marine Permian Biostratigraphy and Biochronology, *The Geological Society of London Special Publications*, **265**: 15−38.

Steiner, M.B., Ogg, J.G., Zhang, Z., and Sun, S., 1989, The Late Permian/ Early Triassic magnetic polarity time scale and plate motions of south China. *Journal of Geophysical Research*, **94**: 7343−7363.

Sugiyama, K., 1997, Triassic and Lower Jurassic radiolarian biostratigraphy in the siliceous claystone and bedded chert units of the southeastern Mino Terrane, Central Japan. *Bulletin of the Mizunami Fossil Museum*, **24**: 79−193.

Sun, Y.D., Joachimski, M.M., Wignall, P.B., Yan, C., Chen, Y., Jiang, H., et al., 2012, Lethally hot temperatures during the Early Triassic Greenhouse. *Science*, **338**: 366−370.

Sun, Z.M., Hounslow, M.W., Pei, J., Zhao, L., Tong, J.N., and Ogg, J.G., 2007, Magnetostratigraphy of the West Pingdingshan section, Chaohu, Anhui Province: relevance for base Olenekian GSSP selection. *Albertiana*, **36**: 22−32.

Szurlies, M., 2004, Magnetostratigraphy: the key to global correlation of the classic Germanic Trias − case study Volpriehausen Formation (Middle Buntsandstein), Central Germany. *Earth and Planetary Science Letters*, **227**: 395−410.

Szurlies, M., 2007, Latest Permian to Middle Triassic cyclo-magnetostratigraphy from the Central European Basin, Germany: implications for the geomagnetic polarity timescale. *Earth and Planetary Science Letters*, **261**: 602−619.

Tanner, L.H., 2010a, The Triassic isotope record. In: Lucas, S.G. (ed), The Triassic Timescale, *The Geological Society of London Special Publications*, **334**: 103−118.

Tanner, L.H., 2010b, Cyclostratigraphy record of the Triassic: a critical examination. In: Lucas, S.G. (ed), The Triassic Timescale, *The Geological Society of London Special Publications*, **334**: 119−137.

Tanner, L.H., 2018, *The Late Triassic World: Earth in a Time of Transition*. Topics in Geobiology, Springer Publications, Vol. 46: 805 pp. https://doi.org/10.1007/978-3-319-68009-5

Tong, J.N., Zakharov, Y.D., Orchard, M.J., Yin, H.F., and Hansen, H.J., 2004, Proposal of the Chaohu section as the GSSP candidate of the I/O boundary. *Albertiana*, **29**: 13−28.

Tong, J.N., Chu, D.L., Liang, L., Shu, W.C., Song, H.J., Song, T., et al., 2019, Triassic integrative stratigraphy and timescale of China. *Science China Earth Sciences*, **62**: 189−222. https://doi.org/10.1007/s11430-018-9278-0.

Tozer, E.T., 1965, *Lower Triassic stages and ammonoid zones of Arctic Canada*, Geological Survey of Canada Paper, 65-12. pp. 1−14.

Tozer, E.T., 1967, *A standard for Triassic Time*. Geological Survey of Canada Bulletin, **156**: 104 pp.

Tozer, E.T., 1984, The Trias and its Ammonites: the evolution of a time scale. *Geological Survey of Canada Miscellaneous Report*, **35**: 171 pp.

Tozer, E.T., 1986, Definition of the Permian−Triassic (P−T) boundary: the question of the age of the *Otoceras* beds. *Memorie della Società Geologica Italiana*, **34**: 291−301.

Tozer, E.T., 1990, How many Rhaetians? *Albertiana*, **8**: 10−13.

Tozer, E.T., 1994, Age and correlation of the *Otoceras* beds at the Permian−Triassic Boundary. *Albertiana*, **14**: 31−37.

Traverse, A., 2007, *Paleopalynology*, second ed. Dordrecht: Springer, 813 pp.

Trotter, J.A., Williams, I.S., Nicora, A., Mazza, M., and Rigo, M., 2015, Long-term cycles of Triassic climate change: a new $\delta^{18}O$ record from conodont apatite. *Earth and Planetary Science Letters*, **415**: 165–174.

Van Veen, P.M., 1995, Time calibration of Triassic/Jurassic microfloral turnover, eastern North America—Comment. *Tectonophysics*, **245**: 93–95.

Van Veen, P., Hochuli, P.A., and Vigran, J.O., 1998, Arctic spores/pollen. Column for Triassic chart of Mesozoic and Cenozoic sequence chronostratigraphic framework of European basins, by Hardenbol, J., Thierry, J., Farley, M.B., Jacquin, Th., de Graciansky, P.-C., and Vail, P.R. (coordinators). In: de Graciansky, P.-C., Hardenbol, J., Jacquin, Th., and Vail, P.R. (eds), Mesozoic-Cenozoic Sequence Stratigraphy of European Basins, *SEPM Special Publication*, **60**: Chart 8.

Veizer, J., and Prokoph, A., 2015, Temperatures and oxygen isotopic composition of Phanerozoic oceans. *Earth-Science Reviews*, **146**: 92–104.

Veizer, J., Ala, D., Azmy, K., Bruckschen, P., Buhl, D., Bruhn, F., et al., 1999, $^{87}Sr/^{86}Sr$, $\partial^{13}C$ and $\partial^{18}O$ evolution of Phanerozoic seawater. *Chemical Geology*, **161**: 59–88 Datasets available at: < http://www.science.uottawa.ca/geology/isotope_data/ >.

Veizer, J., Godderis, Y., and François, L.M., 2000, Evidence for decoupling of atmospheric CO_2 and global climate during the Phanerozoic eon. *Nature*, **408**: 698–701.

Visscher, H., 1992, The new STS Triassic stage nomenclature. *Albertiana*, **10**: 1–2.

Visscher, H., and Brügman, W.A., 1981, Ranges of selected palynomorphs in the Alpine Triassic of Europe. *Review of Palaeobotany and Palynology*, **34**: 115–128.

Visscher, H., Van Houte, M., Brugman, W.A., and Poort, R.J., 1994, Rejection of a Carnian (Late Triassic) 'pluvial event' in Europe. *Review of Palaeobotany and Palynology*, **83**: 217–226.

von Alberti, F.A., 1834, *Beitrag zu einer Monographie des Bunter Sandsteins, Muschelkalks und Keupers und die Verbindung dieser Gebilde zu einer Formation.* Stuttgart and Tübingen: Verlag der J.G. Cottaíshen Buchhandlung 326 pp.

von Gümbel, C.W., 1861, *Geognostische Beschreibung des bayerischen Alpengebirges und seines Vorlands.* Perthes: Gotha 950 pp.

von Salis, K., 1998, Calcareous nannofossils. Column for Triassic chart of Mesozoic and Cenozoic sequence chronostratigraphic framework of European basins, by Hardenbol, J., Thierry, J., Farley, M.B., Jacquin, Th., de Graciansky, P.-C., and Vail, P.R. (coordinators). In: de Graciansky, P.-C., Hardenbol, J., Jacquin, Th., Vail, P.R. (eds), Mesozoic-Cenozoic Sequence Stratigraphy of European Basins, *SEPM Special Publication*, **60**: Chart 8.

Vörös, A., Szabó, I., Kovács, S., Dosztály, L., and Budai, T., 1996, The Felsöörs section: a possible stratotype for the base of the Ladinian Stage. *Albertiana*, **17**: 25–40.

Vrielynck, B., 1998, Conodonts. Column for Triassic chart of Mesozoic and Cenozoic sequence chronostratigraphic framework of European basins, by Hardenbol, J., Thierry, J., Farley, M.B., Jacquin, Th., de Graciansky, P.-C., and Vail, P.R. (eds), Mesozoic-Cenozoic Sequence Stratigraphy of European Basins, *SEPM Special Publication*, **60**: Chart 8.

Vuks, V.J., 2000, Triassic foraminifers of the Crimea, Caucasus, Mangyshlak and Pamirs (biostratigraphy and correlation). *Zentralblatt für Geologie und Paläontologie, I*, **11-12**: 1353–1365.

Vuks, V.J., 2007, New data on the Late Triassic (Norian-Rhaetian) foraminiferans of the western Pre-caucasus (Russia). *New Mexico Museum of Natural History and Science Bulletin*, **41**: 411–412.

Waagen, W., and Diener, C., 1895. Untere Trias. In: Mojsisovics, E., von Waagen, W., Diener, C. (eds), Entwurf einer Gliederung der pelagischen Sedimente des Trias-Systems, *Sitzungberichte Akademie Wissenschaften Wien*, **104**: 1271–1302.

Wallman, K., 2001, The geological carbon cycle and the evolution of marine $\partial^{18}O$ values. *Geochimica et Cosmochimica Acta*, **65**: 2469–2485.

Wang, X.D., Cawood, P.A., Zhao, H., Zhao, L.S., Grasby, S.E., Chen, Z. Q., et al., 2018, Mercury anomalies across the end Permian mass extinction in South China from shallow and deep water depositional environments. *Earth and Planetary Science Letters*, **496**: 159–167. https://doi.org/10.1016/j.epsl.2018.05.044.

Wang, X.D., Cawood, P.A., Zhao, H., Zhao, L.S., Grasby, S.E., Chen, Z. Q., and Zhang, L., 2019, Global mercury cycle during the end-Permian mass extinction and subsequent Early Triassic recovery. *Earth and Planetary Science Letters*, **513**: 144–155. https://doi.org/10.1016/j.epsl.2019.02.026.

Ward, P.D., Haggart, J.W., Carter, E.S., Wilbur, D., Tipper, H.W., and Evans, T., 2001, Sudden productivity collapse associated with the Triassic-Jurassic boundary mass extinction. *Science*, **292**: 1148–1151.

Warrington, G., 2002, Triassic spores and pollen. In: Jansonius, J., McGregor, D.C. (eds), Palynology: principles and applications. 2nd ed., *American Association of Stratigraphic Palynologists Foundation*, **2**: 755–766.

Weems, R.E., and S.G. Lucas. 2015. A revision of the Norian conchostracan zonation in North America and its implication for Late Triassic North American history. In: Sullivan, R.M., Lucas, S.G. (eds), Fossil Record 4, *New Mexico Museum of Natural History and Science Bulletin*, **67**: 303–318.

Whiteside, J.H., Olsen, P.E., Kent, D.V., Fowell, S.J., and Et-Touhami, E., 2007, Synchrony between the Central Atlantic magmatic province and the Triassic–Jurassic mass-extinction event? *Palaeogeography, Palaeoclimatology, Palaeoecology*, **244**: 345–367.

Wignall, P.B., 2001, Large igneous provinces and mass extinctions. *Earth-Science Reviews*, **53**: 1–33.

Wignall, P.B., 2007, The End-Permian mass extinctions – how bad did it get? *Geobiology*, **5**: 303–309.

Wignall, P., 2015, *The Worst of Times: How Life on Earth Survived Eighty Million Years of Extinctions.* Princeton University Press 224 pp.

Wignall, P.B., and Hallam, A., 1992, Anoxia as a cause of the Permo-Triassic mass extinction: facies evidence from northern Italy and the western United States. *Palaeogeography, Palaeoclimatology, Palaeoecology*, **93**: 21–46.

Wing, S.L., and Sues, H.-D., 1992, Mesozoic and early Cenozoic terrestrial ecosystems. *In* Behrensmeyer, A.K., Damuth, J.D., DiMichele, W.A., Potts, R., Sues, H.-D., and Wing, S.L. (eds), *Terrestrial Ecosystems through Time: Evolutionary Paleoecology of Terrestrial Plants and Animals.* Chicago, IL: University of Chicago Press, 324–416.

Wotzlaw, J.-F., Guex, J., Bartolini, A., Gallet, Y., Krystyn, L., McRoberts, C.A., et al., 2014, Towards accurate numerical calibration of the Late Triassic: high-precision U-Pb geochronology constraints on the duration of the Rhaetian. *Geology*, **42**: 571–574.

Wotzlaw, J.-F., Brack, P., and Storck, J.-C., 2018, High-resolution stratigraphy and zircon U–Pb geochronology of the Middle Triassic Buchenstein Formation (Dolomites, northern Italy): precession-forcing of hemipelagic carbonate sedimentation and calibration of the Anisian-Ladinian boundary interval. *Journal of the Geological Society*, **175**: 71–85. https://doi.org/10.1144/jgs2017-052.

Xie, S.C., Pancost, R.D., Wang, Y.B., Yang, H., Wignall, P.B., Luo, G.M., et al., 2010, Cyanobacterial blooms tied to volcanism during the 5 m.y. Permo-Triassic biotic crisis. *Geology*, **38**: 447–450.

Xu, G., Hannah, J.L., Stein, H.J., Bingen, B., Yang, G., Zimmerman, A., et al., 2009, Re–Os geochronology of Arctic black shales to evaluate the Anisian-Ladinian boundary and global faunal correlations. *Earth and Planetary Science Letters*, **288**: 581–587.

Yin, H.F., Zhang, K.X., Tong, J.N., Yang, Z.Y., and Wu, S.B., 2001, The Global Stratotype Section and Point (GSSP) of the Permian-Triassic boundary. *Episodes*, **24**: 102–114.

Yin, H.F., Tong, J.N., and Zhang, K.X., 2005, A review of the global stratotype section and point of the Permian-Triassic boundary. *Acta Geologica Sinica*, **79**: 715–728.

Yin, H.F., Feng, Q., Xie, S., Yu, J., He, W., Liang, H., et al., 2007a, Recent achievements on the research of the Paleozoic–Mesozoic transitional period in South China. *Frontiers of Earth Science in China*, **1**: 129–141.

Yin, H.F., Yang, F.Q., Yu, J.X., Peng, Y.Q., Zhang, S.X., and Wang, S. Y., 2007b, An accurately delineated terrestrial Permian–Triassic Boundary and its implication. *Science in China, Series D: Earth Sciences*, **50**: 1281–1292.

Yu, J.X., Broutin, J., Chen, Z.Q., Shi, X., Li, H., Chu, D.L., and Huang, Q.S., 2015, Vegetation changeover across the Permian-Triassic Boundary in Southwest China. *Earth-Science Reviews*, **149**: 203–224. https://doi.org/10.1016/j.earscirev.2015.04.005.

Zaffani, M., Jadoul, F., and Rigo, M., 2018, A new Rhaetian $\delta^{13}C_{org}$ record: carbon cycle disturbances, volcanism, End-Triassic mass Extinction (ETE). *Earth-Science Reviews*, **178**: 92–104.

Zakharov, Y.D., 1994, Proposals on revision of the Siberian standard for the Lower Triassic and candidate stratotype section and point for the Induan–Olenekian boundary. *Albertiana*, **14**: 44–51.

Zakharov, Y.D., Shigata, Y., Popov, A.M., Sokarev, A.N., Buryi, G.I., Golozubov, V.V., et al., 2000, The candidates of global stratotype of the boundary of the Induan and Olenekian stages of the Lower Triassic in Southern Primorye. *Albertiana*, **24**: 12–26.

Zakharov, Y.D., Shigeta, Y., Popov, A.M., Buryi, G.E., Oleinikov, A.V., Dorukhovskaya, E.A., et al., 2002, Triassic ammonoid succession in South Primorye: 1. Lower Olenekian *Hedenstroemia bosphorensis* and *Anasibirites nevolini* Zones. *Albertiana*, **27**: 42–64.

Zakharov, Y.D., Shigeta, Y., and Igo, H., 2009, Correlation of the Induan-Olenekian boundary beds in the Tethys and Boreal realm: evidence from conodont and ammonoid fossils. *Albertiana*, **37**: 20–27.

Zapfe, H., 1974, Die Stratigraphie der Alpin-Mediterranen Trias. *Schriftenreihe der Erdwissenschaftlichen Kommissionen, Österreichische Akademie der Wissenschaften*, **2**: 137–144.

Zhang, L., Orchard, M.J., Brayard, A., Algeo, T.J., Zhao, L.S., Chen, Z.-Q., et al., 2019, The Smithian/Spathian boundary (late Early Triassic): a review of ammonoid, conodont, and carbon-isotopic criteria. *Earth-Science Reviews*, **195**: 7–36. https://doi.org/10.1016/j.earscirev.2019.02.014.

Zhang, Y., Li, M.S., Ogg, J.G., Montgomery, P., Huang, C., Chen, Z.Q., et al., 2015, Cycle-calibrated magnetostratigraphy of middle Carnian from South China: implications for Late Triassic time scale and termination of the Yangtze Platform. *Palaeogeography, Palaeoclimatology, Palaeoecology*, **436**: 135–166. http://dx.doi.org/10.1016/j.palaeo.2015.05.033.

Zhang, Y., Ogg, J.G., Franz, M., Bachmann, G.H., Szurlies, M., Röhling, H.-G., et al., 2020, Late Triassic magnetostratigraphy from the Germanic Basin and the global correlation of the Carnian Pluvial Episode. *Earth and Planetary Science Letters*, **541**, article #116275: 15 pp. https://doi.org/10.1016/j.epsl.2020.116275.

Zhao, L.S., Orchard, M.J., Tong, J.N., Sun, Z.M., Zuo, J., Zhang, S., et al., 2007, Lower Triassic conodont sequence in Chaohu, Anhui Province, China and its global correlation. *Palaeogeography, Palaeoclimatology, Palaeoecology*, **252**: 24–38. http://dx.doi.org/10.1016/j.palaeo.2006.11.032.

Zhao, L.S., Tong, J.N., Sun, Z.M., and Orchard, M.J., 2008, A detailed Lower Triassic conodont biostratigraphy and its implications for the GSSP candidate of the Induan-Olenekian boundary in Chaohu, Anhui Province. *Progress in Natural Scienc*, **18**: 79–90.

Zhao, L.S., Chen, Y.L., Chen, Z.Q., and Cao, L., 2013, Uppermost Permian to lower Triassic conodont zonation from Three George area, South China. *Palaios*, **28**: 509–522. http://dx.doi.org/10.2110/palo.2012.p12-107r.

Zhu, Z.C., Liu, Y.Q., Kuang, H.W., Benton, M.J., Newell, A.J., Xu, H., et al., 2019, Altered fluvial patterns in North China indicate rapid climate change linked to the Permian-Triassic mass extinction. *Science Reports*, **9**: 16818. https://doi.org/10.1038/s41598-019-53321-z.

Chapter 26

S.P. Hesselbo, J.G. Ogg and M. Ruhl
With contributions by L.A. Hinnov and C.J. Huang

The Jurassic Period

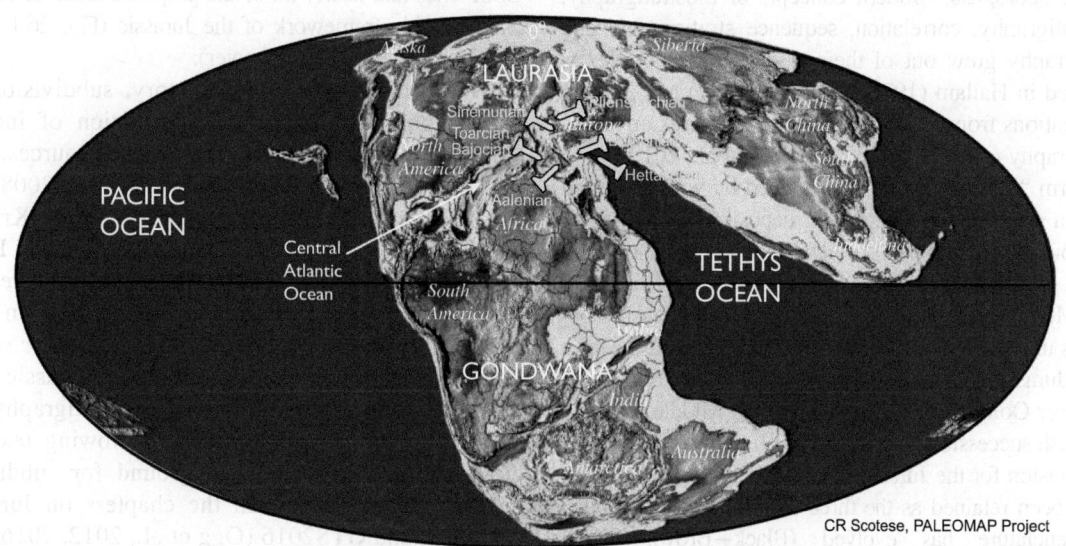

158.4 Ma Jurassic

CR Scotese, PALEOMAP Project

Chapter outline

26.1 **History and subdivisions**   956
  26.1.1 Overview of Jurassic   956
  26.1.2 Subdivisions of the Lower Jurassic   956
  26.1.3 Subdivisions of the Middle Jurassic   965
  26.1.4 Subdivisions of the Upper Jurassic   970
26.2 **Jurassic stratigraphy**   975
  26.2.1 Marine biostratigraphy   975
  26.2.2 Terrestrial biostratigraphy   981

26.2.3 Physical stratigraphy   982
26.3 **Jurassic time scale**   992
  26.3.1 Constraints from radioisotopic
     dates   992
  26.3.2 Jurassic age model   994
  26.3.3 Summary   1004
Acknowledgments   1005
Bibliography   1005

Abstract

Ammonites underwent an evolutionary diversification after the mass extinction of the end Triassic induced by the formation of a Large Igneous province (LIP), and this group provides the most useful marine biostratigraphy. Only two levels within the Jurassic are relatively well determined using U−Pb dating from single zircons in ash beds, at the base Hettangian and the Pliensbachian−Toarcian boundary. Otherwise the Lower Jurassic is scaled using astrochronology and the Middle and Upper Jurassic scaled from Pacific seafloor spreading rates correlated to magnetic reversals. LIP activity during the Early Jurassic (Triassic−Jurassic boundary and Toarcian) perturbed global environments to extents not evidenced since the end Permian, and age relationships allow for a strong causal connection between these LIP eruptions and mass extinctions caused by major paleoenvironmental change, including ocean anoxia. Breakup of the supercontinent Pangea dominated paleogeography and paleoceanography and created shallow seaways that form sources and traps for hydrocarbons.

Geologic Time Scale 2020. DOI: https://doi.org/10.1016/B978-0-12-824360-2.00026-7

Calcareous planktonic algae diversified and migrated from shallow seaways to open oceans to set the stage for the beginning of modern oceanic biogeochemical cycling; calcareous nannofossils provide additional widely used correlation tools.

26.1 History and subdivisions

26.1.1 Overview of Jurassic

The immense wealth of fossils, particularly the ammonites with their diversity of ornamented spirals, in the Jurassic strata of Britain and northwest Europe were a magnet for innovative geologists. Modern concepts of biostratigraphy, chronostratigraphy, correlation, sequence stratigraphy, and paleogeography grew out of their studies, for example, as summarized in Hallam (1975) and Page (2017) and in various publications from the International Congresses on Jurassic Stratigraphy (most recently Villaseñor et al., 2019).

The term "Jura Kalkstein" was applied by von Humboldt (1799) to a series of carbonate shelf deposits exposed in the mountainous Jura Region of northernmost Switzerland. Even though he recognized that these strata were distinct from the German Muschelkalk (Middle Triassic), he erroneously considered his unit to be older. Brongniart (1829) coined the term "Terrains Jurassiques" when correlating this "Jura Kalkstein" to the Lower Oolite Series (now assigned to Middle Jurassic) of the British succession. von Buch (1839) established a three-fold subdivision for the Jurassic. The basic framework of von Buch has been retained as the three Jurassic series, although the nomenclature has evolved (Black–Brown–White, Lias–Dogger–Malm, and currently Lower–Middle–Upper).

d'Orbigny (1842–1851, 1852) grouped the Jurassic ammonite and other fossil assemblages of France and England into 10 main divisions, which he termed "étages" (stages). Seven of d'Orbigny's stages are used today, but none of them has retained its original meaning. Simultaneously, Quenstedt (1848) subdivided each of the three Jurassic series of von Buch of the Swabian Alps of Southwestern Germany into six lithostratigraphic subdivisions characterized by ammonites and other fossils (Geyer and Gwinner, 1979). Quenstedt had denoted his subdivisions by Greek letters (*alpha–zeta*). Alfred Oppel, his pupil, was the first to successfully correlate Jurassic units among England, France, and southwestern Germany. Oppel (1856–1858) modified d'Orbigny's stage framework and further subdivided the Jurassic into biostratigraphic zones. His basic philosophy of defining the bases of Jurassic stages and substages according to ammonite zones in marginal-marine sections was retained at the Colloque du Jurassique à Luxembourg 1962 (Maubeuge, 1964; see also Morton, 1974) during which the majority of the current suite of 11 Jurassic stages were formalized.

Establishing the precise limits of those stages with Global Boundary Stratotype Section and Points (GSSPs) has been more difficult and is not yet complete. The traditional subdivision of the Jurassic was mainly based on shallow marine deposits of the northwest European basin (England to southwest Germany). Hiatuses or other complications commonly punctuate these "stratotypes" and their boundaries, and many of the ammonite taxa represent only certain paleoclimate realms in the Jurassic paleogeography. In particular, a pronounced provinciality during the Late Jurassic has inhibited efforts to assign GSSPs. Establishing reliable high-resolution correlations to tropical (Tethyan), Pacific, deep sea, terrestrial, and other settings have commonly remained tenuous. Nevertheless, these European basins contain all of the ratified GSSP sites and nearly all of the proposed ones for the chronostratigraphic framework of the Jurassic (Fig. 26.1 and also Fig. 26.11 at end of this chapter).

Detailed reviews of the history, subdivisions, biostratigraphic zonations, and correlation of individual Jurassic stages are compiled in several sources, including Arkell (1933, 1956), Morton (1974, 2008), Cope et al. (1980a,b), Harland et al. (1982, 1990), Krymholts et al. (1982), Burger (1995), and Groupe Français d'Étude du Jurassique (1997). An extensive online library of Jurassic stratigraphy articles with an emphasis on Eurasia is maintained at *www.jurassic.ru/epubl. htm*. The International Commission on Jurassic Stratigraphy website is https://jurassicdotstratigraphydotorg. wordpress.com/. Portions of the following text, especially on the historical background for subdivisions, have been modified from the chapters on Jurassic in GTS2012 and GTS2016 (Ogg et al., 2012, 2016).

26.1.2 Subdivisions of the Lower Jurassic

A marine transgression in the area of the Laurasian Seaway (modern European and Circum North Atlantic area) during the latest Triassic and Early Jurassic deposited widespread clay-rich calcareous deposits. These distinctive strata in southwest Germany were called the Black Jurassic (*schwarzer Jura*) by von Buch (1839) and named "Lias" in southern England by Conybeare and Phillips (1822). This series was subdivided into three stages (Sinemurian, Liasian, and Toarcian) by d'Orbigny (1842–1851, 1852). Oppel (1856–1858) replaced the Liasian with a Pliensbachian Stage, and Renevier (1864) separated the lower Sinemurian as a distinct Hettangian Stage. Widespread hiatuses or condensation horizons mark the bases of the classical Sinemurian, Pliensbachian, and Toarcian stages in their originating regions.

26.1.2.1 Triassic–Jurassic Boundary

The original Sinemurian Stage of d'Orbigny (1842–1851, 1852) extended to the base of the Jurassic. Indeed, the original extent of the Lower Jurassic tentatively included beds

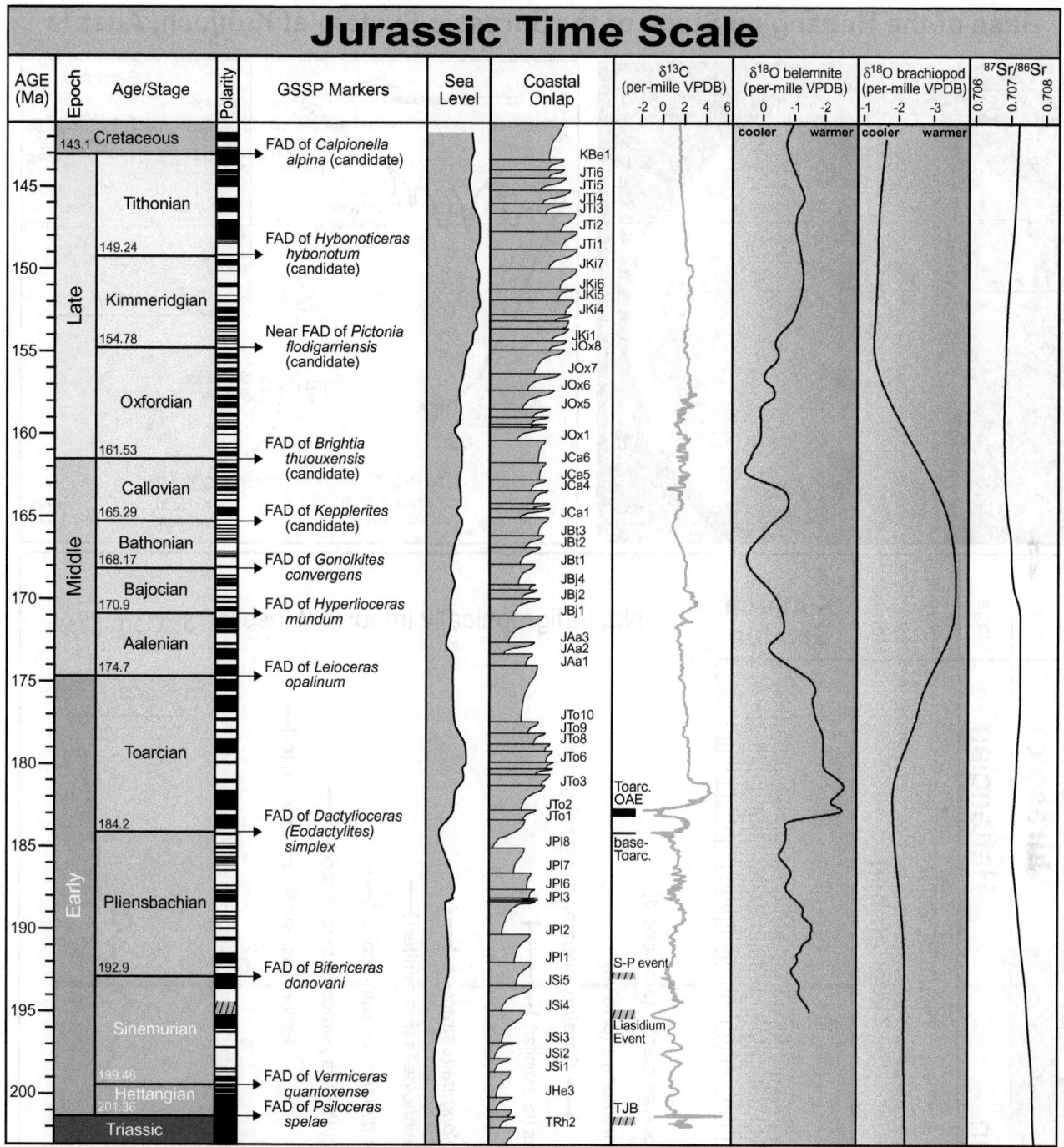

Jurassic Time Scale

FIGURE 26.1 Jurassic overview. Main markers or candidate markers for GSSPs of Jurassic stages are an FAD of ammonite taxa as detailed in the text. ("Age" is the term for the time equivalent of the rock-record "Stage.") Magnetic polarity scale is a composite from many sources and is enlarged in Fig. 26.10. Schematic sea-level, coastal onlap curves, and sequence nomenclature are modified from Haq (2017). See also review in Simmons et al. (2020, Ch. 13: Phanerozoic eustasy, this book). The carbon isotope curve with major excursions is a synthesis by Cramer and Jarvis (2020, Ch. 11: Carbon isotope stratigraphy, this book) from numerous studies. The oxygen isotope curve and implied sea-surface temperature trends are the statistical-mean curves by Grossman and Joachimski (2020, Ch. 10: Oxygen isotope stratigraphy, this book) from a synthesis of numerous belemnite and brachiopod carbonate studies. Strontium isotope curve is modified from McArthur et al. (2020b, Ch. 7: Strontium isotope stratigraphy, this book). The vertical scale of this diagram is standardized to match the vertical scales of the period-overview figure in all other Phanerozoic chapters. The vertical scale of this diagram is standardized to match the vertical scales of the first stratigraphic summary figure in all other Phanerozoic chapters. *FAD,* First appearance datum; *GSSP,* Global boundary Stratotype Section and Point; *S—P Event,* Sinemurian—Pliensbachian boundary event; *TJB,* Triassic—Jurassic boundary event; *T-OAE,* Toarcian Oceanic Anoxic Event; *VPDB,* Vienna PeeDee Belemnite $\delta^{13}C$ and $\delta^{18}O$ standard; *VSMOW,* Vienna Standard Mean Ocean Water $\delta^{18}O$ standard. The vertical scale of this diagram is standardized to match the vertical scales of the first stratigraphic summary figure in all other Phanerozoic chapters.

Base of the Hettangian Stage of the Jurassic System at Kuhjoch, Austria

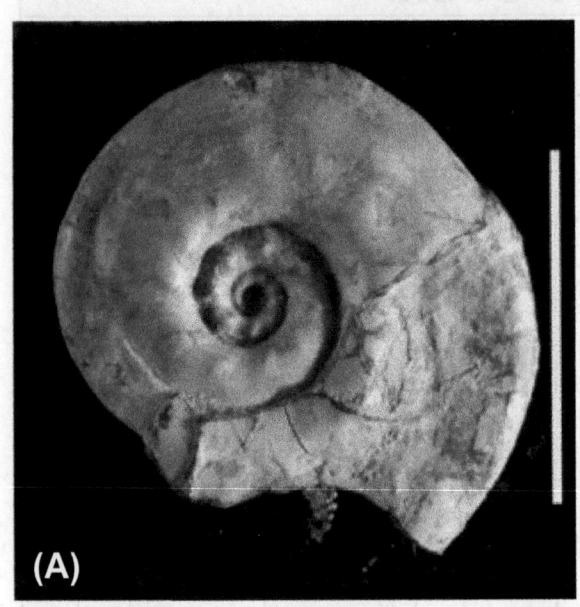

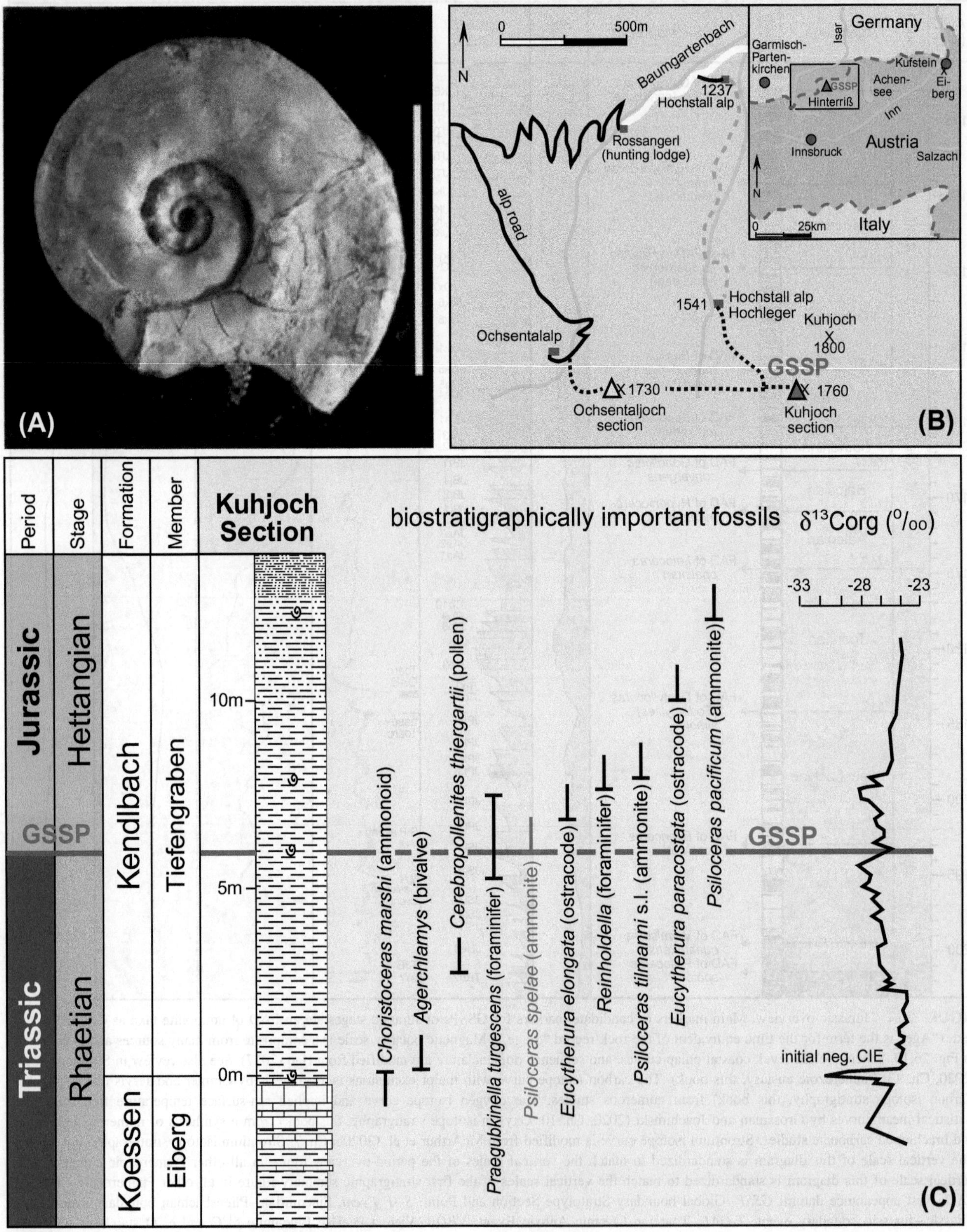

(A)

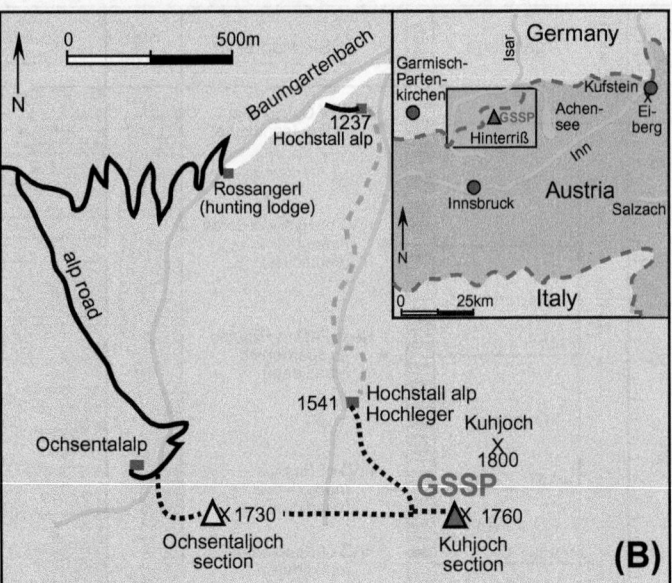

(B)

| Period | Stage | Formation | Member | Kuhjoch Section | biostratigraphically important fossils | δ¹³Corg (⁰/oo) |

that are now assigned to the Rhaetian of the uppermost Triassic (Bonebed of southwest Germany, portions of Penarth Beds in England, Rhätische Gruppe of German and Austrian Alps, etc.). Oppel (1856−1858) redefined the base of the Jurassic as the lowest ammonite assemblage characterized by *Psiloceras planorbis* species that overlay the Bonebed. He referred to characteristic coastal sections in southern England, including Lyme Regis in Dorset and Watchet in Somerset.

The end-Triassic mass extinction terminated many groups of marine life, including the conodonts, whose distinctive phosphatic jaw elements constitute a primary zonation for much of the Paleozoic and Triassic, and the majority of ammonoids. Indeed, in the few regions with continuous deposition, there is an interval devoid of either typical latest Triassic taxa (e.g., conodonts or *Choristoceras* ammonoids) or earliest Jurassic forms (e.g., *Psiloceras* ammonites). In successions within the Laurasian Seaway, the stratigraphic record is commonly characterized by hiatuses, due largely to base-level changes that took place during the transition from wlacustrine to marine environments across the Triassic−Jurassic boundary, with superimposed more rapid sea-level fluctuations, in a series of semi-isolated sedimentary basins (e.g., Hesselbo and Jenkyns, 1998; Hallam and Wignall, 1999). Ammonoid diversity was already very low in late Rhaetian time (*Choristoceras marshi* Zone), and the Hettangian genus *Psiloceras* must be derived from the Triassic genera of the family Discophyllitidae, which lived mainly in the open sea (Hillebrandt, 1997). The informal base of the Jurassic (Hettangian Stage) in southern England was the initial influx and preservation of these *Psiloceras* ammonites (*Psiloceras planorbis* species) during the early stages of the transgression; but this local phenomenon was deemed unsuitable for an international definition. Species within the genus *Psiloceras* are ubiquitous from the eastern Pacific and Tethys to the European Boreal province; therefore one option was to utilize the earliest forms for a base-Jurassic definition. However, the earliest form, *Psiloceras spelae*, first occurs during an inferred global sea-level low point that caused an extended gap in shallow seas; therefore this taxon is only found in rare complete sections (e.g., Peru, Alps, and Nevada) (Bloos, 2008; Hillebrandt and Krystyn, 2009).

Other nonammonoid possibilities that were considered for defining a precise base-Jurassic level that could be globally correlated included a major turnover of siliceous radiolarian microfossils or a negative excursion in carbon isotopes. After a major international effort to correlate environmental and biostratigraphic events associated with the end-triassic extinctions and the extensive eruption of the Central Atlantic Magmatic Province (CAMP) at ∼201 Ma (e.g., review by Hesselbo et al., 2007a), it was decided to place the base of the Jurassic at the initial stages of biological recovery from the end-Triassic extinction as marked by the earliest forms of *Psiloceras* ammonites.

Detailed paleomagnetic and bio- and chemostratigraphic studies (Deenen et al., 2011, 2010; Whiteside et al., 2010; Dal Corso et al., 2014), complemented by high-precision U−Pb radioisotopic dating, demonstrated a coincidence between the onset of the CAMP eruptions and the end-Triassic mass extinctions, and a delay of only about 100 kyr until the recovery at the beginning of the Jurassic (e.g., Schoene et al., 2010; Blackburn et al., 2013; Davies et al., 2017).

26.1.2.2 Hettangian

Renevier (1864) proposed the Hettangian Stage to encompass the *Psiloceras planorbis* and *Schlotheimia angulatus* ammonite zones as interpreted by Oppel. The stage was named after a quarry near Hettange-Grande village in Lorraine (northeastern France), 22 km south of Luxembourg, although the strata in this locality are primarily sandstone with no fossils in the lower part.

The GSSP for the base of the Hettangian Stage was ratified in 2010 as the Kuhjoch section within the Northern Calcareous Alps of Austria (Fig. 26.2) (Hillebrandt et al., 2013). The GSSP level, 5.80 m above the base of the Tiefengraben Member of the Kendlbach Formation, corresponds to the local lowest occurrence of the ammonite subspecies *Psiloceras spelae tirolicum* (Hillebrandt and Krystyn, 2009). Other markers include the lowest occurrences of the widely distributed continental palynomorph *Cerebropollenites thiergartii* (Kuerschner et al., 2007), of the aragonitic foraminifer *Praegubkinella turgescens* and of the ostracod *Cytherelloidea buisensis* (Hillebrandt et al., 2008). The $\delta^{13}C_{org}$ record shows an "initial" negative excursion near the boundary, at the transition from the underlying Kössen to the Kendlbach formations, and a shift to more positive $\delta^{13}C_{org}$ at the GSSP level; and this carbon isotope signature provides a primary method for high-resolution correlation to other sections (e.g., Ruhl et al., 2009, 2011; Deenen et al., 2010). This Austrian section has some problems, including (1) a fairly limited exposed stratigraphic extent, (2) sparse ammonite species resulting in ranges being difficult to delimit, (3) a lack of magnetostratigraphy zones (magnetostratigraphic studies were attempted, but without

FIGURE 26.2 GSSP for the base of Jurassic (base of Hettangian Stage) at Kuhjoch section, Northern Calcareous Alps, Austria. The GSSP level coincides with the lowest occurrence of ammonite *Psiloceras spelae* (new subsp. *tirolicum* Hillebrandt and Krystyn, 2009); marking the beginning of a new marine ecosystem after the major end-Triassic extinctions. The brief negative excursion in carbon isotopes is approximately simultaneous with those end-Triassic extinctions and main eruption phase of the Central Atlantic Magmatic province. *GSSP*, Global Boundary Stratotype Section and Point. Photograph provided by Axel von Hillebrandt.

Base of the Sinemurian Stage of the Jurassic System at East Quantoxhead, West Somerset, SW England.

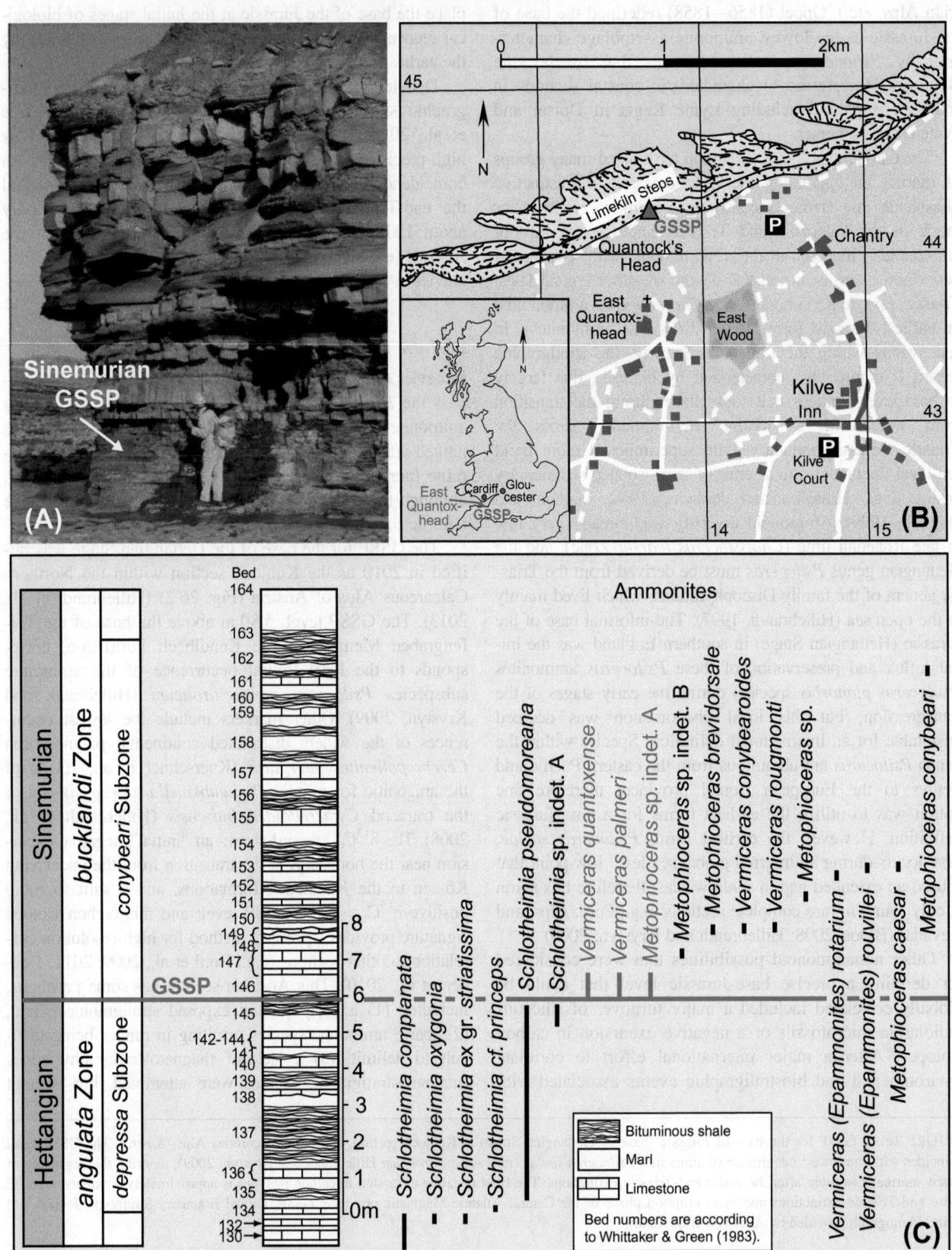

success), (4) no identified cyclic sedimentation to provide estimates of elapsed durations among events, and (5) a possible tectonic break (thrust fault) about 3.8 m below the GSSP that may have truncated part of the palynological and $\delta^{13}C_{org}$ record in the top of the Schattwald beds of the Tiefengraben Member (e.g., Bonis et al., 2009; Ruhl et al., 2009; Warrington, 2010; Palotai et al., 2017).

Palynological and $\delta^{13}C_{org}$ data from exposures at both sides of Kuhjoch ("kuh" = cow, "joch" = col; thus "Kuhjoch" means "cow-col," or "col of cows") have subsequently shown that the "TPo" palynological zone was indeed missing in the original Kuhjoch West section (Bonis et al., 2009) and that this level is also marked by an abrupt 1‰−2‰ negative shift in $\delta^{13}C_{org}$ (Ruhl et al., 2009). Work on the "Kuhjoch East" section, about 10 m east of "Kuhjoch West," showed the TPo Zone to be present and to be marked by a gradual 1‰−2‰ negative shift in $\delta^{13}C_{org}$. A composite palynological framework and $\delta^{13}C_{org}$ record for Kuhjoch, combining data from Kuhjoch East and West, are published in Hillebrandt et al. (2013) and Korte et al. (2019), which form the basis for Fig. 26.2. The TPo Zone in the top of the Schattwald and time-equivalent beds of the Tiefengraben Member (Kendlbach Fm.) appear to be missing from many of the studied Triassic−Jurassic boundary successions in the Eiberg Basin, possibly because this stratigraphic interval in the Tiefengraben Member is most easily sheared, compared to the under- and overlying carbonate rocks of the Eiberg Mbr. (Kössen Fm.) and Breitenberg Mbr. (top Kendlbach Fm.) during alpine uplift and deformation.

The assignment of the base-Jurassic GSSP is only slightly older than the long-standing informal base of the Hettangian Stage placed at the initial entry of *P. planorbis* ammonites into southern England. Cyclostratigraphy of the main Hettangian reference sections in Britain showed that this downward extension, as measured relative to carbon isotope signatures near the Triassic−Jurassic boundary, is in the order of 100 kyr (Ruhl et al., 2010; Xu et al., 2017a). This paleomagnetically and biostratigraphically constrained cyclostratigraphy indicates that the Hettangian spans about 2 Myr in the Somerset (United Kingdom) sections; a duration consistent with radioisotopic dating constraints from Peru (Schaltegger et al., 2008; Schoene et al., 2010; Wotzlaw et al., 2014), and astrochronology and magnetostratigraphy from the Hartford Basin, United States (Kent et al., 2017). Later work, however, suggests that the Hettangian in the Somerset sections might span ∼ 3.2 Myr, with the duration of the pre-*Planorbis* (*tilmanni*) Zone being at least 520 kyr (Weedon et al., 2018, 2019). This important difference of interpretation is further discussed in Section 26.2.

There are no accepted groupings into substages of the three Hettangian ammonite zones (*Psiloceras planorbis*, *Alsatites liasicus*, and *Schlotheimia angulata*).

26.1.2.3 Sinemurian

26.1.2.3.1 History, definition, and boundary stratotype

The Sinemurian Stage was named by d'Orbigny (1842−1851, 1852) after the town of Semur-en-Auxois (*Sinemurum Briennense castrum* in Latin) in the Cote d'Or department in the Burgundy region of east-central France. After the establishment of the Hettangian Stage incorporated its former lower ammonite zones (Renevier, 1864), the base of the revised Sinemurian was assigned to the proliferation of the Arietitidae ammonite group, particularly the lowest occurrence of the early genera *Vermiceras* and *Metophioceras* (base of *Metophioceras conybeari* Subzone of the *Arietites bucklandi* Zone). However, that informal stage boundary was not defined by a generally accepted species or assemblage (Sinemurian Boundary Working Group, 2000), and a stratigraphic gap exists between the Hettangian and Sinemurian throughout most of northwest Europe.

Sedimentation was continuous across the boundary interval in rapidly subsiding troughs in western Britain. Therefore the boundary GSSP was placed in interbedded limestone and claystone at a coastal exposure near East Quantoxhead, Kilve, Somerset, England (Page et al., 2000; Sinemurian Boundary Working Group, 2000; Bloos and Page, 2002) (Fig. 26.3). The GSSP at 0.9 m above the base of Bed 145 coincides with the lowest occurrence of arietitid ammonite genera *Vermiceras* and *Metophioceras*. This level is near the highest occurrence of the ammonite genus *Schlotheimia* that is characteristic of the uppermost Hettangian. This turnover of ammonite genera is a global event for correlating the boundary interval (Bloos and Page, 2002).

26.1.2.3.2 Upper Sinemurian substage

The Sinemurian Stage has two substages. The Colloque du Jurassique à Luxembourg 1962 (Maubeuge, 1964) assigned the base of an upper stage, called Lotharingian, named by Haug (1910) after the Lorraine region of France, to the base of the *Caenisites turneri* ammonite zone. However, current usage follows Oppel (1856−1858) in assigning the base of the Lotharingian Substage as the base of the overlying *Asteroceras obtusum* Zone (e.g., Krymholts et al., 1982; Groupe Français d'Étude du Jurassique, 1997). The lower substage does not have a secondary name.

FIGURE 26.3 GSSP for the base of the Sinemurian Stage at East Quantoxhead section, Somerset, southwest England. The GSSP level coincides with the lowest occurrence of *Vermiceras quantoxense* ammonite (base of *Metophioceras conybeari* Subzone of the *Arietites bucklandi* Zone). *GSSP*, Global Boundary Stratotype Section and Point. Photograph provided by Kevin Page.

Base of the Pliensbachian Stage of the Jurassic System at Wine Haven, Yorkshire, UK.

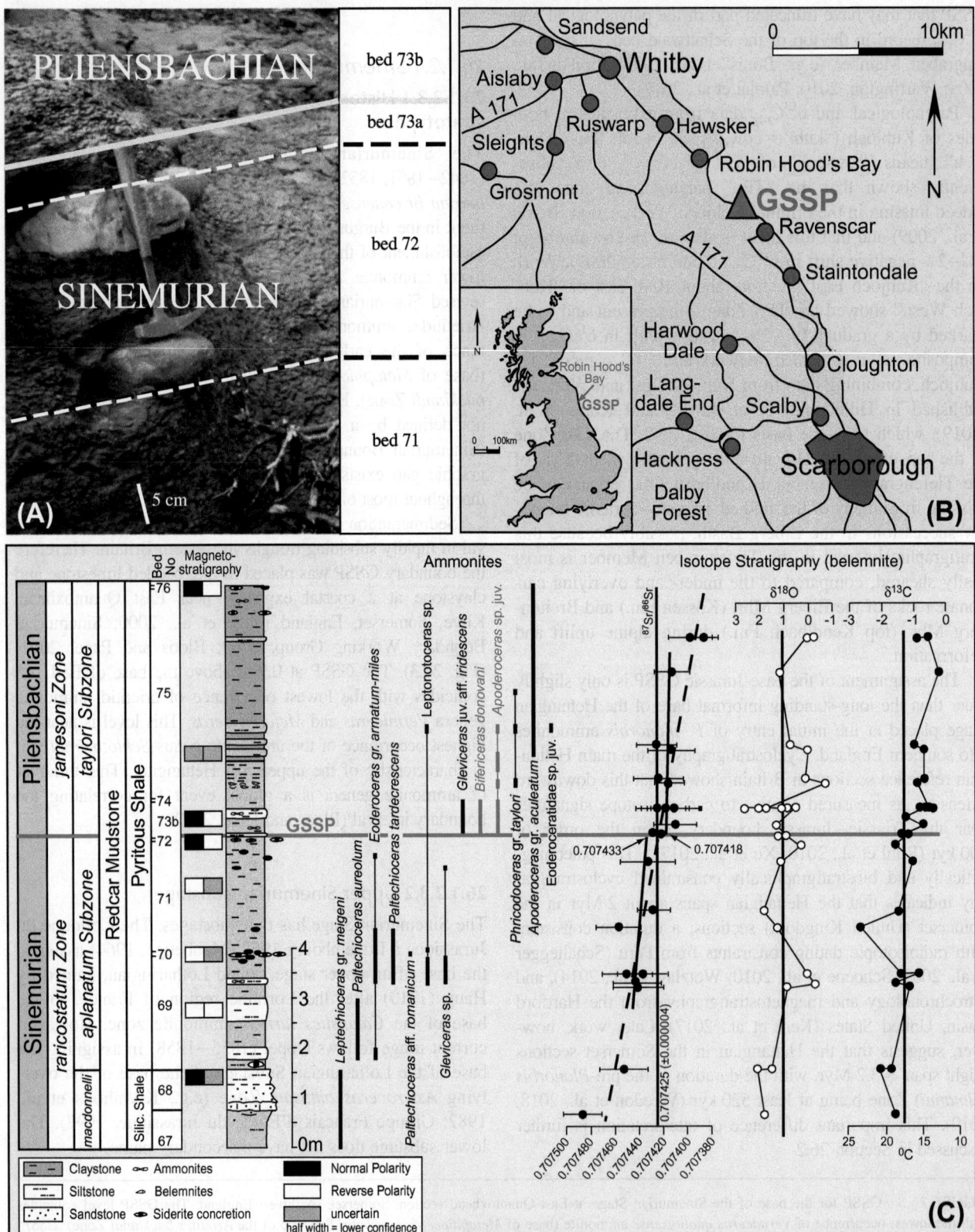

FIGURE 26.4   GSSP for the base of the Pliensbachian Stage at Wine Haven section, Yorkshire, east England. The GSSP level coincides with the lowest occurrence of ammonite *Bifericeras donovani* (base of *Uptonia jamesoni* Zone). *GSSP*, Global Boundary Stratotype Section and Point. Photograph provided by Kevin Page.

26.1.2.4 Pliensbachian

26.1.2.4.1 History, definition, and boundary stratotype

The Pliensbachian Stage was proposed by Oppel (1856–1858) to replace the Liasian stage of d'Orbigny, which lacked a geographic base. The stage was named after the outcrops along the Pliensbach stream near the village of Pliensbach (Geppingen, 35 km southeast of Stuttgart) in the Baden-Württemberg district of Germany. Even though this section has an unconformity to the underlying Sinemurian, the lowest ammonite subzone in this section (*Phricodoceras taylori* Subzone of the *Uptonia jamesoni* Zone) was used as the base of the Pliensbachian Stage (e.g., Dean et al., 1961).

At this level the Psiloceratoidea ammonites that dominated the lower stages of Hettangian and Sinemurian disappear, and the Eoderoceratoidea superfamily diversifies to dominate the NE European fauna of the Pliensbachian Stage (Meister et al., 2003). This faunal event occurs globally. However, a sedimentary or paleontological gap between the Pliensbachian and Sinemurian sequences is a common feature. Of 27 regions considered by the Pliensbachian boundary working group, only a single candidate in north Yorkshire, England, was satisfactory for a potential GSSP (Meister et al., 2003, 2006).

At the coastal section of Wine Haven at Robin Hood's Bay, Yorkshire, the GSSP in the Pyritous Shales Mbr. of the Redcar Fm. is within a claystone bed, 6 cm above a nodule layer (Fig. 26.4). This level coincides with the lowest ammonite occurrences of *Bifericeras donovani* species and *Apoderoceras* genus (Meister et al., 2006). Secondary markers include a brief reversed polarity magnetozone within the uppermost Sinemurian and a broadly defined negative carbon isotope excursion peaking in the lowermost Pliensbachian (Meister et al., 2006; Korte and Hesselbo, 2011; Armendáriz et al., 2012; Franceschi et al., 2014; Silva and Duarte, 2015; Gómez et al., 2016; Price et al., 2016a; Peti et al., 2017; Danisch et al., 2019).

26.1.2.4.2 Upper Pliensbachian substage

The Pliensbachian Stage has two substages. The lower substage of Carixian was named by Lang (1913) after Carixia, the Latin name for Charmouth in England. The upper substage of Domerian was named by Bonarelli (1894, 1895) after the type section in the Modelo formation at Monte Domaro in the Lombardian Alps of Northern Italy.

The Colloque du Jurassique à Luxembourg 1962 (Maubeuge, 1964) assigned the boundary between the Carixian and Domerian substages to the base of the *Amaltheus margaritatus* ammonite zone, at the appearance of the *Amaltheus* genus (typically *Amaltheus stokesi*).

26.1.2.5 Toarcian

26.1.2.5.1 History, definition, and boundary stratotype

The Toarcian Stage was defined by d'Orbigny (1842–1851, 1852) at the Vrines quarry, 2 km northwest of the village of Thouars (Toarcium in Latin) in the Deux-Sèvres region of west-central France. The thin-bedded succession of blue–gray marl and clayey limestone spans the entire Toarcian with 27 ammonite horizons grouped into 8 ammonite zones (Gabilly, 1976).

The Pliensbachian–Toarcian boundary interval is the first phase of a major extinction event in western Europe among rhynchonellid brachiopods, ostracod faunas, benthic foraminifera, and bivalves and marks a turnover in ammonites and belemnites. The base of the Toarcian is marked by a surge of Dactylioceratide *Dactylioceras* (*Eodactylites*) ammonites and extinction of the boreal amaltheid family. The base-Toarcian extinction event has been shown to be a global phenomenon (Caruthers et al., 2013). This episode is followed approximately one ammonite zone higher by a second extinction phase that coincides with a widespread anoxic event and a pronounced extinction among benthic invertebrates, such as bivalves and brachiopods (e.g., review in Wignall and Bond, 2008). Seawater strontium isotope ratios, which had been declining since the Hettangian, reach a minimum at the Pliensbachian–Toarcian boundary and start to increase throughout the early Toarcian (Jones et al., 1994a; Rocha et al., 2016; McArthur et al., 2020a).

At Thouars and throughout northwest Europe, the base of the Toarcian strata is an important flooding surface above a major sequence boundary and is associated with condensation or gaps in the *Dactylioceras tenuicostatum* ammonite zone. This widespread tendency to hiatus necessitated selection of candidate GSSPs in the Mediterranean region where stratigraphic gaps are less pronounced (Rocha et al., 2016). The selected primary marker of the Toarcian GSSP was the lowest occurrence of a diversified *Dactylioceras* (*Eodactylites*) ammonite fauna [(*Eodactylites*) *simplex* horizon, sensu Goy et al., 1996].

The GSSP is at Ponto do Trovao, Peniche, western Portugal, at the base of Bed 15e (Rocha et al., 2016) (Fig. 26.5). The GSSP level coincides with the lowest occurrence of the ammonite *Dactylioceras* (*Eodactylioceras*) *simplex*. This Tethyan *Dactylioceras* (*Eodactylioceras*) fauna interval correlates with the NW European *Protogrammoceras paltus* horizon and is succeeded by an "English" *Orthodactylites* Succession (Rocha et al., 2016). The lowest occurrence of the nannofossil *Zeugrhabdotus erectus* closely follows the boundary and is an important biostratigraphic marker (Ferreira et al., 2019). Strontium isotope analyses from the GSSP yield a $^{87}Sr/^{86}Sr$ value of 0.707073 ± 0.000002 (McArthur et al., 2020a) and the carbon isotope stratigraphy

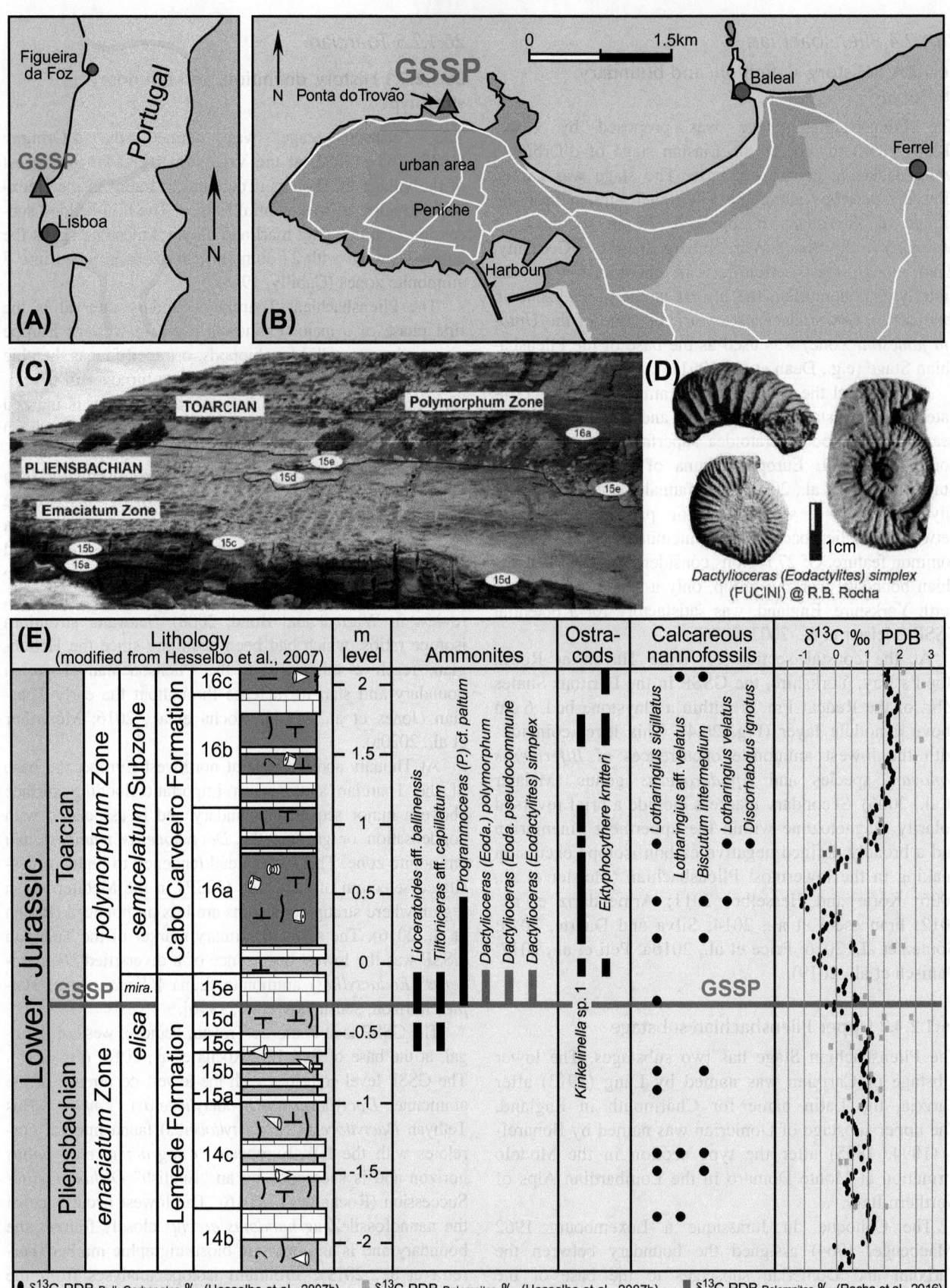

Base of the Toarcian Stage of the Jurassic System at Ponta do Trovao, Peniche, Portugal

FIGURE 26.5 GSSP for the base of the Toarcian Stage at Ponta do Trovão, Peniche, Lusitanian Basin, Portugal. The GSSP level coincides with the lowest occurrence of the ammonite *Dactylioceras (Eodactylites) simplex*, which cooccurs with *Dactylioceras (Eodactylites) pseudocommune* and *Dactylioceras (Eodactylites) polymorphum*. GSSP, Global Boundary Stratotype Section and Point.

is also well known here (Hesselbo et al., 2007b). A complementary section spanning the uppermost Pliensbachian and lowermost Toarcian ammonite zones at Almonacid de la Cuba in the Iberian Range of central-eastern Spain has detailed magnetostratigraphy and isotope stratigraphy (carbon, oxygen, and strontium) (Comas-Rengifo et al., 2010a, b). A potential shortcoming of the GSSP is condensation around the base of the Toarcian, although lack of reworking of macrofossils has been taken to indicate that minimal missing section is there (Rocha et al., 2016).

26.1.2.5.2 Toarcian substages

There is no agreement on the number of substages of the Toarcian. A binary subdivision following the division by Oppel (1856–1858) places a substage boundary at the base of the *Haugia variabilis* ammonite zone, at the appearance of abundant Phymatoceratinae group of ammonites, particularly the *Haugia* genus (e.g., Krymholts et al., 1982; Burger, 1995). An alternative three-substage division (e.g., Groupe Français d'Étude du Jurassique, 1997) groups the *Haugia variabilis* and underlying *Hildoceras bifrons* zones into a Middle Toarcian and places the limit of an Upper Toarcian at the base of the *Grammoceras thouarsense* Zone.

26.1.3 Subdivisions of the Middle Jurassic

Black Lower Jurassic clays (*Schwarzer Jura*) in southwestern Germany are overlain by strata containing clayey sandstone and brown-weathering ferruginous oolite. This lithologic change to the Brown Jurassic (*Brauner Jura*) of von Buch (1839) has been retained as the base of the Middle Jurassic (base of Aalenian). The Middle Jurassic in southern England is characterized by shallow marine carbonates of the Lower Oolite group of Conybeare and Phillips (1822), which comprised an expanded "Bathonian" stage of d'Omalius d'Halloy (1843). The lower portion of d'Omalius d'Halloy's Bathonian was classified as a separate "Bajocian" stage by d'Orbigny (1842–1851, 1852). In turn, Mayer-Eymar (1864) further removed the lower portion of d'Orbigny's Bajocian into a distinct "Aalenian" Stage.

Oppel (1856–1858) assigned the upper limit of his Middle Jurassic series or "Dogger" at the base of the Kellaway Rock of England, or the base of the Callovian Stage of d'Orbigny (1842–1851, 1852). This Callovian stage was shifted into the Middle Jurassic series by the Colloque du Jurassique à Luxembourg 1962 (Maubeuge, 1964) as preferred by Arkell (1956).

The bases of the Aalenian through Bathonian stages have been marked by GSSPs in expanded sections of rhythmic alternations of limestone and marl. The placement of a base for the Callovian Stage has been hindered by a ubiquitous condensation or hiatuses in strata of northwest Europe and elsewhere.

26.1.3.1 Aalenian

26.1.3.1.1 History, definition, and boundary stratotype

The Aalenian Stage was proposed by Mayer-Eymar (1864) for the lowest part of the "*Brauner Jura*" in the vicinity of Aalen at the northeastern margin of the Swabian Alps (Baden-Württemberg, southwestern Germany) where iron ore was mined from the associated ferruginous oolite sandstones (Dietl and Etzold, 1977). His Aalenian stratotype truncated the Bajocian Stage of d'Orbigny (1842–1851, 1852) at the base of the *Sonninia sowerbyi* ammonite zone.

The biostratigraphic placement of the base of the Middle Jurassic was the evolution of the ammonite subfamilies Grammoceratinae and Leioceratinae, especially the first occurrence of species of the genus *Leioceras*, which evolved from *Pleydellia*. The Aalenian GSSP in the Fuentelsaz section in the Iberian Chain of Spain corresponds to this ammonite marker (Goy et al., 1994, 1996; Cresta et al., 2001). The base of marl Bed FZ107 in the Fuentelsaz section of alternating marl and limestone coincides with the first occurrence of *Leioceras opalinum* (base of *L. opalinum* Zone) (Fig. 26.6). The magnetostratigraphy of Fuentelsaz correlates to a composite magnetic pattern derived from other sections in Europe. A secondary reference section for basal Aalenian is at Wittnau near Freiburg in south Germany (Ohmert, 1996).

26.1.3.1.2 Middle and Upper Aalenian substages

The four ammonite zones of the Aalenian are grouped into three substages: the Lower Aalenian is the *Leioceras opalinum* Zone, the Middle Aalenian comprises the *Ludwigia murchisonae* and *Brasilia bradfordensis* zones, and the Upper Aalenian is the *Graphoceras concavum* Zone.

26.1.3.2 Bajocian

26.1.3.2.1 History, definition, and boundary stratotype

The Bajocian Stage was named by d'Orbigny (1842–1851, 1852) after the town of Bayeux in Normandy (Bajoce in Latin). The abandoned quarries are now overgrown, and the nearby coastal cliff section of Les Hachettes indicates that most of the early Bajocian is absent at a hiatus and erosional surface. Indeed, the late Bajocian is largely condensed into a 15-cm-thick layer (Rioult, 1964). Ammonite lists of d'Orbigny indicate that he erroneously assigned species of upper Toarcian to the lower Bajocian and vice versa. This confusion was one reason for Mayer-Eymar (1864) to distinguish a new Aalenian Stage between the Toarcian and Bajocian.

The Colloque du Jurassique à Luxembourg 1962 (Maubeuge, 1964) defined the Bajocian Stage to begin at the base of the *Sonninia sowerbyi* ammonite zone and extend to

Base of the Aalenian Stage of the Jurassic System at Fuentelsaz, Spain

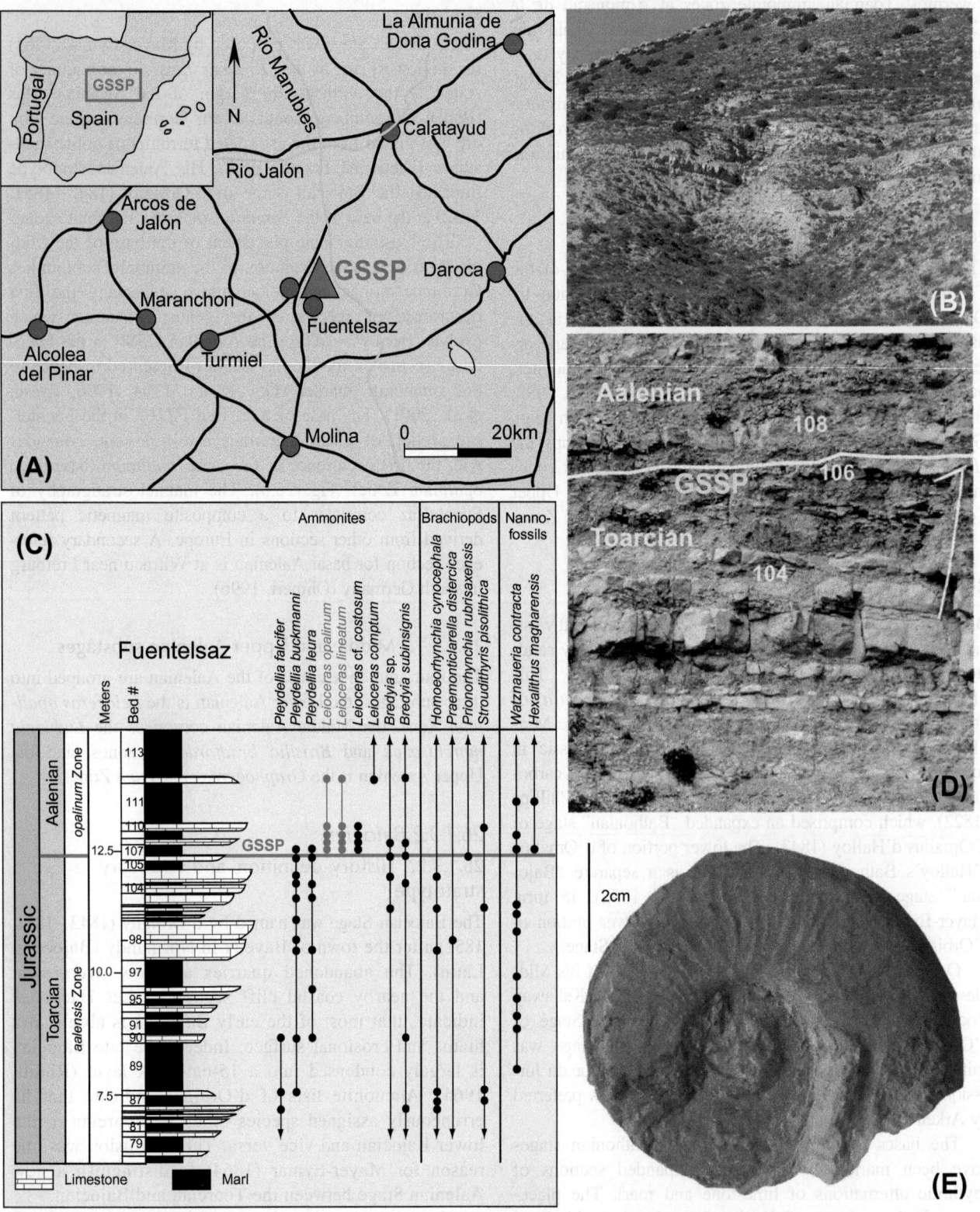

FIGURE 26.6 GSSP for the base of the Aalenian Stage at Fuentelsaz Section, Castilian Branch of the Iberian Range, Spain. The GSSP level coincides with the lowest occurrence of ammonite *Leioceras opalinum* (base of *L. opalinum* Zone). *GSSP*, Global Boundary Stratotype Section and Point. Photos provided by Maria Soledad Ureta.

the top of the *Parkinsonia parkinsoni* Zone. However, the holotype of the *Sonninia sowerbyi* index species was later discovered to be a nucleus of a large Sonniniidae (*Papilliceras*) from the overlying *Otoites sauzei* Zone (Westermann and Riccardi, 1972). Therefore the basal ammonite zone of the Bajocian was redefined to be the *Hyperlioceras discites* Zone, with the zonal base marked by the lowest occurrence of the ammonite genus *Hyperlioceras* (*Toxolioceras*), which evolved from *Graphoceras* (both in ammonite family Graphoceratidae).

Two sections recorded this ammonite datum with supplementary biostratigraphic and magnetostratigraphic data—Murtinheira at Cabo Mondego in Portugal (GSSP selected at the first occurrence datum of *Hyperlioceras mundum* and related species *H. furcatum*, *Bratinsina aspera*, and *B. elegantula*) and Bearreraig Bay on the Isle of Skye in Scotland (selected as an Auxiliary Stratotype Point) (Pavia and Enay, 1997). The GSSP of Cabo Mondego is at the base of Bed AB11 in a rhythmic alternation of gray limestone and marl (Henriques, 1992; Henriques et al., 1994; Pavia and Enay, 1997) (Fig. 26.7). The GSSP level is just below the base of a normal polarity magnetozone (Pavia and Enay, 1997).

26.1.3.2.2 Upper Bajocian substage

The base of the Upper Bajocian is the base of the *Strenoceras* (*Strenoceras*) *niortense* ammonite zone. In older literature the base was assigned as the base of a *Strenoceras subfurcatum* Zone, until it was recognized by Dietl (1981) that the holotype of the index species belongs to *Garantiana* and had originated from the overlying zone; therefore this *Stren. subfurcatum* zone became invalid. The base of the Upper Bajocian corresponds to a major turnover of ammonite genera.

26.1.3.3 Bathonian

26.1.3.3.1 History, definition, and boundary stratotype

The Bathonian Stage was named by Jean Julien d'Omalius d'Halloy (1843) after the town of Bath (Bathonium in Latin) in southwest England, where oolitic limestones are exposed in a number of quarries. But this stratotype succession is incomplete and lacks adequate characterization by ammonites (Torrens, 1965). The lower half of this original Bathonian was reclassified in the system of d'Orbigny (1842–1851, 1852) as a Bajocian Stage exposed in Normandy, but d'Orbigny did not specify a revised stratotype for the truncated Bathonian, nor provide an unambiguous lower boundary. Indeed, d'Orbigny's original description suggests that he had included the equivalent of the present "Lower Bathonian" substage within his Bajocian stratotype (Rioult, 1964). A century of confusion ended when the base of the Bathonian Stage was defined by the Colloque du Jurassique à Luxembourg 1962 (Maubeuge, 1964) as the base of the *Zigzagiceras zigzag* ammonite zone.

The basal Bathonian is well developed in cyclic alternations of marl and limestone in southeastern France. The Bathonian GSSP was placed at the base of limestone bed RB071 at Ravin du Bès, Bas Auran near Digne, Basses-Alpes, France (Innocenti et al., 1988; Fernández-López, 2007; Fernández-López et al., 2009) (Fig. 26.8). This level coincides with the lowest occurrence of ammonites *Gonolkites convergens* (base of *Zigzagiceras zigzag* Zone) and *Morphoceras parvum*. The GSSP level is just below the local lowest occurrence of calcareous nannofossil *Watznaueria barnesiae* (base of zone NJT11).

An auxiliary section for the Bathonian Stage at Cabo Mondego, near the Bajocian GSSP site, provides complementary data about the biochronostratigraphic subdivision and ammonite succession of the Sub-Mediterranean *M. parvum* Subzone and the Northwest European *Gon. convergens* Subzone. Neither section has yet yielded an unambiguous cyclostratigraphy or magnetostratigraphy.

26.1.3.3.2 Middle and Upper Bathonian substages

The Bathonian is generally divided into three substages, with the base of the Middle Bathonian placed at the base of the *Procentes progracilis* ammonite zone.

A divergence of ammonite assemblages in the upper Middle Bathonian has resulted in different basal definitions for an Upper Bathonian substage in each province. In the Sub-Mediterranean province (Tethyan province), a Middle/Upper Bathonian boundary is assigned to the base of the *Hecticoceras* (*Prohecticoceras*) *retrocostatum* Zone. In the northwest European province (Boreal province), a substage boundary is commonly assigned to the base of the *Procerites* (*Procerites*) *hodsoni* Zone, which is a significantly older level (Groupe Français d'Étude du Jurassique, 1997).

26.1.3.4 Callovian

26.1.3.4.1 History, definition, and boundary stratotype

The Callovian Stage was named by d'Orbigny (1842–1851, 1852) after Kellaways (various spellings in the geological literature) in Wiltshire, 3 km northeast of Chippenham, England, because he considered "*Calloviensis*" to be a derivative of the place name. The "Kelloways Stone" contains abundant cephalopods, including *Ammonites calloviensis* (called *Sigaloceras calloviensis* in current taxonomy). Oppel (1856–1858) placed the base of his "Kelloway gruppe" at the base of the *Macrocephalites macrocephalus* Zone, or at the lithologic base of the Upper Cornbrash (currently the upper part of the *Clydoniceras discus* Subzone of the uppermost Bathonian). At this level, ammonites of the genus *Macrocephalites* replace *Clydoniceras*, but the Upper Cornbrash is a condensed deposit "representing but a fraction of the time-intervals involved" (Cope et al., 1980b).

Base of the Bajocian Stage of the Jurassic System at Murtinheira at Cabo Mondego, Portugal

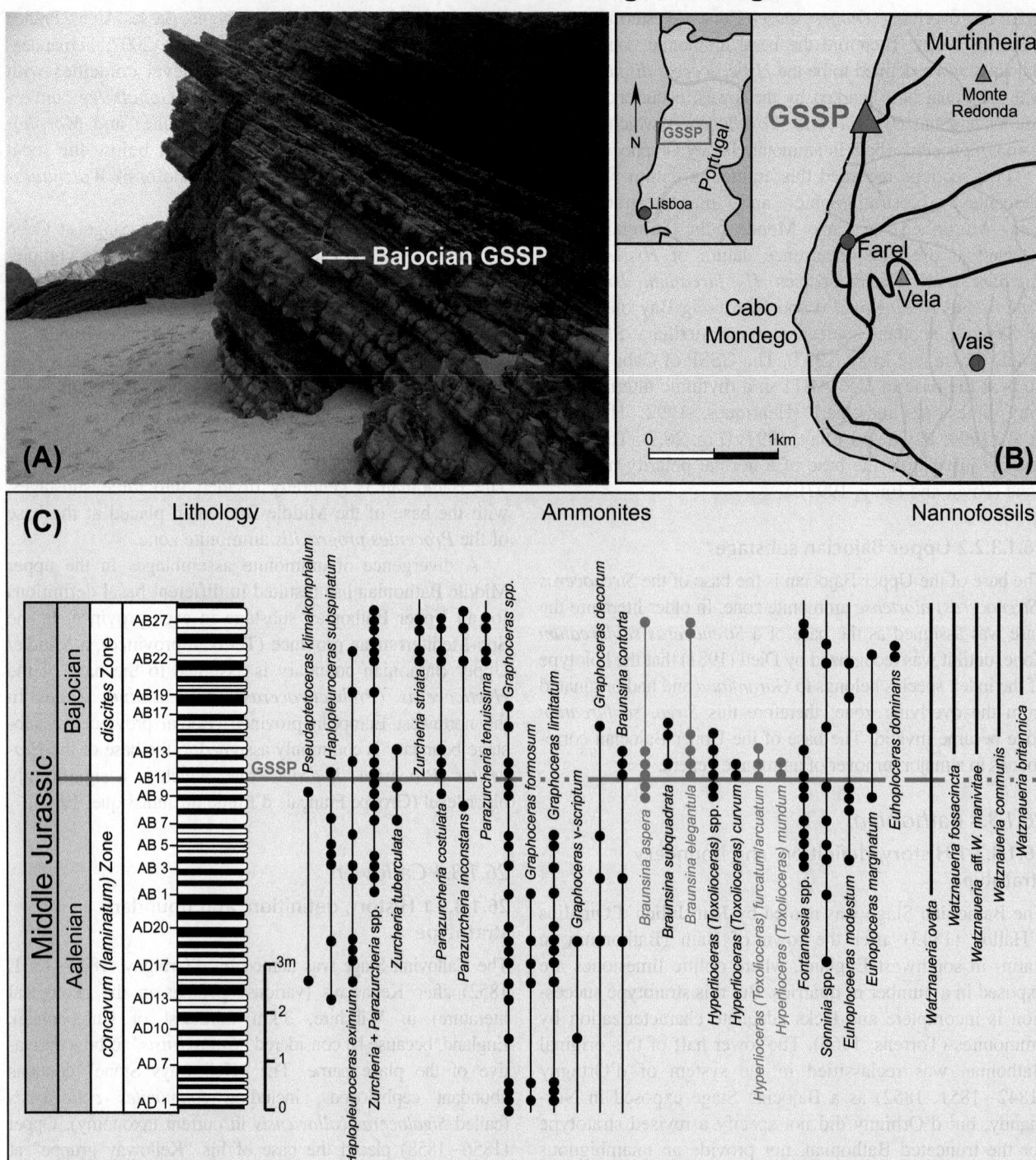

FIGURE 26.7   GSSP for the base of the Bajocian Stage at Cabo Mondego on the Atlantic coast of Portugal. The GSSP level coincides with the lowest occurrence of ammonite *Hyperlioceras* (base of *Hyperlioceras discites* Zone). *GSSP*, Global Boundary Stratotype Section and Point. Photograph provided by Maria Helena Henriques.

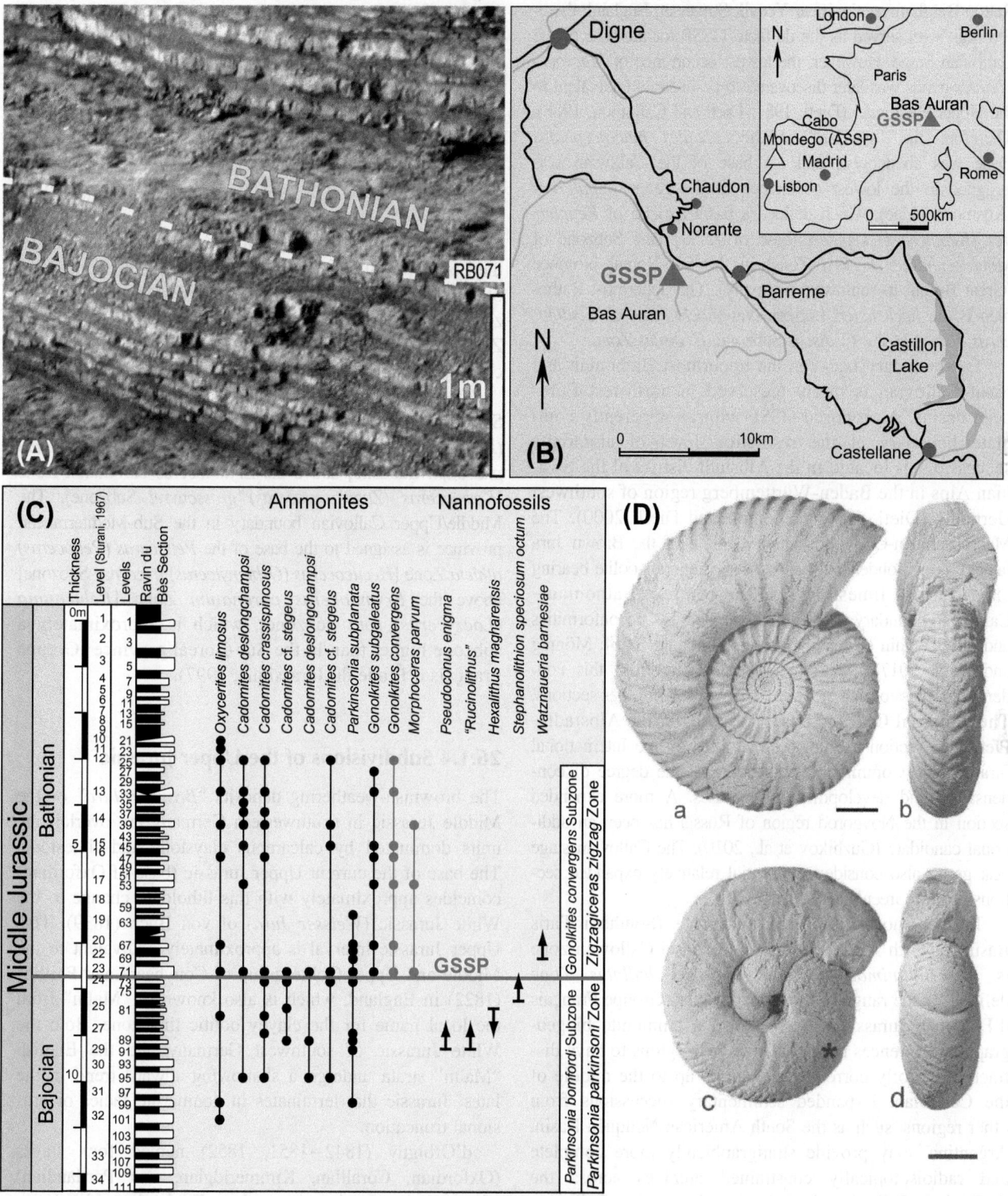

Base of the Bathonian Stage of the Jurassic System at the Ravin du Bès Section, SE France

FIGURE 26.8 GSSP for the base of the Bathonian Stage at Ravin du Bès, Bas Auran area, Alpes de Haute Provence, France. The GSSP level coincides with the lowest occurrence of ammonite *Gonolkites convergens* (base of *Zigzagiceras zigzag* Zone). *GSSP*, Global Boundary Stratotype Section and Point. Photograph provided by Sixto Fernández-López.

Callomon (1964, 1999) noted that the base of the *M. macrocephalus* Subzone in "standard chronostratigraphy" was initially defined as the base of Bed 4 by Arkell (1954) at the Sutton Bingham section near Yeovil, Somerset, England; therefore this level served as the de facto GSSP for the base of the Callovian Stage. However, the lowest occurrence of *Macrocephalites* genus was later discovered to be in strata equivalent to the Upper Bathonian (Dietl, 1981; Dietl and Callomon, 1988). Therefore the "standard" *Macrocephalites macrocephalus* Zone was abandoned, and the base of the Callovian was assigned to the lowest occurrence of the genus *Kepplerites* (Kosmoceratidae), which defined a basal horizon of *Kepplerites (Kepplerites) keppleri* (base of *K. keppleri* Subzone of *Macrocephalites herveyi* Zone) in the Sub-Boreal province (Great Britain to southwest Germany). The uppermost Bathonian is the *hochstetteri* horizon (var. *hochstetteri* of *Clydoniceras discus*) of the *C. discus* Subzone, *C. discus* Zone.

The boundary between the uppermost Bathonian and basal Callovian is rarely preserved in northwest European basins. A proposed GSSP with an apparently complete boundary at the resolution level of ammonite successions is located in the Albstadt district of the Swabian Alps in the Baden-Württemberg region of southwest Germany (Dietl, 1994; Callomon and Dietl, 2000). The Macrocephalen-Oolith formation (Unit ε of the Brown Jura facies) is a condensed facies of ferruginous-oolite-bearing clay to marly limestone, and the compact Bathonian–Callovian boundary interval is bounded by unconformities and may contain minor hiatuses (e.g. Mönnig, 2014; Mönnig and Dietl, 2017). The magnetostratigraphy from this condensed succession has not yet been verified in other sections. The potential GSSP at base of Bed 6a in the Albstadt–Pfeffingen section has not been adopted by the International Stratigraphic Commission on account of the degree of condensation and development of hiatuses. A more expanded section in the Novgorod region of Russia has been an additional candidate (Guzhikov et al., 2010). The Callovian stage task group also considered potential relatively expanded sections in East Greenland (Mönnig, 2014).

In the Sub-Mediterranean province (Southern Paris Basin to North Africa and Italy), the basal Callovian zone is the *Bullatimorphites (Kheraiceras) bullatus* Zone defined by the range of the index species (Groupe Français d'Étude du Jurassique, 1997). Strong ammonite biogeographic differences required these two regions to have distinct and poorly correlated zonations up to the middle of the Callovian. Expanded sedimentary successions from other regions, such as the South American Neuquén Basin, Argentina, may provide stratigraphically more complete and radioisotopically constrained archives across the Bathonian–Callovian boundary and for the Callovian Stage (e.g., Kamo and Riccardi, 2009). Detailed biostratigraphic or other correlation at a zonal or subzonal level between such regions and Europe is yet to be established.

26.1.3.4.2 Middle and Upper Callovian substages

The Callovian Stage is generally divided into three substages. The substage boundaries correspond to two important changes in ammonite fauna, but ammonite provincialism and utilization of different faunal successions led to different placements within each province that do not necessarily coincide (Groupe Français d'Étude du Jurassique, 1997).

In the Sub-Boreal province the Lower/Middle Callovian substage boundary is placed at the base of the *Kosmoceras (Zugokosmoceras) jason* Zone (base of *Kosmoceras (Zugokosmoceras) medea* Subzone), above the *Sigaloceras (Sigaloceras) calloviense* Zone [*Sigaloceras (Catasigaloceras) enodatum* Subzone]. In the Sub-Mediterranean province, the Lower/Middle Callovian boundary is placed at the base of the *Reineckeia anceps* Zone (*Reineckeia stuebeli* Subzone) above the *Macrocephalites (Dolikephalites) gracilis* Zone (*Indosphinctes patina* Subzone). These two levels are considered approximately coeval.

The Middle/Upper Callovian substage boundary in the Sub-Boreal province is assigned to the base of the *Peltoceras (Peltoceras) athleta* Zone [*Kosmoceras (Lobokosmoceras) phaeinum* Subzone] above the *Erymnoceras coronatum* Zone [*Kosmoceras (Zugokosmoceras) grossouvrei* Subzone]. The Middle/Upper Callovian boundary in the Sub-Mediterranean province is assigned to the base of the *Peltoceras (Peltoceras) athleta* Zone [*Hecticoceras (Orbignyiceras) trezeense* Subzone] above the *Erymnoceras coronatum* Zone [*Rehmannia (Loczyceras) rota* Subzone], which is approximately a subzone higher than in the Sub-Boreal province (Groupe Français d'Étude du Jurassique, 1997).

26.1.4 Subdivisions of the Upper Jurassic

The brownish-weathering deposits "*Brauner Jura*" of the Middle Jurassic in southwestern Germany are overlain by units dominated by calcareous claystone and limestone. The base of the current Upper Jurassic (base of Oxfordian) coincides approximately with this lithologic change to the White Jurassic (*Weisser Jura*) of von Buch (1839). This Upper Jurassic interval is approximately equivalent to the Middle and Upper Oolite group of Conybeare and Phillips (1822) in England, which is also known as "Malm" from the local name for the clayey oolitic limestone. Both the White Jurassic of southwest Germany and the English "Malm" strata undergo a shallowing-upward trend in the latest Jurassic that terminates in nonmarine facies or erosional truncation.

d'Orbigny (1842–1851, 1852) named four stages (Oxfordian, Corallian, Kimmeridgian, and Portlandian) after reference sections in southern England, and he designated the base of the Cretaceous as a "Purbeck Stage" followed by a "Neocomian Stage." Oppel (1856–1858) eliminated d'Orbigny's Corallian and Portlandian stages to extend the Kimmeridgian stage to the base of the Purbeckian (also considered to be Cretaceous). Oppel had left an interval "unassigned" between his Oxfordian and Kimmeridgian groups (his *Diceras arietina* Zone, approximately equivalent to the Upper Calcareous Grit Formation of Dorset, England). Later, Oppel (1865) created a new uppermost Jurassic stage, the "Tithonian," in the Mediterranean region that encompassed the upper part of his previous "Kimmeridgian group" and extended through the former Purbeckian to the base of the Neocomian stage. However, Oppel did not specify the limits or reference sections for this Tithonian stage concept. This confusing situation was further distorted when the "Berriasian" Stage of Coquand (1871) came into common use to designate the lowermost Cretaceous, even though it overlapped with the original concept of Oppel's Tithonian Stage.

The combination of (1) shuffling of Upper Jurassic stage nomenclature, (2) imprecise definitions, (3) a pronounced faunal provincialism during the majority of the Late Jurassic that precluded precise correlation even within northwest Europe, and (4) widespread hiatuses in the reference sections, resulted in a proliferation of regional stage and substage nomenclature. Finally, after a century of debate, the Colloque du Jurassique à Luxembourg 1962 (Maubeuge, 1964) voted to "return to the original sense of this stage [Oxfordian] as defined by A. d'Orbigny and given precision by W.J. Arkell," and to discontinue the regional usage of a "Purbeckian" stage, because it was primarily a local facies. However, the controversy over other Upper Jurassic stage definitions and the placement of the Jurassic/Cretaceous boundary led the Colloque to "refer the question back for consultation among interested specialists." These issues have yet to be fully resolved.

During the 1980s and 1990s, the International Subcommission on Jurassic Stratigraphy decided that the Upper Jurassic consists of three stages: Oxfordian, Kimmeridgian, and Tithonian. Through a fortunate episode of biogeography, an interregional biostratigraphic definition of the base of the Oxfordian is well established with ammonites. Similarly, the definition of the base of the Kimmeridgian is now settled. However, it has proven difficult to correlate potential definitions for the base of the Tithonian stage.

26.1.4.1 Oxfordian

26.1.4.1.1 History, definition, and boundary stratotype

The Oxfordian Stage of d'Orbigny (1842−1851, 1852), named after the city of Oxford in south-central England with reference to the Oxford Clay Fm., was overlain by his "Coralline Stage." Oppel (1856−1858) incorporated the majority of that "Corallian Stage" into an expanded "Oxfordian Group." In an error of correlation, he simultaneously

considered the base of his Oxfordian to be the lithologic base of the Oxford Clay where it sits on the Kelloway Rock in Yorkshire (now considered to be approximately the lower/middle Callovian boundary) and the biostratigraphic top of the *Peltoceras athleta* ammonite zone (now considered to be the middle of Upper Callovian). He also left "unassigned" a suite of strata between his "Oxfordian" and the overlying Kimmeridgian Stage.

Ammonites across the Callovian/Oxfordian boundary interval were studied by Arkell (1939, 1946), who placed this boundary at the lowest occurrence of *Vertumniceras mariae* (now placed in the *Quenstedtoceras* genus) above *Quenstedtoceras lamberti*, or essentially at the base of the Oxford Clay Fm. This is similar to the historical usage in southwest Germany, where the Upper or "White" Jurassic begins just above the Lamberti Knollen Bed. The Colloque du Jurassique à Luxembourg 1962 (Maubeuge, 1964) selected Arkell's biostratigraphic definition for the base of the Oxfordian Stage. This Colloque also assigned the upper limit of the Oxfordian as the top of the *Ringsteadia pseudocordata* ammonite zone in the Boreal province.

The main biostratigraphic event is the "Boreal Spread," an extensive expansion of the Boreal Cardioceratidae ammonites from their Arctic province across Europe and mixing with Mediterranean province fauna rich in Phylloceratidae ammonites (Page et al., 2010a). A current prime candidate for the Oxfordian GSSP is a coastal section at Redcliff Point near Weymouth, Dorset, Southern England (Melendez and Page, 2015; Page 2017). The GSSP in this clay-rich section would be placed to coincide with the base of the *Cardioceras scarburgense* Subzone of the *Quenstedtoceras mariae* Zone, but there have been different opinions of which specific ammonite horizon to use for that subzone definition. In GTS2012 the working definition had been the proposed lowest occurrence of ammonite *Cardioceras redcliffense* in a coastal section in Dorset, England (Page et al., 2010b), within the youngest part of a normal polarity magnetozone correlated to marine magnetic anomaly M37n (Ogg et al., 2010a, 2011). However, that *Card. redcliffense* horizon is difficult to correlate and the taxon might be a morphotype of the index species of the uppermost Callovian subzone (Pellenard et al., 2014a). Another concern is that the Redcliff exposure is in an episodically active landslip that lacks easy-to-recognize lithological markers near the boundary.

The Oxfordian working group is currently (as of 2019) favoring a slightly higher level with a brief, marked faunal turnover of the disappearance of several Callovian genera and the appearance of new species, especially *Hecticoceras (Brightia) thuouxensis* followed by the lowest *Cardioceras scarburgense*. The ammonite biostratigraphy across this *Q. lamberti* to *Q. mariae* transition interval has been studied in expanded dark claystone successions in southeast France (e.g., Fortwengler and Marchand, 1994; Page et al., 2010b; Pellenard et al., 2014a) where two sections have been recommended as alternative basal Oxfordian GSSPs (at the Thuoux and Saint-Pierre d'Argençon sections; Pellenard et al., 2014a). These French sections have proved suitable for some other stratigraphic correlation methods that match or complement Redcliff, such as nannofossils, dino-flagellates, and cyclostratigraphy (Pellenard et al., 2014a). In the candidate GSSP at Thuoux, the *Hect (B.) thuouxensis* level also coincides with the first appearance of dinocyst *Wanaea fimbriata*. Even though the candidate GSSP in France has been unsuitable for paleomagnetism, the compar-ison of ammonite assemblages enables approximate correla-tion to the magnetostratigraphy of the Dorset section at the base of the local *Cardioceras woodhamense* biohorizon (e.g., Kiselev et al., 2013), which would project this bound-ary as the lower part of Chron M36Br; and this magnetostra-tigraphic level is used in GTS2020 to estimate the age of this candidate level for placing the Callovian—Oxfordian boundary. Recognition of this level with other biostrati-graphic groups is not yet fully established; but the level is immediately followed by a widespread surge in primitive planktonic foraminifers (Oxford et al., 2002).

Another candidate GSSP section at Dubki in European Russia is excellent for the basal Oxfordian but is poor for the uppermost Callovian (Kiselev et al., 2013).

26.1.4.1.2 Middle and Upper Oxfordian substages

The base of a Middle Oxfordian Substage is commonly placed at the base of the ammonite *Perisphinctes (Aris-phinctes) plicatilis* Zone of the Tethyan and Sub-Boreal faunal provinces, which is coeval with the base of the *Car-dioceras densiplicatus* Zone of the Boreal province.

Beginning with the Middle Oxfordian, faunal differenti-ation among faunal realms became more pronounced and has inhibited standardization and correlation of regional ammonite zones. In addition, even though regional zonal nomenclatures have commonly remained constant, the assigned biostratigraphic boundaries of these "standard" zonal units have been undergoing redefinitions (e.g., Głowniak, 1997; Groupe Français d'Étude du Jurassique, 1997; Głowniak et al., 2008; among others).

In the Sub-Mediterranean province (Tethyan province), the base of an Upper Oxfordian Substage is commonly assigned to the base of the *Perisphinctes (Dichotomoceras) bifurcatus* Zone, although the definition of the zonal base is not always consistent among regional studies. In the Sub-Boreal province, the Middle/Upper substage boundary is assigned as the base of the regional *Perisphinctes cautis-nigrae* Zone; and the Boreal province uses the base of the regional *Amoeboceras serratum* Zone. Cross-regional ammonite markers and magnetostratigraphic studies sug-gest that these three levels are within about 250 kyr of each other (base of *Am. serratum* is youngest; base of *Per. cautisnigrae* is oldest).

26.1.4.2 Kimmeridgian

26.1.4.2.1 History, definition, and boundary stratotype

The Kimmeridgian Stage was named by d'Orbigny (1842—1851, 1852) after the coastal village of Kimmeridge in Dorset, England, where the spectacular cliffs of black Kimmeridge Clay expose a continuous record of that inter-val. Oppel (1856—1858) expanded the Kimmeridgian downward by incorporating a portion of d'Orbigny's for-mer "Corallian stage," but, rather than assign a boundary between the Oxfordian and Kimmeridgian, he left the inter-vening Upper Calcareous Grit formation "unassigned." Oppel initially indicated that the Kimmeridgian "group" would continue upward to the base of the Purbeck (his base Cretaceous), but later, Oppel (1865) inserted a Titho-nian Stage as the uppermost Jurassic stage. Therefore nei-ther of the boundaries of the Kimmeridgian Stage was initially adequately defined.

The Oxfordian/Kimmeridgian boundary was defined by Salfeld (1914) after studying the Perisphinctidae ammonite Succession from the boundary interval. He proposed a *Ringsteadia anglica* Zone (now called *Ringsteadia pseudo-cordata* Zone) in the uppermost Oxfordian and that the base of the Kimmeridgian would correspond to the appear-ance of *Pictonia*. The Colloque du Jurassique à Luxembourg 1962 (Maubeuge, 1964) fixed the base of the Kimmeridgian as the base of the *Pictonia baylei* Zone in the Sub-Boreal province. This level is considered equivalent to the base of the *Amoeboceras bauhini* Zone of the Boreal realm. How-ever, due to faunal provincialism that began in the middle Oxfordian, the ammonite zonations of Great Britain (Boreal and Sub-Boreal provinces) could not be correlated to the Sub-Mediterranean province (Tethyan province). The Col-loque du Jurassique à Luxembourg 1962 (Maubeuge, 1964) indicated that the base of the Kimmeridgian was equivalent to the base of the *Sutneria platynota* Zone of the Sub-Mediterranean province, but uncertainty remained on the potential diachroneity of the base Kimmeridgian between the (Sub-)Boreal and Sub-Mediterranean regions.

The presumed equivalence was later demonstrated to be incorrect from comparisons of dinoflagellate cyst assemblages (Brenner, 1988; Poulson, in Melendez and Atrops, 1999) and rare incursions of ammonites from the Boreal province into the Sub-Mediterranean successions in Poland and in the Swa-bian Alps (Atrops et al., 1993; Matyja and Wierzbowski, 1997, 2002, 2006a,b; Schweigert and Callomon, 1997). These constraints on the correlation among faunal provinces near the Oxfordian—Kimmeridgian boundary imply that much of the upper *Epipeltoceras bimammatum* Zone and *Idoceras planula* Zone of the Sub-Mediterranean province of the Tethyan province, which had constituted the entire upper half of that regional "Upper Oxfordian," should be coeval with the lowest Kimmeridgian (*Amoeboceras bauhini* Zone and *Amoeboceras bayi* horizon of the *Amoebites kitchini* Zone) of the Boreal province (Matyja et al., 2006). This offset in the regional "Oxfordian/Kimmeridgian" boundary placements by one-and-a-half ammonite zones (∼1.3 Myr) has been demonstrated by comparing magnetostratigraphy of Great Britain (Sub-Boreal and Boreal provinces) with different regions in the Sub-Mediterranean province (Przybylski et al., 2010a). In order to allow a uniform Oxfordian/Kimmeridgian boundary that corresponds to an ammonite zonal boundary in the different provinces, which had been a major cause of the postponed decision, Wierzbowski and Matyja (2014) enhanced correlations to recommend the elevation of the *Epi. hypselum* Subzone of the lower *Epi. bimammatum* Zone to zonal status. The temporal offset had created a dilemma in selecting a GSSP for the base of the Kimmeridgian Stage, because an initial choice must be made between faunal provinces and the corresponding definition of the Oxfordian/Kimmeridgian boundary (reviewed in Atrops, 1999; Melendez and Atrops, 1999; Wierzbowski, 2001, 2002, 2010). It was decided by the Oxfordian−Kimmeridgian boundary working group to retain a definition similar to Salfeld (1914) of the appearance of *Pictonia* in the Sub-Boreal province.

An intertidal coastal section in dark-gray claystone at Flodigarry, Staffin Bay, Isle of Skye (northwest Scotland) has yielded detailed Sub-Boreal and Boreal ammonite Successions and a magnetostratigraphy (e.g., Sykes and Callomon, 1979; Wierzbowski et al., 2006, 2015, 2016, 2018; Matyja et al., 2006; Przybylski et al., 2010a) and is now the site proposed for the GSSP, specifically in the upper part of Bed 35 of the Staffin Shale Fm., 1.25 ± 0.01 m below the base of Bed 36 (cf. Morton and Hudson, 1995). The proposed GSSP level is the base of the Sub-Boreal ammonite *Pict. baylei* Zone and the base of the Densicostata Subzone marked by the base of the *flodigarriensis* horizon (at the first appearance of *Pictonia flodigarriensis*) and, independently, the base of the Boreal ammonite *Am. bauhini* Zone. The point is located about 0.15 m above the base of the reversed-polarity magnetozone F3r, which may correlate with the older part of marine magnetic anomaly M26r (Przybylski et al., 2010a). Dinoflagellate cysts are complementary markers for the boundary (Riding and Thomas, 1997; Poulsen and Riding, 2003), and strontium isotope analyses of belemnites from the proposed GSSP yield a low Sr isotope value of 0.70687 at the boundary (Wierzbowski et al., 2017). The proposed GSSP level has also yielded an Re−Os radioisotope age (of 154.1 ± 2.2 Ma, Selby 2007).

26.1.4.2.2 Upper Kimmeridgian substage

In the Sub-Boreal province the base of an Upper Kimmeridgian Substage is the ammonite *Aulacostephanoides mutabilis* Zone, whereas in the Sub-Mediterranean province, it is typically assigned to the base of the *Aspidoceras*

acanthicum Zone. Magnetostratigraphy suggests that the Sub-Mediterranean placement is about 0.3 Myr younger than the Sub-Boreal assignment (Ogg et al., 2010a,b). An alternative assignment in the Boreal province, the base of the *Aulacostephanus eudoxus* ammonite zone, is one ammonite zone higher than the Sub-Boreal substage placement, or about 0.8 Myr younger.

26.1.4.3 Tithonian
26.1.4.3.1 History, definition, and boundary stratotype

In an enlightened departure from the lithology-based stratotype concept, Oppel (1865) defined the Tithonian Stage solely on biostratigraphy. In mythology, Tithon is the spouse of Eos (Aurora), goddess of dawn, and Oppel used this name in a poetic allusion to the dawn of the Cretaceous. He referenced sections in Western Europe, from Poland to Austria, for the characteristics of the Tithonian.

The base of Oppel's Tithonian was placed at the top of the Kimmeridgian *Aulacostephanus eudoxus* ammonite zone, which can be recognized in both the Sub-Boreal and Sub-Mediterranean faunal provinces. Later, Neumayr (1873) established the *Hybonoticeras beckeri* Zone above the *A. eudoxus* in the Sub-Mediterranean province and assigned it to the Kimmeridgian Stage.

Neumayr's revised placement of the Kimmeridgian−Tithonian boundary corresponded closely with the boundary between the "Portlandian" and Kimmeridgian Stages as initially assigned by d'Orbigny (1842−1851, 1852), who had assigned "des *Ammonites giganteus* et *Irius*" as Portlandian index fossils. However, d'Orbigny did not visit England, and he inadvertently combined fossil assemblages from outcrops at Boulogne in France with a name derived from a "type section" on the Isle of Portland in England. The "*Ammonite irius*" is one representative of the *Gravesia* genus, which has the lowest occurrence in the basal *Hybonoticeras hybonotum* Zone of the revised Tithonian. Accordingly, the *Gravesia gravesiana* ammonite zone was assigned as the basal zone of the British Portlandian [Tithonian] Stage by Salfeld (1913). Following Salfield's oral presentation to the Geological Society of London, it was noted that the chronostratigraphic term "Kimmeridgian" only partially encompassed the "Kimmeridgian Clay" Formation; therefore it was recommended that Salfield "*should invent a dual nomenclature—one for the stratigraphical and another for the zoological sequence*" and replace "Kimmeridgian Stage" by a new name (in Salfeld, 1913). Unfortunately, this enlightened recommendation was not pursued, and a confusing equivalence of a "Kimmeridgian Stage" with the "Kimmeridge Clay" formation and an associated lifting the base of d'Orbigny's "Portlandian" Stage became common usage in England, but a stratigraphically lower Kimmeridgian−Tithonian boundary was used

elsewhere in Europe. In Britain the Kimmeridge Clay Formation was arbitrarily subdivided into a lower and upper
member at the approximate Kimmeridgian–Tithonian
boundary level at the lowest occurrence of *G. gravesiana* at
the "Maple Ledge" bed (reviewed in Cox and Gallois,
1981). The "Tithonian" was formally adopted as the name
of the uppermost stage of the Jurassic by a vote of the International Commission on Stratigraphy in September, 1990.

The Second Colloquium on the Jurassic System, held in
Luxembourg in 1967, recommended that the top of the Kimmeridgian be assigned to the base of the *Gravesia gravesiana*
Zone (Resolution du deuxième Colloque du Jurassique à Luxembourg, 1970). However, Cope (1967) subdivided the lowermost Tithonian portion of the Upper Kimmeridge Clay into
several ammonite zones based on successive species of his
reconstituted *Pectinatites* genus and abandoned Salfield's two
Gravesia zones. Cope raised the upper limit of the uppermost
Kimmeridgian *Aulacostephanus autissiodorensis* Zone to the
base of his new *Pectinatites (Virgatosphinctoides) elegans*
Zone, thereby effectively lifting the associated biostratigraphic
division between Lower and Upper Kimmeridge clay. Cox
and Gallois (1981) note that the top of the international
Kimmeridgian Stage now falls within the middle of
Cope's expanded *A. autissiodorensis* Zone in the Sub-
Boreal province; therefore they suggested reinstating a trun-
cated *G. gravesiana* Zone below the *P. (V.) elegans* Zone.

In addition to this complex history of changing concepts,
the extreme provincialism of macrofossils and microfossils
has proven a daunting challenge to find a suitable GSSP definition that is amenable to global correlation. The Kimmeridgian–Tithonian boundary Interval in the Tethyan Faunal
realm is marked by the simultaneous lowest occurrence of
the ammonites *Hybonoticeras* aff. *hybonotum* and *Glochi-
ceras lithographicum* (at the base of the *H. hybonotum*
Zone), immediately followed by the lowest occurrence of
the *Gravesia* genus. Candidate sections in southeast France
for the GSSP include a thick pelagic limestone succession
just below a castle on the hill of Crussol overlooking the
Rhône River just west of Valence (Atrops, 1982, 1994) and
at a quarry within a military training area at Canjuers (Var
district) (Atrops et al., unpubl.). The base of *H. hybonotum*
Zone at Crussol is at the base of normal polarity zone
M22An (Ogg et al., 2010b). The Canjuers Quarry did not
yield a magnetostratigraphy, but its ammonite succession is
better established. Other potential GSSP sections include the
Swabian region of southern Germany (Olóriz and Schweigert, 2010) and, indeed, in the Kimmeridge Clay at Kimmeridge Bay, which is well characterized by multiple
stratigraphic methods (Morgans-Bell et al., 2001; Huang
et al., 2010a), although, in that case, not using an ammonite
as a primary marker. Given the complexities regarding provincialism in many faunal groups at this time, a case may
be made to consider a magneto- or chemostratigraphic event
as primary marker for the base of the Tithonian, with the

lowest occurrence of specific guide fossils for each province
as auxiliary correlation tools.

26.1.4.3.2 Upper Tithonian substage

In the Tethyan faunal domain, the base of an Upper Tithonian Substage is informally assigned to a major turnover in
ammonite assemblages at the base of the *Micracanthoceras
microcanthum* Zone. This level is approximately where
calpionellid microfossils become important in the biostratigraphic correlation of pelagic limestone and is at the base
of normal polarity chron M20n.

26.1.4.3.3 Bolonian–Portlandian and Volgian regional stages of Europe

The century of controversy over the subdivision and
nomenclature for the uppermost Jurassic and lowermost
Cretaceous stages, coupled with markedly distinct ammonite assemblages in different regions of Europe, led to
extensive usage of regional stages.

The Volgian Stage in western Russia was established by
Nikitin (1881), capped by a Ryazanian Horizon (Bogoslovsky, 1897), and later extended downward (Resolution of the
All-Union Meeting on the unified scheme of the Mesozoic
stratigraphy for the Russian Platform, 1955) (reviewed in
Krymholts et al., 1982). The Volgian is zoned by ammonite
assemblages that are extensively distributed in the Boreal
faunal realm; therefore it became the standard in the northern high latitudes outside of Britain. However, the zonations
are different among the Russian Platform, northern Siberia,
subpolar Urals, and Spitsbergen regions (e.g., Rogov and
Zakharov, 2009). In 1996 the Russian Interdepartmental
Stratigraphic Committee resolved to equate the Lower
and Middle Volgian (*Ilowaiskya klimovi* through *Epivir-
gatites nikitini* ammonite zones of Russian platform) to
the Tithonian Stage, assign the Upper Volgian (*Kachpur-
ites fulgens* through *Craspedites nodiger* ammonite
zones) to the lowermost Cretaceous, and use only Tithonian and Berriasian as chronostratigraphic units in the
Russian geologic time scale (Rostovtsev and Prozorowsky, 1997). In the Russian Platform region of the Boreal
Realm, the boundary between Lower and Middle "Volgian" regional substages is assigned at the base of the
Dorsoplanites panderi ammonite zone.

The "Bolonian" and "Portlandian" have been promoted as
"secondary standard stages" for usage in southern English
regional geology, especially in Dorset (e.g., Cope, 1993;
Taylor et al., 2001) but have not been adopted and are not
supported by the ICS. In the Sub-Boreal faunal province in
Britain, the base of "Portlandian" regional stage, named by
d'Orbigny (1842–1851) after the Isle of Portland in Dorset, is
placed at the base of the *Progalbanites albani* ammonite zone
at the base of the Portland Sand formation. The "Bolonian"
was named by Blake (1881) after the Boulogne-sur-Mer section at Pas-de-Calais, northwest France. The "Bolonian" of

Britain is equivalent to the upper Kimmeridgian clay formation between the Kimmeridgian—Tithonian boundary and the Portland Sand, or the *Pectinatites (Virgatosphinctoides) elegans* through *Virgatopavlovia fittoni* ammonite zones. The overlying "Portlandian" is equivalent to the Portland Group in Dorset, or *Progalbanites albani* through approximately the *Titanites anguiformis* ammonite zones (usage varies).

The approximate correlation and scaling of ammonite zonations among the Boreal (Volgian), Sub-Boreal (Bolonian, regional upper Kimmeridgian), and Tethyan (Tithonian) stages has been only partially resolved through applying magnetostratigraphy (e.g., Houša et al., 2007; Ogg et al., 2010b) and detailed intercorrelation webs of macrofossil biostratigraphy (e.g., Rogov and Zakharov, 2009; Rogov, 2010a,b) (Fig. 26.9). The base of the Volgian in Russia appears to coincide exactly with the base of the Tithonian in western Europe (Rogov and Price, 2010).

26.1.4.3.4 Top of Tithonian (base of Cretaceous)

In contrast to most geological systems, there are no major "global events" encoded within the uppermost Jurassic and lowermost Cretaceous. Indeed, it has been nearly impossible to find any significant widespread biostratigraphic, geochemical, or other marker for interregional correlation within this interval (Price et al., 2016b; Wimbledon, 2017). Currently, the microfossil turnover of calpionellids from *Crassicollaria* and large *Calpionella* to small *Calpionella alpina* that defines the base of the *C. alpina* Subzone of the *Calpionella* Zone has been identified as a widespread marine biostratigraphic marker and has been selected by the Berriasian working group as the primary marker for the base of the stage and thereby the base of the system. This bioevent is thought to occur within magnetic Chron M19n.2n (Wimbledon et al., 2013). Various calcareous nannofossil FADs bracket the *alpina* Zone, and correlations are also supported by calcareous dinoflagellate cysts (Reháková, 2000). Presently the prime candidate for the GSSP is Tré Maroua (Hautes-Alpes), which is about 30 km southeast of Gap in the Vocontian Basin of southern France in the Vocontian Basin, France (Wimbledon et al., 2020). Nevertheless, discussions continue with respect to the more radical step of moving the base Cretaceous to the base Valanginian; and supporting arguments are most recently summarized by Granier (2019) and Énay (2020).

In GTS2020, we utilize the middle of Chron M19n.2n for assigning the numerical age to the base of the Cretaceous.

26.2 Jurassic stratigraphy

The ammonite successions of Europe have historically served as global primary standard for the Jurassic. Other bio-, chemo-, and magnetostratigraphic proxies and methods are typically calibrated to these standard European ammonite zones (Fig. 26.10). An extensive compilation and intercorrelation of Jurassic stratigraphy of European basins was coordinated by Hardenbol et al. (1998), but later workers have greatly enhanced or revised portions of these comprehensive chart series. Correlation outside of these European basins, especially to terrestrial realms, is relatively uncertain but improving especially where major environmental change events leave a biological or geochemical imprint in the record. Due to pronounced climate—oceanic provincialism, the late Jurassic interval is most challenging for global correlations (e.g., summary by Zeiss, 2003).

26.2.1 Marine biostratigraphy

26.2.1.1 Ammonites

Alfred Oppel (1856—1858) developed the concept of a biostratigraphic zone and used ammonites to define two-thirds of his 33 Jurassic zones. Jurassic ammonite zonations have undergone constant revision since Oppel, and the Jurassic is currently subdivided into 70—80 zones and typically 160—170 subzones in each faunal realm. Reviews of the development, definitions, and intercorrelation of European ammonite zonations are in Thierry (1998) (including correlation charts), Krymholts et al. (1982), and Groupe Français d'Étude du Jurassique (1997). The precision of correlation of the northwest European standard zones to the regional ammonite zones of Western North America (e.g., Pálfy et al., 1997, 2000) and of Russia (e.g., Zakharov et al., 1997; Rogov, 2004, 2010a,b; Rogov and Zakharov, 2009) and of other regions depends on the varying provinciality of genera and index species. Globally, up to seven suborders of Order Ammonoidea are recognized; and these range through up to 20 biogeographic provinces and subprovinces (Page, 2008; Schweigert, 2015).

Only a single genus, the *Psiloceras* of basal Jurassic age, appears to have survived the end-Triassic mass extinction of the ammonoids. This single form, indeed probably only a single *P. spelae* species, rapidly diversified into over 20 ammonite genera during the Hettangian (Guex, 1995; Guex et al., 2004). A multistage "Early Toarcian crisis" caused a lesser bottleneck, followed by a rapid recovery (e.g., Dera et al., 2010). In the Bajocian the first Perisphinctoidea evolved and became the globally most important ammonites of the late Jurassic (Page, 2008). Belemnites, the cousins of the ammonites, have local Jurassic zonations that provide correlation to the stage or substage level (e.g., Combemorel, 1998).

Nomenclature for many of the Jurassic ammonite zones can be misleading. According to the International Stratigraphic Code, biostratigraphic zones are named according to the genus and species of associated taxa (e.g., *Kotetishvilia compressissima* Zone begins with the lowest occurrence of that ammonite taxon) combined with the appropriate term for the kind of

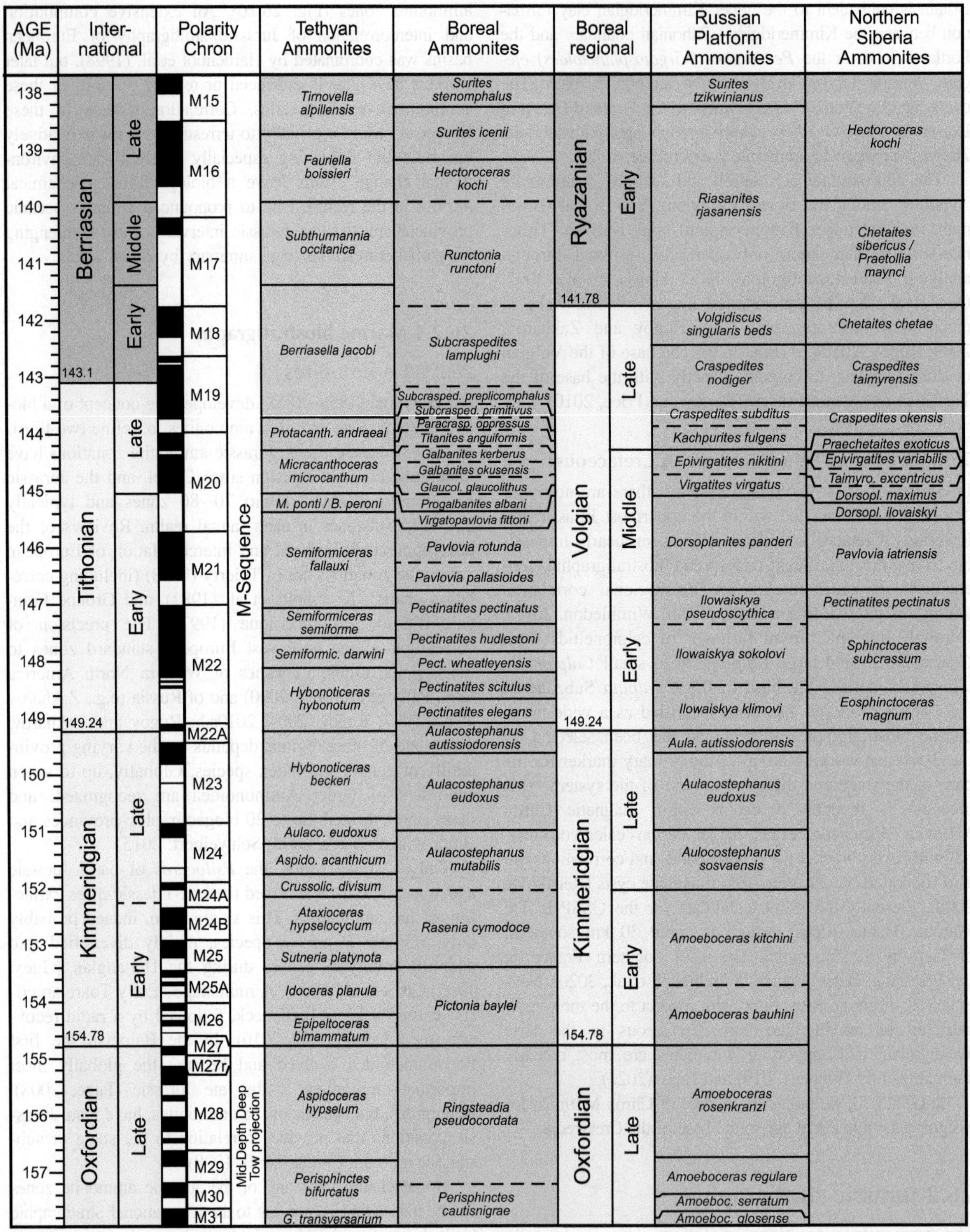

FIGURE 26.9 Interregional correlation of latest Jurassic through earliest Cretaceous ammonite zones and regional stages. Magnetostratigraphic correlations to the marine magnetic anomaly M-sequence and cycle stratigraphy provide the reference numerical scale and absolute durations for Tethyan ammonite zones (e.g., Ogg et al., 2010a; Boulila et al., 2008b, 2010a,b; Pruner et al., 2010). Sub-Boreal ammonite calibrations to Tethyan zone and/or M-sequence incorporates

biostratigraphic unit. This system is followed for most Paleozoic—Triassic and Cretaceous ammonoid zonations. In contrast, most Jurassic ammonite specialists working in northwest Europe advocate and apply "Standard Chronozones," in which the nomenclature is not directly associated with the temporal range of the name-giving species (e.g., Callomon, 1985, 1995). Therefore a standard Eudoxus Zone (capitalized, nonitalics, no genus designation) of the Kimmeridgian in southern England has a base defined at bed E1 at a quarry near Westbury, Wiltshire (Birkelund et al., 1983) and continues to the base of the Autissiodorensis Zone, which was assigned as the top of the Flats Stone Band bed at beach exposures of the Kimmeridge Clay near the village of Kimmeridge, Dorset (Cox and Gallois, 1981; tables in Cox, 1990). In some cases the "name-species" may not appear until quite high in its "Standard ChronoZone," for example, the lowest occurrence of the ammonite *Cardioceras cordatum* is in the upper third of the "Cordatum Zone." Unfortunately, the basal limits of many of these "standard chronozones" remain unstandardized among regions and specialists and have changed through the decades, which often creates correlation problems for nonammonite workers. For example, the base of the Kimmeridgian is accepted as the base of the Baylei Zone, but, there had been significant disagreement whether this Zone begins with the lowest occurrence of the ammonite *Pictonia flodigarriensis* or another taxon. Both systems—a regional designation of Jurassic standard chronozones versus a nomenclature based on ammonite biozones—have defenders (mainly Jurassic ammonite specialists in the former case) and critics.

For clarity in our charts and tables (e.g., Tethyan zones in Table 26.3), we have included the genera of the ammonite "index" species, but with a caution that these are not always the "guide" species of the named zone.

26.2.1.2 Fish and marine reptiles

Ray-finned fish of the primitive teleost group, to which most living bony fish belong, occur in the Jurassic, but there is no established zonation. The earliest large filter feeders, pachycormids up to 9 m long, are known from the Middle Jurassic into the Cretaceous (Calvin, 2010; Friedman et al., 2010).

The most famous dwellers of the Jurassic seas were abundant large marine reptiles (Sauropterygia) comprising long-necked plesiosaurs (near lizard), shorter necked pliosaurs (more lizard) and the streamlined dolphin-like ichthyosaurs

(fish lizard). The Callovian and Late Jurassic assemblages included the pliosaur, *Liopleurodon*, the world's largest known carnivore of over 15 m in length (Page, 2005).

26.2.1.3 Other marine macrofauna

Brachiopod zonations for northwest Europe and for the northern part of the Tethyan province provide important markers within individual basins and approach the resolution of ammonite zones in some stages of the Jurassic (e.g., Alméras et al., 1997; Laurin, 1998). The *Buchia* genus of bivalves is used for correlations within the Oxfordian—Hauterivian strata, especially within the circum-Boreal realm of Eurasia and western North America, and several buchiid zones have been calibrated to ammonite zonations (e.g., Jeletzky, 1965; Surlyk and Zakharov, 1982; Zakharov, 1987; Sha and Fürsich, 1994; Zakharov et al., 1997). However, the correlation potential of brachiopods and the slower evolving successions of bivalves and gastropods are compromised by their benthic habits, which can be reflected in ecological-facies associations and provincialism (reviewed in Hallam, 1976; Cope et al., 1980b).

Ostracods are small (0.2—1.5 mm) crustaceans with calcified bivalved shells. They are a major constituent of shallow marine and brackish benthic faunas, and their evolving assemblages enable a biostratigraphic resolution approaching provided by ammonite zones, especially within portions of the Lower and Middle Jurassic (e.g., reviews by Cox, 1990; Collin, 1998).

26.2.1.4 Foraminifers and calpionellids

Jurassic benthic Foraminifera datums and zonations have been developed for both calcareous and agglutinated forms. Compilations are available for the British and North Sea region (e.g., Copestake et al., 1989), for larger benthic Foraminifera in the Tethyan domain (Peybernes, 1998), and smaller benthic Foraminifera in European basins (Ruget and Nicollin, 1998). Planktonic foraminifera appeared in the *Variabilis* Zone of Toarcian, and evolved and spread in Bajocian—Bathonian (genus *Globuligerina* with eight taxa). The genus *Conoglobigerina* with two taxa appeared in mid-Oxfordian. The group flourished in Kimmeridgian and became virtually extinct in Tithonian (Gradstein et al., 2017). For details see Chapter 3G, Planktonic foraminifera, this volume.

Calpionellids are vase-shaped pelagic microfossils of uncertain origin, which appeared in the late Tithonian

---

magnetostratigraphy (e.g., Ogg et al., 1994; Przybylski et al., 2010a,b), cycle stratigraphy for scaling of Kimmeridgian—Tithonian Succession (e.g., Huang, 2018), and biostratigraphic correlations (e.g., Wierzbowski et al., 2016; Głowniak et al., 2008; Wimbledon et al., 2011). Boreal calibrations of Russian Platform to Northern Siberia and to Sub-Boreal zones are mainly biostratigraphic (e.g., Rogov and Zakharov, 2009; Rogov, 2010b; Rogov and Price, 2010; Harding et al., 2011) supplemented with magnetostratigraphy for the Late Volgian of Northern Siberia (Houša et al., 2007). However, the reference sections for Boreal zonations and for the Portlandian—Ryazanian of the Sub-Boreal realm commonly have ammonite levels separated by stratigraphic gaps; therefore this lack of continuity precludes exact interregional correlations (e.g., charts in Harding et al., 2011). In addition, some ammonite zones are poorly defined (e.g., Cope, 2007, recommended abandoning the *Paracraspedites oppressus* Zone of England). This summary diagram is modified from Ogg et al., 2010b.

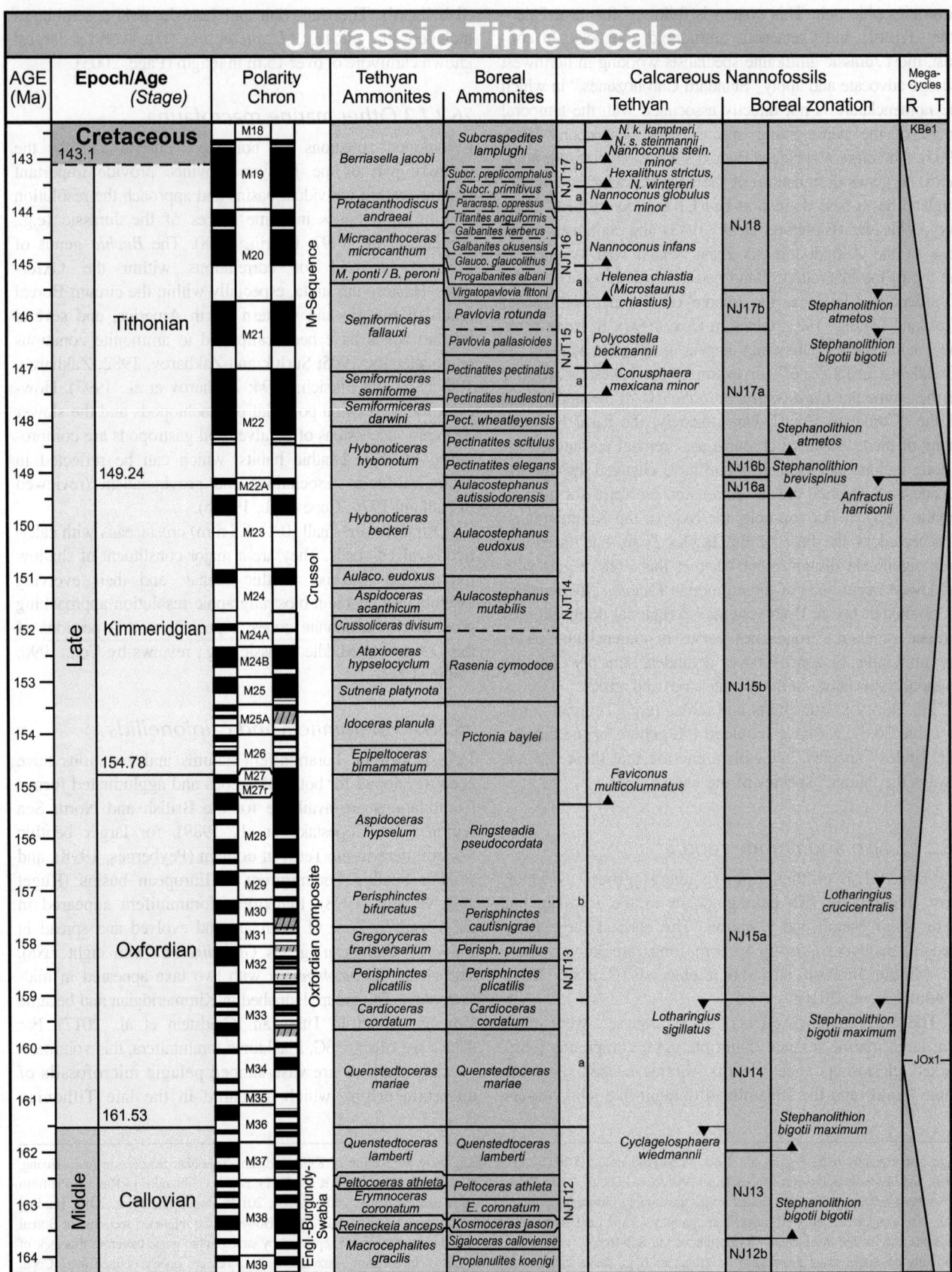

FIGURE 26.10   Summary of age model for epoch/series and age/stage boundaries of the Jurassic with selected marine biostratigraphic zonations. ("Age" is the term for the time equivalent of the rock-record "stage.") Potential definitions of Oxfordian, Kimmeridgian, Tithonian, and Berriasian are indicated; but the final decisions will be made by the International Commission on stratigraphy (see www.stratigraphy.org). Marine biostratigraphy columns are representative

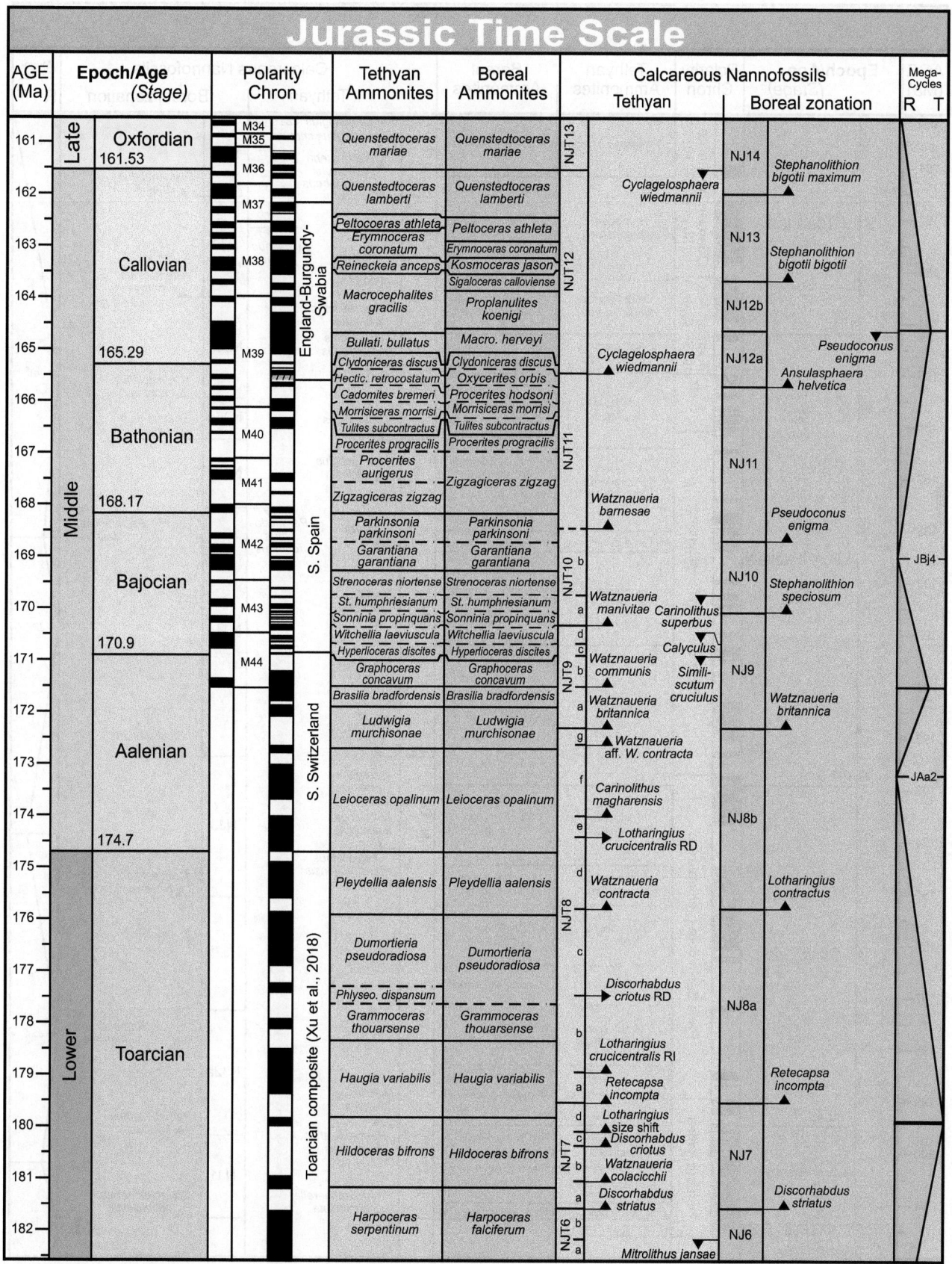

FIGURE 26.10 (Continued)

ammonoid zones for the Tethyan (Sub-Mediterranean province) and Boreal (Sub-Boreal for Late Jurassic) realms. For detailed ammonite zonations in Late Jurassic, see expanded scale in Fig. 26.9. Calcareous nannofossil zones and Boreal zone-markers are a composite from Casellato (2010), Mattioli and Erba (1999)

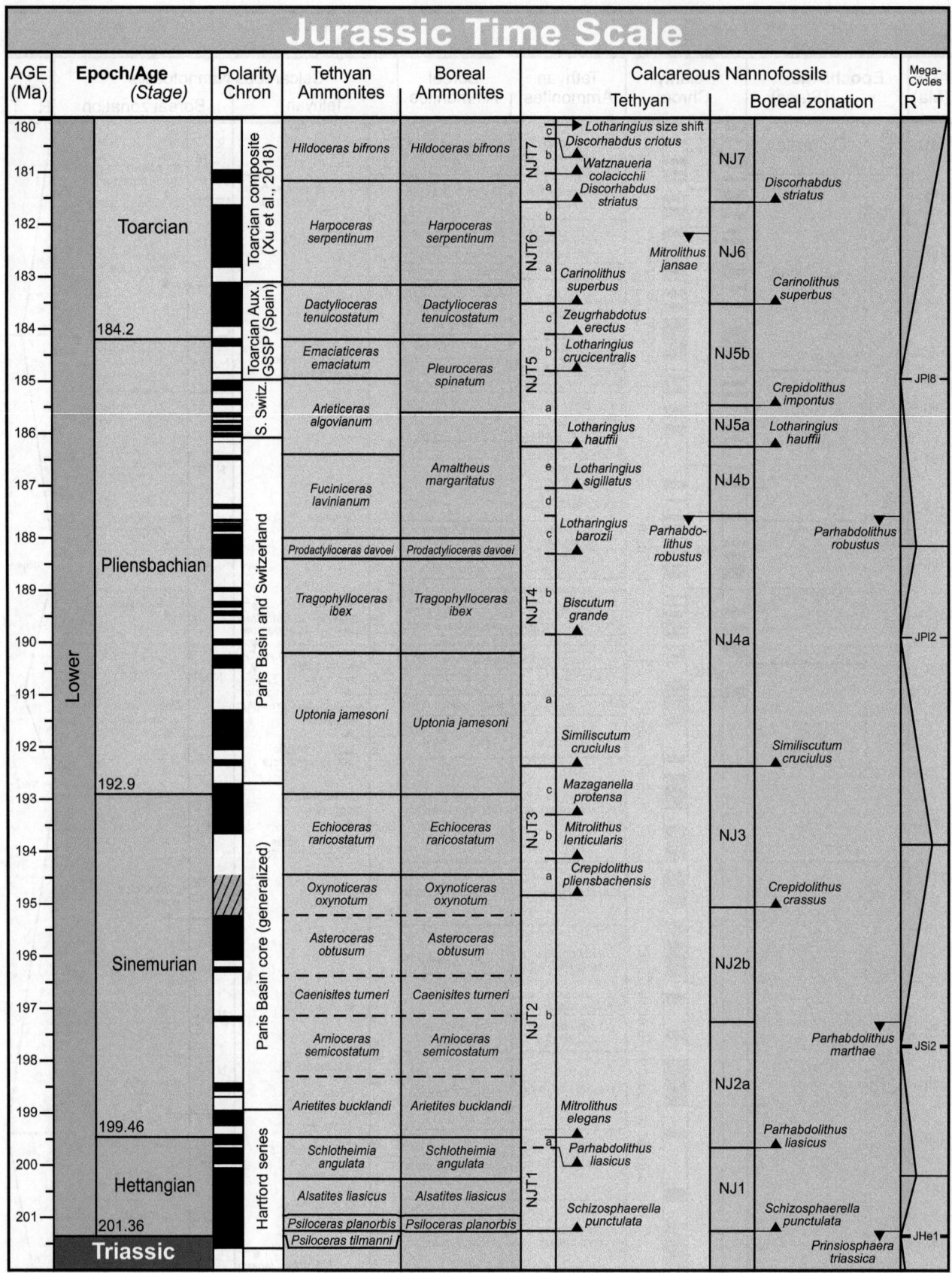

FIGURE 26.10 (Continued)

and Bown (1998) and Ferreira et al. (2019). The mega-cycles of sea-level trends are from Haq (2017). Additional Jurassic zonations, geochemical trends, sea-level curves, etc. are compiled in the internal datasets within the *TimeScale Creator* visualization package (https://timescalecreator.org).

and continued until the middle of the Early Cretaceous (Remane, 1985). They provide important Upper Jurassic and Lower Cretaceous correlation markers, especially in pelagic carbonates of the Tethyan—Atlantic seaway (reviewed by Remane, 1998).

26.2.1.5 Calcareous nannofossils

The Jurassic history of calcifying pelagic algae, which form calcareous nannofossils, consists of a more or less gradual diversification and shift in environmental preference from epicontinental seas to open ocean, accompanied by a steady increase in accumulation rate (summarized in de Kaenel et al., 1996; Bown et al., 2004; Erba, 2006; Suchéras-Marx et al., 2019). The earliest known calcareous nannofossils appeared in the late Triassic; but most of the initial robust types vanished during the end-Triassic extinctions. A single surviving coccolithophorid algal species diversified near the Hettangian/Sinemurian boundary; and the major radiation of Jurassic placolith coccoliths (plates from coccolithophores) took place during Pliensbachian to early Toarcian time. A second major pulse of diversification within the Aalenian—Bajocian boundary interval included the genus *Watznaueria*, which would dominate most assemblages during the rest of the Jurassic and entire Cretaceous (e.g., Erba, 1990; Olivero, 2008; Ferreira et al., 2019). A major overturn of Tethyan nannofossil assemblages took place in the late Tithonian. The Late Jurassic is marked by the appearance of nannoconids, a heavily calcified form, that are often regarded as the first effective carbonate-producers in the Mesozoic oceans and contributed to the nannofossil-rich limestones that characterize the Cretaceous pelagic realm (Erba, 2006; Casellato et al., 2008). The onset of calcareous nannoplankton, their size, and general morphology and the diversification trends may be responses to the chemistry of the ocean-atmosphere system, especially pCO_2 and Mg/Ca ratio (Erba, 2006).

Jurassic nannofossil zonations and markers in the Boreal/Sub-Boreal realm are calibrated to ammonite zones in northwestern Europe (e.g., Bown et al., 1988; Bown, 1998; von Salis et al., 1998). Nannofossil datums in the Tethyan/sub-Mediterranean realm are partially calibrated to ammonite zones (e.g., de Kaenel et al., 1996; Mattioli and Erba, 1999; Mailliot et al., 2006; Ferreira et al., 2019) and to Late Jurassic magnetic polarity zones (e.g., Bralower et al., 1989; Channell et al., 2010; Casellato, 2010).

26.2.1.6 Dinoflagellate cysts

Organic-walled cysts of dinoflagellates are an important correlation tool for the North Sea, where their datums are correlated to ammonite zones of the Boreal realm (e.g., Woollam and Riding, 1983; Riding and Ioannides, 1996; Ioannides et al., 1998; Wiggan et al., 2017). Several of these markers also occur in the Tethyan realm, but the ranges and correlation to ammonite zones are not as well established (Habib and Drugg,

1983; Ioannides et al., 1998; Kuerschner et al., 2007). Independent dinocyst zonations have been developed for the Jurassic of Australia (Helby et al., 1987), for the upper Jurassic of New Zealand (Wilson, 1984) and for other basins.

26.2.1.7 Radiolarians

Siliceous radiolarians are a major component of Jurassic pelagic sediments deposited under high productivity conditions, but their tests are rarely preserved jointly with aragonitic ammonite shells. Detailed radiolarian zonations for the Middle and Late Jurassic have been developed for the western margin of North America (e.g., Pessagno et al., 1993, 2009), for Japan (Matsuoka and Yao, 1986; Matsuoka, 1992), for the former Tethyan seaways exposed in Europe (Baumgartner, 1987; INTERRAD Jurassic-Cretaceous Working Group, 1995; De Wever, 1998), and for the North Sea region (Dyer and Copestake, 1989; Dyer, 1998).

These zonations can be partially correlated to each other; however, the calibration of the radiolarian assemblages to standard geologic stages and to the ammonite-based stratigraphic reference framework of Europe has been challenging and controversial (e.g., Pessagno and Meyerhoff Hull, 1996). An example is the divergent correlations for the radiolarian assemblage overlying basalt at Ocean Drilling Program (ODP) Site 801, which provided a key age control on the Callovian—Oxfordian portion of the marine magnetic anomaly M-sequence and global spreading rates. The radiolarian assemblages from the basal and interpillow sediments were originally interpreted as "late Bathonian to early Callovian" (Shipboard Scientific Party, 1990; Matsuoka, 1992; reviewed in Ogg et al., 1992) but were reinterpreted as equivalent to "middle Oxfordian" of western North America by Pessagno and Meyerhoff Hull (1996). In contrast the same radiolarian assemblages were assigned to "Bajocian" (Bartolini et al., 2000; Bartolini and Larson, 2001) based partly on the zonal calibrations developed by the INTERRAD Jurassic-Cretaceous Working Group (1995) for the Mediterranean region. Possible contributing factors to this divergence are diachroneity of radiolarian or ammonite datums and ranges among basins, errors in taxonomy assignments, imprecise correlation of radiolarian markers to regional ammonite stratigraphy, and miscorrelation of ammonite assemblages among paleogeographic provinces (Pessagno and Meyerhoff Hull, 1996; Kiessling et al., 1999; reviewed in Pessagno et al., 2009).

26.2.2 Terrestrial biostratigraphy

26.2.2.1 Dinosaurs

Dinosaurs are the celebrities amongst Jurassic fauna and the largest land animals that ever lived. Even though they dominated the terrestrial herbivores and carnivores, the biostratigraphic ranges of these enormous creatures are not well established relative to marine-based stages. A set of six generalized "land-vertebrate faunachrons" (LVFs) can be partially correlated across Pangea (Lucas, 2009).

Crocodile-like crurotarsan archosaurs were the main competitors of dinosaurs during the Late Triassic, but their mass extinction in latest Triassic enabled the dinosaurs to flourish and diversify (e.g., Benson, 2018). This crurotarsan extinction either coincides with the major mass extinction of marine life induced by environmental impacts of the CAMP flood basalt eruptions (e.g., Whiteside et al., 2010) or began slightly earlier (Lucas et al., 2011). The Jurassic begins with the appearance of the crocodylomorph *Protosuchus*, defining the base of the Wassonian LVF. The Dawan LVF begins with the appearance of theropod *Megapnosaurus* (*Syntarsus*), a 3-m-long long bipedal dinosaur, in approximately mid-Sinemurian.

No dinosaur genera are shared among any of the continents through most of the middle Jurassic, despite the presumed high mobility of the enormous sauropod dinosaurs (Lucas, 2009). A vaguely calibrated Dashanpuan LVF of possible Bajocian age and a Tuojiangian LVF of possible Bathonian–Callovian age are characterized by stegosaur and sauropod assemblages.

The sparse record of dinosaurs in Lower and Middle Jurassic strata contrasts with the rich fossil deposits of the upper Jurassic, such as the Morrison Formation of western North America. This Kimmeridgian–Tithonian interval included the carnivores of *Allosaurus* and *Ceratosaurus*, the huge sauropod *Brachiosaurus*, and moderate-sized ornithopods (iguanodontids and hypsilophodontids). The Comobluffian LVF, named from a Morrison excavation site, is coeval with the Ningjiangouan LVF of China (Lucas, 2009). These deposits can be partially correlated to the Kimmeridgian stage through magnetostratigraphy (e.g., Steiner et al., 1987), which is consistent with U–Pb ages of 152–156 Ma (Trujillo et al., 2006, 2008).

Flying over this Jurassic park were pterosaurs, which had evolved during the Late Triassic. The first bird, the renowned *Archaeopteryx* fossils from lagoonal settings in southern Germany, appeared in the Tithonian (Brusatte et al., 2015). Theropod ancestors to the future birds are known from the Oxfordian (Hu et al., 2009; Stone, 2010; Choiniere et al., 2010).

26.2.2.2 Plants, pollen, and spores

Terrestrial ecosystem stratigraphy is increasingly well calibrated relative to marine records due to the palynological study of marine sedimentary archives with well-established ammonite biostratigraphy, chemo- and magnetostratigraphy (e.g., Correia et al., 2019). Sporomorph records suggest that the land plant ecosystems did change significantly during the interval of the end-Triassic mass extinctions in the marine realm (Kuerschner et al.,

2007; Bonis et al., 2009, 2010; Lindström et al., 2012), and palynologists recognize first and last occurrences of individual species, as well as changing relative abundances of dominant pollen and spores, as stratigraphic markers for intra- and interbasinal correlation. For example, the pollen *Cerebropollenites thiergartii* has its first appearance close to the first occurrence of the ammonite species *Psiloceras spelae tirolicum* in the base-Jurassic GSSP at Kuhjoch, Austria (Hillebrandt et al., 2013). Thus the occurrence of this pollen is suggested to be a terrestrial marker for the base of the Jurassic also in continental settings or terrestrial depositional environments. Furthermore, some regional macroplant records indicate up to an 85% decline in species richness on the local scale, partly because some plant types are not well represented by the sporomorph record (Mander et al., 2010). Cycads, ginkgos, and conifers have diverse assemblages that respond to local climate (e.g., review by van Konijnenburg-van Cittert, 2008, for United Kingdom); but only a few regional spore-pollen zones/ranges have been published for selected time intervals (e.g., Srivastava, 1987, for Normandy-Germany; Morris et al., 2009, for offshore Norway). Transitional forms between gymnosperms and the angiosperm plants which diversify in the Cretaceous are interpreted from floral assemblages from North China that are below an ash bed dated as ~162 Ma, hence are early Kimmeridgian or older (Chang et al., 2009).

26.2.3 Physical stratigraphy

26.2.3.1 Magnetostratigraphy

The Jurassic magnetic polarity patterns are primarily compiled from ammonite-bearing sections in Europe, with the Sinemurian and the Callovian stages remaining the main uncertain intervals. The Late Jurassic M-sequence of marine magnetic anomalies has been correlated via magnetostratigraphy to ammonite zones. Polarity chrons prior to the Callovian do not have a corresponding marine magnetic anomaly sequence to provide an independent nomenclature or scaling system. Incorporating abbreviations derived from the corresponding ammonite zone or stage is one option (e.g., Ogg, 1995); with a similar philosophy applied at the series-level for the Triassic compilation by Hounslow and Muttoni (2010). However, until the polarity pattern spanning each stage has been adequately verified, a standardized nomenclature system is premature.

26.2.3.1.1 Early Jurassic through Aalenian magnetic polarity scale

The Hettangian Stage is dominated by normal polarity in the Paris Basin (Yang et al., 1996), England (Hounslow et al., 2004; Deenen et al., 2010; Hüsing et al., 2014), and in the Hartford Basin of eastern North America (Kent et al., 1995,

2017; Kent and Olsen, 2008), with only three brief well-documented reversed polarity zones ("H24r–H26r" of Kent and Olsen, 2008; "AQ1r–AQ3r" of Hüsing et al., 2014). Cyclostratigraphy of the Hartford Basin (United States) sequence enabled assignment of the polarity ages relative to the radioisotope-dated CAMP basalts (Kent and Olsen, 2008; Blackburn et al., 2013), which suggested that the uppermost reversed polarity zone "H26r" is at the base of Sinemurian. Subsequent chemo-, magneto-, and cyclostratigraphic work on the biostratigraphically well-constrained Lower Jurassic sedimentary record of the Bristol Channel Basin in Somerset (United Kingdom), including the base Sinemurian GSSP at Kilve, confirmed these three short reversals do straddle the Hettangian–Sinemurian boundary (Hüsing et al., 2014; Xu et al., 2017a).

The Sinemurian in outcrop sections is dominated by reversed polarity magnetization with relatively brief normal polarity zones (e.g., Channell et al., 1984; Steiner and Ogg, 1988). A detailed magnetostratigraphic study of a Paris Basin core was interpreted as indicating numerous additional brief polarity changes (Yang et al., 1996); but the absence of such features in outcrop studies suggests that most of these were single-sample anomalies or other artifacts. The approximate ammonite-zoned stratigraphy of that borehole (Montcornet) provides better biostratigraphic constraints on the main features of the Sinemurian polarity pattern, and a generalized column from that study is used for the GTS2020 scale.

The lower Pliensbachian (Carixian Substage) is also dominated by reversed polarity. The main polarity patterns are consistent between the major studies in southern Switzerland (Horner, 1983, and Horner and Heller, 1983) and the Montcornet Borehole (Moreau et al., 2002). However, a detailed study in central Italy (Speranza and Parisi, 2007) is difficult to correlate to the first two compilations, and this contrast may be a combination of different ammonite subzonal schemes, hiatuses and redeposited intervals, variable sedimentation rates, and other factors. Therefore the polarity patterns within each ammonite zone according to the synthesis by Moreau et al. (2002) are used in GTS2020, based on the midpoints of their reinterpreted ammonite zone boundaries. Cyclostratigraphic analyses of the biostratigraphically constrained composite magnetostratigraphy for the Pliensbachian Stage at the stratigraphically expanded and complete Llanbedr (Mochras Farm) Borehole was used to obtain durations for individual proposed Pliensbachian magnetochrons (Ruhl et al., 2016). Additional paleomagnetic studies from this borehole and/or biostratigraphically well-constrained outcrops are, however, essential to fully constrain the evolution of the Pliensbachian magnetic field and polarity record. For example, Speranza and Parisi (2007) have interpreted two unusually long-duration intervals of transitional or intermediate polarity during the earliest Pliensbachian, but these await confirmation.

An upper Pliensbachian (Domerian Substage) polarity pattern was obtained from a detailed study in southern Switzerland (Horner, 1983; Horner and Heller, 1983). However, those studies did not tabulate how the magnetostratigraphy links to the outcrop sections, thereby precluding a detailed correlation to the observed ammonite zones. Later work on the upper Pliensbachian biomagnetostratigraphy using the Montcornet borehole, Paris Basin, (Moreau et al., 2002), combined with integrated bio- and magnetostratigraphic studies on a Pliensbachian–Toarcian boundary succession in southern Spain (Comas-Rengifo et al., 2010b), have established that the boundary is marked by a short magnetochron of "normal" polarity, pre- and succeeded by short magnetochrons of "reversed" polarity. The base-Toarcian *mirabile* Subzone of the *tenuicostatum* Zone in the southern European realm is thus marked by an interval of reversed polarity (Comas-Rengifo et al., 2010b).

A magnetic polarity record for the entire Toarcian Stage was calibrated in the biostratigraphically complete and expanded Llanbedr (Mochras Farm) borehole (Xu et al., 2018a,b). Combined with the earlier Toarcian magnetic polarity data from Switzerland and Spain (Horner and Heller, 1983; Galbrun et al., 1990; Osete et al., 2007; Comas-Rengifo et al., 2010b), this has allowed construction of a composite magnetic polarity time scale spanning the entire Toarcian Stage (Xu et al., 2018b). The Llanbedr (Mochras Farm) borehole also has the potential to provide a cyclostratigraphic scale for the Toarcian Stage (Hesselbo et al., 2013). Proportional scaling of magnetozones within ammonite subzones with the implicit assumption of a linear correlation between biozone thickness and time in this Llanbedr (Mochras Farm) sedimentary archive was used provisionally by Xu et al. (2018b) to establish relative durations for the individual Toarcian magnetochrons as displayed on the GTS2020 chart.

Details on the magnetostratigraphy at the Toarcian–Aalenian boundary and Aalenian–Bajocian boundary come from the respective GSSPs at Fuentelsaz in Spain (Goy et al., 1996; Cresta et al., 2001) and Capo Mondego in Portugal (Henriques et al., 1994; Pavia and Enay, 1997).

26.2.3.1.2 M-sequence of marine magnetic anomalies and Bajocian through Late Jurassic magnetic polarity scale

The M-sequence of marine magnetic anomalies provides a reference pattern and chron nomenclature for the magnetostratigraphy from upper Jurassic through Lower Cretaceous fossiliferous sections. The most commonly used M-sequence is derived from the Pacific spreading centers as a combination of the Hawaiian and Japanese magnetic lineations with a slowly varying spreading rate (see Ogg, 2020, Ch. 5: Geomagnetic polarity time scale, this book). The oldest magnetic anomaly that is documented in all ocean basins is M25.

Numerous high-resolution magnetostratigraphy studies of Kimmeridgian and Tithonian strata have calibrated ammonites, calpionellids, and calcareous nannofossils from the Sub-mediterranean faunal province and DSDP cores from the central Atlantic to polarity Chrons M25 through M18 (e.g., Ogg et al., 1984, 1991; Lowrie and Ogg, 1986; Ogg, 1988; Bralower et al., 1989; Speranza et al., 2005; Channell et al., 2010; Pruner et al., 2010; Przybylski et al., 2010a). These calibrations constrain the relative duration of each ammonite zone within the Kimmeridgian and Tithonian stages. In turn, orbital-climate cycles recorded in Kimmeridgian and Tithonian strata enable computation of actual durations for some of the zones and polarity chrons, thereby a determination of oceanic spreading rates for the associated marine magnetic anomalies (e.g., Ogg et al., 2010b, 2012, 2020). The succession and relative widths of the modeled marine magnetic anomalies from M18r (base of Cretaceous) through M25r of the "classic" Hawaiian magnetic lineation M-sequence (e.g., Larson and Hilde, 1975) have been an excellent match to these cycle magnetic composites, whereas a later model that averaged three Pacific lineation sets (Tominaga and Sager, 2010) seems to be slightly distorted. Calibration of boreal sections to this magnetic time scale has been partially achieved for the Kimmeridgian and Tithonian (e.g., Ogg et al., 1994, 2010b; Houša et al., 2007; Przybylski et al., 2010a). Reversed polarity Chron M25r is early Kimmeridgian.

Marine magnetic anomalies older than Chron M25 are both closer spaced and lower in amplitude. Magnetic profiles over pre-M26 oceanic crust in the Pacific (Cande et al., 1978; Handschumacher et al., 1988) have been supported and extended by deep-tow surveys (Sager et al., 1998; Tivey et al., 2006; Tominaga et al., 2008, 2015) indicating a possible set of between 50 and 100 polarity chrons (semiarbitrarily grouped into clusters with a nomenclature of "M26" through "M44"). Pacific IODP Site 801 is within the marine anomaly cluster "M42," and the interpreted Bajocian−Bathonian age of Site 801 implies that marine magnetic anomalies M26 through M44 should span the Oxfordian, Callovian, Bathonian, and probably the Bajocian stages.

However, unlike the standard sea-surface polarity model of M0r through M25r, the interpretation of the deep-tow surveys is more ambiguous. A direct modeling of the deep-tow signals emphasizes the narrow paleomagnetic intensity fluctuations near the oceanic crust, thereby creating a model of numerous short-duration, low-amplitude polarity intervals. Alternatively, if the deep-tow data are projected to a mid-depth level, then close-spaced narrow-width fluctuations are diminished and the wider width features are emphasized. Therefore both geophysics models

for pre-M25r marine magnetic anomaly interpretations are published, with the hypothesis that the actual reversal history of the Earth's geomagnetic field is perhaps between these alternatives (e.g., Tominaga et al., 2008). Indeed, the outcrop-based magnetostratigraphy for the Oxfordian through Bajocian seems consistent with an intermediate model for such pre-M25 marine magnetic anomalies.

Magnetostratigraphic studies in Oxfordian strata with sub-Mediterranean ammonite zonation in Spain and Poland yielded a frequency of reversal patterns that seemed consistent with the extended Pacific model (e.g., Steiner et al., 1986; Juárez et al., 1994, 1995; Ogg and Gutowski, 1996); but correlations were ambiguous. Scaling of Oxfordian ammonite zones and subzones by cycle stratigraphy (Boulila et al., 2008a, 2010a) enabled a better approximation of the durations of the polarity zones observed within each of those subzones. Therefore a composite magnetostratigraphy compiled from approximately 20 Oxfordian sections in Europe with partial cycle scaling verified the main features of the deep-tow pattern from M26 through M37 (Przybylski et al., 2010b; Ogg et al., 2010a). Chron M37n of that marine magnetic anomaly model was correlated to a Callovian−Oxfordian boundary using the lowest occurrence of *Cardioceras redcliffense* in GTS2012, but the current working definition for the base Oxfordian using the lowest occurrence of *Cardioceras woodhamense* would correlate to the lower part of Chron M36Br (Ogg et al., 2010a, 2011).

Magnetostratigraphy investigations of the Callovian stage in England, France, and Poland are generally dominated by normal polarity with relatively few reversed polarity intervals (e.g., Ogg et al., 1990; Channell et al., 1990; Belkaaloul et al., 1995, 1997; Callomon and Dietl, 2000; Guzhikov et al., 2010; Gipe, 2013, Gipe et al., 2013). These reversed polarity clusters are mainly in the Bathonian−Callovian boundary interval, within the *Macrocephalites gracilis* ammonite zone, two in the *Erymnoceras coronatum* Zone, and in the latest Callovian. Similar normal polarity dominance is interpreted for marine magnetic anomaly clusters M37−upper M39; and the reversed polarity clusters are correlated to similar features in the mid-depth anomaly model. The implied span of Callovian-aged Pacific crust encompassed by anomalies M37−M39 is approximately half of the span of the Oxfordian-aged crust (intervals of anomalies M26−M37).

A suite of Bathonian and Bajocian ammonite-rich sections in Spain display a complex magnetic polarity pattern of frequent reversals within a dominance of reversed polarity (Steiner et al., 1987). These features are consistent with the interpreted high-frequency and reversed polarity bias of pre-M39n deep-tow marine magnetic anomalies. ODP Site 801 is on deep-tow marine magnetic anomaly M42; but the relationship of its radiolarian assemblages to geologic stages is disputed (see discussion in Jurassic radiolarians). However, the model that the lower units of oceanic crust at Site 801 formed during late Bajocian is consistent with the general match of magnetostratigraphy to trends of marine magnetic anomalies. As with the Oxfordian, a reliable correlation to the deep-tow record will require extensive cycle stratigraphy or other means of adjusting the Callovian—Bajocian magnetostratigraphy records for variable sedimentation rates and hiatuses. Once this is achieved, then the ages and spreading rates for the earliest history of the Pacific plate can be determined. As in GTS2012, the suggested correlation and corresponding estimated age model in GTS2020 for this Bathonian—Bajocian interval is based on a visual fit of the main trends in dominance by normal or reversed polarity of the magnetostratigraphy to the mid-depth deep-tow model. At this point, one can only conclude that the main trends in relative polarity dominance during the Bathonian—Bajocian appear to be consistent with the mid-depth deep-tow marine magnetic surveys of Pacific crust of this same age span.

26.2.3.2 Geochemical stratigraphy

Geochemical anomalies are increasingly used for (inter) regional and global stratigraphic correlation between marine, and also terrestrial, sedimentary archives. Hitherto, the main geochemical anomalies of the Jurassic that have been important for interregional correlation, have occurred in strata comprising the Triassic—Jurassic boundary and the lower Toarcian. A comprehensive compilation of Jurassic chemostratigraphic trends and excursions by Jenkyns et al. (2002) was enhanced by a detailed synthesis of the geochemical signatures associated with Jurassic—Cretaceous anoxic events (Jenkyns, 2010). Biostratigraphically well-constrained sedimentary archives are increasingly studied to document long- and short-term changes in geochemical proxy records, allowing for detailed chemostratigraphic correlation, as well as improved understanding of, for example, the magnitude and rate of change in global (bio-)geochemical cycling.

Chemostratigraphic analyses of sedimentary archives can, in principle, utilize any organic or inorganic geochemical proxy data that can be retained in sediments and that reflects a regional or global change, including (but not limited to) changes in stable isotope ratios, elemental concentrations and ratios, mineral occurrence and properties, molecular biomarker abundance and ratios, kerogen type, etc.

Importantly, the powerful aspect of chemostratigraphy or sedimentary geochemistry is that changes in sedimentary geochemical proxy records (1) often happen at rates that are significantly quicker than the fossil turnover rate of biozones and subzones, allowing for stratigraphic correlation at a temporally higher resolution than, for example, biostratigraphy and (2) are often paced at orbital (or Milankovitch) periodicities, allowing for direct and detailed constraints on the rate of change in the studied geochemical proxy, sedimentation rates, and relative time in the

sedimentary archive. Chemostratigraphy, just like magneto- and cyclostratigraphy and geochronology, should, however, always be used in conjunction with biostratigraphy, or independent biostratigraphic age constraints.

26.2.3.2.1 Carbon isotopes

Changes in the carbon isotope ratio of organic and inorganic substrates in sedimentary archives can reflect changes in the magnitudes of fluxes of carbon between the global exogenic carbon reservoirs and/or the $\delta^{13}C$ value thereof, and thus reflect changes (or perturbations) in the global exogenic carbon cycle. During major global change events, rapid negative shifts (excursions) in the global $\delta^{13}C$ signature are likely often linked to the release of ^{13}C-depleted carbon into the integrated ocean-atmosphere system. Conversely, rapid positive trends (or excursions) in the global $\delta^{13}C$ signature are likely linked to drawdown and sequestration of ^{12}C-enriched organic matter, from the ocean-atmosphere system into the sedimentary realm. Other factors may impact on the size of fluxes between exogenic carbon reservoirs, and the $\delta^{13}C$ values thereof, such as changes in the volume of global biomass, changing organic fractionation or carbon shuttling. Rapid, short-term changes as well as long-term trends in the global $\delta^{13}C$ signature can be used for temporally high-resolution stratigraphic correlation of sedimentary archives within a basin, regionally and globally, including marine—terrestrial correlation.

The Triassic—Jurassic transition is marked by a major carbon cycle perturbation, and associated excursion in $\delta^{13}C$ records. The end-Triassic mass extinction coincides with a negative carbon isotope excursion that may be linked to the initiation of widespread CAMP flood basalt volcanism, oceanic productivity collapse, and/or the release of methane (e.g., Pálfy et al., 2001; Hesselbo et al., 2002, 2007a; Ruhl et al., 2009, 2011; Ogg and Chen, 2020). The brief latest Rhaetian "initial" negative carbon isotope excursion preceded the base of the Jurassic by 100—200 kyr and is separated from the following earliest Hettangian "main" negative carbon isotope excursion by a brief return to more positive $\delta^{13}C$ values (Hesselbo et al., 2002). Global $\delta^{13}C$ values remain relatively negative throughout the Hettangian and Lower Sinemurian, as observed in $\delta^{13}C$ records from marine basins in the United Kingdom, Austria, Denmark, the United States, and South America (Kuerschner et al., 2007; Ruhl et al., 2010; Whiteside et al., 2010; Bartolini et al., 2012; Lindström et al., 2012; Xu et al., 2017a; Yager et al., 2017; Korte et al., 2019). Some marine sedimentary archives however show a $+5‰$ shift to more positive $\delta^{13}C$ values in the mid to late Hettangian, followed by a stabilization at about $+5‰$ during the late Hettangian through Sinemurian (e.g., Williford et al., 2007; van de Schootbrugge et al., 2008; Hesselbo et al., 2020b), possibly due to local carbon cycling, a change in dominant carbon fractionating processes and/or local diagenesis.

The early Toarcian exhibits even larger geochemical excursions associated with the Toarcian Oceanic Anoxic Event (T-OAE). This time period is marked by the widespread burial and sequestration of carbon in marine and continental sedimentary basins, under dysoxic to anoxic, or even euxinic (sulphidic) water columns (Jenkyns, 2010), leading to the formation of organic carbon-rich marine deposits, such as the Posidonienschiefer of Germany and the Jet Rock of England.

A significant 4‰ to 7‰ negative carbon isotope excursion, possibly one of the largest carbon isotope perturbations of the Phanerozoic, is directly associated with the T-OAE. It had a duration of between 200 kyr and >1 Myr based on the assignment of orbital (or Milankovitch) periodicities to the observed facies and geochemical oscillations (Suan et al., 2008; Kemp et al., 2011; Huang and Hesselbo, 2014; Boulila and Hinnov, 2017). The onset of this excursion is just below the base of the *Harpoceras serpentinum* Zone and reaches its lowest values in the lower *Harp. serpentinum* Zone. The onset is also marked by a step-wise shift to more negative values that are possibly orbitally paced and of stratigraphic correlation value (Kemp et al., 2005; Hesselbo and Pienkowski, 2011).

The negative carbon isotope excursion is superimposed on a much longer positive excursion in global $\delta^{13}C$ values, which originates in the upper Pliensbachian *Pl. spinatum* Zone and is maintained until halfway through the Toarcian *Harp. bifrons* Zone (Jenkyns et al., 2002; Hermoso et al., 2013; Xu et al., 2018a). The observed negative excursion has been attributed to the release of ^{13}C-depleted biogenic and/or thermogenic methane from ocean-floor clathrates or dyke-/sill-intruded subsurface (organic-rich) shales, respectively (Hesselbo et al., 2000; Kemp et al., 2005; McElwain et al., 2005; Svensen et al., 2007). It is commonly thought to coincide also with the initial phase of eruptions of the large Karoo-Ferrar Igneous province in Africa−Antarctica, at ~183 Ma (e.g., Pálfy and Smith, 2000; Kerr, 2000; Kemp et al., 2005), forming a likely additional source of carbon from the degassing of the emplaced basalts. Although earlier studies were centered around the northwest European realm, the early Toarcian carbon isotope excursion has been observed in geographically widespread marine sedimentary archives, on both hemispheres, and in both organic and inorganic substrates (Suan et al., 2011; Al-Suwaidi et al., 2016; Fu et al., 2016; Them et al., 2017b).

A major transgression evident from the upper *Dacty. tenuicostatum* Zone and through the lower *H. serpentinum* Zone (e.g., Hallam, 1981; Jenkyns, 1988) and negative steps in the carbon isotope excursion partially coincide with flooding surfaces in the sea-level rise (Hesselbo and Pienkowski, 2011).

A smaller 2‰ to 3‰ negative carbon isotope excursion was furthermore observed globally at the Pliensbachian−Toarcian boundary (Hesselbo et al., 2007b; Littler et al., 2010), with a 200−300 kyr duration (Martinez et al., 2017). Due to common condensation in the boundary interval, this basal episode is not as well studied as the major phase of the T-OAE that follows it, but it may have been linked to initial phases of Karoo-Ferrar eruption.

Oxygen-poor conditions likely contributed to the bivalve extinction pulses already during the earliest Toarcian (e.g., Wignall and Bond, 2008).

More recently, many more Jurassic perturbations in the global $\delta^{13}C$ signature, reflecting changes in the global carbon cycle, have been reported, some shorter in duration and/or magnitude, such as the Liasidium Event in the Sinemurian *Ox. oxynotum* Zone (Riding et al., 2012; Hesselbo et al., 2020a), or much longer in duration, such as the prolonged negative $\delta^{13}C$ excursion at the Sinemurian−Pliensbachian boundary (Jenkyns et al., 2002; Korte and Hesselbo, 2011; Price et al., 2016a; Storm et al., 2020). Similarly, positive and negative carbon isotope excursions have been reported for the Middle Jurassic lower Aalenian and lower Bajocian stages and the Aalenian−Bajocian, Bajocian−Bathonian, and Bathonian−Callovian boundaries (e.g., Bartolini et al., 1996, 1999; Jenkyns et al., 2002; Morettini et al., 2002; Sandoval et al., 2008; Silva et al., 2019).

The upper Jurassic Oxfordian contains significant positive and negative excursions in $\delta^{13}C$ (e.g., Padden et al., 2002; Jenkyns et al., 2002; Gröcke et al., 2003; Weissert and Erba, 2004), of which a relatively brief, but pronounced positive peak within the upper *Perisphinctes plicatilis* ammonite zone might serve as useful global correlation horizon (e.g., Pearce et al., 2005; Głowniak and Wierzbowski, 2007; Wierzbowski, 2015). This positive excursion in carbon isotopes is within a broad elevated ^{13}C enrichment that is coeval with major carbon-rich sediment deposits that became the hydrocarbon source rocks of Saudi Arabia, the North Sea, Siberia, and other regions. Smaller changes are recognized throughout the Kimmeridgian−Tithonian (e.g., Bartolini et al., 1999; Morgans-Bell et al., 2001). The Jurassic−Cretaceous boundary interval has generally low $\delta^{13}C$ values but lacks any carbon isotope excursions to aid in global correlation of the system boundary (e.g., Price et al., 2016b).

26.2.3.2.2 Oxygen-stable isotopes and other climate proxies

The Jurassic is an interval of general warmth, without extensive glacial deposits, but significant regional or global cooling events or colder periods, possibly even marked by the establishment of small icecaps, may have occurred (Rogov and Zakharov, 2010; Korte and Hesselbo, 2011; Korte et al., 2015; Ruebsam et al., 2019).

Oxygen isotope ($\delta^{18}O$) records possibly reflecting marine temperature trends are patchy and heavily biased toward records from Europe and Russia (e.g., Veizer et al., 1999, 2000; Jenkyns et al., 2002; Zakharov et al., 2006; Korte et al., 2015). However, another indicator of temperature, the occurrences of glendonites, pseudomorphs after the hydrated

$CaCO_3$ polymorph ikiaite, occurring within siliciclastic deposits, are favored in methane-rich polar marine sediments (Morales et al., 2017) and provide support to interpretations from oxygen isotopes in marine macrofossil shells suggesting reduced seawater temperatures. Glendonite occurrences are observed in mid to high-latitude settings during a few Jurassic intervals (Rogov and Zakharov, 2010). Thus it appears that the average climates were overall warm (lighter $\delta^{18}O$ values), but with possible cold intervals in the Late Pliensbachian and in the Aalenian−Bajocian.

During the late Pliensbachian a possible cold interval coincides with a positive carbon isotope excursion, suggesting carbon drawdown and diminished atmospheric pCO_2 (e.g., Korte and Hesselbo, 2011). Cooler temperatures inferred for the Aalenian appear to have culminated in the Bajocian, on the basis of peak glendonite abundance. There are only rare reports of glendonites in the Bathonian through lower Callovian (Rogov and Zakharov, 2010). No cold-climate markers are found to support a postulated "cold snap" and glacial-induced sea-level regression at the Callovian−Oxfordian boundary (e.g., Dromart et al., 2003; Pellenard et al., 2014b), although the middle Callovian may have been an intermediate climate optimum followed by a relative cooling (e.g., Cecca et al., 2005; Zakharov et al., 2006). No glendonites are recorded from the upper Jurassic, where there is a trend toward more equable climates—the Boreal/Tethyan temperature difference of c. 7°C−9°C in the middle Oxfordian appears to decrease toward the end-Jurassic (Zák et al., 2011). Late Jurassic Greenhouse conditions reached a maximum during the Kimmeridgian (e.g., Frakes et al., 1992; Zakharov et al., 2006; Sellwood and Valdes, 2008; Wierzbowski, 2015), coinciding with a peak in global sea level. In Britain the humid Kimmeridgian was followed by aridity through the end of the Jurassic (e.g., Hesselbo et al., 2009). These general temperature trends are consistent with other paleoclimate indicators (Jenkyns et al., 2002).

Modeling of Jurassic climates suggests that atmospheric carbon dioxide levels were at least $2-4 \times$ the present level (reviewed by Sellwood and Valdes, 2008). Elevated carbon dioxide may have played a role in the relatively shallow carbonate-compensation depth (CCD) in the oceans (e.g., Ogg et al., 1992). An explosion in organic marine primary producers, with species richness in dinoflagellates rapidly increasing at the Aalenian−Bajocian boundary and peaking in the Kimmeridgian, may have significantly increased carbon drawdown from the ocean-atmosphere system, reducing global warming and or increasing the dissolved and particulate organic and inorganic carbon pools. Increased productivity by calcareous plankton during the Tithonian, especially the robust nannoconid types of nannoplankton, contributed to the lowering of the CCD and onset of the chalk (*creta*) deposits that characterize the Lower Cretaceous in all ocean basins. A decrease in pCO_2 during the Tithonian may have been a major factor in this evolutionary surge of calcifying plankton (e.g., Erba, 2006; Casellato et al., 2008; Ferreira et al., 2019).

26.2.3.2.3 Strontium and Osmium isotope ratios

Marine $^{87}Sr/^{86}Sr$ through the Jurassic progressively decreased from its end-Norian peak (at 0.70795) to an intermediate low (at 0.70708) at the end of the Pliensbachian, indicating a minimum in contribution of nonradiogenic ^{86}Sr from weathering of continental crust or ancient sediments relative to the supply of radiogenic ^{87}Sr from hydrothermal alteration of mid-ocean ridge basalts or weathering of continental flood basalts (Jones et al., 1994a,b; McArthur et al., 2000, 2001; Jones and Jenkyns, 2001; McArthur et al., 2020b, Ch. 7: Strontium isotope stratigraphy, this book). The initial stages of the downturn in marine $^{87}Sr/^{86}Sr$ from the latest Norian peak (0.70795) spanned the major flood basalt outpouring of the Central Atlantic Magmatic Province (CAMP) along the future central Atlantic seaway, at the Triassic−Jurassic transition, but the interpretation of the impact of these CAMP eruptions and major environmental shifts upon the rate of change in the oceanic strontium isotope ratios depends on estimates for the duration of the Rhaetian and Hettangian stages and the filtering of Sr isotope ratios derived from different fossil types (Cohen and Coe, 2007; McArthur, 2008, see also discussion in Ruhl et al., 2016). Similarly, a detailed understanding of the effect of the early Toarcian Karoo−Ferrar volcanic province on the rates and timing of the reversal in strontium isotope trends depends on estimates of the elapsed duration of the earliest Toarcian ammonite zone (e.g., Kemp et al., 2005; Cohen and Coe, 2007; McArthur et al., 2008; Jenkyns, 2010; Huang and Hesselbo, 2014).

Previously, including for GTS2012, the assumed linear changes in $^{86}Sr/^{87}Sr$ (and therefore assumed stability in the supply of Sr isotopes from the weathering of continental crust relative to hydrothermal alteration of mid-ocean ridge basalts) through the Hettangian, Sinemurian, Pliensbachian, and Toarcian stages, were used to estimate and scale the Early Jurassic time scale (Weedon and Jenkyns, 1999; McArthur et al., 2016). Radioisotopic and astrochronologic age constraints on Lower Jurassic sedimentary successions are incompatible with this assumption, suggesting that changes in the relative contribution of continental crustal weathering versus hydrothermal alteration of ocean-island basalts caused nonlinear changes in the isotopic composition of the ocean strontium reservoir (Ruhl et al., 2016 and references therein).

Rising strontium isotope ratios during the early Toarcian, likely in response to Karoo−Ferrar large igneous province (LIP) volcanism induced elevated weathering rates of continental crustal material, resulted in a sustained plateau (0.70730) through the Aalenian. Strontium isotope ratios again decreased through the Bajocian to middle Callovian, with a possible shoulder spanning the Bajocian−Bathonian boundary (Jones et al., 1994b).

During the early Oxfordian, marine $^{87}Sr/^{86}Sr$ reached its lowest ratio (0.70686) throughout the entire Phanerozoic (McArthur et al., 2001). This pronounced episode may indicate a major pulse of seafloor hydrothermal activity (Jones et al., 1994b; Jones and Jenkyns, 2001), which is supported by the formation and expansion of new spreading centers, as Pangea continued its breakup, and by interpretations of other geochemical, deep-sea sediment and spreading rate evidence (e.g., Ogg et al., 1992). The only Phanerozoic minimum that approached this ultralow $^{87}Sr/^{86}Sr$ ratio was at the end of the Middle Permian. After the middle Oxfordian, the strontium isotope ratio began a long-term increase that peaked in the Barremian of the Early Cretaceous.

In contrast to strontium the osmium has a short oceanic residence time that enables recording of pronounced excursions in response to the chemical weathering of the fresh basalts of the end-Triassic and early Toarcian flood basalts (e.g., Cohen et al., 2004; Cohen and Coe, 2007), although estimates of the magnitude of the global signal can be distorted by basinal effects. Recent comparison of the evolution of the $^{187}Os/^{188}Os$ signature through lower Toarcian in the more restricted Cleveland Basin (Yorkshire, United Kingdom), and the more open marine Cardigan Bay Basin (Wales, United Kingdom) and Bighorn Creek (Alberta, western Canada) shows a significant positive excursion, albeit with different magnitudes between localities, suggesting significantly elevated global continental weathering rates during the T-OAE (Cohen et al., 2004; Percival et al., 2016; Them et al., 2017a).

Increasing $^{187}Os/^{188}Os$ values from the lower Pliensbachian at the base-Pliensbachian GSSP at Robin Hood's Bay (Yorkshire, United Kingdom) suggests increasing continental weathering at that time (Porter et al., 2013), which is supported by a ~ 2 Myr plateau in $^{87}Sr/^{86}Sr$ values superimposed on the long-term Early Jurassic Negative Trend (Jenkyns et al., 2002; Ruhl et al., 2016).

The limited dataset for middle—upper Jurassic seawater Os isotopic compositions also shows a rapid change from low $^{187}Os/^{188}Os$ (with a value of c. 0.25) in the Callovian to more radiogenic values (of c. 0.6) in the Kimmeridgian—Tithonian, possibly suggesting the progressive supply of nonradiogenic crustal material into the late Jurassic ocean (Cohen et al., 1999; Selby, 2007).

Except for the Aalenian and Oxfordian time-intervals marked by prolonged and relatively stable $^{87}Sr/^{86}Sr$ values, the rapidly changing ratios of marine $^{87}Sr/^{86}Sr$ can enable global correlation, at times at the ammonite subzone level (McArthur et al., 2001).

26.2.3.3 Cyclostratigraphy

Astronomically forced cyclostratigraphy is an important tool in measuring Jurassic geologic time and establishing "floating" astronomical time scales. The signatures of orbital eccentricity forcing of the precession index dominates much of Jurassic cyclostratigraphy; therefore most Jurassic cyclostratigraphic scales have a resolving power of the long-eccentricity cycle (405-kyr periodicity) or short-eccentricity (100-kyr). Obliquity (~ 37-kyr) forcing had a lesser, although still important role (e.g., as recently summarized by Hinnov, 2018). In GTS2020, we not only largely follow the recent review and synthesis of Huang (2018) but also incorporate studies that have been published subsequently.

26.2.3.3.1 Early and Middle Jurassic cycle stratigraphy

26.2.3.3.1.1 Hettangian The cyclostratigraphic-calibrated length of the Hettangian Stage is currently a matter of some debate. From detailed chemostratigraphic and magnetostratigraphic analysis of coastal outcrops in Somerset United Kingdom, which include the base-Sinemurian GSSP, Ruhl et al. (2010), Deenen et al. (2010), and Hüsing et al. (2014) proposed a duration for the Hettangian of 1.9 Myr (assuming a precise correlation to the base-Hettangian GSSP in Austria using carbon isotope stratigraphy), principally by recognition and counting of the 100-kyr short eccentricity cycle. This duration finds agreement from cycle analysis in the magnetostratigraphy-calibrated Hartford Basin lacustrine successions of the eastern United States (Kent et al., 2017). These interpretations from Somerset are supported by analysis of interpreted 405-kyr long eccentricity cycles from the same dataset (Huang, 2018). However, this interpretation is challenged by Weedon et al. (2018, 2019) who argue that all the SW UK coastal sections contain multiple stratigraphic gaps at different levels, detectable by high-resolution ammonite biostratigraphy, and thus they proposed a composite cyclostratigraphic floating astrochronology, with a minimum duration for the Hettangian of 4.1 Myr. In addition to the occurrence of hiatuses, the principal difference between interpretations is the amount of time assigned to the first zone of the Hettangian, the *Ps. tilmanni* Zone, with the implication that the basal Jurassic in Somerset is strongly condensed. In the absence of stratigraphically well-constrained radioisotopic dates or floating astrochonologies, we here conservatively retain the shorter calibration for the Hettangian but recognize that this may change if and when supporting data for a long Hettangian are obtained.

26.2.3.3.1.2 Sinemurian Apart from the basal ammonite zone, which has a partial cyclostratigraphic calibration (Ruhl et al., 2010; Xu et al., 2017a; Weedon et al., 2019), there are as yet few cyclostratigraphic studies of the Sinemurian Stage, with the longest analyzed dataset being an oceanic ribbon chert succession at Inuyama, Japan (Ikeda et al., 2017). This succession does not yield reliable correlation markers to locations elsewhere for the Sinemurian.

26.2.3.3.1.3 Pliensbachian Analysis of elemental time series (principally calcium and iron content) measured from the Llanbedr (Mochras Farm) borehole, West Wales, provides a floating astrochronology for the Pliensbachian Stage (Ruhl et al., 2016). This marine hemipelagic succession contains a fairly abundant ammonite fauna and so allows durations of the standard ammonite biozones to be estimated. As the cyclostratigraphic interpretation is ultimately based on the strongly fluctuating carbonate content of the mudstone sequence, there is the possibility that diagenetic carbonate precipitation may skew results particularly in the uppermost ammonite zone. Corroboration and verification from additional data are still strongly desirable, but here we use the estimated duration for the stage of 8.7 Myr. This interpretation is compatible with the minimum duration of 6.67 Myr for the Pliensbachian from precession-dominated cyclicity in the Belemnite Marls of southern England (Weedon and Jenkyns, 1999), combined with linear strontium isotope trends, and cycle stratigraphic data from Robin Hood's Bay in northeast England (van Buchem et al., 1994) and Breggia Gorge in southern Switzerland (Weedon, 1989). This duration is also consistent with the minimum estimate of 5 Myr derived from precession-dominated strata in northern Italy (Hinnov and Park, 1999).

26.2.3.3.1.4 Toarcian There have been many attempts to provide a cyclostratigraphic-derived duration for the T-OAE, but as yet there is no consensus whether the dominant cyclicity in the analyzed records is from ~20-kyr precession, ~35-kyr obliquity, or 100-kyr eccentricity (e.g., Kemp et al., 2005; Suan et al., 2008; Boulila et al., 2014, Huang and Hesselbo, 2014; Ruebsam et al., 2014; Boulila and Hinnov, 2017; Thibault et al., 2018; Boulila et al., 2019), resulting in strongly discrepant results of 200 kyr to >1 Myr. Here we favor the 100-kyr interpretation of these cycles that define a step-like motif in many isotopic records for the T-OAE negative carbon isotope excursion (on the basis that this interpretation requires less extreme rates of sedimentation change in continuous successions) and therefore a long duration for the anoxic event itself. The duration of the interval from the Pliensbachian–Toarcian boundary to the T-OAE has also been a focus of attention (Marṭinez et al., 2017; Ait-Itto et al., 2018).

For estimation of the duration of the whole stage, time series of magnetic susceptibility (Huret et al., 2008) and portable-XRF-derived Ti content from the Sancerre borehole, Paris Basin, France, has been used, which yielded a minimum duration of 8.3 Myr from 405-kyr long eccentricity cycles (Boulila et al., 2014, 2019), a result largely but not entirely confirmed by Huang (2018). The analysis of Huang results in a Toarcian stage that is longer by about a million years on interpretation of condensation and even possible missing cycles.

Carbonate cycles interpreted as obliquity-dominated oscillations within the Sogno formation in Northern Italy yielded an 11.37 ± 0.05 Myr duration for the combined Toarcian and Aalenian stages (Hinnov and Park, 1999; Hinnov et al., 1999).

In view of the considerable incompatibilities between interpretations, we do not use cyclostratigraphy to scale ammonite zones for the Toarcian, although expect this approach to be successful in future iterations of the time scale. The floating astrochronology for the earliest Toarcian (Martinez et al., 2017; Ait-Itto et al., 2018) is, however, used here to link single crystal ID-TIMS U–Pb ages from Peru (Sell et al., 2014) to the stage boundary.

26.2.3.3.1.5 Aalenian The Aalenian stage of Sogno, Italy (Hinnov and Park, 1999) display cycles in the precession–eccentricity range. Tuning the longer period to 405 kyr yields the total duration of 3.85 Myr (Huret et al., 2008; Huang, 2018) used in GTS2020. The estimated durations for the ammonite zones are: *Leio. opalinum* (2.0 Myr), *Ludw. murchisonae* (0.8 Myr), *Bras. bradfordensis* (0.4 Myr), and *Graph. concavum* (0.43 Myr), although placements of some of these ammonite zonal boundaries are indirectly assigned according to calibrated nannofossil datums.

26.2.3.3.1.6 Bajocian–Callovian Cycle stratigraphy has been interpreted only for short intervals within the Bajocian, Bathonian, and Callovian stages. A bedded chert succession in the Lombardy basin was analyzed by Ikeda et al. (2016) who derived an estimate of ~4 Myr for the Bajocian and Bathonian combined, although the biostratigraphic markers at the top of the analyzed low-sedimentation-rate sections are imprecise. In contrast a duration of about 4 Myr was proposed for the Lower Bajocian alone from the French Subalpine Basin (Suchéras-Marx et al., 2013); such a long length for the Bajocian is at present difficult to reconcile with ages based on the M-series model for the Bajocian–Aptian. The Gnaszyn Formation (Poland) suggests that the Lower Bathonian spans a minimum of 2 Myr based on tuning to cycles interpreted as 405-kyr (Ziólkowski and Hinnov, 2009). Potential durations from the well-defined lithologic oscillations across the Bathonian GSSP, Ravin du Bès Section, SE France (Fernández-López et al., 2009) suggest ammonite zone durations for *P. parkinsoni* (~0.6 Myr) of uppermost Bajocian, and *Zig. zigzag* (~0.6 Myr) and *Procerites aurigerus* (~0.2 Myr) of lower Bathonian.

Analysis of boreholes to test feasibility of a radioactive-waste storage into upper Callovian clay-rich deposits in the Paris Basin indicated that the upper Callovian spanned 0.9 ± 0.1 Myr and the middle Callovian *Erym. coronatum* Zone spanned a similar 0.9 ± 0.1 Myr (Huret, 2006; Lefranc et al., 2008). Deng et al (2015) analyzed a fluvio–lacustrine succession from NW China spanning the Callovian–Kimmeridgian interval and evidencing a regular cyclicity, but with relatively limited independent age constraint. Lack of consistent results for the Bajocian—Bathonian, combined with the lack of a calibrated astronomical time scale for the Callovian, led Huang (2018) to state that it was not yet possible to compile an astronomical time scale for the Middle Jurassic.

26.2.3.3.2 Late Jurassic cycle stratigraphy

The majority of the late Jurassic has been scaled by cycle stratigraphy, and the Boreal and Tethyan datasets have been merged using magnetostratigraphic intercorrelations. The intervals of cycle-scaled magnetozones were correlated to the M-sequence of marine magnetic anomalies, thereby supporting a spreading rate model. A linear fit to these cycle-determined spreading rates was projected to the pre-Oxfordian mid-depth-tow record of marine magnetic anomalies, which, in turn, was used as a reference magnetic polarity scale to correlate the main features of the Bajocian through Callovian magnetostratigraphy.

26.2.3.3.2.1 Oxfordian Cycle stratigraphy of the ammonite-zoned Lower-to-Middle Oxfordian Terres Noires formation in the Vocontian Basin (SE France) indicated a dominant 405-kyr signal, and a total span for the lowermost four ammonite zones of 4.1 Myr (Boulila et al., 2008a, 2010a). A long span for the *Quen. mariae* ammonite zone is verified by independent cycle stratigraphy analysis of French boreholes (Lefranc et al., 2008).

Middle Oxfordian through Lower Kimmeridgian Carbonate platform successions with rare ammonite levels in the Swiss and French Jura Mountains records major depositional sequences in response to sea-level changes (e.g., Gygi et al., 1998; Gygi, 2000). Small-scale oscillations within these sequences were interpreted as a record of short-term 100-kyr eccentricity orbital-climate cycles to give an estimate of durations for the associated ammonite zones (summarized in Strasser, 2007), although it is probable that there are "missing beats" at some emergent surfaces of sequence boundaries.

26.2.3.3.2.2 Kimmeridgian—Tithonian Cycle stratigraphy at La Méouge (France) yielded durations for the Lower Kimmeridgian ammonite zones (Boulila et al., 2008b, 2010b) that are quite similar to magnetostratigraphy-derived estimates. The Kimmeridge Clay Formation of Dorset (southern England), which spans the Kimmeridgian and Lower Tithonian, has been subject of several cyclostratigraphic studies (e.g., Waterhouse, 1995; Weedon et al., 1999; Huang et al., 2010b). The Kimmeridge Clay was continuously cored with calibrations to individual ammonite zones (e.g., Cope, 2009), and a record of its cyclicity was derived from multiple proxies (Morgans-Bell et al., 2001; Weedon et al., 2004). This dataset was further

refined and tuned to 405- and 100-kyr eccentricity signals (Huang et al., 2010b). The middle part of the *Aul. autissiodorensis* ammonite zone through the top of the *Virg. fittoni* Zone spans 3.72 Myr. This duration incorporates a refinement based on correlation to French magnetostratigraphy sections, which implies that there are three 405-kyr cycles within an ambiguous cycle stratigraphy interval (either two or three 405-kyr cycles) of the *Pect. elegans—Pect. wheatleyensis* ammonite zones in the Kimmeridgian clay (Fig. 2 of Huang et al., 2010b; Ogg et al., 2010b). This total span is similar to the 3.91 Myr estimate for the same interval based only on 38-kyr obliquity tuning (Weedon et al., 2004). Adjustment after reanalysis by Huang (2018) yielded a minimum duration for the Kimmeridgian of 3.64 Myr and the early Tithonian of 3.85 Myr, which is reflected in GTS2020. Upon incorporating potential correlation of ammonite zones between Sub-Boreal and Boreal realms to correlate the lower part of the Volgian Hekkingen Formation and the Kimmeridgae Clay Formation, the composite cyclicity of the two formations indicates a duration of 6.03 Myr for the Lower and Middle Volgian and possibly a 5.2-Myr duration for the Upper Volgian (Huang et al., 2010b). These spans for the Volgian do not contradict a ~7-Myr duration for the entire Tithonian Stage based on radioisotopic U—Pb dates from zircons in volcanic ashes in the Neuquén Basin of Argentina (Lena et al., 2019a).

Detailed cyclostratigraphic, magnetostratigraphic, and biostratigraphic analysis of the Vaca Meurta Formation of the Neuquén Basin, Argentina has also been undertaken and impacts on the inferred age for the J/K boundary (summarized in Kietzmann et al., 2018, and discussed further later).

26.2.3.4 Sequence stratigraphy

Jurassic sea-level trends have been compiled for various intervals in different basins. Examples include: Britain and North Sea (e.g., Partington et al., 1993; Hesselbo, 2008), Switzerland—France (e.g., Gygi et al., 1998; Gygi, 2000; Colombié and Rameil, 2007), Poland (Pienkowski, 2004), Greenland (Surlyk, 1990), Russia (Sahagian et al., 1996; Pinous et al., 1999), and Arabia (Sharland et al., 2001, 2004; Haq and al-Qahtani, 2005; Al-Husseini and Matthews, 2006). The regional trends have been compared on a global scale to construct global sequence stratigraphy scales and eustatic curves (e.g., Arkell, 1956; Hallam, 1981, 2001; Haq et al., 1988; Hardenbol et al., 1998). These results have been synthesized by Haq (2017), and this compilation is used in GTS2020. See also the review in Simmons et al. (2020, Ch. 13: Phanerozoic eustasy, this book).

The main Jurassic sea-level trend is a progressive sea-level rise from the latest Triassic until the late Kimmeridgian (peaking in the *A. eudoxus* ammonite zone). A major sea-level fall trend through the Tithonian reaches a minimum in the late Berriasian. Superimposed on this main cycle are several major sequences. Assignments of even fine-scale sequences depend on interpretation models for the response of sediment facies (other than obvious emergent soils or flooding surfaces) to relative sea-level changes. Therefore interpretations vary among stratigraphers for assigning small-scale sequences within a given region. For example, the same relative enrichment in carbonate within Kimmeridgian basinal successions can be interpreted as a sea-level maximum-flooding episode (e.g., Boulila et al., 2010b, 2011) or as a sea-level lowstand (e.g., Mattioli et al., 2011) depending on whether the carbonate is considered to be mainly pelagic or mainly export wafted from adjacent carbonate platforms. There are innumerable other examples of such diametrically opposite interpretations, applied particularly to the offshore marine facies successions.

Many of these fine-scale sequences appear to correspond to 405-kyr long-eccentricity-induced orbital-climate cycles. For example, the main sequences interpreted from carbonate-clay changes in the Lower Oxfordian of southern France and from sand-influxes and hiatuses in the lower-to-middle Oxfordian of the Dorset Coast (England) represent these 405-kyr orbital-climate oscillations (Boulila et al., 2010a; Ogg et al., 2010a). In this particular case a lowstand exposure in Dorset corresponds to an episode of carbonate-enrichment in the basinal successions of SE France. It remains speculative whether these periodic fine-scale sequences represent major eustatic sea-level oscillations (e.g., storage of water in high-latitude or high-altitude glaciers) or if they are mainly advances/retreats of local coastlines in response to 405-kyr climate cycles altering regional weathering styles and runoff (and oceanic productivity). High-resolution correlation and systematic interpretation of such sequences are required among different paleogeographic realms, which has not yet proven possible. High-resolution elemental records that can be used as proxies for grain size changes in distal environments, such as those of Hermoso et al. (2013) or Thibault et al. (2018), promise a method that can better integrate inferred sea-level changes with multiple other stratigraphic techniques.

26.2.3.5 Other major stratigraphic events
26.2.3.5.1 Large igneous provinces
26.2.3.5.1.1 Central Atlantic Magmatic province The
CAMP may be the geographically largest known LIP of flood basalts and sills (e.g., Marzoli et al., 1999; Hames et al., 2000). The eruptive centers of this c. 10-million km^2 volcanic outpouring extended from central-Northern South America across Eastern North America and the western Sahara to Spain. The recalibrated radioisotopic dates center on ~201.5 Ma, just prior to the base of the Jurassic (Table 26.1), with onset of the earliest magmatic activity at 201.635 ± 0.029 Ma for the Kakoulima-layered mafic

intrusion in Guinea (Davies et al., 2017). Stratigraphic correlations from palynology, ammonoid biostratigraphy, carbon isotope stratigraphy, and magnetostratigraphy, as well as U−Pb dates from ash beds, firmly establish that the onset of CAMP igneous activity is the same age as the end-Triassic mass extinction and therefore implicate volcanic processes as the prime causal agent for the extinction (e.g., Olsen et al., 2003; Hesselbo et al., 2002; Wignall and Bond, 2008; Jourdan et al., 2009; Ruhl et al., 2009; Blackburn et al., 2013; Wotzlaw et al., 2014; Yager et al., 2017; Korte et al., 2019; Panfili et al., 2019). Sedimentary mercury has also been reported recently associated with this boundary and may be useful as a stratigraphic marker for LIP activity (e.g., Thibodeau et al., 2016; Percival et al., 2017).

26.2.3.5.1.2 Karoo−Ferrar The Karoo Basalts in South
Africa and the Ferrar igneous rocks in Antarctica and Australia are part of a volcanic province with a total original volume of ~5-million km^3 (White, 1997). The numerous radioisotopic dates (K−Ar, Ar−Ar, U−Pb) span from about 185 to 179 Ma with a peak in the Karoo Basalts at around 183 Ma (e.g., as reviewed in Moulin et al., 2017). Magnetic reversals have been documented from lava sequences in the Karoo (Hargreaves, et al., 1997; Moulin et al., 2017) and, given ID-TIMS U−Pb age constraints from marine sequences (Sell et al., 2014), provide the means to integrate diverse chemostratigraphic, biostratigraphic, and magnetostratigraphic data (Xu et al., 2018a,b).

As is the case for CAMP, the eruption of the Karoo−Ferrar LIP is firmly established as coincident with, and therefore a prime agent for causing, the early T-OAE suite of organic-rich strata, major geochemical anomalies (e.g., the largest negative excursion in carbon-13 during the Mesozoic), and faunal extinctions (e.g., Jenkyns 1988, Hesselbo et al., 2000, Caruthers et al., 2013, Them et al., 2017b). The T-OAE began at the base of the *Harp. serpentinum* ammonite zone. In contrast to CAMP, timing of peak magmatic activity, and its association with climatic, environmental, biological, and geochemical change at that time, is not fully constrained.

26.2.3.5.2 Major bolide impacts
There are few unambiguously documented impacts of Jurassic age. A modest iridium anomaly was reported from near the palynology-defined Triassic−Jurassic boundary in the Eastern United States; and associated features, such as the fern spike and apparent suddenness of the terrestrial extinctions, had suggested a possible impact relationship (Olsen et al., 2003), but extensive follow-up research has provided no additional support for this hypothesis.

The Puchezh-Katunki impact structure in Volga Federal District of Russia has a relatively modest diameter of 40−80 km and has yielded Ar−Ar ages of 192−196 Ma (i.e., Late Sinemurian—Early Pliensbachian); palynological evidence from the immediate sediment infill supports this age assignment (Holm-Alwmark et al., 2019). Morokweng impact structure in the Kalahari Desert of South Africa has a similar diameter (70 km) with an age from U—Pb and Ar—Ar analyses of 145.2 ± 0.8 Ma (Koeberl et al., 1997; Hart et al., 1997). This age implies an impact event in the latest Jurassic that does not correspond to any significant environmental perturbation.

26.3 Jurassic time scale

The primary reference scales for most stage boundaries and other events in Jurassic stratigraphy are the ammonite zones for the Tethyan and Sub-Boreal faunal realms (Fig. 26.10). The age model for these ammonite zones requires integration of several constraints, in approximate order of importance:

1. Selected radioisotopic dates, especially for Hettangian, Toarcian, and Berriasian.
2. Durations on ammonite zones derived from cycle stratigraphy, especially in Hettangian—Sinemurian, Toarcian—Aalenian, and Oxfordian—Tithonian.
3. Calibrations of ammonite zones via magnetostratigraphy to a spreading rate model calibrated to a cubic spline fit for the early Pacific Plate; these magnetostratigraphic calibrations are fairly well constrained in Oxfordian through Tithonian, and preliminary correlation for Bajocian through Callovian based on a visual comparison to the deep-tow geomagnetic survey pattern.
4. Proportional scaling of ammonite zones according to relative numbers of subzones in the intervals that did not have adequate constraints by one of the above methods.

In turn, the assignment of numerical ages for most of the other Jurassic stratigraphic events is according to their calibrations to this age model for the ammonite zonations.

A partial database of the main radioisotopic dates, cyclostratigraphy-derived durations for Hettangian through Aalenian, and potential constraints on Oxfordian through Berriasian Stage boundary ages (prior to a revised spline fit of the M-sequence) are summarized in Table 26.1. The assumptions, scaling methods, and the derived age model for each Jurassic Stage are summarized in Table 26.2, and those for each Tethyan ammonite zone in Table 26.3.

26.3.1 Constraints from radioisotopic dates

GTS2020 principally uses radioisotopic dates from U—Pb methods only from single zircon analyses (no multigrain) by CA-TIMS, which must follow treatments of annealing followed by chemical abrasion. As for GTS2012, it is still the case that few U—Pb dates for the Jurassic have been acquired by these methods, and those that have are clustered around the key major environmental change events of

the end-Triassic mass extinction and the T-OAE. Nearly all of the U—Pb dates compiled by Pálfy et al. (2000) had utilized multigrain analyses without these preprocessing techniques, and while they do not define GTS2020 they are useful as "secondary guides."

The Triassic—Jurassic boundary has a relatively precise radioisotopic date of 201.36 ± 0.17 Ma, based on CA-ID-TIMS constraints from dated ammonoid-bearing strata in Peru (samples: LM4-90 and LM4-100/101) and a similar age from the former GSSP candidate section in New York Canyon, Nevada (Schoene et al., 2010), as recalculated by Wotzlaw et al. (2014) using updated EARTHTIME tracers.

This methodology has provided significantly improved apparent precision, with radioisotopic uncertainties of <0.1% for the Triassic—Jurassic transition. Uncertainties on the accuracy are, however, not well constrained. For example, *Psiloceras spelae spelae* in the Pucara Basin first appeared midway through the ∼5-m-thick section between the samples LM4-90 and LM4-100/101 that were used for radioisotopic dating. Significant facies changes may, however, correspond to significant variations in sedimentation rates here. Theoretically, the age of the first appearance of *Psil. spelae spelae* in the studied succession of the Pucara Basin should be bracketed between 201.53 and 201.19 Ma (a stratigraphic uncertainty of 340 kyr). Furthermore, the official base of the Jurassic is placed at the lowest occurrence of the ammonite subspecies *Psiloceras spelae tirolicum* in the GSSP at Kuhjoch (Austria), while the base of the Jurassic in the Pucara Basin (Peru) is, however, assigned as the lowest appearance of *Ps. spelae spelae*. Although these two ammonite subspecies probably had their lowest recorded occurrences close in time to one another, there may be some (so far unconstrained) diachroneity among these different horizons in the studied sections. Thus with the advancement of radioisotopic dating methodology, constraints on the ages of stage or period boundaries and events are now increasingly hampered by (bio-)stratigraphic limitations that can be an order of magnitude larger than the uncertainty on the radioisotopic date itself. Such hard-to-quantify uncertainties are prevalent throughout the Paleozoic and Mesozoic time scales.

The Hettangian—Sinemurian boundary is constrained by an U—Pb date of 199.43 ± 0.10 Ma from the boundary interval in Peru (Schaltegger et al., 2008; Guex et al., 2012) but again affected by classical stratigraphic uncertainty.

There are now a number of high-precision U—Pb dates from marine strata around the Pliensbachian—Toarcian boundary and within the Toarcian. These include a radioisotopic date of 185.5 ± 0.2 Ma from the Pliensbachian *Am. margaritatus* Zone equivalent strata of the East Tributary of Bighorn Creek, Alberta, Canada (Them et al., 2017b). Importantly, new radioisotopic dates from the upper Pliensbachian in Oregon, United States (Lena et al., 2019b), assigned to the *margaritatus* and *spinatum* zones have good ammonite-based constraints, albeit primarily for the North American zonal

TABLE 26.1 Initial constraints on the age model for the Jurassic.

Stage or *ammonite zone*	Radioisotopic date or projected ages from cyclostratigraphy (Ma)	Equivalent 405-kyr cycle number (E)	Implied stage duration (Myr)	Reference and notes on radioisotopic dates and cyclostratigraphic ages
Base Berriasian	143.1 ± 0.6			Assigned from Cretaceous age model (Gale et al., 2020, Ch. 27: The Cretaceous Period, this volume).
Tithonian			5.8	
Base Tithonian	148.9 ± 0.7			Initial projection from Cretaceous age model (this volume) prior to a revised cubic-spline fit of M-sequence (Ogg, 2020, Ch. 5: Geomagnetic polarity time scale, this book).
Kimmeridgian	154.1 ± 2.2		5.9	Re−Os date directly from GSSP: Selby (2007).
Base Kimmeridgian	154.8 ± 0.8			Initial projection from Cretaceous age model (this volume) prior to a revised cubic-spline fit of M-sequence.
Oxfordian	156.1 ± 0.89		6	Ar−Ar date from *transversarium* Zone (Pellenard et al., 2013).
Base Oxfordian	160.8 ± 1.0			Initial projection from Cretaceous age model (this volume) before revision of M-sequence fit and the choice of a different ammonite horizon (see Oxfordian, Section 26.1.4.1.1, this chapter).
Callovian	164.6 ± 0.2	405 ± 3	3.1	Kamo and Riccardi (2009).
Base Callovian	163.9 ± 1.1			Initial projection from Cretaceous age model (this volume) prior to a revised cubic-spline fit of M-sequence.
Bathonian			2.1	
Base Bathonian	166.0 ± 1.2	410		Initial projection from Cretaceous age model (this volume) prior to a revised cubic-spline fit of M-sequence.
Bajocian	168.72 ± 1.73		4.9	Ar−Ar dating (Koppers et al., 2003, G3) of basalts at ODP Site 801, Pigafetta Basin, western Pacific; drilled on marine magnetic anomaly M42n.4r on revised deep-tow nomenclature. A biostratigraphic assignment as Bajocian has large uncertainty.
Base Bajocian	170.9	423		Huang (2018) reanalysis of Huret (2006) and Hinnov and Park (1999); Cyclostratigraphic duration of 4 Myr for Aalenian, anchored to base-Toarcian Radioisotopic date.
Aalenian			3.8	
Base Aalenian	174.7	432		Huang (2018) reanalysis of Boulila et al. (2014); Cyclostratigraphic estimated duration of 9.5 Myr, anchored to base Toarcian.
Toarcian			9.5	
thouarsense	180.4 ± 0.4	439 ± 2		Sell et al. (2014); correlation to European zonation uncertain.
bifrons	182.1 ± 0.1	445 ± 3		Sell et al. (2014).
bifrons	182.0 ± 0.1	445 ± 3		Sell et al. (2014).
bifrons-variabilis	180.9 ± 0.4	444 ± 5		Mazzini et al. (2010). Corrected Corfu et al. (2016); stratigraphic position from Al-Suwaidi et al. (2016).
bifrons-variabilis	181.8 ± 0.2	444 ± 5		Mazzini et al. (2010). Corrected Corfu et al. (2016); stratigraphic position from Al-Suwaidi et al. (2016).
tenuicostatum-bifrons	183.0 ± 2.0	450 ± 7		Re−Os age from van Acken et al. (2019). Amalgamated age from stratigraphically widely separated levels.
serpentinum	185.7 ± 0.4	450 ± 3		Leanza et al. (2013). Reported to be Pliensbachian, but Fig. 6 of Leanza et al. (2013) is Hoelderi Zone (Toarcian) based on ammonites (Al-Suwaidi et al., in prep.). Other U-Pb ages reported by Leanza et al. (2013) are of uncertain stratigraphic position.
tenuicostatum	183.2 ± 0.1	454 ± 3		Riccardi and Kamo (2012).
Top *tenuicostatum*	183.2 ± 0.3	453 ± 1		Sell et al. (2014) = Best constraint on To−Pl boundary.
Base Toarcian	184.2	456		Sell et al. (2014); Ait-Itto et al. (2018) use 1 Myr duration for *tenuicostatum* Zone from cyclostratigraphy in Morocco.

(Continued)

TABLE 26.1 (Continued)

Stage or *ammonite zone*	Radioisotopic date or projected ages from cyclostratigraphy (Ma)	Equivalent 405-kyr cycle number (E)	Implied stage duration (Myr)	Reference and notes on radioisotopic dates and cyclostratigraphic ages
Pliensbachian			8.7	
margaritatus	185.5 ± 0.2	462 ± 4		Them et al. (2017b).
jamesoni-davoei	186.8 ± 1.5	471 ± 7		Panä et al. (2018).
Base Pliensbachian	192.9	477		Ruhl et al. (2010), Huang (2018); based on cyclostratigraphic duration of the Pliensbachian from Llanbedr (Mochras Farm) borehole of 8.7 Myr.
Sinemurian			6.6	
angulata-bucklandi	199.4 ± 0.1	493 ± 4		From the *Badouxia canadensis* beds, Peru. Ruhl et al. (2010), Huang (2018).
Base Sinemurian	199.5	493		Based on cyclostratigraphy anchored to base Hettangian using a value of 1.9 Myr (1.8 Myr from base of *planorbis* Zone and 0.1 Myr for *tilmanni* Zone. Note that Hettangian calibration of Weedon et al., 2018 (3.9 Myr) is conservatively not adopted here.
Hettangian			1.9	
Base *tilmanni*	201.4 ± 0.2	498 ± 1		Schoene et al (2010); Wotzlaw et al. (2014).
Base Hettangian	201.4 ± 0.2	498		

Selected radioisotopic dates (with 2-sigma uncertainties) are from sedimentary outcrops and from flood basalt deposits of uppermost Triassic and Jurassic. Dates are from ID-TIMS U−Pb single zircon analyses, unless otherwise stated. Argon−argon ages have been adjusted to an FC monitor standard of 28.201 Ma. There are many other studies of the flood basalt episodes or deposits; see references for reviews and syntheses. Equivalent 405-kyr cycle number [E] relative to present (Laskar et al., 2011) are either as implied by the radiometric date or by relative placement from biostratigraphic correlation combined with scaling between stage boundaries. GSSP, Global Boundary Stratotype Section and Point; ODP, Ocean Drilling Program.

scheme. These dates range from 186.96 ± 0.07 Ma in the *kunae* Zone (i.e., lower *margaritatus* Zone) to 184.02 ± 0.05 which lies above the highest ammonites of the *carlottense* Zone (i.e., upper *spinatum* Zone) and may be in the lowermost Toarcian.

A high-precision U−Pb radioisotopic date of 183.22 ± 0.25 Ma is reported from the strata equivalent to the top *Dacty. tenuicostatum* Zone in Peru (Sell et al., 2014), which presently provides the best biostratigraphic-calibrated numerical constraint on the Pliensbachian−Toarcian boundary. Other precise ages are provided from detrital zircons from Tibet by Fu et al. (2016), and also by Mazzini et al. (2010), Riccardi and Kamo (2012), Leanza et al. (2013), and Sell et al. (2014) from Toarcian levels in South America, but these provide less constraint because of difficulties in correlation to the standard ammonite zonation and age reversals in the ash sequences analyzed (see discussion in Corfu et al., 2016; Sell et al., 2016).

The Bajocian/Bathonian boundary is ∼169 Ma, based on ^{40}Ar/^{39}Ar dating of lower volcanic units in ODP Site 801 (Koppers et al., 2003), although there is a large uncertainty on the biostratigraphic age of these basalts. The age for the Bathonian−Callovian boundary is at ∼165 Ma, based on a U−Pb date of 164.6 ± 0.2 Ma from an ash bed in Argentina with ammonites approximating the Bathonian−Callovian boundary (Kamo and Riccardi, 2009).

In the Late Jurassic, recent work has yielded high-precision U−Pb radioisotopic dates from both detrital zircons

and ash beds in the Neuquén Basin, Argentina and at Mazetepec, Mexico, providing age constraints on both the base Tithonian and base Berriasian (J/K boundary) (Lena et al., 2019a; Aguirre-Urreta et al., 2019; see Table 27.2 of Chapter 27: The Cretaceous Period, this volume).

Increasingly, bio- and chemostratigraphic studies, integrated with dates obtained by the ^{187}Re/^{188}Os method from black shales, are of sufficient precision to be useful as secondary guides for time scale calibration, particularly in time intervals barren of ash beds. With this an age of 154.1 ± 2.2 Ma (Selby, 2007) was obtained for the proposed base-Kimmeridgian GSSP on the Isle of Skye, and an age of 183.0 ± 2.0 Ma from the *H. serpentinum* ammonite zone in the Posidonienschiefer at Dormettingen Quarry, Germany (van Acken et al., 2019), and 180.3 ± 3.2 Ma for an *H. serpentinum* Zone equivalent lacustrine sedimentary succession in the Sichuan Basin, China (Xu et al., 2017b).

26.3.2 Jurassic age model

Portions of the Jurassic age model are interpolation and extrapolations from cycle-derived durations for stage/zones, from matching magnetostratigraphy of outcrops to the spreading rate model for the M-sequence of marine magnetic anomalies, and from the relative numbers of ammonite subzones.

The main radioisotopic controls are at the basal limits of the Hettangian, the Sinemurian, and the Toarcian stages, and the consistency of the cycle stratigraphy−scaled M-sequence

TABLE 26.2 Jurassic age model for stages and ammonite zones.

Stage	Age (Ma)	Estimated uncertainty (2-sigma)	Duration	Calibration (brief)	Stage primary marker and GSSP or working definition; calibration	Method to compute basal age of stage	Method for scaling ammonite zones within stage
CRETACEOUS (Berriasian)	143.1	0.6		mid-Chron M19n.2n	Pending GSSP marker is the base of calpionellid *Calp. alpina*.	Mainly based on cycle-scaled durations of Berriasian through Barremian stages relative to base-Aptian radiometric date	
Tithonian	149.2	0.7	6.1	Base of Chron M22An	Not yet defined. Used base of Chron M22An, which is nearly coeval with base of *H. hybonotum* ammonite zone at Crussol Mountain, France.	Tithonian through Bajocian from magnetostrat correlations to a spline fit of M-sequence spreading rates (including Deep-Tow extension) as constrained by the assigned base-Berriasian age and the approximate stage durations of Berr, Tith, Kimm and Oxf. from cyclostratigraphy	Tethyan zones have magnetostratigraphic placements; Sub-Boreal zones (England) have a mixture of cycle durations and magnetostratigraphic placements.
Kimmeridgian	154.8	0.8	5.5	Base of Chron M26r	Base of *P. baylei* ammonite zone (Isle of Skye), using lowest occurrence of *P. flodigarriensis* ammonite (to be submitted for ratification) which is essentially coeval with base of Chron M26r at the GSSP.	(same)	Tethyan and Sub-Boreal zones have magnetostratigraphic placements; and some of their cycle-derived durations are constraints on the Pacific M-sequence spreading model.
Oxfordian	161.5	1.0	6.8	25% up in Chron M36Br (mid-depth)	Not yet defined, but will be base of *Q. mariae* ammonite zone. Used base of *Brightia thuouxensis* biohorizon projected to Redcliff = lowest part of BB19 reversed polarity zone at a thin N-subzone = c. 25% up in M36Br on Deep-Tow magnetic model.	[same]	Tethyan and Sub-Boreal zones and subzones have magnetostratigraphic placements; and some of their cycle-derived durations are constraints on the Pacific M-sequence spreading model.
Callovian	165.3	1.1	3.8	Base of Chron M39n.3n (mid-depth)	Not yet defined, so base of *B. bullatus* ammonite zone used in GTS2020. GSSP is not yet decided. In magnetostratigraphy from the Albstadt–Pfeffingen GSSP candidate (Swabia, Germany), this level is the base of a brief normal polarity zone, which is interpreted to be the base of marine magnetic anomaly M39n.3n in the Deep-Tow extension (mid-depth projection) to M-sequence.	(same)	Most Tethyan and Sub-Boreal zones (and some subzones) have magnetostratigraphy with possible correlations to the mid-depth Deep-Tow extension to the Pacific M-sequence spreading model.
Bathonian	168.2	1.2	2.9	Base of Chron M42n.1n (mid-depth)	Base of *Z. zigzag* ammonite zone at Ravin du Bès-Bas Auran (France). In magnetostratigraphy of Spain, this level is the base of a normal polarity zone, which is interpreted to be the base of marine magnetic anomaly M42n.1n in the Deep-Tow extension (mid-depth projection) to M-sequence.	(same)	Most Tethyan zones have magnetostratigraphy (Spain) with possible correlations to the mid-depth Deep-Tow extension to the Pacific M-sequence spreading model. Lower Bathonian zones scaled proportionally to the relative number of subzones, due to uncertainty in limits for *P. aurigerus* Zone in reference section.

(Continued)

TABLE 26.2 (Continued)

Stage	Age (Ma)	Estimated uncertainty (2-sigma)	Duration	Calibration (brief)	Stage primary marker and GSSP or working definition; calibration	Method to compute basal age of stage	Method for scaling ammonite zones within stage
Bajocian	170.9	0.8	2.7	80% up in Chron M44n.1r (mid-depth)	Base of *H. discites* ammonite zone at Cabo Mondego (Portugal). In combined magnetostratigraphy of Spain and Switzerland, this level is in the uppermost part (c. 80% up) of a reversed-polarity-dominated zone, interpreted to be marine magnetic anomaly M44n.1r in the Deep-Tow extension (mid-depth projection) to M-sequence.	Implied by revised Aalenian stage duration (3.8 Myr) from cyclostratigraphy relative to base Aalenian	Most Tethyan zones have magnetostratigraphy (Spain) with possible correlations to the mid-depth Deep-Tow extension to the Pacific M-sequence spreading model. However, portions of uppermost Bathonian and the Middle and Lower Bathonian scaled proportionally to the relative number of subzones, due to ambiguities in deep-two models.
Aalenian	174.7	0.8	3.8	Cycle duration	Base of *L. opalinum* ammonite zone at Fuentelsaz (Spain).	Implied by Toarcian stage duration (9.5 Myr) from cyclostratigraphy relative to base Toarcian	Durations of Aalenian ammonite zones from cycle stratigraphy (France, Italy).
Toarcian	184.2	0.3	9.5	U–Pb radioisotope date	Base of *D. tenuicostatum* ammonite zone; but GSSP not yet decided (probably in either Portugal or Spain).	High-resolution U–Pb radioisotopic date and cycle duration of basal ammonite zone	Durations of Toarcian ammonite zones from cycle stratigraphy (England).
Pliensbachian	192.9	0.3	8.7	Cycle duration	Base of *U. jamesoni* ammonite zone at Robin Hood's Bay (England).	Pliensbachian stage duration (8.7 Myr) from cyclostratigraphy relative to base Toarcian	Durations of Pliensbachian ammonite zones from cycle stratigraphy (England).
Sinemurian	199.5	0.3	6.6	Cycle duration of Hettangian	Base of *A. bucklandi* ammonite zone at East Quantoxhead (England).	Implied by Hettangian stage duration (1.8 Myr) from cyclostratigraphy relative to base Hettangian	Proportional to the relative number of subzones within stage duration.
Hettangian	201.4	0.2	1.9	U–Pb radioisotope date	Lowest occurrence of the ammonite *Psiloceras spelae* at Kuhjoch (Austria).	Bracketed by high-resolution U–Pb radioisotopic ages	Durations of Hettangian ammonite zones from cycle stratigraphy (England).

Summary of the derivation of the age models used to interpolate the stage boundaries and for the primary scaling of Tethyan ammonite zones (tabulated in Table 26.3) within each stage. Values are rounded to nearest 0.1 Myr. Details are given in the text. *GSSP*, Global Boundary Stratotype Section and Point.

TABLE 26.3 Jurassic Age Model for Tethyan Ammonite Zones.

Stage/Age	Tethyan Zone	Basal Age (Ma)	Notes on calibration and zonation
CRETACEOUS basal working definition (base calpionellid *C. alpina*) is near base of this zone	*Berriasella jacobi*	143.45	Dash at about Chron M19n.2n.3; due to probable distortions in sedimentation in the compact reference sections—At Puerto Escano (Pruner et al., 2010), base is about M19n.2n.1. At Sierra Gorda (Ogg et al., 1984), this zone's base is at Chron M19n.2n.55 (+/− 0.05); uncertainty in magnetostratigraphic correlation of top of the Tithonian *Durangites* Zone is ± 0.05 of polarity Chron M19n.2n = ~0.04 Myr.
	Durangites	144.03	Placed at Chron M20n.1n.8: At Puerto Escano (Pruner et al., 2010), is about M20n.1n.65. Higher at Sierra Gorda (Ogg et al., 1984) at M19r.1 (+/− 0.2); which is only about 0.1 Myr higher. Uncertainty in magnetostratigraphic correlation to base of the Tithonian *Durangites* Zone is ± 0.2 of polarity chron M19r = approx. 0.08 Myr.
	Micracanthoceras microcanthum	145.05	Base Chron M20n (+/− 0.1) at both Sierra Gorda (Ogg et al., 1984) and Puerto Escano (Pruner et al., 2010). The zones of *P. transitorius* and "*Simplisphinctes*" are combined into a *microcanthum* Zone in some schemes (used here; following diagram from F. Oloriz); but separated as zones in the main reference section. Base of Calpionellid Zone A1 is just above base of *microcanthum* Zone.
	Micracanthoceras ponti/Burckhardticeras peroni	145.38	Chron M20r.5 (+/− 0.1).
	Semiformiceras fallauxi	147.07	Chron M22n.95 (+/− 0.05).
	Semiformiceras semiforme	147.60	Chron M22n.6 (+/− 0.1); *Semiformiceras semiforme* (one of *Haploceras (Volanites) verruciferum*).
	Semiformiceras darwini	148.12	Chron M22n.25 (+/− 0.05); *Semiformiceras darwini* (=zone of *Virgatosimoceras albertinum*).
Tithonian base	*Hybonoticeras hybonotum*	149.24	Base Chron M22An (+/− 0.1).
	Hybonoticeras beckeri	150.78	Chron M23r.2r.1 (+/− 0.05); subzones given equal duration.
	Aulacostephanus eudoxus	151.22	Full name = *Aulacostephanus (Pseudomutabilis) eudoxus*. Base = Chron M24r.1r.8 (+/− 0.1).
	Aspidoceras acanthicum	151.75	Base = Chron M24r.2r.6 (+/− 0.1).
	Crussoliceras divisum	152.15	Assigned as Chron M24Ar.6 (+/− 0.1).
	Ataxioceras hypselocyclum	152.98	Full name = *Ataxioceras (A.) hypselocyclum*. Assigned as Chron M25n.8 (+/− 0.1).
	Sutneria platynota	153.42	Chron M25r.1 (+/− 0.1). Cycles = >0.40 Myr duration.
	Idoceras planula	154.23	Base = base Chron M26n.1r.
Kimmeridgian base (base of a Boreal ammonite zone) is near base of this redefined Tethyan zone	*Epipeltoceras bimammatum*	154.78	REDEFINED as base of *Epipeltoceras bimammatum* Zone by Wierzbowski and Matyja (2014) in order to fit to Boreal placement of the base of Kimmeridgian. In their revised Bobrowniki section, the boundary is very close to base of reversed-polarity interval in magstrat of Przybylski et al. (2010a) = M26r (used here). Span is about 35% of former *bimammatum* Zone, which assumed a missing 0.4 Myr "missed beat" added to Strasser (2007) estimate of 1.2 Myr, for entire *bimammatum* scaling.
	Aspidoceras hypselum	156.80	Redefined as separate zone (upgraded subzone of former longer *E. bimammatum* Zone). Full name of subzone = *Aspidoceras (Euaspidoceras) hypselum*. Base = M28Ar.25. Span of full *hypselum-bimammatum*: Strasser (2007) had as 1.2 Myr; but we assumed a missing 0.4 Myr "missed beat"; plus another 0.1 Myr, for entire *bimammatum* scaling. Both subzones are a merged *E. hypselum* Subzone in Aguilon = >scaling done for both together. Assigned as lower 55% of *bimammatum* Zone based on magstrat.

(Continued)

TABLE 26.3 (Continued)

Stage/Age	Tethyan Zone	Basal Age (Ma)	Notes on calibration and zonation
	Perisphinctes bifurcatus	157.70	Full name = *Perisphinctes (Dichotomoceras) bifurcatus*. Base = Chron M29r.5 (approx. due to hiatus in Aguilon, Spain, magnetostratigraphy reference section). Only 1.5 cycles of 100 kyr known in lower part of zone. Total duration for *bifucatus* zone of 0.9 Myr, based partly on scaling relative to M-sequence.
	Gregoryceras transversarium	158.35	Base = approx. base Chron M30An. Cycle duration = 0.65 (Boulila et al., 2010a).
	Perisphinctes plicatilis	159.10	Full name = *Perisphinctes (Arisphinctes) plicatilis*. Approx. base of Chron M32r on short-wavelength DeepTow = average of Spain and Dorset.
	Cardioceras cordatum	159.66	Full name = *Cardioceras (C.) cordatum*. Base = approx. Chron M33Bn.5.
Oxfordian base	Quenstedtoceras mariae	161.53	Using Callov-Oxf working group preference that Base Oxfordian = base of Scarburgense Subzone as redefined in England as base of *Q. woodhamense* Biohorizon at Redcliff (not the *redcliffense* horizon) = would be "0" level on that magstrat reference section (3 m higher than base redcliffense) = > lowest part of BB19 reversed polarity zone at a thin N-subzone = c. 25% up in M36Br on DeepTow magnetic model [Fig. 5, then Fig. 7 composite, then Fig. 8 in Ogg et al. (2010a)] on British Oxf) = > 0.4 Myr higher than base *redcliffense* as used in GTS2012. Full name = *Quenstedtoceras (Q.) mariae*. Cycle duration (Boulila et al., 2008a) is ~2 Myr.
	Quenstedtoceras lamberti	162.45	Full name = *Quenstedtoceras (Lamberticeras) lamberti*. Base had been Chron M37n.2n.3 (+/− 0.2); but is in R-zone in England. Cycle strat (Huret, 2006) implies *lamberti* that is about 0.76 Myr duration.
	Peltocoeras athleta	162.65	Base = within lower Chron M38n.1n in mid-depth DeepTow model. Cycle strat implies very brief zone (0.1 Myr in Huret, 2006).
	Erymnoceras coronatum	163.30	Chron M38n.4n.7 in mid-depth DeepTow model.
	Reineckeia anceps	163.47	Chron M38n.4n.1 in mid-depth DeepTow model. Implied short-duration fit cycles (0.2 Myr) of Huret (2006).
	Macrocephalites gracilis	164.68	Full name = *Macrocephalites (Dolikephalites) gracilis*. Base = Chron M39n.2n.6 in mid-depth DeepTow model.
Callovian base	Bullatimorphites bullatus	165.29	Full name = *Bullatimorphites (Kheraiceras) bullatus*. Base = Chron M39.3n in mid-depth DeepTow model (assumed small-N at Albstadt proposed GSSP for base Callovian is this level).
	Clydoniceras discus	165.50	Full name = *Clydoniceras (C.) discus* [Late-Middle Bathonian ammonite subzones (or 1.5 s.z. if single-subzone) = 10 units given equal duration from base *subcontractus* to base Callovian].
	Hecticoceras retrocostatum	165.83	Ful name = *Hecticoceras (Prohecticoceras) retrocostatum*. (See note on *C. discus* for scaling method.)
	Cadomites bremeri	166.04	Full name = *Cadomites (C.) bremeri*. (See Note on *C. discus* for scaling method.)
	Morrisiceras morrisi	166.20	Full name = *Morrisiceras (M.) morrisi*. *M. morrisi* Zone given a single-zone (1.5 s.z.) duration. (See Note on *C. discus* for scaling method.)
	Tulites subcontractus	166.36	Full name = *Tulites (T.) subcontractus*. Base placed as Chron M40n.2n.8, which equates magnetostratigraphy-zone "Subc-R" with Chron M40n.1r. Zone given 1.5 s.z. duration. (See Note on *C. discus* for scaling method.)
	Procerites progracilis	166.97	Full name = *Procerites (P.) progracilis*. [Early Bathonian ammonite subzones (total of 6) given equal duration from base *subcontractus* to base Bathonian.]
	Procerites aurigerus	167.57	Full name = *Procerites (Siemiradzkia) aurigerus*. Base may be at base of Chron M41n.3n; but magnetostrat had undifferentiated *zigzag* and *aurigenus* zones = > subzone scaling was done. (See note on *P. progracilis* for scaling method.)

TABLE 26.3 (Continued)

Stage/Age	Tethyan Zone	Basal Age (Ma)	Notes on calibration and zonation
Bathonian Base	*Zigzagiceras zigzag*	168.17	Full name = *Zigzagiceras (Z.) zigzag*. Assigned as base of Chron M42n.1n in mid-depth DeepTow model.
	Parkinsonia parkinsoni	168.72	Full name = *Parkinsonia (P.) parkinsoni*. [Late Bajocian ammonite subzones (6 of them in *parkinsoni* and *garantiana* zones) are given equal duration.]
	Garantiana garantiana	169.27	Full name = *Garantiana (G.) garantiana*. Calibrated as base of Chron M42n.4n in mid-depth DeepTow model.
	Strenoceras niortense	169.74	Full name = *Strenoceras (S.) niortense*. [Early-Middle Bajocian ammonite subzones (14 of them) are given equal duration.]
	Stephanoceras humphriesianum	170.09	Full name = *Stephanoceras (S.) humphriesianum*. (See note at *S. niortense* for scaling method.)
	Sonninia propinquans	170.32	(See note at *S. niortense* for scaling method.)
	Witchellia laeviuscula	170.67	(See Note at *S. niortense* for scaling method.)
Bajocian Base	*Hyperlioceras discites*	170.90	Full name = *Hyperlioceras (H.) discites*. Assigned as Chron M44n.1r.8 of mid-depth DeepTow model (uppermost-Aalenian Long-reversed polarity zone is interpreted as this Chron, and the onset of the long normal polarity zone "Disc-N2" interpreted as Chron M44n.1n).
	Graphoceras concavum	171.50	Duration from Huang (2018); but shifted here by 0.1 Myr to partly adjust its magnetostratigraphy (which, also, might have uncertainties) to fit base of Chron M44n.1r in DeepTow model.
	Brasilia bradfordensis	171.90	Duration from Huang (2018). Genus is "*Ludwigia*" in some schemes. *Ludwigia murchisonae* Zone of S. Switzerland also includes a *bradfordensis* zone (this follows Arkell's usage).
	Ludwigia murchisonae	172.70	Cycle strat duration from Huang (2018).
Aalenian Base	*Leioceras opalinum*	174.70	Cycle strat duration from Huang (2018).
	Pleydellia aalensis	175.91	Cycle strat duration from Huang (2018).
	Dumortieria pseudoradiosa	177.29	Dash base. Cycle strat duration from Huang (2018).
	Phlyseogrammoceras dispansum	177.63	Dash base. Cycle strat duration from Huang (2018).
	Grammoceras thouarsense	178.34	Cycle strat duration from Huang (2018). Dash base—exact position of zonal limits with respect to cycles is not precise.
	Haugia variabilis	179.82	Cycle strat duration from Huang (2018).
	Hildoceras bifrons	181.17	Cycle strat duration from Huang (2018).
	Harpoceras serpentinum	183.16	Onset of main OAE = base of *H. serpentinum* (see Hesselbo et al., 2007a about problems with interregional correlation); the preferred duration is from Huang (2018). In GTS2016 the age (and duration of underlying stage) was constrained to 183.2 ± 0.3 Ma (Sell et al., 2014) for the uppermost *D. tenuicostatum* Zone.
Toarcian Base	*Dactylioceras tenuicostatum*	184.20	Full name = *Dactylioceras (Orthodactylites) tenuicostatum*. Base-Toarcian FIXED as 184.2 Ma based on constraints from radiometric dates above/below. Duration based on percent-up in Mochras borehole.
	Emaciaticeras emaciatum	184.95	Based on relative correlation to *Pleuroceras spinatum* Zone of Sub-Boreal realm, which has a cycle-scaled duration (Ruhl et al., 2016).
	Arieticeras algovianum	186.40	Based on relative correlation to *Amaltheus margaritatus* Zone of Sub-Boreal realm, which has a cycle-scaled duration (Ruhl et al., 2016).

(Continued)

TABLE 26.3 (Continued)

Stage/Age	Tethyan Zone	Basal Age (Ma)	Notes on calibration and zonation
	Fuciniceras lavinianum	188.00	Base is same as base of *Amaltheus margaritatus* Zone of Sub-Boreal realm, which has a cycle-scaled duration (Ruhl et al., 2016).
	Prodactylioceras davoei	188.40	Same as zone of Sub-Boreal realm, which has a cycle-scaled duration (Ruhl et al., 2016).
	Tragophylloceras ibex	190.20	Same as zone of Sub-Boreal realm, which has a cycle-scaled duration (Ruhl et al., 2016).
Pliensbachian Base	*Uptonia jamesoni*	192.90	Same as zone of Sub-Boreal realm, which has a cycle-scaled duration (Ruhl et al., 2016).
	Echioceras raricostatum	194.44	Sinemurian ammonite subzones (17 of them) are given equal duration.
	Oxynoticeras oxynotum	195.22	Sinemurian ammonite subzones are given equal duration.
	Asteroceras obtusum	196.37	Sinemurian ammonite subzones are given equal duration.
	Caenisites turneri	197.14	Sinemurian ammonite subzones are given equal duration.
	Arnioceras semicostatum	198.30	Sinemurian ammonite subzones are given equal duration.
Sinemurian Base	*Arietites bucklandi*	199.46	Basal age from Hettangian cycle duration relative to U−Pb dating of base Hettangian.
	Schlotheimia angulata	200.26	Duration from cycles in England (Ruhl et al., 2010; Xu et al., 2017a).
	Alsatites liasicus	200.96	Duration from cycles in England (Xu et al., 2017a).
	Psiloceras planorbis	201.26	Assigned as 0.1 Myr above base of *P. spelae* (Xu et al., 2017a).
JURASSIC (Hettangian) Base	*Psiloceras spelae*	201.36	Basal "new" zone (was also called "*Psiloceras tilmanni*") has duration of 0.1 Myr (Xu et al., 2017a).

Placement of the base Callovian through base-Berriasian stage boundaries is not yet formalized by the International Commission on Stratigraphy (see ICS summary table at www.stratigraphy.org or GSSP table at https://timescalefoundation.org/gssp/ for details). Usage of zones/subzones varies among Tethyan regions, and this suite is mainly for the Sub-Mediterranean province. Two-decimal ages are given to show relative durations among the zones; and uncertainties on those ages are equal to or greater than the uncertainties on the boundaries of its stage (Table 26.4). The column of Notes is informal remarks on calibrations and different zonal usages. (Details for subzone age models; and for calibrations of other biostratigraphic zonations can be found in pop-up windows for those items within the *TimeScale Creator* database; https://timescalecreator.org.) *GSSP*, Global boundary Stratotype Section and Point; *OAE*, Oceanic Anoxic Event.

pattern of Middle Jurassic through Early Cretaceous, with secondary guides from radioisotopic dates in the Kimmeridgian and Tithonian. This Jurassic age model still requires further testing by much greater acquisition of radioisotopic dates with precise biostratigraphic placements, verification from multiple independent cycle stratigraphy analyses, and obtaining reliable pre-Oxfordian marine magnetic anomaly reference scales.

26.3.2.1 Early Jurassic through base Bajocian—cyclostratigraphy scaling combined with radioisotopic dates

26.3.2.1.1 Hettangian

The Triassic/Jurassic boundary age is 201.4 ± 0.2 Ma based on U−Pb dates from ash beds within the Aramachay formation of the Pucara Group in northern Peru (Levanto section) (Schaltegger et al., 2008, Schoene et al., 2010; Wotzlaw et al., 2014). The stage boundary age is supported by carbon isotope stratigraphy and an age model for the succession based on Bayesian-Monte Carlo analysis of the U−Pb ages (Yager et al., 2017).

Each of the Hettangian ammonite zones of Britain is cycle scaled for a total of 1.8 Myr (Ruhl et al., 2010; Huang, 2018), with an additional 0.1 Myr added for the basal (*Ps. tilmanni*) zone to account for the correlation of the ammonite-barren "pre-*Planorbis*" interval of Britain to the first occurrence of the precursor species to *Ps. spelae tirolicum* (Hillebrandt and Krystyn, 2009) that marks the base-Jurassic GSSP; the precise correlation is based mainly on carbon isotope stratigraphy shown in Korte et al. (2019). A cyclostratigraphic analysis of four different sections in southern Britain, together with new ammonite horizon correlations, led Weedon et al. (2018, 2019) to propose a duration of 3.7 Myr for the Hettangian; pending confirmation from independent studies, this duration is not used in GTS2020, as discussed in Section 26.2.3.

Further U−Pb ages from throughout the Hettangian of the Utcubamba Valley, Pucara Basin, northern Peru (Guex et al., 2012) provide secondary constraints on the Hettangian time scale.

26.3.2.1.2 Sinemurian

The Hettangian cycle stratigraphy projects the base Sinemurian as 199.4 Ma, which is consistent with the 199.4 ± 0.1 Ma for this boundary interval in Peru (Schaltegger et al., 2008; Guex et al., 2012). However, that radioisotopic date may require adjustment to the EarthTime U−Pb standards and techniques for processing zircons.

In GTS2012 the Sinemurian/Pliensbachian boundary was extrapolated by assuming a linear decline in the cycle-scaled Sr trend from base Sinemurian through base Toarcian (LOWESS fit of McArthur et al., 2001), whereas in GTS2020 the same boundary is assigned an age based on cyclostratigraphic analysis of the Pliensbachian and a radioisotopic date for the earliest Toarcian.

The Sinemurian ammonite zones are scaled proportionally to the relative number of component subzones (total of 17 subzones distributed within the 6 ammonite zones); there is currently no complete cyclostratigraphy for the Sinemurian.

26.3.2.1.3 Pliensbachian

The Sinemurian/Pliensbachian boundary is assigned an age of 192.9 Myr based on cyclostratigraphic extrapolation downward from the Pliensbachian/Toarcian boundary, using the cycle stratigraphy of Ruhl et al. (2016) and Huang (2018) which gives a total duration for the Pliensbachian stage of 8.7 Myr. The relative durations of the standard ammonite zones for the Pliensbachian are scaled according to cyclostratigraphic analysis of core from the Llanbedr (Mochras Farm) borehole, Wales, United Kingdom (Page *in* Copestake and Johnson, 2014; Ruhl et al., 2016). Ц−Pb radioisotopic dates from western Canada reported by Pană et al. (2018) provide secondary constraints on the Pliensbachian time scale but interpretation of age spectra was problematic. Tethyan ammonite zones for the upper Pliensbachian were correlated to these Sub-Boreal zones using estimates from published charts (e.g., Groupe Français d'Étude du Jurassique, 1997; Hardenbol et al., 1998).

26.3.2.1.4 Toarcian

In GTS2012 an age of 182.7 Ma (revised as 183.7 Ma in GTS2016) for the base Toarcian was based on the mean age of the reversed polarity initial eruptive episode of Karoo−Ferrar flood basalts and earliest sills, with the rationale that gas emissions from these caused the marked

negative carbon isotope anomaly and mass extinction at the Pliensbachian−Toarcian boundary (e.g., McElwain et al., 2005; Svensen et al., 2007, 2012; Xu et al., 2018a). However, U−Pb radioisotopic dates available from ash beds in the Palquilla Stratigraphic section near Tacna in southern Peru provide a more rigorous constraint (Sell et al., 2014, 2016; Corfu et al., 2016); an age assigned to the top of the *D. tenuicostatum* Zone at that location is here extrapolated down to the base Toarcian using the Moroccan cyclostratigraphic estimate of 1 Myr for the duration of the *Dacty tenuicostatum* Zone from Ait-Itto et al. (2018).

The Toarcian spans ∼9.5 Myr based on the reanalyses by Huang (2018) of cyclostratigraphic data from Boulila et al. (2014), in which additional 405-kyr cycles are suggested. However, the ammonite zonation is relatively poorly defined in the Sancerre borehole and so is not used here for internal scaling.

The Llanbedr (Mochras Farm) borehole, Wales, United Kingdom, provides a direct calibration of magnetozones, standard ammonite zones, and foraminifer zones (Copestake and Johnson, 2014; Xu et al., 2018b) for the Toarcian Stage, which is scaled for GTS2020 based principally on stratigraphic thickness (Xu et al., 2018b) pending publication of cyclostratigraphic analysis of the same core. The duration for the *Dacty. tenuicostatum* Zone inferred on this basis is the same as determined independently from the cyclostratigraphy of Ait-Itto et al. (2018).

Additional CA-ID-TIMS dates from Argentina and Canada of Mazzini et al. (2010), Corfu et al. (2016), and Them et al. (2017a,b) provide secondary constraints on the time scale but suffer from weak biostratigraphic placement with respect to the European ammonite zonation.

26.3.2.1.5 Aalenian

The Toarcian span of 9.5 Myr from the cycle stratigraphy in of the Sancerre borehole (Boulila et al., 2014; Huang, 2018) projects the base Aalenian at 174.7 Ma.

The Aalenian ammonite zones are scaled from cycle stratigraphy, and the stage spans a total of 3.8 Myr from the Huang (2018) reanalysis of Hinnov and Park (1999) and Huret et al. (2008) to derive a 405-kyr cycle calibration. Therefore the Aalenian−Bajocian boundary is assigned an age of 170.9 Ma.

26.3.2.1.6 Bajocian

The Aalenian−Bajocian boundary is ∼80% up within a relatively long reversed polarity magnetozone (Henriques et al., 1994; Pavia and Enay, 1997). Based on the estimated 170.9 Ma age of this boundary derived from cycle stratigraphy relative to the base Toarcian (184.2 Ma) and the projected M-sequence of marine magnetic anomalies, this reversed polarity magnetozone is correlated to the later portion of polarity Chron M44r.

The magnetostratigraphy of Bajocian reference sections (Steiner et al., 1987) is used to partially constrain the age model for the Bajocian. For example, the base of the *Gar. garantiana* ammonite zone is at the beginning of a relatively long normal polarity magnetozone correlated to Chron M42n.4n; therefore it is assigned an age of 169.3 Ma according to the M-sequence age model (Fig. 26.10). Where the magnetostratigraphic correlations or limits of ammonite boundaries are less precise, the relative durations of Bajocian ammonite zones are apportioned according to relative numbers of component subzones.

26.3.2.2 Base Bathonian through Late Jurassic—Calibration to Pacific spreading rate model

The progressive efforts of numerous investigators have compiled the magnetostratigraphy patterns from nearly every ammonite zone of the Sub-Boreal and Tethyan realms and correlated these to the M-sequence of marine magnetic anomalies. Therefore the Jurassic stage boundaries from Bathonian through Tithonian and the ages of most ammonite zones in the Bajocian through Tithonian are derived from outcrop-based magnetostratigraphic calibration to the M-sequence of Pacific marine magnetic anomalies. Several of the Oxfordian and Kimmeridgian ammonite zones have durations computed from cycle stratigraphy (e.g., Boulila et al., 2008a,b, 2010a,b), which, in turn, are important controls on the Pacific spreading rate model for that M-sequence. In turn, the assigned ages are from the spreading rate model for that M-sequence as partially calibrated to the few radioisotopic dates. Details are in Ogg (2020, Ch. 5: The geomagnetic polarity time scale) and in Agterberg et al., (2020, Ch. 14A: Geomathematical and statistical procedures) in this book.

The spreading rate model is a smoothed spline fit based on (i) cycle-scaled estimated durations for the Kimmeridgian through Barremian stages with the base of Chron M0r at the Barremian Aptian boundary assigned as 121.4 Ma, (ii) the associated magnetostratigraphic correlations for the beginnings of those stages, and (iii) the U-Pb radioisotopic date of 168.46 ± 1.7 Ma for marine magnetic anomaly M42n.4r at ODP Site 801 (Koppers et al., 2003). The spline fit implies that the spreading rate had a gradual acceleration from the Bajocian that peaked in the late Tithonian, slowed during the Berriasian, then remained relatively constant to the early Aptian. The lack of radioisotopic dates on magnetic polarity zones in this interval implies that the resulting age model for the pre-Berriasian marine magnetic anomalies ultimately hinges on a fairly simplistic spreading rate model of a gradual increase in Pacific spreading rates from Bajocian through Barremian.

Therefore the uncertainties on the extrapolated boundary ages from the spline fit of the M-sequence polarity scale incorporate the ± 0.6 Myr uncertainty on the base Berriasian (c. M19n.2n.5) as given by the Cretaceous age model (Chapter 27: The Cretaceous Period, this volume), the c. 0.8 Myr uncertainty for the base Bajocian as given by the cyclostratigraphic durations of Toarcian—Aalenian relative to the base Toarcian, and an estimate of the statistical uncertainties on the smoothed trend in spreading rate slope. Therefore all uncertainties on these extrapolated boundary ages are estimated as ± 1 Myr. Of course, this assumes that the magnetostratigraphic correlations are accurate, which is a factor that is difficult to quantify meaningfully for each independent stage boundary.

However, the uncertainties on the durations of the individual stages are much less. A crucial factor is the reliability of the biostratigraphic age model for the older portion of the M-sequence. The projection of the Pacific spreading rate model to pre-Oxfordian marine magnetic anomalies interpreted from deep-tow data still requires verification of the polarity signatures, further cycle stratigraphic work on Bajocian—Bathonian—Callovian zones/chrons, and additional direct radioisotopic dating. If the magnetostratigraphic correlations to the M-sequence are supported by future studies, then the uncertainty on the duration of each individual stage is probably only about 10% of its total current estimated span, based on the cycle-scaled spreading rates in Kimmeridgian—Oxfordian and the consistency of the base-Bajocian age "projected downward from base Oxfordian" with the cycle-scaled span of the Toarcian and Aalenian that gives the same base-Bajocian age "projected upward from base Toarcian."

26.3.2.2.1 Bathonian

The Bajocian—Bathonian boundary is at the base of a normal polarity magnetozone that is tentatively correlated with Chron M42n.1n. This has a projected age of 168.2 Ma according to the M-sequence age model and is supported by the 168.7 ± 1.7 Ma date (adjusted from Koppers et al., 2003) for the marine magnetic anomaly M42n.4r as drilled at ODP Site 801 that yielded microfossil assemblages interpreted as being within the Bajocian—Bathonian boundary interval. This radioisotopic date was used as the main constraint on the older portion of the M-sequence in GTS2020, although our magnetostratigraphic correlation of the reversed polarity dominance at the Bajocian—Bathonian boundary (e.g., Steiner et al., 1987; Steiner and Ogg, 1988) implies a correlation to the latter-most part of Chron M42n with a projected age of 168.2 Ma.

As with the Bajocian, some of the Bathonian ammonite zones are correlated via magnetostratigraphy to the M-sequence polarity pattern, therefore, have extrapolated numerical ages. Other intervals necessitated scaling according to relative numbers of component ammonite subzones.

These extrapolated ages fit well with the 167−163 Ma range of secondary radioisotopic date guides.

26.3.2.2.2 Callovian

The base of the Callovian is within a thin normal polarity−dominated interval that is tentatively correlated with Chron M39n.3n of the M-sequence pattern based on a composite magnetostratigraphy from upper Bathonian through Callovian sections in France, England, and Swabia in Germany (Gipe, 2013; Gipe et al., 2013), hence, has a projected age of 165.3 Ma (with ∼1 Myr uncertainty). This is consistent with the 164.6 ± 0.2 Ma age on the Bathonian−Callovian boundary from Argentina (Kamo and Riccardi, 2009), although the ammonites from that section lack cosmopolitan markers.

Each Callovian ammonite zone has tentative correlations to the M-sequence, some of which incorporate estimated durations from cycle stratigraphy of those zones (e.g., according to Huret et al. (2008), the *Rein. anceps* Zone spans only about 0.2 Myr and *Pelt. athleta* only about 0.1 Myr).

26.3.2.2.3 Oxfordian

The Callovian−Oxfordian boundary, using the placement of the *Cardioceras woodhamense* biohorizon in the magnetostratigraphy of Dorset, is within the lower part of magnetozone M36Br (Gipe, 2013; Gipe et al., 2013), which has a projected M-sequence age of 161.5 Ma. The projected 6.7-Myr span of the Oxfordian to 154.8 Ma is broadly consistent with currently the only reported radioisotopic date from within that stage—an Ar/Ar date of 156.1 ± 0.9 Ma from a volcanic ash layer that is considered to be perhaps equivalent to Middle Oxfordian horizons in another section (Pellenard et al., 2013).

Each Oxfordian ammonite zone has its age derived from a merger of cycle-derived and magnetostratigraphic correlation to the M-sequence (e.g., Boulila et al., 2010a; Ogg et al., 2010a; Przybylski et al., 2010a). The Oxfordian spans 6.7 Myr, of which 2.0 Myr is within the basal *Quen. mariae* ammonite zone according to cycle stratigraphy from SE France (Boulila et al., 2008a).

26.3.2.2.4 Kimmeridgian

The base of the Kimmeridgian, using the Boreal/Sub-Boreal placement at the base of the *P. baylei* ammonite zone, corresponds to the base of Chron M26r with a projected age of 154.8 Ma. This is consistent with the Re−Os radioisotope age of 154.1 ± 2.2 Ma at the proposed GSSP level at Skye (Selby, 2007) and is within the uncertainty on the recalibrated ^{40}Ar/^{39}Ar date (156.3 ± 3.4 Ma) obtained from within oceanic basalts drilled at DSDP Site 765 on marine anomaly M26r (Ludden, 1992).

Each Kimmeridgian ammonite zone has its age derived from magnetostratigraphic correlation to the M-sequence.

The implied durations of zones are generally consistent with the cycle-derived durations from cycle stratigraphy (e.g., Boulila et al., 2008b, 2010b; Strasser, 2007), although the span of a couple of the ammonite zones implied "missing cycle beats" at sequence boundaries. For example, the *Idoceras planula* Zone of basal Kimmeridgian was estimated as ∼0.35 Myr by Strasser (2007), but the magnetostratigraphy implies a duration of 0.75 Myr, implying that a full 400-kyr long eccentricity cycle is probably missing at the regional Oxf-8 sequence boundary below its transgressive base.

26.3.2.2.5 Tithonian

The base of the Tithonian is assigned here as the base of Chron M22An, which is essentially coeval with the historical boundary in Tethyan ammonite stratigraphy. The corresponding M-sequence age is 149.2 Ma. This is consistent with an Early Tithonian U−Pb age of 147.11 ± 0.08 Ma from an ash bed in the Neuquén Basin (Lena et al., 2019a), however, Aguirre-Urreta et al., 2019 noted a possible incompatibility with younger detrital zircons (143 Ma) from a lower stratigraphic position at the same location.

Each Tithonian ammonite zone of the Sub-Mediterranean faunal realm has its age derived from magnetostratigraphic correlation to the M-sequence (Fig. 26.9). The ages of Sub-Boreal/Boreal zones are estimated from a combination of rare correlation levels to the Sub-Mediterranean succession, from magnetostratigraphy (e.g., Ogg et al., 1994; Houša et al., 2007), and from scaling by cycle stratigraphy (e.g., Huang et al., 2010b; Huang, 2018).

The proposed definition for the Jurassic/Cretaceous boundary is the middle of chron M19n.2n, which is adopted for herein (cf. Wimbledon, 2017; Wimbledon et al., 2020). The age model used for the Cretaceous (Gale et al., 2020, Ch. 27: The Cretaceous Period, this volume) now projects this age as 143.1 Ma, largely as a result of the cyclostratigraphic estimates of durations of Lower Cretaceous stages relative to revised radioisotopic dating estimate for the base of the Aptian Stage. The implied duration of the Tithonian herein is also compatible with the combined cyclostratigraphic and magnetostratigraphic calibrations of Kietzmann et al. (2018), but here we reinterpret their M19n−M18n magnetostratigraphic assignment to be the single long Chron M19n of the M-sequence. In contrast, Lena et al. (2019a) propose a younger age (∼140.8 Ma) for the Jurassic/Cretaceous boundary based on radioisotopic dates from a series of ash beds in Mexico and Argentina, where biostratigraphic ages are interpreted from nannofossils and calpionellids, but without use of magnetostratigraphic correlations. If this younger age estimate were used, then a significantly shorter Berriasian Stage would be required, and the adjustment to the interpolated M-sequence spreading rates would result in slightly younger ages for the bases of Kimmeridgian and Tithonian stages, and slightly longer durations for the Bajocian through Callovian stages.

26.3.3 Summary

As with all time scales, this Jurassic compilation is a work in progress. The combination of extensive cycle stratigraphy studies and enhanced methods of radioisotopic dating has greatly refined the age model for the Jurassic compared to the previous GTS2012 compilation (Table 26.4), but several intervals lack adequate constraints. The principal anchors for the current version are ID-TIMS U–Pb dates on single zircon populations derived from locations where a good biostratigraphic control can be demonstrated, and these are focused at the major environmental change events of the Triassic–Jurassic boundary and the T-OAE. Other Lower-to-lower-Middle Jurassic stage boundaries and internal scaling of stages are derived principally from astrochronology. From the Middle Jurassic (base Bajocian) to the base Cretaceous, the stage boundary and ammonite zone ages are primarily based on magnetostratigraphy correlation to a spline fit M-sequence spreading model constrained by only a few radioisotopic dates with relatively high analytical and biostratigraphic uncertainties but augmented by a few intervals within these stages that have cyclostratigraphic scaling of biozones. The GTS2020 ages for the Lower and Middle Jurassic stage boundaries are similar to those of GTS2012, while the ages for the Late Jurassic stages have become younger as a result principally of revised M-sequence scaling based on a much younger base-Aptian age assignment and a proposed cyclostratigraphy that expanded the Valanginian–Hauterivian interval (Table 26.4). The magnetostratigraphic calibration of the

Middle Jurassic awaits an improved pre-M29 marine magnetic model, confirmation of the polarity pattern for each ammonite zone, and additional cycle stratigraphy analyses of the durations of the component ammonite zones in multiple independent sections.

It is also clear that uncertainty in this Jurassic time scale is underquantified: the ID-TIMS U–Pb dating now has high analytical precision, but there are larger uncertainties regarding biostratigraphic correlation of local dated horizons to the European-defined ammonite zones, accumulation rates, reworking of zircons, lead-loss, etc. For example, the interpretation of U–Pb dates from the Jurassic–Cretaceous boundary interval requires further work to refine the age assignment to the potential GSSP. Another factor is that interpretations of astrochronology can result in contradictory age models for the same sedimentary successions; and those uncertainties are complex and difficult to quantify without more extensive radioisotopic datasets. Relatively new age-dating techniques applied directly to the best characterized marine sedimentary successions offer a possibility of future progress.

However, the merger of radioisotopic dating and high-resolution astronomical tuning with the integrated stratigraphy of marine and volcanic successions has firmly established the coincidence in time between the two major episodes of environmental change at the Triassic–Jurassic boundary and during the earliest Toarcian with the massive eruption of the two LIPs of the North Atlantic and the South Atlantic, respectively.

TABLE 26.4 Jurassic Age Model of GTS2020 and comparison with GTS2012.

Base of Stage	GTS 2012			GTS 2020			Status
	Age (Ma)	*uncertainty*	duration	Age (Ma)	*uncertainty*	duration	
Berriasian	145.0	0.8	4.8	**143.1**	0.6	5.4	GSSP pending
Tithonian	152.1	0.9	7.1	**149.2**	0.7	6.1	No GSSP yet
Kimmeridgian	157.3	1.0	5.2	**154.8**	0.8	5.6	GSSP pending
Oxfordian	163.5	1.1	6.2	**161.5**	1.0	6.7	No GSSP yet
Callovian	166.1	1.2	2.6	**165.3**	1.1	3.8	No GSSP yet
Bathonian	168.3	1.3	2.2	**168.2**	1.2	2.9	
Bajocian	170.3	1.4	2.0	**170.9**	0.8	2.7	
Aalenian	174.1	1.0	3.8	**174.7**	0.8	3.8	
Toarcian	182.7	0.7	8.6	**184.2**	0.3	9.5	
Pliensbachian	190.8	1.0	8.1	**192.9**	0.3	8.7	
Sinemurian	199.3	0.3	8.5	**199.5**	0.3	6.6	
Hettangian	201.3	0.2	2.0	**201.4**	0.2	1.9	

Uncertainties and durations are in Myr. Values are rounded to nearest 0.1 Myr.

GSSPs of the Jurassic Stages, with location and primary correlation criteria

Stage	GSSP Location	Latitude, Longitude	Boundary Level	Correlation Events	Reference
Tithonian	*candidates are Mt. Crussol or Canjuers (SE France) and Swabia, Germany*			*Near bases of Hybonoticeras hybonotum ammonite zone, of Gravesia genus, and of polarity chron M22An*	
Kimmeridgian	*candidate is Flodigarry (Isle of Skye, NW Scotland)*			*Ammonite, FAD of Pictonia flodigarriensis (base of Pict. baylei Zone) of Sub-Boreal realm, and base of polarity chron M26r*	
Oxfordian	*candidates are Thuoux and Saint-Pierre d'Argençon sections, SE France*			*Ammonite, FAD of Brightia thouxensis (base of Quenstedtoceras mariae Zone)*	
Callovian	*candidates are Pfeffingen, Swabian Alb, SW Germany, and Novgorod region, Russia*			*Ammonite, FAD of Kepplerites (Kosmoceratidae); defines base of Macrocephalites herveyi Zone in UK and SW Germany*	
Bathonian	Ravin du Bès, Bas-Auran area, Alpes de Haute Provence, France	43°57'38"N 6°18'55"E*	base of limestone bed RB071	Ammonite, FAD of *Gonolkites convergens*	Episodes **32**/4, 2009
Bajocian	Murtinheira Section, Cabo Mondego, Portugal	40°11'57"N 8°54'15"W*	base of Bed AB11 of the Murtinheira Section	Ammonite, FAD of *Hyperlioceras mundum*	Episodes **21**/1, 1997
Aalenian	Fuentelsaz, Spain	41°10'15"N 1°50'W	base of Bed FZ 107 in Fuentelsaz Section	Ammonite, FAD of *Leioceras opalinum*	Episodes **24**/3, 2001
Toarcian	Ponta do Trovao, Peniche, Lusitanian Basin, Portugal	39°22'15"N 9°23'07"W	base of micritic limestone bed 15e	Ammonite, FAD of *Dactylioceras (Eodactylioceras) simplex*	Episodes **39**/3, 2016
Pliensbachian	Robin Hood's Bay, Yorkshire Coast, England	41°10'15"N 1°50'W	base of Bed 73b at Wine Haven, Robin Hood's Bay	Ammonite, FAD of *Bifericeras donovani*	Episodes **29**/2, 2006
Sinemurian	East Quantoxhead, SW England	51°11'27.3"N 3°14'11.2"W*	0.90 m above the base of Bed 145	Ammonite, FAD of *Vermiceras quantoxense*	Episodes **25**/1, 2002
Hettangian	Kuhjoch Section, Karwendel Mtns, Austria	47°29'02"N 11°31'50"E	5.80 m above base of the Tiefengraben Mbr. of the Kendelbach Formation	Ammonite, FAD of *Psiloceras spelae*	Episodes **36**/3, 2013

* according to Google Earth

FIGURE 26.11 Ratified GSSPs and potential primary markers under consideration for defining the Jurassic stages *(status as of January 2020)*. Details of each GSSP are available at http://www.stratigraphy.org, https://timescalefoundation.org/gssp/, and in the *Episodes* publications. *GSSP*, Global Boundary Stratotype Section and Point.

Acknowledgments

We thank colleagues within the Jurassic Stratigraphic Community for their lively and fruitful discussion over many decades of the diverse topics covered in this chapter.

Additional or alternative zonal schemes are available through the *TimeScale Creator* visualization datapacks (https://timescalecreator.org). Documentation for post-2011 advances in formalizing GSSPs, zonal schemes, interregional correlations, and eventual consensus on the best numerical scalings will be found in the newsletters of the subcommission on Jurassic Stratigraphy.

Micha Ruhl and Steve Hesselbo gratefully acknowledge funding from Shell while at the University of Oxford, and Hesselbo acknowledges NERC grant (NE/N018508/1) on Early Jurassic Earth System and Time Scale. James Ogg's participation was partly enabled by a distinguished visiting professorship at Chengdu University of Technology (Chengdu) and China University of Geosciences (Wuhan), China.

Bibliography

Agterberg, F.P., da Silva, A.C., and Gradstein, F.M., 2020, Chapter 14A - Geomathematical and statistical procedures. *In* Gradstein, F.M., Ogg, J.G., Schmitz, M.D., and Ogg, G.M. (eds), The *Geologic Time Scale 2020*. Vol. **1** (this book). Elsevier, Boston, MA.

Aguirre-Urreta, B., Naipauer, M., Lescano, M., López-Martínez, R., Pujana, I., Vennari, V., et al., 2019, The Tithonian chrono-biostratigraphy of the Neuquén Basin and related Andean areas: a review and update. *Journal of South American Earth Sciences*, **92**: 350–367.

Ait-Itto, F., Martinez, M., Price, G.D., and Ait Addi, A., 2018, Synchronization of the astronomical time scales in the Early Toarcian: a link between anoxia, carbon-cycle perturbation, mass extinction and volcanism. *Earth and Planetary Science Letters*, **493**: 1−11.

Al-Husseini, M., and Matthews, R.K., 2006, Stratigraphic note: orbital calibration of the Arabian Jurassic second-order sequence stratigraphy. *GeoArabia*, **11**: 161−170.

Alméras, Y., Boullier, A., and Laurin, B., 1997, Brachiopodes. *In* Groupe Français d'Étude du Jurassique (ed), Biostratigraphie du Jurassique ouest-européen et méditerranéen: zonations parallèles et distribution des invertébrés et microfossiles. *Bulletin des Centres de Recherches Exploration—Production Elf-Aquitaine, Mémoire*, **17**: 169−195.

Al-Suwaidi, A.H., Hesselbo, S.P., Damborenea, S.E., Manceñido, M.O., Jenkyns, H.C., Riccardi, A.C., et al., 2016, The Toarcian oceanic anoxic event (Early Jurassic) in the Neuquén Basin, Argentina: a reassessment of age and carbon isotope stratigraphy. *Journal of Geology*, **124**: 171−193.

Arkell, W.J., 1933, *The Jurassic System in Great Britain*. Oxford: Clarendon Press, p. 681.

Arkell, W.J., 1939, The ammonite succession at the Woodham Brick Company's pit, Akeman Street Station, Buckinghamshire, and its bearing on the classification of the Oxford Clay. *Quarterly Journal of the Geological Society*, **95**: 135−221.

Arkell, W.J., 1946, Standard of the European Jurassic. *Geological Society of America Bulletin*, **57**: 1−34.

Arkell, W.J., 1954, Three complete sections of the Cornbrash. *Proceedings of the Geological Association London*, **65**: 115.

Arkell, W.J., 1956, *Jurassic Geology of the World*. Edinburgh: Oliver and Boyd, p. 806.

Armendáriz, M., Rosales, I., Bádenas, B., Aurell, M., García-Ramos, J.C., and Piñuela, L., 2012, High-resolution chemostratigraphic records from Lower Pliensbachian belemnites: Palaeoclimatic perturbations, organic facies and water mass exchange (Asturian basin, northern Spain). *Palaeogeogr. Palaeoclimatol. Palaeoecol.*, **333−334**: 178−191.

Atrops, F., 1982, La sous-famille des Ataxioceratinae (Ammonitina) dans le Kimméridgien inférieur du Sud-Est de la France: systématique, évolution, chronostratigraphie des genres *Orthosphinctes* et *Ataxioceras. Documents des Laboratoires de Géologie Lyon*, **83**: 463.

Atrops, F., 1994, upper Oxfordian to Lower Kimmeridgian ammonite successions and biostratigraphy of the Crussol and Châteauneuf d'Oze sections. *In* Atrops, F., Fortwengler, D., Marchand, D., and Melendez, G. (eds), *IV Oxfordian & Kimmeridgian Working Group Meeting, Lyon and SE France Basin, Guide Book and Abstracts*, 50−60, 106−111.

Atrops, F., 1999, Report of the Oxfordian-Kimmeridgian boundary working group. *International Subcommission on Jurassic Stratigraphy Newsletter*, **27**: 34.

Atrops, F., Gygi, R., Matyja, B.A., and Wierzbowski, A., 1993, The *Amoeboceras* faunas in the Middle Oxfordian − lowermost Kimmeridgian, Submediterranean succession, and their correlation value. *Acta Geologica Polonica*, **43**: 213−227.

Bartolini, A., and Larson, R.L., 2001, The Pacific microplate and the Pangea supercontinent in the Early to Middle Jurassic. *Geology*, **29**: 735−738.

Bartolini, A., Baumgartner, P.O., and Hunziker, J., 1996, Middle and Late Jurassic carbon stable-isotope stratigraphy and radiolarite sedimentation of the Umbria-Marche Basin (central Italy). *Eclogae Geologicae Helvetiae*, **89**: 811−844.

Bartolini, A., Baumgartner, P.O., and Guex, J., 1999, Middle and Late Jurassic radiolarian palaeoecology versus carbon-isotope stratigraphy. *Palaeogeography, Palaeoclimatology, Palaeoecology*, **145**: 43−60.

Bartolini, A., Larson, R., Baumgartner, P.O., and the ODP Leg 185 Scientific Party, 2000, *Bajocian Radiolarian Age of the Oldest Oceanic Crust In Situ (Pigafetta Basin, western Pacific, ODP Site 801. Leg 185)*. La Grande-Motte, France: European ODP Forum, abstracts, p. 32.

Bartolini, A., Guex, J., Spangenberg, J., Schoene, B., Taylor, D., Schaltegger, U., et al., 2012, Disentangling the Hettangian carbon isotope record: implications for the aftermath of the end-Triassic mass extinction. *Geochemistry, Geophysics, Geosystems*, **13**: Q01007.

Baumgartner, P.O., 1987, Age and genesis of Tethyan Jurassic radiolarites. *Eclogae Geologicae Helvetiae*, **80**: 831−879.

Belkaaloul, K.N., Aissaoui, D.M., Rebelle, M., and Sambet, G., 1995, Magnetostratigraphic correlation of the Jurassic carbonates from the Paris Basin: implications for petroleum exploration. *Journal of the Geological Society, London*, **95**: 173−186.

Belkaaloul, K.N., Aissaoui, D.M., Rebelle, M., and Sambet, G., 1997, Resolving sedimentological uncertainties using magnetostratigraphic correlation: an example from the Middle Jurassic of Burgundy, France. *Journal of Sedimentary Research*, **67**: 676−685.

Benson, R.B.J., 2018, Dinosaur macroevolution and macroecology. *Annual Review of Ecology, Evolution, and Systematics*, **49**: 379−408.

Birkelund, T., Callomon, J.H., Clausen, C.K., Nøhr Hansen, H., and Salinas, I., 1983, The Lower Kimmeridge Clay at Westbury, Wiltshire, England. *Proceedings of the Geological Association*, **94**: 289−309.

Blackburn, T.J., Olsen, P.E., Bowring, S.A., McLean, N.M., Kent, D.V., Puffer, J., et al., 2013, Zircon U-Pb geochronology links the end-Triassic extinction with the central Atlantic Magmatic province. *Science*, **340**: 941−945, https://doi.org/10.1126/science.1234204.

Blake, J.F., 1881, On correlation of the Kimmeridge and Portland rocks of England with those of the continent. Part 1. The Paris Basin. *Quarterly Journal of the Geological Society of London*, **37**: 497−587.

Bloos, G., 2008, *Comment on the T-J Boundary Level at the FAD of the Earliest Psiloceratid, P. spelae*. [Unpublished document accompanying ICS voting package on base-Jurassic GSSP from the secretary of Triassic/Jurassic Boundary Working Group.]

Bloos, G., and Page, K.N., 2002, Global stratotype section and point for base of Sinemurian Stage (Lower Jurassic). *Episodes*, **25**: 22−28.

Bodin, S., Mattioli, E., Fröhlich, S., Marshall, J.D., Boutib, L., Lahsini, S., et al., 2010, Toarcian carbon isotope shifts and nutrient changes from the Northern margin of Gondwana (High Atlas, Morocco, Jurassic): palaeoenvironmental implications. *Palaeogeography, Palaeoclimatology, Palaeoecology*, **297**: 377−390.

Bogoslovsky, N.L., 1897, The Ryazanian horizon: fauna, stratigraphic relationships, and probable age. *Materials on Geology of Russia (St. Petersburg)*, **18**: 157.

Bonarelli, G., 1894, Contribuzione alla conoscenza del Giura-Lias Lombardo. *Atti della Regia Accademia delle Scienze di Torino*, **30**: 63−78.

Bonarelli, G., 1895, Fossili domeriani della Brianza. *Rendiconti del Reale Istituto. Lombardo di Scienze e Lettere (Milano), Serie 2*, **28**: 326−347.

Bonis, N.R., Kürschner, W.M., and Krystyn, L., 2009, A detailed palynological study of the Triassic−Jurassic transition in key sections of the Eiberg Basin (Northern Calcareous Alps, Austria). *Review of Palaeobotany and Palynology*, **156**: 376−400.

Bonis, N.R., Ruhl, M., and Kürschner, W.M., 2010, Milankovitch-scale palynological turnover across the Triassic-Jurassic transition at St. Audrie's Bay, SW UK. *Geological Society of London Journal*, **167**: 877−888.

Boulila, S., and Hinnov, L.A., 2017, A review of tempo and scale of the early Jurassic Toarcian OAE: implications for carbon cycle and sea level variations. *Newsletters on Stratigraphy*, **50**: 363−389.

Boulila, S., Hinnov, L.A., Collin, P.Y., Huret, E., Galbrun, B., and Fortwengler, D., 2008a, Astronomical calibration of the Lower Oxfordian (Terres Noires, Vocontian Basin, France): consequences of revising Late Jurassic time scale. *Earth and Planetary Science Letters*, **276**: 40−51.

Boulila, S., Galbrun, B., Hinnov, L., and Collin, P.Y., 2008b, Orbital calibration of the Early Kimmeridgian (Southeastern France): implications for geochronology and sequence stratigraphy. *Terra Nova*, **20**: 455−462.

Boulila, S., Galbrun, B., Hinnov, L.A., Collin, P.-Y., Ogg, J.G., Fortwengler, D., et al., 2010a, Milankovitch and sub-Milankovitch forcing of the Oxfordian (Late Jurassic) Terres Noires Formation (SE France) and global implications. *Basin Research*, **22**: 712−732.

Boulila, S., de Rafélis, M., Hinnov, L.A., Gardin, S., Galbrun, B., and Collin, P.-Y., 2010b, Orbitally forced climate and sea-level changes in the Paleoceanic Tethyan domain (marl-limestone alternations, Lower Kimmeridgian, SE France). *Palaeogeography, Palaeoclimatology, Palaeoecology*, **292**: 57−70.

Boulila, S., Gardin, S., de Rafélis, M., Hinnov, L.A., Galbrun, B., and Collin, P.-Y., 2011, Reply to the comment on "Orbitally forced climate and sea-level changes in the Paleoceanic Tethyan domain (marl-limestone alternations, Lower Kimmeridgian, SE France).". *Palaeogeography, Palaeoclimatology, Palaeoecology*, **306**: 252−257.

Boulila, S., Galbrun, B., Huret, E., Hinnov, L.A., Rouget, I., Gardin, S., et al., 2014, Astronomical calibration of the Toarcian Stage: implications for sequence stratigraphy and duration of the early Toarcian OAE. *Earth and Planetary Science Letters*, **386**: 98−111.

Boulila, S., Galbrun, B., Sadki, D., Gardin, S., and Bartolini, A., 2019, Constraints on the duration of the early Toarcian T-OAE and evidence for carbon-reservoir change from the High Atlas (Morocco). *Global and Planetary Change*, **175**: 113−128.

Bown, P.R. (ed), 1998. Calcareous Nannofossil Biostratigraphy. Chapman & Hall. London.

Bown, P.R., Cooper, M.K.E., and Lord, A.R., 1988, A calcareous nannofossil biozonation for the early to mid Mesozoic. *Newsletters on Stratigraphy*, **20**: 91−114.

Bown, P.R., Lees, J.A., and Young, J.R., 2004, Calcareous nannoplankton evolution and diversity through time. *In* Thierstein, H.R., and Young, J.R. (eds), *Coccolithophores: From Molecular Processes to Global Impact*. Berlin: Springer-Verlag, 481−508.

Bralower, T.J., Monechi, S., and Thierstein, H.R., 1989, Calcareous nannofossil zonation of the Jurassic-Cretaceous boundary interval and correlation with the geomagnetic polarity timescale. *Marine Micropaleontology*, **14**: 153−235.

Brenner, W., 1988, Dinoflagellaten aus dem Unteren Malm (Oberer Jura) von Süddeutschland; Morphologie, Ökologie, Stratigraphie. *Tübinger Mikropaläontologische Mitteilungen*, **6**: 116.

Brongniart, A., 1829, *Tableau des Terrains qui Composent L'écorce du Globe ou Essai sur la Structure de la Partie Connue de la Terre*. Paris: F.G. Levrault, p. 435.

Brusatte, S.L., O'Connor, J.K., and Jarvis, E.D., 2015, The origin and diversification of birds. *Current Biology*, **25**: R888−R898.

Burger, D., 1995, Timescales 8. Jurassic. Calibration and development correlation charts and explanatory notes. *Australian Geological Survey Organization Record*, **37**: 1−30.

Callomon, J.H., 1964, Notes on the Callovian and Oxfordian Stages. *In* Mautbeuge, P.L. (ed), *Colloque du Jurassique à Luxembourg 1962*. Luxembourg: Publication de l'Institut Grand-Ducal, Section des Sciences Naturelles, Physiques et Mathématiques, 269−291.

Callomon, J.H., 1985, Biostratigraphy, chronostratigraphy and all that—again. *In* Michelsen, O., and Zeiss, A. (eds), International Symposium on Jurassic Stratigraphy, Erlangen, 1984. *Copenhagen: Geological Survey of Denmark*, **Vol. 3**: 611−624.

Callomon, J.H., 1995, Time from fossils: S. S. Buckman and Jurassic high-resolution geochronology. *In* Le Bas, M.J. (ed), Milestones in Geology. *Geological Society of London, Memoir*, **16**: 127−150.

Callomon, J.H., 1999, Report of the Bathonian − Callovian Boundary Working Group. *International Subcommission on Jurassic Stratigraphy Newsletter*, **26**: 53−60.

Callomon, J.H., and Dietl, G., 2000, On the proposed basal boundary stratotype (GSSP) of the Middle Jurassic Callovian Stage. *In* Hall, R.L., and Smith, P.L. (eds), Advances in Jurassic Research. *GeoResearch Forum*, **6**: 41−54.

Calvin, L., 2010, On giant filter feeders. *Science*, **327**: 968−969.

Cande, S.C., Larson, R.L., and LaBrecque, J.L., 1978, Magnetic lineations in the Pacific Jurassic Quiet Zone. *Earth and Planetary Science Letters*, **41**: 434−440.

Caruthers, A.H., Smith, P.L., and Gröcke, D.R., 2013, The Pliensbachian−Toarcian (Early Jurassic) extinction, a global multi-phased event. *Palaeogeography, Palaeoclimatology, Palaeoecology*, **386**: 104−118.

Casellato, C.E., 2010, Calcareous nannofossil biostratigraphy of upper Callovian-lower Berriasian successions from the southern Alps, North Italy. *Rivista Italiana di Paleontologia e Stratigrafia*, **116**: 357−404.

Casellato, C.E., Andreini, G., Erba, E., and Parisi, G., 2008, Calcareous nannofossil and Calpionellid calcification events across Tithonian − Berriasian time interval and low latitudes paleoceanographic implications. *Journal of Nannoplankton Research*, ISSN 1210-8049: 33−33.

Cecca, F., Martin Garin, B., Marchand, D., Lathuiliere, B., and Bartolini, A., 2005, Paleoclimatic control of biogeographic and sedimentary events in Tethyan and peri-Tethyan areas during the Oxfordian (Late Jurassic). *Palaeogeography, Palaeoclimatology, Palaeoecology*, **222**: 10−32.

Chang, S.-C., Zhang, H., Renne, P.R., and Fang, Y., 2009, High-precision $^{40}Ar/^{39}Ar$ age constraints on the basal Lanqi Formation and its implications for the origin of angiosperm plants. *Earth and Planetary Science Letters*, **30**: 212−221.

Channell, J.E.T., Lowrie, W., Pialli, P., and Venturi, F., 1984, Jurassic magnetic stratigraphy from Umbrian (Italian) land sections. *Earth and Planetary Science letters*, **68**: 309−325.

Channell, J.E.T., Massari, F., Benetti, A., and Pezzoni, N., 1990, Magnetostratigraphy and biostratigraphy of Callovian−Oxfordian limestones from the Trento Plateau (Monti Lessini, Northern Italy). *Palaeogeography, Palaeoclimatology, Palaeoecology*, **79**: 289−303.

Channell, J.E.T., Casellato, C.E., Muttoni, G., and Erba, E., 2010, Magnetostratigraphy, nannofossil stratigraphy and apparent polar wander for Adria-Africa in the Jurassic−Cretaceous boundary interval. *Palaeogeography, Palaeoclimatology, Palaeoecology*, **293**: 51−75.

Choiniere, J.N., Xu, X., Clark, J.M., Forster, C.A., Guo, Y., and Han, F., 2010, A basal alvarezsauroid theropod from the early Late Jurassic of Xinjiang, China. *Science*, **327**: 571−574.

Cohen, A.S., and Coe, A.L., 2007, The impact of the central Atlantic Magmatic province on climate and on the Sr- and Os-isotope evolution of seawater. *Palaeogeography, Palaeoclimatology, Palaeoecology*, **244**: 374–390.

Cohen, A.S., Coe, A.L., Bartlett, J.M., and Hawkesworth, C.J., 1999, Precise Re–Os ages of organic-rich mudrocks and the Os isotope composition of Jurassic seawater. *Earth and Planetary Science Letters*, **167**: 150–173.

Cohen, A.S., Coe, A.L., Harding, S.M., and Schwark, L., 2004, Osmium isotope evidence for the regulation of atmospheric CO_2 by continental weathering. *Geology*, **32**: 157–160.

Cohen, A.S., Coe, A.L., and Kemp, D.B., 2007, The Late Palaeocene-Early Eocene and Toarcian (Early Jurassic) carbon isotope excursions: a comparison of their time scales, associated environmental changes, causes and consequences. *Journal of the Geological Society of London*, **164**: 1093–1108.

Collin, J.-P., 1998, Ostracodes. Columns for Jurassic chart of Mesozoic and Cenozoic sequence chronostratigraphic framework of European basins. *In* de Graciansky, P.-C., Hardenbol, J., Jacquin, Th, and Vail, P.R. (eds), Mesozoic-Cenozoic Sequence Stratigraphy of European Basins. *SEPM Special Publication.*

Colombié, C., and Rameil, N., 2007, Tethyan-to-boreal correlation in the Kimmeridgian using high-resolution sequence stratigraphy (Vocontian Basin, Swiss Jura, Boulonnais, Dorset). *International Journal of Earth Science (Geol. Rundsch.)*, **96**: 567–591.

Comas-Rengifo, M.J., Gómez, J.J., Goy, A., Osete, M.L., and Palencia-Ortas, A., 2010a, The base of the Toarcian (Early Jurassic) in the Almonacid de la Cuba section (Spain): ammonite biostratigraphy, magnetostratigraphy and isotope stratigraphy. *Episodes*, **33**: 15–22.

Comas-Rengifo, M.J., Arias, C., Gómez, J.J., Goy, A., Herrero, C., Osete, M.L., et al., 2010b, A complementary section for the proposed Toarcian (Lower Jurassic) global stratotype: the Almonacid De La Cuba Section (Spain). *Stratigraphy and Geological Correlation*, **18**: 133–152.

Combemorel, R., 1998, Belemnites. Columns for Jurassic chart of Mesozoic and Cenozoic sequence chronostratigraphic framework of European basins. *In* de Graciansky, P.-C., Hardenbol, J., Jacquin, Th, and Vail, P.R. (eds), Mesozoic-Cenozoic Sequence Stratigraphy of European Basins. *SEPM Special Publication.*

Conybeare, W.D., and Phillips, W., 1822, *Outlines of the Geology of England and Wales, with an Introduction Compendium of the General Principles of that Science, and Comparative Views of the Structure of Foreign Countries, Part 1.* London: William Phillips, p. 470.

Cope, J.C.W., 1967, The paleontology and stratigraphy of the lower part of the upper Kimmeridge Clay of Dorset. *Bulletin of the British Museum. Natural History, Geology Series*, **15**: 3–79.

Cope, J.C.W., 1993, The Bolonian Stage: an old answer to an old problem. *Newsletters on Stratigraphy*, **28**: 151–156.

Cope, J.C.W., 2007, Drawing the line: the history of the Jurassic–Cretaceous boundary. *Proceedings of the Geologists' Association*, **119**: 105–117.

Cope, J.C.W., 2009, Correlation problems in the Kimmeridge Clay Formation (upper Jurassic, UK): lithostratigraphy versus biostratigraphy and chronostratigraphy. *Geological Magazine*, **146**: 266–275.

Cope, J.C.W., Getty, T.A., Howarth, M.K., Morton, N., and Torrens, H.S., 1980a, *A correlation of Jurassic rocks in the British Isles. Part One: Introduction and Lower Jurassic, Geological Society of London, Special Report*, 14. p. 73.

Cope, J.C.W., Duff, K.L., Parsons, C.F., Torrens, H.S., Wimbledon, W.A., and Wright, J.K., 1980b, A correlation of Jurassic rocks in the British Isles. Part Two: Middle and upper Jurassic. *Geological Society of London, Special Report*, **15**: 109.

Copestake, P., and Johnson, B., 2014, Lower Jurassic Foraminifera from the Llanbedr (Mochras Farm) borehole, North Wales, UK. *Monograph of the Palaeontographical Society*, **167**: 1–403.

Copestake, P., Johnson, B., Morris, P.H., Coleman, B.E., and Shipp, D.J., 1989, Jurassic. *In* Jenkyns, D.G., and Murray, J.W. (eds), *Stratigraphical Atlas of Fossil Foraminifera.* Chichester: The British Micropalaeontological Society, 125–272.

Coquand, H., 1871, Sur le Klippenkalk du département du Var et des Alpes-Maritimes. *Bulletin de la Société Géologique de France*, **28**: 232–233.

Corfu, F., Svensen, H., and Mazzini, A., 2016, Comment to paper: Evaluating the temporal link between the Karoo LIP and climatic–biologic events of the Toarcian Stage with high-precision U–Pb geochronology by Bryan Sell, Maria Ovtcharova, Jean Guex, Annachiara Bartolini, Fred Jourdan, Jorge E. Spangenberg, Jean-Claude Vicente, Urs Schaltegger in Earth and Planetary Science Letters 408 (2014) 48–56. *Earth and Planetary Science Letters*, **434**: 349–352.

Correia, V.F., Riding, J.B., Henriques, M.H., Fernandes, P., Pereira, Z., and Wiggan, N.J., 2019, The Middle Jurassic palynostratigraphy of the northern Lusitanian Basin, Portugal. *Newsletters on Stratigraphy*, **52**: 73–96.

Cox, B.M., 1990, A review of Jurassic chronostratigraphy and age indicators for the UK. *In* Hardman, R.F.P., and Brooks, J.R.V. (eds), Tectonic Events Responsible for Britain's Oil and Gas Reserves. *Geological Society Special Publication*, **55**: 169–190.

Cox, B.M., and Gallois, R.W., 1981, *The Stratigraphy of the Kimmeridge Clay of the Dorset Type Area and Its Correlation With Some Other Kimmeridgian Sequences.* Institute of Geological Sciences, 80-4. Natural Environment Research Council, 44 pp.

Cramer, B.D., and Jarvis, I., 2020, Chapter 11 - Carbon isotope stratigraphy. *In* Gradstein, F.M., Ogg, J.G., Schmitz, M.D., and Ogg, G.M. (eds), *The Geologic Time Scale 2020.* Vol. **1** (this book). Elsevier, Boston, MA.

Cresta, S., Goy, A., Ureta, S., Arias, C., Barrón, E., Bernard, J., et al., 2001, The global stratotype section and point (GSSP) of the Toarcian-Aalenian boundary (Lower-Middle Jurassic). *Episodes*, **24**: 166–175.

Dal Corso, J., Marzoli, A., Tateo, F., Jenkyns, H.C., Bertrand, H., Youbi, N., et al., 2014, The dawn of CAMP volcanism and its bearing on the end-Triassic carbon cycle disruption. *Journal of the Geological Society*, **171**: 153–164.

Danisch, J., Kabiri, L., Nutz, A., and Bodin, S., 2019, Chemostratigraphy of Late Sinemurian – Early Pliensbachian shallow-to deep-water deposits of the central High Atlas Basin: Paleoenvironmental implications. *Journal of African Earth Sciences*, **153**: 239–249.

Davies, J.H.F.L., Marzoli, A., Bertrand, H., Youbi, N., Ernesto, M., and Schaltegger, U., 2017, End-Triassic mass extinction started by intrusive CAMP activity. *Nature Communications*, **8**: 15596.

de Kaenel, E., Bergen, J.A., and von Salis Perch-Nielsen, K., 1996, Jurassic calcareous nannofossil biostratigraphy of western Europe. Compilation of recent studies and calibration of bioevents. *Bulletin de la Société géologique de France*, **167**: 15–28.

De Wever, P., 1998, Radiolarians. Columns for Jurassic chart of Mesozoic and Cenozoic sequence chronostratigraphic framework of European basins. *In* de Graciansky, P.-C., Hardenbol, J., Jacquin, Th, and Vail, P.R. (eds), Mesozoic-Cenozoic Sequence Stratigraphy of European Basins. *SEPM Special Publication.*

Dean, W.T., Donovan, D.T., and Howarth, M.K., 1961, The Liassic ammonite zones and subzones of the North West European province. *Bulletin of the British Museum of Natural History, Geology*, **4**: 435–505.

Deenen, M.H.L., Ruhl, M., Bonis, N.R., Krijgsman, W., Kürschner, W.M., Reitsma, M., et al., 2010, A new chronology for the end-Triassic mass extinction. *Earth and Planetary Science Letters*, **291**: 113–125.

Deenen, M.H.L., Krijgsman, W., and Ruhl, M., 2011, The quest for chron E23r at Partridge Island, bay of Fundy, Canada: CAMP emplacement postdates the end-Triassic extinction event at the North American craton. *Canadian Journal of Earth Sciences*, **48**: 1282–1291.

Deng, S., Wang, S., Yang, Z., Lu, Y., Li, X., Hu, Q., et al., 2015, Comprehensive study of the middleeupper Jurassic Strata in the Junggar Basin, Xinjiang. *Acta Geoscientica Sinica*, **36**: 559–574.

Dera, G., Neige, P., Dommergues, J.-L., Fara, E., Laffont, R., and Pellenard, P., 2010, High-resolution dynamics of Early Jurassic marine extinctions: the case of Pliensbachian–Toarcian ammonites (Caphalopoda). *Journal of the Geological Society*, **167**: 21–33.

Dietl, G., 1981, Zur systematischen Stellung von *Ammonites subfurcatus* ZIETEN und deren Bedeutung für *subfurcatum*-Zone (Bajocium, Mittlerer Jura). *Stuttgarter Beiträge zur Naturkunde, Serie B (Geologie und Paläontologie)*, **81**: 1–11.

Dietl, G., 1994, Der hochstetteri-Horizont – ein Ammonitenfaunen-Horizont (Discus-Zone, Ober-Bathonium, Dogger) aus dem Schwäbischen Jura. *Stuttgarter Beiträge zur Naturkunde, Serie B (Geologie und Paläontologie)*, **202**: 1–39.

Dietl, G., and Callomon, J.H., 1988, Der Orbis-Oolith (Ober-Bathonium, Mittl. Jura) von Sengenthal/Opf. Fränk. Alb, und siene Bedeutung für Korrelation und Gliederung der Orbis-Zone. *Stuttgarter Beiträge zur Naturkunde, Serie B (Geologie und Paläontologie)*, **142**: 1–31.

Dietl, G., and Etzold, A., 1977, The Aalenian at the type locality. *Stuttgarter Beiträge zur Naturkunde, Serie B (Geologie und Paläontologie)*, **30**: 13.

d'Omalius d'Halloy, J.J., 1843, *Précis élémentaire de Géologie*. Paris: Bruxelles-Paris, p. 636.

d'Orbigny, A., 1842, *Paléontologie français. Description zoologique et géologique de tous les animaux mollusques et rayonnés fossiles de France. Terrains oolitiques ou jurassiques. I. Céphalopodes*. Paris: Masson, p. 642.

d'Orbigny, A., 1852, *Cours élémentaire de paléontologie et de géologie stratigraphique*, Vol. 2. Paris: Masson, pp. 383–847.

Dromart, G., Garcia, J.-P., Picard, S., Atrops, F., Lécuyer, C., and Sheppard, S.M.F., 2003, Ice age at the Middle–Late Jurassic transition? *Earth and Planetary Science Letters*, **213**: 205–220.

Dyer, R., 1998, Radiolarians: North Sea central and Viking Graben. Columns for Jurassic chart of Mesozoic and Cenozoic sequence chronostratigraphic framework of European basins. *In* de Graciansky, P.-C., Hardenbol, J., Jacquin, Th, and Vail, P.R. (eds), *Mesozoic-Cenozoic Sequence Stratigraphy of European Basins. SEPM Special Publication.*

Dyer, R., and Copestake, P., 1989, A review of Late Jurassic to earliest Cretaceous radiolaria and their biostratigraphic potential to petroleum exploration in the North Sea. *In* Batten, D.J., and Keen, M.C. (eds), *northwest European Micropalaeontology and Palynology*. Chichester: British Micropalaeontological Society, 214–235.

Énay, R., 2020, The Jurassic/Cretaceous system boundary is an impasse. Why do not go back to Oppel's 1865 original an historic definition of the Tithonian? *Cretaceous Research*, **106**: 104241, https://doi.org/10.1016/j.cretres.2019.104241.

Erba, E., 1990, Calcareous nannofossil biostratigraphy of some Bajocian sections from the Digne area (SE France). *Memorie Descrittive della Carta Geologica d'Italia*, **XL**: 237–255.

Erba, E., 2006, The first 150 million years history of calcareous nannoplankton: Biosphere–geosphere interactions. *Palaeogeography, Palaeoclimatology, Palaeoecology*, **232**: 237–250.

Ferreira, J., Mattioli, E., Sucherás-Marx, B., Giraud, F., Duarte, L.V., Pittet, B., et al., 2019, Western Tethys Early and Middle Jurassic calcareous nannofossil biostratigraphy. *Earth-Science Reviews*, **197**: 102908, https://doi.org/10.1016/j.earscirev.2019.102908.

Fernández-López, S.R., 2007, Ammonoid taphonomy, palaeoenvironments and sequence stratigraphy at the Bajocian/Bathonian boundary on the Bas Auran area (Subalpine Basin, south-eastern France). *Lethaia*, **40**: 377–391.

Fernández-López, S.R., Pavia, G., Erba, E., Guiomar, M., Henriques, M.H., Lanza, R., et al., 2009, The Global Boundary Stratotype Section and Point (GSSP) for base of the Bathonian Stage (Middle Jurassic), Ravin du Bès Section, SE France. *Episodes*, **32**: 222–248.

Fortwengler, D., and Marchand, D., 1994, Nouvelles unites biochronologiques de la Zone à Mariae (Oxfordien inférieur). *Geobios, Mémoire Spécial*, **17**: 203–209.

Frakes, L.A., Francis, J.E., and Syktus, J.I., 1992, *Climate Modes of the Phanerozoic*. Cambridge: Cambridge University Press, p. 286.

Franceschi, M., Dal Corso, J., Posenato, R., Roghi, G., Masetti, D., and Jenkyns, H.C., 2014, Early Pliensbachian (early Jurassic) C-isotope perturbation and the diffusion of the Lithiotis Fauna: insights from the western Tethys. *Palaeogeography, Palaeoclimatology, Palaeoecology*, **410**: 255–263.

Fu, X., Wang, J., Feng, X., Wang, D., Chen, W., Song, C., et al., 2016, Early Jurassic carbon-isotope excursion in the Qiangtang Basin (Tibet), the eastern Tethys: implications for the Toarcian Oceanic anoxic event. *Chemical Geology*, **442**: 62–72.

Friedman, M., Shimada, K., Martin, L.D., Everhart, M.J., Liston, J., Maltese, A., et al., 2010, 100-million-year dynasty of giant planktivorous bony fishes in the Mesozoic seas. *Science*, **327**: 990–993.

Gabilly, J., 1976, *Le Toarcien à Thouars et dans le centre-ouest de la France: biostratigraphie, evolution de la faune (Harpoceratinae-Hildoceratinae)*. Paris: Centre National Recherche Scientifique, p. 217.

Galbrun, B., Gabilly, J., and Rasplus, L., 1988, Magnetostratigraphy of the Toarcian stratotype sections at Thouars and Airvault (Deux-Sevres, France). *Earth and Planetary Science Letters*, **87**: 453–462.

Galbrun, B., Baudin, F., Fourcade, E., and Rivas, P., 1990, Magnetostratigraphy of the Toarcian Ammonitico Rosso limestone at Iznalloz, Spain. *Geophysical Research Letters*, **17**: 2441–2444.

Gale, A.S., Mutterlose, J., and Batenburg, S., 2020, Chapter 27 – The Cretaceous Period. *In* Gradstein, F.M., Ogg, J.G., Schmitz, M.D., and Ogg, G.M. (eds), *The Geologic Time Scale 2020*. Vol. **2** (this book). Elsevier, Boston, MA.

Geyer, O.F., and Gwinner, M.P., 1979, Die Schwäbische Alb und ihr Vorland. *Sammlung Geologischer Führer*, **67**: 271.

Gipe, R.A., 2013, *Callovian (upper Middle Jurassic) Magnetostratigraphy: A Composite Polarity Pattern From France, Britain and Germany, and Its Correlation to the Pacific Marine Magnetic Anomaly Model*. Purdue University Open Access Theses. Paper 36, 107 pp. http://docs.lib.purdue.edu/open_access_theses/36.

Gipe, R.A., Ogg, J.G., and Coe, A.L., 2013, Magnetostratigraphy of the Callovian (upper Middle Jurassic) from outcrops in France, England and Germany and calibration of the Pacific M-sequence of magnetic anomalies. *Geological Society of America Abstracts with Programs*, **45** (7), 811, https://gsa.confex.com/gsa/2013AM/webprogram/Paper231332.html.

Głowniak, E., 1997, Middle Oxfordian ammonites. *International Subcommission on Jurassic Stratigraphy Newsletter*, **25**: 45–46.

Głowniak, E., and Wierzbowski, H., 2007, Comment on "The mid-Oxfordian (Late Jurassic) positive carbon-isotope excursion recognised from fossil wood in the British Isles" by C.R. Pearce, S.P. Hesselbo, A.L. Coe. *Palaeogeography Palaeoclimatology Palaeoecology*, **248**: 247–251.

Głowniak, E., Matyja, B.A., and Wierzbowski, A., 2008, Upgraded subdivision of the Oxfordian and Kimmeridgian as a consequence of recent correlations. *The 5th International Symposium IGCP 506. Hammamet – Tunisia, 28–31 March 2008, Abstract Volume*, 34–35.

Gómez, J.J., Comas-Rengifo, M.J., and Goy, A., 2016, Palaeoclimatic oscillations in the Pliensbachian (early Jurassic) of the Asturian basin (northern Spain). *Climate of the Past*, **12**: 1199–1214.

Goy, A., Ureta, M.S., Arias, C., Canales, M.L., Garcia Joral, F., Herrero, C., et al., 1994, The Fuentelsaz section (Iberian Range, Spain), a possible stratotype for the base of the Aalenian Stage. *Miscellanea del Servizio Geologico Nazionale*, **5**: 1–31.

Goy, A., Ureta, M.S., Arias, C., Canales, M.L., Garcia Joral, F., Herrero, C., et al., 1996, Die Toarcium/Aalenium-Grenze im Profil Fuentelsaz (Iberische Ketten, Spanien). *In* Ohmert, W. (ed) Die Grenzziehung Unter-/Mitteljura (Toarciuim/Aalenium) bei Wittnau und Fuentelsaz. Beispiele interdisziplinarer geowissenschaftlicher Zusammenarbeit. *Geologisches Landesamt Baden-Wurttemberg, Informationen*, **8**: 43–52.

Gradstein, F.M., Gale, A., Kopaevich, L., Waskowska, A., Grigelis, A., Glinskikh, L., et al., 2017, The planktonic foraminifera of the Jurassic. Part II: Stratigraphy, palaeoecology and palaeobiogeography. *Swiss Journal of Palaeontology*, **136**: 259–271, https://doi.org/10.1007/s13358-017-0123-y.

Granier, B., 2019, JK2018: International Meeting around the Jurassic/Cretaceous Boundary Chairperson's Report. *Volumina Jurassica*, **17**: 1–6.

Gröcke, D.R., Price, G.D., Ruffell, A.H., Mutterlose, J., and Baraboshkin, E., 2003, Isotopic evidence for Late Jurassic–Early Cretaceous climate change. *Palaeogeography, Palaeoclimatology, Palaeoecology*, **202**: 97–118.

Grossman, E.L., and Joachimski, M.M., 2020, Chapter 10 - Oxygen isotope stratigraphy. *In* Gradstein, F.M., Ogg, J.G., Schmitz, M.D., and Ogg, G.M. (eds), *The Geologic Time Scale 2020*. Vol. 1 (this book). Elsevier, Boston, MA.

Groupe Français d'Étude du Jurassique, 1997, Biostratigraphie du Jurassique ouest-européen et méditerranéen: zonations parallèles et distribution des invertébrés et microfossiles. *Bulletin des Centres de Recherches Exploration- Production Elf-Aquitaine, Mémoire*, **17**: 422.

Guex, J., 1995, Ammonites hettangiennes de la Gabbs Valley Range (Nevada). *Mémoires Géologie (Lausanne)*, **27**: 130.

Guex, J., Bartolini, A., Atudorei, V., and Taylor, D., 2004, High-resolution ammonite and carbon isotope stratigraphy across the Triassic – Jurassic boundary at New York Canyon. *Earth and Planetary Science Letters*, **225**: 29–41.

Guex, J., Schoene, B., Bartolini, A., Spangenberg, J., Schaltegger, U., O'Dogherty, L., et al., 2012, Geochronological constraints on post-extinction recovery of the ammonoids and carbon cycle perturbations during the Early Jurassic. *Palaeogeography, Palaeoclimatology, Palaeoecology*, **346–347**: 1–11.

Guzhikov, A.Yu, Pimenov, M.V., Malenkina, S.Yu, Manikin, A.G., and Astarkin, S.V., 2010, Paleomagnetic, petromagnetic, and terrigenous–mineralogical studies of upper Bathonian–Lower Callovian sediments in the Prosek Section, Nizhni Novgorod Region. *Stratigraphy and Geological Correlation*, **18**: 42–62.

Gygi, R.A., 2000, Integrated stratigraphy of the Oxfordian and Kimmeridgian (Late Jurassic) in northern Switzerland and adjacent southern Germany. *Memoires of the Swiss Academy of Sciences*, **104**: 152.

Gygi, R.A., Coe, A.L., and Vail, P.R., 1998, Sequence stratigraphy of the Oxfordian and Kimmeridgian stages (Late Jurassic) in northern Switzerland. *In* de Graciansky, P.-C., Hardenbol, J., Jacquin, Th, and Vail, P.R. (eds), Mesozoic-Cenozoic Sequence Stratigraphy of European Basins. *SEPM Special Publication*, **60**: 527–544.

Habib, D., and Drugg, W.S., 1983, Dinoflagellate age of Middle Jurassic-Early Cretaceous sediments in the Blake-Bahama Basin. *Initial Reports of the Deep Sea Drilling Project*, **76**: 623–638.

Hallam, A., 1975, *Jurassic Environments*. Cambridge: Cambridge University Press, p. 269.

Hallam, A., 1976, Stratigraphic distribution and ecology of European Jurassic bivalves. *Lethaia*, **9**: 245–259.

Hallam, A., 1981, A revised sea-level curve for the Early Jurassic. *Journal of the Geological Society of London*, **138**: 735–743.

Hallam, A., 2001, A review of the broad pattern of Jurassic sea-level changes and their possible causes in the light of current knowledge. *Palaeogeography, Palaeoclimatology, Palaeoecology*, **167**: 23–37.

Hallam, A., and Wignall, P.B., 1999, Mass extinctions and sea-level changes. *Earth-Science Reviews*, **48**: 217–250.

Hames, W.E., Renne, P.R., and Ruppel, C., 2000, New evidence for geologically instantaneous emplacement of earliest Jurassic central Atlantic magmatic province basalts on the North American margin. *Geology*, **28**: 859–862.

Handschumacher, D.W., Sager, W.W., Hilde, T.W.C., and Bracey, D.R., 1988, Pre-Cretaceous evolution of the Pacific plate and extension of the geomagnetic polarity reversal time scale with implications for the origin of the Jurassic "Quiet Zone". *Tectonophysics*, **155**: 365–380.

Haq, B.U., 2017, Jurassic sea-level variations: a reappraisal. *GSA Today*, **28**.

Haq, B.U., and al-Qahtani, A.M., 2005, Phanerozoic cycles of sea-level change on the Arabian Platform. *GeoArabia*, **10**: 127–160.

Haq, B.U., Hardenbol, J., and Vail, P.R., 1988, Mesozoic and Cenozoic chronostratigraphy and eustatic cycles. *In* Wilgus, C.K., Hastings, B.S., Kendall, G.St.C., Posamentier, H.W., Ross, C.A., and van Wagoner, J.C. (eds), Sea-Level Changes: An Integrated Approach. *SEPM Special Publication*, **42**: 71–108.

Hardenbol, J., Thierry, J., Farley, M.B., Jacquin, Th, de Graciansky, P.-C., and Vail, P.R., 1998, Mesozoic and Cenozoic sequence chronostratigraphic framework of European basins. *In* de Graciansky, P.-C., Hardenbol, J., Jacquin, Th, and Vail, P.R. (eds), Mesozoic-Cenozoic Sequence Stratigraphy of European Basins. *SEPM Special Publication*, **60**: 3–13.

Harding, I.C., Smith, G.A., Riding, J.B., and Wimbledon, W.A.P., 2011, Inter-regional correlation of Jurassic/Cretaceous boundary strata based on the Tithonian-Valanginian dinoflagellate cyst biostratigraphy of the Volga Basin, western Russia. *Review of Palaeobotany and Palynology*, **167**: 82–116.

Hargreaves, R.B., Rehacek, J., and Hooper, P.R., 1997, Paleomagnetism of the Karoo igneous rocks in South Africa. *South African Journal of Geology*, **100**: 195–212.

Harland, W.B., Cox, A.V., Llewellyn, P.G., Pickton, C.A.G., Smith, A.G., and Walters, R., 1982, *A Geologic Time Scale*. Cambridge: Cambridge University Press, p. 131.

Harland, W.B., Armstrong, R.L., Cox, A.V., Craig, L.E., Smith, A.G., and Smith, D.G., 1990, *A geologic time scale 1989*. Cambridge: Cambridge University Press, p. 263.

Hart, R.J., Andreoli, M.A.G., Tredoux, M., Moser, D., Ashwal, L.D., Eide, E. A., et al., 1997, Late Jurassic age for the Morokweng impact structure, southern Africa. *Earth and Planetary Science Letters*, **147**: 25−35.

Haug, E., 1910, *Traité de Géologie. II. Les périodes géologiques. Fasicule 2. Jurassique et Crétacé*. Paris: Armand Collin, pp. 929−1396.

Helby, R., Morgan, R., and Partridge, A.D., 1987, A palynological zonation of the Australian Mesozoic. *Association of Australasian Palaeontologists Memoir*, **4**: 94.

Henriques, M.H., 1992, *Biostratigrafia e paleontologia (Ammonoidea) do Aaleniano em Portugal (Sector Setentrional da Bacia Lusitaniana)*. Unpublished Ph.D Thesis, Centro de Geociências da Universidade de Coimbra, Instituto Nacional de Investigacao Cientifica, 301 pp.

Henriques, M.H., Gardin, S., Gomes, C.R., Soares, A.F., Rocha, R.B., Marques, J.F., et al., 1994, The Aalenian-Bajocian boundary at Cabo Mondego (Portugal). *In* Cresta, S., and Pavia, G. (eds), Proceedings of the 3rd International Meeting on Aalenian and Bajocian Stratigraphy. *Miscellanea del Servizio Geologico Nazionale*, **5**: 63−77.

Hermoso, M., Minoletti, F., and Pellenard, P., 2013, Black shale deposition during Toarcian super-greenhouse driven by sea level. *Climate of the Past*, **9**: 2703−2712.

Hesselbo, S.P., 2008, Sequence stratigraphy and inferred relative sea-level change from the onshore British Jurassic. *Proceedings of the Geologists' Association*, **9**: 19−34.

Hesselbo, S.P., and Jenkyns, H.C., 1998, British Lower Jurassic sequence stratigraphy. *In* de Graciansky, P.-C., Hardenbol, J., Jacquin, Th, and Vail, P.R. (eds), Mesozoic-Cenozoic Sequence Stratigraphy of European Basins. *SEPM Special Publication*, **60**: 562−581.

Hesselbo, S.P., and Pienkowski, G., 2011, Stepwise atmospheric carbon-isotope excursion during the Toarcian Oceanic Anoxic Event (Early Jurassic, Polish Basin). *Earth and Planetary Science Letters*, **301**: 365−373.

Hesselbo, S.P., Gröcke, D.R., Jenkyns, H.C., Bjerrum, C.J., Farrimond, P., Morgans, B.H.S., et al., 2000, Massive dissociation of gas hydrate during a Jurassic oceanic anoxic event. *Nature*, **406**: 392−395.

Hesselbo, S.P., Robinson, S.A., Surlyk, F., and Piasecki, S., 2002, Terrestrial and marine extinctions at the Triassic-Jurassic boundary synchronized with major carbon-cycle perturbation: a link to initiation of massive volcanism? *Geology*, **30**: 251−254.

Hesselbo, S.P., McRoberts, C.A., and Pálfy, J., 2007a, Triassic−Jurassic boundary events: problems, progress, possibilities. *Palaeogeography, Palaeoclimatology, Palaeoecology*, **244**: 1−10.

Hesselbo, S.P., Jenkyns, H.C., Duarte, L.V., and Oliveira, L.C.V., 2007b, Carbon isotope record of the Early Jurassic (Toarcian) Oceanic Anoxic Event from fossil wood and marine carbonate (Lusitanian Basin, Portugal). *Earth and Planetary Science Letters*, **253**: 455−470.

Hesselbo, S.P., Deconinck, J.-F., Huggett, J.M., and Morgans-Bell, H.S., 2009, Late Jurassic palaeoclimatic change from clay mineralogy and gamma-ray spectrometry of the Kimmeridge Clay, Dorset, UK. *Journal of the Geological Society of London*, **166**: 1123−1133.

Hesselbo, S.P., Bjerrum, C.J., Hinnov, L.A., MacNiocaill, C., Miller, K.G., Riding, J.B., et al., 2013, Mochras borehole revisited: a new global standard for Early Jurassic earth history. *Scientific Drilling*, **16**: 81−91.

Hesselbo, S.P., Hudson, A.J.L., Huggett, J.M., Leng, M.J., Riding, J.B., and Ullmann, C.V., 2020a, Palynological, geochemical, and mineralogical characteristics of the Early Jurassic Liasidium Event in the Cleveland Basin, Yorkshire, UK. *Newsletters on Stratigraphy*, **53** (2), 191−211.

Hesselbo, S.P., Korte, C., Ullmann, C.V., and Ebbesen, A., 2020b, Carbon and oxygen isotope records from the southern Eurasian Seaway following the Triassic-Jurassic boundary: parallel long-term enhanced carbon burial and seawater warming. *Earth-Science Reviews*, **203**: 103131.

Hillebrandt, A.v., 1997, Proposal for the Utcubamba Valley sections in northern Peru. *International Subcommission on Jurassic Stratigraphy Newsletter*, **24**: 21−25.

Hillebrandt, A.v., and Krystyn, L., 2009, On the oldest Jurassic ammonites of Europe (Northern Calcareous Alps, Austria) and their global significance. *Neues Jahrbuch für Geologie und Paläontologie Abhandlungen*, **253**: 163−195.

Hillebrandt, A.v., Krystyn, L., and Kürschner, W.M., 2008, A candidate GSSP for the base of the Jurassic in the Northern Calcareous Alps (Kuhjoch section; Karwendel Mountains, Tyrol, Austria). *International Subcommission on Jurassic Stratigraphy, Triassic/Jurassic Boundary Working Group Ballot 2008*, 44.

Hillebrandt, A.v., Krystyn, L., Kürschner, W.M., Bonis, N.R., Ruhl, M., Richoz, S., et al., 2013, The Global Stratotype Sections and Point (GSSP) for the base of the Jurassic System at Kuhjoch (Karwendel Mountains, Northern Calcareous Alps, Tyrol, Austria). *Episodes*, **36**: 162−198.

Hinnov, L.A., 2018, Cyclostratigraphy and Astrochronology in 2018. *Stratigraphy and Time Scales*, vol. 3. Elsevier, pp. 1−80.

Hinnov, L.A., and Park, J.J., 1999, Strategies for assessing Early−Middle (Pliensbachian−Aalenian) Jurassic cyclochronologies. *Philosophical Transactions of the Royal Society of London, Series A*, **357**: 1831−1859.

Hinnov, L.A., Park, J.J., and Erba, E., 1999, Lower-Middle Jurassic rhythmites from the Lombard Basin, Italy: a record of orbitally-forced cycles modulated by long-term secular environmental changes in West Tethys. *In* Hall, R.L., and Smith, P.L. (eds), Advances in Jurassic Research 2000; Proceedings of the Fifth International Symposium on the Jurassic System, 1998. *GeoResearch Forum*, **6**: 437−453.

Holm-Alwmark, S., Alwmark, C., Ferrière, L., Lindström, S., Meier, M.M.M., Scherstén, A., et al., 2019, An Early Jurassic age for the Puchezh-Katunki impact structure (Russia) based on ^{40}Ar/^{39}Ar data and palynology. *Meteoritics & Planetary Science*, **54**: 1764−1780.

Horner, F., 1983, *Palaeomagnetismus von Karbonatsedimenten der Suedlichen Tethys: Implikationen für die Polaritaet des Erdmagnetfeldes im untern Jura und für die tektonik der Ionischen Zone Grieschenlands*. PhD. thesis, ETH Zürich, 139 pp.

Horner, F., and Heller, F., 1983, Lower Jurassic magnetostratigraphy at the Breggia Gorge (Ticino, Switzerland) and Alpi Turati (Como, Italy). *Geophysics Journal of the Royal Astronomical Society*, **73**: 705−718.

Hounslow, M.K., and Muttoni, G., 2010, The geomagnetic polarity timescale for the Triassic: linkage to stage boundary definitions. *In* Lucas, S.G. (ed), The Triassic Timescale. The Geological Society. *London: Special Publication*, **334**: 61−102.

Hounslow, M.W., Posen, P.E., and Warrington, G., 2004, Magnetostratigraphy and biostratigraphy of the upper Triassic and lowermost Jurassic succession, St. Audrie's Bay, UK. *Palaeogeography, Palaeoclimatology, Palaeoecology*, **213**: 331−358.

Houša, V., Pruner, P., Zakharov, V.A., Kostak, M., Chadima, M., Rogov, M.A., et al., 2007, Boreal−Tethyan correlation of the Jurassic−Cretaceous boundary interval by magneto- and biostratigraphy. *Stratigraphy and Geological Correlation*, **15**: 297−309.

Hu, D., Hou, L., Zhang, L., and Xu, X., 2009, A pre-*Archaeopteryx* troodontid theropod from China with long feathers on the metatarsus. *Nature*, **461**: 640–643.

Huang, C., 2018, Astronomical time scale for the Mesozoic. *Stratigraphy and Time Scales*, Vol. 3. Elsevier, pp. 81–150.

Huang, C., and Hesselbo, S.P., 2014, Pacing of the Toarcian Oceanic Anoxic Event (Early Jurassic) from astronomical correlation of marine sections. *Gondwana Research*, **25**: 1348–1356.

Huang, C., Hesselbo, S.P., and Hinnov, L.A., 2010a, Astrochronology of the late Jurassic Kimmeridgian Clay (Dorset, England) and implications for Earth system processes. *Earth and Planetary Science Letters*, **289**: 242–255.

Huang, C., Hinnov, L.A., Swientek, O., and Smelror, M., 2010b, Astronomical tuning of Late Jurassic-Early Cretaceous sediments (Volgian-Ryazanian stages), Greenland-Norwegian seaway. *AAPG Annual Convention, New Orleans, LA, Abstract.*

Huret, E., 2006, *Analyse cyclostratigraphique des variations de la susceptibilité magnétique des argilites callovo-oxfordiennes de l'Est du Bassin de Paris: application à la recherche de hiatus sédimentaires.* Ph.D thesis, Université Pierre et Marie Curie, Paris, 321 pp.

Huret, E., Hinnov, L.A., Galbrun, B., Collin, P.-Y., Gardin, S., and Rouget, I., 2008, Astronomical calibration and correlation of the Lower Jurassic, Paris and Lombard basins (Tethys). *33rd International Geological Congress, Oslo, Norway, Abstract.*

Hüsing, S.K., Beniest, A., van der Boon, A., Abels, H.A., Deenen, M.H.L., and Krijgsman, W., 2014, Astronomically-calibrated magnetostratigraphy of the Lower Jurassic marine successions at St. Audrie's Bay and East Quantoxhead (Hettangian-Sinemurian; Somerset, UK). *Palaeogeography, Palaeoclimatology, Palaeoecology*, **403**: 43–56.

Ikeda, M., Bôle, M., and Baumgartner, P.O., 2016, Orbital-scale changes in redox condition and biogenic silica/detrital fluxes of the Middle Jurassic Radiolarite in Tethys (Sogno, Lombardy, N-Italy): possible link with glaciation? *Palaeogeography, Palaeoclimatology, Palaeoecology*, **457**: 247–257.

Ikeda, M., Tada, R., and Ozaki, K., 2017, Astronomical pacing of the global silica cycle recorded in Mesozoic bedded cherts. *Nature Communications*, **8**: 15532.

Ioannides, I., Riding, J., Stover, L.E., and Monteil, E., 1998, Dinoflagellate Cysts. Columns for Jurassic chart of Mesozoic and Cenozoic sequence chronostratigraphic framework of European basins. *In* de Graciansky, P.-C., Hardenbol, J., Jacquin, Th, and Vail, P.R. (eds), Mesozoic-Cenozoic Sequence Stratigraphy of European Basins. *SEPM Special Publication60*: SEPM Special Publication.

Innocenti, M., Mangold, C., Pavia, G., and Torrens, H.S., 1988, A proposal for the formal ratification of the boundary stratotype of the Bathonian stage based on a Bas Auran section (S.E. France). *In* Rocha, R.B., and Soares, A.F. (eds), *2nd International Symposium on Jurassic Stratigraphy*, **1**: 333–346.

INTERRAD Jurassic-Cretaceous Working Group, 1995, Middle Jurassic to Lower Cretaceous Radiolaria of Tethys: occurrences, systematics, biochronology. *Mémoires de. Géologie Lausanne*, **23**: 1172.

Jeletzky, J.A., 1965, Late upper Jurassic and early Lower Cretaceous fossil zones of the Canadian western Cordillera, British Columbia. *Geological Survey of Canada Bulletin*, **103**: 70.

Jenkyns, H.C., 1988, The early Toarcian (Jurassic) anoxic event: stratigraphic, sedimentary and geochemical evidence. *American Journal of Science*, **288**: 101–151.

Jenkyns, H.C., 2010, Geochemistry of oceanic anoxic events. *Geochemistry, Geophysics, Geosystems*, **11**: Q03004, https://doi.org/10.1029/2009GC002788.

Jenkyns, H.C., Jones, C.E., Gröcke, D.R., Hesselbo, S.P., and Parkinson, D.N., 2002, Chemostratigraphy of the Jurassic System: applications, limitations and implications for palaeoceanography. *Journal of the Geological Society*, **159**: 351–378.

Jones, C.E., and Jenkyns, H.C., 2001, Seawater strontium isotopes, oceanic anoxic events, and seafloor hydrothermal activity in the Jurassic and Cretaceous. *American Journal of Science*, **301**: 112–149.

Jones, C.E., Jenkyns, H.C., and Hesselbo, S.P., 1994a, Strontium isotopes in Early Jurassic seawater. *Geochimica et Cosmochimica Acta*, **58**: 1285–1301.

Jones, C.E., Jenkyns, H.C., Coe, A.L., and Hesselbo, S.P., 1994b, Strontium isotopic variations in Jurassic and Cretaceous seawater. *Geochimica et Cosmochimica Acta*, **58**: 3061–3074.

Jourdan, F., Marzoli, A., Bertrand, H., Cirilli, S., Tanner, L., Kontak, D.J., et al., 2009, $^{40}Ar/^{39}Ar$ ages of CAMP in North America: implications for the Triassic–Jurassic boundary and the 40K decay constant bias. *Lithos*, **110**: 167–180.

Juárez, M.T., Osete, M.L., Meléndez, G., Langereis, C.G., and Zijderveld, J.D.A., 1994, Oxfordian magnetostratigraphy of Aguilón and Tosos sections (Iberian Range, Spain) and evidence of a pre-Oligocene overprint. *Physics of the Earth and Planetary Interiors*, **85**: 195–211.

Juárez, M.T., Osete, M.L., Meléndez, G., and Lowrie, W., 1995, Oxfordian magnetostratigraphy in the Iberian Range. *Geophysical Research Letters*, **22**: 2889–2892.

Kamo, S.L., and Riccardi, A.C., 2009, A new U-Pb zircon age for an ash layer at the Bathonian-Callovian boundary, Argentina. *GFF*, **131**: 177–182.

Katz, M.E., Wright, J.D., Miller, K.G., Cramer, B.S., Fennel, K., and Falkowski, P.G., 2005, Biological overprint of the geological carbon cycle. *Marine Geology*, **217**: 323–338.

Kemp, D.B., Coe, A.L., Cohen, A.S., and Schwark, L., 2005, Astronomical pacing of methane release in the Early Jurassic period. *Nature*, **437**: 396–399.

Kemp, D.B., Coe, A.L., Cohen, A., and Weedon, G.P., 2011, Astronomical forcing and chronology of the early Toarcian (Early Jurassic) Oceanic Anoxic Event in Yorkshire, UK. *Paleoceanography*, **26**: PA002122.

Kent, D.V., and Olsen, P.E., 2008, Early Jurassic magnetostratigraphy and paleolatitudes from the Hartford continental rift basin (eastern North America): testing for polarity bias and abrupt polar wander in association with the central Atlantic Magmatic province. *Journal of Geophysical Research*, **113**: B06105, https://doi.org/10.1029/2007JB005407.

Kent, D.V., Olsen, P.E., and Witte, W.K., 1995, Late Triassic-Early Jurassic geomagnetic polarity sequence and paleolatitudes from drill cores in the Newark rift basin, eastern North America. *Journal of Geophysical Research*, **100**: 14965–14998.

Kent, D.E., Olsen, P.E., and Muttoni, G., 2017, Astrochronostratigraphic polarity time scale (APTS) for the Late Triassic and Early Jurassic from continental sediments and correlation with standard marine stages. *Earth-Science Reviews*, **166**: 153–180.

Kerr, R.A., 2000, Did volcanoes drive ancient extinctions? *Science*, **289**: 1130–1131.

Kiessling, W., Scasso, R., Zeiss, A., Riccardi, A., and Medina, F., 1999, Combined Radiolaria-ammonite stratigraphy for the Late Jurassic of the Antarctic Peninsula: implications for radiolarian stratigraphy. *Geodiversitas*, **21**: 687–713.

Kietzmann, D.A., Iglesia Llanos, M.P., and Kohan Martinez, M., 2018, Astronomical calibration of the upper Jurassic − Lower Cretaceous in the Neuquén Basin, Argentina: a contribution from the southern hemisphere to the geologic time scale. *Stratigraphy and Time Scales*, vol. 3. Elsevier, pp. 327−355.

Kiselev, D., Rogov, M., Glinskikh, L., Guzhikov, A., Pimenov, M., Mikhailov, A., et al., 2013, Integrated stratigraphy of the reference sections for the Callovian-Oxfordian boundary in European Russia. *Volumina Jurassica*, **11**: 59−96.

Koeberl, C., Armstrong, R.A., and Reimold, W.U., 1997, *Morokweng, South Africa: a large impact structure of Jurassic−Cretaceous Boundary age*, Geology, 25. , pp. 731−734.

Koppers, A.A.P., Staudigel, H., and Duncan, R.A., 2003, High resolution ^{40}Ar/^{39}Ar dating of the oldest oceanic basement basalts in the western Pacific basin. *Geochemistry, Geophysics, Geosystems*, **4**: 8914, https://doi.org/10.1029/2003GC000574.

Korte, C., and Hesselbo, S.P., 2011, Shallow-marine carbon- and oxygen-isotope and elemental records indicate icehouse-greenhouse cycles during the Early Jurassic. *Paleoceanography*, **26**: PA4219.

Korte, C., Hesselbo, S.P., Ullmann, C.V., Dietl, G., Ruhl, M., Schweigert, G., et al., 2015, Jurassic climate mode governed by ocean gateway. *Nature Communications*, **6**, https://doi.org/10.1038/ncomms10015.

Korte, C., Ruhl, M., Pálfy, J., Ullmann, C.V., and Hesselbo, S.P., 2019, Chemostratigraphy across the Triassic−Jurassic boundary. *In* Sial, A. N., Gaucher, C., Ramkumar, M., and Ferreira, V.P. (eds), Chemostratigraphy Across Major Chronological Boundaries, Geophysical Monograph. *The American Geophysical Union; John Wiley & Sons, Inc. AGU Books*, **240**: 185−210.

Krymholts, G.Ya., Mesezhnikov, M.S., and Westermann, G.E.G. (eds), 1982, Zony iurskoi sistemy v SSSR. *Interdepartmental Stratigraphic Committee of the USSR Transactions*, **10**. *In* Russian. English translation by Vassiljeva, T.I., 1988. The Jurassic Ammonite Zones of the Soviet Union. *Geological Society of America, Special Paper*, **223**: 116.

Kuerschner, W.M., Bonis, N.R., and Krystyn, L., 2007, Carbon-isotope stratigraphy and palynostratigraphy of the Triassic−Jurassic transition in the Tiefengraben section − Northern Calcareous Alps (Austria). *Palaeogeography, Palaeoclimatology, Palaeoecology*, **244**: 257−280.

Lang, W.D., 1913, The Lower Pliensbachian − "Carixian" − of Charmouth. *Geological Magazine*, **5**: 401−412.

Larson, R.L., and Hilde, T.W.C., 1975, A revised time scale of magnetic reversals for the Early Cretaceous and Late Jurassic. *Journal of Geophysical Research*, **80**: 2586−2594.

Laskar, J., Fienga, A., Gastineau, M., and Manche, H., 2011, La2010: a new orbital solution for the long-term motion of the Earth. *Astronomy & Astrophysics*, **532**: A89.

Laurin, B., 1998, Brachiopods. Columns for Jurassic chart of Mesozoic and Cenozoic sequence chronostratigraphic framework of European basins. *In* de Graciansky, P.-C., Hardenbol, J., Jacquin, Th, and Vail, P.R. (eds), Mesozoic-Cenozoic Sequence Stratigraphy of European Basins. *SEPM Special Publication60*: SEPM Special Publication.

Lefranc, M., Beaudoin, B., Chilès, J.P., Guillemot, D., Ravenne, C., and Trouiller, A., 2008, Geostatistical characterization of Callovo-Oxfordian clay variability from high-resolution log data. *Physics and Chemistry of the Earth*, **33**: S2−S13.

Lena, L., López-Martínez, R., Lescano, M., Aguirre-Urreta, B., Concheyro, A., Vennari, V., et al., 2019a, High-precision U-Pb ages in the early Tithonian to early Berriasian and implications for the numerical age of the Jurassic/Cretaceous boundary. *Solid Earth*, **10**: 1−14.

Lena, L.F.De, Taylor, D., Guex, J., Bartolini, A., Adatte, T., van Acken, D., et al., 2019b, The drivig mechanisms of the carbon cycle perturbations in the late Pliensbachian (Early Jurassic). *Scientific Reports*, **9**: 18430.

Leanza, H.A., Mazzini, A., Corfu, F., Llambías, E.J., Svensen, H., Planke, S., et al., 2013, The Chachil Limestone (Pliensbachian−earliest Toarcian) Neuquén Basin Argentina:U-Pb age calibration and its significance on the Early Jurassic evolution of southwestern Gondwana. *Journal of South American Earth Science*, **42**: 171−185.

Lindström, S., van de Schootbrugge, B., Dybkjær, K., Pedersen, G.K., Fiebig, J., Nielsen, L.H., et al., 2012, No causal link between terrestrial ecosystem change and methane release during the end-Triassic mass extinction. *Geology*, **40**: 539−542.

Littler, K., Hesselbo, S.P., and Jenkyns, H.C., 2010, A carbon-isotope perturbation at the Pliensbachian−Toarcian boundary: evidence from the Lias Group, NE England. *Geological Magazine*, **147**: 181−192.

Lowrie, W., and Ogg, J.G., 1986, A magnetic polarity time scale for the Early Cretaceous and Late Jurassic. *Earth and Planetary Science Letters*, **76**: 341−349.

Lucas, S.G., 2009, Global Jurassic tetrapod biochronology. *Volumina Jurassica*, **VI**: 99−108.

Lucas, S.G., Tanner, L.H., Donohoo-Hurley, L.L., Geissman, J.W., Kozur, H.W., Heckert, A.B., et al., 2011, Position of the Triassic-Jurassic boundary and timing of the end-Triassic extinctions on land: data from the Moenave Formation on the southern Colorado Plateau, USA. *Palaeogeography, Palaeoclimatology, Palaeoecology*, **302**: 194−205.

Ludden, J., 1992, Radiometric age determinations for basement from Sites 765 and 766. Argo Abyssal Plain and Northwestern Australia. *Proceedings of the Ocean Drilling Program, Scientific Results*, **123**: 557−559.

Mailliot, S., Mattioli, E., Guex, J., and Pittet, B., 2006, The Early Toarcian anoxia, a synchronous event in the western Tethys? An approach by quantitative biochronology (Unitary Associations), applied on calcareous nannofossils. *Palaeogeography, Palaeoclimatology, Palaeoecology*, **240**: 562−586.

Mander, L., Kürschner, W.M., and McElwain, J.C., 2010, An explanation for conflicting records of Triassic-Jurassic plant diversity. *Proceedings of the National Academy of Science*, **107**: 15351−15356.

Martinez, M., Krencker, F.N., Mattioli, E., and Bodin, S., 2017, Orbital chronology of the Pliensbachian−Toarcian transition from the central High Atlas Basin (Morocco). *Newsletters on Stratigraphy*, **50**: 47−69.

Marzoli, A., Renne, P.R., Piccirillo, E.M., Ernesto, M., Bellieni, G., and DeMin, A., 1999, Extensive 200-million-year-old continental flood basalts of the central Atlantic Magmatic Province. *Science*, **284**: 616−618.

Matsuoka, A., 1992, Jurassic and Early Cretaceous radiolarians from Leg 129. Sites 800 and 801. western Pacific Ocean. *Proceedings of the Ocean Drilling Program, Scientific Results*, **129**: 203−211.

Matsuoka, A., and Yao, A., 1986, A newly proposed radiolarian zonation for the Jurassic of Japan. *Marine Micropaleontology*, **11**: 91−105.

Mattioli, E., and Erba, E., 1999, Synthesis of calcareous nannofossil events in tethyan Lower and Middle Jurassic successions. *Rivista Italiana di Paleontologia e Stratigrafia*, **105**: 343−376.

Mattioli, E., Colombié, C., Giraud, F., Olivier, N., and Pittet, B., 2011, Comment on "Orbitally forced climate and sea-level changes in the Paleoceanic Tethyan domain (marl-limestone alternations, Lower

Kimmeridgian, SE France)." by S. Boulila, M. de Rafélis, L.A. Hinnov, S. Gardin, B. Galbrun, and P.-Y. Collin [Palaeogeography, Palaeoclimatology, Palaeoecology 292 (2010) 57-70]. *Palaeogeography, Palaeoclimatology, Palaeoecology*, **306**: 249–251.

Matyja, B.A., and Wierzbowski, A., 1997, The quest for a unified Oxfordian/Kimmeridgian boundary: implications of the ammonite succession at the turn of the Bimammatum and Planula Zones in the Wielun Upland, central Poland. *Acta Geologica Polonica*, **47**: 77–105.

Matyja, B.A., and Wierzbowski, A., 2002, Boreal and Subboreal ammonites in the Submediterranean uppermost Oxfordian in the Bielawy section (northern Poland) and their correlation value. *Acta Geologica Polonica*, **52**: 411–421.

Matyja, B.A., and Wierzbowski, A., 2006a, Syborowa Góra (upper Oxfordian: amoeboceras layer of lower Bimammatum Zone). *In* Wierzbowski, A., Aubrecht, R., Golonka, J., Gutowski, J., Krobicki, M., Matyja, B.A., Pieńkowski, G., and Uchman, A. (eds), *Jurassic of Poland and adjacent Slovakian Carpathians, Field Guide — 7th International Congress on the Jurassic System, Kraków, September 6–18. 2006*. Warszawa: Polish Geological Institute, 141–143.

Matyja, B.A., and Wierzbowski, A., 2006b, Quarries at Raciszyn and Lisowice, upper Oxfordian ammonite succession (upper Bimammatum to Planula Zones). *In* Wierzbowski, A., Aubrecht, R., Golonka, J., Gutowski, J., Krobicki, M., Matyja, B.A., Pieńkowski, G., and Uchman, A. (eds), *Jurassic of Poland and adjacent Slovakian Carpathians, Field Guide — 7th International Congress on the Jurassic System, Kraków, September 6–18. 2006*. Warszawa: Polish Geological Institute, 163–165.

Matyja, B.A., Wierzbowski, A., and Wright, J.K., 2006, The Subboreal/Boreal ammonite succession at the Oxfordian/Kimmeridgian boundary at Flodigarry, Staffin Bay (Isle of Skye), Scotland. *Transactions of the Royal Society of Edinburgh, Earth Science*, **96**: 309–318.

Maubeuge, P.L. (ed), 1964. Colloque du Jurassique à Luxembourg 1962. Publication de l'Institut Grand-Ducal, Section des Sciences Naturelles, Physiques et Mathématiques. Luxembourg,.

Mayer-Eymar, C., 1864, *Tableau synchronistique des terrains jurassiques*. Zürich: J. Höfer.

Mazzini, A., Svensen, H., Leanza, H.A., Corfu, F., and Planke, S., 2010, Early Jurassic shale chemostratigraphy and U-Pb ages from the Neuquén Basin (Argentina): implications for the Toarcian Oceanic Anoxic Event. *Earth and Planetary Science Letters*, **297**: 633–645.

McArthur, J.M., 2008, Comment on 'The impact of the central Atlantic Magmatic Province on climate and on the Sr- and Os-isotope evolution of seawater' by Cohen, A. S. and Coe, A. L. 2007. Palaeogeography, Palaeoclimatology, Palaeoecology, 244. 374–390. *Palaeogeography, Palaeoclimatology, Palaeoecology*, **263**: 146–149.

McArthur, J.M., Donovan, D.T., Thirlwall, M.F., Fouke, B.W., and Mattey, D., 2000, Strontium isotope profile of the early Toarcian (Jurassic) Oceanic Anoxic Event, the duration of ammonite biozones, and belemnite paleotemperatures. *Earth and Planetary Science Letters*, **179**: 269–285.

McArthur, J.M., Howarth, R.J., and Bailey, T.R., 2001, Strontium isotope stratigraphy: LOWESS Version 3: best fit to the marine Sr-isotope curve for 0–509 Ma and accompanying look-up table for deriving numerical age. *Journal of Geology*, **109**: 155–170.

McArthur, J.M., Algeo, T.J., van de Schootbrugge, B., Li, Q., and Howarth, R.J., 2008, Basinal restriction, black shales, Re–Os dating, and the early Toarcian (Jurassic) oceanic anoxic event. *Paleoceanography*, **23**: PA4217, https://doi.org/10.1029/2008PA001607.

McArthur, J.M., Steuber, T., Page, K.N., and Landman, N.H., 2016, Sr-Isotope Stratigraphy: assigning time in the Campanian, Pliensbachian, Toarcian, and Valanginian. *Journal of Geology*, **124**: 569–586.

McArthur, J.M., Page, K., Duarte, L.V., Thirlwall, M.F., Li, Q., Weis, R., et al., 2020a, Sr-isotope stratigraphy (^{87}Sr/^{86}Sr) of the lowermost Toarcian of Peniche, Portugal, and its relation to ammonite zonations. *Newsletters on Stratigraphy*, **53** (3), 297–312.

McArthur, J.M., Howarth, R.J., Shields, G.A., and Zhou, Y., 2020b, Chapter 7 - Strontium isotope stratigraphy. *In* Gradstein, F.M., Ogg, J.G., Schmitz, M.D., and Ogg, G.M. (eds), *The Geologic Time Scale 2020. Vol. 1* (this book). Elsevier, Boston, MA.

McElwain, J.C., Murphy, J.W., and Hesselbo, S.P., 2005, Changes in carbon dioxide during an oceanic anoxic event linked to intrusion of Gondwana coals. *Nature*, **435**: 479–483.

Meister, C., Blau, J., Dommergues, J.-L., Feistburkhardt, S., Hart, M., Hesselbo, S.P., et al., 2003, A proposal for the Global Boundary Stratotype Section and Point (GSSP) for the base of the Pliensbachian Stage (Lower Jurassic). *Ecologae Geologicae Helvetiae*, **96**: 275–298.

Meister, C., Aberhan, M., Blau, J., Dommergues, J.-L., Feist-Burkhardt, S., Hailwood, E.A., et al., 2006, The Global Boundary Stratotype Section and Point (GSSP) for the base of the Pliensbachian Stage (Lower Jurassic), Wine Haven, Yorkshire, UK. *Episodes*, **29**: 93–106.

Melendez, G., and Atrops, F., 1999, Report of the Oxfordian-Kimmeridgian Boundary Working Group. *International Subcommission on Jurassic Stratigraphy Newsletter*, **26**: 67–74.

Melendez, G., and Page, K.N., 2015, Report of the Oxfordian Task Group (OTG) Meeting: the section of Ham Cliff, Redcliff Point, Weymouth, UK (11–14th June 2014). *Volumina Jurassica*, **13**: 159–164.

Mönnig, E., 2014, Report of the Callovian stage task group, 2013. *Volumina Jurassica*, **12**: 197–200.

Mönnig, E., and Dietl, G., 2017, The systematics of the ammonite genus *Kepplerites* (upper Bathonian and basal Callovian, Middle Jurassic) and the proposed basal boundary stratotype (GSSP) of the Callovian Stage. *Neues Jahrbuch für Geologie und Paläontologie, Abhandlungen*, **286**: 235–287.

Morales, C., Rogov, M., Wierzbowski, H., Ershova, V., Suan, G., Adatte, T., et al., 2017, Glendonites track methane seepage in Mesozoic polar seas. *Geology*, **45**: 503–506.

Moreau, M.-G., Bucher, H., Bodergat, A.-M., and Guex, J., 2002, Pliensbachian magnetostratigraphy: new data from Paris Basin (France). *Earth and Planetary Science Letters*, **203**: 755–767.

Morettini, E., Santantonio, M., Bartolini, A., Cecca, F., Baumgartner, P. O., and Hunziker, J.C., 2002, Carbon isotope stratigraphy and carbonate production during the Early-Middle Jurassic: examples from the Umbria-Marche-Sabina Apennines (central Italy). *Palaeogeography, Palaeoclimatology, Palaeoecology*, **184**: 251–273.

Morgans-Bell, H.S., Coe, A.L., Hesselbo, S.P., Jenkyns, H.C., Weedon, G.P., Marshall, J.E.A., et al., 2001, Integrated stratigraphy of the Kimmeridge Clay Formation (upper Jurassic) based on exposures and boreholes in south Dorset, UK. *Geological Magazine*, **138**: 511–539.

Morris, P.H., Cullum, A., Pearce, M.A., and Batten, D.J., 2009, Megaspore assemblages from the Are Formation (Rhaetian-Pliensbachian) offshore mid-Norway, and their value as field and regional stratigraphic markers. *Journal of Micropalaeontology*, **28**: 161–181.

Morton, N. (ed), 1974. The definition of standard Jurassic Stages. Colloque du Jurassique à Luxembourg 1967; Generalities, methods. Mémoires du Bureau de Recherches Géologiques et Minières **75**: 83–93.

Morton, N., 2008, The International Subcommission on Jurassic Stratigraphy. *Proceedings of the Geologists' Association*, **119**: 97–103.

Morton, N., and Hudson, J., 1995, Field Guide to the Jurassic of the Isles of Raasay and Skye, Inner Hebrides, NW Scotland. *In* Taylor, P.D. (ed), *Field Geology of the British Jurassic*. Geological Society of London, 209–280.

Moulin, M., Fluteau, F., Courtillot, V., Marsh, J., Delpech, G., Quidelleur, X., et al., 2017, Eruptive history of the Karoo lava flows and their impact on early Jurassic environmental change. *Journal of Geophysical Research: Solid Earth*, **122**: 738–772.

Neumayr, M., 1873, Die Fauna des Schichten mit *Aspidoceras acanthicum*. *Abhandlungen der Kais.-Königl. Geologischen Reichsanstalt*, **5**: 141–257.

Nikitin, S.N., 1881, Jurassic formations between Rogbinsk, Mologa, and Myshkin. *Materials on the Geology of Russia*, **10**: 194.

Ohmert, W. (ed), 1996. Die Grenzziehung Unter-/Mitteljura (Toarciuim/Aalenium) bei Wittnau und Fuentelsaz. Beispiele interdisziplinarer geowissenschaftlicher Zusammenarbeit. *Geologisches Landesamt Baden-Württemberg, Informationen* **8**: 53.

Ogg, J.G., 1988, Early Cretaceous and Tithonian magnetostratigraphy of the Galicia margin (Ocean Drilling Program Leg 103). *Proceedings of the Ocean Drilling Program, Scientific Results*, **103**: 659–682.

Ogg, J.G., 1995, Magnetic polarity time scale of the Phanerozoic. *In* Ahrens, T.J. (ed), *Global Earth Physics: A Handbook of Physics Constants*. American Geophysical Union Reference Shelf, Vol. 1, p. 240–270.

Ogg, J.G., 2012, Geomagnetic polarity time scale. *In* Gradstein, F.M., Ogg, J.G., Schmitz, M., and Ogg, G. (eds), *The Geologic Time Scale 2012*. Elsevier Publ, 85–113.

Ogg, J.G., 2020, Chapter 5 – Geomagnetic polarity time scale. *In* Gradstein, F.M., Ogg, J.G., Schmitz, M.D., and Ogg, G.M. (eds), *The Geologic Time Scale 2020*. Vol. **1** (this book). Elsevier, Boston, MA.

Ogg, J.G., and Gutowski, J., 1996, Oxfordian and lower Kimmeridgian magnetic polarity time scale. *In* Riccardi, A.C. (ed), Advances in Jurassic Research. *GeoResearch Forum*, **1–2**: 406–414.

Ogg, J.G., Steiner, M.B., Oloriz, F., and Tavera, J.M., 1984, Jurassic magnetostratigraphy, 1. Kimmeridgian–Tithonian of Sierra Gorda and Carcabuey, southern Spain. *Earth and Planetary Science Letters*, **71**: 147–162.

Ogg, J.G., Wieczorek, J., Steiner, M.B., and Hoffmann, M., 1990, Jurassic magnetostratigraphy, 4. Early Callovian through Middle Oxfordian of Krakow Uplands (Poland). *Earth and Planetary Science Letters*, **104**: 289–303.

Ogg, J.G., Hasenyager, R.W., Wimbledon, W.A., Channell, J.E.T., and Bralower, T.J., 1991, Magnetostratigraphy of the Jurassic–Cretaceous boundary interval—Tethyan and English faunal realms. *Cretaceous Research*, **12**: 455–482.

Ogg, J.G., Karl, S.M., and Behl, R.J., 1992, Jurassic through Early Cretaceous sedimentation history of the central Equatorial Pacific and of Sites 800 and 801. *Proceedings of the Ocean Drilling Program, Scientific Results*, **129**: 571–613.

Ogg, J.G., Hasenyager II, R.W., and Wimbledon, W.A., 1994, Jurassic-Cretaceous boundary: Portland-Purbeck magnetostratigraphy and possible correlation to the Tethyan faunal realm. *Géobios*, **17**: 519–527.

Ogg, J.G., Coe, A.L., Przybylski, P.A., and Wright, J.K., 2010a, Oxfordian magnetostratigraphy of Britain and its correlation to Tethyan regions and Pacific marine magnetic anomalies. *Earth and Planetary Science Letters*, **289**: 433–448.

Ogg, J.G., Hinnov, L.A., Huang, C., and Przybylski, P.A., 2010b, Late Jurassic time scale: integration of ammonite zones, magnetostratigraphy, astronomical tuning and sequence interpretation for Tethyan, Sub-boreal and Boreal realms. *Earth Science Frontiers*, **17**: 81–82.

Ogg, J.G., Coe, A.L., Przybylski, P.A., and Wright, J.K., 2011, *Magnetostratigraphy of the Redcliff Section, Dorset.* [submitted to Oxfordian Working Group for base-Oxfordian GSSP voting documentation coordinated by Guillermo Melendez, chair.]

Ogg, J.G., Hinnov, L.A., and Huang, C., 2012, Jurassic. *In* Gradstein, F.M., Ogg, J.G., Schmitz, M., and Ogg, G. (eds), *The Geologic Time Scale 2012*. Elsevier Publ, 731–791.

Ogg, J.G., Ogg, G.M., and Gradstein, F.M., 2016, *A Concise Geologic Time Scale 2016*. Elsevier Publ, p. 234.

Ogg, J.G., and Chen, Z.Q., 2020, Chapter 25 – The Triassic Period. *In* Gradstein, F.M., Ogg, J.G., Schmitz, M.D., and Ogg, G.M. (eds), *The Geologic Time Scale 2020*. Vol. **2**: Elsevier. Boston, MA.

Olivero, D. (with contribution of E. Mattioli), 2008, The Aalenian-Bajocian (Middle Jurassic) of the Digne area. *In* Mattioli, E. (special ed), *Guidebook for the post-congress fieldtrip in the Vocontian Basin, SE France, 11–13 September 2008. 12th Meeting of the International Nannoplankton Association, Lyon, 7–10 September 2008.* Available at http://paleopolis.rediris.es/cg/CG2008_BOOK_01/index.htm.

Olóriz, F., and Schweigert, G., 2010, Working Group on the Kimmeridgian-Tithonian boundary. *International Commission on Jurassic Stratigraphy Newsletter*, **36**: 5–12.

Olsen, P.E., Kent, D.V., Et Touhami, M., and Puffer, J.H., 2003, Cyclo-, magneto-, and bio-stratigraphic constraints on the duration of the CAMP event and its relationship to the Triassic-Jurassic boundary. *American Geophysical Union Geophysical Monograph*, **136**: 7–32.

Oppel, C.A., 1856, Die Juraformation Englands, Frankreichs und des Südwestlichen Deutschlands. *Württemberger Naturforschende Jahreshefte*, **12-14**: 857.

Oppel, C.A., 1865, Die Tithonische Etage. *Zeitschrift der Deutschen Geologischen Gesellschaft, Jahrgang*, **17**: 535–558.

Osete, M.L., Gialanella, P.R., Gómez, J.J., Villalaín, J.J., Goy, A., and Heller, F., 2007, Magnetostratigraphy of early-middle Toarcian expanded sections from the Iberian Range (central Spain). *Earth and Planetary Science Letters*, **259**: 319–332.

Oxford, M.J., Gregory, F.J., Hart, M.B., Henderson, A.S., Simmons, M.D., and Watkinson, M.P., 2002, Jurassic planktonic foraminifera from the United Kingdom. *Terra Nova*, **14**: 205–209.

Padden, M., Weissert, H., Funk, H., Schneider, S., and Gansner, C., 2002, Late Jurassic lithological evolution and carbon-isotope stratigraphy of the western Tethys. *Eclogae Geologicae Helvetiae*, **95**: 333–346.

Page, K.N., 2005, Jurassic. *In* Selley, R.C., Cocks, L.R.M., and Plimer, I.R. (eds), *Encyclopedia of Geology*. Amsterdam: Elsevier, Vol. **1–5**: 353–360.

Page, K.N., 2008, The evolution and geography of Jurassic ammonoids. *Proceedings of the Geologists' Association*, **119**: 35–57.

Page, K.N., 2017, From Oppel to Callomon (and beyond): building a high-resolution ammonite-based biochronology for the Jurassic System. *Lethaia*, **50**: 336–355.

Page, K.N., Bloos, G., Bessa, J.L., Fitzpatrick, M., Hesselbo, S., Hylton, M., et al., 2000, East Quantoxhead, Somerset: a candidate Global Stratotype Section and Point for the base of the Sinemurian Stage (Lower Jurassic). *GeoResearch Forum*, **6**: 163−171.

Page, K.N., Meléndez, G., and Wright, J.K., 2010a, The ammonite faunas of the Callovian-Oxfordian boundary interval in Europe and their relevance to the establishment of an Oxfordian GSSP. *Volumina Jurassica*, **7**: 89−99.

Page, K.N., Meléndez, G., Hart, M.B., Price, G., Wright, J.K., Bown, P., et al., 2010b, Integrated stratigraphical study of the candidate Oxfordian Global Stratotype Section and Point (GSSP) at Redcliff Point, Weymouth, Dorset, UK. *Volumina Jurassica*, **7**: 101−111.

Pálfy, J., and Smith, P.L., 2000, Synchrony between Early Jurassic extinction, oceanic anoxic event, and the Karoo-Ferrar flood basalt volcanism. *Geology*, **28**: 747−750.

Pálfy, J., Parrish, R.R., and Smith, P.L., 1997, A U-Pb age from the Toarcian (Lower Jurassic) and its use for time scale calibration through error analysis of biochronologic dating. *Earth and Planetary Science Letters*, **146**: 659−675.

Pálfy, J., Smith, P.L., and Mortensen, J.K., 2000, A U-Pb and ^{40}Ar/^{39}Ar time scale for the Jurassic. *Canadian Journal of Earth Science*, **37**: 923−944.

Pálfy, J., Demeny, A., Haas, J., Hetenyi, M., Orchard, M.J., and Veto, I., 2001, Carbon isotope anomaly and other geochemical changes at the Triassic-Jurassic boundary from a marine section in Hungary. *Geology*, **29**: 1047−1050.

Palotai, M., Palfy, J., and Sasvari, A., 2017, Structural complexity at and around the Triassic-Jurassic GSSP at Kuhjoch, Northern Calcareous Alps, Austria. *International Journal of Earth Science (Geologische Rundschau)*, **106**: 2475−2487, https://doi.org/10.1007/s00531-017-1450-4.

Pană, D.I., Poulton, T.P., and Heaman, L.M., 2018, U-Pb zircon ages of volcanic ashes integrated with ammonite biostratigraphy, Fernie Formation (Jurassic), western Canada, with implications for Cordilleran-Foreland basin connections and comments on the Jurassic time scale. *Bulletin of Canadian Petroleum Geology*, **66**: 595−622.

Panfili, G., Cirilli, S., Dal Corso, J., Bertrand, H., Medina, F., Youbi, N., et al., 2019, New biostratigraphic constraints show rapid emplacement of the central Atlantic Magmatic Province (CAMP) during the end-Triassic mass extinction interval. *Global and Planetary Change*, **172**: 60−68.

Partington, M.A., Copestake, P., Mitchener, B.C., and Underhill, J.R., 1993, Biostratigraphic calibration of genetic stratigraphic sequences in the Jurassic−lowermost Cretaceous (Hettangian to Ryazanian) of the North Sea and adjacent areas. *In* Parker, J.R. (ed), Petroleum Geology of Northwest Europe: Proceedings of the 4th Conference, London, *The Geological Society*, pp. 371−386.

Pavia, G., and Enay, R., 1997, Definition of the Aalenian−Bajocian Stage boundary. *Episodes*, **20**: 16−20.

Pearce, C.R., Hesselbo, S.P., and Coe, A.L., 2005, The mid-Oxfordian (Late Jurassic) positive carbon-isotope excursion recognized from fossil wood in the British Isles. *Palaeogeography, Palaeoclimatology, Palaeoecology*, **221**: 343−357.

Pellenard, P., Nomade, S., Martire, L., De Oliveira Ramalho, F., Monna, F., and Guillou, H., 2013, The first ^{40}Ar-^{39}Ar date from Oxfordian ammonite-calibrated volcanic layers (bentonites) as a tie-point for the Late Jurassic. *Geological Magazine*, **150**: 1136−1142.

Pellenard, P., Fortwengler, D., Marchand, D., Thierry, J., Bartolini, A., Boulila, S., et al., 2014a, Integrated stratigraphy of the Oxfordian global stratotype section and point (GSSP) candidate in the Subalpine Basin (SE France). *Volumina Jurassica*, **12**: 1−44.

Pellenard, P., Tramoy, R., Pucéat, E., Huret, E., Martinez, M., Brueau, L., et al., 2014b, Carbon cycle and sea-water palaeotemperature evolution at the Middle-Late Jurassic transition, eastern Paris Basin (France). *Marine and Petroleum Geology*, **53**: 30−43, https://doi.org/10.1016/j.marpetgeo.2013.07.002.

Percival, L.M.E., Cohen, A.S., Davies, M.K., Dickson, A.J., Hesselbo, S.P., Jenkyns, H.C., et al., 2016, Osmium-isotope evidence for two pulses of increased continental weathering linked to Early Jurassic volcanism and climate change. *Geology*, **44**: 759−762.

Percival, L.M.E., Ruhl, M., Hesselbo, S.P., Jenkyns, H.C., Mather, T.A., and Whiteside, J., 2017, Mercury evidence for pulsed volcanism during the end-Triassic mass extinction. *Proceedings of the National Academy of Sciences of the United States of America*, **114**: 7929−7934.

Pessagno Jr., E.A., and Meyerhoff Hull, D., 1996, Once upon a time in the Pacific: chronostratigraphic misinterpretation of basal strata at ODP Site 801 (central Pacific) and its impact on geochronology and plate tectonics models. *GeoResearch Forum*, **1-2**: 79−92.

Pessagno Jr., E.A., Blome, C.D., Hull, D.M., and Six, W.M., 1993, Jurassic Radiolaria from the Josephine ophiolite and overlying strata, Smith River subterrane (Klamath Mountains), northwestern Californian and southwestern Oregon. *Micropaleontology*, **39**: 93−166.

Pessagno Jr., E.A., Cantú-Chapa, A., Mattinson, J.M., Meng, X., and Kariminia, S.M., 2009, The Jurassic-Cretaceous boundary: new data from North America and the Caribbean. *Stratigraphy*, **6**: 185−262.

Peti, L., Thibault, N., Clémence, M.-E., Korte, C., Dommergues, J.-L., Bougeault, C., et al., 2017, Sinemurian−Pliensbachian calcareous nannofossil biostratigraphy and organic carbon isotope stratigraphy in the Paris Basin: calibration to the ammonite biozonation of NW Europe. *Palaeogeogr. Palaeoclimatol. Palaeoecol.*, **468**: 142−161.

Peybernes, B., 1998, Larger benthic foraminifera. Columns for Jurassic chart of Mesozoic and Cenozoic sequence chronostratigraphic framework of European basins. *In* de Graciansky, P.-C., Hardenbol, J., Jacquin, Th, and Vail, P.R. (eds), Mesozoic-Cenozoic Sequence Stratigraphy of European Basins. *SEPM Special Publication*.

Pienkowski, G., 2004, The epicontinental Lower Jurassic of Poland. *Polish Geological Institute, Special Publication*, **12**: 1−152.

Pinous, O.V., Sahagian, D.L., Shurygin, B.N., and Nikitenko, B.L., 1999, High-resolution sequence stratigraphic analysis and sea-level interpretation of the middle and upper Jurassic strata of the Nyurolskaya depression and vicinity (southeastern West Siberia, Russia). *Marine and Petroleum Geology*, **16** (3), 245−257.

Porter, S.J., Selby, D., Suzuki, K., and Gröcke, D., 2013, Opening of a trans-Pangaean ma-rine corridor during the Early Jurassic: insights from osmium isotopes across the Sinemurian−Pliensbachian GSSP, Robin Hood's Bay, UK. *Palaeogeography, Palaeoclimatology, Palaeoecology*, **375**: 50−58.

Poulsen, N.E., and Riding, J.B., 2003, The Jurassic dinoflagellate cyst zonation of Subboreal northwest Europe. *In* Ineson, J.R., and Surlyk, F. (eds), The Jurassic of Denmark and Greenland. *Geological Survey of Denmark and Greenland Bulletin*, **1**: 115−144.

Price, G.D., Baker, S.J., Van De Velde, J., and Clémence, M.E., 2016a, High-resolution carbon cycle and seawater temperature evolution during the Early Jurassic (Sinemurian-Early Pliensbachian). *Geochemistry, Geophysics, Geosystems*, **17**: 3917–3928.

Price, G.D., Fozy, I., and Pálfy, J., 2016b, Carbon cycle history through the Jurassic–Cretaceous boundary: a new global $\delta^{13}C$ stack. *Palaeogeography, Palaeoclimatology, Palaeoecology*, **451**: 46–61.

Pruner, P., Houša, V., Olóriz, F., Košťák, M., Krs, M., Man, O., et al., 2010, High-resolution magnetostratigraphy and the biostratigraphic zonation of the Jurassic/Cretaceous boundary strata in the Puerto Escaño section (southern Spain). *Cretaceous Research*, **31**: 192–206.

Przybylski, P.A., Ogg, J.G., Wierzbowski, A., Coe, A.L., Hounslow, M. W., Wright, J.K., et al., 2010a, Magnetostratigraphic correlation of the Oxfordian-Kimmeridgian Boundary. *Earth and Planetary Science Letters*, **289**: 256–272.

Przybylski, P.A., Głowniak, E., Ogg, J.G., Ziółkowski, P., Sidorczuk, M., Gutowski, J., et al., 2010b, Oxfordian magnetostratigraphy of Poland and its Sub-Mediterranean correlations. *Earth and Planetary Science Letters*, **289**: 417–432.

Quenstedt, F., 1848, *Der Jura*. Tübingen: H. Laupp, p. 103.

Reháková, D., 2000, Calcareous dinoflagellates and calpionellid bioevents versus sea-level fluctuations recorded in the West Carpathians (Late Jurassic/Early Cretaceous pelagic environments). *Geologica Carpathica*, **51**: 229–243.

Remane, J., 1985, Calpionellids. *In* Bolli, H.M., Saunders, J.B., and Perch Nielsen, K. (eds), *Plankton Stratigraphy*. Cambridge: Cambridge University Press, 555–572.

Remane, J., 1998, Calpionellids. Columns for Jurassic chart of Mesozoic and Cenozoic sequence chronostratigraphic framework of European basins. *In* de Graciansky, P.-C., Hardenbol, J., Jacquin, Th, and Vail, P.R. (eds), Mesozoic-Cenozoic Sequence Stratigraphy of European Basins. *SEPM Special Publication*.

Renevier, E., 1864, *Notices géologique et paléontologiques sur les Alpes Vaudoises, et les régions environnantes. I. Infralias et Zone à Avicula contorta (Ét. Rhaetien) des Alpes Vaudoises, Bulletin de la Société Vaudoise des Sciences Naturelles*, 8. , pp. 39–97.

Riccardi, A.C., and Kamo, S., 2012, A new U-Pb zircon age for an ash layer at the Pliensbachian-Toarcian boundary, Argentina. *In Unearthing Our Past and Future—Resourcing Tomorrow: 34th International Geological Congress (Brisbane, Australia), Abstracts*. p. 752.

Riding, J.B., and Ioannides, N.S., 1996, Jurassic dinoflagellate cysts. *Bulletin de la Société géologique de France*, **167**: 3–14.

Riding, J.B., and Thomas, J.E., 1997, Marine palynomorphs from the Staffin Bay and Staffin Shale formations (Middle-upper Jurassic) of the Trotternish Peninsula. *Scottish Journal of Geology*, **33**: 59–74.

Riding, J.B., Leng, M.J., Kender, S., Hesselbo, S.P., and Feist-Burkhardt, S., 2012, Isotopic and palynological evidence for a new Early Jurassic environmental perturbation. *Palaeogeography, Palaeoclimatology, Palaeoecology*, **374**: 16–27.

Rioult, M., 1964, Le stratotype du Bajocien. *In* Maubeuge, P.L. (ed), *Colloque du Jurassique à Luxembourg 1962*. Luxembourg: Publication de l'Institut Grand-Ducal, Section des Sciences Naturelles, Physiques et Mathématiques, 239–258.

Rocha, R.B., da, Mattioli, E., Duarte, L.V., Pittet, B., Elmi, S., Mouterde, R., et al., 2016, Base of the Toarcian Stage of the Lower Jurassic defined by the Global Boundary Stratotype Section and Point (GSSP) at the Peniche section (Portugal). *Episodes*, **39**: 460–481.

Rogov, M.A., 2004, Ammonite-based correlation of the Lower and Middle (*Pandieri* Zone) Volgian substages with the Tithonian Stage. *Stratigraphy and Geological Correlation*, **12** (7), 35–57.

Rogov, M.A., 2010a, New data on ammonites and stratigraphy of the Volgian Stage of Spitzbergen. *Stratigraphy and Geological Correlation*, **18**: 505–531.

Rogov, M.A., 2010b, A precise ammonite biostratigraphy through the Kimmeridgian-Volgian boundary beds in the Gorodischi section (Middle Volga area, Russia), and the base of the Volgian Stage in its type area. *Volumina Jurassica*, **8**: 103–130.

Rogov, M.A., and Price, G.D., 2010, New stratigraphic and isotope data on the Kimmeridgian-Volgian boundary beds of the Subpolar Urals, western Siberia. *Geological Quarterly*, **54**: 33–40.

Rogov, M.A., and Zakharov, V.A., 2009, Ammonite- and bivalve-based biostratigraphy and Panboreal correlation of the Volgian Stage. *Science in China Series D: Earth Sciences*, **52**: 1890–1909.

Rogov, M.A., and Zakharov, V.A., 2010, Jurassic and Lower Cretaceous glendonite occurrences and their implication for Arctic paleoclimate reconstructions and stratigraphy. *Earth Science Frontiers*, **17**: 345–347.

Rostovtsev, K.O., and Prozorowsky, V.A., 1997, Information on resolutions of standing commissions of the Interdepartmental Stratigraphic Committee (ISC) on the Jurassic and Cretaceous systems. *International Subcommission on Jurassic Stratigraphy Newsletter*, **24**: 48–49.

Ruget, C., and Nicollin, J.-P., 1998, Smaller benthic foraminifera. Columns for Jurassic chart of Mesozoic and Cenozoic sequence chronostratigraphic framework of European basins. *In* de Graciansky, P.-C., Hardenbol, J., Jacquin, Th, and Vail, P.R. (eds), Mesozoic-Cenozoic Sequence Stratigraphy of European Basins. *SEPM Special Publication*.

Ruhl, M., Kürschner, W.M., and Krystyn, L., 2009, Triassic-Jurassic organic carbon isotope stratigraphy of key sections in the western Tethys realm (Austria). *Earth and Planetary Science Letters*, **281**: 169–187.

Ruhl, M., Deenen, M.H.L., Abels, H.A., Bonis, N.R., Krijgsman, W., and Kürschner, W.M., 2010, Astronomical constraints on the duration of the early Jurassic Hettangian stage and recovery rates following the end-Triassic mass extinction (St Audrie's Bay/East Quantoxhead, UK). *Earth and Planetary Science Letters*, **295**: 262–276.

Ruhl, M., Bonis, N.R., Reichart, G.-J., Sinninghe Damsté, J.S., and Kürschner, W.M., 2011, Atmospheric methane injection caused end-Triassic mass extinction. *Science*, **33**: 430–434.

Ruhl, M., Hesselbo, S.P., Hinnov, L., Jenkyns, H.C., Xu, W., Storm, M., et al., 2016, Astronomical constraints on the duration of the Early Jurassic Pliensbachian Stage and global climatic fluctuations. *Earth and Planetary Science Letters*, **455**: 149–165.

Ruebsam, W., Münzberger, P., and Schwark, L., 2014, Chronology of the Early Toarcian environmental crisis in the Lorraine Sub-Basin (NE Paris Basin). *Earth and Planetary Science Letters*, **404**: 273–282.

Ruebsam, W., Mayer, B., and Schwark, L., 2019, Cryosphere carbon dynamics control early Toarcian global warming and sea level evolution. *Global and Planetary Change*, **172**: 440–453.

Sager, W.W., Weiss, C.J., Tivey, M.A., and Johnson, H.P., 1998, Geomagnetic polarity reversal model of deep-tow profiles from the Pacific Jurassic Quiet Zone. *Journal of Geophysical Research – Solid Earth*, **103**: 5269–5286.

Sahagian, D., Pinous, O., Olferiev, A., and Zakharov, V., 1996, Eustatic curve for the Middle Jurassic-Cretaceous based on Russian Platform and Siberian stratigraphy: zonal resolution. *American Association of Petroleum Geologists Bulletin*, **80**: 1433–1458.

Salfeld, H., 1913, Certain upper Jurassic strata of England. *The Quarterly Journal of the Geological Society of London*, **69**: 423−430.

Salfeld, H., 1914, Die Gliederung des Oberen Jura in Nordwest Europa. *Neues Jahrbuch für Mineralogie, Geologie und Palaeontologie*, **32**: 125−246.

Sandoval, J., O'Dogherty, L., Aguado, R., Bartolini, A., Bruchez, S., and Bill, M., 2008, Aalenian carbon-isotope stratigraphy: calibration with ammonite, radiolarian and nannofossil events in the western Tethys. *Palaeogeography, Palaeoclimatology, Palaeoecology*, **267**: 115−137.

Schaltegger, U., Guex, J., Bartolini, A., Schoene, B., and Ovtcharova, M., 2008, Precise U−Pb age constraints for end-Triassic mass extinction, its correlation to volcanism and Hettangian post-extinction recovery. *Earth and Planetary Science Letters*, **267**: 266−275.

Schweigert, G., 2015, Chapter 14. Ammonoid biogeography in the Jurassic. *In* Klug, C., Korn, D., De Baets, K., Kruta, I., and Mapes, R.H. (eds), *Ammonoid Paleobiology: From Macroevolution to Paleogeography*. Springer Publ, 389−402.

Schoene, B., Guex, J., Bartolini, A., Schaltegger, U., and Blackburn, T.J., 2010, Correlating the end-Triassic mass extinction and flood basalt volcanism at the 100 ka level. *Geology*, **38**: 387−390.

Schweigert, G., and Callomon, W.J., 1997, Der bauhini-Faunenhorizont unde seine Bedeutung für die Korrelation zwischen tethyalem und subborealem Oberjura. *Stuttgarter Beiträge zur Naturkunde, Serie B (Geologie und Paläontologie)*, **247**: 1−69.

Scotese, C.R., 2014, Atlas of Jurassic Paleogeographic Maps, PALEOMAP Atlas for ArcGIS, volume 3, The Jurassic and Triassic, Maps 32-42, Mollweide Projection, PALEOMAP Project, Evanston, IL., 2014.

Selby, D., 2007, Direct rhenium-osmium age of the Oxfordian-Kimmeridgian boundary, Staffin Bay, Isle of Skye, UK and the Late Jurassic geologic timescale. *Norwegian Journal of Geology*, **87**: 291−299.

Sell, B., Ovtcharova, M., Guex, J., Bartolini, A., Jourdan, F., Spangenberg, J.E., et al., 2014, Evaluating the temporal link between the Karoo LIP and climatic−biologic events of the Toarcian Stage with high-precision U−Pb geochronology. *Earth and Planetary Science Letters*, **408**: 48−56.

Sell, et al., 2016, Response to comment on "Evaluating the temporal link between the Karoo LIP and climatic−biologic events of the Toarcian Stage with high-precision U−Pb geochronology. *Earth and Planetary Science Letters*, **434**: 353−354.

Sellwood, B.W., and Valdes, P.J., 2008, Jurassic climates. *Proceedings of the Geologists' Association*, **119**: 5−17.

Sha, J., and Fürsich, F.T., 1994, Bivalve faunas of eastern Heilongjiang, northeastern China. II. The Late Jurassic and Early Cretaceous buchiid fauna. *Beringeria*, **12**: 3−93.

Sharland, P.R., Archer, R., Casey, D.M., Davies, R.B., Hall, S.H., Heward, A.P., et al., 2001, Arabian Plate sequence stratigraphy. *GeoArabia Special Publication*, **2**: 371.

Sharland, P.R., Casey, D.M., Davies, R.B., Simmons, M.D., and Sutcliffe, O.E., 2004, Arabian Plate sequence stratigraphy − revisions to SP2. *GeoArabia*, **9**: 199−214.

Shipboard Scientific Party, 1990, Site 801: Pigafetta Basin, western Pacific. *Proceedings of the Ocean Drilling Program, Initial Reports*, **129**: 91−170.

Silva, R.L., and Duarte, L.V., 2015, Organic matter production and preservation in the Lusitanian Basin (Portugal) and Pliensbachian climatic hot snaps. *Global and Planetary Change*, **131**: 24−34.

Silva, R.L., Duarte, L.V., Wach, G.D., Morrison, N., and Campbell, T., 2019, Oceanic organic carbon as a possible first-order control on the carbon cycle during the Bathonian−Callovian. *Global and Planetary Change*, **184**: 103058, https://doi.org/10.1016/j.gloplacha.2019.103058.

Simmons, M.S., Miller, K.G., Ray, D.C., Davies, A., van Buchem, F.S.P., and Gréselle, B., 2020, Chapter 13 − Phanerozoic eustasy. *In* Gradstein, F.M., Ogg, J.G., Schmitz, M.D., and Ogg, G.M. (eds), *The Geologic Time Scale 2020*. Vol. **1** (this book). Elsevier, Boston, MA.

Sinemurian Boundary Working Group (G. Bloos, coordinator), 2000, Submission of East Quantoxhead (West Somerset, SW England) as the GSSP for the base of the Sinemurian Stage. *International Commission on Stratigraphy Document*, 14 pp.

Speranza, F., and Parisi, G., 2007, High-resolution magnetic stratigraphy at Bosso Stirpeto (Marche, Italy): anomalous geomagnetic field behavior during early Pliensbachian (early Jurassic) times? *Earth and Planetary Science Letters*, **256**: 344−359.

Speranza, F., Satolli, S., Mattioli, E., and Calamita, F., 2005, Magnetic stratigraphy of Kimmeridgian−Aptian sections from Umbria-Marche (Italy): new details on the M-polarity sequence. *Journal of Geophysical Research*, **110**: B12109, https://doi.org/10.1029/2005JB003884.

Srivastava, S.K., 1987, Jurassic spore-pollen assemblages from Normandy (France) and Germany. *Geobios*, **20**: 5−79.

Svensen, H., Corfu, F., Polteau, S., Hammer, Ø., and Planke, S., 2012, Rapid magma emplacement in the Karoo Large Igneous Province. *Earth and Planetary Science Letters*, **325−326**: 1−9.

Steiner, M.B., and Ogg, J.G., 1988, Early and Middle Jurassic magnetic polarity time scale. *In* Rocha, R.B., and Soares, A.F. (eds), *Second International Symposium on Jurassic Stratigraphy*. Lisbon: Instituto Nacional de Investigação Científica, 1097−1111.

Steiner, M.B., Ogg, J.G., Melendez, G., and Sequieros, L., 1986, Jurassic magnetostratigraphy, 2. Middle-Late Oxfordian of Aguilon, Iberian Cordillera, northern Spain. *Earth and Planetary Science Letters*, **76**: 151−166.

Steiner, M.B., Ogg, J.G., and Sandoval, J., 1987, Jurassic magnetostratigraphy, 3. Bajocian-Bathonian of Carcabuey, Sierra Harana and Campillo de Arenas, (Subbetic Cordillera, southern Spain). *Earth and Planetary Science Letters*, **82**: 357−372.

Stone, R., 2010, Bird-dinosaur link firmed up, and in brilliant color. *Science*, **327**: 570−571.

Strasser, A., 2007, Astronomical time scale for the Middle Oxfordian to Late Kimmeridgian in the Swiss and French Jura Mountains. *Swiss Journal of Geosciences*, **100**: 407−429.

Storm, M.S., Hesselbo, S.P., Jenkyns, H.C., Ruhl, M., Ullmann, C.V., Xu, W., et al., 2020, Orbital pacing and secular evolution of the Early Jurassic carbon cycle. *Proceedings of the National Academy of Sciences*, **117**: 3974−3982.

Suan, G., Pittet, B., Bour, I., Mattioli, E., Duarte, L.V., and Mailliot, S., 2008, Duration of the Early Toarcian carbon isotope excursion deduced from spectral analysis−Consequence for its possible causes. *Earth and Planetary Science Letters*, **267**: 666−679.

Suan, G., Nikitenko, B.L., Rogov, M.A., Baudin, F., Spangenberg, J.E., Knyazev, V.J., et al., 2011, Polar record of Early Jurassic massive carbon injection. *Earth and Planetary Science Letters*, **312**: 102−113.

Suchéras-Marx, B., Giraud, F., Fernandez, V., Pittet, B., Lecuyer, C., Olivero, D., et al., 2013, Duration of the Early Bajocian and the associated δ^{13}C positive excursion based on cyclostratigraphy. *Journal of the Geological Society*, **170**: 107−118.

Suchéras-Marx, B., Mattioli, E., Allemand, P., Giraud, F., Pittet, B., Plancq, J., et al., 2019, The colonization of the oceans by calcifying pelagic algae. *Biogeosciences*, **16**: 2501–2510.

Surlyk, F., and Zakharov, V.A., 1982, Buchiid bivalves from the upper Jurassic and Lower Cretaceous of East Greenland. *Palaeontology*, **25**: 727–753.

Svensen, H., Planke, S., Chevallier, L., Malthe-Sørenssen, A., Corfu, F., and Jamtveit, B., 2007, Hydrothermal venting of greenhouse gases triggering Early Jurassic global warming. *Earth and Planetary Science Letters*, **256**: 554–566.

Sykes, R.M., and Callomon, J.H., 1979, The *Amoeboceras* zonation of the Boreal upper Oxfordian. *Palaeontology*, **22**: 839–903.

Surlyk, F., 1990, A Jurassic sea-level curve for East Greenland. *Palaeogeography, Palaeoclimatology, Palaeoecology*, **78**: 71–85.

Taylor, S.P., Sellwood, B., Gallois, R., and Chambers, M.H., 2001, A sequence stratigraphy of Kimmeridgian and Bolonian stages, Wessex-Weald Basin. *Journal of the Geological Society of London*, **158**: 179–192.

Them, T.R., Gill, B.C., Selby, D., Gröcke, D.R., Friedman, R.M., and Owens, J.D., 2017a, Evidence for rapid weathering response to climatic warming during the Toarcian Oceanic Anoxic Event. *Scientific Reports*, **7** (5003), 10.

Them, T.R., Gill, B.C., Caruthers, A.H., Gröcke, D.R., Tulsky, E.T., Martindale, R.C., et al., 2017b, High-resolution carbon isotope records of the Toarcian Oceanic Anoxic Event (Early Jurassic) from North America and implications for the global drivers of the Toarcian carbon cycle. *Earth and Planetary Science Letters*, **459**: 118–126.

Thibault, N., Ruhl, M., Korte, C., Ullmann, C.V., Kemp, D.B., Gröcke, D.R., et al., 2018, The wider context of the Lower Jurassic Toarcian oceanic anoxic event in Yorkshire coastal outcrops, UK. *Proceedings of the Geologists' Association*, **129**: 372–391.

Thibodeau, A., Ritterbush, K., Yager, J., West, A.J., Ibarra, Y., Bottjer, D.J., et al., 2016, Mercury anomalies and the timing of biotic recovery following the end-Triassic mass extinction. *Nature Communications*, **7**: 11147, https://doi.org/10.1038/ncomms11147.

Thierry, J., 1998, Ammonites. Columns for Jurassic chart of Mesozoic and Cenozoic sequence chronostratigraphic framework of European basins. *In* de Graciansky, P.-C., Hardenbol, J., Jacquin, Th, and Vail, P.R. (eds), Mesozoic-Cenozoic Sequence Stratigraphy of European Basins. *SEPM Special Publication*, **60**: 776–777.

Tivey, M.A., Sager, W.W., Lee, S.-M., and Tominaga, M., 2006, Origin of the Pacific Jurassic Quiet Zone. *Geology*, **34**: 789–792.

Tominaga, M., and Sager, W.W., 2010, Revised Pacific M-anomaly geomagnetic polarity timescale. *Geophysical Journal International*, **182**: 203–232.

Tominaga, M., Sager, W.W., Tivey, M.A., and Lee, S.-M., 2008, Deep-tow magnetic anomaly study of the Pacific Jurassic Quiet Zone and implications for the geomagnetic polarity reversal time scale and geomagnetic field behavior. *Journal of Geophysical Research*, **113**: B07110, https://doi.org/10.1029/2007JB005527.

Tominaga, M., Tivey, M.A., and Sager, W.W., 2015, Nature of the Jurassic magnetic quiet Zone. *Geophysical Research Letters*, **42**: 8367–8372, https://doi.org/10.1002/2015GL065394.

Torrens, H.S., 1965, Revised zonal scheme for the Bathonian stage of Europe. *Reports of the Seventh Carpato-Balkan Geological Association Congress*, Vol. 2, p. 47–55.

Trujillo, K.C., Chamberlain, K.R., and Strickland, A., 2006, Oxfordian U/Pb ages from SHRIMP analysis for the upper Jurassic Morrison Formation of southeastern Wyoming with implications for biostratigraphic correlations. *Geological Society of America Abstracts with Programs*, **38** (6), 7.

Trujillo, K.C., Chamberlain, K.R., and Bilbey, S.A., 2008, U/Pb age for a pipeline dinosaur site; new techniques and stratigraphic challenges. *Geological Society of America Abstracts with Programs*, **40** (1), 42.

van Acken, D., Tütken, T., Daly, J.S., Schmid-Röhl, A., and Orr, P.J., 2019, Rhenium-osmium geochronology of the Toarcian Posidonia Shale, SW Germany. *Palaeogeography, Palaeoclimatology, Palaeoecology*, **534**: 109294, https://doi.org/10.1016/j.palaeo.2019.109294.

van Buchem, F.S.P., McCave, I.N., and Weedon, G.P., 1994, Orbitally induced small-scale cyclicity in a siliciclastic epicontinental setting (Lower Lias, Yorkshire, UK). *In* Boer, P.L., and Smith, D.G. (eds), Orbital Forcing and Cyclic Sequences. *International Association of Sedimentologists Special Publication*, **19**: 345–366.

van de Schootbrugge, B., Payne, J.L., Tomasovych, A., Pross, J., Fiebig, J., Benbrahim, M., et al., 2008, Carbon cycle perturbation and stabilization in the wake of the Triassic-Jurassic boundary mass-extinction event. *Geochemistry, Geophysics, Geosystems*, **9**: Q04028, https://doi.org/10.1029/2007GC001914.

van Konijnenburg-van Cittert, J.H.A., 2008, The Jurassic fossil plant record of the UK area. *Proceedings of the Geologists' Association*, **119**: 59–72.

Veizer, J., Ala, D., Azmy, K., Bruckschen, P., Buhl, D., Bruhn, F., et al., 1999, ^{87}Sr/^{86}Sr, ∂^{13}C and ∂^{18}O evolution of Phanerozoic seawater. *Chemical Geology*, **161**: 59–88, http://www.science.uottawa.ca/geology/isotope_data/.

Veizer, J., Godderis, Y., and François, L.M., 2000, Evidence for decoupling of atmospheric CO_2 and global climate during the Phanerozoic eon. *Nature*, **408**: 698–701.

Villaseñor, A.B., Oloriz-Sáez, F., and Rosales-Domínguez, M.C. (eds), 2019. Updating the Knowledge on the Jurassic in the American Region. *Journal of South American Earth Sciences*.

von Buch, L., 1839, *Über den Jura in Deutschland*. Berlin: Der Königlich Preussischen Akademie der Wissenschaften, 87 pp.

von Humboldt, F.W.H.A., 1799, *Über die Unterirdischen Gasarten und die Mittel ihren Nachtheil zu Vermindern*. Wiewag: Ein Beitrag zur Physik der Praktischen Bergbaukunde. Braunschweig, 384 pp.

von Salis, K., Bergen, J., and De Kaenel, E., 1998, Calcareous nannofossils. Columns for Jurassic chart of Mesozoic and Cenozoic sequence chronostratigraphic framework of European basins. *In* de Graciansky, P.-C., Hardenbol, J., Jacquin, Th, and Vail, P.R. (eds), Mesozoic-Cenozoic Sequence Stratigraphy of European Basins. *SEPM Special Publication*.

Warrington, G., 2010, The Hettangian GSSP. *International Commission on Jurassic Stratigraphy Newsletter*, **36**: 5–12.

Waterhouse, H.K., 1995, High-resolution palynofacies investigation of Kimmeridgian sedimentary rocks. *In* House, M.R., and Gale, A.S. (eds), Orbital Forcing Timescales and Cyclostratigraphy. *Geological Society Special Publication*, **85**: 75–114.

Weedon, G.P., 1989, The detection and illustration of regular sedimentary cycles using Walsh power spectra and filtering, with examples from the Lias of Switzerland. *Journal of the Geological Society of London*, **146**: 133–144.

Weedon, G.P., and Jenkyns, H.C., 1999, Cyclostratigraphy and the Early Jurassic timescale: data from the Belemnite Marls, southern England. *Geological Society of America Bulletin*, **111**: 1823–1840.

Weedon, G.P., Jenkyns, H.C., Coe, A.L., and Hesselbo, S.P., 1999, Astronomical calibration of the Jurassic time-scale from cyclostratigraphy in British mudrock formations. *Philosophical Transactions of the Royal Society of London, Series A*, **357**: 1787–1813.

Weedon, G.P., Coe, A.L., and Gallois, R.W., 2004, Cyclostratigraphy, orbital tuning and inferred productivity for the type Kimmeridge Clay (Late Jurassic), southern England. *Journal of the Geological Society*, **161**: 655–666.

Weedon, G.P., Jenkyns, H.C., and Page, K.N., 2018, Combined sea-level and climate controls on limestone formation, hiatuses and ammonite preservation in the Blue Lias Formation, South Britain (uppermost Triassic – Lower Jurassic). *Geological Magazine*, **155**: 1117–1149.

Weedon, G.P., Page, K.N., and Jenkyns, H.C., 2019, Cyclostratigraphy, stratigraphic gaps and the duration of the Hettangian Stage (Jurassic): insights from the Blue Lias Formation of southern Britain. *Geological Magazine*, **156**: 1469–1509.

Weissert, H., and Channell, J.E.T., 1989, Tethyan carbonate carbon isotope stratigraphy across the Jurassic-Cretaceous boundary: an indicator of decelerated global carbon cycling? *Paleoceanography*, **4**: 483–494.

Weissert, H., and Erba, E., 2004, Volcanism, CO_2 and palaeoclimate: a Late Jurassic-Early Cretaceous carbon and oxygen isotope record. *Journal of the Geological Society*, **161**: 695–702.

Westermann, G.E.G., and Riccardi, A.C., 1972, Middle Jurassic ammonite fauna and biochronology of the Argentine-Chilean Andes. 1. Hildoceratacea. *Palaeontologographica, A*, **140**: 1–116.

White, R.S., 1997, Mantle plume origin for the Karoo and Ventersdorp flood basalts, South Africa. *South African Journal of Geology*, **100**: 271–282.

Whiteside, J.H., Olsen, P.E., Eglinton, T.I., Montluçon, D., Brookfield, M.E., and Sambrotto, R.N., 2010, Compound-specific carbon isotopes from Earth's largest flood basalt province directly link eruptions to the end-Triassic mass extinction. *Proceedings of the National Academy of Sciences of the United States of America*, **107**: 6721–6725.

Wierzbowski, A., 2001, Kimmeridgian Working Group. *International Subcommission on Jurassic Stratigraphy Newsletter*, **28**: 15–16.

Wierzbowski, A., 2002, Kimmeridgian Working Group. *International Subcommission on Jurassic Stratigraphy Newsletter*, **29**: 18–19.

Wierzbowski, A., 2010, Kimmeridgian. *International Commission on Jurassic Stratigraphy Newsletter*, **36**: 5–12.

Wierzbowski, H., 2015, Seawater temperatures and carbon isotope variations in central European basins at the Middle–Late Jurassic transition (Late Callovian–Early Kimmeridgian). *Palaeogeography, Palaeoclimatology, Palaeoecology*, **440**: 506–523.

Wierzbowski, A., and Matyja, B.A., 2014, Ammonite biostratigraphy in the Polish Jura sections (central Poland) as a clue for recognition of the uniform base of the Kimmeridgian Stage. *Volumina Jurassica*, **12** (1), 45–98.

Wierzbowski, A., Coe, A.L., Hounslow, M.W., Matyja, B.A., Ogg, J.G., Page, K.N., et al., 2006, A potential stratotype for the Oxfordian/Kimmeridgian boundary: Staffin Bay, Isle of Skye, U.K. *Volumina Jurassica*, **4**: 17–33.

Wierzbowski, A., Smoleń, J., and Iwańczuk, J., 2015, The Oxfordian and Lower Kimmeridgian of the Peri-Baltic Synclise (north-eastern Poland): stratigraphy, ammonites, microfossils (foraminifers, radiolarians), facies and palaeogeographic implications. *Neues Jahrbuch fuür Geologie und Paläontologie, Abhandlungen*, **277**: 63–104.

Wierzbowski, A., Atrops, F., Grabowski, J., Hounslow, M., Matyja, B.A., Olóriz, F., et al., 2016, Towards a consistent Oxfordian-Kimmeridgian global boundary: current state of knowledge. *Volumina Jurassica*, **14**: 14–49.

Wierzbowski, H., Anczkiewicz, R., Pawlak, J., Rogov, M.A., and Kuznetsov, A.B., 2017, Revised Middle–upper Jurassic strontium isotope stratigraphy. *Chemical Geology*, **466**: 239–255.

Wierzbowski, A., Matyja, B.A., and Wright, J.K., 2018, Notes on the evolution of the ammonite families Aulacostephanidae and Cardioceratidae and the stratigraphy of the uppermost Oxfordian and lowermost Kimmerdigian in the Staffin Bay sections (Isle of Skye, northern Scotland). *Volumina Jurassica*, **16**: 27–50.

Wiggan, N.J., Riding, J.B., and Franz, M., 2017, Resolving the Middle Jurassic dinoflagellate radiation: the palynology of the Bajocian of Swabia, southwest Germany. *Review of Palaeobotany and Palynology*, **238**: 55–87.

Wignall, P.B., and Bond, D.P.G., 2008, The end-Triassic and Early Jurassic mass extinction records in the British Isles. *Proceedings of the Geologists' Association*, **119**: 73–84.

Williford, K.H., Ward, P.D., Garrison, G.H., and Buick, R., 2007, An extended organic carbon-isotope record across the Triassic-Jurassic boundary in the Queen Charlotte Islands, British Columbia, Canada. *Palaeogeography, Palaeoclimatology, Palaeoecology*, **244**: 290–296.

Wilson, G.J., 1984, New Zealand Late Jurassic to Eocene dinoflagellate biostratigraphy. *Newsletters on Stratigraphy*, **13**: 104–117.

Wimbledon, W.A.P., 2017, Developments with fixing a Tithonian/Berriasian (J/K) boundary. *Volumina Jurassica*, **15**: 181–186.

Wimbledon, W.A.P., Casellato, C.E., Reháková, D., Bulot, L.G., Erba, E., Gardin, S., et al., 2011, Fixing a basal Berriasian and Jurassic-Cretaceous (J-K) boundary – is there perhaps some light at the end of the tunnel? *Rivista Italiana di Paleontologia e Stratigrafia*, **117**: 295–307.

Wimbledon, W.A.P., Reháková, D., Pszczółkowski, A., Casellato, C., Halásová, E., Frau, C., et al., 2013, A preliminary account of the bio- and magnetostratigraphy of the upper Tithonian – Lower Berriasian interval at Le Chouet, Drôme (SE France). *Geologica Carpathica*, **64**: 437–460.

Wimbledon, W.A.P., et al., 2020, Progress with selecting a GSSP for the Berriasian Stage (Cretaceous) – illustrated by sites in France and Italy. *XIVth Jurassica & Workshop of the ICS Berriasian Group 2019*, pp. 186–187.

Woollam, R., and Riding, J.B., 1983, Dinoflagellate cyst zonation of the English Jurassic. *Report of the Institute of Geological Sciences*, **83**: 41.

Wotzlaw, J.-F., Guex, J., Bartolini, A., Gallet, Y., Krystyn, L., McRoberts, C.A., et al., 2014, Towards accurate numerical calibration of the Late Triassic: high-precision U-Pb geochronology constraints on the duration of the Rhaetian. *Geology*, **42**: 571–574.

Xu, W., Ruhl, M., Hesselbo, S.P., Riding, J.B., and Jenkyns, H.C., 2017a, Orbital pacing of the Early Jurassic carbon cycle, black-shale formation and seabed methane seepage. *Sedimentology*, **64**: 127–149.

Xu, W., Ruhl, M.R., Jenkyns, H.C., Hesselbo, S.P., Riding, J.B., Selby, D., et al., 2017b, Carbon sequestration in an expanding lake system during the Toarcian Oceanic Anoxic Event. *Nature Geoscience*, **10**: 129–134.

Xu, W., Ruhl, M., Jenkyns, H.C., Leng, M.J., Huggett, J.M., Minisini, J. M., et al., 2018a, Evolution of the Toarcian (Early Jurassic) carbon-cycle and global climatic controls on local sedimentary processes

(Cardigan Bay Basin, UK). *Earth and Planetary Science Letters*, **484**: 396–411.

Xu, W., MacNiocaill, C., Ruhl, M., Jenkyns, H.C., Riding, J.B., and Hesselbo, S.P., 2018b, Magnetostratigraphy of the Toarcian Stage (Lower Jurassic) of the Llanbedr (Mochras Farm) borehole, Wales: basis for a global standard and implications for volcanic forcing of palaeoenvironmental change. *Journal of the Geological Society*, **175**: 594–604.

Yager, J.A., West, A.J., Corsetti, F.A., Berelson, W.M., Rollins, N.E., Rosas, S., et al., 2017, Duration of and decoupling between carbon isotope excursions during the end-Triassic mass extinction and central Atlantic Magmatic Province emplacement. *Earth and Planetary Science Letters*, **473**: 227–236.

Yang, Z., Moreau, M.-G., Bucher, H., Dommergues, J.-L., and Trouiller, A., 1996, Hettangian and Sinemurian magnetostratigraphy from the Paris Basin. *Journal of Geophysical Research*, **101**: 8025–8042.

Zák, K., Kosták, M., Man, O., Zakharov, V.A., Rogov, M.A., Pruner, P., et al., 2011, Comparison of carbonate C and O stable isotope records across the Jurassic/Cretaceous boundary in the Tethyan and Boreal realms. *Palaeogeography, Palaeoclimatology, Palaeoecology*, **299**: 83–96.

Zakharov, V.A., 1987, The Bivalve *Buchia* and the Jurassic-Cretaceous boundary in the Boreal Province. *Cretaceous Research*, **8**: 141–153.

Zakharov, V.A., Bogomolov, Yu.I., Il'ina, V.I., Konstantinov, A.G., Kurushin, N.I., Lebedeva, N.K., et al., 1997, Boreal zonal standard and biostratigraphy of the Siberian Mesozoic. *Russian Geology and Geophysics*, **38**: 965–993.

Zakharov, Y.D., Smyshlyaeva, O.P., Shigeta, Y., Popov, A.M., and Zonova, T.D., 2006, New data on isotopic composition of Jurassic-Early Cretaceous cephalopods. *Progress in Natural Science*, **16**: 50–67.

Zeiss, A., 2003, The upper Jurassic of Europe: its subdivision and correlation. *Geological Survey of Denmark and Greenland Bulletin*, **1**: 75–114.

Ziólkowski, P., and Hinnov, L.A., 2009, Cyclostratigraphy of Bathonian using magnetic susceptibility - preliminary report. *Geologia*, **35** (3/1), 115.

A.S. Gale, J. Mutterlose and S. Batenburg
With contributions by F.M. Gradstein, F.P. Agterberg, J.G. Ogg and M.R. Petrizzo

Chapter 27

The Cretaceous Period

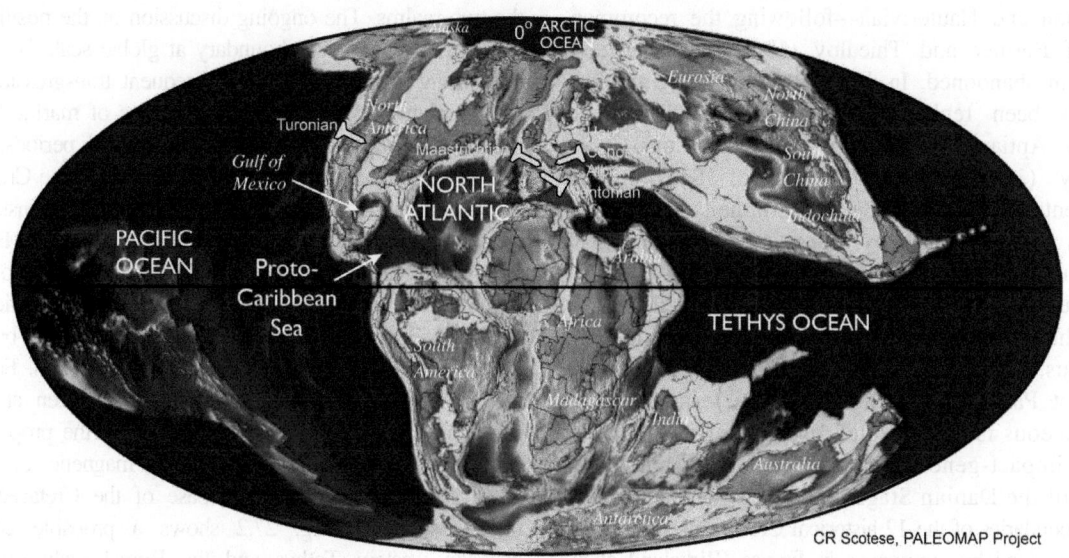

96.6 Ma Cretaceous

CR Scotese, PALEOMAP Project

Chapter outline

27.1 History and subdivisions   1023
   27.1.1 Subdivisions of the Lower Cretaceous   1024
   27.1.2 Subdivisions of the Upper Cretaceous   1034
27.2 Cretaceous stratigraphy   1043
   27.2.1 Marine biostratigraphy   1043
   27.2.2 Terrestrial biostratigraphy   1051

27.2.3 Physical stratigraphy   1051
27.3 Cretaceous time scale   1058
   27.3.1 Previous scales   1059
   27.3.2 Cretaceous numerical age model
      of GTS2020   1061
Bibliography   1068

Abstract

The breakup of the former Pangea supercontinent culminated in the modern drifting continents. Increased rifting caused the establishment of the Atlantic Ocean in the middle Jurassic and significant widening in Cretaceous. An explosion of calcareous nannoplankton and foraminifers in the warm seas created massive chalk deposits. A surge in submarine volcanic activity enhanced supergreenhouse conditions in the middle Cretaceous with high CO_2 concentrations. Angiosperm plants bloomed on the dinosaur-dominated land during late Cretaceous. The Cretaceous dramatically ended with an asteroid impact, which resulted in a mass extinction.

27.1 History and subdivisions

The stratigraphic unit of the "Terrain Crétacé" was established by d'Omalius d'Halloy (1822) to encompass the Chalk (*creta* in Latin) that characterizes a major unit of strata in the Paris Basin and across much of northern Europe. Smith (1815) had already mapped four stratigraphic units between the "lower clay" (Eocene) and the "Portland Stone" (Jurassic). These units were grouped by Conybeare and Phillips (1822) into two main divisions—a higher Chalk and the dominantly clastic formations beneath. This twofold division, adopted in England and France at an early

Geologic Time Scale 2020. DOI: https://doi.org/10.1016/B978-0-12-824360-2.00027-9

1023

stage, has persisted as the two Cretaceous series and epochs, Lower and Upper. Attempts to introduce a Middle Cretaceous in the 1980s were unsuccessful.

D'Orbigny (1840) divided the Cretaceous fossil assemblages of France into five divisions, which were termed "étages" (stages): Neocomian, Aptian, Albian, Turonian and Senonian. Later D'Orbigny (1849−1850) added "Urgonian" between the Neocomian and Aptian and "Cenomanian" stage between the Albian and Turonian. The term "Neocomian" had been coined by Thurmann (1836) for strata in the vicinity of Neuchâtel, Switzerland. This Neocomian interval has subsequently been subdivided into three stages—the Berriasian, Valanginian and Hauterivian—following the recommendation of Barbier and Thieuloy (1965), and the term Neocomian abandoned. In this definition the "Urgonian" stage has been replaced by a "Barremian" and an expanded Aptian stage. The "Senonian," named by d'Orbigny (1842) after the town of Sens in the Départements of Yonne and Seine et Marne is also no longer used, being replaced by four stages, the Coniacian, Santonian, Campanian and Maastrichtian.

The uppermost units of the chalk of Denmark, which had traditionally been included in the Cretaceous, are now classified as a Danian Stage of the lowermost Paleogene (Cenozoic). The termination of the Cretaceous is now defined by a major mass extinction and impact-generated iridium anomaly horizon at the base of the Danian Stage.

The boundaries of the 12 historical Cretaceous stages were primarily defined by ammonites in France (Birkelund et al., 1984; Kennedy, 1984). Refined recognition or proposed definitions of the basal boundaries of these stages have encompassed other globally identifiable criteria, including geomagnetic reversals, carbon-isotope excursions, and microfossil levels (Fig. 27.1; and also Fig. 27.12 later in this chapter).

Historical usage, coupled with the expertise of members of the various boundary working groups, has dictated a preferential selection of boundary stratotypes [Global Boundary Stratotype Section and Points (GSSPs)] for most stages and substages to be within Western European basins. A major problem in Cretaceous chronostratigraphy was, and still is, the correlation of regional biostratigraphic events and associated stage boundary definitions to other paleogeographic and paleoceanographic areas. At the time of completion of this chapter (October 2019) only 6 of the 12 Cretaceous stages had ratified GSSPs (Hauterivian, Albian, Cenomanian, Turonian, Santonian, and Maastrichtian).

27.1.1 Subdivisions of the Lower Cretaceous

The subdivisions of the Berriasian through Aptian stages of the Lower Cretaceous are based on exposures in southeast France and adjacent northwest Switzerland. The Vocontian

Basin preserves a nearly continuous record, either as clay and limestone−marl successions (basin center) or carbonate dominated deposits (basin margin). The marine fauna and microflora thriving in the warm ocean of the northern Tethyan Realm only rarely extended into the Boreal Realm of Northwest Europe, Greenland, Siberia, and other northern regions. The northern higher latitudes were characterized by taxonomically different organisms, causing a distinctive faunal and floral provincialism throughout most of the Early Cretaceous. The segregation, which culminated in the latest Jurassic to earliest Cretaceous, was exacerbated by a global sea-level low stand and resulted in the application of different stage names for the Jurassic/Cretaceous boundary interval in the two realms. The ongoing discussion on the positioning of the Jurassic/Cretaceous boundary at global scale is related to this strong provincialism. A subsequent transgression in the Valanginian allowed a limited exchange of marine biota via marine gateways during certain short-lived periods, but the Tethyan and Boreal Realms persisted through the Cretaceous.

Traditionally, the primary markers that were considered for defining stage and substage boundaries of the Berriasian through Aptian were lowest or highest occurrences of ammonite species. In an effort to achieve global correlation, these macrofossil datums have been replaced or enhanced by magnetostratigraphic, geochemical, or microfossil events. For example, the base of the Albian stage is now taken at the first occurrence of a planktonic foraminifer and the proposed base of the Aptian Stage will probably be a magnetic reversal. The base of the Berriasian Stage (base of the Cretaceous) is a microfossil level. Fig. 27.2 shows a probable correlation between western Tethys and the Boreal realm, using this information.

27.1.1.1 The Jurassic/Cretaceous boundary and the Berriasian Stage

The Cretaceous is the only Phanerozoic system that does not yet have an accepted global boundary definition, despite over a dozen international conferences and working group meetings dedicated to the issue since the 1970s (e.g., reviews by Zakharov et al., 1996; Cope, 2007; Wimbledon, 2017). Difficulties in assigning a global Jurassic/Cretaceous boundary are the product of historical usage, the lack of any major faunal change between the latest Jurassic and earliest Cretaceous, a pronounced provincialism of marine fauna and flora and a concentration of previous studies on often endemic ammonites (Fig. 27.2). Another problem is the occurrence of widespread hiatuses or condensations in many European and Russian epicontinental successions caused by the long-term "Purbeckian regression."

The Tithonian Stage of the latest Jurassic was defined by Oppel (1865) to include all deposits in the Mediterranean area that lie between a restricted Kimmeridgian and

"Valanginian," but no representative section or upper limit were designated. Coquand (1871) coined the "Berriasian" for a limestone succession near the village of Berrias (Ardèche, southeast France) (reviewed in Rawson, 1983). This unit was originally conceived as a subdivision of the Valanginian, subsequently often referred to as "Infra-Valanginian." It was only during a colloquium in Lyon during 1963 that the Berriasian was formally established as a stage below the Valanginian. This Berriasian Stage overlapped to some degree with the original concept of the Tithonian Stage. The historical lower boundary of the Berriasian Stage lacks any significant faunal change, indeed the basal part of the Berrias stratotype lacks any diagnostic ammonites (Cope, 2007). The Colloque sur la limite Jurassique-Crétacé (1975) voted to define the base of the Berriasian at the base of the *Berriasella jacobi* ammonite subzone. This was a shift downward from the 1963 vote to use the base of the *Pseudosubplanites grandis* ammonite subzone (reviewed in Wimbledon et al., 2011).

Some workers raised the Jurassic/Cretaceous boundary to the beginning of the current Valanginian Stage (e.g., Zakharov et al., 1996). An alternative suggestion assigns the lower part of the historical Berriasian Stage to the Jurassic, and defines the Jurassic/Cretaceous boundary at the base of the "middle Berriasian" *Subthurmannia occitanica* ammonite zone (Remane, 1991). This proposal takes intervals with reworked sediments below that zone into account (Hoedemaeker et al., 2003). Using the *S. occitanica* zone has the advantage in that it can be correlated to the southern part of the Boreal Realm where it approximately coincides with the base of the *Runctonia runctoni* ammonite zone.

The current Berriasian Working Group of the International Commission on Cretaceous Stratigraphy has worked to integrate regional ammonite zonations, calpionellid zones, calcareous nannofossil datums, palynomorphs, and magnetostratigraphy (e.g., Schnabl et al., 2015; Wimbledon, 2017; summarized in Fig. 27.2). Their emphasis is on identifying a GSSP level near markers of both regional and global significance.

The absence of the ammonite species *B. jacobi* (=*Strambergella jacobi*) in the lower part of the nominal *B. jacobi* Subzone rules this taxon out as a GSSP marker (Frau et al., 2016; Wimbledon, 2017). It further appears that there are no ammonite-zone boundaries that are synchronous among the main European regions within the basal Cretaceous transition interval.

The lowest occurrence of the small, globular calpionellid species *Calpionella alpina*, which marks the base of the *C. alpina* Subzone, has been documented as a useful event for dating deep-shelf to pelagic limestone sequences. This event, which falls into the middle of magnetozone M19n.2n, has been recorded from southern Europe, Arabia, Iran, and Argentina. The base of the *C. alpina* Subzone has therefore been proposed as

the primary Tithonian–Berriasain boundary marker (Wimbledon, 2017).

Calcareous nannofossils underwent a rapid diversification in the interval under discussion. The lowest occurrences of several species bracket the base of the *C. alpina* Subzone. *Hexalithus strictus* (=*H. geometricus*) predates and *Nannoconus steinmannii minor* postdates the base of the *C. alpina* Subzone. Other nannofossils taxa can be used as secondary markers (Casellato, 2010).

Magnetostratigraphy has proven to be a reliable method for placing biostratigraphic events into a common framework (e.g., Channell et al., 2010). In the GTS2012, the base of Chron M18r was used a temporary assignment for the Jurassic/Cretaceous boundary. However, this datum, even though it can be recognized in both marine and terrestrial deposits in tropical to boreal settings, does not coincide with any consistent biotic marker, and the same is true of the base of M19n.2n. The Berriasian Working Group therefore voted to adopt the turnover in calpionellids from *Crassicollaria* to *Calpionella* as the primary marker for the base of the Berriasian Stage. This datum coincides with the base of the *C. alpina* Subzone, which falls in the middle of Chron M19n.2n and may correspond to the base of the *Arctoteuthis tehamaensis* belemnite zone of Siberia. U–Pb radioisotopic dates zircons in volcanic ash beds from Argentina and Mexico suggested an age of 140.2 ± 0.1 Ma for the base of the Berriasian (Lena et al., 2019). However, this does not fit with regional magnetostratigraphy and with the duration of the Berriasian Stage as indicated by cyclostratigraphy (see later).

In October 2019, following a decade of international collaboration, the Berriasian Working Group of ICS submitted a proposal to define the GSSP for the Berriasian Stage and the Cretaceous System at Bed 14 in the lower section at Tré Maroua (Le Saix, Hautes-Alps), which is about 30 km southeast of Gap in southern France (Wimbledon et al., 2019, 2020). The GSSP level would coincide with the appearance of small, orbicular *C. alpina* calpionellids (base of the Calpionella Zone). This section yielded macrofossils (mainly ammonites) and abundant, well-preserved microfossils in the context of magnetozones M20n–M17n. As of April 2020, this formal proposal was still under consideration by the ICS and IUGS.

27.1.1.1.1 Portlandian–Purbeckian and Volgian–Ryazanian

d'Orbigny (1849–1850) introduced the Portlandian and Purbeckian stages for the Portland Limestone and the nonmarine Purbeck facies of England and northern France. This usage has been largely discontinued, as both facies types are of only regional relevance (Cope, 2007).

The marked provinciality found across the Jurassic/Cretaceous boundary interval lead to the establishment of

Cretaceous Time Scale

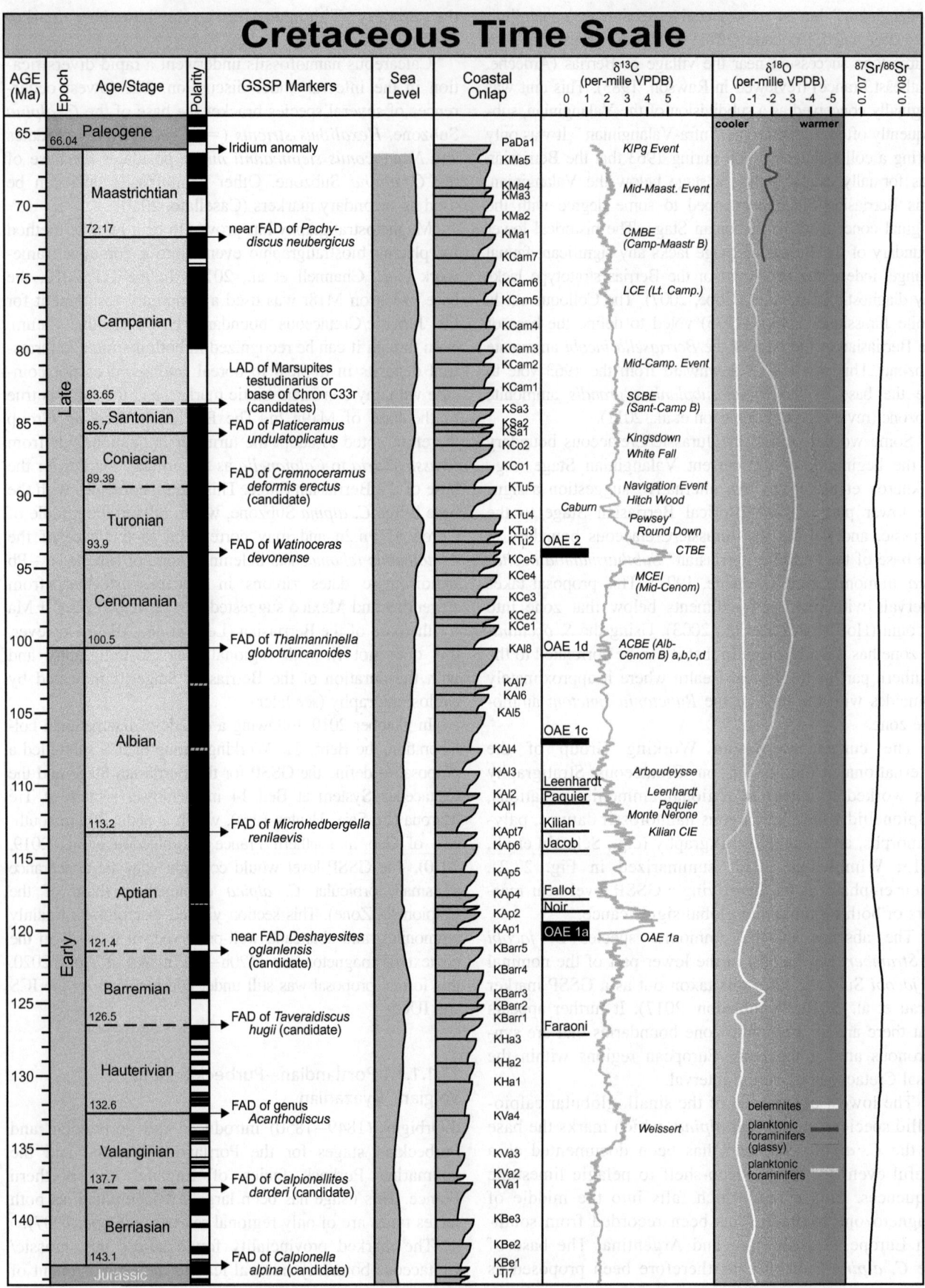

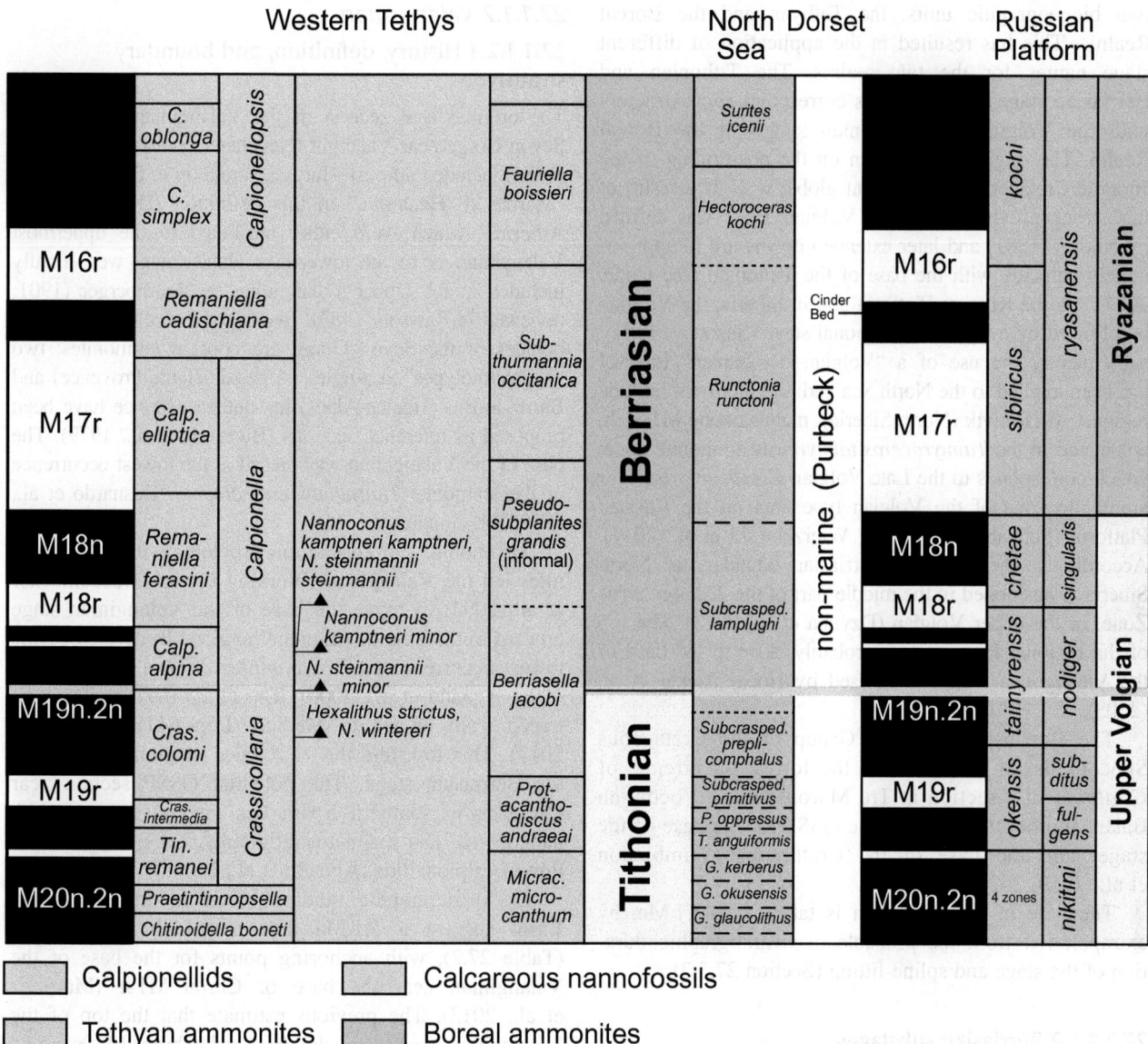

FIGURE 27.2 Possible correlation of the Jurassic–Cretaceous boundary from Western Tethys into the Boreal Realm. Relative placement of selected index fossils for defining the Jurassic–Cretaceous boundary among different paleogeographic regions of the Western Tethys, the Sub-Boreal (North Sea, Dorset), the Boreal (Nordvik, Russian Platform). Yellow line is the proposed GSSP (Global Boundary Stratotype Section and Point). Modified from Wimbledon (2017) and Wimbledon et al. (2019).

FIGURE 27.1 Cretaceous overview. Main markers or candidate markers for GSSPs of Cretaceous stages are currently a combination of calpionellids (Berriasian, Valanginian), ammonites (Hauterivian, Barremian, Aptian, Turonian, Maastrichtian), polarity chron (alternate candidates for Aptian and Campanian), planktonic foraminifers (Albian, Cenomanian), inoceramus bivalve (Coniacian, Santonian), and microcrinoids (Campanian); details are in text and Fig. 27.12. ("Age" is the term for the time equivalent of the rock-record "Stage.") Magnetic polarity column is the C- and M-sequence pattern as scaled in GTS2020 (Ogg, 2020, Ch. 5: Geomagnetiic polarity time scale, this book). Schematic sea-level, coastal onlap curves, and sequence nomenclature are modified from Haq (2014). See also review in Simmons et al. (2020, Ch. 13: Phanerozoic eustasy, this book). The $\delta^{13}C$ curve with major named-excursions (and widespread anoxic events) from Cramer and Jarvis (2020, Ch. 11: Carbon isotope stratigraphy, this book) had compiled numerous studies, including Sprovieri et al. (2006, 2013), Herrle et al. (2004, 2015), Gale et al. (2011), Thibault et al. (2012, 2016), and Jarvis et al. (2006). The oxygen isotope curve and implied sea-surface temperature trends is the statistical mean curve from a synthesis of numerous belemnite and planktonic foraminifera studies (see Grossman and Joachimski, 2020, Ch. 10: Oxygen isotope stratigraphy, this book), although some temperature trends derived from other proxies have more detailed excursions. Strontium isotope curve is modified from McArthur et al. (2020, Ch. 7: Strontium isotope stratigraphy, this book). The vertical scale of this diagram is standardized to match the vertical scales of the period-overview figure in all other Phanerozoic chapters. *GSSP*, Global Boundary Stratotype Section and Point; *VPDB*, Vienna PeeDee Belemnite $\delta^{13}C$ standard; *VSMOW*, Vienna Standard Mean Ocean Water $\delta^{18}O$ standard.

two biogeographic units, the Tethyan and the Boreal Realms. This has resulted in the application of different stage names for the two realms. The Tithonian and Berriasian stages of the Tethys correspond approximately with the Volgian and Ryazanian stages of the Boreal Realm. The ongoing discussion on the positioning of the Jurassic/Cretaceous boundary at global scale is a result of this strong provincialism. The Volgian Stage was defined by Nikitin (1881) and later extended downward to approximately coincide with the base of the Tithonian (see Cope, 2007). On the Russian Platform and in Siberia, the Volgian is followed by a Ryazanian regional stage (Sazonov, 1951); subsequently the use of a "Volgian–Ryazanian" interval has been applied to the North Sea and high northern latitude regions. At Nordvik (North Siberia), magnetozone M19n.2n is situated in the *Taimyroceras taimyrensis* ammonite zone, which corresponds to the Late Volgian *Craspedites nodiger* ammonite zone of the Volgian type area on the Russian Platform (Schnabl et al., 2015; Wierzbowski et al., 2017). Accordingly, the Tithonian/Berriasian boundary in North Siberia is positioned in the middle part of the *T. taimyrensis* Zone, in the upper Volgian (Dzyuba et al., 2013). The top of the regional Ryazanian is probably close to the base of the Valanginian Stage, as revised by Hoedemaeker et al. (2003).

The Berriasian Working Group of the Cretaceous Subcommission recommends the lowest occurrence of *C. alpina* in a section at Tré Maroua in the Vocontian Basin, southeast France, as the GSSP for the base of the stage, and the base of the Cretaceous (Wimbledon et al., 2019, 2020).

The base of the Berriasian is taken at 143.1 Ma, by extrapolation from the probable cyclostratigraphic duration of the stage and spline-fitting (Section 27.3.2).

27.1.1.1.2 Berriasian substages

No formal substages or boundary stratotypes have yet been recommended for the Berriasian Stage. The four ammonite zones of the Berriasian of the Mediterranean area have been used as informal substages (e.g., Reboulet and Atrops, 1999).

The base of the Middle Berriasian *S. occitanica* ammonite zone correlates to the middle portion of polarity-Chron M17r. This level falls near the base of the *Chetaites sibiricus* ammonite zone of North Siberia (Dzyuba et al., 2013).

The base of the Upper Berriasian Substage is currently placed at the base of the *Subthurmannia boissieri* ammonite zone. This event approximates the base of Zone D of the calpionellid zonation and lies in the middle of magnetic polarity zone M16r (M16r.5). It falls amidst the *Heteroceras kochi* ammonite zone of the Russian Platform and North Siberia (Dzyuba et al., 2013).

27.1.1.2 Valanginian
27.1.1.2.1 History, definition, and boundary stratotype

The original type section of the Valanginian stage is the Seyon Gorge near Valangin (Neuchâtel, Switzerland). Desor (1854) included all post–Jurassic strata up to the base of the "Marnes d' Hauterive" in this definition. The "Marnes à *Astieria*," which were either attributed to the uppermost Valanginian or to the lowermost Hauterivian, were finally included in the Upper Valanginian by Baumberger (1901; reviewed in Rawson, 1983). Because the shallow-water carbonates of the Seyon Gorge are poor in ammonites, two "hypostratotypes" at Angles (Alpes-de-Haute-Provence) and Barret-le-Bas (Hautes-Alpes) in southeast France have been proposed as reference sections (Busnardo et al., 1979). The base of the Valanginian was placed at the lowest occurrence of the ammonite *Thurmanniceras otopeta* (Busnardo et al., 1979; Birkelund et al., 1984).

Taxonomic and correlation problems with the ammonites led the Valanginian Working Group (Hoedemaeker et al., 2003) to move the base of the Valanginian stage upward to the base of Calpionellid Zone E, defined by the lowest occurrence of *Calpionellites darderi* (Bulot et al., 1996; Aguado et al., 2000). The *C. darderi* level can be traced from France to Mexico (López-Martínez et al., 2017). This transfers the *T. otopeta* ammonite zone into the Berriasian stage. The potential GSSP section near Caravaca in southern Spain has integrated ammonoid, nannofossil, and magnetostratigraphy, but poor preservation of calpionellids (Aguado et al., 2000).

Cyclostratigraphic studies based in the Vocontian Basin suggest a 5.1 Ma duration of the Valanginian (Table 27.2), with anchoring points for the base of the Valanginian near the base of Chron M14r (Martinez et al., 2013). The previous estimate that the top of the Valanginian was near the base of Chron M10n (e.g., Weissert et al., 1998) as used in GTS2012 is not consistent with this astronomical-tuned duration, therefore is no longer used as a constraint. Corresponding absolute ages are 137 ± 1 Ma for the base of the Valanginian and 131.3 ± 0.25 Ma for the base of the Hauterivian, based on U–Pb ages (Aguirre-Urreta et al., 2015, 2017, 2019).

Correlation of the base of Valanginian from the Tethys into the Boreal Realm is problematic, both areas are characterized by a distinctive faunal and floral provincialism. In the Tethys the interval of the *Tirnovella pertransiens* Zone is marked by the presence of the ammonite genus *Platylenticeras*, which is quite common in northern Europe. The base of the *T. pertransiens* Zone is therefore correlated with the lowest occurrence of the genus *Platylenticeras* in northern Europe (England, Germany; Mutterlose et al., 2014). $^{87}Sr/^{86}Sr$ values of 0.707300–0.707355 are typical for the Lower Valanginian of both realms.

The base of the Valanginian is taken at 137.7 Ma based on cyclostratigraphical tuning of the duration of the stage anchored to the base of the Barremian (Martinez et al., 2015).

27.1.1.2.2 Upper Valanginian substage

In the Tethys the base of the Upper Valanginian is traditionally placed at the base of the distinctive *Saynoceras verrucosum* ammonite zone. Rare specimens of *S. verrucosum* have also been recognized in the West European Province of the Boreal Realm, allowing for correlation. In the Boreal Realm *S. verrucosum* first appears in the upper part of the *Prodichotomites hollwedensis* Zone. This observation and the record of earliest specimens of boreal *Prodichotomites* from the top of the Lower Valanginian in southeast France (Thieuloy, 1977; Kemper et al., 1981) support a placement of the base of the upper Valanginian in northern Europe in the middle of the *P. hollwedensis* ammonite zone.

The Weissert isotope excursion started and peaked in the *S. verrucosum* Zone (135–134 Ma; Martinez et al., 2013, 2015). ^{87}Sr/^{86}Sr values increase from 0.707355 to 0.707379 near the Valanginian/Hauterivian boundary.

Potential boundary stratotypes in southeast France (section at Vergol) and in the Betic Cordillera of Spain are being considered (Bulot et al., 1996; Subcommission on Cretaceous Stratigraphy, 2009).

27.1.1.3 Hauterivian

27.1.1.3.1 History, definition, and boundary stratotype

The original definition of the Hauterivian Stage in the area of Hauterive (Neuchâtel, Switzerland) by Renevier (1874) includes from bottom to top three lithological units: the "Marnes à *Astieria*" (later transferred to the Valanginian), the "Marnes d' Hauterive à *Ammonites Radiates*," and the "Pierre Jaune de Neuchâtel" (see Rawson, 1983). In the Tethys the base of the Hauterivian is recognized by the lowest occurrence of the ammonite species *Acanthodiscus radiatus* (Reboulet, 1996; Mutterlose et al., 1996). This event coincides with the lowest occurrence of the ammonite genus *Acanthodiscus*, which accordingly marks the base of the Hauterivan. In the La Charce GSSP section (southeast France, Fig. 27.3), the presence of *Acanthodiscus* in bed number 189 marks the base of the Hauterivian Stage (Mutterlose et al., in press).

In the Boreal Realm the early Hauterivian shows major influxes of ammonite faunas from the Tethys, events which can be used for correlation. These indicate that the traditional definition of the base of the Hauterivian in the Boreal Realm by the lowest occurrence of *Endemoceras*

amblygonium is nearly coeval with the suggested GSSP level in the Tethyan Realm (Mutterlose et al., 1996). The lowermost part of the boreal *E. amblygonium* Zone corresponds to the uppermost part of the *Criosarasinella furcillata* ammonite zone of the La Charce section. The characterization of the base of the Hauterivian by ammonites is extremely difficult outside Europe and North Africa. In Argentina the base of the *Holcoptychites neuquensis* ammonite zone is correlated with the base of the *A. radiatus* Zone, based on rare ammonite taxa occurring both in Europe and Argentina (Aguirre-Urreta et al., 2005). Magnetostratigraphy had suggested that the base of the Hauterivian was near the top of Chron M10Nn (e.g., Weissert et al., 1998; Sprovieri et al., 2006), but those studies were not directly calibrated to the bases of ammonite zones. The cycle-scaled duration of the Hauterivian was calculated at 5.9 ± 0.4 Ma (Martinez et al., 2015), attributing the base of the Hauterivian an absolute age of 132 ± 1 Ma and the top an age of 126 ± 1 Ma (Table 27.2). Published ^{87}Sr/^{86}Sr data (McArthur et al., 2007; Meissner et al., 2015) suggest Sr-values of >0.707380 for the lowermost Hauterivian for the Tethys and the Boreal Realm.

The base of the Hauterivian is taken at 132.6 Ma based on cyclostratigraphic projection down from the base of the Berriasian (Martinez et al., 2015) and spline fitting. For details see Section 27.3.2.

27.1.1.3.2 Upper Hauterivian substage

The base of the Upper Hauterivian substage in the Tethyan Realm is defined by the highest occurrence of Neocomitinae ammonites and the lowest occurrence of the ammonite species *Subsaynella sayni*. None of these events can be recognized in the Boreal Realm, where the base of the Upper Hauterivian is approximated by the lowest occurrence of the ammonite *Simbirskites*. A correlation tie-point is provided by the Tethyan ammonite *Crioceratites duvali*. Its presence in both realms correlates the base of the Upper Hauterivian of the Tethyan Realm to the upper part of the Boreal *Simbirskites inversum* ammonite zone of the Boreal Realm. The highest occurrence of the calcareous nannofossil species *Cruciellipsis cuvillieri* may serve as an alternative marker for the base of the Upper Hauterivian (Mutterlose et al., 1996). It has been reported from the middle part of the *S. sayni* ammonite zone (Bergen, 1994) and is near the base of magnetic polarity zone M8r in Italy (Channell et al., 1995a). In the Boreal Realm this datum equates the middle part of the *Simbirskites staffi* ammonite zone (Möller et al., 2015).

The La Charce section has also been proposed as a candidate for the Lower/Upper Hauterivian substage boundary section.

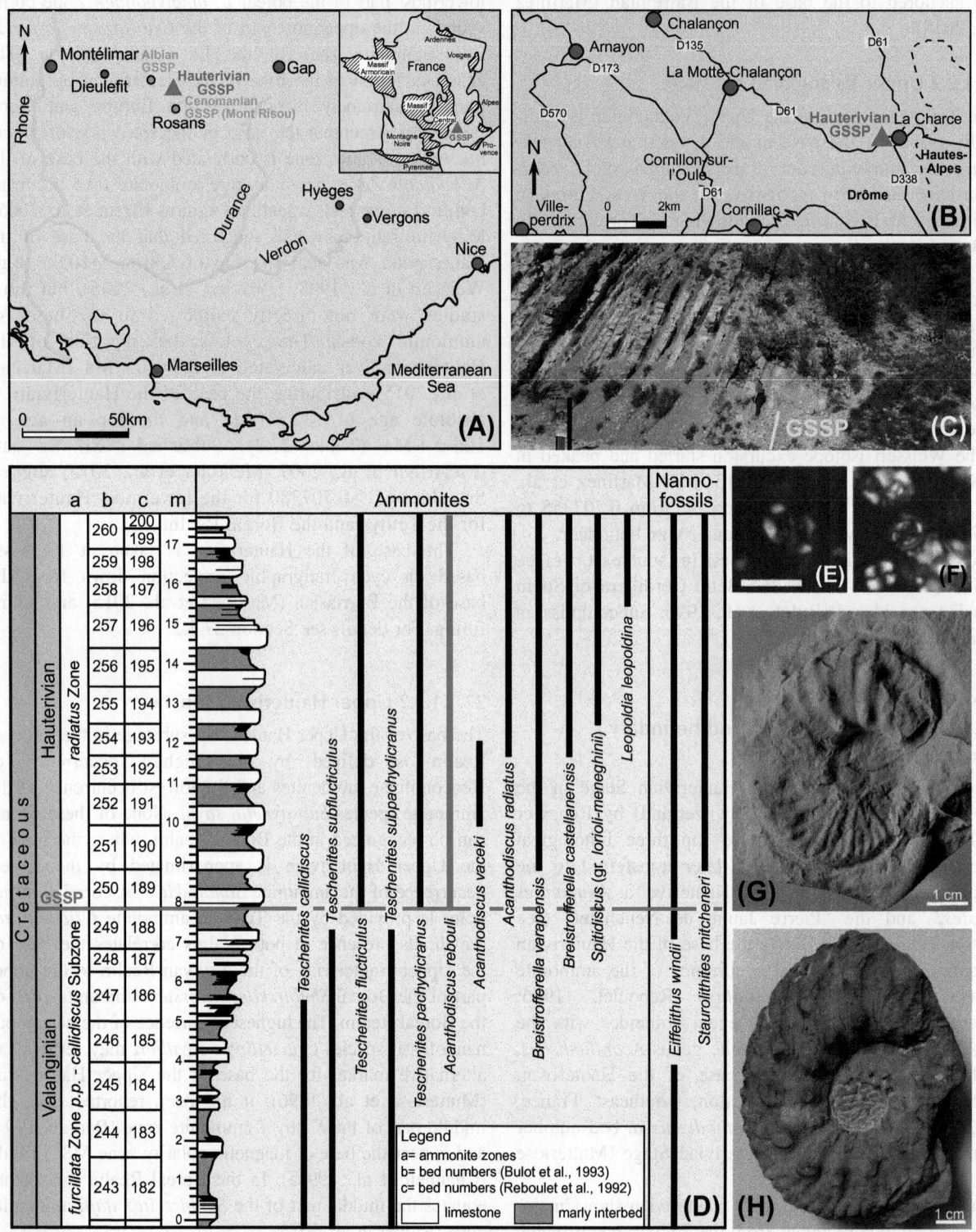

FIGURE 27.3   GSSP for the Hauterivian Stage, in the commune of La Charce, Drôme, southern France. (A and B) Location maps. (C) Photograph of outcrop, with GSSP marked at the base of Bed 189 (photograph by Andy Gale). (D) Stratigraphy and distribution of key ammonites and nannofossils. (E) *Eiffelithus windi*, (F) *Staurolithites mitcheneri*, (G and H) *Acanthodiscus radiatus*. Genus *Acanthodiscus* is the GSSP marker for the base of the Hauterivian. *GSSP*, Global Boundary Stratotype Section and Point.

27.1.1.4 Barremian

27.1.1.4.1 History, definition, and boundary stratotype

The original concept for the Barremian stage is based on marine strata cropping out in the area near Barrême (Alpes-de-Haute-Provence, southeast France). Without giving a specific type locality, Coquand (1861) quoted various belemnites and ammonites for the Barremian, which also encompass parts of the current Upper Hauterivian substage. A more precise definition of the Barremian Stage goes back to Kilian (1888), which has later been redefined by Busnardo (1965) who used the roadside exposures of Angles in southeast France as the type locality.

In the Tethys the base of the Barremian is marked by the lowest occurrence of the ammonite species *Taveraidiscus hugii* (Reboulet and Atrops, 1999). A candidate for the global boundary stratotype is at Río Argos near Caravaca, Spain (Company et al., 1995; Rawson et al., 1996a). Correlation to the Boreal Realm by dinoflagellate cysts indicates that the base of the Barremian Stage approximates the base of the *Hoplocrioceras rarocinctum* ammonite zone (Leereveld, 1995; Rawson et al., 1996a). This correlation is supported by belemnite based $^{87}Sr/^{86}Sr$-isotope data, which show values of ~0.707475 for the lowermost Barremian of both realms (Mutterlose et al., 2014).

In Italy the boundary interval falls in the uppermost part of magnetic polarity zone M4n (at approximately Chron M4n.8 = "M5n.8"; Channell et al., 1995a). Absolute ages for the base of the Barremian of 126 ± 1 Ma and of 121.2 ± 0.5 Ma for the base of the Aptian (Table 27.2) are based on U−Pb zircons (Aguirre-Urreta et al., 2015) resp. Ar^{40}/Ar^{39} sanidines (He et al., 2008). Bodin et al. (2006) calculated a duration of 4.5 Ma for the Barremian, based on phosphorous burial rates.

The base of the Barremian is taken at 126.5 Ma, based on intercalibration of radioisotopic dates from Argentina and cyclostratigraphy (Martinez et al., 2015). For details see Section 27.3.2.

27.1.1.4.2 Upper Barremian substage

In the Tethys region the boundary between the Lower and Upper Barremian is placed at the lowest occurrence of the ammonite *Toxancyloceras vandenheckei* (Rawson et al., 1996a; Reboulet and Atrops, 1999). The proposed substage GSSP is near Caravaca, Spain (Company et al., 1995). This substage boundary level is within the uppermost part of magnetic polarity zone M3r (at approximately Chron M3r.8) in Italy (Channell et al., 1995a). $^{87}Sr/^{86}Sr$-isotope data suggest that in the Boreal Realm the base of the Upper Barremian lies in the upper third of the *Paracrioceras elegans* ammonite zone.

27.1.1.5 Aptian

27.1.1.5.1 History, definition, and boundary stratotype

The Aptian Stage was a vague designation by d'Orbigny (1840) for strata containing "Upper Neocomian" fauna and named after the village of Apt (Vaucluse province, southeast France). The extent and nomenclature of the Aptian and underlying stages have undergone major revisions (reviewed by Moullade et al., 2011). The French sections are poor in ammonites; therefore the classical marker for the base of the Aptian was the lowest occurrence of the deshayesitid ammonite *Prodeshayesites* in northwest Europe (Rawson, 1983; Moullade et al., 1998a,b; Hoedemaeker et al., 2003). However, the local lowest occurrence of this ammonite genus is commonly associated with a major transgression in earliest Aptian and virtually no ammonite-bearing section exists which represents a continuous and complete Barremian−Aptian boundary interval (Erba et al., 1996). Therefore the proposed primary marker for base of the Aptian Stage is the beginning of magnetic polarity-Chron M0r. A quite different boundary concept was proposed by Moullade et al. (2011), who on historical grounds recommended converting the lower three ammonite zone of the current Aptian into a Bedoulian Stage, and beginning the Aptian at the base of *Dufrenoyia furcata* zone (marked by lowest occurrence of *Dufrenoyia* ammonite genus) with a GSSP near Roquefort-La-Bédoule in southeastern France. Reboulet et al. (2011) reject the usage of Bedoulian (see Section 27.1.1.5.2). The GTS2020 scale retains the current Aptian/Barremian usage with its working boundary definition as the base of Chron M0r.

The proposed global boundary stratotype in pelagic limestone at Gorgo a Cerbara in central Italy has an integrated stratigraphy of paleomagnetics, biostratigraphy (calcareous nannofossils, planktonic foraminifers, radiolarians, dinoflagellates) carbon-isotope chemostratigraphy, and cycle stratigraphy (Erba et al., 1996, 1999; Channell et al., 2000; Jenkyns, 2017). However, recently Frau et al. (2018) have reidentified ammonites found beneath, within, and above magnetochron M0r as belonging to the *Martelites sarasini* Zone, traditionally placed within the Barremian, rather than Aptian *Prodeshayesites* as previously recorded.

An important event about 1 Myr after the boundary is Oceanic Anoxic Event 1a (OAE1a), marked by widespread organic-rich shale (e.g., Selli Level in Italy, Goguel level in southeast France, Fischschiefer in northwestern Germany) and the initial, sharp, negative excursion at the base of a longer positive carbon-13 isotopic excursion (Weissert and Lini, 1991; Mutterlose and Böckel, 1998; Li et al., 2008; Jenkyns 2017).

The Aptian Working Group of the Cretaceous Subcommission is likely to recommend taking the maximum negative value of the basal excursion of OAE1a in the Gorgo a Cerbara section in central Italy as GSSP for base of the Aptian Stage (STRATI meeting, Milan, July 2019). However, for our study, the base of the Aptian is taken at 121.4 Ma, at the base of Chron M0, on the basis of dates from Svalbard (Zhang et al. 2019). A modified version of the proposed calibrations and durations of latest Barremian through earliest Aptian Tethyan and Boreal ammonite zones relative to Chron M0r and carbon-isotope excursions (Frau et al., 2018; Luber et al., 2019; Martinez et al., 2020; C. Frau, pers. commun., June 2020) is used the GTS2020 summary diagrams of that interval. For details see Section 27.3.2.

27.1.1.5.2 Substages of Aptian

Two substages are recommended for subdividing the Aptian Stage. The Lower Aptian has traditionally been subdivided using the ammonite lineage *Deshayesites–Dufreynoyia*, but there is currently considerable disagreement as to the application of species names to *Deshayesites*, such that the zonation provided here is provisional (Reboulet and Atrops, 1999; Bersac and Bert, 2012). The Lower/Upper substage boundary is taken at the base of the *Epicheloniceras martinoides* ammonite zone in the standard Tethyan zonation which is approximately equivalent to the base of the *Tropaeum bowerbanki* ammonite zone of the Boreal zonations (Reboulet et al., 2011). At this level, there is an important change in ammonite fauna in both the Tethyan and Sub-Boreal realms.

This level coincides with a traditional boundary between a lower "Bedoulian" and an upper "Gargasian" stage, which were based on sections at Cassis-La Bédoule (Bouches du Rhône province; near Marseilles) and at Gargas (near Apt) in southern France, although historical usage of "Bedoulian" was only for the lower portion of this Lower Aptian (Moullade et al., 2011). A comprehensive synthesis has been compiled for the integrated stratigraphy of the type Bedoulian (Moullade et al., 1998a,b; Ropolo et al., 1998) and type Gargasian (Moullade et al., 2004, 2005; and later articles in *Notebooks on Geology*). The lowest occurrence of planktonic foraminifer *Praehedbergella luterbacheri* is just above the substage boundary followed by the lowest *Globigerineloides ferreolensis* (Moullade et al., 2005), and the lowest calcareous nannofossil *Eprolithus floralis* (base of Zone NC7) slightly precedes the boundary (Moullade et al., 1998b).

In France, an additional uppermost substage of "Clansayesian" was added when Breistroffer (1947) moved the thin "Clansayes" horizon from Albian into the underlying Aptian. However, this substage was not considered useful because (1) the reference sections at Gargas (near Apt) and Clansayes are not suitable for correlation purposes (Rawson, 1983); (2) the rationale for a separate "Clansayesian" substage is questioned (e.g., Owen, 1996a; Casey et al., 1998); and (3) the newly

defined Albian GSSP boundary probably includes part of the Clansayesian interval (Kennedy et al., 2017).

The IUGS Lower Cretaceous Ammonite Working Group or "Kilian Group" voted to "abandon the terms Bedoulian, Gargasian and Clansayesian as they are not recognized internationally, but mainly used in France ... Moreover, the type sections of these French substages do not offer good prospects (low number and/or bad preservation of ammonoids) ..." (reported in Reboulet et al., 2011; see also Reboulet and Atrops, 1999).

27.1.1.6 Albian

27.1.1.6.1 History, definition, and boundary stratotype

The Albian Stage was named after the Aube region (the Roman name was *Alba*) in northeast France (d'Orbigny, 1842). Exposures of Albian clays still exist in the Aube, and the stratigraphy and paleontology of the stratotype has been thoroughly reviewed (Amédro and Matrion, 2007; Amédro et al., 2019; Colleté, 2010).

Prior to 1947, the base of the Albian was assigned as the base of the *Nolaniceras nolani* ammonite zone. Breistroffer (1947) moved this zone and the overlying *Hypacanthoplites jacobi* ammonite zone into an expanded uppermost Aptian, thereby placing the base of the Albian in the Northwest European faunal province of the Boreal realm at the base of the *Leymeriella tardefurcata* ammonite zone with a basal level characterized by *Leymeriella schrammeni*. However, this boundary interval in Western Europe is marked by endemic ammonites, and the occurrences of the earliest *Leymeriella* ammonite species (e.g., *L. schrammeni*) are restricted to northern Germany (Casey, 1996; Hart et al., 1996; Kennedy et al., 2000a; Mutterlose et al., 2003). The successive species *Leymeriella germanica* and *L. tardefurcata* provide an excellent Boreal–Tethyan correlation, because they occur both in northern Germany and the Vocontian Basin in southeast France (Kennedy et al., 2000a), but occur significantly higher than the Aptian–Albian boundary as now defined.

The ratified GSSP for the base of the Albian is the lowest occurrence of the planktonic foraminifer *Microhedbergella miniglobularis* within the thin organic-rich Kilian level, at Pre-Guittard, Arnayon, Drôme, France (Petrizzo et al., 2012; Kennedy et al., 2014, 2017; Fig. 27.4). This marks a major turnover in planktonic foraminiferans, with the extinction of larger *Hedbergella* and *Paraticinella*, and their replacement by small, smooth *Microhedbergella* species. This same turnover has been recorded in the Atlantic and Indian Oceans (Petrizzo et al., 2011). The GSSP falls some meters above the first occurrence of circular *Praediscosphaera columnata* (nannofossil) and coincides with a small negative excursion in $\delta^{13}C$. The GSSP falls within the ammonite zone of *H.*

Base of the Albian Stage of the Cretaceous System, Col de Pré-Guittard Section, Arnayon, Drôme, France.

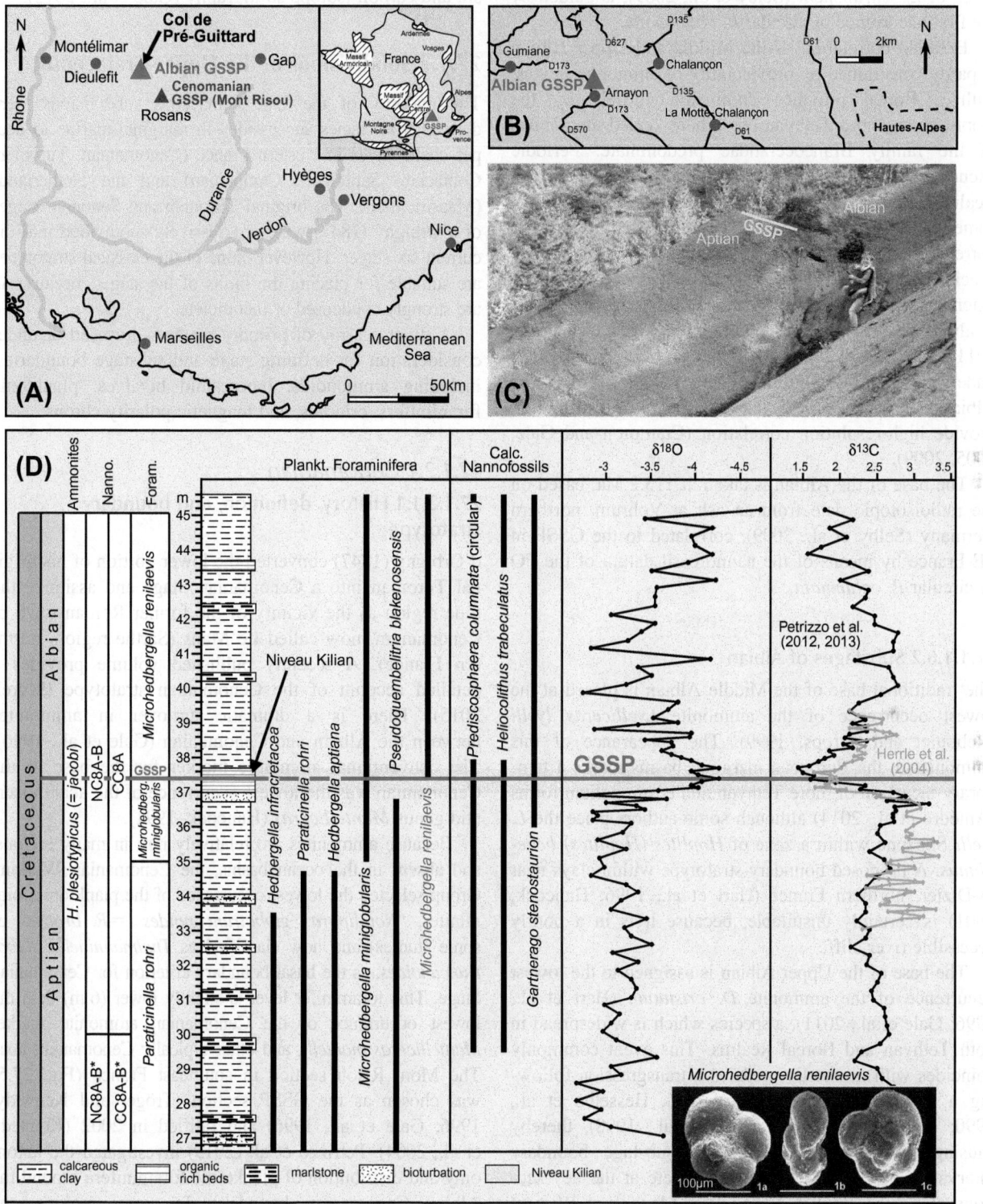

FIGURE 27.4   GSSP for the base of the Albian Stage, Col de Pré-Guittard, Arnayon, Drôme, southeast France. (A and B) Location maps. (C) Photograph of the outcrop, indicating the GSSP within the organic-rich Niveau Kilian. (D) Stratigraphy of the outcrop, and the GSSP marker, the planktonic foraminifer *Microhedbergella renilaevis*. Note also the minor negative carbon isotope excursion at this level. Photograph by Andy Gale. *GSSP*, Global Boundary Stratotype Section and Point.

jacobi. Correlation to the northern German succession at Vohrum, where a dateable ash lies beneath the first occurrence of *L. schrammeni* (Selby et al., 2009), is guided by the first occurrence of circular *P. columnata*.

Detailed correlation of the Middle and Upper Albian is partly constrained by provinciality of ammonite faunas, with a Boreal province dominated entirely by the Hoplitaceae, and a Tethyan one where keeled ammonites of the family Brancoceratidae predominate. Periodic excursions of the Tethyan Ammonites into the Boreal Realm result in the intercalation of the two faunas at some levels, providing valuable evidence for interregional correlation (Owen, 1996a). Thus the basal Upper Albian species brancoceratid species *Dipoloceras cristatum* extends far northwards from the Tethys into the Boreal Realm, and provides an important marker (Gale et al., 2011). Inoceramid bivalves of the genus *Actinoceramus* underwent rapid evolution in the Middle and Upper Albian and, because the species are global in distribution, provide high-resolution correlation (Crampton and Gale, 2005, 2009).

The base of the Albian is taken at 113.2 Ma, based on the radioisotopic date from an ash at Vohrum, northern Germany (Selby et al., 2009), correlated to the GSSP in SE France by means of the nannofossil datum of the FO of circular *P. columnata*.

27.1.1.6.2 Substages of Albian

The traditional base of the Middle Albian is placed at the lowest occurrence of the ammonite *Lyelliceras lyelli* (Reboulet and Atrops, 1999). The appearance of this ammonite in the European marginal basins marks a temporary incursion of more Tethyan and cosmopolitan forms (Amédro et al., 2014) although some authors place the *L. lyelli* Subzone within a zone of *Hoplites (Hoplites) benettianus*. A proposed boundary stratotype within clays near St-Dizier, northern France (Hart et al., 1996; Hancock, 2001) is certainly unsuitable, because it is in a poorly accessible river cliff.

The base of the Upper Albian is assigned to the lowest occurrence of the ammonite *D. cristatum* (Hart et al., 1996; Gale et al., 2011), a species which is widespread in both Tethyan and Boreal Realms. This event commonly coincides with or shortly predates a transgression following a major sequence boundary (e.g., Hesselbo et al., 1990; Amédro, 1992; Hardenbol et al., 1998), thereby causing the Middle/Upper Albian substage boundary interval to be condensed and incomplete at the key sections along the English Channel/La Manche at Wissant (Pas-de-Calais province, northwest France) and at Folkestone (Kent, United Kingdom) (Hart et al., 1996; Gale and Owen in Young et al., 2010). Therefore an expanded basinal clay-rich section at Col de Palluel in

southeast France was extensively studied as a substage GSSP candidate and tied to cyclostratigraphy, nannofossil, and carbon-isotope stratigraphy (Gale et al., 2011).

27.1.2 Subdivisions of the Upper Cretaceous

The majority of the Late Cretaceous subdivisions were derived from facies successions in marginal-marine to deeper chalk facies in western France (Cenomanian, Turonian, Coniacian, Santonian, Campanian) and the Netherlands (Maastrichtian). The original Turonian and Senonian stages of d'Orbigny (1847) were progressively subdivided into the current six stages. However, none of the classical stratotypes are suitable for placing the limits of the stages, because all are strongly condensed or incomplete.

A diverse array of primary markers are used or under consideration for defining stage and substage boundaries, including ammonoids, inoceramid bivalves, planktonic foraminifers, crinoids, and magnetic polarity chrons.

27.1.2.1 Cenomanian

27.1.2.1.1 History, definition, and boundary stratotype

d'Orbigny (1847) converted the lower portion of his original Turonian into a Cenomanian Stage and assigned the type region as the vicinity of the former Roman town of Cenomanum, now called Le Mans (Sarthe region, northern France). A recently published volume provides a detailed account of the Cenomanian stratotype (Morel, 2015). There is a dramatic turnover in ammonites between the Albian and Cenomanian (Gale et al., 1996). The conventional ammonite marker for the base of the Cenomanian was the lowest occurrence of the acanthoceratid genus *Mantelliceras* (Hancock, 1991).

Because ammonites are relatively rare in many regions, and absent in the ocean basins, the Cenomanian Working Group selected the lowest occurrence of the planktonic foraminifer, "*Rotalipora*" *globotruncanoides* (=*R. brotzeni* of some studies) and now classified as *Thalmanninella globotruncanoides*, as the basal boundary criterion for Cenomanian Stage. This foraminifer level is slightly lower (6 m) than the lowest occurrence of the Cenomanian ammonite marker *Mantelliceras mantelli*, and other typically Cenomanian taxa. The Mont Risou section in southeast France (Fig. 27.5) was chosen as the GSSP section (Tröger and Kennedy, 1996; Gale et al., 1996) and ratified in 2002 (Kennedy et al., 2004). Petrizzo et al. (2015) investigated the taxonomy and distribution of planktonic foraminifera around the Albian–Cenomanian boundary. In many regions, the Albian–Cenomanian boundary interval is coincident with a widespread hiatus and condensation associated with a major sequence boundary (e.g., Tröger and Kennedy, 1996; Hardenbol et al., 1998; Robaszynski, 1998; Gale et al.,

Base of the Cenomanian Stage of the Cretaceous System, Mont Risou, Hautes-Alpes, France.

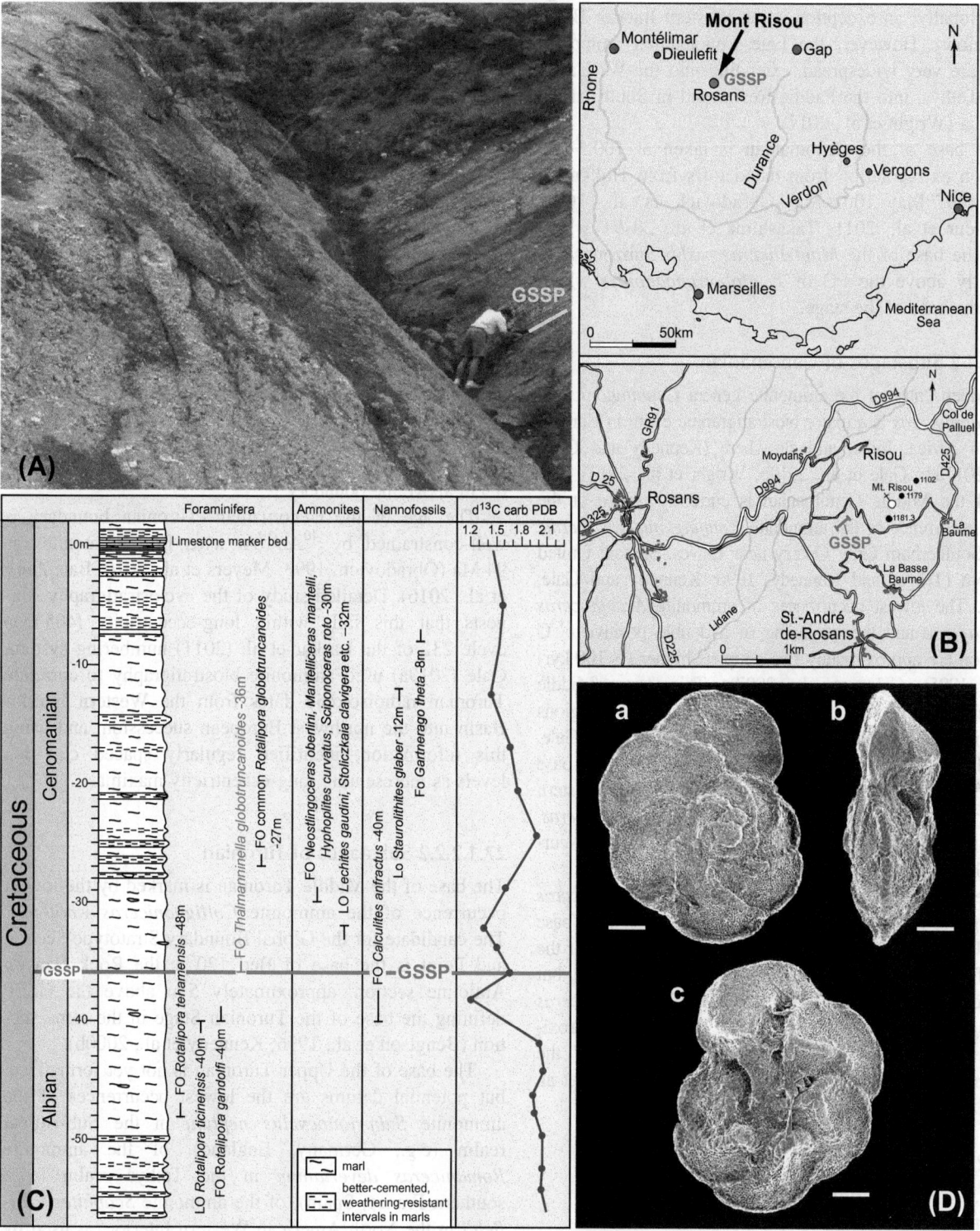

FIGURE 27.5 GSSP for base of the Cenomanian Stage at Mont Risou, Haute Alpes, southeast France. (A) Photo of section. (B) Location maps. (C) Log of section, to show upper part of Marnes Bleues Formation. The GSSP level is the lowest occurrence of the planktonic foraminifera *Thalmanninella globotruncanoides*. (D) Photograph of the marker foraminifer (provided by Atsushi Ando) is the metatype of *T. globotruncanoides* from the type locality (Ando and Huber, 2007). *GSSP*, Global Boundary Stratotype Section and Point.

2019b). Cenomanian ammonites show relatively little endemism (Wright et al., 2017), and most of the standard zones for the Lower and Middle Cenomanian can be recognized globally; an exception is the Western Interior Basin (see below). However, the Late Cenomanian ammonite faunas are very widespread, extending into the WI Basin, across Tethys, into the Pacific Realm and in South Africa and India (Wright et al., 2017).

The base of the Cenomanian is taken at 100.5 Ma, based on extrapolation from dated tuffs from Hokkaido, Japan (99.7 Ma; 100.8 Ma—Obradovich et al., 2002; Quidelleur et al., 2011; Takashima et al., 2019), which fall at the base of the *Mantelliceras saxbii* Subzone, significantly above the FO of *T. globotruncanoides* which marks the base of the stage.

27.1.2.1.2 Substages of Cenomanian

The sudden entry of the ammonite genera *Cunningtoniceras* and *Acanthoceras* is a major biostratigraphic event in Europe, northern Africa, India, and elsewhere (Kennedy and Gale, 2015, 2017a,b; Gale et al., 2019b; Wright et al., 2017). The base of the Middle Cenomanian is currently placed at the lowest occurrence of the ammonite *Cunningtoniceras inerme* at the Southerham Grey Quarry near Lewes, Sussex, United Kingdom (Tröger and Kennedy, 1996; Kennedy and Gale, 2017a). The lowest occurrences of ammonite *Acanthoceras rhotomagense* and the beginning of a double positive δ^{13}C excursion are approximately five couplets higher (~ 100 kyr) (Gale, 1995; Gale et al., 2007). This Lower/Middle Cenomanian boundary interval is missing over large regions due to its coincidence with a major sequence boundary (Gale, 1995; Hardenbol et al., 1998; Robaszynski, 1998). The base of the Middle Cenomanian can be identified in the Western Interior Basin from the FO of the ammonite *Conlinoceras*, associated with the Mid-Cenomanian carbon-isotope excursion (Gale et al., 2007).

The replacement of ammonites of the genus *Acanthoceras* by the genus *Calycoceras* is commonly used to mark the base of the Upper Cenomanian (Hancock, 1991). A marker for the base of the Upper Cenomanian has not yet been selected, but the placement will probably be at the base of the *Calycoceras guerangeri* Zone (Wright et al., 2017). This zone is approximately coeval with *Dunveganoceras pondi* Zone of the Western Interior, of which the base is used by Cobban et al. (2006) for their Upper/Middle substage boundary.

27.1.2.2 Turonian
27.1.2.2.1 History, definition, and boundary stratotype

The concept of the Turonian Stage has undergone continual redefinitions (reviewed in Bengtson et al., 1996). The Turonian proposed by d'Orbigny (1847) in 1842 was later divided by him into a lower Cenomanian Stage and an upper Turonian Stage. The name is derived from the Touraine region of France (Turones and Turonia of the Romans), and d'Orbigny (1852) clarified his later definition by selecting a type region lying between Saumur (on the Loire river) and Montrichard (on the Cher river). A detailed stratigraphical and paleontological account of the Turonian stratotype is provided by Amédro et al. (2018). In this region the lower part of the Turonian contains the ammonite *Mammites nodosoides*, and its lowest occurrence was formerly considered to be the marker for the base of the Turonian Stage (e.g., Harland et al., 1990). After considering several potential placements, the Turonian Working Group placed the base of the Turonian at the lowest occurrence of the ammonite *Watinoceras devonense* (two ammonite zones below the *M. nodosoides* Zone) near the global OAE2 (Bengtson et al., 1996). The GSSP is at Rock Canyon Anticline, east of Pueblo (Colorado, west-central United States) (Kennedy and Cobban, 1991; Bengtson et al., 1996; Kennedy et al., 2000b, 2005) and was ratified in 2003 (Fig. 27.6). The maximum major carbon-isotope peak associated with the OAE2 occurs 0.5 m above the boundary.

The age of the Cenomanian−Turonian boundary is well-constrained by ^{40}Ar/^{39}Ar ages from bentonites as 94 Ma (Obradovich, 1993; Meyers et al., 2010; Batenburg et al., 2016). Detailed study of the cyclostratigraphy suggests that this falls within long-eccentricity (405 kyr) cycle 232 of the Laskar et al. (2011) numbering system. Gale (2019a) used ammonite biostratigraphy to correlate Turonian radioisotopic dates from the Western Interior Basin into the northwest European succession, and using this information, identified regularly spaced clay-rich levels as representing long-eccentricity maxima.

27.1.2.2.2 Substages of Turonian

The base of the Middle Turonian is marked by the lowest occurrence of the ammonite *Collignoniceras woollgari*. The candidate for the Global Boundary Stratotype Section and Point is the base of Bed 120 in the Rock Canyon Anticline section, approximately 5 m above the GSSP defining the base of the Turonian Stage in the same section (Bengtson et al., 1996; Kennedy et al., 2000b).

The base of the Upper Turonian is not yet formalized, but potential datums are the lowest occurrences of the ammonite *Subprionocyclus neptuni* in the Sub-Boreal realm (e.g., Germany, England), of the ammonite *Romaniceras deverianum* in the Tethys realm (e.g., southern France, Spain), of the ammonite *Scaphites whitfieldi* in the North American Western Interior (as used by Cobban et al., 2006, or of an inoceramid bivalve, *Inoceramus perplexus* (=*Mytiloides costellatus* in some studies) (Bengtson et al., 1996; Wiese and Kaplan, 2001). The best of these suggestions is probably the first

Base of the Turonian Stage of the Cretaceous System at Pueblo, Colorado, USA.

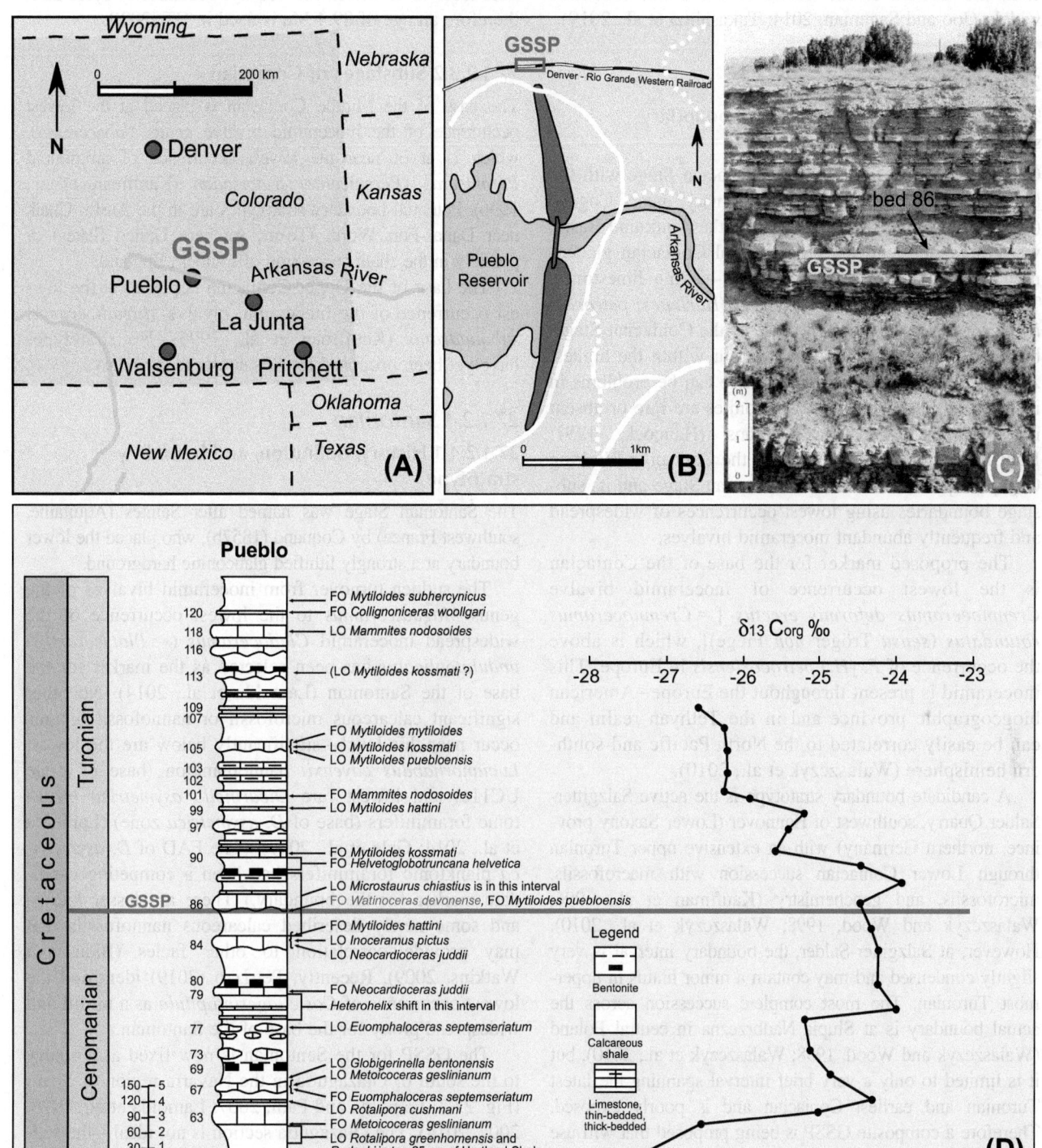

FIGURE 27.6   GSSP for base of the Turonian Stage near Pueblo, Colorado, United States. (A and B) Location maps. (C) Photograph of outcrop. (D) Stratigraphic log of the lower Bridge Creek Limestone Formation. The GSSP level coincides with the lowest occurrence of the ammonite *Watinoceras devonense*. Photograph provided by Jim Kennedy. *GSSP*, Global Boundary Stratotype Section and Point.

occurrence of the widespread species *I. perplexus*, which is close to a small positive carbon-isotope positive excursion, the Caburn Event (Jarvis et al., 2006) identified widely (Joo and Sageman, 2014; Takashima et al., 2019).

27.1.2.3 Coniacian

27.1.2.3.1 History, definition, and boundary stratotype

Coquand (1857a,b) defined the Coniacian Stage with the type locality at Richemont Seminary near Cognac (Charente province, northern part of the Aquitaine Basin, western France). In this region, basal Coniacian glauconitic sands overlie Turonian rudistid-bearing limestones. The entry of ammonoid *Forresteria (Harleites) petrocoriensis* was taken to mark the base of the Coniacian Stage, but this species is now taken to occur within the highest zone of the Turonian. However, there can be problems in identifying this species, and ammonites are rare or absent in important Coniacian sections (Hancock, 1991; Kauffman et al., 1996). Therefore the Coniacian Working Group proposes to define the Coniacian Stage and its substage boundaries using lowest occurrences of widespread and frequently abundant inoceramid bivalves.

The proposed marker for the base of the Coniacian is the lowest occurrence of inoceramid bivalve *Cremnoceramus deformis erectus* [= *Cremnoceramus rotundatus* (*sensu* Tröger *non* Fiege)], which is above the occurrence of *F. (H.) petrocoriensis* in Europe. This inoceramid is present throughout the Europe—American biogeographic province and in the Tethyan realm and can be easily correlated to the North Pacific and southern hemisphere (Walaszczyk et al., 2010).

A candidate boundary stratotype is the active Salzgitter-Salder Quarry, southwest of Hannover (Lower Saxony province, northern Germany) with an extensive upper Turonian through Lower Coniacian succession with macrofossils, microfossils, and geochemistry (Kauffman et al., 1996; Walaszczyk and Wood, 1998; Walaszczyk et al., 2010). However, at Salzgitter-Salder, the boundary interval is very slightly condensed and may contain a minor hiatus in uppermost Turonian. The most complete succession across the actual boundary is at Słupia Nadbrzeżna in central Poland (Walaszczyk and Wood, 1998; Walaszczyk et al., 2010), but it is limited to only a very brief interval spanning the latest Turonian and earliest Coniacian and is poorly exposed. Therefore a composite GSSP is being proposed that will use both of these reference sections.

The lowest occurrence of inoceramid bivalve *C. deformis erectus* correlates approximately to the base of the *Scaphites preventricosus* ammonite zone of the North American Western Interior basin (Cobban et al., 2006). Bentonites from this ammonite zone and the uppermost Turonian zone have yielded ^{40}Ar/^{39}Ar ages indicating a

boundary age of approximately 89.8 Ma (Obradovich, 1993; Siewert, 2011; Sageman et al., 2014). The boundary falls close to or within the 405-kyr eccentricity cycle 221, therefore an age of 89.4 Ma is used in GTS2020.

27.1.2.3.2 Substages of Coniacian

The base of the Middle Coniacian is placed at the lowest occurrence of the inoceramid bivalve genus *Volviceramus*, which is at or near the lowest occurrence of ammonoid *Peroniceras (Peroniceras) tridorsatum* (Kauffman et al., 1996). Potential boundary stratotypes are in the Austin Chalk near Dallas-Fort Worth (Texas, southern United States) or possibly in the chalk succession of southern England.

The base of the Upper Coniacian is placed at the lowest occurrence of the inoceramid bivalve *Magadiceramus subquadratus* (Kauffman et al., 1996). No stratotypes have yet been proposed for this substage boundary.

27.1.2.4 Santonian

27.1.2.4.1 History, definition, and boundary stratotype

The Santonian Stage was named after Saintes (Aquitaine, southwest France) by Coquand (1857b), who placed the lower boundary at a strongly lithified glauconitic hardground.

The sudden turnover from inoceramid bivalves of the genus *Magadiceramus* to the lowest occurrence of the widespread inoceramid *Cladoceramus* (= *Platyceramus*) *undulatoplicatus* has been selected as the marker for the base of the Santonian (Lamolda et al., 2014). No other significant calcareous microfossil or nannofossil datums occur near this level—significantly below are the lowest *Lucianorhabdus cayeuxii* nannoplankton (base of Zone UC11c) and lowest rare *Dicarinella asymetrica* planktonic foraminifers (base of *D. asymetrica* zone) (Lamolda et al., 2014; Gale et al., 2007). (The FAD of *D. asymetrica* planktonic foraminifers had been a competing candidate marker for the boundary.) There are lesser known and some newly described calcareous nannofossils that may provide correlation to other facies (Blair and Watkins, 2009). Recently, Petrizzo (2019) identified the lowest occurrence of *Costellagerina pilula* as a useful and widespread proxy for the base of the Santonian.

The GSSP for the Santonian is now fixed at a quarry to the south of Olazagutia in the Navarra region of Spain (Fig. 27.7; Lamolda and Paul, 2007; Lamolda et al., 1996, 2007, 2014). The Olazagutia section is not ideal—the sediment is strongly lithified and the biostratigraphic record may be incomplete. The Olazagutia section was however ratified as the GSSP in 2012 by the Subcommission on Cretaceous Stratigraphy.

The lowest occurrence of the inoceramid *C. undulatoplicatus* boundary marker is just below the base of the *Clioscaphites saxitonianus* ammonite zone of the North

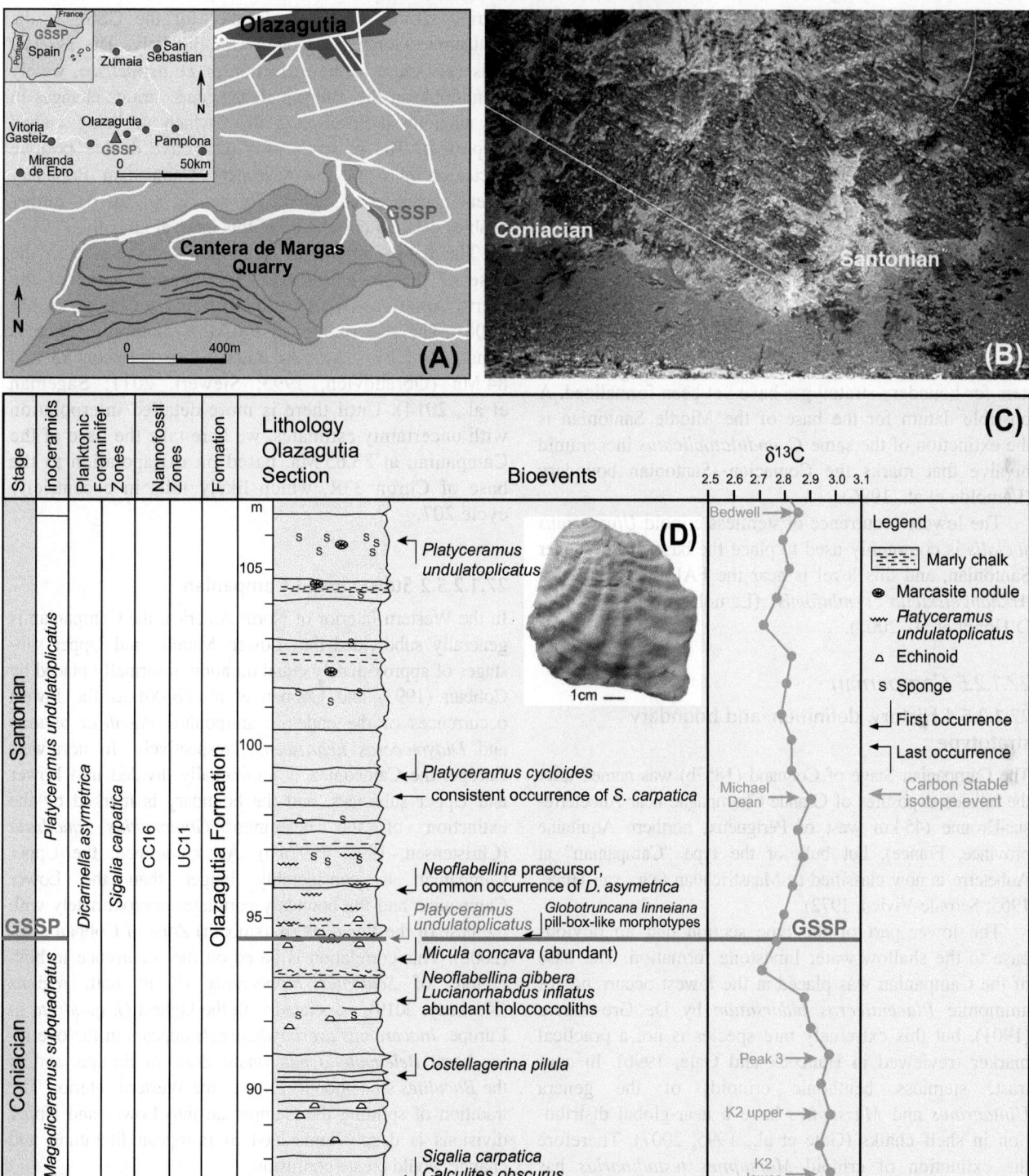

Base of the Santonian Stage of the Cretaceous System in the Olazagutia Section, Spain

FIGURE 27.7   GSSP for the base of the Santonian Stage, Olazagutia, Navarra, north Spain. (A) Map of region. (B) Photograph of GSSP section in quarry face. (C) Log of section, with GSSP marked, at level of lowest occurrence of the inoceramid bivalve *Platyceramus undulatoplicatus* (D). *GSSP*, Global Boundary Stratotype Section and Point.

American Western Interior (Walaszczyk and Cobban, 2007). A set of ^{40}Ar/^{39}Ar ages from bracketing bentonites indicate a boundary age near 86 Ma (Obradovich, 1993; Siewert, 2011; Sageman et al., 2014). A high-resolution δ^{13}C record from Seaford Head in Sussex, United Kingdom, provided an orbital tuning for the Santonian (Thibault et al., 2016) and permits detailed correlation to the standard Tethyan succession at Gubbio. The base of the stage is taken at 85.7 Ma, based upon radioisotopic dates from beds yielding *C. undulatoplicatus* in Texas and the Western Interior Basin of the United States (Appendix 2), approximately at the level of eccentricity cycle 221, equivalent to cycle Sa1 of Thibault et al. (2016).

27.1.2.4.2 Substages of Santonian

The traditional Santonian has three substages, but no markers for boundary stratotypes have yet been formalized. A possible datum for the base of the Middle Santonian is the extinction of the same *C. undulatoplicatus* inoceramid bivalve that marks the Coniacian–Santonian boundary (Lamolda et al., 1996).

The lowest occurrence of stemless crinoid *Uintacrinus socialis* is commonly used to place the base of the Upper Santonian, and this level is near the FAD of nannofossil *Arkhangelskiella cymbifomis* (Lamolda et al., 1996; D'Hondt et al., 2007).

27.1.2.5 Campanian

27.1.2.5.1 History, definition, and boundary stratotype

The Campanian Stage of Coquand (1857b) was named after the hillside exposures of Grande Champagne near Aubeterre-sur-Dronne (45 km west of Périgueux, northern Aquitaine province, France), but bulk of the type "Campanian" at Aubeterre is now classified as Maastrichtian (e.g., van Hinte, 1965; Séronie-Vivien, 1972).

The lower part of the type section had no obvious base to the shallow-water limestone formation. The base of the Campanian was placed at the lowest occurrence of ammonite *Placenticeras bidorsatum* by De Grossouvre (1901), but this extremely rare species is not a practical marker (reviewed in Hancock and Gale, 1996). In contrast, stemless benthonic crinoids of the genera *Uintacrinus* and *Marsupites* have a near-global distribution in shelf chalks (Gale et al., 1995, 2007). Therefore the extinction of crinoid *Marsupites testudinarius* has been an informal boundary marker for the base of the Campanian Stage (Hancock and Gale, 1996), although the occurrence of this taxon is restricted to certain paleoenvironments. The base of the traditional Campanian is probably within the lower portion of reversed-polarity Chron C33r. Therefore the Campanian Working Group is

presently considering using the beginning of Chron C33r as the primary boundary definition, thereby enabling global recognition in pelagic, continental, and other nonshallow-marine settings. A likely locality for the GSSP is the Bottacione Gorge at Gubbio, Umbria, Italy. The reversal falls very close to the extinction of *D. asymetrica*, widely identifiable in the Tethyan Realm, and various changes in the nannofossil *Broinsonia*, all of which are being studied at present. The presence of a distinctive double positive excursion in δ^{13}C, the Santonian–Campanian Boundary Event (Jarvis et al., 2006; Thibault et al., 2016), further enables detailed correlation of this interval globally.

The base-Campanian is generally correlated to the base of the *Scaphites leei III* ammonite zone of the North American Western Interior (e.g., Cobban et al., 2006). The age of the base of this ammonite zone is constrained by ^{40}Ar/^{39}Ar dates to be between 83 and 84 Ma (Obradovich, 1993; Siewert, 2011; Sageman et al., 2014). Until there is more detailed interpolation with uncertainty estimates, we here take the base of the Campanian at 83.65 Ma, based on extrapolation to the base of Chron 33R, which likely falls in eccentricity cycle 207.

27.1.2.5.2 Substages of Campanian

In the Western Interior of North America, the Campanian is generally subdivided into Lower, Middle, and Upper substages of approximately equal duration, informally placed by Cobban (1993) and Cobban et al. (2006) as the lowest occurrences of the endemic ammonites *Baculites obtusus* and *Didymoceras nebrascense*, respectively. In northwest Europe, the Campanian is traditionally divided into Lower and Upper substages, and the boundary is marked by the extinction of the belemnite *Gonioteuthis quadrata* (Christensen, 1990, 1997a,b). As so defined, the Upper Campanian is considerably longer than the Lower Campanian and the boundary correlates approximately with the base of the *Baculites* sp. (smooth) Zone of Cobban et al. (2006). This correlation is based on the occurrence in both regions of *Scaphites hippocrepis* III in both regions (Kennedy, 2019), associated with the highest *G. quadrata* in Europe. *Inoceramus azerbaydjanensis* appears in the overlying basal *Belemnitella mucronata* Zone in Europe, and in the *Baculites* sp. (smooth) Zone in the Western Interior. The tradition of splitting the Campanian into Lower and Upper divisions is deeply embedded in European literature, and change would create confusion.

Campanian correlation has been significantly improved by the discovery of a sharp, short negative excursion in δ^{13}C, the Late Campanian Event (Voigt et al., 2012), which lies beneath and in the lower part of the range of *Radotruncana calcarata*. This has been discovered as far afield as Tibet (Wendler et al., 2011).

27.1.2.6 Maastrichtian
27.1.2.6.1 History, definition, and boundary stratotype

The Maastrichtian Stage was introduced by Dumont (1849) for the "Calcaire de Maastricht" with a type locality at the town of Maastricht (southern Netherlands near border with Belgium). The stratotype was fixed by the Comité d'étude du Maastrichtian as the section of the Tuffeau de Maastrict exposed in the ENCI company quarry at St. Pietersberg on the outskirts of Maastricht, but this local coarse carbonate facies would correspond only to part of the upper Maastrichtian in current usage (reviewed in Rawson et al., 1978; Odin and Lamaurelle, 2001). A revised concept of Maastrichtian Stage was based on belemnites in the white chalk facies. Accordingly, the base of the stage was assigned to the lowest occurrence of belemnite *Belemnella lanceolata*, with a reference section in the chalk quarry at Kronsmoor [50 km northwest of Hamburg, north Germany (e.g., Birkelund et al., 1984; Schönfeld et al., 1996)]. The lowest occurrence of ammonoid *Hoploscaphites constrictus* above this level provided a secondary marker. Comparison of strontium-isotope stratigraphy and indirect correlations by ammonoids indicate that this level is approximately equivalent to the base of the *Baculites eliasi* ammonoid zone of the North American Western Interior (Landman and Waage, 1993; McArthur et al., 1992).

However, belemnites, including *B. lanceolata* are largely restricted to Boreal chalks and are absent in the Tethyan faunal realm, where the ammonoid *Pachydiscus neubergicus* has a much wider geographical distribution (reviewed in Hancock, 1991). Therefore the Maastrichtian Working Group recommended the base of the Maastrichtian to be taken at the lowest occurrence of ammonoid *P. neubergicus* (Odin et al., 1996). In retrospect, this was a poor decision, because the occurrence of *P. neubergicus* is strongly diachronous across its geographic range and uncommon in most localities.

The ratified Maastrichtian GSSP boundary is in an abandoned quarry near the village of Tercis les Bains in southwest France, at 90 cm beneath a coincident lowest occurrence of *P. neubergicus* and *H. constrictus* ammonoids (Fig. 27.8; Odin, 1996, 2001; Odin and Lamaurelle, 2001). The GSSP level was selected, bizzarely as the arithmetic mean of 12 biohorizons with potential correlation potential, including ammonoids, dinoflagellate cysts, planktonic and benthic foraminifers, inoceramid bivalves, and calcareous nannofossils (Odin and Lamaurelle, 2001). The history, stratigraphy, paleontology, and intercontinental correlations are compiled in a large special (and outrageously expensive) volume (Odin and Lamaurelle, 2001).

In many ways the GSSP was poorly chosen; the ammonite marker is inappropriate, the section lacks key planktonic foraminifers above the occurrence of *R.*

calcarata, and remnant paleomagnetism is weak or absent. The abandoned quarry section is growing over, and a tree now covers the GSSP marker. However, the section at Tercis provides a detailed carbon-isotope record, which permits correlation with both Boreal chalks and the deep-water Tethan succession at Gubbio (Voigt et al., 2012). Some distance above a well-marked late Campanian negative event, the carbon-isotope values decline by about 0.3 ppt, the Campanian–Maastrichtian Boundary Event, then rise slightly to a minor peak. The fine details of this curve can be matched precisely in the Boreal chalk successions of northern Germany, Denmark and the United Kingdom, and the Tethyan section at Gubbio, Italy, which allows the level of the Tercis GSSP to be equated with the base of the *Belemnella obtusa* Zone in the Boreal chalks.

This correlation is also aided by the use of inoceramid bivalves, because the boundary marker falls just beneath the first occurrence of *Endocosta typica* at Tercis (Walaszczyk et al., 2002). The base of the *E. typica* zone is equated with the base of the *Baculites baculus* Zone in the North American ammonite succession (Cobban et al., 2006). The age for the base of this ammonite zone is 72.1 Ma according to the spline-fit of dates from bracketing bentonites.

This agreement between the independent sets of correlations implies that the base-Maastrichtian is essentially equivalent to the base of the *B. obtusa* belemnite zone, the base of *E. typica* inoceramid zone, the base of *B. baculus* ammonite zone of North American, and approximately to the base of nannofossil zone UC17. The magnetostratigraphic placement is Chron C32n.2n.88, and the age for all of these base-Maastrichtian levels is approximately 72.2 Ma.

27.1.2.6.2 Upper Maastrichtian substage

The Maastrichtian is commonly divided into two substages in the Boreal chalks, based on belemnite occurrences in northern Europe (Germany, Poland, Denmark, the United Kingdom). The Lower Maastrichtian includes the zones of *B. obtusa*, *Belemnella sumensis*, *Belemnella cimbrica*, and *Belemnella fastigata*, the Upper Maastrichtian includes the zones of *Belemnitella junior* and *Belemnitella casimirovensis* (Christensen, 1990, 1997a,b). It is not clear how this zonation relates to the occurrences of ammonites in other successions. In the Western Interior Basin, the lowest occurrence of ammonoid *Hoploscaphites birkelundi* (formerly *H.* aff. *nicolleti*) is an informal marker for the base of the Upper Maastrichtian (Landman and Waage, 1993; Cobban, 1993; Cobban et al., 2006).

As indicated by this review, the main chronostratigraphic markers for the boundaries of Cretaceous stages, as defined by ratified GSSPs or as the informal working

Base of the Maastrichtian Stage of the Cretaceous System at Tercis les Bains, Landes, France

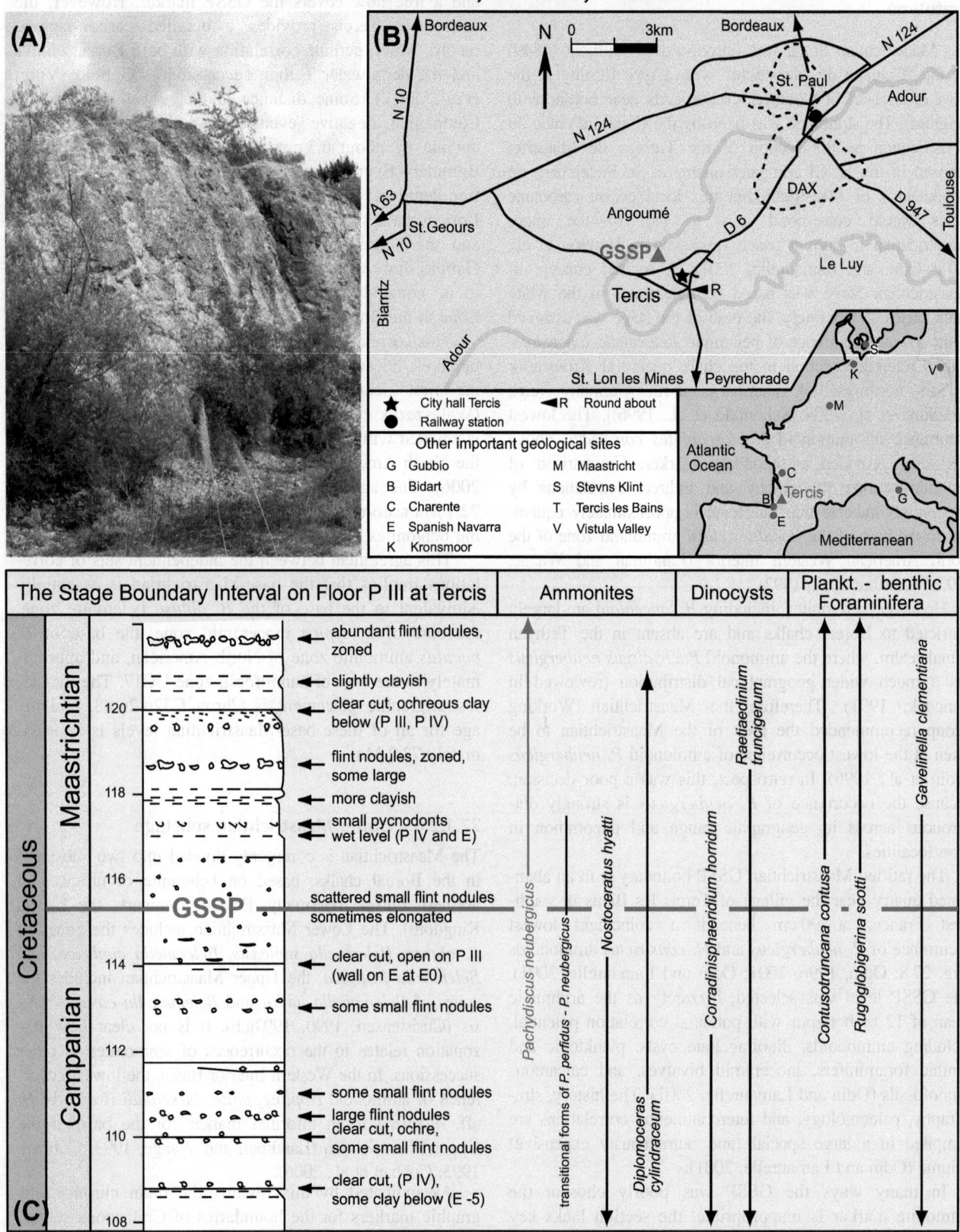

FIGURE 27.8 GSSP for base of the Maastrichtian Stage at Tercis, Landes, southwest France (A and B). The GSSP is situated 90 cm below the lowest occurrence of the ammonite *Pachydiscus neubergicus* (C). Photograph provided by Andy Gale. *GSSP*, Global Boundary Stratotype Section and Point.

TABLE 27.1 Main biostratigraphic or other chronostratigraphic markers for ratified Global Boundary Stratotype Section and Points (GSSPs) (R) and potential GSSPs for the Callovian through Maastrichtian time scale segment.

base of Stage	Ratified GSSP (R) or potential GSSP main marker(s)
Danian (Paleogene base)	Iridium anomaly **(R)**, within Chron C29r
Maastrichtian	near FAD *P. neubergicus* **(R)**, projected to about 88% in subchron C32n.2n
Campanian	base Chron C33r; base of *S. leei* III Zone; LAD *M. testidunarius*
Santonian	FAD *P. undulatoplicatus* **(R)**
Coniacian	FAD *C. deformis erectus*
Turonian	FAD *W. devonense* **(R)**
Cenomanian	FAD *T. globotruncanoides*, just below *M. mantelli* Zone **(R)**
Albian	FAD *M. renilaevis*, base Kilian OAE **(R)**
Aptian	base of Chron M0r, ~FAD *D. oglanlensis* Zone
Barremian	FAD *T. hugii*, upper part of Chron M5n, 0.7 Myr above Faraoni OAE
Hauterivian	FAD *A. radiatus* Zone **(R)**
Valanginian	FAD *C. darderi*, base of calpionellid Zone E, lower part of Chron M14r
Berriasian (Cretaceous base)	base of *C. alpina*; mid Chron 19n.2n.2
Tithonian	near bases of *H. hybonotum* Zone and of Chron M22An
Kimmeridgian	Base of *P. flodigarriensis* Zone, base of Chron M26r
Oxfordian	FAD of *B. thuouxenis*, within lower part of Chron M36Br
Callovian	FAD of *Kepplerites*, within Chron M39n.3n

This is the interval used in the spline fits of Figs. 27.11 and 27.12. *OAE*, Oceanic Anoxic Event. Full name and explanation of each chronostratigraphic marker is in the appropriate text for that stage.

definitions used in GTS2020 are a combination of ammonites, inoceramids, planktonic foraminifers and calpionellids, geomagnetic polarity chrons, and other events (Table 27.1).

27.2 Cretaceous stratigraphy

As in the Jurassic, the ammonite and other macrofossil zones of European and North American basins had provided the traditional standards for subdividing Cretaceous stages. With the advent of petroleum and scientific drilling, microfossil datums from pelagic successions are commonly used for global correlations, especially when augmented by magnetostratigraphy. Cycle stratigraphy has now enabled scaling of many of these zonations, and detailed carbon-isotope curves are becoming a major method for interregional correlation.

27.2.1 Marine biostratigraphy

Ammonites dominate the historical zonation of the Lower Cretaceous of Europe, but ammonites are of restricted occurrence in the European Upper Cretaceous Chalk facies, and

belemnites, inoceramid bivalves and pelagic or benthic crinoids (e.g., *Marsupites*) provide important information. Buchiid bivalves, belemnites, and brachiopods are used for correlation within the Lower Cretaceous within the Boreal Realm. Important microfossil biostratigraphic zonations include planktonic foraminifers, calcareous nannoplankton, dinoflagellate cysts, and calpionellids.

27.2.1.1 Ammonites

Ammonites, despite various limitations, provide the primary reference scale for the majority of the marine Cretaceous sequences in all paleogeographic regions. The Lower Cretaceous standard zonation is the Tethyan ammonite succession of the western Mediterranean region with its revised and enhanced zonal schemes developed by the Kilian working group (e.g., Reboulet et al., 2006, 2009, 2011, 2018). The Upper Cretaceous succession of the Western Interior Basin of the United States, locally yields abundant ammonites on which a standard zonation (Fig. 27.9) is based, and abundant radioisotopically dated volcanic ash beds in the region form the basis for the Late Cretaceous age model (Cobban et al., 2006; Sageman et al., 2014). However, many of the Western Interior

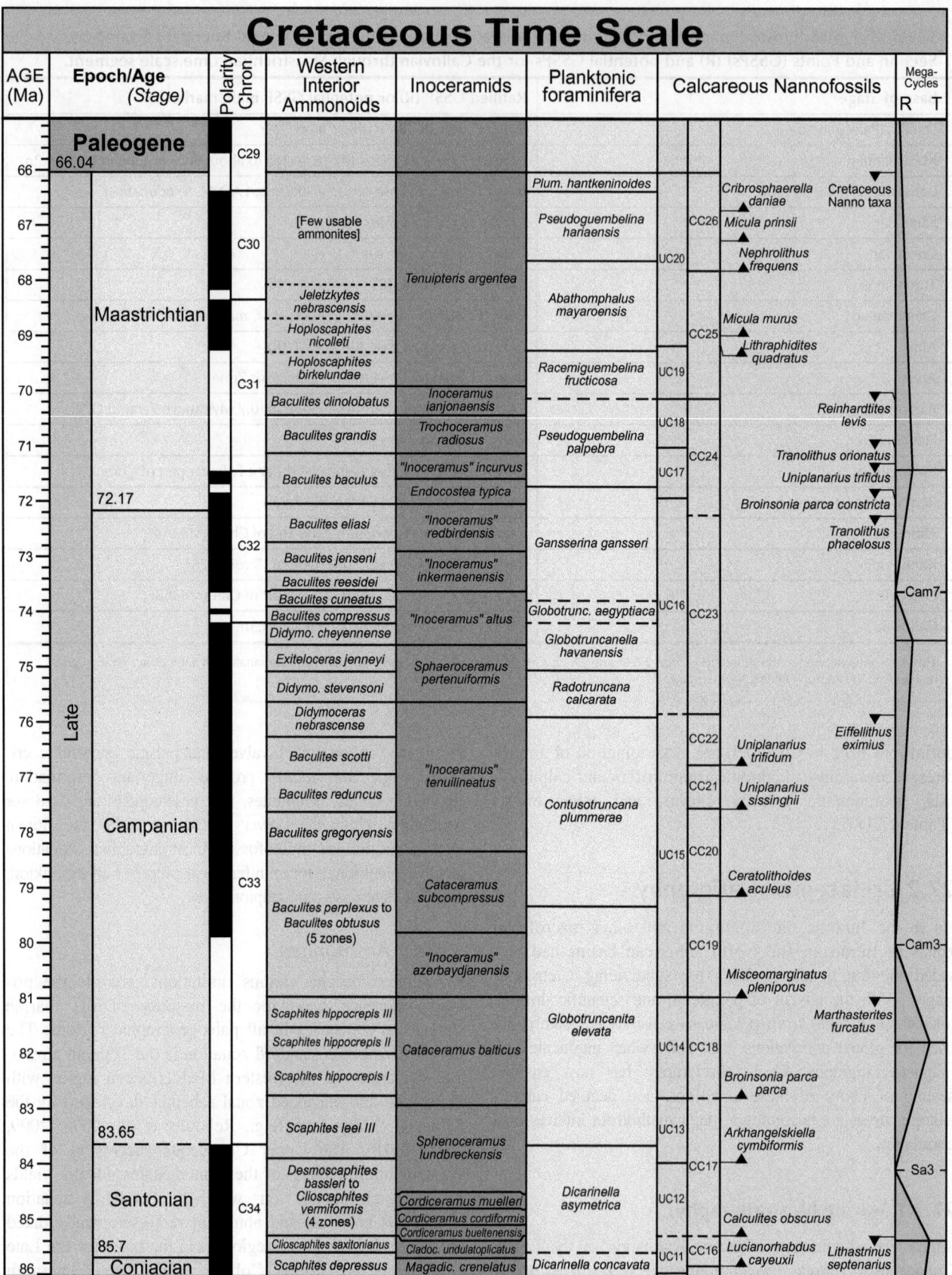

FIGURE 27.9 Cretaceous integrated time scale. Summary of numerical ages of epoch/series and age/stage boundaries of the Cretaceous with selected marine biostratigraphic zonations and principal trends in sea level. ("Age" is the term for the time equivalent of the rock-record "stage.").

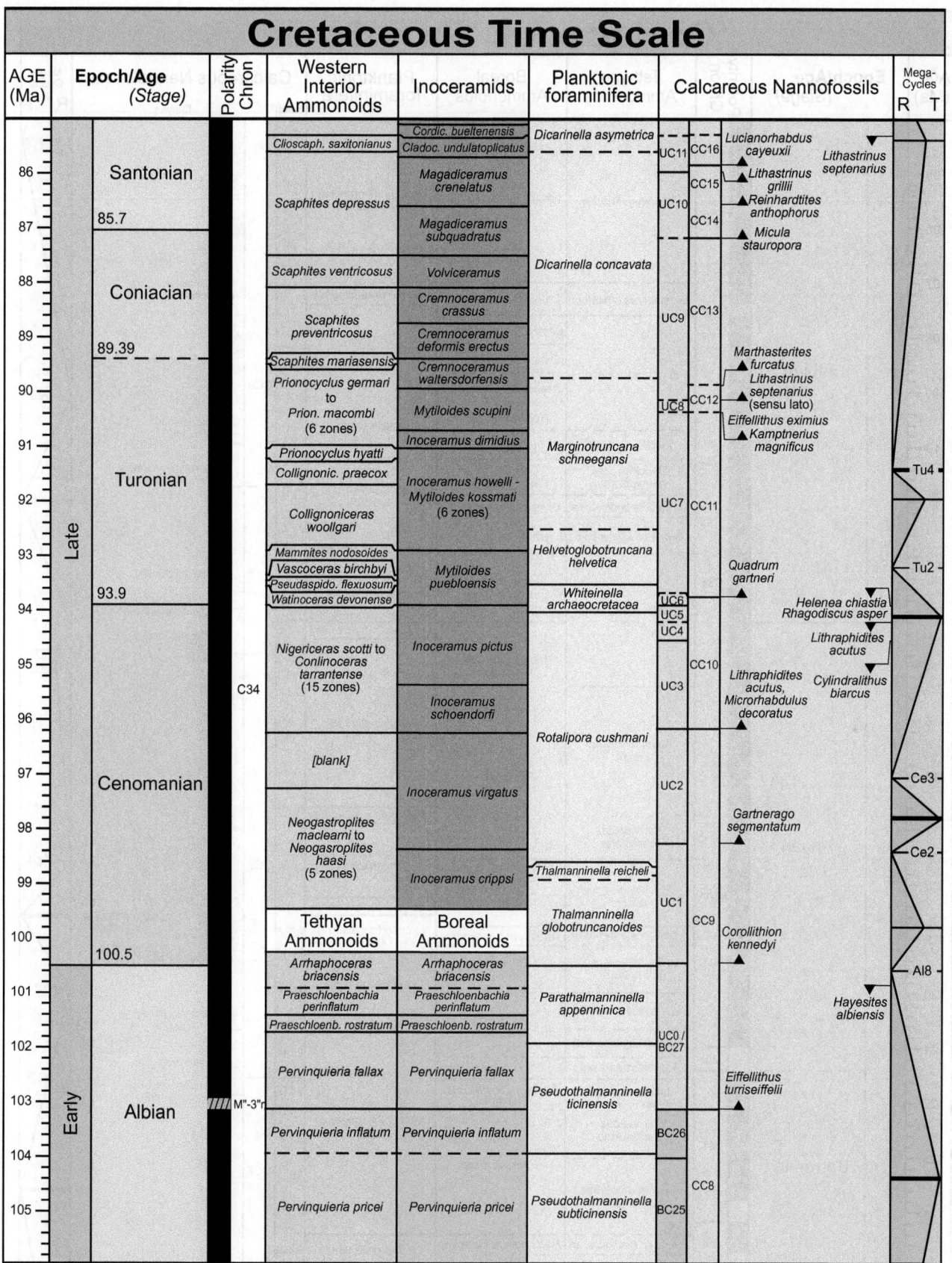

FIGURE 27.9   (Continued)

Selected marine macrofossil biostratigraphy columns for Early Cretaceous are ammonoid zones for the Tethyan realm (Sub-Mediterranean province; Reboulet et al., 2018) and Sub-Boreal realm (Mutterlose et al., 2014). Marine macrofossil zones for Late Cretaceous are ammonoids of the Western Interior

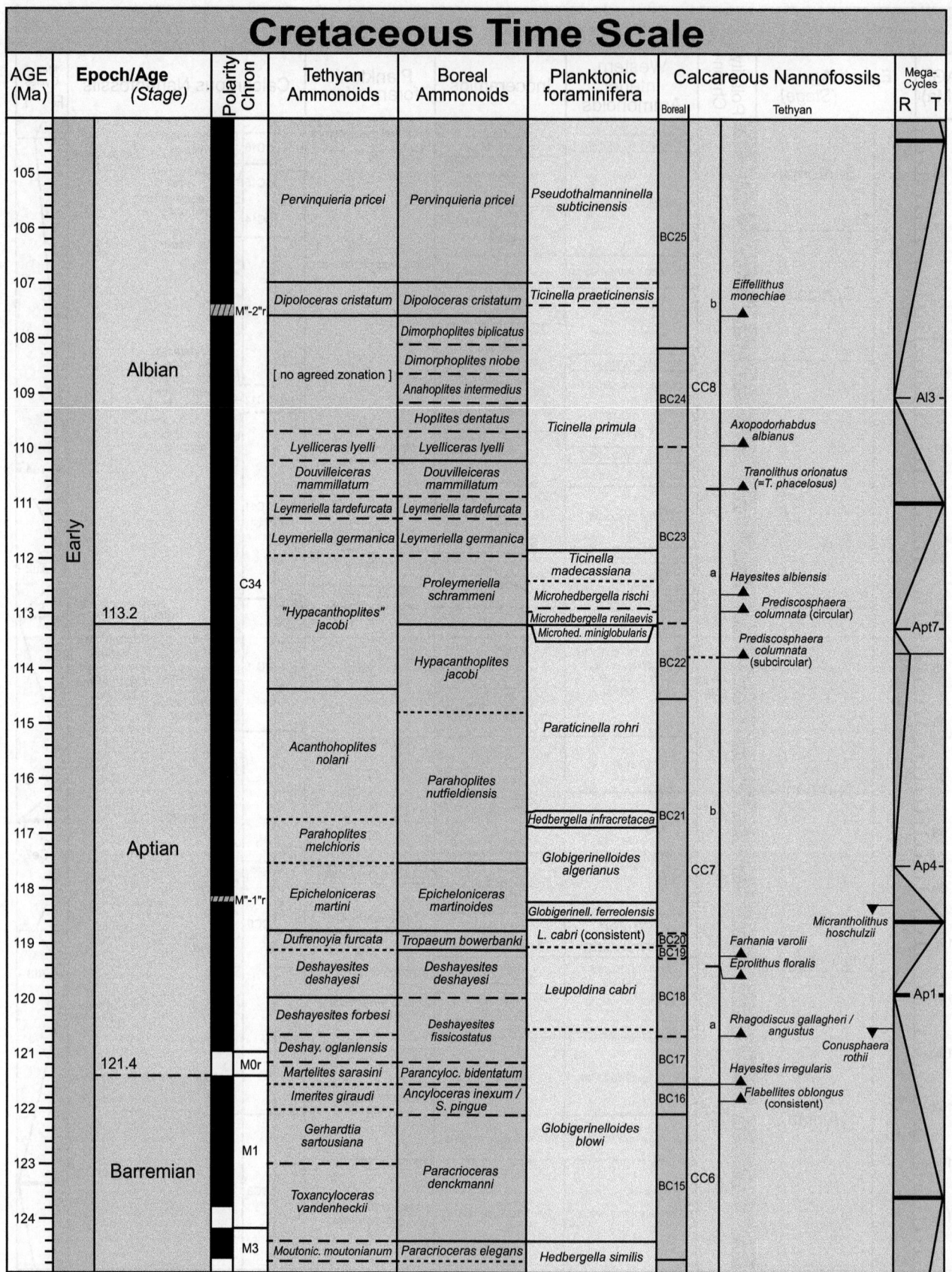

FIGURE 27.9 (Continued)

of United States (a full list of zone names is in Table 27.3) and inoceramids of North America and Europe (Cobban et al., 2006; with partial modification by Walaszczyk, pers. comm., 2019). Selected microfossil zones are planktonic foraminifers (composite from Coccioni and Premoli Silva, 2015; Petrizzo et al., 2012;

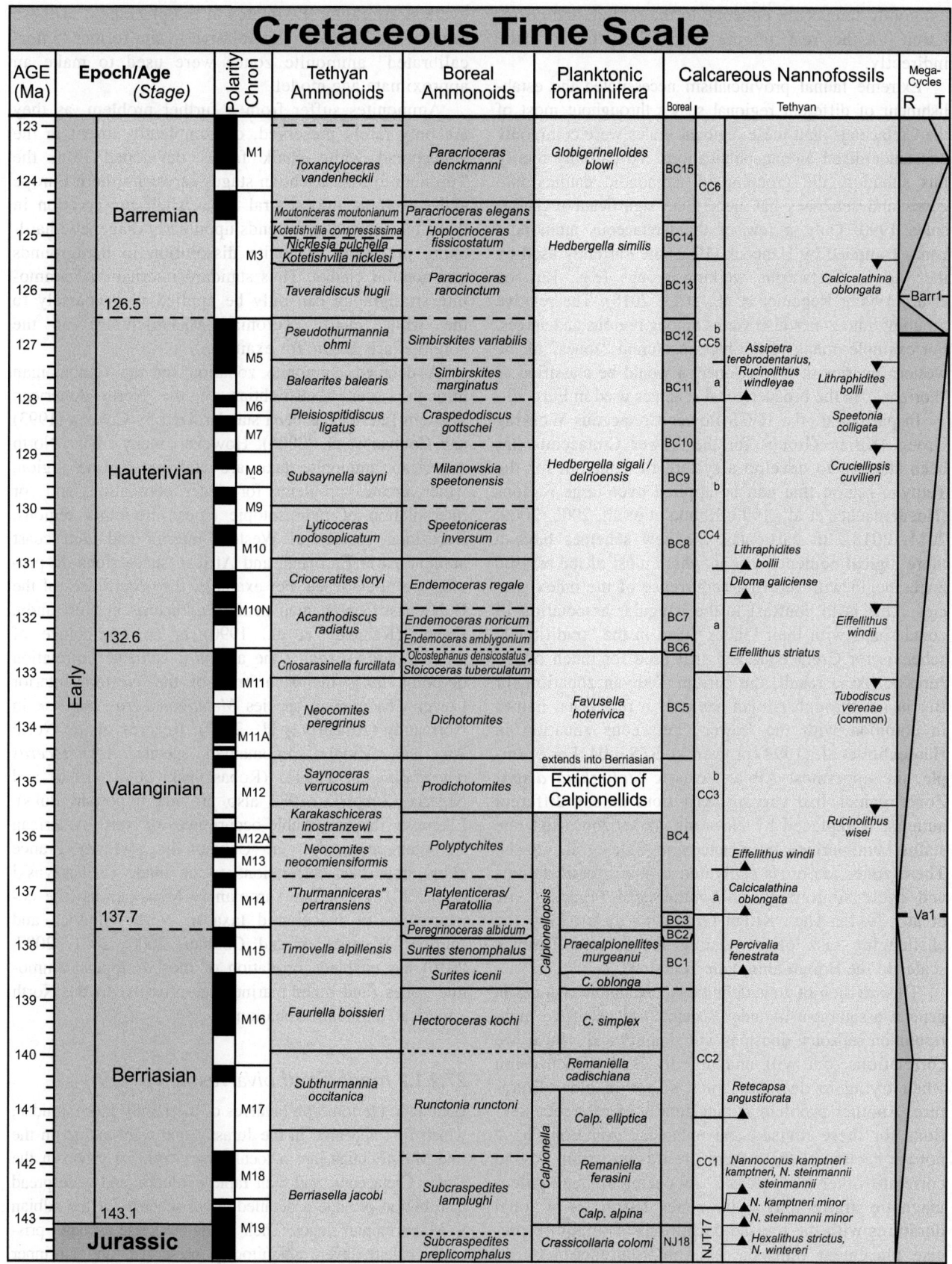

Cretaceous Time Scale

FIGURE 27.9 (Continued)

and other sources, including Huber and Petrizzo, pers. comm., 2019) and early Cretaceous calpionellid zones (Remane, 1998). Early Cretaceous calcareous nannofossil zones are Boreal (BC; Bown et al., 1998) and Tethyan (CC with selected zone/subzone markers; with calibrations compiled from Bergen, 1994; Bralower et al., 1995; and other sources, including Watkins, pers. comm., 2019).

ammonite faunas are endemic to the region, and correlation to the rest of the world has to be made indirectly.

Extreme faunal provincialism necessitated the establishment of different regional scales throughout most of the Cretaceous, and these regional scales were commonly nonstandardized among publications. To partially rectify this situation, the grouping of ammonoid datums into zones and subzones has undergone significant revisions since 1990. Only a few of the Cretaceous ammonoid zones compiled by Hancock (1991) are currently used by the various Cretaceous working groups (e.g., Rawson et al., 1996b; Reboulet et al., 2011, 2018). The relative grouping into zones also varies among regions and stages. For example, many of the high-resolution "zones" of the western interior of North America would be classified as "horizons" in the broader zonal schemes used in Europe.

In particular, the IUGS Lower Cretaceous Working Group (Kilian Group) for the Lower Cretaceous has been striving to develop a systematic zonation for the Tethyan region that can be applied over large regions (Hoedemaeker et al., 1993; Reboulet et al., 2006, 2009, 2011, 2018). In particular, the new schemes have a more logical nomenclature in which most of the revised zones begin with the first occurrence of the index species. This is in contrast to the irregular association of zonal spans with their "index" taxa in the "traditional" schemes for Cretaceous and still used for much of the Jurassic. As a result, the current Tethyan zonation for Berriasian through Albian has only a few zonal names in common with the Lower Cretaceous zonation in Hardenbol et al. (1998) or even in GTS2004. For example, the uppermost Albian "classic *Stoliczkaia dispar* Zone," which had varying definitions among different authors, is replaced by elevating its subzones to zone status (summarized by Kennedy in Gale et al., 2011). These zones are cross-calibrated to microfossil datums and cycle stratigraphy in southeastern France (Gale et al., 2011). The "Kilian Group" is undertaking the challenging task of correlating the revised Tethyan scales to the Boreal ammonite zonations.

This revision of zone definitions, taxonomic changes in general assignment of index species, and extensive high-resolution subzonal divisions will simplify and standardize correlations, but will undoubtedly lead to confusion when trying to decipher zonal scales in older literature. Another problem is that details of reference sections for these revised and enhanced zonations have not yet been published; therefore it is uncertain how to correlate other microfossil, geochemical, and paleomagnetic stratigraphy to the zones. Estimates of zonal durations with the revised definitions also require precise placement onto the existing reference sections that had been used for strontium-isotope trends and cycle stratigraphy. Estimates of the placement of these renamed/redefined zones relative to the former "intercalibrated" ammonite zones were used to make an approximate age model.

Ammonites suffer from a further problem, as they are only rarely preserved, or completely absent in the widespread white chalk facies developed from the Turonian to Maastrichtian stages across northern Europe and eastwards into central Asia. Their preservation in these facies usually depends upon early diagenetic hardening preceding aragonite dissolution in hardgrounds and nodular chalks. Thus standard international ammonite stratigraphy can only be applied very coarsely to the white chalk (Turonian—Maastrichtian) of the Anglo—Paris Basin, for example.

A detailed ammonite zonation for the Cenomanian through lower Maastrichtian of the North American Western Interior has been standardized by Cobban (1993) and Cobban et al. (2006). However, many of the North American ammonite taxa are endemic to the region, which creates problems for wider correlation, and for interpolation of radioisotopic dates into other regions. Occasional records of Western Interior and Gulf coast ammonites in European and African successions do provide useful evidence. For example, the occurrences of the Gulf coast Cenomanian genus *Budaiceras* in Normandy, France (Kennedy et al., 1990) in an assemblage of *Mantelliceras dixoni* Zone age is a valuable correlation tie-point, as is the occurrence of the Western Interior Lower Cenomanian species *Metengonoceras teigense* in Normandy (Amédro et al., 2002). Records of the North American Middle Cenomanian species *Acanthoceras amphibolum* in Tunisia (Robaszynski et al., 1990) and Nigeria (Zaborski, 1985 also provide important links). Likewise, the remarkable occurrences of North American Turonian ammonites at Uchaux in southern France allow important correlations to be made (Robaszynski et al., 2014). For the Turonian to Maastrichtian the co-ocurrences of inoceramid taxa in North America and Europe (Walaszczyk and Cobban, 2000, 2001, 2007, 2016) has enabled correlation of most European ammonite zones (and other marine macrofossils) to the North American ammonite zonation.

27.2.1.2 Inoceramid bivalves

Benthonic pteriomorph bivalves of the family Inoceramidae, which first appeared in the Jurassic and went extinct in the mid-Maastrichtian, are a locally rock-forming group in the Upper Cretaceous, and their rapid evolution and widespread distribution provide a detailed zonal scheme for the Albian to Maastrichtian stages. Their shells comprise an outer prismatic calcite layer, ubiquitously preserved, and an inner nacreous one composed of aragonite, which is usually not normally preserved in sandstones and limestones. They thus partially compensate for the absence of ammonites in sediments such as chalk, in which dissolution on or in the sea-floor has removed taxa which possess aragonitic shells (see above). Inoceramids occupied a considerable diversity of habitats, from deep-water dysaerobic muds, chalks, to shallow marine sandstones, but were largely absent from carbonate platforms. The exceptionally widespread occurrence of species, often effectively global, and relatively low endemism (except in New Zealand) makes them excellent fossils for interregional correlation. Inoceramid stratigraphy had advanced considerably over the last decade, largely thanks to taxonomic revision and increased collecting. The Albian genus *Actinoceramus* provides high-resolution global correlation across the Middle−Upper Albian transition (Crampton and Gale, 2005, 2009). Important additional references are Walaszczyk and Cobban (2000, 2001, 2007, 2016), and Walaszczyk et al. (2001, 2002, 2010, 2016). A standard zonation is presented here based on this work.

27.2.1.2.1 Planktonic microcrinoids

Cretaceous microcrinoids belonging to the Roveacrinida were tiny, pelagic organisms that were common from the Albian to Maastrichtian in open marine sediments, locally occurring in rock-forming abundance. Although species have been used locally to characterize intervals, it has recently become apparent that they are far more diverse and widely distributed than previously thought (Gale, 2016, 2017, 2019a,b). Their rapid evolution and widespread distribution offer considerable potential for interregional correlation, and a series of zones have been proposed for the Albian to Campanian interval (Gale, 2017, 2019a; see also Ch. 3K: Cretaceous microcrinoids (this book)).

27.2.1.2.2 Other marine macrofauna

Workers on chalk facies successions have also employed zonations based upon diverse calcitic fossils, including benthonic crinoids, echinoids, and brachiopods. Most of these schemes have only local (basin-scale) application, but Campanian and Maastrichtian microbrachiopod zones can be applied for high-resolution correlation across northern Europe (Surlyk, 1984). The stemless crinoids of the genera *Uintacrinus* and *Marsupites* have short ranges and global distribution and are very useful fossils in interregional correlation (Gale et al., 2007).

The correlation of shallow-water carbonate platforms is achieved using larger benthonic foraminifers, calcareous algae, and rudist bivalves. The latter group is highly diverse and of considerable value in correlation (Masse and Philip, 1998; Steuber et al., 2016). These groups were strongly affected by OAEs, such that in the latest Cenomanian.

Belemnites have been used quite widely for stratigraphical correlation in the Cretaceous. The belemnite assemblages of the Kimmeridgian to Barremian interval show a distinctive provincialism. The Tethys was characterized by representatives of the Duvaliidae and Belemnopseidae, the Boreal Realm was dominated by Cylindroteuthididae (e.g., Combemorel and Christensen, 1998; Mutterlose et al., 1996; Mutterlose and Böckel, 1998; Dzyuba et al., 2013). This paleobiogeographic situation changed entirely in the Aptian, when the boreal Cylindroteuthididae and the Tethyan Duvaliidae went extinct. They were globally replaced by *Neohibolites*, which experienced a major radiation.

For the Tethys no precise zonation schemes are available, the stratigraphic resolution of specific taxa is on the scale of a substage. Detailed belemnite zonation schemes exist for the Boreal−Atlantic and Boreal−Arctic Provinces of the Boreal Realm. For the Berriasian−Hauterivian succession of the Boreal−Atlantic Province (northwest Europe) six belemnite zones, based on different species of the genera *Acroteuthis* and *Hibolithes*, have been recognized. The Barremian can be subdivided into five belemnite zones, defined by species of a biometrically well-defined evolutionary lineage. The genera *Oxyteuthis* and *Aulacoteuthis* allow thereby a detailed biozonation of the Barremian, which otherwise is difficult to subdivide. The Berriasian−Hauterivian assemblages of the Boreal−Arctic province are characterized by taxa attributed to *Cylindroteuthis*, *Arctoteuthis* and *Simobelus*. The Aptian−Albian of the Boreal Realm can successfully be subdivided into nine belemnite zones using species of *Neohibolites*.

Belemnites of the family Belemnitellidae occur from the Cenomanian to Maastrichtian stages and are used extensively used in Boreal Upper Cretaceous chalk successions of northern Europe, and some species also occur less frequently in North America. The family displayed considerable provinciality, and Central European and Central Russian subprovinces are recognized, the latter extending far eastwards into central Asia (Christensen, 1990, 1996, 1997a,b). A few species show short-term extensions into northern Tethys. Belemnites are particularly important in the biostratigraphy of the Boreal chalks of Campanian and Maastrichtian range, with the genera *Gonioteuthis*, *Belemnitella* and *Belemnella* displaying evolutionary changes which permit development of a refined zonation (Combemorel and Christensen, 1998).

27.2.1.2.3 Planktonic foraminifera

Planktonic foraminifera originated in late Early Jurassic (e.g., Hart, 1980; Caron, 1983; Gradstein et al., 2017) and only underwent radiation and geographic spreading from mid-Cretaceous onward. In general, planktonic foraminifera from Berriasian to Barremian have a scattered geographic and stratigraphic record. A progressive increase in abundance and diversification of genera and species is observed from the mid-Barremian (Aguado et al., 2014) to the Aptian (Premoli Silva and Sliter, 1999). The major turnover observed across the Aptian—Albian boundary interval represents the most dramatic event in the Cretaceous evolutionary history of planktonic foraminifera after the mass extinction at the Cretaceous/Paleogene (K/Pg) Boundary (e.g., Leckie et al., 2002; Huber and Leckie, 2011; Petrizzo et al., 2012, 2013; Kennedy et al., 2014). During the early to middle Albian planktonic foraminifera diversified rapidly with the continuous increase in morphological complexity and the appearance of newly evolving lineages all characterized by novel morphological and wall texture features.

The Early Cretaceous biozonal scheme currently adopted is based on the work by Moullade (1966) implemented and/or modified according to subsequent biostratigraphical studies or syntheses by Longoria (1977), Sigal (1977), Salaj (1984), Caron (1985), Gorbatchik (1986), Banner et al. (1993), Coccioni and Premoli Silva (1994), Robaszynski and Caron (1995), BouDagher-Fadel et al. (1997), Moullade et al. (2005), and Premoli Silva et al. (2018).

The subsequent pattern of evolutionary changes in the planktonic foraminifera from the Cenomanian to the Turonian corresponds to increased speciation and enlargement in test size of trochospiral keeled and unkeeled taxa and mirror the overall trend of rising sea level and global warming and increase of the density gradient within the surface water. After the maximum Late Cretaceous global Warmth registered in the early Turonian, and a relatively stasis from the late Turonian to the early—middle Coniacian, planktonic foraminifera underwent a major compositional changes in the late Coniacian—Santonian marked by high rates of species diversification and the appearance of newly evolved keeled, biserial, and multiserial taxa (e.g., Wonders, 1980; Caron and Homewood, 1983; Hart, 1999; Premoli Silva and Sliter, 1999). The radiation in the late Coniacian—Santonian time interval is followed by extinctions of some keeled taxa in the latest Santonian—earliest Campanian. In general, the Coniacian—Santonian time interval represents the transition from the mid-Cretaceous extreme greenhouse to more temperate climatic conditions in the Campanian—Maastrichtian that determined the onset of climatic bioprovinces. As a consequence, distinct biozonation schemes for the Tethyan, Boreal, Austral, and Transitional Provinces have been developed (e.g., Caron, 1985; Nederbragt, 1990; Huber, 1992; Premoli Silva and Sliter, 1999; Robaszynski and Caron, 1995; Li et al., 1999; Petrizzo, 2003; Campbell et al., 2004).

The Late Cretaceous biozonal scheme currently adopted is based on Caron (1966), Pessagno (1967), Robaszynski et al. (1984, 1990, 2000), Robaszynski and Caron (1979), Masters (1977), Robaszynski and Caron (1995), Premoli

Silva and Sliter (1995) and implemented and/or modified according to studies by Tur et al. (2001), Petrizzo (2000, 2001, 2003, 2019), Bellier and Moullade (2002), Lamolda et al. (2007), Petrizzo and Huber (2006), González-Donoso et al. (2007), Huber et al. (2008), Gale et al. (2011), Petrizzo et al. (2011, 2015, 2017), Pérez-Rodríguez et al. (2012), Elamri and Zaghbib-Turki (2014), Coccioni and Premoli Silva (2015), Haynes et al. (2015), Huber et al. (2017).

The taxonomy of Early and Late Cretaceous planktonic foraminifera is currently under revision using a more systematic and evolutionary framework by the Mesozoic Planktonic Foraminiferal Working Group that has produced a taxonomic database (pforams@mikrotax) available online at http://www.mikrotax.org (see Huber et al., 2016 for further details).

27.2.1.2.4 Benthonic foraminifera and calpionellids

A detailed suite of correlations of smaller and larger benthic foraminifera datums to ammonoid zones is partially established (e.g., Magniez-Jannin, 1995; Arnaud-Vanneau and Bilotte, 1998), although these zonations are mainly applicable to European basins.

Calpionellids are enigmatic pelagic microfossils with distinctive vase-shaped tests in thin-section. Calpionellids appeared in the Tithonian and vanished in the latest Valanginian or earliest Hauterivian (Remane, 1985), and their abundance in carbonate-rich shelf to basinal settings within the Tethyan realm enables biostratigraphic correlation prior to the increase in the diversity of planktonic foraminifera. Six standard zones (Allemann et al., 1971) with finer subdivisions (e.g., Remane et al., 1986) provide the basic framework for interregional correlation (summarized by Remane, 1998). There are variations upon this basic framework (e.g., Pruner et al., 2010), but a nomenclature using lettered-zones is a common system.

27.2.1.2.5 Calcareous nannofossils

The Cretaceous Period was named for the immense chalk formations that blanket much of northwestern Europe, and the main components of this chalk are calcareous nannofossils. Following their rapid surge in abundance at the end of the Jurassic (e.g., Caselleto et al., 2010), calcareous nannofossils remained ubiquitous throughout the Cretaceous and Cenozoic in all oceanic settings and sediments above the carbonate dissolution depth.

Calibration of major calcareous nannofossil datums to ammonoid zones or magnetic polarity zones is established for several intervals in the Tethyan and Boreal Realms (e.g., Lower Cretaceous of Tethyan realm by Bergen, 1994; Bralower et al., 1995; Erba et al., 1999; Channell et al., 2000, 2010); Lower Cretaceous of Subboreal—Boreal realm by Bown et al., 1998; Upper Cretaceous by Burnett et al., 1998;

von Salis, 1998; Huber et al., 2008; Corbett et al., 2014). The calibrations for the zonations are a synthesis of selected zonal scales and markers.

27.2.1.2.6 Organic and siliceous microfossils

Organic-walled cysts of dinoflagellates have been correlated directly to ammonoid zones in the Tethyan realm for Berriasian through Turonian, and in the Boreal realm for Berriasian through Aptian (compiled by Monteil and Foucher, 1998). In some intervals, these widespread dinoflagellate cysts have aided to resolve uncertainties in interregional correlations [e.g., earliest Cretaceous between Sub-Boreal and Boreal realms by Harding et al., 2011; Barremian−Aptian between Tethyan and Austral realms by Oosting et al., 2006).

Siliceous radiolarians (pelagic sediments), charophytes (brackish-water algae tests), and calcareous algae have a relatively lower resolution set of datums and zones compared to other Cretaceous microfossil groups (e.g., respective syntheses by De Wever, 1998; Riveline, 1998; Masse, 1998). Siliceous diatoms evolved in the Jurassic but did not undergo a major evolutionary radiation until the mid-Cretaceous, especially after the Cenomanian−Turonian boundary (e.g., Round et al., 1990; Sinninghe Damsté et al., 2004).

27.2.2 Terrestrial biostratigraphy

It is beyond the scope of this chapter to make a thorough review of terrestrial biostratigraphy, and only a few highlights will be mentioned. Authoritative books on this topic include Lucas et al. (1998), Woodburne (2004), and Kemp (2005).

Dinosaurs, the most renowned group of Cretaceous vertebrates, provide only a broad biostratigraphy (Lucas, 1997). Early Cretaceous sauropods were smaller, but ornithopods (such as *Iguanodon*) were larger than their Jurassic cousins. Stegosaurs, iguanodontids, hypsilophodontid ornithopods and sauropods (except in South America) were nearly extinct by the end of the Early Cretaceous. The rapid diversification of angiosperms (flowering plants) displaced gymnosperms in the mid-Cretaceous and was probably a major factor in the evolution of the suite of hadrosaurid ornithopods, ceratopsians and ankylosaurs browsers. This suite and their tyrannosaurid and coelurosaurian theropod predators were dramatically terminated at the end of the Cretaceous.

The Yixian Formation in China is famous for the well-preserved Jehol Biota (plants, birds, terrestrial life). Radioisotope dating suggests a Barremian to early Aptian age for the main Yixian Formation fossil localities. The Jehol Biota and other localities indicate that placental and marsupial mammals appeared in the later part of the Early Cretaceous. The North American mammal age

"Judithian" began in early Campanian (e.g., Jinnah et al., 2009), and the earliest NALMA of "Aquilan" is possible of Santonian age.

27.2.3 Physical stratigraphy

27.2.3.1 Magnetostratigraphy

27.2.3.1.1 Cretaceous portion of M-sequence

The M-sequence of marine magnetic anomalies formed from the Late Jurassic to the earliest Aptian. Several biomagnetostratigraphic studies have correlated Early Cretaceous calpionellid, calcareous microfossil, and dinoflagellate datums to the M-sequence polarity chrons (e.g., Channell and Grandesso, 1987; Channell et al., 1987, 1993, 1995b, 2000, 2010; Ogg, 1987, 1988; Speranza et al., 2005; Pruner et al., 2010). Correlation of Tethyan ammonite zones to the M-sequence have been achieved for the Berriasian spanning Chrons M18−M15 (Galbrun, 1984), the Berriasian−Valanginian boundary interval spanning M15−M13 (Ogg et al., 1988; Aguado et al., 2000), and less-precise calibrations for portions of the Hauterivian−Barremian interval of Chron M10N to M1 (Cecca et al., 1994; Channell et al., 1995a; Sprovieri et al., 2006). Cycle-calibrated durations of polarity zones in some of these sections indicate a fairly constant spreading rate for the Hawaiian magnetic lineations during the Valanginian through Barremian (e.g., Sprovieri et al., 2006). Polarity zone M0r is a primary marker associated with the proposed GSSP at the base of the Aptian. Andesite volcanic beds in the lower Yixian Formation of China, which are interpreted as eruptions during a reversed-polarity episode, have an Ar−Ar date of 122.0 ± 0.5 Ma (He et al., 2008), which may indicate a correlation to either Chron M0r or M1r in the GTS2020 age model. When coupled with a spreading rate model for the Pacific magnetic lineations within each individual stage (see Ogg, 2020, Ch. 5: Geomagnetic polarity time scale, this book), these correlations partly constrain the relative duration of each ammonite zone within the Berriasian, Hauterivian and Barremian stages.

Correlation of Boreal ammonite zones to the M-sequence has been directly achieved only for the equivalent of uppermost Tithonian and Berriasian in Siberia (Houša et al., 2007) and indirectly for the equivalent of the Berriasian Stage in the Purbeck beds of southern England (Ogg et al., 1991, 1994).

27.2.3.1.2 Reported brief subchrons within Aptian and Albian

An extended 40-Myr normal-polarity Chron C34n or "Cretaceous Normal Polarity Superchron" spans the early Aptian through middle Santonian. Brief reversed-polarity chrons have been reported from three intervals—middle Aptian, middle Albian, and mid-late Albian—especially within drilling cores of deep-sea sediments. However, none of these proposed subchrons have been unambiguously interpreted from marine magnetic anomaly surveys nor have M″-2r″ or M″-3r″ been verified in outcrop sections. The following summary is revised from GTS2012; and the reported possible placement relative to microfossil zones is illustrated in Fig. 27.9. Ryan et al. (1978) proposed a negative numbering for these three "pre-M0r" reversed-polarity events or clusters of events:

1. M″-1r″ in middle Aptian with a biostratigraphic age near the base of the *Globigerinelliodes algerianus* planktonic foraminifer zone (Pechersky and Khramov, 1973; Jarrard, 1974; VandenBerg et al., 1978; Keating and Helsley, 1978a,b,c; Hailwood, 1979; VandenBerg and Wonders, 1980; Lowrie et al., 1980; Tarduno et al., 1989; Ogg et al., 1992). This subchron has also been called the "ISEA" event from an Italian outcrop sample code (Tarduno et al., 1989) and has an estimated duration of less than 100,000 years (Tarduno, 1990). Based on cycle stratigraphy, the base of this foraminifer zone is at approximately 122 Ma in the GTS2012.
2. M″-2r″ set of Middle Albian events near the boundary of the *Biticinella breggiensis* and *Ticinella primula* planktonic foraminifer zones (Keating and Helsley, 1978a; Jarrard, 1974; VandenBerg and Wonders, 1980; Tarduno et al., 1992; Shipboard Scientific Party, 1998; Ogg and Bardot, 2001).
3. M″-3r″ set in Late Albian (Green and Brecher, 1974; Jarrard, 1974; Hailwood, 1979), which may occur at the end of the *Praediscosphaera cretacea* or within the *Eiffellithus turriseiffeli* nannoplankton zones (Tarduno et al., 1992).

Another reversed-polarity event, possibly near the Aptian–Albian boundary, has been reported within basalt flows with a radioisotopic date of 113.3 ± 1.6 Ma (Gilder et al., 2003) but has not yet been verified by other studies.

27.2.3.1.3 Cretaceous portion of C-sequence

Polarity Chrons C33r through lower C29r have been correlated to microfossil and nannofossil datums from basal Cenomanian to the base of the Cenozoic (e.g., Alvarez et al., 1977; Lowrie and Alvarez, 1977; Huber and Leckie, 2011). This polarity time scale has been partly calibrated in the North American Western Interior seaway to regional ammonoid zones and an array of ^{40}Ar/^{39}Ar dates from bentonites (e.g., Obradovich, 1993; Hicks and Obradovich, 1995; Hicks et al., 1995, 1999; Lerbekmo and Braman, 2002), and Chrons C32–C29 have been scaled from cycle stratigraphy in ocean-drilling cores (Herbert et al., 1995; Husson et al., 2011; Thibault et al., 2012) (see tables in Chapter 5: Geomagnetic polarity time scale, and Speijer et al., 2020, Ch. 28: The Paleogene Period, this book). The base of Chron C33r was reported as being within

the regional "upper Santonian" of England (Montgomery et al., 1998), but this polarity reversal is now being considered to mark the base of the Campanian. The ages on the Campanian–Maastrichtian portion of the C-sequence constrain the synthetic age-distance model for the magnetic anomalies of the South Atlantic (revised from Cande and Kent, 1992, 1995; see Ogg, 2020, Ch. 5: Geomagnetic polarity time scale, this book).

27.2.3.2 Geochemical stratigraphy

27.2.3.2.1 Carbon stable isotopes and carbon-enrichment episodes

Several significant excursions in the carbon cycle, called OAEs, punctuate the Cretaceous stratigraphic record, of which the early Aptian OAE1a and latest Cenomanian OAE2 events are the most significant global events (Jenkyns, 2010). Many of the major excursions (>1.5 per mil) in δ^{13}C are associated with widespread organic-rich sediments and changes in facies indicating drowning of carbonate platforms, and many appear to be preceded by or coincide with the eruption of major flood basalts (large igneous provinces: LIPs) that provided a source of the excess carbon (e.g., reviews and syntheses by Weissert et al., 1998; Jenkyns, 1999, 2010, 2017; Larson and Erba, 1999; Erba, 2004; Wortmann et al., 2004). In some cases, relatively elevated concentrations of mercury in black shales are considered a fingerprint of coincident volcanic activity (e.g., Scaife et al., 2017). The middle Cretaceous concentrations of black shales, which are the source rocks for over a quarter of current hydrocarbon deposits, were originally considered as indications of widespread oceanic anoxia (e.g., Schlanger and Jenkyns, 1976; Jenkyns, 2017). Enhanced oceanic productivity and carbon flux to the seafloor, perhaps in response to ultra-greenhouse conditions, as well as improved preservation of organic matter, are considered as first-order controls (e.g., Hochuli et al., 1999; Leckie et al., 2002; Jenkyns, 2010). The nomenclature for the main events is partly derived from distinctive European organic-rich horizons and some have an OAE numbering (Fig. 27.1).

1. *Weissert Event.* The Late Valanginian positive δ^{13}C excursion of ~2 per mil has an onset in the mid-Valanginian (base of *S. verrucosum* ammonite zone), peaks in the following *Neocomites peregrinus* ammonite zone, and returns to background levels at the beginning of the Hauterivian (Lini et al., 1992; Channell et al., 1993; Weissert et al., 1998; Martinez et al., 2015). The excursion approximately coincides with the early phases of eruption of the Paraná–Etendeka LIP of South America and Namibia (Erba et al., 2004; Martinez et al., 2015). The onset of this excursion is heralded by four thin organic carbon-rich layers ("Barrande" layers

B1–B4) in the Vocontian basin of southeastern France (Reboulet, 2001; Reboulet et al., 2003).

2. *Faraoni Event.* A latest Hauterivian organic-enrichment episode during the *Pseudothurmannia catulloi* ammonite subzone (middle of *Pseudothurmannia ohmi* Zone) is documented by a pair of organic-rich sediments in Mediterranean and Atlantic pelagic sections (Bodin et al., 2006). It coincides with only a relatively minor positive $\delta^{13}C$ excursion (e.g., Baudin et al., 1997, 1999; Coccioni, 2003; Föllmi et al., 2006).

3. *Selli Event (OAE1a)* with a complex $\delta^{13}C$ signature that begins in the Lower Aptian *Deshayesites forbesi* ammonite zone has been extensively studied. The organic-rich "Selli" event (called Goguel level in SE France) in the lower portion of the *Leopoldina cabri* foraminiferal zone is typically preceded by a sharp negative followed by a major positive excursion in $\delta^{13}C$ (e.g., Menegatti et al., 1998; Weissert and Lini, 1991; Weissert et al., 1998; Mutterlose and Böckel, 1998; Moullade et al., 1998b; Hochuli et al., 1999; Larson and Erba, 1999; Leckie et al., 2002; Renard et al., 2005) and a global nannoconid crisis, represented by a major decrease of these rock-forming calcareous nannofossils, precedes and continues through the main level of black shale (e.g., Erba, 1994, 2004). The beginning of this OAE1a Selli event appears to be synchronous with the main eruptive phase of the massive Ontong Java Plateau flood basalts in the eastern equatorial Pacific. The major positive-isotope excursion continues into the early Late Aptian. A relatively minor organic-rich "Noir" bed at the base of the Upper Aptian *E. martinoides* ammonite zone in the reference sections in southeastern France occurs at the level of the top of the main positive-isotope peak, and another minor "Fallot" organic-rich level occurs at a minimum in $\delta^{13}C$ near the top of that ammonite zone (Föllmi et al., 2006).

4. *OAE1b cluster.* A complex multiphase set of mostly negative $\delta^{13}C$ excursions spanning the Aptian–Albian boundary interval contains widespread organic-rich layers in the reference sections of southeastern France— the Jacob, Kilian, Paquier (called Urbino in Italy), Leenhardt, and l'Arbouysse Events (Bréheret, 1988; Weissert and Bréheret, 1991; Herrle et al., 2004; Gale et al., 2011). The confusing designation of OAE1b has been applied either to all or to only part of this interval, especially the Paquier level. Cyclostratigraphic studies have quantified the placement of the Jacob, Kilian, and Pacquier/Urbino events as approximately 36, 31, and 26.5 long-eccentricity 405-kyr cycles prior to the base-Cenomanian, respectively (e.g., Fiet et al., 2001; Grippo et al., 2004; Huang et al., 2010; Gale et al., 2011).

5. *Jassines Event* (Gale et al., 2011), Middle Albian, possibly correlative with OAE1c of authors, which may or may not correspond with the organic-rich

"Amadeus" layer (after Mozart) or "Toolebuc" level (e.g., Leckie et al., 2002; Coccioni, 2003).

6. *Albian–Cenomanian Boundary Event* (Jarvis et al., 2006; Gale et al., 2011), a series of four positive $\delta^{13}C$ excursions called the Albian/Cenomanian Boundary Event (the lower part coinciding locally with a set of organic-rich "Breistroffer" or "Pialli" layers). The record of this event extends from the *perinflatum* Subzone of the Albian to the *M. mantelli* Zone of the Cenomanian: it has been referred to as OAE1d (Gambacorta et al., 2015).

7. *Middle Cenomanian Event.* A minor but distinctive, brief double positive peak is widely recognized in the *C. inerme* and basal *A. rhotomagense* zones in Europe and North America (Jenkyns et al., 1994; Paul et al., 1999; Coccioni and Galeotti, 2003; Gale et al., 2007; Zheng et al., 2016; Eldrett et al., 2015) and is also found in Tibet (Li et al., 2006). This interval is locally represented by a thin, laminated dark marl in Tethyan sections (Gale, 1995).

8. *OAE2 (Bonarelli Event, Thomel Event).* Cenomanian–Turonian boundary excursion spans the *Metoicoceras geslinianum* to *W. devonense* ammonite zones, with the peak in uppermost Cenomanian (e.g., Schlanger et al., 1987; Jenkyns et al., 1994; Kerr, 1998; Jenkyns, 1999, 2010; Jarvis et al., 2011; Gale et al., 2019a). The associated organic-rich levels are named Bonarelli in central Italy, Thomel in southern France, and Bahloul in Tunisia. The excursion has a distinctive form, comprising an initial rapid rise in $\delta^{13}C$ values, locally with minor positive excursions (the precursor events), then a brief fall, which coincides with the southerly migration of a Boreal fauna (the Plenus Cold Event; Gale and Christensen, 1995; Zheng et al., 2013; O'Connor et al., in press), and subsequent rise to a plateau in the upper *geslinianum* and *juddii* zones, and fall into the lowermost Turonian.

Above the Cenomanian–Turonian boundary, detailed study of the English chalk provided a very high-resolution carbon-isotope succession up to the level of the lower Campanian, including numerous minor excursions, many of which were named (Gale et al., 2008; Jarvis et al., 2006). The discovery that the same excursions are present in the Western Interior Basin (Joo and Sageman, 2014) and in Japan (Takashima et al., 2019) significantly increases their correlative value. The carbon-isotope stratigraphy of the Campanian and Maastrichtian was described in detail by Voigt et al. (2012) from sections in central Italy, northern Germany, southwest France, and eastern England.

9. *Holywell Event* (Jarvis et al., 2006). A short, but distinctive positive excursion in the low Turonian *Fagesia catinus* Zone, which goes against the trend of overall falling values after OAE2. This is recorded widely, from the United Kingdom (Jarvis et al.,

2006), western France (Kennedy and Gale, 2015), southern France (Gale et al., 2019a), the Western Interior of the United States (Joo and Sageman, 2014), and Japan (Takashima et al., 2019).

10. *Round Down Event* (Jarvis et al., 2006). A minor positive carbon-isotope excursion (+0.4 ppt) in the mid-Turonian, marking the base of a significant overall fall in $\delta^{13}C$ values, falls approximately at the base of the *Romaniceras ornatissimum* ammonite zone.

11. *Pewsey Event* (Gale et al., 2008; Jarvis et al., 2006). A minor positive $\delta^{13}C$ excursion low in the Upper Turonian (*S. neptuni* Zone) set against an overall falling trend. It is identified in Germany (Wiese and Kaplan, 2001), the Western Interior Basin (Joo and Sageman, 2014), and Japan (Takashima et al., 2019).

12. *Hitch Wood Event* (Gale et al., 2008; Jarvis et al., 2006). A major positive carbon-isotope excursion in the Upper Turonian, with the peak in the upper *S. neptuni* Zone. The event is also present in northern Germany (Wiese, 1999), Spain and Italy (Stoll and Schrag, 2000), Japan (Takashima et al., 2019), and the Western Interior Basin (Joo and Sageman, 2014), and Japan.

13. *Navigation Event*. A broad negative $\delta^{13}C$ excursion whose minimum falls in the uppermost Turonian *Forresteria petrocoriense* Zone (Jarvis et al., 2006) at the level of the Navigation Marls in the United Kingdom succession. Recorded possibly in the Western Interior Basin (Joo and Sageman, 2014) and Japan (Takashima et al., 2019).

14. *White Fall Event* (Jarvis et al., 2006). There is an overall rise in values through the Lower and Middle Coniacian, which culminate in a peak within the lower part of the range of the inoceramid *Volviceramus involutus*. Recorded in the Western Interior Basin (Joo and Sageman, 2014).

15. *Horseshoe Bay Event* (Jarvis et al., 2006). A minor but distinctive positive event that occurs beneath the FO of *Uintacrinus*. It is also identified in northern Germany and at Gubbio (Thibault et al., 2016).

16. *Buckle Event* (Jarvis et al., 2006). A minor but distinctive negative $\delta^{13}C$ excursion that occurs at the base of the *U. socialis* Zone in England and Germany and can also be identified at Gubbio, Italy (Thibault et al., 2016).

17. *Santonian—Campanian Boundary Event* (Jarvis et al., 2006; Thibault et al., 2016). A double positive $\delta^{13}C$ excursion of nearly 1‰ that occurs within and just above the *Marsupites* zones. Also identified in the WI Basin (Joo and Sageman, 2014), in northern Germany and at Gubbio (Thibault et al., 2016) and in Japan (Takashima et al., 2019).

18. *Late Campanian Event* (Voigt et al., 2012). A distinctive negative excursion, partly coterminous with the *R. calcarata* Zone, has been found extensively in Europe, from Germany and Denmark, the Trunch borehole, the United Kingdom, Tercis, France, Gubbio, Italy, and Tibet (Li et al., 2006).

19. *Campanian—Maastrichtian Boundary Event* (Voigt et al., 2010, 2012). A prominent negative carbon-isotope shift lasting approximately 2.5 Myr, with a detailed structure, which has been recorded extensively across northwest Europe (Germany, Denmark, the United Kingdom), at Gubbio, Italy, and is recorded from both the Pacific and Atlantic Oceans.

20. *Mid-Maastrichtian Event*. Lower Maastrichtian values climb to a double $\delta^{13}C$ peak, close to the base of magnetochron C31n, between the LO of *Rheinhardites laevis* and the FO of *Lithrapidites quadratus* (Voigt et al., 2012).

21. *K—Pg Event*. The K—Pg boundary is associated with a short-lived negative $\delta^{13}C$ excursion (Voigt et al., 2012).

OAE3, initially described by Scholle and Arthur (1980), has been described from the Atlantic region (Wagerich, 2012) and identified in the Upper Coniacian of the Western Interior Basin by Joo and Sageman (2014), but it is not a discrete event. In addition to the excursions listed above, Jarvis et al. (2006) named numerous minor carbon-isotope excursions within the Cenomanian to Campanian interval, the lateral extent of which is not known. Increasing numbers of studies on carbon-isotope stratigraphy will extend and refine the record in the near future.

27.2.3.2.2 Oxygen stable isotopes and other climate proxies

In the Cenozoic, $\delta^{18}O$ provides an accurate record of oceanic temperature, but problems of diagenetic alteration of foraminiferan tests (Pearson et al., 2001; Wilson et al., 2002; see review by Pearson, 2012) mean that values have commonly been reset to cooler values, especially in older sediments. Work on exceptionally preserved "glassy" material demonstrates the great value of these (Huber et al., 1995, 2002). Although the actual temperature values derived from oxygen isotopes may be contentious, the trends are not, and when compared with data derived from the organic molecule TEX_{86}, provide an accurate record of Cretaceous climate trends (O'Brien et al., 2017). Additional evidence from cool-water indicators, such as pseudomorphs after ikaite, called glendonites, provide additional evidence of colder periods.

Early work on bulk oxygen isotopes from the English chalk (Jenkyns et al., 1994) and phosphate from Israeli shark teeth (Kolodny and Raab, 1988) demonstrated a clear temperature pattern through the Late Cretaceous, with warming up to the Cenomanian—Turonian boundary, followed by gradual progressive cooling, which accelerated in the Maastrichtian. This pattern is now corroborated by data from TEX_{86}, and a much more extensive

$\delta^{18}O$ dataset extending back to the Aptian (O'Brien et al., 2017). The Early Cretaceous story is more problematic; $\delta^{18}O$ shows a progressive warming through the Albian and Cenomanian from low temperatures in the Late Aptian, whereas TEX_{86} demonstrates more variability, including a warm peak in the Early Aptian.

The occurrences of ikaite pseudomorphs in the Cretaceous have been taken as evidence of cool periods, because the mineral only forms at very low temperatures (Price, 1999). These occurrences are concentrated in the Valanginian, Aptian, Albian and Maastrichtian, and have been taken as supporting evidence for high-latitude glaciation during these intervals. This is supported by evidence for larger magnitude sea-level changes during these times (Ray et al., 2019).

27.2.3.2.3 Strontium and osmium isotope ratios

The marine $^{87}Sr/^{86}Sr$ record displays a progressive rise from the Berriasian to a maximum of 0.707493 in the Barremian *P. elegans* ammonite zone (see McArthur et al., 2020, Ch 7: Strontium isotope stratigraphy, this book). If one assumes that this was a quasilinear trend through the Valanginian and Hauterivian, then the relative duration of each ammonite zone and its subzones can be estimated from the reference sections in southeastern France (McArthur et al., 2007).

The Strontium-isotope ratio decreases to a pronounced minimum of 0.707220 just before the Aptian/Albian boundary, rises sharply during the early Albian, flattens to a broad maximum through the mid-to-late Albian, then declines to a late Turonian minimum of approximately 0.707275. From this Turonian minimum, it rises to 0.707830 at the end of the Cretaceous (Jones and Jenkyns, 2001).

These $^{87}Sr/^{86}Sr$ trends enable global correlation and relative dating (e.g., McArthur et al., 1993), except on the cusps of reversals [Aptian/Albian boundary, Late Turonian, middle Barremian (sensu lato) or where the curve is relatively flat (e.g., mid-to-late Albian)]. For example, the age of the Campanian—Maastrichtian was calibrated in GTS2004, using $^{87}Sr/^{86}Sr$ curve correlations between Kronsmoor, Germany, and the US Western Interior (chapter on Cretaceous in GTS2004).

27.2.3.3 Cyclostratigraphy

The ideal Cretaceous time scale encompasses a full record of the sedimentary expressions of every long (405 kyr) eccentricity cycle, the only available stable tuning target from the astronomical solution beyond 52 Ma (Laskar et al., 2011; Zeebe, 2017), with additional tie-points provided by the identification of very long (~1.2 or ~2.4 Myr) eccentricity minima in the geologic record. Currently, the construction of a fully 405 kyr tuned time scale for the Cretaceous is limited by (1) the availability of cyclic successions and the resolution of

cyclostratigraphic studies; (2) the difficulty in determining phase relationships between orbital forcing and sedimentary response; and (3) the availability of radioisotopic dates and difficulties in correlating cyclic successions to available numeric age control. Despite these challenges, there is increasing consistency between cyclostratigraphically obtained durations and available radioisotopic date control, as well as between cyclostratigraphic studies. This section presents the state-of-the-art of the Cretaceous astronomically tuned time scale with a selection of cyclostratigraphic studies used for the scaling of the GTS2020 bio-magnetostratigraphy. It is also important to note that lithological time-series are preferable to those based on $\delta^{13}C$, because lithology directly reflects climatic events, with no phase difference or lag time.

Cyclostratigraphy is only really of value in time scale construction when it is fully integrated with other well established stratigraphical data including magnetostratigraphy, biostratigraphy, and chemostratigraphy (Hilgen et al., 2014), and constrained by high-resolution radioisotopic dates.

27.2.3.3.1 Berriasian—Barremian Stages

Many important studies of cyclostratigraphy of this interval are based on the conspicuously rhythmic limestone—marl successions developed in the Vocontian Basin, SE France, and eastern Spain (e.g., Cotillon, 1987; Cotillon et al., 1980; Sprenger and ten Kate, 1993; Fiet et al., 2006; Martinez et al., 2012, 2013, 2015). However, problems exist with the Berriasian Stage in SE France, because of penecontemporaneous slumping (Pasquier and Strasser, 1997). In addition, the paper by Sprenger and ten Kate (1993) only studied approximately the uppermost third of what would now be considered to be Berriasian. The detailed studies of Martinez et al. (2012, 2013, 2015) identified a complete sequence of 405-kyr eccentricity cycles for the Valanginian and Hauterivian stages, which were fully integrated with ammonite, nannofossil, and carbon-isotope stratigraphy. The proposed durations from these studies for the Valanginian-Hauterivian stages and of Tethyan ammonite zones within those stages is used in GTS2020 (Fig. 27.9). Intercalibration to sections in Argentina with radioisotopic dates provide evidence that the Hauterivian—Barremian boundary fell at 126.07 ± 0.25 Ma. There is currently no reliable study on the cyclostratigraphical duration of the Berriasian Stage, but several studies provide a likely duration of 4—5 Myr (e.g., Kietzmann et al., 2015). Similarly, the entire Barremian Stage currently remains unstudied in detail, but there is evidence from cycle counting that the stage lasted approximately 5.1 Myr (Fiet et al., 2006). Cyclostratigraphy studies on Spanish reference sections have enabled a preliminary scaling of Tethyan ammonite zones of the Barremian (Martinez et al., 2020), and a modifed version is used in GTS2020 (Fig. 27.9).

The cyclostratigraphic dataset from the Greenland—Norwegian sea used by Huang (2018) to calibrate the

Tithonian to Barremian interval remains unpublished, and the log does not show any primary bio- or chemostratigraphical data. Therefore we do not consider it in the present study.

27.2.3.3.2 Aptian—Aptian Stages

Much of the tuning of Aptian—Albian time has been based on the Piobbco core, drilled into pelagic and hemipelagic strata in the Umbria—Marche Apennines, central Italy (Herbert et al., 1995; Grippo et al., 2004). Grippo et al. (2004) used image-processed photos to generate spectra, and identified 26 405-kyr cycles through the Albian, with older levels providing only poor spectra (their Fig. 15). Gale et al. (2011) correlated the 405-kyr cycles into the succession of the Vocontian Basin with the aid of planktonic foraminiferans and used this to calibrate the duration of ammonite zones for the Albian and is used in GTS2020 (Fig. 27.9). The presence of 32 405-kyr eccentricity cycles in the Albian of the Piobbco core matches well radioisotopic estimates for the duration of the stage.

Huang et al. (2010) studied the lower part of the Piobbico core, and from the 405-kyr cycles which they identified estimated the base of the Aptian Stage (base of Chron M0r) to fall at 125.45 Ma. This date was subsequently modified by the changed estimate for the age of the base-Cenomanian and used in GTS2012 (Ogg et al., 2012) to place the base of the Aptian at 126.3 Ma. However, three independent lines of evidence cast doubt on this conclusion. The first is the placement of the Hauterivian—Barremian boundary, based on radioisotopic dates, at 126.02 Ma (Martinez et al., 2015). The second is new radioisotopic dates from the Svalbard, which place the Barremian—Aptian boundary at 121—122 Ma (Midtkandal et al., 2016). The actual date by Midtkandal et al. (2016) of 123.10 ± 0.3 Ma has now been revised by Zhang et al. (2019) to be of mid-Barremian age in the upper half of M1r, casting severe doubt on a 126 Ma age of Chron M0r. In addition, new spreading rate estimates change the date of the base Aptian (base of Chron M0r) to 121.5 Ma (Malinverno et al., 2012). There is evidently a problem with the cyclostratigraphical interpretation of the lower part of the Piobbco core. Ghirardi et al. (2014) applied cyclostratigraphy analysis to the middle Aptian interval of sections in the Vocontian Basin of SE France which have ammonite zones and carbon-isotope stratigraphy (Herrle et al., 2004; with revised ammonite zones by Luber et al., 2017, 2019), and this scaling is used for the middle Aptian in GTS2020 (Fig. 27.9).

The duration of Oceanic Event OAE1a was estimated using cyclostratigraphy at 1.0—1.3 Myr (Li et al., 2008) with a rapid, short-lived initial negative excursion lasting between 27 and 44 kyr. The duration was further constrained by Malinverno et al. (2010) to 1.1 ± 0.11 Myr.

27.2.3.3.3 Cenomanian—Turonian Stages

Reliable radioisotopic dates for the base of the Cenomanian (99.8 ± 0.4 Ma; Obradovich et al., 2002, Japan), assigned on more evidence to 100.5 ± 0.14 Ma, and the base of the Turonian (93.9 Ma; Meyers et al., 2012) imply that the duration of the stage was near 6 Myr. Cyclostratigraphic estimates of the duration of the Cenomanian based on the pelagic succession in the Italian Appennines vary from 6.0 ± 0.5 (Herbert et al., 1995) to 5.4 Myr (Sprovieri et al., 2013). The estimate of Gale (1995) of 4.45 Myr was based on sections that are now known to contain multiple hiatuses. Gale et al. (1999) estimated the Middle and Late Cenomanian to span 3 Myr.

Several cyclostratigraphic studies (Batenburg et al., 2016; Gale et al., 1999; Eldrett et al., 2015) estimate the duration between the onset of the carbon-isotope excursions accompanying the mid-Cenomanian event (MCE) and OAE2 at 1.9—2.0 Myr. Duration estimates for the OAE2 itself are more problematic, as many sedimentary environments with rhythmic deposition experienced a strong change in oxygenation, resulting in potential hiatuses in the lithologic record. Nevertheless, recent duration estimates between the start of the carbon-isotope excursion and the Cenomanian—Turonian boundary range from 430 to 560 kyr with most estimates around 500 kyr (Batenburg et al., 2016; Eldrett et al., 2015; Ma et al., 2014; Meyers et al., 2012; Sageman et al., 2006; Takashima et al., 2019; Voigt et al., 2008). The combined estimated duration between the MCE and the Cenomanian—Turonian boundary of ~2.4 Myr agrees well with radioisotopic date constraints on the first positive peak of the MCE of 96.21 ± 0.36 Ma (Batenburg et al., 2016), and the Cenomanian—Turonian boundary of 93.90 ± 0.15 Ma (Meyers et al., 2012). The Cenomanian—Turonian boundary falls near the maximum of 405-kyr cycle 232 (Batenburg et al., 2016) of the La2011 astronomical solution (Laskar et al., 2011). Gale (2019b) tentatively identified 405-kyr cycles in the Anglo—Paris Basin, using repetitive marly levels in the chalk succession, guided by radioisotopic dates correlated in from the Western Interior of the United States with ammonites. The upper limit of the Turonian Stage was discussed by Ma et al. (2019) on the basis of cyclostratigraphic analysis of the Iona (89.52 ± 0.17 Ma) and Libsack (89.75 ± 0.38 Ma) cores. Sprovieri et al. (2013) obtained a similar duration of the Turonian stage in the Gubbio section based on time-series analysis of $\delta^{13}C$ data.

27.2.3.3.4 Coniacian—Santonian Stages

Borehole-resistivity and lithostratigraphy of the Niobrara Formation of the Western Interior Basin tied to North American ammonite zones in coeval outcrops have been tuned to short- and long-eccentricity cycles (Locklair and Sageman, 2008; Sageman et al., 2014). The Coniacian stage spans 3.4 ± 0.13 Myr, and the Santonian spans 2.39 ± 0.15 Myr. In the southern English chalk, Thibault et al. (2016) tuned a long time-series of carbon-isotope data through the upper Coniacian to lower Campanian and identified five 405 kyr in the Santonian, giving it a duration of approximately 2.25 Myr. Thibault et al. (2016) also matched the record to the La2011 solution in an attempt to anchor the floating time scale.

27.2.3.3.5 Campanian and Maastrichtian Stages

The Campanian is relatively poorly constrained by cyclostratigraphic studies, due to the long length of the stage, the paucity of rhythmic records, and difficulties in correlation due to provincialism of faunas and an absence of major climatic events. The long cyclostratigraphic record of the Western Interior (Sageman et al., 2014) extends four 405-kyr cycles upward into the Campanian, and the Umbria–Marche Basin record 12 (Sprovieri et al., 2013). From the Maastrichtian downward, the 405 kyr interpretation of Site 762C extends 6 405-kyr cycles into the Campanian (Husson et al., 2011) and a composite of German successions spans the topmost 14 405-kyr cycles (Voigt and Schonfeld, 2010), leaving a gap in the middle Campanian. Rhythmic limestone–marl alternations from the Mississippi embayment (O'Connor et al., 2020) span the middle Campanian, but the succession is condensed near the upper and lower stage boundaries. Nevertheless, their estimated duration of the Campanian of 12 Myr (O'Connor et al., 2020) is in good agreement with existing estimates. For the Maastrichtian, independent cyclostratigraphic studies (Batenburg et al., 2012, 2014, 2018; Husson et al., 2011; Thibault et al., 2012; Wu et al., 2014) are in good agreement, supporting a duration of the Maastrichtian stage of 6.1 Myr. The top of the Maastrichtian is anchored to the intercalibrated K/Pg boundary, placed at 66.04 (Renne et al., 2013). Kuiper et al. (2008) identified the K/Pg as representing a 405-kyr minimum, which can be correlated to the base of 405-kyr eccentricity cycle 163 in the La2011 solution (Laskar et al., 2011).

27.2.3.4 Sequence stratigraphy

Cretaceous marginal-marine to deep-shelf successions in Europe and North America record an abundance of basinal and regional transgressions and regressions. At the largest scale, the Cretaceous strata encompass a single transgressive–regressive cycle (the "North Atlantic" cycle of Jacquin and de Graciansky, 1998). The lower boundary is a widespread unconformity during Late Berriasian, the transgression peaked in the Early Turonian, and average sea levels continued to decrease into the Paleocene.

The common features from an extensive suite of compilations edited by de Graciansky et al. (1998) were assembled in a comprehensive synthesis and systematically numbered from the base of each stage (Hardenbol et al., 1998). Coeval emergent horizons recorded by Aptian–Albian seamount carbonate platforms in the central Pacific Ocean imply that some of these depositional sequences reflect global eustatic sea-level oscillations (Röhl and Ogg, 1996). Detailed analysis of facies successions and biostratigraphic constraints across the Arabian Plate yielded a detailed sequence stratigraphy for the Cretaceous (e.g., Sharland et al., 2001, 2004; Simmons et al., 2007; van Buchem et al., 2011), in which many of the main features appear to coincide with the European–American-derived sequence scale. These compilations and additional reference sections were synthesized and recalibrated by Haq (2014) into a revised eustatic and coastal onlap curve for the Cretaceous (used in Fig. 27.1). The magnitude and cause of these sea-level changes during the presumed "supergreenhouse" of the mid-Cretaceous remain controversial (e.g., Miller et al., 2003, 2004), and many features require additional verification and documentation in multiple reference sections. See also review in Simmons et al. (2020, Ch. 13: Phanerozoic eustasy, this book).

Details of the Cretaceous sequence stratigraphy are continually undergoing refinement, but the major global oscillations have probably been identified. The main large-scale deepening and shallowing trends from the sequence stratigraphy charts (Hardenbol et al., 1998; Haq, 2014) are summarized in Fig. 27.9.

27.2.3.5 Other major stratigraphic events

27.2.3.5.1 Large igneous provinces

At least five major LIPs formed during the Cretaceous. Most of these appear to be associated with major distortions in the global carbon budget as indicated by excursions in carbon isotopes, widespread organic-rich shales or "OAE" episodes, increased oceanic carbonate dissolution, and other changes in climate and oceanic chemistry (e.g., Larson and Erba, 1999; Jones and Jenkyns, 2001; Bice et al., 2002). Age constraints and possible feedbacks from these episodes are reviewed by Wignall (2001) and Courtillot and Renne (2003); and selected ages are summarized in Tables 27.2 and 27.3.

1. *Paraná–Etendeka, ~136–133 Ma.* The early stages of rifting of the South Atlantic were accompanied by extrusion of a LIP onto South America (Paraná flood basalts) and smaller fields in Namibia (Etendeka Traps). The full suite of volcanic activity may have had multiple pulses that spanned over 10 Myr, but the main pulse of tholeiitic volcanism appears to be between 135

and 133 Ma (e.g., Stewart et al., 1996; Gibson, 2006; Gibson et al., 2006, dates are after adjustment to a $^{40}Ar/^{39}Ar$ FCS (Fish Canyon sanidine) monitor age of 28.20 Ma). The central Paraná volcanic suite spans a minimum of a triplet of polarity zones (normal-reversed-normal) (Mena et al., 2006). The onset of the northern and western Paraná volcanics from U–Pb dating of baddeleyite/zircon is 134.3 ± 0.8 Ma (Janasi et al., 2011). The onset and main eruptive phase of the Paraná–Etendeka flood basalts coincides with the late Valanginian "Weissert" C-13 positive excursion (e.g., Weissert et al., 1998; Erba et al., 2004), which has a GTS2020 age assignment of ~135 to 133 Ma (Fig. 27.1).

2. *Ontong Java Plateau–Manihiki Plateau, ~125–123 Ma.* During the middle of the early Aptian, the largest series of volcanic eruptions of the past quarter-billion years built the Ontong Java Plateau and Manihiki Plateau in the western equatorial Pacific. A series of deep-sea drilling legs documented that the multikilometer-thick series of volcanic flows forming the Ontong Java Plateau occurred during a short time span at ~125–122 Ma (e.g., Mahoney et al., 1993; Chambers et al., 2004). A controversial theory is that its initiation may have been caused by a bolide impact (IIngle and Coffin, 2004). The upper portions of the basalt flows are interbedded with pelagic sediments of the lower portion of the *L. cabri* foraminifer zone (Mahoney et al, 2001; Sikora and Bergen, 2004). A cascade of environmental effects from the eruption of the Ontong Java Plateau is the suspected culprit for the organic-rich deposits associated with the Early Aptian "OAE1a" or "Selli" episode which was followed by a large positive carbon-isotope excursion (e.g., Larson, 1991; Tarduno et al., 1991; Larson and Erba, 1999). However, the c. 120 Ma age for the main OAE1a in the revised GTS2020 age model is difficult to reconcile with the published dates for those Ontong Java Plateau basalts; therefore those radioisotopic dates should be reexamined.

3. *Kerguelen Plateau–Rajmahal Traps, ~118 Ma.* The Kerguelen Plateau in the southern Indian Ocean is the second largest oceanic plateau after the Ontong Java Plateau. The peak of construction of the southern and largest portion may have been simultaneous with the eruption of the formerly adjacent Rajmahal Traps of eastern India at ~118 Ma (reviewed in Wignall, 2001; Courtillot and Renne, 2003). This episode probably contributed to the broad carbon-isotope excursion that characterizes the late Aptian. A second eruptive episode at ~83 Ma enlarged the Kerguelen Plateau and constructed the Broken Ridge (Courtillot and Renne, 2003).

4. *Caribbean–Colombian Province, ~90 Ma.* The Caribbean–Colombian volcanic province is a large Pacific Ocean plateau that was later emplaced between the North and South American plates. Its eruption may have contributed to the end-Cenomanian OAE2 carbon-isotope excursion (e.g., Kerr, 1998); however, its apparent average age of 89.5 ± 0.3 Ma coincides with the Turonian–Coniacian boundary (Courtillot and Renne, 2003).

5. *Deccan Traps, ~66–65 Ma.* The Deccan Traps cover most of central India. Their eruption peak at ~66 Ma coincides with the catastrophic termination of the Cretaceous (Courtillot and Renne, 2003).

In addition to these LIPs, pulses of large-scale volcanism constructed the Shatsky Rise in the central Pacific during the earliest Cretaceous (Mahoney et al., 2005), the Madagascar traps at ~87 Ma (Bryan and Ernst, 2008), and the Sierra Leone Rise in the central Atlantic at about 73 Ma (Ernst and Buchan, 2001; Large Igneous Provinces Commission, 2011).

27.2.3.5.2 Major bolide impacts

Five impact craters with diameters greater than 40 km are currently documented from the Cretaceous (details from Jourdan et al., 2009; Earth Impact Database, 2018; unless otherwise noted).

1. *Morokweng* crater (~70 km diameter; or perhaps originally over 100 km) in the Kalahari desert of South Africa has a "recommended" age from U–Pb and $^{40}Ar/^{39}Ar$ analyses of 145.2 ± 0.8 Ma (Koeberl et al., 1997; Reimond et al., 1999; Jourdan et al., 2009). This coincides with the Jurassic–Cretaceous boundary interval, although there is no unambiguous evidence of a wave of extinctions of this age.

2. *Mjølnir* crater (~40 km) offshore northern Norway has an estimated age of c. 142.0 ± 2.6 Ma (Earth Impact Database, 2018), and biostratigraphy of regional coring that indicates the impact was near the Volgian–Ryazanian boundary interval (Smelror et al., 2001; Tsikalas, 2005).

3. *Tookoonoka* crater (~50 km) in west Queensland, Australia, has a poorly constrained age estimated as 128 ± 5 Ma.

4. *Kara* crater (~65 km) in the northern Urals of Russia is dated as 70.3 ± 2.2 Ma.

5. The immense *Chicxulub* crater (~170 km) in Yucatan, Mexico, that dramatically terminated the Mesozoic Era at 66.0 Ma.

27.3 Cretaceous time scale

In this section first the GTS2012 is summarized, after which the data and methods are presented that construct the current geologic time scale. For the first time a rather detailed Early Cretaceous dataset is available to spline the

M-sequence, also taking cyclostratigraphic stage durations into account in the geomathematical analysis. The Albian through Maastrichtian scale relies on direct dating of stage and period boundaries and cyclostratigraphic duration of stages.

27.3.1 Previous scales

The numerical time scale for the Cretaceous consists of three main "primary scales"—(1) Tethyan ammonite zones of Berriasian through Barremian are calibrated to the marine magnetic anomaly M-sequence age model scaled with a spline-fit, (2) microfossil and ammonite zones in Aptian–Albian are calibrated to cycle stratigraphy that is constrained by radioisotopic dates, and (3) North American ammonite zones of Cenomanian through early Maastrichtian that have an abundance of interbedded bentonites with radioisotopic dates are scaled with cycle stratigraphy and a spline-fit. Most other Cretaceous events are assigned ages according to their calibration to these primary biostratigraphy scales or via direct calibrations to the M-sequence or C-sequence chrons.

Cretaceous time scales have been composed, using this philosophy and methods by Gradstein et al. (2004, 2012) and by Ogg et al. (2008, 2012, 2016). The GTS2012 scaling is summarized below, after which the current numerical Cretaceous model will be outlined.

27.3.1.1 Constraints from radioisotopic dates used in GTS2012

Compared with any other Phanerozoic interval, the Late Cretaceous has the highest concentration of radioisotopic dates derived from bentonites interbedded with fossiliferous limestone. In contrast, the Early Cretaceous has only a few well-constrained radioisotopic dates; and a selected subset of the main constraints is briefly reviewed here.

The Tithonian–Berriasian boundary (base of Cretaceous) is constrained by a $^{40}Ar/^{39}Ar$ date of 145.5 ± 0.8 Ma on reversed-polarity sills intruding earliest Berriasian sediments on the Shatsky Rise of the Pacific (Mahoney et al., 2005). A Berriasian–Valanginian age older than ~ 138 Ma is supported by U–Pb dates from calcareous nannofossil–zoned sediments in the Great Valley of California (Shimokawa, 2010).

Deposits in the Neuquén Basin of Argentina that are correlated to the basal ammonite zone of the Upper Hauterivian substage yielded a U–Pb SHRIMP age of 132.7 ± 1.3 Ma (Aguirre-Urreta et al., 2008). Basalts from Resolution Guyot in the Pacific with reversed-polarity magnetization interpreted as Chron M5r or M3r of Barremian yielded a $^{40}Ar/^{39}Ar$ age from whole rock incremental heating of 128.4 ± 2.1 Ma (Pringle and Duncan, 1995a), but this method and its biostratigraphy constraints are not considered to be conclusive.

Several age dates indicate that the Aptian began at approximately 126 Ma. The basaltic basement on the Ontong Java Plateau, which is interbedded and overlain by limestone assigned to the *L. cabri* foraminifer zone (Mahoney et al., 2001; Sikora and Bergen, 2004), yielded an average $^{40}Ar/^{39}Ar$ age of 124.3 ± 1.8 Ma (Chambers et al., 2004). As noted above, the eruption of this massive igneous province is considered to be a causal factor in the OAE1a or "Selli" episode of carbon-rich oceanic sediment accumulation and carbon-isotope excursion that begins at about 0.5 Myr after the end of Chron M0r (hence about 0.9 Myr after the beginning of the Aptian) and spans about 1.1 Myr (Larson and Erba, 1999; Malinverno et al., 2010). Dating of reversed-polarity basalts at MIT Guyot in the Pacific yielded an age of 125.4 ± 0.2 Ma (Pringle and Duncan, 1995b; as recomputed to FCS of 28.201 Ma). The overlying basalt is a transition into a well-developed soil that is overlain by transgressive marine sediments containing early Aptian nannofossils and then capped by a thick shallow-water carbonate platform. Initially, the reversed-polarity zone had been interpreted to be the uppermost portion of polarity Chron M1r of middle Barremian, but the nannofossil criteria and the character of the carbon-isotope values from the overlying sediments are consistent with interpreting this zone as basal Aptian and therefore uppermost Chron M0r (Pringle et al., 2003). Chron M0r spans 0.4 Myr (e.g., Herbert et al., 1995), therefore its base, which is a proposed marker for the base of the Aptian Stage, was projected in GTS2012 as approximately 126.0 Ma. This is consistent with a U–Pb date of 124.07 ± 0.17 from calcareous nannofossil–zoned sediments of Early Aptian in the Great Valley of California (Shimokawa, 2010) and with an Ar–Ar date of 124.3 ± 1.8 Ma from basalts of Ontong Java Plateau (Chambers et al., 2004) that are considered to be coeval with the onset of Early Aptian OAE1a anoxic horizon (Table 27.1 in GTS2012).

In contrast, a volcanic episode of reversed polarity within continental deposits of the Yixian Formation of China yielded a 122.0 ± 0.5 date from $^{40}Ar/^{39}Ar$ step-heating method (He et al., 2008). He et al. (2008) had interpreted this reversed-polarity event to be during Chron M0r, thereby implying that the base of the Aptian was younger than 123 Ma! In the publication that had used the date cited above, an alternative interpretation, assuming that the polarity was a primary magnetization, was that this eruption might have captured the brief M″-1r″ or ISEA event of mid-Early Aptian (c. 122 Ma). But the revised dating for the base of the Aptian as c. 121.4 Ma in the GTS2020 age model would imply that the original interpretation by He et al. (2008) is viable.

A volcanic ash within the basal Albian strata in northwest Germany yielded a U–Pb age of 113.1 ± 0.3 Ma (Selby et al., 2009). The top of the basal Cenomanian ammonite subzone has a $^{40}Ar/^{39}Ar$ date of 99.8 ± 0.4 Ma (Obradovich et al., 2002).

From the lowest Cenomanian through lower Maastrichtian, there are nearly 45 radioisotopic-dated horizons from the Western Interior Basin of North America (detailed in Appendix 2 of GTS2012). A Late Cretaceous time scale for North American ammonoid zones was initially calibrated by Obradovich (1993) from his extensive suite of $^{40}Ar/^{39}Ar$ dates on bentonites using a multigrain analysis of sanidine grains at the USGS lab. He added additional dates from sections calibrated to Campanian–Maastrichtian magnetostratigraphic (Hicks et al., 1995, 1999), then summarized and enhanced the sets using a refined ammonite zonation for the Western Interior basin (Cobban et al., 2006). The collections of sanidine grains from Obradovich's separates have been undergoing single-crystal analyses to narrow the uncertainties at the University of Wisconsin at Madison, and this group has analyzed additional horizons and applied U–Pb dating to zircons from the same levels (e.g., Siewert, 2011; Meyers et al., 2012). Supporting U–Pb and Ar–Ar results have been contributed by Quidelleur et al. (2011) from Japanese sections, including verification that the base of the Cenomanian is at approximately 100.0 Ma.

The youngest ammonite-zoned age from the Western Interior suite is ~ 69.9 Ma in the earliest Maastrichtian. The Maastrichtian–Danian boundary (base of Cenozoic) is well-dated as ~ 66.0 Ma.

27.3.1.2 Direct spline-fitting of radioisotopic date suite as used in GTS2012

An initial spline-fit incorporated the majority of the Cretaceous and Jurassic suite of radioisotopic dates (Appendix 2 of GTS2012) and their biochronostratigraphic assignments with uncertainties to a merged scale of North American Western Interior ammonite zones (Late Cretaceous) and Tethyan ammonite zones (basal Cenomanian through Jurassic). Some Cretaceous radioisotope dates were omitted that were difficult to correlate to the primary ammonite zonal scale (e.g., the Kneehills Tuff within *Triceratops* dinosaur beds). The spline-fit processed by Øyvind Hammer used the methods described in Chapter 14 on Geomathematics in the GTS2012 volume). The first spline run with a smoothing factor of 1.5 indicated that 11 dates did not pass the chi-squared distribution test (mainly in Jurassic); therefore a second spline run relaxed the smoothing factor to 0.975. The computed stage boundaries are listed under "Spline #1" in Table 27.2 of GTS2012.

However, the relaxed-fit for this spline did not fully utilize the detailed high-precision radioisotope dates with excellent ammonite zone placements for the Late Cretaceous. As a result, many of those dates were no longer being assigned by that spline-fit into the computed age ranges for those zones. Therefore a second hybrid spline was constructed—

the Late Cretaceous had a more strict spline-fit model, and the Early Cretaceous through Jurassic had the less restrictive smoothing factor. The resulting estimates for Cretaceous stage boundaries were listed under Spline #2 in Table 27.2 of GTS2012. The main change was for the poorly constrained base of Hauterivian, which changed by 0.4 Myr between these two spline-fit versions.

This Spline #2 also yielded a set of smoothly varying durations for Upper Cretaceous ammonite zones. However, in many intervals of the Upper Cretaceous, this smoothly varying scaling can be enhanced by using the durations for individual ammonite zones based on cycle stratigraphy. For example, the cycle-scaled duration of the *Pseudaspidoceras flexuosum* ammonite zone is 0.45 Myr (Meyers et al., 2012), but the spline-fit projected it as 0.84 Myr, twice the actual duration. In intervals of the Lower Cretaceous, improved scalings of ammonite zones can incorporate calibrations to the Pacific M-sequence, relative scalings from linear strontium-isotope trends (e.g., McArthur et al., 2007) or correlations to cycle-scaled microfossil zones (e.g., Grippo et al., 2004; Huang et al., 2010; Gale et al., 2011).

Therefore the spline-fit was enhanced by incorporating these other sets of stratigraphic studies or integrated stratigraphy (e.g., the Bayesian statistical method by Meyers et al., 2012, that optimizes the merger of radioisotopic dates with the cycle stratigraphy in the Turonian–Cenomanian boundary interval). In most cases the modification of the ages for those stage boundaries in the Upper Cretaceous based on the enhanced calibrations of the ammonite zones was less than 0.3 Myr between the spline-fit versions. An issue, not addressed by the geomathematics analysis in GTS2012, is that subjective adjusting of the geomath numerical solution also weakens the estimation of uncertainties. The latter is largely avoided in GTS2020.

27.3.1.3 Cretaceous age model used in GTS2012

The primary standards for Cretaceous calibrations are ammonite zones of the Tethyan realm (Berriasian through Albian) and North America Western Interior (Cenomanian through mid-Maastrichtian) (Table 27.3 of GTS2012). The age model for the Berriasian through Barremian ammonite zones was mainly derived from biostratigraphic correlations to the M-sequence of marine magnetic anomalies, plus intervals with cycle stratigraphy and/or linearization of strontium-isotope trends (Table 27.2 of GTS2012). The Aptian–Albian ammonite zones were scaled according to their correlation to microfossil and nannofossil datums, which, in turn, are scaled by cycle stratigraphy. A spline-fit (Spline #2, as explained above) of numerous radioisotopic dates from volcanic ash horizons interbedded with Western

Interior ammonites with adjustments for cycle stratigraphy of some intervals provided a high-resolution age model for the Cenomanian through early Maastrichtian. The late Maastrichtian correlations rely on microfossil datums calibrated to a spline- and cycle-fit of C-Sequence marine magnetic anomalies.

Only three Cretaceous stages (Maastrichtian, Turonian and Cenomanian) in 2012 had ratified GSSPs, therefore the age model for the other stages were based on selected working definitions (e.g., base of Aptian placed at base of magnetic polarity Chron M0r; base of Albian assigned to the lowest occurrence of a nannofossil in the cycle-scaled reference section) (Table 27.2 of GTS2012).

Details on the subjective merging of the spline results with cyclo- and other high-resolution stratigraphy are in the Cretaceous chapter of GTS2012, and the interested reader is kindly referred to that intricate text. That GTS2012 Cretaceous time scale model is contrasted to the GTS2020 age model later.

27.3.2 Cretaceous numerical age model of GTS2020

The GTS2020 numerical time scale for the Cretaceous follows the concept of GTS2012 with the three "primary scales"—(1) Tethyan ammonite zones of Berriasian through Barremian are calibrated to the M-sequence age model or to cyclostratigraphy in French reference sections, (2) several ammonite and microfossil zones in Aptian—Albian are calibrated to cycle stratigraphy that is constrained by radioisotopic dates, and (3) North American ammonite zones of Cenomanian through early Maastrichtian that have an abundance of interbedded bentonites with radioisotopic dates are scaled with cycle stratigraphy. This latter segment of the Cretaceous has a more mature cyclostratigraphy with more overlapping sections than the pre-Cenomanian Cretaceous. Geomathematical methodology is used for final Early and Late Cretaceous time scaling, as visualized in Figs. 27.10 and 27.11 showing the GTS2020 cubic spline fits for these chronostratigraphic intervals. This practice is new for Early Cretaceous, which for the first time now has a rather detailed and more stable dataset, as explained later.

Six Cretaceous stage boundaries now have a ratified GSSP, that is, Maastrichtian, Santonian, Turonian, Cenomanian, Albian and Hauterivian (Fig. 27.1, Table 27.1, Fig. 27.12), of which only the Hauterivian is within the geomagnetic M-sequence. Nevertheless, there is a detailed understanding of the M-sequence definition of the Berriasian through Aptian stages, as shown in Table 27.2, which also lists the definitions of

pre-Berriasian stages with M-sequence assignment and its km marine.

The Jurassic—Cretaceous boundary and base of Berriasian is in the middle of Chron M19n (Wimbledon et al., 2019, 2020), the Valanginian base is in the lower part of Chron M14r, the base of Hauterivian may be near Chron M10n and the base of Barremian is in the upper part of Chron M5n. Base Aptian is at the very end of the M-sequence at the older limit of Chron M0r at 0 km on the Hawaiian magnetic anomaly profile. The complete radioisotopic, cyclostratigraphic, and M-sequence dataset is in Table 27.2; it extends into marine magnetic anomaly M42 of potential Bajocian age to take advantage of that splined M-sequence segment also. Radioisotopic dates were reviewed earlier, with specific reference to new literature in GTS2016, and details are compiled in Appendix 2 in this book.

The critical factor for the Early Cretaceous scale in GTS2020 is that a rather high-resolution U—Pb radioisotopic, geomagnetic and cyclostratigraphic dataset is now available for the Tithonian through Barremian, not known during construction of GTS2012 (Kietzmann et al., 2018; Lena et al., 2019; Vennari et al., 2014; Aguirre-Urreta et al., 2017, 2019; Zhang et al., 2019). Important cyclostratigraphic constraints on duration of the Valanginian—Hauterivian are from Martinez et al. (2015).

The new information includes:

1. Durations of Tithonian through Barremian stages are slightly over 5 Myr per stage.
2. He et al. (2008) age near 122 Ma likely is earliest Aptian M0r or in latest Barremian M1r.
3. The Aptian Stage is about 5 Myr shorter than thought previously (see Table 27.2 for numerical details).

Together, this new dataset in Table 27.2, with emphasis on modern U—Pb dates makes it likely that the Ar/Ar age date near 146 Ma of Mahoney et al. (2005) on M18, the dates near 125 Ma on the oceanic Ontong Java Plateau and the MIT Guyot in the Pacific (Appendix 2 and Cretaceous chapter of GTS2012), and the interpretation of 62.5 long-eccentricity cycles from base-Cenomanian to top of M0r (Huang et al., 2010) should be reevaluated.

Spline-fitting as applied to Paleozoic periods in GTS2020 was also used to help obtain the GTS2020 Cretaceous time scale, as outlined in detail in Ch14A: Geomathematical and statistical procedures (this book). The resulting smoothing spline if applied only to the array of radioisotopic dates with their estimated placement relative to the Hawaiian magnetic anomaly profile, which is shown in Fig. 27.10, is approximately a straight line representing approximately constant seafloor spreading for that Hawaiian profile from Kimmeridgian through Barremian. The deviations from this spline are small.

TABLE 27.2 Stratigraphic–radioisotopic data set, aligned to the mid-km M-sequence, used for the Lower Cretaceous time scale.

Stage	radioisotopic date, 2-sigma	interpolated age with cycles	duration from 405-kyr cycles	magnetic polarity-chron assignment	mid-km M-sequence	Reference
base-Cenomanian	100.5 ± 0.14					GTS2012 and this chapter
			Albian duration 12.45 ± 0.5			
base Albian	113.10 ± 0.3					GTS2012 and this chapter
early Aptian	121.20 ± 1			several ages dates in Appendix 2		
near base Aptian	122.01 ± 0.52			no biostrat; whole M0r	minus 4.9 ± 4.9	He et al. (2008); this chapter
			Aptian duration 8.1 ± 0.5			
base Aptian				near Chron M0r	0 ± 2.5	
mid-Barremian	123.10 ± 0.3			upper half of M1r	55.8 ± 2.2	Zhang et al. (2019); this chapter
Barremian	125.45 ± 0.43			entire M1r	68.5 ± 6.1	Pringle and Duncan, (1995b)
			Barremian duration 5.00 ± 0.5			
base Barremian		126.07 ± 0.25		upper part of Chron M5n	120 ± 5	
Hauterivian	127.24 ± 0.25 (2σ approx.)			upper Hauterivian, mid-M5n	125.5 ± 5.5	Return to Agrio
Hauterivian	130.39 ± 0.25 (2σ approx.)			uppermost M10N	213.5 ± 4.6	Aguirre-Urreta et al. (2017)
			Hauterivian duration 5.93 ± 0.5			
base Hauterivian	131.29 ± 0.25			in Chron M10n	198 ± 4	Aguirre-Urreta et al. (2019)
			Valanginian duration 5.06 ± 0.5			
base Valanginian		137.05 ± 1.0		chron M14r.3	347 ± 9	Martinez et al. (2015)
Berriasian	139.24 ± 0.16			uppermost M17r	447.8 ± 9.9	Lena et al. (2019)
	139.55 ± 0.18			M17r	457.7 ± 9.9	Vennari et al. (2014)
	139.96 ± 0.17			M18n–M17r boundary interval	476.9 ± 9.3	Lena et al. (2019)
	140.34 ± 0.18			M18r–M18n boundary interval	491.8 ± 5.8	
	140.51 ± 0.16			base M18n to lower M17r	476.3 ± 19.6	Lena et al. (2019) with revised Chron Appendix 2
	142.04 ± 0.17			lower M18r	503 ± 2.1	

TABLE 27.2 (Continued)

Stage	radioisotopic date, 2-sigma	interpolated age with cycles	duration from 405-kyr cycles	magnetic polarity-chron assignment	mid-km M-sequence	Reference
	146.48 ± 1.63			M18r—M18n boundary interval	500 ± 5.1	Mahoney et al. (2005)
			Berriasian duration 5.27 ± 0.5			Kietzmann et al. (2018)
base Berriasian				mid M19n.2n	526.3 ± 14	Berriasian Working Group, ICS
Tithonian	147.11 ± 0.18			early Tithonian; upper M22r	684.7 ± 4.3	Lena et al. (2019)
			Tithonian duration 5.67 ± 0.5			Kietzmann et al. (2018)
base Tithonian				base chron M22An	701 ± 2	
base Kimmeridgian	154.1 ± 2.10			base M26r	845.2 ± 2.5	Selby et al. (2009), GTS2016
			Kimmeridgian duration 5.20 ± 0.5			GTS2012 cycles/ spreading rate
Oxfordian	155.60 ± 0.89			M24Bn—M30r	851 ± 67.7	AB4-Pell
			Oxfordian duration 5.80 ± 0.5			GTS2012 cycles/ spreading rate
base Oxfordian				lower part of M36Br	1013.69 ± 1.05	GTS2012
			Callovian duration 3.00 ± 0.5			GTS2012 cycles/ spreading rate
base Callovian				within M39n.3n (mid-depth scale)	1097.9	GTS2012
Bathonian	164.64 ± 0.27			M39n.3n	1097.9 ± 9.4	Kamo and Riccardi (2009)
Bajocian	168.35 ± 1.31			mid Bajocian, mid-M42-base M43r	1193 ± 18	Appendix 2
Bajocian	171.48 ± 1.22			mid Bajocian, mid-M42-base M43r	1193 ± 18	Appendix 2

Spline-fit of this data set is in Fig. 27.10; for details see text.

Kent and Gradstein (1985) in their Cretaceous and Jurassic geochronology study for the Decade of North America Geology publications had also suggested that a constant and linear spreading of the M-sequence was a reasonable template to interpolate the Oxfordian through Barremian time scale. Despite use of few tie-points for the age versus M-sequence km plot, the Jurassic—Cretaceous boundary was interpolated by Kent and Gradstein (1985) at 144 Ma, only slightly older than the final estimate of 143.1 ± 0.6 Ma in GTS2020, albeit using a different definition for that system boundary.

In Table 14A.5 of Chapter 14 in this book, stage durations according to the spline-curve are compared with Milankovitch-based duration estimates. Although the spline ages are probably unbiased estimates of the true ages, the Milankovitch cycle durations are probably better stage duration estimates. For a description of the technique to incorporate cycle durations in the stage age

TABLE 27.3 Stratigraphic—radioisotopic data set with cycle age and C-chron assignments, partially aligned to the mid-km C-sequence, used for the Upper Cretaceous time scale.

Stage (base)	radioisotopic date, 2-sigma	name of UK bentonite	cycle and cycle-age assignment	stratigraphy-C-chron assignment	mid-km C-sequence	Reference
Danian	66.04 ± 0.05					This chapter
			66.31 ± 0.5	top of 30N	1371.84	Batenburg et al. (2012)
			69.19 ± 0.5	top of 31N	1409.56	Batenburg et al. (2012)
			70.08 ± 0.5	*B. clinolobatus* Zone; mid-31R		
			70.65 ± 0.5	top of 32N	1481.12	Batenburg et al. (2012); GTS2012
Maastrichtian			72.5 ± 0.5	*B. eliasi* Zone; mid-32N		This chapter
	74.85 ± 0.43		184.5 74.3 ± 0.5	*E. jenneyi* Zone; top 33N	1549.41	This chapter; Appendix 2
			75.92 ± 0.5	*R. calcarata* Zone; upper 33N		This chapter; Appendix 2
			76.62 ± 0.5	*B. scotti* Zone; mid-33N		This chapter; Appendix 2
			197 79.9 ± 0.5	base 33N	1732.76	This chapter
			~197 80.62 ± 0.4	*B. obtusus* Zone; beneath base 33N	1723.76	This chapter; Appendix 2
Campanian	83.27 ± 0.11		~207	*S. leei III* Zone; base 33R	1862.32	S1019 in Wang et al. (2016)
	84.43 ± 0.15			*D. bassleri* Zone		Sageman et al. (2014)
			207.5	34N		
Santonian	85.66 ± 0.19		86.08 ± 0.35	top C. *undulatoplicatus* Zone		This chapter; Appendix 2
Coniacian	89.87 ± 0.18	Lewes Marl	222			Appendix 2; Gale (2019a)
	89.37 ± 0.15	Caburn Marl	224			Appendix 2; Gale (2019a)
	91.07 ± 0.28	Southerham Marls	225			
	91.15 ± 0.26	Glynde Marls	226			
	93.67 ± 0.31	Lulworth Marl	229			
Turonian	93.79 ± 0.26	C-T bentonite	232 93.65 ± 0.5			Batenburg et al. (2016)
mid-Cenomanian	96.12 ± 0.31	Thatcher Bentonite	238 96.5 ± 0.5			Batenburg et al. (2016)
	99.7 ± 0.3			~1 subzone above GSSP		Takashima et al. (2019)
Cenomanian	100.5 ± 0.14					GTS2012 and this chapter

Spline-fit for this data set is in Fig. 27.11. For details see text. *GSSP*, Global Boundary Stratotype Section and Point.

estimates the reader is referred to Chapter 14, Geomathematics (this book). The corrected cycle durations then can be used to estimate stage base age

estimates, combined with the constant-spreading-rate estimates for the durations of the Kimmeridgian through Callovian. These revised Callovian through Albian age

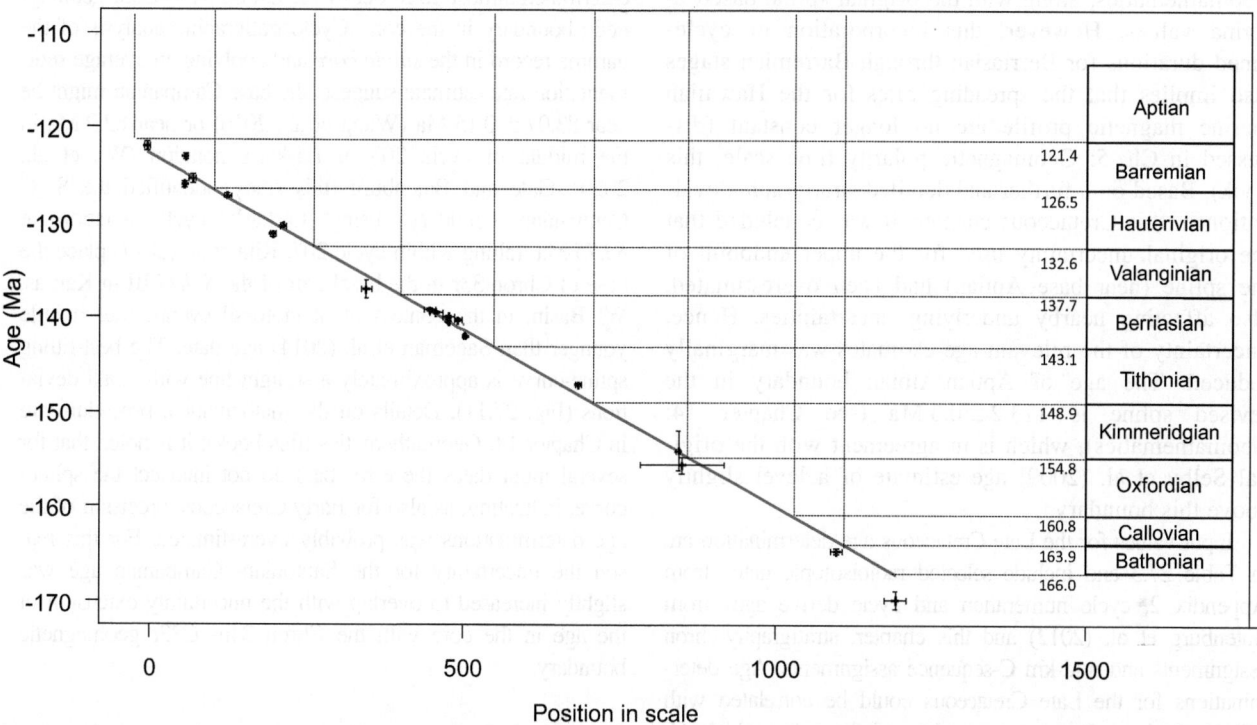

FIGURE 27.10 Early Cretaceous and Late Jurassic spline-fit and geologic time scale. Spline-fit errors are small, if present and may be invisible. For details see text in this chapter and in Agterberg et al. (2020, Ch. 14A: Geomathematics and statistical procedures (this book)). [Note that the basal ages for Jurassic stages used in GTS2020 used a different spline fit with incorporation of cyclostratigraphy (see Hesselbo et al. (2020, Ch. 26: The Jurassic Period, this volume))].

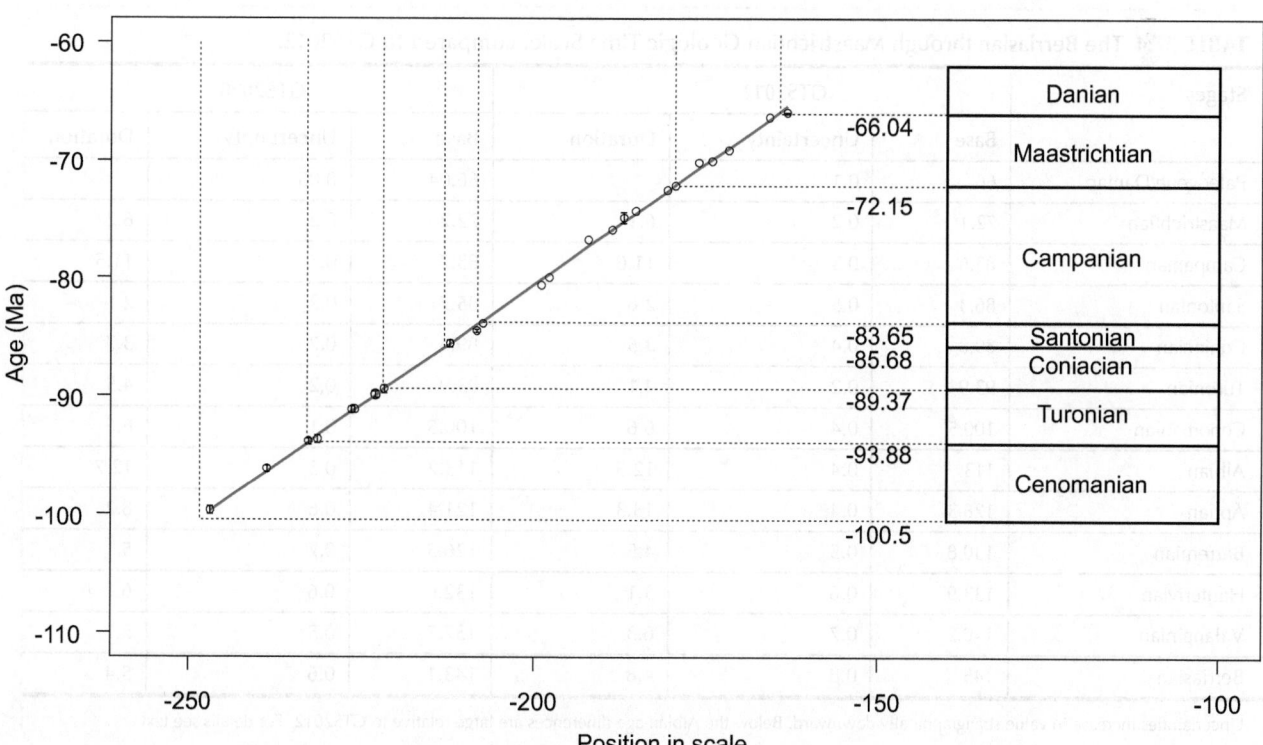

FIGURE 27.11 Late Cretaceous spline-fit and the Late Cretaceous time scale. Spline-fit uncertainties are small, if present, and may be invisible. For details see text in this chapter and in Agterberg et al. (2020, Ch. 14A: Geomathematics and statistical procedures, this book).

estimates are in Table 14A.6 of Chapter 14, Geomathematics, along with the original spline-based 2-sigma values. However, this incorporation of cycle-tuned durations for Berriasian through Barremian stages also implies that the spreading rates for the Hawaiian marine magnetic profile are no longer constant (discussed in Ch. 5: Geomagnetic polarity time scale, this book). Based on a further and detailed stratigraphic evaluation in this Cretaceous chapter, it was concluded that the original uncertainty used for the upper endpoint of the spline (near base Aptian) had been overestimated, also affecting nearby underlying uncertainties. Hence, uncertainty of the relevant age estimates was marginally reduced. The age of Aptian/Albian boundary in the revised spline is 113.2 ± 0.3 Ma (see Chapter 14: Geomathematics), which is in agreement with the original Selby et al. (2009) age estimate of a level slightly above this boundary.

Input values for the Late Cretaceous age determination are in Table 27.3 and include selected radioisotopic dates from Appendix 2, cycle numeration and cycle derive ages from Batenburg et al. (2012) and this chapter, stratigraphy-chron assignments and mid-km C-sequence assignments. Age determinations for the Late Cretaceous could be correlated with Milankovitch cycles, but a number of dates for which C-sequence and cycle input were available do not have reported 2-sigma values, and all dates were given equal weights for the spline-fitting. Wang et al. (2016) published U/Pb age dates on a bentonite in cored nonmarine sediments at 1019 m depth in the Songliao Basin, China. An age of 83.27 ± 0.11 Ma was determined almost 25 m below the Chron 34n–C33r geomagnetic boundary in the core. Cyclostratigraphic analysis of the gamma record in the *single core* and applying an average sedimentation rate estimate suggests the base Campanian might be near 83.07 ± 0.15 Ma (Wang et al., 2016) or near 82.9 Ma in the middle of cycle 205 of Laskar's notation (Wu et al., 2020). Gale and Batenburg (this study) identified the S–C Campanian boundary, using Laskar's cycle notation at *83.75* Ma, falling within cycle 207. Kita et al. (2017) place the base of Chron 33r in the basal part of the *S. leei* III in Kansas, WI Basin, in the context of nannofossil events, thus a little younger than Sageman et al. (2014) age date. The best-fitting spline-curve is approximately a straight line with small deviations (Fig. 27.11). Details on the mathematical procedure are in Chapter 14, Geomathematics (this book); it is noted that for several input dates the error bars do not intersect the spline-curve, indicating, as also for Early Cretaceous, precision of the age determinations was probably overestimated. For this reason the uncertainty for the Santonian–Campanian age was slightly increased to overlap with the uncertainty extension of the age in the core with the Chron 34n–C33r geomagnetic boundary.

27.3.2.1 Age of stage boundaries

The base Cretaceous and base Berriasian interpolate at 143.1 ± 0.6 Ma, 1.9 Myr younger than in GTS2012, and with a smaller uncertainty (Table 27.4). The Berriasian

TABLE 27.4 The Berriasian through Maastrichtian Geologic Time Scale, compared to GTS2012.

Stages	GTS2012			GTS2020		
	Base	Uncertainty	Duration	Base	Uncertainty	Duration
Paleogene/Danian	66	0.1		66.04	0.05	
Maastrichtian	72.1	0.2	6.1	72.2	0.2	6.2
Campanian	83.6	0.3	11.6	83.7	0.5	11.5
Santonian	86.3	0.5	2.6	85.7	0.2	2
Coniacian	89.8	0.4	3.5	89.4	0.2	3.7
Turonian	93.9	0.2	4.1	93.9	0.2	4.5
Cenomanian	100.5	0.4	6.6	100.5	0.1	6.6
Albian	113	0.4	12.5	113.2	0.3	12.7
Aptian	126.3	0.4	13.3	121.4	0.6	8.2
Barremian	130.8	0.5	4.5	126.5	0.7	5.1
Hauterivian	133.9	0.6	3.1	132.6	0.6	6.1
Valanginian	140.2	0.7	6.3	137.7	0.5	5.1
Berriasian	145	0.8	4.8	143.1	0.6	5.4

Uncertainties increase in value stratigraphically downward. Below the Albian age differences are large relative to GTS2012. For details see text.

GSSPs of the Cretaceous Stages, with location and primary correlation criteria

Stage	GSSP Location	Latitude, Longitude	Boundary Level	Correlation Events	Reference
Maastrichtian	Tercis les Bains, Landes, France	43°40'46.1"N 1°06'47.9"W*	level 115.2 on platform IV of the geological site at Tercis les Bains	Mean of 12 biostratigraphic criteria of equal importance. Near ammonite FAD of *Pachydiscus neubergicus*	Episodes **24**/4, 2001
Campanian	*candidates are in Italy and in Texas*			*Crinoid, LAD of Marsupites testudinarius or base of Chron C33r*	
Santonian	Olazagutia, Northern Spain	42°52'5.3"N 2°11'40"W	94.4 m in the eastern border of the Cantera de Margas quarry	Inoceramid bivalve, FAD *Platyceramus undulatoplicatus*	Episodes **37**/1, 2014
Coniacian	*candidates are in Poland (Slupia Nadbrzena) and Germany (Salzgitter)*			*Inoceramid bivalve, FAD of Cremnoceramus deformis erectus*	
Turonian	Pueblo, Colorado, USA	38°16'56"N 104°43'39"W*	base of Bed 86 of the Bridge Creek Limestone Member	Ammonite, FAD of *Watinoceras devonense*	Episodes **28**/2, 2005
Cenomanian	Mont Risou, Hautes-Alpes, France	44°23'33"N 5°30'43"E	36 m below the top of the Marnes Bleues Formation on the south side of Mont Risou	Foraminifer, FAD of *Thalmanninella globotruncanoides*	Episodes **27**/1, 2004
Albian	Col de Pré-Guittard Section, Drôme, France	44°29'47"N 5°18'41"E	37.4 m above the base of the Marnes Bleues Formation and 40 cm above the base of the Kilian Niveau	Foraminifer, FAD of *Microhedbergella renilaevis*	Episodes **40**/3, 2017
Aptian	*candidate is Gorgo a Cerbara, Umbria-Marche, central Italy*			*Base of Chron M0r; near ammonite, FAD of Deshayesites oglanlensis*	
Barremian	*candidate is Río Argos near Caravaca, Murcia province, Spain*			*Ammonite, FAD of Taveraidiscus hugii*	
Hauterivian	La Charce Section, Drôme Province, southeast France	44°28'10"N 5°26'37.4"E	base of Bed 189 of La Charce Section	Ammonite, FAD of genus *Acanthodiscus*	
Valanginian	*candidate is near Caravaca (S. Spain)*			*Calpionellid, FAD of Calpionellites darderi*	
Berriasian	*Tré Maroua, SE of Gap, southeast France*			*Calpionellid, FAD of Calpionella alpina*	

* according to Google Earth

FIGURE 27.12 Ratified and potential GSSPs and primary correlation markers for the Cretaceous stages (status as of January 2020). Details of each GSSP are available at http://www.stratigraphy.org, https://timescalefoundation.org/gssp/, and in the *Episodes* publications. *GSSP*, Global Boundary Stratotype Section and Point.

Stage measures 5.4 Myr in duration, 1 Myr longer than in GTS2012. The Valanginian Stage started at 137.7 ± 0.5 Ma and lasted 5.1 Myr. In GTS2012 base Valanginian was 140.2 ± 0.7 Ma in age and lasted 6.3 Myr. The Hauterivian Stage ranges from 132.6 ± 0.6 to 126.5 ± 0.7 Ma, lasting 6.1 Myr. In GTS2012 the stage had a rather short duration of 3.1 Myr, between 133.9 and 130.8 Ma. The Barremian Stage lasted 5.1 Myr from 126.5 ± 0.7 to 121.4 ± 0.6 Ma. In GTS2012, its top was at 126.3 Ma, a large numerical difference. In GTS2012 the Aptian lasted 13.3 Myr from 126.3 until 113 Ma, being the longest stage in the Cretaceous.

With a duration of 8.2 Myr from 121.4 ± 0.6 until 113.2 ± 0.3 the GTS2020 Aptian is only the third longest stage in the Cretaceous, after the Albian with 12.7 Myr and Campanian with 11.5 Myr durations. The proposed revised interpretation of 405-kyr cycles within the Aptian (Zhang et al., 2019) yields an estimated duration of 8.1 ± 0.5 Myr, virtually identical to the current geomathematical estimate in GTS2020. The Albian lasted 12.7 Myr from 113.2 to 100.5 ± 0.1 Ma for base-Cenomanian. The 405-kyr cyclostratigraphic duration estimate for the Albian of c. 12.45 ± 0.5 Myr (Grippo et al., 2004) is close to the spline-fit result.

Base Turonian has a direct age date of 93.9 ± 0.2, making the Cenomanian 6.6 Myr long. The Turonian Stage lasted for 4.5 Myr until base Coniacian at 89.4 ± 0.2 Ma; the latter stage lasted 3.7 Myr. The Santonian Stage ranges from 85.7 to 83.6 Ma, being the shortest Cretaceous stage with its 2.1 Myr duration. The second longest stage in the Cretaceous, the Campanian Stage lasted from 83.7 to 72.2 Ma, for a duration of 11.5 Myr. The Maastrichtian Stage started at 72.2 Ma and ended at the K/T boundary precisely dated at 66.04 Ma, with a duration of 6.2 Myr.

The Early Cretaceous stretches between 143.1 and 100.5 Ma, with a total duration of 42.6 Myr, and the Late Cretaceous ranges from 100.5 to 66.04 Ma, with a total duration of 34.5 Myr. The total duration of the Cretaceous Period itself is 77.1 Myr from 143.1 to 66.04 Ma.

The GTS2020 geochronologic scale for the Cretaceous Period differs substantially from that in GTS2012 for the boundary ages of the Berriasian through Aptian stages, and much less so for the seven younger stages. Particularly for the mid-Cretaceous, the puzzling age discrepancy with oceanic basement Ar—Ar dates needs investigation.

Bibliography

Agterberg, F.P., da Silva, A.C., and Gradstein, F.M., 2020, Chapter 14A - Geomathematical and statistical procedures. *In* Gradstein, F.M., Ogg, J.G., Schmitz, M.D., and Ogg, G.M. (eds), The *Geologic Time Scale 2020.* **Vol. 1** (this book). Elsevier, Boston, MA.

Aguado, R., Company, M., and Tavera, J.M., 2000, The Berriasian/Valanginian boundary in the Mediterranean region: new data from the Caravaca and Cehegin sections, SE Spain. *Cretaceous Research,* **21**: 1–21.

Aguado, R., de Gea, G.A., and O'Dogherty, L., 2014, Integrated biostratigraphy (calcareous nannofossils, planktonic foraminifera, and radiolarians) of an uppermost Barremian—lower Aptian pelagic succession in the Subbetic Basin (southern Spain). *Cretaceous Research,* **51**: 153–173.

Aguirre-Urreta, M.B., Rawson, P.F., Concheyro, G.A., Bown, P.R., and Ottone, E.G., 2005, Lower Cretaceous (Berriasian-Aptian) biostratigraphy of the Neuquén Basin. *In* Veiga, G., Spalletti, L., Howell, J.A., and Schwarz, E. (eds), The Neuquén Basin: A Case Study in Sequence Stratigraphy and Basin Dynamics. *Geological Society of London, Special Publication,* **252**: 57–81.

Aguirre-Urreta, M.B., Pazos, P.J., Lazo, D.G., Fanning, C.M., and Litvak, V.D., 2008, First U/Pb SHRIMP age of the Hauterivian stage, Neuquen Basin, Argentina. *Journal of South American Earth Sciences,* **26**: 91–99.

Aguirre-Urreta, B., Lescano, M., Schmitz, M.D., Tunik, M., Concheyro, A., Rawson, P.F., et al., 2015, Filling the gap: New precise Early Cretaceous radioisotopic ages from the Andes. *Geological Magazine,* **152** (3), 557–564.

Aguirre-Urreta, M.B., Schmitz, M., Lescano, M., Tunik, M., Rawson, P.F., Concheyro, A., et al., 2017, A high precision U-Pb radioisotopic age for the Agrio Formation, Neuquén Basin, Argentina: implications for the chronology of the Hauterivian Stage. *Cretaceous Research,* **75**: 193–204.

Aguirre-Urreta, M.B., Martinez, M., Schmitz, M., Lescano, M., Omarini, J., Tunik, M., et al., 2019, Interhemispheric radio-astrochronological calibration of the time scales from the Andean and the Tethyan areas in the Valanginian—Hauterivian (Early Cretaceous). *Gondwana Research,* **70**: 104–132.

Alvarez, W., Arthur, M.A., Fischer, A.G., Lowrie, W., Napoleone, G., Premoli Silva, I., et al., 1977, Upper Cretaceous—Palaeocene magnetic stratigraphy at Gubbio, Italy—V. Type section for the Late Cretaceous—Palaeocene geomagnetic reversal time scale. *Geological Society of America Bulletin,* **88**: 383–389.

Allemann, F., Catalano, R., Fares, F., and Remane, J., 1971, Standard calpionellid zonation (Upper Tithonian-Valanginian) of the Western Mediterranean province. *Proceedings II Planktonic Conference, Roma,* **1971**: 1337–1340.

Amédro, F., 1992, L'Albien du bassin Anglo-Parisien: ammonites, zonation phylétique, séquences. *Bulletin des Centres de Recherches Exploration − Production Elf-Aquitaine,* **16**: 187–233.

Amédro, F., and Matrion, B., 2007, Une coupe lithologique synth.tique dans l'Albien-type de l'Aube, France. *Bull. Inf. Géol. Bass. Paris,* **44**: 7–23.

Amédro, F., Cobban, W.A., Breton, G., and Rogron, P., 2002, *Metengonoceras teigenense* Cobban et Kennedy, 1989: une ammonite exotique d'origine Nord-Américaine dans le Cénomanien inférieur de Basse-Normandie (France). *Bulletin trimestriel de la Société géologique de Normandie et des amis du Muséum du Havre,* **87**: 5–25 5 pls (for 2000).

Amédro, F., Matrion, B., Magniez-Jannin, F., and Touch, R., 2014, La limite Albien inférieur- Albien moyen dans l'Albien type de l'Aube (France): ammonites, foraminiferes, séquences. *Revue de Paléobiologie,* **33**: 159–279.

Amédro, F., Matrion, B., Robaszynski, F., 2018. *Stratotype Turonien.* Muséum nationale d'histoire naturelle, Biotope, Meze, Paris, p. 416 (Patrimonie géologique, 8).

Amédro, F., Deconinck, J.-F., and Matrion, B., 2019, L'Albien type de l'Aube (France): Première description litho-biostratigraphique de lat totalité des Argiles tégulines de Courcelles. *Bulletin Inf. Géol. Bass. Paris,* **56** (2): 7–22.

Amodio, S., Ferreri, V., D'Argenio, B., Weissert, H., and Sprovieri, M., 2008, Carbon-isotope stratigraphy and cyclostratigraphy of shallow-marine carbonates: the case of San Lorenzello, Lower Cretaceous of southern Italy. *Cretaceous Research,* **29**: 803–813.

Ando, A., and Huber, B.T., 2007, Taxonomic revision of the Late Cenomanian planktonic foraminifera *Rotalipora greenhornensis* (Morow, 1934). *Journal of Foraminiferal Research,* **37**: 160–174.

Ando, A., Huber, B.T., MacLeod, K.G., Ohta, T., and Khim, B.-K., 2009, Blake Nose stable isotopic evidence against the mid-Cenomanian glaciation hypothesis. *Geology,* **37**: 451–454, https://doi.org/10.1130/G25580A.1.

Arnaud-Vanneau, A., and Bilotte, M., 1998, Larger benthic foraminifera. Columns for Jurassic chart of Mesozoic and Cenozoic sequence chronostratigraphic framework of European basins, by Hardenbol, J., Thierry, J., Farley, M.B., Jacquin, T., de Graciansky, P.-C., and Vail, P.R. (coordinators). *In* de Graciansky, P.-C., Hardenbol, J., Jacquin, T., and Vail, P.R. (eds), Mesozoic-Cenozoic Sequence Stratigraphy of European Basins. *SEPM Special Publication,* **60**: 763–781.

Banner, F.T., Copestake, P., and White, M.R., 1993, Barremian-Aptian Praehedbergellidae of the North Sea area: a reconnaissance. *Bulletin of the Natural History Museum: Geology Series,* **49**: 1–30.

Barron, E.J., 1983, A warm, equable Cretaceous: the nature of the problem. *Earth Science Review,* **19**: 305–338.

Bartolocci, P., Beraldini, M., Cecca, F., Faraoni, P., Marini, A., and Pallini, G., 1992, Preliminary results on correlation between Barremian ammonites and magnetic stratigraphy in Umbria-Marche Apennines (Central Italy). *Palaeopelagos*, **2**: 63−68.

Barbier, R., and Thieuloy, J.P., 1965, Étage Valanginien. *Mémoires du Bureau de Recherche Géologique et Minière*, **34**: 79−84.

Batenburg, S., Sprovieri, M., Gale, A.S., Hilgen, F.J., Lirer, F., and Laskar, J., 2012, Cyclostratigraphy and astronomical tuning of the Maastrichtian at Zumaia (Basque country, northern Spain). *Earth and Planetary Science Letters*, **359**: 264−278.

Batenburg, S., Gale, A.S., Sprovieri, M., Hilgen, F.J., Thibault, N., Boussaha, M., et al., 2014, An astronomical timescale for the Maastrichtian based on the Sopelana and Zumaia sections (Basque country, northern Spain). *Journal of the Geological Society*, **171**: 165−180.

Batenburg, S.J., De Vleesschouwer, D., Sprovieri, M., Hilgen, F.J., Gale, A.S., Singer, B.S., et al., 2016, Orbital control on the timing of oceanic anoxia in the Late Cretaceous. *Climate of the Past*, **12**: 1195−2009. https://doi.org/10.5194/cp-12-1995-2016.

Batenburg, S.J., Friedrich, O., Moriya, K., Voigt, S., Cournède, C., Moebius, I., et al., 2018, Late Maastrichtian carbon isotope stratigraphy and cyclostratigraphy of the Newfoundland Margin (Site U1403, IODP Leg 342). *Newsletters on Stratigraphy*, **51** (2), 245−260.

Baudin, F., Faraoni, P., Marini, A., and Pallini, G., 1997, Organic matter characterization of the "Faraoni Level" from Northern Italy (Lessini Mountains and Trento Plateau): comparison with that from Umbria-Marche Appennines. *Palaeopelagos*, **7**: 41−51.

Baudin, R., Bulot, L.G., Cecca, F., Coccioni, R., Gardin, S., and Renard, M., 1999, Un équivalent du "Niveau Faraoni" dans le Bassin du Sud-Est de la France, indice possible d'un événement anoxique fini-hauterivien étendu à la téthys méditerranéenne. *Bulletin de la Société géologique de France*, **170**: 487−498.

Baumberger, E., 1901, *Über Facies und Transgressionen der unteren Kreide am Nordrande der mediterrano-helvetischen. Bucht im westlichen Jura.* Wissenschaftliche Beilagen Berichte Töchterschule Basel, 1−44.

Bellier, J.P., and Moullade, M., 2002, Lower Cretaceous planktonic foraminiferal biostratigraphy of the western North Atlantic (ODP Leg 171B), and taxonomic clarification of key index species. *Revue de Micropaléontologie*, **45** (1), 9−26.

Bengtson, P., Cobban, W.A., Dodsworth, P., Gale, A.S., Kennedy, W.J., Lamolda, M.A., et al., 1996, The Turonian stage and substage boundaries. *Bulletin de l'Institut Royal des Sciences Naturelles de Belgique, Sciences de la Terre*, **66** (Suppl.), 69−79.

Bergen, J.A., 1994, Berriasian to early Aptian calcareous nannofossils from the Vocontian trough (SE France) and Deep Sea Drilling Site 534: new nannofossil taxa and a summary of low-latitude biostratigraphic events. *Journal of Nannoplankton Research*, **16**: 59−69.

Bersac, S., and Bert, D., 2012, Ontogenesis, variability and evolution of the Lower Greensand Deshayesitidae (Ammonoidea, Lower Cretaceous, southern England); reinterpretation of literature data; taxonomic and biostratigraphic implications. *Annales de la Muséum d'histoire naturelle, Nice*, **27**: 197−270.

Bice, K.L., Bralower, T.J., Duncan, R.A., Huber, B.T., Leckie, R.M., Sageman, B.B., 2002. Cretaceous climate-ocean dynamics: future directions for IODP. In: A JOI/USSSP and NSF Sponsored Workshop. Available from: < http://www.whoi.edu/ccod/CCOD_report.html > .

Birkelund, T., Hancock, J.M., Hart, M.B., Rawson, P.F., Remane, J., Robaszynski, R., et al., 1984, Cretaceous stage boundaries—proposals. *Geological Society of Denmark Bulletin*, **33**: 3−20.

Blair, S.A., and Watkins, D.K., 2009, High-resolution calcareous nanno-fossil biostratigraphy for the Coniacian/Santonian Stage boundary, Western Interior Basin. *Cretaceous Research*, **30**: 367−384.

Bodin, S., Godet, A., Föllmi, K.B., Vermeulen, J., Arnaud, H., Strasser, A., et al., 2006, The late Hauterivian Faraoni oceanic anoxic event in the western Tethys: evidence from phosphorus burial rates. *Palaeogeography, Palaeoclimatology, Palaeoecology*, **235**: 245−264.

Bornemann, A., Norris, R.D., Friedrich, O., Beckmann, B., Schouten, S., Sinninghe Damsté, J., et al., 2008, Isotopic evidence for glaciation during the Cretaceous supergreenhouse. *Science*, **319**: 189−192.

BouDagher-Fadel, M.K., Banner, F.T., and Whittaker, J.E., 1997, *Early Evolutionary History of Planktonic Foraminifera.* Chapman & Hall, pp. 1−269.

Bown, P.R., Rutledge, D.C., Crux, J.A., and Gallagher, L.T., 1998, Lower Cretaceous. *In* Bown, P.R. (ed), *Calcareous Nannofossil Biostratigraphy.* British Micropalaeontology Society Publication Series. London: Chapman & Hall, 86−131.

Bralower, T.J., Sliter, W.V., Arthur, M.A., Leckie, R.M., Allard, D.J., and Schlanger, S.O., 1993, Dysoxic/anoxic episodes in the Aptian-Albian (Early Cretaceous). *In* Pringle, M.S., Sager, W.W., Sliter, W.V., and Stein, S. (eds), The Mesozoic Pacific: Geology, Tectonics, and Volcanism. *American Geophysical Union Geophysics Monograph*, **77**: 5−37.

Bralower, T.J., Leckie, R.M., Slliter, W.V., and Thierstein, H.R., 1995, An integrated Cretaceous microfossil biostratigraphy. *In* Berggren, W.A., Kent, D.V., and Hardenbol, J. (eds), Geochronology, Time Scales and Global Stratigraphic Correlations: A Unified Temporal Framework for a Historical Geology. *SEPM Special Publication*, **54**: 65−79.

Bralower, T.J., Fullagar, P.D., Paull, C.K., Dwyer, G.S., and Leckie, R.M., 1997, Mid-Cretaceous strontium-isotope stratigraphy of deep-sea sections. *Geological Society of America Bulletin*, **109**: 1421−1442.

Bréheret, J.-G., 1988, Épisodes de sedimentation riche en matiere organique dans les marnes bleues d'âge aptien et albien de la partie pélagique du bassin Vocontien. *Bulletin de la Société géologique de France*, **8** (IV), 349−356.

Breistroffer, M., 1947, Sur les zones d'ammonites dans l'Albien de France et d'Angleterre. *Travaux du Laboratoire de Géologie de la Faculté des Sciences de l'Université de Grenoble*, **26**: 17−104.

Bryan, S.E., and Ernst, R.E., 2008, Revised definition of Large Igneous Provinces (LIPS). *Earth-Science Reviews*, **86**: 175−202.

Bulot, L.G., Blanc, E., Thieuloy, J.-P., and Remane, J., 1993, La limite Berriasien-Valanginien dans le Sud-Est de la France: données biostrati-graphiques nouvelles. *Comptes rendus de l'Académie des Sciences, Paris, Série II. Sciences de la Terre et des Planetes*, **317**: 387−394.

Bulot, L.G., Blanc, E., Company, M., Gardin, S., Hennig, S., Hoedemaeker, P.J., et al., 1996, The Valanginian Stage. *Bulletin de l'Institut Royal des Sciences Naturelles de Belgique, Sciences de la Terre*, **66** (Suppl.), 11−18.

Burnett, J.A., Gallagher, L.T., and Hampton, M.J., 1998, Upper Cretaceous. *In* Bown, P.R. (ed), *Calcareous Nannofossil Biostratigraphy.* British Micropalaeontological Society Publication Series. London: Chapman & Hall, 132−199.

Busnardo, R., 1965, Le Stratotype du Barrémien. 1.—Lithologie et Macrofaune, *and* Rapport sur l'étage Barrémien. *Mémoires du Bureau de Recherches Géologiques et Minières*, **34**: 101−116 and 161-169.

Busnardo, R., Thieuloy, J.-P., and Moullade, M., 1979, Hypostratotype Mesogéen de l'Étage Valanginien (sud-est de la France). *Les Stratotypes Français*, **6**: 143 pp.

Campbell, R.J., Howe, R.W., and Rexilius, J.P., 2004, Middle Campanian–lowermost Maastrichtian nannofossil and foraminiferal biostratigraphy of the northwestern Australian margin. *Cretaceous Research*, 25 (6), 827–864.

Cande, S.C., and Kent, D.V., 1992, A new geomagnetic polarity time scale for the Late Cretaceous and Cenozoic. *Journal of Geophysical Research*, 97: 13917–13951.

Cande, S.C., and Kent, D.V., 1995, Revised calibration of the geomagnetic polarity timescale for the Late Cretaceous and Cenozoic. *Journal of Geophysical Research*, 100: 6093–6095.

Caron, M., 1966, Globotruncanidae du Crétacé supérieur du synclinal de la Gruyère (Préalpes medians, Suisse). *Revue de Micropaléontologie*, 9: 68–93.

Caron, M., 1983, La spéciation chez les Foraminifères planctiques: une réponse adaptée aux contraints de l'environnement. *Zitteliana*, 10: 671–676.

Caron, M., 1985, Cretaceous planktonic foraminifera. *In* Bolli, H.M., Saunders, J.B., and Perch-Nielsen, K. (eds), *Plankton Stratigraphy*, Cambridge University Press, Cambridge, 17–86.

Caron, M., and Homewood, P., 1983, Evolution of early planktic foraminifers. *Marine Micropaleontology*, 7 (6), 453–462.

Casellato, C.E., 2010, Calcareous nannofossil biostratigraphy of upper Callovian-lower Berriasian successions from the Southern Alps, North Italy. *Rivista Italiana di Paleontologia e Stratigrafia*, 116: 357–404.

Casey, R., 1961, The stratigraphical palaeontology of the Lower Greensand. *Palaeontology*, 3: 487–621.

Casey, R., 1996, Lower Greensand ammonites and ammonite zonation. *Proceedings of the Geologists' Association*, 107: 69–76.

Casey, R., Bayliss, H.M., and Simpson, M.E., 1998, Observations on the lithostratigraphy and ammonite succession of the Aptian (Lower Cretaceous) Lower Greensand of Chale Bay, Isle of Wight, UK. *Cretaceous Research*, 19: 511–535.

Cecca, F., Pallini, G., Erba, E., Premoli-Silva, I., and Coccioni, R., 1994, Hauterivian-Barremian chronostratigraphy based on ammonites, nannofossils, planktonic foraminifra and magnetic chrons from the Mediterranean domain. *Cretaceous Research*, 15: 457–467.

Chambers, L.M., Pringle, M.S., and Fitton, J.G., 2004, Phreatomagmatic eruptions on the Ontong Java Plateau: an Aptian ^{40}Ar/^{39}Ar age for volcaniclastic rocks at ODP Site 1184. *In* Fitton, J.G., Mahoney, J.J., Wallace, P.J., and Sanders, A.D. (eds), Origin and Evolution of the Ontong Java Plateau. *Geological Society of London, Special Publication*, 229: 325–331.

Chang, S.-C., Zhang, H., Renne, P.R., and Fang, Y., 2009, High-precision ^{40}Ar/^{39}Ar age of the Jehol Biota. *Paleogeography, Paleoecology, Palaeoclimatology*, 280: 94–104.

Channell, J.E.T., and Grandesso, P., 1987, A revised correlation of magnetozones and calpionellid zones based on data from Italian pelagic limestone sections. *Earth and Planetary Science Letters*, 85: 222–240.

Channell, J.E.T., Bralower, T.J., and Grandesso, P., 1987, Biostratigraphic correlation of M-sequence polarity chrons M1 to M22 at Capriolo and Xausa (S. Alps, Italy). *Earth and Planetary Science Letters*, 85: 203–221.

Channell, J.E.T., Erba, E., and Lini, A., 1993, Magnetostratigraphic calibration of the Late Valanginian carbon isotope event in pelagic limestones from Northern Italy and Switzerland. *Earth and Planetary Science Letters*, 118: 145–166.

Channell, J.E.T., Erba, E., Muttoni, G., and Tremolada, F., 2000, Early Cretaceous magnetic stratigraphy in the APTICORE drill core and adjacent outcrop at Cismon (Southern Alps, Italy), and correlation to the proposed Barremian-Aptian boundary stratotype. *Geological Society of America Bulletin*, 112: 1430–1443.

Channell, J.E.T., Cecca, F., and Erba, E., 1995a, Correlations of Hauterivian and Barremian (Early Cretaceous) stage boundaries to polarity chrons. *Earth and Planetary Science Letters*, 134: 125–140.

Channell, J.E.T., Erba, E., Nakanishi, M., and Tamaki, K., 1995b, Late Jurassic-Early Cretaceous time scales and oceanic magnetic anomaly block models. *In* Berggren, W.A., Kent, D.V., and Hardenbol, J. (eds), Geochronology, Time Scales and Global Stratigraphic Correlations: A Unified Temporal Framework for a Historical Geology. *SEPM Special Publication*, 54: 51–63.

Channell, J.E.T., Casellato, C.E., Muttoni, G., and Erba, E., 2010, Magnetostratigraphy, nannofossil stratigraphy and apparent polar wander for Adria-Africa in the Jurassic–Cretaceous boundary interval. *Palaeogeography, Palaeoclimatology, Palaeoecology*, 293: 51–75.

Christensen, W.K., 1990, Upper Cretaceous belemnite stratigraphy of Europe. *Cretaceous Research*, 11: 371–386.

Christensen, W.K., 1996, A review of the Upper Campanian and Maastrichtian belemnite biostratigraphy of Europe. *Cretaceous Research*, 17: 751–766.

Christensen, W.K., 1997a, The Late Cretaceous belemnite family Belemnitellidae: taxonomy and evolutionary history. *Bulletin of the Geological Society of Denmark*, 44: 59–88.

Christensen, W.K., 1997b, Palaeobiogeography and migration in the Late Cretaceous belemnite family Belemnitellidae. *Acta Geologica Polonica*, 42: 457–495.

Clarke, L.J., and Jenkyns, H.C., 1999, New oxygen evidence for long-term Cretaceous climatic change in the Southern Hemisphere. *Geology*, 27: 699–702.

Cobban, W.A., 1993, Diversity and distribution of Late Cretaceous ammonites, Western Interior, United States. *In* Caldwell, W.G.E., and Kauffman, E.G. (eds), Evolution of the Western Interior Basin. *Geological Association of Canada Special Paper*, 39: 435–451.

Cobban, W.A., Walaszczyk, I., Obradovich, J.D., and McKinney, K.C., 2006, *A USGS zonal table for the Upper Cretaceous Middle Cenomanian-Maastrichtian of the Western Interior of the United States based on Ammonites, Inoceramids, and Radiometric Ages*, U.S. Geological Survey Open-File Report, 1250–2006, 45 p.

Coccioni, R., 2003. Cretaceous anoxic events: the Italian record. Abstract volume of the Séance spécialisée de la Société géologique de France. In: Paléocéanographie du Mésozoïque en réponse aux forçages de la paléogéographie et du paléoclimat, 10–11 June 2003, Paris. p. 10.

Coccioni, R., and Galeotti, S., 2003, The mid-Cenomanian Event: prelude to OAE 2. *Palaeogeography, Palaeoclimatology, Palaeoecology*, 190: 427–440, https://doi.org/10.1016/S0031-0182(02)00617-X.

Coccioni, R., and Premoli Silva, I., 1994, Planktonic foraminifera from the Lower Cretaceous of Rio Argos sections (southern Spain) and biostratigraphic implications. *Cretaceous Research*, 15 (6), 645–687.

Coccioni, R., and Premoli Silva, I., 2015, Revised Upper Albian–Maastrichtian planktonic foraminiferal biostratigraphy and magnetostratigraphy of the classical Tethyan Gubbio section (Italy). *Newsletters on Stratigraphy*, 48/1: 47–90.

Colin, J.-P., and Babinot, J.F., 1998, Ostracodes. Columns for Jurassic chart of Mesozoic and Cenozoic sequence chronostratigraphic framework of European basins, by Hardenbol, J., Thierry, J., Farley, M.B., Jacquin, T., de Graciansky, P.-C., and Vail, P.R. (coordinators), and chart supplements. *In* de Graciansky, P.-C., Hardenbol, J., Jacquin, T., and Vail, P.R. (eds), Mesozoic-Cenozoic Sequence Stratigraphy of European Basins. *SEPM Special Publication*, 60: 763–781.

Colleté, C., (ed.) 2010. *Stratotype Albien*. Muséum national d'Histoire naturelle. BRGM Orléans. Paris, 332 p.

Colloque sur la limite Jurassique-Crétacé, 1975, *Mémoires du Bureau de Recherches Géologiques et Minières*, **86**: 386–393 (summary).

Combemorel, R., and Christensen, W.K., 1998, Belemnites. Columns for Jurassic chart of Mesozoic and Cenozoic sequence chronostratigraphic framework of European basins, by Hardenbol, J., Thierry, J., Farley, M.B., Jacquin, T., de Graciansky, P.-C., and Vail, P.R. (coordinators). *In* de Graciansky, P.-C., Hardenbol, J., Jacquin, T., and Vail, P.R. (eds), Mesozoic-Cenozoic Sequence Stratigraphy of European Basins. *SEPM Special Publication,* **60**: 763–781.

Company, M., Sandoval, J., and Tavera, J.M., 1995, Lower Barremian ammonite biostratigraphy in the Subbetic Domain (Betic Cordillera, southern Spain). *Cretaceous Research*, **16**: 243–256.

Conybeare, W.D., and Phillips, W., 1822, *Outlines of the Geology of England and Wales, With an Introduction Compendium of the General Principles of That Science, and Comparative Views of the Structure of Foreign Countries. Part 1*. London: William Phillips 470 p.

Cope, J.C.W., 2007, Drawing the line: the history of the Jurassic-Cretaceous boundary. *Proceedings of the Geologists' Association*, **119**: 105–117.

Coquand, H., 1857a, Notice sur la formation crétacée du département de la Charente. *Bulletin de la Société géologique de France, Série 2*, **14**: 55–98.

Coquand, H., 1857b, Position des *Ostrea columba* et *biauriculata* dans le groupe de la craie inférieure. *Bulletin de la Société géologique de France, Série 2*, **14**: 745–766.

Coquand, H., 1861, Sur la convenance d'établir dans le groupe inférieur de la formation crétacée un nouvel étage entre le Néocomien proprement dit (couches à *Toxaster complanatus* et à *Ostrea couloni*) et le Néocomien supérieur (étage Urgonien de d'Orbigny). *Mémoires de la Société d'Emulation de Provence*, **1**: 127–139.

Coquand, H., 1871, Sur le Klippenkalk du département du Var et des Alpes-Maritimes. *Bulletin de la Société géologique de France*, **28**: 232–233.

Corbett, M.J., Watkins, D.K., and Pospichal, J.J., 2014, Quantitative analysis of calcareous nannofossil bioevents of the Late Cretaceous (Late Cenomanian-Coniacian) Western Interior Seaway and their reliability in established zonation schemes. *Marine Micropaleontology*, **109**: 30–45, https://doi.org/10.1016/j.marmicro.2014.04.002.

Cotillon, P., 1987, Bed-scale cyclicity of pelagic Cretaceous successions as a result of world-wide control. *Marine Geology*, **78**: 109–123.

Cotillon, P., Ferry, S., Gaillard, C., Jautée, E., Latreille, G., and Rio, M., 1980, Fluctuations des paramètres du milieu marin dans le domaine vocontien (France, Sud-Est) au Crétacé inférieur: mise en évidence par l'étude des formations marno-calcaires alternantes. *Bulletin de la Société Géologique de France*, **22**: 735–744.

Courtillot, V.E., and Renne, P.R., 2003, On the ages of flood basalt events. *Comptes Rendus Geoscience*, **335**: 113–140.

Cramer, B.D., and Jarvis, I., 2020, Chapter 11 - Carbon isotope stratigraphy. *In* Gradstein, F.M., Ogg, J.G., Schmitz, M.D., and Ogg, G.M. (eds), *The Geologic Time Scale 2020*. **Vol. 1** (this book). Elsevier, Boston, MA.

Crampton, J.S., and Gale, A.S., 2005, A plastic boomerang: speciation and intraspecific variation in the Cretaceous bivalve *Actinoceramus*. *Paleobiology*, **31**: 559–577.

Crampton, J., and Gale, A.S., 2009, Taxonomy of the *Actinoceramus sulcatus* lineage (Late Cretaceous; Bivalvia, Inoceramidae). *Journal of Paleontology*, **83**: 89–109.

D'Halloy, J.G.J., 1822, Observations sur un essai de cartes géologiques de la France, des Pays-Bas, et des contrées voisines. *Annales de Mines*, **7**: 353–376.

D'Hondt, A.V., 1998, Inoceramids. Columns for Jurassic chart of Mesozoic and Cenozoic sequence chronostratigraphic framework of European basins, by Hardenbol, J., Thierry, J., Farley, M.B., Jacquin, T., de Graciansky, P.-C., and Vail, P.R. (coordinators). *In* de Graciansky, P.-C., Hardenbol, J., Jacquin, T., and Vail, P.R. (eds), Mesozoic-Cenozoic Sequence Stratigraphy of European Basins. *SEPM Special Publication,* **60**: 763–781.

D'Hondt, A.V., Lamolda, M.A., and Pons, J.M., 2007, Stratigraphy of the Coniacian-Santonian transition, meeting organised by the Santonian Working Group of the Subcommission on Cretaceous Stratigraphy (coordinators) *Cretaceous Research*, **28** (1), 142.

d'Omalius d'Halloy, J.G.J., 1822, Observations sur un essai de cartes géologiques de la France, des Pays-Bas, et des contrées voisines. *Annales de Mines*, **7**: 353–376.

d'Orbigny, A., 1840, *Paléontologie française. Terrains crétacés. 1. Céphalopodes*. Paris, 622 p.

d'Orbigny, A., 1842, *Paléontologie française. Terrains crétacés. 2. Gastropodes*. Paris: Masson 456 p.

d'Orbigny, A., 1847, *Paléontologie française. Terrains crétacés. 4. Brachiopodes*. Paris: Masson 390 p.

d'Orbigny, A., 1849–1850, *Paléontologie française. Terrains jurassique. 1. Céphalopodes*. Paris: Bertrand.

d'Orbigny, A., 1852, *Cours élémentaire de paléontologie et de géologie stratigraphique*, Vol. 2. Paris: Masson, pp. 383–847.

De Grossouvre, A., 1901, *Recherches sur la Craie supérieure. Partie 1: stratigraphie générale. Mémoires pour servir à l'explication de la carte géologique détaillée de la France*. Paris: Imprimerie Nationale 1013 p.

De Wever, P., 1998, Radiolarians columns for Jurassic chart of Mesozoic and Cenozoic sequence chronostratigraphic framework of European basins, by Hardenbol, J., Thierry, J., Farley, M.B., Jacquin, T., de Graciansky, P.-C., and Vail, P.R. (coordinators). *In* de Graciansky, P.-C., Hardenbol, J., Jacquin, T., and Vail, P.R. (eds), Mesozoic-Cenozoic Sequence Stratigraphy of European Basins. *SEPM Special Publication,* **60**: 763–781.

Desor, E., 1854, Quelques mots sur l'étage inférieur du groupe néocomien (étage Valanginien). *Bulletin de la Société des Sciences Naturelles de Neuchâtel*, **3**: 172–180.

Desor, E., and Gressly, A., 1859, Études géologiques sur le Jura neuchâtelois. *Bulletin de la Société des Sciences Naturelles de Neuchâtel*, **4**: 1–159.

Douglas, R.G., and Savin, S.M., 1975, Oxygen and carbon isotope analyses of Tertiary and Cretaceous microfossils from Shatsky Rise and other sites in the North Pacific Ocean. *Initial Reports of the Deep Sea Drilling Project*, **32**: 509–520.

Dumont, A., 1849, Rapport sur la carte géologique du Royaume. *Bulletin de l'Académie royal des Sciences, des Lettres et des Beaux-Arts de Belgique*, **16**: 351–373.

Dzyuba, O.S., Izokh, O.P., and Shurygin, B.N., 2013, Carbon isotope excursions in Boreal Jurassic–Cretaceous boundary sections and their correlation potential. *Palaeogeography, Palaeoclimatology, Palaeoecology*, **381–382**: 33–46.

Earth Impact Database, 2018. Maintained by the Planetary and Space Science Centre, University of New Brunswick. Available from: < www.passc.net/EarthImpactDatabase/ > .

Elamri, Z., and Zaghbib-Turki, D., 2014, Santonian-Campanian biostratigraphy of the Kalaat Senan area (west-central Tunisia). *Turkish Journal of Earth Sciences*, **23** (2), 184–203.

Eldrett, J.S., Ma, C., Bergman, S.C., Lutz, B., Gregory, F.J., Dodsworth, P., et al., 2015, An astronomically calibrated stratigraphy of the Cenomanian, Turonian and earliest Coniacian from the Cretaceous Western Interior Seaway, USA: implications for global chronostratigraphy. *Cretaceous Research*, **56**: 316–344.

Erba, E., 1994, Nannofossils and superplumes: the Early Aptian nannoconid crisis. *Paleoceanography*, **9**: 483–501.

Erba, E., 2004, Calcareous nannofossils and Mesozoic oceanic anoxic events. *Marine Micropaleontology*, **52**: 85–106.

Erba, E., 2006, The first 150 million years history of calcareous nannoplankton: biosphere–geosphere interactions. *Palaeogeography, Palaeoclimatology, Palaeoecology*, **232**: 237–250.

Erba, E., Aguaro, R., Avram, E., Barboschkin, E.J., Bergen, J., Bralower, T.J., et al., 1996, The Aptian Stage. *Bulletin de l'Institut Royal des Sciences Naturelles de Belgique, Sciences de la Terre*, **66** (Suppl.), 31–43.

Erba, E., Channell, J.E.T., Claps, M., Jones, C., Larson, R., Opdyke, B., et al., 1999, Integrated stratigraphy of the Cismon Apticore (southern Alps, Italy): a "reference section" for the Barremian-Aptian interval at low latitudes. *Journal of Foraminiferal Research*, **29**: 371–391.

Erba, E., Bartolini, A., and Larson, R.I., 2004, Valanginian Weissert oceanic event. *Geology*, **32**: 149–152.

Erbacher, J., Thurow, J., and Littke, R., 1996, Evolution patterns of radiolaria and organic matter variations: a new approach to identify sea-level changes in mid-Cretaceous pelagic environments. *Geology*, **24**: 499–502.

Ernst, R.E., and Buchan, K.L., 2001, Large mafic magmatic events through time and links to mantle plume-heads. *In* Ernst, R.E., and Buchan, K.L. (eds), Mantle Plumes: Their Identification Through Time. *Geological Society of America Special Paper*, **352**: 483–575.

Falzoni, F., Petrizzo, M.P., Jenkyns, H.C., Gale, A.S., and Tsikos, H., 2016, Planktonic foraminiferal biostratigraphy and assemblage composition across the Cenomanian-Turonian boundary interval at Clot Chevalier (Vocontian Basin, SE France). *Cretaceous Research*, **59**: 69–97.

Fiet, N., 2000, Calibrage temporal de l'Aptien et des sous-étages associés par une approche cyclostratigraphique appliquée à la série pélagique de Marches-Ombrie (Italie centrale). *Bulletin de la Société géologique de France*, **171**: 103–113.

Fiet, N., and Gorin, G., 2000, Lithological expression of Milankovitch cyclicity in carbonate-dominated, pelagic, Barremian deposits in central Italy. *Cretaceous Research*, **21**: 457–467.

Fiet, N., Beaudoin, B., and Parize, O., 2001, Lithostratigraphic analysis of Milankovitch cyclicity in pelagic Albian deposits of central Italy: implications for the duration of the stage and substages. *Cretaceous Research*, **22**: 265–275.

Fiet, N., Quidelleur, X., Parize, O., Bulot, L.G., and Gillot, P.Y., 2006, Lower Cretaceous stage durations combining radiometric data and orbital chronology: towards a more stable relative time scale? *Earth and Planetary Science Letters*, **246**: 407–417.

Föllmi, K.B., Godet, A., Bodin, S., and Linder, P., 2006, Interactions between environmental change and shallow water carbonate buildup along the northern Tethyan margin and their impact on the Early Cretaceous carbon isotope record. *Paleoceanography*, **21**: article #PA4211, 16 pp, https://doi.org/10.1029/2006PA001313.

Frau, C., Bulot, L.G., Réhaková, D., and Wimbledon, W.A.P., 2016, The revision of the ammonite index species *Berriasella jacobi* Mazenot,

1939 and its consequences for the biostratigraphy of the Berriasian Stage. *Cretaceous Research*, **66**: 94–114.

Frau, C., Bulot, L.G., Delanoy, G., Moreno-Bedmar, J.A., Masse, J.-P., Tendil, A.J.-B., and Lanteaume, C., 2018, The Aptian GSSP candidate at Gorgo a Cerbara (central Italy): an alternative explanation of the bio-, litho- and chemostratigraphic markers. *Newsletters in Stratigraphy*, **51**: 311–326.

Galbrun, B., 1984, Magnétostratigraphie de la limitè Jurassique-Crétacé. Proposition d'une échelle de polarité à partir du stratotype du Berriasien (Berrias, Ardèche, France) et la Sierra de Lugar (Province de Murcie, Espagne). *Mémoires des Sciences della Terre*, **38**: 95 p.

Gale, A.S., 1995, Cyclostratigraphy and correlation of the Cenomanian Stage in Western Europe. *In* House, M.R., and Gale, A.S. (eds), Orbital Forcing Timescales and Cyclostratigraphy. *Geological Society Special Publication*, **85**: 177–197.

Gale, A.S., 2016, Roveacrinida (Crinoidea, Articulata) from the Santonian-Maastrichtian (Upper Cretaceous) of England, the US Gulf Coast (Texas, Mississippi) and southern Sweden. *Papers in Palaeontology*, **2**: 489532, https://doi.org/10.1002/spp2.1050.

Gale, A.S., 2017, An integrated microcrinoid zonation for the Lower Campanian chalk of southern England, and its implications for correlation. *Cretaceous Research*, **87**: 312352, https://doi.org/10.1016/j.cretres.2017.02.002.

Gale, A.S., 2019a, Correlation, age and significance of Turonian Chalk Hardgrounds in southern England and northern France: the roles of tectonics, eustasy and condensation. *Cretaceous Research*, **1103**: article #104164, https://doi.org/10.1016/j.cretres.2019.06.010.

Gale, A.S., 2019b, Microcrinoids (Echinodermata: Articulata: Roveacrinida) from the Cenomanian-Santonian chalk of the Anglo-Paris Basin: taxonomy and biostratigraphy. *Revues de Paléobiologie*, **38**: 379–533.

Gale, A.S., Montgomery, P., Kennedy, W.J., Hancock, J.M., Burnett, J. A., and McArthur, J.M., 1995, Definition and global correlation of the Santonian-Campanian boundary. *Terra Nova*, **7**: 611–622.

Gale, A.S., Kennedy, W.J., Burnett, J.A., Caron, M., and Kidd, B.E., 1996, The Late Albian to Early Cenomanian succession at Mont Risou near Rosans (Drôme, SE France): an integrated study (ammonites, inoceramids, planktonic foraminifera, nannofossils, oxygen and carbon isotopes). *Cretaceous Research*, **17**: 515–606.

Gale, A.S., Young, J.R., Shackleton, N.J., Crowhurst, S.J., and Wray, D. S., 1999, Orbital tuning of Cenomanian marly chalk successions: towards a Milankovitch time-scale for the Late Cretaceous. *Philosophical Transactions of the Royal Society of London, Series A*, **357**: 1815–1829.

Gale, A.S., Hardenbol, J., Hathway, B., Kennedy, W.J., Young, J.R., and Phansalkar, V., 2002, Global correlation of Cenomanian (Upper Cretaceous) sequences: evidence for Milankovitch control on sea level. *Geology*, **30**: 291–294.

Gale, A.S., Kennedy, J.W., Lees, J.A., Petrizzo, M.R., and Walaszczyk, I., 2007, An integrated study (inoceramid bivalves, ammonites, calcareous nannofossils, planktonic foraminifera, stable carbon isotopes) of the Ten Mile Creek section, Lancaster, Dallas County, north Texas, a candidate Global boundary Stratotype Section and Point for the base of the Santonian Stage. *Acta Geologica Polonica*, **57**: 113–160.

Gale, A.S., Hancock, J.M., Kennedy, W.J., Petrizzo, M.R., Lees, J.A., Walaszczyk, I., et al., 2008, An integrated study (geochemistry, stable oxygen and carbon isotopes, nannofossils, planktonic foraminifera, inoceramid bivalves, ammonites and crinoids) of the Waxahachie Dam Spillway section, north Texas, a possible boundary stratotype for the base of the Campanian Stage. *Cretaceous Research*, **29**: 131–167.

Gale, A.S., Bown, P., Caron, M., Crampton, J., Crowhurst, S.J., Kennedy, W.J., et al., 2011, The uppermost Middle and Upper Albian succession at the Col de Palluel, Hautes-Alpes, France: an integrated study (ammonites, inoceramid bivalves, planktonic foraminifera, nannofossils, geochemistry, stable oxygen and carbon isotopes, cyclostratigraphy). *Cretaceous Research*, **32**: 59–130.

Gale, A.S., Jenkyns, H.C., Tsikos, H., van Breugel, Y., Bottini, C., Erba, E., et al., 2019a, High-resolution bio- and chemostratigraphy of an expanded record of Oceanic Anoxic Event 2 (Late Cenomanian–Early Turonian) at Clot Chevalier, near Barrême, SE France (Vocontian Basin, SE France). *Newsletters on Stratigraphy*, **52**: 97–129, https://doi.org/10.1127/nos/2018/0445.

Gale, A.S., Kennedy, W.J., and Walaszczyk, I., 2019b, Upper Albian, Cenomanian and Lower Turonian stratigraphy, ammonite and inoceramid bivalve faunas from the Cauvery Basin, Tamil Nadu, South India. *Acta Geologica Polonica*, **69**: 161–338.

Gale, A.S., and Christensen, W.K., 1995, Occurrence of the belemnite *Actinocamax plenus* (Blainville) in the Cenomanian of SE France and its significance. *Bulletin of the Geological Society of Denmark*, **43**: 68–77.

Gambacorta, G., Jenkyns, H.C., Russo, F., Tsikos, H., Wilson, P.A., Faucher, G., et al., 2015, Carbon- and oxygen-isotope records of mid-Cretaceous Tethyan pelagic sequences from the Umbria-Marche and Belluno Basina (Italy). *Newsletters on Stratigraphy*, **48/3**: 299–323.

Gambacorta, G., Malinverno, A., and Erba, E., 2019, Orbital forcing of carbonate versus siliceous productivity in the late Albian-late Cenomanian (Umbria-Marche Basin, central Italy). *Newsletters on Stratigraphy*, **52/2**: 197–220.

Gerasimov, P.A., and Mikhailov, N.R., 1966, Volgian Stage and general stratigraphic scale of the Upper Series of the Jurassic System. *Isvestia Akademia Nauka, S.S.S.R., Geology Series*, **2**: 118–138 [In Russian].

Ghirardi, J., Deconinck, J.-F., Pellenard, P., Martinez, M., Bruneau, L., Amiotte-Suchet, P., and Pucéat, E., 2014, Multi-proxy orbital chronology in the aftermath of the Aptian Ocean Anoxic Event 1a: Palaeoceanographic implications (Serre Chaitieu section, Vocontian Basin, SE France). *Newsletters in Stratigraphy*, **47/3**: 247–262.

Gibson, F., 2006. Timescales of Plume-Lithosphere Interactions in LIPs: ^{40}Ar/^{39}Ar Geochronology of Alkaline Igneous Rocks From the Paraná-Etendeka Large Igneous Province. Available from: < www.largeigneousprovinces.org/06nov.html >.

Gibson, S.A., Thompson, R.N., and Day, J.A., 2006, Timescales and mechanisms of plume–lithosphere interactions: ^{40}Ar/^{39}Ar geochronology and geochemistry of alkaline igneous rocks from the Paraná–Etendeka large igneous province. *Earth and Planetary Science Letters*, **256**: 1–17.

Gilder, S., Chen, Y., Cogne, J.P., Tan, X.D., Courtillot, V., Sun, D.J., et al., 2003, Paleomagnetism of Upper Jurassic to Lower Cretaceous volcanic and sedimentary rocks from the western Tarim basin and implications for inclination shallowing and absolute dating of the M-0 (ISEA?) chron. *Earth and Planetary Science Letters*, **206**: 587–600.

Giraud, F., 1995, Recherche des périodicités astronomiques et des fluctuations du niveau marin à partir de l'étude du signal carbonaté des séries pélagiques alternants. *Documents des Laboratoires de Géologie Lyon*, **134**: 279 p.

Giraud, F., Beaufort, L., and Cotillon, P., 1995, Periodicities of carbonate cycles in the Valanginian of the Vocontian Trough: a strong obliquity control. *In* House, M.R., and Gale, A.S. (eds), Orbital Forcing Timescales and Cyclostratigraphy. *Geological Society Special Publication*, **85**: 143–164.

González-Donoso, J.M., Linares, D., and Robaszynski, F., 2007, The rotaliporids, a polyphyletic group of Albian-Cenomanian planktonic foraminifera: emendation of genera. *Journal of Foraminiferal Research*, **37** (2), 175–186.

Gorbatchik, T.N., 1986. *Jurassic and Early Cretaceous planktonic foraminifera of the south of the USSR* (in Russian). AN SSSR,"Nauka", pp. 1–239. (*Jurskie i Rammelovye planktonye foraminifery Juga SSSR*. Akademia NAUK SSSR. Moskovskoi Oschestvo Prirody).

Graciansky, P.-C., Hardenbol, J., Jacquin, T., and Vail, P.R. (eds.), 1998. *Mesozoic-Cenozoic Sequence Stratigraphy of European Basins*. SEPM Special Publication, **60**: 786 pp.

Gradstein, F.M., Ogg, J.G., and Smith, A.G., 2004, (coordinators) *A Geologic Time Scale 2004*. Cambridge: Cambridge University Press 589 p.

Gradstein, F.M., Ogg, J.G., Schmitz, M.D. and Ogg, G.M. (eds.), 2012. *The Geologic Time Scale* 2012. Elsevier Publ. Co., 1144 pp.

Gradstein, F.M., Gale, A., Kopaevich, L., Waskowska, A., Grigelis, A., Glinskikh, L., et al., 2017, The planktonic foraminifera of the Jurassic. Part II: Stratigraphy, palaeoecology and palaeobiogeography. *Swiss Journal of Paleontology*, **136**: 259–271.

Grant, S.F., Coe, A.L., and Armstrong, H.A., 1999, Sequence stratigraphy of the Coniacian succession of the Anglo-Paris Basin. *Geological Magazine*, **136**: 17–38.

Graciansky, P.-C., Hardenbol, J., Jacquin, T., and Vail, P.R. (eds), 1998. Mesozoic-Cenozoic Sequence Stratigraphy of European Basins. p. 786.

Green, K.A., and Brecher, A., 1974, Preliminary paleomagnetic results for sediments from Site 263, Leg 27. *Initial Reports, Deep Sea Drilling Project*, **27**: 405–413.

Grippo, A., Fischer, A.G., Hinnov, L.A., Herbert, T.M., and Premoli Silva, I., 2004, Cyclostratigraphy and chronology of the Albian stage (Piobbico core, Italy). *In* D'Argenio, B., Fischer, A.G., Premoli Silva, I., Weissert, H., and Ferreri, V. (eds), Cyclostratigraphy: Approaches and Case Histories. *SEPM Special Publication*, **81**: 57–81.

Grossman, E.L., and Joachimski, M.M., 2020, Chapter 10 - Oxygen isotope stratigraphy. *In* Gradstein, F.M., Ogg, J.G., Schmitz, M.D., and Ogg, G.M. (eds), *The Geologic Time Scale 2020*. Vol. **1** (this book). Elsevier, Boston, MA.

Hailwood, E.A., 1979, Paleomagnetism of Late Mesozoic to Holocene sediments from the Bay of Biscay and Rockall Plateau, drilled on IPOD Leg 48. *Initial Reports, Deep Sea Drilling Project*, **48**: 305–339.

Hancock, J.M., 1991, Ammonite scales for the Cretaceous System. *Cretaceous Research*, **12**: 259–291.

Hancock, J.M., 2001, A proposal for a new position for the Aptian/Albian boundary. *Cretaceous Research*, **22**: 677–683.

Hancock, J.M., and Gale, A.S., 1996, The Campanian Stage. *Bulletin de l'Institut Royal des Sciences Naturelles de Belgique, Sciences de la Terre*, **66** (Suppl.), 103–109.

Hancock, J.M., Peake, N.B., Burnett, J., Dhondt, A.V., Kennedy, W.J., and Stokes, R., 1993, High Cretaceous biostratigraphy at Tercis, SW France. *Bulletin de l'Institut Royal des Sciences Naturelles de Belgique, Sciences de la Terre*, **63**: 133–148.

Haq, B.U., 2014, Cretaceous eustasy revisited. *Global and Planetary Change*, **113**: 44–58.

Hardenbol, J., Thierry, J., Farley, M.B., Jacquin, T., de Graciansky, P.-C., Vail, P.R., et al., 1998, Mesozoic and Cenozoic sequence chronostratigraphic framework of European basins, 763–781, and chart supplements. *In* de Graciansky, P.-C., Hardenbol, J., Jacquin, T., and Vail, P.R. (eds), Mesozoic-Cenozoic Sequence Stratigraphy of European Basins. *SEPM Special Publication,* **60**: 3–13.

Harding, I.C., Smith, G.A., Riding, J.B., and Wimbledon, W.A.P., 2011, Inter-regional correlation of Jurassic/Cretaceous boundary strata based on the Tithonian-Valanginian dinoflagellate cyst biostratigraphy of the Volga Basin, western Russia. *Review of Palaeobotany and Palynology*, **167** (1–2), 82–116.

Harland, W.B., Armstrong, R.L., Cox, A.V., Craig, L.E., Smith, A.G., and Smith, D.G., 1990, *A Geologic Time Scale 1989*. Cambridge: Cambridge University Press 263 p.

Hart, M.B., 1980, A water depth model for the evolution of the planktonic Foraminiferida. *Nature*, **286**: 252–254.

Hart, M.B., 1999, The evolution and biodiversity of Cretaceous planktonic Foraminiferida. *Geobios*, **32** (2), 247–255.

Hart, M., Amédro, F., and Owen, H., 1996, The Albian stage and substage boundaries. *Bulletin de l'Institut Royal des Sciences Naturelles de Belgique, Sciences de la Terre*, **66** (Suppl.), 45–56.

Haynes, S.J., Huber, B.T., and Macleod, K.G., 2015, Evolution and phylogeny of mid-Cretaceous (Albian–Coniacian) biserial planktic foraminifera. *Journal of Foraminiferal Research*, **45** (1), 42–81.

He, H.Y., Pan, Y.X., Tauxe, L., Qin, H.F., and Zhu, R.X., 2008, Toward age determination of the M0r (Barremian–Aptian boundary) of the Early Cretaceous. *Physics of the Earth and Planetary Interiors*, **169**: 41–48.

Hennebert, M., Robaszynski, F., and Goolaerts, S., 2009, Cyclostratigraphy and chronometric scale in the Campanian – Lower Maastrichtian – the Abiod Formation at Ellès, central Tunisia. *Cretaceous Research*, **30**: 325–338.

Herbert, T.D., 1992, Paleomagnetic calibration of Milankovitch cyclicity in Lower Cretaceous sediments. *Earth and Planetary Science Letters*, **112**: 15–28.

Herbert, T.D., D'Hondt, S.L., Premoli-Silva, I., Erba, E., and Fischer, A.G., 1995, Orbital chronology of Cretaceous-Early Palaeocene marine sediments. *In* Berggren, W.A., Kent, D.V., and Hardenbol, J. (eds), Geochronology, Time Scales and Global Stratigraphic Correlations: A Unified Temporal Framework for a Historical Geology. *SEPM Special Publication,* **54**: 81–94.

Herrle, J.O., and Mutterlose, J., 2003, Calcareous nannofossils from the Aptian-Lower Albian of southeast France: palaeoecological and biostratigraphic implications. *Cretaceous Research*, **24**: 1–22.

Herrle, J.O., Kössler, P., Friedrich, O., Erlenkeuser, H., and Hemleben, C., 2004, High-resolution carbon isotope records of the Aptian to Lower Albian from SE France and the Mazagan Plateau (DSDP Site 545): a stratigraphic tool for paleoceanographic and paleobiologic reconstruction. *Earth and Planetary Science Letters*, **218**: 149–161.

Herrle, J.O., Schroder-Adams, C.J., Davis, W., Pugh, A.T., Galloway, J.M., and Fath, J., 2015, Mid-Cretaceous High Arctic stratigraphy, climate, and Oceanic Anoxic Events. *Geology*, **43**: 403–406.

Hesselbo, S.P., Coe, A.L., and Jenkyns, H.C., 1990, Recognition and documentation of depositional sequences from outcrop: an example from the Aptian and Albian on the eastern margin of the Wessex Basin. *Journal of the Geological Society*, **147**: 549–559.

Hesselbo, S.P., Ogg, J.G., and Ruhl, M., 2020, Chapter 26 – The Jurassic Period. *In* Gradstein, F.M., Ogg, J.G., Schmitz, M.D., and Ogg, G.M. (eds), *The Geologic Time Scale 2020*. Vol. **2** (this book). Elsevier, Boston, MA.

Hicks, J.F., and Obradovich, J.D., 1995, Isotopic age calibration of the GRTS from C33N to C31N: Late Cretaceous Pierre Shale, Red Bird section, Wyoming, USA. *Geological Society of America, Abstracts with Programs*, **27**: A174.

Hicks, J.F., Obradovich, J.D., and Tauxe, L., 1995, A new calibration point for the Late Cretaceous time scale: the ^{40}Ar/^{39}Ar isotopic age of the C33r/C33n geomagnetic reversal from the Judith River Formation (Upper Cretaceous), Elk Basin, Wyoming, USA. *Journal of Geology*, **103**: 243–256.

Hicks, J.F., Obradovich, J.D., and Tauxe, L., 1999, Magnetostratigraphy, isotopic age calibration and intercontinental correlation of the Red Bird section of the Pierre Shale, Niobrara County, Wyoming, USA. *Cretaceous Research*, **20**: 1–27.

Hilgen, F.J., Hinnov, L.A., Aziz, H.A., Abels, H.A., Batenburg, S., Bosmans, J.H.C., et al., 2014, Stratigraphic continuity and fragmentary sedimentation: the success of cyclostratigraphy as part of integrated stratigraphy. *In* Smith, D.G., Bailey, R.J., Burgess, P.M., and Fraser, A.J. (eds), *Strata and Time: Probing the Gaps in Our Understanding*. Geological Society of London, Special Publication, p. 404.

Hochuli, P.A., Menegatti, A.P., Weissert, H., Riva, A., Erba, E., and Premoli Silva, I., 1999, Episodes of high productivity and cooling in the early Aptian Alpine Tethys. *Geology*, **27**: 657–660.

Hoedemaeker, P.J., and Leereveld, H., 1995, Biostratigraphy and sequence stratigraphy of the Berriasian-lowest Aptian (Lower Cretaceous) of the Río Argos succession, Caravaca, SE Spain. *Cretaceous Research*, **16**: 195–230.

Hoedemaeker, P.J., Company, M.R., Aguirre Urreta, M.B., Avram, E., Bogdanova, T.N., Bujtor, L., et al., 1993, Ammonite zonation for the Lower Cretaceous of the Mediterranean region; basis for the stratigraphic correlation within IGCP, Project 262. *Revista Española de Paleontología*, **8**: 117–120.

Hoedemaeker, P.J., Reboulet, S., Aguirre-Urreta, M.B., Alsen, P., Atrops, F., Barragan, R., et al., 2003, Report on the 1st international workshop of the IUGS Lower Cretaceous Ammonite Working Group, the 'Kilian Group' (Lyon, 11 July 2002). *Cretaceous Research*, **24**: 89–94.

Houša, V., Pruner, P., Zakharov, V.A., Kostak, M., Chadima, M., Rogov, M.A., et al., 2007, Boreal–Tethyan correlation of the Jurassic–Cretaceous boundary interval by magneto- and biostratigraphy. *Stratigraphy and Geological Correlation*, **15** (3), 297–309.

Huang, C., 2018, Astronomical timescale for the Mesozoic. *Stratigraphy Timescales*, **3**: 1–70.

Huang, Z., Ogg, J.G., and Gradstein, F.M., 1993, A quantitative study of Lower Cretaceous cyclic sequences from the Atlantic Ocean and the Vocontian Basin (SE France). *Paleoceanography*, **8**: 275–291.

Huang, C., Hinnov, L.A., Fischer, A.G., Grippo, A., and Herbert, T., 2010, Astronomical tuning of the Aptian Stage from Italian reference sections. *Geology*, **38**: 899–902.

Huber, B.T., 1992, Paleobiogeography of Campanian-Maastrichtian foraminifera in the southern high latitudes. *Palaeogeography, Palaeoclimatology, Palaeoecology*, **92** (3–4), 325–360.

Huber, B.T., and Leckie, R.M., 2011, Planktic foraminiferal species turnover across deep-sea Aptian/Albian boundary sections. *Journal of Foraminiferal Research*, **41**: 53–95.

Huber, B.T., and Petrizzo, M.R., 2014, Evolution and taxonomic study of the Cretaceous planktonic foraminifer genus *Helvetoglobotruncana* Reiss, 1957. *Journal of Foraminiferal Research*, **44**: 40–57.

Huber, B.T., Hodell, D.A., and Hamilton, C.P., 1995, Middle-Late Cretaceous climate of the southern high latitudes: stable isotopic evidence for minimal equator-to-pole thermal gradients. *Geological Society of America Bulletin*, **107**: 1164–1191.

Huber, B.T., Norris, R.D., and MacLeod, K.G., 2002, Deep sea paleotemperature record of extreme warmth during the Cretaceous. *Geology*, **30**: 123–126.

Huber, B.T., MacLeod, K.G., and Tur, N.A., 2008, Chronostratigraphic framework for Upper Campanian-Maastrichtian sediments on the Blake Nose (subtropical North Atlantic). *Journal of Foraminiferal Research*, **38**: 162–182.

Huber, B.T., Petrizzo, M.R., Young, J.R., Falzoni, F., Gilardoni, S.E., Bown, P.R., et al., 2016, Pforams@mikrotax: a new online taxonomic database for planktonic foraminifera. *Micropaleontology*, **62** (6), 429–438.

Huber, B.T., Petrizzo, M.R., Watkins, D.K., Haynes, S.J., and MacLeod, K.G., 2017, Correlation of Turonian continental margin and deep-sea sequences in the subtropical Indian Ocean sediments by integrated planktonic foraminiferal and calcareous nannofossil biostratigraphy. *Newsletters on Stratigraphy*, **50**: 141–185.

Husson, D., Galbrun, B., Laskar, J., Hinnov, L.A., Thibault, N., Gardin, S., et al., 2011, Astronomical calibration of the Maastrichtian. *Earth and Planetary Science Letters*, **305**: 328–340.

Ingle, S., and Coffin, M.F., 2004, Impact origin for the greater Ontong Java Plateau? *Earth and Planetary Science Letters*, **218**: 123–134.

Jacquin, T., and de Graciansky, P.-C., 1998, Major transgressive-regressive cycles: the stratigraphic signature of European basin development. *In* de Graciansky, P.-C., Hardenbol, J., Jacquin, T., and Vail, P.R. (eds), Mesozoic-Cenozoic Sequence Stratigraphy of European Basins. *SEPM Special Publication*, **60**: 15–29.

Janasi, V.A., Freitas, V.A., and Heaman, L.H., 2011, The onset of flood basalt volcanism, Northern Paraná Basin, Brazil: a precise U–Pb baddeleyite/zircon age for a Chapecó-type dacite. *Earth and Planetary Science Letters*, **302**: 147–153.

Jarrard, R.D., 1974, Paleomagnetism of some Leg 27 sediment cores. *Initial Reports, Deep Sea Drilling Project*, **27**: 415–423.

Jarvis, I., Gale, A.S., Jenkyns, H.C., and Pearce, M.A., 2006, Secular variation in Late Cretaceous carbon isotopes: a new $\delta^{13}C$ carbonate reference curve for the Cenomanian–Campanian (99.6–70.6 Ma). *Geological Magazine*, **143**: 561–608.

Jarvis, I., Lignum, J.S., Gröcke, D.R., Jenkyns, H.C., and Pearce, M.A., 2011, Black shale deposition, atmospheric CO_2 drawdown, and cooling during the Cenomanian-Turonian Oceanic Anoxic Event. *Paleoceanography*, **26**: PA3201, https://doi.org/10.1029/2010PA002081.

Jenkyns, H.C., 1999, Mesozoic anoxic events and palaeoclimate. *Zentralblatt für Geologie und Paläontologie, Teil I*, **7–9**: 943–949.

Jenkyns, H.C., 2010, Geochemistry of oceanic anoxic events. *Geochemistry, Geophysics, Geosystems*, **11**: article # Q03004, https://doi.org/10.1029/2009GC002788.

Jenkyns, H.C., 2017, Transient cooling episodes during Cretaceous Oceanic Anoxic Events with special reference to OAE1a (Early Aptian). *Philosophical Transactions of the Royal Society*, **A376**: 20170073.

Jenkyns, H.C., Gale, A.S., and Corfield, R.M., 1994, Carbon- and oxygen-isotope stratigraphy of the English Chalk and the Italian Scaglia and its palaeoclimatic significance. *Geological Magazine*, **131**: 1–34.

Jinnah, Z.A., Roberts, E.M., Deino, A.L., Larsen, J.S., Link, P.K., and Fanning, C.M., 2009, New $^{40}Ar/^{39}Ar$ and detrital zircon U-Pb ages for the Upper Cretaceous Wahweap and Kaiparowits formations on the Kaiparowits Plateau, Utah: implications for regional correlation, provenance, and biostratigraphy. *Cretaceous Research*, **30**: 287–299.

Jones, C.E., and Jenkyns, H.C., 2001, Seawater strontium isotopes, oceanic anoxic events, and seafloor hydrothermal activity in the Jurassic and Cretaceous. *American Journal of Science*, **301**: 112–149.

Joo, Y.J., and Sageman, B.G., 2014, Cenomanian to Campanian carbon isotope chemostratigraphy from the Western Interior Basin, USA. *Journal of Sedimentary Research*, **84**: 529–542.

Jourdan, F., Renne, P.R., and Reimold, W.U., 2009, An appraisal of the ages of terrestrial impact structures. *Earth and Planetary Science Letters*, **286**: 1–13.

Kamo, S.L., and Riccardi, A.C., 2009, A new U–Pb zircon age for an ash layer at the Bathonian–Callovian boundary, Argentina. *GFF*, **131**: 177–182.

Kennedy, W.J., and Gale, A.S., 2017b, The Ammonoidea of the Lower Chalk. Part 7. *Palaeontographical. Society (Monograph)*, 461–571.

Kietzmann, D.A., Iglesia Llanos, M.P., and Kohan Martínez, M., 2018, Astronomical calibration of the Tithonian-Berriasian in the Neuquén Basin, Argentina: A contribution from the Southern Hemisphere to the geologic time scale. *In* Montenari, M. (ed), Stratigraphy and Timescales. **3**: 327–355.

Kolodny, Y., and Raab, M., 1988, Oxygen isotopes in phosphatic fish remains from Israel; paleothermometry of tropical Cretaceous and Tertiary shelf waters. *Palaeogeography, Palaeoclimatology, Palaeoecology*, **64**: 59–67.

Kuiper, K.F., Deino, A., Hilgen, F.J., Krijgsman, W., Renne, P.R., and Wijbrans, J.R., 2008, Synchronizing rock clocks of Earth history. *Science*, **320**: 500–504.

Kauffman, E.J., Kennedy, W.J., and Wood, C.J., 1996, The Coniacian stage and substage boundaries. *Bulletin de l'Institut Royal des Sciences Naturelles de Belgique, Sciences de la Terre*, **66** (Suppl.), 81–94.

Keating, B.H., and Helsley, C.E., 1978a, Magnetostratigraphic studies of Cretaceous age sediments from Sites 361, 363, and 364. *Initial Reports, Deep Sea Drilling Project*, **40**: 459–467.

Keating, B.H., and Helsley, C.E., 1978b, Magnetostratigraphic studies of Cretaceous sediments from DSDP Site 369. *Initial Reports, Deep Sea Drilling Project*, **41** (Suppl.), 983–986.

Keating, B.H., and Helsley, C.E., 1978c, Paleomagnetic results from DSDP Hole 391C and the magnetostratigraphy of Cretaceous sediments from the Atlantic Ocean floor. *Initial Reports, Deep Sea Drilling Project*, **44**: 523–528.

Kemp, T.S., 2005, *The Origin and Evolution of Mammals*. Oxford: Oxford University Press 342 pp.

Kemper, E., Rawson, P.F., and Thieuloy, J.P., 1981, Ammonites of Tethyan ancestry in the early Lower Cretaceous of north-west Europe. *Palaeontology*, **24**: 251–311.

Kennedy, W.J., 1984, Ammonite faunas and the 'standard zones' of the Cenomanian to Maastrichtian Stages in their type areas, with some proposals for the definition of the stage boundaries by ammonites. *Geological Society of Denmark Bulletin*, **33**: 147–161.

Kennedy, W.J., 2019. The Ammonoidea of the Upper Chalk. Part 1. *Monographs of the Palaeontographical Society*, **173**: 112 pp.

Kennedy, W.J., and Cobban, W.A., 1991, Stratigraphy and inter-regional correlation of the Cenomanian-Turonian transition in the Western Interior of the United States near Pueblo, Colorado. A potential boundary stratotype for the base of the Turonian Stage. *Newsletters on Stratigraphy*, **24**: 1–33.

Kennedy, W.J., and Gale, A.S., 2015, Ammonites from the Albian and Cenomanian (Cretaceous) of Djebel Mhrila, Tunisia. *Revues de Paléobiologie*, **34** (2), 235–361.

Kennedy, W.J., and Gale, A.S., 2016, Late Turonian ammonites from northwestern Aquitaine, France. *Cretaceous Research*, **58**: 265–296.

Kennedy, W.J., and Gale, A.S., 2017a, Trans-Tethyan correlation of the Lower-Middle Cenomanian boundary interval; southern England (Southerham, near Lewes Sussex) and Douar el Khiana, northeastern Algeria. *Acta Geologica Polonica*, **67**: 75–108.

Kennedy, W.J., Juignet, P., and Girard, J., 1990, *Budaiceras hyatti* (Shattuck, 1903), a North American index ammonite from the Lower Cenomanian of Haute Normandie, France. *Neues Jahrbuch für Geologie und Paläontologie Abhandlungen*, **1990**: 525–535.

Kennedy, W.J., Christensen, W.K., Hancock, J.M., 1995. Defining the base of the Maastrichtian and its substages. Internal report for the Maastrichtian Working Group. In: *Second International Symposium on Cretaceous Boundaries, Brussels, 8–16 September 1995*, 13 p. [Unpublished].

Kennedy, W.J., Gale, A.S., Bown, P.R., Caron, M., Davey, R.J., Gröcke, D., et al., 2000a, Integrated stratigraphy across the Aptian-Albian boundary in the Marnes Bleues at the Col de Pré-Guittard, Arnayon (Brôme), and at Tartonne (Alpes-de-Haute-Provence), France: a candidate Global Boundary Stratotype Section and Boundary Point for the base of the Albian Stage. *Cretaceous Research*, **21**: 591–720.

Kennedy, W.J., Walaszczyk, I., and Cobban, W.A., 2000b, Pueblo, Colorado, USA, candidate Global Boundary Stratotype Section and Point for the base of the Turonian Stage of the Cretaceous and for the Middle Turonian substage, with a revision of the Inoceramidae (Bivalve). *Acta Geologica Polonica*, **50**: 295–334.

Kennedy, W.J., Gale, A.S., Lees, J.A., and Caron, M., 2004, Definition of a Global Boundary Stratotype Section and Point (GSSP) for the base of the Cenomanian Stage, Mont Risou, Hautes-Alpes, France. *Episodes*, **27**: 21–32.

Kennedy, W.J., Walaszczyk, I., and Cobban, W.A., 2005, The Global Boundary Stratotype Section and Point for the base of the Turonian Stage of the Cretaceous: Pueblo, Colorado, U.S.A. *Episodes*, **28** (2), 93–104.

Kennedy, W.J., Gale, A.S., Huber, B., Petrizzo, M.R., and Jenkyns, H.C., 2014, Integrated stratigraphy across the Aptian-Albian boundary in the Marnes Bleue, at the Col de Pre-Guittard, Arnayon (Drome) revisited, a candidate Global Boundary Stratotype Section and new proposal of a Boundary Point for the base of the Albian Stage: the first occurrence of the planktic foraminiferan *Microhedbergella renilaevis* Huber and Leckie, 2011. *Cretaceous Research*, **51**: 248–259.

Kennedy, W.J., Gale, A.S., Huber, B.T., Petrizzo, M.R., Bown, P., and Jenkyns, H.C., 2017, The Global Boundary Stratotype Section and Point (GSSP) for the base of the Albian Stage, of the Cretaceous, the Col de Palluel section, Arnayon, Drôme, France. *Episodes*, **40**: 177–188.

Kent, D.V., and Gradstein, F.M., 1985, A Cretaceous and Jurassic geochronology. *Bulletin of the Geological Society of America*, **96**: 1419–1427.

Kerr, A.C., 1998, Oceanic plateau formation: a cause of mass extinction and black shale deposition around the Cenomanian-Turonian boundary? *Journal of the Geological Society*, **155**: 619–626.

Kietzmann, D.A., Palma, R.M., and Iglesia Llanos, M.P., 2015, Cyclostratigraphy of an orbitally-driven Tithonian–Valanginian carbonate ramp succession, Southern Mendoza, Argentina: implications for the Jurassic–Cretaceous boundary in the Neuquén Basin. *Sedimentary Geology*, **315**: 29–46.

Kilian, W., 1888, *Description Géologique de la Montagne de Lure (Basses-Alpes)*. Paris: Université Masson, These de Doctorate 458 pp.

Kita, Z.A., Watkins, D.K., and Sageman, B.S., 2017, High-resolution calcareous nannofossil biostratigraphy of the Santonian/Campanian Stage boundary, Western Interior Basin, USA. *Cretaceous Research*, **69**: 49–55.

Koeberl, C., Armstrong, R.A., and Reimold, W.U., 1997, Morokweng, South Africa: a large impact structure of Jurassic–Cretaceous boundary age. *Geology*, **25**: 731–734.

Kuhnt, W., and Moullade, M., 2007, The Gargasian (middle Aptian) of La Marcouline section at Cassis–La Bédoule (SE France): stable isotope record and orbital cyclicity. *Carnets de Géologie*, **2007** (02), 1–9.

Kuhnt, W., Holbourn, A., Gale, A., Chellai, E.H., and Kennedy, W.J., 2009, Cenomanian sequence stratigraphy and sea-level fluctuations in the Tarfaya Basin (SW Morocco). *Geological Society of America Bulletin*, **121**: 1695–1710.

Lamolda, M.A., and Paul, C.R.C., 2007, Carbon and Oxygen Stable Isotopes across the Coniacian-Santonian boundary at Olazagutia (Navarra Province), Spain. *Cretaceous Research*, **28**: 18–29.

Lamolda, M.A., Hancock, J.M., Burnett, J.A., Collom, C.J., Christensen, W.K., Dhondt, A.V., et al., 1996, The Santonian Stage and substages. *Bulletin de l'Institut Royal des Sciences Naturelles de Belgique, Sciences de la Terre*, **66** (Suppl.), 95–102.

Lamolda, M.A., Peryt, D., and Ion, J., 2007, Planktonic foraminiferal bioevents in the Coniacian/Santonian boundary interval at Olazagutia, Navarra province, Spain. *Cretaceous Research*, **28** (1), 18–29.

Lamolda, M.A., Paul, C.R.C., and Peryt, D., 2014, The Global Boundary Stratotype and Section Point for the base of the Santonian Stage, "Cantera de Margas", Olazagutia, northern Spain. *Episodes*, **37**: 2–13.

Landman, N.H., and Waage, K.M., 1993, Scaphitid ammonites of the Upper Cretaceous (Maastrichtian) Fox Hills Formation in South Dakota and Wyoming. *Bulletin of the American Museum of Natural History*, **215**: 257.

Large Igneous Provinces Commission (of the International Association of Volcanology and Chemistry of the Earth's Interior), 2011. LIP Record. Available from: < http://largeigneousprovinces.org/record > [and subpages].

Larson, R.L., 1991, Latest pulse of the Earth: evidence for a mid Cretaceous super plume. *Geology*, **19**: 547–550.

Larson, R.L., and Erba, E., 1999, Onset of the mid-Cretaceous greenhouse in the Barremian-Aptian: igneous events and the biological, sedimentary, and geochemical responses. *Paleoceanography*, **14**: 663–678.

Laskar, J., Fienga, A., Gastineau, M., and Manche, H., 2011, La2010: a new orbital solution for the long-term motion of the Earth. *Astronomy Astrophysics*, **532**: A89.

Leckie, R.M., Bralower, T.J., and Cashman, R., 2002, Oceanic anoxic events and plankton evolution: biotic response to tectonic forcing during the mid-Cretaceous. *Paleoceanography*, **17**: 1041, https://doi.org/10.1029/2001PA000623.

Leereveld, H., 1995, Dinoflagellate cysts from the Lower Cretaceous Río Argos succession (SE Spain). *LPP Contributions Series*, **2**: 175.

Lena, L., López-Martínez, R., Lescano, M., Aguire-Urreta, B., Concheyro, A., Vennari, V., et al., 2019, High-precision U–Pb ages in the early Tithonian to early Berriasian and implications for the numerical age of the Jurassic–Cretaceous boundary. *Solid Earth*, **10**: 1–14.

Lerbekmo, J.F., 1989, The stratigraphic position of the 33-33r (Campanian) polarity chron boundary in southeastern Alberta. *Bulletin of Canadian Petroleum Geology*, **37**: 43–47.

Lerbekmo, J.F., and Braman, D.R., 2002, Magnetostratigraphic and biostratigraphic correlation of late Campanian and Maastrichtian marine and continental strata from the Red Deer River Valley to the Cypress Hills, Alberta, Canada. *Canadian Journal of Earth Sciences*, **39**: 539–557.

Li, L., Keller, G., and Stinnesbeck, W., 1999, The Late Campanian and Maastrichtian in northwestern Tunisia: palaeoenvironmental inferences from lithology, macrofauna and benthic foraminifera. *Cretaceous Research*, **20** (2), 231–252.

Li, X., Jenkyns, H.C., Wang, C., Hu, X., Chen, X., Wei, Y., et al., 2006, Upper Cretaceous carbon- and oxygen-isotope stratigraphy of hemipelagic carbonate facies, southern Tibet, China. *Journal of the Geological Society*, **163**: 375–382.

Li, X., Bralower, T.J., Montañez, I.P., Osleger, D.A., Arthur, M.A., Bice, D.M., et al., 2008, Toward an orbital chronology for the early Aptian Oceanic Anoxic Event (OAE1a, ~120 Ma). *Earth and Planetary Science Letters*, **271**: 88–100.

Lini, A., Weissert, H., and Erba, E., 1992, The Valanginian carbon isotope event: a first episode of greenhouse climate conditions during the Cretaceous. *Terra Nova*, **4**: 374–384.

Locklair, R.E., and Sageman, B.B., 2008, Cyclostratigraphy of the Upper Cretaceous Niobrara Formation, Western Interior, U.S.A.: a Coniacian–Santonian orbital timescale. *Earth and Planetary Science Letters*, **269**: 539–552.

Longoria, J.F., 1977, Bioestratigrafía del Cretàcico Inferior basada en microfòsiles planctònicos. *Boletin de la Sociedad Geològica Mexicana*, **38/1**: 2–17.

López-Martínez, R., Aguirre-Urreta, B., Lescano, M., Concheyro, A., Vennari, V., and Ramos, V.A., 2017, Tethyan calpionellids in the Neuquén Basin (Argentine Andes), their significance in defining the Jurassic/Cretaceous boundary and pathways for Tethyan-Eastern Pacific connections. *Journal of South American Earth Sciences*, **78**: 116–125.

Lowrie, W., and Alvarez, W., 1977, Late Cretaceous geomagnetic polarity sequence: detailed rock and palaeomagnetic studies of the Scaglia Rossa limestone at Gubbio, Italy. *Geophyics Journal of the Royal Astronomical Society*, **51**: 561–581.

Lowrie, W., Alvarez, W., Premoli-Silva, I., and Monechi, S., 1980, Lower Cretaceous magnetic stratigraphy in Umbrian pelagic carbonate rocks. *Geophysical Journal, Royal Astronomical Society*, **60**: 263–281.

Luber, T.L., Bulot, L.G., Redfern, J., Frau, C., Arantegui, A., and Masrour, M., 2017, A revised ammonoid biostratigraphy for the Aptian of NW Africa: Essaouira-Agadir Basin, Morocco. *Cretaceous Research*, **79**: 12–34.

Luber, T.L., Bulot, L.G., Redfern, J., Nahim, M., Jeremiah, J., Simmons, M., et al., 2019, A revised chronostratigraphic framework for the Aptian of the Essaouira-Agadir Basin, a candidate type section for the NW African Atlantic Margin. *Cretaceous Research*, **93**: 292–317.

Lucas, S.G., 1997, *Dinosaurs, the Textbook*, Second ed. Dubuque, IA: Wm. C. Brown Publishers: 336 pp.

Lucas, S.G., Kirkland, J.I., and Estep, J.W. (eds), 1998. Lower and Middle Cretaceous terrestrial ecosystems. *New Mexico Museum of Natural History and Science Bulletin* **14**: 330 pp.

Lucas, S.G., Sulliva, R.M., and Spielmann, J.A., 2012, Cretaceous vertebrate biochronology, North American Western Interior. *Journal of Stratigraphy*, **36** (2), 426–461.

Ma, C., Meyers, S.R., Sageman, B.B., Singer, B.S., and Jicha, B.R., 2014, Testing the astronomical timescale for oceanic anoxic event 2, and its extension into Cenomanian strata of the Western Interior Basin, USA. *Bulletin of the Geological Society of America*, **126**: 974–989.

Ma, C., Meyers, S.R., and Sageman, B.B., 2019, Testing Late Cretaceous astronomical solutions in a 15 million year astrochronological record from North America. *Earth and Planetary Science Letters*, https://doi.org/10.1016/j.epsl.2019.01.053.

Magniez-Jannin, F., 1995, Cretaceous stratigraphic scales based on benthic foraminifera in West Europe (biochronohorizons). *Bulletin de la Société géologique de France*, **166**: 565–572.

Mahoney, J.J., Storey, M., Duncan, R.A., Spencer, K.J., and Pringle, M.S., 1993, Geochemistry and age of the Ontong Java Plateau. *In* Pringle, M.S. (ed), The Mesozoic Pacific: Geology, Tectonics, and Volcanism. *American Geophysical Union Geophysical Monographs*, **77**: 233–262.

Mahoney, J.J., Fitton, J.G., Wallace, P.J., et al., 2001. Basement drilling of the Ontong Java Plateau Sites 1183-1187. *Proceedings of the Ocean Drilling Program, Initial Reports*, **192**: College Station, TX (Ocean Drilling Program) http://www-odp.tamu.edu/publications/192_IR/192ir.htm.

Mahoney, J.J., Duncan, R.A., Tejada, M.L.G., Sager, W.W., and Bralower, T.J., 2005, Jurassic-Cretaceous boundary age and mid-ocean-ridge-type mantle source for Shatsky Rise. *Geology*, **33**: 185–188.

Malinverno, A., Erba, E., and Herbert, T.D., 2010, Orbital tuning as an inverse problem: chronology of the early Aptian oceanic anoxic event 1a (Selli Level) in the Cismon APTICORE. *Paleoceanography*, **25**: article #PA2203, https://doi.org/10.1029/2009PA001769.

Malinverno, A., Hildebrandt, J., Tominaga, M., and Channell, J.E., 2012, M-sequence geomagnetic polarity time scale (MHTC12) that steadies global spreading rates and incorporates astrochronology constraints. *Journal of Geophysical Research: Solid Earth*, **117** (B6), article #B06104, 17 pp, https://doi.org/10.1029/2012JB009260.

Martinez, M., Pellenard, P., Deconinck, J.-F., Monna, F., Riquier, L., Boulila, S., et al., 2012, An orbital floating time scale of the Hauterivian/Barremian GSSP from a magnetic susceptibility signal (Rio Argos, Spain). *Cretaceous Research*, **36**: 106–115.

Martinez, M., Deconinck, J.F., Pellenard, P., Reboulet, S., and Riquier, L., 2013, Astrochronology of the Valanginian stage from reference sections (Vocontian Basin, France) and palaeoenvironmental implications for the Weissert Event. *Palaeogeography, Palaeoclimatology, Palaeoecology*, **376**: 91–102.

Martinez, M., Deconinck, J.F., Pellenard, P., Riquier, L., Company, M., Reboulet, S., et al., 2015, Astrochronology of the Valanginian–Hauterivian stages (Early Cretaceous): chronological relationships between the Paraná–Etendeka large igneous province and the Weissert and the Faraoni events. *Cretaceous Research*, **131**: 158–173.

Martinez, M., Aguardo, R., Company, M., Sandoval, J., and O'Dogherty, L., 2020, Integrated astrochronology of the Barremian Stage (Early Cretaceous) and its biostratigraphic subdivisions. EGU General Assembly, Online 4−8 May 2020: https://presentations.copernicus.org/EGU2020/EGU2020-8923_presentation.pdf.

Masse, J.-P., 1998, Calcareous algae. Columns for Jurassic chart of Mesozoic and Cenozoic sequence chronostratigraphic framework of European basins, by Hardenbol, J., Thierry, J., Farley, M.B., Jacquin, T., de Graciansky, P.-C., and Vail, P.R. (coordinators). *In* de Graciansky, P.-C., Hardenbol, J., Jacquin, T., and Vail, P.R. (eds), Mesozoic-Cenozoic Sequence Stratigraphy of European Basins. *SEPM Special Publication*, **60**: 763−781.

Masse, J.-P., and Philip, J., 1998, Rudists. Columns for Jurassic chart of Mesozoic and Cenozoic sequence chronostratigraphic framework of European basins, by Hardenbol, J., Thierry, J., Farley, M.B., Jacquin, T., de Graciansky, P.-C., and Vail, P.R. (coordinators). *In* de Graciansky, P.-C., Hardenbol, J., Jacquin, T., and Vail, P.R. (eds), Mesozoic-Cenozoic Sequence Stratigraphy of European Basins. *SEPM Special Publication*, **60**: 763−781.

Masters, B.A., 1977, Mesozoic planktonic foraminifera. A world-wide review and analysis. *Oceanic Micropaleontology*, **1**: 301−731.

McArthur, J.M., Kennedy, W.J., Gale, A.S., Thirlwall, M.F., Chen, M., Burnett, J., et al., 1992, Strontium isotope stratigraphy in the late Cretaceous: international correlation of the Campanian/Maastrichtian boundary. *Terra Nova*, **4**: 332−345.

McArthur, J.M., Thirlwall, M.F., Chen, M., Gale, A.S., and Kennedy, W.J., 1993, Strontium isotope stratigraphy in the late Cretaceous: numerical calibration of the Sr isotope curve, and international correlation for the Campanian. *Paleoceanography*, **8**: 859−873.

McArthur, J.M., Kennedy, W.J., Chen, M., Thirlwall, M.F., and Gale, A.S., 1994, Strontium isotope stratigraphy for the Late Cretaceous: direct numerical age calibration of the Sr-isotope curve for the U.S. Western Interior Seaway. *Palaeogeography, Palaeoclimatology, Palaeoecology*, **108**: 95−119.

McArthur, J.M., Janssen, N.M.M., Reboulet, S., Leng, M.J., Thirlwall, M.F., and van de Schootbrugge, B., 2007, Palaeotemperatures, polar ice-volume, and isotope stratigraphy (Mg/Ca, d^{18}O, d^{13}C, ^{87}Sr/^{86}Sr): the Early Cretaceous (Berriasian, Valanginian, Hauterivian). *Palaeogeography, Palaeoclimatology, Palaeoecology*, **248**: 391−430.

McArthur, J.M., Howarth, R.J., Shields, G.A., and Zhou, Y., 2020, Chapter 7 - Strontium isotope stratigraphy. *In* Gradstein, F.M., Ogg, J.G., Schmitz, M.D., and Ogg, G.M. (eds), *The Geologic Time Scale 2020*. Vol. 1 (this book). Elsevier, Boston, MA.

Meissner, P., Mutterlose, J., and Bodin, S., 2015, Latitudinal temperature trends in the northern hemisphere during the Early Cretaceous (Valanginian-Hauterivian). *Palaeogeography, Palaeoclimatology, Palaeoecology*, **424**: 17−39.

Mena, M., Orgeira, M.J., and Lagorioi, S., 2006, Paleomagnetism, rock-magnetism and geochemical aspects of early Cretaceous basalts of the Paraná Magmatic Province, Misiones, Argentina. *Earth Planets Space*, **58**: 1283−1293.

Menegatti, A.P., Weissert, H., Brown, R.S., Tyson, R.V., Farrimond, P., Strasser, A., et al., 1998, High-resolution δ^{13}C stratigraphy through the early Aptian "Livello Selli" of the Alpine Tethys. *Paleoceanography*, **13**: 530−545.

Mesozoic Planktonic Foraminiferal Working Group (Huber, B.T., coordinator), 2006. Mesozoic Planktonic Foraminiferal Taxonomic Dictionary. [Unpublished.].

Meyers, S.R., Siewert, S.E., Singer, B.S., Sageman, B.B., Condon, D., Obradovich, J.D., et al., 2010. Reducing error bars through the inter-calibration of radioisotopic and astrochronologic time scales for the Cenomanian/Turonian Boundary Interval, Western Interior Basin, USA. In: *American Geological Union Fall Meeting, San Francisco, CA, 13−17 December 2010, Abstract*. Available from: < http://www.agu.org/meetings/fm10/waisfm10.html >.

Meyers, S.R., Siewert, S.E., Singer, B.S., Sageman, B.B., Condon, D., Obradovich, J.D., et al., 2012, Intercalibration of radioisotopic and astrochronologic time scales for the Cenomanian-Turonian Boundary interval, Western Interior Basin, USA. *Geology*, **40**: 7−10.

Midtkandal, I., Svensen, H.H., Planke, S., Corfu, F., Polteau, S., Torsvik, T.H., et al., 2016, The Aptian (Early Cretaceous) anoxic event (OAE1a) in Svalbard, Barents Sea and the absolute age of the Barremian-Aptian boundary. *Palaeogeography, Palaeoclimatology, Palaeoecology*, **463**: 126−135, https://doi.org/10.1016/j.palaeo.2016.09.023.

Miller, K.G., Barrera, E., Olsson, R.K., Sugarman, P.J., and Savin, S.M., 1999, Does ice drive early Maastrichtian eustasy? *Geology*, **27**: 783−786.

Miller, K.G., Sugarman, P.J., Browning, J.V., Kominz, M.A., Hernàndez, J.C., Olsson, R.K., et al., 2003, Late Cretaceous chronology of large, rapid sea-level changes: glacioeustasy during the greenhouse world. *Geology*, **31**: 585−588.

Miller, K.G., Sugarman, P.J., Browning, J.V., Kominz, M.A., Olsson, R., K., Feigenson, M.D., et al., 2004, Upper Cretaceous sequences and sea-level history, New Jersey coastal plain. *Geological Society of America Bulletin*, **116**: 368−393.

Mitchell, R.N., Bice, D.M., Montanari, A., Cleaveland, L.C., Christianson, K.T., Coccioni, R., et al., 2008, Oceanic anoxic cycles? Orbital prelude to the Bonarelli Level (OAE 2). *Earth and Planetary Science Letters*, **267**: 1−16.

Möller, C., Mutterlose, J., and Alsen, P., 2015, Integrated stratigraphy of Lower Cretaceous sediments (Ryazanian-Hauterivian) from North-East Greenland. *Palaeogeography, Palaeoclimatology, Palaeoecology*, **437**: 85−97.

Monteil, E., and Foucher, J.-C., 1998, Dinoflagellate cysts. Columns for Jurassic chart of Mesozoic and Cenozoic sequence chronostratigraphic framework of European basins, by Hardenbol, J., Thierry, J., Farley, M.B., Jacquin, T., de Graciansky, P.-C., and Vail, P.R. (coordinators). *In* de Graciansky, P.-C., Hardenbol, J., Jacquin, T., and Vail, P.R. (eds), Mesozoic-Cenozoic Sequence Stratigraphy of European Basins. *SEPM Special Publication*, **60**: 763−781.

Montgomery, P., Hailwood, E.A., Gale, A.S., and Burnett, J.A., 1998, The magnetostratigraphy of Coniacian-Late Campanian chalk sequences in southern England. *Earth and Planetary Science Letters*, **156**: 209−224.

Morel, N. (ed), 2015. *Stratotype Cénomanien*. Muséum National d'Histoire Naturelle. Biotope, Meze, BRGM Orléans. Paris, 384 p.

Moullade, M., 1966, Etude stratigraphique et micropaléontologique du Crétacé inférieur de la «Fosse Vocontienne». *Documents des Laboratoires de Géologie de Lyon*, **15**: 1−369.

Moullade, M., Tronchetti, G., and Masse, J.-P. (eds), 1998a. Le stratotype historique de l'Aptien inférieur (Bédoulien) dans la région de Cassis-La Bédoule (S.E. France). *Géologie Méditerranéenne* **XXV** (3−4), 298.

Moullade, M., Masse, J.-P., Tronchetti, G., Kuhnt, W., Ropolo, P., Bergen, J.A., et al., 1998b, Le stratotype historique de l'Aptien (région de Cassis-La Bédoule, SE France): synthèse stratigraphique. *Géologie Méditerranéenne*, **XXV** (3–4), 289–298.

Moullade, M., Tronchetti, G., Kuhnt, W., Renard, M., and Bellier, J.-P., 2004, The Gargasian (Middle Aptian) of Cassis-La Bédoule (Lower Aptian historical stratotype, SE France): geographic location and lithostratigraphic correlations. *Carnets de Géologie* Notebooks on Geology, Letter, 2004/02, 24 p. Available from, http://paleopolis. rediris.es/cg/CG2004_L02_MM_etal/index_v1.html.

Moullade, M., Tronchetti, G., and Bellier, J.-P., 2005, The Gargasian (Middle Aptian) strata from Cassis-La Bédoule (Lower Aptian historical stratotype, SE France): planktonic and benthic foraminiferal assemblages and biostratigraphy. *Carnets de Géologie* Notebooks on Geology, *Article*, **2005/02**, 20 p. Available from, http://paleopolis. rediris.es/cg/CG2005_A02/index.html.

Moullade, M., Granier, B., and Tronchetti, G., 2011, The Aptian Stage: back to fundamentals. *Episodes*, **34**: 148–156.

Mutterlose, J., and Böckel, B., 1998, The Barremian-Aptian interval in NW Germany: a review. *Cretaceous Research*, **19**: 539–568.

Mutterlose, J., Autran, G., Baraboschkin, E.J., Cecca, F., Erba, E., Gardin, S., et al., 1996, The Hauterivian Stage. *Bulletin de l'Institut Royal des Sciences Naturelles de Belgique, Sciences de la Terre*, **66** (Suppl.), 19–24.

Mutterlose, J., Bornemann, A., Luppold, F.W., Owen, H.G., Ruffell, A., Weiss, W., et al., 2003, The Vöhrum section (northwest Germany) and the Aptian/Albian boundary. *Cretaceous Research*, **24**: 203–252.

Mutterlose, J., Bodin, S., and Fähnrich, W., 2014, Strontium-isotope stratigraphy of the Early Cretaceous (Valanginian–Barremian): implications for Boreal–Tethys correlation and paleoclimate. *Cretaceous Research*, **50**: 252–263.

Mutterlose, J., Rawson, P.F., Reboulet, S., Baudin, F., Bulot, L., Emmanuel, L., et al., in press. The Global Boundary Stratotype Section and Point (GSSP) for the base of the Hauterivian Stage (Lower Cretaceous), La Charce, southeast France. *Episodes*. https:// doi.org/10.18814/epiiugs/2020/020072.

Myers, S.R., Siewert, S.E., Singer, B.S., Sageman, B.B., Condon, D.J., Obradovitch, J.D., et al., 2012, Integration of radioisotopic and astrochronologic timescales for the Cenomanian-Turonian boundary, Western Interior Basin, USA. *Geology*, **40**: 7–10.

Nederbragt, A.J., 1990. *Biostratigraphy and Paleoceanographic Potential of the Cretaceous Planktic Foraminifera Heterohelicidae* (Doctoral dissertation). Centrale Huisdrukkerij Vrije Universiteit.

Niebuhr, B., 2005, Geochemistry and time-series analyses of orbitally forced Upper Cretaceous marl–limestone rhythmites (Lehrte West Syncline, northern Germany). *Geological Magazine*, **142**: 31–55.

Nikitin, S.N., 1881, Jurassic Deposits between Rybinsk, Mologa, and Myshkin. *Materialy dlya geologii Rossii*, **X**: 201–331 [In Russian].

Oboh-Ikuenobe, F.E., Benson, D.G., Scott, R.W., Holbrook, J.M., Evetts, M.J., and Erbacher, J., 2007, Re-evaluation of the Albian-Cenomanian boundary in the U.S. Western Interior based on dinoflagellate cysts. *Review of Palaeobotany and Palynology*, **144**: 77–97.

Oboh-Ikuenobe, F.E., Holbrook, J.M., Scott, R.W., Akins, S.L., Evetts, M.J., Benson, D.G., et al., 2008, Anatomy of epicontinetal flooding: late Albian-early Cenomanian of the southern U.S. Western Interior Basin. *In* Pratt, B.R., and Homden, C. (eds), Dynamics of Epeiric Seas. *Geological Association of Canada Special Publication*, **48**: 201–227.

Obradovich, J.D., 1993, A Cretaceous time scale. *In* Caldwell, W.G.E., and Kauffman, E.G. (eds), Evolution of the Western Interior Basin. *Geological Association of Canada Special Paper*, **39**: 379–396.

Obradovich, J.D., Matsumoto, T., Nishida, T., and Inoue, Y., 2002, Integrated biostratigraphic and radiometric scale on the Lower Cenomanian (Cretaceous) of Hokkaido, Japan. *Proceedings of the Japan Academy, Series B-Physical and Biological Sciences*, **78** (6), 149–153.

O'Brien, C.L., Robinson, S.A., Pancost, R.D., Damste, J.S.S., Schouten, S., Lunt, D.J., et al., 2017, Cretaceous sea-surface temperature evolution: constraints from TEX$_{86}$ and planktonic foraminiferal oxygen isotopes. *Earth-Science Reviews*, **172**: 224–247.

O'Connor, L.K., Batenburg, S.J., Robinson, S.A., Jenkyns, H.C., 2020. An orbitally paced, near-complete record of Campanian climate and sedimentation in the Mississippi embayment, USA. *Newsletters on Stratigraphy*, article #534, https://doi.org/10.1127/nos/2020/0534.

O'Connor, L.K., Remmelzwaal, S., Robinson, S.A., Batenburg, S.J., Jenkynss, H.C., Parkinson, I., et al., in press. Deconstructing the Plenus Cold Event (Cenomanian, Cretaceous). *Paleoceanography*.

Odin, G.S., 1996, Le site de Tercis (Landes). Observations stratigraphiques sur le Maastrichtien. Arguments pour la localisation et la corrélation du Point Stratotype Global de la limite Campanien – Maastrichtien. *Bulletin de la Société géologique de France*, **167**: 637–643.

Odin, G.S., Hancock, J.M., Antonescu, E., Bonnemaison, M., Caron, M., Cobban, W.A., et al., 1996, Definition of a Global Boundary Stratotype Section and point for the Campanian/Maastrichtian boundary. *Bulletin de l'Institut Royal des Sciences Naturelles de Belgique, Sciences de la Terre*, **66** (Suppl.), 111–117.

Odin, G.S., (ed.), 2001. *The Campanian-Maastrichtian Stage Boundary: Characterization at Tercis les Bains (France) and Correlation with Europe and other Continents*. Development in Paleontology and Stratigraphy, **19**: 881 pp.

Odin, G.S., and Lamaurelle, M.A., 2001, The global Campanian-Maastrichtian Stage boundary. *Episodes*, **24**: 229–238.

Ogg, J.G., Ogg, G.M., and Gradstein, F.M., 2016, *A Concise Geologic Time Scale 2016*, 234p. Elsevier.

Ogg, J.G., 1987, Early Cretaceous magnetic polarity time scale and the magnetostratigraphy of DSDP Sites 603 and 534, western Central Atlantic. *Initial Reports of the Deep Sea Drilling Project*, **93**: 849–888.

Ogg, J.G., 1988, Early Cretaceous and Tithonian magnetostratigraphy of the Galicia margin (Ocean Drilling Program Leg 103). *Proceedings of the Ocean Drilling Program, Scientific Results*, **103**: 659–682.

Ogg, J.G., 2020, Chapter 5 - Geomagnetic polarity time scale. *In* Gradstein, F.M., Ogg, J.G., Schmitz, M.D., and Ogg, G.M. (eds), *The Geologic Time Scale 2020*. Vol. **1** (this book). Elsevier, Boston, MA.

Ogg, J.G., Bardot, L., 2001. Aptian through Eocene magnetostratigraphic correlation of the Blake Nose Transect (Leg 171B), Florida Continental Margin. *Proceedings of the Ocean Drilling Program, Scientific Results*, **171B**. 59 p. Available from: < http://www-odp. tamu.edu/publications/171B_SR/chap_09/chap_09.htm > .

Ogg, J.G., and Lowrie, W., 1986, Magnetostratigraphy of the Jurassic-Cretaceous boundary. *Geology*, **14**: 547–550.

Ogg, J.G., and Smith, A.G., 2004, *In* Gradstein, F.M., Ogg, J.G., and Smith, A.G. (eds), *A Geologic Time Scale 2004*. Cambridge: Cambridge University Press, p. 589.

Ogg, J.G., Steiner, M.B., Oloriz, F., and Tavera, J.M., 1984, Jurassic magnetostratigraphy, 1. Kimmeridgian-Tithonian of Sierra Gorda and Carcabuey, southern Spain. *Earth and Planetary Science Letters*, **71**: 147–162.

Ogg, J.G., Steiner, M.B., Company, M., and Tavera, J.M., 1988, Magnetostratigraphy across the Berriasian-Valanginian stage boundary (Early Cretaceous) at Cehegin (Murcia Province, southern Spain). *Earth and Planetary Science Letters*, **87**: 205–215.

Ogg, J.G., Hasenyager II, R.W., Wimbledon, W.A., Channell, J.E.T., and Bralower, T.J., 1991, Magnetostratigraphy of the Jurassic–Cretaceous boundary interval—Tethyan and English faunal realms. *Cretaceous Research*, **12**: 455–482.

Ogg, J.G., Kodama, K., and Wallick, B.P., 1992, Lower Cretaceous magnetostratigraphy and paleolatitudes off northwest Australia, ODP Site 765 and DSDP Site 261, Argo Abyssal Plain, and ODP Site 766, Gascoyne Abyssal Plain. *Proceedings of the Ocean Drilling Program, Scientific Results*, **123**: 523–548.

Ogg, J.G., Hasenyager II, R.W., and Wimbledon, W.A., 1994, Jurassic-Cretaceous boundary: Portland-Purbeck magnetostratigraphy and possible correlation to the Tethyan faunal realm. *Géobios, Mémoire Spécial*, **17**: 519–527.

Ogg, J.G., Ogg, G., and Gradstein, F.M., 2008, *The Concise Geologic Time Scale*. Cambridge: Cambridge University Press 177 pp.

Ogg, J.G., Hinnov, L.A., and Huang, C., 2012, Cretaceous. *In* Gradstein, F.M., Ogg, J.G., Schmitz, M., and Ogg, G. (eds), *The Geologic Time Scale 2012*. Elsevier, 793–853.

Ogg, J.G., Ogg, G.M., and Gradstein, F.M., 2016. *A Concise Geologic Time Scale 2016*. Elsevier, 234 p.

Oosting, A.M., Leereveld, H., Dickens, G.R., Henderson, R.A., and Brinkhuis, 2006, Correlation of Barremian-Aptian (mid-Cretaceous) dinoflagellate cyst assemblages between the Tethyan and Austral realms. *Cretaceous Research*, **27**: 762–813.

Oppel, C.A., 1865, Die Tithonische Etage. *Zeitschrift der Deutschen Geologischen Gesellschaft, Jahrgang*, **17**: 535–558.

Owen, H.G., 1996a. Boreal and Tethyan late Aptian to late Albian ammonite zonation and palaeobiogeography. In: Spaeth, C. (ed), Jost Wiedmann Memorial Volume, Proceedings of the 4th International Cretaceous Symposium, Hamburg, 1992. Mitteilung aus dem Geologisch-Paläontologischen Institut der Universität Hamburg, vol. 77. pp. 461–481.

Owen, H.G., 1996b, "Uppermost Wealden facies and Lower Greensand Group (Lower Cretaceous) in Dorset, southern England: correlation and palaeoenvironment" by Ruffell & Batten (1994) and "The Sandgate Formation of the M20 Motorway near Ashford, Kent and its correlation" by Ruffell & Owen (1995)": reply. *Proceedings of the Geologists' Association*, **107**: 74–76.

Owen, H.G., 2002, The base of the Albian Stage; comments on recent proposals. *Cretaceous Research*, **23**: 1–13.

Paul, C.R.C., Lamolda, M.A., Mitchell, S.F., Vaziri, M.R., Gorostidi, A., and Marshall, J.D., 1999, The Cenomanian-Turonian boundary at Eastbourne (Sussex, UK): a proposed European reference section. *Palaeogeography, Palaeoclimatology, Palaeoecology*, **150**: 83–121.

Pearson, P.N., 2012, Oxygen isotopes in Foraminifera: an overview and historical review. *In* Ivany, L.C., and Huber, B.T. (eds), Reconstructing Earth's Deep Time Climate. *The Paleontological Society Papers*, **18**: 1–38.

Pearson, P.N., Ditchfield, P.W., Singano, J., Harcourt-Brown, K.G., Nicholas, C.J., Olsson, R.K., et al., 2001, Warm tropical sea surface temperatures in the Late Cretaceous and Eocene epochs. *Nature*, **413**: 481–487.

Pechersky, D.M., and Khramov, A.N., 1973, Mesozoic paleomagnetic scale of the U.S.S.R. *Nature*, **244**: 499–501.

Pérez-Rodríguez, I., Lees, J.A., Larrasoaña, J.C., Arz, J.A., and Arenillas, I., 2012, Planktonic foraminiferal and calcareous nannofossil biostratigraphy and magnetostratigraphy of the uppermost Campanian and Maastrichtian at Zumaia, northern Spain. *Cretaceous Research*, **37**: 100–126.

Pessagno, E.A., 1967, Upper Cretaceous planktonic foraminifera from the western Gulf Coastal Plain. *Palaeontographica Americana*, **5**: 245–445.

Petrizzo, M.R., 2000, Upper Turonian–lower Campanian planktonic foraminifera from southern mid–high latitudes (Exmouth Plateau, NW Australia): biostratigraphy and taxonomic notes. *Cretaceous Research*, **21** (4), 479–505.

Petrizzo, M.R., 2001, Late Cretaceous planktonic foraminifera from Kerguelen Plateau (ODP Leg 183): new data to improve the Southern Ocean biozonation. *Cretaceous Research*, **22** (6), 829–855.

Petrizzo, M.R., 2003, Late Cretaceous planktonic foraminiferal bioevents in the Tethys and in the Southern Ocean record: an overview. *Journal of Foraminiferal Research*, **33** (4), 330–337.

Petrizzo, M.R., 2019, A critical evaluation of planktonic foraminiferal biostratigraphy across the Coniacian-Santonian boundary interval in Spain, Texas, and Tanzania, *Geologic Problem Solving with Microfossils IV*, SEPM Special Publication, **111**: 186–198, http://dx.doi.org/10.2110/sepmsp.111.04.

Petrizzo, M.R., and Huber, B.T., 2006, Biostratigraphy and taxonomy of late Albian planktonic foraminifera from ODP Leg 171B (western North Atlantic Ocean). *Journal of Foraminiferal Research*, **36** (2), 166–190.

Petrizzo, M.R., Falzoni, F., and Premoli Silva, I., 2011, Identification of the base of the lower-to-middle Campanian *Globotruncana ventricosa* Zone: comments on reliability and global correlations. *Cretaceous Research*, **32**: 387–405.

Petrizzo, M.R., Huber, B.T., Gale, A.S., Barchetta, A., and Jenkyns, H.C., 2012, Abrupt planktonic foraminiferal turnover across the Niveau Kilian at Col de Pré-Guittard (Vocontian Basin, southeast France): new criteria for defining the Aptian/Albian boundary. *Newsletters on Stratigraphy*, **45**: 55–74.

Petrizzo, M.R., Huber, B.T., Gale, A.S., Barchetta, A., and Jenkyns, H.C., 2013, Erratum. Abrupt planktic foraminiferal turnover across the Niveau Kilian at Col de Pré-Guittard (Vocontian Basin, southeast France): new criteria for defining the Aptian/Albian boundary. *Newsletters on Stratigraphy*, **46** (1), 93.

Petrizzo, M.R., Caron, M., and Premoli-Silva, I., 2015, Remarks on the identification of the Albian/Cenomanian boundary and taxonomic clarification of the planktonic foraminifera index species *globotruncanoides*, *brotzeni* and *tehamaensis*. *Geological Magazine*, **152**: 521–536.

Petrizzo, M.R., Jiménez Berrocoso, Á., Falzoni, F., Huber, B.T., and Macleod, K.G., 2017, The Coniacian–Santonian sedimentary record in southern Tanzania (Ruvuma Basin, East Africa): planktonic foraminiferal evolutionary, geochemical and palaeoceanographic patterns. *Sedimentology*, **64** (1), 252–285.

Premoli Silva, I., and Sliter, W.V., 1995, Cretaceous planktonic foraminiferal biostratigraphy and evolutionary trends from the Bottaccione section, Gubbio, Italy. *Palaeontographia Italica*, **82**: 1–89.

Premoli Silva, I., and Sliter, W.V., 1999, Cretaceous paleoceanography: evidence from planktonic foaminiferal evolution. *In* Barrera, E., and Johnson, C.C. (eds), Evolution of the Cretaceous Ocean-Climate System. *Geological Society of America Special Paper*, **332**: 301–328.

Premoli Silva, I., Soldan, D.M., and Petrizzo, M.R., 2018, Upper Hauterivian-upper Barremian planktonic foraminiferal assemblages from the Arroyo Gilico section (Southern Spain). *Journal of Foraminiferal Research*, **48** (4), 314–355.

Price, G.D., 1999, The evidence for and implications of polar ice during the Mesozoic. *Earth Science Reviews*, **48**: 183–210.

Pringle, M.S., and Duncan, R.A., 1995a, Radiometric ages of basaltic lavas recovered sites 865, 866, and 869. *Proceedings of the Ocean Drilling Program, Scientific Results*, **143**: 277–283.

Pringle, M.S., and Duncan, R.A., 1995b, Radiometric ages of basaltic lavas recovered at Lo-En, Wodejebato, MIT, and Takuyo-Daisan Guyots, northwestern Pacific Ocean. *Proceedings of the Ocean Drilling Program, Scientific Results*, **144**: 547–557.

Pringle, M.S., Chambers, L., Ogg, J.G., 2003. Synchronicity of volcanism on Ontong Java and Manihiki plateaux with global oceanographic events? In: American Geophysical Union and European Union of Geophysics Conference, Nice, France, May 2003, Abstract.

Pruner, P., Houša, V., Olóriz, F., Koštak, M., Krs, M., Man, O., et al., 2010, High-resolution magnetostratigraphy and the biostratigraphic zonation of the Jurassic/Cretaceous boundary strata in the Puerto Escaño section (southern Spain). *Cretaceous Research*, **31**: 192–206.

Pucéat, E., Lécuyer, C., Sheppard, S.M.F., Dromart, G., Reboulet, S., and Grandjean, P., 2003, Thermal evolution of Cretaceous Tethyan marine waters inferred from oxygen isotope composition of fish tooth enamels. *Paleoceanography*, **18**: 1029, https://doi.org/10.1029/2002PA00823.

Quidelleur, X., Paquette, J.L., Fiet, N., Takashima, R., Tiepolo, M., Desmares, D., et al., 2011, New U-Pb (ID-TIMS and LA-ICPMS) and $^{40}Ar/^{39}Ar$ geochronological constraints of the Cretaceous geologic time scale calibration from Hokkaido (Japan). *Chemical Geology*, **286**: 72–83.

Rawson, P.F., 1983, The Valanginian to Aptian Stages—current definitions and outstanding problems. *Zitteliania*, **10**: 493–500.

Rawson, P.F., 1990, Event stratigraphy and the Jurassic-Cretaceous boundary. *Transactions of the Institute of Geology and Geophysics, Academy Sciences USSR, Siberian Branch*, **699**: 48–52 [In Russian with English summary].

Rawson, P.F., Curry, D., Dilley, F.C., Hancock, J.M., Kennedy, W.J., Neale, J.W., et al., 1978, A correlation of Cretaceous rocks in the British Isles. *Geological Society of London Special Report*, **9**: 70.

Rawson, P.F., Avram, E., Baraboschkin, E.J., Cecca, F., Company, M., and Delanoy, G., 1996a, The Barremian Stage. *Bulletin de l'Institut Royal des Sciences Naturelles de Belgique, Sciences de la Terre*, **66** (Suppl.), 25–30.

Rawson, P.F., D'Hondt, A.V., Hancock, J.M., and Kennedy, W.J. (eds), 1996b. Proceedings of the second international symposium on Cretaceous Stage Boundaries, Brussels, 8–16 September 1995. *Bulletin de l'Institut Royal des Sciences Naturelles de Belgique, Sciences de la Terre* **66** (Suppl.), 117 p.

Ray, D.C., van Buchem, F.S.P., Baines, G., Davies, A., Gréselle, B., Simmons, M.D., et al., 2019, The magnitude and cause of short-term eustatic Cretaceous sea-level change: a synthesis. *Earth Science Reviews*, **197**: article #102901: 20 pp, https://doi.org/10.1016/j.earscirev.2019.102901.

Reboulet, S., 1996, L'évolution des ammonites du Valanginien-Hauterivien inférieur du bassin vocontien et de la plate-forme provençale (sud-est de la France): relations avec la stratigraphie séquentielle et implications biostratigraphiques. *Documents des Laboratoires de Géologie Lyon*, **137**: 370 p.

Reboulet, S., 2001, Limiting factors on shell growth, mode of life and segregation of Valanginian ammonoid populations: evidence from adult-size variations. *Geobios*, **34**: 423–435.

Reboulet, S., and Atrops, F., 1999, Comments and proposals about the Valanginian-Lower Hauterivian ammonite zonation of south-eastern France. *Ecologae Geologicae Helvetiae*, **92**: 183–197.

Reboulet, S., Mattioli, E., Pittet, B., Baudin, B., Olivero, D., and Proux, O., 2003, Ammonoid and nannoplankton abundance in Valanginian (early Cretaceous) limestone-marl alternations from the southeast France Basin: carbonate dilution or productivity? *Palaeogeography, Palaeoclimatology, Palaeoecology*, **201**: 113–139.

Reboulet, S., Hoedemaeker, P.J., Aguirre-Urreta, M.B., Alsen, P., Atrops, F., Baraboshkin, E.Y., et al., 2006, Report on the 2nd international meeting of the IUGS Lower Cretaceous ammonite working group, the "Kilian Group", Neuchâtel, Switzerland, 8 September 2005. *Cretaceous Research*, **27**: 712–715.

Reboulet, S., Klein, J., Barragán, R., Company, M., González-Arreola, C., Lukeneder, A., et al., 2009, Report on the 3rd international meeting of the IUGS Lower Cretaceous Ammonite Working Group, the "Kilian Group", Vienna, Austria, 15th April 2008. *Cretaceous Research*, **30**: 496–502.

Reboulet, S., Rawson, P.F., Moreno-Bedmar, J.A., Aguirre-Urreta, M.B., Barragán, R., Bogomolov, Y., et al., 2011, Report on the 4th international meeting of the IUGS Lower Cretaceous Ammonite Working Group, the "Kilian Group" (Dijon, France, 30th August, 2010). *Cretaceous Research*, **32**: 786–793.

Reboulet, S., Szives (reporters), O., Aguirre-Urreta, B., Barragán, R., Company, M., Frau, C., et al., 2018, Report on the 6th international meeting of the IUGS Lower Cretaceous Ammonite Working Group, the Kilian Group (Vienna, Austria, 20th August 2017). *Cretaceous Research*, **91**: 100–110.

Reimond, W.U., Koeberl, C., Brandstäter, F., Kruger, F.J., Armstrong, R.A., and Bootsman, C., 1999, Morokweng impact structure, South Africa: geologic, petrographic, and isotopic results, and implications for the size of the structure. *In* Dressler, B.O., and Sharpton, V.L. (eds), Large Meteorite Impacts and Planetary Evolution II. *Geological Society of America Special Paper*, **339**: 61–90.

Remane, J., 1985, Calpionellids. *In* Bolli, H.M., Saunders, J.B., and Perch Nielsen, K. (eds), *Plankton Stratigraphy*. Cambridge: Cambridge University Press, 555–572.

Remane, J., 1991, The Jurassic-Cretaceous boundary: problems of definition and procedure. *Cretaceous Research*, **12**: 447–453.

Remane, J., 1998, Calpionellids. Columns for Jurassic chart of Mesozoic and Cenozoic sequence chronostratigraphic framework of European basins, by Hardenbol, J., Thierry, J., Farley, M.B., Jacquin, T., de Graciansky, P.-C., and Vail, P.R. (coordinators). *In* de Graciansky, P.-C., Hardenbol, J., Jacquin, T., and Vail, P.R. (eds), Mesozoic-Cenozoic Sequence Stratigraphy of European Basins. *SEPM Special Publication*, **60**: 763–781.

Remane, J., Bakalova-Ivanova, D., Borza, K., Knauer, J., Nagy, I., Pop, G., et al., 1986, Agreement on the subdivision of the Standard Calpionellid Zones defined at the IInd Planktonic Conference Roma 1971. *Acta Geologica Hungarica*, **29**: 5–14.

Renard, M., de Rafélis, M., Emmanuel, L., Moullade, M., Masse, J.-P., Kuhnt, W., et al., 2005. Early Aptian $\partial^{13}C$ and manganese anomalies from the historical Cassis-La Bédoule stratotype sections (S.E. France): relationship with a methane hydrate dissociation event and stratigraphic implications. In: Carnets de Géologie/Notebooks on Geology, Article, 2005/04, 18 p. Available from: < http://paleopo-lis.rediris.es/cg/CG2005_A04/index.html > .

Renevier, E., 1874, Tableau des terrains sédimentaires formés pendant les époques de la phase organique du globe terrestre avec leurs représentants en Suisse et dans les régions classiques, leurs syno-nymes et les principaux fossiles de chaque étage. *Bulletin de la Société vaudoise des Sciences naturelles*, **13**: 218–252.

Renne, P.R., Mundil, M., Balco, G., Min, K., and Ludwig, K.R., 2010, Joint determination of ^{40}K decay constants and ^{40}Ar*/^{40}K for the Fish Canyon sanidine standard, and improved accuracy for ^{40}Ar/^{39}Ar geo-chronology. *Geochimica et Cosmochimica Acta*, **74**: 5349–5367.

Renne, P.R., Deino, A.L., Hilgen, F.J., Kuiper, K.F., Mark, D.F., Mitchell, W.S., et al., 2013, Time scales of critical events around the Cretaceous-Paleogene Boundary. *Science*, **339**: 684, https://doi.org/10.1126/science.1230492.

Riveline, J., 1998, Charophytes. Columns for Jurassic chart of Mesozoic and Cenozoic sequence chronostratigraphic framework of European basins, by Hardenbol, J., Thierry, J., Farley, M.B., Jacquin, T., de Graciansky, P.-C., and Vail, P.R. (coordinators). *In* de Graciansky, P.-C., Hardenbol, J., Jacquin, T., and Vail, P.R. (eds), Mesozoic-Cenozoic Sequence Stratigraphy of European Basins. *SEPM Special Publication*, **60**: 763–781.

Robaszynski, F., 1998, Planktonic foraminifera. Columns for Jurassic chart of Mesozoic and Cenozoic sequence chronostratigraphic framework of European basins, by Hardenbol, J., Thierry, J., Farley, M.B., Jacquin, T., de Graciansky, P.-C., and Vail, P.R. (coordina-tors). *In* de Graciansky, P.-C., Hardenbol, J., Jacquin, T., and Vail, P.R. (eds), Mesozoic-Cenozoic Sequence Stratigraphy of European Basins. *SEPM Special Publication*, **60**: 763–781.

Robaszynski, F., and Caron, M.C., 1979, Atlas of mid-Cretaceous planktonic foraminifera (Boreal Sea and Tethys). *Cahiers de Micropaléontologie*, **1**: 1–185.

Robaszynski, F., and Caron, M., 1995, Foraminiféres planctoniques du crétacé: commentaire de la zonation Europe-Mediterranée. *Bulletin de la Société géologique de France*, **166**: 681–692.

Robaszynski, F., Caron, M., Gonzalez Donoso, J.M., and Wonders, A.H., 1984, The European working group on Planktonic Foraminifera, Atlas of Late Cretaceous globotruncanids. *Revue de Micropalèontologie*, **26**: 145–305.

Robaszynski, F., Caron, M., Dupuis, C., Amédro, F., Gonzalez-Donoso, J.M., Linares, D., et al., 1990, A tentative integrated stratigraphy in the Turonian of Central Tunisia: formations, zones and sequential stratigraphy in the Kalaat Senan area. *Bulletin des Centres Recherches Exploration-Production Elf-Aquitaine*, **14**: 213–384.

Robaszynski, F., Gonzalez Donoso, J.M., Linares, D., Amédro, F., Caron, M., Dupuis, C., et al., 2000, Le Crétacé supérieur de la région de Kalaat Senan, Tunisie Centrale. Litho-biostratigraphie intégrée: zones d'ammonites, de foraminifères planctoniques et de nannofossiles du Turonien supérieur au Maastrichtien. *Bulletin des Centres de Recherche et d'Exploration-Production d'Elf-Aquitaine*, **22**: 359–490.

Robaszynski, F., Amédro, F., Devalque, C., Matrion, B., 2014, Le Turonien des Massifs d'Uchaux et de la Ceze (S.E. France). Memoires de la Classe des Sciences Academie Royale de Belgique, Brussels. 197 p. 48 pls.

Röhl, U., and Ogg, J.G., 1996, Aptian-Albian sea level history from guyots in the western Pacific. *Paleoceanography*, **11**: 595–624.

Ropolo, P., Conte, G., Gonnet, R., Masse, J.-P., and Moullade, M., 1998, Les faunes d'ammonites du Barrémien supérieur/Aptien inférieur (Bédoulien) dans la région stratotypique de Cassis-La Bédoule (S.E. France): état des connaissances et propositions pour une zonation par ammonites du Bédoulien type. *Géologie Méditerranéene*, **25**: 167–175.

Roth, P.H., 1989, Ocean circulation and calcareous nannoplankton evo-lution during the Jurassic and Cretaceous. *Palaeogeography, Palaeoclimatology, Palaeoecology*, **74**: 111–126.

Round, F.E., Crawford, R.M., and Mann, D.G., 1990, *The Diatoms: Biology and Morphology of the Genera*. Cambridge: Cambridge University Press, 747 p.

Ryan, W.B.F., Bolli, H.M., Foss, G.N., Natland, J.H., Hottman, W.E., and Foresman, J.B., 1978, Objectives, principal results, operations, and explanatory notes of Leg 40, South Atlantic. *Initial Reports, Deep Sea Drilling Project*, **40**: 5–20.

Sageman, B.B., Meyers, S.R., and Arthur, M.A., 2006, Orbital time scale and new C-isotope record for Cenomanian-Turonian boundary stra-totype. *Geology*, **34**: 125–128.

Sageman, B.B., Singer, B.S., Meyers, S.R., Walaszczyk, I., Seiwert, S.E., Condon, D.J., et al., 2014, Integrating ^{40}Ar/^{39}Ar, U-Pb, and astronomical clocks in the Cretaceous Niobrara Formation, Western Interior Basin, USA. *Bulletin of the Geological Society of America*, **126**: 956–973, https://doi.org/10.1130/B30929.1.

Salaj, J., 1984, Foraminifers and detailed microbiostratigraphy of the boundary beds of the Lower Cretaceous stages in the Tunisian Atlas. *Geologicky Zbornik*, **35**: 583–599.

Sazonov, N.T., 1951, On some little-known ammonites of the Lower Cretaceous. *Byulleten' Moskovskogo Obshchestva Ispytatelei Prirody*, **56**: 1–176 [In Russian].

Scaife, J.D., Ruhl, M., Dickson, J.A., Mather, T.A., Jenkyns, H.C., Percival, L.M.E., et al., 2017, Sedimentary mercury enrichments as a marker for submarine large igneous province volcanism? Evidence from the Mid-Cenomanian event and Oceanic Anoxic Event 2 (Late Cretaceous). *Geochemistry, Geophysics, Geosystems*, **18**: 4253–4275, https://doi.org/10.1002/2017GC007153.

Schlanger, S.O., and Jenkyns, H.C., 1976, Cretaceous oceanic anoxic events: causes and consequences. *Geology en Mijnbouw*, **55**: 179–184.

Schlanger, S.O., Arthur, M.A., Jenkyns, H.C., and Scholle, P.A., 1987, The Cenomanian-Turonian oceanic anoxic event, I. Stratigraphy and distribution of organic carbon-rich beds and the marine ^{13}C excur-sion. *In* Brooks, J., and Fleet, A.J. (eds), Marine Petroleum Source Rocks. *Geological Society of London, Special Publication*, **26**: 371–399.

Schnabl, P., Pruner, P., and Wimbledon, W.A.P., 2015, A review of magnetostratigraphic results in the Tithonian-Berriasian of Nordvik (Siberia) and possible biostratigraphic constraints. *Geologica Carpathica*, **66**: 489–498.

Scholle, P.A., and Arthur, M., 1980, Carbon isotope fluctuations in Cretaceous pelagic limestones: potential stratigraphic and petroleum exploration tool. *Bulletin of the American Association of Petroleum Geologists*, **64**: 67–87.

Schönfeld, J., Schulz, M.-G., McArthur, J.M., Burnett, J., Gale, A., Hambach, U., et al., 1996, New results on biostratigraphy, geochemis-try and correlation from the standard section for the Upper Cretaceous white chalk of northern Germany (Lägerdorf–Kronsmoor–Hemmoor). In: Spaeth, C. (ed), *Jost Wiedmann Memorial Volume; Proceedings of the 4th International Cretaceous Symposium, Hamburg, 1992.* Mitteilung aus dem Geologisch-Paläontologischen Institut der Universität Hamburg, vol. 77. pp. 545–575.

Scotese, C.R., 2014. *Atlas of Late Cretaceous Paleogeographic Maps, PALEOMAP Atlas for ArcGIS, Volume 2, The Cretaceous, Maps 16–22, Mollweide Projection,* PALEOMAP Project, Evanston, IL.

Scott, R.W., 2007, Calibration of the Albian/Cenomanian boundary by ammonite biostratigraphy: U.S. Western Interior. *Acta Geologica Sinica,* **81**: 940–948.

Scott, R.W., 2009, Uppermost Albian biostratigraphy and chronostratigraphy. *Carnets de Géologie/Notebooks on Geology, Article,* **2009** (03), 15.

Scott, R.W., 2011. CRET1 Chronostratigraphic Database. Available from: < http://precisionstratigraphy.com >.

Scott, R.W., Oboh-Ikuenobe, F.E., Benson, D.G., and Holbrook, J.M., 2009, Numerical age calibration of the Albian/Cenomanian boundary. *Stratigraphy,* **6**: 17–32.

Selby, D., Mutterlose, J., and Condon, D.J., 2009, U-Pb and Re-Os Geochronology of the Aptian/Albian and Cenomanian/Turonian stage boundaries: implications for timescale calibration, osmium isotope seawater composition and Re-Os systematics in organic-rich sediments. *Chemical Geology,* **265**: 394–409.

Séronie-Vivien, M., 1972, Contribution à l'étude de Sénonien en Aquitaine septentrionale, ses stratotypes: Coniacien, Santonien, Campanien. *Les Stratotypes Français,* **2**: 195.

Sharland, P.R., Archer, R., Casey, D.M., Davies, R.B., Hall, S.H., Heward, A.P., et al., 2001, Arabian Plate Sequence Stratigraphy. *GeoArabia Special Publication,* **2**: 372.

Sharland, P.R., Casey, D.M., Davies, R.B., Simmons, M.D., and Sutcliffe, O.E., 2004, Arabian Plate sequence stratigraphy. *GeoArabia: Middle East Petroleum Geosciences,* **9** (2), 199–214.

Shimokawa, A., 2010, *Zircon U-Pb Geochronology of the Great Valley Group: Recalibrating the Lower Cretaceous Time Scale* (M.S. thesis). University of North Carolina at Chapel Hill. 46 p. [See also: Shimokawa, A., Coleman, D.S., Bralower, T.J., 2010. Recalibrating the Lower Cretaceous time scale with U-Pb zircon ages from the Great Valley Group. Geological Society of America *Abstracts with Programs.* Vol. **42**, No. 5, p. 393. https://gsa.confex.com/gsa/2010AM/webprogram/Paper182413.html].

Shipboard Scientific Party, 1998, Site 1049: paleomagnetism section (authored by Ogg, J.G., Bardot, L., and Foster, J.). In: *Proceedings Ocean Drilling Program, Initial Reports,* **171B**. pp. 70–75.

Shipboard Scientific Party, 2004, Explanatory notes: biostratigraphy. In: *Proceedings of the Ocean Drilling Program, Initial Reports,* **207**. Available from: < http://www-odp.tamu.edu/publications/207_IR >.

Siewert, S.E., 2011, *Integrating ^{40}Ar/^{39}Ar, U-Pb and Astronomical Clocks in the Cretaceous Niobrara Formation.* M.S. thesis, University of Wisconsin at Madison, 74 pp.

Sigal, J., 1977, Essai de zonation du Crétacé méditerranéen à l'aide des foraminiféres planctoniques. *Géologie Méditerranéenne,* **4/2**: 99–108.

Sikora, P., and Bergen, J., 2004, Lower Cretaceous Biostratigraphy of Ontong Java sites from DSDP Leg 30 and ODP Leg 192. *In* Fitton, J.G., Mahoney, J.J., Wallace, P.J., and Sanders, A.D. (eds), Origin and Evolution of the Ontong Java Plateau. *Geological Society of London, Special Publication,* **229**: 83–111.

Simmons, M.D., Sharland, P.R., Casey, D.M., Davies, R.B., and Sutcliffe, O.E., 2007, Arabian Plate sequence stratigraphy: potential implications for global chronostratigraphy. *GeoArabia: Middle East Petroleum Geosciences,* **12** (4), 101–130.

Simmons, M.S., Miller, K.G., Ray, D.C., Davies, A., van Buchem, F.S.P., and Gréselle, B., 2020, Chapter 13 - Phanerozoic eustacy. *In* Gradstein, F.M., Ogg, J.G., Schmitz, M.D., and Ogg, G.M. (eds), *The Geologic Time Scale 2020.* Vol. **1** (this book). Elsevier, Boston, MA.

Sinninghe Damsté, J.S., Muyzer, G., Abbas, B., Rampen, S.W., Massé, G., Allard, W.G., et al., 2004, The rise of rhizosolenid diatoms. *Science,* **304**: 584–587.

Smelror, M., Kelley, S.R.A., Dypvik, H., Mørk, A., Nagy, J., and Tsikalas, F., 2001, Mjølnir (Barents Sea) meteorite impact ejecta offers a Volgian–Ryazanian boundary marker. *Newsletters on Stratigraphy,* **38**: 129–140.

Smith, W., 1815. A Delineation of the Strata of England and Wales, With Parts of Scotland.

Speijer, R.P., Pälike, H., Hollis, C.J., Hooker, J.J., and Ogg, J.G., 2020, Chapter 28 – The Paleogene Period. *In* Gradstein, F.M., Ogg, J.G., Schmitz, M.D., and Ogg, G.M. (eds), *The Geologic Time Scale 2020.* Vol. **2** (this book). Elsevier, Boston, MA.

Speranza, F., Satolli, S., Mattioli, E., and Calamita, F., 2005, Magnetic stratigraphy of Kimmeridgian–Aptian sections from Umbria-Marche (Italy): new details on the M-polarity sequence. *Journal of Geophysical Research,* **110**: B12109, https://doi.org/10.1029/2005JB003884.

Sprenger, A., and ten Kate, W.G., 1993, Orbital forcing of calcilutite-marl cycles in southeast Spain and an estimate for the duration of the Berriasian Stage. *Geological Society of America Bulletin,* **105**: 807–818.

Sprovieri, M., Coccioni, R., Lirer, F., Pelosi, N., and Lozar, F., 2006, Orbital tuning of a lower Cretaceous composite record (Maiolica Formation, central Italy). *Paleoceanography,* **21**: PA4212, https://doi.org/10.1029/2005PA001224.

Sprovieri, M., Sabatino, N., Pelosi, N., Batenburg, S.J., Coccioni, R., Iavarone, M., et al., 2013, Late Cretaceous orbitally-paced carbon isotope stratigraphy from the Bottacione Gorge (Italy). *Palaeogeography, Palaeoclimatology, Palaeoecology,* **379–380**: 81–94.

Steuber, T., Scott, R.W., Mitchell, S.F., and Skelton, P.S., 2016, Part N, revised, volume 1, Chapter 26C: stratigraphy and diversity dynamics of Jurassic–Cretaceous Hippuritida (rudist bivalves). *Treatise Online,* **81**: 1–17 7 g., 1 table.

Stewart, K., Turner, S., Kelley, S., Hawkesworth, C., Kirstein, L., and Mantovani, M., 1996, 3-D ^{40}Ar-^{39}Ar geochronology in the Parana continental flood basalt province. *Earth and Planetary Science Letters,* **143**: 95–109.

Stoll, H.M., and Schrag, D.P., 2000, High-resolution stable isotope records from the Upper Cretaceous rocks of Italy and Spain: Glacial episodes in a greenhouse planet? *Geological Society of America Bulletin,* **112**: 308–319.

Strasser, A., Hillgartner, H., and Pasquier, J.B., 2004, Cyclostratigraphic timing of sedimentary processes: an example from the Berriasian of the Swiss and French Jura Mountains. *In* D'Argenio, B., Fischer, A.G., Premoli Silva, I., Weissert, H., and Ferreri, V. (eds), Cyclostratigraphy: Approaches and Case Histories. *SEPM Special Publication,* **81**: 135–151.

Stratotype Albien. Muséum national d'Histoire naturelle. *In* Colleté, C. (ed), Paris: Biotope, Meze, BRGM OrléansBiotope, Meze, BRGM Orléans. Paris, 332 p.

Subcommission on Cretaceous Stratigraphy, 2009. Annual Report 2009. 12 p. Available from: < http://www2.mnhn.fr/hdt203/media/ISCS/ICS2009_Report_Creta.pdf > .

Surlyk, F., 1984, The Maastrichtian Stage in NW Europe and its brachiopod zonation. *Bulletin of the Geological Society of Denmark*, 33: 217–223.

Swisher III, C.C., Wang, Y.Q., Wang, X.L., Xu, X., and Wang, Y., 1999, Cretaceous age for the feathered dinosaurs of Liaoning, China. *Nature*, 400: 58–61.

Takashima, R., Nishi, H., Yamanaka, T., Orihashi, Y., Tsujino, Y., Quidelleur, X., et al., 2019, Establishment of Upper Cretaceous bio- and carbon-isotope stratigraphy in the northwest Pacific Ocean and radiometric ages around the Albian/Cenomanian, Coniacian/Santonian and Santonian/Campanian boundaries. *Newsletters on Stratigraphy*, https://doi.org/10.1127/nos/2019/0472.

Takashima, R., Nishi, H., Hayashi, K., Okada, H., Kawahata, H., Yamanaka, T., Fernando, A.G., and Mampuku, M., 2009, Litho-, bio- and chemostratigraphy across the Cenomanian–Turonian boundary (OAE 2) in the Vocontian Basin of southeastern France. *Palaeogeography, Palaeoclimatology, Palaeoecology*, 273: 61–74.

Tarduno, J.A., 1990, A brief reversed polarity interval during the Cretaceous Normal Polarity Superchron. *Geology*, 18: 638–686.

Tarduno, J.A., Sliter, W.V., Bralower, T.J., McWilliams, M., Premoli-Silva, I., and Ogg, J.G., 1989, M-sequence reversals recorded in DSDP Sediment Cores from the Western Mid-Pacific Mountains and Magellan Rise. *Geological Society of America Bulletin*, 101: 1306–1319.

Tarduno, J.A., Sliter, W.V., Kroenke, L., Leckie, R.M., Mayer, H., Mahoney, J.J., et al., 1991, Rapid formation of Ontong Java Plateau by Aptian mantle volcanism. *Science*, 254: 399–403.

Tarduno, J.A., Lowrie, W., Sliter, W.V., Bralower, T.J., and Heller, F., 1992, Reversed polarity characteristic magnetizations in the Albian Contessa Section, Umbrian Apennines, Italy: implications for the existence of a Mid-Cretaceous Mixed Polarity Interval. *Journal of Geophysical Research*, 97 (B1), 241–271.

Tavera, J.M., Aguado, R., Company, M., Oloriz, F., 1994. Integrated biostratigraphy of the Durangites and Jacobi zones (J/K boundary) at the Puerto Escano section in southern Spain (Province of Cordoba). In: Cariou, E., Hantzpergue, P. (eds), *Third International Symposium on Jurassic Stratigraphy, Poitiers, France, 22–29 September 1991*. Geobios, Mémoire Special, 17 (1): 469–476.

Thierry, J., 1998, Ammonites. Columns for Jurassic chart of Mesozoic and Cenozoic sequence chronostratigraphic framework of European basins, by Hardenbol, J., Thierry, J., Farley, M.B., Jacquin, T., de Graciansky, P.-C., and Vail, P.R. (coordinators). *In* de Graciansky, P.-C., Hardenbol, J., Jacquin, T., and Vail, P.R. (eds), *Mesozoic-Cenozoic Sequence Stratigraphy of European Basins. SEPM Special Publication*, 60: 776–777.

Thibault, N., Husson, D., Harlou, R., Gardin, S., Galbrun, B., Huret, E., et al., 2012, Astronomical calibration of upper Campanian–Maastrichtian carbon isotope events and calcareous plankton biostratigraphy in the Indian Ocean (ODP Hole 762C): implication for the age of the Campanian–Maastrichtian boundary. *Palaeogeography, Palaeoclimatology, Palaeoecology*, 377–378: 52–71.

Thibault, N., Jarvis, I., Voigt, S., Gale, A.S., Attree, K., and Jenkyns, H. C., 2016, Astronomical calibration and global correlation of the Santonian (Cretaceous) based on the marine carbon isotope record. *Paleoceanography* article #PA002941, 31: 847–865, https://doi.org/10.1002/2016PA002941.

Thieuloy, J.-P., 1977, La zone à Callidiscus du Valanginien supérieur vocontien (Sud-Est de la France). Lithostratigraphie, ammonitofaune, limite Valanginien-Hauterivien, corrélations. *Géologie Alpine*, 53: 83–143.

Thurmann, J., 1836, Lettre à ME de Beaumont. *Bulletin de la société géologique de France, Serie 1*, 7: 207–211.

Tröger, K.-A., and Kennedy, W.J., 1996, The Cenomanian Stage. *Bulletin de l'Institut Royal des Sciences Naturelles de Belgique, Sciences de la Terre*, 66 (Suppl.), 57–68.

Tsikalas, F., 2005. Mjølnir Impact Crater. Geophysics Research Group, Department of Geology, University of Oslo. Available from: < http://folk.uio.no/ftsikala/mjolnir/ > .

Tur, N.A., Smirnov, J.P., and Huber, B.T., 2001, Late Albian–Coniacian planktic foraminifera and biostratigraphy of the northeastern Caucasus. *Cretaceous Research*, 22: 719–734.

van Buchem, F.S.P., Simmons, M.D., Droste, H.J., and Davies, R.B., 2011, Late Aptian to Turonian stratigraphy of the eastern Arabian Plate – depositional sequences and lithostratigraphic nomenclature. *Petroleum Geosciences*, 17: 211–222.

VandenBerg, J., Klootwijk, C.T., and Wonders, A.A.H., 1978, Late Mesozoic and Cenozoic movements of the Italian peninsula: further paleomagnetic data from the Umbrian sequence. *Geological Society of America Bulletin*, 89: 133–150.

VandenBerg, J., and Wonders, A.A.H., 1980, Paleomagnetism of Late Mesozoic pelagic limestones from the Southern Alps. *Journal of Geophysical Research*, 85: 3623–3627.

van Hinte, J.E., 1965, The type Campanian and its planktonic Foraminifera. *Proceedings of the Koninklijke Nederlandse Akademie van Wetenschappen, Series B*, 68: 8–28.

Vennari, V.V., Lescano, M., Naipauer, M., Aguirre-Urreta, B., Concheyro, A., Schaltegger, U., et al., 2014, New constraints on the Jurassic–Cretaceous boundary in the High Andes using high-precision U–Pb data. *Gondwana Research*, 26: 374–385.

Voigt, S., and Schonfeld, J., 2010, Cyclostratigraphy of the reference section for the Cretaceous white chalk of northern Germany, Lägerdorf–Kronsmoor – a late Campanian–early Maastrichtian orbital time scale. *Palaeogeography, Palaeoclimatology, Palaeoecology*, 287: 67–80.

Voigt, S., Erbacher, J., Mutterlose, J., Weiss, W., Westerhold, T., Wiese, F., et al., 2008, The Cenomanian – Turonian of the Wunstorf section – (North Germany): global stratigraphic reference section and new orbital time scale for Oceanic Anoxic Event 2. *Newsletters on Stratigraphy*, 43: 65–89.

Voigt, S., Friedrich, O., Norris, R.D., and Schönfeld, J., 2010, Campanian–Maastrichtian carbon isotope stratigraphy: shelf-ocean correlation between the European shelf sea and the tropical Pacific Ocean. *Newsletters on Stratigraphy*, 44: 57–72.

Voigt, S., Gale, A.S., Jung, C., and Jenkyns, H.C., 2012, Global correlation of Upper Campanian-Maastrichtian successions using carbon-isotope stratigraphy: development of a new Maastrichtian timescale. *Newsletters on Stratigraphy*, 45: 25–53.

von Salis, K., 1998, Calcareous nannofossils. Columns for Jurassic chart of Mesozoic and Cenozoic sequence chronostratigraphic framework of European basins, by Hardenbol, J., Thierry, J., Farley, M.B., Jacquin, T., de Graciansky, P.-C., and Vail, P.R. (coordinators). *In* de Graciansky, P.-C., Hardenbol, J., Jacquin, T., and Vail, P.R. (eds), Mesozoic-Cenozoic Sequence Stratigraphy of European Basins. *SEPM Special Publication,* **60**: 763−781.

Wagerich, M., 2012, "OAE3"-regional Atlantic organic carbon burial during the Coniacian-Santonian. *Climate of the Past,* **8**: 1447−1455, https://doi.org/10.5194/cp-8-1447-2012.

Walaszczyk, I., and Cobban, W.A., 2007, Inoceramid fauna and biostratigraphy of the upper Middle Coniacian−lower Middle Santonian of the Pueblo Section (SE Colorado, US Western Interior). *Cretaceous Research,* **28**: 132−142.

Walaszczyk, I., and Cobban, W.A., 2016, Inoceramid bivalves and biostratigraphy of the upper Albian and Cenomanian of the United States Western Interior Basin. *Cretaceous Research,* **59**: 30−68.

Walaszczyk, I., and Wood, C.J., 1998, Inoceramid and biostratigraphy at the Turonian/Coniacian boundary; based on the Salzgitter-Salder Quarry, Lower Saxony, Germany, and the Słupia Nadbrzeżna section, Central Poland. *Acta Geologica Polonica,* **48**: 395−434.

Walaszczyk, I., and Wood, C.J., 2018, Inoceramid bivalves from the Coniacian (Upper Cretaceous) of the Staffhorst shaft (Lower Saxony, Germany) − stratigraphical significance of a unique section. *Cretaceous Research,* **87**: 226−240, http://dx.doi.org/10.1016/j.cretres.2017.07.001.

Walaszczyk, I., Odin, G.S., and Dhondt, A.V., 2002, Inoceramids from the Upper Campanian and Lower Maastrichtian of the Tercis section (SW France), the Global Stratotype Section and Point for the Campanian − Maastrichtian boundary; taxonomy, biostratigraphy and correlation potential. *Acta Geologica Polonica,* **52**: 269−305.

Walaszczyk, I., Wood, C.J., Lees, J.A., Peryt, D., Voigt, S., and Wiese, F., 2010, The Salzgitter-Salder Quarry (Lower Saxony, Germany) and Słupia Nadbrzeżna river cliff section (central Poland): a proposed candidate composite Global Boundary Stratotype Section and Point for the base of the Coniacian Stage (Upper Cretaceous). *Acta Geologica Polonica,* **60**: 445−477.

Walaszczyk, I., Dubicka, Z., Olszewska-Nejbert, D., and Remin, Z., 2016, Integrated biostratigraphy of the Santonian through Maastrichtian (Upper Cretaceous) of Poland. *Acta Geologica Polonica,* **66**: 313−350.

Walaszczyk, I., and Cobban, W.A., 2000, Inoceramid faunas and biostratigraphy of the Upper Turonian − Lower Coniacian of the Western Interior of the United States. *Special Papers in Palaeontology,* **64**: 1−118.

Walaszczyk, I., Cobban, W.A., and Harries, P.J., 2001, Inoceramids and inoceramid biostratigraphy of the Campanian and Maastrichtian of the United States Western Interior Basin. *Revues de Paléobiologie,* **20**: 117−234.

Wang, T.T., Ramezani, J., Wang, C.S., Wu, H.C., He, H.Y., and Bowring, S.A., 2016, High-precision U−Pb geochronologic constraints on the Late Cretaceous terrestrial cyclostratigraphy and geomagnetic polarity from the Songliao Basin, Northeast China. *Earth and Planetary Science Letters,* **446**: 37−44.

Weissert, H., and Bréheret, J.G., 1991, A carbonate-isotope record from Aptian-Albian sediments of the Vocontian Trought (SE France). *Bulletin de la Société géologique de France,* **162**: 1133−1140.

Weissert, H., and Lini, A., 1991, Ice Age interludes during the time of Cretaceous greenhouse climate? *In* Muller, D.W., McKenzie, J.A., and Weissert, H. (eds), *Controversies in Modern Geology.* London: Academic Press, 173−191.

Weissert, H., Lini, A., Föllmi, K.B., and Kuhn, O., 1998, Correlation of Early Cretaceous carbon isotope stratigraphy and platform drowning

events: a possible link? *Palaeogeography, Palaeoclimatology, Palaeoecology,* **137**: 189−203.

Wendler, I., 2013, A critical evaluation of carbon isotope stratigraphy and biostratigraphic implications for Late Cretaceous global correlation. *Earth Science Reviews,* **126**: 116−146.

Wendler, I., Wendler, J., Grafe, K.-U., Lehmann, J., and Willems, H., 2009, Turonian to Santonian carbon isotope data from the Tethys Himalaya, southern Tibet. *Cretaceous Research,* **30**: 961−979.

Wendler, I., Willems, H., Grafe, K.-U., Ding, L., and Luo, H., 2011, Upper Cretaceous inter-hemispheric correlation between the Southern Tethys and the Boreal: chemo- and biostratigraphy and paleoclimatic reconstructions from a new section in the Tethys Himalaya, S-Tibet. *Newsletters on Stratigraphy,* **44/2**: 137−171.

Wierzbowski, H., Anczkiewicz, R., Pawlak, J., Rogov, M.A., and Kuznetsov, A.B., 2017, Revised Middle−Upper Jurassic strontium isotope stratigraphy. *Chemical Geology,* **466**: 239−255.

Wiese, F., 1999, Stable isotope data (^{13}C, ^{18}O) from the Middle and Upper Turonian (Upper Cretaceous) of Liencres (Cantabria, northern Spain) with a comparison to northern Germany (Söhlde & Salzgitter-Salder). *Newsletters on Stratigraphy,* **37**: 37−62.

Wiese, F., and Kaplan, U., 2001, The potential of the Lengerich section (Münster Basin, northern Germany) as a possible candidate Global boundary Stratotype Section and Point (GSSP) for the Middle/Upper Turonian boundary. *Cretaceous Research,* **22**: 549−563.

Wignall, P.B., 2001, Large igneous provinces and mass extinctions. *Earth-Science Reviews,* **53**: 1−33.

Wilson, P.A., Norris, R.D., and Cooper, M.J., 2002, Testing the Cretaceous greenhouse hypothesis using glassy foraminiferal calcite from the core of the Turonian tropics on Demerara Rise. *Geology,* **30**: 607−610.

Wimbledon, W.A.P., 2017, Developments with fixing a Tithonian/Berriasian (J/K) boundary. *Volumina Jurassica,* **XV**: 181−186.

Wimbledon, W.A.P., Casellato, C.E., Reháková, D., Bulot, L.G., Erba, E., Gardin, S., et al., 2011, Fixing a basal Berriasian and Jurassic-Cretaceous (J-K) boundary − is there perhaps some light at the end of the tunnel? *Rivista Italiana di Paleontologia e Stratigrafia,* **117**: 295−307.

Wimbledon, W.A.P., et al., 2019. Proposal of Tré Maroua as the GSSP for the Berriasian Stage (Cretaceous System), Made on Behalf of the Berriasian Working Group of the International Subcommission on Cretaceous Stratigraphy (ICS). 81 p. 24 Figures.

Wimbledon, W.A.P., Reháková, D., Svobodová, A., Schnabl, P., Pruner, P., Elbra, T., et al., 2020, Fixing a J/K boundary: a comparative account of key Tithonian-Berriasian profiles in the departments of Drôme and Hautes-Alpes, France. *Geologica Carpathica,* **71**: 24−46.

Wissler, L., Weissert, H., Buonocunto, F.P., Ferreri, V., and D'Argenio, B., 2004, Calibration of the Early Cretaceous time scale; a combined chemostratigraphic and cyclostratigraphic approach to the Barremian−Aptian interval, Campania Apennines and Aouthern Alps (Italy). *In* D'Argenio, B., Fischer, A.G., Premoli Silva, I., Weissert, H., and Ferreri, V. (eds), Cyclostratigraphy: Approaches and Case Histories. *SEPM Special Publication,* **81**: 123−133.

Wonders, A.A.H., 1980. Middle and Late Cretaceous Planktonic Foraminifera of the Western Mediterranean Area (Doctoral dissertation). Utrecht University.

Woodburne, M.O., 2004, *Late Cretaceous and Cenozoic Mammals of North America: Biostratigraphy and Geochronology.* New York: Columbia University Press 376 p.

Wortmann, U.G., Herrle, J.O., and Weissert, H., 2004, Altered carbon cycling and coupled changes in Early Cretaceous weathering patterns: evidence from integrated carbon isotope and sandstone records of the western Tethys. *Earth and Planetary Science Letters*, **220**: 69–82.

Wright, C.W., Kennedy, W.J., and Gale, A.S., 2017, The Ammonoidea of the Lower Chalk. Part 7. *Palaeontographical Society (Monograph)*, **171**: 461–571.

Wu, H.C., Zhang, S.H., Hinnov, L.A., Jiang, G.Q., Yang, T.S., Li, H.H., et al., 2014, Cyclostratigraphy and orbital tuning of the terrestrial Upper Santonian-Lower Danian in Songlaio Basin, northeastern China. *Earth and Planetary Science Letters*, **407**: 82–95.

Wu, H.C., Hinnov, L.A., Zhang, S.H., Jiang, G.Q., Chu, R., Yang, T.S., et al., 2020. Continental geological evidence for two Solar System chaotic events in the Late Cretaceous. *Proceedings of the National Academy of Sciences (USA)*, in press.

Young, J.Y., Gale, A.S., Knight, R.I., and Smith, A.B. (eds), 2010. *Fossils of the Gault Clay*. p. 342.

Zaborski, P.M.P., 1985, Upper Cretaceous ammonites from the Calabar region, south-east Nigeria. *Bulletin of the British Museum (Natural History), Geology Series*, **39**: 72.

Zakharov, V.A., Bown, P., and Rawson, P.F., 1996, The Berriasian Stage and the Jurassic-Cretaceous boundary. *Bulletin de l'Institut Royal des Sciences Naturelles de Belgique, Sciences de la Terre*, **66** (Suppl.), 7–10.

Zeebe, R.E., 2017, Numerical solutions for the orbital motion of the solar system over the past 100 Myr: limits and new results. *The Astronomical Journal*, **154** (5), 193.

Zhang, Y., Ogg, J.G., Minguez, D.A., Hounslow, M., Olaussen, S., Gradstein, F.M., and Esmeray-Senlet, S., 2019, Magnetostratigraphy of U/Pb-dated boreholes in Svalbard, Norway, implies that the Barremian–Aptian Boundary (beginning of Chron M0r) is 121.2 ± 0.4 Ma. In: AGU Annual Meeting (San Francisco, 9–13 December 2019). #GP44A-06. < https://agu.confex.com/agu/fm19/meetingapp.cgi/Paper/577991 >.

Zheng, X.-Y., Jenkyns, H.C., Gale, A.S., Henderson, G., and Ward, D.J., 2013, Nd-isotope evidence for changes in ocean circulation associated with transient cooling in the European shelf sea during OAE 2 (Cenomanian–Turonian). *Earth and Planetary Science Letters*, **375**: 338–348.

Zheng, X.-Y., Jenkyns, H.C., Gale, A.S., Ward, D.J., and Henderson, G.M., 2016, Recurrent climate-driven reorganizations of ocean circulation in the European epicontinental sea during the mid-Cretaceous; evidence from Nd isotopes. *Geology*, **44**: 151–154.

The Paleogene Period

52.2 Ma Paleogene

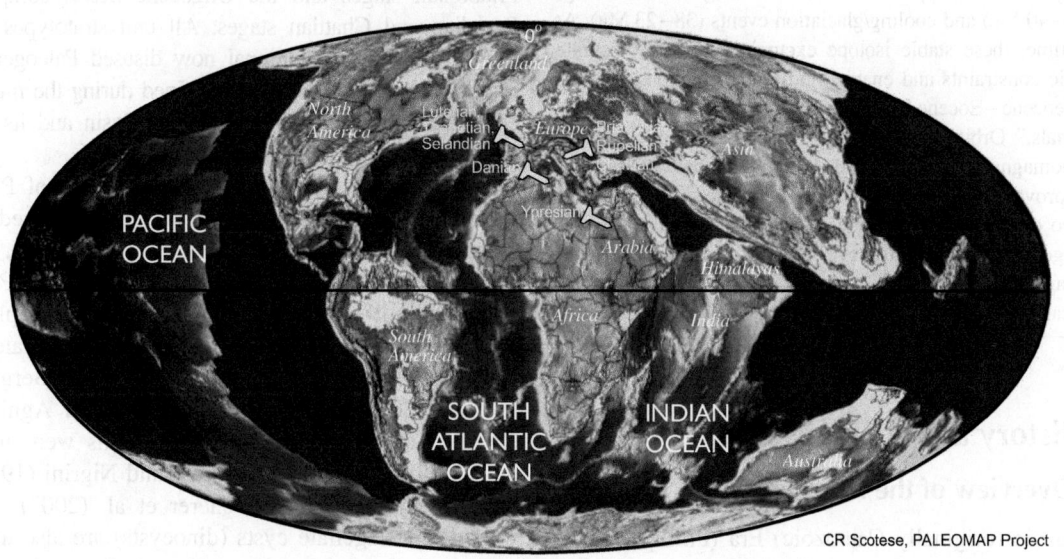

CR Scotese, PALEOMAP Project

Chapter outline

28.1 History and subdivisions   1088
    28.1.1 Overview of the Paleogene   1088
    28.1.2 Danian   1090
    28.1.3 Selandian   1092
    28.1.4 Thanetian   1092
    28.1.5 Ypresian   1095
    28.1.6 Lutetian   1097
    28.1.7 Bartonian   1097
    28.1.8 Priabonian   1099
    28.1.9 Rupelian   1101
    28.1.10 Chattian   1103
28.2 Paleogene biostratigraphy   1105

28.2.1 Marine biostratigraphy   1105
28.2.2 Terrestrial biostratigraphy   1113
28.2.3 Physical stratigraphy   1116
28.3 Paleogene time scale   1124
    28.3.1 Astronomical age model   1124
    28.3.2 Limits of validity of solar system orbital
        solutions   1124
    28.3.3 Updates to magnetic reversal ages since
        the Geologic Time Scale 2012   1125
    28.3.4 Ages of Paleogene stages   1126
Acknowledgments   1127
Bibliography   1127

Abstract

All Paleocene and Oligocene stages (Danian, Selandian, Thanetian, resp. Rupelian, and Chattian) have formally ratified definitions and so have the Ypresian, Lutetian, and Priabonian stages of the Eocene. We anticipate that the Global Boundary Stratotype Section and Point (GSSP) for the Bartonian Stage still requires more research before all stages of the Paleogene (66−23 Ma) are formally defined. Paleogene marine microfossil groups (planktonic and larger benthic foraminifera, calcareous nannofossils, radiolarians, organic-walled dinoflagellate cysts) provide robust zonation schemes for regional to global correlation and are integrated within the magneto-biochronological

Geologic Time Scale 2020. DOI: https://doi.org/10.1016/B978-0-12-824360-2.00028-0

framework. Since land mammal faunas are also increasingly being studied with an integrated magnetostratigraphic and/or chemostratigraphic and geochronologic approach, their age calibrations have considerably been improved since GTS2012. Stable isotope analysis and XRF (X-ray fluorescence) scanning have become key tools in Paleogene high-resolution stratigraphy, correlation, and time scale construction. Stable oxygen and carbon isotope records also provide insight into trends in paleoclimate and carbon cycling, such as the warming trend starting in the middle Paleocene and culminating during the Early Eocene Climatic Optimum, and the subsequent cooling leading to a change from greenhouse to icehouse conditions at the onset of the Oligocene. Numerous short-term isotope excursions mark high climatic variability, expressed in hyperthermal (transient global warming) events (62–40 Ma) and cooling/glaciation events (38–23 Ma). At the same time, these stable isotope excursions provide accurate stratigraphic constraints and enable land-sea correlations, such as for the Paleocene–Eocene Thermal Maximum, the "Mother of all hyperthermals." Orbital tuning of sedimentary cycles, calibrated to the geomagnetic polarity and biostratigraphic scales, has greatly improved the resolution of the Paleogene time scale over the last two decades. We now have astronomical age control for almost all geomagnetic polarity reversals, but differences between published age models still persist through the "Eocene astronomical time scale gap" spanning Chrons C20r through C22n.

28.1 History and subdivisions

28.1.1 Overview of the Paleogene

The Cenozoic (originally Cainozoic) Era (Phillips, 1841) derives its name from the new (kainos) biota, compared to those of the Mesozoic Era. The Cenozoic Era is subdivided into the Paleogene (*palaios* = old, *genos* = birth), Neogene, and Quaternary periods.

The Paleogene Period (and System) is subdivided into three Epochs (and Series) (Fig. 28.1): the Paleocene, Eocene, and Oligocene, referring again to the evolution of the biota (eos = dawn and oligos = little). Lyell (1833) introduced the term Eocene for the oldest Cenozoic rocks in western Europe in which he recognized 3.4% of extant mollusk species. Later, Beyrich (1854), based on marine sequences and their fossil content observed in northern Germany, separated the Oligocene from the Miocene and Eocene. Naumann (1866) combined the Eocene and Oligocene in his *Paleogen Stufe*, as opposed to the *Neogen Stufe* (Hörnes, 1853) which included sequences containing Miocene, Pliocene, and Pleistocene faunas. Finally, Schimper (1874), based on paleobotanic studies in the Paris Basin and other west-European basins, distinguished the Paleocene from the Eocene. Only after a long period of opposition, the Paleocene eventually became accepted as the lowermost epoch of the Cenozoic (e.g., Mangin, 1957).

Larger benthic foraminifera and particularly the nummulites, played an important role in the development of Paleogene biostratigraphy and French-speaking stratigraphers such as Renevier (1874) and Gignoux (1950) used the term *Nummulitique* as an equivalent of Paleogene, but now this term has been abandoned in stratigraphic practice.

The three Paleogene series are formally subdivided into nine stages (Fig. 28.1) as decided by the International Subcommission on Paleogene Stratigraphy at the 1989 International Geological Congress (IGC) in Washington (Jenkins and Luterbacher, 1992). The Paleocene Series consists of the Danian, Selandian, and Thanetian stages, the Eocene Series of the Ypresian, Lutetian, Bartonian, and Priabonian stages and the Oligocene Series comprises the Rupelian and Chattian stages. All unit stratotypes of these stages, as well as additional now disused Paleogene stages (e.g., Montian, Cuisian), were defined during the mid-19th to early-20th century in the North Sea Basin and its southern extension, the Paris Basin (Pomerol, 1981).

Over the last few decades, the practice of Paleogene marine biostratigraphy has been largely based on the magneto-biochronology of Berggren et al., 1985, 1995) in which the geomagnetic polarity time scale (GPTS) was integrated with calcareous microfossil datums (planktonic foraminifera and calcareous nannofossils). Updates to this integrated time scale have been provided by Berggren and Pearson (2005), Wade et al. (2011), and Agnini et al. (2014). Low-latitude radiolarian datums were integrated into the time scale by Sanfilippo and Nigrini (1998a) and diatoms were added by Scherer et al. (2007). Organic-walled dinoflagellate cysts (dinocysts) are also an important group for Paleogene stratigraphy but owing to wide geographic variation in species distribution only regional zonations have been established (e.g., Bijl et al., 2013b). Over the last three decades, the understanding of the functioning of the solar system has greatly improved, and detailed geologic time series can be correlated to the orbital "metronome." This has resulted in the development of astronomically calibrated time scales that cover almost the entire Cenozoic Era.

Originally, the definition and characterization of the classic unit stratotypes were needed to reconstruct the outline of Paleogene geological developments in northwestern Europe. Developing a high-resolution time scale for the evolution of our planet, however, requires accurate chronostratigraphy and geochronology of well-defined levels that can be correlated between widely separated areas. Therefore, since the 1980s the concept of the Global Boundary Stratotype Section and Point (GSSP) has strongly guided Paleogene stratigraphic research, in attempting to precisely define boundary stratotypes of the stages at particular physical levels in well-documented and uninterrupted exposed sequences. A major concern in the definition of the Paleogene GSSPs was that the regions of the classic unit stratotypes could not provide sufficiently continuous sequences to define the boundaries of

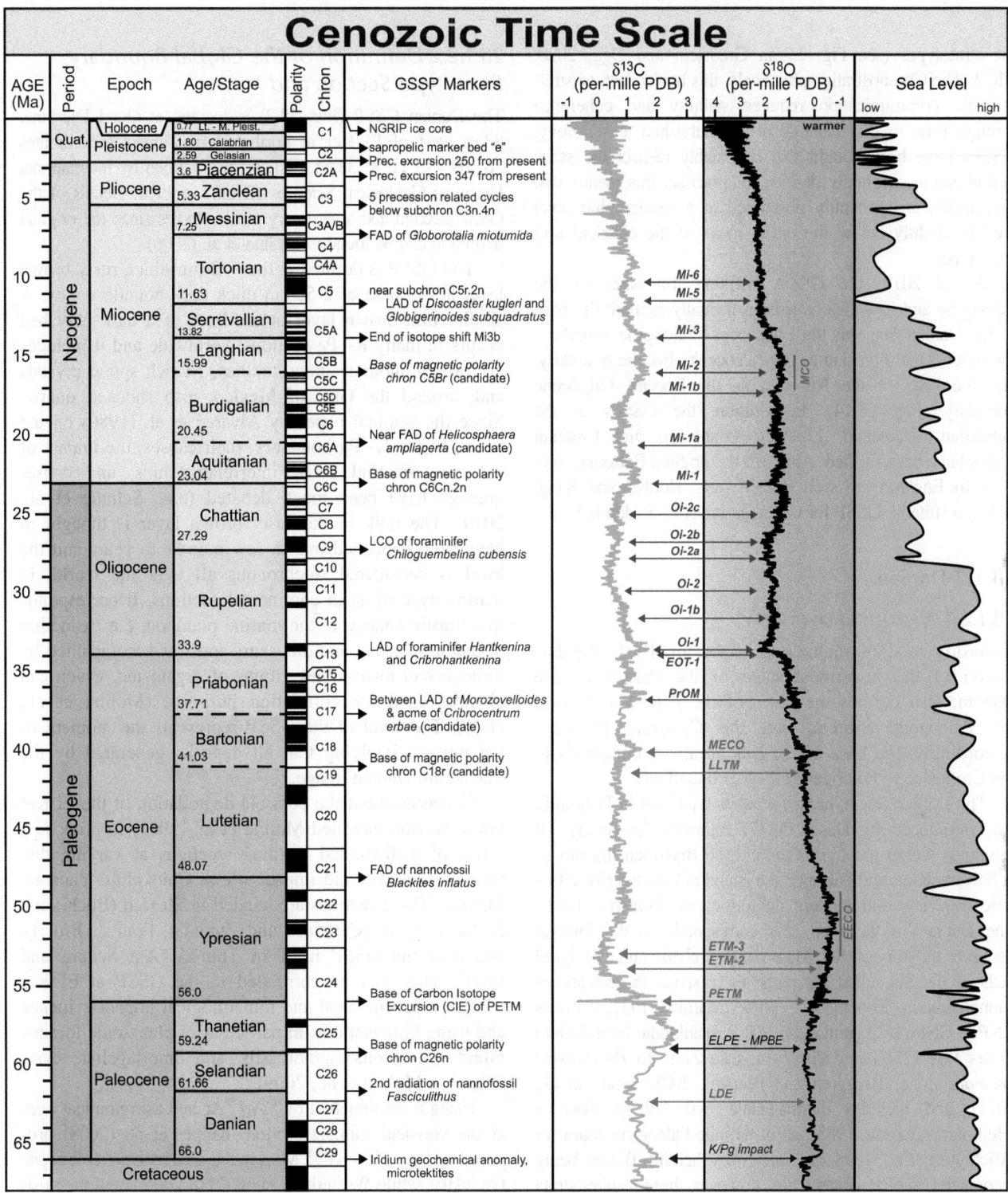

FIGURE 28.1   Integrated chronostratigraphy of the Cenozoic (66 Ma to Recent). "Age" is the term for the time equivalent of the rock-record "Stage." Main markers or candidate markers for GSSPs of Paleogene, Neogene, and Quaternary stages are detailed in this chapter, Chapter 29 (The Neogene Period) and Chapter 30 (The Quaternary Period), respectively. A summary of all basic Paleogene GSSP data is given in Fig. 28.14. *Prec. excursion 250 from Present* = orbital precession cycle excursion #250 before present. Benthic foraminiferal δ¹³C and δ¹⁸O curves are derived from a nine- or five-point moving window of recalibrated data from Cramer et al. (2009), and additional details are in Fig. 28.13. Named isotopic excursions and other events are discussed in this chapter and in Grossman and Joachimski (2020, Ch. 10: Oxygen isotope stratigraphy, this book). Schematic sea-level curve is modified from Hardenbol et al. (1998) and does not show Quaternary high-frequency oscillations. See also review in Simmons et al. (2020, Ch. 13: Phanerozoic eustasy, this book). *EECO*, Early Eocene Climatic Optimum; *ELPE*, early Late Paleocene Event; *EOT*, Eocene−Oligocene Transition; *ETM*, Eocene Thermal Maximum; *FAD*, First Appearance Datum; *GSSP*, Global Boundary Stratotype Section and Point; *K/Pg*, Cretaceous−Paleogene Boundary; *LAD*, Last Appearance Datum; *LCO*, Last Common Occurrence; *LDE*, Latest Danian Event; *LLTM*, Late Lutetian Thermal Maximum; *MCO*, Miocene Climatic Optimum; *MECO*, Mid-Eocene Climatic Optimum; *Mi*, Miocene Oxygen Isotopic Maximum; *MPBE*, Mid-Paleocene Biotic Event; *Oi*, Oligocene Oxygen Isotopic Maximum; *PDB*, PeeDee Belemnite Standard; *PETM*, Paleocene−Eocene Thermal Maximum; *PrOM*, Priabonian Oxygen Isotopic Maximum.

the stratotypes (see Fig. 2.2 in Gradstein and Ogg, 2020, Ch. 2: The Chronostratigraphic Scale, this book). Yet, as stratigraphic communication requires stability and coherence through time in the meaning of the published terminology, GSSPs have been sought that reasonably respect the stage definitions as originally defined. In practice, this means that the GSSPs are generally positioned at a stratigraphic level well to slightly below the oldest rocks of the classical unit stratotypes.

As of 2019, the GSSPs defining the bases of the Paleogene and its series have been formally ratified: the base of the Danian Stage as the Cretaceous—Paleogene boundary, the base of the Ypresian as the Paleocene—Eocene boundary, and the base of the Rupelian as the Eocene—Oligocene boundary (Fig. 28.14). In addition, the GSSPs of the Selandian, Thanetian, Lutetian, Priabonian and Chattian stages have been ratified. Although the original Bartonian sections in England are well studied (e.g., Hooker and King, 2019), a suitable GSSP for this stage is yet to be defined.

28.1.2 Danian

28.1.2.1 Historical overview

According to stratigraphic rules (Salvador, 1994), the definition of the lowermost stage of the Paleogene, the Danian, also defines the bases of the Paleocene Series, the Paleogene System, and the Cenozoic Erathem. Accordingly, the base of the Danian also corresponds to the Cretaceous—Paleogene (K—Pg) boundary.

The Danian Stage, named after its type area in Denmark, was introduced by Desor (1847) following his studies of echinoids within the *Cerithium* and the Bryozoan limestones at Stevens Klint and Fakse on the island of Zealand (for a historic overview and modern definition, see Pomerol, 1981). The Danian in the type area corresponds to the interval between the top of the Maastrichtian chalk and the basal beds of the Selandian. Its range there spans the calcareous nannoplankton Zones NP1—NP4 (Martini, 1971; = Zones CNP1—CNP6 of Agnini et al., 2014), planktonic foraminifera Zones P0—P2 (*Guembelitria cretacea* Zone to *Praemurica uncinata* Zone; Berggren and Pearson, 2005; Wade et al., 2011), and includes dinoflagellate cyst Viborg Zone 1 (Heilmann-Clausen, 1988), all of definite Paleogene character (Michelsen et al., 1998, and references therein). Before being recognized as of Paleogene age, however, these Danian strata were originally considered as the uppermost stage of the Cretaceous (Desor, 1847). Consequently, a supposedly higher stage at the base of the Paleogene was introduced by Dewalque (1868): the Montian (after Mons, Belgium). The Montian, however, has lost its significance because of poor stratotype conditions and because it correlates with the upper part of the Danian Stage in its present definition (Pomerol, 1981; De Geyter et al., 2006).

28.1.2.2 Definition of the Global Boundary Stratotype Section and Point

The Danian GSSP (Fig. 28.2) is located at Oued Djerfane, 8 km west of El Kef in northwestern Tunisia (coordinates 36° 09′13.2″N, 8° 38′54.8″E) and was ratified by International Union of Geological Sciences (IUGS) in 1991. Details of the GSSP Section and a summary of the studies since the original definition can be found in Molina et al. (2006).

The GSSP is defined at the ~2-mm-thick rusty brown layer at the base of a 50-cm-thick dark boundary clay. A similar thin brown layer at the base of a dark clay bed occurs in many K—Pg sections worldwide and it includes an iridium anomaly, microtektites, Ni-rich spinel crystals and, around the Gulf of Mexico, also shocked quartz. Since the seminal paper by Alvarez et al. (1980) on the events at the K—Pg boundary, their causes, the impact of an extraterrestrial body, interrelationships, and consequences have been much debated (e.g., Schulte et al., 2010). The bulk of the rusty brown layer is thought to have been deposited over a few months to years and the level is considered isochronous all over the world, in marine as well as in continental sections. It corresponds to a drastic change in the marine plankton, the extinction of the ammonites, and dinosaurs and was accompanied by turnovers in many other groups of organisms, which can also be used for correlation purposes (Molina et al., 2006). The level of the GSSP represents the moment of the impact, implying that all deposits generated by the impact are Danian in age.

Concerns about the possible degradation of the El Kef GSSP section have led Molina et al. (2009) to propose a series of well-studied auxiliary sections at varying distances to the asteroid impact site at Chicxulub, Yucatan, Mexico. These sections are located in Mexico (Bochil and Mulato), Spain (Caravaca and Zumaia), France (Bidart), and near the GSSP itself in Tunisia (Aïn Settara and Ellès). They can be correlated to the GSSP at El Kef using the geochemical and mineralogical impact evidence and using biozonations, in particular of planktonic foraminifera, calcareous nannofossils, and dinoflagellate cysts (Fig. 3 in Molina et al., 2009).

Using a combination of ^{40}Ar/^{39}Ar and astronomical ages at the classical Zumaia section, Kuiper et al. (2008) proposed an age of ~65.95 Ma for the Cretaceous—Paleogene boundary, while Westerhold et al. (2008) correlated the position of the GSSP exactly in the middle of (geomagnetic polarity) Chron C29r. The astronomical age calibration by Dinarès-Turell et al. (2014) now dates the boundary at 66.00 Ma. The astronomically calibrated duration of the Danian, approximated in terms of the GPTS from the middle of Chron C29r to 30 precession cycles above the base of Chron C26r, is 4.34 Myr (66.00—61.66 Ma), according to the age model of Dinarès-Turell et al. (2014).

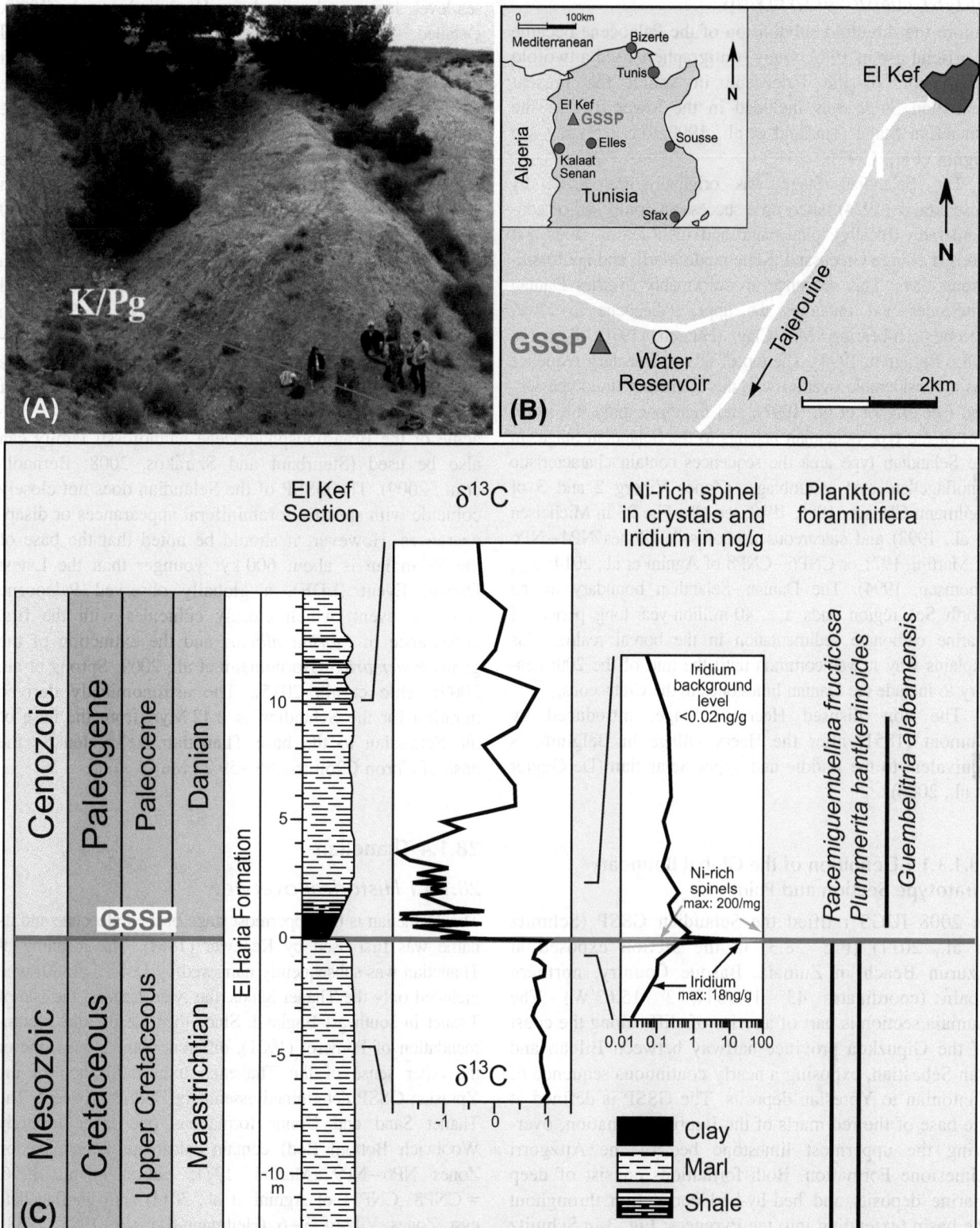

Base of the Danian Stage at El Kef, Tunisia

FIGURE 28.2 Danian GSSP located at El Kef in Tunisia (B). Photograph of the outcrop (A) from Ogg et al. (2008) and stratigraphic data (C) from Keller and Lindinger (1989), Robin and Rocchia (1998), and Molina et al. (2006). *GSSP*, Global Boundary Stratotype Section and Point.

28.1.3 Selandian

28.1.3.1 Historical overview

Before the threefold subdivision of the Paleocene became in official use in 1989, many stratigraphers used a twofold subdivision of the Paleocene in which the present Selandian Stage was included in the lower part of the Thanetian Stage (Harland et al., 1990; for a review see Bignot et al., 1997).

The Selandian Stage was originally introduced by Rosenkrantz (1924) based on a succession composed of conglomerates (locally), glauconitic and fossiliferous sands and marls (Lellinge Greensand, Kerteminde Marl), and nonfossiliferous clays. This sequence unconformably overlies Danian limestones and underlies the upper Paleocene to lower Eocene ash-bearing Mo–Clay (Hansen, 1979; Pomerol, 1981; Berggren, 1994). The top of this sedimentary sequence has a considerable overlap with the Thanetian stratotype section (see Bignot et al., 1997), and therefore, only the lower part of the type Selandian belongs to the Selandian Stage. In the Selandian type area the sequences contain characteristic dinoflagellate cyst assemblages (Zones Viborg 2 and 3 of Heilmann-Clausen, 1985, 1988; see also Fig. 23 in Michelsen et al., 1998) and calcareous nannofossils (Zones NP4–NP6 of Martini, 1971 or CNP6–CNP8 of Agnini et al., 2014; e.g., Thomsen, 1994). The Danian–Selandian boundary in the North Sea region ends a c. 40-million-year-long period of marine carbonate sedimentation in the boreal realm. This explains why it was common until the mid of the 20th century to include the Danian limestones in the Cretaceous.

The now disused Heersian Stage, introduced by Dumont (1851) after the Heers village in Belgium, is equivalent to the middle and upper Selandian (De Geyter et al., 2006).

28.1.3.1.1 Definition of the Global Boundary Stratotype Section and Point

In 2008 IUGS ratified the Selandian GSSP (Schmitz et al., 2011) (Fig. 28.3) in the section exposed at Itzurun Beach in Zumaia, Basque Country, northern Spain (coordinates 43° 17.98′N, 2° 15.63′W). The Zumaia section is part of a series of cliffs along the coast of the Gipuzkoa province halfway between Bilbao and San Sebastian, exposing a nearly continuous sequence of Santonian to Ypresian deposits. The GSSP is defined at the base of the red marls of the Itzurun Formation, overlying the uppermost limestone bed of the Aitzgorri Limestone Formation. Both formations consist of deep marine deposits and bed-by-bed correlation throughout the basin (extending into the Pyrenees; Fig. 3 in Schmitz et al., 2011), suggests that sedimentation across the Danian/Selandian boundary was continuous. The change from the Aitzgorri Limestone Formation to the marls of

the Itzurun Formation is considered the deepwater correlative conformity of the major Sel1 sequence boundary and sea-level low as described in Hardenbol et al. (1998). Detailed stratigraphic studies show that the lithological change at the Selandian GSSP correlates with the facies shift from Danian limestones to Selandian detrital sediments in the type area of Denmark, suggesting a response to the same regional sea-level event (Schmitz et al., 1998).

The GSSP occurs approximately at the top of the lower third of Chron C26r and comparison with the Bjala section in Bulgaria indicates that the GSSP occurs at 30 precession cycles (about 615 kyr) above the base of Chron C26r, providing an astronomical age of 61.66 Ma (Dinarès-Turell et al., 2010; Schmitz et al., 2011). A tool for global marine correlation of the base of the Selandian Stage is provided by the second radiation of the calcareous nannofossil *Fasciculithus* (base Zone NP5), which is situated one precession cycle (21 kyr) below the base of the Selandian. For regional correlations the top of the acme of the Braarudosphaeraceae nannofossil family can also be used (Steurbaut and Sztrákos, 2008; Bernaola et al., 2009). The GSSP of the Selandian does not closely coincide with specific foraminiferal appearances or disappearances. However, it should be noted that the base of the Selandian is about 600 kyr younger than the Latest Danian Event (LDE), a globally observed Paleocene warming event, which closely coincides with the first appearance of *Igorina albeari* and the extinction of the genus *Praemurica* (Bornemann et al., 2009; Sprong et al., 2009; Jehle et al., 2015). The astronomically derived duration for the Selandian is 2.42 Myr, from the base of the Selandian to the base Thanetian, delineated by the base of Chron C26n (61.66–59.24 Ma).

28.1.4 Thanetian

28.1.4.1 Historical overview

The Thanetian is the uppermost stage of the Paleocene and its name was first used by Renevier (1874). The meaning of Thanetian was subsequently narrowed by Dollfus (1880) who included only the Thanet Sands, the type-strata on the Isle of Thanet in southeast England. Since then, despite the recommendation of Pomerol (1981), different authors used one or the other sense of the Thanetian until ratification of the Ypresian GSSP determined essentially Dollfus' concept. The Thanet Sand and Upnor formations (the latter formerly Woolwich Bottom Bed) contain calcareous nannoplankton Zones NP6–NP9 (Martini, 1971; Ellison et al., 1996; = CNP8–CNP11 of Agnini et al., 2014) and dinoflagellate cyst Zones Viborg 3–6 (Heilmann-Clausen, 1985, 1988), D4b–D5a (Costa and Manum, 1988), and Ama-Ahy (Powell et al., 1996). These two formations span Chrons C26n, C25 and the lower part of C24r (Ali and Jolley, 1996).

Base of the Selandian Stage at Zumaia, Spain

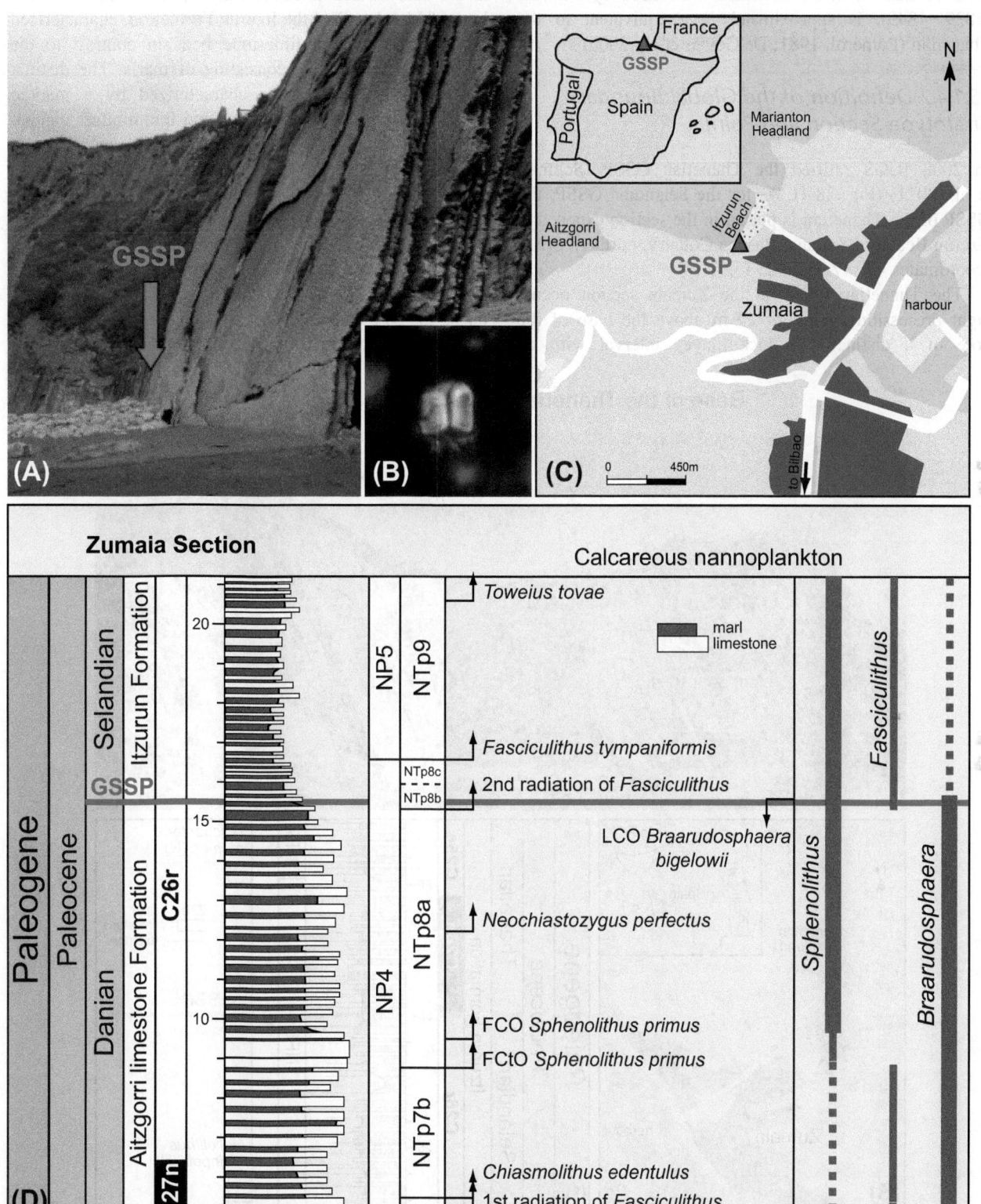

FIGURE 28.3 Selandian GSSP located at Zumaia, Basque Country, Spain (C). The stratigraphic data (D) are based on Schmitz et al. (2011); The photograph of the outcrop (A) by courtesy of Eustoquio Molina and the image of *Fasciculithus tympaniformis* (B) by courtesy of Simonetta Monechi. *GSSP*, Global Boundary Stratotype Section and Point.

The now disused Landenian Stage, named after the town of Landen in Belgium and introduced by Dumont (1839, 1849), is approximately age equivalent to the Thanetian (Pomerol, 1981; De Geyter et al., 2006).

28.1.4.2 Definition of the Global Boundary Stratotype Section and Point

In 2008 IUGS ratified the Thanetian GSSP (Schmitz et al., 2011) (Fig. 28.4). As for the Selandian GSSP, the GSSP of the Thanetian is placed in the section exposed at Itzurun Beach in Zumaia, Basque Country, northern Spain (coordinates 43° 17.98′N, 2° 15.63′W).

The Thanetian GSSP in the Zumaia section occurs eight precession cycles, or 2.8 m above the base of the core of a distinct 1-m-thick clayey interval with an increased magnetic susceptibility. The GSSP is positioned at the base of Chron C26n, which is c. 6.5 m above the base of Member B of the Itzurun Formation, characterized by regular indurated limestone beds in contrast to the underlying Member A, consisting of marls. The distinct 1-m-thick clay horizon is characterized by a marked change in calcareous nannofossil and foraminifera content and is known as the short-lived Mid-Paleocene Biotic Event (MPBE) (Bernaola et al., 2007; Hilgen et al., 2015). The MPBE is located c. 4.5 m above the first occurrence of *Heliolithus kleinpelli*, the marker of Zone NP6, and within planktonic foraminifera Zone P4.

A major increase in the abundance of the dinoflagellate *Alisocysta gippingensis* is considered a useful event for recognizing the base of the Thanetian within the North Sea Basin, while the last occurrence of *Palaeoperidinium pyrophorum* is

Base of the Thanetian Stage at Zumaia, Spain

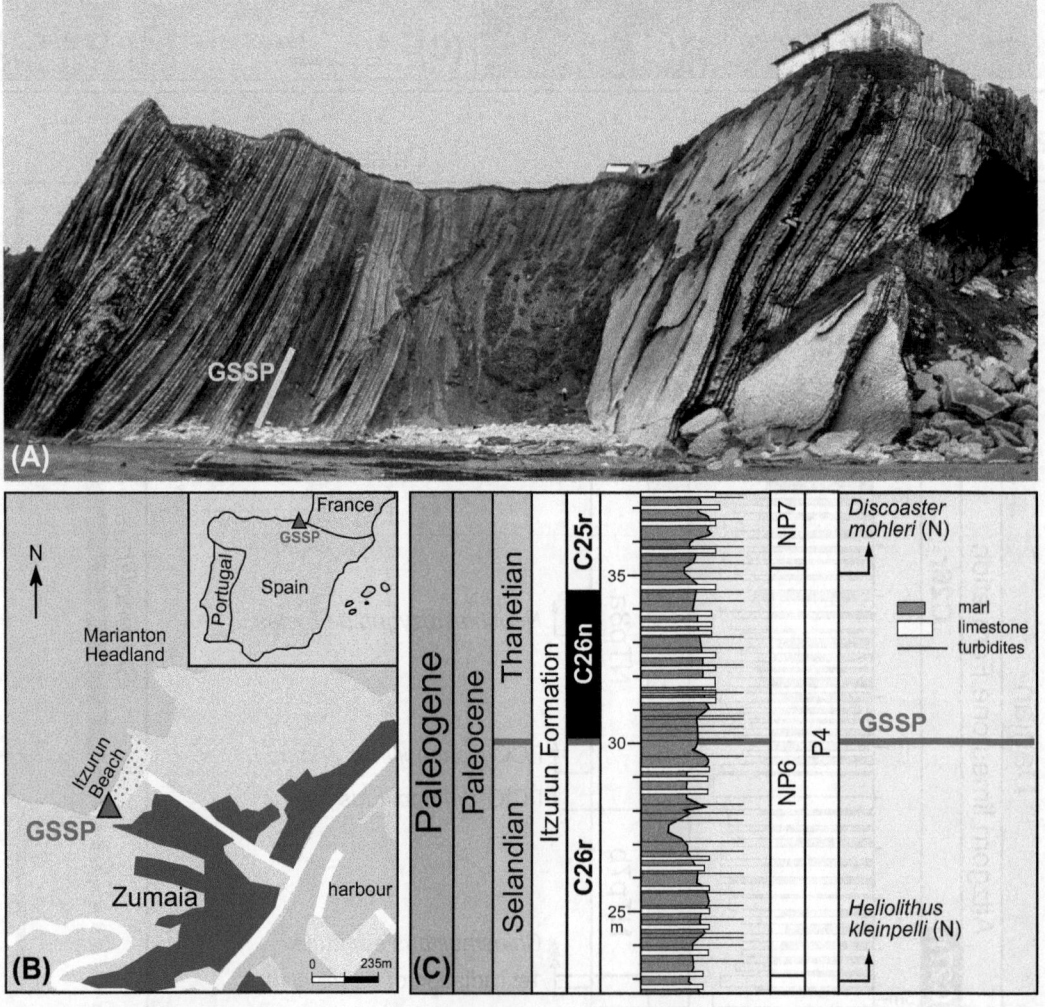

FIGURE 28.4 Thanetian GSSP located at Zumaia, Basque Country, Spain (B). The photograph of the outcrop (A) by courtesy of Victoriano Pujalte. The stratigraphic data (C) are based on Schmitz et al. (2011). Note that the position of the GSSP was accidentally positioned too high in the sequence (to the left) in figure 4 of Schmitz et al. (2011) and accordingly also in Fig. 28.4 in Vandenberghe et al. (2012); the position of the GSSP is corrected here. *GSSP*, Global Boundary Stratotype Section and Point.

useful for interregional correlation (Schmitz et al., 2011). The astronomically derived duration for the Thanetian, approximated by the base of Chron C26n to the base of the Ypresian as marked by a large negative carbon isotope excursion (CIE), is 3.24 Myr (59.24−56 Ma).

28.1.5 Ypresian

28.1.5.1 Historical overview

The Ypresian is the lowermost stage of the Eocene Series and accordingly, the base of the Ypresian also defines the base of the Eocene (Salvador, 1994). The Ypresian Stage was originally introduced by Dumont (1849) to include strata of open marine clays to fine-sands, situated between the terrestrial to marginally marine Landenian and the marine Brusselian in Belgium. Dumont (1851) later assigned the upper, sandier part of this stage to a separate Paniselian Stage (Pomerol, 1981). The Paniselian (named after the Mont-Panisel hill near Mons, Belgium), however, is a poorly defined term and became obsolete as a chronostratigraphic term (De Geyter et al., 2006), leaving the Ypresian as it was originally defined by Dumont (1849). The term Ypresian is derived from the town of Ieper (Dutch spelling) or Ypres (French translation).

The Ypresian sediments in the type area are well constrained by dinocyst and calcareous nannofossil biostratigraphy, sequence-stratigraphy, and magnetostratigraphy (e.g., Ali et al., 1993; Vandenberghe et al., 1998, 2004; Steurbaut, 2006).

Cuisian and Ilerdian are regional stage names that were used in parallel with the Ypresian (Pomerol, 1981). The stratotype of the Cuisian, defined in the Paris Basin, was considered age-equivalent with the Ypresian, and accordingly, it was considered redundant (Pomerol, 1981). In the Tethyan realm the Ilerdian is a regional Mediterranean stage (Hottinger and Schaub, 1960; Pomerol, 1981) corresponding to an important phase in the evolution of larger benthic foraminifera which is not represented in northwestern European basins. Integrated stratigraphic studies have shown that the base of the Ilerdian correlates with the base of the Ypresian (Pujalte et al., 2009; Scheibner and Speijer, 2009; Zamagni et al., 2012).

The biotic and abiotic events that occurred in the Paleocene−Eocene boundary interval, their interrelations, and possible causes have been intensively studied by the Paleocene−Eocene Boundary Working Group (e.g., reviews in Aubry et al., 1998; Aubry et al., 2003, 2007) and many others. Although numerous biostratigraphic events occur in this boundary interval, the onset of a pronounced negative CIE associated with the Paleocene−Eocene Thermal Maximum (PETM), was selected as the best criterion for recognition and global correlation of the Paleocene−Eocene boundary (Fig. 28.5).

Based on cyclostratigraphic analysis, Westerhold et al. (2008) placed the base of the PETM at Chron C24r.36 (=36% up in Chron C24r). Furthermore, addressing the "discrepancy at 50 Ma" in GTS2012, Westerhold et al. (2017) used a new compilation of ODP data from Leg 208 Walvis Ridge and Leg 207 Demarara Rise to assess the duration of the Ypresian. They provided a combined age model of magnetic reversals from the base of Chron C24n to base of Chron C19r, and arrived at an age of 56.0 Ma for the base of the PETM.

The onset of the CIE has the advantage of being applicable in both marine and terrestrial successions (Stott et al., 1996). The negative CIE is a primary expression of the PETM reflecting a massive injection of light carbon (^{12}C) into the ocean and atmosphere that caused a 100−200 kyr period of pronounced global warming, a transient climatic condition commonly referred to as hyperthermal (Thomas et al., 2000). A sudden release of methane due to dissociation of subseafloor clathrates due to deep-water warming and/or submarine landslides (Dickens et al., 1997) remains a plausible source of isotopically light carbon although other sources have been proposed (e.g., Higgins and Schrag, 2006). The CIE coincides with major paleontological events: a deep-sea benthic foraminiferal extinction event, diversification in planktonic foraminifera, calcareous nannofossils and larger benthic foraminifera, a global acme of the dinocyst genus *Apectodinium*, a decline of coral reefs, and on land a mammal dispersal event, which in North America defines the base of the Wasatchian Land Mammal Age (LMA) (see overviews in Aubry et al., 2007; Sluijs et al., 2007; McInerney and Wing, 2011; Speijer et al., 2012).

28.1.5.2 Definition of the Global Boundary Stratotype Section and Point

The GSSP of the Ypresian is located in the Dababiya Section on the east bank of the upper Nile Valley south of Luxor in Egypt (coordinates 25° 30′N, 32° 31′52″E) (Dupuis et al., 2003; Aubry et al., 2007). The "golden spike" is positioned at the base of a 63-cm-thick dark gray clayey interval, the Dababiya Quarry Bed 1, which underlies the 2-m-thick, partly phosphatic and laminated, Dababiya Quarry beds 2−5. Together these beds constitute the Dababiya Quarry Member in the lower part of the Esna Shale Formation.

Unfortunately, there is an increasing concern about the stability of Ypresian GSSP in global correlation. It was realized already early on that the Dababiya Quarry Beds pinch out laterally (Dupuis et al., 2003) and that this was due to the filling up of an erosional channel structure in the Esna shales directly underlying the base of the Ypresian (Ernst et al., 2006; Schulte et al., 2011;

Base of the Ypresian Stage at Dababiya, Egypt

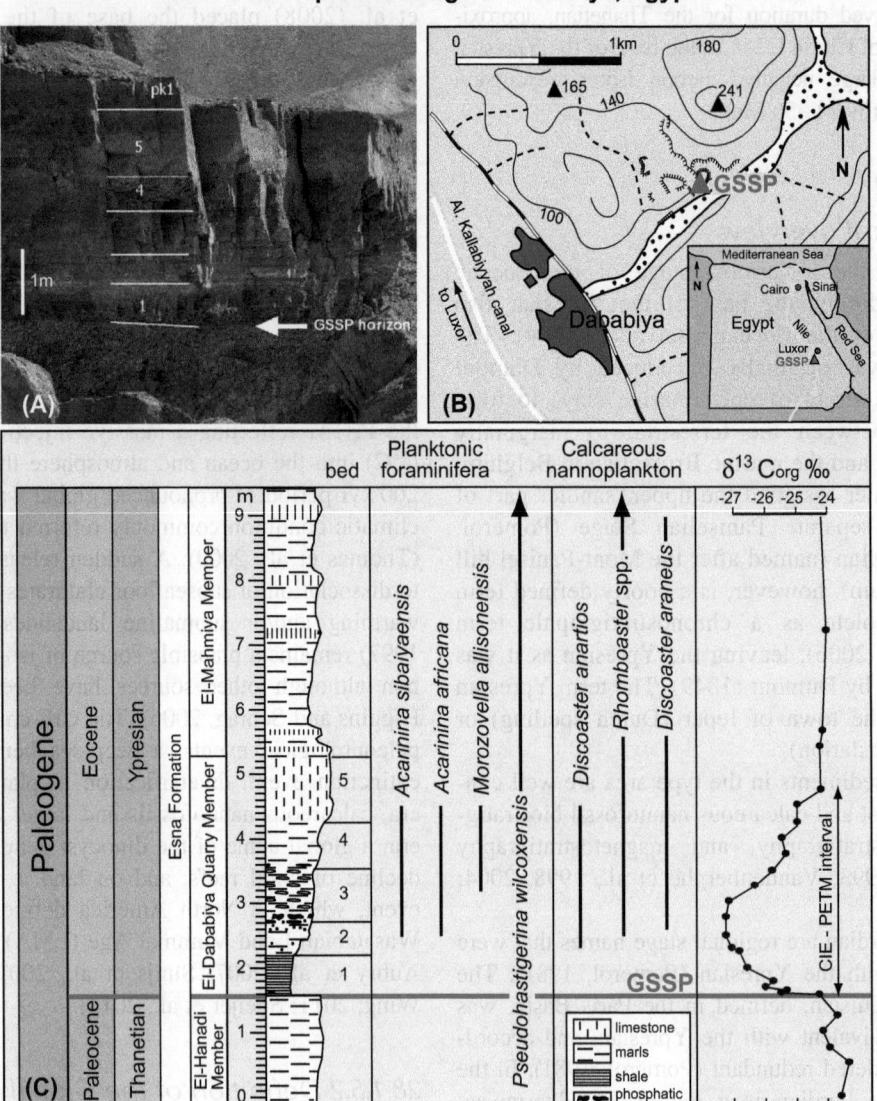

FIGURE 28.5 Ypresian GSSP located at Dababiya, Egypt (B). The photograph of the outcrop (A), the map (B), and the stratigraphic data (C) are based on Aubry et al. (2007). *GSSP*, Global Boundary Stratotype Section and Point.

Khozyem et al., 2014). Consequently, the Dababiya sequence is not conformable across the GSSP. An interval lacking carbonate at the base of the Dababiya Quarry beds is also problematic. Because of these limitations, the search for continuous sections spanning the Paleocene–Eocene boundary goes on. An expanded bathyal PETM section at Forada, northern Italy (Giusberti et al., 2007; Luciani et al., 2007), was suggested as an auxiliary stratotype section and point for the base of the Ypresian (Boscolo-Galazzo et al., 2019).

The base of the Ypresian Stage is about 1 Myr older than the base of the Ypresian stratotype in the Ieper area in Belgium. In fact, the basal Ypresian in its historical type area correlates with the base of the fluvio-lagoonal

upper part of the former Landenian Stage (Steurbaut, 2006). The CIE is situated near the base of the Sparnacian in the adjacent Paris Basin, where sediments are mainly nonmarine (e.g., Aubry et al., 2003). Because the Sparnacian of the Paris Basin is inferred to correlate with the lower part of the Ypresian Stage, the Sparnacian has become obsolete (Pomerol, 1981). The astronomically revised magnetic reversal ages from Francescone et al. (2019) indicate a duration for the Ypresian of 7.93 Myr (56.00–48.07 Ma). This is slightly longer than when using an age of 48.30 Ma for the basal Lutetian (Westerhold et al., 2017) and approximately 400 kyr shorter than by using 47.8 Ma for the base of the Lutetian in GTS2012 (Vandenberghe et al., 2012).

28.1.6 Lutetian

28.1.6.1 Historical overview

The Lutetian was defined in the Paris Basin and named after Lutetia, the Roman name for Paris. According to its author, De Lapparent (1883), the Lutetian stratotype is typified by the *Calcaire grossier* of the Paris Basin. As no stratotype section was indicated by De Lapparent, a neostratotype was selected near Creil, approximately 50 km north of Paris, at St. Leu d'Esserent and St. Vaast-les-Mello (Pomerol, 1981). The type-Lutetian contains characteristic larger benthic foraminifera (*Nummulites laevigatus*, *Orbitolites complanatus*), palynomorphs, and calcareous nannoplankton. In open marine sequences, the base of the Lutetian used to be delineated by the first occurrence of the planktonic foraminifer *Hantkenina nuttalli* (or the former P9/P10 zonal boundary (e.g., Berggren et al., 1995; Luterbacher et al., 2004). However, this species does not appear at the base of Chron C21r, but at the base of Chron C20r, about 3 Myr later (Wade et al., 2011). The base of Chron C21r boundary was another traditional criterion used to define the base of the Lutetian (Molina et al., 2011).

Blondeau et al. (1965) discussed the correlation of the Lutetian with now obsolete Belgian stages, such as Brusselian, Ledian, Wemmelian (De Geyter et al., 2006) and the Bracklesham "Beds" of the Hampshire Basin, United Kingdom.

28.1.6.2 Definition of the Global Boundary Stratotype Section and Point

The Lutetian GSSP was ratified in 2011 (Molina et al., 2011) (Fig. 28.6). It is defined in the cliffs of the Gorrondatxe (or Azkorri) beach section near Getxo village (province of Biscay, Basque Country, northern Spain) at coordinates 43° 22′ 46.47″N and 3° 00′51.61″W. The Gorrondatxe section is largely composed of hemipelagic marls and limestones, but also contains intercalated distal thin-bedded siliciclastic turbidites. These tabular-shaped turbidites are considered not to have significantly scoured the hemipelagic marls. The GSSP level is situated at 167.85 m in the Gorrondatxe section at a dark marly level, below a prominent 15-cm-thick turbidite. This marly level is thought to reflect a global maximal flooding event (Molina et al., 2011). The GSSP level is identified by the first occurrence of the nannofossil *Blackites inflatus* which is very close to the base of Zone CNE8 (Agnini et al., 2014).

The first occurrence of *B. inflatus* lies close to the base of Lutetian stratotype. Cyclostratigraphic analysis reveals that this event lies 39 precession cycles above the base of Chron C21r, which equates to 819 kyr (Molina et al., 2011). Based on this datum, the GSSP is placed at Chron C21r.6. This GSSP level is in agreement with the base of the Lutetian stratotype. A further discussion of an integrated biomagneto- and astrochronology is given in Westerhold et al. (2017). The astronomically derived duration of the Lutetian is now 7.04 Myr (48.07−41.03 Ma) using the base of Chron C18r for delineating the base of the Bartonian.

28.1.7 Bartonian

28.1.7.1 Historical overview

The Bartonian (Mayer-Eymar, 1858) refers to the marine Barton clays and sands of the Hampshire Basin, southern England (now Barton Clay and Becton Sand formations), exposed in the coastal section between Highcliff and Barton-on-Sea (Prestwich, 1847; Pomerol, 1981). The Barton Clay Formation contains rich and diverse dinoflagellate cyst assemblages and calcareous microfossils, mainly calcareous nannoplankton, indicative Zones NP16−17 (upper CNE14−CNE15) (Martini, 1971; Pomerol, 1981; Hooker and King, 2019). These biozones are indicative of a middle Eocene age for the type Bartonian.

There have been two different concepts for the base of the Bartonian in the type area: one corresponding to the base of the Barton Beds as defined by Keeping (1887) and another corresponding to the base of the Barton Clay Formation (Hooker, 1986). Keeping (1887) proposed to place the base of the Bartonian Stage at a distinct horizon rich in *Nummulites prestwichianus* near the base of the Barton Beds (Pomerol, 1981). In the type section at Barton-on-Sea, this horizon is just 3 m above the base of the Barton Clay Formation, but further east in the Hampshire Basin (e.g., on the Isle of Wight) the part of the Barton Clay Formation below the *N. prestwichianus* bed increases up to 26 m in thickness (Dawber et al., 2011; Hooker and King, 2019).

Cavelier and Pomerol (1986), following Hooker (1986), proposed placing the base of the Bartonian Stage at the base of the Barton Clay Formation, tying the chronostratigraphic base to the base of the lithostratigraphic unit. This level has since been abandoned to serve for base unit Bartonian in favor of the *N. prestwichianus* bed. The *N. prestwichianus* bed is situated in the upper part of Zone NP16 (upper CNE14), which agrees with the magnetobiochronology of the Alum Bay sequence in the Isle of Wight where the base of the Bartonian is situated in the middle of Chron C18r (Dawber et al., 2011; Hooker and King, 2019).

In France, a number of poorly defined regional (sub) stages overlap with various parts of the Bartonian, such as the Auversian, Marinesian, and Biarritzian (Pomerol, 1981; Cavelier and Pomerol, 1986; Schuler et al., 1992), but their usage has now been abandoned.

Base of the Lutetian Stage at Gorrondatxe, Spain

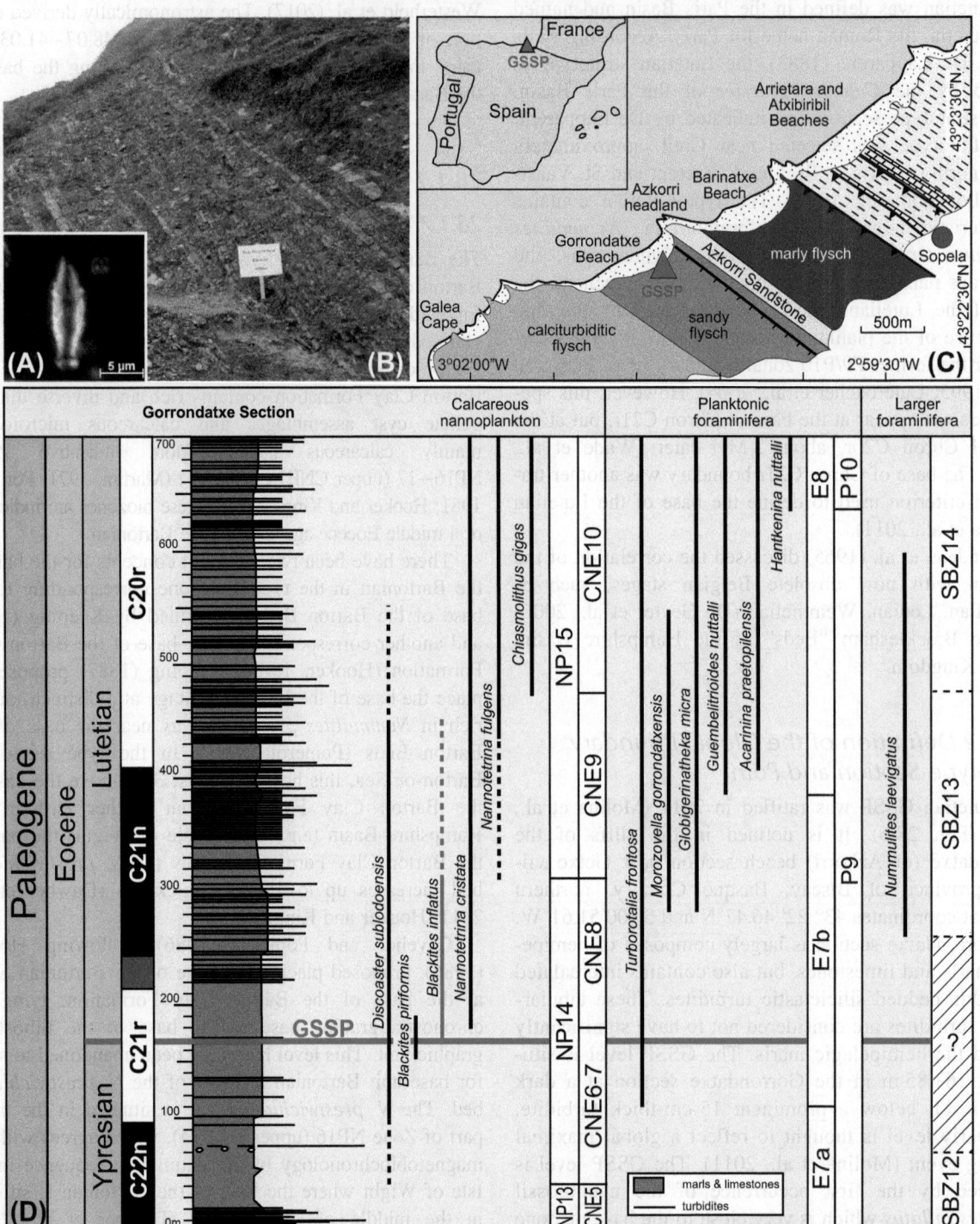

FIGURE 28.6 Lutetian GSSP located at Gorrondatxe, Basque Country, Spain (C). Photograph of the outcrop (B) courtesy of Eustoquio Molina. Stratigraphic data and section (D) based on Molina et al. (2011). Image of *Blackites inflatus* (A) by courtesy of Simonetta Monechi. *GSSP*, Global Boundary Stratotype Section and Point.

28.1.7.2 Toward definition of the Global Boundary Stratotype Section and Point

A formal proposal for a Bartonian GSSP has yet to be made. The Contessa Highway section in Italy provides a well-documented interval of Lutetian and Bartonian strata (Jovane et al., 2007; Jovane et al., 2010; Dinarès-Turell, 2019). Several distinctive marker beds and microfossil events occur near the base of Chron C18r, providing various means for correlation. Stable isotope records may also be used for constraining and correlating the future base of the Bartonian. In the Contessa Highway section, distinct oxygen and CIEs start about 6 m above the base of Chron C18r, within Subchron C18n.2n, and are accompanied by a significant planktonic foraminiferal turnover. These isotopic excursions are similar to the Middle Eocene Climatic Optimum (MECO) as observed in the Indian-Atlantic sector of the Southern Ocean (Bohaty and Zachos, 2003) and are also identified in the Barton Clay Formation of the Hampshire Basin (Dawber et al., 2011; Hooker and King 2019).

Payros et al. (2015) carried out a detailed biomagnetostratigraphic study on the Lutetian-Bartonian transition in the Oyambre section (northern Spain) in order to evaluate its suitability for serving as a GSSP. The Oyambre section provides a well-exposed and continuous sequence of deep-water sediments with numerous biostratigraphic markers but unfortunately yields a poor magnetostratigraphic record in the interval spanning Chron C19n.

In the absence of a formal proposal for the Bartonian GSSP, we retain the definition of the base Bartonian as it was defined in previous time scales (e.g., Hardenbol and Berggren, 1978; Luterbacher et al., 2004; Fluegeman, 2007; Vandenberghe et al., 2012), namely, at the base of Chron C18r. At Alum Bay (Hampshire Basin), this datum lies between the two levels for base Bartonian previously proposed by Keeping (1887) and Cavelier and Pomerol (1986), respectively, as outlined above (Hooker and King, 2019). The astronomically derived duration for the Bartonian is 3.32 Myr (41.03−37.71 Ma), using the Tiziano bed near the base of Subchron C17n.2n as approximation for the basal Priabonian age (Agnini et al., 2019; Galeotti et al., 2019).

28.1.8 Priabonian

28.1.8.1 Historical overview

The Upper Eocene Priabonian Stage (Munier Chalmas and De Lapparent, 1893) has its historical type section in Priabona, northern Italy (Roveda, 1961; Hardenbol, 1968; Barbin, 1988). Hardenbol (1968) discussed the history of the usage of the term Priabonian since its introduction by Suess (1868). In the propositions at the end of the Colloque sur l'Eocène (BRGM, 1968), several parastratotype sections have been designated in the same area in addition to the type section.

Bryozoan marls and limestones, as well as beds with small nummulites are included in the shallow water facies of the Priabonian type area (Barbin, 1988). The type section also contains rare planktonic foraminifera such as *Turborotalia cerroazulensis* assigned to upper Zone E15 and Zone E16 and spans calcareous nannofossil Zones NP19−20 (CNE18−CNE20) with *Isthmolithus recurvus* (Roth et al., 1971). This stratigraphic assignment is supported by the recognition of the dinoflagellate cyst *Melitasphaeridium pseudorecurvatum* Zone in the historical type section (Brinkhuis and Biffi, 1993).

In earlier integrated time scales, the Bartonian/Priabonian boundary was placed at the first occurrence of *Chiasmolithus oamaruensis*, marker of the base of Zone NP18 (Berggren et al., 1985, 1995). Concerns about the diachronous appearance of *C. oamaruensis*, however, led to a provisional placement of the base Priabonian at the base of Subchron C17n.1n in GTS2004 and GTS2012.

The Ludian is a former stage in the stratigraphy of the Paris Basin that can be attributed to the Priabonian (Schuler et al., 1992).

28.1.8.2 Definition of the Global Boundary Stratotype Section and Point

Agnini et al. (2011) presented an integrated biomagnetostratigraphy of the Alano section, near Alano di Piave in the Venetian southern Alps of Italy in which the base of the Tiziano volcanic ash bed was proposed to delineate the Priabonian GSSP (Fig. 28.7). In 2019 a formal proposal for definition and placement of the GSSP of the Priabonian in the Alano section was submitted to the ICS (Agnini et al., 2019). The Alano section (45° 54′51.10″N and 11° 55′4.87″E) is located about 50 km NE from the historical Priabona stratotype and just 8 km NE of the Possagno section, which was already considered as deep-water parastratotype section (BRGM, 1968). Calcareous microfossils are abundant in the hemipelagic marls of the Alano section. Comparison of the paleomagnetic reversals and biochronology with those of ODP Site 1052, Blake Nose (Western Atlantic) enables a detailed correlation with the interval from the upper part of Chron C18r to the basal part of Chron C16r, allowing an accurate age control of the section. The Tiziano bed is positioned immediately above the first occurrence of *C. oamaruensis*, within Subchron 17n.2n (Agnini et al., 2011, 2019). The Tiziano and other ash beds were recently also radiometrically dated and calibrated to a new astrochronological model for the Alano sequence (Galeotti et al., 2019). This yielded a close match between the radiometric age of 37.789 ± 0.024 Ma and an astrochronological age of 37.71 ± 0.01 Ma for the Tiziano bed. Furthermore, a lateral equivalent of the Tiziano ash bed (VL + 3 ash bed) is observed in the Varignano section, some 80 km west of Alano di Piave, yielding a correlative

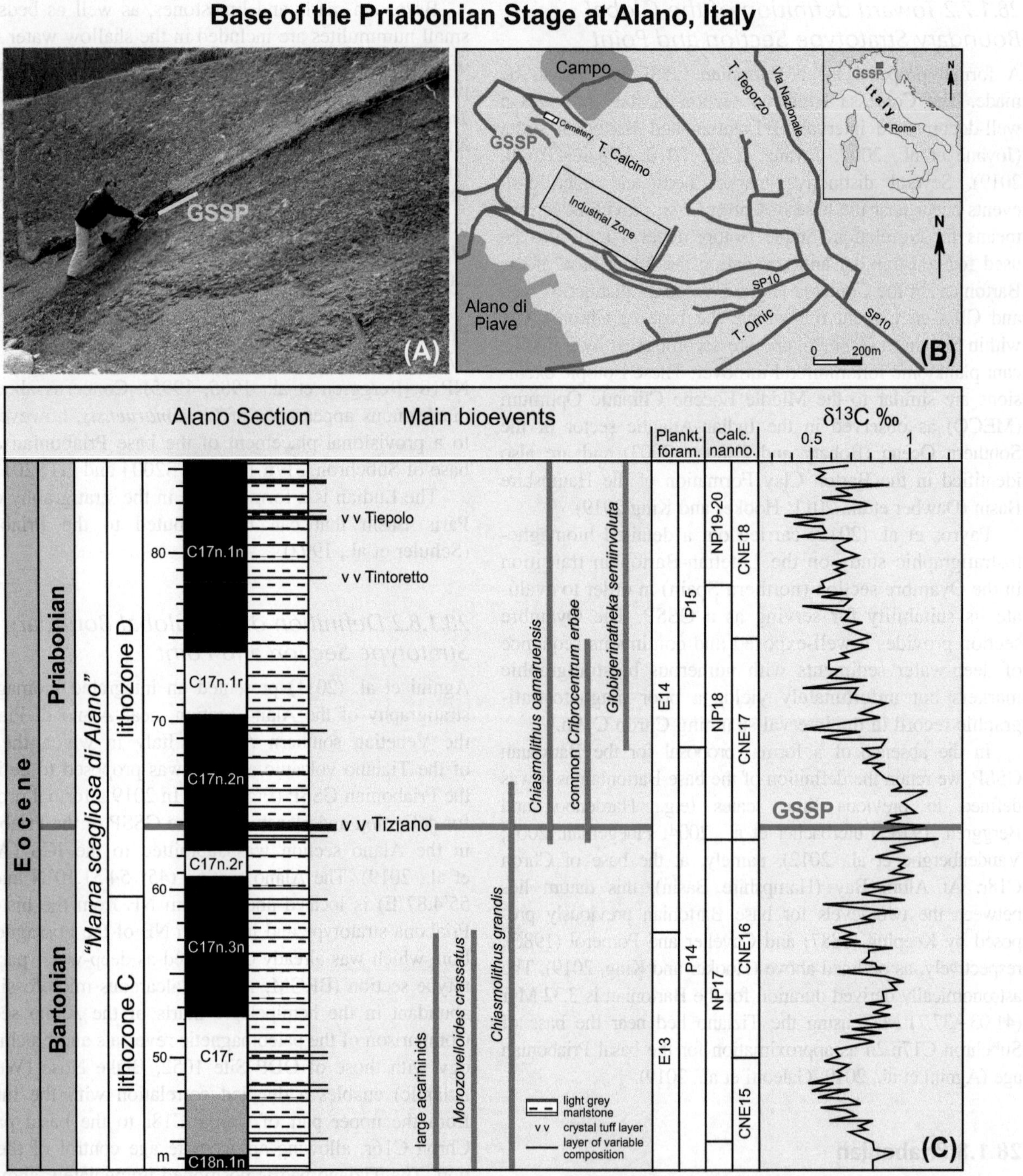

Base of the Priabonian Stage at Alano, Italy

FIGURE 28.7 Priabonian GSSP located at Alano di Piave, Italy (B). Photograph of the outcrop (A) by courtesy of Claudia Agnini. Stratigraphic data (C) are based on Agnini et al. (2019), in which the data of Fornaciari et al. (2010), Spofforth et al. (2010), Agnini et al. (2011), Galeotti et al. (2019), and Luciani et al. (2019) are summarized. GSSP, Global Boundary Stratotype Section and Point.

deepwater sequence, containing beds with redeposited larger benthic foraminifera throughout the sequence (Luciani et al., 2019). The latter enables direct comparison with the Priabona stratotype.

Several criteria enable correlation of the Tiziano bed to other areas (Agnini et al., 2011, 2019): (1) The base of

Subchron C17n.2n is about 1.5 m below the Tiziano bed; (2) the base of the acme of *Cribrocentrum erbae* defines the base of Zone CNE17, predating the Tiziano bed by ~ 30 kyr, and provides a reliable biostratigraphic marker (Fornaciari et al., 2010; Agnini et al., 2011, 2019; Galeotti et al., 2019). In addition, the first rare occurrence of

C. oamaruensis is observed just below the Tiziano bed. However, its rarity and diachrony across latitudes preclude the use of this taxon for accurate stratigraphic correlation (Agnini et al., 2019).

The nearly coeval and global extinctions of *Morozovelloides* and large acarininids have been proposed as an alternative to delineate the base of the Priabonian (Wade et al., 2012; Cotton et al., 2017). It coincides with a turnover in radiolarians in the middle of Subchron C17n.3n, facilitating correlation with noncalcareous deep-sea sediments. In the Alano section, this level is well documented about 6 m below the Tiziano bed, predating it by about 250 kyr. Yet, it cannot be linked to a distinct bed in the marl-dominated sequence (Agnini et al., 2011; Wade et al., 2012; Galeotti et al., 2019), which makes it less suitable for delineating a stable position of the GSSP.

In larger foraminiferal stratigraphy the base of shallow benthic zone SBZ19 has generally been considered to correspond to the base of the Priabonian (e.g., Serra-Kiel et al., 1998). However, this level now turns out to roughly correlate with the base of Subchron 16n.2n (Cotton et al., 2017; Papazzoni et al., 2017). Furthermore the Tiziano equivalent ash bed VL + 3 at Varignano is situated within SBZ18 at the base of Subchron C17n.2n (Luciani et al., 2019) and thus, the base of the Priabonian stratotype is about 1.5 Myr younger than the proposed GSSP and the formally defined Priabonian Stage would comprise the entire historical stratotype.

The astronomically derived duration for the Priabonian, based on the proposed GSSP in the Alano di Piave section (Agnini et al., 2019; Galeotti et al., 2019) is 3.81 Myr (37.71−33.9 Ma).

28.1.9 Rupelian

28.1.9.1 Historical overview

The Rupelian is the lowermost stage of the Oligocene Series, and accordingly, the base of the Rupelian also defines the base of the Oligocene (Salvador, 1994). The term Rupelian was introduced by Dumont (1849) for the Boom Clay deposits exposed in a series of brickyards along the Rupel and Scheldt rivers in Belgium. These clay deposits constituted the upper Rupelian of Dumont while a lower Rupelian part, consisting of the *Nucula comta* Clay sandwiched between two sand deposits, was exposed in northeastern Belgium. In its outcrop area, much of the Boom Clay has been eroded but it becomes thicker and stratigraphically more complete in the subsurface of northern Belgium and the Netherlands (Vandenberghe et al., 2001, 2014; Van Simaeys and Vandenberghe, 2006). These Rupelian clays form a characteristic and continuous deposit in a large part of the North Sea region. Distinct bedding patterns reflect

mainly 41-kyr obliquity cycles (Abels et al., 2007). Several characteristic layers of septaria (carbonate concretions) are widely recognized in the clay and explain the name *Septarienton* (septaria clay) in Germany. Many macro- and microfossil groups have been studied in the stratotype area (references in Pomerol, 1981; Van Simaeys and Vandenberghe, 2006). Calcareous nannoplankton Zones NP23 and NP24 (=CNO_3−CNO_4) and planktonic foraminifera Zones O1−O4 (Berggren and Pearson 2005; Wade et al., 2011) were recognized in the clay and also a benthic foraminiferal and a dinoflagellate cyst zonation was established in the Rupelian stratotype area (Van Simaeys and Vandenberghe, 2006).

In the concept of a threefold subdivision of the Oligocene by Beyrich (1854) the historical Rupelian stratotype as defined by Dumont in 1849 was considered as the middle Oligocene, between the lower Oligocene Latdorfian and the upper Oligocene Chattian (Pomerol, 1981). In the present twofold subdivision of the Oligocene the upper Oligocene Chattian remained unchanged, but the lower Oligocene Rupelian now encompasses both the German Latdorfian, its lateral correlative Tongrian in Belgium and the overlying historical Rupelian. Consequently, the present base Rupelian is older than the base of the historical stratotype of the Rupelian Boom Clay and correlates with a level within Zone NP21 (=base CNO1) (Vandenberghe et al., 2003, 2004).

Two events that are often associated with the start of the Oligocene in fact occurred shortly after the onset. These events are the Oi-1 event, the most significant oxygen isotope cooling event (Zachos et al., 1996; De Man et al., 2004a; Berggren et al., 2018) and the *Grande Coupure*, referring to a major turnover and dispersal event in the mammal record in response to the formation of new land bridges (Woodburne and Swisher, 1995; Vandenberghe et al., 2003; Hooker, 2010a).

Before the Rupelian was officially adopted by the ICS as the standard stage of the lower Oligocene, former overlapping stages such as the Stampian and Sannoisian were used, particularly in the Paris Basin, but have now been abandoned (Cavelier and Pomerol, 1977, 1983; Pomerol, 1981).

28.1.9.2 Definition of the Global Boundary Stratotype Section and Point

The GSSP of the Rupelian is situated in the Massignano section, 10 km southeast of Ancona, on the Adriatic coast of Italy (Premoli Silva et al., 1988; Premoli Silva and Jenkins, 1993) (Fig. 28.8). It is defined in an abandoned quarry on the east side of the Ancona−Sirolo road near Massignano, at coordinates 43° 32′58.2″N and 13° 36′03.8″E. The Massignano section covers a continuous, 23-m-thick upper Eocene to lower Oligocene sequence of open marine marls and calcareous marls. The GSSP is located at the base of a greenish-gray

Base of the Rupelian Stage at Massignano, Italy.

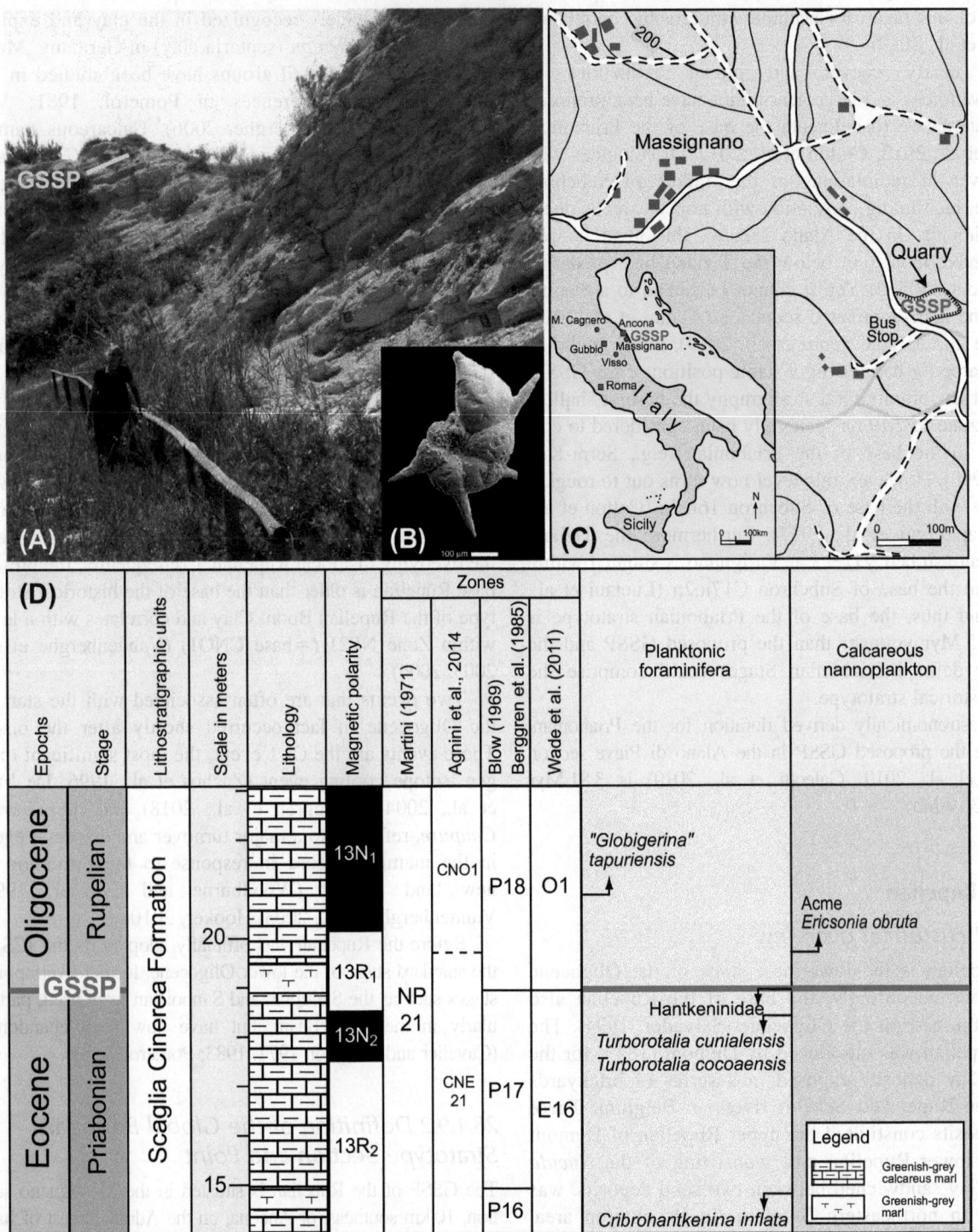

FIGURE 28.8 Rupelian GSSP located at Massignano, Italy (C). Photograph of the outcrop (A) by courtesy of Eustoquio Molina. Image of *Hantkenina alabamensis* (B) by courtesy of Brian Huber. Stratigraphic data (D) are based on Coccioni et al. (1988) and Premoli Silva and Jenkins (1993). *GSSP*, Global Boundary Stratotype Section and Point.

0.5-m-thick marl bed, 19 m above the base of the section. The key marker of the GSSP is the extinction of the hantkeninid planktonic foraminifera (five species) at the top of Zone E16, which lies within nannofossil Zone NP21 (=top CNE20) and Subchron 13r.1r. It must be noted, however, that a normal-polarity subchron (C13r.2n) is positioned just below the GSSP in the Massignano section and is absent in the marine magnetic C-Sequence km model. Therefore, the GSSP position projected to the GPTS is better expressed as Chron C13r.86 (86% up from the base of Chron C13r). K—Ar and Ar—Ar dating of biotite grains from 4.3 m below the base of the GSSP give an age of 34.6 ± 0.3 Ma, suggesting a numerical age of about 34 Ma for the GSSP itself (Premoli Silva and Jenkins, 1993). The selection of this GSSP for the base of the Oligocene honors the original meaning of the Oligocene as defined by Beyrich (1854) who considered the Latdorfian transgression as the start of the Oligocene (see also Brabb, 1968).

A detailed overview of Oligocene chronostratigraphy and planktonic foraminiferal biostratigraphy was recently provided by Berggren et al. (2018). The astronomically derived duration for the Rupelian is 6.61 Myr (33.9—27.29 Ma), using a level 15% up from the base of Chron C9n as approximation for the base of the Chattian Stage (Coccioni et al., 2018).

28.1.10 Chattian

28.1.10.1 Historical overview

The upper Oligocene Chattian was introduced and named after the *Kasseler Meeressande* in Hessen, Germany, by Fuchs (1894), who recognized the similarity to the Doberg strata near Bünde (Westphalia, Germany). Much later, Görges (1957) selected the section at Doberg as the stratotype of which a detailed litholog was provided in Pomerol (1981). More recently, the stratigraphy of the Rupelian—Chattian transition in various parts of the North Sea Basin was discussed by Van Simaeys et al. (2004) and Śliwińska et al. (2012).

The most distinct biostratigraphic marker characterizing the base of the Chattian in the North Sea Basin is the blooming of the benthic foraminifer *Asterigerina guerichi*, known as the *Asterigerina* Horizon (e.g., Indans, 1965). This biohorizon is recognized over the entire North Sea Basin and hence provides a suitable regional correlation marker to define the base of the Chattian (De Man et al., 2004b).

The base of the *Asterigerina* Horizon falls within calcareous nannofossil Zone NP24 (=upper CNO4), adopting the first occurrence of *Helicosphaera recta* as the substitute marker of the base of this biozone (Van Simaeys et al., 2004) and, at least in the subsurface of northern Denmark, it continues within the lower part of Zone NP25 (=CNO5) (Śliwińska et al., 2012). The lower unit A of the type Chattian (Anderson, 1961) falls within Zone NP24, whereas

the middle and the upper units, B and C, are situated in Zone NP25, in the North Sea Basin defined by the first occurrence of *Pontosphaera enormis*. Chattian A and B make up the Eochattian and Chattian C corresponds to the Neochattian as defined by Hubach (1957).

Dinocyst assemblages are diverse and well preserved in both the Rupelian and the Chattian of the North Sea Basin (Köthe, 1990). The first occurrence of *Artemisiocysta cladodichotoma* coincides with the onset of the *A. guerichi* bloom at the base of the Chattian in the North Sea Basin (Van Simaeys et al., 2004; Śliwińska et al., 2012). The lower Chattian is also characterized by the *Svalbardella* dinocyst event, occurring in the middle and upper part of Chron C9n (Van Simaeys et al., 2005a,b) and correlating with the mid-Oligocene Oi-1b deep-sea cooling event as described by Miller et al. (1998). Although Van Simaeys et al. (2005a,b) linked the *Svalbardella* and Oi-2b event to the sequence boundary below the type Chattian, further bio- and magnetostratigraphic correlation and the observation that the *Svalbardella* event overlaps with the *A. guerichi* bloom, rather suggest a correlation of the Oi-2a cooling event with the unconformity at the base of the Chattian (Śliwińska et al., 2014; Coccioni et al., 2018).

28.1.10.2 Definition of the Global Boundary Stratotype Section and Point

The GSSP of the Chattian was ratified by the IUGS at the meeting in Cape Town in 2016 (Coccioni et al., 2018). It is defined in the Monte Cagnero section in the pelagic succession of the Umbria-Marche Basin near Urbania in central Italy (coordinates 43° 38′47.81″N and 12° 28′03.83″E (Fig. 28.9). The Monte Cagnero section consists of a 86-m-thick sequence of rhythmically bedded blue-gray marls, calcareous marls, and marly limestones of the Scaglia Cinerea Formation. Rare thin biotite-rich volcanoclastic layers provide a direct means for radio-isotopic calibration of the magnetostratigraphic framework (Coccioni et al., 2008, 2018).

The GSSP is positioned at the 197 m level, 30 cm below the base of a prominent calcareous bed within a relatively thick marly interval. This level coincides with the last common occurrence (LCO) of the planktonic foraminifera *Chiloguembelina cubensis*, defining also the base of planktonic foraminiferal Zone O5 (Berggren and Pearson, 2005; Wade et al., 2011; Berggren et al., 2018). Note, however, that strong doubts remain on global synchronicity of the LCO of *C. cubensis* (e.g., King and Wade, 2017). The GSSP level falls within the upper part of calcareous nannofossil Zone NP24 [slightly above the base of Zone CNO5 (Agnini et al., 2014)] and is now correlated with the lower part of Chron C9n, instead of Chron C10n as previously thought (Coccioni et al., 2008; Wade et al., 2011). Two volcanoclastic biotite-rich layers

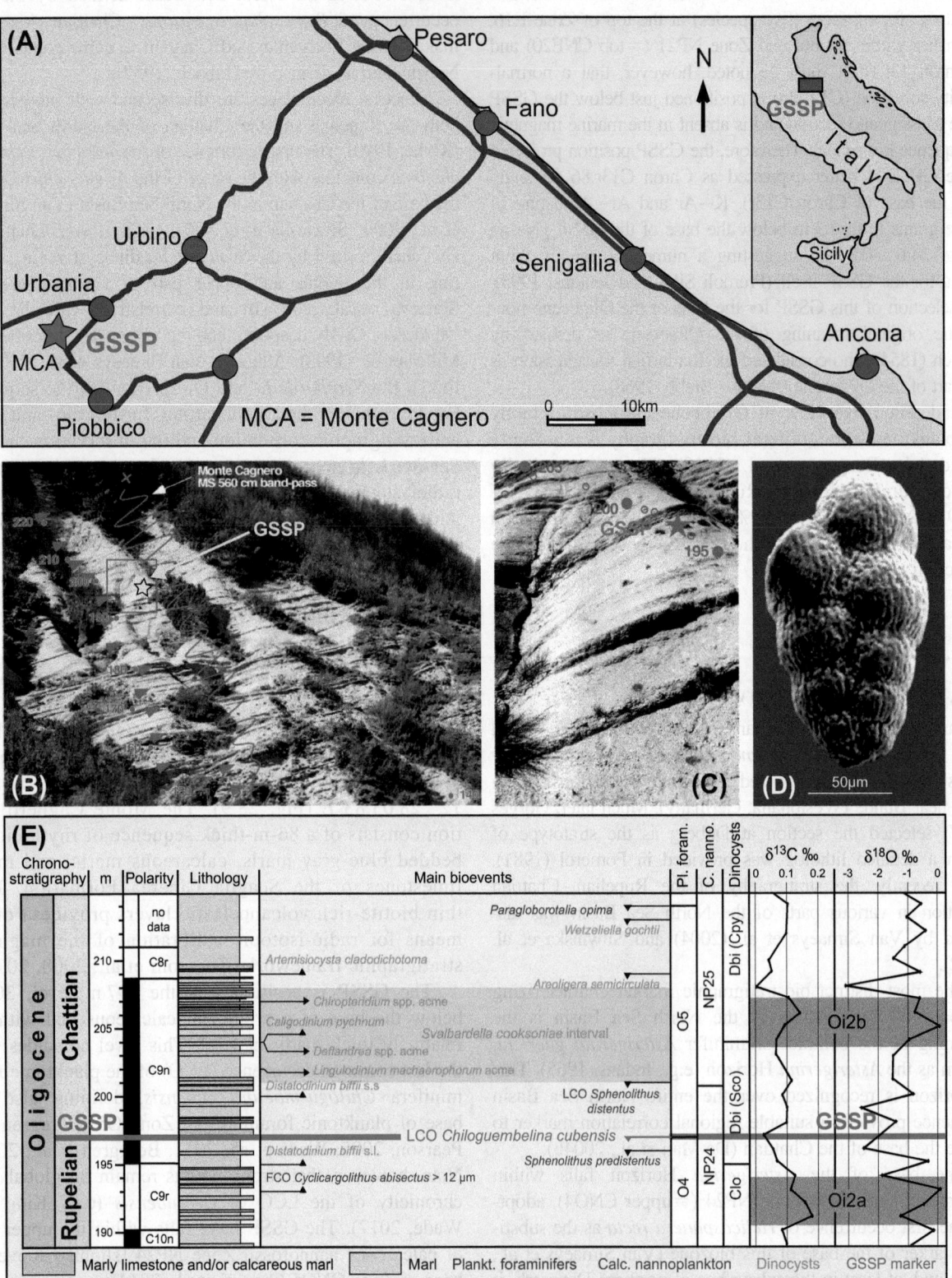

FIGURE 28.9 The Chattian GSSP located at Monte Cagnero near Urbania, Italy (A). Photographs of the outcrop (B, C) and image of *Chiloguembelina cubensis* (D) by courtesy of Rodolfo Coccioni. Stratigraphic data (E) are based on Coccioni et al. (2018). *GSSP*, Global Boundary Stratotype Section and Point.

are situated about 13 m above the GSSP, providing independent ages of 27.0 ± 0.2 Ma and 27.5 ± 0.2 Ma, respectively. Two different astronomical tuning models of the cyclic bedding provide ages of the Chattian GSSP of 27.82 and 27.41 Ma, respectively. Sr isotope and glauconite dates of the historical Rupelian and Chattian stratotype areas bracket the Rupelian—Chattian boundary between about 29 and 27 Ma (De Man et al., 2010) and the hiatus of the Rupelian—Chattian unconformity in the type area spans about 1.7 Ma (Coccioni et al., 2018). Accordingly, the GSSP level at Monte Cagnero is consistent with the original position of the boundary between the historical stratotypes. In Coccioni et al. (2018), the base of the Chattian correlates to 15% up from the base of Chron C9n and is assigned an astronomically derived age of 27.29 Ma, close to option 2 in Coccioni et al. (2018). Since the base of the Neogene is at 23.04 Ma, the total duration of the Chattian is 4.25 Myr.

28.2 Paleogene Biostratigraphy

The biostratigraphy of Paleogene marine successions is well established by means of several microfossil groups. Calcareous nannofossils and planktonic foraminifera are the two most widely used groups for biostratigraphic purposes in open marine deposits, especially in low- and mid-latitude settings where deposition occurs above the carbonate compensation depth (CCD). In areas where calcareous remains do often not preserve well (below CCD or in restricted basins) alternative biostratigraphic schemes have been established based on organic microfossils, notably organic-walled dinoflagellate cysts (dinocysts), or siliceous microfossils such as radiolarians and diatoms. Dinocysts can be particularly useful for biostratigraphy in shallow siliciclastic deposits. Larger benthic foraminifera are a key stratigraphic tool in low- to mid-latitude carbonate platforms. Calibration between the various biostratigraphic schemes can be achieved in areas where pelagic and continental margin taxa co-occur, such as in the Pyrenees, Spain, and New Zealand, for the Paleocene—Eocene transition. The Paleogene is also marked by rapid diversification of land mammals, following the extinction of the dinosaurs at the K—Pg boundary, yielding various terrestrial zonations for the different continents.

Fig. 28.10 shows the correlation of Paleocene stages, magnetostratigraphy, and standard marine zonations based on lower-to-mid-latitude planktonic foraminifera, calcareous nannoplankton, and larger benthic foraminifera. The main changes in these zonations compared to GTS2012 concern revised stratigraphic calibrations of the Shallow Benthic Zones (SBZ) and inclusion of a new calcareous nannoplankton zonation scheme (Agnini et al., 2014). Additional biozonation schemes are provided for radiolarians and dinocysts (Fig. 28.11). These biostratigraphic charts particularly benefitted from the stratigraphic synthesis of northwestern

European basins (King, 2016), providing a new dinocyst zonation and from studies on the radiolarian distribution in the Pacific Ocean (Kamikuri et al., 2012, 2013). Finally, based on numerous new integrated paleontological and chronostratigraphic studies of continental deposits worldwide, a fully revised synthesis of LMAs for North and South America, Europe, and Asia is provided here (Fig. 28.12).

28.2.1 Marine Biostratigraphy

28.2.1.1 Planktonic Foraminifera

The application of planktonic foraminifera to Paleogene stratigraphy (Fig. 28.10) has been primarily developed in two areas, the southern part of central Asia (e.g., Subbotina, 1953) and the Caribbean (e.g., Bolli, 1957). In the late 1950s and 1960s, planktonic foraminifera became a key biostratigraphic tool and assumed even greater importance with the advent of scientific ocean drilling. Paleogene planktonic foraminiferal zonations have been erected by many authors (e.g., Bolli, 1966; Blow, 1979; Toumarkine and Luterbacher, 1985; Berggren et al., 1985, 1995; Olsson et al., 1999; Berggren and Pearson, 2005; Pearson et al., 2006; Wade et al., 2011, 2018). The planktonic foraminiferal zonation used here is based on Wade et al. (2011) and is generally applicable in open-marine, low-to-middle-latitude sequences. Many marker species in this zonation are absent at higher southern latitudes, resulting in different and less detailed Austral and Circum—Antarctic zonation schemes (e.g., Jenkins, 1985; Huber and Quillévéré, 2005).

The Paleocene recovery of planktonic foraminifera from their near demise at the K—Pg boundary led to strong diversification in various lineages, enabling a detailed biostratigraphic subdivision. Recently, after a more than 25-year long endeavor, the Paleogene Planktonic Foraminifera Working Group achieved the finalization of a trilogy of atlases of Paleogene planktonic foraminiferal taxonomy, phylogeny, and biostratigraphy. The working group first provided a state-of-the-art atlas of Paleocene planktonic foraminifera (Olsson et al., 1999) in which the widely used Paleogene P-zonation developed by Berggren et al. (1995) was maintained. Several new lineages developed during the early Paleocene, such as *Subbotina*, *Parasubbotina*, *Praemurica*, and *Globanomalina*, providing several biostratigraphic marker species (Olsson et al., 1999). The late early Paleocene (Zones P2—P3; $\sim 61-62$ Ma) is marked by the appearance and first diversification of the symbiont bearing genera *Acarinina*, *Igorina*, and *Morozovella* (Norris, 1996; Quillévéré et al., 2001). During the late Paleocene to early Eocene, tropical to subtropical open-marine assemblages are commonly dominated by representatives of *Morozovella* and *Acarinina*. These genera provide numerous biozonal markers starting at the base of Zone P3 (first occurrence *Morozovella angulata*; at ~ 62.5 Ma) up to middle Eocene Zone E10 (last occurrence *Morozovella aragonensis* at ~ 43.5 Ma).

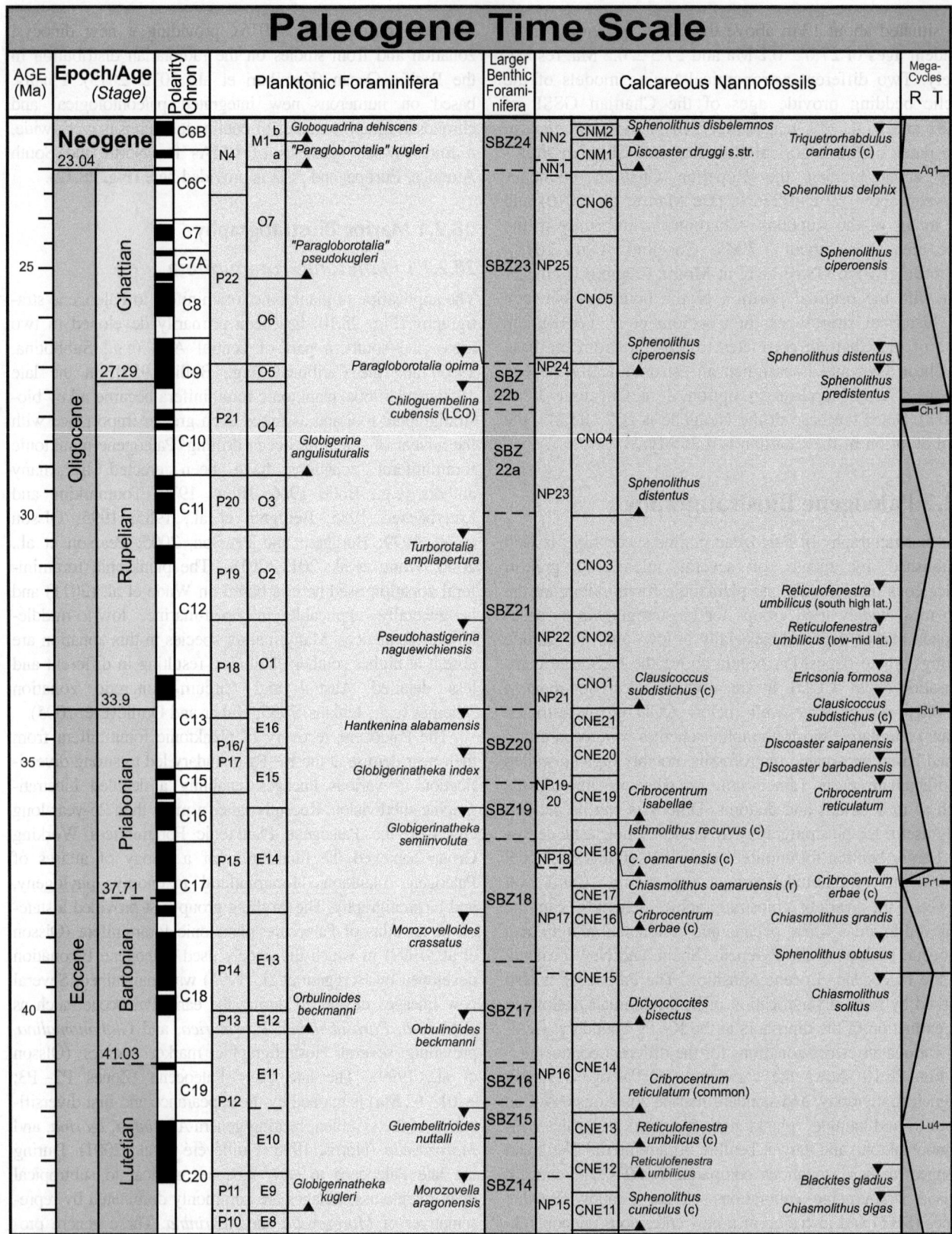

FIGURE 28.10 Paleogene chronostratigraphy, biozonation schemes of marine calcareous microfossils, and transgressive—regressive cycles. Planktonic foraminiferal biostratigraphy is based on Wade et al. (2011). Tethyan SBZ is modified from Cahuzac and Poignant (1997) and Serra-Kiel et al. (1998). Modifications of
the approximate positions of the boundaries between SBZs are mainly based on improved correlations through calcareous nannofossils and magnetostratigraphy
(e.g., Molina et al., 2003; Payros et al. 2009; Pujalte et al., 2009; Scheibner and Speijer, 2009; Rodriguez-Pinto et al., 2012, 2013; Costa et al., 2013). Calcareous
nannofossil biostratigraphy is based on the CN-zonation scheme of Agnini et al. (2014) and the correlation with the NP-zonation of Martini (1971). The main
Paleogene transgressive—regressive trends are modified from Hardenbol et al. (1998). *NP*, Nannoplankton; *SBZ*, Shallow Benthic zonation.

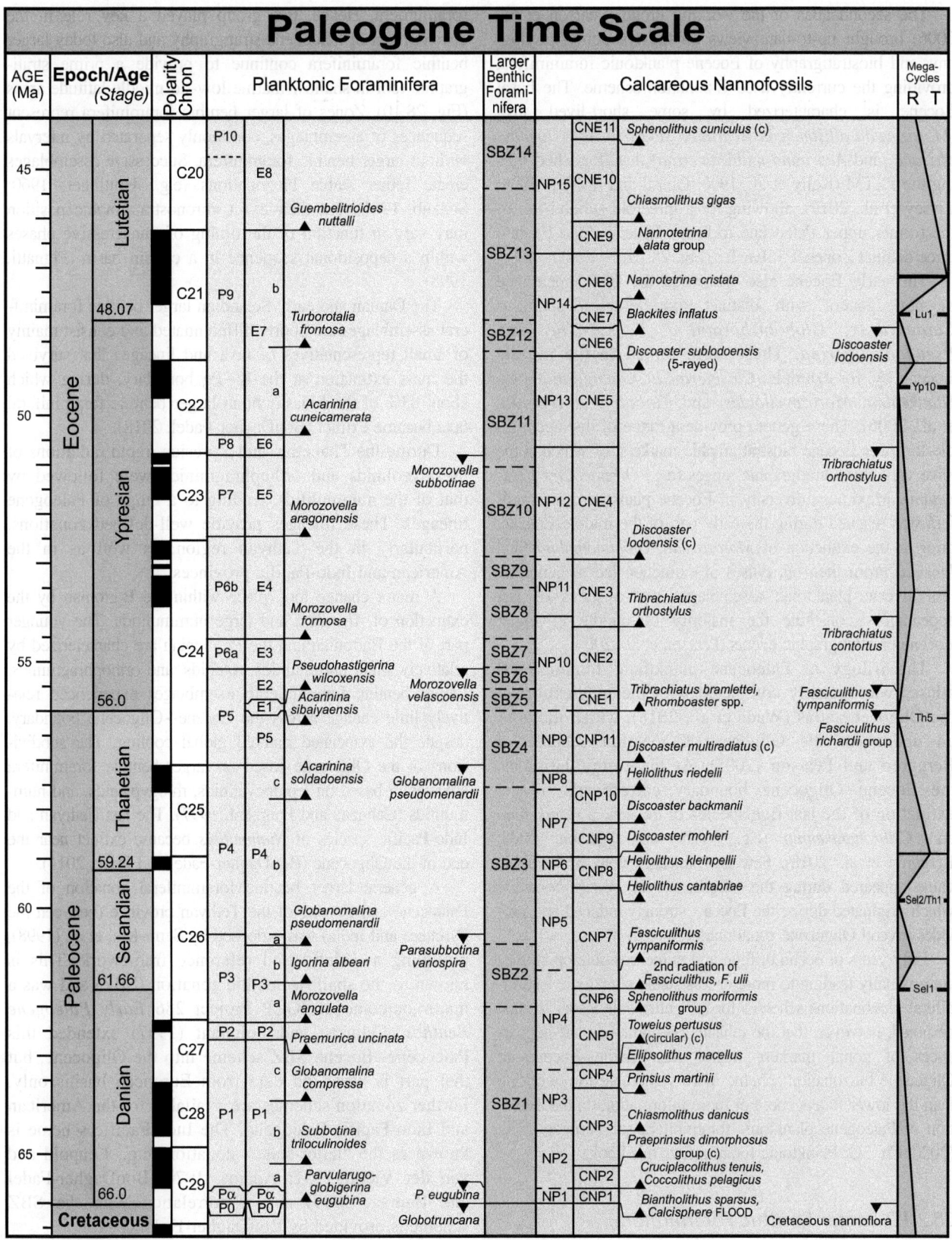

Paleogene Time Scale

FIGURE 28.10 (Continued)

The second atlas of the working group (Pearson et al., 2006) brought up-to-date views on the taxonomy, phylogeny, and biostratigraphy of Eocene planktonic foraminifera, providing the currently used E-zonation scheme. The basal Eocene is characterized by some short-lived taxa (*Morozovella allisonensis*, *Acarinina sibaiyaensis*, *Acarinina africana*, and *Acarinina multicamerata*) that flourished during the PETM (Kelly et al., 1996; Guasti and Speijer, 2008; Livsey et al., 2019), allowing for a threefold subdivision of the former upper Paleocene to lower Eocene Zone P5 into three distinct Zones, P5, E1, E2 (Fig. 28.10).

The early Eocene also witnessed the first appearance of new genera with distinct morphologies, such as *Catapsydrax*, *Globoturborotalita*, *Jenkinsella*, and *Pseudohastigerina*. These were joined in the middle Eocene by, for example, *Clavigerinella*, *Globigerinatheka*, *Hantkenina*, *Morozovelloides*, and *Turborotalia* (Pearson et al., 2006). These genera provide a range of distinct middle to upper Eocene biostratigraphic markers, of which some have very short stratigraphic ranges (e.g., *Orbulinoides beckmanni*). Maximum diversity in Eocene planktonic foraminifera was reached during the early part of the middle Eocene, prior to the extinction of *Morozovella*, *Globanomalina*, and *Igorina*. From then on, pulses of extinction led to generally less diverse planktonic assemblages and a series of last appearances constitute the majority of middle to upper Eocene biostratigraphic events (Pearson et al., 2006).

The trilogy of Paleogene planktonic foraminiferal atlases was recently completed with the publication of the Oligocene atlas (Wade et al., 2018), which includes an update of the Oligocene O-zonation scheme of Berggren and Pearson (2005). At low-to-mid-latitudes, the Eocene–Oligocene boundary corresponds to the extinction of the last five species of the genera *Hantkenina* and *Cribrohantkenina* (e.g., Wade and Pearson, 2008; Berggren et al., 2018). Few new lineages with stratigraphic value appeared during the Oligocene, yet *Paragloborotalia* which originated during the Eocene, strongly radiated and provides several Oligocene zonal markers.

Fifty years of ocean drilling and numerous outcrop studies are gradually leading to more or less stable planktonic foraminiferal biozonations schemes for the entire Paleogene. It must be noted, however, that the reliance on numerous last appearances of zonal markers makes the middle Eocene to Oligocene biozonation scheme more vulnerable to reworking than the lower Paleogene. For more information on the evolution of Paleogene planktonic foraminifera see Petrizzo et al. (2020, Ch. 3G: Planktonic foraminifera, this book).

28.2.1.2 Larger Benthic Foraminifera

Many of the west European centers in which the science of stratigraphy evolved during the 18th and 19th centuries are located on Paleogene strata rich in larger benthic foraminifera. Hence, this group played a key role in the development of Paleogene stratigraphy and also today larger benthic foraminifera continue to provide a prime stratigraphic tool in shallow-marine low- to middle-latitude areas (Fig. 28.10). Zones of larger benthic foraminifera represent sequences of assemblages, commonly separated by intervals without larger benthic foraminifera. Successive assemblages create rather stable biozonations (e.g., Hottinger, 1960; Schaub, 1981), but their exact chronostratigraphic position may vary in function of the timing of transgressive phases within a depositional sequence in a certain basin (Pignatti, 1998).

The Danian and early Selandian larger benthic foraminiferal assemblages are poorly differentiated and consist mainly of small representatives of taxa and lineages that survived the mass extinction at the K–Pg boundary, during which about 83% of the Maastrichtian larger benthic foraminiferal taxa became extinct (BouDagher-Fadel, 2018).

During the Thanetian and Ypresian, rapid radiations of the alveolinids and orthophragminids were followed by that of the nummulitids, leading to a range of Paleogene lineages. These lineages provide well-defined zonations, particularly in the Tethyan region, as well as in the American and Indo-Pacific provinces.

A major change took place within the Bartonian by the extinction of *Alveolina* and large nummulitids. The younger part of the Bartonian and the Priabonian are characterized by relatively small nummulitids, rotaliids, and orthophragminids. Larger benthic foraminiferal assemblages experienced relatively little change across the Eocene–Oligocene boundary, despite the associated marked global cooling. The subdivisions of the Oligocene based on larger benthic foraminifera are mainly based on lepidocyclinids, miogypsinids, and nummulitids (Cahuzac and Poignant, 1997). The last Tethyan and Indo-Pacific species of *Nummulites* became extinct near the end of the Oligocene (BouDagher-Fadel and Price, 2014).

A general larger benthic foraminiferal zonation of the Paleocene and Eocene of the Tethyan province (between the Pyrenees and India) was proposed by Serra-Kiel et al. (1998), providing a widely used reference framework. This is known as the shallow benthic zonation (SBZ) and was a major outcome of IGCP Project 286 *Early Paleogene Benthos*. Cahuzac and Poignant (1997) extended this Paleocene–Eocene SBZ scheme into the Oligocene, but that part is based on data from European basins only. Further zonation schemes are available for the American and Indo-Pacific Paleogene. The Indo-Pacific scheme is known as the "letter-stage" zonation (e.g., Leupold and van der Vlerk, 1931; Adams, 1970; BouDagher-Fadel and Banner, 1999) and a correlation with the SBZ scheme is provided by BouDagher-Fadel (2018).

The correlation between the planktonic zonations and those based on larger benthic foraminifera shown in Fig. 28.10 are partly based on observations in sections where both groups co-occur (e.g., Molina et al., 2003) and partly through magnetostratigraphic data (mostly from Spain; e.g., Burbank et al., 1992; Molina et al., 1992; Serra-Kiel et al., 1994; Payros et al., 2009; Rodriguez-Pinto et al., 2012, 2013; Costa et al., 2013) and calibration with the GPTS. It should be noted, however, that larger benthic foraminiferal successions are generally discontinuous and somewhat diachronous. Consequently, it is not possible to accurately define the ages of the biozonal boundaries with great confidence. This is particularly true for the Paleocene and the Oligocene. Accordingly, shallow benthic zonal boundaries in Fig. 28.10 are indicated by dashed lines to indicate this uncertainty and inaccuracy.

For the Paleocene and Eocene, we largely followed the correlations between the SBZ scheme and the magnetostratigraphy and the partially indirect correlation with the oceanic planktonic schemes as indicated by Serra-Kiel et al. (1998) and modified this scheme to the updated magnetobiostratigraphic schemes. Throughout the Eocene, new data (e.g., Payros et al., 2009; Rodriguez-Pinto et al., 2012, 2013; Costa et al., 2013) led to improvements in the biomagnetostratigraphic calibration. Furthermore, the basal Ilerdian SBZ5 was firmly correlated with the basal Ypresian and thus the classical Ilerdian Stage correlates entirely with the lower Ypresian (Pujalte et al., 2009; Scheibner and Speijer, 2009). This is applicable in the Tethyan region between the Mediterranean and Pakistan (e.g., Afzal et al., 2011; Zamagni et al., 2012; Drobne et al., 2014) and it indicates that the well-known larger benthic foraminiferal turnover between SBZ4 and SBZ5 is in some way related to climate change during and before the PETM (e.g., Orue-Extebarria et al., 2001; Scheibner et al., 2005). Yet, this is questioned for the eastern part of the Tethys (e.g., Zhang et al., 2013; BouDagher-Fadel et al., 2015). The correlation scheme presented here corresponds fairly well with the synthetic scheme of Papazzoni et al. (2017), except for the age of SBZ5.

28.2.1.3 Smaller Benthic Foraminifera

Smaller benthic foraminifera are often useful for correlation of both shallow-and deep-water Paleogene sedimentary strata. Deep-marine shales in many petroleum basins are rich in deep-water agglutinated foraminifera (DWAF), which holds particularly for Paleogene sequences from around the North Atlantic, the Carpathians (flysch sediments), the Caribbean and New Zealand (e.g., Hornibrook et al., 1989; Gradstein et al., 1994; Kaminski and Gradstein, 2005). In general, deep-marine, carbonate-poor, high-latitude sediments harbor diverse agglutinated benthic assemblages.

There is no major taxonomic turnover of DWAF at the K–Pg boundary, nor in the Paleogene itself, although the PETM is marked by distinct faunal shifts, leading to temporary widespread blooms of *Repmanina charoides* (e.g., Kaminski et al., 1996; Galeotti et al., 2004).

There are over 150 stratigraphically useful cosmopolitan Paleogene deeper water agglutinated benthic foraminifera,

and many taxa have stratigraphic ranges that vary slightly or even markedly from basin to basin. However, correlation of wells within a basin can be accomplished with local zonations (e.g., Hornibrook et al., 1989; Gradstein et al., 1994; Kaminski and Gradstein, 2005). These zonations are particularly useful in regions where calcareous planktonic microfossils are generally rare and only present in narrow stratigraphic intervals.

In carbonate-rich open marine sequences, smaller calcareous benthic foraminifera rarely provide improved stratigraphic constraints over calcareous plankton (planktonic foraminifera and calcareous nannoplankton. Yet, deepwater calcareous assemblages clearly evolve from an essentially Cretaceous (Velasco-type) fauna persisting throughout the Paleocene through a Paleogene fauna during most of the Eocene and to a fauna transitional to the modern fauna during the Oligocene (Thomas 2007). The most important turnover of the calcareous benthic fauna took place at the base of the PETM (see Section 28.2.3). The benthic deep-sea fauna experienced a severe reduction in diversity and composition with a ~40% extinction of many Velasco-type taxa (e.g., Thomas and Shackleton, 1996; Thomas, 2007), including the long-lived and widely distributed *Gavelinella beccariiformis*. Whereas the Paleocene Velasco-type fauna is quite uniform globally, the Ypresian (postextinction) fauna is highly variable, largely consisting of survivors from the Paleocene (Arreguin-Rodriguez et al., 2018). The extinction of *Nuttallides truempyi*, a survivor from the Late Cretaceous, is a useful deep-sea marker for the onset of the Oligocene (Thomas, 2007). Berggren and Miller (1989) provided zonation schemes based on bathyal and abyssal benthic foraminifera.

28.2.1.4 Calcareous Nannoplankton

The recognition that calcareous nannofossils are useful for biostratigraphic correlation is generally credited to Bramlette and coworkers (Bramlette and Riedel, 1954; Bramlette and Sullivan, 1961; Bramlette and Wilcoxon, 1967). Following their pioneer efforts, intensive taxonomic and biostratigraphic studies have been carried out (e.g., Bystricka, 1965; Hay et al., 1967) that form the basis of current calcareous nannofossil zonal schemes for the Paleogene. Until a few years ago, the most widely used schemes in low-to-mid-latitude stratigraphic studies were the NP-zonation of Martini (1971) and the CP-zonation of Okada and Bukry (1980). Martini's zonation mainly relied on land sequences from temperate areas, whereas the zonation of Okada and Bukry resulted from analyses of low-latitude oceanic records.

Subsequent high-resolution studies (e.g., Romein, 1979; Perch-Nielsen, 1985; Varol, 1989; Aubry, 1996) redefined and subdivided the NP- and CP-zones. For example, a fourfold subdivision was proposed for lower Eocene Zone

NP10 (Aubry, 1996). Most recently, Agnini et al. (2014) found that several of the zonal and subzonal markers in the NP- and CP-zonation schemes are unreliable and introduced a new Paleogene CN (Calcareous Nannofossil) zonation for low-to-middle latitudes, based on more robust zonal markers. This zonation consists of 11 Paleocene zones (CNP1−11), 21 Eocene zones (CNE1−21), and 6 Oligocene zones (CNO1−6). This zonation and key bioevents are primarily based on the magnetobiochronology of the IODP Expedition 320, Pacific Equatorial Age Transect (PEAT) (Pälike et al., 2010) and turn out robust in other regions too (Agnini et al., 2014). Fig. 28.10 shows both the NP and CN zonations.

Because calcareous nannoplankton distribution is related to the thermal structure of the upper ocean, the presence and stratigraphic ranges of marker species can differ from low to high latitudes. Therefore, regional Paleogene zonations have been proposed for some higher latitude regions (e.g., Wei and Pospichal, 1991; Steurbaut, 1998; Varol, 1998).

Following the biotic crisis at the K−Pg boundary, calcareous nannoplankton underwent a major diversification in the early−middle Paleocene, giving rise to several lineages of biostratigraphic marker taxa (e.g., *Fasciculithus*, *Chiasmolithus*, *Sphenolithus*, *Discoaster*, and *Helicosphaera*). Several radiation pulses appear to have been linked to warming events: the LDE correlates with the first diversification of *Fasciculithus* and related genera, followed by a second pulse during the earliest Selandian (Monechi et al., 2013). The PETM is characterized by diversification among *Discoaster* and the *Rhomboaster−Tribrachiatus* lineage and extinction of large *Fasciculithus* (e.g., Angori and Monechi, 1996; Aubry, 1998).

While early Paleogene calcareous nannoplankton evolution seems to reflect an increasing temperature trend and widespread oligotrophic conditions, late Paleogene calcareous nannoplankton evolution is marked by a progressive decline in diversity and a low rate of originations, influenced by climatic deterioration and eutrophication (Aubry, 1992, 1998). Pronounced taxonomic turnover occurred near the middle−late Eocene boundary and a sharp diversity decline in *Discoaster* took place in the latest Eocene (Monechi, 1986; Nocchi et al., 1988). Several nannofossil taxa became extinct during the early Oligocene. Diversity continued to decrease through the Oligocene, particularly in high latitudes. In contrast the genera *Sphenolithus* and *Helicosphaera* radiated in low latitudes.

Extremely diverse and well-preserved tropical Paleogene nannofossil assemblages were recovered from clay-rich sediments in Tanzanian drill cores (Bown, 2005; Bown and Dunkley Jones, 2006). The recovery of exceptionally rich assemblages from clay-rich coastal sediments suggests that the tiniest and most fragile species are generally selectively removed from the assemblage in oceanic sediments, either during or after deposition. This discovery also means that Paleogene nannofossil diversity may be significantly underestimated, although the overall trends and phases of radiation during the Paleogene are unlikely to be severely affected by selective preservation. For more information, see Watkins and Raffi (2020, Ch. 3F: Calcareous nannofossils, this book).

28.2.1.5 Radiolaria

Riedel (1957) was the first to realize the potential of radiolarians for biostratigraphy based on studies of deep-sea sediment cores. Subsequent studies of radiolarian assemblages in cores collected by the Deep Sea Drilling Project led to the establishment of a low-latitude Cenozoic zonation (Riedel and Sanfilippo, 1970, 1971, 1978; Foreman, 1973; Sanfilippo and Riedel, 1973; Sanfilippo et al., 1985; Nishimura, 1992; Sanfilippo and Nigrini, 1998a). The datums that underpin the Paleogene zones have been mainly calibrated indirectly to the GPTS via co-occurring calcareous microfossil biostratigraphy, but later studies have included direct calibration to magnetochrons (e.g., Kamikuri et al., 2012).

Alternative zonation schemes covering various parts of the Paleogene have been developed for higher latitude regions, such as for the Norwegian and Greenland Seas (Bjørklund, 1976), the Antarctic region and Southern Ocean (Petrushevskaya, 1975; Takemura and Ling, 1997; Funakawa and Nishi, 2005), boreal Russia (Kozlova, 1999), and the Southwest Pacific (Hollis, 1993, 1997, 2002). Sanfilippo and Nigrini (1998a) compiled data from DSDP/ODP Legs 1−135, creating an integrated chart with a composite chronology of radiolarian events tied to numerical ages for the low-latitude Pacific, Indian and Atlantic Oceans. They also introduced codes for the zonation (RP1−RP22 for the Paleogene). A nearly complete lower Paleocene to middle Eocene radiolarian record in ODP Hole 1051 A, western North Atlantic, enabled a well-resolved succession of 200 bioevents for low-latitude Zones RP6 to RP16 (Sanfilippo and Blome, 2001). More recently, the IODP Expedition 320, PEAT, provided an additional continuous radiolarian succession from lower Eocene to lower Miocene with direct calibration to the GPTS (Kamikuri et al., 2012; Kamikuri et al., 2013). The present low-latitude zonation (Fig. 28.11), spanning the middle Paleocene to upper Oligocene is largely based on the new calibrations of Kamikuri et al. (2012). Delayed first occurrences and early last occurrences of several low-latitude index species have been noted in South Pacific Paleogene studies. For example, nine species first occur at the Paleocene−Eocene boundary in New Zealand (Hollis, 2006), of which five already appear in the late Paleocene in low-latitudes (Sanfilippo and Nigrini, 1998b; Sanfilippo and Blome, 2001). This seems

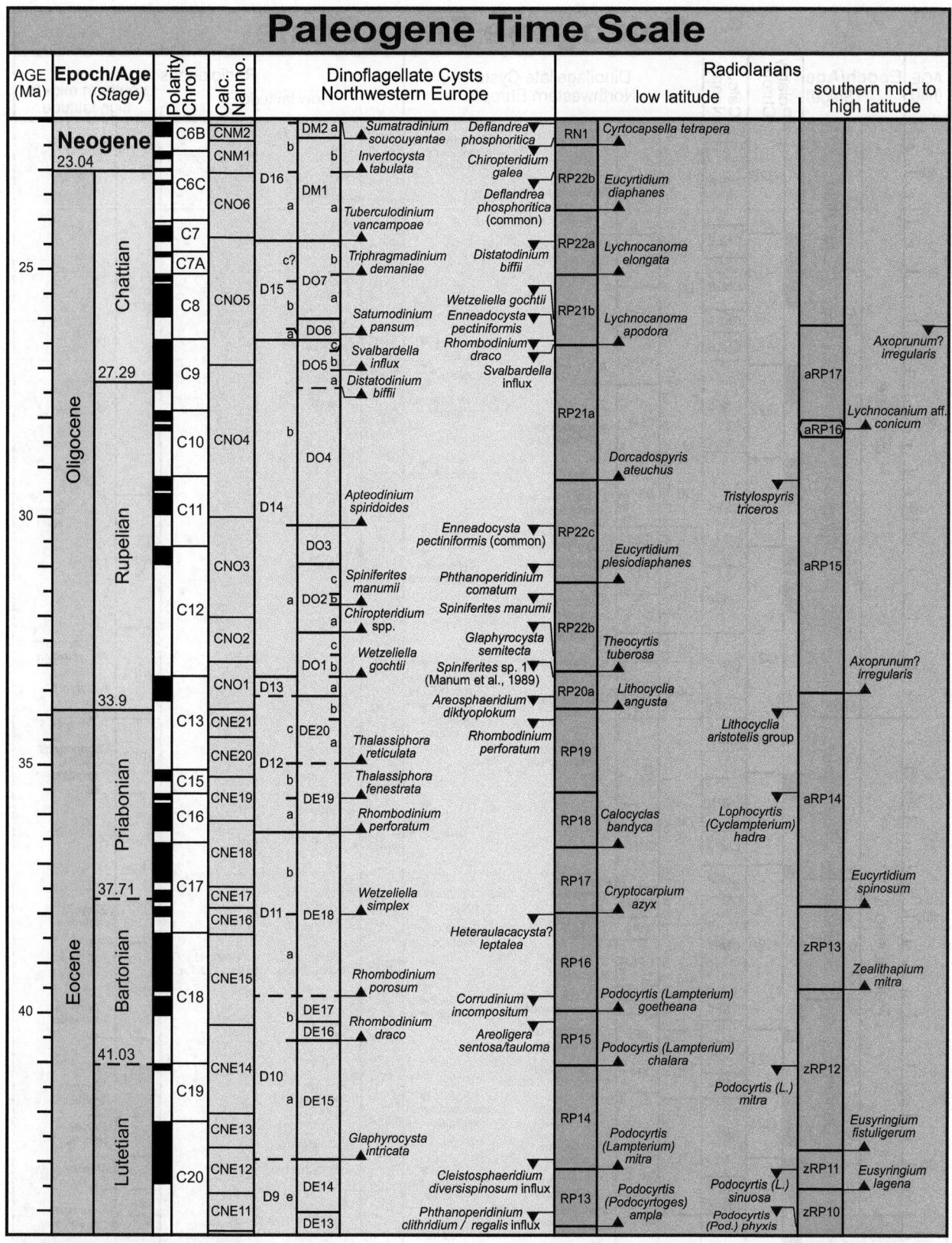

FIGURE 28.11 Paleogene dinoflagellate cyst and radiolarian biostratigraphy calibrated against magnetostratigraphy and calcareous nannofossil biostratigraphy. Dinoflagellate cyst stratigraphy for northwestern Europe was revised from the compilation in GTS2004 by A. J. Powell and H. Brinkhuis (D-zonation) and supplemented by the new zonation scheme (DP, DE, DO, and DM codes) and zonal markers compiled by King (2016).

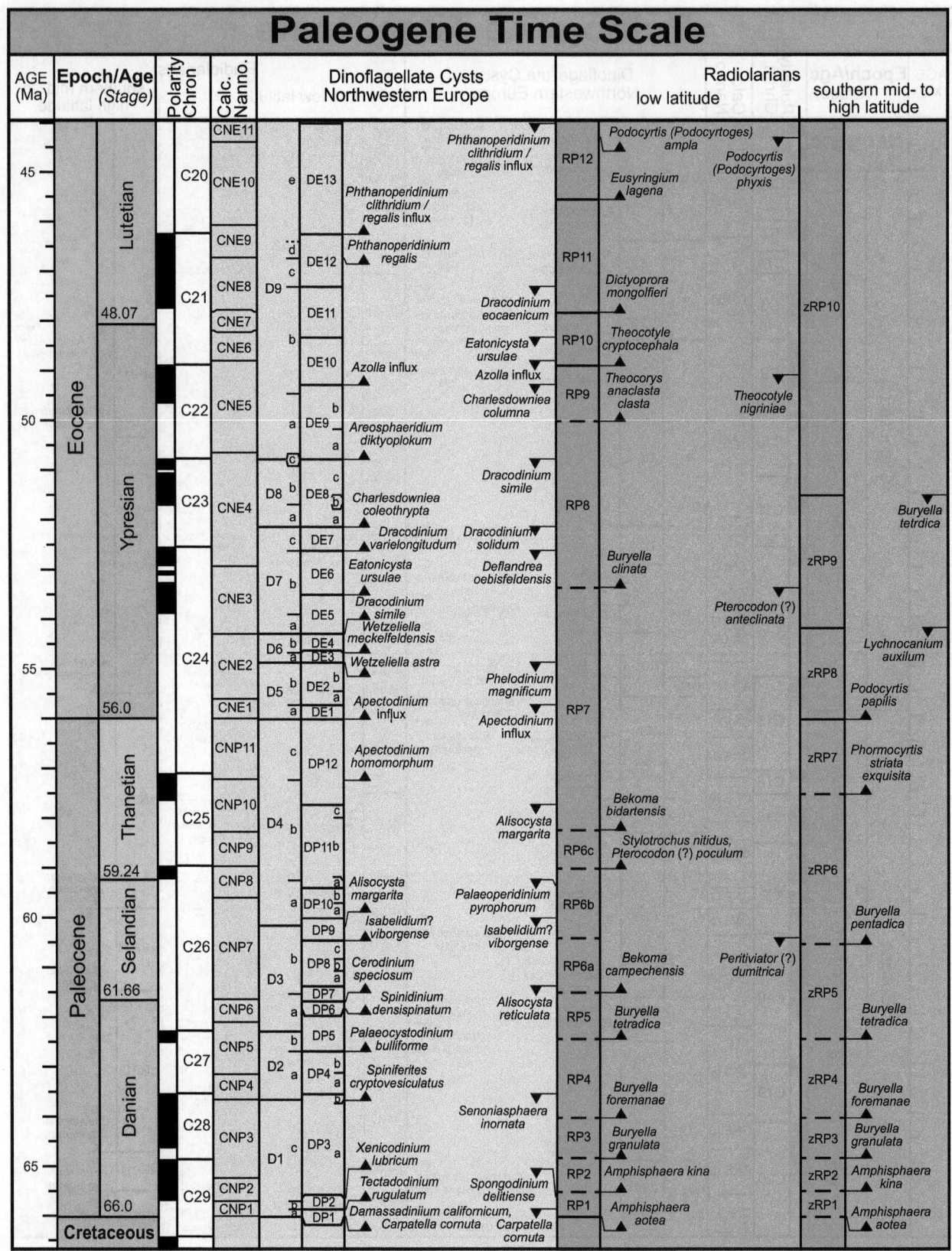

FIGURE 28.11 (Continued) The positioning of zonal markers that are used in both zonation schemes is based on the latter compilation. Most subzonal markers are not indicated. For an extensive review of the validity of all markers, see the appendix in King (2016). Paleogene radiolarian biozonations for low latitudes (RP zones), southwest Pacific (zRP zones) and Southern Ocean (aRP zones) have been recalibrated following the magnetobiochronology of Kamikuri et al. (2012), Dallanave et al. (2015), and Funakawa and Nishi (2005), respectively. Revised definitions for zRP zones are provided by Hollis et al. (2017) and Hollis et al. (2020).

to reflect poleward expansion of the ranges of warm-water taxa during episodes of global warming. Recently, Hollis et al. (2017, 2020) have reviewed and revised the southern mid-latitude zonation for the Paleogene and improved correlation with low latitude and Southern Ocean zones. The revised definitions for southern mid-latitude zones (zRP1−14) are applied in Fig. 28.11.

28.2.1.6 Dinoflagellate cysts

Applications of fossil dinocysts in global Paleogene biostratigraphy and paleoecology have been reviewed in detail in several papers, including Williams and Bujak (1985), Powell (1992), Williams et al. (2004), and Fensome and Williams (2004). These compilations indicate that the stratigraphic range of a given Paleogene dinocyst is rarely synchronous world-wide. Many authors have demonstrated climatic and environmental control on the stratigraphic distribution of taxa in the Paleogene (e.g., Brinkhuis, 1994; Wilpshaar et al., 1996; Bujak and Brinkhuis, 1998; Crouch et al., 2003). Several studies also revealed a differentiation of northern and southern hemisphere assemblages and/or endemic Antarctic assemblages (e.g., Wrenn and Hart, 1988; Wilson et al., 1998; Guerstein et al., 2002; Brinkhuis et al., 2003). Williams et al. (2004) recognized the need to accommodate both latitudinal and hemispherical control of dinocyst assemblages in Paleogene distribution charts. Accordingly, these authors give ranges for low, mid, and high latitudes in both northern and southern hemispheres.

Information concerning Paleogene dinocyst distribution is most comprehensive for the mid-latitudes of the northern hemisphere. This is a reflection of the more intense study of assemblages from these regions, notably from northwestern Europe and the greater North Sea Basin. Although first-order calibration of dinocyst events against magnetostratigraphy is limited, age control is provided through biostratigraphic analysis of numerous outcrops, supplemented by a large volume of largely unpublished subsurface data. The succession of dinocyst events offshore of northwestern Europe is fairly well documented in the public domain (e.g., Mudge and Bujak, 1996; Powell et al., 1996; Mangerud et al., 1999; King, 2016). Powell and Brinkhuis provided a first synthesis for northwestern Europe in GTS2004 (Luterbacher et al., 2004). Their zonation scheme (D1−D16), based on index dinocyst events, was largely maintained in GTS2012 (Vandenberghe et al., 2012) and is also presented here with minor modifications (Fig. 28.11) It must be noted, however, that a new voluminous compilation of Cenozoic stratigraphic data from the greater North Sea Basin by King (2016) suggests that quite a few marker taxa previously used in GTS2004 and GTS2012 are not as reliable as previously thought. Accordingly, King (2016) provides an alternative and more detailed zonation scheme (DP1−12, DE1−20, DO1−7) for the entire Paleogene of northwestern Europe.

Over the last two decades, high-latitude ocean drilling (ODP Leg 189, Tasman Plateau; IODP Expedition 302, Lomonosov Ridge, Arctic Sea; IODP Expedition 318, Wilkes Land Margin, East Antarctica) has also been providing a wealth of new high-latitude paleoenvironmental and stratigraphic data on the Paleogene (e.g., Sluijs et al., 2006, 2009; Warnaar et al., 2009; Bijl et al., 2018), leading to a much better global coverage of dinocyst distribution. Accordingly, there is an increasing amount of regionally well-calibrated Paleogene dinocyst datums (e.g., Bijl et al., 2013b), yet a general synthesis outlining latitudinally well-constrained dinocyst biozonations is not yet available. For more information on their evolution, see Fensome and Munsterman (2020, Ch. 3I: Dinoflagellates, this book).

28.2.2 Terrestrial biostratigraphy

The various continental masses had distinctive land mammal faunas during the Paleogene. These faunas underwent rapid evolution and have been widely used for correlation of nonmarine strata. Intercontinental correlation has often proved problematic owing to endemism, except during geologically brief periods of faunal interchange facilitated by paleogeographic features such as land bridges. Because of the often laterally discontinuous nature of continental strata, occurrences may be in isolated exposures whose stratigraphic relationships are unknown. The stratigraphically and geographically extensive sequences in western North America are notable exceptions.

Broad biostratigraphic units known as LMAs form separate series for North America (NALMA), Europe (ELMA), Asia (ALMA), and South America (SALMA). These can stand independently when correlation to standard chronostratigraphy is uncertain.

Owing to endemism, smaller biostratigraphic−biochronological units vary from having continent-wide applicability to only local use. These may be conventional biozones or, commonly in Europe, reference levels (MP). Reference levels purport to order superpositionally isolated faunas according to evolutionary grade, avoiding the problem of fixing boundaries (Schmidt-Kittler, 1987). Accordingly, the highly discontinuous and point-based European MP reference levels are not separated from each other by lines in Fig. 28.12.

Calibration of mammalian biostratigraphic−biochronological systems to the GPTS is most extensively documented in North America (e.g., Janis, 1998; Woodburne, 2004; Janis et al., 2008; Secord, 2008; Smith

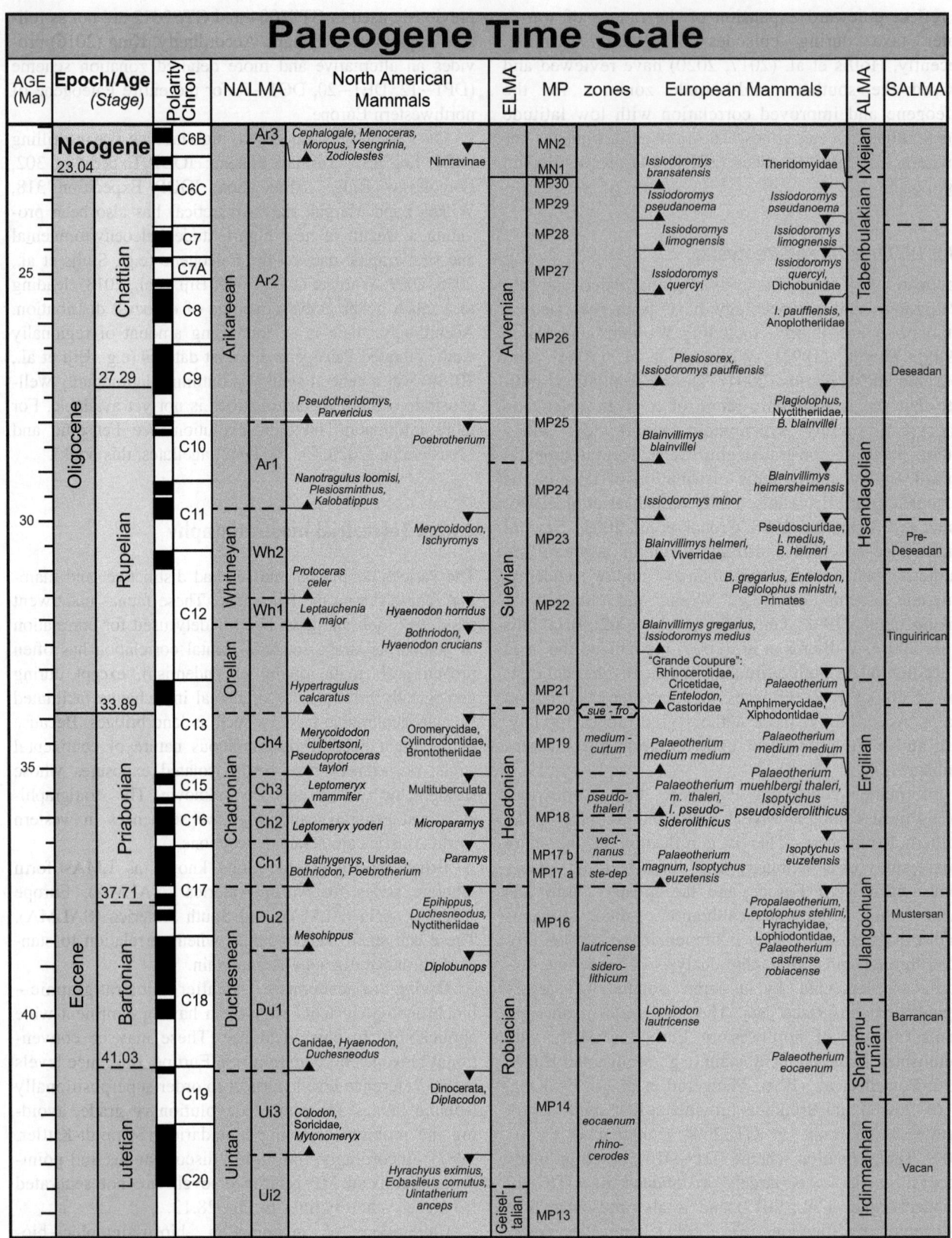

FIGURE 28.12 Paleogene LMAs, mammal bioevents, and zonal markers calibrated against magnetostratigraphy. This scheme provides a major update from earlier schemes in GTS2004 and GTS2012. For the sources used in these earlier compilations, we refer to the captions to Fig. 20.4 (GTS2004) and Fig. 28.10 (GTS2012). Some of the more recent works from these earlier compilations as well as new studies since GTS2012 included in the present revision are listed here. NALMA: Woodburne (2004), Clyde et al. (2007), Janis et al. (2008), Secord (2008), Smith et al. (2008), Gunnell et al. (2009), Kelly et al. (2012), Murphey et al. (2018); ELMA: Smith and Smith (2003), Hooker (2010b, 2015), Comte et al. (2012), Yans et al. (2014b), Lenz et al. (2015), Noiret et al. (2016), Steurbaut et al. (2016), Heilmann-Clausen (2018); ALMA: Wang et al. (2007, 2010, 2019), Clyde et al. (2008, 2010), Kraatz and Geisler (2010), Ting et al. (2011); SALMA: Dunn et al. (2013), Clyde et al. (2014), Woodburne et al. (2014a,b), Krause et al. (2017).

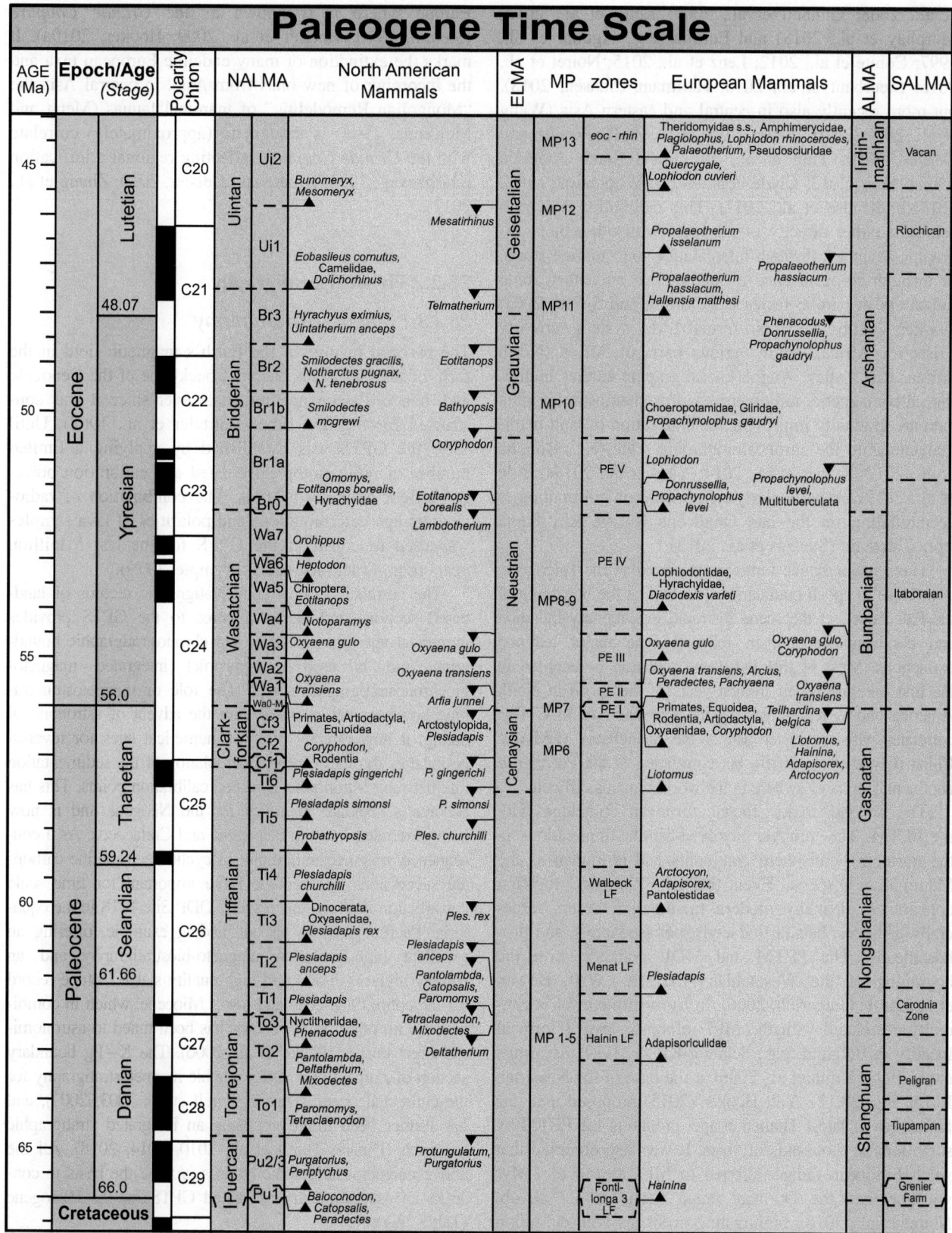

FIGURE 28.12 (Continued) For Europe and North America, the first and last selected appearances are given. They are intended to reflect a balance between those recording important faunal changes, used in biostratigraphy and intercontinental correlation, and those highlighting intercontinental diachronism. *LMAs*, Land Mammal Ages; *NALMA*, North America Land Mammal Ages; *SALMA*, South America Land Mammal Ages; *ALMA*, Asia Land Mammal Ages; *ELMA*, Europe Land Mammal Ages; *MP*, European Reference Levels.

et al., 2008; Gunnell et al., 2009; Kelly et al., 2012; Murphey et al., 2018) and Europe (e.g., Aguilar et al., 1997; Comte et al., 2012; Lenz et al., 2015; Noiret et al., 2016; Steurbaut et al., 2016; Heilmann-Clausen, 2018), but more recently also in central and eastern Asia (Wang et al., 2007, 2010, 2019; Clyde et al., 2010; Kraatz and Geisler, 2010; Ting et al., 2011) and South America (Dunn et al., 2013; Clyde et al., 2014; Woodburne et al., 2014a,b; Krause et al., 2017). This calibration has been achieved either directly or via links with other biostratigraphic schemes, through intercalation with marine strata, or through co-occurrence of mammalian and other zonal indicators in paralic facies (e.g., Smith and Smith, 2003; Hooker, 2010b, 2015). No formal LMA system currently exists for Africa, yet in various parts of Africa (North Africa, Rift Valley, Angola), stratigraphic studies including paleomagnetic, radiometric and/or chemostratigraphic data are gradually improving the calibration of land mammal faunas to the chronostratigraphic scale (e.g., Roberts et al., 2012; Kocsis et al., 2014; Yans et al., 2014a; Solé et al., 2019). A particularly rich source of information is accumulating on the late Oligocene Rukwa Rift fauna from Tanzania (Stevens et al., 2013)

Three major faunal turnovers occurred in the Paleogene, at or close to epoch boundaries. The first, at the beginning of the Paleocene, set the scene for rapid evolutionary radiation and continental endemism following the major tetrapod extinctions. Most of this Paleocene radiation is recorded in the first one and a half million years of the epoch in North America and is represented by the Puercan NALMA. This correlates with the lower part of the Shanghuan ALMA in China (Clyde et al., 2010), but correlation of the Puercan to continental strata elsewhere in the world remains difficult.

The second major faunal turnover coincides with the PETM. The turnover involved similar innovations in the northern hemisphere continents and is known as the Mammalian Dispersal Event (MDE). This marks the first appearance of many modern mammalian orders, especially primates, bats, artiodactyls, perissodactyls, and proboscideans. The PETM and MDE coincide with the beginning of the Wasatchian NALMA (Wa0; Bowen et al., 2001; Gingerich, 2006). In Europe this level is generally correlated with the MP7 reference level (Dormaal locality in Belgium; e.g., Schmidt-Kittler, 1987; Escarguel et al., 1997: Smith et al., 2006) at the base of the Neustrian LMA; Fig. 28.12). Yet, Hooker (2015) proposed that this fauna was of latest Thanetian age, predating the PETM by some tens of thousands of years. It was also observed that south European faunas referred to MP7 are up to 2 Myr younger than the Dormaal fauna (Yans et al., 2014b; Noiret et al., 2016), highlighting problems with definition of the MP reference levels (Hooker 1998).

The third major faunal turnover, in the earliest Oligocene, was less widespread, but well represented in Europe where it is known as the *Grande Coupure* (Stehlin, 1910; Hooker et al., 2009; Hooker, 2010a). It marks the extinction of many endemic European taxa and the incoming of new ones from Asia. In central Asia the "Mongolian Remodeling" of mammal faunas (Meng and McKenna, 1998) is thought to (approximately) correlate with the *Grande Coupure*, reflecting regional aridification (Dashzeveg, 1993; Kraatz and Geisler, 2010; Zhang et al., 2012).

28.2.3 Physical stratigraphy

28.2.3.1 Magnetostratigraphy

The reversal history of the Earth's magnetic field in the form of the GPTS has been the backbone of the Cenozoic and, hence, Paleogene time scale ever since it was constructed for the first time (Heirtzler et al., 1968). Until 1990 the GPTS was established by applying a limited number of radio-isotopically dated age calibration points in seafloor anomaly profiles. The combination of radio-isotopic age determinations and polarities of lava samples was used to construct the GPTS for the last 5 million years (e.g., Mankinen and Dalrymple, 1979).

The correlation of magnetostratigraphic records of land-based sections and deep-sea cores to the GPTS provides numerical ages for bioevents and chronostratigraphic boundaries, and is used to construct integrated magneto-biochronostratigraphic scales. The role of magnetostratigraphy, however, has changed with the advent of astronomical dating; it now directly provides numerical ages for reversal boundaries through (linear) interpolation of the sedimentation rate between astronomically dated calibration points. This has become a standard approach for the Neogene and is now being extended into the Paleogene and Cretaceous. As a consequence, magnetostratigraphy of cyclic deep marine carbonate successions has become more important for time scale construction itself. In this respect, ODP Site 1218 in the equatorial Pacific provides an outstanding example, offering an excellent high-resolution magneto-biostratigraphy and an equally high-resolution and high-quality stable isotope record for the entire Oligocene and lower Miocene, which in combination with other proxy records has been tuned to astronomical target curves (Pälike et al., 2006). The K–Pg boundary section of Zumaia provided a reliable magnetostratigraphy for the entire Paleocene (Dinarès-Turell et al., 2003, 2007), and has further been improved using an integrated stratigraphic approach (Dinarès-Turell et al., 2010, 2014, 2018). All the above magnetostratigraphic records provide the basis to construct an astronomically calibrated GPTS for the Paleogene (Table 28.1).

Furthermore, magnetostratigraphic records remain important for generating age models for sedimentary successions based on their correlation to the GPTS. Paleogene examples

TABLE 28.1 Cenozoic magnetic polarity chrons with comparison of age models used in Cande and Kent (1995), GTS2004 (Gradstein et al., 2004), GTS2012 (Gradstein et al., 2012), GTS2016 (Ogg et al., 2016), and GTS2020 (this book).

Magnetochron (Base)	CK1995	GTS2004	GTS2012	GTS2016	GTS2020	Sources of calibration used in GTS2020
	0.000	0.000	0.000	0.000	0.000	
C1n (Brunhes)	0.780	0.781	0.781	0.773	0.773	no change from GTS2016
C1r.1r (Matuyama)	0.990	0.988	0.988	1.008	1.008	
C1r.1n (Jaramillo)	1.070	1.072	1.072	1.076	1.076	
C1r.2r		1.173	1.173	1.189	1.189	
C1r.2n (Cobb Mountain)		1.185	1.185	1.221	1.221	
C1r.3r	1.770	1.778	1.778	1.775	1.775	
C2n (Olduvai)	1.950	1.945	1.945	1.934	1.934	
C2r.1r	2.140	2.128	2.128	2.120	2.120	
C2r.1n (Feni)	2.150	2.148	2.148	2.155	2.155	
C2r.2r (Matuyama continued)	2.581	2.581	2.581	2.610	2.610	
C2An.1n (Gauss)	3.040	3.032	3.032	3.032	3.032	no change from GTS2012/16
C2An.1r (Keana)	3.110	3.116	3.116	3.116	3.116	
C2An.2n	3.220	3.207	3.207	3.207	3.207	
C2An.2r (Mammoth)	3.330	3.330	3.330	3.330	3.330	
C2An.3n (Gauss continued)	3.580	3.596	3.596	3.596	3.596	
C2Ar (Gilbert)	4.180	4.187	4.187	4.187	4.187	
C3n.1n (Cochiti)	4.290	4.300	4.300	4.300	4.300	
C3n.1r	4.480	4.493	4.493	4.493	4.493	
C3n.2n (Nunivak)	4.620	4.631	4.631	4.631	4.631	
C3n.2r	4.800	4.799	4.799	4.799	4.799	
C3n.3n (Sidufjall)	4.890	4.896	4.896	4.896	4.896	
C3n.3r	4.980	4.997	4.997	4.997	4.997	
C3n.4n (Thvera)	5.230	5.235	5.235	5.235	5.235	
C3r (Gilbert lower part)	5.875	6.033	6.033	6.033	6.023	Drury et al. (2017)
C3An.1n	6.122	6.252	6.252	6.252	6.272	
C3An.1r	6.256	6.436	6.436	6.436	6.386	
C3An.2n	6.555	6.733	6.733	6.733	6.727	
C3Ar	6.919	7.140	7.140	7.140	7.104	
C3Bn	7.072	7.212	7.212	7.212	7.214	
C3Br.1r	7.135	7.251	7.251	7.251	7.262	
C3Br.1n	7.170	7.285	7.285	7.285	7.305	
C3Br.2r	7.341	7.454	7.454	7.454	7.456	
C3Br.2n	7.375	7.489	7.489	7.489	7.499	
C3Br.3r	7.406	7.528	7.528	7.528	7.537	
C4n.1n	7.533	7.642	7.642	7.642	7.650	
C4n.1r	7.618	7.695	7.695	7.695	7.701	
C4n.2n	8.027	8.108	8.108	8.108	8.125	
C4r.1r	8.174	8.254	8.254	8.254	8.257	

(Continued)

TABLE 28.1 (Continued)

Magnetochron (Base)	CK1995	GTS2004	GTS2012	GTS2016	GTS2020	Sources of calibration used in GTS2020
C4r.1n	8.205	8.300	8.300	8.300	8.300	no change from GTS2012/16
C4r.2r	8.631	8.769	8.771	8.771	8.771	
C4An	8.945	9.098	9.105	9.105	9.105	
C4Ar.1r	9.142	9.312	9.311	9.311	9.311	
C4Ar.1n	9.218	9.409	9.426	9.426	9.426	
C4Ar.2r	9.482	9.656	9.647	9.647	9.647	
C4Ar.2n	9.543	9.717	9.721	9.721	9.721	
C4Ar.3r	9.639	9.779	9.786	9.786	9.786	
C5n.1n	9.880	9.934	9.937	9.937	9.937	
C5n.1r	9.920	9.987	9.984	9.984	9.984	
C5n.2n	10.839	11.040	11.056	11.056	11.056	
C5r.1r	10.943	11.118	11.146	11.146	11.146	
C5r.1n	10.991	11.154	11.188	11.188	11.188	
C5r.2r	11.343	11.554	11.592	11.592	11.592	
C5r.2n	11.428	11.614	11.657	11.657	11.657	
C5r.3r	11.841	12.014	12.049	12.049	12.049	
C5An.1n	11.988	12.116	12.174	12.174	12.174	
C5An.1r	12.096	12.207	12.272	12.272	12.272	
C5An.2n	12.320	12.415	12.474	12.474	12.474	
C5Ar.1r	12.605	12.730	12.735	12.735	12.735	
C5Ar.1n	12.637	12.765	12.770	12.770	12.770	
C5Ar.2r	12.705	12.820	12.829	12.829	12.829	
C5Ar.2n	12.752	12.878	12.887	12.887	12.887	
C5Ar.3r	12.929	13.015	13.032	13.032	13.032	
C5AAn	13.083	13.183	13.183	13.183	13.183	
C5AAr	13.252	13.369	13.363	13.363	13.363	
C5ABn	13.466	13.605	13.608	13.608	13.608	
C5ABr	13.666	13.734	13.739	13.739	13.739	
C5ACn	14.053	14.095	14.070	14.070	14.070	
C5ACr	14.159	14.194	14.163	14.163	14.163	
C5ADn	14.607	14.581	14.609	14.609	14.609	
C5ADr	14.800	14.784	14.775	14.775	14.775	
C5Bn.1n	14.888	14.877	14.870	14.870	14.870	
C5Bn.1r	15.034	15.032	15.032	15.032	15.040	Kochhann et al. (2016)
C5Bn.2n	15.155	15.160	15.160	15.160	15.186	
C5Br	16.014	15.974	15.974	15.974	15.994	
C5Cn.1n	16.293	16.268	16.268	16.268	16.261	
C5Cn.1r	16.327	16.303	16.303	16.303	16.351	
C5Cn.2n	16.488	16.472	16.472	16.472	16.434	
C5Cn.2r	16.556	16.543	16.543	16.543	16.532	
C5Cn.3n	16.726	16.721	16.721	16.721	16.637	
C5Cr	17.277	17.235	17.235	17.235	17.154	
C5Dn	17.615	17.533	17.533	17.533	17.466	
C5Dr.1r	17.825	17.717	17.717	17.717	17.634	
C5Dr.1n	17.853	17.740	17.740	17.740	17.676	

TABLE 28.1 (Continued)

Magnetochron (Base)	CK1995	GTS2004	GTS2012	GTS2016	GTS2020	Sources of calibration used in GTS2020
C5Dr.2r	18.281	18.056	18.056	18.056	18.007	interpolated from GTS2012
C5En	18.781	18.524	18.524	18.524	18.497	Liebrand et al. (2016)
C5Er	19.048	18.748	18.748	18.748	18.636	
C6n	20.131	19.722	19.722	19.722	19.535	
C6r	20.518	20.040	20.040	20.040	19.979	
C6An.1n	20.725	20.213	20.213	20.213	20.182	interpolated from GTS2012
C6An.1r	20.996	20.439	20.439	20.439	20.448	
C6An.2n	21.320	20.709	20.709	20.709	20.765	Liebrand et al. (2016)
C6Ar	21.768	21.083	21.083	21.083	21.130	interpolated from GTS2012
C6AAn	21.859	21.159	21.159	21.159	21.204	
C6AAr.1r	22.151	21.403	21.403	21.403	21.441	
C6AAr.1n	22.248	21.483	21.483	21.483	21.519	
C6AAr.2r	22.459	21.659	21.659	21.659	21.691	Liebrand et al. (2016)
C6AAr.2n	22.493	21.688	21.688	21.688	21.722	interpolated from GTS2012
C6AAr.3r	22.588	21.767	21.767	21.767	21.806	
C6Bn.1n	22.750	21.936	21.936	21.936	21.985	Beddow et al. (2018)
C6Bn.1r	22.804	21.992	21.992	21.992	22.042	
C6Bn.2n	23.069	22.268	22.268	22.268	22.342	
C6Br	23.353	22.564	22.564	22.564	22.621	
C6Cn.1n	23.535	22.754	22.754	22.754	22.792	
C6Cn.1r	23.677	22.902	22.902	22.902	22.973	
C6Cn.2n	23.800		23.030	23.030	23.040	
C6Cn.2r	23.999		23.233	23.233	23.212	
C6Cn.3n	24.118		23.295	23.295	23.318	
C6Cr	24.730		23.962	23.962	24.025	
C7n.1n	24.781		24.000	24.000	24.061	
C7n.1r	24.835		24.109	24.109	24.124	
C7n.2n	25.183		24.474	24.474	24.459	
C7r	25.496		24.761	24.761	24.654	
C7An	25.648		24.984	24.984	24.766	
C7Ar	25.823		25.099	25.099	25.099	no change from GTS2012/16
C8n.1n	25.951		25.264	25.264	25.264	
C8n.1r	25.992		25.304	25.304	25.304	
C8n.2n	26.554		25.987	25.987	25.987	
C8r	27.027		26.420	26.420	26.420	
C9n	27.972		27.439	27.439	27.439	
C9r	28.283		27.859	27.859	27.859	
C10n.1n	28.512		28.087	28.087	28.087	
C10n.1r	28.578		28.141	28.141	28.141	
C10n.2n	28.745		28.278	28.278	28.278	
C10r	29.401		29.183	29.183	29.183	
C11n.1n	29.662		29.477	29.477	29.477	
C11n.1r	29.765		29.527	29.527	29.527	
C11n.2n	30.098		29.970	29.970	29.970	
C11r	30.479		30.591	30.591	30.591	

(Continued)

TABLE 28.1 (Continued)

Magnetochron (Base)	CK1995	GTS2004	GTS2012	GTS2016	GTS2020	Sources of calibration used in GTS2020
C12n	30.939		31.034	31.034	30.977	Westerhold et al. (2014) (PEAT)
C12r	33.058		33.157	33.157	33.214	
C13n	33.545		33.705	33.705	33.726	
C13r	34.655		34.999	35.102	35.102	Westerhold et al. (2014) (PEAT), used in GTS2016
C15n	34.940		35.294	35.336	35.336	
C15r	35.343		35.706	35.580	35.580	
C16n.1n	35.526		35.892	35.718	35.718	
C16n.1r	35.685		36.051	35.774	35.774	
C16n.2n	36.341		36.700	36.351	36.351	
C16r	36.618		36.969	36.573	36.573	
C17n.1n	37.473		37.753	37.385	37.385	
C17n.1r	37.604		37.872	37.530	37.530	
C17n.2n	37.848		38.093	37.781	37.781	
C17n.2r	37.920		38.159	37.858	37.858	
C17n.3n	38.113		38.333	38.081	38.081	
C17r	38.426		38.615	38.398	38.398	
C18n.1n	39.552		39.627	39.582	39.582	
C18n.1r	39.631		39.698	39.666	39.666	
C18n.2n	40.130		40.145	40.073	40.073	
C18r	41.257		41.154	41.030	41.030	
C19n	41.521		41.390	41.180	41.180	
C19r	42.536		42.301	42.124	42.196	Westerhold et al. (2017) (Site 1258)
C20n	43.789		43.432	43.426	43.450	Dinarès-Turell et al. (2018)
C20r	46.264		45.724	45.724	46.235	Westerhold et al. (2017) (Site 1258)
C21n	47.906		47.349	47.349	47.760	Francescone et al. (2019)
C21r	49.037		48.566	48.566	48.878	
C22n	49.714		49.344	49.344	49.666	
C22r	50.778		50.628	50.628	50.767	
C23n.1n	50.946		50.835	50.835	50.996	
C23n.1r	51.047		50.961	50.961	51.047	
C23n.2n	51.743		51.833	51.833	51.724	
C23r	52.364		52.620	52.620	52.540	
C24n.1n	52.663		53.074	53.074	52.930	
C24n.1r	52.757		53.199	53.199	53.020	
C24n.2n	52.801		53.274	53.274	53.120	
C24n.2r	52.903		53.416	53.416	53.250	
C24n.3n	53.347		53.983	53.983	53.900	
C24r	55.904		57.101	57.101	57.101	Hilgen et al. (2010), used in GTS2012/2016
C25n	56.391		57.656	57.656	57.656	
C25r	57.554		58.959	58.959	58.959	
C26n	57.911		59.237	59.237	59.237	

TABLE 28.1 (Continued)

Magnetochron (Base)	CK1995	GTS2004	GTS2012	GTS2016	GTS2020	Sources of calibration used in GTS2020
C26r	60.920		62.221	62.221	62.278	Dinarès-Turell et al. (2014)
C27n	61.276		62.517	62.517	62.530	
C27r	62.499		63.494	63.494	63.537	
C28n	63.634		64.667	64.667	64.645	
C28r	63.976		64.958	64.958	64.862	
C29n	64.745		65.688	65.688	65.700	
K/PG	65.000			66.060	66.001	
C29r				66.410		

The GTS2020 scale completes the astronomical-tuned age model for nearly all chrons, whereas earlier scales required interpolations from spreading rate models for the South Atlantic C-sequence synthetic profile of marine magnetic anomalies; and the main additional astronomical-tuning studies are indicated in the right column. See text and Ogg (2020, Ch. 5: Geomagnetic polarity time scale, this book) for additional details. *PEAT*, Pacific Equatorial Age Transect.

are the magnetostratigraphic framework established for the Bighorn Basin in North America to unravel faunal, climate and basin evolution (Clyde et al., 2007), for the Tibetan Plateau to constrain climate and uplift history (Dupont-Nivet et al., 2007; Wang et al., 2008), for Cenozoic ocean drilling cores to constrain planktonic foraminiferal biochronology (Wade et al., 2011), and for Antarctic cores to reconstruct glaciation history (Florindo et al., 2005).

In the Southern Hemisphere, the New Zealand geologic time scale has been recalibrated to the GPTS (Raine et al., 2015). Calibration is mainly based on foraminiferal and nannofossil datums for New Zealand's Paleogene series and stages, but recent advances have been made in direct calibration to magnetochrons (e.g., Dallanave et al., 2015).

28.2.3.2 Chemostratigraphy and climatic evolution

Carbon and oxygen stable isotope analyses (shown in conventional notation here, $\delta^{13}C$ and $\delta^{18}O$) of bulk carbonate or foraminiferal shells are routinely performed in stratigraphic studies of Paleogene sequences. Compilations of the oxygen and carbon isotopic composition of the shells of deep-sea benthic foraminifera (e.g., Savin, 1977; Miller et al., 1987; Zachos et al., 2001; Zachos et al., 2008; Cramer et al., 2009; Cramwinckel et al., 2018) provide a long-term history of the evolution of climate and the carbon cycle through the Paleogene (Fig. 28.13). Furthermore, the Paleogene includes numerous short-term (<200 kyr) climate events. These fluctuations provide a key tool for stratigraphic correlation between marine successions and occasionally also between marine and terrestrial successions, notably for the base of the Ypresian.

Following variable climatic conditions in the early Paleocene, oxygen isotopes show a progressive warming trend from the middle Paleocene to early Eocene. The warmest long-term time interval of the whole Cenozoic was recorded during the Early Eocene Climatic Optimum (EECO), between about 49 and 53 Ma (e.g., Zachos et al., 2008; Westerhold et al., 2018b; Hollis et al., 2019). The $\delta^{18}O$ record indicates that the termination of the EECO marks the start of a long-term climatic cooling trend for the Eocene, punctuated by warming phases, such as the MECO. The Eocene–Oligocene transition (EOT) is marked by further cooling and growth of large ice-sheets on Antarctica. The $\delta^{18}O$ record suggests that global temperatures rose again in the late Oligocene. These long-term trends are generally considered to be related to tectonic phenomena. North Atlantic rifting and associated volcanism at the end of the Paleocene and the beginning of the Eocene could explain the early Eocene warm period, while reduction in seafloor spreading rates during the later Eocene is associated with the cooling. The opening of the Tasmanian and Drake passages is thought to have initiated circum–Antarctic ocean circulation, which led to sudden cooling at the beginning of the Oligocene (Zachos et al., 2001). However, there is debate over the timing of the opening of these gateways (Sijp et al., 2011; Bijl et al., 2013a) and others have alternatively suggested Antarctic ice sheet expansion was more directly linked to a decrease in atmospheric CO_2 (DeConto and Pollard, 2003).

Superimposed on these general climatic trends are short-lived hyperthermal and cooling/glaciation events of some tens to a few hundred thousand years. The main events that aid stratigraphic correlation are summarized here from old to young (Fig. 28.13). A global 2°C–3°C deep-sea warming marks the LDE, at the base of Chron C26r at 62.3 Ma (Bornemann et al., 2009; Westerhold et al., 2011; Dinarès-Turell et al., 2012; Deprez et al., 2017; Barnet et al., 2019). The LDE also appears to closely correlate with land mammal fauna transitions across the globe (Torrejonian to Tiffanian in North America; Shanghuan to Nongshanian in eastern Asia; onset of *Carodnia* fauna in South America;

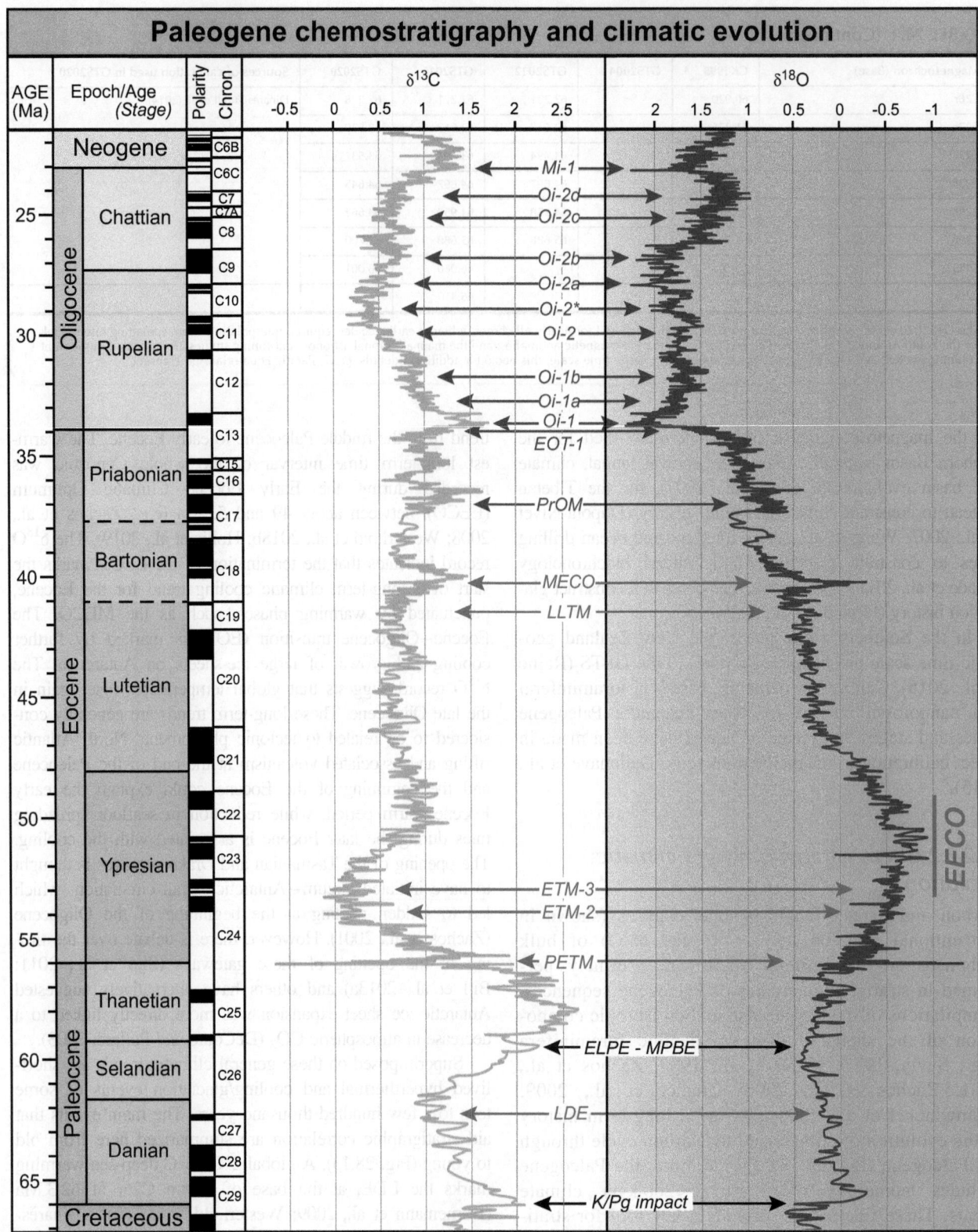

FIGURE 28.13   Positions of main Paleogene climatic events against the generalized oxygen and carbon isotope curves for the Paleogene. The stable isotope curves are based on Cramer et al. (2009); original carbon and oxygen isotope values are recalibrated to GTS2020 and the curves shown are obtained by a nine-point moving window or a five-point window through major excursions. Isotopic events in blue reflect cooling/glaciation events and those in red are warming events (hyperthermals). The references for the naming and positioning of the climatic events are discussed in the text (Section 28.2.3).

Clyde et al., 2008, 2010; Woodburne et al., 2014a) suggesting that this warming period may have played a role in rejuvenating land mammal faunas worldwide. Note that the early Danian Dan-C2 event (Quillévéré et al., 2008) is not included in Fig. 28.13, because high-resolution records from the Pacific and Atlantic Oceans do not provide evidence of deep-sea warming during this event (Westerhold et al., 2011; Barnet et al., 2019).

The Early Late Paleocene Event (ELPE) is represented by a clay-rich interval in North Pacific and South Atlantic deep-sea cores (Westerhold et al., 2011; Littler et al., 2014; Barnet et al., 2019) and is linked to prominent changes among the plankton (Bralower et al., 2002; Petrizzo, 2005). Known as the MPBE, ELPE has also been reported from outcrops of marine strata in Europe (Bernaola et al., 2007; Coccioni et al., 2012) and nonmarine sections in Spain and Argentina (Hyland et al., 2015; Pujalte et al., 2017). The position of the event is situated close to the top of Chron C26r at 59.5 Ma (Bernaola et al., 2007; Westerhold et al., 2011). Although the ELPE may indeed represent a global phenomenon and is associated with a negative $\delta^{18}O$ excursion at South Atlantic ODP Site 1262 (Littler et al., 2014; Barnet et al., 2019), it is still not fully resolved whether it is indeed a global warming event (Westerhold et al., 2011; Littler et al., 2014).

The PETM, at the base of the Eocene, is characterized by the highest global temperatures of the Cenozoic and can be considered as the "Mother of all hyperthermals" as the term was first proposed to characterize this remarkable global warming event (Thomas and Zachos, 2000; Thomas et al., 2000). The PETM is associated with a 2‰–3‰ negative CIE marking the Paleocene–Eocene boundary (Kennett and Stott, 1991; Aubry et al., 2007). The isotopic anomaly developed within a few thousand years in both the deep and shallow ocean indicating that the entire oceanic carbon reservoir experienced a rapid change in isotopic composition. It took ~170 kyr before $\delta^{13}C$ returned to background levels (e.g., Röhl et al., 2007; Zeebe and Lourens, 2019). This oceanic perturbation is also marked by biotic turnovers and variations in the CCD, phenomena also observed with other warming events (e.g., Lourens et al., 2005; Zachos et al., 2010; D'Haenens et al., 2014). The $\delta^{13}C$ anomaly has been used for marine–terrestrial correlations (Koch et al., 1992), indicating that the Clarkforkian–Wasatchian land-mammal turnover and the major extinction event among benthic foraminifera in the deep sea at 56 Ma were virtually coeval and triggered by the same global environmental perturbation (e.g., McInerney and Wing, 2011).

At 54 Ma, just below the top of Chron C24r, ETM-2, and the closely spaced CIEs H1 and H2 reflect another global warming event that shows some similarities with the PETM in isotopic and microfaunal characteristics, although is much less pronounced (Lourens et al., 2005; D'haenens et al.,

2014). Both events are sometimes referred to as hyperthermals. Another early Eocene hyperthermal, which has been labeled variously as ETM-3, X-event, and K (Agnini et al., 2009) occurs at 53 Ma, within Subchron C24n.1n. The K, H1, and H2 codes are derived from the analysis by Cramer et al. (2003), who identified almost 20 short-term negative CIEs (labeled A to L) in deep-sea sediment cores within the interval of Chrons C24n–C25n. This sequence of orbitally tuned CIEs has been extended into the middle Eocene and nomenclature tied directly to magnetochrons, so that the uppermost CIEs that lie within Chron 21r are labeled C21rH1–C21rH6 (Sexton et al., 2011; Westerhold et al., 2018b). The link between CIEs and astronomical cycles provides excellent opportunities for high-resolution stratigraphic correlation of lower and middle Eocene sediments (Lauretano et al., 2016; Westerhold et al., 2017). For example, the EECO has now been defined as the interval of low benthic foraminiferal $\delta^{18}O$ values that lies between CIEs C24n.2rH1 (=J event of Cramer et al. (2003)) and C22nH5 at 53.26–49.14 Ma (Westerhold et al., 2018b; Hollis et al., 2019).

The Late Lutetian Thermal Maximum, within Chron C19r at 41.5 Ma, is currently considered to represent the last hyperthermal of the Paleogene (Intxauspe-Zubiaurre et al., 2018; Westerhold et al., 2018a). MECO, the climatic optimum of the early Bartonian, is a much longer warming period (~750 kyr), but its termination is reminiscent of a typical hyperthermal as it comprises parallel negative excursions in $\delta^{18}O$ and $\delta^{13}C$ over <100 kyr (Edgar et al., 2010). It is also associated with biotic changes and oscillation in CCD. It ranges from the upper part of Chron C18r to Subchron C18n.2n at 40–40.75 Ma (Bohaty et al., 2009; Edgar et al., 2010; Luciani et al., 2010; Moebius et al., 2015).

The long-term cooling trend following the MECO is punctuated by cooling events lasting <200 kyr. One of these is the Priabonian Oxygen Isotope Maximum event, situated in middle Chron C17n.1n around 37 Ma, which may reflect short-lived ice growth in East Antarctica (Scher et al., 2014). The EOT is marked by several cooling and glaciation pulses, notably EOT-1, during the latest Eocene (in Chron C13r, ~34 Ma) and Oi1 in the earliest Oligocene (Chron C13n, ~33.5 Ma). It is generally accepted that these cooling events and the transition from greenhouse to icehouse conditions, is associated with the build-up of permanent ice sheets on Antarctica (e.g., Miller et al., 1991). The coding and timing of the Oligocene cooling events used here are based on the scheme of Wade and Pälike (2004), emended by Boulila et al. (2011). The Oligocene cooling events have been shown to be related to 405, 127 and 96-kyr eccentricity cycles and to 1.2-Myr obliquity cycles (Pälike et al., 2006).

The waxing and waning of an ice mass on Antarctica is inferred to also have had an influence on eustatic sea level (Fig. 28.1). Eustasy estimates for the Oligocene Oi events, based on Pleistocene calibration and sea-level changes associated with these events, are estimated to be in the order of 50 to 65 m (Wade and Pälike, 2004). It is more complicated to reconstruct eustatic change in sea level from $\delta^{18}O$ fluctuations during the earlier Paleogene because there is a lack of direct evidence for the presence of continental ice. Further elaboration on the link between $\delta^{18}O$ curves, eustatic sea-level change and sequence stratigraphic records from marginal basins is provided by Vandenberghe et al. (2012). See also review in Simmons et al. (2020, Ch. 13: Phanerozoic eustasy, this book).

The major transgressive—regressive cycles recognized in the Paleogene (Fig. 28.10) are based on a combination of T—R facies cycles and major sequence boundaries as proposed in Hardenbol et al. (1998), and adding the major cooling event in the earliest Oligocene. Note that the positions of the sequence boundaries are correlated to the current GPTS by means of the—now somewhat outdated—planktonic foraminiferal biozonation in Hardenbol et al. (1998).

28.3 Paleogene time scale

For GTS 2012/2016 the Paleogene time scale was constructed in two fundamentally different ways. The traditional way is to use the GPTS as backbone: This GPTS is constructed using seafloor anomaly profiles in combination with a limited number of radioisotopically dated age control points, as in Cande and Kent (1992, 1995). The alternative method is astronomical dating. This provides direct numerical ages for magnetic reversals and bioevents by tuning cyclic sedimentary successions to astronomical target curves. Both methods were presented independently from one another for GTS2012. Note though, that astronomical dating critically depends on selected radio-isotopic ages for first-order tuning to 405-kyr eccentricity. Here we report on updates to the astronomically calibrated GPTS based on several new astronomical calibration publications that were published since GTS2012 but refrain from updating the radioisotopically derived ages. For the present time scale, we are instead at a point where almost the entire Paleogene sequence can be provided through astronomically derived ages. A detailed discussion of the radio-isotopic age model construction is provided by Vandenberghe et al. (2012).

28.3.1 Astronomical age model

The numerical calibration for the Paleogene time scale has undergone significant revisions since its predecessors in GTS2012 (Vandenberghe et al., 2012) and GTS2004 (Luterbacher et al., 2004). In particular, there have been

concerted efforts to exploit the information obtained by the International Ocean Discovery Program, and its predecessors, to inform a recalibration of the GPTS and biostratigraphic events toward one where now most of the Paleogene can be supported by astronomical age calibrations at least from one drill site. Absolute and relative age uncertainties persist, particularly around Lutetian to Ypresian Chrons C20—C22, that are likely to be further revised as more data become available.

28.3.2 Limits of validity of solar system orbital solutions

Before detailing the updates in the present astronomical age model for the Paleogene, the existing limitations in astronomical integrations that form the basis for astronomical age models need to be reiterated (see Laskar, 2020, Ch. 4: Astrochronology, this book). These limitations relate to (1) the chaotic nature of the solar system, primarily affecting long-term orbital properties, and (2) uncertainties in parameters like tidal dissipation, dynamical ellipticity, and tidal friction that affect the Earth—Moon system and need to be reconstructed from the geologic record. For the first limitation the phase and period of the ~405-kyr long-eccentricity cycle are nearly identical for the different available astronomical solutions throughout the Cenozoic and stays within one full cycle over 250 Myr (Laskar et al., 2004, 2011a). In contrast, the short eccentricity cycles (~80—130 kyr periods) agree back to 50~54 Ma in phase but then rapidly diverge, thus not providing a reliable tuning target prior to this age range, in particular for astronomical solutions that include the major asteroids (Laskar et al., 2011b; Zeebe, 2017). Similarly, the long-term amplitude modulations of eccentricity and obliquity (~2.4 and ~1.2 Myr periods) are also affected by the chaotic diffusion in astronomical solutions beyond ~50 Ma, and while efforts have been made to compare geological data with available astronomical solutions (Laskar et al., 2004, 2011a,b; Westerhold et al., 2017), it is easy to change initial conditions of a single model to obtain a multitude of diverging astronomical solutions beyond ~54 Ma (Zeebe, 2017). As a consequence, a detailed pattern matching based astronomical target other than through comparison with the phasing of the 405-kyr eccentricity cycle cannot be used to inform the geologic time scale at present beyond ~50—53 Ma.

Currently, no published obliquity and precession solutions exist beyond Laskar et al. (2004), as efforts to decode tidal dissipation and dynamical ellipticity throughout the Cenozoic have not yet been completed (Pälike and Shackleton, 2000; Lourens et al., 2001; Zeeden et al., 2014).

TABLE 28.2 Paleogene age model of GTS2020 and comparison with GTS2012 and GTS2016.

Epoch base	Age/Stage base	GTS 2012		GTS 2016		GTS 2020		Comments
		Age (Ma)	duration	Age (Ma)	duration	Age (Ma)	duration	
Miocene	Aquitanian	23.03		23.10		23.04		
Oligocene	Chattian	28.74	5.71	28.09	4.99	27.29	4.25	Ratified 2018 at higher level
	Rupelian	34.23	5.49	33.90	5.81	33.90	6.61	
Eocene	Priabonian	37.75	3.52	37.99	4.09	37.71	3.81	
	Bartonian	41.15	3.40	41.03	3.04	41.03	3.32	No GSSP yet.
	Lutetian	47.82	6.67	47.84	6.81	48.07	7.04	
	Ypresian	55.84	8.02	55.96	8.12	56.00	7.93	
Paleocene	Thanetian	59.12	3.28	59.24	3.28	59.24	3.24	
	Selandian	61.66	2.54	61.61	2.37	61.66	2.42	
	Danian	65.95	4.29	66.04	4.43	66.00	4.34	

28.3.3 Updates to magnetic reversal ages since the Geologic Time Scale 2012

The GTS2012 Paleogene time scale (Vandenberghe et al., 2012) included cyclostratigraphic studies, which aimed to tune the Paleogene on the 405-kyr eccentricity scale, using new astronomical solutions (La2004: Laskar et al., 2004; La2010: Laskar et al., 2011a). These studies were then used in constructing an astronomical time scale for the Paleogene. Since then several additional astronomical solutions for the eccentricity and orbital inclination have been published (La2011: Laskar et al., 2011b; Zeebe, 2017), and some key cyclostratigraphic studies have covered various parts of the Paleogene allowing us now to provide a first fully astronomically derived Paleogene geomagnetic polarity reversal table (Table 28.1). Ages straddling the Oligocene–Miocene boundary have been revised by Beddow et al. (2018) using sediments recovered from the equatorial Pacific during IODP Expedition 320, keeping the new basal age of Subchron C6Cn.2n (23.04 Ma) close to those from GTS2012 and 2016 (23.03 Ma), but revise older magnetochron ages by up to ~70 kyr to 24.124 Ma at the base of Subchron C7n.1r.

Liebrand et al. (2016) used sediments from Walvis Ridge (ODP Expedition 208) to reevaluate the time interval from 17.1 to 30.1 Ma from ODP Site 1264, with new ages that are largely consistent with previous GTS2012 ages. Here we use the revised ages from Liebrand et al. (2016) for the basal ages of Chrons C7n through C7An (24.46–24.77 Ma), which show some more significant deviations to GTS2012. For the main remaining part of the Oligocene down to the base of Chron C11r (30.59 Ma) we keep the ages from GTS2012 which were

derived from ODP Site 1218 in Pälike et al. (2006), as subsequent revisions are largely within error of astronomical age models (± precession or obliquity cycles), as ODP 1218 has a finely defined magnetostratigraphy.

One consequence of this decision is that we arrive at an age of 27.29 Ma for the base of the Chattian at Chron C9n.14., which is closer to the "MCA Tuning 2" age of 27.41 Ma of Coccioni et al. (2018) than their preferred "MCA Tuning 1" with an age of 27.82 Ma. Since there are remaining uncertainties about the utility of the primary foraminiferal marker *C. cubensis* (King and Wade, 2017), we use the position of the Chattian GSSP with respect to the magnetostratigraphy as the tool to convert the magnetic reversal ages to the basal Chattian age.

For the interval from the base of Chron C12n through the base of Chron C19n, we use the revised ages from Westerhold et al. (2014), using sediments from IODP Expedition 320. This age model does not significantly change the age for the Eocene–Oligocene boundary correlated to Chron C13r.14 (33.89 Ma). For Chrons C13n through C19n these ages were already included in the GTS2016 update.

We then use again astronomical age calibrations from the Walvis Ridge ODP Site 1258, presented by Westerhold et al. (2017), for the basal ages of Chrons C19r and C20r, but using a basal age for Chron C20n of 43.45 as provided by Dinarès-Turell et al. (2018) (i.e., c. 50 kyr younger than that from ODP 1258 of Westerhold et al., 2017).

For the basal ages of Chrons C21n through C24n (47.76–53.90) we use the revised ages from Francescone et al. (2019) based on records from the Scaglia Rossa and Scaglia Variegata formations of the

Umbria-Marche Basin in central Italy. Their study revealed a satisfactory match to coeval deep marine records from ODP Sites 1258 and 1263, and it also evaluated in detail marine magnetic anomaly profiles from major oceanic basins characterized by high seafloor spreading rates that were used to independently test the revised ages. Importantly, theses ages result in significant adjustments to the durations and absolute ages of Chrons C20r through C22n, the stratigraphic interval referred to as the "Eocene astronomical time scale gap," and for which several alternative age models with quite different durations exist (e.g., Boulila et al., 2018; Vahlenkamp et al., 2020).

For the basal ages for Chrons C24r−C26n we maintain those published in Hilgen et al. (2010), while for the remainder of Cenozoic geomagnetic polarity sequence down to Chron C29n and base of the Paleocene we use ages from Dinarès-Turell et al. (2014)

Vandenberghe et al. (2012) provided a critical evaluation of the cyclostratigraphic interpretation of Westerhold et al. (2008) and indicated that 25 rather than 24 × 405-kyr cycles are present in the Paleocene

(Hilgen et al., 2010). The results presented here are essentially consistent with those in GTS2012 in terms of the total duration of the Paleocene incorporating 25 eccentricity cycles of 405 kyr.

28.3.4 Ages of Paleogene stages

For GTS2020, cyclostratigraphy and astrochronology are used to directly calibrate magnetic reversal ages. The complete C-sequence results are tabulated in Table 28.1 and the stage boundary ages are summarized in Table 28.2 with a comparison to earlier time scales. In the Paleocene, the base Danian is 66.00 Ma (was 65.95 ± 0.05 Ma in GTS2012), base Selandian 61.66 Ma (was 61.45 ± 0.13 Ma), and base Thanetian 59.24 Ma (was 59.12 ± 0.13 Ma). For the Eocene, the base Ypresian is 56.00 Ma (was 55.84 ± 0.03 Ma), base Lutetian is 48.07 Ma (was 47.82 ± 0.2 Ma), base Bartonian is 41.03 Ma (was 41.15 ± 0.52 Ma), and base Priabonian is 37.71 Ma (was 37.75 ± 0.52 Ma). In the Oligocene Epoch, base Rupelian is 33.9 Ma (was 34.23 ± 0.33 Ma) and base Chattian is 27.29 Ma (was 28.74 ± 0.13 Ma).

GSSPs of the Paleogene Stages, with location and primary correlation criteria					
Stage	**GSSP Location**	**Latitude, Longitude**	**Boundary Level**	**Correlation Events**	**Reference**
Chattian	Monte Cagnero, Urbania, Umbria-Marche, Italy	43°38'47.81"N 12°28'03.83"E*	at meter level 197	Foraminifer, LCO of *Chiloguembelina cubensis*	Episodes **41**/1, 2018
Rupelian	Massignano, near Ancona, Italy	43°31'58.2"N 13°36'03.8"E*	Base of a 0.5 m thick, greenish-grey marl bed, 19 m above base of section	Foraminifer, LAD of *Hantkenina* and *Cribrohantkenina*	Episodes **16**/3, 1993
Priabonian	Alano di Piave, Venetian southern Alps,NE Italy	45°54'51.10"N 11°55'4.87"E	Base of Tiziano tuff bed; 63.57 m leve in section	Foraminifer, LAD of *Morozovelloides* and FAD of nannofossil *Cribrocentrum erbae*	
Bartonian	*Candidate section at Gubbio, central Italy*			*provisional: base of magnetic polarity chronozone C18r*	
Lutetian	Gorrondatxe beach section near Getxo village, W Pyrenees, Spain	43°22'46.47"N 3°00'51.61"W	167.85 m above the base of the Gorrondatxe section in a dark marly level	Calcareous nannofossil, FAD of *Blackites inflatus*	Episodes **34**/2, 2011
Ypresian	Dababiya, near Luxor, Egypt	25°30'04.70"N* 32°31'19.14"E*	Base of Bed 1 in DBH subsection	Base of Carbon Isotope Excursion (CIE)	Episodes **30**/4, 2007
Thanetian	Zumaia section, northern Spain	43°17'58.4"N* 02°15'39.1"W*	6.5 m above the base of Member B of the Itzurun Formation	Base of magnetic polarity chronozone C26n	Episodes **34**/4, 2011
Selandian	Zumaia section, northern Spain	43°17'57.1"N* 02°15'39.6"W*	Base of the red marls of the Itzurun Formation	2nd radiation of nannofossil *Fasciculithus*	Episodes **34**/4, 2011
Danian	Oued Djerfane, west of El Kef, Tunisia	36°09'13.2"N 8°38'54.8"E	Reddish layer at the base of the 50 cm thick, dark boundary clay	Iridium geochemical anomaly	Episodes **29**/4, 2006
* according to Google Earth					

FIGURE 28.14 Paleogene GSSP locations, boundary levels, primary marker fossils and correlation events, and key references (status as of January 2020). *GSSP*, Global Boundary Stratotype Section and Point. *Details of each GSSP are available at http://www.stratigraphy.org, https://timescale-foundation.org/gssp/, and in the Episodes publications.*

Acknowledgments

This chapter is a major revision and update of N. Vandenberghe, F.J. Hilgen, F.J. and R.P. Speijer (2012), which in turn was a major revision and update of H.P. Luterbacher, J.R. Ali, H. Brinkhuis, F.M. Gradstein, J.J. Hooker, S. Monechi, J.G. Ogg, J. Powell, U. Röhl, A. Sanfilippo, and B. Schmitz (2004). We are grateful to all previous contributors as well as to C. Agnini, P.D. Gingerich, C. Heilmann-Clausen, M. Jehle, D. Kroon, D.K. Munsterman, K. Pascher, A. Payros, I. Raffi, T. Smith, N.J. Stevens, B. Wade, and Y. Wang for advice and exchange of information which stimulated up-to-date improvements of this chapter. We also greatly appreciate the efforts of G.M. Ogg in drafting all figures into perfection.

Bibliography

Abels, H.A., Van Simaeys, S., Hilgen, F.J., De Man, E., and Vandenberghe, N., 2007, Obliquity dominated glacio-eustatic sea-level change in the early Oligocene: evidence from the shallow marine siliciclastic Rupelian stratotype (Boom Formation, Belgium). *Terra Nova*, **19**: 65–73.

Adams, C.G., 1970, A reconsideration of the East Indian letter classification of the Tertiary. *Bulletin of the British Museum (Natural History)*, **19**: 87–137.

Afzal, J., Williams, M., Leng, M.J., Aldridge, R., and Stephenson, M.H., 2011, Evolution of Paleocene to early Eocene larger benthic foraminifer assemblages of the Indus Basin, Pakistan. *Lethaia*, **44**: 299–320.

Agnini, C., Macri, P., Backman, J., Brinkhuis, H., Fornaciari, E., Giusberti, L., et al., 2009, An early Eocene carbon cycle perturbation at ~52.2 Ma in the Southern Alps: chronology and biotic response. *Paleoceanography*, **24: PA2209**, 1–14.

Agnini, C., Fornaciari, E., Giusberti, L., Grandesso, P., Lanci, L., Luciani, V., et al., 2011, Integrated bio-magnetostratigraphy of the Alano section (NE Italy): a proposal for defining the middle-late Eocene boundary. *Geological Society of America Bulletin*, **123**: 841–872.

Agnini, C., Fornaciari, E., Raffi, I., Catanzariti, R., Pälike, H., Backman, J., et al., 2014, Biozonation and biochronology of Paleogene calcareous nannofossils from low and middle latitudes. *Newsletters on Stratigraphy*, **47**: 131–181.

Agnini, C., Backman, J., Boscolo Galazzo, F., Condon, D.J., Fornaciari, E., Galeotti, S., et al., 2019, *Proposal for the Global Standard Stratotype-Section and Point (GSSP) for the Base of the Priabonian Stage at the Alano Section (Italy)*. Submitted to the International Commission on Stratigraphy.

Aguilar, J.-P., Legendre, S., and Michaux, J. (eds), 1997, *Actes du Congres BiochroM'97—Biochronologie Mammalienne du Cénozoïque en Europe et Domaines Reliés*. Montpellier: Mémoires et Traveaux de l'Institut de Montpellier de l'École Pratique des Hautes Études, 21 Institut de Montpellier, 181 p.

Ali, J.R., and Jolley, D.W., 1996, Chronostratigraphic framework for the Thanetian and lower Ypresian deposits of southern England. In: Knox, R.W.O.B., Corfield, R.M., and Dunay, R.E. (eds), Correlation of the Early Paleogene in Northwest Europe, *Geological Society Special Publication*, **101**: 129–144.

Ali, J.R., King, C., and Hailwood, E.A., 1993, Magnetostratigraphic calibration of early Eocene depositional sequences in the southern North

Sea Basin. In: Hailwood, E.A., and Kidd, R.B. (eds), High Resolution Stratigraphy, *Geological Society Special Publication*, **70**: 99–125.

Alvarez, L.W., Alvarez, W., Asaro, F., and Michel, H.V., 1980, Extraterrestrial cause for the Cretaceous-Tertiary extinction. *Science*, **208**: 1095–1108.

Anderson, A.J., 1961, Gliederung und paläogeographische Entwicklung der Chattischen Stufe (Oberoligozän) im Nordseebecken. *Meyniana*, **10**: 118–146.

Angori, E., and Monechi, S., 1996, High-resolution calcareous nannofossil biostratigraphy across the Paleocene-Eocene boundary at Caravaca (southern Spain). In: Aubry, M.-P., and Benjamini, C. (eds), Paleocene-Eocene Boundary Events in Space and Time, *Israel Journal of Earth Sciences*, **44**: 197–206.

Arreguin-Rodriguez, G.J., Thomas, E., D'haenens, S., Speijer, R.P., and Alegret, L., 2018, Early Eocene deep-sea benthic foraminiferal faunas: recovery from the Paleocene-Eocene Thermal Maximum extinction in a greenhouse world. *PLoS One*, **13**: e0193167.

Aubry, M.-P., 1992, Late Paleogene calcareous nannoplankton evolution: a tale of climatic deterioration. In: Prothero, D.R., and Berggren, W.A. (eds), *Eocene-Oligocene Climatic and Biotic Evolution*. Princeton, NJ: Princeton University Press, pp. 272–309.

Aubry, M.-P., 1996, Towards an upper Paleocene-lower Eocene high resolution stratigraphy based on calcareous nannofossil stratigraphy. In: Aubry, M.-P., and Benjamini, C. (eds), Paleocene-Eocene Boundary Events in Space and Time, *Israel Journal of Earth Sciences*, **44**: 239–253.

Aubry, M.-P., 1998, Early Paleogene calcareous nannoplankton evolution: a tale of climatic amelioration. In: Aubry, M.-P., Lucas, S.G., and Berggren, W.A. (eds), *Late Paleocene-Early Eocene Climatic and Biotic Events in the Marine and Terrestrial Records*. New York: Columbia University Press, pp. 158–203.

Aubry, M.-P., Lucas, S.G., and Berggren, W.A. (eds), 1998, *Late Paleocene-Early Eocene Climatic and Biotic Events in the Marine and Terrestrial Records*. New York: Columbia University Press, 513 p.

Aubry, M.-P., Berggren, W., Van Couvering, J.A., Ali, J., Brinkhuis, H., and Cramer, B., et al., 2003, Chronostratigraphic terminology at the Paleocene/Eocene boundary. In: Wing, S.L., Gingerich, P.D., Schmitz, B., and Thomas, E. (eds), Causes and Consequences of Globally Warm Climates in the Early Paleogene, *Geological Society of America Special Paper*, **369**: 551–566.

Aubry, M.-P., Ouda, K., Dupuis, C., Berggren, W.A., Van Couvering, J.A., Ali, J., et al., 2007, The Global Standard Stratotype-section and Point (GSSP) for the base of the Eocene Series in the Dababiya section (Egypt). *Episodes*, **30**: 271–286.

Barbin, V., 1988, The Eocene-Oligocene transition in shallow-water environment: the Priabonian Stage type area (Vicentin, northern Italy). In: Premoli Silva, I., Coccioni, R., and Montanari, A. (eds), *The Eocene-Oligocene Boundary in the Marche-Umbria Basin (Italy)*. Ancona: International Subcommission on Paleogene Stratigraphy, pp. 163–171.

Barnet, J.S.K., Littler, K., Westerhold, T., Kroon, D., Leng, M.J., Bailey, I., et al., 2019, A high-fidelity benthic stable isotope record of Late Cretaceous-Early Eocene climate change and carbon cycling. *Paleoceanography and Paleoclimatology*, **34**: 672–691.

Beddow, H.M., Liebrand, D., Wilson, D.S., Hilgen, F.J., Sluijs, A., Wade, B.S., et al., 2018, Astronomical tunings of the Oligocene-Miocene transition from Pacific Ocean Site U1334 and implications for the carbon cycle. *Climate of the Past*, **14**: 255–270, https://doi.org/10.5194/cp-14-255-2018.

Berggren, W.A., 1994, In defense of the Selandian. *GFF*, **116**: 44–46.

Berggren, W.A., and Miller, K.G., 1989, Cenozoic bathyal and abyssal calcareous benthic foraminiferal zonation. *Micropaleontology*, **35**: 308–320.

Berggren, W.A., and Pearson, P.N., 2005, A revised tropical to subtropical Paleogene planktonic foraminiferal zonation. *Journal of Foraminiferal Research*, **35**: 279–298.

Berggren, W.A., Kent, D.V., and Flynn, J.J., 1985, Paleogene geochronology and chronostratigraphy. In: Snelling, N.J. (ed), The Chronology of the Geological Record, *Geological Society London, Memoir*, **10**: 141–195.

Berggren, W.A., Kent, D.V., Swisher, C.C., III, and Aubry, M.-P., 1995, A revised Cenozoic geochronology and chronostratigraphy. In: Berggren, W.A., Kent, D.V., Aubry, M.-P., Hardenbol, and J. (eds), Geochronology, time scales and Global Stratigraphic Correlation, *SEPM Special Publication*, **54**: 129–212.

Berggren, W.A., Wade, B.S., and Pearson, P.N., 2018, Oligocene chronostratigraphy and planktonic foraminiferal biostratigraphy: historical review and current state-of-the-art. In: Wade, B.S. et al. (eds), Atlas of Oligocene Planktonic Foraminifera, *Cushman Foundation for Foraminiferal Research Special Publication*, **46**: 29–54.

Bernaola, G., Martín-Rubio, M., and Baceta, J.I., 2009, New high resolution calcareous nannofossil analysis across the Danian/Selandian transition at the Zumaia section: comparison with South Tethys and Danish sections. *Geologica Acta*, **7**: 79–92.

Bernaola, G., Baceta, J.I., Orue-Etxebarria, X., Alegret, L., Martin-Rubio, M., Arostegui, J., et al., 2007, Evidence of an abrupt environmental disruption during the mid-Paleocene biotic event (Zumaia section, western Pyrenees). *Geological Society of America Bulletin*, **119**: 785–795.

Beyrich, E., 1854, Über die Stellung der hessischen Tertiärbildungen. *Berichte der Verhandlungen der königlichen Preussischen Akademie der Wissenschaften*, **1854**: 640–666.

Bignot, G., Curry, D., and Pomerol, C., 1997, Le dossier selandien. *Bulletin d'information des géologues du Bassin de Paris*, **34**: 3–11.

Bijl, P.K., Bendle, J.A.P., Bohaty, S.M., Pross, J., Schouten, S., Tauxe, L., et al., Expedition 318 Scientists, 2013a, Eocene cooling linked to early flow across the Tasmanian Gateway. *Proceedings of the National Academy of Sciences of the United States of America*, **110**: 9645–9650.

Bijl, P.K., Sluijs, A., and Brinkhuis, H., 2013b, A magneto- and chemostratigraphically calibrated dinoflagellate cyst zonation of the early Paleogene South Pacific Ocean. *Earth-Science Reviews*, **124**: 1–31.

Bijl, P.K., Houben, A.J.P., Hartman, J.D., Pross, J., Salabarnada, A., Escuttia, C., et al., 2018, Paleoceanography and ice sheet variability offshore Wilkes Land, Antarctica—Part 2: Insights from, Oligocene-Miocene dinoflagellate cyst assemblages. *Climate of the Past*, **14**: 1015–1033.

Bjørklund, K.R., 1976, Radiolaria from the Norwegian Sea, Leg 38 of the Deep Sea Drilling Project. In: Talwani, M., Udintsev, G., et al. (eds), *Initial Reports of the Deep Sea Drilling Project, Leg 38*. Washington, DC: U.S. Government Printing Office, pp. 1101–1168.

Blondeau, A., Cavelier, C., Feugueur, L., and Pomerol, C., 1965, Stratigraphie du Paléogène du Bassin du Paris en relation avec les bassins avoisinants. *Bulletin de la Société Géologique de France*, **7**: 200–221.

Blow, W.H., 1969, Late middle Eocene to Recent planktonic foraminiferal biostratigraphy. In: Brönnimann, P., and Renz, H.H. (eds), *Proceedings of the First International Conference on Planktonic Microfossils (Geneva, 1967)*. Leiden: E.J. Brill, pp. 199–422.

Blow, W.H., 1979, *The Cainozoic Globigerinida; A Study of the Morphology, Taxonomy, Evolutionary Relationships and the Stratigraphical Distribution of some Globigerinida (Mainly Globigerinacea)*, Vols. 1–3. Leiden: E.J. Brill, 1413 p.

Bohaty, S.M., and Zachos, J.C., 2003, Significant Southern Ocean warming event in the late middle Eocene. *Geology*, **31**: 1017–1020.

Bohaty, S.M., Zachos, J.C., Florindo, F., and Delaney, M.L., 2009, Coupled greenhouse warming and deep-sea acidification in the middle Eocene. *Paleoceanography*, **24**: PA2207.

Bolli, H.M., 1957, The genera Globigerina and Globorotalia in the Paleocene-Lower Eocene Lizard Springs formation of Trinidad, B.W.I. In: Loeblich, A.R. (ed), Studies in Foraminifera. *U.S. National Museum Bulletin*, **215**: 61–81.

Bolli, H.M., 1966, Zonation of Cretaceous to Pliocene marine sediments based on planktonic foraminifera. *Boletin Informativo—Asociación Venezolana de Geología, Minería y Petróleo*, **9**: 3–32.

Bornemann, A., Schulte, P., Sprong, J., Steurbaut, E., Youssef, M., and Speijer, R.P., 2009, Latest Danian carbon isotope anomaly and associated environmental change in the southern Tethys (Nile basin, Egypt). *Journal of the Geological Society*, **166**: 1135–1142.

Boscolo-Galazzo, F., Capraro, L., Fornaciari, E., Giusberti, L., and Luciani, V., 2019, The Forada section (northeastern Italy): a candidate Auxiliary Boundary Stratigraphic Section and Point (ASSP) for the base of the Ypresian. *STRATI 2019, Third International Congress on Stratigraphy, 2–5 July, 2019, Milano, Italy, Abstract Book*, 307 p.

BouDagher-Fadel, M.K., 2018, Evolution and Geological Significance of Larger Benthic Foraminifera., second ed. London: UCL Press, 693 p.

BouDagher-Fadel, M.K., and Banner, F.T., 1999, Revision of the stratigraphic significance of the Oligocene-Miocene "letter-stages". *Revue de Micropaléontologie*, **42**: 93–97.

BouDagher-Fadel, M.K., and Price, G.D., 2014, The phylogenetic and palaeogeographic evolution of the nummulitoid larger benthic foraminifera. *Micropaleontology*, **60**: 483–508.

BouDagher-Fadel, M.K., Price, G.D., Hu, X., and Li, J., 2015, Late Cretaceous to early Paleogene foraminiferal biozones in the Tibetan Himalayas, and a pan-Tethyan foraminiferal correlation scheme. *Stratigraphy*, **12**: 67–91.

Boulila, S., Galbrun, B., Miller, K.G., Pekar, S.F., Browning, J.V., Laskar, J., et al., 2011, On the origin of Cenozoic and Mesozoic "third-order" eustatic sequences. *Earth-Science Reviews*, **109**: 94–112.

Boulila, S., Vahlenkamp, M., De Vleeschouwer, D., Laskar, J., Yamamoto, Y., Pälike, H., et al., 2018, Towards a robust and consistent middle Eocene astronomical timescale. *Earth and Planetary Science Letters*, **486**: 94–107, https://doi.org/10.1016/j.epsl.2018.01.003.

Bowen, G.J., Koch, P.L., Gingerich, P.D., Norris, R.D., Bains, S., and Corfield, R.M., 2001, Refined isotope stratigraphy across the continental Paleocene-Eocene boundary on Polecat Bench in the northern Bighorn Basin. In: Gingerich, P.D. (ed), Paleocene-Eocene Stratigraphy and Biotic Change in the Bighorn and Clarks Fork Basins, Wyoming, *Papers on Paleontology*, **33**: 73–88.

Bown, P.R., 2005, Palaeogene calcareous nannofossils from the Kilwa and Lindi areas of coastal Tanzania (Tanzania Drilling Project 2003-4). *Journal of Nannoplankton Research*, **27**: 21–95.

Bown, P.R., and Dunkley Jones, T., 2006, New Palaeogene calcareous nannofossil taxa from coastal Tanzania: Tanzania Drilling Project Sites 11–14. *Journal of Nannoplankton Research*, **28**: 17–34.

Brabb, E.E., 1968, Comparison of the Belgian and German Oligocene sequences for the purpose of selecting a stratotype. *Mémoires du Bureau de Recherches Géologiques et Minières*, **69**: 77–82.

Bralower, T.J., Premoli Silva, I., and Malone, M.J., Scientific Participants of Leg 198, 2002, New evidence for abrupt climate change in the Cretaceous and Paleogene; An Ocean Drilling Program expedition to Shatsky Rise, Northwest Pacific. *GSA Today*, **12**: 4–10.

Bramlette, M.N., and Riedel, W.R., 1954, Stratigraphic value of discoasters and some other microfossils related to recent coccolithophores. *Journal of Paleontology*, **28**: 385–403.

Bramlette, M.N., and Sullivan, F.R., 1961, Coccolithophorids and related nannoplankton of the early Tertiary in California. *Micropaleontology*, **7**: 129–188.

Bramlette, M.N., and Wilcoxon, J.A., 1967, Middle Tertiary calcareous nannoplankton of the Cipero Section, Trinidad. *Tulane Studies in Geology and Paleontology*, **5**: 93–131.

BRGM, 1968, Colloque sur l'Eocène, Vols. I–III. *Mémoires Bureau de Recherches Géologique et Minière, 58(I), 59(II), 69(III)*. Orléans: BRGM.

Brinkhuis, H., 1994, Late Eocene to early Oligocene dinoflagellate cysts from the Priabonian type-area (Northeast Italy); biostratigraphy and paleoenvironmental interpretation. *Palaeogeography, Palaeoclimatology, Palaeoecology*, **107**: 121–163.

Brinkhuis, H., and Biffi, U., 1993, Dinoflagellate cyst stratigraphy of the Eocene/Oligocene transition in central Italy. *Marine Micropaleontology*, **22**: 131–183.

Brinkhuis, H., Sengers, S., Sluijs, A., Warnaar, J., and Williams, G.L., 2003, Latest Cretaceous to earliest Oligocene, and Quaternary dinoflagellate cysts from ODP Site 1172, East Tasman Plateau. In: Exon, N.F., Kennett, J.P., and Malone, M.J. (eds), *Proceedings of the Ocean Drilling Program, Scientific Results, Leg 189*. College Station, TX: Ocean Drilling Program, pp. 1–48.

Bujak, J.P., and Brinkhuis, H., 1998, Global warming and dinocyst changes across the Paleocene/Eocene epoch boundary. In: Aubry, M.-P., Lucas, S.G., and Berggren, W.A. (eds), *Late Paleocene-early Eocene Climatic and Biotic Events in the Marine and Terrestrial Records*. New York: Columbia University Press, pp. 277–295.

Burbank, D.W., Puigdefàbregas, C., and Muñoz, J.A., 1992, The chronology of the Eocene tectonic and stratigraphic development of the eastern Pyrenean foreland basin, northeast Spain. *Geological Society of America Bulletin*, **104**: 1101–1120.

Bystricka, H., 1965, Der stratigraphische Wert von Discoasteriden im Palaeogen der Slowakei. *Geologiske Sbornik (Bratislava)*, **16**: 7–10.

Cahuzac, B., and Poignant, A., 1997, Essai de biozonation de l'Oligo-Miocène dans les bassins européens à l'aide des grands foraminifères néritiques. *Bulletin de la Société Géologique de France*, **168**: 155–169.

Cande, S.C., and Kent, D.V., 1992, A new geomagnetic polarity time scale for the late Cretaceous and Cenozoic. *Journal of Geophysical Research*, **97**: 13917–13951, https://doi.org/10.1029/92jb01202.

Cande, S.C., and Kent, D.V., 1995, Revised calibration of the Geomagnetic Polarity time scale for the Late Cretaceous and Cenozoic. *Journal of Geophysical Research*, **100**: 6093–6095, https://doi.org/10.1029/94jb03098.

Cavelier, C., and Pomerol, C., 1977, Proposition d'une échelle stratigraphique standard pour le Paléogène. *Newsletters on Stratigraphy*, **6**: 56–65.

Cavelier, C., and Pomerol, C., 1983, Echelle de corrélation stratigraphique du Paléogène; stratotypes, étages standards, biozones, chimiozones et anomalies magnétiques. *Géologie de la France*, **3**: 261–262.

Cavelier, C., and Pomerol, C., 1986, Stratigraphy of the Paleogene. *Bulletin de la Société Géologique de France, Huitième Série*, **2**: 255–265.

Clyde, W.C., Hamzi, W., Finarelli, J.A., Wing, S.L., Schankler, D., and Chew, A., 2007, Basin-wide magnetostratigraphic framework for the Bighorn Basin, Wyoming. *Geological Society of America Bulletin*, **119**: 848–859.

Clyde, W.C., Tong, Y., Snell, K.E., Bowen, G.J., Ting, S., Koch, P.L., et al., 2008, An integrated stratigraphic record from the Paleocene of the Chijiang Basin, Jiangxi Province (China): implications for mammalian turnover and Asian block rotations. *Earth and Planetary Science Letters*, **269**: 553–563.

Clyde, W.C., Ting, S., Snell, K.E., Bowen, G.J., Tong, Y., Koch, P.L., et al., 2010, New paleomagnetic and stable isotope results from the Nanxiong Basin, China: implications for the K/T boundary and the timing of Paleocene mammalian turnover. *Journal of Geology*, **118**: 131–143.

Clyde, W.C., Wilf, P., Iglesias, A., Slingerland, R.L., Barnum, T., Bijl, P.K., et al., 2014, New age constraints for the salamanca formation and lower Rio Chico Group in the western San Jorge Basin, Patagonia, Argentina: implications for Cretaceous-Paleogene extinction recovery and land mammal age correlations. *Geological Society of America Bulletin*, **126**: 289–306.

Coccioni, R., Monaco, P., Monechi, S., Nocchi, M., and Parisi, G., 1988, Biostratigraphy of the Eocene-Oligocene boundary at Massignano (Ancona, Italy). In: Premoli Silva, I., Coccioni, R., and Montanari, A. (eds), *The Eocene-Oligocene Boundary in the Marche-Umbria Basin (Italy)*. Ancona: International Subcommission on Paleogene Stratigraphy, pp. 59–80.

Coccioni, R., Marsili, A., Montanari, A., Bellanca, A., Neri, R., Bice, D.M., et al., 2008, Integrated stratigraphy of the Oligocene pelagic sequence in the Umbria-Marche basin (northeastern Apennines, Italy): a potential Global Stratotype Section and Point for the Rupelian/Chattian boundary. *Geological Society of America Bulletin*, **120**: 487–511.

Coccioni, R., Bancala, G., Catanzarit, R., Fornaciari, E., Frontalini, F., Giusberti, L., et al., 2012, An integrated stratigraphic record of the Palaeocene-lower Eocene at Gubbio (Italy): new insights into the early Palaeogene hyperthermals and carbon isotope excursions. *Terra Nova*, **24**: 380–386.

Coccioni, R., Montanari, A., Bice, D., Brinkhuis, H., Deino, A., Frontalini, F., et al., 2018, The Global Stratotype Section and Point (GSSP) for the base of the Chattian Stage (Paleogene System, Oligocene Series) at Monte Cagnero, Italy. *Episodes*, **41**: 17–32, https://doi.org/10.18814/epiiugs/2018/v41i1/018003.

Comte, B., Sabatier, M., Marandat, B., and Vianey-Liaud, M., 2012, Les rongeurs de Chery-Chartreuve et Rocourt Saint-Martin (est du Bassin de Paris: Aisne, France): Leur place parmi les faunes de l'Eocene moyen d'Europe. *Palaeovertebrata*, **37**: 167–271.

Costa, E., Garces, M., Lopez-Blanco, M., Serra-Kiel, J., Bernaola, G., Cabrera, L., et al., 2013, The Bartonian-Priabonian marine record of the eastern Pyrenean foreland basin (NE Spain): a new calibration of the larger foraminifers and calcareous nannofossil calibration. *Geologica Acta*, **11**: 177–193.

Costa, L.L., and Manum, S.B., 1988, Dinoflagellate cysts: the description of the international zonation of the Paleogene (D1–D15) and the Miocene (D16–D20). *Geologisches Jahrbuch, Reihe A*, **100**: 321–330.

Cotton, L.J., Zakrevskaya, E.Y., van der Boon, A., Asatrayan, G., Hayrapetyan, F., Israyelyan, A., et al., 2017, Integrated stratigraphy of the Priabonian (upper Eocene) Urtsadzor section, Armenia. *Newsletters on Stratigraphy*, **50**: 269–295.

Cramer, B.S., Wright, J.D., Kent, D.V., and Aubry, M.-P., 2003, Orbital climate forcing of $\delta^{13}C$ excursions in the late Paleocene-early Eocene (chrons C24n-C25n). *Paleoceanography*, **18**: 1–21 PA1097.

Cramer, B.S., Toggweiler, J.R., Wright, J.D., Katz, M.E., and Miller, K.G., 2009, Ocean overturning since the Late Cretaceous: inferences from a new benthic foraminiferal isotope compilation. *Paleoceanography*, **24**: PA4216.

Cramwinckel, M.J., Huber, M., Kocken, I.J., Agnini, C., Bijl, P.K., Bohaty, S.M., et al., 2018, Synchronous tropical and polar temperature evolution in the Eocene. *Nature*, **559**: 382–386.

Crouch, E.M., Brinkhuis, H., Visscher, H., Adatte, T., and Bolle, M.-P., 2003, Late Paleocene-early Eocene dinoflagellate cyst records from the Tethys; further observations on the global distribution of Apectodinium. In: Wing, S.L., Gingerich, P.D., Schmitz, B., and Thomas, E. (eds), Causes and Consequences of Globally Warm Climates in the Early Paleogene, *Geological Society of America, Special Paper*, **369**: 113–131.

Dallanave, E., Agnini, C., Bachtadse, V., Muttoni, G., Crampton, J.S., Strong, C.P., et al., 2015, Early to middle Eocene magneto-biochronology of the southwest Pacific Ocean and climate influence on sedimentation: insights from the Mead Stream section, New Zealand. *Geological Society of America Bulletin*, **127**: 643–660.

Dashzeveg, D., 1993, Asynchronism of the main mammalian faunal events near the Eocene-Oligocene boundary. *Tertiary Research*, **14**: 141–149.

Dawber, C.F., Tripati, A.K., Gale, A.S., MacNiocaill, C., and Hesselbo, S.P., 2011, Glacioeustasy during the middle Eocene? Insights from the stratigtraphy of the Hampshire Basin, UK. *Palaeogeography, Palaeoclimatology, Palaeoecology*, **300**: 84–100.

DeConto, R.M., and Pollard, D., 2003, Rapid Cenozoic glaciation of Antarctica induced by declining atmospheric CO_2. *Nature*, **421**: 245–249.

De Geyter, G., De Man, E., Herman, J., Jacobs, P., Moorkens, T., and Steurbaut, E., et al., 2006, Disused Paleogene regional stages from Belgium; Montian, Heersian, Landenian, Paniselian, Bruxellian, Laekenian, Ledian, Wemmelian and Tongrian. In: Dejonghe, L. (ed), Chronostratigraphic Units Named From Belgium and Adjacent Areas, *Geologica Belgica*, **9**: 203–213.

De Lapparent, A., 1883, *Traité de Géologie*. Paris: Savy, 280 p.

De Man, E., Ivany, L., and Vandenberghe, N., 2004a, Stable oxygen in isotope record of the Eocene-Oligocene transition in the southern North Sea Basin; positioning the Oi-1 event. In Vandenberghe, N. (ed), Symposium on the Paleogene Preparing for Modern Life and Climate, *Netherlands Journal of Geosciences*, **83**: 193–197.

De Man, E., Van Simaeys, S., De Meuter, F., King, C., and Steurbaut, E., 2004b, Oligocene benthic foraminiferal zonation for the southern North Sea Basin. *Bulletin de l'Institut Royal des Sciences Naturelles de Belqique, Sciences de la Terre*, **74**: 177–195.

De Man, E., Van Simaeys, S., Vandenberghe, N., Harris, W.B., and Wampler, J.M., 2010, On the nature and chronostratigraphic position of the Rupelian and Chattian stratotypes in the southern North Sea basin. *Episodes*, **33**: 3–14.

Deprez, A., Jehle, S., Bornemann, A., and Speijer, R.P., 2017, Differential response at the seafloor during Palaeocene and Eocene warming events at Walvis Ridge, Atlantic Ocean (ODP Site 1262). *Terra Nova*, **29**: 71–76.

Desor, F., 1847, Sur le terrain danien, nouvel étage de la Craie. *Bulletin de la Société Géologique de France*, **4**: 179–182.

Dewalque, C., 1868, *Prodrome d'une description géologique de la Belgique*. Bruxelles and Liège: Librairie Polytechnique de Decq, 442 p.

D'haenens, S., Bornemann, A., Claeys, P., Röhl, U., Steurbaut, E., and Speijer, R.P., 2014, A transient deep-sea circulation switch during Eocene Thermal Maximum 2 (ETM 2). *Paleoceanography*, **29**: 1–19 PA2567.

Dickens, G.R., Castillo, M.M., and Walker, J.C.G., 1997, A blast of gas in the latest Paleocene; simulating first-order effects of massive dissociation of oceanic methane hydrate. *Geology*, **25**: 259–262.

Dinarès-Turell, J., 2019, Coherent new orbital tuning of the middle Eocene Contessa section (Umbrian Apennines, Italy) and significance for the Bartonian Stage GSSP. *STRATI 2019, Third International Congress on Stratigraphy, 2–5 July, 2019, Milano, Italy, Abstract Book*, 309. p.

Dinarès-Turell, J., Baceta, J.I., Pujalte, V., Orue-Etxebarria, X., Bernaola, G., and Lorito, S., 2003, Untangling the Palaeocene climatic rhythm; an astronomically calibrated early Palaeocene magnetostratigraphy and biostratigraphy at Zumaia (Basque Basin, northern Spain). *Earth and Planetary Science Letters*, **216**: 483–500.

Dinarès-Turell, J., Baceta, J.I., Bernaola, G., Orue-Extebarria, X., and Pujalte, V., 2007, Closing the Mid-Palaeocene gap: toward a complete astronomically tuned Palaeocene Epoch and Selandian and Thanetian GSSPs at Zumaia (Basque basin, W Pyrenees). *Earth and Planetary Science Letters*, **262**: 450–467.

Dinarès-Turell, J., Stoykova, K., Baceta, J.L., Ivanov, M., and Pujalte, V., 2010, High-resolution intra- and interbasinal correlation of the Danian-Selandian transition (Early Paleocene): the Bjala section (Bulgaria) and the Selandian GSSP at Zuamaia (Spain). *Palaeogeography, Palaeoclimatology, Palaeoecology*, **297**: 511–533.

Dinarès-Turell, J., Pujalte, V., Stoykova, K., Baceta, J.I., and Ivanov, M., 2012, The Palaeocene "top chron C27n" transient greenhouse episode: evidence from marine pelagic Atlantic and peri-Tethyan sections. *Terra Nova*, **24**: 477–486.

Dinarès-Turell, J., Westerhold, T., Pujalte, V., Röhl, U., and Kroon, D., 2014, Astronomical calibration of the Danian stage (Early Paleocene) revisited: settling chronologies of sedimentary records across the Atlantic and Pacific Oceans. *Earth and Planetary Science Letters*, **405**: 119–131, https://doi.org/10.1016/j.epsl.2014.08.027.

Dinarès-Turell, J., Martinez-Braceras, N., and Payros, A., 2018, High-resolution integrated cyclostratigraphy from the Oyambre Section (Cantabria, N Iberian Peninsula): constraints for orbital tuning and correlation of middle Eocene Atlantic deep-sea records. *Geochemistry, Geophysics, Geosystems*, **19**: 787–806, https://doi.org/10.1002/2017gc007367.

Dollfus, G.F., 1880, Essai sur l'étendue des terrains tertiaires dans le bassin Anglo-Parisien. *Bulletin de la Société Géologique de Normandie*, **6**: 584–605.

Drobne, K., Jez, J., Cosovic, V., Ogorelec, B., Stenni, B., Zakrevskaya, E., et al., 2014, Identification of the Paleocene-Eocene boundary based on larger foraminifers of the Palaeogene Adriatic carbonate platform. *In* Rocha, R., et al. (ed), *STRATI 2013*, Springer Geology, 89–93.

Drury, A.J., Westerhold, T., Frederichs, T., Tian, J., Wilkens, R., Channell, J.E.T., et al., 2017, Late Miocene climate and time scale reconciliation: accurate orbital calibration from a deep-sea perspective. *Earth and Planetary Science Letters*, **475**: 254–266, https://doi.org/10.1016/j.epsl.2017.07.038.

Dumont, A., 1839, Rapport sur les traveaux de la carte géologique pendant l'année 1839. *Bulletins de l'Académie Royale des Sciences, des Lettres et des Beaux*, **6**: 464–485.

Dumont, A., 1849, Rapport sur la carte géologique de la Belgique. *Bulletins de l'Académie Royale des Sciences, des Lettres et des Beaux*, **16**: 351–373.

Dumont, A., 1851, Sur la position géologique de l'argile rupelienne et sur le synchronisme des formations teriaires de la Belgique, de l'Angleterre et du Nord de la France. *Bulletins de l'Académie Royale des Sciences, des Lettres et des Beaux*, **18**: 179−195.

Dunn, R.E., Madden, R.H., Kohn, M.J., Schmitz, M.D., Strömberg, C.A.E., Carlini, A.A., et al., 2013, A new chronology for middle Eocene- early Miocene South American Land Mammal Ages. *Geological Society of America Bulletin*, **125**: 539−555.

Dupont-Nivet, G., Krijgsman, W., Langereis, C.G., Abels, H.A., Dai, S., and Fang, X., 2007, Tibetan plateau aridification linked to global cooling at the Eocene-Oligocene transition. *Nature*, **445**: 635−638.

Dupuis, C., Aubry, M.-P., Steurbaut, E., Berggren, W.A., Ouda, K., Magioncalda, R., et al., 2003, The Dababiya Quarry section; lithostratigraphy, clay mineralogy, geochemistry and paleontology. In Ouda, K., Aubry, M.-P. (eds), The Upper Paleocene-Lower Eocene of the Upper Nile Valley; Part 1, Stratigraphy. *Micropaleontology*, **49** (Suppl. 1): 41−59.

Edgar, K.M., Wilson, P.A., Sexton, P.F., Gibbs, S.J., Roberts, A.P., and Norris, R.D., 2010, New biostratigraphic, magnetostratigraphic and isotopic insights into the Middle Eocene Climatic Optimum in low latitudes. *Palaeogeography, Palaeoclimatology, Palaeoecology*, **297**: 670−682.

Ellison, R.A., Ali, J.R., Hine, N.M., and Jolley, D.W., 1996, Recognition of Chron C25n in the upper Paleocene Upnor Formation of the London Basin, UK. In: Knox, R.W.O.B., Corfield, R.M., and Dunay, R.E. (eds), Correlation of the Early Paleogene in Northwest Europe, *Geological Society, London, Special Publications*, **101**: 185−193.

Ernst, S.R., Guasti, E., Dupuis, C., and Speijer, R.P., 2006, Environmental perturbation in the southern Tethys across the Paleocene/ Eocene boundary (Dababiya, Egypt); foraminiferal and clay mineral records. *Marine Micropaleontology*, **60**: 89−111.

Escarguel, G., Marandat, B., and Legendre, S., 1997, Sur l'âge numérique des faunes de mammifères du Paléogène d'Europe occidentale, en particulier celles de l'Éocène inférieur et moyen. In: Aguilar, J.-P., Legendre, S., and Michaux, J. (eds), *Actes du Congrès BiochroM'97—Biochronologie Mammalienne du Cénozoïque en Europe et Domaines Reliés*. Mémoires et Traveaux de l'Institut de Montpellier de l'École Pratique des Hautes Études 21. Montpellier: Institut de Montpellier, pp. 443−460.

Fensome, R.A., and Williams, G.L., 2004, The Lentin and Williams Index of Fossil Dinoflagellates, 2004 ed. Houston, TX: American Association of Stratigraphic Palynologists, 909 p.

Fensome, R.A., and Munsterman, D.K., 2020, Chapter 3I - Dinoflagellates. *In* Gradstein, F.M., Ogg, J.G., Schmitz, M.D., and Ogg, G.M. (eds), *The Geologic Time Scale 2020*. **Vol. 1** (this book). Elsevier, Boston, MA.

Florindo, F., Wilson, G.S., Roberts, A.P., Sagnotti, L., and Verosub, K.L., 2005, Magnetostratigraphic chronology of a late Eocene to early Miocene glacimarine succession from the Victoria Land basin. *Global and Planetary Change*, **45**: 207−236.

Fluegeman, R.H., 2007, Unresolved issues in Cenozoic chronostratigraphy. *Stratigraphy*, **4**: 109−116.

Foreman, H.P., 1973, Radiolaria of Leg 10 with systematics and ranges for the families Amphipyndacidae, Artostrobiidae and Theorperidae. *In* Worzel, J.L., Bryant, W., et al. (eds), *Initial Reports of the Deep Sea Drilling Project, Leg 10*. Washington, DC: U.S. Government Printing Office, pp. 407−474.

Fornaciari, E., Agnini, C., Catanzariti, R., Rio, D., Bolla, E.M., and Valvasoni, E., 2010, Mid-latitude calcareous nannofossil biostratigraphy and biochronology across the middle to late Eocene transition. *Stratigraphy*, **7**: 229−264.

Francescone, F., Lauretano, V., Bouligand, C., Moretti, M., Sabatino, N., Schrader, C., et al., 2019, A 9 million-year-long astrochronological record of the early-middle Eocene corroborated by seafloor spreading rates. *Geological Society of America Bulletin*, **131**: 499−520, https://doi.org/10.1130/B32050.1.

Fuchs, T., 1894, Tertiärfossilien aus den kohlenführenden Miozänablagerungen der Umgebung von Krapina und Radobog und über die Stellung der sogenannten "Aquitanischen Stufe". *Mitteilungen aus dem Jahrbuche der Königlichen Ungarischen Geologischen Anstalt*, **10**: 161−175.

Funakawa, S., and Nishi, H., 2005, Late middle Eocene to early Oligocene radiolarian biostratigraphy in the Southern Ocean (Maud Rise, ODP Leg 113, Site 689). *Marine Micropaleontology*, **54**: 213−247.

Galeotti, S., Kaminski, M.A., Coccioni, R., and Speijer, R.P., 2004, High-resolution deep-water agglutinated foraminiferal record across the Paleocene/Eocene transition in the Contessa Road section (central Italy). In: Bubik, M., and Kaminski, M.A. (eds), Proceedings of the Sixth International Workshop on Agglutinated Foraminifera, *Grzybowski Foundation Special Publication*, **8**: 83−103.

Galeotti, S., Sahy, D., Agnini, C., Condon, D., Fornaciari, E., Francescone, F., et al., 2019, Astrochronology and radio-isotopic dating of the Alano di Piave section (NE Italy), candidate GSSP for the Priabonian Stage (late Eocene). *Earth and Planetary Science Letters*, **525**: 1−10, https://doi.org/10.1016/j.epsl.2019.115746.

Gignoux, M., 1950, *Géologie stratigraphique*, fourth ed. Paris: Masson, 735 p.

Gingerich, P.D., 2006, Environment and evolution through the Paleocene-Eocene thermal maximum. *Trends in Ecology & Evolution*, **21**: 246−253.

Giusberti, L., Rio, D., Agnini, C., Backman, J., Fornaciari, E., Tateo, F., et al., 2007, Mode and tempo of the Paleocene-Eocene Thermal Maximum in an expanded section from the Venetian pre-Alps. *Geological Society of America Bulletin*, **119**: 391−412.

Görges, 1957, Die Mollusken der oberoligozänen Schichten des Doberges bei Bünde in Westfalen. *Paläontologische Zeitschrift*, **31**: 116−134.

Gradstein, F.M., Kaminski, M.A., Berggren, W.A., Kristiansen, I.L., D'Ioro, M.A., 1994, Cenozoic biostratigraphy of the North Sea and Labrador Shelf. *Micropaleontology*, **40** (Suppl.), 152 p.

Gradstein, F.M., Ogg, J.G., and Smith, A.G. (eds), 2004. *A Geologic Time Scale 2004*. Cambridge University Press, Cambridge, 589 pp.

Gradstein, F.M., Ogg, J.G., Schmitz, M.D., and Ogg G.M. (co-ordinators), 2012, *The Geologic Time Scale 2012*. Elsevier Publ., 1176 pp.

Gradstein, F.M., and Ogg, J.G., 2020, Chapter 2 - The chronostratigraphic scale. *In* Gradstein, F.M., Ogg, J.G., Schmitz, M.D., and Ogg, G.M. (eds), *The Geologic Time Scale 2020*. **Vol. 1** (this book). Elsevier, Boston, MA.

Grossman, E.L., and Joachimski, M.M., 2020, Chapter 10 - Oxygen isotope stratigraphy. *In* Gradstein, F.M., Ogg, J.G., Schmitz, M.D., and Ogg, G.M. (eds), *The Geologic Time Scale 2020*. **Vol. 1** (this book). Elsevier, Boston, MA.

Guasti, E., and Speijer, R.P., 2008, *Acarinina multicamerata* n. sp. (Foraminifera): a new marker for the Paleocene-Eocene Thermal Maximum. *Journal of Micropalaeontology*, **27**: 5−12.

Guerstein, G.R., Chiesa, J.O., Guler, M.V., and Camacho, H.H., 2002, Bioestratigrafia basada en quistes de dinoflagelados de la Formacion Cabo Pena (Eoceno terminal—Oligoceno temprano), Tierra del Fuego, Argentina. *Revista Española de Micropaleontología*, **34**: 105−116.

Gunnell, G.F., Murphey, P.C., Stucky, R.K., Townsend, K.E.B., Robinson, P., Zonneveld, J.-P., et al., 2009, Biostratigraphy and biochronology of the latest Wasatchian, Bridgerian and Uintan North American Land Mammal "Ages". *Museum of Northern Arizona Bulletin*, **65**: 279–330.

Hansen, J.M., 1979, Age of the Mo-Clay Formation. *Bulletin of the Geological Society of Denmark*, **27**: 89–91.

Hardenbol, J., 1968, The 'Priabonian' type section (a preliminary note). *Mémoires du Bureau de Recherche Géologique et Minière*, **58**: 629–635.

Hardenbol, J., and Berggren, W.A., 1978, A new Paleogene numerical time scale. In: Cohee, G.V., Glaessner, M.F., and Hedberg, H.D. (eds), *American Association of Petroleum Geologists, Studies in Geology*, **6**: 213–234.

Hardenbol, J., Thierry, J., Farley, M.B., Jacquin, T., De Graciansky, P.-C., and Vail, P.R., 1998, Mesozoic–Cenozoic sequence chronostratigraphic framework of European basins. In: De Graciansky, P.-C., Hardenbol, J., Jacquin, T., Vail, P.R., and Farley, M.B. (eds), Sequence Stratigraphy of European Basins, *SEPM Special Publication*, **60**: 3–14.

Harland, W.B., Armstrong, R.L., Cox, A.V., Graig, L.A., Smith, A.G., Smith, D.G., 1990, *A Geologic time scale 1989*. Cambridge: Cambridge University Press, 263 p.

Hay, W.W., Mohler, H.P., Roth, P.H., Schmidt, R.R., and Boudreaux, J.E., 1967, Calcareous nannoplankton zonation of the Cenozoic of the Gulf Coast and Caribbean-Antillean area and transoceanic correlation. *Transactions of the Gulf Coast Association of Geological Societies*, **17**: 428–480.

Heilmann-Clausen, C., 1985, Dinoflagellate stratigraphy of uppermost Danian to Ypresian in the Viborg 1 Borehole, central Jylland, Denmark. *Danmarks Geologiske Undersøgelse*, **A7**: 1–89.

Heilmann-Clausen, C., 1988, The Danish Sub-Basin: Paleogene dinoflagellates. *Geologisches Jahrbuch, Reihe A*, **100**: 339–343.

Heilmann-Clausen, C., 2018, Observations of the dinoflagellate Wetzeliella in Sparnacian facies (Eocene) near Epernay, France, and a note on tricky acmes of Apectodinium. *Proceedings of the Geologists' Association*, On-line (in press): https://doi.org/10.1016/j.pgeola.2018.06.001.

Heirtzler, J.R., Dickson, G.O., Herron, E.M., Pitman, W.C., and Le Pichon, X., 1968, Marine magnetic anomalies, geomagnetic field reversals, and motions of the ocean floor and continents. *Journal of Geophysical Research*, **73**: 2119–2139.

Higgins, J.A., and Schrag, D.P., 2006, Beyond methane; towards a theory for the Paleocene-Eocene thermal maximum. *Earth and Planetary Science Letters*, **245**: 523–537.

Hilgen, F.J., Kuiper, K.F., and Lourens, L.J., 2010, Evaluation of the astronomical time scale for the Paleocene and earliest Eocene. *Earth and Planetary Science Letters*, **300**: 139–151.

Hilgen, F.J., Abels, H.A., Kuiper, K.F., Lourens, L.J., and Wolthers, M., 2015, Towards a stable astronomical time scale for the Paleocene: aligning Shatsky Rise with the Zumaia–Walvis Ridge ODP Site 1262 composite. *Newsletters on Stratigraphy*, **48**: 91–110.

Hollis, C.J., 1993, Latest Cretaceous to late Paleocene radiolarian biostratigraphy; a new zonation from the New Zealand region. In: Lazarus, D.B., and De Wever, P.D. (eds), Interrad VI, *Marine Micropaleontology*, **21**: 295–327.

Hollis, C.J., 1997, Cretaceous-Paleocene Radiolaria from eastern Marlborough, New Zealand. *Institute of Geological and Nuclear Sciences. Monograph*, **17**: 1–152.

Hollis, C.J., 2002, Biostratigraphy and paleoceanographic significance of Paleocene radiolarians from offshore eastern New Zealand. *Marine Micropaleontology*, **46**: 265–316.

Hollis, C.J., 2006, Radiolarian faunal change across the Paleocene-Eocene boundary at Mead Stream, New Zealand. *Eclogae Geologicae Helvetiae*, **99**: S79–S99.

Hollis, C.J., Pascher, K.M., Kamikuri, S.-I., Nishimura, A., Suzuki, N., and Sanfilippo, A., 2017, Towards an integrated cross-latitude event stratigraphy for Paleogene radiolarians. *Radiolaria Newsletter of the International Association of Radiolarists*, **40**: 288–289.

Hollis, C.J., Dunkley Jones, T., Anagnostou, E., Bijl, P.K., Cramwinckel, M.J., Cui, Y., et al., 2019, The DeepMIP contribution to PMIP4: methodologies for selection, compilation and analysis of latest Paleocene and early Eocene climate proxy data, incorporating version 0.1 of the DeepMIP database. *Geoscientific Model Development*, **12**: 3149–3206.

Hollis, C.J., Pascher, K.M., Kamikuri, S.-i, Nishimura, A., Suzuki, N., and Sanfilippo, A., 2020 (In press), Towards and integrated cross-latitude event stratigraphy for Paleogene radiolarians. *Stratigraphy*, in press.

Hooker, J.J., 1986, Mammals from the Bartonian (middle/late Eocene) of the Hampshire basin, southern England. *Bulletin of the British Museum (Natural History)*, **39**: 191–478.

Hooker, J.J., 1998, Mammalian faunal change across the Paleocene-Eocene transition in Europe. In: Aubry, M.-P., Lucas, S.G., and Berggren, W.A. (eds), *Late Paleocene-Early Eocene Climatic and Biotic Events in the Marine and Terrestrial Records*. New York: Columbia University Press, pp. 428–450.

Hooker, J.J., 2010a, The 'Grande Coupure' in the Hampshire Basin, UK: taxonomy and stratigraphy of the mammals on either side of this major Palaeogene faunal turnover. In: Whittaker, J.E., and Hart, M.B. (eds), Micropalaeontology, Sedimentary Environments and Stratigraphy: A Tribute to Dennis Curry (1921–2001), *The Micropalaeontological Society, Special Publications*, **4**: 147–215.

Hooker, J.J., 2010b, The mammal fauna of the Eocene Blackheath Formation of Abbey Wood, London. *Monograph of the Palaeontographical Society, London*, **164**: 1–162.

Hooker, J.J., 2015, A two-phase Mammalian Dispersal Event across the Paleocene-Eocene transition. *Newsletters on Stratigraphy*, **48**: 201–220.

Hooker, J.J., and King, C., 2019, The Bartonian unit stratotype (S. England): assessment of its correlation problems and potential. *Proceedings of the Geologists' Association*, **130**: 157–169.

Hooker, J.J., Grimes, S.T., Mattey, D.P., Collinson, M.E., and Sheldon, N.D., 2009, Refined correlation of the late Eocene-early Oligocene Solent Group and timing of its climate history. In: Koeberl, C., and Montanari, A. (eds), Late Eocene Earth: Hothouse, Icehouse, and Impacts, *Geological Society of America Special Paper*, **452**: 179–195.

Hörnes, M., 1853, Mitteilung an Prof. Bronn gerichtet, Wien, 3. Okt, 1853. *Neues Jahrbuch für Mineralogie, Geognosie, Geologie und Petrefaktenkunde*, 806–810.

Hornibrook, N.D., Brazier, R.C., Strong, C.P., 1989, Manual of New Zealand Permian to Pleistocene foraminiferal biostratigraphy. *New Zealand Geological Survey, Paleontological Bulletin*, **56**, 175 p.

Hottinger, L., 1960, Recherches sur les alvéolines du Paléocène et de l'Éocène. *Schweizerische Paläontologische Abhandlungen*: **75–76**, 243.

Hottinger, L., and Schaub, H., 1960, Zur Stufeneinteilung des Paläozäns und des Eozäns–Einführung der Stufen Ilerdien und Biarritzien. *Eclogae Geologicae Helvetiae*, **53**: 453–480.

Hubach, H., 1957, Das Oberoligozän des Dobergs bei Bünde in Westfalen. *Berichte der Naturhistorischen Gesellschaft*, **103**: 5–69.

Huber, B.T., and Quillévéré, F., 2005, Revised Paleogene planktonic foraminiferal biozonation for the Austral Realm. *Journal of Foraminiferal Research*, **35**: 299–314.

Hyland, E.G., Sheldon, N.D., and Cotton, J.M., 2015, Terrestrial evidence for a two-stage mid-Paleocene biotic event. *Palaeogeography, Palaeoclimatology, Palaeoecology*, **417**: 371–378.

Indans, J., 1965, Nachweis des Asterigerinen-Horizontes im Oberoligozän des Dobergs bei Bunde/Wstf. *Neues Jahrbuch für Geologie und Paläontologie*, **123**: 20–24.

Intxauspe-Zubiaurre, B., Martinez-Braceras, N., Payros, A., Ortiz, S., Dinarès-Turell, J., and Flores, J.-A., 2018, The last Eocene hyperthermal (Chron C19r event, ~41.5 ma): chronological and paleoenvironmental insights from a continental margin (Cape Oaymbre, N Spain). *Palaeogeography, Palaeoclimatology, Palaeoecology*, **505**: 198–216.

Janis, C.M. (ed), 1998, *Evolution of Tertiary Mammals of North America. Terrestrial Carnivores, Ungulates and Ungulate-like Mammals*, Vol. 1. Cambridge: Cambridge University Press, 691 p.

Janis, C.M., Gunnell, G.F., and Uhen, M.D. (eds), 2008, *Evolution of the Tertiary Mammals of North America, Volume 2: Small Mammals, Xenarthrans, and Marine Mammals*. Cambridge: Cambridge University Press, 795 p.

Jehle, S., Bornemann, A., Deprez, A., and Speijer, R.P., 2015, The impact of the Latest Danian Event on planktic foraminiferal faunas. *PLoS One*, **10**: e0141644.

Jenkins, D.G., 1985, Southern mid-latitude Paleocene to Holocene planktic foraminifera. In: Bolli, H.M., Saunders, J.B., and Perch-Nielsen, K. (eds), Plankton Stratigraphy. Cambridge: Cambridge University Press, pp. 263–282.

Jenkins, D.G., and Luterbacher, H.P., 1992, Paleogene stages and their boundaries: introductory remarks. *Neues Jahrbuch für Geologie und Paläontologie*, **186**: 1–5.

Jovane, L., Florindo, F., Coccioni, R., Dinarès-Turell, J., Marsili, A., Monechi, S., et al., 2007, The middle Eocene climatic optimum event in the Contessa Highway section, Umbrian Apennines, Italy. *Geological Society of America Bulletin*, **119**: 413–427.

Jovane, L., Sprovieri, M., Florindo, F., Coccioni, R., and Marsili, A., 2010, Astronomic calibration of the middle Eocene Contessa Highway section. *Earth and Planetary Science Letters*, **298**: 77–88.

Kamikuri, S.-I., Moore, T.C., Ogane, K., Suzuki, N., Pälike, H., and Nishi, H., 2012, Early Eocene to early Miocene radiolarian biostratigraphy for the low-latitude Pacific Ocean. *Stratigraphy*, **9**: 1–77.

Kamikuri, S.-I., Moore, T.C., Lyle, M., Ogane, K., and Suzuki, N., 2013, Early and middle Eocene radiolarian assemblages in the eastern equatorial Pacific Ocean (IODP Leg 320 Site U1331): faunal changes and implications for paleoceanography. *Marine Micropaleontology*, **98**: 1–13.

Kaminski, M.A., and Gradstein, F.M. (eds), 2005, Atlas of Paleogene Cosmopolitan Deep-water Agglutinated Foraminifera. *Grzybowski Foundation Special Publication*, 10, 548 p.

Kaminski, M.A., Kuhnt, W., and Radley, J.D., 1996, Palaeocene-Eocene deep water agglutinated foraminifera from the Numidian Flysch (Rif, northern Morocco); their significance for the palaeoceanography of the Gibraltar gateway. *Journal of Micropalaeontology*, **15**: 1–19.

Keeping, H., 1887, On the discovery of the *Nummulina elegans* zone at Whitecliff Bay, Isle of Wight. *Geological Magazine*, **4**: 70–72.

Keller, G., and Lindinger, M., 1989, Stable isotope, TOC and CaCO$_3$ record across the Cretaceous/Tertiary boundary at El Kef, Tunisia. *Palaeogeography, Palaeoclimatology, Palaeoecology*, **73**: 243–265.

Kelly, D.C., Bralower, T.J., Zachos, J.C., Premoli Silva, I., and Thomas, E., 1996, Rapid diversification of planktonic foraminifera in the tropical Pacific (ODP Site 865) during the late Paleocene thermal maximum. *Geology*, **24**: 423–426.

Kelly, T.S., Murphey, P.C., and Walsh, S.L., 2012, New records of small mammals from the middle Eocene Duchesne River Formation, Utah, and their implications for the Uintan-Duchesnean North American Land Mammal Age transition. *Paludicola*, **8**: 208–251.

Kennett, J.P., and Stott, L.D., 1991, Abrupt deep-sea warming, palaeoceanographic changes and benthic extinctions at the end of the Palaeocene. *Nature*, **353**: 225–229.

Khozyem, H., Adatte, T., Keller, G., Tantawy, A.A., and Spangenberg, J.E., 2014, The Paleocene-Eocene GSSP at Dababiya, Egypt—Revisited. *Episodes*, **37**: 78–86.

King, C., 2016, A Revised Correlation of Tertiary Rocks in the British Isles and Adjacent Areas of NW Europe. In: Gale, A.S., and Barry, T.L. (eds), *Geological Society Special Report*, **27**, 719 p.

King, D.J., and Wade, B.S., 2017, The extinction of *Chiloguembelina cubensis* in the Pacific Ocean: implications for defining the base of the Chattian (upper Oligocene). *Newsletters on Stratigraphy*, **50**: 311–339, https://doi.org/10.1127/nos/2016/0308.

Koch, P.L., Zachos, J.C., and Gingerich, P.D., 1992, Correlation between isotope records in marine and continental carbon reservoirs near the Palaeocene/Eocene boundary. *Nature*, **358**: 319–322.

Kochhann, K.G.D., Holbourn, A., Kunht, W., Channell, J.E.T., Lyle, M.W., Shackford, J.K., et al., 2016, Eccentricity pacing of eastern equatorial Pacific carbonate dissolution cycles during the Miocene Climatic Optimum. *Paleoceanography*, **31** (9), 1176–1192, https://doi.org/10.1002/2016PA002988.

Kocsis, L., Gheerbrant, E., Mouflih, M., Cappetta, H., Yans, J., and Amaghzaz, M., 2014, Comprehensive stable isotope investigation of marine biogenic apatite from the late Cretaceous—early Eocene phosphate series of Morocco. *Palaeogeography, Palaeoclimatology, Palaeoecology*, **394**: 74–88.

Köthe, A., 1990, Paleogene dinoflagellates from northwest Germany—Biostratigraphy and paleoenvironment. *Geologisches Jahrbuch, Reihe A*, **118**: 3–111.

Kozlova, G.E., 1999, *Paleogene Boreal Radiolarians From Russia (translated from Russian)*. St Petersburg: All-Russia Petroleum Research Exploration Institute (VNIGRI), 323 p.

Kraatz, B.P., and Geisler, J.H., 2010, Eocene-Oligocene transition in central Asia and its effects on mammalian evolution. *Geology*, **38**: 111–114.

Krause, J.M., Clyde, W.C., Ibanez-Meija, M., Schmitz, M., Barnum, T., Bellosi, E.S., et al., 2017, New age constraints for early Paleogene strata of central Patagonia, Argentina: implications for the timing of South American Land Mammal Ages. *Geological Society of America Bulletin*, **129**: 886–903.

Kuiper, K.F., Deino, A., Hilgen, F.J., Krijgsman, W., Renne, P.R., and Wijbrans, J.R., 2008, Synchronizing rock clocks of Earth history. *Science*, **320**: 500–504.

Laskar, J., 2020, Chapter 4 - Astrochronology. *In* Gradstein, F.M., Ogg, J.G., Schmitz, M.D., and Ogg, G.M. (eds), *The Geologic Time Scale 2020*. **Vol. 1** (this book). Elsevier, Boston, MA.

Laskar, J., Robutel, P., Joutel, F., Gastineau, M., Correia, A.C.M., and Levrard, B., 2004, A long-term numerical solution for the insolation quantities of the Earth. *Astronomy & Astrophysics*, **428**: 261–285, https://doi.org/10.1051/0004-6361:20041335.

Laskar, J., Fienga, A., Gastineau, M., and Manche, H., 2011a, La2010: a new orbital solution for the long-term motion of the Earth. *Astronomy & Astrophysics*, **532**, https://doi.org/10.1051/0004-6361/201116836.

Laskar, J., Gastineau, M., Delisle, J.B., Farres, A., and Fienga, A., 2011b, Strong chaos induced by close encounters with Ceres and Vesta. *Astronomy & Astrophysics*, **532**, https://doi.org/10.1051/0004-6361/201117504.

Lauretano, V., Hilgen, F.J., Zachos, J.C., and Lourens, L.J., 2016, Astronomically tuned age model for the early Eocene carbon isotope events: a new high-resolution $\delta^{13}C_{benthic}$ record of ODP Site 1263 between 49 and 54 Ma. *Newsletters on Stratigraphy*, **49**: 383–400.

Lenz, O.K., Wilde, V., Mertz, D.F., and Riegel, W., 2015, New palynology-based astronomical and revised $^{40}Ar/^{39}Ar$ ages for the Eocene maar lake of Messel (Germany). *International Journal of Earth Sciences*, **104**: 873–889.

Leupold, W., and van der Vlerk, I.M., 1931, The Tertiary. *Leidsche Geologische Mededelingen*, **5**: 611–648.

Liebrand, D., Beddow, H.M., Lourens, L.J., Pälike, H., Raffi, I., Bohaty, S.M., et al., 2016, Cyclostratigraphy and eccentricity tuning of the early Oligocene through early Miocene (30.1-17.1Ma): *Cibicides mundulus* stable oxygen and carbon isotope records from Walvis Ridge Site 1264. *Earth and Planetary Science Letters*, **450**: 392–405, https://doi.org/10.1016/j.epsl.2016.06.007.

Littler, K., Röhl, U., Westerhold, T., and Zachos, J.C., 2014, A high-resolution benthic stable-isotope record for the south Atlantic: implications for orbital-scale changes in late Paleocene-early Eocene climate and carbon cycling. *Earth and Planetary Science Letters*, **401**: 18–30.

Livsey, C.M., Babila, T.L., Robinson, M.M., and Bralower, T.J., 2019, The planktonic foraminiferal response to the Paleocene-Eocene thermal maximum on the Atlantic coastal plain. *Marine Micropaleontology*, **146**: 39–50.

Lourens, L.J., Wehausen, R., and Brumsack, H.J., 2001, Geological constraints on tidal dissipation and dynamical ellipticity of the Earth over the past three million years. *Nature*, **409**: 1029–1033, https://doi.org/10.1038/35059062.

Lourens, L.J., Sluijs, A., Kroon, D., Zachos, J.C., Thomas, E., Röhl, U., et al., 2005, Astronomical pacing of late Palaeocene to early Eocene global warming events. *Nature*, **435**: 1083–1087.

Luciani, V., Giusberti, L., Agnini, C., Backman, J., Fornaciari, E., and Rio, D., 2007, The Paleocene-Eocene Thermal Maximum as recorded by Tethyan planktonic foraminifera in the Forada section (northern Italy). *Marine Micropaleontology*, **64**: 189–214.

Luciani, V., Giusberti, L., Agnini, C., Fornaciari, E., Rio, D., and Spofforth, D.J.A., 2010, Ecological and evolutionary response of Tethyan planktonic foraminifera in the middle Eocene climatic optimum (MECO) from the Alano section (NE Italy). *Palaeogeography, Palaeoclimatology, Palaeoecology*, **292**: 82–95.

Luciani, V., Fornaciari, E., Papazzoni, C.A., Dallanave, E., Giusberti, L., Stefani, C., et al., 2019, Integrated stratigraphy at the Bartonian-Priabonian transition: correlation between shallow benthic and calcareous plankton zones (Varignano section, northern Italy). *Geological Society of America Bulletin*, **131**: 26.

Luterbacher, H.P., Ali, J.R., Brinkhuis, H., Gradstein, F.M., Hooker, J.J., Monechi, S., et al., 2004, The Paleogene Period. *In* Gradstein, F.M., Ogg, J.G., and Smith, A.G. (eds), *A Geologic time scale 2004*. Cambridge: Cambridge University Press, pp. 384–408.

Lyell, C., 1833, *Principles of Geology: being an Attempt to Explain the Former Changes of the Earth's Surface, by Reference to Causes Now in Operation*, Vol. III. John Murray: London, 398 p.

Mangerud, G., Dreyer, T., Soyseth, L., Martinsen, O., and Ryseth, A., 1999, High-resolution biostratigraphy and sequence development of the Palaeocene succession, Grane Field, Norway. *In* Jones, R.W., and Simmons, M.D. (eds), *Biostratigraphy in Production and Development Geology*, Geological Society, London, Special Publications, **152**: 167–184.

Mangin, J.-P., 1957, Remarques sur le terme Paléocène et sur la limite Crétacé-Tertiaire. *Comptes Rendus des Séances de la Société Géologique de France*, **14**: 319–321.

Mankinen, E.A., and Dalrymple, G.B., 1979, Revised geomagnetic polarity time scale for the interval 0–5m.y. BP. *Journal of Geophysical Research*, **84**: 615–626.

Martini, E., 1971, Standard Tertiary and Quaternary calcareous nannoplankton zonation. In: Farinacci, A. (ed), *Proceedings of the Second Planktonic Conference, Rome, 1970; Tecnoscienza, Roma*, pp. 739–785.

Mayer-Eymar, K., 1858, Versuch einer neuen Klassifikation der Tertiär-Gebilde Europa's. *Verhandlungen der Schweizer Naturforschenden Gesellschaft für die Gesamten Naturwissenschaften Trogen*, **42**: 165–199.

McInerney, F.A., and Wing, S.L., 2011, The Paleocene-Eocene thermal maximum: a perturbation of carbon cycle, climate, and biosphere with implications for the future. *Annual Review of Earth and Planetary Sciences*, **39**: 489–516.

Meng, J., and McKenna, M.C., 1998, Faunal turnovers of Paleogene mammals from the Mongolian Plateau. *Nature*, **394**: 364–367.

Michelsen, O., Thomsen, E., Danielsen, M., Heilmann-Clausen, C., Jordt, H., and Laursen, G., 1998, Cenozoic sequence stratigraphy in the eastern North Sea. In: De Graciansky, P.-C., Hardenbol, J., Jacquin, T., and Vail, P.R. (eds), Mesozoic and Cenozoic Sequence Stratigraphy of European Basins, *SEPM Special Publication*, **60**: 91–118.

Miller, K.G., Fairbanks, R.G., and Mountain, G.S., 1987, Tertiary oxygen isotope synthesis, sea level history, and continental margin erosion. *Paleoceanography*, **2**: 1–19.

Miller, K.G., Mountain, G.S., Browning, J.V., Kominz, M., Sugarman, P.J., Christie-Blick, N., et al., 1998, Cenozoic global sea level, sequences, and the New Jersey transect: results from coastal plain and continental slope drilling. *Reviews of Geophysics*, **36**: 569–601.

Miller, K.G., Wright, J.D., and Fairbanks, R.G., 1991, Unlocking the ice house; Oligocene-Miocene oxygen isotopes, eustasy, and margin erosion. *Journal of Geophysical Research, B, Solid Earth and Planets*, **96**: 6829–6848.

Moebius, I., Friedrich, O., Edgar, K.M., and Sexton, P.F., 2015, Episodes of intensified biological productivity in the subtropical Atlantic Ocean during the termination of the Middle Eocene Climatic Optimum (MECO). *Paleoceanography*, **30**: 1041–1058.

Molina, E., Canudo, J.I., Guernet, C., McDougall, K., Ortiz, N., Pascual, J.O., et al., 1992, The stratotypic Ilerdian revisited: integrated stratigraphy across the Paleocene/Eocene boundary. *Revue de Micropaléontologie*, **35**: 143–156.

Molina, E., Angori, E., Arenillas, I., Brinkhuis, H., Crouch, E.M., Luterbacher, H., et al., 2003, Correlation between the Paleocene/Eocene boundary and the Ilerdien at Campo, Spain. *Revue de Micropaléontologie*, **46**: 95–109.

Molina, E., Alegret, L., Arenillas, I., Arz, J.A., Gallala, N., Hardenbol, J., et al., 2006, The Global Boundary Stratotype Section and Point for the base of the Danian Stage (Paleocene, Paleogene, "Tertiary", Cenozoic) at El Kef, Tunisia; original definition and revision. *Episodes*, **29**: 263–273.

Molina, E., Alegret, L., Arenillas, I., Arz, J.A., Gallala, N., Grajales-Nishimura, J.M., et al., 2009, The Global Boundary Stratotype Section and Point for the base of the Danian Stage (Paleocene, Paleogene, "Tertiary", Cenozoic): auxillary sections and correlation. *Episodes*, **32**: 84–95.

Molina, E., Alegret, L., Apellaniz, E., Bernaola, G., Caballero, F., Dinares-Turell, J., et al., 2011, The Global Stratotype Section and Point (GSSP) for the base of the Lutetian Stage at the Gorrondatxe section, Spain. *Episodes*, **34**: 86–108.

Monechi, S., 1986, Calcareous nannofossil events around the Eocene-Oligocene boundary in the Umbrian Apennines (Italy). *Palaeogeography, Palaeoclimatology, Palaeoecology*, **57**: 61–69.

Monechi, S., Reale, V., Bernaola, G., and Balestra, B., 2013, The Danian/Selandian boundary at Site 1262 (South Atlantic) and in the Tethyan region: biomagnetostratigraphy, evolutionary trends in fasciculiths and environmental effects of the Latest Danian Event. *Marine Micropaleontology*, **98**: 28–40.

Mudge, D.C., and Bujak, J.P., 1996, An integrated stratigraphy for the Paleocene and Eocene of the North Sea. In: Knox, R.W.O.B., Corfield, R.M., and Dunay, R.E. (eds), Correlation of the Early Paleogene in Northwest Europe, *Geological Society Special Publication*, **101**: 91–113.

Munier Chalmas, E., and De Lapparent, A., 1893, Note sur le nomenclature des terrains sedimentaires. *Bulletin de la Société Géologique de France*, **3**: 438–493.

Murphey, P.C., Kelly, T.S., Chamberlain, K.R., Tsukui, K., and Clyde, W.C., 2018, Mammals from the earliest Uintan (middle Eocene) Turtle Bluff Member, Bridger Formation, southwestern Wyoming, USA, part 3: Marsupialia and a reevaluation of the Bridgerian-Uintan North American Land mammal Age transition. *Palaeontologia Electronica*, **21**: 25A, 1–52.

Naumann, C.F., 1866, *Lehrbuch der Geognosie*, Vol. 3. Leipzig: Engelmann.

Nishimura, A., 1992, Paleocene radiolarian biostratigraphy in the northwest Atlantic at Site 384, Leg 43 of the Deep Sea Drilling Project. *Micropaleontology*, **38**: 317–362.

Nocchi, M., Parisi, G., Monaco, P., Monechi, S., and Madile, M., 1988, Eocene and early Oligocene micropaleontology and paleoenvironments in SE Umbria, Italy. *Palaeogeography, Palaeoclimatology, Palaeoecology*, **67**: 181–244.

Noiret, C., Steurbaut, E., Tabuce, R., Marandat, B., Schnyder, J., Storme, J.-Y., et al., 2016, New bio-chemostratigraphic dating of a unique early Eocene sequence from southern Europe results in precise mammalian biochronological tie-points. *Newsletters on Stratigraphy*, **49**: 469–480.

Norris, R.D., 1996, Symbiosis as an evolutionary innovation in the radiation of Paleocene planktic foraminifera. *Paleobiology*, **22**: 461–480.

Ogg, J.G., 2020, Chapter 5 - Geomagnetic polarity time scale. *In* Gradstein, F.M., Ogg, J.G., Schmitz, M.D., and Ogg, G.M. (eds), *The Geologic Time Scale 2020*. **Vol. 1** (this book). Elsevier, Boston, MA.

Ogg, J.G., Ogg, G., and Gradstein, F.M., 2008, *The Concise Geologic Time Scale*. Cambridge: Cambridge University Press, 177 p.

Ogg, J.G., Ogg, G.M., and Gradstein, F.M., 2016, *A Concise Geologic Time Scale 2016*. Elsevier, 234 pp.

Okada, H., and Bukry, D., 1980, Supplementary modification and introduction of code numbers to the low-latitude coccolith biostratigraphic zonation (Bukry, 1973; 1975). *Marine Micropaleontology*, **5**: 321–325.

Olsson, R.K., Hemleben, C., Berggren, W.A., and Huber, B.T., 1999, Atlas of Paleocene Planktonic Foraminifera. *Smithsonian Contributions to Paleobiology*, 85, 252 p.

Orue-Extebarria, X., Pujalte, V., Bernaola, G., Apellaniz, E., Baceta, J.I., Payros, A., et al., 2001, Did the Late Paleocene thermal maximum affect the evolution of larger foraminifers? Evidence from calcareous plankton of the Campo Section (Pyrenees, Spain). *Marine Micropaleontology*, **41**: 45–71.

Pälike, H., and Shackleton, N.J., 2000, Constraints on astronomical parameters from the geological record for the last 25Myr. *Earth and Planetary Science Letters*, **182**: 1–14, https://doi.org/10.1016/S0012-821x(00)00229-6.

Pälike, H., Norris, R.D., Herrle, J.O., Wilson, P.A., Coxall, H.K., Lear, C.H., et al., 2006, The heartbeat of the Oligocene climate system. *Science*, **314**: 1894–1898, https://doi.org/10.1126/science.1133822.

Pälike, H., Lyle, M., Nishi, H., Raffi, I., Gamage, K., Klaus, A., and the Expedition 320/321 Scientists, 2010, *Proceedings of the Integrated Ocean Drilling Program*, **320/321**. Tokyo, Integrated Ocean Drilling Program Management International, Inc.

Papazzoni, C.A., Cosovic, V., Briguglio, A., and Drobne, K., 2017, Towards a calibrated larger foraminifera biostratigraphic zonation: celebrating 18 years of the application of shallow benthic zones. *Palaios*, **32**: 1–5.

Payros, A., Tosquella, J., Bernaola, G., Dinarès-Turell, J., Orue-Extebarria, X., and Pujalte, V., 2009, Filling the North European early/middle Eocene (Ypresian/Lutetian) boundary gap: insights from the Pyrenean continental to deep-marine record. *Palaeogeography, Palaeoclimatology, Palaeoecology*, **280**: 313–332.

Payros, A., Dinarès-Turell, J., Monechi, S., Orue-Extebarria, X., Ortiz, S., Apellaniz, E., et al., 2015, The Lutetian/Bartonian transition (middle-Eocene) at the Oyambre section (northern Spain): implications for standard chronostratigraphy. *Palaeogeography, Palaeoclimatology, Palaeoecology*, **440**: 234–248.

Pearson, P.N., Olsson, R.K., Huber, B.T., Hemleben, C., and Berggren, W.A., 2006, Atlas of Eocene Planktonic foraminifera. *Cushman Foundation Special Publication*, **41**, 514 p.

Perch-Nielsen, K., 1985, Cenozoic calcareous nannofossils. In: Bolli, H.M., Saunders, J.B., and Perch-Nielsen, K. (eds), *Plankton Stratigraphy*. Cambridge: Cambridge University Press, pp. 427–554.

Petrizzo, M.R., 2005, An early late Paleocene event on Shatsky Rise, northwest Pacific Ocean (ODP Leg 198): evidence from planktonic foraminiferal assemblages. In: Bralower, T.J., Premoli Silva, I., Malone, M.J., et al. (eds), *Proceedings of the Ocean Drilling Program, Scientific Results, Leg 198*. College Station, TX: Ocean Drilling Program, pp. 1–29.

Petrizzo, M.R., Wade, B.S., and Gradstein, F.M., 2020, Chapter 3G - Planktonic foraminifera. *In* Gradstein, F.M., Ogg, J.G., Schmitz, M.D., and Ogg, G.M. (eds), *The Geologic Time Scale 2020*. **Vol. 1** (this book). Elsevier, Boston, MA.

Petrushevskaya, M.G., 1975, Cenozoic radiolarians of the Antarctic, Leg 29, DSDP. In: Kennett, J.P., Houtz, R.E., et al. (eds), *Initial Reports of the Deep Sea Drilling Project, Leg 29*. Washington, DC: U.S. Government Printing Office, pp. 541–675.

Phillips, J., 1841, Figures and Descriptions of the Palaeozoic Fossils of Cornwall, Devon and East Somerset. London: Longman, Brown, Green and Longmans, 231 p.

Pignatti, J.S., 1998, The philosophy of larger foraminiferal biozonation—a discussion. *Slovenska Akademija Znanosti in Umetnosti, Razred za Naravoslovne Vede*, **34**: 15–20.

Pomerol, C. (ed), 1981, Stratotypes of Paleogene Stages. *Bulletin d'information des géologues du Bassin de Paris, Mémoire Hors Série*: **2**, 301 p.

Powell, A.J., 1992, A Stratigraphic Index of Dinoflagellate Cysts. London: Chapman and Hall, 290 p.

Powell, A.J., Brinkhuis, H., and Bujak, J.P., 1996, Upper Paleocene-lower Eocene dinoflagellate cyst sequence biostratigraphy of Southeast England. In: Knox, R.W.O.B., Corfield, R.M., and Dunay, R.E. (eds), Correlation of the Early Paleogene in Northwest Europe, *Geological Society Special Publication*, **101**: 145–183.

Premoli Silva, I., Coccioni, R., and Montanari, A. (eds), 1988, *The Eocene-Oligocene Boundary in the Marche-Umbria Basin (Italy)*. Ancona: International Subcommission on Paleogene Stratigraphy, 268 p.

Premoli Silva, I., and Jenkins, D.G., 1993, Decision on the Eocene-Oligocene boundary stratotype. *Episodes*, **16**: 379–382.

Prestwich, J., 1847, On the probable age of the London Clay, and its relations to the Hampshire and Paris Tertiary systems. *Quarterly Journal of the Geological Society of London*, **3**: 354–377.

Pujalte, V., Schmitz, B., Baceta, J.I., Orue-Etxebarria, X., Bernaola, G., Dinarès-Turell, J., et al., 2009, Correlation of the Thanetian-Ilerdian turnover of larger foraminifera and the Paleocene-Eocene thermal maximum: confirming evidence from the Campo area (Pyrenees, Spain). *Geologica Acta*, **7**: 161–175.

Pujalte, V., Apellaniz, E., Caballero, F., Monechi, S., Ortiz, S., Orue-Extebarria, X., et al., 2017, Integrative stratigraphy and climatic events of a new lower Paleogene reference section from the Betic Cordillera: Rio Gor, Granada Province, Spain. *Spanish Journal of Palaeontology*, **32**: 185–206.

Quillévéré, F., Norris, R.D., Moussa, I., and Berggren, W.A., 2001, Role of photosymbiosis and biogeography in the diversification of early Paleogene acarininids (planktonic Foraminifera). *Paleobiology*, **27**: 311–326.

Quillévéré, F., Norris, R.D., Kroon, D., and Wilson, P.A., 2008, Transient ocean warming and shifts in carbon reservoirs during the early Danian. *Earth and Planetary Science Letters*, **265**: 600–615.

Raffi, I., Wade, B.S., and Pälike, H., 2020, Chapter 29 - The Neogene Period. *In* Gradstein, F.M., Ogg, J.G., Schmitz, M.D., and Ogg, G.M. (eds), *The Geologic Time Scale 2020*. **Vol. 2** (this book). Elsevier, Boston, MA.

Raine, J.I., Beu, A.G., Boyes, A.F., Campbell, H.J., Cooper, R.A., Crampton, J.S., et al., 2015, Revised calibration of the New Zealand Geological time scale: NZGT2015/1. *GNS Science Report*, **2012/39**: 1–53.

Renevier, E., 1874, Tableau de terrains sédimentaires formés pendant les époques de la phase organique du globe terrestre. *Bulletin de la Société Vaudoise des Sciences Naturelles*, **13**: 218–252.

Riedel, W.R., 1957, Radiolaria: a preliminary stratigraphy. In: Petterson, H. (ed), *Reports of the Swedish Dea-Sea Expedition, 1947–1948*, Vol. 6. Göteborg: Elanders Boktryckeri Aktiebolag, pp. 59–96.

Riedel, W.R., and Sanfilippo, A., 1970, Radiolaria, leg 4, Deep Sea Drilling Project. In: Bader, R.G. et al. (eds), *Initial Reports of the Deep Sea Drilling Project*. Washington, DC: U.S. Government Printing Office, pp. 503–575.

Riedel, W.R., and Sanfilippo, A., 1971, Cenozoic Radiolaria from the western tropical Pacific, Leg 7. In: Winterer, E.L. et al. (eds), *Initial Reports of the Deep Sea Drilling Project, Leg 7*. Washington, DC: U.S. Government Printing Office, pp. 1529–1672.

Riedel, W.R., and Sanfilippo, A., 1978, Stratigraphy and evolution of tropical Cenozoic radiolarians. *Micropaleontology*, **24**: 61–96.

Roberts, E.M., Stevens, N.J., O'Connor, P.M., Dirks, P.G.H.M., Gottfried, M.D., Clyde, W.C., et al., 2012, Initiation of the western branch of the East African Rift coeval with the eastern branch. *Nature Geoscience*, **5**: 289–294.

Robin, E., and Rocchia, R., 1998, Ni-rich spinel at the Cretaceous-Tertiary boundary of El Kef, Tunisia. *Bulletin de la Société Géologique de France*, **169**: 365–372.

Rodriguez-Pinto, A., Pueyo, E.L., Serra-Kiel, J., Samso, J.M., Barnolas, A., and Pocovi, A., 2012, Lutetian magnetostratigraphic calibration of larger foraminifer zonation (SBZ) in the Southern Pyrenees: the Isuela section. *Palaeogeography, Palaeoclimatology, Palaeoecology*, **333–334**: 107–120.

Rodriguez-Pinto, A., Pueyo, E.L., Serra-Kiel, J., Barnolas, A., Samso, J. M., and Pocovi, A., 2013, The upper Ypresian and Lutetian in San Pelegrin section (Southwestern Pyrenean basin): magnetostratigraphy and larger foraminifera correlation. *Palaeogeography, Palaeoclimatology, Palaeoecology*, **370**: 13–29.

Röhl, U., Westerhold, T., Bralower, T.J., and Zachos, J.C., 2007, On the duration of the Paleocene-Eocene thermal maximum (PETM). *Geochemistry, Geophysics, Geosystems*, **8**: 1–13.

Romein, A.J.T., 1979, Lineages in early Paleogene calcareous nannoplankton. *Utrecht Micropaleontological Bulletins*, **22**: 231.

Rosenkrantz, A., 1924, De københavnske grønsandslag og deres placering i den danske lagraekke. *Meddelelser fra Dansk Geologisk Forening*, **6**: 1–39.

Roth, P.H., Baumann, P., and Bertolino, V., 1971, Late Eocene-Oligocene calcareous nannoplankton from central and northern Italy. In: Farinacci, A. (ed), *Proceedings of the Second Planktonic Conference, Rome, 1970*. Roma: Tecnoscienza, pp. 1069–1097.

Roveda, V., 1961, Contributo allo studio di alcuni macroforaminiferi di Priabona. *Rivista Italiana di Paleontologia*, **67**: 153–213.

Salvador, A. (ed), 1994, *International Stratigraphic Guide; A Guide to Stratigraphic Classification, Terminology*, and Procedure. Boulder, CO: Geological Society of America, 214 p.

Sanfilippo, A., and Blome, C.D., 2001, Biostratigraphic implications of mid-latitude Paleocene-Eocene radiolarian faunas from Hole 1051A, Ocean Drilling Program Leg 171B, Blake Nose, western North Atlantic. In: Kroon, D., Norris, R.D., and Klaus, A. (eds), Western North Atlantic Paleogene and Cretaceous Palaeoceanography, *Geological Society, London, Special Publications*, **183**: 185–224.

Sanfilippo, A., and Nigrini, C., 1998a, Code numbers for Cenozoic low latitude radiolarian biostratigraphic zones and GPTS conversion tables. *Marine Micropaleontology*, **33**: 109–156.

Sanfilippo, A., and Nigrini, C., 1998b, Upper Paleocene-lower Eocene deep-sea radiolarian stratigraphy and the Paleocene/Eocene series boundary. In: Aubry, M.-P., Lucas, S.G., and Berggren, W.A. (eds), *Late Paleocene-Early Eocene Climatic and Biotic Events in the Marine and Terrestrial Records*. New York: Columbia University Press, pp. 244–276.

Sanfilippo, A., and Riedel, W.R., 1973, Cenozoic radiolaria (exclusive of theoperids, artostrobiids and amphipyndacids) from the Gulf of Mexico, DSDP Leg 10. In: Worzel, J.L., and Bryant, W. (eds), *Initial Reports of the Deep Sea Drilling Project, Leg 10*. Washington, DC: U.S. Government Printing Office, pp. 475–611.

Sanfilippo, A., Westberg-Smith, M.J., and Riedel, W.R., 1985, Cenozoic Radiolaria. In: Bolli, H.M., Saunders, J.B., and Perch-Nielsen, K.

(eds), *Plankton Stratigraphy*. Cambridge: Cambridge University Press, pp. 631–712.

Savin, S.M., 1977, The history of the Earth's surface temperature during the past 100 million years. *Annual Review of Earth and Planetary Sciences*, **5**: 319–355.

Schaub, H., 1981, Nummulites et Assilines de la Téthys paléogène: Taxinomie, phylogenèse et biostratigraphie. *Schweizerische Palaeontologische Abhandlungen*, **104–106**: 236.

Scheibner, C., and Speijer, R.P., 2009, Recalibration of the shallow-benthic zonation across the Paleocene-Eocene boundary: the Egyptian record. *Geologica Acta*, **7**: 195–214.

Scheibner, C., Speijer, R.P., and Marzouk, A.M., 2005, Turnover of larger Foraminifera during the Paleocene-Eocene thermal maximum and paleoclimatic control on the evolution of platform ecosystems. *Geology*, **33**: 493–496.

Scher, H.D., Bohaty, S.M., Smith, B.W., and Munn, G.H., 2014, Isotopic interrogation of a suspected late Eocene glaciation. *Paleoceanography*, **29**: 628–644.

Scherer, R.P., Gladenkov, A.Y., and Barron, J.A., 2007, Methods and applications of Cenozoic marine diatom biostratigraphy. In: Starratt, S. (ed), *Pond Scum to Carbon Sink: Geological and Environmental Applications of the Diatoms*. Paleontological Society Papers 13. Bethesda, MD: The Paleontological Society, pp. 61–83.

Schimper, W.P., 1874, *Traité de Paléonologie végétale*, Vol. III. Paris: J. B. Baillière et Fils, 896 p.

Schmidt-Kittler, N. (ed), 1987, International Symposium on Mammalian Biostratigraphy and Paleoecology of the European Paleogene, Mainz, February 18th–21st, 1987. *Münchner Geowissenschaftliche Abhandlungen, Reihe A*, 10, 312 p.

Schmitz, B., Molina, E., and Von Salis, K., 1998, The Zumaya section in Spain: a possible global stratotype section for the Selandian and Thanetian Stages. *Newsletters on Stratigraphy*, **36**: 35–42.

Schmitz, B., Pujalte, V., Molina, E., Monechi, S., Orue-Extebarria, X., Speijer, R.P., et al., 2011, The Global Stratotype Sections and Points for the bases of the Selandian (middle Paleocene) and Thanetian (upper Paleocene). *Episodes*, **34**: 220–243.

Schuler, M., Cavelier, C., Dupuis, C., Steurbaut, E., Vandenberghe, N., Riveline, J., et al., 1992, The Paleogene of the Paris and Belgian basins; standard stages and regional stratotypes. *Cahiers de Micropaléontologie*, **7**: 29–92.

Scotese, C.R., 2014, Atlas of Paleogene Paleogeographic Maps (Mollweide Projection), Maps 8–15, Vol. 1. *The Cenozoic, PALEOMAP Atlas for ArcGIS*, PALEOMAP Project. Evanston, IL. https://www.academia.edu/11099001/Atlas_of_Paleogene_Paleogeographic_Maps.

Schulte, P., Alegret, L., Arenillas, I., Arz, J.A., Barton, P.J., Bown, P.R., et al., 2010, The Chicxulub asteroid impact and mass-extinction at the Cretaceous-Paleogene boundary. *Science*, **327**: 1214–1218.

Schulte, P., Scheibner, C., and Speijer, R.P., 2011, Fluvial discharge and sea-level changes controlling black shale deposition during the Paleocene-Eocene Thermal Maximum in the Dababiya Quarry section, Egypt. *Chemical Geology*, **285**: 167–183.

Secord, R., 2008, The Tiffanian Land-Mammal Age (Middle and Late Paleocene) in the Northern Bighorn Basin, Wyoming. *University of Michigan Papers on Paleontology*, **35**, 192 p.

Serra-Kiel, J., Canudo, J.I., Dinarès-Turell, J., Molina, E., Ortiz, N., Pascual, J.O., et al., 1994, Cronoestratigrafía de los sedimentos marinos del Terciario inferior de la Cuenca de Graus-Tremp (Zona Central Surpirenaica). *Revista de la Sociedad Geológica de España*, **7**: 273–297.

Serra-Kiel, J., Hottinger, L., Caus, E., Drobne, K., Ferrandez, C., Jauhri, A.K., et al., 1998, Larger foraminiferal biostratigraphy of the Tethyan Paleocene and Eocene. *Bulletin de la Société Géologique de France*, **169**: 281–299.

Sexton, P.F., Norris, R.D., Wilson, P.A., Pälike, H., Westerhold, T., Röhl, U., et al., 2011, Eocene global warming events driven by ventilation of oceanic dissolved organic carbon. *Nature*, **471**: 349–352.

Sijp, W.P., England, M.H., and Huber, M., 2011, Effect of the deepening of the Tasman Gateway on the global ocean. *Paleoceanography*, **26**: PA4207.

Simmons, M.S., Miller, K.G., Ray, D.C., Davies, A., van Buchem, F.S.P., and Gréselle, B., 2020, Chapter 13 - Phanerozoic eustacy. *In* Gradstein, F.M., Ogg, J.G., Schmitz, M.D., and Ogg, G.M. (eds), *The Geologic Time Scale 2020*. **Vol. 1** (this book). Elsevier, Boston, MA.

Śliwińska, K.K., Abrahamsen, N., Beyer, C., Brunings-Hansen, T., Thomsen, E., Ulleberg, K., et al., 2012, Bio- and magnetostratigraphy of Rupelian-mid Chattian deposits from the Danish land area. *Review of Palaeobotany and Palynology*, **172**: 48–69.

Śliwińska, K.K., Heilmann-Clausen, C., Thomsen, E., Rocha, R., Pais, J., Kullberg, J.C., et al., 2014, Correlation between the type Chattian in NW Europe and the Rupelian-Chattian candidate GSSP in Italy. *In* Rocha, R. et al. (eds), *STRATI 2013*. Springer Geology, pp. 283–286.

Sluijs, A., Schouten, S., Pagani, M., Woltering, M., Brinkhuis, H., Sinninghe Damsté, J.S., et al., the Expedition 302 Scientists, 2006, Subtropical Arctic Ocean temperatures during the Palaeocene-Eocene thermal maximum. *Nature*, **441**: 610–613.

Sluijs, A., Bowen, G.J., Brinkhuis, H., Lourens, L., and Thomas, E., 2007, The Palaeocene-Eocene Thermal maximum super greenhouse: biotic and geochemical signatures, age models and mechanisms of global change. In: Williams, M., Haywood, A.M., Gregory, F.J., and Schmidt, D.N. (eds), *Deep-Time Perspectives on Climate Change: Marrying the Signal from Computer Models and Biological Proxies*. The Micropalaeontological Society, Special Publications. London: The Micropalaeontological Society, pp. 323–349.

Sluijs, A., Schouten, S., Donders, T.H., Schoon, P.L., Röhl, U., Reichart, G.-J., et al., 2009, Warm and wet conditions in the Arctic region during Eocene Thermal maximum 2. *Nature Geoscience*, **2**: 777–780.

Smith, T., and Smith, R., 2003, Terrestrial mammals as biostratigraphic indicators in upper Paleocene-lower Eocene marine deposits of the southern North Sea Basin. In: Wing, S.L., Gingerich, P.D., Schmitz, B., and Thomas, E. (eds), Causes and Consequences of Globally Warm Climates in the Early Paleogene, *Geological Society of America Special Paper*, **369**: 513–520.

Smith, T., Rose, K.D., and Gingerich, P.D., 2006, Rapid Asia-Europe-North America geographic dispersal of earliest Eocene primate Teilhardina during the Paleocene-Eocene Thermal Maximum. *Proceedings of the National Academy of Sciences of the United States of America*, **103**: 11223–11227.

Smith, M.E., Carroll, A.R., and Singer, B.S., 2008, Synoptic reconstruction of a major ancient lake system: Eocene Green River Formation, western United States. *Geological Society of America Bulletin*, **120**: 54–84.

Solé, F., Noiret, C., Desmares, D., Adnet, S., Taverne, L., De Putter, T., et al., 2019, Reassessment of historical sections from the Paleogene margin of the Congo Basin reveals an almost complete absence of Danian deposits. *Geoscience Frontiers*, **10**: 1039–1063.

Speijer, R.P., Scheibner, C., Stassen, P., and Morsi, A.M.M., 2012, Response of marine ecosystems to deep-time global warming: a synthesis of biotic patterns across the Paleocene-Eocene thermal maximum (PETM). *Austrian Journal of Earth Sciences*, **105**: 6–16.

Spofforth, D.J.A., Agnini, C., Pälike, H., Rio, D., Bohaty, S., Fornaciari, E., et al., 2010, Organic carbon burial following the Middle Eocene Climatic Optimum (MECO) in the central–western Tethys. *Paleoceanography*, **25**: PA3210. doi: 2009PA001738.

Sprong, J., Speijer, R.P., and Steurbaut, E., 2009, Biostratigraphy of the Danian/Selandian transition in the southern Tethys, with special reference to Lowest Occurrence of planktic foraminifera *Igorina albeari*. *Geologica Acta*, **7**: 63–77.

Stehlin, H.G., 1910, Remarques sur les faunules de mammifères des couches éocènes et oligocènes du Bassin de Paris. *Bulletin de la Société Géologique de France*, **9**: 488–520.

Steurbaut, E., 1998, High-resolution holostratigraphy of middle Paleocene to early Eocene strata in Belgium and adjacent areas. *Palaeontographica Abteilung A: Palaeozoologie-Stratigraphie*, **247**: 91–156.

Steurbaut, E., 2006, Ypresian. In: Dejonghe, L. (ed), Current Status of Chronostratigraphic Units Named From Belgium and Adjacent Areas, *Geologica Belgica*, **9**: 73–93.

Steurbaut, E., and Sztrákos, K., 2008, Danian/Selandian boundary criteria and North Sea Basin-Tethys correlations based on calcareous nannofossil and foraminiferal trends in SW France. *Marine Micropaleontology*, **67**: 1–29.

Steurbaut, E., De Coninck, J., and Van Simaeys, S., 2016, Micropaleontological dating of the Premontre mammal fauna (MP10, Premontre Sands, EECO, early late Ypresian, Paris Basin). *Geologica Belgica*, **19**: 273–280.

Stevens, N.J., Seiffert, E.R., O'Connor, P.M., Roberts, E.M., Schmitz, M.D., Krause, C., et al., 2013, Palaeontological evidence for an Oligocene divergence between Old World monkeys and apes. *Nature*, **497**: 611–614.

Stott, L.D., Sinha, A., Thiry, M., Aubry, M.-P., and Berggren, W.A., 1996, Global $\delta^{13}C$ changes across the Paleocene-Eocene boundary; criteria for terrestrial-marine correlations. In: Knox, R.W.O.B., Corfield, R.M., and Dunay, R.E. (eds), Correlation of the Early Paleogene in Northwest Europe, *Geological Society Special Publication*, **101**: 381–399.

Subbotina, N.N., 1953, Fossil Foraminifera of the USSR: Globigerinids, Hantkeninids and Globorotaliids (Translated from Russian). *Trudy Vsesyuznogo Nauchno-Issledovatel'skogo Geologo-Razvedochnogo Instituta (VNIGRI)*, **76**: 296.

Suess, E., 1868, Über die Gliederung des vicentinischen Tertiärgebirges. *Sitzungsberichte der Kaiserlichen Akademie der Wissenschaften in Wien*, **58**: 265–279.

Takemura, A., and Ling, H.-Y., 1997, Eocene and Oligocene radiolarian biostratigraphy from the Southern Ocean: correlation of ODP Legs 114 (Atlantic Ocean) and 120 (Indian Ocean). *Marine Micropaleontology*, **30**: 97–116.

Thomas, E., 2007, Cenozoic mass extinctions in the deep sea; what perturbs the largest habitat on Earth?. *In* Monechi, S., Coccioni, R., and Rampino, M.R. (eds), *Large Ecosystem Perturbations: Causes and Consequences*. Geological Society of America Special Paper, **424**: 1–23.

Thomas, E., and Shackleton, N.J., 1996, The Paleocene-Eocene benthic foraminiferal extinction and stable isotope anomalies. *In* Knox, R.W.O.B., Corfield, R.M., and Dunay, R.E. (eds), *Correlation of the Early Paleogene in Northwest Europe*. Geological Society Special Publication, **101**: 401–441.

Thomas, E., and Zachos, J.C., 2000, Was the late Paleocene thermal maximum a unique event? *In* Schmitz, B., Aubry, M.-P., and Zachos, J.C. (eds), *Early Paleogene Warm Climates and Biosphere Dynamics*. GFF, **122**: 169–170.

Thomas, E., Zachos, J.C., and Bralower, T.J., 2000, Deep-sea environments on a warm earth; latest Paleocene-early Eocene. *In* Huber, B.T., MacLeod, K.G., and Wing, S.L. (eds), *Warm Climates in Earth History*. Cambridge: Cambridge University Press, pp. 132–160.

Thomsen, E., 1994, Calcareous nannofossil stratigraphy across the Danian-Selandian boundary in Denmark. *GFF*, **116**: 65–67.

Ting, S., Tong, Y., Clyde, W.C., Koch, P.L., Meng, J., Wang, Y., et al., 2011, Asian early Paleogene chronology and mammalian faunal turnover events. *Vertebrata PalAsiatica*, **49**: 1–28.

Toumarkine, M., and Luterbacher, H., 1985, Paleocene and Eocene planktic foraminifera. In: Bolli, H.M., Saunders, J.B., and Perch-Nielsen, K. (eds), Plankton Stratigraphy. Cambridge: Cambridge University Press, pp. 87–154.

Vahlenkamp, M., De Vleeschouwer, D., Batenburg, S.J., Edgar, K.M., Hanson, E., Martinez, M., et al. and Expedition 369 Scientific Participants, 2020, A lower to middle Eocene astrochronology for the Mentelle Basin (Australia) and its implications for the geologic time scale. *Earth and Planetary Science Letters*, **529**: XX.

Van Simaeys, S., De Man, E., Vandenberghe, N., Brinkhuis, H., and Steurbaut, E., 2004, Stratigraphic and palaeoenvironmental analysis of the Rupelian-Chattian transition in the type region; evidence from dinoflagellate cysts, Foraminifera and calcareous nannofossils. *Palaeogeography, Palaeoclimatology, Palaeoecology*, **208**: 31–58.

Van Simaeys, S., Brinkhuis, H., Pross, J., Williams, G.L., and Zachos, J.C., 2005a, Arctic dinoflagellate migrations mark the strongest Oligocene glaciations. *Geology*, **33**: 709–712.

Van Simaeys, S., Munsterman, D., and Brinkhuis, H., 2005b, Oligocene dinoflagellate cyst biostratigraphy of the southern North Sea Basin. *Review of Palaeobotany and Palynology*, **134**: 105–128.

Van Simaeys, S., and Vandenberghe, N., 2006, Rupelian. In: Dejonghe, L. (ed), Current Status of Chronostratigraphic Units Named From Belgium and Adjacent Areas, 9, *Geologica Belgica*, **9**: 95–101.

Vandenberghe, N., Hager, H., van den Bosch, M., Verstraelen, A., Leroi, S., Steurbaut, E., et al., 2001, Stratigraphic correlation by calibrated well logs in the Rupel Group between North Belgium, the Lower-Rhine area in Germany and southern Limburg and the Achterhoek in the Netherlands. In: Vandenberghe, N. (ed), Contributions to the Paleogene and Neogene Stratigraphy of the North Sea Basin, *Aardkundige Mededelingen*, **11**: 69–84.

Vandenberghe, N., Laga, P., Steurbaut, E., Hardenbol, J., and Vail, P.R., 1998, Tertiary sequence stratigraphy at the southern border of the North Sea Basin in Belgium. In: de Graciansky, P.C., Hardenbol, J., Jacquin, T., and Vail, P.R. (eds), Mesozoic and Cenozoic Sequence Stratigraphy of European Basins, *SEPM Special Publication*, **60**: 119–154.

Vandenberghe, N., Brinkhuis, H., and Steurbaut, E., 2003, The Eocene/Oligocene boundary in the North Sea area; a sequence stratigraphic approach. In: Prothero, D.R., Ivany, L.C., and Nesbitt, E.A. (eds), *From Greenhouse to Icehouse: The Marine Eocene-Oligocene Transition*. New York: Columbia University Press, pp. 419–437.

Vandenberghe, N., Van Simaeys, S., Steurbaut, E., Jagt, J.W.M., and Felder, P.J., 2004, Stratigraphic architecture of the Upper Cretaceous and Cenozoic along the southern border of the North Sea Basin in Belgium. In: Vandenberghe, N. (ed), Symposium on the Paleogene Preparing for Modern Life and Climate, *Netherlands Journal of Geosciences*, **83**: 155–171.

Vandenberghe, N., Hilgen, F.J., and Speijer, R.P., 2012, The Paleogene Period. *In* Gradstein, F.M., Ogg, J.G., Schmitz, M.D., and Ogg, G.M. (eds), *The Geologic time scale 2012*. Amsterdam: Elsevier, pp. 855–921.

Vandenberghe, N., De Craen, M., and Wouters, L., 2014, The Boom Clay Geology, from Sedimentation to Present-day Occurrence: a review. *Memoirs of the Geological Survey of Belgium*, **60**, 76 pp.

Varol, O., 1989, Palaeocene calcareous nannofossil biostratigraphy. In: Crux, J.A., and Van Heck, S.E. (eds), *Nannofossils and Their Applications*. Chichester: Ellis Horwood, pp. 267–310.

Varol, O., 1998, Palaeogene. In: Bown, P.R. (ed), *Calcareous Nannofossil Biostratigraphy*. British Micropalaeontological Society Publication Series. London: Chapman and Hall, pp. 200–224.

Wade, B., and Pälike, H., 2004, Oligocene climate dynamics. *Paleoceanography*, **19**: PA4019.

Wade, B.S., and Pearson, P.N., 2008, Planktonic foraminiferal turnover, diversity fluctuations and geochemical signals across the Eocene/Oligocene boundary in Tanzania. *Marine Micropaleontology*, **68**: 244–255.

Wade, B.S., Pearson, P.N., Berggren, W.A., and Pälike, H., 2011, Review and revision of Cenozoic planktonic foraminiferal biostratigraphy and calibration to the geomagnetic polarity and astronomical time scale. *Earth-Science Reviews*, **104**: 111–142.

Wade, B.S., Premec Fucek, V., Kamikuri, S.-I., Bartol, M., Luciani, V., and Pearson, P.N., 2012, Successive extinctions of muricate planktonic foraminifera (*Morozovelloides* and *Acarinina*) as a candidate for marking the base of the Priabonian. *Newsletters on Stratigraphy*, **45**: 245–262.

Wade, B.S., Olsson, R.K., Pearson, P.N., Huber, B.T., and Berggren, W.A. (eds), 2018, Atlas of Oligocene Planktonic Foraminifera. *Cushman Foundation for Foraminiferal Research Special Publication*, **46**, 524 p.

Wang, Y., Meng, J., Ni, X., and Li, C., 2007, Major events of Paleogene mammal radiation in China. *Geological Journal*, **42**: 415–430.

Wang, C., Zhao, X., Liu, Z., Lippert, P.C., Graham, S.A., Coe, R.S., et al., 2008, Constraints on the early uplift history of the Tibetan plateau. *Proceedings of the National Academy of Sciences of the United States of America*, **105**: 4987–4992.

Wang, Y., Meng, J., Beard, C.K., Li, Q., Ni, X., Gebo, D.L., et al., 2010, Early Paleogene stratigraphic sequences, mammalian evolution and its response to environmental changes in Erlian Basin, Inner Mongolia, China. *Science China Earth Sciences*, **53**: 1918–1926.

Wang, Y., Li, Q., Bai, B., Jin, X., Mao, F., and Meng, J., 2019, Paleogene integrative stratigraphy and timescale of China. *Science China Earth Sciences*, **62**: 1–23.

Warnaar, J., Bijl, P.K., Huber, M., Sloan, L., Brinkhuis, H., Röhl, U., et al., 2009, Orbitally forced climate changes in the Tasman sector during the Middle Eocene. *Palaeogeography, Palaeoclimatology, Palaeoecology*, **280**: 361–370.

Watkins, D.K., and Raffi, I., 2020, Chapter 3F - Calcareous nannofossils. *In* Gradstein, F.M., Ogg, J.G., Schmitz, M.D., and Ogg, G.M. (eds), *The Geologic Time Scale 2020*. **Vol. 1** (this book). Elsevier, Boston, MA.

Wei, W., and Pospichal, J.J., 1991, Danian calcareous nannofossils succession at Site 738 in the southern Indian Ocean. In: Barron, J.A., Larsen, B., et al. (eds), *Proceedings of the Ocean Drilling Program, Scientific Results, Leg 119*. College Station, TX: Ocean Drilling Program, pp. 495–512.

Westerhold, T., Röhl, U., Raffi, I., Fornaciari, E., Monechi, S., Reale, V., et al., 2008, Astronomical calibration of the Paleocene time. *Palaeogeography, Palaeoclimatology, Palaeoecology*, **257**: 377–403.

Westerhold, T., Röhl, U., Donner, B., McCarren, H.K., and Zachos, J.C., 2011, A complete high-resolution Paleocene benthic stable isotope record for the central Pacific (ODP Site 1209). *Paleoceanography*, **26**: PA2216, https://doi.org/10.1029/2010PA002092.

Westerhold, T., Röhl, U., Pälike, H., Wilkens, R., Wilson, P.A., and Acton, G., 2014, Orbitally tuned timescale and astronomical forcing in the middle Eocene to early Oligocene. *Climate of the Past*, **10**: 955–973.

Westerhold, T., Röhl, U., Frederichs, T., Agnini, C., Raffi, I., Zachos, J.C., et al., 2017, Astronomical calibration of the Ypresian timescale: implications for seafloor spreading rates and the chaotic behavior of the solar system? *Climate of the Past*, **13**: 1129–1152, https://doi.org/10.5194/cp-13-1129-2017.

Westerhold, T., Röhl, U., Donner, B., Frederichs, T., Kordesch, W.E.C., Bohaty, S.M., et al., 2018a, Late Lutetian Thermal maximum— crossing a thermal threshold in Earth's climate system? *Geochemistry, Geophysics, Geosystems*, **19**: 73–82.

Westerhold, T., Röhl, U., Donner, B., and Zachos, J.C., 2018b, Global extent of early Eocene hyperthermal events: a new Pacific benthic foraminiferal isotope record from Shatsky Rise (ODP Site 1209). *Paleoceanography and Paleoclimatology*, **33**: 626–642.

Williams, G.L., and Bujak, J.P., 1985, Mesozoic and Cenozoic dinoflagellates. In: Bolli, H.M., Saunders, J.B., and Perch-Nielsen, K. (eds), Plankton Stratigraphy. Cambridge: Cambridge University Press, pp. 847–964.

Williams, G.L., Brinkhuis, H., Pearce, M.A., Fensome, R.A., and Weegink, J.W., 2004, Southern Ocean and global dinoflagellate cyst events compared: index events for the Late Cretaceous-Neogene. In: Exon, N.F., Kennett, J.P., and Malone, M.J. (eds), *Proceedings of the Ocean Drilling Program, Scientific Results, Leg 189*. College Station, TX: Ocean Drilling Program, pp. 1–98.

Wilpshaar, M., Santarelli, A., Brinkhuis, H., and Visscher, H., 1996, Dinoflagellate cysts and mid-Oligocene chronostratigraphy in the central Mediterranean region. *Journal of the Geological Society London*, **153**: 553–561.

Wilson, G.S., Roberts, A.P., Verosub, K.L., Florindo, F., and Sagnotti, L., 1998, Magnetobiostratigraphic chronology of the Eocene-Oligocene transition in the CIROS-1 core, Victoria Land margin, Antarctica; implications for Antarctic glacial history. *Geological Society of America Bulletin*, **110**: 35–47.

Woodburne, M.O. (ed), 2004. *Late Cretaceous and Cenozoic Mammals of North America; Biostratigraphy and Geochronology*. New York: Columbia University Press, 391 p.

Woodburne, M.O., and Swisher, C.C., III, 1995, Land mammal high-resolution geochronology, intercontinental overland dispersals, sea level, climate, and variance. *In* Berggren, W.A., Kent, D.V., Aubry, M.-P., and Hardenbol, J. (eds), *Geochronology, time scales and Global Stratigraphic Correlation*, SEPM Special Publication, **54**: 335–365.

Woodburne, M.O., Goin, F.J., Bond, M., Carlini, A.A., Gelfo, J.N., Lopez, G.M., et al., 2014a, Paleogene land mammal faunas of South America—a response to global climatic changes and indigenous floral diversity. *Journal of Mammalian Evolution*, **21**: 1–73.

Woodburne, M.O., Goin, F.J., Raigemborn, M.S., Heizler, M., Gelfo, J.N., and Oliveira, E.V., 2014b, Revised timing of the South American early Paleogene land mammal ages. *Journal of South American Earth Sciences*, **54**: 109–119.

Wrenn, J.H., and Hart, G.F., 1988, Paleogene dinoflagellate cyst biostratigraphy of Seymour Island, Antarctica. In: Feldmann, R.M., and Woodburne,

M.A. (eds), Geology and Palaeontology of Seymour Island, Antarctic Peninsula, *Geological Society of America, Memoir*, **169**: 321–447.

Yans, J., Amagzhaz, M., Bouya, B., Cappetta, H., Iacumin, P., Kocsis, L., et al., 2014a, First carbon isotope chemostratigraphy of the Ouled Abdoun phosphate Basin, Morocco; implications for dating and evolution of earliest African placental mammals. *Gondwana Research*, **25**: 257–269.

Yans, J., Marandat, B., Masure, E., Serra-Kiel, J., Schnyder, J., Storme, J.-Y., et al., 2014b, Refined bio- (benthic foraminifera, dinoflagellate cysts) and chemostratigraphy (δ^{13}C) of the earliest Eocene at Albas-le-Clot (Corbières, France): implications for mammalian biochronology in southern Europe. *Newsletters on Stratigraphy*, **47**: 331–353.

Zachos, J.C., Quinn, T.M., and Salamy, S., 1996, High resolution (104yr) deep-sea foraminiferal stable isotope records of the Eocene-Oligocene climate transition. *Paleoceanography*, **11**: 251–266.

Zachos, J., Pagani, M., Sloan, L., Thomas, E., and Billups, K., 2001, Trends, rhythms, and aberrations in global climate 65Ma to present. *Science*, **292**: 686–693.

Zachos, J.C., Dickens, G.R., and Zeebe, R.E., 2008, An early Cenozoic perspective on greenhouse warming and carbon cycle dynamics. *Nature*, **451**: 279–283.

Zachos, J.C., McCarren, H., Murphy, B., Röhl, U., and Westerhold, T., 2010, Tempo and scale of late Paleocene and early Eocene carbon

isotope cycles: implications for the origin of hyperthermals. *Earth and Planetary Science Letters*, **299**: 242–249.

Zamagni, J., Mutti, M., Ballato, P., and Kosir, A., 2012, The Paleocene-Eocene thermal maximum (PETM) in shallow-marine successions of the Adriatic carbonate platform (SW Slovenia). *Geological Society of America Bulletin*, **124**: 1071–1086.

Zeebe, R.E., 2017, Numerical Solutions for the orbital motion of the solar system over the past 100 Myr: limits and new results. *Astronomical Journal*, **154**, https://doi.org/10.3847/1538-3881/aa8cce.

Zeebe, R.E., and Lourens, L.J., 2019, Solar System chaos and the Paleocene-Eocene boundary age constrained by geology and astronomy. *Science*, **365**: 926–929.

Zeeden, C., Hilgen, F.J., Hüsing, S.K., and Lourens, L.L., 2014, The Miocene astronomical time scale 9–12 Ma: new constraints on tidal dissipation and their implications for paleoclimatic investigations. *Paleoceanography*, **29**: 296–307. https://doi.org/10.1002/2014pa002615.

Zhang, R., Kravchinsky, V.A., and Yue, L., 2012, Link between global cooling and mammalian transformation across the Eocene-Oligocene boundary in the continental interior of Asia. *International Journal of Earth Sciences*, **101**: 2193–2200.

Zhang, Q., Willems, H., and Lin, D., 2013, Evolution of the Paleocene-early Eocene larger benthic foraminifera in Tethyan Himalaya of Tibet, China. *International Journal of Earth Sciences*, **102**: 1427–1445.

I. Raffi, B.S. Wade and H. Pälike,
With contributions by A.G. Beu, R. Cooper, M.P. Crundwell,
W. Krijgsman, T. Moore, I. Raine, R. Sardella and Y.V. Vernyhorova

Chapter 29

The Neogene Period

14.9 Ma Neogene

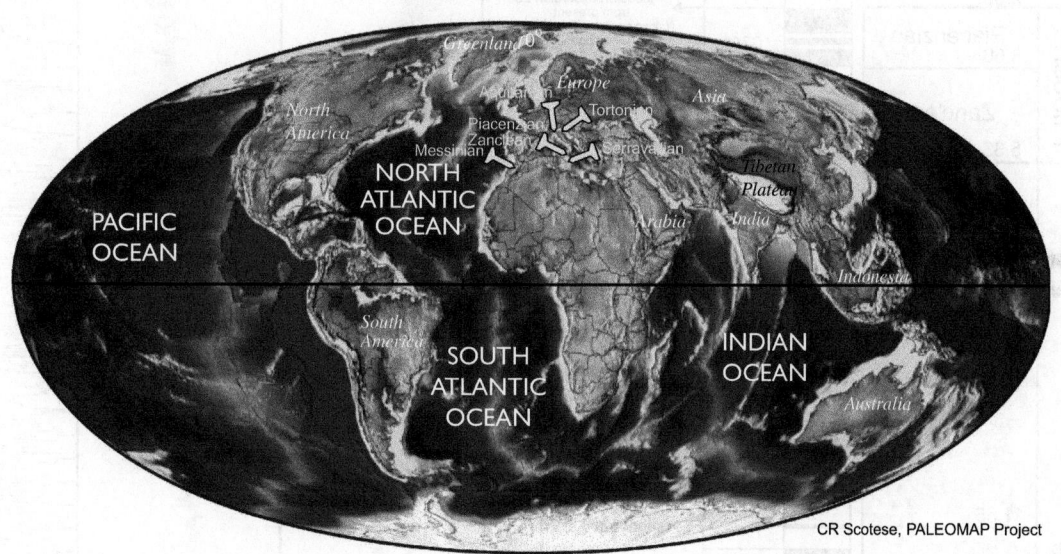

CR Scotese, PALEOMAP Project

Chapter outline

29.1 History and subdivisions	1142	
29.1.1 Overview of Neogene	1142	
29.1.2 Subdivisions of the Miocene	1143	
29.1.3 Subdivisions of the Pliocene	1150	
29.1.4 Regional stages	1150	
29.2 Neogene stratigraphy	1162	
29.2.1 Biostratigraphy	1162	
29.2.2 Terrestrial biostratigraphy	1191	
29.2.3 Physical stratigraphy	1193	
29.3 Neogene astronomically tuned time scale	1194	
29.3.1 Toward a fully astronomically tuned Neogene polarity time scale	1194	
29.3.2 GTS2012 and GTS2020	1198	
29.4 Summary	1199	
Acknowledgments	1200	
Bibliography	1200	

Abstract

The Neogene oceans and continents were mosaicked to form a paleogeography similar to today and exposed to the warm conditions of the mid Neogene to the cooling toward the glacial Quaternary. Antarctic ice sheets stabilized, then Northern Hemisphere ice sheets grew and thickened. Tectonics continued to shape the continents and ocean floor. High rising mountains, such as the Himalaya, altered atmospheric patterns and climate, and land bridges were exposed, as at the end of the Miocene when the Mediterranean was isolated and nearly completely desiccated; or in the early Pliocene when the emerging Isthmus between South and North America influenced the oceanic circulation in the Northern Hemisphere. Tectonic and climate changes influenced the evolution of fauna and flora, and species were forced to adapt or became extinct.

Geologic Time Scale 2020. DOI: https://doi.org/10.1016/B978-0-12-824360-2.00029-2
© 2020 Elsevier B.V. All rights reserved.

29.1 History and subdivisions

29.1.1 Overview of Neogene

The Neogene Period/System presently includes the Miocene Epoch/Series, with the six classical Stages: Aquitanian, Burdigalian, Langhian, Serravallian, Tortonian and Messinian, and the Pliocene Epoch/Series, with the two stages Zanclean and Piacenzian (Fig. 29.1).

In the last few years, several relevant points of discussion have developed involving the Neogene and regarding

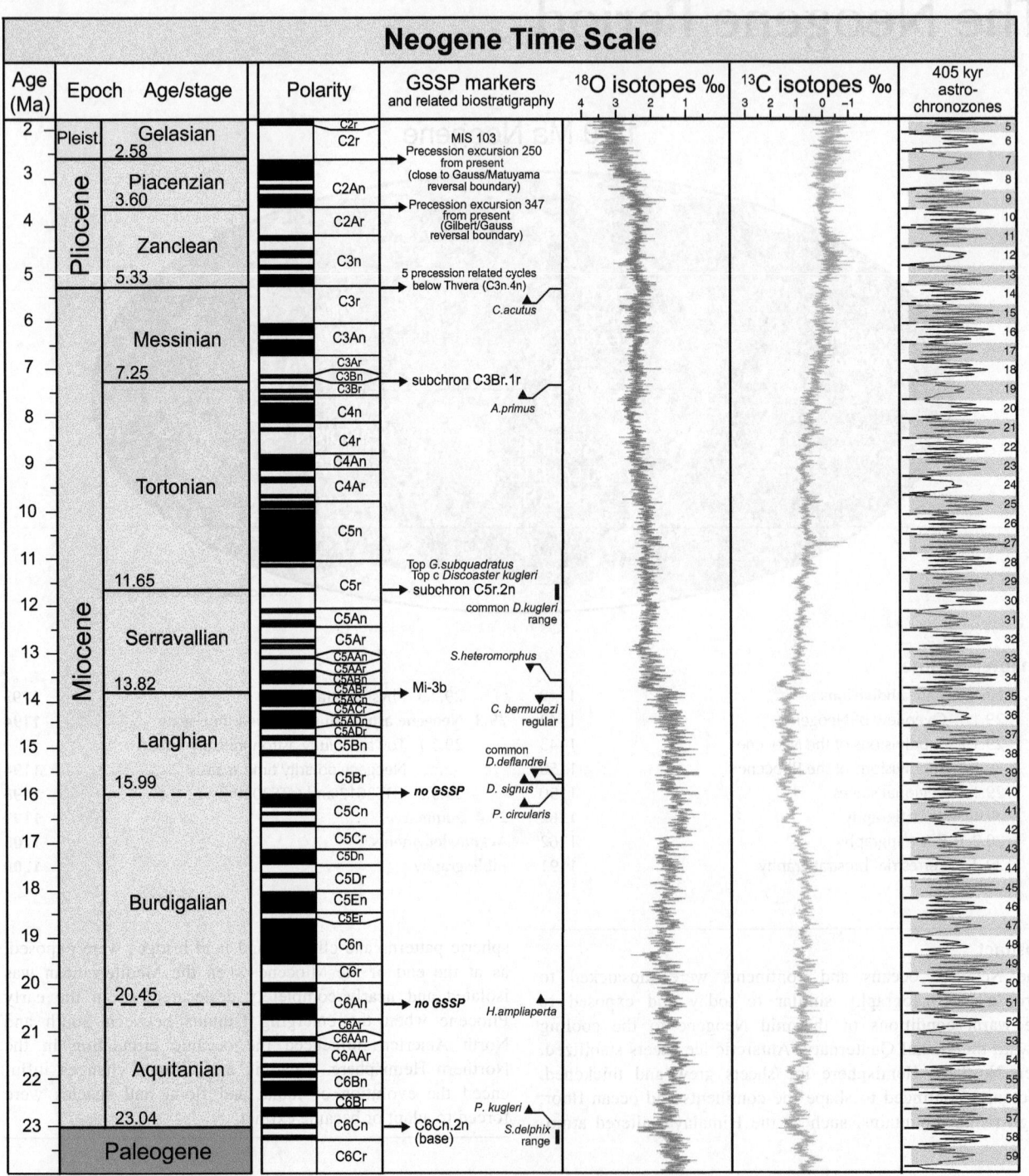

FIGURE 29.1 Neogene overview. Main markers for GSSPs of Neogene stages, together with correlative biohorizons. Oxygen and carbon isotope stack records are from De Vleeschouwer et al. (2017) and De Vleeschouwer et al. (2020, submitted). Astronomically age-calibrated chronozones (Milankovitch chronozones") are defined on the basis of 405-kyr long-eccentricity-related cycles from cycle 7−57 for the Neogene. *GSSP*, Global Boundary Stratotype Section and Point.

the chronostratigraphic approach and philosophy: the rethink of the concept of Global Boundary Stratotype Section and Point (GSSP) versus unit stratotype (Hilgen et al., 2006) and the formal definition of unit-stratotypes and (astro)chronozones (Hilgen et al., 2019). Hilgen et al. (2019) recently introduced a revised nomenclature of the astrochronozones originally proposed in Wade and Pälike (2004) and Wade et al. (2011). The (astro)chronozones imply that cycles used for the tuning can be formally defined as chronostratigraphic units independent of the standard hierarchy in global chronostratigraphy. The proposal by Hilgen et al. (2019) has been intended as a way to conform the standard geologic time scale (GTS) and Global Chronostratigraphic Scale with the progress in integrated high-resolution stratigraphy and astronomical dating. The proposal has been discussed within the International Subcommission on Stratigraphic Classification which assessed its worth. We apply astronomically age calibrated chronozones (Milankovitch chronozones) to the Neogene system/period, although they are not formally defined yet, and incorporate these in the standard chronostratigraphic scale. The scheme reported in Fig. 29.1 provides the definition of 405-kyr cycles with numerically coded "zones." The numbers derive by counting the large-scale 405-kyr eccentricity-related cycles back from the recent, relying on the stability of the 405-kyr cycle (Wade and Pälike, 2004; Wade et al., 2011; Laskar et al., 2011; Westerhold et al., 2012). Thus in the Neogene, Cycles 7–57 are identified (Fig. 29.1) which correspond to Neogene astrochronozones.

Astrobiochronology and tuned ages of reversal boundaries conform the Astronomically Tuned Neogene Time Scale (ATNTS) 2012 or 2016 except for C3r–C4r.1r, which have been updated according to Drury et al. (2017), C5Bn.1r–C5Dr.1n (updated according to Kochhann et al., 2016), C5En–C6AAr.2r (modified according to Liebrand et al., 2016), and C6Bn.1n–C7n.1r which were adjusted by Beddow et al. (2018).

Recently, the use of formal subseries was proposed (Head et al., 2017), but there was no consensus by the three subcommissions (Quaternary, Neogene, and Paleogene) (see Pearson et al., 2017; Finney and Bown, 2017). Therefore the status quo has been maintained, and the uses of the terms early, middle, late as well as the chronostratigraphic terms lower, middle, upper remain informal and should be lower case. The informal subdivisions are intended as in Berggren et al. (1995a), with

early/lower Miocene corresponding to the Aquitanian and Burdigalian stages;

middle/middle Miocene corresponding to the Langhian and Serravallian stages;

late/upper Miocene corresponding to the Tortonian and Messinian stages;

early/lower Pliocene corresponding to the Zanclean stage; and

late/upper Pliocene corresponding to Piacenzian stage.

29.1.2 Subdivisions of the Miocene

In this section the established GSSPs of the Miocene stages are briefly reported. Little has changed compared to what was already reported at the time of GTS2012, and the Langhian and Burdigalian GSSPs are yet to be defined (Fig. 29.1; see also Fig. 29.13 at the end of this chapter). We will provide updated details for these latter stages, while further details on Neogene GSSPs are found in GTS2004 (Chapter 21, Lourens et al., 2004) and GTS2012 (Chapter 29, Hilgen et al., 2012) to which the reader should refer.

29.1.2.1 Aquitanian and the Oligocene–Miocene boundary

The Aquitanian GSSP—the Oligocene–Miocene boundary—was established, more than two decades ago, in the Lemme–Carrosio section in Northern Italy (Steininger et al., 1997; Fig. 29.2) as the result of an ad hoc Working Group (WG) established at the International Geological Congress of Sidney, in 1976. The lengthy activity, undertaken during the preceding two decades, is the result of the difficulty of finding a suitable outcrop section for this important boundary. The final choice fell upon a section, the Lemme–Carrosio, of prevailing middle bathyal, massive and laminated siltstones, and the GSSP was defined at the base of Chron C6Cn.2n, despite the poorly defined magnetostratigraphic record obtained in the section. The biostratigraphic events that approximate the boundary are the calcareous nannofossil biohorizon Top (T) *Sphenolithus delphix*, and planktonic foraminiferal biohorizon Base (B) *Paragloborotalia kugleri*. The corresponding biostratigraphic location for the Oligocene/Miocene boundary resulted at CNO6/CNM1 zonal boundary of the low–mid latitude calcareous nannofossil zonation of Backman et al. (2012) and O7/M1 zonal boundary of the (sub)tropical planktonic foraminiferal zonation of Wade et al. (2011).

Recalibration of the poorly defined age of the boundary at 23.8 Ma (Berggren et al., 1995a) resulted in a new age of 22.9 Ma (Shackleton et al., 2000), subsequently retuned to La2004 solution at 23.03 Ma (see Neogene Chapter 29 in GTS2012; Hilgen et al., 2012). More recently, Beddow et al. (2018) astronomically tuned the carbonate content and benthic foraminiferal $\delta^{13}C$ records at equatorial Pacific Site U1334 and derived an age of 23.040 Ma for the base of Chron C6Cn.2n.

In the deep-sea sections studied for astronomical tuning and age recalibration at the Oligocene–Miocene transition [Ocean Drilling Program (ODP) Sites 522, 926, 929, Raffi, 1999; Shackleton et al., 2000; ODP Site 1090, Billups et al., 2004; Channell et al., 2003; IODP (Integrated ODP) Site U1334, Beddow et al., 2016, 2018] the boundary shows clear magnetostratigraphic and isotope characterization. The Mi1 $\delta^{18}O$ increase (Miller et al., 1991), a related $\delta^{13}C$ increase (Zachos et al., 2001), and a major sea-level fall (Miller et al., 2005) occur

Base of the Aquitanian Stage of the Miocene Series at Lemme-Carrosio, Italy.

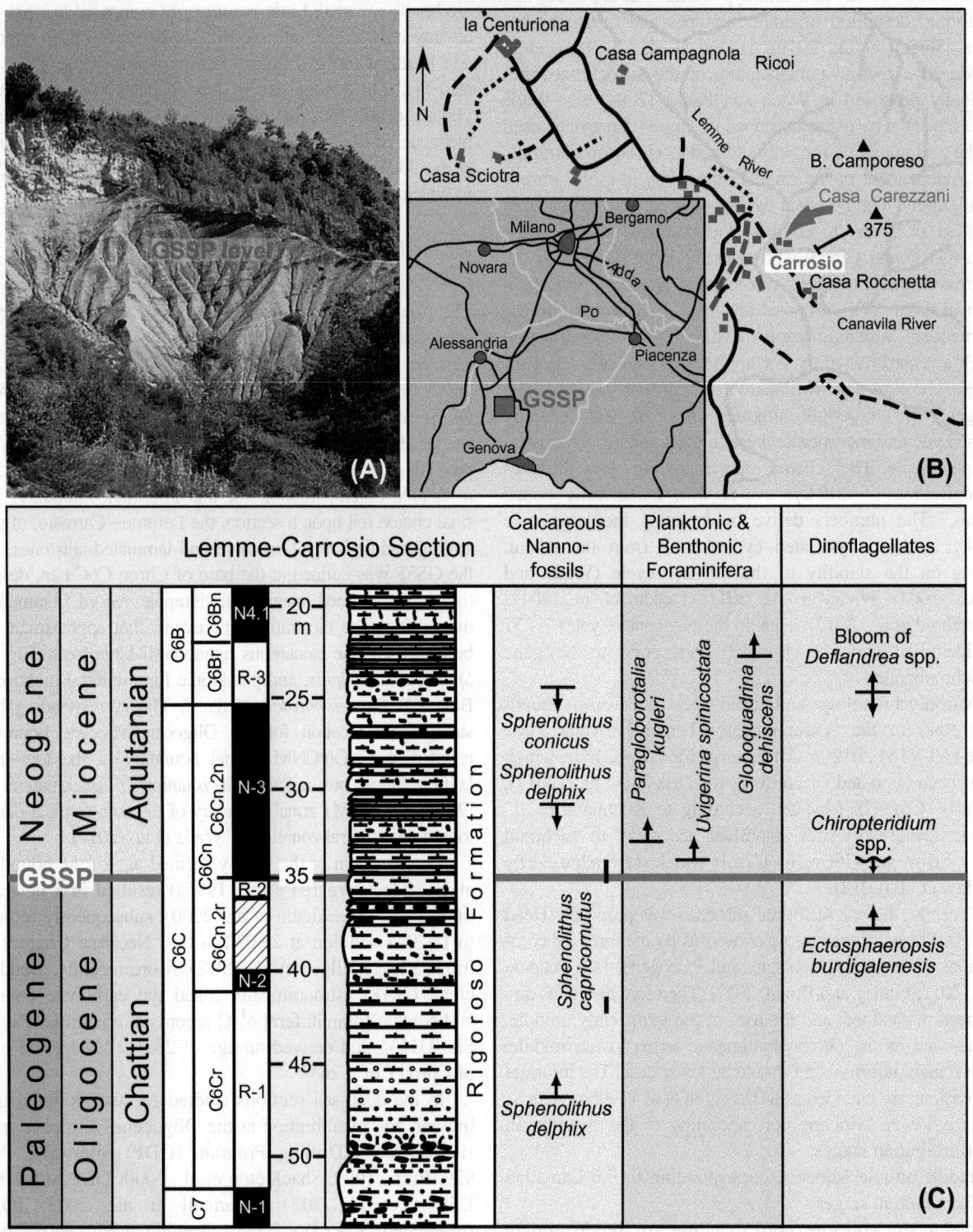

FIGURE 29.2   GSSP for the Aquitanian Stage and Oligocene/Miocene Boundary at Lemme–Carrosio, North Western Italy, showing section (A), location (B), and lithologic log and magneto-biostratigraphic data (C). *GSSP*, Global Boundary Stratotype Section and Point. Calcareous nannofossil data are from Fornaciari and Rio, 1996.

before the boundary in the latest Oligocene (Beddow et al., 2016). The isotopic and clear magnetostratigraphic features are lacking in the Lemme−Carrosio GSSP section preventing precise global correlation: this would suggest the need of replacement of the GSSP section and proposal of a new stratotype section possibly in an (I) ODP core (see Miller and Wright, 2017, for further detail and discussion on this topic).

29.1.2.2 Burdigalian

The criteria of recognition and the age of the Aquitanian/Burdigalian boundary are still an open problem (see discussion in Lourens et al., 2004; Hilgen et al., 2012). As discussed in the Neogene chapter of GTS2012 (Hilgen et al., 2012), among the different criteria that have been proposed so far, the calcareous nannofossil biohorizon B *Helicosphaera ampliaperta* has been considered most suitable for placing (approximating) the Aquitanian/Burdigalian boundary (Fig. 29.1). This choice was made based on the astronomical age of the event (20.44 Ma as calibrated in the Ceara Rise reference sections), and in the presence of *H. ampliaperta* in the historical stratotype of the Burdigalian (e.g., Müller in Bizon and Muller, 1979; Fornaciari and Rio, 1996). It should be noted, however, that although the biohorizon is recognized as useful and reliable in the Mediterranean, it has a low degree of reliability in the open ocean. The other criteria associated to the Aquitanian/Burdigalian boundary are T *Paragloborotalia kugleri* (Berggren et al., 1995a), dated at 21.12 Ma (see Table 29.3, later in this chapter); the top of Chron C6An (Berggren et al., 1995a), dated at 19.979 Ma (Liebrand et al., 2016); and B *Sphenolithus belemnos* (Haq et al., 1987), dated at 19.01 Ma (Backman et al., 2012).

The problem of the Aquitanian/Burdigalian boundary remains, and no significant upgrades have been obtained by the WG on the Langhian and Burdigalian GSSPs, established within the Subcommission on Neogene Stratigraphy (SNS) in 2015. The intent of having the Burdigalian GSSP defined in an astronomically tuned deep-marine section in the Mediterranean, coherently with the other defined Miocene GSSPs, collides with the major issue of the lack of suitable sections in the area. However, recent work by Fabbrini et al. (2019) has identified B *Helicosphera ampliaperta* within Chron C6An.2n at the Contessa Section (Italy) that is a possible candidate for the Burdigalian GSSP.

The alternative to have this boundary defined in an (I) ODP core in the open ocean represents a serious option to be considered. Possible candidates are (1) ODP Site 1264 (Walvis Ridge, Leg 208; Zachos et al., 2004), with excellent carbonate and benthic isotope records that have been tuned to 100-kyr short eccentricity (Liebrand et al., 2016); (2) ODP Site 1090 (Billups et al., 2004) with very good magnetostratigraphy and relatively high-resolution stable isotope records

tuned to obliquity, but at a high-latitude location so standard calcareous plankton biostratigraphy is prevented; (3) IODP Site U1337 (eastern equatorial Pacific, Exp. 320/321; Pälike et al., 2010) with a high-resolution benthic foraminiferal and bulk carbonate stable isotope records (Holbourn et al., 2015); or (4) ODP Sites 925/926/929 (Ceara Rise, Leg 154; Curry et al., 1995) with high-resolution benthic foraminiferal stable isotope records based on a tuned cyclostratigraphy (Shackleton and Hall, 1997; Pälike et al., 2006a; N. Shackleton, unpublished), but no magnetostratigraphy. All these records compose an almost complete astrochronology framework for the Miocene which evidently can enable to defining remaining GSSPs and replacing amended ones (Hilgen et al., 2019). At this time the choice of defining a GSSP in the deep-sea (IODP/ODP) core archives, searching for global chronostratigraphic correlation tools, still finds objections within the stratigraphic community.

29.1.2.3 Langhian

The limits affecting some historical Miocene stratotype sections, for example, poor microfossil content and/or unsuitable sedimentary setting and/or the presence of gap in sedimentation, characterize also the historical Langhian stratotype. It was designated in Northern Italy (Piedmont Basin) and represented by the "Pteropod marls" (following Mayer-Eymar, 1868) now corresponding to the Cessole Formation in the homonymous Cessole Section (Cita and Premoli Silva, 1968). The *Praeorbulina* datum represented the criterion for recognizing the base of the Langhian (see Rio et al., 1997), also taking into account its proximity to top of Chron C5Cn. However, with terrigenous and turbiditic sediments in its lower part, this historical stratotype is not suitable for defining the GSSP, and thus the lower−middle Miocene boundary. Moreover, questions are open as regards to the choice of the "best" criteria (e.g., paleomagnetic reversal, biostratigraphic datum) for approximating the boundary, and different options have been discussed by the Langhian GSSP WG in the recent past. Turco et al. (2011a) suggested that the *Praeorbulina* datum was unreliable as a guiding criterion for recognizing the Langhian GSSP due to its taxonomic ambiguity that resulted in different magnetostratigraphic positions, ranging from the upper part of Chron C5Br, close to the C5Bn.2n/C5Br magnetic boundary (at about 15.2 Ma) to Subchron C5Cn.1n in the Mediterranean (at about 16.2 Ma) and within an interval equivalent to Chron C5Cr in Atlantic Ocean (at about 16.8 Ma). Moreover, the revision of the Langhian historical unit stratotype (Iaccarino et al., 2011) revealed that the placement of *Praeorbulina* datum is not consistent among different authors, and taxonomic issues regarding the definition of *Praeorbulina* are ongoing.

For avoiding an uncertain biostratigraphic horizon, the choice of a paleomagnetic reversal seems more feasible and reliable. The proposals under discussion include (1) the top of C5Br, which could be difficult to detect though, being followed by a short normal magnetozone (C5Bn.2n); (2) the base of Chron C5Br (younger end of C5Cn), which could be a better choice because it is a more distinct paleomagnetic signal (base of a long reversed magnetozone following a short normal magnetozone). We have provisionally placed the Langhian GSSP to coincide with this latter paleomagnetic reversal, dated astronomically at 15.99 Ma (according to GTS2020, slightly older than GTS2012) (Fig. 29.1).

As for the designation of the Langhian GSSP section, the interest has been focused on two Mediterranean sections that have been examined in detail to establish their suitability: the downward extension of the La Vedova beach section in central–northern Italy and St. Peter's Pool section on Malta Island (central Mediterranean). In both the sections the studies dealt with integrated magnetobiostratigraphy, stable isotope stratigraphy, cyclostratigraphy, and astronomical tuning (Foresi et al., 2011; Turco et al., 2011b; Turco et al., 2017). Although St. Peter's Pool section has better preservation of the calcareous plankton, the higher quality magneto–cyclostratigraphic record of La Vedova Section indicates this section for a proposal to Langhian GSSP definition, as resulted from discussion within the SNS Working group during the STRATI 2019 Meeting (Milan, Italy, July 1–5, 2019). A direct link of the Burdigalian/Langhian boundary to the open ocean benthic foraminiferal isotope record is appropriate, taking into account that it falls within a critical paleoclimatic interval, the Miocene Climatic Optimum. For this reason, searching for an auxiliary boundary stratotype in an (I)ODP core would be advisable. Potential candidates are deep-sea Pacific cores from IODP Exp. 320/321, Sites U1337 and U1338 (Holbourn et al., 2014, 2015).

29.1.2.4 Serravallian

The Serravallian GSSP was established in the Ras-il-Pellegrin section on Malta Island (Hilgen et al., 2009), with the boundary defined between the "Globigerina Limestone" and the "Blue Clay Formation" (at top "transitional bed"; Fig. 29.3). The choice was not in line with the GSSP guide for selecting the best criterion (Remane et al., 1996), being coincident with a formation boundary. However, it was a justified selection because the GSSP level corresponds to a global climatic signal, provided by the positive $\delta^{18}O$ shift associated with Mi-3b (Miller et al., 1991, 1996) and the carbon isotope excursion CM6 (Woodruff and Savin, 1991; Holbourn et al., 2014), and is clearly recorded in the bulk carbonate record of Ras-il-Pellegrin section (Fig. 29.3). The Mi-3b $\delta^{18}O$ shift has been precisely recorded in several deep-sea records and represents a clear correlation to the open ocean of the

Serravallian GSSP as defined in a Mediterranean section. The validity of this GSSP section is supported by the soundness of integrated stratigraphic correlations to other Mediterranean and open ocean sections. The biostratigraphic events that approximate the boundary are the calcareous nannofossil biohorizon T common *Sphenolithus heteromorphus* that follows the GSSP, and planktonic foraminiferal biohorizon T *Clavatorella bermudezi* a rare but distinctive secondary marker within Zone M7. The biostratigraphic position corresponds to CNM7/CNM8 zonal boundary of the low–mid latitude calcareous nannofossil zonation of Backman et al. (2012).

The reader is referred to GTS2012 (Chapter 29, Hilgen et al., 2012) for further details on the integrated stratigraphic data on the Serravallian GSSP interval.

29.1.2.5 Tortonian

The GSSP of the Tortonian Stage was established in the Monte dei Corvi sction (northern Italy) (Hilgen et al., 2003, 2005; Hüsing et al., 2007) and formally defined at the midpoint of the sapropel of small-scale sedimentary cycle 76 (Fig. 29.4). It is correlated to the short normal subchron C5r.2n and slightly precedes the $\delta^{18}O$ increase of the Mi-5 event of Miller et al. (1991). The calcareous nannofossil biohorizon Top common *Discoaster kugleri* that corresponds to the CNM10/CNM11 zonal boundary of the low–mid latitude calcareous nannofossil zonation of Backman et al. (2012) approximates the GSSP level. In terms of (sub)tropical planktonic foraminiferal zonation, the GSSP roughly corresponds to the B *Globoturborotalita nepenthes* that marks the M10/M11 zonal boundary (Wade et al., 2011), although diachroneity of the marker events between the open ocean and Mediterranean makes biostratigraphic correlation difficult.

29.1.2.6 Messinian

The GSSP of the Messinian Stage has been formally defined in the Oued Akrech section (Morocco–Atlantic side of the Mediterranean) at the base of the reddish marl of basic cycle OA15 (Fig. 29.5; Hilgen et al., 2000a, 2000b), located within the Subchron C3Br.1r and astronomically calibrated at 7.246 Ma. The calcareous nannofossil biohorizon B *Amaurolithus delicatus* approximates the GSSP and indicates the lower part of Zone CNM17 of the low–mid latitude calcareous nannofossil zonation of Backman et al. (2012). In terms of planktonic foraminiferal biostratigraphy, the GSSP falls approximately midway through (sub)tropical Subzone M13b (*Globorotalia plesiotumida/Globorotalia lenguaensis* concurrent-range subzone) of Wade et al. (2011). The regular occurrence of *Globorotalia miotumida* (=*conomiozea*) group closely coincides with the

Base of the Serravallian Stage of the Miocene Series in the Ras il Pellegrin Section, Fomm Ir-Rih Bay, Malta.

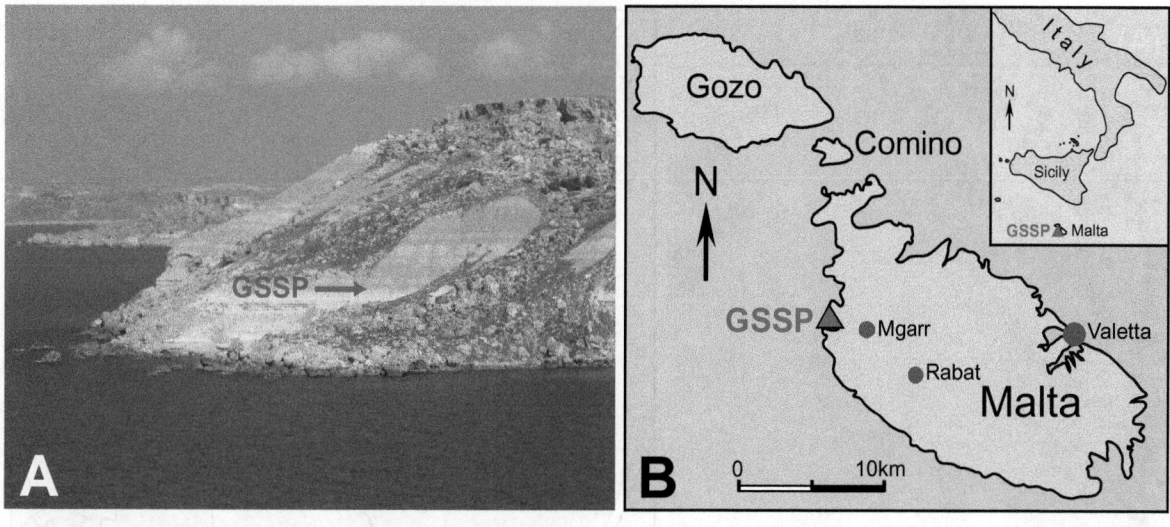

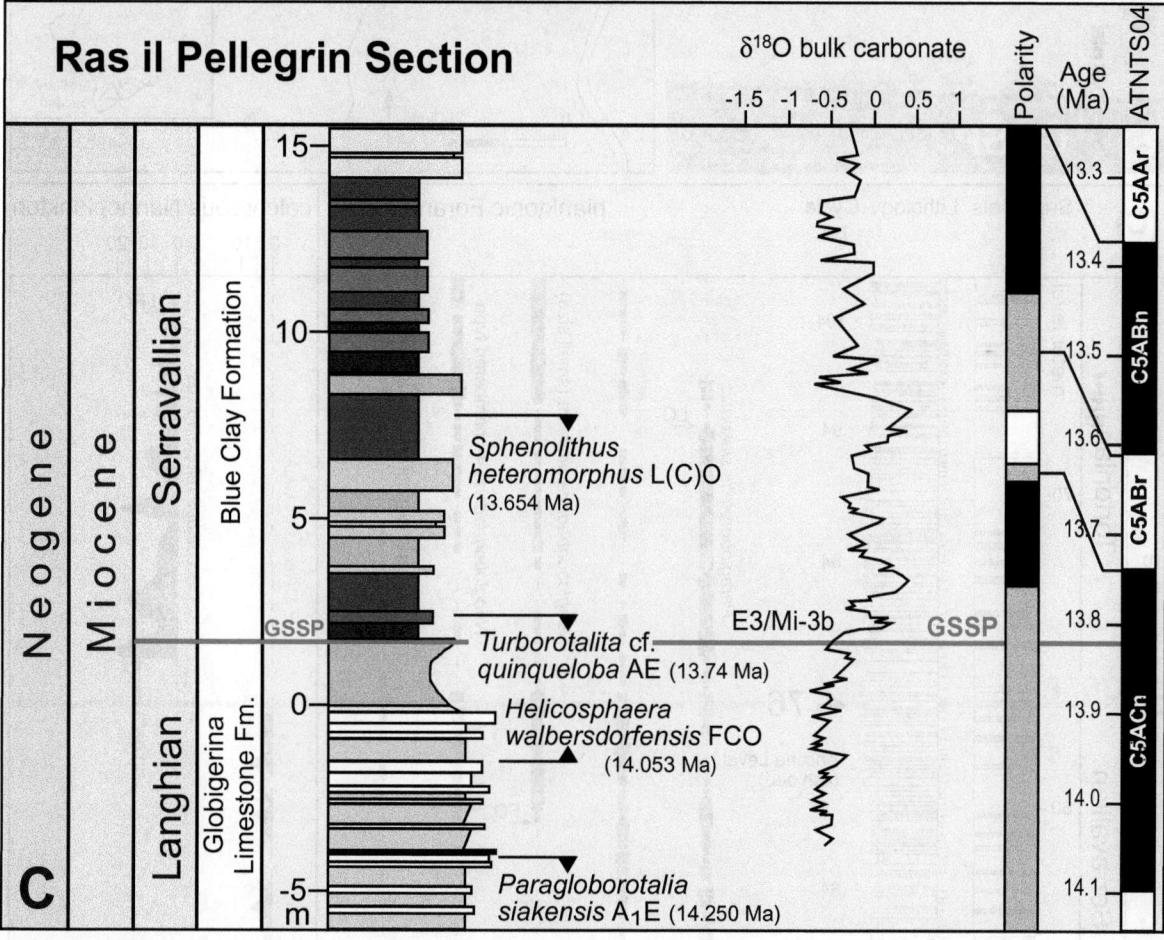

FIGURE 29.3 GSSP for the Serravallian stage at Ras-il-Pellegrin on Malta Island, showing section (A), location (B), and lithologic log and magneto-biostratigraphic and oxygen isotope data (C). E3/Mi-3b indicates oxygen isotope main shift, labeled following Flower and Kennett (1993) and Miller et al. (1996). Generic assignments of foraminifera have been updated. A_1E, First Acme End; *AE*, Acme End; *FCO*, first common occurrence; *GSSP*, Global Boundary Stratotype Section and Point; *L(C)O*, Last Common Occurrence; *LO*, Last Occurrence.

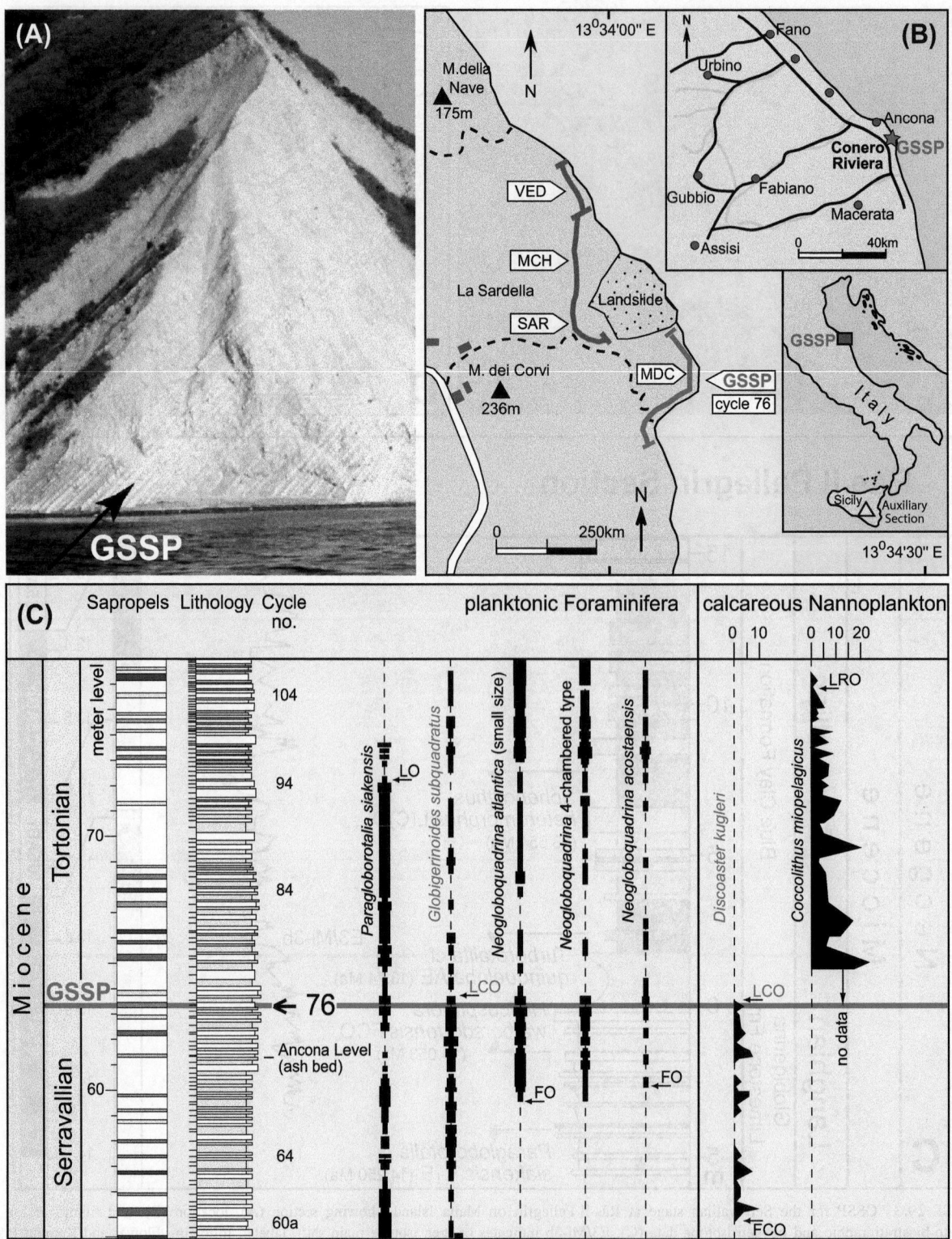

FIGURE 29.4 GSSP for the Tortonian Stage at Monte dei Corvi, central–Eastern Italy, showing section (A), location (B), and lithologic log and magneto-biocyclostratigraphic data (C). *FCO*, First Common Occurrence; *FO*, First Occurrence; *GSSP*, Global Boundary Stratotype Section and Point; *LCO*, Last Common Occurrence; *LO*, Last Occurrence; *LRO*, Last Rare Occurrence.

Base of the Messinian Stage of the Miocene Series of the Neogene System at Oued Akrech, Morocco

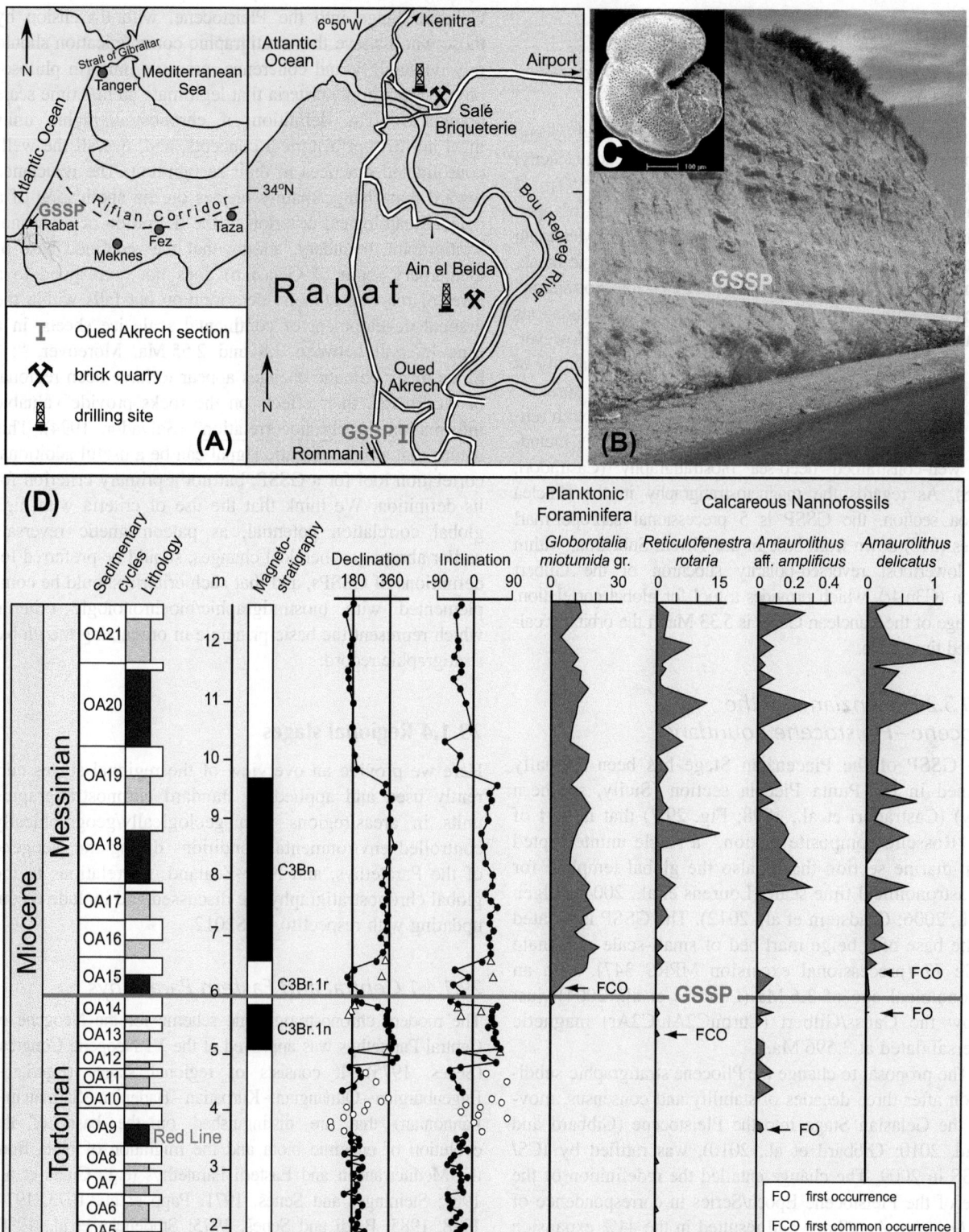

FIGURE 29.5 GSSP for the Messinian Stage at Oued Akrech (Morocco), showing location (A), section (B), the marker species *Globorotalia miotumida* (C), and lithologic log and magneto-biostratigraphic data (D). Acronyms as specified in Fig. 29.3. *GSSP*, Global Boundary Stratotype Section and Point.

Messinian boundary as defined in the GSSP section and in Mediterranean area.

29.1.3 Subdivisions of the Pliocene

29.1.3.1 Zanclean and the Miocene–Pliocene boundary

The GSSP of the Zanclean Stage—the Miocene/Pliocene boundary—was placed at Eraclea Minoa section (Sicily, southern Italy) and located at the base of the "Trubi" marl Formation that represents the reestablishment of open marine conditions in the Mediterranean after the "Messinian Salinity Crisis" (Fig. 29.6). The location at a lithologic break, from nonmarine sands and marls of "Arenazzolo Formation" to marine Trubi marls, does not respect the recommendations for the choice of a GSSP (Remane et al., 1996). However, Van Couvering et al. (2000) clearly explained the validity of this choice that it is a "historically justified boundary horizon" that "can be correlated by multiple lines of evidence with reliable precision" (p. 180, Van Couvering et al., 2000), including well-constrained deep-sea biostratigraphy (Castradori, 1998). As regards the magnetostratigraphy in the Eraclea Minoa section, the GSSP is 5 precessional sapropel-marl cycles (510) down from base of the Thvera Subchron, within the lowermost reversed-polarity subchron of the Gilbert Chron (C3n.4r), which provides a tool for global correlation. The age of the Zanclean GSSP is 5.33 Ma in the orbitally calibrated time scale.

29.1.3.2 Piacenzian and the Pliocene–Pleistocene boundary

The GSSP of the Piacenzian Stage has been formally defined in the Punta Piccola section (Sicily, southern Italy) (Castradori et al., 1998; Fig. 29.7) that is part of the "Rossello composite section," a single uninterrupted deep-marine section that is also the global template for the astronomical time scale (Lourens et al., 2004; Hilgen et al., 2006; Gradstein et al., 2012). The GSSP is located at the base of a beige marl bed of small-scale carbonate cycle 77 (precessional excursion MPRS 347), with an astronomical age of 3.6 Ma (Lourens et al., 2004), just below the Gauss/Gilbert (ChronC2An/C2Ar) magnetic reversal dated at 3.596 Ma.

The proposal to change the Pliocene stratigraphic subdivision after three decades of stability and consensus, moving the Gelasian Stage into the Pleistocene (Gibbard and Head, 2010; Gibbard et al., 2010), was ratified by ICS/IUGS in 2009. The change entailed the redefinition of the base of the Pleistocene Epoch/Series in correspondence of the Gelasian Age/Stage and resulted in the 44% expansion of the Pleistocene (Van Couvering et al., 2009). In Chapter 29 of GTS2012 on the Neogene Period (Hilgen et al., 2012) the new chronostratigraphy was explained, although the authors favored the previous chronostratigraphic position for the Gelasian. Controversial reactions by the scientific community accompanied the shift of the Gelasian Stage into the Pleistocene, with dissension by those who believe that stratigraphic communication should rely on stability and coherence, preserve uniform philosophy, and not apply criteria that legitimate ad hoc time scale boundaries. The definition of chronostratigraphic units must not disrupt historical concepts, and, overall the well-consolidated practices in their recognition. The reluctance toward that change mainly resides on the application of a paleoclimatological criterion in the definition of a chronostratigraphic boundary, adding that the redefined base of Quaternary (base of Gelasian) does not correspond to a level of relevant climatic deterioration but falls within the gradual development of continental scale ice sheets, in a time interval between 2.8 and 2.55 Ma. Moreover, "... Since many climatic changes appear to have been regional or worldwide, their effects on the rocks provide valuable information for chronocorrelation" (Salvador, 1994). This means that a paleoclimatic signal can be a useful additional correlation tool for a GSSP, but not a primary criterion for its definition. We think that the use of criteria with high global correlation potential, as paleomagnetic reversals and/or abrupt geochemical changes, should be preferred for definitions of GSSPs, and that such criteria should be complemented with biostratigraphic/biochronologic criteria, which represent the basic principle in organizing the global stratigraphic record.

29.1.4 Regional stages

Here we provide an overview of the regional stages currently used and applied as standard chronostratigraphic units in areas/regions with geologically/geographically controlled environmental conditions during the Neogene of the Paratethys, and New Zealand. Correlations to the global chronostratigraphy are discussed and include recent updating with respect to GTS2012.

29.1.4.1 Central and Eastern Paratethys

The modern chronostratigraphic scheme for the Neogene of Central Paratethys was approved at the VI Neogene Congress (Senes, 1975). It consists of regional stages (Egerian–Eggenburgian–Ottnangian–Karpatian–Badenian–Sarmatian–Pannonian) that are distinguished on the basis of the evolution of endemic biota and the migration of biota from the Mediterranean and Eastern Paratethys (e.g., Cicha et al., 1967; Steininger and Seneš, 1971; Papp et al., 1973, 1974, 1978, 1985; Báldi and Seneš, 1975; Studencka et al., 1998) (Fig. 29.8). For the early and middle Miocene, calcareous nannofossils provide the most useful biostratigraphy, and reference to the zonations of Backman et al. (2012) and Martini

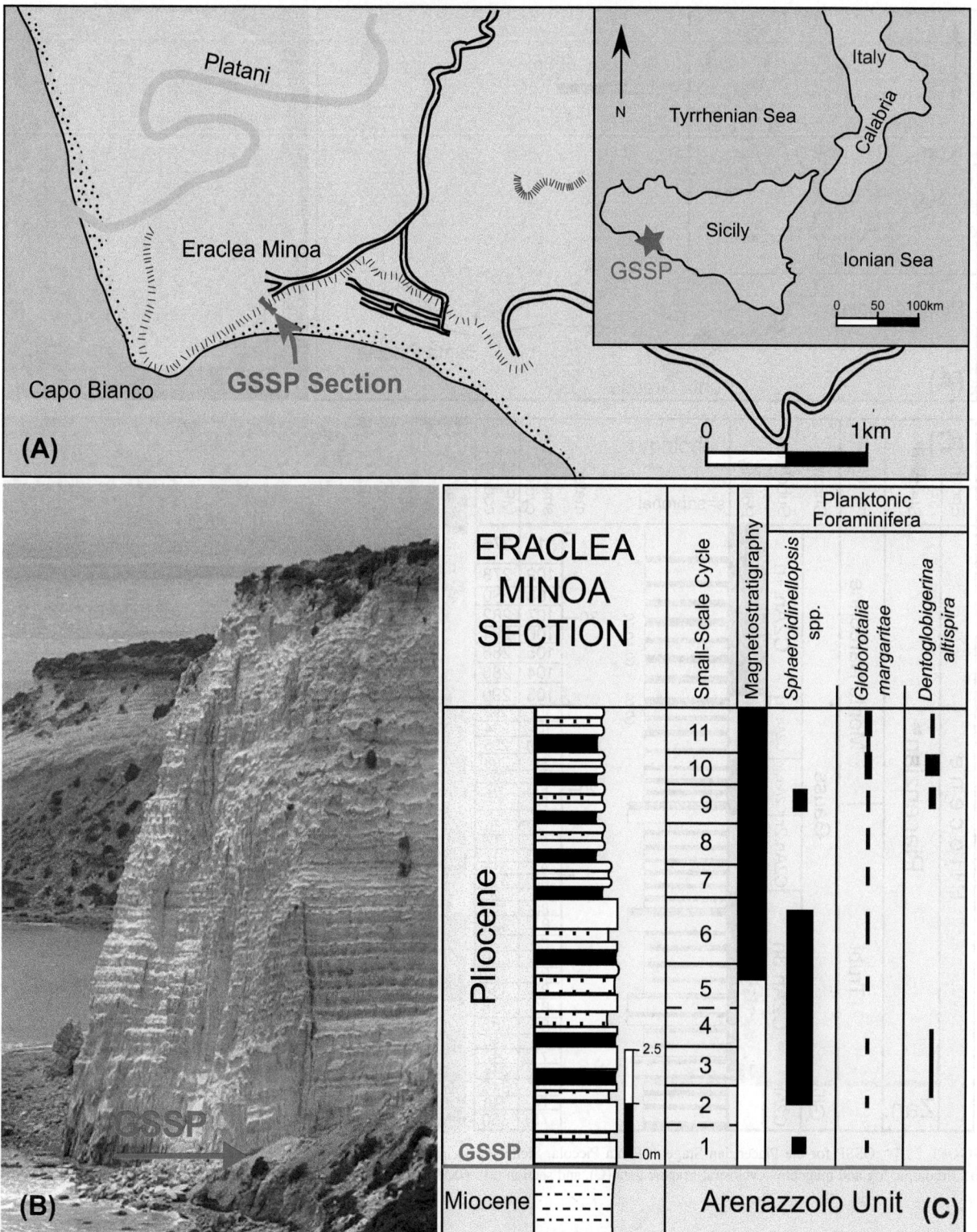

FIGURE 29.6   GSSP for the Zanclean Stage and base of Pliocene Series at Eraclea Minoa, Sicily (Italy), showing location (A), section (B), lithologic log and magneto-biostratigraphic data (C). Generic assignments of foraminifera have been updated. *GSSP*, Global Boundary Stratotype Section and Point.

Base of the Piacenzian Stage of the Pliocene Series of the Neogene System at Punta Piccola, Sicily, Italy

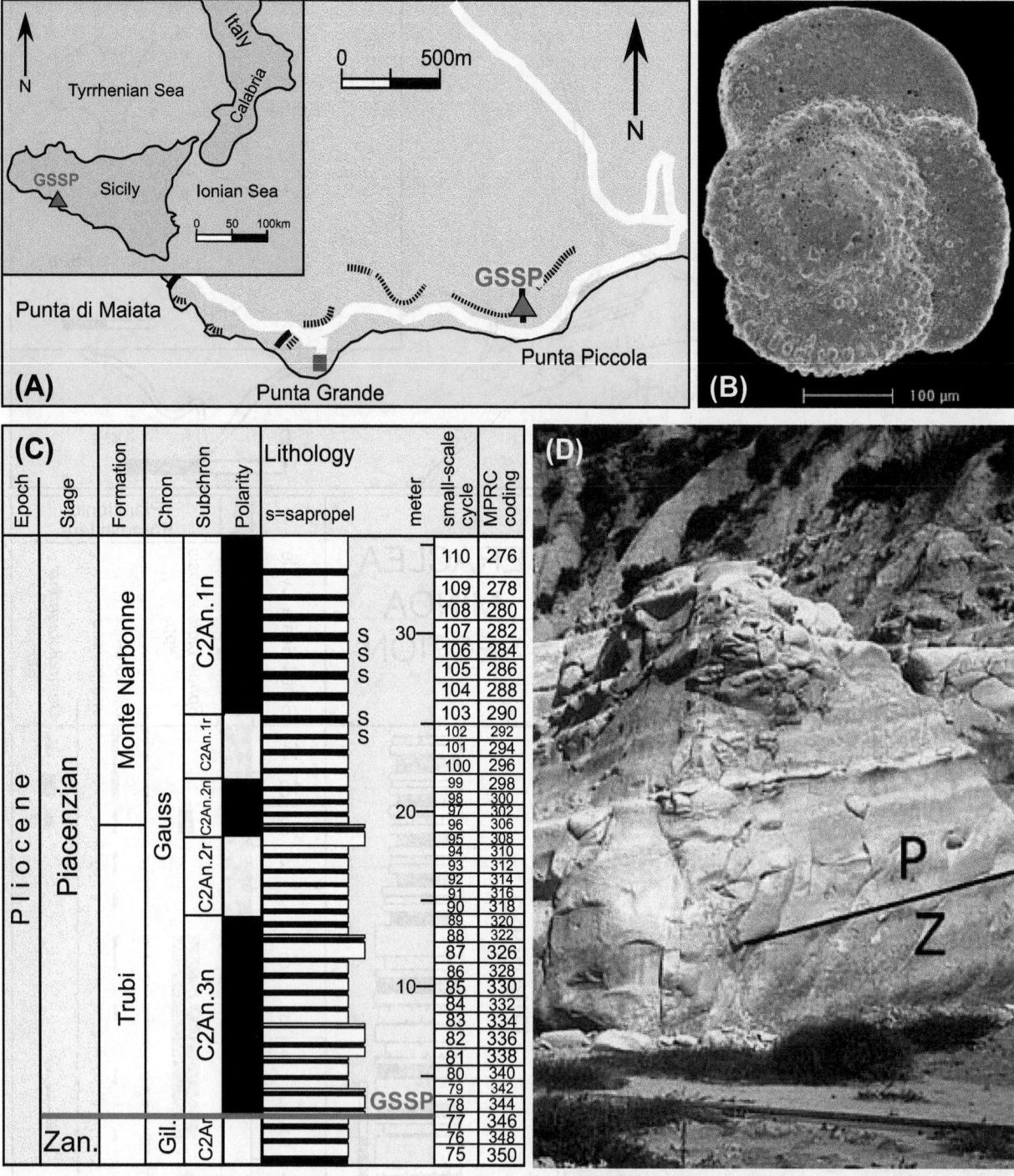

FIGURE 29.7   GSSP for the Piacenzian Stage at Punta Piccola, Sicily (Italy), showing location (A), the marker species *Globorotalia crassaformis* (B), lithologic log and magneto—cyclostratigraphic data (C), and section (D). *GSSP*, Global Boundary Stratotype Section and Point.

(1971) are reported next. Both absolute (isotopic, radiometric) and relative (biomagnetostratigraphic) chronologies are used to provide age constraints for the Neogene time scale of Central Paratethys (e.g., Hardenbol et al., 1998; Rögl et al., 2007; Piller et al., 2007; Hohenegger et al., 2014; Kováč et al., 2018). In addition, climatic changes, sea-level variations, and the migration of planktonic and benthic organisms between the Central and Eastern Paratethys have been taken into account to validate the accuracy of correlations (e.g., Holcova, 2005, 2013; Holcova et al., 2015, 2018; Sant et al., 2017a, 2017b; Kováč et al., 2018).

The Paleogene/Neogene boundary is located within the Egerian Stage of the Central Paratethys (Fig. 29.8). Nannofossil biostratigraphy is mostly used to identify this boundary, through the biohorizons B *Discoaster druggii* and B *Helicosphaera scissura* that occur within Zone CNM1 and mark the boundary of the NN1/NN2 zones (Holcova, 2002, 2005, 2017; Kováč et al., 2018). The exact determination of the Paleogene/Neogene boundary is generally problematic because Zone NN1 is difficult to recognize in Paratethys. An additional marker for this boundary is the larger foraminiferal datum B *Miogypsina gunteri* (e.g., Rögl, 1998; Piller et al., 2007).

The Eggenburgian Stage falls within the nannofossil Zone NN2 and roughly corresponds to Zone CNM4 and to the lower Burdigalian (Fig. 29.8). Its base is correlated with the sequence boundary Aq2 of Hardenbol et al. (1998) in terms of sequence stratigraphy and sea-level

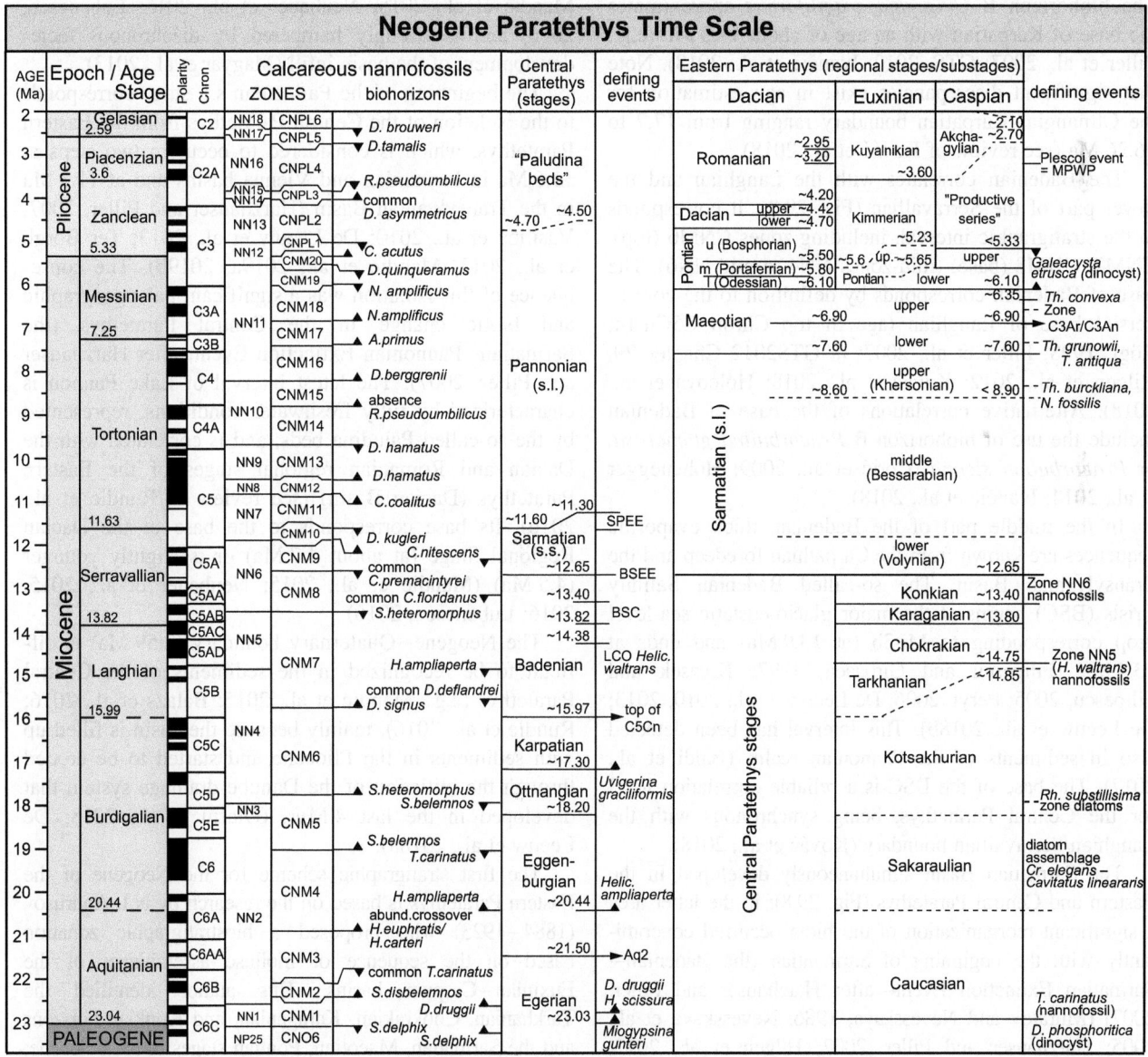

FIGURE 29.8 Regional Neogene stages for the Paratethys with defining events and correlations to standard global stages and low-to-middle-latitude calcareous nannofossil biostratigraphy. Subdivisions are shown both for the Central and Eastern Paratethys.

changes, dated at about 21.5 Ma (Kovač et al., 1999, 2018; Piller et al., 2007; in GTS2012, Chapter 29, Hilgen et al., 2012). The biohorizon B *H. ampliaperta* that occurs within Zone CNM4 (NN2) close to the Aquitanian/Burdigalian boundary (about 20.44 Ma) is recognized within the Eggenburgian and is considered a reliable event in the Central Paratethys (e.g., Holcova, 2005, 2017; Piller et al., 2007; Grunert et al., 2015).

The Ottnangian, corresponding to upper CNM5–lower CNM6 zones (upper NN3–lower NN4 zones), correlates with the Mid Burdigalian. Its base is placed at about 18.2 Ma and corresponds to the middle part of Chron C5En (Kováč et al., 2004; Piller et al., 2007; Pipper et al., 2017).

The Karpatian correlates with the upper part of Burdigalian and corresponds to Zone CNM6 (Zone NN4). The biohorizon B *Uvigerina graciliformis* approximates the base of Karpatian with an age of about 17.3 Ma (e.g., Piller et al., 2007; GTS, 2012; Ivančič et al., 2018). Note that significant discrepancies exist in age estimation for the Ottnangian/Karpatian boundary ranging from 17.7 to 16.26 Ma (see review of Kováč et al., 2018).

The Badenian correlates with the Langhian and the lower part of the Serravallian (Fig. 29.8). It corresponds to the stratigraphic interval, including zones CNM6 (top), CNM7, CNM8 (base) (mid zones NN4 to mid NN6). The base of Badenian corresponds by definition to the controversial base of Langhian (age of top Chron C5Cn.1n; Rögl, 1998; Piller et al., 2007; in GTS2012 Chapter 29, Hilgen et al., 2012; Kováč et al., 2018; Holcova et al., 2018). Alternative correlations of the base of Badenian include the use of biohorizon B *Praeorbulina glomerosa*, or *Praeorbulina sicana* (Ćorić et al., 2009; Hohenegger et al., 2014; Ivančič et al., 2018).

In the middle part of the Badenian, thick evaporitic sequences are known from the Carpathian foredeep and the Transylvanian Basin. The so-called Badenian Salinity Crisis (BSC) begins at the major glacio-eustatic sea-level drop corresponding to Mi-3b (at 13.9 Ma) and ends at 13.4 Ma (Filipescu and Girbacea, 1997; Krezsek and Filipescu, 2005; Peryt, 2006; De Leeuw et al., 2010, 2013; De Leeuw et al., 2018b). This interval has been detected also in sediments of the Pannonian realm (Baldi et al., 2017). The base of the BSC is a reliable correlation level for the Central Paratethys, being synchronous with the Langhian/Serravallian boundary (Kováč et al., 2018).

The Sarmatian Basin simultaneously developed in the Eastern and Central Paratethys (Fig. 29.8): in the latter area a significant reorganization of the biota occurred concomitantly with the beginning of Sarmantian (the Badenian–Sarmatian Extinction Event—after Harzhauser and Piller, 2007) (Muratov and Nevesskaya, 1986; Nevesskaya et al., 2005; Harzhauser and Piller, 2007; Hilgen et al., 2012; Ter Borgh et al., 2014). The base of the Sarmatian of the Central Paratethys (the Sarmatian s.s.) is dated at about

12.65 Ma (Paulissen et al., 2011; Palcu et al., 2015; Mandic et al., 2019a, 2019b).

The Pannonian stage *sensu* Lörenthey (1900) or "Pannonian sensu lato (s.l.)" includes the upper Miocene–Pliocene sediments of the Central Paratethys (Lake Pannon) and is subdivided into two units, Pannonian *sensu* Stevanovic (1951) or "Pannonian sensu stricto (s.s.)" and Pontian *sensu* Andrussov (1887). Moreover, the Pannonian is also subdivided in other (alternative) chronostratigraphic units (see review in Sacchi et al., 1997; Sacchi and Horvath, 2002). The chronostratigraphic framework for the Pannonian s.l. as well as the correlation with the coeval units of the Eastern Paratethys and the Mediterranean is still under discussion (Sacchi et al., 1997; Sacchi and Horvath, 2002; Leever et al., 2011; Csato et al., 2015; Mandic et al., 2015; Neubauer et al., 2015; Lubenescu, 2016) and is strongly hampered by diachronous facies development of the basin infill (Magyar et al., 2013).

The beginning of the Pannonian s.l. stage corresponds to the isolation of the Central Paratethys from the Eastern Paratethys, which is considered to occur in two steps at 11.6 Ma in Pannonian and Vienna basins and at 11.3 Ma in the Transylvanian Basin (Harzhauser and Piller, 2007; Vasiliev et al., 2010; De Leeuw et al., 2013; Ter Borgh et al., 2013; Mandic et al., 2019a, 2019b). The consequence of this isolation was a significant paleogeographic and biotic change in the Central Paratethys (the Sarmatian–Pannonian Extinction Event; after Harzhauser and Piller, 2007). The latest interval of Lake Pannon is characterized by fully freshwater conditions, represented by the so-called Paludina beds, and is correlated with the Dacian and Romanian regional stages of the Eastern Paratethys (Dacian Basin) (see review in Rundic et al., 2016). Its base corresponds to the base of the Dacian Regional Stage (at about 4.7 Ma) or is slightly younger (4.5 Ma) (Mandic et al., 2015; Neubauer et al., 2015, 2016; Lubenescu, 2016).

The Neogene–Quaternary boundary (2.59 Ma) is difficult to be recognized in the sediments of the Central Paratethys (e.g., Mandic et al., 2015; Balazs et al., 2016; Rundic et al., 2016), mainly because the basin is filled up with sediments in the Pliocene and started to be eroded through the initiation of the Danube drainage system that developed in the last 4 Myr. (Olariu et al., 2018; De Leeuw et al., 2018a).

The first stratigraphic scheme for the Neogene of the Eastern Paratethys is based on the research by N.I. Andrusov (1884–1923) who proposed a biostratigraphic zonation based on the sequence of mollusc assemblages of the Euxinian–Caspian basins. This author identified the Tarkhanian, Chokrakian, Karaganian, and Konkian horizons and the Sarmatian, Maeotian, Pontian stages in the Miocene, the Kimmerian, Kuyalnikian (Apsheronian), and Chaudian (Bakunian) stages in the Pliocene (Andrusov, 1961a, 1961b).

Later, the Sakaraulian and Kotsakhurian horizons (beds with fauna) were added to the Miocene by Davitashvili (1933). In 1955, A.G. Eberzin proposed a unified stratigraphic scheme of the Neogene deposits of Southern USSR, subsequently approved by the Interdepartmental Stratigraphic Committee of the USSR (ISC USSR) in 1956 (Eberzin, 1959). In 1973 the USSR Neogene Commission of the ISC converted the horizons of the lower and middle Miocene (Sakaraulian, Kotsakhurian, Tarkhanian, Chokrakian, Karaganian, and Konkian) in regional stages and, subsequently, indicated the Caucasian as the lowest regional stage of the Neogene of the Eastern Paratethys (Nevesskaya et al., 1975a, 1975b). This regional chronostratigraphic scheme was approved at the VI Neogene Congress in Bratislava and accepted by the ISC USSR in 1983 (Nevesskaya et al., 1975a, 1975b, 1984; Muratov and Nevesskaya, 1986). It consists of distinct regional stages that reflect the geological history of the region, from North-Eastern Bulgaria and Eastern Romania in the West to the foothills of the Kopetdag in the East (Nevesskaya et al., 1975a, 1975b, 1984). For the Neogene of the Eastern Paratethys, the described chronostratigraphic scheme is still valid (Fig. 29.8).

In the Eastern Paratethys the Paleogene/Neogene boundary is difficult to be recognized. It is defined according to calcareous nannofossil and/or dinocysts biostratigraphy only in wells of the Northern Black Sea Region (Chernobaevka Formation) and the Kerch Peninsula (Arabatska Formation) (Portniagina, 1980; Andreyeva-Grigorovich and Gruzman, 1989; Andreyeva-Grigorovich, 2004; Vernyhorova and Ryabokon, 2018), and in several outcrops in Ciscaucasia (Alkun Formation) (Beluzhenko et al., 2018) and Caucasus (Uplistsikhe layers) (Minashvili and Ananiashvili, 2017). This boundary is recognized through dinoflagellate cysts at the extinction of the genera *Chiropteridium*, *Wetzeliella*, *Rhombodinium*. The boundary interval contains also some *Deflandrea* species, mainly *Deflandrea phosphoritica*, *D. phosphoritica* var. *spinulosa* (Portniagina, 1980; Andreyeva-Grigorovich, 2004; Filippova et al., 2010, 2013, 2015; Beluzhenko et al., 2014, 2018; Vernyhorova and Ryabokon, 2018). The Paleogene/Neogene boundary can also be recognized for the occurrence of *Triquetrorhabdulus carinatus*, the calcareous nannofossil marker species of Zone NN1, and *Triquetrorhabdulus. milowii* (Beluzhenko et al., 2018). Moreover, the appearance of *T. carinatus* and the presence of *Spenolithus ciperoensis* permit the identification of the upper Oligocene—lower Miocene transition (Minashvili and Ananiashvili, 2017), corresponding to zones CNO6—CNM1 (NP25—NN1) biostratigraphic interval.

In the Eastern Paratethys the lower Miocene comprises three regional stages: Caucasian, Sakaraulian, and Kotsakhurian that are still poorly dated by means of biostratigraphy. The boundaries are generally determined inadequately through lithological changes and rare paleontological data (see review in Vernyhorova and Ryabokon, 2018).

The Caucasian was originally defined as an analog of Aquitanian Stage (Nevesskaya et al., 1975a, 1975b, 1984). The Sakaraulian is correlated with the Eggenburgian, according to the presence of some benthic foraminifera (*Bulimina tumidula*, *Caucasinella elongata*, *Trochammina depressa*), molluscs (e.g., *Plagiocardium abundans*), ostracods (e.g., *Neomonoceratina helvetica*), therefore, to the lower part of the Burdigalian (Nevesskaya et al., 1975a, 1975b, 1984; Muratov and Nevesskaya, 1986). The Sakaraulian deposits of the Crimean Peninsula (part of the Arabatska Formation) contain sediments with diatom assemblages, including the taxa *Craspedodiscus elegans*, *Cavitatus jouseanus* (=*Sindera jouseana*), which are similar to the assemblages of the *Melosira hispanica* Zone from the Eggenburgian that, in turn, is correlated to lower Burdigalian [nannofossil Zone NN2 (=Zone CNM4); Olshtynskaya, 1996; Olshtynska, 2001].

The Kotsakhurian Regional Stage is correlated with the Ottnangian and Karpatian-lowermost Badenian of Central Paratethys, by the presence of some brackish-water molluscs (e.g., *Rzehakia*, *Congeria*, and *Melanopsis*) and, consequently, with the upper part of the Burdigalian (Nevesskaya et al., 1975a, 1975b, 1984; Muratov and Nevesskaya, 1986) and part of the Langhian (Palcu et al., 2019a, 2019b). The Kotsakhurian deposits in the Northern Black Sea region and northern part of the Crimean Peninsula (Karha beds) contain diatom assemblages of the *Rhaphoneis substillisima* Zone and are correlated with the Ottnangian (Olshtynskaya, 1996; Olshtynska, 2001).

The Tarkhanian Regional Stage is still a matter of debate (Fig. 29.8). It is correlated to the lower Miocene (e.g., Konenkova and Bogdanovich, 1994; Barg and Ivanova, 2001), the middle Miocene (e.g., Andreyeva-Grigorovich and Savytskaya, 2000; Krasheninnikov et al., 2003; Golovina, 2012), or to the lower and middle Miocene simultaneously (e.g., Nevesskaya et al., 2003). The different correlations of the Tarkhanian mainly depend on different interpretations of the nannofossil data. Based on the presence of *Helicosphaera ampliaperta* and recognition of Zone NN4 (=CNM6 − lower CNM7), the Tarkhanian was correlated with the Karpatian of the Central Paratethys (Konenkova and Bogdanovich, 1994; Studencka et al., 1998; Barg and Ivanova, 2001). Based on the presence of *Sphenolithus heteromorphus* and recognition of Zone NN5 (=CNM7), the Tarkhanian was correlated with the lower Badenian of the Central Paratethys and the middle Miocene (Rögl, 1998; Andreyeva-Grigorovich and Savytskaya, 2000; Krasheninnikov et al., 2003; Golovina, 2012). According to magnetostratigraphic data (sections near Belaya and Pshekha rivers of Ciscaucasia), the base of Tarkhanian is dated at about 14.85 Ma (Palcu et al., 2019a, 2019b). This age is in accordance with the presence of *Sphenolithus heteromorphus* and *Helicosphaera waltrans* that are used as biostratigraphic markers of the Zone

NN5 (=Zone CNM7) in the Central Paratethys (Rögl and Muller, 1976; Theodoridis, 1984; Ćorić and Švabenská, 2004; Ćorić et al., 2007), and with the appearance of *P. glomerosa* (at an age <15.2 Ma), concomitantly with the Badenian flooding of the Carpathian foredeep (Sant et al., 2017a, 2017b; Kovač et al., 2007, 2017; Golovina, 2012; Palcu et al., 2019a, 2019b).

The middle Miocene in the Eastern Paratethys also includes the regional stages Chokrakian, Karaganian, Konkian, and the regional substages Volhynian (lower Sarmatian s.l.) and part of the middle Sarmatian s.l. (Bessarabian) (Nevesskaya et al., 2003). The Chokrakian is correlated with part of the Badenian, with a base about 14.75 Ma in age (in sections near Belaya and Pshekha rivers of Ciscaucasia) (Palcu et al., 2019a, 2019b). The Karaganian corresponds to the BSC (e.g., Muratov and Nevesskaya, 1986; Peryt et al., 2004; Peryt, 2006), with a base about 13.9 Ma in age (Palcu et al., 2017, 2019a, 2019b).

The Konkian is correlated with the upper Badenian and corresponds to the biostratigraphic interval between Zones CNM8–CNM11 (=Zones NN6–NN7), based on nannofossil assemblages that were found in different sections of the Eastern Paratethys (Muratov and Nevesskaya, 1986; Radionova et al., 2012; Palcu et al., 2017). The base of Konkian is dated at about 13.4 Ma (Palcu et al., 2017, 2019a, 2019b).

The Sarmatian Regional Stage of the Eastern Paratethys or Sarmatian (s.l.) is subdivided into three regional substages: lower (Volhynian), middle (Bessarabian), upper (Khersonian) (Nevesskaya et al., 1975a, 1975b, 1984; Muratov and Nevesskaya, 1986). The beginning of the Sarmatian is characterized by the unification of all Paratethys basins (see Palcu et al., 2019a, 2019b; Simon et al., 2018). During the Sarmatian the Pannonian Basin became isolated first, followed later by the Dacian Basin, and developed as a solo paleogeographic entity separated from the rest the Euxinian–Caspian part of the Eastern Paratethys (Fig. 29.8) (see review in Muratov and Nevesskaya, 1986; Van Baak et al., 2017; Palcu et al., 2019a, 2019b).

The base of the Sarmatian (s.l.) is dated at about 12.65 Ma through magnetostratigraphic data (Ter Borgh et al., 2014; Palcu et al., 2017, 2019a, 2019b). A different radiometric age of 13.6 Ma (±0.89) was previously obtained from the Sarmatian deposits in the Western Near-Black Sea Region (Moldavian Plate), the Kerch Peninsula, and the Taman Peninsula (Chumakov et al., 1984). The base of the middle Sarmatian s.l. (Bessarabian) is still poorly dated, but it is generally placed at an age older than 11.3 Ma (Piller et al., 2007; Filipescu et al., 2011). The base of the upper Sarmatian s.l. (Khersonian) is estimated to be younger than 8.9 Ma, based on the occurrence of oceanic diatoms *Thalassiosira burckliana* and *Nitzschia fossilis* in the lowermost Khersonian deposits of Ciscaucasia (Euxinian) (Radionova et al., 2012). The first occurrence

of these taxa is at 8.9 Ma, according to tropical diatom zonation (Barron and Baldauf, 1995). In the Dacian Basin the base of upper Sarmatian s.l. is tentatively estimated as occurring at 8.6 Ma (Palcu et al., 2019a, 2019b).

The Maeotian is subdivided into lower and upper regional substages (Nevesskaya et al., 1975a, 1975b, 1984; Muratov and Nevesskaya, 1986), separated by a major phase of slope instability named "the Intra Maeotian Event" (Palcu et al., 2019a, 2019b). The Maeotian of the Dacian and the Euxinian (Ciscaucasia) basins corresponds to polarity Chrons C3Ar and C3B, dated at about 7.6–6.1 Ma; the lower–upper Maeotian boundary is recognized at the paleomagnetic reversal C3Ar/C3An at an age of 6.8–6.9 Ma (e.g., Filippova and Trubikhin, 2009; Palcu et al., 2019a, 2019b; Rybkina et al., 2015). This chronology is in agreement with the diatom assemblages present in the lowermost Maeotian deposits that, besides Paratethyan endemic species, include the oceanic taxa *Thalassiosira grunowii* and *T. antiqua* that first occur at 7.8 and 7.7 Ma, respectively (Barron and Baldauf, 1995). The lowermost upper Maeotian deposits also contain the open marine species *Nitzschia miocenica* (first occurrence at 7.1 Ma; Barron, 2003) and *Thalassiosira convexa* (first occurrence at 6.7 Ma; Barron, 2003) (e.g., Radionova et al., 2012; Popov et al., 2016). The "transition beds" between the Maeotian and Pontian successions in Ciscaucasia comprise sediments with *Actinocyclus octonarius* and oceanic diatom index species from the *T. convexa* Zone (e.g., Radionova and Golovina, 2011; Radionova et al., 2012), dated at 6.35–6.1 Ma according to cyclostratigraphy (Rostovtseva and Rybkina, 2014, 2017).

The base of the Pontian Regional Stage of the Eastern Paratethys is dated at about 6.1 Ma, corresponding to the upper part of Chron C3An.1n (Filippova and Trubikhin, 2009; Krijgsman et al., 2010; Vasiliev et al., 2011; Radionova et al., 2012; Rostovtseva and Rybkina, 2014, 2017; Popov et al., 2016; Van Baak et al., 2016, 2017; Grothe et al., 2018).

The Pontian of the Euxinian–Caspian part of the Eastern Paratethys is subdivided into lower and upper regional substages (Nevesskaya et al., 1975a, 1975b, 1984; Muratov and Nevesskaya, 1986), both of which correspond to Chron C3r (Filippova and Trubikhin, 2009; Radionova et al., 2012; Rostovtseva and Rybkina, 2014, 2017; Van Baak et al., 2016). In the late Pontian the Euxinian and the Caspian basins became isolated from each other and had their individual development up to the end of the Neogene (Muratov and Nevesskaya, 1986). The upper Pontian substage of the Euxinian Basin is subdivided in Portaferian and Bosphorian regional beds, containing specific molluscs fauna (Nevesskaya et al., 1975a, 1975b, 1984; Muratov and Nevesskaya, 1986). The age of the lower/upper Pontian boundary in the Euxinian Basin

(Ciscaucasia) is still under discussion with two different dates of 5.65 Ma (Rostovtseva and Rybkina, 2014, 2017) and 5.8 Ma (Vasiliev et al., 2011; Chang et al., 2014).

The occurrence of the dinocyst *Galeacysta etrusca* is recorded in the time interval between 6.1−5.23 Ma in the Pontian of the Euxinian−Caspian Basin, whereas it occurs within 5.37−5.33 Ma interval in the Mediterranean: this indicates that Pontian fauna migrate to the Mediterranean during the final stage of the Messinian Salinity Crisis (Filippova and Trubikhin, 2009; Radionova et al., 2012; Grothe et al., 2014, 2018). This is also revealed by Paratethyan ostracods and molluscs that appear in the Lago Mare facies at the top of Messinian in the Mediterranean (Gliozzi et al., 2007; Guerra-Merchan et al., 2010; Stoica et al., 2016).

Two alternative chronologies also exist for the top of the Pontian in the Euxinian Basin (see review in Van Baak et al., 2017): the different ages vary between 5.23 (Filippova and Trubikhin, 2009; Radionova et al., 2012; Rostovtseva and Rybkina, 2014, 2017) and ~5.6 Ma (Krijgsman et al., 2010; Vasiliev et al., 2011) (Fig. 29.8).

The Pontian Regional Stage in the Dacian Basin is subdivided into lower (Odessian), middle (Portaferrian), and upper (Bosphorian) regional substages. The magnetostratigraphic ages for the substage boundaries have been established for the Maeotian/Odessian at 6.1 Ma, the Odessian/Portaferrian at 5.8 Ma, the Portaferrian/Bosphorian at 5.5 Ma, and the Pontian/Dacian at 4.7 Ma (Vasiliev et al., 2004, 2010; Krijgsman et al., 2010). The simultaneous onset of the Pontian Stage in the whole Eastern Paratethys is supported by the first occurrence of dinocysts *Galeacysta etrusca* and *Caspidinium rugosum* in the Dacian Basin, at about 6.0 Ma (Stoica et al., 2013; Grothe et al., 2018).

The Euxinian, Caspian, and Dacian basins of the Eastern Paratethys had their own development during Pliocene time (Fig. 29.8). The isolation of these basins and the high endemism of their biota significantly complicate interregional correlations. The Pliocene of the Euxinian Basin comprises the Kimmerian and the Kuyalnikian regional stages that in the Caspian Basin correspond to the "Productive Series" and the Akchagylian Regional Stage, respectively (Muratov and Nevesskaya, 1986; Nevesskaya et al., 2005).

The Kimmerian is subdivided into lower and upper regional substages (Muratov and Nevesskaya, 1986). The Pontian/Kimmerian boundary is erosional in the Ciscaucasia, and the base of Kimmerian is dated at about 5.23 Ma (e.g., Nevesskaya et al., 2005; Radionova et al., 2012; Rostovtseva and Rybkina, 2014) or alternatively at 5.6 Ma (Krijgsman et al., 2010; Vasiliev et al., 2011). The base of the Kuyalnikian (Ciscaucasia) is dated at 3.6 Ma and it is concomitant with the Gilbert/Gauss paleomagnetic reversal (Nevesskaya et al., 2005).

The base of the "Productive Series" (Caspian Basin) is indirectly dated at about 5.33 Ma (Van Baak et al., 2013, 2016). The Akchagylian Regional Stage is still a matter

of strong debate (see Krijgsman et al., 2019, and references therein), and two alternative chronologies have been proposed: a long Akchagylian (3.6−1.8 Ma) and a short Akchagylian (2.7−2.1 Ma).

The Pliocene in the Dacian Basin comprises the upper Pontian (Bosphorian) substage, the Dacian Regional Stage, and the Romanian Regional Stage (see review in Van Baak et al., 2015). The base of Dacian is dated at ~4.7 Ma (Vasiliev et al., 2004; Krijgsman et al., 2010; Stoica et al., 2013; Van Baak et al., 2015). The boundary between the lower and upper Dacian regional substages is placed within Chron C3n.1r at an age of about 4.42 Ma (Jorissen et al., 2018). The Romanian is comprised between 4.2 and 1.8 Ma (Van Baak et al., 2015).

The Mid-Pliocene Warm Period, lasting from 3.3 to 2.9 Ma (see review in Haywood et al., 2016), is recognized in the freshwater Romanian Basin on the basis of occurrence of species of molluscs Lymnocardiinae that indicates increased salinity (the Plescoi Event, after Pană et al., 1968). The chronology of this event indicates the 3.2−2.95 Ma interval, coincident with the boundary of the middle/upper Romanian regional substages (Van Baak et al., 2015).

The Neogene−Quaternary boundary (2.588 Ma) is not yet properly detected within the Kuyalnikian Regional Stage but probably correlates with the basal Akchagylian in the South Caspian Basin (Richards et al., 2018; Van Baak et al., 2019) and is located within the Romanian Stage of the Dacian Basin (Vasiliev et al., 2004; Van Baak et al., 2015).

29.1.4.2 New Zealand

In Fig. 29.9 the Neogene regional stages of New Zealand are reported and represent the most up-to-date understanding of the Neogene stratigraphy in New Zealand. Part of the information derives form an ongoing work (by M. Crundwell) devoted to the compilation of detailed range charts for locally age-calibrated Neogene planktonic foraminiferal and bolboform events in the New Zealand region that is summarized in Table 29.1 and Fig. 29.9. Since the contribution in GTS2012 (Chapter 29, Hilgen et al., 2012), a revised New Zealand time scale has been published (Raine et al., 2015a, 2015b), representing an update of the review volume edited by Cooper (2004). This revision made little difference to dating and correlation of New Zealand stages to the Neogene time scale: only the base of the Waipipian Stage (early Piacenzian) slightly increased in age, from 3.6 to 3.7 Ma. This change resulted from the provisional adoption of nannofossil T *Reticulofenestra pseudoumbilicus* as the defining biostratigraphic criterion for the base of the Mangapanian Stage (Raine et al., 2015a, 2015b). Work by Crundwell and McDonnell (2014) to locate a suitable SSP (Stratigraphic boundary Section and Point) for the base of the Mangapanian Stage pointed out the Rangitikei River section, in the southern part of North Island, as a potential location for the SSP. In this section the base of the mid part of the interval with dextral coiling of *Truncorotalia crassaformis* is suggested as the criterion for recognizing the base of stage (Table 29.1), thus resulting in a change of the age of the base of the Mangapanian Stage from 3.0 to 2.93 Ma.

New Zealand regional stages were originally adopted because of the extreme endemism of New Zealand marine invertebrates and benthic foraminifers used for dating when mapping rocks on land. Uplift resulting from New Zealand position astride the Australian–Pacific plate boundary has provided extensive outcropping areas and thick sections of Neogene and Quaternary marine rocks on land. Traditionally, these sediments were stratigraphically subdivided using macrofossils, many of which provide useful biostratigraphy in shallow-water rocks: note that the New Zealand Neogene molluscan fauna consists of more than 6000 species. Based on this biostratigraphy, it was obviously impossible to recognize, at the opposite end of the Earth, the standard European stages. Since correlation by calcareous nannofossils, planktonic foraminifers, bolboforms and dinoflagellates have become a standard, recognition of international biostratigraphic events also became a routine. However, the local Neogene planktonic foraminiferal events listed in Table 29.1 show that even they are more restricted (regional) than would have been expected from previous foraminiferal studies. The chronology of Neogene planktonic foraminiferal events in the New Zealand area derives by correlation with the thick Neogene succession cored at Site 1123 (ODP Leg 181; located in the offshore of the eastern South Island) in which most of the last 20 Myr of sediments have been calibrated to magnetostratigraphy (Shipboard Scientific Party, 1999). Thoroughgoing examination of the phylogeny of planktonic foraminifers (Crundwell and Nelson, 2007; Crundwell, 2014, 2015a, 2015b, 2018) has shown that several taxa require revised classification, and several local endemic taxa are now recognized. These taxa are not all documented yet (work in progress by M. Crundwell), but the events shown in Table 29.1 are defined and the related critical taxa are discussed next.

The Oligocene–Miocene boundary falls within the Waitakian Regional Stage and is recognized using T planktonic foraminifer *Globoturborotalita euapertura*. The Waitakian–Otaian boundary, located within the mid-Aquitanian (at 21.7 Ma), corresponds to the base of the benthic foraminifer *Ehrenbergina marwicki* in the Bluecliffs section (South Canterbury; Morgans et al., 1999), due to the absence of planktonic foraminiferal events in this lower Miocene interval. Planktonic foraminiferal events are used instead to recognize almost all other the stage boundaries in the younger part of Miocene up to the early Pliocene.

Subdivisions within the Altonian Stage have boundaries defined as follows (Table 29.1): the base of the Altonian

Stage is recognized by B *Globoconella praescitula*; the base of middle Altonian corresponds to B *Globoconella zealandica*, the base of the upper Altonian to B *Globoconella miozea* (Fig. 29.9).

B *Praeorbulina curva*, which is used to recognize the base of the Clifdenian Stage (Fig. 29.9), is often difficult to recognize, especially in wells. Therefore the turnover from predominantly dextral to sinistral coiling specimens of *G. miozea* is used as secondary marker. The shift in coiling lasted approximately 100 kyr, and the top of the coiling shift (in which <20% specimens show dextral coiling) is considered as the proxy for the base of Clifdenian.

The base of the Lillburnian Stage (Fig. 29.9) is defined by B *Orbulina suturalis*. T *Fohsella peripheronda* was previously used for recognizing the base of the Waiauan Stage. However, due to a younger age recalibration of this event, it has been suggested to provisionally utilize T *Globoconella conica* for locating the base of the stage, whereas T *Fohsella peripheroronda* is proposed as a marker for the base of a new subdivision, upper Lillburnian.

B *Bolboforma subfragoris* (s.l.) is proposed as a marker for the base of the upper Waiauan Stage, as well as the regional proxy for the base of the late Miocene. The base of the Tongaporutuan Stage is defined by the base of the "Kaiti Coiling Zone" that is the lowest interval in the late Miocene in which assemblages of *Globoconella miotumida* have 20% or more dextrally coiled specimens. T *Globoquadrina dehiscens* is used to subdivide the Tongaporutuan into lower and upper substages (Table 29.1; Fig. 29.9).

The base of the Kapitean Stage was originally defined by the Base of the pectinid bivalve *Sectipecten wollastoni* at Kapitea Creek (Westland). *Sectipecten* is part of a complex of scallops (Pectinidae) of the subfamily Mesopeplinae that evolved rapidly during late Miocene and Pliocene time in New Zealand (Beu, 1995). These bivalves provide a clear means for correlating shallow-water rocks, such as barnacle-rich limestone, which are extensively outcropping on land, although they cannot be used to correlate to deeper water sediments. Adding that the base of the Kapitea Creek succession has condensed sedimentation, and that ambiguity exists in the interpretation of the magnetostratigraphic record between 4.30 and 7.45 Ma at ODP Site 1123, the definition and chronology for base of Kapitean, as well as for base of Opoitian, are complicated. Consequently, for identification of the base of Kapitean planktonic foraminifers belonging to *G. miotumida–conomiozea* complex have been used in several ways. Moreover, the ages reported in Table 29.1 for stage bases have been obtained by interpolation, assuming a constant sedimentation rate for ODP Site 1123 (as explained by Crundwell and Nelson, 2007). T *Globoconella conomiozea* (s.s.), as defined in Cooper (2004), is suggested as the defining criterion for the base of the Kapitean Stage, with an interpolated age of 6.96 Ma. However, the ambiguous paleomagnetic data at ODP Site

TABLE 29.1 Regional Neogene stages for New Zealand and defining events.

New Zealand GTS		Biostratigraphic markers and New Zealand Stages	GTS2020 (Ma)	References
Wanganui	Haweran (Wq)	Base *Hirsutella hirsuta* MIS-1 subzone (proxy base Holocene 0.0117 Ma)	0.01	Crundwell (work in progress)
		Base *Hirsutella hirsuta* MIS-5e subzone (proxy base Late Pleistocene 0.126 Ma)	0.13	Crundwell (work in progress)
		Base Haweran Stage (Wq): Rangitawa tephra 0.34 ± 0.012 Ma ITPFT date	0.34	Cooper (2004), Raine et al. (2015a, 2015b)
	Castlecliffian (Wc)	Base common *Truncorotalia truncatulinoides* (proxy base Middle Pleistocene 0.781 Ma)	0.78	Work in prep.
		Base Castlecliffian Stage (Wc): Base Ototoka tephra c. 1.63 Ma	**1.63**	Cooper (2004), Raine et al. (2015a, 2015b)
	Nukumaruan (Wn)	Top upper dextral coiling zone *Truncorotalia crassaformis* (>50% dextral; proxy base Wn)	2.37	Crundwell (work in progress)
		Base Nukumaruan Stage (Wn): B *Zygochlamys delicatula*; secondary marker B *Truncorotalia crassula* s.s.	**2.4**	Cooper (2004), Crundwell (work in progress)
	Mangapanian (Wm)	Base *Truncorotalia viola* (proxy base Gelasian 2.58 Ma)	2.51	Crundwell (work in progress)
		Proposed base Mangapanian Stage (Wm): Base upper dextral coiling zone *Truncorotalia crassaformis*	**2.93**	Crundwell (work in progress)
	Waipipian (Wp)	Top *Truncorotalia crassaconica* (proxy base Wm)	2.98	Crundwell (work in progress)
		Top middle dextral coiling zone *Truncorotalia crassaformis* (proxy base Piacenzian 3.60 Ma)	3.63	Crundwell (work in progress)
		Proposed base Waipipian Stage (Wp): base middle dextral coiling zone *Truncorotalia crassaformis*	**3.67**	Crundwell (work in progress)
	Opoitian (Wo)	**Intra-Opotian (proposed base upper Wo)**: B *Globoconella pseudospinosa*	4.57	Crundwell (work in progress)
		Proposed base Opoitian Stage (Wo): B *Globoconella puncticulata* s.s. (anagentic population >95% unkeeled; proxy base Zanclean 5.333 Ma)	**5.15***	Crundwell (work in progress) after Raine et al. (2015a, 2015b)
Taranaki	Kapitean (Tk)	**Intra-Kapitean (base upper Tk)**: B *Globoconella sphericomiozea* s.s. (anagenetic population ≥ 5% unkeeled)	5.58*	Crundwell (2004)
		Base Kapitean Stage (Tk): B *Globoconella conomiozea* s.s. (anagenetic population ≥ 10% <4.5 chambers); shallow-water proxy LO *Sectipecten wollastoni*	**6.96***	Raine et al. (2015a, 2015b), Cooper (2004)
	Tongaporutuan (Tt)	Top middle *Truncorotalia juanai* acme (proxy base Messinian 7.246 Ma)	7.29*	Cooper (2004)
		Intra-Tongaporutuan (base upper Tt): T *Globoquadrina dehiscens*	8.96	Cooper (2004)
		Base Tongapoutuan (Tt) Stage: Base Kaiti Coiling Zone (lowest dextral coiling zone *Globoconella miotumida*)	11.04	Cooper (2004)
Southland	Waiauan (Sw)	**Proposed intra-Waiauan (base upper Sw)**: B *Bolboforma subfragoris* s.l.	**11.67**	Raine et al. (2015a, 2015b) after Crundwell and Nelson (2007)
		Base *Globoturborotalita nepenthes* (proxy base Tortonian 11.80 Ma)	11.8	Wade, Table 29.3 (this work)
		Base Waiauan Stage (Sw): immediately above T *Globoconella conica*	**12.98**	Cooper (2004)

(Continued)

TABLE 29.1 (Continued)

	Lillburnian (Sl)	Top *Fohsella peripheroronda* (proposed marker upper Sl, and proxy base Serravallian)	13.91	Crundwell (work in progress), Wade, Table 29.3 (this work)
		Base Lillburnian Stage (Sl): B *Orbulina suturalis*	**15.12**	Cooper (2004), Wade, Table 29.3 (this work)
	Clifdenian (Sc)	**Intra-Clifdenian (base middle Sc)**: B *Praeorbulina circularis* (population)	**~15.40**	Cooper (2004)
		Intra-Clifdenian (base middle Sc): B *Praeorbulina glomerosa* (population)	**~15.70**	Cooper (2004)
		Base Clifdenian Stage (Sc): B *Praeorbulina curva* (population; proxy base Langhian)	**15.97**	Cooper (2004)
Pareora	Altonian (Pl)	*Globoconella miozea* (coiling transition, <20% Dextral; proxy base Sc)	16.02	Cooper (2004)
		Intra-Altonian (proposed base uppermost Pl): T *Globoconella zealandica*	16.39	Crundwell (work in progress) after Gradstein et al. (2012)
		Intra-Altonian (base upper Pl): B *Globoconella miozea* s.s. (marker base upper Pl)	**16.7**	Cooper (2004)
		Top *Globoconella praescitula* s.s.	16.7	after Gradstein et al. (2012)
		Intra-Altonian (base middle Pl): B *Globoconella zealandica* (marker base middle Pl)	17.18	Raine et al. (2015a, 2015b), Wade, Table 29.3 (this work)
		Top *Globoconella incognita*	17.18	Crundwell (this work), Wade, Table 29.3 (this work)
		Base Altonian Stage (Pl): B *Globoconella praescitula*	**18.26**	After Cooper (2004)
	Otaian (Po)	Top *Dentoglobigerina binaiensis*	19.09	Wade, Table 29.3 (this work)
		Base Burdigalian	**20.44**	Hilgen et al. (2012)
		Base *Globoconella incognita*	20.98	Wade, Table 29.3 (this work)
		Intra-Otaian (Base upper Po): B *Globoconella incognita*	20.98	Morgans (personal communication)
		Base Otaian (Po) Stage: B *Ehrenbergina marwicki*	**21.7**	Raine et al. (2015a, 2015b)
Landon	Waitakian (Lw)	Top *Ciperoella ciperoensis*	22.9	Lourens et al. (2004)
		Base *Trilobatus trilobus* s.l.	22.96	Lourens et al. (2004)
		Intra-Waitakian (Base upper Lw): immediately above T *Globoturborotalita euapertura*	23.03	Morgans (personal communication)
		Base Aquitanian	**23.04**	Hilgen et al. (2012)
		Top *Globoturborotalita euapertura*	23.04	Gradstein et al. (2012) Indian Ocean
		Base common *Trilobatus primordius*	23.5	Gradstein et al. (2012) South Atlantic
		Base *Globoquadrina dehiscens*	25.2	Cooper (2004)
		Base Waitakian (Lw) Stage: Base *Globoquadrina dehiscens* (Otiake, Waitaki Valley)	**25.2**	Raine et al. (2015a, 2015b)

Some planktonic foraminifera generic assignments have been updated as per Wade et al. (2018). Ages have been updated as per Wade (this study, Table 29.3). *Note*: This table uses different generic assignments to Table 29.3, pending formal revision by the Neogene Planktonic Foraminifera Working Group. *GTS*, Geologic Time Scale; *T*, top; *B*, base.

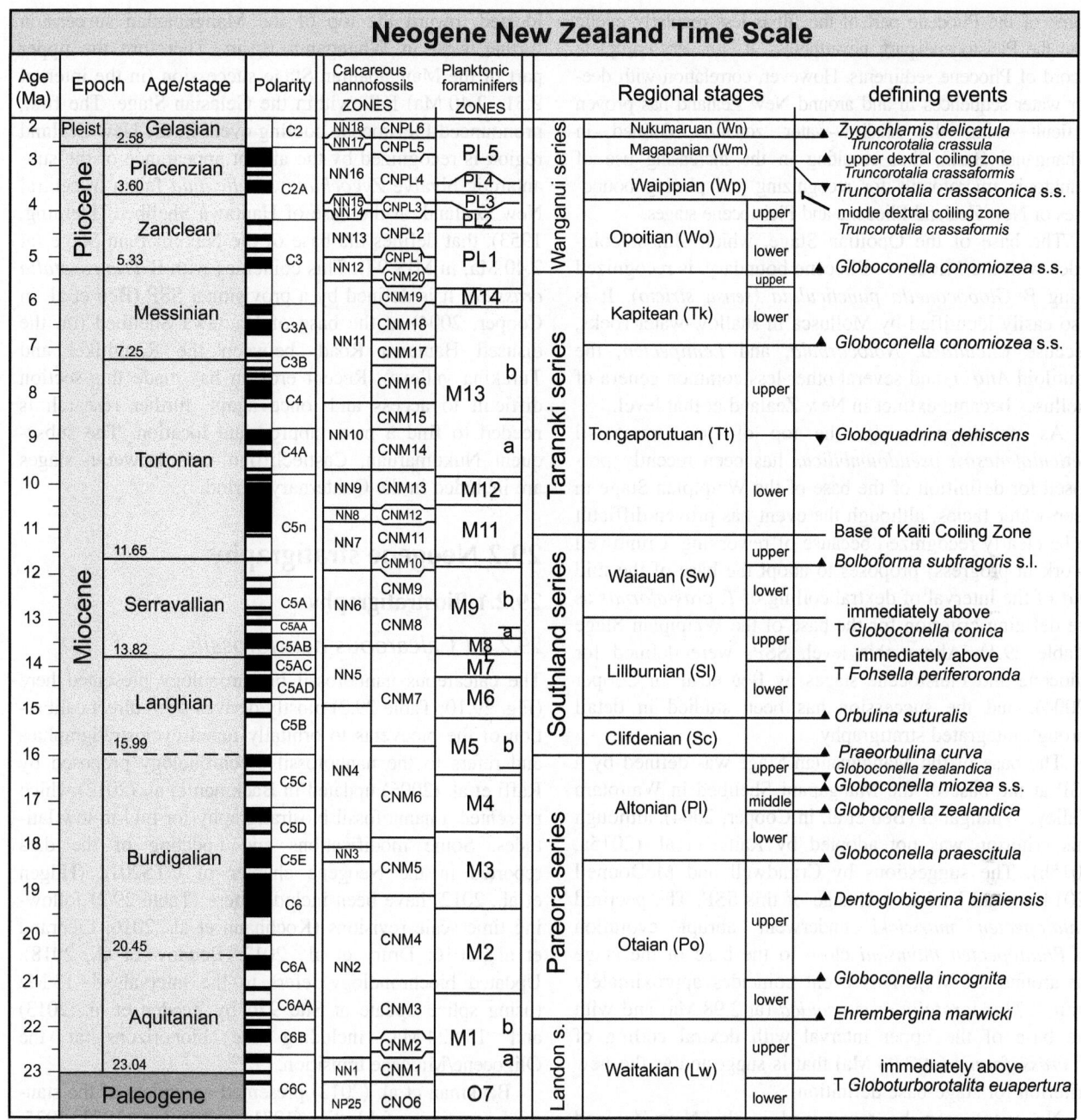

FIGURE 29.9 Regional Neogene stages for New Zealand and defining events and correlations to standard global stages and low-to-calcareous nannofossil and planktonic foraminifera biostratigraphies. This figure uses different planktonic foraminiferal generic assignment to Fig. 29.10 and Table 29.3, pending formal revision by the Neogene planktonic foraminifera working group.

1123 indicate that this event could fall close to the base of either Chron C3An.1n (6.03–6.25 Ma) or Chron C3An.2n (6.44–6.73 Ma). The interpolated age of 6.96 Ma corresponds to a reversed interval rather than either of these expected normal intervals, suggesting that further refinements are needed (ages affected by uncertainty due to magnetic polarity ambiguity are identified by asterisks in Table 29.1).

Regional SSPs have not been defined yet for Miocene or early Pliocene stages in New Zealand, because suitable

localities and criteria require further detailed research. Almost all New Zealand Pliocene and Pleistocene stages were originally defined in sections in Whanganui Basin (southwestern North Island). Here the combination of long-lasting and continued gradual basin subsidence and glacio-marine sea-level fluctuations produced a 4-km-thick basin fill of cyclic shallow-water sediments, deposited throughout Pliocene and Pleistocene time. A recent review by Pillans (2017) of the Quaternary part of the basin highlighted its unique features.

Some of the Pliocene part of the fill is less regularly cyclic than the Pleistocene part; nevertheless, it is a very complete record of Pliocene sediments. However, correlation with deeper water sequences in and around New Zealand has proven difficult using the shallow-water zonation defined in Whanganui Basin, thus resulting in the increasing use of planktonic microfossils for recognizing and placing boundaries of New Zealand Pliocene and Pleistocene stages.

The base of the Opoitian Stage, which almost coincides with the Miocene–Pliocene boundary, is recognized using B *Globoconella puncticulata* (*sensu stricto*). It is also easily identified by Mollusca in shallow-water rocks, because *Cucullaea*, *Notocorbula*, and *Lentipecten*, the nautiloid *Aturia*, and several other less common genera of molluscs became extinct in New Zealand at that level.

As stated previously, the top of the nannofossil *Reticulofenestra pseudoumbilicus* has been recently proposed for definition of the base of the Waipipian Stage in deep-water facies, although the event has proven difficult to be clearly recognized because of reworking. Crundwell (work in progress) proposes to adopt the base of the mid part of the interval of dextral coiling of *T. crassaformis* as the defining criterion for the base of the Waipipian Stage (Table 29.1). Above this level, SSPs were defined for Pliocene and Pleistocene stages by Beu et al. in Cooper (2004), and the succession has been studied in detail through integrated stratigraphy.

The base of the Mangapanian Stage was defined by a SSP at the base of the Mangapani Shellbed in Waitotara Valley, Whanganui (Beu et al. in Cooper, 2004), although this criterion was not adopted by Raine et al. (2015a, 2015b). The suggestions by Crundwell and McDonnell (2014) would lead to a change of this SSP. The pectinid *Phialopecten marwicki* underwent abrupt evolution to *Phialopecten thomsoni* close to the base of the stage (at around 3.0 Ma). This event coincides approximately with T *Truncorotalia crassaconica* (at 2.98 Ma) and with the base of the upper interval with dextral coiling of *T. crassaformis* (at 2.93 Ma) that is suggested as the new criterion for stage-base definition.

No events can be recognized in the New Zealand region at the base of the Gelasian Stage (base of the Quaternary Period and Pleistocene Epoch, 2.588 Ma); a closer regional planktonic foraminiferal event is B *Truncorotalia viola* at 2.51 Ma. Taking into account the position of the proxy event used for recognition of base Gelasian, the Gauss–Matuyama paleomagnetic reversal (Naish et al., 1997; Turner et al., 2005, fig. 10; Pillans, 2017), the base of Gelasian falls within the upper part of Mangaweka Mudstone in the Rangitikei River section, 280 m below the base of the Nukumaruan Stage Succession (base of "Hautawa Shellbed"). This stratigraphic level corresponds to a position between the Parihauhau and Te Rama shellbeds (Fleming, 1953),

located toward the top of the Mangapanian succession further west in Whanganui Basin. Therefore the upper part of the Mangapanian Stage succession (in the interval 2.51–2.40 Ma) falls within the Gelasian Stage. The first, pronounced Pleistocene cooling event in the New Zealand region is recognized by the abrupt appearance of the subantarctic bivalve *Zygochlamys delicatula* fauna in central New Zealand, at the base of Hautawa Shellbed (Fleming, 1953), that defines the base of the Nukumaruan Stage (at 2.40 Ma, in MIS 97). This coincides with B *Truncorotalia crassula*. It is defined by a provisional SSP (Beu et al. in Cooper, 2004) at the base of Hautawa Shellbed (on the disused Hautawa Road, between the Rangitikei and Turakina valleys). Recent erosion has made this section difficult to access and, once again, further research is needed to find a more appropriate location. The subsequent Nukumaruan, Castlecliffian, and Haweran stages are included in the Quaternary Period.

29.2 Neogene stratigraphy

29.2.1 Biostratigraphy

29.2.1.1 Calcareous nannofossils

The calcareous nannofossil biochronology presented here (Fig. 29.10; Table 29.2) mostly derives from direct calibration of the bioevents to orbitally tuned cyclostratigraphies and refers to the nannofossil biochronology proposed by Raffi et al. (2006) updated in Backman et al. (2012) which presented a nannofossil biostratigraphy for mid-to-low latitudes. Some modifications and updating of the data reported in the Neogene chapter of GTS2012 (Hilgen et al., 2012) have been included here (Table 29.2) following time scale revisions (Kochhann et al., 2016; Liebrand et al., 2016; Drury et al., 2017; Beddow et al., 2018). Updated biochronology refers to the intervals 8–13 Ma (using splice update at Site 926 by Zeeden et al., 2013) and 15–24 Ma, including the biohorizons at the Oligocene/Miocene transition.

Backman et al. (2012) presented an update of the standard zonations of Martini (1971) and Bukry (1973, 1975; in Okada and Bukry, 1980). They provided ages for 59 biohorizons in the Miocene–Pliocene interval and included data obtained during the past three decades, from primarily low and middle latitude deep-sea sediments as well as from marine on-land sections in the Mediterranean, refining a previous updated version of nannofossil astrobiochronology (Raffi et al., 2006). Twenty-five of these biohorizons define zonal boundaries and few (five) biohorizons are calibrated using magnetostratigraphy (Backman et al., 2012; Raffi et al., 2006). Twenty-four biozones span the 20.45 Myr time interval of the Neogene, resulting in an average duration of about 0.85 million years per biozone. The span of duration of biozones varies from 0.21 to 2.25

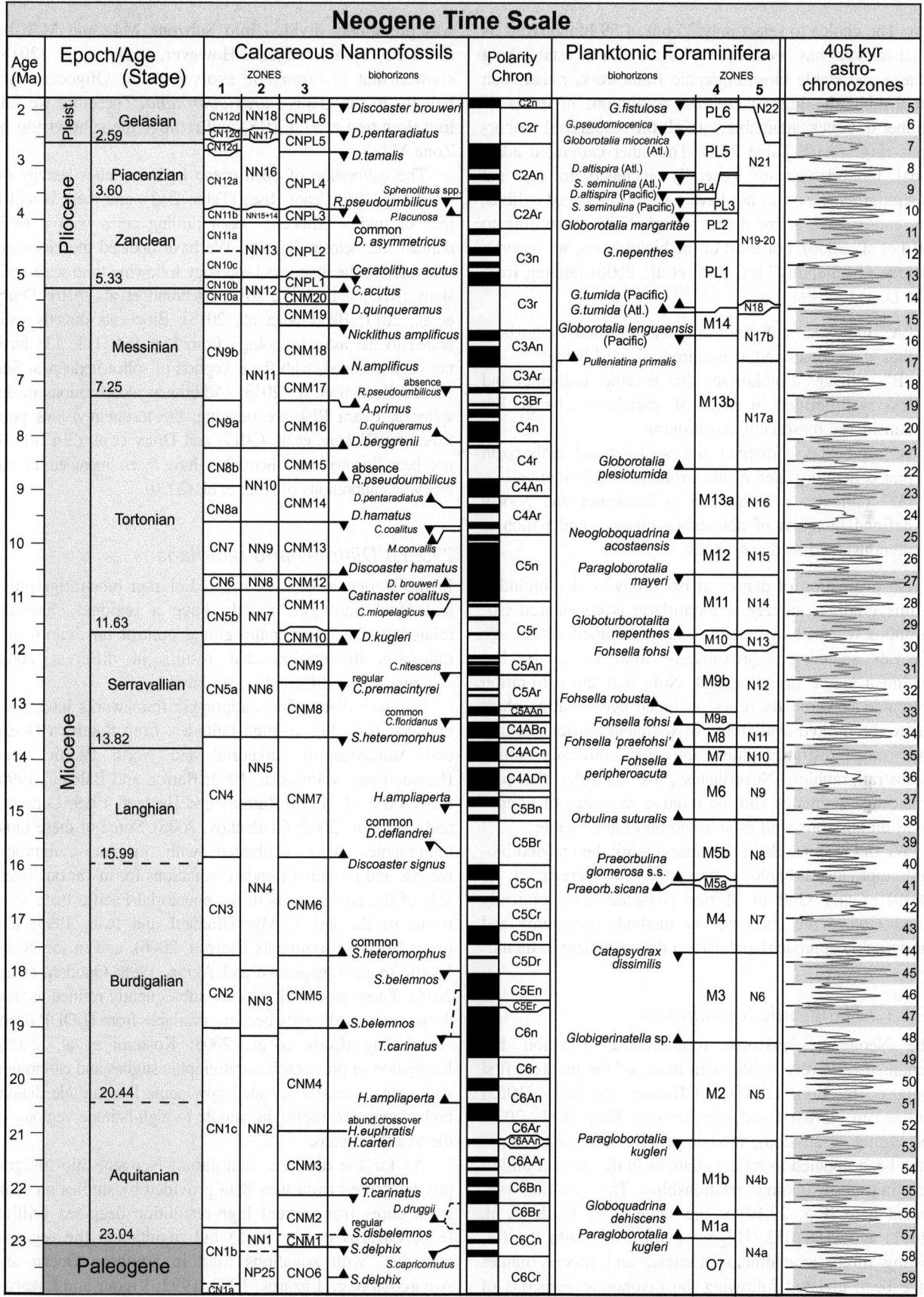

FIGURE 29.10 Calcareous nannofossils and planktonic foraminifera biostratigraphy and biochronology. Marker biohorizons are selected from Tables 29.2 and 29.3, in which additional biohorizons and related notes are also reported. 405-kyr eccentricity-related cycles define astronomically age-calibrated chronozones (Milankovitch chronozones). Zonations: **1**—Okada and Bukry (1980); **2**—Martini (1971); **3**—after Backman et al. (2012) with emended CNM2; **4**—after Wade et al. (2011) with emended M14; **5**—after Blow (1969).

Myr. The choice to select only 25 out of 59 biohorizons as zonal markers has been made with the aim of establishing a basic and stable biostratigraphic framework, rather than establishing the highest possible resolution, in terms of number of zones, attainable with all the recognized biohorizons (Fig. 29.10; Table 29.2). The other calibrated additional biohorizons are generally included in regional biostratigraphies, or do not have a well-established calibration and/or have a low degree of reliability (according to Raffi et al., 2006). For most of the biohorizons, we report a "degree of reliability" (cf. Raffi et al., 2006) ranging from A to D, as follows:

- "A" indicates a distinct and well-defined biohorizon that is demonstrated as isochronous worldwide;
- "B" indicates a biohorizon that is rather indistinct and less well-defined in terms of abundance change but considered reasonably isochronous;
- "C" indicates a distinct and well-defined biohorizon that is demonstrated as diachronous worldwide; and
- "D" indicates a biohorizon as indistinct and poorly defined in terms of abundance change, and which is demonstrated as diachronous.

An evaluation of degree of isochrony is also included in this grading, namely, a biohorizon is considered isochronous when its age estimate is constrained within age intervals spanning approximately from 20 to 100 kyr (within a single orbital cycle). Note that the calibration presented here mostly relies on tuned cyclostratigraphies due to the limited availability of sediment sections having both magnetostratigraphies and astronomically tuned cyclostratigraphies. Nevertheless, we consider the presented age estimates and the relative sequence of nannofossil biohorizons well established and rather stable.

We did not include in the present work the updated biostratigraphy/biochronology presented by Bergen et al. (2019) as "BP Gulf of Mexico Neogene Astronomically tuned Time Scale" because our methods (taxonomy and stratigraphic approach) and dataset do not adhere to theirs.

29.2.1.2 Planktonic foraminifera

The Neogene planktonic foraminiferal zonation has remained relatively stable, with many of the markers first proposed by Bolli (1957) and Banner and Blow (1965) retained in current zonal schemes (see King et al., 2020, for further discussion), though generic and zonal names have been updated to reflect changes in the current understanding of taxonomic relationships. The zonation presented here (Fig. 29.10) is the same as in Wade et al. (2011) and GTS2012 (Hilgen et al., 2012) with the following minor amendments. Generic and species names have been updated following the taxonomic revisions of Spezzaferri et al. (2015) and Wade et al. (2018). Zone M4

was previously divided into Subzone M4a and M4b by the B *Fohsella birnageae*. However, Leckie et al. (2018) showed that *F. birnageae* evolved in the Oligocene and belongs to the genus *Paragloborotalia*. Therefore we follow Berggren et al. (2018) and remove the subdivision of Zone M4.

The calibration of open ocean horizons relies heavily on ODP Leg 154 (Ceara Rise) (Table 29.3), and there is still a lack of middle Miocene ocean drilling cores with a well-defined magnetostratigraphy. We have updated the bioevents calibrated to the magnetochronology following time scale revisions (Kochhann et al., 2016; Liebrand et al., 2016; Drury et al., 2017; Beddow et al., 2018). Bioevents directly calibrated to the astrochronology from Leg 154, 111, 138 have not been changed, with the exception of splice updates at Site 926 by Zeeden et al. (2013). Additional modifications to the splice at Ceara Rise are ongoing, the recent revisions published by Wilkens et al. (2017) and Drury et al. (2017) have not been incorporated here, but have been included in the bioevent calibrations of King et al. (2020).

29.2.1.3 Diatoms and radiolarians

In the Neogene, diatom and radiolarian biostratigraphies and biochronologies mostly have a regional character related to the latitudinal/regional control on diatom and radiolaria distribution that results in different zonal schemes in the different ocean basins.

Neogene diatom biostratigraphic frameworks have been obtained from low to high latitudes, from Southern Ocean and Antarctica to equatorial and North Pacific (e.g., Harwood and Maruyama, 1992; Barron and Baldauf, 1995; Harwood et al., 1998; Ramsay and Baldauf, 1999; Censarek and Gersonde, 2002; Gladenkov, 2006). Some of these biostratigraphies were calibrated with magnetostratigraphic records and provided standard zonations for in various intervals of the Neogene, as in the equatorial Pacific from sediments of the last 13 Myr (Baldauf and Iwai, 1995) and lower Miocene sediments (Barron, 2006), and in the North Pacific (e.g., Yanagisawa and Akiba, 1998; Onodera et al., 2016). These biozonations were subsequently refined as new deep-marine sediments became available from (I)ODP deep-sea drilling (Cody et al., 2008; Koizumi et al., 2009). Integration of previous biostratigraphic studies and calibration with recent datasets revealed how some low-latitude diatom biohorizons are useful in middle-to-high-latitude regions of the world's oceans.

As for the diatoms, radiolarian Neogene biostratigraphy benefitted from new data provided by studies on sediment cores from recent high-resolution deep-sea drilling (e.g., Lazarus et al., 2015) but maintained the regional character, with zonations used in Southern Ocean and Antarctica (e.g., Lazarus, 1990, 1992; Vigour and Lazarus, 2002), and North Pacific (e.g., Morley and Nigrini, 1995;

TABLE 29.3 Neogene–Quaternary calcareous nannofossil biochronology, mostly derived from direct calibration to orbitally tuned cyclostratigraphy (Raffi et al., 2006; Backman et al., 2012; this study).

CHRONO-STRATIGRAPHY		CALCAREOUS NANNOFOSSIL ZONES from Low to Middle Latitudes			BIOHORIZON (DATUM)	AGE (Ma)	CALIBRATION		BIOHORIZON and CALIBRATION REFERENCES	COMMENTS
Stage/Age	Basal age (Ma)	Okada and Bukry (1980) CN Zones	Martini (1971) NN Zones	After Backman et al. (2012) CNPL – CNM Zones		GTS2020	Magnetochron	DSDP/ODP Site/Hole		
					T = Top/LAD; B = Base/FAD; X = Crossover in abundance		(various authors, see ref's)			
Tarantian (Lt. Pleist.)		CN15	NN21	CNPL11	X Gephyrocapsa spp./Emiliania huxleyi	0.09	MIS 5a-5b		Thierstein et al. (1977); Rio et al. (1990); Berggren et al. (1995b)	Diachrony of this abundance cross-over, spanning most of the last glacial cycle. Age of 0.07 Ma at mid-latitudes
	0.126									
Ionian (M. Pleist.)		CN14b/CN15	NN20/NN21		B Emiliania huxleyi	0.29	MIS 8	653 A	Thierstein et al. (1977); Rio et al. (1990)	Ubiquitous species, often forming blooms. Reliability: B
		CN14a/CN14b	NN19/NN20	CNPL10/CNPL11	T Pseudoemiliania lacunosa	0.43	MIS 12	926 C	Thierstein et al. (1977); Backman et al. (2012)	Slightly older calibration (0.47 Ma) in the Mediterranean. Reliability: A
	0.781									
		CN14a		CNPL10	T common Reticulofenestra asanoi	0.91	MIS 23; Chron C1r.1r	926 C	Wei (1993); Raffi (2002)	Isochronous event in equatorial and mid-latitude Atlantic, Mediterranean, mid-latitude Pacific. Reliability: A
				CNPL9/CNPL10	T absence Gephyrocapsa spp. (≥4 μm)	1.06	MIS 29 ÷ MIS 26; Chron C1r.1 n	926B	Gartner (1977); Raffi (2002)	Slightly diachronous in Mediterranean (~0.96 Ma). It corresponds to the top of the "Small Gephyrocapsa zone" of Gartner (1977). Reliability: A
					B common Reticulofenestra asanoi	1.14	MIS 34 ÷ MIS 32 Chron Cn1r.2r	926 C	Wei (1993); Raffi (2002)	Slightly diachronous in Mediterranean (~1.04 Ma). Reliability: D
Calabrian		CN13b	NN19	CNPL9	T Helicosphaera sellii (Atlantic)	1.24	MIS 42 ÷ MIS 38 upper Chron C1r.3r	607	Raffi et al. (1993); references in Raffi et al. (2006)	Isochronous between tropical Pacific and equatorial and mid-latitude Atlantic. Diachronous in eastern Mediterranean (occurring at 1.34 Ma). Reliability: C
				CNPL8/CNPL9	T Gephyrocapsa spp. (> 5.5 μm)	1.25	MIS 38-37; upper Chron C1r.3r	926 C	Raffi et al. (1993); references in Raffi et al. (2006)	It corresponds to the base of a small Gephyrocapsa spp. dominance interval ("Small Gephyrocapsa zone" of Gartner, 1977). Reliability: A
				CNPL8	B Gephyrocapsa spp. (> 5.5 μm)	1.59	MIS 56 ÷ MIS 54; mid Chron C1r.3r	926 C	Raffi (2002); references in Raffi et al. (2006)	Astronomically tuned in eastern Mediterranean at 1.61 Ma. Reliability: B
		CN13a			T Calcidiscus macintyrei	1.60	MIS 58 ÷ 56; lower Chron C1r.3r.	926 C	Raffi (2002); references in Raffi et al. (2006)	Diachronous in eastern Mediterranean (occurring at 1.66 Ma). Reliability: C
				CNPL7/CNPL8	B Gephyrocapsa spp. (≥4 μm)	1.71	MIS 60 ÷ 59; lower Chron C1r.3r (above top of Chron C2n-Olduvai).	926B	Raffi (2002); Raffi et al. (2006)	Occurring slightly earlier (at 1.67 and 1.69 Ma) in Pacific ODP Legs 111 and 138. Reliability: A
	1.806			CNPL7						

(Continued)

TABLE 29.3 (Continued)

Stage	CN	NN	CNPL	CN	NN	Bioevent	Age (Ma)	Magnetochron	Core	References	Comments
Gelasian	CN12d/CN13a	NN18/NN19	CNPL6/CNPL7	CN12d	NN18	T *Discoaster brouweri*	1.93	lowermost Olduvai; = base of Chron C2n.	926 A	Curry et al. (1995); Lourens et al. (2004)	Slightly earlier occurrence (at 1.95 Ma) in Mediterranean, and earlier (2.06 Ma) in Pacific (ODP Leg 138). Reliability: B
	CN12d	NN18	CNPL6			B common *Discoaster triradiatus*	2.16	~C2r.2r.6 (early Matuyama)	926 A	Curry et al. (1995); Lourens et al. (2004)	Astronomically tuned in eastern Mediterranean as occurring slightly later. Reliability: A
	CN12c/CN12d	NN17/NN18	CNPL5/CNPL6			T *Discoaster pentaradiatus*	2.39	~Chron C2r.3r.25	926 A	Curry et al. (1995); Lourens et al. (2004)	Earlier occurrence (at 2.51 Ma) in Mediterranean. Reliability: C
	CN12b/CN12c	NN16/NN17	CNPL5			T *Discoaster surculus*	2.53	~Chron C2r.3r.1	926 C	Curry et al. (1995); Lourens et al. (2004)	Reliability: A
2.588	CN12b										
Piacenzian	CN12a/CN12b		CNPL4/CNPL5			T *Discoaster tamalis*	2.76	~Chron C2An.1 n.6 (near top of Gauss)	926 C	Curry et al. (1995); Lourens et al. (2004)	Earlier occurrence (at 2.87 Ma) in equatorial Pacific (ODP Leg 138). Reliability: C
	CN12a	NN16	CNPL4			T *Sphenolithus* spp.	3.61	base Chron C2An.3 n (base of Gauss)	925B	Curry et al. (1995); Lourens et al. (2004)	Slightly earlier occurrences (at 3.65 Ma and 3.70 Ma) in equatorial Pacific (ODP Leg138) and in eastern Mediterranean, respectively. Reliability: C
3.600											
	CN12a/CN11b	NN15/NN16	CNPL3/CNPL4			T *Reticulofenestra pseudoumbilicus*	3.82	~Chron C2Ar.7 (uppermost part of upper-reversed interval of Gilbert)	926 A	Curry et al. (1995); Lourens et al. (2004)	Reliability: A
	CN11b	NN15 + NN14	CNPL3			T *Amaurolithus tricorniculatus*	~3.93	-			Occurrence at mid-way between Base common *D. asymmetricus* and Top *R. pseudoumbilicus*. Not an assigned age. Reliability: D
	CN11a/CN11b	NN13/NN14	CNPL2/CNPL3			B common *Discoaster asymmetricus*	4.04	~Chron C2n.1 n.8 (near top of Cochiti, or in upper Gilbert)	926 C	Backman et al. (2012)	Reliability: B
Zanclean	CN10c/CN11a	NN13	CNPL2			T *Amaurolithus primus*	4.50	within sub-Chron C3n.2r		Raffi and Flores (1995); Raffi et al. (1995); Lourens et al. (2004)	Astronomically tuned in Pacific (ODP Leg 138), Occurrence between Nunivak and Sidufjall subchrons. Reliability: D
	CN10c		CNPL1/CNPL2			T *Ceratolithus acutus*	5.04	-	926B	Backman and Raffi (1997); Backman et al. (2012)	Reliability: B
	CN10b/CN10c	NN12/NN13				B *Ceratolithus rugosus*	5.08	~Chron C3n.4 n.5 (Thvera)	926 C	Backman and Raffi (1997); Backman et al. (2012)	Relatively indistinct biohorizon. Reliability: D
	CN10b	NN12	CNPL1			T *Ceratolithus atlanticus*	5.22	-	926 A	Backman and Raffi (1997); Backman et al. (2012)	
						T *Triquetrorhabdulus rugosus*	5.23	Chron C3r.85 (within uppermost part of Gilbert)	926 A	Backman and Raffi (1997); Backman et al. (2012)	Reliability: B
						B *Ceratolithus larrymayeri*	5.33	-	926 A	Backman and Raffi (1997); Backman et al. (2012)	Regional marker for base of Pliocene. Astronomically tuned in equatorial Atlantic, also occurring in the eastern Mediterranean (Raffi, unpublished)
5.333											
Messinian	CN10a/CN10b	NN11/NN12	CNM20/CNPL1			B *Ceratolithus acutus*	5.36	uppermost Chron C3r (Gilbert)	926 A	Backman and Raffi (1997); Backman et al. (2012)	Reliability: B
	CN9d/CN10a	NN11/NN12	CNM19/CNM20			T *Discoaster quinqueramus*	5.53	middle of Chron C3r (Gilbert)	926 C	Backman and Raffi (1997); Backman et al. (2012)	Reliability: A

Epoch	CN Zone	NN Zone	CNM	Bioevent	Age (Ma)	Chron	Holes	References	Comments
	CN9c/CN9d		**CNM18/CNM19**	T *Nicklithus amplificus*	**5.98**	lowermost Chron C3r (Gilbert)	926 C	Backman and Raffi (1997); Backman et al. (2012)	Reliability: A
	CN9c		CNM18	X *Nicklithus amplificus/Triquetrorhabdulus rugosus*	**6.79**	-	925, 926	Backman and Raffi (1997); Zeeden et al. (2013)	
	CN9b/CN9c		**CNM17/CNM18**	B *Nicklithus amplificus*	**6.82**	upper part Chron C3Ar	844B, 845 A, 926B	Raffi and Flores (1995); Backman and Raffi (1997); Backman et al. (2012)	Reliability: A
	CN9b	NN11	CNM17	T absence *Reticulofenestra pseudoumbilicus*	**7.10**	Chrons C3Bn/C3Ar transition	844B, 845 A, 926B	Raffi and Flores (1995); Backman and Raffi (1997); Backman et al. (2012)	Reliability: B
7.246	**CN9a/CN9b**		**CNM16/CNM17**	B *Amaurolithus primus* (B *Amaurolithus* spp.)	**7.45**	Chron C3Br.2r	926 A	Raffi and Flores (1995); Backman and Raffi (1997); Zeeden et al. (2013)	Isocronous occurrence in tropical Pacific, Atlantic and Indian oceans and Mediterranean. Reliability: A
				T *Minylitha convallis*	**7.78**	upper Chron C4n.2 n		Raffi et al. (2006)	Diachronous occurrence between equatorial Pacific, Atlantic and Mediterranean (between 7.79 and 8.68 Ma). Reliability: D
	CN9a		CNM16	B common *Discoaster surculus*	**7.88**	within Chron C4n.2 n	710, 844	Raffi et al. (1995); Raffi et al. (2006)	Magnetostratigraphic calibration in tropical Indian and Pacific Oceans. Reliability: C
				B *Discoaster quinqueramus*	**~8.10**	Base Chron C4n	844	Raffi et al. (1995)	Not well-defined record of appearance. Reliability: D
	CN8/CN9a	**NN10/NN11**	**CNM15/CNM16**	B *Discoaster berggrenii*	**8.29**	Chron C4r	844 C, 926B/C	Raffi et al. (2006); Zeeden et al. (2013)	Diachronous occurrence between Atlantic and equatorial Pacific (8.5 Ma). Reliability: D
			CNM15	B *Discoaster loeblichii*	**~8.70**	Base Chron C4r	844	Raffi et al. (1995)	Variable occurrence of taxon in different oceanic areas. Similar magnetostartigraphic position in equatorial Pacific and mid lat. Atlantic
Tortonian	CN8	NN10	**CNM14/CNM15**	B absence *Reticulofenestra pseudoumbilicus*	**8.80**		926B	Raffi et al. (2006); Backman et al. (2012)	The interval of absence of the nominate taxon (also "R. pseudoumbilicus paracme") has been observed in different ocean basins (from tropical Indian, Pacific and Atlantic oceans to Mediterranean)
			CNM14	B common *Discoaster pentaradiatus*	**9.37**	~Chron C4Ar	844	Raffi et al. (1995); Raffi et al. (2003)	Astronomically tuned in eastern Mediterranean. Reliability: C
	CN7/CN8	**NN9/NN10**	**CNM13/CNM14**	T *Discoaster hamatus*	**9.61**	~Chron C4Ar.2r	926B	Backman and Raffi (1997); Zeeden et al. (2013)	Astronomically calibrated later occurrence in eastern Mediterranean (at 9.53 Ma). Reliability: C
				T *Catinaster calyculus*	**9.62**	~Chron C4Ar.2 n	926B	Backman and Raffi (1997); Zeeden et al. (2013)	Reliability: D
				T *Catinaster coalitus*	**9.67**		926B	Backman and Raffi (1997); Zeeden et al. (2013)	Reliability: D
	CN7	NN9	CNM13	B *Minylitha convallis*	**9.75**	-	926	Backman and Raffi (1997); Raffi et al. (2006); Backman et al. (2012)	Later occurrence in eastern Mediterranean (9.61 Ma).
				B *Discoaster neohamatus*	**10.54**	-	926B	Backman and Raffi (1997); Zeeden et al. (2013)	Later occurrence in eastern Mediterranean (at 9.87 Ma). Reliability: C
	CN6/CN7	**NN8/NN9**	**CNM12/CNM13**	B *Discoaster hamatus*	**10.57**	-	926B	Backman and Raffi (1997); Zeeden et al. (2013)	Later occurrence in eastern Mediterranean (at 10.18 Ma). Reliability: C

(Continued)

TABLE 29.3 (Continued)

Stage	CN	NN	CNM	Calcareous nannofossil event	Age (Ma)	Chron	Site	References	Remarks
	CN6	NN8		B Discoaster bellus gr.	10.64	lower C5n	926 A	Raffi et al. (1995); Zeeden et al. (2013)	Later occurrence in eastern Mediterranean (at ~10.4 Ma). Reliability: C
			CNM12	B common Helicosphaera stalis	10.72	-		Fornaciari et al. (1996); Hilgen et al. (2000)	Astronomically tuned in eastern Mediterranean; unreliable biohorizon in oceanic areas
				T common Helicosphaera walbersdorfensis	10.74	-		Theodoridis (1984); Hilgen et al. (2000)	Astronomically tuned in eastern Mediterranean; unreliable biohorizon in oceanic areas
				B Discoaster brouweri	10.78	-	926 A	Backman and Raffi (1997); Raffi et al. (2006); Zeeden et al. (2013)	Astronomically tuned also in eastern Mediterranean (at 10.73 Ma).
				B Catinaster calyculus	10.80	-	926 A	Backman and Raffi (1997); Zeeden et al. (2013)	Astronomically tuned in Atlantic (Leg 154, Sites 925 and 926).
	CN5b/ CN6	NN7/ NN8	CNM11/ CNM12	B Catinaster coalitus	10.89	lower C5n	926 A	Backman and Raffi (1997); Zeeden et al. (2013)	Astronomically tuned also in eastern Mediterranean (at 10.73 Ma).
			CNM11	T Coccolithus miopelagicus	11.04	within Chron C5n.2 n	926 A	Raffi et al. (1995); Backman and Raffi (1997); Zeeden et al. (2013)	Astronomically tuned also in eastern Mediterranean concomitantly with equatorial Atlantic; occurrence at higher stratigraphic level in equatorial Pacific, where it has been calibrated to magnetostratigraphy (at ~10.61 Ma).
	CN5b	NN7	CNM10/ CNM11	T common Discoaster kugleri	11.61	middle of Chron C5r.2 n	926 A	Raffi et al. (1995); Backman and Raffi (1997); Backman et al. (2012); Zeeden et al. (2013)	Astronomically tuned also in eastern Mediterranean (Hilgen et al., 2000), indicated as "primary marker" for approximating the Tortonian GSSP. Reliability: A
11.608			CNM10	T Cyclicargolithus floridanus	~12.0	~uppermost Chron C5An-lowermost Chron C5r		Raffi et al. (1995)	Not well-defined biohorizon, calibrated to magnetostratigraphy Reliability: D
	CN5a/ CN5b	NN6/ NN7	CNM9/ CNM10	B common Discoaster kugleri	11.89	~lower Chron C5r.3r	926 A	Raffi et al. (1995); Backman and Raffi (1997); Backman et al. (2012); Zeeden et al. (2013)	Astronomically tuned also in eastern Mediterranean. Reliability: A
			CNM9	T Coronocyclus nitescens	12.45	-	926 A	Raffi et al. (2006); Turco et al. (2002); Zeeden et al. (2013)	Earlier occurrence in equatorial Pacific (at 12.25 Ma).
			CNM8/ CNM9	T regular Calcidiscus premacintyrei	12.57	-	926b	Raffi et al. (1995); Raffi et al. (2006); Backman et al. (2012)	Astronomically tuned also in eastern Mediterranean. Reliability: A
				B common Triquetrorhabdulus rugosus	12.67	~Chron C5Ar.1r	926 A	Raffi et al. (1995); Raffi et al. (2006)	Magnetostratigraphic position as recorded at Site 845
Serravallian	CN5a	NN6	CNM8	B Calcidiscus macintyrei	13.16		926 A	Raffi et al. (1995); Turco et al. (2002); Zeeden et al. (2013)	Taxon with taxonomic ambiguities with a biostratigraphic position probably controlled also by biogeographic factors. Reliability: D
				T common Cyclicargolithus floridanus	13.33	within Chron C5AAr	926 A	Raffi et al. (1995); Raffi et al. (2006); Zeeden et al. (2013)	Astronomically tuned in eastern Mediterranean, isochronous with the occurrence in equatorial Atlantic (Sites 925, 926; 13.33 Ma). Reliability: A
	CN4/ CN5a	NN5/ NN6	CNM7/ CNM8	T Sphenolithus heteromorphus	13.60	~Chron C5ABr.6	926A-B	Backman and Raffi (1997); Turco et al. (2002); Zeeden et al. (2013)	"Primary marker" for approximating the base of Serravallian GSSP. Reliability: A

Stage / Boundary	CN	NN	CNM/CNO	Event	Age (Ma)	Chron	Site	Reference	Notes / Reliability
13.82	CN4	NN5							
Langhian	CN3/CN4	NN4/NN5	CNM7	T *Helicosphaera ampliaperta*	14.86	mid part of Chron C5Bn	925D	Curry et al. (1995); Backman et al. (2012)	Reliability: D
	CN3	NN4		T common *Discoaster deflandrei*	15.80	mid part of Chron C5Br	925D	Rio et al. (1990); Backman et al. (2012)	Reliability: A
			CNM6/CNM7	B *Discoaster signus*	15.85	mid part of Chron C5Br	925D	Rio et al. (1990); Backman et al. (2012)	This event occurs close to the distinct decrease in abundance (Tc) of *D. deflandrei* in the tropical Indian, Pacific and Atlantic oceans, and in the mid-latitude South Atlantic. Reliability: A
15.99			CNM6			base of C5Br			
	CN2/CN3	NN3/NN4	CNM5/CNM6	B common *Sphenolithus heteromorphus*	17.65	Chron C5Dr.1 n	926, 925	Backman et al. (2012); this study	Reliability: B
			CNM5	T *Sphenolithus belemnos*	17.94	-	926B	Backman et al. (2012)	Reliability: B
Burdigalian	CN1c/CN2	NN2	CNM4/CNM5	B *Sphenolithus belemnos*	19.01	upper part of Chron C6n.1 n	926B	Backman et al. (2012)	Reliability: B
			CNM4	T *Triquetrorhabdulus carinatus*	19.18	-	1218 A	Backman et al. (2012)	Diachronous event, calibrated at ~18.31 in the Equatorial Atlantic (Ceara Rise). Reliability: D
	CN1c			B *Helicosphaera ampliaperta*	20.43	-	926B	Curry et al. (1997); Backman et al. (2012)	Candidate as provisional "primary marker" for approximating the Burdigalian GSSP. Reliability: B
20.44			CNM3/CNM4	X *Helicosphaera euphratis*/*Helicosphaera carteri*	20.98	-	926B	Fornaciari (1996); Backman et al. (2012); this study	Reliability: A
Aquitanian	CN1a-b/CN1c	NN2	CNM2/CNM3	T common *Triquetrorhabdulus carinatus*	22.15	upper part Chron C6Bn.2 n	1218 A	Backman et al. (2012); this study	Reliability: A
		NN1/NN2	CNM2	B *Discoaster druggi* s.s.	22.68	-	1218 A	Lyle et al. (2002); this study	Occurring later (close to base Chron C6Bn, ~22.32 Ma) in South Atlantic (Rio Grande Rise; Site 516; Florindo et al., 2015). Reliability: D
			CNM1/CNM2	B regular *Sphenolithus disbelemnos*	22.90	upper part Chron C6Cn.1r	926, 929, 522, 1218 A	Rio et al. (1990); Fornaciari (1996); Backman et al. (2012); this study	Earlier scatter occurrence at mid latitude S Atlantic (Site 1264). Reliability: B
	CN1a-b	NN1	CNM1	T *Sphenolithus capricornutus*	23.11		1264	Raffi (1999); Liebrand et al. (2016)	Associated with *S. delphix* range at some locations, not useful for biostratigraphic correlations. Reliability: C
Chattian			CNO6/CNM1	T *Sphenolithus delphix*	23.11	Chron C6Cn.3 n	1264	Raffi (1999); Liebrand et al. (2016)	Astronomically tuned at Site 1264 (mid lat. South Atlantic), at similar stratigraphic position at Site 1218 (equatorial Pacific). Reliability: A
				T *Helicosphaera recta*				Backman et al. (2012)	The biohorizon is close to B *S. delphix*, but the nominate taxon shows irregular distribution in oceanic sediments. Reliability: D
23.04		NP25	CNO6	B *Sphenolithus delphix*	23.73	uppermost part of Chron C6Cr	1264	Raffi (1999); Liebrand et al. (2016)	Astronomically tuned at Site 1264 (mid lat. South Atlantic), it results diachronous with equatorial Pacific (Site 2018) and equatorial Atlantic (sites 926 and 929). Reliability: C

Modifications and updating of the data reported in the Neogene Chapter of GTS2012 (Hilgen et al., 2012) are included, following time scale revisions. DSDP, Deep Sea Drilling Project; MIS, Marine Isotope Stage; ODP, Ocean Drilling Program.

TABLE 29.3 Neogene–Quaternary planktonic foraminifera biochronology, with bioevents from Wade et al. (2011) and GTS2012 with minor amendments.

CHRONOSTRATIGRAPHY Stage/Age	STANDARD TROPICAL-SUBTROPICAL BIOZONE — After Blow (1969) Indo-Pacific	After Blow (1969) Atlantic	Wade et al. (2011) with emended M14 Indo-Pacific	Wade et al. (2011) with emended M14 Atlantic	BIOHORIZON (DATUM) T = Top/LAD; B = Base/FAD (see legend for others)	AGE (Ma) This study	CALIBRATIONS Magnetochron/Marine Isotope Stage	Implied age (Ma) (regional)	BIOHORIZON and CALIBRATION REFERENCES	COMMENTS
Tarantian (Lt. Pleist.)					T *Globorotalia flexuosa*	0.07	~C1n.913 (late Brunhes); MIS 4	NA: 0.07;	Joyce et al. (1990)	Calibrated in the Gulf of Mexico to ~C1n.913 = Same chron-age calibration as Berggren et al. (1995b) = late Brunhes (Stage 4); 0.069 Ma
					**T *Globigerinoides ruber* (pink) [Indo-Pac.]	0.12	~C1n.85 (late Brunhes)	I: 0.12; SP: 0.12	Thompson et al. (1979)	Calibrated to magnetics in the South Pacific and Indian Ocean to ~C1n.85 (0.12 Ma) in Wade et al. (2011), based on Thompson et al. (1979).
(0.126)			PT1b	PT1b	B *Globigerinella calida*	0.22	C1n.715 (late Brunhes); MIS 7	SP: 0.22	Chaproniere et al. (1994)	Calibrated in the South Pacific to C1n.715 = Same chron-age calibration as Berggren et al. (1995b) = late Brunhes (Stage 7); 0.22 Ma (Chaproniere et al., 1994)
					**B *Globigerinoides ruber* (pink) [Med.]	0.33	C1n.57 (middle Brunhes)	M: 0.33	Lourens (unpublished) in Lourens et al. (2004)	Cycle-calibrated in the Eastern Mediterranean
Ionian (M. Pleist.)					B *Globorotalia flexuosa*	0.40	C1n.486 (mid-Brunhes); MIS 11	NA: 0.40	Joyce et al. (1990)	Calibrated in the Gulf of Mexico to C1n.486 = Same chron-age calibration as Berggren et al. (1995b) = mid-Brunhes (Stage 11); 0.401 Ma
					B *Globorotalia hirsuta*	0.45	C1n.42 (mid-Brunhes); MIS 12	SA: 0.45	Pujol and Duprat (1983)	South Atlantic calibration to C1n.42 = Same chron-age calibration as Berggren et al. (1995b) = mid-Brunhes (stage 12); 0.45 Ma.
			PT1b/ PT1a	PT1b/ PT1a	T *Globorotalia tosaensis*	0.61	EP: C1n.165 (early Brunhes); SP: C1r.1r (latest Matuyama)	SP: 0.91; EP: 0.61	EP: Mix et al. (1995); SP: Chaproniere et al. (1994)	Astronomically tuned from equatorial Pacific ODP Legs 111 and 138. Berggren et al. (1995b) had assigned as early Brunhes; 0.65 Ma (=C1n.165); and noted that *Globorotalia tosaensis* LAD is placed earlier, at latest Matuyama (0.91 Ma) by Caproniere et al. (1994).
	N22	N22			B *Globorotalia hessi*	0.74	C1n.04 (basal Brunhes)	SP: 0.74	Chaproniere et al. (1994)	Calibrated in the South Pacific to C1n.04 = Same chron-age calibration as Berggren et al. (1995b) = basal Brunhes; 0.74 Ma.
(0.781)					***X *Pulleniatina* coiling change random to dextral [Pacific]	0.79	C1r.1r (upper)	EP: 0.79	Pearson (1995), Wade et al. (2011)	Astronomical age by Wade et al. (2011), based on ODP Hole 144-871 A (Marshall Islands) in Pearson (1995). Uppermost C1r.1r
					**B *Globorotalia excelsa* [Med.]	1.02	mid-C1r.1.n (Middle Jaramillo)	M: 1.02	Glaçon et al. (1990), Channell et al. (1990)	Calibrated to the middle of Jaramillo subchron in the Tyrrhenian Sea, ODP Leg 107.
Calabrian			PT1a	PT1a	T *Globigerinoides obliquus*	1.30	–	EA: 1.30	Chaisson and Pearson (1997)	Astronomically tuned from equatorial Atlantic ODP Sites 925 and 926.
					T *Neogloboquadrina acostaensis*	1.58	–	EA: 1.58	Bickert et al. (1997a), Chaisson and Pearson (1997), Curry et al. (1995)	Astronomically tuned from equatorial Atlantic ODP Sites 925 and 926.
					T *Globoturborotalita apertura*	1.64	–	EA: 1.64	Bickert et al. (1997a), Chaisson and Pearson (1997), Curry et al. (1995)	Astronomically tuned from equatorial Atlantic ODP Sites 925 and 926.
(1.806)					**B *Globigerina cariacoensis* [Med.]	1.78	Top C2n (Top Olduvai)	M: 1.78	Glaçon et al. (1990), Channell et al. (1990)	Mediterranean datum. Same chron-age calibration as Berggren et al. (1995b)

(Continued)

Stage	N-zone	Zone	Zone	Datum	Age	Chron	Age (EA/EP)	References	Remarks
	N22/N21	PT1a/PL6	PT1a/PL6	**T Globigerinoidesella fistulosa**	1.88	EP: Chron C2n (Olduvai)	EA: 1.88; EP: (1.77)	EA: Chaisson and Pearson (1997). EP: Jenkins and Houghton (1989)	EA: Astronomically tuned from Leg 154 (925, 926). EP: Astronomically tuned as 1.77 Ma from Leg 111 (677) in Jenkins and Houghton (1989) but this was based on widely spaced core catcher samples according to Wade et al. (2011). Genus updated as per Spezzaferri et al. (2015).
				**B Globorotalia truncatulinoides [Atl.]	1.93	SP: C2r/C2An (Gauss/Matuyama)	M: 2.00; EA: 1.93; SP: 2.61	EA: Bickert et al. (1997a), Chaisson and Pearson (1997); M: Lourens et al. (1996a), Zijderveld et al. (1991), Hilgen (1991a); SP: Dowsett (1988)	Highly diachronous marker for the base of N22 (Dowsett, 1988). The astronomically calibrated equatorial Atlantic age is favoured here.
Gelasian		PL6		T Globigerinoides extremus	1.98	-	EA: 1.98	Chaisson and Pearson (1997)	Astronomically tuned from ODP Sites 925 and 926.
				B Pulleniatina finalis	2.04	-	EA: 2.04	Chaisson and Pearson (1997)	Astronomically tuned from ODP Sites 925 and 926.
				**T Globorotalia exilis [Atl.]	2.09	~C2r.2 n.5 (Reunion)	EA: 2.09	Berggren et al. (1985)	Atlantic Ocean only. Astronomically tuned from ODP Sites 925 and 926. Berggren et al. (1995b) assign to Reunion Subchron; 2.15 Ma = C2r.2 n.5.
				**B Pulleniatina (reappearance) [Atl.]	2.26	Early Matuyama (below Reunion)	EA: 2.26	Chaisson and Pearson (1997)	Astronomically tuned from ODP Sites 925 and 926.
			PL6	T Globorotalia pertenuis	2.30	-	EA: 2.30; EP: (2.63)	EA: Bickert et al. (1997a), Chaisson and Pearson (1997); Curry et al. (1995); EP: Keigwin (1982)	Astronomically tuned from ODP Sites 925 and 926. Wade et al. (2011) show both the 2.30 Ma age from Lourens et al. (2004) and a 2.60 Ma age according to Berggren et al. (1995b) who assigned it to top Gauss Subchron (=C2An.1 n.95) in the South Atlantic and East equatorial Pacific based on Keigwin (1982). The better constrained astronomically calibrated age is here retained.
	N21			T Globoturborotalita woodi	2.30	-	EA: 2.30	Chaisson and Pearson (1997)	Astronomically tuned from ODP Sites 925 and 926.
		PL6/PL5		**T Globorotalia pseudomiocenica [Indo-Pac.]	2.31	Early Matuyama (below Reunion)	EP: 2.31	Berggren et al. (1995a)	Note mistake in GTS2012 Appendix 3.1, where 2.39 Ma should have read 2.30 Ma (now updated to revised magnetochronology). "This zone is the same as Zone PL6 (Indo-Pacific) of Berggren et al. (1995a). This zone is specific to the Indo-Pacific realm because Globorotalia pseudomiocenica evolved into G. miocenica over the interval of Chron C2An.3 n to Chron C2An.2 n (~3.5–3.2 Ma) in the Atlantic realm (DSDP Site 502, Colombia Basin; Keigwin, 1982), whereas it persisted into younger biostratigraphic level in the Indo-Pacific realm (Berggren et al.: 166)" (Wade et al., 2011).
			PL6/PL5	**T Globorotalia miocenica [Atl.]	2.39	~C2r.3r.8 (Early Matuyama, below Reunion)	EA: 2.39	Bickert et al. (1997a), Chaisson and Pearson (1997). Magnetochron according to Berggren et al. (1995b)	Astronomically tuned from ODP Sites 925 and 926. Essentially same as Berggren et al. (1995b), who assigned to early Matuyama (below Reunion Subchron); = C2r.3r.8.
		PL5	PL5	T Globorotalia limbata [Atl.]	2.39	-	EA: 2.39; EP: 2.24	EA: Bickert et al. (1997a), Chaisson and Pearson (1997); Curry et al. (1995); EP: Mix et al. (1995), Shackleton et al. (1995a,b)	Wade et al. (2011) prefer the LAD (2.39 Ma) at equatorial Atlantic Sites 925 and 926. Astronomically tuned later in the east equatorial Pacific as 2.24 Ma from ODP Legs 111 and 138.

| 2.588 | N22/N21 | | | | | | | | |

TABLE 29.3 (Continued)

CHRONO-STRATIGRAPHY		STANDARD TROPICAL-SUBTROPICAL BIOZONE				BIOHORIZON (DATUM)	AGE (Ma)	CALIBRATIONS		BIOHORIZON and CALIBRATION REFERENCES	COMMENTS
Stage/Age	Basal age (Ma)	After Blow (1969)		Wade et al. (2011) with emended M14		T = Top/LAD; B = Base/FAD (see legend for others)	This study	Magnetochron/Marine Isotope Stage	Implied age (Ma) (regional)		
		Indo-Pacific	Atlantic	Indo-Pacific	Atlantic						
						T Globoturborotalita decoraperta	2.75	-	EA: 2.75	Chaisson and Pearson (1997)	Astronomically tuned from ODP Sites 925 and 926.
						***X Globorotalia crassaformis (S coiling) [Med.]	2.92	C2An.1 n (late Gauss)	M: 2.92	Langereis and Hilgen (1991), Hilgen (1991b)	Mediterranean coiling change D-> S. C2An.1 n.25 = Same chron-age calibration as Berggren et al. (1995b) = C2An.1 n (late Gauss); 2.92 Ma.
						T Globorotalia multicamerata	2.98	C2An.1r (Kaena)	EA: 2.98	Chaisson and Pearson (1997), Berggren et al. (1983, 1985, 1995b), Pujol (1983)	Astronomically tuned from ODP Sites 925 and 926. Berggren et al. (1995b) assign as C2An.1r (Kaena); = C2Ar.1r.3. Berggren et al. (1995b) magnetochron based on Pujol (1983) and Berggren et al. (1983), Rio Grande Rise DSDP 72, South Atlantic.
						***X Globorotalia crassaformis (D coiling) [Med.]	3.00	C2An.1 n (late Gauss)	M: 3.00	Langereis and Hilgen (1991), Hilgen (1991b)	Mediterranean coiling change. C2An.1 n.1 = Same chron-age calibration as Berggren et al. (1995b) = C2An.1 n (late Gauss); 3.00 Ma.
Piacenzian		N21		PL5/PL4		**T Dentoglobigerina altispira [Atl.]	3.13	M: C2An.2 n (late Mammoth to early Kaena); SA: C2An.1r (Kaena)	M: 3.17; EA: 3.13; EP: 3.47	M: Lourens et al. (1996a), Zachariasse et al. (1989), Hilgen (1991b); EA: Chaisson and Pearson (1997), Curry et al. (1995), Tiedemann and Franz (1997); SA: Berggren et al. (1995b); EP: Mix et al. (1995), Shackleton et al. (1995a,b)	Astronomically tuned from equatorial Atlantic ODP Sites 925 and 926. This LAD is reported as slightly earlier (3.17 Ma) in eastern Mediterranean and much earlier (3.47 Ma) at Pacific ODP Legs 111 and 138. Berggren et al. (1995b) had similar assignment in the South Atlantic—C2An.1r (Kaena); 3.09 Ma. However, 0.08 Myr earlier in Mediterranean in late Mammoth to early Kaena C2An.2 n; or about 3.17 Ma, equivalent to a calibration of about C2Ar.1r.3.
				PL4/PL3		**T Sphaeroidinellopsis seminulina [Atl.]	3.16	M: top Mammoth; SA: C2An.1r.0 (base Kaena)	M: 3.19; EA: 3.16; EP: 3.59	M: Lourens et al. (1996a), Zachariasse et al. (1989), Hilgen (1991b); EA: Chaisson and Pearson (1997), Curry et al. (1995), Tiedemann and Franz (1997); SA: Berggren et al. (1995b); EP: Mix et al. (1995), Shackleton et al. (1995a)	Astronomically tuned as 3.16 Ma from Atlantic ODP Sites 925 and 926. Wade et al. (2011) used this age. This LAD is reported as slightly earlier in astronomically tuned eastern Mediterranean and much earlier (3.59 Ma) at astronomically tuned Pacific ODP Legs 111 and 138. Berggren et al. (1995b) assigned as base Kaena (3.12 Ma = C2An.1r.0) and noted that "Sphaeroidinellopsis seminulina" LAD occurs about 0.1 Myr earlier (top of Mammoth, or 3.21 Ma) in the Mediterranean.
						***X Globorotalia crassaformis (S coiling) [Med.]	3.16	C2An.2 n (late Mammoth, early Kaena)	M: 3.16	Langereis and Hilgen (1991), Hilgen (1991b)	Mediterranean coiling change. Same chron-age calibration as Berggren et al. (1995b) = C2An.2 n (late Mammoth, early Kaena); = C2An.2 n.5.
				PL3		**B Globorotalia inflata [S. Atl.]	3.24	NA: C2r.1r - C2r.2r; M: early Matuyama; SA: C2An.2r.7 (top Mammoth)	NA: 1.99 - 2.19; M: 2.09; SA: 3.24	NA: Anthonissen (2008), Flower (1999), Weaver and Clement (1987); M: SA: Berggren et al. (1983, 1985), Pujol (1983); SA: M: Lourens et al. (1996a), Zijderveld et al. (1991), Hilgen (1991a)	Astronomically tuned in the eastern Mediterranean. Same chron-age calibration as Berggren et al. (1985) in the South Atlantic as top Mammoth, 3.25 Ma (=C2An.2r.7). Berggren et al. (1995b) note that this FAD occurs about 1 Myr later (early Matuyama, or about 2.09 Ma) in Mediterranean and North Atlantic. In Berggren et al.'s (1995b) table 5 the South Atlantic age was apparently accidentally deleted but should be top of Mammoth chron, or about 3.25 Ma (as in earlier draft of this paper and in Berggren et al. (1985)).
						***X Globorotalia crassaformis (D coiling) [Med.]	3.26	C2An.2r.6 (mid-Mammoth)	M: 3.26	Langereis and Hilgen (1991), Hilgen (1991b)	Mediterranean coiling change. Same chron-age calibration as Berggren et al. (1995b) = mid-Mammoth, 3.26 Ma (=C2An.2r.6).

Stage	Zone	N-Zone	Event	Age (Ma)	Magnetochron	Regional ages	References	Comments
			**T *Globorotalia* cf. *crassula* [N. Atl.]	3.29	C2An.2r.3 (lower Mammoth)	NA: 3.29	Weaver and Clement (1987)	North Atlantic LAD. Same chron-age calibration as Berggren et al. (1995b), lower Mammoth (=C2An.2r.3).
		N21/N19-N20	B *Globigerinoidesella fistulosa*	3.33	C2An.2r.0 (base Mammoth)	M: 3.33	Hays et al. (1969)	Wade et al. (2011) astrochronology age implies same chron-age calibration as Berggren et al. (1995b), originally form Hays et al. (1969), of C2An.2r (base Mammoth) with an age of 3.33 Ma (=C2An.2r.0). Genus updated as per Spezzaferri et al. (2015).
		N21/N19-N20	B *Globorotalia tosaensis*	3.35	~C2An.3 n.92	EP: 3.35	Hays et al. (1969)	Wade et al. (2011) astrochonology age implies same chron-age calibration as Berggren et al. (1995b) at just below Mammoth, [3.35 Ma in CK95 = C2An.3 n.92]. Berggren et al. (1995b) magnetochron assignment according to Hays et al. (1969).
			**T *Pulleniatina* (disappearance) [Atl.]	3.41	EP: ~C2An.3 n.5	EA: 3.41; EP: 3.45	Chaisson and Pearson (1997), Curry et al. (1995), Tiedemann and Franz (1997), Saito et al. (1975), Keigwin (1982)	Astronomically tuned from equatorial Atlantic ODP Sites 925 and 926. Berggren et al. (1995b) have similar calibration as C2An.3 n (3.45 Ma [a] = C2An.3 n.5). Berggren et al. (1995b) magnetochron designation was based on Saito et al. (1975) and Keigwin (1982).
	PL5/PL4		**T *Dentoglobigerina altispira* [Pac.]	3.47	M: C2An.2 n (late Mammoth to early Kaena); SA: C2An.1r (Kaena)	M: 3.17; EA: 3.13; EP: 3.47	M: Lourens et al. (1996a), Zachariasse et al. (1989), Hilgen (1991b); EA: Chaisson and Pearson (1997), Curry et al. (1995), Tiedemann and Franz (1997); SA: Berggren et al. (1995b); EP: Mix et al. (1995), Shackleton et al. (1995a,b)	Astronomically tuned from equatorial Atlantic ODP Sites 925 and 926. This LAD is reported as slightly earlier in eastern Mediterranean and much earlier (3.47 Ma) at Pacific ODP Legs 111 and 138. Berggren et al. (1995b) had similar assignment in the South Atlantic—C2An.1r (Kaena); 3.09 Ma. However, 0.08 Myr earlier in late Mammoth to early Kaena C2An.2 n; or about 3.17 Ma, equivalent to a calibration of about C2Ar.1r.3.
	PL4	N19–20	B *Globorotalia pertenuis*	3.52	EP: ~C2An.3 n.5	EA: 3.52; EP: 3.45	EA: Chaisson and Pearson (1997), Curry et al. (1995), Tiedemann and Franz (1997); EP: Keigwin (1982)	Astronomically tuned from equatorial Atlantic ODP Sites 925 and 926. Berggren et al. (1995b) assigned as C2An.3 n (3.45 Ma [a] = C2An.3 n.5) in the Caribbean and East equatorial Pacific (Keigwin 1982)
	PL4/PL3	N19–20	**T *Sphaeroidinellopsis seminulina* [Pac.]	3.59	M: top Mammoth; SA: C2An.1r.0 (base Kaena)	M: 3.17; EA: 3.16; EP: 3.59	M: Lourens et al. (1996a), Zachariasse et al. (1989), Hilgen (1991b); EA: Chaisson and Pearson (1997), Curry et al. (1995), Tiedemann and Franz (1997); SA: Berggren et al. (1995b); EP: Mix et al. (1995), Shackleton et al. (1995a)	Astronomically tuned from equatorial Atlantic ODP Sites 925 and 926. Wade et al. (2011) used this 3.16 Ma age. This LAD is reported as slightly earlier in astronomically tuned eastern Mediterranean and much earlier (3.17 Ma) at astronomically tuned Pacific ODP Legs 111 and 138. Berggren et al. (1995b) assigned as base Kaena (3.12 Ma = C2An.1r.0) and noted that "*Sphaeroidinellopsis seminulina*" LAD occurs about 0.1 Myr earlier (top of Mammoth, or 3.21 Ma) in the Mediterranean.
3.600			**T *Pulleniatina primalis* [Pac.]	3.66	~C3Ar.9 (late Gilbert reversal)	EP: 3.66	Keigwin (1982)	Wade et al. (2011) astrochronology implies same chron-age calibration as Berggren et al. (1995b) = late Gilbert reversed (3.65 Ma [a] = C3Ar.9) from the Pacific study of Keigwin (1982)
Zanclean	PL3		**B *Globorotalia miocenica* [Atl.]	3.77	C2An.3 n	EA: 3.77	Chaisson and Pearson (1997), Curry et al. (1995), Tiedemann and Franz (1997), Berggren et al. (1983), Pujol (1983)	Astronomically tuned from Atlantic ODP Sites 925 and 926. Berggren et al. (1995b) assigned as C2An.3 n (3.55 Ma [a] = C2An.3 n.1). Magnetochron assignment according to Berggren et al. (1983) and Pujol (1983).
Zanclean			T *Globorotalia plesiotumida*	3.77	-	EA: 3.77	Chaisson and Pearson (1997), Curry et al. (1995), Tiedemann and Franz (1997)	Astronomically tuned from equatorial Atlantic ODP Sites 925 and 926.

(Continued)

TABLE 29.3 (Continued)

CHRONO-STRATIGRAPHY		STANDARD TROPICAL-SUBTROPICAL BIOZONE				BIOHORIZON (DATUM)	AGE (Ma)	CALIBRATIONS		BIOHORIZON and CALIBRATION REFERENCES	COMMENTS
		After Blow (1969)		Wade et al. (2011) with emended M14		T = Top/LAD; B = Base/FAD (see legend for others)	This study	Magnetochron/ Marine Isotope Stage	Implied age (Ma) (regional)		
Stage/Age	Basal age (Ma)	Atlantic	Indo-Pacific	Indo-Pacific	Atlantic						
				PL3/PL2	PL3/PL2	T *Globorotalia margaritae*	3.85	M: late Gilbert; EP: Base C2An.3 n (Gauss/Gilbert)	M: 3.81; EA: 3.85; EP: 3.58	M: Lourens et al. (1996a), Hilgen (1991b), Langereis and Hilgen (1991), Berggren et al. (1995b); EA: Chaisson and Pearson (1997), Curry et al. (1995), Tiedemann and Franz (1997); EP: Saito et al. (1975)	Cycle-calibrated in Eastern Mediterranean as 3.85 Ma. This LAD is reported slightly earlier (3.85 Ma) at astronomically tuned equatorial Atlantic Sites 925 and 926. This calibration was used by Wade et al. (2011) and is retained here. Berggren et al. (1995b) assigned as Gauss/Gilbert boundary in the eastern equatorial Pacific (3.58 Ma [a,o] = Base C2An.3 n) based on Saito et al. (1975). They reported this event to have occurred 0.2 Myr earlier in Mediterranean (late Gilbert, or 3.79 Ma). The LCO occurs in the Mediterranean just above Cochiti Subchron (3.96 Ma).
						X *Pulleniatina* coiling sinistral to dextral	4.08	EP: Just above Cochiti	EA: 4.08; EP: 3.95	EA: Chaisson and Pearson (1997), Curry et al. (1995), Tiedemann and Franz (1997); EP: Saito et al. (1975), Keigwin (1982)	Astronomically tuned from equatorial Atlantic ODP Sites 925 and 926. Berggren et al. (1995b) assigned this event to just above the Cochiti subchron in the equatorial Pacific based on Saito et al. (1975) and Keigwin (1982).
				PL2	PL2	**T *Pulleniatina spectabilis* [Pac.]	4.21	C3n.1 n.8 (Top Cochiti)	EP: 4.21	Hays et al. (1969)	Pacific only. Same chron-age calibration as Berggren et al. (1995b) as top Cochiti Subchron, (=C3n.1 n.8). Berggren et al. (1985) wrote: "...recorded at Site 503 (eastern Pacific) at about 4 Ma but this is a lone occurrence and probably not a true FAD (Keigwin 1982)". *P. spectabilis* LAD near top of Cochiti subchron according to Hays et al. (1969) with an estimated age of 3.9 Ma at Pacific sites.
						B *Globorotalia crassaformis* sensu lato	4.31	NA: C3n.2 n; M: Base C2An.3 n (Gauss/Gilbert); SP: C3n.2 n.85	NA: 4.50; M: 3.58; EA: 4.31; SP: 4.50	NA: Anthonissen (2009a), Bylinskaya (2005), Weaver and Clement (1987); M: Hilgen (1991b), Langereis and Hilgen (1991); SP: Chaproniere et al. (1994)	Astronomically tuned from equatorial Atlantic ODP Sites 925 and 926. Berggren et al. (1995b) assigned as Nunivak Subchron (4.50 Ma [a,n] = C3n.2 n.85) in the South Pacific based on Chaproniere et al. (1994). Berggren et al. (1995b) reported it to occur about 1 Myr later (Gilbert/Gauss boundary, or 3.58 Ma) in the Mediterranean based on Langereis and Hilgen (1991) and Hilgen (1991b). North Atlantic age similar to South Pacific.
				PL2/PL1	PL2/PL1	T *Globoturborotalita nepenthes*	4.37	EP: C3n.1 n.8 (Top Cochiti)	EA: 4.37; EP: 4.20	EA: Chaisson and Pearson (1997), Curry et al. (1995), Tiedemann and Franz (1997); EP: Hays et al. (1969), Saito et al. (1975), Berggren et al. (1983), Pujol (1983)	Astronomically tuned from Atlantic ODP Sites 925 and 926. Berggren et al. (1995b) assign as top Cochiti Subchron; 4.20 Ma [a] = C3n.1 n.8 based on sites in the equatorial Pacific.
						B *Globorotalia exilis*	4.45	-	EA: 4.45	Chaisson and Pearson (1997), Curry et al. (1995), Tiedemann and Franz (1997)	Astronomically tuned from equatorial Atlantic ODP Sites 925 and 926.
				PL1		T *Sphaeroidinellopsis kochi*	4.53		EA: 4.53	Chaisson and Pearson (1997), Curry et al. (1995), Tiedemann and Franz (1997)	Astronomically tuned from equatorial Atlantic ODP Sites 925 and 926. Mistake in Lourens et al. (2004) where "Bottom" should have read "Top".

Stage	Zonation (N)	Zonation (PL/M)	Datum	Age (Ma)	Calibration	Other ages	References	Comments
			T *Globorotalia cibaoensis*	**4.61**		SA: 4.61	SA: Poore et al. (1983)	C3n.2 n (Nunivak). South Atlantic DSDP Site 357 (Rio Grande Rise). T. *G. cibaoensis* was previously used to "subdivide Zone PL1 and had a calibration of 4.6 Ma in Berggren et al. (1995a). However, Chaisson and Pearson (1997) reported a much younger LAD for this species which was adopted by Lourens et al. (2004) to give an astronomical age of 3.23 Ma" (Wade et al., 2011). The much younger LAD at Ceara Rise is not confirmed by recent IODP expeditions (e.g., Expedition 363, Rosenthal et al., 2018), so the calibration of Berggren et al. (1995a) is retained here, and also in Wade et al. (2011).
	N19 – 20/ N18		T *Globigerinoides seiglei*	**4.72**	SA: C3n.4 n (Thvera); EP: ~C3n.2r.5 (above Sidufjall)	SA: 5.0 – 5.2; EP: 4.72	SA: Keigwin (1982); EP: Berggren et al. (1983); Pujol (1983); Leonard et al. (1983)	Same chron-age calibration as Berggren et al. (1995b), above Sidufjall subchron at 4.7 Ma [a] = C3n.2r.5 in the equatorial Pacific. LAD G. seiglei observed in Hole 502 (Caribbean) within Thvera Subchron (Keigwin, 1982).
	N18/ N17b	PL1/ M14	B *Sphaeroidinella dehiscens* sensu lato	**5.53**	EP: C3n.4 n.13 (basal Thvera)	EA: 5.53; EP: 5.20	EA: Chaisson and Pearson (1997), Shackleton and Crowhurst (1997); EP: Hays et al. (1969), Saito et al. (1975)	Astronomically tuned from equatorial Atlantic ODP Sites 925 and 926. Berggren et al. (1995b) assign as early Gilbert reversed; 5.2 Ma = C3n.4 n.13 in the equatorial Pacific.
	N18		**B *Globorotalia tumida* [Pac.]	**5.57**	EP: ~C3r.45 (early Gilbert)	EA: 5.72; I: (5.57); SP: (5.57); EP: 5.57	EA: Chaisson and Pearson (1997), Shackleton and Crowhurst (1997); I: Srinivasan and Sinha (1992); EP: Shackleton et al. (1995a,b), Saito et al. (1975), Keigwin (1982); SP: Hodell and Kennett (1986)	Astronomically tuned from equatorial Atlantic ODP Sites 925 and 926. *Globorotalia tumida* FAD is reported as later (5.57 Ma) at astronomically tuned Pacific ODP Legs 111 and 138. Berggren et al. (1995a) dated this event in Hole 588 and found it to be synchronous in other SW Pacific sites (586B,587,590) and in Indian Ocean (Holes 114,219,237,238). This FAD is at a comparable position, but without paleomagnetic control, at Hole 806B (Ontong-Java Plateau). They assign as C3r (early Gilbert); 5.6 Ma = C3r.45
			**B *Globorotalia pliozea* [S. Pac.]	**5.66**	SP: ~C3r.45 (early Gilbert)	SP: 5.66	Srinivasan and Sinha (1992)	Same chron-age calibration as Berggren et al. (1995b) = C3r (early Gilbert); [Datum levels in C3r scaled proportionally to match Berggren's relative spacing.]. "*Globorotalia pliozea*" FAD is dated in Hole 588 (SW Pacific).
			**B *Globorotalia sphericomiozea* [S. Pac.]	**5.66**	SP: ~C3r.45 (early Gilbert)	SP: 5.66	Srinivasan and Sinha (1992), Hodell and Kennett (1986)	Same chron-age calibration as Berggren et al. (1995b) = C3r (early Gilbert); [Datum levels in C3r scaled proportionally to match Berggren's relative spacing]. "*Globorotalia sphericomiozea*" FAD is dated in Hole 588 (SW Pacific).
Messinian	N18/ N17b	PL1/ M14-M13b	**B *Globorotalia tumida* [Atl.]	**5.72**	EP: ~C3r.45 (early Gilbert)	EA: 5.72; I: (5.57); SP: (5.57); EP: 5.57	EA: Chaisson and Pearson (1997), Shackleton and Crowhurst (1997); I: Srinivasan and Sinha (1992); EP: Shackleton et al. (1995a,b), Saito et al. (1975), Keigwin (1982); SP: Hodell and Kennett (1986)	Astronomically tuned from equatorial Atlantic ODP Sites 925 and 926. *Globorotalia tumida* FAD is reported as later (5.57 Ma) at astronomically tuned Pacific ODP Legs 111 and 138. Berggren et al. (1995a) dated this event in Hole 588 and found it to be synchronous in other SW Pacific sites (586B,587,590) and in Indian Ocean (Holes 114,219,237,238). This FAD is at a comparable position, but without paleomagnetic control, at Hole 806B (Ontong-Java Plateau). They assign as C3r (early Gilbert); 5.6 Ma = C3r.45
	N17	M14	B *Turborotalita humilis*	**5.81**	-	EA: 5.81	Chaisson and Pearson (1997), Shackleton and Crowhurst (1997)	Astronomically tuned from equatorial Atlantic ODP Sites 925 and 926.
	N17b	M14-M13b	T *Globoquadrina dehiscens*	**5.91**	SP: C3r.15; EP: ~C3r.7	I: (5.91); SP: 5.91; EP: (5.50)	EP: Hays et al. (1969), Saito et al. (1975), Berggren et al. (1983), Pujol (1983); SP: Srinivasan and Sinha, Hodell and Kennett (1986)	Astrochronology age by Wade et al. (2011) is same chron-age calibration as Berggren et al. (1995b) = C3r (early Gilbert), 5.8 Ma = C3r.15. "*Globoquadrina dehiscens*" LAD is calibrated in ODP Hole 588 in SW Pacific, and is synchronous in nearby Holes 586B,587,590 and in Indian Ocean. Wade et al. (2011) clarify: "Hodell and Kennett (1986) have shown the LAD of G. dehiscens to be diachronous, and the extinction appears to occur earlier in higher latitudes in comparison to tropical sites."

5.333

(Continued)

TABLE 29.3 (Continued)

CHRONO-STRATIGRAPHY		STANDARD TROPICAL-SUBTROPICAL BIOZONE				BIOHORIZON (DATUM)	AGE (Ma)	CALIBRATIONS		BIOHORIZON and CALIBRATION REFERENCES	COMMENTS
		After Blow (1969)		Wade et al. (2011) with emended M14							
Stage/Age	Basal age (Ma)	Indo-Pacific	Atlantic	Indo-Pacific	Atlantic	T = Top/LAD; B = Base/FAD (see legend for others)	This study	Magnetochron/Marine Isotope Stage	Implied age (Ma) (regional)		
						B *Globorotalia margaritae*	6.08	NA: ~C3An.1 n.5; (M: ~C3r.8 mid-Thvera); SP: C3An.2 n	NA: 5.99; M: (5.10); EA: 6.08; SP: (6.46)	NA: Weaver and Clement (1987); M: Langereis and Hilgen (1991) EA: Chaisson and Pearson (1997), Shackleton and Crowhurst (1997); SP: Chaproniere et al. (1994)	Astronomically tuned from Atlantic ODP Sites 925 and 926. Extensive notes in Berggren et al. (1995a), who place the global FAD in C3An (mid) at 6.0 Ma (=C3An.1 n.5) in the North Atlantic. The FAD is considered diachronous between the SW Pacific and Indian Ocean, observed to occur above (Ontong-Java Hole 806B) and below (Tonga Platform Hole 840) the FAD of *G. tumida* (Srinivasan and Sinha, 1992). Calibrated to C3An.2 n at South Pacific Tonga Platform. The FAD is nearly 1 Myr later in the Mediterranean (about 5.32 Ma), where its initial occurrence is recorded immediately above base Zanclean in mid-Thvera [~C3r.8] and first common occurrence (FCO) only shortly thereafter at 5.07 Ma (Langereis and Hilgen, 1991).
				M14/M13b		**T *Globorotalia lenguaensis* [Pac.]	6.14	SA: upper C4An; EP: C3An.1 n; SP: ~C3An.1 n.5	NA: 8.58; EA: 9.00; SP: 6.14; EP: 6.14	NA: Zhang et al. (1993), Aubry (1993); EA: Shackleton and Crowhurst (1997), Turco et al. (2002); SA: Poore et al. (1983); SP: Chaproniere et al. (1994); EP: Chaisson and Leckie (1993)	Pacific calibration: Same scaling as Berggren et al. (1995a) - LAD is directly calibrated to C3An.1 n in ODP Hole 840 (Tonga Platform, SW Pacific) and Hole 806B (Ontong-Java Plateau, equatorial West Pacific). This LAD occurs just below FAD of *P. primalis* at ODP 840 and just above it at ODP 806B. The corresponding age is about 6.0 Ma (Cande and Kent, 1995). Chron-age assignment = C3An.1 n at 6.0 Ma = C3An.1 n.5.
						B *Globigerinoides conglobatus*	6.20	C3r.25 (lowermost Gilbert)	EA: 6.20	Chaisson and Pearson (1997), Shackleton and Crowhurst (1997), Berggren et al. (1983), Pujol (1983), Poore et al. (1983)	Astronomically tuned from equatorial Atlantic ODP Sites 925 and 926. Magnetochron assignment in South Atlantic DSDP Leg 72 according to Berggren et al. (1985).
						X *Neogloboquadrina acostaensis* coiling sinistral to dextral	6.37	EP: C3An.2 n	NA: 6.37; M: 6.35; EP: 6.4	NA: Krijgsman et al. (unpublished) in Lourens et al. (2004); M: Krijgsman et al. (1999), Hilgen and Krijgsman (1999), Sierro et al. (2001); EP: Srinivasan and Sinha (1992)	Astronomically tuned from the Eastern Mediterranean (6.35 Ma) and Morocco (6.37 Ma) in Lourens et al. (2004). Chron-age calibration of Berggren et al. (1995b) = C3An.2 n.5.
				M13b		**T *Globorotalia miotumida* (conomiozea) [temperate]	6.52	NA: C3An.2 n	NA: 6.4 – 6.7; M: 6.52	NA: Weaver and Clement (1987), Clement and Robinson (1986); M: Krijgsman et al. (1999), Hilgen and Krijgsman (1999), Sierro et al. (2001)	Cycle-calibrated in Eastern Mediterranean. North Atlantic magnetochron designation based on DSDP Leg 94 (Hole 609, 609B).
		N17b/N17a	N17b/N17a			B *Pulleniatina primalis*	6.57	SP: C3An.2 n.5	SP: 6.57	I and SP: Srinivasan and Sinha (1992)	Wade et al. (2011) astrochronology age projects same chron-age of Berggren et al. (1995b) = C3An.2 n; 6.4 Ma (=C3An.2 n.5). "*Pulleniatina primalis*" FAD occurs at this age in the SW Pacific region (Sites 587,588,590; and at 806B, but without paleomagnetic calibration, and Indian Ocean (Sites 214,219,238). FAD is slightly above FAD of *G. lenguaensis* at Hole 840.
		N17a	N17a			**T *Globorotalia nicolae* [Med.]	6.72	-	M: 6.72	M:Krijgsman et al. (1999), Hilgen and Krijgsman (1999), Sierro et al. (2001)	Cycle-calibrated in Eastern Mediterranean

Stage	Zone	Zone	Zone	Zone	Zone	Event	Age (Ma)	Chron	Age estimate	References	Remarks
						X *Neogloboquadrina acostaensis* coiling dex. to sin.	**6.76**	C3Ar.8	SP: 6.76	SP: Srinivasan and Sinha (1992)	Wade et al. (2011) astrochronology age projects as same with chron-age of Berggren et al. (1995a) = C3An.2r; 6.6 Ma = C3Ar.8.
						**B *Globorotalia nicolae* [Med.]	**6.83**	-	M: 6.83	M: Krijgsman et al. (1999), Hilgen and Krijgsman (1999), Sierro et al. (2001)	Cycle-calibrated in Eastern Mediterranean.
						X *Neogloboquadrina atlantica* coiling dextral to sinistral	**6.97**	C3Ar.3	NA: 6.97	NA: Weaver and Clement (1987), Clement and Robinson (1986), Spiegler and Jansen (1989)	Wade et al. (2011) astrochronology age same as chron-age of Berggren et al. (1995a) = C3Ar; 6.8 Ma = C3Ar.3. *Neogloboquadrina atlantica* (D to S) was calibrated to magnetics in North Atlantic DSDP Hole 609, 611 and Norwegian Sea Hole 642.
7.246						**B *Globorotalia miotumida* (conomiozea) [temperate]	**7.89**	NA: C4n.1r	NA: 7.71; M: 7.89	NA: Anthonissen (2009a), Hodell et al. (2001), Weaver and Clement (1987), Clement and Robinson (1986); M: Krijgsman et al. (1995), Hilgen et al. (1995)	Cycle-calibrated in the Eastern Mediterranean. Calibrated to magnetostratigraphy in North Atlantic DSDP Leg 94 Holes 611 C, 610E. Anthonissen (2009a) compared this event (according to Flower, 1999) in North Atlantic Site 982 (Rockall Plateau) to the orbitally tuned isotope record of ODP Site 982 in Hodell et al. (2001), with a re-calibrated age to 7.71 Ma for the northeast Atlantic.
						B *Candeina nitida*	**8.43**	EP: (C4r.1r.7)	EA: 8.43; EP: (8.12)	EA: Chaisson and Pearson (1997), Shackleton and Crowhurst (1997); EP: Chaisson and Leckie (1993), Berggren et al. (1995a)	Astronomically tuned from equatorial Atlantic ODP Sites 925 and 926. Berggren et al. (1995a) assign an age for "*Candeina nitida*" FAD as recorded in Hole 806B (Ontong-Java Plateau), with correction of estimates made by Berggren et al. (1995a). See the explanation of the mistake in the Chaisson and Leckie (1993) age model (under the remarks for *FAD C. cibaoensis* in Berggren et al., 1995a). Berggren et al. (1995a) assign as C4r.1r.7.
						**B *Neogloboquadrina humerosa* [temperate]	**8.56**	EP: C4r.2r.5	NA: (9.52); EP: 8.56	NA: Anthonissen (2009a), Spezzaferri (1998), Israelson and Spezzaferri (1998); EP: Ryan et al. (1974), Berggren et al. (1985, 1995a)	Wade et al. (2011) astrochronology age projects as same with chron-age for the equatorial Pacific in Berggren et al. (1995a), based on Berggren et al. (1985). They assign as C4r.2r; 8.5 Ma = C4r.2r.5. In the North Atlantic the FAD of *N. humerosa* was recorded in ODP Hole 918D (Irminger Basin) by Spezzaferri (1998) with a strontium isotope value according to Israelson and Spezzaferri (1998), recalibrated to McArthur et al. (2001) strontium isotope age model in Anthonissen (2009a), 9.52 Ma.
	M13b/M13a	M13b/M13a	N17a/N16	N17a/N16	N17a/N16	**B *Globorotalia plesiotumida*	**8.58**	EP: C4.2r.9	NA: (9.30); EA: 8.58; EP: 8.30	NA: Zhang et al. (1993); Aubry (1993); EA: Chaisson and Pearson (1997), Shackleton and Crowhurst (1997); EP: Chaisson and Leckie (1993)	Astronomically tuned from equatorial Atlantic ODP Sites 925 and 926. Berggren et al. (1995a) note that "*Globorotalia plesiotumida*" FAD is recorded in Hole 806B (Ontong-Java Plateau, equatorial Pacific), but its FAD was difficult to define owing to questionable lower occurrences. Berggren et al. (1995a) assigned the FAD based on its lowest questionable occurrence, but its effective FCO may be about 0.5 Myr higher (coeval with FAD of *Globorotalia cibaoensis*). In the Gulf of Mexico, the FAD is recorded in upper part of Nannofossil Zone NN10, similar to the lower Pacific occurrences, and overlaps the upper part of *Discoaster bollii* range with the FAD slightly below FAD of nannofossil *Minylitha convallis* FAD (Zhang et al., 1993; Aubry, 1993). A similar FAD level is in Buff Bay, Jamaica (Berggren, 1993; Aubry, 1993). Berggren et al. (1995a) assign the age as (C4r.2r; (8.3) = C4.2r.9 based on the equatorial Pacific.
Tortonian	M13a	M13a	N16	N16		B *Globigerinoides extremus*	**8.83**	EP: C4.2r.9	EA: 8.827; EP: 8.30	EA: Chaisson and Pearson (1997), Shackleton and Crowhurst (1997); EP: Chaisson and Leckie (1993), Berggren et al. (1995a)	Astronomically tuned from equatorial Atlantic ODP Site 926. Recorded in Hole 806B (Ontong-Java Plateau, equatorial Pacific), with correction of Chaisson and Leckie (1993) estimates made by Berggren et al. (1995a). See the explanation of the mistake in the Chaisson and Leckie (1993) age model (under the remarks for FAD *C. cibaoensis* in Berggren et al., 1995a). Berggren et al. (1995a) had at (C4r.2r); (8.3) = C4.2r.9. Astronomical age updated as per Zeeden et al. (2013).

(Continued)

TABLE 29.3 (Continued)

CHRONO-STRATIGRAPHY		STANDARD TROPICAL-SUBTROPICAL BIOZONE				BIOHORIZON (DATUM)	AGE (Ma)	CALIBRATIONS		BIOHORIZON and CALIBRATION REFERENCES	COMMENTS
		After Blow (1969)		Wade et al. (2011) with emended M14				Magnetochron/ Marine Isotope Stage	Implied age (Ma)		
Stage/Age		Atlantic	Indo-Pacific	Indo-Pacific	Atlantic	T = Top/LAD; B = Base/FAD (see legend for others)	This study		(regional)		
Basal age (Ma)						**T Globorotalia lenguaensis [Atl.]	9.00	SA: upper C4An; EP: C3An.1 n; SP: ~C3An.1 n.5	NA: 8.58; EA: 9.001; SP: 6.14; EP: 6.14	NA: Zhang et al. (1993); Aubry (1993); EA: Shackleton and Crowhurst (1997); Turco et al. (2002); SA: Poore et al. (1983); SP: Chaproniere et al. (1994); EP: Chaisson and Leckie (1993)	Atlantic calibration: Astronomically tuned in equatorial Atlantic ODP Site 926, age updated here as per Zeeden et al. (2013). In Hole 519 (S. Atl.), calibrated to upper C4An (about 8.7 Ma) in Poore et al. (1983). In E68-136 (Gulf of Mexico), the LAD occurs about 10 m above the FAD of *Globorotalia plesiotumida* (=base of N17) and within range of *Minylitha convallis* and above the LAD of *Discoaster bollii*. In contrast, in E66-73, the LAD is at the same level as FAD of *G. plesiotumida* (8.58 Ma), and below the FAD of *M. convallis* and LAD of *D. bollii* (Zhang et al., 1993; Aubry, 1993). Following Turco et al. (2002), Lourens et al. (2004) significantly revised the LAD of *G. lenguaensis* to 8.97 Ma. This revised age is significantly older than reported in Berggren et al. (1995a) (derived from the Tonga Plateau, SW Pacific). This revision, Wade et al. (2011) argued would place this event within the M13a Subzone, "inconsistent with the established order of bioevents". In the Gulf of Mexico, however, Zhang et al. (1993) found the Top of *G. lenguaensis* near the same horizon as the Base *G. plesiotumida* (8.52 Ma), (not Top as reported in GTS2012) more consistent with the older age suggested by Turco et al. (2002). Wade et al. (2011) argued that they retained the Berggren et al. (1995a) age for stability, pending further investigation. In contrast to Wade et al. (2011) this study chooses to honour both calibrations and recognises significant diachrony between the Atlantic and Pacific ages of LAD of *G. lenguaensis*. The Atlantic event is herein no longer used as a zonal marker between M13b and M14 and the resulting zone in the Atlantic scheme is therefore a M14-M13b composite, pending further investigation.
						**B Neogloboquadrina pachyderma [Ind.]	9.37	I: C4Ar.1 n.5	I: 9.37	I: Berggren (1992)	Same chron-age calibration as Berggren et al. (1995a), calibrated in Holes 748, 751 (Kerguelen Plateau, south Indian Ocean) = C4Ar.1 n; > 9.2 Ma = C4Ar.1 n.5.
						B Globorotalia cibaoensis	9.44	EP: (C4n.2 n)	EA: 9.44	EA: Chaisson and Pearson (1997), Shackleton and Crowhurst (1997); EP: Chaisson and Leckie (1993); Berggren et al. (1995a)	Astronomically tuned from equatorial Atlantic ODP Sites 925 and 926. Recorded in Hole 806B (Ontong-Java Plateau, equatorial Pacific), with correction of Chaisson and Leckie (1993) estimates made by Berggren et al. (1995a). See the explanation of the mistake in the Chaisson and Leckie (1993) age model (under the remarks for FAD *G. cibaoensis* in Berggren et al., 1995a). Berggren et al. (1995a) record FAD *G. cibaoensis* in C4n.2 n.
						B Globorotalia juanai	9.69	EP: (C4r.1r.7)	EA: 9.69; EP: (8.1)	EA: Chaisson and Pearson (1997), Shackleton and Crowhurst (1997); EP: Chaisson and Leckie (1993); Berggren et al. (1995a)	Astronomically tuned from Atlantic ODP Sites 925 and 926. Berggren et al. (1995a) assign an age for "*Globorotalia juanai*" FAD as recorded in Hole 806B (Ontong-Java Plateau), with correction of estimates made by Berggren et al. (1995a). See the explanantion of the mistake in the Chaisson and Leckie (1993) age model (under the remarks for FAD *G. cibaoensis* in Berggren et al., 1995a). Berggren et al. (1995a) assign as C4r.1r.7.

			Datum event	Age (Ma)	Magnetochron	Age (regional)	References	Comments
N16/N15	M13a/M12	M13a/M12	**B Neogloboquadrina acostaensis [(sub)tropical]	9.89	NA: C5n.2 n.05	NA: 10.9; M: 11.781; EA: 9.89	NA: Berggren et al. (1985), Miller et al. (1985, 1991); EA: Chaisson and Pearson (1997), Shackleton and Crowhurst (1997), Turco et al. (2002)	Astronomically tuned in the South Atlantic and Mediterranean. This FAD is younger in equatorial Atlantic Sites 925 and 926. The FAD is reported older (11.78 Ma) in cycle-calibrated Mediterranean (Hilgen et al., 2000). Wade et al. (2011) expound further: "The cyclostratraphic age of the LO of *Neogloboquadrina acostaensis* (9.83 Ma) is derived from Ceara Rise (Chaisson and Pearson, 1997). This calibration was adopted by Lourens et al. (2004) and is significantly younger (1.07 Myr) than in Berggren et al. (1995a) (10.90 Ma) and would move this event from early Subchron C5n.2 n to Subchron C5n.1 n. Turco et al. (2002) noted the diachrony of the LO of *Neogloboquadrina acostaensis* between low latitudes and the Mediterranean. The age used in Berggren et al. (1995a) is calibrated to the magnetostratigraphy at Site 563 (Miller et al., 1985) and the discrepancy in calibrated ages may be due to further diachrony between the tropical and subtropical Atlantic Ocean, however, we note that the order of bioevents is consistent between Ceara Rise and Site 563. "Berggren et al. (1995a) note that "*Neogloboquadrina acostaensis*" FAD is calibrated in Holes 558 and 563 and 608 in the North Atlantic at the base of chron C5n = C5n.2 n; 10.9 Ma (=C5n.2 n.05). Astronomical age updated as per Zeeden et al. (2013).
			**T Globorotalia partimlabiata [Med.]	9.94	-	M: 9.94	M: Hilgen et al. (2005), Hilgen et al. (2000a)	Cycle-calibrated in Eastern Mediterranean. Age updated as per magnetochronology.
N15		M12	T Globorotalia challengeri	9.98	-	EA: 9.978	EA: Shackleton and Crowhurst (1997), Turco et al. (2002)	Astronomically tuned from equatorial Atlantic ODP Sites 925 and 926. Age updated as per Zeeden et al. (2013).
			***T Neogloboquadrina nympha [Ind.]	10.20	I: C5n.2 n.8	I: 10.20	I: Berggren (1992)	Same chron–age calibration as Berggren et al. (1995a)—'*Neogloboquadrina nympha*' LAD is calibrated in Holes 748, 751 (Kerguelen Plateau, south Indian Ocean) = C5n.2 n; 10.1 Ma = C5n.2 n.8.
N15/N14	M12/M11	M12/M11	***T Paragloborotalia mayeri [(sub)tropical]	10.54	NA: C5r.2r.2	NA: 11.44; M: 12.07; EA: 10.54	NA: Berggren et al. (1985), Miller et al. (1985, 1991); M: Hilgen et al. (2005), Hilgen et al. (2000a); EA: Shackleton and Crowhurst (1997), Turco et al. (2002)	Here we use the marker *P. mayeri* [following Berggren et al. (1995a) and Wade et al. (2011)] but continue to lump as *P. siakensis/mayeri* due to the ongoing taxonomic controversy (see Leckie et al. 2018 for discussion). Astronomically tuned from Atlantic ODP Site 926 and cycle tuned in the Eastern Mediterranean as "Top *Paragloborotalia siakensis*" in Lourens et al. (2004) (GTS2004) and Hilgen et al. (2000). This is about 1 Myr "above" the projected chron-age of Berggren et al. (1995a) from North Atlantic Sites 558, 563 and 608. LAD of "*N. mayeri*" was observed in close juxtaposition with FAD of "*N. acostaensis*", essentially eliminating Foraminifer Zone N15 (=M12). This has been varyingly interpreted as due to a hiatus in the North Atlantic (Aubry, 1993) or diachrony. Therefore, Berggren et al. assignment was C5r.2r; 11.4 Ma = C5r.2r.2 based on the North Atlantic sites. This LAD also defined their base of Medit. zone Mt9. Wade et al. (2011) further clarify: "The extinction of *Paragloborotalia mayeri* has been recalibrated to 10.53 Ma as per Chaisson and Pearson (1997) [given as siakensis in Turco et al. (2002)]. This is significantly younger (870 kyr) than the previous reported age of 11.40 Ma in Berggren et al. (1995a). The interpolated age would place this event mid-C5n.2 n rather than C5r.2r. The age used in Berggren et al. (1995a) is calibrated to the magnetostratigraphy at Site 563 (Miller et al., 1985) and this discrepancy may be due to diachrony between the tropical and subtropical Atlantic Ocean. Hilgen et al. (2000a) noted the diachrony in the extinction of *P. mayeri* between the tropical Atlantic Ocean and the Mediterranean and diachrony with higher latitudes was suggested by Miller et al. (1991)." The older "Top *P. mayeri*" event in cycle-tuned temperate Mediterranean [12.07 Ma in Lourens et al. (2004)] is here used as a separate regional event. Age updated as per Zeeden et al. (2013).

(Continued)

TABLE 29.3 (Continued)

CHRONO-STRATIGRAPHY			STANDARD TROPICAL-SUBTROPICAL BIOZONE				BIOHORIZON (DATUM)	AGE (Ma)	CALIBRATIONS			BIOHORIZON and CALIBRATION REFERENCES	COMMENTS
Stage/Age			After Blow (1969)		Wade et al. (2011) with emended M14		T = Top/LAD; B = Base/FAD (see legend for others)	This study	Magnetochron/ Marine Isotope Stage	Implied age (Ma)	(regional)		
			Indo-Pacific	Atlantic	Indo-Pacific	Atlantic							
Basal age (Ma)							**B Neogloboquadrina acostaensis** [temperate]	10.57	NA: C5n.2 n.05	NA: 10.92; M: 10.57; EA: 9.83		NA: Berggren et al. (1985), Miller et al. (1985); M: Chaisson and Pearson (1997); EA: Chaisson and Crowhurst (1997), Turco et al. (2002)	Astronomically tuned in the equatorial Atlantic and Eastern Mediterranean. This FAD is younger (9.83 Ma) in equatorial Atlantic Sites 925 and 926. This FAD is reported older (10.57 Ma) in cycle-calibrated Eastern Mediterranean. Wade et al. (2011) expound further: "The cyclostratigraphic age of the LO of Neogloboquadrina acostaensis (9.83 Ma) is derived from Ceara Rise (Chaisson and Pearson, 1997). This calibration was adopted by Lourens et al. (2004) and is significantly younger (1.07 Myr) than in Berggren et al. (1995a) (10.90 Ma) and would move this event from early Subchron C5n.2 n to Subchron C5n.1 n. Turco et al. (2002) noted the diachrony of the LO of Neogloboquadrina acostaensis between low latitudes and the Mediterranean. The age used in Berggren et al. (1995a) is calibrated to the magnetostratigraphy at Site 563 (Miller et al., 1985) and the discrepancy in calibrated ages may be due to further diachrony between the tropical and subtropical Atlantic Ocean, however, we note that the order of bioevents is consistent between Ceara Rise and Site 563." Berggren et al. (1995a) note that "Neogloboquadrina acostaensis" FAD is calibrated in Holes 558 and 563 and 608 in the North Atlantic at the base of chron C5n = C5n.2 n; 10.9 Ma (=C5n.2 n.05). The cycle tuned age from the Mediterranean is here preferred.
	N14	N14	M11	M11			B Globorotalia limbata	10.64	-	EA: 10.64		EA: Chaisson and Pearson (1997), Shackleton and Crowhurst (1997)	Astronomically tuned from equatorial Atlantic ODP Sites 925 and 926.
							T Cassigerinella chipolensis	10.92	-	EA: 10.916		EA: Shackleton and Crowhurst (1997), Turco et al. (2002)	Astronomically tuned from equatorial Atlantic ODP Site 926. Age updated as per Zeeden et al. (2013).
							B Globoturborotalita decoraperta	11.16	-	EA: 11.16		EA: Chaisson and Pearson (1997), Shackleton and Crowhurst (1997)	Astronomically tuned from equatorial Atlantic ODP Site 926. Age updated as per Zeeden et al. (2013), which is significantly younger than the age of 11.49 Ma in Lourens et al. (2004).
							B Globoturborotalita apertura	11.18	-	EA: 11.18		EA: Chaisson and Pearson (1997), Shackleton and Crowhurst (1997)	Astronomically tuned from equatorial Atlantic ODP Sites 925 and 926.
							B Globorotalia challengeri	11.23	-	EA: 11.231		EA: Shackleton and Crowhurst (1997), Turco et al. (2002)	Astronomically tuned from equatorial Atlantic ODP Site 926. Age updated as per Zeeden et al. (2013).
							B regular Globigerinoides obliquus	11.27	-	M: 11.58; EA: 11.273		M: Hilgen et al. (2000a, 2005); EA: Shackleton and Crowhurst (1997), Turco et al. (2002)	Astronomically tuned from equatorial Atlantic ODP Sites 925 and 926 and cycle tuned in the Eastern Mediterranean. The equatorial Atlantic age is here favoured. Age updated as per Zeeden et al. (2013).
							T Globigerinoides subquadratus	11.57	-	M: 11.539; EA: 11.567		EA: Shackleton and Crowhurst (1997), Turco et al. (2002)	Astronomically tuned from equatorial Atlantic ODP Site 926. Tortonian GSSP coincides almost exactly with the Last Common Occurrences of the planktonic foraminifer Globigerinoides subquadratus in the Eastern Mediterranean (Hilgen et al. 2000). Age updated as per Zeeden et al. (2013).
11.608													

Stage	Zone (N)	Zone (N)	Zone (M)	Zone (M)	Bioevent	Age (Ma)	Chron	Regional ages (Ma)	References	Comments
Serravallian	N14/N13	N14/N13	M11/M10	M11/M10	B *Globoturborotalita nepenthes* [Med.]	11.67	NA: C5r.3r.3	NA: 11.80; EA: 11.667	NA: Berggren et al. (1985, 1993), Miller et al. (1991, 1994); EA: Shackleton and Crowhurst (1997), Turco et al. (2002)	Astronomically tuned from equatorial Atlantic ODP Site 926. In the North Atlantic Berggren et al. (1995a) notes that the FAD occurs in North Atlantic Sites 563 and 608 and basal Buff Bay Fm., Jamaica = C5r.3r; 11.8 Ma = C5r.3r.3. The astronomically calibrated age is here favoured. Age updated as per Zeeden et al. (2013).
	N13	N13	M10	M10	**B *Neogloboquadrina* group [Med.]	11.78	-	M: 11.78	M: Hilgen et al. (2000a, 2005)	Cycle-calibrated in Eastern Mediterranean.
	N13/N12	N13/N12	M10/M9b	M10/M9b	T *Fohsella fohsi*	11.93	NA: top C5An.1 n	NA: 11.90; EA: 11.927	NA: Berggren et al. (1985); EA: Chaisson and Pearson (1997), Shackleton and Crowhurst (1997)	Astronomically tuned "*F. fohsi* s.l." from equatorial Atlantic ODP Sites 925 and 926. Berggren et al. (1995a) note that "*Globorotalia fohsi robusta*" LAD is calibrated in North Atlantic Hole 563 = C5An.1 n; 11.9 Ma = top C5An.1 n. Age updated as per Zeeden et al. (2013).
					**T *Globorotalia panda* [Ind.]	11.93	I: C5r.3r.3	I: 11.93	I: Berggren (1992), Wright and Miller (1992), Berggren et al. (1985)	Same chron-age calibration as Berggren et al. (1995a): The LAD is calibrated in Hole 747 A (Kerguelen Plateau, south Indian Ocean); with magnetic stratigraphy reinterpreted in Wright and Miller (1992) = C5r.3r; 11.8 Ma = C5r.3r.3. Age updated as per magnetochronology.
					**T *Paragloborotalia mayeri* sensu stricto [Med.]	12.07	-	M: 12.07	M: Hilgen et al. (2005)	Cycle-calibrated in the eastern Mediterranean. See the comment for "T *Paragloborotalia mayeri/siakensis*".
			M9b	M9b	**T *Tenuitella selleyi*, *T. pseudoedita*, *T. minutissima*, *T. clemenciae* [Ind.]	12.37	I: C5An.2 n.5	I: 12.37	I: Li et al. (1992)	Same chron-age calibration as Berggren et al. (1995a)—"*Tenuitella selleyi*" through "*clemenciae*" simultaneous LADs are calibrated in Hole 747 A (Kerguelen Plateau, south Indian Ocean) = Chron C5An.2 n (C5An.2 n.5).
					B *Globorotalia lenguanensis*	12.43	-	EA: 12.426	EA: Chaisson and Pearson (1997), Shackleton and Crowhurst (1997)	Astronomically tuned from equatorial Atlantic ODP Site 926. Age updated as per Zeeden et al. (2013).
	N12	N12			**B *Paragloborotalia partimlabiata* [Med.]	12.77	-	M: 12.77	M: Hilgen et al. (2005)	Cycle-calibrated in Eastern Mediterranean.
					B *Sphaeroidinellopsis subdehiscens*	13.04		EA: 13.04	EA: Shackleton and Crowhurst (1997), Shackleton et al. (1999), Turco et al. (2002)	Astronomically tuned from equatorial Atlantic ODP Site 926. Age updated as per Zeeden et al. (2013).
			M9b/M9a	M9b/M9a	B *Fohsella robusta*	13.13	NA: C5An.2 n.5	NA: 12.36; EA: 13.13	EA: Chaisson and Pearson (1997), Shackleton et al. (1999); NA: Turco et al. (2002); NA: Berggren (1992), Wright and Miller (1992), Berggren et al. (1985)	Astronomically tuned from equatorial Atlantic ODP Sites 925 and 926. Berggren et al. (1995a) note that "*Globorotalia fohsi robusta*" FAD is based on revised interpretation of magnetostratigraphy of North Atlantic Hole 563, in turn based on stable isotope studies in Wright and Miller (1992) = Chron C5An.2 (C5An.2 n.5). Berggren et al. (1995a) age updated as per magnetochronology.
	N12/N11	N12/N11	M9a	M9a	T *Riveroinella martinezpicoi*	13.30		EA: 13.301	EA: Shackleton and Crowhurst (1997), Shackleton et al. (1999), Turco et al. (2002)	Astronomically tuned from equatorial Atlantic ODP Site 926. Age updated as per Zeeden et al. (2013). Genus updated as per Wade et al. (2018).
			M9a/M8	M9a/M8	B *Fohsella fohsi*	13.40	NA: C5Ar.2r.8	NA: 12.71; EA: 13.398	NA: Berggren et al. (1985), Wright and Miller (1992); EA: Chaisson and Pearson (1997), Shackleton and Crowhurst (1997), Shackleton et al. (1999)	Astronomically tuned from equatorial Atlantic ODP Site 926. Berggren et al. (1995a) assign "*Globorotalia fohsi* s.str." FAD is based on reinterpretation of Wright and Miller (1992) magnetostratigraphy at Hole 563, in turn based on stable isotope studies at Holes 747 A and 608 = C5Ar.n1–2 (undiff.); 12.7 Ma = placed here as middle or Chron C5Ar.2r.8. Updated as per Zeeden et al. (2013).

(Continued)

TABLE 29.3 (Continued)

CHRONOSTRATIGRAPHY Stage/Age	Basal age (Ma)	After Blow (1969) Indo-Pacific	After Blow (1969) Atlantic	Wade et al. (2011) with emended M14 Indo-Pacific	Wade et al. (2011) with emended M14 Atlantic	BIOHORIZON (DATUM) T = Top/LAD; B = Base/FAD (see legend for others)	AGE (Ma) This study	CALIBRATIONS Magnetochron/ Marine Isotope Stage	CALIBRATIONS Implied age (Ma) (regional)	BIOHORIZON and CALIBRATION REFERENCES	COMMENTS
		N11	N11	M8	M8	B Neogloboquadrina nympha	13.49	NA: C5ABn.5; I: C5AB (base)	NA: 13.49; I: 13.73	NA: Wright and Miller (1992); I: Berggren (1992)	Same chron-age calibration as Berggren et al. (1995a)—The FAD is recorded in Holes 747 A, 751, Indian Ocean Kerguelen Plateau assigned to polarity chron C5AB (base). In North Atlantic Hole 608, the FAD occurs in sample assigned to polarity chron C5AAr, just above C5ABn => averaged here as Chron C5ABn.5.
		N11/N10	N11/N10	M8/M7	M8/M7	B Fohsella "praefohsi"	13.77	(NA: C5Ar.2r.8)	NA: (12.71); EA: 13.77	NA: Berggren et al. (1985), Wright and Miller (1992); EA: Shackleton et al. (1999), Turco et al. (2002)	Astronomically tuned from Atlantic ODP Sites 925 and 926. Berggren et al. (1995a) assigned "Globorotalia praefohsi" FAD to a much younger level, based on reinterpretation of magnetostratigraphy at North Atlantic Hole 563, in turn based on stable isotope studies at Holes 747 A and 608 (Wright and Miller, 1992) = C5Ar.n1–2 (undif.) = placed here at Chron C5Ar.2r.8. Note this North Atlantic FAD age-assignment was nearly identical to FAD of "Gt. fohsi s.str."
						**T Globorotalia praescitula [S. Atl.]	13.78	I: C5An.1 n	EA: 13.779; I: 11.94	EA: Shackleton et al. (1999), Turco et al. (2002); I: Berggren (1992), Wright and Miller (1992), Berggren et al. (1985)	Astronomically tuned from equatorial Atlantic ODP Site 926, updated here as per Zeeden et al. (2013). Berggren et al. (1995a) listed LAD as much younger in C5An.1 n or 11.9 Ma; therefore the cycle-calibration needs comparison to other sites. Berggren et al. calibration was in Hole 747 A (Kerguelen Plateau, south Indian Ocean); with magnetic stratigraphy reinterpreted in Wright and Miller (1992) = Chron C5An.1 n. In the northeastern Atlantic (Site 982 and 918) "Top regular Globorotalia praescitula/zealandica group" has an age of ca. 13.8 Ma in Anthonissen (2009b).
	13.82					T regular Clavatorella bermudezi	13.86	-	EA: 13.858	EA: Shackleton et al. (1999), Turco et al. (2002)	Astronomically tuned from equatorial Atlantic ODP Site 926. Astronomical age updated as per Zeeden et al. (2013). Note that Wade et al. (2011) list this LAD without "regular" as "LAD C. bermudezi" following Pearson and Chaisson (1997) at 303.56 mcd in Hole 926 A with an age of 13.8 Ma. However, detailed analysis of Hole 926 A by Turco et al. (2002) resulted in a significantly higher "LAD C. bermudezi" at 268.74 mcd and an age of 12.0 Ma according to Lourens et al. (2004) astrochronology. Turco et al. (2002) did identify a "LRO" last regular occurrence of C. bermudezi at 299.34 mcd, approximating the Pearson and Chaisson observation. We therefore retain the "regular" for this event at 13.86 Ma, appreciating that the LAD at 12.0 Ma may not be a useful correlation level due to very rare abundances of this taxon.
Langhian		N10	N10	M7	M7	T Globorotalia archeomenardii	13.86		EA: 13.858	EA: Shackleton et al. (1999), Pearson and Chaisson (1997)	Astronomically tuned from equatorial Atlantic ODP Site 926. Age updated as per Zeeden et al. (2013).
						T Fohsella peripheroronda	13.91	NA: base C5ADn	NA: 14.62; EA: 13.909	NA: Berggren et al. (1985a); EA: Shackleton et al. (1999), Pearson and Chaisson (1997)	Astronomically tuned from equatorial Atlantic ODP Site 926. Berggren et al. (1995a) used to define base of Mediterranean Zone M7, with age assignment of C5ADn (lower); 14.6 Ma = base C5ADn = nearly 1 Myr older than the equatorial Atlantic assignment. Age update as per Zeeden et al. (2013).
						B Globorotalia praemenardii	13.99	-	EA: 13.99	EA: Shackleton et al. (1999), Pearson and Chaisson (1997)	Astronomically tuned from equatorial Atlantic ODP Site 926. Age updated as per Zeeden et al. (2013).

Stage	N-zone	M-zone	Calcareous plankton event	Age (Ma)	Polarity chron	Regional age (Ma)	References	Comments
	N10/N9	M7/M6	B *Fohsella peripheroacuta*	14.06	(SA: base C5ADr)	EA: 14.059	EA: Shackleton et al. (1999), Pearson and Chaisson (1997); SA: Berggren et al. (1985), Ryan et al. (1974)	Astronomically tuned from equatorial Atlantic ODP Site 926. Age updated as per Zeeden et al. (2013).
	N9	M6	B *Clavatorella bermudezi*	14.63	-	EA: 14.63	EA: Shackleton et al. (1999), Pearson and Chaisson (1997), King et al. (2020)	Astronomically tuned from equatorial Atlantic ODP Sites 925 and 926. Was 14.89 Ma in GTS2004. Updated to 14.63 Ma as per King et al. (2020). Note mistake in Wade et al. (2011) which gives an age of 15.73 Ma.
			T *Globigerinatella insueta*	14.66		EA: 14.66	EA: Shackleton et al. (1999), Pearson and Chaisson (1997)	Astronomically tuned from equatorial Atlantic ODP Sites 925 and 926.
	N9/N8	M6/M5b	B *Orbulina suturalis*	15.12	SA: mid-C5Bn.2 n	SA: 15.12	SA: Poore et al. (1983), Berggren et al. (1985)	Lourens et al. (GTS2004) used FAD *Orbulina universa* FAD at 14.73 Ma based on astronomically tuned from Atlantic ODP Sites 925 and 926. Berggren et al. (1995a) used *Orbulina suturalis* (age of 15.09 Ma, based on mid-C5Bn.2 n; which had 15.1 Ma in South Atlantic DSDP Leg 73) to define base of Zone M6. Wade et al. (2011) argued for retaining the age estimate from Berggren et al. (1995a) due to "the rarity of *Orbulina* at the beginning of its range at Ceara Rise (Pearson and Chaisson, 1997)."
			**T *Globorotalia miozea* [Ind.]	15.87	I: C5Br.15	I: 15.87	I: Li et al. (1992)	Same chron-age calibration as Berggren et al. (1995a); with LAD calibrated in Hole 747 A, Kerguelen Plateau, Indian Ocean; as within lower polarity chron C5B.1 n = C5Br (lower); 15.9 Ma = C5Br.15.
			B *Praeorbulina circularis*	15.98	NA: C5C.1 n.95	NA: 15.98	NA: Berggren et al. (1985), Wade et al. (2011)	Calibrated to C5Cn.1 n (upper) (Berggren et al. 1995a). Berggren et al. (1985) writes: "sequential appearance of praeorbulinid taxa occurs in "normal" (DSDP Hole 563) and expanded (DSDP Hole 558) sequence of anomaly 5 C correlative in North Atlantic."
		M5b	B *Globigerinoides diminutus*	16.07	(SA: C5Cn.1 n.7)	SA: (16.07)	SA: Berggren et al. (1983, 1985)	Same chron-age calibration as Berggren et al. (1985a): Berggren et al. (1985) writes: "FAD *Globigerinoides diminutus* occurs in Hole 516 [South Atlantic DSDP Leg 72] about 15 m below FAD *Orbulina suturalis* in an interval with no paleomagnetic data and about 5 m above FAD *P. sicana* in mid-part of Chron C5CN (Berggren et al., 1983)" = (C5Cn.2 n); (16.1 Ma) = (C5Cn.1 n.7). Further investigation at other sites is required to test this calibration.
	N8		B *Globorotalia archeomenardii*	16.26	-	EA: 16.26	EA: Shackleton et al. (1999), Pearson and Chaisson (1997)	Age estimated from equatorial Atlantic ODP Sites 925 and 926 (Lourens et al., 2004)
Burdigalian		M5b/M5a	B *Praeorbulina glomerosa sensu stricto*	16.27	SA: C5Cn.1 n.8	EA: 16.27	EA: Shackleton et al. (1999), Pearson and Chaisson (1997); SA: Berggren et al. (1983, 1985)	Age estimated from Atlantic ODP Sites 925 and 926 (Lourens et al., 2004). Berggren et al. (1995a) uses to define base of Medit. Zone Mt5b, and assign to C5Cn.1 n (upper) in South Atlantic Hole 516 DSDP Leg 72; 16.1 Ma = C5Cn.1 n.8. Berggren et al. (1985) writes: "FAD *Praeorbulina glomerosa* occurs in Hole 516 [South Atlantic DSDP Leg 72] about 15 m below FAD *Orbulina suturalis* in an interval with no paleomagnetic data and about 5 m above FAD *P. sicana* in mid-part of Chron C5CN (Berggren et al., 1983)"
		M5a	B *Praeorbulina curva*	16.29	NA: base C5Cn.1 n	NA: 16.29	NA: Berggren et al. (1985), Wade et al. (2011)	Wade et al. (2011) astrochronology age projects as same chron-age calibration as Berggren et al. (1995a) = mid-C5Cn. Berggren et al. (1985) writes: "sequential appearance of praeorbulinid taxa occurs in "normal" (DSDP Hole 563) and expanded (DSDP Hole 558) sequence of anomaly 5C correlative in North Atlantic."

TABLE 29.3 (Continued)

Stage/Age	Basal age (Ma)	After Blow (1969) Indo-Pacific	After Blow (1969) Atlantic	Wade et al. (2011) with emended M14 Indo-Pacific	Wade et al. (2011) with emended M14 Atlantic	BIOHORIZON (DATUM) T = Top/LAD; B = Base/FAD (see legend for others)	AGE (Ma) This study	Magnetochron/Marine Isotope Stage	Implied age (Ma) (regional)	BIOHORIZON and CALIBRATION REFERENCES	COMMENTS
		N8/N7	N8/N7	M5a/M4	M5a/M4	**B** *Praeorbulina sicana*	**16.39**	SA: C5Cn.2 n.55	EA: 16.39	SA: Berggren et al. (1983, 1985), Wade et al. (2011)	Age of 16.97 Ma was estimated from Atlantic ODP Sites 925 and 926; in Lourens et al. (2004)—however this implied revised calibration to polarity chrons would seem to invert Medit. zones Mt4b and Mt5a of Berggren et al. (1995a). Berggren et al. (1995a) assign age as Chron C5Cn.2 n (mid) and write: "LAD *P. sicana* occurs in interval of no paleomagnetic data about 20 m above FAD *P. sicana* which occurs in anom. 5 C correlative in Hole 516 (Berggren et al., 1983)." Wade et al. (2011) write: "*Praeorbulina* taxa are rare at Ceara Rise (Pearson and Chaisson, 1997) and therefore were not included in the revised calibration, and we have retained the ages reported in Berggren et al. (1995a)". We tentatively follow the Wade et al. (2011) assessment until further sites are investigated.
						T *Globorotalia incognita*	16.39	I: C5Cn.2 n.5	I: 16.39	I: Berggren (1992)	Same chron-age calibration as Berggren et al. (1995a)—'*Globorotalia incognita*' LAD is calibrated in Hole 747 C, Kerguelen Plateau = Chron C5Cn.2 n (mid).
		N7	N7	M4	M4	B *Globorotalia miozea*	16.62	I: C5Cn.3 n.1	I: 16.62	I: Berggren (1992)	Same chron-age calibration as Berggren et al. (1995a). They note the FAD is recorded in Hole 751, Kerguelen Plateau, south Indian Ocean = C5Cn.3 n (lower); 16.7 Ma = placed at C5Cn.3 n.1, and used to define base of Medit. Zone Mt4.
						B *Globorotalia zealandica*	17.18	I: C5Dn.9	I: 17.18	I: Berggren (1992), Li et al. (1992)	Wade et al. (2011) astrochronology age projects as same chron-age calibration as Berggren et al. (1995a)—'*Globorotalia zealandica*' is recorded in Indian Ocean Kerguelen Plateau Hole 747 A (Li et al. 1992); and observed in lower part of polarity zone C5Dn in Hole 751 (Berggren 1992) = Chron C5Dn (upper) (C5Dn.9). Wade et al. (2011): Mistake in Berggren et al. (1995a) where "LAD *G. zealandica*" should have read "FAD *G. zealandica*"
		N7/N6	N7/N6	M4/M3	M4/M3	**T** *Catapsydrax dissimilis*	17.54	NA: C5Dr.9	NA: 18.3; EA: 17.54	NA: Berggren et al. (1985); EA: Shackleton et al. (1999), Pearson and Chaisson (1997)	Age estimated from Atlantic ODP Sites 925 and 926; in GTS2004 Neogene foraminifer appendix. Berggren et al. (1995a) calibrate LAD as recorded in Holes 558, 563, North Atlantic; and observed at top of polarity zone C5Dr in Hole 608 in interval of strong dissolution = placed at C5Dr.9.
						B *Globigerinatella insueta sensu stricto*	17.59	–	EA: 17.59	EA: Shackleton et al. (1999), Pearson and Chaisson (1997)	Age estimated from Atlantic ODP Sites 925 and 926; in GTS2004 Neogene foraminifer appendix—Note that this is much higher FAD than estimated calibrations by Berggren et al. (1995a); which results in Zone M3 being nearly non-existent. Berggren et al. (1995a)—FAD has an inferred correlation to base of Chron C5En.
		N6	N6	M3	M3	B *Globorotalia praescitula*	18.22	SA: C5En.5	SA: 18.22	SA: Berggren et al. (1983)	Wade et al. (2011) astrochronology age projects as same chron-age calibration as Berggren et al. (1995a), who used to define base of Medit. Zone Mt3—FAD is recorded in Indian Ocean Holes 747 A, 748B, 751, North Atlantic Hole 608; and is recorded in lowest part of polarity zone C5En in South Atlantic Hole 516 F = placed at Chron C5En.5.
						T *Dentoglobigerina binaiensis*	19.09	–	EA: 19.09	EA: Shackleton et al. (1999), Pearson and Chaisson (1997)	Age estimated from Atlantic ODP Sites 925 and 926.

	N6/N5	N6/N5	M3/M2	M3/M2	Biohorizon (datum)	Age	Calibration	Region: Age	References	Age estimated from / notes
20.44					B *Globigerinatella* sp.	19.30	-	EA: 19.30	EA: Shackleton et al. (1999), Pearson and Chaisson (1997)	Age estimated from Atlantic ODP Sites 925 and 926.
					B *Globigerinoides altiaperturus*	**19.97**	M: base C6r	M: 19.97	M: Berggren et al. (1983), Pujol (1983), Montanari et al. (1991)	Wade et al. (2011) astrochronology age projects as same chron-age calibration as Berggren et al. (1995a)—FAD is recored in Hole 516 F, South Atlantic; and is within in older part of polarity chron C6r in Contessa Highway section = set as base of Chron C6r.
	N5	N5	M2	M2	T *Tenuitella munda*	20.83	I: C6Ar.8	I: 20.83	I: Li et al. (1992), Wright and Miller (1992), Berggren et al. (1995a)	Wade et al. (2011) astrochronology age projects as same chron-age calibration as Berggren et al. (1995a)—LAD is recorded in Hole 747, Kergulen Plateau; with the magnetostratigraphy reinterpreted in Wright and Miller (1992) = ChronC6Ar.8.
					B *Globorotalia incognita*	20.98	I: C6Ar.4	I: 20.98	I: Berggren (1992, 1995a), Wright and Miller (1992)	Same chron-age calibration as Berggren et al. (1995a)—The FAD is recorded in Hole 747 A; with magnetostratigraphy reinterpreted in Wright and Miller (1992) = placed at Chron C6Ar.4.
					T *Ciperoella angulisuturalis*	20.99	SA: C6Ar.4	SA: 20.99	SA: Berggren et al. (1983, 1995a), Wright and Miller (1992)	Same chron-age calibration as Berggren et al. (1995a)—''*Globoturborotalita angulisuturalis*' LAD is recorded in Hole 516 F; with magnetostratigraphy reinterpreted in Wright and Miller (1992) = placed at Chron C6Ar.4.
	N5/N4b	N5/N4b	M2/M1b	M2/M1b	T *Paragloborotalia kugleri*	**21.12**	M: C6Ar.6	M: 21.54; EA: 21.12	M: Montanari et al. (1991); EA: Shackleton et al. (1999), Pearson and Chaisson (1997); SA: Berggren et al. (1983), Pujol (1983)	Age estimated from Atlantic ODP Sites 925 and 926; in GTS2004 Neogene foraminifer appendix. Berggren et al. (1995a) use this level as base of Medit. Zone Mt2, and note that LAD is recorded in Hole 516 F, South Atlantic, and at base of polarity zone C6An in Contessa Highway section = placed at Chron C6Ar.6.
	N4b	N4b	M1b	M1b	T *Paragloborotalia pseudokugleri*	21.31	SA: C6Ar.4	EA: 21.31	EA: Shackleton et al. (1999), Pearson and Chaisson (1997); SA: Berggren et al. (1983, 1995a), Wright and Miller (1992)	Age estimated from Atlantic ODP Sites 925 and 926; in GTS2004 Neogene foraminifer appendix. Berggren et al. (1995a) have similar age—LAD is recorded in Hole 516 F; with magnetostratigraphy reinterpreted in Wright and Miller (1992) = placed at Chron C6Ar.4.
Aquitanian					T *Dentoglobigerina globularis*	22.03	SA: C6Bn.1r.3	SA: 22.03	SA: Berggren et al. (1983)	Same chron-age calibration as Berggren et al. (1995a)—''*Globoquadrina globularis*'' LAD is recorded in Hole 516 F, South Atlantic = placed at Chron C6Bn.1r.3.
	N4b/N4a	N4b/N4a	M1b/M1a	M1b/M1a	B *Globoquadrina dehiscens*	**22.50**	M: C6Br.5	M: 22.50	NA: Berggren et al. (1985); M: Montanari et al. (1991); SA: Berggren et al. (1983), Pujol (1983), Poore et al. (1983)	Wade et al. (2011) astrochronology age projects same chron-age calibration as Berggren et al. (1995a)—The FAD is recorded in South Atlantic Hole 516 F, North Atlantic Holes 558, 563; and Contessa Highway Section = placed at Chron C6Br.5.
			M1a	M1a	T *Ciperoella ciperoensis*	22.90		EA: 22.90	EA: Shackleton et al. (1999), Pearson and Chaisson (1997)	Age estimated from Atlantic ODP Sites 925 and 926; in Lourens et al. (2004).
	N4a		M1a		B *Trilobatus trilobus sensu lato*	22.96		EA: 22.96	EA: Shackleton et al. (1999), Pearson and Chaisson (1997)	Age estimated from Atlantic ODP Sites 925 and 926; in Lourens et al. (2004)
23.04			M1a/O7	M1a/O7	B *Paragloborotalia kugleri*	**22.96**	SA: base C6Cn.2 n	NA: 23.04; EA: 22.96	NA: Berggren et al. (1985); EA: Shackleton et al. (1999), Pearson and Chaisson (1997); SA: Berggren et al. (1983)	Age estimated from Atlantic ODP Sites 925 and 926; in GTS2004 Neogene foraminifer appendix. Berggren et al. (1995a) notes that "Globorotalia kugleri" FAD is recorded in Holes 558, 563 (North Atlantic) and 516 F (South Atlantic) = base of Chron C6Cn.2 n.

Bioevents calibrated to the magnetochronology have been updated following time scale revisions, whereas most of the bioevents directly calibrated to the astrochronology have not been changed. Bold fonts indicate the marker and age for the base of zones. ODP, Ocean Drilling Program DSDP, Deep Sea Drilling Program; MIS, Marine Isotope Stage; FAD/LAD, First/Last appearance datum. Abbreviations in Biohorizon (datum) column: X = coiling change. Abbreviations in Calibration column: NA = North Atlantic (including Gulf of Mexico), M = Mediterranean, EA = equatorial Atlantic, SA = South Atlantic, I = Indian Ocean, SP = South Pacific, EP = equatorial Pacific.

TABLE 29.4 Diatom primary biostratigraphic markers (used for defining the zonal boundaries in Fig. 29.11) and selected additional biohorizons.

	Diatom biostratigraphic and additional markers	Age (Ma)	Zone (base)	Reference
T	*Nitzschia jouseae*	2.71	*Rhizosolenia praebergonii* b	1, 2, 3
B	*Rhizosolenia praebergonii*	3.13	*Rhizosolenia praebergonii* a	1, 2, 3
B	*Asteromphalus elegans*	4.21		1, 3
B	*Nitzschia jouseae*	4.92	*Nitzschia jouseae*	1, 3
	Miocene—Pliocene boundary	**5.33**		
B	*Shionodiscus oestrupii*	5.69		1, 3
T	*Thalassiosira miocenica*	5.69	*Thalassiosira convexa* c	1, 3
T	*Thalassiosira praeconvexa*	6.61	*Thalassiosira convexa* b	1, 3
B	*Thalassiosira miocenica*	6.91		1, 3
B	*Thalassiosira convexa var.aspinosa*	6.91	*Thalassiosira convexa* a	1, 3
B	*Thalassiosira praeconvexa*	7.08	*Fragilariopsis miocenica* b	1, 3
B	*Fragilariopsis miocenica*	7.60	*Fragilariopsis miocenica* a	1, 3
T	*Thalassiosira burkliana*	7.84	*Nitzschia porteri* b	1, 3
T	*Thalassiosira yabei*	8.44	*Nitzschia porteri* a	1, 3
B	*Thalassiosira burkliana*	9.09	*Thalassiosira yabei* b	1, 3
B	*Coscinodiscus loeblichii*	9.30		1, 3
T	*Actinocyclus moronensis*	9.77	*Thalassiosira yabei* a	1, 3
T	*Craspedodiscus coscinodiscus*	11.21	*Actinocyclus moronensis*	3
T	*Coscinodiscus gigas var.diorama*	11.21		4
B	*Hemidiscus cuneiformis*	11.67		1
B	*Coscinodiscus temperi delicatus*	12.01	*Craspedodiscus coscinodiscus*	3, 4, 5
B	*Coscinodiscus gigas var.diorama*	12.86	*Coscinodiscus gigas var. diorama*	3
T	*Araniscus lewisianus*	13.01		1, 3, 6
B	*Actinocyclus ellipticus var. spiralis*	14.16	*Araniscus lewisianus*	4
T-1	*Annellus californicus*	14.91	*Cestodiscus peplum* b	1
B	*Actinocyclus ingens*	15.25		4
B	*Cestodiscus peplum*	16.15	*Cestodiscus peplum* a	6
T	*Azpeitia bukryi*	17.54	*Crucidenticula sawamurae* b	3, 6
B	*Crucidenticula sawamurae*	18.42	*Crucidenticula sawamurae* a	3, 7
T	*Actinocyclus radionovae*	19.05	*Nitzschia maleinterpretaria* b	7
T	*Craspedodiscus elegans*	19.37		6, 7
B	*Nitzschia maleinterpretaria*	19.70	*Nitzschia maleinterpretaria* a	7
B	*Coscinodiscus lewisianus robustus*	19.85		7
T	*Bogorovia veniamini*	20.50	*Craspedodiscus elegans*	6, 7
B	*Thalassiosira fraga*	20.97		4
T	*Azpeitia oligocenica*	21.01	*Rossiella fennerae* c	8
B	*Actinocyclus hajosiae*	22.08	*Rossiella fennerae* b	7
T	*Thalassiosira primalabiata*	22.60		1, 6, 7
	Oligocene—Miocene boundary	**23.03**		
T	*Rocella gelida*	24.17	*Rossiella fennerae* a	7, 8

References for biozones and age estimates: (1) Burckle, 1978; (2) Baldauf and Iwai, 1995; (3) Barron, 1992; (4) Barron, 1983; (5) Burckle et al., 1982; (6) Barron, 1985a; (7) Barron, 2006; and (8) Barron, et al., 1985.

TABLE 29.5 Radiolaria primary biostratigraphic markers (used for defining the zonal boundaries in Fig. 29.11) and selected additional biohorizons.

	Radiolaria biostratigraphic and additional markers	Age (Ma)	Zone (base)
T	*Pterocanium prismatium*	2.077	RN13
T	*Anthocyrtidium jenghisi*	2.790	RN12b
B	Cycladophora davisiana	2.889	
T	*Stichocorys peregrina*	2.902	RN12a
B	Lamprocyrtis heteroporos	3.231	
T	*Phormostichoartus fistula*	3.957	RN11b
T	Spongaster pentas	3.977	
T	*Phormostichoartus doliolum*	4.032	RN11a
T	*Didymocyrtis penultima*	4.264	RN10
B	Spongaster tetras	4.264	
B	Pterocanium prismatium	4.729	
T	Solenosphaera omnitubus	5.319	
	Miocene–Pliocene boundary	**5.332**	
B	Spongaster pentas	6.169	
B	Didymocyrtis tetrathalamus	6.599	
trans	*Stichocorys delmontensis > S. peregrina*	7.750	RN9
B	Stichocorys peregrina	7.840	
B	Solenosphaera omnitubus	8.250	
T	Diartus hughesi	8.392	RN8
B	Diartus penultima	8.508	
trans	*Diartus petterssoni > D. hughesi*	8.760	RN7
B	Diartus hughesi	8.992	
B	Didymocyrtis antepenultima	10.010	
T	Cyrtocapsella cornuta	11.855	
B	*Diartus petterssoni*	12.111	RN6
T	Calocycletta robusta	13.348	
Bc	Calocycletta caepa	13.946	
T	Dorcadospyris dentata	14.660	
trans	*D. dentata > Dorcadospyris alata*	14.780	RN5
B	Dorcadospyris alata	15.075	
B	Carpocanopsis cristata	15.797	
B	*Calocycletta costata*	17.490	RN4
B	Dorcadospyris dentata	17.724	
B	*Stichocorys wolffii*	18.567	RN3
T	Dorcadospyris praeforcipata	19.766	
B	Stichocorys delmontensis	20.677	
T	*Theocyrtis annosa*	21.384	RN2
B	Calocycletta virginis	21.391	
B	Calocycletta cornuta	22.260	
B	*Cyrtocapsella tetrapera*	22.347	RN1

(Continued)

TABLE 29.5 (Continued)

	Radiolaria biostratigraphic and additional markers	Age (Ma)	Zone (base)
B	*Eucyrtidium diaphanes*	22.953	
	Oligocene–Miocene boundary	**23.03**	
B	*Dorcadospyris cyclacantha*	23.292	

Zonation refers to low-latitude oceanic basins. Reference and calibration of biohorizons: Lourens et al. (2004), Nigrini et al. (2005), Sanfilippo and Nigrini (1998) T = top; B = base; trans = transition.

TABLE 29.6 Dinoflagellates primary biostratigraphic markers and additional biohorizons, selected from King (2016), with "DM" and "DP" zones defined on the basis of the most reliable events, which are better defined and widespread, and useful for correlation and dating.

	Dinoflagellate biostratigraphic and additional markers	Zone (base)	Age (Ma) or stratigraphic position
B	common *Habibacysta tectata, Filisphaera filifera, Bitectatodinium tepikiense*	DPL1	~2.5
	PIACENZIAN/GELASIAN boundary		2588
T	*Barssidinium spp.*		~2.75
T	*Inverstocysta lacrimosa + Melitasphaeridium choanophorum*	DP4	~2.8
T	*Impagidinium solidum*		
T	*Operculodinium tegillatum*	DP3	Within Piacenzian
	ZANCLEAN/ PIACENZIAN boundary		3.6
T	*Batiacasphaera minuta*		Close to CNPL4/CNPL3 boundary
T	*Corrudinium devernaliae*		Close to CNPL4/CNPL3 boundary
T	*Reticulatosphaera actincoronata*	DP2	~4.3
T	*Hystichokolpoma rigaudie*		
T	*Selenopemphix armageddonensis*		
T	*Barssidinium evangelinae*	DP1	Close to Messin./Zancl. boundary
T	*Operculodinium piaseckii*		
	MIOCENE/PLIOCENE boundary		5333
T	*Labyrinthodinium truncatum*	DM9	~8.4
T	*Gramocysta verricula*		
B	*Selenopemphix armageddonensis*		
T	*Palaecystodinium golzowense*	DM8	
B	*Amiculosphaera umbraculum*		~8.9 or earlier
B	*Gramocysta verricula*		
	SERRAVALLIAN/TORTONIAN boundary		11.63
T	*Cannosphaeropsis passio*	DM7	~11.9
T	*Cerebrocysta poulseni*		
B	*Cannosphaeropsis passio*	DM6b	~12.3
B	*Spiniferites (Achomosphaera) andalousiense*		
T	*Unipontidinium aquaeductum*	DM6a	
T	*Palaecystodinium miocaenicum*	DM5c	Close to Lang./Serraval. boundary
B	*Habibacysta tectata*	DM5b	

TABLE 29.6 (Continued)

	Dinoflagellate biostratigraphic and additional markers	Zone (base)	Age (Ma) or stratigraphic position
	LANGHIAN/SERRAVALLIAN		13.82
B	*Unipontidinium aquaeductum*	DM5a	
T	*Distatodinium paradoxum*		
B	*Labyrinthodinium truncatum*	DM4	Tentatively at 15.2
	BURDIGALIAN/LANGHIAN		*15.99*
B	*Sumatridinium druggii*	DM3b	
T	*Cordosphaeridium cantharellum*	DM3a	
T	*Caligodinium amiculum*		Close to Aquit./Burdigal. boundary
B	*Hystrichosphaeropsis obscura*	DM2c	close to Aquit./Burdigal. boundary
	AQUITANIAN/BURDIGALIAN		20.44
T	abundant *Homotryblium plectilum*	DM2b	
T	*Chiropteridium galea* (= T *Chiropteridium* genus, within NN2)	DM2a	
T	common *Deflandrea phosphoritica*	DM1b	
T	*Distatodinium biffii*	DM1a	
	OLIGOCENE/MIOCENE boundary		23.0400

Motoyama, 1996; Motoyama et al., 2004). The radiolarian low-latitude biostratigraphy compiled by Sanfilippo and Nigrini (1998) has been used in the past as the standard low-latitude zonal scheme, with an extraregional perspective. Nigrini et al. (2005), in the study of equatorial Pacific ODP Leg 199 sites, set a new standard in refining the biohorizons of numerous radiolarian taxa. Since then, the radiolarian biostratigraphic works have been either refinements on Nigrini et al. (2005), or they have taken the improving zonal schemes and calibrated biohorizons in high southern and northern latitudes or in marginal seas successions, as those in the Japan Sea, in high-to-mid-latitude North Pacific and in subarctic areas (e.g., Kamikuri, 2017; Kamikuri et al., 2017).

In Fig. 29.11, diatom and radiolarian datums (Tables 29.4 and 29.5) and zonation for the low-latitude oceanic basins are reported and correlated to magnetostratigraphy and nannofossil biostratigraphy. The diatom zonal scheme mainly refers to biostratigraphic studies in the Pacific deep-sea sections, as those from ODP Legs 138, 145, 199 (e.g., Akiba, 1986; Akiba and Yanagisawa, 1986; Baldauf and Iwai, 1995; Barron and Gladenkov, 1995; Barron, 2003, 2006; Barron et al., 2004; Burckle, 1978; Burckle and Trainer, 1979; Burckle et al., 1982). For the Neogene radiolarians, the additional datums, besides the marker taxa, reported in Fig. 29.11 have been selected among 102 Neogene datums (from Nigrini et al., 2005), choosing evenly spaced in time datums, first appearances

whenever possible, and taxa easy to recognize, fairly abundant, and usually well preserved.

29.2.1.4 Dinoflagellates

Neogene zonations based on dinoflagellate cysts (dinocysts) have a widely known regional character of which the many works published, even in recent times, give evidence (e.g., De Schepper and Head, 2009; Williams and Manum, 1999; Dybkjær and Piasecki, 2010; Köthe, 2012; King, 2016; Schreck et al., 2017; Boyd et al., 2018). Despite the absence of widely applicable dinocyst zonations, when calcareous plankton microfossils are rare or absent, as in sediments from inner neritic or restricted deep-water environments, dinocyst biostratigraphy has been successfully applied, as shown by studies in northern latitude areas where stratigraphically important taxa occur, as in the North Atlantic and Nordic Seas, also in marginal marine settings. Dinocyst biostratigraphy is used to get a better age control in the Paratethys areas, for example, in Eastern Paratethys at the Paleogene/Neogene transition and in the stratigraphic interval of the Pontian regional stage (see details in Section 29.1.4.1).

Age assessments of dinoflagellate events mostly result from comparison with and correlation to nannofossil and foraminiferal biostratigraphies and magnetostratigraphy-calibrated successions (e.g., Schreck et al., 2012; King, 2016; Boyd et al., 2018). The dinoflagellate events and

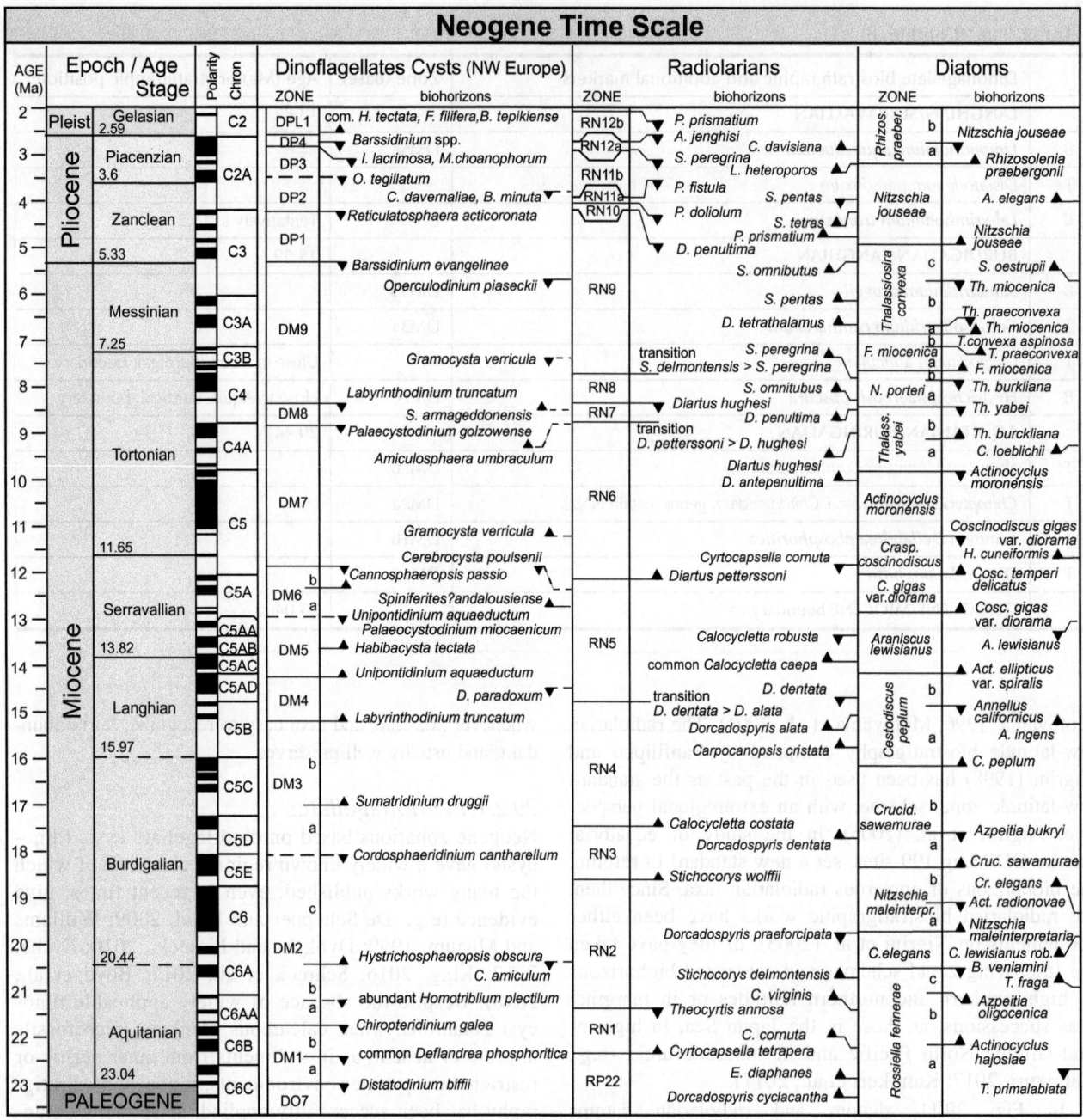

FIGURE 29.11 Neogene dinoflagellate cyst zonation for northwestern Europe (after King, 2016), and diatom and radiolarian datums and zonations for the low-latitude oceanic basins, with their estimated correlation to magnetostratigraphy and calcareous nannofossil zones.

zonation shown in Fig. 29.11 refer to biostratigraphic synthesis of dinocyst data from the Cenozoic sedimentary basins of NW Europe ("Atlantic margin" province and North Sea province; King, 2016). In this synthesis, it is emphasized that significant regional differences exist in the stratigraphic ranges of many dinoflagellate taxa, both within the North Sea Basin, and between this basin and other areas, and reliable interregional standard biostratigraphy for dinocysts has not been developed (King, 2016). However, the availability of more complete

database from deeper water sections in the North Sea, the North Atlantic, and the Nordic seas (e.g., Dybkjær and Piasecki, 2010) provided a series of dinocyst biohorizons recognizable in the North Sea Basin that have been calibrated more or less precisely to the standard geochronologic scale (through calibration to other fossil group zonations). The reader is kindly referred to King (2016) for further details of the selected dinocyst events, those that are better defined and widespread, and useful for correlation and dating. In this new Cenozoic

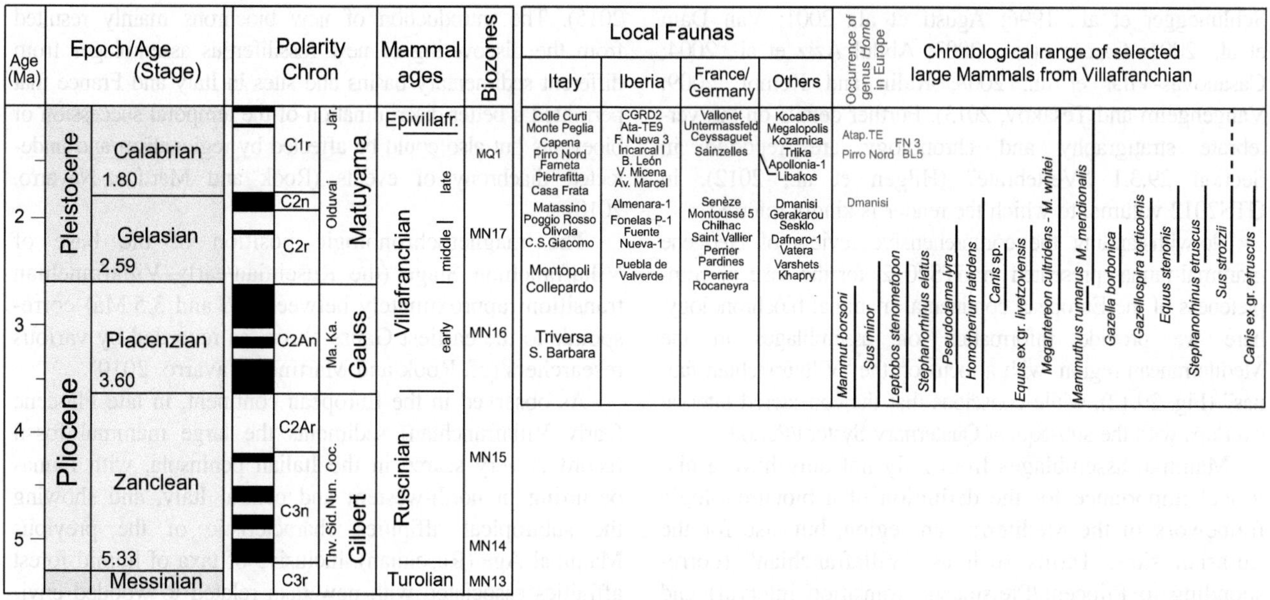

Figure 29.12 data:

Age (Ma)	Epoch	Age (Stage)	Polarity Chron	Mammal ages	Biozones
	Pleistocene	Calabrian 1.80	C1r (Jar.); Matuyama (Olduvai) C2n	Epivillafr. / late	MQ1
2		Gelasian 2.59	C2r (Matuyama)	Villafranchian middle	MN17
3	Pliocene	Piacenzian 3.60	C2An (Gauss, Ma. Ka.)	Villafranchian early	MN16
4		Zanclean	C2Ar; C3n (Gilbert, Thv. Sid. Nun. Coc.)	Ruscinian	MN15
5		5.33			MN14
		Messinian	C3r	Turolian	MN13

Local Faunas:

	Italy	Iberia	France/Germany	Others
(top)	Colle Curti, Monte Peglia, Capena, Pirro Nord, Fameta, Pietrafitta, Casa Frata	CGRD2, Ata-TE9, F. Nueva 3, Incarcal I, B. León, V. Micena, Av. Marcel	Vallonet, Untermassfeld, Ceyssaguet, Sainzelles	Kocabas, Megalopolis, Kozarnica, Trlika, Apollonia-1, Libakos
	Matassino, Poggio Rosso, Olivola, C.S.Giacomo	Almenara-1, Fonelas P-1, Fuente Nueva-1	Senèze, Montoussé 5, Chilhac, Saint Vallier, Perrier Pardines, Perrier Rocaneyra	Dmanisi, Gerakarou, Sesklo, Dafnero-1, Vatera, Varshets, Khapry
	Montopoli Collepardo	Puebla de Valverde		
	Triversa, S. Barbara			

Occurrence of genus *Homo* in Europe: Atap.TE, FN 3, Pirro Nord, BL5, Dmanisi

Chronological range of selected large Mammals from Villafranchian: *Mammut borsoni*, *Sus minor*, *Leptobos stenometopon*, *Stephanorhinus elatus*, *Pseudodama lyra*, *Homotherium latidens*, *Canis* sp., *Equus* ex gr. *livetzovensis*, *Megantereon cultridens - M. whitei*, *Mammuthus rumanus M. meridionalis*, *Gazella borbonica*, *Gazellospira torticornis*, *Equus stenonis*, *Stephanorhinus etruscus*, *Sus strozzii*, *Canis* ex gr. *etruscus*

FIGURE 29.12 Large Mammal biochronologic scheme, including the Ruscinian and Villafranchian stages. Selected taxa are reported, including the occurrence of genus *Homo* in Europe. Modified from Rook and Martìnez Navarro (2010); Martínez-Navarro et al. (2015).

dinoflagellate cyst zonation, "DM" and "DP" zones are defined based on the most reliable events, complemented by additional events considered regionally consistent but rare and/or more environmentally controlled (see also Fig. 29.11 and Table 29.6).

29.2.2 Terrestrial biostratigraphy

29.2.2.1 Mammals

In the terrestrial realm, biochronology represents the preferred conceptual method, compared to chronostratigraphy, for relating biological events to the GTS (e.g., see discussion in Lindsay, 2003). As for continental mammalian faunas, until now the organization of geologic time based on evidences provided by these faunas has been plagued by interpretative and semantic confusion. During the second half of the last century, knowledge on the Neogene and Quaternary terrestrial mammalian faunas notably increased and, the chronological setting has substantially enhanced, overall as far as small mammals are concerned. Discontinuity in the continental sedimentary record, the lack of deposits formed in a regime of virtually continuous sedimentation, anisotropy of paleoenvironmental conditions, taphonomic and sampling biases are all responsible for the fact that the stratigraphic order of the lowest and highest occurrences of taxa remains (stratigraphic data) within a given geographical area does not necessarily reflect the temporal order of their actual first/last appearances (paleobiologic events) in time. Therefore in order to relate biological events to the GTS in the terrestrial realm, a "biochronologic" theoretical approach, based on temporal inferences, has been used instead of the operational "biostratigraphic" approach, based on stratigraphic evidence. Accordingly, in the terrestrial realm a biochronologic unit is a span of time defined by the first/last appearance paleobiologic events (as well as by the exclusive cooccurrences of some taxa), while a biostratigraphic unit is a body of rock strata defined by its fossil content (stratigraphic data based on known fossil occurrences).

Section 29.3.1 "Vertebrate" in the GTS2012 volume (Hilgen et al., 2012) provides an exhaustive review on Neogene sedimentary sequences that are magnetochronologically and/or radioisotopically calibrated, and a comprehensive picture of worldwide database of Neogene mammal faunas. That chapter summarized an improved mammal-based chronology and correlations developed from studies on continuous continental sedimentary sections that had cyclostratigraphic patterns used for astrochronological calibration (e.g., Abdul Aziz et al., 2004; Hordijk and De Bruijn, 2009), which provided a standard dating precision in the order of ∼0.1 Myr. Chronology of mammal faunas is reasonably well established in Europe and North America, as a result of a longer history of mammal paleontology, whereas it is still not complete in other continents. In European mammal chronology, the two different approaches, discussed earlier, have been used for subdividing the Neogene time using vertebrate paleontology, the "stratigraphical" and "faunal" methods (see Lourens et al., 2004; Hilgen et al., 2012), and provided relevant results. Important mammal-related magnetostratigraphic records are available extending across different intervals of the Miocene and Pliocene, from the Oligocene/Miocene transition to late Miocene in Spain, Switzerland, southern Germany, and eastern Europe (e.g., Levèque, 1993; Krijgsman et al., 1996;

Schlunegger et al., 1996; Agustí et al., 2001; Van Dam et al., 2001; Garcés et al., 2003; Abdul Aziz et al., 2004; Casanovas-Vilar et al., 2008; Kälin and Kempf, 2009; Vangengeim and Tesakov, 2013). Further details on the vertebrate stratigraphy and chronology are reported in Section 29.3.1 "Vertebrate" (Hilgen et al., 2012), in GTS2012 volume, to which the reader is kindly referred.

Notwithstanding the comprehensive review of Neogene mammal-faunas presented in GTS2012, for the sake of completeness of the European continental mammal biochronology, here we provide information on assemblages in the Mediterranean region, with a focus on the "Villafranchian faunas" (Fig. 29.12), while conscious that the considered interval overlaps with the subsequent Quaternary System/Period.

Mammal assemblages from Italy not only have a historical importance for the definition of a biochronologic framework of the Mediterranean region, but also for the Eurasian sites. Terms such as "Villafranchian" (corresponding to Pliocene/Pleistocene transition interval) and "Galerian" (middle Pleistocene) come from Italian sites and are still generally used by vertebrate paleontologists interested in the "Old World" terrestrial faunas and continental stratigraphers in southern Europe. At the Pliocene/Pleistocene transition, the geographical position of Italy represented a crossroad between Europe and Africa, between East and West; therefore the Italian fossil record can be considered of special interest for the study of the Plio−Pleistocene terrestrial ecosystems.

The study of the Italian Pliocene−Pleistocene mammal faunas was once the scientific interest of the valuable and internationally known research activity of Augusto Azzaroli, starting from the 1960s through the 1980s. The Upper Valdarno Basin record (Tuscany, central Italy) is the basis for taxonomic definition of much of the early-to-late Villafranchian (late Pliocene to early Pleistocene) species and genera, as well as for the definition of subdivisions of the continental biochronologic scale, that were the bulk of Azzaroli's activity (Azzaroli, 1977; Azzaroli et al., 1982, 1988). The continuous interest in the study of Upper Valdarno Basin fossil mammals contributed to the increasing knowledge and refinement of the biochronologic subdivision of the Villafranchian (Azzaroli et al., 1982, 1988) (Fig. 29.12), and, in a broader sense, to the reconstruction of biotic and abiotic events that characterized the Plio−Pleistocene interval of central Italy. "Faunal Units" (FUs) were recognized in subsequent levels delineating the biochronologic unit "Villafranchian Mammal Age" (Azzaroli, 1977) that corresponds to a timespan from late Pliocene to most of the early Pleistocene, between 3.5 and 1.1−1.0 Ma. Since the initial introduction of FUs for the subdivision in early, middle, and late Villafranchian, Italian paleontologists have proposed some new FUs and/or new Land-Mammal Ages or revised others (e.g., Torre, 1987; Sala et al., 1992; Caloi and Palombo, 1996; Gliozzi et al., 1997; Petronio and Sardella, 1999; Palombo, 2004; Bellucci et al.,

2015). The introduction of new biochrons mainly resulted from the discovering of new fossiliferous assemblages from different sedimentary basins and sites in Italy and France that permitted a better discrimination of the temporal succession of bioevents but also could be affected by regionalism and undetected diachrony of events (Rook and Martínez-Navarro, 2010).

The magnetochronologic position of the base of Villafranchian Stage (the Ruscinian/early Villafranchian transition; approximately between 3.6 and 3.5 Ma) corresponds to the earliest Gauss chron, as recorded by various researchers (cf. Rook and Martínez-Navarro, 2010).

As observed in the European continent, in late Pliocene (early Villafranchian) sediments the large mammal fossil record is very scarce in the Italian peninsula, with faunas occurring in north-western and central Italy, and showing the subtropical affinities characteristic of the previous Mammal Age (Ruscinian), inclusive of taxa of humid forest affinities associated with new taxa related to wooded environment (Triversa FU, Azzaroli, 1977; Azzaroli et al., 1988; Pradella and Rook, 2007; Collepardo local fauna, Bellucci et al., 2019; Fig. 29.12) As regards the early Villafranchian faunal records in Europe, they are present in France, in the Central Massif (the site of Vialette) where one of the first records in Europe of *Canis* and *Equus* is found (Lacombat et al., 2008); in Spain (the site of Villarroya), in which one of the oldest records of Megantereon *cultridens* in the continent is present (Palmqvist et al., 2007); in Eastern Europe (within the Dacic Basin in Romania and in Bulgaria) with a fauna characterized by the appearance of a primitive form of the African origin modern elephant *Mammuthus rumanus*, the ancestor of *Mammuthus meridionalis* (Lister and van Essen, 2003; Markov and Spassov, 2003).

The early/middle Villafranchian transition approximates the base of Gelasian Stage that is the Pliocene/Pleistocene boundary at 2.59 Ma. The first middle Villafranchian large mammal assemblage is recorded in central Italy (Montopoli fauna, in Lower Valdarno Basin, Tuscany; Azzaroli, 1977; Azzaroli et al., 1988) and is well known in the literature because it corresponds to a distinct signal of environmental change represented by the disappearance of the dispersals of a primitive species of the genus *Mammuthus*, the monodactyl horse (*Equus* cf. *livenzovensis*), the large deer *Eucladoceros tegulensis*, and *Gazella borbonica*. The Montopoli FU corresponds to the MN16b unit in the European MN subdivision (Fig. 29.12) (Gliozzi et al., 1997) and occurs at the Gauss/Matuyama transition (Lindsay et al., 1980). Faunas of similar age, with archaic true horses, are known in former Soviet Union area as Ukraine, Tajikista, and southern Adyrgan in eastern Kazakhstan (Vangengeim et al., 2005).

In the middle and late Villafranchian, namely, the early Pleistocene, major episodes in mammalian fauna reorganization occurred and continued at the Villafranchian/

Galerian Mammal Age transition (Epivillafranchian after Bellucci et al., 2015), in parallel with increased climatic instability. This renewal, occurred from middle-to-late Villafranchian, is attained mainly by dispersal events, which followed in phases around the mid part of early Pleistocene (at the Gelasian/Calabrian transition; cf. Torre et al., 1992). This faunal turnover intensely influenced mammal assemblages and involved carnivores and herbivores, resulting in the disappearance of most of the early Villafranchian species, largely among herbivores, while new carnivores and herbivores progressively appeared (Sala et al., 1992; Torre et al., 1992). Characterization of these early Pleistocene faunas comes from fossiliferous sites located in France (St. Vallier FU; Guérin et al., 2004), in Bulgaria and Greece (e.g., Spassov, 2000; Koufos, 2001; Kostopoulos and Athanassiou, 2005), and in Italy (Costa San Giacomo FU, Latium) (Bellucci et al., 2012, 2014). The faunal composition of middle Villafranchian units is characterized by some important first occurrences, for example, those of *Stephanorhinus etruscus*, *Equus stenonis*, the spiral horned antelope *Gazellospira torticornis*, and the first occurrence of *Canis* cf. *etruscus* (Rook and Torre, 1996). The occurrence of *C.* cf. *etruscus* in Italy, together with the finding of *Canis* sp. in the early Villafranchian in France, suggests that the so-called Wolf-event of Azzaroli (1983) started earlier than originally assumed, and that the expansion of wolf-like dogs across Eurasia was a diachronous event (Sotnikova and Rook, 2010).

Besides the faunal turnovers recorded in the late Villafranchian that deeply affected the carnivore and herbivore components of the mammal assemblages, the genus *Homo* dispersal occurred as another important faunal turnover in the Mediterranean European area. The earliest *Homo* out of Africa was recorded at the site of Dmanisi (Caucasus, Georgia) associated with the latest occurrence of *Canis etruscus*, whereas the oldest occurrence of early *Homo* in Western Europe was recorded in southern Italy (at Pirro Nord site), southern Spain (Fuente Nueva-3 and Barranco Leon-5) (cf. Rook and Martínez-Navarro, 2010, and references therein).

In the time interval corresponding to the Epivillafranchian (former Villafranchian/Galerian transition, 1.2−0.9 Ma), the European mammal faunas primarily included survivors from the latest Villafranchian, specifically the large mammal assemblages included Villafranchian taxa together with newcomers, mostly from Asia, considered as "transitional" faunas, persisting in the middle Pleistocene (cf. Rook and Martínez-Navarro, 2010; Bellucci et al., 2015).

The current biochronologic scheme (Fig. 29.12) provides very detailed resolution for Italian assemblages even if it is less useful for correlation to other European mammal faunas. Then, the Villafranchian biochronology (Mammal Ages and FUs), so related to the Italian/French record, pertains a regional character that inevitably affects and limits definition

of this biochronologic unit at large scale. However, the insertion into the biochronologic frame of a number of well-dated sites (from Spain to Greece and western former Soviet Union countries) led to an evident improvement in understanding of the biological history (faunal evolution and turnover events) in this timespan (Rook and Martínez-Navarro, 2010). Examples of these correlatable and well-calibrated sedimentary successions are from southern European basins: in southern Spain the Guadix−Baza basin, in which a thick continuous continental succession is representative of latest Miocene, Pliocene, and most of the Pleistocene time intervals, including a relevant record of Villafranchian mammal faunas (e.g., Agustí et al., 2001; Toro et al., 2003; Arribas, 2008); in Bulgaria and Greece, where a fossiliferous sites are important for middle and early late Villafranchian biochronology (e.g., Spassov, 2000, 2003; Koufos, 2001); in areas south of the Caucasus (Georgia) where a late Pliocene to early Pleistocene fossil record of large mammals is present (e.g., Agustí et al., 2009).

29.2.3 Physical stratigraphy

The Neogene Chapter 29 in the GTS2012 volume (Hilgen et al., 2012) provides exhaustive details on recent cross-calibration efforts for astronomically derived and radioisotopic ages, Climate Change and Milankovitch cycles, Sedimentary cycles, and Ice cores, which we kindly refer the reader to for in-depth summaries. Here, we provide updates for the actual magnetostratigraphy forming the basis for Neogene astronomically tuned time scale and discuss more recent works on Geochemical Stratigraphy (δ^{18}O composite curves and splices) and Sequence Stratigraphy.

29.2.3.1 Magnetostratigraphy

We provide a revised GTS2020 geomagnetic polarity time scale, which is discussed next (in Section 29.3.2). We have not yet reevaluated the detailed implications of exploring detailed paired plate-boundary spreading rates as Wilson did for select intervals in GTS2012, and as Francescone et al. (2019) did with an expanded database of spreading pairs for parts of the Paleogene. For a more in-depth discussion on magnetostratigraphic issues, we refer the reader kindly to GTS2012.

29.2.3.2 Geochemical stratigraphy

An extensive set of different geochemical stratigraphic tools and applications is discussed in GTS2012, and we refer the reader there for in-depth information. Here we provide an update on some recently developed composite curves and "mega-splices." Following the pioneering works in generating longer and more detailed stable oxygen isotope time series of Shackleton and Opdyke (1973), Shackleton and Hall (1989), Ruddiman et al. (1987), Raymo et al. (1989), there has been a continuous expansion of new datasets and time-series published, particularly related to the analysis of cores from the Deep Sea Drilling Project (DSDP), ODP, and IODP. Compared to the initial "mega-splice" of oxygen isotope data by Shackleton et al. (1995b) that included data from Vema core V19−30, ODP Site 677, and ODP Site 846, there have been subsequent publications that either assimilate data from different sites into a master time scale, like the frequently used LR04 stack (Lisiecki and Raymo, 2005), the Zachos et al. (2001) isotope compilation (where data were by necessity not astronomically age calibrated), and more recent similar compilations (e.g., Cramer et al., 2009). However, compilations and stack, while clearly useful, have the disadvantage that age model errors can distort and modify the appearance of stable isotope time series. For this reason, De Vleeschouwer et al. (2017) derived a complete stable oxygen isotope "mega-splice" from individual deep-ocean records that were individually astronomically age calibrated, thus perhaps providing the clearest possible picture of climate variability and correlative potential throughout the Neogene and Oligocene [see Fig. 29.1 that also includes a stable carbon isotope counterpart (De Vleeschouwer et al., 2020, submitted)].

29.2.3.3 Sequence stratigraphy

Glacio-eustasy, the expansion and contraction of Antarctic ice sheets, is the primary driver of eustatic fluctuations in the Neogene (Miller et al., 2005). Since GTS2012 several studies have examined the timing and amplitude of Miocene sea level fluctuations, including on the Marion and Queensland Plateaus (Australian margin; John et al., 2011), New Jersey margin (Browning et al., 2013), Durban Basin (SE African margin; Hicks and Green, 2017), and the Maldives (Betzler et al., 2018). The eustatic sea-level variations in the Miocene significantly affected depositional environments of sedimentary successions, causing erosional unconformity surfaces and changes in accommodation space (Browning et al., 2013; Hicks and Green, 2017). In general, there is good agreement with sequence boundaries in multiple records, although the amplitudes and ages vary, probably due to age model uncertainties in shallower water environments (Miller et al., 2017). See also review in Simmons et al. (2020, Ch. 13, Phanerozoic eustasy, this book).

Sequence boundaries indicating glacio-eustasy on the Marion Plateau and New Jersey margin, identified in both seismic data and cores, indicate sea-level fluctuations of >25 m (John et al., 2011; Browning et al., 2013). The timing of sequence boundaries generally corresponds to the global Miocene oxygen isotope maxima (Mi-events) defined by Miller et al. (1991).

Following the late Oligocene lowstand, the O/M boundary was associated with a global sea-level rise of approximately 50 m (Miller et al., 2017). Sea-level rises

of >19 m are also identified at around 20 (Betzler et al., 2018) and 16 Ma (John et al., 2011).

There are multiple lines of evidence that a major sea-level fall (and possible changes in deep-water circulation) occurred close to the base of the Serravallian. A prominent sequence boundary at 13.9 Ma on the Marion Plateau suggests a sea-level fall of 59 ± 6 m (John et al., 2011). The sequence boundary is coeval with an unconformity identified in the Durban Basin by Hicks and Green (2017) and corresponds to a >1‰ oxygen isotope increase associated with Mi-3 (Holbourn et al., 2013). The back-stripped constraint of sea-level fall from the Marion Plateau is larger than estimates from the New Jersey margin (Browning et al., 2013).

From the Maldives, Betzler et al. (2018) report sea-level lowerings at 19.5, 18.5, 17, 15.6, 15.1, 14.2, and 13.1 Ma. These differ to the timing of sea-level fall recorded on the Marion Plateau, at 16.5, 15.4, 14.7, and 13.9 Ma. Betzler et al. (2018) did not find evidence of sea-level fall associated with Mi-3 (13.9 Ma), suggesting eustatic sea-level lowering and deep-sea cooling is out of phase (Betzler et al., 2018).

29.3 Neogene astronomically tuned time scale

29.3.1 Toward a fully astronomically tuned Neogene polarity time scale

The geomagnetic polarity time scale was traditionally based on seafloor anomaly profiles combined with a limited number of radioisotopically dated tie points (Cande and Kent, 1992, 1995). For the Neogene, this approach has been replaced by astronomical dating of sections having a reliable magnetostratigraphy with cycle tuning producing astronomical ages for magnetic reversal boundaries at the same time. In ATNTS2004, uninterrupted astronomical ages for polarity reversals only go back to 12 Ma, with reliable astronomical reversal ages only being available for very short intervals around 16 and 23 Ma (Lourens et al., 2004; Table 29.7). The Neogene polarity time scale in GTS2012 (Hilgen et al., 2012) was still partly based on Wilson's work on seafloor spreading rates (essentially between 16 and 23 Ma). Thus GTS2012 left the ages for chrons C5Bn.1n (14.87 Ma) through C6Cn.2n virtually unchanged from GTS2004, in anticipation of new studies resolving existing discrepancies between spreading rate-based ages and astronomically tuned sections. For GTS2020, we report next the results of more recent studies that result in the first attempt to use fully astronomically tuned ages also from ~15 Ma to the Oligocene−Miocene boundary.

The age calibration of the lower−middle Miocene (and Oligocene) was initially based on ODP Leg 154

TABLE 29.7 GTS2020 Chron Ages and Comparison with GPTS after Cande and Kent (1995), GTS2004, GTS2012, and GTS2016.

Magnetochron (Base)	Cande and Kent (1995)	GTS2004	GTS2012	GTS2016	GTS2020	References
	(Ma)					
	0.000	**0.000**	**0.000**	**0.000**	**0.000**	
C1n (Brunhes)	0.780	0.781	0.781	0.773	0.773	no change from GTS2016
C1r.1r (Matuyama)	0.990	0.988	0.988	1.008	1.008	
C1r.1n (Jaramillo)	1.070	1.072	1.072	1.076	1.076	
C1r.2r		1.173	1.173	1.189	1.189	
C1r.2n (Cobb Mountain)		1.185	1.185	1.221	1.221	
C1r.3r	1.770	1.778	1.778	1.775	1.775	
C2n (Olduvai)	1.950	1.945	1.945	1.934	1.934	
C2r.1r	2.140	2.128	2.128	2.120	2.120	
C2r.1n (Feni)	2.150	2.148	2.148	2.155	2.155	
C2r.2r (Matuyama continued)	2.581	2.581	2.581	2.610	2.610	
C2An.1n (Gauss)	3.040	3.032	3.032	3.032	3.032	no change from GTS2012/16
C2An.1r (Keana)	3.110	3.116	3.116	3.116	3.116	
C2An.2n	3.220	3.207	3.207	3.207	3.207	
C2An.2r (Mammoth)	3.330	3.330	3.330	3.330	3.330	
C2An.3n (Gauss continued)	3.580	3.596	3.596	3.596	3.596	
C2Ar (Gilbert)	4.180	4.187	4.187	4.187	4.187	
C3n.1n (Cochiti)	4.290	4.300	4.300	4.300	4.300	
C3n.1r	4.480	4.493	4.493	4.493	4.493	
C3n.2n (Nunivak)	4.620	4.631	4.631	4.631	4.631	
C3n.2r	4.800	4.799	4.799	4.799	4.799	
C3n.3n (Sidufjall)	4.890	4.896	4.896	4.896	4.896	
C3n.3r	4.980	4.997	4.997	4.997	4.997	
C3n.4n (Thvera)	5.230	5.235	5.235	5.235	5.235	
C3r (Gilbert lower part)	5.875	6.033	6.033	6.033	6.023	Drury et al. (2017)
C3An.1n	6.122	6.252	6.252	6.252	6.272	
C3An.1r	6.256	6.436	6.436	6.436	6.386	
C3An.2n	6.555	6.733	6.733	6.733	6.727	
C3Ar	6.919	7.140	7.140	7.140	7.104	
C3Bn	7.072	7.212	7.212	7.212	7.214	
C3Br.1r	7.135	7.251	7.251	7.251	7.262	
C3Br.1n	7.170	7.285	7.285	7.285	7.305	
C3Br.2r	7.341	7.454	7.454	7.454	7.456	

(Continued)

TABLE 29.7 (Continued)

Magnetochron (Base)	Cande and Kent (1995)	GTS2004	GTS2012	GTS2016	GTS2020	References
	(Ma)					
	0.000	0.000	0.000	0.000	0.000	
C3Br.2n	7.375	7.489	7.489	7.489	7.499	
C3Br.3r	7.406	7.528	7.528	7.528	7.537	
C4n.1n	7.533	7.642	7.642	7.642	7.650	
C4n.1r	7.618	7.695	7.695	7.695	7.701	
C4n.2n	8.027	8.108	8.108	8.108	8.125	
C4r.1r	8.174	8.254	8.254	8.254	8.257	
C4r.1n	8.205	8.300	8.300	8.300	8.300	no change from GTS2012/16
C4r.2r	8.631	8.769	8.771	8.771	8.771	
C4An	8.945	9.098	9.105	9.105	9.105	
C4Ar.1r	9.142	9.312	9.311	9.311	9.311	
C4Ar.1n	9.218	9.409	9.426	9.426	9.426	
C4Ar.2r	9.482	9.656	9.647	9.647	9.647	
C4Ar.2n	9.543	9.717	9.721	9.721	9.721	
C4Ar.3r	9.639	9.779	9.786	9.786	9.786	
C5n.1n	9.880	9.934	9.937	9.937	9.937	
C5n.1r	9.920	9.987	9.984	9.984	9.984	
C5n.2n	10.839	11.040	11.056	11.056	11.056	
C5r.1r	10.943	11.118	11.146	11.146	11.146	
C5r.1n	10.991	11.154	11.188	11.188	11.188	
C5r.2r	11.343	11.554	11.592	11.592	11.592	
C5r.2n	11.428	11.614	11.657	11.657	11.657	
C5r.3r	11.841	12.014	12.049	12.049	12.049	
C5An.1n	11.988	12.116	12.174	12.174	12.174	
C5An.1r	12.096	12.207	12.272	12.272	12.272	
C5An.2n	12.320	12.415	12.474	12.474	12.474	
C5Ar.1r	12.605	12.730	12.735	12.735	12.735	
C5Ar.1n	12.637	12.765	12.770	12.770	12.770	
C5Ar.2r	12.705	12.820	12.829	12.829	12.829	
C5Ar.2n	12.752	12.878	12.887	12.887	12.887	
C5Ar.3r	12.929	13.015	13.032	13.032	13.032	
C5AAn	13.083	13.183	13.183	13.183	13.183	
C5AAr	13.252	13.369	13.363	13.363	13.363	
C5ABn	13.466	13.605	13.608	13.608	13.608	
C5ABr	13.666	13.734	13.739	13.739	13.739	
C5ACn	14.053	14.095	14.070	14.070	14.070	

TABLE 29.7 (Continued)

Magnetochron (Base)	Cande and Kent (1995)	GTS2004	GTS2012	GTS2016	GTS2020	References
	(Ma)					
	0.000	0.000	0.000	0.000	0.000	
C5ACr	14.159	14.194	14.163	14.163	14.163	
C5ADn	14.607	14.581	14.609	14.609	14.609	
C5ADr	14.800	14.784	14.775	14.775	14.775	
C5Bn.1n	14.888	14.877	14.870	14.870	14.870	
C5Bn.1r	15.034	15.032	15.032	15.032	15.040	Kochhann et al. (2016)
C5Bn.2n	15.155	15.160	15.160	15.160	15.186	
C5Br	16.014	15.974	15.974	15.974	15.994	
C5Cn.1n	16.293	16.268	16.268	16.268	16.261	
C5Cn.1r	16.327	16.303	16.303	16.303	16.351	
C5Cn.2n	16.488	16.472	16.472	16.472	16.434	
C5Cn.2r	16.556	16.543	16.543	16.543	16.532	
C5Cn.3n	16.726	16.721	16.721	16.721	16.637	
C5Cr	17.277	17.235	17.235	17.235	17.154	
C5Dn	17.615	17.533	17.533	17.533	17.466	
C5Dr.1r	17.825	17.717	17.717	17.717	17.634	
C5Dr.1n	17.853	17.740	17.740	17.740	17.676	
C5Dr.2r	18.281	18.056	18.056	18.056	18.007	Interpolated GTS2012
C5En	18.781	18.524	18.524	18.524	18.497	Liebrand et al. (2016)
C5Er	19.048	18.748	18.748	18.748	18.636	
C6n	20.131	19.722	19.722	19.722	19.535	
C6r	20.518	20.040	20.040	20.040	19.979	
C6An.1n	20.725	20.213	20.213	20.213	20.182	Interpolated GTS2012
C6An.1r	20.996	20.439	20.439	20.439	20.448	
C6An.2n	21.320	20.709	20.709	20.709	20.765	Liebrand et al. (2016)
C6Ar	21.768	21.083	21.083	21.083	21.130	Interpolated GTS2012
C6AAn	21.859	21.159	21.159	21.159	21.204	
C6AAr.1r	22.151	21.403	21.403	21.403	21.441	
C6AAr.1n	22.248	21.483	21.483	21.483	21.519	
C6AAr.2r	22.459	21.659	21.659	21.659	21.691	Liebrand et al. (2016)
C6AAr.2n	22.493	21.688	21.688	21.688	21.722	Interpolated GTS2012
C6AAr.3r	22.588	21.767	21.767	21.767	21.806	
C6Bn.1n	22.750	21.936	21.936	21.936	21.985	Beddow et al. (2018)
C6Bn.1r	22.804	21.992	21.992	21.992	22.042	
C6Bn.2n	23.069	22.268	22.268	22.268	22.342	
C6Br	23.353	22.564	22.564	22.564	22.621	

(Continued)

TABLE 29.7 (Continued)

Magnetochron (Base)	Cande and Kent (1995)	GTS2004	GTS2012	GTS2016	GTS2020	References
	(Ma)					
	0.000	0.000	0.000	0.000	0.000	
C6Cn.1n	23.535	22.754	22.754	22.754	22.792	
C6Cn.1r	23.677	22.902	22.902	22.902	22.973	
C6Cn.2n (O/M)	23.800	23.030	23.030	23.030	23.040	

References related to the adopted revisions are reported.

(Ceara Rise, equatorial Atlantic) sites lacking a magnetostratigraphy, using especially but not exclusively high-resolution obliquity-dominated records of magnetic susceptibility (Shackleton et al., 1999). Shackleton et al. (2000) also tuned a high-quality stable isotope record from Ceara Rise to arrive at an age of 23.03 Ma (retuned to La2004 of Laskar et al., 2004, see Lourens et al., 2004; Pälike et al., 2006b) for the Oligocene–Miocene boundary, which in GTS2020 has been slightly revised to 23.04 Ma (Beddow et al., 2018).

At Ceara Rise, the interval between 14 and 18 Ma has proved especially problematic due to lack of relevant data. Moreover, no magnetostratigraphy is available as an independent check on the continuity of the succession. In addition, since then, the previous splices from Ceara Rise have been partly adjusted and corrected (Zeeden et al., 2013; Drury et al., 2017; Wilkens et al., 2017), but unfortunately the main problem remains of no available magnetostratigraphy from Ceara Rise ODP Leg 154. We refer the reader to the detailed discussion about age model challenges in GTS2012, Section 29.4.1.1.

29.3.2 GTS2012 and GTS2020

Differences between ATNTS2004 and 2012 were relatively minor, as could be expected from a GTS that is underlain by astronomical tuning. Nevertheless, for GTS2020 there are updates even for the previously fully calibrated part from the present to ∼15 Ma. Here we lay out the main changes in GTS2020 (Table 29.7).

From the present down to C2r.2r (Matuyama continued), we incorporate the minor changes that were introduced in GTS2016 compared to GTS2012 and generally show differences that are smaller than one obliquity cycle. The ages for polarity reversals from the base of C2An.1n (Gauss) through the base of C3n.4n (Thvera) remain unchanged from GTS2012 (and indeed GTS2004). For the basal ages of polarity chrons C3r (Gilbert lower part) through C4r.1r we adopt the revisions given by Drury et al. (2017), but again these are typically only different

by less than an orbital cycle. The basal ages from C4r.1n (8.30 Ma) through C5Bn.1n (14.87 Ma) remain unchanged from GTS2012.

At 15 Ma, we change from a polarity time scale based on spreading rate interpretations to one almost fully supported by direct astronomical tuning: we adopt the calibrated ages from Kochhann et al. (2016), obtained from IODP Expedition 321 equatorial Pacific Site U1335 for the basal magnetic reversals from C5Bn.1r (15.04 Ma) through C5Dr.1n (17.676 Ma). These revised changes show differences of up to 84 kyr younger and 48 kyr older than GTS2012 in this interval. An alternative age model from nearby IODP Site U1337 was provided by Holbourn et al. (2015), but this site does not provide a high-quality magnetostratigraphy in this interval. The ages from Kochhann et al. (2016) appear to be also consistent with new ages resulting from the extension of the La Vedova section near Ancona (Turco et al., 2017; Hilgen, personal communication 2019). In addition, new U/Pb dating of zircon–bearing ash layers from the Columbia River flood basalts (Kasbohm and Schoene, 2018) appear to also support the revised Kochhann et al. (2016) ages, with the exception of the basal age of C5Br, which according to the U/Pb age correlation appears c. 100 kyr older in GTS2020 (15.994 Ma) than the U/Pb age of <15.895 Ma ± 0.019/0.026.

From the basal age of reversal C5Dr.2r (18.50 Ma) through C6AAr.3r (21.81 Ma), we primarily use ages from Liebrand et al. (2016), who performed an eccentricity-based tuning with data from Walvis Ridge ODP Site 1264. This dataset has a less robust magnetostratigraphy but is underpinned by isotopic correlations with other sites. Age differences to GTS2012 are mostly around an obliquity cycle, but with 112 and 187 kyr younger ages than GTS2012 for C5Er and C6n (19.535 Ma). For nine reversal ages in the interval covered by Liebrand et al. (2016), we have to rely on linear interpolation from GTS2012 to GTS2020 (Table 29.7).

New basal reversal ages from C6Bn.1n (21.99 Ma) through C7n.1r (24.12 Ma) were published by Beddow et al. (2018) using data from equatorial Pacific IODP Expedition 320, Site U1334. These also contain the basal age for C6Cn.2n (23.04 Ma), which is our revised age for the Oligocene—Miocene boundary.

TABLE 29.8 GTS2020 ages for Neogene stage boundaries, including the Neogene/Quaternary boundary, and duration of stages.

Stage	Base (Ma)	Duration (Myr)
Gelasian	2.59	
Piacenzian	3.60	0.78
Zanclean	5.33	1.73
Messinian	7.25	1.92
Tortonian	11.65	4.40
Serravallian	13.82	2.17
Langhian	*15.99*	2.17
Burdigalian	*20.45*	4.46
Aquitanian	23.04	2.59

29.4 Summary

- The Astronomically Tuned Neogene Time Scale—ATNTS2020—includes some modifications and updating of tuned ages of reversal boundaries and biochronologic data with respect to the ATNTS2012. These data partially overcome the known problems in the tuning of the lower Miocene (see discussion in Hilgen et al., 2012), providing improved astronomical ages for this time interval. Time scale revisions and updated biochronology refer to the intervals 8—13 and 15—24 Ma, including the Oligocene/Miocene transition, and have been obtained from the ODP Leg 154 and Leg 208 sites and IODP Leg 320/321 sites. The persistent lack of middle and late Miocene ocean drilling cores with a well-defined magnetostratigraphy still leaves a low degree of uncertainty for the time scale of the Neogene, but additional modifications to the splice of reference sections (as at Ceara Rise ODP sites) are ongoing.
- Regarding the incorporation of the stages in ATNTS2020 (Table 29.8; Figure 29.13), it mirrors what is reported for ATNTS2012 and is straightforward for the Serravallian

GSSPs of the Neogene Stages, with location and primary correlation criteria

Stage	GSSP Location	Latitude, Longitude	Boundary Level	Correlation Events	Reference
Piacenzian	Punta Piccola, Sicily, Italy	37°17'20"N 13°29'36"E*	base of the beige marl bed of carbonate cycle an age of 3.6 Ma	precessional excursion 347 from the present with an astrochronological age estimate of 3.6 Ma	Episodes **21/2**, 1998
Zanclean	Eraclea Minoa, Sicily, Italy	37°23'30"N 13°16'50"E	base of the Trubi Formation	Insolation cycle 510 counted from the present with an age of 5.33 Ma	Episodes **23/3**, 2000
Messinian	Oued Akrech, Morocco	33°56'13"N 6°48'45"W	base of reddish layer of sedimentary cycle 15	Foraminifer, regular FAD of *Globorotalia miotumida (conomiozea)* and calc. nannofossil FAD of *Amaurolithus delicatus*	Episodes **23/3**, 2000
Tortonian	Monte dei Corvi Beach, near Ancona, Italy	43°35'12"N 13°34'10"E	mid-point of sapropel layer of basic cycle number 76	Common LAD of calc. nannofossil *Discoaster kugleri* and foraminifer *Globigerinoides subquadratus*	Episodes **28/1**, 2005
Serravallian	Ras il Pellegrin Section, Fomm Ir-Rih Bay, west coast of Malta	35°54'50"N 14°20'10"E	formation boundary between the Globigerina Limestone and Blue Clay	Mi3b oxygen-isotopic event (global cooling episode)	Episodes **32/3**, 2009
Langhian	*candidate: La Vedova (Italy)*			*Base of magnetic polarity chron C5Br*	
Burdigalian	*Potentially in astronomically-tuned ODP core*			*Near FAD of calc. nannofossil Helicosphaera ampliaperta*	
Aquitanian	Lemme-Carrioso Section, Allessandria Province, Italy	44°39'32"N 08°50'11"E	35 m from the top of the section	Base of magnetic polarity chron C6Cn.2n; FAD of calc. nannofossil *Sphenolithus capricornutus*	Episodes **20/1**, 1997

* according to Google Earth

FIGURE 29.13 Ratified and potential GSSPs and primary correlation markers for the Neogene stages (status as of January 2020). Details of each GSSP are in the text and at http://www.stratigraphy.org, https://timescalefoundation.org/gssp/, and in the *Episodes* publications. *GSSP*, Global Boundary Stratotype Section and Point.

GSSP up to the Piacenzian GSSP. Definition of Langhian and Burdigalian GSSPs is still "work in progress." However, significant progress has been made in evaluating sections and guiding criteria for defining the Langhian GSSP in the near future, with the identification of a suitable section in the Mediterranean (La Vedova composite section, Italy) provided with integrated magnetobiostratigraphy, stable isotope stratigraphy, cyclostratigraphy, and robust astronomical tuning. The problem of the lack of Burdigalian GSSP is still unsolved, due to the absence of undoubtedly suitable sections at least in the Mediterranean area, where the other defined Miocene GSSPs are located. The alternative to have this boundary defined in an (I) ODP core in the open ocean represents a serious option to be considered.

- Correlations of Neogene regional stages for the Paratethys area and New Zealand to the global chronostratigraphy are discussed.
- The biochronology of calcareous plankton (nannofossils and foraminifera) has been updated.
- The dinocyst zonation refers to biostratigraphic synthesis of dinocyst data from the Cenozoic sedimentary basins of NW Europe ("Atlantic margin" province and North Sea province).
- Neogene/Quaternary Larger Mammal assemblages in the Mediterranean region, lacking from the previous "Vertebrate" Section in GTS2012 volume, are provided and complement the picture of European continental mammal biostratigraphy/biochronology.

Acknowledgments

We thank Frederik Hilgen and Thomas Westerhold for useful discussion regarding issues related to biochronology linked to astronomical time scale and geomagnetic polarity time scale; and Marcin Latas and David King for comments on the foraminiferal bioevents table.

Bibliography

Abdul Aziz, H., Van Dam, J.A., Hilgen, F.J., and Krijgsman, W., 2004, Astronomical forcing in Upper Miocene continental sequences: implications for the Geomagnetic Polarity Time Scale. *Earth and Planetary Science Letters*, **222**: 243–258.

Agustí, J., Oms, O., and Remacha, E., 2001, Long Plio-Pleistocene terrestrial record of climate change and mammal turnover in southern Spain. *Quaternary Research*, **56**: 411–418.

Agustí, J., Vekua, A., Oms, O., Lordkipanidze, D., Bukshianidze, M., Kiladze, G., et al., 2009, The middle Pliocene site of Kvabebi and the faunal background to early human occupation of southern Caucasus. *Journal of Quaternary Sciences*, **28**: 3275–3280.

Akiba, F., 1986, Middle Miocene to Quaternary diatom bio- stratigraphy in the Nankai Trough and Japan Trench, and modified lower Miocene through Quaternary diatom zones for middle-to-high latitudes of the north Pacific. In: Kagami, H., Karig, D.E., Coulbourn, W.T., et al., (eds), *Initial Reports Deep Sea Drilling Project*, **87**: 393–481, http://dx.doi.org/10.2973/dsdp.proc.87.106.1986.

Akiba, F., and Yanagisawa, Y., 1986, Taxonomy, morphology and phylogeny of the Neogene diatom zonal marker species in the middle-to-high latitudes of the North Pacific. In: Kagami, H., Karig, D.E., Coulbourn, W.T., et al., (eds), *Initial Reports Deep Sea Drilling Project*, **87**: 483–554, http://dx.doi.org/10.2973/dsdp.proc.87.107.1986.

Andreyeva-Grigorovich, A.S., 2004, Obgruntuvannya nizhnoyi granitsi neogenovoyi sistemy Paratetisa ta yiyi korelyatsya za planktonnymi mikroorganizmamy [Substantiation of the lower boundary of the Neogene System in the Paratethys and its correlation by plankton microorganisms]. *Geological Journal of the Academy of Sciences of Ukraine*, **2**: 53–58, (in Ukrainian).

Andreyeva-Grigorovich, A.S., and Gruzman, A.D., 1989, Biostratigraficheskoe obosnovanie granitsyi paleogena i neogena v Tsentralnom i Vostochnom Paratetise [Biostratigraphic substantiation of the Paleogene and Neogene boundary in the Central and Eastern Paratethys]. *Geological Journal of the Academy of Sciences of Ukraine*, **6**: 91–95, (in Russian).

Andreyeva-Grigorovich, A.S., and Savytskaya, N.A., 2000, Nannoplankton of the Tarkhanian deposits of the Kerch Peninsula (Crimea). *Geologica Carpathica*, **5** (116), 399–406.

Andrusov, N.I., 1961a, *Izbrannyie trudyi. Tom I*. Moskva: AN SSSR, p. 12 (Selected Papers, Vol. I) (in Russian).

Andrusov, N.I., 1961b, *Izbrannyie trudyi. Tom II*. Moskva: AN SSSR, p. 614 (Selected Papers, Vol. II) (in Russian).

Andrussov, N., 1887, Geogische Untersuchungen in der westlichen H. alfte der Halbinsel Kertsch. *Memoires de la societe des naturalistes de la Nouvelle-Russie*, **2**: 69–147.

Anthonissen, E.D., 2008, Late Pliocene and Pleistocene biostratigraphy of the Nordic Atlantic region. *Newsletters on Stratigraphy*, **43**: 33–48.

Anthonissen, D.E., 2009, A new Pliocene biostratigraphy for the northeastern north Atlantic. *Newsletters on Stratigraphy*, **43** (2), 91–126.

Arribas, A., 2008, Vertebrados del Plioceno superior terminal en el suroeste deEuropa: Fonelas P-1 y el Proyecto Fonelas. *Instituto Geológico y Minero de España, serie Cuadernos del Museo Geominero*, **10**: 1–607.

Aubry, M.-P., 1993, Neogene allostratigraphy and depositional history of the DeSoto Canyon area, northern Gulf of Mexico. *Micropaleontology*, **39**: 327–366.

Azzaroli, A., 1983, Quaternary mammals and the "End-Villafranchian" dispersal event. A turning point in the history of Eurasia. *Palaeogeography, Palaeoclimatology, Palaeoecology*, **44**: 117–139.

Azzaroli, A., 1977, The Villafranchian stage in Italy and the Plio–Pleistocene boundary. *Giornale di Geologia*, **41**: 61–79.

Azzaroli, A., De Giuli, C., Ficcarelli, G., and Torre, D., 1982, Table of stratigraphic distribution of terrestrial mammalian faunas in Italy from the Pliocene to the early Middle Pleistocene. *Geografia Fisica e Dinamica Quaternaria*, **5**: 55–58.

Azzaroli, A., De Giuli, C., Ficcarelli, G., and Torre, D., 1988, Late Pliocene to early mid-Pleistocene mammals in Eurasia: faunal succession and dispersal events. *Palaeogeography, Palaeoclimatology, Palaeoecology*, **66**: 77–100.

Backman, J., and Raffi, I., 1997, Calibration of Miocene nannofossil events to orbitally tuned cyclostratigraphies from Ceara Rise.

In: Shackleton, N.J., Curry, W.B., Richter, C., and Bralower, T.J. (eds), *Proceedings of the Ocean Drilling Program, Scientific Results,* **154:** 83–99.

Backman, J., Raffi, I., Rio, D., Fornaciari, E., and Palike, H., 2012, Biozonation and biochronology of Miocene through Pleistocene calcareous nannofossils from low and middle latitudes. *Newsletters Stratigraphy,* **45:** 221–244, https://doi.org/10.1127/0078-0421/2012/0022.

Balazs, A., Matenco, L., Magyar, I., Horvath, F., and Cloetingh, S., 2016, The link between tectonics and sedimentation in back-arc basins: new genetic constraints from the analysis of the Pannonian Basin. *Tectonics,* **35:** 1526–1559 <https://doi.org/10.1002/2015TC004109>.

Baldauf, J.G., and Iwai, M., 1995, Neogene diatom biostratigraphy for the eastern equatorial Pacific, Leg 138. In: Pisias, N.G., Mayer, L.A., et al., (eds), *Proceedings of the Ocean Drilling Program, Scientific Results,* **87:** 105–128, http://dx.doi.org/10.2973/odp.proc.sr.138.107.1995.

Báldi, T. and Seneš, J.(eds), 1975. OM Egerien: Die Egerer, Pouzdřaner, Puchkirchener Schichtengruppe und die Bretkaer Formation. Chronostratigraphie und Neostratotypen. Miozan der Zentralen Paratethys, 5. VEDA SAV, Bratislava, 1–553, (in German with English summary).

Baldi, K., Velledits, F., Ćorić, S., Lemberkovics, V., Lőrincz, K., and Shevelev, M., 2017, Discovery of the Badenian evaporites inside the Carpathian Arc: implications for global climate change and Paratethys salinity. *Geologica Carpathica,* **68** (3), 193–206, https://doi.org/10.1515/geoca-2017-0015.

Banner, F.T., and Blow, W.R., 1965, Progress in the planktonic foraminifera biostratigraphy of the Neogene. *Nature,* **208:** 1164–1166.

Barg, I.M., and Ivanova, T.A., 2001, Stratigrafiya i razvitie Ravninnogo Kryima v miotsene [Miocene stratigraphy and geological history of the Crimean Plain Region]. *Stratigraphy. Geological Correlation,* **8** (3), 83–93.

Barron, J.A., 1983, Latest Oligocene through early middle Miocene diatom biostratigraphy of the eastern tropical Pacific. *Marine Micropaleontology,* **7:** 487–515, https://doi.org/10.1016/0377-8398(83)90012-9.

Barron, J.A., 1985a, Late Eocene to Holocene diatom biostratigraphy of the equatorial Pacific Ocean, Deep Sea Drilling Project Leg 85. In: Mayer, L., Theyer, F., Thomas, E., et al., (eds), *Initial Reports of the Deep Sea Drilling Project,* **85:** 413–456, http://dx.doi.org/10.2973/dsdp.proc.85.108.1985.

Barron, J.A., 1985b, Miocene to Holocene planktic diatoms. In: Bolli, H.M., Saunders, J.B., and Perch-Nielsen, K. (eds), *Plankton Stratigraphy,* Cambridge: Cambridge University Press, 763–809.

Barron, J.A., 1992, Neogene diatom datum levels in the Equatorial and North Pacific. In: Isuzaki, K., and Saitu, T. (eds), *The Centenary of Japanese Micropaleontology.* Tokio: Terra Scientific Publ. Co, 413–425.

Barron, J.A., 2003, Planktonic marine diatom record of the past 18m.y.: appearance and extinctions in the Pacific and Southern Oceans. *Diatoms Research,* **118:** 203–224.

Barron, J.A., 2006, Diatom biochronology for the Early Miocene of the Equatorial Pacific. *Stratigraphy,* **2:** 281–309.

Barron, J.A., Keller, G., and Dunn, D.A., 1985, A multiple microfossil biochronology for the Miocene. In: Kennett, J.P. (ed), The Miocene Ocean: Paleoceanography and Biogeography. *Geological Society of America Memoir,* **163:** 21–36.

Barron, J., and Baldauf, J., 1995, Cenozoic marine diatom biostratigraphy and application to paleoclimatology and paleoceanography. In: Blome, C.D., et al., (eds), *Siliceous Microfossils. Paleontological Society Short Courses in Paleontology,* **8:** 107–118.

Barron, J.A., and Gladenkov, A.Y., 1995, Early Miocene to Pleistocene diatom stratigraphy of Leg 145. In: Rea, D.K., Basov, I.A., Scholl, D.W., and Allan, J.F. (eds), *Proceedings of the Ocean Drilling Program, Scientific Results,* **145:** 3–19, http://dx.doi.org/10.2973/odp.proc.sr.145.101.1995.

Barron, J.A., Fourtanier, E., and Bohaty, S.M., 2004. Oligocene and Earliest Miocene Diatom Biostratigraphy of Site 1220, ODP Leg 199, Equatorial Pacific. In: Lyle, M., Wilson, P.W., Janecek, T.R, and Firth, J. (Eds.), *Proceedings of the Ocean Drilling Program, Scientific Results,* **199:** 1–25. http://www-odp.tamu.edu/publications/199_SR/204/204.htm

Beddow, H.M., Liebrand, D., Sluijs, A., Wade, B.S., and Lourens, L.J., 2016, Global change across the Oligocene-Miocene transition: high-resolution stable isotope records from IODP Site U1334 (equatorial Pacific Ocean). *Paleoceanography,* **31:** 81–97, https://doi.org/10.1002/2015PA002820.

Beddow, H.M., Liebrand, D., Wilson, D.S., Hilgen, F.J., Sluijs, A., Wade, B.S., et al., 2018, Astronomical tunings of the Oligocene-Miocene transition from Pacific Ocean Site U1334 and implications for the carbon cycle. *Climate of the Past,* **14** (3), 255–270 <https://doi.org/10.5194/cp-14-255-2018>.

Bellucci, L., Mazzini, I., Scardia, G., Bruni, L., Parenti, F., Segre, A.G., et al., 2012, The site of Coste San Giacomo (Early Pleistocene, Central Italy): palaeoenvironmental analysis and biochronological overview. *Quaternary International,* **267:** 30–39.

Bellucci, L., Bona, F., Corrado, P., Magri, D., Mazzini, I., Parenti, F., et al., 2014, Evidence of late Gelasian dispersal of African fauna at Coste San Giacomo (Anagni Basin, central Italy): early Pleistocene environments and the background of early human occupation in Europe. *Quaternary Science Reviews,* **96:** 72–85.

Bellucci, L., Sardella, R., and Rook, L., 2015, Large mammal biochronology framework in Europe at Jaramillo: the Epivillafranchian as a formal biochron. *Quaternary International,* **389:** 84–89.

Bellucci, L., Biddittu, I., Brilli, M., Conti, J., Germani, M., Giustini, F., et al., 2019, First occurrence of the short-faced bear Agriotherium (Ursidae, Carnivora) in Italy: biochronological and palaeoenvironmental implications. *Italian Journal of Geosciences,* **138** (1), 124–135.

Beluzhenko, E.V., Filippova, N.Y., and Pismennaya, N.S., 2014, Markiruyuschie gorizontyi oligotsen-nizhnemiotsenovyih (maykopskih) otlozheniy Severnogo Kavkaza i Predkavkazya [Marker horizons of the Oligocene – Lower Miocene (Maikop Group) of Northern Caucasus and Ciscaucasia]. *Bulletin of the Moscow Society of Naturalists, Geol. Series,* **89** (1), 21–35, (in Russian).

Beluzhenko, E.V., Filippova, N.Y., and Golovina, L.A., 2018, Alkunskaya svita i granitsa paleogena i neogena na Severnom Kavkaze i Predkavkaze (litologiya, stratifiya, korrelyatsiya) [The Alkun Formation and the Paleogene—Neogene Boundary in the Northern Caucasus and Ciscaucasia (lithology, stratigraphy, correlation)]. *Proceedings of the Research Institute of Geology of Voronezh State University,* **102:** 102, (in Russian).

Bergen, J.A., Truax III, S., de Kaenel, E., Blair, S., Browning, E., Lundquist, J., et al., 2019, BP Gulf of Mexico Neogene

Astronomically-tuned Time Scale (BP GNATTS). *GSA Bullettin*, <https://doi.org/10.1130/B35062.1>.

Berggren, W.A., 1992, Neogene planktonic foraminifer magnetobiostratigraphy of the southern Kerguelen Plateau (Sites 747, 748, and 751). In: Wise Jr., S.W., Schlich, R., et al., (eds), *Proceedings of the Ocean Drilling Program, Scientific Results*, 120: 631–647.

Berggren, W.A., 1993, Neogene planktonic foraminiferal biostratigraphy of eastern Jamaica. *Geological Society of America Memoir*, **182**: 179–217.

Berggren, W.A., Aubry, M.P., and Hamilton, N., 1983, Neogene magnetobiostratigraphy of Deep Sea Drilling Project Site 516 (Rio Grande Rise, South Atlantic). In: Barker, P., et al., (eds), *Initial Reports of the Deep Sea Drilling Project*, **72**: 675–713.

Berggren, W.A., Kent, D.V., and Van Couvering, J.A., 1985a, The Neogene, Part 2. Neogene geochronology and chronostratigraphy. In: Snelling, N.J. (ed), *The Chronology of the Geological Record*. Geological Society of London Memoir, **10**: 211–260.

Berggren, W.A., Hilgen, F.J., Langereis, C.G., Kent, D.V., Obradovich, J.D., Raffi, I., Raymo, M.E., and Shackleton, N.J., 1995b, Late Neogene chronology: new perspectives in high resolution stratigraphy. *Geological Society of America Bulletin*, **107**: 1272–1287.

Berggren, W.A., Kent, D.V., Swisher III, C.C., and Aubry, M.-P., 1995a, A revised Cenozoic geochronology and chronostratigraphy. In: Berggren, W.A., Kent, D.V., Aubry, M.-P., and Hardenbol, J. (eds), Geochronology, Time Scales and Global Stratigraphic Correlation. *SEPM Special Publication*, **54**: 129–212.

Berggren, W.A., Wade, B.S., and Pearson, P.N., 2018, Oligocene chronostratigraphy and planktonic foraminiferal biostratigraphy: historical review and current state-of-the-art. In: Wade, B.S., Olsson, R.K., Pearson, P.N., Huber, B.T., and Berggren, W.A. (eds), *Atlas of Oligocene Planktonic Foraminifera*. Cushman Foundation of Foraminiferal Research, Special Publication, 29–54, No. 46.

Betzler, C., et al., 2018, Refinement of Miocene sea level and monsoon events from the sedimentary archive of the Maldives (Indian Ocean). *Progress in Earth and Planetary Science*, **5**: 5, https://doi.org/10.1186/s40645-018-0165.

Beu, A.G., 1995, Pliocene limestones and their scallops. Lithostratigraphy, pectinid biostratigraphy and paleogeography of eastern North Island late Neogene limestone. *Institute of Geological and Nuclear Sciences Monograph*, **10**: 1–243, 1 map.

Bickert, T., Curry, W.B., and Wefer, G., 1997, Late Pliocene to Holocene (2.6-0 Ma) western equatorial Atlantic deep-water circulation: inferences from benthic stable isotopes. In: Shackleton, N.J., Curry, W.B., Richter, C., and Bralower, T.J. (eds), *Proceedings of the Ocean Drilling Program, Scientific Results*, **154**: 239–254, http://dx.doi.org/10.2973/odp.proc.sr.154.110.1997.

Billups, K., Pälike, H., Channell, J.E.T., Zachos, J.C., and Shackleton, N.J., 2004, Astronomic calibration of the late Oligocene through early Miocene geomagnetic polarity time scale. *Earth and Planetary Science Letters*, **224**: 33–44.

Bizon and Muller, 1979. In Bizon, G., et al., (eds), Report of the working group on Micropaleontology: 7th International Congress of Mediteranean Neogene. *Annales Geologiques Pays Helleniques*, Athens, Tome hors série, fasc. 3: 1335–1364.

Blow, W.H., 1969. Late middle Eocene to Recent planktonic foraminiferal biostratigraphy. In: Bronniman, P., and Renz, H.H. (Eds.),

Proceedings of the First. International Conference on Planktonic Microfossils, Geneva, 1967, 1: 199–422.

Bolli, H.M., 1957, Planktonic foraminifera from the Oligocene–Miocene Cipero and Lengua formations of Trinidad. *B. W. I. United States National Museum Bulletin*, **215**: 97–123.

Boyd, J.L., Riding, J.B., Pound, M.J., De Schepper, S., Ivanovic, R.F., Haywood, A.M., et al., 2018, The relationship between Neogene dinoflagellate cysts and global climate dynamics. *Earth-Science Reviews*, **177**: 366–385, ISSN 0012-8252, https://doi.org/10.1016/j.earscirev.2017.11.018.

Browning, J.V., Miller, K.G., Sugarman, P.J., Barron, J., McCarthy, F.M.G., Kulhanek, D.K., et al., 2013, Chronology of Eocene-Miocene sequences on the New Jersey shallow shelf: implications for regional, interregional, and global correlations. *Geosphere*, **9**: 1434–1456, https://doi.org/10.1130/GES00857.1.

Bukry, D., 1973. Low-latitude coccolith biostratigraphic zonation. In Edgar, N.T., Saunders, J.B., et al.(eds), *Initial Reports Deep Sea Drilling Project*, **15**: 685–703. https://doi.org/10.2973/dsdp.proc.15.116.1973.

Bukry, D., 1975, Coccolith and silicoflagellate stratigraphy, northwestern Pacific Ocean, Deep Sea Drilling Project Leg 32. In: Larson, R.L., Moberly, R., et al., (eds), *Initial Reports Deep Sea Drilling Project*, **32**: 677–701, http://dx.doi.org/10.2973/dsdp.proc.32.124.1975.

Burckle, L.H., 1978, Early Miocene to Pliocene diatom datum level for the equatorial Pacific. *Proceedings Second Working Group Mtg. Biostratigraphic Datum Planes, Pacific Neogene, IGCP Proj. 114. Spec. Publ.- Geological Research and Development Center*, **1**: 25–44.

Burckle, L.H., and Trainer, J., 1979, Middle and late Pliocene diatom datum levels from the central Pacific. *Micropaleontology*, **25** (3), 281–293, https://doi.org/10.2307/1485303.

Burckle, L.H., Keigwin, L.D., and Opdyke, N.D., 1982, Middle and late Miocene stable isotope stratigraphy: correlation to the paleomagnetic reversal record. *Micropaleontology*, **28** (4), 329–334, https://doi.org/10.2307/1485448.

Bylinskaya, M.E., 2005, Range and stratigraphic significance of the *Globorotalia crassaformis* plexus. *Journal of Iberian Geology*, **31**: 51–63.

Caloi, L., and Palombo, M.R., 1996, Latest Early Pleistocene mammal faunas of Italy, biochronological problems. *Il Quaternario*, **8**: 391–402.

Cande, S.C., and Kent, D.V., 1992, A new geomagnetic polarity time scale for the late Cretaceous and Cenozoic. *Journal of Geophysical Research*, **97**: 13917–13951, https://doi.org/10.1029/92jb01202.

Cande, S.C., and Kent, D.V., 1995, Revised calibration of the Geomagnetic Polarity Time Scale for the Late Cretaceous and Cenozoic. *Journal of Geophysical Research*, **100**: 6093–6095, https://doi.org/10.1029/94jb03098.

Casanovas-Vilar, I., Alba, D.M., Moyà-Solà, S., Galindo, J., Cabrera, L., Garcés, M., et al., 2008, Biochronological, taphonomical, and paleoenvironmental background of the fossil great ape *Pierolapithecus catalaunicus* (Primates, Hominidae). *Journal of Human Evolution*, **55**: 589–603.

Castradori, D., 1998, Calcareous nannofossils in the basal Zanclean of Eastern Mediterranean: remarks on paleoceanography and sapropel formation. In: Robertson, A.H.F., Emeis, K.-C., Richter, C., and Camerlenghi, A. (eds), *Proceedings of the Ocean Drilling Program, Scientific Results*, **160**: 113–123, http://dx.doi.org/10.2973/odp.proc.sr.160.005.1998.

Castradori, D., Rio, D., Hilgen, F.J., and Lourens, L.J., 1998, The Global Standard Stratotype-section and Point (GSSP) of the Piacenzian Stage (Middle Pliocene). *Episodes*, **21**: 88–93.

Censarek, B., and Gersonde, R., 2002, Miocene diatom biostratigraphy at ODP Site 689, 690, 1088, 1092 (Atlantic sector of the Southern Ocean). *Marine Micropaleontology*, **45** (3–4), 309–356, <https://doi.org/10.1016/S0377-8398(02)00034-8>.

Chaisson, W.P., and Leckie, R.M., 1993, High-resolution Neogene planktonic foraminifer biostratigraphy of Site 806, Ontong Java Plateau (western equatorial Pacific). In: Berger, W.H., Kroenke, L.W., Mayer, L.A., et al., (eds), *Proceedings of the Ocean Drilling Program, Scientific Results*, **130**: 137–178.

Chaisson, W.P., and Pearson, P.N., 1997, Planktonic foraminifer biostratigraphy at Site 925: middle Miocene–Pleistocene. In: Shackleton, N.J., Curry, W.B., Richter, C., and Bralower, T.J. (eds), *Proceedings of the Ocean Drilling Program, Scientific Results*, **154**: 3–31.

Chang, L., Vasiliev, I., Van Baak, C.G.C., Krijgsman, W., Dekkers, M.J., Roberts, A.P., et al., 2014, Identification and environmental interpretation of diagenetic and biogenic greigite in sediments: a lesson from the Messinian Black Sea. *Geochemistry, Geophysics, Geosystems*, **15**: 3612–3627 <https://doi.org/10.1002/2014GC005411>.

Channell, J.E.T., Rio, D., Sprovieri, R., and Glaçon, G., 1990, Biomagnetostratigraphic correlations from Leg 107 in the Tyrrhenian Sea. In: Kasten, K.A., Mascle, J., et al., (eds), *Proceedings of the Ocean Drilling Program, Scientific Results*, **107**: 669–682.

Channell, J.E.T., Galeotti, S., Martin, E.E., Billups, K., Scher, H., and Stoner, J.S., 2003, Eocene to Miocene magnetic, bio- and chemostratigraphy at ODP site 1090 (subantarctic South Atlantic). *Geol. Soc. Amer. Bull.*, **115**: 607–623.

Chaproniere, G.C.H., Styzen, M.J., Sager, W.W., Nishi, H., Quinterno, P.J., and Abrahamsen, N., 1994, Late Neogene biostratigraphic and magnetostratigraphic synthesis. *Proceedings of the Ocean Drilling Program, Scientific Results*, **135**: 857–877.

Chumakov, I.S., Ganzei, S.S., Byzova, S.L., Dobrynina, V.Y., and Paramonova, N.P., 1984, Geohronologiya sarmata Vostochnogo Paratetisa [Geochronology of the Sarmatian of the Eastern Paratethys]. *Reports of the USSR Academy of Sciences*, **276** (5), 1189–1193, (in Russian).

Cicha, I., Seneš, J., Tejkal, J. (eds), 1967. M3 Karpatien. Die Karpatische Serie und ihr Stratotypus. Chronostratigraphie und Neostratotypen. *Miozan der Zentralen Paratethys*, **1**. SAV: 1–12.

Cita, M.B., and Premoli Silva, I., 1968, Evolution of the planktonic foraminiferal assemblages in the stratigraphical interval between the type Langhian and the type Tortonian and the biozonation of the Miocene of Piedmont. *Giornale di Geologia*, **35**: 1051–1082.

Clement, B.M., and Robinson, R., 1986, The magnetostratigraphy of Leg 94 sediments. In: Ruddimman, W.R., Kidd, R.B., Thomas, E., et al., (eds), *Initial Results of the Deep Sea Drilling Program*, **94**: 635–650.

Cody, R.D., Richard, H., Levy, R.H., Harwood, D.M., and Sadler, P.M., 2008, Thinking outside the zone: high-resolution quantitative diatom biochronology for the Antarctic Neogene. *Palaeogeography, Palaeoclimatology, Palaeoecology*, **260**: 92–121, https://doi.org/10.1016/j.palaeo.2007.08.020.

Cooper, R.A. (ed), 2004. The New Zealand geological time scale. *Institute of Geological and Nuclear Sciences Monograph*, **22**: 1–284.

Ćorić, S., Pavelić, D., Rögl, F., Mandic, O., and Vrabac, S., 2009, Revised Middle Miocene datum for initial marine flooding of North Croatian Basins (Pannonian Basin System, Central Paratethys). *Geologia Croatica*, **62**: 31–43.

Ćorić, S., Švabenská, L., Rögl, F., and Petrová, P., 2007, Stratigraphical position of *Helicosphaera waltrans* Nannoplankton Horizon (NN5, Lower Badenian). *Joannea Geologie und Paläontologie*, **9**: 17–19.

Ćorić, S., and Švabenská, L., 2004, Calcareous nannofossil biostratigraphy of the Grund Formation (Molasse basin, Lower Austria). *Geologia Croatica*, **55** (2), 147–153.

Cramer, B.S., Toggweiler, J.R., Wright, J.D., Katz, M.E., and Miller, K.G., 2009, Ocean overturning since the Late Cretaceous: Inferences from a new benthic foraminiferal isotope compilation. *Paleoceanography*, **24**: article #PA4216, https://doi.org/10.1029/2008PA001683.

Crundwell, M.P, 2004, *New Zealand Late Miocene biostratigraphy and biochronology: studies of planktic foraminifers and bolboforms at oceanic sites 593 and 1123, and selected onland sections.* PhD thesis. University of Waikato, Hamilton, New Zealand. 678pp.

Crundwell, M.P., Beu, A.G., Cooper, R.A., Morgans, H.E.G., Mildenhall, D.C., and Wilson, G.S., 2004, Miocene (Pareora, Southland and Taranaki Series). 164–194 *In:* Cooper, R.A. (ed), *The New Zealand Geological Timescale*. Lower Hutt: Institute of Geological & Nuclear Sciences Limited. Institute of Geological & Nuclear Sciences monograph, **22**: pp. 164–194.

Crundwell, M.P., 2014, Pliocene to Late Eocene foraminiferal and bolboformid biostratigraphy of IODP Hole 317-U1352C, Canterbury Basin, New Zealand. *GNS Science Report 2014/15*, pp. 1–49.

Crundwell, M.P., 2015a, Pliocene and early Pleistocene planktic foraminifera: important taxa and bioevents in ODP Hole 1123B, Chatham Rise, New Zealand. *GNS Science Report 2015/51*, pp. 1–67.

Crundwell, M.P., 2015b, Revised Pliocene and early Pleistocene planktic foraminiferal biostratigraphy, DSDP Site 284, Challenger Plateau, New Zealand. *GNS Science Report 2015/22*, pp. 1–36 + Appendix.

Crundwell, M.P., 2018, *Globoconella pseudospinosa*, n. sp.: a new early Pliocene planktonic foraminifer from the southwest Pacific. *Journal of Foraminiferal Research*, **48**: 288–300.

Crundwell, M.P., and McDonnell, R.N.G., 2014, Report on the Rangitikei River section, near Mangaweka: a possible SSP (Stratotype Section and Point) for the base of the New Zealand Mangapanian Stage. *GNS Science Report 2014/08*, pp. 1–44.

Crundwell, M.P., and Nelson, C.S., 2007, A magnetostratigraphically-constrained chronology for late Miocene bolboformids and planktic foraminifers in the temperate Southwest Pacific. *Stratigraphy*, **4**: 1–34.

Csato, I., Toth, S., Catuneanu, O., and Granjeon, D., 2015, A sequence stratigraphic model for the Upper Miocene-Pliocene basin fill of the Pannonian Basin, eastern Hungary. *Marine and Petroleum Geology*, **66**: 117–134, https://doi.org/10.1016/j.marpetgeo.2015.02.010.

Curry, W.B., Shackleton, N.J., Richter, C., et al., 1995, *Proceedings of the Ocean Drilling Program, Initial Reports*, **154**: College Station, TX: Ocean Drilling Program.

Davitashvili, L.S., 1933, *Obzor mollyuskov tretichnyih i posletretichnyih otlozheniy Kryimsko-Kavkazskoy neftenosnoy provintsii [Overview of mollusks from Tertiary and post-Tertiary sediments of Crimea-Caucasus oil-bearing province.* Leningrad, Moskva: Gosudarstvennoe nauchno-tekhnicheskoe neftianoe izdatelstvo, p. 168, (in Russian).

De Leeuw, A., Bukowski, K., Krijgsman, W., and Kuiper, K.F., 2010, Age of the Badenian salinity crisis; impact of Miocene climate variability on the circum-Mediterranean region. *Geology*, **38**: 715–718, https://doi.org/10.1130/G30982.1.

De Leeuw, A., Filipescu, S., Maţenco, L., Krijgsman, W., Kuiper, K., and Stoica, M., 2013, Paleomagnetic and chronostratigraphic constraints on the Middle to Late Miocene evolution of the Transylvanian Basin (Romania): implications for Central Paratethys stratigraphy and emplacement of the Tisza–Dacia plate. *Global and Planetary Change*, **103**: 82–98, https://doi.org/10.1016/j.gloplacha.2012.04.008.

De Leeuw, A., Morton, A., van Baak, C.G.C., and Vincent, S.J., 2018a, Timing of arrival of the Danube to the Black Sea: provenance of sediments from DSDP Site 380/380A. *Terra Nova*, **30**: 114–124 <https://doi.org/10.1111/ter.12314>.

De Leeuw, A., Tulbure, M., Kuiper, K.F., Melinte-Dobrinescu, M.C., Stoica, M., and Krijgsman, W., 2018b, New ^{40}Ar/^{39}Ar, magnetostratigraphic and biostratigraphic constraints on the termination of the Badenian Salinity Crisis: evidence for tectonic improvement of basin interconnectivity in Southern Europe. *Global and Planetary Change*, **169**: 1–15, <https://doi.org/10.1016/j.gloplacha.2018.07.001>.

De Schepper, S., and Head, M.J., 2009, Pliocene and Pleistocene dinoflagellate cyst and acritarch zonation of DSDP Hole 610A, eastern North Atlantic. *Palynology*, **33**: 179–218.

De Vleeschouwer, D., Vahlenkamp, M., Crucifix, M., and Pälike, H., 2017, Alternating Southern and Northern Hemisphere climate response to astronomical forcing during the past 35 m.y. *Geology*, **45** (4), 275–378, https://doi.org/10.1130/G38663.1.

De Vleeschouwer, D., Drury, A.J., Vahlenkamp, M., Rochholz, F., Liebrand, D., Pälike, H., 2020. High-latitude biomes and rock weathering mediate climate – carbon cycle feedbacks on eccentricity timescales. *Nature Communications* (submitted 2019).

Dowsett, H.J., 1988, Diachrony of Late Neogene microfossils in the southwest Pacific Ocean: application of the graphic correlation method. *Paleoceanography*, **3** (2), https://doi.org/10.1029/PA003i002p00209>.

Drury, A.J., Westerhold, T., Frederichs, T., Tian, J., Wilkens, R., Channell, J.E.T., et al., 2017, Late Miocene climate and time scale reconciliation: accurate orbital calibration from a deep-sea perspective. *Earth and Planetary Science Letters*, **475**: 254–266, <https://doi.org/10.1016/j.epsl.2017.07.038>.

Dybkjær, K., and Piasecki, S., 2010, Neogene dinocyst zonation for the eastern North Sea Basin, Denmark. *Review of Palaeobotany and Palynology*, **161**: 1–29.

Eberzin, A.G., 1959, Shema stratigrafii neogenovyih otlozheniy Yuga SSSR [Stratigraphic scheme of the Neogene deposits of the South of USSR]. *Proceedings of meeting on the development of a unified stratigraphic scale of Tertiary deposits of the Crimea-Caucasus region, Baku, AN AZSSR*, p. 307 (in Russian).

Fabbrini, A., Baldassini, N., Caricchi, C., Foresi, L.M., Sagnotti, L., Dinarès-Turell, J., et al., 2019, In search of the Burdigalian GSSP: new evidence from the Contessa Section (Italy).). *Italian Journal of Geosciences*, **138** (2), 274–295, https://doi.org/10.3301/IJG.2019.07.

Filipescu, S., and Girbacea, R., 1997, Lower Badenian sea-level drop on the western border of the Transylvanian Basin: foraminiferal palaeobathymetry and stratigraphy. *Geologica Carpathica*, **48** (5), 325–334.

Filipescu, S., Wanek, F., Miclea, A., de Leeuw, A., and Vasiliev, I., 2011, Micropaleontological response to the changing paleoenvironment across the Sarmatian-Pannonian boundary in the Transylvanian Basin (Miocene, Oarba de Mure section, Romania). *Geologica Carpathica*, **62** (1), 91–102, https://doi.org/10.2478/v10096-011-0008-9.

Filippova, N., and Trubikhin, V., 2009, K voprosu o korrelyatsii verhnemiotsenovyih otlozheniy Chernomorskogo i Sredizemnomorskogo basseynov [On the correlation of the Upper Miocene sediments Black Sea and Mediterranean basins]. In: Gladenkov, Y. (ed), *Actual Problems of Neogene and Quaternary Stratigraphy and discussion of 33 Intern. Geol. Congress.* 142–152, (in Russian).

Filippova, N.Y., Beluzhenko, E.V., and Golovina, L.A., 2010, Biostratigrafiya alkunskoy svityi (oligotsen–nizhniy miotsen) Severnoy Osetii po mikropaleontologicheskim dannyim (dinotsistyi, nannoplankton, sporyi i pyiltsa) [Micropaleontological Data (Dinocysts, Nannoplankton, Spores and Pollen) and Biostratigraphy of the Alkun Formation, North Ossetia]. *Stratigraphy. Geoogical. Correlation*, **18** (3), 83–106, (in Russian).

Filippova, N.Y., Beluzhenko, E.V., and Golovina, L.A., 2013, Granitsa paleogena i neogena na Severnom Kavkaze i v Predkavkaze [Paleogene/Neogene boundary in the Northern Caucasus and Ciscaucasia]. *"General Stratigraphic Scale of Russia: Current State and Way of Perfection", All-Russian Conf., Moscow, May 23–25, 2013, GIN RAN, Moskva*, pp. 360–362, (in Russian).

Filippova, N.Y., Beluzhenko, E.V., and Golovina, L.A., 2015, O granitse paleogena i neogena i vozraste alkunskoy svityi na Severnom Kavkaze i v Predkavkaze [On the Paleogene – Neogene Boundary and Age of the Alkun Formation in the North Caucasus and Ciscaucasia]. *Stratigraphy. Geological. Correlation*, **23** (6), 70–95, (in Russian).

Finney, S.C., and Bown, P.R., 2017, The status of subseries/subepochs for the Paleocene to Holocene: Recommendations to authors and editors. *Episodes*, **40**: 2–4.

Fleming, C.A., 1953, The geology of Wanganui Subdivision, Waverley and Wanganui sheet districts (N137 and N138). *New Zealand Geological Survey Bulletin*, **52**: 1–362, 2 maps.

Florindo, F., Gennari, R., Persico, D., Turco, E., Villa, G., Lurcock, P.C., Roberts, A.P., Winkler, A., Carter, L., and Pekar, S.F., 2015, New magnetobiostratigraphic chronology and paleoceanographic changes across the Oligocene-Miocene boundary at DSDP Site 516 (Rio Grande Rise, SW Atlantic). *Paleoceanography*, **30**: 659–681, https://doi.org/10.1002/2014PA002734.

Flower, B.P., 1999. Data Report: planktonic foraminifers from the subpolar North Atlantic and Nordic Seas: sites 980–987 and 907. In Raymo, M.E.; Jansen, E.; Blum, P.; Herbert, T.D. (eds), *Proceedings of the Ocean Drilling Program, Scientific Results*, **162**: 1–16, <https://doi.org/10.2973/odp.proc.sr.162.038.1999>.

Flower, B.P., and Kennett, J.P., 1993, Middle Miocene ocean-climate transition: high-resolution oxygen and carbon isotopic records from Deep Sea Drilling Project Site 588A, southwest Pacific. *Paleoceanography*, **8**: 811–843.

Foresi, L.M., Verducci, M., Baldassini, N., Lirer, F., Mazzei, R., Salvatorini, G., et al., 2011, Integrated stratigraphy of St. Peter's Pool section (Malta): new age for the Upper Globigerina Limestone Member and progress towards the Langhian GSSP. *Stratigraphy*, **8** (2–3), 125–143.

Fornaciari, E., 1996, Biocronologia a nannofossili calcarei e streatigrafia ad eventi nel Miocene italiano. *Ph.D. thesis*. University of Padova.

Fornaciari, E., and Rio, D., 1996, Latest Oligocene to early middle Miocene quantitative calcareous nannofossil biostratigraphy in the Mediterranean region. *Micropaleontology*, **42**: 1–36.

Fornaciari, E., Di Stefano, A., Rio, D., and Negri, A., 1996, Mediterranean middle Miocene quantitative calcareous nannofossil biostratigraphy. *Micropaleontology*, **42**: 37–63.

Francescone, F., Lauretano, V., Bouligand, C., Moretti, M., Sabatino, N., Schrader, C., et al., 2019, A 9 million-year-long astrochronological record of the early-middle Eocene corroborated by seafloor spreading rates. *Geological Society of America Bulletin*, **131** (3–4), 499–520, https://doi.org/10.1130/B32050.1.

Garcés, M., Krijgsman, W., Peláez-Campomanes, P., Álvarez Sierra, M.A., and Daams, R., 2003, *Hipparion* dispersal in Europe: magnetostratigraphic constraints from the Daroca area (Spain). *Coloquios de Paleontología, Volumen extraordinario*, **1**: 171–178.

Gartner, S., 1977, Calcareous nannofossil biostratigraphy and revised zonation of the Pleistocene. *Marine Micropaleontology*, **2**: 1–25.

Gibbard, P.L., and Head, M.J., 2010, The newly ratified definition of the Quaternary System/Period and redefinition of the Pleistocene Series/Epoch, and comparison of proposals advanced prior to formal ratification. *Episodes*, **33**: 152–158.

Gibbard, P.L., Head, M.J., Walker, M.J.C., and the Subcommission on Quaternary Stratigraphy, 2010, Formal ratification of the Quaternary System/Period and the Pleistocene Series/Epoch with a base at 2.58 Ma. *Journal of Quaternary Sciences*, **25**: 96–102.

Glaçon, G., et al., 1990, Planktonic foraminiferal events and stable isotope records in the upper Miocene, Site 654. In: Kastens, K., Mascle, J., et al., (eds), *Proceedings of the Ocean Drilling Program, Scientific Results*, **107**: 415–427.

Gladenkov, A.Y., 2006, The Cenozoic diatom zonation and its significance for stratigraphic correlations in the North Pacific. *Paleontological Journal*, **40** (5), 571–583.

Gliozzi, E., Abbazzi, L., Argenti, P., Azzaroli, A., Caloi, L., Capasso Barbato, L., et al., 1997, Biochronology of selected mammals, molluscs and ostracods from the middle Pliocene to the late Pleistocene in Italy. The state of the art. *Rivista Italiana di Paleontologia e Stratigrafia*, **103**: 369–388.

Gliozzi, E., Ceci, M.E., Grossi, F., and Ligios, S., 2007, Paratethyan Ostracod immigrants in Italy during the Late Miocene. *Geobios*, **40**: 325–337, <https://doi.org/10.1016/j.geobios.2006.10.004>.

Golovina, L.A., 2012, K voprosu o stratifii sredne-verhnego miotsena yuga Rossii na osnove izvestkovogo nanoplanktona [On the question of stratigraphy of the Middle–Upper Miocene of southern Russia based on calcareous nannoplankton]. *15th All Russia Micropaleontological Conference on Modern Micropaleontology, Gelendzhik*, pp. 305–308, (in Russian).

Gradstein, F.M., Ogg, J.G., Schmitz, M.D., and Ogg, G.M. (eds), 2012. *The Geologic Time Scale 2012*. Elsevier BV. Oxford.

Grothe, A., Sangiorgi, F., Mulders, Y.R., Vasiliev, I., Reichart, G.-J., Brinkhuis, H., et al., 2014, Black Sea desiccation during the Messinian Salinity Crisis: fact or fiction? *Geology*, **42** (7), 563–566, https://doi.org/10.1130/G35503.1.

Grothe, A., Sangiorgi, F., Brinkhuis, H., Stoica, M., and Krijgsman, W., 2018, Migration of the dinoflagellate *Galeacysta etrusca* and its implications for the Messinian Salinity Crisis. *Newsletters Stratigraphy*, **51** (1), 73–91, https://doi.org/10.1127/nos/2016/0340.

Grunert, P., Auer, G., Harzhauser, M., and Piller, W.E., 2015, Stratigraphic constraints for the upper Oligocene to lower Miocene Puchkirchen Group (North Alpine Foreland Basin, Central Paratethys). *Newsletters Stratigraphy*, **48** (1), 111–133, https://doi.org/10.1127/nos/2014/0056.

Guérin, C., Faure, M., Argant, A., Argant, J., Crégut-Bonnoure, E., Debard, E., et al., 2004, Le gisement pliocène supérieur de Saint-Vallier (Drôme, France): synthèse biostratigraphiqueet paléoécologique. *Geobios*, **37** (Suppl. 1), 349–360.

Guerra-Merchan, A., Serrano, F., Garces, M., Gofas, S., Esu, D., Gliozzi, E., et al., 2010, Messinian Lago-Mare deposits near the Strait of Gibraltar (Malaga Basin, S Spain). *Palaeogeography, Palaeoclimatology, Palaeoecology*, **285**: 264–276, <https://doi.org/10.1016/j.palaeo.2009.11.019>.

Haq, B.U., Hardenbol, J., and Vail, P.R., 1987, Chronology of fluctuating sea levels since the Triassic. *Science*, **235**: 1156–1167.

Hardenbol, J., Thierry, J., Farley, M.B., Jacquin, T., de Graciansky, P.-C., and Vail, P.R., 1998, Mesozoic and Cenozoic Sequence Chronostratigraphic Framework of European Basins. In: de Graciansky, P.-C., Hardenbol, J., Jacquin, T., and Vail, P.R. (eds), *Mesozoic and Cenozoic Sequence Stratigraphy of European Basins*. SEPM Spec. Publ., 3–13.

Harwood, D.M., Maruyama, T., 1992. Middle Eocene to Pleistocene diatom biostratigraphy of Southern Ocean sediments from the Kerguelen Plateau, Leg 120. In Wise, S.W., Schlich, R., et al. (eds), *Proceedings of the Ocean Drilling Program, Scientific Results*, **120**: 683–733. <https://doi.org/10.2973/odp.proc.sr.120.160.1992>.

Harwood, D.M., Bohaty, S.M., and Scherer, R.P., 1998, Lower Miocene diatom biostratigraphy of the CRP-1 drillcore, McMurdo Sound, Antarctica. *Terra Antartica*, **5** (3), 499–514, hdl:10013/epic.28337.d001.

Harzhauser, M., and Piller, W.E., 2007, Benchmark data of a changing sea. Palaeogeography, Palaeobiogeography and events in the Central Paratethys during the Miocene. *Palaeogeography, Palaeoclimatology, Palaeoecology*, **253**: 8–31, https://doi.org/10.1016/j.palaeo.2007.03.031.

Hays, J.D., Saito, T., Opdyke, N.D., and Burckle, L.H., 1969, Pliocene-Pleistocene sediments of the equatorial Pacific: their paleomagnetic biostratigraphic and climatic record. *Geological Society of America Bulletin*, **80**: 1481–1514.

Haywood, A.M., Dowsett, H.J., and Dolan, A.M., 2016, Integrating geological archives and climate models for the mid-Pliocene warm period. *Nature Communications*, **16** (7), 10646, https://doi.org/10.1038/ncomms10646 (1–14).

Head, M.J., Aubry, M.-P., Walker, M., Miller, K.G., and Pratt, B.R., 2017, A case for formalizing subseries (subepochs) of the Cenozoic Era. *Episodes*, **40** (1), 22–27, https://doi.org/10.18814/epiiugs/2017/v40i/017004.

Hicks, N., and Green, A., 2017, A Mid-Miocene erosional unconformity from the Durban Basin, SE African margin: a combination of global eustatic sea level change, epeirogenic uplift, and ocean current initiation. *Marine and Petroleum Geology*, **86**: 798–811.

Hilgen, F.J., 1991a, Astronomical calibration of Gauss to Matuyama sapropels in the Mediterranean and implication for the Geomagnetic Polarity Time Scale. *Earth and Planetary Science Letters*, **104**: 226–244.

Hilgen, F.J., 1991b, Extension of the astronomically calibrated (polarity) timescale to the Miocene-Pliocene boundary. *Earth and Planetary Science Letters*, **107**: 349–368.

Hilgen, F.J., and Krijgsman, W., 1999, Cyclostratigraphy and astrochronology of the Tripoli diatomite formation (pre-evaporite Messinian, Sicily, Italy). *Terra Nova*, **11**: 16–22.

Hilgen, F.J., Krijgsman, W., Langereis, C.G., Lourens, L.J., Santarelli, A., and Zachariasse, W.J., 1995, Extending the astronomical (polarity) time scale into the Miocene. *Earth and Planetary Science Letters*, **136**: 495–510.

Hilgen, F.J., Bissoli, L., Iaccarino, S., Krijgsman, W., Meijer, R., Negri, A., et al., 2000a, Integrated stratigraphy and astrochronology of the Messinian GSSP at Oued Akrech (Atlantic Morocco). *Earth and Planetary Science Letters*, **182**: 237–251.

Hilgen, F.J., Iaccarino, S., Krijgsman, W., Villa, G., Langereis, C.G., and Zachariasse, W.J., 2000b, The Global Boundary Stratotype Section and Point (GSSP) of the Messinian Stage (uppermost Miocene). *Episodes*, **23**: 172–178.

Hilgen, F.J., Abdul Aziz, H., Krijgsman, W., Raffi, I., and Turco, E., 2003, Integrated stratigraphy and astronomical tuning of the Serravallian and lower Tortonian at Monte dei Corvi (Middle-Upper Miocene, northern Italy). *Palaeogeography, Palaeoclimatology, Palaeoecology*, **199**: 229–264.

Hilgen, F.J., Abdul Aziz, H., Bice, D., Iaccarino, S., Krijgsman, W., Kuiper, K., et al., 2005, The Global Boundary Stratotype Section and Point (GSSP) of the Tortonian Stage (Upper Miocene) at Monte dei Corvi. *Episodes*, **28**: 6–17.

Hilgen, F.J., Brinkhuis, H., and Zachariasse, W.J., 2006, Unit stratotypes for global stages: the Neogene perspective. *Earth-Science Reviews*, **74**: 113–125.

Hilgen, F.J., Abels, H.A., Iaccarino, S., Krijgsman, W., Raffi, I., Sprovieri, R., Turco, E., and Zachariasse, W.J., 2009, The Global Stratotype Section and Point (GSSP) of the Serravallian Stage (Middle Miocene). *Episodes*, **32**: 152–166.

Hilgen, F.J., Lourens, L.J., and Van Dam, J.A., 2012, The Neogene Period. In: Gradstein, F.M., Ogg, J.G., Schmitz, M.D., and Ogg (Coordinators), G.M. (eds), *The Geologic Time Scale 2012*, Elsevier Publisher, pp. 409–440, http://dx.doi.org/10.1016/B978-0-444-59425-9.00029-9.

Hilgen, F.J., Lourens, L.J., Pälike, H., and research support team, 2019, Should Unit-Stratotypes and Astrochronozones be formally defined? A dual proposal (including postscriptum). *Newsletters Stratigraphy. (Published online March 2019)*, https://doi.org/10.1127/nos/2019/0514.

Hodell, D.A., and Kennett, J.P., 1986, Late Miocene-Early Pliocene stratigraphy and paleoceanography of the South Atlantic and southwest Pacific oceans: a synthesis. *Paleoceanography*, **1** (3), 285–311.

Hodell, D.A., Curtis, J.H., Sierro, F.J., and Raymo, M.E., 2001, Correlation of Late Miocene to Early Pliocene sequences between the Mediterranean and North Atlantic. *Paleoceanography*, **16** (2), 164–178, <https://doi.org/10.1029/1999PA000487>.

Hohenegger, J., Ćorić, S., and Wagreich, M., 2014, Timing of the Middle Miocene Badenian Stage of the Central Paratethys. *Geologica Carpathica*, **65** (1), 55–66, https://doi.org/10.2478/geoca-2014-0004.

Holbourn, A., Kuhnt, W., Clemens, S., Prell, W., and Andersen, N., 2013, Middle to late Miocene stepwise climate cooling: evidence from a high-resolution deep water isotope curve spanning 8 million years. *Paleoceanography*, **28** (4), 688–699, <https://doi.org/10.1002/2013PA002538>.

Holbourn, A., Kuhnt, W., Lyle, M., Schneider, L., Romero, O., and Andersen, N., 2014, Middle Miocene climate cooling linked to intensification of eastern equatorial Pacific upwelling. *Geology*, **42**: 19–22, https://doi.org/10.1130/G34890.1.

Holbourn, A., Kuhnt, W., Kochhann, K.G.D., Andersen, N., and Meier, K.J.S., 2015, Global perturbation of the carbon cycle at the onset of the Miocene Climatic Optimum. *Geology*, **43**: 123–126, https://doi.org/10.1130/G36317.1.

Holcova, K., 2002, Calcareous nannoplankton from the Eggenburgian stratotype and faciostratotypes (Lower Miocene, Central Paratethys). *Geologica Carpathica*, **53** (6), 381–390.

Holcova, K., 2005, Quantitative calcareous nannoplankton biostratigraphy of the Oligocene/Miocene boundary interval in the northern part of the Buda Basin (Central Paratethys). *Geological Quarterly*, **49**: 263–274.

Holcova, K., 2013, Morphological variability of the Paratethyan Oligocene–Miocene small reticulofenestrid coccolites and its paleoecological and paleogeographical implications. *Acta Palaeontol. Polonica*, **58** (3), 651–668, <https://doi.org/10.4202/app.2009.0006>.

Holcova, K., 2017, Calcareous nannoplankton and foraminiferal response to global Oligocene and Miocene climatic oscillations: a case study from the Western Carpathian segment of the Central Paratethys. *Geologica Carpathica*, **68** (3), 207–228, https://doi.org/10.1515/geoca-2017-0016.

Holcova, K., Hrabovsky, J., Nehyba, S., Hladilov, A.Š., Dolakov, A.N., and Demeny, A., 2015, The Langhian (Middle Badenian) carbonate production event in the Moravian part of the Carpathia Foredeep (Central Paratethys): a multiproxy record. *Facies*, **61**: 1–26, https://doi.org/10.1007/s10347-014-0419-z.

Holcova, K., Dolakova, N., Nehyba, S., and Vacek, F., 2018, Timing of Langhian bioevents in the Carpathian Foredeep and northern Pannonian Basin in relation to oceanographic, tectonic and climatic processes. *Geological Quarterly*, **62** (1), 3–17, https://doi.org/10.7306/gq.1399.

Hordijk, K., and De Bruijn, H., 2009, The succession of rodent faunas from the Mio/Pliocene lacustrine deposits of the Florina-Ptolemais-Servia Basin (Greece). *Hellenic Journal of Geosciences*, **44**: 21–103.

Hüsing, S.K., Hilgen, F.J., Abdul Aziz, H., and Krijgsman, W., 2007, Completing the Neogene geological time scale between 8.5 and 12.5 Ma. *Earth and Planetary Science Letters*, **253**: 340–358, https://doi.org/10.1016/j.epsl.2006.10.036.

Hüsing, S.K., Kuiper, K.F., Link, W., Hilgen, F.J., and Krijgsman, W., 2009, The upper Tortonian-lower Messinian at Monte dei Corvi (northern Apennines, Italy): completing a Mediterranean reference section for the Tortonian Stage. *Earth and Planetary Science Letters*, **282**: 140–157.

Iaccarino, S.M., Di Stefano, A., Foresi, L.M., Turco, E., Baldassini, N., Cascella, A., et al., 2011, High-resolution integrated stratigraphy of the Mediterranean early-middle Miocene: comparison with the Langhian historical stratotype and new perspectives for the GSSP. *Stratigraphy*, **8**: 199–215.

Israelson, C., and Spezzaferri, S., 1998, Strontium isotope stratigraphy from Ocean Drilling Program Sites 918 and 919. In: Larsen, H.C., Saunders, A.D., and Clift, P.D. (eds), *Proceedings of the Ocean Drilling Program, Scientific Results*, **152**: 233–241.

Ivančič, K., Trajanova, M., Ćorić, S., Rožič, B., and Šmuc, A., 2018, Miocene paleogeography and biostratigraphy of the Slovenj Gradec Basin: a marine corridor between the Mediterranean and Central Paratethys. *Geologica Carpathica*, **69** (6), 528–544, https://doi.org/10.1515/geoca-2018-0031.

Jenkins, D.G., and Houghton, S.D., 1989, Neogene planktonic foraminiferal biostratigraphy of ODP Site 677, Panama Basin. *Proceedings of the Ocean Drilling Program, Scientific Results*, **111**: 289–293.

John, C.M., Karner, G.D., Browning, E., Leckie, R.M., Mateo, Z., Carson, B., et al., 2011, Timing and magnitude of Miocene eustacy derived from the mixed siliciclastic-carbonate stratigraphic record of the northeastern Australian margin. *Earth and Planetary Science Letters*, **304**: 455−467.

Jorissen, E.L., De Leeuw, A., van Baak, C.G.C., Mandic, O., Stoica, M., Abels, H.A., et al., 2018, Sedimentary architecture and depositional controls of a Pliocene river-dominated delta in the semi-isolated Dacian Basin, Black Sea. *Sedimentary Geology*, **368**: 1−23, <https://doi.org/10.1016/j.sedgeo.2018.03.001>.

Joyce, J.E., Tjalsma, L.R.C., and Prutzman, J.M., 1990, High-resolution planktic stable isotope record and spectral analysis for the last 5.35 M.Y.: Ocean Drilling Program Site 625 Northeast Gulf of Mexico. *Paleoceanography*, **5**: 507−529.

Kälin, D., and Kempf, O., 2009, High-resolution stratigraphy from the continental record of the Middle Miocene Northern Alpine Foreland Basin of Switzerland. *Neues Jahrbuch für Geologie und Paläontologie Abhandlungen*, **254**: 177−235.

Kamikuri, S., 2017, Late Neogene radiolarian biostratigraphy of the eastern North Pacific ODP Sites 1020/1021. *Paleontological Research*, **21** (3), 230−254, https://doi.org/10.2517/2016PR027.

Kamikuri, S., Itaki, T., Motoyama, I., and Matsuzaki, K.M., 2017, Radiolarian Biostratigraphy from Middle Miocene to Late Pleistocene in the Japan Sea. *Paleontological Research*, **21** (4), 397−421, <https://doi.org/10.2517/2017PR001>.

Kasbohm, J., and Schoene, B., 2018, Rapid eruption of the Columbia River flood basalt and correlation with the mid-Miocene climate optimum. *Science Advances*, **4** (9), https://doi.org/10.1126/sciadv.aat8223.

Keigwin Jr., L.D., 1982, Stable isotope stratigraphy and paleoceanography of Sites 502 and 503. In: Prell, W.L., Gardner, J.V., et al., (eds), *Initial Reports of the Deep Sea Drilling Project*, **68**: 445−453.

King, C., 2016, Biostratigraphy. *In* King, C., Gale, A.S., and Barry, T.L. (eds), A revised correlation of Tertiary rocks in the British Isles and adjacent areas of NW Europe. *London: Geological Society, London, Special Reports*, **27**: 19−57, https://doi.org/10.1144/SR27.3.

King, D.J., Wade, B.S., Liska, R.D., and Miller, C.G., 2020, A review of the importance of the Caribbean region in Oligo-Miocene low latitude planktonic foraminiferal biostratigraphy and the implications for modern biogeochronological schemes. *Earth-Science Reviews*, **202**: article #102968. https://doi.org/10.1016/j.earscirev.2019.102968.

Kochhann, K.G.D., Holbourn, A., Kunht, W., Channell, J.E.T., Lyle, M.W., Shackford, J.K., et al., 2016, Eccentricity pacing of eastern equatorial Pacific carbonate dissolution cycles during the Miocene Climatic Optimum. *Paleoceanography*, **31** (9), 1176−1192, <https://doi.org/10.1002/2016PA002988>.

Koizumi, I., Sato, M., and Matoba, Y., 2009, Age and significance of Miocene diatoms and diatomaceous sediments from northeast Japan. *Palaeogeography, Palaeoclimatology, Palaeoecology*, **272**: 85−98.

Konenkova, I.D., and Bogdanovich, E.I., 1994, Raspredelenie foraminifer i nannoplanktona v tarhan-chokrakskih otlozheniyah urochischa malyiy Kamyishlak (Kerchenskiy poluostrov) [Distribution pattern of foraminifera and nannoplankton within the Tarkhanian-Tschokrakian deposits of the Malij Kamyshlak section (Kerch Peninsula)]. In: Teslenko, Y.V. (ed), *Ancient Biospheres of the Ukraine*. Kiev: National Academy of Sciences of the Ukraine, pp. 95−96, (in Russian).

Kostopoulos, D.S., and Athanassiou, A., 2005, In the shadow of bovids: suids, cervids and giraffids from the Plio-Pleistocene of Greece. *Quaternaire hors-série*, **2**: 179−190.

Köthe, A., 2012, A revised Cenozoic dinoflagellate cyst and calcareous nannoplankton zonation for the German sector of the southeastern North Sea Basin. *Newsletters Stratigraphy*, **45**: 189−220.

Koufos, G.D., 2001, The Villafranchian mammalian faunas and biochronology of Greece. *Bollettino Società Paleontologica Italiana*, **40**: 217−223.

Kovač, M., Holcova, K., and Nagymarosy, A., 1999, Paleogeography, paleobathymetry and relative sea-level changes in the Danube Basin and adjacent areas. *Geologica Carpathica*, **50** (4), 325−338.

Kováč, M., Baráth, I., Harzhauser, M., Hlavatý, I., and Hudáčková, N., 2004, Miocene depositional systems and sequence stratigraphy of the Vienna Basin. *Courier Forschungsinstitut Senckenberg*, **246**: 187−212.

Kovač, M., Andreyeva-Grigorovich, A., Bajraktarević, Z., Brzobohaty, R., Filipescu, S., Fodor, L., et al., 2007, Badenian evolution of the Central Paratethys Sea: paleogeography, climate and eustatic sea-level changes. *Geologica Carpathica*, **58** (6), 579−606.

Kováč, M., Márton, E., Oszczypko, N., Vojtko, R., Hók, J., Králiková, S., et al., 2017, Neogene palaeogeography and basin evolution of the Western Carpathians, Northern Pannonian domain and adjoining areas. *Global and Planetary Change*, **155**: 133−154 <https://doi.org/10.1016/j.gloplacha.2017.07.004>.

Kováč, M., Halasova, E., Hudáčková, N., Holcova, K., Hyžný, M., Jamrich, M., et al., 2018, Towards better correlation of the Central Paratethys regional time scale with the standard geological time scale of the Miocene Epoch. *Geologica Carpathica*, **69** (3), 283−300, https://doi.org/10.1515/geoca-2018-0017.

Krasheninnikov, V.A., Basov, I.A., and Golovina, L.A., 2003, *Vostochnyiy Paratetis: tarhanskiy i konkskiy reigoyarusyi (stratigrafiya, mikropaleontologiya, bionomiya, paleogeograficheskie svyazi) [The Eastern Paratethys: Tarkhanian and Konkian Regional Stages (stratigraphy, micropaleontology, bionomy, paleogeographic links)]*. Moscow: Nauchnyi Mir, p. 194.

Krezsek, C.S., and Filipescu, S., 2005, Middle to late Miocene sequence stratigraphy of the Transylvanian Basin (Romania). *Tectonophysics*, **410** (1−4), 437−463, https://doi.org/10.1016/j.tecto.2005.02.018.

Krijgsman, W., Hilgen, F.J., Langereis, C.G., Santarelli, A., and Zachariasse, W.J., 1995, Late Miocene magnetostratigraphy, biostratigraphy and cyclostratigraphy in the Mediterranean. *Earth and Planetary Science Letters*, **136**: 475−494.

Krijgsman, W., Garcés, M., Langereis, C.G., Daams, R., van Dam, J., van der Meulen, A.J., et al., 1996, A new chronology for the middle to late Miocene continental record in Spain. *Earth and Planetary Science Letters*, **142**: 367−380.

Krijgsman, W., Hilgen, F., Raffi, I., Sierro, F., and Wilson, D., 1999, Chronology, causes and progression of the Messinian salinity crisis. *Nature*, **400**: 652−655.

Krijgsman, W., Stoica, M., Vasiliev, I., and Popov, V.V., 2010, Rise and fall of the Paratethys Sea during the Messinian Salinity Crisis. *Earth and Planetary Science Letters*, **290** (1−2), 183−191, https://doi.org/10.1016/j.epsl.2009.12.020.

Krijgsman, W., Tesakov, A., Yanina, T., Lazarev, S., Danukalova, G., Van Baak, C.G.C., et al., 2019, Quaternary time scales for the Pontocaspian domain: interbasinal connectivity and faunal evolution. *Earth-Science Reviews*, **188**: 1−40, <https://doi.org/10.1016/j.earscirev.2018.10.013>.

Lacombat, F., Abbazzi, L., Ferretti, M.P., Martinez-Navarro, B., Moulle, P.E., Palombo, M.R., et al., 2008, New data on the Early Villafranchian fauna from Vialette (Haute-Loire, France) based on the collection of the Crozatier Museum (Le Puy-en-Velay, Haute-Loire, France). *Quaternary International*, **179**: 64−71.

Langereis, C.G., and Hilgen, F.J., 1991, The Rossello composite: a Mediterranean and global reference section for the early to early late Pliocene. *Earth and Planetary Science Letters*, **104**: 211−225.

Laskar, J., 2004, A comment on 'Accurate spin axes and Solar System dynamics: climatic variations for the Earth and Mars. *Astronomy & Astrophysics*, **416**: 799−800.

Laskar, J., Gastineau, M., Delisle, J.-B., Farrés, A., and Fienga, A., 2011, Strong chaos induced by close encounters with Ceres and Vesta. *Astronomy and Astrophysics*, **532**: 1−4.

Lazarus, D., 1990, Middle Miocene to recent radiolarians from the Weddell Sea, Antarctica, ODP Leg 113. In: Barker, P.F., et al., (eds), *Proceedings of the Ocean Drilling Program, Scientific Results*, **113**: 709−727.

Lazarus, D.B. (1992): Antarctic Neogene radiolarians from the Kerguelen Plateau, Legs 119 and 120. In Wise, S.W., Schlich, R., et al. (eds), *Proceedings of the Ocean Drilling Program, Scientific Results*, **120**: 785−809, <https://doi.org/10.2973/odp.proc.sr.120.192.1992>.

Lazarus, D., Suzuki, N., Caulet, J.-P., Nigrini, C., Goll, I., Goll, R., et al., 2015, An evaluated list of Cenozoic-Recent radiolarian species names (Polycystinea), based on those used in the DSDP, ODP and IODP deep-sea drilling programs. *Zootaxa*, **3999**: 301−333.

Leckie, M.R., Wade, B.S., Pearson, P.N., Fraass, A.J., King, D.J., Olsson, R. K., et al., 2018, Taxonomy, biostratigraphy, and phylogeny of Oligocene and early Miocene *Paragloborotalia* and *Parasubbotina*. In: Wade, B.S., Olsson, R.K., Pearson, P.N., Huber, B.T., and Berggren, W.A. (eds), *Atlas of Oligocene Planktonic Foraminifera—Chapter 5*. Cushman Foundation Special Publication, 125−178, No. 46.

Leever, K.A., Matenco, L., Garcia-Castellanos, D., and Cloetingh, S.A.P.L., 2011, The evolution of the Danube gateway between Central and Eastern Paratethys (SE Europe): insight from numerical modelling of the causes and effects of connectivity between basins and its expression in the sedimentary record. *Tectonophysics*, **502**: 175−195, https://doi.org/10.1016/j.tecto.2010.01.003Tectonophysics.

Leonard, K.A., Williams, D.F., Baruch, B.W., and Thunell, R.C., 1983, Pliocene paleoclimatic and paleoceanographic history of the south Atlantic Ocean: stable isotopic records from leg 72 Deep Sea Drilling Project Holes 516a and 517. In: Barker, P.F., Carlson, R.L., Johnson, D.A., et al., (eds), *Initial Reports Deep Sea Drilling Project*, **72**: 895−906.

Levèque, F., 1993, Correlating the Eocene-Oligocene mammalian biochronological scale from SW Europe with the marine magnetic anomaly sequence. *Journal of the Geological Society*, **150**: 661−664.

Li, Q., Radford, S.S., and Banner, F.T., 1992, Distribution of microperforate tenuitellid planktonic foraminifers in holes 747A and 749B, Kerguelen Plateau. In: Wise Jr., S.W., Schlich, R., et al., (eds), *Proceedings of the Ocean Drilling Program, Scientific Results*, **120**: 569−594.

Liebrand, D., Lourens, L.J., Hodell, D.A., de Boer, B., van de Wal, R.S.W., and Pälike, H., 2011, Antarctic ice sheet and oceanographic response to eccentricity forcing during the early Miocene. *Climate of the Past*, **7** (3), 869−880, https://doi.org/10.5194/cp-7-869-2011.

Liebrand, D., Beddow, H.M., Lourens, L.J., Pälike, H., Raffi, I., Bohaty, S.M., et al., 2016, Cyclostratigraphy and eccentricity tuning of the early Oligocene through early Miocene (30.1−17.1 Ma): *Cibicides mundulus* stable oxygen and carbon isotope records from Walvis

Ridge Site 1264. *Earth and Planetary Science Letters*, **450**: 392−405, <https://doi.org/10.1016/j.epsl.2016.06.007>.

Lindsay, E., 2003, Chronostratigraphy, biochronology, datum events, land mammal ages, stage of evolution, and appearance event ordination. *Bulletin of the American Museum of Natural History*, **279**: 212−230.

Lindsay, E.H., Opdyke, N.O., and Johnson, N.M., 1980, Pliocene dispersal of the horse *Equus* and late Cenozoic mammalian dispersal events. *Nature*, **287**: 135−138.

Lisiecki, L.R., and Raymo, M.E., 2005, A Pliocene-Pleistocene stack of 57 globally distributed benthic $\delta^{18}O$ records. *Paleoceanography*, **20**: PA1003, https://doi.org/10.1029/2004PA001071.

Lister, A.M., and van Essen, H.E., 2003, *Mammuthus rumanus* the earliest mammoth in Europe. In: Petculescu, A., and Stiuca, E. (eds), *Advances in Vertebrate Palaeontology "Hent to Pantha"*. Bucharest: Institute of Speleology of the Romanian Academy, pp. 47−52.

Lörenthey, I., 1900, Foraminifera der pannonischen Stufe Ungarns. *Neues Jahrbuch für Mineralogie, Geologie und Paläontologie*, **2**: 99−107.

Lourens, L.J., Hilgen, F.J., Zachariasse, W.J., Van Hoof, A.A.M., Antonarakou, A., and Vergnaud-Grazzini, C., 1996, Evaluation of the Plio-Pleistocene astronomical time scale. *Paleoceanography*, **11**: 391−413.

Lourens, L.J., Hilgen, F.J., Laskar, J., Shackleton, N.J., and Wilson, D., 2004, The Neogene Period. In: Gradstein, F.M., Ogg, J.G., and Smith, A.G. (eds), *A Geologic Time Scale 2004*. Cambridge: Cambridge University Press, pp. 409−440.

Lubenescu, V., 2016, Biostratigraphic correlations between the Dacian and Pannonian basins from Romania. *Geo-Eco-Marina*, **22**: 161−179, https://doi.org/10.5281/zenodo.889931.

Lyle, M., Wilson, P.A., Janecek, T.R., et al., 2002, *Proceedings of the Ocean Drilling Program, Initial Reports*, 199. College Station, TX: Ocean Drilling Program, http://dx.doi.org/10.2973/odp.proc.ir.199.2002.

Magyar, I., Radivojević, D., Sztano, O., Rastislav Synak, R., Ujszaszi, K., and Pocsik, M., 2013, Progradation of the paleo-Danube shelf margin across the Pannonian Basin during the Late Miocene and Early Pliocene. *Global and Planetary Change*, **103**: 168−173, https://doi.org/10.1016/j.gloplacha.2012.06.007.

Mandic, O., Kurečić, T., Neubauer, T.A., and Harzhauser, M., 2015, Stratigraphic and palaeogeographic significance of lacustrine molluscs from the Pliocene Viviparus beds in central Croatia. *Geologia Croatica*, **68** (3), 179−2017, https://doi.org/10.4154/gc.2015.15.

Mandic, O., Rundic, L., Coric, S., Pezelj, D., Theobalt, D., Sant, K., et al., 2019a, Age and mode of the middle Miocene marine flooding of the Pannonian basin-constraints from Central Serbia. *Palaios*, **34**: 1−25, https://doi.org/10.2110/palo.2018.052.

Mandic, O., Sant, K., Kallanxhic, M.-E., Ćorić, S., Theobalta, D., Grunert, P., et al., 2019b, Integrated bio-magnetostratigraphy of the Badenian reference section Ugljevik in southern Pannonian Basin—implications for the Paratethys history (middle Miocene, Central Europe). *Global and Planetary Change*, **172**: 374−395, <https://doi.org/10.1016/j.gloplacha.2018.10.010>.

Markov, G.N., and Spassov, N., 2003, Primitive mammoths from Northeast Bulgaria in the context of the earliest mammoth migrations in Europe. In: Petulescu, A., and Stiuca, E. (eds), *Advances in Vertebrate Palaeontology "Hen to Pantha"*. Bucharest: Institute of Speleology of the Romanian Academy, pp. 53−58.

Martínez-Navarro, B., Madurell-Malapeira, J., Ros-Montoya, S., Espigares, M.P., Medin, T., Hortola, P., et al., 2015, The Epivillafranchian and the arrival of pigs into Europe. *Quaternary International*, **389**: 131–138.

Martini, E., 1971, Standard Tertiary and Quaternary Calcareous Nannoplankton Zonation. *Proceedings of the II Planktonic Conference, Roma, 1970. Edizioni Tecnoscienza*, **2**: 739–785.

Mayer-Eymar, K. (ed), 1868. *Tableau Synchronistique des Terrains Tertiaires Supérieurs*. fourth ed. Autographie H. Manz. Zürich,

McArthur, J.M., Howarth, R.J., and Bailey, T.R., 2001, Strontium isotope stratigraphy: LOWESS version 3: best fit marine Sr-isotope curve for 0–509 Ma and accompanying look-up table deriving numerical age. *Journal of Geology*, **109**: 155–170.

Miller, K.G., and Wright, J.D., 2017, Success and failure in Cenozoic global correlations using golden spikes: a geochemical and magnetostratigraphic perspective. *Episodes*, **40** (1), 8–21.

Miller, K.G., Aubry, M.-P., Khan, J., Melillo, A.J., Kent, D.V., and Berggren, W.A., 1985, Oligocene to Miocene biostratigraphy, magnetostratigraphy, and isotopic stratigraphy of the western North Atlantic. *Geology*, **13**: 257–261.

Miller, K.G., Wright, J.D., and Fairbanks, R.G., 1991, Unlocking the Ice House: Oligocene-Miocene oxygen isotopes, eustacy and marginal erosion. *Journal of Geophysical Research*, **96**: 6829–6848.

Miller, K.G., Wright, J.D., van Fossen, M.C., and Kent, D.V., 1994, Miocene stable isotopic stratigraphy and magnetostratigraphy of Buff Bay, Jamaica. *Geological Society of America Bulletin*, **106**: 1605–1620.

Miller, K.G., Mountain, G.S., and the Leg 150 Shipboard Party Members of the New Jersey Coastal Plain Drilling Project, 1996, Drilling and dating New Jersey Oligocene-Miocene sequences: ice volume, global sea level, and Exxon records. *Science*, **271**: 1092–1095.

Miller, K.G., Kominz, M.A., Browning, J.V., Wright, J.D., Mountain, G.S., Katz, M.E., et al., 2005, The Phanerozoic record of global sea-level change. *Science*, **310**: 1293–1298.

Miller, K.G., Baluyot, R., Wright, J.D., Kopp, R.E., and Browning, J.V., 2017, Closing an early Miocene astronomical gap with Southern Ocean $\delta^{18}O$ and $\delta^{13}C$ records: implications for sea level change. *Paleoceanography*, **32**: 600–621, https://doi.org/10.1002/2016PA003074.

Minashvili, T., and Ananiashvili, G., 2017, Problem of Stratigraphic Boundary between the Paleogene and Neogene Systems in the Eastern Paratethys. *Bulletin of the Georgian National Academy of Sciences*, **11** (3), 106–111.

Mix, A.C., Le, J., and Shackleton, N.J., 1995, Benthic foraminiferal stable isotope stratigraphy of site 846: 0–1.8 Ma. In: Pisias, N.G., Mayer, L.A., Janecek, T.R., Palmer-Julson, A., and van Andel, T.H. (eds), Proceeding of the Ocean Drilling Program, Scientific Results,. **138**: 839–854.

Montanari, A., Deino, A., Coccioni, R., Langenheim, V.E., Capo, R., and Monechi, S., 1991, Geochronology, Sr isotope analysis, magnetostratigraphy, and plankton stratigraphy across the Oligocene-Miocene boundary in the Contessa section (Gubbio, Italy). *Newsletters Stratigraphy*, **23**: 151–180.

Morgans, H.E.G., Edwards, A.R., Scott, G.H., Graham, I.J., Kamp, P.J.J., Mumme, T.C., et al., 1999, Integrated stratigraphy of the Waitakian–Otaian stage boundary stratotype, early Miocene, New Zealand. *New Zealand Journal of Geology and Geophysics*, **42**: 581–614.

Morley, J.J., and Nigrini, C., 1995, Miocene to Pleistocene radiolarian biostratigraphy of North Pacific Sites 881, 884, 885, 886 and 887. *Proceedings of the Ocean Drilling program, Scientific Results*, **145**: 55–91.

Motoyama, I., 1996, Late Neogene radiolarian biostratigraphy in the subarctic Northwest Pacific. *Micropaleontology*, **42**: 221–262.

Motoyama, I., Niitsuma, N., Maruyama, T., Hayashi, H., Kamikuri, S., Shiono, M., et al., 2004, Middle Miocene to Pleistocene magnetobiostratigraphy of ODP Sites 1150 and 1151, northwest Pacific: sedimentation rate and updated regional geological timescale. *Island Arc*, **13**: 289–305.

Muratov, M.V., and Nevesskaya, L.A. (eds), 1986. Stratigrafiya SSSR. Neogenovaya sistema. Polutom 1 [Stratigraphy of the USSR. Neogene System Semivol. 1]. Nedra. Moskva, p. 420, (in Russian).

Naish, T., Kamp, P.J.J., and Pillans, B., 1997, Recurring global sea-level changes recorded in shelf deposits near the G/M polarity transition, Wanganui Basin, New Zealand: implications for redefining the Pliocene–Pleistocene boundary. *Quaternary International*, **40**: 61–71.

Neubauer, T.A., Georgopoulou, E., Kroh, A., Harzhauser, M., Mandic, O., and Esu, D., 2015, Synopsis of European Neogene freshwater gastropod localities: updated stratigraphy and geography. *Palaeontologia Electronica*, **18.1.3T**: 1–7, palaeo-electronica.org/content/2015/1153-neogene-freshwater-gastropods.

Neubauer, T.A., Harzhauser, M., Mandic, O., Kroh, A., and Georgopoulou, E., 2016, Evolution, turnovers and spatial variation of the gastropod fauna of the late Miocene biodiversity hotspot Lake Pannon. *Palaeogeography, Palaeoclimatology, Palaeoecology*, **442**: 84–95, <https://doi.org/10.1016/j.palaeo.2015.11.016>.

Nevesskaya, L.A., Bogdanovich, A.K., Vyalov, O.S., Zhizhchenko, B.P., Il'ina, L.B., Nossovskiy, M.F., 1975a, Yarusnaya shkala neogenovyih otlozheniy yuga SSSR [A stage stratigraphic scale of Neogene deposits for the South USSR]. In: Senes, J. (ed), *Proceedings of the VIth Congress of the Regional Committee on Mediterranean Neogene Stratigraphy (RCMNS), Bratislava I*, pp. 267–289.

Nevesskaya, L.A., Bogdanovich, A.K., Vyalov, O.S., Zhizhchenko, B.P., Il'ina, L.B., Nossovskiy, M.F., et al., 1975b, Yarusnaya shkala neogenovyih otlozheniy yuga SSSR [A stage stratigraphic scale of Neogene deposits for the South USSR]. *Proceedings of the USSR Academy of Sciences, Geological series*, **2**: 267–289, (in Russian).

Nevesskaya, L.A., Goncharova, I.A., Il'ina, L.B., Paramonova, N.P., Popov, S.V., Bogdanovich, A.K., et al., 1984, Regionalnaya stratigraficheskaya shkala neogena Vostochnogo Paratetisa [Regional stratigraphic scale of the Neogene of Eastern Paratethys]. *Soviet Geology*, **9**: 91–101, (in Russian).

Nevesskaya, L.A., Goncharova, I.A., Il'ina, L.B., Paramonova, N.P., and Khondkarian, S.O., 2003, O stratigraficheskoy shkale neogena Vostochnogo Paratetisa [The Neogene Stratigraphic Scale of the Eastern Paratethys]. *Stratigraphy Geology Correlation*, **11** (2), 3–26, (in Russian).

Nevesskaya, L.A., Kovalenko, E.I., Beluzhenko, E.V., Popov, S.V., Goncharova, I.A., Danukalova, G.A., et al., 2005, Regionalnaya stratigraficheskaya shema Neogena yuga evropeyskoy chasti Rossii [Regional stratigraphic scheme of the Neogene of the south of European part of Russia]. *National Geology*, **4**: 47–59, (in Russian).

Nigrini, C., Sanfilippo, A., and Moore Jr., T.J., 2005, Cenozoic Radiolarian biostratigraphy: a magnetobiostratigraphic chronology of Cenozoic sequences from ODP Sites 1218, 1219, and 1220, Equatorial Pacific. In: Wilson, P.A., Lyle, M., and Firth, J.V. (eds), *Proceedings of the Ocean Drilling Program, Scientific Results*, **199**: 1–76. Available from:, http://www-odptamu.edu/publications/ 199_SR/VOLUME/CHAPTERS/225.PDF.

Ogg, J.G., Ogg, G.M., and Gradstein, F.M., 2016, *A Concise Geologic Time Scale*. Elsevier Publisher, p. 240.

Okada, H., and Bukry, D., 1980, Supplementary modification and introduction of code numbers to the low–latitude coccolith biostratigraphic zonation (Bukry, 1973; 1975). *Marine Micropaleontology*, **5**: 321–325.

Olariu, C., Krezsek, C., and Jipa, D., 2018, The Danube River inception: evidence for a 4 Ma continental scale river born from segmented ParaTethys basins. *Terra Nova*, **30**: 63–71, <https://doi.org/ 10.1111/ter.12308>.

Olshtynska, A.P., 2001, Miocene marine diatom stratigraphy of the Eastern Paratethys (Ukraine). *Geologica Carpathica*, **52** (3), 173–181.

Olshtynskaya, A.P., 1996, Zonalnaya stratigrafiya miotsena tsentralnoy chasti Vostochnogo Paratetisa po diatomovyim vodoroslyam [Zonal stratigraphy of the Miocene of central part the Eastern Paratethys based on diatom algae]. *Geological Journal of the Academy of Sciences of Ukraine*, **3–4**: 88–92, (in Russian).

Onodera, J., Takahashi, K., and Nagatomo, R., 2016, Diatoms, silicoflagellates, and ebridians at Site U134 on the western slope of Bowers Ridge, IODP Expedition 323. *Deep-Sea Research II*, **125–126**: 8–17, https://doi.org/10.1016/j.dsr2.2013.03.025>.

Palcu, D.V., Tulbure, M., Bartol, M., Kouwenhoven, T.J., and Krijgsman, W., 2015, The Badenian–Sarmatian extinction event in the Carpathian foredeep basin of Romania: paleogeographic changes in the Paratethys domain. *Global and Planetary Change*, **133**: 346–358, <https://doi.org/10.1016/j. gloplacha.2015.08.014>.

Palcu, D.V., Golovina, L.A., Vernyhorova, Y.V., Popov, S.V., and Krijgsman, W., 2017, Middle Miocene paleoenvironmental crises in Central Eurasia caused by changes in marine gateway configuration. *Global and Planetary Change*, **158**: 57–71, <https://doi.org/ 10.1016/j.gloplacha.2017.09.013>.

Palcu, D.V., Popov, S.V., Golovina, L.A., Kuiper, K.F., Liu, S., and Krijgsman, W., 2019a, The shutdown of an anoxic giant: magnetostratigraphic dating of the end of the Maikop Sea. *Gondwana Research*, **67** (2019), 82–100, <https://doi.org/10.1016/j.gr.2018.09.011>.

Palcu, D.V., Vesiliev, I., Stoica, M., and Krijgsman, W., 2019b, The end of the Great Khersonian Drying of Eurasia: Magnetostratigraphic dating of the Maeotian transgression in the Eastern Paratethys. *Basin Research*, **31**: 33–58, https://doi.org/10.1111/bre.12307.

Pälike, H., Frazier, J., and Zachos, J.C., 2006a, Extended orbitally forced palaeoclimatic records from the equatorial Atlantic Ceara Rise. *Quaternary Science Reviews*, **25**: 3138–3149, https://doi.org/ 10.1016/j.quascirev.2006.02.011.

Pälike, H., Norris, R.D., Herrle, J.O., Wilson, P.A., Coxall, H.K., Lear, C.H., Shackleton, N.J., Tripatai, A.K., and Wade, B.S., 2006b, The heartbeat of the Oligocene climate system. *Science*, **314**: 1894–1898.

Pälike, H., Lyle, M., Nishi, H., Raffi, I., Gamage, K., Klaus, A., et al., 2010, *Proceedings of the Integrated Ocean Drilling Program*, 320/321. Tokyo: Integrated Ocean Drilling Program Management International, Inc., http://dx.doi.org/10.2204/iodp.proc.320321.101.2010.

Palmqvist, P., Torregrosa, V., Pérez-Claros, J.A., Martínez-Navarro, B., and Turner, A., 2007, A re-evaluation of the diversity of *Megantereon* (Mammalia, Carnivora, Machairodontinae) and the problem of species identification in extinct carnivores. *Journal of Vertebrate Paleontology*, **27**: 160–175.

Palombo, M.R., 2004, Biochronology of the Plio-Pleistocene mammalian faunas of Italian peninsula: knowledge, problems and perspectives. *Il Quaternario*, **17**: 565–658.

Pană, I., Bonig, N., and Botez, R., 1968, New elements in the Levantin fauna in the Buzau region (in Romanian). *Pet. Gaz.*, **XDX**: 699–701.

Papp, A., Rogl, F., Seneš, J. (eds), 1973. M2 Ottnangien. Die Innviertler, Salgotarjaner, Bantapusztaer Schichtengruppe und die Rzehakia Formation. Chronostratigraphie und Neostratotypen. Miozan der Zentralen Paratethys, 3. SAV, Bratislava, 1–841, (in German with English abstract).

Papp, A., Marinescu, F., Seneš, J. (eds), 1974. M5 Sarmatien (sensu E. Suess 1866). Die Sarmatische Schichtengruppe und ihr Stratotypus. Chronostratigraphie und Neostratotypen. *Miozan der Zentralen Paratethys*, **4**. VEDA SAV, Bratislava, 1–707, (in German with English summary).

Papp, A., Cicha, I., Seneš, J., Steininger, F. (eds), 1978. M4 Badenien: (Moravien, Wielicien, Kosovien). Chronostratigraphie und Neostratotypen. *Miozan der Zentralen Paratethys*, **6**. VEDA SAV, Bratislava, 1–594, (in German with English summary).

Papp, A., Jambor, A., Steininger, F.F. (eds), 1985. M6 Pannonien: (Slavonien und Serbien). Chronostratigraphie und Neostratotypen. *Miozan der Zentralen Paratethys*, **7**. Akademiai Kiado, Budapest, 1–636, (in German with English summary).

Paulissen, W.E., Luthi, S.M., Grunert, P., Ćorić, S., and Harzhauser, M., 2011, Integrated high-resolution stratigraphy of a Middle to Late Miocene sedimentary sequence in the central part of the Vienna Basin. *Geologica Carpathica*, **62** (2), 155–169, https://doi.org/10.2478/v10096-011-0013-z.

Pearson, P.N., 1995, Planktonic foraminifer biostratigraphy and the development of pelagic caps on guyots in the Marshall Islands Group. In: Haggerty, J., Premoli Silva, I., Rack, F., and McNutt, M. K. (eds), *Proceedings of the Ocean Drilling program, Scientific Results*, **144**: 21–59.

Pearson, P.N., and Chaisson, W.P., 1997. Late Paleocene to middle Miocene planktonic foraminifer biostratigraphy of the Ceara Rise. In: Shackleton, N.J., Curry, W.B., Richter, C., and Bralower, T.J. (eds), *Proceedings of the Ocean Drilling Program, Scientific Results*, 154: 33-66.

Pearson, P.N., Wade, B.S., Backman, J., Raffi, I., and Monechi, S., 2017, Sub-series and sub-epochs are informal units and should continue to be omitted from the International Chronostratigraphic Chart. *Episodes*, **40** (1), 5–7, https://doi.org/10.18814/epiiugs/2017/v40i1/ 017004.

Peryt, T.M., 2006, The beginning, development and termination of the Middle Miocene Badenian salinity crisis in Central Paratethys. *Sedimentary Geology*, **188–189**: 379–396, https://doi.org/10.1016/j. sedgeo.2006.03.014.

Peryt, T.M., Peryt, D., Jasionowski, M., Poberezhskyy, A.V., and Durakiewicz, T., 2004, Post-evaporitic restricted deposition in the Middle Miocene Chokrakian-Karaganian of East Crimea (Ukraine). *Sedimentary Geology*, **170**: 21–36, https://doi.org/10.1016/j.sedgeo. 2004.04.003.

Petronio, C., and Sardella, R., 1999, Biochronology of the Pleistocene mammal fauna from Ponte Galeria (Rome) and remarks on the middle Galerian faunas. *Rivista Italiana di Paleontologia e Stratigrafia*, **105**: 155–164.

Pillans, B., 2017, Chapter 4. Quaternary stratigraphy of Whanganui Basin – a globally significant archive. In: Shulmeister, J. (ed), *Landscape and Quaternary Environmental Change in New Zealand. Atlantis Advances in Quaternary Science 3*, pp. 141–170, http://dx.doi.org/10.2991/978-94-6239-237-3_4.

Piller, W.E., Harzhauser, M., and Mandic, O., 2007, Miocene Central Paratethys stratigraphy—current status and future directions. *Stratigraphy*, **4** (2–3), 151–168.

Pipper, M., Reichenbacher, B., Kirscher, U., Sant, K., and Hanebeck, H., 2017, The middle Burdigalian in the North Alpine Foreland Basin (Bavaria, SE Germany)—a lithostratigraphic, biostratigraphic and magnetostratigraphic re-evaluation. *Newsletters Stratigraphy*, 1–25, https://doi.org/10.1127/nos/2017/0403.

Poore, R.Z., Tauxe, L., Percival, S.F., Labrecque, J.L., Wright, R., Peterson, N.P., et al., 1983, Late Cretaceous-Cenozoic magnetostratigraphy and biostratigraphy correlations of the South Atlantic Ocean: DSDP Leg 73. *Palaeogeography, Palaeoclimatology, Palaeoecology*, **42**: 127–149.

Popov, S.V., Rostovtseva, Y.V., Fillippova, N.Y., Golovina, L.A., Radionova, E.P., Goncharova, I.A., et al., 2016, Paleontology and stratigraphy of the Middle – Upper Miocene of Taman Peninsula. Part 1. Description of key-sections and benthic fossil groups. *Paleontological Journal (Suppl.)*, **50** (10), 168, https://doi.org/10.1134/S0031030116100014.

Portniagina, L.A., 1980, Palinologiya oligotsenovyih, miotsenovyih i pliotsenovyih otlozheniy akvatorii Azovskogo morya i Kerchenskogo poluostrova [Palynology of Oligocene, Miocene and Pliocene deposits of the Azov Sea and Kerch peninsula]. *Paleontological Reviewer Lviv Universit*, **17**: 83–87, (in Russian).

Pradella, C., and Rook, L., 2007, Mesopithecus (Primates, Cercopithecoidea) from Villafranca d'Asti (Early Villafranchian; NW Italy) and palaeoecological context of its extinction. *Swiss Journal of Geosciences*, **100**: 145–152.

Pujol, C., 1983, Cenozoic planktonic foraminiferal biostratigraphy of the southwestern Atlantic (Rio Grande Rise). In: Barker, P.F., Carlson, R.L., Johnson, D.A., et al., (eds), *Initial Reports of the Deep Sea Drilling Project*, **72**: 623–673.

Pujol, C., and Duprat, J., 1983, Quaternary planktonic foraminifers of the southwestern Atlantic (Rio Grande Rise) Deep Sea Drilling Project Leg 72. In: Barker, P.F., Carlson, R.L., Johnson, D.A., et al., (eds), *Initial Reports of the Deep Sea Drilling Project*, **72**: 601–615.

Radionova, E., and Golovina, L., 2011, Upper Maeotian-Lower Pontian "Transitional Strata" in the Taman Peninsula: Stratigraphic Position and Paleogeographic Interpretation. *Geologica Carpathica*, **2**: 62–100, https://doi.org/10.2478/v10096-011-0007-x.

Radionova, E.P., Golovina, L.A., Filippova, N.Y., Trubikhin, V.M., Popov, S.V., Goncharova, I.A., et al., 2012, Middle-Upper Miocene stratigraphy of the Taman Peninsula, Eastern Paratethys. *Central Europe Journal of Geoscience*, **4** (1), 188–204, <https://doi.org/10.2478/s13533-011-0065-8>.

Raffi, I., 1999, Precision and accuracy of nannofossil biostratigraphic correlation. *Philosophical Transactions of the Royal Society, Series*, **357**: 1975–1993.

Raffi, I., 2002, Revision of the Early-Middle Pleistocene calcareous nannofossil biochronology (1.75–0.85 Ma). *Marine Micropaleontology*, **45**: 25–55.

Raffi, I., and Flores, J.A., 1995, Pleistocene through Miocene calcareous nannofossils from eastern equatorial Pacific Ocean (ODP Leg 138). In: Pisias, N.G., Mayer, L.A., Janecek, T.R., Palmer-Julson, A., and van Handel, T.H. (eds), *Proceedings of the Ocean Drilling Program, Scientific Results*, **138**: 233–286.

Raffi, I., Backman, J., Rio, D., and Shackleton, N.J., 1993, Plio-Pleistocene Nannofossil Biostratigraphy and Calibration to Oxygen Isotope Stratigraphies from Deep Sea Drilling Project Site 607 and Ocean Drilling Program Site 677. *Paleoceanography*, **8** (3), 387–408.

Raffi, I., Rio, D., d'Atri, A., Fornaciari, E., and Rocchetti, S., 1995, Quantitative distribution patterns and biomagnetostratigraphy of middle and late Miocene calcareous nannofossils from equatorial Indian and Pacific Oceans (ODP Legs 115, 130, and 138). In: Pisias, N.G., Mayer, L.A., Janecek, T.R., Palmer-Julson, A., and van Handel, T.H. (eds), *Proceedings of the Ocean Drilling Program, Scientific Results*, **138**: 479–502.

Raffi, I., Mozzato, C., Fornaciari, E., Hilgen, F.J., and Rio, D., 2003, Late Miocene calcareous nannofossil biostratigraphy and astrobiochronology for the Mediterranean region. *Micropaleontology*, **49** (1), 1–26.

Raffi, I., Backman, J., Fornaciari, E., Pälike, H., Rio, D., Lourens, L., et al., 2006, A review of calcareous nannofossil astrobiochronology encompassing the past 25 million years. *Quaternary Science Reviews*, **25**: 3113–3137.

Raffi, I., Agnini, C., Backman, J., Catanzariti, R., and Pälike, H., 2016, A Cenozoic calcareous nannofossil biozonation from low and middle latitudes: a synthesis. *Proc. In 15th INA Conf., Bohol, Philippines. Journal of Nannoplankton Research*, **36**: 121–132.

Raine, J.I., Beu, A.G., Boyes, A.F., Campbell, H.A., Cooper, R.A., Crampton, J.S., et al., 2015a, Revised calibration of the New Zealand geological timescale: NZGT 2015/1. *GNS Science Report 2015/1*, pp. 1–53.

Raine, J.I., Beu, A.G., Boyes, A.F., Campbell, H.A., Cooper, R.A., Crampton, J.S., et al., 2015b, New Zealand geological timescale NZGT 2015/1. *New Zealand Journal of Geology and Geophysics*, **58**: 398–403.

Ramsay, A.S.T., and Baldauf, J.G., 1999, A reassessment of the Southern Ocean biochronology. *Mem. Geological Society*, **18**: 1–122.

Raymo, M.E., Ruddiman, W.F., Backman, J., Clement, B.M., and Martinson, D.G., 1989, Late Pliocene variation in Northern Hemisphere ice sheets and North Atlantic deep water circulation. *Paleoceanography*, **4**: 413–446.

Remane, J., Bassett, M.G., Cowie, J.W., Gohrbandt, K.H., Lane, H.R., Michelson, O., et al., 1996, Revised guidelines for the establishment of global chronostratigraphic standards by the International Commission on Stratigraphy (ICS). *Episodes*, **19**: 77–81.

Richards, K., Van Baak, C.G.C., Athersuch, J., Hoyle, T.M., Stoica, M., Austin, W.E.N., et al., 2018, Palynology and micropalaeontology of the Pliocene-Pleistocene transition in outcrop from the western Caspian Sea, Azerbaijan: potential links with the Mediterranean, Black Sea and the Arctic Ocean? *Palaeogeography, Palaeoclimatology, Palaeoecology*, **511**: 119–143, <https://doi.org/10.1016/j.palaeo.2018.07.018>.

Rio, D., Fornaciari, E., and Raffi, I., 1990a, Late Oligocene through Early Pleistocene calcareous nannofossils from Western Equatorial Indian

Ocean (Leg 115). In: Duncan, R.A., Backman, J., Peterson, L.C., et al., (eds), *Proceedings of the Ocean Drilling Program, Scientific Results,* **115:** 175–253, http://dx.doi.org/10.2973/odp.proc.sr.115.152.1990.

Rio, D., Raffi, I., and Villa, G., 1990b, Pliocene–Pleistocene calcareous nannofossil distribution patterns in the Western Mediterranean. In: Kastens, K., Mascle, J., et al., (eds), *Proceedings of the Ocean Drilling program, Scientific Results,* **107:** 513–533, http://dx.doi. org/10.2973/odp.proc.sr.107.164.1990.

Rio, D., Cita, M.B., Iaccarino, S., Gelati, R., and Gnaccolini, M., 1997, Chapter A5 Langhian, Serravallian, and Tortonian historical strato-types. In: Montanari, A., Odin, G.S., and Coccioni, R. (eds), *Miocene Stratigraphy: An Integrated Approach.* Developments in Palaeontology and Stratigraphy, **15:** 57–87.

Rögl, F., 1998, Palaeogeographic Considerations for Mediterranean and Paratethys Seaways (Oligocene to Miocene). *Annalen des Naturhistorischen Museums Wien,* **99A:** 279–310.

Rögl, F., and Muller, C., 1976, Das Mittelmiozan und die Baden – Sarmat Grenze im Walbersdorf (Burgenland). *Annalen des Naturhistorischen Musem in Wien.,* **80:** 221–232.

Rögl, F., Ćorić, S., Hohenegger, J., Pervesler, P., Roetzel, R., Scholger, R., et al., 2007, Cyclostratigraphy and transgressions at the Early/Middle Miocene (Karpatian/Badenian) boundary in the Austrian Neogene basins (Central Paratethys). *Scripta Facultatis Scientiarum Naturalium Universitatis Masarykianae Brunensis,* **36:** 7–13.

Rook, L., and Martínez-Navarro, B., 2010, Villafranchian: the long story of a Plio-Pleistocene European large mammal biochronologic unit. *Quaternary International,* **219:** 134–144, https://doi.org/10.1016/j. quaint.2010.01.007.

Rook, L., and Torre, D., 1996, The wolf event in western Europe and the beginning of the Late Villafranchian. *Neues Jahrbuch für Geologie und Paläontologie Monatshefte,* **8:** 495–501.

Rosenthal, Y., Holbourn, A.E., Kulhanek, D.K., and the Expedition 363 Scientists, Western Pacific Warm Pool. *Proceedings of the International Ocean Discovery Program,* 363: College Station, TX (International Ocean Discovery Program). https://doi.org/10.14379/iodp.proc.363.101.2018.

Rostovtseva, Y.V., and Rybkina, A.I., 2014, Cyclostratigraphy of Pontian Deposits of the Eastern Paratethys (Zheleznyi Rog Section, Taman Region). *Moscow University Geological Bulletin,* **69** (4), 236–241.

Rostovtseva, Y.V., and Rybkina, A.I., 2017, The Messinian event in the Paratethys: astronomical tuning of the Black Sea Pontian. *Marine and Petroleum Geology,* **80:** 321–332, <https://doi.org/10.1016/j. marpetgeo.2016.12.005>.

Ruddiman, W.F., Backman, J., Baldauf, J., Hooper, P., Keigwin, L., Miller, K., et al., 1987, Leg 94 paleoenvironmental synthesis. In: Ruddiman, W. F., Kidd, R.B., Thomas, E., et al., (eds), *Initial Reports of the Deep Sea Drilling Project,* **44:** 1207–1215.

Rundic, L., Vasic, N., Zivotic, D., Bechtel, A., Knezevic, S., and Cvetkov, V., 2016, The Pliocene Paludina Lake of Pannonian Basin: new evidence from Northern Serbia. *Annales Societatis Geologorum Poloniae,* **86:** 185–209, https://doi.org/10.14241/asgp.2016.003.

Ryan, W.B.F., Cita, M.B., Rawson, M.O., Burckle, L.H., and Saito, T., 1974, A paleomagnetic assignment of Neogene stage boundaries and the development of isochronous datum planes between the Mediterranean, the Pacific and Indian Oceans in order to investigate the response of the world ocean to the Mediterranean 'Salinity Crisis'. *Rivista Italiana di Paleontologia e Stratigrafia,* **80:** 631–688.

Rybkina, A.I., Kern, A.K., and Rostovtseva, Y.V., 2015, New evidence of the age of the lower Maeotian substage of the Eastern Paratethys based on astronomical cycles. *Sedimentary Geology,* **330:** 122–131, <https://doi.org/10.1016/j.sedgeo.2015.10.003>.

Sacchi, M., and Horvath, F., 2002, Towards a new time scale for the Upper Miocene continental series of the Pannonian basin (Central Paratethys). *European Geosciences Union Stephan Mueller Special Publication Series,* **3:** 79–94.

Sacchi, M., Horvath, F., Magyar, I., and Muller, P., 1997, Problems and progress in establishing a Late Neogene Chronostratigraphy for the Central Paratethys. *Neogene Newsletters,* **4:** 37–46.

Saito, T., Burckle, L.H., and Hays, J.D., 1975, Late Miocene to Pleistocene biostratigraphy of equatorial Pacific sediments. In: Saito, T., and Burckle, L. (eds), *Late Neogene Epoch Boundaries.* Micropaleontology Special Paper 1. 226–244.

Sala, B., Masini, F., Ficcarelli, G., Rook, L., and Torre, D., 1992, Mammal dispersal events in the middle and late Pleistocene of Italy and western Europe. *Courier Forschungs-Institut Senckenberg,* **153:** 58–68.

Salvador, A., 1994, *International Stratigraphic Guide—A Guide to Stratigraphic Classification, Terminology and Procedure.* Geological Society of America, pp. 1–207, http://dx.doi.org/10.1130/9780813774022.

Sanfilippo, A., and Nigrini, C., 1998, Code numbers for Cenozoic low latitude radiolarian biostratigraphic zones and GPTS conversion tables. *Marine Micropaleontology,* **33** (1–2), 109–117, https://doi. org/10.1016/S0377-8398(97)00030-3.

Sant, K., Kirscher, U., Reichenbacher, B., Pipperr, M., Jung, D., Doppler, G., et al., 2017a, Late Burdigalian sea retreat from the North Alpine Foreland Basin: new magnetostratigraphic age constraints. *Global and Planetary Change,* **152:** 38–50, <https://doi. org/10.1016/j.gloplacha.2017.02.002>.

Sant, K., Palcu, D.V., Mandic, O., and Krijgsman, W., 2017b, Changing seas in the Early–Middle Miocene of Central Europe: a Mediterranean approach to Paratethyan stratigraphy. *Terra Nova,* **29:** 273–281, https://doi.org/10.1111/ter.12273.

Schlunegger, F., Burbank, D., Matter, A., Engesser, B., and Mödden, C., 1996, Magnetostratigraphic calibration of the Oligocene to Middle Miocene(30-15 Ma) mammal biozones and depositional sequences of the Swiss Molasse Basin. *Eclogae Geologicae Helveticae,* **89:** 753–788.

Schreck, M., Matthiessen, J., and Head, M.J., 2012, A magnetostratigraphic calibration of Middle Miocene through Pliocene dinoflagellate cyst and acritarch events in the Iceland Sea (Ocean Drilling Program Hole 907A). *Review of Palaeobotany and Palynology,* **187:** 66–94.

Schreck, M., Seung-Il, N., Clotten, C., Fahl, K., De Schepper, S., Forwick, M., et al., 2017, Neogene dinoflagellate cysts and acri-tarchs from the high northern latitudes and their relation to sea surface temperature. *Marine Micropaleontology,* **136:** 51–65, https:// doi.org/10.1016/j.marmicro.2017.09.003.

Scotese, C.R., 2014, *Atlas of Neogene Paleogeographic Maps (Mollweide Projection), Maps 1–7, Volume 1, The Cenozoic, PALEOMAP Atlas for ArcGIS,* PALEOMAP Project, Evanston, IL. https://www.acade-mia.edu/11082185/Atlas_of_Neogene_Paleogeographic_Maps.

Senes, J., 1975, Regional stages of the central Paratethys Neogene and the definition of their lower boundaries. In: Senes, J. (ed), *Proceedings of the VIth Congress of the Regional Committee on Mediterranean Neogene Stratigraphy (RCMNS), Bratislava, 1975.* I, pp. 259–265.

Shackleton, N.J., and Crowhurst, S., 1997, Sediment fluxes based on an orbitally tuned time scale 5 Ma to 14 Ma, Site 926. *Proceedings of the Ocean Drilling Program, Scientific Results*, **154**: 69–82.

Shackleton, N.J., and Hall, M.A., 1989, Stable isotope history of the Pleistocene at ODP Site 677. In: Becker, K., Sakai, H., et al., (eds), *Proceedings of the Ocean Drilling Program, Scientific Results,* **111**: 295–316.

Shackleton, N.J., Hall, M.A., 1997. The late Miocene stable isotope record, Site 926. In: Shackleton, N.J., Curry, W.B., Richter, C., and Bralower, T.J. (Eds.), *Proceedings of the Ocean Drilling Program, Scientific Results,* 154: 367–373. https://doi.org/10.2973/odp.proc.sr.154.119.1997

Shackleton, N.J., and Opdyke, N.D., 1973, Oxygen isotope and paleomagnetic stratigraphy of equatorial Pacific core V28–238: oxygen isotope temperatures and ice volumes on a 10^5 year and 10^6 year scale. *Quaternary Research*, **3**: 39–55.

Shackleton, N.J., Baldauf, J.G., Flores, J.-A., Iwai, M., Moore Jr., T.C., Raffi, I., et al., 1995a, Biostratigraphic summary for Leg 138. In: Pisias, N.G., Mayer, L.A., Janecek, T.R., Palmer-Julson, A., and van Andel, T.H. (eds), *Proceedings of the Ocean Drilling program, Scientific Results,* **138**: 517–536.

Shackleton, N.J., Crowhurst, S., Hagelberg, T., Pisias, N.G., and Schneider, D.A., 1995b, A new late Neogene time scale: application to Leg 138 sites. In: Pisias, N.G., Mayer, L.A., Janecek, T.R., Palmer-Julson, A., and van Andel, T.H. (eds), *Proceedings of the Ocean Drilling program, Scientific Results,* **138**: 73–101.

Shackleton, N.J., Crowhurst, S.J., Weedon, G.P., and Laskar, J., 1999, Astronomical calibration of Oligocene-Miocene time. *Philosophical Transactions of the Royal Society*, **357**: 1907–1929.

Shackleton, N.J., Hall, M.A., Raffi, I., Tauxe, L., and Zachos, J.C., 2000, Astronomical calibration age for the Oligocene-Miocene boundary. *Geology*, **28**: 447–450.

Shipboard Scientific Party, 1999, Site 1123: North Chatham Drift – a 20-Ma record of the Pacific Deep Western Boundary Current. In: Carter, R.N., McCave, I.N., Richter, C., Carter, L., et al., (eds), *Proceedings of the Ocean Drilling Program, Initial Reports,* **181**: 1–184, http://www-odp.tamu.edu/publications/181_IR/VOLUME/CHAPTERS/CHAP_07.PDF.

Sierro, F.J., Krijgsman, W., Hilgen, F.J., and Flores, J.A., 2001, The Abad composite (SE Spain): a Mediterranean reference section for the Messinian and the Astronomical Polarity Time Scale (APTS). *Palaeogeography, Palaeoclimatology, Palaeoecology,* **168**: 143–172.

Simmons, M.S., Miller, K.G., Ray, D.C., Davies, A., van Buchem, F.S.P., and Gréselle, B., 2020, Chapter 13 - Phanerozoic eustasy. *In* Gradstein, F.M., Ogg, J.G., Schmitz, M.D., and Ogg, G.M. (eds), *The Geologic Time Scale 2020.* Vol. **1** (this book). Elsevier, Boston, MA.

Simon, D., Palcu, D., Meijer, P., and Krijgsman, W., 2018, The sensitivity of middle Miocene paleoenvironments to changing marine gateways in Central Europe. *Geology*, **47**: 35–38, <https://doi.org/10.1130/G45698.1>.

Sotnikova, M., and Rook, L., 2010, Dispersal of the Canini (Mammalia, Canidae: Caninae) across Eurasia during the Late Miocene to Early Pleistocene. *Quaternary International*, **212**: 86–97.

Spassov, N., 2000, Biochronology and zoogeographic affinities of the Villafranchian Faunas of south Europe. *Historia Naturalis Bulgarica*, **12**: 89–128.

Spassov, N., 2003, The Plio-Pleistocene vertebrate fauna in South-Eastern Europe and the megafaunal migratory waves from the east to Europe. *Revue de Paleobiologie Geneve*, **22**: 197–229.

Spezzaferri, S., 1998, Planktonic foraminifer biostratigraphy and paleoenvironmental implications of Leg 152 sites (East Greenland margin). In: Saunders, A.D., Larsen, H.C., and Wise Jr., S.W. (eds), *Proceedings of the Ocean Drilling Program, Scientific Results,* **152**: 161–189.

Spezzaferri, S., Kucera, M., Pearson, P.N., Wade, B.S., Rappo, S., Poole, C.R., et al., 2015, Fossil and genetic evidence for the polyphyletic nature of the planktonic Foraminifera "Globigerinoides", and description of the new genus *Trilobatus*. *PLoS One*, 1–20, https://doi.org/10.1371/journal.pone.0128108.

Spiegler, D., and Jansen, E., 1989, Planktonic foraminifer biostratigraphy of Norwegian Sea sediments; ODP Leg 104. *Proceedings of the Ocean Drilling Program, Scientific Results,* **104**: 681–696.

Srinivasan, M.S., and Sinha, D.K., 1992, Late Neogene planktonic foraminiferal events of the southwest Pacific and Indian Ocean: a comparison. In: Tsuchi, R., and Ingle Jr., J.C. (eds), *Pacific Neogene: Environment, Evolution and Events*. Tokyo: University of Tokyo Press, pp. 203–220.

Steininger, F., and Seneš, J., 1971, M1 Eggenburgien: Die Eggenburger Schichtengruppe und ihr Stratotypus, *Chronostratigraphie und Neostratotypen. Miozan der Zentralen Paratethys*, **2**. Bratislava: SAV, pp. 1–827, (in German with English abstract).

Steininger, F.F., Aubry, M.-P., Berggren, W.A., Biolzi, M., Borsetti, A.M., Cartlidge, J.E., et al., 1997, The Global Stratotype Section and Point (GSSP) for the base of the Neogene. *Episodes*, **20**: 23–28.

Stevanovic, P.M., 1951, Pontische Stufe im engeren Sinne – Obere Congerienschichten Serbiens und der angrenzenden Gebiete. *Serbische Akademie der Wissenschaften, Sonderausgabe, Mathematisch-Naturwissenschaftliche Klasse, 2,* **187**: 1–351.

Stoica, M., Lazar, I., Krijgsman, W., Vasiliev, I., Jipa, D.C., and Floroiu, A., 2013, Palaeoenvironmental evolution of the East Carpathian fore-deep during the Late Miocene – Early Pliocene (Dacian Basin; Romania). *Global and Planetary Change*, **103**: 135–148, https://doi.org/10.1016/j.gloplacha.2012.04.004.

Stoica, M., Krijgsman, W., Fortuin, A., and Gliozzi, E., 2016, Paratethyan ostracods in the Spanish Lago-Mare: More evidence for interbasinal exchange at high Mediterranean sea level. *Palaeogeography, Palaeoclimatology, Palaeoecology,* **441**: 854–870, <https://doi.org/10.1016/j.palaeo.2015.10.034>.

Studencka, B., Gontsharova, I.A., and Popov, S.V., 1998, The bivalve faunas as a basis for reconstruction of the Middle Miocene history of the Paratethys. *Acta Geologica Polonica*, **48** (3), 285–432.

Ter Borgh, M., Vasiliev, I., Stoica, M., Knežević, S., Maţenco, L., Krijgsman, W., et al., 2013, The isolation of the Pannonian basin (Central Paratethys): New constraints from magnetostratigraphy and biostratigraphy. *Global and Planetary Change,* **103**: 99–118, <https://doi.org/10.1016/j.gloplacha.2012.10.001>.

Ter Borgh, M., Stoica, M., Donselaar, M.E., Matenco, L., and Krijgsman, W., 2014, Miocene connectivity between the Central and Eastern Paratethys: Constraints from the western Dacian Basin. *Palaeogeography, Palaeoclimatology, Palaeoecology,* **41** (2), 45–67, <https://doi.org/10.1016/j.palaeo.2014.07.016>.

Theodoridis, S., 1984, Calcareous nannofossils biozonation of the Miocene and revision of the Helicoliths and Discoasters. *Utrecht Micropaleontological Bulletin*, **32**: 0–268.

Thierstein, H.R., Geitzenauer, K., Molfino, B., and Shackleton, N.J., 1977, Global synchroneity of late Quaternary coccolith datum levels: validation by oxygen isotopes. *Geology*, 5: 400–404.

Thompson, P.R., Bé, A.W.H., Duplessy, J.-C., and Shackleton, N.J., 1979, Disappearance of pink-pigmented Globigerinoides ruber at 120,000 yr BP in the Indian and Pacific oceans. *Nature*, 280: 554–558.

Tiedemann, R., and Franz, S.O., 1997, Deep-water circulation, chemistry, and terrigenous sediment supply in the equatorial Atlantic during the Pliocene, 3.3-2.6 Ma and 5-4.5 Ma. In: Shackleton, N.J., Curry, W.B., Richter, C., and Bralower, T.J. (eds), *Proceedings of the Ocean Drilling Program, Scientific Results,* 154: 299–318.

Toro, I., Agustí, J., and Martínez-Navarro, B., 2003, El Pleistoceno inferior de Barranco León y Fuente Nueva 3, Orce (Granada). Memoria Científica Campañas 1999–2002, *In E.P.G. Arqueología Monografías,* 17. Junta de Andalucía. Consejería de Cultura.

Torre, D., 1987, Pliocene and Pleistocene marine-continental correlations. *Annales Instituti Publici Geologiae Hungarici,* 70: 71–77.

Torre, D., Ficcarelli, G., Masini, F., Rook, L., and Sala, B., 1992, Mammal dispersal events in the early Pleistocene of western Europe. *Courier Forschungs-Institut Senckenberg,* 153: 51–58.

Turco, E., Bambini, A.M., Foresi, L.M., Iaccarino, S., Lirer, F., Mazzei, R., et al., 2002, Middle Miocene high resolution calcareous plankton biostratigraphy at Site 926 (Leg 154, equatorial Atlantic Ocean): paleoecological and paleobiogeographical implications. *Geobios,* 35: 257–276.

Turco, E., Cascella, A., Gennari, R., Hilgen, F.J., Iaccarino, S.M., and Sagnotti, L., 2011a, Integrated stratigraphy of the La Vedova section (Conero Riviera, Italy) and implications for the Burdigalian/Langhian boundary. *Stratigraphy,* 8: 89–110.

Turco, E., Iaccarino, S.M., Foresi, L., Salvatorini, G., Riforgiato, F., and Verducci, M., 2011b, Revisiting the taxonomy of the intermediate stages in the *Globigerinoides* e *Praeorbulina* lineage. *Stratigraphy,* 8: 163–187.

Turco, E., Hüsing, S., Hilgen, F.J., Cascella, A., Gennari, R., Iaccarino, S., et al., 2017, Astronomical tuning of the La Vedova section between 16.3 and 15.0 Ma. Implications for the origin of megabeds and the Langhian GSSP. *Newsletters Stratigraphy,* 50 (1), 1–29, https://doi.org/10.1127/nos/2016/0302.

Turner, G.M., Kamp, P.J.J., McIntyre, A.P., Hayton, S., McGuire, D.M., and Wilson, G.S., 2005, A coherent middle Pliocene magnetostratigraphy, Wanganui Basin, New Zealand. *Journal of the Royal Society of New Zealand,* 35: 197–227.

Van Baak, C.G.C., Vasiliev, I., Stoica, M., Kuiper, K.F., Forte, A.M., Aliyeva, E., et al., 2013, A magnetostratigraphic time frame for Plio-Pleistocene transgressions in the South Caspian Basin, Azerbaijan. *Global and Planetary Change,* 103: 119–134, https://doi.org/10.1016/j.gloplacha.2012.05.004.

Van Baak, C.G.C., Mandic, O., Lazar, I., Stoica, M., and Krijgsman, W., 2015, The Slanicul de Buzau section, a unit stratotype for the Romanian stage of the Dacian Basin (Plio-Pleistocene, Eastern Paratethys). *Palaeogeography, Palaeoclimatology, Palaeoecology,* 440: 594–613, https://doi.org/10.1016/j.palaeo.2015.09.022.

Van Baak, C.G.C., Stoica, M., Grothe, A., Aliyeva, E., and Krijgsman, W., 2016, Mediterranean-Paratethys connectivity during the Messinian salinity crisis. The Pontian of Azerbaijan. *Global and Planetary Change,* 141: 63–81, https://doi.org/10.1016/j.gloplacha.2016.04.005.

Van Baak, C.G.C., Krijgsman, W., Magyar, I., Sztano, O., Golovina, L.A., Grothe, A., et al., 2017, Paratethys response to the Messinian salinity crisis. *Earth-Science Reviews,* 172: 193–223, <https://doi.org/10.1016/j.earscirev.2017.07.015>.

Van Baak, C.G.C., Arjen Grothe, A., Richards, K., Marius Stoica, M., Aliyeva, E., Davies, G.R., et al., 2019, Flooding of the Caspian Sea at the intensification of Northern Hemisphere glaciations. *Global and Planetary Change,* 174: 153–163, <https://doi.org/10.1016/j.gloplacha.2019.01.007>.

Van Couvering, J.A., Castradori, D., Cita, M.B., Hilgen, F.J., and Rio, D., 2000, The base of the Zanclean Stage and of the Pliocene Series. *Episodes,* 23: 179–186.

Van Couvering, J.A., Aubry, M.-P., Berggren, W.A., Gradstein, F.M., Hilgen, F.J., Kent, D.V., et al., 2009, What, if anything, is Quaternary? *Episodes,* 32: 125–126.

Van Dam, J.A., Alcalá, L., Alonso Zarza, A.M., Calvo, J.P., Garcés, M., and Krijgsman, W., 2001, The Upper Miocene mammal record from the Teruel-Alfambra region (Spain): The MN system and continental Stage/Age concepts discussed. *Journal of Vertebrate Paleontology,* 21: 367–385.

Vangengeim, E.A., and Tesakov, A.S., 2013, Late Miocene mammal localities of Eastern Europe and Western Asia: Toward biostratigraphic synthesis. In: Wang, X., Flynn, L.J., and Fortelius, M. (eds), *Asian Neogene Mammal Volume. New York: Columbia University Press.*

Vangengeim, E.A., Pevzner, M.A., and Teakov, A.S., 2005, Ruscinian and lower villafranchian: age of boundaries and position in magnetochronological scale. *Stratigraphy and Geological Correlation,* 13 (5), 530–546.

Vasiliev, I., Krijgsman, W., Langereis, C.G., Panaiotu, C.E., Maţenco, L., and Bertotti, G., 2004, Towards an astrochronological framework for the eastern Paratethys Mio-Pliocene sedimentary sequences of the Focşani basin (Romania). *Earth and Planetary Science Letters,* 227 (3–4), 231–247, https://doi.org/10.1016/j.epsl.2004.09.012.

Vasiliev, I., de Leeuw, A., Filipescu, S., Krijgsman, W., Kuiper, K., Stoica, M., et al., 2010, The age of the Sarmatian–Pannonian transition in the Transylvanian Basin (Central Paratethys). *Palaeogeography, Palaeoclimatology, Palaeoecology,* 297: 54–69, https://doi.org/10.1016/j.palaeo.2010.07.015.

Vasiliev, I., Iosifidi, A.G., Khramov, A.N., Krijgsman, W., Kuiper, K., Langereis, C.G., et al., 2011, Magnetostratigraphy and radio-isotope dating of upper Mioceneelower Pliocene sedimentary successions of the Black Sea Basin (Taman Peninsula, Russia). *Palaeogeography, Palaeoclimatology, Palaeoecology,* 310: 163–175, https://doi.org/10.1016/j.palaeo.2011.06.022.

Vernyhorova, Y.V., and Ryabokon, T.S., 2018, *Maykopskie otlozheniya (oligotsen – nizhniy miotsen) Kerchenskogo poluostrova: istoriya izucheniya, polemika, stratigrafiya [Maikop deposits (Oligocene – lower Miocene) of the Kerch Peninsula: history of the study, controversy, stratigraphy].* Kyiv: NAS of Ukraine, IGS NAS of Ukraine, p. 112, (in Russian).

Vigour, R., and Lazarus, D., 2002, Biostratigraphy of late Miocene–early Pliocene radiolarians from ODP Leg 183 Site 1138. In: Frey, F.A., Coffin, M.F., Wallace, P.J., and Quilty, P.G. (eds), *Proceedings of the Ocean Drilling Program, Scientific Results,* 183: 1–17. [Online]. Available from:, http://wwwodp.tamu.edu/publications/183_SR/VOLUME/CHAPTERS/007.PDF.

Wade, B.S., and Pälike, H., 2004, Oligocene climate dynamics. *Paleoceanography,* 19: PA4019, https://doi.org/10.1029/2004PA001042.

Wade, B.S., Pearson, P.N., Berggren, W.A., and Pälike, H., 2011, Review and revision of Cenozoic tropical planktonic foraminiferal biostratigraphy and calibration to the geomagnetic polarity and astronomical time scale. *Earth-Science Reviews,* 104: 111–142.

Wade, B.S., Olsson, R.K., Pearson, P.N., Huber, B.T. and Berggren, W.A., 2018 (Eds.), *Atlas of Oligocene Planktonic Foraminifera,* Cushman Foundation Special Publication, No. 46, 528 pp.

Weaver, P.R.E., and Clement, B.M., 1987, Synchroneity of Pliocene planktonic foraminiferal datums in the North Atlantic. *Marine Micropaleontology,* **10**: 295–307.

Wei, W., 1993, Calibration of upper Pliocene-lower Pleistocene nannofossil events with oxygen isotope stratigraphy. *Paleoceanography,* **8**: 85–99.

Westerhold, T., Röhl, U., and Laskar, J., 2012, Time scale controversy: accurate orbital calibration of the early Paleogene. *Geochemistry Geophysics Geosystems,* **13**: Q06015.

Wilkens, R.H., Westerhold, T., Drury, A.J., Lyle, M., Gorgas, T., and Tian, J., 2017, Revisiting the Ceara Rise, equatorial Atlantic Ocean: isotope stratigraphy of ODP Leg 154 from 0 to 5 Ma. *Climate of the Past,* **13** (7), 779–793, <https://doi.org/10.5194/cp-13-779-2017>.

Williams, G.L., and Manum, S.B., 1999, Oligocene-early Miocene dinocyst stratigraphy of Hole 985A (Norwegian Sea). In: Raymo, M.E., Jansen, E., Blum, P., and Herbert, T.D. (eds), *Proceedings of the Ocean Drilling Program, Scientific Results,* **162**: 99–109.

Woodruff, F., and Savin, S.M., 1991, Mid-Miocene isotope stratigraphy in the deep sea: high-resolution correlations, paleoclimatic cycles, and sediment preservation. *Paleoceanography,* **6**: 755–806.

Wright, J.D., and Miller, K.G., 1992, Miocene stable isotope stratigraphy, Site 747, Kerguelen Plateau. In: Wise Jr., S.W., Schlich, R., et al., (eds), *Proceedings of the Ocean Drilling Program, Scientific Results.* **120**: 855–866.

Yanagisawa, Y., and Akiba, F., 1998, Refined Neogene diatom biostratigraphy for the northwest Pacific around Japan, with an introduction of code numbers for selected diatom biohorizons. *Journal of the Geological Society of Japan,* **104**: 395–414.

Zachariasse, W.J., Gudjonsson, L., Hilgen, F.J., Langereis, C.G., Lourens, L. J., Verhallen, P.J.J.M., et al., 1989, Late Gauss to Early Matuyama invasions of Neogloboquadrina atlantica in the Mediterranean and associated record of climatic change. *Paleoceanography,* **5** (2), 239–252.

Zachos, J.C., Shackleton, N.J., Revenaugh, J.S., Pälike, H., and Flower, B.P., 2001, Climate response to orbital forcing across the Oligocene-Miocene Boundary. *Science,* **292**: 274–278.

Zachos, J.C., Kroon, D., Blum, P., et al., 2004, *Proceedings of the Ocean Drilling Program, Initial Reports,* **208**, https://doi.org/10.2973/odp.proc.ir.208.2004.

Zachos, J.C., Dickens, G.R., and Zeebe, R.E., 2008, An early Cenozoic perspective on greenhouse warming and carbon-cycle dynamics. *Nature,* **451**: 279–283.

Zeeden, C., Frederik Hilgen, F., Westerhold, T., Lourens, L., Röhl, U., and Bickert, T., 2013, Revised Miocene splice, astronomical tuning and calcareous plankton biochronology of ODP Site 926 between 5 and 14.4 Ma. *Palaeogeography, Palaeoclimtology, Palaecology,* **369**: 430–451, http://dx.doi.org/10.1016/j.palaeo.2012.11.009.

Zhang, J., Miller, K.G., and Berggren, W.A., 1993, Neogene planktonic foraminiferal biostratigraphy of the northeastern Gulf of Mexico. *Micropaleontology,* **39**: 299–326.

Zijderveld, J.D.A., Hilgen, F.J., Langereis, C.G., Verhallen, P.J.J.M., and Zachariasse, W.J., 1991, Integrated magnetostratigraphy and biostratigraphy of the upper Pliocene-lower Pleistocene from the Monte Singa and Crotone areas in Calabria, Italy. *Earth and Planetary Science Letters,* **107**: 697–714.

P.L. Gibbard and M.J. Head

Chapter 30

The Quaternary Period

21ka Quaternary

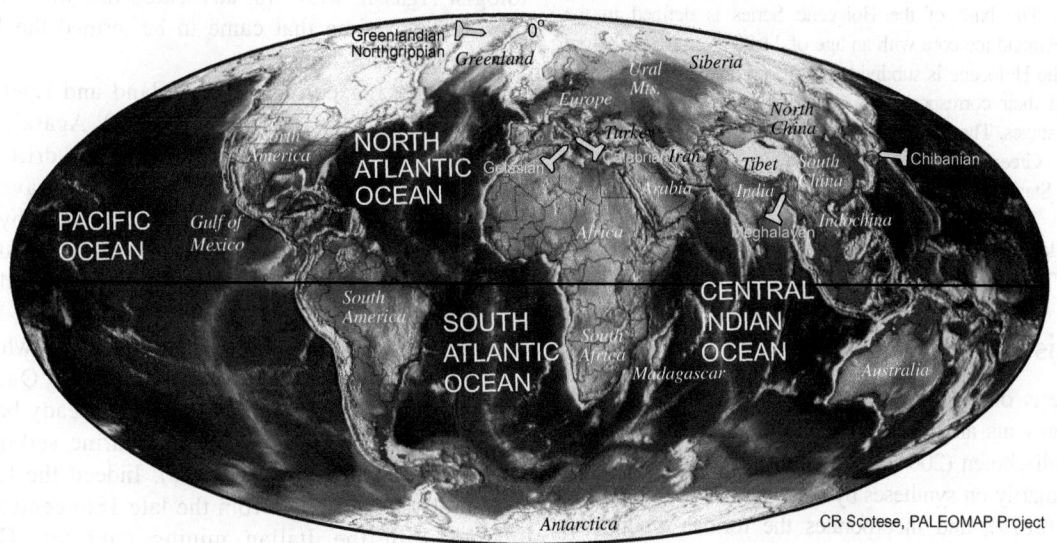

CR Scotese, PALEOMAP Project

Chapter outline

30.1 This chapter 1218
30.2 Evolution of terminology 1218
30.3 The Plio–Pleistocene boundary and
definition of the Quaternary 1219
30.4 Subdivision of the Pleistocene 1223
    30.4.1 Lower Pleistocene Subseries 1223
    30.4.2 Middle Pleistocene Subseries 1225
    30.4.3 Upper Pleistocene Subseries 1227
30.5 Pleistocene/Holocene boundary 1229
30.6 Holocene Series 1232
30.7 Subdivision of the Holocene 1232
    30.7.1 Greenlandian Stage, Lower Holocene
Subseries 1232
    30.7.2 Northgrippian Stage, Middle Holocene
Subseries 1232

30.7.3 Meghalayan Stage, Upper Holocene
Subseries 1234
30.8 Anthropocene Series 1234
30.9 Terrestrial records 1236
    30.9.1 Glacial deposits and climatostratigraphy 1236
    30.9.2 Loess deposits 1238
    30.9.3 Ice cores 1238
    30.9.4 Speleothems 1240
    30.9.5 Long Lake records 1240
    30.9.6 Mammal biochronology and
the Villafranchian Stage 1240
30.10 Ocean–sediment records 1241
30.11 Land–sea correlation 1243
30.12 Quaternary dating methods 1244
Bibliography 1247

Geologic Time Scale 2020. DOI: https://doi.org/10.1016/B978-0-12-824360-2.00030-9

Abstract

The Quaternary System/Period, comprising the Holocene and Pleistocene series/epochs, encompasses the last ~2.6 Myr during which time Earth's climate was strongly influenced by bipolar glaciation, and the genus *Homo* first appeared and diversified. The base of the Quaternary System and Pleistocene Series is defined by the Global Boundary Stratotype Section and Point (GSSP) for the Gelasian Stage at Monte San Nicola in Italy with an age of 2.58 Ma. The Calabrian Stage of the Lower Pleistocene Subseries is defined by a GSSP at the Vrica section, also in Italy, with an age of 1.80 Ma. The Chibanian Stage of the Middle Pleistocene Subseries is defined by a GSSP at the Chiba section in Japan with an age of 0.774 Ma. The Upper Pleistocene subseries is defined in name only with an age of ~129 ka. The base of the Holocene Series is defined in the NGRIP2 Greenland ice core with an age of 11 700 years b2k (before CE 2000). The Holocene is subdivided into lower, middle, and upper subseries and their corresponding Greenlandian, Northgrippian, and Meghalayan stages. The Northgrippian Stage is defined by a GSSP in the NGRIP1 Greenland ice core with an age of 8.2 ka, and the Meghalayan Stage by a GSSP in a speleothem from India with an age of 4.2 ka. These are the only GSSPs ever to be defined in an ice core or a speleothem.

30.1 This chapter

This synthesis of the formal subdivision of the Quaternary System represents an update of earlier versions by Gibbard and van Kolfschoten (2004) and Pillans and Gibbard (2012). It relies primarily on syntheses by Head and Gibbard (2015a) and Head (2019), and incorporates the formal subdivision of the Pleistocene Series into Lower, Middle, and Upper Pleistocene subseries and their corresponding stages, a development ratified in 2020, and incorporates the formal subdivision of the Holocene Series into Lower, Middle, and Upper Holocene subseries and their corresponding stages, a development ratified in 2018 (Walker et al., 2018, 2019a,b). Subseries/subepochs are traditionally used in a formal sense in the Quaternary (Head et al., 2017), and subseries are now ratified for both the Pleistocene and Holocene.

30.2 Evolution of terminology

The classification and interpretation of the youngest stratigraphic successions, variously known as Pleistocene, Holocene, or Quaternary, have been a matter of much debate. During the first two decades of the 19th century, many such successions were attributed to the biblical flood (the "diluvial" theory). This hypothesis could account for unconsolidated sediments that rested unconformable on Tertiary rocks and capped hills, and that commonly contained exotic boulders and the remains of animals, many still extant. A flood origin for the "diluvium" was the accepted paradigm by most eminent geologists of the time, including Buckland and Sedgwick.

During voyages of polar exploration, floating ice had frequently been seen transporting exotic materials, providing an explanation for the boulders in diluvium, and reinforcing the diluvial theory. This explanation led to the adoption of the term "drift" to characterize the sediments. However, geologists working in the Alps and northern Europe had been struck by the extraordinary similarity of the "drift" deposits, and their associated landforms, to those being formed by modern mountain glaciers. Several observers, including Perraudin, Venetz-Sitten, and de Charpentier, proposed that the glaciers had formerly been more extensive, but it was the paleontologist Agassiz who first advocated that this expansion represented a time that came to be termed the Ice Age by Goethe and Schimper.

After having convinced Buckland and Lyell of the validity of his Glacial Theory in 1840, Agassiz's ideas became progressively accepted. The term drift became established for the widespread sands, gravels, and boulder clays thought to have been deposited by glacial ice. Meanwhile, Lyell had already proposed the term Pleistocene in 1839 for the post-Pliocene period closest to the present. He defined this period on the basis of its molluscan fauna, the constituents of which are mostly still extant. However, the term Quaternary (*Quaternaire* or *Tertiaire récent*) had already been proposed in 1829 by Desnoyers for marine sediments in the Seine Basin (Gibbard, 2019). Indeed the term had essentially been in use from the late 18th century, originating with the Italian mining engineer, Giovanni Arduino (1714−95), who distinguished four separate stages or "orders": primary, secondary, tertiary, and a "fourth-order" (*quarto ordine*), comprising the Artesian Alps, the Alpine foothills, the sub-Alpine hills, and the Po Plain, respectively (Schneer, 1969; Ellenberger, 1994; Gibbard, 2019).

Both terms—Pleistocene and Quaternary—became synonymous with the Ice Age and also with the period during which humans evolved. However, unlike the Pleistocene concept, the span of the Quaternary included Lyell's original "recent" that refers to the percentage of living Mollusca. Lyell's "recent" was later replaced by the term Holocene (meaning "wholly recent") that was defined by Gervais (1867−1869) "for the post-diluvial deposits approximately corresponding to the post-glacial period" (Bourdier, 1957). The Holocene Period was originally considered to represent a fifth division or *Quinquennaire* (Parandier, 1891), but this was deemed "excessive"; details are given in Bourdier (1957) and de Lumley (1976).

Because the terms primary and secondary have been abandoned and attempts have been made to suppress the Tertiary (Head et al., 2008a), the continued use of Quaternary has been regarded by some stratigraphers as somewhat archaic. Alternative terms, such as *Anthropogene* (extensively

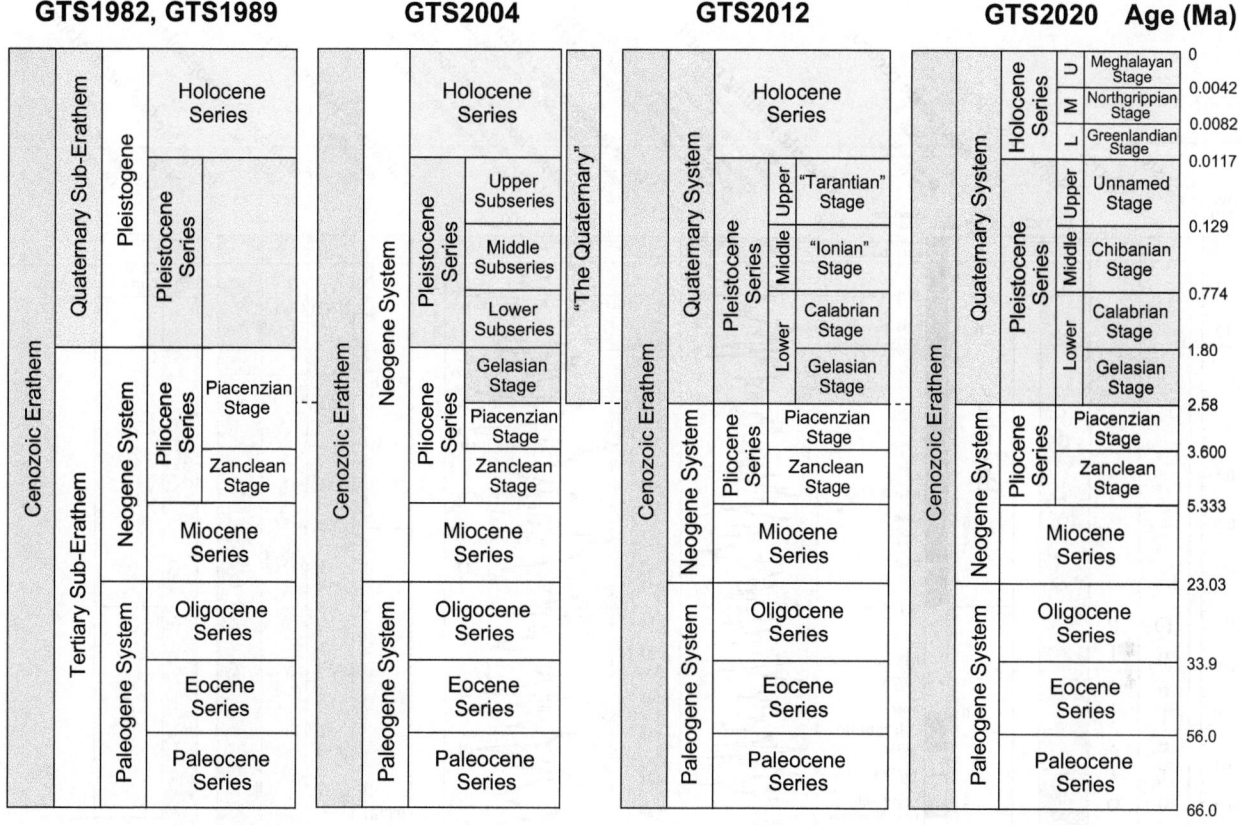

| GTS1982, GTS1989 | GTS2004 | GTS2012 | GTS2020 | Age (Ma) |

FIGURE 30.1 History of Quaternary chronostratigraphy divisions since 1980 (Harland et al., 1982; Harland et al., 1990; Gradstein et al., 2004; Gradstein et al., 2012; this work as based on Head, 2019).

used in the former USSR) or *Pleistogene* (suggested by Harland et al., 1990), have been proposed, but neither has found favor. Subsequently, Gradstein et al. (2004) did not include Quaternary as the youngest period of the Cenozoic Era. Rather, they designated the Quaternary as an informal climatostratigraphic unit, spanning

the interval of oscillating climatic extremes (glacial and interglacial episodes) that was initiated at about 2.6 Ma, therefore encompassing the Holocene and Pleistocene epochs and Gelasian Stage of [the] late Pliocene.

Gibbard and van Kolfschoten (2004)

After exhaustive discussions between the International Commission on Stratigraphy (ICS) and the International Union for Quaternary Research (INQUA), a formal decision on the chronostratigraphic status of the Quaternary was reached in 2009, resulting in the ratification of the Quaternary Period/System and revised base of the Pleistocene Epoch/Series (Gibbard and Head, 2010; Gibbard et al., 2010).

The path leading to the final ratification of the Quaternary is summarized in Fig. 30.1, and a full history of this event is described in Head and Gibbard (2015a).

30.3 The Plio–Pleistocene boundary and definition of the Quaternary

In 1948 at the 18th International Geological Congress in London, an agreement was reached to place the Pliocene–Pleistocene boundary

at the horizon of the first indications of climatic deterioration in the Italian Neogene succession.

King and Oakley (1949)

Furthermore, it was agreed that the boundary should be based on marine faunas and that the Lower Pleistocene should include not only the marine sediments of the Italian Calabrian Stage (a stage introduced by Gignoux, 1910) but also continental deposits of the Villafranchian Stage.

The initial Calabrian boundary was thought to be marked by the first appearance of the cold-water mollusk *Arctica islandica* and the cold-water foraminifer *Hyalinea balthica* (Sibrava, 1978), but Arias et al. (1980) found that *A. islandica* and *H. balthica* both appear at different times, and as it happens earlier than the finally agreed boundary. Subsequently, while various sections in southern Italy competed for the position of stratotype, it became clear that marine microfossils in conjunction with magnetostratigraphy were the best criteria

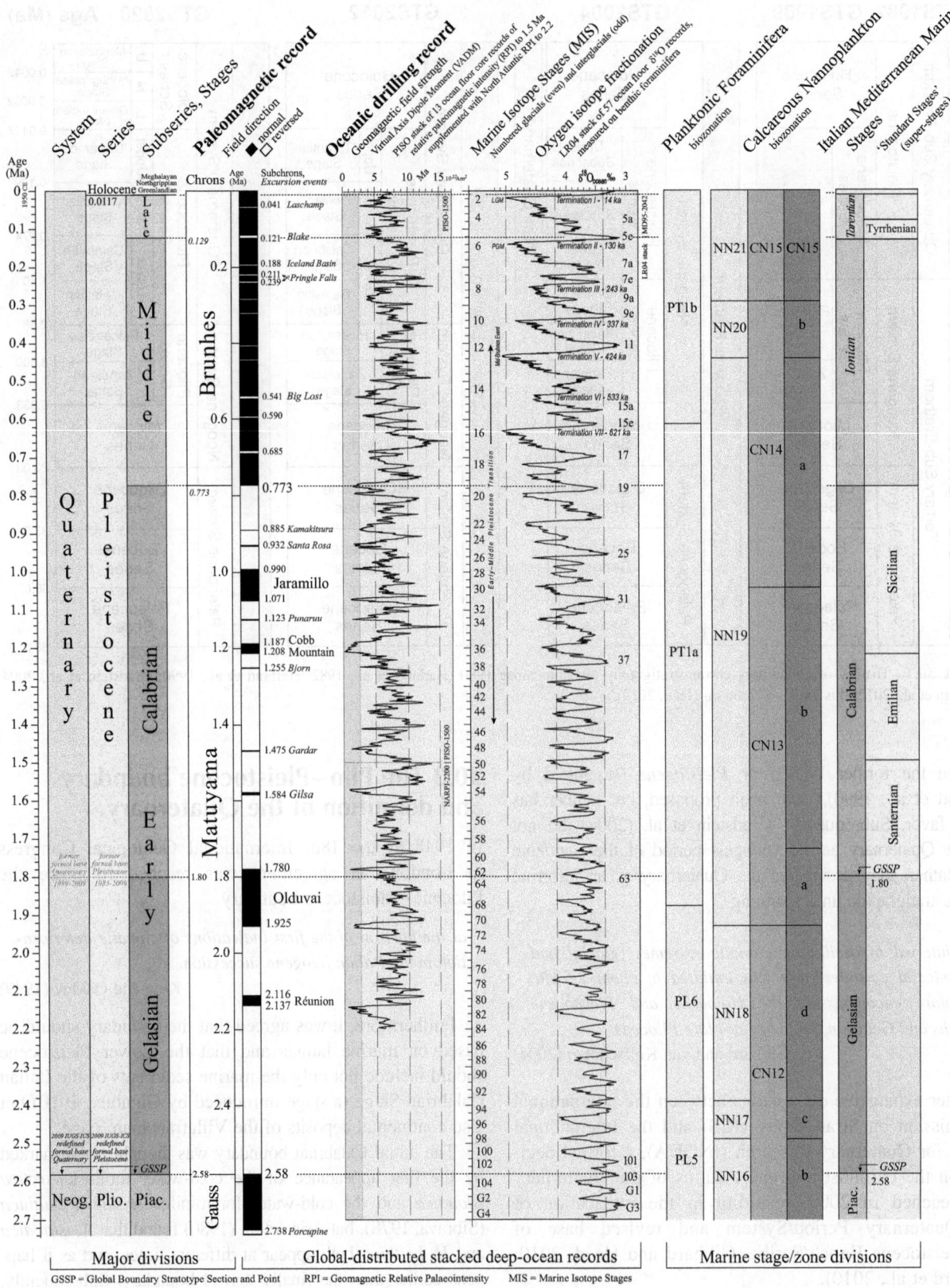

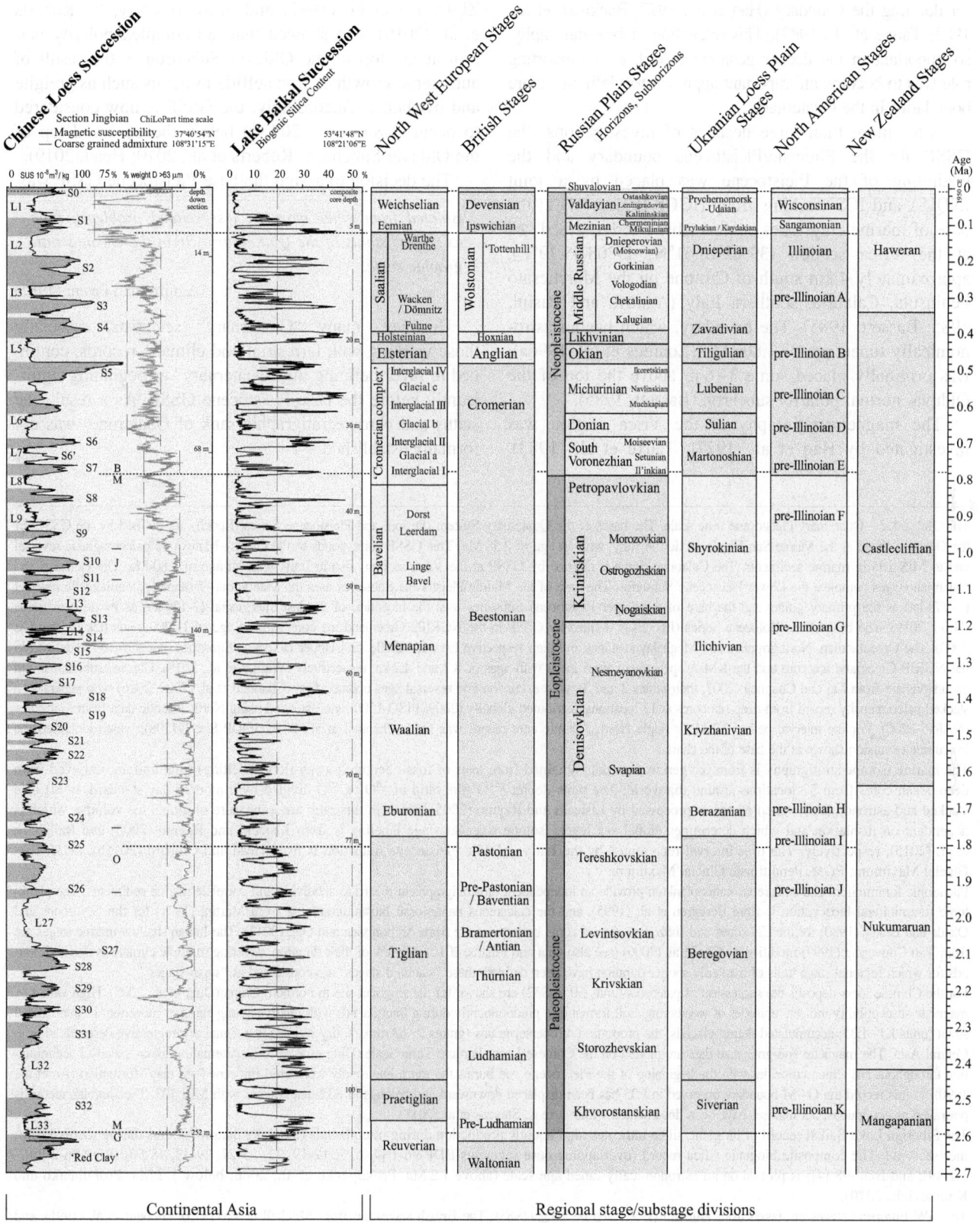

FIGURE 30.2 (Continued)

for defining the boundary (Haq et al., 1977; Backman et al., 1983; Tauxe et al., 1983). This relegation of biostratigraphy, so important in the deeper geologic record, to a supporting role was to become an important approach for defining future boundaries in the Quaternary.

After more than three decades of investigations, the GSSP for the Pliocene/Pleistocene boundary and the beginning of the Pleistocene was placed by a joint INQUA and ICS working group (IGCP Project 41) at the base of marine claystones conformably overlying bed "e" at the Vrica section (39°02′18.61″N, 17°08′05.79″E), approximately 4 km south of Crotone on the Marchesato Peninsula, Calabria, southern Italy (Aguirre and Pasini, 1985; Bassett, 1985). The boundary, which has an astronomically tuned age of 1.806 Ma (Lourens et al., 1996a), was originally placed some 3−6 m above the top of the Olduvai normal polarity subchron (Bassett, 1985).

The magnetostratigraphy of the Vrica section was investigated by Haq et al. (1977), Tauxe et al. (1983), Zijderveld et al. (1991) and, most recently, by Roberts et al. (2010) who showed that the complex polarity pattern at the top of the Olduvai Subchron is the result of authigenic growth of iron sulfide minerals such as greigite and pyrrhotite. Accordingly, the GSSP is now considered to occur ∼8 m (= ∼20 kyr) below the observed top of the Olduvai Subchron (Roberts et al., 2010; Head, 2019).

The decision to assign the base-Pleistocene GSSP was

isolated from other more or less related problems, such as ... the status of the Quaternary within the chronostratigraphic scale.

Aguirre and Pasini (1985)

However, many "Quaternary" scientists, especially those working with terrestrial and climatic records, continued to favor defining the "Quaternary" as beginning significantly before the base-Pleistocene GSSP. As a result, the status and chronostratigraphic rank of Quaternary was not formally established.

FIGURE 30.2   Quaternary/Pleistocene time scale. The bases of the Quaternary System (Period) and Pleistocene Series (Epoch) are defined by the GSSP for the Gelasian Stage at the Monte San Nicola section in Italy, with an age of 2.58 Ma. This GSSP corresponds to the Gauss−Matuyama paleomagnetic reversal and to MIS 103 in marine sediments. The Calabrian Stage is defined by GSSP at the Vrica section, also in Italy, with an age of 1.80 Ma. The Gelasian and Calabrian stages complete the Lower Pleistocene Subseries. The base of the Middle Pleistocene Subseries uses the Matuyama−Brunhes paleomagnetic reversal (∼773 ka) as the primary guide, and the base of the Upper Pleistocene Subseries uses the inception of the last interglacial (∼129 ka) as its primary guide (Head, 2019). The base of the Holocene Series (Epoch) is defined by GSSP in the NGRIP2 Greenland ice core with an age of 11,700 years b2k (before CE 2000). The Greenlandian, Northgrippian, and Meghalayan stages and their respective Lower, Middle, and Upper Holocene Subseries (not shown) are defined in the NGRIP Greenland ice core and the KM-A speleothem from India with ages of 8.2 and 4.2 ka, respectively (Walker et al., 2018). Chrons, subchrons, and excursions are from Laj and Channell (2007, their tables 2 and 3), with excursion and reversal ages updated from Channell et al. (2009, 2016) who generated a stacked paleointensity record from deep-sea cores at 13 locations distributed globally (2009: PISO-1500) and extended using North Atlantic data from four cores (NARPI-2200). For the interval before 2.2 Ma a single North Atlantic core record was used (Channell et al., 2016: IODP Site U1308), which identified the Porcupine excursion shown at the base of the chart.

The marine isotope stratigraphy is from oxygen isotope data obtained from tests of fossil benthic (ocean-floor dwelling) foraminifera, retrieved from deep-ocean cores from 57 locations around the world. The plots depict $\delta^{18}O$ (the ratio of ^{18}O vs ^{16}O divided by a modern-day standard, in ‰) of a stacked and astronomically tuned record as processed by Lisiecki and Raymo (2005). Shifts in this ratio are a measure of global ice volume, which is dependent on insolation and which determines global sea level. Isotope stage/substage labeling is from Lisiecki and Raymo (2005) and Railsback et al. (2015), respectively. The time interval represented by the Early−Middle Pleistocene transition is from Head and Gibbard (2015b). *LGM*, Last Glacial Maximum; *PGM*, Penultimate Glacial Maximum.

Planktonic foraminifera and calcareous nannoplankton provide an independent biostratigraphical means to subdivide and correlate marine sediments. The planktonic foraminiferal biozonation is from Berggren et al. (1995), and the calcareous nannofossil biostratigraphy is from Martini (1971) for the NN zones and Okada and Bukry (1980) for the CN zones and subzones: ages of zonal boundaries are from Anthonissen and Ogg (2012). The Italian shallow marine stages are from Van Couvering (1997) modified by Cita et al. (2006) (see also Cita and Pillans, 2010). In view of their duration, covering multiple climate cycles and periods for which regional stage units of markedly shorter duration have been defined, these "standard stages" are considered as "super stages."

For the Chinese loess deposits the succession of paleosols (units S0 to S32) are shown for the Jingbian site in northern China (Ding et al., 2005). High values of magnetic susceptibility indicate episodes of weathering (soil formation), predominantly during interglacials with relative strong summer monsoon. In intercalated strata (units L1−L33; accumulated during glacials) the proportion of coarser grains (grains > 63 mm, % dry weight) is a signal of progressive desertification in Central Asia. The magnetic and grain-size data are plotted on the Chinese Loess Particle Time Scale (Ding et al., 2002). Alternating loess−paleosol accumulation throughout NE China coincides with the beginning of the Pleistocene and buries the more intensively weathered Pliocene "red clay" formation (An et al., 1990). In this record the G−M boundary observed in L33 has been displaced downward and belongs in S32 that equates with MIS 103. The Jingbian record is portrayed as continuous, but hiatuses have been documented within it (e.g., Stevens et al., 2018).

The Siberian Lake Baikal record of biogenic silica indicates high aquatic production during interglacials (i.e., lake diatom blooms during ice-free summer seasons). The composite biogenic silica record (overlapping core segments BDP-96-1 [4, 6, 7, 12-19, 22, 25-27, 29-33, 48-50]; BDP-96-2 [GC-1,1-49]; and BDP-98 [4]) is plotted on an astronomically tuned age scale (above 1.2 Ma, Prokopenko et al., 2006; below 1.2 Ma, Prokopenko and Khursevich, 2010).

The NW European stages are taken from Zagwijn (1992) and de Jong (1988). The British stages are from Mitchell et al. (1973), Gibbard et al. (1991), and Bowen (1999). The Russian Plain divisional scheme is from Shik (2014) and Shik et al. (2015). The Ukrainian Loess Plain record is based on van Kolfschoten and Gerasimenko (2006). The Russian−Ukrainian Plain schemes recognize the Neopleistocene equivalent to Middle and Late Pleistocene Subseries, and the Eopleistocene equivalent to the Calabrian. The term Paleopleistocene (Tesakov et al., 2015) is the equivalent of the Gelasian. The North American stages are taken from Richmond (unpublished). The New Zealand stages are from Pillans (1991) and Beu (2004) (modified from Cohen and Gibbard, 2019).

The London 1948 IUGS recommendations included the notion that the base-Pleistocene boundary should be placed at the first evidence of climatic cooling. However, the Vrica GSSP boundary level is not the first severe cold-climate oscillation of the late Cenozoic. While the first evidence of continental glaciation in the Northern Hemisphere comes from ice-rafted debris in the Greenland Sea in the mid-Paleogene, around 44 Ma (Tripati et al., 2008), and slightly earlier in the Arctic Ocean (De Schepper et al., 2014), it is now well established that the intensification of Northern Hemisphere glaciation at ~2.7–2.6 Ma is the transition that best characterizes the beginning of the Quaternary/Pleistocene ice ages (Gibbard et al., 2005; Head et al., 2008b). This major cooling takes place at a stratigraphic position close to the base of the Gelasian Stage (Rio et al., 1998), the GSSP for which is defined in marine sediments at Monte San Nicola in Sicily and corresponds to Marine Isotope Stage (MIS) 103 that allows precise recognition in ocean sediments (see discussion in Suc et al., 1997).

This older level corresponds to the Gauss/Matuyama magnetic epoch boundary (~2.59 Ma) as well as being approximately coeval with the base of the Netherlands' terrestrial Praetiglian Stage, the base of the New Zealand Nukumaruan Stage, and the base of the classic Chinese loess succession (Fig. 30.2). The event is clearly defined in the marine oxygen isotope stratigraphy and only slightly postdates the first major influx of ice-rafted debris into the middle latitude of the North Atlantic at around 2.7 Ma (De Schepper et al., 2014). Importantly, a major climatic reorganization across the middle and higher latitudes of the Northern Hemisphere occurs within MIS 104 at ~2.6 Ma, just before the Gauss/Matuyama reversal (Hennissen et al., 2014, 2015; Head, 2019). The fossil mammalian record also shows changes that are obvious near the Gauss/Matuyama reversal marking the boundary between the Early and Middle Villafranchian (Rook and Martínez-Navarro, 2010).

In June 2009 the Executive Committee of the International Union of Geological Sciences (IUGS) formally ratified a proposal by the ICS to define the base of the Quaternary System at the GSSP (see Fig. 30.18 at the end of this chapter) for the Gelasian Stage at Monte San Nicola in Italy (Gibbard and Head, 2010; Gibbard et al., 2010). At the same time the base of the Pleistocene was also lowered to coincide with the Gelasian GSSP. Thus the Gelasian Stage was transferred from the Pliocene Series to the Pleistocene Series (Fig. 30.1). This brought closure to a long and at times contentious process (Head and Gibbard, 2015a).

30.4 Subdivision of the Pleistocene

Two major types of subdivisions have been proposed for the Pleistocene Series: regional and global. A subdivision at regional stage level has been advocated by workers based on sections in elevated shallow marine sediments in Italy (Fig. 30.2). Many other regional schemes exist. In addition, a global subdivision is currently under development, comprising stages and subepochs (subseries), and the first unit to be incorporated is the Gelasian Stage (Fig. 30.3) that now serves as the basal stage of the Pleistocene. At the same time, however, earth scientists concerned with terrestrial and to a lesser extent shallow marine successions have adopted regional subdivision schemes. The regional schemes have found favor despite the difficulties of worldwide correlation. In these schemes, larger, subseries (subepoch) scale units have been adopted. For example, in the former USSR, and particularly in European Russia, the Pleistocene is divided into the Eopleistocene, equivalent to the Early Pleistocene Subseries, and the Neopleistocene, equivalent to the Middle and Late Pleistocene subseries (Anonymous, 1982–1984; Krasnenkov et al., 1997). The Neopleistocene itself has lower, middle, and upper subdivisions, of which the Upper Neopleistocene is exactly equivalent to the Upper Pleistocene (e.g., Tesakov et al., 2015; Head and Gibbard, 2015a and references therein).

A quasiformal tripartite subdivision of the Pleistocene into lower (Early), middle, and upper (Late) has been in use since the 1930s. The first usage of the terms Lower, Middle, and Upper Pleistocene was at the second International Union for Quaternary Research (INQUA) Congress in Leningrad in 1932 (Woldstedt, 1962), although they may have been used in a loose way before this time. Their first use in a formal sense in English was by Zeuner (1935, 1959) and Hopwood (1935) and was based on characteristic assemblages of vertebrate fossils in the European succession. Subepochs (subseries) for the Pleistocene are being defined along with their corresponding stages. They are outlined next.

30.4.1 Lower Pleistocene Subseries

30.4.1.1 Gelasian Stage

The Monte San Nicola section, near Gela, in Sicily, Italy (37°8′45.64″N, 14°12′15.22′E) serves as the GSSP for the Gelasian Stage (Fig. 30.3). It was ratified in August, 1996 (Rio et al., 1998) as, what was then, the uppermost stage of the Pliocene, to represent an interval time beginning with the intensification Northern Hemisphere glaciation. This GSSP was ratified in June, 2009, to serve also as the base of the Quaternary Period and Pleistocene Epoch (Gibbard and Head, 2010).

The GSSP is placed at the base of a marl layer immediately overlying the regionally distinctive sapropelic Nicola bed that is assigned to Mediterranean Precession-Related Cycle (MPRC) 250 (Hilgen, 1991). The Nicola bed is the highest in a cluster of six sapropelic layers occurring at a time of maximum eccentricity, with each layer representing a precession minimum (insolation maximum). The Nicola bed corresponds to the greatest level of summer insolation within this cluster that explains its prominence.

Base of the Gelasian Stage of the Lower Pleistocene Subseries and Pleistocene Series of the Quaternary System at Monte San Nicola, Italy

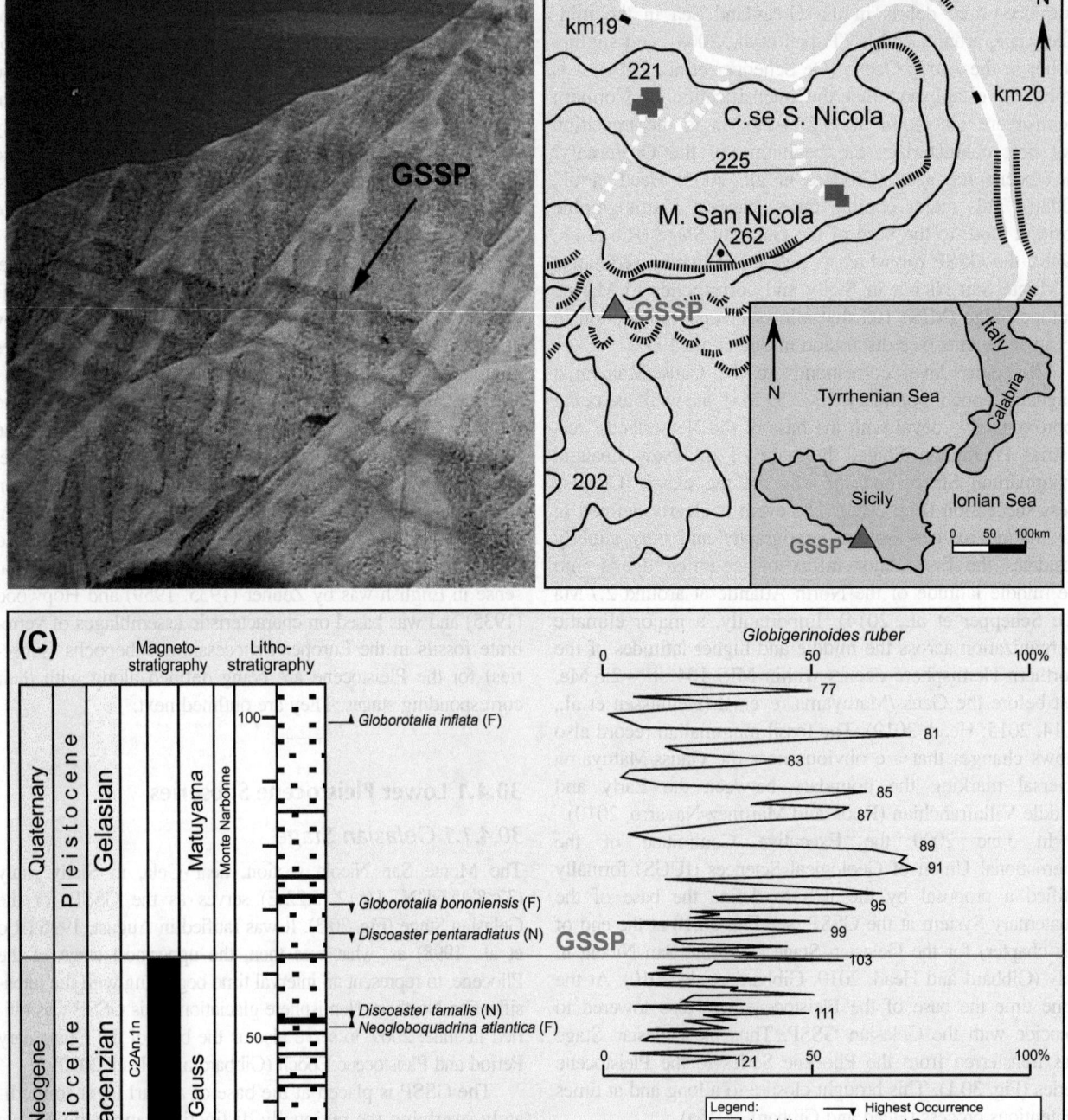

FIGURE 30.3 The GSSP for the Gelasian Stage is defined in marine deposits at Monte San Nicola in Sicily and can be correlated with MIS 103 in the ocean sediments (see Section 30.2). The section is shown with a photograph and stratigraphy. MIS, Marine Isotope Stage. Magnetostratigraphy is modified from Channell et al. (1992) and Head (2019).

The Nicola bed has a midpoint astronomical age of 2.588 Ma (Lourens et al., 1996b; Rio et al., 1998), but the slightly younger GSSP rounds to 2.58 Ma. The Nicola bed and GSSP both occur within MIS 103. The Gauss–Matuyama paleomagnetic reversal serves as a prominent guide to the GSSP both in marine and terrestrial successions. This reversal was initially reported to occur about 1 m (20 kyr) below the GSSP (Rio et al., 1998), which would have placed it in MIS 104. In North Atlantic cores, however, the Gauss–Matuyama boundary occurs within the early part of MIS 103 (Ohno et al., 2012; Lawrence et al., 2013). A reanalysis of the literature indicates that the Gauss–Matuyama reversal may indeed align with the Nicola bed (and therefore the GSSP) at Monte San Nicola (Head, 2019).

30.4.1.2 Calabrian Stage
The Calabrian is defined by the GSSP at the Vrica section (39°02′18.61″N, 17°08′05.79″E) in the province of Crotone, Calabria, southern Italy (Fig. 30.4). The GSSP was initially ratified in May, 1985 to define the base of the Pleistocene (Aguirre and Pasini, 1985; Pasini and Colalongo, 1997), although no corresponding stage was defined at that time (Cita et al., 2008). The Calabrian has a long history of use beginning with its introduction by Gignoux (1910) but was only recently proposed as a formal global stage (Cita et al., 2006, 2008). Following the lowering of the base of the Pleistocene in 2009, an opportunity was seen to subdivide the Lower Pleistocene into two stages, the Gelasian and the newly defined Calabrian (Fig. 30.1). Accordingly, in December 2011, the IUGS ratified the Calabrian Stage with its base defined by the former Pleistocene GSSP at the Vrica section (Cita et al., 2012).

The GSSP is placed at the base of the marl bed immediately overlying sapropelic bed "e," which is assigned to MPRC 176 (Lourens et al., 1996a) with a midpoint astronomical age of 1.806 Ma (Lourens et al., 2004). The GSSP is dated to 1.80 Ma and coincides with the transition from MIS 65–64 (Cita et al., 2012). It occurs ~8 m below the observed top of the Olduvai Subchron, but as noted earlier, diagenetic overprinting obscures the precise position of this polarity reversal (Roberts et al., 2010). Remagnetization may have been caused by sulfidic fluids generated during tectonism along the Calabrian arc (Roberts et al., 2010). Elsewhere, the top of the Olduvai Subchron is consistently placed within MIS 63, which would be consistent with the GSSP occurring ~8 m below it (Head, 2019).

30.4.2 Middle Pleistocene Subseries
The Early–Middle Pleistocene transition marks a fundamental change in Earth's climate state. It is characterized by, interalia, an increase in the amplitude of climate

oscillations, an increasingly asymmetrical waveform, and a shift toward a quasi-100-kyr frequency. The Matuyama/Brunhes paleomagnetic boundary, with an age of ~773 ka (Channell et al., 2010; Head and Gibbard, 2015b), conveniently falls at the approximate midpoint of this transition (Head and Gibbard, 2005, 2015b; Head et al., 2008c).

Participants at the Burg Wartenstein Symposium "Stratigraphy and Patterns of Cultural Change in the Middle Pleistocene," held in Austria in 1973, recommended that

> The beginning of the Middle Pleistocene should be so defined as to either coincide with or be linked to the boundary between the Matuyama Reversed Epoch and the Brunhes Normal Epoch of palaeomagnetic chronology.
> Butzer and Isaac (1975, appendix 2)

A similar recommendation was made by the INQUA Commission on Stratigraphy Working Group on Major Subdivision of the Pleistocene, at the XIIth INQUA Congress in Ottawa in 1987, which placed the Lower/Middle boundary at the Brunhes/Matuyama magnetic reversal (Anonymous, 1988; Richmond, 1996). However, although potential GSSPs in Japan, Italy, and New Zealand were discussed, no decision was reached.

At the 32nd International Geological Congress in Florence in 2004, the Early–Middle Pleistocene boundary Working Group of the ICS Subcommission on Quaternary Stratigraphy (SQS) formally adopted this paleomagnetic reversal as the primary guide for the boundary (Head et al., 2008c), given that it is essentially isochronous and can be recognized in both marine and continental deposits. Indeed a proxy for the reversal can even be recognized in the Antarctic ice core record (Head and Gibbard, 2005, 2015b). It was also decided in Florence that the GSSP should be placed not in a deep-sea core but in a marine section exposed on land (Head et al., 2008c).

Three candidate GSSPs had long been under consideration: the Ideale section of Montalbano Jonico in Basilicata (Nomade et al., 2019), the Valle di Manche in Calabria (Capraro et al., 2017), both in southern Italy, and the Chiba section on the Boso Peninsula, Japan (Okada et al., 2017; Suganuma et al., 2018). Either of the two Italian sections would have defined the Ionian Stage, currently an undefined regional Mediterranean stage (Cita et al., 2006, 2008), as well as the Middle Pleistocene Subseries.

All three sections have detailed marine and terrestrial (pollen) records and highly resolved benthic foraminiferal oxygen isotope records that provide secure astronomical age control. These three sections were scrutinized by the SQS Early–Middle Pleistocene Boundary Working Group, and only the Chiba section was considered to have a reliably preserved Matuyama/Brunhes reversal, as determined by a detailed record of the virtual geomagnetic pole path. The Montalbano Jonico succession has been diagenetically

Base of the Calabrian Stage of the Lower Pleistocene Subseries and Pleistocene Series of the Quaternary System at Vrica, Italy

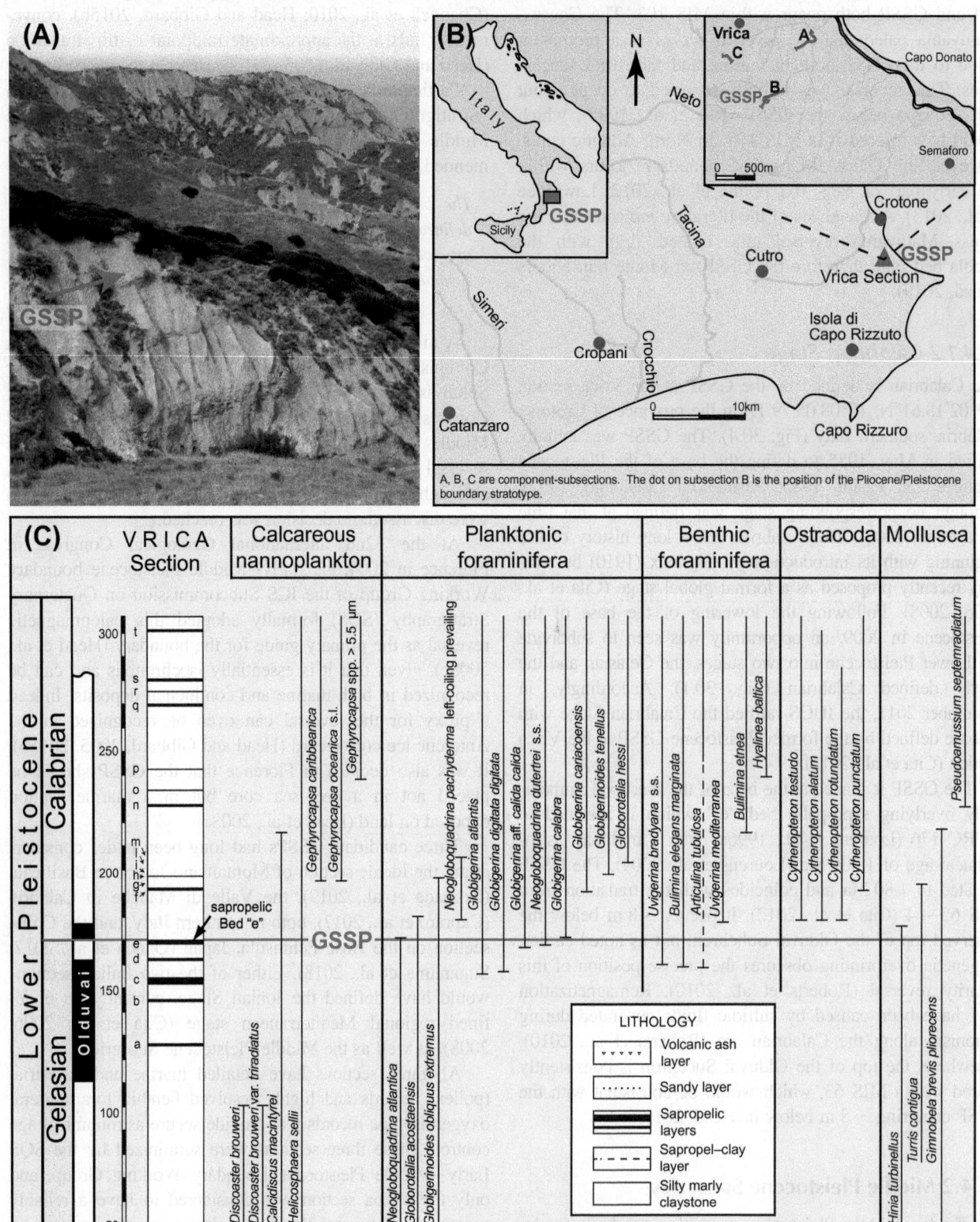

FIGURE 30.4 The GSSP for the Calabrian Stage in the Vrica section, Italy with photograph of the outcrop and stratigraphy. Magnetostratigraphy is from Zijderveld et al. (1991) but as qualified in the text. Biostratigraphy modified from Cita et al. (2012).

altered (Sagnotti et al., 2010), and the Valle di Manche section has a spuriously old Matuyama–Brunhes reversal that most likely is also caused by diagenetic overprinting (Head, 2019). In November 2017 the SQS Early–Middle Pleistocene Boundary Working Group voted to recommend the Chiba section as the GSSP for the Chibanian Stage and Middle Pleistocene Subseries (Head, 2019). These recomendations were approved by the ICS in November, 2019, and ratified in January, 2020.

30.4.2.1 Chibanian Stage

The Chibanian Stage and Middle Pleistocene Subseries are defined by the GSSP at the Chiba section (35°17.66′N, 140°8.79′E) in the Chiba Prefecture, east-central Japan (Figure 30.5). The GSSP was ratified in January, 2020.

The Chiba section is described in Suganuma et al. (2018). Known as the Tabuchi section in some earlier literature (e.g., Kazaoka et al., 2015, 2016), it is exposed along the Yoro River at a point where the Matuyama/Brunhes boundary outcrops. It is one of five sections, all tied by numerous isochronous tephra beds that outcrop along the deeply incised valleys of the Boso Peninsula in east-central Japan. The GSSP is placed at the base of a distinctive and regionally widespread tephra bed, the Ontake-Byakubi tephra (also known as the Byakubi-E tephra bed, or simply Byk-E, which is U–Pb zircon dated to 772.7 ± 7.2 ka; Suganuma et al., 2015). In the Chiba composite section the Matuyama/Brunhes boundary has an astronomical age of 772.9 ka, with a duration of 1.9 kyr for the directional transition. The mid-horizon of its directional transition zone, which is taken as the paleomagnetic boundary itself, is 1.1 m above the Byk-E tephra. This tephra bed and hence the GSSP has an astronomical age of 774.1 ka, occurring just before the end of MIS 19c (Suganuma et al., 2018; Head, 2019).

30.4.3 Upper Pleistocene Subseries

The boundary between the Middle and Upper Pleistocene has yet to be formally defined by a GSSP. Only the name "Upper Pleistocene Subseries" was ratified in January, 2020; and this completes the formal nomenclature of Pleistocene subdivisions at the subseries rank.

The Upper Pleistocene has a long history of use. As long ago as the 2nd INQUA Congress in Leningrad in 1932, a decision was made to define its base at the Last Interglacial (Eemian Stage; Figure 30.6). Later, the lower boundary of the Upper Pleistocene was placed at the base of Marine Isotope Stage 5 (MIS 5), based on a proposal from the INQUA Commission on Stratigraphy (Richmond, 1996). This proposal follows from the assumption that substage 5e of MIS 5 is the oceanic equivalent of the terrestrial northwest European Eemian Stage interglacial (Shackleton, 1977).

However, detailed pollen analyses of deep-sea cores west of Portugal have shown that the base of MIS 5 is some 6 kyr earlier than the base of the Eemian (Shackleton et al., 2003)-see Fig. 30.7. The term Eemian was originally introduced for a warm-temperate marine mollusk fauna in the Netherlands and was later applied across NW Europe to identify the interval of temperate forest cover following the penultimate (Saalian) glaciation, as determined by pollen records (Bosch et al., 2000). The delay between the beginning of MIS 5 and that of the Eemian in NW Europe may partly reflect the slow retreat of Saalian ice sheets but is primarily a discrepancy between differing definitions and climate responses. Marine isotope stage boundaries are defined at numerical midpoints between heavier (glacial) and lighter (interglacial) isotopic values, whereas the expansion of temperate trees in mid-latitudes reflects a threshold crossed as temperatures rise rapidly just before the plateau in isotope values (Fig. 30.7). Aside from the isotope record itself, the MIS 6/5 boundary therefore has little practical means in recognizing and correlating the base of the Upper Pleistocene in most climate records.

Gibbard (2003) and Litt and Gibbard (2008) proposed that, in keeping with the historical association of the boundary with the base of the Eemian, the GSSP should be defined in a high-resolution core succession from the Amsterdam Terminal—which is both the parastratotype and unit-stratotype of the Eemian Stage (Van Leeuwen et al., 2000).

Both the stage and the stage boundary are recognized on the basis of multidisciplinary biostratigraphy, the boundary being placed at the expansion of forest tree pollen above 50% of the total pollen assemblage, the standard practice in northwest Europe (Gibbard, 2003). In particular, the proposed GSSP in the Amsterdam Terminal borehole is based on the steep rise of the *Betula* (birch tree) pollen curve at this point (Litt and Gibbard, 2008). A similar increase in the *Betula* curve is characteristic of many Eemian pollen diagrams across Europe and is thought to be a synchronous response to climatic amelioration, perhaps within a matter of decades as is observed at the transition from the Last Glacial to the Holocene. An age of 127.2 ka is estimated for the base of the Eemian from a varve-dated record at Monticchio in Italy (Brauer et al., 2007), where the chronology is based on a combination of tephrochronology and annual layer counting. Unfortunately, a sharp change in lithology occurs at the base of the Eemian in the Amsterdam Terminal borehole, with the attendant possibility that a brief interval of time is missing. For this and other reasons the proposed GSSP failed to achieve ratification (Head and Gibbard, 2015a).

Cita and Castradori (1994, 1995) proposed the Tarentian Stage (as "Tarentian") for the interval corresponding to the Upper Pleistocene, that is, from the base of the last interglacial to the base of the Holocene, for

Base of the Chibanian Stage of the Pleistocene Series near Chiba, Japan

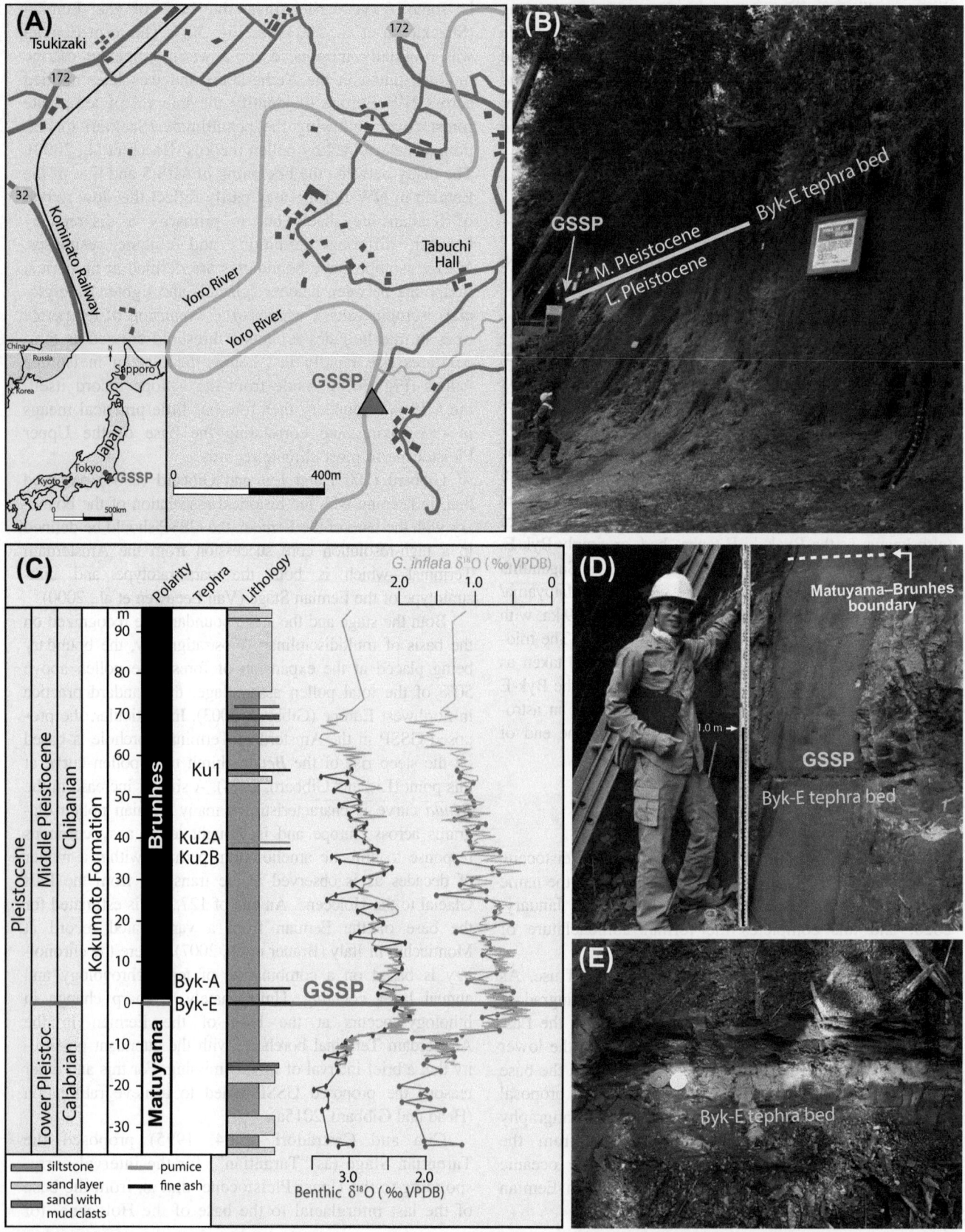

FIGURE 30.5 The GSSP for the Chibanian Stage and Middle Pleistocene Subseries in the Chiba section, Japan with photograph of the outcrop and stratigraphy. (D) shows the detailed position of the GSSP at the base of the Byk-E tephra bed, with the directional midpoint of the Matuyama–Brunhes paleomagnetic reversal also indicated. (E) shows the Byk-E tephra bed in detail. Adapted from Suganuma et al. (in prep.).

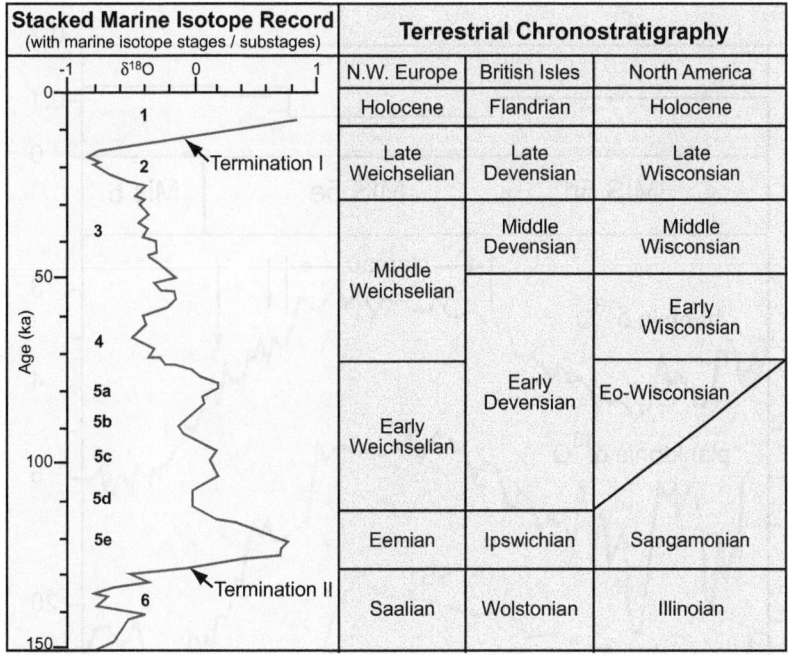

Stacked Marine Isotope Record (with marine isotope stages / substages)	Terrestrial Chronostratigraphy		
δ¹⁸O / Age (ka)	N.W. Europe	British Isles	North America
1	Holocene	Flandrian	Holocene
2 — Termination I	Late Weichselian	Late Devensian	Late Wisconsin
3	Middle Weichselian	Middle Devensian	Middle Wisconsin
4			Early Wisconsin
5a / 5b / 5c / 5d	Early Weichselian	Early Devensian	Eo-Wisconsin
5e	Eemian	Ipswichian	Sangamonian
6 — Termination II	Saalian	Wolstonian	Illinoian

FIGURE 30.6   Marine and continental chronostratigraphy for the past 150 kyr. The stacked marine oxygen isotope record and associated stages and substages are from Martinson et al. (1987), and the terrestrial climato- and chronostratigraphic divisions in northwest Europe and North America are modified from Lowe and Walker (1997).

marine successions in Italy. Thus the Upper Pleistocene would comprise a single stage if the Tarentian were elevated to a global stage (Fig. 30.2). However, the Tarentian has yet to be defined even for regional use (Head and Gibbard, 2015a), although the Fronte Section in Taranto, southern Italy has now been proposed as a potential candidate for a Tarentian GSSP. This section, exposed just above the shoreline, represents a ~9-m thick shallow marine record of the last interglacial (Negri et al., 2015). The suggested level of the GSSP is at the base of the maximum flooding zone, which occurs within the isotopic plateau of MIS 5e. The MIS 6/5 boundary is not represented, and it remains to be seen whether this proposed GSSP level can be correlated unequivocally to the global δ¹⁸O record (Head, 2019).

An ice core in the Antarctic has been suggested as a second potential candidate to define the Upper Pleistocene, with a GSSP placed at the point of an abrupt methane rise (Head, 2016, 2019). This is a distinctive and rapid event (Bazin et al., 2013; Capron et al., 2014) and aligns with maxima for CO_2 and δD in the EPICA Dome C ice core (Pol et al., 2014). It has a gas orbital age of 128.51 ± 1.72 ka in the EPICA Dome C core (Bazin et al., 2013) and appears to be closely related to rising temperatures in the higher northern latitudes. This methane rise aligns closely in age with several North Atlantic proxies for sharply rising temperature, and notably a steep rise in temperate tree pollen on the Portuguese margin (Tzedakis et al., 2018; Head, 2019), which effectively

constitutes the inception of the Eemian. Such a GSSP would therefore offer cross-hemispheric correlatability and tie closely to the original 1932 INQUA proposal that the base of the Upper Pleistocene should be defined at the base of the Eemian Stage.

30.5 Pleistocene/Holocene boundary

In an earlier iteration of this book (Harland et al., 1990), it was stated that "this boundary was thought to correspond to a climatic event around 10,000 radiocarbon years before present (BP)." At the time, the boundary was considered likely to be standardized in a varved lacustrine succession in Sweden (Mörner, 1976). It was originally proposed at the VIII INQUA Congress in Paris in 1969 and was subsequently accepted by the INQUA Holocene Commission in 1982 (Olausson, 1982). The climatic amelioration on which this boundary is identified is well established in a variety of sediments, particularly in northern Europe and North America. In Scandinavia, it corresponds to the following boundaries: European Pollen Zones III−IV, the Younger Dryas−pre-Boreal, and Late Glacial−post-Glacial (Mörner, 1976; Mangerud et al., 1974). However, this boundary definition was not formally ratified by the ICS. Had it been defined precisely at 10,000 (¹⁴C year BP), it would have been the first stratigraphic boundary later than the Proterozoic to be defined chronometrically.

In May, 2008 the Pleistocene−Holocene boundary was defined at a depth of 1492.45 m in the NGRIP2 ice

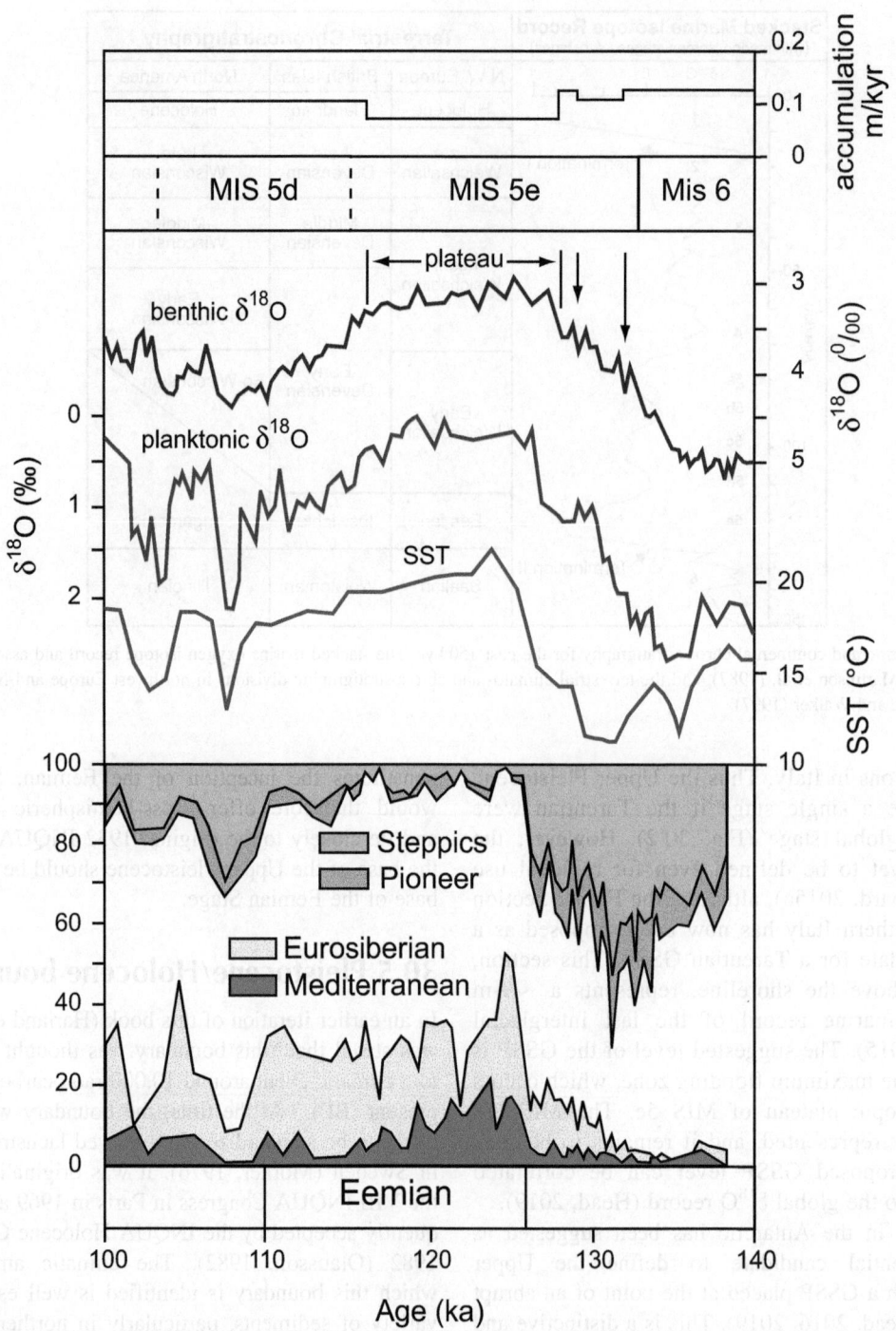

FIGURE 30.7 Marine and continental records of the Last (Eemian) Interglacial in core MD95-2042 off Portugal. The Eemian (regional) Stage, based on the pollen record, begins significantly later than the onset of MIS 5, and it continues well into MIS 5d. The SST reconstruction is based on alkenones (Shackleton et al., 2003). *MIS*, Marine Isotope Stage; *SST*, sea-surface temperature. (After Walker et al., 2008, 2009).

core (75.10°N, 42.32°W) from Greenland, with an age based on annual layer counting of 11,700 years b2k (before CE 2000), with a counting error of 99 years (equivalent to 2σ uncertainty) (Walker et al., 2008, 2009). The boundary is most clearly marked by an abrupt decrease in deuterium excess values, indicating rapid

warming within a period of 1−3 years and over subsequent decades. Other indicators, which change more slowly, include $\delta^{18}O$, dust concentration, annual layer thickness, and a range of chemical species (Fig. 30.8). This was significantly the first GSSP to utilize an ice core. Global auxiliary stratotypes were designated by

Base of the Greenlandian Stage of the Lower Holocene Subseries of the Holocene Series of the Quaternary System in the NGRIP2 ice core, Greenland

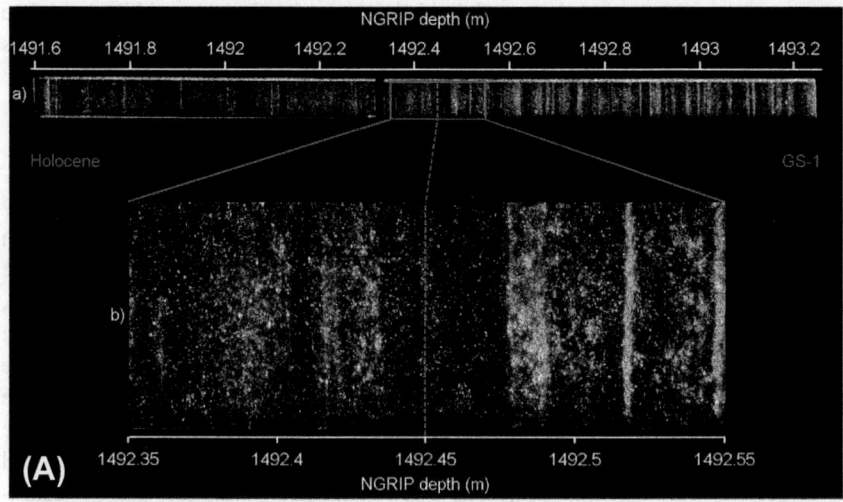

The location of the Pleistocene–Holocene boundary at 1492.45 m is shown in the enlarged lower image.

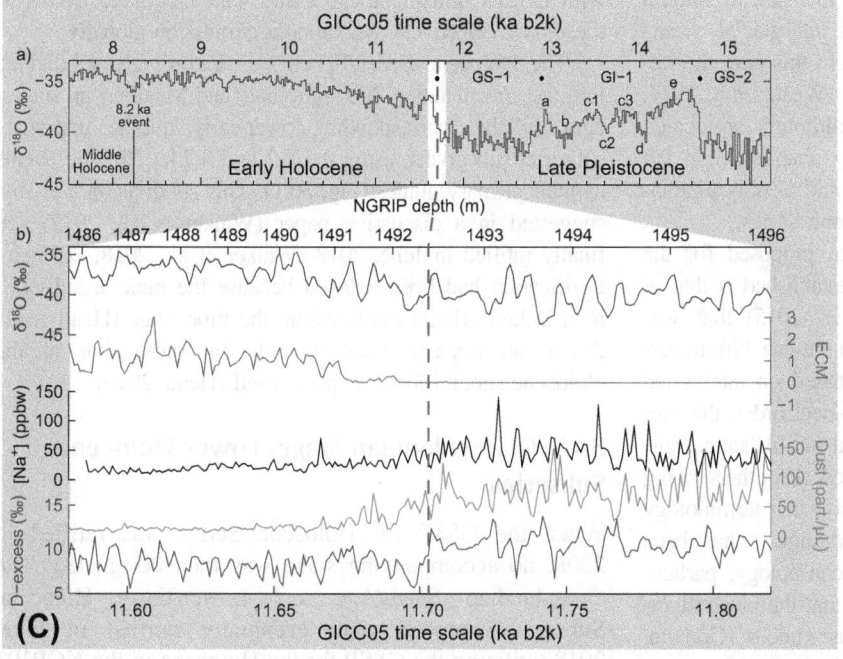

(a) The δ¹⁸O record through the Last Glacial (Greenland Stadial-1)–Interglacial Transition, showing the position of the Pleistocene–Holocene boundary in the NGRIP2 core.

(c) High-resolution multi-parameter record across the Pleistocene–Holocene boundary: δ¹⁸O, electrical conductivity (ECM), Na⁺ concentration, dust content, and deuterium excess.

FIGURE 30.8 Lower Holocene (Greenlandian) GSSP. (A) Visual stratigraphy of the NGRIP2 ice core between 1491.6 and 1493.25 m depth obtained using a digital line scanner. In this photograph the image is "reversed" so that clear ice shows up black, and cloudy ice (which contains impurities such as micrometer-sized dust particles) shows up white. The visual stratigraphy is essentially a seasonal signal and reveals annual banding in the ice. The position of the Pleistocene–Holocene boundary at 1492.45 m is shown in the enlarged lower image (Walker et al., 2009). (B) Location of the NGRIP2 ice core in central Greenland. (C) Paleoenvironmental records. (a) The δ¹⁸O record through the Last Glacial–Interglacial Transition, showing the position of the Pleistocene–Holocene boundary in the NGRIP2 core. The location of the Pleistocene–Holocene boundary at 1492.45 m is shown in the enlarged lower image. (b) High-resolution multiparameter record across the Pleistocene–Holocene boundary: δ¹⁸O, electrical conductivity (ECM), Na⁺ concentration, dust content, and deuterium excess.

Walker et al. (2009) and comprise lacustrine deposits from eastern Canada, Germany, Japan, and New Zealand, and a deep-marine core from the Cariaco Basin, Venezuela.

30.6 Holocene Series

Holocene is the name for the most recent interval of Earth history and includes the present day. Although generally regarded as having begun 10,000 radiocarbon (^{14}C) years, or the last 11,500 calibrated (i.e., calendar) years, BP (i.e., 1950), the base of the Holocene is now assigned an age of 11,700 years b2k (before CE 2000) as described earlier. The term "recent" as an alternative to Holocene has no official validity and should not be used. Sediments accumulating or processes operating at present can be referred to as "modern" or by similar synonyms.

The term Flandrian, derived from marine transgression sediments on the Flanders coast of Belgium (de Heinzelin and Tavernier, 1957), has often been used as a synonym for Holocene (Fig. 30.6). It has been adopted by authors who consider that the present (Holocene) interglacial should have the same stage status as previous interglacial events and thus be included in the Pleistocene. In this case the latter would thus extend to the present day (West, 1968, 1977, 1979; Hyvärinen, 1978). This usage, although advocated particularly in Europe, has been largely abandoned in the last decades (Lowe and Walker, 2015) and is now untenable with formal ratification of the Holocene Series.

Various zonation schemes have been proposed for the Holocene (Flandrian) Epoch. The most established is that of Blytt and Sernander (Lowe and Walker, 2015) that was developed for peat bogs in Scandinavia in the late 19th to earliest 20th centuries. Their terminology, based on interpreted climatic changes, comprised, in chronological order, the pre-Boreal, Boreal, Atlantic, sub-Boreal, and sub-Atlantic. This scheme was refined by the Swede von Post and others, using pollen analysis throughout Europe. Today, this terminology remains in use in northern Europe, although it has been largely displaced by precise numerical chronology, particularly from ^{14}C dating, which has shown that the biostratigraphically defined zone boundaries are diachronous (Godwin, 1975). An attempt to fix these boundaries to precise dates was proposed for northern Europe by Mangerud et al. (1974).

In prehistoric times, as well as later, climatic events have largely served to identify the divisions elaborated by modern ^{14}C, as well as other dating techniques, such as tephrochronology, and dendrochronology as well as human history. Using these techniques, Holocene time can be divided into ultrahigh-resolution divisions. For example, recent developments indicate that cyclic patterns of climate change of durations as short as 200 years or less can be differentiated and potentially used for demonstrating equivalence in peat successions, while ice cores and annually laminated

lacustrine successions allow even more highly resolved correlations. Meanwhile, to provide a formal chronostratigraphic framework at an intermediate scale, the Holocene has been subdivided into early (lower), middle, and late (upper) subepochs (substages) and their respective ages (stages).

30.7 Subdivision of the Holocene

A threefold subdivision of the Holocene gained popularity with the proposal by Mangerud et al. (1974) that for northern Europe, the Flandrian regional stage (equivalent then to the Holocene) should be divided into three substages: Early, Middle, and Late Flandrian. While this subdivisional scheme was shown to have only local or regional applicability, the concept of a tripartite subdivision for the Holocene became widely accepted (Walker et al., 2018). However, defining the increasingly used informal terms of "early", "middle" and "late" Holocene was complicated by the fact that no significant step changes in climate occur on a global scale during the Holocene. The solution was to apply event stratigraphy, with GSSPs utilizing the 8.2 and 4.2 ka climatic events as these have effectively synchronous expression globally.

The Holocene Series/Epoch was accordingly subdivided into the Greenlandian, Northgrippian, and Meghalayan stages/ages and their corresponding lower/early, middle, upper/late subseries/subepochs, using the 8.2 and 4.2 ka climatic events as their primary guides (Fig. 30.1). This subdivision was first suggested in a discussion paper (Walker et al., 2012) and finally ratified in June, 2018 (Walker et al., 2018, 2019a,b). Ratification had been delayed because the rank of subseries then lacked official status within the time scale (Head et al., 2017), an impasse resolved only by ratification of the Holocene subdivisional proposal itself (Head, 2019).

30.7.1 Greenlandian Stage, Lower Holocene Subseries

When the GSSP for Holocene Series was ratified in 2008, no accompanying stage was then designated. The Greenlandian Stage/Age and Lower/Early Holocene Subseries/Subepoch were eventually ratified in June, 2018, utilizing the GSSP for the Holocene in the NGRIP2 Greenland ice core dated at 11,700 yr b2k (before CE 2000), as described earlier (Fig. 30.8).

30.7.2 Northgrippian Stage, Middle Holocene Subseries

The Northgrippian Stage/Age and Middle Holocene Subseries/Subepoch are defined by a GSSP at 1228.67 m depth in the NGRIP1 Greenland ice core (75.10°N, 42.32°W), with an age based on annual ice layer counting of 8236 ± 47 yr b2k (Walker et al., 2018, 2019a,b; Fig. 30.9).

Base of the Northgrippian Stage of the Middle Holocene Subseries of the Holocene Series of the Quaternary System in the NGRIP1 ice core, Greenland

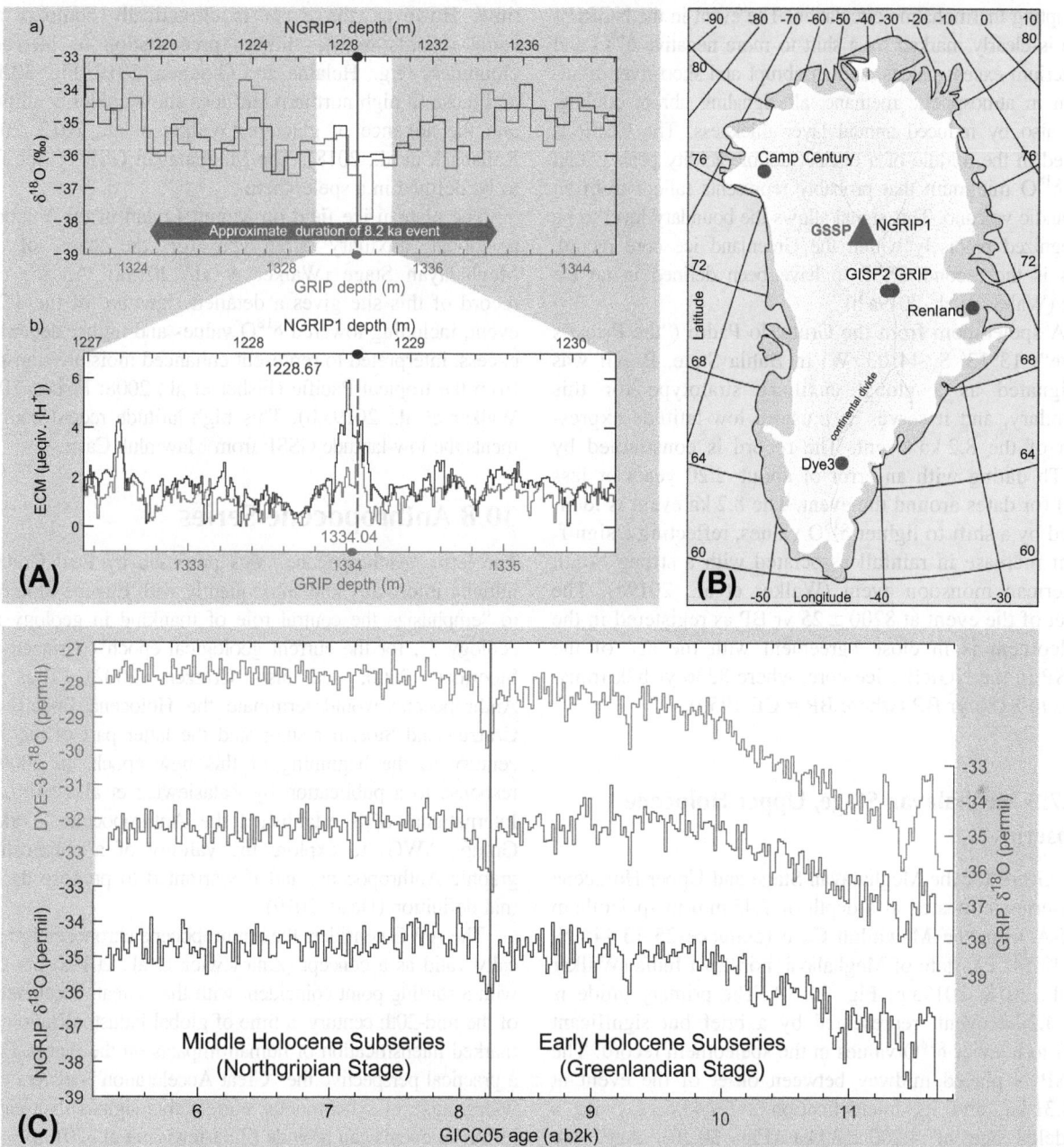

FIGURE 30.9 Middle Holocene (Northgrippian) GSSP. (A) Top (a): water-stable isotope ratios ($\delta^{18}O$) from the GRIP and NGRIP1 ice cores around the 8.2 ka event. The event in NGRIP1 is defined from ~8300 to ~8140 yr b2k. Bottom (b): during the period of low $\delta^{18}O$ values (white bar in the upper panel and expanded here), a distinct acidity double peak is reflected in ECM and can probably be attributed to an Icelandic volcano. It is dated on the GICC05 time scale to 8236 yr b2k (8186 cal. year BP) and is the primary marker for the Early–Middle Holocene (Greenlandian–Northgrippian) boundary (after Walker et al., 2008, 2009). (B) Location of the NGRIP1 ice core in north–central Greenland (NGRIP1 and NGRIP2 were drilled at the same site, with NGRIP 2 penetrating deeper than NGRIP1). (C) Water-stable isotope ratios ($\delta^{18}O$) at 20-year resolution in three Greenland ice core records, DYE-3, GRIP, and NGRIP (NGRIP1 and NGRIP2 combined), over the time interval 11.7–5.3 ka b2k (before CE 2000) on the GICC05 time scale (after Walker et al., 2008, 2009). electrical conductivity measurements (ECM).

It represents a brief global cooling episode—the 8.2 ka climatic events—that seems to have been triggered by catastrophic meltwater release into the North Atlantic, thereby disrupting thermohaline circulation. The event in the NGRIP1 core is clearly marked by a shift to more negative $\delta^{18}O$ and deuterium excess values, and an abrupt and short-lived minimum in atmospheric methane, all signaling abrupt cooling, and also by reduced annual layer thickness. The GSSP is placed in the middle of a distinct double acidity peak within the $\delta^{18}O$ minimum that probably represents fallout from an Icelandic volcano. This signal allows the boundary level to be recognized precisely within the Greenland ice core record. This is the second GSSP to have been defined in an ice core (Walker et al., 2019a,b).

A speleothem from the Gruta do Padre ("the Priest's Cave"; 13°13′ S, 44°03′ W) in Bahia State, Brazil was designated as a global auxiliary stratotype for this boundary, and it serves as a useful low-latitude expression of the 8.2 ka event. The record is constrained by U−Th dating with an error of about ±20 years or less (2σ) for dates around the event. The 8.2 ka event is identified by a shift to lighter $\delta^{18}O$ values, reflecting a significant increase in rainfall associated with a strong South American monsoon event (Walker et al., 2019a). The onset of the event at 8200 ± 25 yr BP as registered in the speleothem is in close agreement with the age of the GSSP in the NGRIP1 ice core, where 8236 yr b2k translates to 8186 yr BP (where BP = CE 1950).

30.7.3 Meghalayan Stage, Upper Holocene Subseries

The GSSP for the Meghalayan Stage and Upper Holocene Subseries is located at a depth of 7.45 mm in speleothem KM-A from the Mawmluh Cave (entrance 25°15′44″ N, 91°42′54″ E), state of Meghalaya, northeast India (Walker et al., 2018, 2019a,b; Fig. 30.10). The primary guide is the 4.2 ka event, represented by a brief but significant shift to heavier $\delta^{18}O$ values in the speleothem record. The GSSP is placed midway between onset of the event at ~4.31 ka, and its intensification at ~4.06 ka, with a modeled age of 4.200 ± 30 ka (Fig. 30.10). Ages are based on U−Th dating and expressed relative to the baseline date of CE 1950, but an age of 4250 yr b2k is preferred to compare directly with the earlier Holocene GSSPs (Walker et al., 2018). The $\delta^{18}O$ record of the KM-A speleothem compares generally but not precisely with two other speleothem records from the Mawmluh Cave (Kathayat et al., 2018), presumably reflecting the complex hydrology in this cave system (Head, 2019).

The 4.2 ka event lasted for two or three centuries and marks an abrupt reduction in precipitation due to a weakening of the monsoon across India and Southeast Asia. In many low- and mid-latitude regions the 4.2 ka event is indeed seen as an aridification episode and has been linked to important human cultural and societal changes at this time. However, the event is climatically complex and some records register higher precipitation or increased cloudiness (e.g., Helama and Oinonen, 2019; Fig. 30.10), and those in high northern latitudes show colder conditions and the advance of glaciers (Walker et al., 2012, 2018; Railsback et al., 2018). The Meghalayan GSSP is the first to be defined in a speleothem.

The plateau ice field on Mount Logan in the Yukon is a global auxiliary stratotype for the base of the Meghalayan Stage (Walker et al., 2019a). An ice core record of this site gives a detailed signature of the 4.2 ka event, including lowered $\delta^{18}O$ values and higher deuterium excess, interpreted to represent enhanced moisture transport from the tropical Pacific (Fisher et al., 2008; Fisher, 2011; Walker et al., 2019a,b). This high-latitude record complements the low-latitude GSSP from Mawmluh Cave.

30.8 Anthropocene Series

The term "Anthropocene" was proposed by Paul Crutzen, initially informally and subsequently with Eugene Stoermer, to "emphasize the central role of mankind in geology and ecology ... for the current geological epoch" (Crutzen and Stoermer, 2000; see also Crutzen, 2002). Thus the Anthropocene would terminate the Holocene Epoch, and Crutzen and Stoermer suggested the latter part of the 18th century as the beginning of this new epoch. In 2009 in response to a publication by Zalasiewicz et al. (2008), the International SQS established the Anthropocene Working Group (AWG) to explore the validity of a chronostratigraphic Anthropocene, and if warranted to propose its formal definition (Head, 2019).

The AWG considers the Anthropocene chronostratigraphically valid as a concept (Zalasiewicz et al., 2019a) but only with a starting point coincident with the "Great Acceleration" of the mid-20th century, a time of global industrialization and marked intensification of human impacts on the planet. From a practical perspective the "Great Acceleration" delivers a far wider range of synchronous, correlatable signals than earlier historical events can provide (Zalasiewicz et al., 2019a).

The currently preferred primary stratigraphic marker for the base of the Anthropocene is the early rise in plutonium-238 representing the fallout from thermonuclear bomb testing that began in 1952. Plutonium-238 has potentially global and rapid dissemination, a long half-life decaying to distinctive daughter products (Waters et al., 2015), and occurs along with an array of other stratigraphic signals associated with the "Great Acceleration." Potential candidate GSSPs will have captured this, and other signals are now being investigated (Waters et al., 2018a,b). The preferred rank is that of series/epoch but this remains open to debate.

Base of the Meghalayan Stage of the Upper Holocene Subseries of the Holocene Series of the Quaternary System in Mawmluh Cave, Meghalaya, India

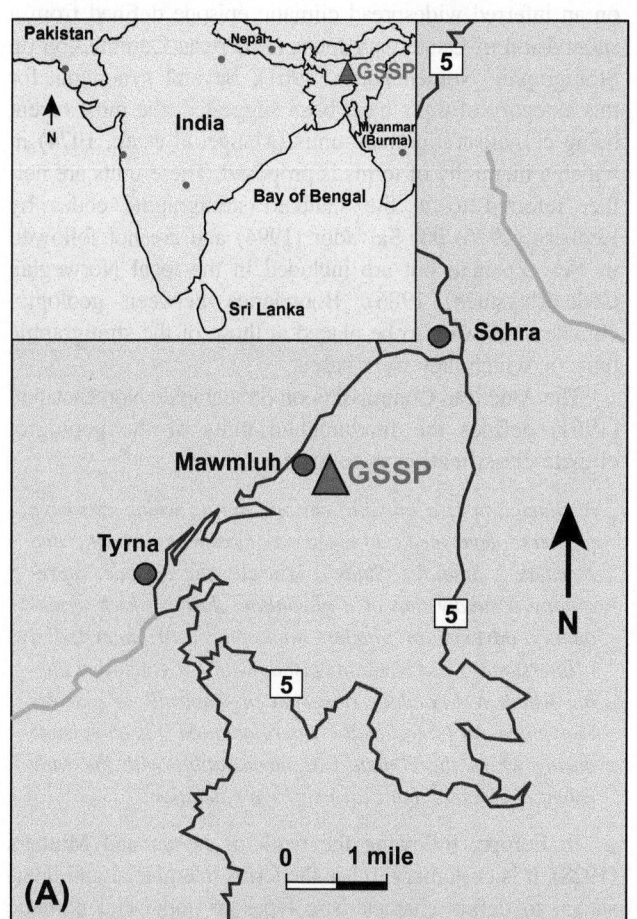

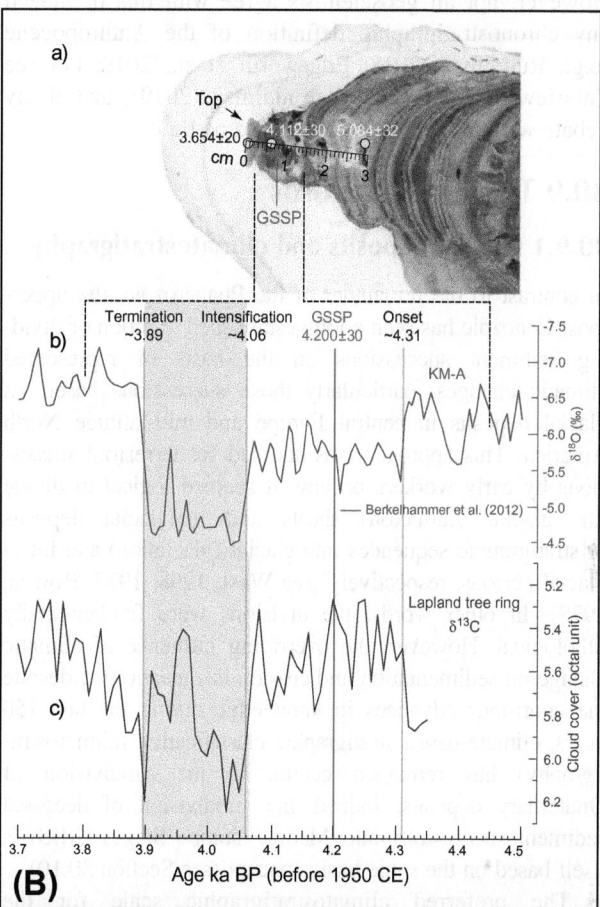

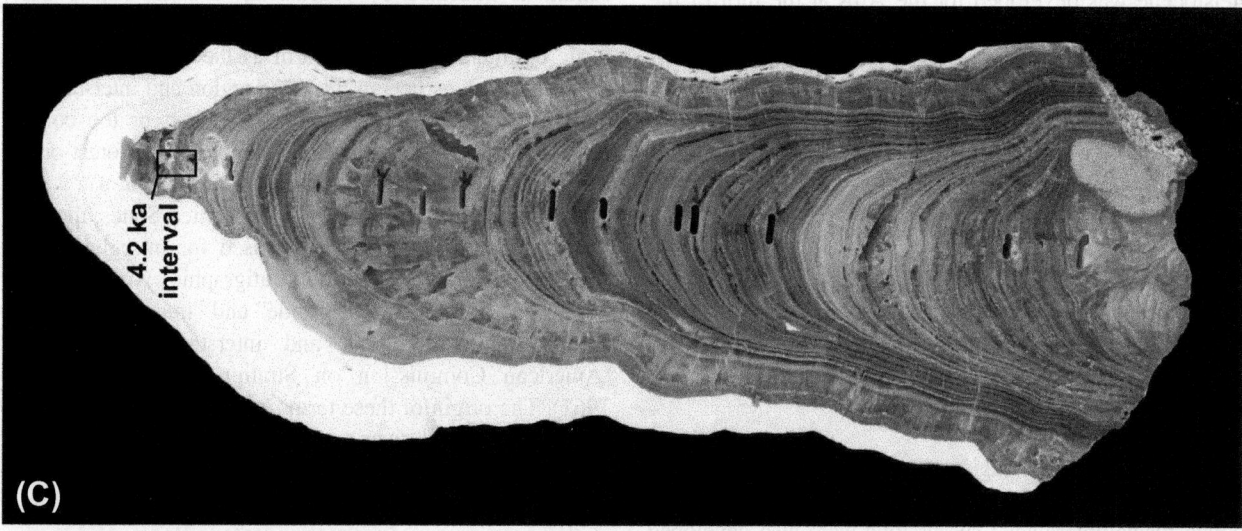

FIGURE 30.10 Late Holocene (Meghalayan) GSSP. (A) Location of the Mawmluh Cave from which the speleothem KM-A was recovered. (B) The speleothem KM-A in detail. Part (a) shows the positions of the GSSP and of the three ^{230}Th dates used to constrain the 4.2 ka interval. Part (b) gives the δ^{18}O record for this speleothem, with the GSSP (4200 ± 30 yr BP, where BP = CE 1950) occupying the midpoint between the onset of aridification (∼4.31 ka) and its intensification (∼4.06 ka). Part (c) for comparison, an inverted δ^{13}C tree ring record from Lapland, northern Finland is shown as a proxy for cloudiness (wetter conditions) and is the most northerly expression yet documented for the 4.2 ka event (Helama and Oinonen, 2019). The gray vertical bar indicates the most intense phase of the 4.2 ka event that might be synchronous for the KM-A and Lapland records (after Head, 2019). (C) The speleothem KM-A in its entirety, length ∼308 mm (after Walker et al., 2019a,b).

The term "Anthropocene" is now widely cited in the geological literature and beyond, and the formal definition of this term would increase its utility (Head, 2019). However, not all geoscientists agree with this or indeed any chronostratigraphic definition of the Anthropocene (e.g., Ruddiman, 2018; Edgeworth et al., 2019; but see Zalasiewicz et al., 2019b; Ruddiman, 2019), and lively debate will probably continue for some time.

30.9 Terrestrial records

30.9.1 Glacial deposits and climatostratigraphy

In contrast to the remainder of the Phanerozoic, the uppermost Cenozoic has seen a long-established tradition of dividing sediment successions on the basis of represented climatic changes, particularly those successions based on glacial deposits in central Europe and mid-latitude North America. This approach was adopted for terrestrial successions by early workers because it seemed logical to divide till (glacial diamicton) sheets and nonglacial deposits or stratigraphic sequences into glacial (glaciation) and interglacial periods, respectively (see West, 1968, 1977; Bowen, 1978). In other words, the divisions were fundamentally lithological. However, the overriding influence of climatic change on sedimentation and erosion has meant that, despite the enormous advances in knowledge during the last 150 years, climate-based stratigraphic classification (climatostratigraphy) has remained central to the subdivision of Quaternary deposits. Indeed, the subdivision of deep-sea sediment successions into Marine Isotope Stages (MIS) is itself based on the same basic concept (see Section 30.10).

The preferred climatostratigraphic scale for the Pleistocene was developed for the Alps at the turn of the century by Penck and Brückner (1909–1911).

For the Alps the succession in increasing age is as follows:

Würm Glacial (Würmian)
Riss-Würm Interglacial
Riss Glacial (Russian)
Mindel—Riss Interglacial
Mindel Glacial (Mindelian)
Günz—Mindel Interglacial
Günz Glacial
Donau—Günz Interglacial
Donau Glacial
? Biber Glacial

For other glaciated regions, including northern Europe, Britain, Russia, North America, and New Zealand, separate named glacial/interglacial successions were also developed (see Figs. 30.2 and 30.6).

Before the impact of the ocean-core isotope sequences, an attempt was made to formalize the climate-based stratigraphic terminology in the American Code of Stratigraphic Nomenclature (American Commission on Stratigraphic Nomenclature, 1961), in which so-called geologic-climate units were proposed. Here a geologic-climate unit is based on an inferred widespread climatic episode defined from a subdivision of Quaternary rocks (American Commission on Stratigraphic Nomenclature, 1961). Several synonyms for this category of units have been suggested, the most recent being climatostratigraphic units (Mangerud et al., 1974) in which a hierarchy of terms is proposed. These units are neither referred to in the standard stratigraphic codes by Hedberg (1976) nor Salvador (1994) and are not followed in New Zealand but are included in the local Norwegian Code (Nystuen, 1986). Boundaries between geologic-climate units were to be placed at those of the stratigraphic units on which they were based.

The American Commission on Stratigraphic Nomenclature (1961) defines the fundamental units of the geologic-climate classification as follows

A Glaciation is a climatic episode during which extensive glaciers developed, attained a maximum extent, and receded. A Stadial ("Stade") is a climatic episode, representing a subdivision of a glaciation, during which a secondary advance of glaciers took place. An Interstadial ("Interstade") is a climatic episode within a glaciation during which a secondary recession or standstill of glaciers took place. An Interglacial ("Interglaciation") is an episode during which the climate was incompatible with the wide extent of glaciers that characterize a glaciation.

In Europe, following the work of Jessen and Milthers (1928), it is customary to use the terms interglacial and interstadial to define characteristic types of nonglacial climatic conditions indicated by vegetational changes (Table 30.1) and interglacial to describe a temperate period with a climatic optimum at least as warm as the present interglacial (Holocene: see earlier) in the same region and interstadial to describe a period that was either too short or too cold to allow the development of temperate deciduous forest or the equivalent of interglacial type in the same region.

In North America, mainly in the United States, the term "interglaciation" is occasionally used for interglacial (see American Commission on Stratigraphic Nomenclature, 1961). Likewise, the terms "stade" and "interstade" may be used instead of stadial and interstadial, respectively (American Commission on Stratigraphic Nomenclature, 1961). The origin of these terms is not certain but the latter almost certainly derives from the French word *stade* (m), which is unfortunate since in French *stade* means (chronostratigraphic) stage (Michel et al., 1997), for example, *stade isotopique marin* = marine isotope stage.

It will be readily apparent that, although in long-standing usage, the glacially based terms are very difficult to apply outside glaciated regions, that is, most of the world. Moreover, as Suggate and West (1969)

TABLE 30.1 Examples of chronostratigraphical substage divisions of interglacial (temperate) stages and related cold (glacial) stages of the Middle and Late Pleistocene.

Chronostratigraphic substages									Vegetation aspect	Characteristic vegetation
Cold stage		e An		e Wo		e De			Early Glacial	Herb dominated
Temperate stage	Cromerian (~750 ka)	Cr IV	Hoxnian (~350 ka)	Ho IV	Ipswichian (~$125 ka)	Ip IV	Flandrian (post-10 ka)		Posttemperate	Birch–pine forest
		Cr III		Ho III		Ip III		Fl III	Late temperate	Mixed deciduous–coniferous forest
		Cr II		Ho II		Ip II		Fl II	Early temperate	Deciduous forest
		Cr I		Ho I		Ip I		Fl I	Pretemperate	Birch–pine forest
Cold stage		l Be		l An		l Wo		l De	Late Glacial	Herb dominated

Modified after West (1968) and West and Turner (1968).
For the Holocene (Flandrian), correlations with the zones of Godwin (1975) are also indicated.
An, Anglian; *Be*, Beestonian; *De*, Devensian; *e*, early; *l*, late; *Wo*, Wolstonian.

recognized, the term glaciation or glacial is particularly inappropriate since modern knowledge indicates that cold rather than glacial climates have tended to characterize the periods intervening between interglacial events. They therefore proposed that the term "cold" stage (chronostratigraphy) be adopted for "glacial" or "glaciation." Likewise, they proposed the use of the term "warm" or "temperate" stage for interglacial, both being based on regional stratotypes. The local nature of these definitions indicates that they cannot necessarily be used across great distances or between different climatic provinces (Suggate and West, 1969; Suggate, 1974; West, 1968, 1977) or indeed across the terrestrial–marine facies boundary. The use of mammalian biostratigraphic data, in particular the evolution of voles, offers the possibility of long-distance correlations between local assemblages. In addition, it is worth noting that the subdivision into glacial and interglacial is mainly applied to the Middle and Late Pleistocene.

Both interglacial and glacial, or temperate and cold, stages have been subdivided into substages and zones. This is achieved in interglacial stages using paleontological, particularly vegetational, assemblages. The cyclic pattern of interglacial vegetation that typifies all known temperate events in Europe was developed as a means of subdividing, comparing, and therefore characterizing temperate events by West (1968, 1977) and Turner and West (1968). In this scheme, temperate (interglacial) event successions are subdivided into four substages: pretemperate, early temperate, late temperate, and posttemperate. Finer-scale zonation schemes are also commonly in use throughout Europe and the former USSR (Table 30.1).

Late Middle- and Late Pleistocene glacial stages have been divided on various bases, but in the Northern Hemisphere the division is based on a combination of vegetation, lithology, and occasionally pedological evidence, often resulting in an unfortunate intermixture of chrono- and climatostratigraphic terminology. Chronological control for Middle and Lower Pleistocene glacial deposits comes from tephrochronology (e.g., Boellstorff, 1978), K/Ar and ^{40}Ar/^{39}Ar and dating of intercalated lavas (e.g., Geirsdottir and Eriksson, 1994; Singer et al., 2004), magnetostratigraphy (e.g., Roy et al., 2004), and cosmogenic nuclide dating (e.g., Balco et al., 2005; Balco and Rovey, 2010).

The last glacial stage (Weichselian, Valdaian, Devensian, Wisconsian, etc.) has particularly been divided into three or four substages (Early Middle or Pleniglacial, Late Weichselian, etc., and Late glacial), using geochronology (mainly ^{14}C). Boundaries are defined at specific dates, especially in the last 30 kyr (Table 30.1).

A widely used event term, the "Last Glacial Maximum," frequently abbreviated to LGM, is used to refer to the maximum global ice volume during the last glacial cycle (CLIMAP, 1981). It corresponds to the trough in the marine isotope record centered on c.18 ^{14}C ka BP (Martinson et al., 1987) and the associated global eustatic sea level low also dated to 18 ^{14}C ka BP (Yokoyama et al., 2000). According to Clark et al. (2009), most ice sheets were at their LGM positions from 26.5 to 19–20 ka cal BP. The LGM has also been assigned chronozone status (23–19 or 24–18 ka cal BP dependent on the dating applied) by Mix et al. (2001) who considered the event should be centered on the calibrated date at 21 ka cal BP. (i.e., LGM *sensu stricto*). However, since the post-MIS 5e (last interglacial, Eemian Stage) last maximum glaciation occurred much earlier in some areas than in others, the term LGM should be used with caution (Ehlers and Gibbard, 2011; Ehlers et al., 2011; Hughes et al., 2013; Hughes and Gibbard, 2015). Moreover, the chronozone has neither been formally defined nor designated a type locality. It is therefore currently of informal status. However, the SQS is currently investigating the possibility of formally defining this and related terms (e.g., Heinrich events).

30.9.2 Loess deposits

In contrast to continental glacial successions, which tend to be fragmentary, thick eolian dust (loess) deposits in Europe and Asia (particularly China) are essentially continuously deposited, much like deep-ocean sediment. In all areas, the loess deposits consist of two major stratigraphic units—loess and paleosols—representing climatically controlled variations in loess accumulation and weathering processes (Fig. 30.11). In the classic sections on the Chinese Loess Plateau, 34 soil-loess couplets have been identified in loess deposits more than 150 m thick, with the Gauss/Matuyama paleomagnetic boundary (~ 2.6 Ma) located within loess L33 (Rutter et al., 1991; Yang and Ding, 2010)—see Figs. 30.2 and 30.12. However, the Gauss/Matuyama equates with MIS 103 ($=$S32), and because paleomagnetic boundaries can be diagenetically displaced many meters in loess deposits (Zhou and Shackleton, 1999; Zhu et al., 2008), it is assumed that the Gauss/Matuyama Boundary in fact relates to S32. Paleosol horizons are characterized by higher magnetic susceptibility, finer grain size, and redder colors than the intervening loess horizons and can be correlated throughout the Loess Plateau.

Underlying the lowermost loess layer (L34) is the so-called red clay, which is still of eolian origin but has been weathered much more strongly than the overlying loess. The boundary between the red clay and L34 is dated to ~ 2.8 Ma and is characterized by a dramatic increase in grain size and loess accumulation rates, interpreted as a major climate shift from long-lasting warm humid conditions to large amplitude cold/dry and warm/humid

FIGURE 30.11 Weinan loess from Shiling Yang. A total number of 34 soil-loess couplets have been identified. Paleosol horizons (S levels) are characterized by higher magnetic susceptibility, finer grain size and redder colors than the intervening loess (L) horizons and can be correlated throughout the Loess plateau.

oscillations (Yang and Ding, 2010). There was also a large increase in aridity over the dust source region.

30.9.3 Ice cores

In the past three decades, the drilling of cores into ice sheets in various parts of the world has revolutionized our records of detailed climatic change. An independent record of Late Pleistocene and Holocene climatic changes has been derived from $\delta^{16}O/\delta^{18}O$ ratios and other measurements in cores through the Greenland and Antarctic ice sheets (Johnsen et al., 1972; Dansgaard et al., 1993; EPICA Community Members, 2004) and from other areas, including temperate and tropical ice caps in Asia, Africa, and South America (Thompson et al., 1989; Thompson et al., 1995, 2002). These have provided spectacularly unrivaled records that allow annual resolution of climatic events. In one case (Kilimanjaro Ice Cap) the ice cores may become the only physical record if present rates of ice cap melting continue—Thompson et al. (2009) have estimated that the ice cap could disappear within the next few decades.

From a stratigraphic point of view, it is the recognition of patterns of a wide range of climatically controlled parameters that provide potentially very high-resolution correlation tools in ice cores. Detailed patterns arise from determination of aerosol particles, dust, trace elements, spores, or pollen grains, etc., that have fallen onto the ice surface and become incorporated into the annual ice layers. They include, for example, dust from wind activity (e.g., Delmonte et al., 2004) or volcanic eruptions (e.g., Zielinski et al., 1997; Davies et al., 2010). Trace gases such as carbon dioxide or methane can be trapped in air bubbles within ice crystals and provide long, continuous records of atmospheric greenhouse gas concentrations (Luthi et al., 2008). In addition, naturally and artificially occurring radioactive isotopes present in the ice layers can be used to provide an independent chronology for dating the ice core successions (e.g., Dreyfus et al., 2008).

The Holocene is well documented by Greenland ice cores, these providing annual layer counting chronologies to ~ 60 ka. Detailed cores from the Greenland Ice Core Project (GRIP) and the Greenland Ice Sheet Project (GISP) have been obtained that provide a record extending to the last interglacial (Eemian), although the oldest ice is deformed (North Greenland Ice Core Project Members, 2004). Chronology comes from a combination of annual layer counting, ice flow modeling, and independent age markers (Andersen et al., 2006; Parrenin et al., 2007; Svensson et al., 2008). Moreover, the Dome C core in Antarctica extends to nearly 800 ka in the Lower Pleistocene (Fig. 30.13; EPICA Community Members, 2004; Luthi et al., 2008; Wolff, 2008). These records have revolutionized our understanding of patterns and

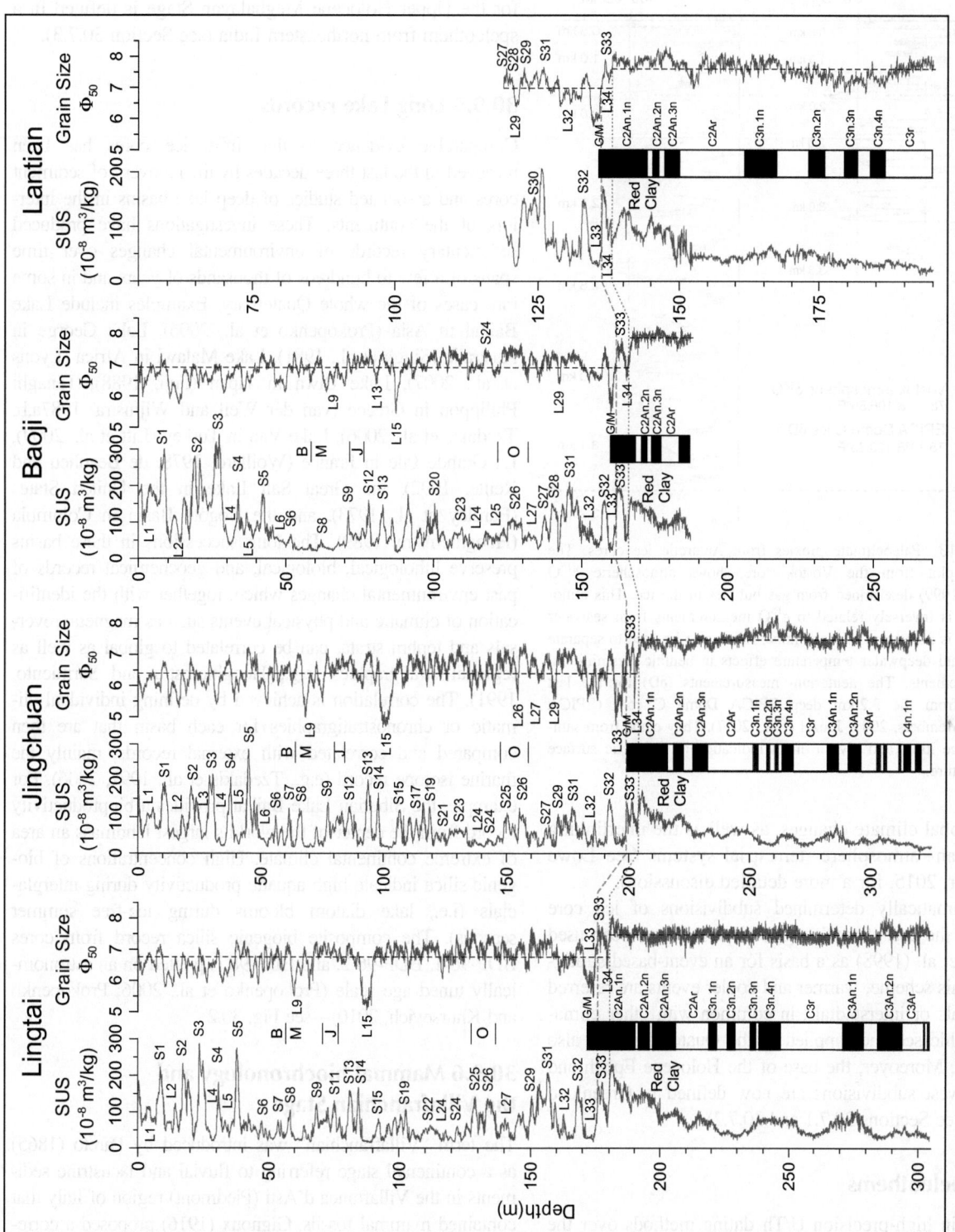

FIGURE 30.12 Magnetic susceptibility (SUS) and grain-size variations in four Chinese loess sections. Soil horizons are characterized by high magnetic susceptibility values that may be used to correlate between sections, in addition to magnetostratigraphy. The Gauss/Matuyama paleomagnetic reversal (~2.6 Ma) is displaced to loess L33 owing to remagnetization but is coeval with S32 (Yang and Ding, 2010).

Antarctic ice drilling record
Oxygen and hydrogen isotope fractionation

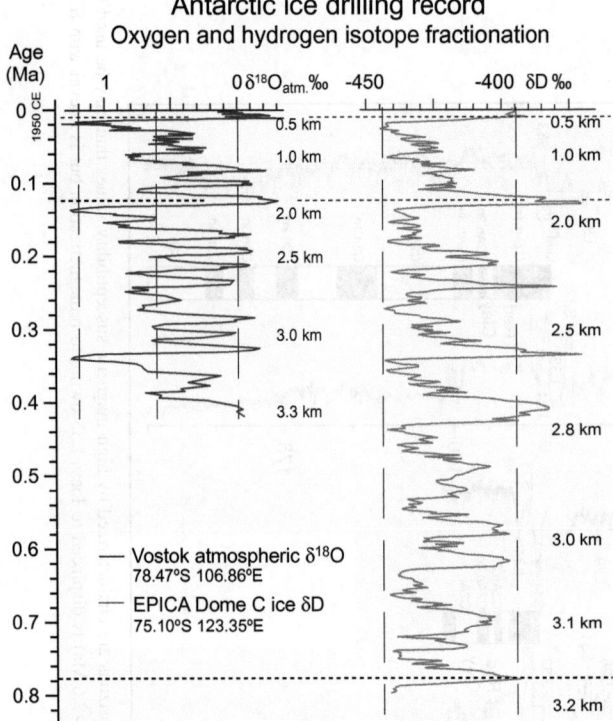

FIGURE 30.13 Paleoclimate proxies from Antarctic ice cores. The 420 ka-long plot from the Vostok core shows atmospheric $\delta^{18}O$ (Petit et al., 1999) determined from gas bubbles in the ice. This atmospheric $\delta^{18}O$ is inversely related to $\delta^{18}O$ measurements from seawater and therefore is a measure of ice volume. It can also be used to separate ice volume and deepwater temperature effects in benthic foraminiferal $\delta^{18}O$ measurements. The deuterium measurements (δD) for the last 800 ka are from the 3.2 km deep EPICA Dome C core (EPICA Community Members, 2004; Jouzel et al., 2007). They come from samples of the ice itself and give a direct indication of Antarctic surface paleotemperature.

rates of global climate changes, as well as the interlinking of the ocean–atmosphere–terrestrial systems (see Lowe and Walker, 2015, for a more detailed discussion).

The climatically determined subdivisions of ice core records for the end of the last glacial period have been used by Björck et al. (1998) as a basis for an event-based stratigraphy. In this scheme, warmer and cooler events are referred to as stadials or interstadials, in common with other climatostratigraphic schemes applied to the Quaternary (see also Fig. 30.14). Moreover, the base of the Holocene Epoch and its two lowest subdivisions are now defined in Greenland ice cores (see Sections 30.7.1 and 30.7.2).

30.9.4 Speleothems

Advances in high-precision U/Th dating methods over the past decade have permitted precise dating of carbonate deposits in caves (speleothems) such as stalagmites and stalactites. In a series of groundbreaking papers, Wang et al. (2001, 2004) and Cheng et al. (2006, 2009) have

demonstrated that the paleoclimate record (principally $\delta^{18}O$) in speleothems is as detailed as that in ice cores and with comparable dating precision (Fig. 30.14). The GSSP for the Upper Holocene Meghalayan Stage is defined in a speleothem from northeastern India (see Section 30.7.3).

30.9.5 Long Lake records

Comparable evidence to that from ice cores has been retrieved in the last three decades by the recovery of sediment cores and associated studies of deep lake basins in the interiors of the continents. These investigations have produced sedimentary records of environmental changes over time spans of a few to hundreds of thousands of years and in some rare cases of the whole Quaternary. Examples include Lake Baikal in Asia (Prokopenko et al., 2006), Lake George in Australia (Singh et al., 1981), Lake Malawi in Africa (Lyons et al., 2007), Lake Biwa in Japan (Fuji, 1988), Tenaghi Philippon in Greece (van der Weil and Wijmstra, 1987a,b; Tzedakis et al., 2006), Lake Van in Turkey (Litt et al., 2009), La Grande Pile in France (Woillard, 1978; de Beaulieu and Reille, 1992), the Great Salt Lake in the United States (Eardley et al., 1973), and the Bogotá Basin in Columbia (Hooghiemstra, 1989). The long successions in these basins preserve lithological, biological, and geochemical records of past environmental changes which, together with the identification of climatic and physical events such as magnetic reversals and tephra strata, can be correlated to global as well as regional stratigraphies (e.g., Hooghiemstra and Sarmiento, 1991). The correlation is achieved by defining individual climatic or chronostratigraphies for each basin that are then compared and correlated with external records, mainly the marine isotope record (e.g., Tzedakis et al., 1997, 2006). For example, the Siberian Lake Baikal provides a bioproductivity record from the center of the world's largest landmass an area of extreme continental climate. High concentrations of biogenic silica indicate high aquatic productivity during interglacials (i.e., lake diatom blooms during ice-free summer seasons). The composite biogenic silica record from cores BDP-96-1, BDP-96-2, and BDP-98 is plotted on an astronomically tuned age scale (Prokopenko et al., 2006; Prokopenko and Khursevich, 2010)—see Fig. 30.2.

30.9.6 Mammal biochronology and the Villafranchian Stage

The term "Villafranchian" was introduced by Pareto (1865) as a continental stage referring to fluvial and lacustrine sediments in the Villafranca d'Asti (Piedmont) region of Italy that contained mammal fossils. Gignoux (1916) proposed a correlation with the marine Calabrian Stage that resulted in the Villafranchian being associated with the basal Pleistocene. However, the Villafranchian is now known to span an interval from about 3.5 to 1.0 Ma (i.e., Upper Pliocene–Lower

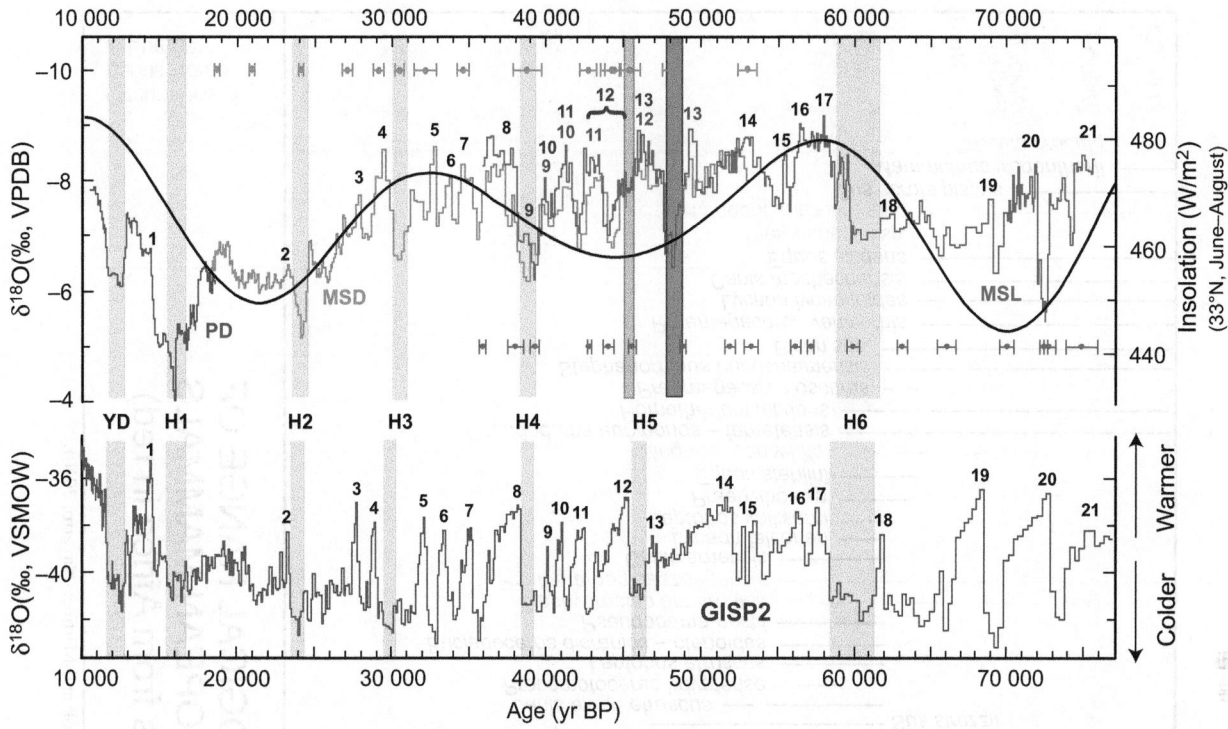

FIGURE 30.14 Comparison between δ^{18}O records from Hulu Cave (China) stalagmites and the GISP2 Greenland ice core (Wang et al., 2001), showing summer insolation at 33°N (the latitude of Hulu Cave). Cold climatic events (YD = Younger Dryas and H1−H6 = Heinrich Events) indicated by vertical bars. Interstadial (warm climate) events are numbered 1−21 in both records.

Pleistocene) and is defined as a Mammal Age, or biochronological unit based on the evolution of large European mammals (Rook and Martínez-Navarro, 2010; Fig. 30.15).

30.10 Ocean−sediment records

Because the span of Quaternary time includes our own, a different order of discrimination is possible and different methods are rapidly developing. The principal development in the Pleistocene time scale depends on the regularity of the climatic cycle that was discovered in 1842 by the French Mathematician Joseph Adhémer, just 5 years after Louis Agassiz announced his theory of a Great Ice Age. This astronomical approach was expanded significantly by James Croll in 1864 and especially by Milutin Milankovitch in 1920 and 1938. However, it was not taken seriously by Quaternary geologists until Zeuner (1945), Emiliani (1955), Broecker et al. (1968) and Evans (1971) were among those to recalculate and relate the astronomical parameters by testing, for example, 42- and 100-kyr cycles against other phenomena, such as the newly established oxygen isotope curve from the oceans. The first rigorous treatment using wide-ranging techniques was by Hays et al. (1976). Isotope studies from the bottom sediments of the world's oceans since then have indicated as many as 103 Quaternary glacial/interglacial cycles (Lisiecki and Raymo, 2005) and have clearly shown that the continental evidence can be very fragmentary.

The marine oxygen isotope scale makes use of the fact that when continental ice builds up as a result of global cooling, the ice is depleted in δ^{18}O relative to the ocean water, leaving the ocean water enriched in δ^{18}O. The oxygen isotope composition of calcareous foraminifera, coccoliths and siliceous diatoms varies in direct proportion to that of the water (see Shackleton and Opdyke, 1973, for discussion of the limitations of isotope stratigraphy). The 16 stages of Emiliani (1955, 1966) obtained from Caribbean and Atlantic sediment cores were extended to 22 by Shackleton and Opdyke (1973) after analysis of the V28-238 core from the equatorial Pacific. Subsequently, another equatorial Pacific core V28-239 (Shackleton and Opdyke, 1976) and an Atlantic core (Van Donk, 1976) extended the reconstruction of glacial−interglacial variability through the Pliocene/Pleistocene boundary. Later developments under the aegis of the Deep Sea Drilling Program resulted in the extension of the isotope record into the Early Pleistocene and Pliocene (Shackleton and Hall, 1989; Ruddiman et al., 1987; Raymo et al., 1989). The succession as shown in Fig. 30.2 is a composite benthic δ^{18}O record from 57 globally distributed sites, containing more than 38,000 individual measurements, to maximize the signal-to-noise ratio (Lisiecki and Raymo, 2005).

As regards nomenclature, the events differentiated in isotope sequences are termed MISs; this term is preferred

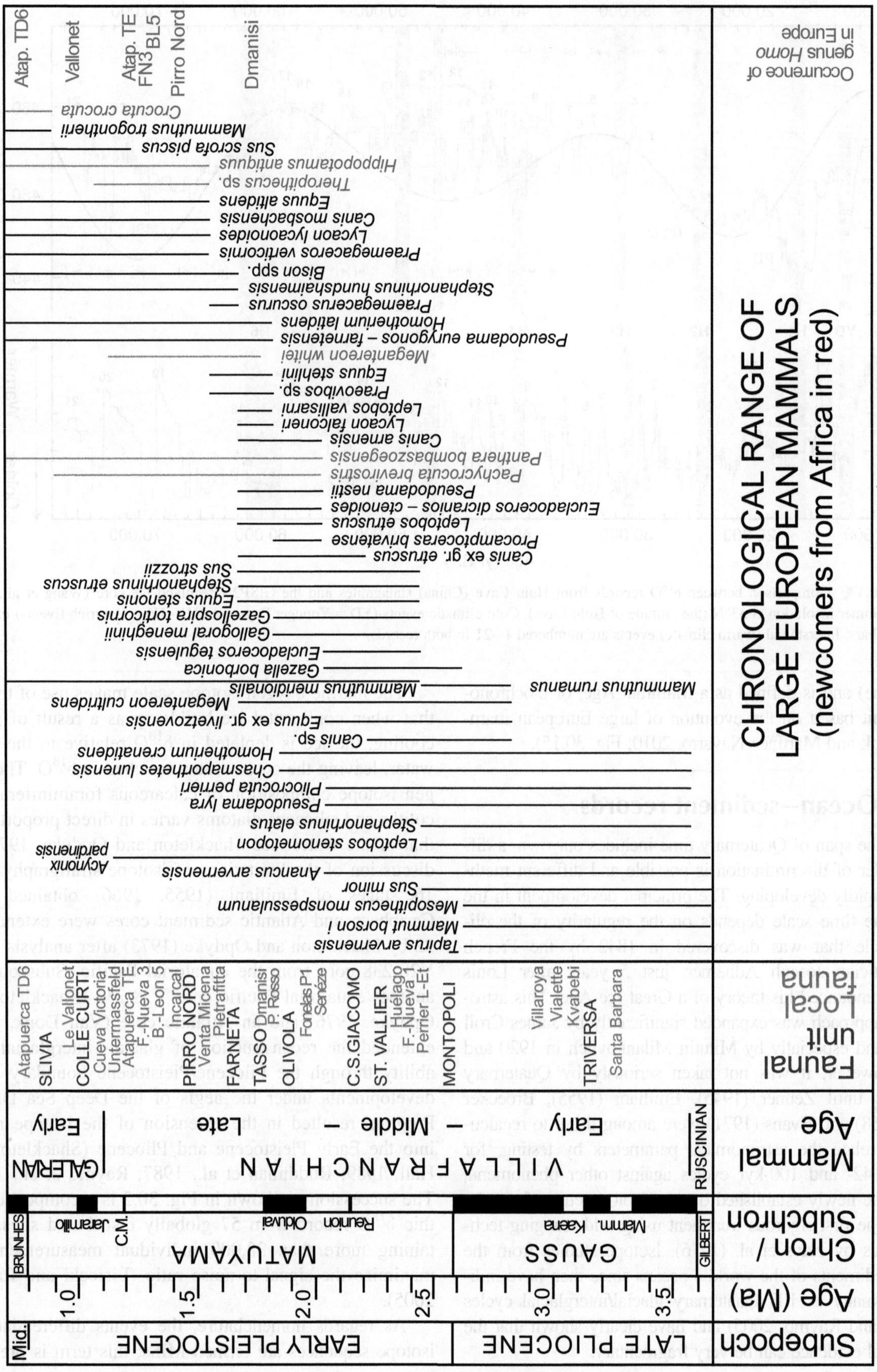

FIGURE 30.15 Chronology of the Villafranchian Mammal Age, with range chart for selected large mammals (Rook and Martínez-Navarro, 2010).

by paleoceanographers to the previously widely used oxygen isotope stages (OIS). This is because of the need to distinguish the isotope stages recognized from those identified from ice core or speleothem records (N. Shackleton, pers. comm.). The stages are numbered from the present-day (MIS 1) backward in time, such that cold-climate or glacial events are assigned even numbers and warm or interglacial (and interstadial) events are given odd numbers. Individual events or substages in MISs are indicated either by lower case letters or in some cases by a decimal system, thus MIS 5 is divided into warm substages 5a, 5c, and 5e, and cold substages 5b and 5d, or 5.1, 5.3, 5.5, and 5.2 and 5.4, respectively, named from the top downward (for discussion, see Railsback et al., 2015). This apparently unconventional top-downward nomenclature originates from Arrhenius' (1952) and Emiliani's (1955) original terminology and reflects the need to identify oscillations down-core from the ocean floor.

The biggest problem with climate-based nomenclature, such as the marine isotope stratigraphy, is where the boundaries should be drawn. Ideally, the boundaries should be placed at a major climate change. However, this is problematic because of the multifactorial nature of climate. But since the events are only recognized through the responses they initiate in depositional systems and biota, a compromise must be agreed. Although there are many places at which boundaries could be drawn, in principle in ocean—sediment cores, they are placed at midpoints between temperature maxima and minima. The boundary points thus defined in ocean records are assumed to be globally isochronous, although a drawback is that temperatures may be very locally influenced and may also show time lag. The extremely slow sedimentation rate of ocean-floor deposits and the relatively rapid mixing rate of oceanic waters argue in favor of the approach. Attempts to date these MIS boundaries are now well established (Martinson et al., 1987).

30.11 Land—sea correlation

In recent years, it has become common to correlate terrestrial successions directly with the MISs established in deep-sea cores. This arises from the need felt to correlate local successions to a regional or global time scale, occasioned by the fragmentary and highly variable nature of many terrestrial successions. The realization that more events are represented in the deep-sea, and indeed ice core, records than were recognized on land, together with the growth in geochronology and tephrochronology, has often led to the replacement of locally established terrestrial time scales. Instead, direct correlations of terrestrial successions to the global isotope scale are advanced, as advocated, for example, by Kukla (1977). The temptation to do this is understandable, but there are serious practical

limitations to this approach (Schlüchter, 1992; Gibbard and West, 2000).

In reality, there are very few means of directly and reliably correlating between the ocean and terrestrial sediment successions. Direct correlation can be achieved using markers that are preserved in both rock successions, such as magnetic reversals or tephra layers, and, rarely, fossil assemblages (particularly pollen). However, this is normally impossible over most of the records and in most geographical areas. Thus these correlations must rely on direct dating or less reliably on the technique of "curve matching," a widely used approach in the Quaternary. The latter can only reliably be achieved where long, continuous terrestrial successions are available, such as Long Lake records (e.g., Tzedakis et al., 1997), but even here it is not always straightforward (e.g., Watts et al., 1995) because of overprinting by local factors. Moreover, the possibility of failure to identify "leads-and-lags" in timing by the matching of curves is very real. In discontinuous successions that typify land and shelf environments, correlations with ocean-basin successions are potentially unreliable, in the absence of fossil groups distributed across the facies boundaries or potentially useful markers.

Increasingly, highly characteristic events are used as a basis for correlation. This event stratigraphy (e.g., Björck et al., 1998; Lowe et al., 1999; Alloway et al., 2007), typically using deposition of a tephra layer or magnetic reversals, can also include geological records of other potentially significant events such as floods, tectonic movements, changes of sea level, and climatic oscillations. Such occurrences, often termed "sub-Milankovitch events," may be preserved in a variety of environmental settings and thus offer important potential tools for high- to very high-resolution cross-correlation.

Of particular importance are the so-called "Heinrich layers" that represent major iceberg-rafting events in the North Atlantic Ocean (Heinrich, 1988; Bond et al., 1992; Bond and Lotti, 1995). These detritus bands can provide important lithostratigraphic markers for intercore correlation in ocean sediments, and the impact of their accompanying sudden coolings (Heinrich events) may be recognizable in certain sensitive terrestrial successions such as ice cores and speleothems (Fig. 30.14). Similarly, the essentially time—parallel periods of abrupt climate change termed "terminations" (Broecker and Van Donk, 1970), seen in marine isotope records (Fig. 30.2), can also be recognized on land as sharp changes in pollen assemblage composition or other parameters, for example, where sufficiently long and detailed successions are available, such as in Long Lake cores (see Tzedakis et al., 1997) and in speleothems (Cheng et al., 2009; see Fig. 30.16). However, their value for correlation may be limited in high-sedimentation-rate successions because these "terminations" are not instantaneous but have durations of several thousand years

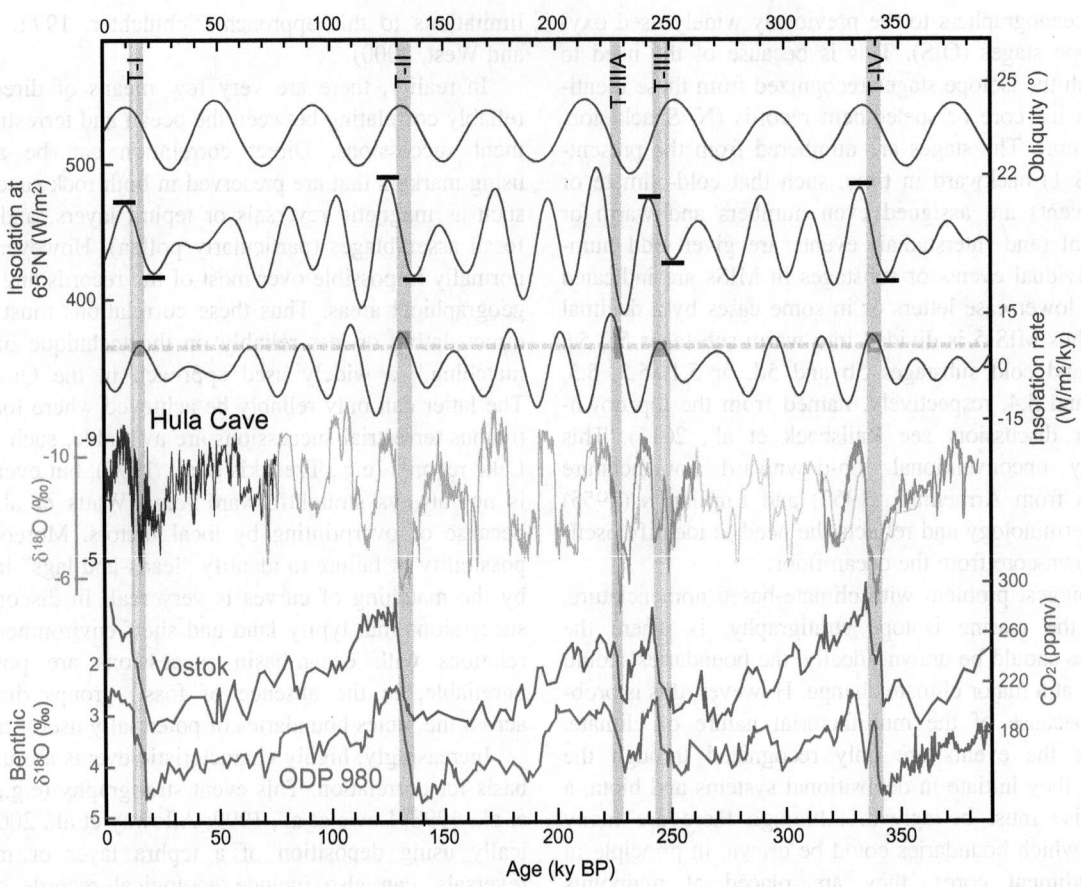

FIGURE 30.16 Correlations between summer insolation at 65°N, $\delta^{18}O$ in Hulu Cave stalagmites (China), Vostok CO_2 record (Antarctica), and benthic $\delta^{18}O$ in ODP 980 core (North Atlantic) (after Cheng et al., 2009).

(Broecker and Henderson, 1998). These matters essentially concern questions of resolution and scale.

Of greater concern for the development of a high-resolution marine—terrestrial correlation is that different proxies respond at different rates and in different ways to climate changes, and these changes themselves may be time transgressive. This has been forcefully demonstrated in marine corps off Portugal by Sanchez-Goñi et al. (1999) and Shackleton et al. (2003) where the MIS 6/5 boundary has been shown to have not been coeval with the Saalian—Eemian Stage boundary on land, as was previously assumed (Fig. 30.7). The same point concerns the MIS 1–2 boundary that predates the Holocene—Pleistocene boundary by some 2000–4000 years. Thus high-resolution land records and low-resolution marine records must be correlated with an eye to the detail since it cannot be assumed that the boundaries recognized in different situations are indeed coeval.

Notwithstanding these problems of detail, which will no doubt be further resolved as new evidence becomes available, it is now generally possible to relate the onshore—offshore successions fairly reliably at a coarse scale at least for the last glacial—interglacial cycle (Upper/Late Pleistocene). This was first proposed by Woillard (1978)

but is now well established (e.g., Tzedakis et al., 1997). For earlier climatic episodes, things are much more complicated; for example, witness the long-standing disagreements over the nature and duration of the northwest European Saalian Stage, already referred to earlier (Litt and Turner, 1993). Questions of whether the Holsteinian/ Hoxnian temperate (interglacial) stage relates to MIS 9 or 11 (e.g., Grün and Schwarcz, 2000; Geyh and Müller, 2005), and thus the immediately preceding Elsterian/ Anglian glacial stage (see Zagwijn, 1992) to MIS 10 or 12 (Turner, 1998; de Beaulieu and Reille, 1995; Litt et al., 2008) or even MIS 8, although now generally resolved, still persist in some quarters. These disagreements, together with precisely how many interglacial-type events, occur within the Saalian Stage, leave much potential for inaccuracy that cannot be resolved by "counting backward" methods. In the absence of reliable dating, these correlations represent little more than a matter of belief.

30.12 Quaternary dating methods

There is a much wider range of geological dating techniques underpinning Quaternary chronology than for earlier

TABLE 30.2 Classification of Quaternary dating methods according to type of method employed.

Sidereal	Isotopic	Radiogenic	Chemical/biological	Geomorphic	Correlation
Historical records	Radiocarbon	Fission track	Amino acid racemization	Weathering rinds	Paleomagnetism
Dendrochronology	K/Ar and Ar/Ar	Luminescence	Obsidian hydration	Soil development	Stable isotopes
Varve chronology	Uranium series	Electron spin resonance	Lichenometry	Geomorphic position	Biostratigraphy
Ice cores	Cosmogenic isotopes		Soil chemistry	Rate of deposition	Lithostratigraphy
	(U + Th)/He			Rate of deformation	Orbital variations
	U/Pb				Tephrochronology
	^{210}Pb				Tektites
					Rock varnish

Source: Modified from Colman, S.M., Pierce, K.L., Birkland, P.W., 1987, Suggested terminology for Quaternary dating methods. *Quaternary Research*, **28**: 314–319.

periods of the geologic time scale. Some of the techniques are solely applied to Quaternary deposits, while others can also be applied to older deposits. Here, we briefly summarize the principles and applications of the less familiar, but nonetheless important, Quaternary dating methods (Table 30.2). The time ranges over which each of the methods is used are shown in Fig. 30.17. General references on Quaternary dating methods include Easterbrook (1988), Rutter and Cato (1995), Wagner (1998), Noller et al. (2000), Walker (2005), and the Encyclopedia of Quaternary Science (Elias, 2007; Elias and Mock, 2013).

Radiocarbon or carbon-14 dating is probably the most familiar and widely used dating method in the Quaternary. Radiocarbon (^{14}C) is a cosmogenic radioactive isotope that is produced in the upper atmosphere by cosmic ray bombardment. It is then mixed through the atmosphere and the oceans and is incorporated into the bodies of animals and plants as well as nonorganic carbonate deposits such as speleothems. After death of an organism or after the ^{14}C is locked into a mineral structure, it undergoes radioactive decay, with a half-life of 5730 years. The method is therefore used for dating a wide range of organic and inorganic materials, such as shell, wood, charcoal, peat, and coral, which are up to about 50,000 years old. Since the amount of ^{14}C in the atmosphere is not constant over time, conversion of radiocarbon age to calendar age utilizes calibration curves from dated tree rings (Reimer et al., 2004).

Luminescence, natural radiation (a, b, g, and cosmic), produces electron and hole pairs that can be trapped at defects in a crystal lattice. Luminescence is emitted when electrons liberated from the traps by heating (thermoluminescence or TL) or exposure to visible-wavelength light (optically stimulated luminescence or OSL) or infrared light (infrared-stimulated luminescence or IRSL) recombine with trapped holes. The greater the radiation dose accrued, the greater the resulting luminescence. Age is found by calibrating the luminescence with known radiation doses, in combination with measurements of the environmental dose rate. The luminescence method dates the last time a sample was heated or exposed to sunlight and is especially used for dating pottery (time of heating) and eolian deposits (last exposure to sunlight). Samples up to about 1 million years old may be dated in some circumstances (e.g., Huntley et al., 1993).

Electron spin resonance (ESR): as with luminescence dating, ESR dating relies on radiation-producing unpaired electrons in crystal lattices, with the concentration of unpaired electrons increasing with time and radiation dose. However, in ESR dating the proportion of trapped electrons is measured directly (using an ESR spectrometer), rather than from the luminescence emitted by the electrons. The method is commonly used for dating carbonates such as corals, shells and speleothems, but arguably its most important application is the dating of fossil teeth (Grün, 2006, 2007).

Cosmogenic nuclides: cosmic ray interactions produce ^{3}He, ^{10}Be, ^{21}Ne, ^{26}Al, and ^{36}Cl in the atmosphere and lithosphere. Accumulation of these nuclides reflects duration of cosmic ray exposure within the upper 1−2 m of Earth's surface and also varies with altitude and latitude. Useful reviews include Gosse and Phillips (2001), Blard et al. (2006), and Balco et al. (2008). The technique is very important for dating moraines and rock surfaces associated with glacial erosion (e.g., Barrows et al., 2002; Balco et al., 2005; Phillips et al., 2009; and Balco and Rovey, 2010).

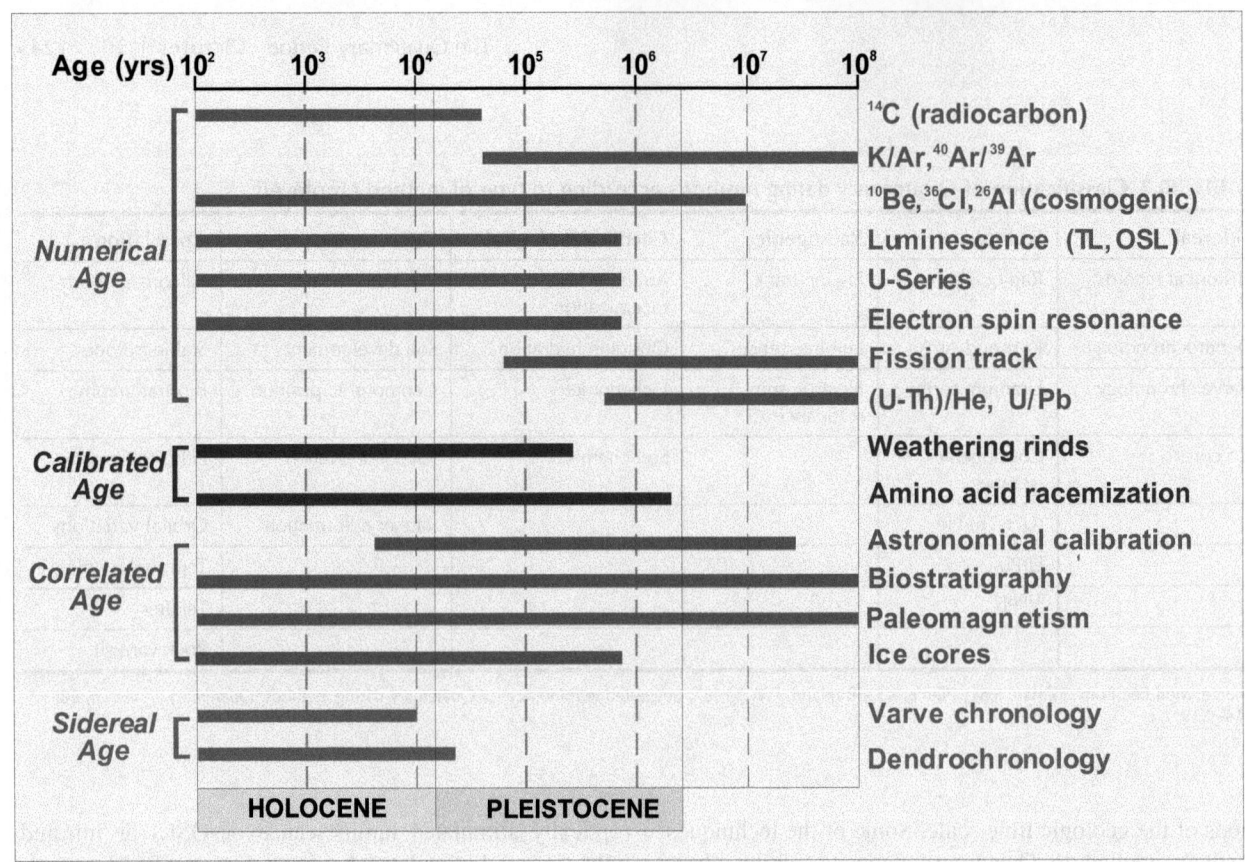

FIGURE 30.17 Quaternary dating methods grouped according to type of age result produced (Colman et al., 1987), showing age ranges over which each method can be applied.

GSSPs of the Quaternary Stages, with location and primary correlation criteria

Stage	GSSP Location	Latitude, Longitude	Boundary Level	Age	Correlation Events	Reference
Anthropocene Series	Informal term awaiting possible formal definition by GSSP			Mid-20th century	Events associated with the "Great Acceleration" of the mid-20th century	Head (2019)
Meghalayan	Mawmluh Cave, Meghalaya, India	25°15'44"N 91°42'54"E	7.45 mm from top of speleothem KM-A	4250 yr bk2 (before CE 2000)	Midpoint of two successive shifts in speleothem δ¹⁸0, 4.2 ka climate event	Walker et al. (2018, 2019a)
Northgrippian	NorthGRIP1 ice core, central Greenland	75.10°N 42.32°W	1228.67 m depth in NGRIP1 ice core	8236 yr b2k, MCE* 47 yr	Volcanic eruption at boundary; 8.2 ka climate event	Walker et al. (2018, 2019a)
Greenlandian	NorthGRIP2 ice core, central Greenland	75.10°N 42.32°W	1492.45 m depth in NGRIP2 ice core	11700 yr b2k, MCE 99 yr	Abrupt decline in deuterium excess values; shift in δ¹⁸0 record; end of the Younger Dryas cold spell	Walker et al. (2008, 2009, 2018, 2019a)
Upper Pleistocene				~129 ka	Onset of last interglacial	Head (2019)
Chibanian	Chiba, Japan	35°17'39.6"N 140°08'47.5"E	Base of Byk-E tephra bed	774 ka	Byk-E tephra; 1.1 m below directional midpoint of Brunhes–Matuyama magnetic reversal (base of Chron 1n)	Suganuma et al. (2018), Head, 2019)
Calabrian	Vrica, Calabria, Italy	39°02'18.61"N 17°08'05.79"E	Base of the marine claystone immediately overlying the sapropelic marker Bed 'e'	1.80 Ma	Bed 'e' assigned to MPRC** 176, coincides with MIS 65–64 transition, GSSP ~8 m below top of Olduvai Subchron	Aguirre & Pasini (1985), Cita et al. (2012)
Gelasian	Monte San Nicola, Sicily, Italy	37°08'45.64"N 14°12'15.22"E	Base of marly layer immediately overlying sapropelic Nicola Bed	2.58 Ma	Nicola Bed assigned to MPRC 250; coincides with MIS 103 and Gauss/Matuyama magnetic reversal (base of Chron 2)	Rio et al. (1998), Gibbard & Head (2010), Gibbard et al. (2010)

* MCE = maximum counting error, reflects the number of "annual" layers identified as uncertain.
** MPRC = Mediterranean Precession-Related Cycle.

FIGURE 30.18 Quaternary GSSP locations, boundary levels, age, primary correlation events, and key references (status as of January 2020). Details of each GSSP are available at http://www.stratigraphy.org, https://timescalefoundation.org/gssp/ and in the *Episodes* publications.

Uranium series: based on decay or accumulation of various parent/daughter isotopes in the U-series decay chains, especially $^{234}U/^{238}U$ and $^{230}Th/^{238}U$, the method is commonly used for dating carbonates such as cave speleothems (Cheng et al., 2009) and fossil corals (Stirling and Andersen, 2009). The method is generally not suitable for dating fossil mollusks because, unlike corals, mollusks do not incorporate U into their structure during growth. Rather, the uranium in mollusk shells is incorporated after death, by diagenetic processes. However, Eggins et al. (2005) have shown that laser ablation multi-collector ICPMS measurements can allow dating of mollusks in some circumstances. Dating of peat has also been attempted (e.g., Geyh and Müller, 2005).

Weathering rinds: the depth of chemical weathering on the surfaces of rocks progressively increases over time, to produce a weathering skin or rind. Has been used for dating glacial moraines and river terraces (e.g., Colman and Pierce, 1981; Knuepfer, 1988).

Amino acid racemization (AAR): living organisms contain dominantly L-amino acids. After death the amino acids undergo diagenetic changes, including racemization that converts L-amino acids into D-amino acids. The D/L ratio increases over time, at a rate dependent on temperature. AAR is ideally used for correlation and dating of coastal deposits containing molluscan shells (e.g., Bowen et al., 1998; Penkman et al., 2013) and can be used to calculate paleotemperatures when age is independently known (e.g., Miller et al., 1997).

Annual layer counting (including dendrochronology, varves, corals, and ice cores): Trees rings, seasonally layered sediments, ice cores, and the carbonate skeletons of various fossils (corals) all contain annual layers that can provide unparalleled precision in dating Quaternary events and deposits (e.g., Lowe and Walker, 2015).

Bibliography

Aguirre, E., and Pasini, G., 1985, The Pliocene-Pleistocene boundary. *Episodes*, **8**: 116–120.

Alloway, B.V., Lowe, D.J., Barrell, D.J.A., Newnham, R.M., Almond, P.C., Augustinus, P.C., et al., 2007, Towards a climate event stratigraphy for New Zealand over the past 30000 years (NZ-INTIMATE project). *Journal of Quaternary Science*, **22**: 9–35.

American Commission on Stratigraphic Nomenclature, 1961, Code of stratigraphic nomenclature. *Bulletin of the American Association of Petroleum Geologists*, **45**: 645–660.

An, Z., Lui, T., Porter, S.C., Kukla, G., Wu, X., and Hua, Y., 1990, The long-term paleomonsoon variation record by the loess-paleosol sequence in central China. *Quaternary International*, **7/8**: 91–95.

Andersen, K.K., Svensson, A., Johnsen, S.J., Rasmussen, S.O., Bigler, M., Rothlisberger, R., et al., 2006, The Greenland Ice Core chronology 2005, 15–42 ka. Part 1: Constructing the timescale. *Quaternary Science Reviews*, **25**: 3246–3257.

Anonymous, 1982–1984, *Stratigraphy of the USSR: Quaternary System*, vol. 1. Moscow: Nedra (1982), 443 pp. Vol. 2 (1984), 556 pp. [In Russian].

Anonymous, 1988, Biostratigraphy rejected for Pleistocene subdivisions. *Episodes*, **11** (3), 228.

Anthonissen, D.E., and Ogg, J.G., 2012, Cenozoic and Cretaceous biochronology of planktonic foraminifera and calcareous nannofossils. *In* Gradstein, F.M., Ogg, J.G., Schmitz, M., and Ogg, G. (eds), *The Geologic Time Scale 2012*. Amsterdam: Elsevier, **vol. 2**: pp. 1083–1127.

Arrhenius, G., 1952, Sediment Cores from the East Pacific. Reports of the Swedish Deep-sea Expedition, 1947-1948 5, Fascicles, 1, 227 p.

Arias, C., Azzaroli, A., Bigazzi, G., and Bonadonna, F., 1980, Magnetostratigraphy and Pliocene–Pleistocene boundary in Italy. *Quaternary Research*, **13**: 65–74.

Backman, J., Shackleton, N.J., and Tauxe, L., 1983, Quantitative nannofossil correlation to open ocean deep-sea sections from Plio-Pleistocene boundary at Vrica, Italy. *Nature*, **304**: 156–158.

Balco, G., and Rovey, C.W.I., 2010, Absolute chronology for major Pleistocene advances of the Laurentide Ice Sheet. *Geology*, **38**: 795–798.

Balco, G., Rovey, C.W.I., and Stone, J.O.H., 2005, The first glacial maximum in North America. *Science*, **307**: 222.

Balco, G., Stone, J.O., Lifton, N.A., and Dunai, T.J., 2008, A complete and easily accessible means of calculating surface exposure ages or erosion rates from ^{10}Be and ^{26}Al measurements. *Quaternary Geochronology*, **3**: 174–195.

Barrows, T.T., Stone, J.O., Fifield, L.K., and Cresswell, R.G., 2002, The timing of the Last Glacial Maximum in Australia. *Quaternary Science Reviews*, **21**: 159–173.

Bassett, M.G., 1985, Towards a common language in stratigraphy. *Episodes*, **8**: 87–92.

Bazin, L., Landais, A., Lemieux-Dudon, B., Kele, H.T.M., Veres, D., Parrenin, F., et al., 2013, An optimized multi-proxy, multi-site Antarctic ice and gas orbital chronology (AICC2012): 120–800 ka. *Climate of the Past*, **9**: 1715–1731.

Berggren, W.S., Hilgen, F.J., Langereis, C.G., Kent, D.V., Obradovich, J.D., Raffi, I., et al., 1995, Late Neogene chronology: new perspectives in high-resolution stratigraphy. *Geological Society of America Bulletin*, **107**: 1272–1287.

Beu, A.G., 2004, Marine mollusca of oxygen isotope stages of the last 2 million years in New Zealand. Part 1: Revised generic positions and recognition of warm-water and cool-water immigrants. *Journal of the Royal Society of New Zealand*, **34**: 111–265.

Björck, S., Walker, M.J.C., Cwynar, L.C., Johnsen, S., Knudsen, K.L., Lowe, J.J., et al., 1998, An event stratigraphy for the Last Termination in the North Atlantic region based on the Greenland ice-core record: a proposal of the INTIMATE group. *Journal of Quaternary Science*, **13**: 283–292.

Blard, P.-H., Bourles, D., Lave, J., and Pik, R., 2006, Applications of ancient cosmic-ray exposures: theory, techniques and limitations. *Quaternary Geochronology*, **1**: 59–73.

Boellstorff, J., 1978, North American Pleistocene stages reconsidered in light of probable Pliocene–Pleistocene continental glaciation. *Science*, **202**: 305–307.

Bond, G.C., and Lotti, R., 1995, Iceberg discharge into the North Atlantic on millenial time scales during the Last Glaciation. *Science*, **267**: 1005–1010.

Bond, G., Heinrich, H., Broecker, W., Labeyrie, L., McManus, J., Andrews, J., et al., 1992, Evidence for massive discharges of icebergs into the North Atlantic Ocean during the last glacial period. *Nature*, **360**: 245−249.

Bosch, J.H.A., Cleveringa, P., and Meijer, T., 2000, The Eemian stage in the Netherlands: history, character and new research. *Geologie en Mijnbouw/Netherlands Journal of Geosciences*, **79** (2/3), 135−145.

Bourdier, F., 1957, Quaternaire. *In* Pruvost, P. (ed), *Lexique Stratigraphique International*. Paris: Europe. Centre National de la Recherche Scientifique, **Vol. 1**: pp. 99−100.

Bowen, D.Q., 1978, *Quaternary Geology*, 1st ed. Oxford: Pergamon Press, p. 221.

Bowen, D.Q., 1999, A revised correlation of the Quaternary deposits in the British Isles. *Geological Society Special Report*, **23**: 174.

Bowen, D.Q., Pillans, B., Sykes, G.A., Beu, A.G., Edwards, A.R., Kamp, P.J.J., et al., 1998, Amino acid geochronology of Pleistocene marine sediments in the Wanganui Basin: a New Zealand framework for correlation and dating. *Journal of the Geological Society*, **155**: 439−446.

Brauer, A., Allen, J.R.M., Mingram, J., Dulski, P., Wulf, S., and Huntley, B., 2007, Evidence for last interglacial chronology and environmental change from Southern Europe. *Proceedings of the National Academy of Sciences, USA*, **104**: 450−455.

Broecker, W.S., and Henderson, G.M., 1998, The sequence of events surrounding Termination II and their implications for the cause of glacial-interglacial CO_2 changes. *Paleoceanography*, **13**: 352−364.

Broecker, W.S., and Van Donk, J., 1970, Insolation changes, ice volumes and the ^{18}O record in deep-sea cores. *Reviews of Geophysics and Space Physics*, **8**: 169−198.

Broecker, W.S., Thurber, D.L., Goddard, J., Ku, T.-L., Matthews, R.K., and Mesolella, K.J., 1968, Milankovitch hypothesis supported by precise dating of coral reefs and deep-sea sediments. *Science*, **159**: 297−300.

Butzer, K.W., and Isaac, G.L. (eds), 1975. *After the Australopithecines*. The Hague, Mouton, p. 991.

Capraro, L., Ferretti, P., Macrì, P., Scarponi, D., Tateo, F., Fornaciari, E., et al., 2017, The Valle di Manche section (Calabria, Southern Italy): a high resolution record of the Early−Middle Pleistocene transition (MIS 21−MIS 19) in the Central Mediterranean. *Quaternary Science Reviews*, **165**: 31−48.

Capron, E., Govin, A., Stone, E.J., Masson-Delmotte, V., Mulitza, S., Otto-Bliesner, B., et al., 2014, Temporal and spatial structure of multi-millennial temperature changes at high latitudes during the last interglacial. *Quaternary Science Reviews*, **103**: 116−133.

Channell, J., Di Stefano, E., and Sprovieri, R., 1992, Calcareous plankton biostratigraphy, magnetostratigraphy and paleoclimatic history of the Plio-Pleistocene Monte San Nicola Section (Southern Sicily). *Bollettino della Societa Paleontologica Italiana*, **31**: 351−382.

Channell, J.E.T., Xuan, C., and Hodell, D.A., 2009, Stacking paleointensity and oxygen isotope data for the last 1.5 Myr (PISO-1500). *Earth and Planetary Science Letters*, **283**: 14−23.

Channell, J.E.T., Hodell, D.A., Singer, B.S., and Xuan, C., 2010, Reconciling astrochronological and 40Ar/39Ar ages for the Matuyama-Brunhes boundary and late Matuyama Chron. *Geochemistry, Geophysics, Geosystems*, **11**: QOAA12. https://doi.org/10.1029/2010GC003203.

Channell, J.E.T., Hodell, D.A., and Curtis, J.H., 2016, Relative paleointensity (RPI) and oxygen isotope stratigraphy at IODP Site U1308: North Atlantic RPI stack for 1.2−2.2 Ma (NARPI-2200) and age of the Olduvai subchron. *Quaternary Science Reviews*, **131**: 1−19.

Cheng, H., Edwards, R.L., Wang, Y., Kong, X., Ming, Y., Kelly, M.J., et al., 2006, A penultimate glacial monsoon record from Hulu Cave and two-phase glacial terminations. *Geology*, **34**: 217−220.

Cheng, H., Edwards, R.L., Broecker, W.S., Denton, G.H., Kong, X., Wang, Y., et al., 2009, Ice age terminations. *Science*, **326**: 248−252.

Cita, M.B., and Castradori, D., 1994, Workshop on marine sections of Gulf of Taranto (Southern Italy) usable as potential stratotypes for the GSSP of the Lower, Middle and Upper Pleistocene. *Il Quaternario*, **7**: 677−692.

Cita, M.B., and Castradori, D., 1995, Workshop on marine sections from the Gulf of Taranto (southern Italy) usable as potential stratotypes for the GSSP of the Lower, Middle and Upper Pleistocene (Bari, Italy, September 29 to October 4, 1994). *Il Quaternario*, **7** (2), 677−692.

Cita, M.B., and Pillans, B., 2010, Global stages, regional stages or no stages in the Plio/Pleistocene? *Quaternary International*, **219**: 6−15.

Cita, M.B., Capraro, L., Ciaranfi, N., Di Stefano, E., Marino, M., Rio, D., et al., 2006, Calabrian and Ionian: a proposal for the definition of Mediterranean stages for Lower and Middle Pleistocene. *Episodes*, **29** (2), 107−114.

Cita, M.B., Capraro, L., Ciaranfi, N., Di Stephano, E., Lirer, F., Maiorano, P., et al., 2008, The Calabrian Stage redefined. *Episodes*, **31**: 408−419.

Cita, M.B., Gibbard, P.L., Head, M.J., and The Subcommission on Quaternary Stratigraphy, 2012, Formal ratification of the base Calabrian Stage GSSP (Pleistocene Series, Quaternary System). *Episodes*, **35** (3), 388−397.

Clark, P.U., Dyke, A.S., Shakun, J.D., Carlson, A.E., Clark, J., Wohlfarth, B., et al., 2009, The Last Glacial Maximum. *Science*, **325**: 710−771.

CLIMAP, 1981, Seasonal reconstructions of the Earth's surface at the last glacial maximum. *Map Series, Technical Report, MC-36*. Geological Society of America, Boulder.

Cohen, K., and Gibbard, P.L., 2019, Global chronostratigraphical correlation table for the last 2.7 million years, version 2019 QI-500. *Quaternary International*, **500**: 20−31.

Colman, S.M., and Pierce, K.L., 1981, Weathering rinds on andesitic and basaltic stones as a Quaternary age indicator, Western United States. *United States Geological Survey Professional Paper 1210*, 41.

Colman, S.M., Pierce, K.L., and Birkland, P.W., 1987, Suggested terminology for Quaternary dating methods. *Quaternary Research*, **28**: 314−319.

Crutzen, P.J., 2002, Geology of mankind. *Nature*, **415**: 23.

Crutzen, P.J., and Stoermer, E.F., 2000, Anthropocene. *Global Change Newsletter*, **41**: 17−18.

Dansgaard, W., et al., 1993, Evidence for general instability of past climate from a 250-kyr ice-core record. *Nature*, **364**: 218−220.

Davies, S.M., Wastegard, S., Abbott, P.M., Barbante, C., Bigler, M., Johnsen, S.J., et al., 2010, Tracing volcanic records in the NGRIP ice-core and synchronising North Atlantic marine records during the last glacial period. *Earth and Planetary Science Letters*, **294**: 69−79.

de Beaulieu, J.-L., and Reille, M., 1992, The last climatic cycle at La Grande Pile (Vosges, France): a new pollen profile. *Quaternary Science Reviews*, **11**: 431−438.

de Beaulieu, J.L., and Reille, M., 1995, Pollen records from the Velay craters: a review and correlation of the Holsteinian Interglacial with isotopic stage 11. *Mededelingen Van de Rijks Geologische Dienst*, **52**: 59−70.

de Heinzelin, J., and Tavernier, R., 1957, Flandrien. *In* Pruvost, P. (ed), *Lexique Stratigraphique International*. Europe, Paris: Centre National de la Recherche Scientifique, **vol. 1**: p. 32.

de Jong, J., 1988, Climatic variability during the past three million years, as indicated by vegetational evolution in northwest Europe and with emphasis on data from The Netherlands. *Philosophical Transactions of the Royal Society B*, **318**: 603−617.

Delmonte, B., Basile-Doelsch, I., Petit, J.R., Maggi, V., Revel-Rolland, M., Michard, A., et al., 2004, Comparing the EPICA and Vostok dust records during the last 220,000 years: stratigraphical correlation and provenance in glacial periods. *Earth-Science Reviews*, **66**: 63−87.

de Lumley, H., 1976, *La Préhistoire Française. Tome 1*. Paris: Editions CNRS, pp. 5−23.

De Schepper, S., Gibbard, P.L., Salzmann, U., and Ehlers, J., 2014, A global synthesis of the marine and terrestrial evidence for glaciation during the Pliocene Epoch. *Earth-Science Reviews*, **135**: 83−102.

Ding, Z.L., Derbyshire, E., Yang, S.L., Yu, Z.W., Xiong, S.F., and Liu, T.S., 2002, Stacked 2.6-Ma grain size record from the Chinese loess based on five sections and correlation with the deep-sea δ^{18}O record. *Paleoceanography*, **17**: 1033. https://doi.org/10.1029/2001PA000725.

Ding, Z.L., Derbyshire, E., Yang, S.L., Sun, J.M., and Liu, T.S., 2005, Stepwise expansion of desert environment across northern China in the past 3.5 Ma and implications for monsoon evolution. *Earth and Planetary Science Letters*, **237**: 45−55.

Dreyfus, G.B., Raisbeck, G.M., Parrenin, F., Jouzel, J., Guyodo, Y., Nomade, S., et al., 2008, An ice core perspective on the age of the Matuyama-Brunhes boundary. *Earth and Planetary Science Letters*, **274**: 151−156.

Eardley, A.J., Shuey, R.T., Gvodetsky, V., Nash, W.P., Picard, M.D., Grey, D.C., et al., 1973, Lake cycles in the Bonneville Basin, Utah. *Geological Society of America Bulletin*, **84**: 211−215.

Easterbrook, D.J. (ed), 1988. *Dating Quaternary Sediments*. Geological Society of America Special Paper, 227, p. 165.

Edgeworth, M., Ellis, E.C., Gibbard, P., Neal, C., and Ellis, M., 2019, The chronostratigraphic method is unsuitable for determining the start of the Anthropocene. *Progress in Physical Geography: Earth and Environment*, **43** (3), 334−344.

Eggins, S.M., Grün, R., McCulloch, M.T., Pike, A.W.G., Chappell, J., Kinsley, L., et al., 2005, In situ U-series dating by laser-ablation multicollector ICPMS: new prospects for Quaternary geochronology. *Quaternary Science Reviews*, **24**: 2523−2538.

Ehlers, J., and Gibbard, P.L., 2011, Quaternary glaciation. *In* Singh, V.P., Singh, P., and Haritashya, U.K. (eds), *Encyclopedia of Snow, Ice and Glaciers*. Dordrecht: Springer, pp. 873−882.

Ehlers, J., Gibbard, P.L., and Hughes, P.D., 2011, Introduction. *In* Ehlers, J., Gibbard, P.L., Hughes, P.D. (eds), *Quaternary Glaciations − Extent and Chronology, Part IV − A Closer Look*. Elsevier, Amsterdam, pp. 1−14.

Elias, S.A. (ed), 2007, *Encyclopedia of Quaternary Science*, **Vol. 4**. Elsevier, Amsterdam.

Elias, S.A., and Mock, C.J. (Eds), 2013, *Encyclopedia of Quaternary Science*. second edition. Elsevier, Amsterdam.

Ellenberger, F., 1994, *Histoire de la Géologie. Tome 2: Technique et documentation*. Lavoisier, Paris, p. 380.

Emiliani, C., 1955, Pleistocene temperatures. *Journal of the Geology*, **63**: 538−578.

Emiliani, C., 1966, Palaeotemperature analysis of Caribbean cores P6304-8 and P6304-9 and a generalized temperature curve for the past 425,000 years. *Journal of the Geology*, **74**: 109−126.

EPICA Community Members, 2004, Eight glacial cycles from an Antarctic ice core. *Nature*, **429**: 623−628.

Evans, P., 1971, Towards a Pleistocene time-scale. *In* Harland, W.B., Francis, H., and Evans, P. (eds), *The Phanerozoic Time-Scale, A Supplement. Part 2*. London: Geological Society of London, 123−356 pp.

Fisher, D.A., 2011, Connecting the Atlantic-sector and the north Pacific (Mt Logan) ice core stable isotope records during the Holocene: the role of El Niño. *Holocene*, **21**: 1117−1124.

Fisher, D., Osterberg, E., Dyke, A., Dahl-Jensen, D., Demuth, M., Zdanowicz, C., et al., 2008, The Mt Logan Holocene − late Wisconsinan isotope record: tropical Pacific − Yukon connections. *Holocene*, **18**: 667−677.

Fuji, N., 1988, Palaeovegetation and palaeoclimatic changes around Lake Biwa, Japan, during the last c. 3 million years. *Quaternary Science Reviews*, **7**: 21−28.

Geirsdottir, A., and Eriksson, J., 1994, Growth of an intermittent ice sheet in Iceland during the Late Pliocene and Early Pliocene. *Quaternary Research*, **42**: 115−130.

Gervais, P., 1867−1869, Zoologie et paleontology générales, *In: Nouvelles Recherches sur les Animaux Vertétebrés et Fossiles*, **Vol. 2**. Paris: Bertrand, p. 263.

Geyh, M.A., and Müller, H., 2005, Numerical 230Th/U dating and a palynological review of the Holsteinian/Hoxnian Interglacial. *Quaternary Science Reviews*, **24**: 1861−1872.

Gibbard, P.L., 2003, Definition of the Middle/Upper Pleistocene boundary. *Global and Planetary Change*, **36**: 201−208.

Gibbard, P.L., 2019, Giovanni Arduino − the man who invented the Quaternary. *Quaternary International*, **500**: 11−19.

Gibbard, P.L., and Head, M.J., 2010, The newly-ratified definition of the Quaternary System/Period and redefinition of the Pleistocene Series/ Epoch, and comparison of proposals advanced prior to formal ratification. *Episodes*, **33**: 152−158.

Gibbard, P.L., and van Kolfschoten, T., 2004, The Pleistocene and Holocene epochs. *In* Gradstein, F.M., Ogg, J.G., and Smith, A.G. (eds), *A Geologic Time Scale 2004*. Cambridge: Cambridge University Press, pp. 441−452.

Gibbard, P.L., and West, R.G., 2000, Quaternary chronostratigraphy: the nomenclature of terrestrial sequences. *Boreas*, **29**: 329−336.

Gibbard, P.L., West, R.G., Zagwijn, W.H. (eds), 1991. Early and early Middle Pleistocene correlations in the southern North Sea Basin. *Quaternary Science Reviews*, **10**: 23−52.

Gibbard, P.L., Smith, A.G., Zalasiewicz, J.A., Barry, T.L., Cantrill, D., Coe, A.L., et al., 2005, What status for the Quaternary? *Boreas*, **34**: 1−6.

Gibbard, P.L., Head, M.J., Walker, M., and The Subcommission on Quaternary Stratigraphy, 2010, Formal ratification of the Quaternary System/Period and the Pleistocene Series/Epoch with a base at 2.58 Ma. *Journal of Quaternary Science*, **25**: 96−102.

Gignoux, M., 1910, Sur la classification du Pliocène et du Quaternarie dans l'Italie du Sud. *Comptes rendus de l'Académie des Sciences*, **150**: 841−844.

Gignoux, M., 1916, L'étage Calabrien (Pliocéne supérieur marin) sur le versant Nord-est de l'Apennin, entre le Monte Gargano et Plaisance. *Bulletin de la Société Géologique de France, Series.*, **4** (14), 324−348.

Godwin, H., 1975, *The History of the British Flora*, 2nd ed. Cambridge: Cambridge University Press, p. 541.

Gosse, J.C., and Phillips, F.M., 2001, Terrestrial in situ cosmogenic nuclides: theory and application. *Quaternary Science Reviews*, **20**: 1475−1560.

Gradstein, F.M., Ogg, J.G., Smith, A.G. (eds), 2004. *A Geologic Time Scale 2004*. Cambridge University Press, p. 589.

Gradstein, F.M., Ogg, J.G., Schmitz, M.D., Ogg, G.M., (eds), *The Geologic Time Scale 2012*, **vols. 1 and 2.** Elsevier. Amsterdam.

Grün, R., 2006, *Direct dating of human remains.* Yearbook of Physical Anthropology, **49**: 2−48.

Grün, R., 2007, Electron spin resonance dating. *In* Elias, S.A. (ed), *Encyclopedia of Quaternary Science.* Amsterdam: Elsevier, **vol. 2**: pp. 1505−1516.

Grün, R., and Schwarcz, H.P., 2000, Revised open system U-series/ESR age calculations for teeth from Stratum C at the Hoxnian Interglacial type locality, England. *Quaternary Science Reviews*, **19**: 1151−1154.

Harland, W.B., Cox, A.V., Llewellyn, P.G., Picton, C.A.G., Smith, A.G., and Walters, R., 1982, *A Geologic Time Scale.* Cambridge: Cambridge University Press, p. xi, 131.

Harland, W.B., Armstrong, R.L., Cox, A.V., Craig, L.E., Smith, A.G., and Smith, D.G., 1990, *A Geologic Time Scale 1989.* Cambridge: Cambridge University Press, p. 263.

Haq, B.U., Berggren, W.A., and Van Couvering, J.A., 1977, Corrected age of the Plio/Pleistocene boundary. *Nature*, **269**: 483−488.

Hays, J.D., Imbrie, J., and Shackleton, N.J., 1976, Variations in the Earth's orbit: pacemaker of the ages. *Science*, **194**: 1121−1132.

Head, M.J., 2016, Defining the Upper Pleistocene Subseries: assessing Antarctic ice cores for a potential global boundary stratotype section and point (GSSP). In *35th International Geological Congress*, Cape Town, South Africa, August 27−September 4, 2016 [Abstract].

Head, M.J., 2019, Formal subdivision of the Quaternary System/Period: present status and future directions. *Quaternary International*, **500**: 32−51.

Head, M.J., and Gibbard, P.L., 2005, Early−Middle Pleistocene transitions: an overview and recommendation for the defining boundary. *In* Head, M.J., and Gibbard, P.L. (eds), Early−Middle Pleistocene Transitions: The Land−Ocean Evidence. *Geological Society of London, Special Publications, 247*: pp. 1−18.

Head, M.J., and Gibbard, P.L., 2015a, Formal subdivision of the Quaternary System/Period: past, present, and future. *Quaternary International*, **383**: 4−35.

Head, M.J., and Gibbard, P.L., 2015b, Early−Middle Pleistocene transitions: linking terrestrial and marine realms. *Quaternary International*, **389**: 7−46.

Head, M.J., Gibbard, P.L., and Salvador, A., 2008a, The Tertiary: a proposal for its formal definition. *Episodes*, **31** (2), 248−250.

Head, M.J., Gibbard, P.L., and Salvador, A., 2008b, The Quaternary: its character and definition. *Episodes*, **31** (2), 234−238.

Head, M.J., Pillans, B., and Farquhar, S., 2008c, The Early−Middle Pleistocene Transition: characterization and proposed guide for the defining boundary. *Episodes*, **31** (2), 255−259.

Head, M.J., Aubry, M.-P., Walker, M., Miller, K.G., and Pratt, B.R., 2017, A case for formalizing subseries (subepochs) of the Cenozoic Era. *Episodes*, **40** (1), 22−27.

Hedberg, H.D., 1976, *International Stratigraphic Guide.* New York: Wiley Interscience, p. 200.

Heinrich, H., 1988, Origin and consequences of cyclic ice-rafting in the Northeast Atlantic Ocean during the past 130000 years. *Quaternary Research*, **29**: 142−152.

Helama, S., and Oinonen, M., 2019, Exact dating of the Meghalayan lower boundary based on high-latitude tree-ring isotope chronology. *Quaternary Science Reviews*, **214**: 178−184.

Hennissen, J.A.I., Head, M.J., De Schepper, S., and Groeneveld, J., 2014, Palynological evidence for a southward shift of the North Atlantic

Current at ∼2.6 Ma during the intensification of late Cenozoic Northern Hemisphere glaciation. *Paleoceanography*, **28**, https://doi.org/10.1002/2013PA002543.

Hennissen, J.A.I., Head, M.J., De Schepper, S., and Groeneveld, J., 2015, Increased seasonality during the intensification of Northern Hemisphere glaciation at the Pliocene−Pleistocene boundary ∼2.6 Ma. *Quaternary Science Reviews*, **129**: 321−332.

Hilgen, F.J., 1991, Extension of the astronomically calibrated (polarity) time scale to the Miocene/Pliocene boundary. *Earth and Planetary Science Letters*, **107**: 349−368.

Hooghiemstra, H., 1989, Quaternary and Upper-Pliocene glaciations and forest development in the tropical Andes: evidence from a long high-resolution pollen record from the sedimentary basin of Bogotá, Columbia. *Palaeogeography, Palaeoclimatology, Palaeoecology*, **72**: 11−26.

Hooghiemstra, H., and Sarmiento, G., 1991, Long continental pollen record from a tropical intermontane basin: Late Pliocene and Pleistocene history from a 540-meter core. *Episodes*, **14**: 107−115.

Hopwood, A.T., 1935, Fossil elephants and Man. *Proceedings of the Geologists' Association*, **46**: 46−60.

Hughes, P.D., and Gibbard, P.L., 2015, A stratigraphical basis for the Last Glacial Maximum (LGM). *Quaternary International*, **383**: 174−185.

Hughes, P.D., Gibbard, P.L., and Ehlers, J., 2013, Timing of glaciation during the last glacial−interglacial cycle: evaluating the meaning and significance of the 'Last Glacial Maximum' (LGM). *Earth-Science Reviews*, **125**: 171−198.

Huntley, D.J., Hutton, J.T., and Prescott, J.R., 1993, The stranded beach-dune sequence of south-east South Australia: a test of thermoluminescence dating, 0-800 ka. *Quaternary Science Reviews*, **12**: 1−20.

Hyvärinen, H., 1978, Use and definition of the term Flandrian. *Boreas*, **7**: 182.

Jessen, K., and Milthers, V., 1928, Stratigraphical and palaeontological studies of interglacial freshwater deposits in Jutland and north-west Germany. *Danmarks Geologisk Undersøgelse*, **II** (48), 379.

Johnsen, S.J., Dansgaard, W., Clausen, H.B., and Langway, C.C., 1972, Oxygen isotope profiles through Antarctic and Greenland ice sheets. *Nature*, **235**: 429−433.

Jouzel, J., Masson-Delmotte, V., Cattani, O., Dreyfus, G., Falourd, S., Hoffmann, G., et al., 2007, Orbital and millennial Antarctic climate variability over the past 800,000 years. *Science*, **317**: 793−796.

Kathayat, G., Cheng, H., Sinha, A., Berkelhammer, M., Zhang, H., Duan, P., et al., 2018, Evaluating the timing and structure of the 4.2 ka event in the Indian summer monsoon domain from an annually resolved speleothem record from Northeast India. *Climate of the Past*, **14**: 1869−1879.

Kazaoka, O., Suganuma, Y., Okada, M., Kameo, K., Head, M.J., Yoshida, T., et al., 2015, Stratigraphy of the Kazusa Group, Boso Peninsula: an expanded and highly-resolved marine sedimentary record from the Lower and Middle Pleistocene of central Japan. *Quaternary International*, **383**: 116−135.

King, W.B.R., and Oakley, K.P., 1949, Definition of the Pliocene-Pleistocene boundary. *Nature*, **163**: 186−187.

Knuepfer, P.L.K., 1988, Estimating ages of late Quaternary stream terraces from analysis of weathering rinds and soils. *Geological Society of America Bulletin*, **100**: 1224−1236.

Krasnenkov, R.V., Iossifova, Yu.I., and Semenov, V.V., 1997, The Upper Don drainage basin – an important stratoregion for climatic stratigraphy of the early Middle Pleistocene (the early Neopleistocene) of Russia. *In* Alekseev, M.N., and Khoreva, I.M. (eds), *Quaternary Geology and Paleogeography of Russia*. Moscow: Geosynthos, pp. 82–96. [In Russian, abstract in English].

Kukla, G.J., 1977, Pleistocene land-sea correlations I. Europe. *Earth-Science Reviews*, **13**: 307–374.

Laj, C., and Channell, J.E.T., 2007, Geomagnetic excursions. Chapter 10. *In* Kono, M. (ed), *Treatise on Geophysics: Volume 5, Geomagnetism*. Amsterdam: Elsevier, pp. 373–416.

Lawrence, K.T., Bailey, I., and Raymo, M.E., 2013, Re-evaluation of the age model for North Atlantic Ocean Site 982 – arguments for a return to the original chronology. *Climate of the Past*, **9**: 2391–2397.

Lisiecki, L.E., and Raymo, M.E., 2005, A Pliocene-Pleistocene stack of 57 globally distributed benthic δ18O records. *Paleoceanography*, **20**: PA1003. https://doi.org/10.1029/2004PA001071.

Litt, T., and Turner, C., 1993, Arbeitsergbenisse der Subkommission für Europäische Quartärstratigraphie: Die Saalesequenz in der Typusregion (Berichte der SEQS 10). *Eiszeitalter und Gegenwart*, **43**: 125–128.

Litt, T., and Gibbard, P.L., 2008, A proposed Global Stratotype Section and Point (GSSP) for the base of the Upper (Late) Pleistocene Subseries (Quaternary System/Period). *Episodes*, **31**: 260–261.

Litt, T., Bettis, E.A., Bosch, A., Dodonov, A., Gibbard, P.L., Jiaqui, L., et al., 2008, A proposal for the Global Stratotype Section and Point (GSSP) for the Middle/Upper (Late) Pleistocene Subseries Boundary (Quaternary System/Period). Subcommission on Quaternary Stratigraphy, Cambridge, p. 35.

Litt, T., Krastel, S., Sturm, M., Kipfer, R., Örcen, S., Heumann, G., et al., 2009, 'PALEOVAN', International Continental Scientific Drilling Program (ICDP): site survey results and perspectives. *Quaternary Science Reviews*, **28**: 1555–1567.

Lourens, L.J., Hilgen, F.J., Raffi, I., and Vergnaud-Grazzini, C., 1996a, Early Pleistocene chronology of the Vrica section (Calabria, Italy). *Paleoceanography*, **11** (6), 797–812.

Lourens, L.J., Antonarakou, A., Hilgen, F.J., Van Hoof, A.A.M., Vergnaud-Grazzini, C., and Zachariasse, W.J., 1996b, Evaluation of the Plio-Pleistocene astronomical timescale. *Paleoceanography*, **11** (4), 391–413.

Lourens, L.J., Hilgen, F.J., Laskar, J., Shackleton, N.J., and Wilson, D., 2004, The Neogene Period. *In* Gradstein, F.M., Ogg, J.G., and Smith, A.G. (eds), *A Geologic Time Scale 2004*. Cambridge: Cambridge University Press, pp. 409–440.

Lowe, J.J., and Walker, M.J.C., 1997, *Reconstructing Quaternary Environments*, 2nd ed. Harlow: Longman, p. 446.

Lowe, J.J., and Walker, M.J.C., 2015, *Reconstructing Quaternary Environments*, 3rd ed. Abingdon: Routledge, p. 538.

Lowe, J.J., Birks, H.H., Brooks, S.J., Coope, G.R., Harkness, D.D., Mayle, F.E., et al., 1999, The chronology of palaeoenvironmental changes during the last Glacial–Holocene transition: towards an event stratigraphy for the British Isles. *Journal of the Geological Society*, **156**: 397–410.

Luthi, D., Le Floc, M., Bereiter, B., Blunier, T., Barnola, J.M., Siegenthaler, U., et al., 2008, High-resolution carbon dioxide concentration record 650,000–800,000 years before present. *Nature*, **453**: 379–382.

Lyons, R.P., Scholz, C.A., King, J.W., Cohen, A.S., and Johnson, T.C., 2007, High amplitude climate variability in a tropical rift-lake: Correlation of drillcore and seismic reflection data in Lake Malawi,

East Africa. *AAPG Annual Meeting*, April 1–4. AAPG Search and Discover Article, Long Beach, California. 90063.

Mangerud, J., Andersen, S.T., Berglund, B.E., and Donner, J.J., 1974, Quaternary stratigraphy of Norden, a proposal for terminology and classification. *Boreas*, **3**: 109–128.

Martini, E., 1971, Standard Tertiary and Quaternary calcareous nanno-plankton zonation. *Proceedings of the Second Planktonic Conference*, 1970, Rome, Tecnoscienza **2**, 739–785.

Martinson, D.G., Pisias, N.G., Hays, J.D., Imbrie, J., Moore, T.C., and Shackleton, N.J., 1987, Age dating and the orbital theory of the ice ages: Development of a high-resolution 0 to 300 000 year chronostratigraphy. *Quaternary Research*, **27**: 1–29.

Michel, J.-P., Fairbridge, R.W., and Carpenter, M.S.N., 1997, *Dictionnaire des Sciences de la Terre*, 3rd ed. Paris and Chichester: Masson and J. Wiley & Sons, p. 346.

Miller, G.H., Magee, J.W., and Jull, A.J.T., 1997, Low-latitude cooling in the Southern Hemisphere from amino acids in emu eggshell. *Nature*, **385**: 241–244.

Mitchell, G.F., Penny, L.F., Shotton, F.W., and West, R.G., 1973, A Correlation of Quaternary deposits in the British Isles. *Geological Society Special Report*, **4**: 99.

Mix, A.C., Bard, E., and Schneider, R., 2001, Environmental processes of the ice age: land, oceans, glaciers (EPILOG). *Quaternary Science Reviews*, **20**: 627–657.

Mörner, N.A., 1976, The Pleistocene/Holocene boundary: proposed boundary stratotypes in Gothenburg, Sweden. *Boreas*, **5**: 193–275.

Negri, A., Amorosi, A., Antonioli, F., Bertini, A., Florindo, F., Lurcock, P.C., et al., 2015, A potential global boundary stratotype section and point (GSSP) for the Tarentian Stage, Upper Pleistocene, from the Taranto area (Italy): results and future perspectives. *Quaternary International*, **383**: 145–157.

Nishida, N., Kazaoka, O., Izumi, K., Suganuma, Y., Okada, M., Yoshida, T., et al., 2016, Sedimentary processes and depositional environments of a continuous marine succession across the Lower–Middle Pleistocene boundary: Kokumoto Formation, Kazusa Group, central Japan. *Quaternary International*, **397**: 3–15.

Noller, J.S., Sowers, J.M., and Lettis, W.R., 2000, Quaternary geochronology. Methods and applications. *American Geophysical Union Reference Shelf*, **4**: 581.

Nomade, S., Bassinot, F., Marino, M., Simon, Q., Dewilde, F., Maiorano, P., et al., 2019, High-resolution foraminifer stable isotope record of MIS 19 at Montalbano Jonico, southern Italy: a window into Mediterranean climatic variability during a low-eccentricity interglacial. *Quaternary Science Reviews*, **205**: 106–125.

North Greenland Ice Core Project Members, 2004, High-resolution record of Northern Hemisphere climate extending into the last interglacial period. *Nature*, **431**: 147–151.

Nystuen, J.P. (Ed.), 1986. Regler og råd for navnsetting av geologiske enheter i Norge. *Norsk Geologisk Tidsskrift*, **66** (Suppl. 1), p. 96.

Ohno, M., Hayashi, T., Komatsu, F., Murakami, F., Zhao, M., Guyodo, Y., et al., 2012, A detailed paleomagnetic record between 2.1 and 2.75 Ma at IODP Site U1314 in the North Atlantic. Geomagnetic excursions and the Gauss–Matuyama transition. *Geochemistry, Geophysics, Geosystems*, **13** (1), Q12Z39.

Okada, H., and Bukry, D., 1980, Supplementary modification and introduction of code numbers to the low-latitude coccolith biostratigraphic zonation (Bukry, 1973; 1975). *Marine Micropaleontology*, **5**: 321–325.

Okada, M., Suganuma, Y., Haneda, Y., and Kazaoka, O., 2017, Paleomagnetic direction and paleointensity variations during the Matuyama—Brunhes polarity transition from a marine succession in the Chiba composite section of the Boso Peninsula, central Japan. *Earth Planets Space*, **69** (45). https://doi.org/10.1186/s40623-017-0627-1.

Olausson, E. (ed), 1982. *The Pleistocene/Holocene boundary in southwestern Sweden*. Sveriges Geologiska Undersökning, **794**, p. 288.

Parandier, H., 1891, Notice géologique et paléontologique sur la nature des terrains traversés par le chemin de fer entre Dijon at Châlons-sur-Saône. *Bulletin de la Société Géologique de France, Series*, **3** (19), 794—818.

Pareto, L., 1865, Note sur les subdivisions que l'on pourrait établir dans les terrains tertiaires de l'Apennin septentrional. *Bulletin de la Société Géologique de France*, **22**: 210—217.

Parrenin, F., Barnola, J.M., Beer, J., Blunier, T., Castellano, E., Chappellaz, J., et al., 2007, The EDC3 chronology for the EPICA Dome C ice core. *Climate of the Past*, **3**: 485—497.

Pasini, G., and Colalongo, M.L., 1997, The Pliocene—Pleistocene boundary-stratotype at Vrica, Italy. *In* Van Couvering, J. (ed), *The Pleistocene Boundary and the Beginning of the Quaternary*. Cambridge, UK: Cambridge University Press, pp. 15—45.

Penck, A., and Brückner, E., 1909—1911, Die Alpen im Eiszeitalter. *Taunitz, Leipz.*, p. 1199.

Penkman, K.E.H., Preece, R.C., Bridgland, D.R., Keen, D.H., Meijer, T., Parfitt, S.A., et al., 2013, An aminostratigraphy for the British Quaternary based on *Bithynia* opercula. *Quaternary Science Reviews*, **61**: 111—134.

Petit, J.R., Jouzel, J., Raynaud, D., Barkov, N.I., Barnola, J.-M., Basile, I., et al., 1999, Climate and atmospheric history of the past 420,000 years from the Vostok ice core, Antarctica. *Nature*, **399**: 429—436.

Phillips, F.M., Zreda, M., Plummer, M.A., Elmore, D., and Clark, D.H., 2009, Glacial geology and chronology of Bishop Creek and vicinity, eastern Sierra Nevada, California. *Geological Society of America Bulletin*, **121**: 1013—1033.

Pillans, B., 1991, New Zealand Quaternary stratigraphy: an overview. *Quaternary Science Reviews*, **10**: 405—418.

Pillans, B., and Gibbard, P., 2012, The Quaternary Period. *In* Gradstein, F.M., Ogg, J.G., Schmitz, M.D., and Ogg, G.M., (eds), *The Geologic Time Scale 2012*. Amsterdam: Elsevier, **vol. 2**: pp. 979—1010.

Pol, K., Masson-Delmotte, V., Cattani, O., Debret, M., Falourd, S., Jouzel, J., et al., 2014, Climate variability features of the last interglacial in the East Antarctic EPICA Dome C ice core. *Geophysical Research Letters*, **41**: 4004—4012.

Prokopenko, A.A., and Khursevich, G.K., 2010, Plio-Pleistocene transition in the continental record from Lake Baikal: Diatom biostratigraphy and age model. *Quaternary International*, **219**: 26—36.

Prokopenko, A.A., Hinnov, L.A., Williams, D.F., and Kuzmin, M.I., 2006, Orbital forcing of continental climate during the Pleistocene: a complete astronomically tuned record from Lake Baikal, SE Siberia. *Quaternary Science Reviews*, **25**: 3431—3457.

Railsback, L.B., Gibbard, P.L., Head, M.J., Voarintsoa, N.R.G., and Toucanne, S., 2015, An optimized scheme of lettered marine isotope substages for the last 1.0 million years, and the climatostratigraphic nature of isotope stages and substages. *Quaternary Science Reviews*, **111**: 94—106.

Railsback, L.B., Liang, F., Brook, G.A., Voarintsoa, N.R.G., Sletten, H.R., Marais, E., et al., 2018, The timing, two-pulsed nature, and variable climatic expression of the 4.2 ka event: a review and high-resolution stalagmite data from Namibia. *Quaternary Science Reviews*, **186**: 78—90.

Raymo, M.E., Ruddiman, W.F., Backman, J., Clement, B.M., and Martinson, D.G., 1989, Late Pliocene variation in northern hemisphere ice sheets and North Atlantic deep water circulation. *Paleoceanography*, **4** (4), 413—446.

Reimer, P.J., Baillie, M.G.L., and Bard, E., 2004, IntCal04 terrestrial radiocarbon age calibration/comparison, 26—0 ka B.P. *Radiocarbon*, **46**: 1029—1058.

Richmond, G.M., 1996, The INQUA-approved provisional Lower-Middle Pleistocene boundary. *In* Turner, C. (ed), *The Early Middle Pleistocene in Europe*. Rotterdam: Balkema, pp. 319—326.

Rio, D., Sprovieri, R., Castradori, D., and Di Stefano, E., 1998, The Gelasian Stage (Upper Pliocene): a new unit of the global standard chronostratigraphic scale. *Episodes*, **21**: 82—87.

Roberts, A.P., Florindo, F., Larrasoana, J.C., O'Regan, M.A., and Zhao, X., 2010, Complex polarity pattern at the former Plio-Pleistocene global stratotype section at Vrica (Italy): remagnetization by magnetic iron sulphides. *Earth and Planetary Science Letters*, **292**: 98—111.

Rook, L., and Martínez-Navarro, B., 2010, Villafranchian: the long story of a Plio-Pleistocene European large mammal biochronological unit. *Quaternary International*, **219**: 134—144.

Roy, M., Clark, P.U., Barendregt, R.W., Glasmann, J.R., and Enkin, R.J., 2004, Glacial stratigraphy and paleomagnetism of late Cenozoic deposits of the north-central United States. *Geological Society of America Bulletin*, **116**: 30—41.

Ruddiman, W.F., Kidd, R.B., Thomas, E., et al., 1987, *Initial Reports Deep Sea Drilling Project, 94*. Washington: U.S. Government Printing Office. http://dx.doi.org/10.2973/dsdp.proc.94.1987.

Ruddiman, W.F., 2018, Three flaws in defining a formal 'Anthropocene'. *Progress in Physical Geography: Earth and Environment*, **42** (4): 451—461.

Ruddiman, W.F., 2019, Reply to Anthropocene Working Group response. *Progress in Physical Geography: Earth and Environment*, **43** (3): 345—351.

Rutter, N.W., and Cato, E.R., 1995, *Dating Methods for Quaternary Deposits*. Geological Association Canada, Geotext, **2**: 308.

Rutter, N., Ding, Z., Evans, M.E., and Liu, T.S., 1991, Baoji-type pedostratigraphic section, Loess Plateau, north-central China. *Quaternary International*, **10**: 1—22.

Sagnotti, L., Cascella, A., Ciaranfi, N., Macri, P., Maiorano, P., Marino, M., et al., 2010, Rock magnetism and palaeomagnetism of the Montalbano Jonico section (Italy): evidence for late diagenetic growth of greigite and implications for magnetostratigraphy. *Geophysical Journal International*, **180**: 1049—1066.

Salvador, A., 1994, *International Stratigraphic Guide*, 2nd ed. Trondheim: International Union of Geological Sciences and Geological Society of America, p. 214.

Sanchez-Goñi, M.F., Eynaud, F., Turon, J.L., and Shackleton, N.J., 1999, High resolution palynological record off the Iberian margin: direct land-sea correlation for the Last Interglacial complex. *Earth and Planetary Science Letters*, **171**: 123—137.

Schlüchter, C., 1992, Terrestrial Quaternary stratigraphy. *Quaternary Science Reviews*, **11**: 603—607.

Schneer, C.J., 1969, Introduction. *In* Schneer, C.J. (ed), *Towards a History of Geology*. The Massachusetts Institute of Technology Press: Cambridge and London, pp. 1—18.

Scotese, C.R., 2014, *Atlas of Neogene Paleogeographic Maps (Mollweide Projection), Maps 1-7, Volume 1, The Cenozoic, PALEOMAP Atlas for ArcGIS*, PALEOMAP Project, Evanston, IL., 2014.

Shackleton, N.J., 1977, The oxygen isotope stratigraphic record of the Late Pleistocene. *Philosophical Transactions of the Royal Society B,* **280**: 169–182.

Shackleton, N.J., and Hall, M.A., 1989, Stable isotope history of the Pleistocene at ODP site 677. *Proceedings of the Ocean Drilling Program, Scientific Results,* **111**: 295–316.

Shackleton, N.J., and Opdyke, N.D., 1973, Oxygen isotope and palaeo-magnetic stratigraphy of the equatorial Pacific core V28-238: oxygen isotope temperatures and ice volumes on a 10^5 and 10^6 year scale. *Quaternary Research,* **3**: 39–55.

Shackleton, N.J., and Opdyke, N.D., 1976, Oxygen isotope and palaeo-magnetic stratigraphy of Equatorial Pacific core V28-239, Late Pliocene to Latest Pleistocene. *In* Cline, R.M., and Hays, J.D. (eds), Investigation of Late Quaternary Paleoceanography and Paleoclimatology. *Geological Society of America Memoir,* **145**: pp. 449–464.

Shackleton, N.J., Sanchez-Goñi, M.F., Pailler, D., and Lancelot, Y., 2003, Marine Isotope Substage 5e and the Eemian Interglacial. *Global and Planetary Change,* **36**: 151–155.

Shik, S.M., 2014, A modern approach to the Neopleistocene stratigraphy and paleogeography of Central European Russia. *Stratigraphy and Geological Correlation,* **22**: 219–230. Translation of: Shik, S.M., 2014. Stratigrafiya. *Geologicheskaya Korrelyatsiya,* **22**: 108–120. (In Russian).

Shik, S.M., Tesakov, A.S., Agadjanian, A.K., Iosifova, Yu.I., Markova, A.K., Pisareva, V.V., et al., 2015, A regional stratigraphical scheme of Eopleistocene and Gelasian (Paleopleistocene) of Central and Southern parts of European Russia. *Bulletin of Regional Intra-Institutional Commission of Central and South of Russian Plains,* **6**: 97–107. (In Russian).

Sibrava, V., 1978, Isotopic methods in Quaternary geology. *In* Cohee, G.V., Glaessner, M.F., and Hedberg, H.D. (eds), Contributions to the Geologic Time Scale. *Studies in Geology,* **6**: 165–169.

Singer, B.S., Ackert, R.P.J., and Guillou, H., 2004, $^{40}Ar/^{39}Ar$ and K-Ar chronology of Pleistocene glaciations in Patagonia. *Geological Society of America Bulletin,* **116**: 434–450.

Singh, G., Opdyke, N.D., and Bowler, J.M., 1981, Late Cainozoic stratigraphy, palaeomagnetic chronology and vegetational history from Lake George, N.S.W. *Journal of the Geological Society of Australia,* **28**: 435–452.

Stevens, T., Buylaert, J.P., Thiel, C., Újvári, G., Yi, S., Murray, A.S., et al., 2018, Ice-volume-forced erosion of the Chinese Loess Plateau global Quaternary stratotype site. *Nature Communications,* **9**: 983–995.

Stirling, C.H., and Andersen, M.B., 2009, Uranium-series dating of fossil coral reefs: extending the sea-level record beyond the last glacial cycle. *Earth and Planetary Science Letters,* **284**: 269–283.

Suc, J.P., Bertini, A., Leroy, S.A.G., and Suballyova, D., 1997, Towards a lowering of the Pliocene/Pleistocene boundary to the Gauss/Matuyama Reversal. *In* Partridge, T.C. (ed), The Plio-Pleistocene Boundary. *Quaternary International,* **40**: pp. 37–42.

Suganuma, Y., Okada, M., Horie, K., Kaiden, H., Takehara, M., Senda, R., et al., 2015, Age of Matuyama–Brunhes boundary constrained by U-Pb zircon dating of a widespread tephra. *Geology,* **43**: 491–494.

Suganuma, Y., Haneda, Y., Kameo, K., Kubota, Y., Hayashi, H., Itaki, T., et al., 2018, Paleoclimatic and paleoceanographic records through Marine Isotope Stage 19 at the Chiba composite section, central Japan: a key reference for the Early–Middle Pleistocene Subseries boundary. *Quaternary Science Reviews,* **191**: 406–430.

Suggate, R.P., 1974, When did the Last Interglacial end? *Quaternary Research,* **4**: 246–252.

Suggate, R.P., and West, R.G., 1969, Stratigraphic nomenclature and subdivision in the Quaternary Working group for Stratigraphic Nomenclature, INQUA Commission for Stratigraphy (unpublished discussion document).

Svensson, A., Andersen, K.K., Bigler, M., Clausen, H.B., Dahl-Jensen, D., Davies, S.M., et al., 2008, A 60000 year Greenland stratigraphic ice core chronology. *Climate of the Past,* **4**: 47–57.

Tauxe, L., Opdyke, N.D., Pasini, G., and Elmi, C., 1983, Age of the Pliocene–Pleistocene boundary in the Vrica section, southern Italy. *Nature,* **304**: 125–129.

Tesakov, A.S., Shik, S.M., Velichko, A.A., Gladenkov, Y, Lavrushin, Y. A., and Yanina, T.A., 2015, Proposed changes in the stratigraphic structure of the Quaternary for the General Stratigraphic Scale of Russia. In: *Proceedings of the All-Russian Scientific Meeting "Stratigraphic and Paleogeographic Problems of the Neogene and Quaternary of Russia (New Materials and Methods)".* GEOS, Moscow, pp. 57–59.

Thompson, L.G., Mosley-Thompson, E., Davis, M.E., Bolzan, J.F., Dai, J., Yao, T., et al., 1989, Holocene-Late Pleistocene climatic ice core records from Qinghai-Tibetan Plateau. *Science,* **246**: 474–477.

Thompson, L.G., Mosley-Thompson, E., Davis, M.E., Lin, P.-N., Henderson, K.A., Cole-Dai, J., et al., 1995, Late Glagial Stage and Holocene tropical ice core records from Huascaran, Peru. *Science,* **269**: 46–50.

Thompson, L.G., Mosley-Thompson, E., Davis, M.E., Henderson, K.A., Brecher, H.H., Zagorodnov, V.S., et al., 2002, Kilimanjaro ice core records: evidence of Holocene climate change in tropical Africa. *Science,* **298**: 589–593.

Thompson, L.G., Brecher, H.H., Mosley-Thompson, E., Hardy, D.R., and Mark, B.G., 2009, Glacier loss on Kilimanjaro continues unabated. *Proceedings of the National Academy of Sciences, United States of America,* **106**: 19770–19775.

Tripati, A.K., Eagle, R.A., Morton, A., Dowdeswell, J.A., Atkinson, K.L., Bahe, Y., et al., 2008, Evidence for glaciation in the Northern Hemisphere back to 44 Ma from ice-rafted debris in the Greenland Sea. *Earth and Planetary Science Letters,* **265**: 112–122.

Turner, C., 1998, Volcanic maars, long Quaternary sequences and the work of the INQUA Subcommission on European Quaternary Stratigraphy. *Quaternary International,* **47/48**: 41–49.

Turner, C., and West, R.G., 1968, The subdivision and zonation of inter-glacial periods. *Eiszeitalter und Gegenwart,* **19**: 93–101.

Tzedakis, P.C., Andrieu, V., Beaulieu, J.L., de, Crowhurst, S., Follieri, M., Hooghiemstra, H., et al., 1997, Comparison of terrestrial and marine records of changing climate of the last 500 000 years. *Earth and Planetary Science Letters,* **150**: 171–176.

Tzedakis, P.C., Hooghiemstra, H., and Pälike, H., 2006, The last 1.35 million years at Tenaghi Philippon: revised chronostratigraphy and long-term vegetation trends. *Quaternary Science Reviews,* **25**: 3416–3430.

Tzedakis, P.C., Drysdale, R.N., Margari, V., Skinner, L.C., Menviel, L., Rhodes, R.H., et al., 2018, Enhanced climate instability in the North Atlantic and southern Europe during the Last Interglacial. *Nature Communications,* **9**: 4235. https://doi.org/10.1038/s41467-018-06683-3.

Van Couvering, J., 1997, Preface, the new Pleistocene. *In* Van Couvering, J. (ed), *The Pleistocene Boundary and the Beginning of the Quaternary*. Cambridge: Cambridge University Press, pp. ii–xvii.

van der Weil, A.M., and Wijmstra, T.A., 1987a, Palynology of the lower part (78–120 m) of the core Tenaghi Philippon II, Middle Pleistocene, Greece. *Review of Palaeobotany and Palynology*, 52: 73–88.

van der Weil, A.M., and Wijmstra, T.A., 1987b, Palynology of the 112.8–197.8 m interval of the core Tenaghi Philippon II, Middle Pleistocene, Greece. *Review of Palaeobotany and Palynology*, 52: 89–117.

Van Donk, J., 1976, A record of the Atlantic Ocean for the entire Pleistocene Epoch. *Geological Society of America Memoirs*, 145: 147–163.

van Kolfschoten, T., and Gerasimenko, N., 2006, The Ukraine Quaternary explored: the Middle and Upper Pleistocene of the Middle Dnieper Area and its importance for East–West European correlation. *Quaternary International*, 149: 1–3.

Van Leeuwen, R.J.W., Beets, D.J., Bosch, J.H.A., Burger, A.W., Cleveringa, P., van Harten, D., et al., 2000, Stratigraphy and integrated facies analysis of the Saalian and Eemian sediments in the Amsterdam-Terminal borehole, the Netherlands. *Netherlands Journal of Geosciences*, 79: 161–198.

Wagner, G.A., 1998, *Age Determination of Young Rocks and Artifacts*. Berlin: Springer, p. 466.

Walker, M., 2005, *Quaternary Dating Methods*. Chichester: Wiley, p. 286.

Walker, M., Johnsen, S., Rasmussen, S.O., Steffensen, J.P., Popp, T., Gibbard, P., et al., 2008, The Global Stratotype Section and Point (GSSP) for the base of the Holocene Series/Epoch (Quaternary System/ Period) in the NGRIP ice core. *Episodes*, 31: 264–267.

Walker, M., Johnsen, S., Rasmussen, S.O., Steffensen, J.P., Popp, T., Gibbard, P., et al., 2009, Formal definition and dating of the GSSP (Global Stratotype Section and Point) for the base of the Holocene using the Greenland NGRIP ice core, and selected auxiliary records. *Journal of Quaternary Science*, 24: 3–17.

Walker, M.J.C., Berkelhammer, M., Björck, S., Cwynar, L.C., Fisher, D.A., Long, A.J., et al., 2012, Formal subdivision of the Holocene Series/ Epoch: a discussion paper by a working group of INTIMATE (Integration of ice-core, marine and terrestrial records) and the Subcommission on Quaternary Stratigraphy (International Commission on Stratigraphy). *Journal of Quaternary Science*, 27: 649–659.

Walker, M., Head, M.J., Berkelhammer, M., Björck, S., Cheng, H., Cwynar, L., et al., 2018, Formal ratification of the subdivision of the Holocene Series/Epoch (Quaternary System/Period): two new Global Boundary Stratotype Sections and Points (GSSPs) and three new stages/subseries. *Episodes*, 41 (4), 213–223.

Walker, M., Head, M.J., Berkelhammer, M., Björck, S., Cheng, H., Cwynar, L., et al., 2019a, Subdividing the Holocene Series/Epoch: formalisation of stages/ages and subseries/subepochs, and designation of GSSPs and auxiliary stratotypes. *Journal of Quaternary Science*, 34 (3), 173–186.

Walker, M., Gibbard, P., Head, M.J., Berkelhammer, M., Björck, S., Cheng, H., et al., 2019b, Formal subdivision of the Holocene Series/Epoch: a summary. *Journal of the Geological Society of India*, 93: 135–141.

Wang, Y.J., Cheng, H., Edwards, R.L., An, Z.S., Wu, J.Y., Shen, C.-C., et al., 2001, A high-resolution absolute-dated late Pleistocene monsoon record from Hulu Cave, China. *Science*, 294: 2345–2348.

Wang, X., Auler, A.S., Edwards, R.L., Cheng, H., Cristalli, P.S., Smart, P.L., et al., 2004, Wet periods in northeastern Brazil over the past 210 ka linked to distant climate anomalies. *Nature*, 432: 740–743.

Waters, C.N., Syvitski, J.P.M., Gałuszka, A., Hancock, G.J., Zalasiewicz, J., Cearreta, A., et al., 2015, Can nuclear weapons fallout mark the beginning of the Anthropocene Epoch? *Bulletin of the Atomic Scientists*, 71: 46–57.

Waters, C.N., et al., 2018a, Global Boundary Stratotype Section and Point (GSSP) for the Anthropocene series: where and how to look for potential candidates. *Earth-Science Reviews*, 178: 370–429.

Waters, C.N., Fairchild, I.J., McCarthy, F.M.G., Turney, C.S.M., Zalasiewicz, J., and Williams, M., 2018b, How to date natural archives of the Anthropocene. *Geology Today*, 34 (5), 182–187.

Watts, W.A., Allen, J.R.M., and Huntley, B., 1995, Vegetation history and palaeoclimate of the last glacial period at Lago Grande di Monticchio, southern Italy. *Quaternary Science Reviews*, 15: 133–153.

West, R.G., 1968, *Pleistocene Geology and Biology*, 1st ed. London: Longmans, p. 377.

West, R.G., 1977, *Pleistocene Geology and Biology*, 2nd ed. London: Longmans, p. 440.

West, R.G., 1979, Further on the Flandrian. *Boreas*, 8: 126.

West, R.G., and Turner, C., 1968, The subdivision and zonation of interglacial periods. *Eiszeitalter und Gegenwart*, 19: 93–101.

Woillard, G.M., 1978, Grande Pile peat bog: a continuous pollen record for the past 140 000 years. *Quaternary Research*, 9: 1–21.

Woldstedt, P., 1962, Über die Bennenung einiger Unterabteilungen des Pleistozäns. *Eiszeitalter und Gegenwart*, 3: 14–18.

Wolff, E.W., 2008, The past 800 ka viewed through Antarctic ice cores. *Episodes*, 31: 216–218.

Yang, S., and Ding, Z., 2010, Drastic climatic shift at ~2.8 Ma as recorded in eolian deposits of China and its implications for redefining the Pliocene- Pleistocene boundary. *Quaternary International*, 219: 37–44.

Yokoyama, Y., Lambeck, K., De Deckker, P., Johnston, P., and Fifield, K., 2000, Timing of the Last Glacial Maximum from observed sea-level minima. *Nature*, 406: 713–716.

Zagwijn, W.H., 1992, The beginning of the Ice Age in Europe and its major subdivisions. *Quaternary Science Reviews*, 11: 583–591.

Zalasiewicz, J., Smith, A., Williams, M., Barry, T.L., Bown, P.R., Brenchley, P., et al., 2008, Are we now living in the Anthropocene? *GSA Today*, 18: 4–8.

Zalasiewicz, J., Waters, C.N., Williams, M., and Summerhayes, C. (eds), 2019a. *The Anthropocene as a Geological Time Unit. A Guide to the Scientific Evidence and Current Debate*. Cambridge. Cambridge University Press, p. i–xiv, 1–361.

Zalasiewicz, J., Waters, C.N., Head, M.J., Poirier, C., Summerhayes, C.P., Leinfelder, R., et al., 2019b, A formal Anthropocene is compatible with but distinct from its diachronous anthropogenic counterparts: a response to W.F. Ruddiman's "three-flaws in defining a formal Anthropocene". *Progress in Physical Geography: Earth and Environment*, 43 (3), 319–333.

Zeuner, F.E., 1935, The Pleistocene chronology of central Europe. *Geological Magazine*, 72: 350–376.

Zeuner, F.E., 1945, The Pleistocene Period: its climate, chronology and faunal successions. *Ray Society Monographs*, 130: 322.

Zeuner, F.E., 1959, *The Pleistocene Period*. London: Hutchinson, p. 447.

Zhou, L.P., and Shackleton, N.J., 1999, Misleading positions of geomagnetic reversal boundaries in Eurasian loess and implications for correlation between continental and marine sedimentary sequences. *Earth and Planetary Science Letters*, **168**: 117−130.

Zhu, Y.M., Zhou, L.P., Mo, D.W., Kaakinen, A., Zhang, Z.Q., and Fortelius, M., 2008, A new magnetostratigraphic framework for late Neogene *Hipparion* Red Clay in the eastern Loess Plateau of China. *Palaeogeography, Palaeoclimatology, Palaeoecology*, **268**: 47−57.

Zielinski, G.A., Mayewski, P.A., Meeker, D., Grönvold, K., Germani, M.S., Whitlow, S., et al., 1997, Volcanic aerosol records and tephrachronology of the Summit, Greenland, ice cores. *Journal of Geophysical Research*, **102**: 26625−26640.

Zijderveld, J.D.A., Hilgen, F.J., Langereis, C.G., Verhallen, P.J.J.M., and Zachariasse, W.J., 1991, Integrated magnetostratigraphy of the upper Pliocene-lower Pleistocene from the Monte Singa and Crotone areas in Calabria, Italy. *Earth and Planetary Science Letters*, **107**: 697−714.

The Anthropocene

Chapter outline

31.1 Origin of the term and history of research as
     a stratigraphic unit   1257
31.2 Lithostratigraphic evidence for the Anthropocene   1258
31.3 Chemostratigraphic indicators of the Anthropocene   1259
31.4 Biostratigraphic indicators of the Anthropocene   1265
31.5 Climatic signals of the Anthropocene   1269
31.6 Anthropocene GSSP possibilities   1269
     31.6.1 Lake deposits   1269
     31.6.2 Marine anoxic basins   1271

31.6.3 Estuaries and deltas   1271
31.6.4 Speleothems   1272
31.6.5 Glacial ice   1272
31.6.6 Corals   1273
31.6.7 Trees   1274
31.6.8 Peat   1274
31.7 Summary   1275
Bibliography   1275

Abstract

The Anthropocene, currently an informal term, being investigated as a potential series/epoch within the Quaternary System/Period. It represents a time when intensified anthropogenic impacts have caused the Earth System to depart from the comparatively stable conditions that characterized the Holocene Epoch. A Holocene/Anthropocene boundary may be best placed to coincide with a marked inflection in a wide array of environmental proxies at about the mid-20th century. A GSSP (Global Boundary Stratotype Section and Point) may be sought among a range of sedimentary environments that offer annual to seasonal resolution.

31.1 Origin of the term and history of research as a stratigraphic unit

The Anthropocene was effectively born as a geological time term in Mexico in 2000. This was at a meeting of the International Geosphere-Biosphere Programme, when atmospheric chemist and Nobel laureate Paul Crutzen proposed the term to convey his meaning that anthropogenic changes to the atmosphere, oceans, and biosphere had intensified. The intensification followed upon the Industrial Revolution, taking the Earth outside of the conditions that had prevailed during the larger part of the Holocene. Crutzen copublished an article first referring to the Anthropocene in 2000 with Eugene Stoermer, a lake ecologist who had independently devised the term and was using it among his colleagues and students. Crutzen subsequently published a single-authored short paper in 2002 in *Nature*. The Anthropocene was expressly indicated as a geological time term, an epoch to follow the Holocene, with a boundary at the beginning of the Industrial Revolution—James Watt's 1784 steam engine invention was suggested as a starting point, reflecting an historical event rather than geological marker.

The Anthropocene began to be widely used and published among the Earth System Science (ESS) community, as a de facto geological time term (e.g., Meybeck, 2003; Steffen et al., 2004). It had not, though, gone through any of the International Commission on Stratigraphy (ICS) procedures for setting up a new unit of the Geological Time Scale and was, and still remains, informal. The growing visibility of the Anthropocene, largely in publications of the ESS community, led to a preliminary analysis by the Stratigraphy Commission of the Geological Society of London, which commenced in 2006. This commission, as a national body, had no power of decision-making over the time scale, which is via the ICS and component subcommissions, but it could compile evidence and make recommendations. 21 out of 22 members of the Stratigraphy Commission were coauthors of a paper (Zalasiewicz et al., 2008) in which they suggested that a stratigraphic case might exist for the Anthropocene, that alignment with the start of the Industrial Revolution had some merit and that the term should be investigated further as a potential new unit of the International Chronostratigraphic Chart.

Geologic Time Scale 2020. DOI: https://doi.org/10.1016/B978-0-12-824360-2.00031-0

This study resulted in an invitation by the Chair of the Subcommission on Quaternary Stratigraphy to set up an Anthropocene Working Group (AWG), which has been in existence since 2009, and which has since been analyzing, and continues to analyze, the case for formalizing the Anthropocene within the Geologic Time Scale (GTS). The AWG developed as a considerably more diverse body than is typical of ICS working groups, as the Anthropocene time interval is one where geological processes overlap not only with a range of human forcings but with an increasingly detailed and sophisticated observational record of both human-driven and Earth processes: hence it includes not only stratigraphers but ES scientists, oceanographers, historians, archeologists, geographers and even an international lawyer, to include consideration of questions of potential wider societal relevance (e.g., Vidas, 2015; Whitmee et al., 2015).

Nevertheless, the AWG focuses on the Anthropocene as a potential chronostratigraphic/geochronologic unit, using standard stratigraphic criteria, so that it can be compared directly with, and on the same terms as, other units of the GTS. This emphasis is important, as since the formation of the AWG, the use of the Anthropocene has increased enormously in the literature (Zalasiewicz et al., 2017b, Fig. 31.1) not least because it has been widely adopted as a concept by disciplines well outside of the Earth sciences, and ranging across the social sciences, humanities and arts (e.g., Biermann, 2014; Clark, 2014; Davis and Turpin, 2015). Interpretation has also expanded, well beyond original ESS meaning and its chronostratigraphic interpretation (which are essentially congruent: see Steffen et al., 2016; Zalasiewicz et al., 2017c) into a broader range of "human-centered"

meanings. These wider meanings of the term are often not consistent with a chronostratigraphic definition, through for instance including a much longer time range (e.g., "early Anthropocene" interpretations stretching well back into the Holocene, and even into the Pleistocene), and/or having globally diachronous and vaguely defined limits reflecting the slow growth and spread of human colonization of the Earth over many millennia (e.g., Smith and Zeder, 2013; Ruddiman, 2013; Bauer and Ellis, 2018).

This account hence focuses on the geological characterization of the Anthropocene within a chronostratigraphic/geochronologic context. This, following the interim conclusions and recommendations of the AWG at the 35th International Geological Congress in Cape Town, 2016 (Zalasiewicz et al., 2017a) may be summarized as possessing "geological reality," as being best considered a new potential epoch (hence, terminating the Holocene), as having a boundary optimally placed in the mid-20th century (reflecting the post-WWII "Great Acceleration" of human population growth, energy use, industrialization and globalization: Steffen et al., 2007, 2015) and best defined via a GSSP, among which the radiogenic "bomb spike" seems most promising as a primary marker. We consider below the lithostratigraphic, chemostratigraphic, and biostratigraphic evidence that underlies such an interpretation.

31.2 Lithostratigraphic evidence for the Anthropocene

Lithostratigraphic criteria for the Anthropocene include a large and growing suite of novel anthropogenic "minerals,"

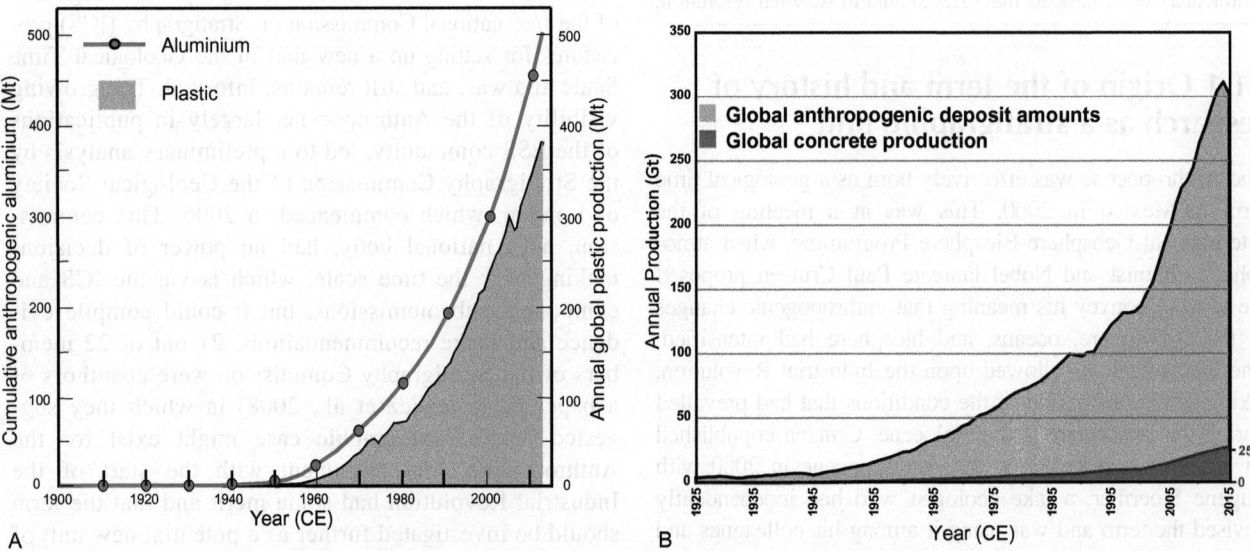

FIGURE 31.1 (A) Cumulative production of plastic and aluminum (from Geyer et al., 2017; Zalasiewicz et al., 2014) and (B) annual global production of total anthropogenic sediment, including concrete shown separately (from Cooper et al., 2018; Waters and Zalasiewicz, 2018).

or more technically "inorganic crystalline mineral-like compounds," as artificial chemical compounds have been specifically excluded from the classification of the International Mineralogical Association (Nickel and Grice, 1998)—although not before 208 anthropogenic minerals had been formally defined and ratified. Nevertheless, the many subsequent novel inorganic crystalline compounds produced by industry and in the laboratories of the materials chemists are minerals in all but formal name. Through the presence of these new long-lived chemical compounds, the Anthropocene may be said to include a considerable and distinctive mineralogical signature.

The context here is the "mineral evolution" of Earth assessed by Hazen et al. (2008). Mineral species on Earth grew from their number in preplanetary asteroids and planetesimals (~ 250) to ~ 1500 on an early Earth with water, some form of plate tectonics and metamorphism. Once life emerged and, crucially, developed oxygenic photosynthesis, the number of minerals rose to >4000 around the Archean–Proterozoic boundary as suites of oxides and hydroxides formed. Further changes in ocean chemistry, climate and biological evolution added only modest numbers of minerals (some 5300 are now recognized).

Early anthropogenic mineral formation included separation of pure metals (rare in nature) and their combination into alloys, beginning with the Bronze and Iron Ages. This process intensified with the Industrial Revolution, and accelerated in post-WWII times. For instance, aluminum, extremely rare in nature in native form, has seen cumulative production of >500 million tons (Fig. 31.1A), almost all since the mid-20th century—more than sufficient to cover the United States land area with a layer of standard kitchen foil—while the production of iron in that time has been ~ 30 times greater (Zalasiewicz et al., 2014).

Such bulk production has been accompanied by a striking diversification of novel anthropogenic mineral-like compounds, such as the tungsten carbide used in ballpoint pens, and novel synthetic garnets used in lasers. Their number now exceeds 200,000 (sources in Hazen et al., 2017, updated), exceeding by more than an order of magnitude the number of "natural" minerals. This explosion in diversity has been accompanied by redistribution of both natural and anthropogenic minerals and mineral-like compounds across the Earth, producing widespread, long-lived and distinctive stratigraphic markers.

A specific novel anthropogenic "mineraloid" is represented by plastics, essentially a post-mid-20th century phenomenon, the production of which is now ~ 380 Mt/year, with total production to 2015 of 8.7 Gt (Fig. 31.1A; Geyer et al., 2017). Most of the plastic produced is either still in use or is in landfills, though the small proportions that have so far leaked into sedimentary systems on land and in the sea have become dispersed widely, and are already effective markers of deposits of recent decades, especially the microplastics now near-ubiquitous in marine sediments (Zalasiewicz et al., 2016a, in press and references therein).

This mineral diversification has been accompanied by a growth in the range and bulk of anthropogenic rock types. Some, like bricks and ceramics, as a transformation of mudrocks, have been made and used throughout much of the Holocene, though their form and composition has evolved and diversified, especially since the Industrial Revolution. Concrete, although intermittently made since Roman times, has since the mid-20th century been produced in increasingly prodigious amounts, and may be said to be a signature rock of the Anthropocene: some 27 billion tons are produced per year at present (Fig. 31.1B) and 500 billion tons have been made to date, the equivalent of a kilo for every square meter of the Earth's surface (Waters and Zalasiewicz, 2018). This ubiquity has made it useful as a generally effective marker, as clasts within the artificial ground that underlies urban areas, for post-1945 deposition (Terrington et al., 2018).

Anthropogenic sedimentation and erosion, associated with forest clearance, farming and urbanization, has developed through the Holocene, though again with major long-term acceleration after WWII. Several studies have indicated that movement of rock and soil by humans now exceeds natural rates of sedimentation and erosion, by an order of magnitude or more (Wilkinson and McElroy, 2007; Hooke et al., 2012). In a study collating global statistics on mineral (and overburden) production and construction, Cooper et al. (2018), estimated annual global anthropogenic sediment production of 316 Gt by 2015 (having risen from ~ 10 Gt in 1950; Fig. 31.1B), ~ 24 times greater than the rate of sediment supply by rivers to the world's oceans. This is generally consistent with an estimate of the total mass of anthropogenically reworked material of some 30 trillion tons, equivalent to about 50 kg/m^2 of the Earth (Zalasiewicz et al., 2016b). Such artificial ground, most concentrated in urban centers, is complemented by strata deposited by natural processes, but with anthropogenic signals, in marine and terrestrial sedimentary settings (including glacial ice). Within such strata, a potential Anthropocene Series may be identified and correlated using a variety of chemostratigraphic and biostratigraphic proxies.

31.3 Chemostratigraphic indicators of the Anthropocene

Human perturbation of chemical cycles has been widespread. Sen and Peuckner-Ehrenbrink (2012) calculated that the anthropogenic fluxes of up to 62 elements now exceed their natural fluxes, via activities such as mining, construction, fossil fuel burning, and deforestation. This has given rise to an array of proxy signals that help characterize the Anthropocene, and that promise to provide the most effective means to define it.

The carbon cycle has been affected by the release into the atmosphere of >600 Gt of carbon, largely from burning of fossil fuels, but with significant components from deforestation and related soil carbon changes, cement manufacture, and other processes [Intergovernmental Panel on Climate Change (IPCC), 2013]. This has caused atmospheric carbon dioxide (CO_2) levels to rise from ~280 ppm in preindustrial times to ~310 ppm by 1950 and then more steeply to ~410 ppm by 2018 (Fig. 31.2A), this latter rise being >100 times faster than that at the Pleistocene—Holocene transition (Waters et al., 2016 and references therein). This ongoing rise is recorded directly in glacial ice, and indirectly as a >2‰ decrease in $\delta^{13}C$ via the Suess effect in such diverse media as foraminifera, tree rings, peat and coral skeletons (see Waters et al., 2018 and references therein; Fig. 31.3). The rise in atmospheric methane (CH_4) levels, similarly recorded in polar ice, has been yet more pronounced than that of CO_2: throughout the Holocene, values fluctuated between 590 and 760 ppb; values began to rise in the early 18th century, and by 1875 were at 800 ppb (Fig. 31.2B), higher than at any time in the last 800 kyr (Waters et al., 2016 and references therein), and have continued rising to their current level (July 2018) of 1850 ppb. Fossil fuel burning has also resulted in a global distribution in recent sediments of fly ash, both as spheroidal carbonaceous particles (Rose, 2015; Swindles et al., 2015), of high preservation potential, and inorganic ash spheres (Rose, 1996); this signal either begins, or increases markedly, in the

mid-20th century (Fig. 31.4), and in itself has been suggested as a potential primary marker for the Anthropocene (Rose, 2015; Swindles et al., 2015).

Levels of chemically bound nitrogen have more than doubled at the Earth's surface, following the invention of the Haber—Bosch process early in the 20th century. Resulting stratigraphic proxy signals include a prominent mid-20th century inflection in $\delta^{15}N$ values in far northern lake deposits (Holtgrieve et al., 2011; Fig. 31.5), likely the result of far-traveled aerosols, which, with associated changes in diatom assemblages, have been suggested as an Anthropocene indicator (Wolfe et al., 2013). Fossil fuel burning has increased the flux of nitrates and nitrogen oxides (NO_x), recorded within Greenland glacial ice (Hastings et al., 2009), but not in Antarctic ice (Fig. 31.5). Surface phosphorus levels have also doubled (Filippelli, 2002) but have not left such a clear signal, as phosphorus only has one stable isotope. Increased flux of phosphorus with nitrogen to coastal waters has, though, caused, from the 1960s, seasonal anoxic "dead zones" to develop, which now cover an area of >250,000 km worldwide (Diaz and Rosenberg, 2008), and are recorded inter alia as changes in dinoflagellate cysts, ostracods, and foraminifera populations (Wilkinson et al., 2014). Perturbations to the sulfur cycle from fossil fuel burning, may be seen, too, as increased sulfates in glacial ice (though not in Antarctica) and speleothems (Mayewski et al., 1990; Fairchild, 2018; Fig. 31.6).

Stratigraphic signals from metal extraction and working have a long history, early signals from Bronze and Iron Age lead smelting being recorded in archives such as European peat-bogs and Arctic ice cores from c.3500 BP, with a more pronounced Roman peak at ~2000 BP. Suggested as potential markers to stratigraphically define an "early" Anthropocene concept (Wagreich and Draganits, 2018), these

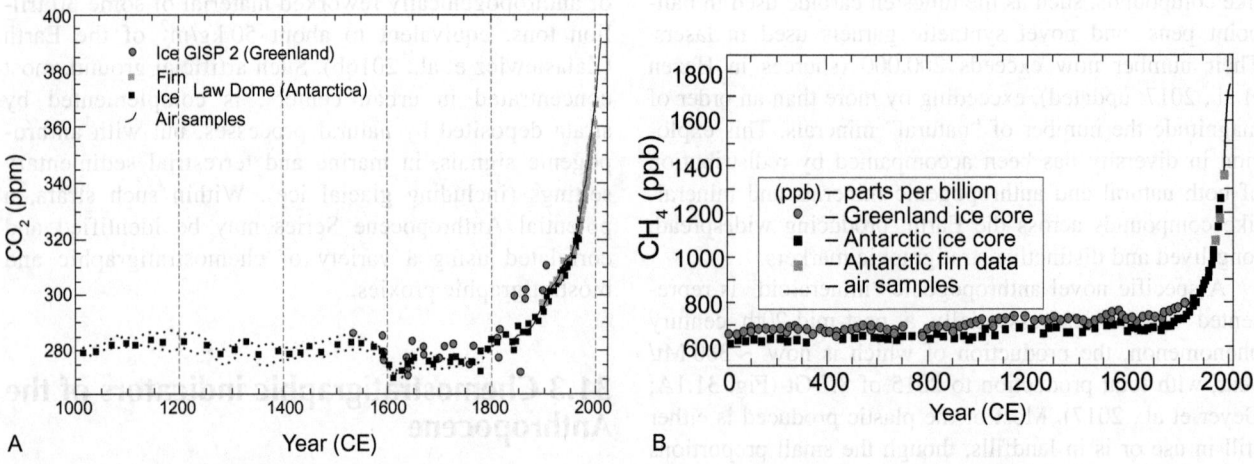

FIGURE 31.2 (A) Atmospheric CO_2 records in ice core from Greenland GISP 2 (Wahlen et al., 1991) and Antarctic Law Dome (Rubino et al., 2013); and (B) Atmospheric CH_4 records in ice core from Greenland GISP 2 (Mitchell et al., 2013) and Antarctic Law Dome (Ferretti et al., 2005).

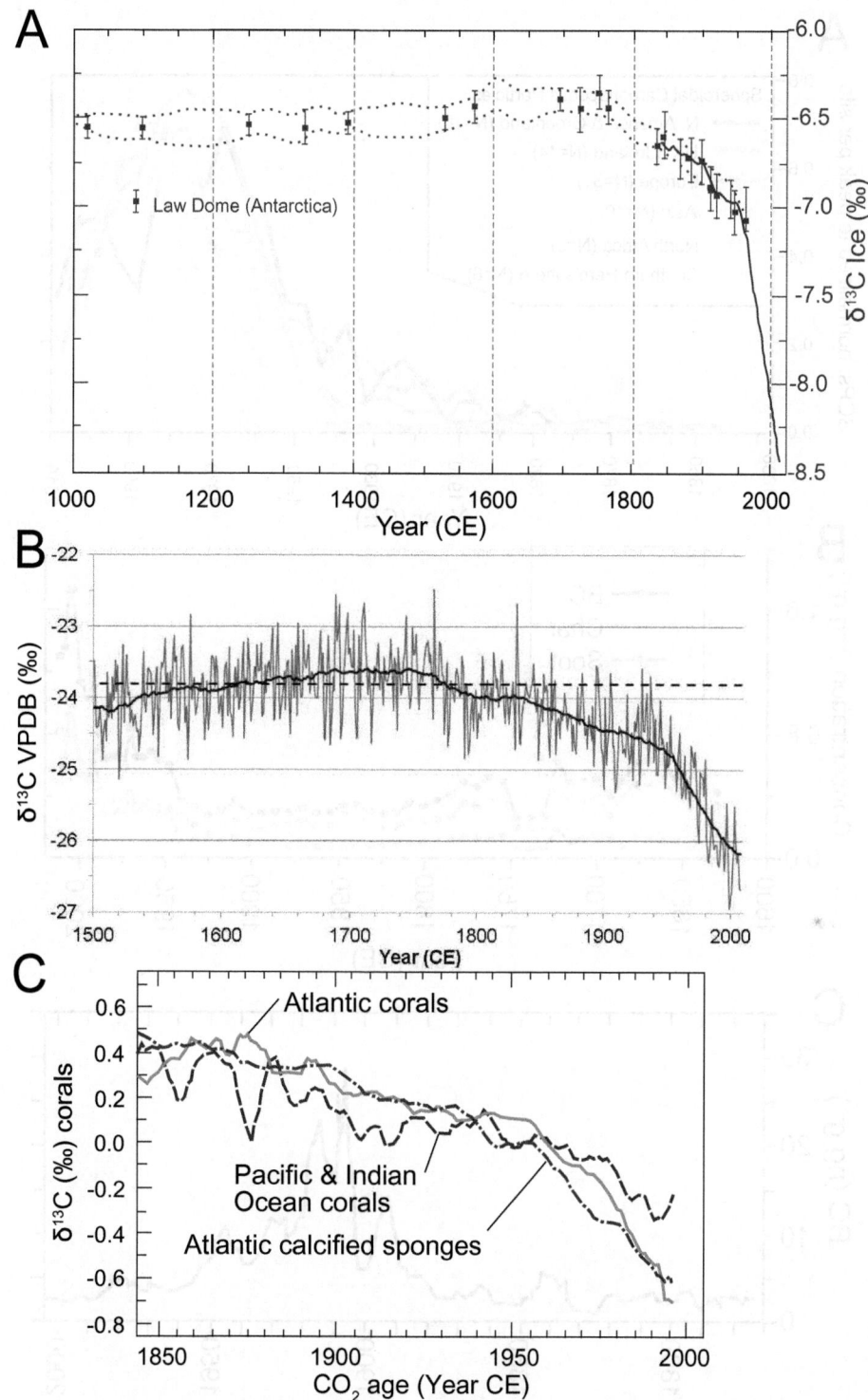

FIGURE 31.3 (A) δ^{13}C variability in the Antarctic Law Dome ice core (Rubino et al., 2013); (B) δ^{13}C variability for a composite tree ring stable isotope chronology developed using *Pinus sylvestris* trees from northern Fennoscandia (Loader et al., 2013). Fine blue line represents annually resolved δ^{13}C variability, thick solid line presents the annual data smoothed with a centrally weighted 51-year moving average. Dashed line represents the mean δ^{13}C value for the "preindustrial" period CE 1500–1799; and (C) changes in δ^{13}C from Atlantic and Pacific/Indian Ocean corals and Atlantic calcified sponges, shown as a 5-year running mean (Swart et al., 2010).

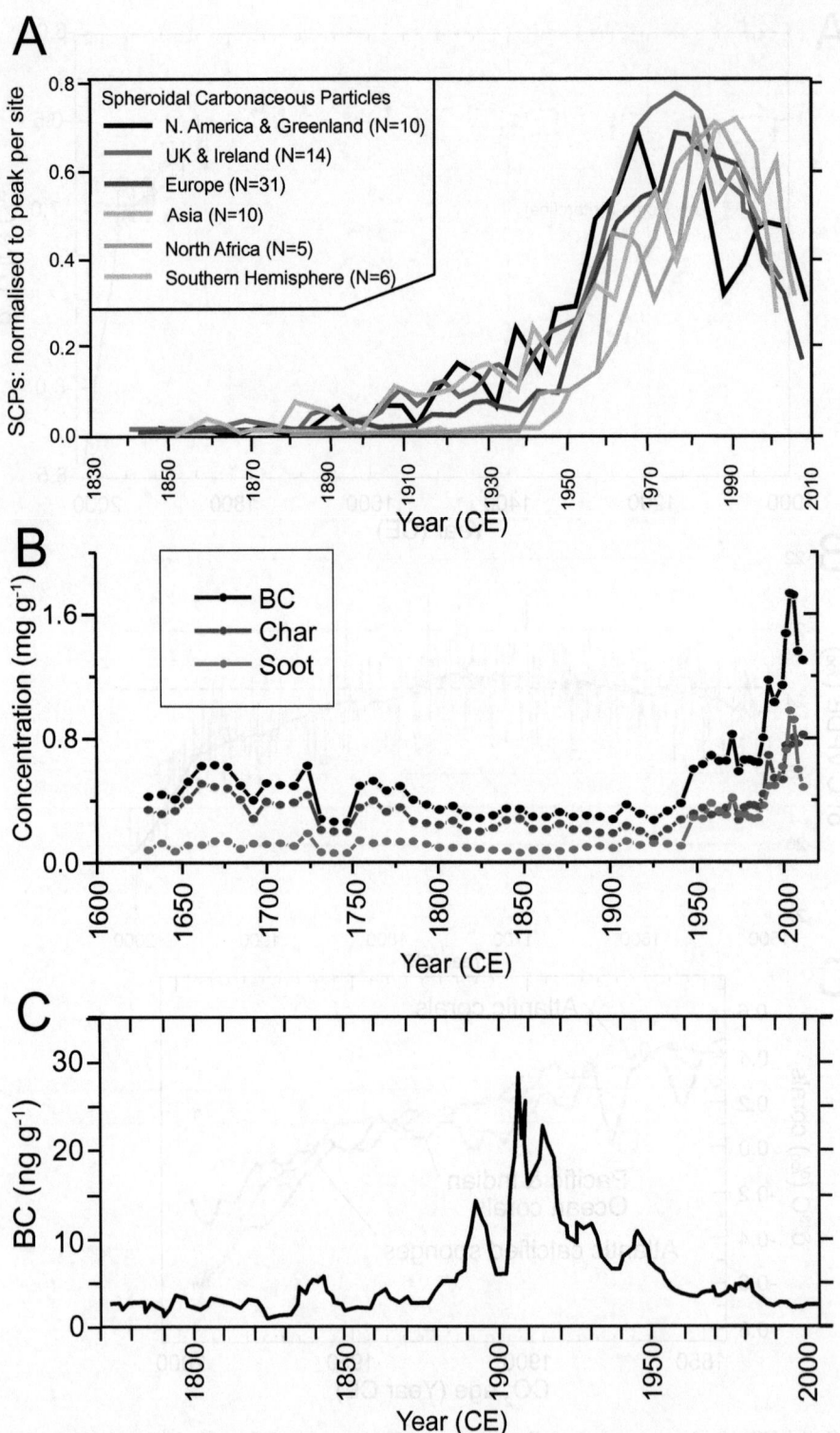

FIGURE 31.4 (A) Global mid-20th century rise and late-20th century spike in spheroidal carbonaceous particles, normalized to the peak value in each lake core (from Waters et al., 2016, modified from Rose, 2015); (B) concentrations of black carbon (BC), char, and soot in the Huguangyan Maar Lake (from Han et al., 2016); (C) BC concentrations (5-year running means) in Greenland ACT2 ice core from CE 1772 to 2003 (McConnell and Edwards, 2008).

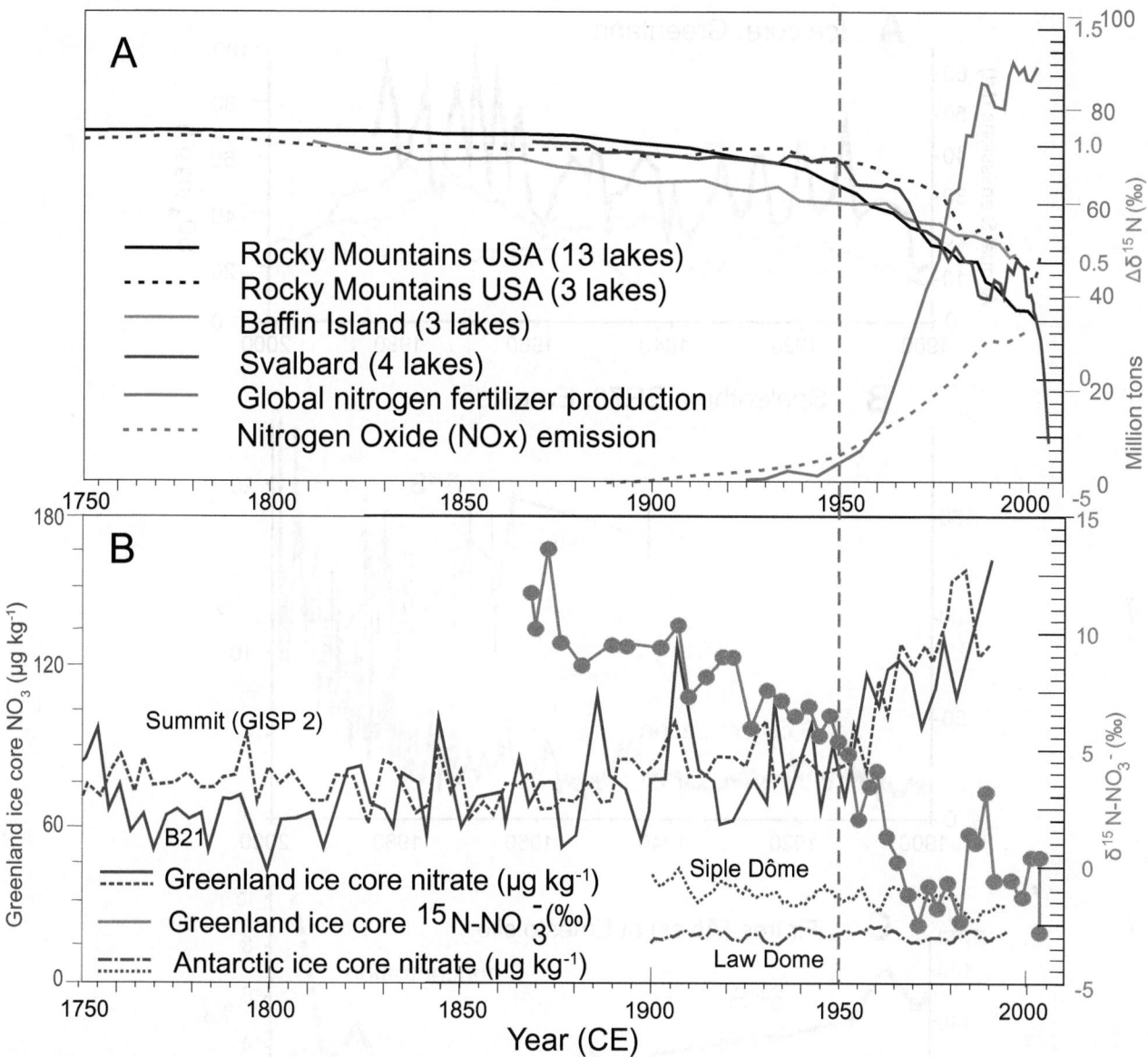

FIGURE 31.5 (A) Relative departures ($\Delta\delta^{15}$N) from mean pre-CE 1900 δ^{15}N in Northern Hemisphere lakes (Wolfe et al., 2013), alongside annual rates of reactive N production from agricultural fertilizer and NO$_x$ emissions from fossil fuel combustion (Holland et al., 2005). (B) Coevolution of ice core NO$_3$ and δ^{15}N for NO$_3$ from GISP 2 Greenland (Hastings et al., 2009) and nitrate concentrations from Antarctic ice core Law Dome and Siple Dôme (Wolff, 2013).

signals are weaker, patchier and less sharply defined than signals associated with and postdating the Industrial Revolution. Greatly increased, and increasingly globalized extraction and use of metals such as lead, zinc and copper then left widespread signals of enhanced concentrations and changed isotope ratios across both hemispheres; complex in detail, an upturn can often be seen around the mid-20th century (Gałuszka and Migaszewski, 2018; Gałuszka and Wagreich, 2019; Fig. 31.7).

More unambiguous recognition of the mid-20th century can be seen among a variety of novel persistent organic pollutants (POPs) such as the organochlorine pesticides DDT, dieldrin and aldrin, and industrial compounds such

as polychlorinated biphenyls (PCBs). Typically disseminated through the atmosphere, they have left clearly detectable residues widely identifying a post-1950 stratal interval both in terrestrial (lake) sediments, glacial ice and in marine sediments (see Gałuszka and Rose, 2019 and references therein; Fig. 31.8).

The sharpest chemical signal associated with the Anthropocene is that from artificial radionuclides, mostly as the "bomb spike" of plutonium, cesium, americium and other radioisotopes and enhanced radiocarbon from atmospheric atomic bomb tests, augmented by accidental releases, such as from Chernobyl and Fukushima (Waters et al., 2015). This

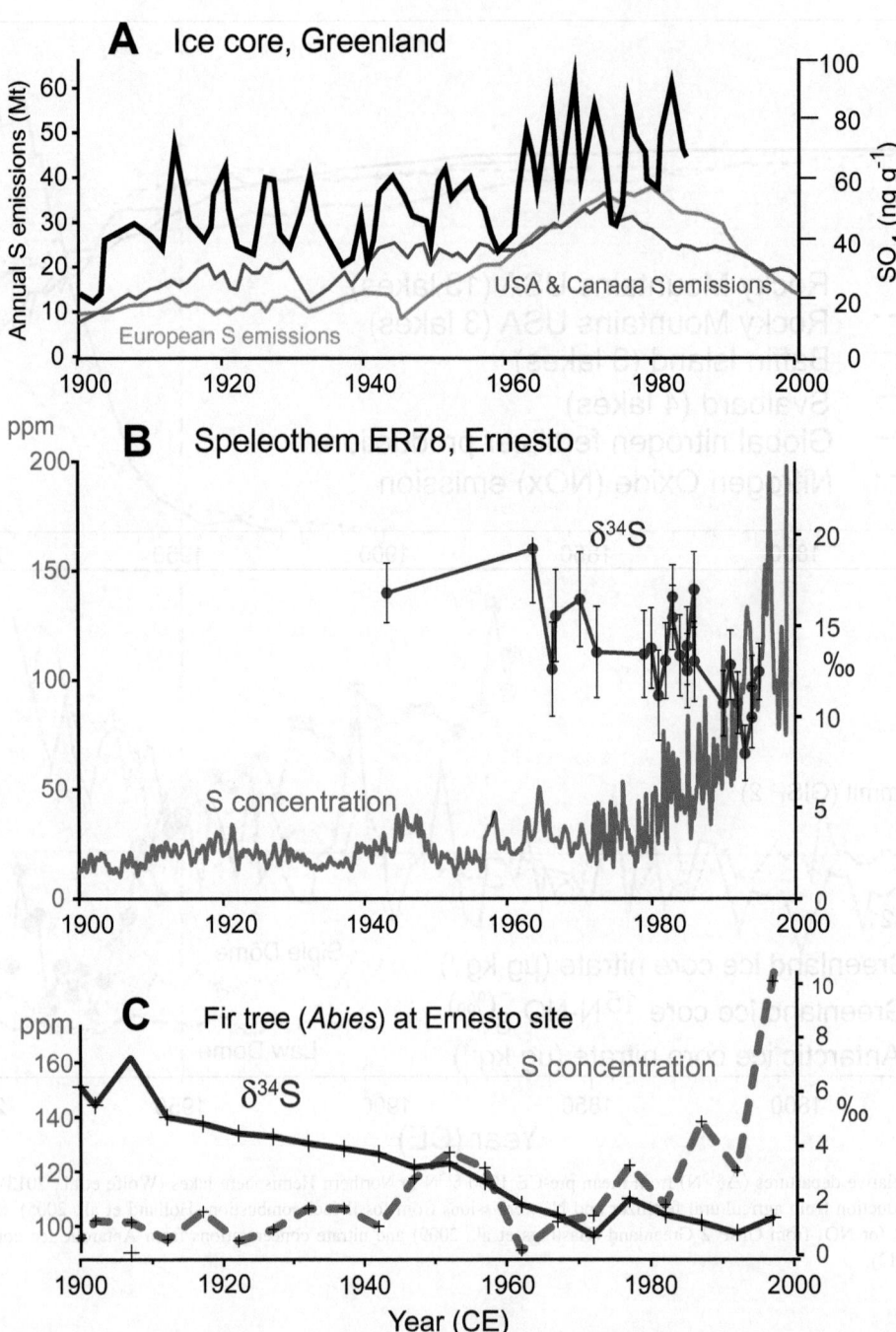

FIGURE 31.6   (A) Sulfate concentrations in Greenland ice core (Mayewski et al., 1990), and sulfur emissions from North America and Europe (Hoesly et al., 2018); (B) S concentration and δ^{34}S from Ernesto cave, Italy (Frisia et al., 2005; Wynn et al., 2010); and (C) S concentrations and δ^{34}S in *Abies alba* (European silver fir tree) from NE Italy (Wynn et al., 2014).

dispersal started with the Trinity bomb test of 1945, followed by wartime use at Hiroshima and Nagasaki, the first of these mooted as a possible Anthropocene Global Standard Stratigraphic Age (GSSA) (Zalasiewicz et al., 2015). Radionuclide dissemination in these was comparatively small and localized, though, and major global spread started with the much larger thermonuclear tests of the early 1950s and

these may form a more reliable primary marker for a potential Anthropocene GSSP (Waters et al., 2015, 2018; Fig. 31.9); the "peak signal" of 1964–65 has also been proposed (Lewis and Maslin, 2015; Turney et al., 2018) although the peak is commonly "smeared out" in many sedimentary archives, and so is not as consistently correlatable (Waters et al., 2018).

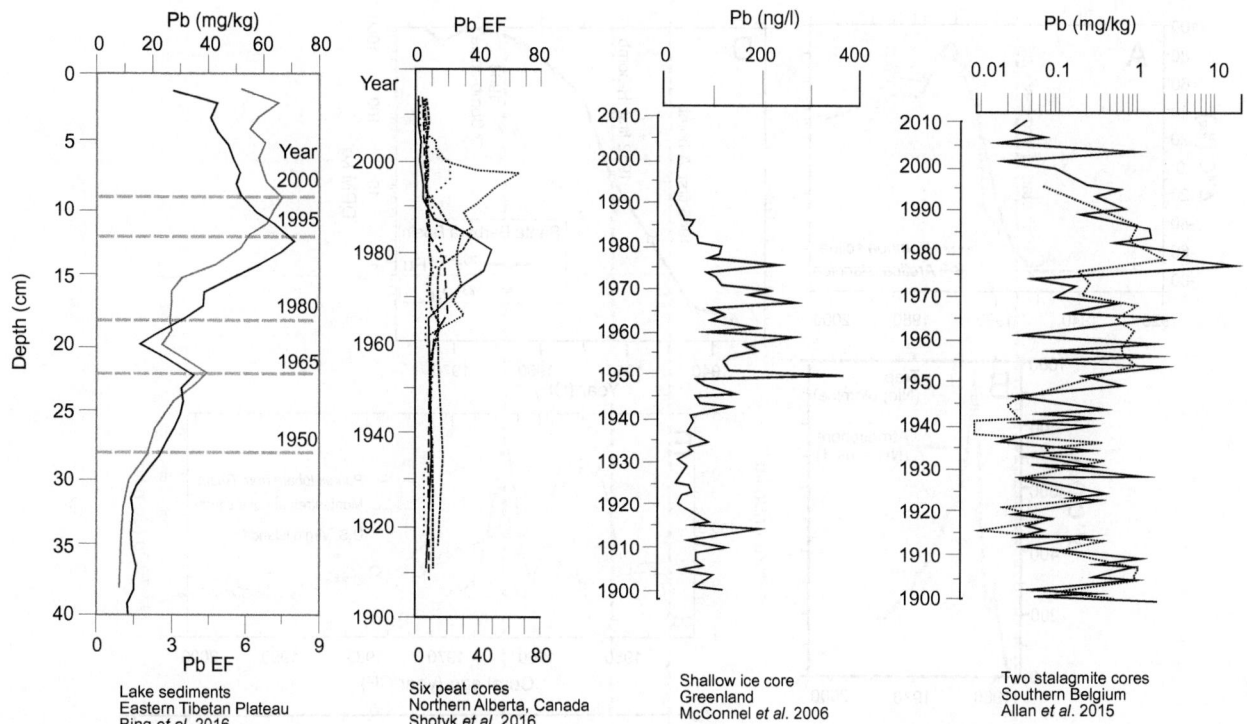

FIGURE 31.7 Examples of temporal changes in lead levels in different natural archives (from Gałuszka and Wagreich, 2019).

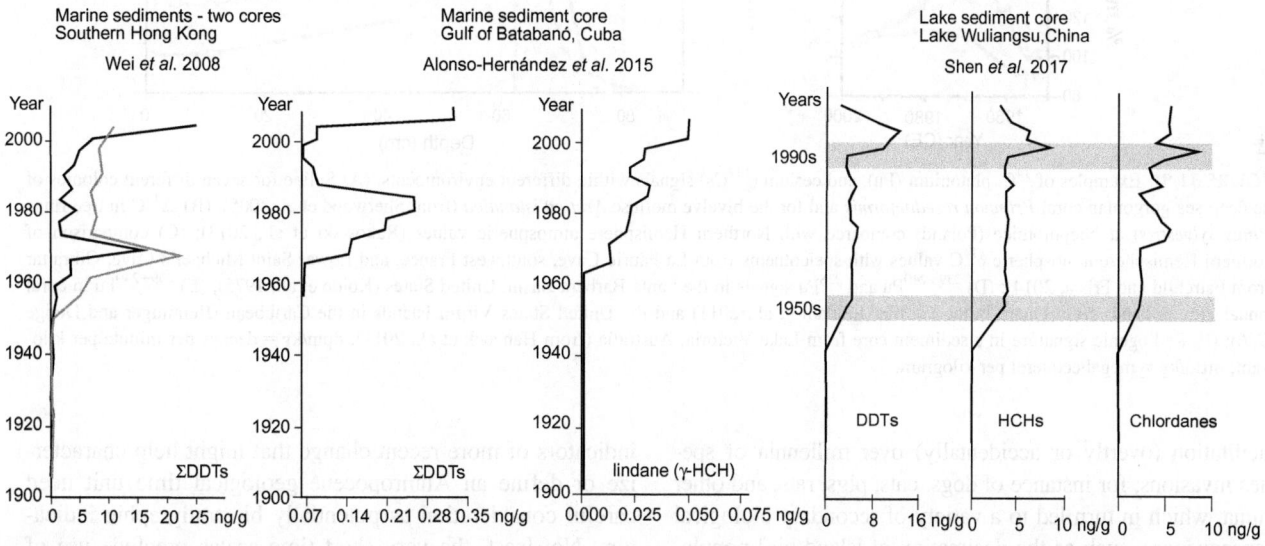

FIGURE 31.8 Examples of temporal changes in organochlorine-pesticide concentrations in dated sediment cores (from Gałuszka and Rose, 2019).

31.4 Biostratigraphic indicators of the Anthropocene

Human impact on the biosphere likely extends back into Late Pleistocene times, with the beginning of extinctions of terrestrial megafauna at ~50 ka, a wave of extinctions that peaked at ~10–12 ka, but that has continued to the present day (Koch and Barnosky, 2006). There has been much debate as to the relative role of human causation versus climate/environment change as extinction drivers, the dominant human role is clear in well-dated examples of island-based extinctions, as on Madagascar and New Zealand, and is evident generally in more recent times (e.g., Barnosky et al., 2011, 2012). Human impact has also included the

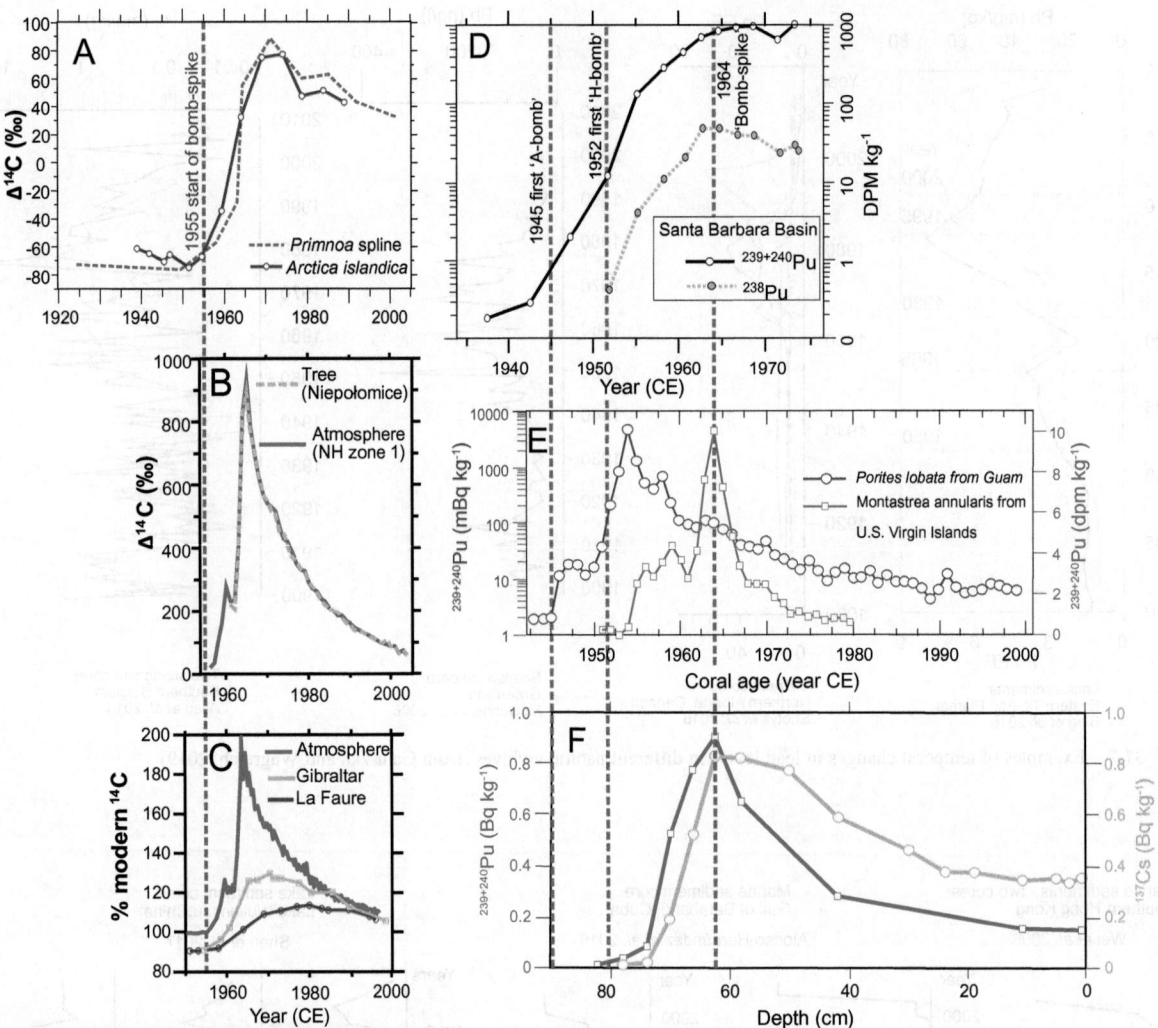

FIGURE 31.9 Examples of ^{14}C, plutonium (Pu), and cesium (^{137}Cs) signals within different environments. (A) Spline for seven different colonies of the deep-sea gorgonian coral *Primnoa resedaeformis* and for the bivalve mollusc *Arctica islandica* (from Sherwood et al. 2005); (B) Δ^{14}C in tree rings (*Pinus sylvestris*) at Niepołomice (Poland) compared with Northern Hemisphere atmospheric values (Rakowski et al., 2013); (C) comparison of Northern Hemisphere atmospheric δ^{14}C values with speleothems from La Faurie Cave, southwest France, and Lower Saint Michael's Cave, Gibraltar (from Fairchild and Frisia, 2014); (D) $^{239+240}$Pu and ^{238}Pu signals in the Santa Barbara Basin, United States (Koide et al., 1975); (E) $^{239+240}$Pu in coral annual growth bands from Guam in the Pacific (Lindahl et al., 2011) and the United States Virgin Islands in the Caribbean (Benninger and Dodge 1986); (F) Radiogenic signature in a sediment core from Lake Victoria, Australia (from Hancock et al., 2011); dpm/kg = decays per minute per kilogram; mBq/kg = megabecquerel per kilogram.

facilitation (overtly or accidentally) over millennia of species invasions, for instance of dogs, cats, pigs, rats, and other fauna, which in turn led to a variety of secondary biological consequences, such as the decimation of island bird populations by rats which in turn affected adjacent reef productivity (Graham et al., 2018). Some of these changes have been exploited biostratigraphically within the Holocene, as in the Santarosean and Santaugustinian land mammal ages of North America of Barnosky et al. (2014), reflecting human entry into North America at ∼14 ka, and introduction of domesticated megafauna 400 years ago respectively.

Given the complexity and longevity of biosphere change since Late Pleistocene times, the (many) biological indicators of more recent change that might help characterize or define an Anthropocene geological time unit need careful consideration as potentially biostratigraphic indicators. Not least, the very short time scales preclude use of "normal" evolutionary changes. However, there have been rapid changes to biological assemblages driven by intensifying anthropogenic pressures over the last century or so, and these do provide biostratigraphic potential.

Among the anthropogenic changes the accelerating rate of species extinctions, now some thousand times higher than background levels (Pimm et al., 2014), has not yet led to a major mass extinction event but threatens one over the course of a few centuries if current rates of loss are sustained (Barnosky et al., 2011, 2014), especially so if projected levels of climate warming [Intergovernmental Panel on Climate Change (IPCC), 2018; and see below] take place. This sets the overall context, but practical Anthropocene biostratigraphic characterization and correlation need to be based on specific taxa and assemblages associated with this overall trend.

A range of potential signals are associated with industry- and agriculture-related chemical pollution, most clearly seen in many lake successions. Indeed, this was the basis for Eugene Stoermer's independent formulation

of the Anthropocene prior to Paul Crutzen's improvisation of the term in 2000. Stoermer specialized in lake diatoms, the patterns of which changed dramatically, as pollution levels increased. By such means of tracking "ecological degradation" (Wilkinson et al., 2014), a post-1950 interval can commonly be widely recognized in lake sedimentary deposits (see also Smol, 2008; Wolfe et al., 2013; Fig. 31.10).

The intensified diaspora of invasive species or "neobiota" over the last century is often associated with increased shipping and, in marine settings, the practice of

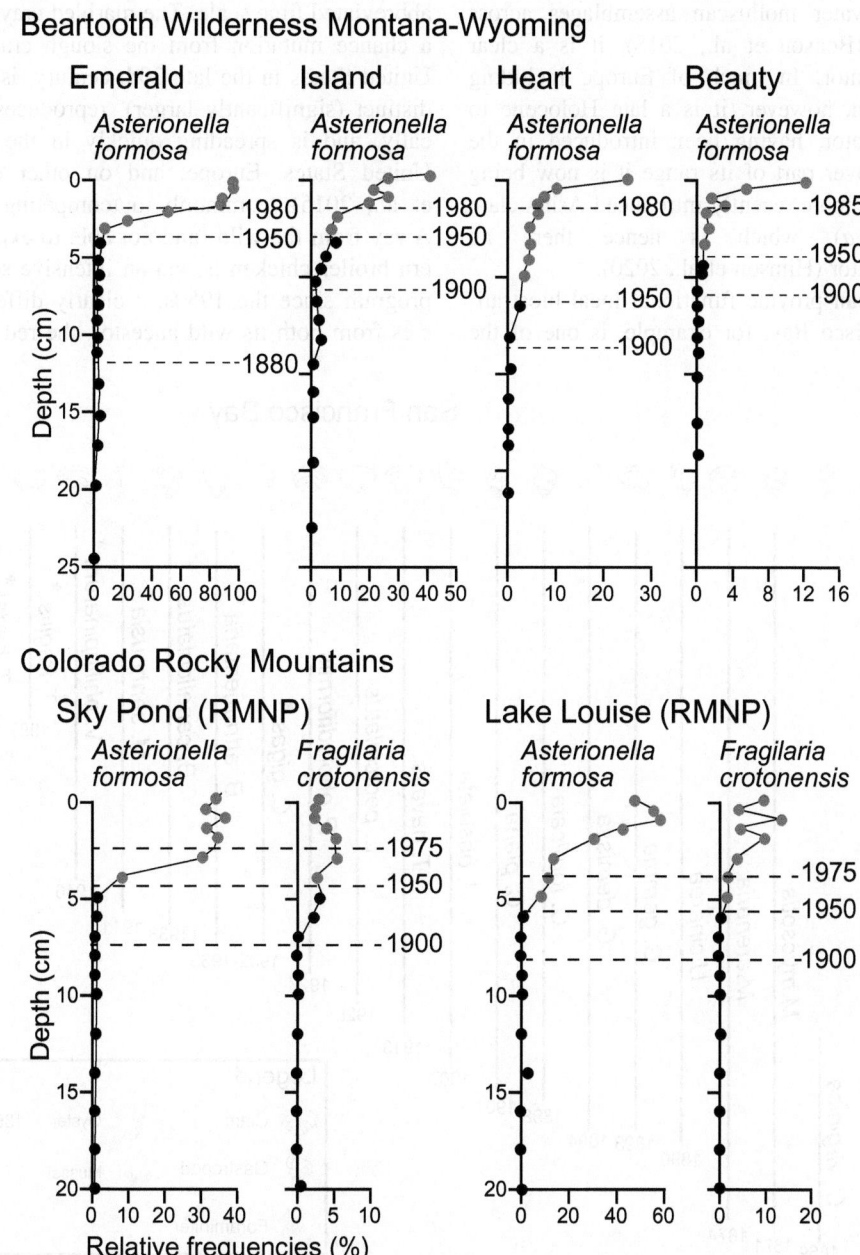

FIGURE 31.10 Replacement of Holocene diatom assemblages by *Asterionella formosa* and/or *Fragilaria crotonensis* in lake sediments from high-altitude sites in the United States since CE ~1950 (Saros et al., 2005).

taking then dumping sediment and water as ballast (Wilkinson et al., 2014). Transformation of much of the Earth's biology in this way has already occurred, as reflected in Samways' (1999) coining of the term Homogenocene, essentially a biology-based synonym for the Anthropocene. Practical biostratigraphic use of this phenomenon may be effected by using observed patterns of species introductions and subsequent spread as a predictive model for their occurrence in recent strata. Thus, for instance, the zebra mussel (*Dreissena polymorpha*), native to the region of the Black Sea and Caspian Sea, was transported to north America in 1988 and now, dominating many freshwater molluscan assemblages across half the continent (Benson et al., 2018), it is a clear Anthropocene indicator. In much of Europe including the United Kingdom, however, it is a late Holocene to Anthropocene indicator, having been introduced in the 19th century. But, over part of its range it is now being out-competed by the more recently introduced Asian clam (*Corbicula fluminea*), which is hence there an Anthropocene indicator (Himson et al., 2020).

Such examples can provide functional local biostratigraphies. San Francisco Bay, for example, is one of the world's most heavily invaded coastal regions, with assemblages commonly dominated by neobiota. Analysis of invasion dates has yielded a local range chart of exotic mollusc and foraminifer taxa across the last century and a half, from which local biozones have been established (Fig. 31.11). Such biozones may be used to constrain an Anthropocene boundary and correlated with comparable local biostratigraphies established elsewhere (Williams et al., 2019) and so provide effective biostratigraphic constraint.

There are few examples of completely novel taxa, but these are not altogether absent, even on such a highly abbreviated time scale. The marbled crayfish, the result of a chance mutation from the slough crayfish of the SW United States in the late 20th century, is morphologically distinct (significantly larger), reproduces parthenogenetically, and is spreading quickly in the wild across the United States, Europe, and on other continents (Vogt et al., 2015), commonly outcompeting native crayfish. Away from the wild (and not able to exist in it) the modern broiler chicken is, via an intensive selective breeding program since the 1950s, a clearly different morphospecies from both its wild ancestor, the red jungle fowl, and

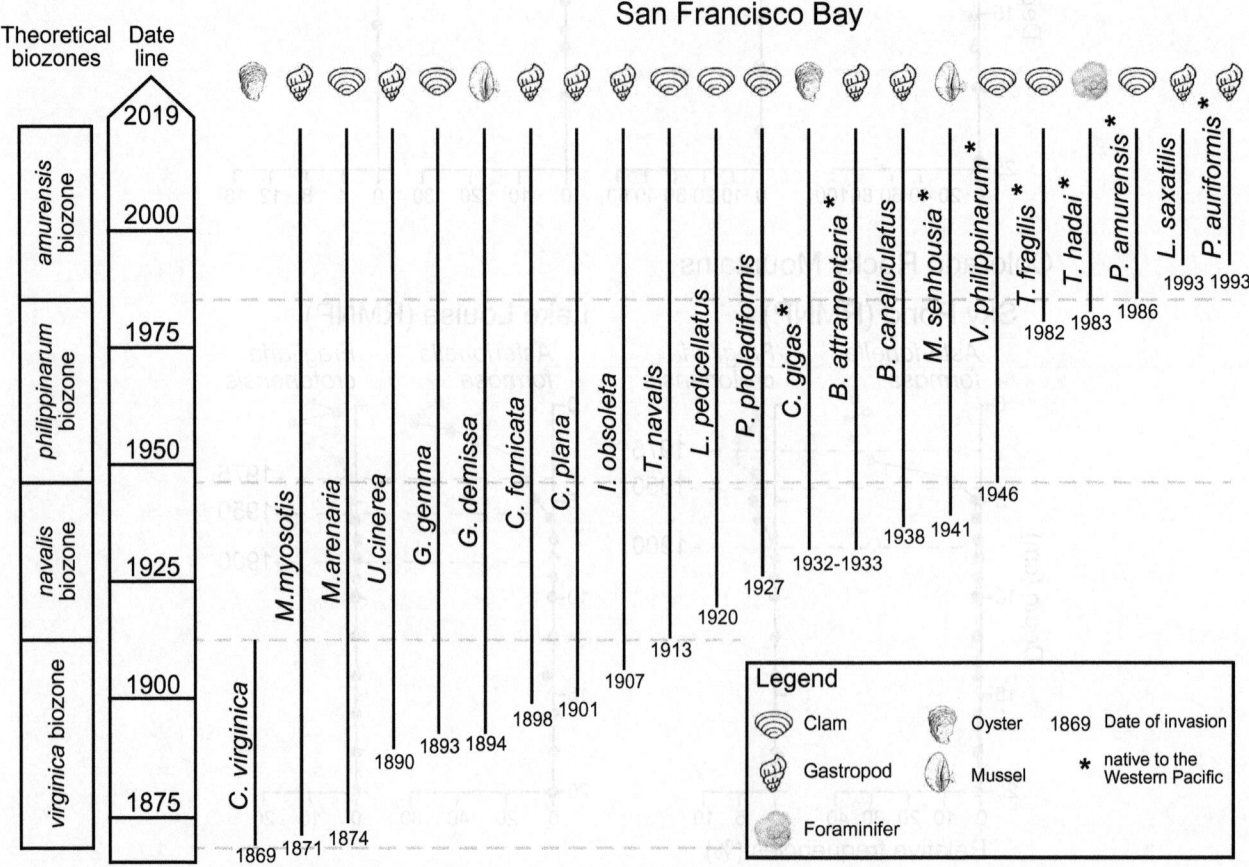

FIGURE 31.11 Biological invasion in San Francisco Bay showing a local range chart of exotic mollusc and foraminifer taxa across the last century and a half, for which theoretical biozones have been suggested.

from earlier domesticated varieties, with sharply distinct skeletal morphology and genetic and isotopic composition (Bennett et al., 2018). Now by an order of magnitude the most common and short-lived bird in the world, with a >20 billion standing population, continually replaced on a ~6-week cycle, its bones, discarded casually and in land-fills, are adding to the biostratigraphic characterization of the Anthropocene.

31.5 Climatic signals of the Anthropocene

The Quaternary Period and its subdivisions, including the recent tripartite subdivision of the Holocene (Walker et al., 2018), are fundamentally based on climate transitions, even if specific boundary levels use other markers, such as paleomagnetic reversals. In the case of the Anthropocene, while there has been sharp change in climate drivers such as greenhouse gas levels, the Earth's radiative balance is not yet in equilibrium (not least as emissions trends remain sharply upward) with most extra heat going into warming the oceans (Yan et al., 2016). Climate and sea level are lagging these changes, although on an upward trajectory (Fig. 31.12), with ~1°C climate warming over preindustrial levels [Intergovernmental Panel on Climate Change (IPCC), 2018], mostly since 1980, and ~20 cm eustatic sea level rise (Church et al., 2013), and as yet no significant deflection of, for instance, oxygen isotope trends in polar ice (Masson-Delmotte et al., 2015). Ultimate equilibrium global temperature and sea level will depend upon carbon emissions [Intergovernmental Panel on Climate Change (IPCC), 2018] but given the long residence time of CO_2 in the atmosphere, a long-term Anthropocene climate state of significantly greater warmth and higher sea levels has already been signaled (Clark et al., 2016).

31.6 Anthropocene GSSP possibilities

The wide array of stratigraphic proxy markers for the Anthropocene, and their equally wide combined distribution across all major sedimentary facies, including glacial ice (and even subglacial sediments, in some instances: Smith et al., 2016) mean that, for practical purposes, an Anthropocene Series, separable from the preceding Holocene, appears capable of wide recognition.

This array of proxy markers clusters around the mid-20th century, hence the optimum boundary level for this potential new geological time unit. Of these markers the sharpest and most unambiguous appears to be the global spread of artificial radionuclides—the beginning of the global "bomb spike" (Waters et al., 2015; Zalasiewicz et al., 2017a), and this might form the primary marker for a proposed Anthropocene GSSP.

The range of environments in which a GSSP may be located has been assessed by Waters et al. (2018) and includes lakes, marine anoxic basins, estuaries and deltas, speleothems, glacial ice, annually banded coral, and tree rings. All of these have the potential for annual (or subannual) resolution within successions that stretch back from the present to at least several centuries back. Peat provides an important terrestrial record, but lacking the annual lamination considered important to ensure the ultrahigh resolution definition considered necessary for a mid-20th century boundary. Anthropogenic deposits are another possibility, but to our knowledge, there are no examples with such continuity and resolution. Within these environments, potential GSSP locations are currently being assessed. The model adopted is that of the Holocene, combining a GSSP with auxiliary stratotypes.

31.6.1 Lake deposits

Lakes cover about 3.7% of the ice-free land surface, notably within boreal and arctic latitudes (Verpoorter et al., 2014) and their bottom sediments commonly comprise continuous, vertically aggraded, high-resolution stratigraphic records. Annual laminae (varves) are common in glacial lakes, characterized by graded summer silt laminae alternating with winter clays, (e.g., Holtgrieve et al., 2011; Wolfe et al., 2013). Varves also occur in hypoxic lakes in which sediment is introduced through seasonal input of clastic or biogenic material and bioturbation is limited. Hypoxia in lake sediments, particularly resulting from increasing nutrient release, has developed extensively since the mid-19th century, about a century earlier than widespread development of hypoxia in coastal zones (Jenny et al., 2016). Hypoxia can change the behavior of trace elements and radionuclides, by either increasing or reducing their mobility, and hence may affect their use as chemostratigraphic markers. Small hypoxic lakes with limited or no fluvial input produce the most suitable environments for preserving clean atmospheric signals, for example, Crawford Lake, Ontario, Huguangyan Maar Lake in China, but also artificial reservoirs, for example, Jasper Ridge, California. Saline lakes develop varves due to seasonal precipitation of evaporite minerals but are more prone to have missing annual laminae.

In addition to lamina counting, independent dating of lake successions using ^{210}Pb is valid for Anthropocene successions, supported by distinct radiogenic spikes. Varved lake sediments can display geochemical signals related to input from local industries, but widespread airborne contaminants, for example, radiogenic fallout (Waters et al., 2015; Fig. 31.9), nitrates and stable nitrogen isotopes (Holtgrieve et al., 2011; Wolfe et al., 2013; Fig. 31.5), fly ash (Rose, 2015; Fig. 31.4), and anthropogenic lead and Pb

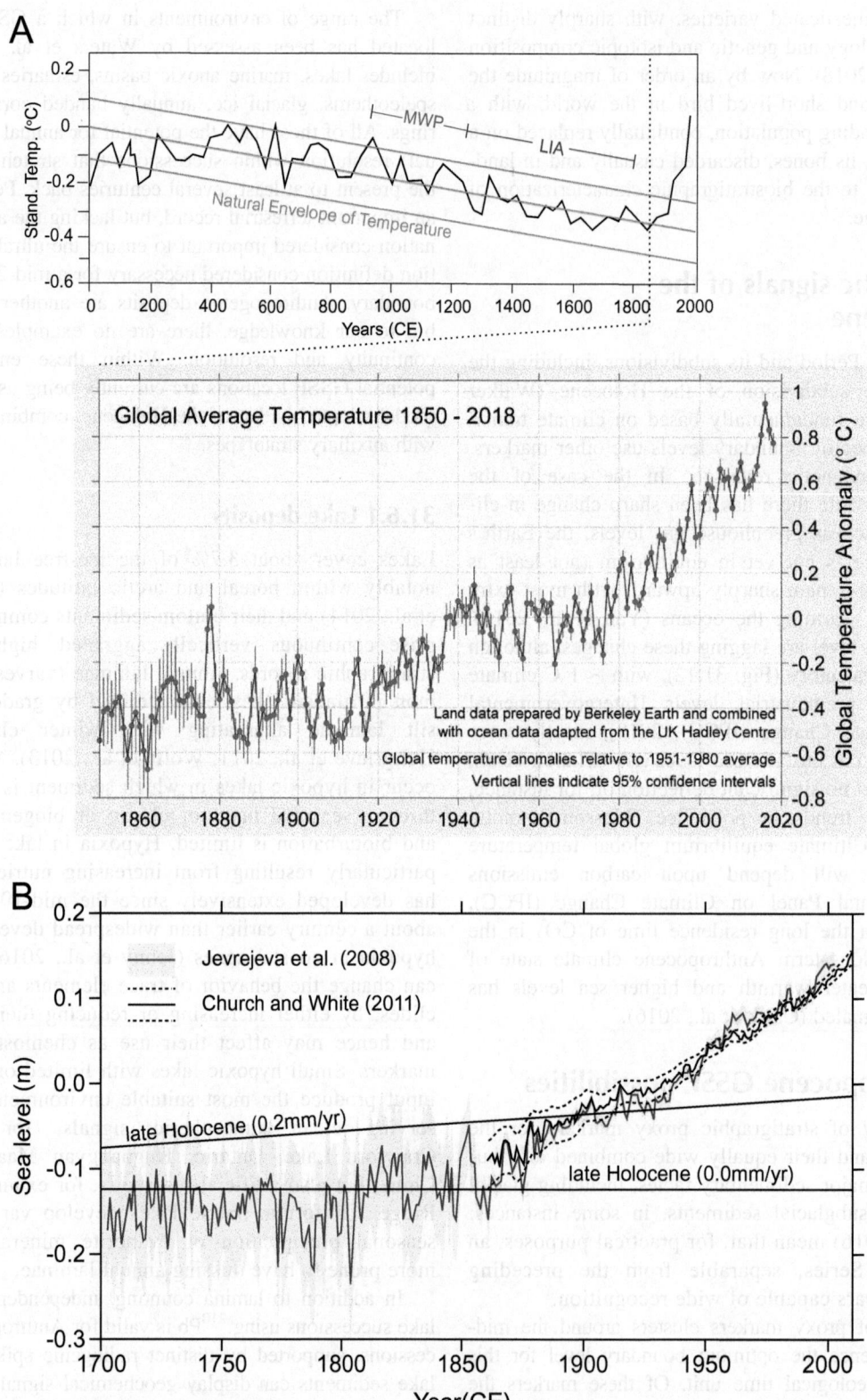

FIGURE 31.12 (A) Upper figure shows standardized global mean temperature for the past 2000 years, represented by 30-year means, showing the natural temperature envelope for the past 2000 years; LIA and the MWP of the Northern Hemisphere are indicated. Lower figure shows global temperature anomalies relative to 1951–80 average from 1850 to 2018 (from Ahmed et al. 2013, redrawn in Waters et al. 2016). (B) Global tide-gauge compilations from 1700 to 2009. *LIA*, Little ice age; *MWP*, Medieval warm period (Jevrejeva et al. 2008; Church and White 2011).

isotopes (Reuer and Weiss, 2002; Fig. 31.7), commonly show patterns across the mid-20th century boundary that are correlatable across continents. Shallow lakes show little settling lag in the signal but may be prone to reworking of signals through erosion within catchment areas.

Lake deposits can show widespread high-density microplastic contamination, including mineral-polyethylene and mineral-polypropylene mixtures, which sink to lake bottoms rapidly (e.g., Corcoran et al., 2015). The microplastics tend to concentrate in nearshore sediments, in low-energy environments close to urban and industrial areas. The timing of onset of deposition of microplastics in lakes remains largely unconstrained in detail; Corcoran et al. (2015) indicate an onset in the Great Lakes of Canada in the 1970s. POPs including chlorinated pesticides (e.g., DDT and its degradation products) show a consistent worldwide rise from the 1950s (Muir and Rose, 2007); many subsequently decline in the youngest lake sediments following regulation or banning of those recognized as environmentally harmful (Fig. 31.8).

Lakes are prone to rapid influx and domination of invasive biota, at the cost of endemic species [e.g., the zebra mussel (*D. polymorpha*) and quagga mussel (*Dreissena rostriformis bugensis*)]. Diatom assemblages in remote Northern Hemisphere lakes show a common pattern of declining Late Holocene benthic diatoms (e.g., *Staurosira* and *Achnanthidium*) and concurrent rise of planktonic diatoms during the Anthropocene, (e.g., *Asterionella formosa* and *Fragilaria crotonensis*), reflecting elevated reactive nitrogen availability and rising temperatures (Wolfe et al., 2013; Fig. 31.10).

31.6.2 Marine anoxic basins

Given that oceans cover about 70% of the Earth's surface, at least one candidate GSSP should be located within marine sediments. However, while many recent marine successions include a range of Anthropocene indicators, few have the level of stratigraphic resolution needed for a GSSP. Deep-ocean oozes typically show very low sedimentation rates—the Anthropocene interval commonly being less than a millimeter thick—preserve few calcareous microfossils, being below the calcite compensation depth, and are typically oxygenated and hence bioturbated, blurring the stratigraphic record. Trawling on the upper continental slope and continental shelf (e.g., Martin et al., 2015) has widely disturbed sediments, that are in any case also prone to bioturbation. Marine anoxic basins, mainly semienclosed basins where the water mass is isolated from the main shelf and typically occur below ~100 m and the reach of storm mixing, potentially offer detailed records of the Holocene–Anthropocene transition. Here, sedimentation accumulation rates are relatively high through a combination of enhanced biological productivity in the photic zone and proximity to terrestrial sediment sources, while

varve preservation is favored by oxygen-deficient bottom waters limiting bioturbation (Schimmelmann et al., 2016). Long-lived marine anoxic basins include the Santa Barbara Basin (California) and Cariaco Basin (Venezuela) and the enclosed Black Sea. Of more recent date are the seasonal marine "dead zones," including the Baltic Sea, the number of which has approximately doubled each decade since the 1960s (Diaz and Rosenberg 2008).

Anthropocene successions in these settings can be dated by lamina counting, by ^{210}Pb dating, by radiogenic fallout signals and land-derived microplastic beads and plastic fibers (e.g., Ivar do Sul and Costa, 2014; Zalasiewicz et al., 2016a). Other important signals include organic and inorganic chemical contamination, effects associated with atmospheric CO_2 increase and warming such as variations in pH, dissolved oxygen content and $\delta^{13}C$ (Fig. 31.3), and resultant biological changes. These signals contrast with latest Holocene strata, for instance in which clinker, from coal burnt to power steamships, was extensively dumped on the sea floor from CE ~1800 to ~1950 (Ramirez-Llodra et al., 2011). The use of Pu isotopes and metal contaminants as markers for the Anthropocene is complicated by the potential decadal time delay for these contaminants to reach the seabed. Typically, the onset of the signal is little affected, but peak signals can show significant lags and prolonged elevated values rather than a distinct peak (Waters et al., 2018; Fig. 31.9).

31.6.3 Estuaries and deltas

Most modern estuaries and deltas developed during the Early Holocene sea-level rise and flooding of river or glacial valleys. Rates of sea level rise during the Mid- and Late Holocene reduced greatly, allowing estuaries to fill with sediments and deltas to build seawards, enhanced by increased sediment flux resulting from early deforestation and agricultural expansion. Subsequently, the trapping of sediment behind major dams, constructed on nearly all major rivers, has greatly reduced sediment flux to the coast since the 1950s (Syvitski and Kettner, 2011). This, along with current and projected absolute sea-level rises, is expected to preserve transgressive, diachronous sequence-stratigraphic systems in these environments. Continuous Anthropocene records are likely to be concentrated in areas of mud deposition in front of major estuaries and in prodelta settings, where moderate sedimentation rates of 1–5 mm/year are typical (Hanebuth et al., 2015). However, this environment is prone to omission surfaces (both through natural erosion and anthropogenic dredging), so many Anthropocene records here will be incomplete.

Many of the geochemical markers and possibilities for independent dating found in lakes and marine basins can also be found in estuaries and deltas, though markers are often enhanced by the common proximity of urban and industrial developments. Estuaries and deltas are particular foci for invasive species (Wilkinson et al., 2014), often associated with marked declines in indigenous species. Widespread species translocation took place during the 19th and 20th centuries through the establishment of global shipping routes, with transfer via hull fouling and in ballast water. Changed species assemblages have also resulted from aquaculture of fish, shellfish, and shrimps in estuaries, especially during the late 20th and early 21st centuries, bringing eutrophication, the spread of pathogens and parasites, and destruction of coastal mangrove systems (Martinez-Porchas and Martinez-Cordova, 2012). San Francisco Bay (Fig. 31.11) exemplifies these complex transitions in biotic assemblages, as well as preserving geochemical signals associated with urbanization and industrialization.

31.6.4 Speleothems

Annually laminated calcareous speleothems (typically stalagmites) occur within natural cave systems of karst environments, in which growth is typically slow, of tens to hundreds of micrometers per year. Speleothems can also occur within artificial tunnels, where degradation of mortar in concrete linings contributes the calcium carbonate; these can have faster growth rates of 10 mm/year.

Speleothems record changing environmental conditions in the atmosphere, soil, and ecosystem through geochemical signals transmitted from the ground surface, through rock, to the underground void. Principal signals that may be used to recognize a potential Holocene−Anthropocene boundary includes shifts in atmospheric ^{14}C (Fig. 31.9) and sulfur (sulfate concentration and $\delta^{34}S$; Fig. 31.6), which relate to global changes to atmospheric chemistry (Fairchild and Frisia, 2014). Superimposed upon these are local, more or less diachronous signals that include (1) variations in growth rates of laminae and $\delta^{18}O$ which relate to air temperature and humidity (Figs. 31.12), (2) $\delta^{13}C$ as an indicator of deforestation and/or introduction of C4 plants, (3) biomarkers such as changes in C27/C31 n-alkane ratios and increase in n-alkanols reflecting the local introduction of agriculture, and (4) shifts in trace elements and isotope ratios (Fairchild, 2018; Fig. 31.7).

Because these signals are transported through plants, soil, and rock before incorporation, they are variably attenuated and delayed (Fairchild and Frisia, 2014, Fairchild, 2018). Speleothems are also insensitive to the large increases in atmospheric CO_2 concentrations and depletion in $\delta^{13}C$ over the past century (Fairchild and Frisia, 2014) and lack Pu isotope determinations. These factors likely exclude a speleothem as GSSP for the Anthropocene (Fairchild, 2018), but one could be chosen as an auxiliary site.

31.6.5 Glacial ice

Extensive, annually resolvable Anthropocene records are present in the continental ice sheets of Antarctica and Greenland. Records are present in mountain glaciers too, but these are prone to loss of annual laminae through seasonal melting and potential mobilization of signals. The Greenland ice sheet is more contaminated by Northern Hemisphere industry, including prior to the 20th century, whereas Antarctic land ice is more pristine and shows a more attenuated but cleaner mid-20th century set of global markers (Waters et al., 2018).

Stratigraphic markers in ice comprise the ice itself and its isotopic compositions; a range of aerosol species that have fallen and been preserved with the snow layers; and trapped air bubbles, formed once fluffy air-filled snow and firn have compacted down into solid ice.

The air bubble record is the most iconic, preserving atmospheric variations in CO_2 (Fig. 31.2A), CH_4 (Fig. 31.2B), and N_2O (Fig. 31.5) that extend far back beyond the instrumental record (Wolff, 2014). However, the Anthropocene interval is mostly firn, where bubbles are not yet closed from connection to the atmosphere and hence their environmental signals—inherently younger than the enclosing ice—are not yet locked in. The depth and hence timing of the firn-to-ice transition is dependent upon the accumulation rate, and so sites with rapid accumulation rates, commonly in more coastal settings, preserve a shorter air/ice lag. In older ice records the nitrogen isotopes in the air bubbles can be used to estimate the correct ages of the air bubbles, facilitating direct comparison with ice of the same age (Parrenin et al., 2013). This correction technique is not really applicable to the Anthropocene record; nevertheless, the fastest accumulating snow, with an ice/air difference of about 30 years, captures the beginning of the sharp upturns in greenhouse gases associated with industrialization and the changes in their $\delta^{13}C$ composition due to the Suess effect (Fig. 31.3) and the ice record can be spliced with direct atmospheric measurements made since 1956 for CO_2 (and later for the other gases) to show the unprecedented scale and speed of the rise in these gases when placed against both a Holocene and a Quaternary context. Thus the marked increase in CO_2 concentration (Fig. 31.2A), $\sim 30\%$ above the highest level of the past 800 kyr (Wolff, 2014), and the change to more depleted $\delta^{13}C$ values (Fig. 31.3), both occur at about CE 1960 in Law Dome ice core (Rubino et al., 2013). CH_4 is now at ~ 1800 ppb, double the highest level of the past 800 kyr (Wolff, 2014), with an initial abrupt rise from CE ~ 1875 in both concentrations and $\delta^{13}C$ values seen in Law Dome (Ferretti et al., 2005; Fig. 31.2B). N_2O concentrations have risen by about 20% in the last 200 years (e.g., Law Dome) and are at levels higher than recorded for the past 800 kyr, with the most rapid rise from CE 1950 onward (Wolff, 2013; Fig. 31.5).

The isotopic composition of the ice itself shows little change across a putative Holocene−Anthropocene boundary, consistent with a climate response that lags greenhouse gas changes at a decadal to multidecadal scale. Greenland δD and $\delta^{18}O$ show no consistent or significant variation across the mid-20th century, though a warming trend is shown by a slight change to less negative $\delta^{18}O$ values from CE $\sim$1850−70 and again from CE 1979 to 2007 (Masson-Delmotte et al., 2015; Fig. 31.12).

Anthropogenic signals derived from aerosols preserved within the ice are not affected by the discrepancy in air bubble age, and these provide the clearest markers for a Holocene/Anthropocene boundary. These include radionuclide signals from nuclear bomb testing (e.g., Pu, ^{14}C), increases in black carbon (BC) and metals such as Pb from industrial activity and automobile emissions, sulfate especially from coal combustion, and nitrate from fertilizers. Silicate dust concentrations in polar snow are exceedingly small and do not appear to show systematic trends across the Holocene−Anthropocene boundary.

The most abrupt anthropogenic signal in ice is radioactive fallout from the atmospheric testing of nuclear devices. This is seen as a marked increase in beta radioactivity (e.g., ^{14}C, ^{90}Sr, and tritium) initially in CE 1954 and peak in CE 1966, followed by slow decay toward natural background levels following the Partial Test Ban Treaty in CE 1963 (Wolff, 2014; Fig. 31.9). The long-lived plutonium radionuclide ^{239}Pu typically has a higher activity in Arctic ice (differing by about a factor of three) than Antarctic ice (Fig. 31.9), with the first detection in polar ice in CE 1953 (Arienzo et al., 2016). The ^{239}Pu bomb-spike (peak signal) occurs in the CE 1962 ice lamina in the Arctic but is not strongly resolved in the Antarctic (Arienzo et al., 2016).

In Greenland, BC is attributed to preindustrial (CE 1772−1860), coal-dominated industrial (CE 1860−1940), and oil-dominated industrial (CE 1940−2003) sources, with peak concentrations in the first decade of the 20th century (McConnell and Edwards, 2008). Lead enrichment in Arctic ice cores started at around 3300−3000 year BP, with an additional significant Roman phase (Hong et al., 1994) and yet higher concentrations in response to the Industrial Revolution from the late 19th century with peaks in the early 20th century reflecting increased coal burning in North America and Europe (McConnell and Edwards, 2008). However, the greatest rises in Pb concentrations are from the 1950s, peaking in the 1960s (Fig. 31.7), associated with emissions of Pb used in alkyl-leaded gasoline (Murozumi et al., 1969; McConnell and Edwards, 2008). At its peak, Pb levels in Greenland ice were above Holocene background levels by a factor of 200, while in Antarctic ice peak values from CE 1950 to 1980 were a factor of five above preindustrial levels (Wolff and Suttie, 1994). Because most combustion of gasoline occurred in the Northern Hemisphere, Greenland ice shows the greatest rise and subsequent fall of Pb signals.

Greenland ice sulfate (SO_4^{2-}) concentrations increased markedly from CE 1900−20 to 1940−80 (Fig. 31.6), culminating in concentrations of a factor of 2−5 above preindustrial values (Fischer et al., 1998), but did not exceed concentrations associated with large volcanic eruptions or

during the Last Glacial Maximum (Wolff, 2014). Greenland ice nitrate concentrations increased by a factor of 2−3 from CE 1950 to 1980 over preindustrial values (Fischer et al., 1998; Wolff, 2013; Fig. 31.5), mainly due to NO_x emissions from fossil fuel combustion, peaking at levels higher than over the past 100 kyr (Wolff 2014). There is an associated decreasing $\delta^{15}N$ trend, beginning about CE 1850, but with greatest rate of change from CE 1950 to 1980 (Fig. 31.5), coincident with a rapid rise in fossil fuel emissions (Hastings et al., 2009). Antarctic ice records no equivalent elevated sulfate and nitrate signatures, being more remote from the main sources of such anthropogenic contamination (Wolff, 2013, 2014).

31.6.6 Corals

Shallow-water coral reefs extend over only about 0.1%−0.2% of the tropical oceans. Cold-water corals occupy a much wider range of latitudes and of water depths, but these are typically smaller and more slowly growing than in tropical reefs. Deeper water corals contain banding that may not necessarily be annual and some proxy signals can be affected by appreciable time lag during settling through the water column (e.g., radiogenic nuclides, $\delta^{13}C$, heavy metals).

Significant anthropogenic stressors of tropical reefs include greater turbidity of marine waters due to increased runoff, rises in pollutants, eutrophication and acidification, and severe overfishing (Hoegh-Guldberg, 2014). Widespread bleaching of tropical coral reefs due to rising sea temperatures commenced in CE 1979 and has subsequently increased in frequency and severity (Hoegh-Guldberg, 2014; Hughes et al., 2017). There has been a 50% reduction in the abundance of reef-building corals over the past 40−50 years, with potential collapse of whole reef systems in the next few decades, as happened during mass extinction events of the geological past (Hoegh-Guldberg, 2014).

The most representative Anthropocene records are in long-lived coral colonies that show annual growth bands that can be independently dated using ^{210}Pb and ^{14}C. Key correlatory signals include Pu radionuclide fallout, the ^{13}C Suess effect, and uptake of pollutants, in particular heavy metals, with proxies of sea-surface temperatures (SSTs) and pH being regionally significant markers.

The Pu signal in corals typically has very high resolution, with only limited evidence of postgrowth mobility (Lindahl et al., 2011). A study in Guam corals shows an initial rise in Pu between CE 1946 and 1951 and a peak signal in CE 1954 (Fig. 31.9), indicating an early local source of the signals from the Pacific Proving Grounds (Lindahl et al., 2011). The Caribbean corals show a more globally representative Pu fallout pattern, with an initial CE 1954 rise and CE 1964 peak (Benninger and Dodge, 1986; Fig. 31.9).

$\delta^{13}C$ in coral skeletons shows a marked decrease from about CE 1955 in Caribbean corals and sclerosponges (the ^{13}C Suess effect), with a more variable and less pronounced rate of change in the Indian and Pacific oceans (Swart et al., 2010; Fig. 31.3). This is attributed to the largest input of anthropogenically derived CO_2 occurring in countries neighboring the Atlantic Ocean (Swart et al., 2010).

Pb signals in corals show strong regional influences. For instance, corals in Bermuda (Kelly et al., 2009) show an initial Pb rise due to Upper Mississippi Valley lead mining in CE 1840–48, followed by a marked increase that began in the late 1940s due to increased consumption of leaded gasoline and peaked in the 1970s (Boyle et al., 2014; Fig. 31.7). Higher $^{206}Pb/^{207}Pb$ ratios seen in Bermuda until the late-1970s are considered sourced from US gasoline, while the markedly lower ratios in the 1980s and 1990s reflect a European source (Kelly et al., 2009; Fig. 31.7).

Elevated SSTs from the 1970s onward have reduced annual band thicknesses in *Porites* colonies in the Great Barrier Reef (De'ath et al., 2009) and are linked with strong decrease in $\delta^{18}O$ (Fig. 31.12) and Sr/Ca ratios in the Caribbean (Hetzinger et al., 2010), though coral proxies have also shown an inception of a warming trend from the mid-19th century. $\delta^{11}B$ isotopic records in corals provide a proxy for seawater pH, with *Porites* coral in the Great Barrier Reef showing a decreasing trend in pH of about 0.2–0.3 U from CE 1940 to 2004, though with marked annual oscillations (Wei et al., 2009).

31.6.7 Trees

A precise dendrochronological time scale has been developed for the Holocene, for instance helping constrain a Medieval Climatic Anomaly from CE 900 to 1200, below average temperatures during the Little Ice Age from CE 1200 to 1850, and warming since CE 1850 (Esper et al., 2002; Wilson et al., 2016; Fig. 31.12). For the Anthropocene, extrapolation of temperature profiles post-CE 1950 is problematic due to increased statistical uncertainty. However, tree rings provide a high-resolution archive of the radiocarbon bomb spike, variations in stable carbon isotopes, sulfur concentrations, and sulfur isotope values.

Trees provide a high-resolution record of the initial decrease in atmospheric $\Delta^{14}C$ by about 20.0‰ between CE 1890 and 1950 in response to the Suess effect of adding low-activity fossil carbon, followed by a marked radiocarbon bomb spike in response to nuclear weapons testing, peaking at double that of pre-CE 1850 times. Annual rings of a pine tree (*Pinus sylvestris*) taken in the Niepołomice area, near Kraków, Poland from CE 1960 to 2003 show a $\Delta^{14}C$ bomb-spike peaking in CE 1964 (Rakowski et al., 2013; Fig. 31.9). In Campbell Island, located in the Southern Ocean, and hence more remote from Northern Hemisphere–dominated thermonuclear

bomb tests, the only alien tree, a Sitka spruce (*Picea sitchensis*), shows peak atmospheric ^{14}C in October–December 1965 (Turney et al., 2018; Fig. 31.9). $^{239+240}Pu$ records in tree rings are less studied, but the distribution of Pu in tree rings seems to show similar trends to those seen in lake sediment cores (Mahara and Kudo, 1995), suggesting plutonium (Pu) as a potentially suitable signal.

Long-term trends of stable carbon isotopic data in tree rings records the isotopic composition of the source atmospheric CO_2 during photosynthesis, with $\delta^{13}C$ values also varying interannually in response to changes in sunlight, soil moisture, precipitation, and relative humidity, with additional species-specific and local environmental effects. The change in tree ring carbon isotopic values resulting from global industrialization is seen as a reduction of atmospheric $\delta^{13}C$ values by ~2‰ since the start of the Industrial Revolution CE ~1820 (the $\delta^{13}C$ Suess effect), with marked inflection at CE ~1950 (Loader et al., 2013; Fig. 31.3). This effect is also observable in ringless trees in the aseasonal tropics and is far greater than anything observed in the tree ring isotope record during the last 1100 years.

Sulfur concentrations within annual rings reflect atmospheric SO_2 pollution; for instance higher post-1950 values were seen in conifers near to Ernesto Cave in NE Italy (Fairchild et al., 2009; Fig. 31.6). However, these trees show considerable noise, including variation between individual trees as the S is mainly sourced from soil–waters rather than directly from the atmosphere (Fairchild et al., 2009). These trees also show a trend in $\delta^{34}S$ toward lighter isotopes throughout the 20th century, with preindustrial values of +7.5‰ and modern values of +0.7‰, though with no significant mid-20th century inflection (Wynn et al., 2014).

31.6.8 Peat

Peat deposits are extensive (Shotyk, 1992; Yu et al., 2010) and provide good environmental archives, particularly those that are ombrotrophic, receiving their nutrients solely from atmospheric inputs with naturally low concentrations of mineral matter and so providing low background values against which anthropogenic inputs may be compared. Bog surface waters are acidic, with varying redox conditions, and so radioisotopes such as ^{137}Cs and metals such as U, Mn, and Fe can be very mobile. Peat bogs are not varved and the most robust age-depth models use a combination of approaches, including ^{14}C and ^{210}Pb.

The Etang de la Gruère, Switzerland, represents the potential of this environment to define the Anthropocene. It has a continuous record of peat accumulation for ~15,000 years, with the first anthropogenically controlled increase in Pb deposition starting at 5320 ^{14}C year BP, thought to be due to forest clearance and the introduction of agriculture

(Shotyk et al., 1998). Increased Pb enrichment and decreased ^{206}Pb/^{207}Pb ratios at 3000 ^{14}C year BP record the first influence of Pb pollution from mining and smelting, but by far the greatest Pb flux occurred in the late 20th century (Fig. 31.7), reaching 1570 times the background value by CE 1979 and associated with greatly decreased ^{206}Pb/^{207}Pb ratios (Shotyk et al., 1998). Reduced Pb contents and increasing Pb isotope ratios in the later decades of the 20th century record the introduction of unleaded gasoline and a reduction in industrial sources of Pb in the region (Shotyk et al., 1998). The highest concentrations of PCBs occur in the peats dating from CE 1960 to 1976 (Fig. 31.8), whereas polycyclic aromatic hydrocarbons (PAHs) peak from CE 1930 to 1951, associated with the greatest use of coal in Switzerland (Berset et al., 2001). The same core has peak records of ^{241}Am activity marking the early 1950s to early 1960s bomb-spike, whereas ^{137}Cs activity is at a maximum in the uppermost living part of the profile, inferred to be sourced by the Chernobyl disaster of CE 1986 (Appleby et al., 1997).

31.7 Summary

Whether the Anthropocene will be formally recognized or not remains moot. Its brevity, ongoing nature, and the novel and geologically unusual nature of many of the signals, together with its "human" associations, have attracted criticism from among both geological (Gibbard and Walker, 2014; Finney and Edwards, 2016) and nongeological (Ruddiman, 2018; Braje, 2016; Bauer and Ellis, 2018) communities. Although these criticisms may be effectively answered (Zalasiewicz et al., 2017b, 2018), the reception to a formal GSSP proposal, once prepared, remains uncertain. Notwithstanding this, it possesses robust geological reality and its strata can be readily correlated worldwide. Its reality is set to be only reinforced in coming years, and so, formal or informal, it deserves a clear, consistent stratigraphic characterization and definition for ongoing use by the wide range of communities for whom it has already become an indispensible term.

Bibliography

Ahmed, M., et al., 2013, Continental-scale temperature variability during the past two millennia. *Nature Geoscience*, **6**: 339–346. https://doi.org/10.1038/ngeo1797.

Appleby, P.G., Shotyk, W., and Fankhauser, A., 1997, Lead-210 age dating of three peat cores in the Jura Mountains, Switzerland. *Water, Air, & Soil Pollution*, **100** (3/4): 223–231. https://doi.org/10.1023/A:1018380922280.

Arienzo, M.M., McConnell, J.R., Chellman, N., Criscitiello, A.S., Curran, M., Fritzsche, D., et al., 2016, A method for continuous ^{239}Pu determinations in Arctic and Antarctic ice cores. *Environmental Science & Technology*, **50** (13): 7066–7073. https://doi.org/10.1021/acs.est.6b01108.

Barnosky, A.D., Hadly, E.A., Bascompte, J., Berlow, E.L., Brown, J.H., Fortelius, M., et al., 2012, Approaching a state-shift in the biosphere. *Nature*, **486**: 52–56. https://doi.org/10.1038/nature11018.

Barnosky, A.D., Holmes, M., Kirchholtes, R., Lindsey, E., Maguire, K.C., Poust, A.W., et al., 2014, Prelude to the Anthropocene: two new North American Land Mammal Ages (NALMAs). *Anthropocene Review*, **1** (3): 225–242. https://doi.org/10.1177/2053019614547433.

Barnosky, A.D., Matzke, N., Tomiya, S., Wogan, G.O.U., Swartz, B., Quental, T.B., et al., 2011, Has the Earth's sixth mass extinction already arrived? *Nature*, **471**: 51–57. https://doi.org/10.1038/nature09678.

Bauer, A.M., and Ellis, E.C., 2018, The Anthropocene divide: obscuring understanding of social-environmental change. *Current Anthropology*, **59** (2): 209–227.

Bennett, C.E., Thomas, R., Williams, M., Zalasiewicz, J., Edgeworth, M., Miller, H., et al., 2018, The broiler chicken as a signal of a human reconfigured biosphere. *Royal Society Open Science*, **5**: 180325. https://doi.org/10.1098/rsos.180325.

Benninger, L.K., and Dodge, R.E., 1986, Fallout plutonium and natural radionuclides in annual bands of the coral *Montastrea annularis*, St. Croix, U.S. Virgin Islands. *Geochimica et Cosmochimica Acta*, **50**: 2785–2797. https://doi.org/10.1016/0016-7037(86)90227-9.

Benson, A.J., Raikow, D., Larson, J., Fusaro, A., and Bogdanoff, A.K., 2018, *Dreissena polymorpha (Pallas, 1771): U.S. Geological Survey, Nonindigenous Aquatic Species Database*. Gainesville, FL. <https://nas.er.usgs.gov/queries/factsheet.aspx?speciesid=5> (revision 13.02.18, access: 20.11.18.)

Berset, J.-D., Kuehne, P., and Shotyk, W., 2001, Concentrations and distribution of some polychlorinated biphenyls (PCBs) and polycyclic aromatic hydrocarbons (PAHs) in an ombrotrophic peat bog profile of Switzerland. *Science of the Total Environment*, **267**: 76–85. https://doi.org/10.1016/S0048-9697(00)00763-4.

Biermann, F., 2014, The Anthropocene: a governance perspective. *Anthropocene Review*, **1** (1): 57–61. https://doi.org/10.1177/2053019613516289.

Boyle, E.A., Lee, J.-M., Echegoyen, Y., Noble, A., Moos, S., Carrasco, G., et al., 2014, Anthropogenic lead emissions in the ocean: the evolving global experiment. *Oceanography*, **27** (1): 69–75.

Braje, T.J., 2016, Evaluating the Anthropocene: is there something useful about a geological epoch of humans? *Antiquity*, **90** (350): 504–512. https://doi.org/10.15184/aqy.2016.32.

Church, J.A., and White, N.J., 2011, Sea-level rise from the late 19th to the early 21st Century. *Surveys in Geophysics*, **32**: 585–602. https://doi.org/10.1007/s10712-011-9119-1.

Church, J.A., Clark, P.U., Cazenave, A., Gregory, J.M., Jevrejeva, S., Levermann, A., et al., 2013, Sea level change. In Stocker, T.F., Qin, D., Plattner, G.K., Tignor, M.M.B., Allen, S.K., Boschung, J., Nauels, A., Xia, Y., Bex, V., Midgley, P.M. (eds), *Climate Change 2013: The Physical Science Basis, Contribution of Working Group I to the Fifth Assessment Report of the Intergovernmental Panel on Climate Change*. pp. 1137–1216. Cambridge University Press, New York. <http://www.ipcc.ch/report/ar5/wg1/>.

Clark, N., 2014, Geo-politics and the disaster of the Anthropocene. *The Sociological Review*, **62** (S1): 19–37. https://doi.org/10.1111/1467-954X.12122.

Clark, P.U., Shakun, J.D., Marcott, S.A., Mix, A.C., Eby, M., Kulp, S., et al., 2016, Consequences of twenty-first-century policy for multimillennial climate and sea-level change. *Nature Climate Change*, **6**: 360–369. https://doi.org/10.1038/nclimate2923.

Cooper, A.H., Brown, T.J., Price, S.J., Ford, J.R., and Waters, C.N., 2018, Humans are the most significant global geomorphological driving force of the 21st century. *Anthropocene Review*. https://doi.org/10.1177/2053019618800234.

Corcoran, P.L., Norris, T., Ceccanese, T., Walzak, M.J., Helm, P.A., and Marvin, C.H., 2015, Hidden plastics of Lake Ontario: Canada and their potential preservation in the sediment record. *Environmental Pollution*, **204**: 17−25. https://doi.org/10.1016/j.envpol.2015.04.009.

Crutzen, P.J., 2002, Geology of Mankind. *Nature*, **415**: 23. https://doi.org/10.1038/415023a.

Crutzen, P.J., and Stoermer, E.F., 2000, The "Anthropocene." *IGBP Global Change Newsletter*, **41**: 17−18.

Davis, H., and Turpin, E., 2015, Art in the Anthropocene: Encounters Among Aesthetics, Politics, Environments and Epistemologies. *Open Humanities Press*, London.

De'ath, G., Lough, J.M., and Fabricius, K.E., 2009, Declining coral calcification on the Great Barrier Reef. *Science*, **323**: 116−119. https://doi.org/10.1126/science.1165283.

Diaz, R.J., and Rosenberg, R., 2008, Spreading dead zones and consequences for marine ecosystems. *Science*, **321**: 926−929. https://doi.org/10.1126/science.1156401.

Esper, J., Cook, E.R., and Schweingruber, F.H., 2002, Low-frequency signals in long tree ring chronologies for reconstructing past temperature variability. *Science*, **295**: 2250−2253. https://doi.org/10.1126/science.1066208.

Fairchild, I.J., 2018, Geochemical records in speleothems. *In* DellaSala, D., and Goldstein, M.I. (eds), *Encyclopedia of the Anthropocene*. Elsevier, Oxford. https://doi.org/10.1016/B978-0-12-809665-9.09775-5.

Fairchild, I.J., and Frisia, S., 2014, Definition of the Anthropocene: a view from the underworld. *In* Waters, C., Zalasiewicz, J., Williams, M., Ellis, M.A., and Snelling, A. (eds), *A stratigraphical basis for the Anthropocene*. Geological Society, Special Publication, London, **395**: 239−254. https//doi.org/10.1144/SP395.7.

Fairchild, I.J., Loader, N.J., Wynn, P.M., Frisia, S., Thomas, P.A., Lageard, J.G.A., et al., 2009, Sulfur fixation in wood mapped by synchrotron X-ray studies: implications for environmental archives. *Environmental Science & Technology*, **43**: 1310−1315. https://doi.org/10.1021/es8029297.

Ferretti, D.F., Miller, J.B., White, J.W.C., Etheridge, D.M., Lassey, K.R., Lowe, D.C., et al., 2005, Unexpected changes to the global methane budget over the last 2,000 years. *Science*, **309**: 1714−1717. https://doi.org/10.1126/science.1115193.

Filippelli, G., 2002, The global phosphorus cycle. *Reviews in Mineralogy and Geochemistry*, **48**: 391−425. https://doi.org/10.2138/rmg.2002.48.10.

Finney, S.C., and Edwards, L.E., 2016, The "Anthropocene" epoch: scientific decision or political statement? *GSA Today*, **26** (2−3): 4−10.

Fischer, H., Wagenbach, D., and Kipfstuhl, J., 1998, Sulfate and nitrate firn concentrations on the Greenland ice sheet. 2. Temporal anthropogenic deposition changes. *Journal of Geophysical Research*, **103**: 21935−21942. https://doi.org/10.1029/98JD01885.

Frisia, S., Borsato, A., Fairchild, I.J., and Susini, J., 2005, Variations in atmospheric sulphate recorded in stalagmites by synchrotron micro-XRF and XANES analyses. *Earth and Planetary Science Letters*, **235**: 729−740. https://doi.org/10.1016/j.epsl.2005.03.026.

Gałuszka, A., and Migaszewski, Z.M., 2018, Chemical signals of the Anthropocene. *In* DellaSala, D., and Goldstein, M.I. (eds), *Encyclopedia of the Anthropocene*. Elsevier, Oxford.

Gałuszka, A., and Rose, N., 2019, Chapter 5.7: organic Ccmpounds. *In* Zalasiewicz, J., Waters, C.N., Williams, M., and Summerhayes, C.P. (eds), *The Anthropocene as a Geological Time Unit*. Cambridge University Press.

Gałuszka, A., and Wagreich, M., 2019, Chapter 5.6: Metals. *In* Zalasiewicz, J., Waters, C.N., Williams, M., and Summerhayes, C.P. (eds), *The Anthropocene as a Geological Time Unit*. Cambridge University Press.

Geyer, R., Jambeck, J.R., and Lavender Law, K., 2017, Production, use, and fate of all plastics ever made. *Science Advances*, **3**: e1700782. https://doi.org/10.1126/sciadv.1700782.

Gibbard, P.L., and Walker, M.J.C., 2014, The term "Anthropocene" in the context of formal geological classification. *In* Waters, C.N., Zalasiewicz, J.A., Williams, M., Ellis, M.A., and Snelling, A.M. (eds), *A stratigraphical basis for the Anthropocene*. Geological Society, Special Publication, London, **395**: 29−37. https//doi.org/10.1144/SP395.1.

Graham, N.A.J., Wilson, S.K., Carr, P., Hoey, A.S., Jennings, S., and McNeil, M.A., 2018, Seabirds enhance coral reef productivity and functioning in the absence of invasive rats. *Nature*, **559**: 250−253. https://doi.org/10.1038/s41586-018-0202-3.

Han, Y.M., Wei, C., Huang, R.-J., Bandowe, B.A.M., Ho, S.S.H., Cao, J.J., et al., 2016. Reconstruction of atmospheric soot history in inland regions from lake sediments over the past 150 years. *Scientific Reports*, **6**: 19151. https://doi.org/10.1038/srep19151.

Hancock, G.J., Leslie, C., Everett, S.E., Tims, S.G., Brunskill, G.J., and Haese, R., 2011, Plutonium as a chronomarker in Australian and New Zealand sediments: a comparison with 137Cs. *Journal of Environmental Radioactivity*, **102**: 919−929. https://doi.org/10.1016/j.jenvrad.2009.09.008.

Hanebuth, T.J.J., Lantzsch, H., and Nizou, J., 2015, Mud depocenters on continental shelves—appearance, initiation times, and growth dynamics. *Geo-Marine Letters*, **35** (6): 487−503. https://doi.org/10.1007/s00367-015-0422-6.

Hastings, M.G., Jarvis, J.C., and Steig, E.J., 2009, Anthropogenic impacts on nitrogen isotopes of ice-core nitrate. *Science*, **324**: 1288. https://doi.org/10.1126/science.1170510.

Hazen, R.M., Papineau, D., Bleeker, W., Downs, R.T., Ferry, J.M., McCoy, T.J., et al., 2008, Mineral evolution. *American Mineralogist*, **93**: 1639−1720. https://doi.org/10.2138/am.2008.2955.

Hazen, R.M., Grew, E.S., Origlieri, M.J., and Downs, R.T., 2017, On the mineralogy of the "Anthropocene Epoch". *American Mineralogist*, **102**: 595−611.

Hetzinger, S., Pfeiffer, M., Dullo, W.-C., Garbe-Schönberg, D., and Halfar, J., 2010, Rapid 20th century warming in the Caribbean and impact of remote forcing on climate in the northern tropical Atlantic as recorded in a Guadeloupe coral. *Palaeogeography, Palaeoclimatology, Palaeoecology*, **296**: 111−124. https://doi.org/10.1016/j.palaeo.2010.06.019.

Himson, S.J., Kinsey, N.P., Aldridge, D.C., Zalasiewicz, J., Williams, M., and Zalasiewicz, J., 2020, Invasive mollusk faunas of the River Thames exemplify biostratigraphic characterization of the Anthropocene. *Lethaia*, **53**: 267−279. https://doi.org/10.1111/let.12355.

Hoegh-Guldberg, O., 2014, Coral reefs in the Anthropocene: persistence or the end of the line? *In* Waters, C.N., Zalasiewicz, J., Williams, M., Ellis, M.A., and Snelling, A. (eds), *A stratigraphical basis for the Anthropocene*. Geological Society, Special Publications, London, **395**: 167−183. https//doi.org/10.1144/SP395.17.

Hoesly, R.M., Smith, S.J., Feng, L.Y., Klimont, Z., Janssens-Maenhout, G., Pitkanen, T., et al., 2018, Historical (1750−2014) anthropogenic emissions of reactive gases and aerosols from the Community Emissions Data System (CEDS). *Geoscientific Model Development*, **11**: 369–408. https://doi.org/10.5194/gmd-11-369-2018.

Holland, E.A., Lee-Taylor, J., Nevison, C., and Sulzman., J.M., 2005, *Global N Cycle: fluxes and N$_2$O Mixing Ratios Originating from Human Activity. Oak Ridge National Laboratory Distributed Active Archive Center.* https://doi.org/10.3334/ORNLDAAC/797.

Holtgrieve, G.W., Schindler, D.E., Hobbs, W.O., Leavitt, P.R., Ward, E.J., Bunting, L., et al., 2011, A coherent signature of anthropogenic nitrogen deposition to remote watersheds of the northern hemisphere. *Science*, **334**: 1545–1548. https://doi.org/10.1126/science.1212267.

Hong, S., Candelone, J.P., Patterson, C.C., and Boutron, C.F., 1994, Greenland ice evidences of hemispheric pollution for lead two millennia ago by Greek and Roman civilizations. *Science*, **265**: 1841–1843. https://doi.org/10.1126/science.265.5180.1841.

Hooke, R.L., Martín-Ducque, J.F., and Pedraza, J., 2012, Land transformation by humans: a review. *GSA Today*, **22** (12): 4–10. https://doi.org/10.1130/GSAT151A.1.

Hughes, T.P., Kerry, J.T., Álvarez-Noriega, M., Álvarez-Romero, J.G., Anderson, K.D., Baird, A.H., et al., 2017, Global warming and recurrent mass bleaching of corals. *Nature*, **543**: 373–377. https://doi.org/10.1038/nature21707.

Intergovernmental Panel on Climate Change (IPCC), 2013, Climate change 2013: the Physical Science Basis. *In* Stocker, T.F., Qin, D., Plattner, G.-K., Tignor, M., Allen, S.K., Boschung, J., Nauels, A., Xia, Y., Bex, V., and Midgley, P.M. (eds), *Contribution of Working Group I to the Fifth Assessment Report of the Intergovernmental Panel on Climate Change.* Cambridge University Press, Cambridge, United Kingdom and New York. 1535 pp.

Intergovernmental Panel on Climate Change (IPCC), 2018, *SR15 The Special Report on Global Warming of 1.5°C.* <http://report.ipcc.ch/sr15/>.

Ivar do Sul, J.A., and Costa, M.F., 2014, The present and future of microplastic pollution in the marine environment. *Environmental Pollution*, **185**: 352–364. https://doi.org/10.1016/j.envpol.2013.10.036.

Jenny, J.-P., Francus, P., Normandeau, A., Lapointe, F., Perga, M.-E., Ojala, A., et al., 2016, Global spread of hypoxia in freshwater ecosystems during the last three centuries is caused by rising local human pressure. *Global Change Biology*, **22**: 1481–1489. https://doi.org/10.1111/gcb.13193.

Jevrejeva, S., Moore, J.C., Grinsted, A., and Woodworth, P.L., 2008, Recent global sea-level acceleration started over 200 years ago? *Geophysical Research Letters*, **35**: L08715. https://doi.org/10.1029/2008GL033611.

Kelly, A.E., Reuer, M.K., Goodkin, N.F., and Boyle, E.A., 2009, Lead concentrations and isotopes in corals and water near Bermuda, 1780–2000. *Earth and Planetary Science Letters*, **283**: 93–100. https://doi.org/10.1016/j.epsl.2009.03.045.

Koch, P.L., and Barnosky, A.D., 2006, Late Quaternary extinctions: state of the debate. *Annual Review of Ecology, Evolution, and Systematics*, **37**: 215–250. https://doi.org/10.1146/annurev.ecolsys.34.011802.132415.

Koide, M., Griffin, J.J., and Goldberg, E.D., 1975, Records of plutonium fallout in marine and terrestrial samples. *Journal of Geophysical Research*, **80**: 4153–4162. https://doi.org/10.1029/JC080i030p04153.

Lewis, S.L., and Maslin, M.A., 2015, Defining the Anthropocene. *Nature*, **519**: 171–180. https://doi.org/10.1038/nature14258.

Lindahl, P., Asami, R., Iryu, Y., Worsfold, P., Keith-Roach, M., and Choi, M.-S., 2011, Sources of plutonium to the tropical Northwest Pacific Ocean (1943–1999) identified using a natural coral archive. *Geochimica et Cosmochimica Acta*, **75**: 1346–1356. https://doi.org/10.1016/j.gca.2010.12.012.

Loader, N.J., Young, G.H.F., Grudd, H., and McCarroll, D., 2013, Stable carbon isotopes from Torneträsk, northern Sweden provide a millennial length reconstruction of summer sunshine and its relationship to Arctic circulation. *Quaternary Science Reviews*, **62**: 97–113. https://doi.org/10.1016/j.quascirev.2012.11.014.

Mahara, Y., and Kudo, A., 1995, Plutonium released by the Nagasaki A-bomb: mobility in the environment. *Applied Radiation and Isotopes*, **46** (11): 1191–1201. https://doi.org/10.1016/0969-8043(95)00161-6.

Martin, J., Puig, P., Palanques, A., and Giamportone, A., 2015, Commercial bottom trawling as a driver of sediment dynamics and deep seascape evolution in the Anthropocene. *Anthropocene*, **7**: 1–15. https://doi.org/10.1016/j.ancene.2015.01.002.

Martinez-Porchas, M., and Martinez-Cordova, L.R., 2012, World aquaculture: environmental impacts and troubleshooting alternatives. *The Scientific World Journal*, Article ID 389623, pp. 9. https://doi.org/10.1100/2012/389623

Masson-Delmotte, V., Steen-Larsen, H.C., Ortega, P., Swingedouw, D., Popp, T., Vinther, B.M., et al., 2015, Recent changes in north-west Greenland climate documented by NEEM shallow ice core data and simulations, and implications for past-temperature reconstructions. *Cryosphere*, **9**: 1481–1504. https://doi.org/10.5194/tc-9-1481-2015.

Mayewski, P.A., Lyons, W.B., Spencer, M.J., Twickler, M.S., Buck, C.F., and Whitlow, S.I., 1990, An ice core record of atmospheric response to anthropogenic sulphate and nitrate. *Nature*, **346**: 554–556.

McConnell, J.R., and Edwards, R., 2008, Coal burning leaves toxic heavy metal legacy in the Arctic. *Proceedings of the National Academy of Sciences of the United States of America*, **105** (34): 12140–12144. https://doi.org/10.1073/pnas.0803564105.

Meybeck, M., 2003, Global analysis of river systems: from Earth System controls to Anthropocene syndromes. *Philosophical Transactions of the Royal Society B*, **358**: 1935–1955. https://doi.org/10.1098/rstb.2003.1379.

Mitchell, L., Brook, E., Lee, J.E., and Sowers, T., 2013, Constraints on the Late Holocene anthropogenic contribution to the atmospheric methane budget. *Science*, **342**: 964–966.

Muir, D.C.G., and Rose, N.L., 2007, Persistent organic pollutants in the sediments of Lochnagar. *In* Rose, N.L. (ed.), Lochnagar: The natural history of a Mountain Lake. *Developments in Paleoenvironmental Research*, **12**: 375–402.

Murozumi, M., Chow, T.J., and Patterson, C.C., 1969, Geochemical concentrations of pollutant lead aerosols, terrestrial dusts and sea salts in Greenland and Antarctic snow data. *Geochimica et Cosmochimica Acta*, **33**: 1247–1294. https://doi.org/10.1016/0016-7037(69)90045-3.

Nickel, E.H., and Grice, J.D., 1998, The IMA Commission on New Minerals and Mineral Names: procedures and guidelines on mineral nomenclature, 1998. *The Canadian Mineralogist*, **36**: 913–926.

Parrenin, F., Masson-Delmotte, V., Köhler, P., Raynaud, D., Paillard, D., Schwander, J., et al., 2013, Synchronous change of atmospheric CO$_2$ and Antarctic temperature during the last deglacial warming. *Science*, **339**: 1060–1063. https://doi.org/10.1126/science.1226368.

Pimm, S.L., Jenkins, C.N., Abell, R., Brooks, T.M., Gittleman, J.L., Joppa, L.N., et al., 2014, The biodiversity of species and their rates of extinction, distribution, and protection. *Science*, **344**: 987. https://doi.org/10.1126/science.1246752.

Rakowski, A.Z., Nadeau, M.-J., Nakamura, T., Pazdur, A., Pawełczyk, S., and Piotrowska, N., 2013, Radiocarbon method in environmental monitoring of CO_2 emission. *Nuclear Instruments and Methods in Physics Research Section B*, **294**: 503–507. https://doi.org/10.1016/j.nimb.2012.07.017.

Ramirez-Llodra, E., Tyler, P.A., Baker, M.C., Bergstad, O.A., Clark, M.R., Escobar, E., et al., 2011, Man and the last great wilderness: human impact on the deep sea. *PLoS One*, **6** (8): 1–25. https://doi.org/10.1371/journal.pone.0022588.

Reuer, M.K., and Weiss, D.J., 2002, Anthropogenic lead dynamics in the terrestrial and marine environment. *Philosophical Transactions of the Royal Society of London, Series A*, **360**: 2889–2904. https://doi.org/10.1098/rsta.2002.1095.

Rose, N.L., 1996, Inorganic ash spheres as pollution tracers. *Environmental Pollution*, **91**: 245–252. https://doi.org/10.1016/0269-7491(95)00044-5.

Rose, N.L., 2015, Spheroidal carbonaceous fly-ash particles provide a globally synchronous stratigraphic marker for the Anthropocene. *Environmental Science & Technology*, **49**: 4155–4162. https://doi.org/10.1021/acs.est.5b00543.

Rubino, M., Etheridge, D.M., Trudinger, C.M., Allison, C.E., Battle, M.O., Langenfelds, R.L., et al., 2013, A revised 1000 year atmospheric $\delta^{13}C$-CO_2 record from Law Dome and South Pole, Antarctica. *Journal of Geophysical Research: Atmospheres*, **118**: 8482–8499. https://doi.org/10.1002/jgrd.50668.

Ruddiman, W.F., 2013, The Anthropocene. *Annual Review of Earth and Planetary Sciences*, **41**: 4.1–4.24.

Ruddiman, W.F., 2018, Three flaws in defining a formal 'Anthropocene'. *Progress in Physical Geography*. https://doi.org/10.1177/0309133318783142.

Samways, M., 1999, Translocating fauna to foreign lands: here comes the Homogenocene. *Journal of Insect Conservation*, **3**: 65–66.

Saros, J.E., Michel, T.J., Interlandi, S.J., and Wolfe, A.P., 2005, Resource requirements of *Asterionella formosa* and *Fragilaria crotonensis* in oligotrophic alpine lakes: implications for recent phytoplankton community reorganizations. *Canadian Journal of Fisheries and Aquatic Sciences*, **62** (7): 1681–1689. https://doi.org/10.1139/f05-077.

Schimmelmann, A., Lange, C.B., Schieber, J., Francus, P., Ojala, A.E.K., and Zolitschka, B., 2016, Varves in marine sediments: a review. *Earth-Science Reviews*, **159**: 215–246. https://doi.org/10.1016/j.earscirev.2016.04.009.

Sen, I.S., and Peuckner-Ehrenbrink, B., 2012, Anthropogenic disturbance of element cycles at the Earth's surface. *Environmental Science & Technology*, **46**: 8601–8609. https://doi.org/10.1021/es301261x.

Sherwood, O.A., Scott, D.B., Risk, M.J., and Guilderson, T.P., 2005, Radiocarbon evidence for annual growth rings in a deep sea octocoral (*Primnoa resedaeformis*). *Marine Ecology Progress Series*, **301**: 129–134. https://doi.org/10.3354/meps301129.

Shotyk, W., 1992, Organic soils. *In* Martini, I.P., and Chesworth, W. (eds), *Weathering, Soils, and Paleosols*. Elsevier, Amsterdam, pp. 203–224 (Ch. 13).

Shotyk, W., Weiss, D., Appleby, P.G., Cheburkin, A.K., Frei, R., Gloor, M., et al., 1998, History of atmospheric lead deposition since 12,370 ^{14}C yr BP from a peat bog, Jura Mountains, Switzerland. *Science*, **281**: 1635–1640. https://doi.org/10.1126/science.281.5383.1635.

Smith, B.D., and Zeder, M.A., 2013, The onset of the Anthropocene. *Anthropocene*, **4**: 8–13. https://doi.org/10.1016/j.ancene.2013.05.001.

Smith, J.A., Andersen, T.J., Shortt, M., Gaffney, A.M., Truffer, M., Stanton, T.P., et al., 2016, Sub-ice-shelf sediments record twentieth century retreat of Pine Island Glacier. *Nature*, **541**: 77–80. https://doi.org/10.1038/nature20136.

Smol, J.P., 2008, *Pollution of Lakes and Rivers: An Environmental Perspective*, second ed. *Wiley-Blackwell*.

Steffen, W., Sanderson, A., Tyson, P.D., Jäger, J., Matson, P.A., Moore III, B., et al., 2004, Global Change and the Earth System: A Planet under Pressure. *The IGBP Book Series*. Springer-Verlag, Berlin, Heidelberg, New York.

Steffen, W., Crutzen, P.J., and McNeill, J.R., 2007, The Anthropocene: are humans now overwhelming the great forces of Nature? *Ambio*, **36**: 614–621. https://doi.org/10.1579/0044-7447(2007)36[614:TAAHNO]2.0.CO;2.

Steffen, W., Broadgate, W., Deutsch, L., Gaffney, O., and Ludwig, C., 2015, The trajectory of the Anthropocene: the Great Acceleration. *Anthropocene Review*, **2** (1): 81–98. https://doi.org/10.1177/2053019614564785.

Steffen, W., Leinfelder, R., Zalasiewicz, J., Waters, C.N., Williams, M., Summerhayes, C., et al., 2016, Stratigraphic and Earth System approaches in defining the Anthropocene. *Earth's Future*, **8**: 324–345. https://doi.org/10.1002/2016EF000379.

Swart, P.K., Greer, L., Rosenheim, B.E., Moses, C.S., Waite, A.J., Winter, A., et al., 2010, The ^{13}C Suess effect in scleractinian corals mirror changes in the anthropogenic CO_2 inventory of the surface oceans. *Geophysical Research Letters*, **37**: L05604. https://doi.org/10.1029/2009GL041397.

Swindles, G.T., Watson, E., Turner, T.E., Galloway, J.M., Hadlari, T., Wheeler, J., and Bacon, K.L., 2015, Spheroidal carbonaceous particles are a defining stratigraphic marker for the Anthropocene. *Scientific Reports*, **5**: 10264. https://doi.org/10.1038/srep10264.

Syvitski, J.P.M., and Kettner, A., 2011, Sediment flux and the Anthropocene. *Philosophical Transactions of the Royal Society A*, **369** (1938): 957–975. https://doi.org/10.1098/rsta.2010.0329.

Terrington, R.L., Silva, É.C.N., Waters, C.N., Smith, H., and Thorpe, S., 2018, Quantifying anthropogenic modification of the shallow geosphere in central London, UK. *Geomorphology*, **319**: 15–34. https://doi.org/10.1016/j.geomorph.2018.07.005.

Turney, C.S.M., Palmer, J., Maslin, M.A., Hogg, A., Fogwill, C.J., Southon, J., et al., 2018, Global peak in atmospheric radiocarbon provides a potential definition for the onset of the Anthropocene Epoch in 1965. *Scientific Reports*, **8**: 3293. https://doi.org/10.1038/s41598-018-20970-5.

Verpoorter, C., Kutser, T., Seekell, D.A., and Transvik, L.J., 2014, A global inventory of lakes based on high-resolution satellite imagery. *Geophysical Research Letters*, **41**: 6396–6402. https://doi.org/10.1002/2014GL060641.

Vidas, D., 2015, The Earth in the Anthropocene – and the world in the Holocene? *European Society of International Law (ESIL) Reflections*, **4** (6): 1–7.

Vogt, G., Falckenhayn, C., Schrimpf, A., Schmid, K., Hanna, K., Panteleit, J., et al., 2015, The marbled crayfish as a paradigm for saltational speciation by autopolyploidy and parthenogenesis in animals. *Biology Open*, **4**: 1583–1594. https://doi.org/10.1242/bio.014241.

Wagreich, M., and Draganits, E., 2018, Early mining and smelting lead anomalies in geological archives as potential stratigraphic markers for the base of an early Anthropocene. *Anthropocene Review*, **5** (2): 177–201. https://doi.org/10.1177/2053019618756682.

Wahlen, M., Allen, D., Deck, B., and Herchenroder, A., 1991, Initial measurements of CO_2 concentrations (1530 to 1940 AD) in air occluded in the GISP 2 Ice Core from central Greenland. *Geophysical Research Letters*, **18**: 1457–1460. https://doi.org/10.1029/91GL01724.

Walker, M.J.C., Berkelhammer, M., Björk, S., Cheng, H., Cwynar, L., Fisher, D., et al., 2018, Formal subdivision of the Holocene Series/Epoch: three proposals by a Working Group of Members of INTIMATE (Integration of Ice-core, Marine and Terrestrial Records) and the Subcommission on Quaternary Stratigraphy. *Episodes*, **41** (4): 213–268.

Waters, C.N., and Zalasiewicz, J., 2018, Concrete: the most abundant novel rock type of the Anthropocene. *In* DellaSala, D., and Goldstein, M.I. (eds), *Encyclopedia of the Anthropocene*. Elsevier, Oxford. https://doi.org/10.1016/B978-0-12-809665-9.09775-5.

Waters, C.N., Syvitski, J.P.M., Gałuszka, A., Hancock, G.J., Zalasiewicz, J., Cearreta, A., et al., 2015, Can nuclear weapons fallout mark the beginning of the Anthropocene Epoch? *Bulletin of the Atomic Scientists*, **71** (3): 46–57. https://doi.org/10.1177/0096340215581357.

Waters, C.N., Zalasiewicz, J., Summerhayes, C., Barnosky, A.D., Poirier, C., Gałuszka, A., et al., 2016, The Anthropocene is functionally and stratigraphically distinct from the Holocene. *Science*, **351** (6269): 137. https://doi.org/10.1126/science.aad2622.

Waters, C.N., Zalasiewicz, J., Summerhayes, C., Fairchild, I.J., Rose, N.L., Loader, N.J., et al., 2018, Global Boundary Stratotype Section and Point (GSSP) for the Anthropocene Series: where and how to look for a potential candidate. *Earth-Science Reviews*, **178**: 379–429. https://doi.org/10.1016/j.earscirev.2017.12.016.

Wei, S., Wang, Y., Lam, J.C.W., Zheng, G.J., So, M.K., Yueng, L.W.Y., et al., 2008, Historical trends of organic pollutants in sediment cores from Hong Kong. *Marine Pollution Bulletin*, **57** (6): 758–766. https://doi.org/10.1016/j.marpolbul.2008.03.008.

Wei, G., McCulloch, M.T., Mortimer, G., Deng, W., and Xie, L., 2009, Evidence for ocean acidification in the Great Barrier Reef of Australia. *Geochimica et Cosmochimica Acta*, **73**: 2332–2346. https://doi.org/10.1016/j.gca.2009.02.009.

Whitmee, S., Haines, A., Beyrer, C., Boltz, F., Capon, A.G., de Souza Dias, B.F., et al., 2015, Safeguarding human health in the Anthropocene epoch: report of the Rockefeller Foundation–Lancet Commission on planetary health. *Lancet*, **386**: 1973–2028. https://doi.org/10.1016/S0140-6736(15)60901-1.

Wilkinson, B.H., and McElroy, B.J., 2007, The impact of humans on continental erosion and sedimentation. *Geological Society of America Bulletin*, **119**: 140–156. https://doi.org/10.1130/B25899.1.

Wilkinson, I.P., Poirier, C., Head, M.J., Sayer, C.D., and Tibby, J., 2014, Microbiotic signatures of the Anthropocene in marginal marine and freshwater palaeoenvironments. *In* Waters, C.N., Zalasiewicz, J.A., Williams, M., Ellis, M.A., and Snelling, A.M. (eds), *A stratigraphical basis for the Anthropocene*. Special Publications, Geological Society, London, **395**: 185–219. https//doi.org/10.1144/SP395.14.

Williams, M., Zalasiewicz, J., Waters, C., Himson, S., Summerhayes, C., Barnosky, A., and Leinfelder, R., 2018, The palaeontological record of the Anthropocene. *Geology Today*, **34** (5): 188–193. https://doi.org/10.1111/gto.12246.

Williams, M., Zalasiewicz, J., Aldridge, D., Waters, C.N, Bault, V., Head, M., and Barnosky, A., 2019, Chapter 3.3: The biostratigraphic signal of the neobiota. *In* Zalasiewicz, J., Waters, C.N., Williams, M., and Summerhayes, C. (eds), *The Anthropocene as a Geological Time Unit*. Cambridge: Cambridge University Press.

Wilson, R., Anchukaitis, K., Briffa, K.R., Büntgen, U., Cook, E., D'Arrigo, R., et al., 2016, Last millennium northern hemisphere summer temperatures from tree rings: Part I: the long term context. *Quaternary Science Reviews*, **134**: 1–18. https://doi.org/10.1016/j.quascirev.2015.12.005.

Wolfe, A.P., Hobbs, W.O., Birks, H.H., Briner, J.P., Holmgren, S.U., Ingólfsson, Ó., et al., 2013, Stratigraphic expressions of the Holocene–Anthropocene transition revealed in sediments from remote lakes. *Earth-Science Reviews*, **116**: 17–34. https://doi.org/10.1016/j.earscirev.2012.11.001.

Wolff, E.W., 2013, Ice sheets and nitrogen. *Philosophical Transactions of the Royal Society B*, **368**: 20130127. https://doi.org/10.1098/rstb.2013.0127.

Wolff, E.W., 2014, Ice sheets and the Anthropocene. *In* Waters, C.N., Zalasiewicz, J.A., Williams, M., Ellis, M.A., and Snelling, A.M. (eds), *A stratigraphical basis for the Anthropocene*. Geological Society, Special Publications, London, **395**: 255–263. https//doi.org/10.1144/SP395.10.

Wolff, E.W., and Suttie, E.D., 1994, Antarctic snow record of southern hemisphere lead pollution. *Geophysical Research Letters*, **21**: 781–784. https://doi.org/10.1029/94GL00656.

Wynn, P.M., Fairchild, I.J., Frisia, S., Spötl, C., Baker, A., Borsato, A., and EIMF, 2010, High-resolution sulphur analysis of speleothem carbonate by secondary ionisation mass spectrometry. *Chemical Geology*, **271**: 101–107. https://doi.org/10.1016/j.chemgeo.2010.01.001.

Wynn, P.M., Loader, N.J., and Fairchild, I.J., 2014, Interrogating trees for isotopic archives of atmospheric sulphur deposition and comparison to speleothem records. *Environmental Pollution*, **187**: 98–105. https://doi.org/10.1016/j.envpol.2013.12.017.

Yan, X.-H., Boyer, T., Trenberth, K., Karl, T.R., Xie, S.-P., Nieves, V., et al., 2016, The global warming hiatus: slowdown or redistribution? *Earth's Future*, **4** (11): 472–482. https://doi.org/10.1002/2016EF000417.

Yu, Z., Loisel, J., Brosseau, D.P., and Beilman, D.W., 2010, Global peatland dynamics since the Last Glacial Maximum. *Geophysical Research Letters*, **37**: L13402. https://doi.org/10.1029/2010GL043584.

Zalasiewicz, J., Williams, M., Smith, A., Barry, T.L., Coe, A.L., Bown, P.R., et al., 2008, Are we now living in the Anthropocene? *GSA Today*, **18** (2): 4–8.

Zalasiewicz, J., Kryza, R., and Williams, M., 2014, The mineral signature of the Anthropocene in its deep-time context. *In* Waters, C.N., Zalasiewicz, J.A., Williams, M., Ellis, M.A., and Snelling, A.M. (eds), *A stratigraphical basis for the Anthropocene*. Geological Society, Special Publication, London, **395**: 109–117. https//doi.org/10.1144/SP395.2.

Zalasiewicz, J., Waters, C.N., Williams, M., Barnosky, A., Cearreta, A., Crutzen, P., et al., 2015, When did the Anthropocene begin? A mid-twentieth century boundary level is stratigraphically optimal. *Quaternary International*, **383**: 196–203. https://doi.org/10.1016/j.quaint.2014.11.045.

Zalasiewicz, J., Waters, C.N., Ivar do Sul, J., Corcoran, P.L., Barnosky, A.D., Cearreta, A., et al., 2016a, The geological cycle of plastics and their use as a stratigraphic indicator of the Anthropocene. *Anthropocene*, **13**: 4–17. https://doi.org/10.1016/j.ancene.2016.01.002.

Zalasiewicz, J., Williams, M., Waters, C.N., Barnosky, A.D., Palmesino, J., Rönnskog, A.-S., et al., 2016b, Scale and diversity of the physical technosphere: a geological perspective. *Anthropocene Review*, **4** (1): 9–22. https://doi.org/10.1177/2053019616677743.

Zalasiewicz, J., Waters, C.N., Summerhayes, C.P., Wolfe, A.P., Barnosky, A.D., Cearreta, A., et al., 2017a, The Working Group on the Anthropocene: summary of evidence and interim recommendations. *Anthropocene*, **19**: 55–60. https://doi.org/10.1016/j.ancene.2017.09.001.

Zalasiewicz, J., Waters, C.N., Wolfe, A.P., Barnosky, A.D., Cearreta, A., Edgeworth, M., et al., 2017b, Making the case for a formal Anthropocene: an analysis of ongoing critiques. *Newsletters on Stratigraphy*, **50**: 205–226. https://doi.org/10.1127/nos/2017/0385.

Zalasiewicz, J., Steffen, W., Leinfelder, R., Williams, M., and Waters, C.N., 2017c, Petrifying Earth process: the stratigraphic imprint of key Earth System parameters in the Anthropocene. *Theory, Culture & Society*, **34**: 83–104. https://doi.org/10.1177/0263276417690587.

Zalasiewicz, J., Waters, C.N., Head, M.J., et al., 2018, The geological and Earth System reality of the Anthropocene: reply to Bauer and Ellis. *Current Anthropology*, **59** (2): 220–223.

Recommended color coding of stages

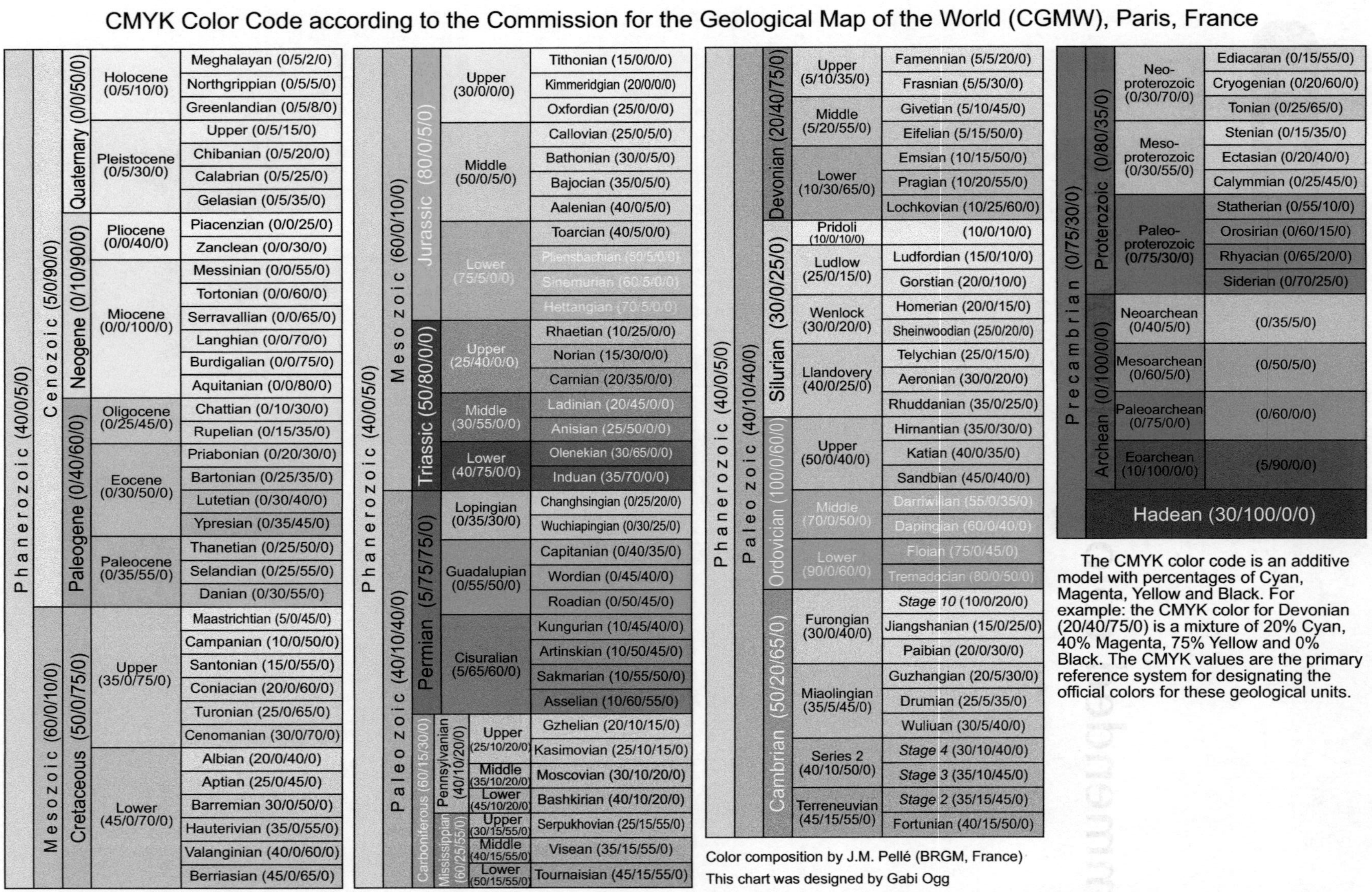

CMYK Color Code according to the Commission for the Geological Map of the World (CGMW), Paris, France

The CMYK color code is an additive model with percentages of Cyan, Magenta, Yellow and Black. For example: the CMYK color for Devonian (20/40/75/0) is a mixture of 20% Cyan, 40% Magenta, 75% Yellow and 0% Black. The CMYK values are the primary reference system for designating the official colors for these geological units.

Color composition by J.M. Pellé (BRGM, France)
This chart was designed by Gabi Ogg

Precambrian (0/75/30/0)

Phanerozoic (40/0/5/0) — Paleozoic (40/10/40/0)

Proterozoic (0/80/35/0)
Series	Stage
Neoproterozoic (0/30/70/0)	Ediacaran (0/15/55/0); Cryogenian (0/20/60/0); Tonian (0/25/65/0)
Mesoproterozoic (0/30/55/0)	Stenian (0/15/35/0); Ectasian (0/20/40/0); Calymmian (0/25/45/0)
Paleoproterozoic (0/75/30/0)	Statherian (0/55/10/0); Orosirian (0/60/15/0); Rhyacian (0/65/20/0); Siderian (0/70/25/0)

Archean (0/100/0/0)
Series	Stage
Neoarchean (0/40/5/0)	(0/35/5/0)
Mesoarchean (0/60/5/0)	(0/50/5/0)
Paleoarchean (0/75/0/0)	(0/60/0/0)
Eoarchean (10/100/0/0)	(5/90/0/0)

Hadean (30/100/0/0)

Early Paleozoic

Devonian (20/40/75/0)
Series	Stage
Upper (5/10/35/0)	Famennian (5/5/20/0); Frasnian (5/5/30/0)
Middle (5/20/55/0)	Givetian (5/10/45/0); Eifelian (5/15/50/0)
Lower (10/30/65/0)	Emsian (10/15/50/0); Pragian (10/20/55/0); Lochkovian (10/25/60/0)

Silurian (30/0/25/0)
Series	Stage
Pridoli (10/0/10/0)	(10/0/10/0)
Ludlow (25/0/15/0)	Ludfordian (15/0/10/0); Gorstian (20/0/10/0)
Wenlock (30/0/20/0)	Homerian (20/0/15/0); Sheinwoodian (25/0/20/0)
Llandovery (40/0/25/0)	Telychian (25/0/15/0); Aeronian (30/0/20/0); Rhuddanian (35/0/25/0)

Ordovician (100/0/60/0)
Series	Stage
Upper (50/0/40/0)	Himantian (35/0/30/0); Katian (40/0/35/0); Sandbian (45/0/40/0)
Middle (70/0/50/0)	Darriwilian (55/0/35/0); Dapingian (60/0/40/0)
Lower (90/0/60/0)	Floian (75/0/45/0); Tremadocian (80/0/50/0)

Cambrian (50/20/65/0)
Series	Stage
Furongian (30/0/40/0)	Stage 10 (10/0/20/0); Jiangshanian (15/0/25/0); Paibian (20/0/30/0)
Miaolingian (35/5/45/0)	Guzhangian (20/5/30/0); Drumian (25/5/35/0); Wuliuan (30/5/40/0)
Series 2 (40/10/50/0)	Stage 4 (30/10/40/0); Stage 3 (35/10/45/0); Stage 2 (35/15/45/0)
Terreneuvian (45/15/55/0)	Fortunian (40/15/50/0)

Mesozoic (60/0/10/0) — Paleozoic (40/10/40/0)

Phanerozoic (40/0/5/0)

Jurassic (80/0/5/0)
Series	Stage
Upper (30/0/0/0)	Tithonian (15/0/0/0); Kimmeridgian (20/0/0/0); Oxfordian (25/0/0/0)
Middle (50/0/5/0)	Callovian (25/0/5/0); Bathonian (30/0/5/0); Bajocian (35/0/5/0); Aalenian (40/0/5/0)
Lower (75/5/5/0)	Toarcian (40/5/0/0); Pliensbachian (50/5/0/0); Sinemurian (60/5/0/0); Hettangian (70/5/0/0)

Triassic (50/80/0/0)
Series	Stage
Upper (25/40/0/0)	Rhaetian (10/25/0/0); Norian (15/30/0/0); Carnian (20/35/0/0)
Middle (30/55/0/0)	Ladinian (20/45/0/0); Anisian (25/50/0/0)
Lower (40/75/0/0)	Olenekian (30/65/0/0); Induan (35/70/0/0)

Permian (5/75/75/0)
Series	Stage
Lopingian (0/35/30/0)	Changhsingian (0/25/20/0); Wuchiapingian (0/30/25/0)
Guadalupian (0/55/50/0)	Capitanian (0/40/35/0); Wordian (0/45/40/0); Roadian (0/50/45/0)
Cisuralian (5/65/60/0)	Kungurian (10/45/40/0); Artinskian (10/50/45/0); Sakmarian (10/55/50/0); Asselian (10/60/55/0)

Carboniferous (60/15/30/0)
Subsystem	Series	Stage
Pennsylvanian (40/10/20/0)	Upper (25/10/20/0)	Gzhelian (20/10/15/0); Kasimovian (25/10/15/0)
	Middle (35/10/20/0)	Moscovian (30/10/20/0)
	Lower (45/10/20/0)	Bashkirian (40/10/20/0)
Mississippian (60/25/55/0)	Upper (30/15/55/0)	Serpukhovian (25/15/55/0)
	Middle (40/15/55/0)	Visean (35/15/55/0)
	Lower (50/15/55/0)	Tournaisian (45/15/55/0)

Cenozoic (5/0/90/0) — Mesozoic (60/0/10/0)

Phanerozoic (40/0/5/0)

Quaternary (0/0/50/0)
Series	Stage
Holocene (0/5/10/0)	Meghalayan (0/5/2/0); Northgrippian (0/5/5/0); Greenlandian (0/5/8/0)
Pleistocene (0/5/30/0)	Upper (0/5/15/0); Chibanian (0/5/20/0); Calabrian (0/5/25/0); Gelasian (0/5/35/0)

Neogene (0/10/90/0)
Series	Stage
Pliocene (0/0/40/0)	Piacenzian (0/0/25/0); Zanclean (0/0/30/0)
Miocene (0/0/100/0)	Messinian (0/0/55/0); Tortonian (0/0/60/0); Serravallian (0/0/65/0); Langhian (0/0/70/0); Burdigalian (0/0/75/0); Aquitanian (0/0/80/0)

Paleogene (0/40/60/0)
Series	Stage
Oligocene (0/25/45/0)	Chattian (0/10/30/0); Rupelian (0/15/35/0)
Eocene (0/30/50/0)	Priabonian (0/20/30/0); Bartonian (0/25/35/0); Lutetian (0/30/40/0); Ypresian (0/35/45/0)
Paleocene (0/35/55/0)	Thanetian (0/25/50/0); Selandian (0/25/55/0); Danian (0/30/55/0)

Cretaceous (50/0/75/0)
Series	Stage
Upper (35/0/75/0)	Maastrichtian (5/0/45/0); Campanian (10/0/50/0); Santonian (15/0/55/0); Coniacian (20/0/60/0); Turonian (25/0/65/0); Cenomanian (30/0/70/0)
Lower (45/0/70/0)	Albian (20/0/40/0); Aptian (25/0/45/0); Barremian (30/0/50/0); Hauterivian (35/0/55/0); Valanginian (40/0/60/0); Berriasian (45/0/65/0)

FIGURE A1.1 CMYK Color Code according to the Commission for the Geological Map of the World (CGMW), Paris, France (except for the new Holocene stages).

RGB Color Code according to the Commission for the Geological Map of the World (CGMW), Paris, France

Precambrian 247/67/112

Proterozoic 247/53/99

Neoproterozoic 254/179/66
- Ediacaran 254/217/106
- Cryogenian 254/204/92
- Tonian 254/191/78

Mesoproterozoic 253/180/98
- Stenian 254/217/154 — 250/167/200
- Ectasian 253/204/138
- Calymmian 253/192/122

Paleoproterozoic 247/67/112
- Statherian 248/117/167 — 248/129/181
- Orosirian 247/104/152
- Rhyacian 247/91/137
- Siderian 247/79/124

Archean 240/4/127
- Neoarchean 249/155/193 — 246/104/178
- Mesoarchean 247/104/169
- Paleoarchean 244/68/159 — 230/29/140
- Eoarchean 218/3/127

Hadean 174/2/126

Explanatory notes:

The RGB color code is an additive model of Red, Green and Blue. Each is indicated on a scale from 0 (no pigment) to 255 (saturation of this pigment). "Devonian (203/140/205)" indicates a mixture of 203 Red, 140 Green and 205 Blue.

The conversion from the reference CMYK values to the RGB codes utilizes Adobe® Illustrator® CS3's color function of "Emulate Adobe® Illustrator® 6.0" (menu Edit / Color Settings / Settings).

ATTENTION: For color conversions using a program other than Adobe® Illustrator®, it is necessary to conserve the reference CMYK, even if the resulting RGB values are slightly different.

Color composition by J.M. Pellé (BRGM, France)
This chart was designed by Gabi Ogg

Phanerozoic 154/217/221 — Paleozoic 153/192/141

Devonian 203/140/55
- **Upper 179/227/238**: Famennian 242/237/197, Frasnian 242/237/173
- **Middle 241/200/104**: Givetian 241/225/133, Eifelian 241/213/118
- **Lower 229/172/77**: Emsian 229/208/117, Pragian 229/196/104, Lochkovian 229/183/90

Silurian 179/225/182
- **Pridoli 230/245/225**: 230/245/225
- **Ludlow 191/230/207**: Ludfordian 217/240/223, Gorstian 204/236/221
- **Wenlock 179/225/194**: Homerian 204/235/209, Sheinwoodian 191/230/195
- **Llandovery 153/215/179**: Telychian 191/230/207, Aeronian 179/225/194, Rhuddanian 166/220/181

Ordovician 0/146/112
- **Upper 127/202/147**: Hirnantian 166/219/171, Katian 153/214/159, Sandbian 140/208/148
- **Middle 77/180/126**: Darriwilian 116/198/156, Dapingian 102/192/146
- **Lower 26/157/111**: Floian 65/176/135, Tremadocian 51/169/126

Cambrian 127/160/86
- **Furongian 179/224/149**: Stage 10 230/245/201, Jiangshanian 217/240/187, Paibian 204/235/174
- **Series 3 166/207/134**: Guzhangian 204/223/170, Drumian 191/217/157, Stage 5 179/212/146
- **Series 2 153/192/120**: Stage 4 179/202/142, Stage 3 166/197/131
- **Terreneuvian 140/176/108**: Stage 2 166/186/128, Fortunian 153/181/117

Phanerozoic 154/217/221

Cenozoic 242/249/29

Quaternary 249/249/127
- Holocene 254/242/224: Meghalayan 254/242/244, Northgrippian 254/242/236, Greenlandian 254/242/229
- Pleistocene 255/242/174: Upper 255/242/211, Chibanian 255/242/199, Calabrian 255/242/186, Gelasian 255/237/179

Neogene 255/230/25
- Pliocene 255/255/153: Piacenzian 255/255/191, Zanclean 255/255/179
- Miocene 255/255/0: Messinian 255/255/115, Tortonian 255/255/102, Serravallian 255/255/89, Langhian 255/255/77, Burdigalian 255/255/65, Aquitanian 255/255/51

Paleogene 253/154/82
- Oligocene 253/192/122: Chattian 254/230/170, Rupelian 254/217/154
- Eocene 253/180/108: Priabonian 253/205/161, Bartonian 253/192/145, Lutetian 252/180/130, Ypresian 252/167/115
- Paleocene 253/167/95: Thanetian 253/191/111, Selandian 254/191/101, Danian 253/180/98

Mesozoic 103/197/202

Jurassic 52/178/201
- Upper 179/227/238: Tithonian 217/241/247, Kimmeridgian 204/236/244, Oxfordian 191/231/241
- Middle 128/207/216: Callovian 191/231/229, Bathonian 179/226/227, Bajocian 166/221/224, Aalenian 154/217/221
- Lower 66/174/208: Toarcian 153/206/227, Pliensbachian 128/197/221, Sinemurian 103/188/216, Hettangian 78/179/211

Triassic 129/43/146
- Upper 189/140/195: Rhaetian 227/185/219, Norian 214/170/211, Carnian 201/155/203
- Middle 177/104/177: Ladinian 201/131/191, Anisian 188/117/183
- Lower 152/57/153: Olenekian 176/81/165, Induan 164/70/159

Permian 240/64/40
- Lopingian 251/167/148: Changhsingian 252/192/178, Wuchiapingian 252/180/162
- Guadalupian 251/116/92: Capitanian 251/154/133, Wordian 251/141/118, Roadian 251/128/105
- Cisuralian 239/88/69: Kungurian 227/135/118, Artinskian 227/123/104, Sakmarian 227/111/92, Asselian 227/99/80

Mesozoic 103/197/202 — Paleozoic 153/192/141

Cretaceous 127/198/78
- Upper 166/216/74: Maastrichtian 242/250/140, Campanian 230/244/127, Santonian 217/239/116, Coniacian 204/233/104, Turonian 191/227/93, Cenomanian 179/222/83
- Lower 140/205/87: Albian 204/234/151, Aptian 191/228/138, Barremian 179/223/127, Hauterivian 166/217/117, Valanginian 153/211/106, Berriasian 140/205/96

Carboniferous 103/165/153
- Pennsylvanian 153/194/181
  - Upper 191/208/197: Gzhelian 204/212/199, Kasimovian 191/208/197
  - Middle 166/199/183: Moscovian 179/203/185
  - Lower 140/190/180: Bashkirian 153/194/181
- Mississippian 103/143/102
  - Upper 179/190/108: Serpukhovian 191/194/107
  - Middle 153/180/108: Visean 166/185/108
  - Lower 140/171/108: Tournaisian 140/176/108

FIGURE A1.2 RGB Color Code according to the Commission for the Geological Map of the World (CGMW), Paris, France (except for the new Holocene stages).

Radioisotopic ages used in GTS2020

M.D. SCHMITZ

GTS 2020 ID	GTS 2012 ID	Period	Epoch	Age	Sample	Locality	Lat-Long	Lithostratigraphy	Age (Ma)	± 2s analytical	± 2s total	Age Type
					Quaternary – not compiled							
					Neogene – not compiled							
			Pliocene									
			Miocene									
		Paleogene										
			Oligocene									
				Chattian								
Pg36					biotite-rich layer; PAC-B2	Pieve d'Accinelli section, northeastern Apennines, Italy	43°35′40.41″N, 12°29′34.16″E	Scaglia Cinerea Fm, 42.3 m above base of section	26.57	0.02	0.04	$^{206}Pb/^{238}U$
				Rupelian								
Pg35	Pg20				biotite-rich layer; MCA-145.8, equivalent to MCA/84-3	Monte Cagnero section (Chattian GSSP), northeastern Apennines, Italy	43°38′47.81″N, 12°28′03.83″E	Scaglia Cinerea Fm, 145.8 m above base of section	31.41	0.03	0.04	$^{206}Pb/^{238}U$
Pg34					biotite-rich layer; MCA-142.8	Monte Cagnero section (Chattian GSSP), northeastern Apennines, Italy	43°38′47.81″N, 12°28′03.83″E	Scaglia Cinerea Fm, 142.8 m above base of section	31.72	0.02	0.04	$^{206}Pb/^{238}U$
			Eocene									
				Priabonian								
Pg33	Pg19				biotite-rich layer; MASS-14.7, equivalent to MAS/86-14.7	Massignano (Oligocene GSSP), near Ancona, northeastern Apennines, Italy	43.5328°N, 13.6011°E	Scaglia Cinerea Fm, 14.7 m above base of section	34.50	0.04	0.05	$^{206}Pb/^{238}U$
Pg32					biotite-rich layer; MASS-12.9	Massignano (Oligocene GSSP), near Ancona, northeastern Apennines, Italy	43.5328°N, 13.6011°E	Scaglia Cinerea Fm, 12.9 m above base of section	34.68	0.04	0.06	$^{206}Pb/^{238}U$
Pg31	Pg18				biotite-rich layer; MASS-12.7, equivalent to MAS/86-12.7	Massignano (Oligocene GSSP), near Ancona, northeastern Apennines, Italy	43.5328°N, 13.6011°E	Scaglia Cinerea Fm, 12.7 m above base of section	34.72	0.02	0.04	$^{206}Pb/^{238}U$
Pg30					biotite-rich layer; MASS-12.9	Massignano (Oligocene GSSP), near Ancona, northeastern Apennines, Italy	43.5328°N, 13.6011°E	Scaglia Variegata Fm, 8.0 m above base of section	35.28	0.03	0.05	$^{206}Pb/^{238}U$
Pg29					biotite-rich layer; MASS-12.9	Massignano (Oligocene GSSP), near Ancona, northeastern Apennines, Italy	43.5328°N, 13.6011°E	Scaglia Variegata Fm, 7.2 m above base of section	35.34	0.03	0.05	$^{206}Pb/^{238}U$
Pg28					biotite-rich layer; MASS-12.9	Massignano (Oligocene GSSP), near Ancona, northeastern Apennines, Italy	43.5328°N, 13.6011°E	Scaglia Variegata Fm, 6.5 m above base of section	35.40	0.03	0.05	$^{206}Pb/^{238}U$
Pg27					biotite-rich layer; MASS-12.9	Massignano (Oligocene GSSP), near Ancona, northeastern Apennines, Italy	43.5328°N, 13.6011°E	Scaglia Variegata Fm, 5.8 m above base of section	35.47	0.03	0.05	$^{206}Pb/^{238}U$
				Bartonian								
				Lutetian								
Pg26	Pg17				Mission Valley ash, SDSNH Loc. 3428	La Mesa, California	~33°N, 117°W	terrestrial facies of Mission Valley Fm	43.35	± 0.50	± 0.50	$^{40}Ar/^{39}Ar$
Pg25	Pg16				DSDP Hole 516F	DSDP Hole 516F	30°16.59′S, 35°17.10′W		46.24	± 0.50	± 0.50	$^{40}Ar/^{39}Ar$
				Ypresian								
Pg24	Pg15				68-0-51, 3497; Blue Point Marker ash	Two-Ocean Plateau and Irish Rock locales, Absaroka volcanic province, western USA	~44°N, 109°W	overlies Aycross Fm	48.41	± 0.21	± 0.21	$^{40}Ar/^{39}Ar$
Pg23	Pg14				CP-1; Continental Peak tuff	Bridger Basin, western USA	42°16′06.2″N, 108°43′7.5″W	Bridger Fm	48.96	± 0.28	± 0.33	$^{40}Ar/^{39}Ar$

Primary radioisotopic age details	Zonal range assignment			Biostratigraphy	Reference
	clade	zonation	Zone		
Weighted mean age of 5 (of 8 with 2 older and 1 younger) single zircon crystal analyses, utilizing CA-TIMS and the ET535 spike.	Plank. foraminifera; calc. nannoplankton		upper Zone O5; lower NP25	upper Zone O5; lower NP25; magnetostratigraphic control at top of Chron C9n.2n	Sahy et al. (2017); Coccioni et al. (2008)
Weighted mean age of 6 (of 7 with 1 older) single zircon crystal analyses, utilizing CA-TIMS and the ET535 spike.	Plank. foraminifera; calc. nannoplankton		middle Zone O2; middle NP23	middle Zone O2; mid NP23; magnetostratigraphic control at upper Chron 12r	Sahy et al. (2017); Coccioni et al. (2008)
Weighted mean age of 4 (of 5 with 1 younger) single zircon crystal analyses, utilizing CA-TIMS and the ET535 spike.	Plank. foraminifera; calc. nannoplankton		middle Zone O2; middle NP23	middle Zone O2; mid NP23; magnetostratigraphic control at upper Chron 12r	Sahy et al. (2017); Coccioni et al. (2008)
Weighted mean age of 3 (of 7 with 4 older) single zircon crystal analyses, utilizing CA-TIMS and the ET535 spike.	Plank. foraminifera; calc. nannoplankton		lowermost NP21, lowermost CP16a	lowermost NP21, lowermost CP16a; magnetostratigraphic control in lower Chron 13r	Sahy et al. (2017); Odin et al. (1991)
Weighted mean age of 3 (of 3) single zircon crystal analyses, utilizing CA-TIMS and the ET535 spike.	Plank. foraminifera; calc. nannoplankton		uppermost NP19/20, uppermost CP15b	uppermost NP19/20, uppermost CP15b; magnetostratigraphic control near base of Chron 13r	Sahy et al. (2017); Odin et al. (1991)
Weighted mean age of 7 (of 10 with 3 older) single zircon crystal analyses, utilizing CA-TIMS and the ET535 spike.	Plank. foraminifera; calc. nannoplankton		uppermost NP19/20, uppermost CP15b	uppermost NP19/20, uppermost CP15b; magnetostratigraphic control near base of Chron 13r	Sahy et al. (2017); Odin et al. (1991)
Weighted mean age of 7 (of 12 with 5 older) single zircon crystal analyses, utilizing CA-TIMS and the ET535 spike.	Plank. foraminifera; calc. nannoplankton		lower NP19/20, lower CP15b	lower NP19/20, lower CP15b; magnetostratigraphic control in C15r	Sahy et al. (2017); Odin et al. (1991)
Weighted mean age of 5 (of 7 with 2 older) single zircon crystal analyses, utilizing CA-TIMS and the ET535 spike.	Plank. foraminifera; calc. nannoplankton		lower NP19/20, lower CP15b	lower NP19/20, lower CP15b; magnetostratigraphic control in C15r	Sahy et al. (2017); Odin et al. (1991)
Weighted mean age of 6 (of 10 with 4 older) single zircon crystal analyses, utilizing CA-TIMS and the ET535 spike.	Plank. foraminifera; calc. nannoplankton		lower NP19/20, lower CP15b	lower NP19/20, lower CP15b; magnetostratigraphic control in C16n.1n	Sahy et al. (2017); Odin et al. (1991)
Weighted mean age of 6 (of 9 with 3 older) single zircon crystal analyses, utilizing CA-TIMS and the ET535 spike.	Plank. foraminifera; calc. nannoplankton		lower NP19/20, lower CP15b	lower NP19/20, lower CP15b; magnetostratigraphic control in C16n.1n	Sahy et al. (2017); Odin et al. (1991)
Laser fusion single-crystal sanidine analyses, originally calibrated to TCs = 28.32 Ma; no analytical data published.	Plank. foraminifera; calc. nannoplankton		NP16, CP14a	coccoliths *Reticulofenestra umbilica* (Levin) and *Discoaster distinctus* Martini; magnetostratigraphic control	Obradovich, unpublished data cited in Walsh et al. (1996)
Laser fusion of biotite and sanidine separates, no analytical data published.	Plank. foraminifera; calc. nannoplankton		NP15, CP10	NP15, CP10; magnetostratigraphic control	Swisher and Montanari, in prep, cited in Berggren et al. (1995) (postscript)
Weighted mean of multi-grain feldspar (46.8% of gas) and multi-grain hornblende (98.7% of gas) incremental heating plateau ages, originally calibrated using FCs = 27.84 Ma.	magnetostratigraphy			magnetostratigraphic control	Hiza (1999); Flynn (1986)
Weighted mean age of 12 (of 16) laser fusion analyses on small multi-grain aliquots of sanidine, using FCs as fluence monitor.	magnetostratigraphy			magnetostratigraphic control	Smith et al. (2008); Clyde et al. (2001)

GTS 2020 ID	GTS 2012 ID	Period	Epoch	Age	Sample	Locality	Lat-Long	Lithostratigraphy	Age (Ma)	± 2s analytical	± 2s total	Age Type
Pg22	Pg15				GR-416; Sixth tuff	Green River Basin, western USA	41°32'31.1"N, 109°28'52.9"W	uppermost Wilkins Peak Member, Green River Fm	49.69	± 0.05	± 0.07	$^{206}Pb/^{238}U$
Pg21	Pg3				Upper Willwood ash; bed B	Bighorn Basin, western USA	42°16'06.2"N, 108°43'7.5"W	Willwood Fm	52.93	± 0.23	± 0.23	$^{40}Ar/^{39}Ar$
Pg20	Pg12				Ash-17	DSDP Site 550, north Atlantic	48°30.91'N, 13°26.37'W		55.48	± 0.12	± 0.12	$^{40}Ar/^{39}Ar$
Pg19	Pg11				SB01-01 bentonite	Longyearbyen section, Spitsbergen, Svalbard Archipeligo	~78°09'10"N, 15°01'38"E	10.9 m above base of the Gilsonryggen Member, Frysjaodden Fm, within the PETM carbon isotope excursion	55.785	± 0.034	± 0.075	$^{206}Pb/^{238}U$
			Paleocene									
				Thanetian								
				Selandian								
Pg18	Pg10				Belt Ash	Southeast Polecat Bench section, northern Bighorn Basin, Wyoming, USA	44°51.5'N, 108°45.1'W	Silver Coulee beds, Fort Union Fm	59.39	± 0.30	± 0.30	$^{40}Ar/^{39}Ar$
				Danian								
Pg17					volcanic tuff in Kiowa core, KJ08-17	Kiowa core, Elbert County Fairgrounds, Kiowa, Colorado	39.35242°N, 104.46642°W	174.52 m below top of core; D1 sequence of fluvial/paludal sandstone, mudstone, and lignite beds	64.52	± 0.03	± 0.08	$^{206}Pb/^{238}U$
Pg16					volcanic tuff in Kiowa core, KJ07-63	Kiowa core, Elbert County Fairgrounds, Kiowa, Colorado	39.35242°N, 104.46642°W	179.26 m below top of core; D1 sequence of fluvial/paludal sandstone, mudstone, and lignite beds	64.63	± 0.05	± 0.09	$^{206}Pb/^{238}U$
Pg15					U coal tephra	Biscuit Butte locality, Garfield County, eastern Montana	47°29'13.1"N, 107°4'7.6"W	base of Lebo Member of the Fort Union Formation	64.66	± 0.05	± 0.09	$^{40}Ar/^{39}Ar$
Pg14					V coal tephra	Biscuit Butte locality, Garfield County, eastern Montana	47°29'21.8"N, 107°4'23.8"W	above the Farrand Channel, upper Tulloch Member of the Fort Union Formation	64.84	± 0.05	± 0.09	$^{40}Ar/^{39}Ar$
Pg13					W coal tephra	Saddle Section locality, Garfield County, eastern Montana	47°30'34.8"N, 107°4'54.9"W	below the Farrand Channel, upper Tulloch Member of the Fort Union Formation	64.91	± 0.05	± 0.10	$^{40}Ar/^{39}Ar$
Pg12					X coal tephra	McGuire Creek locality, McCone County, eastern Montana	47°37'47.5"N, 106°10'12.4"W	capping the Garbani Channel, upper Tulloch Member of the Fort Union Formation	65.29	± 0.06	± 0.11	$^{40}Ar/^{39}Ar$
Pg11					Y coal tephra	Hell Hollow locality, Garfield County, eastern Montana	47.53472°N, 107.1687°W	base of the Garbani Channel, lower Tulloch Member of the Fort Union Formation	65.50	± 0.04	± 0.09	$^{40}Ar/^{39}Ar$
Pg10					volcanic tuff in Kiowa core, KJ08-53	Kiowa core, Elbert County Fairgrounds, Kiowa, Colorado	39.35242°N, 104.46642°W	267.74 m below top of core; D1 sequence of fluvial/paludal sandstone, mudstone, and lignite beds	65.80	± 0.04	± 0.09	$^{206}Pb/^{238}U$
Pg9					MCZ coal, Lerbekmo bentonite (upper)	Lerbekmo (Hell Creek Marina Road) locality, Garfield County, eastern Montana	47.51602°N, 106.9366°W	near base of Tulloch Member of the Fort Union Formation	65.80	± 0.07	± 0.11	$^{40}Ar/^{39}Ar$
Pg8					MCZ coal, McGuire Creek bentonite (lower)	Multiple localities (Lofgren, Z-line, Haxby Road, Lerbekmo, Thomas Ranch) in eastern Garfield and western McCone Counties, eastern Montana	47°31.593'N, 106°56.397'W	near base of Tulloch Member of the Fort Union Formation	65.82	± 0.03	± 0.09	$^{40}Ar/^{39}Ar$
Pg7					IrZ coal, Nirvana bentonite	Multiple localities (Hauso Flats, McKeever Ranch, Nirvana, Hell Hollow) in eastern Garfield and western McCone Counties, eastern Montana	47°31.732'N, 107°12.513'W	base of the Tulloch Member of the Fort Union Formation	65.85	± 0.02	± 0.09	$^{40}Ar/^{39}Ar$
Pg6					Haitian tektites	Beloc, Haiti	18°23'40"N, 72°38'06"W	between levels f and g, in lower part of the marine limestone Beloc Fm	65.83	± 0.06	± 0.10	$^{40}Ar/^{39}Ar$

Primary radioisotopic age details	Zonal range assignment			Biostratigraphy	Reference
	clade	zonation	Zone		
Weighted mean age of 6 single zircon crystal analyses, utilizing CA-TIMS and the ET535 spike.	magnetostratigraphy			magnetostratigraphic control, maximum age for C22n-C22r transition	Machlus et al. (2015); Smith et al. (2004)
Weighted mean age of 16 laser-fusion and 4 five-step laser incremental heating analyses on 1–3 crystal aliquots of sanidine, using FCs as fluence monitor.	NALMA, magnetostratigraphy			Lystitean-Lostcabinian (Wa6–Wa7) North American land mammal age (NALMA) substage boundary; magnetostratigraphic control at base of Chron 24n.1	Smith et al. (2004); Wing et al. (1991); Clyde et al. (1994); Tauxe et al. (1994)
Weighted mean age of 38 laser fusion analyses on single and multi-grain aliquots of sanidine, originally calibrated to FCs = 28.02 Ma.	chemostratigraphy			above the PETM carbon isotope excursion	Storey et al. (2007)
Weighted mean age of five (of 13) single zircon crystal analyses, utilizing CA-TIMS and the ET535 spike.	chemostratigraphy			within the PETM carbon isotope excursion	Charles et al. (2011)
Weighted mean age of 23 laser fusion analyses on 20 grain aliquots of sanidine, originally calibrated to TCs = 28.34 Ma.	magnetostratigraphy			magnetostratigraphic control	Secord et al. (2006)
Weighted mean of 8 (of 12) single zircon crystal analyses, utilizing CA-TIMS and the ET535 spike.	K-Pg event stratigraphy, magnetostratigraphy			with respect to K-Pg boundary defined by iridium layer, disappearance of dinosaurs and plants, and "fern spike" at 302.93 m below top of core; with respect to base of C28n at 175.16 m below top of core	Clyde et al. (2016)
Weighted mean of 6 (of 8) single zircon crystal analyses, utilizing CA-TIMS and the ET535 spike.	K-Pg event stratigraphy, magnetostratigraphy			with respect to K-Pg boundary defined by iridium layer, disappearance of dinosaurs and plants, and "fern spike" at 302.93 m below top of core; with respect to base of C28n at 175.16 m below top of core	Clyde et al. (2016)
Weighted mean age of 173 single crystal sanidine laser fusions, originally calibrated to Renne et al. (2011) optimization.	NALMA, magnetostratigraphy		Torrejonian 1	magnetostratigraphic control at base of C28n	Sprain et al. (2015)
Weighted mean age of 175 single crystal sanidine laser fusions, originally calibrated to Renne et al. (2011) optimization.	NALMA, magnetostratigraphy		Torrejonian 1	magnetostratigraphic control in lower C28r	Sprain et al. (2015)
Weighted mean age of 68 single crystal sanidine laser fusions, originally calibrated to Renne et al. (2011) optimization.	NALMA, magnetostratigraphy		Torrejonian 1	magnetostratigraphic control in upper C29n	Sprain et al. (2015)
Weighted mean age of 87 single crystal sanidine laser fusions, originally calibrated to Renne et al. (2011) optimization.	NALMA, magnetostratigraphy		Puercan 3	magnetostratigraphic control in middle C29n	Sprain et al. (2015)
Weighted mean age of 137 single crystal sanidine laser fusions, originally calibrated to Renne et al. (2011) optimization.	NALMA, magnetostratigraphy		Puercan 2	magnetostratigraphic control in lower C29n	Sprain et al. (2018)
Weighted mean of 8 (of 11) single zircon crystal analyses, utilizing CA-TIMS and the ET535 spike.	K-Pg event stratigraphy, magnetostratigraphy		Puercan 1	with respect to K-Pg boundary defined by iridium layer, disappearance of dinosaurs and plants, and "fern spike" at 302.93 m below top of core; with respect to base of C29n at 268.47 m below top of core	Clyde et al. (2016)
Lerbekmo bentonite, correlated across multiple locations (Flag Butte, Z-line, Haxby Road, Lerbekmo) on the basis of Pb isotopes in feldspars.	NALMA, magnetostratigraphy		Puercan 1	magnetostratigraphic control in middle C29r	Renne et al. (2013); Ickert et al. (2015)
McGuire Creek bentonite, dated in multiple locations (Lofgren, LG11-1; Z-line, ZL12-2; Haxby Road, HX12-1; Lerbekmo, HC-2PR; Thomas Ranch, TR13-3), and correlated on the basis of Pb isotopes in feldspars.	NALMA, magnetostratigraphy		Puercan 1	magnetostratigraphic control in middle C29r	Sprain et al. (2018); Sprain et al. (2015); Ickert et al. (2015); Renne et al. (2013); Swisher et al. (1993)
Nirvana bentonite, dated in multiple locations (Hauso Flats, HF14-1, HF-1PR, HF15-1; McKeever Ranch, MK13-3; Nirvana, NV12-1; Hell Hollow, HH12-1), and correlated on the basis of Pb isotopes in feldspars.	NALMA, magnetostratigraphy		Puercan 1	magnetostratigraphic control in middle C29r	Sprain et al. (2018); Sprain et al. (2015); Ickert et al. (2015); Renne et al. (2013); Swisher et al. (1993)
Weighted mean age of 14 plateau ages on single tektites, originally calibrated to Renne et al. (2011) optimization.	K-Pg event stratigraphy			debris from KT impact event	Renne et al. (2013)

GTS 2020 ID	GTS 2012 ID	Period	Epoch	Age	Sample	Locality	Lat-Long	Lithostratigraphy	Age (Ma)	± 2s analytical	± 2s total	Age Type
Pg5	Pg3				Haitian tektites	Beloc, Haiti	18°23′40″N, 72°38′06″W	from 0.5-thick marl bed in lower part of the marine limestone Beloc Fm	65.92	± 0.14	± 0.18	$^{40}Ar/^{39}Ar$
Pg4	Pg2				Haitian tektites	Beloc, Haiti	18°23′40″N, 72°38′06″W	between levels f and g, in lower part of the marine limestone Beloc Fm	65.84	± 0.16	± 0.18	$^{40}Ar/^{39}Ar$
Pg3	Pg1				C1 glassy melt rock	Chicxulub crater	~20.5°N, 89.8°W	C1 glassy melt rock	65.81	± 0.10	± 0.14	$^{40}Ar/^{39}Ar$
Pg2					volcanic tuff, KJ04-70	"Bowring Pit", West Bijou Creek area, Elbert County, Colorado	39.57059°N, 104.30306°W	1.00 m above K-Pg boundary; D1 sequence of fluvial/paludal sandstone, mudstone, and lignite beds	65.89	± 0.03	± 0.08	$^{206}Pb/^{238}U$
Pg1					volcanic tuff in Kiowa core, KJ10-04	Kiowa core, Elbert County Fairgrounds, Kiowa, Colorado	39.35242°N, 104.46642°W	302.56 below top of core; D1 sequence of fluvial/paludal sandstone, mudstone, and lignite beds	66.02	± 0.02	± 0.08	$^{206}Pb/^{238}U$
		Cretaceous										
			Late									
				Maastrichtian								
K88					volcanic tuff, KJ08-157	"Bowring Pit", West Bijou Creek area, Elbert County, Colorado	39.57059°N, 104.30306°W	0.46 m below K-Pg boundary; D1 sequence of fluvial/paludal sandstone, mudstone, and lignite beds	66.08	± 0.02	± 0.08	$^{206}Pb/^{238}U$
K87					Null coal bentonite	Bug Creek site, Garfield County, eastern Montana	47°40′48.6″N, 106°12′49.6″W	upper Hell Creek Formation	66.08	± 0.10	± 0.13	$^{40}Ar/^{39}Ar$
K86					Null coal bentonite	Thomas Ranch site, Garfield County, eastern Montana	47.66675°N, 106.4258°W	upper Hell Creek Formation	66.14	± 0.30	± 0.32	$^{40}Ar/^{39}Ar$
K85					volcanic tuff in Kiowa core, KJ10-09	Kiowa core, Elbert County Fairgrounds, Kiowa, Colorado	39.35242°N, 104.46642°W	344.50 m below top of core; D1 sequence of fluvial/paludal sandstone, mudstone, and lignite beds	66.30	± 0.03	± 0.08	$^{206}Pb/^{238}U$
K84	K71				AK-476	Strawberry Creek, Alberta	53°17′N, 116°06′W	Kneehills Tuff, 18-cm bentonite, 1.8 m above the upper Kneehills Tuff zone	67.29	± 1.11	± 1.11	$^{40}Ar/^{39}Ar$
K83	K70				91-O-14	Red Bird section, Niobrara County, Wyoming (sec.14, T.38N, R.62W)	43.27°N, 104.27°W	30-cm bentonite bed from unit 112, Red Bird section (Gill and Cobban, 1966), sec. 14, T. 38 N, R. 62 W	70.08	± 0.37	± 0.37	$^{40}Ar/^{39}Ar$
K82	K69				92-O-32	Red Bird section, Niobrara County, Wyoming (sec.14, T.38N, R.62W)	43.27°N, 104.27°W	1.40-m bentonite bed from unit 97, Red Bird section (Gill and Cobban, 1966)	70.66	± 0.65	± 0.66	$^{40}Ar/^{39}Ar$
				Campanian								
K81	K67				Snakebite 1 bentonite	Cruikshank Coolee, north of Herbert, southwestern Saskatchewan	~50.7°N, 107.4°W	Snakebite Member of Bearpaw Fm	73.41	± 0.47	± 0.47	$^{40}Ar/^{39}Ar$
K80	K66				90-O-15, bentonite	Big Horn County, Montana (NE/NE, sec.14, T.1N, R.33E)	45.839°N, 107.573°W	Bearpaw Shale – 22-cm bentonite 1.5 m above base of ammonite zone	74.05	± 0.39	± 0.39	$^{40}Ar/^{39}Ar$
K79	K65				93-O-16, bentonite	Montrose County, Colorado (NE corner SE/SE, sec.17, T.48N, R.7W)	38.4153°N, 107.6522°W	Mancos Shale – Bentonite bed in upper part	74.85	± 0.43	± 0.43	$^{40}Ar/^{39}Ar$
K78	K64				86-O-05, bentonite	Foreman, Little River County, Arkansas	33.695°N, 94.415°W	Anonna Fm, Foreman Quarry	75.92	± 0.39	± 0.39	$^{40}Ar/^{39}Ar$
K77	K63				92-O-13, bentonite	Rio Arriba County, New Mexico (CW1/2, sec.11, T.23N, R.1W)	36.238°N, 106.916°W	Lewis Shale – 15-cm bentonite (possibly equivalent to the Huerfanito bentonite marker bed in the subsurface)	76.62	± 0.51	± 0.51	$^{40}Ar/^{39}Ar$
K76	K62				Judith River Bentonite	Elk Basin (northern Bighorn Basin), Wyoming	~45°N, 109°W	Judith River Bentonite; 22m below C33r/C33n polarity reversal	80.10	± 0.61	± 0.61	$^{40}Ar/^{39}Ar$
K75	K61				RB92-15, bentonite	Red Bird section, Niobrara County, Wyoming (sec.14, T.38N, R.62W)	43.27°N, 104.27°W	Ardmore bentonite bed in the base of the Pierre Shale at the Red Bird section	80.62	± 0.40	± 0.40	$^{40}Ar/^{39}Ar$
K74	K60				Ardmore bentonite	Elk Basin (northern Bighorn Basin), Wyoming	44.954°N, 108.856°W	Ardmore Bentonite; a 4.9 m thick bentonite near the base of the Claggett Shale	81.30	± 0.55	± 0.55	$^{40}Ar/^{39}Ar$

Primary radioisotopic age details	Zonal range assignment			Biostratigraphy	Reference
	clade	zonation	Zone		
Weighted mean age of 52 laser fusions on single or several tektites, originally calibrated to FCs = 27.55 Ma.	K-Pg event stratigraphy			debris from KT impact event	Dalrymple et al. (1993); Izett et al. (1991)
Weighted mean age of 5 plateau ages on single tektites, originally calibrated to FCs = 27.84 Ma.	K-Pg event stratigraphy			debris from KT impact event	Swisher et al. (1992)
Weighted mean age of 5 plateau ages on glassy (andesitic) rock chips, originally calibrated to FCs = 27.84 Ma.	K-Pg event stratigraphy			debris from KT impact event	Swisher et al. (1992)
Weighted mean of 11 (of 14) single zircon crystal analyses, utilizing CA-TIMS and the ET535 spike.	K-Pg event stratigraphy, magnetostratigraphy		Puercan 1	with respect to K-Pg boundary defined by iridium layer, disappearance of dinosaurs and plants, and "fern spike"; within C29r	Clyde et al. (2016)
Weighted mean of 5 (of 9) single zircon crystal analyses, utilizing CA-TIMS and the ET535 spike.	K-Pg event stratigraphy, magnetostratigraphy		Puercan 1	with respect to K-Pg boundary defined by iridium layer, disappearance of dinosaurs and plants, and "fern spike" at 302.93 m below top of core; with respect to base of C29r at 363.89 m below top of core	Clyde et al. (2016)
Weighted mean of 12 (of 14) single zircon crystal analyses, utilizing CA-TIMS and the ET535 spike.	K-Pg event stratigraphy, magnetostratigraphy		Lancian	with respect to K-Pg boundary defined by iridium layer, disappearance of dinosaurs and plants, and "fern spike"; within C29r	Clyde et al. (2016)
Weighted mean age of 57 single crystal sanidine laser fusions, originally calibrated to Renne et al. (2011) optimization.	K-Pg event stratigraphy, magnetostratigraphy		Lancian	magnetostratigraphic control at base of C29r	Sprain et al. (2015)
Weighted mean age of 54 single crystal sanidine laser fusions, originally calibrated to Renne et al. (2011) optimization.	K-Pg event stratigraphy, magnetostratigraphy		Lancian	magnetostratigraphic control at base of C29r	Sprain et al. (2018)
Weighted mean of 6 (of 8) single zircon crystal analyses, utilizing CA-TIMS and the ET535 spike.	K-Pg event stratigraphy, magnetostratigraphy		Lancian	with respect to K-Pg boundary defined by iridium layer, disappearance of dinosaurs and plants, and "fern spike" at 302.93 m below top of core; with respect to base of C29r at 363.89 m below top of core	Clyde et al. (2016)
Multi-grain sanidine laser fusion analysis, originally using TCs = 28.32 Ma as fluence monitor; no analytical data published.	dinosauria	North America	*Triceratops*	*Triceratops* dinosaur zone	Obradovich (1993)
Multi-grain sanidine laser fusion analysis, originally using TCs = 28.32 Ma as fluence monitor.	ammonite	North America	*Baculites clinolobatus* ammonite zone (uppermost)	Top of *Baculites clinolobatus* ammonite zone	Hicks et al. (1999)
Multi-grain sanidine laser fusion analysis, originally using TCs = 28.32 Ma as fluence monitor.	ammonite	North America	*Baculites grandis* ammonite zone.	*Baculites grandis* ammonite zone	Hicks et al. (1999)
Weighted mean age of 10 multi-grain sanidine laser fusion analyses, originally using FCs = 27.84 Ma as fluence monitor.	ammonite	North America	*Baculites reesidei* ammonite zone (uppermost)	top of *Baculites reesidei* ammonite zone	Baadsgaard et al. (1993)
Multi-grain sanidine laser fusion analysis, originally using TCs = 28.32 Ma as fluence monitor.	ammonite	North America	*Baculites compressus* ammonite zone	*Baculites compressus* ammonite zone	Hicks et al. (1999)
Multi-grain sanidine laser fusion analysis, originally using TCs = 28.32 Ma as fluence monitor.	ammonite	North America	*Exiteloceras jenneyi* ammonite zone	*Exiteloceras jenneyi* ammonite zone	Hicks et al. (1999)
Multi-grain sanidine laser fusion analysis, originally using TCs = 28.32 Ma as fluence monitor; no analytical data published.	planktonic foraminifer	Tethyan	*Globotruncanita calcarata*	5m above lowest occurrence of foraminifer *Globotruncanita calcarata*	Obradovich (1993)
Multi-grain sanidine laser fusion analysis, originally using TCs = 28.32 Ma as fluence monitor.	ammonite	North America	*Baculites scotti* ammonite zone	*Baculites scotti* ammonite zone	Hicks et al. (1999)
Multi-grain sanidine laser fusion analysis, originally using TCs = 28.32 Ma as fluence monitor.	ammonite	North America	*Baculites obtusus* to *B. perplexus* ammonite zone	*Baculites obtusus* to *B. perplexus* ammonite zone	Hicks et al. (1995)
Multi-grain sanidine laser fusion analysis, originally using TCs = 28.32 Ma as fluence monitor.	ammonite	North America	*Baculites obtusus* ammonite zone	*Baculites obtusus* ammonite zone	Hicks et al. (1999)
Multi-grain sanidine laser fusion analysis, originally using TCs = 28.32 Ma as fluence monitor.	ammonite	North America	*Baculites obtusus* ammonite zone	*Baculites obtusus* ammonite zone	Hicks et al. (1995)

GTS 2020 ID	GTS 2012 ID	Period	Epoch	Age	Sample	Locality	Lat-Long	Lithostratigraphy	Age (Ma)	± 2s analytical	± 2s total	Age Type
K73	K59				PP-6, upper ash layer	Montagna della Maiella, Valle Tre Grotte, west of Pennapiedimonte, central Apennines, Italy	42°09′N, 14°11′E	Tre Grotte Fm	81.67	± 0.21	± 0.26	$^{206}Pb/^{238}U$
K72	K58				78-O-05, bentonite	Cat Creek Oil field, Petroleum County, Montana	46.880°N, 107.870°W	Eagle Sandstone, thin bentonite in lower part of Scaphites hippocrepis II ammonite zone.	81.84	± 0.11	± 0.22	$^{40}Ar/^{39}Ar$
				Santonian								
K71					S1019, bentonite	Songliao SK-1s core; Songliao Basin, NE China	45°34′14.42″N, 124°40′15.59″E	Bentonite at 1019 m depth in SK-1s borehole; in lower Nenjiang Fm "of deep-water lacustrine gray to black mudstone intercalated with thin marl, shelly limestone, and oil shales"	83.27	± 0.04	± 0.11	$^{206}Pb/^{238}U$
K70	K57				TOM001, volcanic tuff	Horosari Valley, Tomiuchi area, Hokkaido, Japan	42°45′27″N, 142°13′20″E	"Tomiuchibashi Tuff", Kashima Fm, Yazo Group	84.90	± 0.20	± 0.25	$^{206}Pb/^{238}U$
K69	K56				MT08-04; 97-O-04, bentonite	East bank of McDonald Creet, Petroleum County, Montana	47.0044°N, 108.3340°W	Telegraph Creek Fm, 10-cm bentonite	84.41	± 0.14	± 0.24	$^{40}Ar/^{39}Ar$
K68	K55				MT08-04, bentonite	East bank of McDonald Creet, Petroleum County, Montana	47.0044°N, 108.3340°W	Telegraph Creek Fm, 10-cm bentonite	84.43	± 0.09	± 0.15	$^{206}Pb/^{238}U$
K67	K54				70-O-06 (OB93-12), bentonite	bentonite mine, south of Aberdeen, Monroe County, Mississippi	33.775°N, 88.525°W	Bentonite in Tombigbee Sand Member of Eutaw Fm	84.70	± 0.41	± 0.41	$^{40}Ar/^{39}Ar$
K66	K53				91-O-09, bentonite	Head of ravine, 12.9 km west of Shelby, Montana	48.4923°N, 112.0303°W	3-cm-thick bentonite	84.55	± 0.24	± 0.37	$^{40}Ar/^{39}Ar$
K65	K52				MT09-07; 90-O-10, bentonite	Cut bank on east side of Bighorn River, 1.35 miles SE of Hardin, Montana	45.7216°N, 107.5788°W	4-inch-thick bentonite	85.66	± 0.09	± 0.19	$^{40}Ar/^{39}Ar$
K64	K51				92-O-14, bentonite	creek bottom, 1.1 km NW of Whitewright, eastern Grayson County, Texas	33.5200°N, 96.4034°W	Austin Chalk, 91-cm bentonite is immediately above a fossiliferous 0.3–0.6-m-thick hard massive chalk	86.08	± 0.24	± 0.37	$^{40}Ar/^{39}Ar$
				Coniacian								
K63	K50				MT08-1; 91-O-08, bentonite	S. side of Yellowstone river, just south of Billings, Montana	45.7407°N, 108.5119°W	Cody Shale, 30-cm bentonite, 30.5 m above top of Kevin bentonite group in Cody Shale	87.13	± 0.09	± 0.19	$^{40}Ar/^{39}Ar$
K62	K49				MT08-1, bentonite	S. side of Yellowstone river, just south of Billings, Montana	45.7407°N, 108.5119°W	Cody Shale, 30-cm bentonite, 30.5 m above top of Kevin bentonite group in Cody Shale	87.11	± 0.08	± 0.15	$^{206}Pb/^{238}U$
K61	K48				MT08-03; 91-O-13, bentonite	N. bank, Marias River, 0.5 km W of I-25; Toole County, Montana	48.4288°N, 111.8921°W	Marias River Shale, 15-cm bentonite	89.32	± 0.11	± 0.24	$^{40}Ar/^{39}Ar$
K60	K47				MT08-03, bentonite	N. bank, Marias River, 0.5 km W of I-25; Toole County, Montana	48.4288°N, 111.8921°W	Marias River Shale, 15-cm bentonite	89.37	± 0.07	± 0.15	$^{206}Pb/^{238}U$
				Turonian								
K59	K46				MT-09-09, bentonite	ca. 14 km east of main outcrops of Kewvin type section; Toole County, Montana	48.8177°N, 111.8156°W	Possibly marker bed #39 in Ferdig Member of Marias River Fm	89.87	± 0.10	± 0.18	$^{40}Ar/^{39}Ar$
K58	K45				SAN JUAN, bentonite	San Juan County, New Mexico	~36°N, 108°W	Mancos Shale, 40-cm-thick bentonite in Juana Lopez Member	91.24	± 0.09	± 0.23	$^{40}Ar/^{39}Ar$
K57	K44				UT-08-03, bentonite	Ferron sandstone section near Castle Dale, Utah	39.2226°N, 110.9486°W	Mancos Shale, thin bentonite in Juana Lopez Member	91.37	± 0.08	± 0.17	$^{206}Pb/^{238}U$
K56	K43				UT-08-05; 90-O-56, bentonite	Emery Point section, Mesa Butte quadrangle, Emery County, Utah	38.6°N, 111.2°W	Ferron Sandstone, bentonite between Clausen and Washboard members	91.15	± 0.13	± 0.26	$^{40}Ar/^{39}Ar$

Primary radioisotopic age details	Zonal range assignment			Biostratigraphy	Reference
	clade	zonation	Zone		
Weighted mean age of four small multi-grain zircon analyses, utilizing physical abrasion and in-house UNIGE spike (0.1% error in tracer calibration).	ammonite	North America	late Early Campanian (*Sca. hippocrepis III* to *Bac. obtusus* zone interval)	upper part of the *Aspidolithus parcus* (CC 18) calcareous nannoplankton zone; and middle of *Globotruncanita elevata* foraminifer zone	Bernoulli et al. (2004)
Weighted mean age of 43 (of 44) laser fusions on 1–3 grain aliquots of sanidine, using FCs as fluence monitor.	ammonite	North America	*Scaphites hippocrepis II* ammonite zone	lower *Scaphites hippocrepis II* ammonite zone	Sageman et al. (2014); Obradovich (1993)
Six (of 14) youngest single grain zircon analyses yield a weighted mean $^{206}Pb/^{238}U$ age, utilizing CA-TIMS and ET535 spike.	magnetostratigraphy and cyclostratigraphy		Just below base of Chron C33r	Cyclostratigraphy from tuning the interpreted 405kyr (long) eccentricity cycles in the natural-gamma log of the sedimentary record to the La2010a astronomical model. Chron C34n–C33r boundary is at 985.95 m depth in the Songliao SK-1s core.	Wang et al. (2016)
Four (of 5) multigrain chemically-abraded zircon fractions; ID-TIMS.	planktonic foraminifer	Tethyan	Late Santonian to earliest Campanian?	Local *Globotruncana arca* planktonic zone. Within lower *Inoceramus japonicus* zone, which Walaszczyk and Cobban (2006) cite as "Late Santonian evolutionary descendant of *Platyceramus ezoensis*"; although Quidelleur et al. (2011) show as earliest "Campanian".	Quidelleur et al. (2011)
Weighted mean age of two samples, including 64 (of 70) laser fusions on 3–4 grain aliquots of sanidine, using FCs as fluence monitor.	ammonite	North America	*Desmoscaphites bassleri* ammonite zone	*Desmoscaphites bassleri* ammonite zone	Sageman et al. (2014); Obradovich (1993)
Four (of 17) youngest single grain zircon analyses yield a weighted mean $^{206}Pb/^{238}U$ age, utilizing CA-TIMS and ET535 spike.	ammonite	North America	*Desmoscaphites bassleri* ammonite zone	*Desmoscaphites bassleri* ammonite zone	Sageman et al. (2014); Obradovich (1993)
Multi-grain sanidine laser fusion analysis, originally using TCs = 28.32 Ma as fluence monitor; no analytical data published.	ammonite	North America	*Boehmoceras*	Just below *Boehmoceras* fauna of "Uppermost Santonian"	Obradovich (1993)
Weighted mean age of 30 (of 46) laser fusions on 5–7 grain aliquots of sanidine, using FCs as fluence monitor.	ammonite	North America	*Clioscaphites erdmanni* ammonite zone (upper)	Upper part of *Clioscaphites erdmanni* ammonite zone	Sageman et al. (2014); Obradovich (1993)
Weighted mean age of two samples, including 45 (of 51) laser fusions on 7–10 grain aliquots of sanidine, using FCs as fluence monitor.	ammonite	North America	*Clioscaphites vermiformis* ammonite zone (middle)	lower-Upper part of *Clioscaphites vermiformis* ammonite zone	Sageman et al. (2014); Obradovich (1993)
Weighted mean age of one sample, including 19 (of 25) laser fusions on 5 grain aliquots of sanidine, using FCs as fluence monitor.	ammonite	North America	Top of *Clioscaphites saxitonianus* ammonite zone	Top of *Cladoceramus undulatoplicatus* inoceramus bivalve zone, equivalent to top of *Clioscaphites saxitonianus* ammonite zone	Sageman et al. (2014); Obradovich (1993)
Weighted mean age of four samples, including 101 (of 127) laser fusions on 4–5 grain aliquots of sanidine, using FCs as fluence monitor.	ammonite	North America	middle of *Scaphites depressus* ammonite zone	*Protexanites bourgeoisianus (Scaphites depressus)* ammonite zone	Sageman et al. (2014); Obradovich (1993)
Seven (of 11) single grain zircon analyses yield a weighted mean $^{206}Pb/^{238}U$ age, utilizing CA-TIMS and ET535 spike.	ammonite	North America	middle of *Scaphites depressus* ammonite zone	*Protexanites bourgeoisianus (Scaphites depressus)* ammonite zone	Sageman et al. (2014); Obradovich (1993)
Weighted mean age of two samples, including 69 (of 84) laser fusions on 1–10 grain aliquots of sanidine, using FCs as fluence monitor.	ammonite	North America	*Scaphites preventricosus* ammonite zone (upper-middle)	*Forresteria alluaudi–Scaphites preventricosus* ammonite zone (upper ash bed)	Sageman et al. (2014); Obradovich (1993)
Four (of 18) single grain zircon analyses yield a weighted mean $^{206}Pb/^{238}U$ age, utilizing CA-TIMS and ET535 spike.	ammonite	North America	*Scaphites preventricosus* ammonite zone (lower-middle)	*Forresteria alluaudi–Scaphites preventricosus* ammonite zone (lower ash bed)	Sageman et al. (2014); Obradovich (1993)
Weighted mean age of two samples, including 58 (of 63) laser fusions on 8- > 10 grain aliquots of sanidine, using FCs as fluence monitor.	ammonite	North America	*Scaphites nigricollensis* ammonite zone	*Scaphites nigricollensis* ammonite zone	Sageman et al. (2014); Obradovich (1993)
Weighted mean age of two samples, including 85 (of 93) laser fusions on single grain aliquots of sanidine, using FCs as fluence monitor.	ammonite	North America	Top of *Prionocyclus macombi* ammonite zone	Top of *Prionocyclus macombi* ammonite zone; 0.9 m above *P. macombi* and 6 m below *Scaphites warreni*	Sageman et al. (2014); Obradovich (1993)
Five (of 11) single grain zircon analyses yield a weighted mean $^{206}Pb/^{238}U$ age, utilizing CA-TIMS and ET535 spike.	ammonite	North America	Top of *Prionocyclus macombi* ammonite zone	Within *Prionocyclus macombi* ammonite zone; most likely not same level as the San Juan Ar/Ar sample above.	Sageman et al. (2014); Obradovich (1993)
Weighted mean age of two samples, including 83 (of 103) laser fusions on 1–5 grain aliquots of sanidine, using FCs as fluence monitor.	ammonite	North America	*Prionocyclus hyatti* ammonite zone	*Prionocyclus hyatti* ammonite zone	Siewert (2011); Obradovich (1993)

GTS 2020 ID	GTS 2012 ID	Period	Epoch	Age	Sample	Locality	Lat-Long	Lithostratigraphy	Age (Ma)	± 2s analytical	± 2s total	Age Type
K55	K42				90-O-34, bentonite	Laholi Point section, Blue Gap, Navajo Indian Reservation, Arizona	36.1849°N, 109.8836°W	Mancos Shale, bentonite BM-17, 0.16-m-thick in lower calcareous shale member, 25-m above base. Equated to bentonite PCB-20 at top of lower Bridge Creek Limestone Mbr, Pueblo, Colorado; and equivalent bentonite below Bed 97 in USGS #1 Portland core.	93.67	± 0.21	± 0.31	$^{40}Ar/^{39}Ar$
K54	K41				AZLP-08-05 (same bentonite as 90-O-34)	Laholi Point section, Blue Gap, Navajo Indian Reservation, Arizona	36.1849°N, 109.8836°W	Mancos Shale, bentonite BM-17, 0.16-m-thick in lower calcareous shale member, 25-m above base. Equated to bentonite PCB-20 at top of lower Bridge Creek Limestone Mbr, Pueblo, Colorado; and equivalent bentonite below Bed 97 in USGS #1 Portland core.	94.09	± 0.13	± 0.19	$^{206}Pb/^{238}U$
K53	K40				AZLP-08-04; K-07-01C; 90-O-33, bentonite	Laholi Point section, Blue Gap, Navajo Indian Reservation, Arizona; and two equivalent bentonites from other Bridge Creek limestone localities	36.1849°N, 109.8836°W; 38.2803°N, 104.7397°W	Mancos Shale, bentonite BM-15, 0.6-m-thick in lower calcareous shale member, 6.4-m above base. Equated to bentonite PCB-17 in lower Bridge Creek Limestone Mbr, Bed 87 in Turonian GSSP, Pueblo, Colorado; equivalent bentonite below limestone Bed 90 in USGS #1 Portland core.	93.79	± 0.12	± 0.26	$^{40}Ar/^{39}Ar$
K52	K39				AZLP-08-04	Laholi Point section, Blue Gap, Navajo Indian Reservation, Arizona; and two equivalent bentonites from other Bridge Creek limestone localities	36.1849°N, 109.8836°W; 38.2803°N, 104.7397°W	Mancos Shale, bentonite BM-15, 0.6-m-thick in lower calcareous shale member, 6.4-m above base. Equated to bentonite PCB-17 in lower Bridge Creek Limestone Mbr, Bed 87 in Turonian GSSP, Pueblo, Colorado; equivalent bentonite below limestone Bed 90 in USGS #1 Portland core.	94.37	± 0.04	± 0.14	$^{206}Pb/^{238}U$
K51	K38				HKt003	Hakkin river section, Hokkaido, Japan	43°02′44″N, 142°09′26″E	Base of Kakkin Muddy sandstone Member, Saku Fm, Yezo Group	94.30	± 0.30	± 0.35	$^{206}Pb/^{238}U$
				Cenomanian								
K50	K37				K-07-01B; NE-08-01; 90-O-19; 90-O-49	Little Blue River, Thayer County, Nebraska; Four Corners, San Juan County, New Mexico; and one other equivalent bentonite from a Bridge Creek limestone locality	40.1731°N, 97.4463°W; 38.2803°N, 104.7397°W	Greenhorn Limestone; HL-3 marker bed; and Turonian GSSP Bed 80	94.10	± 0.14	± 0.27	$^{40}Ar/^{39}Ar$
K49	K36				NE-08-01	Little Blue River, Thayer County, Nebraska	40.1731°N, 97.4463°W	Greenhorn Limestone; 15–20 cm bed above HL-37	94.01	± 0.04	± 0.14	$^{206}Pb/^{238}U$
K48	K35				Bighorn River Bentonite	Ram River below Ram Falls, Alberta	52° 5′44.00″N, 115°49′3.31″W	Vimy Member of Blackstone Fm; upper prominent 35-cm greenish-grey bentonite in lowest part of member.	94.29	± 0.13	± 0.17	$^{206}Pb/^{238}U$
K47	K34				AZLP-08-02; 90-O-31, bentonite	Lohali Point section, Blue Gap, Navajo Indian Reservation, Arizona; and one equivalent bentonite from other Bridge Creek limestone locality	36.1835°N, 109.8843°W	Mancos Shale; bentonite BM-6, 0.22 m thick in lower calcareous shale member, 6.8 m above base of Mancos Shale.	94.20	± 0.15	± 0.28	$^{40}Ar/^{39}Ar$
K46	K33				AZLP-08-01; 90-O-30, bentonite	Lohali Point section, Blue Gap, Navajo Indian Reservation, Arizona	36.1835°N, 109.8843°W	Mancos Shale, bentonite BM-5, 0.9-m-thick in lower calcareous shale member, 5.7-m above base of Mancos Shale.	94.43	± 0.17	± 0.29	$^{40}Ar/^{39}Ar$
K45	K32				AZLP-08-01, bentonite	Lohali Point section, Blue Gap, Navajo Indian Reservation, Arizona	36.1835°N, 109.8843°W	Mancos Shale, bentonite BM-5, 0.9-m-thick in lower calcareous shale member, 5.7-m above base of Mancos Shale.	94.28	± 0.08	± 0.15	$^{206}Pb/^{238}U$
K44	K31				LP22 biotite	Lohali Point section, Blue Gap, Navajo Indian Reservation, Arizona	36.1835°N, 109.8843°W	Bentonite BM-7; Mancos Shale	95.25	± 1.00	± 1.00	$^{40}Ar/^{39}Ar$
K43	K30				D2315 (OB93-23)	Carbon County, Wyoming	42.13°N, 106.20°W	Frontier Fm	95.39	± 0.18	± 0.37	$^{40}Ar/^{39}Ar$
K42	K29				WY-09-04; 90-O-50	Big Sulphur Draw quadrangle, Natrona County, Wyoming	43.31°N, 106.78°W	Frontier Fm, Soap Creek Bentonite marker	95.53	± 0.09	± 0.25	$^{40}Ar/^{39}Ar$

Primary radioisotopic age details	Zonal range assignment			Biostratigraphy	Reference
	clade	zonation	Zone		
Weighted mean age including 33 (of 41) laser fusions on 5 grain aliquots of sanidine, using FCs as fluence monitor.	ammonite	North America	Top of *Pseudaspidoceras flexuosum* ammonite zone	Top of *Pseudaspidoceras flexuosum* ammonite zone (zone below *V. birchbyi* Zone)	Meyers et al. (2012); Obradovich (1993)
Two youngest (of 11) single grain zircon analyses yield a weighted mean $^{206}Pb/^{238}U$ age, utilizing CA-TIMS and ET535 spike. [Not used in Late Cretaceous spline fit.]	ammonite	North America	Top of *Pseudaspidoceras flexuosum* ammonite zone	Top of *Pseudaspidoceras flexuosum* ammonite zone (zone below *V. birchbyi* Zone)	Meyers et al. (2012); Obradovich (1993)
Weighted mean age of three samples, including 93 (of 107) laser fusions on 5–7 grain aliquots of sanidine, using FCs as fluence monitor.	ammonite	North America	*Watinoceras devonense* ammonite zone (upper half)	upper part of *Watinoceras devonense* ammonite zone	Meyers et al. (2012); Obradovich (1993)
Five youngest (of 7) single grain zircon analyses yield a weighted mean $^{206}Pb/^{238}U$ age, utilizing CA-TIMS and ET535 spike.	ammonite	North America	*Watinoceras devonense* ammonite zone (upper half)	upper part of *Watinoceras devonense* ammonite zone.	Meyers et al. (2012); Obradovich (1993)
Five (of 7) multi-grain chemically-abraded zircon fractions. ID-TIMS.	planktonic foraminifer	Tethyan	Cenomanian-Turonian boundary interval	Between LAD of *Rotalipora cushmani* and FAD of *Marginotruncana schneeganei;* therefore Cenomanian-Turonian boundary interval.	Quidelleur et al. (2011)
Weighted mean age of four samples, including 103 (of 116) laser fusions on 1–3 grain aliquots of sanidine, using FCs as fluence monitor.	ammonite	North America	*Neocardioceras juddii–N. scotti* (lower-middle)	One-third up in undifferentiated *Neocardioceras juddii–Nigericeras scotti* ammonite zone [Meyers et al. (2012) figure]	Meyers et al. (2012), Obradovich (1993)
Six youngest (of 11) single grain zircon analyses yield a weighted mean $^{206}Pb/^{238}U$ age, utilizing CA-TIMS and ET535 spike.	ammonite	North America	*Neocardioceras juddii–N. scotti* (lower-middle)	One-third up in undifferentiated *Neocardioceras juddii–Nigericeras scotti* ammonite zone [Meyers et al. (2012) figure]	Meyers et al. (2012); Obradovich (1993)
Six single zircon grains yielded a weighted mean $^{206}Pb/^{238}U$ age, utilizing CA-TIMS and the ET535 spike	ammonite	North America	*Neocardioceras juddii* ammonite zone	Correlated to "B" bentonite, or Bed 80 in Pueblo section; therefore same as other dated "*N. juddii*" bentonite horizon.	Barker et al. (2011)
Weighted mean age of two samples, including 74 (of 83) laser fusions on single grain aliquots of sanidine, using FCs as fluence monitor.	ammonite	North America	Middle of combined *Vascoceras diartianum-B. clydense* interval	*Euomphaloceras septemseriatum* ammonite zone [middle of undifferentiated *S. gracile* zone (including *B. clydense* ?); on Meyers et al. (2012) figure]	Meyers et al. (2012); Obradovich (1993)
Weighted mean age including 42 (of 46) laser fusions on 3–6 grain aliquots of sanidine, using fluence monitor FCs = 28.201 Ma.	ammonite	North America	Middle of combined *Vascoceras diartianum-B. clydense* interval	*Vascoceras diartianum* portion of *Sciponoceras gracile* ammonite zone [middle of undifferentiated *S. gracile* zone (including *B. clydense* ?); on Meyers et al. (2012) figure]	Meyers et al. (2012); Obradovich (1993)
Two youngest (of 5) single grain zircon analyses yield a weighted mean $^{206}Pb/^{238}U$ age, utilizing CA-TIMS and ET535 spike.	ammonite	North America	Middle of combined *Vascoceras diartianum-B. clydense* interval	*Vascoceras diartianum* portion of *Sciponoceras gracile* ammonite zone [middle of undifferentiated *S. gracile* zone [including *B. clydense* ?); on Meyers et al. (2012) figure]	Meyers et al. (2012); Obradovich (1993)
Plateau of biotite separate, originally using FCs = 28.02 Ma as fluence monitor.	ammonite	North America	Base of combined *Vascoceras diartianum-B. clydense* interval	Near base of undifferentiated *Sciponoceras gracile* ammonite zone (overlying *M. mosbyense* zone)	Quidelleur et al. (2011)
Weighted mean age including 25 (of 33) laser fusions on single sanidine crystals, using fluence monitor FCs = 28.201 Ma.	ammonite	North America	*Dunveganoceras pondi* ammonite zone	*Dunveganoceras pondi* ammonite zone	Ma et al. (2014); Obradovich (1993)
Weighted mean age of three samples, including 60 (of 69) laser fusions on 1->10 grain aliquots of sanidine, using FCs as fluence monitor.	ammonite	North America	*Acanthoceras amphibolum* ammonite zone	*Acanthoceras amphibolum* ammonite zone	Siewert (2011); Obradovich (1993)

GTS 2020 ID	GTS 2012 ID	Period	Epoch	Age	Sample	Locality	Lat-Long	Lithostratigraphy	Age (Ma)	± 2s analytical	± 2s total	Age Type
K41	K28				X bentonite	Burnt Timber Creek, Alberta	51°32′55.91″N, 115°11′42.80″W	Sunkay Member of Blackstone Fm; 30-cm-thick brownish-gray bentonite enclosed by dark marine mudstone.	95.87	± 0.10	± 0.14	$^{206}Pb/^{238}U$
K40	K27				91-O-03; bentonite	Gulch south of Colo. Hwy 76, Pueblo County, Colorado	38.12°N, 104.86°W	Graneros Shale; 7.5-cm-thick bentonite, approximately 0.9 m below its Thatcher Limestone Member	96.12	± 0.19	± 0.31	$^{40}Ar/^{39}Ar$
K39	K26				Bailey Flats core (OB93-26); bentonite	Johnson County, Wyoming	43.6°N, 106.6°W	Basal 7.5 cm of 3-m-thick bentonite, 27 m below lowest occurrence of *C. gilberti* in Frontier Fm	96.56	± 0.45	± 0.45	$^{40}Ar/^{39}Ar$
K38	K25				68-O-09 (OB93-27); bentonite	12.9 km W of Casper, Wyoming; S. side Wyo Hwy 220	42°46′N, 106°26′W	Top of Mowry Shale; Clay Spur Bentonite	97.88	± 0.69	± 0.69	$^{40}Ar/^{39}Ar$
K37	K24				OB93-28; bentonite	N. bank Arrow Creek, Judith Basin County, Montana	47.314°N, 110.487°W	Arrow Creek Member, Colorado Shale; basal 15 cm of Arrow Creek bentonite	99.24	± 0.41	± 0.41	$^{40}Ar/^{39}Ar$
K36	K23				91-O-20 (OB93-29); bentonite	NE of Greybull, Big Horn County, Wyoming	44.56°N, 108.00°W	Thermopolis Shale, 30-cm-thick bentonite ca. 2 m above mudstone unit containing black Mn-nodules, upper part of Shell Creek Shale Member	99.26	± 0.70	± 0.70	$^{40}Ar/^{39}Ar$
K35	K22				91-O-12 (OB93-30); bentonite	Johnson County, Wyoming	43.67°N, 106.85°W	Thermopolis Shale, bentonite 12 m below top	99.46	± 0.59	± 0.59	$^{40}Ar/^{39}Ar$
K34	K21				TNGt006; crystal tuff	Tengunosaw valley, Hokkaido, Japan	43°10′36″N, 142°12′03″E	Hikagenosawa Fm; Yazo Group	97.00	± 0.40	± 0.40	$^{206}Pb/^{238}U$
K33	K17				TNGt005; crystal tuff	Tengunosaw valley, Hokkaido, Japan	43°10′38″N, 142°12′12″E	Hikagenosawa Fm; Yazo Group	99.70	± 1.30	± 1.30	$^{206}Pb/^{238}U$ LA-ICPMS
K32	K20				R8072; crystal tuff	Hotei-zawa, Soeushinai area, Hokkaido, Japan	44°16′48″N, 142°04′56″E	Upper Bed My3; middle part of Yezo Group	99.70	± 0.38	± 0.38	$^{40}Ar/^{39}Ar$
K31	K19				SK069; crystal tuff	Hotei-zawa, Soeushinai area, Hokkaido, Japan	44°16′48″N, 142°04′56″E	Upper Bed My3; middle part of Yezo Group; equivalent to R8072 of Obradovich	99.70	± 0.30	± 0.30	$^{206}Pb/^{238}U$
K30	K18				R8943B; crystal tuff	Kyoei-Sakin-zawa, Soeushinai area, Hokkaido, Japan	44°16′N, 142°07′E	Lower Bed My3; middle part of Yezo Group	99.89	± 0.37	± 0.37	$^{40}Ar/^{39}Ar$
			Early									
				Albian								
K29					TNGt001-8	Tengunosaw valley, Hokkaido, Japan	43°10′45″N, 142°12′48″E	Hikagenosawa Fm; Yazo Group	102.10	± 2.70	± 2.70	$^{206}Pb/^{238}U$ LA-ICPMS
K28	K16				75-O-06 (OB93-31); bentonite	Gully on N. bank of Peace River, short distance below the "Gates", British Columbia	56°06′30″N, 121°49′00″W	Hulcross Fm, bentonite in its upper 50 m.	107.89	± 0.30	± 0.30	$^{40}Ar/^{39}Ar$
K27	K15				Vöhrum tuff	Vöhrum clay pit situated 30 km east of Hannover, 1.6 km SW of Vöhrum, NW Germany	52°19′49″N, 10°11′44″E	Schwicheldt Ton Member, Gault Fm. 2-cm-thick tuff horizon 65 cm above the (local usage) "Aptian/Albian boundary"	113.08	± 0.07	± 0.14	$^{206}Pb/^{238}U$
				Aptian								
K26	K14				OB93-32; bentonite	Otto gott clay pit, NE of Sarstedt, 21 km SE of Hannover, NW Germany	52°20′34″N, 9°57′28″E	Trachytic tuff, attributed to the Waddenzee volcanic center located off the Netherlands coast (Mutterlose and Böckel, 1998) within a succession of clays and marls.	114.84	± 1.30	± 1.30	$^{40}Ar/^{39}Ar$

Primary radioisotopic age details	Zonal range assignment			Biostratigraphy	Reference
	clade	zonation	Zone		
Six single zircon grains yielded a weighted mean $^{206}Pb/^{238}U$ age, utilizing CA-TIMS and the ET535 spike	ammonite	North America	*Acanthoceras amphibolum* ammonite zone	Correlated to "X" bentonite of Frontier Fm in Wyoming; therefore coeval with sample 90-O-50 (*A. amphibolum* ammonite zone). Close to end of regressive phase, just prior to onset of the major transgression culminating in early Turonian most-extensive-flooding of Cretaceous Western Interior basin.	Barker et al. (2011)
Weighted mean age of 48 (of 58) laser fusions on 5 grain aliquots of sanidine, using FCs as fluence monitor.	ammonite	North America	*Conlinoceras tarrantense–C. gilberti* ammonite zone	*Conlinoceras tarrantense* (=*Conlinoceras gilberti*) ammonite zone	Siewert (2011); Obradovich (1993)
Multi-grain sanidine laser fusion analysis, originally using TCs = 28.32 Ma as fluence monitor; no analytical data published.	ammonite	North America	*N. cornutus–base C. gilberti* ammonite zone interval.	27 m below lowest occurrence of *C. gilberti* ammonite	Obradovich (1993)
Multi-grain sanidine laser fusion analysis, originally using TCs = 28.32 Ma as fluence monitor; no analytical data published.	ammonite	North America	*N. cornutus–base C. gilberti* ammonite zone interval.	Top of Mowry Shale	Obradovich (1993)
Multi-grain sanidine laser fusion analysis, originally using TCs = 28.32 Ma as fluence monitor; no analytical data published.	ammonite	North America	*Neogastroplites cornutus* ammonite zone	*Neogastroplites cornutus* ammonite zone	Obradovich (1993)
Multi-grain sanidine laser fusion analysis, originally using TCs = 28.32 Ma as fluence monitor; no analytical data published.	ammonite	North America	*Neogastroplites haasi–N. cornutus* ammonite zones	0.9–4.6 m above mudstone unit containing black Mn-nodules with *Neogastroplites haasi*	Obradovich (1993)
Multi-grain sanidine laser fusion analysis, originally using TCs = 28.32 Ma as fluence monitor; no analytical data published.	ammonite	North America	*Neogastroplites haasi* ammonite zone	Similarity in age to 91-O-20 suggests the *N. haasi* Zone.	Obradovich (1993)
Mean of four (of 9) multi-grain chemically abraded zircon fractions.	planktonic foraminifer	Tethyan	Early Cenomanian	"*Above FAD of Mantelliceras ammonite*" and within range of foraminifer *Thalmanninella globotruncanoides.*"; however, there is a fault below the sampled horizon, therefore is not reliable for scaling relative to underlying strata (Ando, 2016).	Quidelleur et al. (2011)
Eleven spots on 12 zircon grains.	ammonite	Pacific	*Mantelliceras mantelli* ammonite zone (base of upper subzone)	Nearly coeval with local FAD of pelagic forminifer *Thalmanninella globotuncanoides* (primary marker of Albian-Cenomanian boundary).	Quidelleur et al. (2011)
Single or multi-grain sanidine laser fusion analysis, originally using TCs = 28.32 Ma as fluence monitor.	ammonite	Pacific	*Mantelliceras mantelli* ammonite zone (base of upper subzone)	*Mantelliceras saxbii* ammonite subzone (lower part) of *Mantelliceras mantelli* Zone.	Obradovich et al. (2002)
Five multigrain chemically abraded fractions from SK069 were analyzed by ID-IMS.	ammonite	Pacific	*Mantelliceras mantelli* ammonite zone (upper subzone)	Above occurrence of *Mantelliceras saxbii* ammonite	Quidelleur et al. (2011)
Single or multi-grain sanidine laser fusion analysis, originally using TCs = 28.32 Ma as fluence monitor.	ammonite	Pacific	*Mantelliceras mantelli* ammonite zone (top of lower subzone)	Top of *Graysonites woodridgei* ammonite zone (equivalent to European subzone of *Neostlingoceras carcitanense*); upper part. local FAD (but single horizon) of pelagic forminifer *Thalmanninella globotuncanoides* is at this level (Ando 2016); thereby placing a minimum age on the base of Cenomanian.	Obradovich et al. (2002)
Seven spots on eleven zircon grains from TNGt001.8 were analyzed by LA-ICPMS.	planktonic foraminifer	Tethyan	*Ps. subticinensis* foraminifer zone	Above FAD of planktonic foraminifer *Pseudothalmanninella subticinensis*.	Quidelleur et al. (2011)
Multi-grain sanidine laser fusion analysis, originally using TCs = 28.32 Ma as fluence monitor; no analytical data published.	ammonite	North America	*Euhoplites loricatus* ammonite zone	*Pseudopulchellia pattoni* zone of North American Western Interior. Correlated to mid-Middle Albian, *Euhoplites loricatus* zone of Europe.	Obradovich (1993)
Weighted mean of five (of 7) single and 2–3 grain zircon fractions, utilizing CA-TIMS and the ET535 spike	ammonite	Western Europe	*Leymeriella tardefurcata* (lower subzone)	65 cm above the first occurrence of the ammonite *Leymeriella (Proleymeriella) schrammeni anterior*	Selby et al. (2009)
Multi-grain sanidine laser fusion analysis, originally using TCs = 28.32 Ma as fluence monitor; no analytical data published.	ammonite	Western Europe	lower *Acanthohoplites nolani* (or slightly above middle of original *Parahoplites nutfieldiensis*)	*Parahoplites nutfieldiensis* ammonite zone [but definition of this zone was not provided in this paper; but see Singer et al. (2015), below for re-analysis of same sample]	Obradovich (1993)

GTS 2020 ID	GTS 2012 ID	Period	Epoch	Age	Sample	Locality	Lat-Long	Lithostratigraphy	Age (Ma)	± 2s analytical	± 2s total	Age Type
					Re-analysis: OB93-32; bentonite	Otto gott clay pit, Sarstedt, 21 km SE of Hannover, Germany	52°20'34"N, 9°57'28"E	Trachytic tuff, attributed to the Waddenzee volcanic center located off the Netherlands coast (Mutterlose and Böckel, 1998) within a succession of clays and marls.	114.86		± 0.37	⁴⁰Ar/³⁹Ar
					Gott	Otto gott clay pit, Sarstedt, 21 km SE of Hannover, Germany	52°20'34"N, 9°57'28"E	Trachytic tuff, attributed to the Waddenzee volcanic center located off the Netherlands coast (Mutterlose and Böckel, 1998) within a succession of clays and marls.	114.80	± 0.12	± 1.10	⁴⁰Ar/³⁹Ar
K25	K13				PA-140, tuff layer	Bano Nuevo Volcanic Complex, near Cerro Mirador, central Patagonian Cordillera, Argentina	45°15'S 71°34'W	volcanics conformably overlie unconsolidated sands of the Apeleg Fm	120.90	± 1.10	± 1.10	²⁰⁶Pb/²³⁸U ion probe
K24	K12				MC-44, MC-84, MC-86-1; lavas	Bano Nuevo Volcanic Complex, near Cerro Mirador, central Patagonian Cordillera, Argentina	45°17'S 71°28'W	volcanics conformably overlie unconsolidated sands of the Apeleg Fm	122.20	± 1.50	± 1.50	⁴⁰Ar/³⁹Ar
K23					Yixian Basin	Mashenmiao-Zhuanchengzi (MZ) section, Yixian, Liaoning Province, northeast China	41°29'N, 121°03'E,	Basalt flows; terrestrial setting	122.01	± 0.5	± 0.52	⁴⁰Ar/³⁹Ar
K22	K11				144-878A-46M-1, 46M-2, 79R-3; hawaiites	ODP Site 878A (Leg 144), MIT Guyot, western Pacific	27°19.143'N, 151°53.028'E	Seamount basalts below transgressive carbonates	122.18	± 1.43	± 1.43	⁴⁰Ar/³⁹Ar
K21	K10				MC888; tuff	outcrops along McCarty Creek, 2 miles north of Paskenta, Tehama County, Great Valley, California, USA	39.9°N, 122.5°W	Great Valley Group, California, USA	124.07	± 0.16	± 0.24	²⁰⁶Pb/²³⁸U
K20	K9				40R-1, 41R-1, 42R-5, 45R-1; plagioclase phenocrysts from volcaniclastic rocks	ODP Site 1184 (Leg 192), eastern salient of Ontong Java Plateau, western equatorial Pacific	5°0.6653'S, 164°13.9771'E	Four samples from cores into subunit IIE in the lower part of the volcaniclastic succession.	124.32	± 1.80	± 1.80	⁴⁰Ar/³⁹Ar
K19	K8				SG7, SGB25, ML475, KF36, KF53; whole rock basalts	on-shore sections of Ontong Java Plateau basalts, central Malaita, Solomon Islands; equivalent to OJP basalts drilled in ODP Site 807	~9°S, 161°E	Malaita Volcanic Group	125.98	± 2.86	± 2.87	⁴⁰Ar/³⁹Ar
				Barremian								
K18					Svalbard DH	DH-3&7 borehole for U-Pb, DH-2 for C-13 and Magstrat; Svalbard	78°12.1'N, 15°48.9'E	Helvetiafjellet Fm; bentonite at 164.4 m depth in DH7 and coeval one in DH3	123.10	± 0.30	± 0.30	²⁰⁶Pb/²³⁸U
K17	K7				144-878A-89R-4, 91R-3; plagioclase from alkalic basalts	ODP Site 878 (Leg 144), MIT guyot, western Pacific	27°19.143'N, 151°53.028'E	Seamount basalts below transgressive carbonates.	125.45	± 0.41	± 0.43	⁴⁰Ar/³⁹Ar
				Hauterivian								
K16					sample EP 1711–1712	El Portón section, Neuquén Basin, Argentina	37°11'52''S, 69°41'03''W	Interbedded between shales of the upper part of the upper member (Agua de la Mula) of the Agrio Fm	126.97	± 0.04	± 0.15	²⁰⁶Pb/²³⁸U
K15					Agrio del Medio	Neuquén Basin, Argentina	38°20'S, 69°57'W	Interbedded between shales of the upper part of the upper member (Agua de la Mula) of the Agrio Fm	127.42	± 0.03	± 0.15	²⁰⁶Pb/²³⁸U

Primary radioisotopic age details	Zonal range assignment			Biostratigraphy	Reference
	clade	zonation	Zone		
Laser fusions on 7–10 grain aliquots of sanidine, using FCs as fluence monitor. (Note: details given in GSA talk; but not published in detail.)	ammonite	W. Europe	lower *Acanthoplites nolani* (or slightly above middle of original *Parahoplites nutfieldiensis*)	Zone is also called "Nolaniceras nolani". Base of Nolani (hence, top of *P. nutfieldiensis* Zone) is constrained in Germany by 114.86 ± 0.37 Ar/Ar, which verifies and enhances Obradovich's 114 ± 1.3 Ma.	Singer et al. (2015)
Gott K-feldspar sample displays highly reproducible step ages, with a high precision plateau age of 114.85 ± 0.12 Ma, when only analytical uncertainties are considered, using Fish Canyon Tuff age of 28.02 Ma as fluence monitor; not yet converted to new standard of 28.201 Ma (Kuiper et al., 2008)	ammonite	W. Europe	lower *Acanthoplites nolani* (or slightly above middle of original *Parahoplites nutfieldiensis*)	Same level as the other Ar/Ar determinations; slightly above middle of original *Parahoplites nutfieldiensis*	
Weighted mean of 14 (of 17) spot analyses analyzed by SHRIMP II (ANU), using the TEMORA zircon standard.	ammonite	Argentina	*Deshayesites forbesi* through *E. martinoides* zones due to vague biostrat	upper beds of the Apeleg Formation have been dated as early Aptian based on the presence of the ammonite *Tropaeum* or *Australiceras* sp.	Suárez et al. (2010)
Weighted mean of three hornblende 40Ar/39Ar plateau or isochron ages, originally calculated with FCs = 28.02 Ma as the fluence monitor.	ammonite	Argentina	*Deshayesites forbesi* through *E. martinoides* zones due to vague biostrat	upper beds of the Apeleg Formation have been dated as early Aptian based on the presence of the ammonite *Tropaeum* or *Australiceras* sp.	Suárez et al. (2010)
Weighted mean of three whole rock plateau ages; originally calculated with TCs = 28.34 Ma as the fluence monitor.	magnetostratigraphy		Reversed polarity; unknown biostratigraphic age	No biostratigraphy in this terrestrial section that could be correlated to marine standards. Reversed-magnetization basalt flows (10 samples on transect; no boundaries found to normal polarity). Publication assigned it to be Chron M0r based on the reversed polarity.	He et al. (2008)
Weighted mean of three whole rock 40Ar/39Ar isochron ages, originally calculated with TCs = 27.92 Ma as the fluence monitor.	calcareous nannofossil	Tethyan	Within or older than NC6	MIT seamount normally magnetized (above reversed magnetized), upper hawaiite lavas. Platform carbonates 25 m above volcanic basement are assigned to the lower-Aptian *C. litterarius* nannofossil biozone (NC6). Implied to be older than mid-Early Aptian	Pringle and Duncan (1995)
Weighted mean of youngest five (of 7) 1–4 grain zircon fractions, utilizing CA-TIMS and in-house UNC spike (0.1% error in tracer calibration), with analytical error expanded in accommodate geologic scatter.	calcareous nannofossil	Tethyan	lower NC 6B nannofossil zone	*Chiastozygus litterarius* biozone (NC6) — Text says NC6A subzone, as supported by the first occurrence of *C. litterarius* and the last occurrence of *Conusphaera rothii* in MC887, a sample found below MC888. But, LAD of *C. rothii* = base subzone 6B, therefore sample is lower NC6B.	Shimokawa (2010)
Weighted mean of 6 (of 8) laser fusions on 4–5 grain aliquots of plagioclase, sampled from four core intervals, originally calibrated to FCs = 28.02 Ma.	planktonic foraminifer	Tethyan	upper *Leupoldina cabri* foraminifer zone	The oldest sediment overlying basement on the crest of the Ontong Java Plateau occurs within the upper part of the *Leupoldina cabri* planktonic foraminiferal zone.	Chambers et al. (2004)
Weighted mean of 5 whole rock incremental heating plateau ages, originally calculated with FCT-3 biotite = 27.55 Ma.	planktonic foraminifer	Tethyan	*upper Leupoldina cabri* foraminifer zone	The oldest sediment overlying basement on the crest of the Ontong Java Plateau (ODP sites) occurs within the upper part of the *Leupoldina cabri* planktonic foraminiferal zone; but there are no biostratigraphic constraints on these Malaita exposures.	Tejada et al. (2002)
Midtkandal et al. (2016) merged zircon sets from the same bentonite in two boreholes [DH-3 from Corfu et al. (2013)], and used the 10 youngest zircon grains (of 21).	magnetostratigraphy		upper Chron M1r	Indirect palynology correlations. Lowest meters of overlying Carolinefjellet Formation has a negative then positive C-13 excursion which was interpreted as characteristic of earliest Aptian OAE1a. Position of the ash horizon in adjacent borehole magneostratigraphy indicate it is within mid-Chron M1r (Zhang et al., 2019)	Midtkandal et al. (2016)
Weighted average of plagioclase 40Ar/39Ar isochron ages, originally calculated with TCs = 27.92 Ma as the fluence monitor	magnetostratigraphy		Was interpreted as top of M0r; reinterpreted as top M3r	MIT seamount lower alkalic basalts at reversed-upward-to-normal polarity transition; overlain by normal-polarity basalts. Platform carbonates 25 m above volcanic basement are assigned to the lower-Aptian *C. litterarius* nannofossil biozone (NC6). Implied to be older than mid-Early Aptian. Was interpreted as top of Chron M0r; but re-interpretation might be top of M1r or top of M3r.	Pringle and Duncan (1995)
Weighted mean of 6 single zircon grain analyses, utilizing CA-TIMS and the ET535 spike.	ammonite	Argentina	*Sabaudiella riverorum*	Near base of the *Sabaudiella riverorum* regional ammonite zone	Aguirre-Urreta et al. (2019)
Weighted mean of 6 single zircon grain analyses, utilizing CA-TIMS and the ET535 spike.	ammonite	Argentina	upper *Pseudothurmannia ohmi* ammonite zone	Within upper part of regional ammonite zone of *Paraspiticeras groeberi*; which they project to be nearly at end of Hauterivian (uppermost *Pseudothurmannia ohmi* Zone of Tethyan Mediterranean).	Aguirre-Urreta et al. (2015)

GTS 2020 ID	GTS 2012 ID	Period	Epoch	Age	Sample	Locality	Lat-Long	Lithostratigraphy	Age (Ma)	± 2s analytical	± 2s total	Age Type
K14	K6				Caepe Malal; tuff layer	Neuquén Basin, Argentina	37°11'S, 70°23'W	Interbedded between shales of the lowermost part of the upper member (Agua de la Mula) of the Agrio Fm	129.09	± 0.04	± 0.14	$^{206}Pb/^{238}U$
K13					POT 3 tuff	El Portón section, Neuquén Basin, Argentina	37°11'52"S, 69°41'03"W	Pilmatué Member of the lower Agrio Fm	130.39	± 0.04	± 0.16	$^{206}Pb/^{238}U$
				Valanginian								
K12	K5				MC873A; tuff	outcrops along McCarty Creek, 2 miles north of Paskenta, Tehama County, Great Valley, California, USA	39°54'54"N, 122°32'34"W	Great Valley Group	133.51	± 0.22	± 0.29	$^{206}Pb/^{238}U$
K11	K4				XZ0506; rhyolite	Kadong section, south side of Yamzho Iyumco Lake, 28.75N, 90.70E, ca. 51km ESE of Nagarze, Tibet.	52°18'50.2"N, 9°58'05.5"E	Rhyolite is from unit 13 of the upper (volcanic and volcaniclastic) part of Sangxiu Fm	136.00	± 3.00	± 3.00	$^{206}Pb/^{238}U$ ion probe
K10	K3				MC180.5; tuff	outcrops along Kelly Road, Paskenta, Tehama County, Great Valley, California, USA	39°55'N, 122°35'W	Great Valley Group	137.62	± 0.07	± 0.21	$^{206}Pb/^{238}U$
				Berriasian								
K9	K2				CA-6189; rhyolite ignimbrite	southern slope of Lago Norte, 3 km NNW of Lago Norte, Aisén Basin, Patagonia, Argentina	45.1°S, 71.9°W	12 m thick rhyolitic ignimbrite of the Lago Norte Vocanic Complex, overlying a 50 m thick fossiliferous marine succession of Toqui Fm intercalated in the Ibáñez Fm	137.30	± 1.20	± 1.20	$^{206}Pb/^{238}U$ ion probe
K8	K1				GC-670; tuff	Outcrops on south side of Grindstone Creek, west of State Road 306, Glen County, Great Valley, California, USA	39°40'N, 122°35'W	Great Valley Group	138.46	± 0.21	± 0.29	$^{206}Pb/^{238}U$
K7					Lena-Neuquen LL3	Las Loicas; Neuquén Basin, Argentina	35°48'55"S, 70°09'21"W	Vaca Muerta Fm; ca. 54 m level in section	139.24	± 0.05	± 0.16	$^{206}Pb/^{238}U$
K6					Venari-Neuquen	Las Loicas; Neuquén Basin, Argentina	35°48'55"S, 70°09'21"W	Vaca Muerta Fm; ca.56m level in section	139.55	± 0.03	± 0.16	$^{206}Pb/^{238}U$
K5					Lena-Neuquen LL9	Las Loicas; Neuquén Basin, Argentina	35°48'55"S 70°09'21"W	Vaca Muerta Fm; ca. 41 m level in section	139.96	± 0.06	± 0.17	$^{206}Pb/^{238}U$
K4					Lena-Neuquen LL10	Las Loicas; Neuquén Basin, Argentina	35°48'55"S. 70°09'21"W	Vaca Muerta Fm; ca. 31 m level in section	140.34	± 0.06	± 0.18	$^{206}Pb/^{238}U$
K3					Lena-Neuquen MZT-81	Mazatepec, Puebla State; Sierra Madre, Mexico	~20°N, 97°W	Lower Tamaulipas Fm of gray limestone (lower part); ca. 23 m level in section.	140.51	± 0.03	± 0.16	$^{206}Pb/^{238}U$
K2					Lena-Neuquen LL13	Las Loicas; northern Neuquén Basin, Argentina	35°48'55"S, 70°09'21"W	Vaca Muerta Fm; ca. 3 m level in section	142.04	± 0.06	± 0.17	$^{206}Pb/^{238}U$

Primary radioisotopic age details	Zonal range assignment			Biostratigraphy	Reference
	clade	zonation	Zone		
Weighted mean of 5 single zircon grain analyses, utilizing CA-TIMS and the ET535 spike.	ammonite	Argentina	*Subsaynella sayni* ammonite zone (lower)	within the *Spitidiscus riccardii* ammonite zone, correlated with the lower *Subsaynella sayni* zone of the Tethys	Aguirre-Urreta et al. (2015)
Weighted mean of 5 single zircon grain analyses, utilizing CA-TIMS and the ET535 spike.	ammonite	Argentina	*Crioceratites loryi* ammonite zone (lower)	Base of regional *Holcoptychites agrioensis* subzone, *Holcoptychites neuquensis* zone; and they suggest this is approximately coeval to base of *Crioiceratites loryi* zone of Tethyan Mediteranean. Same level as local highest-occurrence of *E. windii* nannofossil.	Aguirre-Urreta et al. (2017)
Weighted mean of youngest eight (of 10) 3–5 grain zircon fractions, utilizing CA-TIMS and in-house UNC spike (0.1% error in tracer calibration), with analytical error expanded in accommodate geologic scatter.	calcareous nannofossil	Tethyan	upper *C. oblongata* (lower NC3b) nannofossil zone (middle Valanginian)	Calcareous nannofossils of *P. fenestrata, R. nebulosus, C. angustiforatus, Cyclagelosphaera deflandrei, Eiffellithus windii, Metadoga mercurius, R. wisei,* and *Tubodiscus verenae* place this sample in the *C. oblongata* (NC3) biozone – lower *E. windii* (lower NC3b) subzone	Shimokawa (2010)
Weighted mean of fifteen spot analyses analyzed by SHRIMP II (Beijing), using the TEMORA zircon standard	calcareous nannofossil	Tethyan	*C. oblongata* (NC3) nannofossil zone	Sample is above a *Calcicalathina oblongata–Speetonia colligata* assemblage of calcareous nannofossils, but minimum age is unconstrained. Assigned here as Valanginian.	Wan et al. (2010)
Weighted mean of six oldest (of 8) 1–2 grain zircon fractions, utilizing CA-TIMS and in-house UNC spike (0.1% error in tracer calibration).	calcareous nannofossil	Tethyan	lowest *C. oblongata* (lower NC3a) nannofossil zone (uppermost Berriasian-lowermost Valanginian)	Calcareous nannofossil occurrences suggest correlation to the *C. angustiforatus* biozone – *P. fenestrata* subzone (NK-2B) at the oldest, but it is not possible to rule out a younger age with the current biostratigraphy; the sample is conservatively placed at the base of the Valanginian in the lower *Calcicalathina oblongata* biozone. [*C. oblongata* begins in lower *T. pertransiens* ammonite zone]	Shimokawa (2010)
Weighted mean of 14 (of 19) spot analyses analyzed by SHRIMP II (ANU), using the TEMORA zircon standard.	ammonite	Argentina	Late Berriasian or younger?	Ammonites (*Groebericeras* Leanza and *Blandfordiceras* Cossman) from the underlying Toqui Formation, the basal unit of the Coihaique Group, indicate early Berriasian and late Berriasian ages.	Suárez et al. (2009)
Weighted mean of five 2–4 grain zircon fractions, utilizing CA-TIMS and in-house UNC spike (0.1% error in tracer calibration), with analytical error expanded in accommodate geologic scatter.	calcareous nannofossil	Tethyan	*Cretarhabdus angustiforatus* biozone (NK-2)	Sample placed in the *Cretarhabdus angustiforatus* biozone (NK-2), supported by the occurrences of *C. angustiforatus* and *Assipetra infracretacea* 1 cm above the sample, and the appearances of *Micrantholithus hoschulzii* and *Rhagodiscus nebulosus* within a meter above the sample.	Shimokawa, 2010
Weighted mean of youngest four (of nine) single zircon grains, utilizing CA-TIMS and the ET2535 or ET535 spike	ammonite; calpionellid	Argentina	upper *A. noduliferum* Zone, *Calpionella* Zone; projected as uppermost Chron M17r	upper *Argentiniceras noduliferum* regional ammonite zone; within Calpionella Zone (calpionellids); therefore, upper Berriasian (Kietzmann et al., 2018).	Lena et al. (2019)
Weighted mean of five single zircon grains, utilizing CA-TIMS and the ET2535 spike	ammonite	Argentina	middle *A. noduliferum* Zone; projected as Chron M17r	middle of Argentiniceras noduliferum regional ammonite zone; therefore, upper Berriasian (Kietzmann et al., 2018).	Vennari et al. (2014)
Weighted mean of youngest four (of 11) single zircon grains, utilizing CA-TIMS and the ET2535 or ET535 spike	ammonite	Argentina	lower *A. noduliferum* Zone; projected as basal Chron M17r-uppermost M18n	lower *Argentiniceras noduliferum* regional ammonite zone; within lower Calpionella Zone (calpionellids); above FO nannofossil *Nannoconus steinmannii minor*; therefore, upper Berriasian (Kietzmann et al., 2018).	Lena et al. (2019)
Weighted mean of youngest four (of eight) single zircon grains, utilizing CA-TIMS and the ET2535 or ET535 spike	ammonite; calpionellid	Argentina	uppermost *B. koeneni* Zone, uppermost *Crassicollaria* Zone; projected as basal Chron M18n-uppermost M18r	uppermost *Substeueroceras koeneni* regional ammonite zone; uppermost Crassicollaria Zone (calpionellids); above FO nannofossil *Rhadodiscus* as per, below FO *Nannoconus wintereri*; therefore, middle Berriasian (Kietzmann et al., 2018).	Lena et al. (2019)
Weighted mean of youngest four (of eight) single zircon grains, utilizing CA-TIMS and the ET2535 or ET535 spike	calpionellid	Tethyan	lower *Calp. elliptica* Zone; projected as between base Chron M18n to lower half M17r.	lower Calpionella elliptica subzone; above FO *Nannoconus kamptneri minor*, below *N. steinmannii steinmanni* FO; therefore, middle Berriasian (Kietzmann et al., 2018).	Lena et al. (2019)
Weighted mean of three single zircon grains, utilizing CA-TIMS and the ET2535 or ET535 spike	ammonite	Argentina	middle(?) *B. koeneni* Zone, lower(?) *Crassicollaria* zone; projected as lower Chron M18r	middle (?) *Substeueroceras koeneni* regional ammonite zone (base not exposed); within lower (?) Crassicollaria Zone (calpionellids); below FO nannofossil *Umbria granulosa*; therefore, middle Berriasian (Kietzmann et al., 2018).	Lena et al. (2019)

GTS 2020 ID	GTS 2012 ID	Period	Epoch	Age	Sample	Locality	Lat-Long	Lithostratigraphy	Age (Ma)	± 2s analytical	± 2s total	Age Type
K1	J19				Shatsky Rise basalt sills	ODP Hole 1213B (Leg 198) on Shatsky Rise, northwest Pacific	31°34.6576'N, 157°17.8621'E	Lithologic Unit IV. Whole-rock = Core 28R-2 (1–6 cm) and 30R-3 (1–6 cm). Feldspars from Core 33R-3 (115–120 cm)	146.48	± 1.62	± 1.63	$^{40}Ar/^{39}Ar$
		Jurassic										
			Late									
				Tithonian								
J40					Lena-Neuquen LYS	Las Loicas; northern Neuquén Basin, Argentina	35°48'55"S, 70°09'21"W	Base of Vaca Muerta Fm overlying non-marine Tordillo Fm	147.11	± 0.08	± 0.18	$^{206}Pb/^{238}U$
				Kimmeridgian								
J39	J18				Staffin black shale	Flodigarry, Staffin Bay, Isle of Skye, U.K. = proposed GSSP for base-Oxfordian	57°39'40"N, 6°14'44"W	Bed 35, which is 1.30 ± 0.1 m below Concretion-rich marker Bed 36, and near the proposed GSSP level (1.25 m below Bed 36); Flodigarry Shale Member, Staffin Fm	154.10	± 2.10	± 2.20	Re-Os
				Oxfordian								
J38					AB4 Pellenard	Kaberlaba section, in the Altopiano di Asiago (Trentino Alto Adige and Veneto regions, Italy	45°50'27"N, 11°29'47"E	Top of middle unit (RAM) of Rosso Ammonitico	155.67	± 0.80	± 0.89	$^{40}Ar/^{39}Ar$
J37					felsic tuff, sample EP03-174-1 (z7850)	6.5 km southwest of Diagonal Mountain, central Bowser Basin, British Columbia	56°41'09"N, 127°22'15"W	upper Hazelton Group, 45 m below the base of the Bowser Lake Group	158.70	± 0.20	± 0.27	$^{206}Pb/^{238}U$
			Middle									
				Callovian								
J36	J15				Ash at Bathonian-Callovian boundary	Chacay Melehué, Neuquén Basin, western Argentina	37°15'S, 70°30'W	"Boundary tuff"; Los Molles Fm	164.64	± 0.20	± 0.27	$^{206}Pb/^{238}U$
				Bathonian								
				Bajocian								
J35	J14				Samples ODP 801C-30R-4 37-41 and 801C-31R-4 138-143	ODP Site 801, Pigafetta basin, western Pacific; marine magnetic anomaly M43r (or M42n.4r on revised deep-tow nomenclature)	18°38.538'N, 156°21.588'E	Tholeiitic basalts (ca. 750 mbsf; this unit extends from ca. 500 mbsf to base of hole at ca. 950 mbsf), oceanic thoeiitic basalt crystalline groundmass	168.72	± 1.71	± 1.73	$^{40}Ar/^{39}Ar$
J34	J13				GUN-F; bentonite	near top of Gunlock section, Washington County, Utah, USA	37.270°N, 113.791°W	Crystal Creek Member, Carmel Fm, Utah, USA	168.35	± 1.30	± 1.31	$^{40}Ar/^{39}Ar$
J33	J12				GUN-4; bentonite	near bottom of Gunlock section, Washington County, Utah, USA	37.270°N, 113.791°W	uppermost Temple Cap Fm, Utah, USA	171.48	± 1.20	± 1.22	$^{40}Ar/^{39}Ar$
				Aalenian								
J32	J11				Eskay Rhyolite	Eskay rhyolite east, Eskay Creek gold mine, Kitimat-Stikine A, Iskut River map area, NW British Columbia	56°39'13.6"N, 130°25'47.4"W	Salmon River Fm, Hazelton Group	174.10	± 0.68	± 0.72	$^{206}Pb/^{238}U$
			Early									
				Toarcian								

Primary radioisotopic age details	Zonal range assignment			Biostratigraphy	Reference
	clade	zonation	Zone		
Weighted mean of whole rock (middle sill), and plagioclase (lower sill) incremental heating plateau ages, originally calculated with FCT biotite = 28.03 Ma as the fluence monitor.	calcareous nannofossil	Tethyan	Younger than NK1 nanno zone of earliest Berriasian	Sills (reversely magnetized) were injected into sediments with Calcareous nannofossil zone NK1 and radiolarian zone *Pseudodictryomitra carpatica* of earliest Berriasian => emplacement of Sills must be younger than earliest Berriasian; and perhaps during Chron M18r.	Mahoney et al. (2005)
Weighted mean of youngest four (of seven) single zircon grains, utilizing CA-TIMS and the ET2535 or ET535 spike	ammonite	Argentina	*V. andesensis* Zone (lowermost Tithonian)	In *Virgatosphinctes andesensis* (same as *Virg. mendozanus* in some scales) regional ammonite zone (base of ammonite recovery at this level of facies change); therefore lowermost Tithonian (Kietzmann et al., 2018)	Lena et al. (2019)
Model 3 isochron age from 9 rock powder aliquots (+1 replicate) sampling a of 10 cm of section, dissolved via CrVI-H2SO4.	ammonite	Subboreal	*Epipeltoceras bimammatum* ammonite zone (middle part)	"subboreal Oxfordian-Kimmeridgian (*Pseudocordata/Baylei* zones) boundary is placed between 1.65–1.47 m and 1.08 m below bed 36—the last occurrence of *Ringsteadia* and the first occurrence of *Pictonia*, respectively. Bed 35 is rich in *Pictonia densicostata*.	Selby (2007)
Plateau of 100% of gas released by step-heating of sanidine separate, originally calibrated to joint optimization scheme of Renne et al. (2011).	ammonite	Tethyan	*Gregoryceras transversarium* ammonite zone or younger	ammonites "chacteristic of the *Gregoryceras transversarium* Biozone, indicates a Middle Oxfordian age for the top of the RAM unit"; and overying upper Rosso Ammon.	Pellenard et al. (2013)
Seven multi-grain zircon fractions combined to yield a weighted mean ^{206}Pb/^{238}U age, utilizing air abrasion and an in-house tracer calibration	ammonite	Tethyan	*Quenstedtoceras lamberti* or *Q. mariae* zones	Upper Hazelton Group contains *Quenstedtoceras* or earliest *Cardioceras* of latest Callovian (*Quenstedtoceras lamberti* Zone) or earliest Oxfordian age (*Q. mariae* Zone), 45 m below the base of the Bowser Lake Group.	Evenchick et al. (2010)
Weighted mean of 6 (of 9) single zircon grains, utilizing CA-TIMS and the ET535 spike	ammonite	Tethyan	*Clydoniceras discus—R. anceps* ammonite zone interval?	Tuff between *L. steinmanni* and *vergarensis* regional ammonite zones. Top of local *Lilloettia steinmanni* ammonite assemblage zone. "A precise age cannot be established. ... Only the overling *Hecticoceras proximum* Zone [which is above a local *Neuqueniceras bodenbenderi* assemblage zone] is chararacterised by an association of Tethyan Oppeliidae indicating the latest early Callovian (Riccardi et al., 1989)."	Kamo and Riccardi (2009)
Weighted mean of two laser incremental heating plateau ages (41%–57% of gas) on acid-leached crystalline groundmass, originally calibrated to FCT-3 biotite = 28.04 Ma.	magnetostratigraphy		Probably Bajocian or Aalenian	Probably the lower massive tholeiitic basalt is mid-Bajocian to early Bathonian [disputed, see Jurassic text; with a much older Early Bajocian or early Aalenian proposed by Bartolini and Larson (2001)]. Important constraint on Jurassic portion (Bajocian-Oxfordian) of M-sequence.	Koppers (2003)
Weighted mean of six single-grain sanidine laser fusion analyses, originally using FCs = 27.84 Ma as fluence monitor.	ammonite	Tethyan	*W. laeviuscula—G. garantiana* ammonite zone interval	Bivalve correlation of Carmel Formation to ammonite-bearing Twin Creek Formation indicates an age of late Early to early Late Bajocian	Kowallis et al. (2001)
Weighted mean of eight single-grain sanidine laser fusion analyses, originally using FCs = 27.84 Ma as fluence monitor.	ammonite	Tethyan	*W. laeviuscula—G. garantiana* ammonite zone interval	Bivalve correlation of Carmel Formation to ammonite-bearing Twin Creek Formation indicates an age of late Early to early Late Bajocian	Kowallis et al. (2001)
Weighted mean of 5 multi-grain zircon fractions, utilizing physical abrasion and in-house UBC spike (0.1% error in tracer calibration). Analytical error expanded in accommodate geologic scatter.	ammonite	Tethyan	uppermost Toarcian—mid-upper Aalenian (*D. pseudoradiosa—B. bradfordensis* zone interval)	Flow-banded rhyolite is overlain by Upper Aalenian strata, correlated to *Erycitoides* cf. *howelli* ammonite zone of western North America zonation) [therefore, could be Lower Aalenian or older?]	Childe (1996)

GTS 2020 ID	GTS 2012 ID	Period	Epoch	Age	Sample	Locality	Lat-Long	Lithostratigraphy	Age (Ma)	± 2s analytical	± 2s total	Age Type
J31					Bed 99	Palquilla section near Tacna in southern Peru.	70.000092°S, 17.616156°W	Pelado Formation	180.35	± 0.39	± 0.44	$^{206}Pb/^{238}U$
J30	J10				PCA-YR-1; volcanic ash	Yakoun River, Queen Charlotte Islands, British Columbia	~53.5°N, 132.2°W	Whiteaves Fm	181.40	± 0.73	± 0.78	$^{206}Pb/^{238}U$
J29	J9				AR08AS-29C; tuff	Serrucho Creek, west of Malargüe city, Mendoza prov., Neuquén Basin, Argentina	35°26′32″S, 69°54′12″W	Upper tuff bed within 19.5-m-thick organic-rich shale of lower Tres Esquinas Fm (21.9 mab in measured section).	180.59	± 0.40	± 0.48	$^{206}Pb/^{238}U$
J28	J8				AR08AS-16; tuff	Serrucho Creek, west of Malargüe city, Mendoza prov., Neuquén Basin, Argentina	35°26′32″S, 69°54′12″W	Lower tuff bed within 19.5-m-thick organic-rich shale of lower Tres Esquinas Fm (13.8 mab in measured section)	181.33	± 0.17	± 0.31	$^{206}Pb/^{238}U$
J27					Bed 28	Palquilla section near Tacna in southern Peru.	70.000092°S, 17.616156°W	Pelado Formation	182.13	± 0.081	± 0.22	$^{206}Pb/^{238}U$
J26					Bed 22	Palquilla section near Tacna in southern Peru.	70.000092°S, 17.616156°W	Pelado Formation	181.99	± 0.13	± 0.24	$^{206}Pb/^{238}U$
J25					Ash bed sample AR08-VT56	Chacay Melehue creek, Vega del Tero, Neuquén Basin, Argentina	37°15′9″S, 70°30′25″W	Chachil Limestone, basal Cuyo Group	182.30	± 0.40	± 0.48	$^{206}Pb/^{238}U$
J24					Multiple samples	Dormettingen Quarry, SW Germany.	48.245°N, 8.761°E	Posidonienschiefer; Multiple samples from sedimentary layers below and between the Unterer Stein, Oberer Stein, Inoceramus Bank, and Nagelkalk horizons.	183.00	± 2.00	± 2.20	Re-Os
J23	J7				EP-89-258; quartz monzonite pluton	Spatsizi River map area, Stikine region, NW British Columbia, Canada	~57°N, 129°W	McEwan Creek pluton	183.23	± 0.62	± 0.67	$^{206}Pb/^{238}U$
J22					Bed 2	Palquilla section near Tacna in southern Peru.	70.000092°S, 17.616156°W	Pelado Formation	183.22	± 0.25	± 0.32	$^{206}Pb/^{238}U$
J21					Ash bed SC1T	St. Clair section, 2 km W of Izee, E. Oregon	44°04′N, 119°24′W	Hyde Formation, Izee terrane, Blue Mountains tectonic province	184.02	± 0.05	± 0.21	$^{206}Pb/^{238}U$
					Pliensbachian							
J20					Ash bed WS13T	Sterrett section, 2 km W of Izee, E. Oregon	44°04′N, 119°24′W	Nicely Formation, Izee terrane, Blue Mountains tectonic province	184.36	± 0.07	± 0.21	$^{206}Pb/^{238}U$
J19					Ash bed SC11T	St. Clair section, 2 km W of Izee, E. Oregon	44°04′N, 119°24′W	Nicely Formation, Izee terrane, Blue Mountains tectonic province	184.55	± 0.04	± 0.21	$^{206}Pb/^{238}U$
J18					Ash bed SC19T	St. Clair section, 2 km W of Izee, E. Oregon	44°04′N, 119°24′W	Nicely Formation, Izee terrane, Blue Mountains tectonic province	185.18	± 0.07	± 0.21	$^{206}Pb/^{238}U$
J17					Ash bed SC28T	St. Clair section, 2 km W of Izee, E. Oregon	44°04′N, 119°24′W	Nicely Formation, Izee terrane, Blue Mountains tectonic province	185.47	± 0.04	± 0.21	$^{206}Pb/^{238}U$
J16	J6				IWE-95-111; rhyolite tuff	ridges west of Skinhead Lake near the town of Granisle, NW British Columbia, Canada (UTM 6085600N 675060E)	54.9°N, 126.5°W	Rhyolite tuff in Hazelton Group	184.75	± 0.60	± 0.63	$^{206}Pb/^{238}U$
J15	J5				GGAJ-92-127; lithic crystal lapilli tuff	South side of Copper Island, Atlin Lake, NW British Columbia, Canada	59.4°N, 133.8°W	Laberge Group	185.68	± 0.19	± 0.33	$^{206}Pb/^{238}U$
J14					bentonite, +2.35 m	East Tributary of Bighorn Creek, Alberta, Canada	51°43′49″N, 115°31′51″W	2.35 m above base of section, in Red Deer Member, Fernie Formation	185.49	± 0.25	± 0.32	$^{206}Pb/^{238}U$
J13					Ash bed SC_47T	St. Clair section, 2 km W of Izee, E. Oregon	44°04′N, 119°24′W	Nicely Formation, Izee terrane, Blue Mountains tectonic province	186.52	± 0.16	± 0.26	$^{206}Pb/^{238}U$
J12					Ash bed Rb_50	Rosebud section, 2 km W of Izee, E. Oregon	44°08′N, 119°19′W	Nicely Formation, Izee terrane, Blue Mountains tectonic province	186.71	± 0.06	± 0.21	$^{206}Pb/^{238}U$

Primary radioisotopic age details	Zonal range assignment			Biostratigraphy	Reference
	clade	zonation	Zone		
Weighted mean of youngest three (of six) single zircon grains, utilizing CA-TIMS and the ET535 spike	ammonite	Tethyan	*Gr. thouarsense* Zone	Above strata containing *Yakounia* ammonites, interpreted as equivalent to *Gr. thouarsense* Standard Zone	Sell et al. (2014); Al-Suwaidi et al. (2016)
Weighted mean of 6 single and multi-grain zircon fractions, utilizing physical abrasion and in-house GSC spike (0.1% error in tracer calibration).	ammonite	Tethyan	upper *Hild. bifrons* to lower *Haugia variabilis*	North American *Crassicosta* ammonite zone is interpreted to correlate to the upper *Hild. bifrons*–lower *Haugia variabilis* ammonite zones of NW Europe	Pálfy et al. (1997); Al-Suwaidi et al. (2016)
Weighted mean of 8 (of 9) multi-grain zircon fractions, utilizing either physical or chemical abrasion and in-house Oslo spike (0.1% error in tracer calibration).	ammonite	Tethyan	*Hild. bifrons* to *H. variabilis* Zone	Andean *Dact. hoelderi* Zone to *P. pacificum* Zone is interpreted to be equivalent to a range from the *Harp. serpentinum* to *Hild. bifrons* Standard Zones	Mazzini et al. (2010); Al-Suwaidi et al. (2016)
Weighted mean of 5 multi-grain zircon fractions, utilizing either physical or chemical abrasion and in-house Oslo spike (0.1% error in tracer calibration).	ammonite	Tethyan	*Hild. bifrons* to *H. variabilis* Zone	Andean *Dact. hoelderi* Zone to *P. pacificum* Zone is interpreted to be equivalent to a range from the *Harp. serpentinum* to *Hild. bifrons* Standard Zones	Mazzini et al. (2010); Al-Suwaidi et al. (2016)
Weighted mean of youngest six (of nine) single zircon grains, utilizing CA-TIMS and the ET535 spike	ammonite	Tethyan	*Hild. bifrons* Zone?	Within strata containing *Porpoceras* ammonites, interpreted as equivalent to *Hild. bifrons* Standard Zone	Sell et al. (2014); Al-Suwaidi et al. (2016)
Weighted mean of youngest three (of 10) single zircon grains, utilizing CA-TIMS and the ET535 spike	ammonite	Tethyan	*Hild. bifrons* Zone?	Within strata containing *Porpoceras* ammonites, interpreted as equivalent to *Hild. bifrons* Standard Zone	Sell et al. (2014); Al-Suwaidi et al. (2016)
Weighted mean of 5 (of 7) single or multi-grain zircon fractions, utilizing either physical or chemical abrasion and in-house Oslo spike (0.1% error in tracer calibration).	ammonite	Tethyan	*Harp. serpentinum*?	Tuff is in the upper Chachil Limestone, and below a level in lower Los Molles Formation with ammonites of *hoelderi* Zone, interpreted to be equivalent to the *Harp. serpentinum* Zone of NW Europe	Leanza et al. (2013); Al-Suwaidi et al. (2016)
The Dormettingen shale suite (n = 14) yields a model 3 isochron age (Excluding the most bioturbated samples 1-BGM, 3- SGS, and 4-AGM).	ammonite	Tethyan	*Dact. tenuicostatum* to *Hild. bifrons* zones	Shale samples included in isochron include horizons spanning the *Dact. tenuicostatum* to *Hild. bifrons* zones	van Acken et al. (2019)
Weighted mean of one zircon and two titanite multi-grain fractions, utilizing physical abrasion and in-house GSC spike (0.1% error in tracer calibration). Analytical error expanded to accommodate geologic scatter.	ammonite	Tethyan	*Dact. tenuicostatum*–*Hild. bifrons* ammonite zone interval?	Intrudes volcanics with intercalated sediments of Early Toarcian age; pluton is perhaps co-magmatic with volcanism.	Evenchick and McNicoll (1993)
Weighted mean of youngest four (of five) single zircon grains, utilizing CA-TIMS and the ET535 spike	ammonite	Tethyan	*Dact. tenuicostatum* Zone/ *Harp. serpentinum* Zone boundary	Lies above *Dactylioceras kanense* and below *Hildaites striatus*	Sell et al. (2014); Al-Suwaidi et al. (2016)
Weighted mean of 7 single grain analyses, excluding 3 older grains; utilizing CA-TIMS and the ET535 spike.	ammonite	Tethyan	*Dact. tenuicostatum*?	Occurs above highest Plienbachian ammonites in section and possibly equivalent to the NW European *tenuicostatum* Zone (Toarcian)	De Lena et al. (2019)
Weighted mean of 4 single grain analyses, excluding 2 older grains; utilizing CA-TIMS and the ET535 spike.	ammonite	North America	*carlottense* Zone	Interpreted to be equivalent to *spinatum* Zone of NW Europe	De Lena et al. (2019)
Weighted mean of 5 single grain analyses, excluding 4 older grains; utilizing CA-TIMS and the ET535 spike.	ammonite	North America	*carlottense* Zone	Interpreted to be equivalent to *spinatum* Zone of NW Europe	De Lena et al. (2019)
Weighted mean of 5 single grain analyses, excluding 4 older grains; utilizing CA-TIMS and the ET535 spike.	ammonite	North America	*carlottense* Zone/*kunae* Zone boundary	Interpreted to be equivalent to *spinatum* Zone *margaritatus* Zone boundary of NW Europe	De Lena et al. (2019)
Weighted mean of 4 single grain analyses, excluding 5 older grains; utilizing CA-TIMS and the ET535 spike.	ammonite	North America	*kunae* Zone	Interpreted to be equivalent to *margaritatus* Zone (*gibbosus* Subzone?) of NW Europe	De Lena et al. (2019)
Weighted mean of 4 multi-grain zircon fractions, utilizing physical abrasion and in-house GSC spike (0.1% error in tracer calibration). Analytical error expanded to accommodate geologic scatter.	ammonite	North America	*Fuc. lavinianum* (lower half)–*Emac. emaciatum* (upper half) zones = *kunae* zone	Interpreted to be equivalent to *margaritatus* Zone of NW Europe	Pálfy et al. (2000)
Weighted mean of 4 multi-grain zircon fractions, utilizing physical abrasion and in-house GSC spike (0.1% error in tracer calibration).	ammonite	North America	*Fuc. lavinianum* (lower half)–*Emac. emaciatum* (upper half) zones=*kunae* Zone	Interpreted to be equivalent to *margaritatus* Zone of NW Europe	Johannson and McNicoll (1997)
Three of five single zircon grain analyses (excluding two older grains) have a weighted mean $^{206}Pb/^{238}U$ age; utilizing CA-TIMS and the ET535 spike.	ammonite	North America	*kunae* Zone	Interpreted to be equivalent to *margaritatus* Zone of NW Europe	Them et al. (2017)
Weighted mean of 6 single grain analyses, excluding 2 older grains; utilizing CA-TIMS and the ET535 spike.	ammonite	North America	*kunae* Zone	Interpreted to be equivalent to *margaritatus* Zone (*subnodosus/gibbosus* subzones boundary?) of NW Europe	De Lena et al. (2019)
Weighted mean of 3 single grain analyses, excluding 3 older grains; utilizing CA-TIMS and the ET535 spike.	ammonite	North America	*kunae* Zone	Interpreted to be equivalent to *margaritatus* Zone (*subnodosus/gibbosus* subzone boundary?) of NW Europe	De Lena et al. (2019)

GTS 2020 ID	GTS 2012 ID	Period	Epoch	Age	Sample	Locality	Lat-Long	Lithostratigraphy	Age (Ma)	± 2s analytical	± 2s total	Age Type
J11					Ash bed SC57T	St. Clair section, 2 km W of Izee, E. Oregon	44°04′N, 119°24′W	Suplee Formation, Izee terrane, Blue Mountains tectonic province	186.82	± 0.04	± 0.21	$^{206}Pb/^{238}U$
J10					Ash bed Rb_22	Rosebud, 2 km W of Izee, E. Oregon	44°08′N, 119°19′W	Suplee Formation, Izee terrane, Blue Mountains tectonic province	186.96	± 0.07	± 0.22	$^{206}Pb/^{238}U$
J9					bentonite, −1.9 m	East Tributary of Bighorn Creek, Alberta, Canada	51°43′44″N, 115°31′51″W	1.9 m below base of section, in Red Deer Member, Fernie Formation	188.58	± 0.25	± 0.32	$^{206}Pb/^{238}U$
J8					Ash bed sample AR08-VT14	Chacay Melehue creek, Vega del Tero, Neuquén Basin, Argentina	37°15′9″S, 70°30′25″W	Chachil Limestone, basal Cuyo Group	185.70	± 0.40	± 0.48	$^{206}Pb/^{238}U$
J7					Ash bed samples C-09-1 and C-09-2	Mirador del Chachil, Neuquén Basin, Argentina	39°1317″S, 70°33′37″W	Chachil Limestone, basal Cuyo Group	186.00	± 0.40	± 0.48	$^{206}Pb/^{238}U$
				Sinemurian								
J6	J4				93JP48; rhyolite flow	southern Telkwa Range, Smithers map area, British Columbia (UTM 6025080N, 622200E)	~56°N, 127°W	Telkwa Fm, Hazelton Group	191.16	± 0.72	± 0.74	$^{206}Pb/^{238}U$
J5	J3				LM4-19B; tuff	Utcubamba valley, northern Peru; on fresh road section from Levanto to Maino.	6°18′26.8″S, 77°53′7.1″W	Ash bed L-19 at ca. 130 m in their section of Aramachay Fm of Pucara Group	199.53	± 0.22	± 0.30	$^{206}Pb/^{238}U$
				Hettangian								
J4					LM4-155/182; volcanic tuff	Utcubamba valley, northern Peru; on fresh road section from Levanto to Maino.	6°18′26.8″S, 77°53′7.1″W	Ash bed at ca. 100 m in their section of Aramachay Fm of Pucara Group	200.35	± 0.13	± 0.22	$^{206}Pb/^{238}U$
J3					LM4-117/118; volcanic tuff	Utcubamba valley, northern Peru; on fresh road section from Levanto to Maino.	6°18′26.8″S, 77°53′7.1″W	Ash bed at ca. 77 m in their section of Aramachay Fm of Pucara Group	200.67	± 0.21	± 0.30	$^{206}Pb/^{238}U$
J2	J2				LM4-100/101; volcanic tuff	Utcubamba valley, northern Peru; on fresh road section from Levanto to Maino.	6°18′26.8″S, 77°53′7.1″W	Ash bed at ca. 68 m in their section, approximately 5 1/2 m above sample LM4-90, in Aramachay Fm of Pucara Group	201.32	± 0.13	± 0.22	$^{206}Pb/^{238}U$
J1	J1				NYC-N10; volcanic tuff	New York Canyon, Nevada, USA	38.5°N, 118.1°W	New York Canyon section [stratigraphy, ammonites, carbon-isotopes in Guex et al. (2004)]; which was candidate for base-Triassic GSSP.	201.33	± 0.17	± 0.27	$^{206}Pb/^{238}U$
		Triassic										
			Late									
				Rhaetian								
T39	T24				LM4-90; volcanic tuff	Utcubamba valley, northern Peru; on fresh road section from Levanto to Maino.	6°18′26.8″S, 77°53′7.1″W	Ash bed 90 at ca. 65 m in their section of Aramachay Fm of Pucara Group.	201.39	± 0.14	± 0.25	$^{206}Pb/^{238}U$
T38	T23				LM4-86; volcanic tuff	Utcubamba valley, northern Peru; on fresh road section from Levanto to Maino.	6°18′26.8″S, 77°53′7.1″W	Ash bed 86 at ca. 59 m level in their section of Aramachay Fm of Pucara Group.	201.51	± 0.15	± 0.24	$^{206}Pb/^{238}U$

Primary radioisotopic age details	Zonal range assignment			Biostratigraphy	Reference
	clade	zonation	Zone		
Weighted mean of 5 single grain analyses, excluding 5 older grains; utilizing CA-TIMS and the ET535 spike.	ammonite	North America	*kunae Zone*	Interpreted to be equivalent to *margaritatus* Zone (*subnodosus/gibbosus* subzones boundary?) of NW Europe	De Lena et al. (2019)
Weighted mean of 4 single grain analyses, excluding 4 older grains; utilizing CA-TIMS and the ET535 spike.	ammonite	North America	*kunae Zone*	Interpreted to be equivalent to *margaritatus* Zone (*subnodosus/gibbosus* subzones boundary?) of NW Europe	De Lena et al. (2019)
Three of five single zircon grain analyses (excluding two older grains) have a weighted mean $^{206}Pb/^{238}U$ age; utilizing CA-TIMS and the ET535 spike.	ammonite	North America	*kunae Zone*	Interpreted to be equivalent to *margaritatus* Zone of NW Europe	Them et al. (2017)
Weighted mean of 2 (of 11) single multi-grain zircon fractions, utilizing either physical or chemical abrasion and in-house Oslo spike (0.1% error in tracer calibration).	ammonite	Andean	*Austromorphites behrendseni*	lower Chachil Limestone contains *Austromorphites behrendseni* Assemblage Zone (interpreted as equivalent to Tethyan *Prodactylioceras davoei* Standard Zone)	Leanza et al. (2013)
Weighted mean of 5 (of 14) single grain analyses, utilizing chemical abrasion and in-house Oslo spike (0.1% error in tracer calibration).	ammonite	Andean	*Austromorphites behrendseni*	lower Chachil Limestone contains *Austromorphites behrendseni* Assemblage Zone (interpreted as equivalent to Tethyan *Prodactylioceras davoei* Standard Zone)	Leanza et al. (2013)
Weighted mean of 3 multi-grain zircon fractions, utilizing physical abrasion and in-house GSC spike (0.1% error in tracer calibration). Analytical error expanded to accommodate geologic scatter.	ammonite	North America	late Sinemurian (*O. oxynotum*−*E. raricostatum* zone interval?)	Sample from 100 m of epiclastic and pyroclastic strata above sediments with ammonite assemblages characteristic of Late Sinemurian *Plesechioceras? harbledownense* ammonite zone of western North America	Pálfy et al. (2000)
Six (of 9) single zircon grain analyses yielded a weighted mean $^{206}Pb/^{238}U$ age, utilizing CA-TIMS and in-house UNIGE spike (0.1% error in tracer calibration).	ammonite	Peruvian	*Badouxia canadensis* ammonite zone	Within *Badouxia canadensis* ammonite beds, with a level with *Angulaticeras* genus about 8 m below and the lowest *Vermiceras densicostatum* and *Metophioceras* occurrence about 12 m above.	Schaltegger et al. (2008), Guex et al. (2012)
Eight single zircon grains analyzed with CA-TIMS and ET2535 spike illustrate excess scatter attributed to a range of growth histories prior to eruption; the authors estimate the eruption age from youngest robust grain analysis.	ammonite	Peruvian	*Kammerkarites* ammonite zone	In upper part of local occurrence of ammonite *Kammerkarites* (middle Hettangian); equivalent to *Saxoceras* zone in NW Europe scale.	Guex et al. (2012)
Seven single zircon grains analyzed with CA-TIMS and ET2535 spike illustrate excess scatter attributed to a range of growth histories prior to eruption; the authors estimate the eruption age from youngest robust grain analysis.	ammonite	Peruvian	*Psiloceras tilmanni* ammonite zone	At the local lowest occurrence of ammonite *Psiloceras tilmanni* (early Hettangian)	Guex et al. (2012)
Fourteen single zircon grains analyzed with CA-TIMS and ET2535 spike illustrate excess scatter attributed to a range of growth histories prior to eruption; originally published as 201.29 ± 0.19/0.28; then revised in Wotzlaw et al. (2014b) as 201.32 ± 0.13 Ma.	ammonite	Peruvian	*Psiloceras spelae* ammonite zone (lower half)	ca. 2.5 m above lowest occurrence of Hettangian genus *Psiloceras (Psiloceras spelae)*; which is presumed to be equivalent to the base-Jurassic GSSP level in Austria.	Schoene et al. (2010); Wotzlaw et al. (2014b)
Seventeen single zircon grains analyzed with CA-TIMS and ET2535 spike. Estimate of eruption age from youngest robust grain at 201.33 ± 0.19 Ma; weighted mean of 9 youngest robust grains is statistically equivalent at 201.45 ± 0.12 Ma.	ammonite	Peruvian	*Psiloceras spelae* ammonite zone (lower half)	1.5 m above the lowest occurrence of Hettangian genus *Psiloceras (Psiloceras spelae)*; which is presumed to be equivalent to the base-Jurassic GSSP level in Austria.	Schoene et al. (2010); Wotzlaw et al. (2014b)
Fourteen single zircon grains analyzed with CA-TIMS and ET2535 spike include Proterozoic xenocrysts, a cluster of 9 equivalent grains and one young analysis attributed to Pb-loss; the authors estimate the eruption age from most precise, young, robust grain, although a weighted mean of nine grains is statistically equivalent.	ammonite	Tethyan	Top-Triassic ammonite zone gap	Mid-way between highest local occurrence of the topmost Triassic ammonite genus *Choristoceras* (*Choristo. crickmayi*, 3 m below) and lowest occurrence of Hettangian genus *Psiloceras* (*Psiloceras spelae*, about 3 m higher) => middle of "extinction interval" of end-Triassic. Tri/Jur boundary projected as about the 64 m level in section (FAD of *Psiloceras spelae*).	Schoene et al. (2010); Wotzlaw et al. (2014b)
Fourteen single zircon grains analyzed with CA-TIMS and ET2535 spike illustrate excess scatter; the authors estimate the eruption age from youngest robust grain, although a weighted mean of youngest three robust grains is statistically equivalent.	ammonite	Tethyan	Top-Triassic ammonite zone gap	One meter above the highest local occurrence of the topmost Triassic ammonite genus *Choristoceras* (*Choristo. crickmayi*, in Bed 84b, 59.5 m level), and 5 m below the Hettangian genus *Psiloceras* (*Psiloceras spelae*, in bed 93b, 64.5 m level) and about 18 m below *Psiloceras tilmanni* (Bed 114-129, lowest at 77.5 m level).	Schoene et al. (2010); Wotzlaw et al. (2014b)

GTS 2020 ID	GTS 2012 ID	Period	Epoch	Age	Sample	Locality	Lat-Long	Lithostratigraphy	Age (Ma)	± 2s analytical	± 2s total	Age Type
T37					LM4-76/77; volcanic tuff	Utcubamba valley, northern Peru; on fresh road section from Levanto to Maino.	6°18′26.8″S, 77°53′7.1″W	Ash bed at ca. 51 m level in their section of Aramachay Fm of Pucara Group.	201.87	± 0.17	± 0.25	$^{206}Pb/^{238}U$
T36					LM4-58/59; volcanic tuff	Utcubamba valley, northern Peru; on fresh road section from Levanto to Maino.	6°18′26.8″S, 77°53′7.1″W	Ash bed at ca. 35 m level in their section of Aramachay Fm of Pucara Group.	202.16	± 0.27	± 0.30	$^{206}Pb/^{238}U$
T35					LM4-50/51; volcanic tuff	Utcubamba valley, northern Peru; on fresh road section from Levanto to Maino.	6°18′26.8″S, 77°53′7.1″W	Ash bed at ca. 29 m level in their section of Aramachay Fm of Pucara Group.	202.38	± 0.16	± 0.25	$^{206}Pb/^{238}U$
T34					LM4-23B; volcanic tuff	Utcubamba valley, northern Peru; on fresh road section from Levanto to Maino.	6°18′26.8″S, 77°53′7.1″W	Ash bed at ca. 9 m level in their section of Aramachay Fm of Pucara Group.	203.71	± 0.11	± 0.24	$^{206}Pb/^{238}U$
				Norian								
T33					LP2010-1d	Utcubamba valley, northern Peru; on fresh road section from Levanto to Maino.	6°18′26.8″S, 77°53′7.1″W	Ash bed at ca.—32 m level in their section of Aramachay Fm of Pucara Group.	205.35	± 0.19	± 0.30	$^{206}Pb/^{238}U$
T32					LP2010-1b	Utcubamba valley, northern Peru; on fresh road section from Levanto to Maino.	6°18′26.8″S, 77°53′7.1″W	Ash bed at ca.—32 m level in their section of Aramachay Fm of Pucara Group.	205.39	± 0.09	± 0.24	$^{206}Pb/^{238}U$
T31					LP2010-3a	Utcubamba valley, northern Peru; on fresh road section from Levanto to Maino.	6°18′26.8″S, 77°53′7.1″W	Ash bed at ca.—39 m level in their section of Aramachay Fm of Pucara Group.	205.70	± 0.15	± 0.27	$^{206}Pb/^{238}U$
				Carnian								
T30	T22				Aglianico tuff bed	Pignola, southern Apennines, southern Italy	40.5°N, 15.8°E	"A 5-cm-thick volcanic ash bed (here named Aglianico) occurs within the hemipelagic to pelagic Calcari con Selce Fm (i.e., cherty limestones) of the Pignola 2 section". Unusual clay-bed (no carbonate) 3 m below is "Carnian pluvial event" that coincides with demise of rimmed carbonate platforms in the Tethys.	230.91	± 0.13	± 0.28	$^{206}Pb/^{238}U$
			Middle									
				Ladinian								
T29	T21				Alpe di Siusi ash bed	Rio Nigra section, Alpe di Siusi/ Seiser Alm area, Dolomites, northern Italy	45°31.0′N, 11°35.0′E	Frommer Member, Fernazza Fm	237.85	± 0.05	± 0.14	$^{206}Pb/^{238}U$
T28	T20				FP2; volcaniclastic sandstone	Flexenpass, eastern Vorarlberg, westernmost Northern Calcareous Alps, Austria	47°08′35″N, 10°10′01″E	Graded volcaniclastic layer in the upper part (44.3 m level in section) of the Reifling Fm	239.34	± 0.28	± 0.38	$^{206}Pb/^{238}U$
T27	T19				Litér, volcanic ash matrix of neptunian dyke	Litér dolomite quarry, Balaton Highlands, Hungary	47°05.9′N, 18°0.4′E	Neptunian dyke into Tagyon Dolomite	238.58	± 0.65	± 0.70	$^{206}Pb/^{238}U$

Primary radioisotopic age details	Zonal range assignment			Biostratigraphy	Reference
	clade	zonation	Zone		
Utilizing CA-TIMS and in-house UNIGE spike (0.1% error in tracer calibration). Originally published as 201.78 ± 0.16 in Guex et al. (2012); then revised by them in Wotzlaw et al. (2014b) as 201.87 ± 0.17 Ma.	ammonite	Tethyan	*Vandaites saximontanum* to *Choristoceras crickmayi*	Between *Vandaites saximontanum* (mid-Rhaetian) and *Choristoceras crickmayi* (late Rhaetian)	Guex et al. (2012); Wotzlaw et al. (2014b)
Utilizing CA-TIMS and in-house UNIGE spike (0.1% error in tracer calibration). Originally published as 202.10 ± 0.27 in Guex et al. (2012); then revised by them in Wotzlaw et al. (2014b) as 202.16 ± 0.27 Ma	ammonite	Tethyan	*Vandaites saximontanum* to *Choristoceras crickmayi*	Between *Vandaites saximontanum* (mid-Rhaetian) and *Choristoceras crickmayi* (late Rhaetian)	Guex et al. (2012); Wotzlaw et al. (2014b)
Utilizing CA-TIMS and in-house UNIGE spike (0.1% error in tracer calibration). Originally published as 202.32 ± 0.16 in Guex et al. (2012); then revised by them in Wotzlaw et al. (2014b) as 202.38 ± 0.16 Ma.	ammonite	Tethyan	*Vandaites saximontanum* to *Choristoceras crickmayi*	Between *Vandaites saximontanum* (mid-Rhaetian) and *Choristoceras crickmayi* (late Rhaetian)	Guex et al. (2012); Wotzlaw et al. (2014b)
Utilizing CA-TIMS and in-house UNIGE spike (0.1% error in tracer calibration). Originally published as 203.62 ± 0.10 in Guex et al. (2012); then revised by them in Wotzlaw et al. (2014b) as 203.71 ± 0.11 Ma.	Ammonite	Tethyan	*Vandaites saximontanum* ammonoid zone	Ca. 5 m above occurrence of *Vandaites saximontanum* (mid-Rhaetian)	Guex et al. (2012); Wotzlaw et al. (2014b)
Youngest six (of 14) single zircon grains, utilizing CA-TIMS and in-house UNIGE spike (0.1% error in tracer calibration).	bivalve	Tethyan	Norian-Rhaetian transition interval	5 m above sample 3a. Occurs "an abundant fauna of *Otapiria*, which are closely related to *Otapiria norica* known from the late Norian *G. cordilleranus* ammonoid zone of northeastern British Columbia (Meyers et al., 2011)." (Wotzlaw et al., 2014b)	Guex et al. (2012); Wotzlaw et al. (2014b)
Date of the youngest zircon (n = 14) consistent with respect to its stratigraphic position relative the other two ash bed samples, utilizing CA-TIMS and in-house UNIGE spike (0.1% error in tracer calibration).	bivalve	Tethyan	Norian-Rhaetian transition interval	5 m above sample 3a. There is "an abundant fauna of *Otapiria*, which are closely related to *Otapiria norica* known from the late Norian *G. cordilleranus* ammonoid zone of northeastern British Columbia (McRoberts, 2011)." (Wotzlaw et al., 2014b)	Guex et al. (2012); Wotzlaw et al. (2014b)
Youngest 8 (of 15) single zircons, utilizing CA-TIMS and in-house UNIGE spike (0.1% error in tracer calibration).	bivalve	Tethyan	Top of *Monotis subcircularis* bivalve zone	Occurs with bivalve *Oxytoma* cf. *O. inaequivalvis*. About 5 m above highest occurrence of "large monotid bivalves, identified as *Monotis subcircularis* ... which occur in hard gray and massive silty limestones at the base of the section. *M. subcircularis* and other large monotid bivalves underwent nearly complete extinction at the top of the Norian, correlative with the top of the *Sagenites quinquepunctatus* ammonoid zone in the Tethys and the top of the *Gnomohalorites cordilleranus* ammonoid zone in the Americas.	Guex et al. (2012); Wotzlaw et al. (2014b)
Eight single zircon grains yield a weighted mean $^{206}Pb/^{238}U$ age utilizing CA-TIMS and in-house MIT spike.	conodont	Tethyan	Assigned in text to either *P. carpathicus* or *E. nodosa* zone of Channell et al. (2003) or to lower and middle part of the *M. nodosus* zone of Orchard (1991).	Just below FAD of conodont *P. carpathicus* and within total range of conodont *M. nodosus*, but they indicate FAD of this taxa varies among sections and workers. Mid-Upper Carnian is suggested by "Lower Carnian conodonts (*Glad.* spp., *N. postkockeli*, *P. inclinata*, *P. tadpole*) abruptly disappear 3 m below the ash bed at a level corresponding to the Carnian pluvial event" and "Carnian-Norian boundary can be placed 6 m above the ash bed with the first occurrence of *M. communisti*."	Furin et al. (2006)
Seven single zircon grains pre-treated by the chemical abrasion (CA-TIMS)	ammonite	Tethyan	*Protrachyceras neumayri* Zone	"close to the *neumayri/regoledanus* subzones"	Mietto et al. (2012)
Nine single zircon grain analyses yielded a weighted mean $^{206}Pb/^{238}U$ age, utilizing CA-TIMS and in-house BGC spike (0.1% error in tracer calibration)	ammonite	Tethyan	*Protrachyceras longobardium* Zone	About 2 m below level with bivalve *Daonella tyrolensis*, and about 16 m above level with *Daonella* cf. *longobardica*. The presence of *Protrachyceras steinmanni* at a similar stratigraphic level in the correlative Bagolino section suggests that this age corresponds to the *P. longobardicum* zone.	Brühwiler et al. (2007); Mietto et al. (2012)
Two (of 5) multi-grain zircon analyses yielded a weighted mean $^{206}Pb/^{238}U$ age, utilizing mechanical	ammonite	Tethyan	*Protrachyceras gredleri* to *P. archelaus* zone	Redeposited tuff in neptunian dyke that contains ammonite assemblage assigned to upper *Protrachyceras gredleri* ammonite zone. Pálfy et al. (2003) schematically	Pálfy et al. (2003)

GTS 2020 ID	GTS 2012 ID	Period	Epoch	Age	Sample	Locality	Lat-Long	Lithostratigraphy	Age (Ma)	± 2s analytical	± 2s total	Age Type
T26					SEC-G	Seceda, NW Dolomites, N. Italy	45°35.9′N, 11°43.5′E	Base of upper Pietra Verde interval of Buchenstein Fm	239.04	± 0.04	± 0.25	$^{206}Pb/^{238}U$
T25					SEC-E	Seceda, NW Dolomites, N. Italy	45°35.9′N, 11°43.5′E	Prominent lapilli layer in Pietra Verde interval of Buchenstein Fm	240.29	± 0.04	± 0.25	$^{206}Pb/^{238}U$
T24					FEO-12	Passo Feudo, Dolomites, N. Italy	46°20′27.0″N, 11°33′32.2″E	Ash layer from "upper part of the Knollenkalke interval suspected to correlate with the middle Pietra Verde interval" at Seceda	240.43	± 0.05	± 0.26	$^{206}Pb/^{238}U$
T23					SEC-D	Seceda, NW Dolomites, N. Italy	45°35.9′N, 11°43.5′E	Lowest prominent ash layer of the middle Pietra Verdi interval, directly overlying the cyclic Knollenkalke interval of Buchenstein Fm	240.58	± 0.04	± 0.25	$^{206}Pb/^{238}U$
T22			*Anisian*									
T21					SEC-B	Seceda, NW Dolomites, N. Italy	45°35.9′N, 11°43.5′E	Tc tuff from basal Knollenkalke interval of Buchenstein Fm	241.71	± 0.05	± 0.29	$^{206}Pb/^{238}U$
T20					FEO-2	Passo Feudo, Dolomites, N. Italy	46°20′27.0″N, 11°33′32.2″E	Ash layer from "basal Plattenkalke interval close to the base of the Buchenstein Fm"	242.01	± 0.05	± 0.26	$^{206}Pb/^{238}U$
T19	T18				MSG.09; volcanic ash	Miniera Val Porina, 820 m elevation, Monte San Giorgio, Lake Lugano, Switzerland	45°54.2′N, 8°57.1′E	Tuff in bituminous shales (Grenzbitumen horizon), Bed 71	242.14	± 0.45	± 0.52	$^{206}Pb/^{238}U$
T18	T15				LAT31; tuff	Cimon Latemar, Dolomites, N. Italy.	46°35′N, 11°22′E	Lower Cyclic Facies (lower-middle)	242.65	± 0.76	± 0.80	$^{206}Pb/^{238}U$
T17	T14				LAT30; tuff	Cimon Latemar, Dolomites, N. Italy.	46°35′N, 11°22′E	Middle Tepee Facies (middle)	242.80	± 0.34	± 0.42	$^{206}Pb/^{238}U$
T16	T12				CHIN-34; volcanic ash	Jinya section, Nanpanjiang Basin, NW Guangxi, south China	24°35′22.0″N, 106°53′13.6″E	Baifeng Fm (shaly base of Fm)	244.60	± 0.43	± 0.50	$^{206}Pb/^{238}U$
T15	T11				CHIN-29; volcanic ash	Jinya section, Nanpanjiang Basin, NW Guangxi, south China	24°35′25.8″N, 106°52′09.7″E	Uppermost part of "Transitional beds" between Luolou and Baifeng Fms	246.80	± 0.39	± 0.47	$^{206}Pb/^{238}U$
T14	T10				GDGB-0; 7-m thick tuffaceous siliciclastic mudstone	Upper Guandao section, Great Bank of Guizhou, Nanpanjiang Basin, South China	25°39′17″N, 106°51′13″E	Luolou Fm, 17.0-m above base of section; 7.0 m above the FOD of *Chiosella timorensis* at 10.0-m	246.30	± 0.14	± 0.29	$^{206}Pb/^{238}U$
T13	T9				GDGB Tuff-110	Lower Guandao section, Great Bank of Guizhou, Nanpanjiang Basin, South China	25°39′17″N, 106°51′13″E	Luolou Fm, 260.25-m above base of section; 17.25 m above the FOD of *Chiosella timorensis* at 243.0-m	246.50	± 0.11	± 0.30	$^{206}Pb/^{238}U$

Primary radioisotopic age details	Zonal range assignment			Biostratigraphy	Reference
	clade	zonation	Zone		
abrasion and in-house NIGL spike (0.1% error in tracer calibration).				show biostratigraphic uncertainty as spanning upper-quarter of *P. gredleri* Zone through lower-quarter of *P. archelaus* Zone.	
Weighted mean of younger cluster (six of 10 analyzed) single-grain zircon analyses using CA-ID-TIMS method.	ammonite	Tethyan	ca. base of *P. archelaus* Zone	"upper part of the Buchenstein succession the correlation is corroborated by the sequential occurrence of three different species or groups of *Daonella* (*D. pichleri*, *D. tyrolensis*, *D. lommeli* group)"	Wotzlaw et al. (2018)
Weighted mean of eight single-grain zircon analyses using CA-ID-TIMS method.	ammonite	Tethyan	ca. middle *P. gredleri* zone	"In the middle Pietra Verde interval representatives of the ammonoid genus *Arpadites* occur in a small interval straddling the accretionary lapilli-bearing horizon"	Wotzlaw et al. (2018)
Weighted mean of younger cluster (three of 8 analyzed) single-grain zircon analyses using CA-ID-TIMS method.	ammonite	Tethyan	ca. middle *P. gredleri* Zone	"In the middle Pietra Verde interval representatives of the ammonoid genus *Arpadites* occur in a small interval straddling the accretionary lapilli-bearing horizon"	Wotzlaw et al. (2018)
Weighted mean of six (of 7) single-grain zircon analyses using CA-ID-TIMS method.	ammonite	Tethyan	ca. base *P. gredleri* Zone	"In the middle Pietra Verde interval representatives of the ammonoid genus *Arpadites* occur in a small interval straddling the accretionary lapilli-bearing horizon"	Wotzlaw et al. (2018)
Weighted mean of younger cluster (five of 10 analyzed) single-grain zircon analyses using CA-ID-TIMS method.	ammonite	Tethyan	ca. lower *N. secedensis* Zone	"characteristic packages of pelagic beds defined at Seceda are still recognizable at Bagolino and Monte Corona (Fig. 5). Moreover, it allows the GSSP equivalent level to be precisely located at Seceda. Based on this bed-by-bed correlation, we place the boundary level between beds D01 and D02", which is ca. 7 m above this dated sample SEC-B.	Wotzlaw et al. (2018)
Weighted mean of six single-grain zircon analyses using CA-ID-TIMS method.	ammonite	Tethyan	middle *R. reitzi* Zone	middle *Reitzi* Zone	Wotzlaw et al. (2018)
Six (of 8) single zircon grain analyses yielded a weighted mean $^{206}Pb/^{238}U$ age, utilizing CA-TIMS and in-house BGC spike (0.1% error in tracer calibration)	ammonite	Tethyan	ca. lower *N. secedensis* Zone	*Daonella* bivalve species are 150 cm below lowest dated sample. Assigned to lowermost *Nevadites secendensis* ammonoid zone by Brack et al. (1996)	Mundil et al. (2010)
Six (of 9) single zircon grain analyses yielded a weighted mean $^{206}Pb/^{238}U$ age, utilizing HF-leaching and in-house BGC spike (0.1% error in tracer calibration).	ammonite	Tethyan	lower N. secedensis Zone [uppermost zone of Anisian]	About 40 m above base of Lower Cyclic Facies, which overlies *Latemarites* ammonites, a genera with range assigned by them to upper *Reitzi* Zone; therefore, sampled level might correspond to lowermost part of *Nevadites secedensis* ammonoid zone.	Mundil et al. (2003)
Ten single zircon grain analyses yielded a weighted mean $^{206}Pb/^{238}U$ age, utilizing CA-TIMS and in-house BGC spike (0.1% error in tracer calibration).	ammonite	Tethyan	upper *R. reitzi* ammonoid zone	uppermost *Reitzi* Zone	Mundil et al. (2003); Brack et al. (2007)
Four (of 7) single zircon grain analyses yielded a weighted mean $^{206}Pb/^{238}U$ age, utilizing CA-TIMS and in-house UNIGE spike (0.1% error in tracer calibration). Uncertainties recalculated in Galfetti et al. (2007).	ammonite	South China	*Balatonites shoshonensis* ammonoid zone	Bracketed by layers containing an ammonoid assemblage diagnostic of the low-palolatitudinal *Balatonites shoshonensis* Zone.	Ovtcharova et al. (2006)
Six (of 8) single zircon grain analyses yielded a weighted mean $^{206}Pb/^{238}U$ age, utilizing CA-TIMS and in-house UNIGE spike (0.1% error in tracer calibration). Uncertainties recalculated in Galfetti et al. (2007).	ammonite	South China	*Acrochordiceras hyatti* ammonoid zone	Poorly preserved *Platycuccoceras*-dominated ammonoid assemblage, indicates *Acrochordiceras hyatti* ammonite zone	Ovtcharova et al. (2006)
Thirteen (of 40) single zircon grain analyses yielded a weighted mean $^{206}Pb/^{238}U$ age, utilizing both physical abrasion and CA-TIMS, and in-house MIT spike.	conodont	South China	*Chiosella timorensis*	"a short distance above the Olenekian boundary". Nearly at base of "Bith-N" polarity zone (in nomenclature of Ogg et al. (2008)]. FAD of conodont *Chiosella timorensis* (working definition of base-Anisian) is about 3 m below this ash layer, and *Ni. germanica* FAD is at about this level.	Ramezani et al. (2007a, b), Lehrmann et al. (2007)
Two of nine single zircon grain analyses (excluding seven older grains) have a weighted mean $^{206}Pb/^{238}U$ age; utilizing CA-TIMS and the ET535 spike.	conodont	South China	*Chiosella timorensis*	Just above LADs of conodonts *Neogondolella regalis* and *Chiosella timorensis*, middle of range of *Neospathodus germanica*, and just above FAD of conodont *Nicoraella kockeli*. Assigned by authors to lowest part of Pelsonian Substage.	Lehrmann et al. (2015)

GTS 2020 ID	GTS 2012 ID	Period	Epoch	Age	Sample	Locality	Lat-Long	Lithostratigraphy	Age (Ma)	± 2s analytical	± 2s total	Age Type
T12	T8				PGD Tuff-3	Lower Guandao section, Great Bank of Guizhou, Nanpanjiang Basin, South China	25°34′42″N, 106°37′08″E	Luolou Fm, 247.9-m above the base of section; 4.9 m above the FOD of Chiosella timorensis at 243.0-m	247.08	± 0.11	± 0.31	$^{206}Pb/^{238}U$
				Early								
				Olenekian								
T11	T7				PGD Tuff-2	Lower Guandao section, Great Bank of Guizhou, Nanpanjiang Basin, South China	25°34′42″N, 106°37′08″E	Luolou Fm, 239.3-m above base of section; 3.7-m below the FOD of Chiosella timorensis at 243.0-m	247.32	± 0.08	± 0.30	$^{206}Pb/^{238}U$
T10	T6				PGD Tuff-1	Lower Guandao section, Great Bank of Guizhou, Nanpanjiang Basin, South China	25°34′42″N, 106°37′08″E	Luolou Fm, 238.8-m above base of section; 4.2-m below the FOD of Chiosella timorensis at 243.0-m	247.46	± 0.05	± 0.29	$^{206}Pb/^{238}U$
T9	T5				CHIN-23; volcanic ash	Jinya section, Nanpanjiang Basin, NW Guangxi, South China	24°36′48.9″N, 106°52′34.0″E	Luolou Fm (upper carbonate unit)	248.12	± 0.37	± 0.45	$^{206}Pb/^{238}U$
				Induan								
T7	T4				CHIN-10; volcanic ash	Jinya section, Nanpanjiang Basin, NW Guangxi, South China	24°36′26.2″N, 106°52′39.6″E	Luolou Fm (upper carbonate unit)	250.55	± 0.47	± 0.54	$^{206}Pb/^{238}U$
T6	T3				CHIN-40; volcanic ash	Jinya section, Nanpanjiang Basin, NW Guangxi, South China	24°35′23.5″ 106°52′49.6″E	Luolou Fm (lower unit of "thin-bedded, dark, laminated, suboxic limestones alternating with dark, organic-rich shales")	251.22	± 0.20	± 0.42	$^{206}Pb/^{238}U$
T5					Bed 34 bentonite; sample MSB34-2	Meishan section D, the GSSP of the base of the Triassic; Changhsing, Zhejiang Province, S. China	31°07′14″N, 119°42′14″E	Bed 34, Yinkeng Fm	251.50	± 0.06	± 0.29	$^{206}Pb/^{238}U$
T4					Bed 33 bentonite; sample MD99-3u	Meishan section D, the GSSP of the base of the Triassic; Changhsing, Zhejiang Province, S. China	31°07′14″N, 119°42′14″E	Bed 33, Yinkeng Fm	251.58	± 0.09	± 0.29	$^{206}Pb/^{238}U$
T3					bentonite; sample PEN-22	Penglaitan section, Laibin area, Guangxi Province, S. China	23°41′8.4″N, 109°18′21″E	lower Ziyun Fm; PTB is placed at the Fmal boundary between the Permian Dalong Fm and the Triassic Ziyun Fm	251.91	± 0.03	± 0.28	$^{206}Pb/^{238}U$
T2					bentonite; sample DGP-21	Dongpan section, southwestern Guangxi Province, S. China	22°16′11.8″N, 107°41′31.3″E	lower Ziyun Fm; PTB is placed at the Fmal boundary between the Permian Dalong Fm and the Triassic Ziyun Fm	251.95	± 0.04	± 0.28	$^{206}Pb/^{238}U$
T1	T2				Bed 28 bentonite; sample MBE0205	Meishan section D, the GSSP of the base of the Triassic; Changhsing, Zhejiang Province, S. China	31°07′14″N, 119°42′14″E	Bed 28, Yinkeng Fm	251.88	± 0.03	± 0.28	$^{206}Pb/^{238}U$
		Permian										
			Lopingian									
				Changhsingian								
P37	P17				Bed 25 bentonite; sample MBE0205	Meishan section D, the GSSP of the base of the Triassic; Changhsing, Zhejiang Province, S. China	31°07′14″N, 119°42′14″E	The base of Bed 25 is the Fmal boundary between the Changhsing Fm and Yinkeng Fm	251.94	± 0.03	± 0.28	$^{206}Pb/^{238}U$
P36					Bed 22 bentonite; sample MZ96(-4.3)	Meishan section D, the GSSP of the base of the Triassic; Changhsing, Zhejiang Province, S. China	31°07′14″N, 119°42′14″E	Bed 22, Changhsing Fm	252.10	± 0.06	± 0.28	$^{206}Pb/^{238}U$

Primary radioisotopic age details	Zonal range assignment			Biostratigraphy	Reference
	clade	zonation	Zone		
Four of six single zircon grain analyses (excluding two older grains) have a weighted mean $^{206}Pb/^{238}U$ age; utilizing CA-TIMS and the ET535 spike.	conodont	South China	*Chiosella timorensis*	Just above LAD of conodont *Neospathodus symmetricus*, in uppermost part of range of *Neospathodus homeri*, in lowermost part of range of *Chiosella timorensis*, and base of range of *Neogondolella regalis*. Assigned by them as about one-third up in Aegean subzone (they assign base-Aegean, hence base-Anisian, to be the FAD of *Chiosella timorensis* at 267.3-m at the Guandao section).	Lehrmann et al. (2015)
Weighted mean of 11 (of 29) single zircon grain analyses, utilizing both physical abrasion and CA-TIMS, and in-house MIT spike.	conodont	South China	*Neospathodus homeri*	Just above PGD Tuff-1; uppermost part of range of conodont *Neospathodus triangularis*, and in upper part of range of *Neospathodus homeri* (uppermost Spathian).	Lehrmann et al. (2006)
Six of six single zircon grain analyses have a weighted mean $^{206}Pb/^{238}U$ age; utilizing CA-TIMS and the ET535 spike.	conodont	South China	*Neospathodus homeri*	Just below PGD Tuff-2; uppermost part of range of conodont *Neospathodus triangularis*, and in upper part of range of *Neospathodus homeri* (uppermost Spathian).	Lehrmann et al. (2015)
Weighted mean of 6 (of 10) single zircon grain analyses, utilizing CA-TIMS and in-house UNIGE spike (0.1% error in tracer calibration). Uncertainties recalculated in Galfetti et al. (2007).	ammonite	Tethyan	*Neopopanoceras haugi*	Low-paleolatitude *Neopopanoceras haugi* zone ammonite fauna (second ammonite zone down from top of Spathian) is correlated with the high-paleolatitude *Keyserlingites subrobustus* zone.	Ovtcharova et al. (2006)
Weighted mean of 6 (of 8) single zircon grain analyses, utilizing CA-TIMS and in-house UNIGE spike (0.1% error in tracer calibration). Uncertainties recalculated in Galfetti et al. (2007).	ammonite	Tethyan	*Tirolites/Columbites*	*Tirolites/Columbites* ammonite assemblage (second ammonite zone from base of Spathian)	Ovtcharova et al. (2006)
Weighted mean of 6 single zircon grain analyses, utilizing CA-TIMS and in-house UNIGE spike (0.1% error in tracer calibration).	ammonite	Tethyan	*Kashmirites densistriatus*	"within the *Kashmirites densistriatus* beds" of early Smithian", which overlie beds with *Hedenstroemia hedenstroemi* ammonites that are assigned to basal-Smithian (basal Olenekian)	Galfetti et al. (2007)
Eleven of fourteen single zircon grain analyses (excluding three older grains) have a weighted mean $^{206}Pb/^{238}U$ age; utilizing CA-TIMS and the ET2535 spike.	conodont	South China	*Clarkina planata* Zone (*Neoclarkina krystyni-Nc. discreta*)	Bed 34, ca. 7 m above the base of the GSSP Bed 27c at Meishan D section	Burgess et al. (2014)
Nine of nine single zircon grain analyses have a weighted mean $^{206}Pb/^{238}U$ age; utilizing CA-TIMS and the ET2535 spike.	conodont	South China	upper *Isarcicela isarcica* Zone?	The base of Bed 33, a 2-cm volcanic clay, is 175-cm above the base of the GSSP Bed 27c at Meishan D section.	Burgess et al. (2014)
Eight of eleven single zircon grain analyses (excluding one younger and two older grains) have a weighted mean $^{206}Pb/^{238}U$ age; utilizing CA-TIMS and the ET2535 spike.	conodont	South China	*Hindeotus parvus* Zone	The bentonite is 0.5-m above the lithologic formation boundary (PTB), and 0.2-m above the FOD of *H. parvus*.	Baresel et al. (2017)
Eight of fourteen single zircon grain analyses (excluding six older grains) have a weighted mean $^{206}Pb/^{238}U$ age; utilizing CA-TIMS and the ET2535 spike.	conodont	South China	*Hindeotus parvus* Zone	The bentonite is 0.3-m above the lithologic formation boundary (PTB) between bed 12 with diverse Permian fauna and Bed 13 with abundant Griesbachian bivalves and ammonoids.	Baresel et al. (2017)
Thirteen of thirteen single zircon grain analyses have a weighted mean $^{206}Pb/^{238}U$ age; utilizing CA-TIMS and the ET2535 spike.	conodont	South China	*Hindeotus parvus* Zone	The base of Bed 28, a 4-cm volcanic clay, is 8-cm above the base of the GSSP Bed 27c at Meishan D section, corresponding to the first appearance of conodont *Hindeodus parvus*.	Burgess et al. (2014)
Sixteen of sixteen single zircon grain analyses have a weighted mean $^{206}Pb/^{238}U$ age; utilizing CA-TIMS and the ET2535 spike.	conodont	South China	*Clarkina meishanensis* Zone	Bed 25, a 5-cm volcanic clay, is the approximate culmination of the late-Permian mass extinction. The base of this former "boundary clay" level is 19-cm below the base of the GSSP Bed 27c at Meishan D section, corresponding to first appearance of conodont *Hindeodus parvus*.	Burgess et al. (2014)
Twelve of twelve single zircon grain analyses have a weighted mean $^{206}Pb/^{238}U$ age; utilizing CA-TIMS and the ET2535 spike.	conodont	South China	*Clarkina changxingensis* Zone	The bentonite in Bed 22 is 4.3-m below the base of the GSSP Bed 27c at Meishan D section.	Burgess et al. (2014)

GTS 2020 ID	GTS 2012 ID	Period	Epoch	Age	Sample	Locality	Lat-Long	Lithostratigraphy	Age (Ma)	± 2s analytical	± 2s total	Age Type
P35					bentonite; sample PEN-28	Penglaitan section, Laibin area, Guangxi Province, S. China	23°41′8.4″N, 109°18′21″E	upper Dalong Fm; PTB is placed at the Fmal boundary between the Permian Dalong Fm and the Triassic Ziyun Fm	252.06	± 0.04	± 0.28	^{206}Pb/^{238}U
P34					bentonite; sample PEN-70	Penglaitan section, Laibin area, Guangxi Province, S. China	23°41′8.4″N, 109°18′21″E	upper Dalong Fm; PTB is placed at the Fmal boundary between the Permian Dalong Fm and the Triassic Ziyun Fm	252.13	± 0.07	± 0.29	^{206}Pb/^{238}U
P33					bentonite; sample PEN-6	Penglaitan section, Laibin area, Guangxi Province, S. China	23°41′8.4″N, 109°18′21″E	upper Dalong Fm; PTB is placed at the Fmal boundary between the Permian Dalong Fm and the Triassic Ziyun Fm	252.14	± 0.08	± 0.29	^{206}Pb/^{238}U
P32					bentonite; sample DGP-17	Dongpan section, southwestern Guangxi Province, S. China	22°16′11.8″N, 107°41′31.3″E	upper Dalong Fm; PTB is placed at the Fmal boundary between the Permian Dalong Fm and the Triassic Ziyun Fm	251.96	± 0.03	± 0.28	^{206}Pb/^{238}U
P31					bentonite; sample DGP-16	Dongpan section, southwestern Guangxi Province, S. China	22°16′11.8″N, 107°41′31.3″E	upper Dalong Fm; PTB is placed at the Fmal boundary between the Permian Dalong Fm and the Triassic Ziyun Fm	251.98	± 0.04	± 0.28	^{206}Pb/^{238}U
P30					bentonite; sample DGP-13	Dongpan section, southwestern Guangxi Province, S. China	22°16′11.8″N, 107°41′31.3″E	upper Dalong Fm; PTB is placed at the Fmal boundary between the Permian Dalong Fm and the Triassic Ziyun Fm	252.10	± 0.04	± 0.28	^{206}Pb/^{238}U
P29					bentonite; sample DGP-12	Dongpan section, southwestern Guangxi Province, S. China	22°16′11.8″N, 107°41′31.3″E	upper Dalong Fm; PTB is placed at the Fmal boundary between the Permian Dalong Fm and the Triassic Ziyun Fm	252.12	± 0.04	± 0.28	^{206}Pb/^{238}U
P28					bentonite; sample DGP-10	Dongpan section, southwestern Guangxi Province, S. China	22°16′11.8″N, 107°41′31.3″E	upper Dalong Fm; PTB is placed at the Fmal boundary between the Permian Dalong Fm and the Triassic Ziyun Fm	252.17	± 0.06	± 0.28	^{206}Pb/^{238}U
P27					Bed 15 bentonite; sample MD99-15(-17.3)	Meishan section D, the GSSP of the base of the Triassic; Changhsing, Zhejiang Province, S. China	31°07′14″N, 119°42′14″E	Bed 15, Changshing Fm	252.85	± 0.11	± 0.29	^{206}Pb/^{238}U
P26					Bed 7 bentonite; sample MD-7b-96(-36.8)	Meishan section D, the GSSP of the base of the Triassic; Changhsing, Zhejiang Province, S. China	31°07′14″N, 119°42′14″E	Bed 7, Changshing Fm	253.45	± 0.08	± 0.29	^{206}Pb/^{238}U
P25					Bed 6 bentonite; sample MZ96(-38.0)	Meishan section D, the GSSP of the base of the Triassic; Changhsing, Zhejiang Province, S. China	31°07′14″N, 119°42′14″E	Bed 6, Changshing Fm	253.49	± 0.07	± 0.28	^{206}Pb/^{238}U
P24	P14				SH27(23); bentonite	Shangsi section, north Sichuan Province, central China	32°20′N, 105°28′E	Bed 23, upper Dalong Fm	253.24	± 0.36	± 0.45	^{206}Pb/^{238}U
P23	P13				Bed 18 bentonite; sample S03-22 (−12.8 m)	Shangsi section, north Sichuan Province, central China	32°20′N, 105°28′E	Bed 18, upper-middle of Dalong Fm	253.60	± 0.08	± 0.29	^{206}Pb/^{238}U
				Wuchiapingian								
P22	P12				Bed 15 bentonite; sample S03-05 (−27.5 m)	Shangsi section, north Sichuan Province, central China	32°20′N, 105°28′E	Bed 15, lower-middle of Dalong Fm	257.79	± 0.14	± 0.31	^{206}Pb/^{238}U
P21					silicified dacitic tuff; sample 13VD82	Druzhba Creek, northeastern periphery of the Okhotsk Massif, Verkhoyansk, Siberia	60.913257°N, 146.817396°E,	middle Druzhba Fm	258.14	± 0.20	± 0.36	^{206}Pb/^{238}U
P20					silicified dacitic tuff; sample 13VD78	Druzhba Creek, northeastern periphery of the Okhotsk Massif, Verkhoyansk, Siberia	60.910240°N, 146.821460°E	lower Druzhba Fm	260.16	± 0.39	± 0.50	^{206}Pb/^{238}U
P19	P11				SH03(8); bentonite	Shangsi section, north Sichuan Province, central China	32°20′N, 105°28′E	Bed 8, lower-middle of Wuchiaping Fm	260.74	± 0.86	± 0.90	^{206}Pb/^{238}U

Primary radioisotopic age details	Zonal range assignment			Biostratigraphy	Reference
	clade	zonation	Zone		
Seven of thirteen single zircon grain analyses (excluding six older grains) have a weighted mean $^{206}Pb/^{238}U$ age; utilizing CA-TIMS and the ET2535 spike.	conodont	South China	*Clarkina changxingensis* Zone	The bentonite is 0.3-m below the lithologic formation boundary (PTB), and 0.6-m below the FOD of *H. parvus*.	Baresel et al. (2017)
Seven of eighteen single zircon grain analyses (excluding eleven older grains) have a weighted mean $^{206}Pb/^{238}U$ age; utilizing CA-TIMS and the ET2535 spike.	conodont	South China	*Clarkina changxingensis* Zone	The bentonite is 0.6-m below the lithologic formation boundary (PTB), and 0.9-m below the FOD of *H. parvus*.	Baresel et al. (2017)
Three of fifteen single zircon grain analyses (excluding twelve older grains) have a weighted mean $^{206}Pb/^{238}U$ age; utilizing CA-TIMS and the ET2535 spike.	conodont	South China	*Clarkina changxingensis* Zone	The bentonite is 1.1-m below the lithologic formation boundary (PTB), and 1.4-m below the FOD of *H. parvus*.	Baresel et al. (2017)
Eleven of twelve single zircon grain analyses (excluding one older grain) have a weighted mean $^{206}Pb/^{238}U$ age; utilizing CA-TIMS and the ET2535 spike.	conodont	South China	*Neogondolella yini* (UAZ1)	The bentonite is 2.7-m below the lithologic formation boundary (PTB) between bed 12 with diverse Permian fauna and Bed 13 with abundant Griesbachian bivalves and ammonoids.	Baresel et al. (2017)
Nine of ten single zircon grain analyses (excluding one younger grain) have a weighted mean $^{206}Pb/^{238}U$ age; utilizing CA-TIMS and the ET2535 spike.	conodont	South China	*Neogondolella yini* (UAZ1)	The bentonite is 3.2-m below the lithologic formation boundary (PTB) between bed 12 with diverse Permian fauna and Bed 13 with abundant Griesbachian bivalves and ammonoids.	Baresel et al. (2017)
Seven of seven single zircon grain analyses have a weighted mean $^{206}Pb/^{238}U$ age; utilizing CA-TIMS and the ET2535 spike.	conodont	South China	*Clarkina changxingensis* Zone	The bentonite is 6.4-m below the lithologic formation boundary (PTB) between bed 12 with diverse Permian fauna and Bed 13 with abundant Griesbachian bivalves and ammonoids.	Baresel et al. (2017)
Eight of eight single zircon grain analyses have a weighted mean $^{206}Pb/^{238}U$ age; utilizing CA-TIMS and the ET2535 spike.	conodont	South China	*Clarkina changxingensis* Zone	The bentonite is 7.3-m below the lithologic formation boundary (PTB) between bed 12 with diverse Permian fauna and Bed 13 with abundant Griesbachian bivalves and ammonoids.	Baresel et al. (2017)
Seven of ten single zircon grain analyses (excluding three older grains) have a weighted mean $^{206}Pb/^{238}U$ age; utilizing CA-TIMS and the ET2535 spike.	conodont	South China	*Clarkina changxingensis* Zone	The bentonite is 9.7-m below the lithologic formation boundary (PTB) between bed 12 with diverse Permian fauna and Bed 13 with abundant Griesbachian bivalves and ammonoids.	Baresel et al. (2017)
Seven of seven single zircon grain analyses have a weighted mean $^{206}Pb/^{238}U$ age; utilizing CA-TIMS and the in-house MIT 535 spike.	conodont	South China	*Clarkina changxingensis* (middle)	The bentonite in Bed 15 is 17.3-m below the base of the GSSP Bed 27c at Meishan D section.	Shen et al. (2011)
Seven of eight single zircon grain analyses (excluding one younger grain) have a weighted mean $^{206}Pb/^{238}U$ age; utilizing CA-TIMS and the in-house MIT 535 spike.	conodont	South China	*Clarkina wangi* (middle)	The bentonite in Bed 7 is 36.8-m below the base of the GSSP Bed 27c at Meishan D section.	Shen et al. (2011)
Seven of seven single zircon grain analyses have a weighted mean $^{206}Pb/^{238}U$ age; utilizing CA-TIMS and the in-house MIT 535 spike.	conodont	South China	*Clarkina wangi* (middle)	The bentonite in Bed 6 is 38.0-m below the base of the GSSP Bed 27c at Meishan D section.	Shen et al. (2011)
Weighted mean of 8 single zircon grain analyses, utilizing CA-TIMS and in-house BGC spike (0.1% error in tracer calibration)	conodont	South China	*Clarkina yini*	The bentonite in Bed 23 is 3.0-m below the extinction horizon at the base of bed 28, and 7.6-m below the FOD of *H. parvus* in bed 30.	Mundil et al. (2004)
Six of seven single zircon grain analyses (excluding one younger grain) have a weighted mean $^{206}Pb/^{238}U$ age; utilizing CA-TIMS and the in-house MIT 535 spike.	conodont	South China	*Clarkina wangi* (middle)	The bentonite in Bed 18 is 12.8-m below the extinction horizon at the base of bed 28.	Shen et al. (2011)
Six of six single zircon grain analyses have a weighted mean $^{206}Pb/^{238}U$ age; utilizing CA-TIMS and the in-house MIT 535 spike.	conodont	South China	*Clarkina transcaucasica*	The bentonite in Bed 15 is 27.5-m below the extinction horizon at the base of bed 28.	Shen et al. (2011)
Two of ten single zircon grain analyses (excluding eight older grains) have a weighted mean $^{206}Pb/^{238}U$ age; utilizing CA-TIMS and the ET2535 spike.	bivalve	NE Asian	*Maitaia tenkensis* Zone, lower Wuchiapingian	Regional correlation of the host Druzhba Formation and the nearby Titan Formation suggest a position within the regional *Maitaia tenkensis* bivalve zone, which is correlated to the lower Wuchiapingian Stage.	Davydov et al. (2018a)
Four of six single zircon grain analyses (excluding two older grains) have a weighted mean $^{206}Pb/^{238}U$ age; utilizing CA-TIMS and the ET2535 spike.	bivalve	NE Asian	*Maitaia tenkensis* Zone, lower Wuchiapingian	Regional correlation of the host Druzhba Formation and the nearby Titan Formation suggest a position within the regional *Maitaia tenkensis* bivalve zone, which is correlated to the lower Wuchiapingian Stage.	Davydov et al. (2018a)
Weighted mean of 5 single zircon grain analyses, utilizing CA-TIMS and in-house BGC spike (0.1% error in tracer calibration)	conodont	South China		The bentonite in Bed 8 is 64.0-m below the extinction horizon at the base of bed 28.	Mundil et al. (2004)

GTS 2020 ID	GTS 2012 ID	Period	Epoch	Age	Sample	Locality	Lat-Long	Lithostratigraphy	Age (Ma)	± 2s analytical	± 2s total	Age Type
P18	P10				SH01(7); bentonite	Shangsi section, north Sichuan Province, central China	32°20′N, 105°28′E	Bed 7, lower-middle of Wuchiaping Fm	259.10	± 1.01	± 1.05	$^{206}Pb/^{238}U$
			Guadalupian									
				Capitanian								
P17					bentonite; sample 10VD-2	Trapper Creek, Cassia County, SE Idaho, USA	42°6.943′N, 114°7.060′W	upper phosphatic Meade Peak Member of the Phosphoria Fm	260.57	± 0.07	± 0.31	$^{206}Pb/^{238}U$
P16					bentonite; sample 12VD-105	Natalkin–Glukharynyi Creek, Ayan-Yuryakh basin, Magadan Province, NE Russia	61°38.566′N, 147°48.506′E	middle Atkan Fm	262.45	± 0.21	± 0.37	$^{206}Pb/^{238}U$
P15					bentonite; sample GM-20	Patterson Hills, Guadalupe Mountains, Texas, USA	31°49′28.4″N, 104°52′32.1″W	20 m above the base of the Rader Limestone	262.58	± 0.45	± 0.53	$^{206}Pb/^{238}U$
				Wordian								
P14	P9				Nipple Hill bentonite	Nipple Hill, Guadalupe Mountains National Park, Texas, USA	31.909°N, 104.789°W	The ash occurs within the undifferentiated lower unit of the Bell Canyon Fm, below the Pinery Limestone Member, and approximately 2 m above the top of the Manzanita Limestone Member of the underlying Cherry Canyon Fm.	265.46	± 0.27	± 0.39	$^{206}Pb/^{238}U$
P13					bentonite; sample GM–29	Monolith Canyon, Guadalupe Mountains, Texas, USA	31°54.740′N, 104°46.943′W	near the base of the South Wells Limestone Fm	266.50	± 0.24	± 0.37	$^{206}Pb/^{238}U$
				Roadian								
P12					bentonite; Ash bed "C", sample PDSE D003	Pingdingshan East Section, Chaohu, Anhui Province, South China	31°38.048′N, 117°49.828′E	the lower part of the Kuhfeng Fm; 1-m above Ash bed "A"	271.04	± 0.10	± 0.33	$^{206}Pb/^{238}U$
P11					bentonite; Ash bed "A", sample PDSE D001	Pingdingshan East Section, Chaohu, Anhui Province, South China	31°38.048′N, 117°49.828′E	base of the Kuhfeng Fm	272.95	± 0.11	± 0.34	$^{206}Pb/^{238}U$
P10					bentonite; sample 13VD83	Druzhba Creek, northeastern periphery of the Okhotsk Massif, Verkhoyansk, Siberia	60.873580°N, 146.867660°E	upper Khuren Fm	273.12	± 0.13	± 0.34	$^{206}Pb/^{238}U$
P9					bentonite; sample 64/2AB-91	Plastovyi Creek, eastern periphery of the Okhotsk Massif, Verkhoyansk, Siberia	60.774157°N, 147.657132°E	middle Khuren Fm	274.00	± 0.12	± 0.34	$^{206}Pb/^{238}U$
			Cisuralian									
				Kungurian								
				Artinskian								
P8	P8				01DES-403; bentonite	Usolka, Dal'ny Tulkas roadcut section, Southern Urals, Russia	53.883829°N, 56.537216°E	12.5 mab of section (base of Bed 9), in Tulkas Fm	288.21	± 0.15	± 0.34	$^{206}Pb/^{238}U$

Primary radioisotopic age details	Zonal range assignment			Biostratigraphy	Reference
	clade	zonation	Zone		
Weighted mean of 6 single zircon grain analyses, utilizing CA-TIMS and in-house BGC spike (0.1% error in tracer calibration)	conodont	South China		The bentonite in Bed 7 is 76.0-m below the extinction horizon at the base of bed 28.	Mundil et al. (2004)
Five of fifteen single zircon grain analyses (excluding ten older grains) have a weighted mean $^{206}Pb/^{238}U$ age; utilizing CA-TIMS and the ET535 spike.	conodont	West Texas	upper Capitanian Stage	*Jinogondolella aserrata, Mesogondolella phosphoriensis, M. rosenkrantzi, M. bitteri* and *M. siciliensis* are reported from the Meade Peak of the Phosphoria and lower-middle Gerster Fm of the Park City Formation in Idaho, Wyoming, Utah and Nevada. *M. siciliensis* is described from Oman in the beds with ammonoid *Timorites* and conodont *M. bitteri* immediately below Wuchiapingian beds (Baud et al., 2012). In the Delaware Basin, *Timorites* ranges from the upper Wordian to the top of the Capitanian.	Davydov et al. (2018b)
Four of eight single zircon grain analyses (excluding four older grains) have a weighted mean $^{206}Pb/^{238}U$ age; utilizing CA-TIMS and the ET535 spike.	bivalve/ ammonoid	NE Asian	lower Capitanian Stage	The Atkan Formation is assigned to the *Maitaia bella* bivalve zone that corresponds to the lower Gizhigian regional stage of northeastern Russia. The latter correlates with the Capitanian of the International Time Scale through the occurrence of the ammonoid *Timorites*, together with Gizhigian bivalves and brachiopods in the Transbaikal region.	Davydov et al. (2016)
Four of six single zircon grain analyses (excluding two younger grains) have a weighted mean $^{206}Pb/^{238}U$ age; utilizing CA-TIMS and the ET2535 spike.	conodont	West Texas	*Jinogondolella postserrata* Zone	Bentonite is found in strata containing *Polydiexodina*, and is correlated to the lower *J. postserrata* Zone	Nicklen (2011)
Seven of seven single zircon grain analyses have a weighted mean $^{206}Pb/^{238}U$ age; utilizing CA-TIMS and the ET535 spike.	conodont	West Texas	upper *Jinogondolella aserrata* Zone	The horizon is 37.2 m below the base of the *Jinogondolella postserrata* conodont zone that defines the base of the Capitanian Stage of the Guadalupian Series.	Ramezani and Bowring (2017)
Eight of thirteen single zircon grain analyses (excluding one younger and four older grains) have a weighted mean $^{206}Pb/^{238}U$ age; utilizing CA-TIMS and the ET2535 spike.	conodont	West Texas	lower *Jinogondolella aserrata* Zone	Lower *Jinogondolella aserrata* Zone	Nicklen (2011)
Eight of ten single zircon grain analyses (excluding two younger grains) have a weighted mean $^{206}Pb/^{238}U$ age; utilizing CA-TIMS and the ET535 spike.	conodont	West Texas	middle *Jinogondolella nankingensis* Zone	Bentonite is found 1.3-m below strata containing *Jinogondolella nankingensis* and *J. aserrata*	Wu et al. (2017)
Seven of thirteen single zircon grain analyses (excluding one younger and five older grains) have a weighted mean $^{206}Pb/^{238}U$ age; utilizing CA-TIMS and the ET535 spike.	conodont	West Texas	middle *Jinogondolella nankingensis* Zone	Bentonite is found 2.3-m below strata containing *Jinogondolella nankingensis* and *J. aserrata*	Wu et al. (2017)
Five of seven single zircon grain analyses (excluding two older grains) have a weighted mean $^{206}Pb/^{238}U$ age; utilizing CA-TIMS and the ET2535 spike.	ammonoid/ conodont	Canadian Arctic/ West Texas	middle *Jinogondolella nankingensis* Zone	The tuff bed occurs within the NE Asian *Kolymia plicata* bivalve zone, which is younger than the Kolymia multiformis bivalve zone, whose type species is co-occurs with the pectinids *"Heteropecten" cf. girtyi* Newell and *"H." cf. gryphus* Newell known from the Word Formation of Texas.	Davydov et al. (2018a)
Three of eight single zircon grain analyses (excluding five older grains) have a weighted mean $^{206}Pb/^{238}U$ age; utilizing CA-TIMS and the ET2535 spike.	ammonoid/ conodont	Canadian Arctic/ West Texas	middle *Jinogondolella nankingensis* Zone	The tuff bed occurs within the upper part of the NE Asian *Kolymia inoceramiformis* bivalve zone, while the ammonoid *Sverdrupites harkeri* occurs immediately below the dated tuff bed. This provides direct correlation to the Roadian Stage in the Canadian Arctic, where the conodont index species of the Roadian Stage *Jinogondolella nankingensis gracilis* and *Sverdrupites harkeri* and other Roadian ammonoids are documented from the Assistant Formation.	Davydov et al. (2018a)
Weighted mean of 8 single zircon grain analyses, utilizing CA-TIMS and the ET535 spike.	conodont	Boreal	*Sweetognathodus "whitei"* Zone	12.3 m above FAD of *Sweetognathus "whitei"*; Burtsevian Substage (Russian Platform) of Artinskian Stage	Schmitz and Davydov (2012)

GTS 2020 ID	GTS 2012 ID	Period	Epoch	Age	Sample	Locality	Lat-Long	Lithostratigraphy	Age (Ma)	± 2s analytical	± 2s total	Age Type
P7	P7				DTR905; bentonite	Usolka, Dal'ny Tulkas roadcut section, Southern Urals, Russia	53.883829°N, 56.537216°E	10.5 mab of section (upper Bed 7), in Tulkas Fm	288.36	± 0.17	± 0.35	$^{206}Pb/^{238}U$
				Sakmarian								
P6	P6				07DTRBed2; bentonite	Usolka, Dal'ny Tulkas roadcut section, Southern Urals, Russia	53.883829°N, 56.537216°E	4.0 mbb of section (upper Bed 2), in Tulkas Fm	290.81	± 0.17	± 0.35	$^{206}Pb/^{238}U$
P5	P5				97USO-91.0; bentonite	Usolka (Krasnousolsky) section, Southern Urals, Russia	53.923777°N, 56.532353°E	91.0 mab of section, in Tulkas Fm	290.50	± 0.17	± 0.35	$^{206}Pb/^{238}U$
P4	P4				97USO-66.2; bentonite	Usolka (Krasnousolsky) section, Southern Urals, Russia	53.923777°N, 56.532353°E	66.2 mab of section (upper Bed 28), in Tulkas Fm	291.10	± 0.18	± 0.36	$^{206}Pb/^{238}U$
				Asselian								
P3	P3				01DES-212; bentonite	Usolka (Krasnousolsky) section, Southern Urals, Russia	53.923777°N, 56.532353°E	41.25 mab of section (upper Bed 18), in Kurkin Fm	296.69	± 0.16	± 0.37	$^{206}Pb/^{238}U$
P2	P2				01DES-202; bentonite	Usolka (Krasnousolsky) section, Southern Urals, Russia	53.923777°N, 56.532353°E	36.0 mab of section (upper Bed 16), in Kurkin Fm	298.05	± 0.46	± 0.56	$^{206}Pb/^{238}U$
P1	P1				01DES-194; bentonite	Usolka (Krasnousolsky) section, Southern Urals, Russia	53.923777°N, 56.532353°E	32.4 mab of section (middle Bed 16), in Kurkin Fm	298.49	± 0.20	± 0.37	$^{206}Pb/^{238}U$
		Carboniferous										
		Pennsylvanian (sub-period)										
				Ghzelian								
Cb47	Cb35				01DES-144; bentonite	Usolka (Krasnousolsky) section, Southern Urals, Russia	53.923777°N, 56.532353°E	30.8 mab of section (lower Bed 16), in Kurkin Fm	299.22	± 0.20	± 0.37	$^{206}Pb/^{238}U$
Cb46	Cb34				97USO-23.3; bentonite	Usolka (Krasnousolsky) section, Southern Urals, Russia	53.923777°N, 56.532353°E	23.3 mab of section (lower Bed 14), in Kurkin Fm	300.22	± 0.16	± 0.35	$^{206}Pb/^{238}U$
Cb45	Cb33				01DES-121; bentonite	Usolka (Krasnousolsky) section, Southern Urals, Russia	53.923777°N, 56.532353°E	21.45 mab of section (upper Bed 12), in Kurkin Fm	301.29	± 0.16	± 0.36	$^{206}Pb/^{238}U$
Cb44	Cb32				01DES-112; bentonite	Usolka (Krasnousolsky) section, Southern Urals, Russia	53.923777°N, 56.532353°E	18.2 mab of section (middle Bed 12), in Kurkin Fm	301.82	± 0.17	± 0.36	$^{206}Pb/^{238}U$
Cb43	Cb31				01DES-63; bentonite	Usolka (Krasnousolsky) section, Southern Urals, Russia	53.923777°N, 56.532353°E	12.0 mab of section, in Kurkin Fm	303.10	± 0.16	± 0.36	$^{206}Pb/^{238}U$
				Kasimovian								
Cb42	Cb30				97USO-2.7; bentonite	Usolka (Krasnousolsky) section, Southern Urals, Russia	53.923777°N, 56.532353°E	11.4 mab of section, in Kurkin Fm	303.54	± 0.18	± 0.39	$^{206}Pb/^{238}U$
Cb41	Cb29				97USO-1.2 & 01DES-63; bentonites	Usolka (Krasnousolsky) section, Southern Urals, Russia	53.923777°N, 56.532353°E	10.65 mab of section, in Kurkin Fm	304.49	± 0.16	± 0.36	$^{206}Pb/^{238}U$
Cb40	Cb28				02VD-5 & 01DES-42; bentonites	Usolka (Krasnousolsky) section, Southern Urals, Russia	53.923777°N, 56.532353°E	9.85 mab of section, in Kurkin Fm	304.83	± 0.16	± 0.36	$^{206}Pb/^{238}U$
Cb39	Cb27				01DES-31 & 08USO-8.4; bentonites	Usolka (Krasnousolsky) section, Southern Urals, Russia	53.923777°N, 56.532353°E	7.8 mab of section, in Kurkin Fm	305.49	± 0.16	± 0.36	$^{206}Pb/^{238}U$
Cb38	Cb26				01DES-371; bentonite	Dal'ny Tulkas quarry section, Southern Urals, Russia	53.884546°N, 56.544646°E	20.15 mab of section, in Kurkin Fm	305.51	± 0.18	± 0.37	$^{206}Pb/^{238}U$
Cb37	Cb25				08USO-7.09; bentonite	Usolka (Krasnousolsky) section, Southern Urals, Russia	53.923777°N, 56.532353°E	6.5 mab of section, in Kurkin Fm	305.95	± 0.17	± 0.37	$^{206}Pb/^{238}U$
Cb36	Cb24				01DES-363; bentonite	Dal'ny Tulkas quarry section, Southern Urals, Russia	53.884546°N, 56.544646°E	19.7 mab of section, in Kurkin Fm	305.96	± 0.17	± 0.36	$^{206}Pb/^{238}U$

Primary radioisotopic age details	Zonal range assignment			Biostratigraphy	Reference
	clade	zonation	Zone		
Weighted mean of 7 single zircon grain analyses, utilizing CA-TIMS and the ET535 spike.	conodont	Boreal	*Sweetognathodus "whitei"* Zone	10.3 m above FAD of *Sweetognathus "whitei"*; Burtsevian Substage (Russian Platform) of Artinskian Stage	Schmitz and Davydov (2012)
Weighted mean of 6 single zircon grain analyses, utilizing CA-TIMS and the ET535 spike.	conodont	Boreal	*Sweetognathodus anceps* Zone	4.2 m below FAD of *Sweetognathus "whitei"*; Sterlitamakian Substage (Russian Platform) of Sakmarian Stage	Schmitz and Davydov (2012)
Weighted mean of 8 single zircon grain analyses, utilizing CA-TIMS and the ET535 spike.	conodont	Boreal	*Sweetognathodus anceps* Zone	Sterlitamakian Substage (Russian Platform) of Sakmarian Stage	Schmitz and Davydov (2012)
Weighted mean of 8 single zircon grain analyses, utilizing CA-TIMS and the ET535 spike.	conodont	Boreal	*Sweetognathodus anceps* Zone	Sterlitamakian Substage (Russian Platform) of Sakmarian Stage	Schmitz and Davydov (2012)
Weighted mean of 9 single zircon grain analyses, utilizing CA-TIMS and the ET535 spike.	conodont	Boreal	*Streptognathodus fusus* Zone	Uskalikian Substage (Russian Platform) of Asselian Stage	Schmitz and Davydov (2012)
Weighted mean of 9 single zircon grain analyses, utilizing both physical abrasion and CA-TIMS and the MIT-1L spike calibrated against the EARTHTIME gravimetric standards.	conodont	Boreal	*Streptognathodus cristellaris–St. sigmoidalis* Zone	4.55 m above FAD of *Streptognathodus isolatus*; Surenian Substage (Russian Platform) of Asselian Stage	Ramezani et al. (2007a, b)
Weighted mean of 14 single zircon grain analyses, utilizing both physical abrasion and CA-TIMS and the MIT-1L spike calibrated against the EARTHTIME gravimetric standards.	conodont	Boreal	*Streptognathodus isolatus* Zone	0.95 m above FAD of *Streptognathodus isolatus*; Surenian Substage (Russian Platform) of Asselian Stage	Ramezani et al. (2007a, b)
Weighted mean of 19 single zircon grain analyses, utilizing both physical abrasion and CA-TIMS and the MIT-1L spike calibrated against the EARTHTIME gravimetric standards.	conodont	Boreal	*Streptognathodus waubaunsensis* Zone	0.65 m below FAD of *Streptognathodus isolatus*; Noginian Substage (Russian Platform) of Gzhelian Stage	Ramezani et al. (2007a, b)
Weighted mean of 7 single zircon grain analyses, utilizing CA-TIMS and the ET535 spike.	conodont	Boreal	*Streptognathodus simplex–St. bellus* Zone	Noginian Substage (Russian Platform) of Gzhelian Stage	Schmitz and Davydov (2012)
Weighted mean of 5 single zircon grain analyses, utilizing CA-TIMS and the ET535 spike.	conodont	Boreal	*Streptognathodus virgilicus* Zone	Pavlovoposadian Substage (Russian Platform) of Gzhelian Stage	Schmitz and Davydov (2012)
Weighted mean of 5 single zircon grain analyses, utilizing CA-TIMS and the ET535 spike.	conodont	Boreal	*Streptognathodus virgilicus* Zone	Pavlovoposadian Substage (Russian Platform) of Gzhelian Stage	Schmitz and Davydov (2012)
Weighted mean of 4 single zircon grain analyses, utilizing CA-TIMS and the ET535 spike.	conodont	Boreal	*Streptognathodus simulator* Zone	Rusavkian Substage (Russian Platform) of Gzhelian Stage	Schmitz and Davydov (2012)
					Schmitz and Davydov (2012)
Weighted mean of 5 single zircon grain analyses, utilizing CA-TIMS and the ET535 spike.	conodont	Boreal	*Streptognathodus firmus* Zone	Rusavkian Substage (Russian Platform) of Gzhelian Stage	Schmitz and Davydov (2012)
Two samples of same ash bed yielded identical weighted mean ages of 304.49 ± 0.16 Ma (six single grains) and 304.42 ± 0.16 Ma (five single grains), utilizing CA-TIMS and the ET535 spike.	conodont	Boreal	*Idiognathodus toretzianus* Zone	Dorogomilovian Substage (Russian Platform) of Kasimovian Stage	Schmitz and Davydov (2012)
Two samples of same ash bed yielded identical weighted mean ages of 304.83 ± 0.16 Ma (four single grains) and 304.82 ± 0.17 Ma (three single grains), utilizing CA-TIMS and the ET535 spike.	conodont	Boreal	*Idiognathodus sagittalis* Zone	Khamovnikian Substage (Russian Platform) of Kasimovian Stage	Schmitz and Davydov (2012)
Two samples of same ash bed yielded identical weighted mean ages of 305.49 ± 0.16 Ma (six single grains) and 305.52 ± 0.19 Ma (six single grains), utilizing CA-TIMS and the ET535 spike.	conodont	Boreal	*Idiognathodus sagittalis* Zone	Khamovnikian Substage (Russian Platform) of Kasimovian Stage	Schmitz and Davydov (2012)
Weighted mean of 4 single zircon grain analyses, utilizing CA-TIMS and the ET535 spike.	conodont	Boreal	*Idiognathodus sagittalis* Zone	Khamovnikian Substage (Russian Platform) of Kasimovian Stage	Schmitz and Davydov (2012)
Weighted mean of 3 single zircon grain analyses, utilizing CA-TIMS and the ET535 spike.	conodont	Boreal	*Streptognathodus subexcelsus* Zone	Krevyakian Substage (Russian Platform) of Kasimovian Stage	Schmitz and Davydov (2012)
Weighted mean of 8 single zircon grain analyses, utilizing CA-TIMS and the ET535 spike.	conodont	Boreal	*Streptognathodus subexcelsus* Zone	Krevyakian Substage (Russian Platform) of Kasimovian Stage	Schmitz and Davydov (2012)

GTS 2020 ID	GTS 2012 ID	Period	Epoch	Age	Sample	Locality	Lat-Long	Lithostratigraphy	Age (Ma)	± 2s analytical	± 2s total	Age Type
					Moscovian							
Cb35	Cb23				01DES-362; bentonite	Dal'ny Tulkas quarry section, Southern Urals, Russia	53.884546°N, 56.544646°E	17.4 mab of section, in Kurkin Fm	307.66	± 0.17	± 0.37	$^{206}Pb/^{238}U$
Cb34	Cb22				06USO-2.0; bentonite	Usolka (Krasnousolsky) section, Southern Urals, Russia	53.923777°N, 56.532353°E	1.4 mab of section, in Zilim Fm	308.00	± 0.18	± 0.37	$^{206}Pb/^{238}U$
Cb33	Cb21				01DES-481; bentonite	Usolka (Krasnousolsky) section, Southern Urals, Russia	53.923777°N, 56.532353°E	1.25 mab of section, in Zilim Fm	308.36	± 0.20	± 0.38	$^{206}Pb/^{238}U$
Cb32	Cb20				01DES-351; bentonite	Dal'ny Tulkas quarry section, Southern Urals, Russia	53.884546°N, 56.544646°E	12.4 mab of section, in Zilim Fm	308.50	± 0.16	± 0.36	$^{206}Pb/^{238}U$
Cb31	Cb19				n1 coal; tonstein	Butovskaya Shaft, Donets Basin	48.0677°N, 37.7918°E	Isaevskaya Fm, C31(N) lithostratigraphic index	307.26	± 0.19	± 0.38	$^{206}Pb/^{238}U$
Cb30	Cb18				m3 coal; tonstein	Zasyadko Shaft, Donets Basin	48.0517°N, 37.7939°E	Lisitchanskaya Fm, C27(M) lithostratigraphic index	310.55	± 0.19	± 0.38	$^{206}Pb/^{238}U$
Cb29	Cb17				l3 coal; tonstein	Krasnolimanskaya Shaft, Donets Basin	48.3594°N, 37.2430°E	Almaznaya Fm, C26(L) lithostratigraphic index	312.01	± 0.17	± 0.37	$^{206}Pb/^{238}U$
Cb28	Cb16				l1 coal; tonstein	Kirov Shaft, Donets Basin	48.1320°N, 38.3548°E	Almaznaya Fm, C26(L) lithostratigraphic index	312.23	± 0.18	± 0.37	$^{206}Pb/^{238}U$
Cb27	Cb15				k7 coal; tonstein	Pereval'skaya Shaft, Donets Basin	48.4460°N, 38.80°E	Kamenskaya Fm, C25(K) lithostratigraphic index	313.16	± 0.17	± 0.37	$^{206}Pb/^{238}U$
Cb26					tonstein; sample Z1	Furst Leopold Coal Mine, Dorsten, Ruhr Basin, Germany	51°40'23"N, 6°58'57"E	10 to 12-m beneath the Aegiranum Marine Band	313.78	± 0.08	± 0.38	$^{206}Pb/^{238}U$
Cb25					Sub-High Main tonstein; sample EH28155	Holme Pierrepont Borehole, Nottinghamshire, England; 181.8-m depth	52°56'47"N, 1°04'13"W	14-m above the Aegiranum Marine Band	314.37	± 0.25	± 0.53	$^{206}Pb/^{238}U$
Cb24	Cb14				k3 coal; tonstein	Pereval'skaya Shaft, Donets Basin	48.4460°N, 38.80°E	Kamenskaya Fm, C25(K) lithostratigraphic index	314.40	± 0.17	± 0.37	$^{206}Pb/^{238}U$
					Bashkirian							
Cb23	Cb13				Bed 32; bentonite	Kljuch section, 5 km W of Kamensk-Ural'sky city, Sverdlov Province, Southern Urals, Russia	56.42414°N, 61.80709°E	Scherbakov Fm, Bed 32	317.54	± 0.17	± 0.38	$^{206}Pb/^{238}U$
Cb22					tonstein; sample T75	Zwartberg Coal Mine, Campine Basin, Belgium	51°01'17"N, 5°31'30"E	between the Wasserfall and Quaregnon marine bands	317.63	± 0.12	± 0.39	$^{206}Pb/^{238}U$
Cb21	Cb12				Bed 9; bentonite	Kljuch section, 5 km W of Kamensk-Ural'sky city, Sverdlov Province, Southern Urals, Russia	56.42414°N, 61.80709°E	Scherbakov Fm, Bed 9	318.63	± 0.22	± 0.40	$^{206}Pb/^{238}U$
Cb20	Cb11				Bed 2; bentonite	Kljuch section, 5 km W of Kamensk-Ural'sky city, Sverdlov Province, Southern Urals, Russia	56.42414°N, 61.80709°E	Scherbakov Fm, Bed 2	319.09	± 0.18	± 0.38	$^{206}Pb/^{238}U$
		Mississippian (sub-period)										
				Serpukhovian								
Cb19					tonstein; sample B9	Oakenclough Brook, Staffordshire, Pennine Basin, England	53°10'10"N, 1°54'54"W	between the Wasserfall and Quaregnon marine bands	324.54	± 0.26	± 0.46	$^{206}Pb/^{238}U$

Primary radioisotopic age details	Zonal range assignment			Biostratigraphy	Reference
	clade	zonation	Zone		
Weighted mean of 7 single zircon grain analyses, utilizing CA-TIMS and the ET535 spike.	conodont	Boreal	*Neognathodus roundyi–Streptognathodus cancellosus* Zone	Peskovian Substage (Russian Platform) of Moscovian Stage	Schmitz and Davydov (2012)
Weighted mean of 4 single zircon grain analyses, utilizing CA-TIMS and the ET535 spike.	conodont	Boreal	*Neognathodus roundyi–Streptognathodus cancellosus* Zone	Myachkovian Substage (Russian Platform) of Moscovian Stage	Schmitz and Davydov (2012)
Weighted mean of 5 single zircon grain analyses, utilizing CA-TIMS and the ET535 spike.	conodont	Boreal	*Neognathodus roundyi–Streptognathodus cancellosus* Zone	Myachkovian Substage (Russian Platform) of Moscovian Stage	Schmitz and Davydov (2012)
Weighted mean of 5 single zircon grain analyses, utilizing CA-TIMS and the ET535 spike.	conodont	Boreal	*Neognathodus roundyi–Streptognathodus cancellosus* Zone	Myachkovian Substage (Russian Platform) of Moscovian Stage	Schmitz and Davydov (2012)
Weighted mean of 9 single zircon grain analyses, utilizing CA-TIMS and the ET535 spike.	conodont	Boreal	*Neognathodus roundyi–Streptognathodus cancellosus* Zone	base C_3a biostratigraphic zone of Donets Basin;	Davydov et al. (2010)
Weighted mean of 10 single zircon grain analyses, utilizing CA-TIMS and the ET535 spike.	conodont	Boreal	*Idiognathodus podolskiensis* Zone	base $C_2{}^mc$ biostratigraphic zone of Donets Basin;	Davydov et al. (2010)
Weighted mean of 6 single zircon grain analyses, utilizing CA-TIMS and the ET535 spike.	conodont	Boreal	*Streptognathodus dissectus* Zone	base $C_2{}^mb$ biostratigraphic zone of Donets Basin;	Davydov et al. (2010)
Weighted mean of 5 single zircon grain analyses, utilizing CA-TIMS and the ET535 spike.	conodont	Boreal	*Streptognathodus dissectus* Zone	base $C_2{}^mb$ biostratigraphic zone of Donets Basin;	Davydov et al. (2010)
Weighted mean of 8 single zircon grain analyses, utilizing CA-TIMS and the ET535 spike.	conodont	Boreal	*Neognathodus uralicus* Zone	base $C_2{}^ma$ biostratigraphic zone of Donets Basin;	Davydov et al. (2010)
Weighted mean of 6 single zircon grain analyses, utilizing CA-TIMS and the ET2535 spike.	ammonoid	NW Europe		Tonstein coal occurs 10–12-m beneath the Aegiranum Marine Band, which defines the base of the Bolsovian (Westphalian C) substage and is thus in the upper Duckmantian (Westphalian B) Substage	Pointon et al. (2012)
Four of nine single zircon grain analyses (excluding two younger and three older grains) have a weighted mean $^{206}Pb/^{238}U$ age; utilizing CA-TIMS and the ET535 spike.	ammonoid	NW Europe		Tonstein coal occurs 14-m above the Aegiranum Marine Band, which defines the base of the Bolsovian (Westphalian C) Substage	Waters and Condon (2012)
Weighted mean of 7 single zircon grain analyses, utilizing CA-TIMS and the ET535 spike.	conodont	Boreal	*Declinognathodus donetzianus* Zone	Tonstein occurs in the base C2ma biostratigraphic zone of Donets Basin, which correlates to the lower Bolvovian Substage of the Westphalian Stage of Western Europe	Davydov et al. (2010)
Weighted mean of 4 single zircon grain analyses, utilizing CA-TIMS and the ET535 spike.	fusulinid	NW Europe		Foraminifera correlate to upper Bashkirian Tashastian Horizon of Urals, and to just below H_6 Limestone of Donets Basin; Cheremshanian Substage (Russian Platform) of Bashkirian Stage	Schmitz and Davydov (2012)
Five of ten single zircon grain analyses (excluding two younger and three older grains) have a weighted mean $^{206}Pb/^{238}U$ age; utilizing CA-TIMS and the ET2535 spike.	ammonoid	NW Europe		Tonstein coal occurs between the Wasserfall and Quaregnon marine bands and thus is in Langsettian (Westphalian A) Substage	Pointon et al. (2012)
Weighted mean of 5 single zircon grain analyses, utilizing CA-TIMS and the ET535 spike.	fusulinid	NW Europe		Foraminifera correlate to middle Bashkirian Ashyabashian Horizon of Urals, and between G_1 and $G_1{}^2$ Limestones of Donets Basin; Prikamian Substage (Russian Platform) of Bashkirian Stage	Schmitz and Davydov (2012)
Weighted mean of 4 single zircon grain analyses, utilizing CA-TIMS and the ET535 spike.	fusulinid	NW Europe		Foraminifera correlate to middle Bashkirian Ashyabashian Horizon of Urals, and between F_1 and $F_1{}^1$ Limestones of Donets Basin; Prikamian substage (Russian Platform) of Bashkirian Stage	Schmitz and Davydov (2012)
Four of nine single zircon grain analyses (excluding five older grains) have a weighted mean $^{206}Pb/^{238}U$ age; utilizing CA-TIMS and the ET2535 spike.	ammonoid	NW Europe		Tuff occurs in the lower Namurian Stage (Arnsbergian substage), E2b2 ammonoid subzone (cycle E2b2ii).	Pointon et al. (2012)

GTS 2020 ID	GTS 2012 ID	Period	Epoch	Age	Sample	Locality	Lat-Long	Lithostratigraphy	Age (Ma)	± 2s analytical	± 2s total	Age Type
Cb18					tonstein; Gabriela coal (seam 365)	Julius Fučík Mine, Petřvald, Moravia-Silesia Region, Czech Republic	49°49.166′N, 18°22.691′E	Jaklovec Member, Ostrava Fm	325.64	± 0.13	± 0.40	$^{206}Pb/^{238}U$
Cb17					tonstein; Eleonara coal (seam 335)	Julius Fučík Mine, Petřvald, Moravia-Silesia Region, Czech Republic	49°49.166′N, 18°22.691′E	Jaklovec Member, Ostrava Fm	325.58	± 0.26	± 0.46	$^{206}Pb/^{238}U$
Cb16	Cb10				tonstein; Ruzeny coal (seam 103)	Upper Silesian Basin, Ostrava, Czech Republic (ton. 106-B2 Karel, 146.92−146.94 m, Staric2 borehole)	~49.8°N, 18.3°E	Lower Hrusov Member, Ostrava Fm	327.58	± 0.17	± 0.39	$^{206}Pb/^{238}U$
Cb15					Main Ostrava Whetstone	Paskov mine, Upper Silesian Basin, Ostrava, Czech Republic	~49.8°N, 18.3°E	boundary between Petrkovice Member and Lower Hrusov Member, Ostrava Fm	327.35	± 0.15	± 0.49	$^{206}Pb/^{238}U$
Cb14	Cb9				tonstein; c11 coal	Yuzhno-Donbasskaya, Shaft No 3, Ugledar; Donets Basin	47.7839°N, 37.2474°E	Samarskaya Fm, C13(C) lithostratigraphic index, D13lower (= C4) limestone	328.14	± 0.20	± 0.40	$^{206}Pb/^{238}U$
Cb13					Bentonite B6; sample BLL1976	BGS Harewood Borehole, 12 km NNE of Leeds, West Yorkshire, England; 304.10-m depth	53°54′N, 1°31′W	upper part of eumorphoceras yatesae (E2a3) Marine Band, Arnsbergian Substage	328.34	± 0.30	± 0.55	$^{206}Pb/^{238}U$
Cb12	Cb8				tonstein; Ludmila coal (043)	Upper Silesian Basin, Ostrava, Czech Republic	~49.8°N, 18.3°E	Petrkovice Member, Ostrava Fm	328.48	± 0.19	± 0.41	$^{206}Pb/^{238}U$
				Visean								
Cb11					bentonite; sample W13	Anhée Sud locality, disused Watrisse Quarry, Namur-Dinant Basin, Belgium	50°16.5′N, 4°54.7′E	Lower Member, Anhee Fm	332.50	± 0.07	± 0.40	$^{206}Pb/^{238}U$
Cb10	Cb7				bentonite; Bed 8	Verkhnyaya Kardailovka section, Southern Urals, Russia	52.28630°N, 58.92426°E	Gusikhin Fm, Bed 21-8	333.87	± 0.18	± 0.39	$^{206}Pb/^{238}U$
Cb9					bentonite; sample W8	Yvoir or Anhée Nord locality, abandon quarry to the north of the village of Anhée, Namur-Dinant Basin, Belgium	50°19.8′N, 4°53.5′E	Poilvache Member, Bonne River Fm	335.59	± 0.19	± 0.44	$^{206}Pb/^{238}U$
Cb8					bentonite; sample W1	Yvoir or Anhée Nord locality, abandon quarry to the north of the village of Anhée, Namur-Dinant Basin, Belgium	50°19.8′N, 4°53.5′E	Seiles/Maizeret Member, Grands Malades Fm	336.22	± 0.06	± 0.40	$^{206}Pb/^{238}U$
Cb7					Krásné Louèky tuffite	Krásné Louèky−Kobylí quarry, Krnov, Moravo-Silesian Basin, Czech Republic	50°7′13.08″N, 17°37′40.30″E	Brantice Member of the Horní Beneöov Fm near the transition to the Moravice Fm	340.05	± 0.10	± 0.41	$^{206}Pb/^{238}U$
Cb6	Cb6				bentonite; C1ve2	Sukhaya Volnovakha, Dokuchaevsk, Tsentral'nyi rudnik, east; Donets Basin	47.73°N, 37.6503°E	lower Styl'skaya Fm, C1A9 lithostratigraphic index	342.01	± 0.19	± 0.41	$^{206}Pb/^{238}U$
Cb5	Cb5				bentonite; C1vc	Sukhaya Volnovakha, Dokuchaevsk, Tsentral'nyi rudnik, east; Donets Basin	47.72°N, 37.6502°E	Skelevatskaya Fm, C1A8 lithostratigraphic index	345.00	± 0.18	± 0.41	$^{206}Pb/^{238}U$
Cb4	Cb4				bentonite; 3/2002	Sukhaya Volnovakha, Dokuchaevsk, Tsentral'nyi rudnik, east; Donets Basin	47.73°N, 37.6503°E	Skelevatskaya Fm, C1A8 lithostratigraphic index	345.17	± 0.18	± 0.41	$^{206}Pb/^{238}U$
				Tournaisian								
Cb3	Cb3				bentonite; 5/2002	Volnovakha River, right bank, near Businova Ravine, Donets Basin, Ukraine	47.6398°N, 37.8926°E	Upper member, Karakubskaya Fm, C1A4 lithostratigraphic index	357.26	± 0.19	± 0.42	$^{206}Pb/^{238}U$
Cb2	Cb2				bentonite (a2); sample DT-13	Apricke section, Ruhr Basin, Nordrhein-Westfalen, Germany.	~50°N, 8°E	Bed 15, Hangenberg Limestone, equivalent to Bed 70 in Hasselbachtal section	358.43	± 0.19	± 0.42	$^{206}Pb/^{238}U$

Primary radioisotopic age details	Zonal range assignment			Biostratigraphy	Reference
	clade	zonation	Zone		
Five of eight single zircon grain analyses (excluding one older and two younger grains) have a weighted mean $^{206}Pb/^{238}U$ age; utilizing CA-TIMS and the ET535 spike.	ammonoid	NW Europe	*Eumorphoceras bisulcatum* Zone	Jaklovec and Poruba Members of the Ostrava Formation assigned to the goniatite E2 (Eumorphoceras bisulcatum) Zone, Arnsbergian Substage of the Namurian Stage of Western Europe.	Jirásek et al. (2018)
Four of seven single zircon grain analyses (excluding three younger grains) have a weighted mean $^{206}Pb/^{238}U$ age; utilizing CA-TIMS and the ET535 spike.	ammonoid	NW Europe	*Eumorphoceras bisulcatum* Zone	Jaklovec and Poruba Members of the Ostrava Formation assigned to the goniatite E2 (Eumorphoceras bisulcatum) Zone, Arnsbergian Substage of the Namurian Stage of Western Europe.	Jirásek et al. (2018)
Weighted mean of 5 single zircon grain analyses, utilizing CA-TIMS and the ET535 spike.	conodont	Boreal	*Lochriea ziegleri* Zone	Tonstein occurs in the Pendleian Substage of the Namurian Stage of Western Europe.	Jirásek et al. (2013)
Weighted mean of 5 single zircon grain analyses, utilizing CA-TIMS and the ET2535 spike.	conodont	Boreal	*Lochriea ziegleri* Zone	Tonstein occurs in the Pendleian Substage of the Namurian Stage of Western Europe.	Jirásek et al. (2013)
Weighted mean of 8 single zircon grain analyses, utilizing CA-TIMS and the ET535 spike.	conodont	Boreal	*Lochriea ziegleri* Zone	middle C1vg2 biostratigraphic zone of Donets Basin; correlative with the Lochriea ziegleri zone of Urals, the Tarussian and lower Steshevian Horizons of the Russian Platform, and the *Eumorphoceras* ammonoid zone (Pendleian and Arnsbergian substages) of the Namurian of Western Europe.	Davydov et al. (2010)
Eleven of seventeen single zircon grain analyses (excluding two older and four younger grains) have a weighted mean $^{206}Pb/^{238}U$ age; utilizing CA-TIMS and the ET535 spike.	ammonoid	NW Europe	*Eumorphoceras yatesae* Zone	Tuff occurs within the upper part of the *Eumorphoceras yatesae* (E2a3) Marine Band of early Arnsbergian age.	Waters and Condon (2012)
Weighted mean of 5 single zircon grain analyses, utilizing CA-TIMS and the ET535 spike.	ammonoid	NW Europe	*Eumorphoceras yatesae* Zone	Pendleian Substage of lower Namurian A (lower Serpukhovian).	Gastaldo et al. (2009)
Five of six single zircon grain analyses (excluding one older grain) have a weighted mean $^{206}Pb/^{238}U$ age; utilizing CA-TIMS and the ET2535 spike.	fusulinid	NW Europe	MFZ14 foraminiferal Zone	near the middle of the MFZ14 foraminiferal zone, in the upper Asbian Substage of northwestern Europe and the upper Lower Warnantian Substage of Belgium	Pointon et al. (2014)
Weighted mean of 4 single zircon grain analyses, utilizing CA-TIMS and the ET535 spike. Twelve other single zircons yielded older ages indicative of reworking of the original pyroclastic deposit.	conodont	Boreal	*Lochriea mononodosa* Zone	1.48 m below the FOD of *Lochriea ziegleri*, in the middle part of the *Lochriea mononodosa* zone, and *Hypergoniatites-Ferganoceras* ammonoid genozone of the Urals.	Schmitz and Davydov (2012)
Three of nine single zircon grain analyses (excluding six older grains) have a weighted mean $^{206}Pb/^{238}U$ age; utilizing CA-TIMS and the ET2535 spike.	fusulinid	NW Europe	MFZ13 foraminiferal zone	near the base of the MFZ13 foraminiferal zone, in the lower Asbian Substage of northwestern Europe and the lowermost Warnantian Substage of Belgium	Pointon et al. (2014)
Seven of eight single zircon grain analyses (excluding one older grain) have a weighted mean $^{206}Pb/^{238}U$ age; utilizing CA-TIMS and the ET2535 spike.	fusulinid	NW Europe	top of the *Eostafella porikensis−Archaediscus gigas* Zone	near the top of the MFZ12 foraminiferal zone, at the top of the Holkerian Substage of NW Europe and the Livian Substage of Belgium	Pointon et al. (2014)
Five of seven single zircon grain analyses (excluding two older grains) have a weighted mean $^{206}Pb/^{238}U$ age; utilizing CA-TIMS and the ET2535 spike.	fusulinid	NW Europe	*Eostafella porikensis−Archaediscus gigas* Zone	Fossils from the Kobyli Quarry include *Globoendothyra* sp., *Eostaffella* sp., *Archaediscus krestovnikovi* Rauser, *Archaediscus* sp., *Endothyra* sp., *Palaeotextulariidae* indet.	Jirásek et al. (2014)
Weighted mean of 7 single zircon grain analyses, utilizing CA-TIMS and the ET535 spike.	fusulinid	NW Europe	base of the *Eostafella porikensis−Archaediscus gigas* Zone	base C1ve2 biostratigraphic zone of Donets Basin; correlative with the MFZ12 foraminiferal zone, at the base of the Holkerian Substage of NW Europe and the Livian Substage of Belgium	Davydov et al. (2010)
Weighted mean of 6 single zircon grain analyses, utilizing CA-TIMS and the ET535 spike.	fusulinid	NW Europe	*Uralodiscus rotundus* Zone	upper C1vc biostratigraphic zone of Donets Basin; correlative with the base of the MFZ10 foraminiferal zone of Western Europe	Davydov et al. (2010)
Weighted mean of 6 single zircon grain analyses, utilizing CA-TIMS and the ET535 spike.	fusulinid	NW Europe	*Uralodiscus rotundus* Zone	middle C1vc biostratigraphic zone of Donets Basin; correlative with the base of the MFZ10 foraminiferal zone of Western Europe	Davydov et al. (2010)
Weighted mean of 8 single zircon grain analyses, utilizing CA-TIMS and the ET535 spike.	fusulinid	NW Europe	*Prochernyshinella disputabilis−Tournayellina beata* Zone	base of C1tb2 biostratigraphic zone of Donets Basin; equivalent to lower Cherepetian Horizon of the eastern Russian Platform, and the MFZ3 foraminiferal zone of Western Europe	Davydov et al. (2010)
Weighted mean of 18 single zircon grain analyses, utilizing CA-TIMS and the ET535 and 2535 spikes.	conodont	Boreal	Lower *Siphonodella duplicata* Zone	Lower part of *S. duplicata* conodont zone, lower Tournaisian	Davydov et al. (2011)

GTS 2020 ID	GTS 2012 ID	Period	Epoch	Age	Sample	Locality	Lat-Long	Lithostratigraphy	Age (Ma)	± 2s analytical	± 2s total	Age Type
Cb1	Cb1				bentonite (a2); Bed 79	Hasselbachtal section, Ruhr Basin, Nordrhein-Westfalen, Germany.	~50.3°N, 8.3°E	Bed 79, Hangenberg Limestone	358.71	± 0.19	± 0.42	$^{206}Pb/^{238}U$
		Devonian										
			Late									
				Famennian								
D27					bentonite; sample A2KQ-2	Kowala Quarry, Holy Cross Mountains, 10 km southwest of Kielce, Poland	50°47'49.1''N, 20°33'55.9''E	Hangenberg Limestone, 30 cm above top of the Hangenberg Black Shale	358.89	± 0.20	± 0.48	$^{206}Pb/^{238}U$
D26					bentonite; sample A2KQ-1	Kowala Quarry, Holy Cross Mountains, 10 km southwest of Kielce, Poland	50°47'49.1''N, 20°33'55.9''E	Wocklum Limestone. 20 cm below the base of the Hangenberg Black Shale	358.97	± 0.11	± 0.43	$^{206}Pb/^{238}U$
D25	D18				bentonite (w3); sample DT-12	Apricke section, Ruhr Basin, Nordrhein-Westfalen, Germany.	~50°N, 8°E	Wocklum Limestone, equivalent to Bed 87 in Hasselbachtal section	359.25	± 0.18	± 0.42	$^{206}Pb/^{238}U$
D24	D17				black shale	Jura Creek, Exshaw, Alberta, Canada	51°04'N, 115°10'W	Exshaw Fm	361.30	± 2.10	± 2.40	Re-Os
D23	D14				pumice tuff	Caldera complex, southern New Brunswick, Canada	~46°N, 66°W	Bailey Rock Rhyolite, which intrudes and overlies the Carrow Fm	362.87	± 0.88	± 0.96	$^{206}Pb/^{238}U$
D22	D13				pumice tuff	Caldera complex, southern New Brunswick, Canada	~46°N, 66°W	Carrow Fm pumiceous tuff, Piskahegan Group	364.08	± 2.22	± 2.25	$^{206}Pb/^{238}U$
D21	D12				WVC754; black shale (0.4%–4% TOC)	Cattaraugus County, western New York, USA	~42°N, 79°W	Dunkirk Fm, ~6.4 m above F-F boundary	367.70	± 2.50	± 2.80	Re-Os
				Frasnian								
D20	D11				WVC785; black shale (0.4%–4% TOC)	Cattaraugus County, western New York, USA	~42°N, 79°W	Hanover Fm, ~2.9 m below F-F boundary	374.20	± 4.00	± 4.20	Re-Os
D19	D10				bentonite; Bed 36	Steinbruch Schmidt quarry, Kellerwald, Germany	~51°N, 9°E?	Kellwasser Horizons	372.36	± 0.05	± 0.41	$^{206}Pb/^{238}U$
D18					bentonite; "Tephra 7.67" (EMC-7.67 m)	Eighteenmile Creek, Erie County, NY, USA; 7.67 m above the base of the Rhinestreet Shale	43.3°N, 78.7°W	Rhinestreet Fm, West Falls Group	375.14	± 0.12	± 0.45	$^{206}Pb/^{238}U$
D17					bentonite; "Tephra 06" (LWG Ash-6)	Little War Gap, East Tennessee, USA	36.5°N, 83.0°W	Belpre Tephra Suite, Lower Dowelltown Member, Chattanooga Shale Fm	375.25	± 0.13	± 0.45	$^{206}Pb/^{238}U$

Primary radioisotopic age details	Zonal range assignment			Biostratigraphy	Reference
	clade	zonation	Zone		
Weighted mean of 9 single zircon grain analyses, utilizing CA-TIMS and the ET2535 spike.	conodont	Boreal	Upper *Siphonodella sulcata* Zone	Upper part of *S. (Eos.) sulcata* conodont zone, basal Tournaisian	Davydov et al. (2011)
Five of six single zircon grain analyses have a weighted mean $^{206}Pb/^{238}U$ age; utilizing CA-TIMS and the ET535 spike.	conodont	NW Europe	middle/upper *costatus-kockeli* Interregnum conodont zone	Above the Hangenberg Black Shale, ~2 m below the Devonian-Carboniferous boundary. Had been assigned as *kockeli* Zone (=new name for Upper *praesulcata* Zone) at Kowala; re-assigned as middle/upper *costatus-kockeli* Interregnum (Becker et al., 2020, this volume)	Myrow et al. (2014)
Six of seven single zircon grain analyses have a weighted mean $^{206}Pb/^{238}U$ age; utilizing CA-TIMS and the ET535 spike.	conodont	NW Europe	Upper *Siphonodella praesulcata* Zone	Below the Hangenberg Black Shale, ~3.5 m below the Devonian-Carboniferous boundary. "Unfortunately, the D/C conodont stratigraphy of Kowala has not yet been fully investigated. Both lower ash beds fall in the upper part of revised *S. (Eos.) praesulcata* Zone, and in the higher *Wocklumeria sphaeroides* Zone (UD VI-D) of the ammonoid zonation." (Becker et al., 2020, this volume)	Myrow et al. (2014)
Weighted mean of 13 single zircon grain analyses, utilizing CA-TIMS and the ET535 and 2535 spikes.	conodont	NW Europe	Upper *Siphonodella praesulcata* Zone	..."..."....."	Davydov et al. (2011)
Isochron age for six samples collected over a 4 m thick interval spanning the Devonian-Carboniferous boundary, processed with CrO3-H2SO4 dissolution	conodont	North America	centered at lower part of revised *S. praesulcata*	Interval between the middle *Palmatolepis expansa* through upper *Siphonodella duplicata* zones. "the Re-Os date comes from an interval stretching from the higher *Bi. costatus* Subzone to the basal *costatus-kockeli-Interregnum* (ckI) sensu Kaiser et al. (2009) (below the Hangenberg Regression)." (Becker et al., 2020, this volume)	Selby and Creaser (2005)
Five multigrain zircon analyses yield a weighted mean $^{206}Pb/^{238}U$ age (sample above or crosscutting spore-bearing bed)	conodont	North America	uppermost *Palmatolepis marginifera* to upper *Palmatolepis expansa* zones	Spore-bearing horizon between the Bailey Rock rhyolite and Carrow Fm pumiceous tuff falls within the *pusillites-lepidophyta* spore zone (FA2d), equivalent to the uppermost *Palmatolepis marginifera* to upper *Palmatolepis expansa* conodont zones (Streel, 2000) [assigned as middle/upper *Bi. costatus* conodont subzone in revised zonal scheme]	Tucker et al. (1998)
Four multigrain zircon analyses yield a weighted mean $^{206}Pb/^{238}U$ age (95% conf. int. including geologic scatter; sample below spore-bearing bed)	conodont	North America	uppermost *Palmatolepis marginifera* to upper *Palmatolepis expansa* zones	..."....."....."	Tucker et al. (1998)
Isochron age for nine samples collected from drillcore, processed with CrO3-H2SO4 dissolution	conodont	North America	*Palmatolepis triangularis* Zone	*Palmatolepis triangularis* conodont zone, ~6.4 m above F-F boundary. Equvalent to "*the Pa. delicatula platys* Zone, which replaced terminologically the former Middle subzone of *Pa. triangularis* Zone" (Becker et al., 2020, this volume)	Turgeon et al. (2007)
Isochron age for eight samples collected from drillcore, processed with CrO₃-H₂SO₄ dissolution.	conodont	North America	*Palmatolepis linguiformis* Zone (MN 13b)	*Palmatolepis linguiformus* conodont biozone, upper Frasnian Stage; ~2.9 m below F-F boundary. "A position of D20 within the lower part of the *Pa. linguiformis* Zone (MN Zone 13b) is accepted with some reservation" (Becker et al., 2020, this volume)	Turgeon et al. (2007)
Eight of eleven single zircon grain analyses have a weighted mean $^{206}Pb/^{238}U$ age; utilizing CA-TIMS and the ET2535 spike.	conodont	NW Europe	middle *Palmatolepis bogartensis* Zone (MN 13a)	Bed 36 bentonite occurs in the middle part of the late *Palmatolepis rhenana* conodont Zone, Upper Frasnian; 2.5 m below the F-F boundary. "middle part of the *Pa. bogartensis* Zone or MN Zone 13a (middle part of the Upper *rhenana* Zone sensu Ziegler and Sandberg 1990)." (Becker et al., 2020, this volume)	Percival et al. (2018)
Six of six single zircon grain analyses have a weighted mean $^{206}Pb/^{238}U$ age; utilizing CA-TIMS and the ET535 spike.	conodont	North America	upper "*Ozarkodina*" *nonaginta* Zone (upper MN 7)	Occurrence of "*Ozarkodina*" *nonaginta* indicates the base of FZ 7 at or near the base of the Rhinestreet Formation, where it occurs in association with *Ag. ancyrogna thoideus* and *Ag. primus*, which do not range above FZ 7 (= lowermost subzone of *Palmatolepis hassi* conodont zone)	Lanik et al. (2016)
Six of nine single zircon grain analyses have a weighted mean $^{206}Pb/^{238}U$ age; utilizing CA-TIMS and the ET535 spike.	conodont	North America	*Palmatolepis housei* Zone [MN 8]	Enclosing strata contain *Ad. nodosa* and *Pa. punctata*, which range from FZ 5 through FZ 9 in the lowermost Dowelltown Member; overlying strata contain *Ag. barba* and *Pa. housei* that are restricted to FZ 8 (= second subzone of *Palmatolepis hassi* conodont zone)	Lanik et al. (2016)

GTS 2020 ID	GTS 2012 ID	Period	Epoch	Age	Sample	Locality	Lat-Long	Lithostratigraphy	Age (Ma)	± 2s analytical	± 2s total	Age Type
D16	D9				bentonite; "Tephra 01" (LWG Ash-1)	Little War Gap, East Tennessee, USA	36.5°N, 83.0°W	Belpre Tephra Suite, Lower Dowelltown Member, Chattanooga Shale Fm	375.55	± 0.10	± 0.44	$^{206}Pb/^{238}U$
			Middle									
				Givetian								
				Eifelian								
D15					Tioga Ash Bed F	Seneca Stone Quarry, Seneca Falls, New York, USA	42°51.3′N, 76°47.2′W	Near base of the Union Springs Fm of the Marcellus Subgroup	390.14	± 0.14	± 0.47	$^{206}Pb/^{238}U$
D14					Tioga Ash Bed B	Seneca Stone Quarry, Seneca Falls, New York, USA	42°51.3′N, 76°47.2′W	Contact between Moorehouse and Seneca Members of the Onendaga Fm	390.82	± 0.18	± 0.48	$^{206}Pb/^{238}U$
			Early									
				Emsian								
D13	D6				Hercules I K-bentonite, sample 12VD-80	Wetteldorf section, Schönecken, Germany	50°10′N, 6°27′E	Heisdorf Fm	394.29	± 0.10	± 0.47	$^{206}Pb/^{238}U$
D12	D5				Volcanic layer	Hans-Platte layer, Eschenbach quarry, Bundenbach (Hunsrück), Germany		Lower Hunsrück Slate	407.75	± 1.08	± 1.16	$^{206}Pb/^{238}U$
				Pragian								
D11	D3				volcaniclastic sandstone, sample 94843525	Cheshire Creek, Limekilns District, Hill End Trough, Eastern Lachlan Orogen, eastern Australia	33°15′20″S, 149°41′54″E	Merrions Fm	411.70	± 0.90	± 1.20	$^{206}Pb/^{238}U$
D10					Milton of Noth Andesite	Rhynie Outlier, NE Scotland	57°20′38″N, 02°49′31″W	Tillybrachty Sandstone and Dryden Flags Fms	411.50	± 1.10	± 1.30	$^{206}Pb/^{238}U$
				Lochkovian								
D9	D2				crystal rich volcaniclastic sandstone, sample 94843520	Hill End Trough, Eastern Lachlan Orogen, eastern Australia	33°03′50″S, 149°37′43″E	Turondale Fm	415.60	± 0.50	± 0.80	$^{206}Pb/^{238}U$
D8					felsic volcanic, sample 93844562B	Cowra Trough, Eastern Lachlan Orogen, eastern Australia	33°04′09″S, 149°36′06″E	Bulls Camp Volcanics	417.70	± 0.50	± 0.80	$^{206}Pb/^{238}U$
D7					K-bentonite, sample H5-1	Smoke Hole section, eastern West Virginia, USA; 84.9 mab	38.82257°N, 79.28615°W	lower Corriganville Limestone Fm of the Helderberg Group	417.22	± 0.21	± 0.50	$^{206}Pb/^{238}U$
D6	D1				Kalkberg' K-bentonite (KKB), sample CV-2; also referenced as Judds Falls Bentonite Bed	US Highway 20, W of Judds Falls, northeast of Cherry Valley, eastern NY, USA; 2.40–2.45 mab	42°49′19″N, 74°43′43″W	Historically described as the Kalkberg Fm, but now interpreted as the New Scotland Fm of the Helderberg Group	417.61	± 0.12	± 0.50	$^{206}Pb/^{238}U$
D5	D1				Kalkberg' K-bentonite (KKB), sample H1-1	US Highway 20, northeast of Cherry Valley, eastern NY, USA; 44.1 mab	42°49′N, 74°44′W	Historically described as the Kalkberg Fm, but now interpreted as the New Scotland Fm of the Helderberg Group	417.68	± 0.21	± 0.52	$^{206}Pb/^{238}U$

Primary radioisotopic age details	Zonal range assignment			Biostratigraphy	Reference
	clade	zonation	Zone		
Seven of fifteen single zircon grain analyses have a weighted mean $^{206}Pb/^{238}U$ age; utilizing CA-TIMS and the ET535 spike.	conodont	North America	*Palmatolepis housei* Zone [*MN 8*]	Enclosing strata contain *Ad. nodosa* and *Pa. punctata*, which range from FZ 5 through FZ 9 in the lowermost Dowelltown Member, and are overlain by strata containing *Ag. barba* and *Pa. housei* that are restricted to FZ 8 (= second subzone of *Palmatolepis hassi* zone)	Lanik et al. (2016)
Eight of ten single zircon grain analyses have a weighted mean $^{206}Pb/^{238}U$ age, utilizing CA-TIMS and the ET535 and 2535 spikes.	conodont	North America	*Polygnathus costatus* Zone	All Tioga ashes are between conodont bearing strata of the *Polygnathus costatus* zone, middle Eifelian.	Harrigan et al. (in review)
Seven of eleven single zircon grain analyses have a weighted mean $^{206}Pb/^{238}U$ age, utilizing CA-TIMS and the ET535 and 2535 spikes.	conodont	North America	*Polygnathus costatus* Zone	All Tioga ashes are between conodont bearing strata of the *Polygnathus costatus* zone, middle Eifelian.	Harrigan et al. (in review)
Nine single zircon grain analyses have a weighted mean $^{206}Pb/^{238}U$ age, utilizing CA-TIMS and the ET535 and 2535 spikes.	conodont	NW Europe	uppermost *Polygnathus patulus* Zone	Uppermost part of the *Polygnathus patulus* conodont zone, uppermost Emsian. The "Hercules I" K-bentonite is situated 13 m below the formal boundary of the Lower and Middle Devonian (GSSP section).	Harrigan et al. (in review)
Ten single zircon analyses include five results, which form a tightly grouped cluster and weighted mean $^{206}Pb/^{238}U$ age.	conodont	NW Europe	lower *Eol. gronbergi* Zone	Tentaculites (dacryoconarids) allow a biostratigraphic assignment to the upper part of the *Nowakia zlíchovensis* dacryoconarid zone.	Kaufmann et al. (2005)
Four of six single zircon grain analyses have a weighted mean $^{206}Pb/^{238}U$ age, utilizing CA-TIMS and an in-house spike.	dacryo-canarid	Australasian	*Nowakia acuaria* Zone	Devoid of fossils; but the Pragian dacryoconarid index, *Nowakia acuaria,* and the brachiopod *Nadiastropha* (sensu stricto) appear in the overlying Limekilns Fm [*N. acuaria* spans ca. base-Pragian through *kitabicus* conodont zone].	Bodorkos et al. (2017); Jagodzinski and Black (1999)
Four air- or chemically abraded zircon fractions yield and equivalent data combined into a weighted mean $^{206}Pb/^{238}U$ age; in-house EARTHTIME-calibrated spike.	palyno-morph	NW Europe	polygonalis–emsiensis Spore Assemblage Biozone = *profunda* through *excavatus* s.str. conodont zones	Spore assemblages indicate an early (but not earliest) Pragian to earliest Emsian zonal assignment. This biostratigraphical age range is potentially further constrained by the presence of *Dictyotriletes subgranifer* (Wellman 2006), indicating a latest Pragian to (?) earliest Emsian age.	Parry et al. (2011)
Six of six single zircon grain analyses combined to procude a weighted mean $^{206}Pb/^{238}U$ age, utilizing CA-TIMS and an in-house spike.	conodont	Australasian	*eurekaensis* (=*postwoschmidti*) through lower *delta* (= through *transitans*) zones	Faunal assemblage assigned to the early Lochkovian *Boucotia australis* brachiopod zone (Garratt and Wright, 1989), itself correlated with the *eurekaensis* conodont zone by Mawson et al. (1989).	Bodorkos et al. (2017); Jagodzinski and Black (1999)
Three of five single zircon grain analyses combined to procude a weighted mean $^{206}Pb/^{238}U$ age, utilizing CA-TIMS and an in-house spike.	conodont	Australasian	*woschmidti* (= *hesperius*) through *eurekaensis* (= *postwoschmidti*) Zone	Bulls Camp Volcanics conformably overlie shales containing FOD of Late Silurian *Monograptus* cf *uniformis,* and "overlain by latitic volcanic rocks devoid of fossils (but which elsewhere overlie conodont-bearing limestones of the *woschmidti* to *eurekaensis* zones) . . . indicate a *woschmidti* to *eurekaensis* Zone age for the Bulls Camp Volcanics."	Bodorkos et al. (2017); Jagodzinski and Black (1999)
Youngest precise analysis of eight single grain zircon analyses, utilizing CA-TIMS and the ET535 spike.	conodont	North America	*L. omoalpha* through *A. trigonicus* zones	above the LOD of *O. elegans detortus* and the FOD of *I. woschmidti woschmidti,* and immediately above the Klonk CIE, indicating a position in the late Middle Lochkovian.	Husson et al. (2016)
Six of eight single grain zircon analyses have a weighted mean $^{206}Pb/^{238}U$ age, utilizing CA-TIMS and the ET535 spike.	conodont	North America	*L. omoalpha* through *A. trigonicus* zones	Conodonts *Wurmiella excavata* and *Ozarkodina planilingua* at 1.75–1.85 mab, *Ancyrodelloides sp.* at 2.35–2.40 mab, and *Pseudooneotodus sp.* at 2.45–2.55 mab indicate a position no lower than the base of the *Caudicriodus postwoschmidti* Zone and no higher than the upper part of the *Ancyrodelloides trigonicus* Zone.	McAdams et al. (2017)
Youngest precise analysis of eight single grain zircon analyses, utilizing CA-TIMS and the ET535 spike.	conodont	North America	*L. omoalpha* through *A. trigonicus* zones	Correlation to regional sections indicate a position a position no lower than the base of the *Caudicriodus postwoschmidti* Zone and no higher than the upper part of the *Ancyrodelloides trigonicus* Zone.	Husson et al. (2016)

GTS 2020 ID	GTS 2012 ID	Period	Epoch	Age	Sample	Locality	Lat-Long	Lithostratigraphy	Age (Ma)	± 2s analytical	± 2s total	Age Type
D4					K-bentonite, sample H2-4	Cobleskill, eastern NY, USA; 18.4 mab	~42.5°N, 72.5°W	Historically described as the Kalkberg Fm, but now interpreted as the New Scotland Fm of the Helderberg Group	417.56	± 0.20	± 0.51	$^{206}Pb/^{238}U$
D3					K-bentonite, sample H2-3	Cobleskill, eastern NY, USA; 15.1 mab	~42.5°N, 72.5°W	Historically described as the Kalkberg Fm, but now interpreted as the New Scotland Fm of the Helderberg Group	417.73	± 0.22	± 0.53	$^{206}Pb/^{238}U$
D2					K-bentonite, sample H2-2	Cobleskill, eastern NY, USA; 12.5 mab	~42.5°N, 72.5°W	Historically described as the Kalkberg Fm, but now interpreted as the New Scotland Fm of the Helderberg Group	417.85	± 0.23	± 0.54	$^{206}Pb/^{238}U$
D1					K-bentonite, sample H2-1	Cobleskill, eastern NY, USA; 9.0 mab	~42.5°N, 72.5°W	Historically described as the Kalkberg Fm, but now interpreted as the New Scotland Fm of the Helderberg Group	418.42	± 0.21	± 0.53	$^{206}Pb/^{238}U$
		Silurian										
			Pridoli									
			Ludlow									
				Ludfordian								
S8					C6 bentonite	Ataky 117 section, Khotyn, Podolia, southwestern Ukraine	48°32.5'N, 26°28.8'E	top of the Pryhorodok Fm	422.91	± 0.07	± 0.49	$^{206}Pb/^{238}U$
S7					M12 bentonite	Malynivtsi 150 section, E. of Khotyn, Podolia, southwestern Ukraine	48°29.8'N, 26°36.1'E	upper Hrynchuk Fm	424.08	± 0.20	± 0.53	$^{206}Pb/^{238}U$
				Gorstian								
			Wenlock									
				Homerian								
S6	S8				WNH15 bentonite	Lion's Mouth Cavern, Wren's Nest Hill, Dudley, England	52.52227°N, 2.09668°W	Upper Quarried Limestone Mbr., Much Wenlock Limestone	427.77	± 0.50	± 0.68	$^{206}Pb/^{238}U$
S5	S7				lower of two bentonites separated by 16 cm; Djupvik bentonite	Djupvik 1 Locality, W. Gotland, Sweden	57°18.9'N, 18°10.1'E	Djupvik Member, Halla Fm	428.06	± 0.48	± 0.68	$^{206}Pb/^{238}U$
S4	S6				upper part of the 30-cm-thick Grötlingbo bentonite	Hörsne 3 Locality, central Gotland, Sweden	57.5°N, 18.5°E	Mulde Brickclay Mbr., Halla Fm	428.47	± 0.54	± 0.72	$^{206}Pb/^{238}U$
				Sheinwoodian								
S3	S4				Ireviken bentonite	Ireviken 1 Locality, Gotland, Sweden	57.8°N, 18.4°E	Lower Visby Fm	431.80	± 0.53	± 0.71	$^{206}Pb/^{238}U$
			Llandovery									
				Telychian								
S2	S3				K-bentonite	Osmundsberg North Quarry, Siljan, Dalarna, south-central Sweden	61°01'03"N, 15°12'04"E	Kallholn Shale	438.74	± 1.11	± 1.20	$^{206}Pb/^{238}U$
				Aeronian								
				Rhuddanian								
S1	S2				Ash	Dob's Linn, Moffat, Scotland, UK	55.44°N, 3.27°W	Birkhill Shales	439.57	± 1.24	± 1.33	$^{206}Pb/^{238}U$

Primary radioisotopic age details	Zonal range assignment			Biostratigraphy	Reference
	clade	zonation	Zone		
Youngest precise analysis of eight single grain zircon analyses, utilizing CA-TIMS and the ET535 spike.	conodont	North America	*L. omoalpha* through *A. trigonicus* zones	Correlation to regional sections indicate a position a position no lower than the base of the *Caudicriodus postwoschmidti* Zone and no higher than the upper part of the *Ancyrodelloides trigonicus* Zone.	Husson et al. (2016)
Youngest precise analysis of eight single grain zircon analyses, utilizing CA-TIMS and the ET535 spike.	conodont	North America	*L. omoalpha* through *A. trigonicus* zones	Correlation to regional sections indicate a position a position no lower than the base of the *Caudicriodus postwoschmidti* Zone and no higher than the upper part of the *Ancyrodelloides trigonicus* Zone.	Husson et al. (2016)
Youngest precise analysis of eight single grain zircon analyses, utilizing CA-TIMS and the ET535 spike.	conodont	North America	*L. omoalpha* through *A. trigonicus* zones	Correlation to regional sections indicate a position a position no lower than the base of the *Caudicriodus postwoschmidti* Zone and no higher than the upper part of the *Ancyrodelloides trigonicus* Zone.	Husson et al. (2016)
Youngest precise analysis of eight single grain zircon analyses, utilizing CA-TIMS and the ET535 spike.	conodont	North America	*L. omoalpha* through *A. trigonicus* zones	Correlation to regional sections indicate a position a position no lower than the base of the *Caudicriodus postwoschmidti* Zone and no higher than the upper part of the *Ancyrodelloides trigonicus* Zone.	Husson et al. (2016)
Eight of eight single grain zircon analyses have a weighted mean $^{206}Pb/^{238}U$ age, utilizing CA-TIMS and the ET535 spike.	graptolite/conodont		Upper *Uncinatograpus spineus–Pseudomonoclimacis latilobus* Zone; upper *Ozarkodina crispa* Zone	Middle *Ozarkodina remscheidensis baccata/Ozarkodina snajdri parasnajdri* Polodian conodont zone places interval in the upper global *Ozarkodina crispa* conodont zone and upper Baltic *Uncinatograpus spineus–Pseudomonoclimacis latilobus* graptolite zone.	Cramer et al. (2015)
Seven out of eight single grain zircon analyses have a weighted mean $^{206}Pb/^{238}U$ age, utilizing CA-TIMS and the ET535 spike.	graptolite/conodont		Upper *Saetograptus leintwardinensis* Zone and upper *Polygnathoides siluricus* Zone	Middle *Ozarkodina crispa* Polodian conodont zone places interval in the upper global *Polygnathoides siluricus* conodont zone and upper Baltic *Saetograptus leintwardinensis* graptolite zone	Cramer et al. (2015)
Six out of eleven single grain zircon analyses have a weighted mean $^{206}Pb/^{238}U$ age, utilizing CA-TIMS and the ET535 spike.	graptolite/conodont		Uppermost *Colonograptus ludensis* Zone	From detailed regional correlation this bentonite is likely only cm's below the base Ludlow GSSP. Therefore, a correlation with the uppermost part of the *Colonograptus ludensis* zone is assigned.	Cramer et al. (2012)
Six out of nine single grain zircon analyses have a weighted mean $^{206}Pb/^{238}U$ age, utilizing CA-TIMS and the ET535 spike.	graptolite/conodont		Upper part of *Colonograptus praedeubeli* to lower part of *C. ludensis* zones	Within the *Kockelella ortus absidata* conodont zone, correlated to a position somewhere from high in the *Colonograptus praedeubeli/deubeli* to low in the C. ludensis graptolite zones	Cramer et al. (2012)
Five out of nine single grain zircon analyses have a weighted mean $^{206}Pb/^{238}U$ age, utilizing CA-TIMS and the ET535 spike.	graptolite/conodont		*Pristiograptus dubius parvus/Gothograptus nassa* Zone	Within the *Ozarkodina bohemica longa* conodont zone and the *Pristiograptus dubius parvus/Gothograptus nassa* graptolite zone.	Cramer et al. (2012)
11 out of 15 single grain zircon analyses from the Ireviken Bentonite yield a weighted mean $^{206}Pb/^{238}U$ age (including analytical error, tracer error, and decay constant error), utilizing CA-TIMS and the ET535 spike.	graptolite		Middle to upper part of the *Cyrtograptus murchisoni* Zone	14 cm above Ireviken Event Datum 2 (base of the Upper *Pseudooneotodus bicornis* conodont zone), which is within cm's of the base Wenlock GSSP.	Cramer et al. (2012)
Four single grain zircon analyses yield a weighted mean $^{206}Pb/^{238}U$ age [including 0.1% tracer uncertainty, Mundil et al. (2004)]	graptolite		*Spirograptus turriculatus* Zone	*Spirograptus turriculatus* Zone, Telychian Stage of the Llandovery Series	Bergstrom et al. (2008)
Six multigrain zircon fractions yield a weighted mean $^{206}Pb/^{238}U$ age.	graptolite		*Coronogr. cyphus* Zone	Exact level uncertain, *Coronogr. cyphus* Zone assigned by Ross et al. (1982) and accepted here.	Tucker et al. (1990), Ross et al. (1982), Toghill (1968)

GTS 2020 ID	GTS 2012 ID	Period	Epoch	Age	Sample	Locality	Lat-Long	Lithostratigraphy	Age (Ma)	± 2s analytical	± 2s total	Age Type
		Ordovician										
			Late									
				Hirnantian								
O42					K-bentonite, sample YC0601	Hirnantian GSSP at Wangjiawan North Section, Yichang, Hubei	30.9841°N, 111.4197°E	039 below base of the Kuanyinchiao Bed	443.20	± 1.60	± 2.73	$^{206}Pb/^{238}U$ ion probe
				Katian								
O41	O16				Ash	Dobbs Linn (Linn Branch), Scotland	55.44°N, 3.27°W	Hartfell Shales	444.88	± 1.07	± 1.17	$^{206}Pb/^{238}U$
O40					Manheim K-bentonite, sample W1204_12.8	Nowadaga Creek, New York, USA	42°59.205'N, 74°48.401'W	Indian Castle Fm, Trenton Group	450.68	± 0.12	± 0.53	$^{206}Pb/^{238}U$
O39					Reedville Calmar K-bentonite	Reedsville, Pennsylvania, USA	52°18'50.2"N, 9°58'05.5"E	Antes Shale	451.20	± 0.13	± 0.50	$^{206}Pb/^{238}U$
O38					Manheim Falls K-bentonite	North Creek, New York, USA	43°5.330'N, 74°55.943'W	Dolgeville Fm, Trenton Group	451.26	± 0.10	± 0.53	$^{206}Pb/^{238}U$
O37					Manheim K-bentonite, sample F1302_1.8	Nowadaga Creek, New York, USA	42°59.724'N, 74°47.286'W	Dolgeville Fm, Trenton Group	451.42	± 0.10	± 0.53	$^{206}Pb/^{238}U$
O36					Chuctanunda K-bentonite	Chuctanunda Creek, New York, USA	42°54.869'N, 74°13.792'W	Flat Creek Fm, Trenton Group	451.71	± 0.13	± 0.53	$^{206}Pb/^{238}U$
O35					Sherman Falls K-bentonite	Flat Creek, New York, USA	42°52.069'N, 74°32.266'W	Flat Creek Fm, Trenton Group	452.82	± 0.08	± 0.53	$^{206}Pb/^{238}U$
				Sandbian								
O34					Shakertown Millbrig K-bentonite	Outcrop on U.S. Highway 68, 1.6 km southwest of intersection with State Route 33, at entrance to Shakertown, Mercer County, KY, USA	37°49'N, 84°45'W	Tyrone Fm	452.86	± 0.29	± 0.59	$^{206}Pb/^{238}U$
O33					K-bentonite, sample Upper Womble	Black Knob Ridge, Atoka, Oklahoma; Katian Global Stratotype Section and Point (GSSP)	34°25'39.08"N, 96°04'3.78"W	Upper Womble Shale	453.16	± 0.24	± 0.57	$^{206}Pb/^{238}U$
O32					K-bentonite, sample LOQ-B	L'Orignal Quarry, Ontario, Canada	45°31'N, 74°22'W	top of the L'Orignal Fm	453.36	± 0.38	± 0.65	$^{206}Pb/^{238}U$
O31					K-bentonite, sample Lower Womble	Black Knob Ridge, Atoka, Oklahoma; Katian Global Stratotype Section and Point (GSSP)	34°25'39.08"N, 96°04'3.78"W	Upper Womble Shale	453.98	± 0.33	± 0.62	$^{206}Pb/^{238}U$
O30					Shakertown Deicke K-bentonite	Outcrop on U.S. Highway 68, 1.6 km southwest of intersection with State Route 33, at entrance to Shakertown, Mercer County, KY, USA	37°48.985'N, 84°45.432'W	Tyrone Fm	453.74	± 0.20	± 0.56	$^{206}Pb/^{238}U$
O29					K-bentonite, sample C-10-11 upper Grimstorp bentonite	Linlandveien road section, Vollen, Oslo, Norway; 7 m above base	59°48.19' N, 10°29.21'E	Arnestad Fm	453.91	± 0.37	± 0.61	$^{206}Pb/^{238}U$

Primary radioisotopic age details	Zonal range assignment			Biostratigraphy	Reference
	clade	zonation	Zone		
Weighted mean of 18 (of 20) zircon spots using SHRIMP II (Beijing), calibrated to standard TEM.	graptolite	South China	top of *Metabolograptus extraordinarius Zone*	uppermost *Metabolograptus extraordinarius* Zone	Hu et al. (2008)
Four multigrain zircon fractions yield a weighted mean $^{206}Pb/^{238}U$ age of 444.88 ± 1.17 Ma	graptolite	Britain	*Paraorthograptus pacificus* Zone	Approximately 4.5 m below Ordov/Sil GSSP, *Paraorthograptus pacificus* Zone	Tucker et al. (1990)
Six of six single grain zircon analyses have a weighted mean $^{206}Pb/^{238}U$ age, utilizing CA-TIMS and the ET535 spike.	graptolite	North America	top of *Diplacanthograptus spiniferus* Zone	top of *Diplacanthograptus spiniferus* Zone	Macdonald et al. (2017)
Eight of eight single grain zircon analyses have a weighted mean $^{206}Pb/^{238}U$ age, utilizing CA-TIMS and the ET2535 spike.	graptolite	North America	lowermost *Diplacanthograptus spiniferus* Zone	Bentonite just above base of the *Diplacanthograptus spiniferus* Zone	Taylor et al. (2015)
Seven of nine single grain zircon analyses (excluding two older grains) have a weighted mean $^{206}Pb/^{238}U$ age, utilizing CA-TIMS and the ET535 spike.	graptolite	Australian	lowermost *Diplacanthograptus spiniferus* Zone	Dlgevilel Fm contains graptolites of the *Orthograptus rudemanni* zone. Manheim bentonite correlated through apatite chemistry to the Calmar K-bentonite (Sell et al., 2015). N.Amer. *O. rudemanni* Zone considered coeval with lower *D. spiniferous* Zone of Australia.	Macdonald et al. (2017)
Six of nine single grain zircon analyses (excluding three older grains) have a weighted mean $^{206}Pb/^{238}U$ age, utilizing CA-TIMS and the ET535 spike.	graptolite	Australian	lowermost *Diplacanthograptus spiniferus* Zone	Dlgevilel Fm contains graptolites of the *Orthograptus rudemanni* zone. Manheim bentonite correlated through apatite chemistry to the Calmar K-bentonite (Sell et al., 2015). N.Amer. *O. rudemanni* Zone considered coeval with lower *D. spiniferous* Zone of Australia.	Macdonald et al. (2017)
Eight of eight single grain zircon analyses have a weighted mean $^{206}Pb/^{238}U$ age, utilizing CA-TIMS and the ET535 spike.	graptolite	Australian	*Diplacanthograptus lanceolatus*	Contains graptolites of the *Corynoides americanus* biozone. *C. americanus* considered coeval with the Australian *Diplacanthograptus lanceolatus* Zone.	Macdonald et al. (2017)
Five of eight single grain zircon analyses (excluding three older grains) have a weighted mean $^{206}Pb/^{238}U$ age, utilizing CA-TIMS and the ET535 spike.	graptolite	Australian	*Diplacanthograptus lanceolatus*	Contains graptolites of the *Corynoides americanus* biozone. *C. americanus* considered coeval with the Australian *Diplacanthograptus lanceolatus* Zone.	Macdonald et al. (2017)
Three of five single grain zircon analyses (excluding two older grains) have a weighted mean $^{206}Pb/^{238}U$ age, utilizing CA-TIMS and the ET2535 spike.	graptolite/ conodont	N. Atlantic/ N. Amer. Midcont.	Upper *Climacograptus bicornis* Zone	*Climacograptus bicornis* Zone/*Phragmodus undatus* Zone	Sell et al. (2013)
Ten of twelve single grain zircon analyses (excluding two younger grains) have a weighted mean $^{206}Pb/^{238}U$ age, utilizing CA-TIMS and the ET2535 spike.	graptolite/ conodont	N. Atlantic/ N. Amer. Midcont.	*Climacograptus bicornis* Zone/*Phragmodus undatus* Zone	Diagnostic graptolites include *C. bicornis, C. bicornis tridentatus, Orthograptus whitfieldi, O. calcaratus* ssp., *Archiclimacograptus modestus, Dicranograptus spinifer, D. contortus, D. arkansasensis, Normalograptus brevis,* and *Nemagraptus gracilis*; the co-occurrence of *I.* cf. *I. superba* and *A. tvaerensis* indicates the *B. alobatus* Subzone of the *A. tvaerensis* Zone.	Sell et al. (2013)
Six of six single grain zircon analyses have a weighted mean $^{206}Pb/^{238}U$ age, utilizing CA-TIMS and the ET535 spike.	graptolite/ conodont	N. Atlantic/ N. Amer. Midcont.	*Climacograptus bicornis* Zone	No biostrat included; but considered to be equivalent to the Millbrig bentonite' which would be near base of *Plectodina tenuis* conodont zone (Fig. 13 in their paper)	Oruche et al. (2018)
Six of eight single grain zircon analyses (excluding two older grains) have a weighted mean $^{206}Pb/^{238}U$ age, utilizing CA-TIMS and the ET2535 spike.	graptolite/ conodont	N. Atlantic/ N. Amer. Midcont.	*Climacograptus bicornis* Zone/*Phragmodus undatus* Zone	Diagnostic graptolites include *C. bicornis, C. bicornis tridentatus, Orthograptus whitfieldi, O. calcaratus* ssp., *Archiclimacograptus modestus, Dicranograptus spinifer, D. contortus, D. arkansasensis, Normalograptus brevis,* and *Nemagraptus gracilis*; the co-occurrence of *I.* cf. *I. superba* and *A. tvaerensis* indicates the *B. alobatus* Subzone of the *A. tvaerensis* Zone.	Sell et al. (2013)
Four of five single grain zircon analyses (excluding one older grain) have a weighted mean $^{206}Pb/^{238}U$ age, utilizing CA-TIMS and the ET2535 spike.	graptolite/ conodont	N. Atlantic/ N. Amer. Midcont.	Upper *Climacograptus bicornis* Zone	*Climacograptus bicornis* Zone/*Phragmodus undatus* Zone	Sell et al. (2013)
Thirteen of sixteen single grain zircon analyses (excluding one younger and two older grains) have a weighted mean $^{206}Pb/^{238}U$ age, utilizing CA-TIMS and in-house spike intercalibrated to ET100 solution.	conodont	North Atlantic	*Amorphognathus tvaerensis* Zone	*Amorphognathus tvaerensis* Zone	Svensen et al. (2015)

GTS 2020 ID	GTS 2012 ID	Period	Epoch	Age	Sample	Locality	Lat-Long	Lithostratigraphy	Age (Ma)	± 2s analytical	± 2s total	Age Type
O28					Vasagard Kinnekulle K-bentonite	Vasagard section along Lœsa brook, Bornholm, Denmark; 0.5−2 m above base	54°52′N, 14°55′W	Skagen Fm	454.41	± 0.17	± 0.53	$^{206}Pb/^{238}U$
O27					Ristikula bed 46 K-bentonite	Ristikula core, Pärnu County, SW Estonia	~58.2°N, 24.8°W	bed 46	454.65	± 0.56	± 0.75	$^{206}Pb/^{238}U$
O26					Arnestad Kinnekulle K-bentonite, sample C-10-9 Arnestad tephra (tuff B)	Linlandveien road section, Vollen, Oslo, Norway; −1 to 0 m below base	59°48.19′N, 10°29.21′E	Arnestad Fm	454.52	± 0.50	± 0.70	$^{206}Pb/^{238}U$
O25					K-bentonite, sample EB16-S-28.63	Sinsen railway cut, Oslo, Norway; 28.63-m above base	59°56.138′N, 10°46.983′E	Arnestad Fm	456.84	± 0.48	± 0.69	$^{206}Pb/^{238}U$
O24					K-bentonite, sample EB16-S-0	Sinsen railway cut, Oslo, Norway; 0-m above base	59°56.138′N, 10°46.983′E	Arnestad Fm	457.66	± 0.65	± 0.83	$^{206}Pb/^{238}U$
O23					K-bentonite	East River Mountain Tunnel, I-77, Mercer County, West Virginia, USA	37°17.0′N, 18°07.5′E	Elway Fm	458.76	± 0.26	± 0.56	$^{206}Pb/^{238}U$
			Middle									
				Darriwilian								
O22	O8				Gritty calcareous ash	Llandindrod, central Wales	~52.2°N, 3.4°W	Llanvirn Series	458.76	± 2.19	± 2.24	$^{206}Pb/^{238}U$
O21	O6				Indurated bentonite	Abereiddy Bay, Wales	~51.9°N, 5.2°W	Lower rhyolitic tuff, Llanrian Volc Fm	462.90	± 1.23	± 1.32	$^{206}Pb/^{238}U$
O20	O5				Ash flow	Arenig Fawr, Wales	~52.9°N, 3.7°W	Serv Fm	465.61	± 1.69	± 1.76	$^{206}Pb/^{238}U$
O19					K-bentonite, sample F1456	Red slate quarry, Giddings Brook Thrust Sheet, eastern New York, USA	43°28.960′N, 73°19.098′W	upper Indian River Fm	464.20	± 0.13	± 0.55	$^{206}Pb/^{238}U$
O18					K-bentonite, sample Mainland 41 m	Mainland Section, Port au Port Peninsula, western Newfoundland, Canada; 41 m above base	~48.5°N, 49.0°W	Cape Cormorant Fm	464.50	± 0.40	± 0.64	$^{206}Pb/^{238}U$
O17					K-bentonite, sample D	West Bay Centre quarry section, Port au Port Peninsula, western Newfoundland, Canada	~48.5°N, 49.0°W	Table Head Group	464.57	± 0.95	± 1.07	$^{206}Pb/^{238}U$
L16					Likhall Bed zircons recovered from limestone	Thorsberg Quarry, Kinnekulle, Sweden	58°34′45″N, 13°25′46″E	Likhall Bed, Taljsten Interval	467.50	± 0.28	± 0.62	$^{206}Pb/^{238}U$
O15	O7				ARG-1 K-bentonite	Cerro Viejo, near Jáchal, San Juan Province, Argentina	30°11′05″S, 68°35′05″W	Lower Member of the Los Azules Fm	465.46	± 3.49	± 3.53	$^{206}Pb/^{238}U$
O14					K-bentonite, sample KB-1	Cerro La Chilca section, San Juan Province, Argentina	30°36′16.9″S, 69°47′41.0″W	Upper San Juan Fm	469.53	± 0.26	± 0.62	$^{206}Pb/^{238}U$

Primary radioisotopic age details	Zonal range assignment			Biostratigraphy	Reference
	clade	zonation	Zone		
Eight of eight single grain zircon analyses have a weighted mean $^{206}Pb/^{238}U$ age, utilizing CA-TIMS and the ET2535 spike.	graptolite/ conodont	Britain	Upper *Diplograptus foliaceus* Zone; *S. cervicornis* Zone	Upper *Diplograptus foliaceus* Zone; *S. cervicornis* Zone	Sell et al. (2013)
Five of five single grain zircon analyses have a weighted mean $^{206}Pb/^{238}U$ age, utilizing CA-TIMS and the ET2535 spike.	chitinozoan		*S. cervicornis* Zone	*S. cervicornis* Zone	Sell et al. (2013)
Eleven of fourteen single grain zircon analyses (excluding one younger and two older grains) have a weighted mean $^{206}Pb/^{238}U$ age, utilizing CA-TIMS and in-house spike intercalibrated to ET100 solution.	conodont	North Atlantic	*Amorphognathus tvaerensis* Zone	*Amorphognathus tvaerensis* Zone	Svensen et al. (2015)
Four of six single grain zircon analyses (excluding one younger and one older grain) have a weighted mean $^{206}Pb/^{238}U$ age, utilizing CA-TIMS and in-house spike intercalibrated to ET100 solution.	graptolite/ conodont	Baltic	*Climacograptus bicornis* graptolite zone *or Amorphognathus tvaerensis* conodont zone	Based on the stratigraphic relationship to the Kinnekulle bentonite suite, these beds are Mid- Late Sandbian in age—*Climacograptus bicornis* graptolite Zone or *Amorphognathus tvaerensis* conodont Zone.	Ballo et al. (2019)
Two of four single grain zircon analyses (excluding one younger and one older grain) have a weighted mean $^{206}Pb/^{238}U$ age, utilizing CA-TIMS and in-house spike intercalibrated to ET100 solution.	graptolite/ conodont	Baltic	Upper *Nemagraptus gracilis* Zone?	Based on the stratigraphic relationship to the Kinnekulle bentonite suite, these beds are Mid Sandbian in age—*Nemagraptus gracilis* graptolite Zone (might be below base of *B. cornis*) or *Amorphognathus tvaerensis* conodont Zone.	Ballo et al. (2019)
Five of five single grain zircon analyses have a weighted mean $^{206}Pb/^{238}U$ age, utilizing CA-TIMS and the ET2535 spike.	condont	N. Atlantic/N. Amer. Midcont.	Upper *Cahabagnathus sweeti* Zone?	Conodont fauna in the Elway Fm immediately below the ERM K-b consists of *Plectodina aculeata?, Pteracontiodus sp. cf. Phragmodus flexuous, Panderodus sp.,* and *Erismodus sp.* conodont fauna in Elway Fm immediately above the ERM K-b consists of *Pl. aculeata, Ph. flexuosus, Pa. sp., Er. sp., Curtognathus sp., . . .* and *Appalachignathus delicatulus.*	Leslie et al. (2012)
Five multigrain zircon fractions yield a weighted mean $^{206}Pb/^{238}U$ age of 458.76 ± 2.24 Ma (95% conf. int. including geologic scatter).	graptolite	Britain	*Didymograptus murchisoni* Zone	*Didymograptus murchisoni* immediately below sampled ash *"considered by Elles to be close to base G. teretiusculus Zone"*	Tucker and McKerrow (1995)
Three multigrain zircon fractions yield a weighted mean $^{206}Pb/^{238}U$ age and corroborating weighted mean $^{207}Pb/^{206}Pb$ age [recalculated using the U decay constant ratio of Mattinson (2010)]	graptolite	Britain	*Didymograptus murchisoni* Zone	Immediately overlying Cyffredin Shale is of *Didymograptus murchisoni* zone age (Tucker and McKerrow 1995).	Tucker et al. (1990)
Two multigrain zircon fractions yield a weighted mean $^{206}Pb/^{238}U$ age.	graptolite	Britain	*Didymograptus artus* Zone	Underlying mudstone contains *Didymograptus artus* Zone graptolites	Tucker et al. (1990)
Six of six single grain zircon analyses have a weighted mean $^{206}Pb/^{238}U$ age, utilizing CA-TIMS and the ET535 spike.	none		?	No fossils have been identified within the Indian River Formation. Sharply overlain by Mount Merino Fm that contains *N. gracilis* graptolite zone in upper part (early Sandbian). Their Fig. 11 suggests that the dated Indian River ashes might be lower Darriwilian.	Macdonald et al. (2017)
Five of five single grain zircon analyses have a weighted mean $^{206}Pb/^{238}U$ age, utilizing CA-TIMS and the ET2535 spike.	graptolite	N. America	*Pterograptus elegans* Zone	*Pterograptus elegans* Zone	Sell et al. (2011)
Three of three single grain zircon analyses have a weighted mean $^{206}Pb/^{238}U$ age, utilizing CA-TIMS and the ET2535 spike.	conodont	N. America	upper *Holmograptus holodentata* Zone	*Holmograptus spinosus* graptolite zone (regional zone equivalent to upper *Holm. lentus* Zone according to Maletz (2009), [D. Goldman 29 Apr 2019 "New conodont information indicates upper *Histiodella holodentata* conodont zone."]	Sell et al. (2011)
Eight of sixteen single grain zircon analyses (excluding six older grains and two younger grains) have a weighted mean $^{206}Pb/^{238}U$ age, utilizing CA-TIMS and the ET2535 spike.	conodont	Argentina	base of *Y. crassus* Zone	The base of the 'Likhall' bed coincides with the boundary between the globally recognized *Lenodus variabilis* and *Yangtzeplacognathus crassus* conodont zones.	Lindskog et al. (2017)
Three multigrain zircon fractions (14 grains total) yield a weighted mean $^{206}Pb/^{238}U$ age of 465.46 ± 3.53 Ma (95% conf. int. including geologic scatter).	graptolite	Austral-asian	*U. sinicus* Subzone of the *U. austrodentatus* Zone	10 graptolite species listed by Mitchell et al. (1998); the *U. sinicus* Subzone of the *U. austrodentatus* Zone. Based on Ortega et al. (2007) this is Da2−L. dentatus graptolite zone.	Huff et al. (1997)
Six of nine single grain zircon analyses (excluding three older grains) have a weighted mean $^{206}Pb/^{238}U$ age, utilizing CA-TIMS and the ET535 spike.	conodont	North Atlantic	*L. variabilis?*	All three dated beds are in the upper San Juan Formation. Conodont range information is not yet available but estimated to be lowermost Darriwilian (*L. variabilis* or *L. antivariabilis*).	Thompson et al. (2012)

GTS 2020 ID	GTS 2012 ID	Period	Epoch	Age	Sample	Locality	Lat-Long	Lithostratigraphy	Age (Ma)	± 2s analytical	± 2s total	Age Type
O13					K-bentonite, sample KBT-10	Talacasto section, San Juan Province, Argentina	31°00'35.5''S, 68°46'12.0''W	Upper San Juan Fm	469.63	± 0.21	± 0.60	$^{206}Pb/^{238}U$
O12					K-bentonite, sample KBT-7	Talacasto section, San Juan Province, Argentina	31°00'35.5''S, 68°46'12.0''W	Upper San Juan Fm	469.86	± 0.33	± 0.65	$^{206}Pb/^{238}U$
				Dapingian								
			Early									
				Floian								
O11					K-bentonite, sample 221598	Olympic 1 well, Broome Platform, Canning Basin, Western Australia; 1165.44−1165.45 m below top of core	18°15'00.8"S, 122°37'10.9"E	Willara Fm	470.18	± 0.13	± 0.55	$^{206}Pb/^{238}U$
O10					K-bentonite, sample KGC2	Knock Airport section, Charlestown Inlier, County Mayo, Ireland; 13-m above base of section	53.914677°N, 8.8075466°W	upper Horan Fm	472.01	± 0.19	± 0.55	$^{206}Pb/^{238}U$
O9					K-bentonite, sample 221599	Olympic 1 well, Broome Platform, Canning Basin, Western Australia; 1239.27−1239.30 m below top of core	18°15'00.8"S, 122°37'10.9"E	Upper Member, Nambeet Fm	471.32	± 0.11	± 0.55	$^{206}Pb/^{238}U$
O8					K-bentonite, sample 221600	Olympic 1 well, Broome Platform, Canning Basin, Western Australia; 1249.31−1249.33 m below top of core	18°15'00.8"S, 122°37'10.9"E	Upper Member, Nambeet Fm	471.78	± 0.13	± 0.56	$^{206}Pb/^{238}U$
O7					K-bentonite, sample 221474	Olympic 1 well, Broome Platform, Canning Basin, Western Australia; 1264.61−1264.62 m below top of core	18°15'00.8"S, 122°37'10.9"E	Upper Member, Nambeet Fm	472.82	± 0.13	± 0.56	$^{206}Pb/^{238}U$
O6					K-bentonite, sample KBT-3N	Talacasto section, San Juan Province, Argentina	31°00'35.5''S, 68°46'12.0''W	Upper San Juan Fm	473.45	± 0.40	± 0.70	$^{206}Pb/^{238}U$
O5					K-bentonite, sample 221477	Olympic 1 well, Broome Platform, Canning Basin, Western Australia; 1335.03−1335.04 m below top of core	18°15'00.8"S, 122°37'10.9"E	Upper Member, Nambeet Fm	477.07	± 0.21	± 0.59	$^{206}Pb/^{238}U$
O4					K-bentonite, sample 221478	Olympic 1 well, Broome Platform, Canning Basin, Western Australia; 1339.56−1339.57 m below top of core	18°15'00.8"S, 122°37'10.9"E	Upper Member, Nambeet Fm	477.03	± 0.16	± 0.57	$^{206}Pb/^{238}U$
				Tremodocian								
O3					K-bentonite, sample 221480	Olympic 1 well, Broome Platform, Canning Basin, Western Australia; 1383.27−1383.28 m below top of core	18°15'00.8"S, 122°37'10.9"E	Upper Member, Nambeet Fm	479.37	± 0.16	± 0.57	$^{206}Pb/^{238}U$
O2	O2				Volcanic sandstone	section McL-6, 4 km downstream of the ford where Bourinot Road crosses McLeod Brook, Cape Breton Island	∼46.0°N, 60.5°W	Chelsey Drive Group	481.13	± 1.12	± 2.76	$^{207}Pb/^{206}Pb$
O1	O1				Crystal-rich volcanic sandstone	Bryn-llin-fawr, Harlech Dome, N. Wales	52°51'35.9"N, 3°47'53.0"W	sequence boundary between Dolgellau and Dol-cyn-afon Fm, Mawddach Group	486.78	± 0.53	± 2.57	$^{207}Pb/^{206}Pb$
		Cambrian										
			Furongian									
				Age 10								
C16	C11				Crystal-rich volcanic sandstone	Ogof-ddu, Criccieth, N. Wales	52°55.138'N, 4°12.803'W	lower Dolgellau Fm, Mawddach Group	488.71	± 1.17	± 2.78	$^{207}Pb/^{206}Pb$

Primary radioisotopic age details	Zonal range assignment			Biostratigraphy	Reference
	clade	zonation	Zone		
Five of seven single grain zircon analyses (excluding two older grains) have a weighted mean $^{206}Pb/^{238}U$ age, utilizing CA-TIMS and the ET535 spike.	conodont	North Atlantic	*L. variabilis?*	*L. antivariabilis* to *L. variabilis* conodont zones, assumed that this same-dated horizon in this section is similar in age to the nearby section for O15 (See comment on O13)	Thompson et al. (2012)
Four of six single grain zircon analyses (excluding two older grains) have a weighted mean $^{206}Pb/^{238}U$ age, utilizing CA-TIMS and the ET535 spike.	conodont	North Atlantic	*L. variabilis?*	*L. antivariabilis* to *L. variabilis* conodont zones, assumed that this same-dated horizon in this section is similar in age to the nearby section for O15 (See comment on O13)	Thompson et al. (2012)
Eight of eight single grain zircon analyses have a weighted mean $^{206}Pb/^{238}U$ age, utilizing CA-TIMS and the ET535 spike.	conodont	Austral-asian	*Jumudontus gananda* Zone	Condonts of the *J. gananda* Zone	Normore et al. (2018)
Four of six single grain zircon analyses (excluding one older and one younger grain) have a weighted mean $^{206}Pb/^{238}U$ age, utilizing CA-TIMS and the ET535 spike.	graptolite	Austral-asian	*C. morsus* Zone	Graptolites sampled at 10-m above base in the section (and 2 m below this dated horizon) include *Pseudisograptus sp.* of the *manubriatus* group, and *Exigraptus uniformis* and *Skiagraptus gnomonicus*, indicating a latest Dapingian (i.e. Yapeenian Ya 2/late Arenig) age.	Herrington et al. (2018)
Seven of seven single grain zircon analyses have a weighted mean $^{206}Pb/^{238}U$ age, utilizing CA-TIMS and the ET535 spike.	conodont	Austral-asian	*O. communis* Zone	Condonts of the *Oepikodus communis* Zone, for which correlation with graptolite biostratigraphy should be *P. fruticosus* graptolite zone.	Normore et al. (2018)
Seven of seven single grain zircon analyses have a weighted mean $^{206}Pb/^{238}U$ age, utilizing CA-TIMS and the ET535 spike.	conodont	Austral-asian	*O. communis* Zone	Condonts of the *Oepikodus communis* Zone, for which correlation with graptolite biostratigraphy should be *P. fruticosus* graptolite zone.	Normore et al. (2018)
Six of six single grain zircon analyses have a weighted mean $^{206}Pb/^{238}U$ age, utilizing CA-TIMS and the ET535 spike.	conodont	Austral-asian	*O. communis* Zone	Condonts of the *Oepikodus communis* Zone, for which correlation with graptolite biostratigraphy should be *P. fruticosus* graptolite zone.	Normore et al. (2018)
Four of four single grain zircon analyses have a weighted mean $^{206}Pb/^{238}U$ age, utilizing CA-TIMS and the ET535 spike.	conodont	Precordiilleran	Uncertain; maybe *base O. evae* Zone	The biostratigraphy is currently questionable; upper Floian; base *Oepikodus evae* conodont zone?	Thompson et al. (2012)
Two of eight single grain zircon analyses have a weighted mean $^{206}Pb/^{238}U$ age, utilizing CA-TIMS and the ET535 spike.	conodont	Austral-asian	*P. oepiki -S. bilobatus* Zone	Conodonts of the *P. oepiki—S. bilobatus* Zone	Normore et al. (2018)
Six of eight single grain zircon analyses have a weighted mean $^{206}Pb/^{238}U$ age, utilizing CA-TIMS and the ET535 spike.	conodont	Austral-asian	*P. oepiki -S. bilobatus* Zone	Conodonts of the *P. oepiki -S. bilobatus* Zone	Normore et al. (2018)
Seven of seven single grain zircon analyses have a weighted mean $^{206}Pb/^{238}U$ age, utilizing CA-TIMS and the ET535 spike.	conodont	Gond-wanan	*P. proteus* Zone	Conodonts of the *P. proteus* Zone	Normore et al. (2018)
Six multigrain zircon fractions yield a weighted mean $^{207}Pb/^{206}Pb$ age of 481.13 ± 2.76 Ma [recalculated using the U decay constant ratio of Mattinson (2010)].	trilobite	Avalonian	Late Tremadocian (Hunnebergian), Late La2 Zone	Trilobite *Peltocare rotundiformis, Hunnegr. cf. copiosus, Adelograptus* of *quasimodo* type	Landing et al. (1997)
Fourteen single zircon grains yield a weighted mean $^{207}Pb/^{206}Pb$ age of 486.78 ± 2.57 Ma [recalculated using the U decay constant ratio of Mattinson (2010)]	graptollite	Avalonian	Base, *R. praeparabola* Zone, base Ordovician	Close to top *Acercare* Zone. Dated ash is 4 m below appearance of *Rhabdinopora*, and 5 m below *R.f. parabola*. It is therefore very close to C/O boundary	Landing et al. (2000)
Nine multigrain zircon fractions to yield a weighted mean $^{207}Pb/^{206}Pb$ age of 488.71 ± 1.17 (analytical) or ± 2.78 Ma (total) [as recalculated using the U decay constant ratio of Mattinson (2010)]	trilobite	Avalonian	Lower *Peltura scarabaeoides* Zone	*Peltura scarabaeoides scarabaeoides* below and *P.s. westergardi* above "indicate the third subzone (*Parabolina lobata* Subzone) of the *Peltura scarabaeoides* Zone in Norway.	Davidek et al. (1998)

GTS 2020 ID	GTS 2012 ID	Period	Epoch	Age	Sample	Locality	Lat-Long	Lithostratigraphy	Age (Ma)	± 2s analytical	± 2s total	Age Type
				Jiangshanian								
				Paibian								
			Maiolingian									
				Guzhangian								
C15	C10				Volcanic ash bed	Taylor Nunatak, Shackleton Glacier, Antarctica	87°19.114'S, 149°26.079'W	Taylor Fm	502.10	± 2.40	± 3.50	$^{207}Pb/^{206}Pb$
				Drumian								
C14					Trieb-1 volcanic ash bed	~650 m southwest of Triebenreuth village; 1.0 m above the base of the lower division of the Triebenreuth Formation, Germany	50°10.658'N, 11°32.965'E	lower Triebenreuth Fm	503.14	± 0.13	± 0.59	$^{206}Pb/^{238}U$
				Wuliuan								
C13	C9				Comley ub volcanic ash bed	200 m south of Comley Quarry, Comley village, Shropshire, England	52°33.670'N, 2°45.756'W	basal Quarry Ridge Grits, basalt Upper Comley Sandstone Fm	509.10	± 0.33	± 0.62	$^{206}Pb/^{238}U$
			Epoch 2									
				Age 4								
C12	C8				SoS-56.1 volcanic ash bed	Somerset Street, Saint John, New Brunswick, Canada	45°16.765'N, 66°3.852'W	9.5 m above the base of the Hanford Brook Fm, middle Somerset St Mbr	508.05	± 1.13	± 2.75	$^{207}Pb/^{206}Pb$
				Age 3								
C11	C7				Section Le-XI volcanic ash bed	Section Le-XI, south of Taliwine n' Aït-Al Mimoun in the upper Lemdad valley, Anti-Atlas, southern Morocco	30°47.822'N, 8°10.480'W	Upper Lemdad Fm	515.56	± 1.03	± 1.16	$^{206}Pb/^{238}U$
C10	C6				Comley lb volcanic ash bed	200 m south of Comley Quarry, Comley village, Shropshire, England	52°33.670'N, 2°45.756'W	Several centimeters below the top of the Green Callavia Sandstone, uppermost Lower Comley Sandstone Fm	514.45	± 0.43	± 0.69	$^{206}Pb/^{238}U$
C9					Purley Shale Formation bentonite	NW corner of Woodlands Quarry, 200 m NNW of Hartshill Green (5 km NW of Nuneaton), Warwickshire, England	52°33.009'N, 1°31.401'W	Seven meters above the base of the Purley Shale Fm, Charnwood Block, Avalon Composite Terrane	517.22	± 0.40	± 0.66	$^{206}Pb/^{238}U$
C8					Mudstone, sample 14CJ-3	Xiaolantian section, Chengjiang County, eastern Yunnan, South China	24°40'53"N, 102°58'50"E	Maotianshan Shale, Yu'anshan Member, Chiungshussuan Fm, South China Block	≤ 518.03	± 0.69	± 0.71	$^{206}Pb/^{238}U$
C7	C5				Cwm Bach 1 volcanic ash bed	Cwm Bach, near Newgale, Pembrokeshire, south Wales	51°51.822'N, 5°8.225'W	closely above only known fossiliferous horizon, Caerfai Bay Shales Fm	519.30	± 0.34	± 0.64	$^{206}Pb/^{238}U$
			Terreneuvian									
				Age 2								
C6					M236 volcanic ash bed	Oud Sdas section, Anti-Atlas, Morocco	30°23.600'N, 8°38.700'W	upper Lie de Vin Fm	520.93	± 0.21	± 0.57	$^{206}Pb/^{238}U$
C5					M234 volcanic ash bed	Oud Sdas section, Anti-Atlas, Morocco	30°23.486'N, 8°38.115'W	lower Lie de Vin Fm	523.17	± 0.22	± 0.57	$^{206}Pb/^{238}U$
C4					M231 volcanic ash bed	Oud Sdas section, Anti-Atlas, Morocco	30°22.229'N, 8°37.676'W	Tifnout Member, upper Adoudou Fm	524.84	± 0.18	± 0.56	$^{206}Pb/^{238}U$

Primary radioisotopic age details	Zonal range assignment			Biostratigraphy	Reference
	clade	zonation	Zone		
Sample TAY-F, two multigrain zircon fractions yield a weighted mean $^{207}Pb/^{206}Pb$ age of 502.1 ± 3.5 Ma [recalculated using the U decay constant ratio of Mattinson (2010)]	trilobite	Gond-wanan	Undillan Stage	Trilobites in carbonate bed, 1 km from dated samples. *Amphoton* cf. *oatesi, Nelsonia* cf. *schesis*, taken to indicate an Undillan, possibly late Floran, age.	Encarnación et al. (1999)
Five single zircon grain analyses have a weighted mean $^{206}Pb/^{238}U$ age utilizing CA-TIMS and the ET535 spike.	trilobite	Gond-wanan	Drumian Stage	Poorly preserved but relatively diverse, trilobite-dominated assemblage of eodiscinids, corynexochids and ptychopariids suggesting a traditional middle Middle Cambrian (middle Celtiberian Series) age are found roughly 40 m above the Triebenreuth volcaniclastic rocks.	Landing et al. (2014)
Seven single zircon grain analyses yield a weighted mean $^{206}Pb/^{238}U$ age of 509.07 ± 0.33 Ma, utilizing CA-TIMS and the ET535 spike.	trilobite	Avalonian	*P. harlani* Zone, upper Stage 4, Series 2	*Paradoxides harlani* and other trilobites in immediately overlying beds indicate the *P. harlani* Biozone of Newfoundland	Harvey et al. (2011)
Eight single zircon grains or small multigrain fractions yield a weighted mean 207Pb/206Pb age of 508.05 ± 2.75 Ma [recalculated using the U decay constant ratio of (Mattinson (2010)]	trilobite	Avalonian	*Protolenus howleyi* Zone Late Branchian, Stage 4, Series 2	*Protolenus* cf. *elegans* Matthew, *Ellipsocephalus* cf. *galeatus* Matthew associated in same bed. Suggests an age for the base of Series 3 and Stage 5 of ~507 Ma.	Landing et al. (1998)
Five single zircon grains yield a weighted mean $^{206}Pb/^{238}U$ age of 515.56 ± 1.16 Ma.	trilobite	Avalonian	*Antatlasia guttapluviae* Zone, Banian Stage; Stage 3	*A. guttapluviae* Zone, based on detailed correlation to section Le-I, 8 km away. The trilobite, *Berabichia vertumnia*, a guide to the *A. guttapluviae* Zone, is 21 m higher in sequence. Lower Botomian.	Landing et al. (1998)
Two single zircon grain analyses yield a weighted mean $^{206}Pb/^{238}U$ age of 514.38 ± 0.43 Ma, utilizing CA-TIMS and the ET535 spike.	trilobite	Avalonian	*Callavia* Zone of upper Stage 3 of Series 2	Dated ash lies in the *Callavia* Zone of upper Stage 3 of Series 2	Harvey et al. (2011)
Five of nine single zircon grains (excluding four older grains) yield a weighted mean $^{206}Pb/^{238}U$ age, utilizing CA-TIMS and the ET535 spike.	trilobite	Avalonian	*Fallotaspis* or *Callavia* Zone of Stage 3 of Series 2	Fossils of the underlying Home Farm Member (Hartshill Sandstone Formation) correlated to faunas of Siberian Tommotian-Atdabanian boundary, and with the *Camenella baltica* Biozone of Cape Breton Island and Newfoundland; fauna some 66 m above the base of the Purley Shale Formation includes *Serrodiscus bellimarginatus* and *Strenuella sabulosa*, correlated with the *sabulosa* Biozone at the base of Stage 4	Wiliams et al. (2013)
Maximum depositional age from youngest single grain analyzed by CA-TIMS with the ET535 spike.	trilobite	South China	*Eoredlichia—Wutingaspis* Zone, Nangaoan Stage (Stage 3)	Mudstone lies below the Chengjiang biota, and above the first occurrence of trilobites of the *Parabadiella* Zone	Yang et al. (2018)
Six single grain zircon analyses have a weighted mean $^{206}Pb/^{238}U$ age, utilizing CA-TIMS and the ET535 spike.	trilobite	Avalonian	*Fallotaspis* Zone of lower Stage 3 of Series 2	Dated ash lies in the *Fallotaspis* Zone of lower Stage 3 of Series 2	Harvey et al. (2011)
Six single grain zircon analyses have a weighted mean $^{206}Pb/^{238}U$ age, utilizing CA-TIMS and the ET535 spike.	carbon isotopes	Morocco	below the Adtabanian—Tommotian boundary	Dated ash lies below the peak of a major positive $\delta^{13}C$ excursion that is correlated with CIE IV of Siberia at the Adtabanian—Tommotian boundary; below the first occurrence of *Fallotaspis* Zone trilobites	Maloof et al. (2010)
Ten of eleven single grain zircon analyses (excluding one older grain) have a weighted mean $^{206}Pb/^{238}U$ age, utilizing CA-TIMS and the ET535 spike.	carbon isotopes	Morocco	lower Tommotian	Dated ash lies below the peak of a major positive $\delta^{13}C$ excursion that is correlated with CIE II of Siberia	Maloof et al. (2010)
Eight single zircon grains (five air-abraded and three chemically abraded) yield a weighted mean $^{206}Pb/^{238}U$ age utilizing the EARTHTIME-calibrated ET535 and MIT-1L spikes.	carbon isotopes	Morocco	Tommotian/Nemakit-Daldynian boundary; late Cordubian	Dated ash lies at the zero-crossing on the descending limb of a major positive $\delta^{13}C$ excursion, and is thus correlated to the Tommotian—Nemakit-Daldynian boundary	Maloof et al. (2010)

GTS 2020 ID	GTS 2012 ID	Period	Epoch	Age	Sample	Locality	Lat-Long	Lithostratigraphy	Age (Ma)	± 2s analytical	± 2s total	Age Type
C3	C4				M223 volcanic ash bed	Oud Sdas section, Anti-Atlas, Morocco	30°22.117'N, 8°36.907'W	Tifnout Member, upper Adoudou Fm	525.34	± 0.18	± 0.56	$^{206}Pb/^{238}U$
				Fortunian								
C2	C3				Volcanic ash bed	Somerset Street, Saint John, New Brunswick, Canada	45°16.765'N, 66°3.852'W	24.3 m above the base of the Chapel Island Formation of the Ratcliffe Brook Group	530.02	± 1.07	± 1.20	$^{206}Pb/^{238}U$
C1					17SWART7 ash 6 volcanic ash bed	Swartkloofberg section, Witputs Subbasin, Nama Basin, southern Namibia	27°26'38.6"S, 16°33'31.4"E	Nomtsas Fm, Nama Group	538.58	± 0.19	± 0.63	$^{206}Pb/^{238}U$
			Ediacaran									
E26					15UNA20 ash 5 volcanic ash bed	Swartpunt section, Witputs Subbasin, Nama Basin, southern Namibia	27°28'27.4"S, 16°41'35.9"E	Urusis Fm, Upper Spitskopf Member, Schwarzrand Subgroup, Nama Group	538.99	± 0.21	± 0.63	$^{206}Pb/^{238}U$
E25					15UNA18 ash 3 volcanic ash bed	Swartpunt section, Witputs Subbasin, Nama Basin, southern Namibia	27°28'27.1"S, 16°41'34.8"E	Urusis Fm, Upper Spitskopf Member, Schwarzrand Subgroup, Nama Group	539.52	± 0.14	± 0.61	$^{206}Pb/^{238}U$
E24					15UNA22 ash 1 volcanic ash bed	Swartpunt section, Witputs Subbasin, Nama Basin, southern Namibia	27°28'22.9"S, 16°41'39.9"E	Urusis Fm, Upper Spitskopf Member, Schwarzrand Subgroup, Nama Group	540.10	± 0.10	± 0.60	$^{206}Pb/^{238}U$
E23	C1				BB5 volcanic ash bed	Oman (3045 m depth, Birba-5 well)	18°09'57.7"N, 55°14'32.7"E	Ara Group, 1 m above base of A4 carbonate unit	541.00	± 0.29	± 0.63	$^{206}Pb/^{238}U$
E22					sample 1.04 volcanic ash bed	Corcal, Corumbá–State of Mato Grosso do Sul, Brazil	19°01.067'S, 57°40.941'W	top of Tamengo Formation, Corumba Group, southern Paraguay Belt	541.85	± 0.77	± 0.97	$^{206}Pb/^{238}U$
E21					sample 1.08 volcanic ash bed	Corcal, Corumbá–State of Mato Grosso do Sul, Brazil	19°01.067'S, 57°40.941'W	top of Tamengo Formation, Corumba Group, southern Paraguay Belt	542.37	± 0.32	± 0.68	$^{206}Pb/^{238}U$
E20	E17				Mkz-11b volcanic ash bed	Oman (2194.4 m depth, Mukhaizna-11 well)	19°22'16.2"N, 56°26'13.1"E	Ara Group, 9 m below top of A3 carbonate unit	542.37	± 0.28	± 0.63	$^{206}Pb/^{238}U$
E19	E15				Minha-1A volcanic ash bed	Oman (3988.3 m depth, Minha-1 well)	18°19'12.0"N, 55°06'19.9"E	Ara Group, 3 m above base of A3 carbonate unit	542.90	± 0.29	± 0.63	$^{206}Pb/^{238}U$
E18	E13				Asala-1 core 21 volcanic ash bed	Oman (3847 m depth, Asala-1 well)	17°54'17.9"N, 54°27'40.8"E	Ara Group, middle of A0 carbonate unit	546.72	± 0.34	± 0.66	$^{206}Pb/^{238}U$
E17	E12				94-N-10B volcanic ash bed	Hauchabfontein, Namibia	24°33'21.3"S, 16°06'57.9"E	Lower Hoogland Member, 270 m above the base of the Kuibis Subgroup, Nama Group	547.32	± 0.31	± 0.65	$^{206}Pb/^{238}U$
E16	E11				JIN04-2 volcanic ash bed	Jijiawan (Jiuqunao) section, 17 km west of Maoping in Yangtze Gorges area, western Hubei Province, South China	30°48'13"N, 111°3'20"E	top of Miaohe member black shale, uppermost Doushantuo Fm	551.09	± 0.84	± 1.02	$^{206}Pb/^{238}U$
E15	E10				Volcanic ash bed	between Medvezhiy and Yeloviy creeks, Zimnie Gory section, White Sea region, Russia	65°28'04.3"N, 39°42'34.2"E	15-m-thick claystone unit in the lower part of sequence B, uppermost Ust-Pineg Fm	552.85	± 0.77	± 2.62	$^{207}Pb/^{206}Pb$
E14					Porto Morrinhos tuff	Porto Morrinhos–State of Mato Grosso do Sul, Brazil	19°30'24.6"S, 57°25'53.4"W	Bocaina Formation, Corumba Group, southern Paraguay Belt	555.18	± 0.34	± 0.70	$^{206}Pb/^{238}U$

Primary radioisotopic age details	Zonal range assignment			Biostratigraphy	Reference
	clade	zonation	Zone		
Twelve single zircon grains (six air-abraded and six chemically abraded) yield a weighted mean $^{206}Pb/^{238}U$ age utilizing the EARTHTIME-calibrated ET535 and MIT-1L spikes.	carbon isotopes	Morocco	Tommotian/Nemakit-Daldynian transition; late Cordubian	Dated ash lies at the peak of a major positive $\delta^{13}C$ excursion that is correlated with a global excursion below the Tommotian−Nemakit-Daldynian boundary	Maloof et al. (2010)
Three multigrain fractions yield a weighted mean $^{206}Pb/^{238}U$ age using physical abrasion and in-house MIT spike.	ichnofossil		Placentian Series, Fortunian Stage	Middle part of trace fossil zone *Rusophycus avalonensis*, Placentian Series	Isachsen et al. (1994)
Three of nine single grain zircon analyses (excluding six older grains) have a weighted mean $^{206}Pb/^{238}U$ age, utilizing CA-TIMS and the ET535 spike.	ichnofossil		basal Cambrian	Above first occurrence of Cambrian advanced bilaterian-sourced trace fossils, represented by *Streptichnus narbonnei* and *Treptichnus c.f. pedum*.	Linnemann et al. (2019)
Three of eleven single grain zircon analyses (excluding eight older grains) have a weighted mean $^{206}Pb/^{238}U$ age, utilizing CA-TIMS and the ET535 spike.	Ediacaran fauna		Nama Assemblage	Below occurrences of Ediacaran rangeomorph/erniettomorph biota, including *Swartpuntia germsi* and *Pteridinium simplex*, Nama Assemblage	Linnemann et al. (2019)
Five of eight single grain zircon analyses (excluding three older grains) have a weighted mean $^{206}Pb/^{238}U$ age, utilizing CA-TIMS and the ET535 spike.	Ediacaran fauna		Nama Assemblage	Below occurrences of Ediacaran rangeomorph/erniettomorph biota, including *Swartpuntia germsi* and *Pteridinium simplex*, Nama Assemblage	Linnemann et al. (2019)
Five of six single grain zircon analyses (excluding one older grain) have a weighted mean $^{206}Pb/^{238}U$ age, utilizing CA-TIMS and the ET535 spike.	Ediacaran fauna		Nama Assemblage	Below occurrences of Ediacaran rangeomorph/erniettomorph biota, including *Swartpuntia germsi* and *Pteridinium simplex*, Nama Assemblage	Linnemann et al. (2019)
Eight single zircon grain analyses have a weighted mean $^{206}Pb/^{238}U$ age, utilizing CA-TIMS and the ET535 spike.	Ediacaran fauna		Nama Assemblage	Simultaneous occurrence of an extinction of Precambrian *Namacalathus* and *Cloudina* (Nama Assemblage) and a negative excursion in carbon isotopes taken as coincident with Cambrian/Ediacaran boundary.	Bowring et al. (2007)
Five of eleven single zircon grain analyses (excluding six older grains) have a weighted mean $^{206}Pb/^{238}U$ age, utilizing CA-TIMS and the ET535 spike.	Ediacaran fauna		Nama Assemblage	The sedimentary succession has yielded macroscopic body fossils including the scyphozoan-like *Corumbella werneri* and *Paraconularia*, along with *Cloudina lucianoi*, in the upper Tamengo Formation, Nama Assemblage	Parry et al. (2017)
Four of eight single zircon grain analyses (excluding 1 older and 3 younger grains) have a weighted mean $^{206}Pb/^{238}U$ age, utilizing CA-TIMS and the ET535 spike.	Ediacaran fauna		Nama Assemblage	The sedimentary succession has yielded macrofossils including the scyphozoan-like *Corumbella werneri* and *Paraconularia*, along with *Cloudina lucianoi*, in the upper Tamengo Formation, Nama Assemblage	Parry et al. (2017)
Sample Mkz-11b just below the Precambrian−Cambrian boundary, yields ten single zircon grain analyses with a weighted mean $^{206}Pb/^{238}U$ age of 542.37 ± 0.63 Ma, utilizing CA-TIMS and the ET535 spike.	Ediacaran fauna		Nama Assemblage	Associated with the highest calcified Ediacaran macrofssils (*Cloudina* and possibly *Namacalathus*), Nama Assemblage	Bowring et al. (2007)
Sample Minha-1A yields eight single zircon grain analyses with a weighted mean $^{206}Pb/^{238}U$ age of 542.90 ± 0.63 Ma, utilizing CA-TIMS and the ET535 spike.	carbon isotopes		Nama Assemblage	Simultaneous occurrence of an extinction of Precambrian *Namacalathus* and *Cloudina* and a negative excursion in carbon isotope, Nama Assemblage	Bowring et al. (2007)
Sample Asala-1 core 21 yields eight single zircon grain analyses with a weighted mean $^{206}Pb/^{238}U$ age of 546.72 ± 0.66 Ma, utilizing CA-TIMS and an EARTHTIME-calibrated spike.	carbon isotopes		Nama Assemblage	Minimum age constraint on +4 per mil carbon isotope peak following the Shuram Excursion, Nama Assemblage	Bowring et al. (2007)
Sample 94-N-10B yields eight single zircon grain analyses with a weighted mean $^{206}Pb/^{238}U$ age of 547.36 ± 0.65 Ma, utilizing CA-TIMS and an EARTHTIME-calibrated spike.	carbon isotopes		Nama Assemblage	Middle of Kuibis Subgroup positive C-isotope excursion. Nama Assemblage Ediacaran macrofossils and Cloudina, Nama Assemblage	Bowring et al. (2007)
Sample JIN04-02 yields two (of ten total) single zircon grain analyses with a weighted mean $^{206}Pb/^{238}U$ age of 551.09 ± 1.02 Ma. A corroborating weighted mean 207Pb/206Pb age of 548.09 ± 2.61 Ma is obtained from all ten zircons [recalculated using the U decay constant ratio of Mattinson (2010)]	Ediacaran fauna		White Sea Assemblage	Ash bed occurs at the top of late acanthomorphic acritarch assemblage and Miaohe biota, and is therefore a minimum age for Doushantuo embryos and small bilaterians, White Sea Assemblage	Condon et al. (2005)
Nineteen single grain and small multigrain zircon fractions yield a weighted mean $^{207}Pb/^{206}Pb$ age of 552.85 ± 2.62 Ma [recalculated using the U decay constant ratio of Mattinson, (2010)]	Ediacaran fauna		White Sea Assemblage	Midpoint of the White Sea occurrence of Ediacaran macrofossils, including *Kimberella*, and *Dickinsonia*, White Sea Assemblage	Martin et al. (2000)
Eight single zircon grain analyses have a weighted mean $^{206}Pb/^{238}U$ age, utilizing CA-TIMS and the ET535 spike.	Ediacaran fauna			Stromatolitic dolostones and phosphorites underlying Nama Assemblage macrofossils	Parry et al. (2017)

GTS 2020 ID	GTS 2012 ID	Period	Epoch	Age	Sample	Locality	Lat-Long	Lithostratigraphy	Age (Ma)	± 2s analytical	± 2s total	Age Type
E13					Park Breccia, monomagmatic volcaniclastic turbidite, sample JNC 912	cutting on the A50 road [same locality as sample CH₂ of Compston et al. (2002)], Charnwood Forest, Leicestershire, England	52°41′38.6″N, 001°16′56.4″W	Park Breccia Member, basal Bradgate Formation, Maplewell Group, Charnian Supergroup	561.85	± 0.66	± 0.89	$^{206}Pb/^{238}U$
E12					vitric tuff, sample JNC 907	southern part of Bardon Hill Quarry, Charnwood Forest, Leicestershire, England	52°42′42.3″N, 001°19′28.9″W	middle Beacon Hill Formation, Maplewell Group, Charnian Supergroup	565.22	± 0.65	± 0.89	$^{206}Pb/^{238}U$
E11					Benscliffe Breccia (pyroclastic flow), sample JNC918	Benscliffe Breccia Member at the "Pillar Rock" type 126 locality in Benscliffe Wood, Charnwood Forest, Leicestershire, England	52°42′26.5″N, 001°14′23.3″W	Benscliffe Breccia Member, basal Beacon Hill Formation, Maplewell Group, Charnian Supergroup	569.08	± 0.73	± 0.94	$^{206}Pb/^{238}U$
E10					volcanic ash bed, sample MPMP33.56	Mistaken Point, eastern Trepassey Bay, Newfoundland	46°37′32.46″N, 53° 9′45.88″W	33.56 m above the base of the Mistaken Point Formation, Conception Group	566.25	± 0.48	± 0.77	$^{206}Pb/^{238}U$
E9					volcanic ash bed, sample Drook-2	Pigeon Cove, eastern Trepassey Bay, Newfoundland	46°41′6.94″N, 53°15′38.03″W	Drook Formation, Conception Group	570.94	± 0.46	± 0.77	$^{206}Pb/^{238}U$
E8					volcanic ash bed, sample OBJP-01	Old Bonaventure, Bonavista Peninsula, Newfoundland	48°17′2.57″N, 53°24′58.26″W	post-glacial strata; Rocky Harbour Formation, Musgravetown Group	579.24	± 0.30	± 0.69	$^{206}Pb/^{238}U$
E7					tuffaceous diamictite, sample OBJP-03	Old Bonaventure, Bonavista Peninsula, Newfoundland	48°17′3.84″N, 53°25′4.08″W	syn-glacial strata; Trinity Diamictite, Rocky Harbour Formation, Musgravetown Group	579.35	± 0.42	± 0.75	$^{206}Pb/^{238}U$
E6					volcanic ash bed, sample B1552-42.2	Old Bonaventure, Bonavista Peninsula, Newfoundland	48°17′3.34″N, 53°25′5.16″W	pre-glacial strata; Rocky Harbour Formation, Musgravetown Group	579.63	± 0.29	± 0.68	$^{206}Pb/^{238}U$
E5					volcanic ash bed, sample NoP-0.9	North Point, St. Mary's Bay, Avalon Peninsula, Newfoundland	46°56′17.87″N, 53°34′30.39″W	post-glacial strata; basal Drook Formation, 0.9 m above the Gaskiers Formation; Conception Group	579.88	± 0.52	± 0.81	$^{206}Pb/^{238}U$
E4					volcanic ash bed, sample GCI-neg6.55	Great Colinet Island, St. Mary's Bay, Avalon Peninsula, Newfoundland	46°57′33.14″N, 53°43′7.82″W	pre-glacial strata; upper Mall Bay Formation, 6.55 m below the base of the Gaskiers Formation; Conception Group	580.90	± 0.53	± 0.82	$^{206}Pb/^{238}U$
E3					volcanic ash bed, sample GCI-neg7.75	Great Colinet Island, St. Mary's Bay, Avalon Peninsula, Newfoundland	46°57′32.69″N, 53°43′8.79″W	pre-glacial strata; upper Mall Bay Formation, 7.75 m below the base of the Gaskiers Formation; Conception Group	580.34	± 0.62	± 0.88	$^{206}Pb/^{238}U$
E2	E2				YG04-2 volcanic ash bed	Jijiawan (Jiuqunao) section, 17 km west of Maoping in Yangtze Gorges area, western Hubei Province, South China	30°48′13″N, 111°03′20″E	9.5 m above base of Doushantuo Fm, 5 m above top of Lower Dolomite Member (Nantuo Cap Carbonate)	632.48	± 0.84	± 1.02	$^{206}Pb/^{238}U$
E1	E1				YG04-15 volcanic ash bed	Wuhe-Gaojiaxi section, south of Sandouping in Yangtze Gorges area, western Hubei Province, South China	30°48′49″N, 111°01′26″E	2.3 m above base of Doushantuo Fm, within the Lower Dolomite Member (cap carbonate)	635.26	± 0.84	± 1.07	$^{206}Pb/^{238}U$
		Cryogenian										
Cr4					ES-1 gray tuffaceous mudstone	Eshan section, eastern Yunnan Province, South China	24°12′15.66″N, 102°28′31.33″E	~20 cm gray tuffaceous mudstone at the top of the Nantuo Fm	634.57	± 0.90	± 1.10	$^{206}Pb/^{238}U$
Cr3	Cr1				NAV.00.2B volcanic ash bed	Navachab section, Damara Belt, Karibib, Namibia	21.98727°S, 15.72989°E	Kachab dropstone interval, ~30 m below the base of the Keilberg cap carbonate, Ghaub Fm, Swakop Group	635.21	± 0.61	± 0.92	$^{206}Pb/^{238}U$
Cr2					dolomitic, fine-grained sandstone; sample R008187	southeast coast of King Island, 100 km northwest of Tasmania	39°59′11.14″S, 144°07′33.34″E	0.7-m-thick transitional bed between the diamictite and lonestone-bearing sandstone and siltstone Cottons Breccia, and the overlying Cumberland Creek Dolostone, Grassy Group	≤ 636.41	± 0.45	± 0.83	$^{206}Pb/^{238}U$
Cr1					DW-1 volcanic ash bed	Duurwater section along Fransfontein Ridge, Namibia	20.20940°S, 15.14693°E	~15 m below the base of the Keilberg cap carbonate, with in the glaciomarine Ghaub Fm, Tsumeb Subgroup, Otavi Group	639.29	± 0.31	± 0.75	$^{206}Pb/^{238}U$

Primary radioisotopic age details	Zonal range assignment			Biostratigraphy	Reference
	clade	zonation	Zone		
Eleven single zircon grains have a weighted mean $^{206}Pb/^{238}U$ age, utilizing CA-TIMS and the ET535 spike.	Ediacaran fauna		Avalon Assemblage	Mercian assemblage of rangiomorphs including *Charnia masoni* and *Bradgatia linfordensis*, Avalon Assemblage	Noble et al. (2015)
Two of five single zircon grain analyses (excluding 3 older grains) have a weighted mean $^{206}Pb/^{238}U$ age, utilizing CA-TIMS and the ET535 spike.	Ediacaran fauna		Avalon Assemblage	Lowest occurrence of the Mercian assemblage represented by a single specimen of *Aspidella*, Avalon Assemblage	Noble et al. (2015)
Two of twelve single zircon grain analyses (excluding 1 younger and 9 older grains) have a weighted mean $^{206}Pb/^{238}U$ age, utilizing CA-TIMS and the ET535 spike.	Ediacaran fauna		Avalon Assemblage	Avalon Assemblage	Noble et al. (2015)
Five single zircon grains have a weighted mean $^{206}Pb/^{238}U$ age, utilizing CA-TIMS and the ET535 spike.	Ediacaran fauna		Avalon Assemblage	Bedding plane exposed at the base of this volcanic ash contains numerous large Ediacaran fossils including characteristic "pizza disc" ivesheadiomorphs, Avalon Assemblage	Pu et al. (2016)
Five single zircon grains have a weighted mean $^{206}Pb/^{238}U$ age, utilizing CA-TIMS and the ET535 spike.	Ediacaran fauna		Avalon Assemblage	Bedding plane exposed at the base of this volcanic ash contains numerous large ivesheadiomorphs and spindle-shaped Ediacaran fossils, Avalon Assemblage	Pu et al. (2016)
Nine of ten single zircon grains (excluding one older analysis) have a weighted mean $^{206}Pb/^{238}U$ age, utilizing CA-TIMS and the ET535 spike.				Minimum age constraint on Gaskiers glaciation	Pu et al. (2016)
Four of six single zircon grains (excluding two older analyses) have a weighted mean $^{206}Pb/^{238}U$ age, utilizing CA-TIMS and the ET535 spike.				Synglacial age constraint	Pu et al. (2016)
Five single zircon grains have a weighted mean $^{206}Pb/^{238}U$ age, utilizing CA-TIMS and the ET535 spike.				Maximum age constraint on Gaskiers glaciation	Pu et al. (2016)
Five single zircon grains have a weighted mean $^{206}Pb/^{238}U$ age, utilizing CA-TIMS and the ET535 spike.				Minimum age constraint on Gaskiers glaciation	Pu et al. (2016)
Nine single zircon grains have a weighted mean $^{206}Pb/^{238}U$ age, utilizing CA-TIMS and the ET535 spike.				Maximum age constraint on Gaskiers glaciation	Pu et al. (2016)
Eight of ten single zircon grains (excluding one younger and one older) have a weighted mean $^{206}Pb/^{238}U$ age, utilizing CA-TIMS and the ET535 spike.				Maximum age constraint on Gaskiers glaciation	Pu et al. (2016)
Three (of nine total) single zircon grain analyses yield a weighted mean $^{206}Pb/^{238}U$ age using physical abrasion and the in-house MIT spike.				Minimum age constraint on top of the Nantuo diamictite (Marinoan glaciation); direct constraint on basal Ediacaran cap carbonate	Condon et al. (2005)
Three (of 18 total) single zircon grain analyses have a weighted mean $^{206}Pb/^{238}U$ age using physical abrasion and the in-house MIT spike.				Minimum age constraint on top of the Nantuo diamictite (Marinoan glaciation); direct constraint on basal Ediacaran cap carbonate (Marinoan deglaciation)	Condon et al. (2005)
Four of seven single zircon grain analyses (excluding two younger and one older grain) have a weighted mean $^{206}Pb/^{238}U$ age, utilizing CA-TIMS and the ET535 spike.				Direct depositional age constraint on the top of the Nantuo diamictite (Marinoan glaciation)	Zhou et al. (2019)
An aliquot of the sample NAV.00.2B analyzed by Prave et al. (2016) yields a weighted mean $^{206}Pb/^{238}U$ age based upon the five youngest single zircon grain analyses, utilizing CA-TIMS and the ET2535 spike.				direct depositional age of the glaciogenic Ghaub Formation and Marinoan glaciation on Congo craton	Hoffmann et al. (2004); Prave et al. (2016)
Seven of eight single zircon grains (excluding one older analysis) have a weighted mean $^{206}Pb/^{238}U$ age, utilizing CA-TIMS and the ET535 spike.				maximum depositional age of the upper most glaciogenic Cottons Breccia, and Marinoan deglaciation in Tasmania	Calver et al. (2013)
Nine of ten single zircon grains (excluding one younger analysis) have a weighted mean $^{206}Pb/^{238}U$ age, utilizing CA-TIMS and the ET2535 spike.				direct depositional age of the glaciogenic Ghaub Formation and Marinoan glaciation on Congo craton; minimum age for the initiation of Marinoan glaciation	Prave et al. (2016)

Bibliography

Aguirre-Urreta, B., Lescano, M., Schmitz, M.D., Tunik, M., Concheyro, A., Rawson, P.F., et al., 2015, Filling the gap: new precise Early Cretaceous radioisotopic ages from the Andes. *Geological Magazine*, **152**: 557–564.

Aguirre-Urreta, B., Martinez, M., Schmitz, M., Lescano, M., Omarini, J., Tunik, M., et al., 2019, Interhemispheric radio-astrochronological calibration of the time scales from the Andean and the Tethyan areas in the Valanginian-Hauterivian (Early Cretaceous). *Gondwana Research*, **70**: 104–132.

Aguirre-Urreta, B., Schmitz, M., Lescano, M., Tunik, M., Rawson, P.F., Concheyro, A., et al., 2017, A high precision U-Pb radioisotopic age for the Agrio Formation, Neuquén Basin, Argentina: Implications for the chronology of the Hauterivian Stage. *Cretaceous Research*, **75**: 193–204.

Al-Suwaidi, A.H., Hesselbo, S.P., Damborenea, S.E., Manceñido, M.O., Jenkyns, H.C., Riccardi, A.C., et al., 2016, The Toarcian Oceanic Anoxic Event (Early Jurassic) in the Neuquén Basin, Argentina: a reassessment of age and carbon isotope stratigraphy. *The Journal of Geology*, **124**: 171–193.

Ando, A., 2016, Recent contributions to the standard Albian/Cenomanian boundary chronology from Hokkaido, Japan: A review for data reintegration and numerical age recalibration. *Cretaceous Research*, **64**: 50–58.

Baadsgaard, H., Lerbekmo, J.F., Wijbrans, J.R., Swisher III, C.C., and Fanning, M., 1993, Multimethod radiometric age for a bentonite near the top of the *Baculites reesidei* Zone of southwestern Saskatchewan (Campanian-Maastrichtian stage boundary?). *Canadian Journal of Earth Sciences*, **30**: 769–775.

Ballo, E.G., Augland, L.E., Hammer, Ø., and Svensen, H.H., 2019, A new age model for the Ordovician (Sandbian) K-bentonites in Oslo, Norway. *Palaeogeography Palaeoclimatology Palaeoecology*, **520**: 203–213.

Baresel, B., Bucher, H., Brosse, M., Cordey, F., Guodun, K., and Schaltegger, U., 2017, Precise age for the Permian–Triassic boundary in South China from high-precision U-Pb geochronology and Bayesian age–depth modeling. *Solid Earth*, **8**: 361–378.

Barker, I.R., Moser, D.E., Kamo, S.L., and Plint, A.G., 2011, High-precision U-Pb zircon ID-TIMS dating of two regionally extensive bentonites: Cenomanian Stage, Western Canada Foreland Basin. *Canadian Journal of Earth Sciences*, **48**: 543–556.

Bartolini, A., and Larson, R.L., 2001, The Pacific microplate and the Pangea supercontinent in the Early to Middle Jurassic. *Geology*, **29**: 735–738.

Baud, A., Richoz, S., Beauchamp, B., Cordey, F., Grasby, S., Henderson, C., Krystyn, L., and Nicora, A., 2012, The Buday'ah Formation, Sultanate of Oman: a Middle Permian to Early Triassic oceanic record of the Neotethys and the late Induan microsphere bloom. *Journal of Asian Earth Sciences*, **43**: 130–144.

Becker, R.T., Marshall, J.E.A., and Da Silva, A.-C., 2020, Chapter 22 – The Devonian Period. *In* Gradstein, F.M., Ogg, J.G., Schmitz, M.D., and Ogg, G.M. (eds), The Geologic Time Scale 2020. **Vol. 2**. Elsevier. Boston, MA.

Bergstrom, S.M., Toprak, F.O., Huff, W.D., and Mundil, R., 2008, Implications of a new, biostratigraphically well-controlled, radio-isotopic age for the lower Telychian Stage of the Llandovery Series (Lower Silurian, Sweden). *Episodes*, **31**: 309–314.

Berggren, W.A., Kent, D.V., Swisher III, C.C., and Aubry, M.-P., 1995, A revised Cenozoic geochronology and chronostratigraphy. *In* Berggren, W.A., Kent, D.V., Aubry, M.-P., and Hardenbol, J. (eds), Geochronology, Time Scales, and Global Stratigraphic Correlation. *SEPM Special Publication*, **54**: 129–212.

Bernoulli, D., Schaltegger, U., Stern, W.B., Frey, M., Caron, M., and Monechi, S., 2004, Volcanic ash layers in the Upper Cretaceous of the Central Apennines and a numerical age for the early Campanian. *International Journal of Earth Sciences*, **93**: 384–399.

Bodorkos, S., Pogson, D.J., and Friedman, R.M., 2017, Zircon U-Pb dating of biostratigraphically constrained felsic volcanism in the Lachlan Orogen via SHRIMP and CA-IDTIMS: implications for the division of Early Devonian time. *Australian Institute of Geoscientists Bulletin*, **65**: 11–14.

Bowring, S.A., Grotzinger, J.P., Condon, D.J., Ramezani, J., Newall, M.J., and Allen, P.A., 2007, Geochronologic constraints on the chron-ostratigraphic framework of the neoproterozoic Huqf Supergroup, Sultanate of Oman. *American Journal of Science*, **307**: 1097–1145.

Brack, P., Mundil, R., Oberli, F., Meier, M., and Rieber, H., 1996, Biostratigraphic and radiometric age data question the Milankovitch characteristics of the Latemar cycles (Southern Alps, Italy). *Geology*, **24**: 371–375.

Brack, P., Rieber, H., Mundil, R., Blendinger, W., and Maurer, F., 2007, Geometry and chronology of growth and drowning of Middle Triassic carbonate platforms (Cernera and Bivera/Clapsavon) in the Southern Alps (northern Italy). *Swiss Journal of Geosciences*, **100**: 327–348.

Brühwiler, T., Hochuli, P.A., Mundil, R., Schatz, W., and Brack, P., 2007, Bio- and chronostratigraphy of the Middle Triassic Reifling Formation of the westernmost Northern Calcareous Alps. *Swiss Journal of Geosciences*, **100**: 443–455.

Burgess, S.D., Bowring, S., and Shen, S.-Z., 2014, High-precision time-line for Earth's most severe extinction. *Proceedings of the National Academy of Sciences*, **111**: pp. 3316–3321.

Calver, C.R., Crowley, J.L., Wingate, M.T.D., Evans, D.A.D., Raub, T.D., and Schmitz, M.D., 2013, Globally synchronous Marinoan deglaciation indicated by U-Pb geochronology of the Cottons Breccia, Tasmania, Australia. *Geology*, **41**: 1127–1130.

Chambers, L.M., Pringle, M.S., and Fitton, J.G., 2004, Phreatomagmatic eruptions on the Ontong Java Plateau: an Aptian $^{40}Ar/^{39}Ar$ age for volcaniclastic rocks at ODP Site 1184. *Geological Society, London, Special Publications*, **229**: 325–331.

Channell, J.E.T., Kozur, H.W., Sievers, T., Mock, R., Aubrecht, R., and Sykora, M., 2003, Carnian-Norian biomagnetostratigraphy at Silickà Brezovà (Slovakia): Correlation to other Tethyan sections and to the Newark Basin. *Palaeogeography, Palaeoclimatology, Palaeoecology*, **191**: 65–109.

Charles, A.J., Condon, D.J., Harding, I.C., Pälike, H., Marshall, J.E.A., Cui, Y., et al., 2011, Constraints on the numerical age of the Paleocene-Eocene boundary. *Geochemistry Geophysics Geosystems*, **12**: Q0AA17, https://doi.org/10.1029/2010GC003426.

Childe, F., 1996, U-Pb geochronology and Nd and Pb isotope character-istics of the Au-Ag-rich Eskay Creek volcanogenic massive sulfide deposit, British Columbia. *Economic Geology*, **91**: 1209–1224.

Clyde, W.C., Zonneveld, J.P., Stamatakos, J., Gunnell, G.F., and Bartels, W.S., 1997, Magnetostratigraphy across the Wasatchian/Bridgerian Nalma Boundary (Early to Middle Eocene) in the Western Green River Basin, Wyoming. *The Journal of Geology*, **105**: 657–670.

Clyde, W.C., Ramezani, J., Johnson, K.R., Bowring, S.A., and Jones, M. M., 2016, Direct high-precision U−Pb geochronology of the end-Cretaceous extinction and calibration of Paleocene astronomical timescales. *Earth and Planetary Science Letters*, **452**: 272−280.

Clyde, W.C., Sheldon, N., Koch, P., Gunnell, G., and Bartels, W., 2001, Linking the Wasatchian/Bridgerian boundary to the Cenozoic Global Climate Optimum: new magnetostratigraphic and isotopic results from South Pass, Wyoming. *Palaeogeography Palaeoclimatology Palaeoecology*, **167**: 175−199.

Clyde, W.C., Stamatakos, J., and Gingerich, P.D., 1994, Chronology of the Wasatchian land-mammal age (early Eocene): magnetostratigraphic results from the McCullough Peaks section, northern Bighorn Basin, Wyoming. *The Journal of Geology*, **102**: 367−377.

Cobban, W., Walaszczyk, I., Obradovich, J., and McKinney, K., 2006, A USGS Zonal Table?for the Upper Cretaceous Middle Cenomanian-Maastrichtian of the Western Interior of the United States based on ammonites, inoceramids, and radiometric ages. *U.S. Geological Survey Open-File Report*, **1250**: 1−50.

Coccioni, R., Marsili, A., Montanari, A., Bellanca, A., Neri, R., Bice, D. M., et al., 2008, Integrated stratigraphy of the Oligocene pelagic sequence in the Umbria-Marche basin (northeastern Apennines, Italy): A potential Global Stratotype Section and Point (GSSP) for the Rupelian/Chattian boundary. *Geological Society of America Bulletin*, **120**: 487−511.

Condon, D., Zhu, M., Bowring, S.A., Wang, W., Yang, A., and Jin, Y., 2005, U-Pb ages from the Neoproterozoic Doushantuo Formation, China. *Science*, **308**: 95−98.

Corfu, F., Polteau, S., Planke, S., Faleide, J.I., Svensen, H., Zayoncheck, A., and Stolbov, N., 2013, *U−Pb geochronology of Cretaceous magmatism on Svalbard and Franz Josef Land, Barents Sea large igneous province: Geological Magazine*, **150**: 1127−1135.

Cramer, B.D., Condon, D.J., Söderlund, U., Marshall, C., Worton, G.J., Thomas, A.T., et al., 2012, U-Pb (zircon) age constraints on the timing and duration of Wenlock (Silurian) paleocommunity collapse and recovery during the 'Big Crisis'. *Geological Society of America Bulletin*, **124**: 1841−1857.

Cramer, B.D., Schmitz, M.D., Huff, W.D., and Bergstrom, S.M., 2015, High-precision U-Pb zircon age constraints on the duration of rapid biogeochemical events during the Ludlow Epoch (Silurian Period). *Journal Of The Geological Society*, **172**: 157−160.

Dalrymple, G., Izett, G., Snee, L., and Obradovich, J.D., 1993, $^{40}Ar/^{39}Ar$ age spectra and total-fusion ages of tektites from Cretaceous—Tertiary Boundary sedimentary rocks in the Beloc Formation, Haiti. *US Geological Survey Bulletin*, **2065**: 1−20.

Davidek, K., Landing, E., Bowring, S.A., Westrop, S., Rushton, A., Fortey, R., et al., 1998, New uppermost Cambrian U-Pb date from Avalonian Wales and age of the Cambrian-Ordovician boundary. *Geological Magazine*, **135**: 305−309.

Davydov, V., Crowley, J., Schmitz, M.D., and Poletaev, V., 2010, High-precision U-Pb zircon age calibration of the global Carboniferous time scale and Milankovitch band cyclicity in the Donets Basin, eastern Ukraine. *Geochemistry Geophysics Geosystems*, **11**: Q0AA04, https://doi.org/10.1029/2009GC00273.

Davydov, V.I., Biakov, A.S., Isbell, J.L., Crowley, J.L., Schmitz, M.D., and Vedernikov, I.L., 2016, Middle Permian U-Pb zircon ages of the 'glacial' deposits of the Atkan Formation, Ayan-Yuryakh anticlinorium, Magadan province, NE Russia: Their significance for global climatic interpretations. *Gondwana Research*, **38**: 74−85.

Davydov, V.I., Biakov, A.V., Schmitz, M.D., and Silantiev, V.V., 2018a, Radioisotopic calibration of the Guadalupian (middle Permian) series − review and updates. *Earth Science Reviews*, **176**: 222−240.

Davydov, V.I., Crowley, J.L., Schmitz, M.D., and Snyder, W.S., 2018b, New U−Pb constraints identify the end-Guadalupian and possibly end-Lopingian extinction events conceivably preserved in the passive margin of North America: implication for regional tectonics. *Geological Magazine*, **155**: 119−131.

Davydov, V.I., Schmitz, M.D., and Korn, D., 2011, The Hangenberg Event was abrupt and short at the global scale: the quantitative integration and intercalibration of biotic and geochronologic data within the Devonian-Carboniferous transition. *Abstracts with Programs - Geological Society of America*, **43**: 128.

De Lena, L.F., Taylor, D., Guex, J., Bartolini, A., Adatte, T., van Acken, D., et al., 2019, The driving mechanisms of the carbon cycle perturbations in the late Pliensbachian (Early Jurassic). *Scientific Reports*, **9**: 1−13.

Encarnación, J., Rowell, A.J., and Grunow, A.M., 1999, A U-Pb age for the Cambrian Taylor Formation, Antarctica: implications for the Cambrian Time Scale. *The Journal of Geology*, **107**: 497−504.

Evenchick, C., and McNicoll, V.J., 1993, U-Pb age for the Jurassic McEwan Creek pluton, north-central British Columbia: regional setting and implications for the Toarcian stage boundary. *Geological Survey of Canada Radiogenic Age and Isotopic Studies*, **7**: 91−97.

Evenchick, C.A., Poulton, T.P., and McNicoll, V.J., 2010, Nature and significance of the diachronous contact between the Hazelton and Bowser Lake groups (Jurassic), north-central British Columbia. *Bulletin of Canadian Petroleum Geology*, **58**: 235−267.

Flynn, J.J., 1986, Correlation and geochronology of middle Eocene strata from the western United States. *Palaeogeography Palaeoclimatology, Palaeoecology*, **55**: 335−406.

Furin, S., Preto, N., Rigo, M., Roghi, G., Gianolla, P., Crowley, J.L., et al., 2006, High-precision U-Pb zircon age from the Triassic of Italy: Implications for the Triassic time scale and the Carnian origin of calcareous nannoplankton and dinosaurs. *Geology*, **34**: 1009−1012.

Galfetti, T., Hochuli, P.A., Brayard, A., Bucher, H., Weissert, H., and Vigran, J.O., 2007, Smithian-Spathian boundary event: Evidence for global climatic change in the wake of the end-Permian biotic crisis. *Geology*, **35**: 291−294.

Garratt, M.J., Wright, A.J., 1989. Late Silurian to Early Devonian biostratigraphy of southeastern Australia. *In:* McMillan, N.J., Embry, A.F., Glass, D.J. (eds), Devonian of the World. Canadian Society of Petroleum Geology, Memoir, 14(III), pp. 647−662.

Gastaldo, R., Purkynova, E., Simunek, Z., and Schmitz, M.D., 2009, Ecological persistence in the Late Mississippian (Serpukhovian, Namurian A) Megafloral Record of the Upper Silesian Basin, Czech Republic. *Palaios*, **24**: 336−350.

Gill, J.R., and Cobban, W.A., 1966. The Red Bird section of the Upper Cretaceous Pierre Shale in Wyoming: United States. Geological Survey, Professional Paper, 393-A, 73 pp.

Guex, J., Bartolini, A., Atudorei, V., Taylor, D., 2004. High-resolution ammonite and carbon isotope stratigraphy across the Triassic−Jurassic boundary at New York Canyon (Nevada). *Earth and Planetary Science Letters*. **225**: 29−41.

Guex, J., Schoene, B., Bartolini, A., Spangenberg, J., Schaltegger, U., O'Dogherty, L., et al., 2012, Geochronological constraints on post-extinction recovery of the ammonoids and carbon cycle perturbations during the Early Jurassic. *Palaeogeography Palaeoclimatology Palaeoecology*, **346-347**: 1–11.

Harrigan, C.O., Schmitz, M.D., Over, D.J., in review, New high-precision U-Pb zircon dates and age-depth modeling to revise the Devonian time scale. *Geological Society of America Bulletin*: in review.

Harvey, T.H.P., Williams, M., Condon, D.J., Wilby, P.R., Siveter, D.J., Rushton, A.W.A., et al., 2011, A refined chronology for the Cambrian succession of southern Britain. *Journal Of The Geological Society*, **168**: 705–716.

He, H., Pan, Y., Tauxe, L., Qin, H., and Zhu, R., 2008, Toward age determination of the M0r (Barremian–Aptian boundary) of the Early Cretaceous. *Physics of the Earth and Planetary Interiors*, **169**: 41–48.

Herrington, R.J., Hollis, S.P., Cooper, M.R., Stobbs, I., Tapster, S., Rushton, A., et al., 2018, Age and geochemistry of the Charlestown Group, Ireland: Implications for the Grampian orogeny, its mineral potential and the Ordovician timescale. *Lithos*, **302-303**: 1–19.

Hicks, J., Obradovich, J., and Tauxe, L., 1999, Magnetostratigraphy, isotopic age calibration and intercontinental correlation of the Red Bird section of the Pierre Shale, Niobrara County, Wyoming, USA. *Cretaceous Research*, **20**: 1–27.

Hicks, J.F., Obradovich, J.D., and Tauxe, L., 1995, A new calibration point for the Late Cretaceous time scale: The ^{40}Ar/^{39}Ar isotopic age of the C33r/C33n geomagnetic reversal from the Judith River Formation (Upper Cretaceous), Elk Basin, Wyoming, USA. *Journal Of Geology*, **103**: 243–256.

Hiza, M.M., 1999. The Geochemistry and Geochronology of the Eocene Absaroka Volcanic Province, Northern Wyoming and Southwest Montana. PhD Thesis, Oregon State University, Corvallis, OR, USA: 261 pp.

Hoffmann, K.-H., Condon, D., Bowring, S.A., and Crowley, J., 2004, U-Pb zircon date from the Neoproterozoic Ghaub Formation, Namibia: constraints on Marinoan glaciation. *Geology*, **32**: 817–820.

Hu, Y., Zhou, J., Song, B., Li, W., and Sun, W., 2008, SHRIMP zircon U-Pb dating from K-bentonite in the top of Ordovician of Wangjiawan Section, Yichang, Hubei, China. *Science in China Series D: Earth Sciences*, **51**: 493–498.

Huff, W.D., Davis, D., Bergstrom, S.M., Krekeler, M.P.S., Kolata, D.R., and Cingolani, C.A., 1997, A biostratigraphically well-constrained K-bentonite U-Pb zircon age of the lowermost Darriwilian Stage (Middle Ordovician) from the Argentine Precordillera. *Episodes*, **20**: 29–33.

Husson, J.N., Schoene, B., Bluher, S., and Maloof, A.C., 2016, Chemostratigraphic and U-Pb geochronologic constraints on carbon cycling across the Silurian-Devonian boundary. *Earth and Planetary Science Letters*, **436**: 108–120.

Hüsing, S., Cascella, A., Hilgen, F., Krijgsman, W., Kuiper, K., Turco, E., et al., 2010, Astrochronology of the Mediterranean Langhian between 15.29 and 14.17 Ma. *Earth and Planetary Science Letters*, **290**: 254–269.

Ickert, R.B., Mulcahy, S.R., Sprain, C.J., Banaszak, J.F., and Renne, P.R., 2015, Chemical and Pb isotope composition of phenocrysts from bentonites constrains the chronostratigraphy around the Cretaceous-Paleogene boundary in the Hell Creek region, Montana. *Geochemistry Geophysics Geosystems*, **16**: 2743–2761.

Isachsen, C.E., Bowring, S.A., Landing, E., and Samson, S.D., 1994, New constraint on the division of Cambrian time. *Geology*, **22**: 496–498.

Izett, G., Dalrymple, G., and Snee, L., 1991, ^{40}Ar/^{39}Ar age of Cretaceous-Tertiary boundary tektites from Haiti. *Science*, **252**: 1539–1542.

Jagodzinski, E., and Black, L., 1999, U–Pb dating of silicic lavas, sills and syneruptive resedimented volcaniclastic deposits of the Lower Devonian Crudine Group, Hill End Trough, New South Wales. *Australian Journal of Earth Sciences*, **46**: 749–764.

Jirásek, J., Hýlová, L., Sivek, M., Jureczka, J., Martínek, K., Sýkorová, I., et al., 2013, The Main Ostrava Whetstone: composition, sedimentary processes, palaeogeography and geochronology of a major Mississippian volcaniclastic unit of the Upper Silesian Basin (Poland and Czech Republic). *International Journal of Earth Sciences*, **102**: 989–1006.

Jirásek, J., Opluštil, S., Sivek, M., Schmitz, M.D., and Abels, H.A., 2018, Astronomical forcing of Carboniferous paralic sedimentary cycles in the Upper Silesian Basin, Czech Republic (Serpukhovian, latest Mississippian): New radiometric ages afford an astronomical age model for European biozonations and substages. *Earth Science Reviews*, **177**: 715–741.

Jirásek, J., Wlosok, J., Sivek, M., Matýsek, D., Schmitz, M., Sýkorová, I., et al., 2014, U-Pb zircon age of the Krásné Loučky tuffite: the dating of Visean flysch in the Moravo-Silesian Paleozoic Basin (Rhenohercynian Zone, Czech Republic). *Geological Quarterly*, **58**: 659–672.

Johannson, G., and McNicoll, V., 1997, New U–Pb data from the Laberge Group, northwest British Columbia: implications for Stikinian arc evolution and Lower Jurassic time scale calibrations. *Radiogenic Age and Isotopic Studies Report 10, Geological Survey of Canada, Current Research*, **1997-F**: 121–129.

Kaiser, S.I., Becker, R.T., Spaletta, C., 2009. High-resolution conodont stratigraphy, biofacies and extinctions around the Hangenberg Event in pelagic successions from Austria, Italy, and France. *In:* Over, D.J. (Ed.), Studies in Devonian Stratigraphy: Proceedings of the 2007 International Meeting of the Subcommission on Devonian Stratigraphy and IGCP 499. Palaeontographica Americana, **63**: 97–139.

Kamo, S.L., and Riccardi, A.C., 2009, A new U–Pb zircon age for an ash layer at the Bathonian–Callovian boundary, Argentina. *GFF*, **131**: 177–182.

Kaufmann, B., Trapp, E., Mezger, K., and Weddige, K., 2005, Two new Emsian (Early Devonian) U-Pb zircon ages from volcanic rocks of the Rhenish Massif (Germany): implications for the Devonian time scale. *Journal Of The Geological Society*, **162**: 363–371.

Kietzmann, D.A., Iglesia Llanos, M.P., and Kohan Martinez, M., 2018, Astronomical calibration of the Upper Jurassic - Lower Cretaceous in the Neuquén Basin, Argentina: a contribution from the Southern Hemisphere to the geologic time scale, *Stratigraphy and Time Scales*, **Vol. 3**. Elsevier, pp. 327–355.

Koppers, A.A.P., 2003, High-resolution ^{40}Ar/^{39}Ar dating of the oldest oceanic basement basalts in the western Pacific basin. *Geochemistry Geophysics Geosystems*, **4**: 8914, https://doi.org/10.1029/2003GC000574.

Kowallis, B.J., Christiansen, E.H., Deino, A.L., Zhang, C., and Everett, B.H., 2001, The record of Middle Jurassic volcanism in the Carmel and Temple Cap Formations of southwestern Utah. *Geological Society of America Bulletin*, **113**: 373–387.

Kuiper, K.F., Deino, A., Hilgen, F.J., Krijgsman, W., Renne, P.R., and Wijbrans, J.R., 2008, Synchronizing rock clocks of earth history. *Science*, **320**: 500–504.

Landing, E., Bowring, S.A., Davidek, K., Rushton, A., Fortey, R., and Wimbledon, W., 2000, Cambrian-Ordovician boundary age and duration of the lowest Ordovician Tremadoc Series based on U-Pb zircon dates from Avalonian Wales. *Geological Magazine*, **137**: 485–494.

Landing, E., Bowring, S.A., Davidek, K.L., and Fortey, R.A., 1997, U-Pb zircon date from Avalonian Cape Breton Island and geochronologic calibration of the Early Ordovician. *Canadian Journal of Earth Sciences*, **34**: 724–730.

Landing, E., Bowring, S.A., Davidek, K.L., Westrop, S.R., Geyer, G., and Heldmaier, W., 1998, Duration of the Early Cambrian: U-Pb ages of volcanic ashes from Avalon and Gondwana. *Canadian Journal of Earth Sciences*, **35**: 329–338.

Landing, E.D., Geyer, G., Buchwaldt, R., and Bowring, S.A., 2014, Geochronology of the Cambrian: a precise Middle Cambrian U−Pb zircon date from the German margin of West Gondwana. *Geological Magazine*, **152**: 28–40.

Lanik, A., Over, D.J., Schmitz, M., and Kirchgasser, W.T., 2016, Testing the limits of chronostratigraphic resolution in the Appalachian Basin, Late Devonian (middle Frasnian), eastern North America: New U-Pb zircon dates for the Belpre Tephra suite. *Geological Society of America Bulletin*, **128**: 1813–1821.

Leanza, H.A., Mazzini, A., Corfu, F., Llambías, E.J., Svensen, H., Planke, S., et al., 2013, The Chachil Limestone (Pliensbachian-earliest Toarcian) Neuquén Basin, Argentina: U-Pb age calibration and its significance on the Early Jurassic evolution of southwestern Gondwana. *Journal of South American Earth Sciences*, **42**: pp. 171–185.

Lehrmann, D.J., Ramezani, J., Bowring, S.A., Martin, M.W., Montgomery, P., Enos, P., et al., 2007, Timing of recovery from the end-Permian extinction: geochronologic and biostratigraphic constraints from south China: comment and reply: Reply. *Geology*, **35**: e136–e137.

Lehrmann, D.J., Ramezani, J., Bowring, S.A., Martin, M.W., Montgomery, P., Enos, P., et al., 2006, Timing of recovery from the end-Permian extinction: geochronologic and biostratigraphic constraints from south China. *Geology*, **34**: 1053–1056.

Lehrmann, D.J., Stepchinski, L., Altiner, D., Orchard, M.J., Montgomery, P., Enos, P., et al., 2015, An integrated biostratigraphy (conodonts and foraminifers) and chronostratigraphy (paleomagnetic reversals, magnetic susceptibility, elemental chemistry, carbon isotopes and geochronology) for the Permian-Upper Triassic strata of Guandao section, Nanpanjiang Basin, south China. *Journal of Asian Earth Sciences*, **108**: 117–135.

Lena, L., López-Martínez, R., Lescano, M., Aguire-Urreta, B., Concheyro, A., Vennari, V., et al., 2019, High-precision U−Pb ages in the early Tithonian to early Berriasian and implications for the numerical age of the Jurassic−Cretaceous boundary. *Solid Earth*, **10**: 1–14.

Leslie, S.A., Sell, B.K., Saltzman, M.R., Repetski, J.E., and Edwards, C.T., 2012, The East River Mountain K-bentonite bed, a central Appalachian marker that closely approximates the Middle-Upper Ordovician (Darriwilian-Sandbian) boundary. *Abstracts with Programs - Geological Society of America*, **44**: 590.

Lindskog, A., Costa, M.M., Rasmussen, C.M.O., Connelly, J.N., and Eriksson, M.E., 2017, Refined Ordovician timescale reveals no link between asteroid breakup and biodiversification. *Nature Communications*, **8**: 1–8.

Linnemann, U., Ovtcharova, M., Schaltegger, U., Gärtner, A., Hautmann, M., Geyer, G., et al., 2019, New high-resolution age data from the Ediacaran-Cambrian boundary indicate rapid, ecologically driven onset of the Cambrian explosion. *Terra Nova*, **31**: 49–58.

Ma, C., Meyers, S.R., Sageman, B.B., Singer, B.S., and Jicha, B.R., 2014, Testing the astronomical time scale for oceanic anoxic event 2, and its extension into Cenomanian strata of the Western Interior Basin (USA). *Geological Society of America Bulletin*, **126**: 974–989.

Macdonald, F.A., Karabinos, P.M., Crowley, J.L., Hodgin, E.B., Crockford, P.W., and Delano, J.W., 2017, Bridging the gap between the foreland and hinterland II: geochronology and tectonic setting of Ordovician magmatism and basin formation on the Laurentian margin of New England and Newfoundland. *American Journal of Science*, **317**: 555–596.

Machlus, M.L., Ramezani, J., Bowring, S.A., Hemming, S.R., Tsukui, K., and Clyde, W.C., 2015, A strategy for cross-calibrating U−Pb chronology and astrochronology of sedimentary sequences: an example from the Green River Formation, Wyoming, USA. *Earth and Planetary Science Letters*, **413**: 70–78.

Mahoney, J.J., Duncan, R.A., Tejada, M.L.G., Sager, W.W., and Bralower, T.J., 2005, Jurassic-Cretaceous boundary age and mid-ocean-ridge−type mantle source for Shatsky Rise. *Geology*, **33**: 185–188.

Maletz, 2009, Imograptus spinosus and the Middle Ordovician (Darriwilian) graptolite biostratigraphy at Les Méchins (Quebec, Canada). *Canadian Journal of Earth Sciences*, **46**: 739–755.

Maloof, A.C., Ramezani, J., Bowring, S.A., Fike, D.A., Porter, S.M., and Mazouad, M., 2010, Constraints on early Cambrian carbon cycling from the duration of the Nemakit-Daldynian−Tommotian boundary $\delta^{13}C$ shift, Morocco. *Geology*, **38**: 623–626.

Martin, M.W., Grazhdankin, D., Bowring, S.A., Evans, D., Fedonkin, M., and Kirschvink, J., 2000, Age of Neoproterozoic bilatarian body and trace fossils, White Sea, Russia: Implications for metazoan evolution. *Science*, **288**: 841–845.

Mattinson, J.M., 2010a, Analysis of the relative decay constants of 235U and 238U by multi-step CA-TIMS measurements of closed-system natural zircon samples. *Chemical Geology*, **275**: 186–198.

Mawson, R., Talent, J.A., Bear, D.S., Brock, G.A., Farrell, J.R., Hyland, K.A., Pyemont, B.D., Sloan, T.R., Sorentino, L., Stewart, M.I., Trotter, J.R., Wilson, G.A., Simpson, A.G., 1989. Conodont data in relation to resolution of stage and zonal boundaries for the Devonian of Australia. *In:* McMillan, N.J., Embry, A.F., Glass, D.J. (eds), Devonian of the World. Canadian Society of Petroleum Geology, Memoir, **14**(III), pp. 485–527.

Mazzini, A., Svensen, H., Leanza, H.A., Corfu, F., and Planke, S., 2010, Early Jurassic shale chemostratigraphy and U−Pb ages from the Neuquén Basin (Argentina): Implications for the Toarcian Oceanic Anoxic Event. *Earth and Planetary Science Letters*, **297**: 633–645.

McAdams, N.E.B., Schmitz, M.D., Kleffner, M.A., Verniers, J., Vandenbroucke, T.R.A., Ebert, J.R., et al., 2017, A new, high-precision CA-ID-TIMS date for the 'Kalkberg' K-bentonite (Judds Falls Bentonite). *Lethaia*, **51**: 344–356.

Meyers, S.R., Siewert, S.E., Singer, B.S., Sageman, B.B., Condon, D.J., Obradovich, J.D., et al., 2012, Intercalibration of radioisotopic and astrochronologic time scales for the Cenomanian-Turonian boundary interval, Western Interior Basin, USA. *Geology*, **40**: 7–10.

Midtkandal, I., Svensen, H.H., Planke, S., Corfu, F., Polteau, S., Torsvik, T.H., et al., 2016, The Aptian (Early Cretaceous) oceanic anoxic event (OAE1a) in Svalbard, Barents Sea, and the absolute age of the Barremian-Aptian boundary. *Palaeogeography Palaeoclimatology Palaeoecology*, **463**: 126–135.

Mietto, P., Manfrin, S., Preto, N., Rigo, M., Roghi, G., Furin, S., et al., 2012, The global boundary stratotype section and point (GSSP) of the Carnian stage (Late Triassic) at Prati di Stuores/Stuores Wiesen section (Southern Alps, NE Italy). *Episodes*, **35**: 414–430.

Mitchell, C.E., Brussa, E.D., and Astini, R.A., 1998. A diverse Da2 fauna preserved within an altered volcanic ash fall, eastern Precordillera, Argentina; implications for graptolite paleoecology. In: Gutierrez-Marco, J.C., Rabano, I. (Eds.), Sixth International Graptolite Conference of the GWG (IPA) and the SW Iberia Field Meeting 1998 of the International Subcommission on Silurian Stratigraphy (ICS-IUGS), Temas Geológico-Mineros 23. Instituto Technologico Geominero de Espana, Madrid, pp. 222–223.

Mundil, R., Ludwig, K., Metcalfe, I., and Renne, P., 2004, Age and timing of the Permian mass extinctions: U/Pb dating of closed-system zircons. *Science*, **305**: 1760–1763.

Mundil, R., Pálfy, J., Renne, P., and Brack, P., 2010, *The Triassic timescale: new constraints and a review of geochronological data*, Geological Society, 334. London: Special Publications, pp. 41–60.

Mundil, R., Zühlke, R., Bechstädt, T., Peterhänsel, A., Egenhoff, S.O., Oberli, F., et al., 2003, Cyclicities in Triassic platform carbonates: synchronizing radio-isotopic and orbital clocks. *Terra Nova*, **15**: 81–87.

Mutterlose, J., and Böckel, B., 1998, The Barremian-Aptian interval in NW Germany: a review. *Cretaceous Research*, **19**: 539–568.

Myrow, P.M., Ramezani, J., Hanson, A.E., Bowring, S.A., Racki, G., and Rakocinski, M., 2014, High-precision U-Pb age and duration of the latest Devonian (Famennian) Hangenberg event, and its implications. *Terra Nova*, **26**: 222–229.

Nicklen, B.L., 2011, Establishing a Tephrochronologic Framework for the Middle Permian (Guadalupian) Type Area and Adjacent Portions of the Delaware Basin and Northwestern Shelf, West Texas and Southeastern New Mexico, USA. University of Cincinnati Dissertation, 134 p.

Noble, S.R., Condon, D.J., Carney, J.N., Wilby, P.R., Pharaoh, T.C., and Ford, T.D., 2015, U-Pb geochronology and global context of the Charnian Supergroup, UK: Constraints on the age of key Ediacaran fossil assemblages. *Geological Society of America Bulletin*, **127**: 250–265.

Normore, L.S., Zhen, Y.Y., Dent, L.M., Crowley, J.L., Percival, I.G., and Wingate, M.T.D., 2018, Early Ordovician CA-IDTIMS U-Pb zircon dating and conodont biostratigraphy, Canning Basin, Western Australia. *Australian Journal of Earth Sciences*, **65**: 1–14.

Obradovich, J.D., 1993, A Cretaceous time scale. *In* Caldwell, W. G. E. and Kauffman, E. G. (eds), *Evolution of the Western Interior Basin*. Geological Association of Canada Special Paper, **39**: 379–396.

Obradovich, J.D., Matsumoto, T., Nishida, T., and Inoue, Y., 2002, Integrated biostratigraphic and radiometric study on the Lower Cenomanian (Cretaceous) of Hokkaido, Japan. Proceedings of the Japan Academy, Series B, **78**: 149–153.

Odin, G., Montanari, A., Deino, A., Drake, R., Guise, P., Kreuzer, H., et al., 1991, Reliability of volcano-sedimentary biotite ages across the Eocene-Oligocene boundary (Apennines, Italy). *Chemical Geology*, **86**: 203–224.

Ogg, J.G., Ogg, G., and Gradstein, F.M., 2008, *The Concise Geologic Time Scale*. Cambridge: Cambridge University Press, p. 177.

Orchard, M.J., 1991, Late Triassic conodont biochronology and biostratigraphy of the Kunga Group, Queen Charlotte Islands, British Columbia. In: Hydrocarbon potential of the Queen Charlotte basin, British Columbia: Geological Survey of Canada, Paper **90-10**: 173–193.

Ortega, G., Albanesi, G.L., and Frigerio, R.S., 2007, Graptolite-conodont biostratigraphy and biofacies of the middle Ordovician Cerro Viejo

succession, San Juan, Precordillera, Argentina. *Palaeogeography, Palaeoclimatology, Palaeoecology*, **245**: 245–264.

Oruche, N.E., Dix, G.R., and Kamo, S.L., 2018, Lithostratigraphy of the upper Turinian − lower Chatfieldian (Upper Ordovician) foreland succession, and a U−Pb ID−TIMS date for the Millbrig volcanic ash bed in the Ottawa Embayment. *Canadian Journal of Earth Sciences*, **55**: 1079–1102.

Ovtcharova, M., Bucher, H., Schaltegger, U., Galfetti, T., Brayard, A., and Guex, J., 2006, New Early to Middle Triassic U−Pb ages from South China: Calibration with ammonoid biochronozones and implications for the timing of the Triassic biotic recovery. *Earth and Planetary Science Letters*, **243**: 463–475.

Pana, D., Poulton, T.P., and Heaman, L.M., 2018, U-Pb zircon ages of volcanic ashes integrated with ammonite biostratigraphy, Fernie Formation (Jurassic), Western Canada, with implications for Cordilleran-Foreland basin connections and comments on the Jurassic time scale. *Journal of the Geological Society*, **66**: 595–622.

Parry, L.A., Boggiani, P.C., Condon, D.J., Garwood, R.J., de M Leme, J., McIlroy, D., et al., 2017, Ichnological evidence for meiofaunal bilaterians from the terminal Ediacaran and earliest Cambrian of Brazil. *Nature Ecology, Evolution*, **1**: 1–10.

Parry, S.F., Noble, S.R., Crowley, Q.G., and Wellman, C.H., 2011, A high-precision U−Pb age constraint on the Rhynie Chert Konservat-Lagerstätte: time scale and other implications. *Journal of the Geological Society*, **168**: 863–872.

Pálfy, J.Ó., Mortensen, J., Smith, P., Friedman, R., McNicoll, V., and Villeneuve, M., 2000, New U-Pb zircon ages integrated with ammonite biochronology from the Jurassic of the Canadian Cordillera. *Canadian Journal of Earth Sciences*, **37**: 549–567.

Pálfy, J.Ó., Parrish, R.R., and Smith, P.L., 1997, A U-Pb age from the Toarcian (Lower Jurassic) and its use for time scale calibration through error analysis of biochronologic dating. *Earth and Planetary Science Letters*, **146**: 659–675.

Pálfy, J.Ó., Parrish, R.R., David, K., and Vörös, A., 2003, Mid-Triassic integrated U−Pb geochronology and ammonoid biochronology from the Balaton Highland (Hungary). *Journal of the Geological Society*, **160**: 271–284.

Pellenard, P., Nomade, S., Martire, L., De Oliveira Ramalho, F., Monna, F., and Guillou, H., 2013, The first ^{40}Ar−^{39}Ar date from Oxfordian ammonite-calibrated volcanic layers (bentonites) as a tie-point for the Late Jurassic. *Geological Magazine*, **150**: 1136–1142.

Percival, L.M.E., Davies, J.H.F.L., Schaltegger, U., De Vleeschouwer, D., Da Silva, A.C., and llmi, K.B.F.X., 2018, Precisely dating the Frasnian−Famennian boundary: implications for the cause of the Late Devonian mass extinction. *Scientific Reports*, **8**: 1–10.

Pointon, M.A., Chew, D.M., Ovtcharova, M., Sevastopulo, G.D., and Crowley, Q.G., 2012, New high-precision U-Pb dates from western European Carboniferous tuffs; implications for time scale calibration, the periodicity of late Carboniferous cycles and stratigraphical correlation. *Journal of the Geological Society*, **169**: 713–721.

Pointon, M.A., Chew, D.M., Ovtcharova, M., Sevastopulo, G.D., and Delcambre, B., 2014, High-precision U−Pb zircon CA-ID-TIMS dates from western European late Viséan bentonites. *Journal of the Geological Society*, **171**: 649–658.

Prave, A.R., Condon, D.J., Hoffmann, K.H., Tapster, S., and Fallick, A.E., 2016, Duration and nature of the end-Cryogenian (Marinoan) glaciation. *Geology*, **44**: 631–634.

Pringle, M., and Duncan, R., 1995, Radiometric ages of basement lavas recovered at Loen, Wodejebato, MIT, and Takuyo-Daisan Guyots, northwestern Pacific Ocean. *Proceedings of the Ocean Drilling Program: Scientific Results*, **144**: 547–557.

Pu, J.P., Bowring, S.A., Ramezani, J., Myrow, P., Raub, T.D., Landing, E., et al., 2016, Dodging snowballs: geochronology of the Gaskiers glaciation and the first appearance of the Ediacaran biota. *Geology*, **44**: 955–958.

Quidelleur, X., Paquette, J.L., Fiet, N., Takashima, R., Tiepolo, M., Desmares, D., et al., 2011, New U–Pb (ID-TIMS and LA-ICPMS) and ^{40}Ar/^{39}Ar geochronological constraints of the Cretaceous geologic time scale calibration from Hokkaido (Japan). *Chemical Geology*, **286**: 72–83.

Ramezani, J., and Bowring, S.A., 2017, Advances in numerical calibration of the Permian timescale based on radioisotopic geochronology. *In* Lucas S. R. and Shen S. Z. (eds), *The Permian Timescale*. Geological Society London Special Publications, SP450: 1–10, 10.1144/SP450.17.

Ramezani, J., Bowring, S.A., Martin, M.W., Lehrmann, D.J., Montgomery, P., Enos, P., et al., 2007a, Timing of recovery from the end-Permian extinction: Geochronologic and biostratigraphic constraints from south China: COMMENT AND REPLY: REPLY. *Geology*, **35**: e137–e138.

Ramezani, J., Schmitz, M.D., Davydov, V., Bowring, S.A., Snyder, W., and Northrup, C., 2007b, High-precision U–Pb zircon age constraints on the Carboniferous–Permian boundary in the southern Urals stratotype. *Earth and Planetary Science Letters*, **256**: 244–257.

Renne, P.R., Balco, G., Ludwig, K.R., Mundil, R., and Min, K., 2011, Response to the comment by WH Schwarz et al. on "Joint determination of 40K decay constants and 40Ar*/40K for the Fish Canyon sanidine standard, and improved accuracy for ^{40}Ar/^{39}Ar geochronology" by PR Renne et al. (2010). *Geochimica et Cosmochimica Acta*, **75**: 5097–5100.

Renne, P.R., Deino, A.L., Hilgen, F.J., Kuiper, K.F., Mark, D.F., Mitchell III, W.S., et al., 2013, Time scales of critical events around the Cretaceous-Paleogene boundary. *Science*, **339**: 684–687.

Riccardi, A.C., Westermann, G.E.G., and Elmi, S., 1989, The Middle Jurassic Bathonian/Callovian ammonite zones of the Argentine/Chilean Andes. *Geobios*, **22**: 553–597.

Ross, R.J., Naeser Jr., C.W., Izett, G.A., Obradovich, J.D., Bassett, M.G., Hughes, C.P., Cocks, L.R.M., Dean, W.T., Ingham, J.K., Jenkins, C.J., Rickards, R.B., Sheldon, P.R., Toghill, P., Whittington, H.B., and Zalasiewicz, J., 1982, Fission-track dating of British Ordovician and Silurian stratotypes. *Geological Magazine*, **119** (2), 135–153.

Sageman, B.B., Singer, B.S., Meyers, S.R., Siewert, S.E., Walaszczyk, I., Condon, D.J., et al., 2014, Integrating ^{40}Ar/^{39}Ar, U-Pb, and astronomical clocks in the Cretaceous Niobrara Formation, Western Interior Basin, USA. *Geological Society of America Bulletin*, **126**: 956–973.

Sahy, D., Condon, D.J., Terry Jr, D.O., Fischer, A.U., and Kuiper, K.F., 2015, Synchronizing terrestrial and marine records of environmental change across the Eocene–Oligocene transition. *Earth and Planetary Science Letters*, **427**: 171–182.

Sahy, D., Condon, D.J., Hilgen, F.J., and Kuiper, K.F., 2017, Reducing disparity in radio-isotopic and astrochronology-based time scales of the Late Eocene and Oligocene. *Paleoceanography*, **32**: 1018–1035.

Schaltegger, U., Guex, J., Bartolini, A., Schoene, B., and Ovtcharova, M., 2008, Precise U–Pb age constraints for end-Triassic mass

extinction, its correlation to volcanism and Hettangian post-extinction recovery. *Earth and Planetary Science Letters*, **267**: 266–275.

Schmitz, M.D., and Davydov, V., 2012, Quantitative radiometric and biostratigraphic calibration of the Pennsylvanian–Early Permian (Cisuralian) time scale and pan-Euramerican chronostratigraphic correlation. *Geological Society of America Bulletin*, **124**: 549–577.

Schoene, B., Guex, J., Bartolini, A., Schaltegger, U., and Blackburn, T.J., 2010, Correlating the end-Triassic mass extinction and flood basalt volcanism at the 100 ka level. *Geology*, **38**: 387–390.

Secord, R., Gingerich, P., Smith, M., Clyde, W., Wilf, P., and Singer, B., 2006, Geochronology and mammalian biostratigraphy of Middle and Upper Paleocene continental strata, Bighorn Basin, Wyoming. *American Journal of Science*, **306**: 211–245.

Sell, B.K., Samson, S.D., Mitchell, C.E., McLaughlin, P.I., Koenig, A.E., and Leslie, S.A., 2015, Stratigraphic correlations using trace elements in apatite from Late Ordovician (Sandbian-Katian) K-bentonites of eastern North America: Geological Society of America Bulletin, **127**: 1259–1274.

Selby, D., 2007, Direct Rhenium-Osmium age of the Oxfordian-Kimmeridgian boundary, Staffin bay, Isle of Skye, UK, and the Late Jurassic time scale. *Norwegian Journal of Geology*, **87**: 291–299.

Selby, D., and Creaser, R.A., 2005, Direct radiometric dating of the Devonian-Mississippian time-scale boundary using the Re-Os black shale geochronometer. *Geology*, **33**: 545–548.

Selby, D., Mutterlose, J., and Condon, D.J., 2009, U–Pb and Re–Os geochronology of the Aptian/Albian and Cenomanian/Turonian stage boundaries: Implications for timescale calibration, osmium isotope seawater composition and Re–Os systematics in organic-rich sediments. *Chemical Geology*, **265**: 394–409.

Sell, B., Ainsaar, L., and Leslie, S., 2013, Precise timing of the Late Ordovician (Sandbian) super-eruptions and associated environmental, biological, and climatological events. *Journal of the Geological Society*, **170**: 711–714.

Sell, B., Ovtcharova, M., Guex, J., Bartolini, A., Jourdan, F., Spangenberg, J.E., et al., 2014, Evaluating the temporal link between the Karoo LIP and climatic–biologic events of the Toarcian Stage with high-precision U–Pb geochronology. *Earth and Planetary Science Letters*, **408**: 48–56.

Sell, B.K., Leslie, S.A., and Maletz, J., 2011, New U-Pb zircon data for the GSSP for the base of the Katian in Atoka, Oklahoma, USA and the Darriwilian in Newfoundland, Canada. *In* Gutiérrez-Marco, J.C., Rábano, I., and García-Bellido, D. (eds), *Ordovician of the World, Cuadernos del Museo Geominero*, **14**: 537–546.

Shen, S., Crowley, J.L., Wang, Y., Bowring, S.A., Erwin, D.H., Cao, C., et al., 2011, Calibrating the end-Permian mass extinction. *Science*, **334**: 1367–1372.

Siewert, S.E., 2011. Integrating ^{40}Ar/^{39}Ar, U-Pb and Astronomical Clocks in the Cretaceous Niobrara Formation. M.S. thesis, University of Wisconsin at Madison: 74 pp.

Singer, B.S., Meyers, S.R., Sageman, B.B., Jicha, B.R., and Condon, D., 2015, Improving Cretaceous time scale uncertainties via multi-collector ^{40}Ar/^{39}Ar dating. *Geological Society of America Abstracts with Programs*, **47** (7), 172, https://gsa.confex.com/gsa/2015AM/webprogram/Paper264460.html.

Shimokawa, A., 2010, Zircon U-Pb Geochronology of the Great Valley Group: recalibrating the Lower Cretaceous Time Scale. *University of North Carolina at Chapel Hill Dissertation*, 46 p.

Smith, M., Carroll, A., and Singer, B., 2008, Synoptic reconstruction of a major ancient lake system: Eocene Green River Formation, western United States. *Geological Society of America Bulletin*, **120**: 54–84.

Smith, M.E., Singer, B.S., and Carroll, A.R., 2004, Discussion and reply: $^{40}Ar/^{39}Ar$ geochronology of the Eocene Green River Formation, Wyoming, Reply. *Geological Society of America Bulletin*, **116**: 253–256.

Sprain, C.J., Renne, P.R., Clemens, W.A., and Wilson, G.P., 2018, Calibration of chron C29r: New high-precision geochronologic and paleomagnetic constraints from the Hell Creek region, Montana. *Geological Society of America Bulletin*, **130**: 1615–1644.

Sprain, C.J., Renne, P.R., Wilson, G.P., and Clemens, W.A., 2015, High-resolution chronostratigraphy of the terrestrial Cretaceous-Paleogene transition and recovery interval in the Hell Creek region, Montana. *Geological Society of America Bulletin*, **127**: 393–409.

Storey, M., Duncan, R.A., and Swisher, C.C., 2007, Paleocene-Eocene thermal maximum and the opening of the northeast Atlantic. *Science*, **316**: 587–589.

Streel, M., 2000, The late Famennian and early Frasnian datings given by Tucker and others (1998) are biostratigraphically poorly constrained. *Subcommission on Devonian Stratigraphy Newsletter*, **17**: 59.

Suárez, M., Demant, A., Cruz, La, De, R., and Fanning, C.M., 2010, $^{40}Ar/^{39}Ar$ and U–Pb SHRIMP dating of Aptian tuff cones in the Aisén Basin, Central Patagonian Cordillera. *Journal of South American Earth Sciences*, **29**: 731–737.

Suárez, M., La Cruz, De, R., Aguirre-Urreta, B., and Fanning, M., 2009, Relationship between volcanism and marine sedimentation in northern Austral (Aisén) Basin, central Patagonia: Stratigraphic, U–Pb SHRIMP and paleontologic evidence, *Journal of South American Earth Sciences*, **27**: pp. 309–325.

Svensen, H.H., Hammer, Ø., and Corfu, F., 2015, Astronomically forced cyclicity in the Upper Ordovician and U–Pb ages of interlayered tephra, Oslo Region, Norway. *Palaeogeography Palaeoclimatology Palaeoecology*, **418**: 150–159.

Swisher III, C.C., Dingus, L., and Butler, R.F., 1993, $^{40}Ar/^{39}Ar$ dating and magnetostratigraphic correlation of the terrestrial Cretaceous-Paleogene boundary and Puercan Mammal Age, Hell Creek-Tullock formations, eastern Montana. *Canadian Journal of Earth Sciences*, **30**: 1981–1996.

Swisher III, C.C., Grajales-Nishimura, J.M., Montanari, A., Margolis, S.V., Claeys, P., Alvarez, W., et al., 1992, Coeval $^{40}Ar/^{39}Ar$ ages of 65.0 million years ago from Chicxulub crater melt rock and Cretaceous-Tertiary boundary tektites. *Science*, **257**: 954–958.

Tauxe, L., Gee, J., Gallet, Y., Pick, T., and Bown, T., 1994, Magnetostratigraphy of the Willwood Formation, Bighorn Basin, Wyoming: new constraints on the location of Paleocene/Eocene boundary. *Earth and Planetary Science Letters*, **125**: 159–172.

Taylor, J.F., Ganis, G.R., Repetski, J.E., Mitchell, C., and Loch, J.D., 2015, Field trip guidebook for the post-meeting field trip: the central appalachians. *In* Taylor, J.F., and Loch, J.D. (eds), *The Ordovician Exposed: 12th International Symposium on the Ordovician System*, 117 p.

Tejada, M., Mahoney, J., Neal, C., Duncan, R., and Petterson, M., 2002, Basement geochemistry and geochronology of Central Malaita, Solomon Islands, with implications for the origin and evolution of the Ontong Java Plateau. *Journal of Petrology*, **43**: 449–484.

Them, T.R., Gill, B.C., Selby, D., Gröcke, D.R.G.X., Friedman, R.M., and Owens, J.D., 2017, Evidence for rapid weathering response to climatic warming during the Toarcian Oceanic Anoxic Event. *Scientific Reports*, **7**: 1–10.

Thompson, C.K., Kah, L.C., Astini, R., Bowring, S.A., and Buchwaldt, R., 2012, Bentonite geochronology, marine geochemistry, and the Great Ordovician Biodiversification Event (GOBE). *Palaeogeography Palaeoclimatology Palaeoecology*, **321-322**: 88–101.

Toghill, P., 1968, The graptolite assemblages and zones of the Brikhill shales (lower Silurian) at Dobb's Linn. *Palaeontology*, **11** (Part 5), 654–668.

Tucker, R., and McKerrow, W., 1995, Early Paleozoic chronology: a review in light of new U-Pb zircon ages from Newfoundland and Britain. *Canadian Journal of Earth Sciences*, **32**: 368–379.

Tucker, R., Bradley, D., Ver Straeten, C., Harris, A., Ebert, J., and McCutcheon, S., 1998, New U–Pb zircon ages and the duration and division of Devonian time. *Earth and Planetary Science Letters*, **158**: 175–186.

Tucker, R.D., Krogh, T.E., Ross, R.J., and Williams, S.H., 1990, Time-scale calibration by high-precision UPb zircon dating of interstratified volcanic ashes in the Ordovician and Lower Silurian stratotypes of Britain. *Earth and Planetary Science Letters*, **100**: 51–58.

Turgeon, S.C., Creaser, R.A., and Algeo, T.J., 2007, Re–Os depositional ages and seawater Os estimates for the Frasnian–Famennian boundary: Implications for weathering rates, land plant evolution, and extinction mechanisms. *Earth and Planetary Science Letters*, **261**: 649–661.

van Acken, D., Tütken, T., Daly, J.S., Schmid-Röhl, A., and Orr, P.J., 2019, Rhenium-osmium geochronology of the Toarcian Posidonia Shale, SW Germany. *Palaeogeography Palaeoclimatology Palaeoecology*, **534**: 109294–12.

Vennari, V.V., Lescano, M., Naipauer, M., Aguirre-Urreta, B., Concheyro, A., Schaltegger, U., et al., 2014, New constraints on the Jurassic–Cretaceous boundary in the High Andes using high-precision U–Pb data. *Gondwana Research*, **26**: 374–385.

Walaszczyk, I., and Cobban, W.A., 2006, Palaeontology and biostratigraphy of the Middle–Upper Coniacian and Santonian inoceramids of the US Western Interior. *Acta Geologica Polonica*, **56** (3), 241–348.

Wan, X., Scott, R., Chen, W., Gao, L., and Zhang, Y., 2010, Early Cretaceous stratigraphy and SHRIMP U-Pb age constrain the Valanginian-Hauterivian boundary in southern Tibet. *Lethaia*, **44**: 231–244.

Wang, T., Ramezani, J., Wang, C., Wu, H., He, H., and Bowring, S.A., 2016, High-precision U–Pb geochronologic constraints on the Late Cretaceous terrestrial cyclostratigraphy and geomagnetic polarity from the Songliao Basin, Northeast China. *Earth and Planetary Science Letters*, **446**: 37–44.

Walsh, S.L., Prothero, D.R., and Lundquist, D.J., 1996, Stratigraphy and paleomagnetism of the middle Eocene Friars Formation and Poway Group, southwestern San Diego County, California. *In* Prothero, D. R., and Emry, R.J. (eds), *The Terrestrial Eocene-Oligocene Transition in North America*. Cambridge: Cambridge University Press, 120–154.

Waters, C.N., and Condon, D.J., 2012, Nature and timing of Late Mississippian to Mid-Pennsylvanian glacio-eustatic sea-level changes of the Pennine Basin, UK. *Journal Of The Geological Society*, **169**: 37–51.

Wellman, C.H., 2006, Spore assemblages from the Lower Devonian 'Lower Old Red Sandstone' deposits of the Rhynie outlier, Scotland. *Transactions of the Royal Society of Edinburgh: Earth Sciences*, **97**: 167−211.

Wiliams, M., Rushton, A.W.A., Cook, A.F., Zalasiewicz, J., Martin, A.P., Condon, D.J., et al., 2013, Dating the Cambrian Purley Shale Formation, Midland Microcraton, England. *Geological Magazine*, **150**: 937−944.

Wing, S.L., Bown, T.M., and Obradovich, J.D., 1991, Early Eocene biotic and climatic change in interior western North America. *Geology*, **19**: 1189−1192.

Wotzlaw, J.-F., Hüsing, S.K., Hilgen, F.J., and Schaltegger, U., 2014a, High-precision zircon U−Pb geochronology of astronomically dated volcanic ash beds from the Mediterranean Miocene. *Earth and Planetary Science Letters*, **407**: 19−34.

Wotzlaw, J.F., Guex, J., Bartolini, A., Gallet, Y., Krystyn, L., McRoberts, C.A., et al., 2014b, Towards accurate numerical calibration of the Late Triassic: High-precision U-Pb geochronology constraints on the duration of the Rhaetian. *Geology*, **42**: 571−574.

Wotzlaw, J.F., Brack, P., and Storck, J.-C., 2018, High-resolution stratigraphy and zircon U−Pb geochronology of the Middle Triassic Buchenstein Formation (Dolomites, northern Italy): precession-forcing of hemipelagic carbonate sedimentation and calibration of the Anisian−Ladinian boundary interval. *Journal of the Geological Society*, **175**: 71−85.

Wu, Q., Ramezani, J., Zhang, H., Wang, T.-T., Yuan, D.-X., Mu, L., et al., 2017, Calibrating the Guadalupian Series (Middle Permian) of South China. *Palaeogeography Palaeoclimatology Palaeoecology*, **466**: 361−372.

Yang, C., Li, X.-H., Zhu, M., Condon, D.J., and Chen, J., 2018, Geochronological constraint on the Cambrian Chengjiang biota, South China. *Journal of the Geological Society*, **175**: 659−666.

Zhang, Y., Ogg, J.G., Minguez, D.A., Hounslow, M., Olaussen, S., Gradstein, F.M., and Esmeray-Senlet, S., 2019, Magnetostratigraphy of U/Pb-dated boreholes in Svalbard, Norway, implies that the Barremian-Aptian Boundary (beginning of Chron M0r) is 121.2 ± 0.4 Ma. In: *AGU Annual Meeting (San Francisco, 9-13 December 2019)*. #GP44A-06. https://agu.confex.com/agu/fm19/meetingapp.cgi/Paper/577991.

Zhou, C., Huyskens, M.H., Lang, X., Xiao, S., and Yin, Q.-Z., 2019, Calibrating the terminations of Cryogenian global glaciations. *Geology*, **47**: 251−254.

Ziegler, W., and Sandberg, C.A., 1990, The Late Devonian Standard Zonation. *Courier Forschungsinstitut Senckenberg*, **121**: 1−115.

Index

Note: Page numbers followed by "*f*" indicate figures and "*t*" indicate tables.

A

Aalenian Stage/Age, 965, 966*f*
Abereiddian (Britain), 652*f*
Acadian (West Avalonia), 586*f*, 610*f*
Adavere (Baltica), 715*f*
Adelaidean (Australia), 586*f*, 610*f*
Aegean Substage, 905*f*, 910, 932*f*, 935*f*
Aeronian Stage/Age, 696−702, 699*f*
Agdzian (Iberia/Morocco), 586*f*
Age 2 (Cambrian), 573−574
Age 3 (Cambrian), 574−575
Age 4 (Cambrian), 575−579
Age 10 (Cambrian), 584−585
Agnostoids, 36*f*, 37, 39−40, 599−600
Aisha-Bibaian (Kazakhstan), 586*f*
Akchagylian (Eastern Paratethys), 1153*f*, 1157
Aksaian (Russia/Siberia; Kazakhstan), 586*f*, 592
Alaunian Substage, 905*f*, 914−915, 921, 929*f*, 933
Albian Stage/Age, 1032−1034, 1033*f*, 1043*t*, 1047*f*
Aleksinian (Russian Platform), 824*f*, 828−829
Alportian (Western Europe), 824*f*, 826−827
Altonian (New Zealand), 1158, 1159*t*, 1161*f*
Amazonian Period (Mars), 449*f*, 455*f*, 459
Amb (Salt Range), 889*f*
Amgan (Russia/Siberia), 586*f*, 593−594, 610*f*
Ammonoids and Ammonites (Ammonoidea)
  Carboniferous, 61−62, 61*f*, 838*f*, 841
  Cretaceous, 62, 67, 1043−1048, 1047*f*
  Devonian, 61−62, 61*f*, 755*f*, 757−758
  Jurassic, 64−67, 65*f*, 975−977, 997*t*
  overview, 61−68
  Paleozoic, 61−62, 61*f*
  Permian, 61−62, 61*f*, 893*f*
  Triassic, 62, 64, 916, 929*f*, 932*f*, 935*f*
Anisian Stage/Age, 909−910, 932*f*
Anthropocene Epoch/Series, 1234−1236, 1257
Aptian Stage/Age, 1031−1032, 1047*f*
Aquitanian Stage/Age, 1143−1145, 1144*f*
Archaeocyathid Extinction Carbon-isotope
  Excursion (AECE), 318*f*, 602−603
Archaeocyathids, 600
Archean Eon, 483−484
Arenig (Britain), 641, 652*f*
Arenigian (Ibero-Bohemia), 652*f*, 653
Argon-Argon (^{40}Ar/^{39}Ar) geochronology, 197−201

Arnsbergian (Western Europe), 824*f*, 826−827
Artinskian Stage/Age, 879−881, 893*f*
Aseri (Baltica), 651−653, 652*f*
Ashgill (Britain), 651, 652*f*
Asselian Stage/Age, 877−879, 878*f*, 893*f*
Astrochronology, 139, 203−204, 718, 769−771, 845−847, 924−925, 988−990, 1038, 1048−1051, 1055−1057
Asturian (Western Europe), 824*f*, 827
Atdabanian (Russia/Siberia; Kazakhstan), 586*f*, 593
Atokan (North America), 824*f*, 831
Aurelucian (Britain), 652*f*
Autunian (Western Europe), 824*f*, 827, 889*f*, 891−894
Ayusokkanian (Russia/Siberia; Kazakhstan), 579, 586*f*, 592

B

Badenian (Central Paratethys), 1153*f*, 1154
Bajocian Stage/Age, 965−967, 968*f*, 980*f*, 1001−1002
Banded iron formations (BIF), 485
Banian (Iberia/Morocco), 586*f*, 610*f*
Barremian Stage/Age, 1031, 1047*f*
Bartonian Stage/Age, 1097−1099
Basal Cambrian Carbon-isotope Excursion (BACE), 318*f*, 602−603
Bashkirian Stage/Age, 819−821, 820*f*
Bathonian Stage/Age, 967, 969*f*, 1002−1003
Batyrbaian (Russia/Siberia; Kazakhstan), 586*f*, 592
Bazovian (NE Siberia), 824*f*
BChron method, 406−407
Begadean (North America), 574−575, 586*f*, 595
Bendigonian (Australasia), 651, 652*f*
Benthic foraminifera
  Carboniferous, 838*f*, 840*f*
  Jurassic, 977
  overview, 88−98
  Permian, 893*f*, 894−895
  Paleogene, 1107*f*, 1108−1109
Berounian (Ibero-Bohemia), 652*f*, 653
Berriasian Stage/Age, 1024−1028, 1047*f*
Bessarabian (Eastern Paratethys), 1153*f*, 1156
Billingen (Baltica), 651−653, 652*f*
Bithynian Substage, 905*f*, 910, 927, 932*f*

Blackhillsian (North America), 652*f*
Bobrikian (Russian Platform), 824*f*
Bolindian (Australasia), 651, 652*f*
Bolorian (Pamir), 889*f*, 894−895
Bolsovia (Western Europe), 827
Boomerangian (Australia), 590
Bootstrap splines, 406
Bosphorian (Eastern Paratethys), 1153*f*, 1156−1157
Botoman (Russia/Siberia; Kazakhstan), 586*f*, 593
Branchian (West Avalonia), 574−575, 586*f*, 610*f*
Bundsandstein (Western Europe), 889*f*
Burdigalian Stage/Age, 1145, 1163*f*
Burrellian (Britain), 652*f*
Burtsevian (Russian Platform), 889*f*

C

C-sequence magnetic polarity scale, 164−173, 166*t*, 1052, 1116−1121, 1194−1198
Caesaraugustan (Iberia/Morocco), 586*f*
Calabrian Stage/Age, 1225, 1226*f*
Calcareous nannofossils
  Cretaceous, 1047*f*, 1050−1051
  Jurassic, 980*f*, 981
  Neogene, 1162−1164, 1163*f*, 1165*t*
  overview, 69−73
  Paleogene, 1107*f*, 1109−1110
  Quaternary, 1222*f*
  Triassic, 917
Callovian Stage/Age, 967−970, 977, 980*f*, 1003
Calorian Period/System (Mercury), 449*f*, 460−461, 466*t*
Calpionellid microfossils, 977−981, 1025, 1027*f*, 1047*f*, 1050
Calymmian Period/System, 316*f*, 328, 482*f*
Cambrian Arthropod Radiation isotope Excursion (CARE), 318*f*, 602−603
Cambrian Period/System
  age model, 607−613
  basal definition, 572*f*, 573
  biostratigraphy, 570*f*, 599−607
  international subdivisions, 566−599, 586*f*
  magnetostratigraphy, 601−602
  regional stages, 567*f*, 585−599
  selected events, 606−607

Cambrian Period/System (*Continued*)
  stable isotope stratigraphy, 289–297, 318*f*, 602–605
  time scale, 570*f*, 607–613, 610*f*
Campanian Stage/Age, 1040, 1047*f*, 1057
Cantabrian (Western Europe), 740*f*, 827
Capitanian Stage/Age, 882–886, 885*f*, 893*f*
Caradoc (Britain), 646, 651, 652*f*
Carbon isotopes
  Cambrian, 318*f*, 567*f*, 602–605, 603*f*
  Carboniferous, 322*f*, 813*f*, 844–845
  Cretaceous, 326*f*, 1027*f*, 1052–1054
  Cryogenian, 317*f*, 502*f*, 506
  Devonian, 321*f*, 735*f*, 755*f*, 766–768
  Ediacaran, 317*f*, 524, 537–540
  Jurassic, 325*f*, 957*f*, 985–986
  Neogene, 327*f*, 1142*f*, 1193–1194
  Ordovician, 319*f*, 633*f*, 657–660, 658*f*
  overview, 309
  Paleogene, 327*f*, 1089*f*, 1121–1124, 1122*f*
  Permian, 323*f*, 877*f*, 895
  Precambrian, 316*f*, 317*f*
  Silurian, 320*f*, 697*f*, 714–716, 715*f*
  Triassic, 324*f*, 905*f*, 922–923
Carboniferous Period/System, 815–816
  age model, 417, 853–855
  basal definition, 747–749, 748*f*, 814*f*, 815
  biostratigraphy, 836–848
  international subdivisions, 811–836, 813*f*
  magnetostratigraphy, 843
  regional stages, 823–836, 824*f*
  stable isotope stratigraphy, 289–297, 295*f*, 322*f*, 813*f*, 843–845
  time scale, 295*f*, 813*f*, 848–855
Carboniferous-Permian spline, 417, 849*t*, 853–855, 854*f*
Carixian Substage, 963
Carnian Pluvial Episode, 914, 923, 932
Carnian Stage/Age, 912–914, 913*f*, 929*f*
Castlemainian (Australasia), 643, 652*f*
Cathedralian (North America), 889*f*
Caucasian (Eastern Paratethys), 1153*f*, 1154–1155
Cautleyan (Britain), 652*f*
Cenomanian Stage/Age, 1034–1036, 1047*f*
Cenozoic Era, 16–18, 17*f*, 28*f*, 1088, 1089*f*
Changhsingian Stage/Age, 886, 888*f*
Chatauquan (North America), 824*f*
Chatfieldian (North America), 652*f*
Chattian Stage/Age, 1103–1105, 1104*f*, 1107*f*
Cheneyan (Britain), 652*f*
Cheremshanian (Russian Platform), 824*f*
Cherepetian (Russian Platform), 824*f*
Chesterian (North America), 824*f*, 830
Chewtonian (Australasia), 651, 652*f*
Chhiddru (Salt Range), 889*f*
Chientangkiangian (China), 652*f*, 653
Chihsian (China), 889*f*, 891
Chitinozoans
  Devonian, 755*f*, 761
  Ordovician, 635*f*, 637*f*, 639*f*, 656, 664*f*
  Silurian, 699*f*, 713
  Overview, 50–55, 51*f*
Chokierian (Western Europe), 825–827, 854*f*

Chokrakian (Eastern Paratethys), 1153*f*, 1154–1156
Chronogram method, 402–403
Chronostratigraphy, 21
Chuanshanian (China), 889*f*, 891
Chuguchanian (NE Siberia), 824*f*
Cincinnatian (North America), 652*f*, 653–654
Cisuralian Epoch/Series, 876–881, 889*f*, 893*f*
Clifdenian (New Zealand), 1158, 1159*t*, 1161*f*
Climate and temperature trends
  Anthopocene, 1269
  Cambrian, 606
  Carboniferous, 295*f*, 813*f*, 815, 845–848
  Cretaceous, 288*f*, 1027*f*, 1054–1055
  Cryogenian, 496–497, 508–510
  Devonian, 294*f*, 735*f*, 755*f*
  Ediacaran, 543
  Jurassic, 288*f*, 957*f*, 986–987
  Neogene, 283–287, 1089*f*, 1193
  Ordovician, 292*f*, 633*f*, 648, 662–664
  Paleogene, 283–287, 1089*f*, 1121–1124, 1122*f*
  Permian, 295*f*, 296*f*, 877*f*, 895
  Precambrian, 297, 298*f*, 481–482
  Quaternary, 283–287, 1236–1240
  Silurian, 293*f*, 697*f*, 715*f*, 717
  Triassic, 288*f*, 296*f*, 905*f*, 914, 922, 924
Conchostracans, 842–843, 919, 929*f*, 932*f*, 935*f*
Coniacian Stage/Age, 1038, 1047*f*
Conodonts
  Cambrian, 57, 570*f*, 601
  Carboniferous, 59, 838*f*, 840*f*, 841
  Devonian, 57–59, 711*f*, 749–757
  Ordovician, 57, 635*f*, 637*f*, 639*f*, 655, 664*f*
  overview, 56–60, 58*f*
  Permian, 59, 893*f*, 894
  Silurian, 57, 699*f*, 713
  Triassic, 59, 916, 929*f*, 932*f*, 935*f*
Constrained optimization methods (CONOP), 314, 361, 428
  Carboniferous-Permian CONOP, 417*f*, 853, 896
  method, 425–439
  Ordovician-Silurian CONOP, 414*f*, 415*f*, 667–674
Copernican Period/System (Moon), 449*f*, 450*f*, 452–453
Cordevolian Substage, 912
Cordubian (Iberia/Morocco), 574–575, 586*f*, 610*f*
Cressagian (Britain), 652*f*
Cretaceous Period/System
  age model, 417–424, 1060–1068
  basal definition, 975, 1024–1028, 1027*f*
  biostratigraphy, 1043–1058, 1043*t*, 1047*f*
  international subdivisions, 1023–1043, 1027*f*, 1067*f*
  magnetostratigraphy, 164–188, 1047*f*, 1051–1052
  regional stages, 1025–1028, 1027*f*
  selected events, 1052–1054, 1057–1058
  stable isotope stratigraphy, 287–289, 326*f*, 1027*f*, 1052–1055

time scale, 1027*f*, 1047*f*, 1058–1068
Cryogenian Period/System
  age model, 501–505
  basal definitions, 512
  biostratigraphy, 505–506
  stable isotope stratigraphy, 508–510
  time scale, 508–510
Cubic smoothing spline, 404–405
Cyclostratigraphy, 139–141, 661–662, 718, 769–771, 770*f*, 845–847, 924–925, 988–990, 1038, 1048–1051, 1055–1057
Cyclothems, 379, 384, 838*f*, 845–847

D

Dacian (Eastern Paratethys), 1153*f*, 1154
Dalaun (China), 824*f*, 832–833
Daldynian-Nemakit (Russia/Siberia), 586*f*, 592, 610*f*
Danian Stage/Age, 1090–1091, 1091*f*, 1107*f*
Dapingian Stage/Age, 637*f*, 643, 644*f*
Darriwilian Stage/Age, 637*f*, 643–646, 645*f*
Dastarian (Pamir), 889*f*
Datsonian (Australia), 591
Delamaran (North America), 595–596
Dengyingxian (South China), 586*f*
Desmoinesian (North America), 824*f*, 831
Devensian (British Isles), 1222*f*, 1229*f*, 1237
Devonian Period/System
  age model, 413–414, 779–781
  basal definition, 734, 737, 737*f*
  biostratigraphy, 749–761, 755*f*
  international subdivisions, 734–749, 735*f*, 736*f*
  magnetostratigraphy, 735*f*, 772
  regional stages, 736*f*
  selected events, 735*f*, 762–766
  stable isotope stratigraphy, 289–297, 294*f*, 321*f*, 735*f*, 766–768
  time scale, 735*f*, 755*f*, 772–781
Devonian spline, 408–409, 413–414, 773, 779, 779*f*
Dewuan (China), 824*f*, 833
Dienerian Substage, 905*f*, 908, 935*f*
Dinoflagellates
  Cretaceous, 100–101, 103, 1051
  Jurassic, 99–101, 103, 981
  Neogene, 104, 965–967, 1188*t*, 1190*f*
  overview, 99–108
  Paleogene, 104, 1041–1043, 1065*f*
  Quaternary, 104
  Triassic, 101, 917
Dinosaurs, 918–919, 981–982, 1031
Dobrotivian (Ibero-Bohemia), 652*f*, 653
Domerian Substage, 963
Dorashamian (Pamir), 886, 889*f*
Dorogomilovian (Russian Platform), 824*f*, 830
Drumian Isotope-Carbon Excursion (DICE), 318*f*, 567*f*, 603*f*, 604
Drumian Stage/Age, 570*f*, 577–579
Duckmantian (Western Europe), 824*f*, 827
Duyunian (South China), 586*f*, 587
Dyeran (North America), 586*f*, 595
Dzhulfian (Pamir), 886, 889*f*

E

Early Cretaceous Epoch, 1024–1034
Early Cretaceous spline, 417–423, 1061, 1065f
Early Devonian Epoch, 735f, 737–741, 755f
Early Imbrian Epoch (Moon), 447–452, 452t
Early Jurassic Epoch, 956–965, 976f
Early Ordovician Epoch, 635–643, 639f
Early Permian (Cisuralian) Epoch, 879, 893f
Early Pleistocene Subepoch, 1227–1229
Early Triassic Epoch, 908–909, 935f
Earth's Moon, 446–453
Eastonian (Australasia), 651, 652f
Ectasian Period/System, 316f, 328, 482f
Ediacaran Period/System
    age model, 543–544, 545f
    basal definition, 522–523, 522f, 525–526
    biostratigraphy, 526–544, 528f
    glaciations, 537
    international subdivisions, 544–547
    stable isotope stratigraphy, 537–543
    time scale, 543–544, 545f
Eemian (NW Europe), 1222f, 1227, 1229
Egerian (Central Paratethys), 1153, 1153f
Eggenburgian (Central Paratethys), 1153–1155, 1153f
Eifelian Stage/Age, 741, 742f, 755f
Emsian Stage/Age, 738–741, 740f, 755f
Eoarchean Era, 483–484
Eocene Epoch/Series, 1089f, 1095, 1107f
Eratosthenian Period (Moon), 449f, 450f, 452
Eustatic sea level. See Sea level and Sequences
Euxinian (Eastern Paratethys), 1153f, 1156–1157

F

Famennian Stage/Age, 735f, 744–747, 746f, 755f
Fassanian Substage, 904, 905f, 912, 932f
Fennian (Britain), 652f
Filippovian (Russian Platform), 881, 889f
Floian Stage/Age, 633f, 639f, 641–643, 642f
Floran (Australia), 590
Foraminifera
    Carboniferous benthic foraminifera, 838f, 840f
    Cretaceous planktonic foraminifera, 1047f, 1049–1050
    Jurassic, 977–981
    Neogene planktonic foraminifera, 1163f, 1164, 1170t
    overviews, 74–98
    Paleogene benthic foraminifera, 1107f, 1108–1109
    Paleogene planktonic foraminifera, 1105–1108, 1107f
    Permian benthic foraminifera, 893f, 894–895
Fortunian Stage/Age, 567f, 570f, 573
Frasnian Stage/Age, 744, 745f, 755f
Furongian Epoch/Series, 567f, 570f, 579–585, 582f
Fusulinids, 89–92, 838f, 840f, 893f, 894–895

G

Gamachian (North America), 652f, 654
Gaskiers glaciation, 544–546
Gelaohean (China), 824f
Gelasian Stage/Age, 1150, 1219f, 1222f, 1223–1225, 1224f
Geologic Time Scale 2020
    overview, 4–11, 6f
    summary, 6f, 7t, 19, 22f
Geomagnetic polarity scale, 161, 1088
    Cambrian, 567f, 601–602
    Carboniferous, 813f, 843
    Cretaceous, 164–188
    C–sequence, 164–173, 166t, 1052, 1116–1121, 1194–1198
    Devonian, 735f, 772
    Jurassic, 173–188, 957f, 982–985
    M–sequence, 173–188, 175t, 980f, 982–985, 1047f, 1051–1052
    Neogene, 1194–1198
    Ordovician, 635f, 637f, 639f
    Paleogene, 1116–1121
    Permian, 877f, 895
    Quaternary, 162t, 1222f
    Silurian, 697f, 713
    Triassic, 905f, 920–921
Gisbornian (Australasia), 646, 651, 652f
Givetian Stage/Age, 735f, 741–744, 743f, 755f
Glaciations
    Late Paleozoic Ice Age, 815, 845–848, 895
    Cryogenian, 231, 490, 495–501, 504–505, 509f, 512–513
    Ediacaran, 537
    Hirnantian, 127–128, 291, 660, 663–664, 717
    Huronian, 485
    Quaternary (Pleistocene), 1222f, 1223, 1236–1238
Global Boundary Stratotype Section and Point (GSSP), 24–31, 26t, 28f, 29f [See each Period for details on their GSSPs]
Gorbiyachinian (Russia/Siberia), 586f, 594
Gorstian Stage/Age, 699f, 706–709, 708f
Graptolites
    Devonian, 44, 45f, 46–48, 711f, 758
    Ordovician, 44, 45f, 635f, 637f, 639f, 655, 664f
    overview, 43, 674–675
    Silurian, 44, 45f, 699f, 700f, 712–713
Greenlandian Stage/Age, 1231f, 1232
Griesbachian Substage, 905f, 908, 935f
GSSP. See Global Boundary Stratotype Section and Point (GSSP)
Guadalupian Epoch/Series, 877f, 881–886, 893f
Gumerovian (Russian Platform), 824f, 828
Guzhangian Stage/Age, 567f, 570f, 579, 580f, 588
Gzhelian Stage/Age, 813f, 822–823, 838f

H

Hadean Eon, 482–483, 482f
Hadrynian (North America), 586f

Haljala (Baltica), 651–653, 652f
Harju (Baltica), 652f
Hastarian (Western Europe), 824f, 825
Hauterivian Stage/Age, 1027f, 1029–1030, 1030f, 1047f
Hesperian Period (Mars), 449f, 458–459
Hessian (North America), 889f
Hettangian Stage/Age, 957f, 959–961, 959f, 988
Hirnantian Stage/Age, 633f, 635f, 648–651, 650f
Holocene Epoch/Series, 1219f, 1229–1234
Homerian Stage/Age, 697f, 699f, 706, 707f
Hot spot analysis, 410–411
Huashibanian (China), 824f, 833
Hunneberg (Baltica), 651–653, 652f

I

Ibexian (North America), 597, 652f, 653
Idamean (Australia), 586f, 591
Illawarra geomagnetic series, 895
Illinoian (North America), 1229f
Illyrian Substage, 905f, 910, 927, 932f
Imbrian Period/System (Moon), 449f, 908
Induan Stage/Age, 905f, 908, 935f
Inoceramids, 1047f, 1048–1051
Ionian Stage/Age, 1219f, 1222f, 1225
Ipswichian (British Isles), 1229f, 1237t
Irenian (Russian Platform), 889f, 890
Irginian (Russian Platform), 889f
Iridium anomaly (base of Cenozoic), 130–132, 1090, 1091f
Issendalenian (Iberia/Morocco), 586f, 610f
Iverian (Australia), 586f, 591
Ivorian (Western Europe), 824f, 825

J

Jaagarahu (Baltica), 715f
Jaani (Baltica), 715f
Jiangshanian Stage/Age, 567f, 570f, 581–584, 583f, 586f, 589
Jinningian (South China), 585–587, 586f, 610f
Jiusian (China), 824f, 833
Julian Substage, 905f, 912, 929f
Jurassic Period/System, 955
    age model, 994–1003
    basal definition, 956–959, 959f
    biostratigraphy, 975–982, 976f
    international subdivisions, 956–975, 957f
    magnetostratigraphy, 976f, 982–985
    regional stages, 974–975, 976f
    selected events, 991–992
    stable isotope stratigraphy, 288f, 325f, 957f, 985–987
    time scale, 957f, 976f, 992–1004
Juuru (Baltica), 652f, 715f

K

Kalabagh (Salt Range), 889f
Kapitean (New Zealand), 1158–1161, 1159t, 1161f

Karaganian (Eastern Paratethys), 1153f, 1154–1156
Karakubian (Russian Platform), 824f
Karpatian (Central Paratethys), 1153f, 1154
Kashirian (Russian Platform), 822, 824f, 829–830
Kasimovian Stage/Age, 813f, 822, 824f, 830, 838f
Kathwai (Salt Range), 855f
Katian Stage/Age, 633f, 635f, 648, 649f
Kaugetuma (Baltica), 715f
Kazanian (Russian Platform), 889f, 890
Keila (Baltica), 651–653, 652f
Keiloran (Australasia), 652f
Khamamitian (NE Siberia), 824f
Khamovnikian (Russian Platform), 824f, 830
Khantaian (Russia/Siberia), 586f, 595
Khatynykhian (NE Siberia), 824f
Khersonian (Eastern Paratethys), 1153f, 1156
Khorokytian (NE Siberia), 824f
Kimmerian (Eastern Paratethys), 1153f, 1154–1155, 1157
Kimmeridgian Stage/Age, 957f, 972–973, 976f, 980f, 990, 1003
Kinderhookian (North America), 824f, 830–831
Kinderscoutian (Western Europe), 824f, 826
Kirinian (NE Siberia), 824f
Kizelian (Russian Platform), 824f, 828
Konkian (Eastern Paratethys), 1153f, 1154–1156
Kosorechian (Russian Platform), 828
Kosvian (Russian Platform), 824f, 828
Kotsakhurian (Eastern Paratethys), 1153f, 1154–1155
Kralodvorian (Ibero-Bohemia), 652f, 1156
Krasnopolyanian (Russian Platform), 824f, 829
Krevyakinian (Russian Platform), 822, 824f, 830
Kubergandian (Pamir), 889f
Kuiperian Period/System (Mercury), 449f, 460–461, 466t
Kukruse (Baltica), 651–653, 652f
Kulyumbean (Russia/Siberia), 594
Kunda (Baltica), 651–653, 652f
Kungurian Stage/Age, 877f, 881, 889f, 890, 893f
Kuressaare (Baltica), 715f
Kuyalnikian (Eastern Paratethys), 1153f, 1157
Kuzel (Western Europe), 824f
Kyglitassian (NE Siberia), 824f

L

Lacian Substage, 905f, 914–915, 921
Ladinian Stage/Age, 910–912, 936t
Lancefieldian (Australasia), 651, 652f
Land Mammal Ages of North America (NALMA), 1113, 1115f
Landon (New Zealand), 1159t
Langhian Stage/Age, 1145–1146
Langsettian (Western Europe), 824f, 827
Languedocian (Iberia/Morocco), 586f, 610f
Large igneous provinces (LIPs)

Central Atlantic Magmatic Province (CAMP), 991
Cretaceous, 1052–1054, 1057–1058
Jurassic, 987, 991
overview, 125, 345–346
Precambrian, 348–350
Siberian Traps, 129, 347
Larger benthic foraminifera (LBF), 88–98
Lasnamagi (Baltica), 652f
Late Cretaceous Epoch, 164–173, 423–424, 1034–1043
Late Cretaceous spline, 422f, 423–424, 1065f
Late Devonian Epoch, 744–749
Late Imbrian Epoch (Moon), 452
Late Jurassic Epoch, 990
Late Ordovician Epoch, 646–651
Late Pleistocene Sub-epoch, 1223
Late Triassic Epoch, 912–915
Lenoxian (North America), 889f
Leonardian (North America), 889f, 890
Lillburnian (New Zealand), 1158, 1159t
Livian (Western Europe), 818, 824f, 825
Llandeilian (Britain), 652f
Llandovery Epoch/Series, 696–703
Llanvirn (Britain), 651
Lochkovian Stage/Age, 737–738
Lomagundi-Jatuli Excursion (LJE), 485
Longmaxian (China), 652f
Longobardian Substage, 642f, 646, 674
Lopingian Epoch/Series, 886–890
Lotharingian Substage, 961
Lousuan (China), 824f
Lower Cretaceous Series, 1024–1034
Lower Devonian Series, 735f, 737–741, 755f
Lower Jurassic Series, 956–965, 976f
Lower Ordovician Series, 635–643, 639f
Lower Permian (Cisuralian) Series, 879, 893f
Lower Pleistocene Sub-series, 1227–1229
Lower Triassic Series, 908–909, 935f
Ludfordian Stage/Age, 709, 710f, 724
Ludlow Epoch/Series, 706–709, 708f
Lunar chronologic units, 452t
Lutetian Stage/Age, 1089f, 1097, 1098f

M

M-sequence magnetic polarity scale, 173, 983–985, 1051
Maastrichtian Stage/Age, 1041–1043
Maeotian (Eastern Paratethys), 1153f, 1156
Magapanian (New Zealand), 1161f
Magnetostratigraphy
Cambrian, 601–602
Carboniferous, 843
Cretaceous, 1051–1052
Devonian, 772
Jurassic, 173–188, 982–985
Neogene, 164–173, 166t, 1193
Ordovician, 633f
overview, 159–161, 161t
Paleogene, 164–173, 166t, 1116–1121
Permian, 879, 882
Quaternary, 1219–1222, 1222f, 1237
Silurian, 713

Triassic, 921
Malevkian (Russian Platform), 824f, 828
Mammals. See Vertebrates
Maokouian (China), 889f, 891
Marinoan glaciation, 497, 504–505
Marjuman (North America), 586f, 596
Mars, 453–460
Marsdenian (Western Europe), 824f, 826–827
Mass extinctions ("Big 5")
End-Cretaceous, 125, 126t, 130–132
End-Permian, 125, 126t, 128–130
End-Triassic, 125, 126t
Late Devonian, 125, 126t
Late Ordovician, 125, 126t, 127–128
Maximum likelihood, 402
Mayan (Russia/Siberia), 586f, 594
Maysvillian (North America), 652f
Medinan (North America), 652f
Meghalayan Stage/Age, 1234
Meishucunian (South China), 586f, 587
Melekessian (Russian Platform), 824f, 829
Melekhovian (Russian Platform), 824f, 830
Meramecian (North America), 824f, 830–831
Mercury, 460–461
Merionethian (West Avalonia), 586f, 610f
Mesoarchean Era, 483–484
Mesoproterozoic Era, 486
Mesozoic Era, 14–16, 15f, 28f
Messinian Stage/Age, 1146–1150
Miaolingian Epoch/Series, 575–579
Middle Devonian Epoch, 734, 741–744
Middle Jurassic Epoch, 988–990
Middle Ordovician Epoch, 637f, 643–646
Middle Pleistocene Sub-epoch, 1225–1227
Middle Triassic Epoch, 921
Midian (Pamir), 889f
Migneintian (Britain), 652f
Mikhailovian (Russian Platform), 824f, 828–829
Milankovitch cycles, 139, 661–662, 718, 769–771, 770f, 845–847, 924–925, 988–990, 1038, 1048–1051, 1055–1057
Mindyallan (Australia), 590–591
Mingxinsi Carbon-isotope Excursion (MICE), 318f, 603f
Miocene Epoch/Series, 1143–1150
Mississippian Subperiod, 815–818, 830, 840f
Missourian (North America), 831–832
Mohawkian (North America), 652f, 653–654
Moliniacian (Western Europe), 818, 823, 824f
Montezuman (North America), 575, 586f, 595
Moon (Earth's), 446–453
Moridunian (Britain), 652f
Morrowan (North America), 824f, 831
Moscovian Stage/Age, 821–822, 829–830
Murgabian (Pamir), 889f
Myachkovian (Russian Platform), 824f, 829–830

N

Nabala (Baltica), 650f, 651–653
NALMA. See Land Mammal Ages of North America (NALMA)

Namurian (Western Europe), 824f, 825
Nangaoan (South China), 586f, 587
Natalian (NE Siberia), 824f
Nealian (North America), 889f
Nectarian Period/System, 447
Neichianshanian (China), 652f
Neoarchean Era, 483–484
Neogene Period/System
    age model, 1143
    basal definition, 1143–1145
    biostratigraphy, 1162–1191
    international subdivisions, 1142–1162,
      1142f
    magnetostratigraphy, 1193
    stable isotope stratigraphy, 285f, 327f,
      1193–1194
    time scale, 1142f, 1194–1199
Neoproterozoic Era, 482f, 495–496
Niuchehean (South China), 584, 586f, 589
Noachian Period/Series (Mars), 457–458
Noginskian (Russian Platform), 824f, 830
Norian Stage/Age, 932–933
Northgrippian, 1232–1234
Nukumaruan (New Zealand), 1159t, 1162
Nyayan (Russia/Siberia), 586f

O

Oandu (Baltica), 652f
Ochoan (North America), 889f, 891
Odessian (Eastern Paratethys), 1153f, 1157
Oeland (Baltica), 652f
Ohesaare (Baltica), 715f
Olenekian Stage/Age, 908–909, 936t
Oligocene Epoch/Series, 1101, 1102f
Opoitian (New Zealand), 1158–1162, 1161f
Ordian (Australia), 586f, 589–590
Ordovician Period/Epoch
    age model, 674–682
    basal definition, 635–643, 641f
    biostratigraphy, 655–657
    international subdivisions, 632–651, 633f
    magnetostratigraphy, 633f
    selected events, 662–664
    stable-isotope stratigraphy, 292f, 319f,
      657–661
    time scale, 633f, 664–682
Ordovician-Silurian spline, 415f, 668–671,
    671f
Oretanian (Ibero-Bohemia), 652f, 653
Orosirian Period/System, 316f, 482f
Osagean (North America), 824f, 830–831
Osmium isotope stratigraphy, 239, 252,
    923–924, 988
Otaian (New Zealand), 1159t, 1161f
Ottnangian (Central Paratethys), 1153f,
    1154–1155
Ovlachanian (NE Siberia), 824f
Oxfordian Stage/Age, 971–972, 990
Oxygen isotope stratigraphy, 279
    Cambrian, 289
    Carboniferous, 281, 290f, 295f, 844
    Cretaceous, 287, 288f, 1054–1055
    Cryogenian, 317f

Devonian, 290f, 294f, 321f, 767
Ediacaran, 317f
Jurassic, 287, 288f, 986–987
Neogene, 283, 285f
Ordovician, 290f, 292f, 297, 661
overview, 279
Paleogene, 283, 285f, 986–987, 1122f
Permian, 290f, 295f, 296f, 895
Precambrian, 297, 298f
Quaternary, 1222f
Silurian, 290f, 293f, 297, 714–716
Triassic, 287, 288f, 296f, 923

P

Paadla (Baltica), 715f
Paibian Stage/Age, 581, 582f, 588
Pakerort (Baltica), 651–653, 652f
Paleoarchean Era, 482f, 483–484
Paleocene Epoch/Series, 1090
Paleogene Period/System
    age model, 1124
    basal definition, 1090–1091, 1091f
    biostratigraphy, 1105–1124
    international subdivisions, 1088–1105,
      1089f
    magnetostratigraphy, 1116–1121
    selected events, 1121–1124
    stable-isotope stratigraphy, 285f, 327f,
      1121–1124
    time scale, 1089f, 1124–1126
Paleoproterozoic, 248, 315–328, 316f,
    484–486
Paleozoic Era, 12–14, 13f, 29f
Pannonian (Central Paratethys), 1153f, 1154,
    1156
Pareora (New Zealand), 1159t, 1161f
Pavlovoposadian (Russian Platform), 824f, 830
Payntonian (Australia), 584, 586f, 591
Pelsonian Substage, 910, 927, 932f
Pendleian (Western Europe), 824f, 826–827
Pennsylvanian Subsystem, 819–823, 838f
Permian Period/System
    age model, 897–898
    basal definition, 877–879, 878f
    biostratigraphy, 894–895
    international subdivisions, 875–890, 877f
    magnetostratigraphy, 882
    stable-isotope stratigraphy, 295f, 296f, 323f,
      895
    time scale, 877f, 896–898
Phanerozoic Eon, 6f, 22f
Piacenzian Stage/Age, 1150, 1152f
Pirgu (Baltica), 651–653, 652f
Placentian (West Avalonia), 586f, 611
Planetary timescale, 443–446, 449f
    Earth's Moon, 446–453, 449f
    Mars, 449f, 453–460
    Mercury, 449f, 460–461
    cratered bodies, 462–474
    Venus, 461–462
Plants, spores and pollen
    Carboniferous, 842
    Cenozoic, 110f, 112f, 118–119

Devonian, 760f, 761
Jurassic, 982
Mesozoic, 110f, 112f, 116–118
Paleozoic, 110–116, 110f, 112f
Precambrian, 109–110, 110f
Pleistocene Epoch/Series, 1223–1229
Pliensbachian Stage/Age, 962f, 963, 1001
Pliocene Epoch/Series, 1150
Podolskian (Russian Platform), 824f, 829–830
Pontian (Eastern Paratethys), 1153f,
    1156–1157
Porkuni (Baltica), 652f, 715f
Portaferrian (Eastern Paratethys), 1153f, 1157
Portlandian (British), 970–971, 973–975
Povolzhian (Russian Platform), 889f
Power law models, 409–410
Pragian Stage/Age, 738, 739f
Pre-Nectarian Period (Moon), 446–447
Pre-Noachian Period (Mars), 453, 456–457
Precambrian
    Archean Eon, 483–484
    chronometric subdivisions (GSSAs), 481,
      482f
    Hadean Eon, 482–483
    International subdivisions, 481–482, 482f
    overview, 481
    Proterozoic Eon, 484–486
Priabonian Stage/Age, 1099–1101, 1100f
Pridoli Epoch/Series, 709, 711f
Prikamian (Russian Platform), 824f, 829
Proterozoic Eon, 482f, 484–486
Protvian (Russian Platform), 824f, 829
Pusgillian (Britain), 652f

Q

Quaternary
    age model, 1244–1247
    basal definition, 1223–1225, 1224f
    international subdivisions, 1222f,
      1223–1229
    magnetostratigraphy, 1222f
    ocean-sediment records, 1241–1243
    stable-isotope stratigraphy, 1245t
    terrestrial records, 1236–1241
    time scale, 1222f, 1244–1247

R

Radaevkian (Russian Platform), 824f, 828–829
Radioisotopic dates
    appendix of GTS2020 dates, 896
    methods and overview, 193
Raikkula (Baltica), 715f
Rakvere (Baltica), 652f, 657–659
Rangerian (North America), 652f
Rawtheyan (Britain), 652f
Redlichiid and Olenellid trilobites Extinction
    Carbon-isotope Excursion (ROECE),
      602–603, 603f
Rhaetian Stage/Age, 915, 936t
Rhenium-Osmium (Re-Os) geochronology,
    201–202
Rhuddanian Stage/Age, 696, 701f

Rhyacian Period/System, 316f, 482f
Richmondian (North America), 652f, 654
Roadian Stage/Age, 882, 883f
Rodinia supercontinent, 486, 510–511
Romanian (Eastern Parathethys), 1153f, 1154, 1157
Rootsikula (Baltica), 715f
Rotliegend (Western Europe), 876, 889f, 891–894
Rupelian Stage/Age, 1101–1103, 1102f
Rusavkinian (Russian Platform), 824f
Ryazanian (Boreal), 974, 976f, 1025–1028

S

Saalian (NW Europe), 1222f, 1227, 1229f
Sakaraulian (Eastern Parathethys), 1153f, 1154–1155
Sakian (Russia/Siberia; Kazakhstan), 581, 586f
Sakmarian Stage/Age, 7t, 323f, 853, 879, 880f
Sandbian Stage/Age, 646–648, 647f
Sangamonian (North America), 1222f
Santonian Stage/Age, 1038–1040, 1039f
Saraninian (Russian Platform), 881, 889f, 890
Sarginian (Russian Platform), 889f, 890, 894–895
Sarmatian (Eastern Parathethys), 1153f, 1154, 1156
Saxonian (Western Europe), 889f, 891–894
Sea level and Sequences
    Cambrian, 383, 605–606, 605f
    Carboniferous, 384, 845–847
    Cenozoic, 385–386
    Cretaceous, 375f, 377f, 385, 1057
    Devonian, 384, 765–766
    Jurassic, 385, 990–991
    Neogene, 375f, 385–386, 1194
    Ordovician, 383, 633f
    overview, 357–359
    Paleogene, 375f, 385–386, 1123–1124
    Permian, 384–385, 877f
    Quaternary, 375f
    Silurian, 383–384, 715f, 716–717
    Triassic, 385, 925
Selandian Stage/Age, 1092, 1093f
Serpukhovian Stage/Age, 7t, 818
Serravallian Stage/Age, 1146, 1147f
Sevatian Substage, 905f, 929f, 933
Severodvinian (Russian Platform), 889f
Severokeltmian (Russian Platform), 824f, 829
Shangsian (China), 824f, 833
Sheinwoodian Stage/Age, 7t, 413t, 699f, 703–706, 705f
Shikhanian (Russian Platform), 880f, 889f
Shiyantou Carbon-isotopeExcursion (SHICE), 602–603, 603f
Shuram Excursion (carbon isotope), 535–536, 538, 540, 547
Siderian, 316f, 482f
Silurian
    age model, 715f
    basal definition, 709
    biostratigraphy, 712–713
    international subdivisions, 696–712, 697f

magnetostratigraphy, 713
    selected events, 320f, 712
    stable-isotope stratigraphy, 293f, 714
    time scale, 718–724, 722f
Silver spikes, 407–408
Sinemurian Stage/Age, 961–962, 961f, 1001
Sjuranian (Russian Platform), 889f
Skullrockian (North America), 586f, 597
Small shelly fossils, 566, 569–571, 570f, 574f
Smithian Substage, 909
Snowball Earth, 201, 328, 497
Sokian (Russian Platform), 889f, 890
Solar system bodies, 443
    Asteroids, 470–473
    Callisto, 466
    Comets, 473–474
    Earth's Moon, 446–453
    Europa, 463
    Ganymede, 463–466
    Io, 463
    Jupiter's moons, 463–466
    Mars, 453–460
    Mercury, 460–461
    Saturn's moons, 466–470
    Venus, 461–462
Solonchanian (NE Siberia), 824f
Southland (New Zealand), 1159t, 1161f
Spathian Substage, 908–909, 922–923
Stage 10 (Cambrian), 584–585
Stage 2 (Cambrian), 573–574
Stage 3 (Cambrian), 574–575
Stage 4 (Cambrian), 575
Stairsian (North America), 652f
Statherian Period/System, 316f, 328, 482f
Stenian Period/System, 316f, 328, 482f, 513
Stephanian (Western Europe), 824f, 825, 827
Steptoean (North America), 581, 586f, 596
Steptoean Positive Isotope-Carbon Excursion (SPICE), 602–603, 603f
Sterlitamakian (Russian Platform), 881, 889f
Steshevian (Russian Platform), 824f, 829
Stratigraphic uncertainties, 403–404, 416–417
Streffordian (Britain), 652f
Strontium 86/87
    Cambrian, 217, 230
    Carboniferous, 229
    Cretaceous, 224t, 228, 1055
    Cryogenian, 231
    Devonian, 229
    Ediacaran, 230, 540–541
    Jurassic, 228, 987–988
    Neogene, 227
    Ordovician, 230, 660–661
    overview, 211–213
    Paleogene, 227–228
    Permian, 228–229
    Precambrian, 212f, 214, 231
    Quaternary, 227
    Silurian, 229, 716
    Triassic, 228, 923–924
Strunian (Western Europe), 736f, 747
Sturtian glaciation, 248–249, 348, 504, 512–513
Sulfur isotopes

Cambrian, 267
Carboniferous, 268–269
Cretaceous, 269
Cryogenian, 317f
Devonian, 268
Ediacaran, 268, 541–543
Jurassic, 269
Neogene, 269–271, 327f
Ordovician, 267
    overview, 259–261
Paleogene, 269–271
Permian, 262f, 269
Precambrian, 265
Silurian, 267–268
Triassic, 269, 923
Sunwaptan (North America), 586f, 596–597
Surenian (Russian Platform), 824f

T

Tangbagouan (China), 824f, 832
Taranaki (New Zealand), 1159t, 1161f
Tarantian Stage/Age, 1219f, 1227–1229
Tarkhanian (Eastern Parathethys), 1153f, 1154–1156
Tarusian (Russian Platform), 818, 829
Tastubian (Russian Platform), 879, 889f
Tatarian (Russian Platform), 889f, 890
Telychian Stage/Age, 703, 704f
Templetonian (Australia), 586f, 590
Terreneuvian Epoch/Series, 573–574, 597–598
Thanetian Stage/Age, 1089f, 1092–1095, 1094f, 1098f
Tithonian Stage/Age, 957f, 973–975, 976f, 980f, 990, 1003
Toarcian Stage/Age, 957f, 963–965, 964f, 980f, 989, 1001
Tommotian (Russia/Siberia; Kazakhstan), 586f, 592–593, 610f
Tongaporutuan (New Zealand), 1158, 1159t, 1161f
Tonian Period/System, 495
Top of Cambrian carbon-isotope Excursion (TOCE), 318f, 328–329, 567f, 585, 603–604, 603f
Topazan (North America), 586f, 596
Tortonian Stage/Age, 1089f, 1142f, 1146, 1148f, 1163f
Tournaisian Stage/Age, 813f, 814f, 815–816, 840f
Toyonian (Russia/Siberia; Kazakhstan), 586f, 593, 610f
Tremadoc (Britain), 641, 651, 652f
Tremadocian Stage/Age, 633f, 635–641, 635f, 641f
Triassic Period/System, 903
    age model, 933–937, 936t
    basal definition, 904–908
    biostratigraphy, 918–920, 929f, 932f, 935f
    international subdivisions, 904–915, 905f
    magnetostratigraphy, 920–922, 929f, 932f, 935f
    selected events, 904

stable-isotope stratigraphy, 905f, 910
time scale, 905f, 926–937, 929f, 932f, 935f
Trilobites, 36–42
Cambrian, 37, 570f, 599–600
overview, 36, 38f
Tuesaian (Kazakhstan), 586f
Tukalandian (Russia/Siberia), 586f, 594–595
Tulean (North America), 652f
Tulian (Russian Platform), 824f, 828–829
Turinian (North America), 652f
Turonian Stage/Age, 1027f, 1036–1038, 1037f, 1047f, 1056
Tuvalian Substage, 905f, 912, 929f, 932

U

U-Pb geochronology, 194–197
U-Pb Isotope Dilution-Thermal Ionization Mass Spectrometry (ID-TIMS), 194
Ufimian (Russian Platform), 876, 890
Uhaku (Baltica), 651–653, 652f
Undillian (Australia), 586f
Ungurian (Kazakhstan), 586f
Upper Cretaceous Series, 164–173, 423–424, 1034–1043
Upper Devonian Series, 744–749
Upper Jurassic Series, 990
Upper Ordovician Series, 646–651
Upper Pleistocene Sub-series, 1223
Upper Triassic Series, 912–915
Upinian (Russian Platform), 824f
Urzhumian (Russian Platform), 889f
Uskalykian (Russian Platform), 889f

V

Valanginian Stage/Age, 1027f, 1028–1029, 1047f, 1055
Varangu (Baltica), 651–653, 652f
Vendian (Russia/Siberia), 497, 522–524, 523f, 547, 586f, 592
Venevian (Russian Platform), 818, 824f, 828–829

Venus, 449f, 461–462
Vereian (Russian Platform), 824f, 829–830
Vertebrates (including Mammals; but See Conodonts for that group)
Carboniferous, 842–843
Cretaceous, 1051
Devonian, 760f, 761
Jurassic, 981–982
Neogene, 1191–1193, 1191f
Paleogene, 1113–1116, 1115f
Quaternary, 1240–1241, 1242f, 1265–1266
Silurian, 699f, 713
Triassic, 918–919, 929f, 932f, 935f
Virgilian (North America), 824f, 832
Viru (Baltica), 652f
Visean Stage/Age, 813f, 816–818, 817f, 824f, 840f
Volgian (Boreal), 974–975, 976f, 990, 1025–1028
Volkhov (Baltica), 652f
Volynian (Eastern Paratethys), 1153f
Vormsi (Baltica), 651–653, 652f
Voznesenian (Russian Platform), 824f
Vyatkian (Russian Platform), 889f

W

Waiauan (New Zealand), 1158, 1159t, 1161f
Waipipian (New Zealand), 1157–1158, 1159t, 1162
Waitakian (New Zealand), 1158, 1159t, 1161f
Wanganui (New Zealand), 1159t, 1161f
Wangcunian (South China), 586f, 588, 605f
Warendan (Australasia), 586f, 591–592, 610f, 652f
Wargal (Salt Range), 882, 889f, 894
Warnantian (Western Europe), 818, 823, 824f, 825
Weichselian (NW Europe), 1222f, 1229f, 1237
Wenlock Epoch/Series, 697f, 699f, 703–706, 705f, 724
Westphalian (Western Europe), 824f, 825, 827
Whiterockian (North America), 652f, 653–654

Whitlandian (Britain), 652f
Wisconsian (North America), 1222f, 1229f, 1237
Wolfcampian (North America), 889f, 890
Wolstonian (British Isles), 1222f, 1229f, 1237t
Wonoka excursion (carbon isotopes), 538, 539f
Wordian Stage/Age, 877f, 882, 884f, 889f, 893f
Wuchiapingian Stage/Age, 877f, 886, 887f, 889f, 893f
Wuliuan Stage/Age, 567f, 570f, 575, 577, 577f, 586f, 587–588

X

Xiaodushanian (China), 824f, 833–834, 889f
Xinchangian (China), 586f, 610f, 652f, 653

Y

Yakhtashian (Pamir), 889f
Yapeenian (Australasia), 651, 652f
Yeadonian (Western Europe), 824f, 826–827
Yiyangian (China), 652f, 653
Younger Dryas, 1229, 1241
Ypresian Stage/Age, 1089f, 1095–1096, 1096f, 1107f

Z

Zanclean Stage/Age, 1142f, 1150, 1151f, 1163f
Zapaltjubian (Russian Platform), 824f
Zechstein (Western Europe), 889f, 891–894, 921
Zhanaarykian (Kazakhstan), 586f
Zhujiaqing Carbon-isotope Excursion (ZHUCE), 318f, 328–329, 567f, 602–603, 603f
Ziganian (Russian Platform), 824f
Zisongian (China), 824f, 889f, 891